1 MONTH OF
FREE
READING

at
www.ForgottenBooks.com

By purchasing this book you are eligible for one month membership to ForgottenBooks.com, giving you unlimited access to our entire collection of over 1,000,000 titles via our web site and mobile apps.

To claim your free month visit:

www.forgottenbooks.com/free1068078

ISBN 978-0-331-68867-2
PIBN 11068078

Forgotten Books is a registered trademark of FB &c Ltd.
Copyright © 2018 FB &c Ltd.
FB &c Ltd, Dalton House, 60 Windsor Avenue, London, SW19 2RR.
Company number 08720141. Registered in England and Wales.

For support please visit www.forgottenbooks.com

DEPARTMENT OF THE INTERIOR

—

. BULLETIN

OF THE

UNITED STATES

GEOLOGICAL SURVEY

No. 178

WASHINGTON
GOVERNMENT PRINTING OFFICE
1901

4. die 13/7

UNITED STATES GEOLOGICAL SURVEY

CHARLES D. WALCOTT, DIRECTOR

THE

EL PASO TIN DEPOSITS

BY

WALTER HARVEY WEED

WASHINGTON

GOVERNMENT PRINTING OFFICE

1901

CONTENTS.

5

ILLUSTRATIONS.

LETTER OF TRANSMITTAL.

DEPARTMENT OF THE INTERIOR,
UNITED STATES GEOLOGICAL SURVEY,
Washington, D. C., March 19, 1901.

SIR: In December, 1900, specimens of tin ore were submitted to the Survey for examination. They proved to contain abundant cassiterite and wolframite in a quartz gangue. They were said to come from the vicinity of El Paso, Texas, and in view of the infrequency of well-authenticated occurrences of tin ore in the United States it was judged wise to have their provenance and manner of occurrence verified at the earliest opportunity. This opportunity presented itself when, in January, 1901, Mr. W. H. Weed had occasion to pass through El Paso, and he was instructed to make a reconnaissance examination of the locality whence the specimens were said to come. The result of his examination is presented briefly in the following report, the immediate publication of which I recommend.

S. F. EMMONS,
Geologist in Charge of Section of Metalliferous Ores.

Hon. CHARLES D. WALCOTT,
Director of United States Geological Survey.

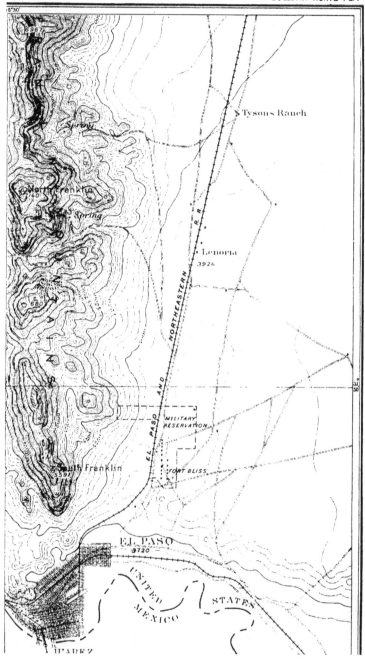

THE EL PASO TIN DEPOSITS.

By W. H. Weed.

LOCATION OF DEPOSITS.

The El Paso tin deposits lie on the east flank of the Franklin Mountains, the southern extension of the Organ or San Andreas Range, about 10 miles north of the city of El Paso. The ores were discovered in 1899 and have been prospected by several open cuts and pits, the deepest of which is about 50 feet below the surface. The property belongs to Judge C. R. Moorhead, of El Paso, to whom I am indebted for many courtesies during my visit to the deposits. The place is distant about 14 miles by wagon road from El Paso, 12 miles of excellent road across the flat mesa being succeeded by 2 miles across the foothills. The White Oaks Railroad crosses the flat 3 or 4 miles east of the property, and the main line of the Southern Pacific lies 10 miles to the south. There is a good spring one-fourth of a mile from the ledges, but there is no large supply of water nearer than the Rio Grande. The mesa is underlain by water, the city of El Paso being supplied from driven wells sunk in the mesa gravels. The mesa is scantily grassed and covered with the usual desert vegetation of small yucca and cactus, while the mountain slopes show cedar bushes, with mesquite, yucca, sotol, and other arid-land plants. The mountains show a very regular crest of bedded rocks surmounting smoother basal slopes of a prevailing red-brown color dotted by green sotol bushes.

GEOLOGICAL STRUCTURE AND FORMATIONS.

The geological structure is very simple and is easily made out, as the mountains are not wooded, but show outcropping edges of the upturned limestones and bare slopes of red granite. The mountain range consists of Cambrian and other Paleozoic limestones, upturned by and resting upon an intrusive mass of coarse-grained granite that forms the central core of the range. This granite is well exposed for a distance of 4 or 5 miles along the eastern side of the mountains, forming the lower half of the mountains proper, and in places extending out to the foothills. The crest of the range consists of steeply tilted, heavily bedded, dark-gray limestones dipping westward. The basal quartzites

11

were observed in the drift seen in arroyos, so that the granite is probably intruded between the base of the Cambrian rocks and the underlying Archean complex.

The eastern foothills consist mainly of limestones, but near the tin deposits these bedded rocks have been cut through and granite now forms the surface, remnants of the limestone cover showing as isolated masses capping the hillocks. These relations are shown in the diagram, fig. 1, which is a rough sketch of the range, representing a cross section at the tin mines. North of the place where this section was made a transverse ridge of the range shows the granite to be sheeted by well-marked planes, dipping eastward at an angle of about 45° to 50°. The granite is very much altered by surface decomposition, and crumbles readily to a coarse sand. No fresh material was observed anywhere on the surface, but fairly good material was obtained from the dump heap of the shaft on the north vein. The granite is sheeted near the veins, the planes of sheeting being parallel to the veins themselves. The general sheeting, however, is in a different direction, the average strike

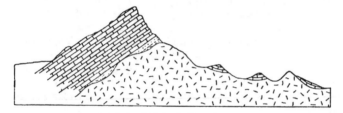

FIG. 1.—Cross section of Franklin Mountains 10 miles north of El Paso, Texas.

being N. 20° E, and the dip 70° SE. A thin section of this granite has been examined under the microscope by Mr. Lindgren, who furnishes the following notes:

The rock is a coarse-grained normal granite. It shows much anhedral quartz with anhedral feldspar, largely microperthite, with some few grains of microcline. A few small flakes of brownish-green hornblende and some small grains of magnetite were also seen. The rock is a soda granite.

White aplite-granite occurs in veinlets and irregular masses intrusive in the granite, but none was observed close to the veins. The mesa is underlain by cemented gravels, which form also the lower slopes of the foothills.

ORES AND VEINS.

The ores consist of cassiterite, or oxide of tin, with wolframite (tungstate of iron and manganese) in a gangue of quartz. Specimens of nearly pure cassiterite weighing several pounds have been found on the surface, and this mineral occurs in the quartz, either alone or associated with wolframite. The most abundant ore is a granular mixture of tin ore and quartz which resembles a coarse granite and corresponds to the greisen ore of European tin deposits. Pyrite occurs rarely in the eastern exposures of the vein, but appears to constitute the bulk of the metallic contents in exposures seen in the westernmost openings. These ores occur in well-defined veins, which run up the slopes nearly at right angles to the direction of the range, the strike being approximately east-west and the veins dipping steeply to the north. Three veins have been discovered, all of which have been exposed by open-cut work and by pits for several hundred feet in length. The most northerly vein is traceable along the surface for a distance of about 1,200 feet. The middle vein lies about 300 feet south of the east end of the northern one, but apparently converges westward toward the northern vein. The southern vein, which is the smallest of the three, lies about 600 feet farther south.

The veins exhibit the usual characters of the European tin veins, notably those of Cornwall, England, their clearly defined fissures

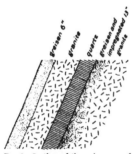

FIG. 2.—Section of tin vein exposed in open cut (north vein).

showing a central core or lead of coarse quartz, sometimes containing tin ore, and flanked on either side by altered rock in which the tin ore replaces the feldspar of the granite. Where this metasomatic replacement is complete the ore shows a mixture of cassiterite, with or without wolframite and quartz. Where the replacement is only partial the greisen ore fades off into the unaltered granite. A cross section of the veins shows, therefore, the same phenomena seen at Cornwall. The diagram, fig. 2, shows an ideal representation of the conditions existing in the veins, and has been drawn from sketches made in the field. The central mass of quartz corresponds to the "leader" of the Cornish veins. It is composed of massive, coarsely crystalline quartz, sometimes showing comb structure, and it is clearly the result of the filling of the open fissure by quartz. The adjacent ore-bearing material is a replacement deposit in which the mineral solutions have substituted ore for the feldspar of the granite by metasomatic action; in other words, the main mass of the ore occurs alongside of a quartz vein, and is due to the alteration of the granite forming the walls of the fissure. In general, the ore passes into the granite by insensible transition and there are no distinct walls.

A thin section of the greisen ore has been examined by Mr. Lindgren, who furnishes the following notes:

The thin section of the tin ore shows it to be a quartz-cassiterite rock. It is a coarsely granular rock consisting of anhedral quartz, with which is intergrown grains of slightly brownish cassiterite. The quartz is full of fluid inclusions and makes up about 75 per cent of the mass. The cassiterite grains are, along the edges, intimately intergrown with quartz. If this is a metasomatic form of the granite a silicification has taken place. The microscope affords no direct evidence, however, that this ore is metasomatic. One small grain of tourmaline and a few flakes of sericite were seen. Neither topaz nor mica occurs in the section, and no remains of feldspar were observed.

The north vein has a course of N. 85½° W. magnetic, as determined from the openings at the east end. At the west end of the workings the course observed, looking back along the outcrop, appears to be N. 80° E. for the northern vein and N. 80° W. for the middle vein; so that if these observations are correct the veins must intersect toward

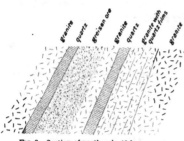

the west. The surveys by the owners of the property show a course N. 85½° W. for the middle and 65° W. for the south vein.

Fig. 3.—Section of north vein (6 feet across)

DEVELOPMENT.

A shaft 35 feet deep has been sunk on the north vein at the eastern end of the vein outcrop. This shaft is about 5 by 10 feet across and shows a very well-defined vein about 5 feet wide, having a dip of about 70° to the north. The sides of the shaft show excellent ore, mostly of the greisen variety, extending down for 8 to 15 feet below the top. At this point a slip crosses the shaft and cuts out the ore. This slip. or fault, is a clay seam but one-fourth to one-half inch in thickness, and seems to have thrown the upper part of the vein to the north. The lower half of the shaft reveals only rusty granite, altered and showing films of quartz, but without recognizable ore. A crosscut south from the bottom of the shaft should reach the vein if the fault is a normal one. In the exposure seen in the upper part of the shaft the ore occurs in bunches in altered granite and lies on the

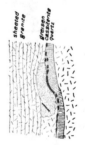

Fig. 4.—Greisen bowlder in tin vein (shaft on middle vein).

north side of a 15-inch streak of sheeted and rusty quartz. A second shaft on the north vein has been sunk at a point about 300 feet west of the one just noted. This shaft is about 25 feet deep. The vein is well exposed at the top, and shows a dip northward, but the shaft passes out of the vein into the sheeted granite, forming the foot wall. A crosscut about 8 feet in length, driven from the bottom of the shaft, cuts the vein, but does not pass through it. The sheeting of the granite seen in this shaft is very pronounced, the rock being divided into plates from one-fourth inch to 12 inches in thickness by planes dipping 61° E. and crossing the vein at 90°. The outcrop of the vein is traceable westward up the slopes by its rusty quartz, and a nearly continuous ledge can be followed. This outcrop has been opened at intervals of a few yards by trenches, which expose the vein and show it to have a thickness of from 2 to 6 feet, with about half this thickness of ore. No samples were, however, taken, and it is uncertain whether the altered granite does not contain a percentage of tin oxide. The most westerly working that could be surely identified as being upon the north vein is a pit 6 feet deep, which shows a 6-foot vein in which the quartz is bluish in color and the tin ore is associated with much pyrite. This point is about 600 or more feet west of the first shaft. West of this point the ledge can not be traced across the slopes, but an opening north a hundred feet higher and a few hundred feet farther west shows a good vein, carrying much pyrite, but devoid of any recognizable tin ore.

The middle vein is developed by a shaft 50 feet deep, which shows a vein having a central leader of quartz 2 feet wide at the top and tapering to 1 foot 4 inches wide at the bottom of the shaft. The dip, as shown by the walls of the shaft, is 70° N. The central quartz mass is spotted with cassiterite, and the altered granite on either side contains recognizable grains of tin oxide.

The south vein lies 500 to 600 feet south of the middle vein. This vein is much narrower than the veins on the north, having an average width of about 1 foot. The strike, as shown near the shaft, is N. 50° W. and the dip 50° N. The vein walls are sometimes defined by a clay selvage one-sixteenth inch wide, but more often show a gradual fading off into the granite.

CONTINUANCE OF VEINS IN DEPTH.

It will be noticed from what has been said that the veins are all well defined at the surface and carry good values in tin ore, but that the ore apparently dies out in depth. Further development is needed to establish the existence of the ore at a greater depth than 50 feet, but it is believed that the veins have been thrown by local slips or faults and will be found by crosscutting from the bottom of the present workings. The character of the fissures and the nature of the ore both indicate that the veins are the result of deep-seated agencies, and are not merely segregations due to descending surface waters. For this reason it is believed that further exploration will develop well-defined tin veins. The absence of topaz in the deposit is noteworthy, for this mineral is commonly associated with cassiterite veins the world over. In other respects the deposits closely resemble the tin veins of Europe, and are clearly due to metasomatic processes. The evidence of a pneumatolitic origin is, however, not conclusive.

DEPARTMENT OF THE INTERIOR

BULLETIN

OF THE

UNITED STATES

GEOLOGICAL SURVEY

No. 179

WASHINGTON
GOVERNMENT PRINTING OFFICE
1902

UNITED STATES GEOLOGICAL SURVEY

CHARLES D. WALCOTT, DIRECTOR

BIBLIOGRAPHY AND CATALOGUE

OF THE

FOSSIL VERTEBRATA OF NORTH AMERICA

BY

OLIVER PERRY HAY

WASHINGTON
GOVERNMENT PRINTING OFFICE
1902

CONTENTS.

LETTER OF TRANSMITTAL.

AMERICAN MUSEUM OF NATURAL HISTORY,
New York, May 6, 1901.

SIR: I have the honor to transmit herewith a Bibliography and Catalogue of the Fossil Vertebrata of North America.

I take pleasure in acknowledging the interest which you have shown at all times in the preparation of this work.

I am, with great respect, your obedient servant,

O. P. HAY.

Hon. C. D. WALCOTT,
   Director of the U. S. Geological Survey,
          Washington, D. C.

7

PREFACE.

The purpose of this bibliography and catalogue is to present a list of all the species of fossil vertebrates which have been described, up to the end of the year 1900, from all that part of the continent of North America lying north of Mexico, and to furnish a guide to the literature relating to all these extinct forms of organized beings. The number of species is so great and the literature describing them is so vast and so widely scattered that even the experts in the science of vertebrate paleontology have great difficulty in making and keeping themselves fully informed as to what has already been done and what is being done. It is hoped that this work will relieve students of much tedious and often unavailing search.

This bulletin may be said to consist of three parts: First, a bibliography which contains the titles of all the books, memoirs, and papers, extended or brief, which the author has been able to consult and which have appeared to him to have a useful bearing on the subject; second, a systematically arranged list of all the species of fossil vertebrates which have been described from the region above indicated; third, an index, alphabetically arranged, of all the systematic names occurring in the volume.

In the bibliography there are given about 4,600 titles. Accompanying each is a statement of the number of pages occupied by the work and the number of plates or figures. If the memoir or paper is found in some serial publication, the name of this is given together with the volume and pages. All of the titles have been obtained through actual examination of the scientific journals. In the case of many of these the search has been made page by page, since the indexes are so often deficient that they can not be wholly relied on. The end in view throughout has been to find all that the paleontologist will need in his work and to give correct citations.

The titles of works and papers are given under the names of the authors, and these authors' names are presented in alphabetical order. Each author's papers are arranged in the sequence of the years of publication. Each title is preceded by the author's name, the year of publication, and a letter of the alphabet, A, B, C, etc., according to the number of papers published during that year, as "Leidy, J., 1868 A," "1868 B" to "1868 J." If an author has issued more than 26

papers in any year, the additional papers are indicated by the letters
AA, BB, etc. Thus we have "Cope, E. D., 1884 JJ." It must be
understood that it is not here implied that the papers of any author
in any year have been issued in the order in which they follow each
other in this bibliography. In the case of an author who, like Professor Cope, was publishing numerous memoirs, papers, and notes in
many journals and Government publications, it is manifestly almost
impossible to ascertain the exact order of issue. If any student
should be under the necessity of determining such order, he will have
to address himself to the task.

How successful the writer has been in collecting a complete list of
species only practical use of the book will prove. What has been
spoken of above as a list of the known species is, however, far more,
for under each species each genus and each group of higher rank
represented by any North American fossil species there has been
included references to the literature appertaining to it—that is, to
put it in another way, each paper and memoir and book cited in the
bibliographical portion has been examined, and the proper citations of
it have been recorded under the species, genera, etc., of the catalogue
of species. An estimate has shown that there are over 40,000 citations
recorded. Nevertheless, it is not improbable that some important
things have escaped attention.

In the catalogue the writer has not usually cited directly the place
of publication of authors' papers, but has referred to them as notated
in the bibliography. If, to give an illustration, the student finds
under *Ceratops* the citation, "Marsh, O. C., 1888 C, p. 477," he must
turn to the list of Professor Marsh's papers for the year 1888 and
run down the list to the letter C. He will find there the complete
title of that paper, place of publication, etc. This method of citation
has been adopted for the following reasons: (1) To economize labor
in the preparation of this book. The writer does not feel that it
could ever have been completed had he been compelled to cite the
journal and volume according to the usual method for all of the approximately 40,000 references herein contained. (2) To bring the size of
this bulletin within limits convenient for use. To refer to papers
in the usual way, even with intelligible abbreviations, would have
increased the size of this volume probably 50 per cent or more.
(3) Because the writer believes that, on the whole, the method here
employed will save the time of the student. If the place of publication alone is given, no clue whatever is afforded to the nature of the
paper. It may or may not contain something of value to him. In
this bulletin he has the complete title. If he is studying the dentition
of a fossil vertebrate, and the paper which he finds cited deals with
the vertebræ, or the carpus, or the habits of the animal in question, he
needs to go no further. The time saved in making one trip to a

library will compensate for any trouble undergone in turning from the catalogue to the bibliography. This method of reference, with some modifications, is in frequent use by scientific writers, though it has probably not been employed before on so large a scale.

Some further explanation may be made regarding the systematic portion of this work. Under each genus and species there will be found, first of all, a reference to the original description. This is followed by the other citations, arranged according to authors in alphabetic order, years of publication, and symbols. If the author cited has employed a generic or specific name different from the one adopted, this is thrown in parenthetically after the citation.

The writer has arranged the classes, orders, families, and genera of our fossil vertebrates according to what has seemed to him to be their natural relationships. This has appeared to him to be preferable to arranging them all alphabetically, although he does not overlook the advantages of the latter method. On the other hand, the species of each genus is here disposed in alphabetical order for the reason that it will save much time in finding them, while attempts at giving them a natural arrangement are often futile.

As regards nomenclature the writer has followed the rules of the American Ornithological Union, because he believes that they meet the requirements of systematic work more completely and consistently than those adopted by any other body of zoologists.

The coming years will surely witness a great increase in our knowledge of the beings which in past ages have peopled our continent. As a result of discoveries and investigations many new forms will be described, while again perhaps many genera and species which are now accepted will be reduced to synonomy. This will, before many years, make obsolete many of the names here adopted, but it is the purpose of this work to facilitate just such discoveries and investigations and changes.

The writer takes pleasure in expressing his obligations to many friends for assistance and encouragement in prosecuting his work on this volume. Encouragement has been afforded by all, or nearly all, of the vertebrate paleontologists of the United States, and as many as have been applied to have given the information desired. Financial assistance has been given by the Smithsonian Institution, and for this acknowledgments are due, especially to its officers, Secretary S. P. Langley and Assistant Secretary Richard Rathbun; and to Hon. C. D. Walcott, director of the United States Geological Survey. Acknowledgments are due to Dr. Theodore Gill for advice with regard to matters pertaining to nomenclature, to scientific literature, and to classification. It must not, however, be inferred that he is responsible for any innovations that have been made. Dr. Leonhard Stejneger, curator of the department of reptiles, United States National Museum, has answered many questions with regard to nomenclature and has given assistance in other ways. The classification of the birds here adopted is, with slight modification, that proposed by him in the Standard Natural History, 1885. Prof. Frederick A. Lucas, assistant curator of the department of comparative anatomy, United States National Museum, has been helpful in many ways, and to him is due the final disposition of certain of the genera and species.

AMERICAN MUSEUM OF NATURAL HISTORY,
New York, May 1, 1901.

BIBLIOGRAPHY AND CATALOGUE OF THE FOSSIL VERTEBRATA OF NORTH AMERICA.

By Oliver Perry Hay.

BIBLIOGRAPHY.

Adam, Walter. 1834 A
On some symmetrical relations of the bones of the megatherium.
Report Brit. Assoc. Adv. Sci., 3d meeting, Cambridge, 1833, pp. 437-440.

Adams, A. Leith. 1874 A
Concluding report on the Maltese fossil elephants.
Report Brit. Assoc. Adv. Sci., 43d meeting, Bradford, 1873, pp. 185-187.

—— 1875 A
On a fossil saurian vertebra (*Arctosaurus osborni*), from the arctic regions.
Proc. Roy. Irish Acad. (2), ii, pp. 177-179, with 4 text figs.

—— 1877 A
A fossil saurian from the arctic regions.
Amer. Jour. Sci. (3), xiii, p. 316.
Describes *Arctosaurus osborni* Adams.

—— 1877 B
Monograph of the British fossil elephants. Part I. Dentition and osteology of *Elephas antiquus*.
Palæontographical Society, xxxi, pp. 1-68; pls. i-v.

—— 1879 A
Monograph of the British fossil elephants. Part II. Dentition and osteology of *Elephas primigenius* (Blumenbach).
Palæontographical Society, xxxiii, pp. 69-146; pls. vi-xv.

—— 1881 A
Monograph of the British fossil elephants. Part III. Osteology of *E. primigenius* and dentition and osteology of *E. meridionalis*.
Palæontographical Society, xxxv, pp. 147-265; pls. xvi-xxviii.

Adams, C. B. 1846 A
Notice of a small Ornithichnite.
Amer. Jour. Sci. (2), ii, pp. 215-216, with 2 figs.

Adams, George I. 1895 A
Two new species of *Dinictis* from the White River beds.
Amer. Naturalist, xxix, pp. 573-578; pl. xxvi and 1 text fig.
Descriptions of *Dinictis fortis* and *D. bombifrons.*

—— 1896 A
The extinct *Felidæ* of North America.
Amer. Jour. Sci. (4), i, pp. 419-444, with pls. x-xii.

—— 1896 B
On the species of *Hoplophoneus.*
Amer. Naturalist, xxx, pp. 46-51; pls. i, ii.
Describes new species *H. insolens* and *H. robustus.*

—— 1897 A
On the extinct *Felidæ.*
Amer. Jour. Sci. (4), iv, pp. 145-149, with 8 figs.
Abstract in Princeton Univ. Bull., ix, pp. 16-17, 1897.

Adloff, Paul. 1898 A
Zur Entwicklungsgeschichte des Nagetiergebisses.
Jenaische Zeitschr. Naturwiss., xxxii, pp. 347-410, with pls. xii-xvi and 4 figs. in the text.

Agassiz, Louis. 1835 A
On the fossil fishes of Scotland.
Report Brit. Assoc. Adv. Sci., 4th meeting, Edinburgh, 1834, pp. 646-647.
Description of new genus *Megalichthys*, type *M. hibberti.*

—— 1841 A
[Note on fossil jaw, on which is based the genus *Phocodon*, type *P. scillæ* Ag.]
Valentin's Repertorium f. Anat. Physiol., vi, p. 286.

13

Agassiz, Louis—Continued.   1843 A

Report on the fossil fishes of the Devonian system or Old Red sandstone.

Report Brit. Assoc. Adv. Sci., 12th meeting, Manchester, 1842, pp. 80–88.

———   1843 B

Recherches sur les poissons fossiles.

Tome i: pp. i–xxxii; 1–188; atlas of 11 pls. (A–K.)

Tome ii: pp. i–xii; pt. 1, 1–310; pt. 2, 1–144; atlas of 146 pls.

Tome iii: pp. i–viii; 1–390; 1–32; atlas of 83 pls.

Tome iv: pp. i–xvi; 1–292; 1–22; atlas of 61 pls.

Tome v: pp. i–xii; 1–160; 1–144; atlas of 95 pls.

This work was issued in livraisons at various times from 1833 to 1844. The dates of publication of the livraisons and of the plates are given in Woodward and Sherborn's Catalogue of British Fossil Vertebrates, pp. xxv–xxix.

Under the genera and species of the present work the actual date of publication, if not 1843, is generally given in parentheses after "1843 B."

———   1844 A

Synoptical table of British fossil fishes, arranged in the order of the geological formations.

Report Brit. Assoc. Adv. Sci., 13th meeting, Cork, 1843, pp. 194–207.

———   1844 B

Monographie des poissons fossiles du vieux grès rouge ou système Dévonien (Old Red sandstone) des Isles Britanniques et de Russie.

Pp. i–xxxvi; 1–171, with pls. A–I; i–xxxiii, Neuchatel (Suisse).

———   1845 A

Report on the fossil fishes of the London clay.

Report Brit. Assoc. Adv. Sci., 14th meeting, York, 1844, pp. 279–310.

———   1846 A

On the ichthyological fossil fauna of the Old Red sandstone.

Edinb. New Philos. Jour., xli, pp. 17–49.

From Monographie des poissons foss. du vieux grès rouge.

———   1848 A

[Letter to Dr. Gibbes regarding *Dorudon*.]

Proc. Acad. Nat. Sci. Phila., iv, p. 4.

———   1849 A

[Remarks on crocodiles of the Green sand of New Jersey and on *Atlantochelys*.]

Proc. Acad. Nat. Sci. Phila., iv, p. 169.

Contains no descriptions.

Agassiz, Louis—Continued.   1850 A

On the fossil remains of an elephant found in Vermont.

Proc. Amer. Assoc. Adv. Sci., 2d meeting, Cambridge, Mass., pp. 100–101.

Said by Professor Agassiz to be found for the first time in this country.

———   1850 B

Remarks on finding femur of *Atlantochelys*.

Proc. Nat. Hist. Soc. Boston, iii, p. 351.

No adequate description and no specific name.

———   1851 A

[Announcement of the finding of the head of a walrus on New Jersey shore.]

Proc. Amer. Assoc. Adv. Sci., 4th meeting, New Haven, 1850, p. 252.

———   1851 B

Report on the vertebrate fossils exhibited to the Association (abstract).

Proc. Amer. Assoc. Adv. Sci., 5th meeting, Washington, D. C., 1851, pp. 178–180.

———   1853 A

Notice of Joseph Leidy's "Memoir of the extinct species of American ox."

Amer. Jour. Sci. (2), xvi, p. 281.

———   1856 A

Notice of the fossil fishes found in California by W. P. Blake.

Amer. Jour. Sci. (2), xxi, pp. 272–275.

Describes 11 new species of Elasmobranchs. The same species were described and figured in Pacific Survey Report, v, pp. 313–316, pl. i.

———   1857 A

Notice of the fossil fishes.

Pacific Survey Report, v, pp. 313–316, with pl. 1.

Contains descriptions and figures of 11 new species of Elasmobranchs.

———   1857 B

Contributions to the natural history of the United States of America. Vol. i, pt. ii, North American *Testudinata*.

Pp. 235–452d, with pls. i–xxvii.

The plates are found in vol. ii. This volume is devoted mostly to the embryology of the *Testudinata*.

———   1862 A

Highly interesting discovery of new sauroid remains.

Amer. Jour. Sci. (2), xxxiii, p. 138.

Refers to the discovery of *Eosaurus* by O. C. Marsh; no name yet proposed.

Agassiz, Louis—Continued.   1862 B
Mastodon tooth in Amador County,
Cal.
Amer. Jour. Sci. (2), xxxiv, p. 135.
Editorial note supposed to have been written by Agassiz.

———   1862 C
Remarks on the megatheroids.
Proc. Nat. Hist. Soc. Boston, ix, p. 101–102.

———   1863 A
Note on the megatherium.
Amer. Jour. Sci. (2), xxxvi, p. 300.

———   1863 B
[Remarks on *Megatherium*.]
Proc. Nat. Hist. Soc. Boston, ix, p. 198.

Ahlborn, Fr.   1896 A
Zur Mechanik des Vogelfluges.
Abhandl. Naturwiss. Verein Hamburg,
xiv, pp. 7–134, with 54 figs. in the text.

Albrecht, Paul.   1879 A
Ueber den Stammbaum der Raubthiere.
Schriften d. phys.-ökon. Gesellsch. Konigsberg, xx, Bericht, pp. 22–23.
Presents an arrangement of the families of the *Carnivora*.

———   1886 A
Über die cetoide Natur der *Promammalia*.
Anatomischer Anzeiger, i, pp. 338–350.

Alessandri, G. De.   1896 A
Avanzi di *Oxyrhina hastalis* del miocene di Alba.
Atti Soc. Ital. Sci. Nat., Milano, xxxvi, pp. 263–269, with pl. i.

Alezais, Henri.   1898 A
Étude anatomique du cobaye (*Cavia cobaya*).
Jour. l'Anat., Physiol. norm. et patholog., xxxiv, pp. 735–756, with 5 text figs.

———   1899 A
Étude anatomique du cobaye (*Cavia cobaya*).
Jour. l'Anat. Physiol., norm. et patholog., xxxv, pp. 333–381.

Allen, Harrison.   1880 A
On some homologies in bunodont dentition.
Proc Acad. Nat. Sci., Phila., pp. 226–228.

———   1882 A
On a revision of the ethmoid bone in the *Mammalia*, with special reference to the description of this bone and of the sense of smelling in the *Cheiroptera*.
Bull. Mus. Comp. Zool., x, pp. 135–164, with pls. i–vii.

Allen, Harrison—Continued.   1886 A
On the tarsus of bats.
Amer. Naturalist, vol. xx, pp. 175–176, with 6 illus.

———   1886 B
On the types of tooth structure in *Mammalia*.
Amer. Naturalist, xx, pp. 296–297.

———   1888 A
The occipito-temporal region in the crania of *Carnivora*.
Science, xi, p. 71.

———   1889 A
Remarks on the pronghorn (*Antilocapra americana*).
Proc. Amer. Philos. Soc., xxvi, pp. 366–367.

———   1891 A
Prof. Joseph Leidy: His labors in the field of vertebrate anatomy.
Science, xviii, pp. 274–276.

———   1893 A
A monograph of the bats of North America.
Bull. U. S. Nat. Mus. No. 43, pp. 198, 38 pls. Washington, D. C., 1893.

Allen, Joel Asaph.   1876 A
Description of some remains of an extinct species of wolf, and an extinct species of deer from the lead region of the Upper Mississippi.
Amer. Jour. Sci. (3), xi, pp. 47–51.

———   1876 B
The American bisons, living and extinct.
Mem. Kentucky Geol. Surv., i pt. ii, pp. i–ix; 1–246, with pls. i–xii.
Mem. Mus. Comp. Zool., iv, No. 10, pp. i–ix; 1–246, with pls. i–xii and map.

———   1876 C
Geographical variation among North American mammals, especially with respect to size.
'Bull. U. S. Geol. and Geog. Surv. Terr., ii, pp. 309–344.

———   1877 A
Monographs of North American Rodentia.
Report U. S. Geol. Surv. Terr., F. V. Hayden, U. S. geologist in charge, vol. xi.
No. ii, *Leporidæ*, pp. 265–378; No. iii, *Hystricidæ*, pp. 379–398; No. iv, *Lagomyidæ*, pp. 399–414; No. v, *Castoroididæ*, pp. 415–426; No. vi, *Castoridæ*, pp. 427–454; No. xi, *Sciuridæ*, pp. 631–940.

Allen, Joel Asaph—Continued.   1877 B
Synoptical list of the extinct *Rodentia* of North America.
Report U. S. Geol. Surv. Terr., F. V. Hayden, U. S. geologist in charge, vol. xi, pp. 943–949.
This list forms a portion of the volume containing J. A. Allen's and E. Coues's monographs of N. A. *Rodentia*.
In this list most of the species are mentioned without comment.

———   1878 A
Description of a fossil passerine bird from the insect-bearing shales of Colorado.
Bull. U. S. Geol. and Geog. Surv. Terr., 1878 iv, pp. 443–445, with pl. i.
See also Nature, vol. xviii, pp. 204–206; Amer. Jour. Sci., 1878, xv, p. 381.
Describes *Palæospiza bella*.

———   1879 A
On the species of the genus *Bassaris*.
Bull. U. S. Geol. and Geog. Surv. Terr., v, pp. 331–340.
Presents the synonomy and the literature bearing on this genus and its species.

———   1880 A
History of North American Pinnipeds; a monograph of the walruses, sea lions, sea bears, and seals of North America.
Pp. i–v; 1–785, with 60 figs. in the text. Washington, 1880.

———   1881 A
Preliminary list of works and papers relating to the mammalian orders *Cete* and *Sirenia*.
Bull. U. S. Geol. and Geog. Surv. Terr., vi, pp. 399–562.

———   1885 A
On an extinct type of dog from Ely Cave, Lee County, Virginia.
Mem. Mus. Comp. Zool. x, No. 2, pp. 1–8, with pls. i–iii.
Describes *Pachycyon robustus*, Allen. This paper is followed (pp. 8–13) by one by N. S. Shaler on the age of the cave.

———   1887 A
Note on Squalodont remains from Charleston, S. C.
Bull. Amer. Mus. Nat. Hist., ii, pp. 35–39, with pls. v, vi.
Describes *Squalodon tiedemani*, sp. nov.

———   1893 A
Descriptions of four new species of *Thomomys*, with remarks on other species of the genus.
Bull. Amer. Mus. Nat. Hist., v. pp. 47–68, with pl. i.

Allen, Joel Asaph—Continued.   1900 A
The North American jumping mice.
Amer. Naturalist, xxxiv, pp. 199–202.

Allen, John H.   1846 A
Some facts respecting the geology of Tampa Bay.
Amer. Jour. Sci. (2), i, pp. 38–42.

Allis, Edward P.   1889 A
The anatomy and development of the lateral line system in *Amia calva*.
Jour. Morphology, ii, pp. 463–540, with pls. xxx–xliii.

———   1899 A
On certain homologies of the squamosal, intercalar, exoccipital, and extrascapular bones of *Amia calva*.
Anatom. Anzeiger, xvi, pp. 49–72.
Appended is a list of 36 papers bearing on the subject.

———   1899 B
A reply to certain of Cole's criticisms of my work on *Amia calva*.
Anatom. Anzeiger, xv, pp. 364–379.

———   1900 A
The premaxillary and maxillary bones and the maxillary and mandibular breathing valves of *Polypterus bichir*.
Anatom. Anzeiger, xviii, pp. 257–289, with 3 figs. in the text.

Alston, Edward R.   1876 A
On the classification of the order *Glires*.
Proc. Zool. Soc., Lond., 1876, pp. 61–98, with pl. iv and 5 figs. in the text.

Alth, Alois v.   1874 A
Ueber die palæozoischen Gebilde Podoliens und deren Versteinerungen.
Abhandl. k. k. geolog. Reichsanstalt, Wien, vii, pp. 1–77, pls. i–iv.
Orphalaspis, *Pteraspis*, *Cyathaspis*, and *Coccosteus*.

———   1886 A
Ueber die Zusammengehörigkeit der den Fischgattungen *Pteraspis*, *Cyathaspis*, und *Scaphaspis* zugeschriebenen Schilder.
Beitrage zur Paläont. Österreich-Ungarns und des Orients, v. Heft. iii, pp. 61–73, with pl. xxiv.

Ameghino, Florentino.   1883 A
Sobre la necesidad de borrar el género *Schistopleurum* y sobre la clasificacion y sinonimia de los Glyptodontes en general.
Bol. Acad. Nac. Ciencias, Córdoba, Argentina, v, pp. 1–34.

Ameghino, Florentino—Cont'd. 1886 A
Contribuciones al conocimiento de
los mamíferos fósiles de los terrenos
terciarios antiguos del Paraná.
Bol. Acad. Nac. Ciencias, Córdoba, Argentina, ix, pp. 1-228.

—— 1889 A
Contribucion al conocimiento de los
mamíferos fósiles de la República
Argentina, obra escrita bajo los auspicios de la Academia Nacional de
Ciencias de la República Argentina
para presentarla á la· Exposicion Universal de Paris de 1889.
Buenos Aires, 1889. Folio, pp. i-xxxiii; 1-1027, with atlas of pls. i-xcviii.

—— 1891 A
Observaciones críticas sobre los
caballos fósiles de la República Argentina.
Revista Argentina de Historia Natural, i,
pp. 4-17; 65-88.

—— 1891 B
Los Plagiaulacídeos argentinos y sus
relaciones zoológicas, geológicas y geográficas.
Revista Argentina de Historia Natural, i,
pp. 38-44.

—— 1891 C
Adicion á la memoria del Dr. H. von
Ihering sobre la distribucion geográfica
de los Creodontes.
Revista Argentina de Historia Natural, i,
pp. 214-219.

—— 1891 D
Determinacion de algunos jalones
para la restauracion de las antiguas conexiones del continente Sud-americano.
Revista Argentina de Historia Natural, i,
pp. 282-288.

—— 1892 A
Répliques aux critiques du Dr. Burmeister sur quelques genres de mammifères fossiles de la République
Argentine.
Bol. Acad. Nac. Ciencias, Córdoba, xii, pp.
437-469; also as separate with pp. 3-35.

—— 1894 A
Enumération synoptique des espèces
mammifères fossiles des formations
éocènes de Patagonie.
Bol.· Acad. Nac. Ciencias, Córdoba, xiii, pp.
259-452, with 66 figs. in the text.

Ameghino, Florentino—Cont'd. 1896 A
Sur l'évolution des dents des mammifères.
Bol. Acad. Nac. Ciencias, Córdoba, Argentina, xiv, pp. 381-517.

—— 1897 A
Mammifères crétacés de l'Argentine.
Deuxième contribution à la connaissance de la faune mammalogique des
couches à Pyrotherium.
Bol. Institut Geogr. Argentino, xviii, pp.
1-117, with 86 figs. in the text.
Many new genera and species described.
Contains brief references to papers by Prof. O. C.
Marsh and Dr. J. L. Wortman on the *Teniodontia.* Reviewed by M. Schlosser in Neues
Jahrb. Min., 1900, ii, pp. 296-305.

—— 1898 A
An existing ground-sloth in Patagonia.
Natural Science, xiii, pp. 324-326.
A translation of the original paper, which
was published in La Plata, Argentina, 1898.

—— 1898 B
Sur l'évolution des dents des mammifères.
Bull. Soc. Géol. France (3), xxvi, pp. 497-499

—— 1899 A
On the primitive type of the plexodont molars of mammals.
Proc. Zool. Soc. Lond., 1899, pp. 555-571,
with 16 figs. in the text.

—— See also **Gervais** and **Ameghino.**

American Geologist.   1888 A
[Editorial notice of a paper by Dr.
Traquair on fossil fishes printed in
Geological Magazine for May, 1888.]
Amer. Geologist, ii, p. 133.

—— 1891 A
New discoveries by Dr. Clarke in the
fish beds of the Devonian.
Amer. Geologist, vii, pp. 143-144.
From reports of meeting of Amer. Assoc.
Adv. Sci.

—— 1891 B
The oldest fish remains known.
Amer. Geologist, vii, pp. 329-330.
Editorial comment on discoveries in Colorado by C. D. Walcott.

—— 1891 C
Discovery of mastodon remains in
the Shenandoah Valley.
Amer. Geologist, vii, p. 335.
Editorial note regarding finding of bones
of the mastodon in Rockingham County, Va.

American Geologist—Cont'd.   1891 D
    [Editorial comment on Dr. G. Baur's
    "Remarks on the reptiles generally
    called *Dinosauria*," published in Amer.
    Naturalist, xxv, pp. 434–454.]
    Amer. Geologist, viii, pp. 55–56.

——————   1891 E
    Supposed Trenton fossil fish.
    Amer. Geologist, viii, pp. 178–180.
    Editorial comment on C. D. Walcott's Lower
    Silurian fishes.

——————   1891 F
    Man and the mammoth.
    Amer. Geologist, viii, pp. 180–183.
    Editorial comment on paper by M. Max.
    Lohest presented before the International Geo-
    logical Congress at Belgium.

——————   1891 G
    "Personal and scientific news."
    Amer. Geologist, viii, pp. 191,196.
    Notes on the finding of various vertebrates
    in Florida by Dr. J. Kost, and of *Megalonyx* in
    Tennessee by Professor Safford.

——————   1892 A
    Prehistoric horses.
    Amer. Geologist, ix, pp. 67–68.
    Editorial comment on Professor Cope's paper
    in Amer. Naturalist for October, 1891.

American Journal of Science.   1828 A
    Teeth of the mastodon.
    Amer. Jour. Sci. (1), xiv, pp. 187–188.
    A notice, taken from the New York Times,
    regarding the finding of teeth of mastodon at
    Sharon, Conn.

——————   1835 A
    Vertebral bone of a mastodon.
    Amer. Jour. Sci., xxvii, pp. 165–166.
    Found in Connecticut.

——————   1845 A
    Large skeleton of the *Zeuglodon* of
    Alabama.
    Amer. Jour. Sci., xlix, p. 218.
    Extract from Mobile Daily Advertiser of
    May 23, 1845. Refers to skeleton found by Dr.
    Koch.

——————   1845 B
    Fossil footmarks found in strata of
    the Carboniferous series in Westmore-
    land County, Pa.
    Amer. Jour. Sci., xlviii, pp. 217–218.

——————   1845 C
    Remarks [on paper by A. T. King].
    Amer. Jour. Sci., xlviii, p. 343.
    See King, A. T. 1845 A.

——————   1846 A
    The mastodon of Newburgh, N. Y.,
    discovered in August, 1845.
    Amer. Jour. Sci. (2), i, pp. 268–270.

American Journal of Science—Con-
    tinued.   1846 B
    Mastodon giganteus.
    Amer. Jour. Sci. (2), ii, pp. 131–133, with
    1 fig. in text.
    An unsigned editorial.

——————   1847 A
    Harlanus, a new genus of fossil pachy-
    derms.
    Amer. Jour. Sci. (2), iii, p. 125.
    Editorial note on paper by R. Owen, in Proc.
    Acad. Phila., iii, pp. 93–96. 1846. (Owen, R.
    1846 A.)

——————   1847 B
    Ornithichnites.
    Amer. Jour. Sci. (2), iii, p. 276.

——————   1852 A
    The *Mastodon giganteus* of North
    America, by John C. Warren, M. D.
    Amer. Jour. Sci. (2), xiv, p. 454.
    An editorial notice of Dr. Warren's book.

——————   1852 B
    On a fossil saurian of the New Red
    sandstone formation of Pennsylvania,
    with some account of that formation,
    etc., by Isaac Lea.
    Amer. Jour. Sci. (2), xiv, p. 451.
    Editorial notice of Lea's paper in Jour.
    Acad. Nat. Sci. Phila. (2), ii, pt. 3.

——————   1853 A
    Notice of the "*Mastodon giganteus*" of
    Dr. J. C. Warren.
    Amer. Jour. Sci., xv, pp. 367–373.
    An editorial notice of Dr. Warren's work on
    the mastodon.

——————   1853 B
    Fossil elephant.
    Amer. Jour. Sci. (2), xv, p. 146.
    Editorial note on occurrence of fossil ele-
    phant at Zanesville, Ohio.

——————   1871 A
    Notice of Cope's "Synopsis of the
    extinct *Batrachia* and *Reptilia* of North
    America."
    Amer. Jour. Sci. (3), i, pp. 220–221.

——————   1872 A
    On two new Ornithosaurians from
    Kansas, by Edward D. Cope.
    Amer. Jour. Sci. (3), iii, pp. 374–375.
    Editorial notice of Cope's paper with above
    title, in Amer. Phil. Soc., xii, p. 420.

——————   1873 A
    [Editorial remarks on Professor
    Cope's "Note on the Cretaceous of
    Wyoming," said to be published in
    Proc. Phil. Soc. Phila.]
    Amer. Jour. Sci. (3), v, p. 231.

American Journal of Science—Continued.   1875 A
   A new mastodon.
     Amer. Jour. Sci. (3), ix, p. 222.
     Editorial notice of announcement by Professor Cope.

—   1876 A
   Fifth annual report of the Geological Survey of Indiana, made during the year 1873.
     Amer. Jour. Sci. (5), xii, p. 307.
     Editorial notice.

—   1882 A
   Mastodons in New Jersey.
     Amer. Jour. Sci. (3), xxiv, pp. 294–295.
     Editorial notice of paper by Professor Lockwood.

American Naturalist.   1869 A
   The egg of the great auk.
     Amer. Naturalist, iii, p. 550.
     A note without signature.

—   1871 A
   Fossil whale in the Drift.
     Amer. Naturalist, v. p. 125.
     Extract from Nature.

—   1871 B
   Fossil walrus.
     Amer. Naturalist, v, p. 316.
     Notice of exhibition of skull of walrus by Professor Newberry.

Ami, Henry M.   1891 A
   On some extinct *Vertebrata* from the Miocene rocks of the north-west territories of Canada recently described by by Professor Cope.
     Science, xviii, p. 53.
     See also Ottawa Naturalist, v, pp. 74–77, for abstract.

—   1898 A
   The mastodon in western Ohio.
     Science (n. s.), vii, p. 80.

Ammon, Ludwig v.   1885 A
   Ueber *Homœosaurus maximiliani.*
     Abhandl. math.-phys. Classe k. bayerischen Akad. Wissensch., xv, pp. 499–528, with pls. i, ii.

—   1896 A
   Ueber neue Stücke von *Ischyodus.*
     Berichte naturw. Vereins Regensburg, 1894–95, Heft v, pp. 252–263, with pls. v, vi.

Anderson, John.   1860 A
   On Dura Den sandstone.
     Report Brit. Assoc. Adv. Sci., 29th meeting, Aberdeen, 1859, p. 97.

—   1861 A
   Report on the excavations in Dura Den.
     Report Brit. Assoc. Adv. Sci., 30th meeting, 1860, pp. 32–34.
     Treats of *Holoptychius*, etc.

Andreæ, A.   1893 A
   Acrosaurus frischmanni H. v. Meyer.
   Ein dem Wasserleben angepasster Rhynchocephale von Solenhofen.
     Bericht Senckenb. naturforsch. Gesellsch., 1893, pp. 21–34, with pls. i, ii.

—   1893 B
   Vorläufige Mittheilung über die Ganoiden (*Lepidosteus* und *Amia*) des Mainzer Beckens.
     Verhandl. naturhist.-medicin. Vereins Heidelberg, (n. s.), v, pp. 7–15, with 2 text figs.
     Contains remarks on North American genera *Lepidosteus, Clastes, Amia, Pappichthys.*

Andrews, Charles W.   1895 A
   Note on a specimen of *Keraterpeton galvani* Huxley, from Staffordshire.
     Geol. Mag. London (4), ii, pp. 81–84, with 1 text fig.

—   1895 B
   Note on a skeleton of a young plesiosaur from the Oxford clay of Peterborough.
     Geol. Mag. London (4), ii, pp. 241–243, with pl. ix and 1 text fig.

—   1895 C
   On the development of the shoulder-girdle of a plesiosaur (*Cryptoclidus oxoniensis,* Phillips sp.), from the Oxford clay.
     Ann. and Mag. Nat. Hist. (6), xv, pp. 333–346, with 4 woodcuts in the text.

—   1895 D
   On the structure of the skull in *Peloneustes philarchus,* a pliosaur from the Oxford clay.
     Ann. and Mag. Nat. Hist. (6), xvi, pp. 242–256, with pl. xiii and 3 figs. in the text.

—   · 1895 E
   The pectoral and pelvic girdles of *Murænosaurus plicatus.*
     Ann. and Mag. Nat. Hist. (6), xvi, pp. 429–434, with 3 figs. in the text.

—   1896 A
   On the structure of the plesiosaurian skull.
     Quart. Jour. Geol. Soc., lii, pp. 246–252, with pl. ix and 2 figs. in the text.

—   1897 A
   Note on the cast of the brain cavity of *Iguanodon.*
     Ann. and Mag. Nat. Hist. (6), xix, pp. 585–591, with pl. xvi and 1 text fig.

Andrews, Edmund.   1875 A
   Dr. Koch and the Missouri mastodon.
     Amer. Jour. Sci. (3), x, pp. 32–34.

Annan, Robert.   1793 A
Account of a large animal found near Hudson River.
Memoirs Amer. Acad. Arts and Sci. (1), ii, pp. 160–164.
Describes what appears to have been the skeleton of a mastodon.

Anonymous.   1834 A
Teeth of the elephant and mastodon.
Amer. Jour. Sci., xxv, pp. 257–258.
Found in Ohio.

——   1837 A
Fossil remains of the elephant, *Elephas primigenius.*
Amer. Jour. Sci., xxxii, pp. 377–379.

Atthey, Thomas.   1868 A
Notes on various species of *Ctenodus*, obtained from the shales of the Northumberland coal field.
Ann. and Mag. Nat. Hist. (4), vol. i, pp. 77–87.
Also in Nat. Hist. Trans. Northumberland and Durham, iii, 1868, pp. 54–66, pls. i–iii. Here the paper is credited to Hancock and Atthey.

——   1868 B
On the occurrence of palatal teeth belonging to the genus *Climaxodus* M'Coy in the Low Main shale of Newsham.
Ann. and Mag. Nat. Hist. (4), ii, pp. 321–323.
Also in Nat. Hist. Trans. Northumberland and Durham, iii, 1868, pp. 306–309.

——   1875 A
On the articular bone and supposed vomerine teeth of *Ctenodus obliquus*, and on *Palæoniscus hancocki*, n. sp., from the Low Main, Newsham, Northumberland.
Ann. and Mag. Nat. Hist. (4), vol. xv, pp. 309–312, pl. xix.

——   1876 A
On *Anthracosaurus russelli* (Huxley).
Ann. and Mag. Nat. Hist. (4), xviii, pp. 146–167, with pls. viii–xi.

——   1877 A
Notes on a paper by R. H. Traquair, M. D., F. G. S., F. R. S., on the structure of the lower jaw in *Rhizodopsis* and *Rhizodus.*
Ann. and Mag. Nat. Hist. (4), vol. xx, pp. 129–130.
See reply to this on p. 244, same volume.
—— See also **Hancock** and **Atthey.**

Atwater, Caleb.   1820 A
On some ancient human bones, etc., with a notice of the bones of the mastodon or mammoth, and of various shells found in Ohio and the West.
Amer. Jour. Sci. (1), ii, pp. 242–246, with pl. ii.

Auld, Robert C.   1892 A
As to the "extinction" of the American horse.
Science, xx, p. 135.

Ayres, Howard.   1893 A
On the genera of the Dipnoi *dipneumones.*
Amer. Naturalist, xxvii, pp. 919–932.

Baer, K. E. von.   1840 A
Untersuchungen über die ehemalige Verbreitung und gänzliche Vertilgung der von Steller beobachteten nordischen Seekuh (*Rytina*, Ill.).
Mém. Acad. Imp. Sci., St.-Pétersbourg (6), Sci. Nat., iii., pp. 53–80.

——   1866 A
Neue Auffindung eines vollständigen Mammuths mit der Haut und den Weichtheilen, im Eisboden Sibiriens, in der Nähe der Bucht des Tas.
Mélanges biolog. Bull. Acad. Imp. Sci. St. Pétersb., v, pp. 644–740, with 1 pl.

——   1867 A
Fortsetzung der Berichte über die Expedition zur Aufsuchung des angekündigten Mammuths.
Mélanges biolog. Bull. Acad. Imp. Sci. St. Pétersb., v, pp. 42–72, with pl.

Baikie, Balfour.   1857 A
On the skull of a manatus from Western Africa.
Proc. Zool. Soc. London, 1857, pp. 29–33, pl. ii.

Baird, Spencer F.   1850 A
On the bone caves of Pennsylvania.
Proc. Amer. Assoc. Adv. Sci., 2d meeting, Cambridge, 1849, pp. 352–355.

——   1857 A
Mammals of North America; the descriptions of species based chiefly on the collection in the museum of the Smithsonian Institution, with 87 pls. of original figs., all illustrating the genera and species and including details of external appearance and osteology.
Philadelphia, 1859, 4to: pp. i–xxxiv; 1–764, with pls. i–lxxxvii, and a number of text figs.

Balfour, F. M.   1881 A
On the development of the skeleton of the paired fins of *Elasmobranchii*, considered in relation to its bearings on the nature of the limbs of the *Vertebrata.*
Proc. Zool. Soc. London, 1881, pp. 652–671, with pls. lvii and lviii and 2 woodcuts in the text.

Balfour, F. M.—Continued. 1882 A
On the structure and development of
Lepidosteus.
Philos. Trans. Roy. Soc. London, clxxiii,
pp. 359-442, pls. 21-29.

Barbour, Edwin H. 1890 A
Remains of the primitive elephant
found in Grinnell, Iowa.
Science, xvi, p. 263.

—— 1890 B
Notes on the palæontological labora-
tory of the United States Geological
Survey under Professor Marsh.
Amer. Naturalist, xxiv, pp. 388-400, with 8
woodcuts in the text.

Barkas, T. P. 1868 A
On *Climaxodus*, or *Pœcilodus*; a pala-
tal tooth from the Low Main coal-shale,
Northumberland.
Geol. Magazine (1), v, pp. 495-497.

—— 1868 B
On the genus *Diplodus.*
Geol. Magazine (1), v, pp. 580-581.

—— 1873 A
Illustrated guide to the fish, amphibian, reptilian, and supposed mammalian remains of the Northumberland
Carboniferous strata.
London, 1873, 8vo., pp. i-viii, 9-117, with a
folio atlas of 10 pls. containing 248 figs.

—— 1881 A
Ctenoptychius, or Kammplatten.
Ann. and Mag. Nat. Hist. (5), viii, pp. 350-
354.

Barkas, W. J. 1874 A
On the microscopical structure of
fossil teeth from the Northumberland
Coal Measures.
Monthly Review of Dental Surgery, ii, pp.
297-300, 344-349, 386-398, 438-445, 482-488, 583-
589, with figs. i-xxxii.

—— 1877 A
On the genus *Ctenodus.*
Jour. and Proc. Roy. Soc. N. S. Wales, x, pp.
99-128, with pls. i-v.
Part I. On the genus *Ctenodus*, a fish found
in the true Coal Measures of Great Britain [pp.
99-109]. Part II. On the microscopic structure
of the mandibular and palatal teeth of *Ctenodus*
[pp. 110-115]. Part III. On the vomerine teeth
of *Ctenodus* [pp. 115-120]. Part IV. On the
dentary, articular, and pterygopalatine bones
of *Ctenodus* [pp. 120-123].

—— 1878 A
On the genus *Ctenodus.*
Jour. and Proc. Roy. Soc. N. S. Wales, xi,
pp. 51-64.
Part V. On the sphenoid cranial bones, oper-
culum, and supposed ear bones of *Ctenodus*
[pp. 51-57]. Part VI. On the scapula?, cora-
coid, ribs, and scales of *Ctenodus* [pp. 58-64].

Barkas, W. J.—Continued. 1878 B
Ctenacanthus, a spine of *Hybodus.*
Jour. and Proc. Roy. Soc. N. S. Wales, xi,
pp. 145-155.

—— 1878 C
On a dental peculiarity of the *Lepidosteidæ.*
Jour. and Proc. Roy. Soc. N. S. Wales, x, pp.
206-207.

Barnston, George. 1863 A
Remarks on the genus *Lutra*, and on
the species inhabiting America.
Canadian Nat. and Geol. (1), viii, pp. 147-
158, with 6 figs. in the text.

Barrett, L. 1859 A
On the atlas and axis of the *Plesio-
saurus.*
Report Brit. Assoc. Adv. Sci., 28th meeting,
Leeds, 1858, pp. 78-80.

Bartlett, John. 1846 A
Letter regarding the finding of zeu-
glodon bones at Natchez, Miss.
Proc. Nat. Hist. Soc. Boston, ii, p. 96.

Barton, Benjamin Smith. 1805 A
Letter to M. Lacépède, of Paris, on
the natural history of America.
Tilloch's Philos. Mag., xii, pp. 97-108; 204-
211.

Bassani, Francesco. 1899 A
La ittiofauna del calcare eocenico di
Gassino in Piemonte.
Atti Accad. Sci. Fis. e Mat. Napoli (2), ix, No.
13, pp. 1-41, with pls. i-iii.

Bate, C. Spence. 1867 A
On the dentition of the common mole
(*Talpa europæa*).
Ann. and Mag. Nat. Hist. (3), xix, pp. 377-
381, with pl. xi.

Baudelot, E. 1873 A
Recherches sur la structure et le développement des écailles des poissons
osseux.
Arch. de Zoologie Expérimentale et Générale, ii, pp. 429-480.

—— 1873 B
Observations sur la structure et le
développement des nageoires des poissons osseux.
Arch. de Zoologie Expérimentale et Générale, ii, pp. xviii-xxiv of "Notes et Revue."
Treats of the structure and development of
the fin rays.

Bauer, Franz. 1898 A
Die Ichthyosaurier des oberen weissen Jura.
Palæontographica, xliv, pp. 283-328, with
pls. xxv-xxvii.
Although describing only European species
this paper contains much of general interest
to students of this group.

Bauer, Franz—Continued.   **1900 A**
Osteologische Notizen über Ichthyosaurier.
Anat. Anzeiger, xviii, pp. 574-588, with 18 figs.
in the text.
Treats of the shoulder girdle, the pelvic girdle, and auditory bones.

Baur, George.   **1883 A**
Der Tarsus der Vögel und Dinosaurier.
Morpholog. Jahrbuch, viii, pp. 417-456, with
pls. xix and xx.
Author's copies of this paper were issued
in 1882 as an "Inaugural-Dissertation zur Er-
langung der Doctorwürde der philosopischen
Facultät der Universität München," pp. 1-44,
pls. xix and xx.

———   **1884 A**
Note on the pelvis in birds and dino-
saurs.
Amer. Naturalist, xviii, pp. 1273-1275.

———   **1884 B**
Dinosaurier und Vögel.   Eine Erwie-
derung an Herrn Prof. W. Dames in
Berlin.
Morpholog. Jahrbuch, x, pp. 446-454.

———   **1884 C**
Ueber das Centrale carpi der Säuge-
thiere.
Morpholog. Jahrbuch, x, pp. 455-457.

———   **1884 D**
Zur Morphologie des Tarsus der Säu-
gethiere.
Morpholog. Jahrbuch, x, pp. 458-461.

———   **1884 E**
Bemerkungen über das Becken der
Vögel und Dinosaurier.
Morpholog. Jahrbuch, x, pp. 613-616.

———   **1884 F**
Der Carpus der Paarhufer.
Morpholog. Jahrbuch, ix, pp. 597-603.

———   **1885 A**
A second phalanx in the third digit
of a carinate bird's wing.
Science, v, p. 355.

———   **1885 B**
A complete fibula in an adult living
carinate bird.
Science, v, p. 375.

———   **1885 C**
Zur Vögel-Dinosaurier-Frage.
Zoolog. Anzeiger, viii, pp. 441-443.

———   **1885 D**
Note on the sternal apparatus in
Iguanodon.
Zoolog. Anzeiger, viii, pp. 561-562.

Baur, George—Continued.   **1885 E**
Zur Morphologie des Carpus and Tar-
sus der Reptilien.
Zoolog. Anzeiger, viii, pp. 631-638.

———   **1885 F**
Ueber das Archipterygium und die
Entwicklung des Cheiropterygium aus
dem Ichthyopterygium.
Zoolog. Anzeiger, viii, pp. 663-666.

———   **1885 G**
On the morphology of the tarsus in
the mammals.
Amer. Naturalist, xix, pp. 86-88.

———   **1885 H**
On the centrale carpi of the mam-
mals.
Amer. Naturalist, xix, pp. 195-196.

———   **1885 I**
The trapezium of the *Camelidæ*.
Amer. Naturalist, xix, pp. 196-197.
See also Morphologisches Jahrbuch, xi, pp.
117-118.

———   **1885 J**
On the morphology of the carpus
and tarsus of vertebrates.
Amer. Naturalist, xix, pp. 718-720.
Contains a "morphological table" and a
list of papers on the carpus and tarsus.

———   **1885 K**
Bemerkungen über den "Astragalus"
und das "Intermedium tarsi" der
Säugethiere.
Morpholog. Jahrbuch, xi, pp. 468-483, with
pl. xxvii.
Cites the literature of the subject and gives
the results of the author's own observations on
the different groups of mammals.

———   **1886 A**
Ueber das Quadratum der Säugetiere.
Biolog. Centralb., vi, pp. 648-658.
Also in Sitzungsber. Gesellsch. Morphol. und
Physiol. München, 1886, pp. 45-57.

———   **1886 B**
Ueber die Morphogonie der Wirbel-
säule der Amnioten.
Biolog. Centralb., vi, pp. 332-342; 353-363.

———   **1886 C**
Der älteste Tarsus (*Archegosaurus*).
Zoolog. Anzeiger, ix, pp. 104-106.
Compare Baur, G., 1886 I.

———   **1886 D**
W. K. Parker's Bemerkungen über
Archæopteryx, 1864, und eine Zusam-
menstellung der hauptsächlichen Lit-
teratur über diesen Vogel.
Zoolog. Anzeiger, ix, pp. 106-109.

Baur, George—Continued.   1886 E

Die zwei Centralia im Carpus von
Sphenodon (*Hatteria*) und die Wirbel
von *Sphenodon* und *Gecko verticillatus*
Laur. (*G. verus* Gray).
Zoolog. Anzeiger, ix, pp. 188–190.

———   1886 F

Bemerkungen über *Sauropterygia* und
Ichthyopterygia.
Zoolog. Anzeiger, ix, pp. 245–252; 823.

———   1886 G

Osteologische Notizen über Reptilien.
Zoolog. Anzeiger, ix, pp. 685–690; 733–743.

———   1886 H

Ueber die Homologien einiger Schä-
delknochen der Stegocephalen und
Reptilien.
Anatom. Anzeiger, i, pp. 348–350.
See correction in Anat. Anzeiger, vol. ii, p.
657.

———   1886 I

The oldest tarsus (*Archegosaurus*).
Amer. Naturalist, xx, pp. 173–174.
Compare Baur, G., 1886 C.

———   1886 J

The proatlas, atlas, and axis of the
Crocodilia.
Amer. Naturalist, xx, pp. 288–293, with 5
illus.

———   1886 K

The intercentrum in *Sphenodon* (*Hatteria*).
Amer. Naturalist, xx, pp. 465–466.

———   1886 L

The ribs of *Sphenodon* (*Hatteria*).
Amer. Naturalist, xx, pp. 979–981.

———   1886 M

Ueber die Kanäle im Humerus der
Amnioten.
Morpholog. Jahrbuch, xii, pp. 299–308.

———   1887 A

Ueber die Abstammung der amnioten
Wirbeltiere.
Biolog. Centralb., vii, pp. 481–493.
Also in Sitzungsbericht der Gesellschaft für
Morphologie und Physiologie, München, 1887,
pp. 44–61.

———   1887 B

Osteologische Notizen über Reptilien.
Fortsetzung ii.
Zoolog. Anzeiger, x, pp. 96–102.

———   1887 C

On the phylogenetic arrangement of
the *Sauropsida*.
Jour. Morphology, i, pp. 93–104.

Baur, George—Continued.   1887 D

On the morphology and origin of the
Ichthyopterygia.
Amer. Naturalist, xxi, pp. 837–840.
Noticed in Geol. Magazine, 1888, v, p. 325.

———   1887 E

On the morphogeny of the carapace
of the *Testudinata*.
Amer. Naturalist, xxi, p. 89.

———   1887 F

Ueber die Ursprung der Extremita-
ten der *Ichthyopterygia*.
Bericht über die xx Versammlung des
oberrhein. geolog. Vereins, pp. 1–4, with 1 plate.
Pagination taken from an author's separate;
the original "Bericht" not seen.

———   1888 A

Unusual dermal ossifications.
Science, xi, p. 144.
Discusses the origin of carapace and plastron
of *Testudinata*.

———   1888 B

Osteologische Notizen über Reptilien.
Fortsetzungen iii, iv, und v.
Zoolog. Anzeiger, xi, pp. 417–424; 592–597;
736–740.

———   1888 C

Beiträge zur Morphogenie des Carpus und Tarsus der Vertebraten. I.
Theil. *Batrachia*. Jena; Verlag von
Gustav Fischer, 1888.
8vo, pp. 1–88, with pls. i–iii.
Appended is a list of books, memoirs, etc.,
80 in number, which treat of the subject discussed by the author.

———   1889 A

Die systematische Stellung von *Dermochelys* Blainv.
Biolog. Centralb., ix, pp. 149–153; 180–191;
618–619.

———   1889 B

Osteologische Notizen über Reptilien.
Fortsetzung vi.
Zoolog. Anzeiger, xii, pp. 40–47.

———   1889 C

Revision meiner Mittheilungen im
Zoologischer Anzeiger, mit Nachträgen.
Zoolog. Anzeiger, xii, pp. 238–243.

———   1889 D

Neue Beiträge zur Morphologie des
Carpus der Säugetiere.
Anatom. Anzeiger, iv, pp. 49–51, with 3 figs.
in the text.

———   1889 E

On the morphology of the ribs and
the fate of the actinosts of the median
fins in fishes.
Jour. Morphology, iii, pp. 463–466.

Baur, George—Continued.   1889 F
On the morphology of the vertebrate skull. I. The otic elements. II. The temporal arches.
Jour. Morphology, iii, pp. 467–474.

——   1889 G
Mr. E. T. Newton on *Pterosauria.*
Geol. Magazine (3), vi, pp. 171–174.
A review of Newton's paper in Phil. Trans. Roy. Soc., vol. 179, p. 303 (1888), on *Scaphognathus.*

——   1889 H
The systematic position of *Meiolania,* Owen.
Ann. and Mag. Nat. Hist. (6), iii, pp. 54–62.

——   1889 I
On "*Aulacochelys,*" Lydekker, and the systematic position of *Anosteira,* Leidy, and *Pseudotrionyx,* Dollo.
Ann. and Mag. Nat. Hist. (6), iii, pp. 273–276.

——   1889 J
On *Meiolania* and some points in the osteology of the *Testudinata;* a reply to Mr. G. A. Boulenger.
Ann. and Mag. Nat. Hist. (6), iv, pp. 37–45, with pl. vi, and 4 figs. in the text.

——   1889 K
Bemerkungen über den Carpus der Proboscidier und der Ungulaten im Allgemeinen.
Morpholog. Jahrbuch, xv, pp. 478–482, with 1 fig. in the text.

——   1890 A
On the characters and systematic position of the large sea lizards, *Mosasauridæ.*
Science, xvi, p. 262.

——   1890 B
The problems of comparative osteology.
Science, xvi, pp. 281–282.

——   1890 C
A review of the charges against the palæontological department of the U. S. Geological Survey and of the defense made by Prof. O. C. Marsh.
Amer. Naturalist, xxiv, pp. 288–304.

——   1890 D
On the classification of the *Testudinata.*
Amer. Naturalist, xxiv, pp. 580–536.

——   1890 E
Professor Marsh on *Hallopus* and other dinosaurs.
Amer. Naturalist, xxiv, pp. 569–571.
Followed by short note by Prof. E. D. Cope.

Baur, George—Continued.   1890 F
The genera of the *Cheloniidæ.*
Amer. Naturalist, xxiv, pp. 486–487.

——   1891 A
The horned saurians of the Laramie formation.
Science, xvii, pp. 216–217.

——   1891 B
Notes on some little-known American fossil tortoises.
Proc. Acad. Nat. Sci. Phila., 1891, pp. 411–430.

——   1891 C
The pelvis of the *Testudinata,* with notes on the evolution of the pelvis in general.
Jour. Morphology, iv, pp. 345–359, with 13 figs. in the text.

——   1891 D
Remarks on reptiles generally called *Dinosauria.*
Amer. Naturalist, xxv, pp. 434–454.
Contains a list of 32 papers cited. A review of this paper by Dames is found in Neues Jahrb. Min., etc., 1893, i, Ref., pp. 545–547.

——   1892 A
Bemerkungen über verschiedene Arten von Schildkröten.
Zoolog. Anzeiger, xv, pp. 155–159.

——   1892 B
On the morphology of the skull in the *Mosasauridæ.*
Jour. Morphology, vii, pp. 1–22, with pls. i and ii.

——   1892 C
On the taxonomy of the genus *Emys.*
Proc. Amer. Philos. Soc., xxx, pp. 40–44; 245.
In the "addition" to his note, on page 245, Dr. Baur made some modifications of his views as expressed on pp. 40–44.

——   1893 A
Notes on the classification and taxonomy of the *Testudinata.*
Proc. Amer. Philos. Soc., xxxi, pp. 210–225.

——   1893 B
Notes on the classification of the *Cryptodira.*
Amer. Naturalist, xxvii, pp. 672–675.

——   1893 C
The discovery of Miocene Amphisbænians.
Amer. Naturalist, xxvii, pp. 998–999.
Describes new genus and species *Hyporhina antiqua;* new family *Hyporhinidæ.*

——   1894 A
Bemerkungen über die Osteologie der Schläfengegend der höheren Wirbeltiere.
Anatom. Anzeiger, x, pp. 315–330.
Includes a list of the titles of 24 papers on the subject discussed.

Baur, George—Continued. 1895 A
Cope on the temporal part of the
skull, and on the systematic position of
the *Mosasauridæ*. A reply.
Amer. Naturalist, xxix, pp. 998–1002.

———— 1895 B
The fins of *Ichthyosaurus*.
Jour. Geology, iii, pp. 238–240.
A review of Prof. Rich. Owen's paper in the
Trans. Geol. Soc., ser. 2, vol. v, pp. 511–514.

———— 1895 C
Die Palatingegend der *Ichthyosauria*.
Anatom. Anzeiger, x, pp. 456–459, with 1 fig.

———— 1895 D
Ueber die Morphologie des Unterkiefers der Reptilien.
Anatom. Anzeiger, xi, pp. 410–415, with 4
figs. Nachtrag, p. 569.
Announces the finding of a hitherto unrec-
ognized bone in the lower jaw of reptiles, the
præspleniale.

———— 1895 E
Das Gebiss von *Sphenodon* (*Hatteria*)
und einige Bemerkungen über Prof.
Rud. Burckhardt's Arbeit über das
Gebiss der *Sauropsidæ*.
Anatom. Anzeiger, xi, pp. 436–439.

———— 1896 A
The paroccipital of the *Squamata* and
the affinities of the *Mosasauridæ* once
more. A rejoinder to Prof. E. D. Cope.
Amer. Naturalist, xxx, pp. 143–147, with pl.
iv.
See also pp. 327–329; 411–412.

———— 1896 B
Bemerkungen über die Phylogenie
der Schildkröten.
Anatom. Anzeiger, xii, pp. 561–570.

———— 1896 C
The *Stegocephali*: A phylogenetic
study.
Anatom. Anzeiger, xi, pp. 657–673, with 8
figs.
Contains description of the osteology of
this group and the titles of 34 papers and
books relating to the subject.

———— 1897 A
Archegosaurus.
Amer. Naturalist, xxxi, pp. 975–980.
A review of Otto Jaekel's paper on "Die Organization von *Archegosaurus*," and a discussion of the composition of the vertebral
column.

———— 1897 B
Ueber die systematische Stellung der
Microsaurier.
Anatom. Anzeiger, xiv, pp. 148–151.

Baur, G., and Case, E. C. 1897 A
On the morphology of the skull of
the *Pelycosauria* and the origin of the
mammals.
Anatom. Anzeiger, xiii, pp. 109–120, with 3
figs.
Abstract in Science (2), v, pp. 592–593.

———— 1899 A
The history of the *Pelycosauria*, with
a description of the genus *Dimetrodon*.
Trans. Amer. Philos. Soc. (2), xx, pp. 1–58,
with pls. i–iii, with 7 figs. in the text.
Appended is a list of titles of works on the
subject, 70 in number.

Bayer, Franz. 1885 A
Ueber die Extremitäten einer jungen
Hatteria.
Sitzungsber. k. Akad, Wissensch., Wien, xc,
i Abth., pp. 237–245, with 1 pl.

Bayle, Émile de. 1855 A
Notice sur le système dentaire de
l' *Anthracotherium magnum*.
Bull. Soc. Geol. France (2), xii, pp. 936–947,
with pl. xxii.

Beard, J. 1890 A
The interrelationships of the *Ichthy-
opsida*.
Anatom. Anzeiger, v, pp. 146–159, 179–188.

Beddard, Frank E. 1888 A
Note on the systematic position of
Monitor.
Anatom. Anzeiger iii, pp. 204–206.

———— 1897 A
Note upon intercentra in the vertebral column of birds.
Proc. Zool. Soc. London, 1897, pp. 465–472,
with 4 figs. in the text.

Belcher, Edward. 1856 A
Notice of the discovery of *Ichthyo-
saurus* and other fossils in the late Arc-
tic exploring expedition, 1852–1854.
Report Brit. Assoc. Adv. Sci., 25th meeting,
Glasgow, 1855, p. 79.
Found, latitude 77° 16' N., longitude, 96° W.

Bell, Robert. 1898 A
On the occurrence of mammoth and
mastodon remains around Hudson Bay.
Bull. Geol. Soc. America, ix, pp. 369–390,
with 1 fig. in the text.
See also Science vii (n. s.), p. 80.

Bemmelen, J. F. van. 1896 A
Bemerkungen zur Phylogenie der
Schildkröten.
Compte-rendu des séances du troisième Con-
grès Internat. de Zoologie, Leyden, 1895, pp.
322–335.

Bemmelen, J. F. van—Cont'd.   1896 B
Bemerkungen über den Schädelbau
von *Dermochelys coriacea.*
Festschrift zum 70sten Geburtstage von Carl
Gegenbaur, ii, pp. 279-285, with pl. i.

———   1899 A
On reptilian affinities in the temporal
region of the monotreme skull.
Proc. Fourth Internat. Congress of Zoology,
Cambridge, England, pp. 162-164.

Bensley, B. A.   1900 A
On the inflection of the angle of the
jaw in the *Marsupialia.*
Science (2), xii, pp. 558-559.

Bensted, W. H.   1861 A
Remains of American *"Missourium"*
associated with flint implements.
The Geologist, iv, pp. 217-221. See also p.
262.
Letter containing extract from Albert
Koch's "Description of the *Missourium*, or
Missouri Leviathan, together with its supposed
habits and Indian tradition." London, E.
Fisher, 1841.

Béraneck, Ed.   1887 A
Ueber das Parietalauge der Reptilien.
Jenaische Zeitschrift, xxi (xiv), pp. 374-
410, with pp. 374-410.

Berthold, Arnold A.   1850 A
Ueber das Backenzahnsystem des
Narwals.
Archiv. f. Anat., Physiol. u. wissensch.
Med., Jahrg. 1850, pp. 386-392, pl. x, figs. 7, 8.

Bettany, G. T.   1873 A
Oreodon remains in the Woodwardian
Museum, Cambridge.
Nature, viii, pp. 309-311.
See also Nature, viii, p. 385.

———   1876 A
On the genus *Merycochœrus* (family
Oreodontidæ), with descriptions of two
new species.
Quart. Jour. Geol. Soc., xxxii, pp. 259-273,
with pls. xvii, xviii.

Beyer, Samuel Walker.   1899 A
Geology of Story County.
Iowa Geol. Survey, ix, pp. 157-239.
Refers to the finding of a tooth of *Elephas
primigenius* at Polk City.

Beyrich, E.   1848 A
Ueber *Xenacanthus decheni* und *Hola-
canthus gracilis,* zwei Fische aus der
Formation des Rothliegenden in Nord
Deutschland.
Bericht Verhandl. k. preuss. Akad. Wis-
sensch. zu Berlin, 1848, pp. 24-33.

Beyrich, E.—Continued.   1850 A
Ueber einige organische Reste der
Lettenkohlenbildung in Thuringen.
Zeitschr. deutsch. geol. Gesellsch., ii, pp.
153-168.

———   1877 A
Ueber einen *Pterichthys* von Gerol-
stein.
Zeitschr. deutsch. geol. Gesellsch., xxix, pp.
751-756, and pl. x.

Bibbins, Arthur.   1895 A
Notes on the palæontology of the
Potomac formation.
Johns Hopkins University Circulars, xv,
No. 121, pp. 17-20, with 2 figs.
Mentions the finding of teeth of certain
dinosaurs.

Bienz, Amié.   1895 A
Dermatemys mavii Gray; eine osteolo-
gische Studie mit Beiträgen zur Kennt-
niss vom Baue der Schildkröten.
Revue suisse de zool. et Ann. Musée d' Hist.
Nat. Genève, iii, pp. 61-135, with pls. ii, iii.
The literature of the subject is presented,
p. 135.

Billings, E.   1863 A
On the remains of the fossil elephant
in Canada.
Canadian Nat. and Geol. (1), viii, pp. 135-
147, with 5 figs. in the text.

Bischoff, Th. Ludwig Wilhelm.   1840 A
Lepidosiren paradoxa. Anatomisch
untersucht und beschrieben.
Leipzig, 1840, 4 to, pp. 1-vi; 1-34, with pls.
i-vii.

Blackmore, Humphrey P.   1874 A
On fossil *Arvicolidæ.*
Proc. Zool. Soc. London, 1874, pp. 460-471,
with 3 figs. of the teeth.

Blainville, H. M. Ducrotay de.   1839 A
Recherches sur l'ancienneté des
Édentés terrestres à la surface de la
terre.
Comptes rendus de l'Acad. Sci. Paris, viii,
pp. 65-69.
See also Ann. Sci. Nat. (2), xi, pp. 113-122.

———   1839 B
Mémoire sur les traces qu'ont lais-
sées à la surface de la terre les Édentés
terrestres.
Comptes rendus de l'Acad. Sci. Paris, viii,
pp. 139-146.

———   1839 C
Recherches sur l'ancienneté des
Édentés terrestres à la su face de la
terre.
Ann. Sci. Naturelles (2), xi, pp. 113-122.

Blainville, H. M. Ducrotay de—Continued.　　　　　　　　1864 A

Ostéographie ou description iconographique comparée du squelette et du système dentaire des mammifères récent et fossiles pour servir de base à la zoologie et à la géologie.

Vols. i-iv, 4to, with pls. vols. i-iv.

Vol. i, *Primates, Chiroptera, Insectivora;* vol. ii, *Carnivora;* vol. iii, *Gravigrada, Ungulagrada;* vol. iv, *Ungulagrada, Bradypus.*

This important work was issued in livraisons from 1839 to 1864. Hence, in the citations found in the catalogue of species, this year, 1864, must be regarded as arbitrary.

Blake, Chas. C.　　　　　　　　1862 A

On a fossil elephant from Texas (*E. texianus*).

The Geologist, v, pp. 57-58, with pl. iv.

—　　　　　　　　1862 B

Elephas texianus v. *columbi.*

The Geologist, vi, pp. 56-60, with 1 text fig.

—　　　　　　　　1863 A

On the geological evidences of horses in the New World.

The Geologist, vii, pp. 24-28.

Blake, William P.　　　　　　　　1855 A

Remains of the mammoth and mastodon in California.

Amer. Jour. Sci. (2), xix, p. 133.

—　　　　　　　　1864 A

Note on fossil remains of the horse and elephant, mingled, at Mare Island, San Francisco Bay.

Proc. California Acad. Nat. Sci., iii, p. 166.
The species not determined.

—　　　　　　　　1867 A

Fossil fish in Great Basin, Nevada.

Proc. California Acad. Nat. Sci., iii, pp. 306-307.
No genera or species designated.

—　　　　　　　　1867 B

Notice of fossil elephants' teeth from the northwest coast.

Proc. California Acad. Nat. Sci., iii, p. 325.
Calls attention to the finding of teeth and tusks of elephants in Alaska; also gives localities in California where such remains have been discovered.

—　　　　　　　　1868 A

Note upon the occurrence of fossil remains of the tapir in California.

Amer. Jour. Sci. (2), xlv, p. 381.

—　　　　　　　　1870 A

On a fossil tooth from Table Mountain.

Amer. Jour. Sci. (2), l, pp. 262-263.

Blake, William P.—Continued.　　　1898 A

Remains of a species of *Bos* in the Quaternary of Arizona.

Amer. Geologist, xxii, pp. 65-72.

—　　　　　　　　1898 B

Bison latifrons and *B. arizonica.*

Amer. Geologist, xxii, pp. 247-248.

—　　　　　　　　1900 A

Remains of the mammoth in Arizona.

Amer. Geologist, xxvi, p. 257.
Records occurrence of *Elephas primigenius*(?) at Yuma.

Blanchard, Émile.　　　　　　　　1857 A

De la détermination de quelques oiseaux fossiles et des caractères ostéologiques des Gallinacés ou Gallides.

Annales Sci. Naturelles (4), Zool., vii, pp. 91-106, with pls. x-xii.

—　　　　　　　　1859 A

Recherches sur les caractères ostéologiques des oiseaux appliquées a la classification naturelle de ces animaux.

Annales Sci. Naturelles (4), Zool., pp. 11-145.

—　　　　　　　　1860 A

Observations sur le système dentaire chez les oiseaux.

Comptes rend. des séance de l'Acad. sciences, Paris, l, pp. 540-542.

Blasius, Wilhelm.　　　　　　　　1884 A

Zur Geschichte der Ueberreste von *Alca impennis* Linn.

Jour. für Ornithologie, xxxii, pp. 58-176.

Blatchley, W. S.　　　　　　　　1898 A

The geology of Lake and Porter counties, Indiana.

22d Ann. Report Depart. Geol. and Natural Resources [of Indiana] for 1897, pp. 25-104, with 9 figs. in the text.
Contains notes on the finding of the mammoth and the mastodon.

Bloch, Leopold.　　　　　　　　1900 A

Schwimmblase, Knochenkapsel und Weber'scher Apparat von *Nemachilus barbatulus* Günther.

Jenaische Zeitschr. Naturwiss., xxxiv, pp. 1-64, with pls. i, ii, and 12 figs. in the text.
Appended is a list of 75 memoirs and papers on the subject.

Blyth, E.　　　　　　　　1837 A

On the osteology of the great auk (*Alca impennis*).

Proc. Zool. Soc. London, 1837, pp. 122-123.

Boas, J. E. V.　　　　　　　　1884 A

Bemerkungen über die Polydactylie des Pferdes.

Morpholog. Jahrbuch, x, pp. 182-184.

Boas, J. E. V.—Continued.   1890 A
Ueber den Metatarsus der Wiederkäuer.
Morpholog. Jahrbuch, xvi, pp. 525–523, with
6 figs. in the text.

Bojanus, L. H.   1827 A
De uro nostrate eiusque sceleto com-
mentatio.
Nova Acta Acad. Cæs. Leop.-Car., xiii, pt.
ii, pp. 413–478, with 5 pls.
Refers (p. 427) *Bison latifrons* to *Urus priscus.*

Born, G.   1876 A
Zum Carpus und Tarsus der Saurier.
Morpholog. Jahrbuch, ii, pp. 1–26, with pl. i.

———   1880 A
Nachträge zu "Carpus und Tarsus."
Morpholog. Jahrbuch, vi, pp. 49–78, with pl. i.
Appended is a list of authors and works
referred to, 14 in number.

Boule, Marcellin.   1899 A
Observations sur quelques équidés
fossiles.
Bull. Soc. Geol., France (3), xxvii, pp. 531–
542, with 22 figs. in the text.

Boulenger, G. A.   1882 A
Catalogue of the *Batrachia Gradientia*
s. *Caudata* and *Batrachia Apoda* in the
collection of the British Museum; second edition.
Pp. i–viii; 1–127, with pls. i–ix.

———   1882 B
Catalogue of the *Batrachia Salientia*
s. *Ecaudata* in the collection of the British Museum; second edition.
Pp. i–xvi; 1–503, with pls. i–xxx, with numerous figs. in the text.

———   1884 A
Synopsis of the families of existing
Lacertilia.
Ann. and Mag. Nat. Hist. (5), xiv, pp. 117–
122.

———   1885 A
Catalogue of the lizards in the British
Museum.
Vol. i, pp. 436, pls. 32. Vol ii, pp. 497, pls.
24. London, 1885.

———   1887 A
Catalogue of the lizards in the British
Museum.
Vol. iii, pp. 575; pls. 40. London, 1887.

———   1888 A
Remarks on a note by Dr. G. Baur on
the Pleurodiran chelonians.
Ann. and Mag. Nat. Hist. (6), ii, pp. 352–354.
Dr. Baur's note occurs in Zoolog. Anzeig.,
Aug., 1888.

Boulenger, G. A.—Continued.   1889 A
Catalogue of the chelonians, rhyn-
chocephalians, and crocodiles in the
British Museum.
London, 1889, pp. 311, 6 pls.

———   1889 B
Remarks in reply to Dr. Baur's article on the systematic position of *Miola-
nia.*
Ann. and Mag. Nat. Hist. (6), iii, pp. 138–141,
with figs. in the text.

———   1891 A
Notes on the osteology of *Heloderma
horridum* and *H. suspectum*, with remarks on the systematic position of the
Helodermatidæ and on the vertebræ of
the *Lacertilia.*
Proc. Zool. Soc. London, 1891, pp. 109–118,
with 6 woodcuts.

———   1891 B
On British remains of *Homœosaurus*,
with remarks on the classification of the
Rhynchocephalia.
Proc. Zool. Soc. London, 1891, pp. 167–172,
with 2 woodcuts.

———   1893 A
Catalogue of the snakes in the British
Museum (Natural History). Vol. i,
containing the families *Typhlopidæ,
Glauconiidæ, Boidæ, Ilysiidæ, Uropeltidæ,
Xenopeltidæ,* and *Colubridæ Aglyphæ,*
part.
Pp. i–xiii; 1–448, with pls. i–xxviii and 26
figs. in the text.

———   1894 A
Catalogue of the snakes in the British
Museum (Natural History). Vol. ii,
containing the conclusion of the *Colubridæ Aglyphæ.*
Pp. i–xi; 1–277, with pls. i–xx, and 25 figs. in
the text.

———   1895 A
Remarks on the value of certain cranial characters employed by Professor
Cope for distinguishing lizards from
snakes.
Ann. and Mag. Nat. Hist. (6), xvi, pp. 366–
367, with 1 fig. in the text.

———   1896 A
Remarks on the dentition of snakes
and on the evolution of the poison fangs.
Proc. Zool. Soc. London, 1896, pp. 614–616.

Boulenger, G. A.—Continued. 1896 B
On the occurrence of Schlegel's gavial
(*Tomistoma schlegelii*) in the Malay pen-
insula, with remarks on the atlas and
axis of the crocodilians.
Proc. Zool. Soc. London, 1896, pp. 628–633,
with 3 figs. in the text.

—— 1896 C
Catalogue of the snakes in the British
Museum (Natural History) Vol. iii,
containing the *Colubridæ* (*Opisthoglyphæ*
and *Proteroglyphæ*), *Amblycephalidæ*,
and *Viperidæ*.
Pp. i–xiv; 1–727, with 37 figs. in the text.

Bouvé, T. T. 1859 A
Statement with regard to *Zeuglodon
cetoides*.
Proc. Nat. Hist. Soc. Boston, vi, pp. 421–422.

—— 1859 B
Communication regarding priority of
describing fossil footprints of the Connecticut Valley.
Proc. Nat. Hist. Soc. Boston, vii, pp. 49–53.

Bowerbank, J. S. 1848 A
Microscopical examination on the
structure of the bones of *Pterodactylus
giganteus* and other fossil animals.
Quart. Jour. Geol. Soc. Lond., iv, pp. 2–10,
with pls. i, ii.

—— 1851 A
On the pterodactyles of the Chalk
formation.
Proc. Zool. Soc. Lond., 1851, pt. xix, pp. 14–
20, with pl. iv.

—— 1852 A
On the probable dimensions of the
great shark (*Carcharias megalodon*) of
the Red Crag.
Report Brit. Assoc. Adv. Sci., 21st meeting,
Ipswich, 1851, p. 54.

—— 1852 B
On the probable dimensions of *Car-
charodon megalodon* from the Crag.
Ann. and Mag. Nat. Hist. (2), ix, pp. 120–123.

Bowers, Stephen. 1889 A
An abstract of a letter regarding the
finding of the remains of fossil vertebrates in Ventura, Cal.
Amer. Geologist, iv, pp. 391–392.

Boyd, C. H. 1881 A
Remains of the walrus (?) in Maine.
Proc. U. S. Nat. Mus., iv, pp. 284–285.

Brainerd, ——. 1852 A
Fossil fishes.
Annals of Science (Cleveland, Ohio), i, pp.
18–20, with 3 text figs.
No specific name is applied to the fishes
described. They are now known as *Gonatodus
brainerdi* (Newb.).

Branco, W. 1885 A
[Review of G. Baur on "Dinosau-
rier und Vögel" and W. Dames on
"Entgegnung an Herrn Dr. Baur"].
Neues Jahrbuch Min. Geol. u. Palæont, 1885,
ii, Referate, pp. 437–441.

—— 1887 A
Beiträge zur Kenntniss der Gattung
Lepidotus.
Abhandl. zur geolog. Specialkarte von
Preussen und der Thüringischen Staaten vii.
Heft 4. pp. 1–85, with pls. i–vii.

—— 1897 A
Die menschenähnlichen Zähne aus
dem Bohnerz der schwäbischen Alb.
Teil ii, Art und Ursache der Reduk-
tion des Gebisses bei Säugern.
Stuttgart, 1897, pp. 1–128.

Brandt, Alexander. 1867 A
Kurze Bemerkungen über aufrecht-
stehende Mammuthleichen.
Bull. Soc. Imp. des Naturalistes de Moscou,
xl, pt. 2, pp. 241–256.

Brandt, Ed. 1868 A
Untersuchungen über das Gebiss der
Spitzmäuse (*Sorex* Cuv.).
Bull. Soc. Imp. Nat. Moscou, xli. pt. 2, pp.
76–95, with pls. i–vi.

—— 1870 A
Untersuchungen über das Gebiss der
Spitzmäuse (*Sorex* Cuv.).
Bull. Soc. Imp. Nat. Moscou, xliii, pt. 2, pp.
1–40.

—— 1888 A
Vergleichend - anatomische Untersuchungen über die Griffelbeine (Ossa
calamiformia) der Wiederkäuer (*Ruminantia*).
Zoolog. Anzeiger xi, pp. 542–548.

Brandt, Johann Friedrich. 1840 A
Beiträge zur Kenntniss der Natur-
geschichte der Vögel mit besonderer
Beziehung auf Skeletbau und verglei-
chende Zoologie.
Mém. Acad. Imp. Sci. St. Pétersbourg (6), Sci.
Nat., iii, pp. 81–237, with pls. i–xviii.

Brandt, Johann Friedrich—C't'd. 1844 A
Notiz über die fossilen Knochen des
Cetotheriums.
Verhandl. d. kais.-russ. mineralog. Gesellsch. St. Petersburg, 1844, pp. 240-244.

—   1848 A
On the position in which the mammoth and rhinoceros have been found in Siberia; from a letter to Baron A. von Humboldt.
Quart. Jour. Geol. Soc., iv, pp. 9-12. (Translations and notices.)

—   1849 A
Symbolæ Sirenologicæ, quibus præcipue *Rhytinæ* historia naturalis illustratur.
Mém. Acad. St. Pétersbourg (6), Sci. Nat., v.; Zool. et Physiol., pp. 1-160, with pls. i-v.

—   1849 B
De *Rhinocerotis antiquitatis* seu *pallasii* structura externa et osteologica observationes e reliquiis quæ in museis Petropolitanis servantur erutæ.
Mém. Acad. Sci. St. Pétersbourg (6), v, pt. ii, pp. 161-416, with pls. i-xxiv.

—   1855 A
Beitrage zur nähern Kenntniss der Säugethiere Russlands.
Mém. Acad. Imp. Sci. St. Pétersbourg (6), vii, pp. 1-365, with pls. i-iii and i-xi.
Abhandlung 3 considers the question of the specific identity of the beaver of America with that of the Old World; also the variation of certain bones of the beaver's skull. Abhandlung 4 treats of the gradual advances in the classification of the rodents, with especial reference to the beaver. Abhandlung 5 deals with stages of craniological development of the rodents and the relationships and the classification to be determined therefrom, with especial reference to the beaver.

—   1862 A
Bemerkungen über die Zahl der Halswirbel der Sirenien.
Mélanges Biol. Bull. Acad Imp. Sci. St. Pétersb., iv, pp. 125-128.

—   1862 B
Einige Wortè über die verschiedenen Entwicklungsstufen der Nasenbeine der Seekühe (*Sirenia*).
Mélanges Biol. Bull. Acad. Imp. Sci. St. Pétersb., iv, pp. 129-132.

—   1862 C
Bemerkungen über die Verbreitung und Vertilgung der *Rhytina*.
Mélanges Biol. Bull. Acad. Imp. Sci. St. Pétersb., iv, pp. 259-268.

Brandt, Johann Friedrich—C't'd. 1866 A
Mittheilungen über die Gestalt und Unterscheinungsmerkmale des Mammuth oder Mamont (*Elephas primigenius*).
Mélanges Biol. Bull. Acad. Imp. Sci. St. Pétersb., v, pp. 567-605, with 1 pl. Also "Ergansungen," pp. 640-644.

—   1866 B
Zur Lebensgeschichte des Mammuth.
Bull. Acad. Imp. Sci. St. Pétersb., x, pp. 111-118.

—   1866 C
Nochmaliger Nachweis der Vertilgung der nordischen oder Steller'schen Seekuh (*Rhytina borealis*).
Bull. Soc. Imp. des Naturalistes de Moscou, xxxix, pp. 572-597.

—   1867 A
Ergänzende Mittheilungen zur Erläuterung der ehemaligen Verbreitung und Vertilgung der Steller'schen Seekuh.
Mélanges Biol. Bull. Acad. Imp. Sci. St. Pétersb., vi, pp. 223-232.
Contains an account of the literature of the subject.

—   1867 B
Untersuchungen über die geographische Verbreitung des Renthiers (*Cervus tarandus* Linn.) in Bezug auf die Würdigung der fossilen Reste derselben.
Verhandl. russ.-kais. mineralog. Gesellschaft zu St. Petersburg (2), ii, pp. 36-132.

—   1867 C
Die geographische Verbreitung des Zubr, oder Bison, des Auerochsen der Neuern (*Bos bison* seu *bonasus*).
Verhandl. russ.-kais. mineral. Gesellsch. zu St. Petersburg (2), ii, pp. 133-185.

—   1867 D
Einige Schlussworte zum Nachweis der Vertilgung der *Rhytina*.
Bull. Soc. Imp. des Naturalistes de Moscou, xl, pt. i, pp. 23-38.
See also pt. ii, pp. 508-543.

—   1868 A
Symbolæ Sirenologicæ. Fasciculi ii et iii. Sireniorum, Pachydermatum, Zeuglodontum et Cetaceorum ordinis osteologia comparata, nec non Sireniorum generum Monographie.
Mém. Acad. Imp. Sci. St. Pétersbourg (7), xii, No. 1, pp. 1-383, with pls. i-ix.

Brandt, Johann Friedrich—C't'd. 1869 A
De Dinotheriorum genere Elephantidorum familiæ adjungendo, nec non de
Elephantidorum generum craniologia
comparata.
Mém. Acad. Imp. Sci. St. Pétersbourg (7),
xiv, No. 1, pp. 1-37.

—— 1869 B
Untersuchungen über die Gattung
der Klippschliefer (*Hyrax*. Herm.),
besonders in anatomischer und ver-
wandtschaftlicher Beziehung nebst Bemerkungen über ihre Verbreitung und
Lebensweise.
Mém. Acad. Imp. Sci. St. Pétersbourg (7),
xiv, pp. i-vi; 1-127, with pls. i-iii.

—— 1873 A
Untersuchungen über die fossilen und
subfossilen Cetaceen Europa's. Mit
Beiträgen von Van Beneden, Cornalia,
Gastaldi, Quenstedt und Paulson nebst
einem geologischen Anhange von Bar-
bot de Marny, G. v. Helmersen, A.
Goebel und Th. Fuchs.
Mém. Acad. Imp. Sci. St. Pétersbourg (7),
xx, No. 1, pp. i-viii; 1-372, with pls. i-xxiv.

—— 1874 A
Ergänzungen zu den fossilen Cetaceen
Europa's.
Mém. Acad. Imp. Sci. St. Pétersbourg (7),
xxi, No. 6, pp. i-iv,; 1-54, with pls. i-v.

—— 1878 A
Tentamen synopseos rhinocerotidum
viventium et fossilium.
Mém. Acad. Imp. Sci. St. Pétersbourg (7),
xxvi, No. 5, pp. i-ii; 1-66, with one plate.

Braus, Hermann.   1898 A
Über die Extremitäten der Selachier.
Verhandl. Anat. Gesellsch., 12th meeting,
Kiel, 1898, pp. 166-179, with 1 page of figs.

Brevoort, J. C.   1859 A
Remains of the American mastodon
found on Long Island, near New York.
Proc. Amer. Assoc. Adv. Sci., 12th meeting,
Baltimore, 1858, pp. 282-284.

Brewer, W. H.   1866 A
Alleged discovery of an ancient hu-
man skull in California.
Amer. Jour. Sci. (2), xliii, p. 424.

Bridge, T. W.   1877 A
The cranial osteology of *Amia calva*.
Jour. Anat. and Physiol., xi, pp. 605-622; pl.
xxiii.

Bridge, T. W.—Continued.   1889 A
Some points in the cranial anatomy
of *Polypterus*.
Proc. Birmingham Philos. Soc., vi, pp. 118-
130, with pls. i-ii.

—— 1895 A
On certain features in the skull of
Osteoglossum formosum.
Proc. Zool. Soc. Lond., 1895, pp. 302-310, with
pl. xxii.

—— 1896 A
The mesial fins of Ganoids and
Teleosts.
Jour. Linn. Soc. Lond., xxv, pp. 530-602,
with pls. xxi-xxiii.

—— 1898 A
On the morphology of the skull in
the Paraguayan *Lepidosiren* and in
other Dipnoids.
Trans. Zool. Soc. Lond., xiv., pp. 325-376,
with pls. xxviii, xxix.

Bridge, T. W., and **Haddon**, A. C.
1889 A
Contributions to the anatomy of
fishes. I. The air bladder and Weber-
ian ossicles in the *Siluridæ*.
Proc. Roy. Soc. Lond., xlvi, pp. 309-328.

—— 1892 A
Contributions to the anatomy of
fishes. II. The air bladder and Weber-
ian ossicles in the Siluroid fishes.
Proc. Roy. Soc. Lond., lii, pp. 139-157.

Bridge, William J.   1879 A
On the osteology of *Polyodon folium*.
Philos. Trans. Royal Soc. of Lond. for 1878,
clxix, pt. ii; pp. 683-733, with pls. lv-lvii.

Briggs, C.   `1838 A
Bones of the mastodon.
Second Annual Report Geol. Surv. Ohio,
pp. 127-129.

Britton, N. L.   1885 A
[Remarks on finding of mastodon
remains east of the Hudson River; followed by remarks by Prof. D. S. Martin and Dr. O. P. Hubbard.]
Trans. N. Y. Acad. Sci., v, pp. 14, 15.

Broadhead, G. C.   1870 A
Fossil horse in Missouri.
Trans. Acad. Sci. St. Louis, iii, pp. xx-xxi.
Reprinted in Amer. Naturalist, iv, p. 60.
Found in Bates County, Mo., 31 feet below
the surface; said by Leidy to belong to an ex-
tinct species.

Broadhead, G. C.—Continued.   1870 B
Bones of large mammals in drift.
Trans. Acad. Sci. St. Louis, iii, pp. xxii,
xxiii.
Tooth of extinct ox found in Missouri; also
in Illinois. Probably *Bos latifrons.*

———   1881 A
The mastodon.
Kansas City Review of Science and Indus-
try, iv, pp. 519-580.
Mentions numerous localities in North
America where the bones of the mastodon
have been found.

Broili, Ferdinand.   1899 A
Ein Beitrag zur Kenntniss von *Eryops megacephalus* (Cope).
Palæontographica, xlvi, pp. 61-84, with pls.
viii-x.

Brongniart, Chas.   1888 A
Sur un nouveau poisson fossile du
terrain houiller de Commentry (Allier),
Pleuracanthus gaudryi.
Bull Soc. Géol. France (3), xvi, pp. 546-550,
with 1 fig. in the text.

———   1888 B
Sur un nouveau poisson fossile du
terrain houiller de Commentry (Allier),
Pleuracanthus gaudryi.
Le Naturaliste, x, pp. 178-180, with 1 fig. in
the text.
The figure gives a restoration of this remark-
able Elasmobranch.

———   1888 C
Études sur le terrain houiller de
Commentry. Faune ichthyologique.
Prèmière partie.
Extracted from Bull. de la Société de
l'Industrie minérale (3), ii, livrais. 4ᵐᵉ, pp. 3-38,
with 15 figs. in the text.
This paper is devoted to the description of
the structure and relationships of the genus
Pleuracanthus, and especially the species *P.
gaudryi.*

Bronn, Heinrich Georg.   1837 A
Lethæa geognostica, oder Abbildungen und Beschreiben der für die Ge-
birgs-Formation bezeichnendsten Versteinerungen.
Vol. i, pp. i-vi; 1-544, with pls. i-xxvii of the
atlas.
Vol. i contains an account of palæontology
up to the period of the chalk.

———   1838 A
Lethæa geognostica, etc.
Vol. ii, pp. 545-1350, with pls. xxviii-xlvii of
the atlas.
Vol. ii contains the palæontology from the
beginning of the Cretaceous to the recent
period.

Brooke, Victor.   1874 A
On Sclater's muntjac and other species of the genus *Cervulus.*
Proc. Zool. Soc. Lond., 1874, pp. 33-42, with
pls. viii, ix, and 5 figs. in the text.

———   1878 A
On the classification of the *Cervidæ,*
with a synopsis of the existing species.
Proc. Zool. Soc. Lond., 1878, pp. 888-928, with
pl. lv and 19 figs. in the text.

Broom, R.   1897 A
On an apparently hitherto undescribed nasal-floor bone in the hairy
armadillo.
Jour. Anat. and Physiol., xxx, pp. 280-282.

Brown, Campbell.   1900 A
Ueber das genus *Hybodus* und seine
systematische Stellung.
Palæontographica, xlvi, pp. 149-174, with
pls. xv, xvi, and 7 figs. in the text.

Browne, Montagu.   1889 A
On a fossil fish (*Chondrosteus*) from
Barrow-on-Soar, hitherto recorded only
from Lyme Regis.
Trans. Leicester Literary and Philos. Soc.,
1889, pp. 16-36, with pls. i, ii, and 3 text figs.

Bru, D. Jean Baptiste.   1804 A
Description des os du *Megatherium*
faité en montant le squelette.
Annales du Muséum, v, pp. 387-400.

Bruce, Adam T.   1883 A
Observations upon the brain casts of
Tertiary mammals.
Contrib. E. M. Mus. Geol. and Arch. Prince-
ton Coll. Bull. No. 3, pp. 36-45, with pl. vii.

Brühl, Bernhard Carl.   1847 A
Die Skeletlehre der Fische.
Vienna, 1847, 8vo., pp. 1-254, with atlas of
19 pls.

Buckland, William.   1824 A
Notice on the *Megalosaurus,* or great
fossil lizard of Stonesfield.
Trans. Geol. Soc. Lond. (2), i, pp. 390-396,
with pls. xl-xliv.

———   1831 A
On the occurrence of the remains of
elephants and other quadrupeds in the
cliffs of frozen mud in Eschscholtz Bay,
within Bering Strait, and in other
distant parts of the shores of the arctic
seas.
Appendix to Beechey's Narrative of a voyage to the Pacific and Behring's Strait to co-
operate with the polar expeditions; performed
on H. M. S. *Blossom,* pp. 593-612, with pls. i-iii.

Buckland, William—Continued. 1837 A
Geology and mineralogy considered
with reference to natural theology.
In two volumes, Philadelphia, 1837.

Vol. i contains the text, consisting of 443
pages. Vol. ii contains the pls. i-lxix, the explanation of the plates, and the index.

Buckley, S. B.   1843 A
Notice of the discovery of a nearly
complete skeleton of the *Zygodon* of
Owen (*Basilosaurus* of Harlan).

Amer. Jour. Sci., xliv, pp. 409–412.
Reprinted in Edinburgh New Philos. Jour.,
vol. xxxv, pp. 77–79, 1843.

————   1846 A
On the Zeuglodon remains of Alabama.

Amer. Jour. Sci. (2), ii, pp. 125–129.
Abstract of this paper printed in Neues
Jahrbuch für Mineralogie, 1847, pp. 510–512.

Büchner, Eug.   1891 A
Die Abbildungen der nordischen
Seekuh (*Rhytina gigas* Zimm.). Mit
besonderer Berücksichtigung neu auf-
gefundener Materialien in seiner Ma-
jestät höchst Eigenen Bibliothek zu
Zarskoje Sselo.

Mém. Acad. Imp. Sci. St.-Pétersbourg (7),
xxxviii, No. 7, pp. 1–24, with 1 pl.

Burckhardt, Rudolph.   1895 A
Das Gebiss der Sauropsiden.

Schwalbe's Morpholog. Arbeiten, v, pp. 341–
385.
Cites the literature of the subject.

————   1900 A
On *Hyperodapedon gordoni*.

Geol. Magazine (4), vii, pp. 486–492; 529–535,
with pl. xix, and 8 text figs.

Burmeister, Hermann.   1864 A
Noticias preliminares sobre las dife-
rentes especies de *Glyptodon* en el Museo
Público de Buenos Aires.

Anales del Museo Púb. Buenos Aires, i, pp.
71–85.

————   1865 A
Bemerkungen über die Arten der
Gattung *Glyptodon* im Museo Público
de Buenos Aires.

Archiv für Anatomie, Physiol., und wissensch. Med., Jahrg., 1865, pp. 317–334, pls. vii,
viii, A.
See also Ann. and Mag. Nat. Hist. (3), xiv,
pp. 81–97.

————   1865 B
Hautpanzer bei *Mylodon*.

Archiv f. Anat. Physiol. u. wissensch. Med.,
Jahrg., 1865, pp. 334–336.

Burmeister, Hermann—Cont'd. 1865 C
Noticias preliminares sobre las dife-
rentes especies de *Glyptodon* en el Museo
Público de Buenos Aires.

Anales Mus. Púb. Buenos Aires, i, pp. 71–85.

————   1866 A
On *Glyptodon* and its allies.

Ann. and Mag. Nat. Hist. (3), xviii, pp. 299–
304.

————   1866 B
Fauna Argentina, Primera parte,
Mamíferos fósiles.

Anales Museo Púb. Buenos Aires, i, pp. 87–
232, with pls. 5–8.

————   1867 A
Fauna Argentina. Primera parte.
Mamíferos fósiles.

Anales Museo Púb., Buenos Aires, i, pp.
233–300, with pls. ix–xiv.

————   1870 A
Ueber das Becken von *Megatherium*.

Verhandl. k. k. zoolog.-bot. Gesellschaft
Wien, xx, pp. 381–388.

————   1871 A
Ueber *Hoplophorus euphractus*.

Archiv f. Anat., Physiol. u. wissensch.
Med., Jahrg. 1871, pp. 164–179, pl. vii A.

————   1871 B
Osteologische Notizen zur Kunde
der Panzerthiere Süd-Amerikas.

Archiv f. Anat., Physiol. u. wissensch.
Med., Jahrg. 1871, pp. 418–429, pl. xi a; pp. 694–
715, pl. xviii.

————   1871 C
On *Saurocetes argentinus*, a new type
of *Zeuglodontidæ*.

Ann. and Mag. Nat. Hist. (4), vii, pp. 51–55.

————   1872 A
Uebersicht der Glyptodonten.

Archiv für Naturgeschichte, xxxviii, i,
pp. 250–264.

————   1873 A
Studien an *Megatherium americanum*.

Archiv f. Anat., Physiol., u. wissensch.
Med., Jahrg. 1873, pp. 626–662, pl. xi.

————   1875 A
Die fossilen Pferde der Pampasfor-
mation, beschrieben von Dr. Herman
Burmeister, Director des Museo Pú-
blico in Buenos Aires. Eine im Auf-
trage der Provinzial-Regierung von
Buenos Aires für die internationale
Ausstellung zu Philadelphia verfasste
Monographie.

Buenos Aires, 1875. Pp. 1–viii; 1–86, with pls.
1–viii. Printed in both Spanish and German.
For supplementary volume see Burmeister,
H., 1889 A.

Burmeister, Hermann—Cont'd. 1874 A
Monografía de los Glyptodontes en
el Museo Público de Buenos Aires.
Anales Mus. Púb. Buenos Aires, ii, pp.
1–412, with pls. i–xiii.
In this monograph are described the
genera *Panochthus*, *Hoplophorus*, *Glyptodon*,
and *Dædicurus*.
This monograph was issued in parts at intervals between 1870 and 1874.

———   1879 A
Neue Beobachtungen an *Dædicurus
giganteus*.
Abhandl. k. Akad. Wissenschaften, Berlin,
1878, pp. 1–23, pls. i, ii.

———   1879 B
Description physique de la République Argentine d'après des observationes personelles et étrangères. Trad-
uit de l'allemand avec le concours de
E. Daireaux. Tome troisième: Animaux vertébrés. Première partie:
Mammifères vivants et èteints. Avec
atlas.
Buenos Ayres, 1879, pp. 1–556.

———   1883 A
Beschreibung des Panzers von *Eutatus seguini*.
Sitzungsber. k.-p. Akad. Wissensch., Berlin, Jahrg,, 1883, pp. 1045–1063, pl. xiii.

———   1888 A
Bericht über *Mastodon andium*.
Sitzungsber. k.-p. Akad. Wissensch., Ber-
lin, Jahrg., 1888, pp. 717–729.

———   1888 B
Ein vollständiger Schädel des Megatherium.
Sitzungsber. k.-p. Akad. Wissensch., Berlin, Jahrg., 1888, pp. 1291–1295.

———   1889 A
Die fossilen Pferde der Pampasform-
ation, beschrieben von Dr. Hermann
Burmeister, Director des Museo Na-
cional in Buenos Aires. Nachtrags-
Bericht; Eine im Auftrage der National-
Regierung für die Ausstellung zu Paris
verfasste Monographie.
Buenos Aires, 1889, pp. i–vii; 1–65, with pls.
ix–xii.
Printed in both Spanish and German. A
supplement to Burmeister H., 1875 A.

Butts, Edward.   1891 A
Recently discovered footprints of the
Amphibian age in the upper Coal-
measure group of Kansas City, Mo.
Kansas City Scientist, v, pp. 17–19, with 4
figs. in the text.

Butts, Edward—Continued.   1891 B
Footprints of new species of amphibi-
ans in the upper Coal-measure group of
Kansas City, Mo.
Kansas City Scientist, v, p. 44, with 2 figs. in
the text.

"C. D."   1838 A
Bones of the mammoth.
Amer. Jour. Sci., xxxviii, p. 201.
Refers to finding of bones of mammoth in
Rochester, N. Y

Calori, Luigi.   1894 A
Sulla composizione dei condili occi-
pitali nelle varie classi di vertebrati e
sull' omologia del terzo condilo occipitale dell' uomo con il condilo occipitale
unico degli uccelli e dei rettili.
Memoirie della R. Accad. Sci. Inst. Bologna
(5), iv, pp. 283–296, with pl.

Camper, Peter.   ·   1788 A
Complementa varia Acad. Imper.
Scient. Petropolitanæ communicanda,
ad clar. ac celeb. Pallas.
Nova Acta Acad. Sci, Imp. Petropolitanæ, ii,
pp. 250–264, with pls. viii, ix.
This paper is divided into the following
heads: Præfatio, 250–252; De cranio bisontis
fossili, 252–254; De bubalo giganteo, 254–257;
De molaribus elephantorum giganteorum et
eorum ossibus, 257–258; De dentibus Hippopota-
morum giganteorum, 258; De alcibus giganteis
Hiberniæ, 258–259; De ossibus mamonteis,
259–264.

Cannon, Geo. L., jr.   1888 A
On the Tertiary *Dinosauria* found in
Denver beds.
Proc. Colorado Sci. Soc., iii, pp. 140–147.
See also "Informal Notes" on pp. 190 and
215 of same volume. These "Informal Notes"
were published in 1889.

———   1890 A
Identification of a Dinosaur from the
Denver group.
Proc. Colorado Sci. Soc., iii, pp. 253–254.
Refers to the finding of *Ornithomimus velox*.

Capellini, Giovanni.   1884 A
Il chelonio veronese (*Protosphargis
veronensis*, Cap.), scoperto nel 1852 nel
cretaceo supèrio presso Sant' Anna di
Alfaedo in Valpolicella.
Memorie R. Accad. Lincei (3), xviii, pp. 291–
320, with pls. i–vi.

Carpenter, W. M.   1838 A
Interesting fossils found in Louisiana.
Amer. Jour. Sci., xxxiv, pp. 201–203, with
3 figs. of teeth of horse.

Carpenter, W. M.—Cont'd.   1839 A
Miscellaneous notices in Opelousas,
Attakapas, etc.
Amer. Jour. Sci., xxxv, pp. 344–346.

———   1846 A
Remarks on some fossil bones re-
cently brought to New Orleans from
Tennessee and from Texas.
Amer. Jour. Sci. (2), i, pp. 244–250, with 4
figs. in the text.

Carte, Alex., and Macalister, Alex.
                                    1869 A
On the anatomy of Balænoptera rostratus.
Philosoph. Trans. Roy. Soc. Lond., clviii,
pp. 201–261, with pls. iv–vii.

Carus, C. G.   1847 A
Resultate geologischer, anatomischer,
and zoölogischer Untersuchungen über
das unter dem Namen Hydrarchos von
Dr. A. C. Koch nach Europa gebrachte
und in Dresden aufgestellte grosse fossile Skelett, in Verbindung mit H. B.
Geinitz, A. F. Günther, und H. G. L.
Reichenbach herausgegeben.
Pp. 16, and 8 pls.; folio, Dresden and Leipzig,
1847.

———   1849 A
Das Kopfskelet des Zeuglodon hydrarchos. Zum erstenmale nach einem
vollständigen Exemplare beschrieben
und abgebildet.
Nova Acta Acad. Cæs. Leop.-Carol. Nat.
Cur., xxii, pp. 373–390, with pls. xxxix A and
xxxix B.

Cary, Austin.   1892 A
A study in foot structure.
Jour. Morphology, vii, pp. 306–316, with pl.
xviii, and 7 figs. in the text.

Case, E. C.   1897 A
Foramina perforating the cranial region of a Permian reptile and on a cast
of its brain cavity.
Amer. Jour. Sci., (4) iii, pp. 321–326, with 4
figs. in the text.
A description of Dimetrodon incisivus.

———   1897 B
. On the osteology and relationships
of Protostega.
Jour. Morphology, xiv, pp. 21–55, with pls.
iv–vi.

———   1898 A
The significance of certain changes
in the temporal region of primitive
Reptilia.
Amer. Naturalist, xxxii, pp. 69–74, with 2
figs. in the text.

Case, E. C.—Continued.   1898 B
The development and geological relations of the vertebrates. Part I, The
fishes; Pt. II, Amphibia; Pt. III, Reptilia.
Jour. Geology, vi, pp. 393–416 (Fishes); pp.
500–523 (Amphibia and Reptilia).

———   1899 A
A redescription of Pariotichus incisivus Cope.
Zoolog. Bulletin, ii, pp. 231–245, with 7 text-
figs.

———   1900 A
The vertebrates from the Permian
bone bed of Vermilion County, Illincis.
Jour. Geology, viii, pp. 698–729, with pls. i–v.
The species of this paper were described by
Cope, but few of them were figured. The original descriptions are reproduced and figures of
many of them are given by Case. Several
bones are figured without being identified.
The species originally belonged to Gurley and
are now owned by the University of Chicago.

——— See also Baur and Case, and Williston and Case.

Chadbourne, P. A.   1871 A
The discovery of the skull of a musk-
ox in Utah.
Amer. Naturalist, v, pp. 315–316.
Extract from Salt Lake Tribune, May 16, 1871.

Chaloner, A. D.   1843 A
Fossil bones of mastodon and ele-
phant.
Proc. Acad. Nat. Sci. Phila., i, pp. 321–322.

Chamberlin, Thomas C.   1900 A
On the habitat of the early verte-
brates.
Jour. Geology, viii, pp. 400–412.

Champley, Robert.   1864 A
The great auk.
Ann. and Mag. Nat. Hist. (3), xiv, pp. 235–236.
Contains a list of the specimens of this bird
and of its skeletons and eggs in the various
museums of the world.
See notes by Dr. J. E. Gray, on p. 319; by
Dr. P. L. Sclater, on p. 320; and on p. 393, by
A. von Pelzeln, all of the same volume.

Chapman, E. J.   1858 A
Mastodon remains. Morpeth, Canada West.
Canadian Jour. Industry, Science, and Art
(2), iii, pp. 356–357.
Announces finding of remains of M. ohioticus
(?) at place named above.

Charlesworth, Edward.   1837 A
Notice of the teeth of Carcharias megalodon in the Red Crag of Suffolk.
Mag. Natural History (2), i, pp. 225–227, with
text fig.

Charlesworth, Edward—C't'd. 1846 A
On the occurrence of the *Mosasaurus*
in the Essex chalk, and on the discov-
ery of flint within the pulp cavities of
the teeth.
Report Brit. Assoc. Adv. Sci., 15th meeting,
Cambridge, 1845, p. 60.
Denies propriety of establishing *Leiodon*.

Charlton, O. C. 1890 A
On the occurrence of mammoth re-
mains in Franklin County, Kans.
Trans. Kansas Acad. Sci., xii, p. 74.

Cheney, T. A. 1872 A
The Chautauqua mastodon.
Amer. Naturalist, vi, pp. 178-179.

Christie, W. J. 1856 A
Extract from a letter regarding the
finding of mastodon bones near Shell
River, British America.
Proc. Nat. Hist. Soc. Boston, v, pp. 265-266.

Clark, J. W. 1871 A
On the skeleton of a Narwhal (*Monodon monoceros*).
Proc. Zool. Soc. Lond., 1871, pp. 42-53, with
2 figs.

Clark, W. B. 1895 A
Contributions to the Eocene fauna of
the middle Atlantic slope.
Johns Hopkins University circulars, xv, pp.
3-6.
A list of the species found, among which are
10 vertebrates.

———   1897 A
Eocene deposits of middle Atlantic
slope.
Bull. U. S. Geol. Surv. No. 141, pp. i-vii; 1-167,
pls. i-xi.
Descriptions of vertebrates, pp. 58-63, pls.
vii, viii.

Clarke, John M. 1885 A
On the higher Devonian faunas of
Ontario County, N. Y.
Bull. U. S. Geol. Surv., No. 16, pp. 1-80, pls.
i-iii.
In this Bulletin are described the four new
species of fishes *Dinichthys newberryi*, *Palæoniscus devonicus*, *Acanthodes ? pristis*, and *Pristacanthus vetustus*.

———   1894 A
New or rare species of fossils from
the horizons of the Livonia salt shaft.
Report State Geol. New York for 1893, ii, pp.
162-168, with pl. 1 and 3 figs. in the text.
Description and figures of *Coccosteus ? halmodeus*.

———   1897 A
The stratigraphic and faunal relations of the Oneonta sandstones and

Clarke, John M.—Continued.
shales, the Ithaca and Portage groups
in central New York.
Fifteenth Ann. Rep. State Geologist, for the
year 1895, vol. i, pp. 31-81.
This paper appeared also in the 49th Ann.
Report of the N. Y. State Museum.
Names *Dinichthys newberryi*, *Acanthodes
priscus*, and *Palæoniscus devonicus* as occurring
in the Portage group.

Clarke, S. F. 1891 A
Embryology of the alligator.
Jour. Morphology, v, pp. 181-206, with pls.
ix-xiii.

Claudius, ——. 1867 A
Das Gehörorgan von *Rhytina stelleri*.
Mém. Acad. Imp. Sci. St. Pétersburg (7), xi,
No. 5, pp. 1-14, with pls. i-iii.

Claypole, E. W. 1883 A
Note on a large fish plate from the
Upper Chemung ? of northern Penn-
sylvania.
Proc. Amer. Philos. Soc., xx, pp. 664-666,
with fig.
Describes and figures *Pterichthys rugosus*.

———   1884 A
Preliminary note on some fossil fishes
recently discovered in the Silurian rocks
of North America.
Amer. Naturalist, xviii, pp. 1222-1226.
New genus and species *Palæaspis americana*
and *P. bitruncata*.

———   1884 B
On some remains of fish from the Up-
per Silurian rocks of Pennsylvania.
The Geolog. Magazine (3), i, pp. 519-521.

———   1885 A
On some fish remains recently dis-
covered in the Silurian rocks of Penn-
sylvania.
Proc. Amer. Assoc. Adv. Sci., 33d meeting,
Phila., 1884, pp. 424-426.
Abstract of paper presented.

———   1885 B
On *Ctenacanthus* and *Gyracanthus*
from the Chemung of Pennsylvania.
Proc. Amer. Assoc. Adv. Sci., 33d meeting,
Phila., 1884, pp. 489-490. Abstract.

———   1885 C
On some remains of fish from the
Upper Silurian rocks of Pennsylvania.
Report Brit. Assoc. Adv. Sci., 54th meeting,
Montreal, 1884, pp. 733-734.

———   1885 D
On the recent discovery of Pteraspi-
dian fish in the Upper Silurian rocks of
North America.
Quart. Jour. Geol. Soc., xli, pp. 48-64, with
8 figs. in the text.

Claypole, E. W.—Continued.   1888 A
  A letter concerning the collection of
  fossil fishes made at Berea, Ohio, by
  Dr. Wm. Clark.
  Amer. Geologist, ii, pp. 62–64.

———   1890 A
  Palæontological notes from Indianapolis (A. A. A. S.), *Pterichthys, Castoroides*, etc.
  Amer. Geologist, vi, pp. 255–260.

———   1891 A
  Megalonyx in Holmes County, Ohio.
  Amer. Geologist, vii, pp. 122–132; 149–158.

———   1892 A
  A new gigantic Placoderm from Ohio.
  Amer. Geologist, x, pp. 1–4, with 1 fig. in the
  text.
  Describes *Gorgonichthys clarki.*

———   1892 B
  Dentition of *Titanichthys* and its
  allies.
  Amer. Geologist, x, p. 193.
  Abstract of paper read before the Geological
  Society of America, 1892.

———   1892 C
  The head of *Dinichthys.*
  Amer. Geologist, x, pp. 199–207, with pls.
  vi–viii.

———   1892 D
  On the structure of the American
  Pteraspidian, *Palæaspis* (Claypole),
  with remarks on the family.
  Quart. Jour. Geol. Soc., xlviii, pp. 542–561,
  with 8 figs. in the text.

———   1893 A
  A new Coccostean—*Coccosteus cuyahogæ.*
  Amer. Geologist, xi, pp. 167–171, with 2 figs.
  in the text.

———   1893 B
  The Cladodont sharks of the Cleveland shale.
  Amer. Geologist, xi, pp. 325–331, with pls.
  vii, viii.

———   1893 C
  Fins of *Palæaspis americana.*
  Amer. Naturalist, xxvii, pp. 375–376.
  Extract from paper in Quarterly Jour.
  Geol. Soc., 1892.

———   1893 D
  The upper Devonian fishes of Ohio.
  The Geol. Magazine (3), x, pp. 443–448, with
  2 figs. in the text.

———   1893 E
  The three great fossil Placoderms of
  Ohio.
  Amer. Geologist, xii, pp. 89–99.
  A popular description of *Dinichthys, Titan-
  ichthys,* and *Gorgonichthys.*

Claypole, E. W.—Continued.   1893 F
  On three new species of *Dinichthys.*
  Amer. Geologist, xii, pp. 275–279, with pl. xii,
  and 1 fig. in the text.
  Descriptions of *D. lincolni, D. clarki,* and *D.
  gracilis.*

———   1893 G
  The fossil fishes of Ohio.
  Report Geol. Surv. of Ohio, vii, pp. 602–619,
  with pls. xxxviii–xliii.

———   1894 A
  Cladodus ? magnificus, a new Sela-
  chian.
  Amer. Geologist, xiv, pp. 137–140, with pl. v.

———   1894 B
  On a new. Placoderm, *Brontichthys
  clarki*, from the Cleveland shale.
  Amer. Geologist, xiv, pp. 379–380, with pl. xii.

———   1894 C
  Structure of the bone of *Dinichthys.*
  Proc. Amer. Micros. Soc., xv, pp. 189–191,
  with 5 figs. in the text.

———   1895 A
  On a new specimen of *Cladodus
  clarki.*
  Amer. Geologist, xv, pp. 1–7, pls. i, ii.

———   1895 B
  The Shaw mastodons.
  Amer. Geologist, xv, pp. 325–326.
  Notes the discovery of three individuals of
  the mastodon at Cincinnati, Ohio. Two of
  these possessed two lower incisors.

———   1895 C
  Recent contributions to knowledge
  of the Cladodont sharks.
  Amer. Geologist, xv, pp. 363–368.

———   1895 D
  The Cladodonts of the upper Devo-
  nian of Ohio.
  The Geolog. Magazine (4), ii, p. 478.
  See also Report Brit. Assoc. Adv. Sci., 1895,
  p. 694.

———   1895 E
  The great Devonian Placoderms of
  Ohio.
  The Geol. Magazine (4), ii, pp. 473–474.
  See also Report Brit. Assoc. Adv. Sci., 1895,
  p. 695.

———   1895 F
  Actinophorous clarki Newb.
  Amer. Geologist, xvi, pp. 20–25, pl. ii.

———   1895 G
  The oldest vertebrate fossil.
  Nature, lii, p. 55.
  This distinction is awarded to *Onchus clintoni.*

———   1895 H
  On the structure of the teeth of the
  Devonian Cladodont sharks.
  Proc. Amer. Micros. Soc., xvi, pp. 191–195,
  with pls. i–iv, containing 7 figs.

Claypole, E. W.—Continued.   1896 A
A new *Titanichthys*.
Amer. Geologist, xvii, pp. 166-169, with pl. x.
Description and figs. of portions of a new species, *T. brevis*.

———   1896 B
The ancestry of the Upper Devonian Placoderms of Ohio.
Amer. Geologist, xvii, pp. 349-360.

———   1896 C
Dinichthys prentis-clarki.
Amer. Geologist, xviii, pp. 199-201, with pl. vii.

———   1896 D
On the teeth of *Mazodus*.
Trans. Amer. Micros. Soc., xviii, pp. 146-148, with pl. of 2 figs.

———   1896 E
On the structure of some Paleozoic spines from Ohio.
Trans. Amer. Micros. Soc., xviii, pp. 151-154, with pl. of 2 figs.

———   1897 A
Man and the *Megalonyx*.
Amer. Geologist, xx, pp. 52-54.

Clift, William.   1835 A
Some account of the remains of the *Megatherium* sent to England from Buenos Ayres, by Woodbine Parish, jr., esq., F. G. S., F. R. S.
Trans. Geol. Soc. Lond. (2), iii, pp. 437-450.

Cocchi, Igino.   1864 A
Monografia dei *Pharyngodopilidæ*, nuova famiglia di pesci labroidi. Studi paleontologica.
Firenze, 1864, pp. 1-88, pls. 1a-va.

Coles, Henry.   1853 A
On the skin of the *Ichthyosaurus*.
Quart. Jour. Geol. Soc. Lond., ix, pp. 79-82, with pl. v.

Collett, John.   1882 A
Recent extinction of the mastodon.
Amer. Naturalist, xvi, pp. 74-75.
Taken from Geological Report of Indiana for 1880.

Collinge, Walter E.   1893 A
The morphology of the sensory canal system in some fossil fishes.
Proc. Birmingham Philos. Soc., ix, pt. 1. Reprint, pp. 1-14, with pls. i, ii.

———   1895 A
The presence of scales in the integument of *Polyodon folium*.
Jour. Anat. and Physiol., xxix, pp. 485-487, with 2 figs. in the text.

Conrad, T. A.   1840 A
On the geognostic position of the *Zeuglodon*, or *Basilosaurus* of Harlan.
Amer. Jour. Sci. (1), xxxviii, pp. 381-382.

———   1869 A
Notes on American fossiliferous strata.
Amer. Jour. Sci. (2), xlvii, pp. 356-364.

———   1871 A
On some points connected with the Cretaceous and Tertiary of North Carolina.
Amer. Jour. Sci. (3), i, pp. 463-469.

Conybeare, William D.   1824 A
Additional notices on the fossil genera *Ichthyosaurus* and *Plesiosaurus*.
Trans. Geol. Soc. Lond. (2), i, pp. 103-123, with pls. xv-xxii.

Cook, Geo. H.   1885 A
Annual report of the State geologist for 1885.
Pp. 1-228.
On page 96 there is a list of 8 species of fossil tracks which have been found in New Jersey.

Cooper, J. Hamilton.   1843 A
On fossil bones found in digging the New Brunswick canal in Georgia.
Proc. Geol. Soc., Lond., iv, pp. 83-84.

Cooper, William.   1824 A
On the remains of the *Megatherium* recently discovered in Georgia.
Annals Lyc. Nat. Hist. of N. Y., i, pp. 114-124, pl. viii.

———   1828 A
Further discovery of fossil bones in Georgia, and remarks on their identity with those of the *Megatherium* of Paraguay.
Annals Lyc. Nat. Hist. of N. Y., ii, pp. 267-270. Abstract in Edinb. New Philos. Jour. 1828, v, pp. 327-329.

———   1831 A
Notices of Big Bone Lick.
Monthly Amer. Jour. of Geol. and Nat. Sci., i, pp. 158-174; 205-217.

———   1833 A
A report on some fossil bones of the *Megalonyx* from Virginia, with a notice of such parts of the skeleton of this animal as have been hitherto discovered, and remarks on the affinities which they indicate.
Annals Lyc. Nat. Hist. N. Y., iii, pp. 166-173.

Cooper, William, **Smith,** J. A., and **DeKay,** J. E.   1831 A
Report of Messrs. Cooper, J. A. Smith, and DeKay to the Lyceum of Natural History on a collection of fossil bones disinterred at Big Bone Lick, Kentucky, in September, 1830, and recently brought to New York.
Reprinted in Edinburgh New Philos. Journal, xi, pp. 352-355, 1831, Amer. Jour. Sci., xx, pp. 370-372, and Monthly Amer. Jour. Geol., i. pp. 43-44.

Cope, E. D.   1864 A
On the characters of the higher groups of *Reptilia Squamata*, and especially of the *Diploglossa.*
Proc. Acad. Nat. Sci. Phila., 1864, pp. 224-231.

————   1865 A
On *Amphibamus grandiceps*, a new batrachian from the Coal-measures.
Proc. Acad. Nat. Sci. Phila., 1865, pp. 134-137.

————   1866 A
On the discovery of the remains of a gigantic Dinosaur in the Cretaceous of New Jersey.
Proc. Acad. Nat. Sci. Phila., 1866, pp. 275-279.

————   1866 B
On the discovery of the remains of a gigantic Dinosaur in the Cretaceous of New Jersey.
Amer. Jour. Sci. (2), xlii, p. 425. Abstract from Proc. Phila. Acad., 1866, pp. 275-279.
Same as preceding.

————   1866 C
Supplement to the description of vertebrates.
Geol. Surv. Illinois, A. H. Worthen, director, ii, pp. 135-141. pl. xxxii, and 1 woodcut.
This paper is supplementary to "Newberry and Worthen, 1866 A."

————   1866 D
Observations on extinct vertebrates of the Mesozoic red sandstone.
Proc. Acad. Nat. Sci. Phila., 1866, pp. 249-250.
Remarks on *Rhytidodon, Clepsisaurus*, and *Mastodonsaurus durus.*

————   1866 E
[Remarks on *Lælaps.*]
Proc. Acad. Nat. Sci. Phila., pp. 316-317.

————   1866 F
On the structures and distribution of the genera of the arciferous *Anura.*
Jour. Acad. Nat. Sci. Phila. (2), vi, pp. 67-112, pl. 25.

Cope, E. D.—Continued.   1866 G
Communication in regard to the Mesozoic sandstone of Pennsylvania.
Proc. Acad. Nat. Sci. Phila., 1866, p. 290.
Notes the finding of *Pterodactylus longispinis*, afterwards described as *Rhabdopelix longispinis.*

————   1866 H
Third contribution to the history of the *Balænidæ* and *Delphinidæ.*
Proc. Acad. Nat. Sci. Phila., xviii, pp. 293-300.
In this paper Harlan's *Delphinus calvertensis* is assigned to *Pontoporia.*

————   1867 A
On *Euclastes*, a genus of extinct *Cheloniidæ.*
Proc. Acad. Nat. Sci. Phila., pp. 39-42.

————   1867 B
[Descriptions of *Eschrichtius cephalus, Rhabdosteus latiradix, Squalodon atlanticus*, and *S. mento.*]
Proc. Acad. Nat. Sci. Phila., 1867, pp. 131-132.

————   1867 C
An addition to the vertebrate fauna of the Miocene period, with a synopsis of the extinct *Cetacea* of the United States.
Proc. Acad. Nat. Sci. Phila., 1867, pp. 138-156.
Species collected in Charles County, Md.

————   1867 D
[Account of extinct reptiles which approach birds.]
Proc. Acad. Nat. Sci. Phila., 1867, pp. 234-235.

————   1867 E
The fossil reptiles of New Jersey.
Amer. Naturalist, i, pp. 23-30.

————   1868 A
On a new large Enaliosaur.
Proc. Acad. Nat. Sci. Phila., 1868, pp. 92-93.
Describes *Elasmosaurus platyurus.*

————   1868 B
On the origin of genera.
Proc. Acad. Nat. Sci. Phila., 1868, pp. 242-300.

————   1868 C
On the genus *Lælaps.*
Amer. Jour. Sci. (2), xlvi, pp. 415-417.

————   1868 D
Note on the fossil reptiles near Fort Wallace.
Page 68 of John L. Le Conte's "Notes on the geology of the survey for the extension of the Union Pacific Railway, E. D., from the Smoky Hill River, Kansas, to the Rio Grande," Philadelphia, Feb., 1868.
Contains description of *Elasmosaurus platyurus* and mention of *Discosaurus carinatus* Cope.

Cope, E. D.—Continued.   1868 E
[Description of new genus and species of *Cheloniidæ*, *Osteopygis emarginatus*.]
Proc. Acad. Nat. Sci. Phila., 1868, p. 147.

——   1868 F
[Remarks on *Palæophis littoralis*
Cope.]
Proc. Acad. Nat. Sci. Phila., 1868, p. 147.

——   1868 G
[Remarks on extinct *Cetacea* from the
Miocene of Maryland.]
Proc. Acad. Nat. Sci. Phila., 1868, pp. 159, 160.

——   1868 H
[Remarks on *Clidastes iguanavus*, *Nec-
toportheus validus*, and *Elasmosaurus*.]
Proc. Acad. Nat. Sci. Phila., 1868, p. 181.

——   1868 I
Second contribution to the history of
the *Vertebrata* of the Miocene period of
the United States.
Proc. Acad. Nat. Sci. Phila., 1868, pp. 184–194.

——   1868 J
Synopsis of the extinct *Batrachia* of
North America.
Proc. Acad. Nat. Sci. Phila., 1868, pp. 208–221.

——   1868 K
On some Cretaceous *Reptilia*.
Proc. Acad. Nat. Sci. Phila., 1868, pp. 233–
242.

——   1868 L
On the origin of genera.
Proc. Acad. Nat. Sci. Phila., 1868, pp. 242–
300.

——   1868 M
[Observations on some extinct rep-
tiles.]
Proc. Acad. Nat. Sci. Phila., 1868, p. 313.

——   1869 A
On the reptilian orders *Pythonomor-
pha* and *Streptosauria*.
Proc. Boston Soc. Nat. Hist., xii, pp. 250–266.

——   1869 B
Synopsis of the extinct *Reptilia* found
in the Mesozoic and Tertiary strata of
New Jersey.
Appendix B of Geology of New Jersey,
George H. Cook, State geologist, Newark, 1868,
pp. 733–742.
The letter of the State geologist to the gov-
ernor (page v.) shows that this volume could
not have been issued before 1869.

——   1869 C
[Remarks on fossil reptiles.]
Proc. Amer. Philos. Soc., xi, p. 16.

Cope, E. D.—Continued.   1869 D
[Remarks on fossil reptiles, *Clidastes
propython*, *Polycotylus latipinnis*, *Orni-
tholarsus immanis*.]
Proc. Amer. Philos. Soc., xi, p. 117.
Reprint in Amer. Jour. Sci. (2), xlviii, p. 278.
Noticed in The Geolog. Magazine, vol. vi, (1869),
p. 476, and vol. vii (1870), p. 427.

——   1869 E
Synopsis of the extinct *Mammalia* of
the cave formations in the United
States, with observations on some *Myri-
opoda* found in and near the same, and
on some extinct mammals of the caves
of Anguilla, W. I., and of other localities.
Proc. Amer. Philos. Soc., xi, pp. 171–192,
with pls. iii–v.
Noticed in Jour. de Zoologie, i, pp. 168–169.

——   1869 F
Second addition to the history of the
fishes of the Cretaceous of the United
States.
Proc. Amer. Philos. Soc., xi, pp. 240–244.

——   1869 G
Third contribution to the fauna of
the Miocene period of the United
States.
Proc. Acad. Nat. Sci. Phila., 1869, pp. 6–12.

——   1869 H
[Remarks on *Holops brevispinus*, *Orni-
tholarsus immanis*, and *Macrosaurus proriger*.]
Proc. Acad. Nat. Sci. Phila., 1869, p. 123.

——   1869 I
[Remarks on *Eschrichtius polyporus*,
Hypsibema crassicauda, *Hadrosaurus
tripos*, and *Polydectes biturgidus*.]
Proc. Acad. Nat. Sci. Phila., 1869, p. 192.

——   1869 J
Remarks on fossils from limestone
caves in Virginia.
Proc. Acad. Nat. Sci. Phila., 1869, p. 8.

——   1869 K
The fossil reptiles of New Jersey.
Amer. Naturalist, iii, pp. 84–91, pl. ii; con-
tinued from Amer. Naturalist, i, p. 30.

——   1869 L
Description of some extinct fishes
previously unknown.
Proc. Boston Soc. Nat. Hist., xii, pp. 310–317.

Cope, E. D.—Continued.　　　1869 M

Synopsis of the extinct *Batrachia, Reptilia,* and *Aves* of North America.

Trans. Amer. Phil. Soc., xiv, pp. i–viii, 1–252, with pls. i–xiv, and 55 woodcuts in the text.

A synopsis appeared in Jour. de Zoologie, i, pp. 182–187, 381–387, 508–510, with pls. xi. Of this memoir only pp. 1–104 were issued in the year 1869; pp. 105–234 appeared in April, 1870, while the remainder was issued in December, 1870. The imprint of the volume is December, 1871.

—　　　　　　　　　　　　　1869 N

Synopsis of the extinct *Mammalia* of New Jersey.

Appendix C of Geology of New Jersey, George H. Cook, State geologist, Newark, 1868, pp. 739–742.

See remark under Cope, 1869, B.

—　　　　　　　　　　　　　1869 O

On two new genera of extinct *Cetacea.*

Canadian Naturalist, iv, pp. 320–321.

Only *Anoplonassa forcipata* is described in this extract from a Boston newspaper.

—　　　　　　　　　　　　　1870 A

On the *Megadactylus polyzelus* of Hitchcock.

Amer. Jour. Sci. (2), xlix, pp. 390–392.

Extract from Cope's "Extinct Reptilia and Aves." Reprinted in Ann. and Mag. Nat. Hist. (4), vol. v, pp. 454–455.

—　　　　　　　　　　　　　1870 B

On *Elasmosaurus platyurus.*

Amer. Jour. Sci. (2), 1, pp. 140–141.

—　　　　　　　　　　　　　1870 C

Additional note on *Elasmosaurus.*

Amer. Jour. Sci. (2), 1, pp. 268–269.

—　　　　　　　　　　　　　1870 D

On some *Reptilia* of the Cretaceous formation of the United States.

Proc. Amer. Philos. Soc., xi, pp. 271–274.

—　　　　　　　　　　　　　1870 E

Verbal communication [at meeting of the Amer. Philos. Soc., Feb. 18, 1870.]

Proc. Amer. Philos. Soc., xi, p. 274.

—　　　　　　　　　　　　　1870 F

Fourth contribution to the history of the fauna of the Miocene and Eocene periods of the United States.

Proc. Amer. Philos. Soc., xi, pp. 285–294.

—　　　　　　　　　　　　　1870 G

On *Adocus,* a genus of Cretaceous *Emydidæ.*

Proc. Amer. Philos. Soc., xi, pp. 295–298.

Cope, E. D.—Continued.　　　1870 H

Observations on the fishes of the Tertiary shales of Green River, Wyoming Territory.

Proc. Amer. Philos. Soc., xi, pp. 380–384.

—　　　　　　　　　　　　　1870 I

Observations on the *Reptilia* of the Triassic formations of the Atlantic regions of the United States.

Proc. Amer. Philos. Soc., xi, pp. 444–446.

Reprinted in Ann. Mag. Nat. Hist., 1870, vi, pp. 498–500.

—　　　　　　　　　　　　　1870 J

[Statements communicated regarding *Liodon perlatus.*]

Proc. Amer. Philos. Soc., xi, pp. 497–498.

—　　　　　　　　　　　　　1870 K

[Remarks on *Adocus syntheticus* and *Lælaps aquilunguis.*]

Proc. Amer. Philos. Soc., xi, p. 515.

—　　　　　　　　　　　　　1870 L

On the *Saurodontidæ.*

Proc. Amer. Philos. Soc., xi, pp. 529–538.

Abstract of in Amer. Naturalist, iv, p. 695.

—　　　　　　　　　　　　　1870 M

On the fishes of a fresh-water Tertiary in Idaho, discovered by Capt. Clarence King.

Proc. Amer. Philos. Soc,, xi, pp. 538–547.

—　　　　　　　　　　　　　1870 N

On the *Adocidæ.*

Proc. Amer. Philos. Soc., xi, pp. 547–553.

—　　　　　　　　　　　　　1870 O

On some species of *Pythonomorpha* from the Cretaceous beds of Kansas and New Mexico.

Proc. Amer. Philos. Soc., xi, pp. 574–584.

—　　　　　　　　　　　　　1870 P

Note on *Saurocephalus,* Harlan.

Proc. Amer. Philos. Soc., xi, p. 608.

Abstract in Amer. Jour. Sci. (3), i, p. 386, 1871.

—　　　　　　　　　　　　　1870 Q

[Observations on fossil reptiles from Kansas.]

Proc. Acad. Nat. Sci. Phila., 1870, p. 132.

—　　　　　　　　　　　　　1870 R

Discovery of a huge whale in North Carolina.

Amer. Naturalist, iv, p. 128.

—　　　　　　　　　　　　　1870 S

Reptilia of the Triassic formations of the United States.

Amer. Naturalist, iv, pp. 562–563.

Cope, E. D.—Continued.   1870 T
Review of G. A. Maack's "Die bis
jetzt bekannten Schildkröten u. d. bei
Kelheim u. Hannover neu aufgefunde-
nen ältesten Arten derselben."
Amer. Jour. Science (2) l. pp. 186–189.
This paper contains a list of species of Ameri-
can turtles known at that date.

————   1871 A
[Remarks on new fossil reptiles from
western Kansas.]
Proc. Acad. Nat. Sci. Phila., 1871, pp. 297–
298.
Names are given of certain fossils, but no
adequate descriptions.

————   1871 B
On the homologies of some of the
cranial bones of the *Reptilia*, and on the
systematic arrangement of the class.
Proc. Amer. Assoc. Adv. Science, 19th meeting, Troy, 1870, pp. 194–247, with 24 figs. in text.
Notice of in Ann. and Mag. Nat. Hist. (4),
vol. vii, pp. 67–68.

————   1871 C
On the extinct tortoises of the Creta-
ceous of New Jersey.
Amer. Naturalist, v, pp. 562–564.

————   1871 D
On the fossil reptiles and fishes of the
Cretaceous rocks of Kansas.
U. S. Geol. Surv. of Wyoming and portions
of contiguous Territories, 2d [4th] annual
report, F. V. Hayden, U. S. Geologist, Washington, 1871, pp. 385–424.

————   1871 E
On the fishes of the Tertiary shales of
Green River, Wyoming Territory.
U. S. Geol. Surv. of Wyoming and portions
of contiguous Territories for 1870, 2d [4th]
annual report, F. V. Hayden, U. S. Geologist,
pp. 425–431.

————   1871 F
The Port Kennedy bone cavern.
Proc. Amer. Philos. Soc., xi, p. 15.

————   ˙1871 G
Supplement to the "Synopsis of the
extinct *Batrachia* and *Reptilia* of North
America."
Proc. Amer. Philos. Soc., xii, pp. 41–52.

————   1871 H
On the occurrence of fossil *Cobitidæ*
in Idaho.
Proc. Amer. Philos. Soc., xii, p. 55.
Discusses position of the genus *Diastichus*,
which was afterwards put among the *Cyprinidæ*.

Cope, E. D.—Continued.   1871 I
Preliminary report on the *Vertebrata*
discovered in the Port Kennedy bone
cave.
Proc. Amer. Philos. Soc., xii, pp. 73–102, with
20 figs. in the text.

————   1871 J
Note of some Cretaceous *Vertebrata* in
the State Agricultural College of Kansas, U. S. A.
Proc. Amer. Philos. Soc., xii, pp. 168–170.

————   1871 K
[Letter to Professor Lesley giving an
account of a journey in the valley of
the Smoky Hill River, in Kansas.]
Proc. Amer. Philos. Soc., xii, pp. 174–176.

————   1871 L
Observations on the extinct Batrachian fauna of the Carboniferous of Linton, Ohio.
Proc. Amer. Philos. Soc., xii, p. 177.

————   1871 M
Observations on the distribution of
certain extinct *Vertebrata* in North
Carolina.
Proc. Amer. Philos. Soc., xii, pp. 210–216.

————   1871 N
Catalogue of the *Pythonomorpha* found
in the Cretaceous strata of Kansas.
Proc. Amer. Philos. Soc., xii, pp. 264–287.

————   1871 O
[Observations on *Sauropleura remex*
and *Œstocephalus amphiuminus.*]
Proc. Acad. Nat. Sci. Phila., 1871, p. 53.

————   1871 P
Cave mammals in Pennsylvania.
Amer. Naturalist, v, p. 58.
Refers to the finding of bones in Port Ken-
nedy cave.

————   1872 A
Bathmodon, a new genus of fossil
mammals.
Amer. Jour. Sci. (3), iii, p. 224.
Relative to paper read before Amer. Philos.
Soc. (Cope, E. D., 1872 J.), correcting date.

————   1872 B
Synopsis of the species of *Chelydrinæ.*
Proc. Acad. Nat. Sci. Phila., 1872, pp. 22–29.

————   1872 C
Observations on the systematic rela-
tions of the fishes.
Proc. Amer. Assoc. Adv. Science, 20th meet-
ing, Indianapolis, 1871, pp. 317–343.
See also Amer. Naturalist, v. pp. 579–593;
Ann. and Mag. Nat. Hist. (4), ix, pp. 155–168;
Trans. Amer. Philos. Soc., xiv, pp. 445–461.

Cope, E. D.—Continued.   1872 D

On the extinct tortoises of the Cretaceous of New Jersey.

Proc. Amer. Assoc. Adv. Science, 20th meeting, Indianapolis, 1871, pp. 344-345.

Abstract of Cope, 1871 D.

—   1872 E

Third account of new *Vertebrata* from
the Bridger Eocene of Wyoming Territory.

Proc. Amer. Philos. Soc., xii, pp. 469-472
(1873).

Issued Aug. 7, 1872, as Pal. Bull. No. 3; reprinted in Proc. Amer. Philos. Soc., Jan., 1873.

—   1872 F

On the geology and palæontology of
the Cretaceous strata of Kansas.

U. S. Geol. Surv. of Montana and portions
of adjacent Territories; 5th annual report, F.
V. Hayden, U. S. Geologist, Washington, D. C.,
1872, pp. 318-349.

—   1872 G

On the vertebrate fossils of the Wasatch strata.

U. S. Geol. Surv. of Montana and portions
of adjacent Territories; 5th annual report, F.
V. Hayden, U. S. Geologist, Washington, 1872;
pp. 350-353.

—   1872 H

On a new Testudinate from the chalk
of Kansas.

Proc. Amer. Philos. Soc., xii, pp. 308-310.

Describes *Cynocercus incisivus*.

—   1872 I

On the families of fishes of the Cretaceous formation in Kansas.

Proc. Amer. Philos. Soc., xii, pp. 327-357.

—   1872 J

On *Bathmodon*, an extinct genus of
ungulates.

Proc. Amer. Philos. Soc., vii, pp. 417-420.

—   1872 K

On two new Ornithosaurians from
Kansas.

Proc. Amer. Philos. Soc., xii, pp. 420-422.

Ornithochirus umbrosus and *O. harpyia* are
described.

—   1872 L

A description of the genus *Protostega*,
a form of extinct *Testudinata*.

Proc. Amer. Philos. Soc., xii, pp. 422-433.

—   1872 M

Description of some new *Vertebrata*
from the Bridger group of the Eocene.

Proc. Amer. Philos. Soc., xii, pp. 460-465.

Issued as Pal. Bull. No. 1, July 29, 1872; reprinted in Proc. Amer. Phil. Soc., January,
1873.

Cope, E. D.—Continued.   1872 N

Second account of new *Vertebrata*
from the Bridger Eocene.

Proc. Amer. Philos. Soc., xii, pp. 466-468.

This forms Pal. Bull. No. 2, and was issued
Aug. 3, 1872; reprinted in Proc. Amer. Philos.
Soc., 1873, January (See Cope, 1873 V).

—   1872 O

On a new genus of *Pleurodira* from
the Eocene of Wyoming.

Proc. Amer. Philos. Soc., xii, pp. 472-477.

Besides the new genus of Pleurodiran turtle
Notomorpha, with 3 species, Cope here describes
the perissodactyle *Notharctus vasaccienus*.

—   1872 P

On the Tertiary coal and fossils of
Osino, Nevada.

Proc. Amer. Philos. Soc., xii, pp. 478-481.

—   1872 Q

On the existence of *Dinosauria* in the
transition beds of Wyoming.

Proc. Amer. Philos. Soc., xii, pp. 481-483
(1873).

This forms Pal. Bull. No. 4, published Aug.
12, 1872; reprinted in Amer. Philos. Soc. Proc.,
Jan., 1873.

—   1872 R

Notices of new *Vertebrata* from the
upper waters of Bitter Creek, Wyoming
Territory.

Proc. Amer. Philos. Soc., xii, pp. 483-486
(1873).

This forms Pal. Bull. No. 6, published Aug.
20, 1872; reprinted Jan., 1873, in Proc. Amer.
Philos. Soc. (See Cope, 1873 V.)

—   1872 S

Second notice of extinct vertebrates
from Bitter Creek, Wyoming.

Proc. Amer. Philos. Soc., xii, pp. 487-488
(1873).

Published in part Aug. 22, 1872, as Pal. Bull.
No. 7; reprinted in Proc. Amer. Philos. Soc.,
Jan., 1873. (See Cope, 1873 V.) The description
of *Anaptomorphus æmulus* forms Pal. Bull. No.
8, Oct., 1872, *fide* Cope, 1873 V.

—   1872 T

On the dentition of *Metalophodon*.

Proc. Amer. Philos. Soc., xii, pp. 542-545.

—   1872 U

On a new vertebrate genus from the
northern part of the Tertiary basin of
Green River.

Proc. Amer. Philos. Soc., xii, p. 554.

Issued also as Pal. Bull. No. 8, Oct. 12, 1872.

Describes *Anaptomorphus æmulus.*

—   1872 V

Descriptions of new extinct reptiles
from the upper Green River Eocene
basin, Wyoming.

Proc. Amer. Philos. Soc., xii, pp. 554-555
(1873).

This constitutes Pal. Bull. No. 9, published
Oct., 1872; reprinted Jan., 1873, *fide* Cope, 1873 V.

Cope, E. D.—Continued. 1872 W
Telegram describing extinct proboscidians from Wyoming.
Palæontological Bull. No. 5.
With conjectural corrections of specific names. Published Aug. 1, 1872, *fide* Cope, 1873 V. Reprinted, with corrections, in Proc. Amer. Philos. Soc., xii, p. 580. (Cope, E. D., 1872 II.)

—— 1872 X
[Description of *Holops pneumaticus*.
Proc. Acad. Nat. Sci. Phila., 1872, pp. 11–12.

—— 1872 Y
List of the *Reptilia* of the Eocene formation of New Jersey.
Proc. Acad. Nat. Sci. Phila., 1872, pp. 14–18.

—— 1872 Z
On an extinct whale from California.
Proc. Acad. Nat. Sci. Phila., 1872, pp. 29–30.
Describes *Eschrichtius davidsonii*.

—— 1872 AA
[Description of *Bathmodon radians*.]
Proc. Acad. Nat. Sci. Phila., 1872, p. 38.
Reprinted in Amer. Jour. Sci. (3), iv, pp. 238–239; Amer. Naturalist, vi, p. 438.

—— 1872 BB
[Description of *Plesiosaurus gulo* and of the turtle afterwards named *Toxochelys latiremis*.]
Proc. Acad. Nat. Sci. Phila., 1872, pp. 127–129.

—— 1872 CC
[Remarks on discoveries recently made by Prof. O. C. Marsh.]
Proc. Acad. Nat. Sci. Phila., 1872, pp. 140–141.

—— 1872 DD
Remarks on *Mastodon* from New Mexico.
Proc. Acad. Nat. Sci. Phila., 1872, p. 142.

—— 1872 EE
Remarks on geology of Wyoming, and on saurodont fishes from Kansas.
Proc. Acad. Nat. Sci. Phila., 1872, pp. 279–281.

—— 1872 FF
The geological age of the coal of Wyoming.
Amer. Naturalist, vi, pp. 669–671.

—— 1872 GG
Food of *Plesiosaurus*.
Amer. Naturalist, vi, p. 439.

—— 1872 HH
The Eocene genus *Synoplotherium*.
Amer. Naturalist, vi, p. 695.

—— 1872 II
Notice of new proboscidians from the Eocene of southern Wyoming.
Proc. Amer. Philos. Soc., xii, p. 580 (1873).
Followed by a note of the secretary, correcting errors in the original, which was issued as Pal. Bull. No. 5. (Cope, E. D., 1872 W.)

Cope, E. D.—Continued 1872 JJ
Carboniferous reptiles of Ohio.
Amer. Naturalist, vi, p. 46.

—— 1872 KK
The proboscidians of the American Eocene.
Amer. Naturalist, vi, pp. 773–774.
The group of which *Eobasileus cornutus* is type.

—— 1872 LL
The armed *Metalophodon*.
Amer. Naturalist, vi, pp. 774–775.
Describes *M. armatus*.

—— 1872 MM
The fish-beds of Osino, Nevada.
Amer. Naturalist, vi, pp. 775–776.
Refers to *Amyzon mentale* and *Trichophanes hians*.

—— 1873 A
Note on the Cretaceous of Wyoming.
Amer. Jour. Sci. (3), v, pp. 230–231.
Slip from the "Proc. Phil. Soc., Philadelphia," published on Feb. 7. Followed by remarks by "Eds." The original not found in the Proc. Amer. Philos. Soc., Phila.

—·— 1873 B
On the new perissodactyles from the Bridger Eocene.
Proc. Amer. Philos. Soc., xiii, pp. 35–36.
Also issued as Pal. Bull. No. 11.

—— 1873 C
On some new extinct *Mammalia* from the Tertiary of the Plains.
Palæont. Bulletin No. 14, pp. 1–2.
Describes *Ælurodon mustelinus* and *Acerotherium megalodus*.

—— 1873 D
Fourth notice of extinct *Vertebrata* from the Bridger and the Green River Tertiaries.
Palæont. Bulletin No. 17, pp. 1–4. "Oct. 25, 1873." (Cope.)

—— 1873 E
On the extinct *Vertebrata* of the Eocene of Wyoming, observed by the expedition of 1872, with notes on the geology.
U. S. Geol. Surv. Terr. for 1872, 6th annual report, F. V. Hayden, U. S. Geologist, Washington, D. C., 1873, pp. 545–649, pls. i–vi.

—— 1873 F
On the short-footed *Ungulata* of the Eocene of Wyoming.
Proc. Amer. Philos. Soc., xiii, pp. 38–74.
For extract see Jour. de Zoologie, ii, 1873, pp. 168–184, pl. vii, followed by remarks by Paul Gervais.

—— 1873 G
On the flat-clawed *Carnivora* of the Eocene of Wyoming.
Proc. Amer. Philos. Soc., xiii, pp. 196–209.

Cope, E. D.—Continued.   1873 H

On the osteology of the extinct tapiroid *Hyrachyus*.

Proc. Am. Philos. Soc., xiii, pp. 212–224.

———   1873 I

[Remarks on additional specimens of *Toxochelys latiremis*.]

Proc. Acad. Nat. Sci. Phila., 1873, p. 10.

———   1873 J

Observations on the structure and systematic position of the genus *Basileus* Cope.

Proc. Acad. Nat. Sci. Phila., 1873, pp. 10–12.

———   1873 K

[Remarks on *Eobasilidæ* and *Bathmodontidæ*.]

Proc. Acad. Nat. Sci. Phila., 1873, p. 102.

———   1873 L

[Remarks on a fossil skull of *Sus scropha*.]

Proc. Acad. Nat. Sci. Phila., 1873, p. 207.

———   1873 M

Some extinct turtles from the Eocene strata of Wyoming.

Proc. Acad. Nat. Sci. Phila., 1873, pp. 277–279. Descriptions of *Trionyx heteroglyptus* and *Plastomenus thomasii*.

———   1873 N

On two new species of *Saurodontidæ*.

Proc. Acad. Nat. Sci. Phila., 1873, pp. 337–339. Describes *Portheus lestrio* and *P. gladius* and the new genus *Daptinus*.

———   1873 O

On some new *Batrachia* and fishes from the Coal-measures of Linton, Ohio.

Proc. Acad. Nat. Sci. Phila., 1873, pp. 340–343.

———   1873 P

On the types of molar teeth.

Proc. Acad. Nat. Sci. Phila., 1873, p. 371.

———   1873 Q

[Remarks on fishes from the Coal-measures of Linton, Ohio.]

Proc. Acad. Nat. Sci. Phila., 1873, pp. 417–419.

———   1873 R

[On *Menotherium lemurinum, Hypisodus minimus, Hypertragulus calcaratus, H. tricostatus, Protohippus*, and *Procamelus occidentalis*.]

Proc. Acad. Nat. Sci., Phila., 1873, pp. 419–420.

———   1873 S

Second notice of extinct *Vertebrata* from the Tertiary of the Plains.

Palæont. Bull. No. 15, pp. 1–6. "Issued Aug. 20, 1873." (Cope.)

Cope, E. D.—Continued.   1873 T

Third notice of extinct *Vertebrata* from the Tertiary of the Plains.

Palæont. Bull. No. 16, pp. 1–8. "Published Aug. 20, 1873." (Cope.)

———   1873 U

On some Eocene mammals obtained by Hayden's Geological Survey of 1872.

Palæont. Bull. No. 12, pp. 1–6. Issued privately March 8, 1873. Not reprinted.

———   1873 V

Palæontological Bulletin. Preliminary.

Pp. 1–2. Gives titles and dates of publication of Palæontological Bulletins Nos. 1–13.

———   1873 W

The gigantic mammals of the genus *Eobasileus*.

Amer. Naturalist, vii, pp. 157–160.

———   1873 X

On some of Professor Marsh's criticisms.

Amer. Naturalist, vii, pp. 290–299. Forms, with slight changes, Palæont. Bull. No. 13.

———   1873 Y

Proboscidians of the American Eocene. Correction.

Amer. Naturalist, vii, p. 49. Makes a correction regarding the teeth of *Eobasileus cornutus*.

———   1873 Z

The *Eobasileus* again.

Amer. Naturalist, vii, p. 181.

———   1873 AA

On Professor Marsh's criticisms.

Amer. Naturalist, vii. Appendix to July number, p. 1. This communication contains only eight lines.

———   1873 BB

On the tusk of *Loxolophodon cornutus*.

Amer. Naturalist, vii, p. 315.

———   1873 CC

Synopsis of new *Vertebrata* from the Tertiary of Colorado, obtained during the summer of 1873.

Pp. 1–19. Washington, Government Printing Office, Oct., 1873. On the title page this paper is said to be extracted from the 7th Annual Report of the U. S. Geol. Surv. Terr., but it does not appear in that volume.

Cope, E. D.—Continued.　　　1873 DD

Some remarkable and gigantic animals.

The Independent, New York, Oct. 30, 1873.
Contains reports of the meeting of the Amer. Assoc. Adv. Sci. at Portland, Me. *Miobasileus, Symborodon*, and *Eobasileus* are referred to. *Symborodon acer, S. bucco, S. altirostris, S. heloceros* are more or less fully described.

——　　　　　　　　　　　1874 A

On some extinct types of horned perissodactyles.

Proc. Amer. Assoc. Adv. Science, 22d meeting, Portland, 1873, pp. 108–109.
Reprinted in Ann. and Mag. Nat. Hist. (4), xiii, pp. 405–406.
Describes *Symborodon torvus, Miobasileus ophryas, Symborodon acer,* and *S. helocerras.*

——　　　　　　　　　　　1874 B

Report on the vertebrate palæontology of Colorado.

Annual Report Geol. and Geog. Surv. Territories for 1873, F. V. Hayden, U. S. Geologist, Washington, D. C., 1874, pp. 427–533, pls. i–viii.
Chapter I. Introduction, pp. 429–430. II. the Cretaceous period, pp. 431–456. III. The Eocene period, pp. 456–461. IV. The Miocene period, pp. 461–518. V. The Loup Fork epoch, pp, 518–532.
An abstract of this paper appeared in Jour. de Zoologie, iv, 1875, pp. 354–359, with a list of the species described.

——　　　　　　　　　　　1874 C

Review of the *Vertebrata* of the Cretaceous period found west of the Mississippi River. Sec. I. On the mutual relations of the Cretaceous and Tertiary formations of the West. Sec. II. List of species of *Vertebrata* from the Cretaceous formations of the West.

Bull. U. S. Geol. and Geog. Surv. of the Terrs,, i, No. 2, 1874, pp. 3–48.

——　　　　　　　　　　　1874 D

Supplementary notices of fishes from the fresh-water Tertiaries of the Rocky Mountains.

Bull. U. S. Geol. and Geog. Surv. of the Terrs., i, No. 2, 1874, pp. 49–51.

——　　　　　　　　　　　1874 E

Abstract of remarks at the meeting of the Amer. Philos. Soc., Jan. 16, 1874.

Proc. Amer. Philos. Soc., xiv, p. 110.
On *Poëbrotherium,* etc.

——　　　　　　　　　　　1874 F

[Observations on age of lignite and and other formations of the West.]

Proc. Acad. Nat. Sci. Phila., 1874, pp. 12–13.

Cope, E. D.—Continued.　　　1874 G

Remarks on *Symborodon, Titanotherium,* and *Eobasileus.*

Proc. Acad. Nat. Sci. Phila., 1874, pp. 89–90.

——　　　　　　　　　　　1874 H

Synopsis of result of work in connection with Hayden's U. S. Geological Survey during 1873.

Proc. Acad. Nat. Sci. Phila., 1874, pp. 116–117.

——　　　　　　　　　　　1874 I

Notes on the Santa Fé marls and some of the contained vertebrate fossils.

Proc. Acad. Nat. Sci. Phila., 1874, pp. 147–152.
Also in Palæont. Bull. No. 18.

——　　　　　　　　　　　1874 J

On a new mastodon and rodent.

Proc. Acad. Nat. Sci. Phila., 1874, pp. 221–223.

——　　　　　　　　　　　1874 K

On the characters of *Symborodon.*

Proc. Acad. Nat. Sci. Phila., 1874, pp. 224–225.

——　　　　　　　　　　　1874 L

Supplement to the extinct *Batrachia* and *Reptilia* of North America. I. Catalogue of the air-breathing *Vertebrata* from the Coal-measures of Linton, Ohio.

Trans. Amer. Philos. Soc., xv, pp. 261–278.
Also issued separately with pp. 1–18.

——　　　　　　　　　　　1874 M

On the homologies and origin of the types of molar teeth of *Mammalia educabilia.*

Jour. Acad. Nat. Sci. Phila. (2), viii, pp. 71–89, with 29 figs. in the text.

——　　　　　　　　　　　1874 N

The succession of life in North America.

Ann. and Mag. Nat. Hist. (4), xiii, pp. 326–331.
Reprinted from The Penn Monthly, v, 1874, pp. 186–145.

——　　　　　　　　　　　1874 O

Report upon vertebrate fossils discovered in New Mexico, with descriptions of new species.

Extract from Appendix FF of the Annual Report of the Chief of Engineers, 1874. Washington, Gov. Print. Office, Nov. 28, 1874, pp. 1–18.
In order to make the pagination of the separata correspond to that of the Engineer's Report add 588.
This paper also forms "Appendix FF₂" (pp 115–130) of "Appendix FF" of the Annual Report of the Chief of Engineers for 1874.

Cope, E. D.—Continued.   1874 P

Report on the stratigraphy and Pliocene vertebrate palæontology of northern Colorado.

Bull. U. S. Geol. and Geog. Survey Terr., i,
Bull. No. 1, ser. 1, pp. 9–28.
Includes also "Supplement. Additions to
the synopsis of new *Vertebrata* from the Territory of Colorado, 1873," pp. 22–28.

                                1874 Q

[Remarks on Professor Marsh's
Brontotherium ingens.]

Proc. Amer. Philos. Soc., xiv, p. 2.
B. ingens is regarded as a synonym of *Symborodon trigonoceras.*

——   1874 R

On Dr. Leidy's "Correction."

Proc. Acad. Nat. Sci. Phila., pp. 224–225.

                                1874 S

[Description of a species of *Ctenodus.*]

Proc. Acad. Nat. Sci. Phila., 1874, pp. 91–92.
This species of *Ctenodus* is without name,
but it was afterwards called *C. ohioensis.*

——   1874 T

[Remarks on skulls of *Eobasileus ga-
leatus* and of a fossil walrus.]

Proc. Amer. Philos. Soc., xiv, pp. 17–18.
The skull of *Eobasileus* was from Wyoming;
that of the fossil walrus from Accomac Har-
bor, Va.

——   1875 A

Note on the genus *Calamadon.*

Amer. Jour. Sci. (3), ix, p. 228.

                                1875 B

The geology of New Mexico.

Amer. Jour. Sci. (3), x, pp. 152–153.

——   1875 C

Systematic catalogue of *Vertebrata* of
the Eocene of New Mexico, collected
in 1874.

Report to the Engineer Department, U. S.
Army, in charge of Lieut. Geo. M. Wheeler.
Washington, April 17, 1875, pp. 5–37.

——   1875 D

Synopsis of the extinct *Batrachia*
from the Coal-measures.

Report of the Geol. Survey of Ohio, Palæontology, ii, 1875, pp. 350–411; pls. xxvi–xlv.

——   1875 E

The *Vertebrata* of the Cretaceous
formations of the West.

Report U. S. Geol. Surv. of the Territories,
ii, pp. 1–303; pls. i–lv, and 10 wood cuts. Washington, D. C., 1875.

Cope, E. D.—Continued.   1875 F

Report on the geology of that part
of northwestern New Mexico examined
during the field season of 1874, by E. D.
Cope, palæontologist and geologist.

Appendix LL of Annual Report of the Chief
of Engineers in annual report upon the geographical explorations and surveys west of the
100th meridian, etc., by Geo. M. Wheeler, 1st
Lieut., Engineers. Washington, D. C., 1875.
Pp. 61–116; pls. (illustrating paleontology) ii,
v, vi.
The pagination here given is that of the
separata. To make this correspond with that
of the complete report 920 must be added.
Same as Cope, 1875 U, with different pagination.

                                1875 G

On the fishes of the Tertiary shales
of the South Park.

Bull. U. S. Geol. and Geog. Surv. Territories,
2d series, No. 1, 1875, pp. 3–5.

                                1875 H

Synopsis of the *Vertebrata* of the
Miocene of Cumberland County, New
Jersey.

Proc. Amer. Philos. Soc., xiv, pp. 361–364.

——   1875 I

On the transition beds of the Sas-
katchawan district.

Proc. Acad. Nat. Sci. Phila., 1875, pp. 9–10.

                                1875 J

The extinct *Batrachia* of Ohio.

Proc. Acad. Nat. Sci. Phila., 1875, p. 16.

——   1875 K

On Green-sand *Vertebrata.*

Proc. Acad. Nat. Sci. Phila., 1875, p. 19.

——   1875 L

On the homologies of the sectorial
tooth of *Carnivora.*

Proc. Acad. Nat. Sci. Phila., 1875, pp. 20–23.

                                1875 M

The feet of *Bathmodon.*

Proc. Acad. Nat. Sci. Phila., 1875, p. 73.

                                1875 N

On fossil lemurs and dogs.

Proc. Acad. Nat. Sci. Phila., 1875, pp. 255–
256.

                                1875 O

On the antelope deer of the Santa Fé
marls.

Proc. Acad. Nat. Sci. Phila., 1875, p. 257.

——   1875 P

On some new fossil *Ungulata.*

Proc. Acad. Nat. Sci. Phila., 1875, pp. 258–
261.
From Palæontological Bulletin No. 19.

Cope, E. D.—Continued.   1875 Q

The phylogeny of the camels.

Proc. Acad. Nat. Sci. Phila., 1875, pp. 261–262.

—   1875 R

The geology of New Mexico.

Proc. Acad. Nat. Sci. Phila. 1875, pp. 263–267.

—   1875 S

On an extinct vulturine bird.

Proc. Acad. Nat. Sci. Phila., 1875, p. 271.

Refers *Cathartes umbrosus* to *Vultur.*

—   1875 T

On fossil remains of reptilia and fishes from Illinois.

Proc. Acad. Nat. Sci. Phila., 1875, pp. 404–411.

Reprinted in Ann. and Mag. Nat. Hist. (4), xvii, 178–184.

According to Prof. Cope (Cope, 1888 BB, p. 286) this paper was not issued until 1876.

—   1875 U

Report on the geology of that part of northwestern New Mexico examined during the field season of 1874, by E. D. Cope, palæontologist and geologist.

From annual report upon the geographical explorations and surveys west of the 100th meridian, in California, etc., by Geo. M. Wheeler; being Appendix LL of the Annual Report of the Chief of Engineers, pp. 981–1017; pls. ii, v, and vi. Pp. 61–97 of separate report LL. Same as Cope, 1875 F, with different pagination.

—   1875 V

Synopsis of the *Vertebrata* whose remains have been preserved in the formations of North Carolina.

Report Geol. Surv. of North Carolina, i, by W. C. Kerr, Appendix B, pp. 29–52, with pls. v–viii.

—   1875 W

Report on the vertebrate fossils from the Fort Union group of Milk River.

British North Amer. Boundary Commission's Report on the geology and resources of the region in the vicinity of the forty-ninth parallel, from the Lake of the Woods to the Rocky Mountains, with lists of plants and animals collected and notes on the fossils; by George Mercer Dawson, geologist and botanist to the commission. Montreal, 1875. Pp. 333–337.

—   1875 X

Check-list of North American *Batrachia* and *Reptilia*, with a systematic list of the higher groups and an essay on geographical distribution; based on the specimens in the U. S. National Museum.

Bull. No. 1, U. S. Nat. Mus., pp. 1–95.

Cope, E. D.—Continued.   1876 A

On the supposed *Carnivora* of the Eocene of the Rocky Mountains.

Palæont. Bull. No. 20, pp. 1–4.

It is stated in this bulletin that it was published Dec. 22, 1875.

—   1876 B

On a gigantic bird from the Eocene of New Mexico.

Proc. Acad. Nat. Sci. Phila., 1876, pp. 10–11.

Reprinted in Amer. Jour. Sci. (3), xii, pp. 306–307, and in Jour. de Zoologie., v, 1876, pp. 264–267.

Describes *Diatryma gigantea.*

—   1876 C

On the *Tæniodonta,* a new group of Eocene *Mammalia.*

Proc. Acad. Nat. Sci. Phila., 1876, p. 39.

—   1876 D

On the geologic age of the vertebrate fauna of the Eocene of New Mexico.

Proc. Acad. Nat. Sci. Phila., 1876, pp. 63–66.

Forms Palæontological Bulletin No. 21.

A translation of this paper appeared in Jour. de Zoologie, v, 1876, pp. 307–311.

—   1876 E

On some supposed lemurine forms of the Eocene period.

Proc. Acad. Nat. Sci. Phila., 1876, pp. 88–89.

—   1876 F

On a new genus of fossil fishes.

Proc. Acad. Nat. Sci. Phila., 1876, p. 113.

—   1876 G

On a new genus of *Camelidæ.*

Proc. Acad. Nat. Sci. Phila., 1876, pp. 144–147.

Describes *Protolabis.*

—   1876 H

Descriptions of some vertebrate remains from the Fort Union beds of Montana.

Proc. Acad. Nat. Sci. Phila., 1876, pp. 248–261.

Issued also as Pal. Bull. No. 22, pp. 1–14.

—   1876 I

On some extinct reptiles and *Batrachia* from the Judith River and Fox Hills beds of Montana.

Proc. Acad. Nat. Sci. Phila., 1876, pp. 340–359.

Issued also as Pal. Bull. No. 23, pp. 1–20.

—   1877 A

A continuation of researches among the *Batrachia* of the Coal-measures of Ohio.

Proc. Amer. Philos. Soc., xvi, pp. 573–578.

Issued as part of Pal. Bull. No. 24.

Cope, E. D.—Continued.	1877 B
On a dinosaurian from the Trias of Utah.
Proc. Amer. Philos. Soc., xvi, pp. 579-584.
Issued as part of Pal. Bull. No. 24.
Describes *Dystrophæus viæmalæ.*

——	1877 C
On a new proboscidian.
Proc. Amer. Philos. Soc., xvi, pp. 584-585.
Also as part of Pal. Bull. No. 24.
Describes *Cænobasileus tremontigerus.* This genus and species were withdrawn in Cope, E. D., 1889 J., p. 207.

——	1877 D
On some new or little known reptiles and fishes of the Cretaceous No. 3 of Kansas.
Proc. Amer. Philos. Soc., xvii, pp. 176-181.
Issued also as part of Pal. Bull. No. 26.

——	1877 E
Descriptions of extinct *Vertebrata* from the Permian and Triassic formations of the United States.
Proc. Amer. Philos. Soc., xvii, pp. 182-193.
Not published until March 1, 1878. Issued as part of Pal. Bull. No. 26. This was "printed Nov. 20, 1877."

——	1877 F
On reptilian remains from the Dakota beds of Colorado.
Proc. Amer. Philos. Soc., xvii, pp. 193-196.

——	1877 G
Report on the geology of the region of the Judith River, Montana, and on vertebrate fossils obtained on or near the Missouri River.
Bull. U. S. Geol. and Geog. Surv., iii, article xix, pp. 565-597; pls. 30-34.

——	1877 H
On a carnivorous dinosaurian from the Dakota beds of Colorado.
Bull. U. S. Geol. and Geog. Surv., iii, article xxxiii, pp. 805-806.
Describes *Lælaps trihedrodon.*

——	1877 I
A contribution to the knowledge of the ichthyological fauna of the Green River shales.
Bull. U. S. Geol. and Geog. Surv., iii, article xxxiv, pp. 807-819.

——	1877 J
On the genus *Erisichthe.*
Bull. U. S. Geol. and Geog. Surv., iii, article xxxv, pp. 821-823.

Cope, E. D.—Continued.	1877 K
Report upon the extinct *Vertebrata* obtained in New Mexico by parties of the expedition of 1874. Chapter xi. Fossils of the Mesozoic periods and Geology of Mesozoic and Tertiary beds. xii. Fossils of the Eocene period. xiii. Fossils of the Loup Fork Epoch.
Geographical surveys west of the one hundredth meridian, First Lieut. Geo. M. Wheeler, Corps of Engineers, U. S. Army, in charge. Vol. iv, Palæontology, Washington, 1877. Pp. 1-370, pls. xxii-lxxxiii.

——	1877 L
On the brain of *Coryphodon.*
Proc. Amer. Philos. Soc., xvi, pp. 616-620; pls. i-ii.

——	1877 M
On the brain of *Procamelus occidentalis.*
Proc. Amer. Philos. Soc., xvii, pp. 49-52; pl. i.

——	1877 N
On the *Vertebrata* of the bone bed in eastern Illinois.
Proc. Amer. Philos. Soc., xvii, pp. 52-63.

——	1877 O
Synopsis of the cold-blooded *Vertebrata* procured by Prof. James Orton during his exploration of Peru in 1876-77.
Proc. Amer. Philos. Soc., xvii, pp. 33-49.
Contains on p. 41 definitions of the subclasses of fishes, including the new class *Hyopomata.*

——	1877 P
On a new species of *Adocidæ* from the Tertiary of Georgia.
Proc. Amer. Philos. Soc., xvii, pp. 82-84.
Issued as part of Palæontological Bulletin No. 25.
Describes *Amiphiemys oxysternum.*

——	1877 Q
On a gigantic saurian from the Dakota epoch of Colorado.
Pal. Bull. No. 25, pp. 5-10.
Describes *Camarasaurus supremus.*

——	1877 R
On *Amphicælias,* a genus of saurians from the Dakota epoch of Colorado.
Pal. Bull. No. 27, pp. 2-3, with date Dec. 10, 1877.

——	1877 S
The lowest mammalian brain.
Amer. Naturalist, xi, pp. 312-313.
Describes brain of *Coryphodon.*

Cope, E. D.—Continued.   1877 T

The discovery of *Lælaps* in Montana.
Amer. Naturalist, xi, p. 311.

————   1877 U

The sea serpents of the Cretaceous
period.
Amer. Naturalist, xi, p. 311.
Brief reference to the discovery of *Elasmosaurus orientalis.*

————   1877 V

The dentition of the herbivorous
Dinosauria.
Amer. Naturalist, xi, pp. 311, 312.

————   1877 W

The largest known saurian.
Amer. Naturalist, xi, p. 629.
Brief description of "*Camæosaurus" supre-
mus.*

————   1877 X

Remains of a huge saurian in Pennsylvania.
Amer. Naturalist, xi, p. 629.
Refers to *Palæoctonus appalachianus,* described in Proc. Amer. Phil. Soc., 1877, xvii, p.
182 (Cope, 1877 E).

————   1877 Y

New vertebrate fossils.
Amer. Naturalist, xi, p. 629.

————   1877 Z

New fossil fishes from Wyoming.
Amer. Naturalist, xi, p. 570.
A list, without descriptions, of fossil fishes
found in Wyoming.

————   1878 A

On the classification of the extinct
fishes of the lower types.
Proc. Amer. Assoc. Adv. Sci., 26th meeting,
Nashville, 1877, pp. 292-300.

————   1878 B

Descriptions of extinct *Batrachia* and
Reptilia from the Permian formation
of Texas.
Proc. Amer. Philos. Soc., xvii, pp. 505-530.
Also part of Pal. Bull. No. 29.

————   1878 C

Descriptions of new *Vertebrata* from
the upper Tertiary formations of the
the West.
Proc. Amer. Philos. Soc., xvii, pp. 219-231.
Issued also as part of Pal. Bull. No. 28.

————   1878 D

On some saurians found in the
Triassic of Pennsylvania, by C. M.
Wheatley.
Proc. Amer. Philos. Soc., xvii, pp. 231-232.
Issued also as part of Pal. Bull. No. 28.

Cope, E. D.—Continued.   1878 E

The saurians of the Dakota epoch.
Amer. Naturalist, xii, pp. 56-57.
Refers to discovery of *Camarasaurus* and
Amphicœlias.

————   1878 F

Descriptions of fishes from the Cretaceous and Tertiary deposits west of
the Mississippi River.
Bull. U. S. Geol. and Geog. Surv. of the
Territories, 1878, vol. iv, pp. 67-77.

————   1878 G

Professor Owen on the *Pythonomorpha.*
Bull. U. S. Geol. and Geog. Surv. Territories,
1878, iv, pp. 299-311.

————   1878 H

Descriptions of new extinct *Vertebrata*
from the upper Tertiary and Dakota
formations.
Bull. U. S. Geol. and Geog. Surv. Territories,
1878, iv, pp. 379-396.

————   1878 I

On the *Vertebrata* of the Dakota epoch
of Colorado.
Proc. Amer. Philos. Soc., xvii, pp. 233-247.
Issued as Pal. Bull. No. 28.

————   1878 J

Clepsydrops in Texas.
Amer. Naturalist, xii, p. 57.
A brief note.

————   1878 K

The affinities of the *Dinosauria.*
Amer. Naturalist, xii, pp. 57-58.
A brief note on the views of Professor Owen
on the *Dinosauria.*

————   1878 L

New artiodactyles of the upper Tertiary.
Amer. Naturalist, xii, p. 58.

————   1878 M

On the saurians recently discovered
in the Dakota beds of Colorado.
Amer. Naturalist, xii, pp. 71-85, with 17 woodcuts in the text.

————   1878 N

A new *Mastodon.*
Amer. Naturalist, xii, p. 128.
Brief notice of *M. campester.*

————   1878 O

The snout fishes of the Kansas chalk.
Amer. Naturalist, xii, pp. 128-129.

————   1878 P

A new genus of *Oreodontidæ.*
Amer. Naturalist, xii, p. 129.
Brief notice of *Ticholeptus.*

Cope, E. D.—Continued. 1878 Q
[Note on fossils obtained by Mr. Russell S. Hill, including bones of *Protostega gigas.*]
Amer. Naturalist, xii, p. 137.

———— 1878 R
A new genus of *Dinosauria* from Colorado.
Amer. Naturalist, xii, pp. 188-189.
Brief notice of discovery of *Hypsirophus.*

———— 1878 S
[Review of Prof. L. Lesquereux's Contributions to the fossil flora of the Western Territories. Part II. The Tertiary flora.]
Amer. Naturalist, xii. pp. 243-246.

———— 1878 T
The structure of *Coryphodon.*
Amer. Naturalist, xii, pp. 324-326.
Reply to Prof. O. C. Marsh, in Nature, vol. xvii, p. 340.

———— 1878 U
A new fauna.
Amer. Naturalist, xii, pp. 327-328.
Notes discovery of *Diadectes, Pariosus, Trimerorhachis,* etc.

———— 1878 V
A new opisthocœlous dinosaur.
Amer. Naturalist, xii, p. 406.
Reprinted in Ann. and Mag. Nat. Hist. 5, vol. ii, p. 194.
Notice of discovery of *Epanterias.*

———— 1878 W
Professor Marsh on Permian reptiles.
Amer. Naturalist, xii, pp. 406-408.
Refers to paper by Professor Marsh in Amer. Jour. Science, May, 1878.

———— 1878 X
The species of *Rhinoceros* of the Loup Fork epoch.
Amer. Naturalist, xii, pp. 488-489.

———— 1878 Y
A new species of *Amphicœlias.*
Amer. Naturalist, xii, pp. 563-564, with 1 woodcut.
Describes *A. fragillimus.*

———— 1878 Z
A new *Diadectes.*
Amer. Naturalist, xii, p. 565.
Describes *D. molaris.*

———— 1878 AA
The vertebræ of *Rachitomus.*
Amer. Naturalist, xii, p. 633.

Cope, E. D.—Continued. 1878 BB
A fossil walrus discovered at Portland, Me.
Amer. Naturalist, xii, p. 633.

———— 1878 CC
The Theromorphous *Reptilia.*
Amer. Naturalist, xii, pp. 829-830.

———— 1878 DD
The homology of the chevron bones.
Amer. Naturalist, xii, p. 319.

———— 1878 EE
Palæontology of Georgia.
Amer. Naturalist, xii, p. 128.
Announces the discovery of fossil reptiles in Cretaceous and Tertiary deposits in Georgia.

———— 1878 FF
Triassic saurians from Pennsylvania.
Amer. Naturalist, xii, p. 58.

———— 1878 GG
A new deer from Indiana.
Amer. Naturalist, iv, p. 189.
Describes *Cariacus dolichopsis.*

———— 1878 HH
Mount Lebanon fishes in Dakota.
Amer. Naturalist, xii, p. 57.

———— 1879 A
The relations of the horizons of extinct *Vertebrata* of Europe and America.
Bull. U. S. Geol. and Geog. Surv. Terr., 1880, v, pp. 33-54.

———— 1879 B
Observations on the faunæ of the Miocene Tertiaries of Oregon.
Bull. U. S. Geol. and Geog. Surv. Terr., 1880, v, pp. 55-69.

———— 1879 C
On the extinct species of *Rhinoceridæ* of North America and their allies.
Bull. U. S. Geol. and Geog. Surv. Terr., 1879-80, v, pp. 227-237.
An adaptation of this paper was published in Amer. Naturalist, xiii, 1879, pp. 771a-771j, with 8 woodcuts.

———— 1879 D
On some characters of the Miocene fauna of Oregon.
Proc. Amer. Philos. Soc., xviii, pp. 63-78.
Also in Pal. Bull. No. 30, pp. 1-16: said to have been issued Dec. 3, 1878.

———— 1879 E
Second contribution to a knowledge of the Miocene fauna of Oregon.
Proc. Amer. Philos. Soc., xviii, pp. 370-376.
Not issued until January, 1880. The same was issued as Pal. Bull. No. 31, pp. 1-7, Dec., 1879.
A notice of this paper with list of species described occurs in Amer. Naturalist, xiv, p. 60.

Cope, E. D.—Continued.   1879 F

On the genera of *Felidæ* and *Canidæ*.
Proc. Acad. Nat. Sci. Phila., 1879, pp. 168–194.
Reprinted in Ann. and Mag. Nat. Hist. (5), vol. v, pp. 36–45, 92–107.
Contains list of genera and species of *Felidæ*, the genera of *Canidæ*, and descriptions of new species.

————   1879 G

The origin of the specialized teeth of the *Carnivora*.
Amer. Naturalist, xiii, pp. 171–173.

————   1879 H

Extinct *Mammalia* of Oregon.
Amer. Naturalist, xiii, p. 181.

————   1879 I

The necks of the *Sauropterygia*.
Amer. Naturalist, xiii, p. 182.

————   1879 J

The scales of *Liodon*.
Amer. Naturalist, xiii, p. 182.
A brief note calling attention to observation of Professor Snow and to his paper in Review of Science and Industry.

————   1879 K

Merycopater and *Hoplophoneus*.
Amer. Naturalist, xiii, p. 197.

————   1879 L

A new genus of *Perissodactyla*.
Amer. Naturalist, xiii, p. 270.
Establishes the genus *Anchisodon*.

————   1879 M

A new genus of *Ichthyopterygia*.
Amer. Naturalist, xiii, p. 271.
Notice of O. C. Marsh's genus, *Sauranodon*.

————   1879 N

The Amyzon Tertiary beds.
Amer. Naturalist, xiii, p. 332.

————   1879 O

A sting ray from the Green River shales of Wyoming.
Amer. Naturalist, xiii, p. 333.
The genus *Xiphotrygon* established.

————   1879 P

American *Aceratheria*.
Amer. Naturalist, xiii, p. 333.

————   1879 Q

The lower jaw of *Loxolophodon*.
Amer. Naturalist, xiii, p. 334.

————   1879 R

New Jurassic *Dinosauria*.
Amer. Naturalist, xiii, pp. 402–404, with 3 wood cuts.
Descriptions of *Camarasaurus leptodirus* and *Hypsirophus seeleyanus*.

Cope, E. D.—Continued.   1879 S

A new *Anchitherium*.
Amer. Naturalist, xiii, pp. 462–463.
Description of *A. præstans*.

————   1879 T

A decade of dogs.
Amer. Naturalist, xiii, p. 530.
A brief note giving a list of ten species of dogs furnished by the Truckee beds in Oregon.

————   1879 U

The cave bear of California.
Amer. Naturalist, xiii, p. 791.
Reprinted in Ann. and Mag. Nat. Hist. (5), vol. v, pp. 260–261; Amer. Jour. Sci., 1880, xix, p. 155.
Describes *Arctotherium simum*.

————   1879 V

On the extinct American rhinoceroses and their allies.
Amer. Naturalist, xiii, pp. 771a–771j, with 8 figs. in the text.

————   1879 W

Palæontological report of the Princeton scientific expedition of 1877.
Amer. Naturalist, xiii, p. 33.
A brief notice of the results of the expedition referred to in the title.

————   1879 X

Scientific news.
Amer. Naturalist, xiii, pp. 798a–798b.
Contains brief notice of excursions to Colorado and descriptions of *Archælurus debilis* and *Hoplophoneus platycopis*.

————   1880 A

Second contribution to the history of the *Vertebrata* of the Permian formation of Texas.
Proc. Amer. Philos. Soc., xix, pp. 38–58.
Issued also as Pal. Bull. No. 32, pp. 1–22.

————   1880 B

On the foramina perforating the posterior part of the squamosal bone of the *Mammalia*.
Proc. Amer. Philos. Soc., xviii, pp. 452–461.

————   1880 C

On the genera of the *Creodonta*.
Proc. Amer. Philos. Soc., xix, pp. 76–82.
Synopsis in Kosmos, x, pp. 299–300.

————   1880 D

On the extinct cats of America.
Amer. Naturalist, xiv, pp. 833–858, with 15 wood cuts in the text.

————   1880 E

Hill's Kansas explorations.
Amer. Naturalist, xiv, p. 141.
Note on fossils collected by Russell S. Hill.

Cope, E. D.—Continued.   1880 F
Notes on sabre-tooths.
Amer. Naturalist, xiv, p. 143.
New genus *Pogonodon* established.

——   1880 G
A new *Hippidium*.
Amer. Naturalist, xiv, p. 223.
H. spectans, from Oregon, described.

——   1880 H
The skull of *Empedocles*.
Amer. Naturalist, xiv, p. 304.

——   1880 I
A new genus of Tapiroids.
Amer. Naturalist, xiv, pp. 382-383.
The genus *Triplopus* characterized.

——   1880 J
The structure of the Permian *Ganocephala*.
Amer. Naturalist, xiv, pp. 383-384.

——   1880 K
A new genus of *Rhinocerontidæ*.
Amer. Naturalist, xiv, p. 540.
The genus *Peraceras* established.

——   1880 L
Extinct *Batrachia*.
Amer. Naturalist, xiv, pp. 609-610.

——   1880 M
The genealogy of the American rhinoceroses.
Amer. Naturalist, xiv, pp. 610-611.
The new genus *Cænopus* proposed.

——   1880 N
The bad lands of the Wind River and their fauna.
Amer. Naturalist, xiv, pp. 745-748.
Two new genera and nine new species proposed.

——   1880 O
The northern Wasatch fauna.
Amer. Naturalist, xiv, p. 908.

——   1880 P
Traquair on *Platysomidæ*.
Amer. Naturalist, xiv, pp. 439-440.
Review of Traquair, 1879 A.

——   1880 Q
The Geological Record.
Amer. Naturalist, xiv, p. 512.
A notice of the Geological Record for 1877, with citations of some errors.

——   1880 R
Marsh on Jurassic *Dinosauria*.
Amer. Naturalist, xiv, p. 302.

——   1880 S
Remarks on fossil vertebrates from California. *Elotherium, Mastodon obscurus*.
Amer. Naturalist, xiv, p. 62.

Cope, E. D.—Continued.   1880 T
The Manti beds of Utah.
Amer. Naturalist, xiv, p. 303.

——   1881 A
On some new *Batrachia* and *Reptilia* from the Permian beds of Texas.
Bull. U. S. Geol. and Geog. Surv. Terr., vi, article ii, pp. 79-82.

——   1881 B
On a wading bird from the Amyzon shales.
Bull. U. S. Geol. and Geog. Surv. Terr., vi, article iii, pp. 83-85.
Describes *Charadrius sheppardianus*.

——   1881 C
On the *Nimravidæ* and *Canidæ* of the Miocene period.
Bull. U. S. Geol. and Geog. Surv. Terr., vi, article iii, pp. 165-181.

——   1881 D
On the *Vertebrata* of the Wind River Eocene beds of Wyoming.
Bull. U. S. Geol. and Geog. Surv. Terr, vi, article viii, pp. 183-202.
Brief notice in Amer. Naturalist, vol. xv, p. 74 (Cope, 1881 CC).

——   1881 E
Review of the *Rodentia* of the Miocene period of North America.
Bull. U. S. Geol. and Geog. Surv. Terr., vi, article xv, pp. 361-386.

——   1881 F
On the *Canidæ* of the Loup Fork epoch.
Bull. U. S. Geol. and Geog. Surv. Terr., vi, article xvi, pp. 387-390.

——   1881 G
The systematic arrangement of the order *Perissodactyla*.
Proc. Amer. Philos. Soc., xix, pp. 377-401.

——   1881 H
On some *Mammalia* of the lowest Eocene beds of New Mexico.
Proc. Amer. Philos. Soc., xix, pp. 484-495.
Issued also as Pal. Bull. No. 33.

——   1881 I
On the origin of the foot structures of the Ungulates.
Amer. Naturalist, xv, pp. 269-273, with 5 figs. in the text.

——   1881 J
On the effect of impacts and strains on the feet of *Mammalia*.
Amer. Naturalist, xv, pp. 542-548, with 11 figs. in the text.

Cope, E. D.—Continued.   1881 K

Catalogue of *Vertebrata* of the Permian formation of the United States.

Amer. Naturalist, xv, pp. 162-164.

———   1881 L

Extinct palæozoic fishes from Canada.

Amer. Naturalist, xv, pp. 252-253.
Notice of a paper read by J. F. Whiteaves.

———   1881 M

A new fossil bird.

Amer. Naturalist, xv, p. 253.
Notice of Dr. J. A. Allen's description of *Palæospiza bella*.

———   1881 N

Mammalia of the Lower Eocene beds.

Amer. Naturalist, xv, pp. 337-338.
Description of *Peryptichus carinidens* and *Deltatherium fundaminis*, new genera and new species.

———   1881 O

The classification of the *Perissodactyla*.

Amer. Naturalist, xv, p. 340.

———   1881 P

Miocene dogs.

Amer. Naturalist, xv, p. 497.
Contains correction of dental formula of *Hyænocyon* and proposal of new genus *Oligobunis*.

———   1881 Q

The *Rodentia* of the American Miocene.

Amer. Naturalist, xv, pp. 586-587.
A catalogue of the known species, together with place of original description and the horizon of occurrence.

———   1881 R

A new *Clidastes* from New Jersey.

Amer. Naturalist, xv, pp. 587-588.
Description of *C. conodon*.

———   1881 S

The temporary dentition of a new Creodont.

Amer. Naturalist, xv, pp. 667-669.
Description of *Triisodon quivirensis* and *Deltatherium absarokæ*.

———   1881 T

A Laramie saurian in the Eocene.

Amer. Naturalist, xv, pp. 669-670.
A new species, *Champsosaurus australis*, found in probably the Puerco.

———   1881 U

Mammalia of the lowest Eocene.

Amer. Naturalist, xv, pp. 829-831.
Descriptions of *Conoryctes comma*, gen. et spec. nov., *Catathlæus rhabdodon*, gen. et spec. nov., and *Mioclænus turgidus*, gen. et spec. nov.

Cope, E. D.—Continued.   1881 V

Eocene *Plagiaulacidæ*.

Amer. Naturalist, xv, pp. 921-922.
Description of *Ptilodus mediævus*, gen. et spec. nov.

———   1881 W

Belodon in New Mexico.

Amer. Naturalist, xv, pp. 922-923.
Descriptions of *B. buceros* and *B. scolopax*.

———   1881 X

A new type of *Perissodactyla*.

Amer. Naturalist, xv, pp. 1017-1018.
Refers *Phenacodus* to the *Perissodactyla*.

———   1881 Y

New genus of *Perissodactyla diplarthra*.

Amer. Naturalist, xv, p. 1018.
Refers *Hyracotherium tapirinum* to a new genus, *Systemodon*.

———   1881 Z

Notes on *Creodonta*.

Amer. Naturalist, xv, pp. 1018-1020.
Describes *Lipodectes penetrans* and *L. petridens*, new species, and establishes the genus *Dissacus*.

———   1881 AA

The Permian formation of New Mexico.

Amer. Naturalist, xv, pp. 1020-1021.
New species described, *Eryops reticulatus* and *Zatrachys apiculus*.

———   1881 BB

Geological news.

Amer. Naturalist, xv, pp. 254, 340, 413, 1023.

———   1881 CC

The *Vertebrata* of the Eocene of the Wind River basin.

Amer. Naturalist, xv, pp. 74-75.
Consists of a notice of the paper Cope 1881 D, and contains description of *Bathyopsis fissidens* Cope.

———   1882 A

The classification of the ungulate *Mammalia*.

Proc. Amer. Philos. Soc., xx, pp. 438-447, with 10 figs. in the text.
Issued also as part of Pal. Bul. No. 35.

———   1882 B

Third contribution to the history of the *Vertebrata* of the Permian formation of Texas.

Proc. Amer. Philos. Soc., xx, pp. 447-461.
Issued also as part of Pal. Bull. No. 35.

———   1882 C

Synopsis of the *Vertebrata* of the Puerco Eocene epoch.

Proc. Amer. Philos. Soc., xx, pp. 461-471.
Also as part of Pal. Bull. No. 35.

Cope, E. D.—Continued. 1882 D

On the systematic relations of the *Carnivora Fissipedia*.

Proc. Amer. Philos. Soc., xx, pp. 471–475. Issued also as part of Pal. Bull. No. 35. Reprinted in Ann. and Mag. Nat. Hist. (5), xii, pp. 112–116.

——— 1882 E

Contributions to the history of the *Vertebrata* of the Lower Eocene of Wyoming and New Mexico, made during 1881. I. The fauna of the Wasatch beds of the basin of the Big Horn River. II. The fauna of the Catathlæus beds, or Lowest Eocene, New Mexico.

Proc. Amer. Philos. Soc., xx, pp. 139–197. Issued also as Pal. Bull. No. 34. For notices of this paper see Amer. Jour. Sci., vol. xxiii, p. 325.

——— 1882 F

On the *Condylarthra*.

Proc. Acad. Nat. Sci. Phila., 1882, pp. 95–97. Reprinted in Ann. and Mag. Nat. Hist. (5), x, pp. 76–79.

——— 1882 G

Contemporaneity of man and Pliocene mammals.

Proc. Acad. Nat. Sci. Phila., 1882, pp. 291–292.

——— 1882 H

On *Uintatherium*, *Bathmodon*, and *Triisodon*.

Proc. Acad. Nat. Sci. Phila., 1882, pp. 294–300.

——— 1882 I

The Tertiary formations of the central region of the United States.

Amer. Naturalist, xvi, pp. 177–196, with 5 wood cuts in the text.

——— 1882 J

The reptiles of the American Eocene.

Amer. Naturalist, xvi, pp. 979–993, with 13 wood cuts in the text.

——— 1882 K

The oldest artiodactyle.

Amer. Naturalist, xvi, p. 71. Reprinted in Ann. and Mag. Nat. Hist. (5), ix, p. 204. Description of *Mioclænus brachystomus*.

——— 1882 L

The characters of the *Tæniodonta*.

Amer. Naturalist, xvi, p. 72. Reprinted in Ann. and Mag. Nat. Hist. (5), ix, pp. 205–206.

——— 1882 M

New forms of *Coryphodontidæ*.

Amer. Naturalist, xvi, p. 73.

Cope, E. D.—Continued. 1882 N

An anthropomorphous lemur.

Amer. Naturalist, xvi, pp. 73–74. Describes *Anaptomorphus homunculus*.

——— 1882 O

Recent extinction of the mastodon.

Amer. Naturalist, xvi, pp. 74–75. Refers to Prof. John Collett's "Geological Report of Indiana for 1880."

——— 1882 P

A new genus of *Tillodonta*.

Amer. Naturalist, xvi, pp. 156–157. Describes *Psittacotherium multifragum*.

——— 1882 Q

Marsh on the classification of the *Dinosauria*.

Amer. Naturalist, xvi, pp. 253–255.

——— 1882 R

New characters of the *Perissodactyla Condylarthra*.

Amer. Naturalist, xvi, p. 334.

——— 1882 S

Mesonyx and *Oxyæna*.

Amer. Naturalist, xvi, p. 334.

——— 1882 T

The rhachitomous *Stegocephala*.

Amer. Naturalist, xvi, pp. 334–335.

——— 1882 U

A second genus of Eocene *Plagiaulacidæ*.

Amer. Naturalist, xvi, pp. 416–417. Description of *Catopsalis foliatus*.

——— 1882 V

Two new genera of the Puerco Eocene.

Amer. Naturalist, xvi, pp. 417–418. Descriptions of *Haploconus lineatus* and *Pantolambda bathmodon*, new genera and new species.

——— 1882 W

The ancestry and habits of *Thylacoleo*.

Amer. Naturalist, xvi, pp. 520–522. Correction of the table of genera of *Plagiaulacidæ*.

——— 1882 X

Notes on Eocene *Mammalia*.

Amer. Naturalist, xvi, p. 522. *Lipodectes penetrans* is a synonym of *Deltatherium fundaminis*. *D. absarokæ* is referred to *Didelphodus*. *Oligotomus osbornianus* is referred to *Ectocion*.

——— 1882 Y

On the *Taxeopoda*, a new order of *Mammalia*.

Amer. Naturalist, xvi, pp. 522–523.

Cope, E. D.—Continued.   1882 Z
A new genus of *Tæniodonta*.
Amer. Naturalist, xvi, pp. 604–605.
Description of *Tæniolabis sulcatus.*

————   1882 AA
New marsupials from the Puerco
Eocene.
Amer. Naturalist, xvi, pp. 684–686.
Polymastodon taöensis, gen. et spec. nov.;
Ptilodus trovessartianus, spec. nov.; *Haploconus
entoconus*, spec. nov.; and *H. gillianus* spec.
nov.

————   1882 BB
Mammalia in the Laramie formation.
Amer. Naturalist, xvi, pp. 830–831.
Description of *Meniscoëssus conquistus*, gen.
et spec. nov.

————   1882 CC
A new form of *Tæniodonta*.
Amer. Naturalist, xvi, pp. 831–832.
Describes *Hemiganus vultuosus*, gen. et spec.
nov.

————   1882 DD
The *Periptychidæ*.
Amer. Naturalist, xvi, pp. 832–833.
Periptychidæ, fam. nov. established; *Hemi-
thlæus kowalevskianus*, gen. et spec. nov., and
Anisonchus coniferus described.

————   1882 EE
Some new forms from the Puerco
Eocene.
Amer. Naturalist, xvi, pp. 833–834.
Mioclænus opisthacus, *M. baldwini*, *Protogonia
plicifera*, and *Dissacus carnifex* described.

————   1882 FF
The recent discoveries of fossil foot-
prints in Carson, Nev.
Amer. Naturalist, xvi, pp. 921–923.
Notice of a paper by Prof. Jos. Le Conte.

————   1882 GG
Two new genera of *Mammalia* from
the Wasatch Eocene.
Amer. Naturalist, xvi, p. 1029.
Genera *Diacodexis* and *Heptodon* estab-
lished.

————   1882 HH
Permian *Vertebrata*.
Amer. Naturalist, xvi, p. 925.
A short note.

————   1882 II
Geological news.
Amer. Naturalist, xvi, pp. 925–926.
Mention of *Orthocynodon* of Scott and
Osborn.

————   1882 JJ
Scientific news.
Amer. Naturalist, xvi, pp. 583–534.

————   1883 A
On a new extinct genus and species
of *Percidæ* from Dakota Territory.
Amer. Jour. Sci. (3), xxv, pp. 414–416.

Cope, E. D.—Continued.   1883 B
The evidence for evolution in the
history of the extinct *Mammalia*.
Science, ii, pp. 272–279.
Address at the Amer. Assoc. Adv. Science,
at Minneapolis, Minn. Reprinted in *Nature*,
vol. xxix, pp. 227–230, 248–250.

————   1883 C
The structure of the skull in *Diclo-
nius mirabilis*, a Laramie dinosaurian.
Science (1), ii, p. 338.
Abstract of paper read before Amer. Assoc.
Adv. Sci.

————   1883 D
The classification of the *Ungulata*.
Proc. Amer. Assoc. Adv. Science, 31st meet-
ing, Montreal, 1882, pp. 477–479.
Abstract of paper read.

————   1883 E
The fauna of the Puerco Eocene.
Proc. Amer. Assoc. Adv. Science, 31st meet-
ing, Montreal, 1882, pp. 479–480.
Abstract of paper read before A. A. A. S.

————   1883 F
On a new basin of White River age
in Dakota.
Proc. Amer. Philos. Soc., xxi, pp. 216–217.

————   1883 G
On the brains of the Eocene *Mamma-
lia*, *Phenacodus* and *Periptychus*.
Proc. Amer. Philos. Soc., xx, pp. 563–565,
with pl. i and ii.
Issued also as part of Pal. Bull. No. 36.

————   1883 H
Fourth contribution to the history of
the Permian formation in Texas.
Proc. Amer. Philos. Soc., xx, pp. 628–636.
Issued also as part of Pal. Bull. No. 36.

————   1883 I
On the mutual relations of the buno-
therian *Mammalia*.
Proc. Acad. Nat. Sci. Phila., 1883, pp. 77–83.
Reprinted in Ann. and Mag. Nat. Hist. (5),
xii, pp. 20–26.

————   1883 J
On the characters of the skull in the
Hadrosauridæ.
Proc. Acad. Nat. Sci. Phila., 1883, pp. 97–107,
with pls. iv–vii.

————   1883 K
On some *Vertebrata* from the Permian
of Illinois.
Proc. Acad. Nat. Sci. Phila., 1883, pp. 108–110.
Diplodus replaced by *Didymodus*; *Thoracodus emydinus* gen. et spec. nov., *Ctenodus heterolophus* sp. nov., *Ctenodus vabasensis* sp. nov.,
described.

Cope, E. D.—Continued.   1883 L

On the fishes of the recent and Pliocene lakes of the western part of the Great Basin, and of the Idaho Pliocene lake.

Proc. Acad. Nat. Sci. Phila., 1883, pp. 134–166, with a map of the region.

———   1883 M

First addition to the fauna of the Puerco Eocene.

Proc. Amer. Philos. Soc., xx, pp. 545–563.
Issued also as part of Pal. Bull. No. 36.
Contains descriptions of 2 new genera and 8 species; also "Note on the Mammalia of the Puerco and the origin of the quadritubercular molar."

———   1883 N

On a new extinct genus of *Sirenia* from South Carolina.

Proc. Acad. Nat. Sci. Phila., 1883, pp. 52–54.
Describes *Dioplotherium manigaulti.*

———   1883 O

The tritubercular type of superior molar tooth.

Proc. Acad. Nat. Sci. Phila., 1883, p. 56.

———   1883 P

Permian fishes and reptiles.

Proc. Acad. Nat. Sci. Phila., 1883, p. 69.

———   1883 Q

On some fossils of the Puerco formation.

Proc. Acad. Nat. Sci. Phila., 1883, pp. 168–170.

———   1883 R

On extinct *Rhinoceri* from the Southwest.

Proc. Acad. Nat. Sci. Phila., 1883, p. 301.

———   1883 S

The extinct *Rodentia* of North America.

Amer. Naturalist, xvii, pp. 43–57, 165–174, 370–381, with 30 wood cuts in the text.

———   1883 T

On the extinct dogs of North America.

Amer. Naturalist, xvii, pp. 235–249, with 14 wood cuts in the text.

———   1883 U

On *Uintatherium* and *Bathmodon.*
Amer. Naturalist, xvii, p. 68.

———   1883 V

The Nevada biped tracks.
Amer. Naturalist, xvii, pp. 69–71, with 3 figs.

Cope, E. D.—Continued.   1883 W

New *Mammalia* from the Puerco Eocene.

Amer. Naturalist, xvii, p. 191.
Notice and abstract of paper read before Amer. Philos. Soc. (Cope, 1883 M.)

———   1883 X

A new fossil Sirenian.

Amer. Naturalist, xvii, p. 309.
Abstract of Cope, E. D., 1883 N.

———   1883 Y

Lydekker on Indian *Mammalia.*
Amer. Naturalist, xvii, pp. 405–406.
Discusses the standing of certain genera of *Rhinocerotidæ.*

———   1883 Z

The ancestor of *Coryphodon.*
Amer. Naturalist, xvii, pp. 406–407.

———   1883 AA

Note on the trituberculate type of molar and the origin of the quadrituberculate.

Amer. Naturalist, xvii, pp. 407–408.

———   1883 BB

The genus *Phenacodus.*
Amer. Naturalist, xvii, p. 535, pl. xii.

———   1883 CC

The structure and appearance of the Laramie dinosaurian.

Amer. Naturalist, xvii, pp. 774–777, pls. xvi–xix.
Description of *Diclonius mirabilis.*

———   1883 DD

The "third trochanter" of the dinosaurs.

Amer. Naturalist, xvii, p. 869.
Notice of observations made by M. Dollo.

———   1883 EE

Some new *Mammalia* of the Puerco formation.

Amer. Naturalist, xvii, p. 968.
Abstract of paper, Cope, E. D., 1883 Q.

———   1883 FF

The progress of the *Ungulata* in Tertiary time.

Amer. Naturalist, xvii, pp. 1056–1057.
Abstract of paper read before the A. A. A. S. in Minneapolis, 1883.

———   1883 GG

A new Chondrostean from the Eocene.

Amer. Naturalist, xvii, pp. 1152, 1153.
Describes *Crossopholis magnicaudatus* gen. et spec. nov.

Cope, E. D.—Continued.   1883 HH
The Carson footprints.
Amer. Naturalist, xvii, p. 1158.

1883 II
Geological notes. Tertiary.
Amer. Naturalist, xvii, pp. 970-971.

1884 A
A Carboniferous genus of sharks still living.
Science, iii, p. 275.
Refers to S. Garman's *Chlamydoselachus.*

1884 B
Pleuracanthus and *Didymodus.*
Science, iii, pp. 645-646.
Reply to Dr. Theo. Gill (Gill, 1884 A.).

1884 C
[On the phylogeny of American artiodactyle mammals.]
Science, iv, p. 339.
Abstract of paper read before Amer. Assoc. Adv. Sci., 1884.

1884 D
[On saurians of the Permian epoch.]
Science, iv, p. 340.
Abstract of paper read before Amer. Assoc. Adv. Sci., 1884.

1884 E
On the trituberculate type of molar tooth in the *Mammalia.*
Proc. Amer. Assoc. Adv. Science, 32d meeting, Minneapolis, 1883, pp. 313-316.
Abstract of paper read.

1884 F
On the structure of the skull in *Diclonius mirabilis,* a Laramie dinosaurian.
Proc. Amer. Assoc. Adv. Sci., 32d meeting, Minneapolis, 1883, pp. 315-316; abstract.

1884 G
The extinct *Mammalia* of the Valley of Mexico.
Proc. Amer. Philos. Soc., xxii, pp. 1-21.
Also as part of Pal. Bull. No. 39.

1884 H
On the structure of the feet in the extinct *Artiodactyla* of North America.
Proc. Amer. Philos. Soc., xxii, pp. 21-27.
Also as part of Pal. Bull. No. 39.

1884 I
Fifth contribution to the knowledge of the fauna of the Permian Formation of Texas and the Indian Territory.
Proc. Amer. Philos. Soc., xxii, pp. 28-47, pl. i.
Also as part of Pal Bull. No. 39.
Besides descriptions of new species, this paper discusses: The posterior foot in Pelycosauria; the structure of the columella auris in *Clepsydrops leptocephalus;* the structure of the quadrate bone in the genus *Clepsydrops;* the articulation of the ribs in *Embolophorus;* the origin of the *Mammalia;* the tarsus; phylogeny of the *Vertebrata.*

Cope, E. D.—Continued.   1884 J
On the distribution of the Loup Fork formation in New Mexico.
Proc. Amer. Philos. Soc., xxi, pp. 306-309.
Issued also as part of Pal. Bull. No. 37.

1884 K
Second addition to the knowledge of the Puerco epoch.
Proc. Amer. Philos. Soc., xxi, pp. 309-324.
Issued also as part of Pal. Bull. No. 37.

1884 L
On the trituberculate type of molar tooth in the *Mammalia.*
Proc. Amer. Philos. Soc., xxi, pp. 324-326.
Issued also as part of Pal. Bull. No. 37.
This paper is practically the same as Cope, 1884 E.

1884 M
Synopsis of the species of *Oreodontidæ.*
Proc. Amer. Philos. Soc., xxi, pp. 503-572.
Issued also as part of Pal. Bull. No. 38.

1884 N
On the structure of the skull in the elasmobranch genus *Didymodus.*
Proc. Amer. Philos. Soc., xxi, pp. 572-590, with 1 pl.
Issued also as part of Pal. Bull. No. 38.

1884 O
The *Vertebrata* of the Tertiary formations of the West. Book I.
Report U. S. Geolog. Survey of the Territories, F. V. Hayden, U. S. geologist in charge, 1884, Washington, iii, pp. i-xxxv, 1-1009; pls. i-lxxva.
This work was not actually issued until the year 1885.
Reviewed in Geol. Magazine, (3) iii, pp. 410-419, 465-477, 512-521.

1884 P
The *Batrachia* of the Permian period of North America.
Amer. Naturalist, xviii, pp. 26-39, with pls. ii-v and 7 woodcuts.

1884 Q
The Loup Fork beds on the Gila River.
Amer. Naturalist, xviii, pp. 58-59.
Mentions finding of *Aphelops fossiger.*

1884 R
On new lemuroids from Puerco formation.
Amer. Naturalist, xviii, pp. 59-62.

1884 S
The *Creodonta.*
Amer. Naturalist, xviii, pp. 255-267, 344-353, 478-485, with 30 woodcuts in the text.

Cope, E. D.—Continued.   1884 T

The history of the *Oreodontidæ*.

Amer. Naturalist, xviii, pp. 280–282.

Synopsis of portion of a paper read before Amer. Philos. Society.

——   1884 U

The skull of a still living shark of the Coal-measures.

Amer. Naturalist, xviii, pp. 412–413.

Refers *Chlamydoselachus anguineus* to *Didymodus*.

——   1884 V

The mastodons of North America.

Amer. Naturalist, xviii, pp. 524–526.

Descriptions of 9 species.

——   1884 W

Marsh on *Diplodocus*.

Amer. Naturalist, xviii, p. 526.

——   1884 X

The pelvisternum of Edentates.

Amer. Naturalist, xviii, pp. 639–640.

Notice of paper by P. Albrecht.

——   1884 Y

The Tertiary *Marsupialia*.

Amer. Naturalist, xviii, pp. 686–697, with 9 illus.

——   1884 Z

Lydekker on extinct *Mammalia* of India.

Amer. Naturalist, xviii, pp. 617–618.

——   1884 AA

The *Condylarthra*.

Amer. Naturalist, xviii, pp. 790–805, 892–906; pls. xxviii–xxx and 28 woodcuts in the text.

——   1884 BB

The *Choristodera*.

Amer. Naturalist, xviii, pp. 815–817.

Remarks on paper by Dr. Lemoine on *Simœdosaurus*.

——   1884 CC

The genus *Pleuracanthus*.

Amer. Naturalist, xviii, p. 818; pl. xxiii.

Plate containing figures of *Didymodus compressus* and *D. platypterus*.

——   1884 DD

Observations on the phylogeny of the *Artiodactyla* derived from American fossils.

Amer. Naturalist, xviii, pp. 1034–1036.

——   1884 EE

The *Amblypoda*.

Amer. Naturalist, xviii, pp. 1110–1121, 1192–1202, with 23 woodcuts in the text.

Concluded in vol. xix, pp. 40–55.

——   1884 FF

The origin of the *Mammalia*.

Amer. Naturalist, xviii, pp. 1136–1137.

Synopsis of paper read before the Amer. Assoc. Adv. Sci., 1884.

Cope, E. D.—Continued.   1884 GG

The structure of the columella auris in *Clepsydrops leptocephalus*.

Amer. Naturalist, xviii, pp. 1253–1256, pl. 38.

——   1884 HH

Note on the phylogeny of the *Vertebrata*.

Amer. Naturalist, xviii, pp. 1255–1257.

——   1884 II

The evidence for evolution in the history of the extinct *Mammalia*.

Nature, xxix, pp. 227–230, 248–250.

Abstract of lecture given before Amer. Assoc. Adv. Sci., 1884.

——   1884 JJ

Genus *Equus*.

Fourteenth Report State Geologist Indiana, pt. ii, pp. 40–41, being an appendix to article by Cope and Wortman.

——   1885 A

The relations between the theromorphous reptiles and the monotreme *Mammalia*.

Proc. Amer. Asso. Adv. Sci., 33d meeting, Phila., 1884, pp. 471–482, with 1 pl.

——   1885 B

On the structure of the feet in the extinct *Artiodactyla* of North America.

Proc. Amer. Assoc. Adv. Sci., 33d meeting, Phila., 1884, pp. 482–490.

——   1885 C

The occurrence of man in the upper Miocene of Nebraska.

Proc. Amer. Assoc. Adv. Sci., 33d meeting, Phila., 1884, p. 593; abstract.

——   1885 D

Second continuation of researches among the *Batrachia* of the Coal-measures of Ohio.

Proc. Amer. Philos. Soc., xxii, pp. 405–408.

Also forming a portion of Pal. Bull. No. 40.

——   1885 E

On the contents of a bone cave in the island of Anguilla (West Indies).

Smithsonian Contributions to Knowledge, xxv, article iii, pp. 1–30, pls. i–iv.

——   1885 F

The *Amblypoda*.

Amer. Naturalist, xix, pp. 40–55, figs. 24–35.

Treats of the *Dinocerata*, and is continued from p. 1202, vol. xviii.

——   1885 G

On the evolution of the *Vertebrata*, progressive and retrogressive.

Amer. Naturalist, xix, pp. 140–148, 234–247, 341–353.

See also Cope, E. D., 1885 AA.

Cope, E. D.—Continued.   1885 H
   The White River beds of Swift Cur-
   rent River, Northwest Territory.
   Amer. Naturalist, xix, p. 163.

———   1885 I
   The position of *Pterichthys* in the
   system.
   Amer. Naturalist, xix, pp. 289-291, with 2
   figs.

———   1885 J
   The oldest Tertiary *Mammalia*.
   Amer. Naturalist, xix, pp. 385-387.

———   1885 K
   The *Lemuroidea* and the *Insectivora*
   of the Eocene period of North America.
   Amer. Naturalist, xix, pp. 457-471, with 18
   woodcuts in the text.

———   1885 L
   The mammalian genus *Hemiganus*.
   Amer. Naturalist, xix, p. 492.

———   1885 M
   Mammals from the Lower Eocene of
   New Mexico.
   Amer. Naturalist, xix, pp. 493-494.

———   1885 N
   The genera of the *Dinocerata*.
   Amer. Naturalist, xix, p. 594.

———   1885 O
   Marsh on the *Dinocerata*.
   Amer. Naturalist, xix, pp. 703-705.

———   1885 P
   Garman on *Didymodus*.
   Amer. Naturalist, xix, pp. 878-879.

———   1885 Q
   The relations of the Puerco and Lar-
   amie deposits.
   Amer. Naturalist, xix, pp. 985-986.

———   1885 R
   The ankle and skin of the dinosaur,
   Diclonius mirabilis.
   Amer. Naturalist, xix, p. 1208; pl. xxxvii,
   figs. 1-3.

———   1885 S
   Pliocene   horses   of   southwestern
   Texas.
   Amer. Naturalist, xix, pp. 1208-1209; pl.
   xxxvii, figs. 4-6.

———   1885 T
   Mr. Lydekker on *Esthonyx*.
   Geol. Magazine (3), ii, pp. 526-527.

———   1885 U
   Palæontological nomenclature.
   Geol. Magazine (3), ii, pp. 572-575.

Cope, E. D.—Continued.   1885 V
   The structure of the columella auris
   in the *Pelycosauria*.
   Memoirs Nat. Acad. Sci., iii, pp. 93-96, with
   5 figs. in the text.

———   1885 W
   The *Batrachia* of the Permian beds
   of Bohemia and the Labyrinthodont
   from the Bijori group (India).
   Amer. Naturalist, xix, pp. 592-594.
   A brief review of works by Fritsch and Lydekker.

———   1885 X
   Polemics in palæontology.
   Amer. Naturalist, xix, pp. 1207-1208.

———   1885 Y
   Eocene paddle-fish and *Gonorhynchi-
   dæ*.
   Amer. Naturalist, xix, pp. 1090-1091.

———   1885 Z
   A contribution to the vertebrate palæontology of Brazil.
   Proc. Amer. Philos. Soc., xxiii, pp. 1-21, with
   pl. i.
   Also as part of Pal. Bull. No. 40, pp. 1-21,
   with pl. i.

———   1885 AA
   The genealogy of the *Vertebrata* as
   learned from palæontology.
   Trans. Vassar Brothers' Institute, iii, pp.
   60-80.
   The substance of this lecture was published
   in the Amer. Naturalist, xix. See Cope, E. D.,
   1885 G.

———   1886 A
   On the structure of the brain and
   auditory apparatus of a theromorphous
   reptile of the Permian epoch.
   Proc. Amer. Assoc. Adv. Science, 34th meeting, Ann Arbor, 1885, pp. 336-341.

———   1886 B
   On two new species of three-toed
   horses from the upper Miocene, with
   notes on the fauna of the Ticholeptus
   beds.
   Proc. Amer. Philos. Soc., xxiii, pp. 357-361.

———   1886 C
   The phylogeny of the *Camelidæ*.
   Amer. Naturalist, xx, pp. 611-624, with 12
   wood cuts in the text.

———   1886 D
   An interesting connecting genus of
   Chordata.
   Amer. Naturalist, xx, pp. 1027-1031.
   Description of *Mycterops ordinatus*.

Cope, E. D.—Continued.   1886 E
The sternum of the *Dinosauria*.
Amer. Naturalist, xx, pp. 153–154, with 2 figs. in the text.

——   1886 F
Corrections of notes on *Dinocerata*.
Amer. Naturalist, xx, p. 155.

——   1886 G
An extinct dog.
Amer. Naturalist, xx, p. 274.
Notice of *Pachycyon robustus* described by Dr. J. A. Allen.

——   1886 H
The vertebrate fauna of the Ticholeptus beds.
Amer. Naturalist, xx, pp. 367–369.

——   1886 I
The *Plagiaulacidæ* of the Puerco epoch.
Amer. Naturalist, xx, p. 451.
Describes *Neoplagiaulax molestus*.

——   1886 J
The long-spined *Theromorpha* of the Permian epoch.
Amer. Naturalist, xx, pp. 544–545.

——   1886 K
Schlosser on the phylogeny of the ungulate *Mammalia*.
Amer. Naturalist, xx, pp. 719–721.
Review of paper by Max Schlosser in Morphologisches Jahrbuch for 1886.
For reply to Professor Cope see Morpholog. Jahrbuch, xii, pp. 575–580.

——   1886 L
Schlosser on *Creodonta* and *Phenacodus*.
Amer. Naturalist, xx, pp. 965–967.
Review of paper by Max Schlosser in Morpholog. Jahrbuch, 1886, p. 287.

——   1886 M
Dollo on extinct tortoises.
Amer. Naturalist; xx, p. 967.
Review of paper by M. Dollo in Bull. Mus. Roy. Belgique, 1886, p. 75.

——   1886 N
A giant armadillo from the Miocene of Kansas.
Amer. Naturalist, xx, pp. 1044–1046.
Describes *Caryoderma snovianum*.

——   1886 O
The batrachian intercentrum.
Amer. Naturalist, xx, pp. 76–77.

——   1886 P
The intercentrum in *Sphenodon*.
Amer. Naturalist, xx, p. 175.

——   1886 Q
Note on *Phenacodus*.
Geol. Magazine (3), iii, pp. 238–239.

Cope, E. D.—Continued.   1886 R
Note on *Erisichthe*.
Geol. Magazine (3) iii, p. 239.

——   1886 S
Edestus and *Pelecopterus*.
Geol. Magazine (3) iii, pp. 141–142.
Refers to statements in Woodward H., 1886 A regarding *Pelecopterus*, with a reply by Mr. W. Davies, F. G. S.

——   1886 T
On two new forms of Polyodont and Gonorhynchid fishes from the Eocene of the Rocky Mountains.
Memoirs Nat. Acad. Sciences, iii, pp. 161–169, with plate containing 6 figures.
Describes *Crossopholis magnicaudatus*, *Notogoneus osculus* and *Priscacara hypsacanthus*.
See notice of this paper in Geol. Magazine, vol. v, 1888, p. 229, by A. S. Woodward.

——   1886 U
The *Vertebrata* of the Swift Current Creek region of the Cypress Hills.
Annual Report of the Geol. and Nat. Hist. Survey of Canada, i, 1885, appendix to article C, pp. 79–85.

——   1886 V
An analytical table of the genera of snakes.
Proc. Amer. Philos. Society, xxiii, pp. 479–499.

——   1886 W
Report on the coal deposits near Zacualtipan, in the State of Hidalgo, Mexico.
Proc. Amer. Philos. Soc., xxiii, pp. 146–151, with 2 text figures.
In this paper are described *Hippotherium peninsulatum*, sp. nov., and *Protohippus castilli*, sp. nov., from Loup Fork shales, near Vera Cruz.

——   1887 A
A contribution to the history of the *Vertebrata* of the Trias of North America.
Proc. Amer. Philos. Soc., xxiv, pp. 209–228, pls. i, ii.

——   1887 B
The classification and phylogeny of the *Artiodactyla*.
Proc. Amer. Philos. Soc., xxiv, pp. 377–400.

——   1887 C
The dinosaurian genus *Cælurus*.
Amer. Naturalist, xxi, pp. 367–369.

——   1887 D
The Mesozoic and Cænozoic realms of the interior of North America.
Amer. Naturalist, xxi, pp. 445–462.

Cope, E. D.—Continued.   1887 E
American Triassic *Rhynchocephalia*.
Amer. Naturalist, xxi, p. 468.
Description of *Typothorax coccinarum*.

———   1887 F
Some new *Tæniodonta* of the Puerco.
Amer. Naturalist, xxi, p. 469.
Description of *Psittacotherium megalodus*.

———   1887 G
The sea-saurians of the Fox Hills
Cretaceous.
Amer. Naturalist, xxi, pp. 563–566.
Description of *Piptomerus megaloporus*, *P. microporus*, *P. hexagonus*, and *Orophosaurus pauciporus*.

———   1887 H
The marsupial genus *Chirox*.
Amer. Naturalist, xxi, pp. 566–567, with 1 woodcut in the text.

———   1887 I
Pavlow on the ancestry of Ungulates.
Amer. Naturalist, xxi, pp. 656–658.

———   1887 J
[Note on *Belodon buceros*.]
Amer. Naturalist, xxi, pp. 659–660.

———   1887 K
Scott and Osborn on White River
Mammalia.
Amer. Naturalist, xxi, pp. 924–926.
Review of above author's paper, "Preliminary account of fossil mammals from the White River formation," etc., in Bull. Mus. Comp. Zool., vol. xiii.

———   1887 L
Marsh on new fossil *Mammalia*.
Amer. Naturalist, xxi, pp. 926–927.
Review of Marsh, 1887 B.

———   1887 M
Scott on *Creodonta*.
Amer. Naturalist, vol. xxi, p. 927.
Review of Scott in Jour. Acad. Nat. Sci., Phila., vol. ix, pp. 155–185.

———   1887 N
The *Perissodactyla*.
Amer. Naturalist, xxi, pp. 985–1007; 1060–1076, with 44 figs. in the text.

———   1887 O
Zittel's Manual of Palæontology.
Amer. Naturalist, xxi, pp. 1014–1019.
A review of the above work of Professor Zittel, and containing a classification of the Teleostomous fishes.

———   1887 P
A saber-toothed tiger from the Loup
Fork beds.
Amer. Naturalist, xxi, pp. 1019–1020.
Description of *Machærodus catocopis*.

Cope, E. D.—Continued.   1887 Q
Thomas on mammalian dentition.
Amer. Naturalist, xxi, pp. 1101–1103.
Review of Thomas, O., in Phil. Trans. Roy. Soc., vol. clxxviii B., pp. 443–486.

———   1887 R
Lydekker, Boulenger, and Dollo on
fossil tortoises.
Geol. Magazine (3), iv, pp. 572–573.

———   1887 S
The origin of the fittest. Essays on
evolution.
8 vo., pp. i–xx; 1–467, with 81 figs. in the text.
New York, 1887.
This work is composed of 21 essays, all of which had been previously published elsewhere.

———   1888 A
Mesozoic Realm.
Amer. Geologist, ii, pp. 261–268.
A portion of the reports of the American committee of the International Congress of Geologists.

———   1888 B
Report of the subcommittee on the
Cenozoic (Interior). Cenozoic Realm.
Amer. Geologist, ii, pp. 285–299.
A portion of the reports of the American committee of the International Congress of Geologists.

———   1888 C
Vertebrate fauna of the Puerco series.
Science, xi, p. 198.
Abstract of paper read before the Nat. Acad. of Sci., 1888.

———   1888 D
The mechanical origin of the sectorial
teeth of the *Carnivora*.
Proc. Amer. Assoc. Adv. Sci., 36th meeting.
New York, 1887, pp. 254–257.

———   1888 E
On the *Dicotylinæ* of the John Day
Miocene of North America.
Proc. Amer. Philos. Soc., xxv, pp. 62–79.

———   1888 F
On the mechanical origin of the den-
tition of the *Amblypoda*.
Proc. Amer. Philos. Soc., xxv, pp. 80–88, with 6 woodcuts in the text.

———   1888 G
On the tritubercular molar in human
dentition.
Jour. Morphology, ii, pp. 7–26, pls. ii, iii.

———   1888 H
On the relation of the hyoid and
otic elements of the skeleton in the
Batrachia.
Jour. Morphology, ii, pp. 297–310, with pls. xxii–xxiv.

Cope, E. D.—Continued.   1888 I

The mechanical causes of the origin
of the dentition of the *Rodentia*.
Amer. Naturalist, xxii, pp. 3–11, with 7 woodcuts in the text.

———   1888 J

Note on the *Marsupialia multituberculata*.
Amer. Naturalist, xxii, pp. 12–13, with 2 figs.
in the text.

———   1888 K

The vertebrate fauna of the Puerco
epoch.
Amer. Naturalist, xxii, pp. 161–163.

———   1888 L

Lydekker's Catalogue of Fossil *Mam-
malia* in the British Museum, Part V.
Amer. Naturalist, xxii, pp. 164–165.
Review of Lydekker R. 1887 A.

———   1888 M

The *Multituberculata* Monotremes.
Amer. Naturalist, xxii, p. 259.
Refers to the discovery of teeth in the young
of *Ornithorhynchus*, and the conclusion is
reached that the *Multituberculata* are Monotremes.

———   1888 N

Glyptodon from Texas.
Amer. Naturalist, xxii, pp. 345–346.
Describes *G. petaliferus*.

———   1888 O

The ossicula auditus of the *Batrachia*.
Amer. Naturalist, xxii, pp. 464–467, pl. vi.

———   1888 P

Notes on *Cælosteus ferox* Newb. and
Titanichthys clarkii Newb.
Amer. Naturalist, xxii, pp. 637–638.

———   1888 Q

Topinard on the latest steps in the
genealogy of man.
Amer. Naturalist, xxii, pp. 660–663.

———   1888 R

Osborn on the Mesozoic *Mammalia*.
Amer. Naturalist, xxii, pp. 723–724, pl. xiv.
Review of Osborn, H. F., 1888 G.

———   1888 S

Lydekker on the *Ichthyosauria* and
Plesiosauria.
Amer. Naturalist, xxii, pp. 724–726.
Review of Lydekker, R., 1888 D.

———   1888 T

Rütimeyer on the classification of
Mammalia, and on American types re-
cently found in Switzerland.
Amer. Naturalist, xxii, pp. 831–835.
Review of Rütimeyer, L. 1888 A.

Cope, E. D.—Continued.   1888 U

The pineal eye in extinct vertebrates.
Amer. Naturalist, xxii, pp. 914–917, pls. xv–
xviii.

———   1888 V

Handbuch der Palæontologie of Zit-
tel.
Amer. Naturalist, xxii, pp. 1018–1019.
Review of Zittel, K. A. 1890 A.

———   1888 W

Schlosser on *Carnivora*.
Amer. Naturalist, xxii, pp. 1019–1020.
Review of Schlosser, M. 1887 B.

———   1888 X

The *Artiodactyla*.
Amer. Naturalist, xxii, pp. 1079–1095, pls.
xxvii, xxviii, and 6 woodcuts in text.

———   1888 Y

Goniopholis in the Jurassic of Colo-
rado.
Amer. Naturalist, xxii, pp. 1106–1107.
Describes *G. lucasii* Cope, transferred from
Amphicotylus.

———   1888 Z

A horned dinosaurian reptile.
Amer. Naturalist, xxii, pp. 1108–1109.
Refers to Marsh, O. C. 1888 C.

———   1888 AA

On the intercentrum of the terrestrial *Vertebrata*.
Trans. Amer. Philos. Soc., xvi, pp. 243–253,
with pl. 1, and 6 figs. in the text.

———   1888 BB

Systematic catalogue of the species of
Vertebrata found in the beds of the
Permian epoch in North America, with
notes and descriptions.
Trans. Amer. Philos. Soc., xvi, pp. 285–297,
with pls. ii and iii, and 1 fig. in the text.
Describes as new *Ectocynodon incisivus*.

———   1888 CC

Synopsis of the vertebrate fauna of
the Puerco series.
Trans. Amer. Philos. Soc., xvi, pp. 298–361,
with pls. iv and v, and 11 figs. in the text.
Contains a list of 106 species belonging to
the Puerco, of which 18 are described as before
unknown.

———   1888 DD

On the shoulder girdle and extremities of *Eryops*.
Trans. Amer. Philos. Soc., xvi, pp. 362–367,
with plate of 4 figs.

Cope, E. D.—Continued.   1888 EE
Schlosser on the Cænozoic marsupials
and *Unguiculata*.
Amer. Naturalist, xxii, pp. 163–164.
A review of Schlosser's paper on "Affen,
Lemuriden," etc.

——   **1889 A**
The age of the Denver formation.
Science, xiii, p. 290.
Discusses "*Bison alticornis*" Marsh.

——   **1889 B**
A review of the North American
species of *Hippotherium*.
Proc. Amer. Philos. Soc., xxvi, pp. 429–458,
with 3 pls.

——   **1889 C**
The mechanical causes of the development of the hard parts of the *Mammalia*.
Jour. Morphology, iii, pp. 137–277, with pls.
ix–xiv, and 93 cuts in the text.

——   **1889 D**
The *Batrachia* of North America.
Bull. U. S. Nat. Mus. No. 34, Washington
D. C., 1889, pp. 1–525; pls. i–lxxxvi.
A work on the living *Batrachia* of N. America.

——   **1889 E**
Catalogue of fossil *Reptilia* and *Batra-
chia* of the British Museum, pt. i, by Dr.
Lydekker.
Amer. Naturalist, xxiii, p. 43.
A review.

——   **1889 F**
The vertebrate fauna of the Equus
beds.
Amer. Naturalist, xxiii, pp. 160–165, February, not of March.
List of species found in the different localities and descriptions of *Alces brevitrabalis*, *A.
semipalmatus*, and *Cariacus ensifer*.

——   **1889 G**
The *Artiodactyla*.
Amer. Naturalist, xxiii, pp. 111–136; pls. iii–
vii; woodcuts in the text, 7–21.
Conclusion of Cope, E. D. 1888 X. The pagi-
nation is that for March, not that for February.

——   **1889 H**
Brongniart and Doederlein on *Xena-
canthini*.
Amer. Naturalist, xxiii, pp. 149–150.

——   **1889 I**
The *Vertebrata* of the Swift Current
River. II.
Amer. Naturalist, xxiii, pp. 151–155.
Genus *Haplacodon* established for *Menodus
angustigenis. Anchitherium westoni, Hypertraga-
lus transversus, Leptomeryx esulcatus, L. mam-
mifer*, and *L. semicinctus* described.

Cope, E. D.—Continued.   1889 J
The *Proboscidia*.
Amer. Naturalist, xxiii, pp. 191–211; 9 wood-
cuts; pls. ix–xvi.
Largely reproduced in Geological Magazine
of London (3), vol. vi, 1889, pp. 438–448.

——   **1889 K**
An intermediate Pliocene fauna.
Amer. Naturalist, xxiii, pp. 253–254.
A short list of fossil vertebrates found in a
lake deposit in Oregon, and a description of
Hippotherium relictum.

——   **1889 L**
Marsh on Cretaceous mammals.
Amer. Naturalist, xxiii, pp. 490–491.
Review of Marsh, O. C. 1889 D and Marsh,
O. C. 1889 F.

——   **1889 M**
On a species of *Plioplarchus* from
Oregon.
Amer. Naturalist, xxiii, pp. 625–626.
Description of *P. septemspinosus*.

——   **1889 N**
On a new genus of Triassic *Dinosau-
ria*.
Amer. Naturalist, xxiii, p. 626.
Genus *Cœlophysis* established for *Cœlurus
longicollis, C. bauri*, and *C. willistoni*.

——   **1889 O**
Vertebrata of Swift Current River.
No. III.
Amer. Naturalist, xxiii, pp. 628–629.

——   **1889 P**
The *Edentata* of North America.
Amer. Naturalist, xxiii, pp. 657–664, with
pls. xxxi, xxxii and 2 woodcuts in the text.

——   **1889 Q**
The horned *Dinosauria* of the Lara-
mie.
Amer. Naturalist, xxiii, pp. 715–717, pls.
xxxiii, xxxiv.
Descriptions of *Monoclonius recurvicornis,
M. sphenocerus*, and *M. fissus*.

——   **1889 R**
Synopsis of the families of *Vertebrata*.
Amer. Naturalist, xxiii, pp. 849–877.

——   **1889 S**
The Silver Lake of Oregon and its
region.
Amer. Naturalist, xxiii, pp. 970–982.
Contains a list of the fossil vertebrates of
the region.

——   **1889 T**
Notes on the *Dinosauria* of the Lara-
mie.
Amer. Naturalist, xxiii, pp. 904–906.
Describes *Pteropelyx grallipes*, gen. et spec.
nov., and *Diclonius pentagonus*.

Cope, E. D.—Continued. 1889 U

Mr. Lydekker on *Phenacodus* and the *Athecæ*.

Nature, xl, p. 296.

———— 1890 A

A. Smith Woodward on *Cœlorhynchus Agassiz*.

Amer. Naturalist, xxiv, pp. 165–166.

———— 1890 B

The homologies of the fins of fishes.

Amer. Naturalist, xxiv, pp. 401–423, pls. xiv–xviii, and 9 woodcuts in the text.

———— 1890 C

Scott and Osborn on the fauna of the Browns Park Eocene.

Amer. Naturalist, xxiv, pp. 470–472.
Review of Scott and Osborn 1890 A.

———— 1890 D

[Note on teeth mentioned by Professor Marsh.]

Amer. Naturalist, xxiv, p. 571.

———— 1890 E

The *Cetacea*.

Amer. Naturalist, xxiv, pp. 599–616, pls. xxxiii, and 8 wood cuts in the text.
Besides the general account of the order there is a list of North American fossil *Cetacea*.

———— 1890 F

The extinct *Sirenia*.

Amer. Naturalist, xxiv, pp. 697–702, with pls. xxv, xxvi, and 3 wood cuts in the text.

———— 1890 G

Note on *Castoroides georgiensis* Moore.

Amer. Naturalist, xxiv, p. 772.
Refers the tooth to *Hippopotamus amphibius*.

———— 1890 H

Newberry's paleozoic fishes of North America.

Amer. Naturalist, xxiv, pp. 844–847.
Review of Newberry, J. C. 1889 A.

———— 1890 I

On two new species of *Mustelidæ* from the Loup Fork Miocene of Nebraska.

Amer. Naturalist, xxiv, pp. 950–952.
Describes *Stenogale robusta* and *Brachypsalis pachycephalus*.

———— 1890 J

On a new dog from the Loup Fork Miocene.

Amer. Naturalist, xxiv, pp. 1067–1068.
Describes *Ælurodon compressus*.

———— 1891 A

On *Vertebrata* from the Tertiary and Cretaceous rocks of the Northwest Territory. I. The species from the Oligo-

Cope, E. D.—Continued.

cene or Lower Miocene beds of the Cypress Hills.

Geol. Surv. Canada. Contributions to Canadian Palæontology, iii, pp. i; 1–25, with pls. i–xiv.

———— 1891 B

On the characters of some paleozoic fishes. I. On a new Elasmobranch from the Permian; *Styptobasis knightiana* gen. et spec. nov., fig 1. II. On new ichthyodorulites; *Hybodus regularis*, sp. nov., fig. 2; *Ctenacanthus amblyxiphias*, sp. nov., fig. 3. III. On the cranial structure of *Macropetalichthys*. IV. On the pectoral limb of the genus *Holonema*, Newb. V. On the paired fins of *Megalichthys nitidus*, Cope. VI. On the non-actinopterygian Teleostomi. VII. On new species of *Platysomidæ*, *Platysomus palmaris* and *P. lacovianus*.

Proc. U. S. Nat. Museum, xiv, pp. 447–463, pls. xxviii–xxxiii.

———— 1891 C

On the structure of certain paleozoic fishes.

Proc. Amer. Assoc. Adv. Science, 39th meeting, Indianapolis, 1890, p. 336.
Abstract of paper describing *Macropetalichthys*, *Megalichthys nitidus*, a species of *Platysomus*, and a species of *Dendrodus*.

———— 1891 D

On two new Perissodactyls from the White River Neocene.

Amer. Naturalist, xxv, pp. 47–49.
Describes *Menodus peltoceras* and *Cænopus simplicidens*.

———— 1891 E

On the non-actinopterygian Teleostomi.

Amer. Naturalist, xxv, pp. 479–481.

———— 1891 F

Catalogue of fossil *Reptilia* and *Batrachia* (*Amphibia*) in the British Museum. Parts II, III, and IV.

Amer. Naturalist, xxv, pp. 644–646.
A review of the above works of Dr. R. Lydekker.

———— 1891 G

A. S. Woodward's Fossil fishes.

Amer. Naturalist, xxv, pp. 646–647.
An editorial notice of A. D. Woodward's Catalogue of fossil fishes in the British Museum.

Cope, E. D.—Continued.   1891 H
On some new fishes from South
Dakota.
Amer. Naturalist, xxv, pp. 654–658.
Describes three new genera (*Gephyura*, *Probalostomus* and *Oligoplarchus*, and five new species.

———   1891 I
Boulenger on *Rhynchocephalia*, *Testudinata*, and *Crocodilia*.
Amer. Naturalist, xxv, pp. 813–814.

———   1891 J
On the skull of the *Equus excelsus*,
Leidy, from the Equus beds of Texas.
Amer. Naturalist, xxv, pp. 912–913.

———   1891 K
The California cave bear.
Amer. Naturalist, xxv, pp. 997–999, with pl. xxi.

———   1891 L
Flower and Lydekker's mammals.
Amer. Naturalist, xxv, pp. 1116–1118.
A review of the above authors' work on mammals.

———   1891 M
Discovery of fish remains in Ordovician rocks.
Amer. Naturalist, xxv, p. 137.
A brief editorial note regarding C. D. Walcott's discovery of fish remains in Colorado.

———   1891 N
Syllabus of lectures on geology and palæontology.
Pp. 1–90, with 60 figs. in the text.   Philadelphia, 1891.

———   1892 A
In the Texas Panhandle.
Amer. Geologist, x, pp. 131–132.

———   1892 B
On a new horizon of fossil fishes.
Proc. Amer. Assoc. Adv. Sci., 40th meeting, Washington, 1891, p. 285.
Abstract of paper regarding fishes described from South Dakota in Amer. Naturalist, 1891, p. 654.

———   1892 C
On the cranial characters of *Equus excelsus*.
Proc. Amer. Assoc. Adv. Sci., 40th meeting, Washington, 1891, p. 285.   Abstract.

———   1892 D
On a new form of *Marsupialia* from the Laramie formation.
Proc. Amer. Assoc. Adv. Sci., 41st meeting, Rochester, 1892, p. 177.
Describes genus *Thlæodon*.

Cope, E. D.—Continued.   E 1892
Report on the palæontology of the
Vertebrata [of Texas].
Geol. Surv. of Texas, 3d annual report, for 1891. Austin, 1892; pp. 251–259.

———   1892 F
The osteology of the *Lacertilia*.
Proc. Amer. Philos. Soc., xxx, pp. 185–221, with pls. ii–vi.

———   1892 G
A contribution to a knowledge of the
fauna of the Blanco beds of Texas.
Proc. Acad. Nat. Sci. Phila., 1892, pp. 226–229.

———   1892 H
On the permanent and temporary
dentitions of certain three-toed horses.
Proc. Acad. Nat. Sci. Phila., 1892, pp. 325–326.

———   1892 I
A hyæna and other *Carnivora* from
Texas.
Proc. Acad. Nat. Sci. Phila., 1892, pp. 326–327.

———   1892 J
A contribution to the vertebrate palæontology of Texas.
Proc. Amer. Philos. Soc., xxx, pp. 123–131.

———   1892 K
On the skull of the dinosaurian *Lælaps incrassatus* Cope.
Proc. Amer. Philos. Soc., xxx, pp. 240–245.

———   1892 L
On some new and little known Palæozoic vertebrates.
Proc. Amer. Philos. Soc., xxx, pp. 221–229, with pls. vii and viii.

———   1892 M
On some points in the kinetogenesis
of limbs.
Proc. Amer. Phil. Soc., xxx, pp. 282–285.

———   1892 N
On the phylogeny of the *Vertebrata*.
Proc. Amer. Philos. Soc., xxx, pp. 278–279, with 2 figs. in the text.

———   1892 O
Remarks on the communication
"Ein fossiler Giftzahn," by Dr. F.
Kinkelin.
Zoolog. Anzeiger, xv, p. 224.
Refers to discoveries of the poison-fangs of serpents earlier than Kinkelin's.

———   1892 P
The age of the staked plains of Texas.
Amer. Nat., xxvi, pp. 49–50.
Refers to occurrence of *Equus simplicidens*, *Mastodon*, and *Testudo turgida*.

Cope, E. D.—Continued. 1892 Q
Fossil *Vertebrata.*
Amer. Nat., xxvi, pp. 89–91.
Abstract of paper read before the American Society of Naturalists at Philadelphia, Dec., 1891.

——— . 1892 R
The homologies of the cranial arches of the *Reptilia.*
Amer. Nat., xxvi, pp. 407–408, with pls. xv–xvii.
Abstract of paper read before U. S. Nat. Acad. Sciences, Apr. 19, 1892. Full paper in Trans. Amer. Philos. Soc., 1892, pp. 11–16.

——— 1892 S
Professor Marsh on extinct horses and other mammals.
Amer. Naturalist, xxvi, pp. 410–412.
Review of Marsh, O. C. 1892 C, and Marsh, O. C. 1892 D.

——— 1892 T
Fourth note on the *Dinosauria* of the Laramie.
Amer. Naturalist, xxvi, pp. 756–758.

——— 1892 U
On a new genus of *Mammalia* from the Laramie formation.
Amer. Naturalist, xxvi, pp. 758–762; pl. xxii.
Describes *Thlæodon padanicus*, belonging to new family *Thlæodontidæ.*

——— . 1892 V
Crook on *Saurodontidæ* from Kansas.
Amer. Naturalist, xxvi, pp. 941–942.
Notice of Crook, A. R. 1892 A.

——— 1892 W
On the permanent and temporary dentitions of certain three-toed horses.
Amer. Naturalist, xxvi, pp. 942–944, pls. xxv, xxvi.

——— 1892 X
American Devonian fishes found in Belgium.
Amer. Naturalist, xxvi, p. 1025.
Notice of paper by M. Lohest in Ann. Soc. Géol. de Belgique, xvi, p. lvii.

——— 1892 Y
A Hyena and other *Carnivora* from Texas.
Amer. Naturalist, xxvi, pp. 1028–1029.
Describes *Borophagus diversidens, Canimartes cusminsii,* and *Felis hillianus.*

——— 1892 Z
On the homologies of the posterior cranial arches in the *Reptilia.*
Trans. Amer. Philos. Soc., xvii, pp. 11–26; with pls. i–v.

Cope, E. D.—Continued. 1892 AA
Vertebrate fauna of the Alachua clays, Florida.
Bull. U. S. Geol. Surv., No. 84, Correlation papers, Neocene, p. 130.

——— 1892 BB
The fauna of the Blanco epoch.
Amer. Naturalist, xxvi, pp. 1058–1059.

——— 1893 A
A preliminary report on the vertebrate paleontology of the Llano Estacado.
From 4th Annual Report of the Geological Survey of Texas, pp. 1–136; pls. i–xxiii.
Notice of, in Amer. Naturalist, vol. xxvii, pp. 811–812.

——— 1893 B
A new extinct species of *Cyprinidæ.*
Proc. Acad. Nat. Sci. Phila., 1893, pp. 19–20.
Describes *Aphelichthys lindahli* from Illinois.

——— 1893 C
Description of the lower jaw of *Tetralelodon shepardii.*
Proc. Acad. Nat. Sci. Phila., 1893, pp. 202–204.

——— 1893 D
Fossil fishes from British Columbia.
Proc. Acad. Nat. Sci. Phila., 1893, pp. 401–402.

——— 1893 E
On the genus *Tomiopsis.*
Proc. Amer. Philos. Soc. xxxi, pp. 317–318.

——— 1893 F
Kansas pterodactyls.
Amer. Naturalist, xxvii, p. 37.
A notice of a paper on the subject by Professor Williston, in Kansas Univ. Quarterly, July, 1892.

——— 1893 G
A remarkable artiodactyle from the White River epoch.
Amer. Naturalist, xxvii, pp. 147–148, pl. i, ii.
Notice of paper Osborn and Wortman 1892 B.

——— 1893 H
Earle on the species of *Coryphodontidæ.*
Amer. Naturalist, xxvii, pp. 250–252.
Review of Earle, Charles, 1892 A.

——— 1893 I
The vertebrate fauna of the Ordovician of Colorado.
Amer. Naturalist, xxvii, pp. 268–269.
Notice of discoveries of fish remains by C. D. Walcott.

——— 1893 J
The genealogy of man.
Amer. Naturalist, xxvii, pp. 321–335, pl. ix, and 9 figs. in the text.

Cope, E. D.—Continued.   1893 K

A new Pleistocene saber-tooth.

Amer. Naturalist, xxvii, pp. 896-897.

Describes *Dinobastis serus.*

——   1893 L

On *Symmorium* and the position of the cladodont sharks.

Amer. Naturalist, xxvii, pp. 999-1001.

——   1893 M

Forsyth Major, and Röse on the theory of dental evolution.

Amer. Naturalist, xxvii, pp. 1014-1016.

——   1893 N

Fritsch's Fauna of the Gaskohle of Bohemia.

Amer. Naturalist, xxvii, pp. 1079-1081, with pls. xix, xx, and 1 fig. in the text.

One plate and the text figure illustrate *Xenacanthus dechrni.*

——   1893 O

Cladodont sharks of the Cleveland shale.

Amer. Naturalist, xxvii, p. 1083.

An editorial notice of Professor Claypole's paper in Amer. Geologist for May, 1893.

——   1893 P

Geology and paleontology, Cenozoic.

Amer. Naturalist, xxvii, p. 473, with pl. xii.

Announces exhibition of mandible of *Tetrabelodon shepardii* and proposal of *Mastodon oligobunis;*

——   1893 Q

The vertebrate paleontology of the Llano Estacado.

Amer. Naturalist, xxvii, pp. 811, 812.

A notice of the author's report (Cope, E. D. 1893 A), on the palæontology of Texas.

——   1893 R

Cary on evolution of foot structure.

Amer. Naturalist, xxvii, pp. 248-250.

——   1894 A

Observations on the geology of adjacent parts of Oklahoma and northwest Texas.

Proc. Acad. Nat. Sci. Phila., 1894, pp. 63-68.

——   1894 B

On the structure of the skull in the plesiosaurian *Reptilia* and on two new species from the upper Cretaceous.

Proc. Amer. Philos. Soc., xxxiii, pp. 109-113, with pl. x and 1 fig. in the text.

Describes *Embaphias circulosus* gen. et sp. nov. and *Elasmosaurus intermedius.*

——   1894 C

Seeley on the fossil reptiles: II. *Pareiasaurus;* VI. The *Anomodontia* and

Cope, E. D.—Continued.

their allies; VII. Further observations on *Pareiasaurus.*

Amer. Naturalist, xxviii, pp. 788-790.

A review of Seeley's papers on the above subjects in Philos. Trans. Roy. Soc., 1888, 1889.

——   1894 D

Scott on the *Mammalia* of the Deep River Beds.

Amer. Naturalist, xxviii, pp. 790-791.

A review of Scott's memoir in Trans. Amer. Philos. Soc., xviii, pp. 55-184.

——   1894 E

Marsh on Tertiary *Artiodactyla.*

Amer. Naturalist, xxviii, pp. 867-869.

A review of Marsh, O. C. 1894 K.

——   1894 F

New and little known Paleozoic and Mesozoic fishes.

Jour. Acad. Nat. Sci. Phila., (2), ix, pp. 427-448; pls. xviii-xx, with 5 figs. in the text.

——   1894 G

On *Cyphornis,* an extinct genus of birds.

Jour. Acad. Nat. Sci. Phila., (2), ix, pp. 449-452; pl. xx, 11-16.

Describes *Cyphornis magnus,* gen. et sp. nov., from Vancouvers Island.

——   1894 H

Extinct *Bovidæ, Canidæ,* and *Felidæ* from the Pleistocene of the Plains.

Jour. Acad. Nat. Sci. Phila., (2), ix, pp. 453-459; pls. xxi-xxii.

——   1894 I

Schlosser on American Eocene *Vertebrata* in Europe.

Amer. Naturalist, xxviii, pp. 594-595.

A review of Schlosser 1894 A.

——   1895 A

Fourth contribution to the marine fauna of the Miocene period of the United States.

Proc. Amer. Philos. Soc., xxxiv, pp. 135-155, with pl. vi.

Note on, in Amer. Naturalist, vol. xxix, p. 572.

——   1895 B

Dean on coprolites.

Amer. Naturalist, xxix, p. 159.

Remarks on Bashford Dean's paper in Trans. N. Y. Acad. Sci., vol. xiii, 1894.

——   1895 C

Baur on the temporal part of the skull, and on the morphology of the skull in the *Mosasauridæ.*

Amer. Naturalist, xxix, pp.558-859; pl. xxxi.

Review of Baur's papers in Anatomischer Anzeiger, x, p. 316, and Amer. Jour. Morphology, 1894, p. 1.

Cope, E. D.—Continued.			1895 D
A batrachian armadillo.
Amer. Naturalist, xxix, p. 998.
Description of *Dissorophus multicinctus.*

——						1895 E
Reply to Dr. Baur's critique on my
paper on the paroccipital bone of the
scaled reptiles and the systematic position of the *Pythonomorpha.*
Amer. Naturalist, xxix, pp. 1008–1005.

——						1895 F
The fossil vertebrates from the fis-
sure at Port Kennedy, Pa.
Proc. Acad. Nat. Sci. Phila., 1895, pp. 446–451.

——						1895 G
The antiquity of man in North
America.
Amer. Naturalist, xxix, pp. 593–599.
Discusses the occurrence of various mam-
mals in caves and other Pleistocene deposits.

——						1895 H
Taylor on box tortoises.
Amer. Naturalist, xxix, pp. 756–757.
Subdivides the genus *Terrapene* into various
genera, among which is *Toxaspis.*

——						1896 A
Criticism of Dr. Baur's rejoinder on
the homologies of the paroccipital bone,
etc.
Amer. Naturalist, xxx, pp. 147–149, with 2
figs. in the text.

——						1896 B
Boulenger on the difference between
Lacertilia and *Ophidia;* and on the
Apoda.
Amer. Naturalist, xxx, pp. 149–152.

——						1896 C
The Paleozoic reptilian order *Cotylo-
sauria.*
Amer. Naturalist, xxx, pp. 301–304, with pl.
vii a.
Describes *Pariotichus aguti.*

——						1896 D
The ancestry of the *Testudinata.*
Amer. Naturalist, xxx, pp. 398–400.
Describes new family *Otocœlidæ,* new
genera *Otocœlus* and *Conodectes,* and new species
O. testudineus and *C. farosus.*

——						1896 E
Permian land *Vertebrata* with carapaces.
Amer. Naturalist, xxx, pp. 986–987, with pls.
xxi, xxii.
Descriptions and figures of *Otocœlus testudineus* and *Dissorhophus articulatus.*

——						1896 F
New and little known *Mammalia* from
the Port Kennedy bone deposit.
Proc. Acad. Nat. Sci. Phila., 1896, pp. 378–394.

Cope, E. D.—Continued.			1896 G
The reptilian order *Cotylosauria.*
Proc. Amer. Philos. Soc., xxxiv, pp. 436–457,
with pls. vii–ix.
Besides the *Cotylosauria,* some new *Batrachia* from the Permian beds of Texas are
described in a supplement, pp. 452–457.

——						1896 H
On some Pleistocene *Mammalia* from
Petite Anse, La.
Proc. Amer. Philos. Soc., xxxiv, pp. 458–468,
with pls. x–xii.

——						1896 I
Second contribution to the history of
the *Cotylosauria.*
Proc. Amer. Philos. Soc., xxxv, pp. 122–139,
with pls. vii–x.
Included in this paper is an "Appendix on
a species of *Trimerorhachis.*" The paper is a
continuation of Cope, 1896 G.

——						1896 J
Sixth contribution to the knowledge
of the marine Miocene fauna of North
America.
Proc. Amer. Philos. Soc., xxxv, pp. 139–146,
with pls. xi, xii.

——						1897 A
Recent papers relating to vertebrate
paleontology.
Amer. Naturalist, xxxi, pp. 315–323.
A notice of papers by Baur, Baur and Case,
Goette, Hay, Wortman, and Marsh.

——						1897 B
The position of the *Periptychidæ.*
Amer. Naturalist, xxxi, pp. 335–336.
Reviewed by M. Schlosser, Neues Jahrb.
Min., 1901, i, p. 149.

——						1897 C
On new Paleozoic *Vertebrata* from
Illinois, Ohio, and Pennsylvania.
Proc. Amer. Philos. Soc., xxxvi, pp. 71–90,
with pls. i–iii.

——						1898 A
[A communication written Oct. 28,
1870, to Prof. B. F. Mudge.]
The University Geological Survey of Kansas,
iv, pp. 29–30.
This letter, now first published, refers to a
collection of fossils which had been submitted
to Prof. Cope for examination.

——						1898 B
Syllabus of lectures on the *Vertebrata.*
Philadelphia, 1898. 8 vo., pp. 1–135, with portrait of the author and 66 figs. in the text; preceded by an essay by Dr. H. F. Osborn on "The
life and works of Cope."

——						1899 A
Vertebrate remains from Port Kennedy bone deposit.
Jour. Acad. Nat. Sci. Phila., (2), xi, pp. 193–
267, with pls. xviii–xxi.

Cope, E. D.—Continued.   1900 A
The crocodilians, lizards, and snakes
of North America.
Ann. Report Smithson. Inst. for 1898, U. S.
Nat. Museum, part ii, pp. 153-1270, with pls.
i-xxxv and 347 figs. in the text.

Cope, E. D., and Wortman, J. L. 1884 A
Post-pliocene vertebrates of Indiana.
14th Report State Geologist of Indiana,
pt. ii, pp. 1-62, with pls. i-vi.

Cornevin, Charles.   1882 A
Nouveaux cas de didactylie chez le
cheval et enterprétation de le polydacty-
lie des Équidés en général.
Assoc. Française pour l'advancement des
Sciences, 10e session, Alger, 1881, pp. 669-675.

Cornuel, J.   1877 A
Description de débris de poissons
fossiles provenant principalement du
calcaire néocomien du département de
la Haute Marne.
Bull. Soc. Géol. France (3), v, pp. 604-626,
with pl. xi.

Cottle, Thomas.   1852 A
Fossil *Pachydermata* in Canada.
Ann. and Mag. Nat. Hist. (2), x, pp. 395-396.
Records the finding of bones and tusk of
Elephas primigenius, Burlington Heights, near
head of Lake Ontario.

——   1853 A
On fossil *Pachydermata* in Canada.
Amer. Jour. Sci. (2), xv, pp. 282-283.
Taken from Ann. and Mag. Nat. Hist. (2), x,
395.

Coues, Elliott.   1866 A
The osteology of the *Colymbus torquatus*, with notes on its myology.
Memoirs Bost. Soc. Nat. Hist., i, pp. 131-172,
with pl. v.

——   1869 A
The great auk.
Amer. Naturalist, iv, p. 57.

——   1872 A
Osteology and myology of *Didelphys
virginiana*.
Memoirs Bost. Soc. Nat. Hist., ii, pp. 41-149,
with 35 figs. in the text.

——   1872 B
Key to North American birds.
Pp. 1-361, with 6 pls. and 238 figs. in the text.
A synopsis of the fossil forms of North Amer-
ican birds is given on pp. 347-350.

——   1875 A
On the cranial and dental characters
of *Mephitinæ*, with descriptions of *Mephitis frontata* n. sp. foss.
Bull. U. S. Geol. and Geog. Surv. Terr., 2d
ser., No. 1, pp. 7-15.

Coues, Elliott—Continued.   1875 B
A critical review of the North American *Saccomyidæ*.
Proc. Acad. Nat. Sci. Phila., xxvii, pp. 272-
327.

——   1875 C
Synopsis of the *Geomyidæ*.
Proc. Acad. Nat. Sci. Phila., xxvii, pp. 130-
138.

——   1876 A
Some account, critical, descriptive,
and historical, of *Zapus hudsonius*.
Bull. U. S. Geol. and Geog. Surv. Terr. (Hayden), i, ser. 2, No. 5, pp. 253-262.
Gives full synonomy and dental and cranial
characters.

——   1877 A
Fur-bearing animals; a monograph
of the North American *Mustelidæ*.
Miscellaneous pubs., No. 8, U. S. Geol. and
Geog. Surv., under F. V. Hayden, pp. i-xiv;
1-348, with pls. i-xx.

——   1877 B
Monographs of North American *Ro-
dentia*.
Report U. S. Geol. and Geog. Surv. Terr., F.V.
Hayden, U. S. geologist in charge, vol. xi.
No. i, *Muridæ*, pp. 1-264, with pls. i-v; No.
vii, *Zapodidæ*, pp. 455-479; No. viii, *Saccomy-
idæ*, pp. 481-542; No. x, *Geomyidæ*, pp. 601-630.

——   1878 A
On consolidation of the hoofs in the
Virginian deer.
Bull. U. S. Geol. and Geog. Surv. Terr., 1878,
iv, pp. 293-294.

——   1878 B
On a breed of solid-hoofed pigs apparently established in Texas.
Bull. U. S. Geol. and Geog. Surv. Terr., 1878,
iv, pp. 296-297.

——   1884 A
Key to North American birds. 2d
edition.
Pp. i-xxx; 1-863, with 561 figs. in the text.
Part ii of this work contains an excellent
account of the anatomy of birds, including the
osteology. Part iv is a "Systematic Synopsis
of the Fossil Birds of America." This synopsis
has remained unchanged in the 3d and 4th
editions and the various reissues of these up to
1896.
See also **Gill and Coues.**

Couper, J. Hamilton.   1842 A
Letter regarding fossil bones found
in Georgia.
Proc. Acad. Nat. Sci. Phila., i, pp. 216-217.
Records finding of bones of *Megatherium*,
Mastodon, a species of *Bos*, etc. A list of these
bones is found on page 190 of the same volume.

Couper, J. Hamilton—Continued.
See note under **Hodson** W. B. 1846 A.

Cox, E. T.   1874 A
Collettosaurus indianaensis Cox.
Fifth Ann. Rep. Geol. Surv. Indiana, made
during the year 1873, pp. 247–248, with 1 pl.

Cragin, F. W.   1888 A
Preliminary description of a new or
little known saurian from the Benton
of Kansas.
Amer. Geologist, ii, pp. 404–407.
Trinacromerum bentonianum is described.

——   1891 A
New observations on the genus *Trin-
acromerum*.
Amer. Geologist, viii, pp. 171–174.

——   1892 A
Observations on llama remains from
Colorado and Kansas.
Amer. Geologist, ix, pp. 257–260.

——   1892 B
A new saber-toothed tiger from the
Loup Fork Tertiary of Kansas.
Science, xix, p. 17.
Describes *Machærodus crassidens.*

——   1894 A
Vertebrates from the Neocomian
shales of Kansas.
Colorado College Studies, v, pp. 69–73, pls. i, ii.
Describes as new species *Plesiosaurus mudgei,
Plesiochelys belvidcrensis, Lamna ? quinquelat-
eralis,* and *Hybodus clarkensis.*

——   1896 A
Preliminary notice of three late Neo-
cene terranes of Kansas.
Colorado College Studies, vi, pp. 53–54.
Records finding of various vertebrate re-
mains, among which are named *Elephas im-
perator* (?), *Megalonyx leidyi, Equus complicatus,
E. curvidens, Auchenia huerfanensis.*

——   1899 A
Goat-antelope from the cave fauna of
Pikes Peak region.
Bull. Geol. Soc. Amer., xi, pp. 610–612, with
pl. lvii.
Presents a description of *Nemorhœdus palmeri*
Crag.

——   1900 A
The capricorns, mammals of an Asiatic
type, former inhabitants of the Pikes
Peak region.
Colorado College Studies, viii, pp. 1–6, with
1 text fig.
A popular account of *Nemorhœdus palmeri*
Crag.

Cramer, Frank.   1895 A
On the cranial characters of the genus
Sebastodes.
Proc. California Acad. Sci. (2), v, pp. 573–610,
with pls. lvii–lxx.
The plates consist of drawings of skulls of
the different species.

Credner, Hermann.   1870 A
Die Kreide von New Jersey.
Zeitschrift deutsch. geol. Gesellschaft, xxii,
pp. 191–251, with map.
Mentions occurrence of *Otodus appendiculatus, Corax heterodon, Oxyrhina mantelli, Lamna
texana, Coprolithus (Macropoma) mantelli, Hyposaurus rogersi, Hadrosaurus foulkii,* and *Mosasaurus mitchelli.*

——   1881 A
Die Stegocephalen (Labyrinthodon-
ten) aus dem Rothliegenden des Plau-
en'schen Grundes bei Dresden.
Zeitschr. deutsch. geol. Gesellsch., xxxiii;
Erster Theil, pp. 298–330, pls. xv–xviii; Zweiter
Theil, pp. 574–603, pls. xxii–xxiv. Describes
Branchiosaurus.

——   1881 B
Ueber einige Stegocephalen (Laby-
rinthodonten) aus dem sächsichen
Rothliegenden.
Sitzungsber. naturf. Gesellsch. Leipzig, 1881,
pp. 1–7.
Contains some remarks on *Branchiosaurus.*

——   1882 A
Die Stegocephalen aus dem Rothlie-
genden des Plauen'schen Grundes bei
Dresden; iii Theil.
Zeitschr. deutsch. geol. Gesellsch., xxxiv,
pp. 213–237, with pls. xii, xiii.
Describes *Pelosaurus* and *Archegosaurus.*

——   1883 A
Die Stegocephalen aus dem Rothlie-
genden des Plauen'schen Grundes bei
Dresden; iv Theil.
Zeitschr. deutsch. geol. Gesellsch., xxxv,
pp. 275–300, with pls. xi, xii.

——   1884 A
[Remarks on *Branchiosaurus ambly-
stomus.*]
Zeitschr. deutsch, geol. Gesellsch., xxxvi,
pp. 685–686.
Regards *Branchiosaurus* as equivalent to
Protriton, but retains former name.

——   1885 A
Die Stegocephalen aus dem Rothlie-
genden des Plauen'schen Grundes bei
Dresden.
Zeitschr. deutsch. geol. Gesellsch., xxxvii,
pp. 694–736, with pls. xxvii–xxix.
Treats of *Melanerpeton, Pelosaurus, Archego-
saurus, Hylonomus,* etc.

Credner, Hermann—Continued. 1886 A
Die Stegocephalen aus dem Rothliegenden des Plauen'schen Grundes bei
Dresden.
Zeitschr. der deutsch. geol. Gesellsch.,
xxxviii, pp. 576–632, with pls. xvi–xix and 13
figs. in the text.
Treats of the development of *Branchiosaurus*.

——   1887 A
Stegocephalen des Rothliegenden.
Two plates, 70 by 90 cm., with "Er-
laüterungen zu den Wandtafeln; Stego-
cephalen des Rothliegenden."
Pp. 1–10, with 3 figs. in the text.
Presents illustrations of the structure of
Branchiosaurus, Pelosaurus, and *Melanerpeton*.

——   1888 A
Die Stegocephalen und Saurier aus
dem Rothliegenden des Plauen'schen
Grundes bei Dresden; vii Theil; *Palæo-
hatteria longicaudata* Crd.
Zeitschr. deutsch. geol. Gesellsch., xl, pp.
490–568, with pls. xxiv–xxvi and 24 text figs.

——   1889 A
Die Stegocephalen und Saurier aus
dem Rothliegenden des Plauen'schen
Grundes bei Dresden; viii Theil.
Kadaliosaurus priscus Cred.
Zeitschr. deutsch. geol. Gesellsch., xli, pp.
319–342, with pl. xv and 5 text figs.

——   1890 A
Die Stegocephalen und Saurier aus
dem Rothliegenden des Plauen'schen
Grundes bei Dresden; ix Theil.
Zeitschr. deutsch. geol. Gesellsch., xlii, pp.
240–277, with pls. ix–xi and 6 text figs.
Describes *Hylonomus* and *Petrobates*.

——   1893 A
Zur Histologie der Faltenzähne
paläozoischer Stegocephalen.
Abhandl. math.-phys. Classe k. sächs.
Gesellsch. Wissensch. Leipzig, xx, pp. 477–552,
with pls. i–iv and 5 text figs.
This important paper treats of the histology
of the genus *Sclerocephalus*.

Crook, Alja Robinson.   1892 A
Ueber einige fossile Knochenfische
aus der mittleren Kreide von Kansas.
Palæontographica, xxxix, pp. 107–124, with
pls. xv–xviii.

Croom, H. B.   1835 A
Some account of the organic remains
found in the marl pits of Lucas Ben-
ners, esq., in Craven County, N. C.
Amer. Jour. Sci., xxvii, pp. 168–171.

Cunningham, John.   1839 A
An account of the footsteps of the
Chirotherium.
Proc. Geol. Soc. Lond., iii, pp. 12–14.

Cunningham, Robt. O.   1871 A
Notes on some points in the osteology
of *Rhea americana* and *Rhea darwini.*
Proc. Zool. Soc. Lond., 1871, pp. 105–110, with
pls. vi and vi a.

Cuvier, Georges.   1799 A
Mémoir sur les espèces d'éléphans
vivants et fossiles.
Mém. de l'Institut Nat. Sci. et Arts, ii, pp.
1–22, pls. ii–vi.

——   1804 A
Sur le Mégalonix, animal de la famille
des Paresseux, mais de la taille de
bœuf, dont les ossemens ont été décou-
vertes en Virginie, en 1796.
Ann. Mus. d'Hist. Nat., v, pp. 358–376, with
pl. xxiii.

——   1804 B
Sur le *Megatherium,* autre animal de
la famille des Paresseux, mais de la
taille du rhinocéros, dont un squelette
fossile presque complet est conservé au
cabinet royal d'historie naturelle à
Madrid.
Ann. Mus. d'Hist. Nat., v, pp. 376–387, with
pls. xxiv, xxv.

——   1806 A
Sur les elephans vivants et fossiles.
Ann. Mus. d'Hist. Nat., viii, pp. 1–58, 93–155,
249–269, with pls. xxxviii–xlv.

——   1806 B
Sur le grand mastodonte.
Ann. Mus. d'Hist. Nat., viii, pp. 270–312, with
pls. xlix–lvi.

——   1806 C
Sur différentes dents du genre des
mastodontes, mais d'espèces moindres
que celles de l'Ohio, trouvées en plusieurs lieux des deux continens.
Ann. Mus. d'Hist. Nat., viii, pp. 401–420, with
pls. lxvi–lxix.

——   1808 A
Sur le grand animal des carrières de
Maestricht.
Ann. Mus. d'Hist. Nat., Paris, xii, pp. 145–178,
with pls. xviii, xix.

——   1808 B
Sur les os fossiles de rumìnans trouvés
dans les terrains meubles.
Ann. Mus. d'Hist. Nat., Paris, xii, pp. 333–398,
with pls. xxxii–xxxiv.
Refers to and figures *Bison latifrons* under
the name Aurochs.

Cuvier, Georges—Continued. 1834 A
Recherches sur les ossemens fossiles,
ou l'on rétablit les caractères de plusieurs animaux dont les révolutions du
globe ont detruit les espèces.
Fourth edition, Paris, 1834-1836, vols. i-x,
octavo, with 2 vols. quarto, containing 260 pls.
First edition was published in 1812, second
edition in 1821, third edition in 1825.

Daily State Journal. 1870 A
Discovery of a mastodon.
Amer. Jour. Sci. (2), i, pp. 422-423.
Extract from Daily State Journal, Spring-
field, Ill.

Dall, William H. 1869 A
[Statement concerning the finding of
the bones of musk ox, buffalo, and
elephant near Yukon River.]
Proc. Boston Soc. Nat. Hist., xiii, p. 136.

———— 1873 A
[Remarks on tusks and bones from
Kotzebue Sound.]
Proc. California Acad. Nat. Sci., iv, pp. 293-
294.

———— 1891 A
Age of the Peace Creek bone beds of
Florida.
Proc. Acad. Nat. Sci. Phila., 1891, p. 121.

Dames, W. 1882 A
On the structure of the head of
Archæopteryx.
Geol. Magazine (2) ix, pp. 566-568.
See also Sitzungsber, k.-p. Akad. Wis-
sensch. Berlin, Jahrg. 1882, pp. 817-819.

———— 1883 A
Ueber *Ancistrodon* Debey.
Zeitschr. deutsch. geol. Gesellsch., xxxv,
pp. 655-670, with pl. xix.

———— 1883 B
Ueber eine tertiäre Wirbelthierfauna
von der westlichen Insel der Birket-
el-Qurûn, im Fajcim (Aegypten).
Sitzungsber. k.-p. Akad. Wissensch. Berlin, pp. 129-153, with pl. iii.

———— 1884 A
Entgegnung an Herrn Dr. Baur.
Morpholog. Jahrbuch, x, pp. 603-612.
Reply to Baur, G. 1884 B.

———— 1884 B
Ueber *Archæopteryx*.
Palæontolog. Abhandl., ii, pp. 117-196 (1-80),
with 1 pl. and 5 woodcuts.

———— 1885 A
[Review of A. von Koenen's "Beiträge zur Kenntniss der Placodermen
des Norddeutschen Oberdevons," pub-

Dames, W.—Continued. 1885 A
lished in Abhandlungen der k. Gesellsch. Wissensch Göttingen, xxx, pp.
1-40.]
Neues Jahrbuch Min., Geol., u. Palæont.,
885, i, Referate, pp. 97-98.

———— 1887 A
Ueber die Gattung *Saurodon*.
Sitz.-Ber. Gesellsch. naturforsch. Freunde
Berlin, 1887, pp. 72-78.

———— 1887 B
Über *Titanichthys pharao*, nov. gen.,
nov. sp., aus der Kreideformation
Aegyptens.
Sitz.-Ber. Gesellsch. naturforsch. Freunde
Berlin, 1887, pp. 69-72, with 4 figs. in the text.
Discusses systematic position of *Enchodus*.

———— 1893 A
Ueber das Vorkommen von Ichthy-
opterygiern im Tithon Argentiniens.
Zeitschr. deutsch. geol. Gesellsch., xlv, pp.
23-33, with pl. i.

———— 1894 A
Die Chelonier der norddeutschen
Tertiärformation.
Palæontolog. Abhandl., vi, pp. 197-220, with
pls. xxvi-xxix.

———— 1894 B
Ueber Zeuglodonten aus Aegypten
und die Beziehungen der Archæoceten
zu den übrigen Cetaceen. I, Beschreibung der Fossilreste; II, Vergleich der
ägyptischen Zeuglodon-Reste mit denen
anderer Gebiete; III, die Stellung der
Zeuglodonten im System der Säugethiere; IV, Ueber den Hautpanzer der
Zeuglodonten.
Palæontolog. Abhandl. (2) i (v), pp. 1-36,
with pls i-vii (xxx-xxxvii).

———— 1895 A
Die Plesiosaurier der süddeutschen
Liasformation.
Abhandl. k. Akad. Wissensch. Berlin, 1895,
phys.-math. Cl., Abhandl., ii, pp. 1-83, pls. i-v,
and 4 figs. in the text.

———— 1895 B
Ueber die Ichthyopterygier der Tri-
asformation.
Sitz.-ber. k.-p. Akad. Wissensch. Berlin, 1895,
pp. 1045-1050.

———— 1897 A
Ueber Brustbein, Schulter-und Beck-
engürtel der *Archæopteryx*.
Sitz-ber. k.-p. Akad. Wissensch. zu Berlin,
1897, pp. 813-834, with 3 figs. in the text.

Dana, E. S.
See **Grinnell** and **Dana.**

Dana, James D.   1875 A
On Dr. Koch's evidence with regard to the contemporaneity of man and the mastodon in Missouri.
Amer. Jour. Sci. (3), ix, pp. 335-346.

——   1875 B
Supplement to the article on Dr. Koch's evidence with regard to the contemporaneity of man and the mastodon.
Amer. Jour. Sci. (3), ix, p. 398.

——   1875 C
Reindeers in southern New England.
Amer. Jour. Sci. (3), x, pp. 353-357.

——   1876 A
The *Vertebrata* of the Cretaceous formations of the West, by E. D. Cope.
Amer. Jour. Sci. (3), xi, pp. 65-66.
Notice and criticism of Professor Cope's work

——   1879 A
Review of J. D. Whitney's "Auriferous Gravels of the Sierra Nevada and California."
Amer. Jour. Sci. (3), xviii, 233-234.

——   1888 A
Notice of Dr. J. S. Newberry's "Descriptions of new fossil fishes," published in Trans. N. Y. Academy of Sciences, vol. vi.
Amer. Jour. Sci. (3), xxxv, p. 498.

——   1896 A
Manual of Geology; treating of the principles of the science with special reference to American geological history, etc.
Fourth edition, pp. 1-1087, with over 1,575 figs. in the text and 2 double-page maps. The first edition of this work was published in 1863; the second, 1876; the third, 1880.

Dareste, Camille.   1849 A
Observations sur l'ostéologie du poisson appelé Triodon macroptère.
Ann. Sci. Naturelles (3), xii, pp. 68-83, with pl. i.

——   1872 A
Études sur les types ostéologiques des poissons osseux.
Comptes rend. Acad. Paris, lxxv, pp. 942-946; 1018-1021; 1086-1089; 1172-1175; 1253-1256.

Davidoff, M.   1880 A
Ueber das Skelett der hinteren Gliedmasse der *Ganoidei holostei* und der physostomen Knochenfische.
Morpholog. Jahrbuch, vi, pp. 125-128.

Davidoff, M.—Continued.   1880 B
Beiträge zur vergleichenden Anatomie der hinteren Gliedmasse der Fische.
Morpholog. Jahrbuch, vi, pp. 433-468, with pls. xxi-xxiii.
Discusses the structure of the ventral fins of the Ganoids and Physostomi.

——   1883 A
Beiträge zur vergleichenden Anatomie der hinteren Gliedmasse der Fische. III Theil, *Ceratodus.*
Morpholog. Jahrbuch, ix, pp. 117-162, with pls. viii and ix.

Davies, A. M.
See **Howes** and **Davies.**

Davies, William.   1872 A
On the rostral prolongations of *Squaloraia polyspondyla*, Ag.
Geol. Magazine (1), ix, pp. 145-150, with pl. iv.

——   1878 A
On the nomenclature of *Saurocephalus lanciformis* of the British Cretaceous deposits, with description and figures of a new species (*S. woodwardii*).
Geol. Magazine, (2), v, pp. 254-262, with pl. viii.

——   1886 A
A note on Professor Cope's "*Edestus* and *Pelecopterus*," etc.
Geol. Magazine (3), iii, pp. 141-142.
A reply to a note by Professor Cope which appears on p. 141 of the same volume.

Davis, Chas. H. S.   1887 A
The *Catopterus gracilis.*
Trans. Sci. Assoc. Meriden, Conn., ii, pp. 21-22, with 1 fig. in the text.

Davis, James W.   1880 A
On the Teleostean affinities of the genus *Pleuracanthus.*
Ann. and Mag. Nat. Hist. (5), v, pp. 349-357.

——   1880 B
On the genus *Pleuracanthus*, Agass., including the genera *Orthacanthus*, Agass. and Goldfuss; *Diplodus*, Agass., and *Xenacanthus*, Beyr.
Quart. Jour. Geol. Soc., xxxvi, pp. 321-336, with pl. xii and 10 text figs.

——   1881 A
On the genera *Ctenoptychius*, Agassiz; *Ctenopetalus*, Agassiz; and *Harpacodus*, Agassiz.
Ann. and Mag. Nat. Hist. (5), viii, pp. 424-427.

Davis, James W.—Continued. 1882 A
On the zoological position of the
genus *Petalorhynchus* Ag., a fossil fish
from the Mountain limestone.
Report Brit. Assoc. Adv. Sci., 51st meeting.
York, 1881, p. 646.

——— 1882 B
Notes on the occurrence of fossil-fish
remains in the Carboniferous limestone
series of Yorkshire.
Proc. Yorkshire Geol. and Polytech. Soc. (2),
viii, pp. 39-63.

——— 1883 A
On the fossil fishes of the Carboniferous limestone series of Great Britain.
Trans. Roy. Dublin Soc. (2), i, pp. 327-548,
with pls. xliii-lxv.

——— 1885 A
Note on *Chlamydoselachus*.
Proc. Yorkshire Geol. and Polytech. Soc. (2),
ix, pp. 98-113, with pl. xl.
*Diplodus, Didymodus, Pleuracanthus, Clado-
dus.*

——— 1886 A
Notes on a collection of fossil-fish re-
mains from the Mountain limestone of
Derbyshire.
Geol. Magazine (3), iii, pp. 148-157.
Discusses relationship of *Xystrodus* to *Tomo-
dus.*

——— 1887 A
Note on a fossil species of *Chlamy-
doselachus*.
Proc. Zool. Soc. Lond., 1887, pp. 542-544.

——— 1887 B
On *Chondrosteus acipenseroides*, Agassiz.
Quart. Jour. Geol. Soc., xliii, pp. 605-616,
with pl. xxiii.

——— 1887 C
The fossil fishes of the chalk of Mount
Lebanon, in Syria.
Trans. Roy. Dublin Soc. (2), iii, pp. 457-636,
with pls. xiv-xxxviii.

——— 1888 A
On fossil-fish remains from the Tertiary and Cretaceo-Tertiary formations
of New Zealand.
Trans. Roy. Dublin Soc. (2), iv, pp. 1-48,
with pls. i-vii.

——— 1890 A
On the fossil fish of the Cretaceous
formations of Scandinavia.
Trans. Roy. Dublin Soc. (2), iv, pp. 363-434,
with pls. xxxviii-xlvi.

Davis, James W.—Continued. 1890 B
On the dentition of *Pleuroplax* (*Pleurodus*), A. S. Woodward.
Ann. and Mag. Nat. Hist. (6), v, pp. 291-
294, with pl. xiii.

——— 1892 A
On the fossil-fish remains of the
Coal-measures of the British Islands.
Part I. *Pleuracanthidæ.*
Trans. Roy. Dublin Soc. (2), iv, pp. 703-748,
with pls. lxv-lxxiii.
*Pleuracanthus, Orthacanthus, Xenacanthus,
Compsacanthus, Diplodus,* and *Didymodus* are
regarded as synonymous.

——— 1894 A
On the fossil-fish remains of the
Coal-measures of the British Islands.
Part II. *Acanthodidæ.*
Trans. Roy. Dublin Soc. (2), v, pp. 249-258,
with pls. xxvii-xxix.

Davison, Charles. 1894 A
On deposits from snowdrift, with especial reference to the origin of the
loess and the preservation of mammoth
remains.
Quart. Jour. Geol. Soc., l, pp. 472-487.

Dawkins, W. Boyd. 1870 A
Fossil mammals in North America.
Nature, ii, pp. 119-120; 232-233.
A review of Dr. Leidy's "The extinct mam-
malian fauna of Dakota and Nebraska."

——— 1871 A
Geology of Oxford.
Nature, v, pp. 145-149, with 6 woodcuts.
A review of Prof. John Phillips's "Geology
of Oxford and the Valley of the Thames."

——— 1872 A
The British Pleistocene *Mammalia.*
Part V. British Pleistocene *Ovidæ,
Oribos moschatus.*
Palæontographical Society, xxv, pp. 1-30, pl.
i-v.

——— 1879 A
On the range of the mammoth in
space and time.
Quart. Jour. Geol. Soc., xxxv, pp. 138-147.

——— 1880 A
The classification of the Tertiary
period by means of the *Mammalia.*
Quart. Jour. Geol. Soc., xxxi, pp. 379-405.

——— 1883 A
On the alleged existence of *Ovibos
moschatus* in the Forest-bed, and on its
range in space and time.
Quart. Journ. Geol. Soc., xxxix, pp. 575-581,
with 1 woodcut.

Dawkins, W. Boyd—Continued. 1889 A
[Address as president of Section C
of the Brit. Assoc. Adv. Sci., 1888, on
the progress of geological inquiry.]
Report. Brit. Assoc. Adv. Sci., 58th meeting,
Bath, 1888, pp. 644–650.

Dawkins, W. B., and **Sanford,** W. A.
                                    1872 A
The British Pleistocene *Mammalia.*
Part iv. British Pleistocene *Felidæ.*
Palæontographical Society, xxv, pp. 177–
194; pls. xxiv–xxv.

Dawson, George M.   1894 A
Notes on the occurrence of mammoth
remains in the Yukon district of Can-
ada and Alaska.
Quart. Jour. Geol. Soc., l, pp. 1–9.
For notice of this paper and for remarks by
Haworth see Nature, xlix, p. 93.

Dawson, J. William.   1855 A
Notice of the discovery of a reptilian
skull in the Coal-measures of Picton.
Quart. Jour. Geol. Soc., xi, pp. 8–9.
The reptile referred to is *Baphetes planiceps.*

——   1860 A
On a terrestrial mollusk, a chilogna-
thous myriapod, and some new species
of reptiles from the Coal-measures of
Nova Scotia.
Quart. Jour. Geol. Soc., xvi, pp. 268–277,
with 29 figs. in the text.

——   1862 A
Note on the discovery of additional
remains of land animals in the Coal-
measures of the South Joggins, Nova
Scotia.
Quart. Jour. Geol. Soc., xviii, pp. 5–7.

——   1863 A
Air breathers of the Coal period.
Amer. Jour. Sci. (2), xxxvi, pp. 430–432.
Abstract of pamphlet with same title, with
illustrations; Bailliere, New York and Lon-
don, 1863; Montreal, Dawson Brothers.
See also Canadian Naturalist, viii, 1863.

——   1863 B
Notice of a new species of *Dendrerpeton,* and of the dermal coverings of certain Carboniferous reptiles.
Quart. Jour. Geol. Soc., xix, pp. 469–473.

——   1863 C
A review of Dr. Dawson's book, "Air
breathers of the Coal period," by S. J.
Mackie.
The Geologist, vi, pp. 434–440.

Dawson, J. William—Continued. 1863 D
The air breathers of the Coal period
of Nova Scotia.
Canadian Naturalist and Geologist, viii, pp.
1–12; 81–92; 159–160; 161–175; 268–295; with pls.
i–vi.

——   1863 E
Note on the footprints of a reptile
from the Coal formation of Cape
Breton.
Canadian Naturalist and Geologist, viii, pp.
430–431, with one fig. in the text.
Name assigned to the tracks, *Sauropus
sydnensis.*

——   1868 A
Acadian geology. The geological
structure, organic remains, and min-
eral resources of Nova Scotia, New
Brunswick, and Prince Edward
Island.
London, 1868; pp. i–xxvii; 1–694, with 232
figs. in the text.

——   1870 A
Note on some new animal remains
from the Carboniferous and Devonian
of Canada.
Quart. Jour. Geol. Soc., xxvi, pp. 166–167.
Also in Canadian Naturalist (2), v, pp. 98–99.
Note on, in Geol. Magazine, vii (1870), p. 87.
Describes briefly *Baphetes minor*

——   1872 A
Note on footprints from the Carbonif
erous of Nova Scotia, in the collection
of the Geological Survey of Canada.
Geol. Magazine (1), ix, pp. 251–253, with 1
woodcut.
Describes *Sauropus unguifer.*

——   1876 A
On a recent discovery of Carbonifer-
ous batrachians in Nova Scotia.
Amer. Jour. Sci. (3), xii, pp. 440–447.

——   1877 A
Lower Carboniferous fishes of New
Brunswick.
Canadian Naturalist (2), viii, pp. 337–340,
with figs. in the text.
Describes the new species *Palæoniscus
(Rhadinichthys) modulus,* and *P. jacksonii.*

——   1877 B
Note on a fossil seal from the Leda
clay of the Ottawa Valley.
Canadian Naturalist (2), viii, pp. 340–341.
The seal is identified as *Phoca grœnlandica.*

——   1878 A
Acadian Geology. The geological
structure, organic remains, and miner-

Dawson, J. William—Continued. 1878 A
al resources of Nova Scotia, New Brunswick, and Prince Edward Island.
Third edition, pp. i-xxvii; 1-694; supplement, pp. 1-103, with 248 figs.

—— 1882 A
On the results of recent explorations of erect trees containing animal remains in the Coal formation of Nova Scotia.
Philos. Trans. Royal Soc., Lond., clxxiii, pt. ii, pp. 621-659; pls. xxxix-xlvii.

—— 1882 B
On the results of recent explorations of erect trees containing reptilian remains in the Coal formation of Nova Scotia.
Canadian Naturalist (2), x, pp. 252-254.
This is a brief abstract of the preceding paper and is reprinted from Nature.

—— 1883 A
On portions of the skeleton of a whale from gravel on the line of the Canadian Pacific railway, near Smiths Falls, Ontario.
Amer Jour. Sci. (3), xxv, pp. 200-202.
Canadian Naturalist (2), x, pp. 385-387.

—— 1890 A
Note on a fossil fish and marine worm found in the Pleistocene nodules of Greens Creek, on the Ottawa.
The Canadian Record of Science, iv, pp. 86-88.
Records the finding of *Cottus uncinatus* (*Artediellus*); also *Mallotus villosus*, *Cyclopterus lumpus*, and *Gasterosteus aculeatus*.

—— 1891 A
Note on *Hylonomus lyelli*, with photographic reproduction of skeleton.
Geol. Magazine (3), viii, pp. 258-259, with pl. viii.

—— 1891 B
On new specimens of *Dendrerpeton acadianum*, with remarks on other Carboniferous amphibians.
Geol. Magazine (3), viii, pp. 145-156, with 4 figs.

—— 1892 A
On the mode of occurrence of remains of land animals in erect trees at the South Joggins, Nova Scotia.
Trans. Royal Soc., Canada, for 1891, ix, sec. iv, pp. 127-128.

—— 1894 A
Some salient points in the science of the earth.
Pp. 499; illus. 46; New York: Harper & Brothers, publishers, 1894.
Chapter x; The oldest air-breathers.

Dawson, J. William—Continued. 1894 B
Preliminary note on recent discoveries of batrachian and other air-breathers in the Coal formation of Nova Scotia.
The Canadian Record of Science, vi, pp. 1-7, with 1 fig. in the text.

—— 1895 A
Synopsis of the air-breathing animals of the Palaeozoic in Canada up to 1894.
Trans. Roy. Soc. of Canada for 1894, xii, sec. iv, pp. 71-78.

—— 1895 B
Note on a specimen of *Beluga catodon*, from the Leda clay, Montreal.
The Canadian Record of Science, vi, pp. 351-354.

—— See also **Lyell** and **Dawson**.

Day, E. C. H. 1864 A
On *Acrodus anningiæ*; with remarks upon the affinities of the genera *Acrodus* and *Hybodus*.
Geol. Magazine (1), i, pp. 57-65, with pls. iii and iv.

"D. L. H." 1838 A
Fossil Fishes.
Amer. Jour. Sci., xxxiv, pp. 198-200.

Dean, Bashford. 1891 A
Pineal foramen of placoderm and catfish.
Nineteenth report Commissioners of Fisheries of the State of New York, pp. 307-363, with pls. i-xiv.
The plates and the explanations thereof are included in the pagination.

—— 1893 A
Report of proceedings of New York Acad. of Sciences, Dec. 12, 1892.
Amer. Naturalist, xxvii, p. 183.
Refers to finding of *Protoceras celer*.

—— 1893 B
On *Trachosteus* and *Mylostoma*; notes on their structural characters.
Trans. N. Y. Acad. Sci., xii, pp. 70-71.

—— 1893 C
Note on the mode of origin of the paired fins.
Trans. N. Y. Acad. Sci., xii, pp. 121-125.

—— 1893 D
On the fin-structures of *Diplurus*.
Trans. N. Y. Acad. Sci., xiii, p. 22.
A very brief abstract.

—— 1894 A
Contributions to the morphology of *Cladoselache* (*Cladodus*).
Jour. Morphology, ix, pp. 87-114, pl. vii.

Dean, Bashford—Continued.	1894 B
A new Cladodont from the Ohio
Waverly, *Cladoselache newberryi*, n. sp.
Trans. N. Y. Acad. Sci., xiii, pp. 115-119,
with pl. i.
Describes *C. newberryi*.

——	1895 A
Fishes, living and fossils; an outline
of their forms and probable relationships.
Pp. i-xiv; 1-300, with frontispiece and 344
figs. in the text; Macmillan & Co., New York
and London.
Review of by A. S. Woodward, Geol. Magazine, 1896, p. 135; and in Nat. Sci., viii, 267.

——	1896 A
On the vertebral column, fins, and
ventral armoring of *Dinichthys*.
Trans. N. Y. Acad. Sci., vol. xv, pp. 157-163,
pls. vii, viii.

——	1896 B
The fin-fold origin of the paired
limbs, in the light of the ptychopterygia
of Palæozoic sharks.
Anatom. Anzeiger, xi, pp. 673-679, with 8 figs.

——	1896 C
Sharks as ancestral fishes.
Nat. Science, viii, pp. 245-253, with 5 figs. in
the text.
Describes and figures *Cladoselache*.

——	1897 A
Note on the ventral armoring of
Dinichthys.
Trans. N. Y. Acad. Sci., xvi, pp. 57-61, with
pls. ii, iii.

——	1897 B
On a new species of *Edestus*, *E.
lecontei*, from Nevada.
Trans. N. Y. Acad. Sci., xvi, pp. 61-69, with
pls. iv, v.
Discusses also its relationships to the other
species.

——	1899 A
Devonian fishes for the American
Museum.
Science (2), x, p. 978.
Refers to a purchase of a collection of Placo-
derms from the Devonian of Ohio.

Deane, James.	1844 A
On the fossil footmarks of Turners
Falls, Mass.
Amer. Jour. Sci., xlvi, pp. 73-77, with 2 pls.
Abstract of, in Quart. Jour. Geol. Soc. Lond.,
i, p. 141.

——	1844 B
On the discovery of fossil footmarks.
Amer. Jour. Sci., xlvii, pp. 381-390.
A protest against Prof. E. Hitchcock's claims
regarding the fossil footmarks of the Connecti-
cut Valley. See Hitchcock, E., 1844 B.

Deane, James—Continued.	1844 C
Answer to the "Rejoinder" of Pro-
fessor Hitchcock.
Amer. Jour. Sci., xlvii, pp. 399-401.
Reply to Hitchcock, E., 1844 B.

——	1845 A
Description of fossil footprints in the
New Red sandstone of the Connecticut
Valley.
Amer. Jour. Sci., xlviii, pp. 158-167, with 1 pl.

——	1845 B
Illustrations of fossil footmarks.
Boston Jour. Nat. Hist., v, pp. 277-284, with 3
figs. in the text.
No species named.

——	1845 C
Notice of a new species of batrachian
footmarks.
Amer. Jour. Sci., xlix, pp. 79-81, with figs.
in the text.
No systematic name given the tracks.

——	1845 D
Fossil footmarks and raindrops.
Amer. Jour. Sci., xlix, pp. 213-214, with 1
page of figs.
No systematic names are applied to the
footmarks.

——	1847 A
Notice of new fossil footprints.
Amer. Jour. Sci. (2), iii, pp. 74-79, with 4 figs.
in the text.

——	1847 B
Fossil footprints.
Amer. Jour. Sci. (2), iv, pp. 448-449.

——	1848 A
Fossil footprints of a new species of
quadruped.
Amer. Jour. Sci. (2), v, pp. 40-41, with 1 fig.
in the text.
In this paper no systematic name is given to
the footprints.

——	1850 A
Fossil footprints of Connecticut River.
Jour. Acad. Nat. Sci. Phila. (2), ii, pp. 71-74,
pls. viii, ix.

——	1856 A
On the sandstone fossils of Connecticut River.
Jour. Acad. Nat. Sci. Phila. (2), iii, pp. 173-
178.

——	1861 A
Ichnographs from the sandstone of
Connecticut River. A memoir upon the
fossil footprints and other impressions
of the Connecticut River sandstone.
4to. Boston, 1861, pp. 1-61, with pls. i-xlvi.
This work was compiled mostly from Dr.
Deane's incomplete papers by T. T. Bouvé.
Accompanying it is an introduction (pp. 2-4),
a biographical notice (pp. 5-12), and a list of

Deane, James—Continued.

Dr. Deane's published papers (pp. 13-14). The book consists mostly of explanations of the plates, and the species have been determined by Mr. Bouvé from Hitchcock's works.

Deecke, W.　　　　　　　　1895 A

Notiz über ein Nothosauriden-Fragment.

Zeitschr. deutsch. geol. Gesellsch., xlvii, pp. 303-306, with 1 fig. in the text.

Deichmüller, J. V.

See **Geinitz** and **Deichmüller.**

De Kay, James E.　　　　　1824 A

Account of the discovery of a skeleton of the *Mastodon giganteum.*

Annals Lyc. Nat. Hist. N. Y., i, pp. 143-147.

A report made by Messrs. De Kay, Van Rensselaer, and Cooper. Records finding of skeleton in Monmouth County, N. J.

———　　　　　　　　　　1828 A

Notes on a fossil skull in the cabinet of the Lyceum, of the genus *Bos*, from the banks of the Mississippi, with observations on the American species of the genus.

Annals Lyc. Nat. Hist. N. Y., ii, pp. 290-291.
See also Edinburgh New Philos. Jour., 1828, v, pp. 326-327.

———　　　　　　　　　　1830 A

On the remains of extinct reptiles of the genera *Mosasaurus* and *Geosaurus* found in the Secondary formation of New Jersey, and on the occurrence of the substance recently named coprolite by Dr. Buckland, in the same locality.

Annals Lyc. Nat. Hist. N. Y., iii, pp. 134-141, pl. iii.

———　　　　　　　　　　1833 A

Observations on a fossil jaw of a species of gavial from west Jersey.

Annals Lyc. Nat. Hist. N. Y., iii, pp. 156-165, pl. iii, figs. 7-11.

———　　　　　　　　　　1842 A

Zoology of New York, or the New York fauna. Part III. Reptiles and Amphibia.

Pp. i-vii; 1-98, with pls. i-xxiii.

———　　　　　　　　　　1842 B

Fossil fishes.

Zoology of New York. Part IV, Fishes, pp. 386-387.

Gives a list of fossil fishes found in the United States and known to De Kay. Localities and references to literature given.

———　　　　　　　　　　1842 C

Mammalia.

Zoology of New York. Part I. *Mammalia*, pp. i-xv; 1-142, with pls. i-xxxiii.

Describes and refers to a number of fossil mammals. Of these a few are figured.

Delbos, Joseph.　　　　　1858 A

Recherches sur les ossements des Carnassiers des cavernes de Sentheim (Haut-Rhin) précédées d'observations sur l'ostéologie de l'ours brun des Pyrénées.

Annales des Sciences Naturelles, Zool. (4), ix, pp. 155-223.

———　　　　　　　　　　1860 A

Recherches sur les ossements des Carnassiers des cavernes de Sentheim (Haut-Rhin): Études sur les ours fossiles.

Annales des Sciences Naturelles, Zool. (4), xiii, pp. 47-108.

———　　　　　　　　　　1860 B

Recherches sur les ossements des Carnassiers des cavernes de Sentheim (Haut-Rhin): Études sur les ours fossiles.

Annales des Sciences Naturelles, Zool., (4), xiv, pp. 5-112.

Delfortrie, E.　　　　　　1874 A

Un *Zeuglodon* dans les faluns du sud-ouest de la France.

Jour. de Zoologie, iii, pp. 25-30, with 2 figs. in the text.

Dennis, J. B. P.　　　　　1861 A

On the mode of flight of the pterodactyles of the Coproldi bed near Cambridge.

Report Brit. Assoc. Adv. Sci., 30th meeting, Oxford, 1860, p. 76.

Depéret, Charles.　　　　1887 A

Recherches sur la succession des faunes de vertébrés miocènes de les vallée du Rhone. Pt. I (pp. 45-118), Stratigraphie paléontologique. Pt. ii (pp. 119-307), Description des vertébrés miocènes du bassin du Rhone.

Archives Mus. d'Hist. Nat. de Lyon, iv, pp. 45-313, with pls. ix-xxv.

———　　　　　　　　　　1892 A

La faune de mammifères miocènes de la Grive-Saint-Alban (Isère).

Archives Mus. d'Hist. Nat. de Lyon, pp. 1-95, pls. i-iv.

Deslongchamps-Eudes, J. A.　1838 A

Mémoire sur le *Poekilopleuron buck-landi*, grand saurien fossile, intermédiaire entre les crocodiles et les lézards.

Mém. Soc. Linn. Normandie, vi, pp. 37-146, with pls. i-viii.

———　　　　　　　　　　1864 A

Mémoires sur les téléosauriens de l'époque jurassique du Department du

Deslongchamps-Eudes, J. A.—Cont'd.
Calvados. Première mémoire contenant l'exposé des caractères géneraux des téléosauriens comparés à ceux des crocodiliens et la description particulière des espèces du lias supérieur.
Mém. Soc. Linn. Normandie, Caen, xiii, No. 3, pp. 1-138, with 9 pls.

Deslongchamps-Eudes, E. 1870 A
Note sur les reptiles fossiles appartenant à la famille des téléosauriens, dont le débris ont été recueillis dans les assises jurassiques de Normandie.
Bull. Soc. Géol. de France, xxvii, pp. 299-351, pls. ii-viii.

Desor, E. 1849 A
Remarks on finding of mastodon at Galena, Mo.
Proc. Boston Soc. Nat. Hist., iii, p. 207.

—— 1852 A
Post-pliocene of the Southern States and its relation to the Laurentian of the North and the deposits of the valley of the Mississippi.
Amer. Jour. Sci. (2), xiv, pp. 49-59.

Destinez, P. 1898 A
Sur deux *Diplodus* et un *Chomatodus* de l'Ampélite à unifère de Chokier, et deux *Cladodus* de Visé.
Ann. Soc. Géol. Belgique, xxiv, pp. 219-223.

Dickeson, M. W. 1845 A
On the geology of the Natchez bluffs.
Proc. 6th meeting Assoc. Amer. Geologists and Naturalists, 1845, pp. 77-79.
In this is described "a curious nondescript quadruped."

—— 1846 A
Remarks on *Megalonyx jeffersoni*.
Proc. Acad. Nat. Sci. Phila., iii, p. 106.

Didelot, Léon. 1875 A
Note sur un *Pycnodus* nouveau du Néocomien moyen (*P. heterodon*).
Bull. Soc. Géol. France (3), iii, pp. 237-256, pl. vi.

Diffenberger, F. R. 1873 A
Elephas americana in Mexico.
Amer. Jour. Sci. (3), vi, p. 62.

Dixon, Frederick. 1850 A
The geology and fossils of the Tertiary and Cretaceous formations of Sussex.
Pp. i-xvi; 1-422, with pls. i-xl.

—— 1878 A
The geology of Sussex, or the geology and fossils of the Tertiary and Cretaceous formations of Sussex; new edi-

Dixon, Frederick—Continued.
tion, revised and augmented, by T. Rupert Jones, Brighton, 1878.
Pp. i-xxiv; 1-469, with pls. i-xxvii.

Dobson, G. E. 1885 A
An attempt to exhibit diagrammatically the several stages of evolution of the *Mammalia*.
Report Brit. Assoc. Adv. Sci., 54th meeting, Montreal, 1884, pp. 768-769, with 1 pl.

—— 1878 A
Catalogue of the *Chiroptera* in the collection of the British Museum.
London, 1878. 8 vo; pp. i-xlii; 1-567, with pls. i-xxix.

—— 1883 A
A monograph of the *Insectivora*, systematic and anatomical. Part I, including the families *Erinaceidæ*, *Centetidæ*, and *Solenodontidæ*. Part II, *Potamogalidæ*, *Chrysochloridæ*, and *Talpidæ* [in part]. Part iii, including pls. xxiii-xxviii.
Pp. 1-172, with pls. i-xxviii. 4to. London. Part i was issued in 1882; pt. ii, in 1883; pt. iii, in 1890. The text was not completed.

Döderlein, Ludwig. 1878 A
Ueber das Skelet des *Tapirus pinchacus*.
Archiv Naturgesch. xliv, i, pp. 37-90.

—— 1889 A
Das Skelet von *Pleuracanthus*.
Zoolog. Anzeiger, xii, pp. 123-127, with 1 text fig.
See also **Steinmann** and **Döderlein.**

Dohrn, Anton. 1886 A
Entstehung und Differenzirung des Zungenbein-und Kiefer-Apparates der Selachier.
Mittheil. Zoolog. Station Neapel, vi, pp. 1-48; 80-92, with pls. i-viii.

Dollo, Louis. 1882 A
Note sur l'ostéologie des *Mosasauridæ*.
Bull. de Musée Roy. d'Hist. nat. de Belgique, i, pp. 55-74, with pls. iv-vi.

—— 1882 B
Première note sur les Dinosauriens de Bernissart.
Bull. de Musée Roy. d'Hist. nat. de Belgique, i, pp. 161-178, with pl. ix.

—— 1882 C
Deuxième note sur les Dinosauriens de Bernissart.
Bull. de Musée Roy. d'Hist. nat. de Belgique, i, pp. 205-211, with pl. xii.
The paper treats especially of the sternum of *Iguanodon* and related genera.

Dollo, Louis—Continued.   1883 A

Note sur la présence chez les oiseaux du "troisième trochanter" des Dinosauriens et sur la function de celui-ci.

Bull. de Musée Roy. d'Hist. nat. de Belgique, ii, pp. 13–19, with pl. i.

————   1883 B

Troisième note sur les dinosauriens de Bernissart.

Bull. de Musée Roy. d'Hist. nat. de Belgique, ii, pp. 85–126, with pls. iii–v.

Treats of *Iguanodon* and its relationships to other *Dinosauria* and to *Aves*. Plate v is a restoration of *Iguanodon bernissartensis*.

————   1883 C

Note sur les restes de dinosauriens rencontrés dans le Crétacé supérieur de la Belgique.

Bull. de Musée Roy. d'Hist. nat. de Belgique, ii, pp. 205–221, with 20 figs. in the text.

————   1883 D

Quatrième note sur les dinosauriens de Bernissart.

Bull. de Musée Roy. d'Hist. nat. de Belgique, ii, pp. 223–252, with pls. ix, x.

This paper is devoted to the anatomy of *Iguanodon bernissartensis*.

————   1883 E

Première note sur les crocodiliens de Bernissart.

Bull. de Musée Roy. d'Hist. nat. de Belgique, ii, pp. 309–340, with pl. xii.

Treats of relationships of the genera of crocodiles.

————   1883 F

Note sur la présence du *Gastornis edwardsii*, Lemoine, dans l'assise inférieure de l'étage landenien, à Mesvin, Près Mons.

Bull. de Musée Roy. d'Hist. nat. de Belgique, ii, pp. 297–305, with pl. ix and 1 text fig.

————   1884 A

Première note sur les chéloniens de Bernissart.

Bull. de Musée Roy. d'Hist. nat. de Belgique, iii, pp. 63–84, with pls. i, ii.

Contains a discussion of the classification of the *Testudinata*.

————   1884 B

Cinquième note sur les dinosauriens de Bernissart.

Bull. de Musée Roy. d'Hist. nat. de Belgique, iii, pp. 129–151, with pls. vi, vii.

Treats especially of the structure and relationship of *Iguanodon*.

Bull. 179——6

Dollo, Louis—Continued.   1884 C

Première note sur le Simœdosaurien d'Erquelinnes.

Bull. de Musée Roy. d'Hist. nat. de Belgique, iii, pp. 151–186, with pls. viii, ix.

Discusses the structure of *Simœdosaurus* and its relation to *Champsosaurus* and *Sphenodon*.

————   1885 A

Sur l'identité des genres *Champsosaurus* et *Simœdosaurus*.

Bull. Soc. géol. de France (3), xiv, pp. 96–96.

————   1885 B

Notes d'ostéologie erpétologique. I. Sur une nouvelle dent de *Craspedon*, dinosaurien du Crétacé supérieur de la Belgique [pp. 309–318, 6 figs. in text]. II. Sur la présence d'un canal basioccipital médian et de deux canaux hypobasilaires chez un genre mosasauriens [pp. 319–331, 8 figs.]. III. Sur la présence d'une interclavicle chez ungenre de mosasauriens et sur la division de ce sous-ordre en familles [pp. 332–335, 2 figs.]. IV. Sur les épiphyses du calcaneum des lacertiliens [pp. 336–338, 4 figs.].

Annales Soc. Sci. de Bruxelles, ix, pp. 309–338, with 20 figs. in text.

————   1886 A

Première note sur les chéloniens du Bruxellien (Éocène moyen) de la Belgique.

Bull. Musée Roy. d'Hist. nat. Belgique, iv, pp. 75–100, with pls. i, ii.

Contains a synopsis of the classification of *Testudinata* and observations on their osteology.

————   1885 C

Première note sur le Hainosaure, mosasaurien nouveau de la craie brune phosphatée de Mesvin-Ciply, Près Mons.

Bull. de Musée Roy. d'Hist. nat. de Belgique, iv, pp. 25–35.

Contains notes on various American genera of *Mosasauridæ*.

————   1887 A

Psephophorus.

Annales Soc. Scientifique de Bruxelles, xi, pp. 139–176.

————   1888 A

On the humerus of *Euclastes.*

Geol. Magazine (3), v, pp. 251–257, with 3 figs.

Dollo, Louis—Continued.   1888 B
Iguanodontidæ et Camptonotidæ.
Comptes rend. Acad. Paris, cvi, pp. 775–777.
Notice of this paper in Geol. Mag., 1888, v, p. 520.

——   1888 C
Première note sur les chéloniens oligocènes et néogènes de la Belgique.
Bull. de Musée Roy. d'Hist. nat. de Belgique, v, pp. 59–96, with pl. iv and a text fig. of a young *Dermochelys coriacea.*
This important paper contains a discussion of the higher groups of *Testudinata*, and especially a consideration of the value of the group *Athecæ.*

——   1888 D
Sur le crâne des mosasauriens.
Bull. Sci. de la France et de la Belgique (3), i (xix), pp. 1–11, with pl. i and 9 figs. in the text.

——   1888 E
Sur la signification du "trochanter pendant" des dinosauriens.
Bull. Sci. de la France et de la Belgique (3), i (xix), pp. 215–224.

——   1890 A
Première note sur les mosasauriens de Maestricht.
Mém. Soc. Belge de Géologie, de Palæont., et d'Hydrologie (Bruxelles), iv, pp. 151–169, with pl. viii.
Contains abundant citations of the literature of the *Mosasauridæ.*

——   1891 A
Nouvelle note sur le Champsosaure, rhynchocephalien adapté à la vie fluviatile.
Mém. Soc. Belge de Géologie, de Paléont., et d'Hydrologie (Bruxelles), v, pp. 147–199, with pls. vi–viii.

——   1892 A
Sur l'origine de la nageoire caudale des *Ichthyosaurus.*
Bull. Soc. Belge Géol., Paléont., Hydrol., vi, Pr. verb., pp. 167–174, with 8 text figs.

——   1892 B
Première note sur les Téléostéens du Crétacé supérieur de la Belgique.
Bull. Soc. Belge Géol., Paléont., Hydrol., vi, pp. 180–189, with 4 text figs.

——   1892 C
Sur un nouveau type de dinosaurien.
Bull. Soc. Belge Géol., Paléont., Hydrol., vi, Pr. verb., pp. 10–12.
The principal groups of dinosaurs are characterized. The new type alluded to appears to be *Triceratops* Marsh.

Dollo, Louis—Continued.   1893 A
Suppression du genre *Leiodon.*
Bull. Soc. Belge Géol., Paléont., Hydrol., vii.
Pr. verb., p. 79.
Regards *Leiodon* as a synonym of *Mosasaurus.*

——   1893 B
Champsosaurus et Pareiasaurus.
Bull. Soc. Belge Géol., Paléont., Hydrol., vii, p. 79.

——   1893 C
Nouvelle note sur l'ostéologie des mosasauriens.
Bull. Soc. Belge Géol., Paléont., Hydrol., vi, pp. 219–259, with pls. iii, iv.

——   1896 A
Sur la phylogénie des *Dipneustes.*
Bull. Soc. Belge Géol., Paléont., Hydrol., ix, pp. 79–128, with pls. v–ix.
Concluded with a list of 123 papers and memoirs bearing on the subject. An abstract appears in Amer. Naturalist, xxx, p. 479.

Domeier, Wilhelm.   1803 A
Nachricht von einem in Nord America gefundenen vollständigen Gerippe eines Vierfüsslers, bisher häufig Mammoth oder Mammut genannt.
Neue Schriften Gesellsch. naturforsch. Freunde zu Berlin, iv, pp. 79–110.

Douglass, Earl.   1900 A
The Neocene lake beds of western Montana and descriptions of some new vertebrates from the Loup Fork. A thesis for the degree of M. S. in the University of Montana, June, 1899. Published by the University.
Pp. 3–27, with pls. i–iv.
Copies of this paper were first distributed about July 1, 1900.

——   1900 B
New species of *Merycochœrus* in Montana, Part i.
Amer. Jour. Sci. (4), x, pp. 428–438, with 3 figs. in the text.
Describes *M. laticeps.* Part ii appeared in same journal January, 1901.

Donald, J. T.   1881 A
Notes on elephant remains from Washington Territory.
The Canadian Naturalist (2), ix, pp. 53–56.

Dugès, Alfredo.   1887 A
Platygonus alemanii, nobis, fósil cuaternario.
La Naturaleza (2), i, pp. 16–18, pls. i, ii.

Duméril, Aug.   1865 A
Histoire naturelle des poissons ou ichthyologie générale. Tome premier,

Duméril, Aug.—Continued.

Élasmobranches, Plagiostomes et Holocéphales ou Chiméres.

Pp. 1–720, with atlas of pls., Paris, 1865.

Dupont, Edward.   1867 A

Human remains in Belgium.

Amer. Jour. Sci. (2), xliii, pp. 121–123; 260–264.

Contains mention of *Elephas primigenius.*

Duralde, Martin.   1804 A

Abstract of a communication from Mr. Martin Duralde relative to fossil bones, etc., of the country of Apelousas, west of the Mississippi, to Mr. William Dunbar, of the Natchez, and by him transmitted to the society.

Trans. Amer. Philos. Soc., vi, pp. 55–58.

No species definitely mentioned except bone of man, and of "elephants."

Duvernoy, G. L.   1851 A

Caractères ostéologiques des genres nouveaux ou des espèces nouvelles de cétacés vivants ou fossiles.

Annales Sci. Nat. (3), xv, pp. 5–71; 381, with pls. i, ii.

——   1855 A

Nouvelles études sur les rhinocéros fossiles.

Arch. Mus. d'Hist. nat., Paris, viii, pp. 1–144, with pls. i–viii.

Earle, Charles.   1891 A

On a new species of *Palæosyops—Palæosyops megarhinus.*

Amer. Naturalist, xxv, pp. 45–47, with 1 fig. in text.

——   1891 B

Palæosyops and allied genera.

Proc. Acad. Nat. Sci. Phila., 1891, pp. 106–117, with 8 figs. in the text.

——   1892 A

Revision of the species of *Coryphodon.*

Bull. Amer. Mus. Nat. Hist., iv, pp. 149–166.

——   1892 B

The variability of specific characters as exhibited by the extinct genus *Coryphodon.*

Science, xx, pp. 7–9, with 2 figs. in the text.

——   1892 C

A memoir upon the genus *Palæosyops* Leidy, and its allies.

Jour. Acad. Nat. Sci. Phila., ix, pp. 267–388, pls. x–xiv.

Reviewed by R. Lydekker, Natural Science, ii, pp. 148–149.

Earle, Charles—Continued.   1893 A

Some points in the comparative anatomy of the tapir.

Science, xxi, p. 118.

Discusses also the names of the tapir.

——   1893 B

On the systematic position of the genus *Protogonodon.*

Amer. Naturalist, xxvii, pp. 377–379.

——   1893 C

The evolution of the American tapir.

The Geol. Magazine (3), x, pp. 391–396.

——   1893 D

The structure and affinities of the Puerco ungulates.

Science, xxii, pp. 49–51.

——   1895 A

On a supposed case of parallelism in the genus *Palæosyops.*

Amer. Naturalist, xxix, pp. 622–626, pl. xxvii.

——   1896 A

Notes on the fossil *Mammalia* of Europe. I. Comparison of the American and European forms of *Hyracotherium.*

Amer. Naturalist, xxx, pp. 131–135.

——   1896 B

Tapirs, past and present.

Science (2), iv, pp. 934–935.

——   1898 A

Relationship of the *Chriacidæ* to the *Primates.*

Amer. Naturalist, xxxii, pp. 261–262.

See also **Osborn** and **Earle;** **Wortman** and **Earle.**

Eastman, Charles R.   1895 A

Beiträge zur Kenntniss der Gattung *Oxyrhina,* mit besonderer Berücksichtigung von *Oxyrhina mantelli* Agassiz.

Palæontographica, xlii, pp. 149–191, with pls. xvi–xviii.

——   1896 A

Observations on the dorsal shields in the Dinichthyids.

Amer. Geologist, xviii, pp. 222–223.

Abstract of paper read before the Amer. Assoc. Adv. Sci., 1896.

——   1896 B

Preliminary note on the relations of certain body plates in the Dinichthyids.

Amer. Jour. Sci. (4), ii, pp. 46–50.

——   1896 C

Review of Bashford Dean's "On the vertebral column, fins, and central armoring of *Dinichthys.*"

Amer. Geologist, xviii, pp. 316–317.

Eastman, Charles R.—Continued. 1896 D
Remarks on *Petalodus alleghaniensis*, Leidy.
Jour. Geology, iv, pp. 174-176.

——— 1897 A
Tamiobatis vetustus; a new form of fossil skate.
Amer. Jour. Sci. (4), iii, pp. 85-90, with pl. i and 1 fig. in the text.

——— 1897 B
On *Ctenacanthus* spines from the Keokuk limestone of Iowa.
Amer. Jour. Sci. (4), iii, pp. 10-13, with 2 figs. in the text.
Describes *C. xiphias* St. J. and W., and *C. acutus* sp. nov.

——— 1897 C
On the characters of *Macropetalichthys*.
Amer. Naturalist, xxxi, pp. 493-499, with pl. xii, and 1 fig. in the text.

——— 1897 D
On the relations of certain plates in the Dinichthyids.
Bull. Mus. Comp. Zool., Harvard, xxxi, pp. 19-43, with pls. i-v.

——— 1898 A
On the occurrence of fossil fishes in the Devonian of Iowa.
Iowa Geol. Surv., vii, pp. 108-116, with pl. iv and 1 fig. in the text.
The new genus *Synthetodus* is established and the new species *Ptyctodus molaris* described.
A notice of this paper, by C. R. Keyes, is found in Amer. Geologist, xxii, pp. 237-239.

——— 1898 B
Dentition of Devonian *Ptyctodontidæ*.
Amer. Naturalist, xxxii, pp. 473-488; 545-560, with 50 text figs.

——— 1898 C
Some new points in Dinichthyid osteology.
Proc. Amer. Assoc. Adv. Sci., xlvii, Boston, pp. 371-372.
See also Science (2), viii, pp. 400-401.

——— 1898 D
Some new points in Dinichthyid osteology.
Amer. Naturalist, xxxii, pp. 747-768, with 6 text figs.
For abstract, see Proc. Amer. Assoc. Adv. Sci., xlvii, Boston, pp. 371-372.

——— 1899 A
Some new American fossil fishes.
Science (2), ix, pp. 642-643.

——— 1899 B
[Notes on fossil fishes found in the Devonian at Milwaukee, Wis.]
Jour. of Geology, vii, pp. 278, 283.
These notes are contained in a paper by Teller and Monroe.

Eastman, Charles R.—Continued. 1899 C
Description of new species of *Diplodus* teeth from the Devonian of northeastern Illinois.
Jour. Geol., vii, pp. 489-493, with pl. vii.
Describes as new *Diplodus priscus* and *D. striatus;* also figures *Phœbodus politus* Newb.

——— 1899 D
Report on the department of vertebrate palæontology.
Annual Report of the assistant in charge of the Mus. Comp. Zool. Harvard College to the president and fellows of Harvard College for 1898-99, pp. 15, 16.
Records the fact that the type of *Titanichthys agassizi* is in Harvard College, and that there is also there a nearly complete skeleton of *Dinichthys pustulosus*.

——— 1899 E
Jurassic fishes from Black Hills of South Dakota.
Bull. Geol. Soc. America, x, pp. 397-406, with pls. xlv-xlviii and 2 text figs.

——— 1899 F
Upper Devonian fish fauna of Delaware County, N. Y.
17th Annual Report of the State Geologist [of New York] for the year 1897, pp. 317-327, with 6 text figs.

——— 1900 A
Fossil Lepidosteids from the Green River shales of Wyoming.
Bull. Mus. Comp. Zool. Harvard Coll., xxxvi, pp. 67-75, with pls. i-ii.

——— 1900 B
Dentition of some Devonian fishes.
Jour. Geology, viii, pp. 32-41, with 7 figs. in the text.

——— 1900 C
Devonische Fischreste aus der Eifel.
Amer. Geologist, xxv, pp. 391-392.
A notice of Huene's paper in Neues Jahrbuch Min. 1900, i, Abh., p. 64.

——— 1900 D
Einige neue Notizen über devonische Fischreste aus der Eifel.
Centralbl. Min., Geolog. und Palæont., 1900, No. 6, pp. 177-178.

——— 1900 E
New fossil bird and fish from the Middle Eocene of Wyoming.
Geol. Magazine (4), vii, pp. 54-58, with pl. iv.
Describes new genus and species of bird, *Gallinuloides wyomingensis*, and fish, *Lepidosteus atrox* Cope.

Edwards, Alphonse Milne.　　　1863 A

Mémoire sur la distribution géologique des oiseaux fossiles et description de quelques espèces nouvelles.

Annales des Sciences Naturelles, Zool. (4), xx, pp. 133-176, pl. iv.

——　　　　　　　　　　　　　　1864 A

Recherches anatomiques, zoologiques, et paléontologiques sur la famille des chevrotains.

Annales des Sciences Naturelles, Zool. (5), ii, pp. 49-167, pls. ii-xii.

——　　　　　　　　　　　　　　1865 A

Rapport sur les notes de MM. Nicolas et Chrétien relatives a des ossemens fossiles provenant de la Vallée de Zacualco.

Arch. Commission Sci. du Mexique, i, pp. 401-407, with 1 pl.

——　　　　　　　　　　　　　　1868 A

Recherches anatomiques et paléontologiques pour servir a l'histoire des oiseaux fossiles de la France.

Vol. i, pp. 1-475, with pls. i-xcvi.

——　　　　　　　　　　　　　　1871 A

Recherches anatomiques et paléontologiques pour servir a l'histoire des oiseaux fossiles de la France.

Vol. ii, pp. 1-632, with pls. xcvii-cc.

Edwards, Arthur M.　　　　　1895 A

Ornithichnites and jaw bone from the Newark sandstone of New Jersey.

Amer. Jour. Sci. (3), l, p. 346.

Edwards, Timothy.　　　　　　1793 A

A description of a horn or bone lately found in the river Chemung, or Tyoga, a western branch of the Susquehanna, about twelve miles from Tyoga Point.

Mem. Amer. Acad., ii, p. 164.

Describes what was apparently the tusk of the mammoth.

Egerton, P. G.　　　　　　　　1837 A

On certain peculiarities in the cervical vertebræ of the *Ichthyosaurus*, hitherto unnoticed.

Trans. Geol. Soc. London (2), v, pp. 187-193, with pl. xiv.

For a preliminary statement see Proc. Geol. Soc. Lond., ii, 1837, p. 192.

——　　　　　　　　　　　　　　1845 A

Description of the mouth of a *Hybodus* found by Mr. Boscawen Ibbetson on the Isle of Wight.

Quart. Jour. Geol. Soc., i, pp. 197-199, with pl. iv.

Egerton, P. G.—Continued.　　1847 A

On the nomenclature of the fossil Chimæroid fishes.

Quart. Jour. Geol. Soc. Lond., iii, pp. 350-353, with pl. xiii and 1 fig. in the text.

——　　　　　　　　　　　　　　1847 B

Fossil fish [of coal-field of eastern Virginia].

Quart. Jour. Geol. Soc. Lond., iii, pp. 275-280.

——　　　　　　　　　　　　　　1848 A

Palichthyologic notes. Supplemental to the works of Professor Agassiz.

Quart. Jour. Geol. Soc. Lond., iv, pp. 302-314, with pl. x and 2 figs. in the text.

This paper is devoted to the consideration of the structure of *Pterichthys* and the description of new species.

——　　　　　　　　　　　　　　1848 B

Observations on Mr. McCoy's paper on "Some fossil fish of the Carboniferous period."

Ann. and Mag. Nat. Hist. (2), ii, pp. 189-190.

——　　　　　　　　　　　　　　1849 A

Palichthyologic notes. No. 2. On the affinities of the genus *Platysomus.*

Quart. Jour. Geol. Soc., v, pp. 329-332.

——　　　　　　　　　　　　　　1849 B

Palichthyologic notes. No. 3. On the *Ganoidei heterocerci.*

Quart. Jour. Geol. Soc. Lond., vi, pp. 1-10, with pl. ii.

——　　　　　　　　　　　　　　1852 A

Lepidotus pectinatus.

Mem. Geol. Surv. United Kingdom, decade 6, no. iii, pp. 1-2, with pl. iii.

——　　　　　　　　　　　　　　1852 B

Ptycholepis minor Egerton.

Mem. Geol. Surv. United Kingdom, decade 6, no. vii, pp. 1-3, pl. vii.

——　　　　　　　　　　　　　　1853 A

Notes on the fossil fish from Albert mine.

Quart. Jour. Geol. Soc. Lond., ix, p. 115.

A brief identification of the fishes collected at Albert mine, New Brunswick, by Charles Lyell.

——　　　　　　　　　　　　　　1853 B

Palichthyologic notes. No. 4. On the affinities of the genera *Tetragonolepis* and *Dapedius.*

Quart. Jour. Geol. Soc. Lond., ix, pp. 274-279, with pl. xi.

——　　　　　　　　　　　　　　1857 A

Palichthyologic notes. No. 9. On some fish remains from the neighborhood of Ludlow.

Quart. Jour. Geol. Soc., London, xiii, pp. 282-289, with pls. ix, x.

Egerton, P. G.—Continued. 1857 B

On the unity of the genera *Pleuracanthus*, *Diplodus*, and *Xenacanthus*, and on the specific distinction of the Permian fossil *Xenacanthus decheni* (Beyrich.)

Ann. and Mag. Nat. Hist. (2), xx, pp. 423-424.

——   1859 A

Palichthyologic notes. No. 12. Remarks on the nomenclature of the Devonian fishes.

Quart. Jour. Geol. Soc., xvi, pp. 119-136.

——   1859 B

On *Chondrosteus*, an extinct genus of the *Sturionidæ*.

Philosoph. Trans. Roy. Soc. Lond., cxlviii, pp. 871-875; pls. 67-70.

——   1861 A

Remarks on the ichthyolites of Farnell road.

Report Brit. Assoc. Adv. Sci., 30th meeting, Oxford, 1860, pp. 77-78.

Description of *Acanthodes mitchelli*.

——   1861 B

British fossils.

Mem. Geol. Sur. United Kingdom, decade 10, pp. 51-75; pls. iv-x.

Descriptions and figs. of *Tristichopterus*, *Acanthodes*, *Climatius*, and *Diplacanthus*.

——   1862 A

On a new species of *Pterichthys* (*P. macrocephalus*, Egerton), from the yellow sandstone of Farlow Co., Salop.

Quart. Jour. Geol. Soc., xviii, pp. 103-106, with pl. iii, figs. 7-9.

——   1866 A

On a new species of *Acanthodes* from the Coal series of Longton.

Quart. Jour. Geol. Soc., xxii, pp. 468-470, with pl. xxiii.

——   1869 A

On two new species of *Gyrodus*.

Quart. Jour. Geol. Soc., xxv, pp. 379-386, with 5 text figs.

——   1871 A

On a new Chimæroid fish from the Lias of Lyme Regis (*Ischyodus orthorhinus ♂*)

Quart. Jour. Geol. Soc., xxvii, pp. 275-279, with pl. xiii.

Eichwald, Eduard.   1840 A

Die Thier- und Pflanzenreste des alten rothen Sandsteins und Bergkalks im Novgorodschen Gouvernement.

Bull. Sci. Acad. Imp. Sci. St. Pétersbourg, vii, pp. 78-91.

Eichwald, Eduard—Continued. 1846 A

Nachtrag zu der Beschreibung der Fische des devonischen Systems aus der Gegend von Pawlowsk.

Bull. Soc. Imp. Nat. Moscou, xix, pt. ii, pp. 277-318.

——   1857 A

Beitrag zur geographischen Verbreitung der fossilen Thiere Russlands.

Bull. Soc. Imp. Nat. Moscou, xxx, pt. ii, pp. 305-354.

Pp. 338-354 are devoted to the fossil vertebrates. Of most interest are some remarks on *Asterolepis* and *Bothriolepis*.

——   1860 A

Lethæa Rossica ou Paléontologie de la Russie.

Vol. i, pp. 1498-1633, with pls. lv-lix.

Only the pages of this volume are quoted here which contain matter bearing on the vertebrates.

——   1866 A

Die *Rhytina borealis* und der *Homocrinus dipentas* in der Lethæa Rossica.

Bull. Soc. Imp. Nat. Moscou, xxxix, pp. 133-162, with pl. viii.

The plate illustrates the structure of *Homocrinus*.

——   1868 A

Lethæa Rossica ou Paléontologie de la Russie.

Vol. ii, pp. 1196-1285, with pls. xxviii-xl, of vol. ii of atlas.

Only those pages of this volume are quoted which contain descriptions of vertebrates.

——   1871 A

Analecten aus der Palæontologie und Zoologie Russlands.

Moscow, 1871, 4to, pp. i-iv; 1-23, with pls. i-iii.

Contains description and figures of *Palæoteuthis marginalis* Eichw., pp. 1-15, pl. i.

Eigenmann, C. H.   1890 A

Description of a fossil species of *Sebastodes*.

Zoe, i, p. 17, with 1 fig.

Describes preopercle of *Sebastodes* (?) rosæ from Tertiary beds at Port Halford, Cal.

Emerton, J. H.   1887 A

The restoration of the skeleton of *Dinoceras mirabile*.

Proc. Boston, Soc. Nat. Hist. xxiii, pp. 342-343.

Emery, Carl,

Die fossilen Reste von *Archegosaurus* und *Eryops* und ihre Bedeutung für die Morphologie des Gliedmassenskelets.

Anatom. Anzeiger, xiv, pp. 202-208, with 7 figs. in the text.

Emery, Carl—Continued. 1897 B
Ueber die Beziehungen des Crossop-
terygiums zu anderen Formen der
Gliedmassen der Wirbeltiere.
Anatom. Anzeiger, xiii, pp. 137-149, with 6
figs. in the text.

——— 1897 C
Beiträge zur Entwicklungsgeschichte
und Morphologie des Hand-und Fuss-
skelets der Marsupialier.
Semon's Zoolog. Forschungsreisen in Aus-
tralien, etc., ii, pp. 371-400, with pls. xxxiii-
xxxvi, with 13 text figs.
On p. 400 is found a record of the literature
examined.

Emmons, Ebenezer. 1845 A
On the supposed *Zeuglodon cetoides* of
Professor Owen.
Amer. Quart. Jour. Agricult. and Sci., ii, pp.
59-63, with 1 page of figs.

——— 1845 B
The *Zeuglodon cetoides* (Owen).
Amer. Quart. Jour. Agricult. and Sci., ii, p.
366.

——— 1846 A
Description of some of the bones of
the *Zeuglodon cetoides* of Professor Owen.
Amer. Quart. Jour. Agricult. and Sci., iii,
pp. 228-231, with pls. i, ii.

——— 1851 A
Description of some of the bones of
the *Zeuglodon cetoides* of Professor Owen.
Pp. 1-6. Year and place of publication unknown. Copy deposited in Smithsonian Institution November 13, 1851.

——— 1856 A
Geological report on the midland
counties of North Carolina. Illustrated
with engravings.
Pp. i-xx; 1-352; pls. 1-8. New York, Geo. P.
Putnam & Co. Raleigh, Henry D. Turner, 1856.
Chap. xl, pp. 298-323, and p. 347 are devoted
to the "animal remains of the Coal Measures
of the Deep and Dan rivers," and to a "notice
of the vertebral remains of the Bristol conglomerate."

——— 1857 A
American geology, containing a statement of the principles of the science,
with full illustrations of the characteristic American fossils, with an atlas and
a geological map of the United States.
Part vi, pp. i-x; 1-152, with pls. i-x, and 114
figs. in the text.

——— 1858 A
· Fossils of the sandstones and slates of
North Carolina.
Proc. Amer. Assoc. Adv. Sci., 11th meeting,
Montreal, 1857, pp. 76-80.

Emmons, Ebenezer—Cont'd. 1858 B
Report of the North Carolina Geological Survey. Agriculture of the eastern
counties, together with descriptions of
the fossils of the marl beds.
Illustrated by engravings. Raleigh, 1858.
Pp. i-xv; 1-314.
Chapters xv-xvii are devoted to descriptions
of the fossil vertebrates, and there are furnished
83 woodcuts in the text.

——— 1860 A
Manual of geology. Second edition.
Pp. i-viii; 1-297, with 218 figs. New York,
1860.

Etheridge, Robert, and **Willett**, Henry. 1889 A
On the dentition of *Lepidotus maxi-
mus*, etc.
Quart. Jour. Geol. Soc., xlv, pp. 356-358,
with pl. xiv.

Evermann, B. W. 1893 A
A skeleton of Steller's sea-cow.
Science, xxi, p. 59.

See also **Jordan** and **Evermann**.

Ewart, J. C. 1894 A
The development of the skeleton of
the limbs of the horse, with observations on polydactyly.
Jour. Anat. Physiol., xxviii, pp. 236-256,
with 3 pages of figs; 342-369, with pl. xii.

——— 1894 B
The second and fourth digits in the
horse; their development and subse-
quent degeneration.
Proc. Roy. Soc. Edinburgh, xx, pp. 185-191.
The author claims to have found these elements in the foetus of the horse.

Eyerman, John. 1886 A
Footprints on the Triassic sandstone
(Jura-Trias) of New Jersey.
Amer. Jour. Sci. (3), xxxi, p. 72.

——— 1889 A
Notes on geology and mineralogy.
Proc. Acad. Nat. Sci. Phila., 1889, pp. 32-35.
Describes footprints from Jura-Trias of New
Jersey.

——— 1890 A
Bibliography of North American ver-
tebrate palæontology for the year 1889.
Amer. Geologist, v, pp. 250-253.

——— 1891 A
Bibliography of North American ver-
tebrate palæontology for the year 1890.
Amer. Geologist, vii, pp. 231-238.

——— 1891 B
A catalogue of the palæontological
publications of Joseph Leidy, M. D.,
LL. D.
Amer. Geologist, viii, pp. 333-342.

Eyerman, John—Continued. 1892 A

Bibliography of North American vertebrate palæontology for the year 1891.

Amer. Geologist, ix, pp. 249-255.

—— 1893 A

Bibliography of North American vertebrate palæontology for the year 1892.

Amer. Geologist, xi, pp. 388-398.

—— 1894 A

Preliminary notice of a new species of *Temnocyon*, and a new genus from the John Day Miocene of Oregon.

Amer. Geologist, xiv, pp. 320, 321.

Descriptions of *T. ferox* and new genus *Hypotemnodon*, type *Temnocyon coryphæus* Cope.

—— 1896 A

The genus *Temnocyon* and a new species thereof, and the new genus *Hypotemnodon*, from the John Day Miocene of Oregon.

Amer. Geologist, xvii, pp. 267-287.

Falconer, Hugh. 1857 A

Description of two species of the fossil mammalian genus *Plagiaulax* from Purbeck.

Quar. Jour. Geol. Soc. Lond., xiii, pp. 261-282, with 17 figs. in the text.

—— 1857 B

On the species of mastodon and elephant occurring in the fossil state in Great Britain. Part I. Mastodon.

Quar. Jour. Geol. Soc. Lond., xii, pp. 307-360, with pls. xi-xii.

Contains diagnoses of the genera *Mastodon* and *Elephas*, the former being divided into the subgenera *Trilophodon* and *Tetralophodon*, the latter genus into *Stegodon*, *Loxodon*, and *Euelephas*.

—— 1862 A

On the disputed affinity of the mammalian genus *Plagiaulax*.

Quar. Jour. Geol. Soc., xviii, pp. 348-369, with 20 figs. in the text.

——. 1865 A

On the species of mastodon and elephant occurring in the fossil state in Great Britain. Part II.

Quar. Jour. Geol. Soc., xxi, pp. 253-332.

—— 1868 A

Palæontological memoirs and notes.

Vol. i, Fauna antiqua sivalensis, pp. 1-590, with pls. i-xxxiv and 16 text figs.; vol. ii, Mastodon, Elephant, Rhinoceros, etc., pp. 1-675, with pls. i-xxxviii and text figs. 1-9. London, 1868.

Compiled by Charles Murchison, with a biographical sketch of Hugh Falconer.

Falconer, Hugh, and Cautley, Proby T. 1846 A

Fauna antiqua sivalensis, being the fossil zoölogy of the Sewalik Hills in the north of India. 1. *Pachydermata*. Elephant and mastodon.

Pp. 1-64, with pls. i-lvi and description of plates forming pp. 1-77.

Farr, Marcus S. 1896 A

Notes on the osteology of the White River horses.

Proc. Amer. Philos. Soc., xxxv, pp. 147-175, with 6 figs. in the text.

Abstract in Princeton Univ. Bull., ix, pp. 17-18, 1897; reviewed by Schlosser in Neues Jahrbuch Mineral., 1899, ii, pp. 316-319.

Farrington, Oliver C. 1899 A

A fossil egg from Dakota.

Pubs. Field Columb. Mus., Geology, i, pp. 193-200, with pls. xx-xxi.

Faujas-Saint-Fond, Barth. 1802 A

Sur deux espèces de bœufs on trouve les crânes fossiles en Allemagne, en France, en Angleterre, dans le nord de l'Amerique et dans d'autres contrées.

Ann. Mus. d'Hist. Nat. Paris, 1802, ii, pp. 188-200, with pls. xliii, xliv.

Contains description and plate of fossil bison (*B. latifrons*).

Featherstonhaugh, G. W. 1831 A

Rhinoceroides alleghaniensis.

Monthly Amer. Jour. Geology and Nat. Sci., i, pp. 10-12, with pl. i.

Felix, Johannes. 1890 A

Beiträge zur Kenntniss der Gattung *Protosphyræna* Leidy.

Zeitschr. deutsch. geol. Gesellsch., xlii, pp. 278-302, with pls. xii-xiv.

Felix, Johannes and Lenk, H. 1891 A

Uebersicht über die geologischen Verhältnisse des mexicanischen Staats Puebla.

Palæontographica, xxxvii, pp. 117-139, with pl. xxx.

Field, Roswell. 1860 A

Ornithichnites, or tracks resembling those of birds.

Amer. Jour. Sci.(2), xxix, pp. 361-363.

—— 1860 B

Ornithichnites.

Proc. Amer. Assoc. Adv. Sci., 13th meeting, Springfield, Mass., 1859, pp. 337-340.

The tracks regarded as belonging to *Reptilia*.

—— 1860 C

Communication on the footmarks of the Connecticut River sandstones.

Proc. Boston Soc. Nat. Hist., vii, p. 316.

Filhol, Henri. 1872 A
Récherches sur les mammifères fossiles des depots de phosphate de Chaux dans les departments du Lot, du Tarn et de Tarn-et-Garone.
Annales sci. géol., iii, art. 7, pp. 1-31, with pls. 13.-1 9.

———— 1874 A
Nouvelles observations sur les mammifères des gisements de phosphates de Chaux (Lémuriens et Pachylémuriens).
Annales sci. géol., v, art. 4, pp. 1-36, with pls. vii and viii.
Devoted especially to descriptions of genera *Necrolemur* and *Palæolemur*. Pl. viii contains 4 figures of *Adapis*.

———— 1876 A
Recherches sur les phosphorites du Quercy. Étude des fossiles qu'on y rencontre et spécialement des mammifères.
Annales sci. géol., vii, art. 7, pp. 1-220, with pls. x-xxxvi.

———— 1877 A
Recherches sur les phosphorites du Quercy. Étude des fossiles qu'on y rencontre et spécialement des mammifères.
Annales sci. géol., viii, art. 1, pp. 1-340, with pls. i-xxviii.

———— 1881 A
Étude des mammifères fossiles de Saint-Gérand-le-Puy (Allier). Seconde partie (1).
Annales sci. géol., xi, pp. 1-86, with pls. i-xx.

———— 1882 A
Étude des mammifères fossiles de Ronzon (Haute-Loire).
Annales sci. géol., xii, art. 3, pp. 1-271, with pls. vi-xxxi.

———— 1882 B
Observations relatives aux caractères ostéologiques de certaines espèces d'*Eudyptes* et de *Spheniscus*.
Bull. Soc. Philomatique de Paris (7), vii, pp. 226-235.

———— 1883 A
Notes sur quelques mammifères fossiles de l'époque miocène.
Archives du Muséum d'Historie Naturelle de Lyon, iii, pp. 1-97, pls. i-v.
1. Observations relatives à divers mammifères fossiles provenant de Saint-Gérand-le-Puy (Allier), pp. 1-42.
2. Observations relatives au carnassiers signalé par Jourdan sous le nom de *Dinocyon thenardi*, pp. 43-53.
3. Observations relatives à divers carnassiers

Filhol, Henri—Continued.
fossiles provenant de la Grive-Saint Alban (Isère), pp. 55-69.
4. Observations relatives aux chiens actuels et aux carnassiers fossiles s'en approchant le plus, pp. 70-97.

———— 1883 B
Observations relatives au mémoire de M. Cope intitulé: Relation des horizons renferment des débris d'animaux vertébrés fossiles en Europe et en Amérique.
Annales sci. géol., xiv, art. 5, pp. 1-51.

———— 1884 A
De la restauration du squelette d'un *Dinoceras*.
Annales sci. géol., xvi, art. 4, pp. 1-10, with pl. ix.
Based especially on Dr. Osborn's study of *Ioxolophodon*, in Osborn, H. F., 1881 A.

———— 1885 A
Observations sur le mémoire de M. Cope intitulé: Relations des horizons renferment des débris d'animaux vertébrés fossiles en Europe et en Amérique.
Annales sci. géol., xvii, art. 5, pp. 1-18, with pl. vi.

———— 1888 A
Études sur les vertébrés fossiles d'Issel (Aude).
Mem. soc. géol. de France (3), v, pp. 1-188, with pls. 1-21.
Devoted principally to the genus *Lophiodon* and its species and to *Pachynolophus*.

———— 1888 B
Caractères de la face du *Machairodus bidentatus*.
Bull. Soc. Philomat. de Paris (7), xii, pp. 129-134.

———— 1891 A
Études sur les mammifères fossiles de Sansan.
Annales sci. géol., France, xxi, pp .1-319, with pls. i-xlvi.

———— 1891 B
De la dentition supérieure de l'*Anthracotherium mimimum*.
Bull. Soc. Philomat., Paris (8), iii, pp. 89-91, with 2 figs. in the text.

———— 1894 A
Observations concernant quelques mammifères fossiles nouveaux du Quercy.
Annales Sci. Nat., Zool. (7), xvi, pp. 129-150, with 21 figs.
Describes remains regarded as belonging to *Dasypodidæ*, *Manidæ*, and *Orycteropodidæ*.

Fisher, G. J.   1859 A
Account of an antler of the reindeer found at Sing Sing, N. Y.
Proc. Acad. Nat. Sci., Phila., xi, p. 194.

Fleischmann, A.   1890 A
Die Stammesverwandtschaft der Nager (*Rodentia*) mit den Beutelthieren (*Marsupialia*).
Sitzungsber. k.-p. Akad. Wissensch., Berlin, 1890, pp. 299–305.

———   1891 A
Die Grundform der Backzähne bei Säugethieren und die Homologie der einzelnen Höcker.
Math. naturwiss. Mittheil. k.-p. Akad. Wissensch., Berlin, 1891, pp. 405–417, with pl. v.

Flores, Eduardo.   1897 A
Sul sistema dentario del genere *Anthracotherium* Cuv.
Bull. soc. geol. Ital., xvi, pp. 92–96.

Flot, L.   1886 A
Description de *Halitherium fossile*, Gervais.
Bull. soc. géol. de France (3), xiv, pp. 483–518, pls. xxvi–xxviii.

———   1886 B
Note sur le *Prohalicore dubaleni*.
Bull. soc. géol. de France (3), xv, pp. 134–138, pl. i.

Flower, W. H.   1864 A
Notes on the skeletons of whales in the principal museums of Holland and Belgium, with descriptions of two species apparently new to science.
Proc. Zool. Soc. Lond., 1864, pp. 384–426, with 17 figs. in the text.

———   1867 A
Description of the skeleton of *Inia geoffrensis* and of the skull of *Pontoporia blainvillii*, with remarks on the systematic position of these animals in the order *Cetacea*.
Trans. Zool. Soc. Lond., vi, pp. 87–116, with pls. xxv–xxviii.

———   1868 A
On the development and succession of the teeth in the armadillos (*Dasypodidæ*).
Proc. Zool. Soc. Lond., 1868, pp. 378–380.

———   1868 B
On the osteology of the cachelot, or sperm whale (*Physeter macrocephalus*).
Trans. Zool. Soc. Lond., vi, pp. 309–372, with pls. lv–lxi and 13 figs. in the text.

Flower, W. H.—Continued.   1869 A
Remarks on the homologies and notation of the teeth of *Mammalia*.
Jour. Anat. and Physiol., iii, pp. 262–263, with 6 figs. in the text.

———   1869 B
On the value of the characters of the base of the cranium in the classification of the order *Carnivora*, and on the systematic position of *Bassaris* and other disputed forms.
Proc. Zool. Soc. Lond., 1868, pp. 4–37, with 15 text figs.

———   1871 A
On the ziphoid whales.
Nature, v, pp. 103–106.

———   1871 B
On the composition of the carpus of the dog.
Jour. Anat. and Physiol., vi, pp. 62–64, with 1 text fig.

———   1872 A
On the recent ziphoid whales, with a description of the skeleton of *Berardius arnouxi*.
Trans. Zool. Soc. Lond., viii, pp. 203–234, with pls. xxvii–xxix.

———   1873 A
Hunterian lectures.   *Artiodactyla.*
Nature, vii, pp. 428–430.
Abstract of three lectures.

———   1874 A
On the structure and affinities of the musk deer (*Moschus moschiferus*).
Proc. Zool. Soc. Lond., 1874, pp. 159–190, with 14 figs. in the text.

———   1876 A
On some cranial and dental characters of the existing species of rhinoceroses.
Proc. Zool. Soc. Lond., 1876, pp. 443–457, with 4 figs. in the text.

———   1876 B
Hunterian lectures on the relations of extinct to existing *Mammalia*.
Nature, xiii, pp. 307–308; 327–328; 350–352; 387–388; 409–410; 449–450; 487–488; 513–514; vol. xiv, p. 11.
Abstract of a course of lectures at the Royal College of Surgeons.

———   1876 C
The extinct animals of North America.
Proc. Roy. Institution, Great Britain, viii, pp. 103–125, with 3 figs. in the text.
A popular lecture on the fossil *Mammalia* of North America.

Flower, W. H.—Continued.   1881 A
Horse. Part I. Zoology and anatomy.
Encyclop. Brit., ed. 9, xii, pp. 172–181, with 6
figs. in the text.
To this article is subjoined a bibliography
of the subject.

—————   1882 A
On the mutual affinities of the ani-
mals composing the order *Edentata*.
Proc. Zool. Soc. Lond., 1882, pp. 358–367.

—————   1882 B
Lemur.
Encyclop. Brit., ed. 9, xiv, pp. 440–445, with 6
text figs.
A portion of this article is devoted to extinct
Lemuroidea.

—————   1883 A
On the arrangement of the orders and
families of existing *Mammalia*.
Proc. Zool. Soc. Lond., 1883, pp. 178–186.

—————   1883 B
On the characters and divisions of the
family *Delphinidæ*.
Proc. Zool. Soc. Lond., 1883, pp. 466–513, with
9 figs. in the text.

—————   1883 C
Mammoth.
Encyclop. Brit., ed. 9, xv, pp. 447–448, with
2 figs. in the text.

—————   1883 D
Mammalia.
Encyclop. Brit., ed. 9, xv, pp. 347–446, with
123 figs. in the text.
This article treats of the anatomy and classi-
fication of the *Mammalia*, together with gen-
eral descriptions of the various orders and
families. The portions on the *Insectivora*,
Chiroptera, and *Rodentia* were furnished by
Dr. G. E. Dobson.

—————   1885 A
An introduction to the osteology of
the *Mammalia*. Third edition, revised
with the assistance of Hans Gadow.
London: Macmillan & Co., 1885, pp. 383, figs.
134.

Flower, W. H. and **Lydekker**, R. 1891 A
An introduction to the study of
mammals living and extinct.
London, pp. 763, figs. 357. 8vo. 1891.

Forbes, Henry O.   1893 A
Observations on the development of
the rostrum of the cetacean genus *Mesoplodon*, with remarks on some of the
species.
Proc. Zool. Soc. Lond. 1893, pp. 216–236, with
pls. xii–xv and 1 page of text figs.

Forshey, C. G.   1846 A
[Communication regarding cranium
of bear found near Natchez, Miss., and
supposed to be that of the polar bear.]
Proc. Bost. Soc. Nat. Hist., ii, p. 163.

Foster, J. W.   1837 A
Miscellaneous observations made
during a tour in May, 1835, to the falls
of the Cuyahoga, near Lake Erie; extracted from the diary of a naturalist.
Amer. Jour. Sci., xxxi, pp. 1–84.
Description of a fossil "*Ovis mammillaris*,"
figs. 19, and of a fossil beaver (*Castoroides*),
figs. 16–18.

—————   1838 A
Organic remains.
2d annual report Geol. Surv. Ohio, pp. 79–83,
with text figs. A–D.
A portion of Foster's report to the State
geologist.

—————   1839 A
Head of *Mastodon giganteum*.
Amer. Jour. Sci., xxxvi, pp. 189–191, with
1 fig.

—————   1849 A
Remarks on geological position of
the mastodon.
Proc. Nat. Hist. Soc. Boston, iii, pp. 111–117.
Accompanied by remarks by Professor
Rogers, Dr. Pickering, and Mr. Desor.

—————   1857 A
On the geological position of the
deposits in which occur the remains of
the fossil elephant of North America.
Proc. Amer. Assoc. Adv. Sci., 10th meeting,
Albany, 1856, pp. 148–169.

—————   1872 A
A new species of elephant.
Nature, vi, p. 443.
Reports the proposal before Amer. Soc. Adv.
Sci., by Foster, of a supposed new species of
fossil elephant from Indiana, *E. mississippien-
sis*. No description given.

Foulke, William P.   1858 A
[Statement respecting fossil bones of
Hadrosaurus.]
Proc. Acad. Nat. Sci. Phila., 1858, pp. 213–
215.

Fraas, Eberhard.   1885 A
[Review of W. Dames: Ueber *Archæopteryx* in Palæontologische Abhandlungen.]
Neues Jahrbuch Min., Geol., und Palæon-
tol., i, 1885, pp. 470–472. Referate.
See Dames, W., 1884 B.

—————   1888 A
Ueber die Finne von *Ichthyosaurus*.
Jahreshefte Ver. f. vaterl. Naturk. Württem-
berg, xliv, pp. 280–303, with pl. vii.

Fraas, Eberhard—Continued. 1889 A
Die Labyrinthodonten der schwäbischen Trias.
Palæontographica, xxxvi, 1–158, with pls. i–xvii.

—— 1891 A
Die Ichthyosaurier der Süddeutschen Trias- und Jura- Ablagerungen.
Pp. 1–81, with pls. i–xiv.
Contains a history of this subject and a discussion of the anatomy of ichthyosaurs.
Pp. 80–81 are occupied by a list of papers and memoirs on the *Ichthyosauria.*

—— 1892 A
The paddles and fins of *Ichthyosaurus.*
Geol. Magazine (3), ix, pp. 516–517, with 2 figs. in the text.

—— 1892 B
Ichthyosaurus numismalis E. Fraas.
Jahreshefte Ver. f. vaterl. Naturk. Würtemberg, xlviii, pp. 22–31.

—— 1894 A
Die Hautbedeckung von *Ichthyosaurus.*
Jahreshefte Ver. f. vaterl. Naturk., Würtemberg, l., pp. 493–497, with pl. v.

—— 1896 A
Neue Selachier-Reste aus dem oberen Lias von Holzmaden in Württemberg.
I. *Hybodus hauffianus* E. Fraas (= *Hybodus reticulatus* Quenstedt, non Agassiz). II. *Palæospinax smith-woodwardii* E. Fraas.
Jahreshefte Ver. f. vaterl. Naturk. Württemberg, liii, Abh., pp. 1–25, with pls. i, ii.

—— 1898 A
Ein neues Exemplar von *Ichthyosaurus* mit Haut-Bekleidung.
Földtani Közlöny, xxviii, pp. 169–173, with pl. ii.

—— 1899 A
Proganochelys quenstedtii Baur (*Psammochelys keuperina* Qu.). Ein neuer Fund der Keuperschildkröte aus dem Stubensandstein.
Jahreshefte Ver. f. vaterl. Naturk. Würtemberg, lv, pp. 401–424, with pls. v–viii, and 5 figs. in the text.

Fraas, Oscar. 1861 A
Ueber *Semionotus* und einige Keuper-Conchylien.
Jahreshefte Ver. f. vaterl. Naturk. Würtemberg. xvii, pp. 81–101.

—— 1867 A
Dyoplax arenaceus, ein neuer Stuttgarter Keuper-Saurier.
Jahreshefte Ver. f. vaterl. Naturk. Würtemberg, xxiii, pp. 109–112, with pl. i.

Franque, Henricus. 1847 A
Nonnulla ad Amiam calvam accuratius cognoscendam.
Pp. 12, with 1 pl.
Inaugural dissertation.

Frenkel, F. 1873 A
Beitrage zur anatomischen Kenntniss des Kreuzbeines der Säugethiere.
Jenaische Zeitschr. f. med. Naturwiss., (1) vii pp. 391–437, with pls. xxi, xxii.

Fricke, Karl. 1875 A
Die fossile Fische aus der Oberen Juraschichten von Hannover.
Palæontographica, xxii, 347–398, pls. xviii–xxii.

Fritsch, Anton. 1876 A
Ueber die Fauna der Gaskohle des Pilsner und Rakonitzer Beckens.
Sitzungsber. k. böhmischen Gesellsch. der Wissensch. in Prag, Jahrgang 1875, pp. 70–79.

—— 1878 A
Die Reptilien und Fische der böhmischen Kreideformation.
Pp. i, ii, 1–46, with 10 pls. and 66 wood cuts. Prag., 1878.

—— 1883 A
Fauna der Gaskohle und der Kalksteine der Permformation Böhmens.
Pp. 1–182, with pls. i–xlviii.
Pp. 1–92 were published in 1879; pp. 93–126, 1880; pp. 127–158, 1881; pp. 159–182, 1883.

—— 1888 A
Ueber die Brustflosse von *Xenacanthus decheni,* Goldf.
Zoolog. Anzeiger, xi, pp. 113–114, with 1 fig. in the text.

—— 1889 A
Ueber *Xenacanthus.*
Zoolog. Anzeiger, xii, pp. 386–387.

—— 1889 B
Fauna der Gaskohle und der Kalksteine der Permformation Böhmens. Zweiter Band. *Stegocephali* (Schluss). *Dipnoi, Selachii* (Anfang).
Pp. 1–114, with 42 pls. and 79 text figs.
Pp. 1–32 are dated 1885; pp. 33–64, 1886; pp. 65–92, 1888; pp. 93–114, 1889.

—— 1890 A
Über Pterygopodien permischer Haifische der Gattungen *Pleuracanthus* und *Xenacanthus.*
Zoolog. Anzeiger, xiii, pp. 318–320, with 1 fig. in text.

—— 1890 B
Preliminary notes on the palæozoic Elasmobranchs, *Pleuracanthus* and *Xenacanthus.*
Geol. Magazine, (3), vii, p. 566.

Fritsch, Anton—Continued.　　　1891 A

Restorations of the Palæozoic Elasmobranch genera *Pleuracanthus* and *Xenacanthus.*

Report Brit. Assoc. Adv. Sci., 60th meeting, Leeds, 1890, p. 822.

—　　　　　　　　　　　　　　1891 B

Über die Xenacanthiden.

Zoolog. Anzeiger, xiv, pp. 21–22 (with restored figure of an old female of *X. decheni*).

—　　　　　　　　　　　　　　1895 A

Fauna der Gaskohle und der Kalksteine der Permformation Böhmens.

Vol. iii, pp. 1–132, pls. 91–132, and text figs. 189–310.

Pp. 1–48 are dated 1890; pp. 49–80, 1893; pp. 81–104, no date, 1894; pp. 105–132, 1895.

This volume is devoted to the genus *Pleuracanthus*, the *Acanthodidæ, Megalichthys, Palæoniscidæ*, etc.

Review of part ii by A. S. Woodward, in Natural Science, ii, pp. 435–438, with 3 figs.

—　　　　　　　　　　　　　　1895 B

Ueber neue Wirbelthiere aus der Permformation Böhmens, nebst einer Uebersicht der aus derselben bekannt gewordenen Arten.

Sitzungsber. k. böhm. Gesellsch. Wissensch., 1895, iii, pp. 1–17, with 1 text fig.

Fry, Edward.　　　　　　　　1846 A

On the osteology of the active gibbon (*Hylobates agilis*).

Proc. Zool. Soc. Lond., 1846, pl. xiv, pp. 11–18.

Fürbringer, Max.　　　　　　1885 A

Über das Schulter- und Ellenbogengelenk bei Vögeln und Reptilien.

Morpholog. Jahrbuch, xi, pp. 118–120.

—　　　　　　　　　　　　　　1885 B

Über die Nervenkanäle im Humerus der Amnioten.

Morpholog. Jahrbuch, xi, pp. 484–486.

—　　　　　　　　　　　　　　1888 A

Untersuchungen zur Morphologie und Systematik der Vögel, zugleich ein Beitrag zur Anatomie der Stütz- und Bewegungsorgane.

Pp. 1–xlix; 1–1712, with 30 pls. Folio, Amsterdam, 1888.

If under the various genera and higher groups citations to particular pages of Fürbringer's work are found, it must be understood only that on those pages some special treatment is given such groups. The same genera and higher groups may be found mentioned anywhere throughout this splendid volume.

—　　　　　　　　　　　　　　1889 A

Untersuchungen zur Morphologie und Systematik der Vögel, zugleich ein

Fürbringer, Max—Continued.

Beitrag zur Anatomie der Stütz- und Bewegungsorgane.

Biolog. Centralbl., ix, pp. 204–217, 385–396, 499–510.

—　　　　　　　　　　　　　　1890 A

Untersuchungen zur Morphologie und Systematik der Vögel, zugleich ein Beitrag zur Anatomie der Stütz- und Bewegungsorgane.

Biolog. Centralbl., x, pp. 48–62, 326–341, 373–377, 491–504, 754–767.

—　　　　　　　　　　　　　　1890 B

Ueber die systematische Stellung der *Hesperornithidæ.*

Monatsschrift deutsch. Verein z. Schutze d. Vogelwelt, xv, pp. 488–513.

—　　　　　　　　　　　　　　1892 A

Untersuchungen zur Morphologie und Systematik der Vögel, etc.

Biolog. Centralbl., xii, pp. 146–157; 722–729.

—　　　　　　　　　　　　　　1892 B

L. Stejneger's Vogelsystem.

Jour. Ornithologie (4), xx, pp. 137–148.

—　　　　　　　　　　　　　　1896 A

Untersuchungen zur Morphologie und Systematik der Vögel, zugleich ein Beitrag zur Anatomie der Stütz- und Bewegungsorgane. (Neunzehntes Stück).

Biolog. Centralbl., xvi, pp. 497–511.

—　　　　　　　　　　　　　　1900 A

Zur vergleichenden Anatomie des Brustschulterapparates und der Schultermuskeln; iv Teil.

Jenaische Zeitschr. f Naturwiss. Jena, xxxiv, pp. 215–718, with pls. xiii–xvii and 141 figs. in the text.

This work deals with the anatomy of the shoulder-girdle of reptiles. Pp. 597–682 were issued separately with the title "Beitrag zur Systematik und Genealogie der Reptilien," pp. 1–88; Inhaltsübersicht pp. 89–91.

Gadow, Hans.　　　　　　　　1877 A

Anatomie des *Phœnicopterus roseus* Pall., und seine Stellung im System.

Jour. Ornithologie, xxv, pp. 382–396.

—　　　　　　　　　　　　　　1885 A

On the anatomical differences in the three species of *Rhea.*

Proc. Zool. Soc. Lond., 1885, pp. 308–322, with 11 woodcuts in the text.

—　　　　　　　　　　　　　　1888 A

Remarks on the numbers and on the phylogenetic development of the remiges of birds.

Proc. Zool. Soc. Lond., 1888, pp. 655–667.

Gadow, Hans—Continued.   1889 A
On the modifications of the first and
second visceral arches, with especial
reference to the homologies of the auditory ossicles.
Philosoph. Trans. Roy. Soc. Lond., clxxix,
pp. 451–485, pls. lxxi–lxxiv.

——   1891 A
Vögel.
Bronn's Klassen und Ordnungen des Thier-
Reichs, vi, pp. 91–1008, with pls. xviii–lix.
Pp. 1–90 of this work, written by E. Selenka,
and published in 1869, are devoted to the osteology of birds. Pages 934–995 treat of the development of the skeleton. On pp. 934–939 is
an important list of works on the development of the skeleton.

——   1892 A
On the classification of birds.
Proc. Zool. Soc. Lond., 1892, pp. 229–256.

——.   1896 A
On the evolution of the vertebral
column of *Amphibia* and *Amniota*.
Philos. Trans. Roy. Soc. Lond. (B), clxxxvii,
pp. 1–57, with 56 figs. in the text.
Appended is a list of 98 memoirs and papers
on the subject.

——   1896 B
[Articles in Newton's Dictionary of
Birds, 1893–1896.]

——   1897 A
Remarks on the supposed relation-
ship of birds and dinosaurs.
Proc. Cambridge Philos. Soc., ix, pp. 204–208.

——   1898 A
A classification of *Vertebrata*, recent
and extinct.
London, Adam and Charles Black, 1898, 8vo,
pp. i–x; 1–82.

——   1899 A
Orthogenetic variation in the shells
of *Chelonia*.
Proc. Cambridge Philosoph. Soc., x, pp. 35–37.

—— and **Abbott**, E. C.   1895 A
On the evolution of the vertebral
column of fishes.
Philosoph. Trans. Royal Soc., Lond.
clxxxvi, pp. 163–221, illustrated by numerous
text-figs.

Garman, Samuel.   1884 A
A peculiar selachian.
Science, iii, pp. 116–117, with 1 woodcut.
Describes *Chlamydoselachus anguineus* Gar-
man.

——   1884 B
The oldest living type of vertebrate,
Chlamydoselachus.
Science, iii, p. 345.
Rejoinder to Prof. E. D. Cope.

Garman, Samuel—Continued.   1884 C
The oldest living type of vertebrate.
Science, iv, p. 484.
Discusses relationships of *Chlamydoselachus;*
also includes a classification of the *Selachii.*

——   1884 D
An extraordinary shark.
Bull. Essex Institute, xvi, pp. 47–52, with
1 page of figs.
Description of new species and genus, *Chlamydoselachus anguineus* Garman, belonging to
the new family *Chlamydoselachidæ.*

——   1885 A
On the frilled shark.
Proc. Amer. Assoc. Adv. Sci., 33d meeting,
Phila., 1884, pp. 537–538.
Refers to *Chlamydoselachus anguineus.*

——   1885 B
Chlamydoselachus anguineus Garm.
A living species of cladodont shark.
Bull. Mus. Comp. Zool., xii, pp. 1–35, with
pls. i–xx.

——   1892 A
The *Discoboli, Cyclopteridæ, Liparop-
sidæ,* and *Liparididæ.*
Mem. Mus. Comp. Zool., xiv, No. 2, pp. 1–96,
pls. i–xiii.

Garrod, A. H.   1873 A
On the affinities of *Dinoceras* and its
allies.
Nature, vii, p. 481.
Places *Dinoceras* among *Artiodactyla.*

——   1873 B
On the order *Dinocerata* (Marsh).
Jour. Anat. Physiol., vii, pp. 267–270, with 1
fig.
A statement of Marsh's results; nothing new
is put forth.

——   1876 A
On some anatomical characters which
bear upon the major divisions of the
passerine birds. Part I.
Proc. Zool. Soc. Lond., 1876, pp. 506–519, with
pls. xlviii–liii.

——   1877 A
Note on the solid-hoofed pigs in the
society's collection.
Proc. Zool. Soc. Lond., 1877, p. 33.

Gaudin, C.
See **Pictet, Gaudin** and **Harpe.**

Gaudry, A.   1872 A
Sur une dent d'*Elephas primigenius*
trouvée par M. Pinard dans l'Alaska.
Comptes rend. Acad. Paris, lxxv, pp. 1281–
1282.

——   1873 A
Sur une dent d'*Elephas primigenius*
trouvée par M. Pinard dans l'Alaska.
Bull. Soc. Géolog. de France (3), i, pp. 123–124.

Gaudry, A.—Continued. 1875 A
Sur la découverte de batraciens dans le terrain primaire.
Bull. Soc. Géolog. de France (3), iii, pp. 299–306, pls. vii, viii.

——— 1875 B
Sur quelques pièces de mammifères fossiles qui ont été trouvées dans les phosphorites du Quercy.
Jour. Zoologie, iv, pp. 518–527, with pl. xviii.

——— 1877 A
Letter to Count de Saporta on North American vertebrate palæontology.
Amer. Naturalist, xi, pp. 184–186.

——— 1878 A
Sur les reptiles des temps primaires.
Comptes rend. Acad. Sci. Paris, lxxxvii, pp. 956–958.
Describes structure of the vertebræ of *Actinodon* and the occurrence of pleurocentra therein.

——— 1878 B
Les enchaînements du monde animal dans les temps géologiques. Mammifères tertiaires.
Paris, 1878. 8°. Pp. 1–296, with 312 figs. in the text.

——— 1883 A
Les enchaînements du monde animal dans les temps géologiques. Fossiles primaires.
Paris, 1883. Pp. 1–319, with 285 figs. in the text.
Pp. 218–250 are devoted to the fishes; pp. 251–288 to the batrachians and the reptiles.

——— 1885 A
Sur les dinocératidés que M. Marsh a recueillis dans l'éocène du Wyoming.
Comptes rend. Acad. Paris, ci, pp. 718–720.

——— 1889 A
Restauration du squelette du *Dinoceras*.
Comptes rend. Acad. Paris, cviii, p. 1292.

——— 1890 A
Remarques sur le nom générique d'*Hipparion*.
Bull. Soc. Géolog. de France (3), xviii, pp. 189–191.

——— 1890 B
Apparences d'inégalité dans le développement des êtres de l'ancien et du nouveau continent.
Comptes rend. Acad. Paris, cx, pp. 482–483.

——— 1890 C
Les enchaînements du monde animal dans les temps géologiques. Fossiles secondaires.
8vo, pp. 1–823, with 408 figs. in the text; Paris, 1898.
Pp. 169–305 of this work are devoted to the *Vertebrata*.

Gaudry, A.—Continued. 1891 A
Similitudes dans la marche de l'évolution sur l'ancien et le nouveau continent.
Bull. Soc. Géolog. de France (3), xix, pp. 1024–1035.

——— 1891 B
Quelques remarques sur les mastodontes à propos de l'animal du Cherichira.
Mém. Soc. Géolog. de France, ii, Mém. No. 8, pp. 1–6, with pls. i, ii.

——— 1892 A
Les Pythonomorphes de France.
Mém. Soc. Géolog. de France. Paléontologie, iii, No. 10, pp. 1–13, with pls. xvii, xviii.

——— 1895 A
Sur les cornes des dinocératidés.
Actes Soc. Scient., Chili, v, pp. cxlii–cxliii.

——— 1897 A
La dentition des ancêtres des tapirs.
Bull. Soc. Géolog. de France (3), xxv, pp. 315–325, with pl. x.
See also Comptes rendus Acad. Sci., cxxv, pp. 755–756.

Gaupp, Ernst. 1891 A
Zur Kenntniss des Primordial-Craniums der Amphibien und Reptilien.
Verhandl. anatom. Gesellsch., 5th meeting, Munich, 1891, pp. 114–120.

——— 1894 A
Zur vergleichenden Anatomie der Schläfengegend am knöchernen Wirbelthiere.
Schwalbe's Morpholog. Arbeiten, iv, pp. 77–131, pls. vi, vii.
Cites the literature of the subject.

——— 1895 A
Mitteilungen zur Anatomie des Frosches.
Anatom. Anzeiger, xi, pp. 1–8, with 5 figs.
This paper treats of the carpus and tarsus of the frog.

——— 1895 B
Ueber die Jochbogen-Bildungen am Schädel der Wirbelthiere.
Jahresber. schles. Gesellsch. vaterl. Cultur, ii Abth., pp. 56–63.

——— 1898 A
Zur Entwicklungsgeschichte des Eidechsenschädels.
Berichte naturforsch. Gesellsch. zu Freiburg i. B., x, pp. 302–316.

——— 1900 A
Das Chondrocranium von *Lacerta agilis*. Ein Beitrag zum Verständnis des Amniotenschädels.
Merkel und Bonnets Anatom. Hefte, xv (Heft xlix), pp. 435–496, with pls. xliii–xlvii, and 23 figs. in the text.

Gazley, Sayres.   1830 A

Notice of osseous remains at Big Bone Lick, Kentucky.

Amer. Jour. Sci. (1), xviii, pp. 139–141.

The above communication is an anonymous one, but the author's name is given in Amer. Jour. Sci., xx, p. 372.

Gegenbaur, Carl.   1864 A

Untersuchungen zur vergleichenden Anatomie der Wirbelthiere. Erstes Heft. Carpus und Tarsus.

Pp. i–vi; 1–127, with 6 pls. Leipzig, 1864.

———   1865 A

Vergleichend - anatomische Bemerkungen über das Fussskelet der Vögel.

Archiv f. Anat. Physiol. u. wissensch. Med., Jahrg. 1865, pp. 450–472, with 4 text figs.

———   1865 B

Untersuchungen zur vergleichenden Anatomie der Wirbelthiere. Zweites Heft: 1. Schultergürtel der Wirbelthiere. 2. Brustflosse der Fische.

Pp. i–iv; 1–176, with 9 pls. Leipzig, 1865.

———   1872 A

Untersuchungen zur vergleichenden Anatomie der Wirbelthiere. Drittes Heft. Das Kopfskelet der Selachier; ein Beitrag zur Erkenntniss der Genese des Kopfskeletes der Wirbelthiere.

Pp. i–x; 1–316, with 22 pls.

———   1872 B

Ueber das Archipterygium.

Jenaische Zeitschr. f. Med. u. Naturwiss., (1), vii, pp. 131–141, with pl. x.

———   1876 A

Zur Morphologie der Gliedmassen der Wirbelthiere.

Morpholog. Jahrbuch, ii, pp. 396–420, with 4 figs. in the text.

———   1887 A

Die Metamerie des Kopfes und die Wirbeltheorie des Kopfskeletes.

Morpholog. Jahrbuch, xiii, pp. 1–114.

Appended is a list of 59 titles of papers bearing on the subject.

———   1895 A

Clavicula und Cleithrum.

Morpholog. Jahrbuch, xxiii, pp. 1–20, with 5 figs. in text.

———   1895 B

Das Flossenskelet der Crossopterygier und das Archipterygium der Fische.

Morpholog. Jahrbuch, xxii, pp. 119–160, with 5 figs. in text.

———   1898 A

Vergleichende Anatomie der Wirbelthiere mit Berücksichtigung der Wir-

Gegenbaur, Carl—Continued.

bellosen. Erster Band. Einleitung, Integument, Skeletsystem, Muskelsystem, Nervensystem und Sinnesorgane.

Leipzig, 1898, 8vo; pp. i–xiv; 1–978, with 619 figs., some colored.

Geinitz, Hans Bruno.   1839 A

Characteristik der Schichten und Petrefacten des sächsischen Kreidegebirges.

Erstes Heft. Der Tunnel bei Obergau, etc., 1839, pp. i–v; 1–29, with pls. i–viii. Zweites Heft. A. Das Land zwischen dem Plauen'schen Grunde bei Dresden und Dohna. B. Fische. Crustaceen, Mollusken, 1840, pp. i–iv; 30–62, with pls. ix–xvi. Drittes Heft. Die sächsischböhmische Schweitz, die Oberlausitz und das Innere von Böhmen, 1842, pp. 63–116; i–xxvi, with pls. xvii–xxiv.

———   1847 A

Koch's *Hydrarchos harlani.*

Neues Jahrbuch f. Mineralogie, 1847, pp. 47–49.

———   1860 A

Zur Fauna des Rothliegenden und Zechsteins.

Zeitschr. deutsch geol. Gesellsch., xii, pp. 467–470.

Refers to the dorsal spine of *Xenacanthus decheni.*

———   1861 A

Dyas, oder die Zechsteinformation und das Rothliegenden. Heft 1. Die animalische Ueberreste der Dyas.

Pp. vii–xviii; 1–342, with 23 pls.

———   1868 A

Die fossilen Fischschuppen aus dem Plänerkalke in Strehlen.

Denkschr. Gesellsch. Natur-u. Heilkunde, Dresden, 1868, pp. 33–48, with pls. i–iv.

———   1875 A

Das Elbthalgebirge in Sachsen. Zweiter Theil. Der mittlere und obere Quader. VI. Würmer, Krebse, Fische und Pflanzen.

Palæontographica, xx, pp. 199–245, with pls. xxxvii–xlvi.

———   1883 A

Die sogenannten Koprolithenlager von Helmstedt, Büddenstedt und Schleweke bei Harzburg.

Abhandl. naturwiss. Gesellsch. Isis in Dresden, i, pp. 3–14, with pl. i, figs. 1–22.

———   1884 A

Ueber ein Graptolithen-führenden Geschiebe mit *Cyathaspis* von Rostock.

Zeitschr. deutsch. geol. Gesellsch., xxxvi, pp. 854–857, with pl. xxx,

Geinitz, Hans Bruno—Cont'd.   1885 A
Ueber Milchzähne der Mammuth,
(*Elephas primigenius*).
Festschrift naturw. Gesellsch. Isis in Dresden.
Feier 50 jährg. Bestehens, pp. 66–74, with pl.
iii.
The figures are those of the very young
animal.

Geinitz, H. B., and **Deichmüller,** J. V.
1882 A
Die Saurier der unteren Dyas von
Sachsen.
Palæontographica, xxix, pp. 1–46, with
9 pls.

Geissler, Gustav.   1895 A
Ueber neue Saurier-Funde aus dem
Muschelkalk von Bayreuth.
Zeitschr. deutsch. geol. Gesellsch., xlvii,
pp. 331–355, with pls. xiii, xiv.
Describes various portions of the skeleton of
Nothosaurus.

Gemmellaro, Gaetano Géorgio. 1857 A
Ricerche sui pesci fossili della Sicilia.
Atti dell' Accademia naturali di Catania
(2), xiii, pp. 279–328, with pls. i–vi.

George, ——.   1869 A
Études zoologiques sur les Hémiones
et quelques autres espèces chevalines.
Annales Sci. Naturelles, Zool. (5), xii, pp.
5–56, pls. 1–4.

Gervais, Paul.   1839 A
[Remarks on *Megatherium.*]
Bull. Soc. Géolog. de France, x, p. 142.

——   1850 A
Mémoire sur la famille des cétacés
Ziphioides et plus particulièrement sur
le *Ziphius longirostris* de la Méditer-
ranée.
Ann. Sci. Naturelles (3), xiv, pp. 1–11.

——   1853 A
Recherches sur l'ostéologie de plusieurs espèces d'amphisbènes, et remarques sur la classification de ces reptiles.
Ann. Sci. Naturelles (3), xx, pp. 298–314, with
pl. xv.

——   1855 A
Recherches sur les mammifères fossiles de l'Amérique méridionale.
Castelnau's Exped. Amer. du Sud., pp. 1–63,
with pls. iv–xiii.

——   1859 A
Zoologie et paléontologie français.
Nouvelles recherches sur les animaux
vertébrés dont on trouve les ossements
enfouis dans le sol de la France et sur

Gervais, Paul—Continued.
leur comparaison avec les espèces
propres aux autres regions du globe.
Deuxième édition, accompagnée d'un
atlas de 84 planches et de figures inter-
calées dans le texte.
Paris, 1859, pp. i–viii; 1–544.

——   1862 A
Sur les *Squalodon.*
Bull. Acad. Roy. de Belgique (2), xiii, pp.
462–469.
Letter addressed to Van Beneden.

——   1871 A
Remarques sur l'anatomie des cétacés
de la division des Balénidés tirées de
l'examen des pièces relatives a ces animaux que sont conservées au Muséum.
Nouv. Arch. du Mus. d'Hist. Nat. Paris, vii,
pp. 65–146, with pls. iii–ix.

——   1872 A
Ostéologie du Sphargis luth (*Sphargis
coriacea*).
Nouv. Arch. du Mus. d'Hist. Nat. Paris, viii,
pp. 199–228, with pls. v–ix.
An abstract of this memoir appeared in
Jour. Zoologie, ii, 1873, pp. 1–4.

——   1873 A
Recherches sur les Édentés tardi-
grades.
Jour. Zoologie, ii, pp. 463–469.

——   1874 A
Form typique des membres chez les
Équidés.
Jour. Zoologie, iii, pp. 300–307.

——   1874 B
Remarques sur les formes cérébrales
propres aux Thalassothériens.
Jour. Zoologie, iii, pp. 570–583.

——   1876 A
Remarques au sujet du genre *Phocodon* d'Agassiz.
Jour. Zoologie, v, pp. 62–70, with 2 figs. in
the text.

——   1876 B
Remarques au sujet du mémoire pré-
cédent [Marsh on *Brontotheriidæ*].
Jour. Zoologie, v, p. 265.

——   1878 A
Sur la dentition des Smilodons.
Comptes rend. Acad. Paris, lxxxvii, pp. 582–
583.

See also **Gervais** and **Ameghino;**
Van Beneden and **Gervais.**

Gervais, H., and Ameghino, F. 1880 A
Les mammifères fossiles de l'Amérique du Sud.
Pp. i–xi; 1–225. Buenos Aires and Paris: 1880. Contains list of about 300 fossil mammals, with diagnoses of new genera and species. About 60 of these species are regarded as new. See Branco, W., Jahresbuch Mineral., i, 1883, p. 300.

Ghigi, Alessandro. 1900 A
Sui denti dei Tapiridi.
Verhandl. anat., Gesellsch., xiv (Pavia), pp. 17–29, with 9 figs. in the text.
These Verhandlungen are found in the Ergänzungsheft of Anatom. Anzeiger, xviii.

Gibbes, Robert W. 1845 A
- Description of the teeth of a new fossil animal found in the Green-sand of South Carolina.
Proc. Acad. Nat. Sci. Phila., 1845, ii, pp. 254–256, pl. i.
A notice of this paper is found in Amer. Jour. Sci., 1845, xlix, p. 216. Describes *Dorudon serratus*.

—   1846 A
On the fossil *Squalidæ* of the United States.
Proc. Acad. Nat. Sci. Phila., iii, pp. 41–43.
Contains a list of the species which this author had identified.

—   1847 A
Description of new species of *Squalidæ* from the Tertiary beds of South Carolina.
Proc. Acad. Nat. Sci. Phila., iii, pp. 266–268.

—   1847 B
On the fossil genus *Basilosaurus*, Harlan (*Zeuglodon* Owen), with a notice of specimens from the Eocene Green-sand of South Carolina.
Jour. Acad. Nat. Sci. Phila., (2), i, pp. 5–15, pls. i–v.

—   1848 A
[Letter from, to Dr. Morton regarding the genus *Dorudon*.]
Proc. Acad. Nat. Sci. Phila., iv, p. 57.

—   1848 B
Monograph of the fossil *Squalidæ*, of the United States.
Jour. Acad. Nat. Sci. Phila., 1848, pp. 139–147, pls. xviii–xxi.

—   1849 A
Monograph of the fossil *Squalidæ*, of the United States.
Jour. Acad. Nat. Sci. Phila., i, pp. 191–206, with pls. xxv–xxvii.

Gibbes, Robert W.—Cont'd. 1850 A
On *Mosasaurus* and other allied genera in the United States.
Proc. Amer. Assoc. Adv. Sci., 2d meeting, 1849, Cambridge, Mass., p. 77.

—   1850 B
New species of fossil *Myliobates*, from the Eocene of South Carolina, and new fossils from the Cretaceous, Eocene, and Pliocene of South Carolina, Alabama, and Mississippi.
Proc. Amer. Assoc. Adv. Sci., 2d meeting, Cambridge, Mass., 1849, pp. 193–194.

—   1850 C
Remarks on the fossil *Equus*.
Proc. Amer. Assoc. Adv. Sci., 3d meeting, Charleston, S. C., 1850, pp. 66–68.
From Ashley River, S. C.; supposed to be taken from Eocene deposits.

—   1850 D
Remarks on the northern *Elephas* of Prof. Agassiz.
Proc. Amer. Assoc. Adv. Sci., 3d meeting, Charleston, S. C., 1850, p. 69.

—   1850 E
Remarks on *Mastodon angustidens*.
Proc. Amer. Assoc. Adv. Sci., 3d meeting, Charleston, S. C., 1850, pp. 69–70.

—   1850 F
Fossils common to several formations.
Proc. Amer. Assoc. Adv. Sci., 3d meeting, Charleston, S. C., 1850, pp. 70–71.

—   1850 G
New species of *Myliobates*, from the Eocene of South Carolina, with other genera not heretofore observed in the United States.
Jour. Acad. Nat. Sci. Phila. (2), i, pp. 299–300.

—   1851 A
A memoir on *Mosasaurus* and three allied new genera, *Holcodus*, *Conosaurus*, and *Amphorosteus*.
Smithsonian Cont. to Knowledge, ii, art. v, pp. 1–13; pl. i–iii.

Gidley, J. W. 1900 A
A new species of Pleistocene horse from the staked plains of Texas.
Bull. Amer. Mus. Nat. Hist., xiii, pp. 111–116, with 5 figs. in the text.
Describes *Equus scotti*.

Giebel, C. G. 1847 A
Fauna der Vorwelt, mit steter Berücksichtigung der lebenden Thiere. Erster Band: Wirbelthiere. Erste Abtheilung: Säugethiere.
Pp. i–xi; 1–283. Leipzig, 1847.

Giebel, C. G.—Continued. 1847 B

Fauna der Vorwelt, mit steter Be-
rücksichtigung der lebenden Thiere.
Erster Band: Wirbelthiere. Zweite
Abtheilung: Vögel und Amphibia.

Pp. i–xi; 1–218. Leipzig, 1847.

——   1847 C

Koch's *Hydrarchos.*

Neues Jahrb. Mineralogie, 1847, pp. 717–721.

——   1847 D

[Letter to Professor Brown.]

Neues Jahrb. Mineralogie, 1847, pp. 819–825.
Contains some observations on *Zeuglodon
cetoides.*

——   1848 A

Fauna der Vorwelt, mit steter Be-
rücksichtigung der lebenden Thiere.
Erster Band: Wirbelthiere. Dritte
Abtheilung: Fische.

Pp. i–xii; 1–467. Leipzig, 1848.

——   1852 A

Ueber die Gebissformel der Spitzmäuse.

Archiv Naturgesch., xviii, pp. 222–227.

——   1857 A

Zur Osteologie der Waschbären
(*Procyon*).

Zeitschr. gesammt. Naturwiss., 1857, pp.
349–373.

——   1859 A

Zur Osteologie der Murmelthiere.

Zeitschrift gesammt. Naturwiss., xiii, pp.
299–309.
Describes osteology of *Arctomys monax.*

——   1866 A

Die Wirbelzahlen am Vogelskelet.

Zeitschr. gesammt. Naturwiss., xxviii, pp.
20–29.
A list showing the number of vertebræ occurring in each of 451 species of birds belonging to various orders and families.

——   1866 B

Osteologie der Klapperschlangen.

Zeitschr. gesammt. Naturwiss., xxviii, pp.
172–180.
The osteology of *Crotalus.*

——   1867 A

Schädel von *Dasypus gigas.*

Zeitschr. gesammt. Naturwiss., xxx, pp. 545–
547.

——   1872 A

Schädel der *Felis concolor F. eyra*, und
F. yaguarundi.

Zeitschr. gesammt. Naturwiss., xl, pp. 431–
432.

Giebel, C. G.—Continued. 1877 A

Das Skelet des westafrikanischen
Crocodilus cataphractus.

Zeitschr. gesammt. Naturwiss. (3), 1877, ii,
pp. 106–113, with pls. 7–10.

——   1879 A

Marsh's neue Mittheilungen über die
jurassischen Dinosaurier Nord-Amerikas.

Zeitschr. gesammt. Naturwiss. (3), iv, pp.
316–318.
A note on some of the genera of Dinosaurs
described by Professor Marsh during the years
1878 and 1879.

——   1880 A

Charakteristik der Hasenschädel.

Zeitschr. gesammt. Naturwiss. (3), v, pp.
318–340, with pls. viii–x.

Gilbert, G. K.   1871 A

Remains of a mastodon.

Proc. Lyc. Nat. Hist. New York, 1871, pp.
220–221.
Describes geological position of mastodon
found at St. Johns, Auglaize County, Ohio.

Gilbert, J. Z.   1898 A

On the skull of *Xerobates* (?) *undata*
Cope.

Kansas Univ. Quarterly, vii, pp. 143–148,
with 4 figs. in the text.

Gill, Theodore.   1861 A

Catalogue of the fishes of the eastern
coast of North America.

Proc. Acad. Nat. Sci. Phila., 1861, pp. 1–63.
This paper has a pagination distinct from
that of the remainder of the volume which
contains it.

——   1862 A

Analytical synopsis of the order of
Squali, and revision of the nomenclature of the genera.

Annals Lyc. Nat. Hist. of New York, vii, pp.
367–408.

——   1865 A

[Description of the genus *Elasmognathus.*]

Proc. Acad. Nat. Sci. Phila., 1865, p. 183.

——   1866 A

Prodrome of a monograph of the
Pinnipedes.

Communications of the Essex Institute, v,
pp. 3–13.

——   1871 A

On the relations of the orders of
mammals.

Proc. Amer. Assoc. Adv. Sci., 19th meeting,
Troy, 1870, pp. 267–270.

Gill, Theodore—Continued.   1871 B
Synopsis of the primary subdivisions of the cetaceans.
Communications of the Essex Institute, vi, pp. 121–126.

——   1872 A
On the characteristics of the primary groups of the class of mammals.
Proc. Amer. Assoc. Adv. Sci., 20th meeting, Indianapolis, 1871, pp. 284–306.
See also Amer. Naturalist, v, pp. 526–533.

——   1872 B
Arrangement of the families of mammals and synoptical tables of characters of the subdivisions of mammals.
Smithsonian Miscel. Coll., No. 230, pp. i–vi, 1–98.

——   1872 C
Arrangement of the families of fishes, or classes *Pisces, Marsipobranchii,* and *Leptocardii.*
Smithsonian Miscel. Coll., No. 247, 1872, pp. i–xlvi; 1–49.

——   1873 A
On the affinities of the sirenians.
Proc. Acad. Nat. Sci. Phila., 1873, pp. 262–273.

——   1873 B
On the homologies of the shoulder-girdle of the dipnoans and other fishes.
Ann. and Mag. Nat. Hist. (4), xi, pp. 173–178.
Extracted from "Arrangement of the families of fishes."

——   1873 C
On the genetic relations of the cetaceans and the methods involved in discovery.
Amer. Naturalist, vii, pp. 19–29.
A statement of the author's views regarding the higher groups and their relations. A reply to Dr. Brandt, of St. Petersburg.

——   1873 D
On the limits of the class of fishes.
Amer. Naturalist, vii, pp. 71–79.

——   1875 A
Synopsis of Insectivorous mammals.
Bull. U. S. Geol. and Geog. Surv. Terr., i, (2), Bull. No. 2, pp. 91–120.

——   1883 A
Nomenclature of the Xiphiids.
Proc. U. S. Nat. Mus., v, pp. 485–486.

——   1883 B
On the classification of the Insectivorous mammals.
Bull. Philos. Soc. Washington, v, pp. 118–120.

——   1883 C
Note on the affinities of the Ephippiids.
Proc. U. S. Nat. Mus., v, pp. 557–560.

Gill, Theodore—Continued.   1884 A
The relations of *Didymodus,* or *Diplodus.*
Science, iii, pp. 429–430.

——   1884 B
The oldest living type of vertebrate, *Chlamydoselachus.*
Science, iii, p. 346.
Contains a classification of the sharks.

——   1884 C
The oldest living type of vertebrates.
Science, iv, p. 524.

——   1884 D
Synopsis of the plectognath fishes.
Proc. U. S. Nat. Mus., vii, pp. 411–427.

——   1888 A
Eutheria and *Prototheria.*
Amer. Naturalist, xxii, pp. 258–259.
Shows that the sense in which Dr. Gill originally used the above terms is not the same as that in which they were used by Professor Huxley.

——   1889 A
On the classification of the mail-cheeked fishes.
Proc. U. S. Nat. Mus., xi, pp. 567–592.
This paper deals with the anatomy and classification of the mail-cheeked fishes, and contains explanations of various anatomical terms; also references to the literature bearing on the group.

——   1891 A
On the relations of *Cyclopteroidea.*
Proc. U. S. Nat. Mus., xiii, pp. 361–376, with pls. xxviii–xxx.

——   1893 A
Families and subfamilies of fishes.
Mem. Nat. Acad. Sci., vi, pp. 125–138.

——   1894 A
The nomenclature of the *Myliobatidæ,* or *Aëtabatidæ.*
Proc. U. S. Nat. Mus., xvii, pp. 111–114.

——   1894 B
The nomenclature of the family *Pœciliidæ,* or *Cyprinodontidæ.*
Proc. U. S. Nat. Mus., xvii, pp. 115–116.

——   1894 C
The differential characters of the *Salmonidæ* and *Thymallidæ.*
Proc. U. S. Nat. Mus., xvii, pp. 117–122.

——   1895 A
Notes on the nomenclature of *Scymnus,* or *Scymnorhinus,* a genus of sharks.
Proc. U. S. Nat. Mus., xviii, pp. 191–193.
Scymnus preoccupied in entomology and to be replaced by *Scymnorhinus.*

Gill, Theodore—Continued.   1896 A
Notes on the synonymy of the *Tor-
pedinidæ*, or *Narcobatidæ*.
Proc. U. S. Nat. Mus., xviii, pp. 161-165.

————— .   1896 B
Note on the nomenclature of the
Pœcilioid fishes.
Proc. U. S. Nat. Mus., xviii, pp. 221-224.

—————   1896 C
Fishes, living and fossil.
Science (2), iii, pp. 909-917.
A review of Dr. Bashford Dean's book,
"Fishes, living and fossil."

—————   1900 A
The earliest use of the names *Sauria*
and *Batrachia*.
Science (2), xii, p. 730.

Gill, T., and Coues, E.   1877 A
Material for a bibliography of North
American mammals.
Report U. S. Geol. Surv. Terr., F. V. Hayden,
xi, appendix B, pp. 951-1081.
This bibliography contains about 3,000 titles
of works on the mammals of North America.

Gilpin, Bernard.   1873 A
Observations on some fossil bones
found in New Brunswick, Dominion of
Canada.
Nova Scotia Inst. of Nat. Sci., iii, pp. 400-
404.
Records finding of cetacean bones, sup-
posed to belong to *Beluga vermontana*.

Goddard, ——.   1841 A
Examination of so-called *Missourium
kochii*.
Proc. Acad. Nat. Sci. Phila., i, pp. 115-116.

Godman, J. D.   1825 A
Description of the os hyoides of the
mastodon.
Jour. Acad. Nat. Sci. Phila., iv, pp. 67-72,
with pl. ii.

—————   1828 A
American natural history.
Vols. i-iii, Philadelphia. 8vo.
Vol. i was published 1825; ii, 1826; iii, 1828.
This work contains descriptions and figures of
some of the earlier known American fossil
vertebrates.

—————   1830 A
Description of a new genus and new
species of extinct mammiferous quadruped.
Trans. Amer. Philos. Soc. (2), iii, pp. 478-485,
pls. xvii, xviii.
Description and figures of *Tetracaulodon
mastodontoideum*, Godm.

Göldi, Emil A.   1883 A
Kopfskelet und Schultergürtel von
Loricaria cataphracta, *Balistes capriscus*,
und *Accipenser ruthenus*.
Zoolog. Anzeiger, vi, pp. 420-422.

—————   1884 A
Kopfskelet und Schultergürtel von
Loricaria cataphracta, *Balistes capriscus*,
und *Accipenser ruthenus*.
Jenaische Zeitschr., xvii (x), pp. 401-451,
with pls. iv-vi.

Goeppert, Ernst.   1895 A
Untersuchungen zur Morphologie der
Fischrippen.
Morphologisches Jahrbuch, xxiii, pp. 145-
217, with pls. xiii-xvi and 21 figs. in text.
Contains a list of papers on the subject.

—————   1895 B
Zur Kenntniss der Amphibienrippen.
Morpholog. Jahrbuch, xxii, pp. 441-448, with
5 figs. in text.

—————   1896 A
Die Morphologie der Amphibienrip-
pen.
Festschrift zum 70sten Geburtstage von Carl
Gegenbaur, i, pp. 395-436, with pls. i, ii, and 10
figs. in the text.
Appended is a list of the works of other
writers on the subject.

Goethe, Johann Wolfgang.   1831 A
Ueber den Zwischenkiefer des Menschen und der Thiere.
Nov. Act. Acad. Caes. Leop.-Carol., xv, pp.
3-48, with pls. i-v.

Goette, Alex.   1878 A
Beiträge zur vergleichenden Morphologie des Skeletsystems der Wirbelthiere. II Die Wirbelsäule und ihre
Anhänge.
Archiv Mikroskop. Anatomie, xv, pp. 442-
541; pls. xxviii-xxxiii.
Treats of the structure and homologies of
the elements of the vertebral column and
homology of ribs, etc., of Ganoids and Elasmobranchs.

—————   1899 A
Ueber die Entwicklung des knöcher-
nen Rückenschildes (Carapax) der
Schildkröten.
Zeitschr. wissensch. Zoologie, lxvi, pp. 407-
434, with pls. xxvii-xxix.

Goldfuss, August.   1845 A
Der Schädelbau des *Mosasaurus*, durch
Beschreibung einer neuen Art dieser
Gattung erläutert.
Nov. Act. Acad. Caes. Leop.-Carol. Nat.
Curiosorum, xxi, pp. 173-200, with pls. vi-ix.
Abstract appeared in Neues Jahrb. Min.,
1847, pp. 122-125.

Goldfuss, August—Continued.   1847 A
Beiträge zur vorweltlichen Fauna des Steinkohlengebirges.
Herausgegeben von dem naturhistorischen Vereine für die preussischen Rheinlande, Bonn, 1847, pp. 1-28, with pls. i-v.

——   1849 A
Description of the *Orthacanthus dechenii*.
Quart. Jour. Geol. Soc., v, pp. 21-23, of translations and notices.
From "Beiträge zur vorweltlichen Fauna des Steinkohlengebirges."

Goode, G. Brown.   1882 A
The taxonomic relations and geographical distribution of the members of the swordfish family.
Proc. U. S. Nat. Mus., iv, pp. 415-433.

Goodrich, E. S.   1894 A
On the fossil *Mammalia* from the Stonesfield slate.
Quart. Jour. Mic. Sci., xxxv, pp. 407-432, with pl. xxvi.
Pp. 425-429 contain a discussion of the primitive mammalian molar.

Gorski, Constantin.   1858 A
Einige Bemerkungen über die Beckenknochen der beschuppten Amphibien.
Archiv f. Anat., Physiol. u. wissensch. Medicin., Jahrg. 1858, pp. 382-389.

Grabbe, A.   1884 A
Beitrag zur Kenntniss der Schildkröten des deutschen Wealden.
Zeitschr. deutsch. geol. Gesellsch., xxxvi, pp. 17-28, with pl. i.
Discusses structure and relations of *Pleurosternon*.

Grant, E.   1842 A
On the structure and history of the mastodontoid animals of North America.
Proc. Geol. Soc. Lond. iii, pp. 770-771.

Grassi, B.   1883 A
Beiträge zur näheren Kenntniss der Entwicklung der Wirbelsäule der Teleostier.
Morpholog. Jahrbuch, viii, pp. 457-473.

Gratacap, L. P.   1886 A
Fish remains and tracks in the Triassic rocks at Weehawken, N. J.
Amer. Naturalist, xx, pp. 243-246, with pls. xii and xiii, and 2 woodcuts in the text.

——   1896 A
Fossils and fossilization.
Amer. Naturalist, xxx, pp. 902-912, 993-1003.

Grateloup, J. P. S.   1840 A
Beschreibung eines fossilen Stückes Kinnlade eines riesigen Saurier-Geschlechtes *Squalodon*, mit *Iguanodon* verwandt, aus dem Meeres-Sande von Léognan bei Bordeaux, pp. 8, 1 pl.
Neues Jahrbuch f. Min., 1841, pp. 830-832.
Abstract from Actes Acad. Roy. Sci. Bordeaux, 1840, p. 208. For original, see same volume Jahrbuch, pp. 567-568, a note from Grateloup, in which he recognizes the Delphinoid nature of this fossil. I have seen only the abstract here referred to.

Gratiolet, Pierre.   1858 A
Note sur un fragment de crâne trouvé à Montrouge, près Paris.
Bull. Soc. Géol. de France (2), pp. 620-625, with pl. v.
Describes *Odobenotherium lartetianum.*

Gray, Asa.   1846 A
Food of the mastodon.
Proc. Bost. Soc. Nat. Hist., ii, pp. 92-98.
Reprinted in Amer. Jour. Sci. (2), iii, p. 436.

Gray, J. E.   1837 A
General arrangement of the *Reptilia*.
Proc. Zool. Soc. Lond., 1837, v, pp. 131-132.

——   1857 A
Observations on the species of the genus *Manatus*.
Proc. Zool. Soc. Lond., 1857, xxv, pp. 59-61.

——   1865 A
Revision of the genera and species of *Mustelidæ* contained in the British Museum.
Proc. Zool. Soc. Lond., 1865, pp. 100-154, with pl. vii, and 3 text figs.

——   1866 A
Catalogue of seals and whales in the British Museum.
Second edition, pp. 402, with 101 woodcuts in text.

——   1867 A
Notice of a new species of American tapir, with observations on the skulls of *Tapirus, Rhinochœrus,* and *Elasmognathus* in the collections of the British Museum.
Proc. Zool. Soc. Lond., 1867, pp. 876-886, with pl. xlii, and 6 text figs.

——   1867 B
Observations on the preserved specimens and skeletons of the *Rhinocerotidæ* in the collection of the British Museum and Royal College of Surgeons, including the descriptions of 3 new species.
Proc. Zool. Soc. Lond., 1865, pp. 1003-1032, with 6 text figs.

Gray, J. E.—Continued. 　　1868 A
Synopsis of the pigs (*Suidæ*) in the
British Museum.
Proc. Zool. Soc. Lond., 1868, pp. 17–49.

———　　　　　　　　　　　　1868 B
Notes on the skulls of the species of
dogs, wolves, and foxes (*Canidæ*) in
the collection of the British Museum.
Proc. Zool. Soc. Lond., 1868, pp. 492–525, with
7 woodcuts in the text.

———　　　　　　　　　　　　1869 A
Catalogue of carnivorous, pachyder-
matous, and edentate *Mammalia* in the
British Museum.
Pp. 398, with 47 woodcuts. London, 1869.

———　　　　　　　　　　　　1869 B
Notes on the families and genera of
of tortoises (*Testudinata*), and on the
characters afforded by the study of
their skulls.
Proc. Zool. Soc. Lond., 1869, pp. 165–225,
with pl. xv, and 20 figs. in the text.

———　　　　　　　　　　　　1869 C
On the bony dorsal shield of the male
Tragulus kanchil.
Proc. Zool. Soc. Lond., 1869, p. 226, with 1 fig.

———　　　　　　　　　　　　1872 A
Description of the younger skull of
Steller's sea bear (*Eumetopias stelleri*).
Proc. Zool. Soc. Lond., 1872, pp. 737–743, with
5 figs.

———　　　　　　　　　　　　1873 A
Notes on the mud tortoises (*Trionyx*,
Geoffroy) and on the skulls of the dif-
ferent kinds.
Proc. Zool. Soc. Lond., 1873, pp. 38–72, with
pl. viii and 13 text figs.
This paper presents a classification of the
Trionychia, together with definitions of the
genera adopted.

———　　　　　　　　　　　　1873 B
Notes on the genera of turtles (*Oia-
copodes*), and especially on their skeletons and skulls.
Proc. Zool. Soc. Lond., 1873, pp. 395–411, with
3 figs. in the text.
In this paper is presented a classification of
the *Chelontidæ* and the *Dermochelyidæ*, together
with definitions of the genera.

———　　　　　　　　　　　　1873 C
On the skulls and alveolar surfaces
of land tortoises (*Testudinata*).
Proc. Zool. Soc. Lond., 1873, pp. 722–728, with
pl. ix.

Green, Jacob. 　　　　　　　1821 A
Some curious facts respecting the
bones of the rattlesnake.
Amer. Jour. Sci. (1), iii, pp. 85–86.

Greene, Francis V. 　　　　　1853 A
Chemical investigation of remains of
fossil *Mammalia*.
Amer. Jour. Sci. (2), xvi, pp. 16–20.
See also Proc. Acad. Nat. Sci. Phila., vi, pp.
292–296.

Grinnell, G. B. 　　　　　　　1881 A
Monograph by Professor Marsh on
the *Odontornithes*, or toothed birds of
North America.
Amer. Jour. Sci. (3), xxi, pp. 255–276, with 9
figs. in the text.

Grinnell, G. B., and **Dana**, E. S. 1876 A
On a new Tertiary lake basin.
Amer. Jour. Sci. (3), xi, pp. 126–128.

Gruber, W. 　　　　　　　　　1859 A
Monographie des canalis supracondy-
loidei humeri und des processus su-
pracondyloidei humeri et femoris der
Säugethiere und des Menschen.
Mém. l'Acad. Imp. Sci. St. Pétersbourg, viii,
pp. 53–128, with pls. i–iii.

Günther, Albert C. 　　　　　　1859 A
Catalogue of the Acanthopterygian
fishes in the collection of the British
Museum.
Vol. i, pp. i–xxxi; 1–524, containing families
Gasterosteidæ, *Berycidæ*, *Percidæ*, *Aphredo-
deridæ.* London, 1859.

———　　　　　　　　　　　　1860 A
Catalogue of the Acanthopterygian
fishes in the collection of the British
Museum.
Vol. ii, pp. i–xxi, 1–548.
This volume contains the *Squamipinnes*,
Triglidæ, *Sciænidæ*, *Sphyrænidæ*, *Scombridæ*,
Carangidæ, *Xiphiidæ*, etc.

———　　　　　　　　　　　　1861 A
Catalogue of the Acanthopterygian
fishes in the collection of the British
Museum.
Vol. iii, pp. i–xxv, 1–586.

———　　　　　　　　　　　　1862 A
Catalogue of the fishes in the British
Museum. Vol. iv, containing *Pomacen-
tridæ*, *Labridæ*, etc.
Pp. i–xxii; 1–534.

———　　　　　　　　　　　　1864 A
Catalogue of the fishes of the British
Museum. Vol. v, containing the fami-
lies *Siluridæ*, *Characinidæ*, etc.
Pp. i–xxii, 1–455, London, 1864.

Günther, Albert C.—Continued. 1866 A
Catalogue of the fishes of the British
Museum. Vol. vi, containing the fami-
lies *Salmonidæ, Percopsidæ, Galaxidæ,
Mormyridæ, Gymnarchidæ, Esocidæ, Um-
bridæ, Scombresocidæ Cyprinodontidæ*.
Pp. i-xv, 1-368, London, 1866.

———   1867 A
Contribution to the anatomy of *Hatteria* (*Rhynchocephalus* Owen).
Philos. Trans. Roy. Soc., London, clvii, pp.
595-629, with pls. xxvi-xxviii.

———   1868 A
Catalogue of the fishes of the British
Museum.
Vol. vii, pp. i-xx, 1-512.
Contains families of the *Physostomi*.

———   1870 A
Catalogue of the fishes of the British
Museum.
Vol. viii, pp. i-xxv, 1-549.
This volume contains the families *Gymno-
tidæ, Symbranchidæ, Murænidæ, Pegasidæ;* also
the *Lophobranchii, Plectognathi, Dipnoi, Gan-
oidei, Chondropterygii, Cyclostomata,* and the
Leptocardii.

———   1871 A
The new ganoid fish (*Ceratodus*) re-
cently discovered in Queensland.
Nature, iv, pp. 406-408, 428-429.

———   1871 B
Ceratodus and its place in the system.
Ann. and Mag. Nat. Hist. (4), vii, pp. 222-227.

———   1872 A
Description of *Ceratodus*, a genus of
Ganoid fishes recently discovered in
rivers of Queensland, Australia.
Philosoph. Trans. Royal. Soc. Lond., clxi,
pp. 511-571; pls. xxx-xliii.

———   1877 A
The gigantic land tortoises (living
and extinct) in the collection of the
British Museum.
Pp. 96; pls. 54. London, 1877.

———   1880 A
An introduction to the study of
fishes.
Edinburgh, 1880, 8vo, pp. i-xvi; 1-720, with
321 figs. in the text.

———   1881 A
Ichthyology.
Encyclop. Brit., 9th ed., xii, pp. 630-696, with
70 figs. in the text.

Günther, Albert C.—Continued. 1886 A
Reptiles.
Encyclop. Brit., 9th ed., xx, pp. 432-445, 465-
473.
Dr. Günther's portion of the article on *Reptilia* treats of the history and literature of the
subject, the various systems of classification,
the general characters of the class, and the
geological and geographical distribution.
There are also lists of the most important
works on the subject.

                                  1888 A
Tortoise.
Encyclop. Brit., 9th ed., xxiii, pp. 455-460,
with 9 figs. in the text.
A popular account of tortoises and a scheme
of classification.

Gürich, Georg.   1891 A
Ueber Placodermen und andere devonische Fischreste im Breslauer min-
eralogischen Museum.
Zeitschr. deutsch. geol. Gesellsch., xliii, pp.
902-911, with 5 figs. in the text.

                                  1891 B
Ueber einen neuen *Nothosaurus* von
Gogolin in Oberschlesien.
Zeitschr. deutsch. geol. Gesellsch., xliii, pp.
967-970, with 2 figs. in the text.

Haacke, Wilhelm.   1888 A
Ueber die Entstehung des Säuge-
tieres.
Biolog. Centralbl., viii, pp. 8-16.

———   1893 A
Ueber die Entstehung des Säuge-
tieres.
Biolog. Centralbl., xiii, pp. 719-732.

Haberlandt, G.   1876 A
Ueber *Testudo præceps* n. sp., die
erste fossile Landschildkröte des
Wiener Beckens.
Jahrbuch k.-k. geol. Reichsanstalt, xxvi,
pp. 243-248, with pl. xvi.
Makes comparisons with Dr. Leidy's *T. cul-
bertsonii.*

Habersham, Joseph.
See note under **Hodson,** W. B.
1846 A.

Haddon, A. C.
See **Bridge** and **Haddon.**

Haeckel, Ernst.   1895 A
Systematische Phylogenie der Wirbelthiere (*Vertebrata*). Dritter Theil
des Entwurfs einer systematischen Phylogenie.
8vo; Berlin, 1895; pp. i-x; 1-660.

Hale, C. S.　　　　　　　　1848 A
　Geology of south Alabama.
　Amer. Jour. Sci. (2), vi., pp. 354–363.

Hall, J.　　　　　　　　1843 A
　Geology of New York. Part iv,
　comprising the survey of the 4th geo-
　logical district.
　Pp. i–xxv; 1–687, with tables containing 74
　figs. of organic remains and 19 pls. 4to.
　Albany, 1843.

————　　　　　　　　1846 A
　On the geological relations of *Casto-*
roides ohioensis.
　Proc. Bost. Soc. Nat. Hist., ii, pp. 167–168.
　Describes skull found near Clyde, N. Y.

————　　　　　　　　1846 B
　Notice of the geological position of
　the cranium of the *Castoroides ohioensis.*
　Jour. Boston Soc. Nat. Hist., v, pp. 385–391.
　This paper constitutes a portion of a joint
　article, by James Hall and Jeffries Wyman, on
　the skull of *Castoroides.* See Wyman, J.,
　1846 B.

————　　　　　　　　1871 A
　Notes and observations on the Cohoes
　mastodon.
　Twenty-first Ann. Report Regents Univ.
　N. Y., on condition of State Cabinet, pp. 99–
　148, with pls. iii–vii.

————　　　　　　　　1887 A
　Note on the discovery of a skeleton
　of an elk, *Elaphus canadensis,* in the
　town of Farmington, Ontario County.
　Sixth Ann. Report State Geologist, New
　York, for 1886, p. 39.

Hallman, E.　　　　　　　　1837 A
　Die vergleichende Osteologie des
　Schläfenbeins; zur Vereinfachung der
　herrschenden Ansichten bearbeitet.
　Hannover, 1837. Pp. 1–130; pls. i–iv.

Hallowell, Edward.　　　　　　1846 A
　Remarks on some bones of a young
　mastodon.
　Proc. Acad. Nat. Sci. Phila., iii, p. 130. See
　also p. 117.

Hambach, G.　　　　　　　　1890 A
　A preliminary catalogue of the fos-
　sils occurring in Missouri.
　Geol. Surv. Missouri, Bull. No. 1, pp. 60–85.
　This contains, on pp. 80 and 81, an alpha-
　betical list of the species of the fossil fishes
　found in Missouri, so tabulated as to show
　their stratigraphical distribution. On p. 82
　are recorded four mammals which have been
　found fossil in the State. Since the list of Mis-
　souri fossils prepared by Chas. R. Keyes was
　published at a later date (1894), but few refer-
　ences are given in the present catalogue to
　Hambach's list. However, this list contains

Hambach, G.—Continued.
　a number of species not before reported from
　Missouri. These are recorded in the present
　Catalogue.

Hammerschmidt, Carl E.　　　1848 A
　Resultate geologischer, anatomischer,
　und zoologischer Untersuchungen über
　Hydrarchos Koch.
　Haidinger's Berichte über die Mittheilungen
　von Freunden der Naturwissenschaften in
　Wien, iii, pp. 322–326.

Hancock, Albany, and Atthey, Thos.
　　　　　　　　　　　　　1868 A
　Notes on the remains of some rep-
　tiles and fishes from the shales of the
　Northumberland coal field.
　Ann. and Mag. Nat. Hist. (4), i, pp. 266–278,
　346–378, with pls. xiv–xvi.
　Also in Nat. Hist. Trans. Northumberland
　and Durham, iii, 1868, pp. 66–120, with pls. i–iii.

————　　　　　　　　1868 B
　Notes on the various species of *Cteno-*
dus obtained from the shales of the
　Northumberland coal field.
　Nat. Hist. Trans. Northumberland and Dur-
　ham, iii, pp. 54–66.

————　　　　　　　　1869 A
　On a new Labyrinthodont amphibian
　from the Northumberland coal field,
　and on the occurrence in the same
　locality of *Anthracosaurus russelli.*
　Nat. Hist. Trans. Northumberland and Dur-
　ham, iii, pp. 310–319.

————　　　　　　　　1869 B
　On the generic identity of *Climaxodus*
　and *Janassa.*
　Ann. and Mag. Nat. Hist. (4), iv, pp. 322–328,
　pl. xii.
　Also in Nat. Hist. Trans. Northumberland
　and Durham, iii, 1869, pp. 330–338.

————　　　　　　　　1870 A
　Note on an undescribed fossil fish
　from the Newsham Coal shale, near
　Newcastle-upon-Tyne.
　Ann. and Mag. Nat. Hist. (4), v, pp. 266–268.
　Describes as new genus *Archichthys,* now re-
　garded as synonym of *Strepsodus.* The genus
　Archichthys founded on portion of a jaw with
　teeth.

————　　　　　　　　1871 A
　A few remarks on *Dipterus* and
　Ctenodus, and on their relationship to
　Ceratodus forsteri Krefft.
　Ann. and Mag. Nat. Hist. (4), vii, pp. 190–
　196, pls. xiii, xiv.
　Also in Nat. Hist. Trans. Northumberland
　and Durham, iv, 1872, pp. 397–407, with pls.
　xiii, xiv.

Hancock, Albany, and Atthey, Thos.—
Continued.   1872 A
Descriptive note on a nearly entire
specimen of *Pleurodus rankinii*, on two
new species of *Platysomus*, a new *Amphicentrum*, with remarks on a few
other fish remains in the Coal-measures
at Newham.
    Ann. and Mag. Nat. Hist. (4), ix, pp. 249–262,
with pls. xvii, xviii.
    Also in Nat. Hist. Trans. Northumberland
and Durham, 1872, iv, pp. 408–423, with pls. xv,
xvi.

Hancock, Albany, and Howse, Richard.
                               1870 A
On *Janassa bituminosa* Schlotheim,
from the marl slate of Midderidge,
Durham.
    Ann. and Mag. Nat. Hist. (4), v, pp. 47–62;
pls. ii, iii.
    Also in Nat. Hist. Trans. Northumberland
and Durham, iii, 1869, pp. 339–357, with pls.
x, xi.

——   1872 A
On *Dorypterus hoffmanni* Germar,
from the marl slates of Midderidge,
Durham.
    Nat. Hist. Trans. Northumberland and Dur-
ham, iv, pp. 243–268, with pls. ix, x.

Hannover, A.   1868 A
Recherches sur la structure et le développement des écailles et des épines
chez les poissons cartilagineux.
    Annales Sci. Naturelles, Zool. (5), ix, pp. 373–
378.

Harlan, R.   1823 A
Observations on fossil elephant teeth,
of North America.
    Jour. Acad. Nat. Sci. Phila., iii, pp. 65–67,
pl. v.

——   1824 A
On a new fossil genus of the order
Enalio Sauri (of Conybeare).
    Jour. Acad. Nat. Sci. Phila., iii, pp. 331–337,
with pl. xii, figs. 1–5.
    In this paper is described the new genus
Saurocephalus, based on the new species *S.
lanciformis*.

——   1824 B
On an extinct species of crocodile not
before described; and some observations on the geology of west Jersey.
    Jour. Acad. Nat. Sci. Phila., iv, pp. 15–24,
with pl. i.
    Describes remains afterwards called *Croco-
dilus macrorhynchus*.

Harlan, R.—Continued.   1825 A
Fauna Americana; being a descrip-
tion of the mammiferous animals inhabiting North America.
    8vo., pp. i–x, 1–318, Philadelphia: Published
by Anthony Finley.

——   1825 B
Notice of the plesiosaurus and other
fossil reliquiæ from the State of New
Jersey.
    Jour. Acad. Nat. Sci. Phila., iv, pp. 231–236.
    The "plesiosaurus" is *Priscodelphinus har-
lani* Leidy.

——   1828 A
Note from R. Harlan, M. D., on the
examination of the large bones disin-
terred at the mouth of the Mississippi
River and exhibited in the city of
Baltimore, Jan. 22, 1828.
    Amer. Jour. Sci., xiv, pp. 186–187.

——   1831 A
Description of the fossil bones of the
Megalonyx discovered in "White Cave,"
Kentucky.
    Jour. Acad. Nat. Sci. Phila., vi, pp. 269–288,
pls. xii–xiv.
    Describes *Megalonyx laqueatus*.
    See for notice of this paper Neues Jahrbuch
Min., 1836, p. 123. Also Amer. Jour. Sci., 1831,
xx, pp. 414–415.

——   1831 B
Description of the jaws, teeth, and
clavicle of the *Megalonyx laqueatus*.
    Monthly Amer. Jour. Geology and Nat. Sci.,
i, pp. 74–76, with pl. iii.

——   1834 A
On some new species of fossil saurians found in America.
    Report Brit. Assoc. Adv. Sci., 3d meeting,
Cambridge, 1833, p. 440.
    A brief abstract, naming only *Ichthyosaurus
missouriensis*.

——   1834 B
Critical notices of various organic re-
mains hitherto discovered in North
America.
    Trans. Geol. Soc. of Pennsylvania, i, pt. i,
pp. 46–112, Philadelphia, 1834.
    See also Harlan, 1835 B; Edinburgh New
Philos. Jour., xvii, pp. 342–362, xvii, pp. 28–40;
Neues Jahrb. Min., 1836, pp. 99–109.

——   1834 C
Notice of fossil bones found in the
Tertiary formation of the State of Louisiana.
    Trans. Amer. Philos. Soc., iv, pp. 397–403, pl.
xx, figs. 1, 2.
    The bones are assigned to the genus *Basilo-
saurus*.

Harlan, R.—Continued.   1834 D

Notice of the discovery of the remains of the *Ichthyosaurus* in Missouri, N. A.

Trans. Amer. Philos. Soc., iv, pp. 405–409, pl. xxx, figs. 3–8.

Describes *Ichthyosaurus missouriensis*, but a note follows which states that the fossil must be assigned to some new genus. This is not named.

————   1834 E

Announcement of the finding of *Ichthyosaurus missouriensis* and *Basilosaurus*.

Bull. Soc. Géol. de France (1), iv, p. 124.

No descriptions are furnished.

————   1835 A

Observations on a large skeleton recently disinterred from the mouth of the Mississippi River.

Med. and Phys. Researches, pp. 76–77.

————   1835 B

Critical notices of various organic remains hitherto discovered in North America.

Med. and Phys. Researches, pp. 253–313.

See also Harlan, 1834 B.

————   1835 C

Description of the fossil bones of the *Megalonyx*, recently discovered in the United States, N. A.

Med. and Phys. Researches, pp. 319–336, with pls. xii–xvi.

————   1835 D

Observations on the fossil bones found in the Tertiary formation in the State of Louisiana.

Med. and Phys. Researches, pp. 337–342.

Treats of *Basilosaurus*.

————   1835 E

Description of the Ichthyosaurian remains recently discovered in the State of Missouri.

Med. and Phys. Researches, pp. 344–348, with 1 pl. of 6 figs.

Treats of *Ichthyosaurus missouriensis*.

————   1835 F

Description of the remains of the *Basilosaurus*, a large fossil marine animal recently discovered in the horizontal limestone of Alabama.

Med. and Phys. Researches, pp. 349–358, with pls. xxvi–xxviii.

————   1835 G

Observations on the fossil elephant teeth of North America.

Med. and Phys. Researches, pp. 359–361, with 1 pl.

The teeth figured are probably those of *E. primigenius*.

Harlan, R.—Continued.   1835 H

Description of a new fossil genus of the order *Enalio Sauri* (of Conybeare).

Med. and Phys. Researches, pp. 362–366, with 1 pl.

————   1835 I

Description of an extinct species of crocodile not before described, and some observations on the geology of west Jersey.

Med. and Phys. Researches, pp. 369–381, with 1 pl. of *Crocodilus macrorhynchus*.

————   1835 J

Notice of plesiosaurian and other fossil reliquiæ from the State of New Jersey.

Med. and Phys. Researches, pp. 382–385, with 1 pl.

————   1839 A

Notice of the discovery of the *Basilosaurus* and the *Batrachiosaurus*.

Proc. Geol. Soc. Lond., iii, pp. 23–24.

Also in Lond. and Edinb. Philos. Mag., xiv, p. 302.

No adequate descriptions are given of the genera proposed.

————   1839 B

[Letter regarding *Basilosaurus* and *Batrachotherium*.]

Bull. Soc. Géol. France, x, pp. 89–90.

See also Neues Jahrb. Min., 1840, p. 741, for abstract.

————   1839 C

On the discovery of the *Basilosaurus* and the *Batrachiosaurus*.

Lond. and Edinb. Philos. Mag. and Jour. Sci., xiv, p. 302.

See also Ann. Sci. Naturelles (2), xii, p. 221.

————   1840 A

A letter from Dr. Harlan addressed to the president [of the Geol. Soc., London] on the discovery of the remains of the *Basilosaurus*, or *Zeuglodon*.

Trans. Geol. Soc. Lond. (2), vi, pp. 67–68.

————   1842 A

Bones of the *Orycterotherium*.

Amer. Jour. Sci., xlii, p. 392.

————   1842 B

Notice of two new fossil mammals from Brunswick canal, Georgia; with observations on some of the fossil quadrupeds of the United States.

Amer. Jour. Sci., xliii, pp. 141–144; pl. iii.

Harlan, R.—Continued.   **1842 C**
Description of the bones of a fossil
animal of the order *Edentata.*
Proc. Amer. Philos. Soc., ii, pp. 109-111.
Describes the new genus and species *Orycte-
rotherium missouriense.*
A notice of this paper was published in Ann.
and Mag. Nat. Hist., 1842, x, p. 72.

——   **1842 D**
Description of a new extinct species
of dolphin from Maryland.
Second Bull. of Proc. Nat. Institute, pp. 195-
196, with 3 large pls.
Describes *Delphinus calvertensis* Harl.

——   **1842 E**
[A list of fossil bones from Bruns-
wick canal, presented by J. Hamilton
Cooper, Darien, Ga.]
Proc. Acad. Nat. Sci. Phila., 1842, pp. 189-190.

——   **1843 A**
Remarks on Mr. Owen's letter to the
editors on Dr. Harlan's new fossil *Mammalia.*
Amer. Jour. Sci., xiv, pp. 208-211.

——   **1843 B**
Description of the bones of a new
fossil animal of the order *Edentata.*
Amer. Jour. Sci. (1), xliv, pp. 69-80, with
pls. i-iii.
Description with figures of *Orycterotherium
missouriense.*

**Harpe, Ph. de la. See Pictet, Gaudin
and Harpe.**

Harrison, J. See Howes and Harrison.

Harrison, R. G.   **1894 A**
Ectodermal or mesodermal origin of
the bones of Teleosts.?
Anatom. Anzeiger, x, pp. 139-143, with 3 figs.
The author denies Klaatsch's view that the
bones are of ectodermal origin.

Hartlaub, Clemens.   **1886 A**
Berträge zur Kenntniss der Manatus-
Arten.
Zoolog. Jahrbücher, i, pp. 1-112, with pls.
i-iv and 13 figs. in the text.
An important work. In the beginning there
is found a list of papers and memoirs on the
subject.

Hartmann, R.   **1893 A**
Ueber die Feliden-Gattung *Machæ-
rodus.*
Sitz.-Ber. Gesellsch. naturforsch. Freunde
Berlin, 1893, pp. 88-94.

Hartt, Ch. Fred.   **1871 A**
Discovery of mastodon remains at
Motts Corners, near Ithaca, N. Y.
Amer. Naturalist, v, pp. 314-315.

Hasse, C.   **1876 A**
Die fossilen Wirbel.
Morpholog. Jahrbuch, ii, pp, 449-477, with
pls. xxx-xxxi.

——   **1877 A**
Die fossilen Wirbel.   Die fossilen
Squatinæ.
Morpholog. Jahrbuch, iii, pp. 328-351, with
pls. xvii-xviii.

——   **1878 A**
Das natürliche System der Elasmobranchier auf Grundlage des Baues und
der Entwicklung der Wirbelsäule.
Zoolog. Anzeiger, i, pp. 144-148; 167-172.

——   **1879 A**
Das natürliche System der Elasmo-
branchier auf Grundlage des Baues und
der Entwicklung ihrer Wirbelsäule.
Eine morphologische und paläontolo-
gische Studie.   Allgemeiner Theil.
Pp. i-vi; 1-76, with 2 pls, and 6 wood cuts.
Jena, 1879.
See for continuation of this work Hasse, C.,
1882 A.

——   **1879 B**
Ueber den Bau und über die Entwicklung des Knorpels bei den Elasmo-
branchiern.
Zoolog. Anzeiger, ii, pp. 325, 351-355, 371-374.

——   **1882 A**
Das natürliche System der Elasmobranchier auf Grundlage des Baues und
der Entwicklung ihrer Wirbelsäule.
Eine morphologische und paläontolo-
gische Studie.   Besonderer Theil.
Pp. 1-285, with 40 pls. Jena., 1882.
For beginning of this work see Hasse, C.
1879 A. For conclusion, Hasse, C., 1885 A.

——   **1885 A**
Das natürliche System der Elasmobranchier auf Grundlage des Baues und
der Entwicklung ihrer Wirbelsäule.
Eine morphologische und paläontolo-
gische Studie.   Ergänzungsheft.
Pp. 1-27, with 1 pl. Jena, 1885.
For earlier parts of this work see Hasse, C.,
1879 A, and Hasse, C., 1882 A.

——   **1893 A**
Allgemeine Bemerkungen über die
Entwicklung und die Stammesge-
schichte der Wirbelsäule.
Anatom. Anzeiger, viii, pp. 288-289.

Haswell, William A.   **1883 A**
On the structure of the paired fins
of *Ceratodus,* with remarks on the general theory of the vertebrate limb.
Proc. Linnean Soc. N. S. Wales (1), vii, pp. 2-
11, with pl. i.

Haswell, William A.—Cont'd.   1884 A
Studies on the Elasmobranch skeleton.
Proc. Linnean Soc. N. S. Wales (1), ix, pp. 71-119, with pls. i and ii.

Hatcher, J. B.   1893 A
The Ceratops beds of Converse County, Wyo.
Amer. Jour. Sci. (3), xlv, pp. 135-144.

——   1893 B
The Titanotherium beds.
Amer. Naturalist, xxvii, pp. 204-221, with 8 figs. in the text.

——   1894 A
A median-horned rhinoceros from the Loup Fork beds of Nebraska.
Amer. Geologist, xiii, pp. 149-150.

——   1894 B
Discovery of *Diceratherium*, the two-horned rhinoceros, in the White River beds of South Dakota.
Amer. Geologist, xiii, pp. 360-361.

——   1894 C
On a small collection of vertebrate fossils from the Loup Fork beds of northwestern Nebraska; with note on the geology of the region.
Amer. Naturalist, xxviii, pp. 236-248, pls. i, ii.
Describes as new *Aelurodon taxoides* and *A. meandrinus*. Treats also of *Aphelops fossiger* and *Teleoceras major*.
Abstract in Princeton Coll. Bull., vi, p. 54, 1894.

——   1895 A
On a new species of *Diplacodon*, with a discussion of the relations of that genus to *Telmatotherium*.
Amer. Naturalist, xxix, pp. 1084-1091, with pls. xxxviii-xxxix, and 2 figs. in the text.

——   1895 B
Discovery, in the Oligocene of South Dakota, of *Eusmilus*, a genus of saber-toothed cats new to North America.
Amer. Naturalist, xxix, pp. 1091-1093.
Description of *Eusmilus dakotensis*.

——   1895 C
The Princeton scientific expedition of 1895.
Princeton Coll. Bull., vii, pp. 95-98.

——   1896 A
Some localities for Laramie mammals and horned dinosaurs.
Amer. Naturalist, xxx, pp. 112-120, with map on pl. iii.

Hatcher, J. B.—Continued.   1896 B
Recent and fossil tapirs.
Amer. Jour. Sci. (4), i, pp. 161-180, with pls. ii-v and 2 figs. in the text.
Notice of this paper by M. Schlosser appears in Neues Jahrb. Min. 1899, ii, pp. 314-316.

——   1897 A
Diceratherium proavitum.
Amer. Geologist, xx, pp. 313-316, with pl. ix.
Regarded as distinct from *Aceratherium tridactylum*.

——   1900 A
The Carnegie Museum palæontological expeditions of 1900.
Science (2), xii, pp. 718-720.
Records finding of specimen of *Platacodon nanus*, and the determination that it is a fish.

——   1900 B
Vertebral formula of *Diplodocus* (Marsh).
Science (2), xii, pp. 828-830.

Hawkins, B. W.   1874 A
On the pelvis of *Hadrosaurus*.
Proc. Acad. Nat. Sci. Phila., 1874, pp. 90-91.

——   1875 A
Pelvis of *Hadrosaurus*.
Proc. Acad. Nat. Sci. Phila., 1875, p. 329.

Hay, O. P.   1895 A
On the structure and development of the vertebral column of *Amia*.
Pubs. Field Columb. Mus., Zool., i, pp. 1-54, pls. i-iii.

——   1895 B
On certain portions of the skeleton of *Protostega gigas*.
Pubs. Field Columb. Mus., Zool., i, pp. 57-62, pls. iv, v.

——   1895 C
Description of a new species of *Petalodus* (*P. securiger*), from the Carboniferous of Illinois.
Jour. Geology, iii, pp. 561-564, with 2 figs. in the text.

——   1896 A
On the skeleton of *Toxochelys latiremis*.
Pubs. Field Columb. Mus., Zool., i, pp. 101-106, pls. xiv, xv.

——   1896 B
The structure and mode of development of the vertebral column.
Science (2), iv, pp. 959-960.

——   1898 A
Protospondyli and *Aetheospondyli* of A. S. Woodward.
Science (2), vii, p. 858.

Hay, O. P.—Continued.   1898 B
Observations on the genus of Cretaceous fishes called by Prof. Cope *Portheus*.
Science (2), vii, p. 646.
Shows that Leidy's *Xiphactinus* is identical with Cope's later published *Portheus*.

____   1898 C
Classification of the Amioid and Lepisosteoid fishes.
Amer. Naturalist, xxxii, pp. 341-349.

____   1898 D
Observations on the genus of fossil fishes called by Professor Cope *Portheus*, by Dr. Leidy *Xiphactinus*.
Zool. Bulletin, ii, pp. 25-54, with 16 figs. in the text.

____   1898 E
Notes on species of *Ichthyodectes*, including the new species *I. cruentus*, and on the related and herein established genus *Gillicus*.
Amer. Jour. Sci. (4), vi, pp. 225-282, with 6 figs. in the text.

____   1898 F
On *Protostega*, the systematic position of *Dermochelys*, and the morphogeny of the chelonian carapace and plastron.
Amer. Naturalist, xxxii, pp. 929-948, with 3 figs. in the text.

____   1899 A
On the names of certain North American fossil vertebrates.
Science (2), ix, pp. 593-594.
The generic name *Wortmania* is proposed for *Hemiganus otariidens* Cope; *Miolabis* for *Procamelus robustus;* *Homogalax* for *Systemodon primævus; Equus eous* for *E. intermedius* Cope.

____   1899 B
On one little known and one hitherto unknown species of *Saurocephalus*.
Amer. Jour. Sci. (4), vii, pp. 299-304, with 5 figs. in the text.

____   1899 C
Notes on the nomenclature of some North American fossil vertebrates.
Science (2), x, pp. 253-254.
New generic name proposed, *Adjidaumo;* new specific names, *Canis (?) marshii, Cynodictis hylactor*.

____   1899 D
Descriptions of two new species of tortoises from the tertiary of the United States.
Proc. U. S. Nat. Mus., xxii, pp. 21-24, with pls. iv-vi.
Describes *Hadrianus schucherti* and *Acherontemys heckmani*.

Hay, O. P.—Continued.   1899 E
On some changes in the names, generic and specific, of certain fossil fishes.
Amer. Naturalist, xxxiii, pp. 783-792.

____   1899 F
A census of the fossil *Vertebrata* of North America.
Science (2), x, pp. 681-684.
This paper purports to give the number of described species and genera belonging to each of the large groups of fossil vertebrates.

____   1899 G
On the nomenclature of certain American fossil vertebrates.
Amer. Geologist, xxiv, pp. 345-349.

____   1900 A
Descriptions of some vertebrates of the Carboniferous age.
Proc. Amer. Philos. Soc., xxxix, pp. 96-123, with pl. vii and 3 text figs.
New species described are *Dittodus lucasi, Cladodus girtyi, Rhizodopsis mazonius, Strepsodus arenosus, S. dawsoni,* and *Elonichthys pettigerus hypsilepis*.

Hay, Robert.   1885 A
Notes on the fossil jaw of bison from the Pliocene of Norton County.
Trans. Kansas Acad. Sci., ix, p. 98.

Haycraft, John Berry.   1891 A
The development of the carapace of the *Chelonia*.
Trans. Roy. Soc., Edinburgh, xxxvi, pp. 335-342, with pl. xxxvi.

Hayden, F. V.   1858 A
Tertiary basin of White and Niobrara rivers.
Proc. Acad. Nat. Sci. Phila., pp. 147-158.
On pp. 157-158 is found a list of all the fossil Tertiary vertebrates known at that time from that region.

____   1858 B
Explorations under the War Department. Explanations of a second edition of a geological map of Nebraska and Kansas, based upon information obtained in an expedition to the Black Hills, under command of Lieut G. K. Warren, Top. Engr., U. S. A.
Proc. Acad. Nat. Sci. Phila., 1858, pp. 139-158.
Contains order of Tertiary beds and a list of fossil vertebrates.
An abstract of this paper appeared in Amer. Jour. Sci., 1858, xxvi, pp. 404-408.

____   1866 A
Discovery of a mastodon tooth near Fort Kearney and another opposite St. Louis.
Proc. Acad. Nat. Sci., Phila., p. 316.
Occupies two and a half lines.

Hayes, Seth.   1895 A
  An examination and description of
  mastodon and accompanying mammalian remains found near Cincinnati,
  June, 1894.
    Jour. Cincinnati Soc. Nat. Hist., xvii, pp.
    217-226.

Haymond, Rufus.   1844 A
  Notice of remains of *Megatherium*,
  Mastodon, and Silurian fossils.
    Amer. Jour. Sci. (1), xlvi, p. 294-296.
    A letter to the senior editor.

Hays, Isaac.   1830 A
  Description of a fragment of the head
  of a new fossil animal discovered in a
  marl pit near Moorestown, N. J.
    Trans. Amer. Philos. Soc. (2), iii, pp. 471-477,
    pl. xvi.
    Proposal of new genus *Saurodon* with species
    S. lanciformis and *S. leanus* (*leax*).

————   1834 A
  Description of specimens of inferior
  maxillary bones of mastodons in the
  cabinet of the American Philosophical
  Society, with remarks on the genus
  Tetracaulodon (Godman), etc.
    Trans. Amer. Philos. Soc. (2), iv, pp. 317-
    330, with pls. xx-xxix.

————   1841 A
  [Remarks on collection of bones,
  mostly of *Mastodon*, exhibited by Dr.
  Koch.]
    Proc. Amer. Phil. Soc., ii, pp. 102-103, 105-
    106.

————   1842 A
  [Review of paper by Prof. Rich.
  Owen on Koch's collection of mam-
  malian remains, especially of the genus
  Tetracaulodon.]
    Proc. Amer. Philos. Soc., ii, pp. 183-184.

————   1842 B
  [On the mastodontoid animals in the
  collection of Mr. Koch.]
    Proc. Amer. Phil. Soc., ii, pp. 264-266.

————   1843 A
  On the family *Proboscidea*, their
  general character and relations, their
  mode of dentition, and geological distribution.
    Proc. Amer. Phil. Soc., iii, pp. 44-48.

————   1844 A
  [Remarks on bones of "*Tetracaulodon*," mastodon, and elephant.]
    Proc. Amer. Philos. Soc., iv, pp. 30-43.

Hays, Isaac—Continued.   1846 A
  [Remarks in regard to recently dis-
  covered mastodons in New York and
  New Jersey.]
    Proc. Amer. Philos. Soc., iv, p. 269.

————   1852 A
  [Remarks on tooth of fossil tapir
  from North Carolina.]
    Proc. Acad. Nat. Sci. Phila., vi, p. 58.

See also **Horner** and **Hays.**

Hazeltine, John.   1835 A
  Fossil tooth.
    Amer. Jour. Sci., xxvii, pp. 166-168.
    Describes a fossil jaw containing teeth which
    appear to have belonged to a young mastodon.
    Found in Chautauqua County, N. Y.

Hébert, Edmund.   1855 A
  Note sur le tibia du *Gastornis parisiensis.*
    Comptes rend. Acad. Sci., Paris, xl, pp. 579-
    582.

————   1855 B
  Note sur le fémur du *Gastornis parisiensis.*
    Comptes rend. Acad. Sci., Paris, xl, pp. 1214-
    1217.

————   1855 C
  Tableau des fossiles de la craie de
  Meudon et description de quelques es-
  pèces nouvelles.
    Mém. Soc. géol. de France, (2), v, pp. 345-
    374, with pls. xxvii-xxix.
    This paper is quoted by A. S. Woodward for
    1854.

————   1856 A
  Recherches sur la faune des premiers
  sédiments tertiaires Parisiens (mammifères pachydermes du genre *Coryphodon*).
    Ann. Sci. Naturelles (4), Zool., vi, pp. 87-136,
    with pls. iii, iv.

Heckel, Joh. Jakob.   1843 A
  Abbildungen und Beschreibungen der
  Fische Syriens, nebst einer neuen Classification und Characteristik sämmtlicher Gattungen der Cyprinen.
    Stuttgart, 1843, pp. 1-258, with pls. i-xxiii.
    Pp. 234-244, with pl. xxiii, are devoted to the
    fossil fishes of the Lebanon. The genera *Pyc-
    nosternix* and *Isodus* are characterized.

————   1850 A
  Ueber die Wirbelsäule fossilen Ganoiden.
    Sitzungsber. k. Akad. Wissensch. Wien.,
    math.-phys. Cl., v, pp. 358-368, with 4 text figs.

Heckel, Joh. Jakob—Cont'd. 1854 A
Ueber den Bau und die Eintheilung
der Pycnodonten, nebst kurzer Beschreibung einiger neuen Arten derselben.
Sitzungsber. k. Akad. Wissensch. Wien.,
math.-phys. Cl., xii, pp. 433-463.

——— 1856 A
Beiträge zur Kenntniss der fossilen
Fische Oesterreichs.
Denkschr. k. Akad. Wissensch. Wien, math.-
naturw. Cl., xi, pp. 187-274, with pls. i-xv.
Also issued separately with pp. 1-88.

Helm, F. 1891 A
On the affinities of *Hesperornis*.
Nature, xliii, p. 368.

Hemstedt, R. 1870 A
Ueber einigen Besonderheiten der
Schädelknochen von *Lepus* und über
das knöcherne Gehörorgan desselben
Genus.
Archiv f. Anat., Physiol. u. wissensch. Med.,
Jahrg. 1870, pp. 437-453, pl. xi.

Hensel, Reinhold. 1859 A
Ueber einen fossilen Muntjac aus
Schlesien.
Zeitschr. deutsch. geol. Gesellsch. xi, pp.
251-279, with pls. x-xii.

——— 1860 A
Ueber *Hipparion mediterraneum*.
Abhandl. k.-p. Akad. Wissensch. Berlin,
1860. pp. 27-121, with pls. i-iv.

——— 1875 A
Zur Kenntniss der Zahnformel für die
Gattung *Sus*.
Nova Acta k. Leop.-Carol. Akad., xxxvii,
No. 5, pp. 1-40, with pl. xxvi.

Hertwig, O. 1874 A
Über Bau und Entwicklung der Pla-
coidschuppen und der Zähne der Selachier.
Jenaische Zeitschr. Naturwiss. viii, pp. 331-
404, with pls. xii-xiii.

——— 1876 A
Über das Hautskelet der Fische.
Morpholog. Jahrbuch, ii, pp. 328-395, with
pls. xxiii-xxviii.
Describes the dermal skeleton of the Silu-
roids and of *Acipenseridæ*.

——— 1879 A
Ueber das Hautskelet der Fische.
Morpholog. Jahrbuch, v, pp. 1-21, with pls.
i-iii.
Describes dermal skeleton of *Lepisosteidæ*
and *Polypteridæ*.

Hertwig, O.—Continued. 1881 A
Ueber das Hautskelet der Fische. III.
Das Hautskelet der *Pediculati*, der *Discoboli*, der Gattung *Diana*, der *Centri-
scidæ*, einiger Gattungen aus der Familie der *Triglidæ* und der Plectognathen.
Morpholog. Jahrbuch, vii, pp. 1-42, with
pls. i-iv.

Hibbert, Samuel. 1836 A
On the fresh-water limestone of Bur-
diehouse, in the neighborhood of Edinburgh, belonging to the Carboniferous
group of rocks, with supplementary
notes on other fresh-water limestones.
Trans. Roy. Soc. Edinburgh, xiii, pp. 169-282,
with pls. v-xii.

Hicks, L. E. 1873 A
Discovery of mastodon remains in
Ohio.
Amer. Jour. Sci. (3), v., p. 79.

Higgins, E. T. 1867 A
On the otolites of fish, and their value
as a test in verifying recent and fossil
species.
Jour. Linn. Soc. Lond., ix, pp., 157-166.

Higley, W. K. 1886 A
A paper on *Elephas primigenius*.
Bull. Chicago Acad. Sci., i, No. x, pp. 123-
127, with 1 pl.

Hilgard, Eugene W. 1860 A
Report on the geology and agriculture
of Mississippi.
Pp. i-xiii, 1-391.
Printed by order of the legislature. Con-
tains remarks on a few fossil vertebrates, and
on page 389 a list of fossil teeth of fishes, by
Dr. Wm. Spillman.

——— 1867 A
On the Tertiary formations of Missis-
sippi and Alabama.
Amer. Jour. Sci. (2), xliii, pp. 29-41.
Contains mention of *Zeuglodon*.

Hilgendorf, F. 1866 A
Ueber das Gebiss der hasenartigen
Nager.
Monatsber. k.-p. Akad. Wissensch. Berlin,
1865 (1866), p. 673.

Hill, F. C. 1881 A
The fossil *Dinocerata* in the E. M.
Museum at Princeton, N. J.
Proc. Amer. Assoc. Adv. Sci., 29th meeting,
Boston, 1880, pp. 524-527, with 1 woodcut.
An abstract of paper read.

——— 1886 A
On the mounting of fossils.
Amer. Naturalist, xx, pp. 353-359, with 8
woodcuts.

Hills, R. C.   1880 A
Note on the occurrence of fossils in
the Triassic and Jurassic beds near San
Miguel, in Colorado.
Amer. Jour. Sci., (3), xix, p. 490.

Hitchcock, Charles H.   1855 A
Impressions (chiefly tracks) on al-
luvial clay in Hadley, Mass.
Amer. Jour. Sci., (2), xix, pp. 391–396.

———   1865 A
Editorial notes.
In Edward Hitchcock's Supplement to the
Ichnology of New England. (Hitchcock, E.,
1858 A).

———   1866 A
Description of a new reptilian bird
from the Trias of Massachusetts.
Annals Lyc. Nat. Hist. New York, viii, pp.
301–302.
Description of tracks of *Tarsodactylus expansus.*

———   1868 A
New American fishes from the Devo-
nian.
Geol. Magazine (1), v, pp. 184–185, with 2
figs. in text.
Reports discovery of *Dinichthys herzeri* by
Dr. Newberry.

———   1868 B
New Carboniferous reptiles and fishes
from Ohio, Kentucky, and Illinois.
Geol. Magazine (1), v, pp. 186–187.
Reports Prof. Newberry's descriptions of
fishes as presented before A. A. A. S., 1867.

———   1869 A
Note on the supposed fossil foot
marks in Kansas.
Amer. Jour. Sci. (2), xlvii, pp. 132–133.

———   1871 A
[Account and complete list of the
Ichnozoa of the Connecticut Valley.]
Walling and Gray's Official Typographical
Atlas of Massachusetts, pp. xx–xxi.
This account is part of a description of the
geology of Massachusetts, and consists principally of a list of the fossil footmarks. This list
is almost identical with the same author's list
of 1889 (1889 B), with the exception of species
published after 1871. I cite this paper, therefore, only in the cases of species, two or three
in number, more particularly mentioned. Few
new facts are presented.

———   1873 A
Footprints in the rocks.
Pop. Sci. Monthly, iii, pp. 428–441, with 7
text-figs.

Hitchcock, Charles H.—Cont'd. 1889 A
[Remarks on A. Wanner's paper on
The discovery of fossil tracks in the
Triassic of York County, Pa.]
Proc. Amer. Assoc. Adv. Sci., 37th meeting,
Cleveland, 1888, p. 186.

———   1889 B
Recent progress in ichnology.
Proc. Boston Soc. Nat. History, xxiv, pp.
117–127.
Contains a list of the American Triassic
species of *Ichnozoa* and descriptions of new
forms. This list is identical with one published by the same author in 1871 (Hitchcock,
C. H., 1871 A), with the exception of a few
species published after the issue of the latter
paper.

Hitchcock, Edward.   1836 A
Ornithichnology. Description of the
footmarks of birds (*Ornithichnites*) on
New Red sandstone in Massachusetts.
Amer. Jour. Sci., xxix, pp. 307–340, with a
folded plate of 24 figs.

———   1837 A
Ornithichnites in Connecticut.
Amer. Jour. Sci., xxxi, pp. 174–175.

———   1837 B
Fossil footsteps in sandstone and
graywacke.
Amer. Jour. Sci., (1), xxxii, pp. 174–176.

———   1841 A
Final report on the geology of Massachusetts.
Vol. ii, pp. 301–831, with pls. xv–li. North-
ampton, 1841.
Pp. 455–525 are devoted to the descriptions of
fossil fishes and footprints, and these are illustrated on pls. xxix–li.

———   1843 A
Description of five new species of
fossil footmarks, from the Red Sandstone of the valley of Connecticut River.
Reports of the 1st, 2d, and 3d meetings of the
Assoc. of Amer. Geol. and Naturalists, pp. 254–
264, with pl. xi.

———   1844 A
Report on ichnolithology, or fossil
footmarks, with a description of several
new species, and the coprolites of birds,
from the valley of the Connecticut
River, and of a supposed footmark
from the valley of the Hudson River.
Amer. Jour. Sci., xlvii, pp. 292–322, with pls.
iii–iv.
Abstract in Neues Jahrb. Min., 1845, pp.
753–757.

Hitchcock, Edward—Continued. 1844 B
Rejoinder to the preceding article of
Dr. Deane.
Amer. Jour. Sci., xlvii, pp. 390–399.
A reply to Deane, J., 1844 B. See Deane, J.,
1844 C.

———   1845 A
Extract of a letter from Prof. E.
Hitchcock, embracing miscellaneous
remarks upon fossil footmarks, the Lin-
colnite, etc., and a letter from Prof.
Richard Owen, on the great birds' nests
of New Holland.
Amer. Jour. Sci., xlviii, pp. 61–65.

———   1845 B
An attempt to name, classify, and
describe the animals that made the
fossil footmarks of New England.
Proc. 6th annual meeting Assoc. Amer. Geologists and Naturalists, New Haven, Conn.,
April, 1845, pp. 23–25.
In this paper the footmarks formerly described under the general terms *Ornithoidichnites* and *Sauroidichnites* are assigned generic
names.

———   1847 A
Description of two new species of
fossil footmarks found in Massachusetts
and Connecticut, or of the animals that
made them.
Amer. Jour. Sci. (2), iv, pp. 46–57, with 3
text figs.

———   1848 A
An attempt to discriminate and describe the animals that made the fossil
footmarks of the United States, and
especially of New England.
Mem. Amer. Acad. Arts and Sci. (2), iii, pp.
129–256, pls. 1–24.

———   1855 A
Shark remains from the Coal formation of Illinois, and bones and tracks
from the Connecticut River sandstone.
Amer. Jour. Sci. (2), xx, pp. 416–417.

———   1856 A
On a new fossil fish and new fossil
footmarks. I. History of the discovery
and general character of the jaw of a
new family of fossil fishes. II. Description of a new and remarkable species
of fossil footmarks from the sandstone
of Turner's Falls, in the Connecticut
Valley.
Amer. Jour. Sci. (2), xxi, pp. 96–100, with 1
fig. in the text.

Hitchcock, Edward—Continued. 1856 B
Account of the discovery of the fossil
jaw of an extinct family of sharks from
the Coal formation.
Proc. Amer. Assoc. Adv. Sci., 9th meeting,
1855, Washington, D. C., pp. 229–230.
A description and figure, without genus and
species, of *Edestus minor* Newb., found in Parke
County, Ind.

———   1858 A
Ichnology of New England. A re-
port on the sandstone of the Connecticut Valley, especially its fossil footmarks.
Boston, 1858. 4to, 220 pp.; with 60 quarto pls.

———   1861 A
Remarks upon certain points in ich-
nology.
Proc. Amer. Assoc. Adv. Sci., 14th meeting,
Newport, 1860, pp. 144–156, with 4 figs. in the
text.

———   1863 A
New facts and conclusions respecting
the fossil footmarks of the Connecticut
Valley.
Amer. Jour. Sci. (2), xxxvi, pp. 46–57.

———   1863 B
Supplement to the ichnology of New
England.
Proc. Amer. Acad. Arts and Sci., vi, pp. 85–92.
In this paper a number of species formerly
described is rejected or recorded as doubtful;
two species are mentioned as new but not described; notes are given on other species, and
there is a general discussion of fossil tracks.

———   1865 A
Supplement to the ichnology of New
England. A report to the government
of Massachusetts in 1863.
Boston, 1865. 4to; pp. i–x, 1–96, pls. i–xx.
Edited by his son, C. H. Hitchcock. Ap-
pendix A (pp. 39, 40) is entitled "Bones of
Megadactylus polyzelus;" Appendix B (pp. 40–93)
consists of a descriptive catalogue of the Hitch-
cock ichnological cabinet at Amherst College.

———   1872 A
Discovery of a tooth of a mastodon
in Massachusetts.
Amer. Jour. Sci. (3), iii, p. 146.

———   1885 A
[Letter to editor of Science announc-
ing discovery of a mastodon's skeleton
near Geneva, N. Y.]
Science, vi, p. 450.

Hitchcock, Fanny R. M.   1887 A
On the homologies of *Edestus.*
Amer. Naturalist, xxi, pp. 847–848.

Hitchcock, Fanny R. M.—C't'd.   1888 A
On the homologies of *Edestus.*
Proc. Amer. Assoc. Adv. Sci., 36th meeting,
New York, 1887, pp. 260-261.
Abstract of paper presented.

————   1888 B
Further notes on the osteology of the
shad (*Alosa sapidissima*).
Annals N. Y. Acad. Sci., iv, pp. 225-228, with
text fig.

Hodson, William B.   1846 A
Memoir on the *Megatherium* and
other extinct gigantic quadrupeds of
the coast of Georgia, with observations
on its geologic features.
Pp. 1-47, with 1 pl. and 2 maps. New York,
1846.
Besides the matter contributed by Hodson
this memoir contains a "Memorandum" by
Dr. Jos. Habersham (pp. 25-30) and "Observations" by J. Hamilton Couper (pp. 31-47).

Hoernes, R.   1875 A
Vorlage von Wirbelthierresten aus
den Kohlenablagerungen von Trifail
in Steiermark.
Verhandl. k. k. geol. Reichsanstalt, 1875-
1876, pp. 310-313.

————   1876 A
Anthracotherium magnum Cuv aus
den Kohlenablagerungen von Trifail.
Jahrbuch k. k. geol. Reichsanstalt, xxvi,
pp. 209-242. with pl. xv.
Contains numerous references to the literature of this subject up to that time.

————   1877 A
Zur Kenntniss der *Anthracotherium
dalmatinum* Meyer.
Verhandl. k. k. geol. Reichsanstalt, 1875-1876,
No. 15, pp. 363-366.

————   1892 A
Zur Kenntniss der Milchbezahnung
der Gattung *Entelodon* Aym.
Sitzungsber Akad. Wissensch. Wien, ci, pp.
17-24, with 1 pl.

Hoffmann, C. K.   1874 A
Amphibien.
Bronn's Klassen und Ordnungen des Thier-
reichs, vi, pp. 1-88, pls. i-xiv.
This portion of Dr. Hoffmann's work is devoted to the osteology of the *Batrachia.* In it
is found a list of 78 works on this subject.

————   1876 A
Beiträge zur Kenntniss des Beckens
der Amphibien und Reptilien.
Niederländisches Archiv Zoologie, iii, pp.
143-194, pls. x, xi, and 15 woodcuts in the text.

Hoffmann, C. K.—Continued.   1878 A
Beiträge zur Anatomie der Wirbelthiere.
Niederländisches Archiv Zoologie, iv, pp.
112-248, pls. ix-xiii.
Discusses the structure of the carpus and
tarsus of reptiles and amphibians, the integument and dermal skeleton of *Testudinata,*
morphology of ribs, etc.

————   1879 A
Beiträge zur vergleichenden Anato-
mie der Wirbelthiere. X; Uber das
Vorkommen von Halsrippen bei den
Schildkröten; XI, Uber das Verhältniss des Atlas und des Epistropheus
bei den Schildkröten; XII, Zur Morphologie des Schultergürtels und des
Brustbeins bei Reptilien, Vögeln, Säugethieren und dem Menschen.
Niederländisches Archiv Zoologie, v, pp.
19-114, with pls. ii-ix.

————   1890 A
Schildkröten.
Bronn's Klassen und Ordnungen des Thier-
reichs, vi, Abtheil. iii, pp. 1-442, with pls.
i-xlviii.
Pp. 1-73 are devoted to the osteology of the
Testudinata. On pp. 1-3 is found a list of the
literature of the subject.

————   1890 B
Eidechsen   und   Wassereidechsen
(*Saurii* und *Hydrosauria*).
Bronn's Klassen und Ordnungen, vi, Abth.
iii, pp. 443-1399, with pls. xlix-cvii.
Pp. 443-610 and pls. xlix-lxxi are devoted
to the osteology of this group. Pp. 1299-1329
are devoted to the palaeontology of the *Reptilia.*

————   1890 C
Schlangen   und   Entwicklungsgeschichte der Reptilien.
Bronn's Klassen und Ordnungen des Thier-
reichs, vi, pt. 3, pp. 1401-2089, with pls. cviii-
clxx.
The description of the skeleton occupies pp.
1420-1448 and pls. cx-cxvi.

Holder, Joseph B.   1883 A
The Atlantic right whales: A contribution embracing an examination of
(1) The exterior characters and osteology of a cisarctic right whale—male;
(2) The exterior characters of a cisarctic right whale—female; (3) The osteology of a cisarctic right whale—sex not
known. To which is added a concise
résumé of historical mention relating
to the present and allied species.
Bulletin Amer. Mus. Nat. Hist., i, pp. 99-138,
with pls. x-xiii.

Holland, W. J.   1900 A
The vertebral formula in *Diplodocus* Marsh.
Science (2), xi, pp. 816-818.

Hollard, Henri.   1853 A
Monographie de la famille des Balistides.
Ann. Sci. Naturelles (3), Zool., xx, pp. 71-114, with pls. i-iii.

———   1854 A
Monographie des Balistides. Part 2.
Ann. Sci. Naturelles (4), Zool., i, pp. 39-72; 303-339.
This monograph is continued in vol. ii, pp. 321-366, 1854, and in vol. iv, pp. 5-27, 1855.

———   1857 A
Étude sur les Gymnodontes, et en particulier sur leur ostéologie et sur les indications qu'elle peut fournier pour leur classification.
Annales des Sci. Naturelles (4), Zool., viii, pp. 275-328.

———   1860 A
Mémoire sur le squelette des poissons plectognathes étudié au point de vue des caractères qu'il peut fournir pour pour la classification.
Annales Sci. Naturelles (4), Zool., xiii, pp. 5-46, pls. ii, iii.

Hollick, Arthur.   1894 A
[Remarks on *Spiraxis* Newb.]
Trans. N. Y. Acad. Sci., xiii, pp. 118-119.

Holmes, F. S.   1849 A
Notes on the geology of Charleston, S. C.
Amer. Jour. Sci., vii, pp. 187-201.

———   1850 A
[Remarks on paper by R. W. Gibbes concerning fossil *Equus* from Eocene of South Carolina.]
Proc. Amer. Assoc. Adv. Sci., 3d meeting, Charleston, S. C., 1850, pp. 68-69.
Remarks on Gibbes. R. W. 1850 C.

———   1850 B
Observations on the geology of Ashley River, South Carolina.
Proc. Amer. Assoc. Adv. Sci., 3d meeting, Charleston, S. C., 1850, pp. 201-204.

———   1858 A
Remains of domestic animals among post-Pliocene fossils in South Carolina.
16 pp., 8vo, Charleston, 1858.
This paper contains extract from a paper by Dr. Leidy and a letter by Professor Agassiz. In this letter Agassiz expresses the belief that "horses, sheep, bulls, and hogs not distin-

Holmes, F. S.—Continued.
guishable at present from the domesticated species" existed in America before the advent of white men.

———   1858 B
Remains of domestic animals among post-Pliocene fossils in South Carolina.
Amer. Jour. Sci. (2), xxv, pp. 442-443.

———   1859 A
[Remarks of Prof. F. S. Holmes on fossils from post-Pliocene of South Carolina, including a letter from Prof. L. Agassiz.]
Proc. Acad. Nat. Sci. Phila., 1859, pp. 177-186.

Holmes, N.   1857 A
[Remarks on contemporaneity of man and the mastodon.]
Trans. Acad. Sci. St. Louis, i, p. 317.

Home, Everard.   1819 A
An account of the fossil skeleton of the *Proteosaurus*.
Philos. Trans. Roy. Soc. Lond., cix, pl. 209-211.
This paper was preceded by others of the same author in the volumes for 1814, 1816, but in these papers no name was applied to the remains.

———   1819 B
Reasons for giving the name *Proteosaurus* to the fossil skeleton which has been described.
Philos. Trans. Roy. Soc. Lond., cix, pp. 212-216, with pl. xv.

Honeyman, D.   1874 A
Skeleton of a whale in the Quaternary of New Brunswick.
Amer. Jour. Sci. (3), vii, p. 597.
Supposed to be *Beluga vermontana*.

———   1874 B
On the Quaternary containing the New Brunswick fossil cetacean.
Amer. Jour. Sci. (3), viii, p. 219.

Hopkins, William.   1855 A
Brief outline or general description of a remarkable fossil, not known to be described, and by some supposed to be an ichthyodorulite.
Proc. Amer. Assoc. Adv. Sci., 8th meeting, Washington, D. C., 1854, pp. 287-290, with 2 woodcuts.

Horner, W. E.   1840 A
Note of the remains of the mastodon, and some other extinct animals collected together in St. Louis, Mo.
Proc. Amer. Philos. Soc., i, pp. 279-282.
This note refers to the collection of Doctor Koch at St. Louis.

Horner, W. E.—Continued.   1841 A
Note on the remains of the mastodon,
and some other extinct animals col-
lected together in St. Louis, Mo.
Amer. Jour. Sci., xl, pp. 56-59. (Extract from
Proc. Amer. Philos. Soc.)
An investigation of Koch's collection.

———   1841 B
Remarks on the dental system of the
mastodon, with an account of some
lower jaws in Mr. Koch's collection,
St. Louis, Mo., where there is a solitary
tusk on the right side.
Proc. Amer. Philos. Soc., i, pp. 307-308.
This is an abstract of "Horner, W. E., 1842 A."

———   1841 C
On the dental system of the mastodon.
Proc. Amer. Philos. Soc., ii, pp. 6-7.
A note of correction of the preceding paper.

———   1842 A
Remarks on the dental system of the
mastodon, with an account of some
lower jaws in Mr. Koch's collection,
St. Louis, Mo., where there is a solitary
tusk on the right side.
Trans. Amer. Philos. Soc., viii, pp. 53-59.

Horner, W. E., and Hays, Isaac.   1842 A
Description of an entire head, and
various other bones of the mastodon.
Trans. Amer. Philos. Soc., viii, pp. 37-48, pls.
i-iv.

Hovey, Edmund O.   1874 A
The largest fossil elephant tooth yet
described.
Proc. Amer. Assoc. Adv. Sci., 22d meeting,
Portland, 1873, p. 112.
Abstract of paper presented.

———   1900 A
The geological and palæontological
collections in the American Museum of
Natural History.
Science (2), xii, pp. 757-760.

Howes, G. B.   1887 A
On the skeleton and affinities of the
paired fins of *Ceratodus*, with observations upon those of *Elasmobranchii*.
Proc. Zool. Soc. Lond., 1887, pp. 3-26, with
pls. i-iii.

———   1887 B
Morphology of the mammalian cora-
coid.
Jour. Anat. Physiol., xxi, pp. 190-198, with
pl. viii.

Howes, G. B.—Continued.   1890 A
Observations on the pectoral fin
skeleton of the living batoid fishes, and
of the extinct genus *Squaloraja*, with
especial reference to the affinities of
the same.
Proc. Zool. Soc. Lond., 1890, pp. 675-688,
with 10 figs. in the text.

———   1893 A
On the mammalian pelvis, with
especial reference to the young of
Ornithorhynchus anatinus.
Jour. Anat. and Physiol., xxvii, pp. 543-556,
with pl. xxviii.

———   1896 A
On the mammalian hyoid, with
especial reference to that of *Lepus, Hyrax*, and *Cholæpus*.
Jour. Anat. and Physiol., xxx, pp. 513-526,
with pl. viii.

——— See also **Hancock** and **Howes.**

Howes, G. B., and Davies, A. M.
                                      1888 A
Observations upon the morphology
and genesis of supernumerary phalan-
ges, with especial reference to those of
the amphibia.
Proc. Zool. Soc., Lond., 1888, pp. 495-511,
with pls. xxiv, xxv.

Howes, G. B., and Harrison, J.   1893 A
On the skeleton and teeth of the
Australian dugong.
Report Brit. Assoc. Adv. Sci., 62d meeting,
Edinburgh, 1892, p. 790.

Howes, G. B., and Ridewood, W.
                                      1888 A
On the carpus and tarsus of the
Anura.
Proc. Zool. Soc. Lond., 1888, pp. 411-182,
with pls. vii-ix.

Howes, G. B., and Swinnerton, H. H.
                                      1900 A
On the development of the skeleton of
the Tuatera, *Sphenodon (Hatteria) punc-
tatus*.
Proc. Zool. Soc. Lond., 1900, pp. 516-517.
An abstract of a paper read before the society.

Howorth, H. H.   1870 A
The extinction of the mammoth.
Report of the Brit. Assoc. Adv. Sci., 39th
meeting, Exeter, 1869, pp. 90-91.
Abstract of paper presented.

———   1881 A
Climate of Siberia in the era of the
mammoth.
Amer. Jour. Sci. (3), xxi, p. 148.
Extracts form Howorth's paper in the Geo-
logical Magazine for December, 1880.

Howorth, H. H.—Continued.   1881 B
The sudden extinction of the mammoth.
Geol. Magazine (2), viii, pp. 309–315.

——   **1892 A**
Did the mammoth live before, during, or after the deposition of the glacial drift.
Geol. Magazine (3), ix, pp. 250–258; 396–405.
See Jukes-Brown, vol. ix, p. 477.

——   **1893 A**
The true horizon of the mammoth.
Geol. Magazine (3), x, pp. 20–27; 161–163; 353–355.

——   **1894 A**
The mammoth age was contemporary with the age of great glaciers.
Geol. Magazine (4), i, pp. 161–167.

Hoy, P. R.   1871 A
Dr. Koch's *Missourium.*
Amer. Naturalist, v, pp. 147–148.

Hubrecht, A. A. W.   1877 A
Beitrag zur Kenntniss des Kopfskelets der Holocephalen.
Niederländisches Archiv fur Zoologie, iii, pp. 255–276, with pl. xvii.
See also Morphologisches Jahrbuch, iii.

——   **1878 A**
Pisces: Fische.
Bronn's Klassen und Ordnungen, vi, Abth. i, pp. 1–86, with pls. i–xii.
Of this work, pp. 1–48, pls. i–vii, appeared in 1876; pp. 49–80, pls. viii–x, in 1878; pp. 81–86, in 1885. Pp. 21–86 are devoted to a consideration of the anatomy of the elasmobranchs.

——   **1896 A**
Die Keimblase von *Tarsius*, ein Hilfsmittel zur schärferen Definition gewisser Säugethierordnungen.
Festschrift zum 70sten Geburtstage von Carl Gegenbaur, ii, pp. 149–178, with pl. i, and 15 figs. in the text.
Compares *Tarsius* with *Anaptomorphus.*

Huene, Friedrich v.   1900 A
Devonische Fischreste aus der Eifel.
Neues Jahrb. Min., Geol. und Pal., 1900, i, Abhandl., pp. 64–66, with 2 text figs.
Records finding of *Dinichthys* and *Rhynchodus.*

Hulke, J. W.   1879 A
Note (3d) on (*Eucamerotus* Hulke) *Ornithopsis* H. G. Seeley=*Bothriospondylus* Owen=*Chondrosteosaurus magnus* Owen.
Quart. Jour. Geol. Soc., xxxv, pp. 752–762, with 4 text figs.

Hulke, J. W.—Continued.   1879 B
Note on *Poikilopleuron bucklandi* of Eudes Deslongchamps (père) identifying it with *Megalosaurus bucklandi.*
Quart. Jour. Geol. Soc. Lond., xxxv, pp. 233–238, with pl. xii.

——   **1880 A**
Supplementary note on the vertebræ of *Ornithopsis,* Seeley=*Eucamerotus,* Hulke.
Quart. Jour. Geol. Soc., xxxvi, pp. 31–34, with pls. iii, iv.

——   **1882 A**
Note on the os pubis and ischium of *Ornithopsis eucamerotus.*
Quart. Jour. Geol. Soc., xxxviii, pp. 372–376, with pl. xiv.

——   **1883 A**
The anniversary address of the President.
Quart. Jour. Geol. Soc., xxxix, Proceedings, pp. 38–65, with 18 figs. in the text.
Discusses the structure of various extinct reptiles, *Enaliosauria* and *Dinosauria,* especially the structure of shoulder girdle and limbs.

——   **1883 B**
An attempt at a complete osteology of *Hypsilophodon foxii.*
Philosoph. Trans. Roy. Soc. London, clxxiii, pp. 1035–1062, pls. lxxi–lxxxii.

——   **1884 A**
The anniversary address of the President [of the Geol. Soc. Lond.].
Quart. Jour. Geol. Soc., vol. xl, Proceedings, pp. 37–57, with 8 text figs.
Discusses structure of certain *Dinosauria,* especially *Iguanodon.*

——   **1885 A**
Note on the sternal apparatus in *Iguanodon.*
Quart. Jour. Geol. Soc., xli, pp. 473–475, with pl. xiv.

——   **1888 A**
Contribution to the skeletal anatomy of the *Mesosuchia,* based on fossil remains from the clays near Peterborough in the collection of A. Leeds, esq.
Proc. Zool. Soc. Lond., 1888, pp. 417–442, with pls. xviii, xix.

——   **1892 A**
On the shoulder girdle in *Ichthyosauria* and *Sauropterygia.*
Proc. Roy. Soc. London, lii, pp. 233–255, with 9 figs. in the text.

Humbert, Alois. See Pictet and Humbert.

Humphry, G. M.   1866 A
On the homologies of the lower jaw,
and the bones connecting it with the
skull in birds, reptiles, and fishes.
Report Brit. Assoc. Adv. Sci., 35th meeting,
Birmingham, 1865, pp. 87–89.

———   1870 A
On the homological relations to one
another of the mesial and lateral fins of
osseous fishes.
Jour. Anat. and Physiol., v, pp. 58–66, pl. ii.

Humphreys, John.   1889 A
The suppression and specialization of
teeth.
Proc. Birmingham Philos. Soc., vi, pp. 137–
161.

See also **Windle** and **Humphreys**.

Hunt, J. G.   1874 A
[Report on food of the mastodon.]
Proc. Boston Soc. Nat. Hist., xvii, pp. 91–92.

Hunter, William.   1769 A
Observations on the bones, commonly
supposed to be elephant's bones, which
have been found near the river Ohio,
in America.
Philos. Trans. Roy. Soc. Lond., 1768, lviii, pp.
34–45, with pl. iv.

Hurst, C. Herbert.   1893 A
The digits in a bird's wing: a study
of the origin and multiplication of errors.
Natural Science, iii, pp. 275–281, with photographic reproduction of wing of Archæopteryx.

———   1895 A
The structure and habits of Archæopteryx.
Natural Science, vi, pp. 112–122; 180–186; 244–
248, with 2 pls. and 2 figs. in the text.

Hutchinson, H. N.   1893 A
Extinct monsters. A popular account
of some of the larger forms of ancient
animal life, with illustrations by J.
Smit and others.
Pp. i–xxii, 1–270, with pls. i–xxvi and 58 text
figs.
Reviewed in Natural Science, ii, pp. 135–143.

Huxley, Thomas H.   1858 A
On Cephalaspis and Pteraspis.
Quart. Jour. Geol. Soc. Lond., xiv, pp. 267–
280, with pls. xiv, xv, showing the microscopical structure of the skeleton.

———   1858 B
On a new species of Plesiosaurus from
Street, near Glastonbury, with remarks
on the structure of the atlas and axis

Huxley, Thomas H.—Cont'd.
vertebræ and of the cranium in that
genus.
Quart. Jour. Geol. Soc. Lond., xiv, pp. 281–
294.

———   1858 C
Observations on the development of
some parts of the skeleton of fishes.
1. On the development of the tail in
Teleostean fishes.
2. On the development of the palato-
pterygoid arc and hyomandibular sus-
pensorium in fishes.
Quart. Jour. Micros. Sci., vii, pp. 33–46.

———   1859 A
Observations on the genus Pteraspis.
Report Brit. Assoc. Adv. Sci., 28th meeting,
Leeds, 1858, pp. 82–83.

———   1859 B
On the dermal armour of Jacare and
Caiman, with notes on the specific and
generic characters of recent Crocodilia.
Jour. Linn. Soc. Lond., iv, pp. 1-28.

———   1859 C
On a new species of Dicynodon (D.
murrayi) from near Colesberg, South
Africa, and on the structure of the skull
in the Dicynodonts.
Quart. Jour. Geol. Soc. Lond., xv, pp. 649-
658, with pls. xxii, xxiii.

———   1861 A
On Pteraspis dunensis (Archæoteuthis
dunensis, Roemer).
Quart. Jour. Geol. Soc. Lond., xvii, pp. 163-
166, with 2 figs. in the text.

———   1861 B
Preliminary essay upon the system-
atic arrangement of the fishes of the
Devonian epoch.
Mem. Geol. Surv. United Kingdom, decade
10, pp. 1–40; pls. 1–3.

———   1861 C
Phaneropleuron andersoni.
Mem. Geol. Surv. United Kingdom, decade
10, pp. 47–49; pl. iii.

———   1863 A
Description of Anthracosaurus russelli,
a new labyrinthodont from the Lanarkshire coal field.
Quart. Jour. Geol. Soc. Lond., xix, pp. 56–68.

———   1864 A
On the cetacean fossils termed "Ziphius," by Cuvier, with a notice of a new
species (Belemnoziphius compressus) from
the Red Crag.
Quart. Jour. Geol. Soc. Lond., xx, pp. 388-
396, with pl. xix.

Huxley, Thomas H.—Cont'd.   1865 A
On the osteology of *Glyptodon*.
Philosoph. Trans. Roy. Soc. Lond., clv, pp. 31-70; pls. 4-9.

———   1866 A
Illustrations of the structure of the Crossopterygian ganoids.
Mem. Geol. Surv. United Kingd., decade 12, pp. 3-45, with pls. ii-x.

———   1867 A
On the classification of birds.
Proc. Zool. Soc. Lond., 1867, pp. 415-472, with 36 figs. in the text.

———   1867 B
Description of vertebrate remains from the Jarrow colliery, part i.
Trans. Roy. Irish Acad., xxiv, pp. 353-369, with pls. xix-xxiii.

———   1868 A
On the animals which are most nearly intermediate between birds and reptiles.
Proc. Roy. Soc. Lond., xvi, pp. 243-248.
Also in Ann. Mag. Nat. Hist. (4), i, 1868, pp. 220-224; Geol. Mag. (1), v, pp. 357-365.

———   1869 A
Triassic *Dinosauria*.
Nature, i, pp. 23-24.
Discusses, among other genera, *Bathygnathus* and *Belodon*.

———   1869 B
On the representatives of the malleus and the incus of the *Mammalia* in the other *Vertebrata*.
Proc. Zool. Soc. Lond., 1869, pp. 391-407, with 8 figs. in the text.

———   1869 C
On *Hyperodapedon*.
Quart. Jour. Geol. Soc., Lond., xxv, pp. 138-152; 157-158, with 5 figs. in the text.

———   1869 D
On the upper jaw of *Megalosaurus*.
Quart. Jour. Geol. Soc. Lond., xxv, pp. 311-314, with pl. xii.

———   1870 A
On the progress of palæontology. Anniversary address delivered before the Geological Society.
Nature, i, pp. 437-443.

———   1870 B
The anniversary address of the president [of the Geol. Soc. Lond.].
Quart. Jour. Geol. Soc., Lond., xxvi; Proceedings, pp. xxix-lxiv.

———   1870 C
Further evidences of the affinity between the Dinosaurian reptiles and birds.
Quart. Jour. Geol. Soc. Lond., xxvi, pp. 12-31.

Huxley, Thomas H.—Cont'd.   1870 D
The classification and affinities of the *Dinosauria*.
Quart. Jour. Geol. Soc. Lond., xxvi, pp. 32-51, with pl. iii.

———   1872 A
A manual of the anatomy of vertebrated animals.
Pp. 1-431, woodcuts 110. New York, D. Appleton & Co., 1872.

———   1874 A
On the structure of the skull and of the heart of *Menobranchus lateralis*.
Proc. Zool. Soc. Lond., 1874, pp. 186-204, with pls. xxix-xxxii.

———   1875 A
Note on the development of the columella auris in the *Amphibia*.
Report Brit. Assoc. Adv. Sci., 44th meeting, Belfast, 1874, pp. 141-142.
Abstract in Nature, vol. xi, 1874, p. 68.

———   1875 B
On *Stagonolepis robertsoni*, and on the evolution of the *Crocodilia*.
Quart. Jour. Geol. Soc. Lond., xxxi, pp. 423-438, with pl. xix.

———   1875 C
Amphibia.
Encyclop. Brit., ed. 9, i, pp. 750-771, with 26 figs. in the text.

———   1876 A
On *Ceratodus forsteri*, with observations on the classification of fishes.
Proc. Zool. Soc. Lond., 1876, pp. 24-59, with 11 woodcuts in the text.

———   1876 B
On the evidence as to the origin of existing vertebrate animals.
Nature, xiii, pp. 388-389, 410-412, 429-430, 467-469, 514-516; xiv, pp. 33-34.
Abstracts of 6 lectures delivered at the Royal School of Mines.

———   1876 C
The direct evidence of evolution; (ii), the negative and favorable evidence.
Popular Sci. Monthly, lvi, pp. 207-223, 285-298, with illus.

———   1880 A
On the cranial and dental characters of the *Canidæ*.
Proc. Zool. Soc. Lond., 1880, pp. 238-288, with 16 woodcuts in the text.

———   1880 B
On the application of the laws of evolution to the arrangement of the *Vertebrata*, and more particularly of the *Mammalia*.
Proc. Zool. Soc. Lond., 1880, pp. 649-661.

Huxley, Thomas H.—Cont'd.   1883 A
On the oviducts of *Osmerus*, with remarks on the relations of the teleostean
with the ganoid fishes.
Proc. Zool. Soc. Lond., 1883, pp. 181–189, with
2 woodcuts.

———   1887 A
Further observations on *Hyperodapedon gordoni*.
Quart. Jour. Geol. Soc., xliii, pp. 675–694, with
pls. xxvi, xxvii.

Hyrtl, Joseph.   1853 A
Ueber normale Quertheilung der
Saurierwirbel.
Sitzungsber. Akad Wissensch. Wien, math.-
nat. Cl., x, pp. 185–192.

———   1854 A
Beitrag zur Anatomie von *Heterotis
ehrenbergii* C. V.
Denkschr. k. Akad. Wissensch. Wien,
math.-nat. Cl., viii, pp. 73–88, with pls. i–iii.
The writer makes comparisons of *Heterotis*
with *Osteoglossum*.

Ihering, H. von.   1891 A
Sobre la distribution geográfica de
los Creodontes.
Revista Argentina de Historia Natural, i,
pp. 209–213.

———   1891 B
Ueber die zoologisch-systematische
Bedeutung der Gehörorgane der Tele-
ostier.
Zeitschr. Wissensch. Zool., liii, pp. 477–514,
with pl. xxxi.
See also Sitzungsbericht Gesellsch. naturf.
Freunde Berlin, 1891, pp. 23–26.

Iwanzon, Nikolai.   1887 A
Der *Scaphirhynchus*. Vergleichend-
anatomische Beschreibung.
Bull. Soc. Imp. des Naturalistes de Moscou
(n. s.), i, pp. 1–41, with pls. i, ii.

Jaccard, Auguste.   See Pictet and Jaccard.

Jackson, C. T.   1851 A
Description of five new species of
fossil fishes.
Proc. Boston Soc. Nat. Hist., iv, pp. 138–142.
Describes *Palæoniscus alberti, P. brownii*, and
P. cairnsii; the others unnamed.

———   1851 B
Descriptions of the fossil fishes of the
Albert coal mine.
Pp. 22–25 of "Report on the Albert coal
mine," etc., Boston, 1851.
Plates not issued with the text and missing
in the copy examined.

Jackson, J. B. S.   1845 A
On the fossil bones of *Mastodon gi-
ganteus* from marl pit in New Jersey.
Proc. Boston Soc. Nat. Hist., ii, pp. 60–62.

Jaeger, Georg Friedr.   1828 A
Ueber die fossile Reptilien welche in
Württemberg aufgefunden worden sind.
Pp. i–vi, 1–48, with pls. i–vi. Stuttgart. 1828.

———   1844 A
Ueber die Stellung und Deutung der
Zähne des Wallrosses.
Müller's Archiv f. Anat., Physiol., etc., 1844,
pp. 70–75.

———   1851 A
Berichtigung einer Angabe Cuviers
über einen Narwhalschädel des Stutt-
gartes Naturalienkabinets, an welchen
beide Stosszähne aus den Zahnhöhlen
hervorragen sollen.
Jahreshefte Vereins vaterl. Naturkunde
Württembergs, vii, pp. 24–32, with 2 figs. on pl. i.

———   1857 A
Bemerkungen über die Veränder-
ungen der Zähne von Säugethieren im
Lauf ihrer Entwicklung, namentlich
(A) bei dem Narwhal (*Monodon monoceros*) und (B) dem Cachalot (*Physeter
macrocephalus*).
Bull. Soc. Imp. Naturalistes Moscou, xxx,
pt. ii, pp. 571–580, with 1 text fig.

Jaekel, O.   1884 A
Die Selachier aus dem oberen Muschelkalk Lothringens.
Abhandl. geol. Specialkarte von Elsass-Loth-
ringen, iii, pp. 275–332, with pls. vii–x.

———   1890 A
Ueber die systematische Stellung und
über fossile Reste der Gattung *Pristro-
phorus*.
Zeitschr. deutsch. geol. Gesellsch., xlii, pp.
86–120, with pls. ii–v, and 7 figs. in the text.
See also Jaekel, O., in Archiv. für Naturgeschichte, lvii, Bd. 1, pp. 15–48.

———   1890 B
Ueber *Phaneropleuron* und *Hemi-
ctenodus*, n. g.
Sitz.-Ber. Gesellsch. naturforsch. Freunde
Berlin, 1890, pp. 1–8, with 2 figs. in the text.

———   1890 C
Ueber die Kiemenstellung und die
Systematik der Selachier.
Sitz.-Ber. Gesellsch. naturforsch. Freunde
Berlin, 1890, pp. 47–57, with 8 figs. in the text.

———   1890 D
Bemerkungen über Flossenstacheln
oder Ichthyodoruliten im Allgemeinen.
Sitz.-Ber. Gesellsch. naturforsch. Freunde
Berlin, 1890, pp. 119–131, with 3 figs. in the text.

Jaekel, O.—Continued.   1891 A
Ueber *Coccosteus.*
Zeitschr. deutsch. geol. Gesellsch., xiii, pp.
773–774.

————   1891 B
Ueber *Menaspis armata* Ewald.
Sitz.-Ber. Gesellsch. naturforsch. Freunde
Berlin, 1891, pp. 115–131.
Discusses, among other matters, the systematic position of the Elasmobranchs.

————   1891 C
Referate über die in den letzten
Jahren erschienenen Arbeiten über
Pleuracanthiden.
Neues Jahrbuch Mineralogie, etc., 1891, ii,
pt. iii, pp. 161–170.
Herein are included reviews of a number of
writers on the genera *Pleuracanthus*, *Xenacanthus*, and *Orthacanthus*. These are Chas. Brongniart, L. Doederlein, E. Koken, and Anton
Fritsch.

————   1891 D
Ueber mikroskopische Untersuchungen im Gebiet der Paläontologie.
Neues Jahrbuch Mineralogie, etc., 1891, i,
Abhandl., pp. 178–196.

————   1892 A
Ueber *Cladodus* und seine Bedeutung
für die Phylogenie der Extremitäten.
Sitz.-Ber. Gesellsch. naturforsch. Freunde
Berlin, 1892, pp. 80–92.

————   1893 A
Ueber die Ruderorgane der Placodermen.
Sitz.-Ber. Gesellsch. naturforsch. Freunde
Berlin, 1893, pp. 178–181.

————   1893 B
[Review of Dr. J. S. Newberry's "The
Palæozoic fishes of North America."]
Neues Jahrbuch Mineral., etc., 1893, i, pp.
174–177.

————   1894 A
Die eocänen Selachier vom Monte
Bolca. Ein Beitrag zur Morphogenie
der Wirbelthiere.
Berlin, 1894, with 8 pls. and 39 text figs., pp.
1–176, 4 to.

————   1894 B
Ueber sogenannten Faltenzähne und
complicirtere Zahnbildungen überhaupt.
Sitz.-Ber. Gesellsch. naturforsch. Freunde
Berlin, 1894, pp. 146–153.

————   1895 A
Organisation der Pleuracanthiden.
Sitz.-Ber. Gesellsch. naturforsch. Freunde,
Berlin, 1895, pp. 69–85, with 2 figs. in the text.

Jaekel, O.—Continued.   1895 B
Ueber eine neue Gebissform fossiler
Selachier.
Sitz.-Ber. Gesellsh. naturforsch. Freunde
Berlin, 1895, pp. 200–202.
Refers to the dentition of *Janassa* and *Petalodus.*

————   1896 A
Ueber die Körperform und Hautbedeckung von Stegocephalen.
Sitz.-Ber. Gesellsch. naturforsch. Freunde
Berlin, 1896, pp. 1–8.

————   1896 B
Die Stammform der Wirbelthiere.
Sitz.-Ber. Gesellsch. naturforsch. Freunde
Berlin, 1896, pp. 107–129.

————   1898 A
Ueber die verschiedenen Rochentypen.
Sitz.-Ber. Gesellsch. naturforsch. Freunde
Berlin, 1898, pp. 44–53.

————   1898 B
Ueber *Hybodus.*
Sitz.-Ber. Gesellsch. naturforsch. Freunde,
1898, pp. 135–146, with 3 figs. in the text.

————   1898 C
Verzeichniss der Selachier des Mainzer Oligocäns.
Sitz.-Ber. Gesellsch. naturforsch. Freunde
Berlin, 1898, pp. 161–169.

————   1899 A
Ueber die Organisation der Petalodonten.
Zeitschr. deutsch. geol. Gesellsch. li, pp.
258–298, with pls. xiv, xv, and 7 text figs.

————   1899 B
Ueber die Zusammensetzung des
Kiefers und Schultergürtels von *Acanthodes.*
Zeitschr. deutsch. geol. Gesellsch., li, Protokolle, pp. 56–60, with 2 text figs.
For more extended paper see Verhandl.
deutsch. zool. Gesellsch., 1899, pp. 249–258, with
2 text figs.

————   1900 A
Ueber die Reste von Edestiden und
die neue Gattung *Helicoprion.*
Neues Jahrb. Min., 1900, ii, 144–148.
A review of Karpinsky, A. 1899 A.

James, Joseph F.   1891 A
The Biological Society of Washington.
Amer. Naturalist, xxv, pp. 298–299.
Contains report of remarks by Profs. H. F.
Osborne and O. C. Marsh on Cretaceous mammals.

Jefferson, Thomas.   1799 A
A memoir on the discovery of certain
bones of a quadruped of the clawed
kind in the western parts of Virginia.
Trans. Amer. Philos. Society, iv, pp. 246–260.
The first account of the bones of *Megalonyx*.
This work is said by C. G. Giebel to have been
issued in 1797.

Jeffries, J. A.   1881 A
On the fingers of birds.
Bull. Nuttall Ornithological Club, vi, pp.
6–11.

————   1882 A
On the claws and spurs on birds'
wings.
Proc. Bost. Soc. Nat. Hist., xxi, pp. 301–306.

————   1883 A
Sternal processes in *Gallinæ*.
Science, ii, p. 622.

————   1883 B
Osteology of the cormorant.
Science, ii, p. 739.

————   1884 A
Osteology of the cormorant.
Science, iii, p. 59.

Johnston, Christopher.   1859 A
Note on Odontology.
Amer. Jour. Dental Science, ix, pp. 337–343.
Gives the name *Astrodon* to certain teeth
found by him at Bladensburg, Maryland. No
description is furnished.

Jordan, D. S.   1891 A
Relations of temperature to vertebræ
among fishes.
Proc. U. S. Nat. Mus., xiv, pp. 107–120.

Jordan D. S., and **Evermann,** B. W.
                    1896 A
The fishes of North and Middle
America. A descriptive catalogue of
the species of fish-like vertebrates found
in the waters of North America, north
of the Isthmus of Panama.
Washington: Government Printing Office,
1896, pt. i, pp. i–lx, 1–1240.
Bull. U. S. Nat. Mus. No. 47.

————   1898 A
The fishes of North America. A de-
scriptive catalogue of the species of fish-
like vertebrates found in the waters of
North America, north of the Isthmus of
Panama.
Part ii, pp. i–xxx, 1241–2183; part iii, pp.
i–xxiv, 2184–3136.

————   1900 A
The fishes of North and Middle
America, etc.
Part iv, pp. i–ci, 3137–3313, with pls. i–cccxcii.

Jourdan, ————   1861 A
Description de restes fossiles de deux
grands mammifères constituant le genre
Rhizoprion (ordre des cétacés, groupe des
Delphinoïdes) et le genre *Dinocyon* (or-
dre des carnassiers), famille des canidés.
Ann. Sci. Naturelles (4), xvi, pp. 367–374,
pl. x.

Karpinsky, A.   1899 A
Ueber die Reste von Edestiden und
die neue Gattung *Helicoprion*.
Verhandl. k. russ. mineralog. Gesellsch., St.
Petersburg (2), xxxvi, No. 2.
In Russian; quoted from Jaekel, 1900 A.

Kaup, J.   1832 A
Ueber zwey Fragmente eines Unter-
kiefers von *Mastodon angustidens* Cuv.,
nach welchen diese Art in die Gattung
Tetracaulodon gehört.
Oken's Isis, xxv, cols. 627–630.

————   1832 B
Ueber *Rhinoceros incisivus* Cuv., und
eine neue Art, *Rhinoceros schleiermacheri*.
Oken's Isis, 1832, cols. 898–904.
Establishes genus *Aceratherium*.

————   1833 A
[Letter addressed to von Leonhard.] .
Neues Jahrbuch Mineral, etc., 1833, p. 327.
Establishes the genus *Hippotherium*, with
Equus gracilis as type.

————   1835 A
Die zwei urweltlichen pferdeartigen
Thiere welche im tertiären Sande bei
Eppelsheim gefunden werden, bilden
eine eigene Unter-Abtheilung der Gattung Pferd, welche in der Zahl der
Fingergliede den Uebergang zur Gattung *Palæotherium* macht und zwischen
diese und Pferd zu stellen ist.
Nova Acta Caes. Leop.-Carol. Akad. Naturforsch., xvii, pp. 173–182, with pl. xii B.

————   1843 A
Bemerkungen über die drei Arten
Mastodon und die drei Arten *Tetracau-
lodon* des Hon. Isaak Hays.
Archiv. Naturgesch., ix, i, pp. 168–175.
See Hays, I., Proc. Amer. Phil. Soc., 1843,
vol. ii, pp. 264–266.

Kehner, F. A.   1876 A
Zur Phylogenie des Beckens.
Verhandl. des naturhist.-medicin. Vereins
zu Heidelberg, v, pp. 346–359; pls. vii, viii.

Kehrer, Gustav.   1886 A
Beiträge zur Kenntniss des Carpus
und Tarsus der Amphibien, Reptilien
und Säuger.
Berichte naturforsch. Gesellsch. zu Freiberg,
Berichte i, pp. 73–88, with pl. iv.

Kersten, Hermann. 1821 A

Capitis *Trichechi rosmari* descriptio osteologica. Dissertatio inauguralis.

Berolini, 8 vo., pp. 5-24, with pls. i-iii.

Keyes, Charles R. 1888 A

On the fauna of the lower Coal measures of central Iowa.

Proc. Acad. Nat. Sci. Phila., 1888, pp. 222-246, with pl. xii.

Notes occurrence of *Petrodus occidentalis* and an unidentified species of *Diplodus*.

——— 1891 A

Fossil faunas in central Iowa.

Proc. Acad. Nat. Sci. Phila., 1891, pp. 242-265. Notes occurrence of *Thrinacodus duplicatus*? and *Deltodus intermedius*.

——— 1894 A

Palæontology of Missouri (part ii).

Missouri Geol. Survey, v, pp. 1-266, with pls. xxxiii-lvi.

The vertebrates are catalogued in chapter xv, pp. 229-239. The synonymy is given, but no descriptions or figures, and no species are mentioned, except the fishes. A list of the fossil vertebrates of Missouri was published in 1890 by G. Hambach (which see), and that contained, with rare omissions, the species which are found in this list of Keyes; also some which are not included in Keyes's list.

Kindle, Edward M. 1898 A

A catalogue of the fossils of Indiana, accompanied by a bibliography of the literature relating to them.

22d Ann. Report Depart. Geol. and Natural Resources [of Indiana] for 1897, pp. 407-514.

The list of fossil vertebrates is found on pp. 484-485.

King, Alfred T. 1844 A

Description of fossil footmarks supposed to be referable to the classes birds, *Reptilia*, and *Mammalia* found in the Carboniferous series in Westmoreland County, Pa.

Proc. Acad. Nat. Sci. Phila., ii, pp. 175-180, and 7 woodcuts.

——— 1845 A

Description of fossil footmarks found in the Carboniferous series in Westmoreland County, Pa.

Amer. Jour. Sci., xlviii, pp. 343-352, with 12 figs.

——— 1845 B

Description of fossil footprints.

Proc. Acad. Nat. Sci. Phila., ii, pp. 299-300. No names are given to the footprints.

——— 1845 C

Footprints.

Amer. Jour. Sci., xlix, pp. 216-217, with figs. Figures the footprints described in King, A. T. 1845 A.

King, Alfred T.—Continued. 1846 A

Footprints in the coal rocks of Westmoreland County, Pa.

Amer. Jour. Sci. (2), i, p. 268, with 2 text figs.

No names are given to these tracks. Lyell regards them as artificial. (Lyell, C. 1846 A. p. 25.)

King, Clarence. 1878 A

Systematic geology.

U. S. Geol. Explor. 40th parallel, Clarence King, geologist in charge. i.

In this volume are found several lists of the fossil vertebrates which characterize the Tertiary formations. No descriptions of the remains are given. The lists are found on the following pages: Jurassic, p. 346; Wasatch, p. 376-377; Bridger, p. 403-405; Uinta, p. 407; White R., p. 411; John Day, p. 424; Loup Fork, p. 430; Equus beds, p. 443. These lists were no doubt furnished by Prof. O. C. Marsh.

King, William. 1850 A

A monograph of the Permian fossils of England. London, 1850. Printed for the Palæontographical Society.

Pp. i-xxxviii; 1-258, with pls. i-xxviii.

Kingsley, J. S., and Ruddick, W. H. 1900 A

The ossicula auditus and mammalian ancestry.

Amer. Naturalist, xxxiii, pp. 219-230, with 3 text figs.

Kiprijanoff, Valerian. 1852 A

Fisch-Ueberreste im Kurskschen eisenhaltigen Sandsteine. I, II.

Bull. Soc. Imp. Naturalistes de Moscou, xxv, pt. 2, pp. 221-226; 483-495, with pls. xii, xiii.

——— 1853 A

Fisch-Ueberreste im Kurskchen eisenhaltigen Sandsteine.

Bull. Soc. Imp. Naturalistes de Moscou, xxvi, pt. 2, pp. 286-294, with pl. ii.

——— 1854 A

Fisch-Ueberreste im Kurskschen eisenhaltigen Sandsteine. V.

Bull. Soc. Imp. Naturalistes de Moscou, xxvii, pt. 2, pp. 273-397, with pls. ii, iii.

——— 1881 A

Studien über die fossilen Reptilien Russlands. Th. I. Gattung *Ichthyosaurus* Konig, aus dem Severischen Sandstein oder Osteolith der Kreidegruppe.

Mém. Acad. Sci. St. Pétersbourg (7), xxviii, No. 8, pp. 1-103, with pls. i-xviii.

A considerable portion of this paper and a number of the plates are devoted to the microscopic structure of the bones of *Ichthyosaurus*.

——— 1882 A

Studien über die fossilen Reptilien Russlands. Theil II, Gattung *Plesio-*

Kiprijanoff, Valerian—Continued.
saurus Conybeare, aus dem Severischen Sandstein oder Osteolith der Kreidegruppe.

Mém. Acad. Imp. Sci. St. Pétersbourg (7), xxx, No. 6, pp. 1–55, with pls. i–xix.

Pp. 34–47, with pls. xv–xix, are devoted to the microscopical structure of the bones of species of *Plesiosaurus*.

Kittary, Modeste. 1850 A
Recherches anatomiques sur les poissons du genre *Acipenser*.

Bull. Soc. Imp. Naturalistes Moscou, xxiii, pt. 2, pp. 389–445, with pls. vi, vii.

Kittl, Ernst. 1887 A
Beiträge zur Kenntniss der fossilen Säugethiere von Maragha in Persien. I. Carnivoren.

Annalen des k. k. naturhistorischen Hofmuseums, Wien, ii, pp. 317–388, with pls. xiv–xviii.

Klaatsch, H. 1890 A
Zur Morphologie der Fischschuppen und zur Geschichte der Hartsubstanzgewebe. I. Die Schuppen der Selachier. II. Die Rhombenschuppen der Ganoiden. III. Die Schuppen der Teliostier. IV. Die Cycloidschuppen der Dipnoer und fossiler Ganoiden. Die Phylogenese der "Cycloidschuppen." V. Die Schuppen von *Ichthyophis*. VI. Die Ableitung der Schuppen. VII. Die Geschichte der Hartsubstanzgewebe.

Morpholog. Jahrbuch, xvi, pp. 97–203; 209–258, with pls. vi–viii.

Appended are references to the works of 48 authors who have treated the subject.

——— 1893 A
Ueber die Wirbelsäule der Dipnoer.

Verhandl. anatom. Gesellschaft, 7ten Versamml., 1893, pp. 130–132.

Discusses the structure of the vertebral column of *Ceratodus* and *Protopterus*, followed by remarks by Hatschek.

——— 1894 A
Ueber die Herkunft der Scleroblasten.

Morphologisches Jahrbuch, xxi, pp. 153–240, with pls. v–ix and 6 figs. in text.

A contribution to knowledge of the genesis of bone.

——— 1896 A
Die Brustflosse der Crossopterygier. Ein Beitrag zur Anwendung der Ar-

Klaatsch, A.—Continued.
chipterygium-Theorie auf die Gliedmassen der Landwirbelthiere.

Festschrift zum 70sten Geburtstage von Carl Gegenbaur, i, pp. 261–391, with pls. i–iv, and 42 text figs.

Appended is a list of 75 papers pertaining to the subject discussed.

Klein, E. F. 1863 A
Beiträge zur Osteologie der Crocodilschädel.

Jahreshefte Vereins vaterl. Naturkunde in Württemberg, xix, pp. 70–100.

——— 1879 A
Beiträge zur Osteologie des Schädels der Knochenfische.

Jahreshefte Vereins vaterl. Naturkunde in Württemberg, xxxv, pp. 66–126, with pl. i.

——— 1881 A
Beiträge zur Osteologie der Fische.

Jahreshefte Vereins vaterl. Naturkunde in Württemberg, xxxvii, pp. 325–360, with pl. ii.

——— 1885 A
Beiträge zur Bildung des Schädels der Knochenfische, ii.

Jahreshefte Vereins vaterl. Naturkunde in Württemberg, xli, pp. 107–261, with pls. ii, iii.

——— 1886 A
Beiträge zur Bildung des Schädels der Knochenfische, iii.

Jahreshefte Vereins vaterl. Naturkunde in Württemberg, xlii, pp. 205–300, with pls. vii, viii.

Klever, Ernst. 1889 A
Zur Kenntniss der Morphogenese des Equidengebisses.

Morpholog. Jahrbuch, xv, pp. 308–330, with pls. xi–xiii.

Appended is a list of the works of 42 authors who have treated the subject of the dentition of the *Equidæ*.

Klippart, John H. 1875 A
Discovery of *Dicotyles* (*Platygonus*) *compressus*, Le Conte.

Proc. Amer. Assoc. Adv. Sci., 23d meeting, Hartford, 1874, pp. 1–6.

Also in Cincinnati Quarterly Journal of Science, 1875, vol. ii, pp. 1–6.

——— 1875 B
Mastodon remains in Ohio.

Cincinnati Quart. Jour. Sci., ii, pp. 151–155.

Klunzinger, C. B. 1871 A
Synopsis der Fische des Rothen Meeres. II Theil.

Verhandl. zool.-botan. Gesellsch. in Wien, xxi, pp. 441–688.

Kneeland, S.   1850 A
The *Manatus* not a cetacean but a
pachyderm.
Proc. Amer. Assoc. Adv. Sci., 3d meeting,
Charleston, 1850, pp. 42–47.

Kner, Rudolph.   1847 A
Ueber die beiden Arten *Cephalaspis
lloydii* und *lewissii*, Agassiz.
Haidinger's Naturwiss. Abhandl. i, pp.
159–168, with pl. v.

———   1860 A
Zur Charakteristik und Systematik
der Labroiden.
Sitz.-Ber. k. Akad. Wissensch. Wien, math.-
nat. Cl., xl. pp. 41–57, with pls. i–ii.

———   1863 A
Ueber enige fossile Fische aus den
Kreide- und Tertiar-schichten von Co-
men und Podsused.
Sitz.-Ber. k. Akad. Wissensch. Wien, math.-
nat. Cl., xlviii, pp. 126–148, with pls. i–iii.

———   1866 A
Betrachtungen über die Ganoiden,
als natürliche Ordnung.
Sitz.-Ber. k. Akad. Wissensch. Wien, liv,
Abth. i, pp. 519–536.

———   1866 B
Die fossile Fische der Asphaltschiefer
von Seefeld in Tirol.
Sitz.-Ber. k. Akad. Wissensch. Wien, math.-
naturw. Cl., liv. Abth. i, pp. 303–334, with pls.
i–vi.
For "Nachtrag" i, see the Sitzungsber. lvi,
Abth. i, pp. 898–909; "Nachtrag" ii, pp. 909–913,
with pls. i–iii.

———   1866 C
Die Fische der bituminösen Schiefer
von Raibl in Kärnthen.
Sitz.-Ber. k. Akad. Wissensch. Wien, liii,
pp. 1–46, with pls. i–vi.

———   1867 A
Ueber *Orthacanthus decheni* Goldf.
oder *Xenacanthus decheni* Beyr.
Sitz.-Ber. k. Akad. Wissensch. Wien, math.-
naturw., Cl., lv. pp. 540–584, with pls. i–x.

———   1868 A
Ueber *Conchopoma gadiforme* nov.
gen. et spec. und *Acanthodes* aus dem
Rothliegenden (der untern Dyas) von
Lebach bei Saarbrücken in Rhein-
preussen.
Sitz.-Ber. k. Akad. Wissensch. Wien, math.-
naturw., Cl., lvii, pt. i, pp. 278–304, with pls.
i–viii.

Knight, ——   1835 A
[On fossil jaw of buffalo.]
Amer. Jour. Sci., xxvii, p. 166.
Opinion expressed on a jaw found at Cha-
tauqua Lake, N. Y.

Knight, W. C.   1895 A
A new Jurassic plesiosaur from Wyo-
ming.
Science (2), ii, p. 449.
Describes *Cimoliosaurus rex* Knight.

———   1898 A
Some new Jurassic vertebrates from
Wyoming.
Amer. Jour. Sci. (4), v, 186, with 2 figs. in
the text.
Describes *Ceratodus robustus* and *C. ameri-
canus.*

———   1898 B
Some new Jurassic vertebrates from
Wyoming.
Amer. Jour. Sci. (4), v, pp. 378–380, with plate
of 3 figs.
Describes and figures *Megalneusaurus rex*, a
new genus based on *Cimoliosaurus rex* Knight.

———   1899 A
The Nebraska Permian.
Jour. Geology, vii, pp. 357–374.

———   1900 A
Some new Jurassic vertebrates.
Amer. Jour. Sci. (4), x, pp. 115–119, with 1
page of figs.

Kober, J.   1882 A
Studien über *Talpa europæa.*
Verhandl. naturf. Gesellsch. in Basel, vii
Theil, pp. 62–119, with pls. i, ii.

———   1884 A
Studien über *Talpa europæa.*
Verhandl. naturf. Gesellsch. in Basel, vii
Theil, pp. 465–485, with pl. vii.

Koch, Albert.   1839 A
The mammoth (mastodon? Eds.).
Amer. Jour. Sci., xxxvi, pp. 198–200.
This article is for the most part an extract
from the *Phila. Presbyterian*, but Leidy (1869
A, p. 398) attributes the authorship to Dr. Koch.

———   1839 B
Remains of the mastodon in Mis-
souri.
Amer. Jour. Sci., xxxvii, pp. 191–192.

———   1840 A
A short description of fossil remains
found in the State of Missouri by the
author.
Written and published by Albert Koch,
proprietor of the St. Louis Museum, St. Louis,
Churchill & Stewart, printers, 1840, pp. 1–8,
with 1 pl.

Koch, Albert—Continued.   1842 A
On the genus *Tetracaulodon*.
Proc. Geol. Soc. Lond., iii, pp. 714–716.

——   1843 A
Description of the *Missourium theristocaulodon* (Koch), or Missouri leviathan (*Leviathan missouriensis*), together
with its supposed habits and Indian
traditions; also comparison on the
whale, crocodile, and missourium with
the leviathan as described in the 41st
chapter of the Book of Job.
Fifth ed., enlarged, Dublin, 1843, pp. 1–28,
with fig. on the front cover.
An edition of this was published as early as
1841.

——   1845 A
Die Riesenthiere der Urwelt, oder das
neuentdeckte *Missourium theristocaulodon*, und die Mastodonten im Allgemeinen und Besondern.
Berlin, 1845, pp. 99, pls. viii.
For review of this paper see Neues Jahrb.
1845, pp. 760–766.

——   1845 B
[Extract from a letter to Dr. Geinitz
on *Zygodon*.]
Neúes Jahrb. f. Mineralogie, 1845, p. 676,
with 3 figs. in the text.

——   1851 A
Entdeckung der Zeuglodonten-Reste.
Haidinger's Berichte Mitthell. Freunden
Naturwissensch. in Wien, vii, pp. 198–199,
203–204.

——   1851 B
Das Skelet des *Zeuglodon macrospondylus*.
Haidinger's Naturwissensch. Abhandl., iv,
and with its own pagination, pp. 58–64, with
pl. vii.

——   1857 A
Mastodon remains in the State of
Missouri, together with evidences of
the existence of man contemporane-
ously with the mastodon.
Trans. Acad. Sci. St. Louis, i, pp. 61–64.

——   1857 B
[Report of Dr. Koch on explorations
in Mississippi and Arkansas for the
purpose of obtaining bones of *Zeuglodon*.]
Trans. Acad. Nat. Sci. St. Louis, i, pp. 17–19.

Koch, Albert—Continued.   1858 A
On the bones of *Mastodon*.
Trans. Acad. Sci., St. Louis, i, pp. 116–117.
Remarks made at meeting of the Academy
and referring to Koch's discoveries.
Besides the papers above given which owe
their authorship to A. Koch the following are
herewith recorded, but have not been examined:
Beschreibung des *Missurian Theristo-
kaulodon* Koch, oder Missuri-Leviathan
(*Leviathan missuriensis*); die vermuthete
Lebensart desselben und indianische
Traditionen über die Ort, wo es ausge-
graben wurde; ferner Vergleichungen
des Wallfisches, des Krokodils und des
Missurium mit dem Leviathan, wie
solcher im 41 Capitel des Buches Hiob
beschrieben wird.
Zeitschrift f. vergleich. Erdkunde, Magde-
burg, 1845, iv, pp. 38–51.

——
Die Mastodontoiden der Urwelt.
With 8 plates. Berlin, 1845.

——
Kurze Beschreibung des Hydrarchos
harlani (Koch), eines riesenmässigen
Meeresgeheuers und dessen Entdeck-
ung in Alabama in Nord-Amerika im
Frühjahr 1845. Dresde[n].

Ueber die Gattung *Zeuglodon* Owen
(*Basilosaurus* Harlan, *Hydrarchus* Koch,
Dorudon Gibbes).
Uebersicht Schlesische Gesellschaft vaterl.
Cultur, 1850, pp. 59–60.

Kölliker, Albert.   1859 A
On the different types in the microscopic structure of the skeleton of osseous fishes.
Proc. Roy. Soc. London, 1859, ix, pp. 656–
668.
Issued in separate form, pp. 1–13.

——   1860 A
Ueber das Ende der Wirbelsäule der
Ganoiden und einige Teleostier.
Pp. 27, with 4 pls., Leipzig, 1860.

——   1860 B
On the structure of the chorda dorsalis of the Plagiostomes and some other
fishes, and on the relation of the proper
sheath to the development of the vertebræ.
Proc. Roy. Soc. Lond., x, pp. 214–222.

Kölliker, Albert—Continued.   1863 A
Weitere Beobachtungen über die Wirbel der Selachier, insbesondere über die Wirbel der *Lumnoidei*, nebst allgemeinen Bemerkungen über die Bildung der Wirbel der Plagiostomen.
Pp. 1–51, with pls. i–v; Frankfort-on-Main, 1864.

Koenen, A. von   1883 A
Beitrag zur Kenntniss der Placodermen des norddeutschen Oberdevon's.
Abhandl. k. Gesellsch. Wissensch. Göttingen, xxx, pp. 1–41, with pls. i–iv.

———   1890 A
Coccosteus decipiens.
Geol. Magazine (3), vii, p. 191.
Discusses question of the existence of a pectoral spine.

———   1895 A
Ueber einige Fischreste des norddeutschen und böhmischen Devons.
Abhandl. k. Gesellsch. Wissensch. Göttingen, xl, math.-phys. Cl., art. 2, pp. 1–37, with pls. i–v.

Koken, Ernst.   1883 A
Die Reptilien der norddeutschen unteren Kreide.
Zeitschr. deutsch. geol. Gesellsch., xxxv, pp. 735–827, with pls. xxiii–xxv.

———   1884 A
Ueber Fisch-Otolithen, insbesondere über diejenigen der norddeutschen Oligocän-Ablagerungen.
Zeitschr. deutsch. geol. Gesellsch., xxxvi, pp. 500–565, with pls. ix–xii.

———   1886 A
[Review of papers by L. Dollo and V. Lemoine.]
Neues Jahrb. Min. Geol. und Paläont., 1886, ii, pp. 289–292, Referate.

———   1887 A
Die Dinosaurier, Crocodiliden und Sauropterygier des norddeutschen Wealden.
Palæontolog. Abhandl., iii, pp. 309–419 (1–111), with 9 pls. and 30 figs. in the text.

———   1888 A
Neue Untersuchungen an tertiären Fisch-Otolithen.
Zeitschr. deutsch. geol. Gesellsch., xl, pp. 274–305, with pls. xvii–xix.
Describes and figures 23 species of otoliths of fishes from the Tertiary deposits of the Southern States of the United States.

Koken, Ernst—Continued.   1888 B
Thoracosaurus macrorhynchus, Bl., aus der Tuffkreide von Maastricht.
Zeitschr. deutsch. geol. Gesellsch., xl, pp. 754–773, with pl. xxxii.

———   1888 C
Eleutherocercus, ein neuer Glyptodont aus Uruguay.
Abhandl. k. Akad. Wissensch. Berlin, 1888, phys.-math. Cl., Abh. i, pp. 1–28, pls. i, ii.

———   1889 A
Ueber *Pleuracanthus* Ag. oder *Xenacanthus* Beyr.
Sitz.-Ber. Gesellsch. naturf. Freunde Berlin, 1889, pp. 77–94, with 6 figs. in the text.

———   1891 A
Neue Untersuchungen an tertiären Fisch-Otolithen, ii.
Zeitschr. deutsch. geol. Gesellsch., xliii, pp. 77–170, with pls. i–x, and 27 figs. in the text.
Description of species, a list of known species of Europe, and a discussion of the significance of the otoliths and of the organ of hearing for the natural classification of fishes.

———   1891 B
[Remarks on the *Physostomi.*]
Sitz.-Ber. Gesellsch. naturf. Freunde Berlin, 1891, pp. 26–28.
Deals especially on relationship of *Cyprinidæ* and *Siluridæ.*

———   1893 A
Beiträge zur Kenntniss der Gattung *Nothosaurus.*
Zeitschr. deutsch. geol. Gesellsch., xlv, pp. 337–377, with pls. vii–xi, and 12 figs. in the text.

———   1896 A
Die Reptilien des norddeutschen Wealden.
Palæontolog. Abhandl., vii (n. s. iii), pp. 117–126 (1–10) with 4 pls. and 1 fig. in the text.

Kollmann, J.   1888 A
Handskelett und Hyperdaktylie.
Verhandl. anatom. Gesellsch., 2ten Versamml., pp. 25–39, with plate of 12 figs.
A paper with the same title is found in Verhandl. Gesellsch. Basel, viii, 1890, pp. 604–626; pl. viii.

Komnek, L. See **Van Beneden** and **Komnek.**

Koninck, L. G. de.   1844 A
Description des animaux fossiles que se trouvent dans le terrain carbonifère de Belgique.
Pp. i–iv; 1–716, with atlas of 60 pls.
Contains (pp. 608–618) descriptions of a few fishes.

Koninck, L. G. de—Continued. 1878 A
Faune du Calcaire carbonifère de la
Belgique. Première partie, poissons et
genre nautile.
Annales Musée nat. Belgique, ii, pp. 9–76,
with pls. i–viii, and figs. in the text.
The pages referred to are devoted to descriptions of the fishes.

Kosmak, George William. 1895 A
Dermal armor of the sturgeon.
Jour. New York Microscop. Soc., xi, pp. 1–21,
with pls. xlvi–xlviii.

Köstlin, Otto. 1844 A
Der Bau des knöchernen Kopfes in
den vier Klassen der Wirbelthiere.
Stuttgart, 1844, 8 vo, pp. i–x; 1–506, with pls.
i–iv.

Kowalevsky, Woldemar. 1873 A
Monographie der Gattung *Anthra-cotherium* Cuv. und Versuch einer
natürlichen Classification der fossilen
Hufthiere.
Palæontographica, xxii, 131–346, with pls.
vii–xvii.

——— 1873 B
Sur l'*Anchitherium aurelianense* Cuv.
et sur l'histoire paléontologique des
chevaux.
Mém. Acad. Imp. Sci. St. Pétersbourg (7),
xx, No. 5, pp. i–iv; 1–73, with pls. i–iii.

——— 1874 A
On the osteology of the *Hyopotamidæ*.
Philosoph. Trans. Roy. Soc. Lond., clxiii,
pp. 19–94; pls. 35–40.

——— 1876 A
Osteologie des Genus *Entelodon* Aym.
Palæontographica, xxii, 415–460, with pls.
xxv–xxvii.

Kramberger, Dragutin. 1881 A
Studien über die Gattung *Saurocephalus* Harlan.
Jahrbuch k.-k. geol. Reichsanstalt Wien.,
xxxi, pp. 371–379.
Saurocephalus and related genera are placed
in family *Scopeloidei* and subfamily *Saurodontidæ*, and a new genus *Solenodon* is described.

Krause, W. 1881 A
Zum Sacralhirn der Stegosaurier.
Biolog. Centralbl., i, p. 461.

Krauss, Ferd. 1858 A
Beiträge zur Osteologie des surinamischen *Manatus*.
Archiv. f. Anat. Physiol. wissensch. Med.,
1858, pp. 390–425.

Krauss, Ferd.—Continued. 1862 A
Beiträge zur Osteologie des surinamischen *Manatus*.
Archiv f. Anat. Physiol., wissensch. Med.,
1862, pp. 415–427; pl. xiii.

——— 1870 A
Beiträge zur Osteologie von *Halicore*.
Archiv f. Anat. Physiol., wissensch. Med.,
1870, pp. 525–614.

Krefft, Gerard. 1870 A
Description of a gigantic amphibian
allied to the genus *Lepidosiren*, from the
Wide Bay district, Queensland.
Proc. Zool. Soc. Lond., 1870, pp. 221–224, with
3 figs.
Description of *Ceratodus forsteri*.

Kükenthal, Willy. 1888 A
Über die Hand der Cetaceen.
Anatom. Anzeiger, iii, pp. 638–646; 912–915,
with 6 figs. in the text.

——— 1889 A
Vergleichend-anatomische und entwicklungsgeschichtliche Untersuchungen an Walthieren.
Denkschr. med.-naturwiss. Gesellsch. zu
Jena., iii, pp. 1–200, with pls. i–xiii.
For conclusion see Kükenthal 1893 C.

——— 1890 A
Über die Hand der Cetaceen.
Anatom. Anzeiger v, pp. 44–52, with 8 figs.
in the text.

——— 1890 B
Über Reste eines Hautpanzers bei
Zahnwalen.
Anatom. Anzeiger, v, pp. 237–240.

——— 1890 C
Ueber die Anpassung von Säugethieren an das Leben im Wasser.
Spengel's Zoolog. Jahrbücher, Abth. f. Systematik, Geog. und Biol. d. Thiere, v. pp. 373–
399.
Translation in Ann. and Mag. Nat. Hist.
(6), vii, 1891, pp. 153–179.

——— 1891 A
Einige Bemerkungen über Säugethierbezahnung.
Anatom. Anzeiger, vi, pp. 364–370.

——— 1891 B
Das Gebiss von *Didelphys*.
Anatom. Anzeiger, vi, pp. 658–666, with 8
figs. in the text.
Abstract in Zoolog. Record for 1891, Mamm.,
p. 10.

——— 1891 C
Ueber die Anpassung von Säugethieren an das Leben im Wasser.
Zoolog. Jahrbücher, Systematik, etc., v, pp.
373–399.

Kükenthal, Willy—Continued.   1892 A
Ueber die Entstehung und Entwick-
lung des Säugetierstammes.
Biolog. Centralbl., xii, pp. 400-413.
A translation of this paper is found in Ann.
and Mag. Nat. Hist. (6), x, 1892, pp. 365-380.

———   1892 B
Ueber den Ursprung und die Ent-
wicklung der Säugetierzähne.
Jenaische Zeitschr., xxvi (xix), pp. 469-489.

———   1893 A
Zur Entwicklung des Handskelettes
des Krokodils.
Morpholog. Jahrbuch, xix, pp. 42-55, with
pl. ii.
The part of the Morpholog. Jahrbuch containing this paper was issued in October, 1892.

———   1893 B
Mittheilungen über den Carpus des
Weisswals. (Die Bildung des Hama-
tums und das Vorkommen von zwei
und drie Centralien.)
Morpholog. Jahrbuch, xix, pp. 56-64, with
pl. iii.

———   1893 C
Vergleichend-anatomische und ent-
wicklungsgeschichtliche   Untersuch-
ungen an Walthieren. II.
Denkschr. med.-naturwiss. Gesellsch. zu
Jena., iii, pp. 224-448, with pls. xiv-xxv.

———   1894 A
Entwicklungsgeschichtliche Untersuchungen am Pinnipediergebisse.
Jenaische Zeitschr., xxviii (xxi), pp. 76-118,
with pls. iii and iv.

———   1895 A
Ueber Rudimente von Hinterflossen
bei Embryonen von Walen.
Anatom. Anzeiger, x, pp. 534-537.

———   1895 B
Zur Dentitionenfrage.
Anatom. Anzeiger, x, pp. 658-659.
Treats of the origin of the mammalian den-
tition.

———   1896 A
Zur Entwicklungsgeschichte des
Gebisses von *Manatus*.
Anatom. Anzeiger, xii, pp. 513-526, with 10
figs. in the text.

———   1897 A
Vergleichend-anatomische und entwicklungsgeschichtliche Untersuchungen an Sirenen.
Semon's Zoolog. Forschungsreisen in Aus-
tralien, etc., iv, pp. 1-7b, with pls. i-v and 47
text figs.

Kunth, A.   1872 A
Ueber *Pteraspis*.
Zeitschr. deutsch. geol. Gesellsch., xxiv, pp.
1-8, with pl. i.

L. P. B.   1885 A
Professor Marsh's Monograph of the
Dinocerata.
Amer. Jour. Sci. (3), xxix, pp. 173-204, with
38 figs in the text.

Lahusen, I.   1880 A
Zur Kenntniss der Gattung *Bothrio-
lepis*, Eichw.
Verhandl. russ-kais. mineralog. Gesellsch.,
St. Petersburg (2), xv, pp. 125-139, with pls. i, ii.

Lambe, Lawrence M.   1898 A
On the remains of mammoths in the
museum of the Geological Survey Department.
Ottawa Naturalist, xii, pp. 136-137.

———   1899 A
On reptilian remains from the Cre-
taceous of northwestern Canada.
Ottawa Naturalist, xiii, pp. 68-70.
See also Summary report of Geol. Surv.
Dept. Canada for year 1898, pp. 182-190.
Account of fossil remains found in districts
of Alberta and Assiniboia.

Landois, ———.   1871 A
Das Gebiss eines jungen Mammuth.
Verhandl. naturhist. Vereines preuss. Rheinlande, etc., xxviii, Corr. Bl., pp. 47-49, with pl.
ix.

———   1881 A
Ueber die Reduktion der Zehen bei
den Säugethieren durch Verkümmer-
ung und Verschmelzung.
Verhandl. naturf. Vereines preuss. Rheinlande, etc., xxxviii, Corr. Bl., pp. 125-129.

Lang, Fr., and **Rütimeyer**, L.   1867 A
Die fossilen Schildkröten von Solo-
thurn.
Neue Denkschriften d. allgem. schweiz.
Gesellsch. f. d. gesammt. Naturwissensch., xxii,
art. 5, pp. 1-48, with pls. i, ii, consisting of pro-
files and sections and a geological map of the
region.

Langdon, F. W.   1883 A
The giant beaver (*Castoroides ohioensis*)
Foster.
Jour. Cincinnati Soc. Nat. Hist., vi, pp. 238-
239.

Lankester, E. R.   1862 A
Restoration of *Pteraspis*.
The Geologist, v, pp. 451-452, with 3 text
figs.

Lankester, E. R.—Continued. 1864 A

On the discovery of the scales of *Pteraspis*, with some remarks on the cephalic shield of that fish.

Quar. Jour. Geol. Soc., xx, pp. 194-197, with pl. xii.

——— 1864 B

Restoration of *Pteraspis*.

The Geologist, vii, p. 136.

——— 1865 A

On the genus *Pteraspis*.

Report Brit. Assoc. Adv. Sci., 34th meeting, Bath, 1864, p. 100. See also p. 58.

Divided into subgenera *Pteraspis, Cyathaspis,* and *Scaphaspis.*

——— 1870 A

On a new *Cephalaspis* discovered in America, etc.

Geol. Magazine, vii, pp. 397-398, with 3 woodcuts in the text.

Describes *C. dawsoni* and refers to the finding of *Machairacanthus sulcatus.*

——— 1873 A

On *Holaspis sericeus*, and on the relationship of the fish-genera, *Pteraspis, Cyathaspis,* and *Scaphaspis.*

Geol. Magazine, x, pp. 241-245, with pl. x.

——— 1873 B

On *Scaphaspis* and *Pteraspis*.

Geol. Magazine, x, pp. 190-192.

Extract from the Academy of Jan. 1, 1873, replying to Kunth.

——— 1873 C

Note on *Holaspis sericeus*.

Geol. Magazine (1), x, pp. 331-332, with 1 woodcut.

——— 1874 A

Magister Schmidt on the shields of *Pteraspis* and *Scaphaspis.*

Geol. Magazine (2), i, p. 288.

See also **Moseley** and **Lankester.**

Lapham, I. A. 1855 A

On the number of teeth of the *Mastodon giganteus.*

Proc. Bost. Soc. Nat. Hist., v, pp. 133-136.

Lartet, Éd. 1837 A

Fossiles provenant du depôt de Sansan.

Compte rendu Acad. Sci., Paris, v, pp. 418, 427.

——— 1866 A

Note sur deux nouveaux siréniens fossiles des terrains tertiaires du bassin de la Garonne.

Bull. Soc. Géol. France, xxiii, pp. 673-686, pl. xiii.

Lataste, Fernand. 1889 A

Considerations sur les deux dentitions des mammifères.

Jour. l'Anat. Physiol., xxv, pp. 200-222.

Numerous references are made to the writings of other authors.

Lathrop, S. P. 1851 A

Mastodon in northern Illinois.

Amer. Jour. Sci. (2), xii, p. 439.

Lawley, R. 1875 A

Monografia del genere *Notidanus.*

Pp. 1-34, with pls. i-iv, Firenze, 1875.

This paper appears to have been published also in Atti Soc. Toscano, iii, 1877.

Lazier and **Parieu.** 1839 A

Note sur la machoire d'un carnassier fossile, nommé Hyénodon *leptorhynchus.*

Ann. Sci. Naturelles, (2), xi, pp. 27-31, with pl. ii.

Lea, Isaac. 1849 A

On reptilian footmarks found in the gorge of Sharp Mountain, near Pottsville, Pa.

Proc. Amer. Phil. Soc., iv, pp. 91-94, with fig. in text.

Noticed in Amer. Jour. of Sci., vol. ix, pp. 124-125.

Describes *Sauropus primævus.*

——— 1850 A

On traces of a fossil reptile (*Sauropus primævus*) found in the Old Red sandstone.

Report Brit. Assoc. Adv. Sci., 19th meeting, Birmingham, 1849, p. 56.

——— 1851 A

Remarks on the bones of a fossil reptilian quadruped.

Proc. Acad. Nat. Sci. Phila., v, pp. 171-172.

Describes *Clepsysaurus pennsylvanicus*, but gives no generic name.

——— 1851 B

Remarks on *Clepsysaurus pennsylvanicus.*

Proc. Acad. Nat. Sci. Phila., v. p. 205.

The generic name is here first given.

——— 1852 A

On the fossil footmarks in the red sandstones of Pottsville, Schuylkill County, Pa.

Trans. Amer. Philos. Soc., x, pp. 307-317, pls. xxxi-xxxiii.

——— 1853 A

Description of a fossil saurian of the New Red sandstone formation of Pennsylvania.

Jour. Acad. Nat. Sci. Phila. (2), pp. 185-202, pls. xvii-xix.

Description of *Clepsysaurus pennsylvanicus.*

Lea, Isaac—Continued.   1853 B

On some new fossil molluscs in the Carboniferous slates of the anthracite seams of the Wilkesbarre coal formation.

Jour. Acad. Nat. Sci. Phila. (2), ii, pp. 203–206, with pl. xx.

Contains, on p. 206, with pl. xx, figs. 4, 5, a description of *Palæoniscus ? leidyiana*.

———   1855 A

Fossil footmarks in the red sandstone of Pottsville, Pa.

Philadelphia, 1855, pp. 1–16, with 1 pl. Elephant folio.

Contains plate showing tracks of *Sauropus primævus* of the natural size.

———   1856 A

[Description of *Centemodon sulcatus*.]

Proc. Acad. Nat. Sci. Phila., 1856, pp. 77–78.

———   1856 B

Reptilian remains in the New Red sandstone of Pennsylvania.

Amer. Jour. Sci. (2), xxii, pp. 122–124.

Repeated on pp. 422–423, same volume. From Proc. Acad. Nat. Sci. Phila., viii, 1856, p. 77.

———   1858 A

[Remarks respecting *Clepsysaurus* and *Bathygnathus borealis*.]

Proc. Acad. Nat. Sci. Phila., 1858, pp. 90–92.

Leboucq, H.   1887 A

La nageoire pectorale des Cétacés au point du vue phylogénique.

Anatom. Anzeiger, ii, pp. 202–208.

———   1888 A

Über das Fingerskelett der Pinnipedier und der Cetaceen.

Anatom. Anzeiger, iii, pp. 530–534.

Also in Verhandl. Anat. Gesellschaft, 1888, pp. 40–44.

———   1896 A

Ueber Hyperphalangie bei den Säugetieren.

Anatom. Anzeiger, xii, Ergänzungsheft, pp. 174–176.

Leche, W.   1877 A

Studien über das Milchgebiss und die Zahnhomologien bei den *Chiroptera*.

Archiv Naturgesch., xliii, i, pp. 353–364.

———   1880 A

Zur Morphologie der Beckenregion bei *Insectivora*.

Morpholog. Jahrbuch, vi, pp. 597–602.

———   1893 A

Studien über die Entwicklung des Zahnsystems bei den Säugethieren.

Morpholog. Jahrbuch, xix, pp. 502–547, with 20 text figs.

Abstract in Zoological Record for 1893, Mamm., p. 8.

Leche, W.—Continued.   1893 B

Nachträge zü Studien über die Entwicklung des Zahnsystems bei der Säugethieren.

Morpholog. Jahrbuch, xx, pp. 113–142, with 12 figs. in the text.

———   1895 A

Zur Entwicklungsgeschichte des Zahnsystems der Säugethiere, zugleich ein Beitrag zur Stammesgeschichte dieser Thiergruppe. I. Theil. Ontogonie.

Bibliotheca Zoologica, Heft 17, pp. 1–160, with 19 pls. and 20 text figs. Stuttgart, 1895.

Pp. 157–160 of this important work are occupied by a list of the authorities which the author has cited.

Reviewed by Max Fürbringer, Morpholog. Jahrbuch, vol. xxiii, p. 592.

———   1896 A

Zur Entwicklungsgeschichte des Zahnsystems der Säugethiere, zugleich ein Beitrag zur Stammesgeschichte dieser Gruppe.

Biolog. Centralbl., xvi, pp. 283–296.

———   1896 B

Die Entwicklung des Zahnsystems der Säugethiere.

Compte rendu des séances du troisième Congres Internat. de Zoologie. Leyden, 1895 pp. 279–289.

———   1897 A

Zur Morphologie des Zahnsystems der Insectivoren.

Anatom. Anzeiger, xiii, pp. 1–11, with 10 figs. in the text.

———   1897 B

Zur Morphologie des Zahnsystems der Insectivoren.

Anatom. Anzeiger, xiii, pp. 513–529, with 7 figs. in the text.

———   1897 C

Untersuchungen über das Zahnsystem lebender und fossiler Halbaffen.

Festschrift zum 70sten Geburtstage von Carl Gegenbaur, iii, pp. 127–166, with pl. i, and 20 figs. in the text.

Le Conte, John L.   1848 A

Notice of five new species of fossil *Mammalia* from Illinois.

Amer. Jour. Sci. (2), v, pp. 102–106, with 3 text figs.

———   1848 B

On *Platygonus compressus;* a new fossil pachyderm.

Mem. Amer. Acad. Arts and Sciences, iii, pp. 257–274, with pls. i–iv.

———   1852 A

Notes on some suilline pachyderms from Illinois.

Proc. Acad. Nat. Sci. Phila., vi, pp. 3–5.

Le Conte, John L.—Continued. 1852 B
Notice of a fossil *Dicotyles* from Missouri.
Proc. Acad. Nat. Sci. Phila., vi, pp. 5–6.
Describes *D. costatus* LeC.

——— 1852 C
[Remarks on *Castoroides ohioensis* from Shawneetown, Illinois.]
Proc. Acad. Nat. Sci. Phila., vi, p. 53.

——— 1852 D
Remarks on some fossil pachyderms from Illinois.
Proc. Acad. Nat. Sci. Phila., vi, pp. 56–57.

Le Conte, Joseph, 1883 A
Carson footprints.
Nature, xxviii, pp. 101–102.

Lee, W. T. 1897 A
[A Mosasauroid from Colorado.]
Amer. Naturalist, xxxi, p. 614.
Said to belong probably to *Clidastes;* no specific name assigned and only a general description.

Lefèvre, Th. 1889 A
Note préliminaire sur les restes de Siréniens recueillis en Belgique.
Zoolog. Anzeiger, xii, pp. 197–200.

Leidy, Joseph. 1847 A
On the fossil horse of America.
Proc. Acad. Nat. Sci. Phila., iii, pp. 262–266, pl. ii.
Describes *Equus americanus.*

——— 1847 B
On a new genus and species of fossil *Ruminantia: Poëbrotherium wilsoni.*
Proc. Acad. Nat. Sci. Phila., iii, pp. 322–326.

——— 1847 C
[Observations on *Equus americanus.*]
Proc. Acad. Nat. Sci. Phila., iii, p. 328.

——— 1848 A
On a new genus and species of fossil *Ruminantia: Poëbrotherium wilsoni.*
Amer. Jour. Sci. (2), v, pp. 276–279.
Extracted from Proc. Phil. Acad., Nov. 1847, p. 322. Reprinted in Ann. and Mag. Nat. Hist. (2) ii, pp. 389–392.

——— 1848 B
On a new fossil genus and species of ruminantoid *Pachydermata: Merycoidodon culbertsonii.*
Proc. Acad. Nat. Sci. Phila., iv, pp. 47–50, with 1 plate of 5 figs.

——— 1849 A
Tapirus americanus fossilis.
Proc. Acad. Nat. Sci. Phila., iv, pp. 180–182.

——— 1850 A
Observations on two new genera of fossil *Mammalia, Eucrotophus jacksoni,* and *Archæotherium mortoni.*
Proc. Acad. Nat. Sci. Phila., v, pp. 90–93.

Leidy, Joseph—Continued. 1850 B
Remarks on *Rhinoceros occidentalis.*
Proc. Acad. Nat. Sci. Phila. v, p. 119.
Contains no description.

——— 1850 C
Descriptions of *Rhinoceros nebrascensis, Agriochærus antiquus, Palæotherium proutii,* and *P. bairdii.*
Proc. Acad. Nat. Sci. Phila., v, pp. 121–122. ·

——— 1851 A
Remarks on *Palæotherium proutii.*
Proc. Acad. Nat. Sci. Phila., v, pp. 170–171.

——— 1851 B
[Description of *Stylemys nebrascensis.*]
Proc. Acad. Nat. Sci. Phila., v, pp. 172–173.

——— 1951 C
Descriptions of *Testudo lata* and *Emys hemispherica.*
Proc. Acad. Nat. Sci. Phila., v, p. 173.

——— 1851 D
Descriptions of fossil ruminant ungulates from Nebraska.
Proc. Acad. Nat. Sci. Phila., v, pp. 237–239.
Describes *Oreodon priscum* and *Cotylops speciosa.*

——— 1851 E
[Remarks on *Oreodon priscus* and *Rhinoceros occidentalis.*]
Proc. Acad. Nat. Sci. Phila., v, p. 276.

——— 1851 F
[Descriptions of two fossil species of *Balæna, B. palæatlantica* and *B. prisca.*]
Proc. Acad. Nat. Sci. Phila., v, pp. 308–309.

——— 1851 G
[Descriptions of a number of fossil reptiles and mammals.]
Proc. Acad. Nat. Sci. Phila., v, pp. 325–328·

——— 1851 H
[Descriptions of fossils from the Green-sand of New Jersey.]
Proc. Acad. Nat. Sci. Phila., v, pp. 329–330.

——— 1851 I
[Remarks on *Aceratherium.*]
Proc. Acad. Nat. Sci. Phila., v, p. 331.

——— 1851 J
[Description of the genus *Arctodon.*]
Proc. Acad. Nat. Sci. Phila., v, p. 278.
No specific name is indicated for the teeth described.

——— 1851 K
Fossil bones from Big Bone Lick, Kentucky.
Proc. Acad. Nat. Sci. Phila., v, p. 140.

——— 1852 A
[Description of a new rhinoceros from Nebraska, *R. americanus.*]
Proc. Acad. Nat. Sci. Phila., vi, p. 2.

Leidy, Joseph—Continued.   1852 B
[Descriptions of *Delphinus conradi* and *Thoracosaurus grandis.*]
Proc. Acad. Nat. Sci. Phila., vi, p. 35.

———   1852 C
[Description of *Pontogeneus priscus.*]
Proc. Acad. Nat. Sci. Phila., vi, p. 52.

———   1852 D
[Description of *Emys culbertsoni.*]
Proc. Acad. Nat. Sci. Phila., vi, p. 34.

———   1852 E
[Remarks on fossil tortoises from Nebraska.]
Proc. Acad. Nat. Sci. Phila., vi, p. 59.

———   1852 F
[Remarks on two crania of extinct species of ox.]
Proc. Acad. Nat. Sci. Phila., vi, p. 71.

———   1852 G
[*Tapirus haysii* named.]
Proc. Acad. Nat. Sci. Phila., vi, p. 106.
Name given to tooth presented by Dr. Hays.

———   1852 H
[Remarks on fossil ox and on extinct *Edentata.*]
Proc. Acad. Nat. Sci. Phila., vi, p. 117.

———   1852 I
[Remarks on *Tapirus haysii.*]   ●
Proc. Acad. Nat. Sci. Phila., vi, p. 148.

———   1852 J
Description of the remains of extinct *Mammalia* and *Chelonia* from Nebraska Territory, collected during the geological survey under the direction of Dr. D. D. Owen.
David D. Owen's "Report of a geological survey of Wisconsin, Iowa, and Minnesota and incidentally a portion of Nebraska Territory," pp. 534-572, pls. ix-xv.

———   1852 K
Description of a new species of crocodile from the Miocene of Virginia.
Jour. Acad. Nat. Sci. Phila. (2), ii, pp. 135-138, pls. xvi.
Description of *Crocodilus antiquus* given.

———   1852 L
Report upon some fossil *Mammalia* and *Chelonia* from Nebraska.
Sixth Annual Report Regents Smithsonian Institution, for 1851, pp. 63-65.

———   1853 A
Memoir on the extinct species of American ox.
Smithsonian Contributions to Knowledge, v, art. 3, pp. 1-20, pls. i-v.
This memoir is said to have been published in Dec., 1852.

Leidy, Joseph—Continued.   1853 B
[Description of *Ursus amplidens.*]
Proc. Acad. Nat. Sci. Phila., vi, p. 303.

———   1853 C
[Observations on extinct *Cetacea.*]
Proc. Acad. Nat. Sci. Phila., vi, pp. 377-378.

———   1853 D
[Remarks on a collection of fossil *Mammalia* from Nebraska.]
Proc. Acad. Nat. Sci. Phila., vi, pp. 392-394.

———   1853 E
Remarks on several fossil teeth.
Proc. Acad. Nat. Sci. Phila., vi, p. 241.
This abstract of remarks contains as names of new species, *Hipparion venustum, Oromys sxopi, Eubradys antiquus,* and *Ereptodon priscus.* No descriptions accompany these names.

———   1853 F
Description of an extinct species of American lion, *Felis atrox.*
Trans. Amer. Philos. Soc., x, pp. 319-321, pl. xxxiv.
This species appears to be first mentioned, without description, in Proc. Amer. Phil. Soc., v. 261, 1853.

———   1853 G
A memoir on the extinct *Dicotylinæ* of America.
Trans. Amer. Philos. Soc., x, pp. 323-343, pls. xxxv-xxxviii.

———   1854 A
The ancient fauna of Nebraska, or a description of remains of extinct *Mammalia* and *Chelonia* from the Mauvais Terres of Nebraska.
Smithsonian Contributions to Knowledge, vi, art. vii, pp. 1-126, pls. i-xxiv.

———   1854 B
Remarks on extinct saurians from Arkansas.
Proc. Acad. Nat. Sci. Phila., vii, p. 72, with pl. ii.
Describes *Brimosaurus grandis.*

———   1854 C
[Remarks on *Sus americanus,* or *Harlanus americanus,* and on other extinct mammals.]
Proc. Acad. Nat. Sci. Phila., vii, pp. 89-90.

———   1854 D
Remarks on a new species of mammal from Nebraska, *Dinictis felina.*
Proc. Acad. Nat. Sci. Phila., vii, p. 127.
Insufficient description.

———   1854 E
Synopsis of extinct *Mammalia,* the remains of which have been discovered in the Eocene formations of Nebraska.
Proc. Acad. Nat. Sci. Phila., vii, pp. 156-157.

Leidy, Joseph—Continued.   1854 F

Description of a fossil apparently indicating an extinct species of the camel tribe.

Proc. Acad. Nat. Sci. Phila., vii, pp. 172–173.

Describes *Camelops kansanus*.

——   1854 G

Remarks on the question of the identity of *Bootherium cavifrons* with *Ovibos moschatus* or *O. maximus*.

Proc. Acad. Nat. Sci. Phila., vii, pp. 209–210.

——   1854 H

On *Bathygnathus borealis*, an extinct saurian of the New Red sandstone of Prince Edward's Island.

Jour. Acad. Nat. Sci. Phila., (2), ii, pp. 327–330, pl. xxxiii.

——   1854 I

Notice of some fossil bones discovered by Mr: Francis A. Lincke, in the banks of the Ohio River, Indiana.

Proc. Acad. Nat. Sci. Phila., vii, pp. 199–201.

These bones were found below Evansville, Ind., and consisted of *Megalonyx jeffersonii*, *Bison americanus*, *Cervus virginianus*, *Equus americanus*, *Tapirus haysii* and *Canis primævus*.

——   1855 A

On *Bathygnathus borealis*, an extinct saurian of the New Red sandstone of Prince Edward's Island.

Amer. Jour. Sci. (2), xix, pp. 444–446.

Extracted from the Jour. of the Acad. of Nat. Sci. Phila., vol. ii, and same as Leidy, J. 1854 H.

——   1855 B

A memoir on the extinct sloth tribe of North America.

Smithsonian Contributions to Knowledge, vii, art. v, pp. 1–68, pls. i–xvi.

——   1855 C

Indications of twelve species of fossil fishes.

Proc. Acad. Nat. Sci. Phila., vii, pp. 395–397.

——   1855 D

Indications of five species, with two new genera, of extinct fishes.

Proc. Acad. Nat. Sci. Phila., vii, p. 414.

——   1856 A

Description of two ichthyodorulites.

Amer. Jour. Sci. (2), xxi, pp. 421–422.

From Proc. Acad. Nat. Sci. Phila., vol. viii, p. 11.

——   1856 B

Notice of some remains of extinct *Mammalia*, recently discovered by Dr.

Leidy, Joseph—Continued.

F. V. Hayden in the Bad Lands of Nebraska.

Amer. Jour. Sci. (2), xxi, pp. 422–428.

From Proc. Phil. Acad., viii, p. 59; same as Leidy, 1856 D.

——   1856 C

Notices of remains of extinct reptiles and fishes discovered by Dr. F. V. Hayden in the Bad Lands of the Judith River, Nebraska Territory.

Amer. Jour. Sci. (2), xxii, pp. 118–120.

Extracted from Proc. Phil. Acad. See Leidy, 1856 F.

——   1856 D

Notice of some remains of extinct *Mammalia*, recently discovered by Dr. F. V. Hayden in the bad lands of Nebraska.

Proc. Acad. Nat. Sci. Phila., viii, p. 59.

Also in Amer. Jour. Sci. (2), xxi, p. 422.

——   1856 E

Description of two ichthyodorulites.

Proc. Acad. Nat. Sci. Phila., viii, pp. 11–12.

Description of *Stenacanthus nitidus* and *Cylindracanthus ornatus*.

——   1856 F

Notice of remains of extinct reptiles and fishes, discovered by Dr. F. V. Hayden in the Bad Lands of the Judith River, Nebraska Territory.

Proc. Acad. Nat. Sci. Phila., viii, pp. 72–73.

——   1856 G

Notice of the remains of a species of extinct seal, from the Post-pliocene deposit of the Ottawa River.

Proc. Acad. Nat. Sci. Phila., viii, pp. 90–91, with pl. iii.

Reprinted in Canadian Naturalist and Geologist, i, p. 238, with pl., 1856.

——   1856 H

Notices of several genera of extinct *Mammalia*, previously less perfectly characterized.

Proc. Acad. Nat. Sci. Phila., viii, pp. 91–92.

——   1856 I

Notices of remains of extinct *Mammalia*, discovered by Dr. F. V. Hayden in Nebraska Territory.

Proc. Acad. Nat. Sci. Phila., viii, pp. 88–90.

——   1856 J

Notice of some remains of extinct vertebrated animals.

Proc. Acad. Nat. Sci. Phila., viii, pp. 163–165.

Leidy, Joseph—Continued   1856 K
Notices of remains of extinct vertebrate animals of New Jersey, collected
by Professor Cook, of the State geological survey, under the direction of Dr.
W. Kitchell.
Proc. Acad. Nat. Sci. Phila., viii, pp. 220-221.

——   1856 L
Notices of extinct vertebrated ani-
mals discovered by Prof. E. Emmons.
Proc. Acad. Nat. Sci. Phila., viii, pp. 255-256.
See also Amer. Jour. Sci., xxiii, 1857, pp. 271-
272.

——   1856 M
Notice of some remains of fishes dis-
covered by Dr. John E. Evans.
Proc. Acad. Nat. Sci. Phila., viii, pp. 256-257.

——   1856 N
Notice of remains of two species of
seals.
Proc. Acad. Nat. Sci. Phila., viii, p. 265.

——   1856 O
Remarks on certain extinct species of
fishes.
Proc. Acad. Nat. Sci. Phila., viii, pp. 301-302.

——   1856 P
Notices of remains of extinct turtles
of New Jersey, collected by Professor
Cook, of the State geological survey,
under the direction of Dr. W. Kitchell.
Proc. Acad. Nat. Sci. Phila., viii, pp. 303-304.

——   1856 Q
Notices of extinct Vertebrata discovered by Dr. F. V. Hayden, during the
expedition to the Sioux country under
the command of Lieut. G. K. Warren.
Proc. Acad. Nat. Sci. Phila., viii, pp. 311-312.

——   1856 R
[Remarks on extinct Dicotylinæ.]
Proc. Acad. Nat. Sci. Phila., 1856, p. 140.

——   1856 S
Descriptions of some remains of fishes
from the Carboniferous and Devonian
formations of the United States.
Jour. Acad. Nat. Sci. Phila. (2), iii, pp. 159-
160.

——   1856 T
Description of some remains of extinct Mammalia.
Jour. Acad. Nat. Sci. Phila. (2), iii, pp. 166-
171, pl. 17.
Describes Camelops kansanus, Canis prima-
vus, Ursus amplidens, Procyon priscus and Anomodon snyderi.

Leidy, Joseph—Continued.   1857 A
Notices of remains of extinct verte-
brated animals discovered by Prof. E.
Emmons.
Amer. Jour. Sci. (2), xxiii, pp. 271-272.
From Proc. Phila. Acad. Nat. Sci., viii, p.
255.

——   1857 B
[Remarks on Dromatherium, Clepsy-
saurus, Omosaurus, etc.]
Proc. Acad. Nat. Sci. Phila., ix, pp. 149-150.

——   1857 C
List of extinct Vertebrata, the remains
of which have been discovered in the
region of the Missouri River, with
remarks on their geological age.
Proc. Acad. Nat. Sci. Phila., ix, pp. 89-91.

——   1857 D
Rectification of the references of certain of the extinct mammalian genera
of Nebraska.
Proc. Acad. Nat. Sci. Phila., ix, pp. 175-176.

——   1857 E
[Remarks on Mosasaurus.]
Proc. Acad. Nat. Sci. Phila., ix, p. 176.

——   1857 F
Notices of some remains of extinct
fishes.
Proc. Acad. Nat. Sci. Phila., 1857, pp. 167-168.

——   1857 G
Notice of remains of the walrus dis-
covered on the coast of the United
States.
Trans. Amer. Philos. Soc., xi, pp. 83-86, with
pl. 4, figs. 1, 2, and pl. 5, fig. 1.

——   1857 H
Descriptions of the remains of fishes
from the Carboniferous limestone of
Illinois and Missouri.
Trans. Amer. Philos. Soc. xi, pp. 87-90, pl.
5, figs. 2-29.

——   1857 I
Remarks on Saurocephalus and its
allies.
Trans. Amer. Philos. Soc., xi, pp. 91-95, with
pl. vi, figs. 8-15.

——   1857 J
Observations on the extinct peccary
of North America; being a sequel to
"A memoir on the extinct Dicotylinæ
of America."
Trans. Amer. Philos. Soc., xi, pp. 97-105, with
pl. vi, figs. 2-7.

Leidy, Joseph—Continued.　　1857 K
　Remarks on the structure of the feet
　of *Megalonyx.*
　Trans. Amer. Philos. Soc., xi, pp. 107-108, pl.
　vi, fig. 1.

———　　　　　　　　　　　　1858 A
　[Remarks concerning *Hadrosaurus.*]
　Proc. Acad. Nat. Sci. Phila., 1858, pp. 215-
　218.

———　　　　　　　　　　　　1858 B
　[Descriptions of *Mastodon mirificus*
　and *Elephas imperator.*]
　Proc. Acad. Nat. Sci Phila., 1858, p. 10.

———　　　　　　　　　　　　1858 C
　[Notice of fossil *Mammalia* from val-
　ley of Niobrara River.]
　Proc. Acad. Nat. Sci. Phila., 1858, p. 11.

———　　　　　　　　　　　　1858 D
　[Remarks on "*Mastodon longirostris*"
　and "*Tapirus mastodontoides*" of Har-
　lan.]
　Proc. Acad. Nat. Sci. Phila., 1858, p. 12.

———　　　　　　　　　　　　1858 E
　Notice of remains of extinct *Verte-*
　brata, from the valley of the Niobrara
　River, collected during the exploring
　expedition of 1857, in Nebraska, under
　the command of Lieut. G. K. Warren,
　U. S. Top. Eng., by Dr. F. V. Hayden.
　Proc. Acad. Nat. Sci. Phila., 1858, pp. 20-29.

———　　　　　　　　　　　　1858 F
　[Description of *Procamelus robustus*
　and *P. gracilis.*]
　Proc. Acad. Nat. Sci. Phila., 1858, p. 89.

———　　　　　　　　　　　　1859 A
　Hadrosaurus foulkii, a new saurian
　from the Cretaceous of New Jersey, re-
　lated to the Iguanodon.
　Amer. Jour. Sci. (2), xxvii, pp. 266-270.
　From Proc. Phil. Acad., 1858, p. 218.

———　　　　　　　　　　　　1859 B
　[Remarks on remains of some extinct
　Vertebrata in Territory of Nebraska.]
　Proc. Amer. Philos. Soc., vii, pp. 10-11.

———　　　　　　　　　　　　1859 C
　[Descriptions of *Xystracanthus arcua-*
　tus and *Cladodus occidentalis.*"]
　Proc. Acad. Nat. Sci. Phila., 1859, p. 3.

———　　　　　　　　　　　　1859 D
　Synonymy of the American *Mosa-*
　saurus.
　Proc. Acad. Nat. Sci. Phila., 1859, pp. 91-92.
　Preceded by remarks on mastodon found in
　Honduras and on the genus *Mosasaurus.*

Leidy, Joseph—Continued.　　1859 E
　[Remarks on fossil vertebrates found
　near Phœnixville, Pa., including new
　genus and species, *Eurydorus serridens.*]
　Proc. Acad. Nat. Sci. Phila., 1859, p. 110.

———　　　　　　　　　　　　1859 F
　[On the occurrence of *Ursus ameri-*
　canus in association with bones of *Mas-*
　todon, Megalonyx, etc.]
　Proc. Acad. Nat. Sci. Phila., 1859, p. 111.

———　　　　　　　　　　　　1859 G
　[Remarks on *Dromatherium sylvestre*
　and *Ontocetus emmonsi.*]
　Proc. Acad. Nat. Sci. Phila., 1859, p. 162.

———　　　　　　　　　　　　1859 H
　[Remarks on finding antler of rein-
　deer at Sing Sing, N. Y.]
　Proc. Acad. Nat. Sci. Phila., 1859, p. 194.

———　　　　　　　　　　　　1859 I
　Remarks on fossils from Bethany,
　Va.; and also from the Green-sand,
　Monmouth County, N. J.
　Proc. Acad. Nat. Sci. Phila., xi, p. 110.
　No descriptions are given and no species are
　named.

———　　　　　　　　　　　　1860 A
　Extinct *Vertebrata* from the Judith
　River and great Lignite formations of
　Nebraska.
　Trans. Amer. Philos. Soc., xi, pp. 139-154,
　with pls. viii-xi.

———　　　　　　　　　　　　1860 B
　Description of vertebrate fossils.
　Holmes's Post-pliocene Fossils of South
　Carolina, pp. 99-122, with pls. xv-xxviii.

———　　　　　　　　　　　　1860 C
　Skull of extinct peccary.
　Proc. Acad. Nat. Sci. Phila., 1860, p. 416.
　This refers to the species found in Gibson
　County, Ind., and afterwards called *Dicotyles*
　nasutus.

———　　　　　　　　　　　　1862 A
　Observations upon the mammalian
　remains found in the crevices of the
　lead-bearing rocks at Galena, Ill.
　J. D. Whitney's Report of a Geolog. Surv. of
　the Upper Mississippi lead region, p. 423.

———　　　　　　　　　　　　1865 A
　Memoir on the extinct reptiles of the
　Cretaceous formations of the United
　States.
　Smithsonian Contributions to Knowledge,
　xiv, article vi, pp. 1-135, pls. i-xx.
　Reviewed in the Geol. Magazine, vol. v,
　1868, p. 432.

Leidy, Joseph—Continued.   1865 B
[Descriptions of *Rhinoceros meridianus*
and *R. hesperius.*]
Proc. Acad. Nat. Sci. Phila., 1865, pp. 176–177.

———   1865 C
Bones and teeth of horses from California and Oregon.
Proc. Acad. Nat. Sci. Phila., 1865, p. 94.
Describes *Equus occidentalis* Leidy.

———   1865 D
Brief review of a memoir on the
Cretaceous reptiles of the United States,
published in the fourteenth volume
of the Smithsonian Contributions to
Knowledge.
Annual Rep. Board Regents Smithsonian
Inst. for 1864, pp. 66–73.
This is a review of the same author's memoir referred to in the title, Leidy, J. 1865 A.

———   1866 A
Drepanodon (Machairodus) occidentalis.
Proc. Acad. Nat. Sci. Phila., 1866, p. 345.
This carnivore from White River is named,
but not described.

———   1866 B
Remarks on fossils presented.
Proc. Acad. Nat. Sci. Phila., xviii, p. 237.
Remarks on Harlan's *Scolopax* and on bird
tibia from Nebraska.

———   1866 C
Remarks on a phalanx of an extinct
reptile.
Proc. Acad. Nat. Sci. Phila., xviii, p. 9.
Reference made to a reptilian phalanx from
Columbus, Miss. Description given, but no
name. May belong to *Hadrosaurus.*

———   1867 A
[Remarks on a skull of *Bison antiquus* from California.]
Proc. Acad. Nat. Sci. Phila., 1867, p. 85.

———   1867 B
[Remarks on fossil skull of *Geomys
bursarius.*]
Proc. Acad. Nat. Sci. Phila., 1867, p. 97.

———   1867 C
[Remarks on a skull of *Castoroides
ohioensis*, found near Charleston, Ill.]
Proc. Acad. Nat. Sci. Phila., 1867, pp. 97–98.

———   1867 D
[Remarks on *Agriochœrus latifrons,
Hippopotamus*, and *Mastodon.*]
Proc. Acad. Nat. Sci. Phila., 1867, p. 32.
The name *Agriochœrus latifrons* is given, but
there is no description.

Leidy, Joseph—Continued.   1868 A
Notice of some vertebrate remains
from Hardin County, Tex.
Proc. Acad. Nat. Sci. Phila., 1868, pp. 174–176.

———   1868 B
Indications of an *Elotherium* in California.
Proc. Acad. Nat. Sci. Phila., 1868, p. 177.
Describes *Elotherium superbum.*

———   1868 C
Notice of some reptilian remains
from Nevada.
Proc. Acad. Nat. Sci. Phila., 1868, pp. 177–178.

———   1868 D
Notice of some remains of horses.
Proc. Acad. Nat. Sci. Phila., 1868, p. 195.

———   1868 E
Notice of some extinct cetaceans.
Proc. Acad. Nat. Sci. Phila., 1868, pp. 196–197.

———   1868 F
Remarks on a jaw fragment of *Megalosaurus.*
Proc. Acad. Nat. Sci. Phila., 1868, pp. 197–200.

———   1868 G
Remarks on *Conosaurus* of Gibbes.
Proc. Acad. Nat. Sci. Phila., 1868, pp. 200–202.

———   1868 H
Notice of American species of *Ptychodus.*
Proc. Acad. Nat. Sci. Phila., 1868, pp. 205–208.

———   1868 I
Notice of some remains of extinct
pachyderms.
Proc. Acad. Nat. Sci. Phila., 1868, pp. 230–233.

———   1868 J
Notice of some remains of extinct
Insectivora from Dakota.
Proc. Acad. Nat. Sci. Phila., 1868, pp. 315–316.

———   1869 A
The extinct mammalian fauna of Dakota and Nebraska, including an account of some allied forms from other
localities, together with a synopsis of
the mammalian remains of North
America.
Jour. Acad. Nat. Sci. Phila. (2), vii, pp. 1–472, with 30 plates.
A synopsis of this memoir was published in
Jour. de Zoologie, 1, 1872, pp. 187–191; 500–508,
with pls. x–xi; ii, pp. 541–545. A list of the
described genera and species is included.

Leidy, Joseph—Continued. 1869 B
Notice of some extinct vertebrates from Wyoming and Dakota.
Proc. Acad. Nat. Sci. Phila., 1869, pp. 63–67.

—— 1870 A
On the *Elasmosaurus platyurus* of Cope.
Amer. Jour. Sci. (2), xlix, p. 392.

—— 1870 B
On *Discosaurus* and its allies.
Amer. Jour. Sci. (2), l, pp. 139–140.
Extracted from Proc. Phil. Acad., 1870, p. 18.

—— 1870 C
[Abstract of remarks made on *Elasmosaurus* at meetings of Acad. Nat. Sci. Phila., Mar. 8 and Apr. 5, 1870.]
Nature, ii, p. 249.

—— 1870 D
[Description of new genus and species, *Megacerops coloradensis.*]
Proc. Acad. Nat. Sci. Phila., 1870, pp. 1–2.

—— 1870 E
[Remarks on *Poicilopleuron valens, Clidastes intermedius, Leiodon proriger, Baptemys wyomingensis,* and *Emys stevensonianus.*]
Proc. Acad. Nat. Sci. Phila., 1870, pp. 3–5.

—— 1870 F
Remarks on *Mylodon ? robustus,* from Central America, and on *Dromatherium sylvestre.*]
Proc. Acad. Nat. Sci. Phila., 1870, pp. 8–9.

—— 1870 G
[Remarks on *Elasmosaurus platyurus* and other fossil vertebrates.]
Proc. Acad. Nat. Sci. Phila., 1870, pp. 9–11.

—— 1870 H
[Remarks on ichthyodorulites and on certain fossil *Mammalia.*]
Proc. Acad. Nat. sci. Phila., 1870, pp. 12–13.

—— 1870 I
Remarks on *Discosaurus* and its allies.
Proc. Acad. Nat. Sci. Phila., 1870, pp. 18–22.

—— 1870 J
[Remarks on fossil vertebrates.]
Proc. Acad. Nat. Sci. Phila., 1870, pp. 66–68.

—— 1870 K
[Remarks on *Elephas americanus* and *Bos americanus* from Kansas and on new cyprinoid fishes from the Rocky Mountain region.]
Proc. Acad. Nat. Sci. Phila., 1870, pp. 69–71.

—— 1870 L
[Remarks on jaw of *Ovibos cavifrons.*]
Proc. Acad. Nat. Sci. Phila., 1870, p. 73.

Leidy, Joseph—Continued. 1870 M
[Remarks on *Mastodon americanus,* etc.]
Proc. Acad. Nat. Sci. Phila., 1870, pp. 96–99.

—— 1870 N
[Description of *Crocodilus elliotti.*]
Proc. Acad. Nat. Sci. Phila., 1870, p. 100.

—— 1870 O
[Remarks on a collection of fossils from the western Territories.]
Proc. Acad. Nat. Sci. Phila., 1870, pp. 109–110.

—— 1870 P
[Remarks on a collection of fossils from Dalles City, Oreg.]
Proc. Acad. Nat. Sci. Phila., 1870, pp. 111–113.

—— 1870 Q
[Descriptions of *Palæosyops paludosus, Microsus cuspidatus,* and *Notharctos tenebrosus.*]
Proc. Acad. Nat. Sci. Phila., 1870, pp. 111–114.

—— 1870 R
[Description of *Graphiodon vinearis* and remarks on *Crocodilus elliotti.*]
Proc. Acad. Nat. Sci. Phila., 1870, p. 122.

—— 1870 S
[Descriptions of *Emys jeanesi, E. haydeni, Baëna arenosa,* and *Saniwa ensidens.*]
Proc. Acad. Nat. Sci. Phila., 1870, pp. 123–124.

—— 1870 T
[Remarks on a collection of fossils from Table Mountain, Cal.]
Proc. Acad. Nat. Sci. Phila., 1870, pp. 125–127.

—— 1870 U
[Description of *Nothosaurops occiduus.*]
Proc. Acad. Nat. Sci. Phila., 1870, p. 74.

—— 1870 V
On fossil bones from Dakota and Nebraska.
Proc. Acad. Nat. Sci. Phila., 1870, pp. 65–66.
Describes a radius and tibia of some undetermined rhinoceros, and records finding of a bone probably belonging to *Ælurodon ferox.*

—— 1870 W
Remarks on Professor Owen's paper on fossil equines from Central and South America.
Proc. Acad. Nat. Sci. Phila., 1870, pp. 126–127.
Contains a list of species of *Protohippus.*

—— 1871 A
Notes on the American mastodon and other fossils.
Amer. Jour. Sci. (3), i, pp. 63–65.
Extracted from Proc. Acad. Nat. Sci. Phila. 1870, pp. 96–99.

Leidy, Joseph—Continued. 1871 B
Remarks on fossil vertebrates from Wyoming.
Amer. Jour. Sci. (3), ii, pp. 372–373.
Extracted from Proc. Acad. Nat. Sci. Phila., Aug. 8, 1871.

——— 1871 C
Report on the vertebrate fossils of the Tertiary formations of the West.
U. S. Geol. Surv. of Wyoming and portions of contiguous Territories, 2d (4th) annual report, F. V. Hayden, U. S. Geologist, pp. 340–370.

——— 1871 D
[Remarks on a collection of fossils from California.]
Proc. Acad. Nat. Sci. Phila., 1871, p. 60.

——— 1871 E
[Remarks on extinct turtles from Wyoming Territory, *Anosteira ornata* and *Hybemys arenarius*.]
Proc. Acad. Nat. Sci. Phila., 1871, pp. 102–103.

——— 1871 F
[Remarks on *Mastodon* and *Equus* from N. Carolina, and on extinct mammals from Wyoming.]
Proc. Acad. Nat. Sci. Phila., 1871, pp. 113–116.

——— 1871 G
[Remarks on remains of *Palæosyops*.]
Proc. Acad. Nat. Sci. Phila., 1871, p. 118.

——— 1871 H
Remarks on a fossil *Testudo* from Wyoming and on supposed fossil turtle eggs.
Proc. Acad. Nat. Sci. Phila., 1871, pp. 154–155.

——— 1871 I
Remarks on donation of fossils from Wyoming.
Proc. Acad. Nat. Sci. Phila., 1871, p. 197.

——— 1871 J
Remarks on *Mastodon*, etc., of California; and note on *Anchitherium*.
Proc. Acad. Nat. Sci. Phila., 1871, pp. 198–199.

——— 1871 K
Remarks on fossil vertebrates from Wyoming.
Proc. Acad. Nat. Sci. Phila., 1871, pp. 228–229.
See Amer. Naturalist, v, pp. 664–666.

——— 1871 L
Notice of some extinct rodents.
Proc. Acad. Nat. Sci. Phila., 1871, pp. 230–232.

——— 1871 M
Remarks on fossils from Oregon.
Proc. Acad. Nat. Sci. Phila., 1871, pp. 247–248.

Leidy, Joseph—Continued. 1872 A
On some new species of fossil *Mammalia* from Wyoming.
Amer. Jour. Sci. (3), iv, pp. 239–240.
From Proc. Acad. Nat. Sci. Phila. Same as Leidy, 1872 G.

——— 1872 B
On the fossil vertebrates of the early Tertiary formation of Wyoming.
U. S. Geol. Surv. of Montana and portions of adjacent Territories; F. V. Hayden, U. S. Geologist, Washington, D. C., 1872, pp. 353–372.

——— 1872 C
Remarks on fossils from Wyoming.
Proc. Acad. Nat. Sci. Phila., 1872, pp. 19–21.

——— 1872 D
Remarks on some extinct mammals.
Proc. Acad. Nat. Sci. Phila., 1872, pp. 37–38.

——— 1872 E
Remarks on some extinct vertebrates.
Proc. Acad. Nat. Sci. Phila., 1872, pp. 38–40.
Descriptions given of *Felis augustus*, *Oligosimus grandævus*, and *Tylosteus ornatus*.

——— 1872 F
Remarks on fossil shark teeth.
Proc. Acad. Nat. Sci. Phila., 1872, p. 166.

——— 1872 G
On some new species of *Mammalia* from Wyoming.
Proc. Acad. Nat. Sci. Phila., 1872, pp. 167–163.

——— 1872 H
Remarks on fossil mammals from Wyoming.
Proc. Acad. Nat. Sci. Phila., 1872, pp. 240–242.

——— 1872 I
Notice of donation of fossils, etc., from Wyoming.
Proc. Acad. Nat. Sci. Phila., 1872, pp. 267–268.

——— 1872 J
Remarks on fossils from Wyoming.
Proc. Acad. Nat. Sci. Phila., 1872, p. 277.
Remarks on *Palæosyops junior*, *Uintacyon edax*, *U. vorax*, *Chameleo pristinus*.

——— 1872 K
Remarks on *Mastodon* from New Mexico.
Proc. Acad. Nat. Sci. Phila., 1872, p. 143.

——— 1872 L
On a new genus of extinct turtles.
Proc. Acad. Nat. Sci. Phila., 1872, p. 162.
New genus *Chisternon* formed for *Baëna undata* Leidy.

——— 1872 M
On some remains of Cretaceous fishes.
Proc. Acad. Nat. Sci. Phila., 1872, pp. 162–163.
Descriptions given of *Otodus divaricatus*, *Oxyrhina extenta*, *Acrodus humilis*, and *Pycnodus faba*.

Leidy, Joseph—Continued.   1873 A
Notice of fossil *Vertebrata* from the
Miocene of Virginia.
Amer. Jour. Sci. (3), v, pp. 311–312.
Extract from Proc. Acad. Nat. Sci. Phila.,
1872.

————   1873 B
Contributions to the extinct vertebrate fauna of the Western Territories.
Report of the U. S. Geological Survey of the
Territories, F. V. Hayden, U. S. geologist in
charge, i, pp. 14–358; pls. i–xxxvii. Washington, 1873.

————   1873 C
Notice of fossil vertebrates from the
Miocene of Virginia.
Proc. Acad. Nat. Sci. Phila., 1873, p. 15.

————   1873 D
Notice of remains of fishes in the
Bridger Tertiary formation of Wyo-
ming.
Proc. Acad. Nat. Sci. Phila., 1873, pp. 97–99.

————   1873 E
Remarks on the occurrence of an ex-
tinct hog in America.
Proc. Acad. Nat. Sci. Phila., 1873, p. 207.

————   1873 F
Remarks on extinct mammals from
California.
Proc. Acad. Nat. Sci. Phila., 1873, pp. 259–260.
Describes *Felis imperialis* and *Auchenia hes-
terna*.

————   1873 G
Remarks on fossil elephant teeth.
Proc. Acad. Nat. Sci. Phila., 1873, pp. 416–417.

————   1874 A
[Remarks on *Thespesius* and *Ischy-
rotherium*.]
Proc. Acad. Nat. Sci. Phila., 1874, pp. 74–75.

————   1874 B
Notice of remains of *Titanotherium*.
Proc. Acad. Nat. Sci. Phila., 1874, pp. 165–166.

————   1874 C
Remarks on fossils presented.
Proc. Acad. Nat. Sci. Phila., 1874, pp. 223–224.

————   1875 A
Remarks on elephant remains.
Proc. Acad. Nat. Sci. Phila., 1875, p. 121.

————   1876 A
Remarks on fossils from the Ashley
phosphate beds.
Proc. Acad. Nat. Sci. Phila., 1876, pp. 80–81.
Abstract in Amer. Jour. Sci. (3), xii, 222–223.

Leidy, Joseph—Continued.   1876 B
Fish remains of the Mesozoic red
shales.
Amer. Jour. Sci. (3), xii, p. 223.
Extracted from Proc. Acad. Nat. Sci. Phila.,
1876. Same as Leidy, 1876 D.

————   1876 C
On *Petalodus*.
Proc. Acad. Nat. Sci. Phila., 1876, p. 9.
Form related to *P. linguifer* found apparently
in Green-sand marl.

————   1876 D
Fish remains of the Mesozoic red
shales.
Proc. Acad. Nat. Sci. Phila., 1876, p. 81.
Reprinted in Amer. Jour. Sci., xii, p. 223.

————   1876 E
Remarks on fossils of the Ashley
phosphate beds.
Proc. Acad. Nat. Sci. Phila., 1876, pp. 86–87.

————   1876 F
Remarks on vertebrate fossils from
the phosphate beds of South Carolina.
Proc. Acad. Nat. Sci. Phila., 1876 pp. 114–115.

————   1877 A
Description of vertebrate remains,
chiefly from the phosphate beds of
South Carolina.
Jour. Acad Nat. Sci. Phila. (2), viii, pp. 209–
261, pls. xxx–xxxiv.

————   1879 A
Fossil remains of a caribou.
Proc. Acad. Nat. Sci. Phila., 1879, pp. 32–33.

————   1879 B
Fossil foot tracks of the anthracite
Coal-measures.
Proc. Acad. Nat. Sci. Phila., 1879, pp. 164–165.

————   1880 A
Bone caves of Pennsylvania.
Proc. Acad. Nat. Sci. Phila., 1880, pp. 346–349.

————   1881 A
Remarks on *Bathygnathus borealis*.
Jour. Acad. Nat. Sci. Phila. (2), viii, pp. 449–
451.

————   1882 A
On an extinct peccary.
Proc. Acad. Nat. Sci. Phila., 1882, pp. 301–302.
Describes *Platygonus vetus*.

————   1884 A
Fossil bones from Louisiana.
Proc. Acad. Nat. Sci. Phila., 1884, p. 22.

————   1884 B
Vertebrate fossils from Florida.
Proc. Acad. Nat. Sci. Phila., 1884, pp. 118–119

Leidy, Joseph—Continued.   1885 A
Rhinoceros and *Hippotherium* from
Florida.
Proc. Acad. Nat. Sci. Phila., 1885, pp. 32–33.

———   1885 B
Remarks on *Mylodon.*
Proc. Acad. Nat. Sci. Phila., 1885, pp. 49–51,
with 6 figs. in the text.

———   1886 A
An extinct boar from Florida.
Proc. Acad. Nat. Sci. Phila., 1886, pp. 37–38,
with 2 figs. in the text.

———   1886 B
Mastodon and llama from Florida.
Proc. Acad. Nat. Sci. Phila., 1886, pp. 11–12.
A notice of this paper, together with a note
by Professor Cope, is found in Amer. Natural-
ist, 1886, xx, p. 755.

———   1886 C
Caries in the mastodon.
Proc. Acad. Nat. Sci. Phila., 1886, p. 38.

———   1887 A
Fossil bones from Florida.
Proc. Acad. Nat. Sci. Phila., 1887, pp. 309–310.

———   1888 A
On a fossil of the puma.
Proc. Acad. Nat. Sci. Phila., 1888, pp. 9–10.

———   1888 B
Toxodon and other remains from
Nicaragua, Central America.
Proc. Acad. Nat. Sci. Phila., 1888, pp. 275–
277, with text figs.

———   1889 A
The saber-tooth tiger of Florida.
Proc. Acad. Nat. Sci. Phila., 1889, pp. 29–31.
Describes *Drepanodon floridanus*, and notes
occurrence of *Auchenia minor.*

———   1889 B
Fossil vertebrates from Florida.
Proc. Acad. Nat. Sci. Phila., 1889, pp. 96–97.

———   1889 C
Notice of some fossil human bones.
Trans. Wagner Free Institute Sci., Phila.,
ii, pp. 9–12.
Contains a notice of the finding of some
human bones in Florida, and also some bones
of *Bison latifrons.*

———   1889 D
Description of mammalian remains
from a rock crevice in Florida.
Trans. Wagner Free Institute Sci., Phila.,
ii, pp. 13–17, with pl. iii, figs. 1, 5–9.

———   1889 E
Description of vertebrate remains
from Peace Creek, Florida.
Trans. Wagner Free Institute Sci., Phila.,
ii, pp. 19–31, with pls. v–viii.

Leidy, Joseph—Continued.   1889 F
Notice of some mammalian remains
from the salt mine of Petite Anse,
Louisiana.
Trans. Wagner Free Institute Sci., Phila.,
ii, pp. 32–40, with pl. v, figs. 1–4, and 9 figs. in
the text.

———   1889 G
On *Platygonus,* an extinct genus allied
to the peccaries.
Trans. Wagner Free Institute Sci., Phila.,
ii, pp. 41–50, with pl. viii, fig. 1.

———   1889 H
Notice and description of fossils in
caves and crevices of the limestone
rocks of Pennsylvania.
Ann. Report Geol. Surv. Penn. for 1887, pp.
1–20, with pls. i, ii.

———   1890 A
Hippotherium and *Rhinoceros* from
Florida.
Proc. Acad. Nat. Sci. Phila., 1890, pp. 182–183,
with 1 fig. in the text.

———   1890 B
Mastodon and capybara of South
Carolina.
Proc. Acad. Nat. Sci. Phila., 1890, pp. 184–185.

———   1890 C
Fossil vertebrates from Florida.
Proc. Acad. Nat. Sci. Phila., 1890, pp. 64–65.

———   1892 A
[List of *Vertebrata* from the Pliocene
beds of Florida.]
Bull. U. S. Geol. Surv., No. 84 (Correlation
papers: Neocene), pp. 129–130.

Leidy, Joseph, and **Lucas,** Fred. A.
1896 A
Fossil vertebrates from the Alachua
clays of Florida; by Joseph Leidy,
M. D., LL. D.; edited by Frederic A.
Lucas.
Trans. Wagner Free Institute Sci., Phila., iv,
pp. vii–xiv, 15–61, with pls. i–xix.

Lemoine, Victor.   1878 A
Recherches sur les ossements fossiles
des terrains tertiaires inférieurs des
environs de Reims. I. Étude du genre
Arctocyon.
Annales Sci. Naturelles, Zool. (6), viii, pp.
1–56, pls. 1–4.

———   1880 A
Sur les ossements fossiles des terrains
tertiaires inférieurs des environs de
Reims.
Assoc. Française pour l'avancement des
Sciences, 8e session, Montpellier, 1879, pp. 585–
594.

Lemoine, Victor—Continued. 1882 A
Sur deux *Plagiaulax* tertiaires, recueillis aux environs de Reims.
Comptes rend. Acad. Sci. Paris, xcv, pp.
1009-1011.

—— 1882 B
Sur l'encéphale de l'*Arctocyon dueil-
lii* et du *Pleuraspidotherium aumonieri*,
mammifères de l'Éocène inférieur des
environs de Reims.
Bull. Soc. Géol. France (3), x, pp. 328-338.

—— 1883 A
Étude sur le *Neoplagiaulax* de la
faune Éocène inférieur des environs de
Reims.
Bull. Soc. Géol. France (3), xi, pp. 249-271,
with pls. v, vi.

—— 1884 A
Du Simœdosaure, reptile de la faune
cernaysienne des environs de Reims.
Comptes rend. Acad. Sci. Paris, xcviii, pp.
697-699.

—— 1884 B
Sur les os de la tête et sur les diverses
espèces du Simœdosaure, reptile de la
faune cernaysienne des environs de
Reims.
Comptes rend. Acad. Sci. Paris, xcviii, pp.
1011-1013.

—— 1885 A
Sur la présence du Simœdosaure dans
les couches éocènes inférieures de
Sézanne.
Bull. Soc. Géol. France (3), xiv, pp. 21-32,
pls. iii-v.

—— 1885 B
Sur les analogies et les différences du
genre Simœdosaure, de la faune cernaysienne des environs de Reims, avec
le genre Champsosaure d'Erquelinnes.
Comptes rend. Acad. Sci. Paris, c, pp. 753-755.

—— 1890 A
Sur les rapports qui paraissent exister
entre les mammifères crétacés d'Amerique et les mammifères de la faune cernaysienne des environs de Riems.
Comptes rend. Acad. Sci. Paris, cx, pp. 480-
482.
See also Bull. Soc. Géol. France (3), xviii,
pp. 321-324, pl. iii.

—— 1891 A
Étude d'ensemble sur les dents des
mammifères fossiles des environs de
Reims.
Bull. Soc. Géol. France (3), xix, pp. 263-
290, with pls. x and xi.

Lemoine, Victor—Continued. 1893 A
Étude sur les os du pied des mammifères de la faune cernaysienne et sur
quelques pièces osseuses nouvelles de
cet horizon paléontologique.
Bull. Soc. Géol. France (3), xxi, pp. 353-368,
pls. ix-xi.

Lenk, H.
See **Felix** and **Lenk**.

Lennox, T. H. 1886 A
The fossil sharks of the Devonian.
Proc. Canadian Institute (3), iii, pp. 120-121.
Makes statement of the finding of a fossil
fin-spine, *Machæracanthus sulcatus*, at St. Marys,
Ontario.

Lepsius, G. R. 1881 A
Halitherium schinzi, die fossile Sirene
des Mainzer Beckens. Eine vergleichend-anatomische Studie.
Abhandl. mittelrhein geol. Vereins, i, pp.
i-vi; 1-200; i-viii, with pls. i-x.
Contains numerous references to the literature of the *Trichechidæ*.

Lesley, J. P. 1889 A
A dictionary of the fossils of Pennsylvania and neighboring States named
in reports and catalogues of the survey.
Compiled for the convenience of the
citizens of the State by J. P. Lesley,
State geologist. 3,000 figures, mostly
facsimile copies of those published by H.
D. Rogers, Hall, Conrad, Vanuxem, Emmons, Logan, Dawson, Billings, Mat-
thews, Hitchcock, Newberry, Meek,
Collett, Worthen, Rominger, D. D.
Owen, Cox, Lyon, Safford, Fontaine,
Lesquereux, Walcott, Leidy, Cope, and
others; and some new species drawn
and described by G. B. Simpson.
Geol. Surv. of Pennsylvania, Report, P₄,
1889. Vol. i, pp. i-xiv; with Errata, pp.
i-xxxi, and list of publications, pp. 1-10. Vol.
ii, pp. 439-914; with additions, corrections,
synonyms, etc., pp. i-x.

—— 1890 A
[Same title as the preceding, of which
it is the conclusion.]
Pp. 915-1283, with critical emendations, additions, synonyms, etc., pp. i-xiv.
The number of pages devoted to corrections
of errors found in this work gives an indication
of the unsatisfactory manner in which it has
been compiled. The figures are usually poorly
reproduced.

—— 1892 A
A summary description of the geology of Pennsylvania in three volumes,

Lesley, J. P.—Continued.
with a new geological map of the State, a map and list of bituminous mines, and many page plate illustrations. Vol. II, describing the Upper Silurian and Devonian formations. By J. P. Lesley, State geologist.
4to., pp. i–xxv; 722–1628.
The figures are usually much reduced and poor. The plates do not follow the order of their numbers.

——— 1895 A
[Same title as the preceding. Vol. III, part I, describing the Carboniferous formation.]
Pp. i–xix; 1629–2153, with pls. ccv–cccxcv. Part iii, describing the bituminous coal fields, E. V. d'Invilliers and the New Red of Bucks and Montgomery counties. Benjamin Lyman. Pp. i–xxiii; 2153–2638, with pls. cccxcvi–dcxi.
There being no descriptions in the abovementioned work of the fossils figured, and many of the figures being very small and indistinct, not all are cited in this Catalogue.

Leuthardt, Franz. 1891 A
Ueber die Reduction der Fingerzahl bei Ungulaten.
Zoolog. Jahrbücher, System., v, pp. 98–146, with pls. i–xxiii.

Levy, Hugo. 1898 A
Berträge zur Kenntniss des Baues und der Entwicklung der Zähne bei den Reptilien.
Jenaische Zeitschr. Naturwiss., xxxii (n. s. xxv), pp. 313–346, with pl. xi.
Contains numerous citations of the literature of the subject.

Lewis, J. L. 1880 A
Fossil remains in southwest Missouri.
Kansas City Review of Science and Industry, iv, p. 207.
Mentions the finding of a tooth of a horse and a large tusk 30 feet below the surface, in Bates County, Mo.

Lewis, H. C. 1882 A
Evidences of the existence of preglacial man.
Proc. Acad. Nat. Sci. Phila., 1882, pp. 292–293.

Leydig, F. 1876 A
Über den Bau der Zehen bei Batrachiern und die Bedeutung des Fersenhöckers.
Morpholog. Jahrbuch, ii, pp. 165–196, with pls. viii–xi.

Lindahl, Josua. 1892 A
Description of a skull of *Megalonyx leidyi*, n. sp.
Trans. Amer. Philos. Soc., xvii, pp. 1–10, with pls. i–x

Lindahl, Josua—Continued. 1897 A
Description of a Devonian ichthyodorulite, *Heteracanthus uddeni*, n. sp., from Buffalo, Iowa.
Jour. Cincinnati Soc. Nat. Hist., xix, pp. 95–98, with pl. vi.

Lindström, G. 1895 A
On the remains of a *Cyathaspis* from the Silurian strata of Gotland.
Bihang k. Svenska Vet.-Akad. Handlingar, xxi, Afd. iv, No. 3, pp. 1–15, with pls. i, ii.
Describes and figures the microscopic structure of the shields and scales of *Cyathaspis*.

Lippincott, James S. 1881 A
An address to the fossil bones in a private museum.
Amer. Naturalist, vol. xv, p. 87.
Reply to J. S. L., p. 38.

Lister, George. 1846 A
Letter to Jeffries Wyman on Koch's *Hydrarchos*.
Proc. Boston Soc. Nat. Hist., ii, pp. 94–95.
Written from Washington County, Ala.

Lockwood, Samuel. 1883 A
A *Mastodon americanus* in a beaver meadow.
Proc. Amer. Assoc. Adv. Sci., 31st meeting, Montreal, 1882, pp. 365–366.

——— 1886 A
The ancestry of *Nasua*.
Amer. Naturalist, xx, pp. 320–325.

Lönnberg, Einar. 1899 A
On some remains of *"Neomylodon listai"* Ameghino, brought home by the Swedish Expedition to Terra del Fuego, 1895–1897.
Svenska Exped. till. Magellansl., ii, No. 7, pp. 149–170, with pls. xii–xiv.

——— 1900 A
On the structure and anatomy of the musk ox (*Ovibos moschatus*).
Proc. Zool. Soc. Lond., 1900, pp. 686–718, with 10 figs. in the text.

Logan, W. E. 1863 A
Superficial geology of Canada.
Geol. Surv. Canada, 1863, pp. 886–930.

Lohest, Maximin. 1884 A
Recherches sur les poissons des terrains paléozoïques de Belgique. Poissons de l'ampélite alunifère, des genres *Campodus*, *Petrodus*, et *Xystracanthus*.
Annales Soc. Géol. Belgique, xi, pp. 295–325, pls. iii, iv.

——— 1888 A
Recherches sur les poissons des terrains paléozoïques de Belgique. Pois-

Lohest, Maximin.—Continued.
sons des Psammites du Condroz, Famennien supérieur.
Annales Soc. Géol. Belgique, xv, pp. 112-208, pls. i-xi.

——　　　　1890 A
De la découverte d'espèces américaines de poissons fossiles dans le Dévonien supérieur de Belgique.
Annales Soc. Géol. Belgique, xvi, Bull. pp. lvii-lix.
The finding of *Dinichthys pustulosus* and *D. terrilli* in Belgium is announced. The former name refers to *D. tuberculatus*, not to *D. pustulosus* Eastman.

Longhi, Prolo.　　　　1899 A
Sopra i resti di un cranio di *Champsodelphis* fossile scoperto nella molassa miocenia del Bellunese.
Atti. Soc. Veneto-Trent. Sci. (2), iii, pp. 323-381, with pls. i-iii.
Treats of the genus *Champsodelphis*, with references to *Squalodon.*

Loomis, Frederic B.　　　1900 A
Die Anatomie und die Verwandtschaft der Ganoid-und Knochen-Fische aus der Kreide-Formation von Kansas, U. S. A.
Palæontographica xlvi, pp. 213-283, with pls. xix-xxvii, and 13 figs. in the text.

Lortet, L.　　　　1892 A
Les reptiles fossiles du bassin du Rhone.
Arch. Mus. d'Hist. Nat. de Lyon, v, pp. 3-139, with pls. i-xii, and 10 figs. in the text.
In this extensive paper are described the genera *Idiochelys, Hydropelta, Eurysternum, Sauranodon, Homœosaurus, Euposaurus, Atoposaurus, Alligatorellus,* etc.

Lortet, Dr. L., and **Chantre,** E.　1879 A
Recherches sur les mastodontes et les faunes mammalogiques que les accompagnent.
Archives Muséum d'Histoire Naturelle de Lyon, ii, pp. 285-311, with pls. i-xvi.
Portion of "Études paléontologiques dans le bassin du Rhone."

Lucas, Frederick A.　　　1888 A
Great auk notes.
The Auk, v, pp. 278-283.

——　　　　1890 A
The expedition to Funk Island, with observations upon the history and anatomy of the great auk.
Ann. Rep. Smith. Inst.; U. S. Nat. Mus., for 1888, pp. 493-529, with pls. lxxii, lxxiii.
Presents also a list of papers relating to the great auk.

Lucas, Frederick A.—Cont'd.　1892 A
On *Carcharodon mortoni* Gibbes.
Proc. Biolog. Soc. of Washington, vii, pp. 151-152.
Shows that *C. mortoni* is a synonym of *C. megalodon.*

——　　　　1895 A
Skeletons of *Zeuglodon.*
Science (2), ii, pp. 42-43.
Refers to the finding of two skeletons by Mr. Chas. Schuchert, and makes some statements of facts about this animal.

——　　　　1895 B
Notes on the osteology of *Zeuglodon cetoides.*
Amer. Naturalist, xxix, pp. 745-746.

——　　　　1896 A
Contributions to the natural history of the Commander Islands. XI. The cranium of Pallas's cormorant.
Proc. U. S. Nat. Mus., xviii, pp. 717-719, with pls. xxxiv and xxxv.
Describes and figures the cranium of *Phalacrocorax perspicillatus.*

——　　　　1897 A
Fossil bison of North America.
Science (2), vi, p. 814.
Abstract of remarks made before Biological Soc. Washington.

——　　　　1898 A
Contributions to palæontology. 1. A new crocodile from the Trias of southern Utah. 2. A new species of *Dinictis* (*D. major*).
Amer. Jour. Sci. (4), vi, p. 399.
The crocodile above described is *Heterodontosuchus ganei.*

——　　　　1898 B
A new snake from the Eocene of Alabama.
Proc. U. S. Nat. Mus., xxi, pp. 637-638, with pls. xlv, xlvi.
This paper describes the new genus and species *Pterosphenus schucherti.*

——　　　　1898 C
Publications of American Museum of Natural History.
Science (2), viii, pp. 96-97.
Contains a reference to figure of model of *Megalosaurus* (*Lælaps*) *aquilunguis.*

——　　　　1898 D
The fossil bison of North America.
Science (2), viii, p. 678.
Abstract of remarks made before Biolog. Soc. Washington. New species *B. occidentalis* described.

Lucas, Frederick A.—Cont'd.   1899 A
The fossil bison of North America.
Proc. U. S. Nat. Mus., xxi, pp. 755-771, with
pls. lxv-lxxxiv and 2 text figs.
Reviewed by Dr. J. A. Allen in Amer. Nat-
uralist, xxxiii, pp. 665-666.

——   1899 B
The characters of *Bison occidentalis*,
the fossil bison of Kansas and Alaska.
Kansas Univ. Quart., viii, A, pp. 17-18, with
pls. viii, ix.

——   1899 C
The nomenclature of the hyoid in
birds.
Science (2), ix, pp. 323-324, with 1 fig.

——   1900 A
The truth about the mammoth.
.McClure's Magazine, xiv, pp. 349-355, Feb.,
with 5 figs. in the text.

——   1900 B
Characters and relations of *Gallinuloides*, a fossil gallinaceous bird from
the Green River shales of Wyoming.
Bull. Mus. Comp. Zool. (Harvard), xxxvi,
pp. 79-84, with pl. i.
Describes *Gallinuloides wyomingensis* Eastman.

——   1900 C
Palæontological notes.
Science (2), xii, pp. 809-810.
Notes on *Thespesius*, and *Basilosaurus
cetoides*.

——   1900 D
A new rhinoceros, *Trigonias osborni*,
from the Miocene of South Dakota.
Proc. U. S. Nat. Mus., xxiii, pp. 221-223, with
2 figs. in the text.

——   1900 E
The pelvic girdle of *Zeuglodon*, *Basilosaurus cetoides* (Owen), with notes on
other portions of the skeleton.
Proc. U. S. Nat. Mus., xxiii, pp. 327-331,
with pls. v-vii.

——   1900 F
A new fossil cyprinoid, *Leuciscus
turneri*, from the Miocene of Nevada.
Proc. U. S. Nat. Mus., xxiii, pp. 333-334,
with pl. viii.
See also **Leidy** and **Lucas; Stejneger** and **Lucas.**

Ludwig, Rudolph.   1877 A
Fossile Crocodiliden aus der Tertiar-
formationen des Mainzer Beckens.
Palæontographica, Suppl., iii, Lief. 4, 5; pp.
1-54, with pls. i-xvi.

Lütken, Chr.   1868 A
On *Xenacanthus* (*Orthacanthus*) de-
chenii Goldfuss.
Geol. Magazine (1), v, pp. 376-380.
A review of various papers by Professor
Kner.

——   1868 B
Professor Kner's Classification of the
Ganoids.
Geol. Magazine (1), v, pp. 429-432.

——   1873 A
Ueber die Begrenzung und Einthei-
lung der Ganoïden.
Palæontographica, xxii, pp. 1-54.

Lund, P. W.   1839 A
Un aperçu des espèces de mammi-
fères fossiles découvertés au Brésil.
Comptes rend. Acad. Sci. Paris, viii, pp.
570-577.
For a translation of this communication see
Annals Nat. Hist. 1839, iii, pp. 210-218.

——   1839 B
Coup d'œil sur les espèces etientes de
mammifères du Brésil; extrait de quelques mémoires présentés à l'Académie
royale des Sciences du Copenhagen.
Annales Sci. Naturelles (2), xi, pp. 214-234.
For abstract see Neues Jahrb. Min., 1840, pp.
120-125.

——   1841 A
Blik paa Brasiliens Dyreverden för
sidste Jordomvæltning.
K. Danske Vidensk. Selskabs Afhandl, viii,
pp. 27-57; 61-86; 217-296, with pls. xiv-xxvii.

——   1842 A
Blik paa Brasiliens Dyreverden för
sidste Jordomvæltning.
K. Danske Vidensk. Selskabs Afhandl, ix,
pp. 137-208, with pls. xxviii-xxxviii.
See also Annales Sci. Naturelles, 1839 (2), xi,
pp. 214-234.

Lydekker, R.   1880 A
Indian Tertiary and Post-tertiary
Vertebrata. Siwalik and Narbada *Proboscidia*. Family i, *Dinotheridæ*. Family ii, *Elephantidæ*.
Palæont. Indica (10), i, pp. 182-294, with
pls. xxix-xlvi.

——   1881 A
Indian Tertiary and Post-tertiary
Vertebrata. Siwalik *Rhinocerotidæ*.
Palæont. Indica (10), ii, pp. 1-62, with pls. i-x.
To this paper is appended a list of the prin-
cipal works that have been published on the
family of *Rhinocerotidæ*.

Lydekker, R.—Continued.　　1882 A

Indian Tertiary and Post-tertiary *Vertebrata.* Siwalik and Narbada *Equidæ.*

Palæont. Indica (10), ii, pp. 67-98, with pls. xi-xv.

Appended is a list of the principal works and memoirs on the osteology and palæontology of *Equus* and *Hippotherium.*

———　　　　　　　　　　　1883 A

Indian Tertiary and Post-tertiary *Vertebrata.* Siwalik selenodont *Suina,* etc.

Palæont. Indica (10), ii, pp. 143-177, with pls. xxiii-xxv.

Besides the descriptions of the Indian species, there is a classification of the *Bunodontia* and the *Selenodontia,* a list of the species of *Hyopotamus,* and references to memoirs relating to the *Anthracotheriidæ, Oreodontidæ,* etc.

———　　　　　　　　　　　1884 A

Indian Tertiary and Post-tertiary *Vertebrata.* Siwalik and Narbada *Carnivora.*

Palæont. Indica (10), ii, pp. 178-355, with pls. xxvi-xlv.

Contains important discussions of the genera of *Carnivora,* with lists of species, and references to authorities.

———　　　　　　　　　　　1884 B

Indian Tertiary and Post-tertiary *Vertebrata.* Additional Siwalik *Perissodactyla* and *Proboscidia.*

Palæont. Indica (10), iii, pp. xi-xxiv; 1-34, with pls. i-v.

Contains discussion of the teeth of elephants and mastodons.

———　　　　　　　　　　　1884 C

Indian Tertiary and Post-tertiary *Vertebrata.* Siwalik and Narbada bunodont *Suina.*

Palæont. Indica (10), iii, pp. 35-104, with pls. vi-xii.

Treats especially of the genera *Sus* and *Hyotherium,* and has appended a list of memoirs.

———　　　　　　　　　　　1885 A

Note on the zoological position of the genus *Microchœrus,* Wood, and its apparent identity with *Hyopsodus,* Leidy.

Quart. Jour. Geol. Soc., xli, pp. 529-531, with 1 woodcut.

———　　　　　　　　　　　1885 B

Catalogue of the fossil *Mammalia* in the British Museum. Part I. Containing the orders *Primates, Chiroptera, Insectivora, Carnivora,* and *Rodentia.*

Pp. i-xxx; 1-268, with 33 figs. in the text. London, 1885.

Lydekker, R.—Continued.　　1885 C

Catalogue of the fossil *Mammalia* in the British Museum. Part II. Containing the order *Ungulata,* suborder *Artiodactyla.*

Pp. i-xxii; 1-324, with 39 figs. in the text. London, 1885.

———　　　　　　　　　　　1885 D

Note on the generic identity of the genus *Esthonyx,* Cope, with *Platychœrops,* Charlesworth (=*Miolophus,* Owen).

Geol. Magazine (3), ii, pp. 360-361.

———　　　　　　　　　　　1885 E

Memoirs on extinct North American vertebrates, by Prof. E. D. Cope.

Geol. Magazine (3), ii, pp. 468-474.

A review of several of Professor Cope's papers.

———　　　　　　　　　　　1886 A

Catalogue of the fossil *Mammalia* in the British Museum. Part III. Containing the order *Ungulata,* suborders *Perissodactyla, Toxodontia, Condylarthra,* and *Amblypoda.*

Pp. i-xvi; 1-186, with 30 figs. in text. London, 1886.

———　　　　　　　　　　　1886 B

Catalogue of the fossil *Mammalia* in the British Museum. Part IV. Containing the order *Ungulata,* suborder *Proboscidea.*

Pp. i-xxiv; 1-235, with 32 figs. in the text. London, 1886.

———　　　　　　　　　　　1886 C

M. Dollo on the evolution of the teeth of herbivorous *Dinosauria.*

Geol. Magazine (3), iii, pp. 274-276, with 10 figs. in the text.

———　　　　　　　　　　　1886 D

Dr. Max Schlosser on the *Ungulata.*

Geolog. Magazine (3), iii, pp. 325-328.

A review of Dr. Schlosser's paper in the Morphologisches Jahrbuch, xii, pp. 1-136.

———　　　　　　　　　　　1886 E

Indian Tertiary and Post-tertiary *Vertebrata.* Siwalik *Crocodilia, Lacertilia,* and *Ophidia.*

Palæont. Indica (10), iii, pp. 209-240, with pls. xxviii-xxxv.

———　　　　　　　　　　　1887 A

Catalogue of the fossil *Mammalia* in the British Museum. Part V. Containing the group *Tillodontia,* the orders *Sirenia, Cetacea, Edentata, Marsupialia, Monotremata,* and Supplement.

Pp. i-xxxv; 1-345, with 55 figs. in the text. London, 1887.

Lydekker, R.—Continued.   1888 A

Note on a new Wealden Iguanodont and other dinosaurs.

Quart. Jour. Geol. Soc., xliv, pp. 46-61, with pl. iii and 3 figs. in the text.

——   1888 B

Catalogue of the fossil *Reptilia* and *Amphibia* in the British Museum. Part I. Containing the orders *Ornithosauria, Crocodilia, Dinosauria, Squamata, Rhynchocephalia,* and *Proterosauria.*

Pp. i-xxviii; 1-309, with 69 figs. in the text. London, 1888.

——   1888 C

Notes on Tertiary *Lacertilia* and *Ophidia.*

Geol. Magazine (3), v, pp. 110-113.

——   1888 D

Note on the classification of the *Ichthyopterygia.*

Geol. Magazine (3), v, pp. 309-314.

——   1888 E

British Museum Catalogue of fossil *Reptilia* and papers on the Enaliosaurians.

Geol. Magazine (3), v., pp. 451-453.

——   1889 A

On a skull of the chelonian genus *Lytoloma.*

Proc. Zool. Soc. Lond., 1889, pp. 60-66, with pls. vi-vii.

——   1889 B

On an apparently new species of *Hyracodontotherium.*

Proc. Zool. Soc. Lond., 1889, pp. 67-69, with 2 woodcuts.

Describes *H. filholi*, from central France.

——   1889 C

On the remains and affinities of five genera of Mesozoic reptiles.

Quart. Jour. Geol. Soc., xlv, pp. 41-59, with 1 pl. and 9 text figs.

——   1889 D

On remains of Eocene and Mesozoic *Chelonia* and a tooth of (?) *Ornithopsis.*

Quart. Jour. Geol. Soc., xlv, pp. 227-246, with pl. viii and 7 woodcuts in the text.

——   1889 E

On certain chelonian remains from the Wealden and Purbeck.

Quart. Jour. Geol. Soc., xlv, pp. 511-518, with 4 figs. in the text.

——   1889 F

Catalogue of the fossil *Reptilia* and *Amphibia* in the British Museum. Part

Lydekker, R.—Continued.

II. Containing the orders *Ichthyopterygia* and *Sauropterygia.*

Pp. i-xxiii; 1-307, with 85 figs. in the text. London, 1889.

——   1889 G

Catalogue of the fossil *Reptilia* and *Amphibia* in the British Museum. Part III. Containing the order *Chelonia.*

Pp. i-xviii; 1-239, with 53 figs. in the text. London, 1889.

——   1889 H

On a Cœluroid dinosaur from the Wealden.

Geol. Magazine (3), vi, pp. 119-121.

——   1889 I

On an ichthyosaurian paddle, showing the contour of the integuments.

The Geol. Magazine (3), vi, pp. 388-390, with 1 fig. in the text.

——   1889 J

[Remarks on Professor Seeley's paper on *Ornithopsis.*]

Quart. Jour. Geol. Soc., xlv., pp. 395-397.

Believes that *Ortiosaurus* should yield to *Cardiodon*, and *Ornithopsis* to *Hoplosaurus* (*Oplosaurus*) or even *Pelorosaurus.*

——   1889 K

Skeleton of *Phenacodus.*

Nature, xl, pp. 57-58, with 1 fig. in the text. The figure is taken from Cope.

——   1889 L

Notes on new and other dinosaurian remains.

Geol. Magazine (3), vi, pp. 352-356, with text figs. A-D.

Contains figures and description of *Arctosaurus osborni* Adams.

——   1890 A

Catalogue of the fossil *Reptilia* and *Amphibia* in the British Museum. Part IV. Containing the orders *Anomodontia, Ecaudata, Caudata,* and *Labyrinthodontia;* and supplement.

Pp. i-xxiii; 1-295, with 66 figs. in the text. London, 1890.

——   1890 B

On the remains of small sauropodous dinosaurs from the Wealden.

Quart. Jour. Geol. Soc. Lond., xlvi, pp. 182-184, with pl. ix and 1 text fig.

Regards the remains as belonging to *Pleurocœlus* Marsh.

——   1891 A

Catalogue of the fossil birds in the British Museum.

Pp. i-xxvii, 1-368, with 75 figs. London, 1891.

Lydekker, R.—Continued.　　1891 B
Professor Osborn on the molars of
the *Perissodactyla.*
Geol. Magazine (3), viii, pp. 317–321, with 3
figs. in the text; p. 384.

———　　　　　　　　　　　　　1891 C
On British fossil birds.
The Ibis (6), iii, pp. 381–410.

———　　　　　　　　　　　　　1892 A
On a remarkable sirenian jaw from
the Oligocene of Italy, and its bearing
on the evolution of the *Sirenia.*
Proc. Zool. Soc. Lond., 1892, pp. 77–88, with
2 woodcuts.

———　　　　　　　　　　　　　1892 B
Recent advances in knowledge of the
ichthyosaurian reptiles.
Natural Science, i, pp. 514–521, with 2 figs. in
the text.

———　　　　　　　　　　　　　1893 A
On a mammalian incisor from the
Wealden of Hastings.
Quart. Jour. Geol. Soc. Lond., xlix, pp.
281–283, with figs. 1, 1a, 2, 2a.

———　　　　　　　　　　　　　1893 B
Some recent restorations of dinosaurs.
Nature, xlviii, pp. 302–304, with 5 figs. in the
text.
See, regarding this paper, O. C. Marsh, Nature, xlviii, p. 437.

———　　　　　　　　　　　　　1893 C
On zeuglodont and other cetacean
remains from the Tertiary of the Cau-
casus.
Proc. Zool. Soc. Lond., 1892, pp. 558–564, with
pls. xxxvi–xxxviii.

———　　　　　　　　　　　　　1893 D
Contributions to a knowledge of the
fossil vertebrates of Argentina; i. The
dinosaurs of Patagonia; II. Cetacean
skulls from Patagonia; III. A study of
extinct Argentine ungulates.
Anales del Museo de la Plata, ii, pp. 1–16,
pls. i–v; pp. 1–13, pls. i–v; pp. 1–91, pls. i–xxxii.

———　　　　　　　　　　　　　1893 E
The restoration of extinct animals.
Natural Science, ii, pp. 135–143, with 2 text
figs.
This is a review of H. N. Hutchinson's
Extinct Monsters.

———　　　　　　　　　　　　　1894 A
Contributions to a knowledge of the
fossil vertebrates of Argentina. Part

Lydekker, R.—Continued.
II. 1. Supplemental observations on
Argentine ungulates. 2. The extinct
edentates of Argentina. 3. On two
extinct carnivores.
Anales del Museo de la Plata, iii, pp. i–vii;
1–118; 1–4; with pls. i, ii; i–lxi.

———　　　　　　　　　　　　　1896 A
[Articles in Newton's Dictionary of
Birds, 1893–1896.]

———　　　　　　　　　　　　　1896 B
A geographical history of mammals.
8vo, pp. i–xii; 1–400, with 82 figs. in the text.

———　　　　　　　　　　　　　1898 A
The deer of all lands. A history of
the family *Cervidæ,* living and extinct.
4to, pp. i–xx; 1–329, with pls. i–xxivand text
figs. 1–80. London, 1898.

———　　　　　　　　　　　　　1898 B
Wild oxen, sheep, and goats of all
lands, living and extinct.
London, 1898. 4to, pp. i–xv; 1–318, with pls.
i–xxvii and 61 figs. in the text.

———　　　　　　　　　　　　　1900 A
The dental formula of the marsupial
and placental *Carnivora.*
Proc. Zool. Soc. Lond., 1899, pp. 922–928, with
pl. lxii.

See also **Flower** and **Lydekker;**
Thomas and **Lydekker.**

Lyell, Charles.　　　　　　1842 A
On the fossil footprints of birds and
impressions of raindrops in the valley
of the Connecticut.
Proc. Geol. Soc. Lond., iii, pp. 793–796.
No species described or figured.

———　　　　　　　　　　　　　1843 A
On the fossil footprints of birds and
impressions of raindrops in the valley
of the Connecticut.
Amer. Jour. Sci., xlv, pp. 394–397.

———　　　　　　　　　　　　　1843 B
On the geological position of the
Mastodon giganteum and associated fos-
sil remains at Bigbone Lick, Kentucky.
Ann. and Mag. Nat. Hist., xii, pp. 125–128.
See also Proc. Geol. Soc. Lond., iv, pp. 36–39,
1843; Amer. Jour. Sci., 1844, xlvi, pp. 320–323.

———　　　　　　　　　　　　　1843 C
On the Tertiary strata of the island
of Martha's Vineyard in Massachusetts.
Proc. Geol. Soc. Lond., iv, pp. 31·33.
Also in Amer. Jour. Sci. (2), xlvi, pp. 318–320.

Lyell, Charles—Continued.   1845 A

On the Miocene Tertiary strata of Maryland, Virginia, and of North and South Carolina.

Quart. Jour. Geol. Soc. Lond., i, pp. 413-429.

Refers to the occurrence of the remains of various species of sharks and of tooth of *Mastodon longirostris*.

———   1846 A

On the evidence of fossil footprints of a quadruped allied to *Cheirotherium* in the Coal strata of Pennsylvania.

Amer. Jour. Sci. (2), ii, pp. 25-29.

———   1846 B

On footmarks discovered in the Coal measures of Pennsylvania.

Quart. Jour. Geol. Soc. Lond., ii, pp. 417-420. No species are named.

———   1846 C

On the newer deposits of the Southern States of North America.

Quart. Jour. Geol. Soc. Lond., ii, pp. 405-410. Refers to the finding of *Zeuglodon* bones in Alabama.

———   1847 A

On the alleged coexistence of man and the megatherium.

Amer. Jour. Sci. (2), iii, pp. 267-269.

———   1847 B

On the structure and probable age of the coal field of the James River, near Richmond, Va.

Quart. Jour. Geol. Soc. Lond., iii, pp. 261-280, with pls. viii and ix.

———   1848 A

On the fossil footmarks of a reptile in the Coal formation of the Allegheny Mountains.

Athenæum, 1848, pp. 166-167.

———   1853 A

On the discovery of some fossil reptilean remains, and a land shell in the interior of an erect fossil tree in the Coal-measures of Nova Scotia, with remarks on the origin of coal fields, and the time required for their formation.

Amer. Jour. Sci. (2), xvi, pp. 33-41. Taken from Proc. Royal Soc. of Great Britain, 1853.

Lyell, Chas., and Dawson, J. W. 1853 A

On the remains of a reptile (*Dendrerpeton acadianum*, Wyman and Owen), and of a land shell discovered in the interior of an erect fossil tree in the coal measures of Nova Scotia.

Quart. Jour. Geol. Soc. Lond., ix, pp. 58-63, with pls. ii-iv.

Maack, G. A.   1865 A

Palæontologische Untersuchungen über noch unbekannte Lophiodonfossilien von Heidenheim am Hahnenkamme in Mittelfranken nebst einer kritischen Betrachtung sämmtlicher bis jetzt bekannten Species des Genus *Lophiodon*.

8vo, Leipzig, 1865, pp. 1-75, with pls. i-xiv.

———   1869 A

Die bis jetzt bekannten fossilen Schildkröten und die im oberen Jura bei Kelheim (Bayern) und Hannover neu aufgefundenen ältesten Arten derselben.

Palæontographica xviii, 193-338, pls. xxxiii-xl.

For review of this work and various corrections see Cope, E. D., 1870 T.

McAdams, William.   1884 A

A new vertebrate from the St. Louis limestone.

Proc. Amer. Assoc. Adv. Sci., 32d meeting, Minneapolis, 1883, p. 269.

Abstract; no generic or specific name.

McCallie, S. W.   1892 A

Remains of the mastodon recently found in Tennessee.

Science, xx, p. 333.

McCoy, Frederick.   1848 A

On some new fossil fish of the Carboniferous period.

Ann. and Mag. Nat. Hist. (2), ii, pp. 1-10; 115-134.

———   1848 B

Reply to Sir Philip Egerton's letter on the *Placodermi*.

Ann. and Mag. Nat. Hist. (2), ii, pp. 277-280.

———   1848 C

On some new ichthyolites from the Scotch Old Red sandstone.

Ann. and Mag. Nat. Hist. (2), ii, pp. 297-312, with text figs.

———   1853 A

On the structure of certain fossil fishes found in the Old Red sandstone of the north of Scotland.

Report Brit. Assoc. Adv. Sci., 22d meeting, Belfast, 1852, p. 55.

Discusses structure of *Holoptychius*.

———   1853 B

On the mode of succession of the teeth in *Cochliodus*.

Report Brit. Assoc. Adv. Sci., 22d meeting, Belfast, 1852, p. 55.

McCoy, Frederick—Continued.   1855 A
Description of the British palæozoic
fossils in the Geological Museum of the
University of Cambridge.
Pp. 1-661, with 26 pls., Cambridge, 1854.
The vertebrate fossils are described on pp.
575-644.

McGee, W. J.   1887 A
Ovibos cavifrons from the Loess of
Iowa.
Amer. Jour. Sci. (3), xxxiv, pp. 217-220.

Mackie, S. J.   1863 A
On a new species of *Hybodus* from
the lower Chalk.
The Geologist, vi, pp. 241-246, pl. xiii.
Discusses the genus *Hybodus*.

———   1863 B
Cestraciont fishes of the Chalk.
The Geologist, vi, pp. 161-163, with pl. ix.
Contains list of species of *Ptychodus* in Brit-
ish Museum.

Mackintosh, H. W.   1878 A
Note on the microscopic structure of
the scale of *Amia calva*.
Proc. Roy. Dublin Soc., i, pp. 93-95, with pls.
i, ii.

McMurrich, J. P.   1884 A
On the osteology of *Amiurus catus*
(L.) Gill.
Zoolog. Anzeiger, vii, pp. 296-299.

———   1884 B
The osteology of *Amiurus catus* (L.)
Gill.
Proc. Canadian Institute, ii, pp. 270-310,
with pl. ii.

Madison (Bishop).   1806 A
The Mammoth, or American elephant.
Med. and Phys. Jour., London, xv, p. 486.
Shows that this animal was herbivorous.

Maggi, Leopoldo.   1898 A
Placche osteodermiche interparietali
degli Stegocefali e rispondenti centri di
ossificazione interparietali dell' uomo.
Rendiconti Reale Instituto Lombardo Sci.
e Lett. (2), xxxi, pp. 211-228, with pl. ii.

———   1898 B
Omologie craniali fra Ittiosauri e feti
dell' uomo e d'altri mammiferi. Ri-
cherche e considerazioni relative all'on-
togenia dei fossili.
Rendiconti Reale Instituto Lombardo Sci. e
Lett. (2), xxxi, pp. 631-641, with pl. ii.

———   1898 C
Il canale cranio-faringeo negli ittios-
auri omologo a quello dell' uomo e
d'altri mammiferi.
Rendiconti Reale Instituto Lombardo Sci. e
Lett. (2), xxxi, pp. 761-771, with pl. iii.

Maggi, Leopoldo—Continued.   1898 D
Le ossa sovraorbitali nei mammiferi.
Rendiconti Reale Instituto Lombardo Sci. e
Lett. (2), xxxi, pp. 1089-1099, with pl. iv.

———   1898 E
Serie di ossicini del tegmen cranii in
alcuni cani (*Canis*), e loro omologhi
ed omotopi in alcuni Storioni (*Acipenser*).
Rendiconti Reale Instituto Lombardo Sci. e
Lett. (2), xxxi, pp. 1473-1492, with pl. v.

Magnus, Hugo.   1869 A
Physiologisch-anatomische Unter-
suchungen über das Brustbein der
Vögel.
Archiv Anat., Physiol. u. wissensch. Med.,
1868, pp. 682-710, with pls. xvi, xvii.
Issued in January, 1869.

Major, C. J. Forsyth.   1873 A
Nagerüberreste aus Bohnerzen Sud-
deutschlands und der Schweiz.
Palæontographica, xxii, 75-130, pls. iii-vi.
Contains remarks on dentition of various
North American genera of mammals.

———   1877 A
Beiträge zur Geschichte der fossilen
Pferde, insbesondere Italiens. Erster
Theil.
Abhandl. schweizer. paläontolog. Gesellsch.,
iv, pp. 1-16, with 4 pls.

———   1880 A
Beiträge zur Geschichte der fossilen
Pferde, insbesondere Italiens. Zweiter
Theil.
Abhandl. schweizer. paläontolog. Gesellsch.,
vii, pp. 1-154, with 3 double plates.
This is the continuation and conclusion of
Major, C. J. F., 1877 A.

———   1893 A
On some Miocene squirrels, with remarks on the dentition and classification of the *Sciurinæ*.
Proc. Zool. Soc. Lond., 1893, pp. 179-215, with
pls. viii-xi.

———   1899 A
On fossil and recent *Lagomorpha*.
Trans. Linn. Soc. Lond. (2), vii, pp. 433-520,
with pls. xxxvi-xxxix.

Malmgren, A. J.   1864 A
Om tandbyggnaden hos Hvalrossen
(*Odobænus rosmarus* L.) och tandom-
bytet hos hans of ödda unge.
Overs. Vetensk.-Akad. Förhandl., xx, pp.
505-522, with pl. vii.

———   1865 A
Ueber den Zahnbau des Walrosses
(*Odobænus rosmarus* L.) und über den
Zahnbau seines ungeborenen Junges.
Archiv Naturgesch., xxxi, i, pp. 182-202.
Translation of Malmgren, 1864 A.

Mandl, L.   1839 A
Recherches sur la structure intime
des écailles des poissons.
Ann. Sci. Naturelles (2), Zool., xi, pp. 337–371,
with pl. ix.

Männer, H.   1899 A
Beiträge zur Entwicklungsgeschichte
der Wirbelsäule bei Reptilien.
Zeit-chr. wissensch. Zool. lxvi, pp. 43–68,
with pls. iv–vii.

Mantell, G. A.   1846 A
Description of footmarks and other
imprints on a slab of New Red sandstone from Turners Falls, Mass., collected by Dr. James Deane, of Greenfield, United States.
Quart. Jour. Geol. Soc., ii, p. 38.

——   1848 A
On the structure of the jaws and teeth
of the Iguanodon.
Philos. Trans. Roy. Soc. Lond., pp. 183–202,
pls. 16–19.

——   1848 B
A brief notice of organic remains recently discovered in the Wealden
formation.
Quart. Jour. Geol. Soc., v, pp. 37–43, with
pl. iii.
Describes remains of *Iguanodon*.

——   1849 A
Additional observations on the osteology of the *Iguanodon* and *Hylæosaurus*.
Philos. Trans. Roy. Soc. Lond., pp. 271–305,
pls. 26–32.
This paper includes "Notes on the vertebral
column of the Iguanodon," by A. G. Melville.

Marck, W. v.   1863 A
Fossile Fische, Krebse und Pflanzen
aus dem Plattenkalk der jüngsten
Kreide in Westphalen.
Palæontographica, xi, pp. 1–83, with pls.
i–xiv.

Marcou, John B.   1887 A
Review of the progress of North
American palæontology for the year
1886.
Amer. Naturalist, xxi, pp. 532–544.

——   1888 A
Review of the progress of North
American palæontology for the year
1887.
Amer. Naturalist, xxii, pp. 679–691.

Marcou, Jules.   1858 A
Geology of North America, with two
reports on the prairies of Arkansas and
Texas, the Rocky Mountains of New
Mexico, and the Sierra Nevada of Cali-

Marcou, Jules—Continued.
fornia, originally made for the United
States Government.
Pp. i–vi, 1–144, with 7 pls. and 3 maps.
The original description of *Ptychodus whip-
plei* occurs on page 33.

Markert, F.   1896 A
Die Flossenstachelen von *Acanthias*.
Ein Beitrag zur Kenntniss der Hart-
substanzgebilde der Elasmobranchier.
Zool. Jahrbücher, Abth. Anat. u. Ontol., ix,
pp. 664–722, with pls. xlvi–xlix and text-
figs. A–K.

Marsh, Dexter.   1848 A
Fossil footprints.
Amer. Jour. Sci. (2) vi, pp. 272–274.
No species are named.

Marsh, O. C.   1862 A
On the saurian vertebræ from Nova
Scotia.
Amer. Jour. Sci. (2), xxxiii, p. 278.
Eosaurus acadianus; name, but no descrip-
tion.

——   1862 B
Description of the remains of a new
Enaliosaurian (*Eosaurus acadianus*)
from the Coal formation of Nova Scotia.
Amer. Jour. Sci. (2), xxxiv, pp. 1–16, with
pls. i, ii.
An abridgment of this description is found
in Canadian Naturalist and Geologist, vol. vii,
pp. 205–213.

——   1863 A
Description of the remains of a new
Enaliosaurian (*Eosaurus acadianus*)
from the Coal formation of Nova Scotia.
Quart. Jour. Geol. Soc., xix, pp. 52–56.
Abstract of Marsh, 1862 B.

——   1867 A
Discovery of additional *Mastodon* re-
mains at Cohoes, N. Y.
Amer. Jour. Sci. (2), xliii, pp. 115–116.

——   1868 A
Notice of a new and diminutive species of fossil horse (*Equus parvulus*)
from the Tertiary of Nebraska.
Amer. Jour. Sci. (2), xlvi, pp. 374–375.
Reprinted in Ann. and Mag. Nat. Hist. (4),
iii, pp. 95–96, 1869; Kosmos, Leipzig, iv, p. 346,
1869. Abstract in Neues Jahrb. Mineral., 1869,
p. 767.

——   1869 A
Notice of new reptilian remains from
the Cretaceous of Brazil.
Amer. Jour. Sci. (2), xlvii, pp. 390–392.
Contains notice of *Thecocampsa squankensis.*
Reprinted in Ann. and Mag. Nat. Hist. (4),
iii, pp. 442–444. Abstract in Neues Jahrb.
Mineral, 1871, p. 112.

Marsh, O. C.—Continued.   1869 B

Notice of some new mosasauroid reptiles from the Green-sand of New Jersey.

Amer. Jour. Sci. (2), xlviii, pp. 392–397.
Abstract in Canadian Naturalist, iv, p. 331, and Geol. Magazine, vii, 1870, pp. 376–377.

———   1869 C

Description of a new and gigantic fossil serpent (*Dinophis grandis*) from the Tertiary of New Jersey.

Amer. Jour. Sci. (2), xlviii, pp. 397–400.
Extract of this paper published in Amer. Naturalist, vol. iv, p. 254.

———   1870 A

Notice of some fossil birds from the Cretaceous and Tertiary formations of the United States.

Amer. Jour. Sci. (2), xlix, pp. 205–217.
Abstracts in Geol. Magazine, vii, 1870, p. 318; Amer. Naturalist, iv, 1870, p. 310. Proc. Acad. Nat. Sci. Phila., xxii, 1870, pp. 5–6.

———   1870 B

Note on the remains of fossil birds.

Amer. Jour. Sci. (2), xlix, p. 272.

———   1870 C

Notice of a new species of gavial from the Eocene of New Jersey.

Amer. Jour. Sci. (2), l, pp. 97–99.
Reprinted in Geol. Magazine, vii, p. 427, 1870.

———   1870 D

Discovery of the Mauvaises Terres formation in Colorado.

Amer. Jour. Sci. (2), l, p. 292.
Reprinted in Canadian Naturalist, v, p. 240, 1871.

———   1870 E

Notice of some new Tertiary and Cretaceous fishes.

Proc. Amer. Assoc. Adv. Sci., 18th meeting, Salem, 1869, pp. 227–230.
Abstract describing briefly *Histiophorus parvulus, Embalorhynchus kinnei, Phyllodus elegans, P. curvidens, Myliobates bisulcus, Dipristis metrsii, Enchodus semistriatus.*

———   1870 F

[Remarks on *Hadrosaurus minor, Mosasaurus crassidens, Liodon laticaudus, Baptosaurus,* and *Rhinoceros matutinus.*]

Proc. Acad. Nat. Sci. Phila., 1870, pp. 2–3.

———   1870 G

[Remarks on *Laornis edvardsianus, Palæotringa littoralis, P. vetus, Telmatornis priscus, T. affinis, Grus haydeni,* and *Puffinus conradii.*]

Proc. Acad. Nat. Sci. Phila., 1870, pp. 5–6.
Names, but no descriptions.

Marsh, O. C.—Continued.   1870 H

New fossil turkey.

Amer. Naturalist, iv, p. 317.
Notice of finding of *Meleagris altus.*

———   1870 I

Announcement of discovery of *Meleagris altus* and of *Dicotyles antiquus.*

Proc. Acad. Nat. Sci. Phila., 1870, p. 11.
No descriptions are furnished.

———   1871 A

Description of some new fossil serpents, from the Tertiary deposits of Wyoming.

Amer. Jour. Sci. (3), i, pp. 322–329.

———   1871 B

Notice of some new fossil reptiles from the Cretaceous and Tertiary formations.

Amer. Jour. Sci. (3), i, pp. 447–459.
Reprinted in Boll. R. Com. Geol. d'Italia, iii, pp. 278–283, 1872. See also Kosmos, vi, pp. 476–479; Pop. Sci. Rev., London, 1871, p. 436, and Neues Jahrb. Mineral, 1871, p. 890.

———   1871 C

Note on a new and gigantic species of pterodactyle.

Amer. Jour. Sci. (3), i, p. 472.

———   1871 D

Notice of some new fossil mammals from the Tertiary formation.

Am. Jour. Sci. (3), ii, pp. 35–44.
Reprinted in Boll. R. Com. Geol. d'Italia, iii, pp. 343–350, 1872.

———   1871 E

Notice of some new fossil mammals and birds from the Tertiary formations of the West.

Amer. Jour. Sci. (3), ii, pp. 120–127.
Reprinted in Boll. R. Com. Geol. d'Italia, iii, pp. 350–353, 1872.

———   1871 F

[Description of *Lophiodon validus.*]

Proc. Acad. Nat. Sci. Phila., 1871, p. 10.

———   1871 G

[Communication on some new reptiles and fishes from the Cretaceous and Tertiary.]

Proc. Acad. Nat. Sci. Phila., 1871, pp. 103–105.

———   1872 A

Discovery of a remarkable fossil bird.

Amer. Jour. Sci. (3), iii, pp. 56–57.
Abstract in Nature, v, p. 348, 1872.
First announcement of *Hesperornis regalis.*

Marsh, O. C.—Continued.   1872 B

Discovery of additional remains of *Pterosauria*, with descriptions of two new species.

Amer. Jour. Sci. (3), iii, pp. 241–248.
Abstract in Nature, v, p. 151, 1872.

——   1872 C

Discovery of the dermal scutes of mosasauroid reptiles.

Amer. Jour. Sci. (3), iii, pp. 290–292.

——   1872 D

Notice of a new species of *Hadrosaurus*.

Amer. Jour. Sci. (3), iii, p. 301.
Describes *H. agilis*.

——   1872 E

Preliminary description of *Hesperornis regalis*, with notices of four other new species of Cretaceous birds.

Amer. Jour. Sci. (3), iii, pp. 360–365.
Abstract in Nature, vi, pp. 90, 94. Reprinted in Ann. and Mag. Nat. Hist., (4), x, pp. 212–217; Boll. R. Com. Geol. d'Italia, iii, pp. 211–217, 1872.

——   1872 F

On the structure of the skull and limbs in mosasauroid reptiles, with descriptions of new genera and species.

Amer. Jour. Sci. (3), iii, pp. 448–464, with 4 pls.
See Kosmos, vi, pp. 447–449; Jour. de Zoologie, ii, p. 583, for abstracts.

——   1872 G

Preliminary description of new Tertiary mammals. Pt. I.

Amer. Jour. Sci. (3), iv, pp. 122–128 and erratum on p. 504.
Abstract of this paper in Neues Jahrb. Mineral., 1872, pp. 990–991.

——   1872 H

Note on *Rhinosaurus*.

Amer. Jour. Sci. (3), iv, p. 147.
Rhinosaurus changed to *Tylosaurus*.

——   1872 I

Preliminary description of new Tertiary mammals. Parts II, III, and IV.

Amer. Jour. Sci. (3), iv, pp. 202–224.
Abstract in Neues Jahrb. Mineral., 1872, pp. 990–991. Advance copies of this paper are said to have been issued Aug. 7, 1872.

——   1872 J

Notice of some new Tertiary and post-Tertiary birds.

Amer. Jour. Sci. (3), iv, pp. 256–262.

——   1872 K

Preliminary description of new Tertiary reptiles. Parts I and II.

Amer. Jour. Sci. (3), iv, pp. 298–309.

Marsh, O. C.—Continued.   1872 L

Note on *Tinoceras anceps*.

Amer. Jour. Sci. (3), iv, p. 322.
Abstract in Pop Sci. Rev., London, 1873, p. 94.

——   1872 M

Notice of a new species of *Tinoceras*.

Amer. Jour. Sci. (3), iv, p. 323.

——   1872 N

Discovery of fossil *Quadrumana* in the Eocene of Wyoming.

Amer. Jour. Sci. (3), iv, pp. 405–406.
See also Geol. Magazine, x, 1873, p. 33, and Amer. Naturalist, vii, pp. 179–180.

——   1872 O

Note on a new genus of carnivores, from the Tertiary of Wyoming.

Amer. Jour. Sci. (3), iv, p. 406.
Reprinted in Geol. Magazine, x, 1873, pp. 33–34.
Describes *Oreocyon latidens*.

——   1872 P

Notice of a new reptile from the Cretaceous.

Amer. Jour. Sci. (3), iv, p. 406.
This paper describes the jaws of *Colonosaurus mudgei* n. g. et n. sp., which were afterwards found to be the jaws of *Ichthyornis dispar*. The paper is noticed in Geol. Magazine, x, 1873, p. 34.

——   1872 Q

Notice of a new and remarkable fossil bird.

Amer. Jour. Sci. (3), iv, p. 344.
Reprinted in Amer. Naturalist, vii, p. 50, and Ann. and Mag. Nat. Hist. (4), xi, p. 80, 1873.
Describes *Ichthyornis dispar*.

——   1872 R

Notice of some remarkable fossil mammals.

Amer. Jour. Sci. (3), iv, pp. 343–344.
Describes *Dinoceras mirabilis* and *Dinoceras lacustris*.

——   1872 S

Discovery of new Rocky Mountain fossils.

Proc. Amer. Philos. Soc., xii, pp. 578–579.
Reprinted in Ann. Sci. Geol., Paris, iii, pp. 99–100.

——   1872 T

New and remarkable fossils.

Amer. Naturalist, vi, pp. 495–497.
Copied from the Yale Courant.

——   1873 A

Notice of a new species of *Ichthyornis*.

Amer. Jour. Sci. (3), v, p. 74.
Description of *Ichthyornis celer*. Abstracts in Nature, vii, p. 310, 1873; Jour. de Zoologie, ii, pp. 40–41.

Marsh, O. C.—Continued. 1873 B

On the gigantic fossil mammals of the order *Dinocerata*.

Amer. Jour. Sci. (3), v, pp. 117-122, with pls. i, ii.

Translated into French in Ann. Sci. Naturelles (5), Zool., xvii. art. 9, pl. 21.

Jour. de Zoologie, ii, pp. 160-168.

For republication, abstracts, etc., see Marsh, 1884 F, p. 227.

—— 1873 C

On a new subclass of fossil birds (*Odontornithes*).

Amer. Jour. Sci. (3), v, pp. 161-162.

Reprinted in Amer. Naturalist, vii, pp. 115-117; Ann. and Mag. Nat. Hist. (4), xi, pp. 233-234. Translated into French in Ann. Sci. Naturelles (5), Zool., xvii, art. 9, pp. 8-11. Noticed in Geol. Magazine, x (1873), p. 115. Remarks on by Troschel in Verhandl. naturh. Vereines preuss. Rheinlande, xxx, Sitzungsber, p. 83. Announces discovery of teeth.

—— 1873 D

Fossil birds from the Cretaceous of North America.

Amer. Jour. Sci. (3), v, pp. 229-230.

—— 1873 E

Note on the dates of some of Professor Cope's recent papers.

Amer. Jour. Sci. (3), v, pp. 235-236.

Also Amer. Naturalist, vii, pp. 303-306.

—— 1873 F

Additional observations on the *Dinocerata*.

Amer. Jour. Sci. (3), v, pp. 293-296.

—— 1873 G

Supplementary note on the *Dinocerata*.

Amer. Jour. Sci. (3), v, pp. 310-311.

—— 1873 H

Notice of new Tertiary mammals.

Amer. Jour. Sci. (3), v, pp. 407-410, 485-488.

Abstract in Nature, viii, p. 76, 1873.

—— 1873 I

New observations on the *Dinocerata*.

Amer. Jour. Sci. (3), v, pp. 300-301.

—— 1873 J

On the gigantic mammals of the American Eocene.

Proc. Amer. Philos. Soc., xiii, pp. 255-256.

—— 1873 K

The fossil mammals of the order *Dinocerata*.

Amer. Naturalist, vii, pp. 146-153, with pls. i and ii.

Published in part in Amer. Jour. Sci. (3), v, pp. 117-122 (Marsh, O. C., 1873 B).

Marsh, O. C.—Continued. 1873 L

On the genus *Tinoceras* and its allies.

Amer. Naturalist, vii, pp. 217-218.

In reply to Prof. E. D. Cope, Amer. Naturalist, vii, pp. 157-160 (Cope, E. D., 1873 W).

—— 1873 M

Tinoceras and its allies.

Amer. Naturalist, vii, pp. 306-308.

—— 1873 N

On some of Professor Cope's recent investigations.

Amer. Naturalist, vii, pp. 51-52.

—— 1873 O

Note on the dates of some of Professor Cope's recent papers.

Amer. Naturalist, vii, p. 173.

—— 1873 P

Reply to Professor Cope's explanation.

Amer. Naturalist, vii, appendix to June number, pp. i-ix.

—— 1874 A

On the structure and affinities of the *Brontotheridæ*.

Amer. Jour. Sci. (3), vii, pp. 81-86, with pls. i and ii.

Abstracts: Nature, ix, p. 227, 1874; Amer. Naturalist, vol. viii, pp. 79-85, 1874; Jour. de Zoologie, iii, pp. 61-62, 1874.

—— 1874 B

Notice of new equine mammals from the Tertiary formation.

Amer. Jour. Sci. (3), vii, pp. 247-258, with 5 figs. in the text.

Reprinted in Ann. and Mag. Nat. Hist. (4), xiii, pp. 397-400. Abstract in Nature, ix, p. 390, 1874.

Contains discussion of origin of modern horse.

—— 1874 C

Notice of new Tertiary mammals. III.

Amer. Jour. Sci. (3), vii, pp. 531-534.

—— 1874 D

Small size of the brain in Tertiary mammals.

Amer. Jour. Sci. (3), viii, pp. 66-67.

Reprints in Ann. and Mag. Nat. Hist., xiv, p. 167, Amer. Naturalist, viii, pp. 503-504, Jour. de Zoologie, iii, pp. 326-327. Abstract in Neues Jahrbuch Mineral., p. 772, 1874, and in Nature, x, p. 273.

—— 1875 A

(An editorial note stating Professor Marsh's claim that Professor Cope's *Calamodon* is identical with Professor Marsh's *Stylinodon*.)

Amer. Jour. Sci. (3), ix, p. 151.

Marsh, O. C.—Continued.   1875 B
Notice of new Tertiary mammals. IV.

Amer. Jour. Sci. (3), ix, pp. 239–250.
A notice of this paper, containing a list of the forms described, may be found in Jour. de Zoologie, iv, 1875, pp. 98–99.

——   1875 C
New order of Eocene mammals.

Amer. Jour. Sci. (3), ix, p. 221.
Reprinted in Amer. Naturalist, ix, pp. 182–183; in Nature, xi, 1875, p. 368; in Ann. and Mag. Nat. Hist. (4), xv, p. 307; Jour. de Zoologie, iv, 1875, pp. 70–71; Pop. Sci. Rev., London, 1875, p. 207.
The new order proposed is denominated *Tillodontia.*

——   1875 D
On the *Odontornithes*, or birds with teeth.

Amer. Jour. Sci. (3), x, pp. 402–408, with pls. ix, x.
Reprinted in Geol. Magazine (2), iii, 1876, pp. 49–53, pl. ii; also in Jour. de Zoologie, iv, 1875, pp. 494–502, pl. xv.

——   1875 E
Odontornithes, or birds with teeth.

Amer. Naturalist, ix, pp. 624–631, with pls. ii, iii.
Same as the preceding paper.

——   1875 F
Mastodon of Otisville, Orange County, N. Y.

Amer. Jour. Sci. (3), ix, p. 483.

——   1876 A
Principal characters of the *Dinocerata.*

Amer. Jour. Sci. (3), xi, pp. 163–168, pls. ii–vi.
Reprinted in Jour. de Zoologie, v, pp. 136–145, pl. iv.
Abstracts, etc., Nature, xiii, p. 374, 1876; Zeitschr. gesammt. Naturwiss. xiv, pp. 31–32, 1876; Neues Jahrb. Mineral., pp. 780–781, 1876; Pop. Sci. Rev., London pp. 326–327, 1876.

——   1876 B
Principal characters of the *Tillodtia.*

Amer. Jour. Sci. (3), xi, pp. 249–252, with pls. viii and ix.
A translation of this paper appeared in Jour. de Zoologie, v, pp. 244–248, pl. xi. Abstract in Nature, xiii, p. 374.

——   1876 C
Principal characters of the *Brontotheridæ.*

Amer. Jour. Sci. (3), xi, pp. 335–340, pls. x–xiii and 2 figs. in text.
A translation of this paper appeared in Jour. de Zoologie, v, 1876, pp. 248–255; pl. xii. Abstract in Nature, xiv, p. 36.

Marsh, O. C.—Continued.   1876 D
On some characters of the genus *Coryphodon.*

Amer. Jour. Sci. (3), xi, pp. 425–428, with 1 page of figs.
Abstracts: Pop. Sci. Rev., London.. 1876, p. 327; Nues Jahrb. Mineral., 1876, p. 781; Archiv. Sci. Phys. et Nat., Geneva, lvi, pp. 273, 274.

——   1876 E
Notice of a new sub-order of *Pterosauria.*

Amer. Jour. Sci. (3), xi, pp. 507–509.
Abstract in Ann. and Mag. Nat. Hist. (4), xviii, pp. 195–196; Jour. de Zoologie, v., pp. 457–458.

——   1876 F
Notice of new *Odontornithes.*

Amer. Jour. Sci. (3), xi, pp. 509–511.
Translated into French in Jour. de Zoologie, v., 1876, pp. 304–306.

——   1876 G
Recent discoveries of extinct animals.

Amer. Jour. Sci. (3), xii, pp. 59–61.
Same, Amer. Naturalist, x, pp. 436–439.
Abstract Neues Jahrb. Mineral., 1876, p. 782.

——   1876 H
Notice of new Tertiary mammals. V.

Amer. Jour. Sci. (3), xii, pp. 401–404.
Abstracts in Arch. Sci. Phys. et Nat., Geneva, xlix, pp. 127–128; Pop. Sci. Rev., London, 1877, pp. 97–98.

——   1876 I
Principal characters of American pterodactyls.

Amer. Jour. Sci. (3), xii, pp. 479–480.

——   1877 A
Characters of *Coryphodontidæ.*

Amer. Jour. Sci. (3), xiv, pp. 81–85, with pl. iv.
Abstracts: Amer. Naturalist, xi, 1877, p. 500; Neues Jahrb. Mineral. 1877, p. 767. Reprint in Jour. de Zoologie, vi, 1877, pp. 380–385.

——   1877 B
Characters of the *Odontornithes*, with notice of a new allied genus.

Amer. Jour. Sci. (3), xiv, pp. 85–87, with pl. v.
Reprinted in Jour. de Zoologie, vi, pp. 385–389.
Baptornis advenus, new genus and species, is described.

——   1877 C
Notice of a new and gigantic dinosaur.

Amer. Jour. Sci. (3), xiv, pp. 87–88.
Reprinted in Jour. de Zoologie, vi, pp. 248–250.
Titanosaurus montanus described.

Marsh, O. C.—Continued. 1877 D
New vertebrate fossils.
Amer. Jour. Sci. (3), xiv, pp. 249–256.
Notice of this paper occurs in Amer. Naturalist, xi, p. 629.

——— 1877 E
Introduction and succession of vertebrate life in America. Vice-president's address before Amer. Assoc. Adv. Science, 1877.
Amer. Jour. Sci. (3), xiv, pp. 337–378.
Proc. Amer. Assoc. Adv. Science, Nashville meeting, 1877, pp. 211–256; published 1878 (Marsh, 1878 G).
For reprints, see Marsh, 1884 F, p. 234.

——— 1877 F
New order of extinct *Reptilia* (*Stegosauria*) from the Jurassic of the Rocky Mountains.
Amer. Jour. Sci. (3), xiv, pp. 513–514.

——— 1877 G
Notice of new dinosaurian reptiles from the Jurassic formation.
Amer. Jour. Sci. (3), xiv, pp. 514–516.

——— 1877 H
Brain of *Coryphodon*.
Amer. Naturalist, xi, p. 375.
A criticism on Professor Cope's paper in same journal, xi, p. 312.

——— 1878 A
New species of *Ceratodus*, from the Jurassic.
Amer. Jour. Sci. (3), xv, p. 76.
Describes *Ceratodus güntheri*.

——— 1878 B
Notice of new dinosaurian reptiles.
Amer. Jour. Sci. (3), xv, pp. 241–244, with 2 figs. in the text.
Abstract in Pop. Sci. Rev., London, 1878, pp. 210–211.

——— 1878 C
Notice of new fossil reptiles.
Amer. Jour. Sci. (3), xv, pp. 409–411,

——— 1878 D
Fossil mammal from the Jurassic of the Rocky Mountains.
Amer. Jour. Sci. (3), xv, p. 459.
Reprinted in Ann. and Mag. Nat. Hist. (5), vol. ii, p. 108; Pop. Sci. Rev., London, 1878, pp. 322–323.
Describes *Dryolestes priscus* Marsh.

——— 1878 E
New pterodactyle from the Jurassic of the Rocky Mountains.
Amer. Jour. Sci. (3), xvi, pp. 233–234.
Abstracts in Neues Jahrb. Mineral., 1878, p. 895; Pop. Sci. Rev., London, 1878, p. 436.
Describes *Pterodactylus montanus*.

Marsh, O. C.—Continued. 1878 F
Principal characters of American Jurassic dinosaurs. Part I.
Amer. Jour. Sci. (3), xvi, pp. 411–416, pls. iv–x.
Abstract in Neues Jahrb. Mineral., 1880, pp. 256–257.

——— 1878 G
The introduction and succession of vertebrate life in America.
Proc. Amer. Assoc. Adv. Science, 26th meeting, Nashville, 1877, pp. 211–258.
Also Pop. Sci. Month., xii, pp. 513–527; 672–697.
Vice-presidential address before Amer. Assoc. Adv. Sci., 1877. Same as Marsh, 1877 E, with footnotes.

——— 1878 H
Brain of a fossil mammal.
Nature, xvii, p. 340.
On the brain of *Coryphodon*.

——— 1879 A
A new order of extinct reptiles (*Sauranodonta*) from the Jurassic formation of the Rocky Mountains.
Amer. Jour. Sci. (3), xvii, pp. 85–86.
Abstract in Ann. and Mag. Nat. Hist. (5), iii, pp. 175–176; Pop. Sci. Rev., London, 1879, p. 206.

——— 1879 B
Principal characters of American Jurassic dinosaurs. Part II.
Amer. Jour. Sci. (3), xvii, 86–92, with pls. iii–x.
Abstract in Neues Jahrb Mineral., 1880, pp. 257–258.

——— 1879 C
The vertebræ of recent birds.
Amer. Jour. Sci. (3), xvii, 266–269, with 5 figs. in text.

——— 1879 D
Polydactyle horses, recent and extinct.
Amer. Jour. Sci. (3), xvii, 499–505, with text figures.
Reprinted in Kosmos, v, pp. 432–438. Abstracts in Pop. Sci. Rev., London, 1879, pp. 318–319; Neues Jahrb. Mineral., 1881, pp. 103–104.

——— 1879 E
Notice of a new Jurassic mammal.
Amer. Jour. Sci. (3), xviii, pp. 60–61.
Reprinted in Ann. and Mag. Nat. Hist. (5), iv, pp. 167–168; Geol. Magazine (2), vi, pp. 371–372.
Describes new genus *Stylacodon*, with new species *S. gracilis*. New family *Stylodontidæ* proposed.

Marsh, O. C.—Continued. 1879 F
Additional remains of Jurassic mammals.
Amer. Jour. Sci. (3), xviii, 215–216, with 1 fig. in text.
Abstract in Pop. Sci. Rev., Lond., 1879, p. 428.

—— 1879 G
Notice of new Jurassic mammals.
Amer. Jour. Sci. (3), pp. xviii, 396–398, with 1 fig. in text.

—— 1879 H
Notice of new Jurassic reptiles.
Amer. Jour. Sci. (3), xviii, pp. 501–505, with pl. iii and 1 text fig.

—— 1879 I
History and methods of palæontological discovery.
Amer. Sci. (3), xviii, pp. 323–359.
Reprinted in Nature, xx, pp. 494–499; 515–521. Kosmos, vi, pp. 339–352, 425–445, 1880; Proc. Amer. Assoc. Adv. Sci., xxviii, p. 1–42, Salem, 1880; Pop. Sci. Month., xvi, pp. 219–236; 363–380.

—— 1880 A
New characters of mosasauroid reptiles.
Amer. Jour. Sci. (3), xix, pp. 83–87, with pl. i and 4 figs. in text.
Abstract in Neues Jahrb. Mineral., 1880, pp. 104–105.

—— 1880 B
The limbs of *Sauranodon*, with notice of a new species.
Amer. Jour. Sci. (3), xix, pp. 169–171, with 1 fig. in text.
For notice of this paper, by C. G. Giebel, see Zeitschr. gesammt. Naturwiss. (3), v, p. 357, 1880; Abstracts in Kosmos, vii, pp. 74–77; Neues Jahrb. Mineral., 1880, pp. 105–106.

—— 1880 C
Principal characters of American Jurassic dinosaurs. Part III.
Amer. Jour. Sci. (3), xix, pp. 253–259, with pls. vi–xi.
Abstract in Neues Jahrb. Mineral., 1880, pp. 106–107.

—— 1880 D
The sternum in dinosaurian reptiles.
Amer. Jour. Sci. (3), xix, pp. 395–396, with pl. xviii.
For synopses of this paper, see Kosmos, vii, p. 317; Neues Jahrb. Mineral., 1881, pp. 104–105.

—— 1880 E
Note on *Sauranodon*.
Amer. Jour. Sci. (3), xix, p. 491.
Notes change of name to *Baptanodon*.

Marsh, O. C.—Continued. 1880 F
Notice of Jurassic mammals representing two new orders.
Amer. Jour. Sci. (3), xx, pp. 235–239, with 1 fig. in the text.
Abstract in Arch. Sci. Phys. et Nat., Geneva, 1880, iv, pp. 420–422.

—— 1880 G
Odontornithes: A monograph on the extinct toothed birds of North America, with 34 plates and 40 woodcuts.
Report of the geological exploration of the 40th parallel, by Clarence King; pp. i–xv; 1–201. Washington, 1880.
For abstracts, see Marsh, O. C., 1896 D. p. 26.

—— 1880 H
List of genera established by Prof. O. C. Marsh, 1862–1879.
Pp. 1–12. No date, but probably issued near beginning of year 1880. For continuation of the list, see Marsh, O. C., 1889 I.

—— 1881 A
Principal characters of American Jurassic dinosaurs. Part IV. Spinal cord, pelvis, and limbs of *Stegosaurus*.
Amer. Jour. Sci. (3), xxi, pp. 167–170, with pls. vi–viii.
See also Kosmos, ix, pp. 319–321; Neues Jahrb. Mineral., 1881, pp. 109–110, 270–271.

—— 1881 B
New order of extinct Jurassic reptiles (*Cœluria*).
Amer. Jour. Sci. (3), xxi, pp. 339–340, with pl. x.
Reprinted in Kosmos, ix, pp. 464–465.

—— 1881 C
Discovery of a fossil bird in the Jurassic of Wyoming.
Amer. Jour. Sci. (3), xxi, pp. 341–342.
Reprinted in Ann. and Mag. Nat. Hist. (5), vii, 1881, pp. 488–489.
Describes *Laopteryx priscus* Marsh.

—— 1881 D
Note on American pterodactyls.
Amer. Jour. Sci. (3), xxi, pp. 342–343.
Abstract in Neues Jahrb. Mineral., 1881, p. 415.

—— 1881 E
Principal characters of American Jurassic dinosaurs. Part V.
Amer. Jour. Sci. (3), xxi, pp. 417–423; pls. xii–xviii, with 7 plates.

—— 1881 F
New Jurassic mammals.
Amer. Jour. Sci. (3), xxi, pp. 511–513.

Marsh, O. C.—Continued.   1881 G
Restoration of *Dinoceras mirabile.*
Amer. Jour. Sci. (3), xxii, pp. 31-32, with pl. ii.
Abstract in Archiv. Sci. Phys. et Nat., Geneva, vi, 1881, pp. 323-324.

——   1881 H
Jurassic birds and their allies.
Amer. Jour. Sci. (3), xxii, pp. 337-340.
Reprinted in Pop. Sci. Month., xx, pp. 312-315; Science (1), ii, pp. 512-513; Geol. Magazine (2), viii, pp. 485-487; Nature, xxv, pp. 22-23; Kosmos, x, pp. 231-234; Ann. and Mag. Nat. Hist. (5), viii, pp. 451-455.

——   1882 A
Classification of the *Dinosauria.*
Amer. Jour. Sci. (3), xxiii, pp. 81-86.
Read before the National Acad. Sci., Nov. 14, 1881.
Reprinted in Kosmos, v, pp. 382-387; Geol. Magazine (2), ix, pp. 80-85; Ann. and Mag. Nat. Hist. (5), ix, pp. 79-84; Nature, xxv, pp. 244-245.

——   1882 B
The wings of pterodactyles.
Amer. Jour. Sci. (3), xxiii, pp. 251-256, with pl. iii.
See also Geol. Magazine (2), ix, pp. 205-210; Kosmos, vi, pp. 103-108; and Nature, xxv, pp. 531-533.
Describes *Rhamphorhynchus phyllurus* Marsh.

——   1882 C
Jurassic birds and their allies.
Report Brit. Assoc. Adv. Sci., 51st meeting, York, 1881, pp. 661-662.
Same as Marsh, O. C., 1881 H.

——   1883 A
Principal characters of American Jurassic dinosaurs. Part VI. Restoration of *Brontosaurus.*
Amer. Jour. Sci., (3), xxvi, pp. 81-85, with pl. i.
See also Geol. Magazine for 1883, pp. 386-388.

——   1883 B
On the supposed human footprints recently found in Nevada.
Amer. Jour. Sci. (3), xxvi, pp. 139-140, with 2 figs. in the text.
Abstract in Nature, xxviii, pp. 370-371; Neues Jahrb. Mineral., 1884, pp. 262-263.

——   1883 C
Birds with teeth.
Third annual report of the U. S. Geol. Surv. to the Secretary of the Interior, 1881-82, by J. W. Powell, director, Washington, 1883, pp. 45-88, with 33 figs. in the text.

Marsh, O. C.—Continued.   1884 A
Principal characters of American Jurassic dinosaurs. Part VII. On the *Diplodocidæ,* a new family of the *Sauropoda.*
Amer. Jour. Sci. (3), xxvii, pp. 160-168, with pls. iii and iv.
Abstract with 3 figures in Science, iii, p. 199; Kosmos, ii, pp. 350-357. Notice by Cope, Amer. Naturalist, xix, p. 67. Reproduced in Geol. Magazine (2), x, pp. 99-107, with 10 figs. in the text.

——   1884 B
Principal characters of American Jurassic dinosaurs. Part VIII. The order of *Theropoda.*
Amer. Jour. Sci. (3), xxvii, pp. 329-340, with pls. viii-xiv.
Abstract with figures in Science, iii, pp. 542-544; Kosmos, ii, pp. 357-365. Reprinted in Geol. Magazine (3), i, pp. 252-262. Notice by Cope in Amer. Naturalist, xix, p. 67.

——   1884 C
A new order of extinct Jurassic reptiles.
Amer. Jour. Sci. (3), xxvii, p. 341, with 1 text fig.

——   1884 D
Principal characters of American Cretaceous pterodactyls. Part I. The skull of *Pteranodon.*
Amer. Jour. Sci. (3), xxvii, pp. 422-426, with pl. xv.
Reprinted in Geol. Magazine (3), i, pp. 345-348.

——   1884 E
On the united metatarsal bones of *Ceratosaurus.*
Amer. Jour. Sci. (3), xxviii, pp. 161-162, with 2 figs. in the text.

——   1884 F
Dinocerata. A monograph of an extinct order of gigantic mammals.
Monographs U. S. Geol. Surv., x, pp. i-xviii; 1-237, with 56 pls. and 200 woodcuts in text. Washington, D. C.
Reviewed in Geol. Magazine, (3), ii, 1885, pp. 212-228, with many figures reproduced.
Abstracts in Amer. Jour. Sci. (3), xxix, pp. 173-204; Geol. Magazine (3), ii, pp. 212-228; Nature, xxxii, pp. 97-99; Pop. Sci. Month., xxviii, pp. 133-134; Neues Jahrb. Mineral., 1886, i, pp. 339-341; Ann. Sci. Géol., Paris, xvii, pp. 1-11.

——   1884 G
The classification and affinities of the dinosaurian reptiles.
Nature, xxxi, pp. 68-69.
Abstract in Science, iv, p. 261.
Read before Brit. Assoc. Adv. Sci., Montreal meeting, 1884.

Marsh, O. C.—Continued.   1885 A
Names of extinct reptiles.
Amer. Jour. Sci. (3), xxix, p. 169.

———   1885 B
The gigantic mammals of the order
Dinocerata.
Fifth Annual Report of the U. S. Geol.
Surv., Washington, D. C., 1885, pp. 243–302, with
100 figs. in the text.

———   1885 C
On American Jurassic mammals.
Report Brit. Assoc. Adv. Sci., 54th meeting,
Montreal, 1884, pp. 734–736.
Illustrated with a geological section.

———   1885 D
On the classification and affinities of
dinosaurian reptiles.
Report Brit. Assoc. Adv. Sci., 54th meeting,
Montreal, 1884, pp. 763–765.
Same as Marsh, O. C., 1884 G.

———   1887 A
American Jurassic mammals.
Amer. Jour. Sci. (3), xxxiii, pp. 326–348,
with pls. vii–x.
See also Geol. Magazine (3), iv, 1887, pp.
242–247, 289–299, pls. vii–ix.

———   1887 B
Notice of new fossil mammals.
Amer. Jour. Sci. (3), xxxiv, pp. 323–331,
with 12 figs. in the text.

———   1887 C
Principal characters of American
Jurassic dinosaurs. Part IX. The
skull and dermal armor of *Stegosaurus.*
Amer. Jour. Sci. (3), xxxiv, pp. 413–417,
with pls. vi–ix.
Reprinted in Geol. Magazine (3), v, pp. 11–15.

———   1888 A
Notice of a new genus of *Sauropoda*
and other new dinosaurs from the
Potomac formation.
Amer. Jour. Sci. (3), xxxv, pp. 89–94, with 9
figs. in the text.

———   1888 B
Notice of a new fossil sirenian from
California.
Amer. Jour. Sci. (3), xxxv, pp. 94–96, with 3
figs. in the text.

———   1888 C
A new family of horned dinosaurs
from the Cretaceous.
Amer. Jour. Sci. (3), xxxvi, pp. 477–478, with
pl. xi.

———   1888 D
The skull and dermal armor of
Stegosaurus.
Geol. Magazine (3), v, pp. 11–15, pls. i–iii,
and 3 figs. in the text.

Marsh, O. C.—Continued.   1889 A
Restoration of *Brontops robustus,* from
the Miocene of America.
Amer. Jour. Sci. (3), xxxvii, pp. 163–165,
with pl. vi.
Reprints in Rep. Brit. Assoc. for 1888, pp.
706–707, 1889; Geol. Magazine (3), vi, pp. 99–101.

———   1889 B
Comparison of the principal forms of
the *Dinosauria* of Europe and America.
Amer. Jour. Sci. (3), xxxvii, pp. 323–331.
Also in Geol. Magazine (3), vi, pp. 204–210.

———   1889 C
Notice of new American dinosaurs.
Amer. Jour. Sci. (3), xxxvii, pp. 331–336,
with figs. 1–5.

———   1889 D
Discovery of Cretaceous *Mammalia.*
Amer. Jour. Sci. (3), xxxviii, pp. 81–92, with
pls. ii–v.
Review by W. Dames in Neues Jahrb. Min-
eral., 1890, pp. 141–143.

———   1889 E
Notice of gigantic horned *Dinosauria*
from the Cretaceous.
Amer. Jour. Sci. (3), xxxviii, pp. 173–175,
with 1 text fig.

———   1889 F
Discovery of Cretaceous *Mammalia,*
pt. ii.
Amer. Jour. Sci. (3), xxxviii, pp. 177–180,
with pls. vii–viii.
For review by W. Dames see Neues Jahrb.
Mineral., 1890, pp. 141–143.

———   1889 G
Skull of the gigantic *Ceratopsidæ.*
Amer. Jour. Sci. (3), xxxviii, pp. 501–506,
with pl. xii.
Reprinted in Geol. Magazine (3), vii, pp. 1–5,
pl. i.

———   1889 H
Restoration of *Brontops robustus* from
the Miocene of America.
Report Brit. Assoc. Adv. Sci., 58th meeting,
Bath, 1888, pp. 706–707.
Same as Marsh, O. C., 1889 A.

———   1889 I
Additional genera established by
Prof. O. C. Marsh, 1880–1889.
Pp. 13–17.—Issued privately, probably toward
end of year 1889. No date. A list of proposed
genera, with derivations and place of publication. For beginning of list see Marsh, O. C.,
1880 H.

———   1890 A
Description of new dinosaurian reptiles.
Amer. Jour. Sci. (3), xxxix, pp. 81–86, with
pl. i.

Marsh, O. C.—Continued. 1890 B

Distinctive characters of the order *Hallopoda.*

Amer. Jour. Sci. (3), xxxix, pp. 415–417, with 1 text figure.

—— 1890 C

Additional characters of the *Ceratopsidæ,* with notice of new Cretaceous dinosaurs

Amer. Jour. Sci. (3), xxxix, 418–426, with pls. v–vii.

—— 1890 D

Notice of new Tertiary mammals.

Amer. Jour. Sci. (3), xxxix, pp. 523–525.

—— 1890 E

Notice of some extinct *Testudinata.*

Amer. Jour. Sci. (3), xl, pp. 177–179, with pls. vii, viii.

—— 1890 F

The skull of the gigantic *Ceratopsidæ.*

Geol. Magazine (3), vii, pp. 1–5, pl. i.

—— 1891 A

A horned artiodactyle (*Protoceras celer*) from the Miocene.

Amer. Jour. Sci. (3), xli, pp. 81–82.

—— 1891 B

The gigantic *Ceratopsidæ,* or horned dinosaurs of North America.

Amer. Jour. Sci. (3), xli, pp. 167–178, pls. i–x. Reprinted in Geol. Magazine (3), viii, pp. 193–199, text figs. 1–19, pls. iv, v; pp. 241–248, text figs. 1–34.

—— 1891 C

Restoration of *Triceratops* [and *Brontosaurus*].

Amer. Jour. Sci. (3), xli, pp. 339–342, with pls. xv, xvi. Reprinted in part in Geol. Magazine (3), viii, pp. 248–250, pl. vii.

—— 1891 D

Restoration of *Stegosaurus.*

Amer. Jour. Sci. (3), xli, pp. 179–181, with pl. ix. Reprinted in Geol. Magazine (3), viii, pp. 385–387, pl. xi.

—— 1891 E

Notice of new vertebrate fossils.

Amer. Jour. Sci. (3), xlii, pp. 265–269.

—— 1891 F

Geological horizons as determined by vertebrate fossils.

Amer. Jour. Sci. (3), xlii, pp. 336–338, with pl. xii. See also Compte rend. Cong. Géolog. Internat., Washington, 1891, pp. 156–159, 1893.

Marsh, O. C.—Continued. 1891 G

On the gigantic *Ceratopsidæ* (or horned lizards) of North America.

Report of Brit. Assoc. Adv. Sci., 60th meeting, Leeds, 1890, pp. 798–795. Same as Marsh, O. C., 1891 B.

—— 1891 H

On the Cretaceous mammals of North America.

Report Brit. Assoc. Adv. Sci., 60th meeting, Leeds, 1890, pp. 853–854.

—— 1891 I

Note on Mesozoic *Mammalia.*

Proc. Acad. Nat. Sci. Phila., 1891, pp. 237–241. Also in Amer. Naturalist, vol. xxv, pp. 611–616. A reply to Osborn, H. F., 1891 A.

—— 1891 J

Restoration of *Stegosaurus.*

Geol. Magazine (3), viii, pp. 385–387, with pl. xi.

—— 1892 A

The skull of *Torosaurus.*

Amer. Jour Sci. (3), xliii, pp. 81–84, with pls. ii, iii.

—— 1892 B

Discovery of Cretaceous *Mammalia.* Part iii.

Amer. Jour. Sci. (3), xliii, pp. 249–262, with pls. v–xi.

—— 1892 C

Recent polydactyle horses.

Amer. Jour. Sci. (3), xliii, pp. 339–355, with 21 figs. in the text.

—— 1892 D

A new order of extinct Eocene mammals (*Mesodactyla*).

Amer. Jour. Sci. (3), xliii, pp. 445–449, with 2 figs. in the text.

—— 1892 E

Notice of new reptiles from the Laramie formation.

Amer. Jour. Sci. (3), xliii, pp. 449–453, with text figs. 1–4.

—— 1892 F

Notes on Triassic *Dinosauria.*

Amer. Jour. Sci. (3), xliii, pp. 543–546, with pls. xv–xvii.

—— 1892 G

Notes on Mesozoic vertebrate fossils.

Amer. Jour. Sci. (3), xliv, pp. 171–176, with pls. ii–v.

—— 1892 H

Restorations of *Claosaurus* and *Ceratosaurus.*

Amer. Jour. Sci. (3), xliv, pp. 343–350, with pls. vi–viii.

Marsh, O. C.—Continued.   1892 I
Restoration of *Mastodon americanus*,
Cuvier.
Amer. Jour. Sci. (3), xliv, p. 350, with pl.
viii.

———   1893 A
A new Cretaceous bird, allied to
Hesperornis.
Amer. Jour. Sci. (3), xlv, pp. 81–82, with 3
figs. in the text.

———   1893 B
The skull and brain of *Claosaurus*.
Amer. Jour. Sci. (3), xlv, pp. 83–86, with
pls. iv, v.

———   1893 C
Restoration of *Coryphodon*.
Amer. Jour. Sci. (3), xlvi, pp. 321–326, with
pls. v, vi.

———   1893 D
Description of Miocene *Mammalia*.
Amer. Jour. Sci. (3), xlvi, pp. 407–412, with
pls. vii–x.

———   1893 E
Restoration of *Mastodon americanus*,
Cuvier.
Geol. Magazine (3), x, p. 164, pl. viii. Same
as Marsh, O. C., 1892 I.

———   1893 F
Restorations of *Anchisaurus*, *Ceratosaurus*, and *Claosaurus*.
Geol. Magazine (3), x, pp. 150–157, pls. vi
and vii, and 1 fig. in the text.

———   1893 G
Restoration of *Anchisaurus*.
Amer. Jour. Sci. (3), xlv, pp. 169–170, with
pl. vi.

———   1893 H
Restoration of *Coryphodon*.
Geol. Magazine (3), x, pp. 482–487, pl. xviii,
and 6 figs. in the text.

———   1893 I
Some recent restorations of dino-
saurs.
Nature, xlviii, pp. 437–438.
A correction of some errors in R. Lydekker's
paper in Nature, xlviii, p. 302.

———   1893 J
Discussion sur la corrélation des
roches clastiques.
Compte rend. Cong. Géol. Internat., Washington, 1891, pp. 156–159.

———   1894 A
Restoration of *Camptosaurus*.
Amer. Jour. Sci. (3), xlvii, pp. 245–246, with
pl. vi.

Marsh, O. C.—Continued.   1894 B
Restoration of *Elotherium*.
Amer. Jour. Sci. (3), xlvii, pp. 407–408, with
pl. ix.

———   1894 C
A new Miocene mammal.
Amer. Jour. Sci. (3), xlvii, p. 409, with 3
figs. in the text.
Describes *Heptacodon curtus*.

———   1894 D
Footprints of vertebrates in the Coal-
measures of Kansas.
Amer. Jour. Sci. (3), xlviii, pp. 81–84, with
pls. ii, iii.

———   1894 E
The typical *Ornithopoda* of the American Jurassic.
Amer. Jour. Sci. (3), xlviii, pp. 85–90, with
pls. iv–vii.

———   1894 F
Eastern division of the Miohippus
beds, with notes on some of the characteristic fossils.
Amer. Jour. Sci. (3), xlviii, pp. 91–94, with
2 figs. in the text.

———   1894 G
Restoration of *Camptosaurus*.
Geol. Magazine (3), i, pp. 193–195, with
pl. vi.
Same as Marsh, O. C., 1894 A.

———   1894 H
Restoration of *Elotherium*.
Geol. Magazine (4), i, pp. 294–295, with
pl. x.
Same as Marsh, O. C., 1894 B.

———   1894 I
Footprints of vertebrates in the Coal-
measures of Kansas.
Geol. Magazine (3), i, pp. 337–339, with
pl. xi.
Descriptions and figures of several species.
The same, in part, as Marsh, O. C., 1894 D.

———   1894 J
A gigantic bird from the Eocene of
New Jersey.
Amer. Jour. Sci. (3), xlviii, p. 344, with 5
figs. in the text.
Describes new genus and species, *Barornis
regens*.

———   1894 K
Miocene artiodactyles from the eastern Miohippus beds.
Amer. Jour. Sci. (3), xlviii, pp. 175–178,
with 7 figs. in the text.

———   1894 L
Description of Tertiary artiodactyles.
Amer. Jour. Sci. (3), xlviii, pp. 259–274, with
34 figs. in the text.

Marsh, O. C.—Continued.　　1894 M

A new Miocene tapir.

Amer. Jour. Sci. (3), xlviii, p. 348.

Describes *Tanyops undans* from South Dakota.

——　　1895 A

The *Reptilia* of the Baptanodon beds.

Amer. Jour. Sci. (3), l, pp. 405–406, with 3 figs. in the text.

Figures and describes hind paddle of *Baptanodon discus* and vertebræ of *B. natans* and of *Pantosaurus* (*Parasaurus*) *striatus*.

——　　1895 B

Restoration of some European dinosaurs, with suggestions as to their place among the *Reptilia*.

Amer. Jour. Sci. (3), l, pp. 407–412, with pls. v–viii, and 1 fig. in the text.

Also in Report Brit. Assoc. Adv. Sci., Ipswich, 1895, pp. 685–688; and Geol. Magazine (4), iii, pp. 1–8, figs. 1, 2 and pls. i–iv.

——　　1895 C

On the affinities and classification of the dinosaurian reptiles.

Amer. Jour. Sci. (3), l, pp. 483–498, with pl. x.

Abstract of paper read before International Congress of Zoologists, Leyden, 1895. See also Marsh, O. C., 1896 D, Marsh, O. C., 1896 A, and Marsh, O. C., 1896 D.

——　　1895 D

Fossil vertebrates.

Johnson's Universal Cyclopædia, viii, pp. 491–498, with double plate containing 16 figs., and 48 figs. in the text.

——　　1896 A

Restoration of some European dinosaurs, with suggestions as to their place among the *Reptilia*.

Geol. Magazine (4), iii, pp. 1–9, pls. i–iv.

See also Marsh, O. C., 1895 C.

——　　1896 B

A new Belodont reptile (*Stegomus*) from the Connecticut River sandstone.

Amer. Jour. Sci. (4), ii, pp. 59–62, with 3 figs. in the text.

Describes *Stegomus arcuatus*.

——　　1896 C

The dinosaurs of North America.

Sixteenth Annual Report U. S. Geol. Surv., 1894–95, part i, pp. 133–244, with pls. ii–lxxxv, and text figs. 1–66.

——　　1896 D

On the affinities and classification of dinosaurian reptiles.

Compte rendu des séances du troisième Congrès International de Zoologie, Leyden, 1895, pp. 196–211, with 1 pl. and 11 figs. in the text.

See also Marsh, O. C., 1895 A and 1895 C.

Marsh, O. C.—Continued.　　1896 E

The geology of Block Island.

Amer. Jour. Sci. (4), ii. pp. 295–298.

——　　1896 F

The Jurassic formation on the Atlantic coast.

Amer. Jour. Sci. (4), ii, pp. 433–447, with 2 figs. in text showing relationships of the strata.

See also Science (2) iv, pp. 805–816, figs. 1–2.

——　　1896 G

Amphibian footprints from the Devonian.

Amer. Jour. Sci. (4), ii, pp. 374–375, with 1 fig. of footprint of *Thinopus antiquus*.

——　　1897 A

Affinities of *Hesperornis*.

Amer. Jour. Sci. (4), iii, pp. 347–348.

——　　1897 B

The *Stylinodontia*, a suborder of Eocene Edentates.

Amer. Jour. Sci. (4), iii, pp. 137–146, with 9 figs. in the text.

——　　1897 C

Vertebrate fossils of the Denver Basin.

Monographs U. S. Geol. Surv., xxvii, pp. 473–527, with pls. xxi–xxxi and text figs. 23–102.

——　　1897 D

Principal characters of the *Protoceratidæ*.

Amer. Jour. Sci. (4), iv, pp. 165–176, with pls. ii–vii, with 7 figs. in the text.

——　　1897 E

Recent observations on European dinosaurs.

Amer. Jour. Sci. (4), iv, pp. 413–416.

——　　1898 A

New species of *Ceratopsia*.

Amer. Jour. Sci. (4), vi, p. 92.

Describes *Triceratops calicornis* and *T. obtusus*.

——　　1898 B

The origin of mammals.

Amer. Jour. Sci. (4), vi, pp. 406–409.

Remarks made at International Congress of Zoologists, Cambridge, Eng., 4th meeting, Proceedings, pp. 71–74. See also Science (2), viii, pp. 963–965.

——　　1898 C

On the families of sauropodous *Dinosauria*.

Amer. Jour. Sci. (4), vi, pp. 487–498.

Also in Report Brit. Assoc. Adv. Sci. for 1898 (1899), pp. 909–910. Geol. Magazine (4), vi, 1899, pp. 157–158.

Matthew, W. D.—Continued. 1898 A
On some new characters of *Clænodon* and *Oxyæna*.
Science (2), viii, p. 880.
A brief abstract.

——— 1898 B
[Remarks on *Mixodectes*.]
Trans. N. Y. Acad. Sci., xvi, pp. 369–370.

——— 1899 A
A provisional classification of the fresh-water Tertiary of the West.
Bull. Amer. Mus. Nat. Hist., xii, pp. 19–75.
Presents lists of mammals occurring in each of the formations of each of the basins.

See also **Wortman and Matthew.**

Maxwell, J. B. 1845 A
[A letter concerning the discovery of mastodon bones near Hackettstown, N. J.]
Proc. Amer. Philos. Soc., iv, pp. 118–121; 127.
Reprinted in Lond. and Edinb. Philos. Mag., 1845, xxvi, pp. 453–456.

Mayer, Paul. 1886 A
Die unpaaren Flossen der Selachier.
Mittheil. Zoolog. Station zu Neapel, vi, pp. 217–285, with pls. 15–19.
Contains also a discussion of the "Halbwirbel der Selachier," p. 261.
Comments on by J. A. Ryder, Amer. Naturalist, xx, p. 142.

Mayer, ———. 1825 A
Die hintere Extremität der Ophidier.
Nova Acta Acad. Caes. Leop.-Car., xii, pp. 819–842, with pls. lxvi, lxvii.

Mayo, Florence. 1888 A
The superior incisors and canine teeth of sheep.
Bull. Mus. Comp. Zool., xiii, pp. 247–258, with pls. i–ii.

Mehnert, Ernst. 1887 A
Untersuchungen über die Entwicklung des Os pelvis der Vögel.
Morpholog. Jahrbuch, xiii, pp. 259–295, with pls. viii–x and 4 figs. in the text.
Contains numerous citations of the literature of the development and homologies of the pelvic bones.

——— 1889 A
Untersuchungen über die Entwicklung des Beckengürtels bei einigen Säugethieren.
Morpholog. Jahrbuch, xv, pp. 97–112, with pl. vi.
Contains numerous references to the literature of the subject.

Mehnert, Ernst—Continued. 1890 A
Untersuchungen über die Entwicklung des Beckengürtels der *Emys lutaria taurica.*
Morpholog. Jahrbuch, xvi, pp. 537–571, with pl. xx.
Contains numerous references to the literature.

Menzbier, M. v. 1887 A
Vergleichende Osteologie der Pinguine in Anwendung zur Haupteintheilung der Vögel.
Bull. Soc. Imp. Natur. Moscou (n. s.), ii, pp. 483–587, with pl. viii.

Mercer, H. C. 1894 A
Reexploration of Hartman's Cave, near Stroudsburg, Pa., in 1892.
Proc. Acad. Nat. Sci. Phila., 1894, pp. 96–104.

——— 1895 A
A preliminary account of the reexploration, in 1894 and 1895, of the "Bone Hole," now known as Irwin's Cave at Port Kennedy, Montgomery County, Pa.
Proc. Acad. Nat. Sci. Phila., 1895, pp. 443–446.

——— 1897 A
The finding of the fossil sloth at Big Bone Cave, Tennessee, in 1896.
Proc. Amer. Philos. Soc., xxxvi, pp. 36–70, with 26 half-tone figs. in the text.

——— 1899 A
The bone cave at Port Kennedy, Pa., and its partial excavation in 1894, 1895, and 1896.
Jour. Acad. Nat. Sci. Phila., xi, pp. 269–286, with 11 text figs. and a chart of the cave.
Besides a description of the cave, a list of the species recovered is given. These are described in Cope, E. D., 1899 A.

Merian, Peter. 1875 A
Ueber einen angeblichen Embryo von *Ichthyosaurus.*
Verhandl. naturf. Gesellsch. Basel, vi, pp. 343–344.

Mermier, Élie. 1896 A
Étude complementaire sur l'*Aceratherium platyodon* de la Mollasse burdigalienne supérieure des environs de Saint-Nazaire en Royans (Drôme).
Ann. Soc. Linn. de Lyon (n. s.), xliii, pp. 225–240, with pls. i, ii.
Additional observations are recorded on pp. 257–260.

Merriam, C. Hart. 1894 A
A new subfamily of murine rodents—the *Neotominæ*—with descriptions of a

Merriam, C. Hart.—Continued.
new genus and species and a synopsis
of the known forms.

Proc. Acad. Nat. Sci. Phila., 1894, pp. 225–252,
with 5 figs. in text.

——— 1895 A

Monographic revision of the pocket
gophers, family *Geomyidæ* (exclusive of
the species of *Thomomys*).

N. A. Fauna, No. 8 (U. S. Dept. Agriculture), pp. 1–259, with pls. 1–19, maps 1–4, and
text figs. 1–71.

——— 1895 B

Revision of the shrews of the American genera *Blarina* and *Notiosorex*.

N. A. Fauna, No. 10 (U. S. Dept. Agriculture), pp. 1–34, with pls. 1–3.

——— 1895 C

Synopsis of the American shrews of
the genus *Sorex*.

N. A. Fauna, No. 10 (U. S. Dept. Agriculture), pp. 57–98, with pls. v–xii.

——— 1896 A

Preliminary synopsis of the American bears.

Proc. Biolog. Soc. Washington, x, pp. 65–83,
with pls. iv–vi, with 17 figs. in the text.
In this paper 10 species of bears are recognized as occurring in North America.

Merriam, J. C. 1894 A
Ueber die Pythonomorphen der Kansas Kreide.

Palæontographica, xli, pp. 1–39, with 4 pls.
and 1 text fig.

——— 1895 A

On some reptilian remains from the
Triassic of northern California.

Amer. Jour. Sci. (3), pp. 55–57, with 2 figs. in
the text.
Describes the new genus and species *Shastasaurus pacificus*.

——— 1896 A

Sigmogomphius lecontei, a new Castoroid rodent from the Pliocene near
Berkeley, Cal.

Bulletin of the department of geology of the
University of California, i, pp. 363–370, with 2
figs. in the text.

——— 1899 A

Report on the expedition to the John
Day fossil fields.

University Chronicle [Univ. of California],
ii, pp. 217–224.
A general account of the region. No species
are described.

Metschnikoff, Olga. 1879 A
Zur Morphologie des Becken- und
Schulterbogens der Knorpelfische.

Zeitschr. wissensch. Zool., xxxiii, pp. 423–
438, with pl. xxiv.

Mettam, A. E. 1895 A
The rudimentary metacarpal and
metatarsal bones of the domestic ruminants.

Jour. Anat. and Physiol., xxix, pp. 244–253,
with 6 figs. in the text.

Meyer, H. v. 1832 A
Cervus alces fossilis (fossiler Elenn).

Nova Acta Acad. Caes. Leop.-Car., xvi, pp.
463–486, with pl. xxxvii.

——— 1834 A

Die fossilen Zähne und Knochen und
ihre Ablagerung in der Gegend von
Georgensmund in Bayern.

4to, pp. i–viii, 1–126, with pls. i–xiv.

——— 1835 A

Ueber fossile Reste von Ochsen, deren
Arten und das Vorkommen derselben.

Nova Acta Acad. Caes. Leop.-Car. xvii, pp.
101–170, with pls. viii–xii A.
Describes *Bos priscus*, and considers briefly
Bos pallasi Dekay and *Bos bombifrons. Bos
priscus* includes *Bison latifrons* of Harlan.

——— 1840 A

[Letter to editor on Grateloup's genus
Squalodon.]

Neues Jahrb. Mineral., 1840, pp. 587–588.

——— 1841 A

[Letters to editor on *Protorosaurus,
Squalodon,* and *Hyotherium.*]

Neues Jahrb. Mineral., 1841, pp. 101–104, 241–
242.

——— 1844 A

Ueber die fossilen Knochen aus dem
Tertiar-Gebilde des Cerro de San Isidro
bei Madrid.

Neues Jahrb. Mineral., 1844, pp. 289–310.
Contains description of *Anchitherium.*

——— 1845 A

Zur Fauna der Vorwelt. Fossile
Säugethiere, Vögel und Reptilien aus
dem Molasse-Mergel von Oeningen.

Pp. i–vi, 1–52, with 12 pls. Frankfurt am
Main, 1845.
Continued in Meyer, H. v. 1855 A, Meyer,
H. v. 1856 B, and Meyer, H. v. 1860 D.

——— 1845 B

Letter directed to Professor Brown.

Neues Jahrb. Mineral, 1845, pp. 308–313.
Contains remarks on *Mosasaurus maximiliani*
and *M. missouriensis.*

Meyer, H. v.—Continued. 1847 A

Die erloschene Cetaceen-Familie der Zeuglodonten, mit *Zeuglodon* und *Squalodon.*

Neues Jahrb. Mineral., 1847, pp. 669–674.

—————— 1851 A

Pterodactylus(Rhamphorhynchus)gemmingi.

Palæontographica, i, pp. 1–21, with pl. v.

—————— 1851 B

Ueber den *Archegosaurus* der Steinkohlenformation.

Palæontographica, i, pp. 209–215, with pl. xxxiii, figs. 15–17.

—————— 1852 A

Ueber die Beschaffenheit des Stosszahnes von *Elephas primigenius* in der Jugend.

Palæontographica, ii, pp. 75–77, with pl. xiv.

—————— 1852 B

Ueber *Chelydra murchisoni* und *Chelydra decheni.*

Palæontographica, ii, pp. 237–247, with pls. xxvi–xxx.

—————— 1854 A

Anthracotherium dalmatinum aus der Braunkohle des Monte Promina in Dalmatien.

Palæontographica, iv, pp. 61–66, with pl. xi.

—————— 1855 A

Zur Fauna der Vorwelt. Die Saurier des Muschelkalkes, mit Rücksicht auf die Saurier aus buntem Sandstein und Keuper.

Pp. i–viii; 1–167, with 70 pls. Frankfurt am Main, 1847–1855.

—————— 1856 A

Ueber den Jugenstand der *Chelydra decheni* aus der Braunkohle des Siebengebirges.

Palæontographica, iv, pp. 56–60.

—————— 1856 B

Zur Fauna der Vorwelt. Saurier aus dem Kupferschiefer der Zechstein-Formation.

Pp. i–vi; 1–28, with 9 pls. Frankfurt am Main, 1856.

—————— 1857 A

Reptilien aus der Steinkohlen-Formation in Deutschland.

Palæontographica, vi, pp. 59–219, with pls. viiia–xxiii.

Meyer, H. v.—Continued. 1858 A

Reptilien aus der Steinkohlen-Formation in Deutschland.

Pp. 1–126, with pls. A; i–xv. Cassel, 1858. Folio.

This memoir is devoted to descriptions and illustrations of *Archegosaurus* and its species.

—————— 1860 A

Rhamphorhynchus gemmingi.

Palæontographica, vii, pp. 79–89; pl. xii.

—————— 1860 B

Frösche aus Tertiär-Gebilden Deutschlands.

Palæontographica, vii, 123–182; pls. xvi–xxii.

—————— 1860 C

Saurier aus der Tuff-Kreide von Mæstricht und Folx-les-Caves.

Palæontographica, vii, pp. 241–244; pl. xxvi.

—————— 1860 D.

Zur Fauna der Vorwelt. Reptilien aus dem lithographischen Schiefer des Jura in Deutschland und Frankreich.

Pp. i–viii; 1–142, with 21 pls. Frankfurt am Main, 1860.

—————— 1861 A

Reptilien aus dem Stubensandstein des obern Keupers.

Palæontographica, vii, pp. 253–346; pls. xxviii–xlvii.

—————— 1861 B

Pterodactylus spectabilis aus dem lithographischen Schiefer von Eichstätt.

Palæontographica, x, pp. 1–10, pl. 1.

—————— 1863 A

Der Schädel des *Belodon* aus dem Stubensandstein des obern Keupers.

Palæontographica, x, pp. 227–246, pls. xxxviii–xlii.

—————— 1864 A

Die diluvialen Rhinoceros-Arten.

Palæontographica, xi, pp. 233–283, with pls. xxxv–xliii.

—————— 1865 A

Der Schädel von *Glyptodon.*

Palæontographica, xiv, pp. 1–18, with pls. i–vii.

—————— 1865 B

Reptilien aus dem Stubensandstein des obern Keupers.

Palæontographica, xiv, pp. 99–104, pls. xxiii–xxix.

—————— 1865 C

Zu *Chelydra decheni* aus der Braunkohle des Siebengebirges.

Palæontographica, xv, pp. 41–47, with pl. ix.

Meyer, H. v.—Continued.   1867 A
Die fossile Reste des Genus *Tapirus.*
Palæontographica, xv, pp. 159-200, with pls.
xxv-xxxii.

————   1867 B
Studien über das Genus *Mastodon.*
Palæontographica, xvii, pp. 1-72, with pls.
i-ix.

Meyer, Otto.   1885 A
Insectivoren und *Galeopithecus* geo-
logisch alte Formen.
Neues Jahrb. Mineral., 1885, ii, pp. 229-230.

Miall, L. C.   1874 A
Report of the committee, consisting
of Professor Phillips, LL. D., F. R. S.,
Professor Harkness, F. R. S., Henry
Woodward, F. R. S., James Thompson,
John Brigg, and L. C. Miall, on the
Labyrinthodonts of the Coal-measures.
Drawn up by L. C. Miall, secretary to
the committee.
Report Brit. Assoc. Adv. Sci., Bradford meet-
ing, 1873, pp. 225-249, with pls. i-iii.

————   1874 B
On the composition and structure of
the bony palate of *Ctenodus.*
Quart. Jour. Geol. Soc., xxx, pp. 772-775,
with pl. xlvii.

————   1875 A
Report of the committee, consisting
of Professor Huxley, LL. D., F. R. S.,
Professor Harkness, F. R. S., Henry
Woodward, F. R. S., James Thompson,
John Brigg, and L. C. Miall, on the
structure and classification of the Laby-
rinthodonts. Drawn up by L. C. Miall,
secretary to the committee.
Report Brit. Assoc. Adv. Sci., Belfast meet-
ing, 1874, pp. 149-192, with pls. iv-vii.
Abstract in Nature, 1874, vol. x, p. 449.

————   1875 B
On the structure of the skull of *Rhizodus.*
Quart. Jour. Geol. Soc., xxxi, pp. 624-627,
with 1 text fig.

————   1875 C
Sur les Labyrinthodontes du terrain
houiller. Deuxième partie.
Jour. de Zool. iv, pp. 19-37.

————   1878 A
Monograph of the Sirenoid and Crossopterygian Ganoids.
Palæontographical Society, xxxii, pp. 1-32;
pls. i, ia, ii-v.

Miall, L. C.—Continued.   1878 B
On the genus *Ceratodus*, with special
reference to the fossil teeth found at
Malédi, Central India.
Palæont. Indica (4), i, pp. 10-17, with pl. iv.

————   1878 C
The skull of the crocodile. A manual for students.
8 vo, pp. 1-50, London, 1878.
Pp. 42-50 contain a translation with annota-
tions of Rathke's Untersuchungen über die Entwicklung und den Körperbau der Krokodile.

————   1880 A
On some bones of *Ctenodus.*
Proc. Yorkshire Zool. and Polytech. Soc. (2),
vii, pp. 289-299, with 12 text figs.

————   1884 A
On a new specimen of *Megalichthys*
from the Yorkshire coal field.
Quart. Jour. Geol. Soc., xl, pp. 347-352, with
6 figs. in the text.

Middendorff, A. Th.   1850 A
Untersuchungen an Schädeln des
gemeinen Landbären als kritische Be-
leuchtung der Streitfrage über die Arten
fossiler Hohlenbären.
Verhandl. russ.-kais. mineralog. Gesellsch.,
St. Petersburg, Jahrg., 1850-51, pp. 7-99, with
text figures.

Middleton, W. G., and Moore, Joseph.   1900 A
Skull of fossil bison.
Proc. Indiana Acad. Sci. for 1899, pp. 178-181,
with plate.

Miller, Gerritt S.   1894 A
On a collection of small mammals
from the New Hampshire mountains.
Proc. Boston Soc. Nat. Hist., xxvi, pp. 177-197,
with pls. iii, iv.

————   1896 A
Genera and subgenera of voles and
lemmings.
N. A. Fauna, No. 12 (U. S. Dept. Agriculture), pp. 1-85, with pls. i-iii and 40 figs. in
the text.

————   1897 A
Revision of the North American bats
of the family *Vespertilionidæ.*
N. A. Fauna, No. 13 (U. S. Dept. Agriculture),
pp. 1-135, with 40 figs. in the text.
This work discusses critically the nomenclature and synonymy of the family *Vespertilionidæ* and its genera and species.

————   1899 A
A new fossil bear from Ohio.
Proc. Biolog. Soc. Washington, xiii, pp. 53-56.
Describes *Ursus procerus.*

Miller, Gerritt S.—Continued. 1899 B
Preliminary list of New York mammals.
Bull. N. Y. State Mus., vi, No. 29, pp. 271–390.
On pages 372–376 is given a list of the fossil species known from the State.

Miller, Hugh. 1859 A
The Old Red sandstone; or new walks in an old field; to which is appended a series of geological papers read before the Royal Physical Society of Edinburgh.
Pages 427, pls. 14. Boston: Gould and Lincoln, 1859.
Original edition published at Edinburgh, 1841.

——— 1873 A
Footprints of the Creator, or the *Asterolepis* of Stromness, with a memoir by Louis Agassiz.
Fifteenth edition. Edinburgh, 1873. William P. Nimmo.

Miller, John. 1872 A
On the so-called hyoid plate of the *Asterolepis* of the Old Red sandstone.
Report Brit. Assoc. Adv. Sci., 41st meeting, Edinburgh, 1871, pp. 106–107.

Miller, S. A. 1877 A
The American Palæozoic fossils: a catalogue of the genera and species, with names of authors, dates, places of publication, groups of rocks in which found, and the etymology and signification of the words, and an introduction devoted to the stratigraphical geology of the Palæozoic rocks.
Cincinnati, Ohio, 1877, pages i–xv; 1–245.
Pages 227–242 are devoted to lists of the genera and species of Palæozoic vertebrates. In 1883 the author published a supplement consisting of pages 247 to 334, in which were additions and corrections to the contents of the original work. The latter and the supplement bound together constitute the second edition, with the original title. As the same author's work of 1889 is fully quoted in this Bibliography and Catalogue, the present one is not quoted.

——— 1889 A
North American geology and palæontology, for the use of amateurs, students, and scientists.
Pp. 1–664, with 1,177 figs. in the text. Cincinnati, Ohio, 1889.
Pp. 586–614 contain a list of the genera and species of Palæozoic fishes, with references to places of publication. The genera are defined. Pp. 614–627 are similarly devoted to the *Batrachia*.

Miller, S. A.—Continued. 1892 A
First appendix, 1892. [To North American geology and palæontology.]
Pp. 665–718.
The names of vertebrates are found on pp. 713–718. A number of new generic names are proposed to replace preoccupied names.

——— 1897 A
Second appendix to North American geology and palæontology.
Pp. 719–793.

Miller, Sylvanus. 1837 A
Retrospective notice of the discovery of fossil mastodon bones in Orange county (N. Y.).
Amer. Jour. Sci., xxxi, pp. 171–172.
Extract from a letter addressed by Sylvanus Miller, esq., to Hon. Dewitt Clinton, in 1815.

Minot, C. S. 1882 A
Is man the highest animal?
Proc. Amer. Assoc. Adv. Sci., 30th meeting, Cincinnati, 1881, pp. 240–242.

Mitchell, Hugh. 1862 A
On the restoration of *Pteraspis*.
The Geologist, v, pp. 404–406, with 3 figs. in the text.

——— 1864 A
Restoration of *Pteraspis*.
The Geologist, vii, pp. 117–118, fig. 1.

Mitchell, James A. 1895 A
The discovery of fossil tracks in the Newark system (Jura-Trias) of Frederick county, Md.
Johns Hopkins University circulars, xv, no. 121, pp. 15–16.
No species are named.

Mitchill, Samuel L. 1818 A
Observations on the geology of North America; illustrated by the description of various organic remains found in that part of the world.
Cuvier's Essay on the theory of the earth, pp. 319–431, with pls. vi–viii. American edition published by Kirk & Mercein, New York.

——— 1824 A
Observations on the teeth of the *Megatherium* recently discovered in the United States.
Annals Lyc. Nat. Hist. N. Y., i, pp. 58–61, pl. vi.

Mitchill, Smith, J. A., and Cooper. 1828 A
Discovery of a fossil walrus in Virginia. Report of Messrs. Mitchill, J. A. Smith, and Cooper on a fossil skull

Mitchell, Smith, J. A., and Cooper—Continued.

sent to Dr. Mitchill by Mr. Cooper, of Accomac County, Va.

Annals Lyc. Nat. Hist. N. Y., ii, pp. 271-272. Notice of in Edinb. New Philos. Jour., 1828, v., p. 325.

Mitivier, M. M. 1892 A

New footprints from the Connecticut Valley.

Proc. Amer. Assoc. Adv. Sci., 40th meeting, Washington, 1891, p. 286. An abstract in 5 lines.

Mivart, St. George. 1864 A

Notes on the crania and dentition of the *Lemuridæ*.

Proc. Zool. Soc. Lond., 1864, pp. 611-648, with 8 figs. in the text.

——— 1865 A

Contributions toward a more perfect knowledge of the axial skeleton in the *Primates*.

Proc. Zool. Soc. Lond., 1865, pp. 545-592, with 13 figs. in the text.

——— 1867 A

Additional notes on the osteology of the *Lemuridæ*.

Proc. Zool. Soc. Lond., 1867, pp. 960-975, with 7 figs. in the text.

——— 1867 B

Notes sur l'ostéologie des Insectivores.

Ann. Sci. Naturelles. (5), Zool. viii, pp. 221-284, with figs. in the text.

——— 1867 C

Notes on the osteology of the *Insectivora*.

Jour. Anat. and Physiol. i, pp. 281-312, with several text figs.; ii, pp. 117-154.

——— 1868 A

Notes sur l'ostéologie des Insectivores.

Ann. Sci. Naturelles, (5), Zool. ix, pp. 311-372, with figs. in the text. A continuation of Mivart, St. G., 1867 B.

——— 1868 B

On the appendicular skeleton of the *Primates*.

Philos. Trans. Roy. Soc. Lond., clvii, pp. 299-429, pls. xi-xiv.

——— 1870 A

On the axial skeleton of the *Urodela*.

Proc. Zool. Soc. Lond., 1870, pp. 260-278, with 19 figs. in the text.

——— 1871 A

On *Hemicentetes*, a new genus of *Insectivora*, with some additional remarks on the osteology of that order.

Proc. Zool. Soc. Lond., 1871, pp. 58-79, with pl. v and 9 text figs.

Mivart, St. George—Continued. 1873 A

On *Lepilemur* and *Cheirogaleus*, and on the zoological rank of the *Lemuroidea*.

Proc. Zool. Soc. Lond., 1873, pp. 484-510, with pl. xliii and 18 woodcuts in the text.

——— 1874 A

On the axial skeleton of the ostrich (*Struthio camelus*).

Trans. Zool. Soc. Lond., viii, pp. 385-451, with 79 figs. in the text.

——— 1875 A

Ape.

Encylop. Brit., ed. 9, ii, pp. 148-169, with 19 figs. in the text.

This article is followed by a bibliography of the subject.

——— 1877 A

On the axial skeleton of the *Struthionidæ*.

Trans. Zool. Soc. Lond., x, pp. 1-52, with 43 figs. in the text.

——— 1878 A

On the axial skeleton of the *Pelecanidæ*.

Trans. Zool. Soc. Lond., x, pp. 315-378, with pls. lv-lxi.

——— 1879 A

Notes on the fins of Elasmobranchs, with considerations on the nature and homologies of vertebrate limbs.

Trans. Zool. Soc. Lond., x, pp. 439-484, with pls. lxxiv-lxxix.

——— 1881 A

The cat; an introduction to the study of the backboned animals, especially mammals.

Pp. i-xxiii; 1-557, with 1 pl. and 208 figs. in the text. 8 vo.

——— 1882 A

On the classification and distribution of the *Æluroidea*.

Proc. Zool. Soc. Lond., 1882, pp. 135-208, with 15 figs. in the text.

——— 1885 A

Notes on the *Pinnipedia*.

Proc. Zool. Soc. Lond., 1885, pp. 484-501. Abstract in Archiv f. Anthrop., xvii, pp. 176-177.

——— 1885 B

On the anatomy, classification, and distribution of the *Arctoidea*.

Proc. Zool. Soc. London, 1885, pp. 340-404. Pp. 397-404 contain valuable tables, which give measurements of various portions of the skeletons, including teeth, of all the genera belonging to the *Arctoidea*.

Mivart, St. George—Continued. 1886 A
Reptiles.
Encyclop. Brit., ed. 9, xx, pp. 445–465, with figs. 3–29.
This portion of the article on *Reptilia* treats of the anatomy of the members of the class, especially of the osteology. Appended is a list of general works on the anatomy.

——— 1892 A
Birds: The elements of ornithology.
Pp. 1–829, with 174 illus., London.

Möbius, K. 1892 A
Die Behaarung des Mammuths und der lebenden Elephanten, vergleichend untersucht.
Sitzungsber. k. p. Akad. Wissensch. Berlin, math. naturw. Cl., 1892, pp. 527–538, with pl. iv.

Mojsisovics, A. 1889 A
Ueber einen seltenen Fall von Polydactylismus beim Pferde.
Anatom. Anzeiger, iv, pp. 255–256.

Molin, Raffaele. 1859 A
Sulla reliquie d' un *Pachyodon* dissoterrate a Libàno due ore nord-est di Belluno in mezzo all' arenaria grigia.
Sitzungsber. Akad. Wissensch. Berlin, math.-nat. Cl., xv, pp. 117–128, with pls. i, ii.

Mollier, S. 1892 A
Zur Entwicklung der Selachierextremitäten.
Anatom. Anzeiger, vii, 1892, pp. 351–365.
Treats of the embryological development of the limbs of *Selachii*.

——— 1893 A
Die paarigen Extremitäten der Wirbelthiere. I. Das Ichthyopterygium.
Merkel and Bonnet's Anatom. Hefte, Ergebnisse, iii, pp. 1–160, with pls. 1–viii and several text figs.

——— 1895 A
Die paarigen Extremitäten der Wirbelthiere. II. Das Cheiropterygium.
Merkel and Bonnet's Anatom. Hefte, Ergebnisse, v, pp. 435–529, with 8 pls.

Monks, Sarah P. 1878 A
The columella and stapes in some North American turtles.
Proc. Amer. Philos. Soc., xvii, pp. 335–337, pls. xvi, xvii.

Moore, Charles. 1857 A
On the skin and food of *Ichthyosauri* and *Teleosauri*.
Report Brit. Assoc. Adv. Sci., 26th meeting, Cheltenham, 1856, pp. 69–70.

Moore, Joseph. 1890 A
A recent find of *Castoroides*.
Amer. Naturalist, xxiv, pp. 767–768.
Found at Richmond, Ind.

Moore, Joseph—Continued. 1890 B
Concerning a skeleton of the great fossil beaver, *Castoroides ohioensis*.
Jour. Cincinnati Soc. Nat. Hist., xiii, pp. 138–169, with 25 figs. in the text.

——— 1891 A
Concerning some portions of *Castoroides ohioensis* not heretofore known.
Proc. Amer. Assoc. Adv. Sci., 39th meeting, Indianapolis, 1890, pp. 265–267.
Abstract of paper presented.

——— 1893 A
The recently found *Castoroides* in Randolph County, Ind.
Amer. Geologist, xii, pp. 67–74, with pl. iii.

——— 1897 A
The Randolph mastodon.
Proc. Indiana Acad. Sci. for 1896, pp. 277–278, with pl.
Records finding of remains of a large mastodon in Randolph County, Ind., and of another near New Paris, Ohio.

——— 1900 A
A cranium of *Castoroides* found at Greenfield, Ind.
Proc. Indiana Acad. Sci. for 1899, p. 171, with pls. i, ii.

Moore, W. D. 1873 A
On footprints in the Carboniferous rocks of western Pennsylvania.
Amer. Jour. Sci. (3), v, pp. 292–293.
? *Chetrotherium reiteri* described.

Moreno, Francisco P. 1892 A
Lijeros apuntes sobre dos géneros de cetáceos fósiles de la República Argentina.
Revista Museo La Plata, iii, pp. 393–400, with pls. x, xi.

Morse, E. S. 1871 A
On the carpal and tarsal bones of birds.
Amer. Naturalist, v, pp. 524–525.

——— 1872 A
On the tarsus and carpus of birds.
Annals Lyc. Nat. Hist. N. Y., x, pp. 141–158, with pls. iv, v.

——— 1880 A
On the identity of the ascending process of the astragalus in birds with the intermedium.
Anniv. Mem. Boston Soc. Nat. Hist., 1880, pp. 3–10, pl. i, and 12 figs. in the text.

——— 1884 A
Man in the Tertiaries.
Amer. Naturalist, xviii, pp. 1001–1012.
Vice-presidential address before section of anthropology A. A. S., 1884.

Morton, S. G. 1830 A
Synopsis of the organic remains of

Morton, S. G.—Continued.

the ferruginous sand formation; with geological remarks.

Amer. Jour. Sci., xvii, pp. 274–296; xviii, pp. 243–250.

Contains brief and unimportant remarks on the few fossil vertebrates known at that time from the Cretaceous rocks.

——   1834 A

Synopsis of the organic remains of the Cretaceous group of the United States, illustrated by 19 plates, to which is added an appendix containing a tabular view of the Tertiary fossils hitherto discovered in North America.

Pp. i–vi, 1–88; Appendix, 1–8; and "Additional observations," pp. 4.

Pls. xi, xii contain figures of teeth of sharks. The names of these, given provisionally by L. Agassiz, are found in the "Additional observations."

——   1835 A

Notice of the fossil teeth of fishes of the United States, the discovery of the Galt in Alabama, and a proposed division of the American Cretaceous group.

Amer. Jour. Sci., xxviii, pp. 276–278.

——   1842 A

Description of some new species of organic remains of the Cretaceous group of the United States; with a tabular view of the fossils hitherto discovered in this formation.

Jour. Acad. Nat. Sci. Phila., viii, pp. 207–227, pls. x, xi.

Contains description and figure of *Ptychodus mortoni*.

——   1844 A

Description of the head of a fossil crocodile from the Cretaceous strata of New Jersey.

Proc. Acad. Nat. Sci. Phila., ii, pp. 82–85, with 1 fig. in the text.

Description of *Crocodilus (Gavialis?) clavirostris*.

——   1844 B

Fossil bones of *Mosasaurus*.

Proc. Acad. Nat. Sci. Phila., ii, pp. 132–133.

Announces exhibition of bone of *M. occidentalis*, but no description is given.

——   1845 A

Description of the head of a fossil crocodile from the Cretaceous strata of New Jersey.

Amer. Jour. Sci., xlviii, pp. 265–267, with 1 fig.

Moseley, H. N., and Lankester, E. R.   1868 A

On the nomenclature of the mammalian teeth and on the dentition of

Moseley, H. N., and Lankester, E. R.— Continued.

of the mole (*Talpa europæa*) and the badger (*Meles taxus*).

Jour. Anat. and Physiol., iii, pp. 72–80, with pl. ii, figs. 5, 6.

Mudge, B. F.   1866 A

Discovery of fossil footmarks in the Liassic (?) formation of Kansas.

Amer. Jour. Sci. (2), xlii, pp. 174–176.

Species not identified. Prof. C. H. Hitchcock regards them as artifacts (Hitchcock, C. H., 1869 A., p. 132).

——   1874 A

Recent discoveries of fossil footprints in Kansas.

Trans. Kansas Acad. Sci., ii, pp. 7–9 (reprint of 1896, pp. 71–74).

Not assigned to genera and species.

——   1875 A

Rare forms of fish in Kansas.

Trans. Kansas Acad. Sci., iii, p. 2? (reprint of 1896, pp. 121–122).

Refers to genus *Agassizodus* and to an undetermined snout fish.

Müller, Aug.   1853 A

Beobachtungen zur vergleichenden Anatomie der Wirbelsäule.

Archiv f. Anat. Physiol. u. wissensch. Med. Jahrg. 1853, pp. 260–316.

Discusses the homologies of the ribs and flesh bones of fishes, the ribs and hæmal arches of the higher vertebrates, the ear bones of the *Cyprinidæ*, etc.

Müller, Johannes.   1831 A

Beiträge zur Anatomie und Naturgeschichte der Amphibien.

Tiedemann and Treviranus's Zeitschr. f. Physiologie, iv, pp. 190–275, with pls. xviii–xxii.

Under *Amphibien* Müller here considers both the groups *Batrachia* and *Reptilia*.

——   1835 A

Vergleichende Anatomie der Myxinoiden, der Cyclostomen mit durchbohrtem Gaumen. Erster Theil, Osteologie und Myologie.

Pp. 1–276, pls. i–ix. Berlin, 1835.

This portion of the anatomy of the Myxinoids appeared in the Abhandl Akad. Wissensch. Berlin, for 1834 (1836), pp. 56–340.

Extensive comparisons are made in this work with the structures of the other fishes. A "Nachtrag" is added in the part of the work on neurology published in 1838.

——   1843 A

Beiträge zur Kenntniss der natürlichen Familien der Knochenfische.

Berichte k. p. Akad. Wissensch. Berlin, 1843, pp. 211–218.

Müller, Johannes—Continued.   1846 A
Fernere Bemerkungen über den Bau
der Ganoiden.
Archiv Naturgesch., xii, pp. 190–208.

———   1846 B
Ueber den Bau der Ganoiden.
Berichte k. p. Akad. Wissensch. Berlin,
1846, pp. 67–85.

———   1846 C
Ueber den Bau und die Grenzen der
Ganoiden und über das natürliche Sys-
tem der Fische.
Abhandl. k.p. Akad. Wissensch. Berlin, 1844,
pp. 117–216, with pls. i–vi.

———   1847 A
Basilosaurus.
Amer. Jour. Sci. (2), iv, pp. 421–422.
Extract from a letter sent by J. Müller to A.
Retzius.
See also reference to a letter from Professor
Müller to Dr. Dunglison in Proc. Amer. Philos.
Soc., iv, p. 338. Date of letter Apr. 20, 1847.

———   1847 B
Ueber die von Herrn Koch in Alabama gesammelten fossilen Knochen-
reste seines *Hydrarchus.*
Archiv Anat. Physiol., u. Wissench. Med.
xiv, pp. 363–377; 878–896.

———   1847 C
Untersuchungen über den Hydrarchos.
Berichte k. p. Akad. Wissensch. Berlin, 1847,
pp. 103–114.
Reprinted in Neues Jahrbuch Min., 1847, pp.
623–631.

———   1847 D
Ueber den Bau des Schädels des
Zeuglodon cetoides Ow.
Berichte k. p. Akad. Wissensch. Berlin, 1847,
p. 160.
A brief note.

———   1847 E
Ueber die Wirbelsäule des *Zeuglodon
cetoides.*
Berichte k. p. Akad. Wissensch. Berlin, 1847
pp. 185–200.

———   1849 A
Ueber die fossilen Reste der Zeuglodonten von Nordamerica, mit Rücksicht auf die europäischen Reste aus
dieser Familie.
Pp. i–iv; 1–38. Folio, with 27 lith. pls.
Berlin, 1849,

———   1851 A
Neue Beiträge zur Kenntniss der
Zeuglodonten.
Monatsber. k. Akad. Wissensch., Berlin,
1851, pp. 236–245.

Müller, J., and **Henle**, J.   1837 A
Ueber die Gattungen der Haifische
und Rochen.
Berichte k. p. Akad. Wissensch. Berlin, i,
pp. 111–118.

———   1837 B
Ueber die Gattungen der Plagiosto-
men.
Arch. Naturgesch. 1837, i, pp. 394–401; 434.

———   1838 A
On the generic characters of cartilag-
inous fishes.
Mag. Nat. Hist. (2), ii, pp. 33–37; 88–92.

———   1841 A
Systematische   Beschreibung   der
Plagiostomen.
Pp. i–xxii; 1–204, with 60 lith. pls. Berlin,
1841.

Murie, James.   1865 A
On the anatomy of a finwhale
(*Physalus antiquorum* Gray) captured
near Gravesend.
Proc. Zool. Soc. Lond., 1865, pp. 206–227,
with 4 woodcuts in text.

———   1870 A
Notes on the anatomy of the prong-
buck, *Antilocapra americana.*
Proc. Zool. Soc. Lond., 1870, pp. 334–368, with
9 woodcuts in the text.

———   1871 A
Researches upon the anatomy of the
Pinnepedia. Part I. On the walrus
(*Trichechus rosmarus* Linn.).
Trans. Zool. Soc. Lond., vii, pp. 411–464, with
pls. li–lv.

———   1872 A
On the form and structure of the
manatee (*Manatus americanus*).
Trans. Zool. Soc. Lond., viii, pp. 127–202,
with pls. xvii–xxvi.

———   1872 B
On the skin, etc., of the *Rhytina*,
suggested by a recent paper of Dr. A.
Brandt's.
Ann. and Mag. Nat. Hist. (4), ix, pp. 306–313,
with pl. xix.

———   1874 A
Researches upon the anatomy of the
Pinnepedia. Part III. Descriptive
anatomy of the sea lion (*Otaria jubata*).
Trans. Zool. Soc. Lond., viii, pp. 501–582, with
pls. lxxv–lxxxii.

Nasmyth, Alexander.   1842 A
On the minute structure of the tusks
of extinct mastodontoid animals.
Proc. Geol. Soc. Lond., iii, pp. 775–780.

Natural Science.   1893 A

The restoration of extinct animals.

Natural Science, ii, pp. 135–143, with 2 figs.

An anonymous paper. Refers to *Triceratops, Ceratosaurus, Hesperornis, Glyptodon*, etc.

——   1894 A

Armored whales.

Natural Science, iv, pp. 174–176.

An unsigned note.

Nature.   1871 A

American notes.

Nature, iii, p. 514.

A note on *Elephas primigenius* in Alaska.

——   1873 A

Notes.

Nature, vii, p. 310.

A note on *Dinoceras mirabilis* and the toothed birds.

——   1873 B

On *Dinoceras mirabilis* (Marsh).

Nature, vii, p. 366, with 1 woodcut.

——   1873 C

[Note on Professor Cope's *Loxolophodon cornutus*, etc.]

Nature, vii, p. 471.

——   1874 A

The "*Brontotheridæ*," a new family of fossil mammals.

Nature, ix, p. 227, figs. 1 and 2.

Editorial remarks on Marsh, O. C., 1874 A.

——   1874 B

A complete specimen of a *Palæotherium*.

Nature, ix, pp. 285–286, with 1 fig.

P. magnum discovered near Paris.

Naumann, Edmund.   1882 A

Ueber japanische Elephanten der Vorzeit.

Palæontographica, xxviii, pp. 1–40, with pls. i–vii.

Negri, Arturo.   1892 A

Trionici eocenici ed oligocenici del Veneto.

Mem. mat. e fisica Soc. Ital. Sci. (3), viii, art. 7, pp. 1–53, with pls. i–v.

On pp. 1–8 are presented the literature bearing on fossil *Trionychidæ* and a list of known species living and extinct.

Nehring, Alfred.   1875 A

Fossile Lemminge und Arvicolen aus dem Diluviallehm von Thiede bei Wolfenbüttel.

Zeitschr. gesammt. Naturwiss., xlv, (n. F., xi), 1875, pp. 1–28, with pl. i.

Discusses the structure of *Arvicola* and *Myodes*, especially that of the teeth.

Nehring, Alfred—Continued.   1886 A

Beiträge zur Kenntniss der *Galictis*-Arten.

Zool. Jahrbücher, Abth. Systemat. Geograph., etc.. i, pp. 177–212, with 3 figs. in the text.

——   1893 A

Ueber pleistocäne Hamster-Reste aus Mittel-und Westeuropa.

Jahrbuch k. k. geol. Reichsanstalt, Wien, xliii, pp. 179–198.

Newberry, J. S.   1853 A

Fossil fishes of the cliff limestone of Ohio.

Annals of Science (Cleveland, Ohio), i, pp. 12–13, text figs. 1, 2.

——   1853 B

On the fossil fishes of the cliff limestone of Ohio.

Annals of Science (Cleveland, Ohio), i, pp. 282–283.

Abstract of paper presented at Cleveland meeting of Amer. Assoc. Adv. Sci., 1853.

The remarks are of a general nature and no species are named. See Newberry, J. S., 1856 A.

——   1856 A

On the fossil fishes of the cliff limestone of Ohio.

Proc. Amer. Assoc. Adv. Sci., 7th meeting, Cleveland, 1853, pp. 166–167.

Same as Newberry, J S., 1853 B.

——   1856 B

Description of several new genera and species of fossil fishes from the Carboniferous strata of Ohio.

Proc. Acad. Nat. Sci. Phila., viii, pp. 96–100.

——   1857 A

New fossil fishes from the Devonian rocks of Ohio.

Amer. Jour. Sci. (2), xxiv, pp. 147–149.

Extracted from Bulletin of National Institute.

——   1857 B

[Letter read changing the genus *Melcolepis* to *Eurylepis*, the former being preoccupied.]

Proc. Acad. Nat. Sci. Phila., 1857, p. 150.

——   1857 C

Fossil fishes from the Devonian of Ohio.

Proc. National Institute, Washington, D. C. (n. s.), i, pp. 119–126, with 1 fig. in the text.

There appears to have been a reprint of this paper with its own pagination.

——   1862 A

Notes on American fossil fishes.

Amer. Jour. Sci. (2), xxxiv, pp. 73–78, with figs. in the text.

Newberry, J. S.—Continued.   1868 A
On some fossil reptiles and fishes from the Carboniferous strata of Ohio, Kentucky, and Illinois.
Proc. Amer. Assoc. Adv. Sci., 16th meeting, Burlington, Vt., 1867, pp. 144–146. Abstract.

———   1868 B
On some remarkable fossil fishes discovered by Rev. H. Herzer, in the Black shale (Devonian) at Delaware, Ohio.
Proc. Amer. Assoc. Adv. Sci., 15th meeting, Buffalo, 1867, pp. 146–147, with two woodcuts.
Describes *Dinichthys herzeri.*

———   1870 A
The geological position of the remains of elephant and mastodon in North America.
Proc. Lyc. Nat. Hist. N. Y., i, pp. 77–84.

———   1870 B
Fossil fishes from the Devonian rocks of Ohio.
Proc. Lyc. Nat. Hist. N. Y., i, pp. 152–153.

———   1870 C
Remarks on the fossil skull of a walrus found at Long Branch.
Proc. Lyc. Nat. Hist. N. Y., i, p. 75.
Recorded as belonging to *Trichechus rosmarus.*

———   1871 A
Fossils from the phosphatic beds of South Carolina.
Proc. Lyc. Nat. Hist. N. Y., i, pp. 240–241.
Brief mention of a few species of vertebrates.

———   1873 A
The classification and geological distribution of our fossil fishes.
Report. Geol. Surv. Ohio, i, Part II, Palæontology, pp. 245–355, with pls. xxiv–xl.

———   1873 B
Notes on the genus *Conchiopsis* Cope.
Proc. Acad. Nat. Sci. Phila., 1873, pp. 425–426.

———   1873 C
Remarks reviewing the history of the class of fishes as traced on the older rocks of North America.
Proc. Lyc. Nat. Hist. N. Y., 2d ser., No. 2, March to June, pp. 25–28.

———   1873 D
Specimens of *Cœlacanthus elegans* Newb. from the Coal-measures of Linton, Ohio.
Proc. Lyc. Nat. Hist. N. Y., 2d ser., No. 2, March to June, pp. 30–32.
Some general remarks on *Cœlacanthus*; no description of *C. elegans.*

Newberry, J. S.—Continued.   1874 A
[Remarks on *Castoroides ohioensis*].
Proc. Lyc. Nat. Hist. N. Y., 2d ser., No. 4, Jan. to June, pp. 92–93.

———   1874 B
Description of the Linton coal bed.
Proc. Lyc. Nat. Hist. N. Y., 2d ser., No. 4, Jan. to June, pp. 134–135.
Contains mention of four genera of fishes found there.

———   1874 C
Account of a second species of *Dinichthys* recently discovered in Ohio.
Proc. Lyc. Nat. Hist. N. Y., 2d ser., No. 4, Jan. to June, pp. 149–151.
Gives a description (not the earliest) of *Dinichthys terrelli.*

———   1874 D
Skull of a walrus (*Trichechus rosmarus*).
Proc. Lyc. Nat. Hist. N. Y., 2d ser., No. 3, Oct. to Dec., 1873, p. 71.
Identification of a skull found in Accomac County, Va., a drawing of which was presented by Dr. R. P. Stevens.

———   1874 E
Remarks on *Cœlacanthus elegans* and on Prof. Cope's genera *Conchiopsis* and *Peplorhina.*
Proc. Lyc. Nat. Hist. N. Y., 2d ser., No. 3, Oct. to Dec., 1873, pp. 76–77.

———   1874 F
[Remarks on skulls of *Dicotyles compressus* found at Columbus, Ohio.]
Proc. Lyc. Nat. Hist. N. Y., 2d ser., No. 3, Oct. to Dec., 1873, pp. 77–78.

———   1875 A
Descriptions of fossil fishes.
Report Geol. Surv. Ohio, ii. Part II, Palæontology, pp. 1–64, with pls. liv–lix.
Contains descriptions and figures of the fossil fishes of the Devonian and the Carboniferous formations of Ohio.

———   1876 A
Descriptions of the Carboniferous and Triassic fossils collected on the San Juan Exploring Expedition under Capt. J. N. Macomb, U. S. Engineers.
Report of the Exploring Expedition from Santa Fe, N. Mex., to the junction of the Grand and Green rivers of the Great Colorado of the West, in 1859, under the command of Capt. J. N. Macomb, Washington, 1876, pp. 135–148, with pl. iii, figs. 1, 1a; 2–2f.
Describes *Deltodus mercurii* Newb., *Ptychodus whipplei* Marcou, and mentions finding of *Lamna texana* Roem., and *Oxyrhina mantelli* Ag.

Newberry, J. S.—Continued.   1877 A
Rhynchodus excavatus.
Geology of Wisconsin, survey of 1873–1877, ii, p. 397.

——   1878 A
Descriptions of new palæozoic fishes.
Ann. N. Y. Acad. Sci., i, pp. 188–192.
Said to have been issued in 1879. Describes *Diplognathus mirabilis, Glyptopomus sayrei, Archæobatis gigas,* gen. nov., *Dinichthys minor, Ctenacanthus compressus, Rhynchodus occidentalis,* and *R. excavatus.*

——   1878 B
Description of new fossil fishes from the Trias of New Jersey and Connecticut.
Ann. N. Y. Acad. Sci., i, pp. 127–128.
Describes *Diplurus longicaudatus* gen. et sp. nov. and *Ptycholepis marshi* sp. nov.

——   1879 A
[Letter to Prof. John Collett, describing a collection of remains of fossil fishes collected in Indiana.]
Report State geologist of Indiana for the years 1876, 1877, 1878. Indianapolis, Ind., 1879, pp. 341–349.

——   1883 A
Some interesting remains of fossil fishes recently discovered.
Trans. N. Y. Acad. Sci., ii, pp. 144–147.
Describes new genus *Mylostoma* and the new species *M. variabilis* and *M. terrelli.*

——   1883 B
The relations of *Dinichthys*, as shown by complete crania recently discovered by Mr. Jay Terrell in the Huron shale of Ohio.
Trans. N. Y. Acad. Sci., iii, p. 20.
A brief abstract.

——   1884 A
On the recent discovery of new and remarkable fossil fishes in the Carboniferous and Devonian rocks of Ohio and Indiana.
Geol. Magazine (3), i, pp. 523–524.

——   1884 B
Ctenacanthus wrighti n. sp.
Thirty-fifth Ann. Rept. N. Y. State Mus. Nat. Hist., p. 206, pl. 16, fig. 12.

——   1885 A
On the recent discovery of new and remarkable fossil fishes in the Carboniferous and Devonian rocks of Ohio and Indiana.
Report Brit. Assoc. Adv. Sci., 54th meeting, Montreal, 1884, pp. 724–725.
Same as Newberry, J. S., 1884 A.

Newberry, J. S.—Continued.   1885 B
Description of some gigantic Placoderm fishes recently discovered in the Devonian of Ohio.
Trans. N. Y. Acad. Sci., v, pp. 25–28.

——   1885 C
Descriptions of some peculiar screwlike fossils from the Chemung rocks.
Ann. N. Y. Acad. Sci., xviii, pp. 217–220, with pl. xviii.
Describes new genus *Spiraxis*, with species *major* and *randalli.*

——   1885 D
Sur les restes de grands poissons fossils récemment découverts dans les roches devoniennes de l'Amerique du Nord.
Compte rend. Cong. Géol. Internat. Berlin, 1885, pp. 273–276,
Cited from a reprint with pp. 11–14.

——   1887 A
The fauna and flora of the Trias of New Jersey and the Connecticut Valley.
Trans. N. Y. Acad. Sci., vi, pp. 124–128.
Contains a list of the fishes found in the Triassic rocks.

——   1887 B
Cœlosteus, a new genus of fishes from the Lower Carboniferous limestones of Illinois.
Trans. N. Y. Acad. Sci., vi, pp. 137–138.
Describes *Cœlosteus ferox* gen. et sp. nov.

——   1887 C
Description of a new species of *Titanichthys.*
Trans. N. Y. Acad. Sci., vi, pp. 164–165.
An abstract describing *Titanichthys clarkii.*

——   1888 A
Fossil fishes and fossil plants of the Triassic rocks of New Jersey and the Connecticut valley.
Mon. U. S. Geol. Surv., xiv, pp. i–xiv, 1–152, with pls. i–xxvi.
For review of by A. S. Woodward, see Nature, xliii, pp. 366–367.

——   1888 B
On the structure and relations of *Edestus*, with a description of a gigantic new species.
Ann. N. Y. Acad. Sci., iv, pp. 113–122, with pls. iv–vi.
Describes *Edestus giganteus* sp. nov.

——   1888 C
Note on a new species of *Rhizodus*, from the St. Louis limestone at Alton, Ill.
Trans. N. Y. Acad. Sci., vii, p. 165.
Describes *Rhizodus anceps* sp. nov.

Newberry, J. S.—Continued.   1888 D

On the fossil fishes of the Erie shale of Ohio.

Trans. N. Y. Acad. Sci., vii, pp. 178–180.

An abstract presenting descriptions of *Cladodus kepleri*, *Actinophorus clarkii*, *Dinichthys curtus*, *D. tuberculatus*.

————   1889 A

The Paleozoic fishes of North America.

Mon. U. S. Geol. Surv., J. W. Powell, Director, xvi, pp. 1–340, pls. i–liii.

Although the imprint of the title-page of this monograph gives 1889 as year of printing, it was not issued until August, 1890.

A review of this monograph by A. S. Woodward appeared in Nature, xliii, p. 146, 1890.

————   1890 A

On the genus *Oracanthus* Ag.

Trans. N. Y. Acad. Sci., ix, pp. 131–133. Abstract.

————   1892 A

American Devonian fishes found in Belgium.

Amer. Naturalist, xxvi, p. 1025.

From Ann. Soc. Géol. Belgique, xvi.

Dinichthys pustulosus (= *D. tuberculatus*) and *D. terrelli* are reported from Belgium. The determinations were made by Dr. Newberry. The paper above quoted is a translation of a paper by Lohest.

————   1897 A

New species and a new genus of American Palæozoic fishes, together with notes on the genera *Oracanthus*, *Dactylodus*, *Polyrhizodus*, *Sandalodus*, *Deltodus*.

Trans. N. Y. Acad. Sci., xvi, pp. 282–304, with pls. xxii–xxvi. Edited by Bashford Dean.

In this paper there are described 16 new species, most of them figured. The new genus *Stenognathus* is proposed, having as type *Dinichthys corrugatus*.

Newberry, J. S., and Worthen, A. H.   1866 A

Descriptions of new species of vertebrates, mainly from the Subcarboniferous limestone and Coal Measures of Illinois.

Geol. Surv. Illinois, A. H. Worthen, Director, ii, 1866, pp. 9–134, pls. i–xiii.

————   1870 A

Descriptions of, fossil vertebrates.

Geol. Surv. Illinois, A. H. Worthen, Director, iv, 1870, pp. 347–374, with pls. i–iv.

Contains descriptions and figures of fossil fishes from the Carboniferous and Subcarboniferous formations of Illinois.

Newton, Alfred.   1861 A

Abstract of Mr. J. Wolley's researches in Iceland respecting the gare-fowl, or great auk (*Alca impennis*, Linn.).

The Ibis (1), iii, pp. 374–399.

————   1868 A

Remarks on Professor Huxley's classification of birds.

The Ibis (2), iv, pp. 85–96.

————   1870 A

On existing remains of the gare-fowl (*Alca impennis*).

The Ibis (2), vi, pp. 257–261.

————   1875 A

Birds.

Encyclop. Brit., ed. 9, iii, pp. 728–778, with figs. 38–49.

This portion of the article on birds is devoted mainly to avian biology. Pages 728–736 deal briefly with fossil birds, subfossil birds, birds recently exterminated, and birds partially exterminated.

————   1879 A

Gare-fowl.

Encyclop. Brit., ed. 9, x, pp. 78–80, with 1 fig.

The author cites the principal literature bearing on the history and structure of this bird.

————   1896 A

A dictionary of birds. By Alfred Newton, assisted by Hans Gadow; with contributions from Richard Lydekker, B. A., F. R. S., Charles S. Roy, M. A., F. R. S., and Robert Shufeldt, M. D., late United States Army.

London, Adam and Charles Black, 1893–1896, pp; i–xii; 1–1088, with numerous text figs.

In the introduction (pp. 1–124) Newton discusses the history of the science of ornithology and schemes of classification. Gadow has articles on the skeleton.

Newton, E. T.   1877 A

On the remains of *Hypsodon*, *Portheus*, and *Ichthyodectes* from British Cretaceous strata, with descriptions of new species.

Quart. Jour. Geol. Soc., xxxiii, pp. 505–523, with pl. xxii and 1 text figs.

————   1878 A

Remarks on *Saurocephalus*, and on the species which have been referred to that genus.

Quart. Jour. Geol. Soc., xxxiv, pp. 786–796.

Contains list of genera and species which have been referred to *Saurocephalus*, etc., and bibliography.

Newton, E. T.—Continued.   1878 B
Description of a new fish from the
lower Chalk of Dover.

Quart. Jour. Geol. Soc., xxxiv, pp. 439-446,
with pl. xix.

Discusses the genera *Portheus*, *Daptinus*, and
Ichthyodectes, and has description and figures
of *Daptinus intermedius*.

————   1878 C
The chimæroid fishes of the British
Cretaceous rocks.

Mem. Geol. Surv. United Kingdom, Mon.
iv, pp. 1-50, with pls. i-xii.

————   1886 A
On the remains of a gigantic species
of bird (*Gastornis klaassenii* n. sp.),
from the lower Eocene beds near Croy-
don.

Trans. Zool. Soc. Lond., xii, pp. 143-160,
with pls. xxviii, xxix.

————   1888 A
Notes on pterodactyls.

Proc. Geologists' Assoc., x, pp. 406-424, with
13 figs. in the text.

————   1889 A
A contribution to the history of
Eocene Siluroid fishes.

Proc. Zool. Soc. Lond., 1889, pp. 201-207,
with pl. xxi.

Contains on pp. 206-207 a list of papers on
fossil Siluroid fishes.

————   1889 B
On the skull, brain, and auditory or-
gan of a new species of pterosaurian
(*Scaphognathus purdoni*) from the
Upper Lias, near Whitby, Yorkshire.

Philos. Trans. Royal Soc. Lond., clxxix, pp.
503-537, pls. lxxvii, lxxviii.

————   1892 A
Note on a new species of *Onychodus*
from the lower Old Red sandstone of
Fairfax.

Geol. Magazine (3), ix, pp. 51-52, with 1 fig.
in the text.

Contains list of six known species of *Onychodus*.

————   1899 A
On the remains of *Amia*, from Oligocene strata in the Isle of Wight.

Quart. Jour. Geol. Soc., lv, pp. 1-10, with pl. i.

Describes two new species and gives list of
known species.

**Nicholson, Henry Allen, and Lydek-
ker, Richard.**   1889 A
A manual of palæontology for the
use of students, with a general intro-

**Nicholson, Henry Allen, and Lydek-
ker, Richard—Continued.**
duction on the principles of palæontology. Third edition.

Vol. ii, pp. i-xi, 889-1624, with figs. 813-1354.

Vol. i of this work treats of the fossil inver-
tebrates; pp. 1475-1560 of vol. ii treats of fossil
plants.

Nickerson, W. S.   1893 A
The development of the scales of
Lepidosteus.

Bull. Mus. Comp. Zool., xxiv, pp. 115-139,
with pls. i-iv.

Nitsche, H.   1891 A
Studien über das Elchwild, *Cervus
alces* L. A. Zahnbildung. B. Geweih-
bildung. C. Bau der Läufe.

Zoolog. Anzeiger, xiv, pp. 181-188, 189-191.

————   1899 A
Ueber die Hirschgeweihe mit mehr
als zwei Stangen, und die Hörner
Wiederkäuer im Allgemeinen.

Proc. 4th Internat. Cong. Zoology, Cambridge, Eng., 1898, pp. 185-187.

Noack, Th.   1894 A
Bemerkungen über die Caniden.

Der Zoolog. Garten, 1894, pp. 165-170; 195-198;
241-246; 260-265.

Nodot, L.   1855 A
Description d'un nouveau genre
d'Édentaté fossile renfermant plusieurs
espèces voisines des glyptodontes et
classification méthodique de treize espèces appartenant a ces deux genres.

Comptes rend. Acad. Sci. Paris, xli, pt. 2,
pp. 335-338.

Describes new genus *Schistopleurum*.

Noetling, Fritz.   1885 A
Die Fauna des samländischen Tertiärs.

Abhandl. zur Specialkarte von Preussen und
den Thüringischen Staaten, vi, pt. iii, pp.
1-107, with pls. i-xi.

Nopcsa, Franz B.   1899 A
Dinosaurierreste aus Siebenbürgen
(Schädel von *Limnosaurus transsylvani-
cus* nov. gen. et spec.).

Denkschr. k. Akad. Wissensch. Wien, math.-
naturw. Cl., lxviii, pp. 555-59, with pls. i-vi.

Contains a list of Hadrosaurs.

Nordmann, Alex.   1862 A
Beiträge zur Kenntniss des Knoch-
enbaues der *Rhytina stelleri*.

Acta Soc. Sci. Fennicæ, vii, pp. 1-38, with
pls. i-v.

Norwood, J. G., and **Owen**, D. D.
                                1846 A
Description of a new fossil fish from
the Palæozoic rocks of Indiana.
Amer. Jour. Sci. (2), i, pp. 367–371, with 2
woodcuts.

———   1846 B
Description of *Macropetalichthys raph-
eidolabis.*
Proc. Boston Soc. Nat. Hist. ii, p. 116.

Oldham, Thomas.   1859 A
On some fossil fish-teeth of the genus
Ceratodus, from Maledi, south of Nag-
pur.
Mem. Geol. Surv. India, i, pp. 295–309, pls.
xiv–xvi.

Orr, Henry.   1885 A
Beitrag zur Phylogenie der Ganoi-
den. Inaugural-Dissertation, Jena.
Pp. 1–37, with 2 figs. in the text.

Orton, Edw.   1891 A
On the occurrence of *Megalonyx jeffer-
soni* in central Ohio.
Bull. Geol. Soc. Amer., ii, p. 635.

Orton, James.   1869 A
The great auk.
Amer. Naturalist, iii, pp. 539–542, with 1 fig.

Osawa, Gakutaro.   1898 A
Verhandl. Anat. Gesellsch., 12th meeting,
Kiel, 1898, pp. 100–105, with 4 figs. in the text.

Osborn, H. F.   1881 A
A memoir upon *Loxolophodon* and
Uintatherium, two genera of the sub-
order *Dinocerata.*
Contrib. E. M. Mus. Geol. and Arch. of the
College of New Jersey [Princeton], i, pp. 1–54,
pls. i–iv, and map.
Notices of this paper, with 2 figs. of *L.
cornutus,* appeared in Amer. Naturalist, xv, p.
888; Amer. Jour. Sci., xxii, p. 235.

———   1883 A
Achænodon, an Eocene bunodont.
Contrib. E. M. Mus. Geol. and Arch.; Princeton Coll., Bull. No. 3, pp. 23–35, with pl. vi, and
8 text figs.

———   1886 A
A new mammal from the American
Triassic.
Science, viii, p. 540, with 1 woodcut.
Describes new genus and species, *Micro-
conodon tenuirostris.*

———   1886 B
Observations upon the Triassic mammals *Dromatherium* and *Microconodon.*
Proc. Acad. Nat. Sci. Phila., 1886, pp. 359–
363, with 4 woodcuts in the text.

Osborn, H. F.—Continued.   1887 A
A pineal eye in the Mesozoic mam-
mals.
Science, ix, p. 92.

———   1887 B
The pineal eye in *Tritylodon.*
Science, ix, p. 114, with 1 woodcut.

———   1887 C
No parietal foramen in *Tritylodon.*
Science, ix, p. 538.

———   1887 D
The origin of the tritubercular type
of mammalian dentition.
Science, x, p. 300.

———   1887 E
The Triassic mammals *Dromatherium*
and *Microconodon.*
Proc. Amer. Philos. Soc., xxiv, pp. 109–111,
with 1 pl.

———   1887 F
On the structure and classification of
the Mesozoic *Mammalia.*
Proc. Acad. Nat. Sci. Phila., 1887, pp. 282–
292, with 3 figs. in the text.
Review of, in Geol. Magazine, 1888, v, p. 132.
Also in Archiv f. Anthrop., xix, pp. 135–141.

———   1887 G
Note on the genus *Athrodon.*
Amer. Naturalist, xxi, p. 1020.
Athrodon preoccupied, *Kurtodon* substituted.

———   1888 A
Additional observations upon the
structure and classification of the Mesozoic *Mammalia.*
Proc. Acad. Nat. Sci. Phila., 1888, pp. 292–301.
with 2 figs. in the text.

———   1888 B
The mylohyoid groove in the Meso-
zoic and recent *Mammalia.*
Amer. Naturalist, xxii, pp. 75–76.

———   1888 C
A review of Mr. Lydekker's arrange-
ment of the Mesozoic *Mammalia.*
Amer. Naturalist, xxii, pp. 232–236.

———   1888 D
Chalicotherium and *Macrotherium.*
Amer. Naturalist, xxii, pp. 728–729.

———   1888 E
The nomenclature of the mammalian
molar cusps.
Amer. Naturalist, xxii, pp. 926–928.

———   1888 F
The evolution of the mammalian
molars to and from the tritubercular
type.
Amer. Naturalist, xxii, pp. 1067–1079, pls. xv,
xvi, and 1 woodcut.
Reviewed by M. Schlosser in Archiv f. Anthrop., xix, pp. 157–159.

Osborn, H. F.—Continued.   1888 G
On the structure and classification of
the Mesozoic *Mammalia.*

Jour. Acad. Nat. Sci. Phila. (2), ix, pp. 186–
265, pls. viii, ix, and 30 figs. in the text.
For review see Nature, xxxviii, pp. 611–614;
Amer. Naturalist, xxii, pp. 723–724. Also, H.
Winge, 1893 C, p. 117.

——   **1888 H**
The characters and phylogeny of the
Amblypoda.

Science (2), vii, p. 226.
Abstract of paper read before Amer.
Morpholog. Society.

——   **1889 A**
The evolution of the mammalian
teeth to and from the tritubercular
type.

Report Brit. Assoc. Adv. Sci., 58th meeting.
Bath, 1888, p. 660.
Abstract of paper read before Brit. Assoc.
Adv. Sci., and published in full in Amer. Nat-
uralist, 1888, p. 1067.

——   **1890 A**
The paleontological evidence for the
transmission of acquired characters.

Science, xv, pp. 110–111.
See Amer. Naturalist, xxiii, pp. 559–566.
Nature, xii, pp. 227–228; Rep. Brit. Assoc. Adv.
Sci. for 1889.
Discusses the evolution of the teeth of *Mam-
malia.*

——   **1890 B**
The palæontological evidence for the
transmission of acquired characters.

Proc. Amer. Assoc. Adv. Sci., 38th meeting,
Toronto, 1889, pp. 273–276.

——   **1890 C**
A review of the Cernaysian *Mammalia.*

Proc. Acad. Nat. Sci. Phila., 1890, pp. 51–62,
with 6 figs. in text.
Review by M. Schlosser in Archiv f. Anthrop, xx, pp. 111–112.

——   **1890 D**
The *Mammalia* of the Uinta formation.   Part III.   The *Perissodactyla.*
Part IV.   The evolution of the ungulate foot.

Trans. Amer. Philos. Soc., xvi, pp. 505–572,
with pls. viii–xi, and 12 figs. in the text.
This forms a portion of a joint paper by Professors Scott and Osborn. See Scott, W. B.,
1890 A, and Scott and Osborn, 1890 A.
For review by R. Lydekker, see Nature, xliii,
p. 177.

——   **1891 A**
A review of the Cretaceous *Mammalia.*

Proc. Acad. Nat. Sci. Phila., 1891, pp. 124–135,
with 8 figs. in the text.
See also Amer. Naturalist, xxv, pp. 595–611,
figs. 1–12.

Osborn, H. F.—Continued.   1891 B
A review of the discovery of the Cretaceous *Mammalia.*

Amer. Naturalist, xxv, pp. 44–45.
Abstract of a review of papers here entered
under titles of Marsh, O. C., 1889 D, and
Marsh, O. C., 1889 F.

——   **1891 C**
A reply to Professor Marsh's "Note
on Mesozoic *Mammalia.*"

Amer. Naturalist, xxv, pp. 775–785, with 1
fig. in the text.
Review by M. Schlosser in Archiv f. Anthrop., xxii, p. 118.

——   **1891 D**
Meniscotheriidæ and *Chalicotheriidæ.*

Amer. Naturalist, xxv, pp. 911–912.

——   **1891 E**
A review of the "Discovery of the
Cretaceous *Mammalia.*"

Amer. Naturalist, xxv, pp. 595–611, with 12
figs. in the text.
This is a review of papers by Professor
Marsh, and is a reprint, with slight changes
and additions, of Osborn, H. F., 1891 A.
Abstract in Archiv f. Anthrop., xxii, pp.
117–118.

——   **1892 A**
Palæonictis in the American Lower
Eocene.

Nature, xlvi, p. 30.

——   **1892 B**
What is *Lophiodon?*

Amer. Naturalist, xxvi, pp. 763–765.

——   **1892 C**
The ancestry of *Chalicotherium.*

Science, xix, p. 276.

——   **1892 D**
A reply to Professor Marsh's "Note
on Mesozoic *Mammalia.*"

Proc. Amer. Assoc. Adv. Sci., 40th meeting,
Washington, 1891, p. 290.
Abstract of Osborn, H. F., 1891 C.

——   **1892 E**
Nomenclature of mammalian molar
cusps.

Amer. Naturalist, xxvi, pp. 436–437.

——   **1892 F**
The contemporary evolution of man.

Amer. Naturalist, xxvi, pp. 455–481, 537–567.
Reviewed by M. Schlosser in Archiv f. Anthrop., xxiii, p. 152.

——   **1892 G**
Is *Meniscotherium* a member of the
Chalicotherioidea?

Amer. Naturalist, xxvi, pp. 506–509, with 4
figs. in the text.

Osborn, H. F.—Continued.   1892 H
Odontogenesis in the ungulates.
Amer. Naturalist, xxvi, pp. 621–623.
A review of paper by Julius Taeker, in
Morpholog. Jahrbuch, xv, p. 308.

———   1892 I
The history and homologies of the
human molar cusps.
Anatom. Anzeiger, vii, pp. 740–747, with 12
text figs.
A review of the contributions of Dr. A.
Fleischmann, Dr. Julius Taeker, and Dr. Carl
Röse.

———   1893 A
The rise of the *Mammalia* in North
America.
Amer. Jour. Sci., (3) xlvi, pp. 379–392; 448–
466, with pl. xi and 5 figs. in the text.
For review of this paper, see Nature, xlix,
pp. 235 and 257.

———   1893 B
Fossil mammals of the Upper Creta-
ceous beds: I. The multituberculates;
II. The trituberculates; III. Faunal
relations of the Laramie mammals.
Bull. Amer. Mus. Nat. Hist., v, pp. 311–330,
with pls. vii, viii, and 3 text figs.
Abstract of, by M. Schlosser, in Archiv f.
Anthrop., xxiv, pp. 126–127.

———   1893 C
Aceratherium tridactylum from the
Lower Miocene of Dakota.
Bull. Amer.Mus. Nat. Hist., v, pp. 85–86.

———   1893 D
The rise of the *Mammalia* in North
America.
Studies from the biological laboratories of
Columbia College; Zoology. Ginn & Co., Bos-
ton, 1893. Pp. 1–45, with 1 pl. and 5 text figs.
Vice-presidential address before the Amer.
Assoc. for the Advance of Science, Madison,
Wis., Aug. 17, 1893. Same as Osborn, 1893 A.
Abstract of, by M. Schlosser, in Archiv f.
Anthrop., xxiv, pp. 123–124.

———   1893 E
The collection of fossil mammals in
the American Museum of Natural History, New York.
Science, xxi, p. 251.

———   1893 F
The *Ancylopoda*, *Chalicotherium*, and
Artionyx.
Amer. Naturalist, xxvii, pp. 118–133, with 4
figs. in the text.
Abstract of, by M. Schlosser, in Archiv f.
Anthrop., xxiv, pp. 125–126.

———   1893 G
Recent researches upon the succes-
sion of the teeth in mammals.
Amer. Naturalist, xxvii, pp. 493–508.
Noticed by M. Schlosser, in Archiv f. An-
throp., xxiv, p. 126.

Osborn, H. F.—Continued.   1893 H
Sur la découverte du *Pulronictis* en
Amérique.
Bull. Soc. Géol. France (3), xx, pp. 434–436,
with 1 fig. in the text.
Describes the new species *P. occidentalis* Osb.

———   1893 I
Protoceras, the new artiodactyle.
Nature, xlvii, pp. 321–322, with 3 figs. in the
text.

———   1893 J
Artionyx, a clawed artiodactyle.
Nature, xlvii, pp. 610–611, with 3 figs. in the
text.

———   1893 K
Foot of *Artionyx*.
Anatom. Anzeiger, viii, p. 368.
Report of meeting of N. Y. Acad. Sci.

———   1893 L
Protoceras celer.
Anatom. Anzeiger, viii, pp. 127–128.
In reports of meeting of N. Y. Acad. Sci.

———   1894 A
A division of the Eutherian mammals
into the *Mesoplacentalia* and *Cenoplacentalia.*
Trans. N. Y. Acad. Sci., xiii, pp. 234–237.

———   1895 A
Fossil mammals of the Uinta Basin.
Bull. Amer. Mus. Nat. Hist., vii, pp. 71–105,
with 17 figs. in the text.
Reviewed by M. Schlosser in Archiv f.
Anthrop., xxv., pp. 181–182.

———   1895 B
Vertebrate paleontology in the Amer-
ican Museum.
Science (2), ii, pp. 178–179.

———   1895 C
The history of the cusps of the human
molar teeth.
Internat. Dental Jour., 1896.
Quoted from reprint, pp. 1–14, with pl. AA
and text figs. A–D.

———   1896 A
Titanotheres of the American Museum
of Natural History.
Amer. Naturalist, xxx, p. 162.
Secretary's report of the session of the New
York Academy of Science, Dec. 9. 1895. Also
found in Anatom. Anzeiger, xi, p. 512 and
Zoolog. Anzeiger, xix, p. 48.

———   1896 B
The cranial evolution of *Titanothe-
rium.*
Bull. Amer. Mus. Nat. Hist., viii, pp.157–197,
with pls. iii and iv and 15 figs. in the text.
Notice of this paper by M. Schlosser in Neues
Jahrb. Mineral., 1899, ii, pp. 319–321, and in
Archiv f. Anthrop., xxv, p. 216.

Osborn, H. F.—Continued.   1896 C
Prehistoric quadrupeds of the Rockies.
The Century Magazine, lii, pp. 705–715, with
8 engravings illustrating restorations of fossil
mammals.

——   1897 A
Lambdotherium not related to *Palæo-syops* or the Titanotheres.
Amer. Naturalist, xxxi, pp. 55–57.

——   1897 B
Reconstruction of *Phenacodus pri-mævus*, the most primitive ungulate.
Amer. Naturalist, xxxi, p. 960.
A synopsis of a paper read before the Brit.
Assoc. Adv. Sci. at Toronto, 1897.
An abstract of this paper is found on p. 684
of the report of the meeting.

——   1897 C
Trituberculy: A review dedicated to
the late Professor Cope.
Amer. Naturalist, xxxi, pp. 993–1016.
Appended is a bibliography of the subject.
The paper is reviewed by M. Schlosser in
Neues Jahrb. Mineral, 1901, i, p. 151.

——   1897 D
Report on the phylogeny of the early
Eocene Titanotheres.
Anatom. Anzeiger, xiii, p. 561.
Report of paper read before New York Bio-
logical Society, Apr. 5, 1897. See also Science
(2), vi, p. 107.

——   1897 E
The origin of the teeth of the *Mam-malia*.
Science (2), v, pp. 576–577.

——   1897 F
The *Ganodonta*, or primitive Eden-
tates with enameled teeth.
Science (2), v, pp. 611–612.

——   1897 G
The Huerfano Lake basin, southern
Colorado, and its Wind River and
Bridger fauna.
Bull. Amer. Mus. Nat. Hist., ix, pp. 247–258.
Brief review in Neues Jahrb. Mineral., 1900,
p. 132.

——   1898 A
Paleontological problems.
Science (2), vii, pp. 145–147.

——   1898 B
Origin of the *Mammalia*.
Science (2), vii, pp. 176–178.
Discussion before Amer. Soc. Naturalists.

——   1898 C
A complete skeleton of *Teleoceras*, the
true rhinoceros from the upper Miocene
of Kansas.
Science (2), vii, pp. 554–557, with 1 fig. in the
text.

Osborn, H. F.—Continued.   1898 D
The life and works of Cope, illustrating the training of a naturalist and the
essential characteristics of a great comparative anatomist.
Preface to second edition of Professor Cope's
Syllabus of Lectures on the *Vertebrata*, pp.
iii–xxxv.

——   1898 E
A complete skeleton of *Teleoceras fos-siger*. Notes upon the growth and
sexual characters of this species.
Bull. Amer. Mus. Nat. Hist., x, pp. 51–59,
with pls iv, iv A.

——   1898 F
A complete skeleton of *Coryphodon radians*.
Bull. Amer. Mus. Nat. Hist., x, pp. 81–91,
with pl. x and 2 figs. in the text.

——   1898 G
Remounted skeleton of *Phenacodus primævus*. Comparison with *Euroto-gonia*.
Bull. Mus. Nat. Hist., x, pp. 159–164, with
pl. xii and 4 text figs.

——   1898 H
Evolution of the *Amblypoda*. Part
I. *Taligrada* and *Pantodonta*.
Bull. Amer. Mus. Nat. Hist., x, pp. 169–218,
with 29 figs. in the text.

——   1898 I
The extinct rhinoceroses.
Mem. Amer. Mus. Nat. Hist., i, pp. 75–164,
with pls. xiiA–xx, with 49 figs. in the text.
Pp. 121–125 of this memoir are devoted to the
bibliography of the rhinoceroses.
A notice of this memoir is found in Natural
Science, 1898, xiii, pp. 151–159.

——   1898 J
Additional characters of the great
herbivorous dinosaur *Camarasaurus*.
Bull. Amer. Mus. Nat. Hist., x, pp. 219–233,
with 13 figs. in the text.

——   1898 K
Paleontological notes.
Science (2), vii, pp. 164–165.

——   1898 L
A placental marsupial.
Science (2), vii, pp. 454–456, with 1 fig. in the
text.

——   1898 M
A complete skeleton of *Coryphodon radians*. Notes upon the locomotion
of this animal.
Science (2), vii, pp. 585–588, with 1 fig. in the
text.

Osborn, H. F.—Continued.   1898 N
Models of extinct vertebrates.
Science (2), vii, pp. 841–845, with text figs.
1–4.

———   1898 O
[Origin of *Mammalia*.]
Science (2), viii, p. 358.
Abstract of remarks at International Congress of Zoologists at Cambridge, England.

———   1898 P
On the presence of a frontal horn in
Aceratherium incisivum Kaup.
Science (2), viii, p. 880.
A brief abstract. *Aceratherium* not a hornless rhinoceros.

———   1898 Q
Casts, models, photographs, and restorations of fossil vertebrates.
Pamphlet issued by Department of Vertebrate Palæontology, Amer. Mus. Nat. Hist., pp. 1–21, with 7 illus.
Remarks on this pamphlet are found in Natural Science, 1898, xiii, p. 230.

———   1898 R
The origin of *Mammalia*.
Amer. Naturalist, xxxii, pp. 309–334, with 14 text figs.
Reviewed in Neues Jahrb. Mineral., 1900, ii, pp. 135–138.

———   1899 A
Frontal horn on *Aceratherium incisivum*. Relation of this type to *Elasmotherium*.
Science (2), ix, pp. 161–162, with pl. i.

———   1899 B
The origin of mammals.
Amer. Jour. Sci. (4), vii, pp. 92–96.

———   1899 C
Additional characters of *Diplodocus*.
Science (2), ix, pp. 315–316.

———   1899 D
A complete mosasaur skeleton, osseous and cartilaginous.
Mem. Amer. Mus. Nat. Hist., i, pp. 167–188, with pls. xxi–xxiii and 13 text figs.
See also, for extract from this paper, Science (2), x, 1899, pp. 919–925, figs. 1–3; for review, Neues Jahrb. Mineral., 1901, i, p. 156.

———   1899 E
A skeleton of *Diplodocus*.
Mem. Amer. Mus. Nat. Hist., i, pp. 191–214, with pls. xxiv–xxviii and 14 text figs.
An extract of this paper was published in Science (2), x, pp. 870–874, with fig. 1.
Reviewed in Neues Jahrb. Mineral., 1901, i, p. 156.

Osborn, H. F.—Continued.   1899 F
Upon the structure of *Tylosaurus dyspelor*, including the cartilaginous sternum.
Science (2), ix, pp. 912–913.
Brief synopsis of paper read before New York Academy of Sciences. See Osborn, H. F., 1899 D.

———   1899 G
The Newburgh mastodon.
Science (2), x, p. 589.

———   1899 H
Fore and hind limbs of carnivorous dinosaurs from the Jurassic of Wyoming. Dinosaur contributions, No. 3.
Bull. Amer. Mus. Nat. Hist., xii, pp. 161–172, with 7 figs. in the text.
Abstract in Neues Jahrb. Mineral., 1901, ii, p. 305.

———   1899 I
[Review of Arthur Smith Woodward's Outlines of vertebrate palæontology for students of zoology.]
Natural Science, xiv, pp. 156–159.

———   1900 A
Intercentra and hypapophyses in the cervical region of mosasaurs, lizards, and *Sphenodon*.
Amer. Naturalist, xxxiv, pp. 1–7, with 4 figs. in the text.
An abstract of this paper was published in Science, x, 1899, p. 896.

———   1900 B
Recent zoöpaleontology.
Science (2), xi, pp. 115–116.

———   1900 C
The geological and faunal relations of Europe and America during the Tertiary period and the theory of the successive invasions of an African fauna.
Science (2), xi, pp. 561–574.

———   1900 D
Recent zoöpaleontology.
Science (2), xii, pp. 767–769, with 1 text fig.
Refers to finding of *Trigonias osborni* Lucas, and *Archelon* Wieland.

———   1900 E
Phylogeny of the rhinoceroses of Europe.
Science (2), xii, p. 885.
An abstract of a paper read before the N. Y. Acad. Sci.

———   1900 F
The angulation of the limbs of *Proboscidia*, *Dinocerata*, and other quadrupeds in adaptation to weight.
Amer. Naturalist, xxxiv, pp. 89–94, with 7 figs. in the text.

Osborn, H. F.—Continued. 1900 G
Reconsideration of the evidence for
a common dinosaur-avian stem in the
Permian.
Amer. Naturalist, xxxiv, pp. 777-799, with
12 figs. in the text.

―― 1900 H
Phylogeny of the rhinoceroses of
Europe.
Bull. Amer. Mus. Nat. Hist., xiii, pp. 229-267,
with 16 figs. in the text.

―― 1900 I
Oxyæna and *Patriofelis* restudied as
terrestrial creodonts.
Bull. Amer. Mus. Nat. Hist., xiii, pp. 269-279,
with pls. xviii, xix, and 8 figs. in the text.

―― 1900 J
Origin of the *Mammalia*. III. Oc-
cipital condyles of reptilian tripartite
type.
Amer. Naturalist, xxxiv, pp. 943-947, with 3
figs in the text.

―― 1900 K
Correlation between Tertiary mammal horizons of Europe and America.
An introduction to the more exact investigation of Tertiary zoögeography.
Preliminary study with third trial
sheet.
Ann. N. Y. Acad. Sci., xiii, pp. 1-72.
Pp. 59-64 are occupied by a bibliography of
the subject; pp. 65-72 by a list of the scientific
publications of the author.

Osborn, H. F. and Earle, Charles.
1895 A
Fossil mammals of the Puerco beds.
Collection of 1892.
Bull. Amer. Mus. Nat. Hist., vii, pp. 1-70,
with 21 figs. in the text.
Reviewed by W. B. Scott in Science (2), i,
p. 660, and by M. Schlosser in Archiv f. Anthrop.,
xxv, pp. 176-179.

**Osborn, H. F., Scott, W. B., and Speir,
Francis, jr.** 1878 A
Palæontological report of the Prince-
ton scientific expedition of 1877.
Contrib. Mus. Geol. and Arch., Princeton
College, No. 1, pp. 1-107, with pls. A, i-x.
Followed by systematic catalogue of the
Eocene vertebrates of Wyoming, pp. 131-146.

Osborn, H. F., and Speir, F. jr. 1879 A
The lower jaw of *Loxolophodon*.
Amer. Jour. Sci., xvii, 304-309, with pl. ii.

Osborn, H. F., and Wortman, J. L.
1892 A
Fossil mammals of the Wahsatch
and Wind River beds. Collection of
1891.
Bull. Amer. Mus. Nat. Hist., iv, pp. 80-147
with pl. iv and 18 figs. in text.
I. Homologies and nomenclature of the
mammalian molar cusps (H. F. O.), p. 84.
II. The classification of the *Perissodactyla*
(H. F. O.), p. 90.
III. The ancestry of the *Felidæ* (J. L. W.),
p. 94.
IV. Taxonomy and morphology of the primates, creodonts, and ungulates; 1, Wahsatch;
2, Wind River (H. F. O.), p. 101.
V. Geological and geographical sketch of
the Big Horn Basin (J. L. W.), p. 135.
VI. Narrative of the expedition (J. L. W.),
p. 144.
For remarks on portions of this paper see
Natural Science, ii, pp. 404-405; also, for ab-
stract, Archiv f. Anthrop., xxiii, pp. 136-138.

―― ―― 1892 B
Characters of *Protoceras* (Marsh), the
new artiodactyl from the Lower Mio-
cene.
Bull. Amer. Mus. Nat. Hist., iv, pp. 351-371,
figs. 1-6.
Abstract of, by M. Schlosser, in Archiv f.
Anthrop., xxiii, pp. 185-136.

―― ―― 1893 A
Artionyx, a new genus of *Ancylopoda*.
Appendix [p. 17]: On the mechanics
of the artiodactyl tarsus.
Bull. Amer. Mus. Nat. Hist., v, pp. 1-18, with
8 text figures.
Remarks on this paper in Natural Science,
ii, p. 326.

―― ―― 1894 A
Fossil mammals of the Lower Miocene
White River beds. Collection of 1892.
Bull. Amer. Mus. Nat. Hist., vi, pp. 199-228,
with pls. ii and iii, and 8 figs. in the text.
Abstract of, by M. Schlosser in Archiv f.
Anthrop., xxiv, pp. 167-168.

―― ―― 1895 A
Perissodactyls of the Lower Miocene
White River beds.
Bull. Amer. Mus. Nat. Hist., vii, pp. 343-375,
with pls. viii-xi and 12 figs. in the text.
Reviewed by M. Schlosser in Archiv f.
Anthrop., xxv, pp. 179-181.

Owen, David Dale. 1842 A
Regarding human footprints in solid
limestone.
Amer. Jour. Sci., xliii, pp. 14-32, with 1 pl.

Owen, David Dale—Continued. 1857 A
Report of the geological survey in Kentucky, made during the years 1854 and 1855.
Amer. Jour. Sci. (2), xxiii, pp. 272–276.
An editorial notice of the above report made by D. D. Owen.

See also Norwood and Owen.

Owen, D. D., Norwood, J. G., and Evans, John. 1850 A
Notice of fossil remains brought by Mr. J. Evans from the Mauvais Terres, or bad lands of White River, 150 miles west of the Missouri.
Proc. Acad. Nat. Sci., v, pp. 66–67.

Owen, Richard. 1832 A
On the anatomy of the weasel-headed armadillo (*Dasypus 6-cinctus* L.).
Proc. Zool. Soc. Lond., pt. ii, 1832, pp. 134–138.

——— 1835 A
On the comparative osteology of the orang and chimpanzee.
Proc. Zool. Soc. Lond., 1835, pt. iii, pp. 80–40.

——— 1838 A
On the osteology of the *Marsupialia*.
Proc. Zool. Soc. Lond., 1838, pt. vi, pp. 120–147.

——— 1838 B
On the dislocation of the tail at a certain point, observable in the skeletons of many *Ichthyosauri*.
Proc. Geol. Soc. Lond., ii, pp. 660–662.

——— 1839 A
Outlines of a classification of the *Marsupialia*.
Proc. Zool. Soc. Lond., 1839, pt. vii, pp. 5–19.

——— 1839 B
Observations on the teeth of the *Zeuglodon, Basilosaurus* of Dr. Harlan.
Proc. Geol. Soc. Lond., iii, pp. 24–28.
Also in Lond. and Edinb. Philos. Mag., xiv, pp. 302–307; Ann. Nat. Hist., 1839, iii, pp. 210–213; Ann. Sci. Naturelles (2), xii, Zool. 1839, pp. 222–229.

——— 1839 C
Description of the *Lepidosiren annectens*.
Trans. Linn. Soc. Lond., xviii, pp. 327–361, with pls. xxiii–xxvii.

——— 1839 D
Observations on the fossils representing the *Thylacotherium prevostii* (Valenciennes), with reference to the doubts of its mammalian and marsupial na-

Owen, Richard—Continued.
tures, and on *Phascolotherium bucklandi*. Pt. I. On the *Thylacotherium*. Pt. II. On the *Phascolotherium*.
Trans. Geol. Soc. Lond. (2), vi, pp. 41–65, with pl. v.

——— 1840 A
Report on British fossil reptiles. Pt. I.
Report of Brit. Assoc. Adv. Sci., 9th meeting, Birmingham, 1839, pp. 43–126.

——— 1840 B
Description of a tooth and part of the skeleton of the *Glyptodon clavipes*, a large quadruped of the Edentate order, to which belongs the tessellated armour described and figured by Mr. Clift in the former volume of the Transactions of the Geological Society, with a consideration of the question whether the *Megatherium* possessed an analagous dermal armour.
Trans. Geol. Soc. Lond. (2), vi, pp. 81–106, with pls. x–xiii.
For preliminary description see Proc. Geol. Soc. Lond., iii, 1840, p. 108.

——— 1840 D
Note on the dislocation of the tail at a certain point observable in the skeletons of many *Ichthyosauri*.
Trans. Geol. Soc. Lond. (2), v, pp. 511–514, with pl. xiii.

——— 1840 E
A description of a specimen of the *Plesiosaurus macrocephalus*, Conybeare, in the collection of Viscount Cole, M. P., D. C. L., F. G. S., etc.
Trans. Geol. Soc. Lond. (2), v, pp. 515–535, with pls. xliii–xlv.

——— 1840 F
Fossil *Mammalia*.
The zoology of the voyage of H. M. S. Beagle. Part i, pp. 13–111, with pls. i–xxxii.

——— 1840 G
Description of the fossil remains of a mammal, a bird, and a serpent from the London clay.
Proc. Geol. Soc. Lond. iii, pp. 162–166.
The genera *Lithornis* and *Palæophis* are established.

——— 1841 A
A description of some of the soft parts, with the integument, of the hind-fin of the *Ichthyosaurus*, indicating the shape of the fin when recent.
Trans. Geol. Soc. Lond. (2), vi, pp. 199–201, with pl. xx.

Owen, Richard—Continued.   1841 B

Description of some ophidiolites (*Palæophis toliapicus*) from the London clay at Sheppey, indicative of an extinct species of serpent.

Trans. Geol. Soc. Lond. (2), vi, pp. 209-210, with pl. xxii.

——   1841 C

Observations on the *Basilosaurus* of Dr. Harlan (*Zeuglodon cetoides* Owen).

Trans. Geol. Soc. Lond. (2), vi, pp. 69-79, with pls. vii-ix.

——   1841 D

On the teeth of species of the genus *Labyrinthodon* (*Mastodonsaurus salamandroides* and *Phytosaurus* (?)) of Jaeger, from the German Keuper and the sandstone of Warwick and Leamington.

Ann. and Mag. Nat. Hist., viii, pp. 58-61.
Taken from Proc. Geol. Soc. Lond., iii, pp. 389-397.

——   1841 E

A description of a portion of the skeleton of the *Cetiosaurus*, a gigantic extinct saurian, occurring in the Oolitic formation of different portions of England.

Proc. Geol. Soc. Lond., iii, pp. 457-462.

——   1841 F

On the structure of fossil teeth from the central or Corn-stone division of the Old Red sandstone, indicative of a new genus of fishes, or fish-like *Batrachia*, for which is proposed the name of *Dendrodus*.

Micros. Journ. and Struct. Record, i, pp. 1-4; 17-20, with 7 figs. in the text.

——   1842 A

Report on British fossil reptiles. Part II.

Report Brit. Assoc. Adv. Sci., 11th meeting, Plymouth, 1841, pp. 60-204.

——   1842 B

On the teeth of a species of the genus *Labyrinthodon* (*Mastodonsaurus* of Jaeger), common to the German Keuper formation and the lower sandstone of Warwick and Leamington.

Trans. Geol. Soc. Lond. (2); vi, pp. 503-513.

——   1842 C

Description of parts of the skeleton and teeth of five species of the genus *Labyrinthodon* . . . with remarks on the

Owen, Richard—Continued.

probable identity of *Cheirotherium* with this genus of extinct batrachians.

Trans. Geol. Soc. Lond., vi, pp. 515-543, with pls. xliii-xlvii.

——   1842 D

Description of the skeleton of an extinct gigantic sloth, *Mylodon robustus*, Owen, with observations on the osteology, natural affinities, and probable habits of the megatheroid quadrupeds in general.

Pp. 1-176, with pls. i-xxiv. London, 1842, 4to.

——   1842 E

Report on the *Missourium* now exhibiting at the Egyptian Hall, with an inquiry into the claims of *Tetracaulodon* to generic distinction.

Proc. Geol. Soc. Lond., iii, pp. 689-695.

——   1843 A

Report on the British fossil Mammalia. Part I. *Unguiculata* and *Cetacea*.

Report Brit. Assoc. Adv. Sci., 12th meeting, Manchester, 1842, pp. 54-74.

——   1843 B

On Dr. Harlan's notice of new fossil *Mammalia*.

Amer. Jour. Sci., xliv, pp. 341-345.
Letter from Professor Owen regarding *Orycterotherium*, *Mylodon*, and *Megatherium*.

——   1844 A

Report on the British fossil Mammalia. Part II. *Ungulata*.

Report Brit. Assoc. Adv. Sci., 13th meeting, Cork, 1843, pp. 208-241.

——   1845 A

Account of various portions of the *Glyptodon*, an extinct quadruped, allied to the armadillo, and recently obtained from the Tertiary deposits in the neighborhood of Buenos Ayres.

Quart. Jour. Geol. Soc., i, pp. 257-262.

——   1845 B

Odontography; or a treatise on the comparative anatomy of the teeth, etc.

Vol. i, text, pp. i-lxxiv, 1-655; Vol. ii, atlas. pls. i, ii, 1-150. London, 1840-1845.
Due to repetitions of numbers on pls., there are altogether 168.
For dates of publication see Woodward and Sherborn, 1890 A.

——   1845 C

Reply to some observations of Prof. Wagner on the genus *Mylodon*.

Ann. and Mag. Nat. Hist. (1), xvi, pp. 100-102.

Owen, Richard—Continued. 1845 D

Description of certain fossil crania discovered by A. G. Bain, esq., in the sandstone rocks at the southeastern extremity of Africa, referable to different species of an extinct genus of *Reptilia* (*Dicynodon*), and indicative of a new tribe or suborder of *Sauria*.

Trans. Geol. Soc., London (2), vii, pp. 59–84, with pls. iii–vi.

—— 1846 A

Observations on certain fossils from the collection of the Academy of Natural Sciences of Philadelphia.

Proc. Acad. Nat. Sci. Phila., iii, pp. 93–96. Reference to this paper is made in Amer. Jour. Sci., iii, 1847, p. 125.

—— 1846 B

A history of British fossil mammals and birds.

Pp. i–xlvi; 1–560. Illustrated by 237 woodcuts. London.

—— 1846 C

[Extracts from letters to Dr. S. G. Morton on *Dorudon* of Gibbes.]

Jour. Acad. Nat. Sci. Phila., i, pp. 9–10. This forms a portion of Dr. Gibbes's paper on *Basilosaurus*, Gibbes, R. W., 1847 B.

—— 1846 D

On the fossil genus *Dorudon*.

Proc. Acad. Nat. Sci., Phila., iii, p. 15. Abstract of letter stating that *Dorudon* of Gibbes is same as *Zeuglodon*.

—— 1847 A

General geological distribution and probable food and climate of the mammoth.

Amer. Jour. Sci. (2), iv., pp. 13–19. Extracted from Owen's British Fossil Mammalia.

—— 1847 B

Report on the archetype and homologies of the vertebrate skeleton. Part I. Special homology. Part II. General homology.

Report Brit. Assoc. Adv. Sci., 16th meeting, Southampton, 1846, pp. 169–340, with 28 woodcuts and 3 tables.

—— 1847 C

Notices of some fossil *Mammalia* of South America.

Report Brit. Assoc. Adv. Sci., 16th meeting, Southampton, 1846, pp. 65–67.

—— 1847 D

Observations on certain fossil bones from the collection of the Academy of Natural Sciences of Philadelphia.

Jour. Acad. Nat. Sci. Phila. (2), i, pp. 18–20. Collection made in digging canal near Brunswick, Ga.

Owen, Richard—Cont. inued. 1847 E

Description of the atlas, axis, and subvertebral wedge-bones in the *Plesiosaurus*, with remarks on the homologies of those bones.

Ann. and Mag. Nat. Hist. (1), xx, pp. 217–225, with 6 figs. in the text.

—— 1848 A

Description of teeth and portions of jaws of two extinct anthracotheroid quadrupeds (*Hyopotamus vectianus* and *H. bovinus*) discovered by the Marchioness of Hastings in the Eocene deposits on the N. W. coast of the Isle of Wight, with an attempt to develop Cuvier's idea of the classification of pachyderms by the number of their toes.

Quart. Jour. Geol. Soc., iv, pp. 104–141, pls. vii–viii, with 18 figs. in text.

—— 1849 A

Notes on remains of fossil reptiles discovered by Prof. Henry Rogers, of Pennsylvania, U. S., in Green-sand formations of New Jersey.

Quart. Jour. Geol. Soc., v, pp. 380–383, with pls. x–xi.

—— 1849 B

Monograph on the fossil *Reptilia*. Part I. *Chelonia*.

Palæontographical Society, ii, pp. 1–76, 38 pls. The above is a joint paper by Professors Owen and Bell.

—— 1849 C

On the development and homologies of the carapace and plastron of the chelonian reptiles.

Philos. Trans. Roy. Soc. Lond., pp. 151–170, with pl. 13, and 9 figs in the text.

—— 1850 A

Monograph of the fossil *Reptilia* of the London clay. Part II. *Crocodilia, Ophidia*.

Palæontographical Society, iii, pt. 2, pp. 1–68, pls. xxix; i–xvi.

—— 1850 B

On the communications between the cavity of the tympanum and the palate in the *Crocodilia* (gavials, alligators, and crocodiles).

Philos. Trans. Roy. Soc. Lond., 1850, pp. 521–527, with pls. xl–xlii.

—— 1851 A

On a new species of pterodactyle (*P. compressirostris*, Owen) from the Chalk, with some remarks on the

Owen, Richard—Continued.
nomenclature of the previously described species.

Proc. Zool. Soc. Lond., 1851, pt. xix, pp. 21–34, with pl. v.

——   1851 B
Monograph on the fossil *Reptilia* of the Cretaceous formations.

Palæontographical Society, v, pp. 1–118, pls. i–xxxvii.

——   1851 C
On the *Megatherium* (*M. americanum* Cuv. & Blum). Part I. Preliminary observations on the exogenous processes of vertebræ.

Philos. Trans. Roy. Soc. Lond., 1851, pp. 719–764; pls. xliv–liii.

——   1852 A
Comparison of the modifications of the osseous structure of the *Megatherium* with that in other known existing and extinct species of the class *Mammalia*.

Amer. Jour. Sci. (2), xiv, pp. 91–97.
Abstract of a memoir read by Professor Owen to the Royal Society of London, cited from Jameson's Edinb. Jour., li, 350.

——   1852 B
On the fossil *Mammalia* from the Eocene fresh-water formation at Hardwell, Hants.

Report Brit. Assoc. Adv. Sci., 21st meeting, Ipswich, 1851, p. 67.

——   1853 A
On the anatomy of the walrus.

Proc. Zool. Soc. Lond., 1853, pt. xxi, pp. 103–106.
Among other things this paper treats of the dentition.

——   1853 B
A monograph of the fossil chelonian reptiles of the Wealden clays and Purbeck limestones.

Palæontographical Society, vii, pp. 1–12, pls. i–ix.

——   1853 C
Notes on the above-described fossil remains.

Quart. Jour. Geol. Soc., ix, pp. 66–87, with pls. ii, iii.
Discusses relationships of the bones discovered by Lyell and Dawson in Nova Scotia, and assigns to them the name *Dendrerpeton acadianum*.

Owen, Richard—Continued.   1853 D
Notice of a batrachoid fossil in British Coal shale.

Quart. Jour. Geol. Soc., ix, pp. 67–70, with pl. ii.
Describes and figures *Parabatrachus* (type *P. colei*), *Megalichthys*, and *Centrodus*.

——   1854 A
On a fossil imbedded in a mass of Pictou coal from Nova Scotia.

Quart. Jour. Geol. Soc., x, pp. 207–208 with pl. ix.

——   1854 B
Monograph on the fossil *Reptilia* of the Wealden formations. Part II. *Dinosauria*.

Palæontographical Society, viii, pp. 1–54, pls. i–xix.

——   1854 C
On some fossil reptilian and mammalian remains from the Purbecks.

Quart. Jour. Geol. Soc., x, pp. 420–433, with 12 figs. in the text.

——   1855 A
Additional remarks on the skull of the *Baphetes planiceps*, Ow.

Quart. Jour. Geol. Soc., xi, pp. 9–10.

——   1855 B
On the *Megatherium* (*M. americanum* Cuv. & Blum). Part II. Vertebræ of the trunk.

Philos. Trans. Roy. Soc. Lond., cxlv., pp. 359–388, pls. xxii–xxvii.

——   1855 C
On the fossil skull of a mammal (*Prorastomus sirenoides*, Owen) from the island of Jamaica.

Quart. Jour. Geol. Soc., xi, pp. 541–543, with pl. xv.

——   1856 A
Monograph on the fossil *Reptilia* of the Wealden formations. Part III. *Megalosaurus bucklandi*.

Palæontograpnical Society, ix, pp. 1–26, pls. i–xii.

——   1856 B
On the *Megatherium* (*M. americanum* Cuv. & Blum). Part III. The skull.

Philos. Trans. Roy. Soc. Lond., cxlvi, pp. 571–589, pls. xxi–xxvi.

——   1856 C
Description of a fossil cranium of the musk-buffalo [*Bubalus moschatus* Owen; *Bos moschatus* (Zimm. & Gmel.) Pallas; *Bos pallasii*, Dekay; *Ovibos pallasii*, H.

Owen, Richard—Continued.
Smith & Bl.] from the "lower-level drift" at Maidenhead, Berkshire.
Quart. Jour. Geol. Soc., xii, pp. 124-131, with 6 figs. in the text.

———
1856 D
Description of the skull of a large species of *Dicynodon* (*D. tigriceps*, Ow.), transmitted from South Africa by A. G. Bain, esq.
Trans. Geol. Soc. Lond. (2), vii, pp. 233-240, with pls. xxix-xxxii.

———
1856 E
On parts of the trunk of the *Dicynodon tigriceps*.
Trans. Geol. Soc. Lond. (2), vii, pp. 241-247, with pls. xxxiii-xxxiv.

———
1856 F
On the affinities of the large extinct bird (*Gastornis parisiensis*) indicated by a fossil femur and tibia discovered in the lowest Eocene formation near Paris.
Quart. Jour. Geol. Soc., xii, pp. 204-217, with pl. iii.
The structure of the tibia and femur in the various orders of birds and a number of genera is discussed.

———
1857 A
On the ruminant quadrupeds and the aboriginal cattle of Great Britain.
Amer. Jour. Sci. (2), xxiii, pp. 132-136.
From Proc. Roy. Inst. Great Britain, May, 1856.

———
1857 B
Monograph on the fossil *Reptilia* of the Wealden formation. Part IV.
Palæontographical Society, x, pp. 1-26, pls. i-xi.
Treats of *Iguanodon* and *Hylæosaurus*.

———
1857 C
Description of a small Lophiodont mammal (*Pliolophus vulpiceps*, Owen) from the London clay, near Harwich.
Quart. Jour. Geol. Soc., xiv, pp. 54-71, with double pls. ii, iii, iv.

———
1857 D
Description of the lower jaw and teeth of an Anoplotheroid quadruped (*Dichobune ovina*, Ow.) of the size of the *Xiphodon gracilis*, Cuv., from the Upper Eocene marl, Isle of Wright.
Quart. Jour. Geol. Soc., xiii, pp. 254-260, with pl. viii.

Owen, Richard—Continued.
1858 A
On the characters, principles of division, and primary groups of the class *Mammalia*.
Amer. Jour. Sci. (2), xxv, pp. 7-18; 177-198.
From Jour. of Proc. Linn. Soc., Lond., 1857.

———
1858 B
Notes and additions [in Buckland's Geology and Mineralogy].
See Buckland, W., 1858 A.

———
1858 C
On the *Megatherium* (*M. americanum*, Cuv. and Blum.) Part IV. Bones of the anterior extremities.
Philos. Trans. Roy. Soc. Lond., cxlviii, pp. 261-278, pls. xviii-xxii.

———
1859 A
On a new genus (*Dimorphodon*) of pterodactyle, with remarks on the geological distribution of flying reptiles.
Report Brit. Assoc. Adv. Sci., 28th meeting, Leeds, 1858, pp. 97-98.

———
1859 B
On remains of new and gigantic species of pterodactyle (*Pter. fittoni* and *Pter. sedgwickii*) from the upper Greensand near Cambridge.
Report Brit. Assoc. Adv. Sci., 28th meeting, Leeds, 1858, pp. 98-103.

———
1859 C
On the gorilla (*Troglodytes gorilla*, Sav.).
Proc. Zool. Soc. London, 1859, pt. xxvii, pp. 1-23.

———
1859 D
Monograph on the fossil *Reptilia*. Including supplement No. 1, Cretaceous *Pterosauria*, and No. 2, Wealden *Crocodilia*.
Palæontographical Society, xi, pp. 1-44, pls. i-xii.

———
1860 A
On the orders of fossil *Reptilia* and their distribution in time.
Report Brit. Assoc. Adv. Sci., 29th meeting, Aberdeen, 1859, pp. 153-166.

———
1860 B
Monograph of the fossil *Reptilia* of the Cretaceous and Purbeck strata, including supplement No. III, Cretaceous *Pterosauria* and *Sauropterygia*. Supplement No. II, *Iguanodon* and Purbeck *Lacertilia*.
Palæontographical Society, xii, pp. 1-39, pls. i-viii.

Owen, Richard—Continued.   1860 C

On the vertebral characters of the order *Pterosauria*, as exemplified in the genera *Pterodactylus* (Cuvier) and *Dimorphodon* (Owen).

Philos. Trans. Roy. Soc. Lond., cxlix, pp. 161–169, with 10 pls.

———   1860 D

On the *Megatherium* (*M. americanum* Cuv. & Blum.). Part. V. Bones of the posterior extremities.

Philos. Trans. Roy. Soc. Lond., vol. cxlix, pp. 809–829, pls. xxxvii–xli.

———   1860 E

Palæontology; or, a systematic summary of extinct animals and their geological remains.

Pp. i–xv, 1–420, with 141 figs., Edinburgh, 1860. A second edition was published in 1861.

———   1860 F

On some reptilian fossils from South Africa.

Quart. Jour. Geol. Soc., xvi, pp. 49–63, with pls. i–iii.

———   1862 A

Description of specimens of fossil *Reptilia* discovered in the Coal-measures of the South Joggins, Nova Scotia, by Dr. J. W. Dawson.

Quart. Jour. Geol. Soc., xviii, pp. 238–244, with pls. ix–x.

———   1863 A

(I) On the Dicynodont *Reptilia*, with a description of some fossil remains brought by H. R. H. Prince Alfred from South Africa, November, 1860. (II) On the pelvis of *Dicynodon*. (III) Notice of a skull and parts of the skeleton of *Rhynchosaurus articeps*.

Philos. Trans. Roy. Soc. Lond., cliii, pp. 455–467, with pls. xix–xxv.

———   1864 A

On the skeleton of the gare-fowl (*Alca impennis*), and the probability of its being an extinct species.

Amer. Jour. Sci. (2), xxxviii, p. 431. Taken from Proc. Zool. Soc. Lond., July 9, 1864.

———   1864 B

Monograph of the fossil *Reptilia* of the Cretaceous formations; including supplement No. II. Cretaceous *Sauropterygia*. Supplement No. III, *Iguanodon*.

Palæontographical Society, xvi, pp. 1–21, pls. i–x.

Owen, Richard—Continued.   1864 C

On the *Archæopteryx* of Von Meyer, with a description of a long-tailed species from the lithographic stone of Solenhofen.

Philos. Trans. Roy. Soc. Lond., cliii, pp. 33–47, pls. i–iv.

———   1865 A

Monograph of the fossil *Reptilia* of the Liassic formations. Part I. *Sauropterygia*.

Palæontographical Society, xvii, pp.1–40, pls. i–xvi.

———   1865 B

Observations on Recherches sur les Squalodons, par P. J. Van Beneden.

Geol. Magazine (1), ii, pp. 405–411.

A review of the above work of Van Beneden, which appeared in Memoirs Royal Acad. of Belgium, 1865.

———   1866 A

On the anatomy of vertebrates. I. Fishes and reptiles.

Pp. i–xlii, 1–650, woodcuts 452. London, 1866, 8vo.

———   1866 B

On the anatomy of vertebrates. II. Birds and mammals.

London, 1866, pp. i–viii, 1–592, with 406 figs. in the text and 3 tables. 8vo.

———   1866 C

Description of the skeleton of the great auk, or garfowl (*Alca impennis* L.).

Trans. Zool. Soc. Lond., v, pp. 317–335, with pls. li and lii.

———   1866 D

Description of part of the lower jaw and teeth of a small Oolitic mammal (*Stylodon pusillus* Ow.).

Geol. Magazine (1), iii, pp. 199–201, with pl. x, figs. 1, 2.

———   1867 A

On the mandible and mandibular teeth of Cochliodonts.

Geol. Magazine (1), iv, pp. 59–63, with pls. iii and iv.

———   1867 B

On the dental characters of genera and species, chiefly of fishes from the Lower Main seam and shales of coal, Northumberland.

Trans. Odontolog. Soc. Great Britain, v, pp. 323–376 with pls. i–xv.

Owen, Richard—Continued.   1868 A
On the anatomy of vertebrates. III. Mammals.
Pp. i-x, 1-915, text figs. 1-614, London, 1868.
Vol. iii treats of the anatomy of the soft portions of the animal body and of the dentary and tegumentary systems. Chapter xi treats of the general conclusions reached.

———   1869 A
On fossil teeth of equines from Central and South America.
Proc. Roy. Soc. Lond., xvii, pp. 267-268.
Equus conversidens described. *E. tau, E. curvidens,* and *E. arcidens* mentioned.

———   1870 A
Monograph of the fossil *Reptilia* of the Liassic formations. Part II. *Pterosauria.*
Palæontographical Society, xxiii, pp. 41-81, pls. xvii-xx.

———   1870 B
On fossil remains of equines from Central and South America referable to *Equus conversidens,* Ow., *Equus tau,* Ow., and *Equus arcidens,* Ow.
Philos. Trans. Roy. Soc. Lond., clix, 559-573, pls. lxi-lxii.

———   1870 C
On remains of a large extinct lama (*Palauchenia magna* Ow.), from Quaternary deposits in the valley of Mexico.
Philos. Trans. Roy. Soc. Lond., clx, pp. 65-77, pls. iv-vii.

———   1871 A
Monograph of the fossil *Mammalia* of the Mesozoic formations.
Palæontographical Society, xxiv, pp. 1-115, pls. i-iv.

———   1872 A
Monograph of the fossil *Reptilia* of Wealden formation. Supplement No. IV. *Dinosauria (Iguanodon).*
Palæontographical Society, xxv, pp. 1-15, pls. i-iii.

———   1873 A
Description of the skull of a dentigerous bird (*Odontopteryx toliapicus,* Ow.) from the London clay of Sheppey.
Quart. Jour. Geol. Soc., xxix, pp. 511-522, with pls. xvi-xvii.

———   1874 A
Monograph on the fossil *Reptilia* of the Wealden and Purbeck formations. Supplement No. V. *Iguanadon.*
Palæontographical Society, xxvii, pp. 1-18, pls. i, ii.

Owen, Richard—Continued.   1874 B
Monograph on the fossil *Reptilia* of the Mesozoic formations. Part I. *Pterosauria (Pterodactylus).*
Palæontographical Society, xxvii, pp. 1-14, pls. i, ii.

———   1875 A
On fossil evidences of a sirenian mammal (*Eotherium ægypticum,* Owen) from the Nummilitic Eocene of the Mokattam Cliffs, near Cairo.
Quart. Jour. Geol. Soc., xxxi, pp. 100-105, with pl. iii.

———   1875 B
On *Prorastomus sirenoides* (Ow.). Part II.
Quart. Jour. Geol. Soc., xxxi, pp. 559-567, with pls. xxviii, xxix.
Reviewed in Geol. Magazine, (2) ii, pp. 422-423.

———   1875 C
Monograph of the fossil *Reptilia* of the Mesozoic formations. Part II. (Genera *Bothriospondylus, Cetiosaurus, Omosaurus.*)
Palæontographical Society, xxix, pp. 15-93, pls. iii-xxii.

———   1876 A
On the existence or not of horns in the *Dinocerata.*
Amer. Jour. Sci. (3), xi, pp. 401-403.

———   1876 B
Evidences of Theriodonts in Permian deposits elsewhere than in South Africa.
Quart. Jour. Geol. Soc., xxxii, pp. 352-363, with 9 text figs.
Discusses *Bathygnathus borealis.*

———   1876 C
Monograph on the fossil *Reptilia* of the Wealden and Purbeck formations. Supplement No. VII. *Crocodilia (Poikilopleuron), Dinosauria (Chondrosteosaurus).*
Palæontographical Society, xxx, pp. 1-7, pls. i-vi.

———   1877 A
On the rank and affinities in the reptilian class of the *Mosasauridæ,* Gervais.
Quart. Jour. Geol. Soc., xxxiii, pp. 682-715, with 24 figs. in the text.

———   1878 A
On the influence of the advent of a higher form of life in modifying the structure of an older and lower form.
Quart. Jour. Geol. Soc., xxxiv, pp. 421-430.

Owen, Richard—Continued.   1878 B
On the affinities of the *Mosasauridæ*, Gervais, as exemplified in the bony structure of the fore fin.
Quart. Jour. Geol. Soc., xxxiv, pp. 748-753, with 4 woodcuts in text.

———   1878 C
Monograph on the fossil *Reptilia* of the Wealden and Purbeck formations. Supplement No. VII. *Crocodilia (Goniopholis, Petrosuchus,* and *Suchosaurus).*
Palæontographical Society, xxxii, pp. 1-15, pls. i-vi.

———   1878 D
On the occurrence in North America of rare extinct vertebrates found fragmentarily in England.
Ann. and Mag. Nat. Hist. (5), ii, pp. 201-223, with pls. x, xi.

———   1879 A
Monograph of the fossil *Reptilia* of the Wealden and Purbeck formations. Supplement, No. IX. *Crocodilia (Goniopholis, Brachydectes, Nannosuchus, Theriosuchus,* and *Nuthetes).*
Palæontographical Society, xxxiii, pp. 1-19, pls. i-iv.

———   1879 B
On the occurrence in North America of rare extinct vertebrates found fragmentarily in England.
Ann. and Mag. Nat. Hist. (5), iv, pp. 53-61, with pl. viii.
A continuation of paper designated Owen R., 1878 D.

———   1880 A
On the occurrence in North America of rare extinct vertebrates found fragmentarily in England.
Ann. and Mag. Nat. Hist. (5), v, pp. 177-181, with pl. viii.
Conclusion of papers designated as Owen, R., 1878 D and Owen R., 1879 B.

———   1881 A
A monograph of the fossil *Reptilia* of the Liassic formations.   Part III. *Ichthyopterygia.*
Palæontographical Society, xxxv, pp. 83-134, pls. xxi-xxxiii.

———   1881 B
On the order *Theriodontia,* with a description of a new genus and species (*Ælurosaurus felinus* Ow.).
Quart. Jour. Geol. Soc., xxxvii, pp. 261-265, with pl. ix.

Owen, Richard—Continued.   1881 C
Description of parts of the skeleton of an Anomodont reptile (*Platypodosaurus robustus* Owen).   Part II. The pelvis.
Quart. Jour. Geol. Soc., xxxvii, pp. 266-271, with pl. x.
Contains remarks by Seely and Twelvetrees.

———   1883 A
On generic characters in the order *Sauropterygia.*
Quart. Jour. Geol. Soc., xxxix, pp. 133-138, with 3 woodcuts in the text.

———   1883 B
On the skull of *Megalosaurus.*
Quart. Jour. Geol. Soc., xxxix, pp. 334-347, with pl. xi and 1 cut in the text.

———   1884 A
On the skull and dentition of a Triassic mammal (*Tritylodon longævus*) from South Africa.
Quart. Jour. Geol. Soc., xl, pp. 146-152, with pl. vi.

———   1886 A
American evidences of Eocene mammals of the "plastic-clay" period.
Report Brit. Assoc. Adv. Sci., 55th meeting, Aberdeen, 1885, p. 1033.   Abstract.

———   1886 B
On a new perissodactyle ungulate from Wyoming.
Geol. Magazine (3), iii, p. 140.
A reference to Woodward, H., 1886 B, concerning *Phenacodus.*

———   1889 A
Monograph on the British fossil *Cetacea* of the Red Crag.
Palæontographical Society, pp. 1-40, with pls. i-v and 14 text figs.
This paper deals especially with the genus *Ziphius,* but a number of other genera are discussed, among them *Belemnoziphius.*

Packard, A. S.   1867 A
Observations on the glacial phenomena of Labrador and Maine, with a view of the recent invertebrate fauna of Labrador.
Mem. Boston Soc. Nat. Hist., i, pp. 210-303, with pls. vii, viii.

———   1868 A
The hairy mammoth.
Amer. Naturalist, ii, pp. 23-35, with 4 woodcuts in the text.

———   1886 A
Geological extinction and some of its apparent causes.
Amer. Naturalist, xx, pp. 29-40.

Palmer, T. S.   1895 A
The earliest name for Steller's sea
cow and dugong.
Science (2), ii, pp. 449–450.
The earliest generic name of the sea cow is
Hydrodamalis Retzius; the earliest specific
name is *gigas* Zimm.

———   1897 A
On the genera of rodents: An at-
tempt to bring up to date the current
arrangement of the order. By Oldfield
Thomas, F. Z. S., Proc. Zool. Soc. London, 1896, pp. 1012–1028.
Science (2), vi, pp. 103–107.
A review of the paper whose title is quoted
above.

———   1899 A
[Review of Dr. E. L. Trouessart's
Catalogus Mammalium.]
Science (2), x, pp. 491–495.
Contains corrections of various generic
names of mammals.

Pander, Christ. H.   1856 A
Monographie der fossilen Fische des
silurischen Systems der russisch-bal-
tischen Gouvernements.
Pp. i–x, 1–91, with 9 pls. St. Petersburg,
1856.

———   1857 A
Ueber die Placodermen des devo-
nischen Systems.
Pp. 1–106, with pls. 1–8, and B. St. Petersburg, 1857. 4°.

———   1858 A
Die Ctenodipterinen des devonischen
Systems.
Pp. i–viii, 1–65, with pls. i–ix. 4°.

———   1860 A
Ueber die Saurodipterinen, Dendro-
dipterinen, Glyptolepiden, und Cheiro-
lepiden des devonischen Systems.
Pp. i–ix, 1–90, with 17 pls. St. Petersburg,
1860.

Pander, C. H., and d'Alton, Ed.   1818 A
Riesenfaulthier (*Megatherium*).
Isis von Oken, 1818, cols. 1083–1086.
A letter addressed to Isis from Madrid.

Panton, J. Hoyes.   1891 A
The mastodon and mammoth in On-
tario, Canada.
Geol. Magazine (3), viii, pp. 504–505.

———   1892 A
The mastodon and mammoth in On-
tario, Canada.
Report Brit. Assoc. Adv. Sci., 61st meeting,
Cardiff, 1891, pp. 654–655.

Paquier, V.   1894 A
Étude sur quelques cétacés du Mio-
cène.
Mém. Soc. Géol. France. Paléontologie, iv,
mém. No. 12, pp. 1–20, with pls. xvii–xviii.

Parker, H. W.   1884 A
[Abstract of an account of the find-
ing of a mammoth in Grinnel, Iowa,
by H. W. Parker; taken from *Daily
Iowa Capital.*]
Science, iv, p. 46.

———   1885 A
Footprints in the rocks of Colorado.
Science, v, pp. 312–313, with 2 woodcuts.

Parker, T. Jeffery.   1887 A
Notes on *Carcharodon rondeletti.*
Proc. Zool. Soc. Lond., 1887, pp. 27–40, with
pls. iv–viii.

———   1890 A
On the presence of a sternum in
Notidanus indicus.
Nature, xliii, p. 142, with 1 fig.

———   1895 A
On the cranial osteology, classifica-
tion, and phylogeny of the *Dinorni-
thidæ.*
Trans. Zool. Soc. Lond., xiii, pp. 373–431,
with pls. lvi–lxii.

Parker, W. K.   1864 A
On the sternal apparatus of birds and
other *Vertebrata.*
Proc. Zool. Soc. Lond., 1864, pp. 339–341.

———   1866 A
On the structure and development
of the skull in the ostrich tribe.
Philos. Trans. Roy. Soc. Lond., clvi, pp. 113–
183, with pls. vii–xv.

———   1866 B
On the osteology of gallinaceous birds
and tinamous.
Trans. Zool. Soc. Lond., v, pp. 149–241, with
pls. xxxiv–xlii.

———   1868 A
A monograph on the structure and
development of the shoulder girdle and
sternum in the *Vertebrata.*
Published by the Ray Society, pp. i–xii,
1–237, with pls. i–xxx.

———   1870 A
On the structure and development of
the skull of the common fowl.
Philos. Trans. Roy. Soc. Lond., clix, pp. 755–
807; pls. lxxx–lxxxvii.

Parker, W. K.—Continued.   1871 A
On the structure and development of
the skull of the common frog (*Rana
temporaria*).
Philos. Trans. Roy. Soc. Lond., clxi, pp. 187–
211, pls. iii–x.

——   1872 A
On the structure and development of
the crow's skull.
Monthly Micros. Jour. Lond., viii, pp. 217–226,
with pls. xxxiv–xxxvi.

——   1873 A
On the development of the skull in
the tit and sparrow-hawk.
Monthly Micros. Jour. Lond., ix, pp. 6–11;
45–50, with pls. ii, v, vi.

——   1874 A
On the structure and development of
the skull in the salmon (*Salmo salar* L.)
Philos. Trans. Roy. Soc. Lond., clxiii, pp. 95–
145; pls. i–viii.

——   1874 B
On the structure and development of
the skull in the pig (*Sus scrofa*).
Philos. Trans. Roy. Soc., clxiv, pp. 289–336,
pls. xxviii–xxxvii.
An analysis of this memoir by E. Alix is
found in Jour. de Zoologie, iv, 1875, pp. 62–69.

——   1875 A
Birds.
Encyclop. Brit., ed. 9, iii, pp. 699–728, woodcuts 1–37.
The remainder of the article on birds is from
the pen of Alfred Newton. Parker deals with
the osteology in particular.

——   1875 B
On ægithognathus birds. (Part I.)
Trans. Zool. Soc. Lond., ix, pp. 289–352, with
pls. liv–lxii.

——   1876 A
On the structure and development of
the bird's skull.
Trans. Linn. Soc. Lond. (2), i, pp. 99–154,
with pls. xx–xxvii.

——   1877 A
On the structure and development of
the skull in the *Batrachia*. Part II.
Philos. Trans. Roy. Soc. Lond., clxvi, pp.
601–669, pls. liv–lxii.

——   1878 A
On the structure and development of
the skull in the sharks and skates.
Trans. Zool. Soc. Lond., x, pp. 189–234, with
pls. xxxiv–xlii.

——   1878 B
On the skull of the ægithognathous
birds.
Trans. Zool. Soc. Lond., x, pp. 251–314, with
pls. xlvi–liv.

Parker, W. K.—Continued.   1879 A
On the structure and development of
the skull in the common snake (*Tropi-
donotus natrix*).
Philos. Trans. Roy. Soc. Lond., clxix, pp.
385–417, pls. xxvii–xxxiii.

——   1879 B
On the evolution of the *Vertebrata*.
Nature, xx, pp. 30–32, 61–64, 81–83, with 4
figs. in the text.

——   1880 A
On the structure and development of
the skull in the *Lacertilia*.
Philos. Trans. Roy. Soc. Lond., clxx, pp. 595–
640, pls. xxxvii–xlv.
See also Proc. Roy. Soc. Lond., 1879, p. 214.

——   1881 A
On the structure and development of
the skull in the *Batrachia*. Part III.
Philos. Trans. Roy. Soc. Lond., clxxii, pp.
1–266, pls. i–iv.

——   1882 A
On the development of the skull in
Lepidosteus osseus.
Philos. Trans. Roy. Soc. Lond., clxxiii, pp.
443–491, pls. xxx–xxxviii.

——   1882 B
On the structure and development of
the skull in sturgeons (*Acipenser ruthenus* and *A. sturio*).
Philos. Trans. Roy. Soc. Lond., clxxiii, pp.
139–185, pls. xii–xviii.

——   1882 C
On the structure and development of
the skull in the urodeles.
Trans. Zool. Soc. Lond., xi, pp. 171–214, with
pls. xxxvi–xli.

——   1882 D
On the morphology of the skull in
the *Amphibia Urodela*.
Trans. Linn. Soc. Lond. (2), ii, pp. 165–212,
with pls. xiv–xxi.

——   1883 A
On the structure and development of
the skull in the *Crocodilia*.
Trans. Zool. Soc. Lond., xi, pp. 263–310, with
pls. lxii–lxxi.

——   1884 A
On the presence of claws in the
wings of birds.
The Ibis (5), vi, pp. 124–126, with 4 figs. in
in the text.

——   1885 A
On mammalian descent: The Hun-
terian lectures for 1884.
Pp. i–xii, 1–229, with addenda and illus.

Parker, W. K.—Continued.　　1885 B
On the structure and development of
the skull in the *Mammalia.*　Part III.
Insectivora.
Nature, xxxi, pp. 377-379.

———　　　　　　　　　　　　1886 A
On the structure and development of
the skull in the *Mammalia.*　Part II.
Philos. Trans. Roy. Soc., Lond., clxxvi, pp.
121-278, pls. xvi-xxxix.

———　　　　　　　　　　　　1887 A
On the morphology of birds.
Nature, xxxv, pp. 331-332.

———　　　　　　　　　　　　1889 A
On the structure and development of
the wing of the common fowl.
Philos. Trans. Roy. Soc. Lond., clxxix, pp.
385-398, pls. lxii-lxv.

———　　　　　　　　　　　　1889 B
On the "manus" of *Phœnicopterus.*
The Ibis (6), i, pp. 183-186, with 2 figs. in the
text.

———　　　　　　　　　　　　1891 A
On the morphology of a reptilian
bird, *Opisthocomus cristatus.*
Trans. Zool. Soc. Lond., xiii, pp. 43-85, with
pls. vii-x.

———　　　　　　　　　　　　1891 B
On the morphology of the *Gallinaceæ.*
Trans. Linn. Soc. Lond., (2), v, pp. 213-244,
with pls. xxii-xxv.

Parsons, F. G.　　　　　　1899 A
The joints of mammals compared
with those of man.
Jour. Anat. and Physiol., xxxiv, pp. 41-68,
with 14 text figs.

———　　　　　　　　　　　　1900 A
The joints of mammals compared
with those of man.　Part II. Joints of
the hind limbs.
Jour. Anat. and Physiol., xxxiv, pp. 301-323,
with 10 figs. in the text.

Patten, William.　　　　　1890 A
On the origin of vertebrates from
arachnids.
Quart. Jour. Micros. Sci., xxxi, pp. 317-378,
with pls. xxiii, xxiv, and 18 text figs.

Paulli, Simon.　　　　　　1899 A
Ueber die Pneumaticität des Schädels bei den Säugethieren. I. Ueber
den Bau des Siebbeins. Ueber die
Morphologie des Siebbeins und die der
Pneumaticität bei den Monotremen
und den Marsupialiern.
Morpholog. Jahrbuch, xxviii, pp. 147-178,
with pl. vii and 16 figs. in the text.

Paulli, Simon—Continued.　　1900 A
Ueber die Pneumaticität des Schädels bei den Säugethieren. Eine Morphologische Studie. II. Ueber die Morphologie des Siebbeins und die der
Pneumaticität bei den Ungulaten und
Probosciden.
Morpholog. Jahrbuch, xxviii, pp. 179-251,
with pls. viii-xiv, and 44 figs. in the text.

———　　　　　　　　　　　　1900 B
Ueber die Pneumaticität des Schädels bei den Säugethieren. Eine Morphologische Studie. III. Ueber die
Morphologie des Siebbeins und die der
Pneumaticität bei den Insectivoren,
Hyracoideen, Chiropteren, Carnivoren, Pinnepedien, Edentaten, Roden-
tiern, Prosimiern, und Primaten, nebst
einer zusamenfassenden Uebersicht
über die Morphologie des Siebbeins
und die der Pneumaticität des Schädels
bei den Säugethieren.
Morpholog. Jahrbuch, xxviii, pp. 483-564,
with pls. xxvii-xxix and 36 figs. in the text.

Pavlow, A.　　　　　　　　1885 A
Notes sur l'histoire géologique des
oiseaux. ·
Bull. Soc. Imp. Natur. Moscou, ix, pp. 100-123.

Pavlow, Marie.　　　　　　1887 A
Études sur l'histoire paléontologique
des ongulés en Amérique et en Europe.
I. Groupe primitif de l'éocène inférieur.
Bull. Soc. Imp. Natur. Moscou (2), i, pp.
343-373, with pl. vii.

———　　　　　　　　　　　　1888 A
Études sur l'histoire paléontologique
des ongulés. II. Le developpement
des *Equidæ.* III. *Rhinoceridæ* et *Tapiridæ.*
Bull. Soc. Imp. Natur. Moscou (2), ii, pp.
135-182, with 2 pls.
For Professor Cope's review of this paper
see Amer. Naturalist, xxii, p. 448; for that of M.
Schlosser, Archiv f. Anthrop., xix, p. 159.

———　　　　　　　　　　　　1890 A
Études sur l'histoire paléontologique
des ongulés. IV. *Hipparion* de la
Russie. V. Chevaux pleistocènes de
la Russie.
Bull. Soc. Imp. Natur. Moscou (2), iii, pp.
653-716, with pls. vii-ix.

———　　　　　　　　　　　　1892 A
Études sur l'histoire paléontologique
des ongulés. VI. Les *Rhinoceridæ* de

Pavlow, Marie—Continued.
la Russie et le developpement des *Rhinoceridæ* en général.

, Bull. Soc. Imp. Natur. Moscou (2), vi, pp. 187-221, with pls. iii-v.
Appended is a list of 105 papers consulted by the authoress.
Reviewed by M. Schlosser in Archiv f. Anthrop,, xxiii, pp. 188-189.

—— 1892 B
Qu'est-ce que c'est que l'*Hipparion?*
Bull. Soc. Imp. Natur. Moscou (2), v, pp. 410-414.

—— 1893 A
Note sur un nouveau crâne d'*Amynodon.*
Bull. Soc. Imp. Natur. Moscou (2), vii, pp. 37-42, with pl. iii.

—— 1894 A
Les mastodontes de la Russie et leurs rapports avec les mastodontes des autres pays.
Mém. Acad. Imp. Sci. St. Pétersbourg (8), i, No. 3, pp. 1-43, with pls. i-iii.

—— 1900 A
Études sur l'histoire paléontologique des ongulés. VII. Artiodactyles anciens.
Bull. Soc. Imp. Natur. Moscou, for 1899, pp. 268-828, with pls. v, vi.
Citations made from reprint with pp. 1-62.

Peale, Rembrandt. 1802 A
A short account of the mammoth.
Tilloch's Philos. Magazine, London, xiv, pp. 162-169.

—— 1802 B
On the differences which exist between the heads of the mammoth and elephant.
Tilloch's Philos. Magazine, London, xiv, pp. 228-229, with pl. v.

—— 1803 A
Account of some remains of a species of gigantic oxen found in America and other parts of the world.
Philos. Magazine, xv, pp. 325-327, with pl. vi.
Account of bones found at Bigbone-lick in Kentucky. The plate figures a portion of the skull and horn core of *Bison latifrons.*

—— 1803 B
An historical disquisition on the mammoth, or great American incognitum, an extinct, immense, carnivorous animal whose remains have been found in North America.
8vo, pp. i-vi, 1-91, with 1 pl. London, 1803.

Pearce, J. Chaning. 1846 A
Notice of what appears to be the embryo of an *Ichthyosaurus* in the pelvic cavity of *Icthyosaurus (communis?).*
Ann. and Mag. Nat. Hist. (1), xvii, pp. 44-46.

Pédroni, P. M. 1844 A
Mémoire sur les poissons fossiles du département de la Gironde.
Actes Soc. Linn. de Bordeaux, xiii, pp. 277-298, with pls. i, ii.
Most of the drawings illustrating this paper are rude.

Perkins, H. C. 1842 A
Notice of fossil bones from Oregon, in a letter to Dr. C. T. Jackson.
Amer. Jour. Sci., xlii, pp. 136-140, with 4 figs.
This paper does not identify with certainty the bones described. May be *Megatherium, Mylodon,* or *Megalonyx,* or even something else. J. A. Allen refers them with doubt to *Bison antiquus* (Allen, J. A., 1876 B, p. 21.) Leidy (1869 A, p. 397) refers some of them to the mammoth.

—— 1844 A
Note to the editors respecting fossil bones from Oregon.
Boston Jour. Nat. Hist., iv, pp. 184-186.
Gives description of a fossil humerus, which is named *Orycterotherium oreponensis.*

Perrin, A. 1895 A
Recherches sur les affinités zoologiques de l'*Hatteria punctata.*
Ann. Sci. Naturelles (7), Zool., xx, pp. 33-102.
The author deals principally with the muscular system, but devotes a few pages (39-46) to the skeleton.

—— 1896 A
Constitution du carpe des anoures.
Bull. Scient. France et Belgique, xxvii, pp. 419-431, pl. xvi.

Peters, Karl. 1869 A
Zur Kenntniss der Wirbelthiere aus den Miocänschichten von Eibiswald in Steiermark. I. Die Schildkrötenreste.
Denkschr. k. Akad. Wissensch. Wien. math.-naturwiss. Cl., xxix, pp. 111-124, pls. i-iii.

—— 1869 B
Zur Kenntniss der Wirbelthiere aus den Miocänschichten von Eibiswald in Steiermark. II. *Amphicyon, Viverra, Hyotherium.*
Denkschr. k. Akad. Wissensch. Wien. math.-naturwiss. Cl., xxix, pp. 189-214, with pls. i-iii.

—— 1870 A
Zur Kenntniss der Wirbelthiere aus den Miocänschichten von Eibiswald in

Peters, Karl—Continued.

Steiermark. III. *Rhinoceros, Anchithe-rium.*

Denkschr. k. Akad. Wissensch. Wien, math.-naturw. Cl. xxx, pp. 29–49, with pls. i–iii.

—— 1895 A

Ueber die Bedeutung des Atlas der Amphibien.

Anatom. Anzeiger, x, pp. 565–574.

Contains titles of 26 papers on the subject.

Peters, Wilhelm.   1839 A

Zur Osteologie der *Hydromedusa maximiliani.*

Archiv Anat. and Physiol., Jahrg. 1839, pp. 280–295, with pl. xiv, figs. 1–4.

—— 1839 B

Ueber die Bildung des Schildkröten-skelet.

Archiv. Anat. and Physiol. u. Wissensch. Med., 1839, pp. 290–295, with pl. xiv, figs. 5–7.

—— 1852 A

Ueber die Gebissformel der Spitz-mäuse.

Archiv f. Naturgesch., xviii, i, pp. 220–277.

—— 1864 A

Ueber das Milchgebiss des Walrosses, *Odobænus rosmarus.*

Monatsber. k. p. Akad. Wissensch. Berlin, 1864, pp. 685–687, with 1 pl.

Philippi, E.   1897 A

Ueber *Ischyodus suevicus* nov. sp. Ein Beitrag zur Kenntniss der fossilen Holo-cephalen.

Palæontographica, xliv, pp. 1–10, with pls. i, ii.

Contains description and figures of *Ischyodus suevicus.*

Pictet, F. J.   1853 A

Traité de paléontologie, ou histoire naturelle des animaux fossiles considé-rés dans leurs rapports zoologiques et géologiques.

Second edition, i, pp. i–xiv; 1–584, with atlas of 110 pls. 4°.

The first edition of the work was published in the year 1844.

—— 1854 A

Traité de paléontologie, etc.

Second edition, ii, pp. 1–297. Only the pages of this volume which treat of *Vertebrata* are in-cluded here. These are devoted to the fishes.

Pictet, F. J., and Campiche, G.   1858 A

Description des fossiles du terrain crétacé des environs de Sante-Croix.

Materiaux pour la paléontologie suisse, ser. ii, art. 2, pp. 1–380, with pls. i–xliii.

Of this work pp. 29–99, inclusive, with pls. i–xii, are devoted to the vertebrates.

Pictet, F. J., Gaudin, C., and Harpe, Ph. de la.   1857 A

Mémoire sur les animaux vertébrés trouvés dans le terrain sidérolitique du Canton de Vaud et appartenant a la faune éocène.

Genève, 1855–1857, pp. 1–120, with pls. i–xiv i 1.

Pictet, F. J., and Humbert, Alois.   1858 A

Monographie des chéloniens de la molasse suisse.

Materiaux pour la paléontologie suisse, ser. i, art. 3, pp. 1–71, with pls. i–xxii.

On pp. 64–66 is found a list of the tortoises of the Swiss molasse.

—— 1866 A

Nouvelles recherches sur les poissons fossiles du Mont Liban.

Genève, 1866, pp. i–vii; 1–115, with pls. i–xix.

Pictet, F. J., and Jaccard, Auguste.   1860 A

Description de quelques débris de rep-tiles et de poissons fossiles trouvés dans l'étage jurassique supérieur (virgalien) du Jura neuchâtelois.

Materiaux pour la paléontologie suisse, ser. iii, art. 1, pp. 1–88, with pls. i–xix.

Plieninger, Th.   1852 A

Belodon plieningeri, H. v. Meyer. Ein Saurier der Keuperformation.

Jahreshefte Ver. f. vaterl. Naturk. Württem-berg, viii, pp. 389–524, with pls. viii–xiii.

Plieninger, Felix.   1895 A

Campylognathus zitteli. Ein neuer Flugsaurier aus dem Lias Schwabens.

Palæontographica, xli, pp. 193–222, with pls. xix and 8 figs. in the text.

This paper contains a very full bibliography of the pterodactyls.

Plummer, John T.   1843 A

Suburban geology, or rocks, soil, and water about Richmond, Wayne County, Ind.

Amer. Jour. Sci. (1), xliv, pp. 281–313, with 14 figs. in the text.

Refers (p. 301) to the finding of a large tusk at Brookville, Ind., and of a mastodon's tooth near Jacksonburg, Ind.

See, also, Amer. Jour. Sci., xl, 1841, p. 149.

Pohlig, Hans.   1887 A

Ueber amerikanische Elephanten-molaren.

Verhandl. Naturf. Ver. preuss. Rheinlande, xliv, Sitzungsber., pp. 117–118.

Recognizes two N. A. species, *E. prim-igenius* and *E. trogontherii.* The latter, the author says, may be identical with *E. columbi.*

198 FOSSIL VERTEBRATA OF NORTH AMERICA. [BULL. 179.

Pohlig, Hans—Continued. 1887 B
Die Spitze eines sehr jugendlichen permanenten Stosszahnes von *Elephas primigenius.*
Verhandl. Naturf. Ver. preuss. Rheinlande, xliv, Sitzungsber., p. 254.

—— 1889 A
Dentition und Kranologie des *Elephas antiquus* Falc, mit Beiträgen über *Elephas primigenius* Blum. und *Elephas meridionalis* Nesti.
Nova Acta Akad. Cæs. Leop.-Car., liii, pp. 1-259, with pls. i-x and 110 figs. in the text.
Reviewed by M. Schlosser in Archiv f. Anthrop., xix, pp., 127-129.

—— 1892 A
Die Cerviden des thüringischen Diluvial-Travertines, mit Beiträgen über andere diluviale und über recente Hirschformen.
Palæontographica, xxxix, pp. 215-264, with pls. xxiv-xxvii.

—— 1892 B
Dentition und Kranologie des *Elephas antiquus* Falc, mit Beiträgen über *Elephas primigenius* Blum. und *Elephas meridionalis* Nesti. Zweiter Abschnitt.
Nova Acta Acad. Cæs. Leop.-Car., lvii, pp. 267-466, with pls. A-E and text figs. 108-152.
Review by M.Schlosser in Archiv f.Anthrop., xxii, pp. 96-97.

—— 1893 A
Le premier crane complet du *Rhinoceros* (*Cænopus*) *occidentalis*, Leidy.
Bull. Soc. Belge Géol., Paléont., Hydrol., vii, Mém., pp. 41-44, with pl. iii.

Pollard, E. C. 1893 A
The succession of teeth in mammals.
Natural Science,·ii, pp. 360-363.

Pollard, H. B. 1892 A
On the anatomy and phylogenetic position of *Polypterus.*
Zool. Jahrbücher, v, pp. 387-428, with pls. xxvii-xxx and 10 figs. in the text.

—— 1894 A
The cirrhostomial origin of the head in vertebrates.
Anatom. Anzeiger, ix, pp. 349-359, with 4 figs. in the text.
Discusses the derivation of the Siluroids from *Amphioxus*, *Myxine*, etc.

—— 1894 B
The suspension of the jaws in fish.
Anatom. Anzeiger, x, pp. 17-25, with 5 figs.

Pollard, H. B.—Continued. 1895 A
. The oral cirri of Siluroids and the origin of the head in vertebrates.
Zool. Jahrbücher; Abth. f. Anat., viii, pp. 379-424, pls. xxiv, xxv.

Pomel, A. 1847 A
Note critique sur les caractères et les limites du genre *Palæotherium.*
Arch. Sci. Phys. et Nat., Geneva, v, pp. 200-207.

—— 1847 B
Sur un nouveau genre de pachydermes fossiles (*Elotherium*) voisin des Hippopotames.
Arch. Sci. Phys. et Nat. Geneva, v, pp. 307-308.

—— 1847 C
Note sur des animaux fossiles découverts dans le départmente de l'Allier.
Bull. Soc. Géol. France (2), iv, pp. 378-385, with pl. iv.

—— 1848 A
Observations paléontologiques sur les hippopotames et les cochons.
Arch. Sci. Phys. et Nat., Geneva, viii, pp. 155-162.
Treats of the species of *Hippopotamus* and of *Sus* then known.

—— 1848 B
Note sur le genre *Hyopotamus* Owen et sur les Anchithériums en général.
Arch. Sci. Phys. et Nat., Geneva, viii, pp. 321-326.

—— 1848 C
Études sur les carnassiers insectivores. I. Insectivores fossiles. II. Classification des insectivores.
Arch. Sci. Phys. et Nat., Genève, ix, pp. 159-165; 244-251.

—— 1848 D
Description de la tête du *Castoroides ohioensis* Foster.
Arch. Sci. Phys. et Nat., Geneva, ix,pp. 165-167.
A review of Wyman, J., 1846 B.

—— 1853 A
Catalogue méthodique et descriptif des vertébrés fossiles découverts dans le bassin hydrographique supérieur de la Loire, et surtout dans la vallée de son affluent principal, l'Allier.
8 vo. Paris, 1853, pp. 1-193.

Portis, Alessandro. 1878 A
Ueber fossile Schildkröten aus dem Kimmeridge von Hannover.
Palæontographica, xxv, pp. 125-140, with pls. xv-xviii.

Portis, Alessandro—Continued.　1882 A
Les chéloniens de la Mollasse Vaudoise conservés dans le Musée Géologique de Lausanne.
Abhandl. schweizer. paläontolog. Gesellsch., ix, pp. 1–78, pls. i–xxix.
Contains 59 titles of papers consulted.

Pouchet, Georges.　　　　　　1866 A
Contribution à l'anatomie des Édentés.
Jour. l'Anat. et Physiol., iii, pp. 113–129, 337–353, with pls. iii, iv, ix, x.
See vol. iv, pp. 35–37, for a letter from Richard Owen to Pouchet on the above memoir.

Pouchet, G., and **Beauregard,** H.　　　　　　1889 A
Recherches sur le cachalot.
Nouv. Arch. du Mus. d'Hist. Nat., Paris (3), i, pp. 1–96, with pls. i–viii.
This portion of the authors' researches is devoted to the osteology of *Physeter macrocephalus.*

Powrie, James.　　　　　　1861 A
Cephalaspides of Forfarshire.
The Geologist, iv, pp. 137–180, with woodcut.

——　　　　　　1864 A
The Scottish *Pteraspis.*
The Geologist, vii, pp. 170–172.

——　　　　　　1864 B
On the fossiliferous rocks of Forfarshire and their contents.
Quart. Jour. Geol. Soc., xx, pp. 413–429, with pl. xx.
Contains remarks on *Acanthodidæ* and *Cephalaspidæ.* New genus *Euthacanthus* (= *Climatius*) is established.

——　　　　　　1867 A
On the genus *Cheirolepis,* from the Old Red sandstone.
Geol. Magazine (1), iv, pp. 147–152, with 2 figs. in the text.

Powrie, James, and **Lankester,** E. Ray.　　　　　　1868 A
A monograph of the fishes of the Old Red sandstone of Britain. Part I. The *Cephalaspidæ,* by E. R. Lankester.
Palæontographical Society, xxi, pp. 1–33, pl. i–v, text figs. 1–12.

——　——　　　　　　1870 A
A monograph of the fishes of the Old Red sandstone of Britain. Part I (concluded). The *Cephalaspidæ.*
Palæontographical Society, xxiii, pp. 1–62, pls. vi–xiv, text figs. 12–33.

Preble, Edward A.　　　　　　1899 A
Revision of the jumping mice of the genus *Zapus.*
N. A. Fauna, No. 15 (U. S. Dept. Agricult.), pp. 1–42, with pl. i.

Preyer, William.　　　　　　1862 A
Ueber *Plautus impennis* Brünn.
Jour. für Ornithologie, x, pp. 110–124, 337–356.

Priem, F.　　　　　　1896 A
Sur des dents de poissons du Crétacé supérieur de France.
Bull. Soc. Géol. France (3), xxiv, pp. 288–295, with pl. ix.

——　　　　　　1897 A
Sur des dents d'Elasmobranches de divers gisements sénoniens (Villedieu, Meudon, Folx-les-caves).
Bull. Soc. Géol. France (3), xxv, pp. 40–56, with pl. i.

——　　　　　　1898 A
Sur des pycnodontes et des squales du Crétacé supérieur du bassin de Paris (Turonien, Sénonien, Montien inférieur).
Bull. Soc. Géol. France (3), xxvi, pp. 229–243, with pl. ii.
Describes new species *Oxlodus attenuatus,* and gives figs. of *Corax pristodontus* and *Scapanorhynchus? (Odontaspis) subulatus.*

Prime, A. J.　　　　　　1845 A
Great American mastodon.
Amer. Quart. Jour. Agricult. and Sci., ii, pp. 203–212, with pl. iv.

Probst, J.　　　　　　1858 A
Ueber das Gebiss des *Notidanus primigenius* Ag.
Jahreshefte Ver. f. vaterl. Naturk. Württemberg, xiv, pp. 124–127, with 10 figs. in the text.

——　　　　　　1859 A
Ueber die Streifung der fossilen Squalidenzähne.
Jahreshefte Ver. f. vaterl. Naturk. Württemberg, xv, pp. 100–102.

——　　　　　　1878 A
Beiträge zur Kenntniss der fossilen Fische aus der Molasse von Baltringen.
Jahreshefte Ver. f. vaterl. Naturk. Württemberg, xxxiv, pp. 112–154, with pl. i.

——　　　　　　1879 A
Beiträge zur Kenntniss der fossilen Fische aus der Molasse von Baltringen.
Jahreshefte Ver. f. vaterl. Naturk. Württemberg, xxxv, pp. 127–191, with pls. ii, iii.
This paper is devoted to descriptions of the teeth of sharks.

——　　　　　　1885 A
Ueber fossile Reste von *Squalodon.* Beitrag zur Kenntniss der fossilen Reste der Meeressäugetiere aus der Molasse von Baltringen.
Jahreshefte Ver. vaterl. f. Naturk. Württemberg, xli, pp. 49–67, with pl. i.

Probst, J.—Continued.   1886 A
Ueber die fossilen Reste von Zahnwalen (Cetodonten) aus der Molasse
von Baltringen.
Jahreshefte Ver. väterl. f. Naturk. Württemberg, xlii, pp. 102-145, with pl. iii.

Prout, H. A.   1846 A
Gigantic *Palæotherium*.
Amer. Jour. Sci. (2), ii, pp. 288-289, with 1 fig.
in the text.

———   1847 A
Description of a fossil maxillary bone
of a *Palæotherium* from near White
River.
Amer. Jour. Sci. (2), iii, pp. 248-250, with 2
figs. in the text.

———   1860 A
[Remarks of Dr. Prout on a fossil
tooth found at King's salt works, near
Abingdon, Va.]
Trans. Acad. Sci. St. Louis, i, pp. 699-700.
On this tooth the genus *Leidyotherium* was
proposed without any specific name. It appears to be quite probable that there was an
error regarding the locality.

Putnam, F. W.   1884 A
Man and the mastodon.
Science, iv, p. 112.
Refers to Plummer, J. T., 1843 A.

———   1885 A
Man and the mastodon.
Science, vi, pp. 375-376.

Pycraft, W. P.   1894 A
The wing of *Archæopteryx*. Part I.
Natural Science, v, pp. 350-360, with pls.
i-iii.

———   1896 A
The wing of *Archæopteryx*.
Natural Science, viii, pp. 261-266, with 3 figs.
in the text.

———   1898 A
Contributions to the osteology of
birds. Part I. *Steganopodes*.
Proc. Zool. Soc. Lond., 1898, pp. 82-101, with
pls. vii, viii, and 8 figs. in the text.

———   1899 A
Contributions to the osteology of
birds. Part III. *Tubinares*.
Proc. Zool. Soc. Lond., 1899, pp. 381-411, with
pls. xxii, xxiii, and 2 text figs.

Quart. Jour. Geol. Sci.   1845 A
Notice of a mastodon recently dis-
covered in North America.
Quart. Jour. Geol. Soc., i, pp. 566-567.
An extract "from an American paper," re-
ferring to the Newburgh, N. Y., mastodon.

Quenstedt, Fr. Aug.   1847 A
Ueber *Lepidotus* im Lias E. Württem-
bergs.
Pp. 1-24, with pls. i, ii. Tübingen, 1847.
Description of *Lepidotus elvensis*, Blv.

———   1850 A
Ueber *Hippotherium* der Bohnenerze.
Jahreshefte Ver. f. vaterl. Naturk. Württem-
berg, vi, pp. 164-185, with pl. i.
This volume, it is stated on the title-page,
was not issued until 1854.

———   1850 B
Die Mastodonsaurier im grünen Keu-
persandsteine Württemberg's sind Ba-
trachier.
Pp. 1-34, with pls. i-iv, Tübingen, 1850.

———   1852 A
Handbuch der Petrefactenkunde.
Pp. 1-792, with 62 pls. Tübingen, 1852.
Pp. 1-258, with pls. 1-19, are devoted to the
Vertebrata. A second edition of this work was
published in 1866. A third edition of 1,239
pages and 100 pls. appeared in 1885.

———   1858 A
Ueber *Pterodactylus liasicus*.
Jahreshefte Ver. f. vaterl. Naturk. Württem-
berg, xiv, pp. 299-310, with pl. ii, fig. 1.

———   1885 A
Handbuch der Petrefactenkunde.
Dritte umgearbeitete und vermehrte
Auflage.
Tübingen, 1885, pp. i-viii, 1-1239, with an
atlas of 100 pls. and explanations and 443 figs.
in the text.
Of this work pp. 19-392 are devoted to the
Vertebrata.

———   1889 A
Psammochelys keuperina.
Jahreshefte Ver. f. vaterl. Naturk. Württem-
berg, xlv, pp. 120-130, with pls. i, ii.

Rapp, Wilhelm.   1852 A
Anatomische Untersuchungen über
die Edentaten. Mit 10 Steindruckta-
feln. 2te Auflage.
Pp. 1-108. Tübingen, 1852.
Treats of the anatomy, including the osteology, of the living *Edentata*.

Rathke, Heinrich.   1853 A
Ueber den Bau und die Entwicklung
des Brustbeins der Saurier.
Pp. 1-26, Königsberg, 1853.
A translation of this work will be found
in Parker, W. K., 1868 A.

Redfield, John Howard.   1837 A
Fossil fishes of Connecticut and Massachusetts, with a notice of an undescribed genus.
Annals Lyc. Nat. Hist. N. Y., iv, pp. 35-40.

Redfield, W. C.　　　　　1838 A
Newly discovered ichnolites.
Amer. Jour. Sci., xxxiii, pp. 201–202.

　　　　　　　　　　　　　1838 B
Fossil fishes in Virginia.
Amer. Jour. Sci., xxxiv, p. 201.

　　　　　　　　　　　　　1839 A
Fossil fishes of the Red sandstone.
Amer. Jour. Sci., xxxvi, pp. 186–187.
Mentions only *Catopterus gracilis* and *Palæoniscus*, species not determined.

　　　　　　　　　　　　　1841 A
Short notices of American fossil
fishes.
Amer. Jour. Sci. (1), xli, pp. 24–28.

　　　　　　　　　　　　　1843 A
Notice of newly discovered fish beds
and a fossil footmark in the Red sandstone formation of New Jersey.
Amer. Jour. Sci. (1), xliv, pp. 134–136, with
1 fig. in text.

　　　　　　　　　　　　　1850 A
On some fossil remains from Broome
County, N. Y.
Proc. Amer. Assoc. Adv. Sci., 2d meeting,
Cambridge, Mass., pp. 255–256.
Vulpes obtained from clay beneath drift.

　　　　　　　　　　　　　1853 A
On the geological age and affinities
of the fossil fishes which belong to the
sandstone formation of Connecticut,
New Jersey, and the coal field near
Richmond, in Virginia.
Annals of Science (Cleveland, Ohio), i, pp.
270–271.
Abstract of communication made to the
Amer. Assoc. Adv. Sci.

　　　　　　　　　　　　　1856 A
On the relations of the fossil fishes of
the sandstone of the Connecticut and
other Atlantic States to the Liassic and
Oolitic periods.
Amer. Jour. Sci. (2), xxii, pp. 357–362.

　　　　　　　　　　　　　1857 A
On the relations of the fossil fishes
of the sandstone of Connecticut and
other Atlantic States to the Liassic and
Jurassic periods.
Proc. Amer. Assoc. Adv. Sci., 10th meeting,
Albany, 1856, pp. 180–188.

Reh, L.　　　　　　　　1894 A
Die Gliedmassen der Robben.
Jenaische Zeitschr. Naturwiss., xxviii (xxi),
pp. 1–44, with pl. i.
Appended is a list of papers, 36 in number,
which bear on the subject discussed.

Reh, L.—Continued.　　　1895 A
Die Schuppen der Säugetiere.
Jenaische Zeitschr. Naturwiss., xxix (xxii),
pp. 167–220, with pl. i.
Appended is a list of papers bearing on the
subject, 90 in number.

Reichenow, Anton.　　　1871 A
Die Fussbildungen der Vögel.
Jour. für Ornithologie, xix, pp. 401–458,
with pl. vi.

Reichert, C. B.　　　　1865 A
Ueber ein Schädel-fragment des
Glyptodon.
Archiv Anat. Physiol. u. wissensch. Med.,
1865, p. 336.

Reinhardt, J.　　　　　1866 A
Ueber den Hautpanzer der mega-
theroiden Thiere.
Archiv Anat. Physiol. u. wissensch. Med.,
1866, pp. 414–415.

　　　　　　　　　　　　　1874 A
Sur les anomalies des vertèbres
sacrées chez les crocodiliens.
Jour. de Zoologie, iii, pp. 308–312.

　　　　　　　　　　　　　1875 A
De i Brasiliens Knoglehuler fundne
Glyplodont-Levninger og en ny, til de
gravigrade Edentater hörende Slægt.
Vidensk. Meddel. naturh. Forening Kjö-
benhavn, 1875, pp. 165–236, with pl. iv.

　　　　　　　　　　　　　1879 A
Beskrivelse af Hovedskallen af et
Kæmpedovendyr, *Grypotherium dar-
winii*, fra La Plata-Landenes plejsto-
cene Dannelser.
Danske Vidensk. Selskab. Skr. (5), xii, pp.
353–374, with pls. i, ii, and résumée in French,
pp. 375–379.

　　　　　　　　　　　　　1880 A
De i brasilianske Knoglehuler fundne
Navlesvin-Arter.
Vidensk. Meddel. naturh. Forening, Kjö-
benhavn, 1880, pp. 3–33, with pl. vii.
Describes as new *Dicotyles stenocephalus* and
compares it with North American species.

Reis, Otto M.　　　　　1888 A
Die Cœlacanthinen, mit besonderer
Berücksichtigung der im Weissen Jura
Bayerns vorkommenden Gattungen.
Palæontographica, xxxv, pp. 1–96, with pls.
i–v.

　　　　　　　　　　　　　1892 A
Ueber die Zurechnung der Acantho-
dier zu den Selachiern.
Sitzungsber. Gesellsch. naturf. Freunde
Berlin, 1892, pp. 153–156.
Followed by remarks in reply by Dr Otto
Jaekel.

Reis, Otto M.—Continued.   1895 A
On the structure of the frontal spine
and the rostro-labial cartilages of
Squaloraja and *Chimæra.*
Geol. Magazine (4), ii, pp. 385-391, pl. xii.

———   1895 B
Illustrationen zur Kenntniss des
Skelets von *Acanthodes bronni* Agassiz.
Abhandl. senckenb. naturforsch. Gesellsch.,
xix, pp. 49-64, with pls. i-vi.

———   1895 C
Palæontologische Beiträge zur Stammesgeschichte der Teleostier.
Neues Jahrb. Mineral., 1895, i, Abhandl., pp.
162-182.

———   1896 A
Ueber *Acanthodes bronni* Agassiz.
Schwalbe's Morpholog. Arbeiten, vi, pp.
143-220, pls. vi, vii, and 2 figs. in the text.
For review of this paper see A. Tornquist, in
Zoolog. Centralbl., i, pp. 649-656.

———   1897 A
Das Skelet der Pleuracanthiden und
ihre systematischen Beziehungen.
Abhandl. senckenb. naturforsch. Gesellsch.,
xx, pp. 57-155, with pl. i.
Reviewed by Tornquist, Zoolog. Centralbl.,
v, 1898, pp. 472-474.

Reissner, E.   1859 A
Ueber die Schuppen von *Polypterus*
und *Lepidosteus.*
Archiv. Anat. Physiol. u. wissensch. Med.,
1859, pp. 259-268, with pl. v, A.

Renevier, E.   1879 A
Les *Anthracotherium* de Rochette.
Bull. Soc. Vaudoise, Sci. Nat. (2), xvi, pp.
140-148, with pls. iv-viii.

Reuss, August Em.   1846 A
Die Versteinerungen der böhmischen
Kreideformation. Abth. I.
Pp. i-iv, 1-58, with i-xiii, Stuttgart. 4to.
Of this "Abtheilung" only pp. 1-13 are de-
voted to the *Vertebrata,* and these are all fishes.

Rhoads, Samuel N.   1894 A
A contribution to the life history of
the Allegheny cave rat, *Neotoma magister.*
Proc. Acad. Nat. Sci. Phila.,1894, pp. 213-221.

———   1895 A
Distribution of the American bison
in Pennsylvania, with remarks on a
new fossil species.
Proc. Acad. Nat. Sci. Phila , 1895, pp. 244-248.
Describes new species, *B. appalachicolus.*

Rhoads, Samuel N.—Cont'd.   1897 A
Notes on living and extinct species
of North American *Bovidæ.*
Proc. Acad. Nat. Sci. Phila., 1897, pp. 483-
502, with pls. xii.

———   1898 A
Notes on the fossil walrus of North
America.
Proc. Acad. Nat. Sci. Phila., 1898, pp.196-201.

Rice, Franklin P.   1885 A
An account of the discovery of a
mastodon's remains in Northborough,
Worcester County, Mass.
Worcester, Mass., 1885. Pp. 3-8, with 1 pl.

Richardson, John.   1854 A
The zoology of the voyage of H. M. S.
Herald. Vertebrates, including fossil
mammals.
Pp. 172; pls. i-xxxiii. London, 1854.

———   1854 B
Extract from the "Zoology of the
Herald" on the mastodon found in the
Swan River region.
Proc. Boston Soc. Nat. Hist., v, pp. 82-84.

———   1855 A
Note on the *Mastodon* (?) and the
Elephas primigenius.
Amer. Jour. Sci. (2), xix, pp. 131-132.

Ridewood, W. G.   1895 A
The teeth of the horse.
Natural Science, vi, pp. 249-258, with figs.
Describes and illustrates the dentition of the
horse from six months before birth to old age.

Riese, Heinrich.   1892 A
Beitrag zur Anatomie des *Tylotriton
rerrucosus.*
Zool. Jahrbücher Abth. Anat. und Ontog.,
pp. 99-154, with pl. 9-11.

Riggs, Elmer S.   1896 A
Hoplophoneus occidentalis.
Kansas Univ. Quarterly, v, pp. 37-52, with
pl. i.

———   1896 B
A new species of *Dinictis* from the
White River Miocene of Wyoming.
Kansas Univ. Quarterly, iv, pp. 237-241, with
1 fig. in the text. Describes *D. paucidens,* sp. nov.

———   1898 A
On the skull of *Amphictis.*
Amer. Jour. Sci. (4), v, pp. 257-259, with 2
figs. in the text.

———   1899 A
The *Mylagaulidæ:* an extinct family
of sciuromorph rodents.
Pubs. Field Columb. Mus., Geol., i, pp. 182-
187, with text figs.

Riggs, Elmer. S.—Continued.   1900 A
Fossil hunting in Wyoming.
Science (2), xi, pp. 233–234.
Mentions Prof. F. A. Lucas's identification of
Labrosaurus with *Antrodemus.*

Riley, Henry, and **Stutchbury**, Samuel.
                                1838 A
A description of various fossil re-
mains of three distinct saurian animals
recently discovered in the Magnesian
Conglomerate near Bristol.
Trans. Geol. Soc. Lond. (2), v, pp. 349–357,
with pls. xxix, xxx.
A brief abstract is given in Neues Jahrb.
Mineral., 1841, pp. 607–609.

Ringueberg, Eug. N. S.   1884 A
A new *Dinichthys* from the Portage
Group of western New York.
Amer. Jour. Sci. (3), xxvii, pp. 476–478, with 2
figs. in the text.
Describes *D. minor*, not *D. minor* of New-
berry; later called *D. ringuebergi* by Newberry.

Roberts, G. E.   1864 A
On some remains of *Bothriolepis.*
Report Brit. Assoc. Adv. Sci., 33d meeting,
Newcastle-on-Tyne, 1863. p. 87.

Rochebrune, A. J.   1880 A
Revision des ophidiens fossiles du
Muséum d'Histoire Naturelle.
Nouv. Arch. du Mus. d'Hist. Nat., Paris (2),
iii, pp. 271–296.

———   1881 A
Mémoire sur les vertèbres des ophidiens.
Jour. l'Anat. et Physiol., xvii, pp. 185–229,
with pls. xiv, xv, and 18 figs. in the text.

Roemer, Ferdinand.   1849 A
Texas: Mit besonderer Rücksicht
auf deutsche Auswanderung und die
physischen Verhältnisse des Landes.
Mit einem naturwissenschaftlichen An-
hange und einer topographisch-geog-
nostischen Karte von Texas.
Pp. i–xv, 1–464, Bonn, 1849.
On pp. 419–420 are found descriptions of
Lamna texana, and the genus *Ancistrodon* Debey.
Otodus appendiculatus, Oxyrhina mantelli and
Corax heterodon are mentioned.

———   1852 A
Die Kreidebildungen von Texas und
ihre organischen Einschlüsse. Mit
einem die Beschreibung von Versteinerungen aus paläozoischen und tertiären
Schichten enthaltenden Anhange und
mit 11 von C. Hohe nach der Natur
auf Stein gezeichneten Tafeln.
Pp. i–vii, 1–100, Bonn, 1852.

Roemer, Ferdinand—Cont'd.   1856 A
Palæoteuthis, eine Gattung nackter
Cephalopoden in devonischen Schich-
ten der Eifel.
Palæontographica, iv, pp. 72–74, with pl. xiii.
Describes *Palæoteuthis dunensis.*

———   1857 A
Ueber Fisch-und Pflanzen-führende
Mergelschiefer des Rothliegenden bei
Klein-Weundorf unweit Löwenberg,
und im Besonderen über *Acanthodes
gracilis,* den am häufigsten in denselben
vorkommenden Fisch.
Zeitschr. deutsch. geol. Gesellsch., ix, pp.
51–84, with pl. iii.

———   1865 A
Ueber das Vorkommen von *Rhizodus
hibberti* Owen (*Megalichthys hibberti*
Agassiz et Hibbert) in den Schiefer-
thonen des Steinkohlengebirges von
Volpersdorf in der Grafschaft Glatz.
Zeitschr. deutsch. geol. Gesellsch., xvii, pp.
272–276, with pl. vi.
The species here called *Rhizodus hibberti* is
renamed *Rhizodopsis robusta* by A. S. Woodward.

———   1890 A
Zur Frage nach dem Ursprunge der
Schuppen der Säugetiere.
Anatom. Anzeiger, viii, pp. 526–532.

———   1892 A
Ueber den Bau und die Entwicklung
des Panzers der Gürteltiere.
Jenaische Zeitschr. Naturwiss., xxvii (xx),
pp. 513–555, with pls. xxiv and xxv.

Röse, C.   1892 A
Zur Phylogenie des Säugetiergebisses.
Biolog. Centralbl., xii, pp. 624–638.
Abstract by M. Schlosser in Archiv f. An-
throp., xxiii, p. 153.

———   1892 B
Ueber Zahnbau und Zahnwechsel
der Dipnoer.
Anatom. Anzeiger, vii, pp. 821–839, with 10
figs. in the text.

———   1892 C
Ueber die Zahnentwicklung der Kro-
kodile.
Verhandl. Anat. Gesellsch., 6th meeting,
Vienna, 1892, pp. 225–226.

———   1892 D
Beiträge zur Zahnentwicklung der
Edentaten.
Anatom. Anzeiger, vii, pp. 495–512, with 14
text figs.

Röse, C.—Continued. 1892 F
Ueber die Zahnentwicklung der Beuteltiere.
Anatom. Anzeiger, vii, pp. 639–650; 693–707, with 23 text figs.

——— 1893 A
Ueber den Zahnbau und Zahnwechsel von *Elephas indicus.*
Schwalbe's Morpholog. Arbeiten, iii, pp. 173–194, pl. x and 2 text figs.
Abstract of, by M. Schlosser, in Archiv f. Anthrop., xxiv, pp. 137–138.
Contains citations of the literature of the subject.

——— 1893 B
Ueber die Zahnentwicklung der *Crocodilia.*
Schwalbe's Morpholog. Arbeiten, iii, pp. 195–228, with 45 figs. in the text.

——— 1893 C
Ueber die Deutung des Milchgebisses der Säugethiere.
Verhandl. deutsch. odontolog. Gesellsch., iv, pp. 308–310, with 1 text fig.
Preceded by a paper on same subject by Dr. M. Schlosser.

——— 1894 A
Ueber die Zahnentwicklung von *Chlamydoselachus anguineus* Garman.
Schwalbe's Morpholog. Arbeiten, iv, pp. 193–206, with 12 figs. in the text.

——— 1894 B
Ueber die Zahnentwicklung der Fische.
Anatom. Anzeiger, ix, pp. 653–662, with 8 figs. in the text.

——— 1895 A
Das Zahnsystem der Wirbelthiere.
Merkel and Bonnet's Anat. Hefte, Ergebnisse, iv, 1894, pp. 542–591.
List of 248 titles of papers bearing on the subject.
Reviewed by M. Schlosser in Archiv f. Anthrop., xxv, pp. 193–195.

——— 1897 A
Ueber die verschiedenen Abänderungen der Hartgewebe bei niederen Wirbeltieren.
Anatom. Anzeiger, xiv, pp. 33–69, with 28 figs. in the text.

Röse, C., and Bartels, O. 1896 A
Ueber die Zahnentwicklung des Rindes.
Schwalbe's Morpholog. Arbeiten, vi, pp. 49–118, with 39 figs. in the text.
Reviewed by M. Schlosser in Archiv f. Anthrop., xxv, pp. 234–235.

Roger, Otto. 1887 A
Ueber die Hirsche.
Correspondenzblatt Vereins Regensb., 1887, Sept., p. 43, with 2 pls.

——— 1889 A
Ueber die Umbildungen des Säugethierskeletes und die Entwicklungsgeschichte der Pferde.
Abhandl. Vereins Regensb., 1889, p. 35.
Reviewed by M. Schlosser in Archiv f. Anthrop., xx, pp. 152–154.

——— 1896 A
Verzeichniss der bisher bekannten fossilen Säugethiere. Neu zusammengestellt von Dr. Otto Roger, kgl. Regierungs-und Kreis-Medizinrath in Augsburg.
Bericht naturwiss. Vereins f. Schwaben und Neuburg (a. V.), xxxii, pp. 1–272.
This is a systematic list of described fossil vertebrates, with references to places of publication of descriptions.

Rogers, Henry D. 1842 A
Report on the *Ornithichnites,* or footmarks of extinct birds in the New Red sandstone of Massachusetts and Connecticut, observed and described by Prof. Hitchcock, of Amherst. Signed by H. ʼ. Rogers, Lardner Vanuxum, Rich· id C. Taylor, Ebenezer Emmons, T. A. Conrad.
Ann. Nat. Hist., viii, pp. 235–238.

——— 1845 A
Remarks upon the bones of *Zeuglodon.*
Proc. Boston Soc. Nat. Hist., ii, p. 79.

——— 1851 A
On the position and characters of the reptilian footprints in the Carboniferous red-shale formation of eastern Pennsylvania.
Proc. Amer. Assoc. Adv. Sci., 4th meeting, New Haven, 1850, pp. 250–251.

——— 1854 A
Communication on the epoch of the mammoth.
Proc. Boston Soc. Nat. Hist., v, pp. 22–23.

——— 1855 A
Remarks on fossil footprints found in Carboniferous sandstones of Pennsylvania.
Proc. Boston Soc. Nat. Hist., v, pp. 182–186.

——— 1865 A
On a peculiar fossil found in the Mesozoic sandstone of the Connecticut

Rogers, Henry D.—Continued.
valley, discovered by Prof. W. B.
Rogers.

> Report Brit. Assoc. Adv. Sci., 34th meeting,
> Bath, 1864, p. 66.
> Zoological position of fossil not determined.

Rohon, J. Victor. 1889 A
Ueber fossile Fische.

> Mém. Acad. Imp. Sci., St. Pétersbourg (7),
> xxxvi, No. 73, pp. 1–17; pls. i, ii.
> Contains remarks on the structure of the
> teeth of *Acanthodes* and *Palæoniscus*, with illus-
> trations.

—— 1889 B
Die Dendrodonten des devonischen
Systems in Russland. Palæontolo-
gische und vergleichend-anatomische
Studie.

> Mém. Acad. Imp. Sci., St. Pétersbourg (7),
> xxxvi, No. 14, pp. 1–52, with pls. i, ii and 1 text
> fig. giving a restoration of *Dendrodus*.

—— 1891 A
Ueber *Pterichthys*.

> Verhandl. russ.-kais. mineralog. Gesellsch.
> St. Petersburg (2), xxviii, pp. 292–315, with
> pl. vii.

—— 1892 A
Ueber devonische Fische vom oberen
Jenissei nebst Bemerkungen über die
Wirbelsäule devonischer Ganoiden.

> Mélanges géol. et paléont. Acad. Imp. Sci.
> St. Pétersbourg, vol. i, pp. 17–34, with 1 pl.
> This paper was published earlier in Bull.
> Acad. St. Pétersbourg (2), i, 1890, pp. 393–410.

—— 1892 B
Holoptychius-Schuppen in Russland.

> Mélanges géol. et paléont. Acad. Imp. Sci.
> St. Pétersbourg, i, pp. 35–56, with 1 pl.

—— 1892 C
Die obersilurische Fische von Oesel.
I Theil, *Thyestidæ* und *Tremataspidæ*.

> Mém. Acad. Imp. Sci. St. Pétersbourg (7),
> xxxviii, No. 13, pp. 1–88, with pls. i–ii.

—— 1893 A
Die obersilurischen Fische von Oesel.
II Theil, *Selachii*, *Dipnoi*, *Ganoidei*,
Pteraspidæ und *Cephalaspidæ*.

> Mém. Acad. Imp. Sci. St. Pétersbourg (7),
> xli, No. 5, pp. 1–124, with pls. i–iii and 29 text
> figs.

—— 1894 A
Metamerie am Primordial-cranium
palæozoischer Fische.

> Zoolog. Anzeiger, xvii, pp. 51–52.
> A brief communication regarding the
> metamery of the primordial skull of *Thyestes*
> and *Tremataspis*.

Rohon, J. Victor—Continued. 1894 B
Zur Kenntnis der Tremataspiden
(Nachtrag zu den Untersuchungen über
"Die obersilurischen Fische von
Oesel").

> Mélanges géol. et paléont. Acad. Imp. Sci.
> St. Pétersbourg, vol. i, pp. 177–201, with 2 pls.

—— 1895 A
Beitrag zur Kenntniss der Gattung
Ptyctodus.

> Verhandl. russ.-kais. mineralog. Gesellsch.
> St. Petersburg (2), xxxiii, pp. 1–16, with pl. i.

—— 1895 B
Die Segmentirung am Primordial-
cranium der obersilurischen Thyesti-
den.

> Verhandl. russ.-kais. mineralog. Gesellsch.
> St. Petersburg (2), xxxiii, pp. 17–64, with pl. ii.

—— 1896 A
Weitere Mittheilungen über die
Gattung *Thyestes*.

> Bull. Acad. Imp. Sci. St. Pétersbourg, iv, pp.
> 223–235, with 1 pl.

—— 1896 B
Beiträge zur Classification der palæozoischen Fische.

> Sitzungsber. k. böhm. Gesellsch. Wissensch.,
> math.-naturw. Cl., 1896, ii, art. xxxvii, pp.
> 1–33, with 8 text figs.

—— 1898 A
Bau der obersilurischen Dipnoer-
Zähne.

> Sitzungsber. k. böhm. Gesellsch. Wissensch.,
> math.-naturw. Cl., 1898, art. xi, pp. 1–17, with
> plate of 11 figs.

Romanovski, G. 1853 A
Ueber eine neue Gattung versteiner-
ter Fisch-Zähne.

> Bull. Soc. Imp. Natur. Moscou, xxvi, pt. i,
> pp. 405–409, with pl. viii.
> Describes genus *Dicrenodus*, type *D. okensis*.

—— 1857 A
Ueber die Verschiedenheit der
beiden Arten: *Chilodus tuberosus* Gieb.
und *Dicrenodus okensis* Rom.

> Bull. Soc. Imp. Natur. Moscou, xxx, pt. i,
> pp. 290–295, with 4 text figs.

—— 1864 A
Description de quelques restes de
poissons fossiles trouvés dans le calcaire
carbonifère du gouvernement de Toula.

> Bull. Soc. Imp. Natur. Moscou, xxxvii, pt.
> 2, pp. 157–170, with pls. iii, iv.

Rosenberg, Emil.   1891 A
Ueber einige Entwicklungsstadien des
Handskelets der *Emys lutaria marsili.*
Morpholog. Jahrbuch, xviii, pp. 1–34, with
pl. i.
In the footnotes are found numerous refer-
ences to the literature bearing on the subject.

Rouault, Marie.   1858 A
Note sur les vertébrés fossiles des
terrains sedimentaires de l'ouest de la
France.
Comptes rend. Acad. Sci. Paris, xlvii, pp.
99–103.
The genus *Nummopalatus* is here established,
p. 101.

**Ruddick, W. H.   See Kingsley and
Ruddick.**

Rütimeyer, L.   1857 A
Ueber schweizerische Anthracotherien.
Verhandl. naturf. Gesellsch. Basel, i, pp. 385–
403.

———   1857 B
Ueber lebende und fossile Schweine.
Verhandl. naturf. Gesellsch. Basel, i, pp. 517–
554.

———   1857 C
Ueber *Anthracotherium magnum* und
hippoideum.
Neue Denkschr. schweizer. Gesellsch. Naturwiss., xv, art. viii, pp. 1–32, with pls. i, ii.

———   1862 A
Beiträge zur Kenntniss der fossilen
Pferde und zur vergleichenden Odon-
tographie der Hufthiere überhaupt.
Verhandl. naturf. Gesellsch. Basel, iii, pp.
558–696, with pls. i–iv.

———   1862 B
Die Fauna der Pfahlbauten der
Schweiz.
Neue Denkschr. schweizer. Gesellsch. Natur-
wiss., ix, pp. 1–248, with pls. i–vi.

———   1865 A
Beiträge zu einer palæontologischen
Geschichte der Wiederkäuer, zunächst
an Linné's Genus *Bos.*
Verhandl. naturf. Gesellsch. Basel, iv, pp.
299–354.

———   1867 A
Die fossilen Schildkröten von Solo-
thurn.
See Lang, Fr., and Rütimeyer, L., 1867 A.

———   1867 B
Versuch einer natürlicher Geschichte
des Rindes, in seinen Beziehungen zu

Rütimeyer, L.—Continued.
den Wiederkauern im Allgemeinen.
Erste Abth.
Neue Denkschr. schweizer. Gesellsch. Naturwiss., xxii, pp. 1–102, with pls. i, ii.

———   1868 A
Versuch einer natürlicher Geschichte
des Rindes, in seinen Beziehungen zu
den Wiederkauern im Allgemeinen.
Zweite Abth.
Neue Denkschr. schweizer. Gesellsch. Naturwiss., xxiii, pp. 1–175, with pls. i–iv, and 25 figs.
in the text.

———   1873 A
Die fossilen Schildkröten von Solo-
thurn und der übrigen Juraformation.
Mit Beiträgen zur Kenntniss von Bau
und Geschichte der Schildkröten im
Allgemeinen.
Neue Denkschr. schweizer. Gesellsch. Natur-
wiss., xxv, art. 2, pp. 1–185, with pls. i–xvii.

———   1874 A
Ueber den Bau von Schale und
Schädel bei lebenden und fossilen
Schildkröten, als Beitrag zu einer
palæontologischen Geschichte dieser
Thiergruppe.
Verhandl. naturf. Gesellsch. Basel., vi, pp.
1–187.

———   1875 A
Weitere Beiträge zur Beurtheilung
der Pferde der Quaternär-Epoche.
Abhandl. schweizer. paläontolog. Gesellsch.,
ii, pp. 1–34, with 3 pls.
Contains discussion of the teeth of living
species of horses.

———   1877 A
Die Rinder der Tertiär-Epoche, nebst
Vorstudien zu einer natürlichen Ge-
schichte der Antilopen. Erster Theil.
Mit drei Doppeltafeln.
Abhandl. schweizer. paläontolog. Gesellsch.,
iv, pp. 1–72.

———   1878 A
Die Rinder der Tertiär-Epoche, nebst
Vorstudien zu einer natürlichen Ge-
schichte der Antilopen. Zweiter Thiel.
Mit vier Doppeltafeln nebst Holzschnit-
ten.
Abhandl. schweizer. paläontolog. Gesellsch.,
v, pp. 1–208.

———   1881 A
Beiträge zu einer natürlichen Geschichte der Hirsche. I Theil.
Abhandl. schweizer. paläontolog. Gesellsch.,
vii, pp. 1–8; viii, pp. 9–95, with 4 pls.
Introduction, with pls. i and ii in vol. vii,
1880.

Rütimeyer, L.—Continued. 1882 A
Studien zu der Geschichte der Hirschfamilie. I. Schädelbau.
Verhandl. naturf. Gesellsch. Basel, vii, pp.
3–61.

———— 1883 A
Beiträge zu einer natürlichen Geschichte der Hirsche.
Abhandl. schweizer. paläontolog. Gesellsch.,
x, pp. 1–120, with pls. v–x.

———— 1884 A
Studien zu der Geschichte der
Hirschfamilie. II. Gebiss.
Verhandl. naturf. Gesellsch. Basel, vii, pp.
399–464.

———— 1888 A
Ueber einzige Beziehungen zwischen den Säugethierstämmen alter
und neuer Welt. Erster Nachtrag zu
der eocänen Fauna von Egerkingen.
Abhandl. schweizer. paläontolog. Gesellsch.,
xv, pp. 1–63, pl. 1.
Reviewed by M. Schlosser in Archiv f. Anthrop., xix, pp. 141–145.

———— 1890 A
Uebersicht der eocänen Fauna von
Egerkingen nebst einer Erwiederung
an Prof. E. D. Cope. Zweiter Nachtrag zu der eocänen Fauna von Egerkingen (1862).
Abhandl. schweizer. paläontolog. Gesellsch.,
xvii, pp. 1–24.
See also Verhandl. naturf. Gesellsch. Basel,
1891, ix, pp. 331–362. For review by M. Schlosser, see Archiv f. Anthrop., xx, pp. 113–114.
Contains a discussion of foot structure.

———— 1891 A
Die eocäne Säugethier-Welt von
Egerkingen. Gesammtdarstellung und
dritter Nachtrag zu den "Eocänen
Säugethieren aus dem Gebiet des
schweizerischen Jura (1862)."
Abhandl. schweizer. paläontolog. Gesellsch.,
xviii, pp. 1–153, with pls. i–viii.
Reviewed by M. Schlosser in Archiv f. An-
throp., xxiii, pp. 139–141.

———— 1892 A
Die eocänen Säugethiere von Egerkingen.
Verhandl. naturf. Gesellsch. Basel, x, pp.
101–129, with figs. A–I in the text.
Discusses the teeth of *Mammalia* and the relationships of the mammals of Egerkingen to
those of the American Eocene.

See also **Lang** and **Rütimeyer.**

Ryder, John A. 1877 A
On the evolution and homologies of
the incisors of the horse.
Proc. Acad. Nat. Sci. Phila., 1877, pp. 152–154.

———— 1877 B
On the laws of digital reduction.
Amer. Naturalist, xi, pp. 603–607.

———— 1878 A
On the mechanical genesis of tooth-
forms.
Proc. Acad. Nat. Sci. Phila., 1878, pp. 45–80,
with 11 figs. in the text.
Reviewed in Amer. Naturalist, xiii, pp. 446–
449.

———— 1879 A
Further notes on the mechanical
genesis of tooth-forms.
Proc. Acad. Nat. Sci. Phila., 1879, pp. 47–51.

———— 1879 B
On the origin of bilateral symmetry
and the numerous segments of the soft
rays of fishes.
Amer. Naturalist, xiii, pp. 41–43.

———— 1885 A
On the morphology and evolution of
the tails of osseous fishes.
Proc. Amer. Assoc. Adv. Sci., 33d meeting,
Phila., 1884, pp. 532–533.
Abstract of paper read.

———— 1885 B
The development of the rays of osse-
ous fishes.
Amer. Naturalist, xix, pp. 200–204, with 5 figs.

———— 1885 C
On the translation forwards of the
rudiments of the pelvic fins of physo-
clist fishes.
Amer. Naturalist, xix, pp. 315–317.

———— 1885 D
On the probable origin, homologies,
and development of the flukes of cetaceans and sirenians.
Amer. Naturalist, xix, pp. 515–519.

———— 1886 A
On the origin of heterocercy and the
evolution of the fins and fin rays of
fishes.
Report U. S. Fish Com. for 1884, pp. 981–1107,
pls. i–xi.

———— 1890 A
The continuity of the primary matrix
of the scales and the actinotrichia of
Teleosts.
Amer. Naturalist, xxiv, pp. 489–491.

Ryder, John A.—Continued.   1892 A
On the mechanical genesis of the
scales of fishes.
Proc. Acad. Nat. Sci. Phila., 1892, pp. 219-224,
with 3 figs. in the text.
See notice of, in Amer. Naturalist, xxvii, p.
391.
Reprinted in Ann. and Mag. Nat. Hist. (6),
xi. pp. 243-248.

Safely, Robert.   1866 A
Discovery of mastodon remains at
.Cohoes, N. Y.
Amer. Jour. Sci. (2), xlii, p. 426.

Safford, J. M.   1853 A.
Tooth of *Getalodus ohioensis*.
Amer. Jour. Sci. (2), xvi, p. 142, with 2 figs.
in the text.
"*Getalodus*" is an evident misprint for *Peta-
lodus*.

————   1892 A
The pelvis of a *Megalonyx* and other
bones from Big Bone Cave, Tennessee.
Bull. Geol. Soc. America, iii, pp. 121-123.

————   1892 B
Exhibition of certain bones of the
Megalonyx not before known.
Proc. Amer. Assoc. Adv. Sci., 40th meeting,
Washington, 1891, p. 289.
Abstract of Safford, 1892 A.

Sagemehl, M.   1883 A
Beiträge zur vergleichenden Ana-
tomie der Fische. I. Das Cranium von
Amia calva L.
Morpholog. Jahrbuch, ix, pp. 177-228, with
pl. x.

————   1884 A
Beiträge zur vergleichenden Anatomie der Fische. III. Das Cranium der
Chariciniden nebst allgemeinen Bemerkungen über die mit einem Weber'-
schen Apparat versehenen Physosto-
menfamilien.
Morpholog. Jahrbuch, x, pp. 1-119, with pls.
i and ii and 1 woodcut.

————   1891 A
Beiträge zur vergleichenden Anatomie der Fische.
Morpholog. Jahrbuch, xvii, pp. 489-595,
with pls. xxviii-xxix.
Contains numerous citations of the literature of the subject.

St. Hilaire, Geoffroy,   1809 A
Sur les tortues molles, nouveau genre
sous le nom de *Trionyx*, et sur la formation des carapaces.
Ann. Mus. d'Hist. Nat. Paris, xiv, pp. 1-20,
with pls. i-v.

St. John, Orestes H.   1870 A
Descriptions of fossil fishes from the
upper Coal-measures of Nebraska.
Proc. Amer. Philos. Soc., xi, pp. 431-437.

————   1872 A
Descriptions of fossil fishes from the
upper Coal-measures of Nebraska.
Final report U. S. Geol. Surv. of Nebraska
and portions of the adjacent Territories, pp.
239-245; pls. iii, iv and vi.

St. John, O. H., and **Worthen**, A. H.
        1875 A
Descriptions of fossil fishes.
Geol. Surv. Illinois, A. H. Worthen, director,
vi, pp. 245-488, with pls. i-xxii.
Contains descriptions and figures of fossil
fishes of the Carboniferous and Subcarboniferous, especially of Illinois.

————   1883 A
Descriptions of fossil fishes. A partial revision of the Cochliodonts and
Psammodonts, including notices of
miscellaneous materials acquired from
the Carboniferous formations of the
United States.
Geol. Surv. Illinois, A. H. Worthen, director,
vii, 1883, pp. 55-264, pls. i-xxvi.

St. John, Samuel.   1851 A
On a specimen of the fossil ox (*Bos
bombifrons*) found in Trumbull County,
Ohio.
Proc. Amer. Assoc. Adv. Sci., v, p. 235. Title
only.
Specimen referred by Leidy (Leidy, 1869 A,
p. 374) to *B. cavifrons*.

Salensky, W.   1899 A
Zur Entwicklungsgeschichte des
Ichthyopterygiums.
Proc. 4th Internat. Congress of Zoology, Cam-
bridge, Eng., 1898, pp. 177-183.

Salter, J. W.   1867 A
On the tracks of *Pteraspis* (?) in the
Upper Ludlow sandstone.
Quart. Jour. Geol. Soc., xxiii, pp. 333-339,
with figs. in the text.

Sanson, A.   1878 A
Détermination spécifique des ossemens fossiles ou anciens de bovidés.
Comptes rend. Acad. Sci. Paris, lxxxvii, pp
756-759.

Saporta, Gaston de.   1877 A
Letters to Dr. F. V. Hayden.
Amer. Naturalist, xi, pp. 187-188.
Expresses gratification at the progress of
palæontology in North America.

Sauvage, H. E. 1870 A
Recherches sur les poissons fossiles des terrains crétacés de la Sarthe.
Ann. Sci. Géol., Paris, ii, art. 7, pp. 1–44, with pls. xvi, xvii (figs. 1–89).

——— 1873 A
De la classification des poissons que composent le famille des Triglides (Joues-cuirassées de Cuvier et Valenciennes).
Comptes rend. Acad. Sci. Paris, lxxvii, pp. 723–726.

——— 1874 A
Revision des espèces du groupe des épinoches.
Nouv. Arch. du Mus. d'Hist. Nat. Paris, x, pp. 5–38, with pl. i.

——— 1874 B
Mémoire sur les dinosauriens et les crocodiliens des terrains jurassiques de Boulogne-sur-mer.
Mém. Soc. Géol. France (2), x, pp. 1–57, with pl. v–x.

——— 1875 A
Note sur les poissons fossiles.
Bull. Soc. Géol. France (3), iii, pp. 631–642; pls. xxii, xxiii.

——— 1875 B
Note sur le genre *Nummopalatus* et sur les espèces de ce genre trouvés dans les terraines tertiaires de la France.
Bull. Soc. Géol. France (3), iii, pp. 613–642, with pls. xxii, xxiii.

——— 1876 A
Notes sur les reptiles fossiles.
Bull. Soc. Géol. France (3), iv, pp. 435–442, pls. xi–xii.
Treats of *Polycotylus, Iguanodon*, etc.

——— 1878 A
Prodrome des Plésiosauriens et Élasmosauriens des formations jurassiques supérieures de Boulogne-sur-mer.
Ann. Sci. Naturelles (6), Zool., viii, art. 13, pp. 1–38, pls. xxvi, xxvii.

——— 1878 B
Description de poissons nouveaux ou imparfaitement connus de la collection du Muséum d'Histoire Naturelle Famille des Scorpénidées, des Platycephalides et des Triglidées.
Nouv. Arch. du Mus. d'Hist. Nat. Paris (2), i, pp. 109–158, with pls. i, ii.

——— 1879 A
Étude sur les poissons et les reptiles des terrains crétacés et jurassiques supérieurs de l'Yonne.
Bull. Soc. Sci. hist. et nat. de l'Yonne, xxxiii, pp. 20–84, with pls. i–viii.

Sauvage, H. E.—Continued. 1887 A
Note sur l'arc pectoral d'un Ichthyosaure du Lias.
Bull. Soc. Géol. France (3), xv, pp. 726–728, pl. xxvii.

——— 1888 A
Poissons du terrain houiller de Commentry; I. Coup d'œil sur la faune ichthyologique du terrain carbonifère et du terrain houiller; II. La faune ichthyologique de Commentry; III. Description des espèces Famille des Palæoniscidées.
Bull. Soc. de l'Industrie Min. (3), ii livrais. 4ᵐᵉ, pp. 39–120.

Savage, J. 1878 A
On mastodon remains in Douglas County [Kansas].
Trans. Kansas Acad. Sci., vi, pp. 10, 11.

Scheidt, Paul. 1894 A
Morphologie und Ontogenie des Gebisses der Hauskatze.
Morpholog. Jahrbuch, xxi, pp. 425–462, with pl. xii.

Schenk, F. 1897 A
Studien über die Entwicklung des knöchernen Unterkiefers der Vögel.
Sitzungsber. k. p. Akad. Wissensch. Wien, math.-naturw. Cl., cvi, Abth. iii, pp. 319–344, with pls. i–v.

Schlosser, M. 1883 A
Über *Chalicotherium*-Arten.
Neues Jahrb. Mineral., 1883, ii, pp. 164–169.

——— 1884 A
Die Nager des europäischen Tertiärs nebst Betrachtungen über die Organization und die geschichtliche Entwicklung der Nager überhaupt.
Palæontographica, xxxi, pp. 9–162, with pls. v–xii.
Nachtrag und Berichtigungen zu "Die Nager des europäischen Tertiärs," xxxi, pp. 323–328.

——— 1884 B
Nachträge und Berichtigungen zu : die Nager des europäischen Tertiärs. Palaeontographica. 31. Band.
Zoolog. Anzeiger, vii, pp. 639–647.

——— 1884 C
Literaturbericht in Beziehung zur Anthropologie mit Einschluss der fossilen und recenten Säugethiere.
Archiv f. Anthrop., xvi, pp. 98–135.

——— 1885 A
Zur Stammesgeschichte der Hufthiere.
Zoolog. Anzeiger, viii, pp. 683–691.

Schlosser, M.—Continued. 1886 A
Zur Stammesgeschichte der Huf-
thiere.
Zoolog. Anzeiger, ix, pp. 252–256, 432–433.
Pp. 432–433 are entitled: Erklärung.

— 1886 B
Beiträge zur Kenntnis der Stammesgeschichte der Hufthiere und Versuch
einer Systematik der Paar- und Unpaar-
hufer.
Morpholog. Jahrbuch, xii, pp. 1–136, with
pls. i–vi.
Abstract in Archiv f. Anthrop., xviii, pp.
131–139.

— 1886 C
Palæontologische Notizen. Ueber
das Verhältnis der Cope'schen Creodonta zu den übrigen Fleischfressern.
Morpholog. Jahrbuch, xii, pp. 287–296.
Abstract in Archiv f. Anthrop., xviii, pp.
130–131.

— 1886 D
Literaturbericht in Beziehung zur
Anthropologie mit Einschluss der fos-
silen und recenten Säugethiere für 1884,
1885.
Archiv f. Anthrop., xvii, pp. 118–194.

— 1887 A
Erwiederung gegen E. D. Cope.
Morpholog. Jahrbuch, xii, pp. 575–580.
Reply to review by Professor Cope in Amer.
Naturalist, 1886, pp. 719–721 (Cope, E. D.,
1886 K).

— 1887 B
Die Affen, Lemuren, Chiropteren, Insectivoren, Marsupialier, Creodonten
und Carnivoren des europäischen Tertiärs und deren Beziehungen zu ihren
lebenden und fossilen aus europäischen
Verwandten.
Beiträge Palæontologie Oesterreich-Ungarns
und des Orients (Mojsisovics und Neumayr),
vi, Heft. i und ii, pp. 1–227, with pls. i–ix.
Abstract in Archiv f. Anthrop., xvii, pp.
279–285.

— 1887 C
Literaturbericht in Beziehung für
Anthropologie mit Einschluss der fos-
silen und recenten Säugethiere für 1886.
Archiv f. Anthrop., xviii, pp. 105–152.

— 1888 A
Ueber die Beziehungen der ausgestor-
benen Säugetierfaunen und ihr Ver-

Schlosser, M.—Continued.
hältnis zur Säugetierfauna der Gegenwart.
Biolog. Centralbl., viii, pp. 582–600, 609–631.
Abstract in Archiv f. Anthrop., xx, pp.
154–160.

— 1888 B
Die Affen, Lemuren, Chiropteren, Insectivoren, Marsupialier, Creodonten
und Carnivoren des europäischen Tertiärs und deren Beziehungen zu ihren
lebenden und fossilen aussereuropä-
ischen Verwandten. Part II.
Beiträge Palæontologie Oesterreich-Ungarns
und des Orients (Mojsisovics und Neumayr),
vii, pp. 1–162.
Abstract in Archiv f. Anthrop., xvii, pp.
285–300, with pl. xii.

— 1889 A
Ueber die Modificationen des Ex-
tremitätenskelets bei den einzelnen
Säugetierstämmen.
Biolog. Centralbl., ix, pp. 684–698, 716–729.

— 1889 B
Literaturbericht für Zoologie in Beziehung zur Anthropologie mit Einschluss der fossilen und recenten Säugethiere für das Jahr 1887 [and 1888].
Archiv f. Anthrop., xix, pp. 78–164.
The notices of the literature for 1887 occupies pp. 78–116; for that of 1888, pp. 117–164.

— 1890 A
Ueber die Deutung des Milchgebisses
der Säugetiere.
Biolog. Centralbl. x, pp. 81–92.
Abstract in Archiv f. Anthrop., xxi, pp. 136–
137.

— 1890 B
Die Affen, Lemuren, Chiropteren,
Insectivoren, Marsupialier, Creodonten
und Carnivoren des europäischen Tertiärs.
Beiträge Palæontologie Oesterreich-Ungarns
und des Orients (Mojsisovics und Neumayr),
viii, pp. 1–107. [387–492.]
Abstract in Zoolog. Record for 1890, Mamm.
p. 14.
Mention is made in this work of many American forms, and their relationships are discussed.

— 1890 C
Die Differenzierung des Säugetiergebisses.
Biolog. Centralbl. x, pp. 238–252; 264–277,
with 7 figs. in the text.
Abstract in Archiv f. Anthrop., xxi, pp. 132–
136.

Schlosser, M.—Continued. 1890 D
Ueber die Modificationen des Extre-
mitätenskelets bei den einzelnen Säugethierstämmen.
Biolog. Centralbl., x, pp. 684-698; 716-729.

——— 1890 E
Literaturbericht für Zoologie in
Beziehung zur Anthropologie mit Ein-
schluss der lebenden und fossilen
Säugethiere für das Jahr 1889.
Archiv f. Anthrop., xx, pp. 113-161.

——— 1891 A
Literaturbericht für Zoologie in
Beziehung zur Anthropologie mit Ein-
schluss der lebenden und fossilen
Säugethiere für das Jahr 1890.
Archiv f. Anthrop., xxi, pp. 97-141.

——— 1892 A
Die Entwickelung der verschiedenen
Säugethierzahnformen im Laufe der
geologischen Perioden.
Verhandl. deutsch. odontolog. Gesellsch.,
iii, pp. 208-230.

——— 1892 B
Literaturbericht für Zoologie in
Beziehung zur Anthropologie mit Ein-
schluss der lebenden und fossilen
Säugethiere für das Jahr 1891.
Archiv f. Anthrop., xxii, pp. 89-130.

——— 1893 A
Ueber die Deutung des Milchgebisses
der Säugethiere.
Verhandl. deutsch. odontolog. Gesellsch.,
iv, pp. 296-307, with 1 text fig.
Followed by Dr. Röse on same subject.

——— 1893 B
Literaturbericht für Zoologie in
Beziehung zur Anthropologie mit Ein-
schluss der lebenden und fossilen
Säugethiere für das Jahr 1892.
Archiv f. Anthrop., xxiii, pp. 111-160.

——— 1894 A
Bemerkungen zu Rütimeyer's "eo-
cäne Säugethierwelt von Egerkingen."
Zoolog. Anzeiger, xvii, pp. 157-162.

——— 1895 A
Literaturbericht für Zoologie in
Beziehung zur Anthropologie mit Ein-
schluss der lebenden und fossilen
Säugethiere für das Jahr 1893 [and
1894].
Archiv f. Anthrop., xxiv, pp. 101-182.
Pp. 101-145 are occupied with the literature
of 1893; pp. 145-182 with that of 1894.

Schlosser, M.—Continued. 1897 A
Literaturbericht für Zoologie in
Beziehung zur Anthropologie mit Ein-
schluss der lebenden und fossilen
Säugethiere für das Jahr 1895 [and
1896].
Archiv f. Anthrop., xxv, pp. 157-244.
Pp. 157-198 are taken up with the literature
of 1895; pp. 198-244 with that for 1896.

——— 1898 A
Literaturbericht für Zoologie in
Beziehung zur Anthropologie mit Ein-
schluss der lebenden und fossilen
Säugethiere für das Jahr 1897.
Archiv f. Anthrop., xxvi, pp. 151-199.

——— 1899 A
Parailurus anglicus and *Ursus böckii*
aus den Ligniten von Baróth-Köpecz,
Comitat Háromszék in Ungarn.
Mitthell. Jahrb. der kgl. ungar. geol. An-
stalt, xiii, pp. 67-95, with pls. x-xii.

——— 1899 B
Uber die Bären und bärenähnlichen
Formen des europäischen Tertiärs.
Palaeontographica, xlvi, pp. 95-147, with pls.
xiii, xiv.

——— 1900 A
Ursus oder *Ursavus* oder *Hyaenarctos?*
Centralbl. Min., Geol., Palaeont., 1900, pp.
261-265; with 3 figs. in the text.

Schlüter, Clemens. 1881 A
Ueber die Fischgattung *Ancistrodon.*
Verhandl. naturf. Vereines preuss. Rhein-
lande, xxxviii, pp. 61-62 with 3 figs. in the text.

Schmidt, Friedrich. 1866 A
Ueber *Thyestes verrucosus* Eichw.,
und *Cephalaspis schrenckii* Pand., nebst
einer Einleitung über das Vorkommen
silurischer Fischreste auf der Insel
Oesel.
Verhandl. russ.-kais. mineralog. Gesellsch.
St. Petersburg (2), i, pp. 217-250, with pls. iv-vi.

——— 1872 A
Wissenschaftliche Resultate der zur
Aufsuchung eines angekündigten Mam-
muthcadavers von der Kaiserlichen
Akademie der Wissenschaften an den
unteren Jenissei ausgesandten Expedition.
Mém. Acad. Imp. Sci. St.-Pétersbourg (7),
xviii, pp. 1-vi; 1-168.
Pp. 28-35 are devoted to "Das Mammuth
und seine Lagerstätte."

Schmidt, Friedrich—Cont'd. 1873 A
Note on *Pteraspis kneri*.
Geol. Magazine, x, pp. 152-153, with 3 figs. in the text.

——— 1873 B
Further remarks on *Pteraspis*.
Geol. Magazine (1), x, pp. 330-331.
The author claims to have found osseous lacunæ in the shield of *Pteraspis*. See, also, reply by E. R. Lankester, this volume, p. 478.

——— 1873 C
Ueber die Pteraspiden überhaupt und über *Pteraspis kneri* aus den obersilurischen Schichten Galiziens insbesondere.
Verhandl. russ.-kais. mineralog. Gesellsch. St. Petersburg (2), viii, pp. 132-152, with pl. v.

——— 1894 A
Ueber *Cephalaspis* (*Thyestes*) *schrencki* Pand. aus dem Obersilur von Rotzeküll auf Oesel.
Mélanges géol. et paléont. Acad. Imp. Sci. St. Pétersbourg, i, pp. 203-210, with 1 pl.

Schmidt, Oscar. 1886 A
The *Mammalia* in their relation to primeval times.
Pp. i-xxii; 1-308, with 51 woodcuts. 8vo. New York, D. Appleton and Co.

Schneider, Anton. 1886 A
Ueber die Flossen der *Dipnoi* und die Systematik von *Lepidosiren* und *Protopterus*.
Zoolog. Anzeiger, ix, pp. 521-524.

Schuchert, Charles. 1900 A
[Account of visits to Alabama, for the purpose of collecting bones of the zeuglodon.]
Proc. U. S. Nat. Mus., xxiii, pp. 326-329.
This account is incorporated in paper by F. A. Lucas (Lucas, F. A., 1900 E).

Schulze, Eilhard. 1894 A
Ueber die Abwärtsbiegung des Schwanztheiles der Wirbelsäule bei *Ichthyosaurus*.
Sitzungsber. k. p. Akad. Wissenscn. Berlin. 1894, pp. 1133-1134.

Schwalbe, G. 1894 A
Ueber Theorien der Dentition.
Verhandl. anatom. Gesellsch., 8th meeting, 1894, pp. 5-45.
Appended is a list of 80 papers on the subject discussed.
Abstract of, by M. S. Schlosser in Archiv f. Anthrop., xxiv, pp. 179-180.

Science. 1885 A
Professor Marsh on the *Dinocerata*.
Science, v, pp. 488-490, with 1 map and 1 woodcut.
A review of Marsh, O. C., 1884 F, with correction in vol. vi, p. 454.

——— 1885 B
The relationships between dinosaurs and birds.
Science, vi, p. 295.
A review of Prof. B. Vetter's article in Festschrift der naturw. Gesellsch. Isis in Dresden, May, 1885, pp. 109-122.

——— 1885 C
[Prof. O. C. Marsh on the brains of Tertiary vertebrates. A notice of his address before the British Association for the Advancement of Science.]
Science, vi, p. 360.

Scott, W. B. 1883 A
Two new Eocene lophiodonts.
Contrib. E. M. Mus. Geol. and Arch.; Princeton Coll. Bull. No. 3, pp. 46-53, with pl. viii.
Describes *Desmatotherium guyotii* and *Dilophodon minusculus*.

——— 1884 A
A new marsupial from the Miocene of Colorado.
Amer. Jour. Sci. (3), xxvii, pp. 442-443, with 1 fig. in the text.

——— 1884 B
[Communication to Princeton Science Club relative to the hind foot of the American *Entelodon*.]
Science, iii, p. 266.

——— 1884 C
[Abstract of paper read before the Amer. Assoc. Adv. Sci., 1884, on the osteology of *Oreodon*.]
Science, iii, p. 342.

——— 1885 A
A fossil elk or moose from the Quaternary of New Jersey (*Cervalces americanus*.)
Science, v, pp. 420-422, with two woodcuts.
Notice of this paper in Amer. Naturalist, xix, p. 495.

——— 1885 B
The osteology of *Oreodon*.
Proc. Amer. Assoc. Adv. Sci., 33d meeting, Phila., 1884, pp. 492-493.
An abstract of paper read.

——— 1885 C
Cervalces americanus, a fossil moose or elk from the Quaternary of New Jersey.
Proc. Acad. Nat. Sci. Phila., 1885, pp. 174-202, with pl. ii and 7 figs. in the text.

Scott, W. B.—Continued. 1885 D
Cervalces americanus.
Fourth annual report E. M. Mus. Geol. and
Arch., Princeton Coll., pp. 4-6, with pl.

——— 1886 A
On some new forms of the *Dinocerata.*
Amer. Jour. Sci. (3), xxxi, pp. 303-307, with
4 figs. in the text.

——— 1886 B
Some points in the evolution of the
horses.
Science, vii, p. 13, with 2 woodcuts.

——— 1887 A
On American elephant myths.
Scribner's Magazine, i, pp. 469-478, with 10
text figs.

——— 1888 A
Origin of American *Carnivora.*
Proc. Amer. Assoc. Adv. Sci., 36th meeting,
New York, 1887, pp. 258-259.
Abstract of paper read.

——— 1888 B
On some new and little-known creo-
donts.
Jour. Acad. Nat. Sci. Phila. (2), ix, pp. 155-
185, pls. v-vii.
Extra copies printed in advance for the au-
thor Oct. 10, 1887.
Reviewed by M. Schlosser in Archiv f.
Anthrop., xviii, pp. 139-140.

——— 1889 A
A comparison of the American and
European Tertiary *Mammalia.*
Princeton Coll. Bull., i, pp. 20-21.

——— 1889 B
Notes on the osteology and systematic position of *Dinictis felina,* Leidy.
Proc. Acad. Nat. Sci. Phila., 1889, pp. 211-244,
with 7 figs. in the text.
Abstract in Princeton Coll. Bull., i, p. 118,
1889.

——— 1889 C
The *Oreodontidæ.*
Princeton Coll. Bull., i, pp. 75-77.

——— 1890 A
The *Mammalia* of the Uinta forma-
tion. Part I. The geological and fau-
nal relations of the Uinta formation.
Part II. The *Creodonta, Rodentia,* and
Artiodactyla.
Trans. Amer. Philos. Soc., xvi, pp. 461-504,
with pls. vii, x, and xi.
This forms a portion of a joint paper by
Professors Scott and Osborn. See Osborn, H. F.
1890 D, and Scott and Osborn, 1890 A.
Notice of, occurs in Princeton Coll. Bull., ii,
p. 46.

Scott, W. B.—Continued. 1890 B
Beiträge zur Kenntniss der *Oreodon-
tidæ.*
Morpholog. Jahrbuch, xvi, pp. 319-395, with
pls. xii-xvi and 10 figs. in the text.
For review of this paper, by R. Lydekker,
see the Geol. Magazine (3), vii, p. 568; by M.
Schlosser, Archiv f. Anthrop., xxi, pp. 137-139.

——— 1890 C
The dogs of the American Miocene.
Princeton Coll. Bull., ii, pp. 37-39.

——— 1891 A
On the osteology of *Poëbrotherium;*
a contribution to the phylogeny of the
Tylopoda.
Jour. Morphology, v. pp. 1-78, pls. i-iii.
An abstract is given by M. Schlosser in Ar-
chiv f. Anthrop., xxii, 1894, pp. 118-120.

——— 1891 B
On the osteology of *Mesohippus* and
Leptomeryx, with observations on the
modes and factors of evolution in the
Mammalia.
Jour. Morphology, v, pp. 301-406, with pls.
xxii, xxiii.
Review in Amer. Geologist, ix, pp. 402-404;
also by Schlosser in Archiv f. Anthrop., xxii,
1894, pp. 120-122.

——— 1891 C
On the mode of evolution in the
Mammalia.
Princeton Coll. Bull., iii, pp. 62-68.

——— 1891 D
The Princeton scientific expedition
of 1891.
Princeton Coll. Bull., iii, pp. 88-91.

——— 1892 A
The evolution of the premolar teeth
in the mammals.
Proc. Acad. Nat. Sci. Phila., 1892, pp. 405-444,
with 8 figs. in the text.
See for abstract, Princeton Coll. Bull., iv,
pp. 74-76; for review by M. Schlosser, Archiv f.
Anthrop., xxiii, p. 155.

——— 1892 B
A revision of the North American
Creodonta, with notes on some genera
which have been referred to that
group.
Proc. Acad. Nat. Sci. Phila., 1892, pp. 291-323.
Reviewed by M. Schlosser in Archiv f. An-
throp., xxiii, pp. 141-142.

——— 1892 C
On some of the factors in the evolu-
tion of the *Mammalia.*
Princeton Coll. Bull., iv, pp. 11-17.

Scott, W. B.—Continued.   1892 D

The genera of American *Creodonta*.
Princeton Coll. Bull., iv, pp. 76–81.

—   1893 A

On a new musteline from the John
Day Miocene.
Amer. Naturalist, xxvii, pp. 658–659.
Description of *Parictis primævus*.

—   1893 B

The mammals of the Deep River
beds.
Amer. Naturalist, xxvii, pp. 659–662.
List of species and description of new species.

—   1893 C

The Princeton scientific expedition
of 1893.
Princeton Coll. Bull., v, pp. 80–84.

—   1894 A

A new insectivore from the White
River beds.
Proc. Acad. Nat. Sci. Phila., 1894, pp. 446–448.

—   1894 B

Notes on the osteology of *Agriochœrus*
Leidy (*Artionyx* O. and W.).
Proc. Amer. Philos. Soc., xxxiii, pp. 243–251,
with 3 text figs.
Notice of this paper in Amer. Naturalist,
xxviii, p. 952, and xxix, p. 368; also in Archiv
f. Anthrop., xxiv, p. 170. Abstract in Princeton
Coll. Bull., vi, pp. 96–100, 1894.

—   1894 C

The manus of a *Hyopotamus*.
Amer. Naturalist, xxviii, pp. 164–165.

—   1894 D

Notes on the osteology of *Ancodus*
(*Hyopotamus*).
Geol. Magazine (4), i, pp. 492–493.

—   1894 E

The structure and relationships of
Ancodus.
Jour. Acad. Nat. Sci. Phila., ix, pp. 461–497,
pls. xxiii, xxiv, with text fig. showing restora-
tion of the skeleton.
Appendix, p. 536, describing the new species
A. rostratus.
Review by M. Schlosser in Archiv f. An-
throp., xxv, pp. 219–221.

—   1894 F

The osteology of *Hyænodon*.
Jour. Acad. Nat. Sci. Phila., ix, pp. 499–535,
with 10 figs. in the text.
Review by M. Schlosser in Archiv f. Anthrop., xxv, pp. 218–219.

—   1895 A

The osteology and relations of *Pro-
toceras*.
Jour. Morphology, xi, pp. 303–374, pls. xx–xxii.
Reviewed by M. Schlosser in Archiv f.
Anthrop., xxv, pp. 217–218.

Scott, W. B.—Continued.   1895 B

A restoration of *Hyænodon*.
Geol. Magazine (4), ii, pp. 441–443, pl. xii A.

—   1895 C

The *Mammalia* of the Deep River
beds.
Trans. Amer. Philos. Soc., xviii, pp. 55–185,
with pls. i–vi.
Abstract in Princeton Coll. Bull., vi, pp. 76–
78, 1894. Noticed by M. Schlosser in Archiv
f. Anthrop., xxiv, pp. 168–170.
This memoir, or reprints of it, appears to
have been issued some time in the year 1894.
The imprint " 1895 " is on the cover of the part
containing it.

—   1895 D

A new Geomyid from the Upper
Eocene.
Amer. Naturalist, xxix, pp. 923–924.
Abstract of paper in the Proc. Nat. Sci.,
Phila. Acad., 1895, describing *Protoptychus
hatcheri;* reviewed by M. Schlosser in Archiv f.
Anthrop., xxv, p. 182.

—   1895 E

Protoptychus hatcheri, a new rodent
from the Uinta Eocene.
Proc. Acad. Nat. Sci. Phila., 1895, pp. 269–286,
with 6 figs. in the text.

—   1895 F

On the *Creodonta*.
Report Brit. Assoc. Adv. Sci., 1895, lxv., pp.
719–720.

—   1896 A

On the osteology of *Elotherium*,
Pomel.
Compte-rendu des séances du troisième Congrès Internat. de Zoologie, Leyden, 1895, pp.
317–319.

—   1896 B

Die Osteologie von *Hyracodon*, Leidy.
Festschrift zum 70sten Geburtstage von Carl
Gegenbaur, ii, pp. 353–384, with pls. i–iii.

—   1897 A

Preliminary notes on the White
River *Canidæ*.
Princeton Univ. Bull., ix, pp. 1–3.

—   1897 B

The osteology of *Hyracodon*.
Princeton Univ. Bull., ix, pp. 11–13.

—   1898 A

The osteology of *Elotherium*.
Trans. Amer. Philos. Soc., xix, pp. 273–324,
with pls. xvii, xviii.

—   1898 B

Notes on the *Canidæ* of the White
River Oligocene.
Trans. Amer. Philos. Soc., xix, pp. 327–415,
with pls. xix, xx.

Scott, W. B.—Continued. 1898 C
Preliminary note on the selenodont
artiodactyls of the Uinta formation.
Proc. Amer. Philos. Soc., xxxvii, pp. 73-81.

———   · 1898 D
Memoir of Edward D. Cope.
Bull. Geol. Soc. America, ix, pp. 401-406.

———   · 1899 A
The selenodont artiodactyls of the
Uinta Eocene.
Trans. Wagner Free Inst. Sci., Phila., vi,
pp. ix-xiii; 1-121, with pls. i-iv.
Note on, by H. F. Osborn in Science (2), xi,
1900 p. 115.

Scott, W. B., and **Osborn,** H. F. 1882 A
Orthocynodon, an animal related to
the rhinoceros, from the Bridger Eocene.
Amer. Jour. Sci. (3), xxiv, pp. 223-225.
Reprinted in Ann. and Mag. Nat. Hist. (5),
x, 1882, pp. 332-334.

——— ———   1883 A
On the skull of the Eocene rhinoceros, *Orthocynodon*, and the relation of
this genus to other members of the
group.
Contrib. E. M. Mus. Geo.. and Arch.; Princeton Coll., Bull. No. 3, pp. 1-22, with pl. v, and
1 fig. in the text.

——— ———   1884 A
On the origin and development of
the rhinoceros group.
Report Brit. Assoc. Adv. Sci., 53d meeting,
Southport, 1883, p. 528.
A brief abstract.

——— ———   1887 A
Preliminary report on the vertebrate
fossils of the Uinta formation, collected
by the Princeton expedition of 1886.
Proc. Amer. Philos. Soc., xxiv, pp. 255-264.
Reviewed by M. Schlosser in Archiv f.
Anthrop., xix, p. 97.

——— ———   1887 B
Preliminary account of the fossil
mammals from the White River formation, contained in the Museum of Comparative Zoology.
Bull. Mus. Comp. Zool., xiii, pp. 151-171,
with pls, i-ii, and 9 figs. in the text.
Reviewed by M. Schlosser in Archiv f.
Anthrop., xix, pp. 91-96.

——— ———   1890 A
The *Mammalia* of the Uinta forma-
tion. Part I. The geological and fau-
nal relations of the Uinta formation.
Part II. The *Creodonta*, by W. B. Scott.
Part III. The *Perissodactyla* by H. F. Os-

Scott, W. B., and **Osborn,** H. F.—Cont'd.
born. Part IV. The evolution of the un-
gulate foot, by H. F. Osborn.
Trans. Amer. Philos. Soc., xvi, pp. 461-572;
pls. vii-xi; text figs. 1-13.
Referred to under designations Scott, W.
B., 1890 A and Osborn, H. F., 1890 D.
Author's reprints of this paper were issued
Aug. 20, 1889. The imprint of Part III of the
Transactions gives the date as 1890.
Reviewed by M. Schlosser in Archiv f. An-
throp., xx, pp. 128-134.

———   1890 B
Preliminary account of the fossil
mammals from the White River and
Loup Fork formations contained in
the Museum of Comparative Zoology.
Part II. *Carnivora* and *Artiodactyla*, by
W. B. Scott. *Perissodactyla*, by Henry
F. Osborn.
Bull. Mus. Comp. Zool., xx, pp. 65-100, with
pls. i-iii and 18 figs. in the text.
Abstract in Zoolog. Record for 1890,
Mamm., p. 15. Reviewed by M. Schlosser in
Archiv f. Anthrop., xx, pp. 114-116.
See also **Osborn, Scott,** and **Speir.**

Seebohm, Henry. 1889 A
An attempt to diagnose the suborders
of the ancient ardeino-anserine assem-
blage of birds by the aid of osteological
characters alone.
The Ibis (6), i, pp. 92-104.

Seeley, Harry G. 1865 A
On the pterodactyle as evidence of a
new subclass of *Vertebrata* (*Saurornia*).
Report Brit. Assoc. Adv. Sci., 34th meeting,
Bath, 1864, p. 69.

———   1866 A
An epitome of the evidence that
pterodactyles are not reptiles, but a
new subclass of vertebrate animals
allied to birds (*Saurornia*).
Ann. and Mag. Nat. Hist. (3), xvii, pp. 321-
331.

———   1870 A
Remarks on Professor Owen's monograph on *Dimorphodon*.
Ann. and Mag. Nat. Hist. (4), vi, pp. 129-152,
with several text figs.

———   1870 B
The *Ornithosauria*, an elementary
study of the bones of pterodactyles,
made from fossil remains found in
the Cambridge upper Greensand, and
arranged in the Woodwardian Museum
of the University of Cambridge.
Cambridge, 1870, 8 vo., pp. i-xiii; 1-135, with
pls. i-xii.
The appendix, pp. 129-132, is devoted to the
bibliography of the subject.

Seeley, Harry G.—Continued. 1871 A
Note on Professor Cope's interpretation of the ichthyosaurian head.
Ann. and Mag. Nat. Hist. (4), vii, pp. 266-268; 389.
This note refers to paper by E. D. Cope. in Amer. Naturalist, October, 1870.

——— 1871 B
Note on some chelonian remains from the London Clay.
Ann. and Mag. Nat. Hist. (4), vii, pp. 227-233, with 1 fig. in the text.
Describes the new genus *Glossochelys*.

——— 1871 C
Additional evidence of the structure of the head in ornithosaurs from the Cambridge upper Greensand; being a supplement to "The *Ornithosauria*."
Ann. and Mag. Nat. Hist. (4), vii, pp. 20-36, with pls. ii, iii.

——— 1874 A
Note on the generic modifications of the plesiosaurian pectoral arch.
Quart. Jour. Geol. Soc., xxx, pp. 436-449, with 13 text figs.

——— 1876 A
On the organization of the *Ornithosauria*.
Jour. Linn. Soc. Lond., xiii, pp. 84-107, with pl. xi.
Review of this paper by "W. D." in Geol. Magazine (2), iv, 1877, p. 124.

——— 1876 B
On the British fossil Cretaceous birds.
Quart. Jour. Geol. Soc., xxxii, pp. 496-512, with pls. xxvi-xxvii.

——— 1879 A
On the *Dinosauria*.
Proc. Geologists' Assoc., vi, pp. 175-185.

——— 1880 A
Report on the mode of reproduction of certain species of *Ichthyosaurus* from the Lias of England and Würtemburg.
Report Brit. Assoc. Adv. Sci., 50th meeting, Swansea, 1880, pp. 68-76, with pl. i.

——— 1880 B
Note on *Psephophorus polygonus*, Meyer, a new type of chelonian reptile allied to the leathery turtle.
Quart. Jour. Geol. Soc., xxxvi, pp. 406-413, with pl. xv.

——— 1887 A
On *Aristosuchus pusillus* (Owen), being further notes on the fossils described by Sir R. Owen as *Poikilopleuron pusillus*.
Quart. Jour. Geol. Soc., xliii, pp. 221-228, with pl. xii.

Seeley, Harry G.—Continued. 1887 B
Mr. Dollo's notes on the dinosaurian fauna of Bernissart.
Geol. Magazine (3), iv, pp. 80-87, 124-130.

——— 1887 C
On the mode of development of the young in *Plesiosaurus*.
Geol. Magazine (3), iv, pp. 562-565.

——— 1888 A
On the mode of development of the young in *Plesiosaurus*.
Report Brit. Assoc. Adv. Sci., 57th meeting, Manchester, 1887, pp. 697-698.

——— 1888 B
The classification of the *Dinosauria*.
Report Brit. Assoc. Adv. Sci., 57th meeting, Manchester, 1887, pp. 698-699.

——— 1888 C
On *Thecospondylus daviesi* (Seeley), with some remarks on the classification of the *Dinosauria*.
Quart. Jour. Geol. Soc., xliv, pp. 79-87, with 5 text fig.

——— 1888 D
Researches on the structure, organization, and classification of the fossil *Reptilia*. II. On *Pareiasaurus bombidens* (Owen), and the significance of its affinities to amphibians, reptiles, and mammals.
Philos. Trans. Roy. Soc. Lond., clxxix, pp. 59-155, with pls. xii-xxi and 14 figs. in the text.

——— 1888 E
Researches on the structure, organization, and classification of the fossil *Reptilia*. III. On parts of the skeleton of a mammal from the Triassic rocks of Klipfontein, Fraserberg, South Africa, (*Theriodesmus phylarchus* Seeley), illustrating the reptilian inheritance in the mammalian hand.
Philos. Trans. Roy. Soc., clxxix, pp. 141-155, with pl. xxvi and 15 figs. in the text.
Reviewed by M. Schlosser in Archiv f. Anthrop,, xix, pp. 145-146.

——— 1888 F
Researches on the structure, organization, and classification of the fossil *Reptilia*. V. On associated bones of a small Anomodont reptile, *Keirognathus cordylus* (Seeley), showing the relative dimensions of the anterior parts of the skeleton and structure of the fore limb and shoulder girdle.
Philos. Trans. Roy. Soc. Lond., clxxix, B, p. 487-501, with pls. lxxv-lxxvi.

Seeley, Harry G.—Continued.   1888 G
On the classification of the fossil animals commonly named *Dinosauria*.
Proc. Roy. Soc. Lond., xliii, pp. 165-171, with 4 figs.

———   1888 H
On the nature and limits of reptilian character in mammalian teeth.
Proc. Roy. Soc. Lond., xliv, pp. 129-141, with 8 text figs.
Reviewed by M. Schlosser in Archiv f. Anthrop., xix, p. 160.

———   1889 A
Researches on the structure, organization, and classification of the fossil *Reptilia*. VI. On the Anomodont *Reptilia* and their allies.
Philos. Trans. Roy. Soc., Lond., clxxx, pp. 215-296, with pls. 9-25.

———   1891 A
The ornithosaurian pelvis.
Ann. and Mag. Nat. Hist. (6), vii, pp. 237-255, with 16 figs. in the text.

———   1891 B
On the shoulder girdle in Cretaceous *Ornithosauria*.
Ann. and Mag. Nat. Hist. (6), vii, pp. 438-445, with 2 figs. in the text.

———   1892 A
Researches on the structure, organization, and classification of the fossil *Reptilia*. VII. Further observations on *Pareiasaurus*.
Philos. Trans. Roy. Soc., clxxxiii, pp. 311-370. with pls. xvii-xxiii and 17 figs. in the text

———   1892 B
The nature of the shoulder girdle and clavicular arch in *Sauropterygia*.
Proc. Roy. Soc. Lond., li, pp. 119-151, with 15 figs. in the text.

———   1893 A
On a reptilian tooth with two roots.
Ann. and Mag. Nat. Hist. (6), xii, pp. 227-230, with 1 text fig.
Refers to Marsh's account of the teeth of *Tricera'ops* and describes and figures the two-rooted tooth of *Nuthetes*.

———   1894 A
On *Euskelesaurus brownii* (Huxley).
Ann. and Mag. Nat. Hist. (6), xiv, pp. 317-340, with 7 figs. in the text.

———   1895 A
Researches on the structure, organization, and classification of the fossil *Reptilia*. Part IX, section 1. On the *Therosuchia*.
Philos. Trans. Roy. Soc. Lond., clxxxv, pt. B. pp. 987-1018, with pl. lxxxviii.

Seeley, Harry G.—Continued.   1895 B
Researches on the structure, organization, and classification of the fossil *Reptilia*. Part IX, section 5. On the skeleton in new *Cynodontia*, from the Karroo rocks.
Philos. Trans. Roy. Soc. Lond., clxxxvi, pp. 59-148, with 83 figs. in text.

———   1895 C
On *Thecodontosaurus* and *Palæosaurus*.
Ann. and Mag. Nat. Hist. (6), xv, pp. 144-163, with 10 figs. in the text.

———   1896 A
On a pyritous concretion from the Lias of Whitby, which appears to show the external form of the body of embryos of a species of *Plesiosaurus*.
Annual Report Yorkshire Philos. Soc. for 1895.

———   1898 A
Origin of *Mammalia*.
Science (2), viii, pp. 357-358.
Abstract of remarks made at International Congress of Zoologists at Cambridge, England, 4th meeting; proceedings, pp. 68-71.

———   1900 A
On an Anomodont reptile, *Aristodesmus rütimeyeri* (Wiedersheim), from the Bunter sandstone, near Basel.
Quart. Jour. Geol. Soc., lvi, pp. 620-645, with 8 figs. in the text.

Segond, L. A.   1864 A
Comparaison morphologique des vertèbres du bassin et du sternum chez les oiseaux.
Jour. de l'Anat., Physiol., etc., i, pp. 602-623.

———   1865 A
Comparaison morphologique des vertèbres du bassin et du sternum chez les oiseaux.
Jour. de l'Anat., Physiol., etc., ii, pp. 36-56.

———   •1873 A
Reptiles et batraciens classés d'après leurs affinités par rapport à cinq types dont les caractères sont empruntés aux parties les moins modifiables du squelette.
Jour. de l'Anat., Physiol., etc., ix, pp. 1-29.

———   1873 B
Des affinités squeletiques des poissons.
Jour. de l'Anat., Physiol., etc., ix, pp. 511-534; 607-627.

Selenka, Emil.   1869 A
Aves. Anatomischer Bau.
Bronn's Klassen und Ordnungen des Thier. Reichs, vi, pp. 1-90, pls.
This article deals with the osteology of birds.

Selwyn, Alfred R. C.   1872 A
On the discovery of reptilian footprints in Nova Scotia.
Geol. Magazine (1), ix, pp. 250-251.

Semon, Richard.   1899 A
Ueber die Entwickelung der Zahngebilde der Dipnoer.
Sitzungsber. Gesellsch. Morphol. u. Physiol. München, xv, pp. 75-85, with 4 text figs.

Serres, M.   1865 A
Deuxième note sur le squelette du *Glyptodon clavipes*.
Comptes rend. Acad. Sci. Paris, lxi, pp. 457-466.

———   1865 B.
Note sur le *Glyptodon ornatus*. De sa carapace et de ses rapports normaux avec le squelette; caractères différentiels des os du bassin avec ceux du *Glyptodon clavipes*.
Comptes rend. Acad. Sci. Paris, lxi, pp. 537-544; 665-670.
Translation in Ann. and Mag. Nat. Hist. (3), xvi, pp. 432-443.

Shaler, N. S.   1871 A
The time of the mammoths.
Amer. Naturalist, iv, pp. 148-166.

———   1871 B
Note on the occurrence of the remains of *Tarandus rangifer*, Gray, at Big Bone Lick, Kentucky.
Proc. Boston Soc. Nat. Hist., xiii, p. 167.

———   1876 A
On the age of the bison in the Ohio Valley.
Mem. Kentucky Geol. Surv., i, pp. 232-236, with 1 diagrammatic section.
This paper forms Appendix II to J. A. Allen's memoir, The American Bisons, and was reprinted in Mem. Mus. Comp. Zool., iv, pp. 232-236.

Shepard, Charles Upham.   1867 A
On the supposed tadpole nests, or imprints made by *Batrachoides nidificans* (Hitchcock), in the red shale of the New Red sandstone of South Hadley, Mass.
Amer. Jour. Sci. (2), xliii, pp. 99-104.

Sherborn, C. D.
See **Woodward,** A. S., and **Sherborn,** C. D.

Shufeldt, R. W.   1881 A
On the ossicle of the antibrachium as found in some North American *Falconidæ*.
Bull. Nuttall Ornithological Club, vi pp. 197-203.

Shufeldt, R. W.—Continued.   1881 B
Osteology of the North American *Tetraonidæ*.
Bull. U. S. Geol. and Geog. Surv., F. V. Hayden, vi, pp. 309-350, with pls. vii-xiv.
Also, with little change in Ann. Rep. U. S. Geol. and Geog. Surv. for 1878 (1883), pp. 653-700, with pls. v-xiii, and 16 figs. in the text.

———   1881 C
Osteology of *Lanius ludovicianus excubiterides*.
Bull. U. S. Geol. and Geog. Surv., F. V. Hayden, vi, pp. 351-359, with pl. xiv.
Also, in Ann. Rep. U. S. Geol. and Geog. Surv. for 1878 (1883), pp. 719-725, with pl.xiv.

———   1881 D
Osteology of *Speotyto cunicularia*, var. *hypogæa*.
Bull. U. S. Geol. and Geog. Surv., F. V. Hayden, vi, pp. 87-117, with pls. i-iii.
Also, with little change, in Annual Report U. S. Geol. and Geog. Surv. for 1878 (1883), pt. i, pp. 598-620, with pls. i-iii.

———   1881 E
Osteology of *Eremophila alpestris*.
Bull. U. S. Geol. and Geog. Surv., F. V. Hayden, vi, pp. 119-147, with pl. iii-iv.
Also in Annual Report U. S. Geol. and Geog. Surv. for 1878 (1883), pp. 627-650, with pl. iv.

———   1883 A
Observations upon the osteology of *Podasocys montanus*.
Jour. Anat. and Physiol., xviii, pp. 86-102, with pl. v.

———   1883 B
Remarks on the osteology of *Phalacrocorax bicristatus*.
Science, ii, pp. 640-642, with 3 figs.
For discussion of this paper see Jeffries, J. A., Science, ii, p. 739, and iii, p. 59.

———   1883 C
Osteology of the *Cathartidæ*.
Twelfth Annual Report U. S. Geol. and Geog. Surv., F. V. Hayden, 1878 (1883), Pt. i, pp. 728-786, with pls. xv-xxiv, and many figs. in the text.

———   1884 A
Osteology of *Numenius longirostris*, with notes upon the skeleton of other American *Limicolæ*.
Jour. Anat. and Physiol., xix, pp. 51-82, with pls. iv and v.

———   1885 A
Contribution to the comparative osteology of the *Trochilidæ, Caprimulgidæ*, and *Cypselidæ*.
Proc. Zool. Soc. Lond., 1885, pp. 886-915, with pls. lviii-lxi and 6 woodcuts in the text.

Shufeldt, R. W.—Continued. 1885 B

The osteology of *Amia calva*, including certain special references to the skeleton of Teleosteans.

Report U. S. Commissioner of Fish and Fisheries, 1883, pp. 747–978, pls. i–xiv.

Part I consists of a translation of Sagemehl's paper on *Amia*. Part II describes osteology of *Amia*, and portions of that of *Micropterus* and *Albula*.

—— 1886 A

Additional notes upon the anatomy of the *Trochilidæ*, *Caprimulgidæ*, and *Cypselidæ*.

Proc. Zool. Soc. Lond., 1886, pp. 501–508, with 6 figs. in the text.

—— 1886 B

Contributions to the anatomy of *Geococcyx californianus*.

Proc. Zool. Soc. Lond., 1886, pp. 466–491, with pls. xliii–xlv.

—— 1886 C

Osteology of *Conurus carolinensis*.

Jour. Anat. and Physiol., xx, pp. 407–425, with pls. x and xi.

—— 1888 A

Contributions to the comparative osteology of arctic and subarctic water birds.

Jour. Anat. and Physiol., xxiii, pp. 1–39, pls. i–v.

—— 1888 B

Osteology of *Porzana carolina* (the Carolina rail).

Jour. Compar. Med. and Surg., ix, pp. 231–248, with 7 figs. in the text.

—— 1888 C

Observations upon the morphology of *Gallus bankiva* of India.

Jour. Compar. Med. and Surg., ix, pp. 343–376, with 30 figs. in the text.

—— 1888 D

On the skeleton in the genus *Sturnella*, with osteological notes upon other North American *Icteridæ* and the *Corvidæ*.

Jour. Anat. and Physiol., xxii, pp. 309–350, with pls. xiv, xv.

—— 1888 E

Observations upon the osteology of the order *Tubinares* and *Steganopodes*.

Proc. U. S. Nat. Mus., xi, pp. 253–315, with 43 figs. in the text.

—— 1889 A

Observations upon the osteology of the North American *Anseres*.

Proc. U. S. Nat. Mus., xi, pp. 215–251, with 20 figs. in the text.

Shufeldt, R. W.—Continued. 1889 B

On the comparative osteology of arctic and subarctic water birds.

Jour. Anat. and Physiol., xxiii, pp. 165–186, pls. vii–xi; 400–427, text figs. 17; 537–558, text figs. 1–8.

This paper is a continuation of Shufeldt, 1888 A.

—— 1889 C

Contributions to the comparative osteology of arctic and subarctic water birds.

Jour. Anat. and Physiol., xxiv, pp. 89–117, pls. vi–viii, and 13 figs. in the text.

A continuation of Shufeldt, 1889 B. Treats of *Fraterculinæ*, *Fratercola*, *Lunda*, *Uria*, and the *Alcidæ*.

—— 1890 A

Contributions to the comparative osteology of arctic and subarctic water birds.

Jour. Anat. and Physiol., xxiv, pp. 169–187, pls. xi, xii, and 2 text figs.; 543–566, with 17 text figs.

—— 1890 B

Contributions to the comparative osteology of arctic and subarctic water birds.

Jour. Anat. and Physiol., xxv, pp. 60–77.

A continuation of Shufeldt, 1890 A.

—— 1890 C

On the affinities of *Hesperornis*.

Nature, xliii, p. 176.

—— 1891 A

On a collection of fossil birds from the Equus beds of Oregon.

Amer. Naturalist, xxv, pp. 359–362.

A general account of Silver Lake, Oregon, and of fossil remains there found.

—— 1891 B

Fossil birds from the Equus beds of Oregon.

Amer. Naturalist, xxv, pp. 818–821.

See also The Auk, 1891, viii, pp. 365–368.

A list of 51 species, with remarks on some of the most interesting.

—— 1891 C

Contributions to the comparative osteology of arctic and subarctic water birds.

Jour. Anat. and Physiol., xxv, pp. 509–525, pls. xi and xii, with 4 figs. in the text.

Continuation and conclusion of Shufeldt, 1890 B.

—— 1892 A

A study of the fossil avifauna of the Silver Lake region, Oregon.

Proc. Amer. Assoc. Adv. Sci. 40th meeting Washington, 1891, p. 286. Abstract.

Shufeldt, R. W.—Continued.   1892 B
Creccoides osbornii.
In Cope's report on palæontology of Texas, 3d Annual Report Geol. Surv. Texas, for 1891. Austin, 1892, pp. 253–255.

———   1892 C
Concerning the taxonomy of the North American *Pygopodes,* based upon their osteology.
Jour. Anat. and Physiol., xxvi, pp. 199–203.

———   1892 D
A study of the avifauna of the Equus beds of the Oregon Desert.
Jour. Acad. Nat. Sci., Phila., ix, pp. 389–425, pls. xv–xvii.

———   1893 A
Comparative osteological notes on the extinct bird *Ichthyornis.*
Jour. Anat. and Physiol., xxvii, pp. 336–342.

———   1894 A
Osteology of certain cranes, rails, and their allies, with remarks upon their affinities.
Jour. Anat. and Physiol., xxix, pp. 21–34, with 3 figs.

———   1894 B
On cases of complete fibulæ in existing birds.
The Ibis (6), vi, pp. 361–366, with 2 figs. in the text.

———   1894 C
Notes on the *Steganopodes* and on fossil birds' eggs.
The Auk, xi, pp. 337–339.

———   1894 D
On the affinities of the *Steganopodes.*
Proc. Zool. Soc. Lond., 1894, pp. 160–162.

———   1897 A
On fossil bird bones obtained by expeditions of the University of Pennsylvania from the bone caves of Tennessee.
Amer. Naturalist, xxxi, pp. 645–650.
All are of existing species and yet living in that region.

———   1897 B
On the feathers of " *Hesperornis.*"
Nature, lvi, p. 30.

———   1899 A
Observations on the classification of birds.
Ann. and Mag. Nat. Hist. (7), iv, pp. 101–111, with 2 figs. in the text.

———   1900 A
The osteology of *Vulpes macrotis.*
Jour. Acad. Nat. Sci. Phila. (2), xi, pp. 395–418, with pls. xxii, xxiii.

Shurtleff, N. B.   1846 A
Remarks on the skeleton of *Mastodon giganteus* found in Newburg, Orange County, N. Y.
Proc. Boston Soc. Nat. Hist., ii, pp. 96–98.

Sickler, F. K. L.   ·   1835 A
Fährten unbekannten Thiere im Sandsteine bei Hildburghausen.
Neues Jahrb. Mineral., etc., 1835, pp. 230–232.
This is followed by remarks on the fossil tracks by H. G. Bronn.

Siebenrock, Friedrich.   1892 A
Zur Kenntniss des Kopfskelettes der Scincoiden, Anguiden, und Gerrhosauriden.
Annalen k. k. naturh. Hofmuseums Wien, vii, pp. 163–196, with pls. xi, xii.

———   1893 A
Das Skelet von *Uroplates fimbriatus* Schneid.
Annalen k. k. naturh. Hofmuseums Wien, viii, pp. 517–536, with pl. xiv and 2 figs. in the text.

———   1893 B
Das Skelet von *Brookesia superciliaris* Kuhl.
Sitzungsber. k. p. Akad. Wissensch. Wien, math.-naturw. Cl., cii, pp. 1–48, with pls. i–iv.

———   1894 A
Das Skelet der *Lacerta simonyi,* und der Lacertidenfamilie überhaupt.
Sitzungsber. k. p. Akad. Wissensch. Wien, math. naturw. Cl., ciii, pp. 205–292, pls. i–iv.

———   1894 B
A contribution to the osteology of the head of *Hatteria.*
Ann. and Mag. Nat. Hist. (6), xiii, pp. 297–311, with pl. xiv.
Translated from Sitzungsber. k. p. Akad. Wissensch. Wien, math. naturw. Cl., cii, Abth. i, 1893, pp. 250–268, pl. i.

———   1895 A
Zu Kenntniss des Rumpfskeletes der Scincoiden, Anguiden und Gerrhosauriden.
Annalen k. k. naturh. Hofmuseums Wien, x, pp. 18–41, with pl. iii and 4 figs. in the text.
Appended is a list of 59 books and papers bearing on the osteology of the *Lacertilia.*

———   1895 B
Das Skelet der *Agamidæ.*
Sitzungsber. k. p. Akad. Wissensch. Wien, math.-naturw. Cl., Abth. i, civ, pp. 1–108, with pls. i–vi.
This work includes a list of 67 papers on the subject treated.

Siebenrock, Friedrich—Cont'd. 1897 A
Das Kopfskelet der Schildkröten.
Sitzungsber. k. p. Akad. Wissensch. Wien.,
math.-naturw. Cl., cvi, pp. 1-84, with pls. i-vi.
This paper includes a list of 61 works on the
subject.

——— 1898 A
Ueber den Bau und die Entwicklung
des Zungenbein-Apparates der Schildkröten.
Annalen k. k. naturh. Hofmuseums Wien,
xiii, pp.424-436, with pls. xvii, xviii and 2 figs.
in the text.
Appended is a list of 22 works bearing on
the subject.

Silliman, Benj., Sr. 1835 A
Vertebral bone of a mastodon.
Amer. Jour. Sci., xxvii, p. 165.

——— 1837 A
Bird tracks at Middletown, Conn.,
in the New Red sandstone.
Amer. Jour. Sci., xxxi, p. 165.
This is an extract from a letter to the editor
from an unnamed correspondent. No species
or genera are named.

Silliman, Benj., Jr. 1868 A
On the existence of the mastodon in
the deep-lying gold placers of California.
Amer. Jour. Sci. (2), xlv, pp. 378-381.

Sirodot, S. 1876 A
Les éléphants du mont Dol; essai
d'organogénie du système des dents
mâchelières du mammouth.
Comptes rend. Acad. Sci., Paris, lxxxii, pp.
734-736, 822-824, 902-905.

——— 1876 B
Les éléphants du mont Dol. Dentition du mammouth, distinction des
molaires inférieure, et supérieurs droits
et gauches.
Comptes rend. Acad. Sci., Paris, lxxxii, pp.
1065-1068.

Sismonda, Eugenio. 1849 A
Descrizione dei pesci e dei crostacei
fossili nel Piemonte.
Mem. della Reale Accad. delle Scienze di
Torino (2), x, pp. 1-88, with pls. i-iii.

Sixta, V. 1900 A
Vergleichend - osteologische Unter-
suchungen über den Bau des Schädels
von Monotremen und Reptilien.
Zoolog. Anzeiger, xxiii, pp. 213-229.
See also J. F. Van Bemmelen in Zoolog.
Anzeiger, xxiii, p. 449.

Skilton, Dr. 1858 A
Communication upon *Equus major*
found near Troy, N. Y.
Proc. Boston Soc. Nat. Hist., vi, pp. 303-304.

Slade, D. D. 1888 A
On certain vacuities or deficiencies in
the crania of mammals.
Bull. Mus. Comp. Zool., xiii, pp. 241-246, with
pls. i and ii.

——— 1890 A
Osteological notes.
Science, xvi, pp. 332-333.
Treats of jugal arch of *Mammalia.*

——— 1891 A
Osteological notes.
Science, xvii, pp. 317-318.
Treats of the jugal arch of the *Primates.*

——— 1891 B
Osteological notes.
Science, xviii, pp. 53-54.
Treats of jugal arch of *Primates.*

——— 1892 A
Osteological notes.
Science, xix, pp. 203-204.
Jugal arch of *Insectivora.*

——— 1892 B
Osteological notes.
Science, xx, p. 46.
Discusses jugal arch of the *Rodentia.*

——— 1893 A
Osteological notes.
Science, xxi, pp. 78-79.
Discusses jugal arch of the *Ungulata.*

Smith, Horace P. 1887 A
Bison latifrons. Leidy.
Jour. Cincinnati Soc. Nat. Hist., x, pp. 19-24,
pl. i.

Smith, J. 1895 A
Notes on a peculiarity in the form of
the mammalian tooth.
Proc. Roy. Soc. Edinburgh, xx, pp. 336-346,
with 1 page of 13 figs.

Smith, J. Augustine. 1846 A
Central cavity of the mastodon.
Amer. Quart. Jour. Agricult. and Sci., iii, pp.
19-22.
The word "central" in the caption of this
paper is probably a misprint for cerebral.

Smith, J. Lawrence. 1844 A
Communication on some fossil bones.
Amer. Jour. Sci. (1), xlvii, pp. 116-117.
On some bones from vicinity of Charleston,
S. C. Species undetermined.

Snow, F. H. 1878 A
On the dermal covering of a mosasauroid reptile.
Trans. Kansas Acad. Sci., vi, pp. 54-58, with
1 pl. and 1 text-fig.

Snow, F. H.—Continued.   1887 A
On the discovery of a fossil bird
track in the Dakota sandstone.
Trans. Kansas Acad. Sci., x, pp. 3–6, with 1
fig. in the text.
Regarded as possibly tracks of *Ichthyornis.*

Sollas, W. J.   1881 A
On a new species of *Plesiosaurus* (*P.
conybeari*) from the lower Lias of Char-
mouth, with observations on *P. megacephalus* Stutchbury and *P. brachy-
cephalus* Owen.
Quart. Jour. Geol. Soc., xxxvii, pp. 440–
481, with pls. xxiii and xxiv and 14 woodcuts.

Speir, Francis.
See **Osborn, Scott, and Speir; Osborn and Speir.**

Stainier, X.   1894 A
Un *Spiraxis* nouveau du Dévonien
belge.
Bull. Soc. Belge Géol., Paléont., Hydrol.,
viii, Mém., pp. 23–28, with 1 text fig.

Stannius, H.   1842 A
Ueber das Gebiss des Lama.
Archiv Anat. Physiol. u. wissensch. Med.,
Jahrg. 1842, pp. 388–389.

———   1842 B
Ueber Gebiss und Schädel des Walrosses unter Berücksichtigung der
Frage, ob die Verschiedenheiten im
Baue des Schädels zur Unterscheidung
mehrerer Arten der Gattung *Trichechus*
berechtigen.
Archiv Anat. Physiol. u. wissensch. Med.,
1842, pp. 390–413.

———   1846 A
Bemerkungen über das Verhältniss
der Ganoïden zu den Clupeïden, insbesondere zu *Butirinus.*
Rostock, 1846, 8vo, pp. 1–20.

———   1854 A
Handbuch der Zootomie von Sie-
bold und Stannius. Zweiter Theil.
Die Wirbelthiere von Hermann Stan-
nius. Erstes Buch. Die Fische.
Zweite Auflage, Berlin, 1854. Pp. 1–279.

———   1856 A
Handbuch der Zootomie von Siebold
und Stannius. Zweiter Theil. Wirbel-
thiere von Hermann Stannius. Zweites
Buch. Zootomie der Amphibien.
Zweite Auflage, Berlin, 1856. Pp. 1–271.

Starks, Edwin Chapin.   1899 A
The osteological characters of the
fishes of the suborder *Percesoces.*
Proc. U. S. Nat. Mus., xxii, pp. 1–10, with pls.
i–iii.
The osteology of *Atherinopsis californiensis,
Mugil cephalus,* and *Sphyræna argentea* is de-
scribed and illustrated.

Stecker, Anton.   1877 A
Zur Kenntniss des Carpus und Tarsus bei *Chamæleon.*
Sitzungsber. k. p. Akad. Wissensch. Wien.
mat.-naturw. Cl., lxxv, pt. i, pp. 7–16, with pls.
i, ii.

Steenstrup, Jap.   1855 A
Et Bidrag til Geirfuglens, *Alca impennis* Lin., Naturhistorie, og særligt til
Kundshaben om dens tidligere Udbred-
ningskreds.
Videnskb. Meddel. naturh. Forening Kjöbenhavn, 1856, pp. 33–118, with 1 map and
1 pl.

Stehlin, H. G.   1893 A
Zur Kenntniss der postembryonalen
Schädelmetamorphosen bei Wiederkauern. Inauguraldissertation, Basel.
4to, Basel, 1893, pp. 1–81, with pls. i–iv.

———   1899 A
Ueber die Geschichte des Suiden-
Gebisses. Erster Theil.
Abhandl. schweizer. paläontolog. Gesell-
sch., xxvi, pp. i–vii; 1–336.

**Steinmann, Gustav, und Döderlein,
Ludwig.**   1890 A
Elemente der Paläontologie.
Pp. i–xix; 1–848 with 1,030 figs. in the text.
Leipzig, Wilhelm Engelmann.
Pp. 516–827 treat on the *Vertebrata.*

Steitz, Auguste.   1870 A
[Exhibition of molar tooth of *Elephas primigenius* from Helena, Montana.]
Trans. Acad. Sci. St. Louis, iii, p. xxxii.

Stejneger, Leonhard.   1883 A
Contributions to the natural history
of the Commander Islands. No. 1.
Notes on the natural history, includ-
ing descriptions of new cetaceans.
Proc. U. S. Nat. Mus., vi, pp. 58–89.
Contains an account of Steller's sea-cow
(*Rytina gigas*) and measurements of skulls es-
tablishing the existence of differences in the
skulls of the two sexes.

———   1884 A
Contributions to the natural history
of the Commander Islands. No. 2.

Stejneger, Leonhard—Continued.

Investigations relating to the date of the extermination of Steller's sea-cow.

Proc. U. S. Nat. Mus., vii, pp. 181–189.

——— 1885 A

Natural history of birds.

The Standard Natural History, edited by John Sterling Kingsley, vol. iv, pp. 1–195, with figs. 1–93; 368–441, with figs. 171–221; 458–547, with figs. 228–273.

——— 1886 A

On the extermination of the great northern sea-cow.

Jour. Amer. Geog. Soc., New York, xviii, pp. 317–328.

——— . 1887 A

How the great northern sea-cow (*Rytina*) became exterminated.

Amer. Naturalist, xxi, pp. 1047–1054.

——— 1893 A

Skeletons of Steller's sea-cow preserved in the various museums.

Science, xxi, p. 81.

Stejneger, L., and **Lucas**, F. A. 1890 A

Contributions to the natural history of the Commander Islands. A.—Contributions to the history of Pallas's cormorant.

Proc. U. S. Nat. Mus., xii, pp. 83–94, with pls. ii–iv.

Steller, Georg Wilhelm. 1751 A

De bestiis marinis.

Novi Commentarii Acad. Scient. Imp. Petropolitanæ, ii, pp. 289–398.

The earliest account of the arctic sea-cow, containing descriptions of its anatomy, habits, etc. The portion of this work devoted to the sea-cow includes pp. 294–300, with pl. xiv.

Stephan, Pierre. 1900 A

Recherches histologiques sur la structure du tissue osseux des poissons.

Bull. Scient. France et Belgique, xxxiii, pp. 281–429, with pls. i–iii.

With a bibliographical index of 140 entries.

Sternberg, Charles. 1899 A

The first great roof.

Popular Science, xxxiii, pp. 126–127, with 1 text fig.

Describes the carapace of *Protostega gigas*.

Stevenson, W. D. 1884 A

The "man-eater shark," *Carcharodon carcharias*.

Amer. Naturalist, xviii, pp. 940–941, with 1 pl. and 1 woodcut.

Stewart, Alban. 1897 A

Restoration of *Oreodon culbertsonii*, Leidy.

Kansas Univ. Quarterly, vi, pp. 13–14, with pl. i.

Stewart, Alban—Continued. 1897 B

Notes on the osteology of *Bison antiquus*.

Kansas Univ. Quarterly, vi, pp. 127–135, with pl. xvii and 1 fig. in the text.

The species here described as *B. antiquus* is now assigned to *B. occidentalis* Lucas.

——— 1898 A

A contribution to the knowledge of the ichthyic fauna of the Kansas Cretaceous.

Kansas Univ. Quarterly, vii, pp. 21–29, with pls. i and ii.

——— 1898 B

Individual variations in the genus *Xiphactinus* Leidy.

Kansas Univ. Quarterly, vii, A, pp. 115–119, with pls. vii–x.

——— 1898 C

Some notes on the genus *Saurodon* and allied species.

Kansas Univ. Quarterly, vii, A, pp. 177–186, with pls. xiv–xvi.

——— 1898 D

A preliminary description of seven new species of fish from the Cretaceous of Kansas.

Kansas Univ. Quarterly, vii, A, pp. 191–196, with pl. xvii and text figs. 1–3.

——— 1899 A

A preliminary description of the opercular and other cranial bones of *Xiphactinus* Leidy.

Kansas Univ. Quarterly, viii, A, pp. 19–21.

——— 1899 B

Pachyrhizodus minimus, a new species of fish from the Cretaceous of Kansas.

Kansas Univ. Quarterly, viii, A, pp. 37–38, with 1 fig. in the text.

——— 1899 C

Notice of three new Cretaceous fishes, with remarks on the *Saurodontidæ* Cope.

Kansas Univ. Quarterly, viii, A, pp. 107–112.

——— 1899 D

Notes on the osteology of *Anogmius polymicrodus* Stewart.

Kansas Univ. Quarterly, viii, A, pp. 117–121, with pl. xxxi.

——— 1899 E

Leptichthys, a new genus of fishes from the Cretaceous of Kansas.

Amer. Geologist, xxiv, pp. 78–79.

——— 1900 A

Teleosts of the Upper Cretaceous.

Univ. Geol. Surv., Kansas, vi, pp. 257–403, with pls. xxxiii–lxxiii, and 6 figs. in the text.

Stewart, Thos. P.   1828 A
Mammoth.
Amer. Jour. Sci. (1), xiv, pp. 188–189.

Stickler, L.   1899 A
Ueber den microscopischen Bau der
Faltenzähne von *Eryops megacephalus*
Cope.
Palæontographica, xlvi, pp. 85–94, with pls.
xi, xii.

Stirrup, Mark.   1893 A
The true horizon of the mammoth.
Geol. Magazine (3), x, pp. 107–111.

———   1894 A
The true horizon of the mammoth.
Geol. Magazine (4), i, pp. 80–82.

Stock, Thos.   1881 A
On some British specimens of the
"Kammplatten" or "Kammleisten"
of Professor Fritsch.
Ann. and Mag. Nat. Hist. (5), viii, pp. 90–95,
with pl. vi.

———   1881 B
On the discovery of a nearly entire
Rhizodus in the Wardie shales.
Geol. Magazine (2), viii, pp. 77–78.

———   1882 A
Further observations on Kammplatten, and note on *Ctenoptychius pectinatus*.
Ann. and Mag. Nat. Hist. (5), ix, pp. 253–257,
with pl. viii.

———   1882 B
Notice of some discoveries recently
made in Carboniferous vertebrate palæontology.
Nature, xxvii, p. 22.
Records finding of "*Diplodus*" teeth with
spines quite different from those of *Pleuracanthus*. Also *Tristychius* associated with *Hybodus*.

———   1883 A
On the structure and affinities of the
genus *Tristychius* Agass.
Ann. and Mag. Nat. Hist. (5), xii, pp. 177–
190, with pl. vii.

Strauch, Alexander.   1890 A
Bemerkungen über die Schildkröten-
sammlung im Zoologischen Museum
der K. Akademie der Wissenschaften
zu St. Petersburg.
Mém. Acad. Imp. Sci. St. Pétersbourg (7),
xxxviii, No. 2, pp. 1–43.

Struckmann, C.   1880 A
Ueber die Verbreitung des Renthiers
in der Gegenwart und in älterer Zeit
nach Maasgabe seiner fossilen Reste

Struckmann, C.—Continued.
unter besonderer Berücksichtigung der
deutschen Fundorte.
Zeitschr. deutsch. geol. Gesellsch., xxxii,
pp. 728–773.

Strüver, Johannes.   1864 A
Die fossilen Fische aus dem Keuper-
sandstein von Coburg.
Zeitschr. deutsch. geol. Gesellsch., xvi, pp.
303–330, with pl. xiii.
Contains discussion of *Semionotus* and *Dictyopyge*.

Struthers, John.   1871 A
On some points in the anatomy of a
great fin-whale (*Balænoptera musculus*).
Jour. Anat. and Physiol., i, pp. 107–125.
Discusses the rudimentary hind limbs, the
pelvic bones, ribs, sternum, and cervical vertebræ.

———   1872 A
On the cervical vertebræ and their
articulations in fin-whales.
Jour. Anat. and Physiol., vii, pp. 1–55, pls.
i, ii.

———   1881 A
On the bones, articulations, and
muscles of the rudimentary hind limbs
of the Greenland right-whale (*Balæna
mysticetus*).
Jour. Anat. and Physiol., xv, pp. 141–176,
301–321, pls. xiv–xvii.

———   1885 A
On the rudimentary hind limb of
Megaptera longimana.
Amer. Naturalist, xix, pp. 124–125.
Abstract of paper read at Montreal before
Brit. Assoc. Adv. Sci., 1884.

———   1885 B
On the development of the foot of
the horse.
Nature, xxxii, pp. 560–561.
A short note.

———   1888 A
On some points in the anatomy of a
Megaptera longimana.
Jour. Anat. and Physiol., xxii, pp. 240–282,
441–460, 629–654, pls. x–xii; xxiii, pp. 124–163,
pl. vi; 306–335, 356–373.
The last two installments of this paper
belong to the year 1889. The first installment,
not above cited, is found in vol. xxiii, pp. 109–
124, and deals with the soft portions of the
anatomy.

———   1893 A
On the rudimentary hind limb of a
great fin-whale (*Balænoptera musculus*)
in comparison with those of the hump-

Struthers, John—Continued.

back whale and the Greenland right-whale.

Jour. Anat. and Phys., vol. xxvii, with pls. xvii–xx.

———   1893 B

On the development of the bones of the foot of the horse, and of digital bones generally; and on a case of polydactyly in the horse.

Jour. Anat. and Physiol., xxvii, pp. 51–62, with pl. 1.

———   1895 A

Carpus of the Greenland right-whale (*Balæna mysticetus*) and of fin-whales.

Jour. Anat. and Physiol., xxix, pp. 145–187, with pls. ii–iv.

Stutchbury, S.

See **Riley** and **Stutchbury**.

Swallow, G. C.   1866 A

Notice of remains of the horse in the altered drift of Kansas.

Trans. Acad. Sci. St. Louis, ii, p. 418.

The species not distinguishable by the writer from the common horse.

Swinnerton, H. H.

See **Howes** and **Swinnerton**.

T. R. D.   1839 A

Fossil shells and bones (*Megatherium*) near Savannah, Ga.

Amer. Jour. Sci., i, 1839, p. 380.

T. R. J.   1889 A

Human relics and bones of mastodon found in association.

Geol. Magazine (3), vi, p. 192.

Said to have been found in Attica, Wyoming County, N. Y.

Taeker, Julius.   1892 A

Zur Kenntnis der Odontogenese bei Ungulaten.

Pp. 27, with 4 pls. Dorpat, 1892.

Teller, Friedrich.   1884 A

Neue Anthracotherienreste aus Süd-steiermark und Dalmatien.

Beiträge Palæontologie Osterreich-Ungarns und des Orients (Mojsisovics und Neumayer), iv, pp. 45–134, with pls. xi–xiv.

On pages 47, 48 is a list of papers which treat of the genus *Anthracotherium*.

———   1891 A

Ueber den Schädel eines fossilen Dipnoërs, *Ceratodus sturii*, nov. spec., aus den Schichten der oberen Trias der Nordalpen.

Abhandl. k. k. geol. Reichsanstalt Wien., xv, Heft 3, pp. 1–39, with pls. i–iv.

On pp 4–5 is found a long list of works on the genus *Ceratodus*.

Bull. 179——15

Thacher, James K.   1877 A

Median and paired fins, a contribution to the history of vertebrate limbs.

Trans. Connecticut Acad. Arts and Sci., iii, pp. 281–310, pls. xlix–lx.

———   1887 B

Ventral fins of ganoids.

Trans. Connecticut Acad. Arts and Sci., iv, pp. 233–242, pls. i, ii.

Thévenin, Armand.   1896 A

Mosasauriens de la craie grise de Vaux-Éclusier près Péronne (Somme).

Bull. Soc. Géol. France (3), xvii, pp. 900–911, with pls. xxix, xxx, and 8 figs. in the text.

———   1896 B

Nouveaux mosasauriens trouvés en France.

Comptes rend. Acad. Sci. Paris, cxxiii, pp. 1319–1320.

Thilenius, G.   1894 A

Ueber Sesambeine fossiler Säugetiere.

Anatom. Anzeiger, x, pp. 42–48, with 2 figs.

Thomas, Oldfield.   1888 A

On the homologies and succession of the teeth in the *Dasyuridæ*, with an attempt to trace the history of the evolution of mammalian teeth in general.

Philos. Trans. Roy. Soc. Lond., clxxviii, pp. 443–462; pls. 27, 28.

Reviewed by M. Schlosser in Archiv f. Anthrop., xix, pp. 115–116.

———   1888 B

On a new and interesting annectant genus of *Muridæ*, with remarks on the relations of the Old and New-World members of the family.

Proc. Zool. Soc. Lond., 1888, pp. 130–135, with pl. v.

———   1890 A

Remarks on Dr. Schosser's "Ueber die Deutung des Milchgebisses der Säugetiere."

Biolog. Centralbl., x, pp. 216–219.

———   1892 A

On the species of *Hyracoidea*.

Proc. Zool. Soc. Lond., 1892, pp. 50–76, with pl. iii.

———   1892 B

Notes on Dr. W. Kükenthal's discoveries in mammalian dentition.

Ann. and Mag. Nat. Hist. (6), ix, pp. 308–313.

———   1896 A

On the genera of rodents; an attempt to bring up to date the current arrangement of the order.

Proc. Zool. Soc. Lond., 1896, pp. 1012–1028.

See for remarks on this paper Dr. T. S Palmer in Science, vi, p. 103.

Thomas, O., and Lydekker, R. 1897 A
On the number of grinding teeth possessed by the manatee.
Proc. Zool. Soc. Lond., 1897, pp. 595-600, with pl. xxxvi.

Thompson, D'Arcy W. 1886 A
On the hind limb of *Ichthyosaurus*, and on the morphology of vertebrate limbs.
Jour. Anat. and Physiol., xx, pp. 532-585, with 2 text figs.

———   1890 A
On the systematic position of *Zeuglodon*.
Studies from the Museum of Zoology, University Coll., Dundee, No. 9, pp. 1-8, with 10 text figs.

———   1890 B
On the systematic position of *Hesperornis*.
Studies from the Museum of Zoology, University Coll., Dundee, No. 10, pp. 1-15, with 17 text figs.

Thompson, James. 1868 A
Note on the spines of *Gyracanthus*.
Trans. Geol. Soc. Glasgow, iii, pp. 130-133.

———   1870 A
On teeth and dermal structure associated with *Ctenacanthus*.
Report Brit. Assoc. Adv. Sci., 39th meeting, Exeter, 1869, pp. 102-103.
Regards *Ctenacanthus*, *Cladodus*, and *Diplodus* as representing one form.

———   1871 A
On *Ctenacanthus hybodoides*, Egerton.
Trans. Geol. Soc. Glasgow, iv, pp. 59-62, with pls. ii, iii.

Thompson, J. Arthur. 1897 A
As regards dentitions—a notice of some new facts and theories.
Proc. Scottish Micros. Soc., ii, pp. 36-50.

Thompson, Zadock. 1850 A
An account of some fossil bones found in Vermont in making excavations for the Rutland and Burlington Railroad.
Amer. Jour. Sci. (2), ix, pp. 256-263, with 2 pages of cuts, 13 in number.

———   1853 A
History of Vermont, natural, civil, and statistical.
Appendix, pp. 5-64.
Contains description and figures of *Beluga vermontana* and an account of the finding of some remains of *Elephas primogenius?*.

Thompson, Zadock—Continued. 1859 A
A Vermont whale.
Edinb. New Philos. Jour., x, p. 299.
Records finding of *Beluga vermontana* at Charlotte, Vt., in 1849.

Tims, H. W. Marett. 1896 A
Notes on the dentition of the dog.
Anatom. Anzeiger, xi, pp. 537-546, with 5 figs.
Contains a list of papers written on the subject discussed, the teeth of mammals, with especial reference to the existence of "premilk" teeth.

———   1896 B
On the tooth-genesis in the *Canidæ*.
Jour. Linn. Soc. Lond., Zool., xxv, pp. 445-480, with 8 figs. in the text.
Contains numerous references to the literature.

Todd, Albert. 1876 A
[Exhibition of tooth identified as that of *Mastodon angustidens*, found in Pike County, Mo.]
Trans. Acad. Sci. St. Louis, iii, p. cxcii.

Toll, Eduard. 1895 A
Die fossilen Eislager und ihre Beziehungen zu den Mammuthleichen.
Mém. Acad. Imp. Sci. St. Pétersbourg (7), xliii, No. 13, pp. 1-86, with pls. i-vii.

Tomes, Charles S. 1898 A
A manual of dental anatomy, human and comparative.
Fifth edition, 8vo, Philadelphia, 1898, pp. i-viii; 1-596, with figs. 1-273.

Tomes, John. 1850 A
On the structure of the dental tissues of the order *Rodentia*.
Philos. Trans. Roy. Soc. Lond., 1850, pp. 529-567, with pls. xliii-xlvi.

Topinard, Paul. 1892 A
De l'évolution des molaires et prémolaires chez les Primates et en particulier chez l'homme.
L'Anthropologie, Paris, 1892, pp. 641-710, with 8 figs.
Reviewed by M. Schlosser in Archiv f. Anthrop., xxiii, pp. 157-159.

Tornier, Gustav. 1886 A
Fortbildung und Umbildung des Ellbogengelenks während der Phylogenesis der einzelnen Säugethiergruppen.
Morpholog. Jahrbuch, xii, pp. 407-413, with 2 woodcuts.

———   1888 A
Die Phylogenese des terminalen Segmentes der Säugethier-Hintergliedmassen.

Tornier, Gustav—Continued.
Morpholog. Jahrbuch, xiv, pp. 223-328, with pls. xi, xii.

"Die Begriffe, 'Fuss,' und 'terminales Segment der hinteren Gliedmassen' werden von mir als gleichwerthig angesehen."

——   **1890 A**
Die Phylogenese des terminalen Segmentes der Säugethier-Hintergliedmassen.
Morpholog. Jahrbuch, xvi, pp. 401-488, with pls. xvii, xviii.

Toula, Franz.   **1892 A**
Zwei neue Säugethierfundorte auf der Balkanhalbinsel.
Sitzungsber. Cl. k. Akad. Wissensch. Wien. math.-naturw. ci, Abth. i, pp. 608-615, with plate.

Describes *Menodus ? rumeticus*, which is compared with *M. proutti* (Leidy).

——   **1897 A**
Phoca vindobonensis n. sp. von Nussdorf in Wien.
Beiträge Palæontologie Oesterreich-Ungarns und des Orients (Mojsisovics und Neumayer), xi, pp. 47-70, with pls. ix-xi.

This paper contains a bibliography and a list of the described fossil species of seals.

Toula, Franz, and Kail, J. A.   **1885 A**
Ueber einen Krokodil-Schädel aus den Tertiärablagerungen von Eggenburg in Niederrösteriech.
Denkschr. k. Akad. Wissensch. Wien, pp. 229-356, with 3 pls. and 3 text figs.

Traquair, R. H.   **1867 A**
Description of *Pygopterus greenocki* (*Agassiz*), with notes on the structural relations of the genera *Pygopterus, Amblypterus*, and *Eurynotus*.
Trans. Roy. Soc. Edinburgh, xxiv, pp. 701-713, with pl. xlv.

Description of species now referred by A. S. Woodward to *Nematoptychius*.

——   **1870 A**
On the cranial osteology of *Polypterus*.
Jour. Anat. and Physiol., v, pp. 166-183, pl. vi.

——   **1871 A**
Notes on the genus *Phaneropleuron* (Huxley), with a description of a new species from the Carboniferous formation.
Geol. Magazine (1), viii, pp. 529-535, with pl. xiv and 1 fig. in the text.

For supplementary remarks see Geol. Magazine (1), ix, 1872, p. 271.

Traquair, R. H.—Continued.   **1873 A**
On a new genus of fossil fish of the order *Dipnoi*.
Geol. Magazine (1), x, pp. 552-555, with pl. xiv and 1 woodcut.

Describes new genus and species, *Ganorhynchus woodwardi.*

——   **1875 A**
On the structure and systematic position of the genus *Cheirolepis*.
Ann. and Mag. Nat. Hist. (4), xv, pp. 237-248, with pl. xvii.

——   **1875 B**
On some fossil fishes from the neighborhood of Edinburgh.
Ann. and Mag. Nat. Hist. (4), xv, pp. 258-268, with pl. xvi.

Describes and figures *Nemoptychius greenocki, Wardichthys*, n. g., and *Rhizodus hibberti.*

——   **1875 C**
On the structure and affinities of *Tristichopterus alatus* Egerton.
Trans. Roy. Soc. Edinburgh, xxvii, pp. 383-396.

——   **1877 A**
On the Agassizian genera *Amblypterus, Palæoniscus, Gyrolepis*, and *Pygopterus*.
Quart. Jour. Geol. Soc., xxxiii, pp. 548-578.

——   **1877 B**
The Ganoid fishes of the British Carboniferous formations. Part I. *Palæoniscidæ.*
Palæontographical Society, xxxi, pp. 1-60, pls. i-vii.

——   **1877 C**
On the structure of the lower jaw in *Rhizodopsis* and *Rhizodus.*
Ann. and Mag. Nat. Hist. (4), xix, pp. 299-305.

——   **1878 A**
On the genera *Dipterus* Sedgw. & Murch., *Palædaphus* Van Beneden and De Koninck, *Holodus* Pander, and *Cheirodus* M'Coy.
Ann. and Mag. Nat. Hist. (5), ii, pp. 1-17, with pl. iii.

Describes the structure of *Dipterus* and places the genus in the *Dipnoi. Palædaphus* also belongs to the *Dipnoi.*

——   **1879 A**
On the structure and affinities of the *Platysomidæ.*
Trans. Roy. Soc. Edinburgh, xxix, pp. 343-391, with pls. iii-vi.

Review of this paper in Geol. Magazine (2), vii, pp. 318-321.

Traquair, R. H.—Continued.   1881 A
Report on the fossil fishes collected
by the Geological Survey of Scotland in
Eskdale and Liddesdale. Part I. *Ganoidei.*
Trans. Roy. Soc. Edinburgh, xxx, pp. 15–71,
with pls. i–vi.

——   1881 B
On the cranial osteology of *Rhizodopsis.*
Trans. Roy. Soc. Edinburgh, xxx, pp. 167–
179, with 3 figs. in the text.

——   1884 A
Remarks on the genus *Megalichthys*
Agassiz, with description of a new spe-
cies.
Geol. Magazine (3), i, pp. 115–121, plate v.
See also Proc. Roy. Phys. Soc. Edinburgh,
viii, pp. 67–77, with pl. iv.

——   1884 B
Description of a fossil shark (*Ctenacanthus costellatus*) from the Lower Carboniferous rocks of Eskdale, Dumfries-
shire.
Geol. Magazine (3), i; pp. 3–8, with pl. ii.

——   1884 C
Notes on the genus *Gyracanthus*
Agassiz.
Ann. and Mag. Nat. Hist. (5), xiii, pp. 37–48.

——   1885 A
On a specimen of *Psephodus magnus*
Agassiz, from the Carboniferous lime-
stone of East Kilbridge, Lanarkshire.
Geol. Magazine (3), ii, pp. 337–344.

——   1885 B
On a specimen of *Psephodus magnus*
Agassiz, from the Carboniferous lime-
stone of East Kilbridge, Lanarkshire.
Trans. Geol. Soc. Glasgow, vii, pp. 392–402,
with pl. xvi.

——   1886 A
On *Harpacanthus*, a new genus of
Carboniferous selachian spines.
Ann. and Mag. Nat. Hist. (5), xviii, pp. 493–
496, with 2 figs. in the text.

——   1887 A
Notes on *Chondrosteus acipenseroides*
Agassiz.
Geol. Magazine (3), iv. pp. 248–257, with 5 figs.
Also in Proc. Roy. Phys. Soc. Edinburgh, ix,
pp. 349–361, 1887.

——   1888 A
Notes on Carboniferous *Selachii.*
Geol. Magazine (3), v, pp. 81–86; 101–104.
See also Proc. Roy. Phys. Soc. Edinburgh,
ix, pp. 412–421.
Discusses *Cladodontidæ, Orodontidæ, Cochlio-
dontidæ, Petalodontidæ,* and *Oracanthus.*

Traquair, R. H.—Continued.   1888 B
Notes on the nomenclature of the
fishes of the Old Red sandstone of Great
Britain.
Geol. Magazine (3), v, pp. 507–517.

——   1888 C
On the structure and classification of
the *Asterolepidæ.*
Ann. and Mag. Nat. Hist. (6), ii, pp. 485–504,
with pls. xvii, xviii.
Also in Proc. Roy. Phys. Soc. Edinburgh, x,
pp. 23–46, pls. i, ii.

——   1889 A
Homosteus Asmuss compared with
Coccosteus Agassiz.
Geol. Magazine (3), vi, pp. 1–8, pl. i.
Also in Proc. Roy. Phys. Soc. Edinburgh, x,
pp. 47–57, with pl. iii.

——   1889 B
Note on the genera *Tristycanthus* and
Ptychacanthus.
Geol. Magazine (3), vi, pp. 27–28.

——   1889 C
On the systematic position of the
"Dendrodont" fishes.
Geol. Magazine (3), vi, pp. 490–492.

——   1890 A
Notes on the Devonian fishes of Scaumenac Bay and Campbelltown, in Canada.
Geol. Magazine (3), vii, pp. 15–22.

——   1890 B
On *Phlyctænius*, a new genus of *Coccosteidæ.*
Geol. Magazine (3), vii, pp. 55–60, with pl. iii.
See also Proc. Roy. Phys. Soc. Edinburgh, x,
pp. 227–285, 1891, for reprint, with *Phlyctænaspis*
substituted in the title for *Phlyctænius.*

——   1890 C
On the supposed pectoral limb in
Coccosteus decipiens.
Geol. Magazine (3), vii, p. 285.
A reply to Prof. V. Koenen, Geol. Magazine,
vii, p. 191.

——   1890 D
On the structure of *Coccosteus decipiens*
Agassiz.
Ann. and Mag. Nat. Hist. (6), v, pp. 125–136,
with pl. x.
Also in Proc. Roy. Phys. Soc. Edinburgh,
1889–90, pp. 211–224, pl. x.

——   1890 E
Note on *Phlyctænius*, a new genus of
Coccosteidæ.
Geol. Magazine (3), vii, p. 144.
A short note proposing *Phlyctænaspis* to re-
place *Phlyctænius*, regarded as being preoccu-
pied by *Phlyctænium* of Zittel.

Traquair, R. H.—Continued. 1890 F
List of the fossil *Dipnoi* and *Ganoidei* of Fife and the Lothians.
Proc. Roy. Soc. Edinburgh, xvii, pp. 385–400.

——— 1891 A
Review of A. S. Woodward's "Catalogue of the fossil fishes in the British Museum."
Geol. Magazine (3), viii, pp. 123–129.

——— 1891 B
On the structure of *Cocrosteus decipiens*, Agassiz.
Proc. Roy. Phys. Soc. Edinburgh, x, pp. 211–224, with pl. x.
See Traquair, R. H., 1890 D.

——— 1893 A
Notes on the Devonian fishes of Campbelltown and Scaumenac Bay.
Geol. Magazine (3), x, pp. 145–149, with 1 text fig.
Treats of *Protodus jexi*, *Doliodus problematicus*, *Cheiracanthus costellatus*, *Cephalaspis campbelltonensis*, *Cephalaspis jexi*, and *Phlyctænaspis acadica*.

——— 1893 B
Fauna der Gaskohle und der Kalksteine des Perm-formations Böhmens.
Geol. Magazine (3), x, pp. 175–178.
Review of Prof. Fritsch's work with title given above.

——— 1893 C
Notes on the Devonian fishes of Campbelltown and Scaumenac Bay in Canada. No. 3.
Geol. Magazine (3), x, pp. 262–267.
Describes *Scaumenacia curta*, *Coccosteus canadensis*, *Gyrolepis quebecensis*, and *Eusthenopteron foordii*.
See notice of this paper in Amer. Naturalist, vol. xxvii, p. 817.

——— 1893 D
Notes on the Devonian fishes of Campbelltown and Scaumenac Bay in Canada. Parts 1 and 2.
Proc. Roy. Phys. Soc. Edinburgh, xii, pp. 111–125.
Contains descriptions of the species referred to under Traquair, R. H., 1893 A, and Traquair, R. H., 1893 C.

——— 1894 A
On a new species of *Diplacanthus*, with remarks on the Acanthodian shoulder-girdle.
Geol. Magazine (4), i, pp. 254–257.

——— 1894 B
Notes on Palæozoic fishes, No. 1.
Ann. and Mag. Nat. Hist. (6), xiv, pp. 368–374, with pl. ix and 1 text fig.
Describes and figures an English species of *Phlyctænaspis*, a new one of *Acanthaspis*, and one of *Eurylepis*.

Traquair, R. H.—Continued. 1894 C
A monograph of the fishes of the Old Red sandstone of Britain. Part II, No. 1. The *Asterolepidæ*.
Palæontographical Society, xlviii, pp. 63–90, with pls. xv–xviii, and text figs. 34–50.

——— 1896 A
Review of A. S. Woodward's "Catalogue of the fossil fishes in the British Museum. Part III."
Geol. Magazine (4), iii, pp. 124–127.

——— 1899 A
Report on fossil fishes collected by the Geological Survey of Scotland in the Silurian rocks of the south of Scotland.
Trans. Roy. Soc. Edinburgh, xxxix, pp. 827–864, with pls. i–v and 18 text figs.

——— 1900 A
Notes on *Drepanaspis gmündenensis* Schlüter.
Geol. Magazine (4), vii, pp. 153–159, with 3 figs. in the text.

——— 1900 B
The bearings of fossil ichthyology on the problem of evolution.
Geol. Magazine (4), vii, pp. 463–470, 516–524.

Trautschold, H. 1874 A
Fischreste aus dem Devonischen des Gouvernements Tula.
Nouv. Mém. Soc. Imp. Natur. Moscou, xiii, pp. 263–276, with pls. xxvi, xxvii.
Figures and describes various species of *Cladodus*, *Orodus*, *Helodus*, *Psammodus*, etc.

——— 1874 B
Die Kalkbrüche von Mjatschkowa. Eine Monographie des oberen Bergkalks.
Nouv. Mém. Soc. Imp. Natur. Moscou, xiii, pp. 277–326, with pls. xxviii–xxxi.

——— 1879 A
Ueber eine Ichthyosaurus-Flosse aus dem Moskauer Kimmeridge.
Verhandl. russ.-kais. mineralog. Gesellsch. St. Petersburg (2), xii, pp. 168–173, with pl. v.

——— 1879 B
Die Kalkbrüche von Mjatschkowa. Eine Monographie des oberen Bergkalks. Schluss.
Nouv. Mém. Soc. Imp. Natur. Moscou, xiv, pp. 1–82, with pls. i–vii.

——— 1880 A
Ueber *Tomodus* Agassiz.
Bull. Soc. Imp. Natur. Moscou, lv, pt. 2, pp. 139–140.

——— 1880 B
Ueber *Bothriolepis panderi* Lahusen.
Bull. Soc. Imp. Natur. Moscou, lv, pt. 2, pp. 169–179, with 1 pl. and 5 text figs.

Trautschold, H.—Continued.   1880 C
Ueber *Dendrodus* und *Coccosteus.*
Verhandl. russ.-kais. mineralog. Gesellsch.
St. Petersburg (2), xv, pp. *139-156, with pls.
iii-x.

—   1883 A
Ueber *Edestus* und einige andere
Fischreste des Moskauer Bergkalks.
Bull. Soc. Imp. Natur. Moscou, lvii, pt. 2, pp.
160-174, with pl. v and 3 text figs.

—   1886 A
Ueber das Genus *Edestus.*
Bull. Soc. Imp. Natur. Moscou, lxi, pt. i, pp.
94-99.

—   1888 A
Ueber *Edestus protopirata*, Trd.
Zeitschr. deutsch. geol. Gesellsch., xl, pp.
750-758, with 2 figs. in the text.

—   1889 A
Ueber *Coccosteus megalopteryx*, Trd.,
C. obtusus und *Cheliophorus verneuili*
Ag.
Zeitschr. deutsch. geol. Gesellsch., xii, pp.
35-48, with pls. iii-vi.

—   1889 B
Ueber vermeintliche Dendrodonten.
Zeitschr. deutsch. geol. Gesellsch., xii, pp.
621-634, with pls. xxiii-xxv.

—   1890 A
Ueber *Protopirata centrodon*, Trd.
Bull. Soc. Imp. Natur. Moscou (n. s.), iv,
pp. 317-321, with 1 fig. in the text.
The genus *Edestus* is considered.

Treuenfels, Paul.   1896 A
Die Zähne von *Myliobatis aquila.*
Inaugural dissertation at Univ. Basel. Breslau, 1896, 8vo., pp. 1-34, with pls. i, iii.

Troost, G.   1835 A
On the localities in Tennessee in
which bones of the gigantic mastodon
and *Megalonyx jeffersonii* are found.
Trans. Geol. Soc. Penn., i, pp. 139-146; 236-243.

Troschel, F. H.   1861 A
Ueber den Unterkiefer der Schlangen
und über die fossile Schlange von Rott.
Archiv Naturgesch., xxvii, i, pp. 326-360,
with pl. x.

Trouessart, E. L.   1889 A
Mammifères.
Annuaire Géol. Universel, v. 1886, Paris.
Pp. 915-970.
Généralités, pp. 915-916; Analyse des principaux mémoires relatifs à la classe des mammifères en général, pp. 916-942; Résumé systématique espèces nouvelles et travaux relatifs aux differents ordres, familles, etc., pp. 942-970.

Trouessart, E. L.—Continued.   1892 A
The fiction of the American horse
and the truth on this disputed point.
Science, xx, p. 188.

—   1897 A
Catalogus Mammalium tam viventium quam fossilium.   Nova editio.
Berolini, 1897, pp. 665-1264, 8vo.
Contains the orders *Primates, Prosimiæ
Chiroptera, Insectivora.*

—   1898 A
Catalogus Mammalium tam viventium quam fossilium.   Nova editio.
Berolini, 1898, pp. 665-1264, 8vo.
Contains the orders *Sirenia, Cetacea, Edentata, Marsupialia, Allotheria, Monotremata.*

—   1899 A
Catalogus Mammalium tam viventium quam fossilium.
Berolini, 1899, pp. 1265-1469, 8vo.
Appendix (Addenda et Corrigenda) and
alphabetical index.

True, Frederick W.   1889 A
A review of the family *Delphinidæ.*
Bull. U. S. Nat. Mus., No. 36, Washington,
D. C., 1889, pp. 191, with 47 pls.

—   1891 A
The puma, or American lion: *Felis
concolor* of Linnæus.
Ann. Rep. Smithson. Inst., 1889; Rep. Nat.
Mus., pp. 591-608, with pl. xciv.

—   1896 A
A revision of the American moles.
Proc. U. S. Nat. Mus., xix, pp. 1-111, with pls.
i-v and 46 text figs.

Tscherski, J. D.   1892 A
Beschreibung der Sammlung posttertiärer Säugethiere.
Mém. Acad. Imp. Sci., St. Pétersbourg (7),
xl, No. 1, pp. i-v, 1-511, with pls. i-vi

Tullberg, Tycho.   1899 A
Ueber das System der Nagethiere:
eine phylogenetische Studie.
8vo, pp. i-v; 1-514; A1-A18, pls. i-lvii.
Upsala, 1899.
Pp. 501-514 are occupied by a list of works
dealing with the subject.

Tuomey, M.   1847 A
Discovery of the cranium of the
zeuglodon.
Amer. Jour. Sci. (2), iv, pp. 283-285, with 1
fig. in the text.

—   1847 B
Notice of the discovery of a cranium
of the zeuglodon.
Proc. Acad. Nat. Sci. Phila. iii, pp. 151-153,
with 2 woodcuts.

Tuomey, M.—Continued. 1847 C
Notice of the discovery of a cranium
of the zeuglodon (*Basilosaurus*).
Jour. Acad. Nat. Sci. Phila. (2), i, pp. 16–17.

———— 1850 A
[Remarks on a fossil reptile from
Alabama, belonging to the genus *Leiodon*. Followed by remarks by Prof.
L. Agassiz.]
Proc. Amer. Assoc. Adv. Sci., 3d meeting,
Charleston, S. C., 1850, p. 74.

———— 1858 A
[Descriptions and figures of *Ctenacanthus elegans*, *Cladodus newmani*, *C. magnificus*.]
Second Biennial Report Geol. Alabama, by
M. Tuomey, A. M., edited by J. W. Mallet,
Ph. D. Pp. i–xix, 1–292.
Appended are lists of Cretaceous and Tertiary fossils, including a number of sharks, pp.
251–268. The species named above are described on pp. 38–40.

Turner, George. 1799 A
Memoir on the extraneous fossils
denominated mammoth bones; principally designed to show that they are
the remains of more than one species
of nondescript animal.
Trans. Amer. Philos. Soc., iv, pp. 510–518.

Turner, H. N., jr. 1848 A
Observations relating to some of the
foramina at the base of the skull in
Mammalia, and on the classification of
the order *Carnivora*.
Proc. Zool. Soc. Lond., 1848, pt. xvi, pp. 63–88.

———— 1850 A
On the generic subdivisions of the
Bovidæ, or hollow-horned ruminants.
Proc. Zool. Soc. Lond., 1850, pt. xviii, pp.
164–178.

———— 1872 A
Some observations on the dentition
of the narwhal (*Monodon monoceros*).
Jour. Anat. and Physiol., vii, pp. 75–79.

———— 1873 A
Note on a bidental skull of a narwhal.
Jour. Anat. and Physiol., viii, pp. 133–134.

———— 1876 A
Additional note on the dentition of
the narwhal (*Monodon monoceros*).
Jour. Anat. and Physiol., x, p. 516.

Turner, William. 1899 A
On the development and morphology
of the marsupial shoulder girdle.
Trans. Roy. Soc. Edinburgh, xxxix, pp. 749–770, with pls. i, ii.

Ubaghs, Casimir. 1875 A
La *Chelonia hoffmanni*.
Ann. Soc. Géol. Belgique, ii, pp. 197–205.

Udden, J. A. 1891 A
Megalonyx beds in Kansas.
Amer. Geologist, vii, pp. 340–345, with 3
woodcuts in the text.

———— 1899 A
Dipterus in the American Middle
Devonian.
Jour. Geology, vii, pp. 494–495, with 1 fig. in
the text.

Underwood, Lucien. 1890 A
A bison at Syracuse, N. Y.
Amer. Naturalist, xxiv, pp. 963–964.
Followed by a note by Professor Cope identifying the specimen as *Bos americanus*.

Ussow, S. A. 1898 A
Die Entwicklung der Cycloid-
Schuppe der Teliostier.
Bull. Soc. Imp. Natur. Moscou (n. s.), xi,
1897, pp. 339–366, with pls. vi, vii.

———— 1900 A
Zur Anatomie und Entwicklungsgeschichte der Wirbel der Teleostier.
Bull. Soc. Imp. Natur. Moscou (n. s.), xv,
1900, pp. 175–240, with pls. i–iv.

Vaillant, Léon. 1883 A
Remarques sur les affinités naturelles
des familles composant le sous-ordre des
poissons malacoptérygiens abdominaux.
Ann. Sci. Naturelles (6), Zool., xv, art. 7, pp.
1–13.

———— 1891 A
Sur la signification taxonomique du
genre *Emys* C. Duméril.
Ann. Sci. Naturelles (7), Zool., xii, pp. 51–63.

———— 1894 A
Essai sur la classification générale des
chéloniens.
Ann. Sci. Naturelles (7), Zool., xvi, pp. 331–345.

Van Beneden, P. J. 1861 A
Un mammifère nouveau du Crag
d'Anvers.
Bull. Acad. Roy. Belgique (2), xii, pp. 22–28.
Describes *Squalodon antverpiensis*, and holds
that *Squalodon* and *Zeuglodon* belong to different families.

Van Beneden, P. J.—Cont'd.   1861 B
Sur le développement de la queue des poissons plagiostomes.
Bull. Acad. Roy. Belgique (2), xi, pp. 287–293, with 1 pl.
Published also as a separate, with pp. 3–8.

——   1865 A
Recherches sur les ossements provenant du Crag d'Anvers, les Squalodons.
Mém. Acad. Roy. Belgique, xxxv, art. iii, pp. 1–85, with pls. i–iv, and many text figs.
On pp. 11–12 of this memoir is found a list of the works which have been published on the Zeuglodonts and Squalodonts.

——   1868 A
De la composition du bassin des cétacés.
Bull. Acad. Roy. Belgique (2), xxv, pp. 428–433, with 1 pl.

——   1872 A
Les Balenidés fossiles d'Anvers.
Jour. de Zoologie, i, pp. 407–419.
Extract from Bull. Acad. Roy. Belgique(2), xxxiv, pp. 6–20, 1872.

——   1872 B
Les baleines fossiles d'Anvers.
Bull. Acad. Roy. Belgique (2), xxxiv, pp. 6–20.

——   1875 A
Le squelette de la baleine fossile du Musée de Milan.
Bull. Acad. Roy. Belgique (2), xl, pp. 736–758, with 1 pl.
Describes and figures *Plesiocetus cortesi.*

——   1877 A
Description des ossements fossiles des environs d'Anvers. Première partie: Les phoques ou les amphitériens.
Ann. Mus. Roy. d'Hist. Nat. Belgique, i, pp. 1–88, with pls. i–xviii, and 17 figs. and 8 maps in the text.

——   1877 B
Note sur un cachalot nain du Crag d'Anvers (*Physeterula dubusii*).
Bull. Acad. Roy. Belgique (2), xliv, pp. 851–856, with 1 pl.

——   1880 A
Description des ossements fossiles des environs d'Anvers. Deuxième partie: Cétacés, genres *Balænula, Balæna,* et *Balænotus.*
Ann. Mus. Roy. d'Hist. Nat. Belgique, iv, pp. 1–83, with pls. i–xxxix.
This work contains a very full bibliography of the fossil *Cetacea.*

——   1880 B
Les mysticètes à court fanons des sables des environs d'Anvers.
Bull. Acad. Roy. Belgique (2), l, pp. 11–26.

Van Beneden, P. J.—Cont'd.   1882 A
Description des ossements fossiles des environs d'Anvers. Troisième partie: Cétacés, genres *Megaptera, Balænoptera, Burtinopsis,* et *Erpetocetus.*
Ann. Mus. Roy. d'Hist. Nat. Belgique, vii, pp. 1–90, with pls. xl–cix.
Part iv, describing the genus *Plesiocetus,* is found in vol. ix of the Annales. Pt. v, describing *Amphicetus, Heterocetus,* etc., is in vol. xiii. These genera have not been found in North America.

——   1882 B
Une baleine fossile de Croatie appartenant au genre Mésocète.
Mém. Acad. Roy. Sci. Lettres et Beaux-Arts, xliv, art. 2, pp. 3–29, with pls. i–ii.

Van Beneden, P. J., and Gervais, P.   1880 A
Ostéographie des cétacés vivants et fossiles, comprenant la description et l'iconographie du squelette et du système dentaire de ces animaux; ainsi que des documents relatifs à leur histoire naturelle.
Paris, 1880, pp. i–viii, 1–634, with atlas of 64 pls.

Van Beneden, P. J. and Koninck, L. G. de.   1864 A
Notice sur le *Palædaphus insignis.*
Bull. Acad. Roy. Belgique (2), xvii, pp. 143–151, with pls. i, ii.

Van Rensselaer, Jeremiah.   1826 A
Notice of a recent discovery of the fossil remains of the mastodon.
Amer. Jour. Sci. (1), xi, pp. 246–250.

——   1827 A
On the fossil remains of the mastodon lately found in Ontario County, N. Y.
Amer. Jour. Sci. (1), xii, pp. 380–381.

——   1828 A
On the fossil tooth of an elephant found near the shore of Lake Erie, and on the skeleton of a mastodon lately discovered on the Delaware and Hudson Canal.
Amer. Jour. Sci. (1), xiv, pp. 31–33.

Van Wijhe, J. W.   1882 A
Ueber das Visceralskelet und die Nerven des Kopfes der Ganoiden und von *Ceratodus.*
Niederländisches Archiv Zoologie, v, pp. 205–320, with pls. xv and xvi.
Besides the subjects of the title, this paper treats of the "otic bones."

Vasseur, Gaston.   1874 A
Sur le pied de derrière de l'*Hyænodon parisiensis.*
Comptes rend. Acad. Sci. Paris, lxxviii, pp. 1446–1447.

Vasseur, Gaston—Continued. 1875 A
Sur le cubitus de *Coryphodon oweni.*
Bull. Soc. Géol., France (3), iii, pp. 181-186, pl. iii.

Vetter, Benjamin. 1881 A
Die Fische aus dem lithographischen Schiefer im Dresdener Museum.
Mittheil. k. min., geol., u. præhist. Mus. Dresden, Heft 4, pp. i-viii, 1-118, with pls. i-iii.

—— 1885 A
Ueber die Verwandschaftsbeziehungen zwischen Dinosauriern und Vögeln.
Festschrift naturw., Gesellsch. Isis, Dresden, Feier 50jährg. Bestehens, pp. 109-122.

Vogt, Carl. 1880 A
Archæopteryx macrura, an intermediate form between birds and reptiles.
The Ibis (4), iv, pp. 434-456, with pl. xiii, and 4 figs. in the text.

Voigt, Fr. S. 1835 A
Thier-Fährten in Hildburghauser Sandsteine (*Palæopithecus*).
Neues Jahrb. Mineral., 1835, pp. 322-326.

Vrolik, A. J. 1873 A
Ueber die Verknöcherung und die Knochen des Schädels der *Teleostei.*
Niederländisches Archiv. Zoologie, i, pp. 219-318, with pls. xviii-xxii.
I. Synonymie der Schädelknochen bei den *Teleostei,* pp. 221-230, with a table.
II. Ueber die Verknöcherung des Teleostierschädels, pp. 231-260.
III. Die Knochen des Teleostierschädels, pp. 261-290, with a table.
IV. Die Verknöcherung des Schläfenbeins der Säugethiere, pp. 291-318.
This paper also gives consideration to the "otic bones." In opposition to the views of Vrolik, see Van Wijhe, J. W., 1882 A.

Wagner, Andreas. 1851 A
Beiträge zur Kenntniss der in den lithographischen Schiefern abgelagerten urweltlichen Fische.
Abhandl. k. bayer. Akad. Wissensch., math.-phys. Cl., vi, pp. 1-80, with pls. i-iv.

—— 1851 B
Beschreibung einer neuen Art von *Ornithocephalus,* nebst kritischer Vergleichung der in der k. palæontologischen Sammlung zu München aufgestellten Arten aus dieser Gattung.
Abhandl. k. bayer. Akad. Wissensch., math.-phys. Cl., vi, pp. 127-193, pls. v-vi.

—— 1853 A
Beschreibung einer fossilen Schildkröte und etlicher anderer Reptilien-

Wagner, Andreas—Continued.
überreste aus den lithographischen Schiefern und dem Grünsandsteine von Kelheim.
Abhandl. k. bayer. Akad. Wissensch., math.-phys. Cl., vii, pp. 241-264, with pls. iv-vi.
Describes *Platychelys oberdorferi, Idiochelys,* and *Homœosaurus.*

—— 1861 A
On a new fossil reptile supposed to be furnished with feathers.
Ann. and Mag. Nat. Hist. (3), ix, pp. 261-267 (1862).
Translated from the Sitzungsber. Akad. Wissensch. München, 1861, p. 146.

—— 1861 B
Monographie der fossilen Fische aus den lithographischen Schiefern Bayerns. Erste Abtheilung. Plakoiden und Pyknodonten.
Abhandl. k. bayer. Akad. Wissensch., math.-phys. Cl., ii. Abth., ii, pp. 277-352, with pls. iv-vii.

—— 1863 A
Monographie der fossilen Fische aus den lithographischen Schiefern Bayerns. Zweite Abtheilung.
Abhandl. k. bayer. Akad. Wissensch., math.-phys. Cl., ix, pp. 613-748, with pls. ii-vii.
Also issued as reprint with pp. 43-138.

Wagner, George. 1898 A
On some turtle remains from the Fort Pierre.
Kansas Univ. Quarterly, vii, A, pp. 201-205, with 2 figs. in the text.

—— 1899 A
On *Tetracaulodon* (*Tetrabelodon*) *shepardii* Cope.
Kansas Univ. Quarterly, viii, A, pp. 99-103, with pls. xxiv, xxv.

Wailes, B. L. C. 1854 A
Report on the agriculture and geology of Mississippi, embracing a sketch of the social and natural history of the State.
Pp. i-xx; 1-371.
Published by order of the legislature.
Contains remarks on a few fossil vertebrates and a list, by Dr. Leidy, of those found in Mississippi.

Walcott, C. D. 1892 A
Notes on the discovery of a vertebrate fauna in Silurian (Ordovician) strata.
Bull. Geol. Soc. America, iii, pp. 153-172, pls. 3-5.
Describes *Dictyorhabdus priscus, Astraspis desiderata,* and *Eryptichius americanus.*
See Neues Jahrb. Mineral., 1895, p. 162, for Otto Jaekel's review; Amer. Naturalist, xxvii, p. 268, for notice by Cope.

Walcott, C. D.—Continued.   1900 A

Correspondence relating to collection of vertebrate fossils made by the late Prof. O. C. Marsh.

Science (2), xi, pp. 22–23.

This letter, directed to Prof. S. P. Langley, contains a partial list of the types of species described by Professor Marsh, and now in the National Museum.

Walker, S. T.   1883 A

On the origin of the fossil bones discovered in the vicinity of Tise's Ford, Florida.

Proc. U. S. Nat. Mus., vi, pp. 427–428.

Walter, Ferdinand.   1887 A

Das Visceralskelet und seine Muskulatur bei den einheimischen Amphibien und Reptilien.

Jenaische Zeitschr. Naturwiss., xxi (xiv), pp. 1–45, with pls. i–iv.

Wanner, Atreus.   1889 A

The discovery of fossil tracks, algae, etc., in the Triassic of York County, Pa.

Ann. Report Geol. Surv. Penn., for 1887, pp. 21–27, with pls. iv–xiii.

No names are given to the tracks figured.

———   1892 A

Fossil tracks in the Trias of York County, Pa.

Proc. Amer. Assoc. Adv. Sci., 39th meeting, Washington, 1891, p. 286. A brief abstract.

Ward, Henry A.   1864 A

Notice of the *Megatherium cuvieri*, the giant fossil ground-sloth of South America. Presented to the University of Rochester by Hiram Sibley, esq.

Pp. 1–34, with 1 large pl.

Has, in addition, a brief description of *Megalonyx jeffersoni*.

Ward, Henry L.   1887 A

The pelvis of the dugong.

Science, ix, p. 586, with 1 woodcut.

Ward, John.   1890 A

The geological features of the North Staffordshire coal fields, their organic remains, their range and distribution; with a catalogue of the fossils of the Carboniferous system of North Staffordshire.

Trans. North Staffordshire Mining and Mechan. Engineers, x, pp. 1–189, with index, pp. i–ix, and pls. i–ix.

Warren, John C.   1846 A

On the osteology and dentition of some North American mastodons.

Ann. and Mag. Nat. Hist. (1), xvii, pp. 145–150.

A letter addressed to Prof. R. Owen and sent by him to the editors.

———   1850 A

On the *Mastodon angustidens*.

Proc. Amer. Assoc. Adv. Sci., 2d meeting, 1849, Cambridge, Mass., pp. 93–94.

———   1852 A

[Remarks on a supposed tooth of *Mastodon angustidens*.]

Proc. Boston Soc. Nat. Hist., iv, pp. 129–131.

———   1852 B

[Remarks on the supposed food of *Mastodon giganteus*.]

Proc. Boston Soc. Nat. Hist., iv, p. 154.

———   1852 C

The *Mastodon giganteus* of North America.

Pp. i–viii; 1–219, 4to, with vignette and pls. i–xxvi and 1 large additional pl. Boston, 1852, John Wilson & Son.

A second edition of this work was published in 1855.

———   1854 A

Remarks on some fossil impressions in the sandstone rocks of Connecticut River.

8vo; pp. 1–54, with 1 pl. and 2 text figs. Boston, 1854.

A popular account of the impressions, footmarks, etc., of the sandstones referred to.

———   1855 A

Supernumerary tooth in *Mastodon giganteus*.

Amer. Jour. Sci. (2), xix, pp. 349–353.

———   1855 B

The *Mastodon giganteus* of North America.

2d edition, 260 pp., 4to, with 31 pls.; Boston, 1855.

Contains 3 new plates and some additions to text of the first edition, which appeared in 1852.

———   1855 C

Supernumerary tooth in jaw of *Mastodon*.

Proc. Boston Soc. Nat. Hist., v, pp 146–150.

———   1855 D

Brief history of the zeuglodon.

Proc. Boston Soc. Nat. Hist., v, pp. 91–92.

Warren, John C.—Continued.　1856 A
New remarkable gigantic fossils and
footmarks.
Proc. Boston Soc. Nat. Hist., v, pp. 298–306.

Waterhouse, G. R.　　　　1838 A
On the skull and dentition of the
American badger (*Meles labradoria*).
Proc. Zool. Soc. Lond., 1838, pt. vi, pp. 153–
154.

———　　　　　　　　　　1839 A
On the skulls and dentition of the
Carnivora.
Proc. Zool. Soc. Lond., 1839, pt. vii, pp. 135–
137.

Weber, Max.　　　　　　1886 A
Studien über Säugethiere. Ein Beitrag
zur Frage nach dem Ursprung der Ce-
taceen.
Jena, 1886, 8vo, pp. i–vi; 1–252, with pls.
i–iv.
Reviewed by M. Schlosser in Archiv f.
Anthrop., xviii, pp. 143–146.

———　　　　　　　　　　1887 A
Über die cetoide Natur der *Promam-
malia*.
Anatom. Anzeiger, ii, pp. 42–55.

———　　　　　　　　　　1888 A
Anatomisches uber Cetaceen. I.
Ueber den Carpus der Cetaceen; II.
Ueber den Magen der Cetaceen.
Morpholog. Jahrbuch, xiii, pp. 616–653, with
pls. xxvii–xxviii, and 2 woodcuts.

———　　　　　　　　　　1893 A
Bemerkungen über den Ursprung
der Haare und über Schuppen bei
Säugethieren.
Anatom. Anzeiger, viii, pp. 413–423.

———　　　　　　　　　　1893 B
Zur Frage nach dem Ursprung der
Schuppen der Säugethiere.
Anatom. Anzeiger, viii, pp. 649–651.

———　　　　　　　　　　1898 A
Studien über Säugethiere.
Pp. i–iv, 1–152, with pls. i–iv. 8vo, Jena, 1898.
Part II of this work is entitled "Anatomische
Bemerkungen über Elephas," pp. 135–152, pl. iv.

Weil, Richard.　　　　　1899 A
Development of the ossicula audita
in the opossum.
Ann. N. Y. Acad. Sci., xii, pp. 103–107, with
pls. ii, iii.

Weithofer, Anton.　　　　1888 A
Beiträge zur Kenntniss der Fauna von
Pikermi bei Athen.
Beiträge Paläontologie Oesterreich-Ungarns
(Mojsisovics and Neumayer), vi, pp. 225–292,
with pls. x–xix.
Reviewed by M. Schlosser in Archiv f.
Anthrop. xix, pp. 146–148.

———　　　　　　　　　　1888 B
Einige Bemerkungen ·über den Car-
pus der Proboscidier.
Morpholog. Jahrbuch, 1888, pp. 507–516.

———　　　　　　　　　　1890 A
Die fossilen Proboscidier des Arno-
thales in Toskana.
Beiträge Paläontologie Oesterreich-Ungarns
(Mojsisovics and Neumayer), viii, pp. 108–240,
with pls. i–xv.
Review in Archiv f. Anthrop., xx, pp. 116–
118.

———　　　　　　　　　　1893 A
Proboscidiani fossili di Valdarno in
Toscano.
Memoire per servire alla descrizione della
carta geologica d'Italia, iv, pt. 2, pp. 1–152, with
pls. 1–xv.

Wellburn, Edgar D.　　　1900 A
On the genus *Megalichthys*, Agassiz;
its history, systematic position, and
structure.
Proc. Yorkshire Geol. and Polytech. Soc. (2),
xiv, pp. 52–71, with pls. xiii–xvii, xix.

———　　　　　　　　　　1900 B
On *Rhadinichthys monensis*, Egerton,
and its distribution in the Yorkshire
coal field.
Geol. Magazine (4), vii, pp. 260–263, with 1
fig. in the text.

Weller, Stuart.　　　　　1899 A
A peculiar Devonian deposit in north-
eastern Illinois.
Jour. Geology, vii, pp. 483–488, with 3 text
figs.
Reports occurrence of *Ptychodus calceolus*,
Diplodus priscus, and *D. striatus*.

West, H. H.　　　　　　1877 A
Wonderful discoveries in the Sandstone rocks of Colorado.
Western Review of Science and Industry, i,
pp. 564–565 (Kansas City, Mo.).
A description of the finding of some huge
reptilian bones near Morrison, Colo. Copied
from the Colorado Springs Gazette.

Weyhe, ——.　　　　　　1875 A
Uebersicht der Säugethiere nach
ihren Beckenformen.
Zeitschr. gesammt. Naturwiss., xlv (n. F.
xi), 1875, pp. 97–123.

Wheatley, Charles M.   1861 A
Remarks on the Mesozoic red sandstone of the Atlantic slope, and notice
of the discovery of a bone bed therein,
at Phœnixville, Pa.
Amer. Jour. Sci. (2), xxxii, pp. 41-48.

————   1871 A
Notice of the discovery of a cave in
eastern Pennsylvania, containing remains of post-Pliocene fossils, including those of mastodon, tapir, megalonyx, mylodon, etc.
Amer. Jour. Sci. (3), i, pp. 235-237, with 2 figs.
showing the location of the cave.

————   1871 B
The bone cave of eastern Pennsyl-
vania.
Amer. Jour. Sci. (3), i, pp. 384-3a5.
Contains a list of the species found up to
that time in Port Kennedy bone cave.

Wheeler, William.   1878 A
A fossil tusk found in Franklin
County, [Kansas].
Trans. Kansas Acad. Sci., vi, p. 11.
Supposed to be mastodon.

Wheeler, William Morton.   1899 A
George Baur's life and writings.
Amer. Naturalist, xxxiii, pp. 15-30, with
portrait.
This paper ends with a list of Dr. Baur's
scientific writings, 144 in number.

Whipple, S. H.   1844 A
Notice of bones of mastodon.
Proc. Amer. Philos. Soc., iv, pp. 35-36.
A letter accompanying a collection of mas-
todon bones from Benton County, Mo.

White, C. A. .   1883 A
On the commingling of ancient faunal
and modern floral types in the Laramie group.
Amer. Jour. Sci. (3), xxvi, pp. 120-123.

————   1887 A
On the interrelations of contempo-
raneous fossil faunas and floras.
Amer. Jour. Sci. (3), xxxiii, pp. 364-374.

White, Philip J.   1895 A
The existence of skeletal elements
between the mandibular and hyoid
arches in *Hexanchus* and *Læmargus*.
Anatom. Anzeiger, xi, pp. 57-60, with 3 figs.

Whiteaves, J. F.   1880 A
On a new species of *Pterichthys*,
allied to *Bothriolepis ornata*, from the
Devonian rocks of the north side of
the Baie des Chaleurs.
Amer. Jour. Sci. (3), xx, 132-136.

Whiteaves, J. F.—Continued.   1881 A
On some remarkable fossil fishes
from the Devonian rocks of Scaumenac
Bay, Province of Quebec.
Amer. Jour. Sci. (3), xxi, pp. 494-496.
Also in Ann. and Mag. Nat. Hist. (5), viii,
159-162.
*Diplacanthus striatus? Phaneropleuron cur-
tum, Eusthenopteron foordii, Glyptolepis microlepidotus, Glyptolepis* ——, and *Cheirolepis canadensis* are described.

————   1881 B
On a new species of *Pterichthys,* allied
to *Bothriolepis ornata* Eichwald, from
the Devonian rocks of the north side
of the Baie des Chaleurs.
Canadian Naturalist (2), x, pp. 23-27.

————   1881 C
On some remarkable fossil fishes
from the Devonian rocks of Scaumenac
Bay, P. Q., with descriptions of a new
genus and three new species.
Canadian Naturalist (2), x, pp. 27-35.

————   1881 D
Description of a new species of
Psammodus from the Carboniferous
rocks of the island of Cape Breton.
Canadian Naturalist (3), x, p. 36.
Describes *Psammodus bretonensis.*

————   1881 E
On some fossil fishes, crustacea, and
mollusca from the Devonian rocks at
Campbellton, N. B., with descriptions
of five new species.
Canadian Naturalist (2), x, pp. 98-101, with
1 fig. in the text.
Describes *Onccosteus acadicus, Cephalaspis
campbelltonensis, Ctenacanthus latispinosus* and
Homacanthus sp. indet.

————   1883 A
Recent discoveries of fossil fishes in
the Devonian rocks of Canada.
Proc. Amer. Assoc. Adv. Sci., 31st meeting,
Montreal, 1882, pp. 353-356.

————   1883 B
Recent discoveries of fossil fishes in
the Devonian rocks of Canada.
Amer. Naturalist, xvii, pp. 158-164.

————   1887 A
Illustrations of the fossil fishes of
the Devonian rocks of Canada.
Trans. Roy. Soc. Canada, iv, sec. iv, 1886,
pp. 101-110; pls. vi-x.

————   1889 A
Illustrations of the fossil fishes of the
Devonian rocks of Canada.   Part II.
Trans. Roy. Soc. Canada, vi, sec. iv, 1888,
pp. 77-96; pls v-x.

Whiteaves, J. F.—Continued.   1889 B
On some fossils from the Hamilton
formation of Canada, with a list of the
species at present known from that
formation and province. 3. The fossils
of the Triassic rocks of British Columbia. 4. On some Cretaceous fossils from
British Columbia, the Northwest Territory, and Manitoba.
Geol. and Nat. Hist. Survey of Canada.
Contrib. to Canad. Palæont. i, part ii, art. 2, pp.
91–196, with pls. xii–xxvi.

————   1892 A
The fossils of the Devonian rocks of
the islands, shores, or immediate
vicinity of Lakes Manitoba and Winni-
pegosis.
Geol. Surv. Canada. Contrib. to Canad.
Palæont., i, part iv, art. 6, pp. 255–359, with
pls. xxxiii–xlvii.
On pages 353 and 354, pls. xlvi and xlvii,
certain fossil fishes are described and figured.

————   1898 A
Note on a fish tooth from the Upper
Arisaig series of Nova Scotia.
Canadian Record of Science, vii, pp. 461–462,
with 1 text fig.
Also in Report Brit. Assoc. Adv. Sci. for
1897, Toronto meeting, pp. 656–657.
Describes *Dendrodus arisaigensis* from rocks
supposed to be of Silurian, possibly Devonian,
age.

————   1898 B
On some additional or imperfectly
understood fossils from the Hamilton
formation of Ontario, with a revised
list of the species therefrom.
Geol. Surv. Canada. Contrib. to Canad.
Palæont., i, pp. 361–436, with pls. xlviii–l.

————   1899 A
The Devonian system in Canada.
Science (2), x, pp. 402–412; 430–438.
Also in Proc. Amer. Assoc. Adv. Sci., xlviii
(Columbus), pp. 198–223.

Whitfield, R. P.   1888 A
Evidence confirmatory of *Mastodon
obscurus* Leidy as an American species.
Proc. Amer. Assoc. Adv. Sci., 36th meeting,
New York, 1887, pp. 252–253.

————   1890 A
Observations on a fossil fish from the
Eocene beds of Wyoming.
Bull. Amer. Mus. Nat. Hist., iii, pp. 117–120,
with pl. iv.

————   1891 A
Mastodon remains on New York Is-
land.
Science, xviii, p. 342.

Whitfield, R. P.—Continued.   1900 A
Note on the principal type specimen
of *Mosasaurus maximus* Cope, with illustrations.
Bull. Amer. Mus. Nat. Hist., xiii, pp. 25–29,
with pls. iv, v.

Whitfield, R. P. and **Hovey**, E. O. 1900 A
A catalogue of the types and figured
specimens in the palæontological collection of the geological department, American Museum of Natural History.
Bull. Amer. Mus. Nat. Hist., xi, pp. 190–356.
On page 348 there is noted the presence of
five types of fossil fishes in the museum.

Whitney, J. D.   1867 A
Notice of a human skull recently taken
from a shaft near Angels, Calaveras
County.
Amer. Jour. Sci. (2), xliii, pp. 265–267.
Taken from Proc. California Acad. Nat. Sci.,
iii, pp. 277.

————   1879 A
The animal remains, not human, of
the auriferous gravel series.
Mem. Mus. Comp. Zool., vi, pp. 239–258.
This forms sec. iv of chap. iii of Whitney's
"The auriferous gravels of the Sierra Nevada
of California."
The descriptions of the species have all been
published by Dr. Leidy elsewhere, but those of
the equine animals are reproduced.

Whittlesey, Charles.   1848 A
Notes upon the drift and alluvium
of Ohio and the West.
Amer. Jour. Sci. (2), v, pp. 205–217.

————   1869 A
On the evidences of the antiquity of
man in the United States.
Proc. Amer. Assoc. Adv. Sci., 17th meeting,
Chicago, 1868, pp. 268–288.

Wiedersheim, Robert.   1876 A
Die ältesten Formen des Carpus und
Tarsus der heutigen Amphibien.
Morpholog. Jahrbuch, ii, pp. 421–434, with
pl. xxix.
See also "Nachträgliche Bemerkungen" in
vol. iii, pp. 152–154, with 5 figs.

————   1877 A
Das Kopfskelet der Urodelen.
Morpholog. Jahrbuch, iii, pp. 352–448, with
pls. xix–xxiii and 1 woodcut; pp. 459–548, with
pls. xxiv–xxvii and 5 woodcuts.

————   1878 A
Labyrinthodon rütimeyeri. Ein Beitrag zur Anatomie von Gesammtskelet
und Gehirn der triassischen Labyrin-
tho donten.
Abhandl. schweizer. paläontolog. Gesellsch.,
v, pp. 1–56, with 3 pls.

Wiedersheim, Robert—Cont'd. 1880 A
Das Skelet von *Pleurodeles waltlii.*
Jenaische Zeitschr. Naturwiss., xiv (vii), pp.
1-38, with pl. l.

—— 1880 B
Das Skelet und Nervensystem von
Lepidosiren annectens (Protopterus).
Jenaische Zeitschr. Naturwiss., xiv (vii), pp.
155-192, with pls. vii, viii.

—— 1881 A
Zur Palæontologie Nord-Amerikas.
Biolog. Centralbl., i, pp. 359-372.

—— 1884 A
Die Stammesentwicklung der Vögel.
Biolog. Centralbl., iii, pp. 654-668.

—— 1889 A
Über die Entwicklung des Schulter-
und Beckengürtels.
Anatom. Anzeiger, iv, pp. 428-441.

—— 1890 A
Weitere Mitteilungen über die Ent-
wicklungsgeschichte des Schulter- und
Beckensgürtels.
Anatom. Anzeiger, v, pp. 18-26.

Wiegmann, Fr. Aug. 1838 A
Ueber das Gebiss des Wallrosses.
Archiv Naturgesch., iv, Bd. l, pp. 112-130.

Wieland, George R. 1896 A
Archelon ischyros: a new gigantic
cryptodire testudinate from the Fort
Pierre Cretaceous of South Dakota.
Amer. Jour. Sci. (4), ii, pp. 399-412, with pl.
vi and 19 figs. in the text.
Describes under above name the carapace
and other bones of a new turtle.

—— 1898 A
The Protostegan plastron.
Amer. Jour. Sci. (4), v, pp. 15-20, with pl. ii
and 2 figs. in the text.

—— 1899 A
The terminology of vertebral centra.
Amer. Jour. Sci. (4), viii, pp. 163-164.

—— 1900 A
The skull, pelvis, and probable rela-
tionships of the huge turtles of the
genus *Archelon* from the Fort Pierre
Cretaceous of South Dakota.
Amer. Jour. Sci. (4), ix, pp. 237-251, with
pl. ii and 6 text figs.
Noticed by Osborn, Science, xii, 1900, p. 2.

—— 1900 B
Some observations on certain well-
marked stages in the evolution of the
testudinate humerus.
Amer. Jour. Sci. (4), ix, pp. 413-424, with 23
text. figs.

Wilckens, M. 1884 A
Uebersicht über die Forschungen
auf dem Gebiete der Paläontologie der
Haustiere. 1. Die pferdeartigen Tiere
des Tertiärs.
Biolog. Centralbl., iv, pp. 137-154; 183-188.

—— 1884 B
Uebersicht über die Forschungen
auf dem Gebiete der Paläontologie der
Haustiere. 2. Die Pferde des Diluviums.
Biolog. Centralbl.. iv, pp. 294-310; 327-344.
Contains Lydekker's list of Pliocene and
recent *Equidæ.*

—— 1885 A
André Sanson, Sur les Equidés quar-
ternaires.
Biolog. Centralbl., v, pp. 184-187.

—— 1885 B
Uebersicht über die Forschungen
auf dem Gebiete der Paläontologie der
Haustiere. 5. Die schweinartigen Tiere
(Suiden).
Biolog. Centralbl., v, pp. 208-222; 233-241;
263-270; 295-308; 332.

—— 1885 C
Uebersicht über die Forschungen
auf dem Gebiete der Paläontologie der
Haustiere. 6. Die Kamelartigen Tiere
(Cameliden).
Biolog. Centralbl., v, pp. 418-434.

—— 1885 D
Uebersicht über die Forschungen auf
dem Gebiete der Paläontologie der
Haustiere. 7. Die hundartigen Tiere
(Caniden) des Tertiärs.
Biolog. Centralbl., v, pp. 459-468; 489-499;
518-529.

—— 1885 E
Uebersicht über die Forschungen auf
dem Gebiete der Paläontologie der
Haustiere. 3. Die Abstammung des
Rindes und die tertiären Formen des-
selben.
Biolog. Centralbl., iv, pp. 749-766.
Reviews of Wilckens's various papers by M.
Schlosser in Archiv f. Anthrop., xvii, pp. 187-
191; xviii, pp. 146-150.

—— 1888 A
Beitrag zur Kenntniss des Pferdege-
bisses mit Rücksicht auf die fossilen
Equiden von Maragha in Persien.
Nova Acta Caes. Akad., Leop.-Car., lii, pp.
257-284, with pls. ix-xvi.

Wilder, B. G. 1871 A
Mastodon remains in central New
York.
Amer. Jour. Sci. (3), ii, p. 58.
A brief note.

Wilder, B. G.—Continued. 1880 A
Some recent American papers in comparative anatomy.
Science (John Michels, editor), i, pp. 322–323.

Wildermuth, H. A. 1877 A
Der feinere Bau der lufthaltigen Vogelknochen nebst Beiträgen zur Kenntniss ihrer Entwicklung.
Jenaische Zeitschr. Naturwiss., xi (o. s.), pp. 537–550, with pl. xxxvi.

Willemoes-Suhm, Rudolph. 1869 A
Ueber *Cœlacanthus* und einige verwandte Gattungen.
Palæontographica, xvii, pp. 73–88, with pls. x–xi.

Williams, Henry S. 1882 A
Note on some fish-remains from the Upper Devonian rocks in New York State.
Proc. Amer. Assoc. Adv. Sci., 30th meeting, Cincinnati, 1881, pp. 192–193.
Describes *Dipterus ithacensis.*

——— 1887 A
On the fossil faunas of the Upper Devonian, the Genesee section, New York.
Bull. U. S. Geol. Surv. No. 41, pp. 1–104, with pls. i–iv.

——— 1891 A
On the plates of *Holonema rugosa.*
Proc. Amer. Assoc. Adv. Sci., 39th meeting, Indianapolis, 1890, p. 337.

——— 1892 A
The scope of palæontology and its value to geologists.
Proc. Amer. Assoc. Adv. Sci., 41st meeting, Rochester, 1892, pp. 149–170.
Vice-presidential address before Amer. Assoc. Adv. Sci., 1892.

——— 1893 A
On the ventral plates of the carapace of the genus *Holonema* of Newberry.
Amer. Jour. Sci. (3), xlvi, 285–288, with figs. in the text.

Williams, Herbert Upham. 1886 A
Notes on the fossil fishes of the Genesee and Portage black shales.
Bull. Buffalo Soc. Nat. Sci., v, pp. 81–84, with 1 pl.

Williamson, W. C. 1837 A
On the affinity of some fossil scales of fish from the Lancashire Coal measures with those of the recent *Salmonidæ.*
Lond. and Edinb. Philos. Mag. and Jour. Sci. (3), xi, pp. 300–301, with pl. ii.
The scales here described and figured are now referred to *Rhizodopsis sauroides.*

Williamson, W. C.—Continued. 1849 A
On the microscopic structure of the scales and dermal teeth of some Ganoid and Placoid fishes.
Philos. Trans. Roy. Soc. Lond., 1849, pp. 435–475; pls. 40–43.

Williston, S. W. 1878 A
American Jurassic dinosaurs.
Trans. Kansas Acad. Sci., vi, pp. 42–46.

——— 1890 A
Structure of the plesiosaurian skull.
Science, xvi, p. 262.

——— 1890 B
Structure of the plesiosaurian skull.
Science, xvi, p. 290.

——— 1890 C
Note on the pelvis of *Cumnoria (Camptosaurus.)*
Amer. Naturalist, xxiv, pp. 472–478.

——— 1890 D
A new plesiosaur from the Niobrara Cretaceous of Kansas.
Trans. Kansas Acad. Sci., xii, pp. 174–178, with 2 figs. in the text.
Describes and figures *Cimoliosaurus (Elasmosaurus) snowi.*
Noticed by Cope, Amer. Naturalist, xxv, p. 653.

——— 1891 A
Kansas mosasaurs.
Science, xviii, p. 345.

——— 1891 B
The skull and hind extremity of *Pteranodon.*
Amer. Naturalist, xxv, pp. 1124–1126.

——— 1892 A
Kansas pterodactyls. Part I.
Kansas Univ. Quarterly, i, pp. 1–13, pl. i.
Contains list of species, with description of the osteology of *Pteranodon* and *Nyctodactylus.*
A brief abstract of this paper is found in Amer. Naturalist, xxvii, p. 37.

——— 1893 A
An interesting food habit of the plesiosaurs.
Trans. Kansas Acad. Sci., xiii, pp. 121–122, with plate.
Refers to the occurrence of pebbles in the stomachs of plesiosaurs.

——— 1893 B
Kansas pterodactyls. Part II.
Kansas Univ. Quarterly, ii, pp. 79–81, with 1 fig. in the text.

——— 1893 C
Kansas mosasaurs. Part II. Restoration of *Clidastes.*
Kansas Univ. Quarterly, ii, pp. 83–84, with pl. iii.

Williston, S. W.—Continued.   1894 A

A food habit of the plesiosaurs.

Amer. Naturalist, xxviii, p 50.

Extract from Trans. Kansas Acad. Sci., xiii, p. 121.

———   1894 B

Vertebrate remains from the lowermost Cretaceous.

Kansas Univ. Quarterly, iii, pp. 1–4, with pl. i.

———   1894 C

Restoration of *Platygonus*.

Kansas Univ. Quarterly, iii, pp. 23–39, with pls. vii, viii, and 7 text figs.

Notice of, by M. Schlosser, in Archiv f. Anthrop., xxiv, p. 171.

———   1894 D

A new turtle from the Benton Cretaceous.

Kansas Univ. Quarterly, iii, pp. 5–18, with pls. ii–iv, with 9 text-figs.

Describes as belonging to new family, genus, and species, *Desmatochelys lowii*.

———   1894 E

Restoration of *Aceratherium fossiger* Cope.

Kansas Univ. Quarterly, ii, pp. 289–290, pl. viii.

———   1894 F

A new Dicotyline mammal from the Kansas Pliocene.

Science, xxiii, p. 164.

Describes *Dicotyles leptorhinus* Williston.

———   1895 A

New or little-known extinct vertebrates.

Kansas Univ. Quarterly, iii, pp. 165–176.

Abstract by M. Schlosser in Archiv f. Anthrop., xxv, p. 183.

———   1895 B

Note on the mandible of *Ornithostoma*.

Kansas Univ. Quarterly, iv, p. 61.

———   1896 A

On the dermal coverings of *Hesperornis*.

Kansas Univ. Quarterly, v, pp. 53–54, with pl. ii.

———   1896 B

On the skull of *Ornithostoma*.

Kansas Univ. Quarterly, iv, pp. 195–197, with pl. i.

Ornithostoma is adopted instead of *Pteranodon* Marsh.

———   1897 A

Restoration of *Ornithostoma* (*Pteranodon*).

Kansas Univ. Quarterly, vi, pp. 35–51, with pl. ii, and 4 figs. in the text.

Williston, S. W.—Continued.   1897 B

Notice of some vertebrate remains from the Kansas Permian.

Kansas Univ. Quarterly, vi, pp. 53–56, with 2 figs. in the text.

See also Science (2), v, p. 895, and Trans. Kansas Acad. Sci., xv, pp. 120–122.

———   1897 C

A new plesiosaur from the Kansas Comanche Cretaceous.

Kansas Univ. Quarterly, vi, p. 57.

Description of *Plesiosaurus gouldii* n. sp.

———   1897 D

Brachysaurus, a new genus of mosasaurs.

Kansas Univ. Quarterly, vi, pp. 95–96, with pl. viii.

Description of *Brachysaurus overtoni* gen. et sp. nov.

———   1897 E

On the extremities of *Tylosaurus*.

Kansas Univ. Quarterly, vi, pp. 99–102, with pls. ix–xii, and 1 fig. in the text.

One of the plates shows the scaled condition of the skin of *Tylosaurus*.

———   1897 F

Range and distribution of the mosasaurs, with remarks on synonymy.

Kansas Univ. Quarterly, vi, pp. 177–185, with pl. xx.

———   1897 G

A new labyrinthodont from the Kansas Carboniferous.

Kansas Univ. Quarterly, vi, pp. 209–210, with pl. xxi.

———   1897 H

The Kansas Niobrara Cretaceous.

Univ. Geol. Surv. Kansas, ii, pp. 235–246.

Contains a brief account of the genera of vertebrates found in this group of deposits.

———   1897 I

The Pleistocene of Kansas.

Univ. Geol. Surv. Kansas, ii, pp. 299–308, with pls.

———   1897 J

Restoration of Kansas Mosasaurs.

Kansas Univ. Quarterly, vi, A, pp. 107–110, with pl. xiii.

———   1898 A

Restoration of Kansas Cretaceous animals.

Kansas Univ. Quarterly, vii.

Frontispiece restoration of *Clidastes* is given.

———   1898 B

Birds.

Univ. Geol. Surv. Kansas, iv, pp 43–53, with pls. v–viii.

This is a synopsis of the fossil birds of Kansas.

Williston, S. W.—Continued.   1898 C
Dinosaurs.
Univ. Geol. Surv. Kansas, iv, pp. 67-70, with pl. ix.
A brief exposition of the characters of the group and a plate of *Claosaurus annectens*, to illustrate the structure of the Kansan species, *C. agilis*.

————   1898 D
Crocodiles.
Univ. Geol. Surv. Kansas, iv, pp. 75-78, with 2 figs. in the text.

————   1898 E
Mosasaurs.
Univ. Geol. Surv. Kansas, iv, pp. 83-221, with pls. x-lxxii and 5 figs. in the text.

————   1898 F
The sacrum of *Morosaurus*.
Kansas Univ. Quarterly, vii, pp. 173-175, with 2 figs. in the text.

————   1898 G
Miocene Edentates.
Science (2), viii, p. 132.
Shows that Cope's *Caryoderma snovianum* is a *Testudo*, possibly *T. undata* Cope.

————   1898 H
Editorial notes.
Kansas Univ. Quarterly, vii, A, p. 235.
Note on nuchal fringe and scales of *Platecarpus*.

————   1898 I
The Pleistocene of Kansas.
Trans. Kansas Acad. Sci., xv, pp. 90-94.
Presents an annotated list of the Pleistocene vertebrates found in Kansas. A reprint of Williston, S. W., 1897 I.

————   1898 J
Notice of some vertebrate remains from the Kansas Permian.
Trans. Kansas Acad. Sci., xv, pp. 120-122.

————   1898 K
Saber-toothed cats.
Pop. Sci. Monthly, liii, pp. 348-351, with 1 text-fig.

————   1899 A
Some additional characters of the mosasaurs. Contributions from the palæontological laboratory, No. 42.
Kansas Univ. Quarterly, viii, A, pp. 39-41.
Describes the episternum and nuchal fringe in *Platecarpus*. Also shows that the food is principally fishes.

————   1899 B
A new genus of fishes from the Niobrara Cretaceous.
Kansas Univ. Quarterly, viii, pp. 113-115, with pl. xxvi.

Williston, S. W.—Continued   1899 C
The red-beds of Kansas.
Science (2), ix, p. 221.
Records the finding of *Eryops megacephalus* in the above beds.

————   1899 D
A new species of *Sagenodus* from the Kansas Coal-measures.
Kansas Univ. Quarterly, viii, pp. 175-181, with pls. xxviii, xxxv-xxxvii.

————   1899 E
Notes on the coraco-scapula of *Eryops* Cope.
Kansas Univ. Quarterly, viii, pp. 185-186, with pls. xxvii, xxix, xxx.

————   1900 A
Some fish-teeth from the Kansas Cretaceous.
Kansas Univ. Quarterly, ix, pp. 27-42, with pls. vi-xiv.

————   1900 B
Cretaceous fishes [of Kansas]. Selachians and Pycnodonts.
Univ. Geol. Surv. Kansas, vi, pp. 237-256, with pls. xxiv-xxxii.
This is a reprint of the preceding paper.

Williston, S. W., and **Case**, E. C.   1892 A
Kansas mosasaurs.
Kansas Univ. Quarterly, i, pp. 15-32, with pls. ii-vi.
Gives list of species and describes the osteology of *Clidastes velox* and *C. westii* n. sp.

————   1898 A
Turtles. Introduction, and *Desmatochelys*, by S. W. Williston. *Toxochelys*, by E. C. Case.
Univ. Geol. Surv. Kansas, iv, pp. 349-387, with pls. lxxiii-lxxxiv, with 1 series of text figs.

Willett, Henry.
See **Etheridge** and **Willett**.

Wilson, J. T., and **Hill**, J. P.   1897 A
Observations upon the development and succession of the teeth in *Perameles*, together with a contribution to the discussion of the homologies of the teeth in marsupial animals.
Quart. Jour. Micros. Sci. (2), xxxix, pp. 427-588, with pls. xxv-xxxii.

Wilson, Thomas.   1892 A
Man and the *Mylodon*.
Amer. Naturalist, xxvi, pp. 628-631.
Presents results of chemical analysis of bones of man and *Mylodon*, found together.

Winchell, Alex. 1863 A
Description of elephantine molars in the museum of the University of Michigan.
Canadian Naturalist, 1863, viii, pp. 398–400.

——— 1864 A
Notice of the remains of a mastodon recently discovered in Michigan.
Amer. Jour. Sci. (2), xxxviii, pp. 223–224.
See also Canadian Naturalist, 1893, p. 398.

Winckler, T. C. 1869 A
Des tortues fossiles conservées dans le Musée Teyler et dans quelques autres musées.
Arch. Mus. Teyler, ii, pp. i–ii; 1–151, with pls. i–xxxiii.

——— 1871 A
Mémoire sur le *Cœlacanthus harlemensis.*
Arch. Mus. Teyler, iii, pp. 101–116, with pl. iv.
Notice and review of, by W. D[awkins] in Geol. Magazine, 1872, pp. 81–32.

——— 1873 A
Le *Plesiosaurus dolichodeirus* Conyb.
Arch. Mus. Teyler, iii, pp. 219–233, with pl. vi, and 10 figs. in the text.

——— 1874 A
Le *Pterodactylus kochii* Wagn. du Musée Teyler.
Archiv. Mus. Teyler, iii, pp. 377–387, with pl. vii.

——— 1874 B
Mémoire sur des dents de poissons du terrain bruxellian.
Arch. Mus. Teyler, iii, pp. 295–304, with pl. vii.

Windle, B. C. A., and Humphreys, John. 1887 A
On man's lost incisors.
Report Brit. Assoc. Adv. Sci., 56th meeting, Birmingham, 1886, pp. 688–691.
Abstract; full paper in Jour. Anat. and Physiol., xxi, p. 84.

Winge, Herluf. 1888 A
Jordfundne og nulevende Gnavere (*Rodentia*) fra Lagoa Santa, Minas Geraes, Brasilien.
E Museo Lundii, i, Art. iii, pp. 1–200, with pls. i–viii.
Abstract of, by M. Schlosser, in Archiv f. Anthrop., xix, pp. 89–90.

——— 1893 A
Jordfundne og nulevende Flagermus (*Chiroptera*) fra Lagoa Santa, Minas

Winge, Herluf—Continued.
Geraes, Brasilien. Med Udsigt over Flagermusenes indbyrdes Slægtskab.
E Museo Lundii, ii, pp. 1–65, with pls. 1, ii. Kjöbenhavn, 1893.
Abstract of, by M. Schlosser, in Archiv f. Anthrop., xxiv, pp. 142–143.

——— 1893 B
Chauves-souris fossiles et vivantes de Lagoa Santa, Minas Geraes, Brésil.
E Museo Lundii, ii, pp. 67–92.

——— 1893 C
Jordfundne og nulevende Pungdyr (*Marsupialia*) fra Lagoa Santa, Minas Geraes, Brasilien. Med Udsigt over Pungdyrenes Slægtskab.
E Museo Lundii, ii, pt. ii, pp. 1–132, with pls. i–iv.
Abstract of, by M. Schlosser, in Archiv f. Anthrop., xxiv, pp. 143–145.

——— 1895 A
Jordfundne og nulevende Aber (*Primates*) fra Lagoa Santa, Minas Geraes, Brasilien. Med Udsigt over Abernes indbyrdes Slægtskab.
E Museo Lundii, ii, pt. iii, pp. 1–45, with pls. i–ii.
Abstract of, by M. Schlosser, in Archiv f. Anthrop., xxv, pp. 238–239.

——— 1895 B
Jordfundne og nulevende Rovdyr (*Carnivora*) fra Lagoa Santa, Minas Geraes, Brasilien. Med Udsigt over Rovdyrenes indbyrdes Slægtskab.
E Museo Lundii, ii, pt. iv, pp. 1–103, with pls. i–viii. Kjöbenhavn, 1895.
Notice of this paper by M. Schlosser in Neues Jahrb. Mineral., 1899, ii, pp. 144–147, and in Archiv f. Anthrop., xxv, pp. 239–241.

Winslow, C. F. 1857 A
Letter on finding human remains and those of elephant and mastodon in California.
Proc. Boston Soc. Nat. Hist., vi, p. 278.

——— 1868 A
On human remains along with those of the mastodon in the drift of California.
Amer. Jour. Sci. (2), xlvi, pp. 407–408.
Taken from Proc. Boston Soc. Nat. Hist., vi, 1857, p. 278.

Wislizenus, A. 1858 A
Was man contemporaneous with the mastodon?
Trans. Acad. Sci. St. Louis, i, pp. 168–171.

Wistar, Caspar. 1799 A

A description of the bones deposited by the president in the museum of the society, and represented in the annexed plates.

Trans. Amer. Philos. Soc., iv, pp. 526–531, with 2 pls.

Description of bones of *Megalonyx* deposited by the president of the society, Thomas Jefferson.

——— . 1818 A

An account of two heads found in the morass called the Big Bone Lick and presented to the society by Mr. Jefferson.

Trans. Amer. Philos. Soc. (2), i, pp. 375–380, with pls. x, xi.

Mentions bones of mammoth and mastodon, and figures bones of extinct deer and ox.

Woodhull, Alfred A. 1872 A

On the elephant in Colorado.

Amer. Jour. Sci. (3), iii, p. 374.

Reports occurrence of *E. americanus* in Colorado.

Woodward, A. S. 1884 A

Chapter on fossil sharks and rays.

Hardwicke's Science-Gossip, xx, pp. 172–174; 227–230; 267–272, with figs. 99–101; 129–143; 161–177.

——— 1885 A

On the literature and nomenclature of British fossil *Crocodilia*.

Geol. Magazine (3), ii, pp. 496–510.

Contains a list of papers on the subject.

——— 1885 B

Chapters on fossil sharks and rays.

Hardwicke's Science-Gossip, xxi, pp. 106–109; 154–156; 226–229; 270–272, with figs. 71–84; 101–111; 155–161; 181–188.

These articles form a continuation of the paper 1884 A. 1886 E is the conclusion.

——— 1886 A

On the relations of the mandibular and hyoid arches in a Cretaceous shark (*Hybodus dubrisiensis* Mackie).

Proc. Zool. Soc Lond.,1886, pp. 218–224, pl. xx.

——— 1886 B

Note on the presence of a columella (epipterygoid) in the skull of *Ichthyosaurus*.

Proc. Zool. Soc. Lond., 1886, pp. 405–408, woodcuts 1–4.

——— 1886 C

On the palæontology of the selachian genus *Notidanus*.

Geol. Magazine (3), iii, pp. 205–217; 253–259, with pl. vi and 2 figs. in the text.

Woodward, A. S.—Continued. 1886 D

The history of fossil crocodiles.

Proc. Geologists' Assoc., ix, pp. 288–344, with 25 figs. in the text.

——— 1886 E

Chapters on fossil sharks and rays.

Hardwicke's Science-Gossip, xxii, pp. 4–6, with figs. 1, 2.

This forms the conclusion of the papers of this author designated 1884 A and 1885 B.

——— 1887 A

On the presence of a canal system, evidently sensory, in the shields of Pteraspidian fishes.

Proc. Zool. Soc. Lond.,1887, pp. 478–481, with 1 woodcut.

——— 1887 B

On the dentition and affinities of the selachian genus *Ptychodus* Agassiz.

Quart. Jour. Geol. Soc., xliii, pp. 121–131, with pl. x.

——— 1887 C

On "leathery turtles," recent and fossil, and their occurrence in British Eocene deposits.

Proc. Geologists' Assoc., x, pp. 2–14.

Contains numerous references to the literature of the subject.

——— 1888 A

[Note on J. S. Newberry's paper "On the structure and relations of *Edestus*, with a description of a gigantic new species," published in Ann. N. Y. Acad. Sci., v, 1888, pp. 1–10.]

Geol. Magazine (3), v, p. 374.

——— 1888 B

Note on Prof. W. Dames's "Die Gattung *Saurodon*," published in Sitzungsb. Ges. naturf. Freunde, Berlin, 1887.

Geol. Magazine (3), v, p. 156.

——— 1888 C

M. Charles Brongniart on *Pleuracanthus*.

Geol. Magazine (3), v, pp. 422–425.

Review of paper published in Comptes rend. Acad. Sci. Paris, April 23, 1888.

——— 1888 D

Note on the occurrence of a species of *Onychodus* in the lower Old Red sandstone passage beds of Ledbury, Herefordshire.

Geol. Magazine (3), v, pp. 500–501.

Woodward, A. S.—Continued.   1888 E

Dr. Anton Fritsch on *Ctenodus* and other Palæozoic fishes.

Geol. Magazine (3), v, pp. 523–526.

A review of Fauna der Gaskohle, etc., Bd. ii, Heft 3, pp. 65–92.

——   1888 F

Notes on the determination of the fossil teeth of *Myliobatis*, with a revision of the English Eocene species.

Ann. and Mag. Nat. Hist. (6), i, pp. 36–47, with pl. i.

——   1888 G

On the fossil fish-spines named *Cœlorhynchus*, Agassiz.

Ann. and Mag. Nat. Hist. (6), ii, pp. 223–226.

——   1888 H

On some remains of the extinct selachian *Asteracanthus* from the Oxford clay of Peterborough, preserved in the collection of Alfred N. Leeds, esq., of Eyebury.

Ann. and Mag. Nat. Hist. (6), ii, pp. 336–342, with pl. xii.

Strophodus probably referable to *Asteracanthus.*

——   1888 I

A synopsis of the vertebrate fossils of the English Chalk.

Proc. Geologists' Assoc., x, pp. 273–338, with pl. i.

——   1888 J

On a head of *Hybodus delabechei*, associated with dorsal fin-spines, from the Lower Lias of Lyme Regis, Dorsetshire.

Annual Report Yorkshire Philos. Soc., 1888, pp. 58–61, with pl. i.

——   1889 A

A comparison of the Cretaceous fish-fauna of Mount Lebanon with that of the English Chalk.

Report Brit. Assoc. Adv. Sci., 58th meeting, Bath, 1888, p. 678.

——   1889 B

Remarks on *Sclerorhynchus atavus.*

Proc. Zool. Soc. Lond., 1889, pp. 449–451, with 1 woodcut.

Describes an extinct saw-fish from Mount Lebanon.

——   1889 C

Note on *Bucklandium diluvii*, König, a siluroid fish from the London clay of Sheppey.

Proc. Zool. Soc. Lond., 1889, pp. 208–210, with pl. xxii.

Woodward, A. S.—Continued.   1889 D

Catalogue of the fossil fishes in the British Museum. Part I. Containing the *Elasmobranchii.*

London, 1889, pt. 1, pp. i–xlvii; 1–474, pls. i–xvii, woodcuts 13.

See notice of this volume in Geol. Magazine, vi, 1889, p. 366.

——   1889 E

Prof. Dr. Von Zittel on palæichthyology.

Geol. Magazine (3), vi, pp. 125–130; 177–181; 227–282.

A review of Dr. Zittel's Handbuch der Palæontologie.

——   1889 F

Palæichthyological notes. I. On the so-called *Hybodus keuperinus*, Murch. & Strickl. II. On *Diplodus moorei*, sp. nov., from the Keuper of Somersetshire. III. On a symmetrical Hybodont tooth from the Oxford clay of Peterborough. IV. On a maxilla of *Saurichthys* from the Rhætic of Aust Cliff, near Bristol.

Ann. and Mag. Nat. Hist. (6), iii, pp. 297–302, with pl. xiv.

——   1889 G

Acanthodian fishes from the Devovian of Canada.

Ann. and Mag. Nat. Hist. (6), iv, pp. 183–184.

——   1889 H

On the palæontology of sturgeons.

Proc. Geologists' Assoc., xi, pp. 24–44, with pl. i, and 13 figs. in the text.

——   1890 A

Vertebrate palæontology in some American and Canadian museums.

Geol. Magazine (3), vii, pp. 390–395; 455–460.

——   1890 B

A new theory of *Pterichthys.*

Ann. and Mag. Nat. Hist. (6), vi, pp. 314–316.

Notice of paper by W. Patten in Quart. Jour. Micros. Sci., xxxi, 1890, pp. 317–378.

——   1890 C

A synopsis of the fossil fishes of the English lower Oolites.

Proc. Geologists' Assoc., xi, pp. 285–306, with pl. iii.

——   1890 D

The fossil fishes of the Hawkesbury series at Gosford, New South Wales.

Mem. Geol. Surv. N. S. Wales, Palæont., No. 4, pp. 1–55, with pls. i–x. Sydney, 1890.

Abstract in Ann. and Mag. Nat. Hist. (6), vi, p. 423.

Woodward, A. S.—Continued. 1891 A
Catalogue of the fossil fishes in the
British Museum. Part II. Containing
the *Elasmobranchii* (*Acanthodii*), *Holocephali, Ichthyodorulites, Ostracodermi,
Dipnoi,* and *Teleostomi* (*Crossopterygii*),
and chondrostean *Actinopterygii.*
Pp. i-xliv; 1-567, pls. i-xvi and 57 text figs.

——— 1891 B
Dr. Anton Fritsch on Palæozoic
elasmobranch fishes.
Geol. Magazine (3), viii, pp. 375-378, with 2
figs.
A review of Dr. Fritsch's Fauna der Gaskohle, etc.

——— 1891 C
Armored palæozoic sharks.
Geol. Magazine (3), viii, pp. 422-425.
A review of three of Dr. Otto Jaekel's papers
on Elasmobranchs.

——— 1891 D
Review of " Pineal fontanelle of Placoderms and catfish " by Bashford
Dean in 19th Report of the Commissioners of Fisheries of N. Y., 1891.
Geol. Magazine (3), viii, p. 518.

——— 1892 A
On the Lower Devonian fish-fauna
of Campbelltown, New Brunswick.
Geol. Magazine (3), ix, pp. 1-6, pl. 1.
Describes *Protodus jexi* gen. et spec. nov.,
Diplodus problematicus sp. nov., *Gyracanthus
incurvus, Acanthodes semistriatus, Cephalaspis
campbelltonensis, Phlyctænaspis acadica.*

——— 1892 B
Palæozoic fishes.
Geol. Magazine (3), ix, pp. 233-235.
Review of Professor Cope's "On the characters of some Palæozoic fishes" and Rohon's
"Ueber *Pterichthys.*"

——— 1892 C
Further contributions to knowledge
of the Devonian fish-fauna of Canada.
Geol. Magazine (3), ix, pp. 481-485; pl. xiii,
and 1 fig. in text.
Treats of *Phlyctænaspis acadica, Diplacanthus
horridus,* sp. nov., *Coccosteus canadensis,* sp.
nov., and *Bothriolepis.*

——— 1892 D
On the skeleton of a Chimæroid fish
(*Ischyodus*) from the Oxford clay of
Christian Malford, Wiltshire.
Ann. and Mag. Nat. Hist. (6), ix, pp. 94-96.

——— 1892 E
The evolution of fins.
Natural Science, i, pp. 28-35, with 8 figs. in
the text.

Woodward, A. S.—Continued. 1892 F
The forerunners of the backboned
animals.
Natural Science, i, pp. 596-602, with 10 figs.

——— 1892 G
The evolution of shark's teeth.
Natural Science, i, pp. 671-675, with 12 figs.
in the text.

——— 1893 A
On the cranial osteology of the Mesozoic ganoid fishes, *Lepidotus* and *Dapedius.*
Geol. Magazine (3), x, pp. 413-414.
See Woodward, A. S., 1893 D.

——— 1893 B
The fossil fishes of the British Coalmeasures.
Geol. Magazine (3), x, pp. 72-74.
A review of James W. Davis's paper on *Pleuracanthidæ* in Trans. Roy. Dublin Soc., iv, pp.
703-748, 1892.

——— 1893 C
Note on a case of subdivision of the
median fin in a dipnoan fish.
Ann. and Mag. Nat. Hist. (6), xi, pp. 241-242, with 1 fig. in the text.
Describes method of separation of the anal
fin of *Phaneropleuron curtum* from the caudal
by fusion of the axonosts.

——— 1893 D
On the cranial osteology of the
Mesozoic ganoid fishes, *Lepidotus* and
Dapedius.
Proc. Zool. Soc. Lond., 1893, pp. 559-565, with
pls. xlix and l, and 6 figs. in the text.

——— 1893 E
Note on the evolution of the scales
of fishes.
Natural Science, iii, pp. 448-450, with figures.

——— 1894 A
On the affinities of the Cretaceous
fish *Protosphyræna.*
Ann. and Mag. Nat. Hist. (6), xiii, pp.
510-512.
Shows the close relationships existing between *Protosphyræna* and *Hypsocormus.*

——— 1894 B
Note on a tooth of *Oxyrhina* from the
Red Crag of Suffolk.
Geol. Magazine (4), i, pp. 75-76, with 2 figs.
in the text.
Refers to the finding of *Oxyrhina crassa.*

——— 1895 A
Catalogue of the fossil fishes in the
British Museum. Part III. Contain-

Woodward, A. S.—Continued.

ing the actinopterygian *Teleostomi* of the order *Chondrostei* (concluded), *Protospondyli, Aëtheospondyli,* and *Isospondyli* (in part).

Pp. i–xliii, 1–544, pls. i–xviii, text figs. 45.
Review of, by R. H. Traquair in Geol. Magazine, 1896, p. 124.

———   1895 B
The problem of the primeval sharks.
Natural Science, vi, pp. 38–43, with 3 figs. in the text.

———   1895 C
On two deep-bodied species of the clupeoid genus *Diplomystus.*
Ann. and Mag. Nat. Hist. (6), xv, pp. 1–3, with pl. i, figs. 1–4.
Describes and figures *Diplomystus longicostatus* and describes *D. birdi* n. sp.

———   1895 D
On the fossil fishes of the Upper Lias of Whitby. Part I.
Proc. Yorkshire Geol. and Polytech. Soc. (2), xiii, pp. 25–42, with pls. iii–v.
Leptolepis saltviciensis, Pholidophorus germanicus, Eugnathus fasciculatus, and *Ptycholepis bollensis* are described and figured.

———   1895 E
The fossil fishes of the Talbragar beds (Jurassic?).
Mem. Geol. Surv. N. S. Wales, Palæont., No. 9, pp. 1–29, with pls. i–vi.

———   1896 A
Review of Bashford Dean's "Fishes, living and fossil."
Geol. Magazine (3), iii, pp. 135–139.

———   1896 B
On some extinct fishes of the Teleostean family *Gonorhynchidæ.*
Proc. Zool. Soc. Lond., 1896, pp. 500–504, with pl. xviii.

———   1896 C
On some remains of the pycnodont fish *Mesturus,* discovered by Alfred N. Leeds, esq., in the Oxford Clay of Peterborough.
Ann. and Mag. Nat. Hist. (6), xvii, pp. 1–15, with pls. i–iii.

———   1897 A
On the fossil fishes of the Upper Lias of Whitby. Part II.
Proc. Yorkshire Geol. and Polytech. Soc. (2), xiii, pp. 155–170, with pls. xix–xxi.
Describes *Caturus* sp. indet. and 3 species of *Pachycormus.*

Woodward, A. S.—Continued.   1898 A
On the fossil fishes of the Upper Lias of Whitby. Part III.
Proc. Yorkshire Geol. and Polytech. Soc. (2), xiii, pp. 325–337, with pls. xlvi–xlviii.
Describes and figures species of *Lepidotus,* and treats of their osteology.

———   1898 B
Outlines of vertebrate palæontology for students of zoology. By Arthur Smith Woodward, assistant keeper of the department of geology in the British Museum, Cambridge.
University Press, 1898, pp. i–xxiv; 1–470, with 228 figs. in the text.
Reviewed by H. F. Osborn in Natural Science, xiv, 1899, pp. 156–159.

———   1899 A
Notes on teeth of sharks and skates from the English Eocene formations.
Proc. Geologists' Assoc., xvi, pp. 1–14, with pl. i.

———   1899 B
On the fossil fishes of the Upper Lias of Whitby. Part IV.
Proc. Yorkshire Geol. and Polytech. Soc. (2), xiii, pp. 455–472, with pls. lxviii, lxix, and 13 text figs.
Describes and figures *Belonorhynchus acutus, B. brevirostris,* and *Gyrosteus mirabilis.*

———   1900 A
On some fish remains from the Parana formation, Argentine Republic.
Ann. and Mag. Nat. Hist. (7), vi, pp. 1–7, with pl. i.

———   1900 B
On a new ostracoderm (*Euphanerops longævus*) from the Upper Devonian of Scaumenac Bay, Province of Quebec, Canada.
Ann. and Mag. Nat. Hist. (7), v, pp. 416–418, with pl. x, figs. 1–1b.

———   1900 C
On a new species of *Deltodus* from the Lower Carboniferous (Yoredale rocks) of Yorkshire.
Ann. and Mag. Nat. Hist. (7), v, pp. 419–420, with pl. x, figs. 2–2b.

———   1900 D
Doctor Traquair on Silurian fishes.
Geol. Magazine (4), vii, pp. 66–72, with 4 figs. in the text.
A review of Traquair's Report on fossil fishes, etc., in Trans. Roy. Soc. Edinburgh, xxxix, pp. 827–864, pls. i–v, and 18 text-figs.

Woodward, A. S. and Sherborn, C. D.
1890 A
A catalogue of British fossil *Vertebrata*.
Pp. i–xxxv, 1–396. 8vo, London.
Table of contents: Introduction, p. v. Dates of publication of Agassiz's "Poisson fossiles," p. xxv. Dates of publication of Owen's "Odontography," p. xxx. Table of the stratigraphical distribution of British fossil *Vertebrata*, p. xxx. Catalogue of British fossil *Vertebrata*, pp. 1–311. Addenda and Corrigenda, p. 395.

Woodward, Henry.
1869 A
Man and the mammoth; being an account of the animals associated with early man in prehistoric times.
Geol. Magazine (1), vi, pp. 58–72.

———
1871 A
Seeley on the *Ornithosauria*.
Nature, iv, p. 100.
A review of H. G. Seeley's two works, here designated as Seeley, H. G., 1869 A, and Seeley, H. G., 1870 A.

———
1874 A
New facts bearing on the inquiry concerning forms intermediate between birds and reptiles.
Report Brit. Assoc. Adv. Sci., 43d meeting, Bradford, 1873, pp. 98–94.

———
1874 B
New facts bearing on the inquiry concerning forms intermediate between birds and reptiles.
Quart. Jour. Geol. Soc., xxx, pp. 8–15.

———
1884 A
Ueber *Archæopteryx* von W. Dames.
Geol. Magazine (3), i, pp. 418–424, with pl. xiv.
A review of W. Dames's work in Palæont. Abhandl., ii, 1884, pp. 117–196.

———
1885 A
On an almost perfect skeleton of *Rhytina gigas* (*Rhytina stelleri*, Steller's sea-cow), obtained by Mr. Robert Damon, F. G. S., from the Pleistocene peat-bogs of Bering Island.
Quart. Jour. Geol. Soc., xli, pp. 457–472, with 5 figs. in the text.
Contains a list of works on *Sirenia*.

———
1885 B
Iguanodon mantelli Meyer.
Geol. Magazine (3), ii, pp. 10–15, with pl. i.

———
1885 C
On "wingless birds," fossil and recent; and a few words on birds as a class.
Geol. Magazine (3), ii, pp. 308–318.

Woodward, Henry—Continued. 1885 D
On the fossil *Sirenia* in the British Museum (Natural History), Cromwell Road, S. W.
Geol. Magazine (3), ii, pp. 412–425, with 2 figs. in the text.

———
1886 A
On a remarkable ichthyodorulite from the Carboniferous series, Gascoyne, Western Australia.
Geol. Magazine (3), iii, pp. 1–7, pl. i, and 5 figs. in the text.
Describes *Edestus davisii*.

———
1886 B
Prof. E. D. Cope on a new type of perissodactyl ungulate from the Wasatch Eocene of Wyoming Territory, United States of North America.
Geol. Magazine (3), iii, pp. 49–52, with pl. ii.
Extracts from Professor Cope's descriptions of *Phenacodus*.

———
1886 C
On "flightless birds," commonly called "wingless birds," fossil and recent; and a few words on birds as a class.
Proc. Geologists' Assoc., ix, pp. 352–376, with pls. i and ii, and 20 figs. in the text.

———
1894 A
Remarks on paper read by Dr. G. M. Dawson on elephant remains in Alaska.
Quart. Jour. Geol. Soc., l, p. 9.

———
1895 A
Note on the reconstruction of *Iguanodon* in the British Museum (Natural History), Cromwell Road.
Geol. Magazine (4) ii, pp. 289–292, pl. x, and 2 figs. in the text.

Woodward, M. F.
1892 A
On the milk dentition of *Procavia* (*Hyrax*) *capensis* and of the rabbit (*Lepus cuniculus*), with remarks on the relation of the milk and permanent dentition of the *Mammalia*.
Proc. Zool. Soc. Lond., 1892, pp. 38–49 with pl. ii.

———
1893 A
Contributions to the study of mammalian dentition. Part I. On the development of the teeth of the *Macropodidæ*.
Proc. Zool. Soc. Lond., 1893, pp. 450–473, with pls. xxxv–xxxvii.
Appended is a list of references to papers by other authors.
Review by M. Schlosser in Archiv f. Anthrop., xxiv, p. 145.

Woodward, M. F.—Continued. 1894 A
On the milk dentition of the *Rodentia*, with a description of a vestigial milk incisor in the mouse (*Mus musculus*).
Anatom. Anzeiger, ix, pp. 619–631, with 3 figs. in the text.

——   **1896 A**
Contributions to the study of mammalian dentition. Part II. On the teeth of certain *Insectivora*.
Proc. Zool. Soc. Lond., 1896, pp. 557–594, with pls. xxiii–xxvi.
Appended is a list of references to 82 papers bearing on the subject.
An abstract in Archiv f. Anthrop., xxv, pp. 242–243.

——   **1896 B**
On the teeth of the *Marsupialia*, with especial reference to the pre-milk dentition.
Anatom. Anzeiger, xiii, pp. 281–291.
Review by M. Schlosser in Archiv f. Anthrop., xxv, pp. 243–244.

Woodworth, J. B.   1895 A
Three-toed dinosaur tracks in the Newark group at Avondale, N. J.
Amer. Jour. Sci. (3), l, pp. 481–482.

——   **1900 A**
Vertebrate footprints on Carboniferous shales of Plainville, Mass.
Bull. Geol. Soc. America, xi, pp. 449–454, with pl. xl and 2 text-figs.
Records finding of batrachian footprints.

——   **1900 B**
Glacial origin of older Pleistocene in Gay Head cliffs, with note on fossil horse of that section.
Bull. Geol. Soc. Amer., xi, pp. 455–460, with pls. xli, xlii.
Records finding of an astragalus of *Equus* in reputed Miocene beds.

Woolworth, Samuel.   1847 A
Description of a tooth of the *Elephas americanus*.
Amer. Jour. Agricult. and Sci., vi, pp. 31–37, with 2 figs. in the text.

Worthen, A. H.   1857 A
On the occurrence of fish remains in the Carboniferous limestone of Illinois.
Proc. Amer. Assoc. Adv. Sci., 10th meeting, Albany, 1856, pp. 189–191.

——   **1871 A**
Vertical range of the mammoth and mastodon.
Amer. Naturalist, v, pp. 606–607.

See also **St. John** and **Worthen**; **Newberry** and **Worthen**.

Wortman, J. L.   1882 A
Remarks on *Ursus amplidens*.
Proc. Acad. Nat. Sci. Phila., 1882, pp. 286–288.

——   **1882 B**
On the origin and development of the existing horses.
Kansas City Review of Science and Industry, v, pp. 719–726, with figs. 1–6; vi, pp. 67–75, figs. 7–22.

——   **1883 A**
L'origine du cheval.
Revue Scientifique (3), v, pp. 705–714, with text-figs. 125–144.

——   **1883 B**
Remarks on *Galera macrodon*.
Amer. Naturalist, xvii, p. 1001.
Remarks made before Acad. Nat. Sci. Phila., April 26, 1883. These remarks do not appear in the "Proceedings" for that month.

——   **1885 A**
Cope's Tertiary *Vertebrata*.
Amer. Jour. Sci. (3), xxx, pp. 295–299.

——   **1886 A**
The comparative anatomy of the teeth of the *Vertebrata*.
8vo, pp. 351–504, with pls. i–vi and text figs. 189–268.
Reprinted from "The American System of Dentistry." Contains many figures of North American fossil vertebrates.

——   **1890 A**
Human and animal remains.
Amer. Naturalist, xxiv, pp. 592–594.
Contains directions for collecting and preserving fossil remains.

——   **1893 A**
On the divisions of the White River, or Lower Miocene, of Dakota.
Bull. Amer. Mus. Nat. Hist., v, pp. 95–105.
Contains a list of the genera characteristic of each of the different beds.

——   **1893 B**
A new theory of the mechanical evolution of the metapodial keels of *Diplarthra*.
Amer. Naturalist, xxvii, pp. 421–434, with 5 figs. in the text.

——   **1894 A**
Osteology of *Patriofelis*, a Middle Eocene creodont.
Bull. Amer. Mus. Nat. Hist., vi, pp. 129–164, with pl. i and 5 figs. in the text.
Reviewed by M. Schlosser in Archiv f. Anthrop., xxv, pp. 184–185.

Wortman, J. L.—Continued.   1894 B
On the affinities of *Leptarctus primus*
of Leidy.
Bull. Amer. Mus. Nat. Hist., vi, pp. 229–231.

———   1895 A
On the osteology of *Agriochœrus.*
Bull. Amer. Mus. Nat. Hist., vii, pp. 145–178,
with pl. i and 24 figs. in the text.
Reviewed by M. Schlosser in Archiv f. An-
throp., xxv, p. 185.

———   1896 A
Species of *Hyracotherium* and allied
perissodactyls from the Wasatch and
Wind River beds of North America.
Bull. Amer. Mus. Nat. Hist., viii, pp. 81–110,
with pl. ii and 18 figs. in the text.
Remarks on, by M. Schlosser in Neues Jahrb.
Mineral., 1899, ii, pp. 139–142 and in Archiv f.
Anthrop., xxv, pp. 221–222.

———   1896 B
Psittacotherium, a member of a new
and primitive suborder of the *Edentata.*
Bull. Amer. Mus. Nat. Hist., viii, pp. 259–262.
The suborder *Ganodonta* established.

———   1897 A
[Abstract of remarks on *Ganodonta,*
made before New York Academy of
Sciences, Dec. 7, 1896.]
Science (2), v, p. 71.

———   1897 B
The *Ganodonta* and their relationship to the *Edentata.*
Bull. Amer. Mus. Nat. Hist., ix, pp. 59–110,
with 36 figs. in the text.
Abstracts of this and Wortman, J. L., 1896 B,
by M. Schlosser in Archiv f. Anthrop., xxvi,
pp. 179–181.

———   1898 A
The extinct *Camelidæ* of North America, and some associated forms.
Bull. Amer. Mus. Nat. Hist., x, pp. 93–142,
with pl. xi and 23 text figs.

———   1899 A
Restoration of *Oxyæna lupina* Cope,
with descriptions of certain new species
of Eocene creodonts.
Bull. Amer. Mus. Nat. Hist., xii, pp. 139–148,
with pl. vii and 8 figs. in the text.
See also **Cope and Wortman; Os-**
born and Wortman.

Wortman, J. L., and **Earle,** Charles.
1893 A
Ancestors of the tapir from the Lower
Miocene of Dakota.
Bull. Amer. Mus. Nat. Hist., v, pp. 159–180,
with 7 text figs.
Notice in Amer. Naturalist, xxviii, p. 466.

Wortman, J. L., and **Matthew,** W. D.
1899 A
The ancestry of certain members of
the *Canidæ,* the *Viverridæ* and *Procy-*
onidæ.
Bull. Amer. Mus. Nat. Hist., xii, pp. 109–139,
with pl. vi and 10 text figs.

Wright, Albert A.   1893 A
On the ventral armor of *Dinichthys.*
Report Geol. Surv. Ohio, vii, pp. 620–626 with
pl. xliv and 2 text figs.

———   1894 A
The ventral armor of *Dinichthys.*
Amer. Geologist, xiv, pp. 313–320, with pl.
ix and 2 text figs.

Wright, R. R.   1884 A
The relationship between the air
bladder and auditory organ in *Amiurus.*
Zoolog. Anzeiger, vii, pp. 248–252.

Wylie, T. A.   1859 A
Teeth and bones of *Elephas primige-*
nius, lately found near the western fork
of White River, in Monroe County, Ind.
Amer. Jour. Sci. (2), xxviii, pp. 283–284.

Wyman, Jeffries.   1845 A
Communication on skeleton of *Hy-*
drarchos sillimani.
Proc. Boston Soc. Nat. Hist., ii, pp. 65–68.
Refers the skeleton shown by Dr. Koch to
Dorudon and probably to *Zeuglodon.*

———   1846 A
[Remarks on *Castoroides ohioensis*
found in Wayne County, N. Y.]
Proc. Boston Soc. Nat. Hist., ii, pp. 138–139.

———   1846 B
An anatomical description of the
cranium [of *Castoroides ohioensis*].
Boston Jour. Nat. Hist., v, pp. 391–401, with
pls. xxxvii–xxxix.
This constitutes a portion of a joint paper
by the above author and Dr. James Hall. See
Hall, James, 1846 B.

———   1850 A
Notice of fossil bones from the neighborhood of Memphis, Tenn.
Amer. Jour. Sci. (2), x, pp. 56–64, with 5 figs.
in the text.

———   1850 B
Notice of remains of vertebrated animals found at Richmond, Va.
Amer. Jour. Sci. (2), x, pp. 228–235, with 1
page of figs.

———   1850 C
Remarks on teeth of fossil fishes from
Richmond, Va.
Proc. Boston Soc. Nat. Hist., iii, pp. 246–247.
Teeth belonging to genus *Phyllodus.*

Wyman, Jeffries—Continued.   1850 D
Remarks on finding of bones of *Megalonyx*, *Castor*, and *Castoroides* at Memphis, Tenn.
Proc. Boston Soc. Nat. Hist., iii, pp. 280–281.

——   1850 E
Remarks on zeuglodon bones shown by Dr. Durkee.
Proc. Boston Soc. Nat. Hist., iii, pp. 328–329.

——   1853 A
Description of the interior of the cranium and of the form of the brain of *Mastodon giganteus*.
Amer. Jour. Sci. (2), xv, pp. 48–55, with 3 outline figs.

——   1853 B
[Remarks on mastodon tooth found about 40 miles west of Chicago, and on teeth of mammoth.]
Proc. Boston Soc. Nat. Hist., iv, pp. 376–378.

——   1853 C
Notes on the reptilian remains.
Quar. Jour. Geol. Soc., ix, pp. 64–66, with pls. ii, iii.
Discusses the structure and relationships of the bones of *Dendrerpeton acadianum*.

——   1855 A
Notice of fossil bones from the red sandstone of the Connecticut Valley.
Amer. Jour. Sci. (2), xx, pp. 394–397.
No systematic names employed.

——   1855 B
Fossil footprints.
Proc. Boston Soc. Nat. Hist., v. p. 256. Also in Amer. Jour. Sci. (2), xxi, 1856, p. 446.

——   1855 C
Description of a portion of the lower jaw of *Mastodon andium* of Cuvier; also of a tooth and fragment of the femur of a mastodon, brought from Chile by Lieut. J. M. Gilliss, U. S. N.
Exec. Doc. No. 121, 33d Cong., 1st sess.: The U. S. Naval Astron. Exped. to the Southern Hemisphere during the years 1849–1852, Lieut. J. M. Gilliss, supt., vol. ii, pp. 275–281, with pls. xii, xiii.

——   1856 A
Account of some fossil bones collected in Texas.
Proc. Boston Soc. Nat. Hist., vi, pp. 51–55.
Describes some bones of *Elephas, Mastodon,* and *Megatherium*.

——   1857 A
Notes on the teeth of an elephant discovered near Zanesville, Ohio.
Proc. Amer. Assoc. Adv. Sci., 10th meeting, Albany, 1856, pp. 169–172.

Wyman, Jeffries—Continued.   1857 B
On a batrachian reptile from the Coal formation.
Proc. Amer. Assoc. Adv. Sci., 10th meeting, Albany, 1856, pp. 172–173.
For more complete descriptions and names of the species, see Wyman, J. 1858 A.

——   1858 A
On some remains of batrachian reptiles discovered in the Coal formation of Ohio, by Dr. J. S. Newberry and C. M. Wheatley.
Amer. Jour. Sci. (2), xxv, pp. 158–163, with 2 figs. in the text.
Abstract in Zeitschr. gesammt. Naturwiss., xiii, pp. 71–72, 1859.

——   1860 A
Exhibition of skull of capybara found fossil on the southwestern frontier of the United States.
Proc. Boston Soc. Nat. Hist., vii, p. 350.

——   1862 A
Observations upon the remains of extinct and existing species of *Mammalia* found in the crevices of the lead-bearing rocks and in the superficial accumulations within the lead region of Wisconsin, Iowa, and Illinois.
J. D. Whitney's Report Geol. Surv. Upper Miss. Lead Region, pp. 421–423.

——   1867 A
Remarks on finding of bones of great auk on Goose Island, Casco Bay.
Proc. Boston Soc. Nat. Hist., xi, pp. 301–303.

——   1872 A
The osteology and myology of *Didelphys virginiana*, with an appendix on the brain.
Mem. Boston Soc. Nat. Hist., ii, pp. 41–154, with 36 figs. in the text.

Yates, L. G.   1874 A
Fossil elephant and mastodon in California.
Amer. Jour. Sci. (3), viii, p. 143.

——   1874 B
Letter relating to mammalian fossils in California.
Proc. Acad. Nat. Sci. Phila., 1874, pp. 18–21.

Young, John.   1866 A
On the affinities of *Platysomus* and allied genera.
Quart. Jour. Geol. Soc., xxii, pp. 301–317, with pls. xx, xxi, and 4 figs. in the text.
Restriction and definition of *Platysomus* Ag.

Young, John—Continued. 1866 B
Notice of new genera of Carbonifer-
erous *Glyptodipterines.*
Quart. Jour. Geol. Soc., xxii, pp. 596-608,
with 8 figs. in the text.
Defines *Rhizodus* Owen.

Zigno, A de. 1887 A
Quelques observations sur les siréniens fossiles.
Bull. Soc. Géol. France (3), xv, pp. 728-732,
pl. xxvii.

Zilliken, J. E. 1879 A
Die Entstehung des Kamelhöckers.
Kosmos, vi, pp. 143-145.

Zittel, Karl A. 1882 A
Ueber Flugsaurier aus dem lithographischen Schiefer Bayerns.
Palæontographica, xxix, 47-81, with pls. x-
xiii (i-iv).

——— 1886 A
Ueber *Ceratodus.*
Sitzungsber. k. p. Akad. Wissensch. Mün-
chen., 1886, pp. 258-251, with plate.

——— 1890 A
Handbuch der Palæontologie. I.
Abth. Palæozoologie. III Band. *Verte-*
brata (*Pisces, Amphibia, Reptilia, Aves*).
Pp. i-xii; 1-900, with 719 illus. Munich and
Leipsic, 1887-1890.
Pp. i-xii; 1-256 were issued in 1887; pp. 257-
436 in 1888; pp. 437-632 in 1889; pp. 633-900 in
1890.

Zittel, Dr. Karl A.—Continued. 1893 A
The geological development, descent,
and distribution of the *Mammalia.*
Geol. Magazine (3), x, pp. 401-412, 455-468,
501-514.

——— 1893 B
Handbuch der Palæontologie. I.
Abth. Palæozoologie. IV Band. *Ver-*
tebata (*Mammalia*).
Pp. i-xi, 1-799, with 590 figs. in the text.
Munich and Leipsic, 1891-1893.

——— 1896 A
Fossil fishes in the British Museum.
Natural Science, viii, pp. 408-413.
A review of the first three volumes of A. S.
Woodward's "Catalogue of the fossil fishes in
the British Museum."

Zograf, N. 1896 A
Note sur l'odontographie des Ganoï-
dei Chondrostei.
Ann. Sci. Naturelles (8) Zool., i, pp. 197-
219, with pls. iv, v.

Zwick, Wilhelm. 1897 A
Beiträge zur Kenntniss des Baues
und der Entwicklung der Amphibien-
gliedmassen, besonders von Carpus und
Tarsus.
Zeitschr. wissensch. Zool., lxiii, pp. 62-114,
with pls. iv, v.
A list of the principal memoirs and papers
dealing with the subject is appended.

ADDENDUM.

Ballou, William H. 1898 A
The serpent-like sea saurians.
Pop. Sci. Month., liii, pp. 209-225, with 2 pls.
and 6 text figs.

Baur, G. 1892 D
Der Carpus der Schildkröten.
Anatom. Anzeiger, vii, pp. 206-211, with figs.
1-4.

Cope, E. D. 1877 AA
Verbal communication on a new locality of the Green River shales containing fishes, insects, and plants in a good
state of preservation.
Palæont. Bull. No. 25, p. 1.
This appears to have been a communication
made to the American Philosophical Society,
but it does not appear in the proceedings. It
contains a list of the fishes obtained, some
general remarks on their relationships, but no
descriptions. "Published August 3, 1877."

Lucas, F. A. 1890 B
Description of some bones of Pallas
cormorant (*Phalacrocorax perspicillatus*).
Proc. U. S. Nat. Mus., xii, pp. 88-94, with
pls. ii-iv.

——— 1890 C
The expedition to Funk Island, with
observations upon the history and anatomy of the great auk.
Rep. U. S. Nat. Mus., 1887-1888, pp. 493-529,
with pls. lxxi-lxxiii.

Rohon, J. V. 1899 A
Die Devonischen Fische von Timan in
Russland.
Sitzungsber. k. böhm. Gesellsch. Wissensch.,
math.-naturw. Cl., 1899, art. viii, pp. 1-77, with
45 text figs.

——— 1899 B
Ueber Parietalorgane und Paraphysen.
Sitzungsber. k. böhm. Gesellsch. Wissensch.,
math.-naturw. Cl., 1899, art. xxxiii, pp. 1-15,
with 6 text figs.

TABULAR KEY TO CATALOGUE.

Class ELASMOBRANCHII.

Subclass	Superorder	Order	Suborder	Superfamily	Family	Page.
Plagiostomata	Ichthyotomi	Pleuracanthides			Pleuracanthidæ	268
					Cladodontidæ	267
		Pleuropterygia			Cladoselachidæ	272
		Acanthodii			Acanthodidæ	273
					Diplacanthidæ	274
	Aristoselachii	Selachii	Squali	Petalodontoidea	Petalodontidæ	275
					Psammodontidæ	284
					Cochliodontidæ	286
					Orodontidæ	295
				Heterodontoidea	Heterodontidæ	296
				Hexanchoidea	Hexanchidæ	300
					Lamnidæ	301
				Galeoidea	Scyllorhinidæ	310
					Galeidæ	310
				Squaloidea	Somniosidæ	314
			Rajæ	Pachyura	Tæniobatidæ	315
					Rajidæ	315
					Pristidæ	316
					Ptychodontidæ	317
				Masticura	Dasyatidæ	318
					Myliobatidæ	319
					Mantidæ	321
Holocephala				Chimæroidea	Ptychodontidæ	322
					Chimæridæ	323

ICHTHYODORULITES (p. 328).

Class PISCES.

Aspidoganoidei		Heterostraci		Pteraspidae	889
		Osteostraci		Cephalaspidae	840
		Antiarcha		Euphaneropidae	841
				Asterolepidae	842
	Placodermi	Arthrodira		Coccosteidae	844
				Macropetalichthyidae	849
				Asterosteidae	849
				Phyllolepidae	849
				Mylostomatidae	850
Arygostei	Dipnoi	Ctenodipterini		Dipteridae	851
				Phaneropleuridae	852
		Sirenoidei		Ctenodontidae	858
				Ceratodontidae	855
	Crossopterygia	Rhipidistia		Holoptychiidae	856
				Megalichthyidae	859
				Osteolepidae	861
		Actinistia		Onychodontidae	862
				Coelacanthidae	863
Teleostomi	Actinopteri	Chondrostei		Palaeoniscidae	865
				Platysomidae	869
				Dictyopygidae	870
				Acipenseridae	871
				Polyodontidae	872
		Pycnodonti		Pycnodontidae	871
		Holostei	Ginglymodi	Lepidotidae	876
				Lepidosteidae	871
			Halecomorphi	Isopholidae	871
				Macrosemiidae	874
				Pachycormidae	874
		Nematognathi		Amiidae	890
				Siluridae	892

Class PISCES—Continued.

Subclass	Superorder	Order	Suborder	Superfamily	Family	Page.
Teleostomi	Actinopteri	Isospondyli			Pholidophoridae	383
					Chirocentridae	388
					Stratodontidae	387
					Pachyrhizodontidae	387
					Enchodontidae	388
					Albulidae	390
					Clupeidae	390
					Salmonidae	392
					Gonorhynchidae	392
					Osteoglossidae	393
		Plectospondyli			Cyprinidae	394
		Phthinobranchii	Hemibranchii		Dercetidae	396
					Gasterosteidae	397
			Lophobranchii (not represented).		Syngnathidae	
					Hippocampidae	
		Mesichthyes	Haplomi		Lucidae	396
					Poeciliidae	398
			Synentognathi (not represented).		Esocidae	
					Hemirhamphidae	
					Scombresocidae	399
		Percomorphi	Percesoces		Mugilidae	399
					Sphyraenidae	400
			Squamipinnes		Chaetodontidae	400
			Pharyngognathi		Labridae	401
					Pomacentridae	402
				Berycoidea	Berycidae	402
				Scombroidea	Xiphiidae	403
					Trichiuridae	403
					Carangidae	403

			Family	No.
Teleostomi	Actinopteri	Percomorphi	Percoidea	Aphredoderidæ	403
				Percidæ	404
				Sparidæ	406
			Sciænoidea	Sciænidæ	405
			Scorpænoidea	Scorpænidæ	407
		Pareloplitæ	Cottoidea	Cottidæ	406
			Cyclopteroidea	Cyclopteridæ	407
		Plectognathi	Scleroderml	Balistidæ	407
			Ostracoderml	Ostraciidæ	
			Gymnodontes	Diodontoidea	Diodontidæ	408

Class BATRACHIA.

			Family	No.
Stegocephali	Microsauria		Protritonidæ	410
			Molgophidæ	411
			Hylonomidæ	412
			Ptyonidæ	414
			Tuditanidæ	415
			Diplocaulidæ	416
	Apocospondyli		Dendrerpetontidæ	417
			Sauropleuridæ	419
			Archegosauridæ	419
			Cricotidæ	420
			Anthracosauridæ	421
			Eryopidæ	421
			Mastodonsauridæ	423
Urodela			Genera of uncertain position	424
Salientia			Ranidæ	425

Class REPTILIA.

Subclass	Superorder	Order	Suborder	Superfamily	Family	Page
		Cotylosauria			Diadectidæ	426
					Pariotichidæ	428
					Parelasauridæ	429
		Chelydosauria			Otocœlidæ	429
		Anomodontia			Dicynodontidæ	430
					Clepsydropidæ	431
		Pelycosauria			Bolosauridæ	434
			Athecæ		Dermochelyidæ	436
				Pleurodira	Pleurosternidæ	437
					Pelomedusidæ	438
					Plesiochelyidæ	439
			Thecophora		Protostegidæ	440
					Thalassemydidæ	442
				Cryptodira	Toxochelyidæ	442
					Desmatochelyidæ	443
		Testudines			Chelonidæ	444
					Adocidæ	446
					Chelydridæ	446
					Anosteiridæ	446
					Emydidæ	447
					Testudinidæ	449
					Plastomenidæ	452
			Trionychia		Trionychidæ	453
		Plesiosauria			Plesiosauridæ	455
					Elasmosauridæ	457
		Rhynchocephalia			Champsosauridæ	460
		Ichthyosauria			Proteosauridæ	462
					Baptanodontidæ	463

464 474 474 476 476 478 478 479 480 480 482 484 486 487 491 492 493 494 494 495 497 498 501 502 503 505 505 506 507 508 509 511 511 515 516

Mosasauridæ

Iguanidæ

Anguidæ

Varanidæ

Amphisbænidæ

Chamæleonidæ

Palæophidæ

Boidæ

Colubridæ

Morosauridæ

Atlantosauridæ

Diplodocidæ

Megalosauridæ

Anchisauridæ

Cœluridæ

Ceratosauridæ

Ornithomimidæ

Hallopidæ

Stegosauridæ

Nodosauridæ

Ceratopsidæ

Camptosauridæ

Iguanodontidæ

Nanosauridæ

Ornithocephalidæ

Pteranodontidæ

Nyctosauridæ

Phytosauridæ

Stegosauroidea

Ceratopsoidea

Iguanodontoidea

Pythonomorpha

Sauria

Rhiptoglossa

Serpentes

Opisthocœlia

Theropoda

Orthopoda

Pterosauri

Parasuchia

Aëtosauria

Eusuchia

Squamata

Dinosauria

Pterosauri

Loricata

Class AVES

Subclass	Superorder	Order	Suborder	Superfamily	Family	Page
Saururæ	Odontotormæ	Ornithopappi			Archaeopterygidae	518
		Pteropappi			Ichthyornithidae	519
					Apatornithidae	520
	Odontolcæ	Dromæopappi			Hesperornithidae	521
	Dromæognathæ	Gastornithes			Gastornithidae	522
Eurhipiduræ	Euornithes	Cecomorphæ		Colymboidea	Colymbidae	523
				Alcoidea	Alcidae	524
				Laroidea	Laridae	525
				Procellarioidea	Procellariidae	526
		Grallæ		Scolopacoidea	Charadriidae	526
					Scolopacidae	526
				Gruoidea	Gruidae	527
					Rallidae	528
		Chenomorphæ		Anatoidea	Anatidae	529
				Phoenicopteroidea	Phoenicopteridae	531
		Herodii		Ardeoidea	Ardeidae	531
		Steganopodes		Pelecanoidea	Pelecanidae	532
					Sulidae	532
					Phalacrocoracidae	533
		Gallinæ	Alectoropodes		Tetraonidae	534
					Phasianidae	534
					Gallinuloididae	535
		Raptores	Cathartides		Cathartidae	535
			Accipitres		Falconidae	536
			Striges		Strigidae	536
		Passeres			Corvidae	537
					Icteridae	537
					Fringillidae	537

Class MAMMALIA.

Prototheria
- Protodonta — Dromatheriidæ 555
- Allotheria
  - Bolodontidæ 557
  - Plagiaulacidæ 558

Eutheria
- Didelphia
  - Marsupialia
    - Polyprotodontia
      - Stagodontidæ 561
      - Cimolestidæ 565
      - Triconodontidæ 566
      - Amphitheriidæ 568
      - Paurodontidæ 568
      - Dryolestidæ 569
      - Didelphidæ 570
- Monodelphia
  - Bruta
    - Tæniodonta
      - Conoryctidæ 572
      - Stylinodontidæ 573
    - Xenarthra
      - Megatheriidæ 575
      - Glyptodontidæ 580
      - Dasypodidæ 581
  - Sirenia
    - Prorastomidæ 582
    - Halitheriidæ 582
    - Trichechidæ 583
    - Hydrodamalidæ 583
  - Cete
    - Zeuglodontes
      - Basilosauridæ 584
    - Odontocete
      - Squalodontidæ 586
      - Platinistidæ 588
      - Delphinidæ 590
      - Physeteridæ 593
    - Mysticete
      - Balænidæ 595
  - Ungulata
    - Condylarthra
      - Phenacodontidæ 598
      - Meniscotheriidæ 602
      - Mioclænidæ 605
      - 606

Class MAMMALIA—Continued.

Subclass	Superorder	Order	Suborder	Superfamily	Family	Page.
Eutheria	Monodelphia	Ungulata	Perissodactyla	Equoidea	Equidae	608
				Tapiroidea	Lophiodontidae	624
					Tapiridae	628
				Brontotheroidea	Brontotheriidae	629
				Rhinocerotoidea	Hyracodontidae	638
					Amynodontidae	641
					Rhinocerotidae	642
			Artiodactyla		Pantolestidae	648
				Pantolestoidea	Homacodontidae	649
					Helohyidae	650
					Eohyidae	650
				Anthracotheroidea	Anthracotheriidae	651
				Suoidea	Suidae	653
				Cameloidea	Agriochoeridae	662
					Camelidae	673
				Traguloidea	Tragulidae	681
					Cervidae	681, 686
				Bovoidea	Antilocapridae	686
					Bovidae	690
		Ancylopoda		Chalicotheroidea	Chalicotheriidae	693
		Amblypoda	Taligrada		Periptychidae	696
			Pantodonta		Pantolambdidae	697
					Coryphodontidae	700
			Dinocerata		Bathyopsidae	701
					Tinoceridae	707
		Proboscidea			Elephantidae	715
		Tillodontia			Esthonychidae	716
					Anchippodontidae	

No.	Family	Suborder	Order
718	Sciuridae	Sciuromorpha	Glires (Simplicidentata)
721	Mylagaulidae	Sciuromorpha	
721	Castoridae	Sciuromorpha	
728	Ischyromyidae	Sciuromorpha	
726	Muridae	Myomorpha	
730	Geomyidae	Myomorpha	
782	Dipodidae	Myomorpha	
782	Hystricidae	Hystricomorpha	
782	Erethizontidae	Hystricomorpha	
784	Castoroididae	Hystricomorpha	
784	Dasyproctidae	Hystricomorpha	
785	Caviidae	Hystricomorpha	
785	Ochotonidae	Lagomorpha	Duplicidentata
737	Leporidae	Lagomorpha	
738	Talpidae		Insectivora
738	Soricidae		
742	Leptictidae		
744	Vespertilionidae		Chiroptera
746	Oxyclaenidae		Creodonta (Ferae)
747	Arctocyonidae		
749	Triisodontidae		
751	Mesonychidae		
754	Proviverridae		
755	Viverravidae		
756	Ambloctonidae		
759	Hyaenodontidae		
762	Uintacyonidae		
764	Ursidae	Arctoidea	Fissipedia
765	Procyonidae	Arctoidea	
769	Mustelidae	Arctoidea	
776	Canidae	Cynoidea	
777	Hyaenidae	Aeluroidea	
783	Felidae	Aeluroidea	
783	Otariidae		Pinnipedia
784	Odobenidae		
786	Phocidae		
788	Mixodectidae		Prosimiae (Primates)
788	Limnotheridae		

Monodelphia — Eutheria

CATALOGUE.

Class ELASMOBRANCHII Bonaparte.

Bonaparte, C. L. 1832–1841, Icon. Fauna Ital., iii, Introd.

Agassiz, L. 1843 B, i, p. 170; iii, p. 73. ("Placoides.")
    1844 A. ("Placoides.")
    1845 A. ("Placoides.")

Balfour, F. M. 1881 A.

Beard, J. 1890 A.

Braus, H. 1898 A.

Bridge, W. J. 1879 A.

Brühl, B. C. 1847 A.
    1896 A, p. 583.

Calori, L. 1894 A.

Cope, E. D. 1872 C. (Selachii.)
    1877 O, p. 41.
    1878 A.
    1884 N, p. 579.
    1885 G.
    1887 S, pp. 319, 320, 324.
    1889 R, pp. 853, 854.
    1890 B.
    1891 N, pp. 10, 15.
    1898 B, pp. 17, 21.

Dean, B. 1895 A.
    1896 B.

Dohrn, A. 1886 A.

Duméril, A. 1865 A.

Eastman, C. R. 1898 A, p. 108.

Fritsch, A. 1889 B, p. 95. (Selachii.)

Gadow, H. 1898 A, p. 5.

Gadow and Abbott 1895 A.

Garman, S. 1884 C.

Gegenbaur, C. 1865 B, pp. 77, 132.
    1872 A.
    1876 A.
    1895 B.
    1898 A.

Gill, T. 1861 A, p. 22.
    1862 A.
    1884 B.
    1884 C.
    1893 A.

Goeppert, E. 1895 A.
    1895 B.

Goette, A. 1878 A.

Günther, A. 1870 A, p. 348. (Chondropterygii.)
    1872 A, p. 554. (Chondropterygii.)
    1880 A, p. 313. (Chondropterygii.)
    1881 A, p. 685. (Chondropterygii.)

Haeckel, E. 1895 A, p. 238. (Selachii.)

Hannover, A. 1868 A.

Hasse, C. 1876 A.
    1877 A.
    1878 A.
    1879 A.
    1879 B.
    1882 A.
    1893 A.

Haswell, W. A. 1884 A.

Hay, O. P. 1899 F, p. 682.

Hertwig, O. 1874 A.

Howes, G. B. 1890 A.

Hubrecht, A. A. W. 1878 A.

Huxley, T. H. 1872 A, p. 111.
    1876 A.

Jaekel, O. 1890 A.
    1890 C.
    1890 D.
    1891 B.
    1891 D. (Selachii.)
    1894 A.

Klaatsch, H. 1890 A, p. 103.
    1896 A.

Köstlin, O. 1844 A.

Markert, F. 1896 A, p. 664.

Marsh, O. C. 1877 E.

Mayer, P. 1886 A.

Metschnikoff, O. 1879 A, p. 424.

Mivart, St. G. 1879 A.

Mollier, S. 1892 A.
    1893 A.

Müller, J. 1846 C, p. 203. (Elasmobranchii, seu Selachii.)

Müller and Henle 1841 A.

Newberry, J. S. 1873 A.
    1875 A.

Newberry and Worthen 1866 A.

Nicholson and Lydekker 1889 A, p. 923.

Osborn, H. F. 1899 I, p. 158.

Owen, R. 1847 B.
    1860 E, p. 99.
    1866 A. (Plagiostomi.)

Parker, W. K. 1878 A.

Pictet, F. J. 1854 A, p. 225. ("Placoides.")

Pollard, H. B. 1894 B.

Reis, O. M. 1896 A, p. 202.

Röse, C. 1892 A.
    1897 A, p. 62.

Ryder, J. A. 1885 B.
    1886 A.
    1892 A.

St. John and Worthen 1875 A.
    1883 A.

Stannius, H. 1854 A.

Steinmann and Döderlein 1890 A, p. 529.

Thacher, J. K. 1877 A, p. 284.
    1877 B.

Van Beneden, P. J. 1861 B

Wiedersheim, R. 1889 A, p. 428.

Woodward, A. S. 1884 A.
    1885 B.
    1886 E.
    1889 D.
    1891 A, p. xi.
    1891 C.
    1895 B.
    1898 B, p. 17.

Worthen, A. H. 1857 A.

Wortman, J. L. 1886 A, p. 370.

Zittel, K. A. 1890 A, p. 60. (Selachii.)

Subclass PLAGIOSTOMATA Müller.

Müller, J. 1835 A.
Gill, T. 1861 A., p. 23. (Plagiostomi.)
Günther, A. 1870 A, p. 352.
    1871 A, p. 428.
    1880 A, p. 313.
Howes, G. B. 1887 A. (Plagiostomi.)

Huxley, T. H. 1872 A, p. 120. (Plagiostomi.)
Müller, J. 1846 C, p. 203. (Plagiostomi.)
Owen, R. 1846 B, p. 23. (Plagiostomi.)
Stannius, H. 1854 A. (Plagiostomi.)
Zittel, K. A. 1890 A, p. 64. (Plagiostomi.)

Superorder *ICHTHYOTOMI* Cope.

Cope, E. D. 1884 U.
Brongniart, C. 1888 A. (Pterygacanthidæ, as subclass.)
    1888 B, p. 180. (Pterygacanthidæ.)
    1888 C, p. 36. (Pterygacanthidæ.)
Cope, E. D. 1884 B.
    1884 N, p. 581.
    1884 N, p. 589. (Xenacanthini.)
    1884 HH.
    1885 G. 1887 S, p. 825.
    1889 H.
    1889 R, p. 854.
    1891 N, p. 15.
    1892 N, p. 280.
    1893 L.
    1893 N, p. 1080.
    1894 F, p. 430.
    1898 B, pp. 22, 23.

Davis, J. W. 1892 A, p. 704.
Fritsch, A. 1889 B, p. 96. (Xenacanthides.)
    1891 B. (Ichthyotomi.)
    1895 A, p. 33.
Gadow, H. 1898 A, p. 5. (Proselachii.)
Garman, S. 1884 C.
    1885 B, p. 29.
Gill, T. 1884 A. (Xenacanthini.)
    1884 B. (Xenacanthini.)
    1884 C.
Haeckel, E. 1895 A, p. 240. (Proselachii.)
Kehner, F. A. 1876 A, p. 348. ("Proselachier.")
Koken, E. 1889 A, p. 87.
Nicholson and Lydekker, R. 1889 A, p. 926.
Steinmann and Döderlein, 1890 A, p. 541. (Proselachii.)
Woodward, A. S. 1889 D, p. 1.
    1898 B, pp. 17, 32.

Order PLEURACANTHIDES Haeckel.

Haeckel, E. 1895 A, p. 236.

PLEURACANTHIDÆ.

Brongniart, C. 1888 B.
    1888 C.
Brown, C. 1900 A, p. 174.
Cope, E. D. 1889 R, p. 854.
    1894 F, p. 436.
    1898 B, p. 23.
Davis, J. W. 1880 A.
    1880 B.
    1892 A.
Fritsch, A. 1889 B, p. 96. (Xenacanthidæ.)
    1891 B. (Xenacanthidæ.)
    1895 A, p. 33. (Xenacanthidæ.)
Geinitz, H. B. 1860 A, p. 468. (Xenacanthi.)
    1861 A, p. 22. (Xenacanthi.)

Gegenbaur, C. 1895 B (Xenacanthini.)
Hasse, C. 1878 A, p. 169.
Jaekel, O. 1890 D, p. 124. (Xenacanthini.)
    1894 A, p. 64. (Xenacanthini.)
    1895 A. (Pleuracanthidæ.)
Nicholson and Lydekker 1889 A, p. 926.
Reis, O. M. 1897 A.
Steinmann and Döderlein 1890 A, p. 541.
Traquair, R. H. 1900 B, p. 516.
Woodward, A. S. 1885 B, p. 155.
    1889 D, p. 1.
    1892 E.
    1898 B, p. 33.
Zittel, K. A. 1890 A, p. 88. (Xenacanthidæ.)

PLEURACANTHUS Agassiz. Type *P. lævissimus* Agassiz.

Agassiz, L. 1843 B (1837), iii, p. 66.
    1844 A.
Barkas, T. P. 1873 A, p. 17.
Beyrich, E. 1848 A. (Xenacanthus, type *X. decheni*.)
Brongniart, C. 1888 A.
    1888 B.
    1888 C.
Brown, C. 1900 A, p. 172.
Case, E. C. 1898 B, p. 403.
Cope, E. D. 1884 B.
    1884 N, p. 589.

Cope, 1889 H.
    1890 B, pl. xiv.
    1893 N, pl. xxix.
Davis, J. W. 1880 A. (Pleuracanthus, Xenacanthus.)
    1880 B. (Pleuracanthus, Xenacanthus.)
    1885 A.
    1892 A, p. 705.
Dean, B. 1894 A. (Xenacanthus.)
    1895 A, p. 83, figs. 90-90b.
    1896 C, p. 246.
Döderlein, L. 1889 A.

Egerton, P. G. 1857 B. (Pleuracanthus, Xenacan-
thus.)
Emery, C. 1897 B.
Fritsch, A. 1888 A, fig. (Xenacanthus.)
    1889 A. (Xenacanthus.)
    1889 B, p. 99.
    1890 A, fig. (Pleuracanthus, Xenacanthus.)
    1890 B.
    1891 A.
    1891 B. (Xenacanthus.)
    1895 A (1890), p. 3 (Pleuracanthus); p. 21
      (Xenacanthus).
Gegenbaur, C. 1898 A.
Geinitz, B. 1860 A, p. 468.
    1861 A, p. 22.
Gill, T. 1884 A. (Xenacanthus.)
    1884 C. (Xenacanthus.)
Haeckel, E. 1895 A, p. 241.
Hasse, C. 1879 A, p. 65.
Jaekel, O. 1890 D, p. 125. (Xenacanthus.)
    1891 C, p. 168. (Xenacanthus.)
    1894 A, p. 103. (Orthacanthus.)
    1895 A. (Pleuracanthus, Xenacanthus.)
    1899 A, p. 295.
Kner, R. 1867 A. (Pleuracanthus, Xenacanthus.)
    1868 A, p. 279. (Xenacanthus.)
Koken, E. 1889 A.
Lütken, C., 1868 A. (Xenacanthus.)
Miller, S. A. 1889 A, p. 607.
Mollier, S. 1893 A, p. 114.
Nicholson and Lydekker, 1889 A, p. 926.
Owen, R. 1860 E, p. 108.
Pictet, F. J. 1854 A, p. 273 (Xenacanthus); p. 294
    (Pleuracanthus).
Quenstedt, F. A. 1885 A, p. 289.
Reis, O. M. 1896 A.
    1897 A.
Roemer, F. 1857 A, p. 60.
Steinmann and Döderlein 1890 A, p. 541.

Ward, J. 1890 A, p. 134.
Woodward, A. S. 1888 C.
    1889 D, pp. xix, 2.
    1889 F, p. 300.
    1891 B, p. 377, figs. 1, 2.
    1892 E, figs. 2, 3, 7.
    1893 B.
    1896 A, p. 138.
    1898 B, p. 33.
Zittel, K. A. 1890 A, p. 89.

Pleuracanthus arcuatus Newb.

Newberry, J. S. 1856 B, p. 100.
Lesley, J. P. 1889 A, p. 506, figure (Orthacanthus);
    p. 700 (Pleuracanthus).
    1895 A, pl. dlxxi. (Orthacanthus.)
Newberry, J. S. 1873 A, p. 332, pl. xl, fig. 4. (Ortha-
    canthus.)
Nicholson and Lydekker, 1889 A, p. 939.
Woodward, A. S. 1889 D, p. 8.
    Coal-measures; Ohio.

Pleuracanthus biserialis Newb.

Newberry, J. S. 1856 B, p. 100.
Woodward, A. S. 1889 D, p. 9.
    Coal-measures; Ohio.

Pleuracanthus dilatatus Newb.

Newberry, J. S. 1856 B, p. 100.
Woodward, A. S. 1889 D, p. 9.
    Coal-measures; Ohio.

Pleuracanthus quadriseriatus (Cope).

Cope, E. D. 1877 E, p. 192. (Orthacanthus.)
Case, E. C. 1900 A, p. 700, pl. i, figs. 3a, 3b. [P. (Or-
    thacanthus).]
Cope, E. D. 1881 K, p. 163. (Orthacanthus.)
Woodward, A. S. 1889 D, p. 9.
    Permian; Illinois.

ORTHACANTHUS Agassiz.   Type *O. cylindricus* Agassiz.

Agassiz, L. 1843 B, iii, pl. xlv, figs. 7–9.
Beyrich, E. 1848 A, p. 28.
Brongniart, C. 1888 B, p. 180. (Syn. of Pleuracan-
thus.)
Cope, E. D. 1893 N, p. 1080.
Davis, J. W. 1880 A, p. 349.
    1880 B, p. 321.
    1892 A, p. 705. (Syn. of Pleuracanthus.)
Fritsch, A. 1889 B, p. 99.
    1890 B.
    1891 A.
    1891 B.
    1895 A (1890), p. 33.
Goldfuss, A. 1849 A.
Jaekel, O. 1891 C, p. 167.
    1894 A, p. 103.
    1895 A, p. 83.
Kner, R. 1867 A.
Koken, E. 1889 A.
Lütken, C. 1868 A.
Miller, S. A. 1889 A, p. 604.
Newberry, J. S. 1889 A, p. 214.

Nicholson and Lydekker 1889 A, p. 926. (Syn. of
    Pleuracanthus.)
Pictet, F. J. 1854 A, p. 295.
Reis, O. M. 1897 A.
Ward, J. 1890 A, p. 137.
Woodward, A. S. 1889 D, p. 2. (Syn. of Pleuracan-
thus.)
    1891 B, p. 377.
    1896 A, p. 135.
Zittel, K. A. 1890 A, p. 90.

Orthacanthus gracilis Newb.

Newberry, J. S. 1875 A, p. 56, pl. lix, fig. 7.
Case, E. C. 1900 A, p. 701, pl. i, fig. 4. [Pleura-
    canthus (Orthacanthus).]
Cope, E. D. 1881 K, p. 163.
    1888 BB, p. 285.
Lesley, J. P. 1889 A, p. 506, fig.
Miller, S. A. 1889 A, p. 604, fig. 1154.
    1895 A, pl. D lxxi.
Woodward, A. S. 1889 D, p. 9 (Orthacanthus?).
    Permian; Illinois.

Compsacanthus Newb. Type *C. lævis* Newb.

Newberry, J. S. 1856 B, p. 100.
Davis, J. W. 1880 A, p. 355.
  1883 A, p. 354.
  1892 A, p. 705. (Syn. of Pleuracanthus.)
Miller, S. A. 1889 A, p. 592.
Newberry, J. S. 1873 A, p. 331.
Woodward, A. S. 1889 D, p. 9. (Syn. of Pleuracanthus.)
Zittel, K. A. 1890 A, p. 117.

Compsacanthus lævis Newb.

Newberry, J. S. 1856 B, p. 100.
Miller, S. A. 1889 A, p. 592, fig. 1114.
Newberry, J. S. 1873 A, p. 332, pl. xl, figs. 5, 5a.
  1889 A, p. 204.
Woodward, A. S. 1889 D, p. 9. (Pleuracanthus.)
  Coal-measures; Ohio.

DIACHANODUS Garman. Type *Didymodus? compressus* Cope, = *D. texensis* Cope; not *Diplodus compressus* of Newberry.

Garman, S. 1885 B, p. 30.
Cope, E. D. 1883 K, p. 108 (Didymodus in part, to replace Diplodus, preoccupied by Rafinesque).
  1884 A. (Didymodus.)
  1884 B. (Didymodus.)
  1884 N. (Didymodus.)
  1884 U. (Didymodus.)
  1884 CC. (Didymodus.)
  1885 P. (Didymodus.)
  1894 F, p. 436. (Didymodus).
Davis, J. W. 1885 A. (Didymodus.)
  1892 A, p. 709. (Syn. of Pleuracanthus.)
Fritsch, A. 1895 A (1890), p. 46. (Orthacanthus.)
Garman, S. 1884 B. (Diplodus.)
  1885 A. (Diplodus.)
Gill, T. 1884 A. (Diplodus.)
Goldfuss, A. 1847 A, p. 23, pl. v, figs. 9–11. (Orthacanthus.)
Jaekel, O. 1894 A, p. 108. (Didymodus.)
Koken, E. 1889 A, pp. 87 (Pleuracanthus); p. 91 (Triacranodus).
Woodward, A. S. 1889 D, p. 15 (Diacranodus).
  1890 A, p. 394. (Didymodus.)
  1898 B, p. 32. (Didymodus.)

Zittel, K. A. 1890 A, p. 90. (Didymodus.)

Diacranodus platypternus (Cope).

Cope, E. D. 1884 N, p. 587, figs. 8, 9. (Didymodus.)
  1884 CC, p. 818, pl. xxiii. (Didymodus.)
  1888 BB, p. 285. (Didymodus.)
Garman, S. 1885 B, p. 30. (Diacranodus.)
  Permian; Texas.

Diacranodus texensis (Cope).

Cope, E. D. 1883 K, p. 108. (Didymodus? compressus Newb.)
Case, E. C. 1900 A, p. 701, pl. i, figs. 5a–5d. [Pleuracanthus (Didymodus).]
Cope, E. D. 1884 N, pp. 573, 575.(Diplodus compressus Cope, not of Newberry.)
  1884 CC, p. 818, pl. xxiii. (Didymodus compressus.)
  1885 D, p. 406. (Pleuracanthus compressus.)
  1888 BB, p. 285. (Didymodus.)
Garman, S. 1885 B, p. 30. (Diacranodus compressus.)
Woodward, A. S. 1889 D, p. 15. (Didymodus.)
  Permian; Texas, Illinois.

DITTODUS Owen. Type, none assigned; *D. divergens* Owen = *Diplodus gibbosus* Agassiz may be taken.

Owen, R. 1867 B, p. 325, pl. i.
  Unless otherwise indicated, the following writers designate this genus by the name *Diplodus*. This name is, however, preoccupied.
Agassiz, L. 1843 B, iii, p. 204. (Diplodus, type *D. gibbosus*. Name preoccupied by Rafinesque, 1810.)
  1844 A.
Barkas, T. P. 1868 B.
  1874 A, pp. 344, 346.
Cope, E. D. 1883 K, p. 108. (Didymodus in part.)
  1884 A.
  1884 B.
  1884 N.
Davis, J. W. 1880 A, p. 349.
  1880 B, p. 321.
  1885 A. (Syn. of Pleuracanthus.)
  1892 A, p. 705. (Syn. of Pleuracanthus.)
Destinez, P. 1898 A, p. 219.
Eastman, C. R. 1899 C, p. 483.
Egerton, P. G. 1857 B.
Garman, S. 1884 B.

Gill, T. 1884 A.
Hasse, C. 1879 A, p. 67.
Hay, O. P. 1899 E, p. 791. (Dittodus.)
Jaekel, O. 1895 A, p. 78.
Kner, R. 1867 A.
Koken, E. 1889 A.
Miller, S. A. 1889 A, p. 596.
  1892 A, p. 714. (Dissodus, type *D. gibbosus*.)
Newberry, J. S. 1873 A, pp. 285, 333, 334.
  1875 A, p. 44.
Owen, R. 1867 B, p. 325 (Dittodus); p. 359 (Aganodus); p. 346 (Ochlodus); p. 363 (Pternodus).
Pictet, F. J. 1854 A, p. 259.
Stock, T. 1882 B.
Thompson, J. 1871 A. (Dittodus.)
Ward, J. 1890 A, p. 138.
Woodward, A. S. 1885 B, p. 155.
  1889 D, pp. 2, 12. (Pleuracanthus, Diplodus.)
  1889 F, p. 299.
  1898 B, p. 33. (Diplodus.)
Woodward and Sherborn 1890 A, p. 66.
Zittel, K. A. 1890 A, p. 90.

Dittodus acinaces (Dawson).

Dawson, J. W., 1860 B, suppl. p. 46, fig. 43. (Diplodus.)
    1868 D, p. 11. (Diplodus.)
    1868 A, p. 211, fig. 58. (Diplodus.)
    1878 A, p. 211, fig. 58. (Diplodus.)
    1894 A, p. 269. (Diplodus.)
Miller, S. A. 1892 A, p. 714. (Dissodus.)
Woodward, A. S. 1889 D, p. 13. (Diplodus.)
    Coal-measures; Nova Scotia.

Dittodus compressus (Newb.).

Newberry, J. S., 1856 A, p. 99. (Diplodus.)
Cope, E. D. 1877 N, p. 53. (Diplodus.)
    1881 K, p. 163. (Diplodus.)
Hay, O. P. 1900 A, p. 16.
Kindle, E. M. 1898 A, p. 484.
Miller, S. A. 1892 A, p. 714. (Dissodus.)
Newberry, J. S. 1873 A, pp. 334, 335. (Diplodus.)
    1875 A, p. 44, pl. lviii, fig. 2. (Diplodus.)
    1889 A, p. 214. (Diplodus.)
Newberry and Worthen 1866 A, p. 60, pl. iv, fig. 2. (Diplodus.)
St. John, O. 1872 A, p. 240, pl. iv, figs. 19a, 19b. (Diplodus.)
St. John and Worthen 1870 A, p. 432. (Diplodus.)
Woodward, A. S. 1889 D, p. 12. (Diplodus.)
    Coal-measures; Indiana, Ohio.

Dittodus gibbosus (Binney).

Binney, E. W., 1841, Trans. Manchester Geol. Soc., i, p. 169, pl. v, figs. 17-18. (Diplodus.)
Agassiz, L. 1843 B, iii, p. 204, pl. 226, fig. 1. (Diplodus gibbosus; not figs. 2-5, *fide* Woodward); p. 205, pl. 226, figs. 6-8. (Diplodus minutus.)
Cope, E. D. 1884 N, p. 573. (Diplodus.)[1]
Davis, J. W. 1892 A, p. 725. (Pleuracanthus lævissimus.)
Hancock and Atthey 1868 A, p. 370. (Diplodus.)
Miller, S. A. 1892 A, p. 714. (Dissodus.)
Owen, R. 1867 B, p. 325, pl. i. (Dittodus parallelus.)[2]
Woodward, A. S. 1889 D, p. 10. (Diplodus.)
    Coal-measures; England; Permian; Texas.

Dittodus gracilis (Newb.).

Newberry, J. S., 1856 B, p. 99. (Diplodus.)

Miller, S. A. 1892 A, p. 714. (Dissodus.)
Newberry, J. S. 1873 A, pp. 334, 336. (Diplodus.)
    1875 A, p. 44, pl. lviii, figs. 3, 3a. (Diplodus.)
    1889 A, p. 214. (Diplodus.)
Woodward, A. S. 1889 D, p. 13. (Diplodus.)
    Coal-measures; Ohio.

Dittodus latus (Newb.).

Newberry, J. S., 1856 B, p. 99. (Diplodus.)
Destinez, P. 1898 A, p. 220.
Hay, O. P. 1900 A, p. 96.
Kindle, E. M. 1898 A, p. 484.
Miller, S. A. 1892 A, p. 714. (Dissodus.)
Newberry, J. S. 1873 A, pp. 334, 336. (Diplodus.)
    1875 A, p. 44, pl. lviii, figs. 1-1b. (Diplodus.)
Newberry and Worthen 1866 A, p. 59, pl. iv, figs. 1-1e. (Diplodus.)
Woodward, A. S. 1889 D, p. 12. (Diplodus.)
    Coal-measures; Indiana, Ohio, Illinois, Belgium.

Dittodus lucasi Hay.

Hay, O. P., 1900 A, p. 97, fig. 1.
    Coal-measures; Illinois.

Dittodus penetrans (Dawson).

Dawson, J. W., 1860 B, suppl., p. 50, fig. 42. (Diplodus.)
    1868 A, p. 211, fig. 57. (Diplodus.)
    1878 A, p. 211, fig. 57. (Diplodus.)
Miller, S. A. 1892 A, p. 714. (Dissodus.)
Woodward, A. S. 1889 D, p. 14. (Diplodus.)
    Coal-measures; Nova Scotia.

Dittodus priscus (Eastman).

Eastman, C. R., 1899 C, p. 490, pl. vii, figs. 1, 2. (Diplodus.)
Weller, S. 1899 A. (Diplodus.)
    Devonian: Illinois.

Dittodus striatus (Eastman).

Eastman, C. R., 1899 C, p. 490, pl. vii, figs. 3, 4. (Diplodus.)
Weller, S. 1899 A. (Diplodus.)
    Devonian: Illinois.

PROTODUS A. S. Woodward.   Type *P. jexi* A. S. Woodward.

Woodward, A. S., 1892 A, p. 1.
Eastman, C. R. 1899 C, p. 489.
Miller, S. A. 1892 A, p. 717.
Traquair, R. H. 1893 A, p. 145.
    1893 D, p. 111.

Protodus jexi (A. S. Woodward).

Woodward, A. S., 1892 A, p. 2, pl. 1. fig. 1.
Traquair, R. H. 1893 A, p. 145.
    Lower Devonian; New Brunswick.

THRINACODUS St. J. and Worth.   Type *T. nanus* St. J. and Worthen.

St. John and Worthen, 1875 A, p. 289.
Cope, E. D. 1884 B. (Thrinacodus.)
    1884 N. (Pleuracanthus.)
Davis, J. W. 1892 A, p. 705. (Syn. of Pleuracanthus.)

Gill, T. 1884 A.
Koken, E. 1889 A, p. 87. (Pleuracanthus.)
Miller, S. A. 1889 A, p. 613.
Zittel, K. A. 1890 A, p. 90. (Syn. of Diplodus.)
    Probably not distinct from *Dittodus*.

---

[1] Cope's Texan specimens may belong to some other species.
[2] See, however, Woodward, A. S. 1891 A, p. 354.

Thrinacodus bicornis Newb.

Newberry, J. S., 1879 A, p. 344.
Kindle, E. M. 1898 A, p. 485.
Woodward, A. S. 1889 D, p. 14.   (Diplodus.)
  Subcarboniferous (St. Louis); Indiana.

Thrinacodus duplicatus (Newb. and Worth.).

Newberry and Worthen, 1866 A, p. 61, pl. iv, figs. 3, 3a.   (Diplodus.)
Destines, P. 1898 A, p. 220.   (Diplodus.)
Keyes, C. R. 1891 A, p. 264.   (Thrinacodus duplicatus?)
Lesley, J. P. 1890 A, p, 1186, figures.
  1895 A, pl. dlxxii.
St. John and Worthen 1875 A, p. 289.   (Thrinacodus.)

Woodward A. S. 1889 D, p. 14.   (Diplodus.)
  Subcarboniferous (Keokuk); Illinois, Belgium.   ·

Thrinacodus incurvus (Newb. and Worth.).

Newberry and Worthen, 1866 A, p. 62, pl. iv, figs. 4, 4a.   (Diplodus.)   .
Lesley, J. P. 1890 A, p. 1186.
St. John and Worthen 1875 A, p. 289.
Woodward, A. S. 1889 D, p. 14.   (Diplodus.)
  Subcarboniferous (Keokuk); Illinois.

Thrinacodus nanus St. J. and Worth.

St. John and Worthen, 1875 A, p. 289, pl. v, figs. 1, 2.
Woodward A. S. 1889 D, p. 14.   (Diplodus.)
  Subcarboniferous (Kinderhook); Iowa.

DOLIODUS Traquair.   Type *Diplodus problematicus* A. S. Woodward.

Traquair, R. H., 1898 A, p. 145.
Eastman, C. R. 1899 C, p. 489.
Traquair, R. H. 1898 D, p. 112.

Doliodus problematicus (A. S. Woodward).

Woodward, A. S. 1892 A, p. 2, pl. 1, fig. 2.   (Diplodus.)

Miller, S. A. 1892 A, p. 714.   (Dissodus.)
Traquair, R. H. 1898 A, p. 145.
  1898 D, p. 112.
  Lower Devonian; New Brunswick.

CLADODONTIDÆ.

Brown, C. 1900 A, p. 174.
Claypole, 1893 B.
  1895 C.
  1895 D.
Cope, E. D. 1893 L.
  1894 F, p. 427.
  1898 B, p. 23.

Garman, S. 1885 A ("Cladodonts").
  1885 B.   ("Cladodonts.")
Gill, T. 1884 B, p. 346.   (Hybodontidæ, in part.)
Nicholson and Lydekker 1889 A, p. 927.
Woodward, A. S. 1889 D, p. 16.

CLADODUS Agassiz.   Type *C. mirabilis* Agassiz.

Agassiz, L. 1843 B, iii, p. 196.
  1844 A.
Barkas, W. J. 1873 A, p. 19.
  1874 A, pp. 386, 392, 438.
Brown, C. 1900 A, p. 172.
Claypole, E. W. 1893 B.
  1893 D.
  1895 C.
  1895 H.
Cope, E. D. 1893 L.
  1893 O.
  1894 F.
Davis, J. W. 1883 A, p. 372.
  1885 A.
Dean, B. 1894 A.
Eastman, C. R. 1899 C, p. 489.
  1900 B, p. 35.
Emery, C. 1897 B, p. 147.
Gadow, H. 1898 A, p. 5.
Garman, S. 1885 B.
Hasse, C. 1879 A, p. 69.
Jaekel, O. 1892 A.
  1899 A, p. 296.
Koken, E. 1889 A, p. 92.
Koninck, L. G. de 1878 A, p. 26.
Kükenthal, W. 1892 A.

Lesley, J. P. 1889 A, p. 130.
McCoy, F. 1855 A, p. 619.
Miller, S. A. 1889 A, p. 590.
Newberry and Worthen 1866 A, p. 20.
Nicholson and Lydekker 1889 A, p. 927.
Pictet, F. J. 1854 A, p. 258.
Reis, O. 1896 A, p. 212.
St. John and Worthen 1875 A, p. 432.
Steinmann and Döderlein 1890 A, p. 546.
Thompson, J. 1870 A.
  1871 A.
Traquair, R. H. 1888 A, p. 81.
  1900 B, p. 517.
Trautschold, H. 1874 B, p. 286.
Woodward, A. S. 1884 A, p. 267.
  1889 D, p. 16.
Woodward and Sherborn 1890 A, p. 34.
Zittel, K. A. 1890 A, p. 67.

Cladodus acuminatus Newb.

Newberry, J. S. 1856 B, p. 99.
Miller, S. A. 1889 A, p. 590, fig. 1110.
Newberry, J. S. 1875 A, p. 45, pl. lviii, fig. 4.
Woodward, A. S. 1889 D, p. 23.
  Coal-measures; Ohio.

Cladodus alternatus St. J. and Worth.

St. John and Worthen 1875 A, p. 265, pl. ii, figs. 14-18.
Koninck, L. G. de 1878 A, p. 29.
Woodward, A. S. 1889 D, pp. 22, 23.
    Subcarboniferous (Kinderhook, Burlington); Iowa.

Cladodus angulatus Newb. and Worth.

Newberry and Worthen 1866 A, p. 24, pl. i, figs. 7-8a.
Woodward, A.S. 1889 D, p. 21.   (Syn. of C. robustus.)
    Subcarboniferous (Keokuk); Illinois.

Cladodus bellifer St. J. and Worth.

St. John and Worthen 1875 A, p. 270, pl. iv, fig. 10.
Koninck, L. G. de 1878 A, p. 27, pl. iii, fig. 4.
Sauvage, H. E. 1888 A, p. 47.
Woodward, A. S. 1889 D, p. 23.
    Subcarboniferous (Burlington); Iowa.

Cladodus carinatus St. J. and Worth.

St. John and Worthen 1875 A, p. 279, pl. iv, figs. 6, 7.
Woodward, A. S. 1889 D, p. 23.
    Coal-measures; Illinois, Iowa.

Cladodus claypolei Hay.

Hay, O. P. 1899 E, p. 783.
Claypole, 1894 A, p. 137, pl. v.   (C. magnificus, not of Tuomey, 1858.)
    Devonian (Cleveland shale); Ohio.

Cladodus concinnus Newb.

Newberry, J. S. 1875 A, p. 48, pl. lviii, fig. 8.
    1889 A, p. 170, pl. xxi.
Woodward, A. S. 1889 D, p. 23.
    Devonian (Cleveland shale); Ohio.

Cladodus coniger Hay.

Hay, O. P. 1899·E, p. 783.
Miller, S. A. 1892 A, p. 714.   (C. carinatus.)
Newberry, J. S. 1889 A, p. 108.   (C. carinatus, not of St. John and Worthen.)
    Devonian (Chemung); Pennsylvania.

Cladodus costatus Newb. and Worth.

Newberry and Worthen 1866 A, p. 27, pl. i, figs. 13, 13a.
Woodward, A. S. 1889 D, p. 23.
    Subcarboniferous (Chester); Illinois.

Cladodus deflexus Newb. and Worth.

Newberry and Worthen 1870 A, p. 355, pl. iii, figs. 3, 3a.
Woodward, A. S. 1889 D, p. 23.
    Subcarboniferous (Burlington); Illinois.

Cladodus eccentricus St. J. and Worth.

St. John and Worthen 1875 A, p. 272, pl. iv, fig. 4.
Keyes, C. R. 1894 A, p. 229.
Woodward, A. S. 1889 D, p. 23.
    Subcarboniferous (St. Louis); Illinois, Missouri.

Cladodus elegans Newb. and Worth.

Newberry and Worthen 1870 A, p. 354, pl. iv, fig. 9.
Keyes, C. R. 1894 A, p. 229.
Woodward, A. S. 1889 D, p. 23.
    Subcarboniferous (St. Louis); Missouri.

Cladodus euglypheus St. J. and Worth.

St. John and Worthen 1875 A, p. 274, pl. iv, figs. 1-3.
Keyes, C. R. 1894 A, p. 229.
Woodward, A. S. 1889 D, p. 24.
    Subcarboniferous (St. Louis); Illinois, Missouri, Iowa.

Cladodus exiguus St. J. and Worth.

St. John and Worthen 1875 A, p. 261, pl. iii, figs. 13-15.
Woodward, A. S. 1889 D, p. 24.
    Subcarboniferous (Kinderhook); Iowa.

Cladodus exilis St. J. and Worth.

St. John and Worthen 1875 A, p. 256, pl. i figs. 1-6.
Woodward, A. S. 1889 D, p. 24.
    Subcarboniferous (Kinderhook); Iowa.

Cladodus ferox Newb. and Worth.

Newberry and Worthen 1866 A, p. 26, pl. i, fig. 11.
Hambach, G. 1890 A, p. 80.
Woodward, A. S. 1889 D, p. 24.
    Subcarboniferous (St. Louis); Missouri.

Cladodus fulleri St. J. and Worth.

St. John and Worthen 1875 A, p. 276, pl. iv, fig. 9
Woodward, A. S. 1889 D, p. 24.
    Coal-measures; Illinois, Iowa.

Cladodus girtyi Hay.

Hay, O. P. 1900 A, p. 98, fig. 2.
    Coal-measures; Colorado.

Cladodus gomphoides St. J. and Worth.

St. John and Worthen 1875 A, p. 269, pl. iv, figs. 12-16.
Woodward, A. S. 1889 D, p. 24.
    Subcarboniferous (Burlington); Iowa.

Cladodus gracilis Newb. and Worth.

Newberry and Worthen 1866 A, p. 30, pl. i, fig. 17.
Kindle, E. M. 1898 A, p. 484.
Trautschold, H. 1874 B, p. 287.
Woodward, A. S. 1889 D, p. 24.
    Coal-measures; Indiana.

Cladodus grandis Newb. and Worth.

Newberry and Worthen 1866 A, p. 29, pl. i, figs. 15, 15a.
Woodward, A. S. 1889 D, p. 21.
    Subcarboniferous (Chester); Illinois.

Cladodus hertzeri Newb.

Newberry, J. S. 1875 A, p. 46, pl. lviii, fig. 5.
Woodward, A. S. 1889 D, p. 24.
    Subcarboniferous (below Berea Grit); Ohio.

Cladodus intercostatus St. J. and Worth.

St. John and Worthen 1875 A, p. 267, pl. iv, fig. 11.
Woodward, A. S. 1889 D, p. 24.
Subcarboniferous (Burlington): Iowa, Illinois.

Cladodus ischypus Newb. and Worth.

Newberry and Worthen 1870 A, p. 354, pl. iv, figs. 6, 6a.
Woodward, A. S. 1889 D, p. 24.
Subcarboniferous (St. Louis): Missouri.

Cladodus keokuk St. J. and Worth.

St. John and Worthen 1875 A, p. 268.
Woodward, A. S. 1889 D, p. 24.
Subcarboniferous (Keokuk): Iowa, Illinois.

Cladodus lamnoides Newb. and Worth.

Newberry and Worthen 1866 A, p. 30, pl. i, fig. 16.
Trautschold, H. 1874 B, p. 286, text figure and pl. xxviii, fig. 3; pl. xxix, fig. 2.
Woodward, A. S. 1889 D, pp. 22, 24.
Subcarboniferous (Keokuk); Illinois. Also reported from Russia.

Cladodus magnificus Tuomey.

Tuomey, M. 1858 A, p. 39, figs. C, Cb.
Newberry, J. S. 1889 A, p. 216.
Newberry and Worthen 1866 A, p. 24, pl. i, figs. 6, 6a.
Subcarboniferous; Alabama, Illinois, Iowa.

Cladodus micropus Newb. and Worth.

Newberry and Worthen 1866 A, p. 21, pl. i, figs. 2, 2b.
Woodward, A. S. 1889 D, p. 21. (Syn. of C. robustus.)
Subcarboniferous (Keokuk); Illinois.

Cladodus monroei Eastman.

Eastman, C. R. 1900 B, p. 36, fig. 2.
Devonian (Hamilton); Ohio.

Cladodus mortifer Newb. and Worth.

Newberry and Worthen 1866 A, p. 22, pl. i, fig. 5.
Newberry, J. S. 1897 A, p. 285, pl. xxii, figs 2, 2a.
St. John, O. 1870 A, p. 431.
1872 A, p. 239, pl. iii, figs. 6a–b; pl. vi, figs. 13a–13d.
Woodward, A. S. 1889 D, p. 24. (Syn. of C. occidentalis.)
Coal-measures; Illinois, Indiana, Iowa, Nebraska, Kansas.

Cladodus newmani Tuomey.

Tuomey, M. 1858 A, p. 39, fig. B.
Newberry, J. S. 1889 A, p. 216.
Subcarboniferous; Alabama.

Cladodus occidentalis Leidy.

Leidy, J. 1859 C, p. 3.
1873 B, pp. 311, 352, pl. xvii, figs. 4–6.
Woodward, A. S. 1889 D, p. 24.
Coal-measures; Kansas.

Cladodus pandatus St. J. and Worth.

St. John and Worthen 1875 A, p. 278, pl. iv, fig. 8.
Woodward, A. S. 1889 D, p. 25.
Coal-measures; Illinois.

Cladodus parvulus Newb.

Newberry, J. S. 1875 A, p. 48, pl. lviii, figs. 9, 9a.
Woodward, A. S. 1889 D, p. 25.
Devonian; Ohio.

Cladodus pattersoni Newb.

Newberry, J. S. 1875 A, p. 47, pl. lviii, figs. 6, 6a.
1889 A, p. 171.
Woodward, A. S. 1889 D, p. 25.
Subcarboniferous (Waverly); Ohio, Kentucky.

Cladodus politus Newb. and Worth.

Newberry and Worthen 1866 A, p. 27, pl. i, fig. 12.
Woodward, A. S. 1889 D, p. 25.
Subcarboniferous (Chester); Illinois.

Cladodus praenuntius St. J. and Worth.

St. John and Worthen 1875 A, p. 270, pl. iv, fig. 17.
Woodward, A. S. 1889 D, p. 25.
Subcarboniferous (Burlington); Iowa.

Cladodus raricostatus St. J. and Worth.

St. John and Worthen 1875 A, p. 271, pl. iv, fig. 18.
Woodward, A. S. 1889 D, p. 25.
Subcarboniferous (Keokuk); Iowa.

Cladodus rivi-petrosi Claypole.

Claypole, E. W. 1893 B, p. 328, pl. viii.
1893 D, p. 445, fig. 1.
Devonian (Cleveland shale); Ohio.

Cladodus robustus Newb. and Worth.

Newberry and Worthen 1866 A, p. 20, pl. i, figs. 1, 1a.
Woodward, A. S. 1889 D, pp. 21, 25. (C. robustus, micropus, angulatus, turritus).
Subcarboniferous (Keokuk); Illinois.

Cladodus romingeri Newb.

Newberry, J. S. 1875 A, p. 49.
1889 A, p. 177, pl. xxvii, fig. 10.
Woodward, A. S. 1889 D, p. 25.
Subcarboniferous (Waverly); Michigan.

Cladodus spinosus Newb. and Worth.

Newberry and Worthen 1866 A, p. 22, pl. i, figs. 3, 3a.
Dana, J. D. 1896 A, p. 644, fig. 1021.
Hambach, G. 1890 A, p. 80.
Woodward, A. S. 1889 D, p. 22.
Subcarboniferous (St. Louis); Missouri.

Cladodus splendens Newb.

Newberry, J. S. 1897 A, p. 284, pl. xxii, fig. 1.
Subcarboniferous (Kinderhook); Iowa.

Cladodus springeri St. J. and Worth.

St. John and Worthen 1875 A, p. 259, pl. ii, figs. 1–13
Destinez, P. 1898 A, p. 222.
Koninck, L. G. de 1878 A, p. 28, pl. iii, figs. 5, 6.
Sauvage, H. E. 1883 A, p. 47.

Woodward, A. S. 1889 D, p. 22.
Subcarboniferous (Kinderhook); Iowa. Also reported from Belgium.

Cladodus stenopus Newb. and Worth.

Newberry and Worthen 1866 A, p. 23, pl. i, figs. 4, 4a.
Woodward, A. S. 1889 D, p. 25.
Subcarboniferous (St. Louis); Illinois.

Cladodus striatus Agassiz.

Agassiz, L. 1843 B, iii, p. 197 pl. xxii b, figs. 14–17.
Davis, J. W. 1883 A, p. 375, pl. xlix, figs. 12, 13 (C. striatus); p. 374, pl. xlix, figs. 10, 11 (C. elongatus); p. 379, pl. xlix, fig. 19 (C. curtus); p. 380, pl. xlix, fig. 20 (C. hornei).
Eastman, C. R. 1900 B, p. 36.
Traquair, R. H. 1888 A. p. 81.
Woodward, A. S. 1889 D, p. 19.
Subcarboniferous; England, Ireland: Devonian (Corniferous); Ohio (if correctly identified).

Cladodus subulatus Newb.

Newberry, J. S. 1875 A, p. 47, pl. lviii, fig. 7.
Woodward, A. S. 1889 D, p. 25.
Devonian (Cleveland shale); Ohio.

Cladodus succinctus St. J. and Worth.

St. John and Worthen 1875 A, p. 265, pl. iii, figs 8–12.
Koninck, L. G. de 1878 A, p. 29.
Woodward, A. S. 1889 D, pp. 22, 25.
Subcarboniferous (Kinderhook); Iowa.

Cladodus terrelli Newb.

Newberry, J. S. 1889 A, p. 170, pl. xxvii, figs. 5–7.
Dean, B. 1894 A, p. 107. (C. terrelli?).
Devonian (Cleveland shale); Ohio.

Cladodus tumidus Newb.

Newberry, J. S. 1889 A, p. 172, pl. xxvii, figs. 8–9.
Devonian (Cleveland shale); Ohio.

Cladodus turritus Newb. and Worth.

Newberry and Worthen 1866 A, p. 28, pl. i, fig. 14.
Woodward, A. S. 1889 D, p. 21. (Sy.a. of C.robustus).
Subcarboniferous (Keokuk); Illinois.

Cladodus vanhornei St. J. and Worth.

St. John and Worthen 1875 A, p. 273, pl. iv, fig. 5.
Woodward, A. S. 1889 D, p. 25.
Subcarboniferous (St. Louis); Illinois.

Cladodus wachsmuthi St. J. and Worth.

St. John and Worthen 1875 A, p. 268, pl. iii, figs. 1–7.
Davis, J. W. 1883 A, p. 376.
Koninck, L. G. de 1878 A, p. 29.
Woodward, A. S. 1889 D, pp. 22, 26. (C. vachsmuthi.)
Subcarboniferous (Kinderhook); Iowa.

Cladodus zygopus Newb. and Worth.

Newberry and Worthen 1866 A, p. 25, pl. i, figs. 9, 9a.
Woodward, A. S. 1889 D, p. 26.
Subcarboniferous (Chester); Illinois.

PHŒBODUS St. J. and Worth.   Type *P. sophiæ* St. J. and Worth.

St. John and Worthen 1875 A, p. 251.
Eastman, C. R. 1899 C, p. 489.
Garman, S. 1885 B, p. 31 (Pternodus, type P. springeri; not of Owen, 1867).
Miller, S. A. 1889 A, p. 588 (Bathychilodus): p. 606 (Phœbodus).
St. John and Worthen 1875 A, p. 252. (Bathycheilodus, type B. macisaacsii St. J. and Worth.)
Woodward, A. S. 1889 D, p. 27.

Phœbodus macisaacsii (St. J. and Worth.).

St. John and Worthen 1875 A, p. 252, pl. i, figs. 12–13d. (Bathycheilodus.)
Miller, S. A. 1889 A, p. 588, fig. 1103. (Bathychilodus).
Woodward, A. S. 1889 D, p. 27.
Middle Devonian; Iowa.

Phœbodus politus Newb.

Newberry, J. S. 1889 A, p. 173, pl. xxvii, figs. 27–2na.
Eastman, C. R. 1899 C, p. 491, pl. vii, fig. 5.
Devonian (Cleveland shale); Ohio, Illinois.

Phœbodus sophiæ St. J. and Worth.

St. John and Worthen 1875 A, p. 251, pl. i, figs. 14–14d.
Lesley, J. P. 1889 A, p. 635, figures.
Miller, S. A. 1889 A, p. 606, fig. 1158.
Newberry, J. S. 1889 A, p. 174.
Woodward, A. S. 1889 D, p. 27.
Middle Devonian, Iowa.

Phœbodus springeri (St. J. and Worth.).

St. John and Worthen 1875 A, p. 255, pl. i, figs. 7–7d. (Pristicladodus.)
Garman, S. 1885 B, p. 31. (Pternodus.)
Lesley, J. P. 1889 A, p. 745, figures. (Pristicladodus springeri, P. armatus.)
Miller, S. A. 1889 A, p. 608, fig. 1162. (Pristicladodus.)
St. John and Worthen 1875 A, p. 256, pl. i, figs. 8–11c. (Pristicladodus springeri, P. armatus.)
Woodward, A. S. 1889 D, p. 27. (Phœbodus.)
Subcarboniferous (Kinderhook); Iowa.

LAMBDODUS St. J. and Worth.   Type *L. costatus* St. J. and Worth.

St. John and Worthen 1875 A, p. 280.
Miller, S. A. 1889 A, p. 600.
Nicholson and Lydekker 1889 A, p. 928.
Woodward, A. S. 1889 D, p. 27.

Lambdodus calceolus St. J. and Worth.

St. John and Worthen 1875 A, p. 281, pl. v., fig. 5.
Keyes, C. R. 1894 A, p. 230.

St. John and Worthen 1875 A, p. 282, pl. i, figs. 6–6c (var. robustus).

Woodward, A. S. 1889 D, p. 27 (L. calceolus and L. robustus).

Subcarboniferous (Burlington); Iowa, Illinois: (Keokuk); Illinois, Missouri, Iowa.

Lambdodus costatus St. J. and Worth.

St. John and Worthen 1875 A, p. 280, pl. v, fig. 3.

Keyes, C. R. 1894 A, p. 229.

Miller, S. A. 1889 A, p. 600, fig. 1140.

Subcarboniferous (Burlington, Keokuk); Iowa, Illinois, Missouri.

Lambdodus hamulus St. J. and Worth.

St. John and Worthen 1875 A, p. 283, pl. v, figs. 25a–25c. (L. hamatus, on explanation of plate.)

Woodward, A. S. 1889 D, p. 27.

Subcarboniferous (Chester); Illinois.

Lambdodus reflexus St. J. and Worth.

St. John and Worthen 1875 A, p. 284, pl. v, figs. 25, 25a.

Woodward, A. S. 1889 D, p. 27.

Subcarboniferous (Chester); Illinois.

Lambdodus transversus St. J. and Worth.

St. John and Worthen 1875 A, p. 282, pl. v, figs. 4a, 4b.

Woodward, A. S. 1889 D, p. 27.

Subcarboniferous (St. Louis); Illinois.

CARCHAROPSIS Agassiz.   Type *C. prototypus* Agassiz = *Pristicladodus dentatus* McCoy.

Agassiz, L. 1843 B, iii, p. 318.   (Definition in footnote.)

Davis, J. W. 1883 A, p. 381; p. 384.   (Pristicladodus.)

Eichwald, E. 1860 A, p. 1604.   (Dicrenodus.)

Günther, A. 1880 A, p. 319.

Hasse, C. 1878 A, p. 168.   (Pristicladodus.)

McCoy, F. 1855 A, p. 642.

Miller, S. A. 1889 A, p. 589 (Carcharopsis); p. 608. (Pristicladodus).

1892 A, p. 714.   (Dicrenodus.)

Nicholson and Lydekker 1889 A, p. 944 (Carcharopsis); 928 (Dicrenodus).

Pictet, F. J. 1854 A, p. 239.

Romanovsky, G. 1853 A, p. 408, pl. viii.   (Dicrenodus, type *D. okensis.*)

St. John and Worthen 1875 A, p. 253 (Pristicladodus); p. 254 (Carcharopsis).

Woodward, A. S. 1886 A, p. 268.

1889 D, p. 28.   (Dicrenodus.)

Woodward and Sherborn 1890 A, p. 63.   (Dicrenodus.)

Zittel, K. A. 1890 A, p. 81 (Carcharopsis); p. 66 (Pristicladodus).

Carcharopsis wortheni Newb.

Newberry and Worthen 1866 A, p. 69, pl. iv, figs. 14 14a.

Dana, J. D. 1896 A, p. 644, fig. 1020.

Davis, J. W. 1883 A, p. 383.   (Pristicladodus.)

Miller, S. A. 1889 A, p. 589, fig. 1106.

St. John and Worthen 1875 A, p. 255.

Woodward, A. S. 1889 D, p. 29.   (Dicrenodus.)

Subcarboniferous; Alabama.

HYBOCLADODUS St. J. and Worth.   Type *H. plicatilis* St. J. and Worth.

St. John and Worthen 1875 A, p. 284.

Miller, S. A. 1889 A, p. 600.

Nicholson and Lydekker 1889 A, p. 928.

Woodward, A. S. 1889 D, p. 29.

Hybocladodus compressus (Newb. and Worth).

Newberry and Worthen 1866 A, p. 78, pl. v, fig. 1. (Helodus.)[1]

St. John and Worthen 1875 A, p. 287, pl. v. fig. 8.

Woodward, A. S. 1889 D, p. 29.

Subcarboniferous (Burlington); Iowa.

Hybocladodus intermedius St. J. and Worth.

St. John and Worthen 1875 A, p. 287, pl. v, fig. 11.

Woodward, A. S. 1889 D, p. 29.

Subcarboniferous (Keokuk); Iowa, Illinois.

Hybocladodus nitidus St. J. and Worth.

St. John and Worthen 1875 A, p. 288, pl. v, fig. 7.

Woodward, A. S. 1889 D, p. 29.

Subcarboniferous (Chester); Illinois.

Hybocladodus plicatilis St. J. and Worth.

St. John and Worthen 1875 A, p. 286, pl. v, fig. 9

Miller, S. A. 1889 A, p. 600.

Woodward, A. S. 1889 D, p. 30.

Subcarboniferous (Burlington); Iowa.

Hybocladodus tenuicostatus St. J. and Worth.

St. John and Worthen 1875 A, p. 286, pl. v, fig. 10.

Woodward, A. S. 1889 D, p. 30.

Subcarboniferous (Keokuk); Illinois, Iowa.

SYMMORIUM Cope.   Type *S. reniforme* Cope.

Cope, E. D. 1893 L.

1894 F, p. 427.

Symmorium reniforme Cope.

Cope, E. D. 1893 L.

Brown, C. 1900 A, p. 171, fig. 6c.

Cope, E. D. 1894 F, p. 427, pl. xviii, figs. 1–5.

Coal-measures; Illinois.

[1] Not *H. compressus* Newberry and Worthen 1870 A, p. 360.

MONOCLADODUS Claypole. Type *M. clarki* Claypole.

Claypole, E. W. 1893 B, p. 329.
    1893 G, p. 619.
Cope, E. D. 1893 O. (Syn. of Styptobasis.)
    1894 F, p. 434.

Monocladodus clarki Claypole.
Claypole, E. W. 1893 B, p. 329, pl. viii. •
    1896 D, p. 447, fig. 2.

Claypole, E. W. 1893 G, p. 619, pl. xl, fig. 2.
Cope, E. D. 1893 O.
    Devonian (Cleveland shale); Ohio.

Monocladodus pinnatus Claypole.
Claypole, E. W. 1893 B, p. 330.
Cope, E. D. 1893 O.
    Devonian (Cleveland shale); Ohio.

STYPTOBASIS Cope. Type *S. knightiana* Cope.

Cope, E. D. 1891 B, p. 447.
    1893 O.

Styptobasis aculeata Cope.
Cope, E. D. 1894 F, p. 434, pl. xx, figs. 1–5.
    Coal-measures; Illinois.

Styptobasis knightiana Cope.
Cope, E. D. 1891 B, p. 447, pl. xxviii, fig. 1.
Knight, W. C. 1899 A, p. 374.
Woodward, A. S. 1892 B, p. 233.
    Permian; Nebraska.

Order PLEUROPTERYGIA Dean.

Dean, B. 1894 A, p. 110. | Woodward, A. S. 1898 B, p. 29. (Pleuropterygii.)

CLADOSELACHIDÆ.

Dean, B. 1894 A, p. 111.
    1895 A.
    1896 C.

CLADOSELACHE Dean. Type *Cladodus fyleri* (Newb.).

Dean, B. 1898 C, p. 124.
Brown, C. 1900 A, p. 172.
Case, E. C. 1896 B, p. 402.
Cope, E. D. 1894 F, p. 429.
Dean, B. 1894 A.
    1895 A.
    1896 B.
    1896 C.
Jaekel, O. 1892 A. (Syn. of Cladodus.)
Reis, O. M. 1896 A.
Traquair, R. H. 1900 B, p. 516.
Woodward, A. S. 1892 E. (Cladodus.)
    1895 B, p. 40.
    1898 B, p. 30.

Cladoselache clarki (Claypole.)
Claypole, E. W. 1893 B, p. 327, pl. vii. (Cladodus.)
Dana, J. D. 1896 A, p. 619, fig. 976. (Cladodus.)
Claypole, E. W. 1895 A, pls. i, ii. (Cladodus.)
Dean, B. 1896 B, p. 677, fig. 7.
    Devonian (Cleveland shale); Ohio.

Cladoselache fyleri (Newb.).
Newberry, J. S. 1889 A, pls. xlvi, xlix, figs. 2, 3; no
    description. (Cladodus.)
Claypole, E. W. 1893 B, p. 325. (Cladodus.)
    1893 G, p. 616. (Cladodus.)
    1895 H, p. 191. (Cladodus.)
Dana, J. D. 1896 A, p. 619, fig. 977.
Dean, B 1894 A.
    1894 B, p. 115, pl. ii, figs. 1, 2,
    1895 A, p. 79, fig. 86.
    1896 B.

Jaekel, O. 1892 A, p. 91. (Cladodus.)
Mollier, S. 1893 A, p. 118. (Cladodus.)
Woodward, A. S. 1895 B, p. 40.
    1898 B, p. 31, fig. 24.
    Devonian (Cleveland shale); Ohio.

Cladoselache kepleri (Newb.).
Newberry, J. S. 1888 D, p. 178. (Cladodus.)
Claypole, E. W. 1893 B, p. 327, pl. vii. (Cladodus.)
    1893 G, p. 616. (Cladodus.)
Dean, B. 1894 A, p. 108.
    1896 B, p. 676, fig. 5.
Jaekel, O. 1892 A, p. 82. (Cladodus.)
Newberry, J. S. 1889 A, pls. xliv, xlv. (Cladodus.)
Woodward, A. S. 1889 D, p. 457. (Cladodus.)
    Devonian (Cleveland shale); Ohio.

Cladoselache newberryi Dean.
Dean, B. 1894 B, pl. i, figs. 3–8.
Cope, E. D. 1895 B. (Cladodus.)
Dean, B. 1896 B, p. 677, fig. 5.
Woodward, A. S. 1895 B, p. 41, fig. 2.
    Subcarboniferous (Waverly); Ohio.

Cladoselache sinuatus (Claypole).
Claypole, E. W. 1893 B, p. 327, pl. vii. (Cladodus.)
    1893 G, p. 618, pl. xliii, fig. 6 (Cladodus?
    sinuatus); explan. pl. xliii (Cladodus ro-
    tundiceps).
Dana, J. D. 1896 A, p. 619, fig. 976. (Cladodus.)
    Devonian (Cleveland shale); Ohio.

Order ACANTHODII Owen.

Owen, R. 1860 E, p. 130.
Cope, E. D. 1891 N, pp. 15, 18.
    1893 L.
    1898 B, p. 22.
Fritsch, A. 1895 A (1893), p. 49. ("Tribus Acanthodides.")

Gadow, H. 1898 A, p. 6. (Acanthodi.)
Haeckel, E. 1895 A, pp. 236, 245. (Acanthodini.)
Huxley, T. H. 1861 B, p. 37. (Acanthodidæ.)
Jaekel, O. 1896 B, p. 126.
Woodward, A. S. 1891 A, pp. xi, 1.
    1898 B, p. 35.

ACANTHOËSSIDÆ.[1]

Agassiz, L. 1844 B, p. 32. ("Acanthodiens.")
    1846 A, p. 45. ("Acanthodiens.")
Cope, E. D. 1893 L.
Davis, J. W. 1894 A, p. 249.
Dean, B. 1895 A.
Egerton, P. G. 1866 A.
Fritsch, A. 1895 A (1893), p. 49.
Günther, A. 1881 A, p. 686. (Acanthodini.)
Huxley, T. H. 1861 B, p. 37.
    1872 A, p. 129.
Jaekel, O. 1892 A.
Klaatsch, H. 1890 A, p. 149.
Lütken, C. 1873 A, p. 41.
Newberry, J. S. 1873 A, p. 256.
Nicholson and Lydekker 1889 A, p. 966.

Pictet, F. J. 1854 A, p. 188. ("Acanthodiens.")
Powrie, J. 1864 B, p. 417.
Reis, O. M. 1892 A.
    1895 B.
    1896 A.
    1897 A, p. 87.
Rohon, V. 1898 A, p. 20.
Steinmann and Döderlein 1890 A, p. 564.
Traquair, R. H. 1881 A, p. 18.
    1891 A, p. 124.
    1894 A.
Woodward, A. S. 1891 A, p. 2.
    1898 B, p. 39.
Zittel, K. A. 1890 A, p. 165.

ACANTHOËSSUS[2] Agassiz. Type A. browni Agassiz.

Agassiz, L. 1832, Leonhard and Brown's Zeitschr.
    f. Mineral., p. 149.
    By the following authors this genus is referred to by the name Acanthodes:
Agassiz, L. 1843 B (1834), ii, pt. i, pp. 3, 19.
    (Acanthoëssus changed to Acanthodes.)
    1844 A.
    1844 B, p. 32.
Beyrich, E. 1848 A, p. 24. (Holacanthodes.)
Davis, J. W. 1894 A, p. 249.
Dean, B. 1895 A.
Egerton, P. G. 1861 B.
Fritsch, A. 1895 A (1893), p. 56.
Gadow, H. 1898 A, p. 6.
Jaekel, O. 1899 B, p. 56, figs. 1, 2.
Kner, R. 1868 A, p. 290.
Miller, S. A. 1889 A, p. 586.
Nicholson and Lydekker 1889 A, p. 966.
Pictet, F. J. 1854 A, p. 189.
Quenstedt, F. A. 1852 A, p. 191.
    1885 A, p. 302.
Reis, O. M. 1896 A.
Roemer, F. 1857 A, p. 65.
Rohon, V. 1889 A.
    1898 A, pp. 20, 55.
Steinmann and Döderlein 1890 A, p. 564.
Traquair, R. H. 1888 B, p. 511. (Mesacanthus.)
    1893 B, p. 177.
    1894 A.
Woodward, A. S. 1891 A, p. 1.
    1898 B, p. 39.
Zittel, K. A. 1890 A, p. 166.

Acanthoëssus affinis (Whiteaves).
Whiteaves, J. F. 1887 A, p. 107. (Acanthodes mitchelli?, with A. affinis as alternative.)
Dana, J. D. 1896 A, p. 620, fig. 978. (Acanthodes.)
Traquair, R. H. 1890 A, p. 16. (Mesacanthus affinis.)
Whiteaves, J. F. 1883 A, p. 354. (Acanthodes mitchelli?.)
    1888 B, p. 160. (Acanthodes mitchelli?.)
    1889 A, p. 91, pl. v, figs. 1–1f. (Acanthodes affinis.)
Woodward, A. S. 1891 A, p. 14. (Acanthodes affinis.)
    Upper Devonian; Quebec, Canada.

Acanthoëssus concinnus (Whiteaves).
Whiteaves, J. F. 1887 A, p. 107, pl. x, figs. 1, 1a. (Acanthodes.)
Traquair, R. H. 1890 A, p. 15. (Acanthodes.)
Whiteaves, J. F. 1889 A, pl. v, fig. 2. (Acanthodes).
Woodward, A. S. 1891 A, p. 10. (Acanthodes.)
    Upper Devonian; Quebec, Canada.

Acanthoëssus? pristis (Clarke).
Clarke, J. M. 1885 A, p. 42. (Acanthodes?.)
Woodward, A. S. 1891 A, p. 15. (Acanthodes?)
    Devonian (Genesee); New York.

Acanthoëssus semistriatus (A. S. Woodward).
Woodward, A. S. 1892 A, p. 3, pl. 1, fig. 3. (Acanthodes.)
    Lower Devonian; New Brunswick.

---

[1] By most authors this family is called Acanthodidæ. The change is necessitated by adopting the earlier generic name Acanthoëssus for Acanthodes.

[2] The law of priority, which lies at the base of all codes of nomenclature, requires the restoration of the name first given by Agassiz to this genus.

CHEIRACANTHUS Agassiz.   Type *C. murchisoni* Agassiz.

Agassiz, L. 1843 B (1835), ii, pt. i, p. 125.
1844 B, p. 39.
Gadow, H. 1898 A, p. 6.
McCoy, F. 1855 A, p. 582.
Pictet, F. J. 1854 A, p. 189.
Reis, O. M. 1896 A.
Traquair, R. H. 1898 A.
Woodward, A. S. 1891 A, p. 16.
Woodward and Sherborn, 1890 A, p. 29.
Zittel, K. A. 1890 A, p. 167.

Cheiracanthus costellatus Traquair.

Traquair, R. H. 1884 B, p. 3, pl. ii. (Ctenacanthus.)
Garman, S. 1884 C, p. 484.   (Ctenacanthus.)
Traquair, R. H. 1888 A, p. 81.   (Ctenacanthus costellatus?)
1893 A, p. 146.
1898 D, p. 112.
Woodward, A. S. 1889 D, p. 242.   (Sphenacanthus.)
Subcarboniferous; Scotland: Devonian; Province of Quebec, Canada (if correctly identified).

DIPLACANTHIDÆ.

Cope, E. D. 1891 N, p. 19.
Traquair, R. H. 1894 A.

Woodward, A. S. 1891 A, p. 22.

DIPLACANTHUS Agassiz.   Type *D. crassisimus* Duff.

Agassiz, L. 1844 B. pp. 34, 40.
Dean, B, 1894 A, p. 110.
Egerton, P. G. 1861 B.
Gadow, H. 1898 A, p. 6.
McCoy, F. 1855 A, p. 584.
Pictet, F. J. 1854 A, p. 191.
Reis, O. M. 1896 A, p. 167.
Traquair, R. H. 1888 F, p. 512.   (Rhadinacanthus, type *R. longispinis*.)
1891 A, p. 125.
1894 A.
Woodward, A. S. 1891 A, p. 23.
Woodward and Sherborn 1890 A, p. 65.
Zittel, K. A. 1890 A, p. 167.

Diplacanthus crassisimus Duff.

Duff, P. 1842, Geol. Moray, p. 71, pl. x, fig. 2. (Diplocanthus.)
Agassiz, L. 1844 B, pp. 34, 41, pl. xiv, figs. 1–5 (D. striatus); pp. 34, 42, pl. xiii, figs. 3, 4 (D. striatulus); pp. 34, 43, pl. xiii, figs. 1, 2; pl. xiv, figs. 6, 7 (D. crassispinus).
McCoy, F. 1848 C, p. 301.   (D. gibbus.)
1855 A, p. 584, pl. ii B, fig. 4.   (D. gibbus.)
Traquair, R. H. 1888 B, p. 512.   (D. striatus.)
Whiteaves, J. F. 1881 A, p. 496.   (D. striatus ?.)
1883 A, p. 354.   (D. ?)
1883 B, p. 160.   (D. "sp. undt.")
1883 D, p. 29.   ("Diplacanthus.")
Woodward, A. S. 1891 A, p. 24.
Devonian; Scotland; Canada?.

CLIMATIUS Agassiz.   Type *C. reticulatus* Agassiz.

Agassiz, L. 1844 B, p. 119.
Egerton, P. G. 1861 B, p. 28.
Nicholson and Lydekker 1889 A, p. 968.
Pictet, F. J. 1854 A, p. 298.
Powrie, J. 1864 B, p. 420 (Climatius); p. 425, pl. xx (Euthacanthus, type *E. macnicoli*).
Woodward, A. S. 1891 A, p. 28.
Woodward and Sherborn, 1890 A, p. 36.
Zittel, K. A. 1890 A, p. 168.

Climatius? latispinosus (Whiteaves).

Whiteaves, J. F. 1881 E, p. 99.   (Ctenacanthus.)
Traquair, R. H. 1890 A, p. 21.   (Climatius.)
Whiteaves, J. F. 1883 A, p. 355.   (Ctenacanthus.)
1883 B, p. 164.   (Ctenacanthus.)
1889 A, p. 96, pl. x, figs. 3–3b.   (Ctenacanthus.)
Woodward, A. S. 1889 G, p. 183.   (Climatius.)
1891 A, p. 33.   (Climatius?.)
1892 A, p. 8.   (Climatius.)
Lower Devonian; New Brunswick.

Superorder *EUSELACHII*,[1] *nom. nov.*

Order SELACHA Bonaparte.

Bonaparte, C. L., 1832–1841, Iconograph. Fauna Ital., iii, Introd.
The form *Selachii* of this word is employed by the following writers unless otherwise indicated. The content of the term also varies greatly with the different authors.
Agassiz, L. 1843 B (1838), iii, p. 78. ("Placoides.")
Beard, J. 1890 A.

Cope, E. D. 1872 C, p. 326.
1877 O, p. 41.
1878 A, p. 292.
1884 HH.
1885 G.
1889 R, p. 854.
1891 N, p. 15.
1896 B, p. 24.

---

[1] This superordinal name is employed to include all *Elasmobranchii* except the *Ichthyotomi*. The living sharks and rays may be regarded as constituting the types of the group. It is quite probable that a more complete knowledge of the Paleozoic families will cause them to be removed from association with the yet living families.

Dohrn, A. 1886 A.
Emery, C. 1897 B.
Gadow, H., 1898 A, p. 5.
Garman, S. 1884 C.
Gegenbaur, C. 1895 B.
Gill, T. 1884 B.
      1884 C.
Goeppert, E. 1895 A.
      1895 B.
Günther, A. 1881 A, p. 685.
Haeckel, E., 1895 A, p. 242.   (Plagiostomi.)
Hasse, C. 1893 A.
Haswell, W. A. 1884 A, p. 112.   (Selachoidei.)
Howes, G. B. 1890 A.
Jaekel, O. 1890 A.
      1890 C.
      1894 A.
Jordan and Evermann 1896 A, p. 15.
Kehner, F. A. 1876 A.
Klaatsch, H. 1894 A.

Köstlin, O. 1844 A, p. 437.   ("Haifische.")
Leidy, J. 1871 C, p. 369.
Mayer, P. 1886 A.
Mollier, S. 1893 A, p. 6.
Newberry and Worthen 1866 A.
Owen, R. 1866 A, p. 13.
Parker, W. K. 1878 A.
St. John and Worthen 1875 A.
      1883 A.
Sagemehl, M. 1888 A, p. 224.
Steinmann and Döderlein 1890 A, p. 543.
Tomes, C. S. 1898 A, p. 240.   (Selachia.)
Stephan, P. 1900 A.   ("Sélaciens.")
Wiedersheim, R. 1889 A, p. 428.
Woodward, A. S. 1886 A.
      1889 D, p. 30.
      1892 G.
      1895 B.
      1898 B, p. 89.
Zittel, K. A. 1890 A, p. 60.

Suborder SQUALI.[1]

Duméril, A. 1865 A, p. 309.   ("Squales ou Pleuro-
   trèmes.")
Gegenbaur, C. 1898 A.
Gibbes, R. W. 1848 A.
      1847 A.
      1848 B.
Gill, T. 1861 A, p. 23.
      1862 A.
Günther, A. 1870 A, p. 358.   (Selachoidei.)
      1880 A, p. 314.   (Selachoidei.)
Haeckel, E. 1895 A, p. 242.   (Squalacei.)

Müller, J. 1846 C, p. 203.   (Squalidæ.)
Müller and Henle 1841 A.
Owen, R. 1845 B, p. 26.   ("Squaloids.")
      1847 B, p. 255.
      1860 E, p. 106.   (Squalidæ.)
Pictet, F. J. 1854 A, p. 234.   ("Squalidiens.")
Stannius, H. 1854 A.   (Squalidæ.)
Steinmann and Döderlein 1890 A, p. 545.
Woodward, A. S. 1889 D, pp. xxxix, 157.   (Aster-
   ospondyli.)

Superfamily PETALODONTOIDEA, nom. nov.

PETALODONTIDÆ.

Newberry and Worthen, 1866 A, p. 31.
Cope, E. D. 1891 A, p. 16.
      1898 B, p. 24.
Davis, J. W. 1883 A, pp. 331, 483.
Jaekel, O. 1895 B.
      1898 A, p. 49 (with subfamilies Janassinæ,
   Polyrhizodontinæ, and Pristodontinæ.)
      1899 A, p. 258.
      1900 A, p. 147.

Koninck, L. G. de 1878 A, p. 29.
Newberry, J. S. 1889 A, p. 185.
Newberry and Worthen 1866 A, p. 46.
Nicholson and Lydekker 1889 A, p. 927.
Traquair, R. H. 1888 A, p. 85.
Woodward, A. S. 1885 B, pp. 156, 226.
      1889 D, p. 35.
      1898 B, p. 50.
Zittel, K. A. 1890 A, p. 96.

JANASSA Münster.   Type Trilobites bituminosus Schlotheim.

Münster, G. 1839, Beiträge Petrefakt., pt. 1, p. 67.
Agassiz, L. 1843 B, iii, p. 375.
Atthey, T. 1868 B.
Barkas, T. P. 1868 A.   (Climaxodus.)
Cope, E. D. 1877 N, p. 52.   (Strigilina, with S. lin-
   guæformis Cope = J. strigilina Cope as type.)
Davis, J. W. 1882 B, p. 60.
Hancock and Atthey 1869 A.
Hancock and Howes 1870 A.
Jaekel, O. 1894 A, p. 65.
      1895 B, p. 201.
      1896 A, p. 48.
      1899 A, p. 258, pls. xiv, xv, and text figs.

McCoy, F. 1848 A, p. 128 (Climaxodus, with C.
   imbricatus as type.)
      1855 A, p. 620.
Miller, S. A. 1889 A, p. 591 (Climaxodus); p. 600.
   (Janassa.)
Nicholson and Lydekker 1889 A, p. 929.
Pictet, F. J. 1854 A, p. 270 (Climaxodus); p. 282
   (Janassa.)
Woodward, A. S. 1885 B, p. 227.
      1888 E.
      1889 D, p. 34.
      Zittel, K. A. 1890 A, p. 96.

---

[1] Additional literature bearing on this group may be looked for under the heads "Elasmobranchii"
and "Selachii."

Janassa brevis (Newb.).

Newberry. J. S. 1856 B, p. 100. (Climaxodus.)
Miller, S. A. 1889 A, p. 591.
Woodward, A. S. 1889 D, p. 39. (Janassa.)
  Coal-measures; Ohio.

Janassa gurleiana Cope.

Cope. E. D. 1877 E, p. 191. (Strigilina.)
Case, E. C. 1900 A, p. 700, pl. i, figs. 2a–2c. (J. gurleyana.)
Cope, E. D. 1881 K, p. 163.
  1888 BB, p. 285.
Woodward, A. S. 1889 D, p. 39.
  Permian; Illinois.

Janassa ordiana Cope.

Cope, E. D. 1881 K, p. 163.
  1888 BB, p. 285.
Woodward, A. S. 1889 D, p. 39.
  Permian; Texas.

Janassa strigilina Cope.

Cope, E. D. 1881 K, p. 163.
Case, E. C. 1900 A, p. 699, pl. i, figs. 1a–1c.
Cope, E. D. 1877 E, p. 191. (Strig. linguaeformis.)
  1877 N, p. 52. (Strigilina linguaeformis Cope, not of Atthey.)
  1888 BB, p. 285.
Miller, S. A. 1889 A, p. 600. (J. linguiformis.)
Woodward, A. S. 1889 D, p. 38.
  Permian; Illinois.

PELTODUS Newb. and Worth. Type *P. unguiformis* Newb. and Worth.

Newberry and Worthen 1870 A, p. 362.
Jaekel, O. 1899 A, p. 278.
Miller, S. A. 1889 A, p. 605.
St. John and Worthen 1875 A, p. 409.
Woodward, A. S. 1889 D, pp. 34, 39. (Syn. of Janassa.)

Peltodus ? plicomphalus St. J. and Worth.

St. John and Worthen 1875 A, p. 411, pl. xiii, fig. 9.
Lesley, J. P. 1889 A, p. 614, figures.
Woodward, A. S. 1889 D, p. 39. (Janassa.)
  Subcarboniferous (Chester); Illinois.

Peltodus pulvinulus Newb. and Worth.

Newberry and Worthen 1870 A, p. 362.
  Subcarboniferous; Illinois.

Peltodus quadratus St. J. and Worth.

St. John and Worthen 1875 A, p. 410, pl. xiii, figs. 6, 7.
Keyes, C. R. 1894 A, p. 232.
Lesley, J. P. 1889 A, p. 614, figures.
Woodward, A. S. 1889 D, p. 39. (Janassa.)
  Subcarboniferous (St. Louis); Illinois, Missouri.

Peltodus transversus St. J. and Worth.

St. John and Worthen 1875 A, p. 412, pl. xiii, fig. 8.
Lesley, J. P. 1889 A, p. 614, figure.
Woodward, A. S. 1899 D, p. 39. (Janassa.)
  Coal-measures; Illinois.

Peltodus unguiformis Newb. and Worth.

Newberry and Worthen 1870 A, p. 363, pl. ii, fig. 7.
Woodward, A. S. 1889 D, p. 39. (Janassa.)
  Coal-measures; Illinois.

TANAODUS St. J. and Worth. Type, none named.

St. John and Worthen 1875 A, p. 367.
Jaekel, O. 1899 A, p. 278.
Koninck, L. G. de 1878 A, p. 52.
Miller, S. A. 1889 A, p. 612.
Woodward, A. S. 1889 D, pp. 34, 39. (Syn. of Janassa.)

Tanaodus angularis (Newb. and Worth.).

Newberry and Worthen 1866 A, p. 55, pl. iii, figs. 16, 16a. (Chomatodus angularis in text; C. angulatus on plate.)
St. John and Worthen 1875 A, p. 368. (T. angulatus.)
Woodward, A. S. 1889 D, p. 48. (Petalodus.)
  Coal-measures; Illinois.

Tanaodus bellicinctus St. J. and Worth.

St. John and Worthen 1875 A, p. 376, pl. xi, figs. 14–16, 25.
Lesley, J. P. 1890 A, p. 1156, figures.
Woodward, A. S. 1889 D, p. 39. (Janassa.)
  Subcarboniferous (Chester); Illinois.

Tanaodus depressus St. J. and Worth.

St. John and Worthen 1875 A, p. 378, pl. xi, figs. 11–13.
Lesley, J. P. 1890 A, p. 1156, figures.
Woodward, A. S. 1889 D, p. 39. (Janassa.)
  Subcarboniferous (Chester); Illinois.

Tanaodus gracillimus (Newb. and Worth.).

Newberry and Worthen 1866 A, p. 51, pl. iii, figs. 12, 12a. (Chomatodus.)
Miller, S. A. 1889 A, p. 613, fig. 1176.
St. John and Worthen, 1775 A, p. 368.
Woodward, A. S. 1889 D, p. 48. (Petalodus.)
  Subcarboniferous (Burlington); Iowa.

Tanaodus grossiplicatus St. J. and Worth.

St. John and Worthen 1875 A, p. 375, pl. xi, fig. 26.
Lesley, J. P. 1890 A, p. 1156, figures.
Woodward, A. S. 1889 D, p. 39. (Janassa.)
  Subcarboniferous (Chester); Illinois.

Tanaodus multiplicatus (Newb. and Worth.).

Newberry and Worthen 1866 A, p. 57, pl. iii, figs. 18, 18a. (Chomatodus.)
Koninck, L. G. de 1878 A, p. 53, pl. vi, figs. 10, 11.
Lesley, J. P. 1890 A, p. 1157, figures.
St. John and Worthen, 1875 A, p. 368.
Sauvage, H. E. 1888 A, p. 47.
Woodward, A. S. 1889 D, p. 49. (Petalodus.)
  Subcarboniferous (Burlington); Iowa. Also in Subcarboniferous of Belgium.

Tanaodus obscurus (Leidy).

Leidy, J. 1857 H, p. 89, pl. v, figs. 22, 23. (Chomatodus.)
St. John and Worthen, 1875 A, p. 368.
Woodward, A. S. 1889 D, p. 228. (Chomatodus.)
  Coal-measures; Illinois.

Tanaodus polymorphus St. J. and Worth.

St. John and Worthen 1875 A, p. 380, pl. xi, figs. 17-19, 24.
Lesley, J. P. 1890 A, p. 1157, figures.
Woodward, A. S. 1889 D, p. 39. (Janassa.)
  Subcarboniferous (Chester); Illinois.

CYMATODUS Newb. and Worth.

Newberry and Worthen 1870 A, p. 363.
Miller, S. A. 1889 A, p. 594.
Woodward, A. S. 1889 D, pp. 39, 458.

THORACODUS Cope.

Cope, E. D. 1883 K, p. 108.
Miller, S. A. 1889 A, p. 613.
Woodward, A. S. 1889 D, p. 39. (=Janassa?.)
  Probably a synonym of Janassa.

FISSODUS St. J. and Worth.

St. John and Worthen 1875 A, p. 413.
Jaekel, O. 1895 B, p. 201. (Syn. of Petalodus.)
  1899 A, p. 289.
Miller, S. A. 1889 A, p. 597.
Nicholson and Lydekker 1889 A, p. 929.
Woodward, A. S. 1889 D, p. 40.

Fissodus bifidus St. J. and Worth.

St. John and Worthen 1875 A, p. 414, pl. xiii, figs. 1, 2.

CHOLODUS St. J. and Worth.

St. John and Worthen 1875 A, p. 415.
Miller, S. A. 1899 A, p. 590.
Woodward, A. S. 1889 D, p. 40.

Tanaodus praenuntius St. J. and Worth.

St. John and Worthen 1875 A, p. 371, pl. xi, figs. 6-10.
Keyes, C. R. 1894 A, p. 231.
Lesley, J. P. 1890 A, p. 1158, figures.
Woodward, A. S. 1889 D, p. 39. (Janassa.)
  Subcarboniferous (St. Louis); Illinois, Iowa, Missouri.

Tanaodus pumilis St. J. and Worth.

St. John and Worthen 1875 A, p. 369, pl. xi, figs. 1-5.
Lesley, J. P. 1890 A, p. 1158, figures.
Woodward, A. S. 1889 D, p. 39. (Janassa.)
  Subcarboniferous (St. Louis); Illinois, Iowa.

Tanaodus sculptus St. J. and Worth.

St. John and Worthen 1875 A, p. 373, pl. xi, figs. 20-23.
Keyes, C. R. 1894 A, p. 231.
Lesley, J. P. 1890 A, p. 1158, figures.
Woodward, A. S. 1889 D, p. 39. (Janassa.)
  Subcarboniferous (St. Louis); Illinois, Iowa, Missouri.

Tanaodus sublunatus St. J. and Worth.

St. John and Worthen 1875 A, p. 368, pl. xi, fig. 27.
Lesley, J. P. 1890 A, p. 1159, figures.
Woodward, A. S. 1889 D, p. 39. (Janassa.)
  Subcarboniferous (St. Louis); Illinois.

Type *C. oblongus*, Newb. and Worth.

Cymatodus oblongus Newb. and Worth.

Newberry and Worthen 1870 A, p. 364, pl. iv, figs. 7, 7a.
Woodward, A. S. 1889 D, pp. 39, 458.
  Coal-measures; Illinois.

Type *T. emydinus* Cope.

Thoracodus emydinus Cope.

Cope, E. D. 1883 K, p. 108.
Case, E. C. 1900 A, p. 702.
Miller, S. A. 1889 A, p. 613. (T. eurydinus.)
Woodward, A. S. 1889 D, p. 39.
  Permian; Illinois.

Type *F. bifidus* St. J. and Worth.

Miller, S. A. 1889 A, p. 598, fig. 1131.
Woodward, A. S. 1889 D, p. 40.
  Subcarboniferous (Chester); Illinois.

Fissodus tricuspidatus St. J. and Worth.

St. John and Worthen 1875 A, p. 415, pl. xiii, fig. 3.
Woodward, A. S. 1889 D, p. 40.
  Subcarboniferous (Chester); Illinois.

Type *C. inaequalis* St. J. and Worth.

Cholodus inaequalis St. J. and Worth.

St. John and Worthen 1875 A, p. 416, pl. xiii, figs. 4, 5.
Miller, S. A. 1889 A, p. 590, fig. 1106.
  Coal-measures; Illinois, Iowa.

PETALORHYNCHUS Newb. and Worth.   Type *Petalodus psittacinus* McCoy.

Newberry and Worthen 1866 A, p. 32.
Davis, J. W. 1882 A.
    1882 B, p. 60.
    1883 A, pp. 485, 516.
Jaekel, O. 1895 B.   (Syn. of Petalodus.)
    1899 A, p. 286.
Miller, S. A. 1889 A, p. 606.
Nicholson and Lydekker 1889 A, p. 929.
Woodward, A. S. 1889 D, p. 40.
Zittel, K. A. 1890 A, p. 97.

Petalorhynchus distortus St. J. and Worth.

St. John and Worthen 1875 A, p. 406, pl. xii, figs. 7, 8.
Keyes, C. R. 1894 A, p. 232.
Lesley, J. P. 1889 A, p. 626, figures.
Woodward, A. S. 1889 D, p. 42.
    Subcarboniferous (St. Louis); Illinois, Iowa, Missouri.

Petalorhynchus pseudosagittatus St. J. and Worth.

St. John and Worthen 1875 A, p. 405, pl. xii, figs. 1–4.

Keyes, C. R. 1894 A, p. 232.
Lesley, J. P. 1889 A, p. 626, figures.
Woodward, A. S. 1889 D, p. 42.
    Subcarboniferous (St. Louis); Illinois, Missouri, Iowa.

Petalorhynchus spatulatus St. J. and Worth.

St. John and Worthen 1875 A, p. 406, pl. xii, figs. 5, 6,
Lesley, J. P. 1889 A, p. 626, figures.
Woodward, A. S. 1889 D, p. 42.
    Subcarboniferous (St. Louis); Illinois, Iowa?

Petalorhynchus striatus Newb. and Worth.

Newberry and Worthen 1866 A, p. 40, pl. ii, figs. 8–8b.
Lesley, J. P. 1889 A, p. 626, figures.
    1895 A, pl. dlxxi.
Woodward, A. S. 1889 D, p. 42.
    Subcarboniferous (Burlington); Illinois.

PETALODUS Owen.   Type *P. hastingsiæ* Agassiz.

Owen, R. 1845 B, p. 61, pl. xxii.
Barkas, W. J. 1874 A, p. 536.
Davis, J. W. 1882 A.
    1882 B, p. 58.
    1883 A, p. 484.
Hancock and Atthey, 1869 A.
Hancock and Howes, 1870 A, pp. 54, 61.
Jaekel, O. 1895 B.
    1899 A, p. 286.
Koninck, L. G. de 1878 A, p. 49.
Leidy, J. 1855 D (Sicarius, type *S. extinctus*).
    1856 S, p. 161.
McCoy, F. 1855 A, p. 635.
Miller, S. A. 1889 A, p. 606.
Newberry and Worthen 1866 A, pp. 31, 46.
Nicholson and Lydekker 1889 A, p. 930.
Pictet, F. J. 1854 A, p. 270.
Trautschold, H. 1874 B, p. 298.
Woodward, A. S. 1885 B, pp. 156, 226.
    1889 D, p. 42.
Zittel, K. A. 1890 A, p. 97.

Petalodus curtus Newb. and Worth.

Newberry and Worthen 1870 A, p. 355, pl. iii, fig. 2.
Lesley, J. P. 1889 A, p. 625, figures.
St. John and Worthen 1875 A, p. 394, pl. xii, figs. 12–12c (with ?).
Woodward, A. S. 1889 D, p. 47.
    Subcarboniferous (Keokuk); Illinois, Iowa.

Petalodus hybridus St. J. and Worth.

St. John and Worthen 1875 A, p. 394, pl. xii, fig. 10.
Woodward, A. S. 1889 D, p. 47.
    Subcarboniferous (St. Louis); Illinois.

Petalodus knappi Newb.

Newberry, J. S. 1879 A, p. 345.
Kindle, E. M. 1898 A, p. 485.
Woodward, A. S. 1889 D, p. 47.
    Subcarboniferous (Keokuk); Indiana.

Petalodus linguifer Newb. and Worth.

Newberry and Worthen 1866 A, p. 37, pl. ii, figs. 4–5c.
Hambach, G. 1890 A, p. 81.
Leidy J. 1876 C.
Woodward, A. S. 1889 D, p. 47.
    Subcarboniferous (Chester); Illinois, Missouri (*fide* Hambach).

Petalodus ohioensis Safford.

Safford, J. M. 1853 A, figs. 1, 2.   (Getalodus, *err. typ.*)
Eastman, C. R. 1896 D, p. 174.   (P. alleghaniensis.)
Hambach, G. 1890 A, p. 81.   (P. destructor.)
Hay, O. P. 1895 C.   (P. ohioensis, P. securis.)
Leidy, J. 1855 D.   (Sicarius extinctus.)
    1856 S. p. 161, pl. xvi, figs. 4–6..   (P. alleg.)
    1859 C.   (P. alleg.)
    1873 B, pp. 312, 353, pl. xvii, fig 3.   (P. alleg.)
Lesley, J. P. 1889 A, p. 624, figures.   (P. alleg.)
    1895 A, pl. dlxxi.   (P. alleg., P. destructor.)
Miller, S. A. 1889 A, p. 605, fig. 1157.   (P. alleg.)
Newberry, J. S. 1875 A, p. 52, pl. lviii, figs. 13, 13a, (P. alleg., P. destructor.)
Newberry and Worthen 1866 A, p. 35, pl. ii, figs. 1–3.   (P. destructor.)
St. John, O. 1870 A, p. 433.   (P. destructor.)
    1872 A, p. 240, pl. iii, fig. 5.

St. John and Worthen 1875 A, p. 396. (P. alleg.)
Woodward, A. S. 1889 D, p. 46.
　　Coal-measures; Pennsylvania, Ohio, Illinois,
　　Nebraska, Arkansas, Missouri.

ANTLIODUS Newb. and Worth.　　Type, *A. mucronatus* Newb. and Worth.

Newberry and Worthen 1866 A, p. 33.
Davis, J. W. 1883 A, p. 485.
Koninck, L. G. de 1878 A, p. 51.
Miller, S. A. 1889 A, p. 587.
Newberry, J. S. 1889 A, p. 185.
St. John and Worthen, 1875 A, p. 368.
Woodward, A. S. 1889 D, pp. 42, 48. (Syn. of Peta-
　　lodus.)
Woodward and Sherborn 1890 A, p. 145. (Syn. of
　　Petalodus.)
Zittel, K. A. 1890 A, p. 97.

Antliodus arcuatus Newb.

Newberry, J. S. 1889 A, p.208, pl. xix, figs. 3, 4.
Kindle, E. M. 1898 A, p.484.
　　Subcarboniferous (St. Louis); Indiana.

Antliodus cucullus Newb. and Worth.

Newberry and Worthen 1866 A, p. 41, pl. iii, figs.
　　1–1b.
Woodward, A. S. 1889 D, p. 48. (Petalodus).
　　Subcarboniferous (Keokuk); Illinois.

Antliodus gracilis St. J. and Worth.

St. John and Worthen 1875 A, p. 393, pl. xi, fig. 29.
Woodward, A. S. 1889 D, p. 48. (Petalodus.)
　　Subcarboniferous (Warsaw); Illinois.

Antliodus minutus Newb. and Worth.

Newberry and Worthen 1866 A, p. 43, pl. iii, figs.
　　3–3b.
Kindle, E. M. 1898 A, p. 484.
Koninck, L. G. de 1878 A, p. 52, pl. vi, fig. 9.
Newberry, J. S. 1879 A, p. 341.
Sauvage, H. E. 1888 A, p. 47.
Woodward, A. S. 1889 D, p. 48. (Petalodus.)
　　Subcarboniferous; Europe: (Keokuk); Illi-
　　nois: (St. Louis); Indiana.

Antliodus mucronatus Newb. and Worth.

Newberry and Worthen 1866 A, p. 38, pl. ii, figs.
　　6–6b.
Woodward, A. S. 1889 D, p. 48. (Petalodus.)
　　Subcarboniferous (St. Louis); Illinois.

Antliodus parvulus Newb. and Worth.

Newberry and Worthen 1866 A, p. 38, pl. ii, figs.
　　7–7a.

Petalodus proximus St. J. and Worth.
St. John and Worthen 1875 A, p. 395, pl. xii, fig. 11.
Woodward, A. S. 1889 D, p. 48.
　　Coal-measures; Illinois.

Woodward, A. S. 1889 D, p. 48. (Petalodus.)
　　Subcarboniferous (Burlington, Keokuk);
　　Illinois.

Antliodus perovalis St. J. and Worth.

St. John and Worthen 1875 A, p. 393, pl. xi, fig. 28.
Woodward, A. S. 1889 D, p. 48. (Petalodus.)
　　Subcarboniferous (Warsaw); Illinois.

Antliodus politus Newb. and Worth.

Newberry and Worthen 1866 A, p. 42, pl. iii, figs.
　　2, 2a.
Woodward, A. S. 1889 D, p. 48. (Petalodus.)
　　Subcarboniferous (Keokuk); Illinois.

Antliodus robustus Newb. and Worth.

Newberry and Worthen 1866 A, p. 39, pl. ii, figs.
　　9–9c.
Woodward, A. S. 1889 D, p. 48. (Petalodus.)
　　Subcarboniferous (Chester); Illinois.

Antliodus sarcululus Newb. and Worth.

Newberry and Worthen 1870 A, p. 356, pl. iii, figs.
　　8–8b.
Woodward, A. S. 1889 D, p. 458. (Petalodus.)
　　Subcarboniferous (Burlington); Iowa.

Antliodus similis Newb. and Worth.

Newberry and Worthen 1866 A, p. 41, pl. ii, figs. 10,
　　10a.
Kindle, E. M. 1898 A, p. 484.
Newberry, J. S. 1879 A, p. 346.
Woodward, A. S. 1889 D, p. 48. (Petalodus.)
　　Subcarboniferous (Keokuk); Illinois, Indiana.

Antliodus simplex Newb. and Worth.

Newberry and Worthen 1866 A. p. 44, pl. iii, figs.
　　4, 4a.
Woodward, A. S. 1889 D. p. 48. (Petalodus.)
　　Subcarboniferous (Burlington); Iowa.

Antliodus sulcatus Newb. and Worth.

Newberry and Worthen 1866 A, p. 45, pl. iii, figs.
　　5, 5a.
Woodward, A S. 1889 D, p. 48. (Petalodus.)
　　Subcarboniferous (Keokuk); Illinois.

CHOMATODUS Agassiz.　　Type *C. linearis* Agassiz.

Agassiz, L. 1843 B (1888), iii, p. 108.
　　1844 A.
Davis, J. W. 1883 A, p. 507.
Destinez, P. 1898 A, p. 220.
Koninck, L. G. de 1878 A, p. 45.
Leidy, J. 1857 H, p. 89. (Palæobatis, type *P. in-
　　signis* Leidy.)
McCoy, F. 1855 A, p. 617.

Miller, S. A. 1889 A, p. 590 (Chomatodus); p. 604
　　(Palæobatis).
Newberry, J. S. 1889 A, p. 186.
Newberry and Worthen 1866 A, p. 34.
Owen, R. 1845 B, p. 61.
Pictet, F. J. 1854 A, p. 266.
St. John and Worthen 1875 A, p. 367.

Woodward, A. S. 1889 D, p. 48 (Syn. of Petalodus);
p. 228 (Chomatodus).
Woodward and Sherborn 1890 A, p. 22.
Zittel, K. A. 1890 A, p. 97.

Chomatodus affinis Newb. and Worth.

Newberry and Worthen 1866 A, p. 54, pl. iii, figs. 15, 15a.
Woodward, A. S. 1889 D, p. 48. (Petalodus.)
Subcarboniferous (Keokuk); Illinois.

Chomatodus angustus Newb.

Newberry, J. S. 1879 A, p. 342.
Kindle, E. M. 1898 A, p. 484.
Woodward, A. S. 1889 D, p. 228.
Subcarboniferous (St. Louis); Indiana.

Chomatodus arcuatus St. John.

St. John, O. 1870 A, p. 435.
Destinez, P. 1898 A, p. 221.
St. John, O. 1872 p. 243, pl. vi, figs. 14a, 14b.
St. John and Worthen 1875 A, pl. x, figs. 23a, 23b.
Woodward, A. S. 1889 D, p. 228.
Coal-measures; Iowa, Nebraska, Belgium.

Chomatodus chesterensis St. J. and Worth.

St. John and Worthen 1875 A, p. 363, pl. x, figs. 15–17d.
Woodward, A. S. 1889 D, p. 228.
Subcarboniferous (Chester); Illinois.

Chomatodus comptus St. J. and Worth.

St. John and Worthen 1875 A, p. 356, pl. x, figs. 19–22.
Woodward, A. S. 1889 D, p. 228.
Subcarboniferous (Burlington); Iowa.

Chomatodus costatus Newb. and Worth.

Newberry and Worthen 1866 A, p. 85, pl. v, figs. 17, 17a.
Woodward, A. S. 1889 D, p. 459.
Subcarboniferous (Keokuk); Illinois.

Chomatodus cultellus Newb. and Worth.

Newberry and Worthen 1866 A, p. 52, pl. iii, figs. 13, 13a.
St. John and Worthen 1875 A, p. 359.
Woodward, A. S. 1889 D, p. 48. (Petalodus.)
Subcarboniferous (Chester); Illinois.

Chomatodus elegans Newb. and Worth.

Newberry and Worthen 1866 A, p. 86, pl. v, figs. 18–18b.
Woodward, A. S. 1899 D, p. 228.
Subcarboniferous (Keokuk); Iowa.

Chomatodus inconstans St. J. and Worth.

St. John and Worthen 1875 A, p. 360, pl. x, figs. 5–14.
Woodward, A. S. 1889 D, p. 228.
Subcarboniferous (St. Louis); Iowa.

Chomatodus incrassatus St. J. and Worth.

St. John and Worthen 1875 A, p. 359, pl. x, fig. 18.
Keyes, C. R. 1894 A, p. 231.

Miller, S. A. 1889 A, p. 590, fig. 1109.
Woodward, A. S. 1889 D, p. 458. (Petalodus.)
Subcarboniferous (St. Louis); Illinois, Iowa, Missouri.

Chomatodus insignis (Leidy).

Leidy, J. 1857 H, p. 89, pl. v, figs. 24–26. (Palæobatis.)
Miller, S. A. 1889 A, p. 604. (Palæobatis.)
St. John and Worthen 1875 A, p. 359, pl. x A, figs. 5–5c.
Woodward, A. S. 1889 D, p. 48. (Petalodus.)
Subcarboniferous (St. Louis); Missouri, Illinois.

Woodward regards the identification of Leidy's species with that of St. John and Worthen as doubtful.

Chomatodus loriformis Newb. and Worth.

Newberry and Worthen 1866 A, p. 56, pl. iii, figs. 19, 19a.
Woodward, A. S. 1889 D, p. 49. (Petalodus.)
Subcarboniferous (Keokuk); Illinois.

Chomatodus molaris Newb. and Worth.

Newberry and Worthen 1866 A, p. 56, pl. iii, figs. 17, 17a.
Woodward, A. S. 1889 D, p. 49. (Petalodus.)
Subcarboniferous (Keokuk); Illinois.

Chomatodus newberryi nom. nov.

Kindle, E. M. 1898 A, p. 484. (C. obliquus.)
Newberry, J. S. 1879 A, p. 342. (C. obliquus, not of McCoy, 1848.)
Woodward, A. S. 1889 D, p. 228. (C. obliquus.)
Subcarboniferous (St. Louis); Indiana.

Chomatodus parallelus St. J. and Worth.

St. John and Worthen 1875 A, p. 358, pl. x A. figs. 3, 4.
Keyes, C. R. 1894 A, p. 231.
Woodward, A. S. 1889 D, p. 49. (Petalodus.)
Subcarboniferous (Warsaw); Illinois, Missouri.

Chomatodus pusillus Newb. and Worth.

Newberry and Worthen 1866 A, p. 58, pl. iii, figs. 14–14b.
Kindle, E. M. 1898 A, p. 484.
Woodward, A. S. 1889 D, p. 49. (Petalodus.)
Subcarboniferous (Keokuk); Illinois, Indiana.

Chomatodus selliformis Newb.

Newberry, J. S. 1879 A, p. 341.
Kindle, E. M. 1898 A, p. 484.
Woodward, A. S. 1889 D, p. 228.
Subcarboniferous (St. Louis); Illinois, Indiana.

Chomatodus varsouviensis St. J. and Worth.

St. John and Worthen 1875 A, p. 363, pl. x, figs. 1–4b.
Woodward, A. S. 1889 D, p. 228.
Subcarboniferous (Warsaw); Illinois, Missouri.

LISGODUS St. J. and Worth. Type *L. curtus* St. J. and Worth.

St. John and Worthen 1875 A, p. 363.
Miller, S. A. 1889 A, p. 601.
Woodward, A. S. 1889 D, p. 49.

Lisgodus affinis Newb.

Newberry, J. S. 1879 A, p. 343.
Kindle, E. M. 1896 A, p. 484.
Woodward, A. S. 1889 D, p. 49.
Subcarboniferous (St. Louis): Illinois, Missouri, Indiana.

Lisgodus curtus St. J. and Worth.

St. John and Worthen 1875 A, p. 364, pl. xA, figs. 20–22.
Keyes, C. R. 1894 A, p. 231.
Miller, S. A. 1889 A, p. 601, fig. 1143.

Woodward, A. S. 1889 D, p. 49.
Subcarboniferous (Burlington); Iowa, Illinois, Missouri.

Lisgodus selluliformis St.J.and Worth.

St. John and Worthen, 1875 A, p. 366, pl. xA, fig. 16.
Keyes, C. R. 1894 A, p. 231.
Woodward, A. S. 1889 D, p. 49.
Subcarboniferous (St. Louis); Illinois, Missouri.

Lisgodus serratus St. J. and Worth.

St. John and Worthen 1875 A, p. 365, pl. xA, figs. 17–19.
Woodward, A. S. 1889 D, p. 49.
Subcarboniferous (Burlington); Iowa, Illinois.

CALOPODUS St. J. and Worth. Type *C. apicalis* St. J. and Worth.

St. John and Worthen 1875 A, p. 408.
Miller, S. A. 1889 A, p. 589.
Woodward, A. S. 1889 D, p. 49.

Calopodus apicalis St. J. and Worth.

St. John and Worthen 1875 A, p. 408, pl. xii, figs. 16, 17.

Miller, S. A. 1889 A, p. 589, fig. 1105.
Woodward, A. S. 1889 D, p. 49.
Coal-measures, Iowa.

CTENOPETALUS Davis. Type *Petalodus serratus* Owen.

Davis, J. W. 1881 A, p. 426. (Ex. Agass. MSS.)
1882 B, p. 59.
McCoy, F. 1855 A, p. 626. (Ctenoptychius.)
Miller, S. A. 1889 A, p. 594.
Newberry, J. S. 1889 A, p. 186.
Pictet, F. J. 1854 A, p. 264. (Ctenoptychius.)
Ward, J. 1890 A, p. 141. (Ctenoptychius.)
Woodward, A. S. 1889 D, p. 49. (Syn. of Ctenoptychius.)
Woodward and Sherborn 1890 A, p. 54. (Syn. of Ctenoptychius.)
Zittel, K. A. 1890 A, p. 99.

Ctenopetalus bellulus St. J. and Worth.

St. John and Worthen 1875 A, p. 398, pl. xii, fig. 9.
[C. (Petalodus.)]
Woodward, A. S. 1889 D, p. 52. (Ctenoptychius.)
Subcarboniferous (St. Louis); Iowa, Illinois.

Ctenopetalus limatulus St. J. and Worth.

St. John and Worthen 1875 A, p. 399, pl. xii, fig. 18.

Woodward, A. S. 1889 D, p. 53. (Ctenoptychius.)
Subcarboniferous (Chester); Illinois.

Ctenopetalus medius St. J. and Worth.

St. John and Worthen 1875 A, p. 400, pl. xA, fig. 26.
Woodward, A. S. 1889 D, p. 53. (Ctenoptychius.)
Subcarboniferous (Chester); Illinois.

Ctenopetalus occidentalis St. J. and Worth.

St. John and Worthen 1875 A, p. 401, pl. xii, fig. 14.
Miller, S. A. 1889 A, p. 594, fig. 1120.
Woodward, A. S. 1889 D, p. 53. (Ctenoptychius.)
Coal-measures; Iowa.

Ctenopetalus vinosus St. J. and Worth.

St. John and Worthen 1875 A, p. 396, pl. xii, fig. 13.
Woodward, A. S. 1889 D, p. 54. (Ctenoptychius.)
Subcarboniferous (Keokuk); Iowa.

CTENOPTYCHIUS Agassiz. Type *C. apicalis* Agassiz.

Agassiz, L. 1843 B (1838), iii, p. 99.
1844 A.
Barkas, W. J. 1873 A, p. 18.
1874 A, pp. 439, 482, 533.
Davis, J. W. 1881 A, p. 425.
Jaekel, O. 1895 B, p. 201.
1899 A, p. 286.
Miller, S. A. 1889 A, p. 594 (Ctenoptychius); p. 605 (Peripristis).

Newberry, J. S. 1875 A, p. 52.
1889 A, p. 186.
Newberry and Worthen 1866 A, pp. 33, 73.
Nicholson and Lydekker 1889 A, p. 930.
Owen, R. 1845 B, p. 61.
Pictet, F. J. 1854 A, p. 264.
St. John, O. 1870 A, p. 434. (Peripristis, type *P. semicircularis.*)
1872 A, p. 241. (Peripristis.)

Stock, T. 1882 A.
    1885 B, p. 227.
Woodward, A. S. 1889 D, p. 49.
Woodward and Sherborn 1890 A, p. 54.
Zittel, K. A. 1890 A, p. 99.

Ctenoptychius acuminatus (St. J. and Worth.).

St. John and Worthen 1875 A, p. 402, pl. xA, figs. 6–6d. (Pristodus?)
Lesley, J. P. 1889 A, p. 745, figures. (*Pristodus*)
Woodward, A. S. 1889 D, p. 53.
    Subcarboniferous (Kinderhook); Iowa.

Ctenoptychius cristatus Dawson.

Dawson, J. W. 1868 A, p. 209, fig. 52.
    1878 A, p. 209, fig. 52.
Lesley, J. P. 1895 A, pl. dlxx.
Miller, S. A. 1889 A, p. 594, fig. 1121.
Woodward, A. S. 1889 D, p. 56.
    Coal-measures; Nova Scotia.

Ctenoptychius pertenuis St. J. and Worth.

St. John and Worthen 1875 A, p. 382, pl. xA, fig. 27.

HARPACODUS St. J. and Worth.   Type *H. occidentalis* St. J. and Worth.

St. John and Worthen 1875 A, p. 354. (Ex. Agass. MSS.)
Davis, J. W. 1881 A, p. 426.
    1883 A, p. 514.
Miller, S. A. 1889 A, p. 598.
Newberry, J. S. 1889 A, p. 186.
Woodward, A. S. 1889 D, p. 49. (Syn. of Ctenoptychius.)
Zittel, K. A. 1890 A, p. 99.

Harpacodus compactus St. J. and Worth.

St. John and Worthen 1875 A, p. 355, pl. xA, figs. 1–1b.

POLYRHIZODUS McCoy.   Type *P. magnus* McCoy.

McCoy, F. 1848 A, p. 125.
Davis, J. W. 1883 A, pp. 485–499.
Jaekel, O. 1895 B.
    1899 A, p. 278.
McCoy, F. 1855 A, p. 641.
Miller, S. A. 1889 A, p. 594 (Dactylodus); p. 607 (Polyrhizodus).
Newberry, J. S. 1897 A, p. 294. (Dactylodus.)
Newberry and Worthen 1866 A, p. 33 (Dactylodus, type *D. princeps*); p. 34 (Polyrhizodus).
Nicholson and Lydekker 1889 A, p. 930. (Polyrhizodus, Dactylodus.)
Pictet, F. J. 1854 A, p. 271.
Trautschold, H. 1874 B, p. 294 (Dactylodus); p. 296 (Polyrhizodus).
Woodward, A. S. 1889 D, p. 56.
Zittel, K. A. 1890 A, p. 97.

Polyrhizodus amplus St. J. and Worth.

St. John and Worthen 1875 A, p. 387, pl. xiii, fig. 13.
Keyes, C. R. 1894 A, p. 232.
Lesley, J. P. 1889 A, p. 732, figures.

Woodward, A. S. 1889 D, p. 53.
    Subcarboniferous (Chester); Illinois.

Ctenoptychius semicircularis Newb. and Worth.

Newberry and Worthen 1866 A, p. 72, pl. iv, figs. 18–18b.
Kindle, E. M. 1898 A, p. 484 (Ctenoptychius); p. 485 (Peripristis).
Miller, S. A. 1889 A, p. 605, fig. 1156. (Peripristis.)
Newberry, J. S. 1875 A, p. 52, pl. lviii, fig. 14.
St. John, O. 1870 A. p. 434. (Peripristis.)
    1872 A, p. 242, pl. iii, figs. 3, 4; pl. iv, figs. 20–20b.
Woodward, A. S. 1889 D, p. 53.
    Coal-measures; Ohio, Indiana, Nebraska.

Ctenoptychius stevensoni St. J. and Worth.

St. John and Worthen 1875 A, p. 383, pl. xii, fig. 15.
Woodward, A. S. 1889 D, p. 54.
    Coal-measures; West Virginia.

Woodward, A. S. 1889 D, p. 53. (Ctenoptychius).
    Subcarboniferous (Chester); Illinois.

Harpacodus occidentalis St. J. and Worth.

St. John and Worthen 1875 A, p. 355, pl. xA, figs. 2–2d.
Keyes, C. R. 1894 A, p. 230.
Miller, S. A. 1889 A, p. 599, fig. 1135.
Woodward, A. S. 1889 D, p. 53. (Ctenoptychius.)
    Subcarboniferous (St. Louis); Illinois, Missouri.

Woodward, A. S. 1889 D, p. 59.
    Subcarboniferous (St. Louis); Illinois, Missouri.

Polyrhizodus carbonarius St. J. and Worth.

St. John and Worthen 1875 A, p. 389, pl. xA, figs. 24, 25; pl. xiii, fig. 1.
Lesley, J. P. 1889 A, p. 732, figures.
Woodward, A. S. 1889 D, p. 59.
    Coal-measures; Illinois.

Polyrhizodus concavus (St. J. and Worth.).

St. John and Worthen 1875 A, p. 390, pl. xiii, figs. 17, 18. (Dactylodus.)
Miller, S. A. 1889 A, p. 594, fig. 1123. (Dactylodus.)
Quenstedt, F. A. 1885 A, p. 294, pl. xxiii, fig. 31. (Dactylodus.)
Woodward, A. S. 1889 D, p. 59.
    Subcarboniferous (St. Louis); Illinois.

Polyrhizodus dentatus Newb. and Worth.

Newberry and Worthen 1866 A, p. 50, pl. iii, figs. 10, 10a.
Lesley, J. P. 1889 A, p. 733, figures.
Woodward, A. S. 1889 D, p. 59.
Subcarboniferous (Chester); Illinois.

Polyrhizodus digitatus (Leidy).

Leidy, J. 1857 H, p. 90, pl. v, figs. 27–29. (Ctenoptychius.)
Newberry and Worthen 1866 A, p. 47. (Dactylodus.)
1866 A, p. 47, pl. iii, figs. 7, 7a. (Dactylodus lobatus.)
Woodward, A. S. 1889 D, p. 59. (P. lobatus.)
Subcarboniferous (St. Louis); Illinois, Missouri.

Polyrhizodus excavatus (St. J. and Worth.).

St. John and Worthen 1866 A, p. 392; pl. xiii, fig. 16. (Dactylodus.)
Woodward, A. S. 1889 D, p. 59.
Subcarboniferous (Chester); Illinois.

Polyrhizodus inflexus (Newb. and Worth.).

Newberry and Worthen 1866 A, p. 48, pl. iii, figs. 8, 8a. (Dactylodus.)
Woodward, A. S. 1889 D, p. 59.
Subcarboniferous (Chester); Illinois.

Polyrhizodus latus Newberry.

Newberry, J. S. 1897 A, p. 296, pl. xxiii, figs. 7, 7a. (Dactylodus.)
Subcarboniferous (St. Louis); Illinois.

Polyrhizodus littoni Newb. and Worth.

Newberry and Worthen 1870 A, p. 357, pl. iv, figs. 10, 10a.
Keyes, C. R. 1894 A, p. 232.
Kindle, E. M. 1898 A, p. 485.
Lesley, J. P. 1889 A, p. 733, figures.
Miller, S. A. 1889 A, p. 607, fig. 1160.
Newberry, J. S. 1889 A, p. 209, pl. xix, figs. 5, 6.
Woodward, A. S. 1889 D, p. 59.
Subcarboniferous (St. Louis); Missouri, Indiana.

Polyrhizodus minimus (St. J. and Worth.).

St. John and Worthen 1875 A, p. 391, pl. xiii, fig. 19. (Dactylodus.)
Woodward, A. S. 1889 D, p. 60.
Subcarboniferous (St. Louis); Illinois.

Polyrhizodus modestus Newb.

Newberry, J. S. 1875 A, p. 50, pl. lviii, figs. 10, 10a.
Lesley, J. P. 1889 A, p. 733, figures.
Miller, S. A. 1889 A, p. 607, fig. 1161.
Woodward, A. S. 1889 D, p. 60.
Subcarboniferous; Ohio.

Polyrhizodus nanus St. J. and Worth.

St. John and Worthen 1875 A, p. 386, pl. xiii, fig. 15.
Lesley, J. P. 1889 A, p. 733, figures.
Woodward, A. S. 1889 D, p. 60.
Subcarboniferous (Keokuk); Iowa.

Polyrhizodus piasaensis St. J. and Worth.

St. John and Worthen 1875 A, p. 386, pl. xiii, fig. 12.
Lesley, J. P. 1889 A, p. 734, figures.
Woodward, A: S. 1889 D, p. 60.
Subcarboniferous (Warsaw); Illinois.

Polyrhizodus ponticulus Newb. and Worth.

Newberry and Worthen 1866 A, p. 51, pl. iii, figs. 11, 11a.
Lesley, J. P. 1889 A, p. 734, figures.
Woodward, A. S. 1889 D, p. 60.
Subcarboniferous (Chester); Illinois.

Polyrhizodus porosus Newb. and Worth.

Newberry and Worthen 1866 A, p. 49, pl. iii, figs. 9, 9a.
Lesley, J. P. 1889 A, p. 734, figures.
Woodward, A. S. 1889 D, p. 60.
Subcarboniferous(Burlington); Illinois, Iowa.

Polyrhizodus princeps (Newb. and Worth.).

Newberry and Worthen 1866 A, p. 45, pl. iii, figs. 8–8b. (Dactylodus.)
Hambach, G. 1890 A, p. 80.
Jaekel, O. 1899 A, p. 285. (Dactylodus.)
Newberry, J. S. 1897 A, p. 295, pl. xxiii, figs. 5, 5a, 6. (Dactylodus.)
Woodward, A. S. 1889 D, p. 60.
Subcarboniferous (St. Louis); Illinois, Missouri.

Polyrhizodus rectus (Newb.).

Newberry, J. S. 1897 A, p. 296, pl. xxiii, figs. 8, 9. (Dactylodus.)
Subcarboniferous (St. Louis); Illinois.

Polyrhizodus truncatus Newb. and Worth.

Newberry and Worthen 1870 A, p. 357, pl. iii, figs. 16, 16a.
Woodward, A. S. 1889 D, p. 60.
Subcarboniferous (Burlington); Illinois.

Polyrhizodus williamsi St. J. and Worth.

St. John and Worthen 1875 A, p. 384, pl. xA, fig. 23; pl. xiii, fig. 11.
Keyes, C. R. 1894 A, p. 232.
Lesley, J. P. 1889 A, p. 734, figures.
Woodward, A. S. 1889 A, p. 60.
Subcarboniferous (Keokuk); Missouri, Io

PSAMMODONTIDÆ.

Koninck, L. G. de 1878 A, pp. 29, 3G·
Davis, J. W. 1883 A, pp. 431, 459.
Jaekel, O. 1895 B.
    1896 A, p. 49 (with subfams. Psammondontinæ, Helodontinæ, and Psephodontinæ.

Nicholson and Lydekker 1889 A, p. 934.
St. John and Worthen 1883 A, p. 197.
Woodward, A. S. 1889 D, p. 91.
    1896 B, p. 50.
Zittel, K. A. 1890 A, p. 95.

PSAMMODUS Agassiz.   Type *P. rugosus* Agassiz.

Agassiz, L. 1843 B (1838), iii, p. 108.
    1844 A.
Davis, J. W. 1883 A, p. 459.
Günther, A. 1872 A, p. 519.
Koninck, L. G. de 1844 A, p. 615.
    1878 A, p. 41.
McCoy, F. 1855 A, p. 643.
Miller, S. A. 1889 A, p. 608.
Newberry and Worthen 1866 A, p. 107.
Nicholson and Lydekker 1889 A, p. 935.
Owen, R. 1845 B, p. 59, pls. xvii, xx, xxi.
Pictet, F. J. 1854 A, p. 266.
Quenstedt, F. A. 1885 A, p. 293.
St. John and Worthen 1883 A, p. 197.
Steinmann and Döderlein 1890 A, p. 549.
Trautschold, H. 1874 B, p. 288.
Woodward, A. S. 1889 D, p. 99.
Zittel, K. A. 1890 A, p. 95.

Psammodus angularis Newb. and Worth.

Newberry and Worthen 1866 A, p. 107, pl. xi, figs. 2–2b.
Lesley, J. P. 1889 A, p. 784, figures (P. angularis); p. 787, figures (P. porosus?).
Miller. S. A. 1889 A, p. 609, fig. 1164. (P. porosus.)
Newberry and Worthen, 1866 A, p. 107, pl. xi, figs. 1–1b. (" P. porosus Ag.?").
St. John and Worthen, 1883 A, p. 222.
Woodward, A. S. 1889 D, p. 107.
    Subcarboniferous (Chester); Illinois.

Psammodus antiquus Newb.

Newberry, J. S. 1857 C, p. 124.
    1873 A, p. 265.
Woodward, A. S. 1889 D, p. 459.
    Devonian (Corniferous); Ohio.

Psammodus cælatus St. J. and Worth.

St. John and Worthen 1883 A, p. 217, pl. xviii, fig. 1.
Hambach, G. 1890 A, p. 81.
Lesley, J. P. 1889 A, p. 784, figures.
Woodward, A. S. 1889 D, p. 107.
    Subcarboniferous (St. Louis); Iowa, Missouri (*fide* Hambach).

Psammodus crassidens St. J. and Worth.

St. John and Worthen, 1883 A, p. 218, pl. xviii, figs. 2–6; text figs. a–c.
Lesley, J. P. 1889 A, p. 784, figures (P. crassidens) p. 788, figures (P. rugosus).
    1895 A, pl. ccxxxi.
Miller, S. A. 1889 A, p. 608, fig. 1163.

Newberry and Worthen 1866 A, p. 108, pl. xi, figs. 3, 3a. (P. rugosus, not of Agassiz.)
Woodward, A. S. 1889 D, p. 107.
    Subcarboniferous (St. Louis); Illinois, Iowa.

Psammodus glyptus St. J. and Worth.

St. John and Worthen 1883 A, p.·209, pl. xiv, figs. 5, 6.
Kindle, E. M. 1896 A, p. 485.
Lesley, J. P. 1889·A, p. 786, figures.
    1895 A, pl. ccxxxi.
Newberry, J. S. 1889 A, p. 210, pl. xix, figs. 7, 8.
Woodward, A. S. 1889 D, p. 107.
    Subcarboniferous (Burlington); Illinois.

Psammodus grandis St. J. and Worth.

St. John and Worthen 1883 A, p. 211, pl. xv, figs. 1–3.
Lesley, J. P. 1889 A, p. 786, figures.
    1895 A, pl. ccxxxi.
Woodward, A. S. 1889 D, p. 107.
    Subcarboniferous (Keokuk); Iowa.

Psammodus lovianus St. J. and Worth.

St. John and Worthen 1883 A, p. 207, pl. xiv, figs. 7–9.
Lesley, J. P. 1889 A, p. 786, figures.
    1895 A, pl. ccxxxi.
Woodward, A. S. 1889 D, p. 108.
    Subcarboniferous (Burlington); Illinois, Iowa.

Psammodus plenus St. J. and Worth.

St. John and Worthen 1883 A, p. 213, pl. xvi, figs. 1–4; pl. xvii, figs. 1–4.
Keyes, C. R. 1894 A, p. 235.
Lesley, J. P. 1889 A, p. 786, figures.
    1895 A, pl. ccxxxii.
Woodward, A. S. 1889 D, p. 108.
    Subcarboniferous (St. Louis); Missouri, Illinois, Michigan.

Psammodus reticulatus Newb. and Worth.

Newberry and Worthen 1866 A, p. 109, pl. xi, figs. 5, 5a.
Hambach, G. 1890 A, p. 81.
Koninck, L. G. de 1878 A, p. 42. (P. porosus?).
Lesley, J. P. 1889 A, p. 788, figure.
    1895 A, pl. ccxxxi.
St. John and Worthen 1883 A, p. 224, pl. xix, figs. 3, 5.
Woodward, A. S. 1889 D, p. 108.
    Subcarboniferous (Chester); Illinois, Missouri (*fide* Hambach).

Psammodus springeri St. J. and Worth.

St. John and Worthen 1883 A, p. 202, pl. xx, figs.
  4–11, and 7 text figures.
Lesley, J. P. 1889 A, p. 789, figures.
  1895 A, pl. ccxxxii.
Miller, S. A. 1889 A, p. 609, figs. 1165, 1166.
Woodward, A. S. 1889 D, p. 108.
  Subcarboniferous (Burlington); Iowa, Illinois.

Psammodus tumidus St. J. and Worth.

St. John and Worthen 1883 A, p. 205, pl. xlv, figs.
  1–4.
Lesley, J. P. 1889 A, p. 790, figures.
Woodward, A. S. 1889 D, p. 108.
  Subcarboniferous (Burlington); Illinois, Iowa.

Psammodus turgidus St. J. and Worth.

St. John and Worthen 1883 A, p. 206, pl. xv, fig. 4;
  text-fig. *a.*
Lesley, J. P. 1889 A, p. 790, figures.
  1895 A, pl. ccxxxi.
Miller, S. A. 1889 A, p. 609, fig. 1168.
Woodward, A. S. 1889 D, p. 108.
  Subcarboniferous (division beds of Burlington and Keokuk); Iowa.

Psammodus sp?

Dawson, J W. 1868 A, p. 210, fig. 54.
Lesley, J. P. 1889 A, p. 790, figure.
  Coal-measures; Nova Scotia.

MAZODUS Newb. Type *M. kepleri* Newb.

Newberry, J. S. 1889 A, p. 178.
Claypole, E. W. 1896 D, p. 146.
  1896 E, p. 152.
Miller, S. A. 1892 A, p. 716.

Mazodus kepleri Newb.

Newberry, J. S. 1889 A, p. 180, pl. xxi, figs. 1–3.
Claypole, E. W. 1896 D, p. 146, figs. 1, 2.
  Subcarboniferous (Waverly); Ohio.

ARCHÆOBATIS Newb. Type *A. gigas* Newb.

Newberry, J. S. 1878 A, p. 190.
  1879 A, p. 346.
  1885 A.
Nicholson and Lydekker 1885 A.
Woodward, A. S. 1889 D, p. 108.
  1890 A, p, 391.
Zittel, K. A. 1890 A, p. 96.

Archæobatis gigas Newb.

Newberry, J. S. 1878 A, p. 190.

Kindle, E. M. 1898 A, p. 484.
Newberry, J. S. 1879 A, p. 347.
  1884 A.
  1885 A.
Woodward, A. S. 1889 D, p. 108.
  1890 A, p. 391.
  Subcarboniferous (St. Louis); Indiana.

COPODUS Davis. Type *C. cornutus* Davis.

Davis, J. W. 1883 A, p. 464.
  1883 A, p. 469. (Labodus.)
Eastman, C. R. 1898 A, p. 112.
Miller, S. A. 1889 A, p. 592.
  1892 A, p. 716.
Newberry, J. S. 1889 A, p. 198. (Labodus.)
Nicholson and Lydekker 1889 A, p. 935.
St. John and Worthen 1883 A, p. 227; 6 text-figs.
Woodward, A. S. 1884 A, p. 271.
  1889 D, p. 91.
Woodward and Sherborn 1890 A, p. 43.
Zittel, K. A. 1890 A, p. 96.

Copodus marginatus (Newb.).

Newberry, J. S. 1889 A, p. 198, pl. xix, fig. 9. (Labodus.)

Kindle, E. M. 1898 A, p. 484. (Labodus.)
  Subcarboniferous (St. Louis); Indiana.

Copodus pusillus St. J. and Worth.

St. John and Worthen 1883 A, p. 231, pl. xx, fig. 1.
Woodward, A. S. 1889 D, p. 99.
  Subcarboniferous (Chester); Illinois.

Copodus vanhornii St. J. and Worth.

St. John and Worthen 1883 A, p. 229, pl. xx,
  figs. 2, 3.
Keyes, C. R. 1894 A, p. 235.
Woodward, A. S. 1889 D, p. 99.
  Subcarboniferous (St. Louis); Missouri, Illinois.

Superfamily HETERODONTOIDEA,[1] *nom. nov.*

Gill, T., 1882, in Jordan and Gilbert's Synop. Fishes
N. A., p. 967. (Prosarthri.)
  1884 B, p. 346. (Proarthri.)
  1884 C. (Proarthri.)

1898 A, p. 129. (Prosarthri.)
Jordan and Evermann 1896 A, p. 19. (Proarthri.)
Woodward, A. S. 1889 D, p. ix. ("Proarthri.")

---

[1] This name is proposed in order to conform to the rule adopted by many zoologists, according to which the name of the superfamily is derived from that of a typical family.

COCHLIODONTIDÆ.

Owen, R. 1867 A.
Brown, C. 1900 A, pp. 173, 174.
Cope, E. D. 1891 N, p. 17.
    1898 B, p. 26.
Davis J. W. 1882 B.
    1883 A, pp. 331, 410.
    1890 B.
Koninck, L. G. de 1878 A, p. 29.

Nicholson and Lydekker 1889 A, p. 939.
St. John and Worthen 1883 A, p. 201.
Traquair, R. H. 1888 A, p. 84.
Woodward, A. S. 1884 A, p. 269.
    1889 D, p. 169.
    1892 G.
Zittel, K. A. 1890 A, p. 68.

HELODUS Agassiz.   Type *H. simplex* Agassiz.

Agassiz, L. 1843 B (1838), iii, p. 104.
    1844 A.
Amer. Geologist 1888 A.
Davis, J. W. 1890 B.
Jaekel O. 1898 A, p. 50.
Koninck, L. G. de 1644 A, p. 614.
    1878 A, p. 38.
McCoy, F. 1885, A, p. 630.
Miller, S. A. 1889 A, p. 599.
Newberry and Worthen 1866 A, p. 74.
    1870 A, p. 365.
Nicholson and Lydekker 1889 A, p. 939.
Pictet, F. J. 1854 A, p. 266.
Quenstedt, F. A. 1885 A, p. 294.
St. John and Worthen 1875 A, p. 296.
Traquair, R. H. 1888 A, p. 84.
Trautschold, H. 1874 B, p. 288.
Ward, J. 1890 A, p. 143.
Woodward, A. S. 1884 A, p. 269.
    1889 D, pp. 171, 226.
Woodward and Sherborn 1890 A, p. 92.
Zittel, K. A. 1890 A, p. 75.

Helodus angulatus Newb. and Worth.

Newberry and Worthen 1866 A, p. 83, pl. v, figs. 9–15.
St. John and Worthen 1875 A, p. 362.
Woodward, A. S. 1889 D, p. 221.
    Subcarboniferous (Burlington); Illinois.

Helodus biformis Newb. and Worth.

Newberry and Worthen 1866 A, p. 77, pl. iv, fig. 22.
St. John and Worthen 1875 A, p. 419.
Woodward, A. S. 1889 D, p. 226.
    Subcarboniferous (Kinderhook): Iowa.

Helodus carbonarius Newb. and Worth.

Newberry and Worthen 1866 A, p. 75, pl. iv, figs. 20, 20a.
Woodward, A. S. 1889 D, p. 226.
    Coal-measures: Illinois.

Helodus coniculus Newb. and Worth.

Newberry and Worthen 1866 A, p. 75, pl. iv, figs. 19, 19a.
Kindle, E. M. 1898 A, p. 484.
Newberry, J. S. 1879 A, p. 344.
Wordward, A. S. 1889 D, p. 226.
    Subcarboniferous (Keokuk); Illinois: (St. Louis); Indiana: (Burlington); Iowa.

Helodus consolidatus Newb. and Worth.

Newberry and Worthen 1866 A, p. 87, pl. vi, figs. 1–2a.

Hambach, G. 1890 A, p. 81.
Woodward, A. S. 1889 D, p. 226.
    Subcarboniferous (Keokuk); Illinois, Missouri.

Helodus crenulatus Newb. and Worth.

Newberry and Worthen 1866 A, p. 82, pl. v, figs. 7, 7a.
Hambach, G. 1890 A, p. 81.
Woodward, A. S. 1889 D, p. 226.
    Subcarboniferous (Keokuk); Illinois, Missouri (*fide* Hambach).

Helodus dens-humani Newb. and Worth.

Newberry and Worthen 1866 A, p. 76, pl. iv, figs. 21–21b.
Woodward, A. S. 1889 D, p. 226.
    Subcarboniferous (Keokuk); Illinois.

Helodus denticulatus Newb. and Worth.

Newberry and Worthen 1866 A, p. 81, pl. v, figs. 6–6b.
St. John and Worthen 1875 A, p. 317.   (Helodus?)
Woodward, A. S. 1889 D, p. 226.
    Subcarboniferous (Keokuk); Illinois.

Helodus elytra Newb. and Worth.

Newberry and Worthen 1866 A, p. 78, pl. iv, fig. 23.
St. John and Worthen 1875 A, p. 299.   (Chomatodus.)
    1888 A, p. 96.   (Helodus.)
    Subcarboniferous (Keokuk); Illinois.

Helodus gibberulus Agassiz.

Agassiz, L. 1843 B (1838), iii, p. 106, pl. xii, figs. 1, 2.
Davis, J. W. 1883 A, p. 405, pl. li, fig. 19.   (Lophodus.)
Newberry, J. S. 1889 A. p. 102.
Trautschold, H. 1874 A, p. 269, pl. xxvi, fig. 9.
Woodward, A. S. 1889 D, p. 219.
    Carboniferous; Europe: Subcarboniferous (Waverly); Pennsylvania: ("Mountain limestone"); Illinois, Indiana.

Helodus gibbus Leidy.

Leidy, J. 1857 H, p. 88, pl. v, fig. 18.
Woodward, A. S. 1889 D, p. 226.
    Subcarboniferous (Warsaw); Illinois.

Helodus lævis Newb.

Newberry, J. S. 1879 A, p. 343.
Kindle, E. M. 1898 A, p. 484.
Woodward, A. S. 1889 D, p. 226.
    Subcarboniferous (St. Louis); Indiana.

Helodus limax Newb. and Worth.

Newberry and Worthen 1866 A, p. 80, pl. v, figs. 5–5b.

Woodward, A. S. 1889 D, p. 226.

   Subcarboniferous (Burlington); Iowa.

Helodus politus Newb. and Worth.

Newberry and Worthen 1866 A, p. 79, pl. v, fig. 2.

Woodward, A. S. 1889 D, p. 227.

   Subcarboniferous (Keokuk); Illinois.

Helodus rugosus Newb. and Worth.

Newberry and Worthen 1870 A, p. 359, pl. ii, figs. 10, 10a.

Woodward, A. S. 1889 D, p. 227.

   Coal-measures; Illinois.

      VENUSTODUS St. J. and Worth.

St. John and Worthen 1875 A, p. 344.

Miller, S. A. 1889 A, p. 613.

Woodward, A. S. 1889 D, pp. 218–225.

   1889 F, p. 296.

Venustodus argutus St. J. and Worth.

St. John and Worthen 1875 A, p. 352, pl. ix, figs. 5, 6.

Lesley, J. P. 1890 A, p. 1250, figures.

Woodward, A. S. 1889 D, p. 228.

   Subcarboniferous (Chester); Illinois.

Venustodus robustus St. J. and Worth.

St. John and Worthen 1875 A. p. 345, pl. ix, figs. 15–18.

Lesley, J. P. 1890 A, p. 1251, figures.

Miller, S. A. 1889 A, p. 614, fig. 1177.

Woodward, A. S. 1889 D, p. 228.

   Subcarboniferous (Burlington); Iowa.

Venustodus tenuicristatus St. J. and Worth.

St. John and Worthen 1875 A, p. 348, pl. ix, figs. 19–24.

      XENODUS S. A. Miller.

Miller, S. A. 1892 A, P. 718.

Hay, O. P. 1899 E, p. 784.

Newberry, J. S. 1889 A, p. 67. (Goniodus, not of Agassiz, 1888.)

      PERIPLECTRODUS St. J. and Worth.

St. John and Worthen 1875 A, p. 324.

Miller, S. A. 1889, p. 605.

Woodward, A. S. 1889 D, p. 228.

Periplectrodus compressus St. J. and Worth.

St. John and Worthen 1875 A, p. 326, pl. viii, figs. 26–26c.

Lesley, J. P. 1889 A, p. 622, figures.

Woodward, A. S. 1889 D, p. 229.

   Subcarboniferous (St. Louis); Illinois.

Periplectrodus expansus St. J. and Worth.

Helodus sulcatus Newb. and Worth.

Newberry and Worthen 1866 A, p. 83, pl. v, figs. 16, 16a.

Woodward, A. S. 1889 D, p. 221.

   Subcarboniferous (Keokuk); Illinois.

Helodus undulatus Newb. and Worth.

Newberry and Worthen 1866 A, p. 82, pl. v, figs. 8–8b.

Woodward, A. S. 1889 D, p. 227.

   Subcarboniferous (Keokuk); Illinois.

Helodus wortheni Hay.

Hay, O. P. 1899 E, p. 784.

Newberry and Worthen 1870 A, p. 360, pl. iii, figs. 15, 15a. (Helodus compressus; not H. compressus Newberry and Worthen 1866 A, p. 78, pl. v, fig. 1=Hyboclododus compressus.)

   Subcarboniferous (Burlington); Illinois.

Type *Chomatodus venustus* Leidy.

Keyes, C. R. 1894 A, p. 230.

Lesley, J. P. 1890 A. p. 1251, figures.

Woodward, A. S. 1889 D, p. 229.

   Subcarboniferous (Keokuk): Illinois, Iowa, Missouri.

Venustodus variabilis St. J. and Worth.

St. John and Worthen 1875 A, p. 346, pl. ix, figs. 7–14.

Lesley, J. P. 1890 A, p. 1252, figures.

Woodward, A. S. 1889 D, p. 229.

   Subcarboniferous(Burlington); Illinois,Iowa.

Venustodus venustus (Leidy).

Leidy, J. 1857 H, p. 89, pl. v, figs, 19–21. (Chomatodus.)

Hambach, G. 1890 A, p. 81. (V. leidyi.)

Lesley, J. P. 1890 A, p. 1250, figures. (V. leidyi.)

Miller, S. A. 1889 A, pp. 590, 613, 614.

Newberry and Worthen 1866 A, p. 34.

St. John and Worthen 1875 A, pp. 345, 350, pl. ix, figs. 1–4. (V. leidyi.)

Woodward, A. S. 1889 D, p. 228. (V. leidyi.)

   Subcarboniferous (St. Louis); Illinois, Missouri.

Type *Goniodus hertzeri* Newb.

Xenodus hertzeri (Newb.).

Newberry, J. S. 1889 A, p. 69, pl. xxvii, figs. 11–15. (Goniodus.)

Miller, S. A. 1892 A, p. 718.

   Devonian (Huron Shale); Ohio.

Type *P. warreni* St. J. and Worth.

St. John and Worthen 1875 A, p. 327, pl. viii, figs. 27–27c.

Lesley, J. P. 1889 A, p. 623, figures.

Woodward, A. S. 1889 D, p. 229.

   Subcarboniferous (Chester); Illinois.

Periplectrodus warreni St. J. and Worth.

St. John and Worthen 1875 A, p. 325, pl. viii, figs. 25–25c.

Lesley, J. P. 1889 A, p. 624, figures.

Woodward, A. S. 1889 D, p. 229.

   Subcarboniferous (Burlington); Iowa.

PSEPHODUS Morris and Roberts.   Type *Cochliodus magnus* Portlock.

Morris and Roberts, 1862, Quart. Jour. Geol. Soc.,
   xviii, p. 101.
Amer. Geologist 1888 A.
Davis, J. W. 1882 B p. 54
     1883 A, p. 438.
Jaekel, O. 1896 A, p. 50.
Koninck, L. G. de 1878 A, p. 59.
Miller, S. A. 1889 A, p. 610 (Psephodus); p. 612
   (Tæniodus, in part.)
Newberry and Worthen 1866 A, p. 92. (Aspidodus,
   type *A. convolutus*.)
Nicholson and Lydekker 1889 A, p. 939.
St. John and Worthen, 1883 A, pp. 59, 63; p. 73
   (Tæniodus).
Traquair, R. H. 1885 A.
     1885 B.
     1888 A, p. 84.
Woodward, A. S. 1884 A, p. 270.
     1889 D, p. 176.
     1892 G, p. 674.
Zittel, K. A. 1890 A, p. 72.

Psephodus crenulatus (Newb. and Worth.).

Newberry and Worthen 1866 A, p. 98, pl. viii, figs.
   3–11. (Aspidodus.)
Hambach, G. 1890 A, p. 80. (Aspidodus.)
Kindle, E. M. 1896 A, p. 484. (Aspidotus.)
Newberry, J. S. 1879 A, p. 341. (Aspidodus.)
Newberry and Worthen 1866 A, p. 94, pl. viii, figs.
   12, 22a. (Aspid. convolutus.)
St. John and Worthen 1875 A, p. 419.
     1883 A, pp. 63, 67, 69.
Woodward, A. S. 1889 D, p. 184.
   Subcarboniferous (Chester); Illinois, Ken-
   tucky, Missouri, Indiana.

Psephodus latus St. J. and Worth.

St. John and Worthen 1883 A, p. 72, pl. ii, figs. 1–3.
Keyes, C. R. 1894 A. p. 232.
Lesley, J. P. 1889 A, p. 791, figures.
Woodward, A. S. 1889 D, p. 184.
   Subcarboniferous (St. Louis); Illinois, Mis-
   souri.

Psephodus lunulatus St. J. and Worth.

St. John and Worthen 1883 A, p. 74, pl. ii, fig. 4.
   (P. cunulatus, *errore*).
Lesley, J. P. 1889 A, p. 791, figures.
Woodward, A. S. 1889 D, p. 185.
   Subcarboniferous (Chester); Illinois.

Psephodus obliquus St. J. and Worth.

St. John and Worthen 1883 A, p. 66, pl. i, figs. 1–5
Lesley, J. P. 1889 A, p. 791, figures.
     1895 A, pl. ccxxxii.
Woodward, A. S. 1889 D, p. 184.
   Subcarboniferous (Kinderhook); Iowa.

Psephodus placenta (Newb. and Worth.).

Newberry and Worthen 1866 A, p. 80, pl. v, figs.
   4, 4a. (Helodus.)
Lesley, J. P. 1889 A, p. 792, figures.
     1895 A, pl. ccxxxii.
St. John and Worthen 1883 A, p. 69, pl. ii, figs. 5–8.
Woodward, A. S. 1889 D, p. 184.
   Subcarboniferous (Kinderhook); Iowa.

Psephodus politus Newb.

Newberry, J. S. 1897 A, p. 301, pl. xxiv, figs. 13–23.
   [P. (Helodus); not H. politus of Newb. and
   Worth.]
   Subcarboniferous (Burlington); Illinois.

Psephodus regularis (St. J. and Worth.).

St. John and Worthen 1883 A, p. 77, pl. xiii, fig.
   11. (Tæniodus.)
Kindle, E. M. 1896 A, p. 485. (Tæniodus.)
Lesley, J. P. 1890 A, p. 1173, figs. (Tæniodus.)
     1895 A, pl. ccxxxiv. (Teniodus.)
Woodward, A. S. 1889 D, p. 185.
   Subcarboniferous (Warsaw); Indiana.

Psephodus? reticulatus St. J. and Worth.

St. John and Worthen 1875 A, p. 417, pl. vi, figs.
   19–24.
Leslie, J. P. 1889 A, p. 793, figures.
     1895 A, pl. ccxxxii.
Woodward, A. S. 1889 D, p. 185.
   Subcarboniferous (Kinderhook); Iowa.

Psephodus? symmetricus St. J. and Worth.

St. John and Worthen 1883 A, p. 71, pl. i, figs. 6–7d.
Lesley, J. P. 1889 A, p. 793, figures.
     1895 A, pl. ccxxxii.
Woodward, A. S. 1889 D, p. 185.
   Subcarboniferous (Kinderhook); Iowa.

SANDALODUS Newb. and Worth.   Type, none specified.

Newberry and Worthen 1866 A, p. 102.
Davis, J. W. 1883 A, p. 436.
Koninck, L. G. de 1878 A, p. 62.
Miller, S. A. 1889 A, p. 612.
Newberry and Worthen 1866 A, p. 111. (Trigono-
   dus, type *T. major*.)
Nicholson and Lydekker 1889 A, p. 939.
St. John and Worthen 1883 A, p. 184.
Woodward, A. S. 1889 D, p. 185.

Zittel, K. A. 1890 A, p. 72.

Sandalodus angustus Newb. and Worth.

Newberry and Worthen 1866 A, p. 108, pl. x, fig. 3.
Lesley, J. P. 1890 A, p. 920, figure.
     1895 A, pl. ccxxxii.
Woodward, A. S. 1889 D, p. 187.
   Subcarboniferous (Keokuk); Illinois.

Sandalodus crassus (Newb. and Worth.).

Newberry and Worthen 1866 A, p. 91, pl. viii, figs. 2, 2a. (Cochliodus? crassus, not Sandalodus crassus Newb. and Worth. 1870 = S. spatulatus Newb. and Worth.)

Keyes, C. R. 1894 A, p. 234. (S. lævissimus, S. crassus.

Lesley, J. P. 1889 A, p. 789, figure. (Psammodus semicylindricus.)

    1890 A, p. 921, figure. (S. grandis, S. lævissimus.)

    1895 A, pl. ccxxxii (S. grandis); pl. ccxxxiii (S. lævissimus).

Newberry, J. S. 1897 A, p. 298 (S. grandis.)

Newberry and Worthen 1866 A, p. 105, pl. x, fig. 9 (Sandalodus grandis); p. 109, pl. xi, figs. 4, 4a (Psammodus? semicylindricus); p. 110, pl. xi, figs. 6, 6a (P? rhomboideus),

St. John and Worthen 1883 A, p. 186, pl. xii, figs. 8, 9. (S. lævissimus.)

Woodward, A. S. 1889 D, p. 188. (S. lævissimus.) Subcarboniferous; Illinois, Iowa, Missouri, Indiana.?

Sandalodus ellipticus Newb.

Newberry, J. S. 1897 A, p. 299, pl. xxiv, fig. 2. Subcarboniferous (Kinderhook); Iowa.

Sandalodus? minor (Newb. and Worth.).

Newberry and Worthen 1866 A, p. 112, pl. xi, figs. 7, 7a. (Trigonodus.)

Kindle, E. M. 1898 A, p. 485.

Lesley, J. P. 1890 A, p. 1219, figures. (Trigonodus.)

Newberry, J. S. 1879 A, p. 341. (Trigonodus.)

Woodward, A. S. 1889 D, p. 185. (Psephodus?.) Subcarboniferous (Keokuk); Illinois, (St. Louis,) Indiana.

Sandalodus parvulus Newb. and Worth.

Newberry and Worthen 1866 A, p. 102, pl. x, fig. 1.

Keyes, C. R. 1894 A, p. 233.

Lesley, J. P. 1890 A, p. 921, figures.

St. John and Worthen 1883 A, p. 107, pl. iv, figs. 4-8. (Stenopterodus.)

Woodward, A. S. 1889 D, pp. 190, 217. Subcarboniferous (St. Louis); Illinois, Missouri.

Sandalodus rhomboideus (Newb. and Worth.).

Newberry and Worthen 1866 A, p. 100, pl. ix, fig. 8. (Deltodus.)

Keyes, C. R. 1894 A, p. 234. (S. spatulatus.)

Lesley, J. P. 1889 A, p. 789, figure. (Psammodus.)

    1890 A, p. 922, figures. (S. spatulatus.)

Miller, S. A. 1889 A, p. 612, fig. 1175. (S. spatulatus.)

Newberry, J. S. 1889 A, p. 204, pl. xxi, figs. 6-8. (S. crassus.)

Newberry and Worthen 1866 A, p. 108, pl. x, fig. 2. (S. spatulatus.)

    1870 A, p. 369, pl. iv, figs. 3, 3a. (S. crassus.) (S. spatulatus.)

St. John and Worthen 1883 A, p. 188, pl. xii, fig. 7. (S. spatulatus.)

Woodward, A. S. 1889 D, p. 190. (S. spatulatus.) Subcarboniferous (St. Louis), Illinois, Missouri.

Sandalodus sp?

St. John and Worthen 1883 A, p. 187, pl. xii, figs. 5, 6.

Subcarboniferous (Warsaw); Illinois.

ORTHOPLEURODUS St. J. and Worth.   Type *Sandalodus carbonarius* Newb. and Worth. = *O. angularis* (Newb. and Worth.).

St. John and Worthen 1883 A, pp. 59, 191.

Miller, S. A. 1889 A, p. 604.

Woodward, A. S. 1889 D, p. 188. (Syn. of Sandalodus.)

Zittel, K. A. 1890 A, p. 72.

Orthopleurodus angularis (Newb. and Worth.).

Newberry and Worthen 1866 A, p. 97, pl. ix, fig. 1. (Deltodus.)

Keyes, C. R. 1894 A, p. 234. (O. carbonarius.)

Kindle, E. M. 1898 A, p. 485. (O. carbonarius.)

Lesley, J. P. 1889 A, p. 568, figures. (O. carbonarius.)

Lesley, J. P. 1890 A, p. 920, figures. (Sandalodus carbonarius.)

    1895 A, pl. dlxxi. (O. carbonarius.)

Newberry, J. S. 1889 A, p. 200, pl. xix, fig. 17. (O. carbonarius.)

Newberry and Worthen 1886 A, p. 104, pl. x, figs. 4, 5. (Sandalodus carbonarius.)

St. John, O. 1870 A, p. 436 (Xystrodus? occidentalis); p. 437 (Deltodus? angularis).

St. John, O. 1872 A, p. 244, pl. iv, figs. 18a-18d (Xystrodus occidentalis.)

St. John and Worthen 1883 A, pp. 177, 192, pl. xiii, figs. 6, 8. (O. carbonarius.)

Woodward, A. S. 1889 D, p. 188. (O. carbonarius.) Coal-measures; Indiana, Illinois, Iowa, Missouri, Kansas, and Nebraska.

Orthopleurodus convexus St. J. and Worth.

St. John and Worthen 1883 A, p. 193, pl. xii, figs. 4, 5.

Lesley, J. P. 1889 A, p. 569, figures.

Woodward, A. S. 1889 D, p. 190. (Sandalodus.) Coal-measures; West Virginia.

Orthopleurodus novo-mexicanus St. J. and Worth.

St. John and Worthen 1883 A, p. 195, pl. xiii, figs. 1-3.

Lesley, J. P. 1889 A, p. 569, figures.

Woodward, A. S. 1889 D, p. 190 (Sandalodus). Subcarboniferous; New Mexico.

PLATYODUS Newb.   Type *P. lineatus* Newb.

Newberry, J. S. 1875 A, p. 58.
Miller, S. A. 1889 A, p. 606.
Woodward, A. S. 1889 D, p. 191.

Lesley, J. P. 1889 A, p. 691, figure.
Miller, S. A. 1889 A, p. 607, figure 1159.
Subcarboniferous (Waverly); Kentucky.

Platyodus lineatus Newb,
Newberry, J. S. 1875 A, p. 58, pl. IIx, figures 1, 2.

VATICINODUS St. J. and Worth.   Type *V. vetustus* St. J. and Worth.

St. John and Worthen 1883 A, pp. 59, 80.
Miller, S. A. 1889 A, p. 613.
Woodward, A. S. 1889 D, pp. 186, 190.
Zittel, K. A. 1890 A, p. 69.

Woodward, A. S. 1889 D, p. 191.
Upper Coal-measures; Illinois.

Vaticinodus? simillis St. J. and Worth.

St. John and Worthen 1883 A, p. 86, pl. iv, figs. 17–19.
Lesley, J. P. 1890 A, p. 1249, figures.
Woodward, A. S. 1889 D, p. 191.
Subcarboniferous (Chester); Illinois.

Vaticinodus? carbonarius St. J. and Worth.

St. John and Worthen 1883 A, p. 88, pl. iv, fig. 20.
Lesley, J. P. 1890, A, p. 1247, figure.
Woodward, A. S. 1889 D, p. 190.
Lower Coal-measures; Illinois.

Vaticinodus? simplex St. J. and Worth.

St. John and Worthen 1883 A, p. 84, pl. iv, figs. 22–26.
Keyes, C. R. 1894 A, p. 233.
Lesley, J. P. 1890 A, p. 1250, figures.
Woodward, A. S. 1889 D, p. 191.
Subcarboniferous (St. Louis); Illinois, Missouri, Iowa.

Vaticinodus discrepans St. J. and Worth.

St. John and Worthen 1883 A, p. 83, pl. iii, figs. 2, 3.
Lesley, J. P. 1890 A, p. 1247, figure.
Woodward, A. S. 1889 D, p. 190.
Subcarboniferous (Burlington); Iowa.

Vaticinodus vetustus St. J. and Worth.

St. John and Worthen, 1883 A, p. 82, pl. iii, fig. 1.
Lesley, J. P. 1890 A, p. 1250.
Newberry, J. S. 1897 A, p. 300.
Woodward, A. S. 1889 D, pp. 186, 190. (Sandalodus.)
Subcarboniferous (Kinderhook); Iowa.

Vaticinodus? lepis St. J. and Worth.
St. John and Worthen 1883 A, p. 88, pl. iv, fig. 21.
Lesley, J. P. 1890 A, p. 1249, figure.

PLATYXYSTRODUS Hay.   Type *Cochliodus striatus* McCoy.

Hay, O. P. 1899 E, p. 785.
Davis, J. W. 1886 A, p. 153. (Xystrodus: not of Plieninger, 1860.)
Miller, S. A. 1889 A, p. 614. (Xystrodus.)
Morris and Roberts, 1862, Quart. Jour. Geol. Soc., xviii, p. 101. (Xystrodus, type *Cochliodus striatus* McCoy.)
St. John and Worthen 1883 A, pp. 59, 175. (Xystrodus.)
Woodward, A. S. 1889 D, p. 192. (Xystrodus.)
Zittel, K. A. 1890 A, p. 72. (Xystrodus.)

Woodward, A. S. 1889 D, p. 195. (Xystrodus.)
Subcarboniferous (St. Louis); Illinois, Iowa, Missouri.

Platyxystrodus inconditus (St. J. and Worth.).
St. John and Worthen 1883 A, p. 179, pl. viii, fig. 1. (Xystrodus.)
Hay, O. P. 1899 A, p. 785.
Lesley, J. P. 1890 A, p. 1264, figure. (Xystrodus.)
Woodward, A. S. 1889 D, p. 195. (Xystrodus.)
Subcarboniferous (Keokuk); Illinois, Iowa.

Platyxystrodus bellulus (St. J. and Worth.).
St. John and Worthen 1883 A, p. 183, pl. viii, fig. 3. (Xystrodus.)
Hay, O. P. 1899 E, p. 785.
Lesley, J. P. 1890 A, p. 1264, figs. (Xystrodus.)
Woodward, A. S. 1889 D, p. 195. (Xystrodus.)
Lower Coal-measures; Illinois.

Platyxystrodus simplex (St. J. and Worth).
St. John and Worthen 1883 A, p. 178, pl. viii, figs. 4, 5. (Xystrodus.)
Davis, J. W. 1886 A, p. 153. (Xystrodus.)
Hay, O. P. 1899 E, p. 785.
Lesley, J. P. 1890 A, p. 1265, figure. (Xystrodus.)
Woodward, A. S. 1889 D, p. 195. (Xystrodus.)
Subcarboniferous (Burlington); Illinois, Iowa.

Platyxystrodus imitatus (St. J. and Worth.).
St. John and Worthen 1883 A, p. 180, pl. viii, fig. 2. (Xystrodus.)
Hay, O. P. 1899 E, p. 785.
Keyes, C. R. 1894 A, p. 233. (Xystrodus.)
Lesley, J. P.; 1890 A, p. 1264, figure. (Xystrodus.)

Platyxystrodus verus (St. J. and Worth.).
St. John and Worthen 1883 A, p. 181, pl. viii, figs. 6, 7. (Xystrodus.)
Hay, O. P. 1899 E, p. 785.
Lesley, J. P. 1890 A, p. 1265, figure. (Xystrodus.)
Woodward, A. S. 1889 D, p. 194. (Xystrodus.)
Subcarboniferous (Chester); Illinois.

ICANODUS S. A. Miller.   Type *Tomodus convexus* St. J. and Worth.

Miller, S. A. 1892 A, P. 716.
Davis, J. W. 1883 A, p. 446. (Tomodus, preoccupied by Trautschold, 1879.)
    1886 A, p. 153. (Tomodus.)
Hay, O. P. 1899 E, p. 785.
Miller, S. A. 1889 A, p. 613. (Tomodus.)
St. John and Worthen 1883 A, pp. 59, 171. (Tomodus.)
Woodward, A. S. 1889 D, p. 191. (Tomodus.)
Zittel, K. A. 1890 A, p. 72. (Tomodus.)

Icanodus limitaris (St. J. and Worth).
St. John and Worthen 1883 A, p. 173, pl. xiii, fig.
    12. (Tomodus; called on explanation of plate "Teniodus regularis.")
Hay, O. P. 1899 E, p. 785.
Miller, S. A. 1892 A, p. 716.
Woodward, A. S. 1889 D, p. 192. (Tomodus.)
    Subcarboniferous (Burlington); Iowa.

STENOPTERODUS St. J. and Worth.   Type, none specified; *S. planus* St. J. and Worth.
                                    may be taken.

St. John and Worthen 1883 A, pp. 59, 101.
Miller, S. A. 1889 A, p. 612.
Woodward, A. S. 1889 D, p. 217.

Stenopterodus elongatus St. J. and Worth.
St. John and Worthen 1888 A, p. 106, pl. iv, figs. 1–3.
Woodward, A. S. 1889 D. p. 217.
    Subcarboniferous (Warsaw); Illinois.

Stenopterodus planus St. J. and Worth.
St. John and Worthen 1883 A, p. 102, pl. iv, figs. 9–14.
Woodward, A. S. 1889 D, p. 217.
    Subcarboniferous (Burlington); Illinois, Iowa.

Stenopterodus spindet.
St. John and Worthen 1883 A, p. 105, pl. iv, figs. 15, 16.
    Subcarboniferous (Keokuk); Illinois.

DELTODOPSIS St. J. and Worth.   Type *D. angustus* (Newb. and Worth.).

St. John and Worthen 1883 A, p. 158.
Miller, S. A. 1889 A, p. 596.
Newberry, J. S. 1889 A, p. 186.
Woodward, A. S. 1889 D, p. 217.
Zittel, K. A. 1890 A, p. 72.

Deltodopsis affinis St. J. and Worth.
St. John and Worthen 1883 A, p. 160, pl. xi, fig. 1.
Woodward, A. S. 1889 D, p. 217.
    Subcarboniferous (Warsaw); Illinois.

Deltodopsis angustus (Newb. and Worth.).
Newberry and Worthen 1870 A, p. 368, pl. iii, fig. 7. (Deltodus.)
St. John and Worthen 1883 A, p. 163, pl. xi, figs. 7–10.
Woodward, A. S. 1889 D, pp. 200, 218. (Deltodus.)
    Subcarboniferous (Chester); Illinois.

Deltodopsis? bialveatus St. J. and Worth.
St. John and Worthen 1883 A, p. 169, pl. xi, figs. 15–18. (Including varieties *keokuk* and *convexus*.)
Woodward, A. S. 1889, D, p. 217.
    Subcarboniferous (Burlington); Iowa: (St. Louis); Illinois.

Deltodopsis convolutus St. J. and Worth.
St. John and Worthen 1883 A. p. 165, pl. xi, figs. 11, 12
Woodward, A. S. 1889 D. p. 217.
    Subcarboniferous (Burlington); Iowa, Illinois.

Deltodopsis? exornatus St. J. and Worth.
St. John and Worthen 1883 A, p. 168, pl. xi, fig. 14.
Woodward, A. S. 1889 D, p. 218.
    Subcarboniferous (Warsaw); Illinois.

Deltodopsis? inflexus St. J. and Worth.
St. John and Worthen 1883 A, p. 167, pl. xi, fig. 13.
Woodward. A. S 1889 D, p. 218.
    Subcarboniferous (Keokuk); Illinois.

Deltodopsis sancti-ludovici St. J. and Worth.
St. John and Worthen 1883 A, p. 161, pl. xi, figs. 2–6.
Keyes, C. R. 1894 A, p. 235.
Woodward, A. S. 1889 D, p. 218.
    Subcarboniferous (St. Louis); Illinois, Iowa, Missouri.

DELTODUS Morris and Roberts.   Type *Pœcilodus sublævis* (McCoy).

Morris and Roberts, 1862, Quart. Jour. Geol. Soc., xviii, p. 100. (Ex Agassiz MSS.)
Davis, J. W. 1883 A, p. 427.
Koninck, L. G. de 1878 A, p. 63.
Miller, S. A. 1889 A, p. 595 (Deltodus); p. 612 (Tæniodus).

Newberry, J. S. 1889 A, p. 186.
Newberry and Worthen 1866 A, p. 95.
St. John and Worthen 1883 A, pp. 60, 141 (Deltodus); pp. 59, 75 (Tæniodus).
Traquair, R. H. 1888 A, p. 84.
Trautschold, H. 1879 B, p. 54.

Woodward, A. S. 1884 A, p. 270.
  1889 A, p. 195.
  1892 G, fig. 6.
Woodward and Sherborn 1890 A, p. 60.
Zittel, K. A. 1890 A, p. 73.

Deltodus alatus Newb. and Worth.

Newberry and Worthen 1870 A, p. 368, pl. ii, fig. 6.
Woodward, A. S. 1889 D, p. 199.
  Subcarboniferous (Keokuk); Illinois.

Deltodus cinctulus St. J. and Worth.

St. John and Worthen 1883 A, p. 146, pl. ix, figs. 6, 7.
Keyes, C. R. 1894 A, p. 235.
Kindle, E. M. 1898 A, p. 484.
Woodward, A. S. 1889 D, p. 200.
  Subcarboniferous (Warsaw); Illinois, Missouri, Indiana.

Deltodus cinctus Newb.

Newberry, J. S. 1879 A, p. 344.
Woodward, A. S. 1889 D, p. 200.
  Subcarboniferous (St. Louis); Iowa.

Deltodus cingulatus Newb. and Worth.

Newberry and Worthen 1866 A, p. 99, pl. ix, fig. 6.
Woodward, A. S. 1889 D, p. 200.
  Subcarboniferous (Chester); Illinois.

Deltodus complanatus Newb. and Worth.

Newberry and Worthen 1866 A, p. 98, pl. ix, fig. 4.
Koninck, L. G. de 1878 A, p. 64.
Lesley, J. P. 1890 A, p. 920, figure (Sandalodus);
  p. 1218, figures (Trigonodus major).
  1895 A, pl. ccxxxiii. (Sandalodus.)
Newberry, J. S. 1897 A, p. 298, pl. xxiv, figs. 1-7.
Newberry and Worthen 1866 A, p. 112, pl. xi, 8-9a.
  (Trigonodus major.)
St. John and Worthen 1883 A, p. 184, pl. xii, figs. 1-4. (Sandalodus.)
Woodward, A. S. 1889 D, p. 189. (Sandalodus.)
  Subcarboniferous (Burlington); Iowa.

Deltodus fasciatus Newb. and Worth.

Newberry and Worthen 1870 A, p. 366, pl. ii, fig. 17.
Lesley, J. P. 1890 A, p. 1172, figures. (Teniodus fasciatus.)
  ` 1895 A, pl. ccxxxiv. (Teniodus fasciatus.)
St. John and Worthen 1883 A, p. 76, pl. xiii, fig 9. (Tæniodus fasciatus?)
Woodward, A. S. 1889 D, p. 200.
  Subcarboniferous (Keokuk); Illinois.

Deltodus grandis Newb.

Newberry, J. S. 1879 A, p. 344.
Kindle, E. M. 1898 A, p. 484 (D. grandis?).`
Newberry, J. S. 1897 A, p. 297.
Newberry & Worthen 1866 A, p. 101, pl. ix, figs. 9, 9a.
Woodward, A. S., 1889 D, p. 188. (Syn. of Sandalodus lævissimus.)
  Subcarboniferous (Keokuk); Illinois, Indiana.

Deltodus inornatus Newb.

Newberry, J. S. 1897 A, p. 298, pl. xxiii, figs. 10, 11.
  Subcarboniferous (Kinderhook); Iowa.

Deltodus intermedius St. J. and Worth.

St. John and Worthen 1883 A, p. 158, pl. ix, figs. 14, 15.
Keyes, C. H. 1891 A, p. 265. (D. intermedius?)
Woodward, A. S. 1889 D, p. 200.
  Subcarboniferous (St. Louis); Illinois, Iowa: Coal-measures; Iowa.

Deltodus latior St. J. and Worth.

St. John and Worthen 1883 A, p. 145, pl. ix, figs. 11, 12.
Woodward, A. S. 1889 D, p. 200.
  Subcarboniferous (Keokuk); Illinois, Iowa.

Deltodus littoni Newb. and Worth.

Newberry and Worthen 1870 A, p. 367, pl. iv, figs. 8, 8a.
Keyes, C. R. 1894 A, p. 235.
St. John and Worthen 1883 A, p. 149.
Woodward, A. S. 1889 D, p. 200.
  Subcarboniferous; Missouri.

Deltodus mercurii Newb.

Newberry, J. S. 1876 A, p. 137, pl. iii, figs. 1, 1a.
St. John and Worthen 1883 A, pl. x, figs. 2-2d.
Woodward, A. S. 1889 D, p. 200.
  Coal-measures; New Mexico.

Deltodus obliquus (St. J. and Worth.).

St. John and Worthen 1883 A, p. 78, pl. xiii, fig. 10. (Tæniodus.)
Lesley, J. P. 1890 A, p. 1173, figures. (Teniodus.)
  1895 A, pl. ccxxxiv. (Teniodus.)
Woodward A. S. 1889 D, p. 201.
  Subcarboniferous (Warsaw); Indiana.

Deltodus occidentalis (Leidy).

Leidy, J. 1857 N, p. 88, pl. v, figs. 3-16. (Cochliodus.)
Newberry and Worthen 1866 A, p. 97, pl. ix, figs. 2-3a. (D. stellatus.)
St. John and Worthen 1883 A, p. 150, pl. ix, figs. 9, 10. (D. occidentalis?)
Woodward, A. S. 1889 D, p. 199.
  Subcarboniferous; Illinois.

Deltodus parvus St. J. and Worth.

St. John and Worthen 1883 A, p. 151, pl. ix, figs. 1-5.
Keyes, C. R. 1894 A, p. 235.
Woodward A. S. 1889 D, p. 201.
  Subcarboniferous (St. Louis); Illinois, Iowa, Missouri.

Deltodus planidens Cope.

Cope, E. D. 1894 F, p. 435, pl. xx, fig. 6.
  Carboniferous?; Texas.

Deltodus powellii St. J. and Worth.

St. John and Worthen 1883 A, p. 155, pl. x, fig. 1-1*f*
Woodward, A. S. 1889 D, p. 201.
  Carboniferous; Utah.

Deltodus propinquus St. J. and Worth.

St. John and Worthen 1883 A, p. 156, pl. x, figs. 3, 4.
Woodward A. S. 1889 D, p. 201.
1900 C, p. 420.
Coal-measures; Illinois.

Deltodus spatulatus Newb. and Worth.

Newberry and Worthen 1866 A, p. 100, pl. ix, fig. 7.
Kindle, E. M. 1898 A, p. 484.
Newberry, J. S. 1879 A, p. 346.
1897 A, p. 299, pl. xxiv, figs. 8–11.
St. John and Worthen 1883 A, p. 166.

Woodward, A. S. 1889 D, p. 199.
Subcarboniferous; Illinois, Indiana.

Deltodus trilobus St. J. and Worth.

St. John and Worthen 1883 A, p. 148, pl. ix, fig. 8.
Woodward, A. S. 1889 D, p. 201.
Subcarboniferous (Warsaw); Illinois.

Deltodus undulatus Newb. and Worth.

Newberry and Worthen 1866 A, p. 98, pl. ix, fig. 5.
Woodward, A. S. 1889 D, p. 201.
Subcarboniferous (Keokuk); Iowa.

CHITINODUS St. J. and Worth. Type *C. springeri* St. J. and Worth.

St. John and Worthen 1883 A, pp. 60, 109.
Miller, S. A. 1889 A, p. 590.
Woodward, A. S. 1889 D, p. 217.
Zittel, K. A. 1890 A, p. 70.

Chitinodus antiquus St. J. and Worth.

St. John and Worthen 1883 A, p. 116, pl. vi, fig. 2.
Woodward, A. S. 1889 D, p. 217.
Subcarboniferous (Burlington); Iowa.

Chitinodus latus (Leidy).

Leidy, J. 1857 H, p. 88, pl. v, fig. 17. (Cochliodus.)
Dana, J. D. 1896 A, p. 644, fig. 1018. (Cochliodus nobilis.)
Newberry, J. S. 1897 A, p. 301, pl. xxiv, fig. 24. (Helodus coxanus.)
Newberry and Worthen 1866 A, p. 88, pl. vi, figs. 3–6 [Helodus (Cochliodus) nobilis]; p. 89, pl. vii, figs. 1–4; pl. viii, fig. 1 (Cochliodus nobilis).
1870 A, p. 365 (Cochliodus nobilis).
St. John and Worthen 1883 A, pp. 112, 124.
Woodward, A. S. 1889 D, p. 208. (Cochliodus.)
Zittel, K. A. 1890 A, p. 70. (Chitinodus nobilis.)
Subcarboniferous (Keokuk); Illinois, Iowa.

Chitinodus liratus St. J. and Worth.

St. John and Worthen 1883 A, p. 119, pl. vi, fig. 1.
Woodward, A. S. 1889 D, p. 217.
Subcarboniferous (St. Louis); Illinois.

Chitinodus rugosus Newb. and Worth.

Newberry and Worthen 1866 A, p. 94, pl. viii, fig. 13. (Pœcilodus.)
1866 A, p. 95, pl. viii, fig. 14. (Pœcilodus ornatus.)
St. John and Worthen 1883 A, p. 112. (P. rugosus and P. ornatus to C. rugosus.)
Woodward, A. S. 1889 D, p. 201 (P. ornatus to Deltodus ornatus); p. 205 (P. rugosus); p. 217.
Subcarboniferous (Keokuk); Illinois.

Chitinodus springeri St. J. and Worth.

St. John and Worthen 1883 A, p. 112, pl. vi, figs. 3–15.
Woodward, A. S. 1889 D, p. 217.
Subcarboniferous (Burlington); Illinois, Iowa.

Chitinodus tribulis St. J. and Worth.

St. John and Worthen 1883 A, p. 117, pl. vii, figs. 18–21.
Woodward, A. S. 1889 D, p. 217.
Subcarboniferous (Keokuk); Illinois, Iowa.

PŒCILODUS McCoy. Type *P. jonesii* McCoy.

McCoy, F. 1848 A, p. 129.
Barkas, T. P. 1873 A, p. 19.
Davis, J. W. 1883 A, pp. 413, 441.
McCoy, F. 1855 A, p. 638.
Miller, S. A. 1889 A, p. 607.
Pictet, F. J. 1854 A, p. 270.
Traquair, R. H. 1888 A, p. 84.
Trautschold, H. 1874 B, p. 290.
Woodward, A. S. 1889 D, p. 201.
1892 G, p. 674.
Zittel, K. A. 1890 A, p. 70.

Pœcilodus carbonarius St. J. and Worth.

St. John and Worthen 1883 A, p. 139, pl. viii, figs. 20–21.
Lesley, J. P. 1889 A, p. 727, figures.
Woodward, A. S. 1889 D, p. 204.
Upper Coal-measures; Illinois, Kansas.

Pœcilodus cestriensis St. J. and Worth.

St. John and Worthen 1883 A, p. 135, pl. viii, figs. 15–17.

Lesley, J. P. 1889 A, p. 727, figures.
Woodward, A. S. 1889 D, p. 205.
Subcarboniferous (Chester); Illinois.

Pœcilodus convolutus Newb. and Worth.

Newberry and Worthen 1870 A, p. 366, pl. ii, fig. 9.
Woodward, A. S. 1889 D, p. 205.
Subcarboniferous (Keokuk); Illinois.

Pœcilodus sancti-ludovici St. J. and Worth.

St. John and Worthen 1883 A, p. 132, pl. viii, figs. 8–12.
Keyes, C. R. 1894 A, p. 235.
Lesley, J. P. 1889 A, p. 728, figures.
Woodward, A. S. 1889 D, p. 205.
Subcarboniferous (St. Louis); Illinois, Missouri, Iowa.

Pœcilodus springeri St. J. and Worth.

St. John and Worthen 1883 A, p. 138, pl. viii, fig. 19.
Lesley, J. P. 1889 A, p. 728, figures.

Woodward, A. S. 1889 D, p. 205, 215. (Deltoptychius.)
Subcarboniferous; New Mexico.

Pœcilodus varsoviensis St. J. and
Worth.

St. John and Worthen 1883 A, p. 131, pl. viii, figs.
13, 14.
Lesley, J. P. 1889 A, p. 728, figures.

Woodward, A. S. 1889 D, p. 205.
Subcarboniferous (Warsaw); Illinois.

Pœcilodus wortheni St. John.

St. John, in St. John and Worthen 1883 A, p. 136,
pl. viii, fig. 18.
Lesley, J. P. 1889 A, p. 728, figures.
Woodward, A. S. 1889 D, pp. 205, 215. (Deltopty-
chius.)
Subcarboniferous (Chester); Illinois.

COCHLIODUS Agassiz.   Type *C. contortus* Agassiz.

Agassiz, L. 1843 B (1838), iii, p. 113.
1844 A.
Amer. Geologist 1888 A, p. 133.
Davis, J. W. 1882 B, p. 54.
1883 A, p. 420.
Jaekel, O. 1891 B, p. 128.
Koninck, L. G. de 1878 A, p. 56.
McCoy, F. 1853 B.
1855 A, p. 621.
Miller, S. A. 1889 A, p. 591.
Newberry, J. S. 1889 A, p. 186.
Newberry and Worthen 1866 A, pp. 74, 89.
Nicholson and Lydekker 1889 A, p. 940.
Owen, R. 1845 B, p. 62, pl. xxii, fig. 1.
1860 E, p. 109.
1867 A.
Pictet, F. J 1854 A, p. 267.
St. John and Worthen 1883 A, p. 58.
Steinmann and Döderlein 1890 A, p. 550.
Traquair, R. H. 1888 A, p. 84.
Trautschold H. 1874 B, p. 291.
Woodward, A. S. 1884 A, p. 269.
1889 D, p. 205.
1892 G, fig. 3.
Zittel, K. A. 1890 A, pp. 69, 71.

Cochliodus costatus Newb. and Worth.

Newberry and Worthen 1870 A, p. 364, pl. iii, figs.
10–12, 12a.
St. John and Worthen 1883 A, pp. 167, 171.
Woodward, A. S. 1889 D, p. 208.
Subcarboniferous (Burlington); Iowa.

Cochliodus leidyi St. J. and Worth.

St. John and Worthen 1883 A, p. 127, pl. vii, figs.
11–16.
Woodward, A. S. 1889 D, p. 208.
Subcarboniferous (Chester); Illinois.

Cochliodus obliquus St. J. and Worth.

St. John and Worthen 1883 A, p. 126, pl. vii, fig. 17.
Keyes, C. R. 1894 A, p. 288.
Woodward, A. S. 1889 D, p. 212.
Subcarboniferous (St. Louis); Missouri.

Cochliodus vanhornii St. J. and Worth.

St. John and Worthen 1883 A, p. 121, pl. vii, figs.
1–10.
Keyes, C. R. 1894 A, p. 233.
Woodward, A. S. 1889 D, p. 208.
Subcarboniferous (St. Louis); Illinois, Iowa,
Missouri.

DELTOPTYCHIUS Davis.   Type *Cochliodus acutus* McCoy.

Davis, J. W. 1883 A, p. 432.
Miller, S. A. 1889 A, p. 595.
St. John and Worthen 1883 A, pp. 59, 89.
Woodward, A. S. 1889 D, p. 212.
1892 G, p. 674.
Woodward and Sherborn 1890 A, p. 61.
Zittel, K. A. 1890 A, p. 70.

Deltoptychius expansus St. J. and
Worth.

St. John and Worthen 1883 A, p. 96, pl. v, figs. 9, 13.
Keyes, C. R. 1894 A, p. 233.
Woodward, A. S. 1889 D, pp. 212, 215.
Subcarboniferous (St. Louis); Illinois, Mis-
souri, Iowa.

Deltoptychius nitidus (Leidy).

Leidy, J. 1857 A, p. 87, pl. v. fig. 2 (Cochliodus).
St. John and Worthen 1883 D, p. 99, pl. v, fig. 16.
Woodward, A. S. 1889 D, p. 215.
Subcarboniferous (Chester); Illinois.

Deltoptychius primus St. J. and Worth.

St. John and Worthen 1883 A, p. 93, pl. v, figs. 6–8.
Woodward A. S. 1889 D, pp. 212, 215.
Subcarboniferous (Burlington); Iowa.

Deltoptychius varsoviensis St. J. and
Worth.

St. John and Worthen 1883 A, p. 96, pl. v, figs.
14, 15.
Woodward, A. S. 1889 D, pp. 212, 215.
Subcarboniferous (Warsaw); Illinois.

Deltoptychius wachsmuthi St. J. and
Worth.

St. John and Worthen 1883 A, p. 98, pl. v, figs. 1–5.
Keyes, C. R. 1894 A, p. 233.
Lesley, J. P. 1889 A, p. 198, figure.
1895 A, pl. ccxxix.
Miller, S. A. 1889 A, p. 595, fig. 1124.
Woodward, A. S. 1889 D, pp. 209, 212, 215.
Subcarboniferous (Keokuk); Illinois, Iowa,
Missouri.

ORODONTIDÆ.

Davis, J. W 1882 B.
  1883 A, pp. 331, 396.
Koninck, L. G. de 1878 A, p. 29.
Traquair, R. H. 1888 A, p. 84.

Ward, J. 1890 A, p. 34.
Woodward, A. S. 1884 A, p. 269.
Zittel, K. A. 1890 A, p. 74. (Cestraciontidæ, in part.)

ORODUS Agassiz. Type *O. cinctus* Agassiz.

Agassiz, L. 1843 B (1838), iii, p. 96.
  1844 A.
Davis, J. W. 1882 B, p. 49.
  1883 A, p. 390.
Eastman, C. R. 1900 C, p. 392.
Hasse, C. 1878 A, p. 145.
Jaekel, O. 1890 A, p. 98.
  1899 A.
Koninck, L. G. de. 1844 A, p. 512.
  1878 A, p. 29.
Miller, S. A. 1889 A, p. 608.
Newberry and Worthen 1866 A, p. 62.
Nicholson and Lydekker 1889 A, p. 940.
Owen, R. 1860 E, p. 108.
Pictet, F. J. 1854 A, p. 263.
Quenstedt, F. A. 1865 A, p. 294.
Trautschold, H. 1874 B, p. 292.
Woodward, A. S. 1884 A, p. 269.
  1889 D, p. 230.
Woodward and Sherborn 1890 A, pp. 49, 122.
Zittel, K. A. 1890 A, p. 75.

Orodus alleni St. J. and Worth.

St. John and Worthen 1875 A, pp. 310, 318, pl. vii, fig. 19.
Woodward, A. S. 1889 D, p. 236.
  Lower Coal-measures; Iowa.

Orodus carinatus St. J. and Worth.

St. John and Worthen 1875 A, p. 307, pl. v, figs. 24a-24d.
Woodward, A. S. 1889 D, p. 236.
  Subcarboniferous (Keokuk); Iowa.

Orodus colletti Newb.

Newberry, J. S. 1879 A, p. 348.
Kindle, E. M. 1898 A, p. 484.
Woodward, A. S. 1889 D, p. 236.
  Subcarboniferous (St. Louis); Indiana.

Orodus dædaleus St. J. and Worth.

St. John and Worthen 1875 A, p. 301, pl. vi, figs. 7, 8.
Woodward, A. S. 1889 D, p. 237.
  Subcarboniferous (Kinderhook); Iowa.

Orodus decussatus St. J. and Worth.

St. John and Worthen 1875 A, p. 300, pl. vi, figs. 10-15.
Woodward, A. S. 1889 D, p. 237.
  Subcarboniferous (Kinderhook); Iowa.

Orodus elegantulus Newb. and Worth.

Newberry and Worthen 1866 A, p. 64, pl. iv, figs. 6, 6a.
Kindle, E. M. 1898 A, p. 484.
Lesley, J. P. 1889 A, p. 505, figure.

Newberry, J. S. 1875 A, p. 51, pl. lviii, figs. 12, 12a.
  1879 A, p. 346.
Woodward, A. S. 1889 D, p. 235.
  Subcarboniferous; Iowa, Indiana, Ohio.

Orodus fastigiatus St. J. and Worth.

St. John and Worthen 1875 A, p. 306, pl. vi, figs. 1-3.
Woodward, A. S. 1889 D, p. 237.
  Subcarboniferous (Burlington); Iowa.

Orodus gibbosus (Newb. and Worth.).

Newberry and Worthen 1866 A, p. 79, pl. v, figs. 3-3a. (Helodus.)
Miller, S. A. 1889 A, p. 599, fig. 1137. (Helodus.)
St. John and Worthen 1875 A, p. 305. (Orodus.)
Woodward A. S. 1889 D, p. 238. (O. variocostatus?)
  Subcarboniferous (Keokuk); Illinois, Iowa.

Orodus major St. J. and Worth.

St. John and Worthen 1875 A, p. 302, pl. vii, fig. 10.
Koninck, L. G. de 1878 A, p. 30. (Syn. of O. ramosus.)
Woodward, A. S. 1889 D, p. 237.
  Subcarboniferous (Burlington); Iowa.

Orodus mammillaris Newb. and Worth.

Newberry and Worthen 1866 A, p. 66, pl. iv, figs. 10, 10a.
Dana, J. D. 1896 A, p. 644, fig. 1022.
Woodward, A. S. 1889 D, p. 235.
  Subcarboniferous (Keokuk); Illinois.

Orodus minutus Newb. and Worth.

Newberry and Worthen 1866 A, p. 68, pl. iv, fig. 12.
Woodward, A. S. 1889 D, p. 237.
  Subcarboniferous (Keokuk); Illinois.

Orodus neglectus St. J. and Worth.

St. John and Worthen 1875 A, p. 308, pl. vi, fig. 26.
Woodward, A. S. 1889 D, p. 237.
  Subcarboniferous (St. Louis); Illinois, Iowa.

Orodus ornatus Newb. and Worth.

Newberry and Worthen 1866 A, p. 65, pl. iv, figs. 7-8a.
Kindle, E. M. 1898 A, p. 484.
Newberry, J. S. 1879 A, p. 346.
Woodward, A. S. 1889 D, p. 233.
  Subcarboniferous (Keokuk); Illinois, Indiana.

Orodus ? parallellus St. J. and Worth.

St. John and Worthen 1875 A, p. 295, pl. v, fig. 23.
Woodward, A. S. 1889 D, p. 237.
  Subcarboniferous (Kinderhook); Iowa.

Orodus parvulus St. J. and Worth.

St. John and Worthen 1875 A, p. 309, pl. vi, fig. 6.
Woodward, A. S. 1889 D, p. 237.
  Subcarboniferous (St. Louis); Illinois.

Orodus plicatus Newb. and Worth.

Newberry and Worthen 1866 A, p. 63, pl. iv, fig. 5.
Hambach, G. 1890 A, p. 81.
Woodward, A. S. 1889 D, p. 237.
  Subcarboniferous (St. Louis); Missouri.

Orodus ramosus Agassiz.

Agassiz, L. 1843 B (1838), iii, p. 97, pl. xi, figs. 5–8.
Davis, J. W. 1883 A, p. 390, pl. l, figs. 1–7 (O. ramo-
  sus); p. 399, pl. li, fig. 15 (O. subteres).
Kindle, E. M. 1898 A, p. 484. (O. multicarinatus.)
Koninck, L. G. de 1844 A, p. 613, pl. lv, fig. 2.
  1878 A, p. 30, pl. iv, fig. 1.
Newberry, J. S. 1889 A, pp. 185, 205, pl. xxvii, fig. 29.
Newberry and Worthen 1866 A, p. 62, pl. iv, figs.
  13–13a. (O. multicarinatus.)
Romanovsky, G. 1864 A, p. 157, fig. 2.
Woodward, A. S. 1889 D, p. 231 (O. ramosus); p.
  237 (O. ? multicarinatus).
  Subcarboniferous; Europe, Indiana, Michi-
  gan.

Orodus tuberculatus Newb. and Worth.

Newberry and Worthen 1866 A, p. 66, pl. iv, figs.
  9, 9a.
Woodward, A. S. 1888 D, p. 234.
  Subcarboniferous (Burlington); Illinois,
  Iowa.

Orodus turgidus St. J. and Worth.

St. John and Worthen 1875 A, p. 310, pl. vi, figs.
  4, 5.
Woodward, A. S. 1889 D, p. 238.
  Subcarboniferous (Chester); Illinois.

Orodus variabilis Newb.

Newberry, J. S. 1875 A, p. 50, pl. lviii, figs. 11–11A.
Lesley, J. P. 1889 A, p. 506, figure.
Newberry, J. S. 1873 A, p. 279. (No description;
  syn. ? of Ctenacanthus furcicarinatus.)
Woodward, A. S. 1889 D, p. 238.
  Subcarboniferous (Waverly); Ohio, Ken-
  tucky.

Orodus variocostatus St. J. and Worth.

St. John and Worthen 1875 A, p. 304, pl. viii, figs.
  1–9.
Hambach, G. 1890 A, p. 81.
Woodward, A. S. 1889 D, p. 238.
  Subcarboniferous (Burlington);   Illinois,
  Iowa, Missouri (*fide* Hambach).

Orodus whitei St. J. and Worth.

St. John and Worthen 1875 A, p. 297, pl. vi, fig. 26.
Woodward, A. S. 1889 D, p. 238.
  Subcarboniferous (Kinderhook); Iowa.

LEIODUS St. J. and Worth.   Type *L. calcaratus*, St. J. and Worth.

St. John and Worthen 1875 A, p. 335.
Miller, S. A. 1889 A, p. 600. (Liodus.)
Woodward, A. S. 1889 D, p. 240.

Leiodus calcaratus St. J. and Worth.

St. John and Worthen 1875 A, p. 336, pl. vii, figs.
  11–18.
Miller, S. A. 1889 A, p. 600, fig. 1141. (Liodus.)

Woodward, A. S. 1889 D, p. 240.
  Subcarboniferous (Burlington); Illinois, Iowa.

Leiodus grossipunctatus St. J. and
Worth.

St. John and Worthen 1875 A, p. 337.
Woodward, A. S. 1889 D, p. 240.
  Subcarboniferous (Keokuk); Illinois, Iowa.

CAMPODUS Koninck.   Type *C. agassizianus* Koninck.

Koninck, L. G. de 1844 A, p. 617.
Amer. Geologist 1888 A, p. 133. (Lophodus.)
Davis, J. W. 1882 B, p. 50. (Agassizodus.)
Jaekel, O. 1894 A, p. 137.
Lohest, M. 1884 A, pp. 296, 305.
Miller, S. A. 1889 A, p. 587. (Agassizodus.)
Mudge, B. F. 1875 A. (Agassizodus.)
Newberry and Worthen 1870 A, p. 360. Lophodus,
  type *L. variabilis;* not Lophodus of Roman-
  ovsky.)
Pictet, F. J. 1854 A, p. 267.
St. John and Worthen 1875 A, pp. 297, 311, 316.
  (Agassizodus, to replace Lophodus.)
Woodward, A. S. 1889 D, p. 238.
Zittel, K. A. 1890 A, p. 75.

Campodus basalis (Cope).

Cope, E. D. 1894 F, p. 432, pl. xviii, figs. 6, 7. (Orodus, with Agassizodus as a section.)
  Coal-measures; Illinois.

Campodus corrugatus (Newb. and
Worth.).

Newberry and Worthen 1870 A, p. 358, pl. iii, figs.
  18, 18a. (Orodus.)

St. John and Worthen 1875 A, pp. 316, 328, pl. viii,
  fig. 24. (Agassizodus.)
Woodward, A. S. 1889 D, p. 239. (Campodus.)
  Coal-measures; Illinois, Kansas.

Campodus scitulus (St. J. and Worth.).

St. John and Worthen 1875 A, p. 322, pl. vi, figs.
  16–18c. (Agassizodus.)
Cope, E. D. 1894 F, p. 433. (Orodus, section Agas-
  sizodus.)
Woodward, A. S. 1889 D, p. 239. (Campodus.)
  Coal-measures; Illinois, Iowa.

Campodus variabilis (Newb. and
Worth.).

Newberry and Worthen 1870 A, p. 361, pl. iv, figs.
  4–4b; 5, 5a; 11–11b. (Lophodus.)
Lohest, M. 1884 A, p. 305. (Campodus.)
Miller, S. A. 1889 A, p. 587, fig. 1101. (Agassiz-
  odus.)
St. John and Worthen 1875 A, p. 318, pl. viii, figs.
  1–22. (Agassizodus.)
Woodward, A. S. 1889 D, p. 239. (Campodus.)
  Subcarboniferous; Belgium: Coal-measures;
  West Virginia, Iowa, Kansas.

Campodus virginianus (St. J. and Worth.).

St. John and Worthen 1875 A, p. 321, pl. viii, figs. 23a–23d. (Agassizodus.)

Cope, E. D. 1894 F, p. 433. (Orodus, section Agassizodus.)

Lohest, M. 1884 A, p. 306. (Campodus agassizianus.)

Woodward, A. S. 1889 D, p. 238. (C. agassizianus.)

Coal-measures; West Virginia.

MESODMODUS St. J. and Worth.   Type *M. exsculptus* St. J. and Worth.

St. John and Worthen 1875 A, p. 290.

Miller, S. A. 1889 A, p. 602.

Woodward, A. S. 1889 D, p. 289.

Miller, S. A. 1889 A, p. 602, fig. 1148.

Woodward, A. S. 1889 D, p. 240.

   Subcarboniferous (Kinderhook); Iowa.

Mesodmodus explanatus St. J. and Worth.

St. John and Worthen 1875 A, p. 293, pl. v, figs. 15–17.

Woodward, A. S. 1889 D, p. 239.

   Subcarboniferous (Kinderhook); Iowa.

Mesodmodus exsculptus St. J. and Worth.

St. John and Worthen 1875 A, p. 291, pl. v, figs. 18–22.

Mesodmodus ornatus St. J. and Worth.

St. John and Worthen 1875 A, p. 294, pl. v, figs. 12–14.

Woodward, A. S. 1889 D, p. 240.

   Subcarboniferous (Burlington); Illinois, Iowa.

DESMIODUS St. J. and Worth.   Type *D. tumidus* St. J. and Worth.

St. John and Worthen 1875 A, p. 337.

Miller, S. A. 1889 A, p. 595.

Newberry, J. S. 1889 A, p. 186.

Woodward, A. S. 1889 D, p. 240.

Desmiodus costelliferus St. J. and Worth.

St. John and Worthen 1875 A, p. 341, pl. xA, figs. 10, 11.

Keyes, C. R. 1894 A, p. 230.

Miller, S. A. 1889 A, p. 595, fig. 1125.

Woodward, A. S. 1889 D, p. 240.

   Subcarboniferous (St. Louis); Illinois, Missouri.

Desmiodus ? flabellum St. J. and Worth.

St. John and Worthen 1875 A, p. 343, pl. xA, fig. 15.

Keyes, C. R. 1894 A, p. 230.

Woodward, A. S. 1889 D, p. 240.

   Subcarboniferous (Keokuk); Missouri.

Desmiodus ? ligoniformis St. J. and Worth.

St. John and Worthen 1875 A, p. 342, pl. xA, figs. 12–14.

Keyes, C. R. 1894 A, p. 230.

Woodward, A. S. 1889 D, p. 240.

   Subcarboniferous (Keokuk); Missouri, Iowa.

Desmiodus minusculus (Newb. and Worth.).

Newberry and Worthen 1866 A, p. 67, pl. iv, fig. 11. (Orodus.)

St. John and Worthen 1875 A, p. 339, 340. (Desmiodus.)

Woodward, A. S. 1889 D, p. 237. (Orodus.)

   Subcarboniferous (Keokuk); Illinois.

Desmiodus tumidus St. J. and Worth.

St. John and Worthen 1875 A, pp. 299, 339, pl. xA, figs. 7–9.

Keyes, C. R. 1894 A, p. 230.

Woodward, A. S. 1889 D, p. 240.

   Subcarboniferous (St. Louis); Illinois, Missouri.

PETRODUS McCoy, F.   Type *P. patelliformis* McCoy.

McCoy, F. 1848 A, p. 132.

Davis, J. W. 1883 A, p. 400.

Jaekel, O. 1899 A, p. 278. (Syn.? of Polyrhizodus.)

Koninck, L. G. de 1878 A, p. 36.

Lohest, M. 1884 A.

McCoy, F. 1855 A, p. 637.

Miller, S. A. 1889 A, p. 606.

Newberry and Worthen 1866 A, p. 71. 1870 A, p. 346.

Pictet, F. J. 1854 A, p. 263.

Trautschold, H. 1874 B, p. 298. (Ostinaspis.)

Woodward, A. S. 1889 D, p. 245.

Zittel, K. A. 1890 A, p. 76.

   It is probable that the bodies included in this genus are not teeth but dermal tubercles.

Petrodus acutus Newb. and Worth.

Newberry and Worthen 1866 A, p. 72, pl. iv, fig. 17.

Trautschold, H. 1874 B, p. 209, pl. xxviii, fig. 12. (Ostinaspis.)

Woodward, A. S. 1889 D, p. 247.

   Coal-measures; Illinois.

Petrodus occidentalis Newb. and Worth.

Newberry and Worthen 1866 A, p. 70, pl. iv, figs. 15–16a.
Jaekel, O. 1899 A, p. 283.
Keyes, C. R. 1888 A, p. 243.
Kindle, E. M. 1898 A, p. 485.
Lesley, J. P. 1889 A, p. 628, figures.
Newberry and Worthen 1870 A, p. 346.

Woodward, A. S. 1889 D, p. 246.
    Coal-measures; Illinois, Indiana.

Petrodus ? pustulosus Newb. and Worth.

Newberry and Worthen 1870 A, p. 369, pl. ii, figs. 5, 5a; pl. iii, fig. 6.
Newberry, J. S. 1889 A, p. 178.
Woodward, A. S. 1889 D, p. 247.
    Subcarboniferous (Burlington); Iowa.

STEMMATIAS Hay.   Type *Stemmatodus cheiriformis* St. J. and Worth.

Hay, O. P. 1899 E, p. 784.
Davis, J. W. 1892 A, pp. 721, 726. (Syn. of Pleuracanthus.)
Jaekel, O. 1899 A, p. 278. (Stemmatodus, syn. of Polyrhizodus.)
Lesley, J. P. 1890 A, p. 1048.
St. John and Worthen 1875 A, p. 328. (Stemmatodus; preoccupied by Heckel.)
Ward, J. 1890 A, p. 153. (Stemmatodus.)
Wellburn, E. D. 1899, Proc. Yorksh. Geol. Polytech. Soc., xiii, p. 425. (Stemmatodus.)
Woodward, A. S. 1889 D, pp. 245, 247. (Stemmatodus.)

Stemmatias bicristatus (St. J. and Worth.).

St. John and Worthen 1875 A, p. 331, pl. viii, figs. 32, 33, 35. (Stemmatodus.)
Hay, O. P. 1899 E, p. 784.
Lesley, J. P. 1890 A, p. 1048, figures. (Stemmatodus.)
Woodward, A. S. 1889 D, p. 247. (Stemmatodus.)
    Subcarboniferous (Burlington); Iowa.

Stemmatias bifurcatus (St. J. and Worth.).

St. John and Worthen 1875 A, p. 330, pl. viii, figs. 31a–31c. (Stemmatodus.)
Hay, O. P. 1899 E, p. 784.
Lesley, J. P. 1890 A, p. 1048, figures. (Stemmatodus.)
Woodward, A. S. 1889 D, p. 247. (Stemmatodus.)
    Subcarboniferous (Burlington); Iowa.

Stemmatias cheiriformis (St. J. and Worth.).

St. John and Worthen 1875 A, p. 330, pl. viii, figs. 29, 30. (Stemmatodus.)
Hay, O. P. 1899 E, p. 784.
Lesley, J. P. 1890 A, p. 1048, figures. (Stemmatodus.)

Woodward, A. S. 1889 D, p. 247. (Stemmatodus.)
    Subcarboniferous (Burlington); Iowa.

Stemmatias compactus (St. J. and Worth.).

St. John and Worthen 1875 A, p. 334, pl. viii, figs. 38a–38c. (Stemmatodus.)
Hay O. P. 1899 E, p. 784.
Lesley, J. P. 1890 A, p. 1049, figure. (Stemmatodus.)
Woodward, A. S. 1889 D, p. 247. (Stemmatodus.)
    Subcarboniferous (Chester); Illinois.

Stemmatias keokuk (St. J. and Worth.).

St. John and Worthen 1875 A, p. 334. (Stemmatodus.)
Hay, O. P. 1899 E, p. 784.
Woodward, A. S. 1889 D, p. 247. (Stemmatodus).
    Subcarboniferous (Keokuk); Iowa, Illinois.

Stemmatias simplex (St. J. and Worth.).

St. John and Worthen 1875 A, p. 332, pl. viii, figs. 34, 36, 37. (Stemmatodus.)
Hay, O. P. 1899 E, p. 784.
Lesley, J. P. 1890 A, p. 1049, figures. (Stemmatodus.)
Woodward, A. S. 1889, D, p. 247. (Stemmatodus.)
    Subcarboniferous (Burlington); Iowa.

Stemmatias symmetricus (St. J. and Worth.).

St. John and Worthen 1875 A, p. 333, pl. viii, fig. 28. (Stemmatodus.)
Hay, O. P. 1899 E, p. 784.
Lesley, J. P. 1890 A, p. 1049, figures. (Stemmatodus.)
Woodward, A. S. 1889 D, p. 247. (Stemmatodus.)
    Subcarboniferous (Burlington); Iowa.

HETERODONTIDÆ.

Gill, T. 1862 A. (Heterodontoidæ.)
Agassiz, L. 1843 B (1838), iii, pp. 83, 159. ("Cestraciontes.")
Brown, C. 1900 A, p. 173. (Cestraciontidæ, Hybodontidæ.)
Davis, J. W. 1882 B. (Hybodontidæ.)
    1883 A, pp. 331, 332. (Hybodontidæ.)
Fraas, E. 1896 A, pp. 17, 18. (Hybodontidæ, Acrodontidæ, Cestraciontidæ.)
Fritsch, A. 1889 B, p. 97. (Hybodontidæ.)
Garman, S. 1885 B, p. 31. (Hybodontidæ.)
Gegenbaur, C. 1872 A. (Cestraciodontidæ.)

Gill, T. 1872 C, p. 23. (Heterodontidæ.)
    1884 B, p. 346. (Hybodontidæ, Heterodontidæ.)
Günther, A. 1880 A, p. 328. (Hybodontidæ, Cestraciontidæ.)
    1881 A, p. 685. (Hybodontidæ, Cestraciontidæ.)
Hasse, C. 1878 A. (Hybodontidæ, Cestraciontidæ.)
    1879 A, p. 50 ("Acrodontiden"); p. 68 (Hybodontidæ).
    1882 A, p. 183. (Cestraciontidæ.)
Haswell, W. A. 1884 A. (Cestraciontidæ.)

Howes, G. B. 1887 A.   (Cestracion.)
Huxley, T. H. 1876 A.   (Cestraciontidæ.)
Jaekel, O. 1890 D, p. 121, fig. 2.   (Cestracion.)
    1894 A, p. 137.   (Heterodontus.)
Koninck, L. G. de 1878 A, p. 29.   (Cestraciontidæ.)
Marsh, O. C. 1877 E.   (Cestraciontidæ.)
Mayer, P. 1882 A, p. 277.   (Cestraciontidæ.)
Mivart, St. G. 1879 A, p. 449, pl. lxxvi.   (Cestracion.)
Müller and Henle 1841 A, p. 76.   (Cestraciontes.)
Nicholson and Lydekker 1889 A, p. 940.   (Hybodontidæ, Cestraciontidæ.)
Owen, R. 1860 E, p. 106.   (Cestraciontidæ.)
Pictet, F. J. 1854 A, p. 254 ("Hybodontes"); p. 260 ("Cestraciontes").

Quenstedt, F. A. 1885 A, p. 272 ("Hybodonten"); p. 278 ("Cestracionten").
Reis, O. M. 1897 A, p. 69.   ("Hybodonten").
Stannius, H. 1854 A.   (Cestraciones.)
Steinmann and Döderlein 1890 A, pp. 545, 546.   (Hybodontidæ, Cestraciontidæ.)
Tomes, C. S. 1898 A, p. 244 ("Cestracionts").
Woodward, A. S. 1884 A, p. 267.   (Hybodontidæ.)
    1885 B, p. 106.   (Cestraciontidæ.)
    1889 D, p. 229.   (Cestraciontidæ.)
    1898 B, pp. 42, 44.   (Cestraciontidæ.)
Zittel, K. A. 1890 A, p. 66 (Hybodontidæ); p. 74 (Cestraciontidæ).

HYBODUS Agassiz.   Type *H. plicatilis* Agassiz.

Agassiz, L. 1843 B (1837), pp. 41, 178.
    1843 B, p. 201.   (Sphenonchus, in part.)
    1844 A.
Barkas, W. J. 1874 A, p. 386.
    1877 B, p. 145.
Brown, C. 1900 A, p. 149, pls. xv, xvi.
Day, E. C. H. 1864 A.
Egerton, P. G. 1845 A, pl. iv.
Fraas, E. 1896 A.
Garman, S. 1885 B, p. 31.
Giebel, C. G. 1848 A, p. 311.
Jaekel, O. 1884 A, p. 294.
    1898 B, p. 135 (Hybodus); p. 138 (Orthybodus); p. 143 (Parahybodus).
Mackie, S. J. 1863 A.
Nicholson and Lydekker 1889 A, p. 940.
Owen, R. 1845 B, p. 56, pl. xi.
    1860 E, p. 105.
Pictet, F. J. 1854 A, p. 255.
Quenstedt, F. A. 1852 A, p. 174.
    1885 A, p. 273.
Steinmann and Döderlein 1890 A, p. 546.
Williamson, W. C. 1849 A, p. 465, pl. 43, fig. 33.
Woodward, A. S. 1884 A, p. 268.
    1886 A.
    1888 I, p. 251.

Woodward, A. S. 1888 J, p. 58.
    1889 D, p. 250.
    1889 E.
    1889 F.
    1898 B, p. 44.
Woodward and Sherborn 1890 A, p. 99.
Zittel, K. A. 1890 A, p. 67.

Hybodus clarkensis Cragin.

Cragin, F. W. 1894 A, p. 72, pl. ii, figs. 11-14.
    Cretaceous (Neocomian); Kansas.

Hybodus copei Hay.

Hay, O. P. 1899 E, p. 784.
Cope, E. D. 1891 B, p. 448, pl. xxviii, fig. 2.   (H regularis; preoccupied by Reuss, 1846.)
    Triassic?; Texas.

Hybodus polyprion Agassiz.

Agassiz, L. 1843 B, iii, p. 185, pl. xxiii, figs. 1-15.
Marsh, O. C. 1889 A, p. 230.
Woodward, A. S. 1886 C, p. 257, pl. vi, figs. 1, 2.
    1889 D, p. 268.
    Stonesfield Slate, England: Jurassic; Black Hills, Wyoming.

ACRODUS Agassiz.   Type *A. gaillardoti* Agassiz.

Agassiz, L., 1837, in H. B. Geinitz, Beiträge Kennt. Thüring. Muschelkgeb., p. 21.   (*Fide* A. S. Woodward.)
    1843 B (1838), iii, p. 139 (Acrodus); p. 201 (Sphenonchus, in part).
    844 A.
Day, E. C. H. 1864 A.
Giebel, C. G. 1848 A, p. 325.
Hasse, C. 1882 A, p. 191, pls. xxiv, xxv.
Jaekel, O. 1884 A, p. 311.
    1890 A, p. 93.
    1891 D, p. 191
    1894 A, p. 132.
    1899 A, p. 296.
Nicholson and Lydekker 1889 A, p. 942.
Owen, R. 1845 B, p. 54, pls. xiv-xvi.
    1860 E, p. 109.
Pictet, F. J. 1854 A, p. 261.
Quenstedt, F. A. 1852 A, p. 178.
    1885 A, p. 278.
Steinmann and Döderlein 1890 A, p. 546.

Traquair, R. H. 1888 A, p. 84.
Woodward, A. S. 1885 B, p. 107.
    1889 D, p. 279.
    1889 E.
    1889 F.
Woodward and Sherborn, 1890 A, p. 5.
Zittel, K. A. 1890 A, p. 76.

Acrodus emmonsi Leidy.

Leidy, J. 1872 M, p. 163.
Emmons, E. 1858 B, p. 244, fig. 97.   ("Acrodus.")
Leidy, J. 1873 B, pp. 301, 352.
Woodward, A. S. 1889 D, p. 298.
    Miocene?; North Carolina?.

Acrodus humilis Leidy.

Leidy, J. 1872 M, p. 163.
    1873 B, pp. 300, 352, pl. xxxvii, fig. 5.
Woodward, A. S. 1889 D, p. 298.
    Cretaceous; New Jersey.

Superfamily *HEXANCHOIDEA*,[1] *nom. nov.*

Gill, T. 1882, Jordan & Gilbert's Synop. Fishes
N. A., p. 967. (Opisthrarthri.)
   1884 B, p. 346. (Opisarthri.)
   1893 A, p. 129. (Opistharthri.)
Hasse, C. 1878 A. (Palæonotidani.)
   1879 A, p. 85. (Diplospondyli.)

Haswell, W. A. 1884 A, p. 112. (Palæoselachii.)
Jordan & Evermann 1896 A, p. 17. (Diplospon-
dyli.)
Koken, E. 1889 A, p. 90. (Opisarthri.)
Osborn, H. F. 1899 I, p. 158. (Notidani.)
Woodward, A. S. 1889 D, p. ix. ("Opisthrarthri.")

HEXANCHIDÆ.

Gill, T., 1885, Report Smithsonian Inst. for 1884, p.
618.
   The following authors, unless otherwise in-
dicated, refer to this group under the name
Notidanidæ:
Brown, C. 1900 A, pp. 169, 174.
Duméril, A. 1865 A, p. 430. (Notandani.)
Fritsch, A. 1895 A, p. 46.
Gadow, H. 1889 A. (Heptanchus.)
Gegenbaur, C. 1865 B.
   1872 A.
   1872 B.
Gill, T. 1893 A, p. 129. (Hexanchidæ.)
Günther, A. 1870 A, pp. 355, 397.
   1880 A, p. 324.
   1881 A, p. 685.

Hasse, C. 1879 A, p. 35.
   1882 A, p. 37.
Haswell, W. A. 1884 A, p. 112.
Jaekel, O. 1894 A.
Kölliker, A. 1863 A, p. 1. (Hexanchus.)
Müller and Henle 1841 A, p. 80.
Noetling, F. 1885 A, p. 16.
Reis, O. M. 1897 A, pp. 69, 86.
Sagemehl, M. 1883 A, p. 227.
Semon, R. 1899 A, p. 76.
Stannius, H. 1854 A. (Notidani.)
Steinmann and Döderlein 1890 A, p. 547.
White, P. J. 1895 A.
Woodward, A. S. 1889 D, p. 157.
Zittel, K. A. 1890 A, p. 65.

HEPTRANCHIAS Rafinesque.   Type *Squalus cinereus* Gmelin.

Rafinesque, C. S. 1810, Caratteri di alcuni nuovi
   generi, p. 13.
   Unless indicated otherwise, the following
authors refer to this genus and *Hexanchus*
under the name *Notidanus*.
Agassiz, L. 1843 B, iii, pp. 216, 308.
Brown, C. 1900 A.
Cuvier, G. 1817, Règne Anim, Ed. 1, p. 128.
Duméril, A. 1865 A, p. 432. (Heptanchus.)
Gegenbaur, C. 1872 A, pl. i, figs 1, 2; pl. iii, figs.
   6, 7; pl. iv, figs. 1, 2; pl. vii, figs. 1, 2; pl. x; pl.
   xv, figs. 1, 2; pl. xvi, fig. 3; pl. xviii; pl. xx, figs.
   2, 3, 5, 6, 8; pl. xxi, fig. 5. (Heptanchus and
   Hexanchus.)
   1872 B. (Hexanchus, Heptanchus.)
Gibbes, R. W. 1849 A, p. 195.
Günther, A. 1870 A, p. 397.
   1880 A, p. 325.
Hasse, C. 1882 A, p. 39, pl. vi (Hexanchus); p. 44.
   pl. vi, vii (Heptanchus).
Haswell, W. A. 1884 A, pp. 73, 88. (Heptanchus.)
Hubrecht, A. A. W. 1878 A.
Huxley, T. H. 1876 A.
Kölliker, A. 1860 B. (Heptanchus.)
Lawley, R. 1875 A.
Mayer, P. 1886 A, p. 263. (Heptanchus.)
Mivart, St. G. 1879 A, p. 443, pl. lxxv, fig. 2.
Müller and Henle, 1841, p. 80 (Hexanchus); p. 81
   (Heptanchus).
Parker, T. J. 1890 A.
Pedroni, P. M. 1844 A, p. 281.
Pictet, F. J. 1854 A, p. 243.
Probst, J. 1879 A, p. 156.

Quenstedt, F. A. 1885 A, p. 262.
Rafinesque, C. S., 1810, Caratteri di alcuni nuovi
   generi, p. 14. (Hexanchus.)
Reis, O. M. 1897 A, p. 69.
Stannius, H. 1854 A.
Woodward, A. S. 1886 C.
   1889 D, p. 157.
   1892 G, figs. 9-12.
   1899 A, p. 6.
Zittel, K. A. 1890 A, p. 66.

Heptranchias ? plectrodon (Cope).

Cope, E. D. 1867 C, p. 141. (Notidanus.)
   Tertiary; Maryland.

Heptranchias ? primigenius(Agassiz).

Agassiz, L. 1843 B, iii, p. 218, pl. xxvii, figs. 6-8,
13-17; doubtfully figs. 4, 5. (Notidanus.)
   The following authors refer this species to the
genus *Notidanus*.
Agassiz, L. 1843 B, p. 220, pl. xxvii, figs. 9-12.
   (N. recurvus.)
Bassani, F. 1899 A, p. 25, pl. ii, figs. 13-15.
Cope, E. D. 1867 C, p. 141.
   1875 H, p. 362.
   1875 V, p. 29.
Dana, J. D. 1896 A, p. 901, fig. 1515.
Davis, J. W. 1888 A, p. 33, pl. vi, fig. 6.
Gibbes, R. W. 1849 A, p. 195, pl. xxv, fig. 95.
Lawley, R. 1875 A, p. 20, pl. i, figs. 1-5.
Leidy, J. 1877 A, p. 254.
Noetling, F. 1885 A, p. 17, pl. i, figs. 4, 5.
Pedroni, P. M. 1844 A, p. 281, pl. i, figs. 10, 11.

[1] This name is proposed simply for the purpose of conforming to the rule, adopted by many zoologists, of deriving the superfamily name from the name of a typical family by changing the termina-
tion *idæ* into *oidea.*

CATALOGUE.

Probst, J. 1858 A, p. 124, figs. 1–10.
  1879 A, p. 158, pl. iii, figs. 1–5 (N. primigenius); p. 162, pl. iii, figs. 12–17 (N. recurvus); p. 166, pl. iii, figs. 6–11 (N. d'anconæ).
Quenstedt, F. A. 1852 A, p. 167, pl. xiii, fig. 3.
  1885 A, p. 263, pl. xx, fig. 6.

Woodward, A. S. 1886 C, pp. 216, 253, pl. vi, figs. 19–22.
  1889 D, p. 163.
Wyman, J. 1850 B, p. 234.
Zittel K. A. 1890 A, p. 66, fig. 60.
  Eocene and Miocene; Europe: Eocene; Virginia: Miocene; New Jersey, Maryland, North Carolina.

XIPHODOLAMIA Leidy. Type *X. ensis* Leidy.

Leidy, J. 1877 A, p. 252.
Woodward, A. S. 1889 D, p. 168. (Syn. ? of Notidanus.)
  1899 A, p. 6. (Syn. ? of Notidanus.)

Xiphodolamia ensis Leidy.

Leidy, J. 1877 A, p. 252, pl. xxxiv, figs. 25–30.
Woodward, A. S. 1889 D, p. 168. (Notidanus?)
  1899 A, p. 6.
"Marls" (Cretaceous?); New Jersey.

Superfamily GALEOIDEA,[1] nom. nov.

Gill, T. 1893 A, p. 129. (Asterospondyli, in part.)
Hasse, C. 1879 A, p. 48. (Asterospondyli.)
  1882 A, pp. 186–284. (Asterospondyli.)
Haswell, W. A. 1884 A, p. 113. (Neoselachii, in part.)

Jaekel, O. 1890 A, p. 110. (Asterospondyli.)
  1894 A, p. 58. (Actinospondyli, in part.)
Jordan and Evermann, 1896 A, p. 21. (Galei.)
Woodward, A. S. 1889 D, p. 157. (Asterospondyli, in part.)

LAMNIDÆ.

Davis, J. W. 1890 A, p. 365.
Duméril, A, 1865 A, p. 408. (Lamnæ.)
Gill, T. 1862 A, p. 397. (Lamnoidæ.)
Günther, A. 1870 A, pp. 354, 389.
  1880 A, p. 319.
  1881 A, p. 685.
Hasse, C. 1882 A, p. 214. ("Familie Lamna.")
Haswell, W. A. 1884 A, p. 114.
Jaekel, O. 1894 A, p. 157.
Jordan and Evermann 1896 A, p. 47.
Kiprijanoff, V. 1854 A, p. 373.

Kölliker, A. 1863 A. (Lamnoidei.)
Mayer, P. 1880 A, p. 277.
Müller and Henle 1888 A, p. 36. (Lamnoidea.)
  1841 A, p. 66. (Lamnæ.)
Nicholson and Lydekker 1889 A, p. 944.
Probst, J. 1879 A, p. 127.
Stannius, H. 1854 A. (Lamnoidei.)
Steinmann and Döderlein 1890 A, p. 547.
Woodward, A. S. 1884 A, p. 228.
  1889 D, p. 349.
Zittel, K. A. 1890 A, p. 81.

CARCHARIINÆ.

Duméril, A. 1865 A, p. 416. (Odontaspides.)
Gill, T. 1862 A, p. 398. (Odontaspidoidæ.)
  1893 A, p. 130. (Carchariidæ.)

Jordan and Evermann 1896 A, p. 46. (Carchariidæ.)
Müller and Henle 1841 A, p. 73. (Odontaspides.)

CARCHARIAS Raf. Type *C. taurus* Raf.

Rafinesque, C. S., 1810, Caratteri di alcuni nuovi generi, p. 10.
Agassiz, L. 1843 B (1838), iii, pp. 87, 287, 306. (Odontaspis.)
Davis, J. W. 1890 A, p. 368. (Odontaspis.)
Duméril, A. 1865 A, p. 417. (Odontaspis.)
Günther, A. 1870 A, p. 354. (Odontaspis.)
  1880 A, p. 321. (Odontaspis.)
Hasse, C. 1882 A, pp. 160, 230. (Odontaspis.)
Jordan and Evermann 1896 A, p. 46. (Carcharias.)
Kölliker, A. 1863 A, p. 14, figs. 6, 7. (Odontaspis.)
Müller and Henle 1837 A, p. 114. (Odontaspis.)
  1837 B, p. 397.
  1841 A, p. 73. (Odontaspis.)
Pictet, F. J. 1854 A, p. 251. (Odontaspis.)
Probst, J. 1859 A. (Odontaspis.)
Woodward, A. S. 1884 A, p. 228. (Odontaspis.)
  1889 D, p. 360. (Odontaspis.)
  1899 A, p. 7. (Odontaspis.)

Woodward and Sherborn 1890 A, p. 129. (Odontaspis.)
Zittel, K. A. 1890 A, p. 81. (Odontaspis.)

Carcharias contortidens (Agassiz).

Agassiz, L. 1843 B, iii, p. 294, pl. xxxvii, α, figs. 17–23. [Lamna (Odontaspis).]
Emmons, E. 1858 B, p. 239. [Lamna (Odontaspis).]
Gemmellaro, G. G. 1857 A, p. 320, pl. i, fig. 13; pl. vi α, figs. 18–20. [Lamna (Odontaspis).]
Gibbes, R. W. 1849 A, p. 197, pl. xxvi, fig. 119. [Lamna (Odontaspis).]
Giebel, C. G. 1848 A, p. 363. (Odontaspis.)
Jaekel, O. 1898 C, p. 163. (Odontaspis.)
Probst, J. 1859 A, p. 101. [Lamna (Odontaspis)].
  1879 A, p. 144, pl. ii, figs. 33–39. [Lamna (Odontaspis).]
Sismonda, E. 1849 A, p. 48, pl. ii, figs. 25–28. [Lamna (Odontaspis).]

---

[1] See note under *Hexanchoidea*, p. 300.

Tuomey, M. 1858 A, p. 268. (Lamna.)
Woodward, A. S. 1889 D, p. 366. (Odontaspis.)
Eocene and Miocene; Europe: Eocene; South
Carolina and Alabama: Miocene?; North Carolina.

Carcharias duplex (Agassiz)

Agassiz, L. 1843 B, iii, p. 297, pl. xxxvii a, fig. 1.
[Lamna (Odontaspis).]
Tuomey M. 1858 A, p. 268. (Lamna.)
Woodward, A. S. 1889 I, p. 375. (Odontaspis.)
Said by Tuomey to occur in Tertiary of Alabama. The identification is doubtful.

Carcharias verticalis (Agassiz).

Agassiz, L. 1843 B, iii, p. 294, pl. xxxvii a, figs.
31, 32. (Odontaspis.)
Dames, W. 1883 B, p. 145, pl. iii, figs. 8–10. (Odontaspis.)
Gibbes, R. W. 1849 A, p. 198, pl. xxvi, figs. 124–127. [Lamna (Odontaspis).]
Giebel, C. G. 1848 A, p. 363. (Odontaspis.)
Tuomey, M. 1858 A, p. 268. (Lamna.)
Woodward, A. S. 1889 D, p. 375. (Odontaspis.)
Eocene: Europe, Egypt, South Carolina, Alabama.

LAMNINÆ.

Gill, T. 1862 A, p. 398.

Jordan and Evermann 1896 A, p. 47.

SPHENODUS Agassiz. Type *Sphenodus longidens* Agassiz.

Agassiz, L. 1843 B, iii, p. 288.
Davis, J. W. 1890 A, p. 349.
Quenstedt, F. A. 1852 A, p. 172.
1885 A, p. 270.
Woodward, A. S. 1889 D, p. 349. (Orthacodus, for
Sphenodus Agassiz, regarded as preoccupied by
Sphenodon Gray, 1831.)

Zittel, K. A. 1890 A, p. 81.

Sphenodus rectidens Emmons.

Emmons, E. 1858 B, p. 235, figs. 61, 62.
Cretaceous; North Carolina.

LAMNA Cuvier. Type *Squalus cornubicus* Gmelin.

Cuvier G. 1817, Règne Anim., ed. 1, v, ii, p. 126.
Agassiz, L. 1843 B, iii, pp. 287, 306.
Davis, J. W. 1888 A, p. 15.
1890 A, p. 371.
Duméril, A. 1865 A, p. 404.
Gibbes, R. W. 1849 A, p. 196.
Günther, A. 1870 A, p. 389.
Hasse, C. 1882 A, p. 215, pl. xxviii.
Jordan and Evermann 1896 A, p. 49.
Klunzinger, C. B. 1871 A, p. 669.
Kölliker, A. 1863 A. p. 8, fig. 3.
Müller and Henle 1837 A, p. 114.
1837 B, p. 397.
1841 A, p. 67.
Nicholson and Lydekker 1889 A, p. 944.
Pictet, F. J. 1854 A, p. 249.
Probst J. 1859 A.
Quenstedt, F. A. 1852 A, p. 170.
1885 A, p. 268.
Sismonda, E. 1849 A, p. 45.
Stephan, P. 1900 A, p. 327.
Tomes, C. S. 1898 A, p. 79.
Woodward, A. S. 1884 A, p. 228.
1889 D, p. 392.
Zittel, K. A. 1890 A, p. 82.

Lamna clavata Agassiz.

Agassiz, L. 1856 A, p. 275.
1857 A, p. 316, pl. i, figs. 19–21.
Woodward, A. S. 1889 D, p. 407.
Miocene; California.

Lamna crassidens Agassiz.

Agassiz, L. 1843 B, iii, p. 292, pl. xxxv, figs. 8–21.
Emmons, E. 1858 B, p. 240.

Gemmellaro, G. G. 1857 A, p. 318, pl. vi a, figs. 15, 16.
Gibbes, R. W. 1846 A, p. 42.
1849 A, p. 197, pl. xxvi, figs. 116–118.
Probst, J. 1879 A, p. 373, pl. ii, figs. 64–68.
Woodward, A. S. 1889 D, p. 373. (Odontaspis?)
Wyman, J. 1850 B, p. 233.
Eocene, Europe, North Carolina (Emmons),
South Carolina: Miocene?; Virginia (Wyman).

Lamna cuspidata Agassiz.

Agassiz, L. 1843 B, iii, p. 290, pl. xxxviia, figs.
43–50.
1843 B, iii, p. 291, pl. xxxviia, figs. 51–53
(L. denticulata); p. 293, pl. xxxviia, figs. 27–30
[L. (Odontaspis) hopei); p. 295, pl. xxxviia,
figs. 24–26 [L. (Odontaspis) dubia).
Cope, E. D. 1867 C, p. 142. (L. cuspidata, L. denticulata?, L. hopei.)
1875 E, p. 297.
1875 H, p. 362. (L. cuspidata, L. denticulata.)
1875 V, p. 29.
Geinitz, H. B. 1883 A, p. 5, pl. i, figs. 1–3.
Gibbes, R. W. 1846 A, p. 42.
1849 A, p. 197, pl. xxv, figs. 103–106.
1850 F. (L. cuspidata and L. denticulata.)
Jaekel, O. 1898 C, p. 162.
Leidy, J. 1872 F.
1873 A, p. 304, pl. xviii, figs. 44, 45. (L. cuspidata?).
Lyell, C. 1845 A, p. 415.
1859 A, p. 101. [Lamna (Odontaspis).]
Noetling, F. 1885 A, p. 71, pl. v, figs. 1–3. (Odontaspis hopei.)

Probst, J. 1879 A, p. 149, pl. ii, figs. 59–63. [Lamna (Odontaspis).]
Quenstedt, F. A. 1852 A, p. 171, pl. xiii, fig. 17 (Lamna cuspidata); fig. 15 (L. denticulata).
　　1885 A, p. 268, pl. xx, fig. 27.
Sauvage, H. E. 1882, p. 48, pl. i, figs. 15, 16.
Sismonda E. 1849 A, p. 47, pl. ii, figs. 29–32 (Lamna cuspidata); p. 48, pl. ii, figs. 17–22 (Lamna dubia).
Tuomey, M. 1858 A, p. 268.
Woodward, A. S. 1889 D, p. 368.　(Odontaspis.)
　　Eocene; Europe, Georgia, Alabama, Virginia, Kansas, South Carolina: Miocene; Maryland, North Carolina (Cope).

Lamna elegans Agassiz.

Agassiz, L. 1843 B, iii, p. 289, pl. xxxv, figs. 1–5 (not figs. 6, 7); pl. xxxviia, fig. 59 (not fig. 58).
Bassani, F. 1899 A, p. 13, pl. i, figs. 1–17.　(Odontaspis).
Clark, W. B. 1895 A, p. 4.
　　1897 A, p. 61.
Cope, E. D. 1867 C, p. 142.
　　1875 H, p. 362.
　　1875 V, p. 29.
Dana, J. D. 1896 A, p. 901, fig. 1514.
Davis, J. W. 1888 A, p. 15, pl. iii, fig. 1.　(L. huttoni.)
Dixon, F. 1850 A, p. 208, pl. x, figs. 28–31.
　　1878 A, p. 249, pl. x, figs. 29–31.
Emmons, E. 1858 B, p. 239, figs. 70–71a.
Geinitz, H. B. 1883 A, p. 5, pl. i, figs. 4–6.
Gibbes, R. W. 1846 A, p. 42.
　　1849 A, p. 196, pl. xxv, figs. 96–102.
Leidy, J. 1872 F.
　　1873 B, p. 305.
　　1877 A, p. 254.
Noetling, F. 1885 A, p. 61, pl. iv, figs. 1–9.
Owen, R. 1845 B, p. 32, pl. v, figs. 3, 4; pls. vi, vii.
Sismonda, E. 1849 A, p. 46, pl. ii, figs. 33–35.
Tuomey, M. 1858 A, p. 261.
Woodward, A. S. 1884 A, p. 228.
　　1889 D, p. 361.　(Odontaspis.)
　　1899 A, p. 8, pl. i, figs. 15–18.　(Odontaspis.)
Wyman, J. 1850 B, p. 233.
　　Eocene and Miocene; Europe, New Zealand, Australia, and United States: Eocene; South Carolina, North Carolina, Alabama, and Virginia: Miocene; Maryland.

Lamna gracilis Agassiz.

Agassiz, L. 1843 B, iii, p. 295, pl. xxxvii a, figs. 2–4. [Lamna (Odontaspis).]
Eichwald, E. 1868 A, p. 1222, pl. xxxviii, fig. 10.
Gibbes, R. W. 1849 A, p. 196, pl. xxvi, figs. 128–130. [Lamna (Odontaspis).]
Giebel, C. G. 1847 A, p. 362.
Pictet and Campiche 1858 A, p. 88, pl. xi, figs. 9–18. (Odontaspis.)
Sauvage, H. E. 1879 A, p. 59, pl. iii, figs. 1–12. (Odontaspis.)
Woodward, A. S. 1889 D, p. 359.　(Scapanorhynchus?).
　　Cretaceous; Europe: Eocene, South Carolina

Lamna macrorhiza Cope.

Cope, E. D. 1875 E, p. 297, pl. xlii, figs. 9, 10.

Williston, S. W. 1900 A, p. 38.
　　1900 B, p. 249.
Woodward, A. S. 1889 D, p. 399.
　　Cretaceous (Niobrara); Kansas: Cretaceous of Europe.

Lamna manitobensis Whiteaves.

Whiteaves, J. F. 1889 B, p. 192, pl. xxvi. figs. 6a, 6c.
　　Cretaceous (Niobrara); Manitoba.

Lamna mudgei Cope.

Cope, E. D. 1875 E, p. 297, pl. xlii, figs. 11, 12.
Williston, S. W. 1900 A, p. 38.
　　1900 B, p. 248.
Woodward, A. S. 1889 D, p. 408.
　　Cretaceous (Niobrara); Kansas: ("Greensand"); New Jersey.

Lamna ornata Agassiz.

Agassiz, L. 1856 A, p. 275.
　　1857 A, p. 316, pl. i, fig. 28.
Woodward, A. S. 1889 D, p. 408.
　　Miocene; California.

Lamna ? quinquelateralis Cragin.

Cragin, F. W. 1894 A, p. 72, pl. ii, figs. 9, 10.
Williston, S. W. 1900 A, p. 39.
　　1900 B, p. 250.
　　Cretaceous (Comanche); Kansas.

Lamna rhaphiodon Agassiz.

Agassiz, L. 1843 B, iii, p. 296, pl. xxxvii a, figs. 12–16, not fig. 11. [Lamna (Odontaspis).]
Tuomey, M. 1858 A, p. 268.
　　Reported from the Tertiary of Alabama by Tuomey, but the identification is doubtful. For synonymy and literature see Woodward, A. S. 1889 D, p. 353. See also under Lamna texana.

Lamna subulata Agassiz.

Agassiz, L. 1843 B, iii, p. 296, pl. xxxvii a, figs. 5–7.
Cope, E. D. 1867 C, p. 142.
Geinitz, H. B. 1875 A, p. 209, pl. xxxviii, figs. 29–36.
Hébert, E. 1855 C, p. 355, pl. xxvii, fig. 10.
　　Professor Cope has reported this species with doubt from the Miocene of Charles County, Md. For synonymy and literature of the species see Woodward, A. S. 1889 D, p. 356, under name of Scapanorhynchus ? subulatus.

Lamna texana Roemer.

Roemer, F. 1849 A, p. 419.
Cope, E. D. 1875 E, p. 296.
　　1875 F, p. 63.
　　1875 K, p. 38.
Credner, H. 1870 A, p. 242.
Dana, J. D. 1896 A, p. 843, fig. 1402.
Leidy, J. 1873 B, p. 305, pl. xviii, figs. 46–49. (L. elegans?.)
Newberry, J. S. 1876 A, p. 141.
Roemer, F. 1852 A, p. 29, pl. i, figs. 7a, 7b.
Williston, S. W. 1900 A, p. 40, pl. viii, fig. 4; pl. xiv, fig. 5. (Scapanorhynchus rhaphiodon.)

Williston, S. W. 1900 B, p. 251, pl. xxvi, fig. 4; pl.
　xxxii, fig.5.　(Scapanorhynchus raphiodon.)
Woodward, A. S. 1889 D, p. 353.
　　Cretaceous; Texas, Kansas, Alabama, Missis-
　sippi, New Jersey, Kansas.

Lamna sp. indet.

Williston, S. W. 1900 A, p. 38, pl. xli, figs. 5, 6.
　　1900 B, p. 249, pl. xxx, figs. 5, 6.
　　Cretaceous (Kiowa shales); Kansas.

LEPTOSTYRAX Willist.　Type *L. bicuspidatus* Willist.

Williston, S. W. 1900 A, p. 42.

Leptostyrax bicuspidatus Willist.

Williston, S. W. 1900 A, p. 42.
　　1900 B, p. 253, pl. xxiv, figs. 15, 15a; pl. xxvi,
　　fig. 7.
　　Cretaceous (Mentor beds); Kansas.

OTODUS Agassiz.　Type *O. obliquus* Agassiz.

Agassiz, L. 1843 B, iii, pp. 266, 307.
　　1844 A.
Davis, J. W. 1888 A, p. 14.
　　1890 A, pp. 367, 401.
Gibbes, R. W. 1849 A, p. 198.
Hasse, C. 1882 A, p. 206, pl. xxvii.
Kiprijanoff, V. 1854 A, p. 377.
Nicholson and Lydekker 1889 A, p. 944.
Pictet, F. J. 1854 A, p. 245.
Pictet and Campiche 1858 A, p. 81.
Probst, J. 1879 A, p. 154.
Sismonda, E. 1849 A, p. 38.
Woodward, A. S. 1884 A, p. 229.
　　1889 D, p. 392.　(Syn. of Lamna.)
　　1899 A, p. 10.
Zittel, K. A. 1890 A, p. 83.

Otodus ? apiculatus Agassiz.

Agassiz, L. 1843 B, iii, p. 275, pl. xxxii, figs. 32–35.
Eastman, C. R. 1895 A, p. 178.　(Syn. of Oxyrhina
　　hastalis.)
Gibbes, R. W. 1846 A, p. 42.
　　1849 A, p. 200, pl. xxvi, fig. 147.
Woodward, A. S. 1889 D, p. 387.
　　Eocene; South Carolina.

Otodus appendiculatus Agassiz.

Agassiz, L. 1843 B (1835), iii, p. 270, pl. xxxii, figs.
　　1–25.
Cope, E. D. 1867 C, p. 142.
　　1875 E, p. 296.
　　1875 H, p. 362.
Credner, H. 1870 A, p. 242.
Dana, J. D. 1896 A, p. 843, figs. 1401, 1404.
Davis, J. W. 1890 A, p. 402, pl. xli, figs. 1–11.
Dixon, F. 1850 A, pl. xxx, fig. 25; pl. xxxi, fig. 17.
Emmons, E. 1858 B, p. 240.
Geinitz, H. B. 1875 A, p. 208.
　　1839 A, p. 11, pl. i, fig. 3 (Squalus cornubi-
　　cus); p. 12, pl. i, fig. 5 (Odontaspis rhap-
　　piodon.)
Gibbes, R. W. 1846 A, p. 42.
　　1849 A. p. 199, pl. xxvi, figs. 138–140.
　　1850 F.
Giebel, C. G. 1848 A, p. 42.
Hébert, E. 1855 C, p. 355.
Pictet, F. J. 1854 A, p. 247.

Pictet and Campiche 1858 A, p. 82, pl. x, figs. 3, 4.
Priem, F. 1897 A, p. 40, pl. i, figs. 1–8.　(Lamna.)
Quenstedt, F. A. 1852 A, p. 170, pl. xiii, fig. 8.
　　1885 A, p. 268, pl. xx, figs. 23, 24.
Reuss, A. E. 1846 A, p. 5, pl. iii, figs. 22–29.
Roemer, F. 1849 A, p. 419.
　　1852 A, p. 30, pl. i, figs. 9a, 9b.
Sauvage, H. E. 1870 A, p. 26, figs. 57–59 (O. appendi-
　　culatus); p. 34, figs. 73–75 (Lamna acuminata).
　　1879 A, p. 60, pl. iii, fig. 21 (O. appendiculatus);
　　p. 61, pl. iii, figs. 15–20 (Lamna acuminata).
Spillman, W, in Hilgard, E. W. 1860 A, p. 399.
Williston, S. W. 1900 A, p. xiv, figs. 3–3c.　(Lam-
　　na.)
　　　　1900 B, p. 247, pl. xxvi, figs. 3–3c; pl. xxxi,
　　　　figs. 47–49.　(Lamna.)
　　　　1884 A, p. 229, fig. 131.
Woodward, A. S. 1889 D, p. 393.　(Lamna.)
Wyman, J. 1850 B, p. 234.
　　　　Cretaceous; Europe, Australia, New Jersey,
　　North Carolina, Mississippi: Tertiary; Virginia,
　　Texas, North Carolina.

Otodus crassus Agassiz.

Agassiz, L. 1843 B, iii, p. 271, pl. xxxvi, figs. 29–31.
Gibbes, R. W. 1849 A, p. 200, pl. xxvi, fig. 142.
Kiprijanoff, V. 1854 A, p. 384, pl. ii, figs. 4–20.
Woodward, A. S. 1889 D, p. 400.　(Lamna.)
　　　　Cretaceous; Europe, Alabama.

Otodus divaricatus Leidy.

Leidy, J. 1872 M, p. 162.
Cope, E. D. 1874 C, p. 48.
　　1875 E, p. 295.
　　1873 B, pp. 305, 351, pl. xviii, figs. 26–28.
Williston, S. W. 1900 A, p. 37, pl. vi, figs. 1–1b.
　　(Lamna sulcata.)
　　　　1900 B, p. 248, pl. xxiv, figs. 1–1b.　(L. sulcata.)
Woodward, A. S. 1889 D, p. 398.　(Syn. of Lamna
　　sulcata.)
　　　　Cretaceous; Texas, Kansas.

Otodus levis Gibbes.

Gibbes, R. W. 1847 A, p. 288.
　　1849 A, p. 199, pl. xxvi, fig. 141.
Woodward, A. S. 1889 D, p. 408.　(Lamna.)
　　　　Eocene; South Carolina, New Jersey.

Otodus macrotus Agassiz.

Agassiz, L. 1843 B, iii, p. 273, pl. xxxii, figs. 29–31.
1843 B, iii, pl. xxxv, figs. 6, 7; pl. xxxvii *a*, fig. 58 (Lamna elegans, in part); p. 290, pl. xxxvii *a*, figs. 35–42 (Lamna compressa).
Dixon, F. 1850 A, p. 204, pl. xi, figs. 20, 21. (O. lanceolatus, not of Agassiz.)
1878 A, p. 249, pl. xi, figs. 20, 21. (O. lanceolatus.)
Emmons, E. 1858 B, p. 240.
Gibbes, R. W. 1846 A, p. 42.
1849 A, p. 200, pl. xxvi, figs. 143, 144; p. 197, pl. xxv, figs. 107–112. (Lamna compressa.)
1850 F. (Lamna compressa.)
Winckler, T. C. 1874 B, p. 295.
Woodward, A. S. 1889 D, p. 402. (Lamna.)
1899 A, p. 9, pl. i, figs. 19, 20.
Wyman, J. 1850 B, p. 233.
Eocene; Europe, New Zealand, South Carolina, Virginia.

Otodus obliquus Agassiz.

Agassiz, L. 1843 B, iii, p. 267, pl. xxxi; pl. xxxvi, figs. 22–27.
1843 B, iii, p. 269, pl. xxxvii, figs. 19–23. (O. lanceolatus.)
Bassani, F., 1899 A, p. 18, pl. i, figs. 32–35. (Lamna.)
Clark, W. B. 1895 A, p. 4. (Lamna? obliqua.)
1897 A, p. 61. (Lamna? obliqua.)
Cope, E. D. 1867 C, p. 142.

Dames, W. 1883 B, p. 145, pl. iii, fig. 6.
Davis, J. W. 1890 A, p. 407, pl. xii, fig. 13.
Dixon, F. 1850 A, p. 204, pl. x, figs. 32–35; pl. xv, fig. 11.
1878 A, p. 249, pl. x, figs. 32–35; pl. xv, fig. 11.
Gibbes, R. W. 1846 A, p. 42.
1849 A, p. 199, pl. xxvi, figs. 131–137.
1850 F.
Geinitz, H. B. 1883 A, p. 6, pl. i, figs. 12–18.
Morton, S. G. 1834 A, pl. xi, fig. 1 (Lamna obliqua); fig. 5 (Lamna lanceolata, *fide* Agassiz).
1835 A, p. 277. (Lamna lanceolata.)
1842 A, p. 220. (Lamna lanceolata.)
Noetling, F. 1885 A, p. 84, pl. vi, figs. 4–6. (Carcharodon.)
Woodward, A. S. 1884 A, p. 229. (Otodus.)
1889 D, p. 404. (Lamna? obliqua.)
1899 A, p. 10.
Wyman, J. 1850 B, p. 234. (O. obliquus and O. lanceolatus.)
Upper Cretaceous and Eocene; Europe and Egypt: Eocene; Virginia, New Jersey: Miocene; Virginia, North Carolina (*fide* Cope).

Otodus trigonatus Agassiz.

Agassiz, L. 1843 B, iii, p. 274, pl. xxxvi, figs. 35–37.
?Gibbes, R. W. 1849 A, p. 200, pl. xxvi, figs. 145, 146.
Woodward, A. S. 1889 D, p. 401. (Lamna.)
1899 A, p. 10, pl. i, figs. 23, 24.
Upper Eocene; Bavaria, South Carolina (?).

ISURUS Rafinesque. Type *I. oxyrhynchus* Raf. = *Oxyrhina spallanzanii* Bonap.

Rafinesque, C. S. 1810, Caratteri di alcuni nuovi generi, p. 11.
Unless otherwise stated, this genus is referred to by the following writers under the name *Oxyrhina*. This name is antedated by that of Rafinesque, *Isurus*.
Agassiz, L. 1843 B (1838), iii, pp. 86, 276, 307 (Oxyrhina, type O. spallanzanii).
1844 A.
Davis, J. W. 1890 A, pp. 367, 371.
Duméril, A. 1865 A, p. 407.
Eastman, C. R. 1896 A.
Gibbes, R. W. 1849 A, p. 201.
Hasse, C. 1882 A, p. 230, pl. xxxi.
Jordan and Evermann, 1896 A, p. 47. (Isurus.)
Kölliker, A. 1863 A, p. 11, figs. 4, 5.
Leidy, J. 1873 B, p. 308.
Müller and Henle, 1837 A, p. 114.
1837 B, p. 397.
1841 A, p. 681.
Nicholson and Lydekker, 1889 A, p. 944.
Pictet, F. J. 1854 A, p. 247.
Pictet and Campiche, 1858 A, p. 83.
Probst, J. 1879 A, p. 128.
Quenstedt, F. A. 1852 A, p. 172.
1885 A, p. 270.
Steinmann and Döderlein, 1890 A, p. 547.
Woodward, A. S. 1884 A, p. 229.
1889 D, p. 376.
Zittel, K. A. 1890 A, p. 81.

Isurus crassus (Agassiz).

Agassiz, L. 1843 B, iii, p. 283, pl. xxxvii, fig. 16 (not pl. xxxiv, fig. 14). (Oxyrhina.)

Eastman, C. R. 1896 A, p. 182. (Oxyrhina.)
Gemmellaro, G. G. 1857 A, p. 317, pl. i, fig. 11. (Oxyrhina.)
Gibbes, R. W. 1849 A, p. 202, pl. xxvii, figs. 159, 160. (Oxyrhina.)
Woodward, A. S. 1889 D, p. 389. (Oxyrhina.)
1894 B. (Oxyrhina.)
Eocene; South Carolina, Alabama.

Isurus desorii (Agassiz).

Agassiz, L. 1843 B, iii, p. 282, pl. xxxvii, figs. 8–13. (Oxyrhina.)
1843 B, iii, p. 282, pl. xxxvii, figs. 3–5. (Oxyrhina leptodon.)
Bassani, F. 1899 A, p. 13, pl. ii, figs. 24–38. (Oxyrhina.)
Cope, E. D. 1867 C, p. 142. (Oxyrhina.)
1875 V, p. 29. (Oxyrhina.)
Eastman, C. R. 1895 A, p. 180. (Oxyrhina.)
Emmons, E. 1858 B, p. 236, fig. 67. (Oxyrhina.)
Gemmellaro, G. G. 1857 A, p. 314, pl. vi *a*, figs. 9–11 (O. leptodon); pl. vi *a*, figs. 12, 13 (O. desorii); p. 319, pl. vi *a*, fig. 17 (Lamna lyellii).
Gibbes, R. W. 1847 A, p. 267. (Oxyrhina.)
1849 A, p. 203, pl. xxvii, figs. 169–171 (O. desorii); p. 203, pl. xxvii, figs. 172, 173 (O. wilsonii).
1850 F. (Oxyrhina.)
Noetling, F. 1885 A, p. 50, pl. iii. (O. xiphodon.)
Probst, J. 1879 A, p. 131, pl. ii, figs. 7–13 (O. desorii); p. 141, pl. ii, figs. 69, 75 (Alopecias gigas).
Sismonda, E. 1849 A, p. 40, pl. ii, figs. 7–16. (Oxyrhina.)

Woodward, A. S. 1889 D, p. 382. (Oxyrhina.)
Tertiary; Europe and United States: Eocene;
Alabama, North Carolina, South Carolina:
Miocene; Maryland (*fide* Cope).
For additional references to descriptions of
foreign specimens see Woodward, A. S. 1889 D,
pp. 382, 383.

Isurus hastalis (Agassiz).

Agassiz, L. 1843 B, iii, p. 277, pl. xxxiv, excl. figs.
1, 2, 14. (Oxyrhina.)
1843 B, iii, p. 278, pl. xxxiii, figs. 11–17 (O.
xiphodon); p. 279, pl. xxxvii, figs. 17, 18
(O. trigonodon); p. 279, pl. xxxvii, figs.
14, 15 (O. plicatilis); p. 281, pl. xxxiii, fig.
10 (O. retroflexa); p. 281, pl. xxxvii, figs.
1, 2 (O. quadrans).
Alessandri, G. de 1896 A, p. 263, pl. i. (Oxyrhina.)
Clarke, W. B. 1895 A, p. 4. (Oxyrhina.)
1897 A, p. 61. (Oxyrhina.)
Cope, E. D. 1867 C, p. 142. (O. hastalis, O. xiphodon.)
1875 H, p. 362. (O. xiphodon.)
Eastman, C. R. 1895 A, p. 178. (Oxyrhina.)
Emmons, E. 1858 B, p. 236, figs. 64–66. (O. xiphodon, O. hastilis.)
Gemmellaro, G. G. 1857 A, p. 312, pl. vi, fig. 5
(O. hastalis); p. 313, pl. vi a, figs. 6–8 (O. xiphodon).
Gibbes, R. W. 1846 A, p. 43 (O. xiphodon, O. hastalis); p. 42 (O. retroflexa).
1849 A, pp. 197, 201, pl. xxvi, figs. 148–152
(O. hastalis); p. 201, pl. xxvii, figs. 153, 154
(O. xiphodon); p. 202, pl. xxxvii, figs. 155–
157 (O. plicatilis).
1850 F, p. 70.
Giebel, C. G. 1848 A, p. 357. (Oxyrhina.)
Hitchcock, E., 1835, Geol. Mass., pl. xiii, fig. 37.
(No name or description.)
Leidy, J. 1877 A, p. 254. (O. xiphodon.)
Lyell, Chas. 1843 C, p. 32. (Oxyrhina.)
1845 A, p. 415. (Lamna xiphodon.)
Probst, J. 1879 A, p. 129, pl. ii, figs. 1–6 (O. hastalis); p. 132, pl. ii, figs. 14–19 (O. xiphodon).
Quenstedt, F. A. 1885 A, p. 270, fig. 84. (Oxyrhina.)
Sauvage, H. E. 1875 A, p. 633. (O. xiphodon.)
Sismonda, E. 1849 A, p. 40, pl. i, figs. 41–47 (O. hastalis); p. 42, pl. i, figs. 48–50 (O. plicatilis); figs.
51, 52 (O. xiphodon); p. 43, pl. ii, figs. 1–6 (O. isocelica).
Tuomey, M. 1858 A, p. 268. (O. hastata, O. xiphodon.)
Woodward, A. S. 1889 D, p. 385. (Oxyrhina.)
1900 A, p. 4, pl. i, figs. 6–8. (Oxyrhina.)
Wyman, J. 1850 B, p. 234. (O. hastalis, O. xiphodon.)
Tertiary; Europe, Australia, New Zealand,
and United States: Eocene; Alabama, South
Carolina: Miocene; Virginia, Maryland, South
Carolina, North Carolina: Pliocene?; Argentine
Republic.
For additional references to foreign specimens
see Woodward, A. S. 1889 D, pp. 385, 386, and Eastman, C. R. 1895 A, p. 178.

Isurus mantellii (Agassiz).

Agassiz, L. 1843 B, iii, p. 280, pl. xxxiii, figs. 1–5,
7–9. (Oxyrhina.)
1843 B, iii, p. 292, pl. xxxvii a, fig. 54. (Lamna acuminata.)
Cope, E. D. 1875 E, p. 296. (Oxyrhina extenta.)
Credner, H. 1870 A, p. 242. (Oxyrhina.)
Dana, J. D. 1896 A, p. 843, fig. 1405. (Oxyrhina.)
Davis, J. W. 1890 A, p. 391, pl. xxxix, figs. 1–7.
(Oxyrhina.)
Dixon, F. 1850 A, pl. xxx, fig. 24 (O. mantelli);
pl. xxx, fig. 19 (Lamna acuminata).
Eastman, C. R. 1895 A, pls. xvi–xviii. (Oxyrhina.)
Fritsch, A. 1878 A, p. 7, fig. 12. (Oxyrhina.)
Geinitz, H. B. 1875 A, p. 207, pl. xxxviii, figs. 1–21.
Gibbes, R. W. 1842 A, p. 43. (Oxyrhina.)
1849 A, p. 197, pl. xxv, figs. 113–115 (Lamna acuminata); p. 202, pl. xxvii, fig. 158
(Oxyrhina).
Giebel, C. G. 1848 A, p. 357. (Oxyrhina.)
Hilgard, E. W. 1860 A, p. 389. (Oxyrhina.)
Leidy, J. 1872 M, p. 162. (Oxyrhina extenta.)
1873 B, pp. 302, 351, pl. xviii, figs. 21–25.
(Oxyrhina extenta.)
Morton, S. G. 1834 A, pl. xi, fig. 4. (Lamna mantelli, *fide* Agassiz.)
1835 A, p. 277. (L. mantelli, L. acuminata.)
1842 A, p. 219. (Lamna mantelli and L. acuminata.)
Newberry, J. S. 1876 A, p. 141. (Oxyrhina.)
Quenstedt, F. A. 1852 A, p. 172, pl. xiii, fig. 14.
(Oxyrhina.)
1885 A, p. 270, pl. xx, fig. 37. (Oxyrhina.)
Reuss, A. E. 1846 A, p. 5, pl. iii, figs. 1–6. (Oxyrhina.)
Roemer, F. 1849 H, p. 419. (Oxyrhina.)
1852 A, p. 29, pl. i, figs. 6 a, b. (Oxyrhina.)
Sauvage, H. E. 1870 A, p. 21, figs. 39–41 (Oxyrhina); p. 24, figs. 54–56 (Otodus oxyrhinoides).
Williston, S. W. 1900 A, p. 36, pl. xiii, figs. 41–46;
pl. xiv, figs. 2–2m. (Oxyrhina.)
1900 B, p. 246, pl. xxxi, figs. 41–46; pl. xxxii,
figs. 2–2m.
Woodward, A. S. 1884 A, p. 229, fig. 136. (Oxyrhina.)
1889 D, p. 376. (Oxyrhina.)
Cretaceous; Europe, Alabama, New Jersey,
Virginia, Mississippi, Kansas, Texas.

Isurus minutus (Agassiz).

Agassiz, L. 1843 B, iii, p. 285, pl. xxxvi, figs. 39–47.
(Oxyrhina.)
Cope, E. D. 1867 C, p. 142. (Oxyrhina.)
1875 H, p. 362. (Oxyrhina.)
1875 V, p. 29. (Oxyrhina.)
Eastman, C. R. 1895 A, p. 182. (Oxyrhina.)
Gemmellaro, G. G. 1857 A, p. 316, pl. via fig. 14.
(Oxyrhina.)
Gibbes, R. W. 1849 A, p. 202, pl. xxvii, figs. 161–163;
not fig. 164. (Oxyrhina.)
Sismonda, E. 1849 A, p. 44, pl. ii, figs. 36–39.
Woodward, A. S. 1889 D, p. 391. (Oxyrhina.)
Eocene; South Carolina, Maryland, New Jersey, North Carolina: Miocene; North Carolina
(*fide* Cope).
It is not improbable that the identification of
the American specimens assigned to this species
is wrong. See *Carcharhinus gibbesii*.

Isurus planus (Agassiz).

Agassiz, L. 1856 A, p. 274.   (Oxyrhina.)
   1857 A, p. 315, pl. i, figs. 29, 30.   (Oxyrhina.)
Eastman, C. R. 1895 A, p. 183.   (Oxyrhina.)
Woodward, A. S. 1889 D, p. 391.   (Oxyrhina.)
   Miocene; California.

Isurus sillimanii (Gibbes).

Gibbes, R. W. 1847 A. p. 268.   (Oxyrhina.)
Cope, E. D. 1867 C, p. 142.   (Oxyrhina.)
   1875 V, p. 29.   (Oxyrhina.)
Eastman, C. R. 1895 A, p. 181.   (Oxyrhina.)
Gibbes, R. W.. 1849 A, p. 202, pl. xxvii, figs. 165-168.
   (Oxyrhina.)
   1850 F.   (Oxyrhina.)

Woodward, A. S. 1899 D, p. 391.   (Oxyrhina.)
   Eocene; South Carolina: Miocene; Maryland, North Carolina (fide Cope).

Isurus tumulus (Agassiz).

Agassiz, L. 1856 A, p. 275.   (Oxyrhina.)
   1857 A, p. 316, pl. i, figs. 26, 27, 36, 37, 42-44.
   (Oxyrhina.)
Eastman, C. R. 1895 A, p. 183.   (Oxyrhina.)
Woodward, A. S. 1889 D, p. 391.   (Oxyrhina.)
   Miocene; California.

Isurus sp. indet.

Cope E. D. 1875 E, p. 296.
   1875 F, p. 96.

CARCHARODONTINÆ.

Gill, T. 1893 A, p. 130.

Jordan and Evermann, 1896 A, p. 47.

CARCHARODON A. Smith.   Type *Squalus carcharias* Linn.

Smith, A. in Müller and Henle 1838 A, p. 87.
Agassiz L., 1843 B. iii, pp. 245, 307.
   1844 A.
Davis, J. W. 1888 A, p. 8.
Duméril, A. 1865 A, p. 410.
Gibbes, R. W. 1848 B, p. 142.
   1849 A, p. 191.
Günther, A. 1870 A, p. 391.
   1880 A, p. 320.
Hasse, C. 1882 A, p. 224.
Haswell, W. A. 1884 A, p. 83.
Müller & Henle 1841 A, p. 70.
Kölliker, A. 1863 A, p. 15, figs. 8, 9.
Nicholson & Lydekker 1889 A, p. 945.
Parker, T. J. 1887 A.
Pedroni, P. M. 1844 A, p. 285.
Pictet, F. J. 1854 A, p. 237.
Probst, A. 1879 A, p. 137.
Sismonda, E. 1849 A, p. 33.
Steinmann & Döderlein 1890 A, p. 548.
Woodward, A. S. 1884 A, p. 229.
   1889 D, p. 410.
Zittel, K. A. 1890 A, p. 84.

Carcharodon auriculatus (Blainville).

Blainville, H. M. D., 1818, Nouv. Dict. d'Hist. Nat.,
   xxvii, p. 384.   (Squalus.)
Agassiz, L. 1843 B, iii, p. 254, pl. xxviii, figs. 17-19
   (C. auriculatus); p. 255, pl. xxviii, figs. 20-25
   (C. angustidens); pl. xxx, fig. 3 (C. lanceolatus,
   C. angustidens in text); p. 256, pl. xxxa, figs. 8, 9
   (C. turgidus); p. 257, pl. xxx, fig. 1 (C. lanceolatus); p. 257, pl. xxxa, fig. 14 (C. toliapicus): p.
   258 pl. xxviii, figs. 11-16 (C. heterodon); p. 258,
   pl. xxviii, figs. 8-10 (C. megalotis); p. 259, pl.
   xxviii, fig. 7 (C. disauris).
Bassani, F., 1899 A. p. 22 pl. i, figs. 30-39.
Cope, E. D. 1867 C, p. 142.   (C. angustidens.)
   1875 E, p. 295.   (C. angustidens.)
   1875 H, p. 362.   (C. angustidens.)
   1875 V, p. 29.   (C. angustidens.)
Dana, J. D., 1896 A, p. 901, fig. 1513.   (C. angustidens.)
Davis, J. W. 1880 A, p. 9, pl. i, figs. 4-6; pl. iv, fig.
   22.   (C. angustidens.)

Dixon, F 1850 A, p. 204, pl. xi, fig. 19.   (C. heterodon.)
   1858 A, p. 249, pl. xi, fig. 19.   (C. heterodon.)
Emmons, E. 1858 B, p. 231, fig. 57; p. 233.   (C. crassidens, Emmons, non Sismonda.)
Gemmellaro, G. G. 1857 A, p. 304, pl. v, figs. 6a-8.
   (C. angustidens and var. turgidus.)
Gibbes, R. W. 1846 A, p. 42.   (C. auriculatus, angustidens, heterodon, megalotis, lanceolatus, semiserratus, turgidus.)
   1847 A, p. 267.   (C. acutidens.)
   1848 B, pp. 145, 146, pls. xix-xxi, figs. 10-38.
   (C. angustidens, with varieties auriculatus, heterodon, megalotis, lanceolatus,
   semiserratus, toliapicus, and turgidus.)
   1848 B, p. 146, pl. xxi, figs. 39-44.   (C. acutidens.)
   1850 F.   (C. acutidens, angustidens.)
Giebel, C. G. 1848 A, p. 350 (C. angustidens, C. auriculatus); p. 351 (C. heterodon, lanceolatus,
   toliapicus, turgidus).
Hilgard, E. W. 1860 A, pp. 132, 142.   (C. angustidens.)
Leidy, J. 1876 A, p. 222.   (C. angustidens.)
   1876 E.   (C. angustidens.)
   1877 A, p. 253.   (C. angustidens.)
Morton, S. G. 1834 A, pl. xii, figs. 3, 4, 5.   (Carcharias
   lanceolatus and C. megalotis, fide Agassiz.)
   1835 A, p. 277.   (C. megalotis.)
   1842 A, p. 220.   (Carcharias myalotis.)
Noetling, F. 1885 A, p. 82, pl. vi, figs. 1-3.   (C. angustidens.)
Pedroni, P. M. 1844, A, p. 285.
Rösé, C. 1897 A, p. 35, figs. 5, 6.   (C. angustidens.)
Tuomey, M. 1858 A, p. 268.   (Carcharias angusti-
   dens, auriculatus, heterodon.)
Wailes, B. L. C. 1854 A, p. 277.   (C. angustidens.)
Woodward,.A. S. 1884 A, p. 229, fig. 141.   (C. auricu-
   latus.)
   1889 D, p. 411.   (C. auriculatus.)
Wyman, J. 1850 B, p. 231.   (C. angustidens.)
   Tertiary; Europe, North Africa, Arabia, Vic-
   toria, New Zealand, North America: Eocene;
   Alabama, Mississippi, North Carolina, South
   Carolina: Miocene; Maryland, Virginia.

Carcharodon carcharias (Linn.).

Linnæus, 1758, Syst. Nat., ed. x, p. 235. (Squalus carcharias.)
Agassiz, L. 1843 B, iii, p. 254, pl. xxxa, figs. 5-7. (C. sulcidens.)
Davis, J. W. 1890 A, p. 410, pl. xii, fig. 14. (C. rondeletii.)
Duméril, A. 1865 A. p. 411, pl. vii, fig. 7.
Emmons, E. 1858 B, p. 230, figs. 55, 56. (C. sulcidens.)
Gibbes, R. W. 1846 A, p. 42. (C. sulcidens.)
    1848 A, p. 350. (C. sulcidens.)
    1848 B, p. 147, pl. xxi, figs. 52, 53. (C. sulcidens.)
Giebel, C. G. 1848 A, p. 350. (C. sulcidens.)
Günther, A. 1880 A, p. 320, fig. 114. (C. rondeletii.)
Hasse, C. 1882 A, p. 228, pl. xxxi. (C. rondeletii.)
Jordan and Evermann 1896 A, p. 50. (C. carcharias.)
Kölliker, A. 1863 A, p. 15. (C. rondeletii.)
Müller and Henle, 1841 A, p. 70. (C. rondeletii.)
Stevenson, W. B. 1884 A, with 1 plate and 1 woodcut.
Woodward, A. S. 1884 A, p. 230. (C. sulcidens.)
    1889 D, p. 420. (C. rondeletii.)
    Recent in all temperate and tropical seas: Pliocene; Europe: Miocene and Eocene; South Carolina, North Carolina.

Carcharodon contortidens Emmons.

Emmons, E. 1858 B, p. 233, fig. 60.
    Doubtful species.
    Eocene: North Carolina.

Carcharodon lanciformis Gibbes.

Gibbes, R. W. 1847 A, p. 267.
    1848 B, p. 146, pl. xxi, figs. 46-51.
    1850 F.
Leidy, J. 1877 A, p. 254.
Woodward, A. S. 1889 D, p. 411.
    Phosphate beds; South Carolina.

Carcharodon leptodon Agassiz.

Agassiz, L. 1843 B, iii, p. 259, pl. xxviii, figs. 1-6.
?Gibbes, R. W. 1846 A, p. 42.
?1850 G, p. 300, pl. xlii, figs. 7-9.
Woodward, A. S. 1889 D, p. 421.
    Eocene; South Carolina.
    The identification of the South Carolina specimens with Agassiz's C. leptodon is doubtful.

Carcharodon megalodon Agassiz.

Agassiz, L., in Charlesworth, E. 1837 A, with woodcut.
    1843 B, iii, p. 247, pl. xxix (C. megalodon); p. 250, pl. xxxa, fig. 10 (C. rectidens); 251, pl. xxxa, figs. 11-13 (C. subauriculatus); p. 251, pl. xxx, figs. 2, 4, 6, 7, 8 (C. productus); p. 253, pl. xxx, figs. 9-12 (C. polygyrus).
Bowerbank, J. S. 1852 A. (Carcharias megalodon.)
    1852 B.
Clark, W. B. 1895 A, p. 4. (C. polygyrus.)
    1897 A, p. 61. (C. polygyrus.)

Cope, E. D. 1867 C, p. 142.
    1875 H, p. 362.
    1875 V, p. 29.
Davis, J. W. 1888 A, p. 12, pl. ii, figs. 1-3.
DeKay, J. E. 1842 B, p. 386. (Carcharias polygyrus)
Emmons, E. 1858 B, p. 227, figs. 50, 51; p. 229, figs. 52-54 (C. ferox); p. 232, fig. 59 (C. triangularis).
Gemmellaro, G. G. 1857 A, p. 299, pl. ii, figs. 3a-3c; pl. iii, figs. 1a, 2a, 3a; pl. iv, figs. 1a-3a (C. megalodon); p. 307, pl. iv, fig. 4 (C. latissimus); p. 308 pl. v, figs. 3-5 (C. productus).
Gibbes, R. W. 1846 A, p. 42 (C. megalodon, rectidens); p. 42 (C. polygyrus, productus, mortoni.)
    1847 A, p. 266. (C. mortoni.)
    1848 B, p. 143, pls. xviii, xix, figs. 1-9 (C. megalodon); p. 146, pl. xxi, fig. 45 (C. mortoni.)
    1850 F.
Giebel, C. G. 1848 A, p. 348 (C. megalodon); p. 349 (C. productus, rectidens); p. 350 (C. auriculatus, polygyrus).
Hilgard, E. W. 1860 A, p. 142.
Leidy, J. 1876 A, p. 222.
    1876 E.
    1876 F.
    1877 A, p. 253.
Lucas, F. A. 1892 A, p. 151.
Lyell, C. 1843 C, p. 32. (Carcharias.)
    1845 A, p. 415. (C. megalodon, productus).
Morton, S. G. 1834 A, pl. xii, fig. 2. (Carcharias polygyrus, *fide* Agassiz.)
    1835 A, p. 277. (Carcharias polygyrus.)
    1842 A, p. 220. (Carcharias polygyrus.)
Pedroni, P. M. 1844 A, p. 286, pl. i, figs. 24, 25.
Pictet, F. J. 1854 A, p. 238.
Probst, J. 1879 A, p. 138.
Quenstedt, F. A. 1852 A, p. 169. (Carcharias verus.)
    1885 A, p. 206, fig. 82. (Carcharias verus.)
Sismonda, E. 1849 A, p. 34, pl. i, figs. 8-13 (C. megalodon); p. 35, pl. i, figs. 32, 33 (C. crassidens); p. 37, pl. i, figs. 25-29 (C. productus).
Tuomey, M. 1858 A, p. 268. (Carcharias polygyrus, productus.)
Wailes, B. L. C. 1854 A, p. 276.
Woodward, A. S. 1884 A, p. 229.
    1889 D, p. 415 (C. megalodon); p. 422 (C. mortoni).
    1900 A, p. 4, pl. i, fig. 9.
Zittel, K. A. 1890 A, p. 84, fig. 87.
    Tertiary; Europe, Arabia, East Indies, West Indies, North and South America, Central America, Australia, New Zealand; Eocene and phosphate beds; South Carolina, Alabama, Mississippi, Maryland, Massachusetts, North Carolina, Virginia: Pliocene ; Argentine Republic.

Carcharodon rectus Agassiz.

Agassiz, L. 1856 A, p. 274.
    1857 A, p. 315, pl. i, figs. 39-41.
Bowers, S. 1889 A.
Woodward, A. S. 1889 D, p. 422.
    Miocene; California.

CORAX Agassiz. Type *Galeus pristodontus* Agassiz.

Agassiz, L. 1843 B, iii, p. 224.
Davis, J. W. 1890 A, pp. 367, 411.
Kiprijanoff, V. 1853 A, p. 290.
Pictet, F. J. 1854 A., p. 240.
Pictet and Campiche 1858 A, p. 79.
Priem. F. 1897 A, p. 47.
Reuss, A. E. 1846 A, p. 3.
Williston, S. W. 1900 A, p. 41.
1900 B, p. 252.
Woodward, A. S. 1884 A. p. 227.
1889 D, p. 422.
Zittel, K. A. 1890 A, p. 84.

Corax affinis Agassiz.

Agassiz, L. 1843 B, iii, p. 227, pl. xxvi, fig. 2 (C. affinis); pl. xxvia, figs. 21–24. (C. appendiculatus.)
1843 B, iii, p. 229, pl. xxvia, figs. 51, 57. (C. planus.)
Giebel, C. G. 1848 A, p. 349. (Carcharodon minor.)
Hébert, E. 1855 C, p. 354, pl. xxvii, fig. 9. (Sphyrna plana.)
Woodward, A. S. 1889 D, pp. 427, 428.
Upper Cretaceous; Europe. A specimen in the Enniskillen collection from the Eocene of Alabama may belong to this species.

Corax curvatus Willist.

Williston, S. W. 1900 A, p. 41, pl. xii, figs. 7, 8.
1900 B, p. 253, pl. xxx, figs. 7, 8.
Cretaceous (Benton); Kansas.

Corax falcatus Agassiz.

Agassiz, L. 1843 B, iii, p. 226, pl. xxvi, fig. 14; pl. xxvia, figs. 1–15. (On pl. xxvi as Galeus pristodontus.)
Cope, E. D. 1875 E, p. 295. (Galeocerdo.)
Credner, H. 1870 A, p. 242. (Heterodon.)
Dana, J. D. 1896 A, p. 843, fig. 1408. (C. heterodon.)
Dixon, F. 1850 A, pl. xxx, fig. 18.
Fritsch, A. 1878 A, p. 11, figs. 23, 24. (C. heterodon.)
Geinitz, H. B. 1839 A, p. 11, pl. i, fig. 1 (Galeus pristodontus); p. 11, pl. i, fig. 2. (G. appendiculatus.)
1875 A, p. 210, pl. xl, figs. 2–15. (C. heterodon.)
Giebel, C. G. 1848 A, p. 370. (C. heterodon.)
Hébert, E. 1855 C, p. 354, pl. xxvii, fig. 9. (Sphyrna plana.)
Kiprijanoff, V. 1853 A, p. 293, pl. ii, figs. 1a–1c. (C. heterodon.)
1854 A, p. 293, pl. ii. (C. heterodon.)

Leidy, J. 1873 A, pp. 301, 351, pl. xviii, figs. 29–43.
Pictet and Campiche 1858 A, p. 80, pl. x, figs 1, 2.
Osborn, Scott and Speir 1878 A, p. 105. (Galeocerdo.)
Reuss, A. E. 1846 A, p. 3, pl. iii, figs. 49–71 (C. heterodon); p. 4, pl. iv, fig. 1–3 (C. obliquus).
Roemer, F. 1849 A, p. 419. (C. heterodon.)
1852 A, p. 30, pl. i, fig. 8. (C. heterodon.)
Sauvage, H. E. 1870 A, p. 40, pl. ii, figs. 84, 85.
Tuomey, M. 1858 A, p. 261.
Williston, S. W., 1900 A, p. 41, pl. xiii, figs. 1–40, pl. xiv, figs. 1–11.
1900 B, p. 252, pl. xxxi, figs. 1–40; pl. xxxii, figs. 1–11.
Woodward, A. S. 1884 A, p. 227, fig. 134.
1889 D, p. 424.
Cretaceous; Europe, Alabama, Mississippi, Texas, Kansas, Colorado, New Jersey.

Corax pristodontus Agassiz.

Agassiz, L. 1843 B, iii, p. 224, pl. xxvi, figs. 9–13. (On plate as Galeus pristodontus.)
In Morton, S. G. 1835 A, p. 277. (Galeus pristodontus; no description.)
1843 B, iii, p. 225, pl. xxvi, figs. 4–8; pl. xxvi a, figs. 25–34. (C. kaupii.)
Cope, E. D. 1867 C, p. 141. (Galeocerdo appendiculatus; reference uncertain as to species.)
1875 E, p. 295. (Galeocerdo.)
1877 K, pp. 11, 38. (Galeocerdo.)
De Kay, J. E. 1842 B, p. 387. (Galeus.)
Emmons, E. 1856 B, p. 238, fig. 69. (Galeocerdo.)
Gibbes, R. W. 1846 A, p. 42.
1849 A, p. 192, pl. xxv, fig. 70. (Galeocerdo.)
1850 F. (Galeocerdo.)
Giebel, C. G. 1848 A, p. 370.
Harlan, R. 1835 B, p. 292. (Squalus.)
Hébert, E. 1855 C, p. 353, pl. xxvii, fig. 8.
Morton, S. G. 1834 A, pl. xi, fig. 6. (No description.)
1842 A, p. 219. (Galeus.)
Pictet, F. J. 1854 A, p. 240.
Priem, F. 1897 A, p. 45, pl. i, figs. 18, 19.
1898 A, p. 236, pl. ii, fig. 5.
Spillman, W., in Hilgard, E. W. 1860 A, p. 389. (C. appendiculatus.)
Tuomey, M. 1858 A, p. 261. (Galeus.)
Woodward, A. S. 1884 A, p. 227, fig. 132.
1888 I, p. 293.
1889 D, p. 423.
Zittel, K. A. 1890 A, p. 84, fig. 88.
Upper Cretaceous; Europe, India, Africa, New Jersey; Eocene; Alabama, Mississippi, South Carolina; Miocene; Virginia.

XENODOLAMIA Leidy. Type *X. pravus* Leidy.

Leidy, J. 1877 A, p. 251.
Woodward, A. S. 1889 D, p. 429.

Xenodolamia pravus Leidy.

Leidy, J. 1877 A, p. 251, pl. xxxiv, figs. 33, 34.
Woodward, A. S. 1889 D, p. 429.
Phosphate beds; South Carolina.

Xenodolamia simplex Leidy.

Leidy, J. 1877 A, p. 251, pl. xxxiv, figs. 35, 36.
Woodward, A. S. 1889 D, p. 429.
Phosphate beds; South Carolina.

SCYLLIORHINIDÆ.

Gill, T. 1862 A, p. 406. (Scylliorhinoidæ.)
Duméril, A. 1865 A, p. 312. (Scyllia.)
Gegenbaur, C. 1865 B. (Scylliidæ.)
Günther, A. 1870 A, p. 400. (Scylliidæ.)
1880 A, p. 325. (Scylliidæ.)
Hasse, C. 1878 A. (Scyllia.)
1879 A, p. 52. (Scyllia.)
1882 A, p. 242. (Scyllia.)
Haswell, W. A. 1884 A, p. 115. (Scylliidæ.)
Jaekel, O. 1890 A, p. 94. (Scylliidæ.)
1894 A, p. 159. (Scylliidæ.)

Jordan and Evermann 1896 A, p. 22 (Scylliorhinidæ); p. 25 (Ginglymostomidæ).
Kölliker, A. 1863 A, p. 27. (Scyllium.)
Mayer, P. 1886 A, p. 274. (Scylliidæ.)
Müller and Henle 1841 A, p. 3. (Scyllia.)
Probst, J. 1879 A, p. 169. (Scylliidæ.)
Woodward, A. S. 1889 D, p. 338.
Zittel, K. A. 1890 A (1887), p. 79. (Scylliidæ.)

SCYLLIORHININÆ.

Gill, T. 1862 A, p. 407.

SCYLLIORHINUS Blainville.   Type *S. canicula* Cuvier.

Blainville, H. M. D., 1816, Jour. de Physique, p. 263.
Cuvier, 1817, Règne Anim., ii, p. 124. (Scyllium.)
Günther, A. 1870 A, p. 400. (Scyllium.)
Jordan and Evermann 1896 A, p. 22.
Woodward, A. S. 1889 D, p. 340. (Scyllium.)

Scylliorhinus? gracilis Willist.

Williston, S. W. 1900 A, p. 35, pl. vi, fig. 6. [Scyllium (Lamna?).]
1900 B, p. 245, pl. xxiv, fig. 6. [(Scyl. Lamna)].
Cretaceous (Greenleaf sandstone); Kansas.

Scylliorhinus planidens Willist.

Williston, S. W. 1900 A, p. 35, pl. vi, fig. 7. (Scyllium.)
1900 B, p. 245, pl. xxiv, fig. 7.
Cretaceous (Greenleaf sandstone); Kansas.

Scylliorhinus rugosus Willist.

Williston, S. W. 1900 A, p. 35, pl. vi, fig. 5. (Scyllium.)
1900 B, p. 245, pl. xxiv, fig. 5.
Cretaceous (Greenleaf sandstone); Kansas.

GINGLYMOSTOMINÆ.

Gill, T. 1890 A, p. 130. (Ginglymostominæ.)
1862 A, p. 406. (Ginglymostomoidæ.)
Günther, A. 1870 A, p. 407. (Scylliidæ, in part.)
Hasse, C. 1879 A, p. 51. (Scylliolamnidæ, in part.)
1882 A, p. 193. (Scylliolamnidæ, in part.)

Jordan and Evermann, 1896 A, p. 25. (Ginglymostomidæ.)

GINGLYMOSTOMA Müller and Henle.   Type *Squalus cirratus* Gmelin.

Müller and Henle, 1837 B, p. 396.
Duméril, A. 1865 A, p. 333.
Günther, A. 1870 A, p. 407.
1880 A, p. 326.
Hasse, C. 1879 A, p. 71.
1882 A, pp. 201, 203, pl. xxv.
Klunzinger, C. B. 1871 A, p. 670.
Kölliker, A. 1863 A, p. 5, fig. 1.
Leidy J. 1877 A, P. 250. (Acrodobatis, type *A. serra.*)
Müller and Henle, 1838 A, p. 35.
1841 A, p. 22.
Woodward, A. S. 1889 D, p. 348.
Zittel. K. A. 1890 A, p. 80.

Ginglymostoma obliquum (Leidy).

Leidy, J. 1877 A, p. 250, pl. xxxiv, fig. 14. (Acrodobatis.)
Woodward, A. S. 1889 D, p. 348. (G. serra?)
Eocene; New Jersey.

Ginglymostoma serra (Leidy).

Leidy, J. 1877 A, p. 250, pl. xxxiv, figs. 10-13. (Acrodobatis.)
Woodward, A. S. 1889 D, p. 348.
Eocene; South Carolina.

GALEIDÆ.

Duméril, A. 1865 A, p. 388. (Galei.)
Gegenbaur, C. 1865 B. (Carchariidæ.)
Gill, T. 1862 A, p. 399. (Galeorhinoidæ.)
Günther, A. 1870 A, p. 353. (Carchariidæ.)
1880 A, p. 316. (Carchariidæ.)
1881 A, p. 685. (Carchariidæ.)
Hasse, C. 1882 A, p. 256 ("Fam. Hemigaleus"); p. 268 (Carchariidæ).
Jaekel, O. 1894 A, p. 156. (Carchariidæ.)

Jordan and Evermann 1896 A, p. 27. (Galeidæ.)
Mayer, P. 1886 A, p. 275. (Carchariidæ.)
Müller and Henle 1841 A, p. 7 (Carchariæ); p. 57 (Galei).
Steinmann and Döderlein 1890 A, p. 546. (Carchariidæ.)
Woodward, A. S. 1884 A, p. 227 (Carchariidæ); 1889 D, p. 435 (Carchariidæ.)
Zittel, K. A. 1890 A, p. 85. (Carchariidæ.)

GALEORHININÆ.

Gill, T. 1862 A, pp. 399, 401.
Günther, A, 1870 A, p. 358.   (Carchariina, in part.)

Jordan and Evermann 1896 A, p. 27.

GALEOCERDO Müller and Henle.   Type *G. tigrinus* M. and H.

Müller and Henle 1838 A, p. 36.
Agassiz, L. 1843 B, iii. pp. 230, 304.
Davis, J. W. 1888 A, p. 7.
Duméril, A. 1865 A, p. 393.
Gibbes, R. W. 1849 A, p. 191.
Gill, T. 1862 A, p. 400.
Günther, A. 1870 A, p. 377.
     1880 A, p. 317.
Hasse, C. 1882 A, p. 259, pls. xxxvi, xxxvii.
Jaekel, O. 1894 A, p. 164.
Jordan and Evermann 1896 A, p. 32.
Klunzinger, C. B. 1871 A, p. 663.
Müller and Henle 1837 A, p. 114.
     1841 A, p. 59.
Pedroni, P. M. 1844 A, p. 282.
Pictet, F. J. 1854 A, p. 241.
Steinmann and Döderlein 1890 A, p. 548.
Woodward, A. S. 1884 A, p. 227.
     1889 D, p. 443.
Zittel, K. A. 1890 A, p. 85.

Galeocerdo aduncus Agassiz.

Agassiz, L. 1843 B, iii, p. 231, pl. xxvi, figs. 24-28.
Cope, E. D. 1867 C, p. 141.
     1875 H, p. 362.
     1875 V, p. 29.
     1877 K, p. 38.
Emmons, E. 1858 B, p. 237.
Gibbes, R. W. 1849 A, p. 191, pl. xxv, figs. 54-58.
     1850 F.
Giebel, C. G. 1848 A, p. 369.
Leidy, J. 1877 A, p. 254.
Pedroni, P. M. 1844 A, p. 283, pl. i, figs. 12, 13.
Pictet, F. J. 1854 A, pp. 241, 242.
Quenstedt, F. A. 1852 A, p. 168, pl. xiii, fig. 2
     (Galeus).
     1885 A, p. 264, pl. xx, fig. 8 (Galeus).
Tuomey, M. 1858 A, p. 268 (Galeus).
Woodward, A. S. 1889 D, p. 444.
     1900 A, p. 5, pl. i, figs. 10, 10a.
Zittel, K. A. 1890 A, p. 85, fig. 90.
     Tertiary; Europe: Eocene; North Carolina,
     South Carolina, Alabama: Miocene; Maryland.

Galeocerdo contortus Gibbes.

Gibbes, R. W. 1849 A, p. 193, pl. xxv, fig. 71-74.
Clark, W. B. 1895 A, p. 4.
     1897 A, p. 62.
Cope, E. D. 1867 C, p. 141.
     1875 V, p. 29.
Emmons, E. 1858 B, additions and corrections to
     p. 241, figs. 82a, 83a.
Gibbes, R. W. 1850 F
Jaekel, O. 1896 C, p. 165 (var. hassiæ).
Leidy, J. 1877 A, p. 254.
Pictet, F. J. 1854 A, p. 242.
Woodward, A. S. 1889 D, p. 443.

Wyman, J. 1850 B, p. 234.
     Eocene; North Carolina, South Carolina, Ala-
     bama: Miocene; Virginia and Maryland. Va-
     riety *hassiæ* in Germany.

Galeocerdo crassidens Cope.

Cope, E. D. 1872 I, p. 356.
     1875 E, pp. 243, 395.
Williston, S. W. 1900 A, p. 41.   (Syn.? of Corax fal-
     catus.)
Woodward, A. S. 1889 D, p. 447.
     Cretaceous (Niobrara); Kansas.

Galeocerdo hartwellii Cope.

Cope, E. D. 1872 I, p. 355.
     1875 E, pp. 244, 296.   (G. hartvelii.)
Osborn, Scott and Speir, 1878 A, p. 105.
Williston, S. W. 1900 A, p. 41.   (Syn.? of Corax fal-
     catus.)
     1900 B, p. 252, pl. xxxi, fig. 23.   (Corax hart-
     velii; syn. of C. falcatus.)
Woodward, A. S. 1889 D, p. 447.
     Cretaceous (Niobrara); Kansas, Colorado.

Galeocerdo lævissimus Cope.

Cope, E. D. 1867 C, p. 141.
     Miocene; Maryland.

Galeocerdo latidens Agassiz.

Agassiz, L. 1843 B, iii, p. 231, pl. xxvi, figs. 22, 23;
     (? figs. 20, 21.)
Cope, E. D. 1867 C, p. 141.
     1875 V. p. 29.
Dames, W., 1883 B, p. 142.
Dixon, F. 1850 A, p. 202, pl. xi, figs 22, 23.
     1878 A, p. 248, pl. xi, figs. 22, 23.
? Emmons, E. 1858 B, p. 239, fig. 68.
Gibbes, R. W. 1846 A, p. 42.
     1849 A, p. 192, pl. xxv, figs. 59-62.
     1850 F.
Giebel, C. G. 1848 A, p. 369.
Hilgard, E. W. 1860 A, p. 142.
Tuomey, M. 1858 A, p. 268.   (Galeus.)
Woodward, A. S. 1889 D, p. 444
     Tertiary; Europe and North Africa: Eocene;
     North Carolina, South Carolina, Alabama, and
     Mississippi: Miocene; Maryland.

Galeocerdo minor Agassiz.

Agassiz, L. 1843 B, iii, p. 232, pl. xxvi, figs. 15-19
     (? figs. 20, 21); pl. xxviia, figs. 64-66.
Giebel, C. G. 1848 A, p. 369.
Molin, R. 1860, Sitzungsber. Akad. Wiss. Berlin,
     xl, p. 583.   (Protogaleus.)
Winckler, 1874 B, p. 296, pl. vii, fig. 1.   (G. recti-
     conus.)
Woodward, A. S 1889 D, p. 446.   (Galeocerdo?
     minor.)
     Tertiary; Europe: Eocene; Alabama.   (Wood-
     ward.)

Galeocerdo productus Agassiz.

Agassiz, L. 1856 A, p. 273.
    1857 A, p. 314, pl. i, figs. 1-6.
Woodward, A. S. 1889 D, p. 448.
    Miocene; California.

Galeocerdo subcrenatus Emmons.

Emmons, E. 1858 B, p. 238.
    A doubtful species, without indicated forma-
    tion; North Carolina.

Galeocerdo sp. indet.

Woodward, A. S. 1889 D, p. 456.
    Eocene; South Carolina.

GALEORHINUS Blainville.   Type *Squalus galeus* Linn.

Blainville, H. M. D. 1816, Bull. Sci. Philom., p. 121.
Cuvier G. 1817 Règne Anim., ed. i, p. 117. (Galeus,
    with *Squalus galeus* as type.)
Gegenbaur, C. 1872 A, pl. ii, fig. 2; pl. v, fig. 2; pl.
    viii, fig. 3; pl. xii, fig. 2; pl. xvii, fig. 2; pl. xix.
    fig. 4. (Galeus.)
Gill, T. 1862 A, p. 402.
    1864, Proc. Acad. Nat. Sci. Phila., p. 148.
    (Eugaleus, type *Squalus galeus.*)
Günther, A. 1870 A, p. 379. (Galeus.)
    1880 A, p. 318. (Galeus.)

Hasse, C. 1882 A, p. 263, pl. xxxviii. (Galeus.)
Jaekel, O. 1894 A, p. 166. (Galeus.)
Jordan and Evermann, 1896 A, p. 31. (Galeo-
    rhinus.)
Müller and Henle, 1841 A, p. 57. (Galeus.)
Woodward, A. S. 1889 D, p. 452. (Galeus.)
Zittel, K. A. 1890 A, p. 85. (Galeus.)

Galeorhinus sp. indet.

Woodward A. S. 1889 D, p. 456. (Galeus.)
    Eocene; Alabama.

CARCHARHININÆ.

Duméril, A. 1865 A, p. 340. (Carchariæ.)
Günther, A. 1870 A. p. 353. (Carchariina, in part.)

Jordan and Evermann, 1896 A, p. 28.

CARCHARHINUS Blainville.   Type *C. commersoni*=*C. lamia* (Raf.).

Blainville, H. M. D. 1816, Jour. de Physique, p. 284.
Agassiz, L. 1843 B, iii, pp. 240, 302, 366. (Car-
    charias.)
Cuvier, G. 1817 Règne Anim., ed. i, p. 125. (Car-
    charias, type *C. lamia.*)
Duméril, A. 1865 A, p. 343. (Carcharias.)
Gill, T. 1862 A, p. 401. (Eulamia, type *C. lamia.*)
Günther, A. 1880 A, p. 318. (Carcharias.)
Hasse, C. 1882 A, p. 268, pl. xxxix. (Carcharias.)
Jaekel, O. 1894 A, p. 157. (Carcharias.)
Jordan and Evermann 1896 A, p. 33. (Carcha-
    rhinus.)
Klunzinger, C. B. 1871 A, p. 655. (Carcharias.)
Müller and Henle 1837 A, p. 114. (Carcharias.)
    1841 A, p. 27. (Carcharias.)
Quenstedt, F. A. 1852 A, p. 169. (Carcharias.)
    1885 A, pp. 265, 272. (Carcharias.)
Steinmann and Döderlein 1890 A, p. 548. (Car-
    charias.)
Woodward, A. S. 1889 D, p. 435. (Carcharias.)
Zittel, K. A. 1890 A, (1887) p. 86. (Carcharias.)
Subgenus APRIONODON Gill.
Gill, T. 1862 A, p. 411. (For Aprion of Müller and
    Henle, preoccupied).
Duméril, A. 1865 A, p. 348.
Günther A. 1870 A, p. 361. (Aprion.)
Jaekel, O. 1894 A, p. 162. (Aprion.)
Jordan and Evermann 1896 A, p. 42.
Müller and Henle 1841 A, p. 31. (Aprion.)
Probst, J. 1878 A, p. 121. (Aprion.)
Woodward, A. S. 1889 D, p. 437.
Zittel, K. A. 1890 A, p. 86.

Subgenus PRIONACE Cantor.

Cantor, T, E. 1850 Malayan Fishes, p. 399. (For
    Prionodon of Müller and Henle, preoccupied.)
Agassiz, L. 1843 B, iii, p. 243. (Glyphis.)
Duméril, A. 1865 A, p. 351. (Prionodon.)
Gegenbaur, C. 1872 A, pl. ii, fig. 4; pl. v, fig. 3; pl.
    viii, fig. 4; pl. xviii, fig. 5. (Prionodon.)
Gibbes, R. W. 1849 A, p. 194. (Glyphis.)
Günther, A, 1870 A, p. 363. (Prionodon.)
Hasse, C. 1882 A, p. 271, pl. xxxix. (Prionodon.)
Jaekel, O. 1894 A, p. 162. (Prionodon.)
Jordan and Evermann 1896 A, p. 33.
Müller and Henle 1841 A, p. 36. (Prionodon.)
Pictet, F. J. 1854 A, p. 236. (Glyphis.)
Probst, J. 1878 A, p. 124. (Prionodon.)
Woodward, A. S. 1889 D, p. 439. (Prionodon.)
Zittel, K. A. 1890 A, p. 86. (Prionodon.)

Carcharhinus (Prionace) antiquus
    (Agassiz).

Agassiz, L. 1856 A, p. 273. [Carcharias (Priono-
    don).]
    1857 A, p. 314, pl. i, figs. 15, 16. [Carcharias
    (Prionodon).]
Woodward, A. S. 1889 D, p. 441. [Carcharias
    (Prionodon).]
    Miocene; California

Carcharhinus (Prionace) egertoni
    (Agassiz).

Agassiz, L. 1843 B, iii, p. 228, pl. xxxvi, figs. 6, 7.
    (Corax egertoni.)

Cope, E. D. 1867 C, p. 141. (Galeocerdo.)
  1875 F, p. 63. (Galeocerdo.)
  1875 H, p. 362. (Galeocerdo.)
  1875 V, p. 29. (Galeocerdo.)
Emmons, E. 1858 B, additions and corrections to
  p. 242, fig. 90. (Galeocerdo.)
Gibbes, R. W. 1847 A, p. 268. (Glyphis subulata.)
  1849 A, p. 192, pl. xxv, figs. 66–69 (Galeocerdo
  egertoni); p. 194, pl. xxv, figs. 86, 87 (Gly-
  phis subulata).
  1850 F. (Glyphis subulata.)
Giebel, C. G. 1848 A, p. 371. (Carcharias.)
Woodward, A. S. 1889 D, p. 439. [Carcharias
  (Prionodon).]
Wyman, J. 1850 B, p. 234. (Galeocerdo egertoni,
  Glyphis subulata.)
  Eocene; South Carolina: Miocene: Maryland,
  North Carolina.

Carcharhinus (Aprionodon) gibbesii (Woodward).

Woodward, A. S. 1889 D, p. 437. [Carcharias
  (Aprionodon).]
?Cope, E. D. 1867 C, p. 142. (Oxyrhina minuta.)
Eastman, C. R. 1896 A, p. 182. (Carcharias.)

MICRODUS Emmons. Type M. levis Emmons.

Emmons, E. 1857 A, p. 48.
  It is doubtful what disposition is to be made of
  the tooth described by Emmons under the name
  Microdus levis.

HEMIPRISTIS Agassiz. Type H. serra Agassiz.

Agassiz, L. 1843 B, iii, pp. 237, 303.
Gibbes, R. W. 1849 A, p. 193.
Jaekel, O. 1894 A, p. 168.
Kiprijanoff, V. 1854 A, p. 373.
Klunzinger, C. B. 1871 A, p. 664. (Dirrhizodon,
  type D. elongatus.)
Pedroni, P. M. 1844 A, p. 284.
Pictet, F. J. 1854 A, p. 242.
Probst, J. 1878 A, p. 141.
Quenstedt, F. A. 1852 A, p. 168.
  1885 A, p. 265.
Sismonda, E. 1849 A, p. 82.
Woodward, A. S. 1889 D, p. 448.
Zittel, K. A. 1890 A, p. 85.

Hemipristis heteropleurus Agassiz.

Agassiz, L. 1856 A, p. 274.
  1857 A, p. 315, pl. i, fig. 14.
  Miocene; California.

Hemipristis serra Agassiz.

Agassiz, L. 1843 B, iii, p. 237, pl. xxvii, figs. 18–30.
  1843 B, iii, p. 238, pl. xxvii, figs. 31–33. (H.
  paucidens.)
Cope, E. D. 1867 C, p. 142.
  1875 H, p. 362.
  1875 V, p. 29.
Emmons, E. 1858 B, p. 235, fig. 63. (H. crenula-
  tus.)
Gemmellaro, G. G. 1857 A, p. 296, pl. i, fig. 6a.

Gibbes, R. W. 1846 A, p. 42. (Galeocerdo minor,
  not of Agassiz.)
  1849 A, p. 192, pl. xxv, figs. 63–65 (G. minor);? p. 202, pl. xxvii, fig. 164 (Oxyrhina
  minuta).
  1850 F. (G. minor.)
Leidy, J. 1877 A, p. 254. (Prionodon antiquus.)
  Eocene; South Carolina, Alabama: Miocene;
  Maryland.?

Carcharhinus (Prionace) tenuis (Agassiz).

Agassiz, L. 1843 B, iii, p. 242, pl. xxxa, fig. 16.
  (Carcharias.)
?Gibbes, R. W. 1849 A, p. 191. (Carcharias.)
  ?1850 G, p. 300, pl. xlii, fig. 8. (Carcharias.)
Woodward, A. S. 1889 D p. 442. [Carcharias
  (Prionodon).]
  Senonian; Switzerland: Eocene?; South Caro-
  lina.

Carcharhinus sp. indet.

Woodward, A. S. 1889 D, pp. 455, 456.
  Eocene: Alabama.

Microdus lævis Emmons.

Emmons, E. 1857 A, p. 48 with figure.
  Triassic; North Carolina.

Gibbes, R. W. 1846 A, p. 42. (H. serra, H. paucidens, Lamna hopei.)
  1849 A, p. 193, pl. xxv, figs. 75–85 (H. serra);
  p. 198, pl. xxvi, figs. 120–123 (Lamna
  hopei.)
  1850 F. (H. serra, Lamna hopei.)
Giebel, C. G. 1848 A, p. 368. (H. serra, H. pau-
  cidens.)
Leidy, J. 1877 A, p. 254.
Pedroni, P. M. 1844 A, p. 284, pl. i. fig. 20 (H. ser-
  ra); figs. 21, 22 (H. paucidens).
Pictet, F. J. 1854 A, p. 243.
Probst, J. 1878 A, p. 143, pl. i, figs. 49–57.
Quenstedt, F. A. 1885 A, p. 265, fig. 81.
Sauvage, H. E. 1875 A, p. 634, pl. xxii, fig. 2.
  (Odontaspis sacheri.)
Sismonda, E. 1849 A, p. 33, pl. i, figs. 17, 18.
Woodward, A. S. 1884 A, p. 228, fig. 133.
  1889 D, p. 449.
  1900 A, p. 5, pl. i, figs. 11, 11a.
Wyman, J. 1850 B, p. 234.
Zittel, K. A. 1890 (1887) A, p. 85, fig. 89.
  For additional bibliography see Woodward,
  A. S. 1889 D, p. 449.
  Tertiary; Europe: Eocene: North Carolina,
  South Carolina: Miocene; Virginia: Plio-
  cene?; Argentine Republic.

Hemipristis sp. indet.

Woodward, A. S. 1889 D, p. 456.
  Eocene: South Carolina.

SPHYRNINÆ.

Duméril, A. 1865 A, p. 380.　(Cestraciontes.)
Günther, A. 1870 A, p. 353.　(Zygænina.)
Jordan and Evermann 1896 A, p. 43.　(Sphyrnidæ.)

SPHYRNA Rafinesque.　Type *Squalus zygæna* Linn.

Rafinesque, C. S., 1810, Caratteri di alcuni generi, etc., p. 12.
Agassiz, L. 1843 B, iii, pp. 234, 303, 366, pl. N, figs. 8–10.
Duméril, A. 1865 A, p. 381.　(Cestracion.)
Gegenbaur, C. 1872 A.　(Zygæna.)
Gibbes, R. W. 1849 A, p. 194.
Günther, A. 1870 A, p. 380.　(Zygæna.)
　1880 A, p. 318.　(Zygæna.)
Hasse, C. 1882 A, p. 273.　(Zygæna.)
Hubrecht, A. A. W. 1878 A.　(Zygæna.)
Jaekel, O. 1894 A, p. 165.　(Zygæna.)
Jordan and Evermann 1896 A, p. 43.
Klunzinger, C. B. 1871 A, p. 665.　(Zygæna.)
Mivart, St. G. 1879 A.　(Zygæna.)
Müller and Henle 1841 A, p. 51.
Woodward, A. S. 1889 D, p. 453.
Zittel, K. A. 1890 A, p. 86.

Sphyrna gibbesii, nom. nov.

Emmons, E. 1858 B, additions and corrections to p. 241, fig. 84*a*.　(S. denticulata.)
Gibbes, R. W. 1849 A, p. 196, pl. xxv, fig. 94.　(S. denticulata, not of Agassiz.)
　1850 F.　(S. denticulata.)
　Eocene; South Carolina, North Carolina.

Sphyrna lata Agassiz.

Agassiz, L. 1843 B, iii, p. 235, pl. xxvia, figs. 58, 59.
?Gibbes, R. W. 1849 A, p. 195, pl. xxv, figs. 91–93.
　1850 F.
Woodward, A. S. 1889 D, p. 454.
?Wyman, J. 1850 B, p. 234.
　Tertiary; Virginia: Eocene; South Carolina.
　The identification of the American specimens with Agassiz's species is doubtful.

Sphyrna magna Cope.

Cope, E. D. 1867 C, p. 142.
Woodward, A. S. 1889 D, p. 455.
　Miocene; Virginia.

Sphyrna prisca Agassiz.

Agassiz, L. 1843 B, iii, p. 234, pl. xxvi *a*, figs. 35–50.
Cope, E. D. 1867 C, p. 142.
　1875 H, p. 362.　(Zygæna.)
　1875 V, p. 29.　(Zygæna.)
Gemmellaro, G. G. 1857 A, p. 295, pl. i, fig. 5; pl. vi, fig. 3.
Gibbes, R. W. 1889 A, p. 194, pl. xxv, figs. 88–90.
Pedroni, P. M. 1844 A, p. 284, pl. i, figs. 15, 16.
Woodward, A. S. 1889 D, p. 453.
Zittel, K. A. 1890 A, p. 86, fig. 93.　(S. serrata.)
　Miocene; Europe, Maryland, New Jersey, North Carolina, South Carolina.

Superfamily SQUALOIDEA.

Gill, T. 1893 A, p. 129.　(Tectospondyli, *lapsu calami.*)
Hasse, C. 1879 A, p. 41.　(Cyclospondyli.)
　1882 A, pp. 55–94, pls. viii–x.
Jaekel, O. 1894 A, p. 51.　(Cyclospondyli.)
Jordan and Evermann 1896 A, p. 53.　(Cyclospondyli.)
Woodward, A. S. 1889 D, p. xxvii.　(Cyclospondyli.)

SOMNIOSIDÆ.

Gill, T. 1896 A, p. 192.
Duméril, A. 1865 A, p. 450.　(Scymni.)
Gill, T. 1862 A, p. 405.　(Scymnoidæ, Echinorhinoidæ.)
　1893 A, p. 129.　(Dalatiidæ and Echinorhinidæ.)
Günther, A. 1870 A, p. 417.　(Spinacidæ, in part.)
Hasse, C. 1879 A, p. 43.　(Scymnidæ.)
Jordan and Evermann 1896 A, p. 56 (Dalatiidæ); p. 57 (Echinorhinidæ).
Müller and Henle 1841 A, p. 91.　(Scymni.)
Woodward, A. S. 1889 D, p. 30.　(Spinacidæ, in part.)
Zittel, K. A. 1890 A, p. 87.　(Spinacidæ, in part.)

SCYMNORHINUS Bonaparte.　Type *Scymnus lichia* Cuvier.

Bonaparte, C. S. 1846, Cat. Met. Pesci Europei, p. 16.
Agassiz, L. 1843 B, iii, pp. 305, 367.　(Scymnus.)
Cuvier, G. 1817, Règne Anim., ii, p. 130.　(Scymnus, preoccupied.)
Duméril, A. 1865 A, p. 450.
Gegenbaur, C. 1872 A, pl. i, fig. 3; pl. iv, fig. 3; pl. vii, fig. 3; pl. xi, fig. 1; pl. xvii, fig. 4; pl. xix, fig. 2.　(Scymnus.)
Gill, T. 1895 A.
Hasse, C. 1878 A, p. 168.　(Scymnus.)
　1882 A, p. 65, pl. ix.　(Scymnus.)
Hubrecht, A. A. W. 1878 A.　(Scymnus.)
Kölliker, A. 1863 A, p. 26.　(Scymnus.)

Müller and Henle 1837 A, p. 116.　(Scymnus.)
　1837 B, p. 399.　(Scymnus.)
　1838 A, p. 89.　(Scymnus.)
　1841 A, p. 91.　(Scymnus.)
Woodward, A. S. 1889 D, p. 33.　(Scymnus.)
Zittel, K. A. 1890 A, p. 88.　(Scymnus.)

Scymnorhinus occidentalis (Agassiz).

Agassiz, L. 1856 A, p. 272.　(Scymnus.)
　1857 A, p. 314, pl. i, figs. 9–13.　(Scymnus.)
Woodward, A. S. 1889 D, p. 458.　(Scymnus.)
　Miocene; California.

ECHINORHINUS Blainville.   Type *Squalus spinosus* Gmelin.

Blainville, H. M. D., 1816, Bull. Soc. Philom., p. 121.
Agassiz, L. 1843 B (1838), ill, p. 94. (Goniodus.)
Duméril, A. 1865 A, p. 458.
Günther, A. 1870 A, p. 428.
Hasse, C. 1882 A, p. 73, pl. ix.
Jordan and Evermann 1896 A, p. 57.
Müller and Henle 1838 A, p. 89.

Müller and Henle 1841 A, p. 96.
Woodward, A. S. 1889 D, p. 34.

Echinorhinus blakei Agassiz.
Agassiz, L. 1856 A, p. 272.
    1857 A, p. 313, pl. i, figs. 7, 8, 17.
Miocene of California.

Suborder RAJÆ.

Brühl, B. C. 1847 A.
Duméril, A. 1865 A, p. 468. ("Raies ou Hypotrèmes.")
Gegenbaur, C. 1872 A.
Gill, T. 1893 A, p. 130. (Hypotremi.)
Günther, A. 1870 A, p. 434. (Batoidei.)
    1880 A, p. 335. (Batoidei.)
Haeckel, E. 1895 A, p. 236. (Rajacei.)
Hasse, C. 1879 A, p. 44. (Tectospondyli.)
    1882 A, p. 163. (Tectospondyli.)
Haswell, W. A. 1884 A, p. 115. (Batoidei.)
Jaekel, O. 1890 C.
    1890 D, p. 110. (Tectospondyli.)
    1892 A.

Jaekel, O. 1894 A, p. 40. (Batoidei.)
    1898 A, p. 44. ("Rochen.")
Jordan and Evermann 1896 A, p. 59. (Batoidei.)
Kehner, F. A. 1876 A. (Rail.)
Köstlin, O. 1844 A, p. 425. ("Rochen.")
Müller, J 1846 C. p. 203. (Rajidæ.)
Müller and Henle 1841 A, p. 103.
Parker, W. K. 1878 A. (Rail.)
Steinmann and Döderlein 1890 A, p. 549. (Batoidei.)
Woodward, A. S. 1889 D, p. 30. (Tectospondyli, in part.)
Zittel, K. A. 1890 A, p. 93. (Batoidei.)

Superfamily PACHYURA Gill.

Gill, T. 1872 C, p. 23.
    1882, in Jordan and Gilbert's Synop. N. A. Fishes, p. 36.
    1893 A, p. 130. (Sarcura.)

Jaekel, O. 1894 A, p. 69. (Rhinoraji.)
    1898 A, p. 45. (Rhinoraji.)
Jordan and Evermann 1896 A, p. 59. (Sarcura.)

TAMIOBATIDÆ, nom. nov.

TAMIOBATIS Eastman.   Type *T. vetustus* Eastman.

Eastman, C. R. 1897 A, p. 85.
Müller, S. A. 1897 A, p. 793.
    Relationships to the Rhinobatidæ doubtful.

Tamiobatis vetustus Eastman.
Eastman, C. R. 1897 A, p. 85, pl. i.
Devonian?; Kentucky.

RAJIDÆ.

Agassiz, L. 1843 B, ill, p. 73. ("Raies.")
Duméril, A. 1865 A, p. 525. (Rajæ.)
Günther, A. 1870 A, pp. 435, 455.
    1880 A, p. 340.
    1881 A, p. 686.
Hasse, C. 1878 A, p. 170.
    1879 A, p. 48. (Rajæ.)
    1882 A, p. 163. (Rajæ.)
Haswell, W. A. 1884 A, p. 117.
Howes, G. B. 1890 A.
Hubrecht, A. A. W. 1878 A.
Kölliker, A. 1860 B, p. 212.
Jaekel, O. 1890 A, p. 94.

Jaekel, O. 1891 D, p. 191.
    1894 A, p. 81.
    1898 A, p. 52.
Jordan and Evermann 1896 A, p. 66.
Müller and Evermann 1841 A, p. 132. (Rajæ.)
Nicholson and Lydekker 1889 A, p. 933.
Owen, R. 1845 B, p. 43.
Stannius, H. 1854 A.
Steinmann and Döderlein 1890 A, p. 550.
Stephan, P. 1900 A.
Woodward, A. S. 1889 D, p. 84.
Zittel, K. A. 1890 A, p. 103.

RAJA Linn.   Type *R. batis* Linn.

Linnæus, C., 1858. Syst. Nat., ed. x, p. 231.
Duméril, A. 1865 A, p. 526.
Gegenbaur, C. 1872 A, pl. iii, fig. 2; pl. vi, fig. 4; pl. xiii, fig. 1; pl. xiv, figs. 4, 6; pl. xv, fig. 1.

Günther, A. 1870 A, p. 455.
    1880 A, p. 340.
Hasse, C. 1878 A.
    1882 A, p. 163, pl. xxii.

Hubrecht, A. A. W. 1878 A.
Jordan and Evermann 1896 A, p. 66.
Müller and Henle 1837 A, p. 111.
    1841 A, p. 132, plates.
Stephan, P. 1900 A.
Woodward, A. S. 1889 D, p. 85.
Zittel, K. A. 1890 A, p. 104.

Raja dux Cope.

Cope, E. D. 1867 C, p. 141.
Woodward, A. S. 1889 D, p. 459.
    Miocene; Maryland.

GRYPHODOBATIS Leidy.   Type *G. uncus* Leidy.

Leidy, J. 1877 A, p. 249.
    A genus whose position is doubtful.

Gryphodobatis uncus Leidy.

Leidy, J. 1877 A, p. 249, pl. xxxiv, figs. 8, 9.
Woodward, A. S. 1889 D, p. 156.
    Phosphate beds; South Carolina.

ONCOBATIS Leidy.   Type *O. pentagonus* Leidy.

Leidy, J. 1870 K, p. 70.
Nicholson and Lydekker, 1889 A, p. 934.
Woodward, A. S. 1889 D, p. 90.

Oncobatis pentagonus Leidy.

Leidy, J. 1870 K, p. 70.

Cope, E. D. 1883 L, p. 153.   (Raja.)
Leidy, J. 1871 C, p. 370.
    1873 B, pp. 264, 353, pl. xvii, figs. 18, 19.
Woodward, A. S. 1889 D, p. 90.
    Pliocene; Idaho.

PRISTIDÆ.

Duméril, A. 1865 A, p. 471.   (Pristides.)
Gegenbaur, C. 1872 A.
Günther, A. 1870 A, p. 436.
    1880 A, p. 336.
    1881 A, p. 686.
Hasse, C. 1882 A, p. 121.
Haswell, W. A. 1884 A, p. 116.
Hubrecht, A. A. W. 1878 A.
Jaekel, O. 1894 A, p. 75.

Jordan and Evermann 1896 A, p. 60.
Müller and Henle 1841 A, p. 105.   (Pristides.)
Nicholson and Lydekker 1889 A, p. 930.
Pictet, F. J. 1854 A, p. 274.   (Pristides.)
Woodward, A. S. 1885 B, p. 227.
    1888 B, p. 451.
    1889 D, p. 73.
    1898 B, p. 52.
Zittel, K. A. 1890 A, p. 94.

PRISTIS LATHAM.   Type *Squalus pristis* Linn.= *Pristis antiquorum.*

Latham, John 1794, Trans. Linn. Soc., ii, p. 276.
Agassiz, L. 1843 B, iii, p. 382.
Duméril, A. 1865 A, p. 471.
Gegenbaur, C. 1872 A, pl. iii, fig. 4; pl. xiii, figs.
    4-6; pl. xiv, fig. 2.
Günther, A. 1880 A, p. 347.
Hasse, C. 1882 A, p. 121, pl. xvi.
Hubrecht, A. A. W. 1878 A, pl. ii, figs. 2-5, 8.
Jaekel, O. 1890 D.
    1894 A.
Jordan and Evermann 1896 A, p. 60.
Müller and Henle 1837 A, p. 116.
    1837 B, p. 399.
    1841 A, p. 105.
Nicholson and Lydekker 1889 A, p. 190.
Owen, R. 1845 B, p. 40, pls. viii, ix.
Pictet, F. J. 1854 A, p. 275.
Stannius, H. 1854 A.
Tomes, C. S. 1898 A, p. 246.
Woodward, A. S. 1889 D, p. 73.
    1899 A, p. 3.
Zittel, K. A. 1890 A, p. 94.

Pristis agassizi Gibbes.

Gibbes, R. W. 1847 B, p. 11, pl. i, figs. 6, 7.
Woodward, A. S. 1889 D, p. 75.
    Eocene; South Carolina.

Pristis amblodon Cope.

Cope, E. D. 1869 L, p 312.
    1875 H, p. 362.
Woodward, A. S. 1889 D, p. 458.
    Eocene; New Jersey.

Pristis attenuatus Cope.

Cope, E. D. 1869 F, p. 244.
    1875 V, p. 29.
Woodward, A. S. 1889 D, p. 459.
    Tertiary; North Carolina.

Pristis brachyodon Cope.

Cope, E. D. 1869 A, p. 312.
    1875 V, p. 29.
Woodward, A. S. 1889 D, p. 459.
    Cretaceous (Greensand); Virginia: Miocene:
    North Carolina.

Pristis curvidens Leidy.

Leidy, J. 1855 D.
Woodward, A. S. 1889 D, p. 75.
    Cretaceous (Greensand); New Jersey.

Pristis ensidens Leidy.

Leidy, J. 1855 D.
    1877 A, p. 252, pl. xxxiv, figs. 31, 32.
Woodward, A. S. 1889 D, p. 75.
    Phosphate beds; South Carolina.

Superfamily *MASTICURA* Gill.

Gill, T. 1872 C, p. 22.
   1882, in Jordan and Gilbert's Synop. Fishes
   N. A., p. 36.
   1893 A, p. 130.

Jaekel, O. 1894 A, pp. 64, 115. (Centrobati.)
   1896 A, p. 45. (Centrobati.)
   1899 A, p. 298. (Centrobatidæ.)
Jordan and Evermann 1896 A, pp. 59, 79.

PTYCHODONTIDÆ.

Jaekel, O. 1898 A, p. 51. (Ptychodontinæ.)

HEMIPTYCHODUS Jaekel. Type *Ptychodus mortoni* Mantell.

Jaekel, O. 1894 A, p. 137.

Hemiptychodus mortoni (Mantell).

Mantell, G. A., in Morton, S. G. 1842 A, p. 215, pl. xi, figs. 7, 7a. (Ptychodus.)
Agassiz L. 1843 B, iii, p. 158, pl. xxv, figs. 1–3. (Ptychodus.)
Cope, E. D. 1874 C, p. 48. (Ptychodus.)
   1875 E, p. 294. (Ptychodus.)
Dana, J. D. 1896 A, p. 843, fig. 1406. (Ptychodus.)
De Kay, J. E. 1842 B, p. 386. (Ptychodus.)
Dixon, F. 1878 A, p. 392, pl. xxxi, figs. 6, 7. (Ptychodus.)
Giebel, C. G. 1848 A, p. 334. (Ptychodus.)
Hilgard, E. W. 1860 A, p. 389. (Ptychodus.)

Jaekel, O. 1894 A, p. 137. (Hemiptychodus.)
Leidy, J. 1868 H, p. 205. (Ptychodus.)
   1870 H, p. 12. (Ptychodus.)
   1873 B, pp. 295,352, pl.xviii, figs.1–14. (Ptychodus.)
Morton, S. G. 1834 A, pl. xviii, figs. 1, 2. (No name or description).
Tuomey, M. 1858 A, p. 268. (Ptychodus.)
Williston, S. W. 1900 A, p. 30, pl. vii; pl. viii, fig. 1; pl. ix. (Ptychodus.)
   1900 B, p. 238, pls. xxv, xxvi, fig. 1; pl. xxviii. (Ptychodus.)
Woodward, A. S. 1887 B, p. 130. (Ptychodus.)
   1889 D, p. 149. (Ptychodus.)
   Cretaceous; Alabama, Mississippi, Kansas.

PTYCHODUS Agassiz. Type *P. mammillaris* Agassiz.

Agassiz, L. 1843 B (1835), Feuill., p. 54.
   1843 B (1839), iii, pp. 56, 150.
   1844 A.
Cope, E. D. 1874 C, p. 47. (Sporetodus, type *S. janevaii.*)
   1886 S.
Davis, J. W. 1890 A, p. 375.
Dixon, F. 1850 A, p. 361.
   1858 A, p. 390.
Hasse, C. 1884, Palæontographica, xxxi, p. 9, pl. ii, figs. 16, 17. (Selache davisi.)
Jaekel, O. 1890 A, p. 94.
   1891 D, p. 191.
   1894 A, p. 132.
Geinitz, H. B. 1839 A, pp. 12, 38, 63.
Leidy J. 1868 H.
   1870 H.
Mackie, S. J. 1863 B.
Owen, R. 1845 B, p. 57, pls. xvii–xix.
   1860 E, p. 110.
Pictet, F. J. 1854 A, p. 264.
Priem, F. 1896 A.
Quenstedt, F. A. 1852 A, p. 180.
   1885 A, p. 281.
Sauvage, H. E. 1870 A, p. 15.
Williston, S. W. 1900 A, p. 29.
Woodward, A. S. 1885 B, p. 107.
   1887 B.
   1888 I, p. 294, fig. 1.
   1889 D, p. 122.
Zittel, K. A. 1890 A, p. 78, figs. 77, 78.

Ptychodus anonymus Willist.

Williston, S. W. 1900 A, p. 32, pl. xi, figs. 5–8, 16–18, 20–22, 24.
   1900 B, p.241, pl. xxix, figs. 5–8, 16–18, 20–22, 24.
   Cretaceous (Niobrara, Benton?); Kansas.

Ptychodus decurrens Agassiz.

Agassiz, L. 1843 B (1839), Feuill., p. 54.
   1843 B (1839), iii, p. 154, pl. xxv b, figs. 1, 2, 4, 6–8 (not figs. 3, 5).
Dixon, F. 1850 A, p. 362, pl. xxx, figs. 7, 8; pl. xxxi, fig. 1; pl. xxxii, fig. 5 (P. decurrens); p. 363, pl. xxxi, fig. 9 (P. depressus).
   1858 A, p. 390, pl. xxx, figs. 7, 8; pl. xxxi, fig. 1; pl. xxxii, fig. 5.
Fritsch, A. 1878 A, p. 14, fig. 34.
Kiprijanoff, V. 1852 A, p. 490, pl. xiii, figs. 4, 5.
Owen, R. 1845 B, ii, p. 57, pls. xviii, xix.
Quenstedt, F. A. 1852 A, p. 181, pl. xiii, fig. 59.
   1885 A, p. 281, text fig. 86; pl. xxi, figs. 63, 64.
Reuss, A. E. 1846 A, p. 1, pl. ii, figs. 9, 10.
Sauvage, H. E. 1879 A, p. 53, pl. iii, figs. 25, 26.
Woodward, A. S. 1887 B, pl. x, figs. 1–10, 13.
   1889 D, p. 138.
   1898 B, p. 53, fig. 42.
Zittel, K. A. 1890 A, p. 79, figs. 78 a, b.
   Cretaceous; England, France, Germany; also from Kearns Cañon, Arizona (specimen in U. S. National Museum).

Ptychodus janevaii Cope.

Cope, E. D. 1874 C, p. 47. (Sporetodus.)
   1875 E, pp. 244, 294. (P. janewayii.)
Williston, S. W. 1900 A, p. 33, pl. xii, figs. 9, 10, 11. (P. janewayii.)
   1900 B, p. 242, pl. xxx, figs. 9, 10, 11. (P. janewayii.)
Woodward, A. S. 1889 D, p. 151.
   Cretaceous (Niobrara); Kansas.

Ptychodus mammillaris Agassiz.

Agassiz, L. 1843 B (1835), Feuill., p. 54.
　1843 B (1839), iii, p. 151, pl. xxv b, figs. 11-20
　(P. mammillaris); p. 154, pl. xxv b, figs. 3,
　5 (P. decurrens); p. 155, pl. xxv b, figs. 9,
　10 (P. altior).
Cope, E. D, 1875 E, p. 294.
Davis, J. W. 1890 A, p. 378, pl. xxxviii.
De Kay, J. E. 1842 B, p. 386.
Dixon, F. 1850 A, p. 361, pl. xxx, fig. 6; pl. xxxi,
　fig. 4.
　1878 A, p. 390, pl. xxx, fig. 6; pl. xxxi, fig. 4.
Fritsch, A. 1878 A, p. 14, text fig. 33.
Giebel, C. G. 1848 A, p. 333.
Kiprijanoff, V. 1852 A, p. 487, pl. xii, fig. 3; pl. xiii,
　fig. 3.
Pictet, F. J. 1854 A, p. 265, pl. xxviii, fig. 27.
Quenstedt, F. A. 1852 A, p. 181.
　1885 A, p. 282, pl. xxi, figs. 61, 62. (P de-
　currens.)
Reuss, A. E. 1846 A, p. 2, pl. ii, figs. 11-13.
Sauvage, H. E. 1870 A, p. 16, figs. 86-89.
Woodward, A. S. 1885 B, p. 109, fig. 81.
　1889 D, p. 133.
　Cretaceous; Europe, Delaware.
　For additional citations of literature bearing
　on European specimens, see Woodward, A. S.,
　1889 D, p. 133.

Ptychodus martini Willist.

Williston, S. W. 1900 A, p. 32, pl. x.
　1900 B, p. 240, pl. xxviii.
　Cretaceous (Niobrara); Kansas.

Ptychodus occidentalis Leidy.

Leidy J. 1868 H, p. 207.
Cope, E. D. 1874 C, p. 48.
　1875 E, p. 294.
Leidy, J. 1873 E, pp. 298, 352, pls. xvii, figs. 7, 8;
　pl. xviii, figs. 15-18.
Williston, S. W. 1900 A, p. 33, pl. xi, fig. 4; pl. xii,
　fig. 13.
　1900 B, p. 242, pl. xxix, fig. 4; pl. xxx, fig. 13.
Woodward, A. S. 1889 D, p. 151.
　Cretaceous (Niobrara); Kansas.

Ptychodus papillosus Cope.

Cope, E. D. 1875 E, p. 294.
　1875 F, p. 63.
　Cretaceous; Colorado.

Ptychodus parvulus Whiteaves.

Whiteaves, J. F. 1889 B, p. 191, pl. xxvi, figs. 5 a,
b, c.
　Cretaceous (Niobrara); Manitoba.

Ptychodus polygyrus Agassiz.

Agassiz, L. 1843 B (1835), Feuill., p. 55.
Agassiz, L., in Buckland, W. 1837 A, pl. xxxvii f.
　(No description).
Agassiz, L. 1843 B (1839) B, iii, p. 156, pl. xxv, figs.
　4-11; pl. xxv b, figs. 21-23.
Cope, E. D. 1874 C, p. 48.
　1875 E, p. 294.
Dixon, F. 1850 A, p. 363, pl. xxx, fig. 9; pl. xxxi, fig.
　10 (P. polygyrus); p. xii, pl. xxx, figs. 1, 2 (P.
　latissimus).
　1878 A, p. 391, pl. xxx, figs. 1, 2 (P. latissi-
　mus); p. 392, pl. xxx, fig. 9; pl. xxi, fig. 10
　(P. polygyrus).
Fritsch, A. 1878 A, p. 14, fig. 35.   (P. polygyrus.)
Gibbes, R. W. 1850 A.
　1850 G, pl. 42, figs. 5, 6.
Giebel, C. G. 1848 A, p. 333.
Kiprijanoff, V. 1852 A, p. 494, pl. xiii, fig. 6.'
Leidy, J. 1868 H, p. 208.
　1873 B, p. 352.
Mackie, S. J. 1863 B, pl. ix.
Williston, S. W. 1900 A, p. 31, pl. xi, fig. 9; pl. x,
　fig. 14.
　1900 B, p. 240, pl. xxix, fig. 9; pl. xxx, fig. 14.
Woodward, A. S. 1885 B, p. 109, fig. 82.
　1889 D, p. 143.
　1898 B, p. 52, fig. 41.
　Cretaceous; Alabama, Kansas.

Ptychodus whipplei Marcou.

Marcou, J. 1858 A, p. 33, pl. i, figs. 4, 4a.
Cope, E. D. 1874 C, p. 48.   (P. whippleyi.)
　1875 E, p. 294.   (P. whippleyi.)
　1875 F, p. 63.   (P. whippleyi.)
Leidy, J. 1873 B, p. 300, 352, pl. xviii, figs. 19, 20.
　(P. whippleyi.)
Newberry, J. S. 1876 A, p. 137, pl. iii, figs. 2-2f.   (P.
　whippleyi.)
Osborn, Scott and Speir 1878 A, p. 106.   (P.
　whippleyi.)
Williston, S. W. 1900 A, p. 33, pl. xi, figs. 10-15.
　1900 B, p. 243, pl. xxix, figs. 10-15.   (P.
　whippleyi.)
Woodward, A. S. 1889 D, p. 152.   (P. whippleyi.)
　Cretaceous (Niobrara); Kansas, Colorado,
　New Mexico.

Ptychodus sp. indet.

Williston, S. W. 1900 A, p. 34, pl. xi, figs. 2, 3, pl.
　xiii, fig. 53.
　1900 B, p. 243, pl. xxix, figs. 2, 13; pl. xxxi,
　fig. 53.
　Cretaceous (Benton); Kansas

DASYATIDÆ.

Jordan and Evermann 1896 A, p. 79.
　Unless otherwise stated, the authors below
　cited refer to this family under the name
　Trygonidæ:
Duméril, A. 1865 A, p. 578.   (Trygones.)
Gegenbaur, C. 1865 B.
Gill, T. 1893 A, p. 130.   (Dasybatidæ.)

Günther, A. 1870 A, p. 471.
　1880 A, p. 342.
　1881 A, p. 686.
Hasse, C. 1878 A, p. 169.
　1879 A, p. 47.
　1882 A, p. 136..
Haswell, W. A. 1884 A, p. 117.

Howes, G. B. 1890 A.
Jaekel, O. 1890 C, p. 56.
　　1891 D, p. 191.　("Torpediniden.")
　　1894 A, p. 139.　(Trygoninæ.)
　　1898 A, p. 51.　(Trygoninæ.)

Müller and Henle 1841 A, p. 156.　(Trygones.)
Woodward, A. S. 1889 D, p. 152.
Zittel, K. A. 1890 A, p. 105.

DASYATIS Rafinesque.　Type *D. ujus*=*D. pastinica* Linn.

Rafinesque, C. S., 1810, Caratteri di alcuni nuovi
generi, p. 16.
　　Unless otherwise stated, the following au-
thors refer to this genus under the name
Trygon:
Cope, E. D. 1879 O.　(Xiphotrygon, type *X. acu-
tidens.*)
　　1884 O, p. 49.　(Xiphotrygon.)
Duméril, A. 1865 A, p. 582.
Gegenbaur, C. 1872 A, pl. iii; pl. vi, fig. 6; pl. xi,
　fig. 4; pl. xiii, fig. 2; pl. xiv, fig. 3.
Günther, A. 1880 A, p. 343.
Hasse, C. 1878 A, p. 170.
　　1882 A, p. 142, pl. xix.
Haswell, W. A. 1884 A, p. 100.
Howes, G. B. 1890 A.
Hubrecht, A. A. W. 1878 A.
Jaekel, O. 1890 A, p. 94.
　　1894 A, p. 144.
Jordan and Evermann 1896 A, p. 82.　(Dasyatis.)
Marsh, O. C. 1877 D, p. 261.　(Heliobatis, type
H. radians.)
Müller and Henle 1837 A, p. 117.
　　1838 A, p. 90.
　　1841 A, p. 158.
Newberry, J. S. 1888 B, p. 121.　(Heliobatis.)
Nicholson and Lydekker 1889 A, p. 937.
Woodward, A. S. 1889 D, p. 153.　(Xiphotrygon.)
Zittel, K. A. 1890 A, p. 105.　(Trygon, Xiphotry-
gon.)

Dasyatis carolinensis (Emmons).

Emmons, E. 1858 B, p. 243, figs. 91, 92, 94, 95.
　(Trygon.)
　　1860 A, p. 215, fig. 183°.　(Trygon.)
Tertiary; North Carolina.

Dasyatis hastata? (De Kay).

De Kay, J. E. 1842 B, p. 373, pl. lxv, fig. 214.
　(Trygon.)
Jordan and Evermann 1896 A, p. 84.
Leidy, J. 1860 B, p. 119, pl. xxvii, figs. 1, 2.
Post-pliocene; South Carolina.

Dasyatis radians (Marsh).

Marsh, O. C. 1877 D, p. 256.　(Heliobatis.)
Cope, E. D. 1879 O.　(Xiphotrygon acutidens.)
　　1884 O, p. 50, pl. i, figs. 1-5.　(X. acutidens.)
Jaekel, O. 1894 A, p. 134.　[Trygon (Tæniura.)]
King, C. 1878 A, p. 394.　(Heliobatis.)
Woodward, A. S. 1889 D, p. 155.　(X. acutidens.)
Zittel, K. A. 1890 A, p. 105.　(X. acutidens.)
Eocene (Green R.); Wyoming.

Dasyatis sp. indet.

Leidy, J. 1873 B, p. 358, pl. xxxii, figs. 54, 55.
　(Trygon.)
Miocene; Virginia.

MYLIOBATIDÆ.

Duméril, A. 1865 A, p. 631.　(Myliobatides.)
Gegenbaur, C. 1865 B.
Gill, T. 1893 A, p. 130.　(Aëtobatidæ.)
　　1894 A.　(Aëtobatidæ.)
Günther, A. 1870 A, p. 488.　(Myliobatina.)
　　1880 A, p. 344.
　　1881 A, p. 686.
Hasse, C. 1878 A, p. 170.
　　1879 A, p. 47.
　　1882 A, p. 149.
Haswell, W. A. 1884 A, p. 117.
Jaekel, O. 1891 D, p. 191.
　　1894 A, p. 150.　(Myliobatinæ.)
　　1898 A, p. 51.　(Myliobatinæ.)

Jordan and Evermann, 1896 A, p. 87.
Müller and Henle 1841 A, p. 176.　(Myliobatides.)
Nicholson and Lydekker, 1889 A, p. 935.
Pictet, F. J. 1854 A, p. 279.　(Myliobatides.)
Semon, R. 1899 A, p. 76.
Stannius H. 1854 A.　(Myliobatides.)
Steinmann and Döderlein, 1890 A, p. 549.
Woodward, A. S. 1889 D, p. 109.
Zittel, K. A. 1890 A, p. 99.

MYLIOBATIS Dum.　Type *Raja aquila* Linn.

Duméril, C. 1817, in Cuvier's Règne Anim., ii,p.
137.
　　By many of the following authors this name
is spelled *Myliobates.*
Agassiz, L. 1843 B, iii, p. 317.
Duméril, A, 1865 A, p. 633.
Gibbes, R. W. 1850 G.
Gill, T. 1894 A, p. 113.

Günther, A. 1870 A, p. 488.
　　1880 A, p. 344.
Hasse, C. 1882 A, p. 148, pls. xx, xxi
Howes, G. B. 1890 A.
Hubrecht, A. A. W. 1878 A.
Jaekel, O. 1890 A, p. 94.
　　1894 A, p. 129.
Jordan and Evermann 1896 A, p. 89.

Müller and Henle, 1837 A, p. 117.
1837 B, p. 401.
1841 A, p. 176.
Nicholson and Lydekker 1889 A, p. 935.
Owen, R. 1845 B, p. 46, pls. xxv-xxvii.
Quenstedt, F. A. 1852 A, p. 182.
1885 A, p. 287.
Röse, C. 1897 A, p. 39, figs. 9-15.
Sismonda, E. 1849 A, p. 50.
Steinmann and Döderlein, 1890 A, p. 549.
Tomes, C. S. 1898 A, p. 247.
Treuenfels, P. 1896 A, p. 1, pls. i, ii.
Woodward, A. S. 1888 F.
1889 D, p. 109.
1899 A, p. 3.
Wyman, J. 1850 B, p. 284.
Zittel, K. A. 1890 A, p. 100.

Myliobatis bisulcus Marsh.

Marsh, O. C. 1870 E, p. 229.
Leidy, J. 1877 A, p. 239.
Woodward, A. S. 1889 D, p. 121.
Eocene; New Jersey.

Myliobatis copeanus Clark.

Clark, W. B. 1896 A, p. 4.
1897 A, p. 61, pl. vii, figs. 3a, 3b.
Eocene; Maryland, Virginia.

Myliobatis fastigiatus Leidy.

Leidy, J. 1877 A, p. 239, pl. xxxi, fig. 11.
1876 E, p. 86. (No description).
Woodward, A. S. 1889 D, p. 121.
Eocene; New Jersey.

Myliobatis gigas Cope.

Cope, E. D. 1867 C, p. 140.
Leidy, J. 1877 A, p. 241, pl. xxxiii, fig. 4.
Woodward, A. S. 1889 D, p. 122.
Miocene; Maryland.

Myliobatis glottoides Cope.

Cope, E. D. 1870 F, p. 293.
Eocene; New Jersey.

Myliobatis holmesii Gibbes.

Gibbes, R. W. 1850 G, p. 299, pl. xlii, figs. 1-3.
Cope, E. D. 1867 C, p. 140.
Gibbes, R. W. 1850 B. (No description.)
Leidy, J. 1876 E, p. 86. (M. magister; no description).
1877 A, p. 233, pl. xxxiii, fig. 7 (M. magister);
p. 234 (M. holmesii).
Woodward, A. S. 1889 D, p. 122.
Phosphate beds; South Carolina.

Myliobatis jugosus Leidy.

Leidy, J. 1877 A, p. 240, pl. xxxi, figs. 4, 5.
1876 E, p. 86. (No description.)
Woodward, A. S. 1889 D, p. 122.
Eocene; New Jersey.

Myliobatis leidyi Hay.

Hay, O. P. 1899 E, p. 785.
Leidy, J. 1855 C, p. 395. (M. serratus, not of H. v.
Meyer, 1843.)
1877 A, p. 239, pl. xxxii, fig. 5. (M. serratus.)
Woodward, A. S. 1888 F, p. 38. (M. serratus.)
1889 D, p. 123. (M. serratus.)
Eocene; New Jersey.

Myliobatis mordax Leidy.

Leidy, J. 1877 A, p. 234, pl. xxxiv, fig. 3.
1876 E, p. 86. (No description.)
Woodward, A. S. 1889 D, p. 123.
Phosphate beds; South Carolina.

Myliobatis pachyodon Cope.

Cope, E. D. 1867 C, p. 140.
Leidy, J. 1877 A, p. 242, pl. xxxii, fig. 6.
Woodward, A. S. 1889 D, p. 123.
Miocene; Maryland.

Myliobatis rectidens Cope.

Cope, E. D. 1870 F, p. 294.
Cretaceous; New Jersey.

Myliobatis rugosus Leidy.

Leidy, J. 1855 C, p. 395.
1855 C, p. 396. (M. obesus.)
1877 A, p. 236, pl. xxxi, figs. 6-10; pl. xxxiv,
fig. 44. (M. obesus.)
Woodward, A. S. 1889 D, p. 123.
Eocene; New Jersey.

Myliobatis transversalis Gibbes.

Gibbes, R. W. 1850 G, pl. xlii, fig. 4.
1850 B. (No description.)
Woodward, A. S. 1889 D, p. 124.
Eocene; South Carolina.

Myliobatis vicomicanus Cope.

Cope, E. D. 1867 C, p. 140.
1870 F, p. 294.
Leidy, J. 1877 A, p. 242, pl. xxxiii, fig. 5.
Woodward, A. S. 1889 D, p. 122. (Syn. of M. gigas.)
Miocene; Maryland.

Myliobatis sp. indet.

Leidy, J. 1873 B, p. 354, pl. xxxii, figs. 52, 53.
Miocene; Virginia.

MESOBATIS Leidy.   Type Aëtobatis eximius Leidy.

Leidy J. 1877 A, p. 244.
Woodward, A. S. 1889 D, p. 131.

Mesobatis eximius Leidy.

Leidy, J. 1855 C, p. 396. (Aëtobatis.)
1877 A, p. 244, pl. xxxi, fig. 12.

Woodward, A. S. 1889 D, p. 131.
Phosphate beds; South Carolina.

AËTOBATUS Blainville. Types *A. vulgaris, A. narinari, etc.*

Blainville, H. M. D., 1816, Jour. de Physique, lxxxiii, p. 261.
　Most of the authors here cited spell this name Aëtobatis.
Agassiz, L. 1843 B, iii, p. 325.
Duméril, A. 1865 A, p. 640.
Gill, T. 1894 A, p. 114.
Günther, A. 1870 A, p. 492.
　1880 A, p. 345.
Hasse, C. 1882 A, p. 155, pl xxi.
Jaekel, O. 1894 A, p. 130.
　1899 A, p. 266.
Jordan and Evermann 1896 A, p. 88.
Müller and Henle 1837 A, p. 118.
　1837 B, p. 401.
　1841 A, p. 179.
Pictet, F. J. 1854 A, p. 281.
Quenstedt, F. A. 1852 A, p. 182.
　1885 A, p. 287.
Woodward, A. S. 1889 D, p. 127.
　1899 A, p. 5.
Zittel, K. A. 1890 A, p. 101.

Aëtobatus arcuatus Agassiz.
Agassiz, L. 1843 B, iii, p. 327. (Aëtobatis.)
Cope, E. D. 1867 C, p. 139. (Aëtobatis.)
　1875 V, p. 29. (Aëtobatis.)
Leidy, J. 1877 A, p. 245, pl. xxxi, fig. 14. (Ætobatis.)
Woodward, A. S. 1889 D, 130. (Aëtobatis.)
　Miocene; Europe, Maryland, North Carolina.
　The identification of the American specimens is doubtful.

Aëtobatus perspicuus Leidy.
Leidy, J. 1855 C, p. 396. (Aëtobatis).
　1877 A, p. 244, pl. 31, fig. 13. (Ætobatis).
Woodward, A. S. 1889 D, p. 131. (Aëtobatis).
　Eocene ?; New Jersey.

Aëtobatus profundus Cope.
Cope, E. D. 1867 C, p. 139. (Aëtobatis).
Leidy, J. 1877 A, p. 246, pl. xxxi, fig. 19. (Ætobatis.)
Woodward, A. S. 1889 D, p. 131. (Aëtobatis.)
　Miocene; Maryland.

PLINTHICUS Cope. Type *P. stenodon* Cope.

Cope, E. D. 1869 L, p. 316.
Woodward, A. S. 1889 D, p. 459. (Syn.? of Aëtobatus.)

Woodward, A. S. 1889 D, p. 459.
　Miocene; New Jersey.

Plinthicus stenodon Cope.
Cope, E. D. 1869 L, p. 316.
　1875 H, p. 362.

RHINOPTERA Kuhl. Type *R. marginata* Cuvier.

Kuhl, 1828, in Cuvier's Règne Anim., ed. 2, p. 401.
Agassiz, L. 1843 B, iii, p. 79 (Zygobatis); p. 328 (Zygobates).
Duméril, A. 1865 A, p. 644.
Günther, A. 1880 A, p. 346.
Hasse, C. 1882 A, p. 158, pl. xxi.
Jaekel, O. 1894 A, p. 129.
Müller and Henle 1837 B, p. 401.
　1841 A, p. 181.
Quenstedt, F. A. 1885 A, p. 288. (Zygobatis, Rhinoptera.)
Woodward, A. S. 1885 B, p. 270. (Zygobatis.)
　1889 D, p. 125. (Rhinoptera.)
　1899 A, p. 5. (Rhinoptera or Zygobatis.)

Zittel, K. A. 1890 A, p. 101. (Zygobatis.)

Rhinoptera dubia Leidy.
Leidy, J. 1855 C, p. 396. (Zygobatis.)
　1877 A, p. 247, pl. xxxi, figs. 21–37.
Woodward, A. S. 1889 D, p. 127.
　Phosphate beds; South Carolina.

Rhinoptera, sp. indet.
Agassiz, L. 1856 A, p. 275. (Zygobates.)
　1857 A, p. 316, pl. i, figs. 31–35. (Zygobates.)
Leidy, J. 1871 C, p. 370. (Zygobates.)
　Miocene; California.

MANTIDÆ.

Jordan and Evermann 1896 A, p. 91.
Duméril, A. 1865 A, p. 649. (Cephalopteræ.)
Gill, T. 1893 A, p. 130. (Mobulidæ.)
Günther, A. 1870 A, p. 435. (Ceratoptera.)

Jaekel, O. 1894 A, pp. 115–150. (Ceratopterina.)
　1898 A, p. 51. (Ceratopterinæ.)
Müller and Henle 1838 A, p. 91. (Cephalopteræ.)
　1841 A, p. 184. (Cephalopteræ.)

MANTA Bancroft. Type *M. birostris* Walb.

Bancroft, E. N. 1829, Zool. Jour., iv, p. 444.
Duméril, A. 1865 A, p. 659. (Ceratoptera.)
Günther, A. 1870 A, p. 497. (Ceratoptera.)
　1880 A, p. 347. (Ceratoptera.)
Hasse, C. 1882 A, p. 159, pl. xxi. (Ceratoptera.)
　1885 A, p. 20. (Ceratoptera.)

Jaekel, O. 1894 A. (Ceratoptera.)
Jordan and Evermann 1896 A, p. 92. (Manta.)
Müller and Henle 1837 A, p. 118. (Ceratoptera.)
　1837 B, p. 401. (Ceratoptera.)
　1838 A, p. 91. (Ceratoptera.)
　1841 A, p. 184. (Ceratoptera.)

Zittel, K. A. 1890 A, p. 102. (Ceratoptera.)

Manta unios (Leidy).

Leidy, J. 1876 E, p. 86. (Ceratoptera.)
  1877 A, p. 248, pl. xxxiv, figs. 1, 2. (Ceratoptera.)

Woodward, A. S. 1889 D, p. 131. (Ceratoptera.)
  Phosphate beds; South Carolina.

Subclass HOLOCEPHALA Bonaparte.[1]

Bonaparte, C. L. 1832–41, Icon. Fauna Ital., iii.
  (Pesci), Introd. (Holocephala.)
Bridge, T. W. 1896 A, p. 534.
Cope, E. D. 1872 C.
  .   1877 O, p. 41.
  1878 A, p. 292.
  1884 N, p. 579.
  1884 HH.
  1885 G.
  1887 S, p. 324.
  1889 R, p. 854.
  1891 N, p. 10.
  1898 B, pp. 17, 28.
Duméril, A. 1865 A, p. 663.
Gadow, H. 1898 A, p. 6.
Gadow and Abbott 1895 A.
Gegenbaur, C. 1865 B, pp. 88, 145.
  1898 A.
Günther, A. 1870 A, p. 348.
  1880 A, p. 348.
  1881 A, p. 686.

Haeckel, E. 1895 A, p. 243.
Hasse, C. 1878 A.
  1882 A, p. 25.
Hubrecht, A. A. W. 1877 A.
  1878 A.
Huxley, T. H. 1872 A, p. 120.
  1876 A.
Müller, J. 1835 A.
  1846 C, p. 203.
Nicholson and Lydekker 1889 A, p. 949. (Chimeroidei.)
Owen, R. 1860 E, p. 116.
  1866 A.
Reis, O. M. 1896 A, p. 204.
  1897 A, p. 65.
Stannius, H. 1854 A.
Steinmann and Döderlein 1890 A, p. 550.
Woodward, A. S. 1889 D, p. vi.
  1891 A, pp. xvi, 36.
  1898 B, p. 154.
Zittel, K. A. 1890 A, p. 106.

Superfamily CHIMÆROIDEA.[2]

Cope, E. D. 1889 R, p. 854.
  1891 N, p. 14.
  1898 B, p. 28.
Davies, W. 1872 A.
Dean, B. 1895 A, p. 99.

Newton, E. T. 1878 C.
Woodward, A. S. 1888 I, p. 299.
  ·   1891 A, p. 36.
  1898 B, p. 54.

PTYCTODONTIDÆ.

Woodward, A. S. 1891 A, p. 37.
Cope, E. D. 1891 N, p. 14.
  1898 B, p. 28.
Eastman, C. R. 1898 B.

PTYCTODUS Pander.   Type *P. obliquus* Pander.

Pander, C. H. 1858 A, p. 48.
Cope, E. D. 1871 D, p. 423.
  1872 F, p. 339.
Eastman, C. R. 1898 B, pp. 473, 548.
Miller, S. A. 1889 A, p. 610.
Newberry and Worthen 1866 A, p. 106. (Rinodus, type *R. calceolus.*)
  1870 A, p. 374.
Rohon, J. V. 1895 A.
Steinmann and Döderlein, 1890 A, p. 50.
Woodward, A. S. 1891 A, p. 38.
  1898 B, p. 55.
Zittel, K. A. 1890 A, p. 108.

Ptyctodus calceolus Newb. and Worth.

Newberry and Worthen, 1866 A, p. 106, pl. x, figs. 10–10c. (Rinodus.)

Eastman, C. R. 1898 A, p. 115, fig. 10, A.
  1898 B, p. 476, figs. 1–17.
  1899 B, p. 282.
Lesley, J. P. 1889 A, p. 841, figures.
Miller, S. A. 1889 A, p. 611, fig. 1170.
Newberry, J. S. 1875 A, p. 59, pl. lix, figs. 13, 13a. (Ptyctodus.)
  1889 A, p. 62. (Rhynchodus.)
Newberry and Worthen, 1870 A, p. 374.
Weller, S. 1899 A.
Whiteaves, J. F. 1892 A, p. 353.
  1898 B, p. 410.
Woodward, A. S. 1891 A, p. 39.
  Devonian (Hamilton); Illinois, Iowa, Missouri, Manitoba.

---

[1] This name is by most authors spelled *Holocephali.*
[2] The name of this superfamily is commonly spelled *Chimæroidei.*

Ptyctodus compressus Eastman.

Eastman, C. R. 1898 B, p. 479, figs. 18–27.
Devonian (Hamilton); Wisconsin.

Ptyctodus ferox Eastman.

Eastman, C. R. 1898 B, p. 480, figs. 35–40; p. 552.
1899 B, p. 282.
Devonian (Hamilton); Wisconsin.

RHYNCHODUS Newb.   Type *R. secans* Newb.

Newberry, J. S. 1873 A, pp. 307, 254, 266.
Eastman, C. R. 1900 C, p. 392.
Huene, F. 1900 A, p. 64.
Miller, S. A. 1889 A, p. 611.
Newberry, J. S. 1870 B, p. 153.
    1875 A, p. 58.
    1889 A, p. 45.
Nicholson and Lydekker 1889 A, p. 952.   (Rhyn-
    chodes.)
Woodward, A. S. 1891 A, p. 39.
    1898 B, p. 55.
Zittel, K. A. 1890 A, p, 108.   (Rhynchodes.)

Rhynchodus excavatus Newb.

Newberry, J. S. 1877 A, p. 397.
Eastman, C. R. 1898 B, p. 486.
    1899 B, p. 282.
Newberry, J. S. 1878 A, p. 192.
    1889 A, p. 50, pl. xxix, fig. 1.
Woodward, A. S. 1891 A, p. 39.
Devonian (Hamilton); Wisconsin, Iowa.

Rhynchodus occidentalis Newb.

N. wberry, J. S. 1878 A, p. 192.
Eastman, C. R. 1898 B, p. 485.
Woodward, A. S. 1891 A, p. 39.
Devonian (Hamilton); Iowa.

Rhynchodus secans Newb.

Newberry, J. S. 1873 A, p. 310, pl. xxvi   figs. 1, 1a;
    pl. xxix, figs. 1–2a.
Eastman, C. R. 1898 B, pp. 485, 546.
Lesley, J. P. 1889 A, p. 881, figures.
    1892 A, pl. clix.
Newberry, J. S. 1889 A, p. 47, pl. xxviii, figs. 1–3.
Woodward, A. S. 1891 A, p. 39.
Devonian (Corniferous); Ohio.

Rhynchodus sp. indet.

Whiteaves, J. F. 1892 A, p. 353.   (No description).
Devonian; Manitoba.

PALÆOMYLUS Woodward.   Type *Rhynchodus frangens* Newb.

Woodward, A. S. 1891 A, p. 39.
Eastman, C. R. 1898 B, p. 545.

Palæomylus crassus (Newb.).

Newberry, J. S. 1873 A, p. 312, pl. xxix, fig. 3.
    (Rhynchodus.)
Eastman, C. R. 1898 B, p. 545.
Lesley, J. P. 1889 A, p. 879, figure. (Rhynchodus.)
    1892 A, pl. cl. (Rhynchodus.)
Newberry, J. S. 1889 A, p. 49, pl. xxviii, fig. 4.
    (Rhynchodus.)
Woodward, A. S. 1891 A, p. 40.
Devonian (Corniferous); Ohio.

Palæomylus frangens (Newb.).

Newberry, J. S. 1873 A, p. 311, pl. xxviii, figs. 2,
    2a, 3. (Rhynchodus.)

Eastman, C. R. 1898 B, p. 545.
Lesley, J. P., 1889 A, p. 880, figure. (Rhynchodus.)
    1892 A, pl. clix. (Rhynchodus.)
Miller, S. A, 1889 A, p. 611, fig. 1174. (Rhyncho-
    dus.)
Newberry, J. S. 1889 A, p. 48, pl. xxix, figs. 2, 3.
Woodward A. S. 1891 A, p. 40.
Devonian (Corniferous); Ohio.

Palæomylus greenei (Newb.).

Newberry, J. S. 1889 A, p. 51. (Rhynchodus.)
Eastman, C. R. 1898 B, p. 546, fig. 48.
Woodward, A. S. 1891 A, p. 40.
Devonian (Hamilton); Wisconsin.

CHIMÆRIDÆ.

Agassiz, L. 1843 B, iii, p. 336. ("Chimérides.")
Brühl, B. C. 1847 A. ("Chimæren.")
Davies, W. 1872 A.
Dean, B. 1895 A.
Egerton, P. G. 1847 A.
    1871 A.
Gadow and Abbott 1895 A.
Gegenbaur, C. 1865 B, p. 88.
Günther, A. 1870 A, p. 349.
    1880 A, p. 348.
    1881 A, p. 686.

Hubrecht, A. A. W. 1877 A.
Jordan and Evermann 1896 A, p. 98.
Köstlin, O. 1844 A, p. 417.
Newton, E. T. 1878 C.
Nicholson and Lydekker 1889 A, p. 950.
Reis, O. M. 1895 A.
Stannius, H. 1854 A. (Chimæræ.)
Woodward, A. S. 1891 A, p. 52.
    1898 B, p. 58.
Zittel, K. A. 1890 A, p. 107.

324   FOSSIL VERTEBRATA OF NORTH AMERICA.   [BULL. 179.

CHIMÆRA Linn.   Type *C. monstrosa* Linn.

Linnæus, C., 1758, Syst. Nat., ed. x, p 236.
Agassiz, L. 1843 B, iii, pp. 64, 336, 354.
　1844 A.
Brühl, B. C. 1847 A.
Duméril, A. 1865 A, p. 684.
Egerton, P. G. 1847 A.
Gegenbaur, C. 1898 A.
Goette, A. 1878 A, p. 531.
Hasse, C. 1882 A, p. 26.
Howes, G. B. 1887 A, p. 22.
Hubrecht, A. A. W. 1877 A.
　1878 A.
Huxley, T. H. 1872 A, p. 113
　1876 A. .
Kölliker, A. 1860 B.

Marsh, O. C. 1877 E.
Metschnikoff, O. 1879 A, pl. xxiv, figs. 8–12
Müller, J. 1835 A.
Owen, R. 1845 B, p. 64, pl. xxix.
　1860 E, p. 116.
Parker, W. K. 1871 A, p. 194.
Reis, O. M. 1895 A.
Röse, C. 1895 A, p. 568.
Ryder, J. A. 1886 A.
Stannius, H. 1854 A.
Stephan, P. 1900 A.
Woodward, A. S. 1891 A, p. 91.
Zittel, K. A. 1890 A, p. 112.
　There are no known American fossil species
belonging to this genus.

EDAPHODON Buckland.   Type *E. bucklandi* Agassiz.

Buckland, W., 1838, Proc. Geol. Soc., ii, p. 687.
Agassiz, L. 1843 B, iii, p. 340 (Psittacodon, in part); p. 350 (Edaphodon).
Ammon, L. 1896 A. (On Ischyodus.)
Cope, E. D. 1875 E, p. 283. (Diphrissa.)
　1875 K. (Diphrissa, type *Ischyodus solidulus*)
Egerton, P. G. 1847 A, p. 351.
Leidy, J. 1873 B, p. 306 (Edaphodon), p. 309 (Eumylodus, type *E. laqueatus*).
Marsh, O. C. 1870 E, p. 230. (Dipristis.)
Newton, E. T. 1878 C.
Nicholson and Lydekker 1889 A. p. 951.
Noetling, F. 1885 A, p. 1.
Owen, R. 1860 E, p. 117.
Philippi, E. 1897 A, p. 1. (On Ischyodus.)
Pictet, F. J. 1854 A, p. 232.
Woodward, A. S. 1888 I, p. 300.
　1891 A, p. 73.
Zittel, K. A. 1890 A, p. 111 (Edaphodon); p. 112 (Eumylodus).

Edaphodon divaricatus (Cope).

Cope, E. D. 1869 L, p. 315. (Ischyodus.)
　1875 E, pp. 284, 294. (Ischyodus.)
Woodward, A. S. 1891 A, p. 85.
　Cretaceous; New Jersey.

Edaphodon eocænus (Cope).

Cope, E. D. 1875 E, pp. 285, 288. (Ischyodus.)
Woodward, A. S. 1891 A, p. 85.
　Eocene; New Jersey.

Edaphodon fecundus (Cope).

Cope, E. D. 1875 E, pp. 285, 291. (Ischyodus.)
Woodward, A. S. 1891 A, p. 85.
　Cretaceous; New Jersey.

Edaphodon gaskillii (Cope).

Cope, E. D. 1875 E, pp. 285, 290. (Ischyodus.)
Woodward, A. S. 1891 A, p. 85.
　Cretaceous; New Jersey.

Edaphodon incrassatus (Cope).

Cope, E. D. 1875 E, pp. 285, 289. (Ischyodus.)
Woodward, A. S. 1891 A, p. 85.
　Cretaceous; New Jersey.

Edaphodon laqueatus (Leidy).

Leidy, J. 1873 B, pp. 309, 351, pls. xix, figs. 21, 22; pl. xxxvii, figs. 13, 14. (Eumylodus.)
Cope, E. D, 1875 E, p. 282. (Eumylodus.)
Woodward, A. S. 1891 A, p. 86.
　Cretaceous; Mississippi.

Edaphodon laterigerus (Cope).

Cope, E. D. 1869 F, p. 243. (Ischyodus.)
　1875 E, pp. 284, 288. (Ischyodus.)
Woodward, A. S. 1891 A, p. 85.
　Cretaceous; New Jersey.

Edaphodon latidens (Cope).

Cope, E. D. 1875 E, p. 283. (Diphrissa.)
Woodward, A. S. 1891 A, p. 86.
　Cretaceous; New Jersey.

Edaphodon longirostris (Cope).

Cope, E. D. 1875 E, pp. 284, 287. (Ischyodus.)
Woodward, A. S. 1891 A, p. 85.
　Cretaceous; New Jersey.

Edaphodon miersii (Marsh).

Marsh, O. C. 1870 E, p. 230. (Dipristis.)
Cope, E. D. 1875 E, pp. 285, 292. (Ischyodus.)
Woodward, A. S. 1891 A, p. 85.
　Cretaceous; New Jersey.

Edaphodon mirificus Leidy.

Leidy, J. 1856 K, p. 221.
Cope, E. D. 1875 E, pp. 285, 291. (Ischyodus.)
Leidy, J. 1873 B, pp. 306, 350, pl. xxxvii, figs. 6–12.
Newton, E. T. 1878 C, p. 24.
Woodward, A. S. 1891 A, p. 85.
　Cretaceous; New Jersey.

Edaphodon monolophus (Cope).

Cope, E. D. 1869 L, p. 314. (Ischyodus.)
　1875 E, pp. 285, 289. (Ischyodus.)
Woodward, A. S. 1891 A, p. 85.
　Cretaceous; New Jersey.

Edaphodon smockii (Cope).

Cope, E. D. 1869 L, p. 316. (Ischyodus.)
　1875 E, pp. 284, 288. (Ischyodus.)
Woodward, A. S. 1891 A, p. 85.
　Cretaceous; New Jersey.

Edaphodon solidulus (Cope).

Cope, E. D. 1869 F. p. 244. (Ischyodus.)
  1875 E, p. 283. (Diphrissa.)
  1875 K. (Diphrissa.)
Woodward, A. S. 1891 A, p. 86.
  Cretaceous; New Jersey.

Edaphodon stenobryus (Cope).

Cope, E. D. 1875 E, p. 285. (Ischyodus.)

Woodward, A. S. 1891 A, p. 85.
  Cretaceous; New Jersey.

Edaphodon tripartitus (Cope).

Cope, E. D. 1875 E, pp. 284, 286. (Ischyodus.)
  1869 L, p. 314. (I. mirificus, errore.)
Woodward, A. S. 1891 A, p. 85.
  Cretaceous; New Jersey.

MYLOGNATHUS Leidy.   Type *M. priscus* Leidy.

Leidy, J. 1856 Q, p. 312.
  1860 A, p. 153.
Nicholson and Lydekker 1889 A, p. 951.
Woodward, A. S. 1891 A, p. 86. (Syn. of Edaphodon.)

Mylognathus priscus Leidy.

Leidy, J. 1856 Q, p. 312.
Cope, E. D. 1875 E, p. 293.
Leidy, J. 1860 A, p. 153, pl. xi, figs. 24–30.
Woodward, A. S. 1891 A, p. 86. (Edaphodon?.)
  Eocene (Fort Union); Nebraska.

BRYACTINUS Cope.   Type *B. amorphus* Cope.

Cope, E. D. 1875 E, p. 282.
Nicholson and Lydekker 1889 A, p. 951.
Woodward, A. S. 1891 A, p. 86. (Syn.? of Eda-
  phodon.)

Bryactinus amorphus Cope.

Cope, E. D. 1875 E, p. 282, pl. xlv, fig. 12.
Woodward, A. S. 1891 A, p. 86. (Edaphodon?.)
  Cretaceous; New Jersey.

ISOTÆNIA Cope.   Type *I. neocæsariensis* Cope.

Cope, E. D. 1875 E, p. 293.
Nicholson and Lydekker 1889 A, p. 951.
Woodward, A. S. 1891 A, p. 86. (Syn. ? of Eda-
  phodon.)

Isotænia neocæsariensis Cope.

Cope, E. D. 1875 E, p. 293.
Woodward, A. S. 1891 A, p. 86. (Edaphodon?.)
  Cretaceous; New Jersey.

LEPTOMYLUS Cope.   Type *L. densus* Cope.

Cope, E. D. 1869 L, p. 313.
  1875 E, p. 281.
Nicholson and Lydekker 1889 A, p. 951.
Woodward, A. S. 1891 A, p. 87.

Leptomylus cookii Cope.

Cope, E. D. 1870 H, p. 384.
  1875 E, p. 382.
Woodward, A. S. 1891 A, p. 87.
  Cretaceous; New Jersey.

Leptomylus densus Cope.

Cope, E. D. 1869 L, p. 313.
  1875 E, p. 281.
Woodward, A. S. 1891 A, p. 86.
  Cretaceous; New Jersey.

Leptomylus forfex Cope.

Cope, E. D. 1875 E, p. 281.
Woodward, A. S. 1891 A, p. 87.
  Cretaceous; New Jersey.

SPHAGEPŒA Cope.   Type *S. aciculata* Cope.

Cope, E. D. 1869 F, p. 241.
  1875 E, p. 293.
Woodward, A. S. 1891 A, p. 84.

Sphagepœa aciculata Cope.

Cope, E. D. 1869 F, p. 241.
  1875 E, p. 293.

Woodward, A. S. 1891 A, p. 84.
  Cretaceous; New Jersey.

DICTYORHABDUS Walcott.   Type *D. priscus* Walcott.

Walcott, C. D. 1892 A, p. 165.
  Amer. Geologist, 1891 E. (Palæchimæra; no
  definition.)
    The position of the genus is doubtful.

Dictyorhabdus priscus Walcott.

Walcott, C. D. 1892 A, p. 165, pl. iii, figs. 1–5.

Amer. Geologist 1891· E. (Palæchimæra prisca;
  no description.)
Cope, E. D. 1893 I.
Dana, J. D. 1896 A, p. 509, fig. 697.
  Silurian (Ordovician); Colorado.

MYLEDAPHUS Cope.   Type *M. bipartitus* Cope.

Cope, E. D. 1876 H, p. 260.

Myledaphus bipartitus Cope.

Cope, E. D. 1876 H, p. 260.
  1877 G, p. 574.
  Cretaceous (Fort Union); Montana.
  The relationships of this fossil are doubtful.

HEDRONCHUS Cope.   Type *H. sternbergi* Cope.

Cope, E. D. 1876 H, P. 259.

Hedronchus sternbergi Cope.

Cope, E. D. 1876 H, p. 259.
  1877 G, p. 574.
  1880 Q.

Cretaceous (Ft. Union); Montana.
The relationships of this fossil are extremely problematical.

ICHTHYODORULITES.

Agassiz, L. 1843 B, iii, pp. 1, 67, 69, 212.
  1844 B, p. 111.
Hay, O. P. 1899 F, p. 682.
Jaekel, O. 1890 D.
Koninck, L. G. de 1878 A, p. 65.
Newberry and Worthen 1866 A, pp. 113-127.

Pictet, F. J. 1854 A, p. 229.
St. John and Worthen 1875 A, pp. 420-488
  1883 A, pp. 233-260.
Woodward, A. S. 1891 A, pp. 92-159.
Zittel, K. A. 1890 A, pp. 115-122.

ONCHUS Agassiz.   Type *O. murchisoni* Agassiz.

Agassiz, L. 1843 B (1837), iii, p. 6.
  1843 A.
  1844 A.
  1844 B, p. 117.
Eichwald E. 1860 A, p. 1594.
Hasse, C. 1878 A, p. 144.
  1879 A, p. 59.
McCoy, F. 1855 A, p. 577.
Nicholson and Lydekker 1889 A, p. 940.
Owen, R. 1860 E, p. 100.
Pictet, F. J. 1854 A, p. 283.
Reis, O. M. 1896 A.
Rohon, V. 1893 A, p. 38, with figures.
  1899 A, p. 9.
Traquair, R. H. 1899 A, p. 843.
Woodward, A. S. 1891 A, p. 94.
Zittel, K. A. 1890 A, pp. 64, 67, 166.

Onchus clintoni Claypole.

Claypole, E. W. 1885 D, p. 61, fig. 6.
  1884 A, p. 1225.   (No description).
  1895 G.
Lesley, J. P. 1892 A, p. 774.
Woodward, A. S. 1891 A, p. 96.
Zittel, K. A. 1890 A, p. 317.   (*O. clintonensis*.)
  Silurian (Clinton); Pennsylvania.

Onchus pennsylvanicus Claypole.

Claypole, E. W. 1885 D, p. 61, fig. 5.
  1885 A, p. 426.
  1885 C.
Lesley, J. P. 1892 A, p. 774.
  Upper Silurian; Pennsylvania.

Onchus rectus Eastman.

Eastman, C. R. 1899 F, p. 323, fig. 4.
  Devonian (Catskill); New York.

CTENACANTHUS Agassiz.   Type *C. major* Agassiz.

Agassiz, L. 1843 B (1837), iii, p. 10.
  1843 A.
  1844 A.
  1844 B, p. 119.
American Geologist 1888 A.
Barkas, W. J. 1874 A, p. 386.
  1878 B, p. 145.
Claypole, E. W. 1885 B.
  1895 C.
Davis, J. W. 1883 A, p. 332.
Garman, S. 1884 C.
Gill, T. 1884 C.
McCoy, F. 1855 A, p. 624.
Koninck L. G. de 1878 A, p. 65.

Miller, S. A. 1889 A, p. 593.
Newberry, J. S. 1873 A, pp. 279, 326.
  1873 C, p. 27.
  1889 A, p. 197.
Newberry and Worthen 1866 A, pp. 116, 118.
Nicholson and Lydekker 1889 A, pp. 940, 968.
Pictet, F. J. 1854 A, p. 299.
St. John and Worthen 1875 A, p. 432.
Thompson J. 1870 A.
  1871 A.
Traquair, R. H. 1884 B.
  1888 A, p. 81.
Woodward, A. S. 1884 A, p. 267.
  1891 A, p. 97.

Woodward and Sherborn 1890 A, p. 49.
Zittel, K. A. 1890 A, pp. 67, 116.

Ctenacanthus acutus Eastman.

Eastman, C. R. 1897 B, p. 12, fig. 2.
　　Subcarboniferous (Keokuk); Iowa.

Ctenacanthus amblyxiphias Cope.

Cope, E. D. 1891 B, p. 449, pl. xxviii, fig. 3.
　　Permian; Texas.

Ctenacanthus angulatus Newb. and Worth.

Newberry and Worthen 1866 A, p. 118, pl. xii, figs. 4, 4a.
Woodward, A. S. 1891 A, p. 102.
　　Subcarboniferous (Chester); Illinois.

Ctenacanthus angustus Newb.

Newberry, J. S. 1889 A, p. 181.
Claypole, E. W. 1896 E, p. 151, figs. 1, 2.
Woodward, A. S. 1891 A, p. 102.
　　Devonian (Cleveland shale); Ohio.

Ctenacanthus burlingtonensis St. J. and Worth.

St. John and Worthen 1875 A, p. 426, pl. xv, figs. 6, 7.
Woodward, A. S. 1891 A, p. 102.
　　Subcarboniferous (Burlington); Iowa, Illinois.

Ctenacanthus buttersi St. J. and Worth.

St. John and Worthen 1883 A, p. 241, pl. xxvi, fig. 2.
Woodward, A. S. 1891 A, p. 102.
　　Coal-measures; Illinois.

Ctenacanthus cannaliratus St. J. and Worth.

St. John and Worthen 1883 A, p. 239, pl. xxi, fig. 3.
Woodward, A. S. 1891 A, p. 102.
　　Subcarboniferous (Chester); Illinois.

Ctenacanthus chemungensis Claypole.

Claypole, E. W. 1885 B, p. 490. (Inadequate description.)
　　Devonian (Chemung); Pennsylvania.

Ctenacanthus clarkii Newb.

Newberry, J. S. 1889 A, p. 168, pl. xxvi, figs. 2, 3.
Woodward, A. S. 1891 A, p. 102.
　　Devonian (Cleveland shale); Ohio.

Ctenacanthus compressus Newb.

Newberry, J. S. 1878 A, p. 191.
　　1889 A, p. 168, pl. xxiii, fig. 4.
Woodward, A. S. 1891 A, p. 103.
　　Devonian (Cleveland shale); Ohio.

Ctenacanthus coxianus St. J. and Worth.

St. John and Worthen 1883 A, p. 233, pl. xxi, fig. 1.
Eastman, C. R. 1897 B, p. 13.
Woodward, A. S. 1891 A, p. 103.
　　Subcarboniferous (Keokuk); Iowa.

Ctenacanthus cylindricus Newb.

Newberry, J. S. 1889 A, p. 202, pl. xxxvi, fig. 1.
Eastman, C. R. 1897 B, p. 13.
Woodward, A. S. 1891 A, p. 103.
　　Subcarboniferous (Keokuk); Kentucky.

Ctenacanthus deflexus St. J. and Worth.

St. John and Worthen 1883 A, p. 234 pl. xxii, fig. 1.
Woodward, A. S. 1891 A, p. 103.
　　Subcarboniferous (St. Louis); Illinois.

Ctenacanthus depressus Newb.

Newberry, J. S. 1897 A, p. 290, pl. xxii, fig. 6.
　　Subcarboniferous (Kinderhook); Iowa.

Ctenacanthus elegans Tuomey.

Tuomey, M. 1858 A, p. 38, text-fig. A.
Woodward, A. S. 1891 A, p. 103.
　　Subcarboniferous; Alabama.

Ctenacanthus excavatus St. J. and Worth.

St. John and Worthen 1875 A, p. 428, pl. xv, figs. 4, 5.
Eastman, C. R. 1897 B, p. 13.
Keyes, C. R. 1894 A, p. 236.
Woodward, A. S. 1891 A, p. 103.
　　Subcarboniferous (Keokuk); Missouri, Iowa.

Ctenacanthus? fallax Woodward.

Woodward, A. S. 1891 A, p. 105. (Ex Leidy MS. "Indeterminate.")
Leidy J. 1857 G, pl. v, fig. 30. (No name or description).
　　Subcarboniferous; ——.

Ctenacanthus formosus Newb.

Newberry, J. S. 1873 A, p. 328, pl. 36, figs. 2-2b.
　　1875 A, p. 53, pl. lix, figs. 1-1e.
Woodward, A. S. 1891 A, p. 103.
　　Subcarboniferous (Waverly); Ohio, Kentucky.

Ctenacanthus furcicarinatus Newb.

Newberry, J. S. 1875 A, p. 54, pl. lix. figs. 2-2c.
Koninck, L. G. de 1878, A, p. 68.
Newberry, J. S. 1873 A, p. 279. (No description; teeth =? Orodus variabilis.)
Woodward, A. S. 1891 A, p. 103.
　　Subcarboniferous (Waverly); Kentucky.

Ctenacanthus gemmatus St. J. and Worth.

St. John and Worthen 1875 A, p. 429, pl. xv, figs. 9, 10.
Woodward, A. S. 1891 A, p. 103.
　　Subcarboniferous (St. Louis); Illinois.

Ctenacanthus gradocostatus St. J. and Worth.

St. John and Worthen 1875 A, p. 425, pl. xy, figs. 2, 3.
Woodward, A. S. 1891 A, p. 103.
　　Subcarboniferous (Burlington); Illinois, Iowa.

Ctenacanthus gurleyi Newb.

Newberry, J. S. 1897 A, p. 290.
　　Subcarboniferous (St. Louis); Indiana.

Ctenacanthus harrisoni St. J. and Worth.

St. John and Worthen 1883 A, p. 236, pl. xxiii, fig. 1.
Woodward, A. S. 1891 A, p. 103.
Subcarboniferous (St. Louis); Illinois.

Ctenacanthus keokuk St. J. and Worth.

St. John and Worthen 1875 A, p. 427, pl. xv, fig. 8.
Eastman, C. R. 1897 B, p. 13.
Keyes, C. R. 1894 A, p. 236.
Woodward, A. S. 1891 A, p. 103.
Subcarboniferous (Keokuk); Missouri, Illinois, Iowa.

Ctenacanthus littoni Newb.

Newberry, J. S. 1889 A, p. 201, pl. xxv, fig. 3.
Woodward, A. S. 1891 A, p. 103.
Subcarboniferous (St. Louis); Missouri.

Ctenacanthus marshii Newb.

Newberry, J. S. 1873 A, p. 326, pl. xxvi, figs. 3, 3b.
Woodward, A. S. 1891 A, p. 104.
Coal-measures; Ohio.

Ctenacanthus mayi Newb. and Worth.

Newberry and Worthen 1870 A, p. 372, pl. 11, figs. 2, 2a; also p. 346.
Woodward, A. S. 1891 A, p. 104.
Subcarboniferous (Burlington); Iowa.

Ctenacanthus pellensis St. J. and Worth.

St. John and Worthen 1883 A, p. 237, pl. xxi, fig. 2.
Woodward, A. S. 1891 A, p. 104.
Subcarboniferous (St. Louis); Iowa.

Ctenacanthus pugiunculus St. J. and Worth.

St. John and Worthen 1875 A, p. 430, pl. xxi, fig. 9.
Keyes, C. R. 1894 A, p. 236.
Newberry, J. S. 1897 A, p. 288, pl. xxii, fig. 4. (Oracanthus.)
Woodward, A. S. 1891 A, p. 104.
Subcarboniferous (St. Louis); Missouri, Illinois.

Ctenacanthus randalli Newb.

Newberry, J. S. 1889 A, p. 105.
Woodward, A. S. 1891 A, p. 104.
Devonian (Chemung); Pennsylvania.

Ctenacanthus sculptus St. J. and Worth.

St. John and Worthen 1875 A, p. 421, pl. xiv, fig. 1.

Woodward, A. S. 1891 A, p. 104.
Subcarboniferous (Kinderhook); Iowa.

Ctenacanthus similis St. J. and Worth.

St. John and Worthen 1875 A, p. 431, pl. xv, figs. 11-11d.
Woodward, A. S. 1891 A, p. 104.
Subcarboniferous (Chester); Illinois.

Ctenacanthus speciosus St. J. and Worth.

St. John and Worthen 1875 A, p. 424, pl. xiv, figs. 3, 4.
Davis, J. W. 1883 A, p. 340.
Woodward, A. S. 1891 A, p. 104.
Subcarboniferous (Kinderhook); Iowa.

Ctenacanthus spectabilis St. J. and Worth.

St. John and Worthen 1875 A, p. 420, pl. xv, figs. 1-1e.
Woodward, A. S. 1891 A, p. 104.
Subcarboniferous (Kinderhook); Iowa.

Ctenacanthus triangularis Newb.

Newberry, J. S. 1873 A, p. 329, pl. 36, figs. 1-1b.
Miller, S. A. 1889 A, p. 593, fig. 1118.
Trautschold, H. 1879 B, p. 61, pl. vii, fig. 15.
Woodward, A. S. 1891 A, p. 104.
Subcarboniferous (Waverly); Ohio; also in Russia.

Ctenacanthus varians St. J. and Worth.

St. John and Worthen 1875 A, p. 422, pl. xiv, fig. 12.
Woodward, A. S. 1891 A, p. 104.
Subcarboniferous (Kinderhook); Iowa.

Ctenacanthus vetustus Newb.

Newberry, J. S. 1873 A, p. 326, pl. xxxv, figs. 3-3d.
Claypole, E. W. 1885 B, p. 489.
Woodward, A. S. 1891 A, p. 104.
Devonian (Huron shale); Ohio.

Ctenacanthus wrighti Newb.

Newberry, J. S. 1884 B, p. 206, pl. xvi, figs. 12-14.
1889 A, p. 66, pl. xxvi, fig. 4.
Woodward, A. S. 1891 A, p. 104.
Devonian (Hamilton); New York.

Ctenacanthus xiphias (St. J. and Worth.).

St. John and Worthen 1883 A, p. 241, pl. xxvi, fig. 1. (Acondylacanthus.)
Eastman, C. R. 1897 B, p. 10, fig. 1.
Woodward, A. S. 1891 A, p. 110. (Acondylacanthus.)
Subcarboniferous (Keokuk); Iowa.

ANACLITACANTHUS St. J. and Worth.

St. John and Worthen 1875 A, p. 442.
Miller, S. A. 1889 A, p. 587.

Anaclitacanthus semicostatus St. J. and Worth.

St. John and Worthen 1875 A, p. 443, pl. xvi, fig. 14.

Type *A. semicostatus* St. J. and Worth.

Woodward, A. S. 1891 A, p. 105.
Subcarboniferous (Burlington); Iowa.

EUNEMACANTHUS St. J. and Worth.   Type *E. costatus* (Newb. and Worth.).

St. John and Worthen 1883 A, p. 246.
Miller, S. A. 1889 A, p. 597.

Eunemacanthus costatus (Newb. and Worth.).

Newberry and Worthen 1866 A, p. 120, pl. xii, fig. 5.
(Ctenacanthus? costatus.)

Hambach, G. 1890 A, p. 80.
St. John and Worthen 1883 A, p. 246, pl. xxiii, fig. 2.
Woodward, A. S. 1891 A, p. 105.
Subcarboniferous (St. Louis); Illinois, Missouri (*fide* Hambach).

HOMACANTHUS Agassiz.   Type *H. arcuatus* Agassiz.

Agassiz, L. 1844 B. p. 113.
Newberry and Worthen 1866 A, p. 113.
Nicholson and Lydekker 1889 A, p. 968.
Owen, R. 1860 E, p. 101.
Pictet, F. J. 1854 A, p. 288.
Woodward, A. S. 1891 A, p. 105.

Homacanthus gracilis Whiteaves.

Whiteaves, J. W. 1889 A, p. 96, pl. x, fig. 4.

McCoy, F. 1855 A, p. 682.
Traquair, R. H. 1890 A, p. 21.   (Climatius.)
Whiteaves, J. F. 1881 E, p. 106.   (H. sp. undet.)
1883 A, p. 355.   (H. sp.?.)
1883 B, p. 164.   (H. sp. undet.)
Woodward, A. S. 1891 A, p. 106.
Lower Devonian; New Brunswick.

AMACANTHUS St. J. and Worth.   Type *A. gibbosus* (Newb. and Worth.).

St. John and Worthen 1875 A, p. 464.
Miller, S. A. 1889 A, p. 587.

Amacanthus gibbosus (Newb. and Worth.).

Newberry and Worthen 1866 A, p. 113, pl. xii, fig. 1.
(Homacanthus.)

Keyes, C. R. 1894 A, p. 237.
Miller, S. A. 1889 A, p. 587.
St. John and Worthen 1875 A, p. 464, pl. xxii, fig. 6.
Woodward, A. S. 1891 A, p. 106.
Subcarboniferous (St. Louis); Missouri, Illinois.

MARRACANTHUS St. J. and Worth.   Type *M. rectus* (Newb. and Worth.).

St. John and Worthen 1875 A, p. 465.
Keyes, C. R. 1894 A, p. 237.
Miller, S. A. 1889 A, p. 601.

Marracanthus rectus (Newb. and Worth.).

Newberry and Worthen 1866 A, p. 115, pl. xii, fig. 6.
(Homacanthus?)

Miller, S. A. 1889 A, p. 602, fig. 1147.
St. John and Worthen 1875 A, p. 466, pl. xxii, figs. 7–9.
Woodward, A. S. 1891 A, p. 106.
Subcarboniferous (St. Louis); Missouri, Illinois.

HOPLONCHUS J. W. Davis.   Type *H. elegans* J. W. Davis.

Davis, J. W. 1876, Quart. Jour. Geol. Soc., xxxii, p. 336.
1879, Quart. Jour. Geol. Soc., xxxv, p. 181.
Newberry, J. S. 1889 A, p. 169.
Woodward, A. S. 1891 A, p. 107.

Hoplonchus parvulus Newb.

Newberry, J. S. 1875 A, p. 55, pl. lix, fig. 3.
(Ctenacanthus.)
1889 A, p. 169, pl. xxv, fig. 5.
Woodward, A. S, 1891 A, p. 107.
Devonian (Cleveland shale); Ohio.

ACONDYLACANTHUS St. J. and Worth.   Type *A. gracilis* St. J. and Worth.

St. John and Worthen 1875 A, p. 432.
Davis, J. W. 1883 A, p. 346.
Miller, S. A. 1889 A, p. 586.
Woodward, A. S. 1886 E, p. 4.
1891 A, p. 107.
Woodward and Sherborn 1890 A, p. 4.
Zittel, K. A. 1890 A, p. 117.

Acondylacanthus aequicostatus St. J. and Worth.

St. John and Worthen 1875 A, p. 434, pl. xvi, figs 12, 13.
Woodward, A. S. 1891 A, p. 109.
Subcarboniferous (Keokuk); Iowa.

Acondylacanthus gracilis St. J. and Worth.

St. John and Worthen 1875 A, p. 433, pl. xvi, figs. 8–11.
Miller, S. A. 1889 A, p. 586, fig. 1099.
Woodward, A. S. 1891 A, p. 109.
Subcarboniferous (Kinderhook); Iowa.

Acondylacanthus gracillimus (Newb. and Worth.).

Newberry and Worthen 1866 A, p. 126, pl. xiii, fig. 3. (Ctenacanthus.)
Keyes, C. R. 1894 A, p. 236.   (Ctenacanthus.)

Newberry, J. S. 1889 A, p. 206, pl. xxv, fig. 6. (A. occidentalis.)
Newberry and Worthen 1866 A, p 116, pl. xii, fig. 2. (Leptacanthus ? occidentalis.)
St. John and Worthen 1875 A, p. 435. (A. occidentalis.)
1883 A, p. 238, pl. xxiv, fig. 1. (Ctenacanthus.)
Woodward, A. S. 1891 A, p. 109. (A. occidentalis.)
Subcarboniferous (St. Louis); Missouri, Illinois, Michigan.

Acondylacanthus? mudgianus St. J. and Worth.

St. John and Worthen. 1883 A, p. 244, pl. xxiv, fig. 3.

ASTEROPTYCHIUS McCoy.

McCoy, F. 1848 A, p. 118.
Agassiz, L. 1843 B, iii, p. 176. (Name only).
1844 A.
McCoy, F. 1855 A, p. 615.
Miller, S. A. 1889 A, p. 588.
Newberry, J. S. 1889 A, p. 186.
Woodward, A. S. 1891 A, p. 110.
Zittel, K. A. 1890 A, p. 117.

Asteroptychius bellulus St. J. and Worth.

St. John and Worthen 1875 A, p. 439, pl. xvi, fig. 7.
Woodward, A. S. 1891 A, p. 111.
Coal-measures: Illinois, Iowa.

Asteroptychius elegans Newb.

Newberry, J. S. 1889 A, p. 176, pl. xxv, fig. 4.
Woodward, A. S. 1891 A, p. 111.
Subcarboniferous (Waverly); Michigan.

Asteroptychius gracilis Newb.

Newberry, J. S. 1897 A, p. 293.
Subcarboniferous (St. Louis); Illinois.

Asteroptychius keokuk St. J. and Worth.

St. John and Worthen 1875 A, p. 436, pl. xvi, fig. 2.
Woodward, A. S. 1891 A, p. 111.
Subcarboniferous (Keokuk); Illinois.

Asteroptychius sancti-ludovici St. J. and Worth.

St. John and Worthen 1875 A, p. 437, pl. xvi, figs. 3, 4.

COSMACANTHUS Agassiz.

Agassiz, L. 1844 B, p. 120.
Miller, S. A. 1889 A, p. 598. (Gisacanthus.)
Pictet, F. J. 1854 A, p. 287.
St. John and Worthen 1875 A, p. 440. (Geisacanthus, type G. stellatus.)
Woodward, A. S. 1891 A, p. 111.
Woodward and Sherborn 1890 A, p. 47.
Zittel, K. A. 1890 A, p. 117.

Cosmacanthus bullatus (St. J. and Worth.).

St. John and Worthen 1875 A, p. 441, pl. xvii, figs. 3, 4. (Geisacanthus.)

Woodward, A. S. 1891 A, p. 110.
Coal-measures; Kansas.

Acondylacanthus nuperus St. J. and Worth.

St. John and Worthen 1883 A, p. 242, pl. xxvi, fig. 3.
Woodward, A. S. 1891 A, p. 110.
Coal-measures; Illinois.

Acondylacanthus rectus St. J. and Worth.

St. John and Worthen 1883 A, p. 241, pl. xxvi, fig. 2
Woodward, A. S. 1891 A, p. 109.
Coal-measures; Illinois.

Type A. ornatus McCoy.

Keyes, C. R. 1894 A, p. 236.
Woodward, A. S. 1891 A, p. 111.
Subcarboniferous (St. Louis); Illinois, Missouri.

Asteroptychius tenellus St. J. and Worth.

St. John and Worthen 1883 A, p. 248, pl. xxi, fig. 4.
Woodward, A. S. 1891 A, p. 111.
Coal-measures: Kansas.

Asteroptychius ? tenuis St. J. and Worth.

St. John and Worthen 1875 A, p. 438, pl. xvi, figs. 5, 6.
Woodward, A. S. 1891 A, p. 111.
Subcarboniferous (Chester); Illinois.

Asteroptychius triangularis Newb. and Worth.

Newberry and Worthen 1870 A, p. 370, pl. ii, fig. 4.
Woodward, A. S. 1891 A, p. 111.
Subcarboniferous (Burlington); Iowa.

Asteroptychius vetustus St. J. and Worth.

St. John and Worthen 1875 A, p. 435, pl. xvi, fig. 1.
Woodward, A. S. 1891 A, p. 111.
Subcarboniferous (Kinderhook); Iowa.

Type C. malcolmsoni Agassiz.

Woodward, A. S. 1891 A, p. 112.
Subcarboniferous (Chester); Illinois.

Cosmacanthus stellatus (St. J. and Worth.).

St. John and Worthen 1875 A, p. 440, pl. xxi, fig. 10. (Geisacanthus.)
Keyes, C. R. 1894 A, p. 236. (Geisacanthus.)
Woodward, A. S. 1891 A, p. 113.
Subcarboniferous (St. Louis); Missouri.

BYTHIACANTHUS St. J. and Worth.   Type *B. vanhornei* St. J. and Worth.

St. John and Worthen 1875 A, p. 444.
Miller, S. A. 1889 A, p. 589.
Newberry, J. S. 1889 A, p. 186.

Bythiacanthus ? siderius (Leidy).
Leidy, J. 1870 H, p. 18.   (Asteracanthus.) 1873 B,
     pp. 313, 353, pl. xxxii, fig. 59. (Astera-
     canthus.)
St. John and Worthen 1875 A, p. 445.

Woodward, A. S. 1889 D, p. 320.
     1891 A, p. 113.
Subcarboniferous; Tennessee.

Bythiacanthus vanhornei St. J. and
Worth.
St. John and Worthen 1875 A, p. 445, pl. xvii, fig. 1.
Woodward, A. S. 1891 A, p. 113.
Subcarboniferous (St. Louis); Illinois.

GLYMMATACANTHUS St. J. and Worth.   Type *G. irishii* St. J. and Worth.

St. John and Worthen 1875 A, p. 446.
Miller, S. A. 1889 A, p. 596.
Newberry, J. S. 1889 A, p. 186.

Glymmatacanthus irishii St. J. and
Worth.
St. John and Worthen 1875 A, p. 447, pl. xvii, fig. 2.
Miller, S. A. 1880 A, p. 596, fig. 1133.
Woodward, A. S. 1891 A, p. 113.
Subcarboniferous (Kinderhook); Iowa.

Glymmatacanthus petrodoides St. J.
and Worth.
St. John and Worthen 1883 A, p. 250, pl. xxv. fig. 2.
Woodward, A. S. 1891 A, p. 113.
Subcarboniferous (Chester); Illinois.

Glymmatacanthus rudis St. J. and
Worth.
St. John and Worthen 1883 A, p. 249, pl. xxv, fig. 1.
Woodward, A. S. 1891 A, p. 113.
Subcarboniferous (Keokuk); Iowa.

APATEACANTHUS Woodward.   Type *A. vetustus* (Clarke).

Woodward, A. S. 1891 A, p. 118.

Apateacanthus vetustus (Clarke, J.
M.).
Clarke, J. M. 1885 A, p. 42, pl. i, fig. 7. (Pristacan-
thus.)

Lesley, J. P., 1889 A, p. 744, figure. (Pristacanthus.)
Newberry, J. S. 1889 A, p. 61.
Woodward, A. S. 1891 A, p. 119.
Devonian (Genesee); New York.

COELORHYNCHUS Agassiz.   Type *C. rectus* Agassiz.

Agassiz, L. 1843 B, v. p. 92.
     1844 A.
Cope, E. D. 1890 A.
Gibbes, R. W. 1850 A.
Leidy, J. 1856 A, p. 422. (Cylindracanthus.)
     1856 E, p. 12. (Cylindracanthus; type *C. or-
     natus.*)
Marsh, O. C. 1870 E. p. 228.
Reis, O. M. 1895 C. p. 182.
Williamson, W. C. 1849 A, p. 471, pl. xliii, figs.
     35-37.
Woodward, A. S. 1888 G.
     1891 A, p. 120.

Zittel, K. A. 1890 A, pp. 122, 258.
     This genus probably belongs to the *Sclero-
dermi.*

Coelorhynchus acus Cope.
Cope, E. D. 1870 F, p. 294.
     Eocene; New Jersey.

Coelorhynchus ornatus Leidy.
Leidy, J. 1856 E, p. 12.   (Cylindracanthus.)
     1856 A, p. 422. (Cylindracanthus.)
     1856 O, p. 302. (Coelorhynchus.)
Woodward, A. S. 1891 A, p. 122.
     Eocene; Alabama.

MACHAERACANTHUS Newb.   Type *M. peracutus* Newb.

Newberry, J. S. 1857 C, p. 125.
Fritsch, A. 1895 A (1893), p. 72.
Hopkins, W., 1855 A. (No name.)
Jaekel, O. 1893 B, p. 175.
Miller, S. A. 1889 A, p. 601.
Newberry, J. S. 1857 A, p. 148.
     1862 A, p. 76, fig. 2.
     1873 A, pp. 264, 302.
     1873 C, p. 27.
     1875 A, p. 57.
     1889 A, p. 37.

Nicholson and Lydekker 1889 A, p. 947.
Reis, O. M. 1896 A.
Rouault, M. 1858 A, p. 102. (Machaerius, preoccu-
pied.)
Traquair, R. H. 1893 B, p. 177.
Woodward, A. S. 1891 A, p. 123.
Zittel, K. A. 1890 A, p. 116.

Machaeracanthus major Newb.
Newberry, J. S. 1857 C, p. 125.
Lesley, J. P 1892 A, pl. cl.

Newberry, J. S. 1873, A, pp. 262, 304, pl. xxv, fig. 2.
    1889 A, p. 39, pl. xxix, fig. 4.
Woodward, A. S. 1891 A, p. 124.
Zittel, K. A. 1890 A, p. 116, fig. 124.
    Devonian (Corniferous); Ohio.

Machæracanthus peracutus Newb.

Newberry, J. S. 1857 C, p. 125.
Hopkins, W. 1854 A. (No name).
Miller, S. A. 1889 A, p. 601, fig. 1145.
Newberry, J. S. 1862 A, fig. 2.
    1873 A, p. 305, pl. xxix, fig. 6.
    1889 A, p. 40, pl. xxix, fig. 6.
Woodward, A. S. 1891 A, p. 124.
    Devonian (Corniferous); Ohio.

Machæracanthus sulcatus Newb.

Newberry, J. S. 1857 C, p. 126.
Hall, J. 1843 A, p. 174, fig. 69. ("Ichthyodoru-
    lite.")
Jaekel, O. 1893 B, p. 175.
Lankester, E. R. 1870 A, fig. 3.
Lennox, T. H 1886 A, p. 120.
Lesley, J. P. 1889 A, p. 298, figure. ("Ichthyodoru-
    lite.")
Newberry, J. S. 1873 A, pp. 264, 305.
    1889 A, p. 40, pl. xxix, fig. 4.
Woodward, A. S. 1891 A, p. 124.
    Devonian (Corniferous); Ohio.

GAMPHACANTHUS S. A. Miller.   Type *H. politus* Newb.

Miller, S. A. 1892 A, p. 715.
Hay, O. P. 1899 E, p. 791.
Newberry, J. S. 1889 A, p. 65.   (Heteracanthus,
    type *H. politus;* preoccupied.)
Woodward, A. S. 1891 A, p. 125.   (Heteracanthus.)

Gamphacanthus politus (Newb.).

Newberry, J. S. 1889 A, p. 66, pl. xxi, figs. 4, 5.
    (Heteracanthus.)
Eastman, C. R. 1898 B, p. 552.   (Heteracanthus.)
    1899 B, p. 282.   (Heteracanthus.)

Miller, S. A. 1892 A, p. 715.
Woodward, A. S. 1891 A, p. 125. (Heteracanthus.)
    Devonian (Hamilton); Wisconsin.

Gamphacanthus uddeni (Lindahl).

Lindahl, J. 1897 A, p. 95, pl. vi.   (Heteracanthus.)
Eastman, C. R. 1898 B, p. 557.   (Heteracanthus.)
    1899 B, pp. 278, 282.   (Heteracanthus.)
    Devonian (Hamilton); Iowa, Wisconsin, Illi-
    nois.

ACANTHASPIS Newb.   Type *A. armatus* Newb.

Newberry, J. S. 1875 A, p. 36.
Jaekel, O. 1893 B, p. 175.
Miller, S. A. 1889 A, p. 586.
Newberry, J. S. 1870 B, p. 153.
    1889 A, p. 35.
Traquair, R. H. 1893 A, p. 149.
    1893 D, p 117.
    1894 B, p. 370.
Woodward, A. S. 1891 A, p. 128.
    1892 C, p. 482.
    1892 F, fig. 8.

Woodward, A. S. 1896 B, p. 16.
Zittel, K. A. 1890 A, p. 157.

Acanthaspis armata Newb.

Newberry, J. S. 1875 A, p. 37, pl. iv, figs. 1–6.
Claypole, E. W. 1896 B, p. 354.
Miller, S. A. 1889 A, p. 586, fig. 1097.
Newberry, J. S. 1889 A, p. 36, pl. xxxi; figs. 1–4.
Traquair, R. H. 1894 B, p. 370.
Woodward, A. S. 1891 A, p. 129.
    Devonian (Corniferous); Ohio.

ECZEMATOLEPIS S. A. Miller.   Type *Acantholepis pustulosus* (Newb.).

Miller, S. A. 1892 A, p. 715.
Hay, O. P. 1899 E, p. 791.
Jaekel, O, 1893 B, p. 175.   (Acantholepis.)
Miller, S. A. 1889 A, p. 586.   (Acantholepis.)
Newberry, J. S. 1875 A, p. 38.   (Acantholepis, type
    A. pustulosus; preoccupied.)
    1889 A, p. 33.   (Acantholepis.)
Woodward, A. S. 1891 A, p. 128.   (Acantholepis.)
Zittel, K. A. 1890 A, p. 157.   (Acantholepis.)

Eczematolepis fragilis (Newb.).

Newberry, J. S. 1857 C, p. 126.   (Oracanthus
    fragilis, O. granulatus, O. abbreviatus.)

Eastman, C. R. 1899 B, p. 283.   (Acantholepis.)
Miller, S. A. 1889 A, p. 586, fig. 1098.   (Acantho-
    lepis.)
    1892 A, p. 715.   (Eczematolepis.)
Newberry, J. S. 1875 A, p. 38, pl. lvi, figs. 1–6.
    (Acantholepis pustulosus)
    1875 A, p. 39.   (Oracanthus fragilis, granu-
    latus, abbreviatus.)
    1889 A, p. 34, pl. xxxi, figs. 5–5b.   (Acan-
    tholepis pustulosus.)
Woodward, A. S. 1891 A, p. 129.   (Acantholepis
    pustulosus.)
    Devonian (Corniferous); Ohio, New York.

PHLYCTÆNACANTHUS Eastman.   Type *P. telleri* Eastman.

Eastman, C. R. 1898 B, p. 552.

Phlyctænacanthus telleri Eastman.

Eastman, C. R. 1898 B, p. 551, fig. 49.
    1899 B, p. 283.
    Devonian (Hamilton); Wisconsin, Iowa.

STETHACANTHUS Newb. Type *S. altonensis* (St. J. and Worth.).

Newberry, J. S. 1889 A, p. 198.
Miller, S. A. 1892 A, p. 717.
Woodward, A. S. 1891 A, p. 129.

Stethacanthus altonensis (St. J. and Worth).

St. John and Worthen 1875 A, p. 454, pl. xix, figs. 1–3. (Physonemus.)
Lesley, J. P, 1889 A, p. 641, figures. (Physonemus.)
1890 A, p. 1062.
Newberry, J. S. 1889 A, p. 198, pl. xxiv, figs. 1–3. 1890 A, p. 133.
Woodward, A. S. 1891 A, p. 129.
Subcarboniferous (St. Louis); Illinois.

Stethacanthus compressus Newb.

Newberry, J. S. 1897 A, p. 292, pl. xxiii, figs. 3, 4.
Subcarboniferous (Burlington); Iowa.

Stethacanthus productus Newb.

Newberry, J. S. 1897 A, p. 291, pl. xxiii, figs. 1, 2.
Subcarboniferous (Burlington); Iowa.

Stethacanthus tumidus Newb.

Newberry, J. S. 1889 A, pp. 198, 200, pl. xxv, figs. 1, 2.
Lesley, J. P. 1890 A, p. 1063.
Woodward, A. S. 1891 A, p. 129.
Devonian (Cleveland shale); Ohio, Pennsylvania?.

PHYSONEMUS McCoy. Type *P. arcuatus* McCoy.

McCoy, F. 1848 A, p. 117.
Davis, J. W. 1883 A, p. 368.
McCoy, F. 1855 A, p. 638.
Miller, S. A. 1889 A, p. 606.
Newberry, J. S. 1889 A, p. 195.
1890 A, p. 133.
Pictet. F. J. 1854 A, p. 291.
St. John and Worthen 1875 A, p. 448.
Woodward, A. S. 1891 A, p. 130.
Zittel, K. A. 1890 A, p. 118.

Physonemus carinatus St. J. and Worth.

St. John and Worthen 1875 A, p. 452, pl. xviii, figs. 4, 5.
Woodward, A. S. 1891 A, p. 131.
Subcarboniferous (Kinderhook); Iowa.

Physonemus chesterensis St. J. and Worth.

St. John and Worthen 1875 A, p. 455, pl. xix, fig. 4.
Lesley, J. P. 1889 A, p. 642, figure.
Woodward, A. S. 1891 A, p. 131.
Subcarboniferous (Chester); Illinois.

Physonemus depressus St. J. and Worth.

St. John and Worthen 1875 A, p. 452, pl. xviii, fig. 3.
Lesley, J. P. 1889 A, p. 642, figures.
Woodward, A. S. 1891 A, p. 131.
Subcarboniferous (Kinderhook); Iowa.

Physonemus falcatus St. J. and Worth.

St. John and Worthen 1883 A, p. 252, pl. xxiv, fig. 6.
Keyes, C. R. 1894 A, p. 237.

Lesley, J. P. 1889 A, p. 643, figure.
Woodward, A. S. 1891 A, p. 131.
Subcarboniferous (St. Louis); Missouri.

Physonemus gigas Newb. and Worth.

Newberry and Worthen 1870 A, p. 373, pl. ii, fig. 1.
Davis, J. W. 1883 A, p. 368.
Lesley, J. P. 1889 A, p. 643, figures.
Newberry, J. S. 1889 A, p. 195.
St. John and Worthen 1875 A, p. 449.
Woodward, A. S. 1891 A, p. 132.
Subcarboniferous (Burlington); Illinois.

Physonemus parvulus St. J. and Worth.

St. John and Worthen 1875 A, p. 453, pl. xviii, figs. 11, 12.
Keyes, C. R. 1894 A, p. 237.
Lesley, J. P. 1889 A, p. 643, figures.
Woodward, A. S. 1891 A, p. 132.
Subcarboniferous (Keokuk); Missouri, Illinois.

Physonemus proclivus St. J. and Worth.

St. John and Worthen 1875 A, p. 451, pl. xviii, figs. 1, 2.
Lesley, J. P. 1889 A, p. 644, figures.
Woodward, A. S. 1891 A, p. 132.
Subcarboniferous (Kinderhook); Iowa.

Physonemus stellatus Newb.

Newberry, J. S. 1889 A, p. 200, pl. xxi, fig. 12.
Woodward, A. S. 1891 A, p. 132.
Subcarboniferous (St. Louis); Indiana.

XYSTRACANTHUS Leidy. Type *X. arcuatus* Leidy.

Leidy, J. 1859 C, p. 3.
Koninck, L. G. de 1878 A, p. 75. (Drepanacanthus.)
Jaekel, O. 1899 A, p. 285.
Lohest, M. 1884 A, p. 319.
Miller, S. A. 1889 A, p. 596 (Drepanacanthus); p. 614 (Xystracanthus).
Newberry, J. S. 1889 A, p. 195. (Drepanacanthus.)
Newberry and Worthen 1866 A. p. 120. (Drepanacanthus; type *D. anceps*, Newb. and Worth.)

St. John and Worthen 1875 A, p. 457.
Trautschold, H. 1874 B, p. 297. (Drepanacanthus.)

Xystracanthus acinaciformis St. J. and Worth.

St. John and Worthen 1875 A, p. 459, pl. xx, fig. 2.
Lesley, J. P. 1890 A, p. 1262, figure.
Woodward, A. S. 1891 A, p. 131. (Physonemus.)
Coal-measures; Illinois.

Xystracanthus anceps (Newb. and Worth.).

Newberry and Worthen 1866 A, p. 122, pl. xii, fig. 8. (Drepanacanthus.)
Lesley, J. P. 1890 A, p. 1262, figure.
St. John and Worthen 1875 A, p. 458. (Xystracanthus.)
Woodward, A. S. 1891 A, p. 131. (Physonemus.)
Coal-measures; Illinois.

Xystracanthus arcuatus Leidy.

Leidy, J. 1859 C, p. 3.
1873 B, pp. 312, 353, pl. xvii, fig. 25.
St. John and Worthen 1875 A, p. 457.
Woodward, A. S. 1891 A, p. 131. (Physonemus.)
Coal Measures; Kansas.

Xystracanthus gemmatus (Newb. and Worth.).

Newberry and Worthen 1866 A, p. 123, pl. xiii, fig. 1. (Drepanacanthus.)

BATACANTHUS St. John and Worthen.

St. John and Worthen 1875 A, p. 468.
Miller, S. A. 1889 A, p. 588.
Woodward, A. S. 1891 A, p. 132.

Batacanthus baculiformis St. J. and Worth.

St. John and Worthen 1875 A, p. 469, pl. xxi, figs. 4–8.
Keyes, C. R. 1894 A, p. 238.
Woodward, A. S. 1891 A, p. 133.
Subcarboniferous (Keokuk); Iowa, Missouri, Illinois.

Batacanthus ? necis St. J. and Worth.

St. John and Worthen 1883 A, p. 253, pl. xxv, fig. 4.

ORACANTHUS Agassiz.

Agassiz, L. 1843 B (1837), iii, p. 13.
1844 A.
American Geologist 1888 A.
Davis, J. W. 1883 A, p. 525 (Oracanthus): p. 533 (Phoderacanthus).
Leidy, J. 1856 B, p. 161.
McCoy, F. 1855 A, p. 634.
Miller, S. A. 1889 A, p. 603 (Oracanthus); p. 607 (Pnigeacanthus).
Newberry, J. S. 1889 A, p. 34.
1890 A.
1897 A, p. 286.
Newberry and Worthen 1866 A, p. 122.
Nicholson and Lydekker 1889 A, p. 947.
Pictet, F. J. 1854 A, p. 284.
St. John and Worthen 1875 A, p. 480. (Pnigeacanthus; type *Oracanthus pnigeus*.)
Traquair, R. H. 1888 A, p. 85.
Woodward, A. S. 1891 A, p. 135.
Zittel, K. A. 1890 A, p. 162.

Oracanthus lineatus Newb.

Newberry, J. S. 1897 A, p. 290.
Subcarboniferous (St. Louis); Indiana.

Woodward, A. S. 1891 A, p. 132. (Physonemus.)
Subcarboniferous (Keokuk); Iowa.

Xystracanthus mirabilis St. J. and Worth.

St. John and Worthen 1875 A, p. 458, pl. xx, fig. 1.
Lesley, J. P. 1890 A, p. 1264, figure.
1895 A, pl. dlxxii.
Woodward, A. S. 1891 A, p. 132. (Physonemus.)
Coal-measures; Illinois.

Xystracanthus reversus (St. J. and Worth.).

St. John and Worthen 1875 A, p. 456, pl. xix, figs. 5, 6. (Drepanacanthus.)
Keyes, C. R. 1894 A, p. 237. (Drepanacanthus.)
St. John and Worthen 1883 A, p. 253, pl. xxiv, fig. 5. (Drepanacanthus.)
Woodward, A. S. 1891 A, p. 132. (Physonemus.)
Subcarboniferous (St. Louis); Illinois, Missouri.

Type *B. baculiformis* St. J. and Worth.

Woodward, A. S. 1891 A, p. 133.
Subcarboniferous (Keokuk); Iowa.

Batacanthus stellatus (Newb. and Worth.).

Newberry and Worthen 1866 A, p. 125, pl. xii, fig. 7. (Drepanacanthus ?)
Hambach, G. 1890 A, p. 80.
St. John and Worthen 1875 A, p. 470, pl. xxi, figs. 1–3.
Woodward, A. S. 1891 A, p. 133.
Subcarboniferous (Keokuk); Illinois, Missouri (*fide* Hambach).

Type *O. milleri* Agassiz.

Oracanthus multiseriatus Newb.

Newberry, J. S. 1857 C, p. 126.
Miller, S. A. 1889 A, p. 603.
Subcarboniferous; Ohio.

Oracanthus ? obliquus St. J. and Worth.

St. John and Worthen 1875 A, p. 477, pl. xxii, fig. 16.
Lesley, J. P. 1889 A, p. 501, figures.
Woodward, A. S. 1891 A, p. 139.
Subcarboniferous (Keokuk); Illinois.

Oracanthus pnigeus Newb. and Worth.

Newberry and Worthen 1866 A, p. 117, pl. xii, fig. 3.
Lesley, J. P. 1889 A, p. 724, fig.
Miller, S. A. 1889 A, p. 607. (Pnigeacanthus.)
Newberry, J. S. 1897 A, p. 286.
St. John and Worthen 1875 A, p. 480. (Pnigeacanthus deltoides.)
1883 A, p. 260. (Pnigeacanthus pnigeus).
Woodward, A. S. 1891 A, p. 138.
Subcarboniferous (Keokuk); Iowa.

Oracanthus rectus St. J. and Worth.

St. John and Worthen 1883 A, p. 257, pl xxv, fig. 3.
Lesley, J. P. 1889 A, p. 501, figs.
Woodward, A. S. 1891 A, p. 139.
  Subcarboniferous (Chester); Illinois.

Oracanthus trigonalis (St. J. and Worth.).

St. John and Worthen 1883 A, p. 259, pl. xxiv, fig. 4. (Pnigeacanthus.)
Woodward, A. S. 1891 A, p. 139.
  Subcarboniferous (St. Louis); Illinois.

Oracanthus vetustus Leidy.

Leidy, J. 1855 D.

GYRACANTHUS Agassiz.

Agassiz, L. 1843 B (1837), iii, p. 17.
      1844 A.
Barkas, T. P. 1873 A, p. 15.
Hasse, C. 1879 A, p. 60.
Jaekel, O. 1890 D, p. 123.
McCoy, F. 1855 A, p. 629.
Miller, S. A. 1889 A, p. 596.
Newberry, J. S. 1873 A, pp. 281, 330.
      1875 A, p. 57.
      1889 A, p. 186.
Pictet, F. J. 1854 A, p. 284.
Reis, O. M. 1896 A.
Thompson, J. 1868 A, p. 130.
Traquair, R. H. 1884 C.
      1889 B.
Ward, J. 1890 A, p. 150.
Woodward, A. S. 1885 B, p. 272.
      1891 A, p. 139.
Woodward and Sherborn 1890 A, p. 88.

Gyracanthus alleni Newb.

Newberry, J. S. 1873 A, p. 331, pl. xxxvii, figs. 3–3d.
Claypole, E. W. 1885 B, p. 490.
Miller, S. A. 1889 A, p. 598, fig. 1134.
Traquair, R. H. 1884 C, p. 41.
Woodward, A. S. 1891 A, p. 144.
  Subcarboniferous (Waverly); Ohio.

Gyracanthus compressus Newb.

Newberry, J. S. 1873 A, p. 330, pl. xxxvii, figs. 1–2b.
Woodward, A. S. 1891 A, p. 144.
  Subcarboniferous (Waverly); Ohio, Indiana.

Gyracanthus? **cordatus** St. J. and Worth.

St. John and Worthen 1883 A, p. 251, pl. xxvi, fig. 4.

ERISMACANTHUS McCoy.

McCoy, F. 1848 A, p. 118.
Davis, J. W. 1883 A, p. 364. (Cladacanthus.)
McCoy, F. 1855 A, p. 628.
Miller, S. A. 1889 A, p. 597.
Pictet, F. J. 1854 A, p. 292.
Woodward, A. S. 1886 E, p. 4.
      1889 E, p. 130.
      1891 A, p. 146.
Zittel, K. A. 1890 A, p. 118.

Keyes, C. R. 1894 A, p. 238.
Leidy, J. 1856 S, p. 161, pl. 16, figs. 1–3.
Lesley, J. P. 1889 A, p. 501, figs.
      1895 A, pl. ccxxix.
Newberry, J. S. 1890 A.
      1897 A, p. 285, pl. xxii, fig. 3.
St. John and Worthen 1875 A, p. 476, pl. xxii, fig. 15. (O. consimilis.)
      1875 A, p. 479.
      1883 A, p. 255, pl. xxiv, fig. 2.
Woodward, A. S. 1891 A, p. 139.
  Carboniferous; Missouri Territory: Subcarboniferous (St. Louis); Illinois, Missouri.

Type *G. formosus* Agassiz.

Woodward, A. S. 1891 A, p. 145.
  Subcarboniferous (Keokuk); Iowa.

Gyracanthus duplicatus Dawson.

Dawson, J. W. 1868 A, p. 210, fig. 55.
      1878 A, p. 210, fig. 55.
Lesley, J. P. 1895 A, pl. dlxx.
Newberry, J. S. 1889 A, p. 124.
Traquair, R. H. 1884 C, p. 41.
Woodward, A. S. 1891 A, p. 144.
  Coal-measures; Nova Scotia.

Gyracanthus incurvus Traquair.

Traquair, R. H. 1890 A, p. 21.
Woodward, A. S. 1891 A, p. 144.
      1892 A, p. 2, pl. 1, figs. 4, 5.
  Devonian; Province of Quebec.

Gyracanthus inornatus Newb.

Newberry, J. S. 1889 A, p. 177, pl. xxiii, fig. 5.
Woodward, A. S. 1891 A, p. 145.
  Subcarboniferous (Waverly); Ohio.

Gyracanthus magnificus Dawson.

Dawson, J. W. 1868 A, p. 210, fig. 55a.
      1878 A, p. 210, fig. 55a.
Lesley, J. P. 1895 A, pl. dlxx.
Newberry, J. S. 1875 A, p. 57.
      1889 A, p. 124.
Woodward, A. S. 1891 A, p. 145.
  Subcarboniferous; Cape Breton.

Gyracanthus sherwoodi Newb.

Newberry, J. S. 1889 A, p. 119, pl. xviii, fig. 4.
Woodward, A. S. 1891 A, p. 145.
  Devonian; Pennsylvania.

Type *E. jonesii* McCoy.

Erismacanthus maccoyanus St. J. and Worth.

St. John and Worthen 1875 A, p. 461, pl. xxii, figs. 1–5.
Keyes, C. R. 1894 A, p. 237.
Woodward, A. S. 1891 A, p. 147.
  Subcarboniferous (St. Louis); Missouri, Illinois.

GAMPSACANTHUS St. J. and Worth.   Type *G. typus* St. J. and Worth.

St. John and Worthen 1875 A, p. 470.
Davis, J. W. 1883 A, p. 363.
Miller, S. A. 1889 A, p. 598.
Woodward, A. S. 1891 A, p. 148.
Zittel, K. A. 1890 A, p. 118.

Gampsacanthus? latus St. J. and
Worth.
St. John and Worthen 1875 A, p. 474, pl. xxii, fig. 14.
Keyes, C. R. 1894 A, p. 238.
Woodward, A. S. 1891 A, p. 148.
Subcarboniferous (Keokuk); Missouri.

Gampsacanthus squamosus St. J. and
Worth.
St. John and Worthen 1875 A, p. 473, pl. xxii, fig. 13.

Keyes, C. R. 1894 A, p. 238.
Woodward, A. S. 1891 A, p. 148.
Subcarboniferous (St. Louis); Missouri.

Gampsacanthus typus St. J. and
Worth.
St. John and Worthen 1875 A, p. 472, pl. xxii, fig. 12.
Keyes, C. R. 1894 A, p. 238.
Lesley, J. P. 1895 A, pl. ccxxix.
Miller, S. A. 1889 A, p. 598, fig. 1132.
Woodward, A. S. 1891 A, p. 148.
Zittel, K. A. 1890 A, p. 118, fig. 129.
Subcarboniferous (St. Louis); Missouri, Illinois.

LECRACANTHUS St. J. and Worth.   Type *L. unguiculus* St. J. and Worth.

St. John and Worthen 1875 A, p. 475.
Miller, S. A. 1889 A, p. 600.
Woodward, A. S. 1891 A, p. 148.

Lecracanthus unguiculus St. J. and
Worth.
St. John and Worthen 1875 A, p. 476, pl. xxii, figs. 10, 11.

Keyes, C. R. 1894 A, p. 238.
Woodward, A. S. 1891 A, p. 148.
Subcarboniferous (St. Louis); Missouri, Illinois.

LISTRACANTHUS Newb. and Worth.   Type *L. hystrix* Newb. and Worth.

Newberry and Worthen 1870 A, p. 371.
Koninck, L. G. de 1878 A, p. 75.
Miller, S. A. 1889 A, p. 601.
Newberry, J. S. 1873 A, p. 336.
Woodward, A. S. 1891 A, p. 148.

Listracanthus hildrethi Newb.
Newberry, J. S. 1875 A, p. 56, pl. lix, fig. 6.
Woodward, A. S. 1891 A, p. 149.
Coal-measures; Ohio.

Listracanthus hystrix Newb. and
Worth.
Newberry and Worthen 1870 A, p. 372, pl. ii, figs. 3, 3a
Cope, E. D. 1894 F, p. 433.
Hambach, G. 1890 A, p. 81.
Koninck, L. G. de 1878 A, p. 75, pl. v. fig. 11.
Miller, S. A. 1889 A, p. 601, fig. 1144.
Newberry, J. S. 1873 A, p. 327.
1875 A, p 56, pl. lix, fig. 5.
Woodward, A. S. 1891 A, p. 148.
Coal-measures; Illinois, Ohio, Missouri (*fide* Hambach). Also reported from Belgium.

CYRTACANTHUS Newb.   Type *C. dentatus* Newb.

Newberry, J. S. 1873 A, p. 306.
Miller, S. A. 1889 A, p. 594.
Woodward, A. S. 1891 A, p. 150.

Cyrtacanthus dentatus Newb.
Newberry, J. S. 1873 A, p. 307, pl. xxix, fig. 5.

Miller, S. A. 1889 A, p. 594, fig. 1122.
Newberry, J. S. 1889 A, p. 28. (Cladacanthus.)
Woodward, A. S. 1891 A, p. 150.
Devonian (Corniferous); Ohio.

HARPACANTHUS Traquair.   Type *Tristychius fimbriatus* Stock.

Traquair, R. H. 1886 A, p. 493.
Newberry, J. S. 1889 A, p. 203.
Stock, T. 1882 A. ("Kammplatten.")
1882 B. (Tristychius.)
1883 A. (Tristychius.)
Traquair, R. H. 1888 A, p. 83. (Tristychius.)
Woodward, A. S. 1884 A, p. 268.
1891 A, p. 150.

Harpacanthus fimbriatus (Stock).
Stock, T. 1883 A, p. 177. (Tristychius.)
Newberry, J. S. 1889 A, p. 203, pl. xxi, fig. 11.
Traquair, R. H. 1886 A, p. 493, figs. 1, 2.
Woodward, A. S. 1891 A, p. 150, fig. 11.
Subcarboniferous (St. Louis); Illinois. Also in Scotland.

EDESTUS Leidy.   Type *E. vorax* Leidy.

Leidy, J. 1855 D.
Agassiz, L., in Hitchcock, E. 1856 B.
Cope, E. D. 1870 L, p. 535.
    1872 F, p. 339.
    1875 E, p. 244c.
    1890 H.
Dean, B. 1896 D, p. 61.
Eastman, C. R. 1898 B, p. 554.
Hitchcock, E. 1856 B.
Hitchcock, Fanny R. 1887 A.
    1888 A.
Jaekel, O. 1899 A, p. 297 (Edestus); p. 298 (family Edestidæ).
    1900 A, p. 144.
Leidy, J. 1856 O, p. 301.
    1856 S, p. 157.
Lesley, J. P. 1889 A, *Errata*, p. xxi.
Miller, S. A. 1889 A, p. 597.
Newberry, J. S. 1868 A, p. 145
    1873 A, p. 286.
    1884 A.
    1885 A.
    1888 B.
    1889 A, p. 217.
Newberry and Worthen 1866 A, p. 84.
    1870 A, p. 350.
Nicholson and Lydekker 1889 A, p. 947.
Trautschold, H. 1883 A.
    1886 A.
    1888 A, p. 750.
    1890 A.
Woodward, A. S. 1888 A.
    1891 A, p. 151.
Woodward, H. 1886 A.
Zittel, K. A. 1890 A, p. 119.

Edestus giganteus Newb.

Newberry, J. S. 1888 B, p. 121, pl. vi, fig. 1.
Dana, J. D. 1896 A, p. 681, fig. 1106.
    1888 A.
Dean, B. 1897 B, p. 64, pl. v, fig. 3.
Newberry, J. S. 1889 A, p. 225, pl. xli, fig. 1.
Woodward, A. S. 1888 A.
    1891 A, p. 153.
    Coal-measures; Illinois.

Edestus heinrichsii Newb. and Worth.

Newberry and Worthen 1870 A, p. 350, pl. i, fig. 1.
Dean, B. 1897 B, p. 64, pl. v, fig. 2.

Eastman, C. R. 1898 B, p. 554.
Kindle, E. M. 1898 A, p. 484.
Newberry, J. S. 1879 A, p. 347.
    1888 B, pl. v, figs. 2, 2a, 2b.
    1889 A, p. 215.
Woodward, A. S. 1891 A, p. 152.
Woodward, H. 1886 A, p. 3.
    Coal-measures; Illinois, Indiana.

Edestus lecontei Dean.

Dean, B. 1897 B, p. 62, pl. iv; pl. v, fig. 6.
Jaekel, O. 1900 A, p. 147.
    Carboniferous?; Nevada.

Edestus minor Newb.

Newberry, J. S., in Newberry and Worthen 1866 A, p. 84, pl. iv, fig. 24.
Dana, J. D. 1896 A, p. 681, fig. 1107.
Dean, B, 1897 B, p. 64, pl. v, fig. 4.
Hitchcock, E. 1856 B, text figure. (No name.)
Kindle, E. M. 1898 A, p. 484.
Lesley, J. P. 1889 A, p. 214, figure. (E. vorax.)
Newberry, J. S. 1879 A, p. 348.
    1888 B, pl. v, fig. 1.
    1889 A, p. 215.
Newberry and Worthen 1870 A, p. 350, pl.i, fig. 2. (E. vorax, by error.)
Trautschold, H. 1888 A, p. 753.
Woodward, A. S. 1891 A, p. 153, fig. 12.
Zittel, K. A. 1890 A, p. 119.   (E. vorax.)
    Coal-measures; Indiana.

Edestus vorax Leidy.

Leidy, J. 1855 D.
Dean, B. 1897 B, p. 63.
Leidy, J. 1856 O, p. 301.
    1856 S, p. 159, pl. xv.
    1870 H, p. 13.
Lesley, J. P. 1895 A, pl. ccxxix.
Miller, S. A. 1889 A, p. 597, fig. 1129.
Newberry, J. S. 1868 A, p. 145.
    1879 A, p. 347.
    1888 B, pl. iv, fig. 2.
    1889, A, p. 218.
Woodward, A. S. 1891 A, p. 154.
Woodward, H. 1886 A, p. 3.
    Coal-measures; Arkansas.

EUCTENIUS Traquair.   Type *E. unilateralis* (Barkas).

Traquair, R. H., 1881, Geol. Magazine (2), viii, pp. 36, 334.
Barkas, T. P. 1881 A. (Ctenoptychius.)
Fritsch, A. 1883 A (1880), i, pl. xx. ("Kammplatten.")
Stock, T. 1881 A. ("Kammplatten.")
    1882 A.

Ward, J. 1890 A, p. 152.
Woodward, A. S. 1891 A, p. 155.

Euctenius sp. indet.

Newberry, J. S. 1889 A, p. 228.
Woodward, A. S. 1891 A, p. 156.
    Coal-measures; Ohio.

OESTOPHORUS S. A. Miller.   Type *Sphenophorus lilleyi* Newb.

Miller, S. A. 1892 A, p. 716.
Newberry, J. S. 1889 A, p. 91. (Sphenophorus, type S. *lilleyi*; preoccupied.)

Oestophorus lilleyi (Newb.).

Newberry, J. S. 1889 A, p. 92. (Sphenophorus.)
Hay, O. P. 1899 E, p.791.

Lesley, J. P. 1890 A, p. 977. (Sphenophorus.)
Miller, S. A. 1892 A, p. 716.
Woodward, A. S. 1891 A, p. 157. (Sphenophorus.)
Devonian (Chemung); Pennsylvania.

Oestophorus n. sp. Eastman.

Eastman, C. R. 1899 B, p. 287. (Sphenophorus
n. sp.; no description.)
Devonian; Iowa.

CALLOGNATHUS Newb.   Type *C. regularis* Newb.

Newberry, J. S. 1889 A, p. 69.
Miller, S. A. 1892 A, p. 714.

Callognathus regularis Newb.

Newberry, J. S. 1889 A, p. 70, pl. xxvii, fig. 18.
Woodward, A. S. 1891 A, p. 157.
Devonian (Cleveland shale); Ohio.

Callognathus serratus Newb.

Newberry, J. S. 1889 A, p. 70, pl. xxvii, figs. 16, 17.
Woodward, A. S. 1891 A, p. 157.
Devonian (Cleveland shale); Ohio.

PROSPIRAXIS Williams.   Type *Spiraxis major* Newb.

Williams, H. S. 1887 A, p. 86.
Cope, E. D. 1895 B, p. 159. (Spiraxis.)
Hollick, A. 1894 A, p. 118. (Spiraxis.)
Newberry, J. S. 1885 C, p. 219. (Spiraxis, not of
Adams, 1850.)
Stainier, X. 1894 A. (Spiraxis.)

Prospiraxis major (Newb.).

Newberry, J. S. 1885 C, p. 219, pl. xviii, fig. 1.
(Spiraxis.)
Cope, E. D. 1895 B, p. 159. (Spiraxis.)
Lesley, J. P. 1890 A, p. 994, figures. (Spiraxis.)

Williams, H. S. 1887 A, p. 86. (Spiraxis ? .)
Devonian (Chemung); Ohio, New York?

Prospiraxis randallii (Newb.).

Newberry, J. S. 1885 C, p. 220, pl. xviii, figs 2, 3.
(Spiraxis.)
Cope, E. D. 1895 B, p. 159. (Spiraxis.)
Lesley, J. P. 1890 A, p. 995. (Spiraxis.)
Williams, H. S. 1887 A, p. 86. (Spiraxis?.)
Devonian (Chemung); Pennsylvania New
York. ?

Class PISCES Linnæus.

Linnæus, C., 1758, Syst. Nat., ed. x, p. 239.
Agassiz, L. 1843 B.
    1844 A.
    1845 A.
Baudelot, E. 1873 A.
    1873 B.
Bridge, W. J. 1879 A.
    1896 A.
Cope, E. D. 1872 C.
    1877 O.
    1878 A.
    1885 G.
    1889 R.
    1890 B.
    1891 N.
    1898 B.
Dean, B. 1895 A.
Gadow, H. 1898 A, p. 7. (Teleostomi.)
Gegenbaur, C. 1898 A.
Gill, T. 1861 A. (Telostei, Ganoidei.)
    1872 C.
    1873 B.
    1873 D.
    1893 A. (Teleostomes.)
    1896 C. p. 912.
Grassi, B. 1883 A.
Günther, A. 1880 A.
    1881 A.
Haeckel, E. 1895 A, p. 244 (Ganoides); p. 247 (Os-
teodermi); p.252(Teleostei); p.257 (Dipneusta).
Harrison, R. G. 1894 A. (Teleostei.)
Hertwig, O. 1876 A.
    1881 A.
Hubrecht, A. A. W. 1878 A.

Humphrey, J. 1866 A.
Huxley, T. 1870 A, p. 440.
    1876 A.
Jordan, D. S. 1891 A.
Jordan and Evermann, 1896 A.
    1898 A.
Nicholson and Lydekker 1889 A.
Owen, R. 1847 B.
    1866 A.
Parker, W. K. 1874 A.
    1882 A.
Pictet, F. J. 1854 A, p. 37. ("Teleostiens.")
Pollard, H. B. 1894 B. ("Fishes.")
Rösc, C. 1894 B. ("Fishes.")
Ryder, J. 1879 B.
    1885 A.
    1889 B.
    1892 A.
Stannius, H. 1854 A.
Steinmann and Döderlein 1890 A, p. 551. (Euich-
thyes.)
Woodward, A. S. 1891 A.
    1895 A.
Zittel, K. A. 1890 A.

    The term *Pisces* is employed here in much
the same sense in which it was used originally
by Linnæus; that is, to designate the fish-like
animals, exclusive of the sharks and rays and
the lampreys. There are furthermore included
certain fossil groups and the Dipnoi, not known
to Linnæus. The authors cited above use the
term in various senses, and some employ other
terms to designate this group.

Subclass ASPIDOGANOIDEI Gill.

Gill, T., 1876, Johnson's Univ. Cyclop., iv, p. 185.
Cope, E. D. 1889 R, p. 852. (Ostracodermi, pre-
    occupied.)
    1891 G. (Ostracophori.)
    1891 N, p. 8. (Ostracophori.)
    1898 B, p. 15. (Ostracophori.)
Dean, B. 1895 A. (Ostracodermi.)
Gadow, H. 1898 A, p. 4. (Hypostomata.)
Gill, T. 1896 C, p. 912. (Aspidoganoidei.)
Hay, O. P. 1899 F, p. 682.
Jaekel, O. 1896 B, p. 127. (Ostracodermi.)
Patten, W. 1890 A. (Ostracodermi.)

Rohon, J. V. 1892 C, p. 86. (Protocephali.)
    1894 A, p. 52. (Protocephali.)
    1895 B, p. 16. (Protocephali.)
    1896 A. (Genus Thyestes.)
    1896 B, p. 16. (Protocephali.)
Traquair, R. H. 1900 A, p. 158. (Ostracodermi.)
    1900 B, p. 467. (Ostracodermi.)
Woodward, A. S. 1891 A, pp. xvi, 159. (Ostraco-
    dermi.)
    1892 F, p. 599.
    1898 B, p. 3. (Ostracodermi.)
    1900 D. (Ostracodermi.)

Order HETEROSTRACI Lankester.

Powrie and Lankester 1868 A, pp. 5, 14.
Cope, E. D. 1891 N, p. 9.
    1898 B, p. 15.
Gadow, H. 1898 A, p. 4.
Rohon, J. V. 1895 B, p. 16. (Aspidorhini.)
    1896 A. (Genus Thyestes.)
    1896 B. (Aspidorhini.)

Traquair, R. H. 1899 A, p. 858.
    1900 B, p. 467.
Woodward, A. S. 1891 A, pp. xix, 159.
    1898 B, p. 5.
    1900 D.
Zittel, K. A. 1890 A, p. 144.

PTERASPIDÆ.

Claypole, E. W. 1885 D, p. 56.
    1885 C, p. 734.
    1892 D, p. 542.
Cope, E. D. 1887 O, p. 1015.
Dean, B. 1895 A, figures.
Jaekel, O. 1896 B, p. 128.
Nicholson and Lydekker 1889 A, p. 960.
Patton, W. 1890 A.
Reis, O. M. 1896 A, pp. 212, 213.
Rohon, J. V. 1892 C.
    1894 A.

Rohon, J. V. 1896 B.
    1897 A.
    1899 B, p. 9.
Schmidt, F. 1873 C. ("Pteraspiden.")
Steinmann and Döderlein 1890 A, p. 553.
Traquair, R. H. 1899 A, p. 849.
Woodward, A. S. 1887 A.
    1891 A, p. 159.
    1892 F.
Zittel, K. A. 1890 A, p. 144.

PTERASPIS Kner.   Type *Cephalaspis lloydii* Agassiz, = *P. rostrata* Agassiz.

Kner, R., 1847 A, p. 165.
Agassiz, L. 1843 B, ii, p. 135. (Cephalaspis, in
    part.)
Alth, A. 1874 A, p. 41.
    1886 A.
Claypole, E. W. 1884 A. (Pteraspis, Scaphaspis.)
    1885 A.
Collinge, W. E. 1893 A, p. 2.
Eichwald, E. 1871 A, p. 1, pl. i. (Palæoteuthis,
    type *P. marginalis*.)
Huxley, T. H. 1858 A.
    1859 A.
    1861 A.
    1861 B, p. 38.
    1872 A, p. 128.
Jaekel, O. 1892 A, p. 90.
Kunth, A. 1872 A.
Lankester, E. R. 1862 A.
    1864 A.
    1864 B.
    1865 A. (Pteraspis, Scaphaspis.)
    1873 A.
    1873 B. (Scaphaspis, Pteraspis.)
    1874 A.

Mitchell, H. 1862 A.
    1864 A.
Nicholson and Lydekker 1889 A, p. 960.
Owen, R. 1860 E, p. 122.
Powrie, J. 1864 A.
Powrie and Lankester 1868 A. (Pteraspis and
    Scaphaspis.)
Roemer, F. 1856 A. (Palæoteuthis.)
Rohon, J. V. 1892 C.
    1896 B.
Ryder, J. A. 1890 A, p. 491.
Salter, J. W. 1867 A.
Schmidt, F. 1866 A, p. 217.
    1873 A.
    1873 B.
    1873 C.
Traquair, R. H. 1899 A, p. 850.
    1900 B, p. 466.
Woodward, A. S. 1887 A.
    1891 A, p. 160.
    1892 F, figs. 1, 9.
Zittel, K. A. 1890 A, p. 147.
No American species of this genus are known.

CYATHASPIS Lankester. Type *C. banksi* Huxley and Salter.

Lankester, R. 1865 A, p. 100.
Alth, A. 1874 A, p. 45.
 1886 A.
Claypole, E. W. 1892 D.
Geinitz, H. B. 1884 A, p. 854.
Jaekel, O. 1892 A, p. 90.
Kunth, A. 1872 A.
Lankester, E. R. 1873 B.
Lindström, G. 1895 A.
Matthew, G. F. 1887 B, p. 73. (Diplaspis.)
 1888 A, p. 49. (Diplaspis.)
Miller, S. A. 1889 A, p. 596. (Diplaspis.)
Powrie and Lankester 1868 A, p. 26.
Rohon, J. V. 1896 B, p. 18.
Schmidt, F. 1873 A.
 1873 C.

Traquair, R. H. 1900 B, p. 467.
Woodward, A. S. 1891 A, p. 170.
 1892 F, fig. 2.
Zittel, K. A. 1890 A, p. 148.

Cyathaspis acadica (Matthew).

Matthew, G. F. 1886 A. (Pteraspis ?)
American Geologist 1891 B, p. 329. (Diplaspis.)
Matthew, G. F. 1887 A.
 1887 B, p. 69. (Diplaspis.)
 1888 A, p. 49, pl. iv, figs. 1–4. (Diplaspis.)
 1891 A. (Diplaspis.)
Miller, S. A. 1889 A, p. 596. (Diplaspis.)
Rohon, J. V. 1896 B, p. 19.
Woodward, A. S. 1891 A, p. 173.
Upper Silurian; New Brunswick.

PALÆASPIS Claypole. Type *P. americana* Claypole.

Claypole, E. W. 1884 A, p. 1224.
 1885 A, p. 426.
 1885 C.
 1885 D, pp. 50, 62.
 1892 D.
Collinge, W. E. 1893 A, p. 3.
Lankester, E. R. 1873 A, p. 242. (Holaspis, preoccupied.)
 1873 C.
Lesley, J. P. 1892 A, p. 775.
Miller, S. A. 1889 A, p. 604.
Traquair, R. H. 1900 B, p. 467.
Woodward, A. S. 1891 A, p. 169.
 1898 B, p. 435.
Zittel, K. A. 1890 A, p. 148. (Holaspis.)

Palæaspis americana Claypole.

Claypole, E. W. 1884 A, p. 1224.
 1884 A, p. 1224. (P. bitruncata.)
 1885 A, p. 426. (P. bitruncata.)

Claypole, E. W. 1884 A, p. 1224.
 1885 D, p. 62, fig. 7 (P. americana); fig. 8 (P. bitruncata).
 1892 D, p. 561, fig. 8.
 1893 C.
Dean, B. 1895 A, p. 71.
Lesley, J. P. 1889 A, p. 575, figures. (P. americana, P. bitruncata.)
Matthew, G. F. 1888 A, p. 54.
Rohon, J. V. 1896 B, p. 21.
Traquair, R. H. 1899 A, p. 858.
Woodward, A. S. 1891 A, p. 170.
Zittel, K. A. 1890 A, p. 317.
Devonian (Onondaga); Pennsylvania.

Palæaspis elliptica Claypole.

Claypole, E. W. 1885 C.
 1885 A, p. 426.
Upper Silurian; Pennsylvania.

Order OSTEOSTRACI Lankester.

Powrie and Lankester, 1868 A, p. 14.
Cope, E. D. 1891 N, p. 9.
 1898 B, p. 15.
Gadow, H. 1898 A, p. 4.
Powrie and Lankester, 1870 A, p. 33.
Rohon, J. V. 1892 C, p. 86.
 1895 B, p. 16. (Aspidocephali.)
 1894 A, p. 51. (Aspidocephali.)
 1894 B. (Tremataspidæ.)

Rohon, J. V. 1892 C, p. 86.
 1896 A. (Genus Thyestes.)
 1896 B, p. 21. (Aspidocephali.)
Schmidt, F. 1894 A. (Genus Thyestes.)
Traquair, R. H. 1899 A, p. 857.
 1900 B, p. 467.
Woodward, A. S. 1891 A, pp. xix, 176.
 1898 B, p. 8.

CEPHALASPIDÆ.

Huxley, T. H. 1861 B, p. 38.
Agassiz, L. 1843 A, p. 86. ("Céphalaspides.")
 1844 B, p. 4. ("Céphalaspides.")
 1846 A, p. 87. ("Céphalaspides.")
Cope, E. D. 1886 D.
 1887 O, p. 1015.
Dean, B. 1895 A.
Haeckel, E. 1895 A, p. 286. (Cephalaspid.æ.)
Huxley, T. H. 1872 A, p. 128.
Jaekel, O. 1892 A, p. 90.
 1896 B, p. 128.
Newberry, J. S. 1873 A, p. 256.
Nicholson and Lydekker, 1889 A, pp. 959, 961.

Pictet, F. J. 1854 A, p. 218. ("Céphalaspides.")
Pollard, H. B. 1895 A, p. 419.
Powrie and Lankester, 1868 A.
 1870 A.
Rohon, J. V. 1892 C.
 1896 B, p. 23.
Schmidt, F. 1866 A.
 1873 C.
Steinmann and Döderlein, 1890 A, p. 558.
Woodward, A. S. 1890 B, p. 316.
 1891 A, p. 176.
 1898 B, p. 8.
Zittel, K. A. 1890 A, p. 148.

CEPHALASPIS Agassiz.   Type *C. lyelli* Agassiz.

Agassiz, L. 1843 B (1835), ii, p. 153.
   1843 A.
   1844 A.
   1844 B.   (Pterichthys.)
Alth, A. 1874 A.
   1886 A.
Claypole, E. W. 1885 C.
   1885 D.
   1892 D, p. 556.
Cope, E. D. 1886 D.
   1888 U.
Huxley, T. H. 1858 A.
   1859 A.
   1861 B, p. 38.
   1872 A, p. 128.
Jaekel, O. 1892 A, p. 90.
Kunth, A. 1872 A.
Miller, H. 1859 A, p. 50, pls. x, xiii.
Miller, S. A. 1889 A, p. 589.
Newberry, J. S. 1856 A.
Nicholson and Lydekker, 1889 A, p. 961.
Owen, R. 1860 E, p. 122.
Pander, C. H. 1856 A.
Pictet, F. J. 1854 A, p. 219.
Powrie, J. 1861 A.
Powrie and Lankester, 1868 A.
   1870 A.
Quenstedt, F. A. 1885 A, p. 853.
Rohon, J. V. 1892 C.
   1896 B, p. 23.
Schmidt, F. 1866 A.
Traquair, R. H. 1857 A.
   1899 A, p. 857.
   1900 B, p. 467.
Woodward, A. S. 1891 A, p. 177.
   1892 A, figs. 5, 10.
   1896 B, p. 9.
   1900 D.
Zittel, K. A. 1890 A, p. 149.

Cephalaspis campbelltonensis Whiteaves.

Whiteaves, J. F. 1881 E, J. 98

Claypole, E. W. 1884 A, p. 1223.
Miller, S. A. 1892 A, p. 714.
Rohon, J. V. 1896 B, p. 25.
Traquair, R. H. 1890 A, p. 21.   (C. campbelltonensis; C. whiteavesi.)
   1893 A, p. 146.
   1893 D, p. 118.
Whiteaves, J. F. 1883 A, p. 355.
   1883 B, p. 164.
   1889 A, p. 92, pl. x, fig. 2.
Woodward, A. S. 1891 A, p. 190, pl. ix, fig. 5.
   1892 A, p. 4, pl. i, fig. 6.
   Lower Devonian; New Brunswick.

Cephalaspis dawsoni Lankester.

Lankester, E. R. 1870 A, p. 397, figs. 1, 2.
Claypole, E. W. 1884 A, p. 1223.
Rohon, J. V. 1896 B, p. 25.
Whiteaves, J. F. 1881 E, p. 99.
   1883 B, p. 164.
   1899 A, p. 410.
Woodward, A. S. 1891 A, p. 192, fig. 26.
   Lower Devonian; Gaspé, Canada.

Cephalaspis jexi Traquair.

Traquair, R. H. 1893 A, p. 147.
Rohon, J. V. 1896 B, p. 25.
Traquair, R. H. 1893 D, p. 114.
   Upper Devonian; Scaumenac Bay, Canada.

Cephalaspis laticeps Traquair.

Traquair, R. H. 1890 A, p. 16.
Miller, S. A. 1892 A, p. 714.
Rohon, J. V. 1896 B, p. 25.
Whiteaves, J. F. 1899 A, p. 412.
Woodward, A. S. 1891 A, p. 192.
   Upper Devonian; Scaumenac Bay, Canada.

EUPHANEROPIDÆ

Woodward, A. S. 1900 B, p. 418.

EUPHANEROPS Woodward.   Type *E. longævus* Woodward.

Woodward, A. S. 1900 B, p. 416.

Euphanerops longævus Woodward.

Woodward, A. S. 1900 B, p. 416, pl. x, figs. 1–1b.
   Devonian; Quebec.

Subclass AZYGOSTEI, nom. nov.

Gill, T., 1876, Johnson's Cyclopedia, iv, p. 185.
   (Dipnoi.)

Superorder PLACODERMI McCoy.

McCoy, F. 1848 A, p. 6.
Cope, E. D. 1886 D, p. 1081. (Placoganoidei, Placodermi.)
    1887 O, p. 1019.
    1889 R, p. 866.
Dames, W. 1885 A.
Dean, B. 1891 A.
Egerton, P. G. 1848 B.
Günther, A., 1880 A, p. 351.
    1881 A, p. 686.
Hay, O. P. 1899 F, p. 682.
Haeckel, E. 1895 A, pp. 236, 248.
Huxley, T. H. 1872 A, p. 129.
Jaekel, O. 1896 B, p. 128.
Koenen, A. 1883 A.
    1895 A.

Lütken, C. 1873 A, p. 41.
Marsh, O. C. 1877 E.
McCoy, F. 1848 B.
    1855 A, p. 598. (Placodermata.)
Newberry, J. S. 1873 A, p. 257.
    1875 A, p. 3.
Nicholson and Lydekker 1889 A, 962.
Owen, R. 1860 E, p. 118.
Pander, C. H. 1857 A.
Rohon, J. V. 1892 A, p. 32.
    1899 A, p. 16.
    1899 B, p. 10.
Steinmann and Döderlein 1890 A, p. 551.
Zittel, K. A. 1890 A, p. 151.

Order ANTIARCHA Cope.

Cope, E. D. 1885, I p. 291.
    1886 D, p. 1081.
    1887 S, p. 323.
    1891 N, p. 9.
    1898 B, p. 15.

Gadow, H. 1898 A, p. 4.
Traquair, R. H. 1900 A, p. 159.

ASTEROLEPIDÆ.

Cope, E. D. 1885 I. (Pterichthyidæ.)
    1886 D. (Pterichthyidæ, Bothriolepididæ.)
    1887 O, p. 1015. (Pterichthyidæ, Bothriolepididæ.)
    1887 S, p. 323. (Pterichthyidæ.)
    1889 R, p. 853. (Pterichthyidæ.)
Gürich, G. 1891 A, p. 910. (Pterichthyidæ.)
Haeckel, E. 1895 A, pp. 248, 656. (Asterolepides)
Jaekel, O. 1898 A.
Nicholson and Lydekker 1889 A, p. 962. (Asterolepididæ.)

Rohon, J. V. 1899 A, p. 20.
Steinmann and Döderlein 1890 A, p. 552. (Asterolepididæ.)
Traquair, R. H. 1888 C, pls. xvii, xviii.
    1894 C, pp. 64, 70.
    1900 B, p. 467.
Woodward, A. S. 1891 A, pp. xviii, 208.
    1898 B, p. 12.

ASTEROLEPIS Eichwald.   Type A. ornata Eichwald.

Eichwald, E. 1840 A, p. 79.
Agassiz, L. 1844 B, p. 89 (Asterolepis, in part); p. 111 (Odontacanthus).
Beyrich, E. 1877 A.
Egerton, P. H. 1859 A, p. 121.
Eichwald, E. 1846 A, pt. ii, p. 283.
    1857 A, p. 345.
    1860 A, p. 1504, pl. lvi, fig. 1 (Asterolepis); p. 1521 (Coccosteus).
Jaekel, O. 1896 B, p. 128.
Miller, J. 1872 A.
Newberry, J. S. 1856 A.
    1873 A, p. 328.

Pander, C. H. 1857 A.
Rohon, J. V. 1891 A.
    1899 A, pp. 20, 70.
Traquair, R. H. 1888 B, p. 508.
    1888 C, p. 492.
    1889 A.
    1894 C, p. 71.
Whiteaves J. F. 1880 A, p. 132.
Woodward, A. S. 1891 A, p. 203.
    1898 B, p. 15.
Zittel, K. A. 1890 A, p. 155.
    There are no known American species of this genus.

ASTRASPIS Walcott.   Type A. desiderata Walcott.

Walcott, C. D. 1892 A, p. 166.

Astraspis desiderata Walcott.

Walcott, C. D. 1892 A, p. 166, pl. iii, figs. 6–14; pl. iv, figs. 1–4.

American Geologist 1891 E. (Asterolepis? desiderata.)
Cope, E. D. 1898 I, p. 268.
Dana, J. D. 1896 A, p. 509, fig. 696.
Silurian (Ordovician); Colorado.

PTERICHTHYODES Bleeker.   Type *Pterichthys milleri* Agassiz.

Bleeker, P., 1859, Tentamen, p. 11.
The following authors, unless otherwise
noted, employ the name *Pterichthys* for this
genus. It is, however, preoccupied by Swain-
son, 1839.
Agassiz, L. 1843 A, p. 81.
1844 A.
1844 B, pp. 6, 151.
1846 A, p. 48.
Beyrich, E. 1877 A.
Collinge, W. E. 1893 A, p. 3.
Cope, E. D. 1885 I.
1886 D.
1887 S, p. 323.
1888 U, p. 914.
Dean, B. 1895 A, figures.
Egerton, P. H. 1848 A.
1859 A, pp. 121, 127.
Eichwald, E. 1846 A, p. 283.
Gaudry, A. 1883 A, p. 228.
Günther, A. 1880 A, p. 351, figs. 135, 136.
1862 A.
Gürich, G. 1891 A, p. 910.
Hay, O. P. 1899 E, p. 791. (Pterichthyodes.)
Huxley, T. H. 1861 B, p. 29.
1872 A, p. 129.
Jaekel, O. 1893 A.
1896 B, p. 128.

McCoy, F. 1855 A, p. 598.
Miller, H. 1859 A, p. 42, pls. i, ii.
Miller, S. A. 1889 A, p. 610.
1892 A, p. 716. (Millerichthys.)
Newberry, J. S. 1889 A, p. 110.
Nicholson and Lydekker 1889 A, p. 962.
Owen, R. 1860 E, p. 119.
Patton, W. H. 1890 A, p. 359.
Pictet, F. J. 1854 A, p. 221.
Quenstedt, F. A. 1885 A, p. 354.
Rohon, J. V. 1891 A.
1899 B, p. 10.
Steinmann and Döderlein 1890 A, p. 552.
Traquair, R. H. 1888 B, p. 508.
1888 C, p. 485.
1891 A, p. 125.
1894 C, pp. 63, 90.
Trautschold, H. 1880 B.
Whiteaves, J. F. 1880 A.
1888 B, p. 159.
Woodward, A. S. 1890 B.
1891 A, p. 208.
1892 B, p. 234.
1892 F, fig. 7.
1896 A, p. 139.
1898 B, p. 12.
Zittel, K. A. 1890 A, p. 153.
No American species of this genus are known.

BOTHRIOLEPIS Eichwald.   Type *B. ornata* Eichwald.

Eichwald, E. 1840 A, p. 79.
Agassiz, L. 1844 B, p. 20 (Pamphractus); pp. 97,
151 (Bothriolepis); p. 134 (Placothorax).
Claypole, E. W. 1890 A, p. 257.
Collinge, W. E. 1893 A, p. 3.
Cope, E. D. 1886 D, p. 1081.
1887 O, p. 1015.
1889 R, p. 858.
Eichwald, E. 1846 A, p. 283.
1857 A, p. 345.
1860 A, p. 1511.
Gürich, G. 1891 A, p. 909.
Lahusen, I. 1880 A.
Leidy, J. 1856 E, p. 11. (Stenacanthus.)
1856 S, p. 164. (Stenacanthus.)
Miller, S. A. 1889 A, p. 588.
Newberry, J. S. 1889 A, p. 108.
Nicholson and Lydekker 1889 A, p. 962.
Patton, W. H. 1890 A, p. 359.
Roberts, G. E. 1864 A.
Rohon, J. V. 1899 A, p. 27.
1899 B, p. 10.
Steinmann and Döderlein 1890 A, p. 552.
Traquair, R. H. 1888 B, p. 509.
1888 C, p. 496.
1894 C, p. 63.
Trautschold, H. 1880 B.
Whiteaves, J. F. 1880 A, p. 132.
1881 B, p. 23.
1883 B, p. 159.
1899 A, p. 411.
Woodward, A. S. 1891 A, p. 228.
1892 B, p. 234.
1892 C, p. 484.
1898 B, p. 14.
Woodward and Sherborn 1890 A, p. 18.

Zittel, K. A. 1890 A, p. 157.

Bothriolepis canadensis Whiteaves.

Whiteaves, J. F. 1880 A, p. 132. (Pterichthys.)
Cope, E. D. 1881 L. (Pterichthys.)
1885 I, fig. 1.
1886 D.
1887 S, p. 323, fig. 62.
1888 U, p. 914, pl. xv, figs. 1, 2.
1891 N, p. 9, fig. 2.
1892 X.
1898 B, p. 16, fig. 2.
Dana, J. D. 1894 A, p. 617, fig. 960.
Miller, S. A. 1889 A, p. 588, fig. 1104.
Traquair, R. H. 1888 B, p. 509.
1888 C, p. 496, pl. xviii, fig. 6.
Whiteaves, J. F. 1881 B, p. 26. [Pterichthys
(Bothriolepis).]
1881 C, p. 28. (Pterichthys.)
1883 A, p. 353. (Pterichthys.)
1883 B, p. 158. (Pterichthys.)
1887 A, p. 101, pls. vi-ix. (Pterichthys.)
1889 A, p. 91.
1899 A, p. 411.
Williams, H. S. 1893 A, p. 286, figure.
Woodward, A. S. 1891 A, p. 228.
1892 C, p. 484.
Devonian; Province of Quebec, Canada.

Bothriolepis minor Newb.

Newberry, J. S. 1889 A, p. 112, pl. xx, figs. 6-8.
Cope, E. D. 1892 L, p. 224.
Eastman, C. R. 1899 F, p. 324, fig. 5.
Woodward, A. S. 1891 A, p. 232.
Devonian (Chemung); Pennsylvania.

Bothriolepis nitida (Leidy).

Leidy, J. 1856 E, p. 11. (Stenacanthus.)
Cope, E. D. 1892 L, p. 221. (B. leidyi.)
Eastman, C. R. 1899 F, p. 324. (B. leidyi.)
Leidy, Jos. 1856 A, p. 421. (Stenacanthus.)
   1856 S, p. 163, pl. xvii, fig. 4 (Holoptychius americanus, in part); p. 164, pl. xvi, figs. 7–8 (Stenacanthus).

Newberry, J. S. 1889 A, pp. 109, 111, pl. xviii, fig. 2; pl. xx, figs. 1–5. (B. leidyi.)
Whitfield and Hovey 1900 A, p. 348. (B. leidyi,)
Woodward 1891 A, p. 232. (B. leidyi.)
Devonian (Catskill); Pennsylvania, New York.

Order ARTHRODIRA Cope.

Cope, E. D. 1891 B, p. 454.
   1891 N, p. 14.
   1898 B, p. 29.
Dean, B. 1895 A, p. 129.
Dollo, L. 1896 A.
Eastman, C. R. 1899 A, p. 643. ("Arthrodires.")

Gadow, H. 1898 A, p. 11.
Traquair, R. H. 1891 A, p. 126.
   1900 B, pp. 470, 522.
Woodward, A. S. 1891 A, p. 276.
   1898 B, p. 64.

COCCOSTEIDÆ.

Dean, B. 1897 A, p. 61.
Eastman, C. R. 1896 A.
   1898 D, p. 748.
   1899 A, p. 643.
Steinmann and Döderlein 1890 A, p. 552.
Gürich, G. 1891 A, p. 910.

Jaekel, O. 1896 B, p. 127. ("Coccosteiden.") ·
Koenen, A. 1895 A, p. 3.
Rohon, J. V. 1899 A, p. 31.
Traquair, R. H. 1890 B, p. 55.
Woodward, A. S. 1891 A, p. 277.
   1898 B, p. 65.

COCCOSTEINÆ, nom. nov.

Coccosteus Agassiz.   Type C. decipiens Agassiz.

Agassiz, L. 1844 B, p. 22.
   1843 A. (Definition insufficient; no species).
Alth, A. 1874 A, p. 38.
Claypole, E. W. 1893 A.
   1893 G, p. 615.
Collinge, W. E. 1893 A, p. 4.
Cope, E. D. 1886 D, p. 1030.
Dames, W. 1885 A.
Davis, J. W. 1880 A, p. 356.
Dean, B. 1895 A, p. 130.
Eastman, C. R. 1898 D, p. 748.
Egerton, P. H. 1859 A, p. 127.
Gaudry, A. 1883 A, p. 232.
Günther, A. 1880 A, p. 354, fig. 137.
Gürich, G. 1891 A.
Huxley, T. H. 1861 B, p. 29.
   1872 A, p. 129.
Jaekel, O. 1891 A.
   1899 A, p. 276.
Koenen, A. 1883 A, p. 9.
   1890 A.
   1895 A, p. 3.
Lahusen, I. 1880 A.
McCoy, F. 1855 A, p. 601.
Miller, H. 1859 A, p. 18, pls. iii, ix.
Miller, S. A. 1889 A, p. 591.
Newberry, J. S. 1856 A.
   1875 A, pp. 1–27, 32.
Nicholson and Lydekker 1889 A, p. 964.
Owen, R. 1860 E, p. 123.
   1866 A, p. 196, fig. 127.
Pander, C. H. 1857 A.
Pictet, F. J. 1854 A, p. 220.
Pollard, H. B. 1895 A, p. 419.
Quenstedt, F. A. 1885 A, P. 355.
Rohon, J. V. 1892 A, p. 18.
   1899 A, p. 31.
Ryder, J. A. 1890 A, p. 491.

Steinmann and Döderlein 1890 A, p. 552.
Traquair, R. H. 1888 B, p. 511.
   1889 A.
   1890 C.
   1890 D.
   1891 A, p. 126.
   1891 B.
   1893 A, p. 149.
   1900 A, p. 159.
Trautschold, H. 1880 C, p. 145.
   1889 A.
   1889 B.
Woodward, A. S. 1891 A, p. 278.
   1898 B, p. 60.
Woodward and Sherborn 1890 A, p. 37.
Wright, A. A. 1893 A, p. 621.
Zittel, K. A. 1890 A, p. 158.

Coccosteus canadensis Woodward.

Woodward, A. S. 1892 C, p. 488, pl. xiii, fig. 2.
Traquair, R. H. 1893 C, p. 265.
Devonian; Canada.

Coccosteus cuyahogæ Claypole.

Claypole, E. W. 1893 A.
   1893 G, p. 615, pl. xli, fig. 2.
Devonian (Cleveland shale); Ohio.

Coccosteus macromus Cope.

Cope, E. D. 1892 L, p. 225.
Devonian (Chemung); Pennsylvania.

Coccosteus occidentalis Newb.

Newberry, J. S. 1875 A, p. 32, pl. liv, figs. 2, 2a.
Claypole, E. W. 1893 A, p. 168.
   1893 G, p. 615.
Newberry, J. S. 1889 A, p. 52, pl. xxv, figs. 2, 2a.
Woodward, A. S. 1891 A, p. 293.
Zittel, K. 1890 A, p. 160.
Devonian (Corniferous); Ohio.

LIOGNATHUS Newb. Type *L. spatulatus* Newb.

Newberry, J. S. 1873 A, p. 306.
Miller, S. A. 1889 A, p. 600.
  1892 A, p. 716. (Lispognathus, type *Liogna-thus spatulatus*.)
Newberry, J. S. 1889 A, p. 52. (Syn. of Coccosteus.)
Zittel, K. A. 1890 A, p. 162.

Liognathus spatulatus Newb.

Newberry, J. S. 1873 A, p. 306, pl. xxix, fig. 4.

Claypole, E. W. 1893 A, p. 169. (L. spathulatus.)
  1893 G, p. 615.
Miller, S. A. 1889 A, p. 600, fig. 1142.
  1892 A, p. 716. (Lispognathus).
Newberry, J. S. 1889 A, p. 52. (Syn. of Coccosteus occidentalis.)
Devonian (Corniferous); Ohio.

PHLYCTÆNASPIS Traquair. Type *Coccosteus acadicus* Whiteaves.

Traquair, R. H. 1890 E, p. 144.
Collinge, W. E. 1893 A, p. 5.
Eastman, C. R. 1898 D, p. 755.
Traquair, R. H. 1890 A, p. 20. (Phlyctænius, type *Coccosteus acadicus;* preoccupied by Phlyctænium Zittel.)
  1890 B. (Phlyctænius.)
  1894 B. (Phlyctænaspis.)
  1900 A, p. 159.
Whiteaves, J. F. 1899 A, p. 412.
Woodward, A. S. 1891 A, p. 295.

Phlyctænaspis acadica (Whiteaves).

Whiteaves, J. F. 1881 E, p. 94, fig. 1. (Coccosteus.)
Claypole, E. W. 1893 A, p. 168. (Coccosteus.)
  1893 G, p. 615. (Coccosteus.)

Dana, J. D. 1896 A, p. 616, fig. 964.
Traquair, R. H. 1890 A.
  1890 B, p. 60, pl. iii, figs. 1, 2. (Phlyctænius.)
  1895 A, p. 147, fig. 1.
  1893 D, p. 115.
Whiteaves, J. F. 1883 A, p. 355. (Coccosteus.)
  1883 B, p. 163. (Coccosteus.)
  1889 A, p. 93, pl. ix. (Coccosteus.)
Williams, H. S. 1893 A, p. 286. (Phlyctænius.)
Woodward, A. S. 1891 A, p. 295.
  1892 A, p. 5, pl. 1, figs. 7, 8.
  1892 C, p. 481.
Devonian; New Brunswick.

ASPIDICHTHYS Newb. Type *A. clavatus* Newb.

Newberry, J. S. 1873 A, pp. 269, 322.
Koenen, A. 1883 A, p. 33.
  1895 A, p. 13.
Miller, S. A. 1889 A, p. 587.
Newberry, J. S. 1870 B, p. 153. (Aspidophorus; preoccupied.)
  1889 A, pp. 26, 59, 72.
Whiteaves, J. F. 1892 A, p. 355.
Zittel, K. A. 1890 A, p. 158.

Aspidichthys clavatus Newb.

Newberry, J. S. 1873 A, pp. 257, 269, 273, 323, pl. xxxv, figs. 1, 2.

Miller, S. A. 1889 A, p. 588, fig. 1102.
Newberry, J. S. 1889 A, p. 73, pl. xxxviii, figs. 3, 4.
Whiteaves, J. F. 1892 A, p. 358.
Williams, H. S. 1887 A, p. 43.
Devonian (Cleveland shale); Ohio: (Portage); New York.

Aspidichthys ? notabilis Whiteaves.

Whiteaves, J. F. 1892 A, p. 354, pl. xlvii, figs. 1, 1a.
  1896 B, p. 411, pl. 1, figs. 1, 2.
Devonian, Manitoba.

DINICHTHYINÆ.

Eastman, C. R. 1897 D, p. 84.
Most of the following authors regard this group as of family value:
Claypole, E. W. 1892 B.
Cope, E. D. 1890 H.
Dean, B. 1895 A, p. 130.
  1895 A.

Eastman, C. R. 1896 A.
  1896 B.
  1898 C, p. 872. (Coccosteidæ, in part.)
  1898 D, p. 747. (Dinichthynæ.)
Newberry, J. S. 1885 D, p. 14. (Dinichthydæ.)
Nicholson and Lydekker 1889 A, p. 965.
Woodward, A. S. 1891 A, p. 277. (Coccosteidæ, in part.)

DINICHTHYS Newb. Type *D. hertzeri* Newb.

Newberry, J. S. 1868 B, p. 146.
American Geologist 1891 A.
Claypole, E. W. 1888 A, p. 63.
  1892 A (Dinichthys); p. 1 (Gorgonichthys, type *G. clarki*).
  1892 B.
  1892 C.
  1893 E, p. 97. (Gorgonichthys.)

Claypole, E. W. 1888 A, p. 63.
  1893 G, p. 606 (Dinichthys); p. 614 (Gorgonichthys).
  1894 C, p. 189.
  1895 E, p. 473. (Gorgonichthys.)
  1896 B.
Cope, E. D. 1890 H.
  1891 B, p. 450.
  1892 X.

Dean, B. 1891 A.
1895 A.
1896 A, pls. vii, viii.
1897 A, p. 57, pls. ii, iii.
Eastman, C. R. 1896 A.
1896 B.
1896 C.
1897 D, p. 19.
1898 A.
1896 C, p. 371.
1898 D, p. 747.
1899 A, p. 643.
1900 B, p. 32.
Hay, O. P. 1899 E, 791.
Hitchcock, C. H. 1868 A, p. 184.
Huene, F. 1900 A, p. 64.
Koenen, A. 1895 A, p. 16.
Miller, S. A. 1889 A, p. 596.
1892 A, p. 717.   (Ponerichthys, type *D. ter-rellii* Newb.)
Newberry, J. S. 1870 B, p. 153.
1873 A, pp. 257, 269, 313.
1875 A, p. 3.
1883 B.
1884 A.
1885 A.
1885 B, p. 25.
1885 D, p. 11.
1889 A, p. 135.
1892 A.
Nicholson and Lydekker 1889 A, p. 965.
Pollard, H. B. 1895 A, p. 420.
Reis, O. M. 1896 A, p. 152.
Steinmann and Döderlein 1890 A, p. 553.
Woodward, A. S. 1891 A, p. 300.
1898 B, p. 68.
Wright, A. A. 1894 A, p. 313.
1893 A, p. 620.
Zittel, K. A. 1890 A, p. 160.

Dinichthys canadensis Whiteaves.

Whiteaves, J. F. 1892 A, p. 353, pl. xlvi, fig. 12.
Claypole, E. W. 1896 B, p. 355.
1893 G, p. 606.
Miller, S. A. 1897 A, p. 793.   (Ponerichthys.)
Upper Devonian; Manitoba.

Dinichthys clarki (Claypole).

Claypole, E. W. 1892 A.   (Gorgonichthys.)
1893 E, p. 97.   (Gorgonichthys.)
1893 G, p. 614, pl. xli, fig. 3.   (Gorgonich-thys.)
1895 E, p. 473.   (Gorgonichthys.)
Eastman, C. R. 1900 A, p. 35.
Devonian (Cleveland shale); Ohio.

Dinichthys curtus Newb.

Newberry, J. S. 1888 D, p. 179.
Claypole, E. W. 1893 G, p. 606.
Eastman, C. R. 1900 B, p. 33.
Miller, S. A. 1892 A, p. 717.   (Ponerichthys.)
Newberry, J. S. 1889 A, p. 156, pls. xlviii, fig. 3?;
pl. liii, figs. 1–3.
Woodward, A. S. 1891 A, p. 301.
Wright, A. A. 1893 A, p 623.
Devonian (Cleveland shale); Ohio.

Dinichthys gouldii Newb.

Newberry, J. S. 1885 B, p. 26.
Claypole, E. W. 1893 G, p. 606.
1896 C.
Dean, B. 1896 A, pls. vii, viii.
1897 A, p. 60, pl. ii.
Miller, S. A. 1892 A, p. 717.   (Ponerichthys.)
Newberry, J. S. 1889 A, p. 150, pl. ix, fig. 1; pl. x,
figs. 1, 2.
Woodward, A. S. 1891 A, p. 301.
Devonian (Cleveland shale); Ohio.

Dinichthys gracilis Claypole.

Claypole, E. W. 1893 F, p. 279, text figures.
1893 G, p. 607, pl. xli, fig. 1.
Miller, S. A. 1897 A, p. 793.   (Ponerichthys.)
Devonian (Cleveland shale); Ohio.

Dinichthys halmodeus (Clarke).

Clarke, J. M. 1894 A, p. 162, pl. i, and text figures.
(Coccosteus.)
Eastman, C. R. 1900 B, p. 34.
Devonian; New York.

Dinichthys hertzeri Newb.

Newberry, J. S. 1868 B.
Claypole, E. W. 1893 E, p. 94.
1893 F, p. 277.
1893 G, p. 606.
Dana, J. D. 1896 A, p. 617, figs. 965, 966.
Dean, B. 1891 A, p. 310, pl. i; pl. iii, figs. 3, 4.
Eastman, C. R. 1900 B, p. 34.
Hitchcock, C. H. 1868 A, p. 184, figs. 1, 2.
Lesley, J. P. 1889 A, p. 202, figure.
1892 A, p. 1333.
Miller, S. A. 1889 A, p. 596.
1892 A, p. 717.   (Ponerichthys.)
Newberry, J. S. 1873 A, p. 316, pls. xxx, xxxi.
1875 A, p. 3.
1878 A, p. 189.
1885 D, p. 11.   (*D. stertzeri, errore.*)
1889 A, pp. 64, 338, pl. xxxiii, fig. 2.
Woodward, A. S. 1891 A, pp. 301, 302.
Zittel, K. A. 1890 A, p. 160.
Devonian (Huron shale); Ohio.

Dinichthys intermedius Newb.

Newberry, J. S. 1889 A, p. 152, pl. x, figs. 1, 2; pl.
xlvii, figs. 1, 2, 3?, 4; pls. li, lii.
Claypole, E. W. 1892 C.
1893 G, p. 606.
1894 A, p. 190.
1896 C.
Dean, B. 1891 A, p. 312, pl. ii; pl. iii, figs. 1, 2.
1895 A, p. 133, figs. 133, 136, 144.
Eastman, C. R. 1897 D, p. 19, pl. i, fig. 1; pl. iii,
fig. 1 (?).
1898 C, p. 371.
1898 D, pp. 747, 755, fig. 3.
1900 B, p. 33.
Miller, S. A. 1892 A, p. 717.   (Ponerichthys.)
Woodward, A. S. 1891 A, p. 301.
Wright, A. A. 1893 A, p. 623.
Devonian (Cleveland shale); Ohio.

Dinichthys lincolni Claypole.

Claypole, E. W. 1893 F, p. 277, text figure.
Miller, S. A. 1897 A, p. 793.   (Ponerichthys.)
Devonian (Marcellus); Pennsylvania.

Dinichthys minor Newb. (not of Ringueberg).

Newberry, J. S. 1878 A, p. 191.
Claypole, E. W. 1893 G, p. 606.
Dean, B. 1897 A, p. 58.   (D. minor?)
Miller, S. A. 1892 A, p. 717.   (Ponerichthys.)
Newberry, J. S. 1889 A, p. 149, pl. viii, figs. 1–8.
Devonian (Cleveland shale); Ohio.

Dinichthys newberryi Clarke.

Clarke, J. M. 1885 A, p. 17, pl. i, fig. 1.
    1897 A, p.
Claypole, E. W. 1893 G, p. 606.
Eastman, C. R. 1897 D, p. 30, pl. i, fig. 2; pl. iv.
Newberry, J. S. 1889 A, p. 153.
Lesley, J. P. 1889 A, p. 202, figure.
Woodward, A. S. 1891 A, p. 301.
Devonian (Hamilton); New York.

Dinichthys præmaxillaris nom. nov.

Claypole, E. W. 1893 F, p. 279, pl. xii.   (D. clarki, preoccupied by Gorgonichthys clarki 1892.)
    1893 G, p. 607, pl. xli, fig. 1.   (D. clarki.)
Eastman, C. R. 1900 B, p. 35.   (D. clarki.)
Miller, S. A. 1897 A, p. 793.   (Ponerichthys clarki.)
Devonian (Cleveland shale); Ohio.

Dinichthys ? precursor Newb.

Newberry, J. S. 1889 A, p. 51, pl. xli.
Claypole, E. W. 1893 G, p. 606.   (D. præcursor.)
Eastman, C. R. 1899 F, p. 319.   (Syn. of D. tuberculatus.)
Miller, S. A. 1892 A, p. 717.   (Ponerichthys præcursor.)
Woodward, A. S. 1891 A, p. 302.
Devonian (Corniferous); Ohio.

Dinichthys prentis-clarki. Claypole.

Claypole, E. W. 1896 C, pl. vii.
Miller, S. A. 1897 A, p. 792.   (D. prentis clarkei.)
Devonian (Cleveland shale); Ohio.

Dinichthys pustulosus Eastman.

Eastman, C. R. 1897 D, p. 38, pl. iii, fig. 4 (D. pustulosus); p. 40 (D. mixeri).
    1898 B, p. 556.
    1898 C, p. 371.
    1898 D, p. 748, figs. 1, 2.
    1899 B, pp. 278, 283.
    1900 D, p. 32, fig. 1.
Devonian (Hamilton); Wisconsin, Iowa, Illinois, New York.

Dinichthys ringuebergi Newb.

Newberry, J. S. 1889 A, p. 60.
Claypole, E. W. 1893 G, p. 606.

Eastman, C. R. 1897 D, p. 49.
Lesley, J. P. 1892 A, p. 1333.   (No name.)
Ringueberg, E. N. S. 1884 A.   (D. minor; not of Newberry.)
Woodward, A. S. 1891 A, p. 302.
Devonian (Portage); Ohio.

Dinichthys terrelli Newb.

Newberry, J. S. 1873 A, p. 322, pls. xxxii–xxxiv.
Claypole, E. W. 1893 G, pp. 606, 607, pl. xliii, figs. 1–5.
    1893 E, p. 95.
    1894 C, p. 189, figs. 1–5.
Cope, E. D. 1891 B, p. 450.
Dean, B. 1891 A, p. 310.
    1895 A, p. 137.
    1896 A, pl. viii, fig. 3.
    1897 A, p. 59, pl. iii.
Eastman, C. R. 1896 B.
    1897 D, pl. ii, figs. 1, 5, 6; pl. iii, fig. 2.
    1898 D, pp. 747, 758.
Miller, S. A. 1892 A, p. 717.   (Ponerichthys.)
Newberry, J. S. 1873 A, p. 316, pls. xxxii–xxxiv.
    (D. hertzeri, in part.)
    1874 C, p. 149.
    1875 A, p. 27, charts v, vi.
    1885 D, p. 11.
    1888 D, p. 179.
    1889 A, passim, pl. iv, figs. 1, 2.
    1892 A.
Woodward, A. S. 1890 A, p. 391.
    1891 A, p. 302.
    1891 D.
Wright, A. A. 1893 A, p. 621, pl. xliv; text fig. 1.
    1894 A, pl. ix.
Devonian (Cleveland shale); Ohio.

Dinichthys tuberculatus Newb.

Newberry, J. S. 1888 D, p. 179.
Claypole, E. W. 1893 F, p. 277.
    1893 G, p. 606.
Cope, E. D. 1892 X, p. 1025.   (D. pustulosus.)
Eastman, C. R. 1896 A.
    1897 D, p. 38.
    1899 B, pp. 278, 283.
    1899 F, p. 318.
Lohest, M. 1890 A, p. lvii.   (D. pustulosus; fide Eastman.)
Miller, S. A. 1892 A, p. 717.   (Ponerichthys.)
Newberry, J. S. 1889 A, p. 98, pl. xxxii.
Woodward, A. S. 1891 A, p. 302.
Devonian (Chemung); Pennsylvania: (Corniferous); Ohio: (Hamilton); Wisconsin: (Upper Devonian); Iowa.

Dinichthys sp. indet.

Eastman, C. R. 1899 F, p. 320, fig. 1.
Devonian (Chemung); New York.

Dinichthys sp. indet.

Eastman, C. R. 1899 F, p. 326, fig. 6.
Devonian (Catskill); New York.

STENOGNATHUS Newb.   Type *Dinichthys corrugatus* Newb.

Newberry, J. S. 1897 A, p. 303.

Stenognathus corrugatus Newb.

Newberry, J. S. 1889 A, p. 151, pl. vii, figs. 3, 3a.
(Dinichthys.)
1885 B, p. 26.   (No description.)
1897 A, p. 303, pl. xxiv, figs. 27, 28.   (Stenognathus.)

Woodward, A. S. 1891 A, p. 301.   (Dinichthys).
Devonian (Cleveland shale); Ohio.

TITANICHTHYS Newb.   Type *T. agassizii* Newb.

Newberry, J. S. 1885 B, p. 27.
American Geologist 1891 A.
Claypole, E. W. 1892 A.
    1892 B.
    1893 E, p. 95.
    1893 G, pp. 608, 611.
Cope, E. D. 1891 B, p. 450.
Eastman, C. R. 1896 B.
    1898 D.
    1899 A, p. 643.
Miller, S. A. 1892 A, p. 717.
Newberry, J. S. 1885 D, p. 13.
Woodward, A. S. 1891 A, p. 302.

Titanichthys agassizii Newb.

Newberry, J. S. 1885 B, p. 27.
Claypole, E. W. 1893 G, p. 608, pl. xlii, fig. 5.
Cope, E. D. 1891 B, p. 450.
Dana, J. D. 1888 A, p. 498.
Eastman, C. R. 1896 C, p. 372.
    1898 D, p. 761, fig. 4.
    1899 D, p. 15.
Newberry, J. S. 1889 A, p. 158, pl. l; pl. li, figs. 1, 2; pl. iv, fig. 4.
Woodward, A. S. 1891 A, p. 302.
Devonian (Cleveland shale); Ohio.

TRACHOSTEUS Newb.

Newberry, J. S. 1889 A, p. 166.
Dean, B. 1896 B.
    1895 A, p. 136.
    1897 A, p. 59.
Miller, S. A. 1892 A, p. 717.
Woodward, A. S. 1891 A, p. 311.

DIPLOGNATHUS Newb.   Type *D. mirabilis* Newb.

Newberry, J. S. 1878 A, p. 188.
Dean, B. 1895 A, p. 137.
    1897 A, p. 60.
Newberry, J. S. 1885 A.
    1885 B, p. 27.
    1889 A, p. 159.
Zittel, K. A. 1890 A, p. 162.

Diplognathus mirabilis Newb.

Newberry, J. S. 1878 A, p. 188.

BRONTICHTHYS Claypole.   Type *B. clarki* Claypole.

Claypole, E. W. 1894 B.
    1895 E.
Miller, S. A. 1897 A, p. 792.

Titanichthys attenuatus Wright.
Wright, A. A., in Claypole, E. W. 1893 G, p. 612, pl. xlii, figs. 1, 2.
Devonian (Cleveland shale); Ohio.

Titanichthys brevis Claypole.
Claypole, E. W. 1896 A, pl. x.
Devonian (Cleveland shale); Ohio.

Titanichthys clarkii Newb.
Newberry, J. S. 1887 C, p. 164.
Claypole, E. W. 1888 A, p. 62.
    1893 G, p. 608, pl. xliii, fig. 3.
Cope, E. D. 1888 P.
Dana, J. D. 1888 A, p. 498.
Dean, B. 1895 A, p. 136, fig. 139.
Eastman, C. R. 1896 C, p. 372.
    1898 D, p. 765, fig. 5; fig. 6.?
Newberry, J. S. 1889 A, p. 133, pl. li, figs. 3, 4; pl. iii.
Woodward, A. S. 1891 A, p. 302.
Devonian (Cleveland shale); Ohio.

Titanichthys rectus Claypole.
Claypole, E. W. 1893 G, p. 609, pl. xliii, fig. 5
Devonian (Cleveland shale?); Ohio.

Titanichthys sp. indet.
Claypole, E. W. 1893 G, p. 609, pls. xxxviii, xxxix.

TRACHOSTEUS Newb.   Type *T. clarkii* Newb.

Trachosteus clarkii Newb.
Newberry, J. S. 1889 A, p. 167, pl. xlii, figs. 1–8.
Claypole, E. W. 1888 A, p. 63.
Dean, B. 1895 A, p. 137, fig. 140.
Woodward, A. S. 1891 A, p. 311.
Devonian (Cleveland shale); Ohio.

Dean, B. 1895 A, p. 137, figs. 141–143.
Newberry, J. S. 1884 A.
    1885 B, p. 27.
    1889 A, p. 159, pl. xl, figs. 1–4; pl. xli, figs. 1–3.
Woodward, A. S. 1891 A, p. 312.
Devonian (Cleveland shale); Ohio.

Brontichthys clarki Claypole.
Claypole, E. W. 1894 B, pl. xii.
Devonian (Cleveland shale); Ohio.

MACROPETALICHTHYIDÆ.

Eastman, C. R. 1896 C, p. 372.
  1896 D, p. 748.

MACROPETALICHTHYS Norw. and Owen.

Norwood and Owen 1846 A.
Claypole, E. W. 1896 B, p. 354.
Cope, E. D. 1890 H.
  1891 B, p. 449.
  1891 C.
Dean, B. 1895 A, p. 135.
Eastman, C. R. 1897 C.
  1896 C, p. 372.
  1896 D, p. 748.
  1900 B, p. 36.
  1900 C, p. 392 (Macropetalichthys); p. 391
    (Physichthys).
  1900 D, p. 177. (Physichthys.)
Koenen, A. 1895 A, p. 22.
Meyer, H. v., 1855, Palæontographica, iv, p. 80.
  (Physichthys.)
Miller, S. A. 1889 A, p. 601.
Newberry, J. S. 1846 A, p. 291.
  1857 A, p. 147. (Agassizichthys.)
  1857 C, p. 122. (Agassizichthys.)
  1862 A, p. 75, fig. 1.
  1870 B, p. 152.
  1873 A, p. 290 et passim.
  1889 A, p. 41.
Nicholson and Lydekker 1889 A, p. 976.
Norwood and Owen 1846 B.
Woodward, A. S. 1891 A, p. 303.
  1892 B, p. 233.
Zittel, K. A. 1890 A, p. 164.

Type M. rhapheidolabis Norw. and Owen.

Macropetalichthys rhapheidolabis
  Norwood and Owen.

Norwood and Owen 1846 A, figs. 1, 2.
Kindle, E. M. 1896 A, p. 484.
Miller, S. A. 1889 A, p. 601.
Norwood and Owen 1846 B.
Newberry, J. S. 1853 A, p. 12, fig. 1.
Woodward, A. S. 1891 A, p. 304. (Syn. of M. sulli-
  vanti.)
Devonian (Corniferous); Indiana.
Owen in the original description mentions
  the alternative name, Pterichthys norwoodensis.

Macropetalichthys sullivanti Newb.

Newberry, J. S. 1857 C, p. 122, fig. 1 (Agassizich-
  thys manni); p. 123 (A. sullivanti).
Cope, E. D. 1891 B, pp. 449, 456, pl. xxx, fig. 5.
Eastman, C. R. 1897 C, pl. xii and text fig. 1.
Lesley, J. P. 1892 A, pl. cl.
Miller, S. A. 1889 A, p. 601, fig. 1146.
Newberry, J. S. 1857 A, p. 147. (Agassizichthys
  sullivanti, A manni.)
  1862 A, p. 75. (A sullivanti, A. manni.)
  1873 A, p. 294 et passim, pl. xxv, fig. 1.
  1889 A, p. 28 (M. manni); p. 306, pl. xxxviii,
    figs. 1-2a (M. sullivanti).
Whiteaves, J. W. 1880 A, p. 133.
  1889 B, p. 119.
Woodward, A. S. 1891 A, p. 304.
Devonian (Corniferous); Ohio.

GLYPTASPIS Newb.

Newberry, J. S. 1889 A, p. 157.
Dean, B. 1897 A, p. 59.
Miller, S. A. 1892 A, p. 715.
Newberry, J. S. 1897 A, p. 304.

Type G. verrucosa Newb.

Glyptaspis verrucosa Newb.

Newberry, J. S. 1889 A, p. 158, pl. xiii, figs. 1, 2.
Woodward, A. S. 1891 A, p. 311.
Devonian (Cleveland shale); Ohio.

ASTEROSTEIDÆ.

Woodward, A. S. 1891 A, p. 312.

ASTEROSTEUS Newb.

Newberry, J. S. 1875 A, p. 35.
Eastman, C. R. 1900 B, p. 36.
Miller, S. A. 1889 A, p. 588.
Newberry, J. S. 1889 A, p. 44.
Woodward, A. S. 1891 A, p. 312.
Zittel, K. A. 1890 A, p. 164.

Type A. stenocephalus Newb.

Asterosteus stenocephalus Newb.

Newberry, J. S. 1875 A, p. 36, pl. liv, fig. 1.
Claypole, E. W. 1896 B, p. 354.
Newberry, J. S. 1889 A, p. 45, pl. xxx, fig. 1.
Whitfield and Hovey 1900 A, p. 348.
Woodward, A. S. 1891 A, p. 313.
Devonian (Corniferous); Ohio.

PHYLLOLEPIDÆ.

Woodward, A. S. 1891 A, p. 313.

PHYLLOLEPIS Agassiz.

Agassiz, L. 1844 B, p. 67.
Fritsch, A. 1889 B (1888), p. 89.
Giebel, C. G. 1848 A, p. 227.
Newberry, J. S. 1889 A, p. 97.
Pictet, F. J. 1854 A, p. 144.
Woodward, A. S. 1891 A, p. 313.

Type P. concentrica Agassiz.

Zittel, K. A. 1890 A, p. 180.

Phyllolepis delicatula Newb.

Newberry, J. S. 1889 A, p. 97, pl. xix, fig. 11.
Woodward, A. S. 1891 A, p. 314.
Devonian (Chemung); Pennsylvania.

HOLONEMA Newb.   Type *H. rugosum* Claypole.

Newberry, J. S. 1889 A, p. 92.
Claypole, E. W. 1890 A.
Cope, E. D. 1891 B, p. 456.
  1892 L, p. 221, pl. vii, fig. 3.
Dana, J. D. 1896 A, p. 616, fig. 698.
Dean, B. 1897 A, p. 59.
Miller, S. A. 1892 A, p. 715.
Williams, H. S. 1891 A.
  1893 A, p. 285.
Woodward, A. S. 1891 A, p. 314.
  1892 B, p. 234.

Holonema horridum Cope.

Cope, E. D. 1892 L, p. 222, pl. vii, fig. 1.
Devonian (Chemung); Pennsylvania.

Holonema rugosum (Claypole).

Claypole, E. W. 1883 A, fig. 1. (Pterichthys.)
  1890 A, p. 255. (Pterichthys.)
Cope, E. D. 1891 B, p. 456.
Lesley, J. P., 1889 A, p. 808, figure.
Newberry, J. S. 1889 A, p. 93, pl. xvii, figs. 1–4.
Williams, H. S. 1891 A.
  1893 A, p. 285, with figures.
Woodward, A. S. 1891 A, p. 315.
  Devonian (Chemung); Pennsylvania, New
  York.

MYLOSTOMATIDÆ.

Woodward, A. S. 1891 A, p. 315.

MYLOSTOMA Newb.   Type M. *variabile* Newb.

Newberry, J. S. 1883 A, p. 145.
Dean B. 1893 B.
  1895 A, p. 136.
  1897 A, p. 60.
Miller, S. A. 1892 A, p. 716.
Newberry, J. S. 1884 A.
  1885 A, p. 724.
  1889 A, p. 161.
Woodward, A. S. 1891 A, p. 315.
Zittel, K. A. 1890 A, p. 128.

Mylostoma terrelli Newb.

Newberry, J. S. 1883 A, p. 147.
  1889 A, p. 164, pl. xiv, figs. 1, 2.

Woodward, A. S. 1891 A, p. 316.
  Devonian (Cleveland shale); Ohio.

Mylostoma variabile Newb.

Newberry, J. S. 1883 A, p. 146.
Dean, B. 1895 A, p. 137, fig. 138.
Newberry, J. S. 1889 A, p. 165, pls. xv, xvi.
Woodward, A. S. 1891 A, p. 316.
  Devonian (Cleveland shale); Ohio.

Superorder DIPNOI Müller.

Müller, J. 1846 C, p. 201.
Ayres, H. 1893 A.
Balfour, F. M. 1882 A, p. 432.
Baur, G. 1896 C, p. 670.
Beard, J. 1890 A.
Beyrich, E. 1850 A, p. 153.
Bridge, T. W. 1898 A, p. 325.
Chamberlin, T. C. 1900 A, p. 411. ("Dipnoans.")
.   Cope, E. D. 1872 C.
  1877 O, p. 41.
  1878 A, p. 292.
  1884 N, p. 579.
  1884 H H.
  1885 G.
  1887 S, pp. 184, 824.
  1889 R, pp. 853, 854.
  1890 B.
  1891 N, pp. 10, 14.
  1898 B, pp. 17, 21.
Dean, B. 1895 A. p. 116.
Dollo, L. 1896 A. ("Dipneustes.")
Eastman, C. R. 1898 A, p. 110.
Fritsch, A. 1889 B, p. 65.
Gadow, H., 1898 A, p. 11.
Gadow and Abbott 1895 A.

Gegenbaur, C. 1865 B, pp. 72, 146.
  1895 A.
  1895 B.
  1898 A.
Gill, T. 1861 A, p. 22.
  1873 B.
  1893 A, p. 130.
  1896 C, p. 912.
Goeppert, E. 1895 A.
Günther A. 1870 A, p. 321.
  1871 A.
  1872 A.
  1880 A, p. 355.
  1881 A, p. 686.
Haeckel, E. 1895 A, p. 257. (Dipneusta.)
Hasse, C. 1893 A.
Hay, O. P. 1899 F, p. 682.
Huxley, T. H. 1872 A, p. 145.
Kehner, F. A. 1876 A, p. 351.
Klaatsch, H. 1890 A, p. 209.
  1893 A.
  1896 A, p. 316.
Miall, L. C. 1878 A.
Mollier, S. 1893 A, p. 86.
Müller, J. 1846 C, p. 201.

Newberry, J. S. 1873 A, pp. 249, 252.
Nicholson and Lydekker 1889 A, p. 952.
Orr, H. 1885 A.
Owen, R. 1839 C. (On Lepidosiren annectens.)
Parker, W. K. 1879 B, p. 31.
Röse, C. 1892 B.
1895 A, p. 566.
Rohon, J. V. 1898 A.
Ryder, J. A. 1896 A.
Schneider A. 1886 A.
Semon, R. 1899 A, p. 75.
Stannius, H. 1854 A.

Steinmann and Döderlein 1890 A, p. 553.
Stephan, P. 1900 A. ("Dipnoiques.")
Teller, F. 1891 A.
Thacher, J. K. 1877 A.
Traquair, R. H. 1900 B, p. 521.
Wiedersheim R. 1889 A, p. 480.
Woodward, A. S. 1888 E.
1891 A, pp. 20, 234.
1898 B, p. 60.
Zittel, K. A. 1886 A.   (On Ceratodus.)
1890 A, p. 122.

Order CTENODIPTERINI Pander.

Pander, C. H. 1858 A.
Agassiz, L. 1846 A, p. 44.   ("Diptériens.")
Eichwald, E. 1860 A, p. 1534. ("Cténodiptériens.")

Haeckel, E. 1895 A, p. 263.   (Ctenodipneusta.)
Huxley, T. H. 1861 B, p. 13.   (Ctenododipterini.)
Zittel, K. A. 1890 A, p. 125. (Ctenodipterini.)

DIPTERIDÆ.

Owen, R. 1860 E, p. 129.
Günther, A. 1872 A, p. 554.   (Ctenodipteridæ.)
1880 A, p. 359.   (Ctenodipteridæ.)
Quenstedt, F. A. 1852 A, p. 228.   (Dipterini.)
1885 A, p. 349.   (Dipterini.)

Steinmann and Döderlein 1890 A, p. 555.
Woodward A. S. 1891 A, p. 235.
1898 B, p. 62.   (Ctenodontidæ, part.)

DIPTERUS Sedgw. and Murch.   Type *D. valenciennesi* Sedgw. and Murch.

Sedgwick and Murchison, 1828, Trans. Geol. Soc.
Lond. (2), iii, p. 143.
Agassiz, L. 1843 A.
1843 B, ii, pt. i, p. 3 (1833) (Catopterus, type
C. analis); p. 112 (1835) (Dipterus).
1844 A.
1844 B, pp. 5, 29.   (Polyphractus.)
Atthey, T. 1868 A.
Bridge, T. W. 1898 A, p. 367.
Cope, E. D. 1878 A, p. 298.
Dean, B. 1895 A, p. 121, with figures.
Dollo, L. 1895 A.
Eastman, C. R. 1900 B, p. 37.
Egerton, P. H. 1859 A, p. 125.
Eichwald, E. 1860 A, p. 1535.
Emery, C. 1897 B, p. 147.
Günther, A. 1871 A, p. 428.
1872 A, p. 556.
Hancock and Atthey 1868 A, p. 370.
1868 B, p. 54.
1871 A.
Huxley, T. H. 1861 B, p. 13, figs. 9, 10.
McCoy, F. 1855 A, p. 590.
Miall, L. C. 1878 B.
Miller, H. 1859 A, pl. v.
Miller, S. A. 1889 A, p. 596.
Newberry, J. S. 1875 A, p. 8.
1889 A, p. 87.
Pander, C. H. 1858 A, p. 6, pls. i-v.
Pictet, F. J. 1854 A, p. 192.
Rohon, J. V. 1898 A.
Steinmann and Döderlein 1890 A, p. 555.
Traquair, R. H. 1878 A, p. 1.
1888 B, p. 507.
Woodward, A. S. 1891 A, p. 235.
1893 C.
1896 B, p. 63.

Woodward and Sherborn 1890 A, p. 69.
Zittel, K. A. 1890 A, p. 125, fig. 185.

Dipterus calvini Eastman.

Eastman, C. R. 1900 B, p. 38, fig. 7.
Udden, J. A. 1899 A, p. 494.
Devonian (Cedar Valley limestone); Iowa.

Dipterus contraversus Hay.

Hay, O. P. 1899 E, p. 786.
Eastman, C. R. 1900 B, p. 37.
Newberry, J. S. 1889 A, p. 119, pl. xxvii, fig. 33.
[D. (Ctenodus) radiatus; preoccupied.]
Woodward, A. S. 1891 A, p. 242. [Dipterus
(Ctenodus) radiatus.]
Devonian (Catskill); Pennsylvania.

Dipterus costatus Eastman.

Eastman, C. R. 1900 B, p. 39, fig. 4.
Devonian (Upper); Iowa.

Dipterus flabelliformis Newb.

Newberry, J. S. 1889 A, p. 90, pl. xvi, fig. 21. [D.
(Ctenodus.]
Eastman, C. R. 1900 B, p. 37.
Newberry, J. S. 1897 A, p. 302.   (Ctenodus.)
Woodward, A. S. 1891 A, p. 242.
Devonian (Chemung); Pennsylvania.

Dipterus fleisheri (Newb.).

Newberry, J. S. 1897 A, p. 302, pl. xxiv, fig. 25.
(Ctenodus.)
Eastman, C. R. 1900 B, p. 37.
Devonian ("Catskill"); Pennsylvania.

Dipterus ithacensis Williams.

Williams, H. S. 1882 A.
Woodward, A. S. 1891 A, p. 243.
Devonian (Chemung); New York.

Dipterus minutus Newb.

Newberry, J. S. 1889 A, p. 91, pl. xxvii, fig. 26. [D. (Ctenodus).]
Eastman, C. R. 1900 B, p. 37.
Woodward, A. S. 1891 A, p. 242.
Devonian (Chemung); Pennsylvania.

Dipterus mordax Eastman.

Eastman, C. R. 1900 B, p. 39, figs. 6, 8.
Devonian (Upper); Iowa.

Dipterus nelsoni Newb.

Newberry, J. S. 1889 A., p 89, pl. xxvii, figs. 19, 20. [D. (Ctenodus) nelsoni.]
Dana, J. D. 1896 A, p. 617, fig. 968. (Ctenodus.)
Eastman, C. R. 1899 F, p. 322.
1900 B, p. 37.
Newberry, J. S. 1889 A, p. 90, pl. xxvii, figs. 22, 23. [D. (Ctenodus) levis.]
Williams, H. S. 1887 A, p. 62, pl. iii, fig. 1 (D. nelsoni); pp. 63, 91, pl. iii, fig. 2 (D.? lævis, or D. alleganensis).

Woodward, A. S. 1891 A, p. 242. (D. nelsoni, D. levis.)
Devonian (Chemung); Pennsylvania, New York.

Dipterus sherwoodi Newb.

Newberry, J. S. 1875 A, p. 61, pl. lviii, figs. 17–17b.
Dana, J. D. 1896 A, p. 617, fig. 967.
Eastman, C. R. 1900 B, p. 37.
Lesley, J. P. 1892 A, p. 1458.
Miller, S. A. 1889 A, p. 596, fig. 1128.
Newberry, J. S. 1889 A, p. 118, pl. xxvii, figs. 3, 3b. [D. (Ctenodus).]
Woodward, A. S. 1891 A, p. 242.
Devonian (Catskill); Pennsylvania.

Dipterus uddeni Eastman.

Eastman, C. R. 1900 B, p. 37, figs. 5, 5¹,
Devonian (Cedar Valley limestone); Iowa.

PALÆDAPHUS Van Beneden and de Koninck.   Type *P. insignis* Van Beneden and de Koninck.

Van Beneden and de Koninck, 1864 A, p. 143, pls. i, ii.
Fritsch, A. 1889 B (1888), p. 88.
Miller, S. A. 1889 A, p. 599. (Heliodus.)
Newberry, J. S. 1875 A, p. 62. (Heliodus.)
1889 A, p. 85. (Heliodus.)
Nicholson and Lydekker, 1889 A, p. 967.
Traquair, R. H. 1878 A, p. 12.
Woodward, A. S. 1891 A, p. 243.
Zittel, K. A. 1890 A, p. 128.

Palædaphus lesleyi (Newb.).

Newberry, J. S. 1875 A, p. 64, pl. lviii, fig. 18. (Heliodus.)
Lesley, J. P. 1892 A, p. 1461. (Heliodus.)
Miller, S. A. 1889 A, p. 599, fig. 1136. (Heliodus.)
Newberry, J. S. 1889 A, p. 86, pl. xviii, fig. 3. (Heliodus.)
Woodward, A. S. 1891 A, p. 244.
Devonian (Chemung); Pennsylvania.

CONCHODUS McCoy.   Type *C. ostreaformis* McCoy.

McCoy, F. 1848 A, p. 312.
1855 A, p. 593.
Miller, S. A. 1889 A, p. 592.
Pictet, F. J. 1854 A, p. 269.
Rohon, J. V. 1898 A.
Woodward, A. S. 1891 A, p. 245.

Conchodus plicatus Dawson.

Dawson, J. W. 1868 A, p. 209, fig. 58.
1878 A, p. 209, fig. 58.
Miller, S. A. 1889 A, p. 592, fig. 1115.
Woodward, A. S. 1891 A, p. 255.
Coal-measures; Nova Scotia.

GANORHYNCHUS Traquair.   Type *G. woodwardi* Traquair.

Traquair, R. H. 1873 A.
Miller, S. A. 1892 A, p. 715.
Woodward, A. S. 1891 A, p. 245.

Ganorhynchus beecheri Newb.

Newberry, J. S. 1889 A, p. 95, pl. xix, fig. 2.

Woodward, A. S. 1891 A, p. 246.
Devonian (Chemung); Pennsylvania.

Ganorhynchus oblongus Cope.

Cope, E. D. 1892 L, p. 225.
Devonian (Catskill); Pennsylvania.

PHANEROPLEURIDÆ.

Collenge, W. E. 1893 A, p. 3. (Phaneropleuron.)
Cope, E. D. 1878 A, p. 293. (Phaneropleuron.)
1887 O, p. 1018. (Haplistia.)
1889 R, p. 856. (Haplistia.)
1891 B, p. 460. (Haplistia.)
Günther, A. 1872 A, p. 555.
1880 A, p. 360.
1881 A, p. 687.
Haeckel, E. 1896 A, p. 262.

Huxley, T. H. 1861 B, p. 24. (Phaneropleurini.)
1861 C, (Phaneropleuron.)
Jaekel, O, 1890 B, p. 1.
Klaatsch, H. 1896 A, p. 316.
Steinmann and Döderlein, 1890 A, p. 554.
Traquair, R. H. 1871 A.
1875 C, p. 394.
1893 D, p. 121.
Woodward, A. S. 1891 A, p. 247. (Phaneropleuron.)

SCAUMENACIA Traquair. Type *S. curta* (Whiteaves).

Traquair, R. H. 1898 C, p. 262.
Dollo, L. 1896 A.
Fritsch, A. 1889 B, p. 66 (1888). (Phaneropleuron.)
Traquair, R. H. 1890 F, p. 393. (Phaneropleuron.)
1893 D, p. 118.
Woodward, A. S. 1898 B, p. 71.

Scaumenacia curta (Whiteaves).

Whiteaves, J. F. 1881 C, p. 29. (Phaneropleuron.)
Bridge, T. W. 1898 A, p. 368.
Dana, J. D. 1896 A, p. 617, fig. 969. (Phaneropleuron.)
Jaekel, O, 1890 B, p. 2.
Traquair, R. H. 1890 A, p. 15.
1890 F, p. 393. (Phaneropleuron.)

Traquair, R. H. 1893 C, p. 262.
1900 B, p. 520
Whiteaves, J. F. 1888, A, p. 354. (Phaneropleuron.)
1883 B, p. 160. (Phaneropleuron.)
1881 C, p. 29. (Phaneropleuron.)
1887 A, p. 106, pl. x, figs. 2–2e. (Phanero-pleuron.)
1889 A, p. 91, pl. v, fig. 3. (Phaneropleuron.)
1891 A, p. 495. (Phaneropleuron.)
Woodward, A. S. 1891 A, p. 248. (Phaneropleuron.)
1893 C.
1898 B, p. 22, fig. 20; p. 71. (Phaneropleuron.)
Devonian (Upper); Canada.

Order SIRENOIDEI Müller.

Müller, J. 1843 A, p. 217.
Bischoff, T. L. W. 1840 A, pls. i–viii. (Lepidosiren.)
Bridge, T. W. 1898 A.
Cope, E. D. 1889 R, p. 854.
1898 B, p. 29.
Gadow, H. 1898 A, p. 11.
Gill, T. 1898 A, p. 130.
Günther, A. 1870 A, p. 321.
1872 A, p. 554. (Sirenidæ.)

Günther, A. 1881 A, p. 686. (Sirenidæ.)
Kölliker, A. 1859 A.
Miall, L. C. 1878 A, p. 12.
Müller, J. 1846 C, p. 201.
Owen, R. 1866 A.
Wiedersheim, R. 1880 B. (Lepidosiren.)
Woodward, A. S. 1891 A, p. 235, part.
1898 B, p. 61, part.
Zittel, K. A. 1890 A, p. 130.

CTENODONTIDÆ.

Cope, E. D. 1889 R, p. 854. (Lepidosirenidæ, part.)
Traquair, R. H. 1888 A, p. 61.
1893 D, p. 121.

Woodward, A. S. 1891 A. p. 250.
1898 B, p. 62.

CTENODUS Agassiz. Type *C. cristatus* Agassiz.

Agassiz, L. 1843 B (1838), p. 137.
1844 A.
1844 B, p. 122.
Atthey, T. 1868 A.
1875 A.
Barkas, W. J. 1873 A, pp. 27, 113.
1877 A, pls. i–v.
1878 A.
Bridge, T. W. 1898 A, p. 367.
Collinge, W. E. 1893 A, p. 4.
Cope, E. D. 1878 A, p. 293.
Dollo, L. 1891 K, p. 1, 1.¹⁶²
1893 A
Fritsch, A. 1889 B (1888), pp. 66, 98.
Günther, A. 1871 A, p. 428.
Hancock and Atthey 1868 A, p. 357.
1868 B, p. 54.
1871 A.
1872 A, p. 257.

Jaekel, O. 1890 B, p. 6.
Miall, L. C. 1874 B, pl. xlvii.
1878 B.
1880 A, p. 289.
Miller, S. A. 1889 A, p. 598.
Pictet, F. J. 1854 A, p. 269.
Steinmann and Döderlein 1890 A, p. 555.
Woodward, A. S. 1888 E.
1891 A, p. 250.
Woodward and Sherborn 1890 A, p. 51.
Zittel, K. A. 1890 A, p. 127.

Ctenodus wagneri Newb.

Newberry, J. S. 1889 A, p. 172, pl. xxvii, fig. 30.
1884 A. (No description.)
1885 A, p. 121. (No description.)
Woodward, A. S. 1891 A, p. 255.
Devonian (Cleveland shale); Ohio.

SAGENODUS Owen. Type *S. inæqualis* Owen.

Owen, R. 1867 B, p. 365, pl. xii.
Barkas, W. J., 1874, Month. Rev. Dental Surg., ii, p. 538. (Petalodopsis, type *P. mirabilis*.)
Cope, E. D. 1877 E, p. 192. (Ptyonodus, type *P. vinslovii*.)
1881 K, p. 162. (Ptyonodus.)
1897 C.

Fritsch, A. 1889 B (1888), p. 66. (Ctenodus, in part.)
Gaudry, A. 1883 A, p. 239. (Megapleuron, type *M. rochei*.)
Hancock and Atthey 1868 A, p. 347.
1871 A. (Ctenodus, in part.)
1872 A, p. 272. (Ctenodus, in part.)

Hay, O. P. 1900 A, p. 100.
Jaekel, O. 1890 B, p. 7. (Hemictenodus.)
Williston, S. W. 1899 D, p. 175.
Woodward, A. S. 1890 A, p. 394. (Ptyonodus,
Hemictenodus.)
    1891 A, p. 255. (Sagenodus.)
Woodward and Sherborn 1890 A, p. 50. (Syn. of
Ctenodus.)

Sagenodus angustus (Newb.).

Newberry, J. S. 1897 A, p. 303, pl. xxiv, fig. 26.
(Ctenodus.)
Dean, B., in Newberry, J. S. 1897 A, p.· 303.
(Sagenodus.)
Eastman, C. R. 1900 B, p. 37. (Dipterus.)
Devonian (Chemung); Pennsylvania.

Sagenodus copeanus Williston.

Williston, S. W. 1899 D, p. 178, pls. xxviii, xxxv-
xxxvii.
Coal-measures; Kansas.

Sagenodus dialophus (Cope).

Cope, E. D. 1878 B, p. 528. (Ctenodus.)
    1881 K, p. 162. (Ctenodus.)
    1888 B B, p. 285. (Ctenodus.)
Williston, S. W. 1899 D, p. 176.
Woodward A. S. 1891 A, p. 261. (Sagenodus.)
Permian: Texas.

Sagenodus foliatus Cope.

Cope, E. D. 1897 C, pp. 76, 77, pl. i, fig. 1.
Hay, O. P. 1900 A, p. 107.
Williston, S. W. 1899 D, p. 177.
Coal-measures; Illinois.

Sagenodus fossatus (Cope) :

Cope, E. D. 1877 N, p. 53. (Ctenodus.)
Case, E. C. 1900 A, p. 705, pl. i, figs. 10a, 10b.
Cope, E. D. 1888 B B, p. 285. (Ctenodus.)
Williston, S. W. 1899 D, p. 176.
Woodward, A. S. 1891 A, p. 261.
Permian; Illinois.

Sagenodus gurleyanus (Cope).

Cope, E. D. 1877 N, p. 54. (Ctenodus.)
Case, E. C. 1900 A, p. 704, pl. i, figs. 8a–8c.
Cope, E. D. 1881 K, p. 162. (Ctenodus.)
    1888 B B, p. 285. (Ctenodus.)
Hay, O. P. 1899 E, p. 785.
Miller, S. A. 1889 A, p. 593. (Ctenodus gurlei-
anus.)
Williston, S. W. 1899 D, p. 176.
Woodward, A. S. 1891 A, p. 261.
Permian; Illinois.

Sagenodus heterolophus (Cope).

Cope, E. D. 1883 K, p. 109. (Ctenodus.)
Case, E. C. 1900 A, p. 706.
Williston, S. W. 1899 D, p. 176.
Woodward, A. S. 1891 A, p. 261.
Permian; Illinois.

Sagenodus jugosus Hay.

Hay, O. P. 1900 A, p. 107.
Newberry, J. S. 1875 A, p. 60. (Ctenodus reticulatus; not of Newberry and Worthen, 1870.)

Woodward, A. S. 1891 A, p. 261. (Sagenodus
reticulatus.)
Coal-measures; Ohio.

Sagenodus lacovianus Cope.

Cope, E. D. 1897 C, p. 79, pl. i, fig. 5.
    1897 C, p. 79, pl. i, fig. 4. (S. conchiolepis.)
Hay, O. P. 1900 A, p. 108.
Williston, S. W. 1899 D, p. 177. (S. lacovianus, S.
conchiopsis.)
Coal-measures; Illinois.

Sagenodus occidentalis (Newb. and
Worth.).

Newberry and Worthen 1866 A, p. 19, fig. 2.
(Rhizodus.)
Cope, E. D. 1897 C, p. 75 (S. occidentalis); p. 81,
pl. i, fig. 7 (S. browniae).
Hay, O. P. 1899 E, p. 787. (Megalichthys.)
    1900 A, p. 101.
Miller, S. A. 1889 A, p. 611, fig. 1173. (Rhizodus.)
Newberry, J. S. 1889 A, p. 192. (Rhizodus.)
Newberry and Worthen 1870 A, pl. iv, fig. 1.
(Rhizodus.)
Williston, S. W. 1899 D, p. 177.
Woodward, A. S. 1891 A, p. 348. (Strepsodus?).
Coal-measures; Illinois.

Sagenodus ohionensis (Cope).

Cope, E. D. 1875 D, p. 410, pl. xlv, fig. 2. (Ctenodus.)
    1873 O, p. 341. (Leptophractus obsoletus,
part.)
    1874 S, p. 91. (Ctenodus, without specific
name.)
Newberry, J. S. 1875 A, p. 60. (Ctenodus.)
    1889 A, p. 226; woodcut, fig. 3. (Ctenodus.)
Woodward, A. S. 1891 A, p. 262.
Coal-measures; Ohio.

Sagenodus paucicristatus (Cope).

Cope, E. D. 1877 N, p. 53. (Ceratodus.)
Case, E. C. 1900 A, p. 707, pl. i, figs. 11a, 11b.
Cope, E. D. 1877 E, p. 192. (Ptyonodus.)
    1881 K, p. 162. (Ptyonodus.)
    1888 BB, p. 285. (Ptyonodus.)
Williston, S. W. 1899 D, p. 175.
Woodward, A. S. 1891 A, p. 261.
Permian; Illinois.

Sagenodus periprion (Cope).

Cope, E. D. 1878 K, p. 527. (Ctenodus.)
    1881 K, p. 162. (Ctenodus.)
    1888 B B, p. 285. (Ctenodus.)
Williston, S. W. 1899 D, p. 176.
Woodward, A. S. 1891 A, p. 261.
Permian; Texas.

Sagenodus porrectus (Cope).

Cope, E. D. 1878 B, p. 527. (Ctenodus.)
    1881 K, p. 162. (Ctenodus.)
    1888 B B, p. 285. (Ctenodus.)
Williston, S. W. 1899 D, p. 176.
Woodward, A. S. 1891 A, p. 261.
Permian; Texas.

Sagenodus pusillus (Cope).

Cope, E. D. 1877 E, p. 191. (Ctenodus.)
Case, E. C. 1900 A, p. 705, pl. i, figs. 9a, 9b.
Cope, E. D. 1881 K, p. 162. (Ctenodus.)
    1888 BB, p. 285. (Ctenodus.)
Williston, S. W. 1899 D, p. 176.
Woodward, A. S. 1891 A, p. 261.
    Permian; Illinois.

Sagenodus quadratus (Newb.).

Newberry, J. S. 1873 A, p. 343, pl. xxxix, fig. 8. (Rhizodus.)
Cope, E. D. 1897 C, pp. 76, 77.
Eastman, C. R. 1900 B, p. 37. (Dipterus.)
Hay, O. P. 1900 A, p. 103.
Lesley, J. P. 1889 A, p. 877, figure. (Rhizodus.)
    1895 A, pl. dlxxi. (Rhizodus.)
Newberry, J. S. 1889 A, p. 192. (Rhizodus.)
Williston, S. W. 1899 D, p. 176.
Woodward, A. S. 1891 A, p. 262. (Sagenodus?.)
    Coal-measures; Ohio, Illinois.

Sagenodus quincunciatus Cope.

Cope, E. D. 1897 C, p. 80, pl. i, fig. 6.
Hay, O. P. 1900 A, p. 109.
Newberry and Worthen 1870 A, p. 349, pl. iii, figs. 13, 14. (Rhizodus reticulatus, part.)
Williston, S. W. 1899 D, p. 177.
    Coal-measures; Illinois.

Sagenodus reticulatus (Newb. and Worth.).

Newberry and Worthen 1870 A, p. 349, pl. iii, fig. 9, not figs. 13, 14. (Rhizodus.)
Cope, E. D. 1897 C, p. 78, pl. i, figs. 2, 3 (Sagenodus reticulatus); p. 81, pl. i, fig. 8 (S. magister).
Hay, O. P. 1900 A, p. 106.
Williston, S. W. 1899 D, p. 177. (S. reticulatus, S. magister.)

Woodward, A. S. 1891 A, p. 262. (Sagenodus?.)
    Coal-measures; Illinois.

Sagenodus serratus (Newb.).

Newberry, J. S. 1875 A, p. 59, pl. lviii, figs. 15, 16. (Ctenodus.)
Miller, S. A. 1899 A, p. 593, fig. 1119. (Ctenodus.)
Newberry, J. S. 1889 A, p. 226, pl. xxvii, fig. 31. (Ctenodus.)
Williston, S. W. 1899 D, p. 176.
Woodward, A. S. 1891 A, p. 261. (Sagenodus.)
    Coal-measures; Ohio.

Sagenodus textilis Hay.

Hay, O. P. 1899 E, p. 786.
Cope, E. D. 1897 C, p. 82, pl. i, fig. 9. (S. gurleianus; preoccupied by Cope, 1877.)
Hay, O. P. 1900 A, p. 110.
Williston, S. W. 1899 D, p. 177. (S. gurleianus.)
    Coal-measures; Illinois.

Sagenodus vabasensis (Cope).

Cope, E. D. 1883 K, p. 110. (Ctenodus.)
Case, E. C. 1900 A, p. 704, pl. i, fig. 7.
Cope, E. D. 1888 BB, pp. 285, 288. (Ctenodus.)
Williston, S. W. 1899 D, p. 176.
Woodward, A. S. 1891 A, p. 261.
    Permian; Illinois.

Sagenodus vinslovii (Cope).

Cope, E. D. 1875 T, p. 410. (Ceratodus.)
Case, E. C. 1900 A, p. 703, pl. i, figs. 6a, 6b.
Cope, E. D. 1877 E, p. 192. (Ptyonodus.)
    1877 N, p. 58. (Ceratodus.)
    1881 K, p. 162. (Ptyonodus.)
    1888 BB, p. 285. (Ptyonodus.)
Williston, S. W. 1899 D, p. 176.
Woodward, A. S. 1891 A, p. 262.
    Permian; Illinois.

GNATHORHIZA Cope.   Type *G. serrata* Cope.

Cope, E. D. 1883 H, p. 629.
Miller, S. A. 1889 A, p. 598.
    The systematic position of this genus is very problematical.

Gnathorhiza serrata Cope.

Cope, E. D. 1883 H, p. 629.
    1888 BB, p. 286.
    Permian; Texas.

CERATODONTIDÆ.

Gill, T. 1872 C, p. 22.
Cope, E. D. 1889 R, p. 854.
Gill, T. 1893 A, p. 130.
Günther, A. 1872 A, p. 554. (Sirenidæ, in part.)
Steinmann and Döderlein 1890 A, p. 555.

Quenstedt, F. A. 1885 A, p. 294. (Ceratodontia.)
Van Wijhe, J. W. 1882 A. (Ceratodus.)
Woodward, A. S. 1891 A, p. 264. (Lepidosirenidæ, part.)

CERATODUS Agassiz.   Type *C. latissimus* Agassiz.

Agassiz, L. 1843 B (1838), iii, p. 129.
    1844 A.
Balfour, F. M. 1881 A.
Barkas, W. J. 1877 A.
    1878 A.
Beyrich, E. 1850 A, p. 153.
Bridge, T. W. 1898 A, p. 350.
Bridge, W. J, 1879 A.
Brown, C. 1900 A, p. 172,

Cope, E. D. 1893 N, p. 1051.
Davidoff, M. 1883 A.
Dollo, L, 1896 A.
Gadow, H. 1889 A, p. 458.
Gadow and Abbott 1895 A.
Gegenbaur, C. 1872 B.
    1876 A.
    1895 A, figs. 2, 3.
    1898 A, figures.

Günther, A. 1870 A, p. 323.
  1871 A.
  1871 B.
  1872 A.
  1880 A, p. 357.
Haeckel, E. 1895 A, p. 264.
Haswell, W. A. 1888 A.
Howes, G. B. 1887 A.
Huxley, T. H. 1876 A.
  1876 B, xiii, pp. 389, 411.
  1880 B, p. 660.
Klaatsch, H. 1890 A, p. 209.
  1896 A, p. 316.
Krefft, G. 1870 A.
Miall, L. C. 1878 A, p. 18.
  1878 B.
Mivart, St. G. 1879 A.
Mollier, S. 1893 A, p. 86 *et seq.*
Newberry, J. S. 1875 A, pp. 9–27.
Nicholson and Lydekker 1889 A, p. 958.
Oldham, T. 1859 A.
Pictet, F. J. 1854 A, p. 267.
Pollard, H. B. 1894 B.
Quenstedt, F. A. 1852 A, p. 186.
  1885 A, p. 295.
Reis, O. M. 1897 A, p. 132.
Röse, C. 1892 B.
Salensky W. 1899 A, p. 181.
Semon, R, 1899 A, p. 78.
Stephan, P. 1900 A, p. 307.
Teller, F. 1891 A.
Thacher, J. K. 1877 A.
Tomes, C. S. 1898 A, p. 264.
Traquair, R. H. 1900 B, p. 521.
Van Wijhe, J. W. 1882 A.
Woodward, A. S. 1888 E.
  1891 A, p. 264.
  1898 B, p. 64.

Zittel, K. A. 1886 A, p. 253
  1890 A, p. 182.

Ceratodus americanus Knight.

Knight, W. C. 1898 A, p. 186, fig. 2.
  Jurassic; Wyoming.

Ceratodus eruciferus [1] Cope.

Cope, E. D. 1876 H, p. 259.
  1877 G, p. 574. (Arotus?)
Woodward, A. S. 1891 A, p. 275.
  Cretaceous (Laramie); Montana.

Ceratodus favosus Cope.

Cope, E. D. 1884 I, p. 28.
  1888 BB, p. 286.
Woodward, A. S. 1891 A, p. 274.
  Permian; Texas.

Ceratodus güntheri Marsh.

Marsh, O. C. 1878 A, with figure.
King, C. 1878 A, p. 346.
Marsh, O. C. 1878 G, p. 224.
  1897 C, p. 507, fig. 67.
  Jurassic; Colorado.

Ceratodus hieroglyphus [1] Cope.

Cope, E. D. 1876 H, p. 260.
  1877 G, p. 554. (Arotus.)
  1885 D, p. 408.
Woodward, A. S. 1891 A, p. 275.
  Cretaceous (Laramie); Montana.

Ceratodus robustus Knight.

Knight, W. C. 1898 A, p. 186, fig. 1.
  Jurassic; Wyoming.

SYNTHETODUS Eastman.   Type, *S. trisulcatus* Eastman.

Eastman, C. R. 1898 A, p. 111.

Synthetodus trisulcatus Eastman.

Eastman, C. R. 1898 A, p. 112, pl. iv, figs. 1–10.
  Upper Devonian; Iowa.

Subclass TELEOSTOMI Owen.

Owen, R. 1866 A, p. 7.
  References are here given also to the literature of the groups known as *Ganoidei* and *Teleostei.*
Agassiz, L. 1843 B, ii, p. 1. (Ganoidei.)
Balfour, F. M. 1882 A, p. 431. (Ganoidei.)
Baudelot, E. 1873 A.
  1873 B.
Baur, G. 1889 E. (Teleostei.)
Beard, J. 1890 A. (Teleostei, Ganoidei.)
Chamberlin, T. C. 1900 A, p. 411. ("Teleostomes.")
Cope, E. D. 1872 C, p. 318 (Ganoidei.); p. 396 (Actinopteri).
  1877 N, p. 56. (Hyopomata.)
  1877 O, p. 41. (Hyopomata.)
  1878 A, p. 292. (Hyopomata.)
  1884 I, p. 46. (Hyopomata.)
  1884 N, p. 579. (Hyopomata.)

Cope, E. D. 1884 HH. (Hyopomata.)
  1885 G. (Hyopomata.)
  1887 O, p. 1015.
  1889 R, pp. 854, 855.
  1890 B.
  1891 E.
  1891 N, p. 10.
  1898 B, pp. 18, 30.
Dean, B, 1895 A, p. 139.
Gadow, H. 1898 A, p. 7.
Gadow and Abbott 1895 A. (Ganoidei, Teleostei.)
Gegenbaur, C. 1895 A.
  1898 A.
Gill, T. 1872 C. p. 1.
  1873 B.
  1873 D.
  1893 A, p. 130. ("Teleostomes.")
  1896 C, p. 912.

[1] See the genus of Ichthyodorulites *Euctenius.*

Goeppert, E. 1895 A, p. 158. (Teleostei.)
Günther, A. 1871 A, p. 428.   (Ganoidei, Teleostei.)
    1872 A, p. 554. (Ganoidei.)
    1880 A, p. 373.   (Teleostei.)
    1881 A, p. 688.   (Ganoidei, Teleostei.)
Haeckel, E. 1895 A, p. 244 (Ganoides); p. 252 (Teleostei).
Hallmann, E. 1837 A, p. 52.
Hay, O. P. 1895 A.
    1896 C, p. 342.
Hollard, H. 1860 A.
Huxley, T. H. 1858 C. (Teleostei, Ganoidei.)
    1861 B, p. 29. (Teleostei.)
    1872 A, p. 121 (Ganoidei); p. 130 (Teleostei).
    1883 A. (Teleostei.)
Jordan and Evermann 1896 A, p. 97.
Klaatsch, H. 1890 A, p. 155.
    1894 A.
Klein, E. F. 1879 A.
    1881 A.
    1885 A.

Klein, E. F. 1889 A.
Kner, R. 1866 A.
Kölliker, A. 1859 A. (Teleostei, Ganoidei.)
Köstlin, O. 1844 A.
Lütken, C. 1868 B. (Ganoidei, Teleostei.)
    1873 A. (Ganoidei.)
Miall, L. C. 1878 A. (Ganoidei.)
Müller, J. 1846 C.
Parker, W. K. 1874 A.
Reis, O. M. 1895 C, p. 162.
Röse, C. 1892 A. (Teleostei.)
Steinmann and Döderlein 1890 A, p. 556.
Stephan, P. 1900 A. ("Téléostéens," "Ganoïdes.")
Traquair, R. H. 1877 B. (Ganoidei.)
Ussow, S. A. 1898 A.
Vaillaint, L. 1883 A.
Woodward, A. S. 1891 A, pp. xxi, 316.
    1898 B, p. 69.
Zittel, K. A. 1890 B, p. 133 (Ganoidei); p. 252 (Teleostei).

Superorder CROSSOPTERYGIA.

Baur, G. 1896 C, p. 659.
Bridge, W. J. 1879 A.
Bridge, T. W. 1889 A, p. 118. (Polypterus.)
    1896 A, p. 541. (Polypterus.)
    1898 A, p. 371.
Brown, C. 1900 A, p. 172.
Chamberlin, T. C. 1900 A, p. 411. ("Crossopterygians.")
Cope, E. D. 1872 C, p. 328.
    1877 N, p. 56.
    1877 O, p. 41.
    1878 A, p. 298.
    1885 G, p. 237.
    1887 O, p. 1017. (Crossopterygia and Rhipidopterygia.)
    1887 S, p. 325.
    1889 R, p. 885. (Crossopterygia and Rhipidopterygia.)
    1890 B.
    1891 B, p. 459 (Rhipidopterygia); p. 460 (Crossopterygia.)
    1891 E, p. 480.
    1891 N, p. 19. (Crossopterygia, Rhipidopterygia.)
    1898 B, p. 30. (Crossopterygia, Rhipidopterygia.)
Davidoff, M. 1880 B. (Polypterus.)
Dean, B. 1895 A, p. 148.
Dollo, L. 1896 A, p. 105.
Emery, C. 1879 B.
Gadow, H. 1898 A, p. 7. (Crossopterygii.)
Gegenbaur, C. 1890 B, p. 104, pl. viii, fig. 6. (Polypterus.)
    1895 A.

Gegenbaur, C. 1895 B.
    1898 A.
Gill, T. 1872 C, p. 22.
    1893 A, p. 130.
Goeppert, E. 1895 A, p. 148.
Günther, A. 1872 A, p. 555.
Haeckel, E. 1895 A, p. 249. (Crossopterygii.)
Hay, O. P. 1899 F, p. 682.
Howes, G. B. 1887 A, p. 19.
Huxley, T. H. 1861 B, p. 24. (Crossopterygidæ, as suborder.)
    1866 A.
    1872 A, p. 127. (Crossopterygidæ.)
    1876 A.
Jaekel, O. 1890 B, p. 8.
Klaatsch, H. 1890 A, p. 218.
    1896 A.
Lütken, C. 1873 A, p. 45. (Crossopteri.)
Newberry, J. S. 1873 A, p. 255.
Orr, H. 1885 A.
Osborn, H. F. 1899 I, p. 158.
Pollard, H. B. 1892 A. (Polypterus.)
Reis, O. M. 1897 A, p. 130.
Ryder, J. 1886 A.
Steinmann and Döderlein 1890 A, p. 560.
Traquair, R. H. 1870 A. (Polypterus.)
    1875 C, p. 394.
    1890 B, p. 517. (Crossopterygii.)
Van Wijhe, J. W. 1882 A, p. 251.
Woodward, A. S. 1891 A, p. 316.
    1898 C.
    1898 B, p. 69.
Zittel, K. A. 1890 A, p. 168.
    1896 A, p. 406.

Superorder RHIPIDISTIA Cope.

Cope, E. D. 1887 O, p. 1017.
    1887 O, p. 1018. (Taxistia.)
    1889 R, p. 885 (Rhipidistia); p. 886 (Taxistia).
    1891 B, p. 460 (Rhipidistia); p. 462 (Taxistia).

Cope, E. D. 1891 E. (Rhipidistia, Taxistia.)
    1891 N, p. 19. (Rhipidistia, Taxistia.)
    1898 B, pp. 30, 31. (Rhipidistia, Taxistia.)
Traquair, R. H. 1891 A, p. 127.
Woodward, A. S. 1891 A, p. 318.
    1898 B, p. 71.

HOLOPTYCHIIDÆ.

Collinge, W. E. 1898 A, p. 5.
Cope, E. D. 1889 R, p. 856.
Günther, A. 1880 A, p. 365.
    1881 A, p. 687.
Lütken, C. 1873 A, p. 34.   (Cyclodipterini)
Nicholson and Lydekker 1889 A, p. 968.
Owen, R. 1860 E, p. 132.
Rohon, J. V. 1899 A, p. 41.

Steinmann and Döderlein 1890 A, p. 560.
Traquair, R. H. 1889 C.
    1900 B, p. 518.
Woodward, A. S. 1891 A, p. 321.
    1893 E, p. 450.
    1898 B, p. 72.
Zittel, K. A. 1890 A, p. 176.   (Cyclodipterini.)

HOLOPTYCHIUS Agassiz.   Type *H. nobilissimus* Agassiz.

Agassiz L., 1839, in Murchison's Silur. Syst., p. 599.
    (Holoptychus, misprint.)
    1844 A.
    1844 B.
Anderson, J. 1860 A.
    1861 A.
Collinge, W. E. 1898 A, p. 6.
Egerton, P. H. 1859 A, p. 126.
Emery, C. 1897 B, p. 147.
Huxley, T. H. 1861 B, p. 5, fig. 5.
Jaekel, O. 1894 B, p. 148.
Klaatsch, H. 1890 A, p. 221.
Leidy, J. 1856 S, p. 162.
Lohest, M. 1888 A, p. 121.
McCoy, F. 1853 A.
    1855 A, p. 593.
Miller, H. 1859 A, pl. xiv.
Miller, S. A. 1889 A, p. 599.
Newberry, J. S. 1856 A.
Nicholson and Lydekker 1889 A, p. 968.
Pictet, F. J. 1854 A, p. 145.
Quenstedt, F. A. 1852 A, p. 229.
    1885 A, p. 350.
Rohon, J. V. 1892 B.
    1899 A, p. 41.
Steinmann and Döderlein 1890 A, p. 561.
Traquair, R. H. 1888 B, p. 512.
    1900 B, p. 519.
Williamson, W. C. 1849 A, p. 457, pl. xlii, figs. 21-24.
Woodward, A. S. 1891 A, p. 322.
    1892 E.
Young, J. 1866 B, p. 601.
Zittel, K. A. 1890 A, p. 178.

Holoptychius americanus Leidy.

Leidy, J. 1856 S, p. 163, pl. xvi, figs. 9, 10; pl. xvii, figs. 1-3 (not fig. 4).
Cope, E. D. 1892 L, p. 222.
    1892 X.
    1897 C, p. 72.
DeKay, J. E. 1842 B, p. 385.   (H. nobilissimus.)
Eastman, C. R. 1899 F, pp. 321, 327.
Hall, J. 1843 A, p. 281, text figs. 130.²   (Holoptychus nobilissimus.)
Lesley, J. P. 1889 A, p. 284, figures.   (H. americanus, H. nobilissimus.)
Lohest, M. 1888 A, p. 139.   (H. nobilissimus.)
Miller, S. A. 1889 A, p. 599, fig. 1438.
Newberry, J. S. 1873 A, pp. 271, 272.
    1889 A, pp. 107, 113, pl. xix, figs. 12, 13.
Redfield, W. C. 1841 A, p. 24.   (H. nobilissimus.)
Whitfield and Hovey 1900 A, p. 348.
Williams, H. S. 1887 A, pp. 27, 28.   ("Holoptychius.")

Woodward, A. S. 1891 A, p. 326.
Young, J. 1866 B, p. 599.
    Devonian (Catskill, Ithaca); Pennsylvania, New York.

Holoptychius filosus Cope.

Cope, E. D. 1892 L, p. 228, pl. vii, fig. 3.
    1897 C, p. 72.
    Devonian (Chemung); Pennsylvania.

Holoptychius flabellatus Cope.

Cope, E. D. 1897 C, p. 75, pl. ii, fig. 3.
    Devonian (Catskill?); Pennsylvania.

Holoptychius giganteus Agassiz.

Agassiz, L. 1844 B, pp. 78, 140, pl. xxiv, figs. 3-10.
Cope, E. D. 1892 L, p. 222.
    1897 C, p. 72.
Lohest, M. 1888 A, p. 146, pl. vi, figs. 2, 3; pl. vii, figs. 5, 6.
McCoy, F. 1855 A, p. 594 (H. giganteus); p. 595 (H. princeps).
Newberry, J. S. 1889 A, p. 101, pl. xix, figs. 15, 16.
Woodward, A. S. 1891 A, p. 325.
    Devonian; Europe: (Catskill?): Pennsylvania.

Holoptychius granulatus Newb.

Newberry, J. S. 1889 A, p. 100, pl. xx, fig. 9.
Whitfield and Hovey, 1900 A, p. 348.
Woodward, A. S. 1891 A, p. 331.
    Devonian (Chemung): Pennsylvania?.

Holoptychius hallii Newb.

Newberry, J. S. 1889 A, p. 114, pl. xx, figs. 10, 10a.
Cope, E. D. 1897 C, p. 72.
Woodward, A. S. 1891 A, p. 326.
    Devonian (Catskill): New York.

Holoptychius latus Cope.

Cope, E. D. 1897 C, p. 74, pl. ii, fig. 2.   [Holoptychius (Glyptolepis).]
    Devonian (Catskill?); Pennsylvania.

Holoptychius ? pustulosus Newb.

Newberry, J. S. 1889 A, p. 100, pl. xx, figs. 11, 11a.
Woodward, A. S. 1891 A, p. 331.
    Devonian (Chemung); Pennsylvania.

Holoptychius quebecensis (Whiteaves).

Whiteaves, J. F. 1889 A, p. 77, pl. v, fig. 4.   (Glyptolepis.)
Dana, J. D. 1896 A, p. 618, fig. 970.

Traquair, R. H. 1890 A, p. 17.
   1898 C, p. 265. (Gyrolepis.)
   1898 D, p. 128. (Glyptolepis.)
Whiteaves, J. F. 1881 A, p. 495. (Glyptolepis
  microlepidotus, not of Agassiz.)
   1883 A, p. 354. (Glyptolepis microlepidotus,
    not of Agassiz.)
   1883 B, p. 162. (Glyptolepis.)
   1881 C, pp. 82, 33. (Glyptolepis microlepi-
    dotus and G. "near leptopterus.")
Woodward, A. S. 1891 A, p. 336.
   Upper Devonian; Province of Quebec, Canada.

Holoptychius ? radiatus Newb.

Newberry, J. S. 1889 A, p. 115, pl. xx, figs. 12–14.

Cope, E. D. 1897 C, p. 72.
Whitfield and Hovey 1900 A, p. 348.
Woodward, A. S. 1891 A, p. 331.
   Devonian (Catskill?); Pennsylvania.

Holoptychius serrulatus Cope.

Cope, E. D. 1897 C, p. 71, pl. ii, fig. 1.
   Devonian (Catskill?); Pennsylvania.

Holoptychius tuberculatus Newb.

Newberry, J. S. 1889 A, p. 101, pl. xix, fig. 14.
Cope, E. D. 1897 C, p. 73.
Woodward, A. S. 1891 A, p. 331.
   Devonian (Chemung); Pennsylvania.

ERIPTYCHIUS Walcott.   Type *E. americanus* Walcott.

Walcott, C. D. 1892 A, p. 167.
Chamberlin, T. C. 1900 A, p. 401. (No name.)

Eriptychius americanus Walcott.

Walcott, C. D. 1892 A, p. 167, pl. iv, figs. 5–11.

American Geologist 1891 E. (Holoptychius.)
Cope, E. D. 1893 I.
Dana, J. D. 1896 A, p. 509, fig. 696.
   Lower Silurian (Ordovician); Colorado.

DENDRODUS Owen.   Type *D. biporcatus* Owen.

Owen, R. 1841 F, p. 4.
Agassiz, L. 1844 B, pp. 61, 84, 144. (Lamnodus, Den-
  drodus.)
Eichwald, E. 1860 A, p. 1558.
Lohest, M. 1888 A, p. 118.
McCoy, F. 1855 A, p. 597.
Nicholson and Lydekker 1889 A, p. 969.
Owen, R. 1845 B, pp. 171, 175.
   1860 E, p. 133.
Pander, C. H. 1860 A, pp. 27, 53.
Pictet, F. J. 1854 A, p. 147.
Quenstedt, F. A. 1852 A, p. 229.

Quenstedt, F. A. 1885 A, p. 351.
Rohon, J. V. 1889 B, p. 1–52, pls. i, ii.
Traquair, R. H. 1889 C, p. 490.
Trautschold, H. 1880 C, p. 139.
   1889 B.
Woodward, A. S. 1891 A, p. 338.
Zittel, K. A. 1890 A, p. 177.

Dendrodus arisaigensis Whiteaves.

Whiteaves, J. F. 1898 A, p. 401, fig. 1.
   Silurian or Devonian; Nova Scotia.

APEDODUS Leidy.   Type *A. priscus* Leidy.

Leidy, J. 1856 S, p. 164.
Miller, S. A. 1889 A, p. 587.
Woodward, A. S. 1891 A, p. 322. (Syn.? of Holop-
  tychius.)
Young, J. 1866 B. (Syn. of Rhizodus.)

Apedodus priscus Leidy.

Leidy, J. 1856 S, p. 165, pl. xvii, figs. 5–7.
Newberry, J. S. 1889 A, p. 108.
Woodward, A. S. 1891 A, p. 346.
   Devonian (Catskill); Pennsylvania.

MEGALICHTHYIDÆ nom. nov.

The following authors denominate this family
Rhizodontidæ.
Collinge, W. E. 1898 A, p. 5.
Nicholson and Lydekker 1889 A, p. 969.
Steinmann and Döderlein 1890 A, p. 561.
Traquair, R. H. 1881 B, p. 179.

Traquair, R. H. 1888 B, p. 514.
   1891 A, p. 127.
   1900 B, p. 517.
Woodward, A. S. 1891 A, p. 346.
   1898 B, p. 74.

MEGALICHTHYS Agassiz and Hibbert.   Type *M. hibberti* Agassiz and Hibbert.

Agassiz and Hibbert, in Hibbert, S. 1836 A, p. 202.
   Unless otherwise indicated, the name *Rhi-
  zodus* of Owen is employed for this genus by the
  following writers. *Megalichthys* of Agassiz and
  Hibbert has, however, priority.
Agassiz, L. 1835 A, p. 646. (Megalichthys.)
   1844 B, p. 69.
Atthey, T. 1877 A.

Barkas, T. P. 1873 A, p. 41.
Eichwald, E. 1860 A, p. 1571. (Holoptychius.)
Gaudry, A. 1883 A, p. 236.
Hancock and Atthey 1868 A, p. 346.
Hay, O. P. 1899 E, p. 786.
Klaatsch, H. 1890 A, p. 220.
McCoy, F. 1855 A, p. 612.
Miall, L. C. 1875 B, fig. 1.

Miller, S. A. 1889 A, p. 611.
Newberry, J. S. 1868 A, p. 145.
   1889 A, pp. 192, 340.
Newberry and Worthen 1866 A, p. 18.
Nicholson and Lydekker 1889 A, p. 970.
Owen, R. 1845 B (1840), p. 55, pls. xxxv–xxxvii.
   (Rhizodus; type *Megalichthys hibberti* Agassiz
   and Hibbert.)
Roemer, F. 1865 A.
Steinmann and Döderlein 1890 A, p. 561.
Stock, T. 1881 B, p. 77.
Traquair, R. H. 1875 B.
   1877 C.
   1891 A, p. 127.
Woodward, A. S. 1891 A, p. 342.
   1898 B, p. 74.
Woodward and Sherborn 1890 A, p. 178.
Young, J. 1866 B, p. 599.
Zittel, K. A, 1890 A, p. 182.

Megalichthys anceps (Newb.).

Newberry, J. S. 1888 C. (Rhizodus.)
Dana, J. D. 1888 A. (Rhizodus.)
Newberry, J. S. 1889 A, p. 191, pl. xliii, fig. 1.
   (Rhizodus.)
Woodward, A. S. 1891 A, p. 346. (Rhizodus.)
   Subcarboniferous (St. Louis); Illinois.

Megalichthys angustus Newb.).

Newberry, J. S. 1856 B, p. 99. (Rhizodus.)

Hay, O. P. 1899 E, p. 787.
Hitchcock, C. H. 1868 B. (Rhizodus.)
Lesley, J. P. 1889 A, p. 875, figure. (Rhizodus.)
Newberry, J. S. 1868 A, p. 145. (Rhizodus.)
   1873 A, p. 342, pl. xxxix, fig. 6. (Rhizodus.)
   1889 A, p. 192. (Rhizodus.)
Woodward, A. S. 1891 A, p. 348. (Rhizodus.)
   Coal-measures; Ohio.

Megalichthys incurvus (Newb.).

Newberry, J. S. 1856 B, p. 99. (Rhizodus.)
Hay, O. P. 1899 E, p. 787.
Woodward, A. S. 1891 A, p. 348. (Rhizodus.)
   Coal-measures; Ohio.

Megalichthys lancifer (Newb.).

Newberry, J. S. 1856 B, p. 99. (Rhizodus.)
Dawson, J. W. 1863 D, p. 11. (Rhizodus.)
   1868 A, p. 210, fig. 55a. (R. lancifer?.)
   1878 A, p. 210, fig. 55a. (Rhizodus.)
   1894 A, p. 269. (Rhizodus.)
Hay, O. P. 1899 E, p. 787.
Lesley, J. P. 1889 A, p. 876, figure. (Rhizodus.)
Newberry, J. S. 1873 A, p. 342, pl. xxxix, fig. 9.
   (Rhizodus.)
   1889 A, p. 192. (Rhizodus.)
Woodward, A. S. 1891 A, p. 348. (Rhizodus.)
   Coal-measures; Ohio.

RABDIOLEPIS Emmons.   Type *R. speciosa* Emmons.

Emmons, E. 1857 A, p. 45.
   1860 A, p. 183.

Rabdiolepis speciosa Emmons.

Emmons, E. 1857 A, p. 45, fig. 17.
Cope, E. D. 1875 V, p. 30. (Rhabdolepis.)

Emmons, E. 1860 A, p. 183, fig. 161.
Wheatley, C. M. 1861 A, p. 44. (Radiolepis).
Woodward, A. S. 1891 A, p. 354.
   Triassic (Chatham); North Carolina, Penn-
   sylvania?.

STREPSODUS Young.   Type *Holoptychius sauroides* Binney.

Young, J. 1866 B, p. 602.
Barkas, T. P. 1873 A, p. 32, figs. 113–119 (Strepso-
   dus); pp. 75, 94, figs. 194, 223, 224 (Labyrintho-
   dontosaurus).
Hancock and Atthey 1868 A, p. 353.
   1870 A.
Ward, J. 1890 A, p. 160.
Woodward, A. S. 1891 A, p. 348.
Young J. 1856 B, p. 601, fig. 7. (Dendroptychius.)

Strepsodus arenosus Hay.

Hay, O. P. 1900 A, p. 113, fig. 3.
   Subcarboniferous; Pennsylvania.

Strepsodus dawsoni Hay.

Hay, O. P. 1900 A, p. 114.

Dawson, J. W. 1868 A, p. 210, fig. 55b. (Rhizodus
   lancifer, part.)
   1878 A, p. 210, fig. 55b. (R. lancifer, part.)
   Coal-measures; Nova Scotia.

Strepsodus hardingi (Dawson).

Dawson, J. W. 1868 A, p. 255, fig. 77a–d. (Rhizo-
   dus.)
Dana, J. D. 1888 A. (Rhizodus.)
Dawson, J. W. 1878 A, p. 255, fig. 77a–d. (Rhizo-
   dus.)
Hay, O. P. 1900 A, p. 111.
Lesley, J. P. 1889 A, p. 876, figure. (Rhizodus.)
Woodward, A. S. 1891 A, p. 353.
   Subcarboniferous; Nova Scotia.

RHIZODOPSIS Young.   Type *Holoptychius granulatus* Salter.

Young, J. 1866 B, p. 596.
Barkas, T. P. 1873 A, p. 23, figs. 59–69 (Rhizo-
   dopsis); p. 38, figs 143, 144 (Orthognathus, type
   O. reticulosus).
Hancock and Atthey 1868 A, p. 349.

Nicholson and Lydekker 1889 A, p. 970.
Owen, R. 1867 B, p. 325 (Dittodus, part); p. 354
   (Ganalodus part); p. 366 (Characodus); p. 370
   (Gastrodus).
Roemer, F. 1865 A. (Rhizodus, part.)

Traquair, R. H. 1881 B, p. 169.
Ward, J. 1890 A, p. 161, pl. viii, figs. 5–7.
Williamson, W. C. 1849 A, p. 457, pl. xlii, figs.
   21–23. (Holoptychius, part.)
Woodward, A. S. 1891 A, p. 354.
Young, J. 1866 B, p. 696, fig. 8.

Zittel, K. A. 1890 A, p. 181.

Rhizodopsis ? mazonius Hay.

Hay, O. P. 1900 A, p. 110.
Coal-measures, Illinois.

CŒLOSTEUS Newb.   Type *C. ferox* Newb.

Newberry, J. S. 1887 B.
Cope, E. D. 1890 H.
Miller, S. A. 1892 A, p. 714.
Newberry, J. S. 1889 A, p. 188.
Woodward, A. S. 1891 A, p. 34[6].

Cœlosteus ferox Newb.

Newberry, J. S. 1887 B.
Cope, E. D. 1888 P.
Newberry, J. S. 1889 A, p. 190, pl. xxxv, figs. 1–4.
Woodward, A. S. 1891 A, p. 346.
Subcarboniferous (St. Louis): Illinois.

SAURIPTERIS Hall.   Type *S. taylori* Hall.

Hall, J. 1843 A, p. 282.
   1840, Ann. Rep. Geol. Surv., N. Y., 4th dist.,
      p. 453. (Sauritolepis, nom. nud.)
   1843 A, p. 282, footnote. (Sauritolepis.)
Newberry, J. S. 1873 A, pp. 271, 272. (Holoptychius ? and Bothriolepis?.)
Woodward, A. S. 1890 A, p. 398.
   1891 A, p. 364. (Sauripterus.)

Sauripteris taylori Hall.

Hall, J. 1843 A, p. 282, text figs. 130¹, 131.

Newberry, J. S. 1873 A, pp. 271, 272. (Holopty-
   chius ? and Bothriolepis?.)
   1889 A. p. 113. (Sauripterus.)
Lesley, J. P. 1889 A, p. 94 (Bothriolepis); p. 286,
   figure (Holoptychius); *Errata*, p. xi (Saurip-
   teris).
Whiteaves, J. F. 1880 A, p. 133.
Whitfield and Hovey, 1900 A, p. 348.
Williams, H. S. 1887 A, p. 101. (Sauropterus.)
Woodward, A. S. 1890 A, p. 398.
   1891 A, p. 365.
Devonian (Catskill); Pennsylvania.

EUSTHENOPTERON Whiteaves.   Type *E. foordii* Whiteaves.

Whiteaves, J. F. 1881 A, p. 495.
Miller, S. A. 1889 A, p. 597.
Nicholson and Lydekker 1889 A, p. 970.
Traquair, R. H. 1900 B, p. 517.
Whiteaves, J. F. 1883 A, p. 354.
   1883 B, p. 161.
   1881 C, p. 30.
   1899 A, p. 411.
Woodward, A. S. 1891 A, p. 361.
   1898 B, p. 76.
Zittel, K. A. 1890 A, p. 180.

Eusthenopteron foordii Whiteaves.

Whiteaves, J. F. 1881 A, p. 495.
Bridge, T. W. 1896 A, p. 588.
Cope, E. D. 1892 N, p. 280, fig. 1.

Dana, J. D. 1896 A, p. 618, fig. 971.
Dean, B. 1896 A, p. 152, fig. 154.
Jaekel, O. 1892 A, p. 91.
   1896 A, p. 8.
Traquair, R. H. 1890 A, p. 17.
   1893 C, p. 266.
   1893 D, p. 124.
Whiteaves, J. F. 1883 A, p. 354.
   1883 B, p. 161.
   1881 C, p. 31, text fig.
   1889 A, p. 78, pl. v, fig. 5; pls. vi, vii; p. 91,
      pl. x, fig. 1.
Woodward, A. S. 1891 A, p. 362.
   1898 B, p. 77, fig. 58.
Upper Devonian; Province of Quebec, Canada.

OSTEOLEPIDÆ.

Agassiz, L. 1843 B (1835), ii, p. 117. (Osteolepis.)
   1844 B, p. 50. (Osteolepis.)
Collinge, W. E. 1893 A, p. 5.
Cope, E. D. 1889 R, p. 856.
Huxley, T. H. 1861 B. (Saurodipterini.)
Klaatsch, H. 1896 A, p. 318. ("Rhombodipte-
   rinen.")
Lütken, C. 1873 A, p. 33. (Rhombodipterini.)
McCoy, F. 1855 A, p. 585. (Saurodipteridæ.)
Nicholson and Lydekker 1889 A, p. 971. (Rhom-
   bodipterini.)

Pander, C. H. 1860 A. ("Saurodipterinen.")
Rohon, J. V. 1892 A, p. 29.
   1899 B, p. 13. ("Osteolepiden.")
Steinmann and Döderlein 1890 A, p. 562. (Rhom-
   bodipteridæ.)
Traquair, R. H. 1891 A, p. 127.
   1900 B, p. 517.
Woodward, A. S. 1891 A, p. 367.
   1898 B, p. 77.
Zittel, K. A. 1890 A, p. 183. (Rhombodipterini.)

PARABATRACHUS Owen.   Type *P. colei* Owen.

Owen, R. 1858 D, p. 67.
Unless otherwise indicated the following authorities use for this genus the name *Megalichthys*. Nevertheless, the name *Megalichthys* was originally based on materials belonging to a different genus and species. See p. 359.
Agassiz, L. 1843 B, ii, pt. ii, pp. 89, 154.   (Not Megalichthys of 1836.)
Barkas, T. P. 1873 A, p. 35.
Cope, E. D. 1880 A, p. 52.   (Ectosteorhachis.)
    1883 H, p. 628.   (Ectosteorhachis.)
    1891 B, p. 457.
Gaudry, A. 1883 A, p. 284.
Hay, O. P. 1899 E, p. 788.   (Parabatrachus.)
Huxley, T. H. 1861 B, p. 12.
    1863 A, p. 67.   (Parabatrachus.)
Klaatsch, H. 1890 A, p. 220.
McCoy, F. 1848 A, p. 3.   (Centrodus, type *C. striatulus*, preoccupied.)
Miall, L. C. 1884 A.
Miller, S. A. 1889 A, p. 596.   (Ectosteorachis.)
Newberry, J. S. 1873 A, pp. 256, 343.
    1889 A, p. 215.
Nicholson and Lydekker 1889 A, pp. 970, 972.
Pictet, F. J. 1854 A, p. 179.
Quenstedt, F. A. 1852 A, p. 227.
    1885 A, p. 348.
Roemer, F. 1865 A.
Traquair, R. H. 1881 B, p. 178.
    1884 A.
Wellburn, E. D. 1900 A, p. 52.
Williamson, W. C. 1849 A, p. 456, pl. xlii, figs. 16-19.
Woodward, A. S. 1891 A, p. 378.
    1892 B, p. 233.
Young, J. 1866 B.
Zittel, K. A. 1890 A. p. 184.

Parabatrachus ciceronius (Cope).

Cope, E. D. 1883 H, p. 628.   (Ectosteorhachis.)
    1883 P.   (Ectosteorhachis.)
    1888 BB, p. 286.   (Ectosteorhachis.)
    1891 B, p. 457.   (Megalichthys.)
Hay, O. P. 1899 E, p. 788.
Wellburn, E. D. 1900 A, p. 60.   (Megalichthys.)
Woodward, A. S. 1891 A, p. 388.   (Megalichthys.)
Permian; Texas.

Parabatrachus macropomus (Cope).

Cope, E. D. 1892 L, p. 226, pl. viii.   (Megalichthys.)

Cope, E. D. 1894 F, pl. xix, fig. 1.   (Megalichthys.)
Hay, O. P. 1899 E, p. 788.
Carboniferous; Kansas.

Parabatrachus maxillaris (Agassiz).

This species is known in all scientific works as *Megalichthys hibberti*. The earliest specific name available is, however, *maxillaris*. Most of the authors here cited use the name *M. hibberti*.
Agassiz, L. 1843 B, ii, pt. ii, pp. 89-96, pls; lxiii, lxiii *a*, lxiv (M. hibberti); p. 96 (M. maxillaris).
Cope, E. D. 1880 A, p. 58.
Fleming, J., 1835, Edinb. New Phil. Jour., xix, p. 314, pl. iv, figs. 1-3.   (Ichthyolithus clackmannensis.[1])
Hay, O. P. 1899 E, p. 788.   (P. maxillaris.)
Miall, L. C. 1884 A, p. 347, figs. 1-4, 6.
Newberry, J. S. 1873 A, p. 343, pl. xl, fig. 3. ("Megalichthys.")
    1889 A, p. 215.
Traquair, R. H. 1884 A.
Ward, J. 1890 A, p. 162, pl.
Wellburn, E. D. 1900 A, p. 52, pls. xiv-xvii, xix.
Whiteaves, J. F. 1881 D, p. 36.   (Psammodus bretonensis.)
Williamson, W. C. 1849 A, p. 450, pl. xli, fig. 15; pl. xlii, figs. 16-19.
Woodward, A. S. 1891 A, pp. 378, 388.
Woodward and Sherborn 1890 A, p. 117.
Young, J. 1866 B.
Coal-measures; Great Britain, Cape Breton Island. Scales and bones not distinguishable as different have been found in Ohio and in Nova Scotia.

Parabatrachus nitidus (Cope).

Cope, E. D. 1880 A, p. 52.   (Ectosteorhachis.)
    1888 B B, p. 286.   (E. nitidus.)
    1891 B, p. 457, pl. xxxii, figs. 8, 9.   (Megalichthys.)
    1891 C.   (Megalichthys.)
Hay, O. P. 1899 E, p. 788.
Woodward, A. S. 1891 A, p. 388.   (Megalichthys.)
Permian; Texas.

GLYPTOPOMUS Agassiz.   Type *G. minor* Agassiz.

Agassiz, L. 1844 B, p. 57.
Huxley, T. H. 1861 B, p. 4.
    1866 A, p. 3.
Pictet, F. J. 1854 A, p. 193.
Woodward, A. S. 1891 A, p. 389.

Glyptopomus sayrei Newb.

Newberry, J. S. 1878 A, p. 189.
    1889 A, p. 116, pl. xviii, fig. 1.
Woodward, A. S. 1891 A, p. 390.
Devonian (Catskill); Pennsylvania.

ONYCHODONTIDÆ. ·

Woodward, A. S. 1891 A, p. 391. | Rohon, J. V. 1899 A, p. 60.

---

[1] *Ichthyolithus* was not regarded as a generic name.

ONYCHODUS Newb. Type *O. sigmoides* Newb.

Newberry, J. S. 1857 C, p. 124.
Claypole, E. W. 1896 B, p. 354.
Eastman, C. R. 1900 B, p. 36.
1900 C, p. 392.
Jaekel, O. 1893 B, p. 175.
Miller, S. A. 1889 A, p. 602.
Newberry, J. S. 1857 A, p. 148.
1862 A, p. 77, fig. 3.
1870 B, p. 152.
1873 A, pp. 256, 264, 265, 273, 296.
1888 B, p. 118.
1889 A, p. 58.
Newton, E. T. 1992 A.
Nicholson and Lydekker 1889 A, p. 968.
Rohon, J. V., 1899 A, p. 60.
Woodward, A. S. 1888 D.
1889 B.
1890 A, p. 391.
1891 A, p. 392.
Zittel, K. A. 1890 A, p. 179.

Onychodus ortoni Newb.

Newberry, J. S. 1889 A, p. 71, pl. xix, figs. 1, 1a.
1888 B, p. 118. (Insufficient description.)
Woodward, A. S. 1891 A, p. 393.
Devonian (Huron shale); Ohio.

Onychodus sigmoides Newb.

Newberry, J. S. 1857 C, p. 124. (O. sigmoides,.O. hopkinsi.)
Eastman, C. R. 1899 B. pp. 278, 287.
1899 F, p. 321, figs. 2, 3.
Hitchcock, Fanny R, 1888 A.
Lealey, J. P. 1889 A, p. 497, figures.
1892 A, pl. cl.
Miller, S. A. 1889 A, p. 603, figs. 1150, 1151.
Newberry, J. S. 1862 A, p. 77, fig. 3. (O. hopkinsi.)
1873 A, pp. 265, 299, pl. xxvi, figs. 1-5; pl. xxvii, figs. 1, 2 (O. sigmoides); pp. 270, 302 (O. hopkinsi).
1888 B, p. 118. (O. sigmoides, O. hopkinsi.)
1889 A, p. 56, pl. xxxvi, figs. 1-4; pl. xxxvii, figs. 1-10 (O. sigmoides); p. 99 (O. hopkinsi).
Woodward, A. S. 1891 A, p. 329 (O. sigmoides); p. 393 (O. hopkinsi).
Devonian (Corniferous); Ohio, New York: (Chemung); New York.
O. sigmoides and O. hopkinsi are united on the authority of Dr. C. R. Eastman.

Onychodus sp. indet.

Whiteaves, J. F. 1892 A, p. 356. (No description.)
Devonian; Manitoba.

Order ACTINISTIA Cope.

Cope, E. D. 1887 O, p. 1017.
1889 R, p. 855.
1891 B, p. 460.
1891 E.

Cope, E. D. 1891 N, p. 19.
Woodward, A. S. 1891 A, p. 394.
1898 B, p. 78.

CŒLACANTHIDÆ.

Agassiz, L. 1843 B, 18 ii, pt. ii, p. 168. ("Célacanthes.")
1844 B, p. 59. ("Célacanthes.")
1846 A, p. 45. ("Célacanthes.")
Cope, E. D. 1877 N, p. 56.
1878 A, p. 298.
1889 R, p. 855.
Dollo, L. 1896 A.
Gegenbaur, C. 1895 B.
Günther, A. 1872 A, p. 559.
1880 A, p. 365.
1881 A, p. 687.
Haeckel, E. 1895 A, p. 136. (Cœlacanthini.)
Huxley, T. H. 1861 B, p. 15. (Cœlacanthini.)
1866 A. (Cœlacanthini.)

Huxley, T. H. 1872 A, p. 128. (Cœlacanthini.)
King, W. 1850 A, p. 235.
McCoy, F. 1855 A, p. 589. (Cœlacanthi.)
Newberry, J. S. 1873 A, p. 255.
Nicholson and Lydekker 1889 A, p. 972.
Owen, R. 1860 E, p. 131. (Cœlacanthi.)
Pictet, F. J. 1854 A, p. 141. ("Célacanthes.")
Reis, O. M. 1888 A.
Steinmann and Döderlein 1890 A, p. 562.
Vetter, B. 1881 A, p. 9. (Cœlacanthini.)
Willemoes-Suhm, R. 1869 A.
Woodward, A. S. 1891 A, p. 394.
1892 E.
1898 B, p. 78.
Zittel, K. A. 1890 A, p. 171. (Cœlacanthini.)

CŒLACANTHUS Agassiz. Type *C. granulatus* Agassiz.

Agassiz, L. 1843 B, Feuill (1836), p. 83; ii (1843), pt. ii, p. 170.
1843 B, ii, pt. ii, p. 178. (Hoplopygus.)
1844 A.
Cope, E. D. 1873 O, p. 341. (Conchiopsis, with C. fliferus as type = Cœlacanthus elegans.)
Heckel, J. J. 1850 A, p. 362.
Huxley, T. H. 1861 B, p. 16.
1866 A, p. 8.

Miller, S. A. 1889 A, p. 592.
Newberry, J. S. 1868 A, p. 144.
1873 A, pp. 285, 337.
1873 B. (On Conchiopsis.)
1874 E, p. 76. (On Conchiopsis.)
Nicholson and Lydekker 1889 A, p. 973.
Pictet, F. J. 1854 A, p. 141.
Reis, O. M. 1888 A, p. 71. (Rhabdoderma.)
Vetter, B. 1881 A, p. 9.

Ward, J. 1890 A, p. 168.
Winckler, T. C. 1871 A.
Woodward, A. S. 1891 A, p. 399.
Woodward and Sherborn 1890 A, p. 39.
Zittel, K. A. 1890 A, p. 171.

Cœlacanthus elegans Newb.

Newberry, J. S. 1856 B, p. 98.
Cope, E. D. 1873 O, p. 342. (Conchiopsis filiferus,
    C. anguliferus.)
    1873 Q, p. 418.
Dean, B. 1895 A, p. 153, fig. 155.
Hancock and Atthey 1872, p. 256, pl. xvii, fig. 4.
    (C. lepturus.)
Hay, O. P. 1900 A, p. 116.
Huxley, T. H. 1866 A, pp. 14, 20, pl. v, figs. 1, 2, 4
    (C. elegans); p. 15, pl. ii, figs. 1‑4; pl. iii, figs.
    1‑3; pl. iv, figs. 1‑6 (C. lepturus).
Miller, S. A. 1889 A, p. 592, fig. 1112.
Newberry, J. S. 1873 A, p. 339, pl. xl, figs. 1‑1d.
    1873 B.
    1873 D, p. 30.
    1874 E, p. 76.

Newberry, J. S. 1889 A, p. 213.
Woodward, A. S. 1891 A, p. 403.
    Coal-measures; Ohio.

Cœlacanthus ornatus Newb.

Newberry, J. S. 1856 B, p. 98.
Huxley, T. H. 1866 A, p. 14.
Newberry, J. S. 1873 A, p. 340.
    1889 A, p. 227.
Reis, O. M. 1888 A, p. 5. (Rhabdoderma.)
Woodward, A. S. 1891 A, p. 406.
    Coal-measures; Ohio.

Cœlacanthus robustus Newb.

Newberry, J. S. 1856 B, p. 98.
Hay, O. P. 1900 A, p. 115.
Huxley, T. H. 1866 A, p. 14.
Newberry, J. S. 1873 A, p. 341, pl. xl, figs. 2, 2a.
    1889 A, p. 228.
Reis, O. M. 1888 A, pp. 5, 71. (Rhabdoderma.)
Woodward, A. S. 1891 A, p. 406.
    Coal-measures; Ohio, Illinois.

PEPLORHINA Cope.   Type *P. anthracina* Cope.

Cope, E. D. 1873 O, p. 343.
    1873 Q, p. 418.
    1875 D, p. 409.
Newberry, J. S. 1874 E, p. 76.
Nicholson and Lydekker 1889 A, p. 969.
Woodward, A. S. 1891 A, p. 408.

Peplorhina anthracina Cope.

Cope, E. D. 1873 O, p. 343.
    1873 Q, p. 418.
    1875 D, p. 410, pl. xxxv, fig. 6; pl. xli, figs. 4‑6.
Lesley, J. P. 1889 A, p. 622, figures.
Newberry, J. S. 1873 B.
    1874 E, p. 76.
Woodward, A. S. 1891 A, p. 408.
    Coal-measures; Ohio.

Peplorhina arctata Cope.

Cope, E. D. 1877 N, p. 54.
Case, E. C. 1900 A, p. 707.
Cope, E. D. 1882 B, p. 461. ("Theromorphous
    saurian.")
Woodward, A. S. 1891 A, p. 408.
    Permian; Illinois.

Peplorhina exanthematica Cope.

Cope, E. D. 1873 O, p. 342. (Conchiopsis.)
    1873 Q, p. 418.
Newberry, J. S. 1873 B. (Syn. of P. anthracina.)
    1874 E, p. 76. (Syn. of P. anthracina.)
Woodward, A. S. 1891 A, p. 408.
    Coal-measures; Ohio.

DIPLURUS Newb.   Type *D. longicaudatus* Newb.

Newberry, J. S. 1878 B.
Dean, B. 1893 D.
    1895 A, p. 153.
Newberry, J. S. 1888 A, p. 70.
Nicholson and Lydekker 1889 A, p. 973.
Woodward, A. S. 1891 A, p. 409.
Zittel, K. A. 1890 A, p. 174.

Diplurus longicaudatus Newb.

Newberry, J. S. 1878 B.
Bridge, T. W. 1896 A, p. 588.
Dean, B. 1893 B.
    1895 A, p. 154, fig. 156.
Newberry, J. S. 1888 A, p. 74, pl. xx, figs. 1‑5.
Woodward, A. S. 1891 A, p. 409.
    Triassic; New Jersey, Connecticut.

Superorder ACTINOPTERI Cope.

Cope, E. D. 1872 C, p. 325.
    1877 N, p. 56.
    1877 O, p. 41.
    1878 A, p. 298.
    1887 O, p. 1017. (Actinopterygia.)
    1889 R, p. 855. (Actinopterygia.)
    1891 B, p. 459. (Actinopterygia.)
    1891 N, pp. 19, 21. (Actinopterygia.)

Cope, E. D. 1898 B, pp. 30, 33. (Actinopterygia.)
Gadow, H. 1898 A, p. 7. (Actinopterygia.)
Hay, O. P. 1899 F, p. 682.
Reis, O. M. 1897 A, p. 130. ("Actinopterygier.")
Woodward, A. S. 1891 A, p. 423.
    1895 A.
    1898 B, p. 81. (Actinopterygia.)

Order CHONDROSTEI Müller.

Müller, J. 1846 C.
Agassiz, L. 1843 B (1833), ii, pt. i, pp. 1, 2, 3, 10.
  ("Lepidoides," in part; "Sauroides," in part.)
Beard, J. 1890 A, p. 182.
Bridge, W. J. 1879 A.
Cope, E. D. 1880 P.   (Lysopteri.)
  1872 C, p. 327.
  1877 N, p. 57.
  1877 O, p. 41.
  1878 A, p. 293.
  1887 O, p. 1017.   (Podopterygia, in part.)
  1889 R, pp. 855, 856 (Podopterygia, in part);
    p. 857 (Lysopteri).
  1891 B, p. 459 (Podopterygia, in part); p.
    460 (Lysopteri).
  1891 E.   (Podopterygia, in part; Lysopteri.)
  1891 N, pp. 19, 20.   (Podopterygia.)
  1898 B, pp. 30, 32.   (Podopterygia.)
Egerton, P. G. 1859 B.
Gadow, H. 1898 A, p. 8.
Gegenbaur, C. 1898 A.
Gill, T. 1872 C, p. 22.   (Chondroganoidei.)
  1893 A, p. 130.

Goette, A. 1878 A, p. 442.
Günther, A. 1870 A, p. 332.
  1880 A, p. 360.
  1881 A, p. 687.
Haeckel, E. 1895 A, p. 249.
Huxley, T. H. 1872 A, p. 128.
Jordan and Evermann 1896 A, p. 102.
Klaatsch, H. 1890 A, p. 146.
Ryder, J. A. 1886 A.
Stannius, H. 1854 A.
Thacher, J. K. 1877 A.
  1877 B.
Traquair, R. H. 1879 A, p. 380.   (Acipenseroidei.)
  1877 B.
  1887 A.
Woodward, A. S. 1889 H, p. 43.
  1891 A, pp. v, 423.
  1895 A, p. v.
  1898 B, p. 82.
Zittel, K. A. 1890 A, p. 163 (Chondrostei); p. 186
  (Heterocerci).
Zograf, N. 1896 A.

PALÆONISCIDÆ.

Bridge, W. J. 1879 A, p. 723.
Cope, E. D. 1889 R, p. 858.
  1891 E.
Fritsch, A. 1895 A, pp. 83, 122.
Günther, A. 1880 A, p. 369.
  1881 A, p. 687.
Kosmak, G. W. 1895 A, p. 20.
Nicholson and Lydekker 1889 A, p. 976.
Owen, R. 1860 E, p. 136.
Steinmann and Döderlein 1890 A, p. 565.

Traquair, R. H. 1877 B.
  1891 A, p. 128.
Vogt, C. 1852, Zool. Briefe., ii.   (Palæoniscida.)
Woodward, A. S. 1889 H, p. 43.
  1891 A, p. 425.
  1893 E, p. 450.
  1895 A, p. vi.
  1898 B, p. 82.
Zittel, K. A. 1890 A, p. 188.

GONATODUS Traquair.   Type *G. punctatus* (Agassiz).

Traquair, R. H. 1877 A, p. 555.
  1890 F, p. 399.   (Gonatodus, Drydenius?)
Ward, J. 1890 A, p. 178.
Woodward, A. S. 1891 A, p. 434.

Gonatodus brainerdi (Thomas).

Thomas, W. H. B. 1853, Cleveland Times, Sept. 14,
  as Palæoniscum brainerdi (*fide* J. S. Newberry).
Brainerd, —— 1852 A, p. 18, figs. 1–3.   ("Palæoniscus.")

Hitchcock, C. H. 1868 B, p. 187.   (Palæoniscus; no
  description.)
Lesley, J. P. 1889 A, p. 584, figure.   (Palæoniscus.)
Newberry, J. S. 1862 A, p. 78.   (Palæoniscus.)
  1868 A, p. 145.   (Palæoniscus.)
  1873 A, pp. 281, 346.   (Palæoniscus.)
  1889 A, p. 125, pl. xxxiv, figs. 1, 2.   (Gona-
    todus.)
Woodward, A. S. 1891 A, p. 436.
  Devonian (Berea grit); Ohio.

AMBLYPTERUS Agassiz.   Type *A. latus* Agassiz.

Agassiz, L. 1843 B (1833), ii, pp. 3, 28.
  1844 A.
Eichwald, E. 1860 A, p. 1587.
Fritsch, A. 1895 A, p. 94.
Martin, K. 1873 A, p. 720.
Miller, S. A. 1889 A, p. 567.
Pictet, F. J. 1854 A, p. 181.
Quenstedt, F. A. 1852 A, p. 225.
  1885 A, p. 345.
Steinmann and Döderlein 1890 A, p. 566.
Traquair, R. H. 1867 A.
  1877 A.

Traquair, R. H. 1877 B.
Woodward, A. S. 1891 A, p. 437.
Zittel, K. A. 1890 A, pp. 189, 192.

Amblypterus carolinæ nom. nov.

Cope, E. D. 1875 V, p. 30.   (A. ornatus.)
Emmons, E. 1857 A, p. 44, fig. 16.   (A. ornatus,
  not of Giebel.)
  1860 A, p. 183, fig. 161[1–3].   (A. ornatus.)
Woodward, A. S. 1891 A, p. 448.   (A. ornatus.)
  Triassic; North Carolina.

HAPLOLEPIS S. A. Miller.   Type *E. corrugata* (Newb.).

Miller, S. A. 1892 A, p. 715.
Hay, O. P. 1899 E, p. 791.
Miller, S. A. 1889 A, p. 597 (Eurylepis); p. 602 (Mecolepis).
Newberry, J. S. 1856 B, p. 96. (Mecolepis, preoccupied.)
    1857 B, p. 150. (Eurylepis, preoccupied.)
    1857 B. (Eurylepis.)
    1868 A, p. 144. (Eurylepis.)
    1873 A, p. 347. (Eurylepis.)
    1889 A, p. 212. (Eurylepis.)
Traquair, R. H. 1894 B, p. 373. (Eurylepis.)
Woodward, A. S. 1891 A, p. 448. (Eurylepis.)
Zittel, K. A. 1891 A, p. 192. (Eurylepis.)

Haplolepis corrugata (Newb.).

Newberry, J. S. 1856 B, p. 96. (Mecolepis.)
Miller, S. A. 1892 A, p. 715.
Newberry, J. S. 1857 B. (Eurylepis.)
    1873 A, pp. 138, 350, pl. xxxviii, figs. 4, 4a. (Eurylepis.)
Woodward, A. S. 1891 A, p. 450. (Eurylepis.)
Coal-measures; Ohio.

Haplolepis granulata (Newb.).

Newberry, J. S. 1856 B, p. 97. (Mecolepis granulatus, M. insculptus.)
Miller, S. A. 1892 A, p. 715.
Newberry, J. S. 1857 B, p. 150. (Eurylepis granulatus, E. insculptus.)
    1873 A, p. 352, pl. xxxix, figs. 2–2a, (Eurylepis insculpta.); figs. 5–5a (Eurylepis granulata).
Woodward, A. S. 1891 A, p. 449.
Coal-measures; Ohio.

Haplolepis minima (Newb.).

Newberry, J. S. 1873 A, p. 353, pl. xxxix, fig. 3. (Eurylepis.)
Miller, S. A. 1892 A, p. 715.
Woodward, A. S. 1891 A, p. 450. (Eurylepis.)
Coal-measures; Ohio.

Haplolepis ornatissima (Newb.).

Newberry, J. S. 1856 B, p. 97. (Mecolepis.)
Miller, S. A. 1892 A, p. 715.
Newberry, J. S. 1857 B, p. 150. (Eurylepis.)
    1873 A, p. 352, pl. xxxix, figs. 4, 4a. (Eurylepis.)
Woodward, A. S. 1891 A, p. 450. (Eurylepis.)
Coal-measures; Ohio.

Haplolepis ovoidea (Newb.).

Newberry, J. S. 1856 B, p. 97. (Mecolepis.)
Miller, S. A. 1892 A, p. 715.
Newberry, J. S. 1857 B, p. 150. (Eurylepis.)
    1873 A, p. 351, pl. xxxix, fig. 1. (Eurylepis.)
Woodward, A. S. 1891 A, p. 450. (Eurylepis.)
Coal-measures; Ohio.

Haplolepis serrata (Newb.).

Newberry, J. S. 1856 B, p. 97. (Mecolepis.)
Miller, S. A. 1892 A, p. 715.
Newberry, J. S. 1857 B, p. 150. (Eurylepis.)
Woodward, A. S. 1891 A, p. 450. (Eurylepis.)
Coal-measures; Ohio.

Haplolepis striolata (Newb.).

Newberry, J. S. 1873 A, p. 355. (Eurylepis).
Miller, S. A. 1892 A, p. 715.
Woodward, A. S. 1891 A, p. 450. (Eurylepis.)
Coal-measures; Ohio.

Haplolepis tuberculata (Newb.).

Newberry, J. S. 1856 B, p. 96. (Mecolepis.)
Miller, S. A. 1889 A, p. 597, fig. 1130. (Eurylepis.)
    1892 A, p. 715.
Newberry, J. S. 1857 B, p. 150. (Eurylepis.)
    1873 A, p. 350, pl. xxxviii, figs. 2–3a. (Eurylepis.)
Woodward, A. S. 1891 A, p. 449. (Eurylepis.)
Coal-measures; Ohio.

CHEIROLEPIS Agassiz.   Type *C. traillii* Agassiz.

Agassiz, L. 1843 B (1835), pt. ii, p. 128.
    1844 B, p. 44.
Eichwald, E. 1860 A, p. 1573.
Huxley, T. H. 1861 B, p. 39.
McCoy, F. 1855 A, p. 580. (Chirolepis.)
Miller, S. A. 1889 A, p. 589. (Chirolepis.)
Nicholson and Lydekker 1889 A, p. 977.
Pander, C. H. 1860 A, p. 69.
Pictet, F. J. 1854 A, p. 190.
Powrie, J. 1867 A.
Reis, O. M. 1896 A, p. 187.
Traquair 1875 A.
    1888 B, p. 516.
Woodward, A. S. 1891 A, p. 451.

Woodward and Sherborn 1890 A, p. 31.
Zittel, K. A. 1890 A, p. 188.

Cheirolepis canadensis Whiteaves.

Whiteaves, J. F. 1881 C, p. 33.
Cope, E. D. 1881 L, p. 253. (C. cummingsii?)
Dana, J. D. 1896 A, p. 618, fig. 974.
Traquair, R. H. 1890 A, p. 20.
Whiteaves, J. F. 1881 A, p. 496.
    1883 A, p. 355.
    1883 D, p. 33.
    1889 A, p. 90.
Woodward, A. S. 1891 A, p. 457.
Upper Devonian; Province of Quebec.

RHADINICHTHYS Traquair.   Type *R. ornatissimus* Agassiz.

Traquair, R. H. 1877 A, p. 558.
Collinge, W. E. 1898 A, p. 7.
Traquair, R. H. 1881 A, p. 25.
Ward, J. 1890 A, p. 175.
Wellburn, E. D. 1900 B, p. 260.
Woodward, A. S. 1891 A, p. 461.
Zittel, K. A. 1890 A, p. 193.

Rhadinichthys alberti (Jackson).
Jackson, C. T. 1851 B, p. 22, pl. i, fig. 1. (Palæoniscus.)
Dawson, J. W. 1868 A, p. 231, fig. 62. (Palæoniscus alberti?)
   1877 A, p. 338.
   1878 A, supplement, p. 100. (Palæoniscus.)
Egerton, P. G. 1853 A. (Palæoniscus.)
Jackson, C. T. 1851 A, p. 138. (Palæoniscus.)
Lesley, J. P. 1889 A, p. 584, figure. (Palæoniscus.)
Miller, S. A. 1889 A, p. 611, fig. 1172.
Newberry, J. S. 1889 A, p. 187. (Palæoniscus.)
Traquair, R. H. 1877 A, p. 559.
Woodward, A. S. 1891 A, p. 465.
Subcarboniferous; New Brunswick.

Rhadinichthys cairnsii (Jackson).
Jackson, C. T. 1851 B, p. 23, pl. i, fig. 3. (Palæoniscus.)
Dawson, J. W. 1877 A, p. 339.
   1878 A, supplement, p. 100. (Palæoniscus.)
Egerton, P. G. 1853 A. (Palæoniscus.)
Jackson, C. T. 1851 A, p. 139. (Palæoniscus.)
Newberry, J. S. 1889 A, p. 187. (Palæoniscus.)
Woodward, A. S. 1891 A, p. 465.
Subcarboniferous; New Brunswick.

Rhadinichthys gracilis (Newb. and Worth).
Newberry and Worthen 1870 A, p. 347, pl. iii, fig. 4. (Palæoniscus.)
Woodward, A. S. 1891 A, p. 469.
Coal-measures; Illinois.

Rhadinichthys leidyianus (Lea).
Lea, I. 1853 B, p. 206, pl. xx, figs. 4, 5. (Palæoniscus.)
Amer. Jour. Science 1852 B.
Woodward, A. S. 1891 A, p. 469.
Coal-measures; Pennsylvania.

Rhadinichthys? lineatus (Newb.).
Newberry, J. S. 1856 B, p. 97. (Mecolepis.)
   1857 B. (Eurylepis.)
   1873 A, p. 358, pl. xxxix, figs. 7, 7a. (Eurylepis.)
   1889 A, p. 228. (Rhadinichthys?.)
Woodward, A. S. 1891 A, p. 450. (Eurylepis?.)
Coal-measures; Ohio.

Rhadinichthys modulus (Dawson).
Dawson, J. W. 1877 A, p. 337, fig. 1. [Palæoniscus (Rhadinichthys).]
   1878 A, supplement, p. 99, fig. 18. [P. (Rhadinichthys).]
Newberry, J. S. 1889 A, p. 187. (Palæoniscus.)
Woodward, A. S. 1891 A, p. 466.
Subcarboniferous; New Brunswick.

PYGOPTERUS Agassiz.   Type *P. humboldti* Agassiz.

Agassiz, L. 1843 B (1833), ii, pt. i, p. 10; ii, pt. ii, pp. 74, 152 (1844).
   1844 A.
Barkas, W. J. 1878 C, pp. 205, 207.
McCoy, F. 1855 A, p. 608.
Martin, K. 1873 A, p. 722.
Miller, S. A. 1889 A, p. 611.
Pictet, F. J. 1854 A, p. 179.
Quenstedt, F. A. 1852 A, p. 226.
   1885 A, p. 347.
Traquair, R. H. 1867 A.
   1875 B.
   1877 A.

Woodward, A. S. 1891 A, p. 470.
Zittel, K. A. 1890 A, p. 194.

Pygopterus humboldti[1] Agassiz.
Agassiz, I. 1843 B, ii, pt. i (1833), p. 10 (P. humboldti); pt. ii (1844), p. 74, pls. liv, lv (P. humboldti); p. 76, pls. liii, liiia (P. mandibularis).
Lea, I. 1856 C, p. 78. (P. mandibularis.)
Miller, S. A. 1889 A, p. 611, fig. 1171. (P. mandibularis.)
Wheatley, C. M. 1861 A, p. 44. (P. mandibularis.)
Woodward, A. S. 1891 A, p. 470.
Permian; Europe: Triassic; Pennsylvania?.

PALÆONISCUM Blainv.   Type *P. freieslebeni* Blainv.

Blainville, H. M. D., 1818, Nouv. Dict. d' Hist. Nat., vol. xxvii, p. 320.
   The following authorities, unless otherwise indicated, call this genus *Palæoniscus*. The original form was *Palæoniscum.*
Agassiz, L. 1843 (1833) B, ii, pt. i, pp. 4, 41.
   1844 A.
   1851 B.
Barkas, W. J. 1878 C, p. 207.

Egerton, P. G. 1849 B, p. 4.
Eichwald, E. 1860 A, p. 1581.
Hancock and Athey 1868 A, p. 358, pl. xv, figs. 3–5; pl. xvi, figs. 1, 2.
Koninck, L. G. de 1844 A, p. 608. (Palæoniscum.)
Kosmak, G. W. 1895 A, p. 20.
McCoy, F. 1855 A, p. 605.
Martin, K. 1873 A, p. 700, pl. xxii.
Miller, S. A. 1839 A, p. 604.

---

[1] This species, under the name *P. mandibularis*, was reported with doubt by Lea from the Triassic of Pennsylvania. The scales there found probably belong to some other species.

Newberry, J. S. 1873 A, p. 344.
Nicholson and Lydekker 1889 A, p. 977.
Owen, R. 1845 B, p. 69.
Pictet, F. J. 1854 A, p. 184.
Quenstedt, F. A. 1852 A, p. 223.
1885 A, p. 343.
Redfield, W. C. 1841 A, p. 25.
1856 A, p. 36.
Rohon, J. V. 1889 A, p. 10.
Steinmann and Döderlein 1890 A, p. 566.
Traquair, R. H. 1877 A
Williamson, W. C. 1849 A, p. 445, pl. xl, fig. 7; pl. xli, fig. 1.
Woodward, A. S. 1891 A, p. 476.
Zittel, K. A. 1890 A, p. 190.

Palæoniscum antiquum Williams.

Williams, H. U. 1886 A, p. 84, fig. 2.

Woodward, A. S. 1891 A, p. 485.
Devonian (Portage); New York.

Palæoniscum devonicum Clarke.

Clarke, J. M. 1885 A, pp. 20, 41, pl. i, figs. 2–6.
Lesley, J. P. 1889 A, p. 585, figures.
Woodward, A. S. 1891 A, p. 485.
Devonian (Genesee); New York.

Palæoniscum reticulatum Williams.

Williams, H. U. 1886 A, p. 86, fig. 1.
Woodward, A. S. 1891 A, p. 485.
Devonian (Portage); New York.

Palæoniscum scutigerum Newb.

Newberry, J. S. 1868 A, p. 145. (No description.)
Hitchcock, C. H. 1868 B, p. 186. (No description.)
Coal-measures; Ohio.

TEGEOLEPIS S. A. Miller.   Type *Actinophorus clarkii* Newb.

Miller, S. A. 1892 A, p. 717.
Claypole, E. W. 1895 F. (Actinophorus.)
Hay, O. P. 1899 E, p. 791.
Newberry, J. S. 1888 D, p. 179. (Actinophorus, type *A. clarkii*; preoccupied.)
1889 A, p. 174. (Actinophorus.)
Woodward, A. S. 1891 A, p. 486. (Actinophorus.)

Tegeolepis clarkii (Newb.).

Newberry, J. S. 1888 D, p. 179. (Actinophorus.)
Claypole, E. W. 1895 F. (Actinophorus.)
Newberry, J. S. 1889 A, p. 175, pl. xlix, fig. 1. (Actinophorus.)
Woodward, A. S. 1891 A, p. 487. (Actinophorus.)
Devonian (Cleveland shale); Ohio.

ELONICHTHYS Giebel.   Type *E. germari* Giebel.

Giebel, C. G. 1848 A, p. 249.
Traquair, R. H. 1877 A, p. 553. (Cosmoptychius.)
1877 B, p. 42 (Cosmoptychius); p. 47 (Elonichthys).
Ward, J. 1890 A, p. 171.
Woodward, A. S. 1891 A, p. 487.
Woodward and Sherborn 1890 A, p. 48 (Cosmoptychius); p. 73 (Elonichthys).
Zittel, K. A. 1890 A, p. 190.

Elonichthys browni (Jackson).

Jackson, C. T. 1851 B, p. 22, pl. i, fig. 2. (Palæoniscus.)
Dawson, J. W. 1877 A, p. 339. [Palæoniscus (Elonichthys) brownii.]
1878 A, supplement p. 101. [Palæoniscus (Elonichthys) brownii.]
Jackson, C. T. 1851 A. p. 139. (Palæoniscus.)
Newberry, J. S. 1889 A, p. 187.
Traquair, R. H. 1877 A, p. 553. (Elonichthys.)
Woodward, A. S. 1891 A, p. 501.
Subcarboniferous; New Brunswick.

Elonichthys elegans (Emmons).

Emmons, E. 1860 A, p. 191, fig. 167¹⁻². (Rabdiolepis.)
Leidy, J. 1876 B. (Rhabdolepis elegans?)
1876 D. (Rabdiolepis elegans?)
Mesozoic; N. Carolina, Pennsylvania.

Elonichthys hypsilepis Hay.

Hay, O. P. 1900 A, p. 117 (E. peltigerus?); p. 120 (E. hypsilepis); pl. vii (E. peltigerus hypsilepis).
Coal-measures; Illinois.

Elonichthys macropterus (Agassiz).

Agassiz, L. 1843 B (1833), ii, pt. i, pp. 4, 31; pl. i, figs. 4–7; pl. iii, figs. 1–4. (Amblypterus.)
Goldfuss, A. 1847 A, p. 20, pl. v, figs. 1–8. (Amblypterus.)
Hay, O. P. 1900 A, p. 120. (Syn.? of E. hypsilepis.)
?Newberry and Worthen 1870 A, p. 348. (Amblypterus.)
Traquair, R. H. 1877 A, p. 552.
1877 B, pl. ii, fig. 6. (Rhabdolepis.)
Woodward, A. S. 1891 A, p. 491.
Coal-measures; Europe, Illinois?

Elonichthys peltigerus Newb.

Newberry, J. S. 1856 B, p. 98.
Hay, O. P. 1900 A, p. 117.
Lesley, J. P. 1889 A, p. 585, figure. (Palæoniscus.)
1896 A, pl. dlxx. (Palæoniscus.)
Miller, S. A. 1889 A, p. 605, fig. 1155. (Palæoniscus.)
Newberry, J. S. 1873 A, p. 345, pl. xxxviii, figs. 1–1b. (Palæoniscus.)
Newberry and Worthen 1866 A, p. 17, fig. 1. (Palæoniscus.)
Traquair, R. H. 1877 A, p. 553.
1877 B, p. 47.
Woodward, A. S. 1891 A, p. 489.
Coal-measures; Ohio, Illinois, Indiana.

ACROLEPIS Agassiz.   Type *A. sedgwicki* Agassiz.

Agassiz, L. 1843 B, ii, pt. i, p. 11 (1833); ii, pt. ii,
   p. 79 (1844).
      1844 A.
Barkas, W. J. 1878 C, p. 207.
Eichwald, E. 1860 A, p. 1577.
Fritsch, A. 1895 A, p. 115.
McCoy, F. 1855 A, p. 609.
Martin, K. 1873 A, p. 717.
Miller, S. A. 1889 A, p. 587.
Pictet, F. J. 1854 A, p. 180.
Quenstedt, F. A. 1852 A, p. 226.
      1885 A, p. 346.

Woodward, A. S. 1891 A, p. 501.
Zittel, K. A. 1890 A, p. 191.

Acrolepis ? hortonensis Dawson.

Dawson, J. W. 1868 A, p. 254, fig. 77 *e, f.*
      1877 A, p. 339.   (Palæoniscus jacksoni.)
      1878 A, p. 255, fig. 77 *e, f.* (A. ? hortonensis);
         supplement, p. 101 (Palæoniscus jack-
         soni).
Newberry, J. S. 1889 A, p. 187.
Woodward, A. S. 1891 A, p. 508.
      Subcarboniferous; New Brunswick.

PLATYSOMIDÆ.

Young, J. 1866 A, p. 316.
Bridge, W. J. 1879 A, p. 723.
Cope, E. D. 1880 P.
      1891 E.
Günther, A. 1880 A, p. 370.
      1881 A, p. 687.
Nicholson and Lydekker 1889 A, p. 979.

Steinmann and Döderlein 1890 A, p 567.
Traquair, R. H. 1879 A.
      1891 A, p. 129.
      1896 A, p. 125.
Woodward, A. S. 1891 A, p. 527.   (Platysomatidæ.)
      1898 B, p. 87.
Zittel, K. A. 1890 A, p. 197.

CHIRODUS McCoy.   Type *C. pes-ranæ* McCoy.

McCoy, F. 1848 A, p. 130.
Davis, J. W., 1884, Quart. Jour. Geol. Soc., xl, p.
   620.   (Hemicladodus.)
Miller, S. A. 1889 A, p. 589.
Pictet, F. J. 1854 A, p. 268.
Rohon, J. V. 1896 A.
Traquair, R. H. 1878 A, p. 15.   (Cheirodus.)
Ward, J. 1890 A, p. 183.
Woodward, A. S. 1891 A, p. 585.   (Cheirodus.)

Woodward and Sherborn 1890 A, p. 30.   (Cheiro-
   dus.)
Young, J. 1866 A, p. 306.   (Amphicentrum.)
Zittel, K. A. 1890 A, p. 199.   (Cheirodus.)

Chirodus acutus Newb.

Newberry, J. S. 1856 B, p. 99.   (Cheirodus.)
Miller, S. A. 1889 A, p. 589.
Woodward, A. S. 1891 A, p. 540.   (Cheirodus.)
      Coal-measures; Ohio.

PLATYSOMUS Agassiz.   Type *P. gibbosus* Agassiz.

Agassiz, L. 1843 B (1835), ii, pt. i, pp. 6, 161.
      1844 A.
Collinge, W. E. 1898 A, p. 9.
Egerton, P. G. 1849 A.
Heckel, J. J. 1850 A, p. 365.
King, W. 1850 A, p. 228.
Koninck, L. G. de 1878 A, p. 24.
McCoy, F. 1855 A, p. 614.
Miller, S. A. 1889 A, p. 607.
Newberry, J. S. 1873 A, p. 258.
Nicholson and Lydekker 1889 A, p. 980.
Quenstedt, F. A. 1852 A, p. 227.
      1885 A, p. 347.
Pictet, F. J. 1854 A, p. 208.
Steinmann and Döderlein 1890 A, p. 567.
Traquair, R. H. 1879 A.
Woodward, A. S. 1891 A, p. 541.
Young, J. 1866 A, p. 302.
Zittel, K. A. 1890 A, p. 200.

**Platysomus circularis Newb. and
   Worth.**

Newberry and Worthen 1870 A, p. 347, pl. iv, fig. 2.
Woodward, A. S. 1891 A, p. 550.
      Coal-measures; Illinois.

Platysomus lacovianus Cope.

Cope, E. D. 1891 B, p. 462, pl. xxxi, fig. 11.
      Coal-measures; Illinois.

**Platysomus orbicularis Newb. and
   Worth.**

Newberry and Worthen 1870 A, pl. iii, fig. 1.   (No
   description.)
Woodward, A. S. 1891 A, p. 550.
      Coal-measures; Illinois.

Platysomus palmaris Cope.

Cope, E. D. 1891 B, p. 460, pl. xxxiii, fig. 10.
      Permian; Indian Territory.

TURSEODUS Leidy.   Type *T. acutus* Leidy.

Leidy, J. 1857 F, p. 167.

Turseodus acutus Leidy.

Leidy, J. 1857 F, p. 167.

Wheatley, C. M. 1861 A, p. 44.
Woodward, A. S. 1891 A, p. 527.
      Triassic?; Pennsylvania.

Bull. 179——24

DICTYOPYGIDÆ.

Hay, O. P. 1899 E, p. 789.
Woodward, A. S. 1890 C, p. 1.   (Catopteridæ.)
1890 D, p. 15.   (Catopteridæ.)

Woodward, A. S. 1895 A, pp. vii, 1.   (Catopteridæ.)
1896 B, p. 87.   (Catopteridæ.)
Zittel, K. A. 1890 A, p. 202.   (Stylodontidæ, in part.)

REDFIELDIUS Hay.   Type *Catopterus gracilis* Redfield.

Hay, O. P. 1899 E, p. 789.
The following authors refer to this genus as
Catopterus. This name is, however, preoccupied through Agassiz's use of it for what is now called *Dipterus.*
Egerton, P. G. 1849 B, p. 8.
1853 B, p. 277.
Newberry, J. S. 1888 A, p. 50.
Nicholson and Lydekker 1889 A, p. 983.
Pictet, F. J. 1854 A, p. 187.
Redfield, J. H. 1837 A, p. 39.   (Catopterus, type *C. gracilis*; not Catopterus of Agassiz, 1833.)
Redfield, W. C. 1841 A, p. 26.
1856 A.
1857 A.
Traquair, R. H. 1877 A, p. 562.
Woodward, A. S. 1890 D, p. 15.
1895 A, pp. vii, 1.
1896 B, p. 88.
Zittel, K. A. 1890 A, p. 203.

Redfieldius anguilliformis (W. C. Redfield).

Redfield, W. C. 1841 A, p. 27.   (Catopterus.)
De Kay, J. E. 1842 B, p. 386.   (Catopterus.)
Egerton, P. G. 1849 B, p. 10.   (Catopterus.)
Hay, O. P. 1899 E, p. 789.   (Redfieldius.)
Newberry, J. H. 1888 A, p. 59, pl. xviii, fig. 5.   (Catopterus).
Woodward, A. S. 1895 A, p. 3.   (Catopterus.)
Triassic; Connecticut, New Jersey.

Redfieldius gracilis (J. H. Redfield).

Redfield, J. H. 1837 A, p. 39, pl. 1.   (Catopterus.)
The following authors, unless otherwise noted, put this species in *Catopterus* Redfield.
"D. L. H." 1838 A, p. 199.
Dana, J. D. 1896 A, p. 751, fig. 1170.
Davis, C. H. S. 1887 A.
De Kay, J. E. 1842 B, p. 386.
Egerton, P. G. 1849 B, p. 10.
Hay, O. P. 1899 E, p. 789.   (Redfieldius.)
Hills, R. C. 1880 A.
Hitchcock, E. 1841 A, pp. 440, 460.

Newberry, J. S. 1888 A, p. 55, pl. xvi, figs. 1–3.
Redfield, W. C. 1838 B.
1839 A.
1841 A, p. 27.
1857 A.
Wheatley, C. M. 1861 A, p. 44.
Woodward, A. S. 1895 A, p. 2.
Triassic; Connecticut, Massachusetts, New Jersey.

Redfieldius minor (Newb.).

Newberry J. S. 1888 A, p. 57, pl. xvii, figs. 1–4.   (Catopterus.)
Hay, O. P. 1899 E, p. 789.
Woodward, A. S. 1895 A, p. 3.   (Catopterus.)
Triassic; Connecticut.

Redfieldius ornatus (Newb.).

Newberry. J. S. 1888 A, p. 58, pl. xviii, figs. 3–3b.   (Catopterus.)
Hay, O. P. 1899 E, p. 789.
Woodward, A. S. 1895 A, p. 3.   (Catopterus.)
Triassic; Connecticut.

Redfieldius parvulus (W. C. Redfield).

Redfield, W. C. 1841 A, p. 28.   (Catopterus.)
De Kay, J. E. 1842 B, p. 386.   (Catopterus.)
Egerton, P. G. 1849 B, p. 10.   (Catopterus.)
Hay, O. P. 1899 E, p. 789.
Newberry, J. S. 1888 A, p. 60, pl. xvi, figs. 4, 5.   (Catopterus.)
Woodward, A. S. 1895 A, p. 2.   (Catopterus.)
Triassic; Connecticut.

Redfieldius redfieldi (Egerton.).

Egerton, P. G., in Lyell, C. 1847 B, p. 278.   (Catopterus.)
1849 B, p. 8.   (Catopterus.)
Hay, O. P. 1899 E, p. 789.
Newberry, J. S. 1888 A, p. 53, pl. xv, figs. 1–3.   (Catopterus.)
Woodward, A. S. 1895 A, p. 3.   (Catopterus.)
Triassic; Connecticut, New Jersey.

DICTYOPYGE Egerton.   Type *D. macrura* Redfield.

Egerton, P. G., in Lyell, C. 1847 B, p. 276.
1849 B, p. 8.
1853 B, p. 277.
Emmons, E. 1856 A, p. 52.
Newberry, J. S. 1888 A, p. 61.
Redfield, W. C. 1856 A, p. 361.
Struver, J. 1864 A, p. 322.
Traquair, R. H. 1877 A, p. 563.
Woodward, A. S. 1890 D, p. 16.
1895 A, p. 4.

Woodward, A. S. 1898 B, p. 88.
Zittel, K. A. 1890 A, p. 208.

Dictyopyge macrura (W. C. Redfield).

Redfield, W. C. 1841 A, p. 27.   (Catopterus.)
De Kay, J. E. 1842 B, p. 386.   (Catopterus.)
Egerton, P. G., in Lyell, C. 1847 B, p. 275, pl. viii; pl. ix, fig. 1.
Emmons, E. 1857 A, p. 52, pl. ix, figs. 1, 2.   (Catopterus.)

Newberry, J. S. 1888 A, p. 64, pl. xviii, figs. 1, 2.
Redfield, W. C. 1841 A, p. 27. (Catopterus.)
    1856 A, p. 361. (Catopterus.)
    1857 A, p. 185. (Catopterus.)

Woodward, A. S. 1890 D, p. 16.
    1895 A, p. 4, fig. 1.
    Triassic; Virginia.

ACIPENSERIDÆ.

Agassiz, L. 1843 B (1844), ii, pt. ii, p. 277. ("Acipenserides.")
Bridge, W. J. 1879 A.
    1896 A, p. 534.
Brühl, B. C. 1847 A.
Cope, E. D. 1872 C, p. 320.
Davis, J. W. 1877 B.
Egerton, P. G. 1859 B.
Gegenbaur, C. 1898 A.
Göldi, E. A. 1883 A.
    1884 A.
Goette, A. 1878 A, p. 442.
Günther, A. 1870 A, p. 332.
    1880 A, p. 360.
    1881 A, p. 687.
Hertwig, O. 1876 A, p. 878.

Iwanzow, N. 1887 A. (Scaphorhynchus.)
Jordan and Evermann 1896 A, p. 102.
Klaatsch, H. 1890 A, p. 146.
Kölliker, A. 1859 A. (Acipenserini.)
Köstlin, O. 1844 A, p. 402. ("Stör.")
Kowmak, G W. 1895 A, p. 11.
Müller, J. 1835 A.
Newberry, J. S. 1873 A, p. 256.
Parker, W. K. 1882 B.
Pollard, H. B. 1892 A.
Stannius, H. 1854 A.
Thacher, J. K. 1877 A.
    1877 B.
Woodward, A. S. 1889 H.
    1895 A, p. 42.
Zittel, K. A. 1890 A, p. 163.

ACIPENSER Linn.   Type *A. sturio* Linn.

Linnæus, C., 1758, Syst. Nat., ed. x, p. 237.
Allis, E. P. 1900 A, p. 274.
Bridge, W. J. 1879 A.
Brühl, B. C. 1847 A.
Dean, B. 1895 A.
Gadow and Abbott 1895 A.
Gegenbaur, C. 1865 B, p. 95, pl. vi, fig. 1; pl. viii, figs. 2, 3.
    1895 A, fig. 1.
    1898 A.
Goette, A. 1878 A, p. 442.
Günther, A. 1880 A, p. 361.
Hay, O. P. 1895 A.
Huxley, T. H. 1872 A, p. 121.
Iwanzow, N. 1887 A. (On Scaphorhynchus.)
Jordan and Evermann 1896 A, p. 103.
Kittary, M. 1850 A, p. 389, pls. vi, vii.
Kölliker, A. 1860 B.
Köstlin, O. 1844 A.
Kowmak, G. W. 1895 A, p. 1, pls. xlvi–xlviii.
Maggi, L. 1898 E, fig. 21.

Metschnikoff, O. 1879 A, pl. xxiv, figs. 1–7.
Owen, R. 1866 A.
Parker, W. K. 1868 A.
    1882 B.
Salensky, W. 1899 A, p. 178.
Stannius, H. 1854 A.
Stephan, P. 1900 A.
Thacher, J. K. 1877 B, pl. i, figs. 1, 2.
Van Wijhe, J. W. 1882 A, p. 220.
Williamson, W. C. 1849 A, p. 448, pl. lxii, fig. 11.
Woodward, A. S. 1889 H.
    1895 A, p. 43.
Zograf, N. 1896 A.

Acipenser ornatus Leidy.

Leidy, J. 1873 C.
    1873 A, p. 312.
    1873 B, p. 350, pl. xxxii, fig. 58.
Woodward, A. S. 1895 A, p. 45.
Miocene; Virginia.

POLYODONTIDÆ.

Bridge, W. J. 1879 A.
    1896 A, p. 586.
Browne, M. 1886 A. (Chondrosteus.)
Brühl, B. C. 1847 A. ("Spatularien.")
Collinge, W. E. 1895 A.
Gegenbaur, C. 1865 B, p. 99, pl. viii, fig. 4. (Polyodon.)
    1898 A.
Günther, A. 1870 A, p. 346.
    1880 A, p. 362.
Huxley, T. H. 1872 A, p. 122.
Jordan and Evermann 1896 A, p. 101.

Mivart, St. G. 1879 A, p. 471.
Müller, J. 1835 A. (Spatularia.)
Parker, W. K. 1871 A, p. 195.
Thacher, J. K. 1877 B.
Traquair, R. H. 1877 B.
    1887 A. (Chondrosteus.)
Van Wijhe, J. W. 1882 A, p. 240
Woodward, A. S. 1889 H.
    1895 A, pp. vii, 46.
    1899 B, p. 461. (Gyrosteus)
Zittel, K. A. 1890 A, p. 164.

CROSSOPHOLIS Cope.   Type *C. magnicaudatus* Cope.

Cope, E. D. 1883 G G.   (Crassopholis, *errore*.)
  1885 Y, p. 1090.
  1886 T, p. 161.
  1887 S, p. 325.
Woodward, A. S. 1889 H, p. 29.
  1895 A, p. 46.
  1898 B, p. 94.
Zittel, K. A. 1890 A, p. 165.

Crossopholis magnicaudatus Cope.
Cope, E. D. 1883 GG.
  1885 Y, p. 1090.
  1886 T, p. 161, pl. i, figs. 1–3.
Woodward, A. S. 1895 A, p. 47.
Eocene; Wyoming.

Order PYCNODONTI.

Agassiz, L. 1843 B, ii, pt. ii, p. 181. ("Pycnodontes.")
Cornuel, J. 1877 A, p. 609. ("Pycnodontes.")
Davis, J. W. 1887 C, p. 495. (Pycnodontoidea.)
Didelot, L. 1875 A. ("Pycnodontes.")
Egerton, P. G. 1849 A.
  1869 A, p. 381.
Fricke, K. 1875 A.
Günther, A. 1840 A, p. 366. (Pycnodontoidei.)
Heckel, J. J. 1850 A. ("Pycnodonten.")
  1854 A. ("Pycnodonten.")
  1856 A. ("Pycnodonten.")
Huxley, T. H. 1876 B, xiii, p. 411. (Pycnodonta.)

Lütken, C. 1868 B. (Lepidopleuridæ.)
  1873 A, pp. 30, 45. (Pycnodontes.)
Owen, R. 1845 B, p. 70. ("Pycnodonta.")
  1860 E, p. 160. (Pycnodontes.)
Pictet, F. J. 1854 A, p. 194. (Pycnodontes.)
Pictet and Jaccard 1860 A, p. 42. (Pycnodontes.)
Quenstedt, F. A. 1852 A, p. 209. ("Pleurolepiden.")
  1885 A, p. 325. ("Pleurolepiden.")
Wagner, A. 1861 B, p. 49. (Pycnodontes.)
Young, 1866 A, p. 315. (Lepidopleuridæ. part.)
Zittel, K. A. 1890 A, p. 237. (Pycnodontidæ.)

PYCNODONTIDÆ.

Cope, E. D. 1877 N, p. 61.
  1878 A, pp. 297, 298.
  1885 Z, p. 7.
Davis, J. W. 1887 C, p. 495.
Günther, A. 1880 A, p. 366.
  1881 A, p. 687.
Hancock and Howes 1872 A. ("Pycnodonta.")
Hay, O. P. 1898 A, p. 358.
  1898 C, p. 347.
Heckel, J. J. 1856 A.
Huxley, T. H. 1872 A, p. 129.
King, W. 1850 A, p. 227.
Klaatsch, H. 1890 A, p. 149.
Newberry, J. S. 1873 A, p 258.

Nicholson and Lydekker 1889 A, p. 983.
Owen, R. 1845 B, p. 70. ("Pycnodonts.")
  1860 E, p. 140.
Pictet, F. J. 1854 A, p. 194. (Pycnodontes.)
Steinmann and Doderlein 1890 A, p. 543.
Traquair, R. H. 1879 A, p. 380. (Pycnodontidæ.)
  1896 A, p. 125.
Woodward, A. S. 1895 A, pp. xi, 189. *
  1896 C, p. 1.
  1898 B, p. 101.
Young, J. 1866 A.
Zittel, K. A. 1890 A, p. 236.
  1896 A, p. 411.

TYPODUS Quenstedt.   Type *T. annulatus* Quenstedt.

Quenstedt, F. A., 1858, Der Jura, p. 781.
Egerton, P. G. 1869 A, p. 582. (Mesodon.)
Hay, O. P. 1899 E, p. 789.
Heckel, J. J. 1850 A, p. 365. (Mesodon.)
  1856 A, 202. (Mesodon.)
Pictet, F. J. 1854 A, p. 203. (Mesodon.)
Pictet and Jaccard 1860 A, p. 45. (Mesodon.)
Quenstedt, F. A. 1852 A, p. 330. (Mesodon.)
Wagner, A. 1851 A, p. 56. (Mesodon; type *Gyrodus macropterus* Agassiz.)
  1861 B, p. 69.
Woodward, A. S. 1895 A, p. 199. (Mesodon.)
  1896 C, p. 13. (Mesodon.)
  1898 B, p. 104. (Mesodon.)
Zittel, K. A. 1890 A, p. 247. (Mesodon.)
  The name *Mesodon* was employed by Rafinesque for a mollusk in 1819. According to Mr. A. S. Woodward, *Typodus* is probably a

synonym of *Mesodon*, and is here accepted provisionally as the name of this genus.

Typodus abrasus (Cragin).
Cragin, F. W. 1894 A, p. 72, pl. ii, figs. 18, 19. (Mesodon.)
Hay, O. P. 1899 E, p. 789.
Williston, S. W. 1900 A, p. 29. (Mesodon.)
  1900 B, p. 256. (Mesodon abrasus; Lepidotus? sp.)
Woodward, A. S. 1895 A, p. 532. (Mesodon.)
Cretaceous (Neocomian); Kansas.

Typodus diastematicus (Cope).
Cope, E. D. 1894 F, p. 443, fig. 5. (Mesodon.)
Hay, O. P. 1899 E, p. 789.
Woodward, A. S. 1895 A, p. 213. (Mesodon.)
Cretaceous; Texas.

Typodus dumblei (Cope).

Cope, E. D. 1892 J, p. 128. (Microdon.)
1892 E, p. 256. (Microdon.)
1894 F, p. 444, pl. xx, fig. 7. (Mesodon.)

Hay, O. P. 1899 E, p. 789.
Woodward, A. S. 1895 A, p. 213. (Mesodon.)
Cretaceous; Texas.

CŒLODUS Heckel. Type *C. saturnus* Heckel.

Heckel, J. J. 1856 A, p. 202.
Cope, E. D. 1894 F, p. 447.
Davis, J. W. 1890 A, p. 415.
Priem, F. 1898 A, p. 230.
Quenstedt, F. A. 1885 A, p. 330.
Vetter, B. 1881 A, p. 34.
Woodward, A. S. 1895 A, p. 249.
Woodward and Sherborn 1890 A, p. 41.
Zittel, K. A. 1890 A, p. 248.

Williston, S. W. 1900 A, p. 28, pl. vi, fig. 12.
1900 B, p. 254, pl. xxiv, fig. 12.
Woodward, A. S. 1895 A, p. 257.
Cretaceous (Comanche); Oklahoma.

Cœlodus stantoni Willist.

Williston, S. W. 1900 A, p. 28, pl. vi, fig. 11; pl. viii, fig. 6.
1900 B, p. 255, pl. xxiv, fig. 11; pl. xxvi, fig. 6.
Cretaceous (Kiowa shales); Kansas.

Cœlodus brownii Cope.

Cope, E. D. 1894 F, p. 447, pl. xx, fig. 10.
1894 A, p. 66.

PYCNODUS Agassiz. Type *Zeus platessus* Blainv.

Agassiz, L. 1848 B (1833), i, pl. G, fig. 1. (Name and figure only.)
1843 B, ii, pt. 1 (1833), p. 16; pt. ii (1844), p. 183 (Pycnodus); p. 201 (Periodus).
1844 A.
Bronn, H. G. 1837 A, p. 494, pl. xxv, fig. 3.
Cope, E. D. 1878 A, p. 297.
Cornuel, J. 1877 A, p. 620.
Didelot, L. 1875 A.
Egerton, P. G. 1849 A.
Heckel, J. J. 1850 A, p. 365.
1856 A, p. 204.
Leidy, J. 1873 B, p. 292.
King, W. 1850 A, p. 228.
Nicholson and Lydekker 1889 A, p. 984.
Owen, R. 1845 B, p. 70, pl. xxxiv.
Pictet, F. J. 1854 A, p. 197.
Pictet and Campiche 1858 A, p. 55.
Quenstedt, F. A. 1852 A, p. 212.
1885 A, p. 329.
Vetter, B. 1881 A, p. 25.
Wagner, A. 1851 A, pp. 35, 57.
Woodward, A. S. 1895 A, p. 275.
Zittel, K. A. 1890 A, p. 250.

Pycnodus carolinensis Emmons.

Emmons, E. 1858 B, p. 244, fig. 96.
Cope, E. D. 1875 V, p. 30.
Leidy, J. 1873 B, pp. 294, 350.
Woodward, A. S. 1895 A, p. 279.
Miocene; North Carolina.

Pycnodus phaseolus Hay.

Hay, O. P. 1899 E, p. 788.
Cope E. D. 1875 E, p. 280. (P. faba.)
Leidy, J. 1872 M, p. 163. (P. faba; preoccupied by Meyer, 1847.)
1873 B, pp. 292, 349, pl. xix, figs. 15, 16. (P. faba.)
Woodward, A. S. 1895 A, p. 265. (P. faba.)
Cretaceous; Mississippi, New Jersey.

Pycnodus robustus Leidy.

Leidy, J. 1857 F, p. 168.
1873 B, pp. 293, 350, pl. xxxvii, figs. 18, 19.
Woodward, A. S. 1895 A, p. 266.
Cretaceous; New Jersey.

Pycnodus sp. indet.

Emmons E. 1857 A, p. 50, fig. 25.
Triassic (Chatham); North Carolina.

HADRODUS Leidy. Type *H. priscus* Leidy.

Leidy, J. 1857 F, p. 167.
1873 B, p. 294.
This genus was suspected by Leidy to be reptilian, allied to *Placodus*.

Hadrodus priscus Leidy.

Leidy, J. 1857 F, p. 167.
1873 B, pp. 294, 350, pl. xix, figs. 17-20.
Cretaceous; Mississippi.

URANOPLOSUS Sauvage. Type *U. cotteaui* Sauvage.

Sauvage, H. E. 1879 A, p. 47, pl. i, fig. 1.
Cope, E. D. 1894 F, p. 445.

Uranoplosus arctatus Cope.

Cope, E. D. 1894 F, p. 445, pl. xx, fig. 8.
1894 A, p. 65. (No description.)
Cretaceous (Comanche); Oklahoma.

Uranoplosus flectidens Cope.

Cope, E. D. 1894 F, p. 446, pl. xx, fig. 9.
1894 A, p. 65. (No description.)
Cretaceous (Comanche); Oklahoma.

374   FOSSIL VERTEBRATA OF NORTH AMERICA.   [BULL. 179.

Order HOLOSTEI Müller.

Müller, J. 1846 C, pp. 147, 203.
Gadow, H. 1898 A, p. 8.
Giebel, C. G. 1847 A, p. 141.

Gill, T. 1872 C, p. 21.   (Hyoganoidei.)
Jordan and Evermann 1896 A, p. 107.

Suborder GINGLYMODI Cope.

Cope E. D. 1872 C, p. 328.
Barkas, W. J. 1878 C, p. 203.   (Lepidosteidæ.)
Cope, E. D., 1872 C, p. 330.
   1878 A, p. 295.
   1884 O, p. 52.
   1887 O, p. 1018.
   1889 R, p. 857.
Gill, T. 1872C, p. 21.   (Rhomboganoidei.)
Günther, A. 1881 A, p. 687.   (Lepidosteoidei.)
Hay, O. P. 1898 C, p. 341.   (No name.)
Haeckel, E. 1895 A, p. 236.   (Lepidostones.)
Huxley, T. H. 1861 B, p. 28.   (Lepidosteidæ, as suborder.)

Jordan and Evermann 1896 A, p. 108.   (Rhomboganoidea.)
Lütken, C. 1873 A, pp. 23, 45.   (Lepidosteini.)
Nicholson and Lydekker 1889 A, p. 980.   (Lepidosteoidea.)
Steinmann and Döderlein 1890 A, p. 569.   (Euganoidei.)
Traquair, R. H. 1896 A, p. 126.   (Aëtheospondyli.)
Woodward, A. S. 1893 A, p. 413.   (Lepidosteoidei.)
   1895 A, pp. xxii, 415.   (Aëtheospondyli, in part.)
   1898 B, p. 11.   (Aëtheospondyli, in part.)
Zittel, K. A. 1890 A, p. 201 (Lepidosteidæ); p. 399 (Lepidostei).

LEPIDOTIDÆ.

Owen, R. 1860 E, p. 143.
Cope, E. D. 1877 N, p. 60.
   1878 A, p. 298.
   1889 R, p. 858. (Sphærodontidæ, Lepidotidæ.)
Günther, A. 1880 A, p. 368.   (Sphærodontidæ.)
Hay, O. P. 1898 A, p. 358.   (Semionotidæ.)
   1898 C, p. 347.   (Semionotidæ.)
Nicholson and Lydekker 1889 A, p. 986.

Wagner A. 1863 A, p. 613 (Stylodontes); p. 617 (Sphærodontes).
Williamson, W. C. 1849 A, p. 441.   (Lepidotus.)
Woodward, A. S. 1890 D, p. 30. (Semionotidæ.)
   1895 A, pp. ix, 49.   (Semionotidæ.)
   1895 E, p. 8.   (Semionotidæ.)
   1898 B, p. 95.   (Semionotidæ.)
Zittel, K. A. 1890 A, p. 202 (Stylodontidæ); p. 207 (Sphærodontidæ).

LEPIDOTUS Agassiz.   Type *Cyprinus elvensis* Blainville.

Agassiz, L. 1843 B (1837), ii, pp. 8, 233.
   1844 A.
Barkas, W. J. 1878 C, p. 205.
Branco, W. 1887 A.
Bronn, H. G. 1837 A, 485.
Collinge, W. E. 1893 A, p. 9.
Cornuel, J. 1877 A, p. 620.
Egerton, P. G. 1852 A.
   1869 A, p. 385.
Etheridge, R. 1889 A.
Fricke, K. 1875 A, p. 377.
Hay, O. P. 1898 C, p. 347.
Heckel, J. J. 1850 A, p. 362.
Kner, R. 1866 B, p. 313.
Nicholson and Lydekker 1889 A, p. 986.
Owen, R. 1845 B, p. 69, pls. 30, 31.

Pictet, F. J. 1854 A, p. 160.
Pictet and Jaccard 1860 A, p. 26.
Quenstedt, F. A. 1847 A, pls. i, ii.
   1852 A, p. 195.
   1885 A, p. 307.
Wagner A. 1863 A, p. 618.
Williamson, W. C. 1849 A, p. 441, pl. xl, figs. 3, 4.
Woodward, A. S. 1893 A.
   1893 D, pls. xlix, l.
   1895 A, p. 77.
   1898 A, p. 325.
Zittel, K. A. 1890 A, p. 208.

Lepidotus sp. indet.
Williston, S. W. 1900 A, p. 29.
   Cretaceous (Kiowa shales): Kansas.

ACENTROPHORUS Traquair.   Type *A. varians* (Kirkby).

Traquair, R. H. 1877 A, p. 565.
Fritsch, A. 1895 A, p. 81.
Lutken, C. 1873 A, p. 26, footnote.
Newberry, J. S. 1888 A, p. 67
Traquair, R. H. 1900 B, p. 523.
Woodward, A. S. 1895 A, p. 51.
Zittel, K. A. 1890 A, p. 203.

Acentrophorus chicopensis Newb.
Newberry, J. S. 1888 A, p. 69, pl. xix, figs. 3, 4.
   1887 A, p. 127.   (No description).
Woodward, A. S. 1895 A, p. 54.
   Triassic; New Jersey, Connecticut.

SEMIONOTUS Agassiz.   Type *S. bergeri* Agassiz.

Agassis, L., 1832, Neues Jahrb. Mineral., p. 144.
 1843 B (1837), ii, pt. 1, pp. 8, 222.
Barkas, W. J. 1878 C, p. 205.
Bronn, H. G. 1837 A, p. 485, pl. xxiv, fig. 4.
Egerton, P. G., in Lyell, C. 1847 B.   (Ischypterus;
 type *Palæoniscus fultus*.)
 1849 B, p. 8.   (Ischypterus.)
Fraas, O. 1861 A.
Kner, R. 1866 B, p. 318.
Newberry, J. S. 1888 A, p. 24.
Nicholson and Lydekker 1889 A, p. 982.
Owen, R. 1845 B, p. 69.
Pictet, F. J. 1854 A, p. 163.
Quenstedt, F. A. 1852 A, p. 205.
 1885 A, p. 320.
Redfield, W. C. 1853 A, p. 271.   (Ischypterus.)
 1856 A, p. 362.   (Ischypterus, in part.)
 1857 A.   (Ischypterus, in part.)
Strüver, J. 1864 A.
Traquair, R. H. 1877 A.   (Ischypterus.)
Williamson, W. C. 1849 A, p. 444.
Woodward, A. S. 1895 A, p. 55.
 1898 B, p. 96.
Zittel, K. A. 1890 A, p. 208 (Ischypterus); p. 204
 (Semionotus).

Semionotus agassizii (W. C. Redfield).

Redfield, W. C. 1841 A, p. 26. (Palæoniscus.)
De Kay, J. E. 1842 B, p. 386.
Egerton, P. G. 1849 B, p. 10.   (Ischypterus.)
Giebel, C. G. 1848 A, p. 245.   (Palæoniscus.)
Newberry, J. S. 1888 A, p. 30, pl. iii, fig. 1.
 (Ischypterus.)
Redfield, W. C. 1843 A, p. 185.
Traquair, R. H. 1877 A, p. 559.   (Ischypterus.)
Woodward, A. S. 1891 A, p. 486.
 1895 A, p. 62.
Triassic; New Jersey.

Semionotus alatus (Newb.).

Newberry, J. S. 1888 A, p. 37, pl. viii, figs. 1, 2.
 (Ischypterus.)
 1887 A.   (Ischypterus; no description.)
Woodward, A. S. 1895 A, p. 63.
Triassic; New Jersey.

Semionotus braunii (Newb.).

Newberry, J. S. 1888 A, p. 43, pl. xii, fig. 3; pl. xiii,
 figs. 1, 2.   (Ischypterus.)
Gratacap, L. P. 1886 A, p. 243, figure.   (Palæoniscus
 latus, *errore*.)
Newberry, J. S. 1887 A.   (Ischypterus braunii; no
 description.)
Woodward, A. S. 1895 A, p. 63.
Triassic; New Jersey.

Semionotus elegans (Newb.).

Newberry, J. S. 1888 A, p. 37, pl. vii, fig. 2; pl. x,
 fig. 1; pl. xiv, figs. 1, 2.   (Ischypterus.)
 1887 A.   (Ischypterus, no description.)
Woodward, A. S. 1895 A, p. 63.
Triassic; New Jersey.

Semionotus fultus (J. H. Redfield.)

Redfield, J. H. 1837 A, p. 37.

Agassiz, L. 1843 B, ii, pp. 43, 102, pl. viii, figs. 4, 5.
De Kay, J. E. 1842 B, p. 385.
Egerton, P. G., in Lyell, C. 1847 B, p. 277.
 1849 B, p. 10.   (Ischypterus.)
Fraas, E. 1861 A.
Giebel, C. G. 1848 A, p. 343.   (Palæoniscus.)
Hitchcock, E. 1841 A, p. 459.   (Palæoniscus.)
Newberry, J. S. 1888 A, p. 35, pl. vi, fig. 2; pl. vii,
 fig. 1.   (Ischypterus.)
 1888 A, p. 41, pl. xii, fig. 1.   (I. macropterus.)
Redfield, W. C. 1841 A, p. 25.   (Palæoniscus ma-
cropterus.)
 1843 A, p. 135.   (Palæoniscus.)
 1857 A, p. 184.   (Palæoniscus.)
Traquair, R. H. 1877 A, p. 559.   (Ischypterus.)
Woodward, A. S. 1891 A, p. 486.
 1895 A, p. 58.
Zittel, K. A. 1890 A, p. 208.   (I. macropterus,
 I. fultus.)
Triassic; Massachusetts, Connecticut, New
Jersey.

Semionotus gigas (Newb.).

Newberry, J. S. 1888 A, p. 49, pl. xiv, fig. 3.
 (Ischypterus.)
 1887 A.   (Ischypterus; no description.)
Woodward, A. S. 1895 A, p. 63.
Triassic; New Jersey.

Semionotus lenticularis (Newb.).

Newberry, J. S. 1888 A, p. 39, pl. x, figs. 2, 3.
 (Ischypterus.)
 1887 A.   (Ischypterus no description.)
Woodward, A. S. 1895 A, p. 63.
Triassic; New Jersey.

Semionotus lineatus (Newb.).

Newberry, J. S. 1888 A, p. 40, pl. xi, figs. 1, 2.
 (Ischypterus.)
Woodward, A. S. 1895 A, p. 63.
Triassic; New Jersey.

Semionotus marshii (W. C. Redfield).

Redfield, W. C., in Newberry, J. S. 1888 A, p. 28,
 pl. ii, fig. 1.   (Ischypterus.)
 1856 A, p. 363.   (No description.)
 1857 A, p. 188.   (No description.)
Woodward, A. S. 1895 A, p. 63.
Triassic; Massachusetts.

Semionotus micropterus (Newb.).

Newberry, J. S. 1888 A, p. 2, pl. xii, fig. 2.   (Ischypterus.)
 1887 A.   (Ischypterus; no description.)
Woodward, A. S. 1895 A, p. 63.
Triassic; Connecticut.

Semionotus minutus (Newb.).

Newberry, J. S. 1888 A, p. 48, pl. xiii, figs. 5, 5a.
 (Ischypterus.)
 1887 A.   (Ischypterus; no description.)
Woodward, A. S. 1895 A, p. 63.
Triassic; Connecticut.

Semionotus modestus (Newb.).

Newberry, J. S. 1888 A, p. 38, pl. ix, figs. 1, 3. (Ischypterus.)

1887 A. (Ischypterus; no description.)
Woodward, A. S. 1896 A, p. 63.
Triassic; New Jersey.

Semionotus ovatus (W. C. Redfield).

Redfield, W. C. 1841 A, p. 26.
De Kay, J. E. 1842 B, p. 386.
Egerton, P. G., in Lyell, C. 1847 B, p. 277, pl. ix, fig. 2. ("Tetragonolepis.")
Newberry, J. S. 1888 A, p. 27, pl. i, fig. 1. (Ischypterus.)
Traquair, R. H. 1877 A, p. 559. (Ischypterus.)
Woodward, A. S. 1891 A, p. 486.
1895 A, p. 63.
Triassic; Massachusetts, New Jersey.

Semionotus parvus (W. C. Redfield).

Redfield, W. C., in Newberry, J. S. 1888 A, p. 45, pl. xiii, fig. 4. (Ischypterus.)
Hitchcock, E. 1841 A, pl. xix. (No name.)
Newberry, J. S. 1887 A. (Ischypterus; no description.)
Woodward, A S 1895 A, p. 63.
Triassic; Massachusetts.

Semionotus robustus (Newb.).

Newberry, J. S. 1888 A, p. 36, pl. vi, fig. 1. (Ischypterus.)
1887 A. (Ischypterus; no description.)

Woodward, A. S. 1895 A, p. 63.
Triassic; New Jersey.

Semionotus tenuiceps (Agassiz).

Agassiz, L. 1843 B (1835), ii, pt. i, pp. 159, 303, pl. xivc, figs. 4, 5. (Eurynotus.)
"D. L. H." 1838 A, p. 199. (Palæoniscus latus.)
Deecke, W., Palæontographica, xxxv, p. 14. (Allolepidotus americanus.)
Egerton, P. G. 1849 B, p. 10. (Ischypterus.)
Emmons, E. 1857 A, p. 143, figs. 1, 2 (Ischypterus macropterus); p. 144, pl. ixa, fig. 3; text fig. 112 (Eurinotus tenuiceps, var. ceratocephalus).
1860 A, p. 188, fig. 164. (Eurinotus ceratocephalus.)
Giebel, C. G. 1848 A, p. 286.
Hitchcock, E. 1841 A, p. 459, pl. xxix, figs. 1 and 2. (Eurynotus.)
Newberry, J. S. 1888 A, p. 32, pl. v, figs. 1–3; pl. vii, fig. 3; p. 46, pl. xiii, fig. 3. (S. latus.)
Redfield, J. H. 1837 A, p. 39 (Eurynotus tenuiceps); p. 38, pl. ii (Palæoniscus latus).
Redfield, W. C. 1841 A, p. 26. (Palæoniscus latus.)
1843 A, p. 135. (P. latus.)
1856 A, p. 362. (P. latus.)
Traquair, R. H. 1877 A, p. 559. (Ischypterus.)
Woodward, A. S. 1891 A, p. 531 (S. tenuiceps); p. 486 (Palæoniscus latus).
1895 A, p. 59.
Triassic; Massachusetts, Connecticut, New Jersey.

LEPISOSTEIDÆ.

Beard, J. 1890, A, p. 183.
Brühl, B. C. 1847 A. (Lepidosteus.)
Cope, E. D. 1872 C, p. 330.
1878 A, p. 295.
1884 O, p. 52.
1887 O, p. 1018.
1889 R, p. 857.
Günther, A. 1870 A, p. 328.
1880 A, p. 367.
1881 A, p. 687.
Hay, O. P. 1895 A.
Hertwig, O. 1879 A.
Huxley, T. H. 1861 B, p. 28. (Lepidosteini.)

Jordan and Evermann 1896 A, p. 108.
Klaatsch, H. 1890 A, p. 125.
Kölliker, A. 1859 A. (Lepidosteini.)
Lütken, C. 1868 B, p. 430.
Müller, J. 1846 A, p. 208. (Lepidosteini.)
1846 B, p. 85. (Lepidosteini.)
Newberry, J. S. 1873 A, p. 255.
Nicholson and Lydekker 1889 A, p. 968.
Stannius, H. 1854 A. (Lepidosteini.)
Ussow, S. 1900 A. ("Lepidostien.")
Woodward, A. S. 1895 A, pp. xviii, 140.
Zittel, K. A. 1890 A, p. 222. (Ginglymodi.)

Lepisosteus Lacépède.   Type *L. gavialis=L. osseus.*

Lacépède, 1803, Hist. Nat. Poissons, vol. v, p. 331. Most authors employ the orthography *Lepidosteus.* The original is given above.
Agassiz, L. 1843 B, ii, pt. ii, p. 4. (Lepidosteus.)
Allis, E. P. 1900 A, p. 273.
Andreæ, A. 1893 B. (Lepidosteus, Clastes.)
Balfour, F. M. 1882 A.
Baur, G. 1889 E.
1889 F, p. 469.
Beard, J. 1890 A, p. 182.
Bridge, T. W. 1896 A, p. 540.
Brühl, B. C. 1847 A.
Cope, E. D. 1869 F, p. 242. (Pneumatosteus, type *P. nahunticus.*)
1872 C, p. 319.
1873 E, p. 633. (Clastes, type *C. anax.*)
1875 C, p. 37. (Clastes.)

Cope, E. D. 1875 V, p. 30. (Pneumatosteus.)
1881 D, p. 184. (Clastes.)
1884 O, p. 52 (Pneumatosteus); p. 53 (Clastes).
Davidoff, M. 1880 A.
1880 B.
Eastman, C. R. 1900 A, p. 68.
Gadow and Abbott 1895 A.
Gegenbaur, C. 1865 B, p. 104, pl. vi, fig. 5; pl. viii, fig. 5.
1895 A, p. 6.
Goette, A. 1878 A, p. 453.
Günther, A. 1870 A, p. 328.
1880 A, p. 367.
Hay, O. P. 1895 A.
1896 B, p. 959.
1898 C, p. 345.

Hertwig, O. 1879 A, pls. i, ii.
Huxley, T. H. 1872 A, p. 124.
1883 A.
Klaatsch, H. 1890 A, p. 125.
Kosmak, G. W. 1895 A, p. 8.
Mackintosh, H. W. 1878 A, pl. ii, fig. 6.
Marsh, O. C. 1898 E, p. 567.
Müller, J. 1846 A.
1846 B.
1846 C.
Nickerson, W. S. 1893 A.
Nicholson and Lydekker 1889 A, p. 968. (Pneumatosteus.)
Owen, R. 1845 B, p. 74, pl. xxxv.
Parker, W. K. 1882 A.
Reisner, E. 1859 A, p. 254, pl. v.
Röse, C. 1894 B.
Stannius, H. 1846 A.
1854 A.
Stephan, P. 1900 A.
Thacher, J. K. 1877 B, pl. ii, fig. 3.
Traquair, R. H. 1896 A, p. 126.
1900 B, p. 523.
Van Wijhe, J. W. 1882 A, p. 265.
White C. A. 1883 A. (Clastes.)
Williamson, W. C. 1849 A, p. 439, pl. xl, figs. 1, 2.
Woodward, A. S. 1895 A, p. 440 (Lepidosteus): p. 445 (Clastes).
Zittel, K. A. 1890 A, p. 222. (Lepidosteus, Clastes.)
Zograf, N. 1896 A.

Lepisosteus aganus (Cope).

Cope, E. D. 1877 K, p. 38, pl. xxiii, figs. 10-29 (Clastes.)
Eocene (Wasatch); New Mexico.

Lepisosteus atrox Leidy.

Leidy, J. 1873 D, p. 97.
Cope, E. D. 1873 E, p. 638 (Clastes anax): p. 634 (C. atrox).
1884 O, p. 53, pl. ii, figs. 50-52 (Clastes anax): p. 54, pl. ii, figs. 1, 2 (C. atrox).
Eastman, C. R. 1900 A, p. 69, pl. i, fig. 2; pl. ii. (Lepidosteus.)
1900 E, p. 57. (Lepidosteus.)
Leidy, J. 1873 B, pp. 189, 349, pl. xxxii, figs. 14, 15.
Woodward, A. S. 1895 A, p. 444 (C. atrox); p. 445 (C. anax).
Eocene (Bridger); Wyoming.

Lepisosteus cuneatus (Cope).

Cope, E. D. 1884 O, p. 55, pl. i, fig. 6. (Clastes.)
1880 T, p. 303. (Clastes.)
Eastman, C. R. 1900 A, p. 68. (Lepidosteus.)
1900 E, p. 57. (Lepidosteus.)
Eocene (Manti beds); Utah.

Lepisosteus cycliferus (Cope).

Cope, E. D. 1873 E, p. 634. (Clastes.)
1877 K, p. 40. (Clastes.)
1884 O, p. 54, pl. ii, figs. 25-45. (Clastes.)
Eastman, C. R. 1900 A, p. 68. (Lepidosteus.)
Woodward, A. S. 1895 A, p. 445. (Clastes.)
Eocene (Bridger); Wyoming.

Lepisosteus glaber Marsh.

Marsh, O. C. 1871 G, p. 105.
Cope, E. D. 1873 E, p. 634. (Clastes.)
1884 O, p. 53.
Eastman, C. R. 1900 A, p. 67.
King, C. 1878 A, p. 376 (Clastes?); p. 405.
Woodward, A. S. 1895 A, p. 444.
Eocene (Bridger); Wyoming.

Lepisosteus integer (Cope).

Cope, E. D. 1877 K, p. 41, pl. xxiv, figs. 1-16. (Clastes.)
Eocene (Wasatch); New Mexico.

Lepisosteus nahunticus (Cope).

Cope, E. D. 1869 F, p. 242. (Pneumatosteus.)
1875 V, p. 31. (Pneumatosteus.)
Eastman, C. R. 1900 A, p. 68. (Lepidosteus.)
Miocene; North Carolina.

Lepisosteus notabilis Leidy.

Leidy, J. 1873 D, p. 98.
Cope, E. D. 1877 K, p. 40. (Clastes.)
Eastman, C. R. 1900 A, p. 68. (Lepidosteus.)
Leidy, J. 1873 B, pp. 192, 349, pl. xxxi, figs. 12, 13.
Woodward, A. S. 1895 A, p. 444.
Eocene (Bridger); Wyoming.

Lepisosteus occidentalis Leidy.

Leidy, J. 1856 F, p. 73. (Lepidotus.)
Cope, E. D. 1877 G, p. 574. (Lepidosteus.)
1884 O, p. 52. (Clastes.)
Leidy, J. 1856 C, p. 120. (Lepidotus haydeni; L. occidentalis.)
1856 F, p. 73. (Lepidotus haydeni.)
1860 A, p. 149, pl. xl, figs. 20-23. (Lepidotus.)
Woodward, A. S. 1895 A, p. 125 (Lepidotus haydeni); p. 126 (L. occidentalis).
Cretaceous (Laramie); Wyoming.

Lepisosteus osseus (Linn.).

Linnæus, C. 1758, Syst. Nat., ed. x, p. 313. (Esox.)
Agassiz, L. 1843 B (1836), ii, pt. ii, p. i, pls. B', B''.
Jordan and Evermann 1896 A, p. 110.
Leidy, J. 1860 B, p. 118, pl. xxv, fig. 1. (L. bison?).
Stephan, P. 1900 A, p. 355, fig. 3.
Williamson, W. C. 1849 A, p. 439, pl. xl, figs. 1, 2.
Recent; Vermont to Texas: Post-pliocene: South Carolina.

Lepisosteus simplex Leidy.

Leidy, J. 1873 D, p. 98.
Cope, E. D. 1877 K, p. 40. (Clastes.)
Eastman, C. R. 1900 A, p. 74, pl. i, fig, 1.
Leidy, J. 1873 B, pp. 191, 349, pl. xxxii, figs. 18, 26, 31-43.
Woodward, A. S. 1895 A, p. 444.
Eocene (Bridger); Wyoming.

Lepisosteus whitneyi Marsh.

Marsh, O. C. 1871 G, p. 105.
Eastman, C. R. 1900 A, p. 67.
King, C. 1878 A, p. 405.
Woodward, A. S. 1895 A, p. 445.
Eocene (Bridger); Wyoming.

Suborder HALECOMORPHI Cope.[1]

Cope, E. D. 1872 C, p. 330.
1877 N, p. 58.
1878 A, p. 295.
1887 O, p. 1018.
1889 R, p. 857.
Gill, T. 1872 C, p. 21.   (Cycloganoidei.)
Günther, A. 1880 A, p. 370.
1881 A, p. 688.   (Amioidei.)
Haeckel, E. 1895 A, p. 236.   (Amiacei.)
Hay, O. P. 1898 C, p. 341.   (No name.)
Heckel, J. J. 1850 A.
Huxley, T. H. 1861 B, p. 24.   (Amiadæ, as suborder.)

Steinmann and Döderlein 1890 A, p. 576.   (Amioidei.)
Traquair, R. H. 1896 A, p. 125.   (Protospondyli, in part.)
Wagner, A. 1861 B, p. 43.   (No name.)
Woodward, A. S. 1893 A, p. 413.   ("Amioidei.")
1895 A, p. xxii;   p. 48.   (Protospondyli, in part.)
1898 B, p. 94.   (Protospondyli, i part.)
Zittel, K. A. 1890 A, pp. 222, 899.
1896 A, p. 410.   (Protospondyli.)

ISOPHOLIDÆ.

Hay, O. P. 1899 E, p. 790.
The generic name *Eugnathus*, being preoccupied, must be succeeded by *Isopholis* Zittel. The family name also must be changed accordingly.
Agassiz, L. 1843 B (1844), ii, p. 97.   (Eugnathus.)
Hay, O. P. 1898 C, p. 346.   (Eugnathidæ.)
Nicholson and Lydekker 1889 A, p. 986.   (Eugnathidæ.)

Quenstedt, F. A. 1852 A, p. 204.   (Eugnathus.)
1885 A, p. 319.   (Eugnathus.)
Wagner, A. 1863 A, p. 670.   (Eugnathus.)
Woodward, A. S. 1895 A, pp. xiv, 285.   (Eugnathidæ.)
1898 B, p. 106.   (Eugnathidæ.)
Zittel, K. A. 1890 A, p. 212.   (Saurodontidæ, in part.)

PTYCHOLEPIS Agassiz.   Type *P. bollensis* Agassiz.

Agassiz L., 1832, Neues Jahrb. Mineral., p. 132.   (*Fide* A. S. Woodward.)
1843 B (1844), ii, pt. ii, p. 107.
Bronn, H. G. 1837 A, p. 488.
Egerton, P. G. 1852 B.
Newberry, J. S. 1888 A, p. 65.
Quenstedt, F. A. 1852 A, p. 203.
1885 A, p. 318.
Steinmann, and Döderlein 1890 A, p. 572.
Williamson, W. C. 1849 A, p. 444.

Woodward, A. S. 1895 A, p. 316.
1895 D.
Zittel, K. A. 1890 A, p. 213.

Ptycholepis marshii Newb.

Newberry, J. S. 1878 B.
Lesley, J. P. 1889 A, p. 830, figure.
Newberry, J. S. 1888 A, p. 66, pl. xix, figs. 1-2a.
Woodward, A. S. 1895 A, p. 324.
Triassic; Connecticut.

MACROSEMIIDÆ.

Cope, E. D. 1889 R, p. 858.
Agassiz. L., 1834, Neues Jahrb. Mineral., p. 387.   (Macrosemius.)
Bronn, H. G. 1837 A, p. 492.   (Macrosemius.)

Hay, O. P. 1898 C, p. 347.
Wagner, A. 1851 A, p. 73.   (Macrosemius.)
Woodward, A. S. 1895 A, p. 163.

MACREPISTIUS Cope.   Type *M. arenatus* Cope.

Cope, E. D. 1894 F, p. 441.
Woodward, A. S. 1895 A, p. 173.

Macrepistius arenatus Cope.

Cope, E. D. 1894 F, p. 441, pl. xix, fig. 2.

Cope, E. D. 1894 A, p. 65.
Woodward, A. S. 1895 A, p. 173.
Cretaceous (Trinity); Texas.

PACHYCORMIDÆ.

Woodward, A. S. 1895 A, p. 374.
Cope, E. D. 1875 E, p. 244 A.   (Order Actinochiri; family Pelecopteridæ.)
1877 J, p. 822.   (Erisichtheidæ.)
1878 A, p. 299.   (Actinochiri; Pelecopteridæ.)
1892 V.   (Protosphyrænidæ.)
Crook, A. R. 1892 A, p. 109.   (Protosphyrænidæ.)

Dollo, L. 1892 B, p. 182.   (Protosphyrænidæ.)
Felix, J. 1890 A, p. 302.   (Pelecopteridæ.)
Loomis, F. B. 1900 A, p. 221.
Nicholson and Lydekker 1889 A, p. 995.   (Protosphyrænidæ.)
Stewart, A. 1900 A, p. 362.
Woodward, A. S. 1897 A, p. 158.
1898 B, p. 109.

---

[1] This subordinal name is here employed in a wider sense than was given to it by Cope.

PROTOSPHYRÆNA Leidy.   Type *P. ferox* Leidy.

Leidy, J. 1856 O, p. 302.
Agassiz, L. 1843 B, Feuill (1835), p. 55; lii (1839),
p. 150, pl. xa, figs. 1–4 (Ptychodus, in part); v,
pt. 1 (1844), p. 102 (Saurocephalus).
Cope, E. D. 1871 D, p. 415.
    1872 F, p. 340.
    1872 E E, p. 281.   (Erisichthe, type *E. nitida.*)
    1874 C, p. 41.   (Erisichthe.)
    1875 E, pp. 189, 217, 275 (Erisichthe); p. 244 C
    (Pelecopterus).
    1877 J, p. 821.   (Erisichthe.)
    1878 A, pp. 298, 299.   (Erisichthe and Pele-
    copterus.)
    1878 O.   (Erisichthe.)
    1886 R.
    1886 S.   (Protosphyræna and Pelecopterus.)
    1888 V.
    1892 V.
    1892 W.
Davies, W. 1878 A.
    1886 A.
Felix, J. 1890 A.
Hay, O. P. 1898 D, p. 26.
Hitchcock, Fanny R. 1888 A.   (Pelecopterus.)
Leidy, J. 1857 I.
Loomis, F. B. 1900 A, p. 215.
Newton, E. T. 1878 A.
    1878, in Dixon 1878 A, p. 401.
Nicholson and Lydekker 1889 A, p. 996.
Stewart, A. 1898 A, p. 22.
    1900 A, p. 362.
Woodward, A. S. 1888 I, p. 820.
    1894 A.
    1895 A, p. 399.
    1898 B, p. 111.
Woodward, H. 1886 A, p. 4.   (Pelecopterus.)
Zittel, K. A. 1890 A, p. 78 (Pelecopterus, syn. of
Ptychodus); p. 261 (Pelecopterus); p. 263 (Pro-
tosphyræna).
    1896 A, p. 412.

Protosphyræna angulata Cope.

Cope, E. D. 1872 I, pp. 333, 337.   (Portheus.)
    1872 F, p. 342.   (Portheus.)
    1872 EE, p. 281.   (Erisichthe.)
    1875 E, p. 275.   (Erisichthe.)
    1875 V, p. 32.
Felix, J. 1890 A, p. 293.   (Portheus.)
Loomis, F. B. 1900 A, p. 222.
Newton, E. T. 1878 A, p. 794.
Woodward, A. S. 1895 A, p. 413.
Cretaceous; North Carolina.

Protosphyræna bentoniana Stewart.

Stewart, A. 1898 A, p. 27, pl. i, figs. 2–8b.   (P. ben-
tonia, *crrore typog.*)
    1900 A, p. 365.   (P. bentoniana.)
Cretaceous (Niobrara); Kansas.

Protosphyræna chirurgus (Cope).

Cope, E. D. 1875 E, pp. 244 E, 273, pl. xlviii, fig. 1;
pl. liv, fig. 9.   (Pelecopterus.)
Cretaceous (Niobrara); Kansas.

Protosphyræna dimidiata (Cope).

Cope, E. D. 1878 A, p. 800.   (Erisichthe.)

Cope, E. D. 1877 J, p. 822.   (E. nitida, in part.)
Cretaceous (Niobrara); Kansas.

Protosphyræna gigas Stewart.

Stewart, A. 1899 C, p. 110.
Loomis, F. B. 1900 A, p. 221.
Stewart, A. 1900 A, p. 367.
Cretaceous (Pierre); Kansas.

Protosphyræna gladius (Cope).

Cope, E. D. 1873 N, p. 337.   (Portheus.)
    1874 C, p. 40.   (Portheus.)
    1875 E, pp. 244 E, 273, pl. xliv, fig. 12; pl. lii,
    fig. 3.   (Pelecopterus.)
Cretaceous (Niobrara); Kansas.

Protosphyræna nitida (Cope).

Cope, E. D. 1872 EE, p. 280.   (Erisichthe.)
    1874 C, p. 42.   (Erisichthe.)
    1875 E, pp. 217, 275, pl. xlviii, figs. 3–8.
    (Erisichthe.)
    1877 J, p. 821.   (Erisichthe nitida, in part;
    see P. dimidiata.)
Felix, J. 1890 A, pls. xii–xiv.
Loomis, F. B. 1900 A, p. 227, pl. xix, figs. 6, 7;
    p. 223, text fig. 4, c.
Newton, E. T. 1878 A, p. 794.
Woodward, A. S. 1895 A, p. 409.
Cretaceous (Niobrara); Kansas.

Protosphyræna obliquidens Loomis.

Loomis, F. B. 1900 A, p. 225, pl. xx, figs. 1–4.
Cretaceous (Niobrara); Kansas.

Protosphyræna penetrans (Cope).

Cope, E. D. 1877 J, p. 822.   (Erisichthe.)
Crook, A. R. 1892 A, p. 109.
Felix, J. 1890 A, p. 297, pl. xiv, fig. 1.
Loomis, F. B. 1900 A, p. 234, pl. xix, figs. 1–5; p. 223,
    text fig. 4, b.
Newton, E. T. 1878 A, p. 795.
Stewart, A. 1898 D, p. 192.
    1900 A, p. 369, pl. lxiii; text fig. 4.
Woodward, A. S. 1895 A, p. 409.
Cretaceous (Niobrara); Kansas.

Protosphyræna perniciosa (Cope).

Cope, E. D. 1874 C, p. 41.   (Ichthyodectes.)
    1875 E, pp. 244 D, 273, pl. xliv, fig. 13; pl.
    xlviii, fig. 2; pl. lii, fig. 2.   (Pelecopterus.)
Woodward, A. S. 1895 A, p. 414.
Cretaceous (Niobrara); Kansas.

Protosphyræna recurvirostris Stew-
art.

Stewart, A. 1898 D, p. 191, text fig. 1.
    1900 A, p. 366, text fig. 2.
Cretaceous (Niobrara); Kansas.

Protosphyræna tenuis Loomis.

Loomis, F. B. 1900 A, p. 226, pl. xx, figs. 5–7; p. 223,
    text fig. 4, a.
Cretaceous (Niobrara); Kansas.

Protosphyræna ziphioides (Cope).

Cope, E. D. 1877 J, p. 823.   (Erisichthe.)

Felix, J. 1890 A, p. 297.
Loomis, F. B. 1900 A, p. 222. (Erisichthe xiphiodes.)
Newton, E. T. 1878 A, p. 795.
Woodward, A. S. 1895 A, p. 413.
Cretaceous (Niobrara); Kansas.

Protosphyræna occidentalis Stew.

Stewart, A 1898 A, p. 28, pl. i, figs. 1a, 1b.
1900 A, p. 368, pl. lvii, figs. 1a, 1b.
Cretaceous (Niobrara); Kansas.

AMIIDÆ.

Baur, G. 1889 F, p. 470.
Beard, J. 1890 A, p. 184.
Bridge, T. W. 1877 A.
   1896 A, p. 537.
Cope, E. D. 1872 C, p. 330.
   1877 N, p. 58.
   1878 A, p. 296.
   1887 O, p. 1018.
   1889 R, p. 857.
Günther, A. 1870 A, p. 324.
   1880 A, p. 372.
   1881 A, p. 688.

Hay, O. P. 1898 C, p. 347.
Huxley, T. H. 1872 A, p. 127.
Jordan and Evermann 1896 A, p. 112.
Klaatsch, H. 1890 A, p. 178.
Kölliker, A. 1859 A, p. 7.
Lütken, C. 1873 A, p. 39.
Nicholson and Lydekker 1889 A, p. 988.
Pictet, F. J. 1854 A, p. 134. (Amiades.)
Steinmann and Döderlein 1890 A, p. 577.
Woodward, A. S. 1895 A, pp. xvi, 360.
Zittel, K. A. 1890 A, p. 233. (Halecomorphi.)

AMIOPSIS Kner.   Type *A. prisca* Kner.

Kner, R. 1863 A, p. 126.
Woodward, A. S. 1895 A, p. 371.
Zittel, K. A. 1890 A, p. 23⁴.

Amiopsis ? dartoni Eastman.

Eastman, C. R. 1899 E, p. 406, pl. xlviii, figs. 1, 2.
Jurassic; South Dakota.

AMIA Linn.   Type *A. calva* Linn.

Linnæus C., 1766, Syst. Nat., ed. xii, p. 500.
Allis, E. P. 1889 A.
   1899 A.
   1899 B.
   1900 A.
Andreæ, A. 1893 B.
Baur, G. 1889 F, p. 469.
   1896 C, p. 662.
Beard, J. 1890 A, p. 182.
Bridge T. W. 1877 A.
   1896 A, p. 537.
Cope, E. D. 1872 C, p. 319.
   1879 N.
   1884 O, pp. 742, 745.
   1897 A, p. 321.
Davidoff, M. 1880 A.
   1880 B.
Dean, B. 1895 A.
Franque, H. 1847 A.
Gadow and Abbott 1895 A.
Gegenbaur, C. 1865 B, p. 102, pl. vi, fig. 4; pl. viii, fig. 7.
   1895 A, p. 6.
Goette, A. 1878 A, p. 453.
Günther, A. 1870 A, p. 324.
Hay, O. P. 1895 A.
   1898 C, p. 343.
Hulke, J. W. 1858 A, p. 424.
Huxley, T. H. 1872 A, p. 124.
   1876 A, p. 57.
   1883 A.
Jordan and Evermann 1896 A, p. 112.
Klaatsch, H. 1890 A, p. 178.
   1896 A, p. 323.
Kner, R. 1863 A, p. 126.
Kölliker, A. 1859 A.
   1860 A.
   1863 A, p. 38, figs. 17–19.
Leidy, J. 1873 B, p. 185. (Protamia and Hypamia).

Leidy, J. 1873 D, p. 98. (Protamia and Hypamia).
Lütken, C. 1873 A, p. 39.
Mackintosh, H. W. 1878 A, pls. i, ii.
Marsh, O. C. 1877 E.
Müller, J. 1846 A, p. 201.
   1846 B, p. 78.
   1846 C.
Newton, E. T. 1899 A, p. 1.
Nicholson and Lydekker 1889 A, p. 987.
Owen, R. 1845 B, p. 75.
Parker, W. K. 1882 A, p. 488.
Ryder, J. A. 1886 A, p. 1001.
Sagemehl, M. 1883 A.
   1884 A.
Shufeldt, R. 1885 B.
Stannius, H. 1846 A.
   1854 A.
Stephan, P. 1900 A.
Thacher, J. K. 1877 B, pl. ii, fig. 6.
Traquair, R. H. 1896 A, p. 125.
   1900 B, p. 524.
Van Wijhe, J. W. 1882 A, p. 279.
Woodward, A. S. 1895 A, p. 367.
Zittel, K. A. 1890 A, p. 235.

Amia depressa Marsh.

Marsh, O. C. 1871 G, p. 105.
King, C. 1878 A, p. 405.
Osborn, Scott, and Speir 1878 A, p. 102. (Identification doubtful.)
Woodward, A. S. 1895 A, p. 372.
Eocene (Bridger); Wyoming.

Amia dictyocephala Cope.

Cope, E. D. 1875 G, p. 3.
   1884 O, p. 745, pl. lix, fig. 1.

Woodward, A. S. 1895 A, p. 372.
Eocene (Amyzon shales); Colorado.
"*Amia reticulata*," referred to by Cope in works above cited, appears to be a rejected name for A. *dictyocephala.*

Amia elegans Leidy.

Leidy, J. 1873 D, p. 98. [Amia (Hypamia).]
King, C. 1878 A, p. 405. (Hypamia.)
Leidy, J. 1873 B, pp. 189, 349, pl. xxxii, figs. 19–22.
[Amia (Hypamia).]
Woodward, A. S. 1895 A, p. 373.
Eocene (Bridger); Wyoming.

Amia gracilis Leidy.

Leidy, J. 1873 D, p. 98. [Amia (Protamia).]
1873 B, pp. 188, 348, pl. xxxii, figs. 23, 24.
[Amia (Protamia).]
Woodward, A. S. 1895 A, p. 373.
Eocene (Bridger); Wyoming.

Amia macrospondyla Cope.

Cope, E. D. 1891 A, p. 2, pl. i, fig. 2.
Woodward, A. S. 1895 A, p. 372.
Miocene; N. W. Territory, Canada.

Amia media Leidy.

Leidy, J. 1873 D, p. 98. (Protamia.)
King, C. 1878 A, p. 405.

PAPPICHTHYS Cope.

Cope, E. D. 1873 E, p. 634.
Andreæ, A. 1893 B.
Cope, E. D. 1881 D, p. 184.
1882 E, p. 190.
1884 O, p. 56.
Hay, O. P. 1895 A, p. 12.
Newton, E. T. 1899 A, p. 2.
Nicholson and Lydekker 1889 A, p. 991.
Woodward, A. S. 1895 A, p. 373.
Zittel, K. A. 1890 A, p. 236.

Pappichthys corsoni Cope.

Cope, E. D. 1873 E, p. 636.
1873 E, p. 636. (P. symphysis.)
1884 O, p. 60, pl. iv, figs. 21–36.
Osborn, Scott, and Speir 1878 A, p. 104.
Woodward, A. S. 1895 A, p. 373.
Eocene (Bridger); Wyoming.

Pappichthys lævis Cope.

Cope, E. D. 1873 E, p. 636.

Leidy, J. 1873 B, pp. 188, 348, pl. xxxii, figs. 7–11.
[Amia (Protamia.)]
Woodward, A. S. 1895 A, p. 373.
Eocene (Bridger); Wyoming.

Amia newberriana Marsh.

Marsh, O. C. 1871 G, p. 105.
King, C. 1878 A, p. 405.
Woodward, A. S. 1895 A, p. 372.
Eocene (Bridger); Wyoming.

Amia scutata Cope.

Cope, E. D. 1875 G, p. 3.
1884 O, p. 745, pl. lx, fig. 1.
Osborn, Scott, and Speir 1878 A, p. 102.
Woodward, A. S. 1895 A, p. 372.
Eocene (Amyzon shales); Colorado.

Amia uintaensis Leidy.

Leidy, J. 1873 D, p. 98. [Amia (Protamia).]
King, C. 1878 A, p. 405.
Leidy, J. 1873 B, pp. 185, 348, pl. xxxii, figs. 1–6.
(Protamia.)
Osborn, Scott, and Speir 1878 A, p. 101.
Woodward, A. S. 1895 A, p. 373.
Eocene (Bridger); Wyoming.

Amia whiteavesiana Cope.

Cope, E. D. 1891 A, p. 2, pl.i, fig. 1.
Woodward, A. S. 1895 A, p. 372.
Lower Miocene; N. W. Territory, Canada.

Type *P. plicatus* Cope.

Cope, E. D. 1884 O, p. 58, pl. iii, figs. 2–11.
Osborn, Scott, and Speir 1878 A, p. 104.
Woodward, A. S. 1895 A, p. 373.
Eocene (Bridger); Wyoming.

Pappichthys plicatus Cope.

Cope, E. D. 1873 E, p. 634.
1884 O, p. 59, pl. iii, figs. 12–19; pl. iv, figs. 1–5.
King, C. 1878 A, p. 405.
Osborn, Scott, and Speir 1878 A, p. 104.
Woodward, A. S. 1895 A, p. 373.
Eocene (Bridger); Wyoming.

Pappichthys sclerops Cope.

Cope, E. D. 1873 E, p. 635.
1884 O, p. 57, pl. iii, fig. 1.
Woodward, A. S. 1895 A, p. 373.
Eocene (Bridger); Wyoming.

Order NEMATOGNATHI Gill.

Gill, T. 1861 A, pp. 11, 56.
Allis, E. P. 1900 A, p. 271.
Bridge and Haddon 1892 A.
Brühl, B. C. 1847 A. ("Siluroiden.")
Cope, E. D. 1872 C, p. 330.
1878 A, p. 295.
1884 O, p. 61.
1886 D, p. 1081.
1889 R, p. 857.

Dean, B, 1891 A, p. 315. ("Siluroids.")
Gill, T. 1872 C, pp. xxxviii, 18.
1893 A, p. 132.
1896 C, p. 912. ("Siluroids.")
Göldi, E. A. 1883 A.
1884 A.
Hertwig, O. 1876 A. (Siluroidei.)
Jordan and Evermann 1896 A, p. 114.
McMurrich, J. P. 1884 B.

Müller, J. 1846 B, p. 82. ("Biluroiden.")
Parker, W. K. 1868 A, p. 22. ("Siluroids.")
Pollard, H. B. 1894 A.
    1895 A.

Stannius, H. 1854 A. (Siluroidei.)
Woodward, A. S. 1898 B, p. 119.
Zittel, K. A. 1890 A, p. 260.

SILURIDÆ.

Bridge, W. J. 1879 A.
    1896 A, p. 550.
Bridge and Haddon 1889 A.
    1892 A.
Brühl, B. C. 1847 A. ("Siluroiden.")
Cope, E. D. 1872 C, p. 331.
Dareste, C. 1872 A, p. 945. ("Siluroides.")
Davis, J. W. 1880 A, p. 356.
Gegenbaur, C. 1869 B.
    1898 A.
Gill, T. 1893 A, p. 132.
Göldi, E. A. 1883 A.
    1884 A.
Günther, A. 1864 A, p. 1.
    1880 A, p. 559.
    1881 A, p. 692.
Hertwig, O. 1876 A.
Huxley, T. H. 1861 B, p. 33.
Ihering, H. 1891 B, p. 498.
Jordan, D. S. 1891 A.
Jordan and Evermann 1896 A, p. 115.
Klaatsch, H. 1890 A, p. 184.
Klein, E. F. 1881 A.

Klein, E. F. 1886 A.
Koken, E. 1891 B, p. 26.
Kölliker, A. 1859 A. (Siluroidei.)
McMurrich, J. P. 1884 A.
    1884 B.
Nicholson and Lydekker 1889 A, p. 1001.
Newton, E. 1889 A.
Owen, R. 1845 B, p. 86. ("Siluroids.")
    1886 A, p. 10 et seq.
Pictet, F. J. 1854 A, p. 119. ("Siluriens.")
Pollard, H. B. 1894 A.
    1894 B.
    1895 A.
Ryder, J. A. 1886 A, p. 1003. (Ameiurus.)
Sagemehl, M. 1884 A, p. 1.
Stannius, H. 1854 A. (Siluroidei.)
Stephan, P. 1900 A. ("Siluridés.")
Vaillant, L. 1883 A.
Woodward, A. S. 1889 C.
    1891 D.
Wright, R. R. 1884 A.
Zittel, K. A. 1890 A, p. 260.

RHINEASTES Cope.   Type *R. peltatus* Cope.

Cope, E. D. 1872 R, p. 486.
    1873 E, p. 637 (Phareodon, in part); p. 638
    (Rhineastes, with subgenera Rhineastes
    and Astephus).
    1884 O, p. 62. (With subgenera Rhineastes
    and Astephus.)
    1891 A, p. 2.
Nicholson and Lydekker 1889 A, p. 1002.
Osborn, Scott, and Speir 1878 A, p. 100. (Sp. indet.
    described.)

Rhineastes arcuatus Cope.
Cope, E. D. 1873 E, pp. 638, 641.
    1884 O, pp. 63, 66, pl. ᵥ, fig. 12.
Leidy, J. 1873 D, p. 99. (Pimelodus antiquus;
    no description.)
    1873 B, pp. 193, 348, pl. xxxii, figs. 44–46.
    (Pimelodus antiquus.)
Eocene (Bridger); Wyoming.

Rhineastes calvus Cope.
Cope, E. D. 1873 E, pp. 638, 640.
    1884 O, pp. 63, 65, pl. ᵥ, figs. 3, 4.
Eocene (Bridger); Wyoming.

Rhineastes pectinatus Cope.
Cope, E. D. 1874 D, p. 49.

Cope, E. D. 1874 B, p. 459.
    1875 G, p. 3.
    1884 O, p. 747, pl. ᵥ, fig. 13.
Tertiary (Eocene?); Colorado.

Rhineastes peltatus Cope.
Cope, E. D. 1872 R, p. 486.
    1873 E, p. 639.
    1884 O, p. 63, pl. ᵥ, figs. 1, 2.
Eocene (Bridger); Wyoming.

Rhineastes radulus Cope.
Cope, E. D. 1873 E, p. 639.
    1884 O, p. 67, pl. ᵥ, figs. 14–17.
King, C. 1878 A, p. 405.
Eocene (Bridger); Wyoming.

Rhineastes rhæas Cope.
Cope, E. D. 1891 A, p. 3, pl. i, fig. 3.
Oligocene or Lower Miocene; Northwest
    Territory, Canada.

Rhineastes smithii Cope.
Cope, E. D. 1872 R, p. 486.
    1873 E, p. 639.
    1884 O, pp. 63, 64, pl. ᵥ, figs. 5–11.
Eocene (Bridger); Wyoming.

AMEIURUS Rafinesque.   Type *A. cupreus* Raf.=*A. natalis* Le Sueur.

Rafinesque, C. S., 1820, Ichthyol. Ohioensis, p. 65.
Allis, E. P. 1900 A, p. 271.
Günther, A. 1864 A, p. 98. (Amiurus.)
    1880 A, p. 567. (Amiurus.)
Jordan and Evermann 1896 A, p. 135.

McMurrich, J. P. 1884 A.
    1884 B.
Ryder, J. A. 1886 A, p. 1008.
Wright, R. R. 1884 A.

Ameiurus cancellatus Cope.

Cope, E. D. 1891 A, p. 3, pl. i, figs. 4, 5.
Oligocene; Northwest Territory, Canada.

Ameiurus maconnellii Cope.

Cope, E. D. 1891 A, p. 4, pl. i, figs. 6, 7.
Oligocene; Northwest Territory, Canada.

Order ISOSPONDYLI Cope.

Cope, E. D. 1872 C, p. 332.
  1877 N, p. 60.
  1878 A, pp. 295, 296.
  1884 O, p. 67.
  1887 O, p. 1018.
  1889 R, p. 857.
Gill, T. 1872 C, p. 15.

Jordan, D. S. 1891 A, p. 112.
Jordan and Evermann 1896 A, p. 407.
Traquair, R. H. 1896 A, p. 126.
  1900 B, p. 524.
Woodward, A. S. 1896 A, pp. xix, 446.
  1898 B, p. 113.

PHOLIDOPHORIDÆ.

Eastman, C. R. 1899 E, p. 399.
Traquair, R. H. 1900 B, p. 524.
Woodward, A. S. 1890 D, p. 43.
  1895 A, p. 446.

Woodward, A. S. 1898 B, p. 114.
Zittel, K. A. 1890 A, p. 212. (Saurodontidæ, in part).

PHOLIDOPHORUS Agassiz.   Type *P. bechei* Agassiz.

Agassiz, L., 1832, Neues Jahrb. Mineral., p. 145.
  1843 B (1844), ii, pt. 1, p. 271.
Eastman, C. R. 1899 E, p. 399.
Giebel, C. G. 1848 A, p. 208.
Kner, R. 1866 B, p. 324.
  1866 C, p. 32.
Lütken, C. 1873 A, p. 324.
Quenstedt, F. A. 1852 A, p. 207.
  1885 A, p. 319.
Vetter, B. 1881 A, p. 58.
Wagner, A. 1863 A, p. 27.

Woodward, A. S. 1890 D, p. 43.
  1895 A, p. 447.
  1895 D.
  1898 B, p. 114.
Woodward and Sherborn 1890 A, p. 146.
Zittel, K. A. 1890 A, p. 213.

Pholidophorus americanus Eastman.

Eastman, C. R. 1899 A, p. 642.
  1899 E, p. 398, pls. xlv-xlvii.
Jurassic; South Dakota.

CHIROCENTRIDÆ.

Günther, A. 1868 A, p. 475.
Cope, E. D. 1870 L. (Saurodontidæ.)
  1871 D, p. 414. (Saurodontidæ.)
  1872 F, p. 337. (Saurodontidæ.)
  1872 I, p. 327. (Saurodontidæ.)
  1875 E, p. 183. (Saurodontidæ.)
  1877 G, p. 588. (Saurodontidæ.)
  1888 V, p. 1019. (Saurocephalidæ.)
  1892 V. (Saurocephalidæ.)
Crook, A. R. 1892 A, pp. 111, 121. (Ichthyodectidæ.)
Cuvier and Valenciennes, 1846, Hist. Nat. Poissons, xix, p. 150. (" Chirocentres.")
Günther, A. 1880 A, p. 663.
Loomis, F. B. 1900 A, p. 236.

Nicholson and Lydekker 1889 A, p. 996. (Saurocephalidæ.)
Reis, O. M. 1895 C, p. 182. (Ichthyodectidæ.)
Steinmann and Döderlein 1890 A, p. 580. (Saurocephalidæ.)
Stewart, A. 1898 A, p. 21. (Saurodontidæ.)
  1898 C, p. 179. (Saurodontidæ.)
  1899 C, p. 107. (Saurodontidæ.)
  1900 A, p. 262 (Ichthyodectidæ); p. 310 (Saurodontidæ.)
Woodward, A. S. 1898 B, p. 118. (Saurodontidæ or Ichthyodectidæ.)
Zittel, K. A. 1890 A (1888), p. 262. (Saurocephalidæ.)

ICHTHYODECTINÆ.

Crook, A. R. 1892 A, pp. 111, 121. (Ichthyodectidæ, part.)

Stewart, A. 1899 C, p. 107. (Ichthyodectidæ.)
  1900 A, p. 262. (Ichthyodectidæ.)

PORTHEUS Cope.   Type *P. molossus* Cope.

Cope, E. D. 1871 K, p. 175.
  1872 F, pp. 339, 341.
  1872 I, pp. 330, 331.
  1873 N, p. 338.
  1874 C, p. 39.
  1875 E, pp. 184, 189, 204, 273, pl. lv, fig. 1.
  1892 V.

Crook, A. R. 1892 A, pp. 114, 122.
Günther, A. 1880 A, p. 500.
Hay, O. P. 1898 B, p. 646. (Xiphactinus.)
  1898 D, p. 25, figs. 1-16. (Xiphactinus.)
? Leidy, J. 1870 H, p. 12. (Xiphactinus; type *X. audax.*)
Loomis, F. B. 1900 A, p. 246.

Newton, E. T. 1877 A.
    1878 B.
Nicholson and Lydekker 1889 A. p. 996.
Stewart, A. 1898 B, p. 115. (Xiphactinus.)
    1898 C, pp. 180, 184. (Xiphactinus.)
    1899 A, p. 19. (Xiphactinus.)
    1899 C, p. 107. (Xiphactinus.)
    1900 A, p. 265. (Xiphactinus.)
Woodward, A. S. 1888 I, p. 310.
    1898 B, p. 118.
Zittel, K. A. 1890 A, p. 262.

Portheus ? audax (Leidy).

Leidy, J. 1870 H, p. 12. (Xiphactinus.)
Cope, E. D. 1870 L, p. 533. (Saurocephalus.)
    1871 D, pp. 415, 418. (Saurocephalus.)
    1872 F, pp. 341, 348. (? Xiphactinus.)
    1875 E, p. 276. (Xiphactinus.)
    1898 A (1870), p. 29. (X. audax = Saurocephalus thaumas.)
Hay, O. P. 1898 D, p. 27. (Xiphactinus.)
Leidy, J. 1873 B, pp. 290, 348, pl. xvii, figs. 9, 10. (Xiphactinus.)
Stewart, A. 1898 B, p. 115, figures.
    1900 A, p. 267, pls. xxxiii-xlvii (in part?).
Cretaceous (Niobrara); Kansas.
    Probably identical with one of the other species of Portheus, but it may also belong to some other genus.

Portheus brachygnathus (Stewart).

Stewart, A. 1898 B. (Xiphactinus.)
    1899 C, p. 107. (Xiphactinus.)
    1900 A, p. 293, pl. xlvi, figs. 3, 4. (Xiphactinus.)
Cretaceous (Niobrara); Kansas.

Portheus lestrio Cope.

Cope, E. D. 1873 N, p. 338.
    1875 E, pp. 201, 274, pls. xlii, figs. 1-3; xlvii, fig. 1.
Crook, A. R. 1892 A, p. 123.
Hay, O. P. 1898 D, pp. 27, 52. (Xiphactinus.)
Stewart, A. 1898 B, p. 118, footnote. (Xiphactinus.)
Newton, E. T. 1877 A, p. 509.
Cretaceous (Niobrara); Kansas.

Portheus lowii Stewart.

Stewart, A. 1898 A, p. 24, pl. ii, fig. 2.

Hay, O. P. 1898 D, p. 27. (Xiphactinus.)
Stewart, A. 1900 A, p. 293, pl. xlviii, fig. 2. (Xiphactinus.)
Cretaceous (Benton); Kansas.

Portheus molossus Cope.

Cope, E. D. 1871 K, p. 175.
    1872 F, pp. 324, 338, 342.
    1872 I, pp. 328, 333.
    1874 C, p. 39.
    1875 E, pp. 50, 183, 273, pls. xxxix, xl, figs. 1-9; xli; xliv, figs 5, 10, 11; xlv, figs. 9-11;
Crook, A. R. 1892 A, pp. 114, 122, pl. xviii.
Hay, O. P. 1898 D, pp. 35, 52, figs. 3, 6, 8. (Xiphactinus.)
Newton, E. T. 1877 A, p. 509.
Stewart, A. 1898 B, p. 116, figs. (Syn. of Xiphactinus audax.)
    1900 A, p. 267, part? of pls. xxxiii-xlvii. (Xiphactinus audax.)
Zittel, K. A. 1890 A, p. 263, fig. 268.
Cretaceous (Niobrara); Kansas.

Portheus mudgei Cope.

Cope, E. D. 1874 C, p. 40.
    1875 E, pp. 203, 274.
Crook, A. R. 1892 A, p. 123, pl. xxxix.
Hay, O. P. 1898 D, pp. 27, 52. (Xiphactinus.)
Newton, E. T. 1877 A, p. 509.
Cretaceous (Niobrara); Kansas.

Portheus thaumas Cope.

Cope, E. D. 1870 L, p. 533. (Saurocephalus.)
    1871 D, pp. 418, 422. (Saurocephalus.)
    1872 F, p. 342.
    1872 I, pp. 333, 335.
    1874 C, p. 40.
    1875 E, pp. 196, 273, pl. xliii, figs. 1-4; xliv, figs. 1-4.
Crook, A. R. 1892 A, p. 123.
Hay, O. P. 1898 D, p. 27, figs. 2, 7, 10-12, 15, 16.
Newton, E. T. 1877 A, p. 509. (Xiphactinus.)
Osborn, Scott and Speir 1878 A, p. 99. (Identification doubtful.)
Stewart, A. 1898 B, p. 116, figs. (Syn. of Xiphactinus audax.)
    1900 A, p. 267, part? of pls. xxxiii-xlvii. (Xiphactinus audax.)
Cretaceous (Niobrara); Kansas, Colorado?.

ICHTHYODECTES Cope.   Type I. ctenodon Cope.

Cope, E. D. 1870 L, p. 536.
    1871 D, p. 421.
    1872, F, pp. 340, 343.
    1872, I, pp. 330, 338.
    1875 E, pp. 189, 205, 275.
    1892 V.
Crook, A. R. 1892 A, p. 111.
Loomis, F. B. 1900 A, p. 236.
Newton, E. T. 1877 A, p. 520.
    1878 B.
Nicholson and Lydekker 1889 A, p. 996.
Stewart, A. 1898 C, p. 180.
    1899 A, p. 20.
    1900 A, p. 205.
Woodward, A. S. 1888 I, p. 311.

Woodward, A. S. 1898 B, p. 118.
Zittel, K. A, 1890 A, p. 263.

Ichthyodectes acanthicus Cope

Cope, E. D. 1877 D, p. 177.
Hay, O. P. 1898 E, p. 227.
Stewart, A. 1900 A, p. 301.   •
Cretaceous (Niobrara); Kansas.

Ichthyodectes anaides Cope.

Cope, E. D. 1872 I, p. 339.
    1872 Γ, p. 343.
    1874 C, p. 40.
    1875 E, pp. 206, 274, pl. xliv, figs. 14, 15; pl. xlv, figs. 1-8.

Crook, A. R. 1902 A, pp. 111, 123, pl. xv.
Hay, O. P. 1898 E, p. 226, fig. 2.
Loomis, F. B. 1900 A, p. 244.
Stewart, A. 1900 A, p. 296, pl. xlix, figs. 1-3.
Cretaceous (Niobrara); Kansas.

Ichthyodectes cruentus Hay.

Hay, O. P. 1898 E, p. 225, figs. 1, 3, 4.
Stewart, A. 1900 A, p. 300, pl. l, figs. 8-10b.
Cretaceous (Niobrara); Kansas.

Ichthyodectes ctenodon Cope.

Cope, E. D. 1870 L, p. 536.
1871 D, p. 421.
1872 F, p. 343.
1872 I, pp. 339, 340, 345.
1874 C, p. 40.
1875 E, pp. 207, 241, 274, pl. xlvi, figs. 1-4.
Crook, A. R. 1892 A, p. 123.
Hay, O. P. 1898 E, p. 226.
Loomis, F. B. 1900 A, p. 244, pl. xxiii, figs. 7, 8.
Stewart, A. 1900 A, p. 303, pl. xlix, figs. 5-7; pl. li,
figs. 12, 13.
Cretaceous (Niobrara); Kansas.

Ichthyodectes goodeanus Cope.

Cope, E. D. 1877 D, p. 176.
Hay, O. P. 1898 A, p. 227.
Cretaceous (Niobrara); Kansas.

Ichthyodectes hamatus Cope.

Cope, E. D. 1872 I, pp. 339, 340.
1872 F, p. 343.

Cope, E. D. 1874 C, p. 40.
1875 E, pp. 209, 274, pl. xlvi, fig. 5.
Crook, A. R. 1892 A, p. 123.
Hay, O. P. 1898 E, pp. 225, 226.
Loomis, F. B. 1900 A, p. 243, pl. xxiii, figs. 9a, 9b.
Stewart, A. 1900 A, p. 298, Pl. xlviii, fig. 2; pl. l,
figs. 1-7.
Cretaceous (Niobrara); Kansas.

Ichthyodectes multidentatus Cope.

Cope, E. D. 1872 I, pp. 339, 342.
1872 F, p. 343.
1874 C, p. 41.
1875 E, pp. 212, 275, pl. l, figs. 6, 7
Crook, A. R. 1892 A, p. 123.
Hay, O. P. 1898 E, p. 227.
Loomis, F. B. 1900 A, p. 243.
Cretaceous (Niobrara); Kansas.

Ichthyodectes parvus Stewart.

Stewart, A. 1900 A, p. 302, pl. li, figs. 1-11 (I.
acanthicus, not of Cope); p. 302 (I. parvus).
Cretaceous (Niobrara); Kansas.

Ichthyodectes prognathus Cope.

Cope, E. D. 1870 L, p. 532.   (Saurocephalus.)
1871 D, p. 417.   (Saurocephalus.)
1872 F, pp. 340, 343.
1872 I, p. 340.
1874 C, p. 41.
1875 E, pp. 210, 274, pl. xlvi, figs. 6-10.
Crook, A. R. 1892 A, p. 123.
Hay, O. P. 1898 E, p. 226.
Cretaceous (Niobrara); Kansas.

Stewart, A. 1899 C, p. 107. (Saurodontidæ.) | Stewart, A. 1900 A, p. 310. (Saurodontidæ.)

SAUROCEPHALUS Harlan.   Type S. lanciformis Harlan, not of Agassiz.

Harlan, R. 1824 A.
Agassiz, L. 1843 B (1844), v, p. 101.   (In part.)
Cope, E. D. 1870 L, pp. 529, 530.
1870 P.
1871 A.
1871 D, p. 415.
1872 F, pp. 340, 343.
1872 I, pp. 330, 342.
1873 N, p. 339. (Daptinus, with S. phlebotomus as type.)
1875 E, pp. 189, 215, 275. (Saurodon.)
1888 V. (Saurodon.)
1892 V. (Saurodon.)
1898 A.
Crook, A. R. 1892 A, p. 120.
Davies, W. 1878 A.
Giebel, C. G. 1848 A, p. 88.
Harlan, R. 1834 B, p. 85. (Saurodon.)
1835 B, p. 286.
1835 B, p. 288. (Saurodon.)
1835 H.
Hay, O. P. 1899 B, p. 302.
Hays, I. 1830 A. (Saurodon, with S. leanus as type.)
Kramberger, D, 1881 A.
Leidy, J. 1856 O, p. 302.

Leidy, J. 1857 I, p. 95.
Loomis, F. B. 1900 A, p. 247 (Saurodon); p. 249
(Saurocephalus).
Newton, E. T. 1878 A.
1878 B. (Daptinus.)
Nicholson and Lydekker 1889 A, p. 998 (Saurocephalus); p.999 (Saurodon).
Owen, R. 1845 B, p. 130.
Pictet, F. J. 1854 A, p. 93 (Saurocephalus); p. 94
(Saurodon).
Stewart, A. 1898, A, p. 22. (Saurodon, Daptinus,
Saurocephalus.)
1898 C. (Saurodon, Saurocephalus.)
1900 A, p. 311 (Saurodon); p. 323 (Saurocephalus).
Woodward, A. S. 1888 I, p. 312. (Daptinus.)
Zittel, K. A. 1890 A, p. 264.

Saurocephalus arapahovius Cope.

Cope, E. D. 1872 I, p. 343.
1872 F, pp. 344, 348.
1874 C, p. 41.
1875 E, pp. 216, 275; pl. xlix, fig. 5.
Loomis, F. B. 1900 A, p. 251. (S. arapalovius, syn. of
S. lanciformis.)

Newton, E. T. 1878 A.
Stewart, A. 1898 C, p. 186.
   1900 A, p. 323.
   Cretaceous (Niobrara): Kansas.

Saurocephalus broadheadi (Stewart.)

Stewart, A. 1898 A, p. 24, pl. ii, figs. 1a–1c. (Daptinus.)
Loomis, F. B. 1900 A, p. 252, pl. xxiv, fig. 6; pl. xxv, fig. 1.
Stewart, A. 1898 C, p. 186. (Saurodon.)
   1900 A, p. 313, pl. xlviii, figs. 1a–1c.
   Cretaceous (Niobrara); Kansas.

Saurocephalus dentatus Stewart.

Stewart, A. 1898 A, p. 25, pl. i, figs. 3a, 3b.
Loomis, F. B. 1900 A, p. 251. (Syn. of S. lanciformis.)
Stewart, A, 1898 C, p. 186.
   1900 A, p. 323, pl. lviii, figs. 3, 4.
   Cretaceous (Niobrara); Kansas.

Saurocephalus ferox (Stewart).

Stewart, A. 1898 A, p. 183, pl. xv; pl. xvi, figs. 1–3. (Saurodon.)
   1900 A, p. 319, pls. lvi, lvii. (Saurodon.)
   Cretaceous (Niobrara); Kansas.

Saurocephalus lanciformis Harlan.

Harlan, R. 1824 A, pl. xii. (Not of Agassiz, L. 1844 A.)
Cope, E. D. 1870 L, p. 580.
   1871 D, p. 415.
   1875 E, pp. 216, 275.
Davies, W. 1878 A, p. 260.
Giebel, C. G. 1848 A, p. 89.
Harlan, R. 1834 B, p. 83.
   1835 B, pp. 286, 289.
   1835 H.
Hay, O. P. 1899 B, p. 299, figs. 1–4.
Hays, I. 1830 A, p. 477, pl. xvi, fig. 11.
Hilgard, E. W. 1860 A, pp. 142, 389.
Leidy, J. 1856 O, p. 302.
   1857 I, pl. vi, figs. 8–11.
Loomis, F. B. 1900 A, p. 251, pl. xxv, figs. 2–5.
Morton, S. G. 1835 A, p. 277.
Newton, E. T. 1878 A.
Owen, R. 1845 B, p. 130, pl. lv.
Pictet, F. J. 1854 A, p. 93.
Stewart, A. 1898 C, p. 186.
   Cretaceous (Niobrara): Nebraska, Kansas.

Saurocephalus leanus (Hays).

Hays, I. 1830 A, p. 477, pl. xvi, figs. 1–10. (Saurodon.)
Agassiz, L. 1844 A, p. 205. (Cimolichthys levisiensis, in part.)
Cope, E. D. 1870 L, p. 581. (Saurodon.)
   1871 D, p. 421. (Saurodon.)
   1875 E, p. 275. (Saurodon.)
Davies, W. 1878 A, p. 260.
Giebel, C. G. 1848 A, p. 89. (Saurodon.)
Harlan, R. 1834 B, p. 83.
   1835 B, p. 287 (S. leanus); p. 288 ("S. leae").
Leidy, J. 1856 O, p. 302.
   1857 I, pl. vi, figs. 12–15.
Morton, S. G. 1835 A, p. 277.
Newton, E. T. 1878 A.
Woodward, A. S. 1888 B.
Zittel, K. A. 1890 A, p. 264.
   Cretaceous; New Jersey.

Saurocephalus pamphagus Hay.

Hay, O. P. 1899 B, p. 303, fig. 5.
   Cretaceous (Niobrara); Kansas.

Saurocephalus phlebotomus Cope.

Cope, E. D. 1870 L, p. 580.
   1871 D, p. 416.
   1872 I, p. 343. (Saurodon.)
   1873 N, p. 339. (Daptinus.)
   1874 C, p. 41. (Daptinus.)
   1875 E, pp. 213, 275, pl. xlvii, figs. 3, 4, 6; pl. xlix, figs. 1–4. (Daptinus.)
   1877 G, p. 568. (Saurodon.)
   1888 V.
   1892 V.
Crook, A. R. 1892 A, pp. 120, 123.
Loomis, F. B. 1900 A, p. 248, pl. xxiv, figs. 1–5. (Saurodon,)
Newton, E. T. 1878 B.
Stewart, A. 1898 C, p. 186. (Saurodon.)
   1900 A, p. 312, pl. lvii, figs. 4, 5. (Saurodon.)
Zittel, K. A. 1890 A, p. 264. (Daptinus.)
   Cretaceous (Niobrara); Kansas.

Saurocephalus pygmæus (Loomis).

Loomis, F. B. 1900 A, p. 248, pl. xxiii, fig. 10. (Saurodon.)
   Cretaceous (Niobrara); Kansas.

Saurocephalus xiphirostris (Stewart).

Stewart, A. 1898 C, p. 179, pl. xiv. (Saurodon.)
   1900 A, p. 314, pl. lv. (Saurodon).
   Cretaceous (Niobrara); Kansas.

GILLICUS Hay. Type *Ichthyodectes arcuatus* Cope.

Hay, O. P. 1898 E, p. 230.
Loomis, F. B. 1900 A, p. 237. (Syn. of Ichthyodectes.)
Stewart, A. 1900 A, p. 304.

Gillicus arcuatus (Cope).

Cope, E. D. 1875 E, pp. 198, 2(4, 274, pl. xlvii, figs. 7–9. (Portheus; the figures probably not belonging to this species.)
   1877 D, p. 177. (Ichthyodectes.)
   1892 V, p. 942.

Crook, A. R. 1892 A, p. 112, pl. xvi (I. polymicrodus); p. 123 (Portheus).
Hay, O. P. 1898 D, pp. 30, 42. (I. arcuatus, I. polymicrodus.)
   1898 E, pp. 228, 230. (G. arcuatus, G. polymicrodus.)
Loomis, F. B. 1900 A, p. 242, pl. xxiii, figs. 1–6, text fig. 7. (Ichthyodectes occidentalis.)
Newton, E. T. 1877 A, p. 509.
Stewart, A., p. 307, pls. lii–liv.
   Cretaceous (Niobrara); Kansas.

STRATODONTIDÆ.

Cope, E. D. 1872 I, p. 348.
  1872 F, p. 345.
  1872 I, p. 343. (Pachyrhizodontidæ, part.)
  1875 E, p. 218.
Crook, A. J. 1892 A, p. 109.

Nicholson and Lydekker 1889 A, p. 996. (Pachyrhizodontidæ, part.)
Reis, O. M. 1895 C, p. 170.
Stewart, A 1900 A, p. 326.
Zittel, K. A. 1890 A, p. 268.

STRATODUS Cope.   Type *S. apicalis* Cope.

Cope, E. D. 1872 I, p, 348.
  1872 F, pp. 345, 346.
  1875 E, pp. 219, 226.
Reis, O. M. 1895 C, p. 170.
Stewart, A. 1900 A. p. 327.
Woodward, A. S. 1888 I, p. 314.
Zittel, K. A. 1890 A, p. 269.

Stratodus apicalis Cope.

Cope, E. D. 1872 I, p. 349.

Cope, E. D. 1872 F, pp. 346, 348.
  1874 C, p. 47.
  1875 E, pp. 227, 279, pl. xlix, figs. 6–8.
Stewart, A. 1900 A, p. 328, pl. lx; pl. lxi, fig. 1.
  Cretaceous (Niobrara); Kansas.

Stratodus oxypogon Cope.

Cope, E. D. 1877 D, p. 180.
  Cretaceous (Niobrara); Kansas.

EMPO Cope.   Type *E. nepaholica* Cope.

Cope, E. D. 1872 I, p. 347.
  1872 F, p. 345.
  1872 I, p. 348. (Cimolichthys.)
  1874 C, p. 45.
  1875 E, pp. 219, 228, 279.
Loomis, F. B. 1900 A, p. 267. (Syn. of Cimolichthys.)
Nicholson and Lydekker 1889 A, p. 999.
Stewart, A, 1900 A, p. 330.
Zittel, K. A. 1890 A, p. 268.

Empo contracta Cope.

Cope, E. D. 1874 C, p. 46.
  1875 E, pp. 232, 279, pl. liii, figs. 14–17.
Loomis, F. B. 1900 A, p. 273, pl. xxvii, figs. 8, 9. (Cimolichthys.)
Stewart, A. 1900 A, p. 339.
  Cretaceous (Niobrara); Kansas.

Empo lisbonensis Stewart.

Stewart, A. 1899 C, p. 111.
Loomis, F. B. 1900 A, p. 222. (Syn. of E. merrillii.)
Stewart, A. 1900 A, p. 337, pl. lxi, figs. 10a, 10b.
  Cretaceous (Pierre); Kansas.

Empo merrillii Cope.

Cope, E. D. 1874 C, p. 46.

Cope, E. D. 1875 E, pp. 232, 279, pl. liii, figs. 10–13.
Loomis, F. B. 1900 A, p. 271, pl. xxvii, fig. 7. (Cimolichthys.)
  Cretaceous (Niobrara); Kansas.

Empo nepaholica Cope.

Cope, E. D. 1872 I, p. 347.
  1872 F, p. 345.
  1872 I, p. 351. (Cimolichthys sulcata.)
  1874 C, p. 46. (E. sulcata, E. nepæolica.)
  1875 E, pp. 230, 279, pls. xlix, fig. 9; pl. l, fig. 8; lii, fig. 1; liii, figs. 3–5. (E. nepæolica.)
Loomis, F. B. 1900 A, p. 271, pl. xxvii, figs. 1–3. (Cimolichthys nepæolica)
Stewart, A. 1900 A, p. 332. (E. nepæolica.)
  Cretaceous (Niobrara); Kansas.

Empo semianceps Cope.

Cope, E. D. 1872 I, p. 351. (Cimolichthys.)
  1872 F, p. 326. (Cimolichthys.)
  1874 C, p. 46.
  1875 E, pp. 233, 279, pl. liii, figs. 1, 2, 6–9.
Loomis, F. B. 1900 A, p. 273, pl. xxvii, figs. 4–6. (Cimolichthys.)
Stewart, A. 1900 A, p. 338, pl. lxi, figs. 6–9.
  Cretaceous (Niobrara); Kansas.

PACHYRHIZODONTIDÆ.

Cope, E. D. 1872 I, p. 343.
  1872 F, p. 344.
Loomis, F. B. 1900 A, p. 258. (Salmonidæ, part.)

Nicholson and Lydekker 1889 A, p. 998.
Stewart, A. 1900 A, p. 349. (Salmonidæ, part.)

PACHYRHIZODUS Agassiz.   Type *A. basalis* Agassiz.

Agassiz, L., 1850, Dixon's Geol. Sussex, p. 374.
Cope, E. D. 1872 F, p. 344.
  1872 I, p. 344.
  1875 E, pp. 219, 220, 276.
  1887 D, p. 177.
Loomis, F. B. 1900 A, p. 258.
Nicholson and Lydekker 1889 A, p. 999.
Pictet, F. J. 1854 A, p. 94.

Reis, O. M. 1895 C, p. 170.
Stewart, A. 1900 A, p. 349.
Woodward, A. S. 1888 I, p. 313.
Zittel, K. A. 1890 A, p. 268.

Pachyrhizodus caninus Cope.

Cope, E. D. 1872 I, p. 344.
  1872 F, p. 348.

Cope, E. D. 1874 C, p. 42.
  1875 E, pp. 221, 276, pl. 1, figs. 1–4.
Crook, A. J. 1892 A, p. 109.
Loomis, F. B. 1900 A, p. 262, pl. xxvii, figs. 10–12.
Stewart, A. 1900 A, p. 355, pl. lxx, figs. 3–6.
  Cretaceous (Niobrara); Kansas.

Pachyrhizodus curvatus Loomis.

Loomis, F. B. 1900 A, p. 264 (P. recurvatus); p. 265,
  pl. xxv, figs. 6–8 (P. curvatus).
  Cretaceous (Niobrara); Kansas.

Pachyrhizodus kingii Cope.

Cope, E. D. 1872 I, p. 346.
  1872 F, p. 348.
  1874 C, p. 42.
  1875 E, pp. 223, 276, pl. xlvi, fig. 11.
Loomis, F. B. 1900 A, p. 263.   (Syn. of P. latimentum.)
  Cretaceous (Niobrara); Kansas.

Pachyrhizodus latimentum Cope.

Cope, E. D. 1872 I, p. 346.
  1872 F, p. 348.
  1874 C, p. 42.
  1875 E, pp. 223, 276, pl. 1, fig. 5; pl. li, figs. 1–7.
Loomis, F. B. 1900 A, p. 263, pl. xxvi, figs. 7, 8.

Stewart, A. 1900 A, p. 357, pl. lxviii; pl. lxx, figs.
  9, 10.
  Cretaceous (Niobrara); Kansas.

Pachyrhizodus leptognathus Stewart.

Stewart, A. 1898 D, p. 193, pl. xvii, fig. 1.
Loomis, F. B. 1900 A, p. 264, pl. xxvi, figs. 10–16.
Stewart, A. 1900 A, p. 351, pl. lxix, fig. 1.
  Cretaceous (Niobrara); Kansas.

Pachyrhizodus leptopsis Cope.

Cope, E. D. 1874 C, p. 42.
  1875 E, pp. 225, 276, pl. li, figs. 8–8c.
Loomis, F. B. 1900 A, p. 264.   (P. lepitopsis.)
Stewart, A. 1900 A, p. 354, pl. lxx, fig. 1.
  Cretaceous (Niobrara); Kansas.

Pachyrhizodus minimus Stewart.

Stewart, A. 1899 B, p. 37, figure.
  1900 A, p. 361, text fig. 1.
  Cretaceous (Niobrara); Kansas.

Pachyrhizodus velox Stewart.

Stewart, A. 1898 D, p. 193, pl. xvii, fig. 2.
Loomis, F. B. 1900 A, p. 265.   (P. ferox.)
Stewart, A. 1900 A, p. 353, pl. xlix, fig. 9.
  Cretaceous (Niobrara); Kansas.

CONOSAURUS Gibbes.   Type *C. bowmani* Gibbes.

Gibbes, R. W. 1851 A, p. 9.
Cope, E. D. 1872 F, p. 345.
  1875 E, p. 276.
Leidy, J. 1868 G, p. 202.   (Conosaurops, to replace
  Conosaurus.)
Loomis, F. B. 1900 A, p. 261.
Zittel, K. A. 1890 A, p. 268.   (Syn. of Pachyrhiz-
  odus.)

Conosaurus bowmani Gibbes.

Gibbes, R. W. 1851 A, p. 10, pl. iii, figs. 1–5.
Cope, E. D. 1875 E, p. 276.
Gibbes, R. W. 1850 A.   (Name only.)
Leidy, J. 1868 G, p. 202.   (Conosaurops.)
  Cretaceous (Greensand); New Jersey.

ORICARDINUS Cope.   Type *O. tortus* Cope.

Cope, E. D. 1877 D, p. 177.

Oricardinus sheareri Cope.

Cope, E. D. 1872 I, p. 347.   (Pachyrhizodus.)
  1872 F, p. 348.   (Pachyrhizodus.)
  1874 C, p. 43.   (Pachyrhizodus.)
  1875 E, pp. 225, 276.   (Pachyrhizodus.)
  1877 D, p. 177.   (Oricardinus.)

Loomis, F. B. 1900 A, p. 264.   (Pachyrhizodus
  sheari.)
  Cretaceous (Niobrara); Kansas.

Oricardinus tortus Cope.

Cope, E. D. 1877 D, p. 177.
  Cretaceous (Niobrara); Kansas.

ENCHODONTIDÆ.

Loomis, F. B. 1900 A, p. 267.   (Enchodidæ.)
Nicholson and Lydekker 1889 A, p. 997.

Stewart, A. 1900 A, p. 378.

ENCHODUS Agassiz.   Types *E. faujasi* and *E. halocyon.*

Agassiz, L. 1843 B (1844), v, p. 64.
  1844 A.
Cope, E. D. 1872 F, p. 347.
  1872 I, p. 354.
  1875 E, pp. 219, 238, 301 (Enchodus); pp.
  219, 235, 277 (Phasganodus).
Dames, W. 1887 B, p. 71.
Davis, J. W. 1887 C, p. 521.
Dollo, L. 1892 B, p. 183.
Günther, A. 1880 A, p. 433.
Heckel, J. J. 1843 A, p. 240.   (Isodus.)
Leidy, J. 1873 B, p. 289.

Loomis, F. B. 1900 A, p. 273.
Nicholson and Lydekker 1889 A, p. 996.
Pictet, F. J. 1854 A, p. 81.
Woodward, A. S. 1888 I, p. 315.
  1889 A.
Zittel, K. A. 1890 A, p. 269.

Enchodus amicrodus Stewart.

Stewart, A. 1898 D, p. 193, text-fig. 3.
Loomis, F. B. 1900 A, p. 280.
Stewart, A. 1900 A, p. 379, text fig. p. 378.
  Cretaceous (Niobrara); Kansas.

Enchodus anceps Cope.

Cope, E. D. 1872 I, p. 352. (Cimolichthys.)
  1874 C, p. 44. (Enchodus?)
  1875 E, pp. 236, 277 (Phasganodus); p. 301
  (Enchodus).
Loomis, F. B. 1900 A, p. 277.
  Cretaceous (Niobrara); Kansas.

Enchodus calliodon Cope.

Cope, E. D. 1872 I, p. 354.
  1872 F, pp. 347, 348.
  1874 C, p. 44.
  1875 E, pp. 240, 278, 301.
Loomis, F. B. 1900 A, p. 277.
  Cretaceous (Niobrara); Kansas.

Enchodus carinatus (Cope).

Cope, E. D. 1869 F, p. 241. (Sphyræna.)
  1871 D, p. 424. (Sphyræna.)
  1872 F, p. 347. (Cimolichthys.)
  1875 E, pp. 80, 285, 277. (Phasganodus.)
Loomis, F. B. 1900 A, p. 277.
  Cretaceous (Niobrara); Kansas.

Enchodus dolichus Cope.

Cope, E. D. 1875 E, pp. 239, 278, 300, pl. liv, figs.
  8, 8a.
  1885 Z, p. 3.
Loomis, F. B. 1900 A, p. 279, pl. xxvii, figs. 16, 17.
Stewart, A. 1900 A, p. 377, pl. lxx, fig. 12.
  Cretaceous (Niobrara); Kansas.

Enchodus ferox Leidy.

Leidy, J. 1855 C, p. 397.
Cope, E. D. 1875 E, p. 277.
Emmons, E. 1860 A, p. 214, fig. 182¹.
Loomis, F. B. 1900 A, p. 277.
Morton, S. G. 1834 A, p. 82, pl. xii, fig. 1. ("Sphyræna.")
  Cretaceous; New Jersey.

Enchodus gentryi (Cope).

Cope, E. D. 1875 H, p. 362. (Phasganodus.)
  Miocene; New Jersey.

Enchodus gladiolus Cope.

Cope, E. D. 1872 I, p. 353. (Cimolichthys.)
  1874 C, p. 43. (Phasganodus?.)
  1875 E, pp. 285, 277 (Phasganodus); 301, pl.
  xlii, fig. 7 (Enchodus).
  Cretaceous (Niobrara); Kansas.

Enchodus oxytomus Cope.

Cope, E. D. 1875 E, p. 278.
Loomis, F. B. 1900 A, p. 277.
  Cretaceous; New Jersey.

Enchodus parvus Stewart.

Stewart, A. 1898 D, p. 192, text fig. 2.
Loomis, F. B. 1900 A, p. 280. (Syn. of E. shumardii.)
Stewart, A. 1900 A, p. 378, text fig.
  Cretaceous (Niobrara); Kansas.

Enchodus petrosus Cope.

Cope, E. D. 1874 C, p. 44.
  1875 E, pp. 239, 278, pl. liv, figs. 4–6.
Loomis, F. B. 1900 A, p. 278, pl. xxvii, figs. 13–15.
Stewart, A. 1900 A, p. 376, pl. lxx, fig. 11.
  Cretaceous (Niobrara) Kansas.

Enchodus pressidens Cope.

Cope, E. D. 1869 F, p. 241.
  1875 E, p. 277.
Loomis, F. B. 1900 A, p. 277.
  Cretaceous; New Jersey.

Enchodus semistriatus Marsh.

Marsh, O. C. 1870 E, p. 230.
Cope, E. D. 1875 E, p. 277. (Phasganodus.)
Loomis, F. B. 1900 A, p. 277.
  Cretaceous; New Jersey.)

Enchodus shumardii Leidy.

Leidy, J. H. 1856 M, p. 257.
Cope, E. D. 1874 C, p. 44.
  1875 E, p. 279.
Leidy, J. 1873 B, pp. 289, 347, pl. xvii, fig. 20.
Loomis, F. B. 1900 A, p. 280, pl. xxvii, figs. 18, 19.
Stewart, A. 1900 A, p. 375.
Whiteaves, J. F. 1889 B, p. 194, pl. xxvi, figs. 7 a, b, c.
  Cretaceous; Dakota, Manitoba.

Enchodus tetræcus Cope.[1]

Cope, E. D. 1875 E, p. 278.
Loomis, F. B. 1900 A, p. 277 (E, tetracus).
Stewart A. 1900 A, p. 375. (E. tetracus.)
  Cretaceous; Delaware, New Jersey.

Enchodus sp. indet.

Cope, E. D. 1871 D, p. 424.
  Cretaceous (Niobrara); Kansas.

TETHEODUS Cope. Type *T. pephredo* Cope.

Cope, E. D. 1874 C, p. 43.
  1875 E, pp. 219, 287, 277.
Loomis, F. B. 1900 A, p. 278. (Syn. of Enchodus.)
Zittel, K. A. 1890 A, p. 270.

Tetheodus pephredo Cope.

Cope, E. D. 1874 C, p. 43.

Cope, E. D. 1875 E, pp. 287, 277, pl. liv, figs. 1–3.
Loomis, F. B. 1900 A, p. 278. (T. pephero; syn. of Enchodus petrosus.)
  Cretaceous (Niobrara); Kansas.

---

[1] Agassiz (1843 B (1844), v, p. 65, pl. xxv c, figs. 1, 2; pl. xxxiii, figs. 2–4) reports *E. halocyon* from Cretaceous of Maryland. Its original locality is Europe. The determination was probably erroneous.

CIMOLICHTHYS Leidy. Type *C. lewesiensis* Leidy.

Leidy, J. 1856 O, p. 302.
Agassiz, L. 1843 B, v, pt. i, p. 102. (Saurocephalus, Saurodon, not of Harlan and Hays.)
Cope, E. D. 1870 L, p. 530.
  1871 D, p. 415.
  1872 F, p. 346.
  1875 E, pp. 219, 235, 277, 300. (Phasganodus, in part.)
Dames, W. 1887 A. (Phasganodus=Saurodon Agassiz.)
Dixon, F. 1878 A, p. 411.
Leidy, J. 1857 F, p. 167. (Phasganodus, type *P. dirus.*)
  1873 B, p. 347. (Phasganodus.)
Loomis, F. B. 1900 A, p. 267.

Woodward, A. S. 1888 I, p. 316.
Zittel, K. A. 1890 A, p. 269 (Cimolichthys); p. 270 (Phasganodus).

Cimolichthys dirus (Leidy.)

Leidy, J. 1857 F, p. 167. (Phasganodus.)
Cope, E. D. 1874 C, p. 43. (Phasganodus.)
  1875 E, p. 277. (Phasganodus.)
Dames, W. 1887 A, p. 74. (Saurodon.)
Leidy, J. 1873 B, pp. 289, 347, pl. xvii, figs. 23, 24. (Phasganodus.)
Stewart, A. 1900 A, p. 376, pl. lxx, fig. 14. (Enchodus.)
  Cretaceous (Fox Hills); North Dakota.

ALBULIDÆ.

Günther, A. 1868 A, pp. 468, 469. (Albulina.)
Jordan and Evermann 1896 A, p. 410.

Loomis, F. B. 1900 A, p. 252.

SYNTEGMODUS Loomis. Type *S. altus* Loomis.

Loomis, F. B. 1900 A, p. 252.

Syntegmodus altus Loomis.

Loomis, F. B. 1900 A, p. 253, pl. xxii, fig. 9.
  Cretaceous (Niobrara); Kansas.

APSOPELIX Cope. Type *A. sauriformis* Cope.

Cope, E. D. 1871 D, p. 423.
  1875 E, p. 241.
Zittel, K. A. 1890 A, p. 312.
A genus of uncertain position.

Apsopelix sauriformis Cope.

Cope, E. D. 1871 D, p. 424.
  1874 C, p. 47.
  1875 E, pp. 242, 279, pl. xlii, fig. 6.
  Cretaceous (Benton); Kansas.

CYCLOTOMODON Cope. Type *C. vagrans* Cope.

Cope, E. D. 1876 F, p. 113.
Systematic position doubtful.

Cyclotomodon vagrans Cope.

Cope, E. D. 1876 F, p. 113.
  Phosphate deposit; South Carolina.

CLUPEIDÆ.

Agassiz, L. 1843 B (1844), v, p. 96. ("Halécoides," in part.)
  1845 A, p. 292.
Bridge, T. W. 1896 A, p. 557.
Brühl, B. C. 1847 A. ("Clupeen.")
Cope, E. D. 1872 C, p. 333.
  1884 O, p. 67.
Dareste, C. 1872 A, p. 1087.
Grassi, B. 1883 A.
Günther, A. 1868 A, p. 381.
  1880 A, p. 655.
  1881 A, p. 693.
Hitchcock, Fanny R. 1888 B.
Jordan, D. S. 1891 A.
Jordan and Evermann 1896 A, p. 417.
Klein, E. F. 1879 A, p. 106.

Klein, E. F. 1886 A.
Kölliker, A. 1859 A. (Clupeinæ.)
Marck, W. 1863 A, p. 41. (Clupeoidei.)
Müller, J. 1846 C, p. 790.
Nicholson and Lydekker 1889 A, p. 994.
Owen, R. 1845 B, p. 135.
  1866 A, p. 10, *et seq.*
Parker, W. K. 1868 A, p. 56.
Quenstedt, F. A. 1885 A, p. 367. (Clupacei.)
Stannius, H. 1846 A. ("Clupeiden.")
  1854 A.
Steinmann and Döderlein 1890 A, p. 579.
Stephan, P. 1900 A. ("Clupéidés.")
Vaillant, L. 1883 A.
Vrolik, A. J. 1873 A, p. 267.
Zittel, K. A. 1890 A, p. 271.

DIPLOMYSTUS Cope. Type *D. dentatus* Cope.

Cope, E. D. 1877 I, p. 808.

A few references are here included to the literature of the genus *Clupea*, which is closely related to *Diplomystus* and to which most of the species of the latter were originally assigned.
Agassiz L. 1843 B (1844), v, p. 115. (Clupea.)
Cope, E. D. 1884 O, p. 73.
 1885 Z, p. 4.
Günther, A. 1868 A, p. 412. (Clupea.)
Jordan and Evermann 1896 A, p. 421. (Clupea.)
Müller, J. 1843 A, p. 216.
Nicholson and Lydekker 1889 A, p. 996.
Pictet, F. J. 1854 A, p. 113. (Clupea.)
Quenstedt, F. A. 1885 A, p. 367. (Clupea.)
Reis, O. M. 1895 C, p. 167. (Clupea.)
Woodward, A. S. 1895 C, p. 1.
Zittel, K. A. 1890 A, p. 276. (Clupea, Diplomystus.)

Diplomystus altus (Leidy).

Leidy, J. 1873 B, pp. 196, 347, pl. xvii, fig. 2. (Clupea.)
Cope, E. D. 1877 I, p. 811.
 1877 Z, p. 570.
 1877 AA.
 1884 O, p. 79.
King, C. 1878 A, p. 394. (Clupea.)
Eocene (Green River); Wyoming.

Diplomystus analis Cope.

Cope, E. D. 1877 I, p. 809.
 1877 Z, p. 570.
 1877 AA.
 1884 O, p. 75, pl. vii, fig. 4; pl. viii, fig. 3; pl. x, fig. 2.
Eocene (Green River); Wyoming.

Diplomystus dentatus Cope.
Cope, E. D. 1877 I, p. 808.
 1877 Z, p. 570.
 1877 AA.
 1884 O, P. 74, pl. x, fig. 1.
Eocene (Green River); Wyoming.

Diplomystus humilis (Leidy).
Leidy, J. 1856 M, p. 256. (Clupea.)
Cope, E. D. 1870 H, p. 382. (Clupea pusilla, C. humilis.)
 1871 E, p. 429. (Clupea pusilla, C. humilis.)
 1877 I, p. 811.
 1877 Z, p. 570.
 1877 AA.
 1884 O, p. 77, pl. ix, fig. 8; pl. x, fig. 4.
King, C. 1878 A, p. 394. (Clupea humilis, C. pusilla.)
Leidy, J. 1871 C, p. 369. (Clupea humilis, C. pusilla.)
 1872 B, pp. 353, 372. (Clupea.)
 1873 B, pp. 196, 347, pl. xvii, fig. 1. (Clupea.)
Eocene (Green River); Wyoming.

Diplomystus pectorosus Cope.
Cope, E. D. 1877 I, p. 810.
 1877 Z, P. 570.
 1877 AA.
 1884 O, p. 76, pl. x, fig. 3.
Eocene (Green River); Wyoming.

Diplomystus theta Cope.
Cope, E. D. 1874 D, p. 51. (Clupea.)
 1874 B, p. 461. (Clupea.)
 1877 Z, p. 570.
 1884 O, p. 77.
Eocene (Green River); Wyoming.

SPANIODON Pictet. Type *S. blondellii* Pictet.

Pictet, F. J., 1850, Poiss. foss. du Liban, p. 34.
Cope, E. D. 1878 F, p. 69.
 1878 H H, p. 57.
Davis, J. W. 1887 C, p. 587.
Pictet, F. J. 1854 A, p. 115.

Zittel, K. A. 1890 A, p. 274.

Spaniodon simus Cope.
Cope, E. D. 1878 F, p. 69.
Cretaceous (Niobrara); S. Dakota.

SARDINIUS Marck. Type *Osmerus cordieri* Agassiz.

Marck, W., 1858, Zeitschr. deutsch. geol. Gesellsch., x, p. 245.
Cope, E. D. 1878 A, p. 70.
Davis, J. W. 1887 C, p. 567.
Marck, W. 1863 A, p. 43.
Zittel, K. A. 1890 A, p. 275.

Sardinius? blackburnii Cope.
Cope, E. D. 1891 H, p. 654.
 1892 B.
Oligocene?; S. Dakota.

Sardinius lineatus Cope.
Cope, E. D. 1878 F, p. 71.
Cretaceous (Niobrara); S. Dakota.

Sardinius nasutulus Cope.
Cope, E. D. 1878 F, p. 70.
Cretaceous (Niobrara); S. Dakota.

Sardinius percrassus Cope.
Cope, E. D. 1878 F, p. 72.
Cretaceous (Niobrara); S. Dakota.

SALMONIDÆ.

Agassiz, L. 1843 B (1844), v, p. 96. (Halécoides, in
    part.)
    1845 A, p. 306.
Bridge, T. W. 1896 A, p. 550.
Bruch, 1862, Osteol. Rheinlachs.
Brühl, B. C. 1847 A.
Cope, E. D. 1872 C, p. 333.
Dareste, C. 1872 A, p. 1175.
Gill, T. 1894 C, p. 117.
Grassi, B. 1883 A.
Günther, A. 1866 A, p. 1.
    1880 A, p. 630.
    1881 A, p. 693.
Jordan and Evermann 1896 A, p. 460 (Salmonidæ);
    p. 519 (Argentinidæ).
Klein, E. F. 1879 A, p. 107.

Kölliker, A. 1859 A. (Salmones.)
Müller, J. 1843 A, p. 214. (Salmones.)
Nicholson and Lydekker 1889 A, p. 994.
Owen, R. 1845 B, p. 141. ("Salmonoids.")
    1866 A.
Parker, W. K. 1868 A, p. 55.
    1874 A.
Quenstedt, F. A. 1885 A, p. 365. (Salmonei.)
Reis, O. M. 1895 C, p. 170.
Stannius, H. 1854 A. (Salmones.)
Stephan, P. 1900 A. (Salmones.)
Ussow, S. 1900 A. ("Salmoniden.")
Vaillant, L. 1883 A.
Vrolik, A. J. 1873 A, p. 248.
Zittel, K. A. 1890 A, p. 279.

RHABDOFARIO Cope.   Type *R. lacustris* Cope.

Cope, E. D. 1870 M, p. 545.
Zittel, K. A. 1890 A, p. 280.

Rhabdofario lacustris Cope.

Cope, E. D. 1870 M, p. 546.
    Pleistocene; Idaho.

LEPTICHTHYS Stewart.   Type *L. agilis* Stewart.

Stewart, A. 1899 E, p. 78.
    1900 A, p. 872.

Leptichthys agilis Stewart.

Stewart, A. 1899 E, p. 78.
    1900 A, p. 372, pl. lxxii, fig. 1.
    Cretaceous; Kansas.

MALLOTUS Cuvier.   Type *M. villosus* (Müller).

Cuvier, G., 1829, Règne Anim., ed. ii, vol. ii, p. 305.
Agassiz, L. 1843 B (1844), v., p. 98.
Günther, A. 1866 A, p. 170.
    1880 A, p. 647.
Jordan and Evermann 1896 A, p. 520.
Zittel, K. A. 1890 A, p. 280.

Mallotus villosus (Müller).

Müller, 1877, Prod. Zool. Dan., p. 245. (Clupea.)
Agassiz, L. 1843 B (1844), v., p. 98, pl. lx.

Dawson, J. W. 1877 B, p. 341.
    1890 A, p. 86.
Günther, A. 1866 A, p. 170.
Jordan and Evermann 1896 A, p. 521.
Logan, W. E. 1863 A, p. 916, fig. 494.
Quenstedt, F. A. 1885 A, p. 365.
Woodward, A. S. 1889 E, p. 230.
Zittel, K. A. 1890 A, p. 280, fig. 290.
    Living in northern seas: Pleistocene; Canada.

GONORHYNCHIDÆ.

Cope, E. D. 1872 C, p. 333.
Gill, T. 1872 C, p. 16.
Günther, A. 1868 A, p. 373.
    1880 A, p. 652.

Günther, A. 1881 B, p. 693.
Nicholson and Lydekker 1889 A, p. 999.
Woodward, A. S. 1896 B.
Zittel, K. A. 1890 A, p. 285. (Ganorhynchidæ.)

NOTOGONEUS Cope.   Type *N. osculus* Cope.

Cope, E. D. 1885 Y, p. 1091.
    1886 T, p. 163.
Nicholson and Lydekker 1889 A, p. 999.
Woodward, A. S. 1896 B, p. 500.
Zittel, K. A. 1890 A, p. 285.

Notogoneus osculus Cope.

Cope, E. D. 1885 Y, p. 1091.
    1886 T, p. 163, pl. i. figs. 4, 5.
Whitfield, R. P. 1890 A, p. 117.
Woodward, A. S. 1896 B, p. 500, pl. xviii, figs. 1, 2.
    Eocene (Green River); Montana.

OSTEOGLOSSIDÆ.

Bridge, T. W. 1895 A, p. 302. (Osteoglossum.)
Cope, E. D. 1872 C, p. 333.
    1884 O, pp. 67, 68.
Gegenbaur, C. 1898, p. 356. (Osteoglossum.)
Günther, A. 1868 A, vii, p. 377.
Hyrtl, J. 1854 A, p. 73.
Loomis, F. B. 1900 A, p. 229. (Plethodidæ, part.)
Müller, J. 1843 A, p. 217. (Osteoglossum.)

Müller, J. 1846 A. (Osteoglossum.)
    1846 C. (Osteoglossum.)
Nicholson and Lydekker 1889 A, p. 999.
Owen, R. 1845 B, p. 137, pl. xlviii. (Osteoglossum.)
Reis, O. M. 1895 C, p. 169.
Zittel, K. A. 1890 A, p. 281.

PHAREODUS Leidy.   Type *P. acutus* Leidy.

Leidy, J. 1873 D, p. 99.
Cope, E. D. 1871 E, p. 430. (Osteoglossum.)
    1873 E, p. 637. (Phareodon, in part.)
    1877 I, p. 807. (Dapedoglossus, type *D. testis.*)
    1878 Z, p. 570. (Dapedoglossus.)
    1884 O, p. 68. (Dapedoglossus.)
Leidy, J. 1873 B, p. 193.
Nicholson and Lydekker 1889 A, p. 999. (Dapedoglossus.)
Zittel, K. A. 1890 A, p. 281. (Dapedoglossus.)

Phareodus acutus Leidy.

Leidy, J. 1873 D, p. 99.
Cope, E. D. 1873 E, p. 637 (Phareodon acutus); p. 638 (P. sericeus).
    1884 O, pp. 69, 72, pl. v, figs. 18–20. (Dapedoglossus.)
King, C. 1878 A, p. 405.
Leidy, J. 1873 B, pp. 193, 349, pl. xxxii, figs. 47–51.
Eocene (Bridger); Wyoming.

Phareodus æquipinnis (Cope).

Cope, E. D. 1878 F, p. 77. (Dapedoglossus.)
    1884 O, p. 73, pl. vii, fig. 2. (Dapedoglossus.)
Eocene (Green River); Wyoming.

Phareodus encaustus (Cope).

Cope, E. D. 1871 E, p. 430. (Osteoglossum.)
    1877 I, p. 808. (Dapedoglossus.)
    1877 AA (Dapedoglossus).
    1878 Z, p. 570. (Dapedoglossus.)
    1884 O, p. 70, pl. vi, fig. 1. (Dapedoglossus.)
Leidy, J. 1872 B, p. 372. (Osteoglossum.)
Eocene (Green River); Wyoming.

Phareodus testis Cope.

Cope, E. D. 1877 I, p. 807.
    1877 AA. (Dapedoglossus.)
    1878 Z, p. 570.
    1884 O, p. 71, pl. vii, fig. 1; pl. viii, figs. 1, 2.
Eocene (Green River); Wyoming.

ANOGMIUS Cope.   Type *A. contractus* Cope.

Cope, E. D. 1871 J, p. 170.
    1872 I, p. 354.
    1875 E, pp. 220 A, 240.
    1877 D, p. 178.
Loomis, F. B. 1900 A, p. 229 (Thryptodus, type *T. zitteli*); p. 235 (Pseudothryptodus, type *P. intermedius*); p. 254 ("Agnomius," a syn. of Osmeroides).
Sauvage, H. E. 1879 A, p. 52. (Syn. ? of Ischyrhiza.)
Stewart, A. 1900 A, p. 349.
Zittel, K. A. 1890 A, p. 268. (Syn. of Pachyrhizodus.)
    Without doubt, close comparison will much reduce the number of species of this genus.

Anogmius aratus Cope.

Cope, E. D. 1877 G, p. 585.
Stewart, A. 1900 A, p. 340.
Cretaceous (Niobrara); Kansas.

Anogmius contractus Cope.

Cope, E. D. 1871 J, p. 170.
    1872 I, p. 354.
    1875 E. p. 240 (Anogmius); 220 A (Pachyrhizodus).
Cretaceous (Niobrara); Kansas.

Anogmius evolutus Cope.

Cope, E. D. 1877 D, p. 179.
Loomis, F. B. 1900 A, p. 257, pl. xxvi, figs. 5, 6. (Osmeroides.)

Stewart, A. 1898 D, p. 196. (Beryx multidentatus.)
    1900 A, p. 347, pl. lxv, figs. 8, 9, 10; pl. lxvii.
Cretaceous (Niobrara); Kansas.

Anogmius favirostris Cope.

Cope, E. D. 1877 D, p. 178.
Loomis, F. B. 1900 A, p. 256. (Osmeroides.)
Cretaceous (Niobrara); Kansas.

Anogmius intermedius (Loomis).

Loomis, F. B. 1900 A, p. 236, pl. xxii, figs. 3–8. (Pseudothryptodus.)
Cretaceous (Niobrara); Kansas.

Anogmius polymicrodus Stewart.

Stewart, A. 1898 D, p. 195. (Beryx?.)
Loomis, F. B. 1900 A, p. 256, pl. xxvi, figs. 1–4. (Osmeroides.)
Stewart, A. 1899 A, p. 20.
    1899 D, p. 117, pl. xxxi.
    1900 A, p. 342, pl. lxiv; pl. lxv, figs. 1–7; pl. lxvi; pl.lxvii, fig. "phar."
Cretaceous (Niobrara); Kansas.

Anogmius rotundus (Loomis).

Loomis, F. B. 1900 A, p. 235, pl. xxii, figs. 1, 2. (Thryptodus.)
Cretaceous (Niobrara); Kansas.

Anogmius zitteli (Loomis).

Loomis, F. B. 1900 A, p. 234, pl. xxi, figs. 1–10. (Thryptodus.)
Cretaceous (Niobrara); Kansas.

INCERTÆ SEDIS.

PROBALLOSTOMUS Cope   Type *P. longulus* Cope.

Cope, E. D. 1891 H, p. 655.

Proballostomus longulus Cope.

Cope, E. D. 1891 H, p. 655.

Cope, E. D. 1892 B, p. 285.
Oligocene?; S. Dakota.

Order PLECTOSPONDYLI.

Cope, E. D. 1872 C, p. 332.
1878 A, p. 295.
Gill, T. 1893 A, p. 131.

Jordan and Evermann 1896 A, p. 160.
Woodward, A. S. 1898 B, p. 119.

CYPRINIDÆ.

Bloch, L. 1900 A. ("Cobitiden," "Cypriniden.")
Bridge, T. W. 1896 A, p. 547.
Brühl, B. C. 1847 A. ("Cyprinen.")
Cope, E. D. 1871 H. (>Cobitidæ.)
1871 I, p. 99. (>Cobitidæ.)
1872 C, p. 332. (Catostomidæ, Cyprinidæ, Cobitidæ.)
1883 L, p. 161. (>Cobitidæ.)
Dareste, C. 1872 A, p. 934. ("Cyprinoïdes.")
Gill, T. 1872 C, pp. 17, 18. (Catostomidæ, Cyprinidæ, Cobitidæ.)
1893 A, pp. 131, 132. (Catostomidæ, Cyprinidæ, Cobitidæ.)
Grassi, B. 1883 A.
Günther, A. 1868 A, p. 3.
1880 A, p. 587.
1881 A, p. 692.
Ihering, H. 1891 B.
Jordan, D. S. 1891 A.
Jordan and Evermann 1896 A, p. 161 (Catostomidæ); p. 199 (Cyprinidæ).

Klein, E. F. 1879 A, p. 110.
Kölliker, A. 1859 A. (Cyprinoïdei.)
Koken, E. 1891 B, p. 26.
Leidy, J. 1871 C, p. 369.
Müller, A. 1853 A.
Müller, J. 1846 C, p. 183. (Cyprinoïdei.)
Owen, R. 1845 B, p. 144. ("Cyprinoids.")
1866 A.
Parker, W. K. 1868 A.
Pictet, F. J. 1854 A, p. 102. (Cyprinoïdes.)
Quenstedt, F. A. 1885 A, p. 361.
Sagemehl, M. 1884 A, p. 1.
1891 A. (Cyprinidæ, Cobitidæ.)
Segond, L. A. 1873 B, p. 529. ("Cyprinoïdes.")
Stannius, H. 1854 A. (Cyprinoïdei.)
Stephan, P. 1900 A.
Vaillant, L. 1883 A.
Vrolik, A. J. 1873 A, p. 265.
Zittel, K. A. 1890 A, p. 282. (Cyprinoïdæ.)

CATOSTOMINÆ.

Cope, E. D. 1872 C, p. 332. (Catostomidæ.)
Gill, T. 1872 C, p. 17. (Catostomidæ.)
1893 A, p. 131. (Catostomidæ.)
Günther, A. 1868 A, p. 12.

Günther, A. 1880 A, p. 588. (Catostomina).
Jordan and Evermann 1896 A, p. 161. (Catostomidæ.)
Sagemehl, M. 1891 A.

PROTOCATOSTOMUS Whitfield.   Type *P. constablei* Whitfield.

Whitfield, R. P. 1890 A, p. 120.

Protocatostomus constablei Whitfield.

Whitfield, R. P. 1890 A, p. 120, pl. iv.
Eocene; Wyoming.
See *Notogoneus osculus*, p. 392.

CATOSTOMUS Le Sueur.   Type *C. catostomus* Le Sueur.

Le Sueur, 1817, Jour. Acad. Nat. Sci. Phila., i, p. 89.
Jordan and Evermann 1896 A, p. 173.
Sagemehl, M. 1891 A.

Catostomus batrachops Cope.

Cope, E. D. 1883 L, p. 151.
1889 F.
1889 S, p. 980.
Pleistocene (Equus beds); Oregon.

Catostomus cristatus Cope.

Cope, E. D. 1883 L, p. 160.
Pleistocene (Equus beds); Idaho.

Catostomus labiatus Ayres.

Ayres, W. O. 1855, Proc. Califor. Acad. Sci., i, p. 32.
Cope, E. D. 1883 L, p. 150.
1889 F.
1889 S.
Jordan and Evermann 1896 A, p. 177.
Recent and Pleistocene; Klamath Lake, Oregon.

Catostomus shoshonensis Cope.

Cope, E. D. 1883 L, p. 159.
Pleistocene (Equus beds); Idaho.

AMYZON Cope.　Type *A. mentale* Cope.

Cope, E. D. 1872 P, p. 480.
　1873 E, p. 642.
　1884 O, pp. 742, 748.
Osborn, Scott, and Speir 1878 A, p. 99.
Zittel, K. A. 1890 A, p. 284.

Amyzon brevipinne Cope.
Cope, E. D. 1893 D.
　Tertiary (Amyzon shales); British Columbia.

Amyzon commune Cope.
Cope, E. D. 1874 D, p. 50.
　1874 B, p. 460.
　1875 G, p. 3.
　1884 O, p. 749, pl. v, fig. 21.
Osborn, Scott, and Speir 1878 A, p. 99.
Tertiary (Amyzon shales); Colorado.

Amyzon fusiforme Cope.
Cope, E. D. 1875 G, p. 5.
　1884 O, p. 751.
　Tertiary (Amyzon shales); Colorado.

Amyzon mentale Cope.
Cope, E. D. 1872 P, p. 481.
　1872 MM, p. 775.
　1873 E, p. 643.
　1884 O, p. 749, pl. lix, fig. 2; lx, fig. 2.
　Tertiary (Amyzon shales); Colorado.

Amyzon pandatum Cope.
Cope, E. D. 1875 G, p. 4.
　1884 O, p. 750.
　Tertiary (Amyzon shales); Colorado

CYPRININÆ.

This subfamily is equivalent to the *Cyprinidæ* as recognized by many of the authors referred to under that family.

Brühl, B. C. 1847 A. ("Cyprinen.")
Günther, A. 1880 A, p. 589. (Cyprininæ.)

DIASTICHUS Cope.　Type *D. macrodon* Cope.

Cope, E. D. 1870 M, p. 539.
　1871 H, p. 55.
　1883 L, pp. 154, 158.
Zittel, K. A. 1890 A, p. 284.

Diastichus macrodon Cope.
Cope, E. D. 1870 M, p. 539.
　1883 L, p. 158.
　Late Tertiary; Idaho.

Diastichus parvidens Cope.
Cope, E. D. 1870 M, p. 540.
　1883 L, p. 158.
　Late Tertiary; Idaho.

Diastichus strangulatus Cope.
Cope, E. D. 1883 L, p. 158.
　Late Tertiary; Idaho.

APHELICHTHYS Cope.　Type *A. lindahlii* Cope.

Cope, E. D. 1893 B, p. 19.

Aphelichthys lindahlii Cope.
Cope, E. D. 1893 B, p. 19.
　Pliocene or Pleistocene; S. Illinois.

RUTILUS Raf.　Type *Cyprinus rutilus* Raf.

Rafinesque, C. S., 1820, Ichthyol. Ohioensis, pp. 48, 50.
Cope, E. D. 1870 M, p. 543. (Anchybopsis, type *A. latus*.)
　1878 C, p. 229. (Anchybopsis.)
　1883 L, p. 142. (Myloleucus.)
Günther, A. 1868 A, p. 267. (Leuciscus, in part.)
Jordan and Evermann 1896 A, p. 243.
Zittel, K. A. 1890 A, p. 284. (Anchybopsis.)

Rutilus altarcus (Cope).
Cope, E. D. 1878 C, p. 229. (Anchybopsis.)
　1883 L, p. 146. (Leucus.)
　1889 F. (Leucus.)
　1889 S, p. 980. (Leucus.)
　Pleistocene (Equus); Oregon.

Rutilus condonianus (Cope).
Cope, E. D. 1883 L, p. 156. (Leucus.)
　Pleistocene (Equus); Oregon.

Rutilus gibbarcus (Cope).
Cope, E. D. 1878 C, p. 230. (Alburnops.)
　1878 C, p. 229. (Alburnops breviarcus.)
　1883 L, p. 143. (Myloleucus.)
　1889 F, p. 161. (Myloleucus.)
　1889 S, p. 980. (Myloleucus.)
　Pleistocene (Equus); Oregon.

Rutilus latus (Cope).
Cope, E. D. 1870 M, p. 543. (Anchybopsis.)
　1883 L, pp. 144, 146, 156. (Leucus.)
　Pleistocene (Equus); Idaho.

NOTROPIS Raf.   Type *N. atherinoides* Raf.

Rafinesque, C. S., 1818, Amer. Month. Mag., ii, p. 204.
Jordan and Evermann 1896 A, p. 254.

Notropis angustarcus (Cope).

Cope, E. D. 1878 C, p. 230. (Alburnops.)

Cope, E. D. 1883 L, p. 142. (Cliola.)
1889 F. (Cliola.)
1889 S, p. 980. (Cliola.)
Pleistocene (Equus); Oregon.

LEUCISCUS Cuvier.   Type, none designated.

Cuvier, G., 1817, Règne Anim., ed. i, p. 194.
Cope, E. D. 1870 M, p. 540 (Oligobelus, type *O. arciferus*); p. 541 (Semotilus).
1883 L, p. 156. (Squalius.)
Günther, A. 1868 A, p. 207.
1880 A, p. 598.
Jordan and Evermann 1896 A, p. 228.
Zittel, K. A. 1890 A, p. 282 (Leuciscus); p. 284 (Oligobelus).

Leuciscus arciferus (Cope).

Cope, E. D. 1870 M, p, 541. (Oligobelus.)
1883 I, p. 158. (Squalius.)
Pleistocene: Idaho.

Leuciscus bairdii (Cope).

Cope, E. D. 1870 M, p. 542. (Semotilus.)
1883 L, p. 158. (Squalius.)
Pleistocene; Idaho.

Leuciscus laminatus (Cope).

Cope, E. D. 1870 M, p. 541. (Oligobelus.)
1883 L, p. 157. (Squalius.)
Pleistocene; Idaho.

Leuciscus posticus (Cope).

Cope, E. D. 1870 M, p. 541. (Semotilus.)
1883 L, p. 157. (Squalius.)
Pleistocene; Idaho.

Leuciscus reddingi (Cope).

Cope, E. D. 1883 L, p. 157. (Squalius.)
Pleistocene; Idaho.

Leuciscus turneri Lucas.

Lucas, F. A. 1900 F, p. 338, pl. viii.
Miocene; Nevada.

MYLOCYPRINUS Leidy.   Type *M. robustus* Leidy.

Leidy, J. 1870 K, p. 70.
Cope, E. D. 1870 M, pp. 539, 543.
1883 L, p. 154.
Leidy, J. 1871 C, p. 369.
Peters, 1880, Monatsber. Akad. Wissensch. Berlin, p. 925, figure. (Mylopharyngodon, type *Leuciscus æthiops.*)
Zittel, K. A. 1890 A, p. 284.

Mylocyprinus inflexus Cope.

Cope, E. D. 1883 L, p. 154.
Pleistocene; Idaho.

Mylocyprinus robustus Leidy.

Leidy, J. 1870 K, p. 70.
Cope, E. D. 1870 M, p. 544 (M. kingii, M. robustus); p. 545 (M. longidens).
1883 L, p. 155. (M. robustus, with M. kingii and longidens as subspecies.)
Leidy, J. 1871 C, p. 369.
1873 B, pp. 262, 348, pl. xvii, figs. 11-17.
Pleistocene; Idaho.

Order PHTHINOBRANCHII nom nov.

Interclavicles developed. Elements of the branchial arches more or less reduced. Posttemporal simple. Basis cranii simple. Head more or less produced. Ventral fins abdominal, subabdominal, or wanting. Most families with more or less well-developed dermal armature.

In this order it is intended to unite the *Hemibranchii* and the *Lophobranchii.*

Suborder HEMIBRANCHII Cope.

Cope, E. D., 1871, Trans. Amer. Philos. Soc., xiv, p. 456.
1872 C, p. 338.
Gill, T. 1872 C, p. 13.

Gill, T. 1893 A, p. 137.
Jordan and Evermann 1896, A p. 741.
Ussow, S. 1900 A, p. 227. (Gasterosteiformes.)

DERCETIDÆ.

Cope, E. D. 1878 F, p. 67.
Agassiz, L. 1843 B, ii, pt. ii, p. 258. (Genus Dercetis.)

Günther, A. 1880 A, p. 666. (Hoplopleuridæ.)
Lütken, C. 1873 A, p. 17. (Dercetiformes.)
Marck, W. 1863 A, p. 58. (Dercetiformes.)

Nicholson and Lydekker 1889 A, p. 997.
Pictet, F. J. 1854 A, p. 213 (Hoplopleurides); p. 216, pl. xxxii, figs. 13–16 (Dercetes).
Pictet and Humbert 1866 A, p. 90. (Hoplopleurides.)

Reis, O. M. 1896 C, pp. 169, 180, 182. (Hoplopleuridæ.)
Williston, S. W. 1899 B, p. 115.
1900, in Stewart A. 1900 A, p. 380.
Zittel, K. A. 1890 A, p. 264. (Hoplopleuridæ.)

ICHTHYOTRINGA Cope. Type *I. tenuirostris* Cope.

Cope, E. D. 1878 F, p. 69.
1878 HH, p. 57.
Williston, S. W. 1899 B, p. 115.
1900, in Stewart A. 1900 A, p. 382.

Ichthyotringa tenuirostris Cope.
Cope, E. D. 1878 F, p. 69.
Cretaceous (Niobrara); South Dakota.

LEPTECODON Williston. Type *L. rectus* Willist.

Williston, S. W. 1899 B, p. 113.
Leptecodon rectus Willist.
Williston, S. W. 1899 B, p. 113, pl. xxvi.

Williston, S. W. 1900, in Stewart, A. 1900 A, p. 380, pl. lxxiii.
Cretaceous (Niobrara); Kansas.

TRIÆNASPIS Cope. Type *T. virgulatus* Cope.

Cope, E. D. 1878 F, p. 67.
1878 HH, p. 57.
Williston, S. W. 1899 B, p. 115.
1900, in Stewart, A. 1900 A, p. 382.

Triænaspis virgulatus Cope.
Cope, E. D. 1878 F, p. 67.
Cretaceous (Niobrara); S. Dakota.

LEPTOTRACHELUS Marck. Type *L. armatus* Marck.

Marck, W. 1863 A, p. 59.
Cope, E. D. 1878 F, p. 68.
1878 HH, p. 57.
Davis, J. W. 1887 C, p. 619.
Nicholson and Lydekker 1889 A, p. 997. (Syn. of Dercetes.)
Pictet and Humbert 1866 A, p. 93.

Reis, O. M. 1895 C, p. 169.
Zittel K. A. 1890 A, p. 268.
Leptotrachelus longipinnis Cope.
Cope, E. D. 1878 F, p. 68.
Williston, S. W. 1899 B, p. 115.
1900, in Stewart, A. 1900 A, p. 382.
Cretaceous (Niobrara); S. Dakota.

GASTEROSTEIDÆ.

Cope, E. D. 1872 C, p. 338.
Gill, T. 1893 A, p. 137.
Günther, A. 1859 L, p. 1.
1880 A, p. 504.

Jordan and Evermann 1896 A, p. 742.
Parker, W. K. 1868 A, p. 39.
Sauvage, H. E. 1874 A. ("Épinoches.")

GASTEROSTEUS Linn. Type *G. aculeatus* Linn.

Linnæus, C., 1758, Syst. Nat., ed. x, p. 489.
Günther, A. 1859 A, p. 2.
1880 A, p. 505.
Huxley, T. H. 1868 C, p. 39.
Jordan and Evermann 1896 A, p. 746.
Owen, R. 1845 B, p. 91.
Parker, W. K. 1868 A, p. 39.
Sauvage, H. E. 1874 A, p. 5.

Gasterosteus bispinosus Walbaum.
Walbaum, 1792, Artedi Pisc., p. 450.

Dawson, J. W. 1877 B, p. 341. (G. aculeatus?)
1890 A, p. 86. (G. aculeatus?)
Jordan and Evermann 1896 A, p. 748.
Living; Labrador to New Jersey: Pleistocene; Canada.
It seems most probable that the stickleback found fossil in Canada by Dr. Dawson is the *G. bispinosus*, common on our Eastern coast.

Order MESICHTHYES nom. nov.

Actinopteri with anterior vertebræ unmodified and not connected with auditory apparatus. Branchial arches well developed. Precoracoid arch and interclavicles absent. Ventral fins abdominal.

This order is intended to include the *Haplomi*, the *Synentognathi*, and the *Percesoces*.

Suborder HAPLOMI Cope.

Cope, E. D. 1872 C, pp. 328, 333.
Gill, T. 1872 C, p. 14.

Gill, T, 1893 A, p. 133.
Jordan and Evermann 1896 A, p. 622.

LUCIIDÆ.

Unless otherwise indicated, the family name
Esocidæ is used by the following writers.

Bridge, T. W. 1896 A, p. 545.
Brühl B. C. 1847 A. (Esox.)
Cope, E. D. 1872 C, p. 333.
    1875 E, p. 280. (Ischyrhizidæ.)
Dareste, C. 1872 A, p. 1086.
Gill, T. 1872 C, p. 15.
Grassi, B. 1883 H.
Günther, A. 1866 A, p. 226.
    1880 A, p. 623.
    1881 A, p. 693.
Jordan and Evermann 1896 A, p. 624. (Luciidæ.)
Klein, E. F. 1879, p. 108. (Esox.)
Kölliker, A. 1859 A. (Esoces.)
Müller, A. 1853 A.

Müller, J. 1843 A, p. 215. (Esoces.)
    1846 C, p. 188. (Esoces.)
Nicholson and Lydekker 1889 A, p. 1000.
Owen, R. 1845 B, p. 131. ("Lucioids.")
    1866 A, p. 10 et seq.
Parker, W. K. 1868 A, p. 54.
Pictet, F. J. 1854 A, p. 108.
Quenstedt, F. A. 1885 A, p. 365. (Esocini.)
Reis, O. M. 1895 C, p. 170.
Röse, C. 1897 A, with figures.
Stannius, H. 1854 A. (Esoces.)
Stephan, P. 1900 A. (Esox.)
Vaillant, L. 1883 A.
Vro!ik, A. J. 1873 A, pp. 235, 261.
Woodward, A. S. 1898 B, p. 120.
Zittel, K. A. 1890 A, p. 270.

ISCHYRHIZA Leidy.   Type *I. mira* Leidy.

Leidy, J. 1856 K, p. 221.
Cope, E. D. 1872 I, p. 355.
    1875 E, p. 280.
Leidy, J. 1857 A, p. 272.
Zittel, K. A. 1890 A, p. 270.
    This genus is assigned only provisionally to
    the *Luciidæ*.

Ischyrhiza antiqua Leidy.

Leidy, J. 1856 L, p. 256.
Cope, E. D. 1875 E, p. 280.
    1875 V, p. 32.
    1887 G, p. 566.
Emmons, E. 1858 B, p. 225, figs. 47, 48.
Leidy, J. 1857 A, p. 272.

Leidy, J. 1860 B, p. 120. 'Syn. of I. mira.)
    Cretaceous; North Carolina: (Fox Hills);
    New Mexico.

Ischyrhiza mira Leidy.

Leidy, J. 1856 K, p. 221.
Cope, E. D. 1872 I, p. 355.
    1875 E, p. 280.
Leidy, J. 1860 B, p. 120, pl. xxv, figs. 3–9. (I. an-
    tiqua a synonym.)
    Cretaceous; New Jersey: Miocene?; South
    Carolina.

Ischyrhiza ? radiata Clark.

Clark, W. B. 1895 A, p. 4.
    1897 A, p. 60, pl. vii, figs.
    Eocene; Virginia.

PŒCILIIDÆ.

Cope, E. D. 1872 C, p. 338. (Cyprinodontidæ.)
Gill, T. 1894 B, p. 115. (Pœciliidæ.)
    1896 B, p. 221. (Pœciliidæ.)
Günther, A. 1866 A, p. 299. (Cyprinodontidæ.)
    1880 A, p. 613. (Cyprinodontidæ.)
    1881 A, p. 693. (Cyprinodontidæ.)
Ihering, H. 1891 B, p. 477. (Cyprinodontidæ.)
Jordan and Evermann 1896 A, p. 630.

Kölliker A. 1859 A. (Cyprinodontes.)
Müller, J. 1846 C, pp. 183, 202. (Cyprinodontes.)
Nicholson and Lydekker 1889 A, p. 1000. (Cypri-
    nodontidæ.)
Reis, O. M. 1895 C, p, 170. ("Cyprinodonten.")
Stannius H. 1854 A. (Cyprinodontes.)
Zittel, K. A. 1890 A, p. 281. (Cyprinodontidæ.)

GEPHYURA Cope.   Type *G. concentrica* Cope.

Cope, E. D. 1891 H, p. 654.
    The position of the genus is in doubt.

Gephyura concentrica Cope.

Cope, E. D. 1891 H, p. 654.
    1892 B, p. 285.
    Oligocene ?; South Dakota.

Suborder PERCESOCES Cope.

Cope, E. D., 1871, Trans. Amer. Phil. Soc., xiv, p.
    456.
    1872 C, pp. 335, 337.
Gill, T 1872 C, p. 13.

Gill, T. 1893 A, p. 133.
Jordan and Evermann 1896 A, p. 787.
Starks, E. C. 1899 A, p. 1.

MUGILIDÆ.

Agassiz, L. 1843 B (1844), v, p. 119. ("Mugil-
  oides.")
Cope, E. D. 1872 C, p. 338.
Dareste, C. 1872 A, p. 1175. ("Mugiloides.")
Gill, T. 1872 C, p. 13.
Günther, A. 1861 A, p. 409.
  1880 A, p. 501.
Jordan and Evermann 1896 A, p. 808.
Kölliker, A. 1859 A. (Mugiloidei.)

Nicholson and Lydekker 1889 A, p. 1006.
Owen, R. 1846 B, p. 120. ("Mugiloids.")
  1866 A, p. 11 *et seq.*
Parker, W. K. 1868 A, p. 50.
Stannius, H. 1854 A. (Mugiloidei.)
Starks, E. C. 1899 A, p. 1.
Zittel, K. A. 1890 A, p. 312.
  It is not certain that any of the genera here
  assigned to this family really belong to it.

CLADOCYCLUS Agassiz.   Type *C. lewesiensis* Agassiz.

Agassiz, L. 1843 B (1844), v, p. 103.
Geinitz, H. B, 1868 A, p. 42.
  1875 A. p. 224.
Leidy, J. 1873 B, p. 288.
Loomis, F. B, 1900 A, p. 237. (Syn. of Ichthyo-
  dectes.)
Nicholson and Lydekker 1889 A, p. 1006.
Pictet, F. J. 1854 A, p. 94.
Woodward and Sherborn 1890 A, p. 34.

Cladocyclus occidentalis Leidy.

Leidy, J. 1856 M, p. 256.
  1857 C, p. 90.
  1873 B, pp. 288, 347, pl. xvii, figs. 21, 22; pl.
    xxx, fig. 5.
Loomis, F. B. 1900 A, pp. 237, 242. (Syn. of Ichthyodectes arcuatus.)
Whiteaves, J. F. 1889 B, p. 195, pl. xxvi, figs. 8, 9.
  Cretaceous; South Dakota, Manitoba.

PELECORAPIS Cope.   Type *P. varius* Cope.

Cope, E. D. 1874 C, p. 39.
  1875 E, pp. 182, 273.
  1877 G, p. 587.
Zittel, K. A. 1890 A, p. 312.

Pelecorapis berycinus Cope.

Cope, E. D. 1877 G, p. 587.
  Cretaceous (Pierre); Montana.

Pelecorapis varius Cope.

Cope, E. D. 1874 C, p. 39.
  1875 E, pp. 182, 273.
  1877 G, p. 587.
  Cretaceous (Benton); Kansas.

SYLLÆMUS Cope.   Type *S. latifrons* Cope.

Cope, E. D. 1875 E, pp. 180, 273.
  1877 K, p. 26.
Nicholson and Lydekker 1889 A, p. 1005. (Syn.?
  of Calamopleurus.)
Stewart, A. 1900 A, p. 383.
Woodward, A. S. 1888 I, p. 325.
Zittel, K. A. 1890 A, p. 312. (Scyllæmus.)

Syllæmus latifrons Cope.

Cope, E. D. 1875 E, pp. 181, 273.
  1877 K, p. 27.
Stewart, A. 1900 A, p. 384, pl. lxxii.
  Cretaceous; New Mexico?: Cretaceous (Benton); Kansas.

SPHYRÆNIDÆ.

Agassiz, L., 1843 B, v. p. 93. ("Sphyrænoides.")
  1845 A, p. 291. (Sphyrænoidæ.)
Bridge, T. W. 1896 A, p. 569.
Cope, E. D. 1870 L, p. 529.
Gill, T. 1872 C, p. 12.
Günther, A. 1860 A, p. 334.
  1880 A, p. 499.

Günther, A. 1881 A, p. 690.
Jordan and Evermann 1898 A, p. 822.
Klein, E. F. 1879 A, p. 91.
Nicholson and Lydekker 1889 A, p. 1006.
Quenstedt, F. A. 1885 A, p. 375. (Sphyrænoidei.)
Starks, E. C. 1899 A, p. 1.
Zittel, K. A. 1890 A, p. 312.

DICTYODUS Owen.   Without designated type.

Owen, R., 1839, Rep. Brit. Assoc. Adv. Sci. for 1838,
  p. 142.
Agassiz, L. 1843 B (1844), v, pp. 8, 98. (Sphyræno-
  dus.)
Giebel, C. G. 1848 A, p. 92. (Sphyrænodus.)
Owen, R. 1845 B, pp. 6, 12, pl. liv (Dictyodus),
  p. 128 (Sphyrænodus).
Pictet, F. J. 1854 A, p. 93. (Sphyrænodus.)
Zittel, K. A. 1890 A, p. 312.

Dictyodus silovianus (Cope).

Cope, E. D. 1875 H, p. 362. (Sphyrænodus).
  Miocene; New Jersey.

Dictyodus speciosus (Leidy.)

Leidy, J. 1856 K, p. 221. (Sphyræna.)
Cope, E. D. 1867 C, p. 142. (Sphyræna.)
  1875 H, p. 362. (Sphyrænodus.)
  Miocene; New Jersey.

SPHYRÆNA Bloch and Schn.   Type *Esox sphyræna* Linn.

Bloch and Schneider, 1801, Syst. Ichth., p. 109.
Agassiz L. 1843 B (1844), v, pp. 8, 93.
    1845 A, p. 291.
Günther, A. 1860 A, p. 334.
Jordan and Evermann 1896 A, p. 822.
Owen, R. 1845 B, pp. 4, 6, 126, pl. liii.
Pictet, F. J. 1854 A, p. 92.

Starks, E. C. 1899 A, p. 8.
Zittel, K. A. 1890 A, p. 312.

Sphyræna major Leidy.

Leidy, J. 1855 C, p. 397.
    1877 A, p. 254, pl. xxxiv, figs. 37–41.
    Sands of Ashley River; South Carolina.

Order PERCOMORPHI.

Cope, E. D. 1872 C, p. 341.
    1884 O, p. 79.

Cope, E. D. 1889 R, p. 860.

Suborder SQUAMIPINNES.

Günther, A. 1860 A. p. 1.
Jordan and Evermann 1896 A. p. 781,
    1898 A. p. 1665.

Kölliker, A. 1859 A. (Sqamipennes.)
Zittel, K. A. 1890 A. p 299. (Squamipennes.)

CHÆTODONTIDÆ.

Gill, T. 1872 C, p. 8. (Chætodontidæ, Ephippiidæ.)
Cope, E. D. 1872 C, p. 342.
Günther, A. 1860 A, p. 1. (Squamipinnes, in part.)
    1880 A, p. 397. (Squamipinnes, in part.)
Jordan and Evermann 1898 A, p. 1666 (Ephippidæ); p. 1669 (Chætodontidæ).

Kölliker, A. 1859 A. (Squamipennes.)
Nicholson and Lydekker 1889 A, p. 1013.
Parker, W. K. 1868 A, p. 50.
Zittel, K. A. 1890 A, p. 299. (Squamipennes.)

CHÆTODIPTERUS Lacépède.   Type *C. faber* (Brouss.).

Lacépède, 1802, Hist. Nat. Poiss., iv, p. 503.
Günther, A. 1860 A, p. 60. (Ephippus.)
    1880 A, p. 402, fig. 171. (Ephippus.)
Jordan and Evermann 1898 A, p. 1667.
Nicholson and Lydekker 1889 A, p. 1013. (Ephippus.)
Owen, R. 1866 A, pp. 108, 111. (Ephippus.)
Zittel, K. A. 1890 A, p. 299. (Ephippus.)

Chætodipterus faber (Brouss.).

Broussonet, 1782, Ichth. Decas, 1, v, pl. 4, *fide* Jordan and Evermann. (Chætodon.)
Jordan and Evermann 1898 A, p. 1668.
Leidy, J. 1889 E, p. 31. (Ephippus gigas.)
    Recent; Cape Cod to Rio Janeiro: Pleistocene; Florida.

Suborder PHARYNGOGNATHI.

Cope, E. D. 1872 C, p. 343.
Gill, T. 1872 C, p. 6.
    1893 A, p. 135. (Pomacentroidea, Labroidea).
Ihering, H. 1891 B, p. 508.
Jordan and Evermann 1898 A, p. 1571.
Kner, R. 1860 A, p. 41.

Kölliker, A. 1859 A.
Müller, J. 1846 C, pp. 131, 155, 199.
Parker, W. K. 1868 A, p. 49.
Reis, O. M. 1895 C, p. 163.
Stannius, H. 1854 A.
Zittel, K. A. 1890 B, p. 287.

LABRIDÆ.

Agassiz L. 1845 A, p. 291. (Labroidæ.)
Brühl, B. C. 1847 A.
Cocchi, I. 1864 A. (Pharyngodopilidæ.)
Cope, E. D. 1872 C, p. 343.
Dareste, C. 1872 A, p. 1174. ("Labroides.")
Gill, T. 1872 C, p. 7.
    1893 A, p. 135.
Günther, A. 1862 A, p. 65.
    1880 A, p. 525.
Jordan, D. S. 1891 A.
Jordan and Evermann 1898 A, p. 1571.
Kner, R. 1860 A, p. 41. (" Labroiden.")
Kölliker, A. 1859 A. (Labroidei.)

Müller, J. 1846 C, p. 168. (Labroidei.)
Nicholson and Lydekker 1889 A, p. 1004. (Labridæ, Pharyngodopilidæ.)
Owen, R. 1845 B, p. 108. ("Labroids.")
    1866 A, p. 11 *et seq.* (Cyclo-labridæ.)
Parker, W. K. 1868 A, p. 50.
Reis, O. M. 1895 C.
Sauvage, H. E. 1875 B, p. 615. (Pharyngodopilidæ, Phyllodidæ.)
Zittel, K. A. 1890 A, p. 288.
    In this family are included provisionally the genera *Nummopalatus* and *Phyllodus.*

PROTAUTOGA Leidy.   Type *P. conidens* Leidy.

Leidy, J. 1873 C, p. 15.   (As subgenus of Tautoga.)
 1873 B, p. 347.
Nicholson and Lydekker 1889 A, p. 1004.
Zittel, K. A. 1890 A, p. 289.

Protautoga conidens Leidy.
Leidy, J. 1873 C, p. 15.   [Tautoga (Protautoga.)]
 1873 A, p. 312.   [Tautoga (Protautoga.)]
 1873 B, p. 346, pl. xxxii, figs. 56, 57.   [Tautoga (Protautoga.)]
Miocene; Virginia.

NUMMOPALATUS Roualt.   Type *N. edwardsius* Roualt.

Rouall, M. 1858 A, p. 101.
Cocchi, I. 1864 A, p. 59.   (Pharyngodopilus.)
Cornuel, J. 1877 A, p. 620.
Günther, A. 1880 A, p. 526.
Leidy, J. 1855 C, p. 396.   (Odax.)
Reis, O. M. 1895 C, p. 174.
Sauvage, H. E. 1875 B, p. 613.

Zittel, K. A. 1890 A, p. 289.

Nummopalatus carolinensis (Leidy.)
Leidy, J. 1855 C, p. 396.   (Odax.)
 1877 A, p. 256, pl. xxxiv. figs. 19, 24.   (Pharyngodopilus.)
Phosphate beds; South Carolina.

PHYLLODUS Agassiz.   Type not designated.

Agassiz, L. 1843 B, ii, pt. ii, p. 288.
 1844 A.
Cocchi, I. 1864 A, p. 59.
Cornuel, J. 1877 A, p. 619.
Giebel, C. G. 1848 A, p. 174.
Günther, A. 1880 A, p. 526.
Nicholson and Lydekker 1889 A, p. 1004, fig. 941.
Owen, R. 1845 B, p. 188, pl. xliv, fig. 2; pl. xlvii, figs. 1, 2.
Pictet, F. J. 1854 A, p. 208.
Quenstedt, F. A. 1885 A, p. 381.
Sauvage, H. E. 1875 B, p. 615.
Wyman, J. 1850 B, p. 234.
 1850 C.
Zittel, K. A. 1890 A, p. 289.

Phyllodus curvidens Marsh.
Marsh, O. C. 1870 E, p. 229.
Cope, E. D. 1875 H, p. 362.
Miocene; New Jersey.

Phyllodus elegans Marsh.
Marsh, O. C. 1870 E, p. 228.
Eocene; New Jersey.

Phyllodus hipparionyx Eastman.
Eastman, C. R. 1900, Geol. Surv. Maryland, iv, p. 108.
Wyman, J. 1850 B, p. 234, figs. 9a, 9b.   ("Phyllodus.")
Eocene; South Carolina.

POMACENTRIDÆ.

Gill, T. 1872 C, p. 7.
Günther, A, 1862 A, p. 1.
 1880 A, p. 524.
Jordan and Evermann 1898 A, p. 1543.

Kner, R. 1860 A, p. 42.   ("Pomacentrinen.")
Nicholson and Lydekker 1889 A, p. 1004.
Owen, R. 1866 A, p. 11.   (Ctenolabridæ.)
Zittel, K. A. 1890 A, p. 287.

PRISCACARA Cope.   Type *P. serrata* Cope.

Cope, E. D. 1877 I, p. 816.
 1884 O, p. 92.
Nicholson and Lydekker 1888 A, p. 1004.
Zittel, K. A. 1890 A, p. 288.

Priscacara clivosa Cope.
Cope, E. D. 1878 F, p. 76.
 1877 AA.
 1884 O, p. 96, pl. xiii, fig. 8.
Eocene (Green River); Wyoming.

Priscacara cypha Cope.
Cope, E. D. 1877 I, p. 817.
 1878 Z, p. 570.
 1884 O, p. 94, pl. xiii, fig. 2.
Eocene (Green River); Wyoming.

Priscacara hypsacantha Cope.
Cope, E. D. 1886 T, p. 164, pl. i, fig. vi.
Eocene (Green River); Wyoming.

Priscacara liops Cope.
Cope, E. D. 1877 I, p. 818.
 1877 AA.
 1878 Z, p. 570.
 1884 O, p. 97, pl. xiv, figs. 2, 3.
Eocene (Green River); Wyoming.

Priscacara oxyprion Cope.
Cope, E. D. 1878 F, p. 74.
 1884 O, p. 94, pl. xiv, fig. 5.
Eocene (Green River); Wyoming.

Priscacara pealei Cope.
Cope, E. D. 1878 F, p. 75.
 1884 O, p. 96, pl. viii, fig. 1; pl. xiv, fig. 4.
Eocene (Green River); Wyoming.

Priscacara serrata Cope.
Cope, E. D. 1877 I, p. 816.

Cope, E. D. 1877 AA.
1878 , p. 570.
1884 Ɛ, p. 93, pl. xiii, fig. 1.
1886 T, p. 165.
Eocene (Green River); Wyoming.

Priscacara testudinaria Cope.
Cope, E. D. 1884 O, p. 98, pl. i, fig. 7.
1880 T, p. 303.   (No description.)
Eocene (Manti); Utah.

Superfamily BERYCOIDEA Gill.

Gill, T. 1872 C, p. 10.

Gill, T. 1893 A, p. 133.

BERYCIDÆ.

Bridge, T. W. 1896 A, p. 560.
Cope, E. D. 1872 C, p. 342.
Gill, T. 1872 C, p. 10.
Günther, A. 1859 A, pp. 8, 12.
1880 A, p. 420.

Jordan and Evermann 1896 A, p. 837.
Klein, E. F. 1879 A, p. 80.
Nicholson and Lydekker 1889 A, p. 1010.
Reis, O. M. 1895 C, p. 175.
Zittel, K. A. 1890 A, p. 291.

BERYX Cuvier.   Type B. decadactylus C. and V.

Cuvier, G., 1829, Règne Anim., ed. ii, p. 151.
Agassiz, L. 1844 A.
Bridge, T. W. 1896 A, p. 563.
Dixon, F. 1878 A, p. 398.
Eichwald, E. 1868 A, p. 1196.
Günther, A. 1859 A, p. 12.
1880 A, p. 422.
Huxley, T. H. 1876 B, xiii, p. 388.
Pictet, F. J. 1854 A, p. 49.

Pictet and Humbert 1866 A, pp. 13, 27.
Williamson, W. C. 1849 A, p. 444.
Zittel, K. A. 1896 A, p. 291.

Beryx insculptus Cope.
Cope, E. D. 1869 F, p. 240.
1872 I, p. 357.
1875 E, p. 272, pl. lii, fig. 4.
Cretaceous; New Jersey.

Superfamily SCOMBROIDEA Gill.

Gill, T. 1872 C, p. 8.
1893 A, p. 133.
Kölliker, A. 1859 A.   (Scomberoidei.)

Parker, W. K. 1868 A, p. 50.   ("Scombroids.")
Stannius, H. 1854 A.   (Scomberoidei, as family.)

XIPHIIDÆ.

Agassiz, L. 1845 A, p. 289.   (Xiphioidæ.)
Cope, E. D. 1872 C, p. 342.   (Xiphiadidæ.)
Gill, T. 1872 C, p. 8,
1883 A, p. 485.   (Xiphiidæ, Histiophoridæ.)
Günther, A. 1860 A, p. 511.

Günther, A. 1880 A. p. 431.
Jordan and Evermann 1896 A, p. 893.
Nicholson and Lydekker 1889 A, p. 1010.
Reis, O. M. 1895 C.
Zittel, K. A. 1890 A, p. 300.

ISTIOPHORINÆ.

Gill, T. 1883 A, p. 486.   (Histiophoridæ.)
1893 A, p. 133.   (Histiophoridæ.)
Goode, G. B. 1882 A, p. 417.   (Tetrapturinæ.)

Jordan and Evermann 1896 A, p. 890.   (Istiophoridæ.)

ISTIOPHORUS Lacépède.   Type I. gladifer = I. gladius.

Lacépède, 1802, Hist. Nat. Poissons, iii, p. 374.
Goode, G. B. 1882 A, p. 423.   (Histiophorus.)
Günther, A. 1860 A, p. 512.   (Histiophorus.)
1880 A, p. 431.   (Histiophorus.)
Jordan and Evermann 1896 A, p. 890.
Nicholson and Lydekker 1889 A, p. 1010.   (Histiophorus.)
Owen, R. 1845 B, p. 121.   (Histiophorus.)
Quenstedt, F. A. 1885 A, p. 371.   (Histiophorus.)

Istiophorus antiquus (Leidy).
Leidy, J. 1855 C, p. 397.   (Xiphias.)
Cope, E. D. 1869 L, p. 310.   (Histiophorus.)
Eocene; New Jersey.

Istiophorus homalorhamphus Cope.
Cope, E. D. 1869 L, p. 310.   (Histiophorus.)
Tertiary greensand; New Jersey.

Istiophorus parvulus Marsh.
Marsh, O. C. 1870 E, p. 227.   (Histiophorus.)
Eocene; New Jersey.

Istiophorus robustus (Leidy).
Leidy, J. 1860 B, p. 119, pl. xxvii, figs. 3–5.   (Xiphias.)
Cope, E. D. 1869 L, p. 310.   (Histiophorus.)
Post-pliocene; South Carolina.

EMBALORHYNCHUS Marsh.   Type *E. kinnei* Marsh.

Marsh, O. C. 1870 E, p. 228.   **Embalorhynchus kinnei** Marsh.

                                           Marsh, O. C. 1870 E, p. 228.
                                           Eocene; New Jersey.

TRICHIURIDÆ.

Cope, E. D. 1872 C, p. 342.   Günther, A. 1880 A, p. 453.
Dames, W. 1887 B, p. 71.   Jordan and Evermann 1896 A, p. 888.
Günther, A. 1860 A, p. 342.   Zittel, K. A. 1890 A., p. 302.

TRICHIURUS Linn.   Type *T. lepturus* Linn.

Linnæus, C., 1758, Syst. Nat., ed. x, p. 246.   Jordan and Evermann 1896 A, p. 889.
Günther, A. 1860 A, p. 346.   Leidy, J. 1860 B, p. 121, pl. xxv, fig. 2. (T. lep-
Jordan and Evermann 1896 A, p. 888.   turus fossilis.)
Owen, R. 1845 B, pp. 2, 125, pl. i, fig. 8.   Recent; West Indies to Virginia: Post-plio-
                                                      cene; South Carolina.
Trichiurus lepturus Linn.

Linnæus C., 1758, Syst. Nat., ed. x, p. 246.

CARANGIDÆ.

Gill, T. 1872 C, p. 8.   Nicholson and Lydekker 1889 A, p. 1008.
Cope, E. D. 1872 C, p. 342.   Pictet and Humbert 1866 A, p. 45. ("Carang-
Gill, T. 1893 A, p. 133.   idæs.")
Günther, A. 1860 A, p. 417.   Reis, O. M. 1896 C, p. 177.
      1880 A, p. 440.   Zittel, K. A. 1890 A, p. 304.
Jordan and Evermann 1896 A, p. 895.

PLATAX Cuvier.   Type not specified.

Cuvier and Valenciennes, 1831, Hist. Nat. Poiss.,   Pictet, F. J. 1854 A, p. 67.
   vii, p. 213.   Pictet and Humbert 1866 A, p. 46.
Agassiz, L. 1844 A, p. 207.   Quenstedt, F. A. 1885 A, p. 386.
Davis, J. W. 1887 C, p. 523.   Zittel, K. A. 1890 A, p. 304.
Giebel, C. G. 1848 A, p. 56.
      1860 A, p. 489.   **Platax** sp. indet.
Günther, A. 1880 A, p. 448.
Nicholson and Lydekker 1889 A, p. 1009.   Gibbes, R. W. 1850 A, p. 77.
Owen, R. 1845 B, p. 2, pl. i, fig. 2.   1850 G, p. 300, pl. xlii, figs. 11–13.
                                                       Pliocene; South Carolina.

Superfamily *PERCOIDEA* Gill.

Gill, T. 1893 A, p. 134.

APHREDODERIDÆ.

Cope, E. D. 1870 H, p. 380. (Asineopidæ.)   Günther, A. 1880 A, p. 396.
   1877 AA. (Asineopidæ.)   Jordan and Evermann 1896 A, p. 785.
   1884 O, p. 79 (Aphododeridæ); p. 85 (Asine-   Nicholson and Lydekker 1889 A, p. 1014. (Aphre-
   opidæ).   doderidæ, Asineopidæ.)
Günther, A. 1859 A, p. 271.   Zittel, K. A. 1890 A, p. 296. (Percidæ, in part.)

AMPHIPLAGA Cope.   Type *A. brachyptera* Cope.

Cope, E. D. 1877 I, p. 812.   Cope, E. D. 1877 Z, p. 570.
   1884 O, p. 83.   1877 AA.
                                                   1884 O, p. 84.
Amphiplaga brachyptera Cope.   Eocene (Green River); Wyoming.

Cope, E. D. 1877 I, p. 812.

ASINEOPS Cope.   Type *A. squamifrons* Cope.

Cope, E. D. 1870 H, p. 380.
    1871 E, p. 425.
    1884 O, p. 84.
Leidy, J. 1871 C, p. 369.
Nicholson and Lydekker 1889 A, p. 1014.
Zittel, K. A. 1890 A, p. 299.

Asineops pauciradiatus Cope.

Cope, E. D. 1877 I, p. 813.
    1877 AA.
    1878 Z, p. 570.

Cope, E. D. 1884 O, p. 87, pl. xiv, fig. 1.
    Eocene (Green River); Wyoming.

Asineops squamifrons Cope.

Cope, E. D. 1870 H, p. 381.
    1871 E, p. 426. (A. squamifrons, A. viriden-
        sis.)
    1884 O, p. 85, pl. ix, fig. 5; pl. xi.
King, C. 1878 A, p. 394. (A. squamifrons, A. viri-
        densis.)
Leidy, J. 1871 C, p. 369.
    Eocene (Green River); Wyoming.

TRICHOPHANES Cope.   Type *T. hians* Cope.

Cope, E. D. 1872 P, p. 479.
    1873 E, p. 641.
    18⁷⁹ N.
    1884 O, pp. 79, 742, 752.
Osborn, Scott, and Speir 1878 A, p. 97.

Trichophanes copei. Osb., Sc. and Sp.

Osborn, Scott, and Speir, 1878 A, p. 98.
Cope, E. D. 1884 O, p. 752.
    Oligocene; Colorado.

Trichophanes foliarum Cope.

Cope, E. D. 1878 F, p. 73.
    1884 O, p. 753, pl. lix, fig. 4.
    Oligocene; Colorado.

Trichophanes hians Cope.

Cope, E. D. 1872 P, p. 481.
    1872 MM, p. 775.
    1873 E, p. 642.
    1884 O, P. 753, pl. lix, fig. 3.
    Oligocene; Nevada.

ERISMATOPTERUS Cope.   Type *E. rickseckeri* Cope.

Cope, E. D. 1871 E, p. 427.
    1884 O, p. 80.
Zittel, K. A. 1890 A, p. 296.

Erismatopterus endlichii Cope.

Cope, E. D. 1877 I, p. 811.
    1877 Z, p. 570.
    1877 AA.
    1884 O, p. 82, pl. xii, fig. 5.
    Eocene (Green River); Wyoming.

Erismatopterus levatus Cope.

Cope, E. D. 1870 H, p. 382. (Cyprinodon.)
    1871 E, p. 428.
    1884 O, p. 80, pl. ix, figs. 6, 7.
Leidy, J. 1871 C, p. 369. (Cyprinodon.)
    Eocene (Green River); Wyoming.

Erismatopterus rickseckeri Cope.

Cope, E. D. 1871 E, p. 427.
    1884 O, p. 81, pl. vi, fig. 2.
Leidy, J. 1872 B, p. 372.
    Eocene (Green River); Wyoming.

PERCIDÆ.

Agassiz, L. 1845 A, p. 286.   (Percoidæ.)
Bridge, T. W. 1896 A, p. 564.
Brühl, B. C. 1847 A. (" Percoïden.")
Cope, E. D. 1872 C, p. 342.
    1883 A.
Gill, T. 1872 C, p. 11.
Günther, A. 1859 A, p. 51.
    1880 A, p. 375.
    1881 A, p. 688.

Jordan and Evermann 1896 A, p. 1015.
Kölliker. A. 1859 A. (Percoidei, part.)
Nicholson and Lydekker 1889 A, p. 1012.
Owen, R. 1845 B, p. 88. (" Percoïds.")
    1866 A, p. 11 *et seq.*
Parker, W. K. 1868 A, p. 52.
Reis, O. M. 1895 C.
Segond, L. A. 1873 B, p. 528. (" Percoïdes.")
Zittel, K. A. 1890 A, p. 293.

MIOPLOSUS Cope.   Type *M. abbreviatus* Cope.

Cope, E. D. 1877 I, p. 813.
    1884 O, p. 88.
Nicholson and Lydekker 1889 A, p. 1013.
Zittel, K. A. 1890 A, p. 294.

Mioplosus abbreviatus Cope.

Cope, E. D. 1877 I, p. 813.
    1877 Z, p. 570.
    1877 AA.

Cope, E. D. 1884 O, p. 88.
    Eocene (Green River); Wyoming.

Mioplosus beani Cope.

Cope, L. D. 1877 I, p. 816.
    1877 Z, p. 570.
    1877 AA.
    1884 O, p. 91, pl. xii, fig. 3.
    Eocene (Green River); Wyoming.

Mioplosus labracoides Cope.

Cope, E. D. 1877 I, p. 814.
1877 Z, p. 570.
1877 AA.
1884 O, p. 89, pl. xii, fig. 1.
Eocene (Green River); Wyoming

Mioplosus longus Cope.

Cope, E. D. 1877 I, p. 815.
1877 Z, p. 570.
1877 AA.

Cope, E. D. 1884 O, p. 90, pl. xii, fig. 2.
Eocene (Green River); Wyoming.

Mioplosus multidentatus Cope.

Cope, E. D. 1891 H, p. 657.
1892 B, P. 285.
Tertiary (Oligocene?); South Dakota.

Mioplosus sauvageanus Cope.

Cope, E. D. 1884 O, p. 92.
Eocene (Green River); Wyoming.

PLIOPLARCHUS Cope.   Type *P. whitei* Cope.

Cope, E. D. 1883 A, p. 414.
1884 O, p. 727.
Nicholson and Lydekker 1889 A, p. 1014.
Zittel, K. A. 1890 A, p. 296.

Plioplarchus septemspinosus Cope.

Cope, E. D. 1889 M, p. 625.
Tertiary, below John Day beds; Oregon

Plioplarchus sexspinosus Cope.

Cope, E. D. 1883 A, p. 416.

Cope, E. D. 1884 O, p. 729.
Upper Cretaceous or Lower Tertiary; South
Dakota.

Plioplarchus whitei Cope.

Cope, E. D. 1883 A, p. 414.
1884 O, p. 728, pl. xxiv G, fig. 1.
Upper Cretaceous or Lower Tertiary; South
Dakota.

OLIGOPLARCHUS Cope.   Type *O. squamipinnis* Cope.

Cope, E. D. 1891 H, p. 656.

Oligoplarchus squamipinnis Cope.

Cope, E. D. 1891 H, p. 656.

Cope, E. D. 1892 B, p. 285.
Oligocene?; South Dakota.

SPARIDÆ.

Agassiz, L. 1845 A. p. 287. (Sparoidæ.)
Brühl, B. C. 1847 A. ("Sparoiden.")
Cope, E. D. 1872 C, p. 342.
Jordan and Evermann 1896 A, p. 1343.

Kölliker, A. 1859 A. (Sparoidei.)
Owen, R. 1845 B, p. 91. ("Sparoids.")
Segond, L. A. 1873 B, p. 534. ("Sparoïdes.")
Stannius, H. 1854 A. (Sparoidei.)

CROMMYODUS Cope.   Type *Phacodus irregularis* Cope.

Cope, E. D. 1869 F, p. 243. (Substitute for Phacodus Cope, not of Dixon.)

Crommyodus irregularis Cope.

Cope, E. D. 1869 L, p. 311. (Phacodus.)

Cope, E. D. 1869 F, p. 243.
1875 H, p. 362.
Miocene; New Jersey.

Superfamily SCIÆNOIDEA *Gill.*

Gill, T. 1872 C, p. 10.

SCIÆNIDÆ.

Brühl, B. C. 1847 A. ("Sciænoiden.")
Günther, A. 1860 A, p. 265.
1880 A, p. 426.
Jordan and Evermann 1896 A, p. 1392.
Kölliker, A. 1859 A. (Sciænoidei.)

Owen, R. 1845 B, p. 100. ("Sciænoids.")
1866 A, p. 11 et seq.
Parker, W. K. 1868 A, p. 50.
Segond, L. A. 1873 B, pp. 531, 534. ("Sciénoïdes.")
Stannius, H. 1854 A. (Sciænoidei.)

POGONIAS Lacépède.   Type *P. fasciatus=Labrus cromis* Linn.

Lacépède, 1802, Hist. Nat. Poissons, iii, p. 138.
Günther, A. 1860 A, p. 269.
1880 A, p. 426.
Jordan and Evermann 1896 A, p. 1482.

Pogonias cromis (Linn.).

Linnæus, C., 1758, Syst. Nat., ed. x, p. 479. (Labrus.)
Günther, A. 1880 A, p. 27, fig. 187.

Jordan and Evermann 1896 A, p. 1482.
Leidy, J. H. 1855 C, p. 397.   (P. chromis.)
    Recent; Long Island Sound to N. Mexico
    Sands of Ashley River; South Carolina.

Pogonias multidentatus Cope.
Cope, E. D. 1869 L, p. 310.
    Miocene; Virginia.

Pogonias sp. indet.
Wyman, J. 1850 C, p. 234.
    Eocene?; South Carolina.

Suborder PAREIOPLITÆ Richardson.

Richardson, J., 1836, Fauna bor.-amer., iii, p. 36.
Gill, T. 1889 A, p. 589.   (Acanthopterygii buccis
    loricatis.)

Günther, A. 1860, A, pp. 90, 211.   (Cataphracti.)
Jordan & Evermann, 1896 A, p. 1756.   (Loricati of
    Jenyns, 1835, not Loricata of Merrem, 1820.)

Superfamily SCORPÆNOIDEA Gill.

Gill, T. 1889 A, p. 589.

Gill, T. 1893 A, p. 135.

SCORPÆNIDÆ.

Gill, T. 1860 A, p. 95.   (Scorpænina.)
    1872 C, p. 6.
Günther, A. 1880 A, p. 412.
Sauvage, H. E. 1873 A, p. 726.   (Scorpæni.)

Sauvage, H. E. 1878 B, p. 111.
Segond, L. 1873 B, p. 582.   ("Scorpènes.")
Zittel, K. A. 1890 A, p. 300.

SEBASTODES Gill.   Type *S. paucispinis* (Ayres).

Gill, T., 1861, Proc. Acad. Nat. Sci. Phila., p. 165.
Cramer, F. 1896 A, p. 573.
Gill, T., 1864, Proc. Acad. Nat. Sci. Phila., p. 145.
Jordan and Evermann 1896 A, p. 1765.
Sauvage, H. E. 1878 B, p. 112.

Sebastodes? rosæ Eigenmann.
Eigenmann, C. H. 1890 A, p. 17, fig. 1.
    Tertiary; California.

Superfamily COTTOIDEA Gill.

Gill, T. 1872, C. p. 6.
    1889 A, p. 589.

Gill, T. 1893 A, p. 135.

COTTIDÆ.

Bridge, T. W. 1896 A, p. 571.
Cope, E. D. 1872 C, p. 337.
Dareste, C. 1872 A, p. 1255.   ("Poissons à joues
    cuirassées.")
Gill, T. 1872 C, p. 6.
    1889 A.
    1891 A, p. 362.
Günther, A. 1860 A, p. 152.   (Cottina.)
    1880 A, p. 476.

Jordan and Evermann 1896 A, p. 1879.
Nicholson and Lydekker 1889 A, p. 1006.
Owen, R. 1845 B, p. 90.   ("Cottoids," in part.)
Parker, W. K. 1868 A, p. 43.
Sauvage, H. E. 1878 B, p. 111.   (Scorpænidæ, in
    part; Cottini.)
Segond, L. A. 1873 B, p. 533.   ("Cottes.")
Zittel, K. A. 1890 A, p. 310.

COTTUS Linn.   Type *C. gobio* Linn.

Linnæus, C., 1758, Syst. Nat., ed. x, p. 264.
Günther, A. 1860 A, p. 154.
    1880 A, p. 476.
Jordan and Evermann 1896 A, p. 1941.
Sauvage, H. E. 1878 B, p. 135.

Cottus cryptotremus Cope.
Cope, E. D. 1883 L, pp. 162, 163.
    Pleistocene (Equus); Idaho.

Cottus divaricatus Cope.
Cope E. D. 1883 L, p. 162.
    Pleistocene (Equus); Oregon

Cottus hypoceras Cope.
Cope, E. D. 1883 L, pp. 162, 164.
    Pleistocene (Equus); Oregon

Cottus pontifex Cope.
Cope, E. D. 1883 L, pp. 162, 163.
    Pleistocene (Equus); Oregon.

ARTEDIELLUS Jordan.   Type *Cottus uncinatus* Reinh.

Jordan, D. S., 1885, Cat. Fishes N. A., p. 110.
Jordan and Evermann 1896 A, p. 1905.

Artediellus atlanticus Jord. and Ev.

Jordan and Evermann 1896 A, p. 1906.
Dawson, J. W. 1877 B, p. 341. (Cottus sp.)

Dawson, J.W. 1890 A, p. 86.   (Cottus uncinatus.)
This species is regarded by Jordan and Evermann as being closely related to the European *A. uncinatus*, but probably distinct.
Recent; Labrador to Cape Cod: Pleistocene; Canada.

Superfamily CYCLOPTEROIDEA *Gill*.

Gill, T. 1889 A, p. 589.
Garman, S. 1892 A.   (Discoboli.)
    1891 A, p. 361.
    1896 A, p. 135.

Günther, A. 1861 A, p. 154.   (Discoboli.)
    1880 A, p. 483.   (Discoboli.)
Hertwig, O. 1881 A.   (Discoboli.)
Stannius, H. 1854 A.   (Discoboli.)

CYCLOPTERIDÆ.

Jordan and Evermann 1896 A, p. 2094.
Garman, S. 1892 A, p. 19.

Gill, T. 1872 C, p. 5.
    1891 A, p. 366.

CYCLOPTERUS Linn.   Type *C. lumpus* Linn.

Linnæus, C., 1788, Syst. Nat., ed. xi, i, p. 260.
Garman, S. 1892 A, p. 20.
Gill, T. 1891 A, p. 368.
Günther, A, 1861 A, p. 154.
    1880 A, p. 484.
Jordan and Evermann 1896 A, p. 2096.

Cyclopterus lumpus Linn.

Linnæus, C., 1788, Syst. Nat., ed. xi, i, p. 260.
Dawson, J. W. 1877 B, p. 341.
    1890 A, p. 86.

Garman, S. 1892 A. p. 21, pl. viii, figs. 1–3, 15–17; pl. ix, fig. 2; pl. x, fig. B; pl. xi, fig. 10; pls. xii, xiii.
Gill, T. 1891 A, p. 368, pls. xxvii, xxx.
Günther, A. 1861 A, p. 155.
    1880 A, p. 484, figs. 218, 219.
Jordan and Evermann 1896 A, p. 2096.
Logan, W. E. 1863 A, p. 917.
Recent; eastern and western shores of northern Atlantic: Pleistocene; Ontario, Canada.

Order PLECTOGNATHI.

Cope, E. D. 1872 C, p. 340.
Dareste, C. 1849 A.   ("Plectognathes.")
    1872 A, p. 1019.   ("Plectognathes.")
Gill, T. 1872 C, pp. xl, l.
    1884 D, p. 411.
    1893 A, p. 137.
Günther, A. 1870 A, p. 207.
    1880 A, p. 683.
Haeckel, E. 1895 A, p. 236.
Hertwig, O. 1881 A.
Hollard, H. 1860 A.

Klein, E. F. 1881 A.
    1886 A.
Müller, J. 1846 C, p. 202.
Nicholson and Lydekker 1889 A, p. 1014.
Owen, R. 1866 A, p. 11 *et seq.*
Parker, W. K. 1868 A, p. 33.
Reis, O. M. 1895 C, p. 163.
Stannius, H. 1854 A.
Stephan, P. 1900 A.   ("Plectognathes.")
Zittel, K. A. 1890 A, p. 256.

Suborder SCLERODERMI.

Agassiz, L. 1843 B, ii, pp. 248, 265.   ("Sclérodermes.")
Gill, T. 1872 C, p. 1.
    1884 D, p. 414.
    1893 A, p. 137.
Günther, A. 1870 A, p. 207.
    1880 A, p. 684.

Hollard, H. 1860 A, p. 31.   ("Sclérodermes.")
Jordan and Evermann 1896 A, p. 781.
    1896 A, p. 1697.
Müller, J. 1846 C, p. 122.
Owen, R. 1845 B, p. 82.   ("Scleroderms.")
    1866 A, p. 11.
Zittel, K. A. 1890 A, p. 258.

BALISTIDÆ.

Brühl, B. C. 1847 A.   (Balistes.)
Cope, E. D. 1872 C, p. 20.
Dareste, C. 1872 A, p. 1174.   ("Balistes.")
Gill, T. 1872 C, p. 1.
    1884 D, p. 416.
    1893 A, p. 137.
Hollard, H. 1853 A, p. 71.   ("Balistides.")

Hollard, H. 1854 A, p. 39.   ("Balistides.")
    1860 A, p. 23.   ("Balistides.")
Kölliker, A. 1859 A.   (Balistini.)
Owen, R. 1846 B, p. 82.   (Genus Balistes.)
Parker, W. K. 1868 A, pp. 33, 51.
Stannius, H. 1854 A.   (Balistini.)
Stephan, P. 1900 A.   (Balistides.)

408   FOSSIL VERTEBRATA OF NORTH AMERICA.   [BULL. 179-

GRYPODON Hay.   Type *Ancistrodon texanus* Dames.

Hay, O. P. 1899 E, p. 790.
The following authors employ the name *Ancistrodon*. It is practically preoccupied.
Dames, W. 1883 A, p. 664.
Nicholson and Lydekker 1889 A, p. 1016.
Roemer, F. 1849 A, p. 419.
    1852 A, p. 30.
Schlüter, C. 1881 A, p. 61.
Woodward and Sherborn 1890 A, p. 10.

Zittel, K. A. 1890 A, p. 259.

Grypodon texanus (Dames).
Dames, W. 1883 A, p. 664. (Ancistrodon.)
Hay, O. P. 1899 E, p. 790. (Grypodon.)
Roemer, F. 1849 A, p. 419. (No specific name.)
    1852 A, p. 30, pl. i, fig. 10. (No specific name.)
Woodward, A. S. 1888 I, p. 330. (Ancistrodon.)
    Cretaceous; Texas.

Suborder GYMNODONTES.

Agassiz, L. 1843 B, ii, p. 268. ("Gymnodontes.")
Dareste, C. 1849 A.
Gill, T. 1872 C, p. 1.
    1884 D, p. 418.
    1893 A, p. 138.
Günther, A. 1870 A, pp. 207, 269.

Hollard, H. 1857 A.
    1860 A.
Kölliker, A. 1859 A.
Müller, J. 1846 C, p. 208.
Stannius, H. 1854 A.

Superfamily DIODONTOIDEA Gill.

Gill, T. 1884 D, p. 423.

Gill, T. 1893 A, p. 138.

DIODONTIDÆ.

Cope, E. D. 1872 C, p. 340.
Gill, T. 1884 D, p. 423.
Günther, A. 1870 A, p. 269. (Tetrodontina.)
    1880 A, p. 686. (Gymnodontes.)

Hollard, H. 1857 A. ("Diodoniens.")
Jordan and Evermann 1898 A, p. 1742.
Nicholson and Lydekker 1889 A, p. 1015.
Owen, R. 1845 B, p. 77.

DIODON Linn.   Type *D. hystrix* Linn.

Linnæus, C., 1758, Syst. Nat., ed. x, p. 335.
Agassiz, L. 1843 B, ii, p. 273.
Bridge, T. W. 1896 A, p. 582.
Dareste, C. 1849 A.
Günther, A. 1870 A, p. 306.
    1880 A, p. 689.
Jordan and Evermann 1898 A, p. 1744.
Owen, R. 1845 B, pp. 14, 77, pl. xxxviii, figs. 1, 2;
    pl. xxxix, fig. 2.
    1866 A, p. 39.
Parker, W. K. 1868 A, p. 52.

Pictet, F. J. 1854 A, p. 123.
Zittel, K. A. 1890 A, p. 257.

Diodon vetus Leidy.
Leidy, J. 1855 C, p. 397.
Cope, E. D. 1871 M, p. 210. (D. antiquus.)
    1875 V, p. 31, pl. viii, fig. 6. (D. antiquus.)
Leidy, J. 1877 A, p. 255, pl. xxxiv, figs. 15, 16.
    Phosphate beds; South Carolina: Miocene;
    North Carolina.

Dr. E. Koken, in Zeitschrift deutchen geolog. Gesellsch., vol. xl, pp. 274-305, pls. xvii-xix, has described, from the Tertiary deposits of Alabama and Mississippi, the following list of otolites of fishes, designating the families, in some cases the genera, to which these otolites probably belonged. The word Otolithus being evidently not regarded as a generic name, the specific names have no standing in the system of nomenclature.
On the subject of the otolites of fishes see also Higgins, F. T. 1867 A., and Koken, E. 1891 A.
Otolithus (Carangidarum) americanus Kok., p. 277, pl. xix, figs. 1-3.
Otolithus (Apogonidarum) hospes Kok., p. 278, pl. xviii, fig. 15.
Otolithus (Pagelli) elegantulus Kok., p. 279, pl. xvii, figs. 5, 6.
Otolithus (Sparidarum) insuetus Kok., p. 280, pl. xvii, fig. 9.
Otolithus (Sciaenidarum) radians Kok., p. 280, pl. xix, figs. 7, 8.
Otolithus (Sciaenidarum) gemma Kok., p. 281, pl. xix, figs. 9-13.
Otolithus (Sciaenidarum) eporrectus Kok., p. 282, pl. xviii, figs. 16, 17.
Otolithus (Sciaenidarum) claybornensis Kok., p. 283, pl. xix, figs. 1-4.
Otolithus (Sciaenidarum) intermedius Kok., p. 283, pl. xix, figs. 2, 3.
Otolithus (Sciaenidarum) similis Kok., p. 284, pl. xix, figs. 10, 11, 14.
Otolithus (Sciaenidarum) decipiens Kok., p. 285, pl. xix, figs. 5, 6.
Otolithus (Trachini) lævigatus Kok., p. 286, pl. xviii, figs. 13, 14.
Otolithus (Cottidarum) sulcatus Kok., p. 287, pl. xviii, fig. 12.
Otolithus (Triglae) cor Kok., p. 287, pl. xviii, fig. 10.

Otolithus (*Cepolæ*) *comes* Kok., p. 288, pl. xvii, fig. 12.
Otolithus (*Mugilidarum*) *debilis* Kok., p. 288, pl. xvii, fig. 8.
Otolithus (*Gadidarum*) *meyeri* Kok., p. 289, pl. xviii, figs. 8, 9.
Otolithus (*Gadidarum*) *elevatus* Kok., p. 290, pl. xviii, figs. 4, 5.
Otolithus (*Gadidarum*) *mucronatus* Kok., p. 290, pl. xvii, figs. 10, 11.
Otolithus (*Platessæ*) *sector* Kok., p. 292, pl. xvii, figs. 14–16.
Otolithus (*Solæ*) *glaber* Kok., p. 293, pl. xviii, fig. 3.
Otolithus (*Congeris*) *brevior* Kok., p. 293, pl. xviii, fig. 7.
Otolithus (inc. sedis) aff. *umbonata* Kok., p. 294.

Class BATRACHIA Macartney.

Macartney, J., 1802, Lectures on Comp. Anat., i. tab. 3.

This class of animals is designated by some of the following authors *Batrachia;* by others, *Amphibia.*

Baur, G. 1886 B, p. 360.
   1888 C, p. 2.
   1896 C, p. 654.
Beard, J. 1890 A.
Born, G. 1876 A
   1880 A.
Boulenger, G. A. 1892 A.
   1892 B.
Calori, L. 1894 A.
Cope, E. D. 1868 J.
   1869 M, p. 3.
   1875 D.
   1879 A, p. 33.
   1884 I, p. 44.
   1884 O, p. 100.
   1884 P.
   1884 HH.
   1885 G.
   1887 S, p. 331.
   1888 H.
   1888 O.
   1888 AA, p. 244.
   1889 D.
   1889 R, p. 360.
   1891 N, p. 28.
   1892 D, p. 211.
   1892 N.
   1898 B, p. 43.
Cuvier, G. 1834 A, x, pp. 265, 386, pls. cclii–cclv.
Gadow, H. 1896 A.
   1898 A, p. 12.
Gaupp, E. 1891 A.
   1894 A.
   1900 A.
Gegenbaur, C. 1865 B, p. 52.
   1876 A.
   1898 A.
Gill, T. 1900 A, p. 730.
Goeppert, E. 1895 B.
   1896 A.

Haeckel, E. 1896 A, p. 266.
Hancock and Atthey 1868 A, p. 266.
Hasse, C. 1893 A.
Hay, O. P. 1896 A.
   1899 F, p. 682.
Hoffmann, C. K. 1874 A.
   1876 A.
   1878 A.
Howes, G. B. 1888 A.
   1888 B.
Huxley, T. H. 1867 B.
   1870 A.
   1872 A, p. 149.
   1875 A.
   1875 C.
   1876 B, xiii, pp. 411, 429.
Kehrer, G. 1886 A, p. 73.
Kingsley and Ruddick 1900 A, p. 227.
Köstlin, O. 1844 A.
Levy, H. 1898·A.
Leydig, F. 1876 A.
Lydekker, R. 1890 A, p. 121.
Marsh, O. C. 1877 E.
   1898 B, p. 408.
Müller, J. 1831 A, p. 199. (Nuda.)
Nicholson and Lydekker 1889 A, p. 1018.
Osborn, H. F. 1899 I, p. 156.
Owen, R. 1842 A, p. 181.
   1845 B, p. 187.
   1860 A, pp. 154, 160.
   1866 A.
Parker, W. K. 1868 A, p. 58.
   1879 B, pp. 31, 61.
Peters, K. 1895 A.
Stannius, H. 1856 A. (Amphibia dipnoa.)
Steinmann and Döderlein 1890 A, p. 601.
Walter F. 1887 A.
Wiedersheim, R. 1876 A.
   1887 A.
   1889 A, p. 432.
   1890 A, p. 19.
Woodward, A. S. 1898 B, p. 122.
Zittel, K. A. 1890 A, pp. 337, 432.
Zwick, W. 1897 A.

Order STEGOCEPHALI Cope.

Cope, E. D. 1868 J, p. 209.
Baur, G. 1886 H.
   1886 M, p. 304.
   1889 F, p. 469.
   1894 A.
   1896 C.

Burkhardt, R. 1895 A.
Cope, E. D. 1869 M, p. 6.
   1875 D, pp. 352–356.
   1875 X, p. 10.
   1884 P, p. 26.
   1885 G, p. 243.

Cope, E. D. 1887 S. (Rhachitomi, Embolomeri,
    Stegocephali.)
    1888 V.
    1888 AA, p. 248.
    1889 R, p. 861.
    1891 N, p. 28.
    1892 Z.
    1895 C, p. 856.
    1898 B, p. 48.
Credner, H. 1881 A.
    1882 A.
    1883 A.
    1885 A.
    1886 A.
    1887 A.
    1888 A, p. 555.
    1889 A.
    1890 A.
    1893 A.
Fraas, E. 1889 A, p. 8.
Fritsch, A. 1883 A (1879), p. 68.
Gadow, H. 1896 A, p. 23.
    1898 A, p, 12.
Gaupp, E. 1894 A.
    1895 B, p. 56.
Gegenbaur, C. 1895 A, p. 8.

Gegenbaur, C. 1898 A.
Geinitz and Deichmüller 1882 A.
Goeppert, H. 1896 A, p. 431.
Haeckel, E. 1895 A, p. 273.
Hay, O. P. 1899 F, p. 682.
Jaekel, O. 1896 B, p. 129.
Kehner, F. A. 1876 A.
Lortet, L. 1892 A, p. 125.
Lydekker, R. 1890 A, p. 139.   (Labyrinthodontia.)
Maggi, L. 1898 A.
Miall, L. C. 1874 A.   (Labyrinthodonta.)
    1875 A.   (Labyrinthodonta.)
Nicholson and Lydekker 1889 A, p. 1021.   (Laby-
    rinthodontia.)
Osborn, H. F. 1899 I, p. 157.
    1900 A, p. 4.
Pollard, H. B. 1892 A, p. 408.
Röse, C. 1897 A, p. 48.
Rohon, J. V. 1899 B, p. 13.   ("Stegocephalen.")
Seeley, H. G. 1888 D, p. 95.   (Labyrinthodontia.)
    1892 A, p. 333.   (Labyrinthodontia.)
Steinmann and Döderlein 1890 A, p. 608.
Wiedersheim, R. 1878 A, p. 38.
Woodward, A. S. 1898 B, p. 123.   (Stegocephalia.)
Zittel, K. A. 1890 A, p. 344.

Suborder MICROSAURIA Cope.

Cope, E. D. 1868 J, p. 210.
Baur, G. 1897 B.
Case, E. C. 1898 B, p. 505.
Cope, E. D. 1868 J, p. 210.   (Xenorhachia.)
    1869 M, pp. 6, 8.   (Microsauria, Xenorha-
    chia.)
    1875 D, p. 356.   (Microsauria, Xenorhachia.)
    1882 B, p. 452.
    1888 V.   (Lepospondyli.)
    1889 D, p. 13.   (Stegocephali.)
    1889 R, p. 861.
    1891 N, pp. 28, 30.
    1898 B, pp. 43, 47.
Dawson, J. W. 1892 A.
    1895 A, p. 73.
Fraas, E. 1889 A, p. 8.   (Lepospondyli.)
Fritsch, A. 1883 A (1879), pp. 33, 58.   (Micro-
    sauria aistopoda.)

Fritsch, A. 1889 B (1885), p. 73.
Fürbringer, M. 1900 A, p. 284.
Gadow, H. 1898 A, p. 12.   (Lepospondyli.)
Haeckel, E. 1895 A, p. 268 (Lepospondyli); p. 274
    (Progonamphibia.)
Hay, O. P. 1895 A, p. 45.   (Lepospondyli.)
Lydekker, R. 1890 A, p. 196 (Microsauri); p. 205
    (Aistopoda); p. 205 (Branchiosauria).
Miall, L. C. 1890 A, p. 151.   (Microsauria, Aisto-
    poda, Nectridia, Heleothrepta.)
Nicholson and Lydekker 1889 A, p. 1025.
Osborn, H. F. 1899 I, p. 156.
Wiedersheim, R. 1878 A, p. 39.
Woodward, A. S. 1898 B, p. 127 (Branchiosauria);
    p. 129 (Aistopoda); p. 130 (Microsauria).
Zittel, K. A. 1890 A, p. 370.   (Lepospondyli.)

PROTRITONIDÆ.

Nicholson and Lydekker 1889 A, pp. 1022, 1039.
    Unless otherwise indicated the following
authors refer to this family under the name
Branchiosauridæ.   If *Pelion* really belongs to
this family the family name should be *Peliontidæ*.   References to *Branchiosaurus* are also
added.
Baur, G. 1888 C, p. 63.   (Branchiosaurus.)
Cope, E. D. 1875 X, p. 11.   (Peliontidæ.)
Credner, H. 1881 A.   (Branchiosaurus.)
    1881 B.   (Branchiosaurus.)
    1883 A.   (Branchiosaurus.)
    1884 A.   (Branchiosaurus.)

Credner, H. 1887 A.   (Branchiosaurus.)
Fritsch, A 1876, A, p. 72.   (Branchiosaurus.)
    1883 A (1879), p. 69.
    1883 A (1880), p. 93.
Gadow, H. 1896 A, p. 23.   (Branchiosaurus.)
Gaudry, A. 1883 A, p. 253.   (Protriton, Branchio-
    saurus.)
Hay, O. P. 1895 A, p. 45.
Jaekel, O. 1896 A, p. 1.   (Branchiosaurus.)
Lydekker, R. 1890 A, p. 210.   (Protritonidæ.)
Maggi, L. 1898 A, figs. 1, 3.   (Branchiosaurus.)
Zittel, K. A. 1890 A, p. 370.

SPARODUS Fritsch.   Type *S. validus* Fritsch.

Fritsch, *A.* 1876 A, p. 73.
Dawson, J. W. 1882 A, p. 643, pl. xl, figs. 52-55.
    1895 A, p. 75.
Fritsch, A. 1876 A, p. 76.   (Batrachocephalus, preoccupied.)
    1883 A (1879), p. 84.
Gadow, H. 1896 A, p. 28.
Lydekker, R. 1890 A, p. 212.

Zittel, K. A. 1890 A, p. 375.
    It is doubtful to which family this genus belongs.

Sparodus sp. indet.

Dawson, J. W. 1895 A, p. 75.
    Coal-measures; Nova Scotia.

AMPHIBAMUS Cope.   Type *A. grandiceps* Cope.

Cope, *E. D.* 1865 A.
Baur, G. 1888 C, p. 63.
Cope, E. D. 1866 C.
    1869 M, pp. 7, 8.
Fritsch, A. 1883 A (1879), p. 60.
Miall, L. C. 1875 A, p. 181.
Miller, S. A. 1889 A, p. 618.
Zittel, K. A. 1890 A, p. 375.
    This genus may belong among the *Hylonomidæ.*

Amphibamus grandiceps Cope.

Cope, *E. D.* 1865 A.
    1866 C, p. 135, pl. xxxii and text figure.
    1869 M, p. 8.
Dana, J. D. 1896 A, p. 683, fig. 1108.
Fritsch, A. 1883 A (1880), p. 98, fig. 44.
Hay, O. P. 1900 A, p. 120.
Miller, S. A. 1889 A, p. 618, fig. 1178.
    Coal-measures; Illinois.

PELION Wyman.   Type *P. lyellii* Wyman.

Wyman, *J.*, in Cope, E. D. 1868 J., p. 211.   (To replace *Raniceps* Wyman, preoccupied by Cuvier and Kirby.)
Cope, E. D. 1869 M, p. 9.
    1874 L, p. 269.
    1875 D, p. 389.
    1875 X, p. 11.
    1880 A, p. 50.
Fritsch, A. 1883 A (1879), p. 64.
Huxley, T. H. 1868 A, p. 67.   (Raniceps.)
Miall, L. C. 1875 A, p. 189.
Miller, S. A. 1889 A, p. 624.
Wyman, J. 1858 A, p. 160.   (Raniceps.)
Zittel, K. A. 1890 A, p. 375.

Pelion lyellii Wyman.

Wyman, *J.* 1858 A, p. 160, fig. 1.   (Raniceps.)
Cope, E. D. 1869 M, p. 9.
    1874 L, p. 269.
    1875 D, p. 390, pl. xxvi, fig. 1.
Dana, J. D. 1896 A, p. 683, fig. 1109.
Fritsch, A. 1883 A (1879), p. 94, fig. 45.
Gaudry, A, 1875 A.
Lesley, J. P. 1889 A, p. 613, figure.
Miller, S. A. 1889 A, p. 624, fig. 1190.
Newberry, J. S. 1868 A, p. 144.   (Raniceps.)
Wyman, J., in Cope, E. D. 1868 J, p. 211.
    Coal-measures; Ohio.

MOLGOPHIDÆ.

Cope, *E. D.* 1875 D, p. 357.
    1875 D, p. 357.   (Phlegethontiidæ.)
    1875 X, p. 11.   (Phlegethontiidæ, Molgophidæ.)
Fritsch, A. 1876 A, p. 72.   (Genus Dolichosoma.)
    1883 A (1879), p. 107.   (Aistopoda.)
Lydekker R. 1890 A, p. 205.   (Dolichosomatidæ.)

Miall, L. 1875 A, pp. 151, 171.   (Aistopoda.)
Nicholson and Lydekker 1889 A, p. 1024.   (Dolichosomatidæ.)
Wiedersheim, R. 1878 A, p. 42.   (Phlegethontiidæ, Molgophidæ.)
Zittel, K. A. 1890 A, p. 388.   (Aistopoda.)

PHLEGETHONTIA Cope.   Type *P. linearis* Cope.

Cope, *E. D.* 1871 L.
    1872 JJ.
    1874 L, p. 262.
    1875 D, pp. 357, 366.
Fritsch, A. 1883 A (1879), p. 64.
Miall, L. C. 1875 A, p. 190.
Miller, S. A. 1889 A, p. 624.
Nicholson and Lydekker 1889 A, p. 1024.
Zittel, K. A. 1890 A, p. 388.

Phlegethontia linearis Cope.

Cope, *E. D.* 1871 L.
    1874 L, p. 262.
    1875 D, p. 367, pl. xliii, figs. 1, 2.

Fritsch, A. 1883 A (1879), p. 107, fig. 54.
Lesley, J. P. 1889 A, p. 634, figure.
Miller, S. A. 1889 A, p. 624, fig. 1191.
    Coal-measures; Ohio.

Phlegethontia serpens Cope.

Cope, *E. D.* 1874 L, p. 263.
    1871 L.   (No description.)
    1875 D, p. 367, pl. xxxii, fig. 2.
Fritsch, A. 1883 A (1879), p. 107, fig. 53.
Lesley, J. P. 1889 A, p. 635, figure.
    Coal-measures; Ohio.

MOLGOPHIS Cope.   Type *M. macrurus* Cope.

Cope, E. D. 1868 J, p. 220.
    1869 M, p. 20.
    1874 L, p. 263.
    1875 D, p. 368.
Fritsch, A. 1883 A (1879), p. 63.
Miall, L. C. 1875 A, p. 188.
Miller, S. A. 1889 A, p. 623.
Nicholson and Lydekker 1889 A, p. 1024.
Zittel, K. A. 1890 A, p. 383.

Molgophis brevicostatus Cope.

Cope, E. D. 1875 D, p. 369, pl. xliv, fig. 1.
Fritsch, A. 1883 A (1879), p. 107, fig. 55.
Miller, S. A. 1889 A, p. 623, fig. 1188.
Coal-measures; Ohio.

Molgophis macrurus Cope.

Cope, E. D. 1868 J, p. 220.
    1869 M, p. 21.
    1875 D, p. 368, pl. xliii, fig. 3.
Wyman, J. 1858 A, p. 160, fig. 1. (No name. May
    be not this species.)
    Coal-measures; Ohio.

Molgophis wheatleyi Cope.

Cope, E. D. 1874 L, p. 263.
    1875 D, p. 369, pl. xlv, fig. 1.
    Coal-measures; Ohio.

HYLONOMIDÆ.

Fritsch, A. 1883 A, p. 159.
Lydekker, R. 1890 A, p. 201.

Nicholson and Lydekker 1889 A, p. 1027. (Hylo-
    plesionidæ.)
Steinmann and Döderlein 1890 A, p. 608.

HYLONOMUS Dawson.   Type *H. lyelli* Dawson.

Dawson, J. W. 1860 A, p. 274.
Baur, G. 1897 B.
Cope, E. D. 1871 A, p. 9.
Credner, H. 1885 A, p. 274.
    1890 A.
Dawson, J. W. 1863 A, pp. 430, 446.
    1863 B.
    1863 D, pp. 160, 163, 281.
    1868 A, p. 370.
    1882 A, p. 634.
    1891 B, p. 154.
    1894 A, p. 279.
    1895 A, p. 73.
Fritsch, A. 1883 A (1879), p. 59.
    1883 A (1883), p. 159 (Hylonomus); p. 160
    (Hyloplesion).
Fürbringer, M. 1900 A, pp. 286, 296, 554.
Gadow, H. 1896 A, p. 28.
Huxley, T. H. 1863 A, p. 67.
Lydekker, R. 1890 A, p. 201.
Miall, L. C. 1875 A, p. 173.
Miller, S. A. 1889 A, p. 622.
Nicholson and Lydekker 1889 A, p. 1027.
Steinmann and Döderlein 1890 A, p. 608.
Woodward, A. S. 1898 B, p. 131. (Hyloplesion,
    Hylonomus.)
Zittel, K. A. 1890 A, p. 376.

Hylonomus latidens Dawson.

Dawson, J. W. 1882 A, p. 637, pl. xxxix, figs. 18–22.
    1895 A, p. 74.
Lydekker, R. 1890 A, p. 224.
    Coal-measures; Nova Scotia.

Hylonomus lyelli Dawson.

Dawson, J. W. 1860 A, p. 274, figs. 14–18.
Cope, E. D. 1871 A, p. 9.
Dawson, J. W. 1862 A.
    1863 A, pp. 430, 446.

Dawson, J. W. 1863 C, p. 439.
    1863 D, pp. 167, 281, pls. iv, figs. 23–31; pl. v,
    figs. 1–29.
    1868 A, p. 370, fig. 144.
    1878 A, p. 370; suppl., p. 61.
    1882 A, p. 635, pl. xxxix, figs. 1–14; pl. xlv,
    fig. 140.
    1891 A, pl. viii.
    1891 B, p. 149.
    1894 A, p. 279.
    1894 B, p. 6.
    1895 A, p. 74.
Lesley, J. P. 1889 A, p. 290, figures.
Lydekker, R. 1890 A, p. 222.
Miall, L. C. 1875 A, p. 173.
Owen, R. 1862 A, p. 238, pl. ix, figs. 1–5.
    Coal-measures; Nova Scotia.

Hylonomus multidens Dawson.

Dawson, J. W. 1882 A, p. 637, pl. xxxix, figs 23–26.
    1895 A, p. 74.
    Coal-measures; Nova Scotia.

Hylonomus wymani Dawson.

Dawson, J. W. 1860 A, p. 277, figs. 27–29.
Cope, E. D. 1871 A, p. 9.
Dawson, J. W. 1862 A.
    1863 A, p. 430.
    1863 C, p. 439.
    1863 D, pp. 270, 282, pl. vi, figs. 18–31.
    1868 A, p. 378, figs. 146a–i.
    1878 A, p. 378, figs. 146a–i.
    1882 A, p. 637, pl. xxxix, figs. 15–17.
    1894 A, p. 279.
    1895 A, p. 74.
Lesley, J. P. 1889 A, p. 291, figure.
Lydekker, R. 1890 A, p. 224.
Miall, L. C. 1875 A, p. 174.
Owen, R. 1862 A, p. 240, pl. ix, figs. 11–15.
    Coal-measures; Nova Scotia.

SMILERPETON Dawson.   Type *Hylonomus aciedentatus* Dawson.

Dawson, J. W. 1882 A, p. 634.
   1894 A, p. 279.
   1895 A, p. 74.
Gadow, H. 1896 A, p. 28.
Lydekker, R. 1890 A, p. 224. (Smilerpetum.)
Nicholson and Lydekker 1889 A, p. 1027.
Zittel, K. A. 1890 A, p. 376.

Smilerpeton aciedentatum Dawson.

Dawson, J. W. 1860 A, p. 275, figs. 19–23. (Hylonomus.)
Cope, E. D. 1871 A, p. 9. (Hylonomus.)
Dawson, J. W. 1862 A. (Hylonomus.)
   1863 B, p. 473. (Hylonomus.)
   1863 C, p. 439. (Hylonomus.)

Dawson, J. W. 1863 D, pp. 268, 287, pl. vi, figs. 1–16.
  (Hylonomus.)
   1868 A, p. 376, fig. 145a–t. (Hylonomus.)
   1878 A, p. 376, figs. 145a–t. (Hylonomus.)
   1882 A, p. 638, pl. xl, figs. 28–45.
   1894 A, p. 279.
   1895 A, p. 75.
Lesley, J. P. 1889 A, p. 289, figure. (Hylonomus.)
Lydekker, R. 1890 A, p. 224. (Smilerpetum.)
Miall, L. C. 1875 A, p. 174.
Miller, S. A. 1889 A, p. 622, fig. 1186. (Hylo-
  nomus.)
Owen, R. 1862 A, p. 239, pl. ix, fig. 6.
  Coal-measures; Nova Scotia.

HYLERPETON Owen.   Type *H. dawsonii* Owen.

Owen, R. 1862 A, p. 241.
Dawson, J. W. 1876 A, p. 443.
   1878 A, suppl., p. 58.
   1882 A, p. 634.
   1895 A, p. 74.
Fritsch, A. 1883 A (1879), p. 59.
Huxley, T. H. 1863 A, p. 67.
Lydekker, R. 1890 A, p. 224. (Hylerpetum.)
Miall, L. C. 1875 A, p. 174.
Miller, S. A. 1889 A, p. 622.
Zittel, K. A. 1890 A, p. 376.

Hylerpeton dawsonii Owen.

Owen, 1862 A, p. 241, pl. ix.
Cope, E. D. 1869 M, p. 13.
Dawson, J. W. 1863 A, p. 431.
   1863 B, p. 473.
   1863 C, p. 439.
   1863 D, pp. 272, 282, pl. vi, figs. 32–46.
   1868 A, p. 380, fig. 147.
   1876 A, pp. 442, 443.
   1878 A, p. 380, figs. 147a–d; suppl., p. 58.

Dawson, J. W. 1882 A, p. 639, pl. xli, figs. 62–85.
   1894 B, p. 6.
   1895 A, p. 74.
Gadow, H. 1896 A, p. 28.
Lesley, J. P. 1889 A, p. 289, figure.
Lydekker, R. 1890 A, p. 225.
Miall, L. C. 1875 A, p. 194.
Miller, S. A. 1889 A, p. 622, fig. 1185.
  Coal-measures; Nova Scotia.

Hylerpeton intermedium Dawson.

Dawson, J. W. 1895 A, p. 75.
  Coal-measures; Nova Scotia.

Hylerpeton longidentatum Dawson.

Dawson, J. W. 1876 A, p. 444.
   1878 A, suppl., p. 58.
   1882 A, p. 640, pl. xlii, figs. 86–109.
   1894 B, p. 6.
   1895 A, p. 74.
Lydekker, R. 1890 A, p. 225.
  Coal-measures; Nova Scotia.

FRITSCHIA Dawson.   Type *Hylerpeton curtidentatum* Dawson.

Dawson, J. W. 1882 A. p. 634.
   1895 A, p. 75.
Lydekker, R. 1890 A, p. 225.
Zittel, K. A. 1890 A, p. 381.

Fritschia curtidentata Dawson.

Dawson, J. W. 1876 A, p. 444. (Hylerpeton.)

Dawson, J. W. 1878 A, suppl., p. 59. (Hylerpeton.)
   1882 A, p. 641, pl. xliii, figs. 110–128.
   1895 A, p. 75.
Lydekker, R. 1890 A, p. 225.
  Coal-measures; Nova Scotia.

BRACHYDECTES Cope.   Type *B. newberryi* Cope.

Cope, E. D. 1868 J, p. 214.
   1869 M, p. 14.
   1874 L, p. 268.
   1875 D, pp. 357, 388.
Dawson, J. W. 1891 B, p. 158.
Fritsch A. 1883 A (1879), p. 61.
Miall, L. C. 1875 A, p. 184.
Miller, S. A. 1889 A, p. 619.
Zittel, K. A. 1890 A, p. 377.

Brachydectes newberryi Cope.

Cope, E. D. 1868 J, p. 214.
   1869 M, p. 14.
   1874 L, p. 269.
   1875 D, p. 388, pl. xxvii, fig. 1.
Miall, L. C. 1875 A, p. 184.
Miller, S. A. 1889 A, p. 619, fig 1180.
  Coal-measures; Ohio.

PTYONIIDÆ.

Cope, E. D. 1875 D, p. 357.
1875 X, p. 11.
Fritsch, A. 1888 A (1880), p. 127. (Nectridea.)
Lydekker, R. 1890 A, p. 196. (Urocordylidæ.)

Miall, L. C. 1875 A, p. 168. (Nectridea.)
Nicholson and Lydekker 1889 A, p. 1025. (Urocordylidæ.)
Wiedersheim, R. 1878 A, p. 42.

KERATERPETON Huxley.   Type *K. galvani* Huxley.

Huxley, T. H. 1867 B, p. 354.
Andrews, C. W. 1895 A.
Cope, E. D. 1875 D, p. 371. (Ceraterpeton.)
Fritsch, A. 1883 A (1879), p. 57: p. 136 (1881).
Lydekker, R. 1890 A, p. 197. (Ceraterpetum.)
Maggi, L. 1898 A, fig. 10.
Miall, L. C. 1875 A, p. 170.
Miller, S. A. 1889 A, p. 619. (Ceraterpeton.)
Osborn, H. F. 1896 K, p. 165. (Ceraterpeton.)
Woodward, A. S. 1898 B, p. 132.
Zittel, K. A. 1890 A, p. 379.

Keraterpeton divaricatum Cope.

Cope, E. D. 1885 D, p. 406. (Ceraterpeton.)
Coal-measures; Ohio.

Keraterpeton punctolineatum Cope.

Cope, E. D. 1875 J. (Ceraterpeton punctolineatum.)
1875 D, p. 372 (C. lineopunctatum); pl. xli, fig. 4 (C. punctolineatum).
Coal-measures; Ohio.

Keraterpeton tenuicorne Cope.

Cope, E. D. 1875 D, p. 372 (Ceraterpeton lennicorne, *err. typ.*); pl. xlii, fig. 4 (C. recticorne, *err. typ.*).
1885 D, p. 407. (Ceraterpeton tenuicorne.)
1897 C, p. 85, pl. iii, fig. 2. (Ceraterpeton.)
Osborn, H. F. 1896 K, p. 165. (C. lennicorne.)
Coal-measures; Ohio, Pennsylvania.

OÖSTOCEPHALUS Cope.   Type *O. amphiuminus* Cope=*O. remex* Cope.

Cope, E. D. 1869 J, p. 218.
1869 M, p. 16.
1871 L.
1872 JJ.
1874 L, p. 266.
1875 D, p. 380.
Fritsch, A. 1883 A (1879), p. 57. (Syn.? of Urocordylus.)
Huxley, T. H. 1867 B, p. 359, pl. xx, figs. 1, 2. (Urocordylus.)
Miall, L. C. 1875 A, p. 169. (Syn. of Urocordylus.)
Miller, S. A. 1889 A, p. 623.
Newberry, J. S. 1868 A, p. 144. (Urocordylus.)
Nicholson and Lydekker 1889 A, p. 1025. (Urocordylus.)
Zittel, K. A. 1890 A, p. 381.
This genus is probably identical with *Urocordylus* Huxley.

Oöstocephalus rectidens Cope.

Cope, E. D. 1874 L, p. 268.

Cope, E. D. 1875 D, p. 386, pl. xxvii, figs. 3, 4.
Miall, L. C. 1875 A, p. 170.
Miller, S. A. 1889 A, p. 623, fig. 1189.
Coal-measures; Ohio.

Oöstocephalus remex Cope.

Cope, E. D. 1868 J, p. 219. (Sauropleura.)
1868 J, p. 219. (O. amphiuminus.)
1869 M, pp. 11, 17, fig. 2.
1871 G, p. 41. (O. remex and O. amphiuminus.)
1871 O.
1874 L, p. 268.
1875 D, pp. 366, 381, pl. xxvii, fig. 5?; pl. xxxi, fig. 1; pl. xxxii, fig. 1; pl. xxxiii, fig. 2; pl. xxxiv, fig. 4?.
Fritsch, A. 1883 A (1881), p. 128, fig. 72.
Miall, L. C. 1875 A, p. 170.
Coal-measures; Ohio.

PTYONIUS Cope.   Type not designated.

Cope, E. D. 1874 L, p. 264.
1875 D, p. 373.
Fritsch, A. 1883 A. (1879); p. 57. (Syn.? of Urocordylus.)
Miall, L. C. 1875 A, p. 169. (Syn. of Urocordylus.)
Miller, S. A. 1889 A, p. 625.
Zittel, K. A. 1890 A, p. 381.
This genus is probably identical with *Urocordylus* Huxley.

Ptyonius marshii Cope.

Cope, E. D. 1869 M, p. 24. (Colosteus.)
1871 L, p. 177. (Oestocephalus.)
1874 L, p. 265.
1875 D, p. 375, pl. xxvii, fig. 6; pl. xxvii, figs. 2a, 3.

Lesley, J. P. 1889 A, p. 841, figure.
Miall, L. C. 1875 A, p. 170.
Coal-measures; Ohio.

Ptyonius nummifer Cope.

Cope, E. D. 1875 D, p. 374, pl. xli, figs. 2, 3.
Lesley, J. P. 1889 A, p. 842, figures.
Coal-measures; Ohio.

Ptyonius pectinatus Cope.

Cope, E. D. 1868 J, p. 216. (Sauropleura.)
1869 M, p. 20. (Oestocephalus.)
1874 L, p. 266.
1875 D, p. 377, pl. xxvii, fig. 7; pl. xxviii, figs. 2, 4, 6; pl. xxix, fig. 2; pl. xxx, fig. 2; pl. xxxv, figs. 1-8; pl. xli, fig. 1.

Fritsch, A. 1895 B, p. 8. (P. pectinatus?)
Lesley, J. P. 1889 A, p. 843, figures.
Miall, L. C. 1875 A, p. 170.
  Coal-measures; Ohio; Bohem'a?.

Ptyonius serrula Cope.

Cope, E. D. 1874 L, p. 266.
  1871 L, p. 177. (Oestocephalus serrula; no
    description.)
  1875 D, p. 379, pl. xxviii, fig. 5?; pl. xxx,
    fig. 1.
Lesley, J. P. 1889 A, p. 844, figure.
Miall, L. C. 1875 A, p. 170.

Miller, S. A. 1889 A, p. 625, fig. 1192.
  Coal-measures; Ohio.

Ptyonius vinchellianus Cope.

Cope, E. D. 1874 L, p. 265.
  1871 L. (Oestocephalus winchellianus; no
    description.)
  1875 D, p. 376, pl. xxviii, fig. 1.
Fritsch, A. 1883 A (1881), p. 128, fig. 71.
Lesley, J. P. 1889 A, p. 844, figure. (P. uinchell-
    anus.)
Miall, L. C. 1875 A, p. 170.
  Coal-measures; Ohio.

OTENERPETON Cope.   Type *C. alveolatum* Cope.

Cope, E. D. 1897 C, p. 88.

Otenerpeton alveolatum Cope.

Cope, E. D. 1897 C, p. 88, pl. iii, fig. 1[1].
  Coal-measures; Pennsylvania.

TUDITANIDÆ.

Cope, E. D. 1875 D, p. 397.
  1875 X, p. 11 (Tuditanidæ); p. 12 (Oocytin-
    idæ).
Fritsch, A. 1883 A, p. 173. (Microbrachidæ.)

Nicholson and Lydekker 1889 A, p. 1027. (Micro-
    brachidæ.)
Wiedersheim, R. 1878 A, p. 42.

TUDITANUS Cope.   Genus based on *T. punctulatus* and *T. brevirostris.*

Cope, E. D. 1874 L, p. 271.
  1871 L. (No definition; no described spe-
    cies.)
  1872 JJ. (No definition; no described spe-
    cies.)
  1875 D, p. 391.
  1896 C.
Fritsch, A. 1883 A (1879), p. 65.
  1883 A, p. 173.
Miall, L. C. 1875 A, p. 191.
Miller, S. A. 1889 A, p. 626.
Nicholson and Lydekker 1889 A, p. 1027.
Zittel, K. A. 1890 A, p. 378.

Tuditanus brevirostris Cope.

Cope, E. D. 1874 L, p. 272.
  1875 D, p. 393, pl. xxvi, figs. 3, 4.
Lesley, J. P. 1890 A, p. 1236, figure.
  Coal-measures; Ohio.

Tuditanus huxleyi Cope.

Cope, E. D. 1874 L, p. 274.
  1875 D, p. 397, pl. xxxiv, fig. 2.
Lesley, J. P. 1890 A, p. 1237, figure.
  Coal-measures; Ohio.

Tuditanus longipes Cope.

Cope, E. D. 1874 L, p. 210. (Sauropleura.)·
  1875 D, p. 398, pl. xxvi, fig. 2. (Not pl.
    xxxiv, fig. 1, fide Cope.)
  1896 C.
  1897 C, p. 88.
Lesley, J. P. 1890 A, p. 1257, figures.
  Coal-measures; Ohio.

Tuditanus mordax Cope.

Cope, E. D. 1874 L, p. 274.

Cope, E. D. 1875 D, p. 395.
Miller, S. A. 1889 A, p. 627. (Syn. of Keraterpeton
    punctolineatum.)
  Coal-measures; Ohio.

Tuditanus obtusus Cope.

Cope, E. D. 1868 J, p. 213. (Dendrerpeton.)
  1869 M, pp. 8, 12, fig. 1. (Dendrerpeton.)
  1874 L, p. 274.
  1875 D, p. 396.
  1885 D, p. 407.
  Coal-measures; Ohio.

Tuditanus punctulatus Cope.

Cope, E. D. 1874 L, p. 271.
  1875 D, p. 392, pl. xxxiv, fig. 1. (T. longipes
    on expl. of plate by error.)
  1896 C.
  1897 C, p. 88. (Isodectes.)
  1898 B, p. 61. (Isodectes.)
Fritsch, A. 1883 A, p. 173, fig. 110.
  Coal-measures; Ohio.

Tuditanus radiatus Cope.

Cope, E. D. 1874 L, p. 273.
  1875 D, p. 394, pl. xxvii, fig. 1; pl. xxxiv,
    fig. 3; text, fig. 10.
Lesley, J. P. 1890 A, p. 1238, figure.
Miller, S. A. 1889 A, p. 626, fig. 1194.
  Coal-measures; Ohio.

Tuditanus tabulatus Cope.

Cope, E. D. 1877 A, p. 577.
  Coal-measures; Ohio.

---

[1] On plate and explanation this species is called C. *foveolatum.*

COCYTINUS Cope.   Type *C. gyrinoides* Cope.

Cope, E. D. 1871 L.
1872 JJ.
1874 L, p. 276.
1875 D, p. 360.
1875 X, p. 12.
Fritsch, A. 1883 A (1879), p. 61; p. 173 (1883).
Miall, L. C. 1875 A, p. 185.
Miller, S. A. 1889 A, p. 619.
Nicholson and Lydekker 1889 A, p. 1027.
Zittel, K. A. 1890 A, p. 378.

The family position of this genus is doubtful.

Cocytinus gyrinoides Cope.
Cope, E. D. 1871 L.
1874 L, p. 278.
1875 D, p. 364, pl. xxxix, fig. 4.
Fritsch, A. 1883 A, p. 173, fig. 111.
Miller, S. A. 1889 A, p. 620, fig. 1182.
Coal-measures; Ohio.

DIPLOCAULIDÆ.

Cope, E. D. 1881 K, p. 163. | Cope, E. D. 1882 B, p. 452.

DIPLOCAULUS Cope.   Type *D. salamandroides* Cope.

Cope, E. D. 1877 E, p. 187.
1881 K, p. 163.
1882 B, p. 451.
Miller, S. A. 1889 A, p. 621.
Zittel, K. A. 1890 A, p. 382.

Case, E. C. 1900 A, p. 710.
Cope, E. D. 1882 HH.
1888 BB, p. 286.
1896 G, p. 455, pl. ix.
Permian; Texas.

Diplocaulus limbatus Cope.
Cope, E. D. 1896 G, p. 456.
Permian; Texas.

Diplocaulus magnicornis Cope.
Cope, E. D. 1882 B, p. 453.

Diplocaulus salamandroides Cope.
Cope, E. D. 1877 E. p. 187.
Case, E. C. 1900 A, p. 710, pl. i, figs. 16a–17b.
Cope, E. D. 1881 K, p. 163.
1888 BB, p. 286.
Permian; Illinois.

LEPOSPONDYLOUS GENERA OF UNCERTAIN POSITION.

AMBLYODON Dawson.   Type *A. problematicum* Dawson.

Dawson, J. W. 1882 A, p. 635.

Amblyodon problematicus Dawson.
Dawson, J. W. 1882 A, p. 644, pl. xl, figs. 57–61.

Dawson, J. W. 1895 A, p. 75.
Coal-measures; Nova Scotia.

HYPHASMA Cope.   Type *H. lævis* Cope.

Cope, E. D. 1875 J.
1875 D, p. 387.
Fritsch, A. 1883 A (1881), p. 127.
Miller, S. A. 1889 A, p. 622.
Zittel, K. A. 1890 A, p. 382.

Hyphasma lævis Cope.
Cope, E. D. 1875 J.
1875 D, p. 387, pl. xxxvii, fig. 4.
Miller, S. A. 1889 A, p. 622, fig. 1187.
Coal-measures; Ohio.

EURYTHORAX Cope.   Type *E. sublævis* Cope.

Cope, E. D. 1871 L.
1874 L, p. 275.
1875 D, pp. 357, 401.
Fritsch, A. 1883 A (1879), p. 62.
Miall, L. C. 1875 A, p. 186.
Miller, S. A. 1889 A, p. 622.
Zittel, K. A. 1890 A, p. 378.

Eurythorax sublævis Cope.
Cope, E. D. 1871 L.
1872 JJ.
1874 L, p. 275.
1875 D, p. 402, pl. xlii, fig.
Coal-measures; Ohio.

THYRSIDIUM Cope.   Type *T. fasciculare* Cope.

Cope, E. D. 1875 D, p. 365.
Miller, S. A. 1899 A, p. 626.
Zittel, K. A. 1890 A, p. 382.

Thyrsidium fasciculare Cope.
Cope, E. D. 1875 D, p. 365, pl. xlii, fig. 3.
Coal-measures; Ohio.

PLEUROPTYX Cope.   Type *P. clavatus* Cope.

Cope, E. D. 1875 J.
1875 D, pp. 357, 370.
Miller, S. A. 1889 A, p. 624.

Pleuroptyx clavatus Cope.
Cope, E. D. 1875 J.
1875 D, p. 370, pl. xlii, fig. 1; pl. xliv, fig. 2.
Coal-measures; Ohio.

CERCARIOMORPHUS Cope.   Type *C. parvisquamis* Cope.

Cope, E. D. 1885 D, p. 405.
Subordinal position doubtful.

Cercariomorphus parvisquamis Cope.
Cope, E. D. 1885 D, p. 405.
Coal-measures; Ohio.

Suborder APŒCOSPONDYLI nom. nov.

Baur, G. 1886 K.   (Embolomeri.)
 1886 C, p. 3. (Ganocephala, Rhachitomi, Embolomeri.)
Case, E. C. 1878 B, p. 510 (Rhachitomi); p. 511 (Embolomeri).
Cope, E. D. 1868 J, pp. 210, 221. (Labyrinthodontia and Ganocephala.)
 1875 D, p. 356. (Labyrinthodontia and Ganocephala.)
 1880 A, p. 49. (Labyrinthodontia and Ganocephala.)
 1880 J. (Ganocephala.)
 1880 L, p. 610. (Embolomeri, Ganocephala.)
 1882 T. (Rhachitomi.)
 1884 P, pp. 26, 36. (Embolomeri, Stegocephali.)
 1885 G, p. 243. (Embolomeri.)
 1885 W. (Embolomeri.)
 1887 S, p. 331. (Rhachitomi, Embolomeri.)
 1888 H. (Ganocephala.)
 1888 V. (Ganocephala, Embolomeri, Rhachitomi.)
 1888 AA. (Embolomeri, Rhachitomi.) .
 1889 D, p. 13. (Ganocephala, Embolomeri, Rhachitomi.)
 1889 R, p. 861. (Ganocephala, Embolomeri, Rhachitomi.)
 1891 F, p. 645. (Embolomeri, Rhachitomi, Ganocephali.)
 1891 N, p. 28. (Ganocephali, Rhachitomi, Embolomeri.)
 1896 B, p. 43. (Ganocephala, Rhachitomi, Embolomeri.)
Credner, H. 1882 A. (Stegocephali.)
 1883 A. (Stegocephali.)
Credner, H. 1885 A. (Stegocephali.)

Credner, H. 1886 A. (Stegocephali.)
 1888 A. (Stegocephali.)
 1889 A. (Stegocephali.)
 1890 A. (Stegocephali.)
Dawson, J. W. 1894 A, p. 265. (Labyrinthodontia.)
Fraas, E. 1889 A, p. 8. (Temnospondyli, Stereospondyli.)
Fritsch, A. 1883 A (1879), p. 33. (Labyrinthodontia.)
Gadow, H. 1898 A, p. 12. (Temnospondyli.)
Hoffman, C. K. 1874 A, p. 47. (Labyrinthodontia.)
Huxley, T. H. 1863 A. (Labyrinthodontia.)
 1870 A. (Labyrinthodontia.)
 1872 A, p. 150. (Labyrinthodontia.)
 1876 B, xiii, p. 412. (Labyrinthodontia.)
Lydekker, R. 1890 A, p. 139. (Labyrinthodontia vera.)
Marsh, O. C. 1877 E. (Labyrinthodontia.)
Meyer, H. 1857 A, p. 208. ("Labyrinthodonten.")
Miall, L. C. 1875 C. (Labyrinthodontes.)
Nicholson and Lydekker 1889 A, p. 102.
Owen, R. 1845 B, pp. 182, 196. ("Labyrinthodonts.")
 1860 A. pp. 155, 158, 168. (Ganocephala, Labyrinthodontia.)
 1860 E, p. 182. (Ganocephala.)
 1866 A, p. 14. (Labyrinthodontia and Ganocephala.)
Steinmann and Döderlein 1890 A, p. 611. (Embolomeri.)
Wiedersheim, R. 1878 A. (Ganocephala, Labyrinthodontia.)
Woodward, A. S. 1898 B, p. 182. (Labyrinthodontia.)
Zittel, K. A. 1890 A, p. 384 (Temnospondyli); p. 397 (Stereospondyli)

DENDRERPETONTIDÆ.

Fritsch, A. 1889 B (1885), p. 5.
Cope, E. D. 1875 X, p. 10. (Baphetidæ.)
Dawson, J. W. 1895 A, p. 75. (Dendrerpetonidæ.)

Lydekker, R. 1890 A, p. 170. (Dendrerpetidæ.)
Nicholson and Lydekker 1889 A, p. 1032. (Dendrerpetidæ.)

DENDRERPETON Owen.   Type *D. acadianum* Owen.

Owen, R. 1853 C, p. 66.
Cope, E. D. 1868 J, p. 212.
 1869 M, p. 12.
 1880 A, p. 50.

Credner, H. 1888 A, p. 560.
Dawson, J. W. 1860 A, p. 273.
 1863 A, pp. 431, 444.
 1863 B.

Bull. 179——27

Dawson, J. W. 1863 D, pp. 81, 282.
  1868 A, p. 362.
  1878 A, p. 362; suppl. p. 60.
  1882 A, p. 635.
  1891 B, p. 154.
  1896 A, p. 76.
Fritsch, A. 1883 A (1879), p. 58.
  1889 B, p. 6.
Huxley, T. H. 1863 A, p. 67.
Lydekker, R. 1890 A, p. 170.   (Dendrerpetum.)
Maggi, L. 1896 A, fig. 19.
Miall, L. C. 1875 A, p. 550.
Miller, S. A. 1889 A, p. 620.
Nicholson and Lydekker 1889 A, p. 108.
Zittel, K. A. 1890 A, p. 396.

Dendrerpeton acadianum Owen.

Owen, R. 1853 C.
Cope, E. D. 1868 J, p. 214.
  1869 M, pp. 9, 13.
Dawson, J. W. 1860 A, p. 273, figs. 10–18.
  1862 A.
  1863 A, p. 431.
  1863 B, p. 470.
  1863 C.
  1863 D, pp. 81, 159, 282, pls. iii, vi.
  1868 A, p. 362, fig. 142.
  1876 A, pp. 440, 443, 445.
  1878 A, p. 362, figs. 142a–i; suppl. p. 60.
  1882 A, p. 642, pl. xl, figs. 46–51; pl. xliv, figs. 129, 130, 132–137.
  1891 B, p. 145, figs. 1–3.
  1894 A, p. 272, figure.

Dawson, J.W.1894 B, p. 6.
  1895 A, p. 76.
Fritsch, A. 1883 A (1879), p. 59.
  1889 B (1885), p. 5, fig. 123.
Lesley, J. P. 1889 A, p. 193, figures.
  1895 A, pl. dixxii.
Lydekker, R. 1890 A, p. 223.
Lyell, C. 1853 A, p. 35.
Lyell and Dawson 1853 A, pls. ii, iii.
Marsh, O. C. 1862 B, p. 1.
Miall, L. C. 1875 A, p. 172.
Miller, S. A. 1889 A, p. 620, fig. 1188.
Wyman, J. 1853 C.
  Coal-measures; Nova Scotia.

Dendrerpeton oweni Dawson.

Dawson, J. W. 1863 D, pp. 161, 282, pl. iv.
  1863 A, p. 431.
  1863 B, p. 470.
  1863 C, p. 438.
  1868 A, p. 368, fig. 143.
  1876 A, pp. 443, 444.
  1878 A, p. 368, figs. 143a–g; suppl. p. 60.
  1882 A, p. 643, pl. xliv, figs. 131, 138, 139.
  1891 B, p. 149.
  1895 A, p. 76.
Fritsch, A. 1883 A (1879), p. 59.
Lesley, J. P. 1889 A, p. 195, figures.
  1895 A, pl. dixxii.
Lydekker, R. 1890 A, p. 223.
Miall, L. C. 1875 A, p. 172.
  Coal-measures; Nova Scotia.

BAPHETES Owen.   Type *B. planiceps* Owen.

Owen, R. 1854 A, p. 207.
Dawson, J. W. 1863 D, pp. 8, 282.
  1868 A, p. 359.
  1878 A, p. 328.
  1894 A, p. 266.
  1895 A, p. 76.
Fritsch, A. 1883 A (1879), p. 61.
  1889 B (1885), p. 5.
Huxley, T. H. 1863 A, p. 67.
Miall, L. C. 1875 A, p. 183.
Miller, S. A. 1889 A, p. 618.
Nicholson and Lydekker 1889 A, p. 1083.
Owen, R. 1855 A.
  1860 E, p. 184.
Zittel, K. A. 1890 A, p. 398.
  The family position of this genus is doubtful.

Baphetes minor Dawson.

Dawson, J. W. 1870 A.
  1894 A, p. 268.
  Coal-measures; Nova Scotia.

Baphetes planiceps Owe

Owen, R. 1854 A, pl. ix.
Cope, E. D. 1869 M, p. 25.
Dawson, J. W. 1855 A.
  1863 A, p. 430.
  1863 C.
  1863 D, pp. 9, 282, pl. ii.
  1868 A, p. 359, fig. 141.
  1878 A, pp. 328, 359, figs. 137, 141.
  1894 A, p. 266.
  1895 A, p. 76.
Fritsch, A. 1889 B (1885), p. 5, fig. 124.
Lesley, J. P. 1889 A, p. 80, figures.
Miller, S. A. 1889 A, p. 619, fig. 1179.
Nicholson and Lydekker 1889 A, p. 1083, fig. 962.
Owen R. 1855 A.
  Coal-measures; Nova Scotia.

PLATYSTEGOS Dawson.   Type *P. loricatum* Dawson.

Dawson, J. W. 1895 A, p. 76.

Platystegos loricatum Dawson.

Dawson, J. W. 1895 A, p. 76.
  Coal-measures; Nova Scotia.

SAUROPLEURIDÆ nom. nov.

Cope, E. D. 1875 D, p. 357. (Colosteidæ.)
1875 X, p. 10. (Colosteidæ.)
Wiedersheim R. 1878 A, p. 42. (Colosteidæ.)
The name *Sauropleuridæ* is demanded be-

cause the genus *Colosteus*, from which *Colosteidæ* is derived, is regarded as a synonym of *Sauropleura*.

SAUROPLEURA Cope. Type *S. digitata* Cope.

Cope, E. D. 1868 J, p. 215.
1869 M, p. 22. (Colosteus, type, *C. foveatus*.)
1874 L, p. 269.
1875 D, pp. 373, 402.
1875 D, p. 405. (Colosteus.)
1897 C, p 88.
Fritsch, A. 1883 A (1879), p. 61 (Colosteus); pp. 65, 127 (Sauropleura).
Miall, L. C. 1875 A, p. 185 (Colosteus); p. 190 (Sauropleura).
Miller, S. A. 1889 A, p. 620 (Colosteus); p. 625 (Sauropleura).
Zittel, K. A. 1890 A, p. 379 (Colosteus); p. 382 (Sauropleura).

Sauropleura digitata Cope.

Cope, E. D. 1868 J, p. 216.
1869 M, PP. 16, 242.
1874 L, p. 270.
1875 D, p. 408, pl. xxxivi, fig. 1
1897 C, p. 88.
Dawson, J. W. 1891 B, p. 158.
Fritsch, A. 1883 A (1881), p. 128, fig. 70.
Lesley, J. P. 1890 A, p. 926, figure.
Coal-measures; Ohio.

Sauropleura foveata (Cope).

Cope, E. D. 1869 M, p. 24. (Colosteus.)
1875 D, p. 405, pl. xxxvi, fig. 1.
Coal-measures; Ohio.

Sauropleura latithorax Cope.

Cope, E. D. 1897 C, p. 86, pl. iii, fig. 4.
Coal-measures; Ohio.

Sauropleura newberryi Cope.

Cope, E. D. 1875 D, p. 404, pl. xxxvii, figs. 2, 3; pl. xli, fig. 5.
Lesley, J. P. 1890 A, p. 928, figure.
Coal-measures; Ohio.

Sauropleura pauciradiata Cope.

Cope, E. D. 1874 L, p. 275. (Colosteus.)
1875 D, p. 406, pl. xl, figs. 1, ?. (Colosteus.)
1897 C, p. 86.
Coal-measures; Ohio.

Sauropleura scutellata (Newb.).

Newberry J. S. 1856 B, p. 96. (Pygopterus.)
Cope, E. D. 1869 M, p. 23. (Colosteus crassiscutatus.)
1871 G, p. 41. (Colosteus.)
1873 Q, p. 418. (Colosteus.)
1874 L, p. 275. (Colosteus.)
1875 D, p. 407, pl. xxix, fig. 3; pl. xxxiii, fig. 1; pl. xxxvi, fig. 2. (Colosteus.)
1877 A, p. 578. (Colosteus.)
1897 C, p. 88. (Sauropleura.)
Woodward, A. S. 1891 A, p. 474. ("Doubtful fossil.")
Coal-measures; Ohio.

LEPTOPHRACTUS Cope. Type *L. obsoletus* Cope.

Cope, E. D. 1873 O, p. 340.
1875 D, p. 399.
1882 B, p. 461.
Fritsch, A. 1883 A (1879), p. 62.
Miall, L. C. 1875 A, p. 187.
Miller, S. A. 88 A, p. 62.
Zittel, K. A. 1890 A, p. 378.

Leptophractus lineolatus Cope.

Cope, E. D. 1877 A, p. 576.
Coal-measures; Ohio.

Leptophractus obsoletus Cope.

Cope, E. D. 1873 O, p. 341.
1875 D, p. 400, pl. xxxviii; pl. xxxix, figs. 1, 2, 37.
Woodward, A. S. 1891 A, p. 262. (Sagenodus ohioensis, in part.)
Coal-measures; Ohio.

ARCHEGOSAURIDÆ.

References are here given also to the literature of the genus *Archegosaurus*.
Baur, G. 1886 C. (Archegosaurus.)
1888 C. (Archegosaurus.)
1896 C, p. 662. (Archegosaurus.)
1897 A. (Archegosaurus.)
Cope, E. D. 1882 T. (Trimerorhachidæ.)
1884 P, p. 28. (Trimerorhachidæ.)
1889 R, p. 261. (Trimerorhachidæ, Archegosauridæ.)

Cope, E. D.1891 N, p. 29. (Trimerorhachidæ, Archegosauridæ.)
1898 B, p. 45. (Ganocephali.)
Credner, H. 1882 A, p. 231. (Archegosaurus.)
1885 A, p. 718. (Archegosaurus.)
Emery, C. 1897 A. (Archegosaurus.)
Fraas, E. 1889 A. (Archegosaurus.)
Fritsch, A. 1889 B (1885), p. 13. (Archegosauridæ.)
Gadow, H. 1896 A, p. 21. (Archegosaurus.)
Gaudry, A. 1883 A, p. 261. (Archegosaurus.)

Goldfuss, A. 1847 A, p. 3, pls. i–iii. (Archego-
saurus.)
Huxley, T. H. 1863 A, p. 67. (Archegosauria.)
Jaekel, O. 1896 A, p. 3. (Archegosaurus.)
    1896 B, p. 118. (Archegosaurus.)
Lydekker, R. 1890 A, p. 177. (Archegosauridæ.)
Maggi, L. 1898 A, figs. 4–6. (Archegosaurus.)
Meyer, H. 1851 B, (Archegosaurus.)
    1857 A. (Archegosaurus.)
    1858 A. (Archegosaurus.)
Miall, L. C. 1875 A, p. 166. (Archegosauria.)

Miall, L. C. 1875 C. (Archegosaurus.)
Nicholson and Lydekker 1889 A, p. 1029. (Arche-
gosauridæ.)
Owen, R. 1860 A, p. 156. (Archegosaurus.)
Pictet, F. J. 1853 A, p. 551. (Archegosaurus.)
Quenstedt, F. A. 1852 A, p. 153. (Archegosaurus.)
    1885 A, p. 245. (Archegosaurus.)
Woodward, A. S. 1898 B, p. 133. (Archegosaurus.)
Zittel, K. A. 1890 A, p. 384. (Archegosaurus.)
It is possible that the name Trimerorhachidæ
has priority over Archegosauridæ.

TRIMERORHACHIS Cope.   Type *T. insignis* Cope.

Cope, E. D. 1878 B, p. 524.
Baur, G. 1888 C, p. 6.
Cope, E. D. 1878 U.
    1878 DD.
    1880 A, p. 54.
    1884 P, p. 32.
    1888 H, p. 298.
    1888 U, p. 915.
    1888 AA, p. 248.
Gadow, H. 1896 A, p. 21.
Hulke, J. W. 1888 A, p. 428.
Lydekker, R. 1890 A, p. 189.
Miller, S. A. 1889 A, p. 626.
Nicholson and Lydekker 1889 A, p. 1080.
Zittel, K. A. 1890 A, p. 389.

Trimerorhachis bilobatus Cope.

Cope, E. D. 1883 H, p. 629.
    1884 P, p. 32.
    1888 BB, p. 286.
Permian; Texas.

Trimerorhachis conangulus Cope.

Cope, E. D. 1896 I, p. 187.
Permian; Texas.

Trimerorhachis insignis Cope.

Cope, E. D. 1878 B, p. 524.
    1880 A, p. 54.
    1881 K, p. 163.
    1883 H, p. 630.
    1884 P, p. 32, figs. 3, 4.
    1888 O, p. 466, pl. vi, fig. 1.
    1888 AA, p. 43, fig. 1.
    1888 BB, p. 286.
Lydekker, R. 1890 A, p. 190, fig. 46.
Zittel, K. A. 1890 A, p. 389, figs. 379 a–d.
Permian; Texas.

Trimerorhachis mesops Cope.

Cope, E. D. 1896 G, p. 454.
Permian; Texas.

DISSORHOPHUS Cope.   Type *D. multicinctus* Cope.

Cope, E. D, 1895 D.
Baur, G. 1896 B, p. 569.

Dissorhophus articulatus Cope.

Cope, E. D. 1896 E, pl. xxi.
Permian; Texas.

Dissorhophus multicinctus Cope.

Cope, E. D. 1895 D.
Permian; Texas.

CRICOTIDÆ.

Cope, E. D. 1884 P, p. 38.
    1889 R, p. 861.
    1891 N, p. 30.
    1896 B, p. 46.

Fritsch, A. 1889 B (1885), p. 11. (Diplovertebridæ.)
Lydekker, R. 1890 A, p. 175. (Diplospondylidæ.)

CRICOTUS Cope.   Type *C. heteroclitus* Cope.

Cope, E. D. 1875 T, p. 405.
Baur, G. 1888 C, p. 14.
    1896 C, p. 662.
Cope, E. D. 1878 DD,
    1880 J.
    1880 L, p. 610.
    1884 I, p. 29.
    1884 P, p. 38.
    1885 W.
    1886 O.
    1886 P.
    1888 U, p. 915.

Cope, E. D. 1888 AA.
    1888 BB, p. 286.
    1888 DD, p. 363.
Gadow, H. 1896 A, p. 28.
Hay, O. P. 1895 A, p. 45.
Lydekker, R. 1890 A, p. 175.
Maggi, L. 1898 A, fig. 20.
Miller, S. A. 1889 A, p. 620.
Williston, S. W. 1897 B, figs. 1–4.
    1898 J, p. 120.
Woodward, A. S. 1898 B, p. 136.
Zittel, K. A. 1890 A, p. 394.

Cricotus crassidiscus Cope.

Cope, E. D. 1884 I, p. 28.
1878 B, p. 522. (C. heteroclitus, in part.)
1884 P, p. 39, text fig. 7, pl. v. (See C. heteroclitus.)
1888 AA, p. 246, fig. 3.
1888 BB, p. 286.
Williston, S. W. 1897 B, p. 54.
1898 J, p. 121.
Zittel, K. A. 1890 A, p. 351, fig. 331; p. 395, fig. 384.
(C. heteroclitus; not of Cope.)
Permian; Texas.

Cricotus gibsoni Cope.

Cope, E. D. 1877 E, p. 185.
Case, E. C. 1900 A, p. 709, figs. 15a–15c.
Cope, E. D. 1881 K, p. 163.
1884 P, p. 39.
1888 BB, p. 286.
Permian; Illinois.

Cricotus heteroclitus Cope.

Cope, E. D. 1875 T, p. 405.

Case, E. C. 1900 A, p. 708, pl. i, figs. 12a–12d; 13, 14.
Cope, E. D. 1877 E, p. 186. (C. discophorus.)
1877 N, p. 63.
1878 B, p. 523. (C. discophorus.)
1881 K, p. 163.
1884 I, p. 29.
1884 P, p. 39, fig. 7 and pl. v. (See C. crassidiscus.)
1888 AA, p. 253, pl. i, figs. 7, 8.
1888 BB, p. 286.
Lydekker, R. 1890 A, p. 176, fig. 42.
Permian; Illinois.

Cricotus hypantricus Cope.

Cope, E. D. 1884 I, pp. 29, 30.
1888 AA, p. 253, pl. i, figs. 2–6.
1888 BB, p. 286.
Permian; Texas.

Cricotus sp. indet.

Case, E. C. 1900 A, p. 709, pl. v, figs. 13a–16.
Permian; Illinois.

ANTHRACOSAURIDÆ.

Cope, E. D. 1875 X, p. 10.
Atthey, T. 1876 A, p. 146, pls. viii–xi. (Anthracosaurus.)
Hancock and Atthey 1869 A, p. 318. (Anthracosaurus.)

Huxley, T. H. 1863 A. (Anthracosaurus.)
Lydekker, R. 1890 A, p. 157.
Miall, L. C. 1869 A, p. 159. (Anthracosaurus.)
Nicholson and Lydekker 1889 A, p. 1083.
Zittel, K. A. 1890 A, p. 396. (Gastrolepidoti.)

EOSAURUS Marsh. Type *E. acadianus* Marsh.

Marsh, O. C. 1862 B.
Agassiz, L. 1862 A.
Cope, E. D. 1869 M, p. 30.
Fritsch, A. 1883 A (1879), p. 61.
Huxley, T. H. 1863 A, p. 62.
Lydekker, R. 1890 A, p. 166.
Marsh, O. C. 1863 A.
Miall, L. C. 1875 A, p. 186.
Miller, S. A. 1889 A, p. 621.
Nicholson and Lydekker 1889 A, p. 1037.
Zittel, K. A. 1890 A, p. 400.
The position of Eosaurus in this family is not at all established.

Eosaurus acadianus Marsh.

Marsh, O. C. 1862 B, pls. i, ii.
Cope, E. D. 1869 M, p. 30.
Dawson, J. W. 1863 A, p. 431.
1863 C, p. 440.
1863 D, pp. 275, 283.
1868 A, p. 382, fig. 148.
1878 A, p. 382, fig. 148.
Huxley, T. H. 1863 A, p. 62.
Lesley, J. P. 1895 A, pl. dlxxii.
Lydekker, R. 1890 A, p. 166.
Marsh, O. C. 1862 A. (No description.)
1863 A.
Miller, S. A. 1889 A, p. 621, fig. 1184.
Coal-measures; Nova Scotia.

ERYOPIDÆ.

Cope, E. D. 1882 T.
1882 B, pp. 460, 461.
1884 P, p. 28.
1885 W.
1891 N, p. 29.

Lydekker, R. 1890 A, p. 191.
Nicholson and Lydekker 1889 A, p. 1029.

ERYOPS Cope. Type *E. megacephalus* Cope.

Cope, E. D. 1877 E, p. 188.
Baur G. 1888 C, p. 13.
Baur and Case 1899 A, p. 3, footnote.
Broili, F. 1899 A, p. 84.

Cope, E. D. 1878 B, p. 515 (Epicordylus, type E. erythroliticus); p. 520 (Eryops); p. 521 (Parioxys, type P. ferricolus); p. 526 (Rhachitomus, type R. valens).

Cope, E. D. 1878 U.   (Rhachitomus, Parioxys, Epi-
    cordylus.)
1878 AA.   (Rhachitomus.)
1878 DD.   (Epicordylus.)
1879 A, p. 34.   (Rhachitomus.)
1880 A, pp. 49-51.   (Eryops, Rhachitomus.)
1880 J.
1882 T.
1884 P, p. 38.
1886 O.
1886 P.
1888 AA.
1888 DD.
1894 C, p. 789.
1896 G, p. 439.
Emery C. 1897 A.
Gadow H. 1896 A, p. 21.
Lydekker R. 1890 A, p. 191.
Miller, S. A. 1889 A, p. 622.
Nicholson and Lydekker 1889 A, pp. 1022, 1082.
Seeley, H. G. 1892 A.
Stickler, L. 1899 A, p. 85.
Williston, S. W.1899 E, p. 185.
Zittel, K. A. 1890 A, p. 392.

Eryops erythroliticus Cope.

Cope, E. D. 1878 B, p. 515.   (Epicordylus.)
1878 U.   (Epicordylus.)
1888 AA, p. 253, pl. i, fig. 2.
1888 BB, p. 286.
Permian; Texas.

Eryops ferricolus Cope.

Cope, E. D. 1878 B, p. 521.   (Parioxys.)
1878 U.   (Parioxys.)
1881 K, p. 163.   (Parioxys.)
1884 P, p. 35.
1888 BB, P. 286.
Permian; Texas.

Eryops megacephalus Cope.

Cope, E. D. 1877 E, p. 188.
Baur, G. 1897 A, p. 978.
Broili, F. 1899 A, p. 61, pls. viii-x.
Cope, E. D. 1878 B, p. 520 (E. megacephalus)· p.
    526 (Rhachitomus valens).
1878 U.   (Rhachitomus valens.)
1880 A, p. 51, pls. i-iv.
1881 K, p. 163.
1884 P, p. 29, figs. 5, 6 and pl. iii.
1888 AA, figs. 2, 6.
1888 BB, p. 286.
1888 DD, with plate.
Dana, J. D. 1896 A, p. 686, fig. 1123.
Emery, C. 1897 A, p. 102, figs. 1, 3.
Hulke, J. W. 1888 A, p. 423.   (Rhachitomus
    valens.)
Lydekker, R. 1890 A, p. 191, fig. 47.
Seeley, H. G. 1892 A, pp. 349, 360.
Stickler, L. 1899 A, p. 85, pls. xi, xii.
Williston, S. W. 1899 C, p. 221.
    1899 E, p. 185, pls. xxvii, xxix, xxx.
Zittel, K. A. 1890 A, p. 393, figs. 382, 383.
Permian; Texas, Indian Territory.

Eryops? platypus Cope.

Cope, E. D. 1877 A, p. 574.   (Ichthycanthus.)
Baur, G. 1888 C, p. 16.   (Ichthycanthus.)
Cope, E.D. 1885 Z, p. 9.   (Ichthycanthus.)
    1888 BB, p. 289, fig. 1.   (Eryops.)
Coal-measures; Ohio.

Eryops reticulatus Cope.

Cope, E. D. 1881 AA, p. 1020.
1884 P, p. 34.
1888 BB, p. 286.
Permian; New Mexico.

ICHTHYCANTHUS Cope.   Type *I. ohiensis* Cope.

Cope, E. D. 1877 A, p. 573.
Baur, G. 1888 C. p. 16.
Cope, E. D. 1885 Z, p. 9.
Miller, S. A. 1889 A, p. 622.
Nicholson and Lydekker 1889 A, p. 1082.

Ichthycanthus ohiensis Cope.

Cope, E. D. 1877 A, p. 573.
    1885 Z, p. 9.
Coal-measures; Ohio.

ZATRACHYS Cope.   Type *Z. serratus* Cope.

Cope, E. D. 1878 B, p. 523.
Broili, F. 1899 A, p. 84.   (Zatrachis.)
Cope, E. D. 1882 B, p. 461.
    1882 T.
1884 P, p. 36.
1888 U.
1896 D.
Miller, S. A. 1889 A, p. 627.
Zittel, K. A. 1890 A, p. 390.

Zatrachys apicalis Cope.

Cope, E. D. 1881 AA, p. 1020.
1884 P, p. 36.
1888 BB, p. 286.
1896 E, p. 937.
Permian; New Mexico.

Zatrachys conchigerus Cope.

Cope, E. D. 1896 G, p. 453.
Permian; Texas.

Zatrachys microphthalmus Cope.

Cope, E. D. 1896 G, p. 452.
Permian; Texas.

Zatrachys serratus Cope.

Cope, E. D. 1878 B, p. 523.
1881 K, p. 163.
1884 P, p. 36.
1888 O, p. 466, pl. vi, fig. 2.
1888 BB, pp. 286, 289.
Permian; Texas.

ANISODEXIS Cope. Type *A. imbricarius* Cope.

Cope, E. D. 1882 B, p. 459.
Broili, F. 1899 A, p. 84.
Cope, E. D. 1884 P, p. 36.
Miller, S. A. 1889 A, p. 618.
Zittel, K. A. 1890 A, p. 398.

Anisodexis enchodus Cope.

Cope, E. D. 1885 B, p. 406.

Cope, E. D. 1897 C, p. 88.
Coal-measures; Ohio.

Anisodexis imbricarius Cope.

Cope, E. D. 1882 B, p. 459.
1884 P, p. 36.
1888 BB, p. 286.
Permian; Texas.

ACHELOMA Cope. Type *A. cumminsi* Cope.

Cope, E. D. 1882 B, p. 455.
Broili, F. 1899 A, p. 84.
Cope, E. D. 1884 P, p. 35.
Miller, S. A. 1889 A, p. 617.
Zittel, K. A. 1890 A, p. 398.

Acheloma cumminsi Cope.

Cope, E. D. 1882 B, p. 455.
1884 P, p. 35.
1888 BB, p. 286.
Permian; Texas.

MASTODONSAURIDÆ.

Huxley, T. H. 1863 A, p. 65. (Mastodonsauria.)
Lydekker, R. 1890 A, p. 141.
Miall, L. C. 1875 A, p. 149.

Quenstedt, F. A. 1850 B. ("Mastodonsaurier.")
1885 A, p. 240. (Mastodonsauri.)
Zittel, K. A. 1890 A, p. 401. (Labyrinthodonta.)

MASTODONSAURUS Jaeger. Type *Salamandroides giganteus* Jaeger.

Jaeger, G. F. 1828 A, p. 35.
Cope, E. D. 1898 B, p. 57, pl. 1, fig. 1.
Fraas, E. 1889 A, p. 32.
Huxley, T. H. 1863 A, p. 65.
Jaekel, O. 1896 A, p. 5.
Lydekker, R. 1890 A, p. 142.
Maggi, L. 1898 A, figs. 11, 23.
Miall, L. C. 1875 A, p. 151.
Owen, R. 1841 D, p. 5.
1842 B. (Labyrinthodon.)
1842 C. (Labyrinthodon.)

Pictet, F. J. 1853 A, p. 547.
Quenstedt, F. A. 1850 B.
1852 A, p. 148.
1885 A, p. 240.
Woodward, A. S. 1898 B, p. 137.
Zittel, K. A. 1890 A, p. 404.

Mastodonsaurus sp. indet.

Williston, S. W. 1897 G, pl. xxi.
Carboniferous; Kansas.

EUPELOR Cope. Type *Mastodonsaurus durus* Cope.

Cope, E. D. 1868 J, p. 221.
1898 A, p. 12.
Fritsch, A. 1883 A (1879), p. 62.
Miall, L. C. 1875 A, p. 186.
Zittel, K. A. 1890 A, p. 408.

Eupelor durus Cope.

Cope, E. D. 1866 D, p. 250. (Mastodonsaurus.)

Cope, E. D. 1868 J, p. 221.
1869 M, p. 25.
1887 A, p. 209.
Fritsch, A. 1883 A (1879), p. 62.
Miall, L. C. 1875 A, p. 186.
Triassic; Pennsylvania.

PARIOSTEGUS Cope. Type *P. myops* Cope.

Cope, E. D. 1868 J, p. 211.
1869 M, p. 10.
Fritsch, A. 1883 A (1879), p. 64.
Miall, L. C. 1875 A, p. 189.
Zittel, K. A. 1890 A, p. 189.

Pariostegus myops Cope.

Cope, E. D. 1868 J, p. 212.
1869 M, p. 11.
1875 V, p. 32.
Triassic; North Carolina.

DICTYOCEPHALUS Leidy. Type *D. elegans* Leidy.

Leidy, J. 1856 L, p. 256.
1857 A, p. 272.
Miall, L. C. 1875 A, p. 186.
Zittel, K. A. 1890 A, p. 408.

Dictyocephalus elegans Leidy.

Leidy, J. 1856 L, p. 256.
Cope, E. D. 1868 J, p. 221.
1869 M, p. 25.

Cope, E. D. 1875 V, p. 32.
Emmons E, 1856 A, p. 347. (No name.)
1857 A, p. 59, figs. 31, 32.
1856 A, p. 78.
1860 A, p. 184, fig. 162.
Fritsch, A. 1883 A (1879), p. 61.
Leidy, J. 1857 A, p. 272.
Triassic (Chatham); North Carolina.

Order URODELA.

Baur, G. 1888 C, pp. 36–62.
Boulenger, G. A, 1882 A.
Cope, E. D. 1866 F, p. 97.
    1869 M, p. 5.
    1871 B.
    1875 D, p. 352.
    1888 H, p. 299.
    1888 O.
    1888 R.
    1889 D, p. 29.
    1891 N, pp. 28, 31.
    1898 B, p. 48.
Emery, C. 1897 B.
Gadow, H. 1889 A, p. 459.
    1896 A, p. 2.
    1898 A, p. 13.
Gegenbaur, C. 1864 A.
    1865 B, p. 66.
    1898 A.
Goeppert, E. 1895 B.
    1896 A, p. 398.
Haeckel, E. 1895 A, p. 279.
Hasse, C. 1893 A.
Hay, O. P. 1895 A, p. 46.
    1896 B, p. 969.
    1899 F, p. 682.

Hoffmann, C. K. 1874 A.
    1878 A.
Howes, G. B. 1888 A.
Huxley, T. H. 1872 A.
    1874 A.
    1875 A.
    1876 C.
Kehner, F. A. 1876 A.
Mivart, St. G. 1870 A.
Parker, W. K. 1868 A, p. 56.
    1879 B, pp. 31, 61.
    1881 A, p. 262.
    1882 C.
    1882 D.
Peters, K. 1895 A.
Pollard, H. B. 1892 A, p. 420.
Riese H. 1892 A.
Stannius, H, 1856 A.
Walter, F. 1887 A.
Wiedersheim, R. 1876 A.
    1877 A.
    1880 A.
    1889 A, p. 432.
Zittel, K. A. 1890 A, p. 412.
Zwick, W. 1897 A.

GENERA INCERTÆ SEDIS.

SCAPHERPETON Cope.   Type *S. tectum* Cope.

Cope, E. D. 1876 I, p. 358.
    1882 J, p. 979.
    1884 O, p. 100.
Zittel, K. A. 1890 A, p. 420.

Scapherpeton excisum Cope.

Cope, E. D. 1876 I, p. 357.
    1877 G, p. 574.
    Cretaceous (Laramie); Montana.

Scapherpeton favosum Cope.

Cope, E. D. 1876 I, p. 357·

Cope, E. D. 1877 G, p. 574.
    Cretaceous (Laramie); Montana.

Scapherpeton laticolle Cope.

Cope, E. D. 1876 I, p. 356.
    1877 G, p. 573.
    Cretaceous (Laramie); Montana.

Scapherpeton tectum Cope.

Cope, E. D. 1876 I, p. 355.
    1877 G, p. 573.
    Cretaceous (Laramie); Montana.

HEMITRYPUS Cope.   Type *H. jordanianus* Cope.

Cope, E. D. 1876 I, p. 358.
    1882 J, p. 979.
    1884 O, p. 100.
Zittel, K. A. 1890 A, p. 420.

Hemitrypus jordanianus Cope.

Cope, E. D. 1876 I, p. 358.
    1877 G, p. 574.
    1882 J, p. 979.
    Cretaceous (Laramie); Montana.

Order SALIENTIA Laurenti.

Laurenti, J. N., 1768, Synop. Rept., p. 24.
    This group is usually denominated *Anura*,
and this term is employed by the following
writers unless otherwise indicated. *Salientia*
has priority and is unobjectionable.
Born, G. 1880 A, p. 51.
Boulenger, G. A. 1882 B. (Batrachia salientia.)

Cope, E. D. 1866 F.
    1868 J, p. 209.
    1869 M, p. 6.
    1875 D, pp. 352, 354.
    1888 H, p. 301. (Salientia.)
    1889 D, p. 232. (Salientia.)
    1889 R, p. 861. (Salientia.)

Gadow, H. 1889 A, p. 461.
　1896 A, p. 15.
　1898 A. p. 14.
Gaupp, E, 1895 A.
Gegenbaur, C. 1864 A.
　1865 B, p. 52.
　1898 A.
Goeppert, E. 1896 A, p. 427.
Haeckel, E. 1896 A. p. 261. (Batrachia.)
Hay, O. P. 1899 F, p. 682.
Hoffmann, C. K. 1874 A. (Ecaudata.)
Howes, G. B. 1888 B.
Huxley, T. H. 1872 A.
　1876 C.
Leydig, F. 1876 A.

Meyer, H. 1845 A, p. 18.
Müller, J. 1831 A, p. 209. (Batrachia.)
Nicholson and Lydekker 1889 A, p. 1042. (Ecaudata.)
Parker, W. K. 1868 A, p. 66. (Amphibia anoura.)
　1871 A.
　1877 A. (Batrachia.)
　1879 B, p. 31.
　1881 A.
Perrin, A. 1896 A.
Peters, K. 1895 A.
Stannius, H. 1856 A. (Batrachia.)
Wiedersheim, R. 1889 A, p. 435.
Zittel, K. A. 1890 A, p. 421.
Zwick, W. 1897 A.

RANIDÆ.

Boulenger, G. A. 1882 B, p. 3.
Cope, E. D. 1871 B.
　1889 D, p. 390.
Hoffmann, C. K. 1874 A.
Meyer, H. 1860 B.

Parker, W. K. 1871 A.
　1881 A.
Stannius, H. 1856 A. (Ranina.)
Zittel, K. A. 1890 A, p. 428.

RANA Linn.　　Type *R. temporaria* Linn.

Linnaeus, C., 1758, Syst. Nat., ed. x, p. 210.
Boulenger, G. A. 1882 B, 6.
Cope, E. D. 1889 D, p. 393.
Hoffmann, C. K. 1874 A.
Parker, W. K. 1881 A.

Zittel, K. A. 1890 A, p. 428.
Rana sp. indet.
Wheatley, C. 1871 B.
　Pleistocene; Pennsylvania.

EOBATRACHUS Marsh.　　Type *E. agilis* Marsh.

Marsh, O. C. 1887 A, p. 328. (No definition.)

Eobatrachus agilis Marsh.

Marsh, O. C. 1887 A, p. 328. (No description.)
　1897 C, p. 508. (No description.)
　Jurassic; Wyoming.

Class REPTILIA.

Baur, G. 1885 E.
　1885 B.
　1886 H.
　1892 D. p. 211.
　1895 A.
　1895 D.
　1895 E.
　1896 A.
Bemmelen, J. F. 1899 A, p. 162.
Boulenger, G. A. 1885 A. (Lacertilia.)
　1887 A. (Lacertilia.)
　1889 A. (Testudinata, Crocodilia.)
　1893 A. (Ophidia.)
　1894 A. (Ophidia.)
　1896 C. (Ophidia.)
Burckhardt, R. 1895 A.
Calori L. 1894 A.
Case, E. C. 1898 A.
Cope, E. D. 1864 A.
　1869 M, pp. 3, 26.
　1871 B.
　1885 G.
　1887 S, p. 333
　1888 AA.

Cope, E. D. 1889 R, p. 868.
　1891 N, p. 85.
　1892 N.
　1892 R.
　1898 B, p. 52.
　1900 A, p. 159.
Credner, H. 1888 A.
Cuvier, G. 1834 A, ix, pp. 1–24.
Fürbringer, M. 1886 B.
Gadow, H. 1896 A, p. 23.
　1898 A. p. 17.
Gaupp, E. 1891 A.
　1895 B.
　1900 A.
Gegenbaur, C. 1864 A.
　1865 B, p. 24.
　1898 A.
Günther, A. 1886 A.
Haeckel, E. 1895 A, p. 291.
Hallmann, E. 1837 A, pp. 16, 29.
Hoffmann, C. K. 1876 A.
　1878 A.
Humphrey, G. M. 1866 A.

Huxley, T. H. 1867 A, p. 415.
   1868 A.
   1870 A.
   1872 A, p. 167.
   1876 B, xiii pp. 429, 514.
Kehner, F. A. 1876 A.
Kehrer, G. 1886 A, p. 73.
Kingsley and Ruddick 1900 A, p. 219.
Köstlin, O. 1844 A.
Leidy, J. 1865 A.
Levy, H. 1896 A.
Lydekker, R. 1888 B.
   1889 F.
   1889 G.
   1890 A.
Minner, H. 1899 A, p. 43.
Marsh, O. C. 1877 E.
Meyer, H. 1845 A.
   1855 A.
   1856 B.
   1860 D.
Mivart, St. G. 1886 A.
Nicholson and Lydekker 1899 A, p. 1047.

Osborn, H. F. 1899 I, p. 156.
   1900 J, p. 943.
Owen, R. 1840 A.
   1842 A.
   1845 B, p. 179.
   1847 B, p. 283.
   1860 A.
Parker, W. K. 1868 A, p. 90.
   1880 A.
Pictet, F. J. 1853 A, p. 422
Röse, C. 1895 A.
Seeley, H. G. 1888 H.
   1898 A, p. 357.
Segond, L. A. 1873 A, pp. 1, 24.
Sixta, V. 1890 A.
Stannius, H. 1856 A. (Amphibia monopnoa.)
Walter, F. 1887 A.
Wiedersheim, R. 1889 A, p. 437.
Woodward, H. 1874 A.
   1874 B.
   1898 B, p. 140.
Zittel, K. A. 1890 A, pp. 437, 801.

Order COTYLOSAURIA Cope.

Cope, E. D. 1880 H.
Baur, G. 1885 J. (< Theromorpha.)
   1887 A. (< Theromorpha.)
   1887 C, p. 102. (< Theromorpha.)
   1891 C, p. 354. (< Theromora.)
   1896 B, p. 569.
Baur and Case 1897 A. (< Pareiasauria.)
Burckhardt, R. 1895 A. (< Theromorpha.)
Case, E. C. 1898 A, pp. 70, 73. (< Pareiasauria.)
   1898 B, p. 517.
   1899 A, p. 244.
Cope, E. D. 1878 CC. (< Theromorpha.)
   1880 A. (< Theromorpha.)
   1882 B, p. 448.
   1885 A. (< Theromorpha.)
   1885 G, p. 246. (< Theromorpha.)
   1887 S, p. 333. (< Theromorpha.)
   1889 R, p. 866. (< Theromora.)
   1891 F, p. 645.
   1891 N, p. 38.
   1891 Z, p. 13.
   1894 C.
   1896 C.
   1896 D.
   1896 G.

Cope, E. D. 1896 I.
   1898 B, pp. 54, 60.
   1900 A, p. 159. (Cotylosauria, Theromora.)
Fürbringer, M. 1900 A, pp. 338, 584, 552, 639, 674.
   (Theromorpha.)
Günther, A. 1886 A, p. 442. (< Anomodontia.)
Haeckel, E. 1895 A, p. 306. (< Theromora.)
Hay, O. P. 1899 F, p. 682.
Kükenthal, W. 1892 A. (< Theromorpha.)
Lydekker, R. 1890 A, p. 112. (< Anomodontia, Pariasauria.)
Osborn, H. F. 1898 B. (< Theromora.)
   1899 B, p. 95.
Seeley, H. G. 1889 A, p. 288. (< Pareiasauria.)
   1892 A. (< Pareiasauria.)
   1895 A, p. 998. (Pareiasauria.)
Steinmann and Döderlein 1890 A, p. 622.
   (< Theromorpha.)
Woodward, A. S. 1898 B, p. 145. (Pariasauria, part.)
Zittel, K. A. 1890 A, p. 553 (< Theromorpha); p. 570 (Pareiasauria).
   The *Pareiasauria* are a related but probably distinct group.

DIADECTIDÆ.

Cope, E. D. 1880 A. p. 45.
Baur and Case 1899 A.
Cope, E. D. 1880 H.
   1882 B, p. 448.
   1883 K, p. 108.
   1886 A.
   1888 U.
   1888 BB, p. 288.

Cope, E. D. 1896 G, pp. 439, 441.
   1896 I, p. 130.
   1898 B, p. 60.
Lydekker, R. 1890 A, p. 100.
Marsh, O. C. 1878 C, p. 410. (Nothodontidæ.[1])
Nicholson and Lydekker 1889 A, p. 1051.
Zittel, K. A. 1890 A, p. 581.

---

[1] This family is assigned to the *Diadectidæ* on the authority of Drs. Baur and Case, as above cited.

DIADECTES Cope.　　Type *D. sideropelicus* Cope.

Cope, E. D. 1878 B. p. 505.
　1878 U.
　1880 A, p. 45.
　1880 H.
　1888 U, p. 916, pl. xvi.
　1896 G, p. 441.
　1896 I, pp. 181, 183.
Miller, S. A. 1889 A, p. 620.
Zittel, K. A. 1890 A, p. 582.

Diadectes biculminatus Cope.

Cope, E. D. 1896 I, p. 182, text fig. 3.
　1878 B, p. 505. (" No. 2" of D. sideropelicus.)
Permian; Texas.

Diadectes latibuccatus Cope.

Cope, E. D. 1878 B, p. 505.
　1880 H.
　1881 K, p. 163.　(Empedocles.)
　1883 H, p. 634.　(Empedias.)
　1888 BB, p. 288.　(Empedias.)
　1896 G, p. 442.　(Empedias.)

Cope, E. D. 1896 I, p. 181.
　Permian; Texas.

Diadectes phaseolinus Cope.

Cope, E. D. 1880 A, p. 46.
　1881 K, p. 163.
　1883 H, p. 634.　(Empedias.)
　1886 A.
　1888 U, p. 916, pl. xvi, fig. 3.
　1888 BB, p. 288.　(Empedias.)
　1896 G, p. 442.　(Empedias.)
　1896 I, p. 182, text fig. 2.
Permian; Texas.

Diadectes sideropelicus Cope.

Cope, E. D. 1878 B, p. 505.
　1878 U.
　1881 K, p. 164.
　1888 BB, p. 288.
　1896 I, p. 183.
Lydekker, R. 1890 A, p. 104.
Permian; Texas.

EMPEDIAS Cope.　　Type *E. alatus* Cope.

Cope, E. D. 1883 H, p. 634. (To replace Empedocles, preoccu ied.)
Baur, G. 1897 B p p. 150.
Cope, E. D. 1878 B, pp. 516, 529. (Empedocles.)
　1878 U.
　1880 A, p. 46.　(Empedocles.)
　1880 H.　(Empedocles.)
　1896 G, p. 441.
　1896 I, p. 181.
Lydekker, R. 1890 A, p. 101.
Miller, S. A. 1889 A, p. 621.
Seeley, H. G. 1888 H, p. 187.
　1889 A, p. 284.
　1895 A, p. 1009.
Steinmann and Döderlein 1890 A, p. 623. (Empedocles.)
Zittel, K. A. 1890 A, p. 581.

Empedias alatus Cope.

Cope, E. D. 1878 B, p. 517. (Empedocles.)
　1878 U.　(Empedocles.)
　1881 K, p. 164.　(Empedocles.)
　1888 BB, p. 288.
Permian; Texas.
This species is not included by Professor Cope in either the list contained in Cope 1896 G, or in that of Cope 1896 I.

Empedias fissus Cope.

Cope, E. D. 1883 H, p. 634.
　1888 BB, p. 288.
　1896 G, p. 442.
　1896 I, p. 182, text fig. 4.
Permian; Texas.

Empedias molaris Cope.

Cope, E. D. 1878 Z.　(Diadectes.)
　1880 A, p. 47.　(Empedocles.)
　1880 H.　(Empedocles.)
　1881 K, p. 164.　(Empedocles.)
　1883 H, p. 634.
　1886 A.　(Empedias.)
　1888 BB, p. 288.
　1891 N, p. 38, fig. 17.
　1896 G, p. 442.
　1896 I, p. 181.
Kükenthal, W. 1892 A.
Lydekker, R. 1890 A, p. 102, figs. 20-22.
Seeley, H. G. 1888 H, p. 138, fig. 5.
Wortman, J. L. 1896 A, p. 383.
Zittel, K. A. 1890 A, p. 581, fig. 524.
Permian; Texas.

BOLBODON Cope.　　Type *B. tenuitectus* Cope.

Cope, E. D. 1896 I, p. 134.

Bolbodon tenuitectus Cope.

Cope, E. D. 1896 I, p. 134, text fig. 1.
　Permian; Texas.

CHILONYX Cope.　　Type *Bolosaurus rapidens* Cope.

Cope, E. D. 1888 H, p. 681.
Baur, G. 1897 B, p. 150.
Cope, E. D. 1888 P.

Cope, E. D. 1892 Z, p. 13.
　1896 G, p. 441.
　1896 I, p. 181.

Cope, E. D. 1896 B, p. 57, pl. i, fig. 2.
Lydekker, R. 1890 A, pp. 67, 95, 111.
Miller, S. A. 1889 A, p. 619.
Seeley, H. G. 1896 A, p. 1009.
Zittel, K. A. 1890 A, p. 582.

Chilonyx rapidens Cope.

Cope, E. D. 1878 B, p. 507.   (Bolosaurus.)

Cope, E. D. 1883 H, p. 631.
    1883 P.
    1888 BB, p. 288.
    1892 Z, p. 25, pl. i, fig. 2.
    1894 B, pl. x, fig. 2.
    1896 G, p. 442, pl. viii, fig. 6.
    1896 I, p. 181.
Permian; Texas.

NOTHODON Marsh.   Type *N. lentus* Marsh.

Marsh, O. C. 1878 C, p. 410.
Baur and Case 1899 A, p. 4.
Credner, H. 1888 A, p. 555.
Zittel, K. A. 1890 A, p. 574.

Nothodon lentus Marsh.

Marsh, O. C. 1878 C, p. 410.
Baur and Case 1899 A, p. 4:
Permian; New Mexico.

PARIOTICHIDÆ.

Cope, E. D. 1883 H, p. 631.
Case, E. C. 1899 A, p. 238.
Cope, E. D. 1888 BB, p. 287.
    1896 G, pp. 439, 442.

Lydekker, R. 1890 A, p. 112.
Nicholson and Lydekker 1889 A, p. 1056.
Zittel, K. A. 1890 A, p. 580.

PARIOTICHUS Cope.   Type *P. brachyops* Cope.

Cope, E. D. 1878 B, pp. 508, 529.
Baur, G. 1897 B, p. 150.
Case, E. C. 1899 A, p. 231.
Cope, E. D. 1878 B, pp. 508, 529.   (Ectocynodon,
    type *E. ordinatus* Cope.)
    1879 A, p. 34
    1882 B, p. 450.   (Ectocynodon.)
    1883 H, p. 631.   (Pariotichus, Ectocynodon.)
    1892 Z, p. 14.
    1896 C.
    1896 G, p. 443.
    1896 B, p. 57, fig. 3.
Miller, S. A. 1889 A, p. 621 (Ectocynodon); p. 624
    (Pariotichus).
Nicholson and Lydekker 1889 A, p. 1056. (Parioti-
    chus, Ectocynodon.)
Seeley, H. G. 1896 A, p. 998.
Zittel, K. A. 1890 A, p. 581. (Pariotichus, Ecto-
    cynodon.)

Pariotichus aduncus Cope.

Cope, E. D. 1896 I, p. 135.
Permian; Texas.

Pariotichus aguti Cope.

Cope, E. D. 1882 B, p. 451.   (Ectocynodon.)
    1888 BB, p. 287.   (Ectocynodon.)
    1896 C, pl. vii, a.

Cope, E. D. 1896 G, pp. 444, 445, pl. vii.
Permian; Texas.

Pariotichus brachyops Cope.

Cope, E. D. 1878 B, p. 508.
    1881 K, p. 163.
    1888 BB, p. 287.
    1896 G, p. 445.
Permian; Texas.

Pariotichus incisivus Cope.

Cope, E. D. 1888 BB, p. 290.   (Ectocynodon.)
Case, E. C. 1899 A, p. 231, figs. 1–7.
    1900 A, p. 720.
Cope, E. D. 1896 G, p. 445.
Permian; Texas.

Pariotichus isolomus Cope.

Cope, E. D. 1896 G, pp. 445, 446.
Permian; Texas.

Pariotichus ordinatus Cope.

Cope, E. D. 1878 B, p. 508.   (Ectocynodon.)
    1881 K, p. 163.   (Ectocynodon.)
    1888 BB, p. 287.   (Ectocynodon.)
    1896 G, pp. 446, 447.
Permian; Texas.

ISODECTES Cope.   Type *Pariotichus megalops* Cope.

Cope, E. D. 1896 G, p. 442.
    To this genus Cope has also referred *Tudi-
tanus punctulatus,* which, however, has here
been retained under *Tuditanus.*

Isodectes megalops Cope.

Cope, E. D. 1883 H, p. 630.   (Pariotichus.)

Cope, E. D. 1888 BB, p. 287.   (Pariotichus.)
    1892 Z, p. 25, pl. i, fig. 3.   (Pariotichus.)
    1894 B, pl. x, fig. 3.   (Pariotichus.)
    1896 G, p. 442.
Permian; Texas.

CAPTORHINUS Cope.　Type *C. angusticeps* Cope.

Cope, E. D. 1896 G, p. 443.

Captorhinus angusticeps Cope.

Cope, E. D. 1896 G, p. 443.
Permian; Texas.

PANTYLUS Cope.　Type *P. cordatus* Cope.

Cope, E. D. 1881 A, p. 79.
Baur, G. 1897 B, p. 150.
Cope, E. D. 1882 HH.
　　1883 H, p. 681.
　　1892 Z, p. 14.
　　1896 G, pp. 442, 449.
　　1896 B, p. 57, fig. 4.
Miller, S. A. 1889 A, p. 624.
Seeley, H. G. 1895 A, p. 908.
Zittel, K. A. 1890 A, p. 581.

Pantylus coïcodus Cope.
Cope, E. D. 1896 G, p. 450.
Permian; Texas.

Pantylus cordatus Cope.

Cope, E. D. 1881 A, p. 79.
　　1881 K, p. 163.
　　1888 BB, p. 287.
　　1892 Z, p. 25, pl. i, fig. 4.
　　1894 B, pl. x, fig. 4.
　　1896 G, p. 449.
Permian; Texas.

HYPOPNOUS Cope.　Type *H. squaliceps* Cope.

Cope, E. D. 1896 G, p. 450.

Hypopnous squaliceps Cope.

Cope, E. D. 1896 G, p. 451, pl. viii, figs. 3–5.
Permian; Texas.

HELODECTES Cope.　Type *H. paridens* Cope.

Cope, E. D. 1880 A, pp. 45, 48.
　　1896 G, p. 442.
Miller, S. A. 1889 A, p. 622.
Zittel, K. A. 1890 A, p. 582.

Helodectes isaaci Cope.
Cope, E. D. 1880 A, p. 49.
　　1881 K, p. 164.

Cope, E. D. 1888 BB, p. 288.
Permian; Texas.

Helodectes paridens Cope.

Cope, E. D. 1880 A, p. 48.
　　1881 K, p. 164.
　　1888 BB, p. 288.
Permian; Texas.

PAREIASAURIDÆ.

Cope, E. D. 1896 G, pp. 437, 439.
　　1896 I, p. 137.
Haeckel, E. 1895 A, p. 308.　(Pareosaurida.)
Lydekker, R. 1890 A, p. 112.

Seeley, H. G. 1888 D.　(Pareiasaurus.)
　　1889 A.　(Pareiasaurus.)
　　1892 A.　(Pareiasaurus.)

LABIDOSAURUS Cope.　Type *Pariotichus hamatus* Cope.

Cope, E. D. 1896 I, p. 136.
Baur G. 1897 B, p. 150.

Labidosaurus hamatus Cope.

Cope, E. D. 1896 G, p. 448, pl. viii, figs. 1, 2.　(Pariotichus.)
　　1896 I, p. 136.
Permian; Texas.

Order CHELYDOSAURIA Cope.

Cope, E. D. 1896 B, pp. 54, 61.

Hay, O. P. 1899 F, p. 682

OTOCŒLIDÆ.

Cope, E. D. 1896 D.
Baur, G. 1896 B, p. 569.

Cope, E. D. 1896 I, p. 123.

OTOCŒLUS Cope.   Type *O. testudineus* Cope.

Cope, E. D. 1896 D.
Baur, G. 1896 B, p. 569.
   1897 B, p. 150.
Cope, E. D. 1896 I, p. 123.
   1897 A, p. 315.
Woodward, A. S. 1898 B, p. 149.

Otocœlus mimeticus Cope.

Cope, E. D. 1896 I, p. 128, pl. ix, fig. 1.

Cope, E. D. 1896 E, pl. xxii, fig. 11.   (No description; called on plate O. testudineus.)
   Permian; Texas.

Otocœlus testudineus Cope.

Cope, E. D. 1896 D.
   1896 E, pl. xxi, fig. 2.
   1896 I, p. 124, pls. vii, viii; pl. ix, fig. 2.
   Permian; Texas.

CONODECTES Cope.   Type *C. favosus* Cope.

Cope, E. D. 1896 D.
Baur, G. 1896 B, p. 569.
Cope, E. D. 1896 I, pp. 124, 129.

Conodectes favosus Cope.

Cope, E. D. 1896 D.
   1896 I, p. 129.
   Permian; Texas.

Order ANOMODONTIA Owen.

Owen, R. 1860 A, p. 255.
Cope, E. D. 1878 CC, p. 829.
   1880 A, p. 38.
   1889 R, p. 865.
Fürbringer, M. 1900 A.
Gadow, H. 1898 A, p. 18.
Haeckel, E. 1895 A, p. 314.
Hay, O. P. 1899 F, p. 682.
Lydekker, R. 1890 A, p. 10.
Osborn, H. F. 1898 O, p. 358.
   1900 J, p. 945.

Owen, R. 1881 C, p. 266.
Quenstedt, F. A. 1885 A, p. 218.
Seeley, H. G. 1888 D, p. 101.
   1888 F.
   1889 A.
   1898 A, p. 3o7.
   1900 A, p. 620.
Woodward, A. S. 1898 B, p. 144, in part; p. 157 (Dicynodontia.)
Zittel, K. A. 1890 A, p. 556.

DICYNODONTIDÆ.

Cope, E. D. 1889 R, p. 865.
Huxley, T. H. 1859 C.   ("Dicynodonta.")

Lydekker, R. 1890 A, p. 16.
Osborn, H. F. 1898 B.   (Dicynodontia.)

DICYNODON Owen.   Type *D. lacerticeps* Owen.

Owen, R., 1845, Quart. Jour. Geol. Soc., i, p. 318.
Hulke, J. W. 1892 A, p. 244, figs. 6, 7.
Huxley, T. H. 1859 C.
Lydekker, R. 1890 A, p. 16.
Osborn, H. F. 1898 R, p. 316.
Owen, R. 1845 D.
   1856 D.
   1856 E.
   1860 F.
   1863 A.

Quenstedt, F. A. 1885 A, p. 218.
Zittel, K. A. 1890 A, p. 564.

Dicynodon rosmarus Cope.

Cope, E. D. 1869 M, p. 282.
   1870 I, p. 445.
   An unsatisfactorily described species; its generic identity doubtful.
   Triassic; Pennsylvania.

Order PELYCOSAURIA Cope.

Cope, E. D. 1878 B, p. 528.
Baur, G. 1885 E, p. 637.
   1886 M, p. 302.
   1887 A, p. 488.
   1887 C, p. 102.   (As part of Theromorpha.)
   1897 A, p. 977.
Baur and Case 1897 A.
   1899 A.
Case, E. C. 1898 A, pp. 70, 73.
Cope, E. D. 1878 CC.   (As part of Theromorpha.)
   1880 A.   (As part of Theromorpha.)
   1880 J.

Cope, E. D. 1882 B, p. 448.
   1884 I, pp. 38, 43.
   1885 A.
   1885 G.   (Theromorpha, part.)
   1885 V.
   1886 O.
   1888 U, p. 915.
   1888 AA, p. 244.
   1889 R, p. 866.   (As part of Theromora.)
   1891 F, p. 645.   (Theriodontia.)
   1891 N, p. 35.   (Theriodontia.)
   1892 R.

Cope. E. D. 1894 C.
  1897 A, p. 316.
  1898 B, pp. 55, 67.
Fürbringer, M. 1888 A.
  1900 A.
Gadow, H. 1898 A, p. 18.   (Theriodontia.)
Haeckel, E. 1895 A, p. 308.
Hay, O. P. 1899 F, p. 68.
Lydekker, R. 1890 A, p. 66.   (Theriodontia.)
Marsh, O. C. 1898 B, p. 407.   (Theriodontia.)
Nicholson and Lydekker 1889 A, p. 1057.   (Theriodontia.)

Osborn, H. F. 1898 A, p. 147.   (Theriodontia.)
  1898 B, p. 177.
  1898 R, p. 315.   (Theriodontia.)
  1899 B, p. 95.   (Theriodontia.)
Owen, R. 1876 B.   (Theriodontia.)
  1881 B.   (Theriodontia.)
  1881 C.   (Theriodontia.)
Seeley, H. G. 1889 A, p. 282.
  1895 A, p. 1007.
Steinmann and Döderlein 1890 A, p. 622.
Woodward, A. S. 1898 B, p. 186.
Zittel, K. A. 1890 A, pp. 554, 572.

CLEPSYDROPIDÆ.

Cope, E. D. 1884 I, p. 38.
Baur and Case, 1899 A.
Cope, E. D. 1882 B, p. 450.   (Edaphosauridæ.)
  1883 H, p. 631.   (Edaphosauridæ.)
  1885 A, p. 477.
  1888 BB, p. 287.
  1898 B, p. 67.
Fürbringer, M. 1900 A.

Lydekker, R. 1890 A, p. 104.
Marsh, O. C. 1878 C, p. 410.   (Sphenacodontidæ.)
Nicholson and Lydekker 1889 A, p. 1059.
Zittel, K. A. 1890 A, p. 574.   (Cynodontia.)
  The family *Sphenacodontidæ* proposed by Professor Marsh, is assigned to the *Clepsydrop-idæ* on the authority of Drs. Baur and Case, as cited above.

CLEPSYDROPS Cope.   Type *C. collettii* Cope.

Cope, E. D. 1875 T, p. 407.
Baur and Case 1897 A.
  1899 A.
Case, E. C. 1900 A, p. 711.
Cope, E. D. 1878 B, pp. 510, 529.
  1878 J.
  1878 U.
  1878 DD.
  1879 A, p. 34.
  1881 K, p. 168.
  1885 A.
  1885 O.
  1888 O.
  1888 AA.
  1892 Z, p. 11 (Diopeus, type *C. leptocephalus*);
    p. 14 (Clepsydrops).
Kingsley and Ruddick 1900 A, P. 229.
Lydekker, R. 1890 A, p. 105.
Miller, S. A. 1889 A, p. 619.
Nicholson and Lydekker 1889 A, p. 1059.
Seeley, H. G. 1889 A, pp. 284, 288.
  1895 A, p. 998.
  1895 B, pp. 111, 119.
Zittel, K. A. 1890 A, p. 574.

Clepsydrops collettii Cope.

Cope, E. D. 1875 T, p. 407.
Baur and Case 1899 A, p. 1.
Case, E. C. 1900 A, pp. 711, 720, pl. ii, figs. 1a–3b;
  pl. iv, figs. 7a–7d.
Cope, E. D. 1877 F, p. 196.
  1877 N, p. 61.
  1881 K, p. 163.
  1884 I, p. 38.
  1888 BB, p. 287.
Williston, S. W. 1898 J, p. 121.
  Permian; Illinois, Kansas?.

Clepsydrops leptocephalus Cope.

Cope, E. D. 1884 I, p. 30, pl. 1, figs. 1–5.

Baur and Case 1897 A, p. 116.
  1899 A, p. 13.
Case, E. C. 1900 A, p. 720.
Cope, E. D. 1884 EE, pl. xxxviii, figs. 1, 2.
  1884 GG, p. 1253.
  1885 A, figs. 1, 2.
  1885 V.
  1888 BB, p. 287.
  1892 Z, p. 11, pl. ii, figs. 6, 8.   (Diopeus.)
  1897 A, p. 318.   (Diopeus.)
Zittel, K. A. 1890 A, p. 574.
  Permian; Texas.

Clepsydrops limbatus Cope.

Cope, E. D. 1877 F, p. 196.
Baur and Case 1899 A, p. 3.
Cope, E. D. 1878 B, p. 515.   (Embolophorus?)
  1880 Q.
  Permian; Texas.

Clepsydrops macrospondylus Cope.

Cope, E. D. 1884 I, p. 35.
Baur and Case 1899 A, p. 14.
Cope, E. D. 1888 BB, p. 287.
  Permian; Texas.

Clepsydrops natalis Cope.

Cope, E. D. 1878 B, pp. 509, 529.
Case, E. C. 1900 A, p. 717.
Cope, E. D. 1881 K, p. 163.
  1884 I, pp. 38, 42.
  1884 EE, pl. xxxviii, fig. 6.
  1884 HH, p. 1255.
  1885 A, fig. 6.
  1888 BB, p. 287.
  1892 Z.
Seeley, H. G. 1889 A, p. 290.
Zittel, K. A. 1890 A, p. 574.
  Permian; Texas.

Clepsydrops pedunculatus Cope.

Cope, E. D. 1877 N, p. 62.
Baur and Case 1899 A, p. 2.
Case, E. C. 1900 A, p. 713, pl. ii, figs. 4a-5d.
Cope, E. D. 1881 K, p. 163.
1888 BB, p. 287.
Permian; Illinois.

Clepsydrops vinslovii Cope.

Cope, E. D. 1877 N, p. 61.
Baur and Case 1899 A, p. 2.
Case, E. C. 1900 A, p. 714, pl. ii, figs. 7a-7d.
Cope, E. D. 1881 K, p. 163.
1888 BB, p. 287.
Permian; Illinois.

DIMETRODON Cope.   Type *D. incisivus* Cope.

Cope, E. D. 1878 B, pp. 512, 529.
Ameghino, F. 1896 A, p. 487.
Baur and Case 1897 A.
1899 A.
Case, E. C. 1897 A.
1899 A.
Cope, E. D. 1880 A, p. 42, pl. vi.
1881 K, p. 163.
1886 J.
1888 AA, p. 244.
1888 BB, p. 292.
1888 DD, p. 363.
1889 C, p. 226.
1897 A, p. 315.
Miller, S. A. 1889 A, P. 620.
Nicholson and Lydekker 1889 A, p. 1059.
Osborn, H. F. 1888 G, p. 222, fig. 8.
Seeley, H. G. 1889 A, p. 283.
1895 A, p. 1007.
1895 B, p. 110.
Steinmann and Döderlein 1890 A, p. 624.
Wortman, J. L. 1886 A, p. 383.
Zittel, K. A. 1890 A, p. 576.

Dimetrodon gigas Cope.

Cope, E. D. 1878 U.   (Clepsydrops.)
Baur and Case 1899 A, p. 12.
Cope E. D. 1878 B, p. 515.
1881 K, p. 163.

Cope, E. D. 1888 BB, p. 287.
Permian; Texas.

Dimetrodon incisivus Cope.

Cope, E. D. 1878 U.   (Clepsydrops.)
Baur and Case 1897 A, figures.
1899 A, p. 12, pls. i-iii.
Case, E. C. 1897 A, figs. 1-4.
Cope, E. D. 1878 B, p. 512.
1881 A, p. 81.
1881 K, p. 163.
1886 J.
1888 BB, pp. 287, 292.
Seeley, H. G. 1889 A, p. 283.
Zittel, K. A. 1890 A, p. 576, fig. 519.
Permian; Texas.

Dimetrodon rectiformis Cope.

Cope, E. D. 1878 B, p. 514.
1881 K, p. 163.
1888 BB, p. 287.
Permian; Texas.

Dimetrodon semiradicatus Cope.

Cope, E. D. 1881 A, p. 80.
1881 K, p. 163.   (D. biradicatus.)
1888 BB, p. 287.
Permian; Texas.

NAOSAURUS Cope.   Type *N. claviger* Cope.

Cope, E. D. 1886 J.
Baur and Case 1899 A.
Cope, E. D. 1892 Z, p. 14.
1897 A, p. 316.
Fritsch, A. 1895 B, p. 2.
Lydekker, R. 1890 A, p. 106.
Seeley, H. G. 1889 A, p. 283.
1895 A, p. 998.
Zittel, K. A. 1890 A, p. 576.

Naosaurus claviger Cope.

Cope, E. D. 1886 J.
1888 BB, pp. 287, 293, pl. ii, figs. 1, 2; pl. iii, figs. 1-3.
1892 Z, pl. ii, fig. 7.
Lydekker, R. 1890 A, p. 107.
Osborn, H. F. 1898 N, pp. 842, 844, fig. 4.
1898 Q, p. 7, pl. i, fig. 3

Zittel, K. A. 1890 A, p. 576, fig. 520.
Permian; Texas.

Naosaurus cruciger Cope.

Cope, E. D. 1878 CC.   (Dimetrodon.)
Baur and Case 1899 A, p. 11.
Cope, E. D. 1880 A, p. 44.   (Dimetrodon.)
1881 A, p. 82.   (Dimetrodon.)
1881 K, p. 163.   (Dimetrodon.)
1886 J.
1888 BB, pp. 287, 294, pl. iii, fig. 8.
Lydekker, R. 1890 A, p. 107.
Permian; Texas.

Naosaurus microdus Cope.

Cope, E. D. 1884 I, p. 37.   (Edaphosaurus.)
1886 J.
1888 BB, pp. 287, 294, pl. ii, fig. 3.
Permian; Texas.

EDAPHOSAURUS Cope.   Type *E. pogonias* Cope.

Cope, E. D. 1882 B, p. 448.
    1884 I, p. 37.
    1886 A.
    1892 Z, p. 15.
Miller, S. A. 1889 A, p. 621.
Seeley, H. G. 1889 A, p. 284.
    1895 A, p. 998.

Zittel, K. A. 1890 A, p. 577.

Edaphosaurus pogonias Cope.
Cope, E. D. 1882 B, p. 449.
    1888 BB, p. 287.
    1892 Z, pl. ii, fig. 5.
    Permian; Texas.

* EMBOLOPHORUS Cope.   Type *E. fritillus* Cope.

Cope, E. D. 1878 B, pp. 518, 529.
    1884 I, p. 43.
    1885 A, p. 474, figs. 4, 5.
Lydekker, R. 1890 A, p. 108.
Miller, S. A. 1889 A, p. 621.
Seeley, H. G. 1889 A, p. 284.
    1895 B, p. 111.
Zittel, K. A. 1890 A, p. 577.

Embolophorus dollovianus Cope.
Cope, E. D. 1888 BB, p. 287.

Cope, E. D. 1884 I, p. 43, pl. 1, figs. 4, 5. ("Embolophorus.")
Lydekker, 1890 A, p. 109.   (E. dolloverianus.)
    Permian; Texas.

Embolophorus fritillus Cope.
Cope, E. D. 1878 B, p. 518.
    1881 K, p. 164.
    1884 BB.
Hulke, J. W. 1888 A, p. 423.
Lydekker, R. 1890 A, p. 109.
    Permian; Texas.

ARCHÆOBELUS Cope.   Type *A. vellicatus* Cope.

Cope, E. D. 1877 E, p. 192.
Case, E. C. 1900 A, p. 715.
Miller, S. A. 1889 A, p. 618.
Zittel, K. A. 1890 A, p. 576.

Archæobelus vellicatus Cope.
Cope, E. D. 1877 E, p. 192.

Case, E. C. 1900 A, p. 715, pl. iii, fig. 1.
Cope, E. D. 1877 N, p. 56. ("Species No. 4.")
    1881 K, p. 163.
    1888 BB, p. 287.
    Permian; Illinois.

LYSOROPHUS Cope.   Type *L. tricarinatus* Cope.

Cope, E. D. 1877 E, p. 187.
Case, E. C. 1900 A, p. 714.
Cope, E. D. 1888 BB, p. 287.   (Lysorhophus.)
Miller, S. A. 1889 A, p. 623.
Zittel, K. A. 1890 A, p. 576.

Lysorophus tricarinatus Cope.
Cope, E. D. 1877 E, p. 187.

Baur and Case 1899 A, p. 2.
Case, E. C. 1900 A, pp. 714, 717, pl. ii, figs. 12a-13c.
Cope, E. D. 1881 K, p. 164.
    1888 BB, p. 287.   (Lysorhophus.)
    Permian; Illinois.

THEROPLEURA Cope.   Type *T. retroversa* Cope.

Cope, E. D. 1878 B, p. 519.
Baur and Case 1899 A, p. 11.
Cope, E. D. 1880 A, p. 40.
    1888 BB, p. 287.
Miller, S. A. 1889 A, p. 626.
Nicholson and Lydekker 1889 A, p. 1060.
Seeley, H. G. 1889 A, p. 283.
Zittel, K. A. 1890 A, p. 576.

Theropleura obtusidens Cope.
Cope, E. D. 1880 A, p. 41.
Baur and Case 1899 A, p. 11.
Cope, E. D. 1881 K, p. 164.
    1888 BB, p. 287.
    Permian; Texas.

Theropleura retroversa Cope.
Cope, E. D. 1878 B, p. 519.

Cope, E. D. 1881 K, p. 163.
    1888 BB, p. 287.
    Permian; Texas.

Theropleura triangulata Cope.
Cope, E. D. 1878 B, p. 520.
    1881 K, p. 164.
    1888 BB, p. 287.
    Permian; Texas.

Theropleura uniformis Cope.
Cope, E. D. 1878 B, p. 519.
Baur and Case 1899 A, p. 11.
Cope, E. D. 1880 A, p. 40.
    1881 K, p. 163.
    1888 BB, p. 287.
    Permian; Texas.

OPHIACODON Marsh.   Type *O. mirus* Marsh.

Marsh, O. C. 1878 C, p. 411.
Baur and Case 1899 A, p. 5.
Credner, H. 1888 A, p. 555.
Miller, S. A. 1889 A, p. 624.
Zittel, K. A. 1890 A, p. 574.

Ophiacodon grandis Marsh.

Marsh, O. C. 1878 C, p. 411.

Baur and Case 1899 A, p. 5.   (Eryops?)
Permian; New Mexico.

Ophiacodon mirus Marsh.

Marsh, O. C. 1878 C, p. 411.
Baur and Case 1899 A, p. 5.
Permian; New Mexico.

SPHENACODON Marsh.   Type *S. ferox* Marsh.

Marsh, O. C. 1878 C, p. 410.
Baur and Case 1899 A, p. 4.
Credner, H. 1888 A, p. 555.
Zittel, K. A. 1890 A, p. 574.

Sphenacodon ferox Marsh.

Marsh, O. C. 1878 C, p. 410.
Baur and Case 1899 A, p. 4.
Permian; New Mexico.

BOLOSAURIDÆ.

Cope, E. D. 1878 B, p. 529.
    1881 K, p. 164.
    1883 P.

Cope, E. D. 1888 BB, p. 288.
Lydekker, R. 1890 A, p. 95.
Nicholson and Lydekker 1889 A, p. 1060.

BOLOSAURUS Cope.   Type *B. striatus* Cope.

Cope, E. D. 1878 B, pp. 506, 529.
    1878 U.
    1888 BB, p. 288.
    1890, in Lydekker, R. 1890 A, p. 95.
Miller, S. A. 1889 A, p. 619.
Nicholson and Lydekker 1889 A, p. 1060.
Zittel, K. A. 1890 A, p. 582.

Bolosaurus striatus Cope.

Cope, E. D. 1878 B, p. 506.
    1878 U.
    1881 K, p. 164.
    1888 BB, p. 288.
Permian; Texas.

METAMOSAURUS Cope.   Type *M. fossatus* Cope.

Cope, E. D. 1878 B, p. 516.
    1888 BB, p. 288.
Miller, S. A. 1889 A, p. 623.
Zittel, K. A. 1890 A, p. 582.
The position of this genus is uncertain.

Metamosaurus fossatus Cope.

Cope, E. D. 1878 B, p. 516.
    1881 K, p. 164.
    1888 BB, p. 288.
Permian; Texas.

Order TESTUDINES Shaw.

Shaw, G. 1802, Gen. Zool., iii, p. 5.
    Unless otherwise indicated, the following authors refer to this order under the name *Testudinata.*
Agassiz, L. 1857 B.
Baur, G. 1885 E, p. 632.   (Chelonia.)
    1886 G, pp. 687, 740.
    1887 A, p. 484.
    1887 B.
    1887 C, p. 97.
    1887 E.
    1887 F.
    1888 A.
    1888 B.
    1889 B.
    1889 C.
    1889 F. p. 472.
    1889 H.
    1889 J.
    1890 D.
    1891 B, p. 417.

Baur, G. 1891 C.
    1892 A.
    1892 D, p. 206.
    1893 A.
    1893 B.
    1894 A.
    1895 A.
    1895 D.
    1896 A.
    1896 B.
Bemmelen, J. F. 1896 A.
Bienz, A. 1895 A.
Born, G. 1880 A, p. 68.
Boulenger, G. A. 1889 A.   (Chelonia.)
Calori, L. 1894 A.
Case, E. C. 1897 B.   (Testudines.)
Cope, E. D. 1869 M, p. 27.
    1871 B.
    1873 E, p. 648.
    1875 X, p. 16.
    1882 E, p. 143.

Cope, E. D. 1882 J, p. 986.
  1884 O, p. 111.
  1885 G, p. 246.
  1887 S, p, 334.
  1889 R, p. 863.
  1891 N, p. 39.
  1892 R.
  1892 Z, p. 20.
  1895 E.
  1896 A.
  1896 I.
  1898 B, pp. 55, 92.
  1900 A, p. 160.
Cuvier, G. 1834 A, ix, pp. 347–496, pls. ccxxxix-ccxliii.
Dollo, L. 1884 A.
  1886 A. (Rhynchochelones, or Euchelonia.)
  1888 C.
  1893 C, p. 253. ("Chéloniens.")
Fürbringer, M. 1900 A, pp. 311, 527, 540, 556, 593, 630 (Chelonia.)
Gadow, H. 1889 A, p. 462.
  1896 A, p. 41.
  1898 A, p. 20. (Chelonia.)
  1899 A, p. 35.
Gaupp, E. 1894 A.
  1895 B, p. 58.
Gegenbaur, C. 1864 A.
  1865 B, p. 35.
  1898 A.
Goette, A. 1899 A, p. 407.
Gray, J. E. 1869 B.
  1873 B.
Günther, A. 1877 A.
  1886 A, p. 445.
  1888 A, p. 456. (Chelonia.)
Haeckel, E. 1895 A, p. 316. (Chelonia.)
Hallmann, E. 1837 A, p. 22.
Hay, O. P. 1898 F, p. 929. (Testudines.)
  1899 F, p. 682. (Testudines.)
Haycraft, J. B. 1891 A. (Chelonia.)
Hoffmann, C. K. 1876 A.
  1878 A.
  1879 A.
  1890 A.
Hulke, J. W. 1892 A, p. 248.
Huxley, T. H. 1872 A, p. 170. (Chelonia.)

Kehner, F. A. 1876 A.
Kehrer, G. 1886 A, p. 83.
Köstlin, O. 1844 A.
Lang and Rütimeyer 1867 A.
Leidy, J. 1871 C, p. 365.
Lortet, L. 1892 A.
Lydekker, R. 1889 D. (Chelonia.)
  1889 E. (Chelonia.)
  1889 G. (Chelonia.)
Maack, G. A. 1869 A.
Marsh, O. C. 1877 E. (Chelonia.)
Mehnert, E. 1890 A.
Meyer, H. 1860 D, p. 140.
Mivart, St. G. 1886 A. (Chelonia.)
Monks, S. P. 1878 A.
Müller, J. 1831 A, p. 199. (Testudines.)
Nicholson and Lydekker 1889 A, p. 1082. (Chelonia.)
Owen, R. 1842 A, p. 160. (Chelonia.)
  1849 B. (Chelonia.)
  1849 C. (Chelonia.)
  1860 A, pp. 161–166. (Chelonia.)
  1866 A. (Chelonia.)
Parker, W. K. 1864 A.
  1868 A, p. 133. (Chelonia.)
  1879 B, p. 63.
Peters, W. 1839 B. ("Schildkröten.")
Pictet, F. J. 1853 A, p. 435. ("Chéloniens.")
Portis, A. 1882 A. (Chelonia.)
Rosenberg, E. 1891 A.
Rütimeyer, L. 1873 A.
  1874 A.
Seeley, H. G. 1880 A. (Chelonia.)
  1900 A, p. 641. ("Chéloniens.")
Segond, L. A. 1873 A, p. 2, et seq. (Vernacular name.)
Siebenrock, F. 1897 A.
  1898 A.
Stannius, H. 1856 A. (Chelonia.)
Steinmann and Döderlein 1890 A, p. 630.
Strauch A. 1890 A, p. 9. (Chelonia.)
Tomes, C. 1898 A, p. 267. (Chelonia.)
Wiedersheim, R. 1889 A, p. 438.
Wieland, G. R. 1900 B, p. 413.
Woodward, A. S. 1898 B, p. 170. (Chelonia.)
Zittel, K. A. 1890 A, pp. 449, 500.

Suborder ATHECÆ Cope.

Cope, E. D. 1871 B, p. 235.
Baur, G. 1886 G, p. 687.
  1888 A.
  1889 A.
  1890 A.
Boulenger, G. A. 1889 A, pp. 4, 7.
Case, E. C. 1897 B, p. 21.
Cope, E. D. 1875 X, p. 16.
  1891 I.
  1891 N, p. 39.
  1898 B, p. 62.
Dames, W. 1894 A.
Dollo, L. 1886 A, p. 79.

Dollo, L. 1887 A.
  1888 C.
Gadow, H. 1890 A, p. 21.
Goette, A. 1899 A, p. 422. (Atheca.)
Günther, A. 1888 A, p. 456.
Haeckel, E. 1895 A, pp. 319, 325. (Atheconia.)
Hay, O. P. 1898 F.
Lydekker, R. 1889 G, p. 223. (Athecata.)
Seeley, H. G. 1880 B, p. 412. (Dermochelyidæ.)
Steinmann and Döderlein 1890 A, p. 633.
Strauch, A. 1890 A, p. 38. (Atheca.)
Woodward, A. S. 1887 C, p. 7.

DERMOCHELYIDÆ.[1]

Fitzinger, 1843, Syst. Rept. (Dermatochelyidæ.)
Agassiz, L. 1857 B, p. 320. (Sphargididæ.)
Baur, G. 1886 G, p. 687.
 1887 E.
 1888 A.
 1889 A.
 1889 B, p. 42.
 1889 H, p. 55. (Dermochelys.)
 1889 J, p. 39.
 1890 D, p. 584.
 1893 B, p. 673.
 1896 B.
Bemmelen, J. F. 1896 A, p. 323.
 1896 B, p. 279.
Boulenger, G. A. 1888 A, p. 353.
 1889 A, pp. 4, 7. (Sphargididæ.)
Case, E. C. 1897 B.
Dames, W. 1894 A. (Sphargididæ.)
Dollo, L. 1887 A. (Sphargis.)
 1888 C. (Sphargididæ.)
Fürbringer, M. 1900 A, p. 632. (Sphargidæ.)

Gervais, P. 1872 A. (Sphargididæ.)
Goette, A. 1899 A, p. 422. ("Dermochelyden.")
Gray, J. E., 1825, Ann. Philos., x. (Sphargidæ.)
 1869 B, p. 224. (Sphargididæ.)
Günther, A. 1888 A, p. 456. (Sphargididæ.)
Hay, O. P. 1896 F, p. 931.
Hoffmann, C. K. 1890 A, with plates. (Sphargis.)
Kükenthal, W. 1890 B, p. 240.
 1890 C, p. 162.
Lydekker, R. 1889 G, p. 223.
Nicholson and Lydekker 1889 A, p. 1090.
Seeley, H. G. 1880 B.
Siebenrock, F. 1897 A. (Sphargididæ.)
Strauch, A. 1890 A, p. 38. (Sphargida.)
Vaillant, L. 1894 A. (Sphargididæ.)
Wieland, G. R. 1900 A, p. 237.
 1900 B, p. 413.
Winckler, T. C. 1869 A, p. 60. (Sphargis.)
Woodward, A. S. 1887 C. (Sphargis.)
Zittel, K. A. 1890 A, p. 517.

Suborder THECOPHORA[2] Dollo.

Dollo, L. 1886 A, pp. 79, 91.
Boulenger, G. A. 1889 A, pp. 4, 11.
Case, E. C. 1897 B.
Dollo, L. 1887 A.
 1888 C.
Fraas, E. 1899 A, p. 401. (Proganochelys.)
Gadow, H. 1898 A, p. 20.
Goette, A. 1899 A, p. 426.

Günther, A. 1888 A, p. 457. (Testudinata.)
Haeckel, E. 1895 A, p. 319.
Hay, O. P. 1896 F, p. 940.
Lydekker, R. 1889 G, p. 3. (Testudinata.)
Nicholson and Lydekker 1889 A, p. 1091. (Testudinata.)
Seeley, H. G. 1880 B, p. 412. (Aspidochelyidæ.)
Vaillant, L. 1894 A. (<Craspedota.)

Superfamily PLEURODIRA Cope.

Cope, E. D. 1869 M, p. 156.
Baur, G, 1887 B, p. 101.
 1888 B.
 1889 H.
 1889 J.
 1890 D, p. 535.
 1891 B, p. 420 (Pleurodira); p. 422 (Amphichelydia).
 1891 C, p. 351.
 1896 B, p. 566. (Amphichelydia.)
Boulenger, G. A. 1888 A.
 1889 A, pp. 5, 187.
 1889 B.
Cope, E. D. 1871 B, p. 236.
 1872 D.
 1891 N, p. 39.
 1898 B, p. 63.
Dollo, L. 1886 A.

Duméril and Bibron, 1835, Érp. Gén., ii, p. 175. ("Pleurodères.")
Fraas, E. 1899 A. p. 401. (Proganochelys.)
Fürbringer, M. 1900 A, p. 632.
Gadow, H. 1898 A, p. 21.
Goette, A. 1899 A, p. 430.
Günther, A. 1888 A, p. 458.
Haeckel, E. 1895 A, p. 329. (Pleurodera.)
Hay, O. P. 1898 F, p. 947.
Lydekker, R. 1889 G, p. 158 (Pleurodira); p. 204 (Amphichelydia).
Nicholson and Lydekker 1889 A, p. 1091 (Amphichelydia); p. 1094 (Pleurodira).
Peters, W. 1839 A. (Hydromedusa maximiliani.)
Quenstedt, F. A. 1889 A. (On Psammochelys.)
Siebenrock, F. 1898 A, p. 425.
Steinmann and Döderlein 1890 A, p. 638.
Zittel, K. A. 1890 A, pp. 449, 542.

---

[1] References are given to the literature of this family on account of the bearing it has on opinions regarding the evolution and classification of *Testudines*.

[2] This group as here understood is equivalent to that so called by Dollo, *minus* the *Trionychia*.

PLEUROSTERNIDÆ.

Cope, E. D. 1875 X, p. 17.
Baur, G. 1888 B, p. 420.
    1891 B,'p. 424(Pleurosternidæ); p. 428 (Baënidæ).
Cope, E. D. 1882 E, pp. 143, 145. (Baënidæ.)
    1884 O, p. 111 (Pleurosternidæ); pp. 111, 112 (Baënidæ).
    1896 B, p. 63. (Baënidæ; Pleurosternidæ.)
Dollo, L. 1886 A, p. 78. (Pleurosternidæ; Baënidæ.)

Grabbe, A. 1884 A. (Pleurosternon.)
Günther, A. 1888 A, p. 457. (Baënidæ.)
Lydekker, R. 1889 E. (Pleurosternum.)
    1889 G, p. 205.
Nicholson and Lydekker 1889 A, p. 1092.
Owen, R. 1853 B, p. 21. (Pleurosternon.)
Rütimeyer, L. 1873 A, p. 143. (Pleurosternon.)
Steinmann and Döderlein 1890 A, p. 639.
Woodward, A. S. 1898 B, p. 172. (Pleurosternum.)

COMPSEMYS Leidy. Type *C. victus* Leidy.

Leidy, J. 1856 Q, p. 312.
Baur, G. 1890 D, p. 534.
    1891 B, p. 411.
    1896 B, p. 565.
Cope, E. D. 1869 M, p. 124.
    1875 E, p. 91.
    1875 I.
    1875 W, p. 336.
    1877 F, p. 196.
    1882 E, p. 145.
    1884 O, p. 113.
Leidy, J. 1860 A, p. 152.
Lydekker, R. 1889 G, p. 137.
Marsh, O. C. 1890 E, p. 177. (Glyptops, type *G. ornatus*.)
Nicholson and Lydekker 1889 A, p. 1105.
Zittel, K. A. 1890 A, p. 534.

Compsemys lineolatus Cope.

Cope, E. D. 1874 C, p. 30. (Adocus?.)
    1874 B, p. 454. (Adocus?.)
    1875 E, pp. 92, 263, pl. vi, figs. 11, 12. (Adocus.)
    1877 G, p. 573.
Cretaceous (Laramie); Montana, Colorado, Saskatchewan?.

Compsemys obscurus (Leidy).

Leidy, J. 1856 Q, p. 312. (Emys.)
Cope, E. D. 1869 M, p. 124.
    1874 C, p. 30.

Cope, E. D. 1875 E, p. 261.
    1877 G, p. 573. (Emys.)
Leidy, J. 1860 A, p. 153, pl. xi, fig. 4.
Cretaceous (Laramie); Montana.

Compsemys plicatulus Cope.

Cope, E. D. 1877 F, p. 196.
Baur, G. 1890 D, p. 534.
    1891 B.
    1896 B, p. 565.
Marsh, O. C. 1890 E, p. 176, pl. vii. (Glyptops ornatus.)
    1897 C, p. 507, figs. 63, 64. (Glyptops ornatus.)
Jurassic; Colorado, Wyoming.

Compsemys victus Leidy.

Leidy, J. 1856 Q, p. 312.
Cope, E. D. 1869 M, p. 124.
    1874 B, p. 454.
    1874 C, p. 30.
    1875 E, p. 261, pl. vi, figs. 15, 16.
    1875 I. (C. victus?.)
    1877 G, p. 573.
Leidy, J. 1860 A, p. 152, pl. xi, figs. 5-7.
Marsh, O. C. 1897 C, p. 527.
Cretaceous (Laramie); Montana, Colorado, Saskatchewan?.

Compsemys sp. indet.

Cope, E. D. 1888, CC, p. 301.
Eocene (Puerco); New Mexico.

BAËNA Leidy. Type *B. arenosa* Leidy.

Leidy, J. 1871 C, p. 367.
Baur, G. 1888 B, p. 422.
    1889 H, p. 58.
    1891 B, p. 425.
Cope, E. D. 1873 E, p. 621.
    1882 E, p. 145.
    1882 J p. 990.
    1884 O, pp. 112, 144.
    1886 M.
Günther, A. 1888 A, p. 456.
Leidy, J. 1872 L, p. 162. (Chisternon, type *Baëna undata*.)
    1873 B, p. 160 (Baëna); p. 169 (Chisternon).
    1873 S, p. 123.
Nicholson and Lydekker 1889 A, p. 1093.
Steinmann and Döderlein 1890 A, p. 637.
Zittel, K. A. 1890 A, p. 536.

Baëna arenosa Leidy.

Leidy, J. 1870 S, p. 123.
Baur, G. 1891 B, p. 427.
Cope, E D. 1873 F p. 623.
    1875 C, p. 36.
    1875 F, p. 96.
    1877 K, p. 52, pl. xxiv, fig. 32.
    1882 J, p. 990, fig. 8.
    1884 O, p. 148, pl. xvii, figs. 1, 2.
Dollo, L. 1888 A, p. 11, fig. 11.
King, C. 1878 A, p. 404.
Leidy, J. 1871 C, p. 367. (B. arenosa, B. affinis.)
    1871 K, p. 229.
    1872 B, p. 368.
    1873 B, pp. 161, 341, pl. xiii, figs. 1-5.
Eocene (Bridger); Wyoming: (Wasatch); New Mexico.

Baëna hebraica Cope.

Cope, E. D. 1872 M, p. 463.
    1873 E, p. 621.
    1884 O, p. 146, pl. xix, figs. 1, 2.
    Eocene (Bridger); Wyoming.

Baëna? ponderosa Cope.

Cope, E. D. 1873 E, p. 624.
    1884 O, p. 150, pl. xvii, figs. 3–8.
    Eocene (Bridger); Wyoming.

Baëna undata Leidy.

Leidy, J. 1871 K, p. 228.
Cope, E. D. 1873 E, p. 622.
    1884 O, p. 147, pl. xix, figs. 3–5.
Leidy, J. 1871 B, p. 373.
    1872 B, p. 369.
    1872 L, p. 162. (Chisternon.)
    1873 B, pp. 169, 341, pl. xiv, figs. 1, 2. (Chisternon.)
Osborn, Scott, and Speir 1878 A, p. 96.
    Eocene (Bridger); Wyoming.

POLYTHORAX Cope.   Type *P. missuriensis* Cope.

Cope, E. D. 1876 H, p. 258.
    1882 E, p. 145.
    1884 O, p. 112.
Nicholson and Lydekker 1889 A, p. 1098.

Polythorax missuriensis Cope.

Cope, E. D. 1876 H, p. 258.
    1877 G, p. 573.
    Cretaceous (Laramie); Montana.

PELOMEDUSIDÆ.

Cope, E. D. 1868 B, p. 282.
Baur, G. 1888 B, p. 420. (Pelomedusidæ, Podocnemidæ.)
    1891 B, p. 424. (Bothremydidæ.)

Boulenger, G. A. 1889 A, p. 191.
Lydekker, R. 1889 G, p. 170.
Nicholson and Lydekker 1889 A, p. 1099.
Steinmann and Döderlein 1890 A, p. 639.

BOTHREMYS Leidy.   Type *B. cookii* Leidy.

Leidy, J. 1865 A, pp. 110, 120.
Baur, G. 1889 J, p. 38.
    1891 B, p. 423.
Cope, E. D. 1867 A, p. 40.
    1869 M, pp. 148, 157.
Günther, A. 1888 A, p. 457.
Lydekker, R. 1889 G, p. 174.
Nicholson and Lydekker 1889 A, p. 1099.
Zittel, K. A. 1890 A, p. 547.

Bothremys cookii Leidy.

Leidy, J. 1865 A, pp. 110, 120, pl. xviii, figs. 4–8.
Cope, E. D. 1869 B, p. 735.
    1869 K, p. 89.
    1869 M, p. 157.
    1875 E, p. 263.
Leidy, J. 1865 D, p. 73.
Marsh, G. A. 1869 A, p. 280.
    Cretaceous (Green-sand No. 4); New Jersey.

TAPHROSPHYS Cope.   Type *T. molops* Cope.

Cope, E. D. 1869, K, p. 90.
Baur, G. 1888 B, p. 421.
    1891 B, p. 424.
Cope, E. D. 1869 K, p. 90. (Prochonias, type *Platemys sulcatus* Leidy).
    1869 M, p. 157.
    1873 E, p. 649.
Günther, A. 1888 A, p. 457.
Lydekker, R. 1889 G, p. 174.
Nicholson and Lydekker 1889 A, p. 1099.
Zittel, K. A. 1890 A, p. 547. (Syn. of Bothremys).

Taphrosphys leslianus Cope.

Cope, E. D. 1869 M, pp. 159, 166 (Taphrosphys); p. 165, line 14 (Prochonias).
    1875 E, p. 264.
    Cretaceous (Green-sand); New Jersey.

Taphrosphys longinuchus Cope.

Cope, E. D. 1869 M, pp. 159, 162.
    1875 E, p. 263.
    Cretaceous (Green-sand); New Jersey.

Taphrosphys molops Cope.

Cope, E. D. 1869 M, pp. 158, 159, figs. 43, 44.
    1869 B, p. 735.
    1869 K, p. 89.
    1869 M, pp. 158, 160 (Prochonias enodis); p. 162 (T. molops, var. enodis).
    1870 D, p. 274.
    1875 E, p. 263.
Zittel, K. A. 1890 A, p. 547. (Bothremys.)
    Cretaceous (Green-sand); New Jersey.

Taphrosphys nodosus Cope.

Cope, E. D. 1869 M, pp. 159, 167, pl. i, fig. 16.
    1875 E, p. 264.
    Cretaceous (Green-sand); New Jersey.

Taphrosphys strenuus Cope.

Cope, E. D. 1869 K, pp. 89, 90. (Prochonias.)
    1869 B, p. 735. (T. princeps.)
    1869 K, p. 89. (Prochonias princeps.)
    1869 M, pp. 160, 167 (Prochonias princeps); pp. 159, 166 B (Taphrosphys); p. 167 (Prochonias).

Cope, E. D. 1875 E, p. 264.
1875 V, p. 34.
1878 EE.
Cretaceous (Green-sand): New Jersey,
Georgia, North Carolina.
Cope's species *T. princeps* is last mentioned,
and then only incidentally, in 1869 M, pp. 160,167.
It appears to have been regarded by Cope as a
synonym of *T. strenuus.* In the Cope collection
of reptiles the present writer has found the
femur described on p. 166–B of Cope 1869M, and
it has written on it, probably by Cope himself,
the name "*Prochonias princeps.*"

Taphrosphys sulcatus (Leidy).
Leidy, J. 1856 P, p. 303. (Platemys.)
Cope, E. D, 1869 B, p. 735.
    1869 K, pp. 89, 90. (Prochonias.)
    1869 M, pp. 159,164, text figs. 45,45 *bis* (Taphrosphys); p. 165 (Prochonias).
    1875 E, p. 264.
Leidy, J. 1865 A, pp. 109,120, pl. xix, fig. 4. (Platemys.)
Maack, G. A. 1869 A, p. 281. (Platemys.)
Cretaceous (Green-sand); New Jersey.

PLESIOCHELYIDÆ.

Lydekker, R. 1889 G, p. 183.

PLESIOCHELYS Rütimeyer.   Type *P. solodurensis* Rütimeyer.

Rütimeyer L. 1873 A. p. 48.
Lydekker, R. 1889 D, p. 239.
    1889 G, p. 196.
Maack, G. A. 1869 A, p. 320.   (Stylemys in part;
    not of Leidy.)
Portis, A. 1878 A, p. 131.

Zittel, K. A. 1890 A, p. 544.

Plesiochelys belviderensis Cragin.
Cragin, F. W. 1894 A, p. 71, pl. ii, figs. 1–8.
    Cretaceous (Comanche series); Kansas.

Superfamily CRYPTODIRA Cope.

Cope, E. D. 1869 M, p. 123.
Baur, G. 1889 H.
    1889 J.
    1890 D, p. 535.
    1891 B, p. 419.
    1893 B.
Boulenger, G. A. 1889 A, pp. 4,11.
    1889 B.
Cope, E. D. 1871 B, p. 235.
    1884 O, p. 111.
    1891 N, p. 39.
    1898 B, p. 62.
Dollo, L. 1886 A.

Duméril and Bibron, 1835, Érp. Gén., ii, p, 172.
    ("Cryptodères.")
Fürbringer, M. 1900 A, p. 632.
Gadow, H. 1898, p. 20.
Goette, A. 1899 A, p. 430.
Günther, A. 1888 A, p. 457.
Hay, O. P. 1898 F, p. 947.
Lydekker, R. 1889 G, p. 25.
Nicholson and Lydekker 1889 A, p. 1101.
Williston, S. W., in Williston and Case 1898 A, p.
    367.
Zittel, K. A. 1890, pp. 449.

PROTOSTEGIDÆ.

Cope, E. D. 1873 E, p. 649.
Baur, G. 1888 A.
    1889 A, pp. 184, 191.
    1890 D, p. 584.
Case, E. C. 1897 B.
Cope, E. D. 1875 X, p. 16.

Lydekker, R. 1889 G, p. 228.
Nicholson and Lydekker 1889 A, p. 1089.
Steinmann and Döderlein 1890 A, p. 634.
Wieland, G. R. 1898 A, p. 19.
    1900 A, p. 250.

PROTOSTEGA Cope.   Type *P. gigas* Cope.

Cope, E. D. 1871 K, p. 175.
Baur, G. 1886 G, p. 687.
    1889 A.
    1890 D.
    1896 B.
Capellini, G. 1884 A.
Case, E. C. 1897 B.
Cope, E. D. 1872 F, p. 334.
    1872 L.
    1873 E, p. 648.
    1875 E, pp. 99, 256.
    1884 O, p. 114.

Dollo, L. 1888 C, p. 72.
Goette, A. 1899 A, p. 426.
Hay, O. P. 1895 B.
    1898 F, p. 929.
Lydekker, R. 1889 G, pp. vii, 229.
Marsh, O. C. 1877 E, p. 345.   (Atlantochelys.)
    1877 F. (Atlantochelys.)
Nicholson and Lydekker 1889 A, p. 1089.
Wieland, G. R. 1900 A, pp. 237, 250.
    1900 B, p. 416.
Woodward, A. S. 1887 C, p. 9.
Zittel, K. A. 1890 A, p. 519.

Protostega gigas Cope.

Cope, E. D. 1871 K, p. 175.
Capellini, G. 1884 A.
Case, E. C. 1897 B, pls. iv-vi.
Cope, E. D. 1872 F, pp. 323, 335.
 1872 L, p. 433.
 1875 E, pp. 48, 102, 256, pl. ix, figs. 1-7; pls.
  x-xiii.
 1878 Q.

Dana, J. D. 1876, Man. Geol., ed. ii, p. 466. (Atlantochelys.)
Hay, O. P. 1895 B, pls. iv, v.
Sternberg, C. 1899 A, p. 126, figures.
Wieland, G. A. 1896 A.
 1898 A.
 1900 A, p. 238.
 1900 B, p. 416, figs. 8, 19.
Cretaceous (Niobrara); Kansas.

ARCHELON Wieland.   Type *A. ischyros* Wieland.

Wieland, G. R. 1896 A, p. 399.
Hay, O. P. 1898 F, p. 929. (Protostega.)
Wieland, G. R. 1898 A, p. 15. (Protostega.)

Archelon ischyros Wieland.

Wieland, G. R. 1896 A, p. 399, pl. vi, text figs. 2-19.
Case, E. C. 1897 B, p. 52. (Protostega.)
Wieland, G. R. 1898 A, p. 15, pl. ii, text figs. 1, 2
 (Protostega.)

Wieland, G. R. 1900 A, p. 237, pl. ii, text figs. 1-3; 6.
 1900 B, p. 420, figs. 17, 18.
Williston, S. W. 1897 H, p. 246. (Protostega.)
Cretaceous (Pierre); South Dakota.

Archelon marshii Wieland.

. *Wieland, G. R.* 1900 A, p. 248.
Cretaceous (Pierre); South D kota.

NEPTUNOCHELYS Wieland.   Type *Protostega tuberosa* Cope.

Wieland, G. R. 1900 B, p. 418.

Neptunochelys tuberosa (Cope).

Cope, E. D. 1872 L, p. 433. (Platecarpus, *errore*.)
Baur, G. 1889 A, p. 189.
Cope, E. D. 1869 A, p. 265. (Platecarpus tympaniticus, in part.)
 1869 M, pp. 199, 200. (Platecarpus tympaniticus, in part.)

Cope, E. D. 1872 F, p. 334. (Protostega.)
 1875 E, p. 257. (Protostega.)
Leidy, J. 1865 A, p. 118, pl. viii, figs. 1, 2. (?Hol-
  codus acutidens, in part.)
 1873 B, p. 342. (Atlantochelys.)
Wieland, G. R. 1900 B, p. 418, figs. 9, 10. (Neptunochelys.)
Cretaceous; Mississippi.

ATLANTOCHELYS Agassiz.   Type *A. mortoni* Agassiz.

Agassiz, L., in Leidy, J. 1865 A, p. 43.
 1849 A, p. 169. (No definition.)
 1850 B. (No definition.)
Baur, G. 1889 A, p. 189.
Cope, E. D. 1869 M, p. 185.
Dollo, L. 1888 C, p. 67.
Hay, O. P. 1898 F, p. 930.
Leidy, J. 1873 B, p. 342.
Lydekker, R. 1889 G, p. 229.
Marsh, O. C. 1877 E, p. 345.
 1877 F.

Atlantochelys mortoni Agassiz.

Agassiz, L., in Leidy, J. 1865 A, p. 43, pl. viii, figs. 3-5.

Agassiz, L. 1849 A, p. 169. (No description.)
 1850 B. (No description.)
Baur, G. 1889 A, p. 189. ("Protostega neptunia.")
Cope, E. D. 1869 M, p. 185.
 1872 F, p. 334. (Protostega neptunia.)
 1872 L, p. 433. (Protostega neptunia.)
 1873 E. (Protostega neptunia.)
Hay, O. P. 1898 F, p. 930.
Leidy, J. 1865 A, p. 117. (Mosasaurus mitchelli,
  in part.)
 1873 B, p. 342.
Wieland, G. R. 1900 B, p. 419, figs. 14-16.
Cretaceous; New Jersey.

THALASSEMYDIDÆ.

Baur, G. 1889 A, p. 191. (Thalassemydidæ, Ly-
  tolomidæ.)
Cope, E. D. 1869 M, p. 235. (Propleuridæ.)
 1870 T, p. 137. (Propleuridæ.)
 1871 C, p. 563. (Propleuridæ.)
 1872 B, p. 22. (Propleuridæ.)
 1872 D. (Propleuridæ.)
 1879 A, p. 36. (Propleuridæ.)
 1882 E, pp. 143, 144. (Propleuridæ.)
 1882 J, p. 989. (Propleuridæ.)
 1884 O, pp. 111, 112. (Propleuridæ.)

Cope, E. D. 1887 R. (Propleuridæ.)
Dollo, L. 1884 A, p. 71. (Thalassemydidæ.)
 1886 A, p. 89. (Cheloniidæ, in part.)
 1888 A. (Propleuridæ.)
Lortet, L. 1892 A, p. 24. (Thalassemydidæ.)
Lydekker, R. 1889 A, p. 36. (Propleuridæ.)
 1889 G, p. 25. (Cheloniidæ, in part.)
Nicholson and Lydekker 1889 A, p. 1112. (Cheloniidæ.)
Zittel, K. A. 1890 A, p. 525 (Chelonemydidæ);
  p. 527 (Thalassemydidæ).

OSTEOPYGIS Cope. Type *O. emarginatus* Cope.

Cope, E. D. 1868 E, p. 147.
Baur, G. 1889 A, p. 186.
    1889 B, p. 42.
Case, E. C. 1897 B, pp. 33, 43, 50.
Cope, E. D. 1869 K. p. 88. (Osteopygis; Propleura,
    type *Chelone sopita* Leidy.)
    1869 M, pp. 132, 235 (Osteopygis); pp. 138,
    235 (Propleura); pp. 143, 235 (Catapleura,
    type *C. repanda*).
    1875 E, p. 257.
    1882 E, p. 144. (Osteopygis, Propleura, Cata-
    pleura.)
    1884 O, p. 112. (Osteopygis, Propleura, Cata-
    pleura.)
Lydekker R. 1889 A, p. 61.
    1889 D, p. 233. (Lytoloma.)
    1889 G, p. 52. (Syn.? of Lytoloma.)
Nicholson and Lydekker 1889 A, p. 1113. (Lyto-
    loma.)
Zittel, K. A. 1890 A, p. 526.

Osteopygis chelydrinus Cope.

Cope, E. D. 1869 K, p. 89.
    1868 E, p. 147. (No description.)
    1869 B, p. 735. (No description.)
    1869 M, pp. 135, 138, pl. vii, fig. 8.
    1875 E, p. 259. (Catapleura.)
Cretaceous; New Jersey.

Osteopygis emarginatus Cope.

Cope, E. D. 1869 K, p. 89.
    1869 B, p. 735.
    1868 E, p. 147. (No description.)
    1869 M, pp. 135, 136, 235, pl. vii, fig. 3.
    1875 E, p. 259.
Cretaceous; New Jersey.

Osteopygis erosus Cope.

Cope, E. D. 1875 E, p. 258.
Cretaceous; New Jersey.

Osteopygis platylomus Cope.

Cope, E. D. 1869 K, p. 89.
    1869 B, p. 735.
    1869 M, pp. 135, 137, and 11, figs. 38, 39.
    1875 E, p. 258.
Cretaceous; New Jersey.

Osteopygis ponderosus (Cope).

Cope, E. D. 1871 G, p. 46. (Catapleura.)
    1875 E, p. 259. (Catapleura.)
Cretaceous; New Jersey.

Osteopygis repandus (Cope).

Cope, E. D. 1869 M, p. 142, pl. vii, fig. 2. (Cata-
    pleura.)
    1868 E. (Osteopygis; no description.)
    1869 B, p. 734. (Propleura; no description.)
    1875 E, p. 259. (Catapleura.)
Cretaceous; New Jersey.

Osteopygis sopitus (Leidy).

Leidy, J. 1865 A, pp. 104, 119. (Chelone; not p.
    105, pl. xix, fig. 5 = Lytoloma angusta Cope.)
Cope, E. D. 1868 E, p. 147. (Osteopygis.)
    1869 B, p. 735. (Propleura.)
    1869 K, p. 88. (Propleura.)
    1869 M, pp. 140, 235, pl. vii, figs. 4-7. (Pro-
    pleura.)
    1875 E, p. 258.
Maack, G. A. 1869 A, pp. 238, 283. (Chelone.)
Cretaceous; New Jersey.

PERITRESIUS Cope. Type *Chelone ornata* Leidy.

Cope, E. D. 1869 K, p. 88.
    1869 M, p. 150.
    1882 E, p. 144.
    1884 O, p. 112.
Zittel, K. A. 1890 A, p. 526.
    The systematic position of this genus is
doubtful.

Peritresius ornatus (Leidy.)

Leidy, J. 1856 P, p. 303. (Chelone.)

Cope, E. D. 1869 B, p. 735.
    1869 K, p. 88.
    1869 M, p. 150.
    1875 E, p. 260.
    1878 EE.
Leidy, J. 1865 A, pp. 105, 119, pl. xviii, fig. 10.
    (Chelone.)
Maack, G. A. 1869 A, p. 283. (Chelone.)
Cretaceous; New Jersey, Georgia.

LYTOLOMA Cope. Type *L. angusta* Cope.

Cope, E. D. 1869 K, p. 88.
Baur, G. 1889 B, p. 43.
    1889 J, p. 39. (Euclastes.)
Case, E. C. 1897 B, pp. 44, 50.
Cope, E. D. 1867 A, p. 39. (Euclastes, type *E.
platyops*.)
    1869 M, pp. 131, 144, 235 (Lytoloma); pp. 131,
    148 (Euclastes).
    1882 E, p. 144.
    1884 O, p. 112.
    1887 R.

Dames, W. 1894 A, p. 208.
Dollo, L. 1888 A.
Lydekker, R. 1889 A, p. 61.
    1889 D, p. 233.
    1889 G, p. 51.
Nicholson and Lydekker 1889 A, p. 1113.
Seeley, H. G. 1871 B. (Glossochelys, type *E.
harricensis*.)
Wieland, G. R. 1900 B, p. 416, fig. 6.
Zittel, K. A. 1890 A, p. 526. (Euclastes.)

Lytoloma angusta Cope.

Cope, E. D. 1869 K, p. 89.
    1869 M, p. 145, pl. xi, fig. 1.
    1875 E, p. 257.
Leidy, J. 1865 A, p. 105, pl. xix, fig. 5. (Chelone sopita, in part.)
    Cretaceous (Green-sand No. 5); New Jersey.

Lytoloma jeanesii Cope.

Cope, E. D. 1869 M, p. 145.
    1869 B, p. 735. (Propleura; no description.)
    1875 E, p. 257.
    Cretaceous (Green-sand No. 5); New Jersey.

Lytoloma platyops (Cope.)

Cope, E. D. 1867 A, p. 41. (Euclastes.)
    1869 B, p. 735. (Euclastes.)
    1869 K, p. 89. (Euclastes.)
    1869 M, p. 149, pl. vii, fig. 9. (Euclastes.)
    1875 E, p. 259. (Euclastes.)
    Cretaceous (Green-sand No. 5); New Jersey.

Lytoloma sp. indet.

Clark, W. B. 1895 A, p. 4. (Euclastes.)
    1897 A, p. 69. (Euclastes.)
    Eocene; Maryland.

SYLLOMUS Cope.   Type *S. crispatus* Cope.

Cope, E. D. 1896 J, p. 139.

Syllomus crispatus Cope.

Cope, E. D. 1896 J, p. 139.
    Neocene; Virginia.

TOXOCHELYIDÆ.

Baur, G. 1895 D, p. 569. (Toxochelydæ.)
    1896 B, p. 564.

Hay, O. P. 1896 A, p. 106.

TOXOCHELYS Cope.   Type *T. latiremis* Cope.

Cope, E. D. 1873 I, p. 10.
Baur, G. 1889 H, p. 58.
    1895 D, p. 569.
Case, E. C. 1897 B, p. 28.
    1893, in Williston and Case 1898 A, p. 370.
Cope, E. D. 1872 BB. (No name.)
    1875 E, pp. 98, 300.
    1877 D, p. 176.
Hay, O. P. 1896 A.
    1896 F, p. 935.
Lydekker, R. 1889 G, p. 129.
Nicholson and Lydekker 1889 A, p. 1104.
Wieland, G. R. 1900 B, p. 418.
Zittel, K. A. 1890 A, pp. 526, 534.

Toxochelys brachyrhinus Case.

Case, E. C., in Williston and Case 1898 A, p. 378, pl. lxxxiv.
    Cretaceous; Kansas.

Toxochelys latiremis Cope.

Cope, E. D. 1873 I, p. 10.

Case, E. C., in Williston and Case 1898 A, p. 371, pl. lxxix; pl. lxxx, figs. 1, 2; pl. lxxxi, figs. 1–8, 10–13; pl. lxxxii, figs. 1–3, 6; pl. lxxxiii, figs. 2–4.
Cope, E. D. 1872 BB, p. 129. (Description; no name.)
    1875 E, pp. 98, 260, pl. viii, fig. 12.
    1877 D, p. 176.
Hay, O. P. 1896 A, pls. xiv, xv.
Leidy, J. 1873 B, p. 279, pl. xxxvi, figs. 17–21. (? Cynocercus incisus.)
Wagner, G. 1898 A, p. 201, text figs. 1, 2.
Wieland, G. R. 1900 B, p. 418.
    Cretaceous (Niobrara and Pierre); Kansas.

Toxochelys serrifer Cope.

Cope, E. D. 1875 E, p. 299.
Case, E. C., in Williston and Case 1898 A, p. 379, pl. lxxx, figs. 3–9; pl. lxxxii, figs. 4, 5; pl. lxxxiii, fig. 1.
Hay, O. P. 1896 F, p. 935, figs. 1–3.
    Cretaceous (Niobrara); Kansas.

CYNOCERCUS Cope.   Type *C. incisus* Cope.

Cope, E. D. 1882 H.
    1872 F, p. 335.
    1875 E, p. 96.
Hay, O. P. 1896 A.
Williston, S. W., in Williston and Case 1898 A, p. 368.
Zittel, K. A. 1890 A, p. 526.
    This genus is not improbably identical with *Toxochelys.* If so, *Cynocercus* supersedes *Toxochelys.*

Cynocercus incisus Cope.

Cope, E. D. 1872 H.
    1872 F, p. 335.
    1872 BB, p. 129.
    1874 C, p. 28.
    1875 E, pp. 96, 260, pl. viii, figs. 3–5.
Williston, S. W., in Williston and Case 1898 A, p. 368, fig. 6.
    Cretaceous (Niobrara); Kansas.

DESMATOCHELYIDÆ.

Williston, S. W. 1894 D.

Williston, S. W., in Williston and Case 1898 A, p. 367.

DESMATOCHELYS Willist.   Type *D. lowii* Willist.

Williston, S. W. 1894 D.

    1896, in Williston and Case 1896 A, p. 351.

Desmatochelys lowii Willist.

Williston, S. W. 1894 D.

Wieland, G. R. 1900 B, p. 419, fig. 13.
Williston, S. W., in Williston and Case 1896 A, p. 352. pls. lxxiii–lxxviii.
   Cretaceous (Niobrara); Nebraska, Kansas.

CHELONIIDÆ.

Gray, J. E., 1825, Ann. Phil. (2), x. (Cheloniadæ.)
Agassiz, L. 1857 B, p. 324. (Chelonoidæ.)
Baur, G. 1889 A.
    1889 B.
    1890 F, p. 486.
    1891 C, p. 350.
    1893 B, p. 673.
Bemmelen, J. F. 1896 A.
Boulenger, G. A. 1889 A, p. 180.
Case, E. C. 1897 B.
Cope, E. D. 1869 M, p. 153.
    1871 C, p. 130.
    1872 D.
    1882 E, p. 143.
    1884 O, p. 111.
Dollo, L. 1888 C.

Gadow, H. 1899 A, p. 37.
Goette, A. 1899 A.
Gray, J. E. 1873 B, p. 396.
Günther, A. 1888 A, pp. 457, 458.
Hay, O. P. 1896 A.
    1898 F, p. 931.
Hoffmann, C. K. 1890 A.
Lydekker, R. 1889 G, p. 25.
Nicholson and Lydekker 1889 A, p. 1111.
Owen, R. 1842 A, p. 168.
Pictet, F. J. 1853 A, p. 458. (Chélonées.)
Siebenrock, F. 1897 A.
    1898 A, p. 425.
Steinmann and Döderlein 1890 A, p. 636.
Strauch, A. 1890 A, p. 37.
Zittel, K. A. 1890 A, p. 521.

CHELONIA Brongniart.

Brongniart, A., 1799, Jour. de Science, etc., v, pp. 184, 201.
Agassiz, L. 1857 B, p. 377.
Baur, G. 1889 A.
    1889 F, p. 467.
    1890 F, p. 486.
    1896 B.
Bemmelen, J. F. 1896 B. (Chelone.)
Boulenger, G. A. 1889 A, p. 180. (Chelone.)
Cope, E. D. 1882 E, p. 144.
    1884 O, p. 112.
Dollo, L. 1888 A. (Chelone.)
Fürbringer, M. 1900 A. (Chelone.)
Hoffmann, C. K. 1890 A.
Lydekker, R. 1889 A, p. 60. (Chelone.)
    1889 D. (Chelone.)
    1889 G, p. 27. (Chelone.)
Owen, R. 1842 A, p. 168. (Chelone.)
    1849 B. (Chelone.)

Owen, R. 1849 C. (Chelone.)
    1851 B, p. 1. (Chelone.)
Peters, W. 1859 B.
Pictet, F. J. 1853 A, p. 459.
Segond, L. A. 1873 A, p. 2. ("Tortue franche.")
Stannius, H. 1856 A.
Strauch, A. 1890 A, p. 37. (Chelone.)
Ubaghs, C. 1875 A.
Winckler, T. C. 1869 A.
Zittel, K. A. 1890 A, p. 523.
   No species certainly belonging to this genus have yet been described from North America. For *Chelonia couperi* see Harlan, R. 1842 B, p. 144, pl. iii, figs. 2, 3; but it is probable that the bone there described and figured is not even a portion of a turtle. See also J. H. Couper in Hodson, W. B. 1846 A, p. 46, for mention of this supposed turtle.

PUPPIGERUS Cope.   Type *Chelone grandæva* Leidy.

Cope, E. D. 1869 M, p. 235.
    1882 E, p. 144.
    1884 O, p. 112.
Günther, A. 1888 A, p. 457.
Lydekker, R. 1889 A, p. 61.
    1889 D, p. 254. (Syn. of Lytoloma.)
    1889 G, p. 52. (Syn. of Lytoloma.)
Nicholson and Lyddeker 1889 A, p. 1113.
Zittel, K. A. 1890 A, p. 527.

Puppigerus grandævus (Leidy.)

Leidy, J. 1851 H, p. 329. (Chelonia.)
Cope, E. D. 1869 B, p. 738. (Chelone.)
    1869 M, p. 153 (Chelone); p. 235 (Puppigerus).

Cope, E. D. 1875 H, p. 363.
Leidy, J. 1856 P, p. 303. (Chelone.)
Maack, G. A. 1869 A, p. 283. (Chelone.)
   Miocene; New Jersey.

Puppigerus parvitectus Cope.

Cope, E. D. 1869 M, p. 155. (Chelone parvitecta.)
    1867 C, p. 143. ("Sp. undet.")
    1869 B, p. 738. (Chelone parviscutum: no description.)
    1870 T, p. 138. (Chelone parviscutatus.)
    1872 Y, p. 15.
   Miocene; New Jersey, Maryland.

LEMBONAX Cope.   Type *L. polemicus* Cope.

Cope, E. D. 1869 M, p. 168.
1882 J, p. 989.
Zittel, K. A. 1890 A, p. 524.
The position of this genus is very problematical.

Lembonax insularis Cope.

Cope, E. D. 1872 Y, p. 16.
Eocene; New Jersey.

Lembonax polemicus Cope.
Cope, E. D. 1869 M, p. 168.
Eocene; New Jersey.

Lembonax prophylæus Cope.
Cope, E. D. 1872 Y, p. 15.
Eocene; New Jersey.

ADOCIDÆ.

Cope, E. D. 1870 K.
Baur, G, 1888 B, p. 419.
1891 B, p. 430.
Bienz, A. 1895 A. (On Dermatemydidæ.)
Cope, E. D. 1869 M (1870), p. ii.
1870 N.
1871 C, p. 563.
1871 G, p. 43.
1872 D.

Cope, E. D. 1873 E, p. 649.
1875 X, p. 17.
1882 E, p. 145.
1884 O, pp. 111, 113.
1891 F, p. 645.
Dollo, L. 1886 A, p. 78.
Lydekker, R. 1889 G, p. 129.
Nicholson and Lydekker 1889 A, p. 1106.

ADOCUS Cope.   Type *Emys beatus* Leidy.

Cope, E. D. 1868 K, p. 235.
Baur, G. 1891 B, p. 428.
Cope, E. D. 1869 K, p. 88.
1869 M, pp. 128, 232, and ii.
1870 G.
1870 K.
1873 E, p. 649.
1875 E, pp. 91, 262.
1882 E, p. 145.
1884 O, p. 113.
1891 F, p. 645.
Günther, A. 1888 A, p. 457.
Lydekker, R. 1889 G, p. 129.
Nicholson and Lydekker 1889 A, p. 1106.
Steinmann and Döderlein 1890 A, p. 637.
Zittel, K. A. 1890 A, p. 536.

Adocus agilis Cope.
Cope, E. D. 1868 K, p. 235.
1869 B, p. 734.
1869 M, pp. 233, 234, and ii.
1870 G, p. 296.
1870 N, p. 549.
1871 G, p. 44.
1875 E, p. 262.
Cretaceous (Green-sand No. 5); New Jersey.

Adocus beatus (Leidy).
Leidy, J. 1865 A, pp. 107, 119, pl. xviii, figs. 1–3. (Emys.)
Baur, G. 1891 B, p. 428.
Cope, E. D. 1868 K, p. 235.
1869 B, p. 734.
1869 M, pp. 129, 233.
1870 G, p. 296.

Cope, E. D. 1870 N, p. 547.
1871 G, p. 43.
1875 E, p. 262.
Marsh, O. C. 1890 E, p. 178, pl. vii, fig. 3.   (A. punctatus.)
Cretaceous (Green-sand No. 5); New Jersey.

Adocus pectoralis Cope.
Cope, E. D. 1868 K, p. 236. (Pleurosternum.)
1869 B, p. 734. (Pleurosternum.)
1869 M, p. 130 (Pleurosternum); p. 233 and ii, pl. vii, fig. 1 (Adocus).
1870 G, p. 296.
1871 G, p. 43.
1875 E, p. 262.
Cretaceous (Green-sand No. 5); New Jersey.

Adocus pravus (Leidy).
Leidy, J. 1856 P, p. 303.   (Emys.)
Cope, E. D. 1868 K, p. 235.
1869 B, p. 734.
1869 M, pp. 129, 233, 234.
1870 G, p. 297.
1871 G, p. 44.
1875 E, p. 262.
Leidy, J. 1865 A, pp. 108, 120, pl. xix, fig. 1.   (Emys.)
Maack, G. A. 1869 A, p. 278.   (Emys parva.)
Cretaceous (Green-sand No. 5); New Jersey.

Adocus syntheticus Cope.
Cope, E. D. 1870 N, p. 548.
1870 K.
1871 G, p. 44.
1875 E, p. 262.
Cretaceous (Green-sand No. 5); New Jersey.

AGOMPHUS Cope.   Type *Emys turgidus* Cope.

Cope, E. D. 1871 G, p. 46.
Baur, G. 1888 B, p. 595.
1891 B, p. 429.

Cope, E. D. 1873 E, p. 625.
1877 P.   (Amphiemys, type *A. oxysternum*).
1882 E, p. 145.   (Agomphus, Amphiemys.)

Cope, E. D. 1884 O, p. 113. (Agomphus, Amphiemys.)
  1891 F, p. 645.
Lydekker, R. 1889 G, p. 129.
Nicholson and Lydekker 1889 A, p. 1106.
Zittel, K. A. 1890 A, p. 536. (Agomphus, Amphiemys.)

Agomphus firmus (Leidy).

Leidy, J. 1856 P, p. 303. (Emys.)
Cope, E. D. 1868 K, p. 235. (Adocus.)
  1869 B, p. 734. (Adocus.)
  1869 M, pp. 125, 126. (Emys.)
  1870 G, p. 295. (Adocus.)
  1871 G, p. 46.
  1875 E, p. 262.
Leidy, J. 1865 A, pp. 106, 119, pl. xix, fig. 2. (Emys.)
Maack, G. A. 1869 A, p. 277. (Emys.)
  Cretaceous (Green-sand No. 5); New Jersey.

Agomphus oxysternum Cope.

Cope, E. D. 1877 P. (Amphiemys.)

Baur, G. 1888 B, p. 595.
  1891 B, p. 429.
Cope, E. D. 1878 EE. (Amphiemys.)
  Tertiary?; Georgia.

Agomphus petrosus Cope.

Cope, E. D. 1868 K, p. 236. (Adocus.)
  1869 B, p. 734. (Adocus.)
  1869 M, pp. 125, 126. (Emys.)
  1870 G, p. 295. (Adocus.)
  1871 G, p. 46.
  1873 E, p. 625.
  1875 E, p. 262.
  Cretaceous (Green-sand No. 5); New Jersey.

Agomphus turgidus Cope.

Cope, E. D. 1869 M, pp. 125, 127. (Emys.)
  1871 G, p. 46.
  1875 E, p. 262.
  Cretaceous (Green-sand No. 5); New Jersey.

BASILEMYS gen. nov.   Type *Compsemys variolosus* Cope.

Basilemys imbricarius (Cope).

Cope, E. D. 1876 H, p. 257. (Compsemys.)
  1877 G, p. 573. (Compsemys.)
  Cretaceous (Laramie); Montana.

Basilemys variolosus (Cope).

Cope, E. D. 1876 H, p. 257. (Compsemys.)

Cope, E. D. 1875 E, pp. 91, 261. (C. ogmius.)
  1875 I. (C. ogmius.)
  1875 W, p. 336. (C. ogmius.)
  1877 G, p. 573. (C. variolosus.)
  Cretaceous (Laramie); Saskatchewan, Montana.

ZYGORAMMA Cope.   Type *Z. striatula* Cope.

Cope, E. D. 1870 N, p. 550.
  1871 G, p. 44.
Zittel, K. A. 1890 A, p. 536.

Zygoramma microglypha Cope.

Cope, E. D. 1871 G, p. 44.
  1875 E, p. 263.
  Cretaceous (Green-sand No. 5); New Jersey.

Zygoramma striatula Cope.

Cope, E. D. 1870 N, p. 550.
  1871 G, p. 44.
  1875 E, p. 263.
  Cretaceous (Green-sand No. 5); New Jersey.

BAPTEMYS Leidy.   Type *B. wyomingensis* Leidy.

Leidy, J. 1870 E, p. 5.
Baur, G. 1891 B, p. 429.
Cope, E. D. 1869 M (1870), pp. 233 and ii.
  1871 Q, p. 563.
Leidy, J. 1871 C, p. 367.
Lydekker, R. 1889 G, pp. 129, 143.
Nicholson and Lydekker 1889 A, p. 1106.
Zittel, K. A. 1890 A, p. 536.

Baptemys costilatus (Cope).

Cope, E. D. 1875 C, p. 36. (Dermatemys?.)
  1875 F, p. 96. (Dermatemys.)
  1875 U, p. 1016. (Dermatemys.)
  1877 K, p. 52, pl. xxiv, figs. 17-31. (Dermatemys.)
  1884 O, p. 142. (Dermatemys.)
King, C. 1878 A, p. 877. (Dermatemys.)
  Eocene (Wasatch); N. Mexico.

Baptemys wyomingensis Leidy.

Leidy, J. 1870 E, p. 5.
Baur, G. 1891 B, p. 429.
Cope, E. D. 1869 M (1870), p. 233 (Adocus vyomingensis); p. ii (Baptemys).
  1870 G, p. 297. (Adocus vyomingensis.)
  1873 E, p. 624. (Dermatemys.)
  1882 J, p. 991, fig. 9. (Dermatemys.)
  1884 O, p. 142. (Dermatemys vyomingensis.)
King, C. 1878 A, p. 404.
Leidy, J. 1871 C, p. 367.
  1872 B, p. 367.
  1873 B, pp. 157, 340, pl. xii, pl. xv, fig. 6.
  Eocene (Bridger); Wyoming.

Baptemys sp. indet.

Cope, E. D. 1881 D, p. 184. (Dermatemys.)
  Eocene (Wind River); Wyoming.

HOMOROPHUS Cope.   Type *H. insuetus* Cope.

Cope, E. D. 1870 N, p. 561.
    1882 E, p. 145. (Homorhophus.)
    1884 O, p. 113. (Homorhophus.)
Nicholson and Lydekker 1889 A, p. 1106.
Zittel, K. A. 1890 A, p. 536.

Homorophus insuetus Cope.

Cope, E. D. 1870 N, p. 552.
    1871 G, p. 44.
    1875 E, p. 263.
    Cretaceous (Green-sand No. 5); New Jersey.

CHELYDRIDÆ.

Swainson, 1839, Nat. Hist. and Classif. Fishes, etc.,
    p. 116. (Chelidridæ.)
Agassiz, L. 1857 B, pp. 341, 409. (Chelydroidæ.)
Baur, G. 1889 A, p. 186.
    1889 H.
    1891 B, p. 428.
    1891 C, p. 346.
Bienz, A. 1895 A.
Boulenger, G. A. 1889 A, p. 19.
Cope, E. D. 1871 C, p. 130. (Chelydrinæ.)
    1872 B. (Chelydrinæ.)
    1872 D.
    1882 E, p. 143.
    1884 O, pp. 111, 112.

Dollo, L. 1888 A.
Günther, A. 1888 A, p. 457.
Hay, O. P. 1896 A.
Lydekker, R. 1889 G, p. 133.
Maack, G. A. 1869 A, p. 230.
Meyer, H. 1852 B.
    1856 A.
Nicholson and Lydekker 1889 A, p. 1104.
Rütimeyer, L. 1874 A, p. 16.
Siebenrock, F. 1897 A.
Steinmann and Döderlein 1890 A, p. 635.
Zittel, K. A. 1890 A, p. 532.

CHELYDRA Schweigger.   Type *Testudo serpentina* Linn.

Schweigger, 1814, Prod., p. 23.
Agassiz, L. 1857 B, p. 416.
Baur, G. 1886 G, p. 740.
    1889 B, p. 41.
    1889 D.
    1891 H, p. 58.
    1892 D, p. 207.
Bienz, A. 1895 A.
Boulenger, G. A. 1889 A, p. 20.
Cope, E. D. 1869 M, p. 131.
    1872 D.
    1884 O, P. 112.
    1888 CC, p. 305.
Dollo, L. 1888 A.
Gray, J. E. 1869 B, p. 180.
Hay, O. P. 1896 A.
    1898 F, p. 937.
Hoffmann, C. K. 1890 A.
Lang and Rütimeyer, 1867 A.
Lydekker, R. 1889 G, p. 134.
Meyer, H. 1845 A, p. 12.
    1852 B.
    1856 A.
    1865 C.

Nicholson and Lydekker 1889 A, p. 1104.
Pictet, F. J. 1853 A, p. 453.
Rütimeyer, L. 1874 A, p. 17.
Siebenrock, F. 1898 A, p. 425.
Strauch, A. 1890 A, p. 22.
Wieland, G. R. 1900 B, p. 416, fig. 5.
Winckler, T. C. 1869 A, p. 81.
Zittel, K. A. 1890 A, p. 535.

Chelydra crassa Cope.

Cope, E. D. 1888 CC, p. 306.
    1882 C, p. 461. (Dermatemys sp.; no description.)
    1888 C. (No description.)
    Eocene (Puerco); New Mexico.

Chelydra serpentina (Linn.).

Linnæus, C., 1858, Syst. Nat., ed. x, p. 199.
Agassiz, L. 1857, p. 417, pl.
Boulenger, G. A. 1889 A, p. 20, figs. 3, 4.
Cope, E. D. 1869 A, p. 124.
    Recent: Pleistocene; Maryland.

ACHERONTEMYS Hay.   Type *A. heckmani* Hay.

Hay, O. P. 1899 D, p. 23.

Acherontemys heckmani Hay.

Hay, O. P. 1899 D, p. 23, pl. vi.
    Miocene (Roslyn sandstone); Washington.

ANOSTEIRIDÆ.

Baur, G, 1889 H. (Staurotypidæ or Cinosternidæ,
    in part.)
    1889 I, p. 276. (Pseudotrionychidæ.)

Günther, A. 1888 A, p. 457. (Pseudotrionychidæ,
    in part.)
Lydekker, R. 1889 G, p. 143. (Anosteirinæ.)

ANOSTEIRA Leidy.   Type *A. ornata* Leidy.

Leidy, J. 1871 E, p. 102.
Baur, G. 1889 H, p. 58.
    1889 I, p. 278.
Cope, E. D. 1873 E, pp. 620, 649.   (Anostira.)
    1873 M, p. 278.   (Anostira.)
    1882 E, p. 145.   (Anostira.)
    1882 J, p. 990.   (Anostira.)
    1884 O, pp. 112, 127.   (Anostira.)
Dollo, L. 1886 A, p. 96.   (Anostira.)
Günther, A. 1888 A, p. 457.   (Anostira.)
Leidy, J. 1872 B, p. 270.
    1873 B, p. 174.
Lydekker, R. 1889 G, p. 143.
Nicholson and Lydekker 1889 A, p. 1105.
Zittel, K. A. 1890 A, p. 534.

Anosteira ornata Leidy.

Leidy, J. 1871 E, p. 102.

Cope, E. D. 1873 E, p. 621.
    1882 J, p. 989, fig. 7.
    1884 O, p. 128.
Dollo, L. 1886 A, pl. ii, figs. 7, 8.
King, C. 1878 A, p. 404.
Leidy, J. 1872 B, p. 370.
    1873 B, pp. 174, 341, pl. xvi, figs. 1–6.
Eocene (Bridger); Wyoming.

Anosteira radulina Cope.

Cope, E. D. 1872 V, p. 555.
Baur, G. 1889 I, p. 278.   (A. radulata.)
Cope, E. D. 1873 E, p. 620.
    1884 O, p. 128, pl. xviii, figs. 18, 19.
Eocene (Bridger); Wyoming.

EMYDIDÆ.

Gray, J. E. 1825, Ann. Philos. (2), x.
Agassiz, L. 1857 B, pp. 351, 430.   (Emydoidæ.)
Baur, G. 1891 A, p. 348.
    1892 A, p.159.
    1893 B, p. 675.
Boulenger, G. A. 1889 A, p. 48.   (<Testudinidæ.)
Cope, E. D. 1872 D.
    1882 E, pp. 143, 145.·   (Emydidæ, Cistudin-
        idæ.)
    1884 O, pp. 111, 113.   (Emydidæ, Cistudin-·
        idæ.)
    1899 A, p. 196.   (Terrapenidæ.)
Dollo, L. 1886 A.   (Emydidæ, Cistudinidæ.)

Gray, J. E. 1869 B, p. 175.   (>Cistudinidæ.)
Günther, A 1888 A, p. 457.   (<Testudinidæ.)
Lydekker, R. 1889 G, p. 71.   (<Testudinidæ.)
Mehnert, E. 1890 A.
Nicholson and Lydekker 1889 A, p. 1107.   (<Tes-
    tudinidæ.)
Pictet, F. J. 1853 A, p. 446.   (Emydidea.)
Rosenberg, E. 1891 A.
Rütimeyer, L. 1874 A, pp. 18, 53, 82.
Steinmann and Döderlein 1890 A, p. 637.
Vaillant, L. 1891 A.
Zittel, K. A. 1890 A, p. 637.

EMYS Duméril.   Type *Testudo orbicularis* Linn.

Duméril, 1806, Zool. Anal., p. 76.
Agassiz, L. 1857 B, p. 441.
Baur, G. 1892 A, p. 158.
    1892 C, p. 44 (Terrapene); p. 245 (Emydes).
    1892 D, p. 208.
Boulenger, G. A. 1889 A, p. 111.
Cope, E. D. 1872 M, p. 463.   (Palæotheca, type *P.
    polycypha.*)
    1872 O, p. 474.   (Notomorpha, type *N.
    gravis.*)
    1873 E, p. 625.
    1877 K, p. 53.
    1882 J, p. 992.
    1884 O, pp. 113, 129 (Emys); p. 142 (Noto-
    morpha).
Hoffmann, C. K. 1890 A.
Leidy, J. 1871 C, p. 366.
Lydekker, R. 1889 G, p. 102.
Mehnert, E. 1890 A.
Owen, R. 1842 A, p. 160.
    1849 B.
Pictet, F. J. 1853 A, p. 447.
Pictet and Humbert 1856 A.
Pictet and Jaccard 1860 A, p. 15.
Rosenberg, E. 1891 A.
Strauch, A. 1890 A, p. 14.
Vaillant, L. 1891 A.

Winckler, T. C. 1869 A.
Zittel, K. A. 1890 A, p. 538.

It is not probable that any of the species here
included under *Emys* really belongs to the
genus as it is restricted by modern writers.
The species are more closely related to *Chrys-
emys* and *Clemmys.*

Emys cibollensis Cope.

Cope, E. D. 1877 K, p. 57, pl. xxvii, fig. 4; pl.
    xxviii. figs. 3–6.
    1875 C, p. 36.   (E. stevensoniana.)
    1875 F, p. 96.   (E. stevensonianus.)
    1884 O, p. 131.   (Syn.? of E. euthneta.)
Eocene (Wasatch); New Mexico.

Emys euglypha Leidy.

Leidy, J. 1889 B, p. 97.
    1889 E, p. 27, pl. iv, fig. 1.
    1892 A.
Pliocene (Peace Creek beds); Florida.

Emys euthneta Cope.

Cope, E. D. 1873 E, p. 628.
    1884 O, p. 129, 133, pl. xviii, figs. 34–42.
King, C. 1878 A, p. 376.
Eocene (Wasatch); Wyoming.

Emys haydeni Leidy.

Leidy, J. 1870 S, p. 123.
Cope, E. D. 1884 O, pp. 130, 187.
Leidy, J. 1871 C, p. 366.
    1872 B, p. 367.   (E. wyomingensis, in part.)
    1873 B, pp. 145, 340, pl. ix, fig. 6.   (E. wyomingensis, in part.)
Eocene (Bridger); Wyoming.

Emys latilabiata Cope.

Cope, E. D. 1872 E, p. 471.
    1873 E, pp. 625, 626.
    1884 O, pp. 130, 138.
Eocene (Bridger); Wyoming.

Emys lativertebralis Cope.

Cope, E. D. 1877 K, p. 53, pl. xxvii, figs. 1–3; pl. xxviii, figs. 1, 2.
    1875 C, p. 36.   (E. latilabiatus.)
    1875 F, p. 96.   (E. latilabiatus.)
    1882 J, p. 991, fig. 10.
    1884 O, p. 129.
Eocene (Wasatch); New Mexico.

Emys megaulax Cope.

Cope, E. D. 1873 E, p. 628.
    1873 E, p. 629.   (E. pachylomus.)
    1884 O, p. 132, pl. xviii, figs. 26–33.
King, C. 1878 A, p. 376.   (E. megaulax, E. pachylomus.)
Eocene (Wasatch); Wyoming.

Emys petrolei Leidy.

Leidy, J. 1868 A, p. 176.
Cope, E. D. 1869 M, p. 128.
Leidy, J. 1873 B, pp. 260, 340, pl. ix, fig. 7.
Quaternary; Texas.

Emys polycypha Cope.

Cope, E. D. 1872 M, p. 463.   (Palæotheca.)
    1873 E, pp. 625, 630.
    1884 O, pp. 129, 131, pl. xviii, figs. 20–22.
Eocene (Bridger); Wyoming.

Emys septaria Cope.

Cope, E. D. 1873 E, p. 625.

NOTOMORPHA Cope.

Cope, E. D. 1872 O, p. 474.
    1884 O, p. 142.
Günther, A. 1888 A, p. 457.

Notomorpha gravis Cope.

Cope, E. D. 1872 O, p. 476.

EUPACHEMYS Leidy.

Leidy, J. 1877 A, p. 232.

Eupachemys obtusa Leidy.

Leidy, J. 1877 A, p. 232, pl. xxxiv, figs. 4, 5.

HYBEMYS Leidy.

Leidy, J. 1871 E, p. 103.
    1873 B, p. 174.

Hybemys arenarius Leidy.

Leidy, J. 1871 E, p. 103.

Cope, E. D. 1882 J, p. 992.
    1884 O, pp. 130, 139, pl. xvii, figs. 9–13.
Leidy, J. 1870 E, p. 50.   (E. stevensonianus, in part.)
    1871 B, p. 372.   (E. stevensonianus, in part.)
    1871 C, p. 366.   (E. stevensonianus, in part.)
    1873 B, pp. 140, 340, pl. ix, figs. 2, 3.   (E. vyomingensis, in part.)
Eocene (Bridger); Wyoming.

Emys shaughnessiana Cope.

Cope, E. D. 1884 O, pp. 130, 135, pl. xxiii, figs. 3–8.
    1882 J, p. 992.   (Not adequately defined.)
Eocene (Bridger); Wyoming.

Emys terrestris Cope.

Cope, E. D. 1872 M, p. 463.   (Palæotheca.)
    1873 E, p. 629.
    1884 O, pp. 129, 130, 131, pl. xviii, figs. 23–25.
Eocene (Bridger); Wyoming.

Emys testudinea Cope.

Cope, E. D. 1872 O, p. 475.   (Notomorpha.)
    1873 E, p. 627.
    1877 K, p. 58.
    1884 O, pp. 129, 134, pl. xxiii, figs. 12, 13.
Eocene (Green River); Wyoming.

Emys wyomingensis Leidy.

Leidy, J. 1869 B, p. 66.
Cope, E. D. 1873 E, pp. 625, 626.
    1877 K, p. 53.   (E. vyomingensis.)
    1884 O, pp. 135, pl. xxiii, figs. 9–11.   (E. vyomingensis.)
Leidy J. 1870 E, p. 5.   (E. stevensonianus, in part.)
    1870 S, p. 123.   (E. jeanesi.)
    1871 B, p. 372.   (E. jeanesianus; E. stevensonianus, in part.)
    1871 C, p. 366.   (E. stevensoni, in part; E. jeanesi.)
    1872 B, p. 367.
    1873 B, pp. 140, 340, in part; pl. ix, figs. 4, 5; pl. x, figs. 1, 2.
Osborn, Scott, and Speir 1878 A, p. 95.
Eocene (Bridger); Wyoming.

Type *N. gravis* Cope.

Cope, E. D. 1872 O, p. 477.   (N. garmanii.)
    1873 E, pp. 625, 626.   (Emys.)
    1884 O, p. 143, pl. xxiii, figs. 14–16.
Eocene (Wasatch); Wyoming.

Type *E. obtusa* Leidy.

Leidy, J. 1889 E, p. 29.   ("E. rugosus".)
This species may be identical with *Testudo crassiscutata.*
Phosphate beds; South Carolina.

Type *H. arenarius* Leidy.

King, C. 1878 A, p. 404.
Leidy, J. 1872 B, p. 369.
    1873 B, pp. 174, 340, pl. xv, fig. 9.
Eocene (Bridger); Wyoming.

TRACHEMYS Agassiz.   Type *T. scabra=Testudo scripta* Schoepff.

Agassiz, L. 1857 B, p. 434.
Baur, G. 1892 D, p. 208.
Cope, E. D. 1878 C, p. 228.   (Pseudemys.)
  1878 H, p. 396.   (Pseudemys.)

Trachemys bisornata (Cope).
Cope, E. D. 1878 C, p. 228.   (Pseudemys.)
  Pliocene; Texas.

Trachemys hilli (Cope).
Cope, E. D. 1878 H, p. 395.   (Pseudemys.)
  Miocene (Loup Fork); Kansas.

CLEMMYS Ritgen.   Type *Testudo punctata* Schoepff=*T. guttata* Schneider.

Ritgen, F. A., 1828, Nova Acta Acad. Leop.-Car.,
  xiv, i, p. 272.
Agassiz, L. 1857 B, p. 442 (Nanemys); p. 443 (Cale-
  mys, Glyptemys); p. 444 (Actemys).
Baur, G. 1892 C, p. 44.
Boulenger, G. A. 1889 A, p. 100.
Goette, A. 1899 A.
  For additional synonymy and literature of
  this genus see Boulenger, as cited.

• **Clemmys insculpta** Le C.
Le Conte, J., 1830, Ann. Lyc. New York, iii, p. 112.
  (Testudo.)

Cope, E. D. 1899 A, p. 194.
Boulenger, G. A. 1889 A, p. 107.
  Recent; Eastern U. S.: Pleistocene. (Port
  Kennedy); Pennsylvania.

Clemmys percrassa Cope.
Cope, E. D. 1899 A, p. 194, pl. xviii, figs. 1-1*g*.
  Pleistocene (Port Kennedy); Pennsylvania.

TERRAPENE[1] Merrem.   Type *Testudo carolina* Linn.

Merrem, B. 1820, Tent. Syst. Amphib., p. 27.
Agassiz, L. 1857 B, p. 444.   (Cistudo.)
Baur, G. 1892 A, p. 158.   (Cistudo.)
  1892 C, p. 44 (Cistudo); p. 245 (Terrapene).
Boulenger, G. A. 1889 A, p. 114.   (Cistudo.)
Cope, E. D. 1895 H, p. 757.   (Terrapene; Toxaspis,
  type *T. major*.)
  1899 A, p. 196.   (Toxaspis.)
Fleming, 1822, Phil. Zool., ii, p. 270.   (Cistuda.)
Pictet and Humbert 1856 A, p. 35.   (Cistudo.)
Zittel, K. A. 1890 A, p. 587.   (Cistudo.)

Terrapene anguillulata (Cope).
Cope, E. D. 1899 A, p. 196, pl. xix, fig. 1.   (Toxaspis.)

  Pleistocene (Port Kennedy); Pennsylvania.

Terrapene eurypygia (Cope).
Cope, E. D. 1869 M, p. 124.   (Cistudo.)
  Pleistocene; Maryland.

Terrapene marnochii (Cope).
Cope, E. D. 1878 C, p. 229.   (Cistudo.)
  1885 S.   (Cistudo.)
  1889 F.   (Cistudo.)
  Pleistocene (Equus beds); Texas.

TESTUDINIDÆ.

Gray, J. E., 1825, Ann. Phil. (2), x.
Agassiz, L. 1857 B, pp. 356, 446.   (Testudinina.)
Baur, G. 1889 H.
  1889 J.
  1892 A, p. 159.
  1893 B, p. 677.
Boulenger, G. A. 1889 A, p. 48.   (In part.)
Burckhardt, K. 1895 A.
Cope, E. D. 1882 E, p. 143.
  1884 O, pp. 111, 113.
Dollo, L. 1887 A.

Duméril and Bibron, 1834, Erp. Gén., i, p. 352;
  ii, pp. 1-170.   (Chersites.)
Gray, J. E. 1869 B, p. 166.
  1873 C, p. 722.   (Testudinata.)
Günther, A. 1888 A, p. 457.
Lydekker, R. 1889 G, p. 71.   (In part.)
Nicholson and Lydekker 1889 A, p. 1107.   (In part.)
Siebenrock, F. 1897 A.
Stannius, H. 1856 A.   (Testudinea.)
Steinmann and Döderlein 1890 A, p. 638.
Wieland, G. R. 1900 B, p. 413.
Zittel, K. A. 1890 A, p. 539.   (Chersidæ.)

---

[1] *Terrapene clausa* has been reported by Dr. Leidy (1889 H, p. 6) from a cave in Pennsylvania.
The species accompanying the remains, with few exceptions, yet live in the vicinity.

HADRIANUS Cope. Type *H. octonarius* Cope.

Cope, E. D. 1872 N, p. 468.
Baur, G. 1896 B, p. 568.
Cope, E. D. 1873 E, p. 630.
   1877 K, p. 58.
   1882 E, p. 146.
   1882 J, p. 992.
   1884 O, pp. 113, 140.
Hay, O. P. 1899 D, p. 22.
Leidy, J. 1873 B, p. 137.
Lydekker, R. 1889 G, p. 72.
Zittel, K. A. 1890 A, p. 540. (Syn. of Testudo.)

Hadrianus allabiatus Cope.

Cope, E. D. 1872 E, p. 471.
   1873 E, p. 630.
   1884 O, p. 140, pl. xv, figs. 13-15.
Osborn, Scott, and Speir 1878 A, p. 94.
Eocene (Bridger); Wyoming.

Hadrianus corsoni (Leidy).

Leidy, J. 1871 H. (Testudo.)
Cope, E. D. 1872 M, p. 463. (Testudo hadrianus.)
   1872 N, p. 468. (Hadrianus quadratus.)
   1873 E, p. 631.
   1875 C, p. 36.
   1875 F, p. 96.
   1877 K, p. 58, pl. xxiv, fig. 36.

Cope, E. D. 1882 J, p. 992.
   1884 O, p. 141.
Leidy, J. 1871 B, p. 372. (Emys carteri.)
   1871 K, p. 228. (E. carteri.)
   1872 B, p. 366 (Testudo); p. 367 (E. carteri).
   1872 I, p. 268. (Testudo.)
   1873 B, pp. 132, 339, pl. xl, figs. 1, 2; pl. xv.
   fig. 7; pl. xxix, figs. 2-4; pl. xxx, figs. 1-4.
   (Testudo; on pl. xv, Emys carteri.)
Maack, G. A. 1869 A, p. 278. (Emys carteri.)
Eocene (Bridger); Wyoming: (Wasatch); New Mexico.

Hadrianus octonarius Cope.

Cope, E. D. 1872 N, p. 468.
   1873 E, p. 630.
   1882 J, p. 992, figs. 11-18.
   1884 O, p. 140, pl. xx.
Hay, O. P. 1899 D, p. 23.
Osborn, Scott, and Speir 1878 A, p. 95.
Eocene (Bridger); Wyoming.

Hadrianus schucherti Hay.

Hay, O. P. 1899 D, p. 22, pls. iv, v.
Schuchert, C. 1900 A, p. 328. ·
Eocene (Jackson); Alabama.

STYLEMYS Leidy. Type *S. nebrascensis* Leidy. ·

Leidy, J. 1851 B, p. 172.
Agassiz, L. 1857 B, p. 448. (No name.)
Cope, E. D. 1869 M, p. 123.
   1875 F, p. 74.
   1882 E, p. 145.
   1884 O, pp. 113, 769.
Leidy, J. 1871 C, p. 366.
   1873 B, p. 223.
Lydekker, R. 1889 G, p. 93.
Nicholson and Lydekker 1889 A, p. 1109.
Zittel, K. A. 1890 A, p. 540. (Syn. of Testudo.)

Stylemys nebrascensis Leidy.

Leidy, J. 1851 B, p. 172.
Cope, E. D. 1869 M, p. 124, pl. vii, figs. 10-12. (S. culbertsonii, S. nebrascensis.)
   1874 B, p. 511.
   1884 O, p. 769.
   1891 A, p. 5.
Haberlandt, G. 1876 A. (Testudo culbertsoni.)
Hayden, F. V. 1858 B, p. 158. (Testudo.)
Leidy J. 1851 C, p. 173. (Emys hemispherica, Testudo lata.)
   1851 G, p. 327. (E. oweni.)
   1852 D, p. 34. (E. culbertsonii.)
   1852 E, p. 59. (T. hemispherica, T. nebrascensis, T. oweni, T. culbertsonii.)
   1852 J, p. 567, pl. xii A, figs. 1, 2 (T. nebrascensis); p. 568, pl. xii A, figs. 3, 4 (T. oweni); p. 569, pl. xii (T. culbertsonii); p. 570, pl. xii B, figs. 1, 2 (T. hemispherica): p. 572 (T. lata).

Leidy, J. 1852 L, p. 65. (T. lata, E. hemispherica, E. nebrascensis.)
   1854 A, pp. 103, 115, pl. xxi (T. nebrascensis); pp. 105, 115, pl. xx; pl. xxiv, fig. 3 (T. hemispherica); pp. 106, 115, pl. xxi, pl. xxiv, fig. 4 (T. oweni); pp. 108, 115, pl. xxii, pl. xxiv, fig. 2 (T. culbertsonii): pp. 110, 115, pl. xxiii, pl. xxiv, fig. 1 (T. lata).
   1869 A, p. 26. (Testudo, or Stylemys.)
   1871 C, p. 366..
   1873 B, pp. 224, 339, pl. xix, figs. 7, 9, 10. (Testudo.)
Lydekker, R. 1889 G, p. 94.
Maack, G. A. 1869 A, 245 (T. nebrascensis, hemispherica, oweni); p. 246 (T. lata, culbertsonii).
Marsh, O. C. 1870 D. (Testudo.)
Winckler, T. C. 1869 A, p. 145, pls. xxxi-xxxiii. (T. nebrascensis, culbertsoni, lata, oweni.)
Zittel, K. A. 1890 A, p. 540, fig. 501. (Testudo.)
Oligocene (White River); South Dakota, Colorado.

Stylemys niobrarensis Leidy.

Leidy, J. 1858 E, p. 29. [Testudo (Stylemys.)]
Cope, E. D. 1869 M, p. 124.
   1874 B, p. 531.
Hayden, F. V. 1858 B, p. 158. (Testudo.)
Leidy, J. 1869 A, p. 26. (Testudo, or Stylemys.)
   1871 C, p. 366.
   1873 B, pp. 225, 340, pl. iii, figs. 4-6; pl. xix, figs. 6-8. (Testudo.)

Maack, G A. 1869 A. p. 222.   (Testudo.)
Miocene; South Dakota.

Stylemys oregonensis Leidy.

Leidy, J. 1871 M, p. 248.

Cope, E. D. 1879 B, p. 55.
        1884 O, p. 770.   (Syn. of S. nebrascensis.)
Leidy, J. 1873 B, pp. 226, 340, pl. xv, fig. 10.   (Testudo.)
Miocene (John Day); Oregon.

TESTUDO Linn.   Type *T. græca* Linn.

Linnæus, C., Syst. Nat., ed. x, 1858, p. 197.   (In part.)
Agassiz, L. 1857 B, p. 446.   (Testudo, Xerobates.)
Baur, G. 1889 J.
        1892 A.
Bienz, A. 1895 A, p. 99.
Boulenger, G. A. 1889 A, p. 149.
Cope, E. D. 1874 B, p. 511.
        1878 H, p. 393.   (Testudo, Xerobates.)
        1882 E, p. 146.
        1884 O, pp. 113, 762.   (Testudo, Xerobates.)
        1886 N.   (Caryoderma, type *C. snovianum*.)
        1889 P, p. 662.   (Caryoderma.)
Flower and Lydekker 1891 A, p. 204.   (Caryoderma.)
Fürbringer, M. 1900 A, p. 320.
Gray, J. E. 1869 B, p. 167.
        1873 C, p. 723.   (Xerobates.)
Günther, A. 1877 A.
Hoffmann, C. K. 1890 A.
Leidy, J. 1852 J, p. 566.
        1854 A, pp. 101; 115.
Lydekker, R. 1889 G, p. 71.
Pictet, F. J. 1853 A, p. 443.
Pictet and Humbert 1856 A, p. 17.
Siebenrock, F, 1897 A.
        1898 A.
Strauch, A. 1890 A, p. 11.
Wieland, G. R. 1900 B, p. 415, figs. 1–4.
Zittel, K. A. 1890 A, p. 540.   (Testudo.)
        1898 B, p. 147.   (Caryoderma.)

Testudo amphithorax Cope.

Cope, E. D. 1873 S. p. 6.
        1873 C C, p. 19.
        1874 B, p. 511.
        1884 O, pp. 762, 767, pl. lxi, fig. 4.
Oligocene (White River); Colorado.

Testudo brontops Marsh.

Marsh, O. C. 1890 E, p. 179, pl. viii.
Dana, J. D. 1896 A, p. 901, fig. 1516.
Marsh, O. C. 1897 C, pp. 523, 527, figs. 95, 96.
Oligocene (White River); South Dakota, Colorado.

Testudo crassiscutata Leidy.

Leidy, J. 1889 E, p. 31, pl. vi, figs. 4–7.
        1892 A.
Pliocene; Florida.
This species may be identical with *Eupachemys obtusa*.

Testudo cultrata Cope.

Cope, E. D. 1873 S, p. 6.
        1873 C C, p. 19.
        1874 B, p. 511.
        1884 O, pp. 762, 763, pl. lxiii, figs. 1–3.
        1892 E, p. 256.
Oligocene (White River); Colorado.

Testudo cyclopygia (Cope).

Cope, E. D. 1878 H, p. 394.   (Xerobates.)
Hay, O. P. 1899 G, p. 349.
Miocene (Loup Fork); Kansas.

Testudo gilbertii Hay.

Hay, O. P. 1899 G, p. 349.
Gilbert, J. Z. 1896 A, p. 148, figs. 1–4.   (Xerobates undata.)
Miocene (Loup Fork); Kansas.

Testudo hexagonata Cope.

Cope, E. D. 1893 A, p. 77, pl. xxii, fig. 2.
Pleistocene (Equus beds); Texas.

Testudo klettiana Cope.

Cope, E. D. 1875 F, p. 75.
        1877 K, p. 286, pl. lxvii, fig. 3.
Miocene (Loup Fork); New Mexico.

Testudo laticaudata Cope.

Cope, E. D. 1893 A, p. 75, pl. xxii, fig. 1.
Pleistocene (Equus beds); Texas.

Testudo laticunea Cope.

Cope, E. D. 1873 S, p. 6.
        1873 CC, p. 19.
        1874 B, p. 511.
        1884 O, pp. 762, 765, pl. lxi, fig. 1.
Oligocene (White River); Colorado.

Testudo ligonia Cope.

Cope, E. D. 1873 S, p. 6.
        1873 CC, p. 19.
        1874 B, p. 511.
        1884 O, pp. 762, 766, pl, lxi, figs. 2, 3.
Oligocene (White River); Colorado.

Testudo orthopygia (Cope).

Cope, E. D. 1878 H, p. 393.   (Xerobates.)
Hay, O. P. 1899 G, p. 349.
Miocene (Loup Fork); Kansas.

Testudo pertenuis Cope.

Cope, E. D. 1892 G, p. 226.
        1893 A, pp. 47, 49, fig. 1.
Pliocene (Blanco); Texas.

Testudo quadrata.

Cope, E. D. 1884 O, pp. 762, 764, pl. lxi, fig. 5.
Oligocene (White River); Colorado.

Testudo snoviana (Cope).

Cope, E. D. 1886 N.   (Caryoderma.)
        1889 P, p. 662, pl. xxxii.   (Caryoderma.)
Trouessart, E. L. 1898 A, p. 1136.   (Caryoderma.)
Williston, S. W. 1898 G, p. 132.   (Testudo undata?)
Miocene (Loup Fork; Kansas.

452 FOSSIL VERTEBRATA OF NORTH AMERICA. [BULL. 179.

Testudo turgida Cope.

Cope, E. D. 1892 J, p. 127.
 1892 E, p. 255.
 1892 P.
 1893 A, p. 47.
Pliocene (Blanco); Texas.

Testudo undata Cope.

Cope, E. D. 1875 F, p. 74.

Cope, E. D. 1877 K, p. 288, pl. lxvii, figs. 1, 2.
Hay, O. P. 1899 G, p. 349.
Miocene (Loup Fork); New Mexico.

Testudo sp. indet.

Leidy, J. 1860 B, p. 122, pl. xxviii, figs. 1–4.
Pleistocene; South Carolina.

Suborder TRIONYCHIA.

Unless otherwise indicated, the following authors refer to this group under the name *Trionychoidea.*
Baur, G. 1888 B, p. 786.
 1889 C, p. 240. (Chilotæ.)
 1890 D, p. 536. (Trionychia, Chilotæ.)
 1891 B, pp. 418, 420. (Trionychia.)
Boulenger, G. A. 1889 A, pp. 5, 237.
Cope, E. D. 1891 N, p. 39.
 1898 B, p. 62.
Duméril and Bibron, 1834, Érp. Gén., i, p. 353; ii, p. 461. (Potamites.)
Gadow, H. 1898 A, p. 91. (Trionychoidea.)

Gray, J. E. 1872 A.
Günther, A. 1888 A, p. 457.
Haeckel, E. 1895 A, p. 326. (Diacostalia.)
Lydekker, R. 1889 G, p. 3.
Nicholson and Lydekker 1889 A, p. 1116.
Pictet, F. J. 1853 A, p. 455. (Trionychidea.)
Rütimeyer, L. 1874 A, pp. 13, 51.
Seeley, H. G. 1880 B, p. 412. (Peltochelyidæ.)
Siebenrock, F. 1897 A. (Trionychoidæ.)
Stannius, H. 1856 A. (Trionychoidea.)
Steinmann and Döderlein 1890 A, p. 634.
Vaillant, L. 1894 A. (Mecraspedota.)
Zittel, K. A. 1890 A, p. 513. (Trionychia.)

PLASTOMENIDÆ nom. nov.

Trionychia without fontanelles in the plastron behind the anterior border of the hyoplastron.

PLASTOMENUS Cope. Type *P. thomasii* Cope.

Cope, E. D. 1873 M.
 1873 E, p. 617.
 1875 C, p. 35.
 1875 F, p. 92.
 1875 I.
 1875 W, p. 336.
 1877 K, p. 47.
 1882 E, p. 144.
 1884 O, pp. 112, 122.
Dollo, L. 1886 A, p. 94.
Lydekker, R. 1889 G, p. 7. (Syn. of Trionyx.)
Nicholson and Lydekker 1889 A, p. 1118.
Zittel, K. A. 1890 A, p. 517.

Plastomenus communis Cope.

Cope, E. D. 1875 C, p. 35.
 1875 F, p. 96.
 1877 K, pp. 48, 50, pl. xxv, figs. 1–6. (Except "var. ii," p. 50, which belongs to P. molopinus.)
 1882 C, p. 461.
 1884 O, pp. 123, 126.
 1888 C C, p. 301. (P. communis?)
King C. 1878 A, p 377.
Eocene (Puerco? and Wasatch); New Mexico.

Plastomenus corrugatus Cope.

Cope, E. D. 1875 C, p. 35.
 1875 F, p. 96.
 1875 W, p. 337.
 1877 K, p. 48, pl. xxv, figs. 20–26.
 1884 O, p. 123.
King, C. 1878 A, p. 377.
Eocene (Wasatch); New Mexico.

Plastomenus costatus Cope.

Cope, E. D. 1875 E, pp. 94, 261, pl. viii, fig. 8.
 1875 I.
 1875 W, p. 337.
Cretaceous (Laramie); Saskatchewan.

Plastomenus fractus Cope.

Cope, E. D. 1875 C, p. 35.
 1875 F, p. 96.
 1877 K, p. 49, pl. xxv, figs. 12–19.
 1884 O, p. 123.
Eocene (Wasatch); New Mexico.

Plastomenus insignis Cope.

Cope, E. D. 1874 B, p. 454.
 1874 C, p. 29.
 1875 E, pp. 95, 261, pl. vi, fig. 10.
Cretaceous (Laramie); Colorado.

Plastomenus lachrymalis Cope.

Cope, E. D. 1874 O, p. 15.
 1875 C, p. 35.
 1875 F, p. 96.
 1877 K, p. 48, 51, pl. xxv, fig. 7.
 1884 O, p. 123.
Eocene (Wasatch); New Mexico.

Plastomenus molopinus Cope.

Cope, E. D. 1872 M, p. 461. (Anosteira.)
 1873 E, p. 620.
 1873 M, p. 279.
 1877 K, p. 50, pl. xxv, figs. 5, 6. (Var. ii of P. communis.)
 1884 O, pp. 123, 125, pl. xviii, figs. 9–14.

? Leidy, J. 1873 B, p. 180, pl. xvi, fig. 12.
("Trionyx.")
Eocene (Bridger); Wyoming: (Wasatch);
New Mexico.

Plastomenus œdemius Cope.

Cope, E. D. 1872 M, p. 461. (Anostira.)
1873 E, p. 619.
1873 M.
1877 K, p. 48.
1884 O. pp. 123, 126, pl. xviii, figs. 15-17.
Eocene (Bridger); Wyoming.

Plastomenus punctulatus Cope.

Cope, E. D. 1874 B, p. 453.
1874 C, p. 29.
1875 E, pp. 94, 261, pl. vi, fig. 9.
1877 G, p. 573.
Cretaceous (Laramie); Colorado, South Dakota, Montana.

Plastomenus serialis Cope.

Cope, E. D. 1877 K, pp. 48, 51, pl. xxv, figs. 8-10.
1875 C, p. 35. (P. ? thomasii.)

Cope, E. D. 1884 O, p. 123.
Eocene (Wasatch); New Mexico.

Plastomenus thomasii Cope.

Cope, E. D. 1872 M, p. 462. (Trionyx.)
1873 E, p. 618 (P. thomasii); p. 619 (P. multifoveatus).
1873 M, p. 278 (P. thomasii); p. 279 (P. multifoveatus).
1875 C, p. 35. (P. catenatus.)
1875 F, p. 96. (P. catenatus, P. thomasii.)
1877 K, pp. 49, 51 (P. thomasii); p. 49, pl.
xxv, fig. 11 (P. multifoveatus).
1884 O, pp. 123, 125 (P. thomasii); pp. 123,
124, pl. xviii, figs. 2-8 (P. multifoveatus).
Eocene (Bridger); Wyoming: (Wasatch); New
Mexico.

Plastomenus trionychoides Cope.

Cope, E. D. 1872 M, p. 461. (Anostira.)
1873 E, p. 619.
1'73 M, p. 2'9.
1884 O, p. 123, pl. xviii, fig. 1.
Eocene (Bridger); Wyoming.

TRIONYCHIDÆ.

Agassiz, L. 1857 B, pp. 329, 394.
Baur, G. 1887 B, p. 96.
1888 B, p. 736.
1889 C.
1889 H.
1889 I, p. 273.
1891 C, p. 351.
1893 A, p. 213.
Boulenger, G. A. 1889 A, p. 241.
Cope, E. D. 1869 M, p. 150.
1882 E, p. 143.
1882 J, p. 989.
1884 O, pp. 111, 112.

Cope, E. D. 1889 R, p. 865.
Dollo, L, 1884 A.
1886 A.
1888 A.
Gray, J. E. 1873 C, p. 38.
Günther, A. 1888 A, p. 457.
Hoffmann, C. K. 1890 A.
Lydekker, R. 1889 G, p. 4.
Negri, A. 1892 A.
Siebenrock, F. 1897 A.
1898 A.
Steinmann and Döderlein 1890 A, p. 634.
Strauch, A. 1890 A, p. 33. (Trionychida.)

TRIONYX Geoffroy.   Type *Testudo punctata* Lacépède.

St. Hilaire, G. 1809 A, p. 1.
Baur, G. 1893 A, p. 214.
Boulenger, G. A. 1889 A, p. 242.
Cope, E. D. 1869 M, p. 151.
1873 E, p. 649.
1875 C, p. 34.
1877 K, p. 43.
1884 O, pp. 112, 117.
Fürbringer, M, 1900 A. p. 320.
Gray, J. E. 1869 B, p. 212.
1873 A.
Hoffmann, C. K. 1890 A.
Leidy, J. 1871 C, p. 367.
Lydekker, R. 1889 G, p. 7.
Maack, G. A. 1869 A.
Owen, R. 1842 A, p. 168.
1849 B.
1849 C.
Pictet, F. J. 1853 A, p. 456.
Pictet and Humbert 1856 A.
Seeley, H. G. 1880 B, p. 410.
Siebenrock, F. 1897 A.
1898 A.
Wagler, J., 1830, Nat. Syst. Amphib., p. 134.

Winckler, T. C. 1869 A.
Zittel, K. A. 1890 A. p. 515.
It is highly probable that none of the species
here recorded under *Trionyx* really belongs
under the genus as understood in recent works
on living turtles, especially as the genus was
restricted by Wagler in 1830. See Dr. Baur as
above cited.

Trionyx buiei Cope.

Cope, E. D. 1869 M, pp. 151, 153.
1875 V. p. 34.
Miocene; North Carolina.

Trionyx cariosus Cope.

Cope, E. D. 1875 C, p. 35.
1875 F, p. 95.
1877 K, p. 44, pl. xxvi, figs. 5-10.
1884 O, p. 118.
Eocene (Wasatch); New Mexico.

Trionyx cellulosus Cope.

Cope, E. D. 1867 C, p. 142.
Miocene; Maryland.

Trionyx coalescens (Cope).

Cope, E. D. 1875 E, pp. 93, 261, pl. viii, figs. 6, 7.
(Plastomenus.)
1875 I. (Plastomenus.)
Lambe, L. M. 1899 A, pp. 68, 70. (Plastomenus.)
Cretaceous (Laramie); Saskatchewan.

Trionyx concentricus Cope.

Cope, E. D. 1872 M, p. 461.
1873 E, p. 617.
1884 O, pp. 118, 120, pl. xvi, figs. 3–6.
Eocene (Bridger); Wyoming.

Trionyx foveatus Leidy.

Leidy, J. 1856 F, p. 73.
Baur, G. 1891 B, p. 418.
Cope, E. D. 1869 M, p. 152.
1874 C, p. 29.
1875 E, p. 260.
1877 G, p. 573.
Leidy, J. 1856 C, p. 120.
1856 Q, p. 312.
1860 A, p. 148, pl. xi, figs. 1–3.
Marsh, O. C. 1897 C, p. 527.
Cretaceous (Laramie); Montana, Colorado.

Trionyx guttatus Leidy.

Leidy, J. 1869 B, p. 66.
Cope, E. D. 1869 M, pp. 151, 152.
1873 E, p. 617.
1875 C, p. 35. (T. uintaensis.)
1875 F, p. 96. (T. uintaensis.)
1877 K, p. 46.
1884 O, p. 119.
King, C. 1878 A, p. 404.
Leidy, J. 1871 C p. 367.
1872 B, p. 370.
1873 B, pp. 176, 342, pl. ix, fig. 1.
Eocene (Bridger); Wyoming; (Wasatch); New
Mexico.

Trionyx halophilus Cope.

Cope, E. D. 1869 G, p. 12.
1869 B, p. 735.
1869 M, p. 151, pl. vii, fig. 15.
1875 E, p. 261.
Cretaceous (Green-sand No. 4); Delaware,
New Jersey, Virginia.

Trionyx heteroglyptus Cope.

Cope, E. D. 1873 M.
1873 E, p. 616.
1884 O, pp. 118, 120, pl. xvi, fig. 2.
Eocene (Bridger); Wyoming.

Trionyx leptomitus Cope.

Cope, E. D. 1875 C, p. 35.
1875 F, p. 96.
1877 K, p. 44, pl. xxv, figs. 27–31; pl. xxvi,
figs. 1–4.
1884 O, p. 118.
King, C. 1878 A, p. 377.
Eocene (Wasatch); New Mexico.

Trionyx leucopotamicus Cope.

Cope, E. D. 1891 A, p. 5, pl. i, figs. 8, 9.

Ami, H. M. 1891 A.
Oligocene (White River); N. W. Terr. of British America, North Dakota.

Trionyx lima Cope.

Cope, E. D. 1869 G, p. 12.
1869 K, p. 90.
1869 M, pp. 151, 153, pl. vii, fig. 14.
1875 H, p. 363.
Miocene; New Jersey.

Trionyx mammilaris Cope.

Cope, E. D. 1877 G, p. 573. (No description.)
Cretaceous (Laramie); Montana.

Trionyx pennatus Cope.

Cope, E. D. 1869 G, p. 12.
1869 B, p. 735.
1869 M, p. 152, pl. vii, fig. 13.
Eocene (Green-sand); New Jersey.

Trionyx priscus Leidy.

Leidy, J. 1851 H, p. 329.
Cope, E. D. 1869 B, p. 735.
1869 G, p. 12.
1869 M, pp. 151, 153.
1875 E, p. 260.
Leidy, J. 1865 A, pp. 113, 120, pl. xviii, fig. 9.
Maack, G. A. 1869 A, p. 281.
Cretaceous (Green-sand No. 4); New Jersey.

Trionyx punctiger Cope.

Cope, E. D. 1891 A, p. 5.
1883 F, p. 217 ("Trionyx sp. 2").
Oligocene (White River); South Dakota.

Trionyx radulus Cope.

Cope, E. D. 1875 C, p. 35.
1875 F, p. 96.
1877 K, p. 45, pl. xxvi, figs. 11–16.
1884 O, pp. 118, 119.
King, C. 1878 A, p. 377.
Eocene (Wasatch); New Mexico; (Bridger);
Wyoming.

Trionyx scutumantiquum Cope.

Cope, E. D. 1873 E, p. 617.
1882 J, p. 988, fig. 6.
1884 O, pp. 118, 121, pl. xvi, figs. 1, 1a.
King, C. 1878 A, p. 376.
Eocene (Wasatch); Wyoming.

Trionyx uintaënsis Leidy.

Leidy, J. 1872 I, p. 267.
Cope, E. D. 1884 O, p. 118.
1891 A, p. 5.
Leidy, J. 1873 B, pp. 178, 342, pl. xxix, fig. 1.
Eocene (Bridger); Wyoming.

Trionyx vagans Cope.

Cope, E. D. 1874 B, p. 453.
1874 C, p. 29.
1875 E, pp. 96, 260, pl. vi, figs. 13, 14.
1875 I, p. 1.
1877 G, p. 573.
Leidy, J. 1856 Q, p. 312. (T. foveatus.)
Cretaceous (Laramie); Colorado, Montana,
Saskatchewan.

Trionyx ventricosus Cope.

Cope, E. D. 1877 K, p. 45.
　　　1884 O, p. 118.
　　　Eocene (Wasatch); New Mexico.

Trionyx virginianus Clark.

Clark, W. B. 1895 A, p. 4.
　　　1897 A, p. 59, pl. viii, figs. 1a, 1b.
　　　Eocene; Virginia.

Trionyx sp. indet.
　　　Unnamed specimens of Trionyx have been
　　　reported as follows: Leidy, J. 1873 B, pp. 180,
　　　342, pl. xvi, figs. 11, 12; Eocene, Wyoming.
　　　Leidy, J. 1877 A, p. 233, pl. xxxiv, fig. 3; Phos-
　　　phate beds, South Carolina. Cope, E. D. 1867
　　　C, p. 143; Miocene, Maryland. Cope, E. D. 1888
　　　CC, p. 301; Puerco beds, New Mexico.

AXESTEMYS Hay.　Type *Axestus byssinus* Cope.

Hay, O. P. 1899 G, p. 348.
　　　The following authors call this genus *Axes-
　　tus;* but this name is preoccupied in ento-
　　mology.
Cope, E. D. 1872 M, p. 462.
　　　1873 E, p. 615.
　　　1882 E, p. 144.
　　　1884 O, pp. 112, 116.
Lydekker, R. 1889 G, p. 7.
Nicholson and Lyddeker 1889 A, p. 1118.

Zittel, K. A. 1890 A, p. 517.

Axestemys byssinus (Cope).

Cope, E. D. 1872 M, p. 462. (Axestus byssinus;
　　　misprint.)
　　　1873 E, p. 616. (Axestus.)
　　　1884 O, p. 116, pl. xv, figs. 1–12. (Axestus.)
Hay, O. P. 1899 G, p. 348.
　　　Eocene (Bridger); Wyoming.

Order PLESIOSAURIA.

Andrews, C. W. 1895 C.
　　　1895 D. (Sauropterygia.)
　　　1895 E.
　　　1896 A.
Baur, G. 1886 F. (Sauropterygia.)
　　　1887 A, p. 485. (Sauropterygia.)
　　　1887 C. (Sauropterygia.)
　　　1890 B.
Cope, E. D. 1869 M, p. 31. (Sauropterygia.)
　　　1871 B, p. 235. (Sauropterygia.)
　　　1872 F, p. 327. (Sauropterygia.)
　　　1875 E, p. 69. (Sauropterygia.)
　　　1879 I. (Sauropterygia.)
　　　1885 G, p. 246. (Sauropterygia.)
　　　1887 G, p. 564. (Polycotylinæ.)
　　　1887 S, p. 334. (Sauropterygia.)
　　　1888 S.
　　　1889 R, p. 863.
　　　1891 N, p. 39.
　　　1892 R. (Sauropterygia.)
　　　1892 Z, p. 21. (Sauropterygia.)
　　　1896 B, pp. 55, 66.
　　　1900 A, p. 159.
Dames, W. 1895 A.
Deecke, W. 1895 A. (Nothosaurus.)
Fürbringer, M. 1900 A, pp. 321, 527, 635. (Sauro-
　　　pterygia.)
Gadow, H. 1898 A, p. 23. (Plesiosauri.)

Gegenbaur, C. 1898 A.
Geissler, G. 1895 A. (Nothosaurus.)
Günther, A. 1886 A, p. 444. (Sauropterygia.)
Gürich, G. 1891 B. (Nothosaurus.)
Haeckel, E. 1895 A, p. 335. (Sauropterygia.)
Hay, O. P. 1899 F, p. 682.
Hulke, J. W. 1892 A. (Sauropterygia.)
Huxley, T. H. 1872 A, p. 180.
Koken, E. 1893 A.
Kükenthal, W. 1890 C, p. 176.
Lydekker, R. 1889 F, p. 118. (Sauropterygia.)
Meyer, H. 1855 A. (Nothosaurus.)
Nicholson and Lydekker 1889 A, p. 1067. (Sauro-
　　　pterygia.)
Owen, R. 1840 E.
　　　1847 E.
　　　1860 A, p. 159. (Sauropterygia.)
　　　1860 E, p. 211. (Sauropterygia.)
　　　1866 A, p. 16. (Sauropterygia.)
　　　1883 A. (Sauropterygia.)
Quenstedt, F. A. 1852 A, p. 130. (Plesiosauri.)
　　　1885 A, p. 208. (Plesiosauri.)
Seeley, H. G. 1874 A.
　　　1892 B, p. 121. (Sauropterygia.)
Steinmann and Döderlein 1890 A, p. 625. (Sauro-
　　　pterygia.)
Woodward, A. S. 1898 B, p. 159. (Sauropterygia.)
Zittel, K. A. 1890 A, p. 473. (Sauropterygia.)

PLESIOSAURIDÆ.

Fürbringer, M. 1900 A.
Hoffmann, C. K. 1890 B.
Hulke, J. W. 1892 A.
Huxley, T. H. 1858 B.
Lydekker, R. 1889 F, p. 120.
Nicholson and Lydekker 1889 A, p. 1074.

Owen, R. 1866 A.
Sauvage, H. E. 1878 A.
Seeley, H. G. 1892 B.
Steinmann and Döderlein 1890 A, p. 628.
Zittel, K. A. 1890 A, p. 486.

PLESIOSAURUS Conybeare.   Type *P. dolichodeirus* Conybeare.

Conybeare, W. D., 1821, Trans. Geol. Soc., v, pt. 2,
 p. 560.
Andrews, C. W. 1896 B.
  1895 C.
  1896 A.
Barrett, L. 1859 A.
Buckland, W. 1837 A, i, p. 157; ii, pls. xvi-xix.
Conybeare, W. D. 1824 A, p. 119.
Cope, E. D. 1870 D, p. 274.   (Taphrosaurus, type *P. lockwoodii.*)
  1875 E, pp. 88, 256.
  1879 A, p. 84.
Cuvier, G. 1834 A, x, pp. 445-470, pls. cclix, cclx.
Dames, W. 1895 A.
Fürbringer, M. 1900 A, p. 326.
Hulke, J. W. 1883 A.
  1892 A.
Huxley, T. H. 1872 A, p. 182.
Kiprijanoff, W. 1882 A.
Lydekker, R. 1889 F, pp. viii, 251.
Marsh, O. C. 1877 E.
Newberry, J. S. 1875 A, p. 18.
Owen, R. 1840 A, p. 49.
  1840 E.
  1845 B, p. 275, pls. lxxiii, lxxiv.
  1851 B, p. 58.
  1860 A, p. 160.
  1860 E, p. 233.
  1864 B, p. 1.
  1865 A, p. 1.
  1883 A.
Pictet, F. J. 1853 A, p. 534.
Quenstedt, F. A. 1852 A, p. 130.
  1885 A, p. 208.
Sauvage, H. E. 1878 A.
Seeley, H. G. 1874 A.
  1887 C.
  1888 A.
  1892 B.
  1896 A.

Sollas, W. J. 1882 A.
Steinmann and Döderlein 1890 A, p. 629.
Williston S. W. 1893 A.
  1894 A.
Winckler, T. C. 1873 A.
Woodward, A. S. 1888 I, p. 277.
  1898 B, p. 165.
Zittel, K. A. 1890 A, p. 486.

Plesiosaurus brevifemur Cope.

Cope, E. D. 1875 E, p. 256.
  1869 M, p. 41, figs. 13-15; p. 43, in part.
    (Cimoliasaurus magnus.)
Cretaceous (Green-sand No. 5); New Jersey.

Plesiosaurus gouldii Willist.

Williston, S. W. 1897 C.
Cretaceous (Comanche); Kansas.

Plesiosaurus gulo Cope.

Cope, E. D. 1872 BB.
  1872 GG.
  1874 C, p. 28.
  1875 E, pp. 52, 88, 256.
Cretaceous (Niobrara); Kansas.

Plesiosaurus lockwoodii Cope.

Cope, E. D. 1869 M, p. 40.
  1870 D, p. 274.   (Taphrosaurus.)
  1875 E, p. 256.   (Plesiosaurus.)
Cretaceous (Clay No. 1); New Jersey.

Plesiosaurus mudgei Cragin.

Cragin, F. W. 1894 A, p. 69, pl. i, figs. 1-3, 4?.
Cretaceous (Neocomian); Kansas.

Plesiosaurus shirleyensis Knight.

Knight, W. C. 1900 A, p. 115, figs. A and C.
Jurassic (Shirley); Wyoming.

PANTOSAURUS Marsh.   Type *Parasaurus striatus* Marsh.

Marsh, O. C. 1893 J, p. 159.   (To replace Parasaurus
 Marsh, preoccupied.)
  1891 F, p. 338.   (Parasaurus; no definition.)
  1895 A, p. 406.
Pantosaurus striatus Marsh.
Marsh, O. C. 1893 J, p. 159.

Marsh, O. C. 1891 F, p. 338.   (Parasaurus; no
 description.)
  1895 A, p. 406, fig. 3.
  1897 C, p. 485, fig. 32.
Jurassic; Wyoming.

PIRATOSAURUS Leidy.   Type *P. plicatus* Leidy.

Leidy, J. 1865 A, p. 29.
Zittel, K. A. 1890 A, p. 498.
  A genus of uncertain affinities.
Piratosaurus plicatus Leidy.
Leidy, J. 1865 A, pp. 29, 116, pl. xix, fig. 8.

Cope, E. D. 1869 M, p. 56.
  1875 E, p. 254.
Leidy, J. 1865 D, p. 68.
Cretaceous; Minnesota.

NOTHOSAUROPS Leidy.   Type *N. occiduus* Leidy.

Leidy, J. 1870 U.
Cope, E. D. 1874 C, p. 28.   (Syn. of Plesiosaurus.)
    1875 E, p. 256.   (Syn. of Plesiosaurus.)
Leidy, J. 1873 B, p. 287.   (Nothosaurus.)
Zittel, K. A. 1890 A, p. 600.   (Syn. ? of Champso-
    saurus.)
    A genus of uncertain position.

Nothosaurops occiduus Leidy.
Leidy, J. 1870 U.
Cope, E. D. 1874 C, p. 28.   (Plesiosaurus.)
    1875 E, p. 256.   (Plesiosaurus.)
Leidy, J. 1873 B, pp. 287, 345, pl. xv, figs. 11-13.
    (Nothosaurus.)
    Cretaceous (Laramie); S. Dakota.

MEGALNEUSAURUS Knight.   Type *Cimoliosaurus rex* Knight.

Knight, W. C. 1898 B, p. 378.

Megalneusaurus rex Knight.
Knight, W. C. 1895 A.   (Cimoliosaurus.)
    1898 B, p. 379, figs. 1-3.
    Jurassic; Wyoming.

ELASMOSAURIDÆ

Cope, E. D. 1869 M, p. 47.
    1875 E, pp. 69, 78.

Fürbringer, M. 1900 A, pp. 322, 527.
Seeley, H. G. 1892 B, pp. 134, 151.

POLYCOTYLUS Cope.   Type *P. latipinnis* Cope.

Cope, E. D. 1869 D.
    1869 M, p. 34.
    1871 D, p. 386.
    1872 F, p. 335.
    1875 E, p. 70.
    1887 G, p. 564.
    1888 S.
Lydekker, R. 1889 C, p. 52.   (Syn. of Cimolia-
    saurus.)
Nicholson and Lydekker 1889 A, p. 1077.   (Syn. of
    Cimoliasaurus.)
Sauvage, H. E. 1876 A.
    1878 A, p. 35.

Seeley, H. G. 1892 B, p. 135.
Zittel, K. A. 1890 A, p. 491.   (Syn. of Cimolia-
    saurus.)

Polycotylus latipinnis Cope.
Cope, E. D. 1869 D.
    1869 M, p. 36, pl. i, figs. 1-12.
    1871 D, p. 388.
    1872 F, pp. 320, 335.
    1874 C, p. 27.
    1875 E, pp. 45, 72, 255, pl. vii, figs. 7, 7a
Leidy, J. 1873 B, p. 279.
    Cretaceous (Niobrara); Kansas, Nebraska

PIPTOMERUS Cope.   Type *P. megaloporus* Cope.

Cope, E. D. 1887 G, p. 564.
Lydekker, R. 1889 C, p. 182.
Seeley, H. G. 1892 B, pp. 136, 138.
Zittel, K. A. 1890 A, p. 495.

Piptomerus hexagonus Cope.
Cope, E. D. 1887 G, p. 565.
    Cretaceous (Fox Hills); N. Mexico.

Piptomerus megaloporus Cope.
Cope, E. D. 1887 G, p. 564.
    Cretaceous (Fox Hills); N. Mexico.

Piptomerus microporus Cope.
Cope, E. D. 1887 G, p. 565.
    Cretaceous (Fox Hills); N. Mexico.

URONAUTES Cope.   Type *U. cetiformis* Cope.

Cope, E. D. 1876 I, p. 345.
Lydekker, R. 1889 F, p. 182.
Seeley, H. G. 1892 B, p. 136.
Zittel, K. A. 1890 A, p. 498.

Uronautes cetiformis Cope.
Cope, E. D. 1876 I, p. 346.
    Cretaceous (Fox Hills); Montana.

Uronautes sp. indet.
Cope, E. D. 1887 G, p. 566.
    Cretaceous (Fox Hills); New Mexico.

TRINACROMERUM Cragin.   Type *T. bentonianum* Cragin.

Cragin, F. W. 1888 A.
Cope, E. D. 1894 B, p. 111.
Cragin, F. W. 1891 A.
Lydekker, R. 1889 F, p. 180.   (Syn. of Cimoliasaurus.)
Nicholson and Lydekker 1889 A, p. 1079.   (Trinacromerum.)

Seeley, H. G. 1892 B, pp. 136, 151.
Zittel, K. A. 1890 A, p. 496.

Trinacromerum bentonianum Cragin.

Cragin, F. W. 1888 A.
1891 A.
Cretaceous (Benton); Kansas.

ELASMOSAURUS Cope.   Type *E. platyurus* Cope.

Cope, E. D. 1868 A.
1868 H.
1869 A, p. 266.
1869 K, p. 86.
1869 M, pp. 40, 44, 47, and ii.
1870 C.
1871 D, p. 391.
1872 F, p. 336.
1875 E, pp. 75, 78, 255.
1877 G, p. 577.
1879 I.
1888 S, p. 725.
Hulke, J. W. 1883 A.
1892 A.
Leidy, J. 1851 G, p. 326.   (Discosaurus, type *D. vetustus* Leidy.)
1870 A.   (Discosaurus=Elasmosaurus.)
1870 C.
1870 H, p. 18.
1870 I, p. 18.
Lydekker, R. 1888 E.
1889 C, p. 53.   (Syn. of Cimoliasaurus.)
1889 F, p. 180.   (Syn. of Cimoliasaurus.)
Nicholson and Lydekker 1889 A, p. 1077.   (Syn. of Cimoliasaurus.)
Sauvage, H. E. 1878 A.
Seeley, H. G. 1892 B, p. 134.
Zittel, K. A. 1890 A, p. 491.   (Syn. of Cimoliasaurus.)

Elasmosaurus intermedius Cope.

Cope, E. D. 1894 B, p. 112.
Cretaceous (Pierre); S. Dakota.

Elasmosaurus orientalis Cope.

Cope, E. D. 1868 M.
1869 A, p. 266.
1869 B, p. 733.

Cope, E. D. 1869 K, p. 87.
1869 M, pp. 42, 55, pl. ii, fig. 10.
1870 C.
1875 E, p. 255.
1877 G, pp. 567, 578.
1877 U.
Leidy, J. 1870 I, p. 22.   (Discosaurus.)
Cretaceous; New Jersey, Upper Missouri region.

Elasmosaurus platyurus Cope.

Cope, E. D. 1868 A.
1868 D.
1869 A, p. 266.
1869 K, p. 87.
1869 M, p. 47, text figs. 7-12; pl. ii, figs. 1-9, pl. iii.
1870 B.
1871 C, p. 47.
1871 D, p. 393.
1872 F, pp. 320, 336.
1875 E, pp. 44, 79, 256.
1877 G, p. 578.
1888 S.
Leidy, J. 1870 A.
1870 G, p. 10.
1870 I, pp. 18, 22.
Lydekker, R. 1889 F, p. 181.   (Cimoliasaurus.)
Cretaceous (Niobrara); Kansas.

Elasmosaurus serpentinus Cope.

Cope, E. D. 1877 G, p. 578.
1877 U.
Cretaceous (Niobrara); Nebraska.

Elasmosaurus sp. indet.

Cope, E. D. 1894 B, p. 113.
Cretaceous (Niobrara); Dakota.

CIMOLIASAURUS Leidy.   Type *C. magnus* Leidy.

Leidy, J. 1851 G, p. 325.
Cope, E. D. 1868 H.
1869 A, p. 266.
1869 K, p. 86.
1869 M, p. 40.
1870 B.
1870 C.
1871 D, p. 393.
1875 E, pp. 75, 78, 255.
1888 S, p. 725.
Fürbringer, M. 1900 A, p. 885.
Koken, E. 1896 A, p. 9.
Leidy, J. 1851 G, p. 326.   (Discosaurus, type *D. vetustus.*)
1854 B.   (Brimosaurus, type *B. grandis.*)

Leidy, J. 1865 A, pp. 22, 25, 76, 103, 116.   (Cimoliasaurus, Discosaurus.)
1870 A.   (Discosaurus.)
1870 B.   (Discosaurus, Cimoliasaurus.)
1870 G.   (Brimosaurus.)
1870 H, p. 21.
1870 I.   (Discosaurus.)
Lydekker, R. 1888 E.
1889 C, p. 53.
1889 F, pp. viii, 180.
Nicholson and Lydekker 1889 A, p. 1077.
Seeley, H. G. 1892 B, p. 135.
Williston, S. W. 1890 A.
Woodward, A. S. 1898 B, p. 169.
Zittel, K. A. 1890 A, p. 491.

Cimoliasaurus grandis (Leidy).

Leidy, J. 1854 B, pl. ii, figs. 1-3. (Brimosaurus.)
Cope, E. D. 1869 A, p. 266.
    1869 M, pp. 43, 54.
    1870 C.
    1871 D, p. 400.
    1875 E, p. 255.
Leidy, J. 1870 G, p. 10. (Brimosaurus.)
    1870 I, p. 22. (Discosaurus.)
Cretaceous; Arkansas.

Cimoliasaurus laramiensis Knight.

Knight, W. C. 1900 A, p, 117, figs. B and D.
Jurassic (Shirley); Wyoming.

Cimoliasaurus magnus Leidy.

Leidy, J. 1851 G, p. 325.
Cope, E. D. 1869 A, p. 266.
    1869 B, p. 733.
    1869 K, p. 87.
    1869 M, pp. 42, 43 (part), fig. 16; not figs.
    13-15.
    1870 B.
    1870 C.
    1875 E, p. 255.
Leidy J. 1854 B, pl. ii, figs. 4-6.
    1865 A, pp. 25, 116, pl. v, figs. 13-19; pl. vi,
    figs. 1-18.
    1870 G, p. 10.
    1870 I, p. 22. (Discosaurus.)
Lydekker, R, 1889 F, p. 211.
Cretaceous (Green-sand No. 5); New Jersey.

Cimoliasaurus planior (Leidy).

Leidy, J. 1870 I, pp. 20, 22. (Discosaurus,)
Cope, E. D. 1875 E, p. 255.
Leidy J. 1865 A, p. 23, par. 2, pl. v, figs. 10-12.
    (Discosaurus vetustus.)
        1865 D, p. 68. (Discosaurus vetustus, part.)
Cretaceous; Mississippi.

Cimoliasaurus snowii Williston.

Williston, S. W. 1890 A, p. 262.
Cope, E. D. 1894 B, p. 109, figure.
Williston, S. W. 1890 D, figs. 1, 2.
Cretaceous (Niobrara); Kansas.

Cimoliasaurus vetustus (Leidy).

Leidy, J. 1851 G, p. 326. (Discosaurus.)
Cope, E. D. 1869 A, p. 266. (C. magnus, *fide* Leidy.)
        1869 M, p. 42.
        1870 B, p. 141. (Discosaurus.)
        1875 E, p. 255.
Leidy, J. 1865 A, pp. 22, 116, pl. iv, figs. 13-18, pl.
    v, figs. 1-9. (Discosaurus.)
        1865 D, p. 68, in part. (Discosaurus.)
        1870 I, p. 21. (Discosaurus.)
Lydekker, R. 1889 F, p. 211. (Syn. of C. magnus.)
Cretaceous; Alabama, New Jersey.
    The vertebræ figured on pl. v, figs. 4, 5, 6, of
Leidy 1865 A, must be regarded as the type of
this species.

Cimoliasaurus sp. indet.

Williston, S. W. 1894 B, p. 2, pl. i, figs. 1, 2.
Cretaceous; Kansas.

OROPHOSAURUS Cope.   Type *O. pauciporus* Cope.

Cope, E. D. 1887 G, p. 564.
        1894 B, p. 311.
Lydekker R, 1889 F, p. 182.
Seeley, H. G. 1892 B, p. 136.
Zittel, K. A. 1890 A, p. 491. (Syn. of Cimolia-
saurus.)

Orophosaurus pauciporus Cope.

Cope, E. D. 1887 G, p. 565.
Cretaceous (Fox Hills); New Mexico.

EMBAPHIAS Cope.   Type *E. circulosus* Cope.

Cope, E. D. 1894 B, p. 111.

Embaphias circulosus Cope.

Cope, E. D. 1894 B, p. 111.
Cretaceous (Pierre); South Dakota.

OLIGOSIMUS Leidy.   Type *O. grandævus* Leidy.

Leidy, J. 1872 E, p. 39.
        1873 B, p. 286.
Lydekker, R. 1889 C, p. 180. (Syn. ? of Cimolia-
saurus.)
Seeley, H. G. 1892 B, p. 135.
Zittel, K. A. 1890 A, p. 495. (Oligosomus.)

Oligosimus grandævus Leidy.

Leidy, J. 1872 E, p. 39.
        1873 B, pp. 286, 345, pl. xvi, figs. 18, 19.
Cretaceous? ; Wyoming.

Order RHYNCHOCEPHALIA Günther.

Günther, A. 1867 A, p. 626.
Ammon, L. 1885 A. (Homœosaurus.)
Andreæ, A. 1893 A.
Baur, G. 1885 E, p. 635.
        1886 E.

Baur, G. 1886 G, pp. 685, 733.
        1886 M, p. 304.
        1887 A, p. 486.
        1887 C.
        1889 B, p. 45.

Baur, G. 1894 A, p. 321.
1895 E.
1900 A, p. 160.
Baur and Case 1897 A.
Boulenger, G. A. 1889 A, p 1.
1891 B.
Burckhardt, R. 1895 A.
Case, E. C. 1898 A, p. 71.
Cope, E. D. 1869 M, p. 33.
1871 B.
1878 G, p. 306.
1880 A, p. 38.
1885 G, p. 246.
1887 E.
1887 S, p. 334.
1889 R, p. 863.
1891 N, p. 45.
1892 R.
1898 B, pp. 56, 73.
Credner, H. 1888 A.
Fürbringer, M. 1888 A.

Fürbringer, M. 1900 A, pp. 276, 531, 589.
Gadow, H. 1898 A, p. 18. (Rhynchocephali.)
Gaupp, E. 1894 A.
Gegenbaur, C. 1898 A.
Haeckel, E. 1895 A, p. 305.
Hoffmann, C. K. 1890 B.
Huxley, T. H. 1872 A, p. 194.
Lortet, L. 1892 A, p. 28.
Lydekker, R. 1888 B, p. 291.
Meyer, H. 1856 B. (Protorosaurus.)
1860 D, p. 106. (Homœosaurus.)
Mivart, St. G. 1886 A, p. 448.
Nicholson and Lydekker 1889 A, p. 1131.
Osborn, H. F. 1900 A, p. 1.
Perrin, A. 1895 A.
Siebenrock, F. 1894 B. (Hatteria.)
Steinmann and Döderlein 1890 A, p. 619.
Wagner, A. 1853 A, p. 254.
Woodward, A. S. 1898 B, p. 183.
Zittel, K. A. 1890 A, pp. 451, 583.

SPHENODON Gray.   Type *Hatteria punctata* Gray.

Gray, J. E., 1831, Zool. Miscell., p. 13.
On account of the importance of the genus *Sphenodon* in ascertaining the relationship of the reptiles, the following references are given to the literature of this animal.
Ammon, L. 1885 A. (Hatteria.)
Andreæ, A. 1893 A. (Hatteria.)
Baur, G. 1886 A.
1886 B.
1886 K.
1886 L.
1891 C, p. 345.
1892 D, p. 207.
Bayer, F. 1885 A.
Boulenger, G. A. 1889 A, p. 2.
Burckhardt, R. 1900 A.

Credner, H. 1888 A.
Dollo, L. 1891 A.
Fürbringer, M. 1900 A.
Gadow, H. 1896 A, p. 37.
Günther, A. 1867 A. (Hatteria.)
Hay, O. P. 1898 F, p. 945.
Howes and Swinnerton 1900 A, p. 516.
Huxley, T. H. 1869 C.
1887 A.
Kehrer, G. 1886 A, p. 83.
Osawa, G. 1898 A. (Hatteria.)
Osborn, H. F. 1900 A, p. 1.
Siebenrock, F, 1894 B.
Tomes, C. 1898 A, p. 273.
Zittel, K. A. 1890 A, p. 583.

CHAMPSOSAURIDÆ.

Cope, E. D. 1884 O, p. 106.
Boulenger, G. A. 1891 B, p. 169.
Burckhardt, R. 1900 A, p. 534.
Cope, E. D. 1876 I, p. 350. (Choristodera.)
1882 J, p. 981. (Choristodera.)
1884 BB.
1891 N, p. 45. (Choristodera.)

Cope, E. D. 1898 B, p. 73. (Choristodera.)
Dollo, L. 1884 C, p. 160. (Simœdosauria.)
Lemoine, V. 1884 A.
1884 B. (Simœdosauria.)
1885 A. (Simœdosauria.)
Nicholson and Lydekker 1889 A, p. 1183.
Zittel, K. A. 1890 A, p. 599.

CHAMPSOSAURUS Cope.   Type *C. profundus* Cope.

Cope, E. D. 1876 I, p. 348.
Andreæ, A. 1893 A, p 33.
Baur, G. 1887 C, p. 100.
1895 E.
Cope, E. D. 1883 E.
1884 O, p. 104.
1888 C.
Dollo, L. 1884 C, p. 154. (Simœdosaurus.)
1885 A.
1891 A.
1893 B, p. 79.
Fürbringer, M. 1900 A, p. 291.
Koken, E. 1886 A.

Lemoine, V. 1884 A. (Simœdosaurus.)
1884 B. (Simœdosaurus.)
1885 A. (Simœdosaurus.)
1885 B. (Simœdosaurus.)
1893 A, p. 364. (Simœdosaurus.)
Nicholson and Lydekker 1889 A, p. 1183.
Steinmann and Döderlein 1890 A, p. 622.
Woodward, A. S. 1898 B, p. 189.
Zittel, K. A. 1890 A, p. 600.

Champsosaurus annectens Cope.

Cope, E. D. 1876 I, p. 351.
1877 G, p. 573.
Cretaceous (Laramie); Montana.

Champsosaurus australis Cope.
Cope, E. D. 1881 T.
    1884 O, p. 107, pl. xxiii *b*, figs. 1–4.
    1888 CC, p. 301.
    Eocene (Puerco); N. Mexico.

Champsosaurus brevicollis Cope.
Cope, E. D. 1876 I, p. 352.
    1877 G, p. 573.
    Cretaceous (Laramie); Montana.

Champsosaurus profundus Cope.
Cope, E. D. 1876 I, p. 350.
    1877 G, p. 573.
    Cretaceous (Laramie); Montana.

Champsosaurus puercensis Cope.
Cope, E. D. 1882 E, p. 195.
    1884 O, p. 107, pl. xxiii *b*, figs. 5–10.
    1888 CC, p. 301.
    Eocene (Puerco); N. Mexico.

Champsosaurus saponensis Cope.
Cope, E. D. 1882 E, p. 196.
    1884 O, p. 109, pl. xxiii, figs. 11–22.
    1888 CC, p. 301.
    Eocene (Puerco); N. Mexico.

Champsosaurus vaccinsulensis Cope.
Cope, E. D. 1876 I, p. 353.
    1877 G, p. 573.
    Cretaceous (Laramie); Montana.

RHABDOPELIX Cope.   Type *R. longispinis* Cope.

Cope, E. D. 1869 M, p. 169.
    The position of this genus in the zoological
    system is problematical.

Rhabdopelix longispinis Cope.
Cope, E. D. 1869 M, pp. 170, 245, fig. 46.
    1866 G. (Pterodactylus; no description.)
    1870 I, p. 445.
    Triassic; Pennsylvania.

ISCHYROTHERIUM Leidy.   Type *I. antiquum* Leidy.

Leidy, J. 1856 I, p. 89.
Cope, E. D. 1869 M, p. 38. (Ischyrosaurus.)
    1874 R. (Ischyrosaurus.)
    1875 E, p. 256. (Ischyrosaurus.)
    1884 O, p. 106. (Ischyrosaurus.)
Leidy, J. 1874 A.
Zittel, K. A. 1890 A, p. 498. (Ischyrosaurus.)
    The position of this genus is uncertain.

Ischyrotherium antiquum Leidy.
Leidy, J. 1856 I, p. 89.
Cope, E. D. 1869 M, p 39. (Ischyrosaurus.)
    1874 C, p. 28. (Ischyrosaurus.)
    1875 E, p. 256. (Ischyrosaurus.)
Leidy, J. 1860 A, p. 150, pl. x, figs. 8–17.
    1869 A, p. 414.
Lepsius, G. R. 1881 A, p. 185. (Syn. of Prorasto-
    mus sirenoides.)

Order ICHTHYOSAURIA Blainville.

Agassiz, L. 1843 A, p. 84. (Icththyosauria.)
Bauer, F. 1898 A, p. 283.
Baur, G. 1886 F. (Ichthyopterygia.)
    1887 A. (Ichthyopterygia.)
    1887 C, p. 95. (Ichthyopterygia.)
    1887 D. (Ichthyopterygia.)
    1887 F. (Ichthyopterygia.)
    1890 B. (Ichthyosauria.)
    1891 C, p. 354. (Ichthyosauria.)
    1895 C. (Ichthyosauria.)
    1895 E. (Ichthyosauria.)
Cope, E. D. 1869 M, pp. 26, 27. (Ichthyopterygia.)
    1871 B. (Ichthyopterygia.)
    1887 S, p. 333. (Ichthyopterygia.)
    1888 S. (Ichthyopterygia.)
    1889 R, p. 863. (Ichthyopterygia.)
    1892 R. (Ichthyopterygia.)
    1895 C. (Ichthyopterygia.)
    1898 B, pp. 55, 64. (Ichthyopterygia.)
    1900 A, p. 159. (Ichthyopterygia.)
Dames, W. 1893 A. (Ichthyopterygia.)
    1895 B, p. 1045. ("Ichthyosaurier.")
Fraas, E. 1891 A.
Fürbringer, M. 1900 A. (Ichthyopterygia.)
Gadow, H. 1898 A, p. 24.
Gaudry, A. 1890 A, p. 179.

Gegenbaur, C. 1864 A, p. 31. (Enaliosauria.)
    1865 B, p. 51. (Enaliosauria.)
    1898 A.
Haeckel, E. 1895 A, p. 337. (Ichthyopterygia.)
Hay, O. P. 1899 F, p. 682. (Ichthyosauria.)
Hoffmann, C. K. 1890 B. (Ichthyosauria.)
Hulke, J. W. 1883 A. (Ichthyosauria.)
    1892 A. (Ichthyosauria.)
Huxley, T. H. 1872 A, p. 208. (Ichthyosauria.)
Jaeger, G. F. 1828 A. (Ichthyosauria.)
Conybeare, Trans. Geol. Soc., London (1), v, pt. 2,
    p. 563. (Ichthyosaurus.)
Kükenthal, W. 1890 C, p. 176. (Ichthyosauria.)
    1891 C. (Ichthyosauria.)
    1893 A, p. 54. (Ichthyosauria.)
Lydekker, R. 1888 D. (Ichthyopterygia.)
    1889 F, p. 1. (Ichthyopterygia.)
    1892 B. (Ichthyosauria.)
Marsh, O. C. 1879 A, p. 85. (Sauranodonta.)
    1880 E, p. 491. (Baptanodonta.)
Nicholson and Lydekker 1889 A, p. 1120 (Ich-
    thyopterygia); p. 1124 (Ichthyosauria).
Owen, R. 1838 B, p. 660. (Ichthyosauri.)
    1860 A, p. 159. (Ichthyopterygia.)
    1860 E, p. 198. (Ichthyopterygia.)
    1866 A. (Ichthyopterygia.)

Owen, R. 1881 A. (Ichthyopterygia.)
Quenstedt, F. A. 1852 A, p. 120. (Ichthyosauri.)
 1885 A, p. 194. (Ichthyosauri.)
Rohon, J. V. 1899 B, p. 13. ("Ichthyosaurier.")
Seeley, H. G. 1891 A. (Ichthyosauria.)

Seeley, H. G. 1892 B, p. 119. (Ichthyosauria.)
 1900 A, p. 645. (Ichthyosauria.)
Steinmann and Döderlein 1890 A, p. 640. (Ichthyopterygia.)
Woodward, A. S. 1898 B, p. 176. (Ichthyopterygia.)
Zittel, K. A. 1890 A, p. 451. (Ichthyosauria.)

PROTEOSAURIDÆ nom. nov.

This family has hitherto borne the name *Ichthyosauridæ*. The revival of the generic name *Proteosaurus* for the genus hitherto called *Ichthyosaurus*, requires a corresponding change in the family name. The following writers employ the name *Ichthyosauridæ*.
Bauer, F. 1898 A, p. 283.
 1900 A, p. 574.
Baur, G. 1887 D.

Dames, W. 1895 B, p. 1045.
Fraas, E. 1891 A.
 1892 B.
Fürbringer, M. 1900 A, p. 627.
Günther, A. C. 1886 A, p. 442.
Kiprijanoff, W. 1881 A.
Lydekker, R. 1889 F, p. 6.
Nicholson and Lydekker 1889 A, p. 1124.
Steinmann and Döderlein 1890 A, p. 642.

PROTEOSAURUS Home.   Type *Ichthyosaurus platyodon* Conybeare.

Home, E. 1819 A.
 Unless otherwise indicated, this genus is referred to by the following writers under the name *Ichthyosaurus*. *Proteosaurus*, however, clearly has priority. On this point see R. Lydekker in his "Catalogue Fossil Reptilia," pt. ii, p. vii. Although no species of this genus has as yet been, with certainty, discovered in America, a knowledge of its structure is essential in understanding the genus *Baptanodon;* hence citations of the literature are given.
Bauer, F. 1898 A, p. 283.
 1900 A, p. 574.
Baur, G. 1886 H.
 1887 A, p. 491.
 1887 D.
 1887 F.
 1889 F, p. 467.
 1896 B.
Buckland, W. 1837 A, i, p. 133; ii, pls. vii-xix.
Coles, H. 1853 A.
 1824 A, p. 103.
Cope, E. D. 1871 B.
 1892 Z, p. 17.
Cuvier, G. 1834 A, x, pp. 390-444, pls. cclvi-cclx.
Dames, W. 1893 A.
Dollo, L. 1892 A.
Egerton, P. G. 1837 A.
Fraas, E. 1888 A.
 1891 A.
 1892 A.
 1892 B.
 1894 A.
 1898 A.
Gaudry, A. 1890 C, p. 179.
Home, E. 1819 B. (Proteosaurus.)
Hulke, J. W. 1883 A.
 1892 A.
Huxley, T. H. 1872 A, p. 208.
Jaeger, G. F. 1828 A.
Kiprijanoff, W. 1881 A.
König, 1821, in Conybeare,W. D., 1821, Trans. Geol. Soc. Lond. (1), v. p. 563. (Ichthyosaurus, type *I. communis* Conybeare; not Ichthyosaura Latreille, 1802, Hist Nat. Rept., iv, p. 310, type *Proteus tritonius* Laur.)

Koken, E. 1883 A, p. 737.
Kükenthal, W. 1890 C.
 1891 C, p. 392.
Lydekker, R. 1888 D.
 1889 F, pp. vii, 12.
 1889 I.
 1892 B.
Maggi, L. 1898 B, figs. 1-5.
 1898 C, figs. A, D.
Marsh, O. C. 1877 E.
Merian, P. 1875 A.
Moore, C. 1857 A.
Nicholson and Lydekker 1889 A, p. 1124.
Osborn, H. F. 1900 B, p. 116.
Owen, R. 1838 B, p. 660.
 1840 A, p. 86.
 1840 B.
 1841 A.
 1842 A, p. 193.
 1845 B, p. 275, pls. lxxiii, lxxiiiA.
 1851 B, p. 68.
 1860 E, p. 200.
 1866 A.
 1881 A, p. 83.
Pearce, J. C. 1846 A.
Pictet, F. J. 1853 A, p. 531.
Quenstedt, F. A. 1852 A, p. 120.
 1885 A, p. 195.
Sauvage, H. E. 1887 A.
Schulze, F. 1894 A, p. 1133.
Seeley, H. G. 1871 A.
 1880 A.
 1892 B, p. 119.
Steinmann and Döderlein 1890 A, p. 642.
Thompson, D. A. W. 1886 A.
Trautschold, H. 1879 A.
Woodward, A. S. 1886 B.
 1888 I, p. 278.
 1898 B, p. 179, with figures.
Zittel, K. A. 1890 A, p. 454.

Proteosaurus sp. indet.

Belcher, E. 1856 A. (Ichthyosaurus sp. indet.)
Marsh, O. C. 1896 F, p. 444. (Ichthyosaurus.)
 Jurassic; Exmouth Island (lat. 77° 16′ N., long. 96° W.).

BAPTANODONTIDÆ.

Marsh, *O. C.* 1880 E, p. 491.
Baur, G. 1887 D.
1887 F.

Haeckel, E. 1895 A, p. 340. (Baptanodontia.)
Marsh, O. C. 1879 A, p. 86. (Sauranodontidæ.)

BAPTANODON Marsh. Type *B. natans* Marsh.

Marsh, *O. C.* 1880 E, p. 491. (To replace Sauranodon, preoccupied.)
Bauer, F. 1898 A, p. 324.
Baur, G. 1886 F, p. 245.
1887 D.
1887 F.
Cope, E. D. 1879 M. (Sauranodon.)
Hulke, J. W. 1883 A. (Sauranodon.)
Lydekker, R. 1888 D. (Syn. of Ophthalmosaurus.)
1888 E.
1889 F, p. 6.
Marsh, O. C. 1879 A. (Sauranodon, type *S. natans.*)
1880 B. (Sauranodon.)
1891 F, p. 337.
1897 C, p. 485.
Morse, E. S. 1880 A, p. 7.
Nicholson and Lydekker 1889 A, p. 1126.
Steinmann and Döderlein 1890 A, p. 642.
Wilder, B. G. 1880 A, p. 322. (Sauranodon.)

Zittel, K. A. 1890 A, p. 473.

Baptanodon discus Marsh.

Marsh, *O. C.* 1880 B, p. 171, figure. (Sauranodon.)
1880 E, p. 491.
1884 F, p. 183, fig. 148.
1895 A, p. 405, fig. 1.
1897 C, p. 485, fig. 30.
Wilder, B. G. 1880 A, p. 322. (Sauranodon.)
Jurassic; Wyoming.

Baptanodon natans Marsh.

Marsh, *O. C.* 1879 A, p. 86. (Sauranodon.)
Cope, E. D. 1879 M, p. 271. (Sauranodon.)
Lydekker, R. 1889 F, p. 7, fig. 5.
Marsh, O. C. 1880 B, p. 169. (Sauranodon.)
1880 E, p. 491.
1895 A, p. 406, fig. 2.
1897 C, p. 486, fig. 31.
Jurassic; Wyoming.

CHONESPONDYLUS Leidy. Type *C. grandis* Leidy.

Leidy, *J.* 1868 C, p. 178.
A genus of uncertain position.

Chonespondylus grandis Leidy.

Leidy, *J.* 1868 C, p. 176.
Cope, E. D. 1869 M, p. 29. (Ichthyosaurus.)
Triassic; Nevada.

CYMBOSPONDYLUS Leidy. Type *C. piscosus* Leidy.

Leidy, *J.* 1868 C, p. 178.
A genus of very uncertain position.

Cymbospondylus petrinus Leidy.

Leidy, *J.* 1868 C, p. 178.
Cope, E. D. 1869 M, p. 30.
Triassic: Nevada.

Cymbospondylus piscosus Leidy.

Leidy, *J.* 1868 C, p. 178.
Cope, E. D. 1869 M, p. 30.
Triassic; Nevada.

SHASTASAURUS Merriam. Type *S. pacificus* Merriam.

Merriam, *J. C.* 1895 A, p. 55.
Boulenger, G. A., 1896, Zool. Record; Rept., p. 29. (Schastasaurus.)
Dames, W. 1895 B, p. 1048.
The position of this genus is in doubt.

Shastasaurus pacificus Merriam.

Merriam, *J. C.* 1895 A, p. 55, figs. 1, 2.
Dames, W. 1895 B, p. 1048.
Triassic; California.

Order SQUAMATA Oppel.

Oppel, *M.*, 1811, Ordnungen der Reptilien, p. 14.
Baur, G. 1887 A, p. 482. (Streptostylica.)
1889 F, p. 473.
1892 B, p. 20.
1894 A, p. 327.
1895 A.
1896 A.
Boulenger, G. A. 1891 A, p. 117.
1898 A, p. 1.

Calori, L. 1894 A.
Cope, E. D. 1864 A, p. 224.
1889 R, p. 864.
1891 N, p. 45.
1892 R.
1895 C.
1895 E.
1898 B, pp. 56, 74.
1900 A, pp. 160, 175.

Cuvier, G. 1834 A, x.
Fürbringer, M. 1900 A.
Gray, J. E. 1837 A.
Günther, A. C. 1867 A, p. 625.
Haeckel, E. 1895 A, p. 340.   (Lepidosauria.)
Hay, O. P. 1899 F, p. 682.
Lydekker, R. 1888 B, pp. viii, 248.

Müller, J. 1831 A, p. 199.
Nicholson and Lydekker 1889 A, p. 1137.
Stannius, H. 1856 A.   (Streptostylica.)
Steinmann and Döderlein 1890 A, p. 644.
Woodward, A. S. 1898 B, p. 190.
Zittel, K. A. 1890 A, pp. 450, 602.   (Lepidosauria.)

Suborder PYTHONOMORPHA Cope.

Cope, E. D. 1869 A, p. 258.
Ballou, W. H. 1898 A, p. 212.
Baur, G. 1887 C, p. 99.
Bemmelen, J. F. 1896 A, p. 333.
Boulenger, G. A. 1891 A, p. 117.
Cope, E. D. 1869 M, pp. 27, 29, 175.
    1870 E.
    1870 O.
    1871 B.
    1871 C, p. 400.
    1872 F, pp. 321, 327, 328.
    1872 CC.
    1874 C, p. 30.
    1875 E, pp. 113, 264, pl. xxxvi.
    1878 G.
    1879 A, p. 36.
    1885 G, p. 246.
    1887 S, p. 384.
    1891 N, pp. 45, 50.
    1892 Z, p. 19.
    1895 C, p. 858.
    1895 E, p. 1004.

Cope, E. D. 1896 A.
    1898 B, p. 75.
Dana, J. B. 1896 A, p. 870.
Fürbringer, M. 1900 A, pp. 271, 615, 622.   (Mosasauria.)
Gadow, H. 1898 A, p. 24.
Gaudry, A. 1892 A.
Haeckel, E. 1895 A, p. 349.
Hoffmann, C. K. 1890 B, p. 1320.   (Mosasauria.)
Huxley, T. H. 1872 A, p. 197.   (Mosasauria.)
Lydekker, R. 1888 B, p. 261.
Marsh, O. C. 1877 E.   (Mosasauria.)
Merriam, J. C. 1894 A.
Nicholson and Lydekker 1889 A, p. 1143.
Osborn, H. F. 1899 I, p. 159.
    1900 A, p. 1.   ("Mosasaurs.")
Owen, R. 1878 B, p. 755.
Steinmann and Döderlein 1890 A, p. 645.
Williston, S. W. 1898 E, p. 170.   (Mosasauria.)
Williston and Case 1892 A.
Woodward, A. S. 1898 B, p. 191, with figures.
Zittel, K. A. 1890 A, p. 611.

MOSASAURIDÆ.

Baur, G. 1890 A.
    1890 B.
    1892 B.
Cope, E. D. 1869 A, p. 258.
    1869 M, pp. 182, 183.
    1878 G.
    1895 C.
Dollo, L, 1882 A.
    1885 B.
    1885 C.
    1888 D.
    1890 A.
    1893 B.
Fürbringer, M. 1900 A.
Gaudry, A. 1890 C, p. 202.

Lydekker, R. 1888 B, p. 261.
Marsh, O. C. 1872 F.
    1880 A.
Merriam, J. C. 1894 A.
Nicholson and Lydekker 1889 A, p. 1143.
Owen, R. 1877 A.
    1878 B.
    1879 B.
    1880 A, p. 177.
Quenstedt, F. A. 1885 A, p. 189.   (Mosasauri.)
Thévenin, A. 1896 A.
Williston, S. W. 1891 A.
    1897 F.
    1898 E.
Williston and Case 1892 A, p. 16.

MOSASAURINÆ.

Williston, S. W. 1897 F, pp. 177, 181.
Cope, E. D. 1869 A, p. 258.   (Clidastidæ.)
    1869 M, p. 182.   (Clidastidæ.)

Marsh, O. C. 1876 G, p. 59.   (Edestosauridæ.)
Williston, S. W. 1898 E, pp. 91, 196.

MOSASAURUS Conybeare.   Type M. belgicus Holl.

Conybeare, 1822, in Parkinson's Introduct. Study
    Foss. Organ. Remains, p. 198.
Ballou, W. H. 1898 A, p. 212.
Baur, G. 1892 B.
Buckland, W. 1837 A, ii. p. 167, pl. xx.
Charlesworth, E. 1846 A.
Cope, E. D. 1869 A, p. 260.
    1869 K, p. 85.

Cope, E. D. 1869 M, pp. 185, 186, fig. 47.
    1874 C, p. 31.
    1875 E, p. 128.
    1878 G.
Cuvier, G. 1808 A, p. 145.   ("Animal de Maestricht.")
    1834 A, x, pp. 119–175, pls. cclvi–cclviii
De Kay, J. E. 1830 A.

Dollo, L. 1882 A, p. 56.
1888 D.
1890 A.
1898 A. (Leiodon, a synonym.)
1893 B. (Syn. of Plioplatecarpus.)
Gaudry, A. 1890 A, p. 202.
1892 A, p. 10.
Gibbes, R. W. 1850 A.
1851 A. (Mosasaurus, Holcodus; Amphorosteus, type *A. brumbyi.*)
Goldfuss, A. 1845 A.
Harlan, R. 1834 B, p. 81.
1835 J.
1839 A. (Batrachiosaurus.)
1839 B, p. 90. (Batrachiotherium.)
Holmes, F. S. 1849 A, p. 197.
Leidy, J. 1857 E.
1859 D.
1865 A, p. 30 (Mosasaurus); p. 118 (Baseodon, type *B. reversus*=M. dekayi).
1873 B, p. 270.
Lippincott, J. S. 1881 A.
Lydekker, R. 1888 B, p. 261.
Merriam, J. C. 1894 A.
Meyer, H. 1860 C.
1860 D, p. 98.
Nicholson and Lydekker 1889 A, p. 1144.
Owen, R. 1842 A, p. 144.
1845 B, p. 258.
1851 B, p. 29.
1877 A.
Pictet, F. J. 1853 A, p. 504 (Mosasaurus); p. 506 (Batrachiosaurus, Batrachiotherium).
Quenstedt, F. A. 1852 A, p. 116.
1885 A, p. 189.
Steinmann and Döderlein 1890 A, p. 646.
Thévenin, A. 1896 A.
1896 B.
Williston, S. W. 1897 F.
1898 E, p. 195.
Woodward, A. S. 1898 B, p. 194, with figure.
Zittel, K. A. 1890 A, p. 519.

Mosasaurus brumbyi (Gibbes.)

Gibbes, R. W. 1851 A, p. 10, pl. iii, figs. 10–16. (Amphorosteus.)
Cope, E. D. 1869 A, p. 264.
1869 M, pp. 196, and ii.
1870 O, p. 576.
1871 D, p. 411.
1875 E, p. 270.
Gibbes, R. W. 1850. (Name only.)
Leidy, J. 1865 A, pp. 37,118. (Amphorosteus.)
Lydekker, R. 1888 B, p. 267. (Liodon.)
Cretaceous (Rotten limestone); Alabama.

Mosasaurus carolinensis Gibbes.

Gibbes, R. W. 1851 A, p. 8, pl. ii, figs. 1–3.
Cope, E. D. 1869 M, p. 198. (Syn.? of M. dekayi.)
1875 E, p. 269. (Syn.? of M. dekayi.)
Gibbes, R. W. 1850 A. (No description.)
Leidy, J. 1865 A, pp. 32,117. (Syn. of M. mitchellii.)
Cretaceous; South Carolina.

Mosasaurus copeanus Marsh.

Marsh, O. C. 1869 B, p. 393.

Bull. 179——30

Cope, E. D. 1869 M, p. 198.
1875 E, p. 270.
Marsh, O. C. 1872 F, p. 455.
Cretaceous (Green-sand No. 4); New Jersey.

Mosasaurus couperi Gibbes.

Gibbes, R. W. 1851 A, p. 7, pl. ii, figs. 4, 5.
Cope, E. D. 1869 M, p. 198. (Syn.? of M. dekayi.)
1875 E, p. 269. (Syn.? of M. dekayi.)
Gibbes, R. W. 1850 A. (No description.)
Leidy, J. 1865 A, pp. 32, 117. (Syn. of M. mitchillii.)
Cretaceous; Georgia.

Mosasaurus crassidens Marsh.

Marsh, O. C. 1870 F.
Cope, E. D. 1869 M, p. 198.
1875 E, p. 270.
1875 V, p. 1.
Cretaceous; North Carolina.

Mosasaurus dekayi Bronn.

Bronn, H. G. 1838 A, p. 760.
Cope, E. D. 1867 E. p. 27. (M. mitchilli.)
1869 A, p. 262. (M. mitchilli.)
1869 B, p. 733 (M. mitchilli); p. 734 (Baseodon reversus).
1869 K, p. 86. (M. mitchilli.)
1869 M, pp. 198, 285, (?) pl. xi, fig. 8; text figs. 48, 49.
1875 E, p. 299, pl. xxxvii, fig. 16.
1895 C, p. 859, pl. xxxi, figs. 2, 2a.
De Kay, J. E. 1830 A, p. 135. (Mosasaurus sp.)
1842 A, p. 28, pl. xxii, figs. 57, 58. (M. major.)
Emmons, E. 1858 B, p. 217, figs. 36A, 37. (M. maximiliani.)
Gibbes, R. W, 1850 A.
1851 A, p. 8, pl. i, figs. 2, 6.
Harlan, R. 1835 J. ("Saurian reptile.")
Leidy, J. 1865 A, p. 116, plates referred to, in part (M. mitchilli); p. 118, pl. x, figs. 14, 15 (Baseodon reversus.)
1865 D p. 70, in part.
Mitchill, S. L. 1818 A, pp. 384, 385, 431, pl. viii, fig. 4. ("Saurian.")
Morton, S. G. 1844 B, p. 132. (M. occidentalis; no description.)
Owen, R. 1849 A, p. 382, pl. x, fig. 5. (M. maximiliani.)
Cretaceous; New Jersey, South Carolina?.

Mosasaurus depressus Cope.

Cope, E. D. 1869 A, p. 262.
1869 B, p. 734.
1869 C.
1869 M, pp. 189, 195, pl. xi, fig. 6, text fig. 48.
1875 E, p. 269, pl. xxxvii, fig. 12.
Cretaceous; New Jersey.

Mosasaurus fulciatus Cope.

Cope, E. D. 1869 M, p. 194.
1870 E.
1875 E, p. 269, pl. xxxvii, fig. 13.
Cretaceous; New Jersey.

Mosasaurus horridus Williston.

Williston, S. W. 1898 E, pp. 108, 110, 115, 145, 147, 150, 152, 154, pls. xix–xxi; pl. xxxii.
Ballou, W. H. 1898 A, p. 217.
Williston, S. W. 1897 D, p. 96. (No description.)
Cretaceous (Niobrara); Kansas.

Mosasaurus maximus Cope.

Cope, E. D. 1869 A, p. 262.
1869 M, pp. 188, 189, pl. xi, fig. 7; text figs. 48, 49.
1870 R, p. 86.
1875 E, p. 269, pl. xxxvii, fig. 15.
Marsh, O. C. 1869 B, p. 398.
Whitfield, R. P. 1900 A, p. 25, pls. iv, v.
Cretaceous; New Jersey.

Mosasaurus miersii Marsh.

Marsh, O. C. 1869 B, p. 395.
Cope, E. D. 1869 M, p. 199.
1875 E, p. 270.
Cretaceous; New Jersey.

Mosasaurus minor Gibbes.

Gibbes, R. W. 1851 A, p. 7, pl. i, figs. 3–5.
Cope, E. D. 1869 A, p. 261.
1869 M, pp. 186, 189, 196.
1875 E, p. 270. (Syn.? of *M. acutidens*–Holcodus acutidens.)
Gibbes, R. W. 1850 A. (No description.)
Leidy, J. 1865 A, p. 118. (Holcodus acutidens.)
Cretaceous; Alabama.

Mosasaurus missouriensis (Harlan).

Harlan, R. 1834 D, p. 405, pl. xx. (Ichthyosaurus.)
Baur, G. 1892 B, p. 9. (*M. maximiliani.*)
Cope, E. D. 1869 A, p. 263.
1869 K, p. 86.
1869 M, pp. 189, 195.
1871 D, p. 401.
1875 E, pp. 219, 269.
Gibbes, R. W. 1850 A. (*M. maximiliani.*)
1851 A, p. 6, pl. i, fig. 7. (*M. maximiliani.*)

Goldfuss, A. 1845 A, pls. vi–ix. (*M. maximiliani.*)
1845, Versamml. deutsch. Naturf., Mainz, p. 141. (*M. neovidii.*)
Harlan, R. 1834 A. (Ichthyosaurus missouriensis; no description.)
1834 B, p. 80. (Ichthyosaurus.)
1834 D, p. 405, pl. xxx, figs. 3–8. (Ichthyosaurus.)
1834 E. (Ichthyosaurus.)
1835 B, p. 284. (Ichthyosaurus.)
1835 E. (Ichthyosaurus.)
1839 A. ("Batrachiosaurus.")
1839 B. ("Batrachotherium.")
1839 C. (Batrachiosaurus.)
1842 B, p. 142. (Batrachiotherium.)
Hoffman, C. K. 1890 B, p. 1320. (*M. maximiliani.*)
Leidy, J. 1865 A, p. 117, pl. vii, figs. 15–18; pl. viii, figs. 6, 8; pl. x, figs. 18, 19; pl. xvii, figs. 12, 13. (Determinations doubtful.)
Meyer, H. 1845 B, p. 313. (*M. maximiliani, M. missouriensis.*)
Pictet, F. J. 1858 A, p. 505. (*M. maximiliani.*)
Quenstedt, F. A. 1852 A, p. 118.
Zittel, K. A. 1890 A, p. 621. (*M. maximiliani.*)
Cretaceous; Nebraska.

Mosasaurus oarthrus Cope.

Cope, E. D. 1869 M, pp. 189, 196.
1870 E.
1875 E, p. 269, pl. xxxvii, fig. 17.
Cretaceous; New Jersey.

Mosasaurus princeps Marsh.

Marsh, O. C. 1869 B, p. 392.
Baur, G. 1892 B, p. 4.
Cope, E. D. 1869 M, pp. 188, 192.
1875 E, p. 269. (Syn. *M. dekayi.*)
Cretaceous; New Jersey.

Mosasaurus sp. indet.

Leidy, J. 1865 A, pp. 45, 118, pl. viii, figs. 9, 10.
1873 B, pp. 279, 343, pl. xxxvi, fig. 15.
Harlan, R. 1835 B, p. 285.

CLIDASTES Cope.   Type *C. iguanavus* Cope.

Cope, E. D. 1868 H, p. 181. (Clidastes; Nectoportheus, type *N. validus* Cope.)
1868 K, p. 233.
1869 A.
1869 M, p. 211, figs. 50–52.
1871 D, p. 412.
1871 N, p. 266.
1872 F, pp. 322, 330.
1874 C, p. 31.
1875 E, pp. 128, 130, pl. iv.
1878 G.
Dollo, L. 1885 C, p. 29 (Edestosaurus); p. 30 (Clidastes).
Lydekker, R. 1888 B, p. 272.
Marsh, O. C. 1871 B, p. 447. (Edestosaurus, type *E. dispar* Marsh.)
1872 C, p. 291. (Edestosaurus.)
1872 F, p. 463. (Edestosaurus.)
1872 T, p. 496. (Edestosaurus, Clidastes.)
Merriam, J. C. 1894 A, p. 31.
Nicholson and Lydekker 1889 A, p. 1143.

Owen, R. 1877 A.
Steinmann and Döderlein 1890 A, p. 646.
Williston, S. W. 1891 A.
1896 C.
1897 E.
1897 F.
1897 J, p. 107.
1898 A, figure.
1898 E, p. 195.
Williston and Case 1892 A, p. 17.
Woodward, A. S. 1898 B, p. 195. (Edestosaurus.)
Zittel, K. A. 1890 A, p. 623 (Clidastes); p. 624 (Edestosaurus).

Clidastes cinerearum Cope.

Cope, E. D. 1870 O, p. 583.
1871 D, p. 413.
1871 N, p. 266.
1872 F, pp. 322, 330.
1875 E, pp. 48, 137, 266, pl. xxi, figs. 14–17.
1877 G p. 583.

Marsh, O. C. 1871 B, p. 449.
Williston, S. W. 1891 A.   (Syn.? of C. velox.)
    1898 E, p. 196.
Williston and Case 1892 A, p. 17.
    Cretaceous (Niobrara); Kansas.

Clidastes conodon Cope.

Cope, E. D. 1881 R.
    Cretaceous; New Jersey.

Clidastes dispar (Marsh).

Marsh, O. C. 1871 B, p. 447.   (Edestosaurus.)
Cope, E. D. 1871 N, p. 269.   (Edestosaurus.)
    1872 F, p. 330.   (Edestosaurus.)
    1872 CC, p. 141.   (Edestosaurus.)
    1874 C, p. 33.
    1875 E, p. 266.
    1877 G, p. 583.
Fürbringer, M. 1900 A, p. 272.
Marsh, O. C. 1872 F, p. 463, pl. xi.   (Edestosaurus.)
    1880 A, p. 86, pl. i.   (Edestosaurus.)
Owen, R. 1880 A, pl. viii, fig. 1.   (Edestosaurus.)
Williston, S. W. 1891 A.
    1898 E, pp. 198, 201.   (C. velox in part; C. tortor in part.)
Williston and Case 1892 A, pp. 17, 30.
Woodward, A. S. 1898 B, p. 193, fig. 117.
Zittel, K. A. 1890 A, p. 617, fig. 549.
    Cretaceous (Niobrara); Kansas.

Clidastes iguanavus Cope.

Cope, E. D. 1868 H.
    1868 K, p. 233.
    1869 A, p. 258.
    1869 B, p. 734.
    1869 K, p. 86.
    1869 M, p. 220, pl. v, fig. 3.
    1872 F, p. 330.
    1875 E, pp. 138, 266.
Marsh, O. C. 1871 B, p. 449.
    Cretaceous; New Jersey.

Clidastes intermedius Leidy.

Leidy, J. 1870 E, p. 4.
Cope, E. D. 1869 M, p. 221.
    1871 D, p. 412.
    1875 E, pp. 136, 267.
Marsh, O. C. 1871 B, p. 451.
Leidy, J. 1873 B, pp. 281, 344, pl. xxxiv, figs. 1–5.
    Cretaceous; Alabama.

Clidastes liodontus Merriam.

Merriam, J. C. 1894 A, p. 35.
Williston, S. W. 1898 E, p. 204.
    Cretaceous; Kansas.

Clidastes propython Cope.

Cope, E. D. 1869 D.
    1869 A, p. 258.
    1869 M, p. 221, pl. xii, figs. 1–21.
    1871 B, figs. 15, 19.
    1875 E, p. 265, pl. xxxvii, fig. 1.
    1878 G, p. 303.
Dana, J. D. 1876 A, p. 65.
Marsh, O. C. 1871 B, p. 451.
Zittel, K. A. 1890 A, p. 623, figs. 545, 547, 548, 553.
    Cretaceous; Alabama.

Clidastes pumilis Marsh.

Marsh, O. C. 1871 G, p. 104.

Cope, E. D. 1871 N, p. 266.
    1872 F, pp. 323, 330.
    1874 C, p. 34.
.   1875 E, pp. 48, 266.
Marsh, O. C. 1871 B, p. 452.
Merriam, J. C. 1894 A, p. 35.
Williston, S. W. 1898 E, p. 198.   (Syn. of C. velox.)
Williston and Case 1892 A, p. 17.   (Syn. of C. velox.)
    Cretaceous (Niobrara); Kansas.
    It seems not improbable that this species will yet prove to be the young of C. velox.

Clidastes rex (Marsh).

Marsh, O. C. 1872 F, p. 460, pl. xii.   (Edestosaurus.)
Cope, E. D. 1874 C, p. 33.
    1875 E, p. 266.
Marsh, O. C. 1897 C, p. 527.   (Edestosaurus.)
Williston, S. W. 1898 E, p. 201.   (Syn. of C. tortor.)
Williston and Case 1892 A, pp. 17, 30.
    Cretaceous (Niobrara); Kansas.

Clidastes stenops Cope.

Cope, E. D. 1871 N, p. 268.   (Edestosaurus.)
    1872 F, p. 330.   (Edestosaurus.)
    1874 C, p. 33.
    1875 E, pp. 138, 266, pls. xiv, fig. 2; pl. xvii, figs. 7, 8; pl. xviii, figs. 1–5; pl. xxxvi, fig. 4; pl. xxxvii, figs. 1–3; pl. xxxviii, fig. 3.
    1878 G, p. 299.
Marsh, O. C. 1872 F, p. 464.   (Edestosaurus.)
Williston, S. W. 1898 E, p. 197.
Williston and Case 1892 A, pp. 17, 30.
Zittel, K. A. 1890 A, p. 613, fig. 541.
    Cretaceous (Niobrara); Kansas.

Clidastes tortor Cope.

Cope, E. D. 1871 N, pp. 265, 266.   (Edestosaurus.)
    1872 F, pp. 322, 330.   (Edestosaurus.)
    1872 CC, p. 141.   (Edestosaurus.)
    1874 C, p. 33.
    1875 E, pp. 48, 131, 265, pl. xiv, fig. 1; pl. xvi, figs. 1, 6; pl. xvii, fig. 1; pl. xix, figs. 1–6; pl. xxxvi, fig. 3; pl. xxxvii, fig. 2; pl. xxxviii, fig. 2.
    1877 G, p. 583.
Marsh, O. C. 1872 F, p. 464.   (Edestosaurus.)
Merriam, J. C. 1894 A, p. 34.   (C. medius.)
Williston, S. W. 1897 D, p. 97.
    1898 E, p. 201, pl. xxiii; pl. xxvii, fig. 2; pl. xxviii, fig. 6; pl. xxix, fig. 2; pl. xxxvii; pl. xxxviii; pl. xxxix; pl. xlii, figs. 5, 6; pl. xlvii; pl. liv, figs. 4, 5; pl. lx, fig. 5; pl. lxiv, fig. 1.
Williston and Case 1892 A, p. 17.
    Cretaceous (Niobrara); Kansas.

Clidastes ? validus Cope.

Cope, E. D. 1868 H, p. 181.   (Nectoportheus.)
    1869 A, p. 260.   (Macrosaurus.)
    1869 B, p. 734.   (Macrosaurus.)
    1869 K, p. 86.   (Macrosaurus.)
    1869 M, p. 208.   (Liodon validus, in part); pl. v, fig. 5, (Macrosaurus validus); p. v, expl. of pl. v (Clidastes antivalidus, based on type vertebra of Nectoportheus validus).
    1871 D, p. 414.   (C. antivalidus.)
    Cretaceous; New Jersey.

Clidastes velox (Marsh).

Marsh, O. C. 1871 B, p. 450. (Edestosaurus.)
Ballou, W. H. 1896 A, P. 215, fig. 221.
Cope, E. D. 1871 N, p. 269. (Edestosaurus.)
    1872 F, p. 330. (Edestosaurus.)
    1874 C, p. 33. (C. affinis, C. velox.)
    1875 E, p. 266. (C. velox, C. affinis.)
Leidy, J. 1873 B, pp. 283, 344, pl. xxxiv, figs. 6–11.
    (C. affinis.)
Marsh, O. C. 1872 F, p. 464. (Edestosaurus.)
Merriam, J. C. 1894 A, p. 34, pl. i, fig. 4; pl. iii,
    figs. 6, 7, 9, 10.
Williston, S. W. 1891 A.
    1893 C, pl. iii.
    1897 J, p. 110.
    1898 E, pls. x–xii; pl. xxiv, figs. 6, 7; pl.
    xxvii, figs. 3, 4; pl. xxviii, figs. 1–4; pl.
    xxxi, fig. 6; pl. xxxiii; pl. xxxiv; pl. xl,
    figs. 6, 7; pl. lxii, fig. 3; pl. lxxii, fig. 1.
    (Includes, besides C. velox (Marsh) also C.
    pumilis Marsh, C. affinis Leidy, and C.
    dispar (Marsh).

Williston and Case 1892 A, p. 17, pls. ii, iii.
    Cretaceous (Niobrara); Kansas.

Clidastes westii Williston and Case.

Williston, S. W. 1892 A, p. 29, pls. iv–vi.
    1898 E, p. 206, pl. xxiii; pl. xxvii, fig. 1; pl.
    xxxv; pl. xxxvi; pl. xxxix; pl. lili.
    Cretaceous (Pierre); Kansas.

Clidastes wymani Marsh.

Marsh, O. C. 1871 G, p. 104.
Cope, E. D. 1871 N, p. 266.
    1872 F, p. 330.
    1874 C, p. 33. (C. vymanii.)
    1875 E, p. 266. (C. vymanii.)
Marsh, O. C. 1871 B, p. 451.
    1872 F, p. 464. (Edestosaurus.)
Williston, S. W. 1898 E, p. 208.
Williston and Case 1892 A, pp. 17, 30.
    Cretaceous (Niobrara); Kansas.

BAPTOSAURUS Marsh.   Type *Halisaurus platyspondylus* Marsh.

Marsh, O. C. 1870 F. (To replace Halisaurus, pre-
    occupied by Halosaurus Günther.)
Cope, E. D. 1869 M, pp. 185, 208.
    1874 C, p. 31.
    1875 E, pp. 128, 272.
Dollo, L. 1885 C, p. 28.
Marsh, O. C. 1869 B, p. 395. (Halisaurus, type *H.
platyspondylus*)
Merriam, J. C. 1894 A, p. 36.
Nicholson and Lydekker 1889 A, p. 1142.
Williston, S. W. 1897 F.
    1897 H, p. 244.
    1897 J, p. 107.
    1898 E, p. 207.
Zittel, K. A. 1890 A, p. 624.
    The systematic position of this genus is
' doubtful.

Baptosaurus fraternus Marsh.

Marsh, O. C. 1869 B, p. 397. (Halisaurus.)
Cope, E. D. 1869 M (1870), p. 210.
    1875 E, p. 272.

Marsh, O. C. 1870 F.
    Cretaceous; New Jersey.

Baptosaurus onchognathus Merriam.

Merriam, J. C. 1894 A, p. 37, pl. iv, fig. 9.
Williston, S. W. 1898 E, p. 208, pl. xxv, fig. 6; pl.
    xli, fig. 3.
    Cretaceous (Niobrara); Kansas.

Baptosaurus platyspondylus Marsh.

Marsh, O. C. 1869 B, p. 395. (Halisaurus.)
Cope, E. D. 1869 M (1870), p. 209.
    1875 E, p. 272.
Marsh, O. C. 1869 B, p. 395. (Halisaurus.)
    1870 F.
    Cretaceous; New Jersey.
    This species is said to have been described
at the Salem meeting of the Amer. Assoc. Adv.
Sci., 1869, under the name *Macrospondylus
platyspondylus*, but the name does not occur in
the "Proceedings."

PLATECARPINÆ.

Williston, S. W. 1897 F, pp. 177, 181. | Williston, S. W. 1898 E, pp. 91, 177.

SIRONECTES Cope.   Type *S. anguliferus* Cope.

Cope, E. D. 1874 C, PP. 31, 34.
    1875 E, pp. 128, 139, 267.
Dollo, L. 1885 C, p. 30.
Nicholson and Lydekker 1889 A, p. 1144.
Williston, S. W. 1897 H, p. 245.
    1897 J, p. 107.
Zittel, K. A. 1890 A, p. 624.

Sironectes anguliferus Cope.

Cope, E. D. 1874 C, pp. 31, 34.
    1875 E, pp. 189, 267, pl. xxiii, figs. 16–18; pl.
    xxiv, figs. 1–15.
Williston, S. W. 1898 E, pp. 188, 194. (? Platecarpus planifrons.)
    Cretaceous (Niobrara); Kansas.

HOLCODUS Gibbes.   Type *H. acutidens* Gibbes.

Gibbes, R. W. 1851 A, p. 9.
Cope, E. D. 1869 M, pp. 185, 199.
    1871 N, p. 269,
    1872 F, p. 331.
    1872 CC, p. 141.
Leidy, J. 1865 A, p. 118. (In part).
Marsh, O. C. 1872 C, p. 291.
    1872 F, p. 455, (Syn.? of Mosasaurus.)
Williston, S. W. 1897 F, p. 184. (=? Platecarpus.)
    1898 E, p. 178. (=? Platecarpus.)
Zittel, K. A. 1890 A, p. 620. (Syn. of Mosasaurus.)

Holcodus acutidens Gibbes.
Gibbes, R. W. 1851 A, p. 9, pl. iii, figs. 6-9; not fig. 13.

Cope, E. D. 1869 M, pp. 200, 210.
    1872 CC, p. 141.
    1875 E, p. 270. (Mosasaurus.)
Gibbes, R. W. 1850 A. (H. columbiensis; no description.)
Leidy, J. 1865 A, p. 118 (in part), pl. x, fig. 17.
Marsh, O. C. 1869 B, p. 394.
Williston, S. W. 1898 E, p. 178. (Platecarpus?)
Zittel, K. A. 1890 A, p. 623.
Cretaceous; Mississippi.

PLATECARPUS Cope.   Type *P. tympaniticus* Cope.

Cope, E. D. 1869 A, p. 264.
    Most of the authorities named below include in this genus also the species placed in this work in the genus *Lestosaurus* Marsh.
Cope, E. D. 1869 M, pp. 185, 199.
    1872 CC, p. 141.
    1874 C, pp. 31, 34.
    1875 E, pp. 128, 141, 266.
    1878 G. (In part).
Leidy, J. 1865 A, p. 118. (Holcodus, in part.)
Lydekker, R. 1888 B, p. 269.
Merriam, J. C. 1894 A.
Nicholson and Lydekker 1889 A, p. 1144.
Owen, R. 1877 A.
Steinmann and Döderlein 1890 A, p. 646.
Thévenin, A. 1896 B.

Williston, S. W. 1891 A.
    1897 E.
    1898 E.
Woodward, A. S. 1898 B, p. 194.
Zittel, K. A. 1890 A, p. 622.

Platecarpus tympaniticus Cope.
Cope, E. D. 1869 A, p. 265.
    1869 M, p. 200.
    1874 C, p. 35.
    1875 E, p. 267, pl. xxxvii, fig. 11.
Leidy, J. 1865 A, p. 118 (in part), pl. vii, figs. 4-7; pl. xi, fig. 14. (?Holcodus acutidens.)
Williston, S. W. 1897 F, p. 178.
    1898 E, p. 179.
Cretaceous; Mississippi.

LESTOSAURUS Marsh.   Type *L. simus* Marsh.

Marsh, O. C. 1872 F, p. 454.
Baur, G. 1892 B. (Platecarpus.)
Cope, E. D. 1872 CC, p. 141. (Lestosaurus.)
    1874 C, pp. 31, 34. (Platecarpus.)
    1875 E, pp. 128, 141, 266 (Platecarpus).
    1878 G. (Platecarpus.)
Dollo, L. 1885 C, p. 28 (Lestosaurus); p. 29 (Platecarpus).
Lydekker, R. 1888 B, p. 269. (Platecarpus.)
Marsh, O. C. 1872 T, p. 497.
    1880 A, p. 84.
Merriam, J. C. 1894 A. (Platecarpus.)
Nicholson and Lydekker 1889 A, p. 1144. (Platecarpus.)
Osborn, H. F. 1899 D. (Platecarpus.)
Owen, R. 1877 A. (Platecarpus.)
Steinmann and Döderlein 1890 A, p. 646. (Platecarpus.)
Williston, S. W. 1891 A. (Platecarpus.)
    1897 F. (Platecarpus.)
    1897 J, p. 106. (Platecarpus.)
    1898 E. (Platecarpus.)
    1898 H, p. 235. (Platecarpus.)
    1899 A, p. 40. (Platecarpus.)
Zittel, K. A. 1890 A, p. 622 (Platecarpus).

Lestosaurus clidastoides (Merriam).
Merriam, J. C. 1894 A, p. 30. (Platecarpus.)
Williston, S. W. 1898 E, p. 190. (Platecarpus.)
Cretaceous; Kansas.

Lestosaurus coryphæus (Cope).
Cope, E. D. 1871 N, p. 269. (Holcodus.)
Ballou, W. H. 1898 A, pp. 216, 223, figure. (Platecarpus.)
Baur, G. 1892 B, pls. i, ii. (Platecarpus.)
Cope, E. D. 1872 F, pp. 322, 331. (Holcodus.)
    1874 C, p. 35. (Platecarpus.)
    1875 E, pp. 47, 142, pl. xv, fig. 1; pl. xvi, fig. 3; pl. xvii, fig. 6: pl. xx, figs. 4-7; pl. xxi, figs. 1, 2; pl. xxxvi, fig. 6; pl. xxxvii, fig. 9. (Platecarpus.)
Leidy, J. 1873 B, pp. 276, 344, pl. xxxiv, figs. 12-14; pl. xxxvi, figs. 4-14.
Marsh, O. C. 1872 F, pp. 457, 461. (Edestosaurus.)
Merriam, J. C. 1894 A, p. 29, pl. i, figs. 1, 2; pl. iii, figs. 3-5, 11; pl. iv, figs. 2, 8. (Platecarpus.)
Osborn, H. F. 1899 D, p. 171, text figs. 2, 3. (Platecarpus.)
    1900 A, p. 2, fig. 1. (Platecarpus.)

Wagner, G. 1896 A, p. 203.   (Platecarpus coryphæus?)
Williston, S. W. 1897 J, p. 110.   (Platecarpus.)
    1898 E, pp. 108, 145, 149, 158, 160, 186, pls. xlii-xv; pl. xxii, fig. 2; pl. xxiv, figs. 1-5; pl. xxv, figs. 1-5; pl. xxvi; pl. xxvii, fig.; pl. xxviii, fig. 5; pl. xxix, figs. 3-5; pl. xxxi, figs. 1-5; pl, xlii figs. 3, 4; pl. xlvi, figs. 1-4; pl. l; pl. lii; pl. liv. figs. 2, 3; pl. lv; pl. lvi; pl. lvii, figs. 4-10; pl. lviii; pl. lx, fig. 3; pl. lxi, fig. 3; pl. lxiii; pl. lxiv, fig. 2; pl. lxxii, fig. 2.   (Platecarpus.)
    1899 A, p. 39, pl. xii.   (Platecarpus.)
Woodward, A. S. 1898 B, p. 193, fig. 117.
    Cretaceous (Niobrara, Pierre?); Kansas.

Lestosaurus crassartus (Cope).

Cope, E. D. 1871 N, p. 278.   (Liodon.)
    1872 F, p. 332.   (Liodon.)
    1872 CC, p. 141.   (Liodon.)
    1874 C, p. 36.   (Platecarpus.)
    1875 E, pp. 153, 268, pl. xxvi, figs. 4-12. (Platecarpus.)
Marsh, O. C. 1872 F, p. 462.   (Liodon.)
Williston, S. W. 1898 E, p. 180, pl. xlv, figs. 3, 4. (Platecarpus?.)
    Cretaceous (Pierre); Colorado.

Lestosaurus curtirostris (Cope).

Cope, E. D. 1871 N, p. 273.   (Liodon.)
    1871 A.   (Liodon; no description.)
    1872 F, p. 331.   (Liodon.)
    1872 CC, p. 141.   (Liodon.)
    1874 C, p. 36.   (Platecarpus.)
    1875 E, pp. 150, 268, pl. xiv, fig. 3; pl. xv, fig. 3; pl. xvi, figs. 4, 5; pl. xvii, fig. 2; pl. xviii, figs. 7, 8; pl. xxi, figs. 7-13; pl. xxxvi, figs. 1, 5; pl. xxxvii, fig. 10; pl. xxxviii, fig. 1.   (Platecarpus.)
Lydekker, R. 1888 B, p. 270, fig. 59.   (Platecarpus.)
Marsh, O. C. 1872 F, p. 461.   (Lestosaurus.)
    Cretaceous (Niobrara); Kansas.

Lestosaurus felix Marsh.

Marsh, O. C. 1872 F, p. 457, pl. xiii.
Cope, E. D. 1874 C, p. 35.   (Platecarpus.)
    1875 E, p. 267.   (Platecarpus.)
Williston, S. W. 1898 E, p. 189.   (Platecarpus.)
    Cretaceous (Niobrara); Kansas.

Lestosaurus glandiferus (Cope).

Cope, E. D. 1871 N, p. 276.   (Liodon.)
    1872 F, p. 332.   (Liodon.)
    1874 C, p. 36.   (Platecarpus?)
    1875 E, pp. 156, 268, pl. xxvi, figs. 13, 14. (Platecarpus.)
Williston, S. W. 1898 E, p. 182.   (Platecarpus.)
    Cretaceous (Niobrara, Pierre?); Kansas.

Lestosaurus gracilis Marsh.

Marsh, O. C. 1872 F, p. 460.
Cope, E. D. 1874 C, p. 36.   (Platecarpus.)
    1875 E, p. 268.   (Platecarpus.)
Williston, S. W. 1898 E, p. 187.   (Platecarpus.)
    Cretaceous (Niobrara); Kansas.

Lestosaurus ictericus (Cope).

Cope, E. D. 1870 Q.   (Liodon.)
    1870 O, p. 577.   (Liodon.)
    1871 D, p. 406.   (Liodon.)
    1871 N, p. 272.   (Holcodus.)
    1872 F, p. 331.   (Holcodus.)
    1874 C, p. 35.   (Platecarpus.)
    1875 E, pp. 144, 267, pl. xiv, fig. 4; pl. xv, fig. 2; pl. xvii, figs. 3, 4; pl. xviii, fig. 6; pl. xix, figs. 1-3; pl. xxv, figs. 1-26; pl. xxxvi, fig. 7; pl. xxxvii, fig. 8.   (Platecarpus.)
Marsh, O. C. 1872 F, p. 461.   (Lestosaurus.)
Merriam, J. C. 1894 A, p. 30.   (Platecarpus.)
Williston, S. W. 1898 E, p. 184, pls. xliii, xliv. (Platecarpus.)
    Cretaceous (Niobrara); Kansas.

Lestosaurus latifrons Marsh.

Marsh, O. C. 1872 F, p. 458.
Cope, E. D. 1874 C, p. 36.   (Platecarpus.)
    1875 E, p. 268.   (Platecarpus.)
Williston, S. W. 1898 E, p. 189.   (Platecarpus.)
    Cretaceous (Niobrara); Kansas.

Lestosaurus? latispinus (Cope).

Cope, E. D. 1871 J, p. 169.   (Liodon.)
    1871 N, p. 276.   (Liodon.)
    1872 F, p. 332.   (Liodon.)
    1874 C, p. 38.   (Liodon.)
    1875 E, pp. 155, 268, pl. xxvi, figs. 1-4. (Platecarpus?.)
Williston, S. W. 1898 E. p. 181.   (Platecarpus?.)
    Cretaceous (Pierre); Kansas.

Lestosaurus mudgei (Cope).

Cope, E. D. 1870 Q.   (Liodon.)
    1870 O, p. 581.   (Liodon.)
    1871 D, p. 405.   (Liodon.)
    1871 N, p. 273.   (Holcodus.)
    1872 F, p. 331.   (Holcodus.)
    1874 C, p. 36.   (Platecarpus.)
    1875 E, pp. 157, 268, pl. xvi, fig. 2; pl. xvii, fig. 5; pl. xxvi, figs. 2, 3; pl. xxxvii, fig. 7. (Platecarpus.)
Marsh, O. C. 1872 F, p. 463, pl. xiii.   (Rhinosaurus.)
Williston, S. W. 1898 E,'p. 186.   (Platecarpus.)
    Cretaceous (Niobrara); Kansas.

Lestosaurus oxyrhinus (Merriam).

Merriam, J. C. 1894 A, p. 30.   (Platecarpus.)
Williston, S. W. 1898 E, p. 190.   (Platecarpus.)
    Cretaceous; Kansas.

Lestosaurus planifrons (Cope).

Cope, E. D. 1874 C, p. 31.   (Clidastes.)
    1875 E, pp. 135, 265; pl. xxii, figs. 1-7; pl. xxiii, figs. 1-15; pl. xxxv, fig. 16.   (Clidastes.)
    1878 G, p. 299.   (Clidastes.)
Williston, S. W. 1898 E, pp. 188, 194.   (Platecarpus.)
Williston and Case 1892 A, p. 17.   (Clidastes.)
    Cretaceous (Niobrara); Kansas.

Lestosaurus simus Marsh.

Marsh, O. C. 1872 F, p. 455, pls. x, xi.
Cope, E. D. 1874 C, p. 36.   (Platecarpus.)
    1875 E, p. 268.   (Platecarpus.)

Marsh, O. C. 1880 A, P. 86, pl. i.
1897 C, p. 527.
Merriam, J. C. 1894 A, p. 30. (Platecarpus.)
Owen, R. 1880 A, pl. viii, figs. 2, 3.
Williston, S. W. 1898 E, p. 188, pl. xlvii. (Platecarpus.)
Woodward, A. S. 1898 B, p. 193, fig. 117.
Zittel, K. A. 1890 A, p. 622, figs. 550, 552.
Cretaceous (Niobrara); Kansas.

Lestosaurus tectulus (Cope).

Cope, E. D. 1871 N, p. 271. (Holcodus.)
1872 F, p. 331. (Holcodus.)
· 1874 C, p. 36. (Platecarpus.)
1875 E, pp. 159, 269; pl. xxi, figs. 3-6; pl. xxvii, figs. 5-10. (Platecarpus.)
Williston, S. W. 1898 E, p. 188. (Platecarpus.)
Cretaceous (Pierre); Kansas.

BRACHYSAURUS Willist.   Type *B. overtoni* Willist.

Williston. S. W. 1897 D.
1897 F, p. 178.
1897 H, p. 46.
1898 E, p. 192.

Brachysaurus overtoni Willist.

Williston, S, W. 1897 D, p. 96, pl. viii.
1898 E, pp. 148, 192, pl. xxii; pl. xxx; pl. xl, figs. 10, 11; pl. lxii, fig. 1.
Cretaceous (Pierre); Kansas.

HOLOSAURUS Marsh.   Type *H. abruptus* Marsh.

Marsh, O. C. 1880 A, p. 87.
Dollo L. 1893 C, p. 221.
Williston, S. W. 1897 H, p. 245.   (Syn. of Platecarpus.)
1897 J, p. 107.
1898 E, p. 192.
Zittel, K. A. 1890 A, p. 624.

Holosaurus abruptus Marsh.

Marsh, O. C. 1880 A, p. 87.
1897 C, p. 527.
Williston, S. W. 1898 E, p. 193.
Cretaceous (Niobrara); Kansas.

TYLOSAURINÆ.

Williston, S. W. 1897 F, pp. 177, 180.
Marsh, O. C. 1876 G, p. 59, (Tylosauridæ.)

Williston, S. W. 1895 A, p. 169.   (Tylosauridæ.)
1898 E, pp. 90, 171.

MACROSAURUS Owen.   Type *M. lævis* Owen.

Owen, R. 1849 A, p. 382.
Cope, E. D. 1869 M, p. 200.   (Syn. of Liodon.)
1875 E, p. 270.   (Syn. of Liodon.)
Gibbes, R. W. 1851 A, p. 13.
· Leidy, J. 1865 A, p. 74.
Zittel, K. A. 1890 A, p. 622.   (Syn. of Liodon.)

Macrosaurus lævis Owen.

Owen, R. 1849 A, p. 382, pl. xi, figs. 1-6.
Cope, E. D. 1867 E, p. 26.
1869 A, p. 260.
1869 B, p. 734.
1869 M, pp. 205, in part.   (Liodon.)
1875 E, p. 270.   (Liodon.)
Leidy, J. 1865 A, pp. 74, 118.
1865 D, p. 70.
Owen, R. 1879 B, p. 54.   (Leiodon.)
Cretaceous; New Jersey.

Macrosaurus mitchillii (De Kay).

DeKay, J. E. 1830 A, p. 140, pl. iii, figs. 3, 4. (Geosaurus.)

Cope, E. D, 1869 A, p. 261. (Mosasaurus impar.)
1869 M, pp. 193, 205.   (Liodon.)
1871 G, p. 43.   (Liodon.)
1875 E, p. 270.   (Liodon.)
1875 V, p. 41.   (Liodon.)
De Kay, J. E. 1842 A, p. 28, pl. xxii, figs. 55, 56.   (Geosaurus.)
Emmons, E. 1858 B, p. 224, figs. 45, 46.   (Drepanodon impar.)
Gibbes R. W. 1851 A, p. 9.   (Geosaurus.)
Giebel, C. G. 1847 B, p. 134.   (Geosaurus.)
Harlan, R. 1834 B, p. 82.   (Geosaurus.)
1835 B, p. 285.   (Geosaurus.)
Leidy, J. 1856 L, p. 255.   (Drepanodon impar.) ·
1857 A, p. 271.   (Drepanodon impar.)
1859 B, p. 10.   (Lesticodus impar.)
1859 D, p. 92.   (Geosaurus.)
1835 A, p. 65, pl. xi, figs. 1-4, fide Cope.   (Mosasaurus mitchillii.)
1865 D, p. 70 in part.
Williston, S. W. 1897 F, pp. 182, 183.   (Tylosaurus.)
1898 E, p. 172.   (Tylosaurus.)
Cretaceous; New Jersey.

TYLOSAURUS Marsh.   Type T. *micromus* Marsh.

Marsh, O. C. 1872 H, p. 147.   (To replace Rhinosaurus Marsh and Rhamphosaurus Cope, both preoccupied.)
Since the American species assigned to *Tylosaurus* may prove to be congeneric with those

called by Owen *Leiodon* and by others *Liodon*, references to literature of the latter title is given. *Leiodon* is, however, preoccupied by Swainson, 1839.
Charlesworth, E. 1846 A. (Liodon.)

Cope, E. D. 1869 M, p. 200. (Liodon.)
1871 D, p. 401. (Liodon.)
1871 N, p. 278. (Liodon.)
1872 F, p. 331. (Liodon.)
1872 CC, p. 141. (Rhamphosaurus, to replace Rhinosaurus.)
1874 C, pp. 31, 36, 37. (Liodon, Nectoportheus.)
1875 E, pp. 128, 160, 271. (Liodon, Nectoportheus.)
1878 G. (Liodon.)
1879 J. (Liodon.)
Dollo, L. 1885 C, pp. 28, 30. (Liodon, Tylosaurus.)
Gaudry, A. 1892 A, p. 2. (Leiodon.)
Leidy, J. 1873 B, p. 343.
Lydekker, R. 1888 B, p. 264. (Liodon.)
Marsh, O. C. 1872 C, p. 291. (Liodon.)
1872 F, pp. 455, 461. (Rhinosaurus, type R. micromus.)
1880 A, p. 84.
Merriam, J. C. 1894 A, p. 14,
Nicholson and Lydekker 1889 A, p. 1144.
Osborn, H. F. 1899 D, p. 167.
Owen, R. 1842 A, p. 144. (Leiodon, type L. anceps Owen.)
1845 B, p. 261, pl. lxxii. (Leiodon.)
1851 B, p. 41. (Leiodon.)
1879 B. (Leiodon.)
1880 A. (Leiodon.)
Pictet, F. J. 1853 A, p. 507. (Leiodon.)
Snow, F. H. 1878 A, p. 54. (Liodon.)
Steinmann and Döderlein 1890 A, p. 646. (Liodon.)
Williston, S. W. 1891 A. (Liodon.)
1897 E.
1897 F, p. 177.
1897 J, p. 108.
1898 E, p. 171.
Zittel, K. A. 1890 A, p. 622. (Liodon.)

Tylosaurus congrops (Cope).
Cope, E. D. 1869 M, p. 206. (Liodon.)
1870 D, p. 271. (Liodon.)
1875 E, p. 272. (Liodon.)
Cretaceous; Alabama.

Tylosaurus dyspelor (Cope).
Cope, E. D. 1870 Q. (Liodon.)
1870 O, p. 574. (Liodon.)
1871 D, p. 410. (Liodon.)
1871 J, p. 169. (Liodon.)
1871 K, p. 174, (Liodon.)
1871 N, p. 280. (Liodon.)
1872 F, pp. 321, 333. (Liodon.)
1872 CC, p. 141. (Liodon.)
1874 C, p. 38. (Liodon.)
1875 E, pp. 46, 167, 271, pl. xxvii, fig. 11; pl. xxviii, figs. 1–7; pls. xxix–xxxiii; pl. xxx 1, fig. 1; pl. xxxvii, fig. 5. (Liodon.)
Gibbes, R. W. 1851 A, p. 6. (Liodon.)
Leidy, J. 1873 B, pp. 271, 343, pl. xxxv, figs. 1–11.
Marsh, O. C. 1872 F, p. 463. (Rhinosaurus.)
1872 T, p. 496. (Liodon.)
Merriam, J. C. 1894 A, p. 23, pl. iv, figs. 3, 4.
Osborn, H. F. 1899 D, p. 167, pls. xxi–xxiii, text figs. 1, 7–9, 11–14.
1899 F, p. 913.

Snow, F. H. 1878 A, p. 54. (Liodon.)
Williston, S. W. 1898 E, p. 145 et seq., pl. xliii, figs. 1, 2; pl. xlvi, fig. 4; pl. lix; pl. lxi, figs. 1, 2.
1897 J, p. 110.
Cretaceous (Niobrara); New Mexico, Kansas.

Tylosaurus laticaudus (Marsh).
Marsh, O. C. 1870 F. (Liodon.)
Cretaceous; New Jersey.

Tylosaurus micromus Marsh.
Marsh, O. C. 1872 F, p. 461, pl. xiii. (Rhinosaurus.)
Cope, E. D. 1874 C, p. 36. (Liodon.)
1875 E, p. 271. (Liodon.)
Marsh, O. C. 1880 A, p. 85, fig. 1.
Merriam, J. C. 1894 A, p. 28, pl. i, fig. 3; pl. iii, fig. 8.
Williston, S. W. 1898 E, p. 175.
Cretaceous (Niobrara); Kansas.

Tylosaurus nepæolicus (Cope).
Cope, E. D. 1874 C, p. 37. (Liodon, Rhamphosaurus.)
1875 E, pp. 177, 271, pl. xxxv, figs. 11–15. (Liodon.)
Williston, S. W. 1898 E, p. 176.
Cretaceous (Niobrara); Kansas.

Tylosaurus perlatus (Cope).
Cope, E. D. 1869 M (1870), p. 11 of Errata and Addenda (Liodon); p. 198 ("Mosasaurus brumbyi").
1870 J, p. 497. (Liodon.)
1875 E, p. 271. (Liodon.)
Lydekker, R. 1888 B, p. 267. (Liodon.)
Cretaceous; Alabama.

Tylosaurus proriger (Cope).
Cope, E. D. 1869 H, p. 123. (Macrosaurus.)
Ballou, W. H. 1898 A, pp. 217, 224, figure.
Cope, E. D. 1869 M, p. 202, pl. xii, figs. 22–24. (Liodon.)
1871 A. (Liodon.)
1871 D, p. 401. (Liodon.)
1871 K, p. 175. (Liodon.)
1871 N, p. 271. (Liodon.)
1872 F, pp. 321, 333. (Liodon.)
1872 CC, p. 111. (Rhamphosaurus, provisionally.)
1874 C, p. 38. (Liodon.)
1875 E, pp. 46, 161, 271, pl. xxviii, figs. 8, 9; pl. xxx, figs. 10–14; pl. xxxvi, fig. 2; pl. xxxvii, fig. 6. (Liodon.)
Leidy, J. 1870 E, p. 4. (Liodon.)
1873 B, pp. 274, 344, pl. xxxv, fig. 13; pl. xxxvi, figs. 1–3.
Marsh, O. C. 1872 F, p. 463. (Rhinosaurus.)
Merriam, J. C. 1894 A, p. 23, pl. ii; pl. iii, figs. 1, 2.
Osborn, H. F. 1899 D, p. 169.
Williston, S. W. 1897 E, p. 102, pls. ix, x.
1897 F, pl. xx.
1897 J, p. 110.
1898 E, p. 173, pls. xvi–xviii; pl. xxix, fig. 1; pl. xl, figs. 1–7; pl. xli, fig. 1; pl. xlviii, fig. 7; pl. lv; pl. lx, figs. 1, 2; pls. lxv–lxx; pl. lxxii, fig. 3.
Cretaceous (Niobrara); Kansas.

Tylosaurus rapax sp. nov.

Cope, E. D. 1869 M, p. 187, text fig. 48^b; p. 188, text fig. 49^4; pp. 201, 207. (Liodon validus; not Nectoportheus validus Cope, 1868.)
1870 D, p. 272. (Liodon validus.)
1875 E, p. 271, pl. xxxvii, fig. 4. (Liodon validus.)

Leidy, J. 1865 A, p. 74, in part, pl. iii, figs. 1-2; pl. vii, figs. 19, 20. (Macrosaurus lævis.)
Cretaceous; New Jersey.

Tylosaurus sectorius (Cope).

Cope, E. D. 1871 G, p. 41. (Liodon.)
1875 E, p. 271. (Liodon.)
Cretaceous; New Jersey.

DIPLOTOMODON Leidy.　Type *D. horrificus* Leidy.

Leidy, J. 1868 G, p. 202. (To replace Tomodon, preoccupied.)
Cope, E. D. 1868 C, p. 417. (Tomodon.)
1875 E, p. 272.
Leidy, J. 1865 A, p. 102. (Tomodon, type *T. horrificus*; not Tomodon Duméril, 1853.)
Zittel, K. A. 1890 A, pp. 624, 727.

Diplotomodon horrificus Leidy.

Leidy, J. 1865 A, pp. 102, 119, pl. xx, figs. 7-9. (Tomodon.)
Cope, E. D. 1875 E, p. 272.
Leidy, J. 1865 D, p. 72. (Tomodon.)
1868 G, p. 202.
Cretaceous; New Jersey.

ELLIPTONODON Emmons.　Type *E. compressus* Emmons.

Emmons, E. 1858 B, p. 222.
A genus whose value and systematic position are doubtful.

Elliptonodon compressus Emmons.

Emmons, E. 1858 B, p. 222, figs. 41, 42.
Cope, E. D. 1869 M, p. 186.
Leidy, J. 1865 A, pp. 63, 117.
Cretaceous; North Carolina.

POLYGONODON Leidy.　Type *P. vetus* Leidy.

Leidy, J. 1856 K, p.221.
Cope, E. D. 1869 M, pp. 185, 210.
A genus of uncertain ordinal position.

Polygonodon rectus Emmons.

Emmons, E. 1858 B, p. 218 (Mosasaurus); fig. 37 A (Polygonodon).
Cope, E. D. 1869 M, p. 211.
1875 V, p. 41.
Emmons, E. 1860 A, p. 208, fig. 180^2.
Leidy, J. 1865 A, pp. 76, 118. (Syn. of *P. vetus*.)
Miocene; North Carolina.

Polygonodon vetus Leidy.

Leidy, J. 1856 K, p. 221.
Cope, E. D. 1869 B, p. 734.
1869 M, p. 210.
1875 E, p. 280.
Leidy, J. 1865 A, pp. 76, 118, pl. ix, figs. 12, 13.
1865 D, p. 70.
Cretaceous; New Jersey.

Suborder SAURIA Macartney.

Macartney, J., 1802, Lectures on Comp. Anat., i, tab. 3.
　Unless otherwise indicated, the following authors use the term *Lacertilia* for this order.
Baur, G. 1885 E, p. 634.
　1887 C, p. 99.
　1895 A.
　1896 D.
　1896 A.
Born, G. 1876 A. ("Saurier.")
　1880 A, p. 63. ("Saurier.")
Boulenger, G. A. 1884 A.
　1885 A.
　1887 A.
　1891 A.
　1893 A, p. 1.
　1895 A.
Burckhardt, C. K. 1895 A.
Calori, L. 1894 A.
Cope E. D. 1864 A.
　1859 M, p. 28.

Cope, E. D. 1871 B.
　1878 G, p. 300.
　1882 J, p. 981.
　1887 S, p. 384.
　1891 N, pp. 45, 47.
　1892 F.
　1892 Z, p. 19.
　1895 C, p. 857.
　1895 D.
　1896 A.
　1896 B.
　1898 B, p. 74.
Cuvier, G. 1834 A, x, pp. 1-98.
Dollo, L. 1885 B.
Fürbringer, M. 1900 A, pp. 231, 537, 605, 620.
Gadow, H. 1896 A, p. 29.
　1898 A. p. 24. (Sauria.)
Gaupp, E. 1894 A.
　1895 B.
　1898 A, p. 302.
　1900 A.

Gegenbaur, C, 1864 A.
   1865 B, p. 40.
   1898 A.
Gill, T. 1900 A, p. 730.
Gorski, C. 1858 A.
Günther, A. 1886 A, p. 444.
Haeckel, E. 1895 A. p. 344.
Hoffmann, C. K. 1876 A.
   1878 A.
   1879 A, p. 99.
   1890 B. (Sauril.)
Huxley, T. H. 1870 A.
   1872 A, p. 186.
·   1876 B, xiii, p. 429.
Leidy, J. 1871 C, p. 368.
Levy, H. 1898 A.
Lydekker, R. 1888 B, p. 275.
Marsh, O. C. 1877 E.
Mivart, St. G. 1886 A, p. 448.
Müller, J. 1831 A, p. 238. (Lacertina.)
Nicholson and Lydekker 1889 A, p. 1138.

Osborn, H. F. 1900 A, p. 1
Owen, R. 1842 A, p. 144.
   1860 A, p. 165.
   1866 A.
   1877 A.
Parker, W. K. 168 A, p. 100.
   1879 B, p. 63.
   1880 A.
Rathke, H. 1853 A. ("Saurier.")
Segond, L. A. 1873 A, p. 14. (" Lézard.")
Siebenrock, F. 1892 A.
   1893 A.
   1894 A.
   1895 A.
   1895 B.
Stannius, H. 1856 A. (Sauria.)
Steinmann and Döderlein 1890 A, p. 644.
Tomes, C. S. 1898 A. p. 270. (" Saurians.")
Wiedersheim, R. 1889 A, p. 438.
Zittel, K. A. 1890 A, pp. 450, 602.

IGUANIDÆ.

Gray, J. E., 1827, Philos. Mag. (2), ii, p. 127.
Boulenger, G. A. 1884 A, p. 119.
   1885 A, p. 1.
Cope, E. D. 1864 A, p. 227. (Anolidæ, Iguanidæ.)
   1900 A, p. 220.
Gadow, H. 1898 A. p. 25.

Hoffmann, C. K. 1890 B, p. 1217.
Lydekker, R. 1888 B, p. 277.
Parker, W. K. 1868 A, p. 107. (Iguana.)
Siebenrock, F. 1895 B. (On Agamidæ.)
Stannius, H. 1856 A. (Iguanoidea.)
Zittel, K. A. 1890 A, p. 607.

IGUANAVUS Marsh.   Type *I. exilis* Marsh.

Marsh, O. C. 1872 K, p. 309.
Nicholson and Lydekker 1889 A, p. 1140.

King, C. 1878 A, p. 405.
   Eocene (Bridger); Wyoming.

Iguanavus exilis Marsh.

Marsh, O. C. 1872 K, p. 309.

Iguanavus teres Marsh.

Marsh, O. C. 1892 E, p. 451.
   Cretaceous (Laramie); Wyoming.

CHAMOPS Marsh.   Type *C. segnis* Marsh.

Marsh, O. C. 1892 E., p. 450.
   Provisionally assigned to the *Iguanidæ.*

Chamops segnis Marsh.

Marsh, O. C. 1892 E, p. 450, figs. 2, 3.
   Cretaceous (Laramie); Wyoming.

ANGUIDÆ.

Cope, E. D. 1864 A. pp. 227, 228. (Anguidæ, Gerrhonotidæ.)
Boulenger, G. A. 1884 A, p. 121.
   1885 A, ii, p. 265.
Cope, E. D. 1877 K, p. 42. (Placosauridæ.)
   1882 J, p. 981. (Placosauridæ.)
   1884 O, p. 772. (Gerrhonotidæ.)
   1900 A, p. 488.
Gadow, H. 1898 A. p. 25.

Gaupp, E. 1898 A. (Anguis.)
Hoffmann, C. K. 1890 B. (Anguis.)
Lydekker, R. 1888 B, p. 278.
Marsh, O. C. 1877 E, p. 222. (Glyptosauridæ.)
Nicholson and Lydekker 1889 A, p. 1140.
Parker, W. K. 1868 A, p. 98.
Siebenrock, F. 1892 A.
   1895 A.
Zittel, K. A. 1890 A, p. 607.

PLACOSAURUS Gervais.   Type *P. rugosus* Gervais.

Gervais, P., 1852, Zool. Paléont. France, ed. i, p. 260.
Lydekker, R. 1888 C, p. 110.
Nicholson and Lydekker 1889 A, p. 1140.
Pictet, Gaudin, and Harpe 1857 A, p. 93.
Zittel, K. A. 1890 A, p. 608.

Placosaurus ? sp. indet.

Cope, E. D. 1877 K, p. 42, pl. xxxii, figs. 26–36.
   Eocene; New Mexico.
   The fossil fragments above described Professor Cope referred with doubt to the Placosauridæ.

GLYPTOSAURUS Marsh.   Type *G. sylvestris* Marsh.

Marsh, O. C. 1871 B, p. 456.
Boulenger, G. A. 1891 A, p. 117.
Cope, E. D. 1875 C, p. 34.
Leidy, J. 1872 B, p. 371.
      1873 B, p. 182.
Marsh, O. C. 1871 G, p. 105.
      1872 K, p. 302.
Nicholson and Lydekker 1889 A, p. 1141.
Zittel, K. A. 1890 A, p. 608.

Glyptosaurus anceps Marsh.
Marsh, O. C. 1871 B, p. 458.
      1871 G, p. 105.
Eocene (Bridger); Wyoming.

Glyptosaurus brevidens Marsh.
Marsh, O. C. 1872 K, p. 305.
Eocene (Bridger); Wyoming.

Glyptosaurus nodosus Marsh.
Marsh, O. C. 1871 B, p. 458.
      1871 G, p. 105.
Eocene (Bridger); Wyoming.

Glyptosaurus ocellatus Marsh.
Marsh, O. C. 1871 B, p. 458.
? Leidy, J. 1873 B, p. 183, pl. xvi, figs. 13–17.
Marsh, O. C. 1871 G, p. 105.
      1872 K, p. 306.
Eocene (Bridger); Wyoming.

Glyptosaurus princeps Marsh.
Marsh, O. C. 1872 K, p. 302.
King, C. 1878 A, p. 405.
Eocene (Bridger); Wyoming.

Glyptosaurus rugosus Marsh.
Marsh, O. C. 1872 K, p. 305.
Eocene (Bridger); Wyoming.

Glyptosaurus sphenodon Marsh.
Marsh, O. C. 1872 K, p. 306.
Eocene (Bridger); Wyoming.

Glyptosaurus sylvestris Marsh.
Marsh, O. C.. 1871 B, p. 456.
      1871 G, p. 105.
Eocene (Bridger); Wyoming.

XESTOPS Cope.   Type *Oreosaurus vagans* Marsh.

Cope, E. D. 1873 CC, p. 19. (To replace Oreosaurus
      Marsh, preoccupied.)
Marsh, O. C. 1872 K, p. 308. (Oreosaurus ,type
      O. vagans; not Oreosaurus Peters, 1862.)
Nicholson and Lydekker 1889 A, p. 1141.
Zittel, K. A. 1890 A, p. 608.

Xestops gracilis (Marsh).
Marsh, O. C. 1872 K, p. 307. (Oreosaurus.)
Eocene (Bridger); Wyoming.

Xestops lentus (Marsh).
Marsh, O. C. 1872 K, p. 307. (Oreosaurus.)
King, C. 1878 A, p. 405.
Eocene (Bridger); Wyoming.

Xestops microdus (Marsh).
Marsh, O. C. 1872 K, p. 308. (Oreosaurus.)
Eocene (Bridger); Wyoming.

Xestops minutus (Marsh).
Marsh, O. C. 1872 K, p. 308. (Oreosaurus.)
Eocene (Bridger); Wyoming.

Xestops vagans (Marsh).
Marsh, O. C. 1872 K, p. 308. (Oreosaurus.)
Eocene (Bridger); Wyoming.

SANIWA Leidy.   Type *S. ensidens* Leidy.

Leidy, J. 1870 S, p. 124.
      1871 C, p. 368.
      1872 B, p. 370.
      1873 B, p. 181. (Saniva.)
Marsh, O. C. 1872 K, p. 299.
Nicholson and Lydekker 1889 A, p. 1141. (Saniva.)
Zittel, K. A. 1890 A, p. 608. (Saniva.)

Saniwa ensidens Leidy.
Leidy, J. 1870 S, p. 124.
Cope, E. D. 1873 E, p. 632. (Saniva.)
King, C. 1878 A, p. 405.

Leidy, J. 1871 C, p. 368.
      1872 B, p. 370.
      1873 B, pp. 181, 344, pl. xv, fig. 15; pl. xxvii,
      fig. 35. (Saniva.)
Marsh, O. C. 1871 B, p. 457. (Saniva.)
Eocene (Bridger); Wyoming.

Saniwa major Leidy.
Leidy, J. 1873 B, pp. 182, 345, pl. xv, fig. 14; pl.
      xxvii, figs. 36, 37. (Saniva.)
Eocene (Bridger); Wyoming.

PELTOSAURUS Cope.   Type *P. granulosus* Cope.

Cope, E. D. 1873 S, p. 5.
      1873 CC, p. 19.
      1874 B, p. 512.
      1884 O, p. 771.
Hoffmann, C. K. 1890 B, p. 1324.
Nicholson and Lydekker 1889 A, p. 1140.
Zittel, K. A. 1890 A, p. 608.

Peltosaurus granulosus Cope.
Cope, E. D. 1873 S, p. 5.
      1873 CC, p. 19.
      1874 B, p. 513.
      1874 H.
      1884 O, p. 773, pl. lx, figs. 3–11.
Miocene; Colorado.

Exostinus Cope.   Type *E. serratus* Cope.

Cope, *E. D.* 1878 CC, pp. 16, 19.
    1874 B, p. 513.
    1884 O, p. 775.
Nicholson and Lydekker 1889 A, p. 1140.
Zittel, K. A. 1890 A, p. 610.

Exostinus serratus Cope.

Cope, *E. D.* 1873 CC, p. 16.
    1874 B, p. 514.
    1884 O, p. 776, pl. lx, figs. 12–14.
Oligocene (White River); Colorado.

VARANIDÆ.

Cope, *E. D.* 1864 A, p. 227.
Baur, G. 1890 A, p. 292.
Beddard, F. E. 1888 A.
Boulenger, G. A. 1884 A, p. 122.
    1885 A, p. 308.
    1891 A, p. 117.
Cope, E. D. 1878 G, p. 305.
    1900 A, p. 470.

Dollo, L. 1893 C, p. 252.
Fürbringer, M. 1900 A.
Gadow, H. 1898 A, p. 26.
Hoffmann, C. K. 1900 B, p. 1068.   (Monitoridæ.)
Lydekker, R. 1888 B, p. 281.
Parker, W. K. 1868 A, p. 116.   (Monitor,)
Stannius, H. 1856 A.   (Varani.)
Zittel, K. A. 1890 A, p. 608.

Thinosaurus Marsh.   Type *T. paucidens* Marsh.

Marsh, *O. C.* 1872 K, p. 299.
Boulenger, G. A. 1891 A, p. 115.
Hoffmann, C. K. 1890 B, p. 1324.
Zittel, K. A. 1890 A, p. 609.

Thinosaurus grandis Marsh.

Marsh, *O. C.* 1872 K, p. 301.
Eocene (Bridger); Wyoming.

Thinosaurus agilis Marsh.

Marsh, *O. C.* 1872 K, p. 302.
Eocene (Bridger); Wyoming.

Thinosaurus leptodus Marsh.

Marsh, *O. C.* 1872 K, p. 300.
Cope, E. D. 1873 E, p. 632.
King, C. 1878 A, p. 405.
Eocene (Bridger); Wyoming.

Thinosaurus crassus Marsh.

Marsh, *O. C.* 1892 K, p. 301.
Eocene (Bridger); Wyoming.

Thinosaurus paucidens Marsh.

Marsh, *O. C.* 1872 K, p. 299.
Eocene (Bridger); Wyoming.

Tinosaurus Marsh.   Type *T. stenodon* Marsh.

Marsh, *O. C.* 1872 K, p. 304.
Zittel, K. A. 1890 A, p. 609.
The position of this genus is uncertain.

Tinosaurus stenodon Marsh.

Marsh, *O. C.* 1872 K, p. 304.
Eocene (Bridger); Wyoming.

Tinosaurus lepidus Marsh.

Marsh, *O. C.* 1872 K, p. 308.
Eocene (Bridger); Wyoming.

AMPHISBÆNIDÆ.

Gray, J. E., 1825, Philos. Mag. (2), x.
Baur, G. 1893 C.   (Amphisbænidæ; Hyporhinidæ.)
Boulenger, G. A. 1884 A.
    1885 A, p. 430.
Cope, E. D. 1864 A, p. 226.   (Amphisbænia.)
    1871 B.
    1900 A, p. 682.
Fürbringer, M. 1900 A, pp. 258, 539, 616, 622.   (Amphisbænia; Amphisbænidæ.)

Gadow, 1898 A, p. 27.
Gervais, P. 1853 A, p. 298.   ("Amphisbènes.")
Haeckel, E. 1895 A, p. 349.   (Glyptoderma.)
Hoffmann, C. K. 1890 B, p. 1288.
Müller, J. 1831 A, p. 258.   (Amphisbænoidea).
Owen, R. 1845 B, p. 334.   (Amphisbænæ.)
Parker, W. K. 1868 A, p. 95.   (Amphisbænæ.)
Stannius, H. 1856 A.   (Amphisbænoida.)

Rhineûra Cope.   Type *R. floridana* Cope.

Cope, *E. D.*, 1861, Proc. Acad. Nat. Sci. Phila., p. 75.
Boulenger, G. A. 1885 A, p. 459.
Cope, E. D. 1900 A, p. 684.
Fürbringer, M. 1900 A, p. 265.

Rhineûra hatcherii Baur.

Baur, *G.* 1893 C.
Oligocene (White River); South Dakota.

HYPSORHINA Baur.   Type *H. antiqua* Baur.

Baur, G. 1898 C.

Hypsorhina antiqua Baur.

Baur, G. 1898 C.
   Oligocene (White River); South Dakota.

TYLOSTEUS Leidy.   Type *T. ornatus* Leidy.

Leidy, J. 1872 E, p. 40.
   1873 B, p. 285.
   The position of this fossil is doubtful. It is not improbably a portion of the dermal armor of some Dinosaur.

Tylosteus ornatus Leidy.

Leidy, J. 1872 E, p. 40.
   1873 B, pp. 285, 345. pl. xix, fig. 14: fig. 137.
   Cretaceous?; Montana? (Head of Missouri River.)

ACIPRION Cope.   Type *A. formosum* Cope.

Cope, E. D. 1873 CC, p. 17.
   1874 B, p. 514.
   1884 O, p. 776.

Aciprion formosum Cope.

Cope, E. D. 1873 CC, p. 17.
   1874 B, p. 514.
   1884 O, p. 776, pl. lx, fig. 15.
   Oligocene (White River); South Dakota.

DIACIUM Cope.   Type *D. quinquepedale* Cope.

Cope, E. D. 1873 CC, p. 17.
   1874 B, p. 514.
   1884 O, p. 777.

Diacium quinquepedale Cope.

Cope, E. D. 1873 CC, p. 17.
   1874 B, p. 514.
   1884 O, p. 777, pl. lx, fig. 20.
   Oligocene (White River); Colorado.

CREMASTOSAURUS Cope.   Type *C. carinicollis* Cope.

Cope, E. D. 1873 CC, p. 18.
   1874 B, p. 515.
   1884 O, p. 780.

Cremastosaurus carinicollis Cope.

Cope, E. D. 1873 CC, p. 18.
   1874 B, p. 515. (In part.)
   1884 O, pp. 779, 781, pl. lx, fig. 16.
   Oligocene (White River); Colorado.

PLATYRHACHIS Cope.   Type *P. coloradoensis* Cope.

Cope, E. D. 1873 CC, p. 19.
   1874 B, p. 516.
   1884 O, p. 777.

Platyrhachis coloradoensis Cope.

Cope, E. D. 1873 CC, p. 19.
   1874 B, p. 516.
   1884 O, p. 778, pl. lx, fig. 17.
   Oligocene (White River); Colorado.

Platyrhachis rhambastes Cope.

Cope, E. D. 1884 O, p. 779, pl. lx, fig 18.
   1874 B, p. 516. (Cremastosaurus carinicollis, in part.)
   Oligocene (White River); Colorado.

Platyrhachis unipedalis Cope.

Cope, E. D. 1873 CC, p. 18. (Diacium.)
   1874 B, p. 516. (Cremastosaurus.)
   1884 O, p. 779, pl. lx, fig. 19.
   Oligocene (White River); Colorado.

NAOCEPHALUS Cope.   Type *N. porrectus* Cope.

Cope, E. D. 1872 M, p. 465.
   1873 E, p. 631.
   A genus of uncertain affinities.

Naocephalus porrectus Cope.

Cope, E. D. 1872 M, p. 465.
   1873 E, p. 632.
   Eocene (Bridger); Wyoming.

Suborder RHIPTOGLOSSI Wiegmann.

Wiegmann, A. F. A., 1834, Herp. Mex., p. 13.
Boulenger, G. A. 1884 A, p. 120.   (Rhiptoglossa.)
    1887 A, p. 437.   (Rhiptoglossa.)
    1891 A, p. 117.   (Rhiptoglossa.)
Cope, E. D. 1864 A.   (Rhiptoglossa.)
    1895 C, p. 859.   (Rhiptoglossa.)
    1900 A, p. 208.   (Rhiptoglossa.)
Dollo, L. 1893 C, p. 259.

Fürbringer, M. 1900 A, pp. 265, 586, 619, 622.
    (Chamæleontina.)
Gadow, H. 1898 A, p. 27.   (Chamæleontes.)
Haeckel, E. 1895 A, p. 348.   (Chamæleontes.)
Nicholson and Lydekker 1889 A, p. 1142.   (Rhip-
    toglossa.)
Tomes, C. S. 1898 A, p. 272.   ("Chamæleons.")
Stannius, H. 1856 A.   (Chamæleonidea.)

CHAMÆLEONIDÆ.

Gray, J. E. 1825, Ann. Philos. (2), x, p. 200.
Boulenger, G. A. 1884 A, p. 120.
    1887 A, p. 437.
Cope, E. D. 1895 C, p. 859.

Cope, E. D. 1900 A, p. 209.
Hoffmann, C. K. 1890 B, p. 1280.
Nicholson and Lydekker 1889 A, p. 1142.
Siebenrock, F. 1893 B.

CHAMÆLEO Laurenti.

Laurenti, J. N., 1768, Syn. Rept., p. 45.
Born, G. 1876 A, p. 21.
Boulenger, G. A. 1887 A, p. 438.   (Chamæleon.)
Cope, E. D. 1900 A, p. 211.   (Chamæleon.)
Deslongchamps-Eudes, J. A. 1838 A, p. 96.   ("Ca-
    méléon.")
Fürbringer, M. 1900 A.
Hoffmann, C. K. 1890 B, pp. 450, 1280.
Parker, W. K. 1868 A, p. 122.

Rathke, H. 1853 A, p. 25.
Siebenrock, F. 1893 B.
Stecker, A. 1877 A.   (Chamæleon.)

Chamæleo pristinus Leidy.

Leidy, J. 1872 J, p. 277.
    1873 B, pp. 184, 345, pl. xxvii, figs. 38, 39.
    Eocene; Wyoming.

Suborder SERPENTES Linn.

Linnæus, C., 1758, Syst. Nat., ed. x, p. 194.
    The following authorities, unless otherwise in-
    dicated, refer to this group under the name
    Ophidia.
Baur, G. 1887 C, p. 99.
    .   1895 A.
    1896 A.
Boulenger, G. A. 1893 A.
    .   1895 A.
    1896 A.
    1896 C.
Calori, L. 1894 A.
Cope, E. D. 1864 A.
    1871 B.
    1878 G, p. 300.
    1882 J, p. 981.
    1886 V.
    1887 S, p. 334.
    1891 N, pp. 45, 50.
    1892 Z, p, 20.
    1895 C, p. 857.
    1895 D.
    1896 A.
    1896 B.
    1898 B, p. 75 (Serpentes); p. 79 (Ophidia).
Fürbringer, M. 1900 A.
Gadow, H. 1889 A, p. 471.

.Gadow, H. 1898 A, p. 28.
Gill, T. 1900 A, p. 780.
Günther, A. C. 1866 A, p. 445.
Haeckel, E. 1895 A, p. 353.   (Serpentes.)
Hoffmann, C. K. 1890 C.
Huxley, T. H. 1872 A, p. 200.
Levy, H. 1900 A.
Lydekker, R. 1888 B, p. 249.
Marsh, O. C. 1877 E.
Müller, J. 1831 A, p. 268.
Nicholson and Lydekker 1889 A, p. 1146.
Owen, R. 1845 B, p. 219.
    1850 A, p. 51.
    1860 A, p. 166.   ·
    1866 A.
    1877 A.
Parker, W. K. 1879 A.
    1879 B.
Pictet, F. J. 1853 A, p. 555.   ("Ophidiens.")
Rochebrune, A. T. 1881 A.
Segond, L. A. 1873 A, p. 17.   ("Serpent.")
Stannius, H. 1856 A.
Steinmann and Döderlein 1890 A, p. 646.
Tomes, C, S. 1895 A, p. 273.
Walter, F. 1887 A.
Zittel, K. A. 1890 A, pp. 450, 634.

PALÆOPHIDÆ.

Lydekker, R. 1888 C, p. 118.
    1888 B, p. 256.

Nicholson and Lydekker 1889 A, p. 1147.

PALÆOPHIS Owen.   Type *P. toliapicus* Owen.

Owen, R. 1840 G, p. 165.
Cope, E. D. 1868 K, p. 234.
    1869 M, pp. 227, 299.
Lydekker, R. 1888 C, p. 112.
    1888 B, p. 256.
Marsh, O. C. 1869 C, p. 400. (Dinophis, type *Palæophis littoralis* Cope.)
    1871 A, p. 323. (Dinophis.)
    1877 E, p. 223. (Titanophis, to replace Dinophis, preoccupied.)
Nicholson and Lydekker 1889 A, p. 1148.
Owen, R. 1841 B, p. 209.
    1842 A, p. 150.
    1850 A, p. 56.
Pictet, F. J. 1853 A, p. 556.
Rochebrune, A. T. 1880 A, p. 274.
Zittel, K. A. 1890 A, p. 628. (Palæophis, Titanophis.)

Palæophis grandis (Marsh).

Marsh, O. C. 1869 C. (Dinophis.)
Cope, E. D. 1869 M, p. 228.

Cope, E. D. 1872 Y, p. 14.
    Eocene; New Jersey.

Palæophis halidanus Cope.

Cope, E. D. 1868 K, p. 234.
    1869 B, p. 738.
    1869 M, p. 227, pl. v, fig. 2.
    1872 Y, p. 14.
    1882 J, p. 981, fig. 2.
Marsh, O. C. 1869 C, p. 400. (Dinophis.)
    Eocene; New Jersey.

Palæophis littoralis Cope.

Cope, E. D. 1868 F.
    1868 K, p. 234.
    1869 B, p. 737.
    1869 M, p. 227, pl. v, fig. 1.
    1872 Y, p. 14.
    1882 J, p. 981, fig. 1.
Marsh, O. C. 1869 C, p. 400. (Dinophis.)
    Eocene; New Jersey.

PTEROSPHENUS Lucas.   Type *P. schucherti* Lucas.

Lucas, F. A. 1898 B.

Pterosphenus schucherti Lucas.

Lucas, F. A. 1898 B, p. 637, pls. xlv, xlvi.

Schuchert, C. 1900 A, p. 328.
    Eocene (Jackson); Alabama.

BOIDÆ.

Gray, J. E. 1842, Zool. Miscell., p. 41. (In part.)
Boulenger, G. A. 1893 A, p. 71.
Cope, E. D. 1864 A, p. 230. (Pythonidæ, Boidæ.)
    1886 V, pp. 479, 454.
    1900 A, p. 721.
Gadow, H. 1898 A, p. 28.
Hoffmann, C. K. 1890 C, p. 1758. (Boaidæ.)

Rochebrune, A. T. 1881 A, p. 201. ("Aprotérodontiens.")
Segond, L. A. 1873 A, p. 17. ("Python.")
Stannius, H. 1856 A. (Peropoda.)
Tomes, C. S. 1898 A, p. 373. ("Pythons.")
Zittel, K. A. 1890 A, p. 629. (Boæidæ.)

BOAVUS Marsh.   Type *B. occidentalis* Marsh.

Marsh, O. C. 1871 A, p. 322.
Cope, E. D. 1884 O, p. 108.
Nicholson and Lydekker 1889 A, p. 1147.
Zittel, K. A. 1890 A, p. 629.

Boavus agilis Marsh.

Marsh, O. C. 1871 A, p. 324.
King, C. 1878 A, p. 406.
    Eocene (Bridger); Wyoming.

Boavus brevis Marsh.

Marsh, O. C. 1871 A, p. 324.
King, C. 1878 A, p. 406.
    Eocene (Bridger); Wyoming.

Boavus occidentalis Marsh.

Marsh, O. C. 1871 A, p. 323.
King, C. 1878 A, p. 406.
    Eocene (Bridger); Wyoming.

PROTAGRAS Cope.   Type *P. lacustris* Cope.

Cope, E. D. 1872 E, p. 471.
    1884 O, p. 102.
Nicholson and Lydekker 1889 A, p. 1147.
Zittel, K. A. 1890 A, p. 629. (Syn. ? of Boavus. Marsh.)

Protagras lacustris Cope.

Cope, E. D. 1872 E, p. 471.
    1873 E, p. 682.
    1884 O, p. 108, pl. xxiii, figs. 17, 18.
    Eocene (Bridger); Wyoming.

LITHOPHIS Marsh.   Type *L. sargenti* Marsh.

Marsh, O. C. 1871 A, p. 325.
Nicholson and Lydekker 1889 A, p. 1147.

Lithophis sargenti Marsh.

Marsh, O. C. 1871 A, p. 325.
King, C. 1878 A, p. 406.
    Eocene (Bridger); Wyoming.

LESTOPHIS Marsh.   Type *L. crassus* Marsh.

Marsh, O. C. 1885 A, p. 326. (To replace Limno- | **Lestophis crassus** Marsh.
phis, preoccupied.)
1871 A, p. 326. (Limnophis.) | *Marsh, O. C.* 1871 A, p. 328. (Limnophis.)
Nicholson and Lydekker 1889 A, p. 1147. | King, C. 1878 A, p. 405. (Limnophis.)
| Marsh, O. C. 1885 A.
| Eocene (Bridger); Wyoming.

APHELOPHIS Cope.   Type *A. talpivorus* Cope.

Cope, E. D. 1873 CC, p. 16. | **Aphelophis talpivorus** Cope.
1874 B, p. 518.
1884 O, p. 781. | *Cope, E. D.* 1873 CC, p. 16.
Nicholson and Lydekker 1889 A, p. 1147. | 1874 B, p. 518.
Zittel, K. A. 1890 A, p. 630. | 1884 O, p. 782, pl. lx, fig. 21.
| Oligocene (White River); Colorado.

CALAMAGRAS Cope.   Type *C. murivorus* Cope.

Cope, E. D. 1873 CC, p. 15. | **Calamagras murivorus** Cope.
1874 B, p. 517.
1884 O, p. 784. | *Cope, E. D.* 1873 CC, p. 15. (C. murivorous, C.
Nicholson and Lydekker 1889 A, p. 1147. | truxalis.)
Zittel, K. A. 1890 A, p. 630. | 1874 B, p. 517. (C. murivorus, C. truxalis.)
| 1884 O, p. 784.
| Oligocene (White River); Colorado.

OGMOPHIS Cope.   Type *O. angulatus* Cope.

Cope, E. D. 1884 O, p. 782. | Cope, E. D. 1884 O, p. 783, pl. lviiia, fig. 12.
Nicholson and Lydekker 1889 A, p. 1147. | Oligocene (White River); Colorado.
Zittel, K. A. 1890 A, p. 630.

Ogmophis angulatus Cope. | **Ogmophis oregonensis** Cope.

Cope, E. D. 1873 CC, P. 16. (Calamagras.) | *Cope, E. D.* 1884 O, p. 783, pl. lviiia,.figs. 9–11.
1874 B, p. 518. (Calamagras.) | Miocene (John Day); Oregon.

COLUBRIDÆ.

Gray, J. E. 1849, Cat. Snakes, p. 2. | Cope, E. D. 1886 V, p. 484.
Boulenger, G. A. 1893 A. | 1900 A, p. 732.
1894 A. | Gadow, H. 1898 A, p. 28.
1896 C. | Hoffmann, C. K. 1890 C, p. 1683.
| Stannius, H. 1856 A. (Colubrina.)

COLUBER Linn.

Linnæus, C. 1758, Syst. Nat., ed. x, p. 216. | **Coluber** sp. indet.
Boulenger, G. A. 1894 A, p. 24.
Cope, E. D. 1900 A, p. 825. | Wheatley, C. M. 1871 A.
| Pleistocene; Pennsylvania.

BASCANION Baird and Girard.   Type *Coluber constrictor* Linn.

Baird and Girard, 1853, Cat. N. Amer. Rept., p. 93. | **Bascanion acuminatus** (Cope).
Boulenger, G. A. 1893 A, p. 379. (Zamenis, in part.)
Cope, E. D. 1900 A, p. 787. (Zamenis, in part.) | Cope, E. D. 1899 A, p. 197. (Zamenis.)
| Pleistocene; Pennsylvania.

CROTALIDÆ.

Boulenger, G. A. 1896 C, p. 519. (Crotalinæ.) | Hoffmann, C. K. 1890 C, p. 1796.
Cope, E. D. 1886 V, pp. 480–498. | Rochebrune, A. T. 1881 A, p. 216. ("Crotaliens.")
1892 O. | Stannius, H. 1856 A. (Crotalina.)
1900 A, p. 1130. | Troschel, F. H. 1861 A. (Crotalina.)
Giebel, C. G. 1866 B.

CROTALUS Linn. Type *C. horridus* Linn.

Linnæus, C., 1758, Syst. Nat., ed. x, p. 214.
Boulenger, G. A. 1896 C, p. 572.
Cope, E. D. 1900 A, p. 1149.

Crotalus horridus Linn.

Linnæus, C., 1758, Syst. Nat., ed. x, p. 214.

Cope, E. D. 1900 A, p. 1185.
Green, J. 1821 A.
Wheatley, C. M. 1871 A.
  Pleistocene; Pennsylvania, New Jersey.

NEURODROMICUS Cope. Type *N. dorsalis* Cope.

Cope, E. D. 1873 CC, p. 15.
  1874 B, p. 516.
  1884 O, p. 785.
Hoffmann, C. K. 1890 C, p. 1814. (Neurodromus.)
Nicholson and Lydekker 1889 A, p. 1148.
Zittel, K. A. 1890 A, p. 631.

Neurodromicus dorsalis Cope.

Cope, E. D. 1873 CC, p. 15.
  1874 B, p. 517.
  1884 O, p. 786, pl. lviii, figs. 7, 8.
Oligocene (White River); Colorado.

HELAGRAS Cope. Type *H. prisciformis* Cope.

Cope, E. D. 1883 M, p. 545.
  1883 W.
  1884 O, p. 780.
Nicholson and Lydekker 1889 A, p. 1146.
Zittel, K. A. 1890 A, p. 632.
A genus of uncertain position.

Helagras prisciformis Cope.

Cope, E. D. 1883 M, p. 545.
  1883 W.
  1884 O, p. 731, pl. xxivg. fig. 2.
  1888 CC, p. 302.
Zittel, K. A. 1890 A, p. 632.
  Eocene (Puerco); New Mexico.

CONIOPHIS Marsh. Type *C. precedens* Marsh.

Marsh, O. C. 1892 E. p, 450.
A genus of uncertain affinities.

Coniophis precedens Marsh.

Marsh, O. C. 1892 E, p. 450, figs. 1a–e.
Walcott, C. D. 1900 A, p. 23.
  Cretaceous (Laramie); Wyoming.

Order DINOSAURIA Owen.

Owen, R. 1842 A, p. 102.
Baur, G. 1883 A.
  1884 A.
  1884 B.
  1884 E.
  1885 C.
  1885 E.
  1887 A, p. 487.
  1887 C, p. 101.
  1887 S, p. 333.
  1891 C, p. 354.
  1891 D.
Burckhardt, R. 1895 A.
Cannon, G. L. 1888 A, pp. 140, 190.
Cope, E. D. 1869 M, pp. 33, 36.
  1871 B, p. 234.
  1872 Q.
  1877 G, p. 588.
  1878 K.
  1879 A, p. 38.
  1882 J, p. 981.
  1882 Q.
  1883 J, p. 97.
  1883 DD.
  1885 G, p. 246.
  1886 E.
  1889 R, p. 864.

Cope, E. D. 1891 N, p. 41.
  1892 R.
  1892 Z, p. 17.
  1898 B, pp. 56, 68.
  1900 A, p. 160.
Dames, W. 1884 A.
Dawkins, W. B. 1871 A.
Dollo, L. 1882 B.
  1883 B.
  1883 C.
  1884 B.
  1888 E.
Fürbringer, M. 1888 A.
  1900 A, pp. 347, 535, 651.
Gadow, H. 1897 A, p. 204.
  1898 A, p. 22.
Gegenbaur, C. 1898 A.
Günther, A. 1866 A, p. 442.
Haeckel, E. 1895 A, p. 374.
Hay, O. P. 1899 F, p. 682.
Hoffmann, C. K. 1890 B.
Hulke, J. W. 1888 A.
  1883 B.
  1884 A.
Huxley, T. H. 1868 A, p. 362.
  1870 A, p. 438.
  1870 C.

Huxley, T. H. 1870 D.
  1872 A, p. 223. (Ornithoscelida.)
  1875 B.
  1876 B, xiii, p. 515.
Kehner, F. A. 1876 A.
Lydekker, R. 1888 A.
  1888 B, p. 131.
  1893 E, p. 139.
Marsh, O. C. 1877 E.
  1878 F.
  1879 B.
  1880 D.
  1881 E.
  1881 H.
  1882 A.
  1884 A.
  1884 B.
  1884 G.
  1885 D.
  1889 B.
  1895 B.
  1895 C.
  1896 A.
  1896 C.
  1896 D.

Morse, E. S. 1880 A, p. 3.
Nicholson and Lydekker 1889 A, p. 1151.
Osborn, H. F. 1899 H, p. 161.
  1900 G, p. 777.
Owen, R. 1854 B, p. 1.
  1860 A, p. 164.
  1860 E, p. 257.
  1866 A, p. 18.
  1875 C.
Parker, W. K. 1887 A, p. 331.
Pictet, F. J. 1853 A, p. 466. ("Dinosauriens.")
Quenstedt, F. A. 1852 A, p. 111.
Seeley, H. G. 1879 A.
  1887 B.
  1888 B.
  1888 C.
  1888 G.
Steinmann and Döderlein 1890 A, p. 658.
Tomes, C. S. 1898 A, p. 287.
Vetter, B. 1885 A.
White, C. A. 1883 A.
Wiedersheim, R. 1881 A, p. 366.
Woodward, A. S. 1898 B, p. 196.
Zittel, K. A. 1890 A, pp. 450, 689, 767.

SUBORDER OPISTHOCŒLIA Owen.

Owen, R. 1860 A, p. 164.
  Unless otherwise indicated, the following authors employ the term Sauropoda for this group.
Baur, G. 1883 A.
  1884 A.
  1887 C, p. 101. (Opisthocœlia.)
  1891 D, p. 451. (Cetiosauria.)
Cope, E. D. 1882 Q. (Sauropoda a synonym of Opisthocœlia.)
  1883 J, p. 98. (Opisthocœlia.)
  1889 E.
  1889 R, p. 804. (< Saurischia.)
  1891 N, p. 43. (< Saurischia.)
  1898 B, p. 68. (< Saurischia.)
Dollo, L. 1884 B.
  1892 C, p. 10.
Fürbringer, M. 1900 A.
Gadow, H. 1897 A, p. 204.
  1898 A. p. 22.
Haeckel, E., 1895 A, pp. 382, 386.
Hulke, J. W. 1879 A.
  1880 A.
  1882 A.
Huxley, T. H. 1867 A, p. 415.
  1869 B.
Lydekker, R. 1886 C, p. 275.
  1888 A, p. 53.

Lydekker, R. 1888 B, p. 131.
Marsh, O. C. 1878 F, p. 412.
  1879 B.
  1881 E.
  1882 A, p. 83.
  1883 A, pp. 82, 85.
  1884 A.
  1884 G.
  1885 D.
  1889 B, p. 324.
  1896 C, pp. 164, 241.
  1896 D, p. 207.
  1898 C.
Nicholson and Lydekker 1889 A, p. 1170.
Osborn, H. F. 1898 J, p. 221. (Cetiosauria.)
  1900 G, p. 786. (Cetiosauria.)
Owen, R. 1875 C, p. 27. (Cetiosaurus.)
Science 1885 B.
Seeley, H. G. 1888 B. (< Saurischia.)
  1888 C, p. 86. (Cetiosauria.)
  1888 G, p. 171. (< Saurischia.)
  1892 A, p. 349. (< Saurischia.)
Steinmann and Döderlein 1890 A, p. 661.
Vetter, B. 1885 A, p. 117.
Woodward, A. S. 1898 B, p. 200.
Zittel, K. A. 1890 A, pp. 693, 702.

MOROSAURIDÆ.

Marsh, O. C. 1882 A, p. 83.
Baur, G. 1895 A, p. 1001.
  1896 A, p. 145.
Lydekker, R. 1888 B, p. 133. (Cetiosauridæ.)
  1889 C, p. 59. (Cetiosaurus.)
  1889 J. (On Cardiodon.)
  1890 A, p. 236. (<Cetiosauridæ.)
Marsh, O. C. 1883 A, p. 85.
  1884 A.

Marsh, O. C. 1884 G.
  1885 D.
  1888 A, p. 92. (Pleurocœlidæ.)
  1889 B, p. 324. (Morosauridæ, Cetiosauridæ, Pleurocœlidæ.)
  1896 C, p. 495 (Morosauridæ); p. 496 (Pleurocœlidæ.)
  1896 C, p. 241 (Morosauridæ); p. 242 (Pleurocœlidæ, Cardiodontidæ).

Marsh, O. C. 1896 D, p. 208. (Morosauridæ, Pleurocœlidæ.)
  1896 F, p. 435.
  1897 C, p. 496.
  1896 C, p. 488. (Morosauridæ, Pleurocœlidæ.)
Nicholson and Lydekker 1889 A, p. 1177. (<Ceti-
 osauridæ.)

Owen, R. 1841 E. p. 457. (Cetiosaurus.)
  1859 D, p. 27. (Cetiosaurus.)
Pictet, F. J. 1853 A, p. 493. (Cetiosaurus.)
Steinmann and Döderlein 1890, A, p. 668.
Williston, S. W. 1895, A, p. 169.
Williston and Case 1892 A, p. 16.
Zittel, K. A. 1890 A, p. 709.

MOROSAURUS Marsh.   Type *M. impar* Marsh.

Marsh, O. C. 1878 B, p. 242.
Baur, G. 1883 A, p. 433.
Giebel, C. G. 1879 A, p. 317.
Hulke, J. W. 1888 A, p. 425.
Krause, W. 1881 A.
Lydekker, R. 1888 A, p. 59.
  1888 B, p. 134.
  1889 C, p. 59.
  1890 A, p. 236.
Marsh, O. C. 1878 F, p. 412.
  1879 B.
  1883 A, p. 82.
  1884 A.
  1888 A, p. 90.
  1896 C, p. 495.
  1896 C, p. 181.
  1897 C, p. 496.
Nicholson and Lydekker 1889 A, pp. 1177, 1179.
Osborn, H. F. 1898 J, p. 221.
  1899 H, p. 168.
  1900 G, p. 787.
Steinmann and Döderlein 1890 A, p. 668.
Wiedersheim, R. 1881 A.
Williston, S. W. 1895 A, p. 165.
  1898 F.
Zittel, K. A. 1890 A, p. 709.

Morosaurus agilis Marsh.
Marsh, O. C. 1889 C, p. 334, fig. 3.
  1896 C, pl. xxx, fig. 1; pl. xxxvii, figs. 1, 2.
Osborn, H. F. 1899 H, p. 170, fig. 6.
 Jurassic; Colorado.

Morosaurus grandis Marsh.
Marsh, O. C. 1877 G, p. 515. (Apatosaurus.)

King, C. 1878 A, p. 346. (Apatosaurus.)
Lydekker, R. 1888 B, p. 133, figs. 20, 21.
Marsh, O. C. 1878 F, p. 514, pls. v–vii.
  1879 B, pls. iii, v, vii.
  1880 C, p. 254.
  1883 A, pp. 83, 85.
  1896 C, p. 181, pl. xxx, figs. 2, 3; pl. xxxi,
   pl. xxxii, figs. 4–6; pl. xxxiv, figs. 1, 5–7,
   pl. xxxv, fig. 3; pl. xxxviii· pl. xxxix,
   figs. 4, 5; text figs. 31–33.
  1897 C, p. 497, figs. 49–51.
Williston, S. W. 1898 F, figs. 1, 2.
Zittel, K. A. 1890 A, p. 709, figs. 614–617.
 Jurassic; Wyoming?, Colorado.

Morosaurus impar Marsh.
Marsh, O. C. 1878 B, p. 242.
King, C. 1878 A, p. 346.
 Jurassic; "Rocky Mountain region."

Morosaurus lentus Marsh.
Marsh, O. C. 1889 C, p. 333, fig. 2.
  1896 C, p. 488, fig. 8.
  1896 C, p. 182, pl. xxxii, figs. 1, 2; pl. xxxiii,
   pl. xxxiv, figs. 3, 4; pl. xxxv, fig. 4; pl.
   xxxvi, fig. 1; pl. xxxvii, figs. 3–5; text figs
   34, 63.
  1896 D, p. 20, fig. 8.
 Jurassic; Wyoming.

Morosaurus robustus Marsh.
Marsh, O. C. 1878 F, p. 414, pl. viii.
  1879 H, p. 503.
  1896 C, pl. xxxv, figs. 1, 2.
 Jurassic; Wyoming.

PLEUROCŒLUS Marsh.   Type *P. nanus* Marsh.

Marsh, O. C. 1888 A, p. 90.
Lydekker, R. 1888 B, p. 134.
  1890 A, p. 237.
  1890 B, p. 182, pl. ix.
Marsh, O. C. 1895 C, p. 490.
  1896 C, p. 183.
  1897 C, p. 415.
  1898 C, p. 488.
Nicholson and Lydekker 1889 A, p. 1179.
Zittel, K. A. 1890 A, p. 714.

Pleurocœlus altus Marsh.
Marsh, O. C. 1888 A, p. 92.
 Jurassic (Potomac); Maryland.

Pleurocœlus montanus Marsh.
Marsh, O. C. 1896 C, p. 184, figs. 35–41.
Walcott, C. D. 1900 A, p. 23.
 Formation and locality not given.

Pleurocœlus nanus Marsh.
Marsh, O. C. 1888 A, p. 90, figs. 1–6.
Lydekker, R. 1890 A, p. 238, fig. 52.
Marsh, O. C. 1896 C, pls. xl, xli.
 Jurassic (Potomac); Maryland.

EPANTERIAS Cope.   Type *E. amplexus* Cope.

Cope, E. D. 1878 V, p. 406.
Zittel, K. A. 1890 A, p. 709.

Epanterias amplexus Cope.
Cope, E. D. 1878 V, p. 406.
 Jurassic; Colorado.

CAULODON Cope.   Type *C. diversidens* Cope.

Cope, E. D. 1877 F, p. 198.
Nicholson and Lydekker, 1889 A, p. 1176.
Zittel, K. A. 1890 A, p. 711.
   The position of this genus is doubtful.

Caulodon diversidens Cope.

Cope, E. D. 1877 F, p. 198.
   1878 I, p. 247.

Osborn, H. F. 1898 J, p. 227.
   Jurassic; Colorado.

Caulodon leptoganus Cope.

Cope, E. D. 1878 I, p. 247.
   Jurassic; Colorado.

SYMPHYROPHUS Cope.   Type *S. musculosus* Cope.

Cope, E. D. 1878 I, p. 246.
   1878 H, p. 391.
   1878 M, p. 84.
Zittel, K. A. 1890 A, p. 726.
A genus of doubtful position.

Symphyrophus musculosus Cope.

Cope, E. D. 1878 I, p. 246.
   1878 H, p. 391.
Osborn, H. F. 1898 J, p. 227.
   Jurassic; Colorado.

ASTRODON Johnston.   Type *A. johnstoni* Leidy.

Johnston, C., in Leidy, J. 1865 A, p. 102.
   1859 A, p. 341. (Name; no description.)
Marsh, O. C. 1896 C, p. 242.
   1898 C, p. 488.
Zittel, K. A. 1890 A, p. 717.
A genus of uncertain affinities.

Astrodon johnstoni Leidy.

Leidy, J. 1865 A, pp. 102, 119, pl. xiii, figs. 20–23; pl. x, fig. 0.
Bibbins, A. 1895 A, p. 19, fig. H.   (This species?)
Cope, E. D. 1869 M, p. 99.
Marsh, O. C. 1896 C, p. 164, text fig. 6.
   Jurassic (Potomac); Maryland.

ATLANTOSAURIDÆ.

Marsh, O. C. 1877 G, p. 514.
Cope, E. D. 1877 R, p. 3.   (Camarasauridæ, Amphi-
   coelidæ.)
   1878 I, p. 243.   (Camarasauridæ, Amphicoeli-
   idæ.)
   1883 J, p. 99.   (Camarasauridæ.)
   1898 B, p. 68.   (Cetiosauridæ.)
Günther, A. 1886 A, p. 443.
Lydekker, R. 1888 B, p. 143.
Marsh, O. C. 1878 B, p. 240.
   1882 A, p. 83.
   1883 A, pp. 82, 85.

Marsh, O. C. 1884 A, p. 167.
   1885 D.
   1889 B.
   1893 B.
   1895 C, p. 495.
   1896 C, pp. 166, 241.
   1896 D, p. 208.
   1897 C, p. 494.
   1898 C, p. 487.
Nicholson and Lydekker 1889 A, p. 1173.
Steinmann and Döderlein 1890 A, p. 662.
Zittel, K. A. 1890 A, p. 705.

ATLANTOSAURUS Marsh.   Type *Titanosaurus montanus* Marsh.

Marsh, O. C. 1877 G, p. 514.   (To replace Titano-
   saurus Marsh, preoccupied.)
Cope, E. D. 1878 I, p. 237.
Giebel, C. G. 1879 A, p. 318.
Hulke, J. W. 1879 A.
Lydekker, R. 1888 B, p. 144.
Marsh, O. C. 1877 C.   (Titanosaurus, type *T.
montanus.*)
   1877 E.
   1878 G, p. 224.
   1879 B, p. 88, pls. vi, vii.
   1895 C, p. 495.
   1896 C, p. 166.
   1897 C, p. 485.
Nicholson and Lydekker 1889 A, p. 1173.
Osborn, H. F. 1898 J, p. 219 *et seq.*
Steinmann and Döderlein 1890 A, p. 662.
Zittel, K. A. 1890 A, p. 705.

Atlantosaurus immanis Marsh.

Marsh, O. C. 1878 B, p. 241.

Barbour, E. H. 1890 B, p. 392, figs. 3, 4.
King, C. 1878 A, p. 346.
Lydekker, R. 1888 B, p. 145.
Marsh, O. C. 1879 B, p. 89, pl. vii.
   1896 C, p. 166, pl. xvi.
   1897 C, pp. 488, 526, figs. 34–35a.
Nicholson and Lydekker 1889 A, p. 1174.
   Jurassic; Wyoming, Colorado.

Atlantosaurus montanus Marsh.

Marsh, O. C. 1877 C.   (Titanosaurus.)
Cope, E. D. 1877 Q, p. 6.   (No name.)
   1878 I, p. 237.   (Titanosaurus.)
King, C. 1878 A, p. 346.
Marsh, O. C. 1877 E.
   1879 B, p. 88, pl. vi,
   1896 C, p. 166, pl. xv; pl. xvii, fig. 1.
   1897 C, p. 487, fig. 33.
Osborn, H. F. 1898 J, p. 228, fig. 7 B.
Williston, S. W. 1878 A, p. 43.   (Titanosaurus.)
   Jurassic; Colorado.

APATOSAURUS Marsh.   Type *A. ajax* Marsh.

Marsh, O. C. 1877 G, p. 514.
Giebel, C. G. 1879 A, p. 317.
Marsh, O. C. 1879 B, p. 86.
    1883 A, p. 82.
    1895 C, p. 495.
    1896 C, p. 166.
    1897 C, p. 489.
Nicholson and Lydekker 1889 A, p. 1173.
Osborn, H. F. 1898 J, p. 219 *et seq.*
Zittel K. A. 1890 A, p. 706.

Apatosaurus ajax Marsh.
Marsh, O. C. 1877 G, p. 514.
King, C. 1878 A, p. 346.

Marsh, O. C. 1879 B, p. 87, pl. iv; pl. vi, fig. 1.
    1896 C, p. 168, pl. xviii, figs. 2, 3; text figs. 7, 8.
    1897 C, pp. 490, 526, figs. 37, 38.
Osborn, H. F. 1898 J, p. 228, figs. 7 B, 9.
    Jurassic; Colorado.

Apatosaurus laticollis Marsh.
Marsh, O. C. 1879 B, p. 88, pl. iii, figs. 1, 2.
Barbour, E. H. 1890 B, p. 390, figs. 1, 2.
Marsh, O. C. 1896 C, pl. xviii, fig. 1; pl. xix, fig. 1.
    1897 C, pp. 490, 526, fig. 36.
Zittel, K. A, 1890 A, p. 706, fig. 609.
    Jurassic; Colorado.

BAROSAURUS Marsh.   Type *B. lentus* Marsh.

Marsh, O. C. 1890 A, p. 85, pl. i, figs. 1. 2.
    1895 C, p. 495.
    1896 C, p. 174.
    1898 C, p. 488.
Zittel, K. A. 1890 A, p. 716.

Barosaurus affinis Marsh.
Marsh, O. C. 1899 A, p. 228.   (No description.)
    Jurassic; Wyoming.

Barosaurus lentus Marsh.
Marsh, O. C. 1890 A, p. 85, text figs. 1, 2.
    1896 C, p. 174, text figs. 24–26.
    1899 A, p. 228.
    Jurassic; South Dakota.

BRONTOSAURUS Marsh.   Type *B. excelsus* Marsh.

Marsh, O. C. 1879 H, p. 508.
Fürbringer, M. 1900 A, p. 352.
Hulke, J. W. 1883 A, p. 62.
    1888 A, p. 425.
Lydekker, R. 1888 A, p. 56.
    1898 E, p. 139.
Marsh, O. C. 1880 D, p. 395.
    1881 E, p. 417, pls. xii–xviii.
    1883 A.
    1891 C, p. 341, pl. xvi.
    1895 C, p. 495.
    1896 C, p. 168.
    1897 C, p. 492.
Nicholson and Lydekker 1889 A, p. 1173.
Osborn, H. F. 1898 J, p. 219 *et seq.*
    1899 E, pp. 205, 207, 213.
    1899 H, p. 162, fig. 1 B, p. 171, fig. 7.
    1900 G, p. 788.
Steinmann and Döderlein 1890 A, p. 662.
Woodward, A. S. 1898 B, p. 201.
Zittel, K. A. 1890 A, p. 706.

Brontosaurus amplus Marsh.
Marsh, O. C. 1881 E, p. 421.

Marsh, O. C. 1896 C, p. 173, text figs. 17, 20, 29.
    Jurassic; Wyoming.

Brontosaurus excelsus Marsh.
Marsh, O. C. 1879 H, p. 508.
Fürbringer, M. 1900 A, p. 353, fig. 124.
Geinitz and Deichmüller 1882 A.
Hutchison, H. N. 1893 A, p. 66, pl. iv.
Lydekker, R. 1888 B, p. 144, fig. 23.
    1893 B, p. 302, fig. 1.
Marsh, O. C. 1880 D, p. 395, pl. xviii.
    1881 E, p. 417, pls. xii–xviii.
    1883 A, pl. i.
    1891 C, p. 341, pl. xvi.
    1893 I, p. 437.
    1895 C, pl. x, fig. 2.
    1896 C, p. 169, pls. xx–xxiv; pl. xliii; text figs. 9–16, 21–23.
    1896 D, pl. i, fig. 2.
    1897 C, p. 492, pl. xxi, text figs. 39–43.
Osborn, H. F. 1898 J, 223 *et seq.*, figs. 6, 7 C, 8, 12.
Woodward, A. S. 1898 B, p. 201, figs. 121, 122.
Zittel, K. A. 1890 A, p. 707, figs. 610–613.
    Jurassic; Wyoming, Colorado.

CAMARASAURUS Cope.   Type *C. supremus* Cope.

Cope, E. D. 1877 A.
    1877 W.
    1878 E.
    1878 I, pp. 233, 234.
    1878 M.
    1878 V.
    1879 A, p. 35.
Hulke, J. W. 1879 A.
    1880 A.

Hulke, J. W. 1882 A.
Lydekker, R. 1888 A, p. 54.
Marsh, O. C 1895 C, p. 495.
    1896 C, p 488.
Osborn, H. F. 1898 J, p. 219.
    1900 G, pp. 786, 788.
Owen, R. 1878 D.
Zittel, K. A. 1890 A, p. 711.   (Camarosaurus.)

Camarasaurus leptodirus Cope.

Cope, E. D. 1879 R, p. 403, figs. 1–3.
Jurassic: Colorado.

Camarasaurus supremus Cope.

Cope, E. D. 1877 Q.
1877 W.
1878 E.

Cope, E. D. 1878 I, p. 237, figs. 1–12.
1878 M, figs. 1–12.
1891 N, p. 43, fig. 21.
1896 B, p. 68, fig. 22, A. B.
Marsh, O. C. 1879 B, p. 89.
Osborn, H. F. 1898 J, p. 227 et seq., figs. 10, 11.
Williston, S. W. 1878 A, p. 43.
Zittel, K. A. 1890 A, p. 712, fig. 618.
Jurassic; Colorado.

AMPHICŒLIAS Cope.   Type *A. altus* Cope.

Cope, E. D. 1877 R.
1878 I, p. 242.
1878 M.
Lydekker, R. 1888 A, p. 54.
Marsh, O. C. 1881 E, p. 419. (Syn. of Brontosaurus.)
1896 C, p. 241. (Syn. of Camarasaurus.)
1898 C, p. 488.
Osborn, H. F. 1898 J, p. 219 et seq.
Zittel, K. A. 1890 A, p. 709.

Amphicœlias altus Cope.

Cope, E. D. 1877 R, p. 3.
1878 I, p. 243, figs. 13–15.
1878 M, figs. 13–17.
1879 X, p. 798 a.

Cope, E. D. 1896 B, p. 69, fig. 22, C.
Jurassic; Colorado.

Amphicœlias fragillimus Cope.

Cope, E. D. 1878 Y, figure.
1881 BB, p. 413. (A. fragilissimus; syn. ? of
Cœlurus fragilis.)
1884 W. (A. fragilissimus.)
1896 B, p. 68. (A. fragilissimus.)
Jurassic; Colorado.

Amphicœlias latus Cope.

Cope, E. D. 1878 R, p. 4.
1878 I, p. 245, figs. 17, 18.
Jurassic; Colorado.

DIPLODOCIDÆ.

Marsh, O. C. 1884 A.
1884 G.
1885 D.
1889 B, p. 324.
1896 C, p. 495.

Marsh, O. C. 1896 C, p. 241.
1896 D, p. 206.
1898 C, p. 488.
Nicholson and Lydekker 1889 A, p. 1176.
Zittel, K. A. 1890 A, p. 714.

DIPLODOCUS Marsh.   Type *D. longus* Marsh.

Marsh, O. C. 1878 F, p. 414.
Baur, G. 1891 D, p. 445.
Cope, E. D. 1884 W.
Dollo, L. 1884 B.
Giebel, C. G. 1879 A, p. 317.
Hatcher, J. B. 1900 A, p. 718.
1900 B, p. 828.
Holland, W. J. 1900 A, p. 816.
Lydekker, R. 1888 A, p. 53.
Marsh, O. C. 1880 C, p. 255, pl. vi, figs. 4, 5.
(Stegosaurus armatus, errore.)
1883 A, p. 85.
1884 A.
1896 C, p. 495.
1896 C, p. 175.
1897 C, p. 494.
Nicholson and Lydekker 1889 A, p. 1176.
Osborn, H. F. 1896 J, p. 221.
1899 C, p. 315.
1899 E.
1899 H, p. 165, fig. 1 A.
1900 G, pp. 787, 790, fig. 8.
Steinmann and Döderlein 1890 A, p. 662.

Woodward, A. S. 1896 B, p. 204.
Zittel, K. A. 1890 A, p. 714

Diplodocus lacustris Marsh.

Marsh, O. C. 1884 A, p. 164.
Jurassic; Colorado.

Diplodocus longus Marsh.

Marsh, O. C. 1878 F, p. 414, pl. viii, figs. 3, 4.
Cope, E. D. 1884 W.
Hatcher, J. B. 1900 B, p. 828. ("Diplodocus.")
Holland, W. J. 1900 A, p. 816. ("Diplodocus.")
Lydekker, R. 1888 B, p. 134, fig. 19.
Marsh, O. C. 1884 A, p. 168, pls. iii, iv.
1896 C, p. 175, pls. xxv–xxix, fig. 1; pl.
xxxix, fig. 5, text figs. 27–29.
1897 C, p. 495, figs. 44–48.
Osborn, H. F. 1899 E, p. 191, pls. xxiv–xxviii, text
figs. 1–8, 11–14.
Walcott, C. D. 1900 A, p. 23.
Woodward, A. S. 1898 B, p. 204, fig. 124.
Zittel, K. A. 1890 A, p. 715, figs. 620–622.
Jurassic; Wyoming, Colorado.

Suborder THEROPODA Marsh.

Marsh, O. C. 1881 E, p. 423.
Baur, G. 1883 A, p. 441. (Theropoda.)
　　1887 C, p. 101. (Goniopoda.)
　　1891 D, p. 451. (Megalosauria.)
Cope, E. D. 1866 E, p. 317. (Goniopoda.)
　　1867 D. (>Symphypoda.)
　　1869 M, p. 99 (Goniopoda); p. 120 (Symphy-
　　　　poda).
　　1871 B, p. 234. (Goniopoda, Symphypoda.)
　　1882 Q. (Goniopoda.)
　　1883 J, p. 98. (Goniopoda.)
　　1889 R, p. 864. (<Saurischia.)
　　1891 N, p. 43. (<Saurischia.)
　　1896 B, p. 68. (<Saurischia.)
Dollo, L. 1892 C, p. 12. ("Théropodes.")
Fitzinger, L., 1843, Syst. Rept., p. 35. (Fam. Meg-
　　alosauri.)
Fürbringer, M. 1888 A.
　　1900 A.
Gadow, H., 1897 A, p. 204.
　　1898 A, p. 22.

Günther, A. 1886 A, p. 443.
Haeckel, E. 1896 A, p. 384.
Huxley, T. H. 1870 D, p. 35.
Lydekker, R. 1888 A, p. 59.
　　1888 B, p. 154.
Marsh, O. C. 1882 A, p. 84.
　　1884 B.
　　1884 G.
　　1885 D, p. 764.
　　1889 B, p. 329.
　　1895 C, p. 492.
　　1896 C, pp. 146, 239.
　　1896 D, p. 205.
Nicholson and Lydekker 1889 A, p. 1164.
Seeley, H. G. 1888 B. (<Saurischia.)
　　1892 A, p. 349. (<Saurischia.)
Steinmann and Döderlein 1890 A, p. 659.
Vetter. B. 1885 A, p. 117.
Woodward, A. S. 1898 B, p. 197.
Zittel, K. A. 1890 A, pp. 450, 693, 717.

MEGALOSAURIDÆ.

Huxley, T. H. 1870 D, p. 34.
Baur, G. 1883 A, p. 440.
Cope, E. D. 1868 C, p. 417. (Dinodontidæ.)
　　1869 M, p. 117. (Dinodontidæ.)
　　1889 R, p. 864.
　　1896 B, p. 68.
Dawkins, W. B. 1871 A.
Lydekker, R. 1888 B, p. 157.
Marsh, O. C. 1879 B, p. 89. (Allosauridæ.)
　　1881 E, p. 423. (Allosauridæ.)
　　1882 A, pp. 84, 85. (Megalosauridæ, Labro-
　　　　sauridæ.)
　　1884 B, p. 337. (Megalosauridæ, Labrosaur-
　　　　idæ.)
　　1884 G. (Megalosauridæ, Labrosauridæ.)

Marsh, O. C. 1885 D, p. 764. (Megalosauridæ, La-
　　brosauridæ.)
　　1890 C, p. 424. (Dryptosauridæ.)
　　1895 C, p. 493. (Megalosauridæ, Dryptosaur-
　　　　idæ, Labrosauridæ.)
　　1896 C, pp. 203, 239. (Megalosauridæ, Drypto-
　　　　sauridæ, Labrosauridæ.)
　　1896 D, p. 206. (Megalosauridæ, Dryptosaur-
　　　　idæ, Labrosauridæ.)
Nicholson and Lydekker 1889 A, p. 1167.
Owen, R. 1883 B.
Seeley, H. G. 1894 A.
Steinmann and Döderlein, 1890 A, p. 660.
Zittel, K. A. 1890 A, p. 722.

MEGALOSAURUS Buckland.　Type *M. bucklandi* Meyer.

Buckland, W. 1824 A.
Bronn, H. G. 1837 A, p. 580, pl. xxxiv, fig. 1.
Cope, E. D. 1868 K, p. 239. (Megalosaurus, Pœcil-
　　opleuron.)
　　1869 M, p. 100. (Pœcilopleurum); p. 108
　　　　(Megalosaurus).
Deslongchamps-Eudes, J. A. 1838 A, p. 36. (Pœcil-
　　opleuron, type *P. bucklandi*, in part.)
Gaudry, A. 1890 C, p. 227.
Hulke, J. W. 1879 B.
Huxley, T. H. 1869 D.
　　1870 D.
Lydekker, R. 1888 B, p. 159.
Marsh, O. C. 1895 A, p. 408.

Osborn, H. F. 1900 G, p. 786.
Owen, R., 1842 A, p. 103.
　　1845 B, p. 269.
　　1856 A.
　　1860 E, p. 258.
　　1876 C. (Pœcilopleuron.)
　　1883 B, pl. xi.
Pictet, F. J. 1853 A, p. 467 (Megalosaurus); p.
　　497 (Pœcilopleuron).
Quenstedt, F. A. 1852 A, p. 112.
Zittel, K. A. 1890 A, p. 722.
　　No species of this genus is at present known
from this country.

DRYPTOSAURUS Marsh.　Type *Lælaps aquilunguis* Cope.

Marsh, O. C. 1877 C, p. 88. (To replace *Lælaps*
　　Cope, preoccupied.)
　　Unless otherwise indicated the following
writers refer to this genus under name *Lælaps*.

Baur, G. 1883 A, p. 441.
Cope, E. D. 1866 A. (Lælaps, type *L. aquilunguis*.)
　　1866 E.
　　1867 D.

Cope, E. D. 1866 C.
1866 K, p. 237.
1869 M, p. 100.
1871 B, p. 239.
1876 I, p. 340.
1877 H, p. 805.
1877 T.
1878 K.
1892 Z, p. 17, pl. iii.
Huxley, T. H. 1870 C.
Leidy, J. 1868 F, p. 199.
Lippincott, J. S. 1881 A.
Lydekker, R. 1888 B, pp. 159, 169. (Dryptosaurus.)
Marsh, O. C. 1895 C, p. 498.
1896 C, p. 208.
1896 D.
Morse, E. S. 1880 A, p. 4.
Osborn, H. F. 1898 B, p. 325.
Zittel, K. A. 1890 A, p. 726.

Dryptosaurus aquilunguis (Cope).

Cope, E. D. 1866 A. (Lælaps.)
1866 B. (Lælaps aquilunguis, part; L. macropus, part.)
1866 E. (Lælaps.)
1867 E, p. 27. (Lælaps.)
1868 C. (Lælaps.)
1868 K, p. 237. (Lælaps.)
1869 B, p. 737. (Lælaps.)
1869 K, p. 91, pl. ii. (Lælaps.)
1869 M, pp. 100, 118, 122 J, pls. viii, ix; pl. x,
figs. 1-6. (Lælaps.)
1870 K. (Lælaps.)
1875 E, p. 249. (Lælaps.)
1877 T. (Lælaps.)
Lucas, F. A. 1898 C, p. 97.
Lydekker, R. 1888 B, p. 170.
Marsh, O. C. 1877 C, p. 88.
Osborn, H. F. 1898 N, pp. 843, 844, fig. 3. (Megalosaurus.)
1898 Q, pp. 6, 19, pl. iii. (Megalosaurus.)
Cretaceous; New Jersey.

Dryptosaurus cristatus (Cope).

Cope, E. D. 1876 I, p. 344. (Lælaps.)
1877 G, p. 572. (Lælaps.)
Cretaceous; Montana.

Dryptosaurus explanatus (Cope).

Cope, E. D. 1876 H, p. 249. (Lælaps.)
1877 G, p. 572. (Lælaps.)
Cretaceous; Montana.

Dryptosaurus falculus (Cope).

Cope, E. D. 1876 H, p. 249. (Lælaps.)
1877 G, p. 572. (Lælaps.)
Cretaceous; Montana.

Dryptosaurus hazenianus (Cope).

Cope, E. D. 1876 I, p. 343. (Lælaps.)
1877 G, p. 572. (Lælaps.)
Cretaceous; Montana.

Dryptosaurus incrassatus (Cope).

Cope, E. D. 1876 H, p. 248. (Lælaps.)
1876 I, pp. 340, 341. (Lælaps.)
1877 G, p. 572. (Lælaps.)
1877 T. (Lælaps.)
1892 K, p. 240. (Lælaps.)
1892 Z, p. 17, pl. iii. (Lælaps.)
Hay, O. P. 1899 G, p. 348.
Lambe, L. M. 1899 A, pp. 69, 70. (Lælaps.)
Cretaceous; Montana, British America.

Dryptosaurus kenabekides Hay.

Hay, O. P. 1899 G, p. 348.
Leidy, J. 1856 F, p. 72. (Deinodon horridus, in part.)
1860 A, p. 143 (in part), pl. ix, figs. 21-34.
(Deinodon horridus.)
1868 F, p. 198. (Deinodon horridus.)
Cretaceous (Laramie); Montana.

Dryptosaurus lævifrons (Cope).

Cope, E. D. 1876 I, p. 344. (Lælaps.)
1877 G, p. 572. (Lælaps.)
Cretaceous; Montana.

Dryptosaurus macropus (Cope).

Cope, E. D. 1868 C, p. 417. (Lælaps.)
1866 B. (L. aquilunguis, in part.)
1869 B, p. 737. (Lælaps.)
1869 M, p. 118, figs. 31-34. (Lælaps.)
1875 E, p. 249. (Lælaps.)
Leidy, J. 1865 A, p. 119. (Cœlosaurus antiquus, in
part, fide Cope.)
Cretaceous; New Jersey.

Dryptosaurus trihedrodon (Cope).

Cope, E. D. 1877 H, p. 805. (Lælaps.)
1877 Q. p. 5. (Lælaps.)
1878 M, p. 73. (Lælaps.)
1878 R. (Syn.? of Hypsirophus discurus.)
Jurassic; Colorado.

DEINODON Leidy. Type D. horridus Leidy.

Leidy, J. 1856 F, p. 72.
Cope, E. D. 1866 A, p. 279. (Dinodon, restricted.)
1868 C. (Dinodon.)
1869 M, pp. 100, 116. (Aublysodon; Dinodon
said to be preoccupied.)
1875 E, p. 248. (Aublysodon.)
1876 I, p. 341. (Aublysodon.)
Hay, O. P. 1899 G, p. 346.
Leidy, J. 1856 C, p. 119. (Deinodon.)
1860 A, p. 143, in part. (Deinodon.)
1868 F, p. 198. (Dinodon, restricted; Aublysodon, type A. mirandus.)

Marsh, O. C. 1892 G, p. 174. (Aublysodon.)
Zittel, K. A. 1890 A, p. 726. (Aublysodon.)

Deinodon amplus (Marsh).

Marsh, O. C. 1892 G, p. 174, pl. ii, fig. 5. (Aublysodon.)
Cretaceous (Laramie); Wyoming?

Deinodon cristatus (Marsh).

Marsh, O. C. 1892 G, p. 175, pl. iii, fig. 6. (Aublysodon.)
Cretaceous (Laramie); Wyoming?

Deinodon horridus Leidy.

Leidy, J. 1856 F, p. 72. (In part.)
Cope, E. D. 1866 A, p. 279. (Dinodon horridus, restricted.)
 1868 C, p. 416. (Dinodon.)
 1869 M, p. 120. (Aublysodon.)
 1871 C, p. 120. (Aublysodon.)
 1874 C, p. 26. (Aublysodon.)
 1875 E, p. 249. (Aublysodon.)
 1876 I, p. 341. (Aublysodon.)
 1877 G, p. 572. (Aublysodon.)
Leidy, J. 1856 C, p. 119. (In part.)
 1860 A, p. 143, pl. ix, figs. 36–45; not figs. 21–34; 46–48.
 1868 F, p. 196. (Aublysodon mirandus).

Marsh, O. C. 1892 G, P. 174, pl. iii, fig. 4. (Aublysodon mirandus.)
 Cretaceous (Laramie): Montana.
 The teeth represented on plate ix, of Leidy 1860 A, figs. 21–34, 46–48, are by Cope's restriction of *Deinodon* left without a name. They possibly belong to *Dryptosaurus* and are in this work given the name *Dryptosaurus kenabekides*.

Deinodon lateralis (Cope).

Cope, E. D. 1876 H, p. 248. (Aublysodon.)
 1877 G, p. 572. (Aublysodon.)
 Cretaceous (Laramie); Montana.

ALLOSAURUS Marsh.   Type *A. fragilis* Marsh.

Marsh, O. C. 1877 G, p. 515.
Bibbins, A. 1895 A, p. 19.
Giebel, C. G. 1879 A, p. 318.
Lydekker, R. 1888 B, p. 159.
Marsh, O. C. 1879 B, p. 90.
 1884 B, pls. xi, xii.
 1889 B, p. 329.
 1896 C, p. 163.
Nicholson and Lydekker 1889 A, p. 1168.
Osborn, H. F. 1899 H, p. 161, fig. 1, C. D.; figs. 4, 5.
 1900 G, pp. 779, 785, figs. 2, 4.
Steinmann and Döderlein 1890 A, p. 660.
Williston, S. W. 1878 A, p. 46.
Zittel, K. A. 1890 A, p. 725.

Allosaurus fragilis Marsh.

Marsh, O. C. 1877 G, p. 515.
King, C. 1878 A, p. 346.
Lydekker, R. 1888 B, p. 154, fig. 25.
Marsh, O. C. 1879 B, p. 91, pls. viii, x.
 1884 B, pl. xiii.
 1896 C, p. 163, pl. x, fig. 2; pl. xi, figs. 1, 2 (Allosaurus); pl. xiii, fig. 5 (Labrosaurus).
 1897 C, p. 505, fig. 61.
 1899 A, p. 232, fig. 3.
Zittel, K. A. 1890 A, p. 726, fig. 628. (A. agilis errore.)
 Jurassic; Colorado.

CŒLOSAURUS Leidy.   Type *C. antiquus* Leidy.

Leidy, J. 1865 A, p. 119.
Cope, E. D. 1869 M, p. 119.
Zittel, K. A. 1890 A, p. 727.

Cœlosaurus antiquus Leidy.

Leidy, J. 1865 A, p. 119, pl. iii, fig. 3; pl. xvii, figs. 7–11.

Cope, E. D. 1869 B, p. 737.
 1869 M, pp. 118, 119.
 1875 E, p. 249.
 Cretaceous; New Jersey.

CREOSAURUS Marsh.   Type *C. atrox* Marsh.

Marsh, O. C. 1878 B, p. 243.
 1879 B, p. 90.
 1884 B.
 1896 C, p. 163.
Nicholson and Lydekker 1889 A, p. 1168.
Williston, S. W. 1878 A, p. 46.
Zittel, K. A. 1890 A, p. 725.

Creosaurus atrox Marsh.

Marsh, O. C. 1878 B, p. 243, figs. 1, 2.
King, C. 1878 A, p. 347.
Marsh, O. C. 1879 B, p. 90, pl. x.
 1884 B, pls. ix, xi.
 1896 C, p. 163, pl. xii.
 Jurassic; Colorado.

ANTRODEMUS Leidy.   Type *A. valens* Leidy.

Leidy, J. 1870 E, p. 3.
 1873 B, p. 267. (Poicilopleuron.)
Marsh, O. C. 1879 B, p. 91. (Labrosaurus, type *Allosaurus lucaris* Marsh.)
 1884 B, pl. ix. (Labrosaurus.)
 1896 C, p. 163. (Labrosaurus.)
Riggs, E. S. 1900 A, p. 233.
Zittel, K. A. 1890 A, p. 727. (Labrosaurus.)
 The identification of Marsh's *Labrosaurus* with Leidy's *Antrodemus* and of *L. ferox* with *A. valens* has been made by Mr. F. A. Lucas, of the U. S. National Museum.

Antrodemus lucaris (Marsh).
Marsh, O. C. 1878 B, p. 242. (Allosaurus.)
King, C. 1878 A, p. 346. (Allosaurus.)
Marsh, O. C. 1879 B, p. 91. (Labrosaurus.)
 Jurassic; Colorado.

Antrodemus medius (Marsh).
Marsh, O. C. 1888 A, p. 93. (Labrosaurus.)
 Jurassic (Potomac); Maryland.

Antrodemus sulcatus (Marsh).
Marsh, O. C. 1896 C, p. 270, pl. xiii, fig 1. (Labrosaurus.)
 Formation and locality not given.

Antrodemus valens Leidy.

Leidy, J. 1870 E, p. 3. [Polcilopleuron (Antrodemus).]
Leidy, J. 1873 B, pp. 267, 338 (Polcilopleuron); pl.
  xv, figs. 16–18 (Antrodemus).
Marsh, O. C. 1884 B, p. 333, pl. ix. (Labrosaurus
  ferox.)

Marsh, O. C. 1896 C, p. 163 (Allosaurus); pl. xiii,
  figs. 2–4 (Labrosaurus ferox).
Walcott, C. D. 1900 A, p. 23. (Labrosaurus ferox.)
Zittel, K. A. 1890 A, p. 724.   (Megalosaurus.)
  Jurassic; Colorado.

PARONYCHODON Cope.   Type *P. lacustris* Cope.

Cope, E. D. 1876 H, p. 256.
  1876 I, p. 345.
Osborn, H. F. 1893 B, p. 320.   (Syn.? of Meniscoëssus.)
The position of this genus is very doubtful.

Paronychodon lacustris Cope.

Cope, E. D. 1876 H, p. 256.
  1876 I, p. 350.
  1877 G, p. 572.
Cretaceous (Laramie); Montana.

ZAPSALIS Cope.   Type *Z. abradens* Cope.

Cope, E. D. 1876 I, p. 344.
A genus of doubtful position.

Zapsalis abradens Cope.

Cope, E, D. 1876 I, p. 345
  1877 G, p. 572.
Cretaceous (Laramie); Montana.

PALÆOCTONUS Cope.   Type *P. appalachianus* Cope.

Cope, E. D. 1877 E, p. 182.
  1893 A, p. 15.
A genus of uncertain position.

Palæoctonus appalachianus Cope.

Cope, E. D. 1877 E, p. 182.
  1877 X.
  1878 FF.
  1893 A, p. 15, pl. ii, figs. 2, 3.
Triassic; Pennsylvania.

Palæoctonus dumblianus Cope.

Cope, E. D. 1893 A, p. 16, pl. ii, figs. 5, 6.
Triassic; Texas.

Palæoctonus orthodon Cope.

Cope, E. D. 1893 A, p. 15, pl. ii, fig. 1.
Triassic; Texas.

SUCHOPRION Cope.   Type *S. cyphodon* Cope.

Cope, E. D. 1877 E, p. 185.
A genus of uncertain position.

Suchoprion aulacodus Cope.

Cope, E. D. 1878 E, p. 184.   (Palæoctonus.)
  1878 D, p. 232.
Triassic; Pennsylvania.

Suchoprion cyphodon Cope.

Cope, E. D. 1877 E, p. 185.
  1878 FF.
Triassic; Pennsylvania.

Suchoprion sulcidens Cope.

Cope, E. D. 1878 FF. (No description).
Triassic; Pennsylvania.

TROÖDON Leidy.   Type *T. formosus* Leidy.

Leidy, J. 1856 F, p. 72.
Cope, E. D. 1869 M, p. 120.
Leidy, J. 1856 C, p. 119.
  1860 A, p. 147.
Zittel, K. A. 1890 A, p. 727.
A genus of doubtful position.

Troödon formosus Leidy.

Leidy, J. 1856 F, p. 72.

Cope, E. D. 1869 M, p. 120.
  1874 C, p. 26.
  1875 E, p. 248.
  1877 G, p. 572.
Leidy, J. 1856 C, p. 119.
  1860 A, p. 147, pl. ix, figs. 53–55.
Cretaceous (Laramie); Montana.

ZATOMUS Cope.   Type *Z. sarcophagus* Cope.

Cope, E. D. 1871 M, p. 211
  1875 V, p. 35.
The position of this genus is doubtful.

Zatomus sarcophagus Cope.

Cope, E. D. 1871 M, p. 211.
  1875 V, p. 35.
Emmons, E. 1857 A, p. 62, fig. 34.   (No name).
Mesozoic; North Carolina.

ANCHISAURIDÆ.

Marsh, O. C. 1885 A, p. 169.
Gadow, H. 1897 A, p. 204.
Günther, A. C. 1886 A, p. 443.
Lydekker, R. 1888 B, p. 174.
    1890 A, p. 246.
Marsh, O. C. 1882 A, p. 84. (Amphisauridæ.)
    1884 B, p. 338. (Amphisauridæ.)
    1884 G. (Amphisauridæ.)

Marsh, O. C. 1885 D, p. 765.
    1895 C, p. 498.
    1896 C, pp. 147, 239.
    1896 D, p. 206.
Nicholson and Lydekker 1889 A, p. 1186.
Zittel, K. A. 1890 A, p. 730.

ANCHISAURUS Marsh. · Type *Megadactylus polyzelus* Hitch.

Marsh, O. C. 1885 A. (To replace Amphisaurus Marsh, preoccupied.)
Baur, G. 1883 A, p. 443. (Amphisaurus.)
    1884 B, p. 447. (Amphisaurus.)
Cope, E. D. 1869 M, p. 122 A. (Megadactylus.)
    1870 S. (Megadactylus.)
    1877 Q, p. 6. (Megadactylus.)
    1878 I, p. 236. (Megadactylus.)
    1885 O, p. 705. Megadactylus.)
    1887 C, p. 368. (Megadactylus.)
Dames, W. 1884 A. (Amphisaurus.)
Marsh, O. C. 1882 A, p. 84. (Amphisaurus, to replace Megadactylus, preoccupied.)
    1889 C, p. 331.
    1893 F, p. 151.
    1893 G.
    1895 C, p. 498.
    1896 C, p. 147.
Nicholson and Lydekker 1889 A, pp. 1166, 1170.
Woodward, A. S. 1898 B, p. 198.
Zittel, K. A. 1890 A, p. 730.

Anchisaurus colurus Marsh.

Marsh, O. C. 1891 E, p. 267.
Fürbringer, M. 1900 A, p. 354, fig. 130.
Marsh, O. C. 1892 F, p. 543, pls. xv, xvi.

Marsh, O. C. 1893 C, p. 169, pl. vi.
    1893 F, p. 151, fig. 1.
    1895 C, pl. x, fig. 1.
    1896 C, p. 148, pl. ii, figs. 1–3; pl. iii, figs. 1, 2.
    1896 D, pl. i, fig. 1.
Woodward, A. S. 1898 B, p. 198, fig. 119.
Triassic; Connecticut.

Anchisaurus polyzelus (Hitch.).

Hitchcock, E. 1865 A, p. 39. (Megadactylus.)
Cope, E. D. 1869 M, p. 122 E, pl. xiii. (Megadactylus.)
    1870 A. (Megadactylus.)
Hitchcock, E. 1858 A, p. 187. (No name.)
Marsh, O. C. 1892 F, p. 546, pls. xvi, xvii. (Anchisaurus.)
    1896 C, p. 147, pl. iii, figs. 4, 5. (Anchisaurus.)
Triassic; Massachusetts.

Anchisaurus solus Marsh.

Marsh, O. C. 1892 F, p. 545.
    1893 C, p. 169.
    1893 F, p. 151.
    1896 C, p. 149.
Triassic; Connecticut.

AMMOSAURUS Marsh.   Type *Anchisaurus major* Marsh.

Marsh, O. C. 1891 E, p. 267.
    1892 F, p. 545.
    1893 G, p. 170.
    1895 C, p. 493.
    1896 C, p. 150.

Ammosaurus major Marsh.

Marsh, O. C. 1889 C, p. 331. (Anchisaurus.)
    1891 E, p. 331, fig. 1.
    1892 F, p. 546, pls. xvi, xvii.
    1896 C, p. 147, pl. iii, figs. 3, 6.
Zittel, K. A. 1890 A, p. 730, fig. 636. (Anchisaurus.)
Triassic; Connecticut.

BATHYGNATHUS Leidy.   Type *B. borealis* Leidy.

Leidy, J. 1854 H.
Cope, E. D. 1867 D.
    1870 S.
Huxley, T. H. 1869 A.
Leidy, J. 1855 A.
    1868 F. p, 199.
Marsh, O. C. 1882 A, p. 84.
    1892 C, p. 240.
    1895 C, p. 493.
Meyer, H. 1855 A, p. 161.
Nicholson and Lydekker 1889 A, p. 1166.
Owen, R. 1860 A, p. 163.
Zittel, K. A. 1890 A, p. 731.

Bathygnathus borealis Leidy.

Leidy, J. 1854 H, pl. xxxiii.
Cope, E. D. 1869 M, p. 119.
Dana, J. D. 1896 A, p. 754, fig. 1180.
Dawson, J. W. 1868 A, p. 119, fig. 29.
    1878 A, p. 119, fig. 29.
Lea, I. 1856 B, p. 123.
    1858 A.
Leidy, J. 1855 A.
    1861 A.
Lesley, J. P. 1889 A, p. 80, figures.
Owen, R. 1860 E, p. 252.
    1876 B, p. 359, fig. 9.
Triassic; Prince Edwards Island.

CLEPSYSAURUS Lea.   Type *C. pennsylvanicus* Lea.

Lea, I. 1851 B.
Cope, E. D. 1869 M, p. 122 A
    1870 I, p. 444.
    .1870 S.
    1871 M, p. 210.
Emmons, E. 1856 A, p. 298.   (Clepsisaurus.)
Lea, I. 1853 A, p. 198.
Leidy, J. 1857 B, p. 150.
Marsh, O. C. 1895 C, p. 498.
Nicholson and Lydekker 1889 A, p. 1166.
Zittel, K. A. 1890 A, p. 731.

Clepsysaurus pennsylvanicus Lea.
Lea, I. 1851 B.
Amer. Jour. Science 1852 B.
Cope, E. D. 1866 D, p. 249.
    1869 B, p. 733.
    1869 M, p. 122 A.
    1871 M, p. 210.
    1875 V, p. 85, pl. viii, fig. 4.
Dana, J. D. 1896 A, p. 764, fig. 1181.

Emmons, E. 1856 A, p. 299, figs. A, B, and 21.
    1857 A, p. 67, in part.   (Clepsisaurus.)
    1860 A, p. 175, figs. 158¹⁻⁵, 160⁶.   (Clepsi-
        saurus.)
Lea, I. 1851 A.   (No generic name.)
    1853 A, p. 199, pls. xvii-xix.
    1856 B.
    1858 A.
Leidy, J. 1854 H, p. 327.
    1855 A.
Lesley, J. P. 1889 A, p. 133, figure.
Meyer, H. 1856 A, p. 161.
Wheatley, C. M. 1861 A, p. 44.
    Triassic; Pennsylvania, New Jersey, North
        Carolina.

Clepsysaurus veatleianus Cope.
Cope, E. D. 1877 E, p. 184.
    1878 F F.   (C. wheatleianus.)
Lesley, J. P. 1889 A, p. 184.
    Triassic; Pennsylvania.

THECODONTOSAURUS Riley and Stutchbury.

Riley and Stutchbury 1838 A, p. 352.
Cope, E. D. 1877 E, p. 182.   (Palæosaurus.)
Huxley, T. H. 1870 D, p. 43.   (Thecodontosaur-
    us, Palæosaurus.)
Lydekker, R. 1888 B, p. 174.
    1890 A, p. 246.
Marsh, O. C. 1895 C, p. 498.
Owen, R. 1842 A, p. 154.   (Thecodontosaurus,
    Palæosaurus.)
    1845 B, p. 266.
Riley and Stutchbury 1838 A, p. 352.   (Palæosaur-
    us, with species cylindrodon and platyodon.)
Seeley, H. G. 1896 C.   (Thecodontosaurus, Palæo-
    saurus.)
Zittel, K. A. 1890 A, p. 721 (Thecodontosaurus); p.
    722 (Palæosaurus).

Thecodontosaurus and *Palæosaurus* of Riley
    and Stutchbury are here united. In case
Palæosaurus of these authors is distinct, as
    seems probable, it must be provided with a new
    name, since *Palæosaurus* was used by Geoffroy
    in 1831 in another sense.

Thecodontosaurus fraserianus Cope.
Cope, E. D. 1878 D, p. 232.   (Palæosaurus.)
    1878 FF.
    Triassic; Pennsylvania.

Thecodontosaurus gibbide- Cope.
Cope, E. D. 1878 D, p. 231.
    1878 FF.
    Triassic; Pennsylvania.

ARCTOSAURUS Adams.   Type *A. osborni* Adams.

Adams, A. L. 1875 A, p. 177.
    1877 A.
Lydekker, R. 1890 A, p. 250.
Marsh, O. C. 1895 C, p. 498.
    1896 C, p. 152.
Nicholson and Lydekker 1889 A, p. 1166.
Zittel, K. A. 1890 A, p. 731.
    The position of this genus is doubtful.

Arctosaurus osborni Adams.
Adams, A. L. 1875 A, p. 177, figs. A-D.
    1877 A.
Lydekker, R. 1889 L, p. 352, figs. A-D.
    1890 A, p. 251, fig. 59, A-D.
    Mesozoic; Bathurst Island.

CŒLURIDÆ.

Marsh, O. C. 1881 B.   (Cœluria, Cœluridæ.)
Cope, E. D. 1889 R, p. 864.
Lydekker, R. 1888 B, p. 155.
    1890 A, p. 243.
Marsh, O. C. 1881 E, p. 423.   (Cœluria, Cœluridæ.)
    1882 A, p. 85.   (Cœluria, Cœluridæ.)
    1884 B, p. 338.   (Cœluria, Cœluridæ.)
    1884 G.   (Cœluria, Cœluridæ.)

Marsh, O. C. 1885 D, p. 765.   (Cœluria, Cœluridæ.)
    1895 C, p. 494.   (Cœluria, Cœluridæ.)
    1896 C, p. 240.   (Cœluria, Cœluridæ.
    1896 D, p. 206.
Menzbier, M. 1887 A.   (Cœluria.)
Nicholson and Lydekker 1889 A, p. 1170.
Steinmann and Döderlein 1890 A, p. 661
Zittel, K. A. 1890 A p, 731.

CŒLURUS Marsh.   Type *C. fragilis* Marsh.

Marsh, O. C. 1879 H, p. 504.
Cope, E. D. 1887 A, p. 221.
    1887 C.
Lydekker, R. 1888 B, p. 155.
    1889 H.
Marsh, O. C. 1881 B.
    1884 B.
    1896 C, p. 155.
Nicholson and Lydekker 1889 A. p. 1170.
Seeley, H. G. 1887 A.
    1888 C, P. 83.
Vetter, B. 1885 A. p. 119.
Zittel, K. A. 1890 A, p. 731.

Cœlurus agilis Marsh.

Marsh, O. C. 1884 B, p. 335, pl. x, figs. 3a, 3b.
    1896 C, pl. x, figs. 3, 4.
Jurassic; Colorado.

Cœlurus fragilis Marsh.

Marsh, O. C. 1879 H, p. 504.
Cope, E. D. 1881 BB, p. 413.   (Amphicœlias fragil-
    lissimus a synonym?)
    1887 C, p. 369.
Marsh, O. C. 1881 B, pl. x.
    1884 B, p. 340, pls. xi, xiii.
    1896 C, p. 155, pl. vii.
Seeley, H. G. 1888 C, p. 5.
Zittel, K. A. 1890 A, p. 731, figs. 637-640.
Jurassic; Wyoming.

Cœlurus gracilis Marsh.

Marsh, O. C. 1888 A, p. 94.
Zittel, K. A. 1890 A, p. 732.
Jurassic (Potomac); Maryland.

CŒLOPHYSIS Cope.   Type not indicated.

Cope, E. D. 1889 N.
Lydekker, R. 1888 B, p. 155.   (Tanystropheus.)

Cœlophysis bauri Cope.

Cope, E. D. 1887 C, p. 368.   (Cœlurus.)
    1887 A, p. 226.   (Tanystropheus.)
    1889 N.
Zittel, K. A. 1890 A, p. 733.   (Tanystropheus.)
Triassic; New Mexico.

Cœlophysis longicollis Cope.

Cope, E. D. 1887 C, p. 368.   (Cœlurus.)

Cope, E. D. 1887 A, p. 221.   (Tanystropheus.)
    1889 N.
Zittel, K. A. 1890 A, p. 733.   (Tanystropheus).
Triassic; New Mexico.

Cœlophysis willistoni Cope.

Cope, E. D. 1887 A, p. 227.   (Tanystropheus.)
    1889 N.
Zittel, K. A. 1890 A, p. 733.   (Tanystropheus.)
Triassic; New Mexico.

TICHOSTEUS Cope.   Type *T. lucasanus* Cope.

Cope, E. D. 1877 F, p. 194.
    1878 M, p. 84.
Zittel, K. A. 1890 A, p. 732.
A genus of uncertain position.

Tichosteus æquifacies Cope.

Cope, E.D. 1878 H, p. 392.
Jurassic; Colorado.

Tichosteus lucasanus Cope.

Cope, E. D. 1877 F, p. 194.
    1880 Q.
Osborn, H. F. 1898 J, p. 227.
Jurassic; Colorado.

CERATOSAURIDÆ.

Marsh, O. C. 1884 B, p. 330.
    1884 E.
    1884 G.   (Ceratosauria, Ceratosauridæ.)
    1885 D.   (Ceratosauria, Ceratosauridæ.)
    1892 H.
    1895 C, p. 494.   (Ceratosauria, Ceratosaur-
    idæ.)
    1896 C, p. 240.   (Ceratosauria, Ceratosaur-
    idæ.)

Marsh, O. C. 1896 D, p. 207.   (Ceratosauria, Cerat-
    osauridæ.)
Lydekker, R. 1888 B, p. 157.   (Megalosauridæ, in
    part.)
Steinmann and Döderlein 1890 A, p. 660.
Zittel, K. A. 1890 A, p. 727.

CERATOSAURUS Marsh.   Type *C. nasicornis* Marsh.

Marsh, O. C. 1884 B.
Baur, G. 1884 B, p. 448.
    1890 C, p. 301.
    1891 D, p. 446.
Cope, E. D. 1892 K, p. 244.   (Megalosaurus.)
Dollo, L. 1884 B.
Fürbringer, M. 1888 A, p. 1609.
Gaudry, A. 1890 C, p. 252.

Lydekker, R. 1888 A, p. 58.
    1888 B, p. 157.
    1889 C, p. 45.
    1893 B, p. 302.   (Megalosaurus.)
    1893 E, p. 138.   (Megalosaurus?.)
Marsh, O. C. 1884 E.
    1889 B, p. 330.
    1892 H, p. 347, pl. vii.

Marsh, O. C. 1893 F, p. 152.
    1893 I, p. 437.
    1895 C, p. 494.
    1896 C, p. 156.
    1896 D, p. 203.
    1897 C, p. 503.
Natural Science 1893 A, p. 138.
Nicholson and Lydekker 1889 A, p. 1168.
Seeley, H. G. 1887 A, p. 225.
Steinmann and Döderlein 1890 ., p. 660.
Vetter, B. 1885 A, p. 111.
Woodward, A. S. 1896 B, p. 199.
Zittel, K. A. 1890 A, p. 727.

Ceratosaurus nasicornis Marsh.

Marsh, O. C. 1884 B, p. 330, pls. viii–xi.

Cope, E. D. 1892 K, P. 245. (Megalosaurus.)
    1892 Z, p. 18.
Marsh, O. C. 1884 E, fig. 1.
    1892 H, p. 347, pl. vii.
    1893 F, p. 152, pl. vi.
    1895 C, p. 489, pl. x, fig. 5; text figs. 9, 10.
    1896 C, p. 156, pls. viii–x, fig. 1.; pl. xiv, text
      figs. 1, 64, 65.
    1896 D, p. 202, pl. i, fig. 5; text figs. 9, 10.
    1897 C, p. 503, pl. xxv, text fig. 60.
Menzbier, M. 1887 A, p. 533.
Walcott, C. D. 1900 A, p. 23.
Woodward, A. S. 1898 B, p. 200, fig. 120.
Zittel, K. A. 1890 A, p. 727, figs. 629–635.
    Jurassic; Colorado.

ORNITHOMIMIDÆ.

Marsh, O. C. 1890 A, p. 85.
    1890 C, p. 425.
    1895 C, p. 494.

Marsh, O. C. 1896 C, pp. 203, 240.
    1896 D, p. 207.
Zittel, K. A. 1896 D, p. 207.

ORNITHOMIMUS Marsh. Type *O. velox* Marsh.

Marsh, O. C. 1890 A, p. 84.
Baur, G. 1890 C, p. 303.
Gadow, H. 1897 A, p. 206.
Marsh, O. C. 1895 C, p. 494.
    1896 C, p. 204.
    1897 C, p. 518.
Zittel, K. A. 1890 A, p. 767.

Ornithomimus grandis Marsh.
Marsh, O. C. 1890 A, p. 85.
    1892 E, p. 452.
    1896 C, p. 206.
Zittel, K. A. 1890 A, p. 767.
    Cretaceous (Laramie); Colorado.

Ornithomimus minutus Marsh.
Marsh, O. C. 1892 E, p. 452.
    1896 C, p. 206.
    Cretaceous (Laramie); Colorado.

Ornithomimus sedens Marsh.

Marsh, O. C. 1892 E, p. 451.
    1896 C, p. 205, figs. 49–52.
    Cretaceous (Laramie); Colorado.

Ornithomimus tenuis Marsh.

Marsh, O. C. 1890 A, p. 85.
    Cretaceous (Laramie); Colorado.

Ornithomimus velox Marsh.

Marsh, O. C. 1890 A, p. 84, pl. i, figs. 1–3.
Cannon, J. L. 1890 A.
Dana, J. D. 1896 A, p. 847, fig. 1416.
Marsh, O. C. 1896 C, p. 205, pl. lviii, figs. 1–4.
    1897 C, pp. 518, 527, figs. 84–86.
Zittel, K. A. 1890 A, p. 766, figs. 677, 678.
    Cretaceous (Laramie); Colorado.

HALLOPIDÆ.

Marsh, O. C. 1881 E, p. 423. (Hallopoda, Hallopod-
    idæ.)
Cope, E. D. 1883 J, p. 98. (Hallopoda.)
Günther, A. 1886 A, p. 443.
Marsh, O. C. 1882 A, p. 85. (Hallopoda, Hallopod-
    idæ.)
    1884 B, p. 338.
    1885 D.

Marsh. O. C. 1890 B.
    1895 C, p. 494. (Hallopoda, Hallopidæ.)
    1896 C, p. 240. (Hallopoda, Hallopidæ.)
    1896 D, p. 207. (Hallopoda, Hallopidæ.)
Nicholson and Lydekker 1889 A, p. 1169.
Zittel, K. A. 1890 A, p. 693 (Hallopoda); p. 736
    (Hallopidæ).

HALLOPUS Marsh. Type *Nanosaurus victor* Marsh.

Marsh, O. C. 1881 E, p. 422.
Baur, G. 1890 E.
Marsh, O. C. 1884 B, p. 338.
    1895 C, pp. 488, 494.
    1896 C, p. 153.
    1896 D, p. 201.
Nicholson and Lydekker 1889 A, p. 1169.
Vetter, B. 1885 A, p. 119.
Zittel, K. A. 1890 A, p. 736.

Hallopus victor Marsh.

Marsh, O. C. 1877 D, p. 255. (Nanosaurus.)

Marsh, O. C. 1881 E, p. 422.
    1890 B, p. 417, figure.
    1891 F, p. 337.
    1895 C, p. 484, text figs. 4, 5.
    1896 C, p. 153, pl vi, text figs. 3, 59, 60.
    1896 D, p. 199, figs. 4, 5.
    1897 C, p. 482 figs. 25, 26.
Woodward, A. S. 1898 B, p. 200.
Zittel, K. A. 1890 A, p. 736, fig. 643.
    Jurassic; Colorado.

Suborder ORTHOPODA Cope.

.bre, *E. D.* 1866 E, p. 317.
Baur, G. 1887 A, p. 487. (Orthopoda.)
   1887 C, p. 101. (Orthopoda.)
   1890 B. (Orthopoda.)
   1891 A. (Iguanodontia.)
   1891 B. (Orthopoda.)
   1891 C, p. 354. (Iguanodontia.)
   1891 D, p. 451. (Iguanodontia.)
Cope, E. D. 1869 M, p. 90. (Orthopoda.)
   1871 B, p. 234. (Orthopoda.)
   1882 Q. (Orthopoda.)
   1883 J, p. 98. (Orthopoda.)
   1889 R, p. 864. (Orthopoda.)
   1891 N, p. 43. (Orthopoda.)
   1898 B, p. 68. (Orthopoda.)
Dollo, L. 1888 B. (Iguanodontia.)
Fürbringer, M. 1888 A.

Fürbringer, M. 1900 A.
Gadow, H. 1897 A, p. 204.
   1898 A, p. 22.
Lydekker, R. 1888 B, p. 175. (Ornithopoda.)
Marsh, O. C. 1894 E, p. 90. (Predentata.)
   1895 C, p. 496. (Predentata.)
   1896 C, pp. 186, 242. (Predentata.) .
   1896 D, p. 209. (Predentata.)
Nicholson and Lydekker 1889 A, p.1152. (Ornithopoda.)
Osborn, H. F. 1900 G, p. 786. (Iguanodontia.)
Seeley, H. G. 1888 B. (Ornithischia.)
   1888 G. (Ornithischia.)
Steinmann and Döderlein 1890 A, p. 663. (Orthopoda.)
Zittel, K. A. 1890 A, pp. 692, 786. (Orthopoda.)

Superfamily STEGOSAUROIDEA nom. nov.

This group is called by the following writers the *Stegosauria*.
Baur, G. 1883 A, p. 435.
   1884 A.
Cope, E. D, 1880 R.
   1883 J, p. 98.
Dollo, L. 1892 C, p. 11. ("Stegosauriens.")
Fürbringer, M. 1888 A.
   1900 A, p. 348.
Gadow, H. 1897 A, p. 205. (Stegosauri.)
   1898 A, p. 22. (Stegosauri,)
Haeckel, E. 1895 A, p. 387. (Pachypoda.)
Krause, W. 1881 A.
Lydekker, R. 1888 B, p. 175.
   1889 C, p. 43.
Marsh, O. C. 1877 F, p. 513.
   1878 G, p. 224.
   1880 C.
   1881 A.

Marsh, O. C. 1882 A, p. 83.
   1881 E, p. 423.
   1884 G.
   1885 D.
   1887 C.
   1888 D.
   1889 B, p. 326.
   1891 D, p. 181.
   1891 J, p. 387.
   1895 C, p. 496.
   1896 C, pp. 193, 242.
   1896 D, p. 209.
   1897 C, p. 498.
Science 1885 B.
Steinmann and Döderlein 1890 A, p. 663.
Vetter, B. 1885 A, p. 117.
Wiedersheim, R. 1881 A.
Zittel, K. A. 1890 A, pp. 693, 740.

STEGOSAURIDÆ.

Marsh, O. C. 1880 C, p. 259.
Cope, E. D. 1896 B, p. 70. (Hypsirhophidæ.)
Lydekker, R. 1888 B, p. 176. (Omosauridæ.)
   1889 L.
   1890 A, p. 251.
Marsh, O. C. 1881 E, p. 423.
   1882 A, p. 83.
   1884 B, p. 336.

Marsh, O. C. 1884 G.
   1885 D, p. 764.
   1895 C, p. 496.
   1896 C, p. 242.
   1896 D, p. 209.
Nicholson and Lydekker 1889 A, p. 1161.
Steinmann and Döderlein 1890 A, p. 664.
Zittel, K. A. 1890 A, p. 744.

STEGOSAURUS Marsh.   Type *S. armatus* Marsh.

Marsh, O. C. 1877 F, p. 513.
Cope, E. D. 1878 H, p. 389. (Hypsirophus.)
   1878 R. (Hypsirophus, type *H. discurus.*)
   1888 Z. (Hypsirophus.)
   1892 Z, p. 18. (Hypsirophus.)
Koken, E. 1897 A, p. 51.
Krause, W. 1881 A.
Lydekker, R. 1888 B, p. 177. (Omosaurus, Stegosaurus.)
   1889 F, p. xl.

. Lydekker, R. 1890 A, p. 253.
   1893 B, p. 302. (Hypsirophus.)
Marsh, O. C. 1880 C, p. 253.
   1881 A.
   1883 A, p. 85.
   1887 C.
   1888 D.
   1889 B, p. 327.
   1891 D.
   1891 J.

Marsh, O. C. 1892 G, p. 174.
1893 I, p. 438. (Stegosaurus, Hypsirophus.)
1895 C, p. 497.
1896 C, pp. 186, 242.
1897 C, p. 498.
Nicholson and Lydekker 1889 A, p. 1161.
Owen, R. 1875 C, p. 45. (Omosaurus.)
Steinmann and Döderlein 1890 A, p. 664.
Woodward, A. S. 1898 B, p. 210.
Zittel, K. A. 1890 A, p. 726 (Hypsirophus); p. 744.
(Stegosaurus).

Stegosaurus affinis Marsh.

Marsh, O. C. 1881 A, p. 169.
1896 C, p. 191.
Jurassic; Wyoming?

Stegosaurus armatus Marsh.

Marsh, O. C. 1877 F, p. 513.
King, C. 1878 A, p. 346.
Lydekker, R. 1888 B, p. 179, fig. 32.
Marsh, O. C. 1879 H, p. 504.
1880 C, p. 253 (part), pl. vi (except figs. 4, 5,
which belong to Diplodocus).
1891 D, p. 179.
1896 C, p. 188.
Jurassic; Colorado.

Stegosaurus discurus (Cope).

Cope, E. D. 1878 R. (Hypsirophus.)
1878 H, p. 389. (Hypsirophus.)
Jurassic; Colorado.

Stegosaurus duplex Marsh.

Marsh, O. C. 1887 C, p. 416.
1881 A, p. 168 (pelvis), pl. vi, fig. 1. (S. un-
gulatus.)
1888 D, p. 14.
1891 D, p. 179.
1896 C, p. 193.
Jurassic; Wyoming.

Stegosaurus seeleyanus (Cope).

Cope, E. D. 1879 R. (Hypsirophus.)
Jurassic; Colorado.

Stegosaurus stenops Marsh.

Marsh, O. C. 1887 C, p. 114, pl. vi.
Cope, E. D. 1892 Z, p. 18. (Hypsirophus.)
Lydekker, R. 1888 B, p. 176, fig. 31.
Marsh, O. C. 1888 D, pl. i.
1890 C, p. 426, pl. vii, fig. 2.
1891 D, p. 179.
1896 C, p. 186, pl. xliii, pl. xlvii, fig. 3.
1897 C, p. 499, fig. 52.
Woodward, A. S. 1898 B, p. 212, fig. 131.
Zittel, K. A. 1890 A, p. 745, figs. 650, 654.
Jurassic; Colorado.

Stegosaurus sulcatus Marsh.

Marsh, O. C. 1887 C, p. 415, pl. viii, figs. 4-6.
1888 D, pl. ii, figs. 4-6.
1896 C, p. 193, pl. i, figs. 4-6.
Jurassic; Colorado?

Stegosaurus ungulatus Marsh.

Marsh, O. C. 1879 H, p. 504.
Cope, E. D. 1880 R.
1881 BB, pp. 254, 340. (Hypsirophus.)
Fürbringer, M. 1900 A, p. 351, figs. 121, 126.
Geinitz and Deichmüller 1882 A, p. 15.
Hutchison, H. N. 1893 A, p. 110, pl. xi.
Lydekker, R. 1893 B, p. 304, fig. 3. (Hypsirophus.)
Marsh, O. C. 1881 A (part), pls. vi-viii (except
fig. 1, pl. vi, which belongs to S. duplex).
1887 C, p. 415, pls. vii, viii.
1888 D, pl. ii, figs. 1-3.
1891 D, pl. ix.
1891 J, pl. xi.
1892 G, p. 176, pl. iv.
1894 E, pl. vii, fig. 6.
1896 C, p. 188, pl. xliv, figs. 1, 3, 4; pls. xlv-
xlvii; pl. xlviii, figs. 1, 2; pl. xlix; pl. l,
figs. 1-3; pl. lii.
1896 D, pl. i, fig. 8.
1897 C, p. 498, pl. xxii, text figs. 54-57.
Woodward, A. S. 1898 B, p. 211, fig. 130.
Zittel, K. A. 1890 A, p. 745, figs. 651-653, 655-660.
Jurassic; Wyoming.

DIRACODON Marsh.   Type *D. laticeps* Marsh.

Marsh, O. C. 1881 E, p. 421.
1887 C, p. 416.
1888 D, p. 14.
1891 D, p. 181.
1895 C, p. 497.
1896 C, p. 193.
Nicholson and Lydekker 1889 A, p. 1168.
Zittel, K. A. 1890 A, p. 748.

Diracodon laticeps Marsh.

Marsh, O. C. 1881 E, p. 421.
1887 C, p. 416, pl. ix.
1888 D, pl. iii.
1896 C, p. 193, pl. ii.
Jurassic; Wyoming.

PRICONODON Marsh.   Type *P. crassus* Marsh.

Marsh, O. C. 1888 A, p. 93.
1896 C, p. 187.
Zittel, K. A. 1890 A, p. 748.

Priconodon crassus Marsh.

Marsh, O. C. 1888 A, p. 93, figs. 3-9.
1896 C, pl. xliv, fig. 2.
Walcott, C. D. 1900 A, p. 23.

PALÆOSCINCUS Leidy.   Type *P. costatus* Marsh.

Leidy, J. 1856 F, p. 72.
Cope, E. D. 1869 M, p. 99.
Leidy, J. 1856 C, p. 118.
Marsh, O. C. 1889 B, p. 328.
  1892 G, p. 178.
  1896 C, p. 497.
  1896 C, p. 225.
Zittel, K. A. 1890 A, p. 748.
  The affinities of this genus are uncertain.

Palæoscincus costatus Leidy.

Leidy, J. 1856 F, p. 72.
Cope, E. D. 1869 M, p. 99.

Cope, E. D. 1874 C, p. 26.
  1875 E, p. 248.
  1877 G, p. 572.
Leidy, J. 1856 C, p. 118.
  1860 A, p. 146, pl. ix, figs. 49–61; pl. xi, fig 3.
Marsh, O. C. 1896 C, p. 225.
  Cretaceous (Laramie); Montana.

Palæoscincus latus Marsh.

Marsh, O. C. 1892 G, p. 178, pl. iii, fig. 3.
  1896 C p. 225, pl. lxxxv.
  Cretaceous (Laramie); Wyoming.

DYSTROPHÆUS Cope.   Type *D. viæmalæ* Cope.

Cope, E. D. 1877 B, p. 579.
  1877 K, p. 31.
Marsh, O. C. 1896 C, p. 497.
Zittel, K. A. 1890 A, p. 749.
  A genus of uncertain relationships.

Dystrophæus viæmalæ Cope.

Cope, E. D. 1877 B p. 581.
  1877 K p. 34, text fig. 3.
  1878 M, p. 84.
Marsh, O. C. 1896 C, p. 152.
  Triassic?; Utah.

HYPSIBEMA Cope.   Type *H. crassicauda* Cope.

Cope, E. D. 1869 I.
  1869 M, p. 122G.
  1871 M, p. 211.
  1875 V, p. 36.
Zittel, K. A. 1890 A, p. 748.
  A genus of uncertain position.

Hypsibema crassicauda Cope.

Cope, E. D. 1869 I.
  1869 M, p. 122 G, pl. i, figs. 14, 15.
  1871 M, p. 211.
  1875 E, p. 247.
  1875 V, p. 36, pls. vi, vii.
Nopcsa, F. B. 1899 A, p. 557.
  Cretaceous; New Jersey.

NODOSAURIDÆ.

Marsh, O. C. 1890 C, p. 425.
  1896 C, p. 497.

Marsh, O. C. 1896 C, p. 243.
  1896 D, p. 210.

NODOSAURUS Marsh.   Type *N. textilis* Marsh.

Marsh, O. C. 1889 E, p. 175.
  1896 C, p. 497.
  1896 C, p. 225.
Nicholson and Lydekker 1889 A, p. 1164.
Zittel, K. A. 1890 A, p. 754.

Nodosaurus textilis Marsh.

Marsh, O. C. 1889 E, p, 175, figure.
Lambe, L. M. 1899 A, p. 70. (Identification doubtful.)
Marsh, O. C. 1896 C, p. 225, pl. lxxv, fig. 5.
  Cretaceous; Wyoming, British America.

Superfamily CERATOPSOIDEA nom. nov.

Dollo, L. 1892 C, p. 12. ("Ceratopsiens.")
Fürbringer, M. 1900 A, p. 848. (Ceratopsia.)
Gadow, H. 1897 A, p. 205. (Ceratopsia.)
  1898 A. p. 23. (Ceratopsia.)
Marsh, O. C. 1890 C, p. 418. (Ceratopsia.)

Marsh, O. C. 1895 C, p. 497. (Ceratopsia.)
  1896 C, p. 243. (Ceratopsia.)
  1896 D, p. 210. (Ceratopsia.)
Zittel, K. A. 1890 A, p. 749. (Ceratopsia.)

CERATOPSIDÆ.

Marsh, O. C. 1888 C, p. 473.
Baur, G. 1891 A. (Agathaumidæ.)
    1891 C, p. 354. (Agathaumidæ.)
Cannon, G. L. 1888 A, p. 215.
Cope, E. D. 1889 Q, p. 715. (Agathaumidæ.)
    1889 R, p. 864. (Agathaumidæ.)
    1891 N, p. 43. (Agathaumidæ.)
    1898 B, p. 70. (Agathaumidæ.)
Marsh, O. C. 1889 G.
    1890 A, p. 83.
    1890 C, p. 418.

Marsh, O. C. 1890 F.
    1891 B.
    1891 C.
    1891 G.
    1892 A.
    1895 C, p. 497.
    1896 C, pp. 206, 243.
    1896 D, p. 210.
    1897 C, p. 509.
Nicholson and Lydekker 1889 A, p. 1163.
Steinmann and Döderlein 1890 A, p. 665.

AGATHAUMAS Cope.   Type *A. sylvestris* Cope.

Cope, E. D. 1872 Q.
Amer. Geologist 1891 D.
Baur, G. 1891 A.
    1891 C, p. 354.
    1891 D, p. 450.
Cope, E. D. 1874 B, p. 444 (Agathaumas); pp.
  448, 451 (Polyonax, type *P. mortuarius*).
    1874 C, p. 17 (Agathaumas); p. 24 (Poly-
    onax).
    1875 E, p. 53 (Agathaumas); p. 68 (Poly-
    onax).
    1883 J, p. 99.
    1888 Z. (Polyonax.)
    1889 A. (Polyonax.)
    1889 Q, p. 715. (Polyonax.)
    1889 T, p. 906. (Polyonax.)
    1891 N, p. 43.
    1892 T, p. 758.
Marsh, O. C. 1893 I, p. 438.
    1895 C, p. 497.
    1896 C, p. 217. (Agathaumas; Polyonax.)
Osborn, H. F. 1893 B, p. 326.
Zittel, K. A. 1890 A, p. 751. (Agathaumas, Poly-
  onax.)

Agathaumas mortuarius Cope.

Cope, E. D. 1874 B, pp. 448, 451. (Polyonax.)
    1874 C, p. 24. (Polyonax.)
    1875 E, p. 64, pl. ii, figs. 3–5; pl. iii, figs. 1–4.
    (Polyonax.)
    1889 Q, p. 715.
Cretaceous (Laramie); Colorado.

Agathaumas sylvestris Cope.

Cope, E. D. 1872 Q.
Baur, G. 1891 A.
Cope, E. D. 1872 EE, p. 279.
    1872 FF.
    1873 A.
    1874 B, pp. 442, 445.
    1874 C, pp. 16, 18.
    1875 E, pp. 54, 248, pls. iv, v: pl. vi, figs. 1–4.
    1878 S, p. 245.
    1889 Q, p. 715.
    1891 N, p. 43.
    1892 T, p. 758.
Cretaceous (Laramie); Wyoming.

TRICERATOPS Marsh.   Type *Ceratops horridus* Marsh.

Marsh, O. C. 1889 E, p. 173.
Ameghino, F. 1896 A, p. 391.
Amer. Geologist 1891 D.
Baur, G. 1891 A. (Agathaumus.)
    1891 D, p. 450. (Agathaumus.)
Cope, E. D. 1889 Γ, p. 906. (Polyonax.)
Fürbringer, M. 1900 A, p. 350.
Lydekker, R. 1893 E, p. 138.
Marsh, O. C. 1889 G, p. 502.
    1890 C, p. 418.
    1890 F, p. 2.
    1891 B, p. 167.
    1891 C, p. 339.
    1893 B.
    1895 C, p. 497.
    1896 C, p. 208.
    1897 C, p. 510.
    1898 B, p. 407.
Natural Science 1893 A.
Nicholson and Lydekker 1889 A, p. 1163.
Seeley, H. G. 1893 A, p. 438.
Steinmann and Döderlein 1890 A, p. 665.
Woodward, A. S. 1898 B, p. 213.
Zittel, K. A. 1890 A, pp 750, 752. (Syn. ? of
  Monoclonius.)

Triceratops calicornis Marsh.

Marsh, O. C. 1898 A, p. 92.
Walcott, C. D. 1900 A, p. 28. (T. californis.)
Cretaceous; Wyoming.

Triceratops elatus Marsh.

Marsh, O. C. 1891 E, p. 265.
Cretaceous (Laramie); Wyoming.

Triceratops galeus Marsh.

Marsh, O. C. 1889 E, p. 174.
    1897 C, p. 527.
Cretaceous (Laramie), Colorado.

Triceratops horridus Marsh.

Marsh, O. C. 1889 C, p. 334. (Ceratops.)
    1889 E, p. 173.
    1889 G, p. 502.
    1890 C, p. 426, pl. vi, figs. 7–9.
    1891 B, p. 178, pl. ix.
    1891 G.
    1896 C, p. 208, pl. lxix, figs. 10–12.
    1897 C, p. 527.
Zittel, K. A. 1890 A, p. 752, fig. 665.
Cretaceous (Laramie); Wyoming, Colorado.

Triceratops obtusus Marsh.

Marsh. O. C. 1898 A, p. 92.
Walcott, C. D. 1900 A, p. 23,
  Cretaceous (Laramie); Wyoming.

Triceratops prorsus Marsh.

Marsh, O. C. 1890 A.
Dana, J. D. 1896 A, p. 846, figs. 1411-1413.
Fürbringer, M. 1900 A, p. 351, figs. 123, 125, 128.
Hutchison, H. N. 1893 A, p. 116, pl. xi, text fig. 30.
Lydekker, R. 1893 B, p. 304, fig. 4. (Agathaumus.)
Marsh, O. C. 1890 C, p. 425, pl. v, figs. 3-5; pl. vi,
  figs. 5, 6.
  1891 B, p. 177, pls. i-ix.
  1891 C, p. 339, pl. xv.
  1892 A, p. 84, pl. iii, fig. 6.
  1894 E, pl. vii, fig. 4.   •
  1895 C, pl. x, fig. 10.
  1896 C, p. 208, pl. llx; pl. lx, fig. 4; pl. lxi,
    figs. 1-8; pl. lxiv; pl. lxviii; pl. lxix, figs.
    1-3, 7-9; pls. lxv, lxvi; pl. lxvii, figs. 2-4.
  1896 D, pl. i, fig. 11.
  1897 C, p. 511, pl. xxvii; text figs. 75, 76.

Woodward, A. S. 1898 B, p. 213, fig. 132.
  Cretaceous (Laramie); Wyoming.

Triceratops serratus Marsh.

Marsh, O. C. 1890 A, p. 81.
Dana, J. D. 1896 A, p. 846, fig. 1414.
Marsh, O. C. 1890 C, p. 425, pl. v, fig. 2; pl. vi, figs.
  1-6.
  1891 B, p. 177, pls. i, ii.
  1892 A, p. 84, pl. iii.
  1896 C, p. 208, pl. lx, fig. 3; pl. lxi, figs. 7,
    9, 10.
  1897 C, p. 512, fig. 78.
Woodward, A. S. 1898 B, p. 214, fig. 133.
Zittel, K. A. 1890 A, p. 752, figs. 661, 664.
  Cretaceous (Laramie); Wyoming.

Triceratops sulcatus Marsh.

Marsh, O. C. 1890 C, p. 422.
Walcott, C. D. 1900 A, p. 23.
  Cretaceous (Laramie); Wyoming.

Triceratops sp. indet.

Lambe, L. M. 1899 A, p. 70.
  Cretaceous; British America.

STERRHOLOPHUS Marsh.   Type *Triceratops flabellatus* Marsh.

Marsh, O. C. 1891 C, p. 340.
  1895 C, p. 497.
  1896 C, p. 216.
Woodward, A. S. 1898 B, p. 216.

Sterrholophus flabellatus Marsh.

Marsh, O. C. 1889 E, p. 174. (Triceratops.)
  1889 G, p. 506, pl. xii. (Triceratops.)
  1890 C, p. 425, pl. v, fig. 1; pl. vii, fig. 1.
  (Triceratops.)

Marsh, O. C. 1890 F, pl. i. (Triceratops.)
  1891 B, p. 177, pls. i, vii, ix. (Triceratops.)
  1891 C, p. 340.
  1892 A, p. 84 pl. iii, fig. 2.
  1896 C, p. 208, pl. lx, figs. 1, 2; pl. lxxvi, fig. 1;
    pl. lxxxi.
  1897 C, p. 511, fig. 77.
Woodward, A. S. 1898 B, p. 215, fig. 134.
Zittel, K. A. 1890 A, p. 750, figs. 662, 663.
  Cretaceous (Laramie); Wyoming.

TOROSAURUS Marsh.   Type *T. latus* Marsh.

Marsh, O. C. 1891 E, p. 266.
  1892 A, p. 81, pls. ii, iii.
  1893 I, p. 488.
  1895 C, p. 497.
  1896 C, p. 214.
  1898 B, p. 407.

Torosaurus gladius Marsh.

Marsh, O. C. 1891 E, p. 266.
Dana, J. D. 1896 A, p. 846, fig. 1415.

Marsh, O. C. 1892 A, p. 81, pl. ii.
  1896 C, p. 215, pl. lxii, fig. 2; lxiii, text fig. 54.
  Cretaceous (Laramie); Wyoming.

Torosaurus latus Marsh.

Marsh, O. C. 1891 E, p. 266.
  1892 A, p. 81, pl. ii.
  1896 C, p. 214, pl. lxii, fig. 1.
  Cretaceous (Laramie); Wyoming.

MONOCLONIUS Cope.   Type *M. crassus* Cope.

Cope, E. D. 1876 H, p. 255.
Amer. Geologist, 1891 D.
Baur, G. 1890 E, p. 570.
  1891 A.
  1891 D, p. 460.
Cope, E. D. 1883 J, p. 99.
  1889 Q, p. 716.
  1889 T, p. 906.
Hatcher, J. B. 1896 A.
Marsh, O. C. 1896 C, p. 497.

Marsh, O. C. 1896 C, p. 217.
Zittel, K. A. 1890 A., p. 752. (Syn? of Triceratops.)

Monoclonius crassus Cope.

Cope, E. D. 1876 H, p. 255.
Baur, G. 1891 A.
Cope, E. D. 1877 G, p. 573.
  1880 Q.
  1886 E, fig. 2.
  1889 Q, p. 715, pl. xxxiii, fig. 1.

Seeley, H. G. 1887 B.
Zittel, K. A. 1890 A, p. 752.
Cretaceous (Laramie); Montana.

Monoclonius fissus Cope.

Cope, E. D. 1889 Q, p. 717.
Cretaceous (Laramie): Montana.

Monoclonius recurvicornis Cope.

Cope, E. D. 1889 Q. p. 716, pl. xxxiv, fig. 1.
1877 G, p. 588, pl. xxxiv, figs. 7, 8. ("Dinosaurian.")

Zittel, K. A. 1890 A, p. 752.
Cretaceous (Laramie): Montana.

Monoclonius sphenocerus Cope.

Cope, E. D. 1889 Q. p. 716, pl. xxxiii, fig. 3.
Osborn, H. F. 1898 N, pp. 842, 844, fig. 1. (Agathaumus.)
1898 Q. p. 5, pl. i, fig. 1. (Agathaumus.)
Cretaceous (Laramie); Montana.

CERATOPS Marsh.   Type *C. montanus* Marsh.

Marsh, O. C. 1888 C, p. 477.
Amer. Geologist 1891 D.
Baur, G, 1890 E, p. 570. (Monoclonius.)
1891 A.
1891 D, p. 450. (Monoclonius.)
Cope, E. D. 1889 T, p. 906. (Monoclonius.)
Hatcher, J. B. 1896 A.
Marsh, O. C. 1890 F, p. 4.
1891 C, p. 340.
1896 C, p. 497.
1896 C, p. 216.
Nicholson and Lydekker 1889 A, p. 1163.
Zittel, K. A. 1890 A, p. 751. (Syn.? of Polyonax.)

Ceratops alticornis Marsh.

Marsh, O. C. 1887 B, p. 323, figs. 1, 2. (Bison.)
Barbour, E. H. 1890 B, p. 392, figs. 5, 6.
Cope, E. D. 1889 A. (Polyonax.)

Marsh, O. C. 1889 E, p. 174.
1897 C, pp. 512, 527, figs. 79, 80.
Walcott, C. D. 1900 A, p. 23.
Cretaceous (Laramie); Colorado.

Ceratops montanus Marsh.

Marsh, O. C. 1888 C, p. 477, pl. xi.
1889 B, p. 327.
1890 A, p. 83.
1892 A, p. 84, pl. iii, fig. 4.
1896 C, p. 216, pl. lxiii, fig. 1.
Walcott, C. D. 1900 A, p. 23.
Cretaceous (Laramie); Montana.

Ceratops paucidens Marsh.

Marsh, O. C. 1889 C, p. 336. (Hadrosaurus.)
1890 A, p. 83.
Cretaceous (Laramie); Montana.

CLAORHYNCHUS Cope.   Type *C. trihedrus* Cope.

Cope, E. D. 1892 T. p. 757.

Claorhynchus trihedrus Cope.

Cope, E. D. 1892 T, p. 757.
Cretaceous (Laramie); Montana?.

MANOSPONDYLUS Cope.   Type *M. gigas* Cope.

Cope, E. D. 1892 T, p. 757.

Manospondylus gigas Cope.

Cope, E. D. 1892 T, p. 757.
Cretaceous (Laramie).

Superfamily IGUANODONTOIDEA nom. nov.

This group is referred to by the following
writers under the name *Ornithopoda*, and regarded usually either as an order or a suborder.
Baur, G. 1883 A, p. 436.
1884 A.
1887 C, p. 102.
Cope, E. D. 1882 Q. (Orthopoda.)
1883 J, p. 98.
Dollo, L. 1892 C, p. 11. (Ornithopodes.)
Fürbringer, M. 1888 A.
1900 A.
Gadow, H. 1897 A, p. 205.
1898 A, p. 22.
Haeckel, E. 1895 A, p. 389.
Hulke, J. W. 1883 B. (On Hypsilophodon.)
Lydekker, R. 1888 B, p. 175.
1889 C.

Marsh, O. C. 1881 E, p. 423.
1882 A, p. 84.
1884 G.
1885 D, p. 764.
1887 C, p. 417.
1889 B, p. 328.
1894 E.
1895 C, p. 497.
1896 C, pp. 226, 243.
1896 D, p. 210.-
Nicholson and Lydekker 1889 A, p. 1152.
Science 1885 B.
Steinmann and Döderlein 1890 A, p. 665.
Vetter, B. 1885 A, p. 118.
Woodward, A. S. 1898 B, p. 205.
Zittel, K. A. 1890 A, pp. 693, 754.

CAMPTOSAURIDÆ.

Marsh, O. C. 1881 E, p. 423.   (Camptonotidæ.)
Dollo, L. 1888 B.   (Camptonotidæ.)
Lydekker, R. 1888 B, p. 191.   (Iguanodontidæ, in part.)
Marsh, O. C. 1882 A, p. 84.   (Camptonotidæ.)
    1884 G.   (Camptonotidæ).
    1885 A.
    1885 D.

Marsh, O. C. 1896 C, p. 497.   (Camptosauridæ, Laosauridæ.)
    1896 C, pp. 196, 243.   (Camptosauridæ, Laosauridæ.)
    1896 D, p. 210.   (Camptosauridæ, Laosauridæ.)
Steinmann and Döderlein 1890 A, p. 665.
Zittel, K. A. 1890 A, p. 755.

CAMPTOSAURUS Marsh.   Type *Camptonotus dispar* Marsh.

Marsh, O. C. 1885 A.   (To replace Camptonotus, preoccupied.)
Baur, G. 1883 A, p. 437.   (Camptonotus.)
Dollo, L. 1888 E.   (Camptonotus.)
Lydekker, R. 1888 B, p. 192.
    1889 C, p. 46.
    1890 A, p. 257.
Marsh, O. C. 1879 H, p. 501.   (Camptonotus, type *C. dispar.*)
    1881 A, p. 168.
    1894 A.
    1894 E, p. 85.
    1894 G.
    1896 C, p. 497.
    1896 C, pp. 196, 201.
    1897 C, p. 502.
    1899 A, p. 251.
Nicholson and Lydekker 1889 A, pp. 1152, 1156.
Steinmann and Döderlein 1890 A, p. 665.
White, C. A. 1883 A.
Williston, S. W. 1890 C, p. 472.
Zittel, K. A. 1890 A, p. 755.

Camptosaurus amplus Marsh.

Marsh, O. C. 1879 H, p. 503.   (Camptonotus.)
    1885 A.
    1896 C, p. 196.
Jurassic; Wyoming.?

Camptosaurus dispar Marsh.

Marsh, O. C. 1879 H, p. 501, pl. iii.   (Camptonotus.)
Fürbringer, M. 1900 A, p. 350.
Lydekker, R. 1888 B, p. 192, fig. 36.
Marsh, O. C. 1885 A.
    1892 G, p. 176, pl. v.
    1894 A, pl. vi.
    1894 E, p. 85, pl. v, fig. 1.
    1894 G, pl. vi.
    1896 C, p. 196, pls. liv, lvi.
    1896 D, pl. i, fig. 7.
    1897 C, p. 502, pl. xxiii.
    1899 A, p. 232, fig. 2.
Williston, S. W. 1890 C.   (Cumnoria.)
Zittel, K. A. 1890 A, p. 756, fig. 666, B. C.
Jurassic; Wyoming.

Camptosaurus medius Marsh.

Marsh, O. C. 1894 E, p. 85, pl. iv.
    1896 C, p. 196, pl. liii.
    1897 C, p. 502, figs. 58, 59.
Jurassic; Colorado?.

Camptosaurus nanus Marsh.

Marsh, O. C. 1894 E, p. 85, pl. v, fig. 3.
    1896 C, p. 196, pl. lv, fig. 2.
Walcott, C. D. 1900 A, p. 28.
Jurassic; Wyoming.?

LAOSAURUS Marsh.   Type *L. celer* Marsh.

Marsh, O. C. 1878 B, pp. 244.
Giebel, C. G. 1879 A, p. 317.
Marsh, O. C. 1878 F, p. 415.
    1879 B, pp. 89, 90.
    1879 H, p. 501.
    1894 E, p. 86.
    1895 C, p. 498.
    1896 C, pp. 199, 201.
    1897 C, p. 503.
Nicholson and Lydekker 1889 A, p. 1159.
Williston, S. W. 1878 A, p. 45.
Zittel, K. A. 1890 A, p. 755.

Laosaurus celer Marsh.

Marsh, O. C. 1878 B, p. 244.
King, C, 1878 A, p. 346.
Marsh, O. C. 1896 C, p. 199.
Jurassic; Wyoming.

Laosaurus consors Marsh.

Marsh, O. C. 1894 E, p. 87, pl v, fig. 4; pl. vi, fig. 1; pl. vii, fig. 1.
    1895 C, pl. x, fig. 4.
    1896 C, p. 199, pl. lv, figs. 1, 3; pl. lvii; text fig. 61.
    1896 D, p. 199, pl. i, fig. 4; text fig. 6.
    1897 C, p. 503, pl. xxiv.
Jurassic; Wyoming.

Laosaurus gracilis Marsh

Marsh, O. C. 1878 B, p. 244.
King, C. 1878 A, p. 436.
Marsh, O. C. 1896 C, p. 199.
Jurassic; Colorado.

DRYOSAURUS Marsh.   Type *Laosaurus altus* Marsh.

Marsh, O. C. 1894 E, p. 86.
    1895 C, p. 498.
    1896 C, pp. 198, 201.

Dryosaurus altus Marsh.

Marsh, O. C. 1878 F, p. 415, pls. ix, x.   (Laosaurus.)

Marsh, O. C. 1879 A, pl. viii, fig 1.   (Laosaurus.)
    1892 G, p. 176.   (Laosaurus.)
    1894 E, p. 86, pl. vii, fig. 2.
    1896 C, p. 198, pl. lv, fig. 4.
Zittel, K. A. 1890 A, p. 756, fig. 666 A.
Jurassic; Colorado, Wyoming.

IGUANODONTIDÆ.

Andrews, C. W. 1897 A. (Iguanodon.)
Baur, G. 1883 A, p. 437.
    1885 D. (Iguanodon.)
    1891 D. (Iguanodon.)
Dollo L. 1882 B.
    1882 C.
    1883 A.
    1883 B.
    1883 C.
    1883 D.
    1884 B.
Fürbringer, M. 1900 A, p. 358.
Hulke, J. W. 1883 A.
    1884 A.
    1885 A.
Huxley, T. H. 1870 D.
Lydekker, R. 1888 B, p. 191.
    1889 L.
Mantell, G. A. 1848 A. (Iguanodon.)
    1848 B, p. 40. (Iguanodon.)
    1849 A. (Iguanodon.)
Marsh, O. C. 1884 G.
    1890 C, p. 425. (Claosauridæ.)
    1896 B, p. 411, pl. viii. (Iguanodon.)

Marsh, O. C. 1896 C. p. 498. (Iguanodontidæ, Claosauridæ.)
    1896 C, pp. 230, 244. (Iguanodontidæ, Claosauridæ.)
    1896 D, p. 211. (Iguanodontidæ, Claosauridæ.)
Nicholson and Lydekker 1889 A, p. 1140.
Owen, R. 1842 A, p. 124. (Iguanodon.)
    1845 B, p. 246. (Iguanodon.)
    1851 B, p. 105. (Iguanodon.)
    1857 B, p. 1. (Iguanodon.)
    1860 B, p. 27. (Iguanodon.)
    1864 B, p. 19. (Iguanodon.)
    1872 A. (Iguanodon.)
    1874 A. (Iguanodon.)
Pictet, F. J. 1853 A, p. 470. (Iguanodon.)
Quenstedt, F. A. 1885 A, p. 184. (Iguanodon.)
Sauvage, H. E. 1874 B, p. 12. (Iguanodon.)
Seeley, H. G. 1887 B.
    1891 A, p. 258, fig. 15. (Iguanodon.)
Woodward, A. S. 1898 B, p. 205. (Iguanodon.)
Woodward, H. 1885 B. (Iguanodon.)
    1895 A.
Zittel, K. A. 1890 A, p. 757.

THESPESIUS Leidy.   Type *T. occidentalis* Leidy.

Leidy, J. 1856 Q, p. 311.
Cope, E. D. 1869 M, pp. 91, 98. (Syn. of Hadrosaurus.)
    1871 G, p. 52.
Fürbringer, M. 1900 A, p. 350. (Claosaurus.)
Hatcher, J. B. 1900 A, p. 719.
Leidy, J. 1857 C.
    1860 A, p. 151.
    1865 A, p. 84.
    1870 J, p. 67.
    1874 A, p. 74.
Lucas, F. A. 1900 C, p. 809. (Thespesius.)
Marsh, O. C. 1890 C, p. 423. (Claosaurus, type Hadrosaurus agilis Marsh.)
    1892 G, p. 171. (Claosaurus.)
    1892 H. (Claosaurus.)
    1893 B, p. 83. (Claosaurus.)
    1893 F, p. 153. (Claosaurus.)
    1895 C, p. 498 (Claosaurus.)
    1896 C, p. 219. (Claosaurus.)
    1896 D. (Claosaurus.)
    1897 C, p. 516. (Claosaurus.)
Woodward, A. S. 1898 B, p. 208. (Claosaurus.)
Zittel, K. A. 1890 A, p. 763. (Claosaurus, Thespesius?.)

The genus *Claosaurus* is identified with *Thespesius* on the authority of Mr. F. A. Lucas, of the U. S. National Museum.

Thespesius agilis (Marsh.)

Marsh, O. C. 1872 D. (Hadrosaurus.)
Cope, E. D. 1874 C, p. 20. (Hadrosaurus.)
    1875 E, p. 247. (Hadrosaurus.)
Marsh, O. C. 1877 C. (Hadrosaurus.)
    1890 C, p. 423, fig. 3. (Claosaurus.)
    1896 C, p. 219, fig. 55· (Claosaurus.)
Nopcsa, F. B. 1899 A, p. 557. (Claosaurus.)
Williston, S. W. 1898 C, p. 69.

Zittel, K. A. 1890 A, p. 763, fig. 673. (Claosaurus.)
Cretaceous (Niobrara); Kansas.

Thespesius occidentalis Leidy.

Leidy, J. 1856 Q, p. 311.
Cope, E. D. 1869 M, p. 98. (Hadrosaurus?.)
    1871 G, p. 51. (Hadrosaurus.)
    1874 B, p. 446. (Hadrosaurus.)
    1874 C, p. 20. (Hadrosaurus.)
    1874 P, p. 10. (H. milo; no description.)
    1874 R. (Hadrosaurus.)
    1875 E, pp. 56, 248, pl. iii, figs. 5, 6 (Hadrosaurus); p. 58 (Agathaumus milo a synonym).
Dana, J. D. 1896 A, p. 844, figs. 1408-1411. (Claosaurus annectens,)
Fürbringer, M. 1900 A, p. 351, figs. 123, 125, 128. (Claosaurus annectens.)
Hatcher, J. B. 1893 A, p. 143. (Claosaurus annectens.)
Leidy, J. 1857 C, p. 91.
    1860 A, p. 151, pl. x, figs. 1-5. (Not figs. 6, 7, fide Leidy.)
    1865 A, p. 84.
    1870 J, pp. 67, 68.
Lucas, F. A. 1900 C, p. 809.
Marsh, O. C. 1892 E, p. 453, fig. 4. (Claosaurus annectens.)
    1892 G, p. 171, pls. ii, iii. (C. annectens.)
    1892 H, p. 344, pl. vi. (C. annectens.)
    1893 B, p. 83, pls. iv, v. (C. annectens.)
    1893 F, p. 155, pl. vii. (C. annectens.)
    1894 E, pl. vii, fig. 5. (C. annectens.)
    1895 C, pl. x, fig. 12. (C. annectens.)
    1896 C, p. 219, pls. lxxii-lxxiv; pl. lxxv. fig. 4; pl. lxxvi, fig. 2; pl. lxxvii, fig. 3; pl. lxxviii, fig. 2; pl. lxxix, fig. 5; pl. lxxx. fig. 3. (C. annectens.)

Marsh, O. C. 1896 D, pl. i, fig. 12.   (C. annectens.)
1897 C, p. 517, pl. xxviii; text figs. 81, 82.   (C.
annectens.)
Nopcsa, F. B. 1899 A. P. 556 (Hadrosaurus occidentalis): p. 557 (Claosaurus annectens).

Woodward, A. S. 1898 B, p. 209, fig. 128.   (C. annec-
tens.)
Cretaceous (Laramie); Wyoming, Colorado,
South Dakota.

PTEROPELYX Cope.   Type *P. grallipes* Cope.

Cope, E. D. 1889 T, p. 904.
1892 T, p. 758.   (Claosaurus a synonym.)
Marsh, O. C. 1892 H, p. 346.
1896 C, p. 224.

Pteropelyx grallipes Cope.
Cope, E. D. 1889 T, p. 904.
Cretaceous; (Laramie) Montana.

TRACHODONTIDÆ.

Lydekker, R. 1888 B, p. 241.   (Replacing Hadrosauridæ.)
Cope, E. D. 1869 M, p. 91.   (Hadrosauridæ.)
1883 J.   (Hadrosauridæ.)
1891 N, p. 43.   (Hadrosauridæ.)
1896 R, p. 70.   (Hadrosauridæ.)
Marsh, O. C. 1882 A, p. 84.   (Hadrosauridæ.)
1884 G.   (Hadrosauridæ.)
1885 D, p. 764.   (Hadrosauridæ.)

Marsh, O. C. 1890 C, p. 424.
1895 C, p. 498.
1896 C, p. 244.
1896 D, p. 211.
Nicholson and Lydekker 1889 A, p. 1158.
Nopcsa, F. B. 1899 A. p. 556.   ("Hadrosauriden".)
Steinmann and Döderlein 1890 A, p. 667.
Zittel, K. A. 1890 A, p. 763.   (Hadrosauridæ.)

TRACHODON Leidy.   Type *T. mirabilis* Leidy.

Leidy, J. 1856 F, p. 72.
Baur, G. 1883 A, p. 489.   (Hadrosaurus.)
Cope, E. D. 1869 M, p. 91.   (Hadrosaurus.)
1871 G, p. 52.   (Thespesius.)
1874 B, p. 447.   (Hadrosaurus.)
1874 C, p. 22.   (Hadrosaurus.)
1875 E, p. 247.   (Hadrosaurus.)
1876 H, p. 258.   (Diclonius, type *D. pentagonus*.)
1877 V.   (Diclonius.)
1883 J, p. 99.   (Hadrosaurus.)
1883 CC.   (Diclonius.)
1884 F.   (Diclonius.)
1887 S, p. 338, pls. xv, xvi.
1889 E.   (Hadrosaurus.)
1889 T, p. 904.   (Diclonius.)
1898 B, p. 71.   (Diclonius.)
Dollo, L, 1883 C, p. 211.   (Hadrosaurus.)
1884 B, p. 137, pl. vi, fig. 2.   (Diclonius.)
Foulke, W. P. 1858 A.   (Hadrosaurus.)
Gadow, H. 1897 A, p. 206.   (Hadrosaurus, Diclonius.)
Hawkins, B. W. 1874 A.   (Hadrosaurus.)
1875 A.   (Hadrosaurus.)
Leidy, J. 1856 C, p. 118.   (Trachodon.)
1858 A, p. 215.   (Hadrosaurus, type *H. foulkii*, Leidy.)
1859 A.   (Hadrosaurus.)
1865 A, p. 84.   (Trachodon, Hadrosaurus.)
1865 D, p. 71.   (Hadrosaurus, Trachodon.)
1868 F, p. 199.   (Trachodon, Hadrosaurus.)
1870 J, p. 67.   (Hadrosaurus.)
Lippincott, J. S. 1881 A.   (Hadrosaurus.)
Lydekker, R. 1886 C, p. 275, fig. 3.   (Hadrosaurus.)
1888 B, p. 243.   (Trachodon.)
Marsh, O. C. 1893 B, p. 85.   (Hadrosaurus.)
1895 C, p. 498.
1896 C, p. 224.
Nicholson and Lydekker 1889 A, p. 1153.   (Trachodon.)

Nopcsa, F. B. 1899 A, p. 555.   (Hadrosaurus.)
Osborn, H. F. 1893 B, p. 326.   (Diclonius.)
Owen, R. 1875 C, p. 74.   (Hadrosaurus.)
Steinmann and Döderlein 1890 A, p. 667 (Hadro-
saurus); p. 668 (Diclonius).
Woodward, A. S. 1890 A, p. 393.   (Hadrosaurus.)
Zittel, K. A. 1890 A, p. 763.   (Hadrosaurus.)

Trachodon breviceps Marsh.
Marsh, O. C. 1889 C, p. 335, figs. 4, 5.   (Hadro-
saurus.)
Cope, E. D. 1889 T, p. 906.   (Hadrosaurus.)
Hatcher, J. B. 1896 A, p. 116.   (Hadrosaurus.)
Marsh, O. C. 1890 C, p. 423, figs. 1, 2.   (Hadro-
saurus.)
1896 C, pl. lxxv, figs. 1, 2.   (Trachodon.)
Nopcsa, F. B. 1899 A, p. 557.   (Hadrosaurus.)
Zittel, K. A. 1890 A, p. 765, fig. 676.   (Hadro-
saurus.)
Cretaceous (Laramie); Montana.

Trachodon calamarius (Cope).
Cope, E. D. 1876 H, p. 255.   (Diclonius.)
1877 G, pp. 572, 588.   (Diclonius.)
Cretaceous (Laramie); Montana.

Trachodon cavatus (Cope).
Cope, E. D. 1871 G, p. 50.   (Hadrosaurus.)
1875 E, p. 248.   (Hadrosaurus.)
Nopcsa, F. B. 1899 A, p. 557.   (Hadrosaurus.)
Cretaceous; New Jersey.

Trachodon foulkii Leidy.
Leidy, J. 1858 A.   (Hadrosaurus.)
Cope, E. D. 1867 E, p. 27.   (Hadrosaurus.)
1869 B, p. 736.   (Hadrosaurus.)
1869 M, p. 98.   (Hadrosaurus.)
1875 E, p. 247.   (Hadrosaurus.)
Emmons, E. 1860 A, p. 208, fig. 180⁶.   (Hadrosaurus.)

Leidy, J. 1859 A. (Hadrosaurus.)
1865 A, pp. 76, 118, pl. ii, figs. 9–11; pl. viii,
fig. 13; pl. xii, figs. 1–20; pl. xiii, figs. 1–19,
24–26; pl. xiv, figs. 1–12; pl. xv, figs. 1–6;
pl. xvi, figs. 1–8; pl. xvii, figs. 1–5; pl. xx,
figs. 1, 2. (Hadrosaurus.)
1865 D, p. 70. (Hadrosaurus.)
1868 F, p. 199. (Trachodon.)
1870 J, p. 68. (Hadrosaurus.)
Lydekker, R. 1888 B, p. 224, fig. 50.
Nicholson and Lydekker 1889 A, p. 1153, fig. 1058.
Nopcsa, F. B. 1899 A, p. 556. (Hadrosaurus.)
Zittel, K. A. 1890 A, p. 764, fig. 674.
Cretaceous; New Jersey.

Trachodon longiceps Marsh.

Marsh, O. C. 1890 C, p. 422.
Nopcsa, F. B. 1899 A, p. 557. (Hadrosaurus.)
Cretaceous (Laramie); Wyoming.

Trachodon minor (Marsh).

Marsh, O. C. 1870 F. (Hadrosaurus.)
Cope, E. D. 1869 M (1870), p. 122 J. (Hadrosaurus.)
1875 E, p. 247. (Hadrosaurus.)
Leidy, J. 1870 J, p. 68.
Nopcsa, F. B. 1899 A, p. 557. (Hadrosaurus.)
Cretaceous; New Jersey.

Trachodon mirabilis Leidy.

Leidy, J. 1856 F, p. 72.
Cope, E. D. 1869 M, p. 98. (Hadrosaurus.)
1874 C. pp. 8, 19. (Hadrosaurus.)
1875 E, pp. 56, 247. (Hadrosaurus.)
1877 G, p. 573. (Trachodon.)
1883 C. (Diclonius.)
1883 J. pls. iv–vii. (Diclonius.)
1883 CC. (Diclonius.)
1884 F. (Diclonius.)
1885 R, pl. xxxvii, figs. 1–3. (Diclonius.)

Cope, E. D. 1886 E, fig. 1. (Diclonius.)
1891 N, p. 44, fig. 22. (Diclonius.)
1898 B, p. 71, fig. 23. (Diclonius.)
Lambe, L. M. 1899 A, p. 69.
Leidy, J. 1856 C, p. 118.
1860 A, p. 140, pl. ix, figs. 1–20.
1870 J, p. 68. (Hadrosaurus.)
Marsh, O. C. 1892 H, p. 344. (Hadrosaurus.)
Nopcsa, F. B. 1899 A, p. 556. (Hadrosaurus.)
Osborn, H. F. 1898 N, pp. 842, 844, fig. 2. (Hadrosaurus.)
1898 Q, p. 6, pl. i, fig. 2. (Hadrosaurus.)
Zittel, K. A. 1890 A, p. 764, fig. 675. [Hadrosaurus (Diclonius).] ·
Cretaceous (Laramie); Montana, British America.

Trachodon pentagonus (Cope).

Cope, E. D. 1876 H, p. 253. (Diclonius.)
1877 G, pp. 572, 594. (Diclonius.)
1888 T, p. 905. (Diclonius.)
Cretaceous (Laramie); Montana.

Trachodon perangulatus (Cope).

Cope, E. D. 1876 H, p. 254. (Diclonius.)
1877 G, pp. 572, 588. (Diclonius.)
Cretaceous (Laramie); Montana.

Trachodon tripos (Cope).

Cope, E. D. 1869 I. (Hadrosaurus.)
1869 M, p. 122 I. (Hadrosaurus.)
1870 J, p. 68. (Hadrosaurus.)
1871 M, p. 214. (Hadrosaurus.)
1875 E, p. 247. (Hadrosaurus.)
1875 V, p. 40, pl. v, figs. 1, 2. (Hadrosaurus.)
1878 EE. (Hadrosaurus.)
Leidy, J. 1870 J, p. 68. (Hadrosaurus.)
Nopcsa, F. B. 1899 A, p. 557. (Hadrosaurus.)
Cretaceous; North Carolina, Georgia.

CIONODON Cope.   Type C. arctatus Cope.

Cope, E. D. 1874 P, p. 10. (Cinodon, errore typog.)
1874 B, p. 447.
1874 C, p. 21.
1875 E, pp. 57, 246.
1875 W, p. 334.
1876 H, p. 253.
1877 V.
1883 J, p. 99.
Lydekker, R. 1886 C, p. 276.
Marsh, O. C. 1895 C, p. 498.
Steinmann and Döderlein 1890 A, p. 667.
Zittel, K. A. 1890 A, p. 765.

Cionodon arctatus Cope.

Cope, E. D. 1874 P, p. 10.

Cope, E. D. 1874 B, pp. 448, 449.
1874 C, p. 22.
1874 F.
1875 E, pp. 60, 246, pls. i, ii.
1877 G, p. 597.
Nopcsa, F. B. 1899 A, p. 557.
Cretaceous (Laramie); Colorado.

Cionodon stenopsis Cope.

Cope, E. D. 1875 I.
1875 E, pp. 63, 247.
1875 W, p. 335.
Nopcsa, F. B. 1899 A, p. 557.
Cretaceous (Laramie); Saskatchewan.

DYSGANUS Cope.   Type D. encaustus Cope.

Cope, E. D. 1876 H, p. 250.
1883 J, p. 99.
1890 D.
Zittel, K. A. 1890 A, p. 765.
A genus of uncertain affinities.

Dysganus bicarinatus Cope.

Cope, E. D. 1876 H, p. 252.
1877 G, p. 572.
Cretaceous (Laramie); Montana.

Dysganus encaustus Cope.

Cope, E. D. 1876 H, p. 250.
  1877 G, p. 572.
  Cretaceous (Laramie); Montana.

Dysganus haydenianus Cope.

Cope, E. D. 1876 H, p. 251.

Cope, E. D. 1877 G, pp. 572, 596.
  Cretaceous (Laramie); Montana.

Dysganus peiganus Cope.

Cope, E. D. 1876 H, p. 252.
  1877 G, p. 572.
  Cretaceous (Laramie); Montana.

ORNITHOTARSUS Cope.   Type *O. immanis* Cope.

Cope, E. D. 1869 H, p. 123.
  1869 M, p. 120.
  1883 J, p. 99.
Morse, E. S. 1880 A, p. 7.
Zittel, K. A. 1890 A, p. 765.

Ornithotarsus immanis Cope.

Cope, E. D. 1869 H, p. 123.

Cope, E. D. 1869 D.
  1869 M, p. 121, figs. 35–36a.
  1875 E, p. 248.
  1885 R.
Nopcsa, F. B. 1899 A, p. 577.
  Cretaceous; New Jersey.

PNEUMATOARTHRUS Cope.   Type *P. peloreus* Cope.

Cope, E. D. 1870 I, p. 446.
  1875 E, p. 260. (Pneumatarthrus.)
  1878 I, p. 236. (Pneumatarthrus.)
Zittel, K. A. 1890 A, p. 765. (Syn.? of Ornithotarsus.)

Pneumatoarthrus peloreus Cope.

Cope, E. D. 1870 I, p. 446.

Cope, E. D. 1872 L, p. 433. (Pneumatarthrus.)
  1875 E, p. 260. (Pneumatarthrus.)
Leidy, J. 1865 A, p. 100, pl. xiii, figs. 27, 28. (Young? of Hadrosaurus.)
  Cretaceous; New Jersey.

NANOSAURIDÆ.

Marsh, O. C. 1878 B, p. 244.
  1896 C, p. 498.
  1896 C, p. 244.

Marsh, O. C. 1896 D, p. 211.
Zittel, K. A. 1890 A, p. 766.

NANOSAURUS Marsh.   Type *N. agilis* Marsh.

Marsh, O. C. 1877 D, p. 254.
  1891 F, p. 337.
  1894 E, p. 87.
  1896 C, p. 498.
  1896 C, pp. 199, 201, 202.
  1897 C, p. 483.
Nicholson and Lydekker 1889 A, p. 1059.
Steinmann and Döderlein 1890 A, p. 624.
Williston, S. W. 1878 A, p. 45.
Zittel, K. A. 1890 A, p. 766.

Nanosaurus agilis Marsh.

Marsh, O. C. 1877 D, p. 254.

King, C. 1878 A, p. 346.
Marsh, O. C. 1894 E, pl. vi, fig. 3.
  1896 C, p. 199, text figs. 42, 43.
  1897 C, p. 484, figs. 27, 28.
  Jurassic; Colorado.

Nanosaurus rex Marsh.

Marsh, O. C. 1877 G, p. 516.
King, C. 1878 A, p. 346.
Marsh, O. C. 1894 E, p. 88, pl. vi, fig. 5.
  1896 C, p. 200, text figs. 44–48.
  1897 C, p. 484, fig. 29.
  Jurassic; Colorado.

MACELOGNATHIDÆ.

Marsh, O. C. 1884 C, p. 341.

| Woodward, A. S. 1887 C, p. 8.

MACELOGNATHUS Marsh.   Type *M. vagans* Marsh.

Marsh, O. C. 1884 C, p. 341.
Amer. Geologist, 1891 D.
Baur, G. 1891 D, p. 450. (Cœlurus.)
Marsh, O. C. 1897 C, p. 508.
Nicholson and Lydekker 1889 A, p. 1089.
Zittel, K. A. 1890 A, p. 513.
  For this genus Professor Marsh has established the family *Macelognathidæ* and the order

Macelognatha, and assigned them a place among the *Testudinata.* The relationships of the fossil are problematical.

Macelognathus vagans Marsh.

Marsh, O. C. 1884 C, p. 341, figure.
  1897 C, p. 507, figs. 65, 66.
  Jurassic; Wyoming.

APATODON Marsh.   Type *A. mirus* Marsh.

Marsh, O. C. 1877 D, p. 255.
Baur, G. 1890 C, p. 301.
A doubtful genus of doubtful position.

Apatodon mirus Marsh.

Marsh, O. C. 1877 D, p. 255.
Jurassic; Rocky Mountain region.

BRACHYROPHUS Cope.   Type *B. altarkansanus* Cope.

Cope, E. D. 1878 H, p. 390.
Even the ordinal portion of this genus is doubtful.

Brachyrophus altarkansanus Cope.

Cope, E. D. 1878 H, p. 390.
Jurassic; Colorado.

Order PTEROSAURI Kaup.

Kaup, J. 1834, Isis, p. 315.
Baur, G. 1887 A, p. 487. (Ornithosauria.)
    1887 C, p. 101. (Ornithosauria.)
    1889 G. (Pterosauria.)
    1891 C, p. 355. (Pterosauria.)
Blainville, H. M. D., 1835, Nouv. Ann. Mus., iv, p 238. (Pterodactylia.)
Bonaparte, C. L., 1838, Nuovi Ann. Sci. Nat. Bologna, i, p. 391. (Ornithosauri.)
Cope, E. D. 1869 M, pp. 26, 28, 169, and ii. (Pterosauria.)
    1871 B, p. 234. (Ornithosauria.)
    1885 G, p. 246. (Ornithosauria.)
    1887 S, p. 333. (Ornithosauria.)
    1889 R, p. 863. (Ornithosauria.)
    1891 N, p 41. (Ornithosauria.)
    1892 R. (Pterosauria.)
    1898 B, pp. 56, 67. (Ornithosauria.)
    1900 A, p. 160 (Ornithosauria.)
Fürbringer, M. 1900 A, pp. 355, 527, 660. (Patagiosauria.)
Gadow, H 1898 A, p. 23. (Pterosauria.)
Gaudry, A. 1890 C, p. 235.
Günther, A. 1886 A, p. 443. (Ornithosauria.)
Haeckel, E. 1895 A, p. 390. (Pterosauria.)
Hawkins, B. W. 1874 A. (Ornithosauria.)
Hay. O. P. 1899 F, p. 682.
Huxley, T. H. 1872 A, p. 228. (Pterosauria.)
Kehner, F. A. 1876 A. (Pterosauria.)
Lortet, L. 1892 A, p. 128. (Pterosauria.)
Lydekker, R. 1888 B, p. 2 (Ornithosauria); p. 3 (Pteranodontia, Pterosauria.)
Marsh, O. C. 1876 E. (>Pteranodontia.)
    1876 G. (>Pteranodontia.)
    1876 I. (>Pteranodontia.)
    1877 E. (Pterosauria, Pteranodontia.)
    1881 D. (>Pteranodontia.)
    1882 B. (Pterosauria, Pteranodontia.)
    1884 D. (Pterosauria, Pteranodontia.)

Meyer, H. 1851 A. (Pterosauria.)
    1860 A. (Pterosauria.)
    1860 D. (Pterosauria.)
Newton, E. T. 1888 A. (Ornithosauria.)
    1889 B. (Ornithosauria.)
Nicholson and Lydekker 1889 A, p. 1196 (Ornithosauria); p. 1198 (Pteranodontia); p. 1199 (Pterosauria.)
Owen, R. 1840. (Pterosauria.)
    1851 A. (Pterosauria.)
    1859 A. (Pterosauria.)
    1859 B. (Pterosauria.)
    1860 A, p. 162. (Pterosauria.)
    1860 B. (Pterosauria.)
    1860 C. (Pterosauria.)
    1860 E, p. 244. (Pterosauria.)
    1866 A. (Pterosauria.)
    1870 A. (Pterosauria.)
    1874 B (Pterosauria.)
Pictet, F. J. 1853 A, p. 522. ("Ptérodactyliens.")
Plieninger, F. 1895 A. (Pterosauria.)
Quenstedt, F. A. 1852 A, p. 135. (Pterodactyli.)
    1885 A, p. 219. Pterosauri.)
Seeley, H. G. 1865 A. (Saurornia.)
    1866 A. (Saurornia.)
    1870 A, p. 151. (Ornithosauria.)
    1870 B. (Ornithosauria.)
    1871 C. (Ornithosauria.)
    1876 A. (Ornithosauria.)
    1891 A. (Ornithosauria.)
    1891 B. (Ornithosauria.)
Steinmann and Döderlein, 1890 A, p. 647. (Pterosauria.)
Williston, S. W. 1892 A. (Pterosauria.)
    1897 A (Pterosauria.)
Woodward, A. S. 1898 B, p. 224. (Ornithosauria.)
Woodward, H. 1871 A. (Ornithosauria.)
Zittel, K. A. 1882 A. (Pterosauria.)
    1890 A, pp. 450, 773, 779. (Pterosauria.)

ORNITHOCEPHALIDÆ nom. nov.

This family, unless otherwise indicated, is denominated by the following authors *Pterodactylidæ*.

Bowerbank, J. S. 1851 A.
Cope, E. D. 1893 F.
Fürbringer, M. 1900 A.
Haeckel, E. 1895 A, p. 397. (Pterodactylida.)
Huxley, T. H. 1876 B, xiii, p. 515.
Lydekker, R. 1888 B, p. 8.
Nicholson and Lydekker 1889 A, p. 1199.
Owen, R. 1851 A.

Owen, R. 1859 A.
    1859 D.
    1860 B.
Quenstedt, F. A. 1885 A, p. 219.
Seeley, H. G. 1866 A.
    1870 B. p. 111. (Pterodactylæ.)
    1876 A.
Steinmann and Döderlein 1890 A, p. 650.
Williston, S. W. 1892 A, p. 12. (Pterodactylinæ.)
    1897 A.
Zittel, K. A. 1890 A, p. 791.

CATALOGUE. **507**

ORNITHOCEPHALUS Sömm.   Type *O. antiquus* Sömm.

Sömmerring, S. T., 1812, Denkschr. k. Akad. München, iii, p. 126.

Unless otherwise indicated the following authors refer to this genus under the name *Pterodactylus*. The latter name is usually credited to Cuvier, 1809, Ann. Mus., xiii, p. 424. There, however, we find only the French term *Petro-dactyle*. Even in case this is a misprint for Pterodactyle, it is inadmissable as a generic name. *Pterodactylus* appears to have been used first in 1824 by Cuvier.

Bowerbank, J. S. 1848 A.
    1851 A.
Cope, E. D. 1872 F, p. 337. (Ornithochirus, not of Seeley.)
    1879 A, p. 35.
    1893 F.
Cuvier, G. 1834 A, x, pp. 215–263, pl. ccii.
Dennis, J. B. P. 1861 A.
Hallmann, E. 1837 A, p. 20.
Huxley, T. H. 1872 A, p. 228.
Lortet, L. 1888 B, p. 4.
Lydekker, R. 1888 B, p. 4.
Meyer, H. 1860 D, p. 85.
    1861 B.
Newton, E. T. 1888 A.

Nicholson and Lydekker 1889 A, p. 1200.
Owen, R. 1845 B, p. 273, pl. lxiii A.
    1851 A.
    1851 B, p. 80.
    1859 B.
    1859 D, p. 1.
    1860 A, p. 163.
    1860 B, p. 1.
    1860 C.
    1870 A.
    1874 B.
Pictet, F. J. 1853 A, p. 524.
Quenstedt, F. A. 1858 A.
Seeley, H. G. 1865 A.
    1870 A.
    1870 B, p. 111.
    1871 C.
    1876 A.
    1891 A.
Steinmann and Döderlein 1890 A, p. 650.
Wagner, A. 1851 B, p. 180. (Ornithocephalus.)
Winckler, T. C. 1874 A.
Zittel, K. A. 1890 A, p. 771.

At present no American species is assigned to this genus.

DERMODACTYLUS Marsh.   Type *Pterodactylus montanus* Marsh.

Marsh, O. C. 1881 D, p. 342.
Nicholson and Lydekker 1889 A, p. 1201.
Zittel, K. A. 1890 A, p. 798.

The relationships of this genus are somewhat obscure.

Dermodactylus montanus Marsh.

Marsh, O. C. 1878 E, p. 233. (Pterodactylus.)
    1881 D.
    1897 C, p. 506.
Jurassic; Wyoming.

PTERANODONTIDÆ.

Marsh, O. C. 1876 E, p. 507.
Cope, E. D. 1898 B, p. 67.
Fürbringer, M. 1888 A.
    1900 A.
Marsh, O. C. 1884 D.
    1897 C, p. 509. (Pteranodontia.)

Nicholson and Lydekker 1890 A, p. 1199.
Steinmann and Döderlein 1890, A, p. 650.
Williston, S. W. 1892 A. (Pteranodontinæ.)
    1893 B. (Ornithostomatidæ.)
    1897 A, p. 36. (Ornithostomatinæ.)
Zittel, K. A. 1890 A, p. 798.

PTERANODON Marsh.   Type *P. longiceps* Marsh.

Marsh, O. C. 1876 E, p. 507.

Since the question of the identity of *Pteranodon* Marsh and *Ornithostoma* Seeley has been raised, references to the literature of the latter genus are here included.

Cope, E. D. 1893 F.
Grinnell, G. B. 1881 A, p. 257.
Marsh, O. C. 1876 G, p. 59.
    1876 I.
    1881 D.
    1882 B, p. 258.
    1884 D.
    1897 C, p. 509.
Newton, E. T. 1888 A, p. 418 (Ornithostoma); p. 419 (Pteranodon).
Nicholson and Lydekker 1889 A, p. 1199.
Seeley, H. G. 1871 C, p. 35. (Ornithostoma.)

Seeley, H. G. 1876 B, p. 499. (Ornithostoma.)
    1891 B, p. 442. (Ornithostoma.)
Steinmann and Döderlein 1890 A, p. 651.
Williston, S. W. 1891 B.
    1892 A, pp. 1, 3, pl. i.
    1893 B. (Ornithostoma.)
    1896 B. (Ornithostoma.)
    1896 B, pl. i. (Ornithostomr)
    1897 A. (Ornithostoma.)
Woodward, A. S. 1898 B, p. 229.
Zittel, K. A. 1890 A, p. 799.

Pteranodon comptus Marsh.

Marsh, O. C. 1876 E, p. 509.
Williston, S. W. 1892 A, p. 2.
    Cretaceous (Niobrara); Kansas.

Pteranodon ingens Marsh.

Marsh, O. C. 1872 B, p. 246.   (Pterodactylus.)
Amer. Naturalist, 1872 A.   (Pterodactylus.)
Cope, E. D. 1872 F, p. 337.   (Ornithochirus umbrosus.)
    1872 K, p. 42.   (Ornithochirus umbrosus.)
    1874 C, p. 26.   (Pterodactylus umbrosus, P. ingens.)
    1875 E, p. 249 (Pterodactylus ingens); pp. 48, 65, 247, pl. vii, figs. 1–4 (P. umbrosus).
Marsh, O. C. 1872 S, p. 578.   (Pterodactylus.)
    1872 T, p. 496.   (Pterodactylus.)
    1876 E, p. 508.
    1876 I, p. 480.
    1884 D, p. 426.
Williston, S. W. 1891 B, p. 1124.   (P. umbrosus.)
    1892 A, p. 2.   (P. ingens; P. umbrosus.)
    1893 B, figure.   (Ornithostoma.)
    1897 A, pl. ii, text figs. 1, 2. (Ornithostoma.)
    1897 H, p. 245.   (Ornithostoma.)
Cretaceous (Niobrara); Kansas.

Pteranodon longiceps Marsh.

Marsh, O. C. 1876 E, p. 598.
Lydekker, R. 1888 B, p. 3, fig 1.
Marsh, O. C. 1876 I, p. 480.
    1884 D, pl. xv.
    1897 C, p. 510, figs. 71–74.
Williston, S. W. 1891 B, p. 1124.
    1892 A, p. 2.
Woodward, A. S. 1898 B, p. 229, fig. 140.
Zittel, K. A. 1890 A, p. 799, fig. 694.
Cretaceous; Kansas.

Pteranodon nanus Marsh.

Marsh, O. C. 1881 D, p. 343.
Williston, S. W. 1892 A, p. 2.
Cretaceous; Kansas.

Pteranodon occidentalis Marsh.

Marsh, O. C. 1872 B, p. 241.   (Pterodactylus.)
Cope, E. D. 1871 C.   (Pterodactylus owenii Marsh, not of Seeley.)
    1872 F, p. 337.   (Ornithochirus harpygia.)
    1872 K, p. 471.   (Ornithochirus harpygia.)
    1874 C, p. 26.   (Pterodactylus.)
    1875 E, pp. 48, 66, 249, pl. vii, figs. 5, 6. (Pterodactylus.)
Marsh, O. C. 1871 C, p. 472. (Pterodactylus oweni.)
    1872 S, p. 578.   (Pterodactylus.)
    1872 T, p. 496.   (Pterodactylus.)
    1876 E, p. 508.
    1884 D, p. 423.
    1897 C, p. 527.
Williston, S. W. 1892 A, p. 1.
Cretaceous (Niobrara); Kansas.

Pteranodon velox Marsh.

Marsh, O. C. 1872 B, p. 247.   (Pterodactylus.)
Cope, E. D. 1874 C, p. 26.   (Pterodactylus.)
    1875 E, p. 250.   (Pterodactylus.)
Marsh, O. C. 1872 T, p. 496.   (Pterodactylus.)
    1876 I, p. 480.
Williston, S. W. 1892 A, p. 2.
Cretaceous (Niobrara); Kansas.

NYCTOSAURIDÆ.

Nicholson and Lydekker 1889 A, p. 1199. (Family suggested.)

Williston, S. W. 1897 A, p. 36.   (Nyctodactylinæ.)

NYCTOSAURUS Marsh.   Type *Pteranodon gracilis* Marsh.

Marsh, O. C. 1876 I, p. 480.
Cope, E. D. 1893 F.   (Nyctodactylus.)
Marsh, O. C. 1881 D, p. 343. (Nyctodactylus, to replace Nyctosaurus.)
Newton, E. T. 1888 A, p. 420. (Nyctodactylus.)
Nicholson and Lydekker 1889 A, p. 1199. (Nyctodactylus.)
Williston, S. W. 1892 A, p. 2. (Nyctodactylus.)
    1897 A, p. 39. (Nyctodactylus.)

Williston, S. W. 1897 H, p. 245. (Nyctodactylus.)
Zittel, K. A. 1890 A, p. 799. (Nyctodactylus.)

Nyctosaurus gracilis Marsh.

Marsh, O. C. 1876 E, p. 508.   (Pteranodon.)
    1876 I, p. 480.
    1881 D, p. 343.   (Nyctodactylus.)
Williston, S. W. 1892 A, pp. 3, 5. (Nyctodactylus.)
Cretaceous; Kansas.

Order LORICATA Merrem.

Merrem, B., 1820, Tentamen Syst. Amphib., p. 7.
    Unless otherwise indicated the following authorities apply the name *Crocodilia* to this group.
Baur, G. 1885 E, p. 633.
    1886 G, pp. 689, 738.
    1886 J.
    1887 A, p. 487.
    1887 C, p. 101.
    1889 C, p. 240.
    1891 C, p. 355.
    1894 A, p. 323.
    1896 D.

Boulenger, G. A. 1889 A, p. 273. (Emydosauria.)
    1896 B.
Burckhardt, R. 1896 A.
Cope, E. D. 1869 M, pp. 32, 61.
    1871 R.
    1889 R, p. 863.
    1891 N, p. 45.
    1892 R.
    1898 B, pp. 56, 73. (Loricata.)
    1900 A, p. 161. (Loricata.)
Cuvier, G. 1834 A, ix, pp. 25–344, pls. ccxxix–ccxxxviii.
Delongchamps-Eudes, E. 1870 A.

Dollo, L. 1883 E.
Fürbringer, M. 1888 A.
　　1900 A, pp. 297, 527, 596, 649.
Gadow, H. 1896 A, p. 38.
　　1898 A. p. 18.
Gaupp, E. 1895 B.
Gegenbaur, C. 1864 A, pp. 32, 87.
　　1865 B, p. 32.
Günther, A. 1886 A, p. 444.
Haeckel, E. 1895, p. 358.
Hallmann, E. 1837 A, p. 16.
Hay, O. P. 1899 F, p. 682. (Loricata.)
Hoffmann, C. K. 1876 A.
　　1878 A.
　　1879 A, p. 102.
　　1890 B.
Hulke, J. W. 1888 A.
Huxley, T. H. 1859 B.
　　1870 A, p. 440.
　　1872 A, p. 214.
　　1875 B.
　　1876 B, xliii, pp. 430, 467, 514.
Kehner, F. A. 1876 A.
Koken, E. 1883 A, p. 792.
　　1887 A, p. 84.
　　1888 B.
Kükenthal, W. 1893 A.
Leidy, J. 1871 C, p. 367.
Lortet, L. 1892 A, p. 91.
Ludwig, R. 1877 A.
Lydekker, R. 1888 B, p. 42.

Lydekker, R. 1898 E, p. 189.
Marsh, O. C. 1877 E.
　　1885 D.
Mivart, St. G. 1886 A.
Müller, J. 1831 A, pp. 199, 238.
Nicholson and Lydekker 1889 A, p. 1180.
Owen, R. 1842 A, p. 65.
　　1845 B, p. 285.
　　1850 A.
　　1850 B, p. 521.
　　1860 A, p. 164.
　　1860 E, p. 271.
　　1866 A.
　　1878 A.
　　1879 A.
Parker, W. K. 1868 A, p. 126.
　　1879 B, p. 81.
　　1883 A.
Pictet, F. J. 1853 A, p. 475. ("Crocodiliens.")
Röse, C. 1893 B.
　　1895 A.
Seeley, H. G. 1888 H. p. 131.
Stannius, H. 1856 A.
Steinmann and Döderlein 1890 A, p. 651.
Tomes, C. S. 1898 A. p. 268.
Tornier, G. 1888 A.
Wiedersheim, R. 1889 A, p. 438.
Woodward, A. S. 1886 D.
　　1898 B, p. 216.
Zittel, K. A. 1890 A, pp. 450, 633, 647, 683.

Suborder PARASUCHIA Huxley.

Huxley, T. H. 1875 B, p. 427.
Baur, G. 1894 A, p. 123. (Phytosauria.)
Cope, E. D. 1869 M, pp. 32, 56. (Thecodontia.)
　　1889 R, p. 365.
　　1900 A, p. 161.
Fürbringer, M. 1900 A, p. 302.
Gadow, H. 1898 A. p. 19.
Haeckel, E. 1895 A. p. 364.
Koken, E. 1887 A, p. 96.
　　1888 B, p. 763.

Lydekker, R. 1888 B, p. 123.
　　1890 A, p. 235.
Marsh, O. C. 1895 C, p. 485.
　　1896 B, p. 59. (Belodontia.)
　　1896 D, p. 199. (Belodontia.)
Owen, R. 1860 A, p. 163. (Thecodontia.)
Woodward, A. S. 1898 B, p. 217.
Zittel, K. A. 1890 A, p. 637.

PHYTOSAURIDÆ.

Lydekker, R. 1888 B, p. 123.
Baur, G. 1886 G, p. 740. (Belodontidæ.)
　　1887 A, p. 487. (Belodontidæ.)
　　1887 C, p. 102. (Belodontidæ.)

Cope, E. D. 1898 B, p. 78. (Belodontidæ.)
Nicholson and Lydekker 1889 A, p. 1183. (Phytosauridæ.)

PHYTOSAURUS Jaeger.   Type *P. cylindricodon* Jaeger.

Jaeger, G. F. 1828 A, p. 22.
Baur, G. 1886 G, p. 740. (Belodon.)
　　1887 A, p. 487. (Belodon.)
　　1894 A. (Phytosaurus.)
Cope, E. D. 1866 D. (Rhytidodon, emendation of Rutiodon.)
　　1869 M, p. 56. (Belodon.)
　　1870 I, p. 444. (Belodon.)
　　1879 A, p. 34. (Belodon.)
　　1888 U, p. 916. (Belodon.)
　　1889 E. (Belodon.)
Emmons, E. 1856 A, p. 80. (Rutiodon, type *R. carolinensis.*)

Fürbringer, M. 1900 A, p. 302.
Huxley, T. H. 1869 A.
　　1875 B.
Koken, E. 1887 A, p. 96.
　　1888 B, p. 764.
Lea, I. 1856 B, p. 123. (Centemodon.)
　　1856 A, p. 78. (Centemodon, type *C. sulcatus.*)
Leidy, J. 1856 J, p. 165. (Compsosaurus, type *C. priscus.*)
　　1856 L, p. 256. (Omosaurus, type *O. perplexus.*)
Lydekker, R. 1888 B, p. 123.

Lydekker, R. 1890 A, p, 235.
Marsh, O. C. 1877 E. (Belodon.)
   1895 C, p. 485. (Belodon.)
   1896 B, p. 61. (Belodon; Rhytidodon.)
   1896 D, p. 199. (Belodon.)
Meyer, H. 1842, Neues Jahrb. Mineral., p. 302.
   (Belodon, type *B. plieningeri* Meyer.)
   1855 A, p. 148. (Belodon.)
   1861 A, pls. xxxviii–xlvii. (Belodon.)
   1863 A, pls. xxxviii–xlii. (Belodon.)
   1865 B, pls. xxiii–xxix. (Belodon.)
Nicholson and Lydekker 1889 A, p. 1188.
Pictet, F. J. 1853 A, p. 514.
Plieninger, Th. 1852 A. (Belodon.)
Quenstedt, F. A. 1850 B, p. 23.
   1852 A, p. 110. (Belodon.)
Steinmann and Döderlein 1890 A, p. 652. (Belodon.)
Woodward, A. S. 1886 D, p. 300, fig. 5.
   1898 B, p. 218. (Belodon.)
Zittel, K. A. 1890 A, p. 688. (Belodon.)

Phytosaurus buceros (Cope).

Cope, E. D. 1881 W, p. 922. (Belodon.)
   1887 A, p. 217, pl. ii. (Belodon.)
   1887 J. (Belodon.)
   1888 U, p. 916, pl. xvii, fig. 1. (Belodon.)
   1891 N, p. 46, fig. 23. (Belodon.)
   1898 B, p. 72, fig. 24. (Belodon.)
Triassic, New Mexico.

Phytosaurus carolinensis (Emmons).

Emmons, E. 1856 A, p. 307, fig. 22; pl. viii, fig. 1. (Rutiodon.)
Cope, E. D. 1866 D. (Rhytidodon = Belodon.)
   1869 M, pp. 25, 59. (Belodon.)
   1870 I, p. 444. (Belodon.)
   1871 M, p. 210. (Belodon.)
   1875 V, p. 34. (Belodon.)
   1877 E, p. 185. (Belodon.)
Dana, J. D. 1896 A, p. 754, fig. 1183. (Belodon.)
Emmons, E. 1856 A, p. 308 (Rutiodon, or Clepsisaurus, carolinensis); p. 818, figures (Paleosaurus sulcatus).
   1857 A, p. 82, figs. 52–56. (Rutiodon.)
   1858 A, p. 78. (Rutiodon.)
   1860 A, p. 175, figs. 157, 160 [1]. (Rutiodon.)
Lea, I. 1856 B, p. 123 (Centemodon sulcatus.)
   1856 A, p. 78. (Centemodon sulcatus.)
Leidy, J. 1856 L, p. 256. (Omosaurus perplexus)
   1857 A, p. 272 (Omosaurus perplexus.)
Marsh, O. C. 1896 B, p. 60. (Rhytidodon.)
Wheatley, C. M 1861 A, p. 44. (Centemodon sulcatus.)
Triassic; North Carolina

Phytosaurus leaii (Emmons).

Emmons, E. 1856 A, p. 299. (Clepsisaurus.)
Cope, E. D. 1869 M, p. 59. (Belodon.)
   1870 I, p. 444. (Belodon.)

Cope, E. D. 1871 M. ("Clepsysaurus.")
   1875 V, p. 35. (Belodon.)
Emmons, E. 1857 A, p. 79, text fig. 51; pl. x. (Clepsisaurus.)
Triassic; North Carolina.

Phytosaurus lepturus (Cope).

Cope, E. D. 1869 M. p. 59, pl. xiv, figs. 1–14. (Belodon.)
   1870 I, p. 444.
   1870 S, p. 563.
Triassic; Pennsylvania.

Phytosaurus priscus (Leidy).

Leidy, J. 1856 J, p. 165. (Compsosaurus.)
Dana, J. D. 1896 A, p. 754, fig. 1182. (Belodon.)
Cope, E. D. 1869 M, p. 59, pl. xiv, figs. 15–25. (Belodon.)
   1870 I, p. 444. (Belodon.)
   1871 M, p. 210. (Belodon.)
   1875 V, p 34, pl. viii, fig. 5. (Belodon.)
   1877 E, p. 185. (Belodon.)
Emmons, E. 1856 A, p. 299 (Clepsisaurus pennsylvanicus, in part); p. 318, figures (Palæosaurus carolinensis).
   1857 A, pp. 65–79, figs. 37–50 (Clepsisaurus pennsylvanicus, in part); p. 86, figs. 57, 58, 60 (Palæosaurus carolinensis).
Hills, R. C. 1880 A. (Belodon.)
Leidy, J. 1856 L, p. 255. (Palæosaurus.)
   1857 A, p. 271. (Palæosaurus.)
Triassic; North Carolina.

Phytosaurus rostratus (Marsh).

Marsh, O. C. 1896 B, p. 61, fig. 2. (Rhytidodon.)
   1900 A, p. 23. (Rhytidodon.)
Triassic; North Carolina.

Phytosaurus scolopax (Cope).

Cope, E. D. 1881 W, p. 923. (Belodon.)
   1887 A, p. 221. (Belodon.)
Triassic; New Mexico.

Phytosaurus serridens (Leidy).

Leidy, J 1859 E. (Eurydorus.)
Cope, E D. 1866 D, p. 249 (Belodon.)
   1869 M, pp. 57, 58. (Belodon.)
Wheatley, C. M 1861 A, p. 44. (Eurydorus.)
Triassic; Pennsylvania.

Phytosaurus superciliosus (Cope).

Cope, E. D. 1893 A, p. 12, pl. i, figs. 1–10. (Belodon.)
Triassic; Texas.

Phytosaurus validus (Marsh).

Marsh, O. C. 1893 G, p. 170. (Belodon. No description.)
   1896 B, p. 59. (No description.)
Triassic; Connecticut.

EPISCOPOSAURUS Cope. Type *E. horridus* Cope.

Cope, E. D. 1887 A, p. 213.
1896 I, p. 123.
Marsh, O. C. 1896 B, p. 62.
Nicholson and Lydekker 1889 A, p. 1184.
Zittel, K. A. 1890 A, p, 644.

Episcoposaurus haplocerus Cope.

Cope, E. D. 1892 J, p. 129.

Cope, E. D. 1892 E, p. 257.
Triassic; Texas.

Episcoposaurus horridus Cope.

Cope, E. D. 1887 A, p. 213.
1892 E, p. 257.
Triassic; New Mexico.

Suborder AËTOSAURIA Nich. and Lydek.

Nicholson and Lydekker 1889 A, p. 1182.
Baur, G. 1894 A, p. 323.
Cope, E. D. 1898 B, p. 73. (Pseudosuchia.)
Fraas, O. 1867 A. (Genus Dyoplax.)
Fürbringer, M. 1900 A, p. 305. (Pseudosuchia.)

Gadow, H. 1898 A, p. 19. (Pseudosuchia.)
Haeckel, E. 1895 A, p. 364.
Marsh, O. C. 1895 C, p. 484.
Zittel, K. A. 1890 A, p. 644. (Pseudosuchia.)

AËTOSAURIDÆ.

Baur, J. 1887 C, p. 102. (Aethosauridæ.) | Nicholson and Lydekker 1889 A, p. 1182.

TYPOTHORAX Cope. Type *T. coccinarum* Cope.

Cope, E. D. 1875 R, p. 265.
1875 F, p. 84.
1877 K, p. 29.
1887 A, p. 210.
1887 E.
1896 I, p. 123.
Marsh, O. C. 1896 B, p. 62.
Nicholson and Lydekker 1889 A, pp. 1137, 1183.
Steinmann and Döderlein 1890 A, p. 658.
Zittel, K. A. 1890 A, p. 644.

Typothorax coccinarum Cope.

Cope, E. D. 1875 R, p. 265. (In part.)
1875 A.
1875 F, p. 84.
1877 K, p. 30, pl. xxii, figs. 1-9.
1887 A, p. 210, pl. i.
1887 E.
1896 I, p. 123.
Triassic; New Mexico.

STEGOMUS Marsh. Type *S. arcuatus* Marsh.

Marsh, O. C. 1896 B, p. 59.
Cope, E. D. 1896 I, p. 123.
Regarded by Professor Marsh as belonging to the *Parasuchia*.

Stegomus arcuatus Marsh.

Marsh, O. C. 1896 B, p. 60, pl. i.
Triassic; Connecticut.

Suborder EUSUCHIA Huxley.

Huxley, T. H. 1875 B.
Cope, E. D. 1871 B, p. 235. (Procœlia.)
1889 R, p. 864.
1891 N, p. 45.
1898 B, p. 73.
1900 A, p. 161.
Deslongchamps-Eudes, J. A. 1864 A. ("Téléosauriens.")
Fürbringer, M. 1900 A, p. 306.

Gadow, H. 1898 A, p. 38. (Crocodilia.)
1898 A, p. 19.
Haeckel, E. 1895 A, pp. 364, 367. (Typosuchia.)
Lydekker, R. 1888 B, p. 43.
Nicholson and Lydekker 1889 A, p. 1185.
Owen, R. 1860 A, p. 165. (Procœlia.)
Parker, W. K. 1868 A, p. 126. (Crocodilia.)
Woodward, A. S. 1898 B, p. 223.
Zittel, K. A. 1890 A, p. 647.

CROCODYLIDÆ.

Gray, J. E., 1825, Annals Philos. (2), x, p. 195. (Crocodilidæ.)
The usual way of spelling this name is *Crocodilidæ*.
Bemmelen, J. F. 1896 A, p. 332.
Boulenger, G. A. 1889 A, p. 274.
Calori, L. 1894 A.
Cope, E. D. 1871 B.
1900 A, p. 161.
Deslongchamps-Eudes, J. A. 1864, A. ("Crocodiliens.")
Gaupp, E. 1896 B.

Gegenbaur, C. 1898 A.
Hoffmann, C. K. 1890 B.
Huxley, T. H. 1859 B, p. 5 (Crocodilidæ); p. 2 (Alligatoridæ).
1872 A, p. 221. (Alligatoridæ, Crocodilidæ.)
Lydekker, R. 1888 B, p. 44.
Nicholson and Lydekker 1889 A, p. 1192.
Röse, C. 1892 C.
Steinmann and Döderlein 1890 A, p. 657.
Zittel, K. A. 1890 A, p. 681 (Crocodilidæ); p. 679 (Alligatoridæ).

CROCODYLUM Laurenti.   Type *C. niloticus* Linn.

Laurenti, J. N. 1768, Synop. Rept., p. 53.
The name of this genus is usually spelled *Crocodilus.* The above form is the original one.
Boulenger, G. A. 1889 A, p. 277.
1896 B.
Cope, E. D. 1867 C, p. 143.   (Thecachampsa, type *T. contusor.*)
1869 G, p. 11.   (Thecachampsa.)
1869 M, pp. 62, 231.   (Thecachampsa.)
1870 D, p. 174.   (Thecachampsa.)
1872 V, p. 554.   (Ichthyosuchus.)
1875 C, p. 31.
1882 J, p. 983.   (Crocodilus, Thecachampsa.)
1884 O, p. 151.
1900 A, p. 171.
Cuvier, G. 1834 A, ix, pp. 25–344, with plates.
Gegenbaur, C. 1898 A.
Giebel, C. G. 1877 A, p. 105.
Hoffmann, C. K. 1890 B.
Huxley, T. H. 1859 B, p. 6.
1869 B.
1872 A, p. 217.
1875 B.
Klein, E. F. 1863 A.
Kükenthal, W. 1893 A.
Leidy, J. 1871 C, p. 368.
1872 B, p. 366.
Ludwig, R. 1877 A.
Lydekker, R. 1888 B, p. 53.
Marsh, O. C. 1870 C, p. 98.   (Thecachampsa; not tenable.)
Miall, L. C. 1878 C.
Nicholson and Lydekker 1889 A, p. 1194 (Thecachampsa); p. 1195 (Crocodilus).
Owen, R. 1847 B, p. 283.
1850 B, p. 521.
1866 A.
Parker, W. K. 1883 A.
Segond, L. A. 1873 A, p. 10.   ("Crocodile.")
Steinmann and Döderlein 1890 A, p. 657.
Toula and Kail 1885 A.
Woodward, A. S. 1886 D.
Zittel, K. A. 1890 A, p. 682 (Crocodilus); p. 683 (Thecachampsa).

Crocodylus acer Cope.

Cope, E. D. 1882 J, pp. 982, 986, fig. 2.
1884 O, pp. 152, 154, pl. xxiii, figs. 1, 2.
Zittel, K. A. 1890 A, p. 682, fig. 604.
Eocene (Manti beds); Utah.

Crocodylus affinis Marsh.

Marsh, O. C. 1871 B, p. 454.
Cope, E. D. 1882 J, p. 984, fig. 4; p. 986.
1884 O, pp. 152, 162, pl. xxi, figs. 1–3.
Leidy, J. 1872 B, p. 366.
Eocene (Bridger); Wyoming.

Crocodylus antiquus Leidy.

Leidy, J. 1852 K, pl. xvi.   (Crocodilus.)
Cope, E. D. 1867 C, p. 143.   (Thecachampsa contusor.)
1869 G, p. 12.   (Thecachampsa.)
1869 M, p. 64, fig. 16.   (Thecachampsa.)
1882 J, p. 983.

Emmons, E. 1858 B, p. 215, fig. 35 A.
Eocene; Virginia.

Crocodylus aptus Leidy.

Leidy, J. 1869 B, p. 67.
Cope, E. D. 1844 O, p. 162.
Leidy, J. 1871 C, p. 368.
1872 B, p. 366.
1873 B, p. 126, pl. viii, fig. 2.
Eocene (Wasatch?); Wyoming.

Crocodylus brevicollis Marsh.

Marsh, O. C. 1871 B, p. 456.
King, C. 1878 A, p. 406.
Eocene (Bridger); Wyoming.

Crocodylus chamensis Cope.

Cope, E. D. 1874 O, p. 15.   (Alligator.)
1875 C, p. 33.   (C. heterodon, in part.)
1877 K. p. 67, pl. xxxii, figs. 1–22.
1884 O, p. 166.
Eocene (Wasatch); New Mexico.

Crocodylus clavis Cope.

Cope, E. D. 1872 R, p. 485.
1873 E, p. 612.
1882 J, p. 983, fig. 5; p. 986.
1884 O, pp. 152, 157, pl. xxi, figs. 4–9; pl. xxii.
Osborn, Scott, and Speir 1878 A, p. 93.
Eocene (Bridger); Wyoming.

Crocodylus elliotti Leidy.

Leidy, J. 1870 N.
Cope, E. D. 1873 E, p. 612.
1875 C, p. 33.
1877 K, p. 65, pl. xxxi, figs. 6–17.
1882 J, pp. 983, 986.
1884 O, p. 152.   (Thecachampsa.)
King, C. 1878 A, p. 406.
Leidy, J. 1870 R.
1871 C, p. 368.
1872 B, p. 366.
1873 B, p. 126, pl. viii, figs. 4–8.
Marsh, O. C. 1871 B, p. 454.
Osborn, Scott, and Speir 1878 A, p. 85.
Zittel, K. A. 1890 A, p. 683.   (Thecachampsa.)
Eocene (Bridger); Wyoming, New Mexico.

Crocodylus fastigiatus Leidy.

Leidy, J. 1851 G, p. 327.
Cope, E. D. 1869 M, p. 65.   (Thecachampsa.)
1882 J, p. 986.
Eocene; Virginia.

Crocodylus grinnellii Marsh.

Marsh, O. C. 1871 B, p. 455.
Cope, E. D. 1873 E, p. 613.
Leidy, J. 1872 B, p. 366.
Marsh, O. C. 1871 G, p. 105.
Eocene (Bridger); Wyoming.

Crocodylus grypus Cope.

Cope, E. D. 1875 C, p. 32.
1877 K, p. 63, pl. xxx.
1882 J, p. 983.
King, C, 1878 A, p. 377.
Eocene (Bridger?); New Mexico.

Crocodylus heterodon Cope.

Cope, E. D. 1872 T, p. 544. (Alligator.)
　　1873 E, p. 614. (Alligator.)
　　1875 C, p. 34.
　　1882 J, pp. 983, 986.
　　1884 O, p. 164, pl. xxiii a, figs. 10-18.
King, C. 1878 A, p. 376 (Alligator); p. 377 (Croc-
　　odilus).
Osborn, Scott, and Speir 1878 A, p. 98.
　　Eocene (Wasatch); Wyoming.

Crocodylus humilis Leidy.

Leidy, J. 1856 F, p. 73.
Cope, E. D. 1869 M, p. 82.
　　1874 C, p. 27. (Bottosaurus?)
　　1875 E, p. 254. (Bottosaurus?)
　　1877 G, p. 573.
Leidy, J. 1856 C, p. 120.
　　1860 A, p. 146, pl. xi, figs. 9-19.
Marsh, O. C. 1897 C, p. 527.
　　Cretaceous (Laramie); Montana, Colorado.

Crocodylus liodon Marsh.

Marsh, O. C. 1871 B, p. 454.
Cope, E. D. 1873 E, p. 613.
　　1875 C, p. 33.
　　1877 K, p. 67, pl. xxi, figs. 18-23. (C. liodon?)
　　1884 O, p. 153.
Leidy, J. 1872 B.
Marsh, O. C. 1871 G, p. 104.
　　Eocene; Wyoming, New Mexico.

Crocodylus marylandicus Clark

Clark, W. B. 1895 A, p. 4.
　　1897 A, p. 58, pl. viii, fig. 1.
　　Eocene; Maryland.

Crocodylus parvus Osb., Scott, and Speir.

Osborn, Scott, and Speir 1878 A, p. 91.
　　Eocene (Bridger); Wyoming.

Crocodylus polyodon Cope.

Cope, E. D. 1873 E, p. 614. (Diplocynodus.
　　1882 J, p. 981.
　　1884 O, p. 154, pl. xxiia, figs. 1, 2.
　　Eocene (Bridger); Wyoming.

Crocodylus rugosus (Emmons).

Emmons, E. 1858 B, p. 219, figs. 38, 39. (Polyptych-
　　odon.)
Cope, E. D. 1871 M, p. 210. (Thecachampsa.)
　　1875 V, p. 33, pl. viii, fig. 3. (Theca-
　　champsa.)
Emmons, E. 1860 A, p. 208, fig. 180 ?. (Polyptych-
　　odon.)
Leidy, J. 1865 A, pp. 18, 116. (Polyptychodon.)
　　Eocene; North Carolina.

Crocodylus sericodon Cope.

Cope, E. D. 1867 C, p. 143. (Thecachampsa.)
　　1869 B, p. 738. (Thecachampsa.)
　　1869 G, p. 12. (Thecachampsa.)
　　1869 K, p. 91. (Thecachampsa.)
　　1869 M, p. 64, pl. v, figs. 7, 8. (Theca-
　　champsa.)
　　1875 H, p. 363. (Thecachampsa.)
　　1882 J, p. 984.
　　Miocene; Maryland, New Jersey.

Crocodylus serratus Cope.

Cope, E. D. 1872 Y, p. 171. (Thecachampsa.)
　　1882 J, p. 986.
　　Eocene; New Jersey.

Crocodylus sicarius Cope.

Cope, E. D. 1869 G, p. 12. (Thecachampsa.)
　　1869 K, p. 91. (Thecachampsa.)
　　1869 M, p. 63, pl. v, fig. 6. (Thecachampsa.)
　　1882 J, p. 984.
　　Miocene; Maryland.

Crocodylus solaris Marsh.

Marsh, O. C. 1877 D, p. 254.
　　Pliocene?; South Dakota.

Crocodylus squankensis Marsh.

Marsh, O. C. 1869 A, p. 391. (Thecachampsa.)
Cope, E. D. 1871 C, p. 65. (Thecachampsa.)
　　1872 Y, p. 17. (Thecachampsa.)
　　1882 J, p. 983.
Marsh, O. C. 1870 C, p. 98. (Thecachampsa.)
　　Eocene; New Jersey.

Crocodylus stavelianus Cope.

Cope, E. D. 1885 Q.
　　Eocene (Puerco); New Mexico.

Crocodylus subulatus Cope.

Cope, E. D. 1872 V, p. 554. [Crocodilus (Ichthy-
　　osuchus.)]
　　1873 E, p. 613. (Diplocynodus.)
　　1877 K, p. 60. (Diplocynodus.)
　　1882 J, pp. 983, 986.
　　1884 O, p. 152, pl. xxiiia, figs. 4-9.
　　Eocene (Bridger); Wyoming.

Crocodylus sulciferus Cope.

Cope, E. D. 1872 V, p. 555.
　　1873 E, p. 612.
　　1882 J, p. 983.
　　1884 O, p. 157, pl. xxiiia, fig. 23.
　　Eocene (Bridger); Wyoming.

Crocodylus wheelerii Cope.

Cope, E. D. 1875 C, p. 33.
　　1875 F, p. 96. (C. veelerii.)
　　1877 K, p. 64, pl. xxxi, figs. 1-5
　　1882 J, p. 983.
　　Eocene; New Mexico.

The following references are given to notices of undetermined and undescribed remains of *Crocodylus:*

Cope, E. D. 1881 D, p. 184. Eocene; Wyoming.
　　1882 E, p. 143. Eocene.
Leidy, J. 1865 A, p. 116. Eocene; New Jersey,
　　North Carolina.

Wyman, J. 1850 B, p. 233, fig. 8. Eocene; South
　　Carolina.

LIMNOSAURUS Marsh.   Type *Crocodilus ziphodon* Marsh.

Marsh, O. C. 1872 K, p. 309.

Limnosaurus ziphodon Marsh.

Marsh, O. C. 1871 G, p. 104.   (Crocodilus.)
Cope E. D. 1882 J, p. 986.   (Crocodilus ziphodon.)

King, C. 1878 A, p. 405.
Leidy, J. 1872 B, p. 366.   (Crocodilus.)
Marsh, O. C. 1871 B, p. 453.   (Crocodilus.)
1872 K, p. 309.
Eocene; Wyoming.

POLYDECTES Cope.   Type *P. biturgidus* Cope.

Cope, E. D. 1869 I.
1870 D, p. 271.
1875 V, p. 33.

Polydectes biturgidus Cope.
Cope, E. D. 1869 I.
1870 D, p. 271.
1875 V, p. 33, pl. viii, fig. 2.
Cretaceous?; North Carolina.

ALLIGATOR Cuvier.   Type *A. lucius* Cuvier = *Crocodilus mississippiensis* Daudin.

Cuvier, G., 1807, Ann. Mus., x, p. 30.
Boulenger, G. A. 1889 A, p. 289.
1896 B.
Clarke, S. F. 1891 A.
Cope, E. D. 1873 E, p. 614.
1900 A, pp. 161, 164.
Hoffmann, C. K. 1890 B.
Huxley, T. H. 1859 A, p. 3.
Ludwig, R. 1877 A.
Lydekker, R. 1889 B, p. 44.
Nicholson and Lydekker 1889 A, p. 1196.

Alligator mississippiensis Daudin.
Daudin, F. M., 1802, Hist. Nat. Rept., ii, p. 412.
(Crocodilus.)
Boulenger, G. A. 1889 A, p. 290.
Leidy, J. 1860 B, p. 122, pl. xxvii, figs. 6, 7.
1889 E, p. 31.
1892 A.
Recent and in Peace Creek beds; Florida:
Pleistocene; South Carolina.

BOTTOSAURUS Agassiz.   Type *Crocodilus macrorhyncus* Harlan=*Bottosaurus harlani* Meyer.

Agassiz, L., 1849, Proc. Acad. Nat. Sci. Phila., iv,
p. 169.
Cope, E. D. 1869 K, p. 90.
1869 M, pp. 62, 65.
Leidy, J. 1865 A, p. 12.
Zittel, K. A. 1890 A, p. 679.

Bottosaurus harlani (Meyer).
Meyer, H., 1832, Palæologica, p. 108.   (Crocodilus.)
Cope, E. D. 1867 E, p. 26.
1869 B, p. 736.
1869 M, pp. 65, 231.   (In part.)
1871 G, p. 48.   (B. macrorhynchus.)
1875 E, p. 258.   (B. macrorhynchus.)
1875 V, p. 48.   (B. macrorhynchus.)
De Kay, J. E. 1842 A, p. 27.   (Crocodilus macrorhynchus.)
Gibbes, R. W. 1851 A, p. 7.   (C. macrorhynchus.)
Giebel, C. G. 1847 B, p. 122.   (C. macrorhynchus.)
Harlan, R. 1824 B, pl. i ("Crocodile").
1834 B, p. 76.   (C. macrorhyncus.)

Harlan, R. 1835 B, p. 280.   (C. macrorhyncus.)
1835 I, p. 380, with plate.   (C. macrorhyncus.)
Leidy, J. 1865 A, pp. 12, 115, pl. iv, figs. 19–21; pl.
xvii, figs. 11–14.   (B. harlani, in part.)
1865 D, p. 67.   (B. harlani.)
Cretaceous; New Jersey.

Bottosaurus perrugosus Cope.
Cope, E. D. 1874 C, p. 26.
1874 B, p. 452.
1875 E, pp. 68, 258, pl. vi, figs. 5–8.
Lambe, L. M. 1899 A, p. 69.   (Identification
doubtful.)
Cretaceous; Colorado, British America?.

Bottosaurus tuberculatus Cope.
Cope, E. D. 1869 M, p. 230.
1869 M, p. 65.   (B. harlani, in part.)
1871 G, p. 48.
Cretaceous; New Jersey.

DIPLOCYNODON Pomel.   Type *D. ratelii* Pomel.

Pomel, A. 1847 C, p. 383.
Cope, E. D. 1869 M, p. 62.   (Plerodon.)
1877 K, p. 60.   (Diplocynodus.)
1882 J, p. 986.   (Plerodon.)
Lydekker, R. 1886 E, p. 211.   (Plerodon.)
1888 B, p. 45.
Meyer, H., 1839, Neues Jahrb. Mineral., p. 77.
(Plerodon.)
1857, Neues Jahrb. Mineral., p. 538.   (Pleurodon.)

Nicholson and Lydekker 1889 A, p. 1195.
Zittel, K. A. 1890 A, p. 679.

Diplocynodon sphenops (Cope).
Cope, E. D. 1875 C, p. 31.   (Diplocynodus.)
1877 K, p. 60, pl. xxix.   (Diplocynodus.)
1882 J, p. 986.   (Plerodon.)
King, C. 1878 A, p. 377.   (Diplocynodus.)
Eocene; New Mexico.

GAVIALIDÆ.

Huxley, T. H. 1859 B.
Boulenger, G. A. 1889 A, p. 274. (Crocodilidæ, in part.)

Huxley, T. H. 1872 A, p. 222.
Zittel, K. A. 1890 A, p. 674 (Gavialidæ) ; p. 672 (Rhynchosuchidæ).

GAVIALIS Oppel.   Type *G. gangeticus* Linn.

Oppel, M., 1811, Ordnungen der Reptilien, p. 19.
Boulenger, G. A. 1889 A, p. 275.
Cope, E. D. 1900 A, p. 162.
Lydekker, R. 1888 B, p. 65. (Garialis.)
Owen, R. 1850 B, p. 521.
Toula and Kail, 1885 A.
Zittel, K. A. 1890 A, p. 674.

Gavialis fraterculus Cope.

Cope, E. D. 1869 M, p. 82. (Hyposaurus.)

Cope, E. D. 1869 B, p. 736. (Hyposaurus; no description.)
1875 E, p. 254.
1875 K.
Cretaceous; New Jersey.

Gavialis minor Marsh.

Marsh, O. C. 1870 C.
Cope, E. D. 1872 Y, p. 18. (Thecachampsa?.)
Eocene; New Jersey.

HETERODONTOSUCHUS Lucas.   Type *H. ganei* Lucas.

Lucas, F. A. 1898 A.

Heterodontosuchus ganei Lucas.

Lucas, F. A. 1898 A.
Triassic; Utah.

THORACOSAURUS Leidy.   Type *T. grandis* Leidy = *Gavialis neocesariensis* De Kay.

Leidy, J. 1852 B.
Agassiz, L. 1849 A. (Sphenosaurus, type *Crocodilus clavirostris;* not of Meyer, 1847.)
Cope, E. D. 1869 M, p. 79.
1900 A, p. 161.
Gibbes, R. W. 1851 A, pp. 7, 13.
Koken, E. 1888 B.
Leidy, J. 1865 A, p. 5.
Nicholson and Lydekker 1889 A, p. 1193.
Steinmann and Döderlein 1890 A, p. 657. (Thoracosuchus.)
Zittel, K. A. 1890 A, p. 672.

Thoracosaurus neocesariensis (De Kay).

De Kay, J. E. 1842 A, p. 28, pl. xxii, fig. 59. (Gavialis.)
Agassiz, L. 1849 A. (Sphenosaurus clavirostris.)
Cope, E. D. 1867 E, p. 26. (T. neocæsariensis.)
1869 B, p. 736. (T. neocæsariensis.)
1869 M, pp. 68, 79. (T. neocæsariensis.)
1875 E, p. 250. (T. neocæsariensis.)

De Kay, J. E. 1833 A, pl. iii, figs. 7-11. ("Gavial.")
Gibbes, R. W. 1850 A. (Sphenosaurus clavirostris.)
1851 A, p. 7 (Croc. clavirostris); p. 13 (Croc. basifissus).
Giebel, C. G. 1847 B, p. 122. (Croc. clavirostris.)
Koken, E. 1888 B, p. 757.
Leidy, J. 1852 B. (Thoracosaurus grandis.)
1852 A, p. 135. (Croc. dekayi, C. basifissus.)
1865 A, pp. 5, 115, pl. i, figs. 1-6; pl. ii, figs. 1-3; pl. iii, figs. 5-11.
1865 D, p. 67.
Morton, S. G. 1844 A. [Crocodilus (Gavialis?) clavirostris.]
1845 A. (Croc. clavirostris.)
Owen R. 1849 A, p. 381, pl. x, figs. 1, 2. (Croc. basifissus.)
1860 A, p. 165. (Croc. basifissus.)
Woodward, A. S. 1890 A, p. 398.
Zittel, K. A. 1890 A, p. 673.
Cretaceous; New Jersey.

HOLOPS Cope.   Type *H. brevispinis* Cope.

Cope, E. D. 1869 H, p. 123.
1869 M, pp. 62, 67, 231.
1900 A, p. 162.
Huxley, T. H. 1875 B.
Lydekker, R. 1886 E, p. 211.
Nicholson and Lydekker 1889 A, p. 1193.
Zittel, K. A. 1890 A, p. 673.

Holops basitruncatus (Owen).

Owen, R. 1849 A, p. 381, pl. x, figs. 3, 4. (Crocodilus.)
Cope, E. D. 1869 B, p. 736. (Thoracosaurus tenebrosus.)

Cope, E. D. 1869 H, p. 123. (H. tenebrosus.)
1869 M, pp. 67, 77, fig. 19, p. 231, pl. iv, figs. 8, 9 (H. basitruncatus; called erroneously on plate Bottosaurus harlani); p. 78 (H. tenebrosus).
1875 E, p. 253.
Gibbes, R. W. 1851 A, p. 13. (Crocodilus.)
Leidy, J. 1852 K, p. 135. (Crocodilus.)
1865 A, p. 14 (syn. of Bottosaurus harlani); p. 115, pl. iii, figs. 12-15 (Crocodilus tenebrosus.)
Owen, R. 1860 A. p. 165. (Crocodilus.)
Eocene; New Jersey.

Holops brevispinis Cope.

Cope, E. D. 1869 H, p. 123.
 1869 B, p. 736. (Thoracosaurus; no description.)
 1869 M, p. 69, pl. 1, fig. 13; pl. iv, figs. 4–6.
 1875 E, p. 252.
 Cretaceous; New Jersey.

Holops cordatus Cope.

Cope, E. D. 1869 M, p. 73, fig. 18; pl. iv, fig. 1.
 1875 E, p. 253.
 Cretaceous; New Jersey.

Holops glyptodon Cope.

Cope, E. D. 1869 M, pp. 74, 231.
 1869 B, p. 736. (Thoracosaurus; no description.)
 1875 E, p. 253.
 Cretaceous; New Jersey.

Holops obscurus (Leidy).

Leidy, J. 1865 A, pp. 15, 16, 115, pl. ii, figs. 4, 5. (Crocodilus.)
Cope, E. D. 1867 E, p. 26. (Thoracosaurus.)
 1869 B, p. 736. (Thoracosaurus.)
 1869 H, p. 123.
 1869 M, pp. 68, 75, 231, pl. iv, figs. 1–3.
 1871 C, p. 75.
 1875 E, p. 253.
Leidy, J. 1865 A, pl. i, figs. 7–9.
Lydekker, R. 1886 E, p. 211.
 Cretaceous; New Jersey.

Holops pneumaticus Cope.

Cope, E. D. 1872 X.
 1875 E, p. 250.
 Cretaceous; New Jersey.

GONIOPHOLIDÆ.

Lydekker, R. 1888 B, p. 76. (Goniopholididæ.)
 1890 A, p. 229. (Goniopholididæ.)

Nicholson and Lydekker 1889 A, p. 1189.
Zittel, K. A. 1890 A, p. 676.

GONIOPHOLIS Owen. Type *G. crassidens* Owen.

Owen, R. 1842 A, p. 69.
Cope, E. D. 1878 H p. 391. (Amphicotylus, type *A. lucasii.*)
Lydekker, R. 1888 B, p. 79.
 1890 A, p. 229.
Marsh, O. C. 1877 D, p. 254. (Diplosaurus, type *D. felix.*)
 1877 E. (Diplosaurus.)
 1896 B, p. 62. (Diplosaurus.)
Nicholson and Lydekker 1889 A, p. 1191.
Owen, R. 1878 A.
 1879 A.
Zittel, K. A. 1890 A, p. 676.

Goniopholis felix (Marsh).

Marsh, O. C. 1877 D, p. 254. (Diplosaurus.)
King, C. 1878 A, p. 346. (Diplosaurus.)
Marsh, O. C. 1896 B, p. 61, fig. 2. (Diplosaurus.)
 1897 C, p. 507, fig. 62.

Zittel, K. A. 1890 A, p. 677. (Goniopholis.)
 Jurassic; Colorado.

Goniopholis lucasii Cope.

Cope, E. D 1878 H, p. 391. (Amphicotylus.)
 1888 Y.
Zittel, K. A. 1890 A, p. 677.
 Jurassic; Colorado.

Goniopholis vebbianus (Cope).

Cope, E. D. 1872 H, p. 310. (Hyposaurus.)
 1872 F, p. 327. (Hyposaurus vebbii.)
 1874 C, p. 26. (H. vebbii.)
 1875 E, pp. 52, 67, 250 (H. vebbii); pl. ix, figs. 8–8d (H. vebbianus).
Marsh, O. C. 1877 D, p. 254. (Diplosaurus vebbii.)
Williston, S. W. 1898 D, p. 78. (Hyposaurus.)
Zittel, K. A. 1890 A, p. 677. (Goniopholis vebbii.)
 Cretaceous (Benton); Kansas.

HYPOSAURUS Owen. Type *H. rogersii* Owen

Owen, R. 1849 A, p. 383.
Cope, E. D. 1866 F, p. 111.
 1869 M, p. 80.
 1875 K.
Gibbes, R. W. 1851 A, p. 13.
Lydekker, R. 1888 B, p. 90.
Marsh, O. C. 1877 E.
Nicholson and Lydekker 1889 A, p. 1190.
Zittel, K. A. 1890 A, pp. 672, 676.

Hyposaurus ferox Marsh.

Marsh, O. C. 1871 G, p. 104.
 Cretaceous; New Jersey.

Hyposaurus rogersii Owen.

Owen, R. 1849 A, p. 383, pl. xi, figs. 7–10.

Cope, E. D. 1867 E, p. 26.
 1869 B, p. 736.
 1869 M, p. 80, pl. iv, figs. 10, 11.
 1875 E, p. 250.
Gibbes, R. W. 1851 A, p. 9, pl. iii, fig. 13. (Holcodus acutidens, New Jersey specimen. *fide* Leidy.)
Leidy, J. 1865 A, pp. 18, 116, pl. iii, figs. 4, 16–21; pl. iv, figs. 1–12.
 1865 D, p. 68.
Williston, S. W. 1894 B, p. 3, pl. i, figs. 4, 5. (Identification doubtful.)
 1898 D, p. 76, figs. 3, 4. (Identification doubtful.)
 Cretaceous, New Jersey, Kansas.

PLIOGONODON Leidy.   Type *P. priscus* Leidy.

Leidy, J. 1856 L, p. 255.
 1857 A, p. 271.
Zittel, K. A. 1890 A, p. 620. (Syn.? of Mossasurus.)
 A genus of uncertain affinities.

Pliogonodon priscus Leidy.

Leidy, J. 1856 L, p. 255.

Cope, E. D. 1875, V, p. 33.
Emmons, E. 1858 B, p. 223, figs 43, 44. (P. nobilis.)
Leidy, J. 1857 A, p. 271.
 1865 A, pp. 103, 119.
Miocene?; North Carolina.

Class AVES Linn.

Linnæus, C., 1758, Syst. Nat., ed. x, p. 78.
Ahlborn, F. 1896 A.
Baur, G. 1883 A.
 1884 A.
 1884 B.
 1884 E.
 1885 A.
 1885 B.
 1885 C.
 1886 B.
 1887 A.
Blanchard, É. 1859 A.
 1860 A.
Burckhardt, R. 1896 A.
Calori, L. 1894 A.
Cope, E. D. 1869 M, pp. 4, 236.
 1885 G.
 1889 R, pp. 863, 869.
 1891 N, p. 58.
 1898 B, p. 83.
Coues, E. 1884 A.
Dames, W. 1884 A.
Dollo, L. 1883 B.
Edwards, A. M. 1863 A.
 1869 A.
Fürbringer, M. 1885 A.
 1888 A.
 1889 A.
 1890 A.
 1892 A.
 1896 A.
Gadow, H. 1888 A.
 1889 A, p. 474.
 1891 A.
 1892 A.
 1896 A, p. 42.
 1896 B.
Gaupp, E. 1900 A.
Gegenbaur, C. 1864 A.
 1865 A.
 1898 A.
Giebel, C. G. 1866 A.
Haeckel, E. 1895 A, p. 400.
Hallmann, E. 1837 A, pp. 14, 32.
Hoffmann, C. K. 1879 A.
Humphrey, G. M. 1866 A.
Hurst, C. H. 1893 A.
Huxley, T. H. 1867 A.
 1868 A.
 1870 A.
 1870 C.
 1872 A, pp. 168, 233.
 1876 B, xiii, pp. 468, 514.
Jeffries, J. A. 1881 A.

Jeffries, J. A. 1882 A.
Köstlin, O. 1844 A.
Leidy, J. 1871 C, p. 365.
Lucas, F. A. 1899 C, p. 323.
Lydekker, R. 1891 A.
Magnus, H. 1869 A
Marsh, O. C. 1877 E.
Marshall, W. 1872 A.
Mehnert, E. 1887 A.
Morse, E. S. 1871 A.
 1872 A.
 1880 A.
Newton, A. 1868 A.
 1875 A.
 1896 A.
Nicholson and Lydekker 1889 A, p. 1208.
Osborn, H. F. 1900 G, p. 777.
Owen, R. 1864 C.
 1866 B.
 1873 C.
Parker, W. K. 1864 A.
 1870 A.
 1875 A.
 1875 B.
 1876 A.
 1878 B.
 1879 B, p. 81.
 1884 A.
 1887 A.
 1889 A.
 1891 A.
 1891 B.
Pavlow, A. 1885 A.
Schenk, F. 1897 A.
Segond, L. A. 1864 A.
 1865 A.
Selenka, E. 1869 A.
Shufeldt, R. W. 1885 A.
 1886 A.
 1886 B.
 1886 C.
 1899 A, p. 101.
Steinmann and Döderlein 1890 A, p. 670.
Stejneger, L. 1885 A.
Vetter, B. 1885 A.
Wiedersheim, R. 1884 A.
Wildermuth, H. A. 1877 A.
Woodward, A S. 1898 B, p. 231.
Woodward, H. 1874 A.
 1874 B.
 1885 C.
 1886 C.
Zittel, K. A. 1890 A, pp. 804, 857.

Subclass SAURURÆ Haeckel.

Cope, E. D. 1889 R, p. 869.
1898 B, p. 83.
Fürbringer, M. 1888 A, p. 1519.
Gadow, H. 1892 A, p. 236. (Archornithes.)
1898 A, p. 30. (Archæornithes.)
Haeckel, E. 1895 A, pp. 402, 409.
Huxley, T. H. 1867 A, p. 418.
1872 A, p. 233.
Menzbier, M. 1887 A.

Nicholson and Lydekker 1889 A, p. 1221.
Parker, W. K. 1875 A.
Pavlow, A. 1885 A, p. 113.
Stejneger, L. 1885 A, p. 21.
Woodward, A. S. 1898 B, p. 282. (Subclass Archæ-
ornithes; order Saururæ.)
Woodward, H. 1886 C, p. 370.
Zittel, K. A. 1890 A, p. 820.

Order ORNITHOPAPPI Stejneger.

Stejneger, L. 1885 A, p. 21.
Cope, E. D. 1889 R, p. 869.

Cope, E. D. 1898 B, p. 84.
Haeckel, E. 1895 A, p. 410. (Archæornithes.)

ARCHÆOPTERYGIDÆ.

Cope, E. D. 1898 B, p. 84.
Fürbringer, M. 1888 A, p. 1519.

Huxley, T. H. 1872 A, p. 233.
Nicholson and Lydekker 1889 A, p. 1221.

ARCHÆOPTERYX Meyer.

Meyer, H., 1861, Neues Jahrb. Mineral., p. 679.
Although no species of this genus has been found in North America, the following citations of the literature are given on account of the important information furnished by the genus.
Baur, G. 1884 B, p. 451.
1886 D.
Branco, W. 1885 A.
Dames, W. 1882 A.
1884 A:
1884 B.
1897 A, p. 818.
Edwards, A. M. 1863 A.
1869 A, p. 673.
Fraas, E. 1885 A.
Fürbringer, M. 1888 A, p. 1522.
Gadow, H. 1888 A, p. 166.
1891 A.
1892 A, p. 236.
Gaupp, E. 1894 A.
Haeckel, E. 1895 A, p. 412.
Hurst, C. H. 1893 A.
1895 A.
Huxley, T. H. 1868 A, p. 361.

Marsh, O. C. 1882 C.
Meyer, H., 1861, Palæontographica, i, p. 53.
Newton, E. 1875 A, p. 728.
Nicholson and Lydekker 1889 A, p. 1221.
Owen, R. 1864 C.
Parker, W. K. 1868 A, p. 142.
1875 A.
Pavlow, A. 1885 A.
Pycraft, W. P. 1894 A.
1896 A.
Selenka, E. 1869 A, pl. iii, figs. 4, 5.
Steinmann and Döderlein 1890 A, p. 668.
Stejneger, L. 1885 A, p. 21.
Vetter, B. 1885 A.
Vogt, C. 1880 A.
Wagner, A. 1861 A.
Wiedersheim, R. 1884 A, p. 659.
Woodward, A. S. 1898 B, p. 232, figs. 141, 142.
Woodward, H. 1874 B.
1884 A.
1885 C, p. 311.
1886 C, p. 359, figs. 5, 20.
Zittel, K. A. 1890 A, p. 820.

LAOPTERYX Marsh.   Type L. priscus Marsh.

Marsh, O. C. 1881 C, p. 341.
Fürbringer, M. 1888 A, p. 1478.
Marsh, O. C. 1882 C.
Menzbier, M. 1887 A, p. 581.
Nicholson and Lydekker 1889 A, p. 1229.
Pavlow, A. 1885 A, p. 107.
Stejneger, L. 1885 A, p. 23.
Zittel, K. A. 1890 A, p. 824.
The true position of this genus is problematical.

Laopteryx priscus Marsh.

Marsh, O. C. 1881 C, p. 341.
Coues, E. 1884 A, p. 829.
Fürbringer, M. 1888 A.
Marsh, O. C. 1897 C, p. 508.
Stejneger, L. 1885 A, p. 23.
Jurassic; Wyoming.

Subclass EURHIPIDURA Gill.

Gill, T., 1874, Baird, Brewer, and Ridgway's Land | Gadow, H. 1898 A, p. 30. (Neornithes.)
Birds, i, Introd., p. xiii. | Haeckel, E. 1895 A, p. 402. (Ornithuræ.)
Cope, E. D. 1889 R, p. 869. | Stejneger, L. 1885 A, p. 31.
1891 N, p. 54. | Woodward, A. S. 1898 B, p. 294.
Cope, E. D. 1898 B, p. 84.

Superorder *ODONTOTORMÆ* Marsh.

Marsh, O. C. 1875 E, p. 630. | Marsh, Ò. C. 1880 G, p. 119.
Cope, E. D. 1889 R, p. 869. | 1882 C. (<Odontornithes.)
1891 N, p. 55. | 1883 C, p. 86.
1898 B, p. 84. | Menzbier, M. 1887 A.
Fürbringer, M. 1888 A, p. 1421. | Nicholson and Lydekker 1889 A, p. 1231.
1892 B. | Owen, R. 1878 A, p. 370. (<Odontornithes.)
Grinnell, G. B. 1881 A, p. 258. | Pavlow, A. 1885 A, p. 113.
Lydekker, R. 1891 A, p. 200. | Steinmann and Döderlein 1890 A, p. 673. (Odon-
1896 A, p. 649. | tormæ.)
Marsh, O. C. 1873 C. (<Odontornithes.) | Stejneger, L. 1885 A, p. 23.
1875 D, p. 407. (Ichthyornithes.) | Woodward, A. S. 1898 B, p. 242.
1876 F, p. 511. | Woodward, H. 1886 C, p. 370. (<Odontornithes.)
1876 G, p. 59. | Zittel, K. A. 1890 A, p. 834.
1877 E. (<Odontornithes.)

Order PTEROPAPPI Stejneger.

Stejneger, L. 1885 A, p. 23. | Cope, E. D. 1898 B, p. 88.
Cope, E. D. 1889 R, p. 870. | Gadow, H. 1898 A, p. 31. (Ichthyornithes.)
1891 N, p. 56. | Haeckel, E. 1895 A, p. 410. (Ichthyornithes.)

ICHTHYORNITHIDÆ.

Marsh, O. C. 1873 A. | Lydekker, R. 1891 A, p. 201.
Cope, E. D. 1898 B, p. 88. | Marsh, O. C. 1873 C.
Fürbringer, M, 1888 A, p. 1421. | 1875 E, p. 630.
1896 A, p. 497. | Nicholson and Lydekker 1889 A, p. 1231.
| Steinmann and Döderlein 1890 A, p. 673.

ICHTHYORNIS Marsh.   Type *I. dispar* Marsh.

Marsh, O. C. 1872 Q. | Woodward, A. S. 1898 B, p. 242.
Fürbringer, M. 1888 A. | Woodward, H. 1886 C, p. 361.
1896 A, p. 499. | Zittel, K. A. 1890 A, p. 834.
Grinnell, G. B. 1881 A, figs. 7, 9.
Huxley, T. H. 1876 B, xiii, p. 515
Lydekker, R. 1896 A, p. 649. | **Ichthyornis agilis** Marsh.
Marsh, O. C. 1872 P. (Colonosaurus, type *C.* | *Marsh, O. C.* 1873 D, p. 230. (Graculavus.)
madgei.) | Coues, E. 1884 A, p. 827.
1873 C. | Marsh, O. C. 1880 G, p. 197.
1875 E, p. 630. | Williston, S. W. 1898 B, p. 46.
1877 E. | Cretaceous (Niobrara); Kansas.
1879 C.
1880 G. | **Ichthyornis anceps** Marsh.
1881 H. | *Marsh, O. C.* 1872 E, p. 364. (Graculavus.)
1883 C. | Cope, E. D. 1874 C, p. 17. (Graculavus.)
Menzbier, M. 1887 A. | 1875 E, p. 245 (Graculavus.)
Nicholson and Lydekker 1889 A, pp. 1219, 1231. | Coues, E. 1872 B, p. 350. (Graculavus.)
Seeley, H. G. 1876 B. | 1884 A, p. 827.
Shufeldt, R. W. 1898 A. | Marsh, O. C. 1873 D, p. 229. (Graculavus.)
Steinmann and Döderlein 1890 A, p. 673. | 1880 G, p. 198.
Stejneger, L. 1885 A, p. 26. | Williston, S. W. 1898 B, p. 47.
Thompson, D'A. W. 1890 B. | Cretaceous (Niobrara); Kansas.

Ichthyornis dispar Marsh.

Marsh, O. C. 1872 Q.
Cope, E. D. 1874 C, p. 17.
  1875 E, p. 246.
Coues, E. 1872 B, p. 350.
  1884 A, p. 827.
Dana, J. D. 1896 A, p. 851, fig. 1431.
Marsh, O. C. 1872 P.   (Colonosaurus mudgei.)
  1873 C.
  1873 D, p. 230.
  1875 D, p. 403, pl. ix.
  1875 E, p. 625, pl. ii.
  1880 G, p. 197, pls. xxi–xxvi.
  1883 C, p. 70, figs. 19, 21–24, 28, 29.
Nicholson and Lydekker 1889 A, p. 1231, fig. 1116.
Shufeldt, R. W. 1893 A.
Williston, S. W. 1897 H. p. 245.
  1898 B, p. 46, pl. vii, figs. 1, 2.
Woodward, H. 1874 B.
Zittel, K. A. 1890 A, p. 835, figs. 712, 713.
  Cretaceous (Niobrara); Kansas.

Ichthyornis lentus Marsh.

Marsh, O. C. 1877 D, p. 253.   (Graculavus.)
Coues, E. 1884 A, p. 827.
Marsh, O. C. 1880 G, p. 198.
  Cretaceous (Niobrara); Kansas.

Ichthyornis tener Marsh.

Marsh, O. C. 1880 G, p. 198, pl. xxx, fig. 8.
Coues, E. 1884 A, p. 828.
Williston, S. W. 1898 B, p. 47.
  Cretaceous (Niobrara); Kansas.

Ichthyornis validus Marsh.

Marsh, O. C. 1880 G, p. 196, pl. xxx, figs. 11–14.
Coues, E. 1884 A, p. 828.
  Cretaceous (Niobrara); Kansas.

Ichthyornis victor Marsh.

Marsh, O. C. 1876 F, p. 511.
Coues, E. 1884 A, p. 828.
Dana, J. D. 1896 A, p. 851, fig. 1430.
Grinnell, G. B. 1881 A, figs. 7, 9.
Marsh, O. C. 1880 G, p. 199, pls. xxvii–xxxiv.
  1883 C. p. 74, figs. 25, 27, 30.
  1897 C, p. 527, pl. xxvi, fig. 1.
Shufeldt, R. W. 1893 A.
Stejneger, L. 1885 A, p. 25, fig. 9.   (No specific name.)
Williston, S. W. 1898 B, p. 47.
Woodward, A. S. 1898 B, p. 243, fig. 145.
Zittel, K. A. 1890 A, p. 834, fig. 711.
  Cretaceous (Niobrara); Kansas.

APATORNITHIDÆ.

Fürbringer, M. 1888 A. | Fürbringer, M. 1896 A, p. 497.

APATORNIS Marsh.   Type *Ichthyornis celer* Marsh.

Marsh, O. C. 1873 C.
Fürbringer, M. 1888 A.
  1896 A, p. 497.
Grinnell, G. B. 1881 A.
Lydekker, R. 1896 A, p. 652.
Marsh, O. C. 1880 G, p. 141.
Nicholson and Lydekker 1889 A, p. 1231.
Zittel, K. A. 1890 A, p. 835.

Cope, E. D. 1874 C, p. 17.   (Ichthyornis.)
  1875 E, p. 246.   (Ichthyornis.)
Coues, E. 1884 A, p. 826.
Marsh, O. C. 1873 C.
  1873 D, p. 230.
  1875 D, p. 404.
  1875 E, p. 626.
  1880 G, p. 192, pls. xxviii–xxxiii.
Williston, S. W. 1898 B, p. 47.
  Cretaceous (Niobrara); Kansas.

Apatornis celer Marsh.

Marsh, O. C. 1873 A.   (Ichthyornis.)

Superorder ODONTOLCÆ Marsh.

Marsh, O. C. 1875 D, p. 407.
Cope, E. D. 1889 R, p. 869.
  1891 N, p. 54.
  1898 B, p. 84.
Fürbringer, M. 1888 A.
  1892 B.   (Odontholcæ.)
Gadow, H. 1858 A, p. 31.
Grinnell, G. B. 1881 A, p. 258.
Haeckel, E. 1895 A, p. 416.   (Hesperornithes.)
Lydekker, R. 1891 A, p. 202.
  1896 A, p. 649.
Marsh, O. C. 1873 C.   (<Odontornithes.)
  1875 D, p. 407.
  1875 E, p. 630.

Marsh, O. C. 1876 F.   (<Odontornithes.)
  1876 G, p. 59.
  1877 B.   (<Odontornithes.)
  1878 C, p. 409.   (<Odontornithes.)
  1880 G, p. 5.
  1883 C, p. 86.
Menzbier, M. 1887 A.
Nicholson and Lydekker 1889 A, p. 1224.
Pavlow, A. 1885 A, p. 113.
Steinmann and Döderlein 1890 A, p. 671.
Stejneger, L. 1885 A, p. 27.   (Odontholcæ.)
Woodward, A. S. 1898 B, p. 234.
Wiedersheim, R. 1884 A, p. 664.
Zittel, K. A. 1890 A, p. 826.

Order DROMÆOPAPPI Stejneger.

Stejneger, L. 1885 A, p. 27.
Cope, E. D. 1889 R, 869.

Cope, E. D. 1898 B, p. 87.
Shufeldt, R. W. 1892 C. (Hesperornoides.)

HESPERORNITHIDÆ.

Marsh, O. C. 1872 E, p. 363. (Hesperornidæ.)
Cope, E. D. 1889 R.
 1898 B, p. 87.
Fürbringer, M, 1888 A.
 1890 B.
 1898 A, p. 499.

Lydekker, R. 1891 A, p. 204.
Marsh, O. C. 1876 F, p. 509.
Shufeldt, R. W. 1892 C, p. 202.
Steinmann and Döderlein 1890 A, p. 671.

HESPERORNIS Marsh.   Type *H. regalis* Marsh.

Marsh, O. C. 1872 E, p. 360.
Cope, E. D. 1883 DD.
Dollo, L. 1883 A, p. 16.
Fürbringer, M. 1888 A, p. 1476.
Grinnell, G. B. 1881 A.
Helm, F. 1891 A.
Huxley, T. H. 1876 B, xiii, p. 515.
Lydekker, R. 1891 A, p. 204.
 1891 C.
 1893 E, p. 140.
 1896 A, p. 649.
Marsh, O. C. 1875 D.
 1876 F, p. 509. (Lestornis, type *L. crassipes*.)
 1877 B.
 1877 E.
 1879 C.
 1880 G, p. 5.
 1881 H.
 1882 C.
 1883 C, p. 52.
 1897 A.
Menzbier, M. 1887 A.
Nicholson and Lydekker 1889 A, p. 1224.
Osborn, H. F. 1900 G, p. 783.
Seeley, H. G. 1876 B, p. 510.
Shufeldt, R. W. 1888 A, p. 17.
 1890 A, p. 169.
 1890 C.
 1893 A, p. 336.
 1897 B.
Stejneger, L. 1885 A, p. 25.
Thompson, D'A. W. 1890 B.
Vetter, B. 1885 A.
Wiedersheim, R. 1884 A, p. 664.
Williston, S. W. 1896 A, pl. ii.
 1898 B, p. 47, pls. v–viii.
Woodward, A. S. 1898 B, p. 235.
Woodward, H. 1885 C, p. 312.
 1886 C, p. 361, pl. i.
Zittel, K. A. 1890 A, p. 827.

Hesperornis crassipes Marsh.

Marsh O. C. 1876 F, p. 509. (Lestornis.)

Coues, E. 1884 A, p. 827.
Marsh, O. C. 1880 G, p. 196, pls. vii, xxvii, text
 fig. 40.
Nicholson and Lydekker 1889 A, p. 1225.
Stejneger, L. 1885 A, p. 30. (Lestornis.)
Thompson, D'A. W. 1890 B, p. 9.
Williston, S. W. 1898 B, p. 45.
 Cretaceous (Niobrara); Kansas.

Hesperornis gracilis Marsh.

Marsh, O. C. 1876 F, p. 510.
Coues, E. 1884 A, p. 82.
Marsh, O. C. 1880 G, p. 197.
Williston, S. W. 1898 B, p. 46.
 Cretaceous (Niobrara); Kansas.

Hesperornis regalis Marsh.

Marsh, O. C. 1872 A.
Cope, E. D. 1874 C, p. 17.
 1875 E, p. 245.
Coues, E. 1872 B, p. 350.
 1884 A, p 826.
Dana, J. D. 1896 A, p. 850, figs. 1423–1429.
Grinnell, G. B 1881 A, figs. 1, 3, 5, 6.
Huxley, T. H 1876 C, p. 215.
Marsh, O. C. 1872 E, p. 360.
 1873 D, p. 230
 1875 D, p 404, pl. x.
 1877 B, pl. v
 1880 G, p. 195, pls. i–vi; viii–xvi; xviii–xx;
  text figs. 1, 4–6, 8–14, 21–27, 35, 36.
 1883 C, p. 52, figs. 3–5, 7, 9–14, 17, 18.
 1897 C, p. 527, pl. xxvi, fig. 2.
Nicholson and Lydekker 1889 A, p. 1225, figs.
 1111–1113
Shufeldt, R. W. 1890 A, p. 169.
Stejneger, L. 1885 A, p. 27, figs. 13–16.
Thompson, D'A W 1890 B, figures.
Wiedersheim, R. 1884 A, p. 663.
Williston, S. W. 1898 B, p 45, pls. v, vi.
Woodward A. S. 1898 B, p 235, fig. 143.
Zittel, K. A. 1890 A, p. 816, figs. 704–707.
 Cretaceous (Niobrara); Kansas.

BAPTORNIS Marsh.   Type *B. advenus* Marsh.

Marsh, O. C. 1877 B, p. 86.
Fürbringer, M. 1888 A.
Lydekker, R. 1891 C.
Zittel, K. A. 1890 A, p. 828.

Baptornis advenus Marsh.

Marsh, O. C. 1877 B, p. 86.

Coues, E. 1884 A p. 826.
Marsh, O. C. 1880 G, p. 192, figs. 37–39.
Williston, S. W. 1896 B, p. 46.
Zittel, K. A. 1890 A, p. 828.
Cretaceous (Niobrara); Kansas.

CONIORNIS Marsh.   Type *C. altus* Marsh.

Marsh, O. C. 1893 A, p. 81.

Coniornis altus Marsh.

Marsh. O. C. 1893 A, p. 81, figs. 1–3.
Cretaceous; Montana.

Superorder DROMÆOGNATHÆ *Huxley.*

Huxley, T. H. 1867 A, p. 418.
This suborder, as here recognized, is equivalent to the *Ratitæ* of Merrem *plus* the *Crypturi* (tinamous). It is the *Dromæognathæ* of Stejneger and the *Ratitæ* of Cope, 1896. Unless otherwise indicated, the following authorities are to be understood as employing the term *Ratitæ*.
Cope, E. D. 1889 R, p. 870.
    1891 N, p. 56.
    1896 B, p. 84.
Cunningham, R. O. 1871 A.
Fürbringer, M. 1885 A.
    1888 A.
Gadow, H. 1885 A.
    1888 A.
    1891 A.
    1892 A, p. 237.
    1893 A, p. 30.
Haeckel, E. 1895 A, p. 415.

Huxley, T. H. 1867 A, p. 418.
    1872 A, p. 233.
Menzbier, M. 1887 A.
Mivart, St. G. 1874 A.
    1877 A.
    1892 A, p. 258.
Newton, A. 1868 A.
Nicholson and Lydekker 1889 A, p. 1222.
Parker, W. K. 1866 A. ("Ostrich tribe.")
    1866 B.   (Struthionidæ.)
    1875 A.
    1889 A.
    1895 A.
Pavlow, A. 1886 A, p. 113.
Selenka, E. 1869 A.
Steinmann and Döderlein 1890 A, p. 671.
Stejneger, L. 1885 A, p. 31.   (Dromæognathæ.)
Woodward, H. 1886 C, p. 370.
Zittel, K. A. 1890 A, p. 825.

Order GASTORNITHES.

Cope, E. D. 1889 R, p. 870.
    1896 B, p. 85.

Nicholson and Lydekker 1889 A, p. 1228.
Stejneger, L. 1885 A, p. 54.

GASTORNITHIDÆ.

Dollo, L. 1883 F. p. 297. (Gastornis.)
Edwards, A. M. 1868 A, i, p. 165, pls. xxviii, xxix. (Gastornis.)
Fürbringer, M. 1888 A, p. 1178.
Hébert, E. 1855 A. (Gastornis.)
    1855 B. (Gastornis.)

Lydekker, R. 1891 A, p. 357.
Newton, E. T. 1886 A, p. 143. (Gastornis.)
Nicholson and Lydekker 1889 A, p. 1228.
Owen, R. 1856 F, p. 204, pl. iii. (Gastornis.)
Woodward, A. S. 1898 B, p. 241 (Gastornis.)
Zittel, K. A. 1890 A, p. 836.

DIATRYMA Cope.   Type *D. giganteum* Cope.

Cope, E. D. 1876 B.
    1876 A.
    1877 K, p. 69.
Nicholson and Lydekker 1889 A, p. 1229.
Zittel, K. A. 1890 A, p. 829.

Diatryma giganteum Cope.

Cope, E. D. 1876 B.
    1877 K, p. 70, pl. xxxii, figs. 23–25.
Coues, E. 1884 A, p. 825. (Gastornis.)
Stejneger, L. 1885 A, p. 55.
Eocene; New Mexico.

BARORNIS Marsh.   Type *B. regens* Marsh.

Marsh, O. C. 1894 J.

Barornis regens Marsh.

Marsh, O. C. 1894 J.
Eocene; New Jersey.

Superorder *EUORNITHES* Stejneger.

Stejneger, L. 1885 A, p. 64.
This group is equivalent to that usually called *Carinatæ*, with the tinamous and penguins removed. Unless otherwise indicated, the following authors employ the term *Carinatæ*.
Cope, E. D. 1889 R, p. 870.
1891 N, p. 55. (Euornithes.)
1898 B, pp. 85, 94. (Euornithes.)
Fürbringer, M. 1885 A.
1888 A, p. 1141.
Gadow, H. 1888 A.
1891 A.
1892 A, p. 238.
1898 A, p. 31.
Haeckel, E. 1895 A, p. 413.
Huxley, T. H. 1867 A, p. 418.
1872 A, p. 234.

Lydekker, R. 1891 A, p. 1 (Carinatæ); p. 2 (Euornithes).
Menzbier, M. 1887 A.
Newton, A. 1868 A.
Nicholson and Lydekker 1889 A, p. 1229.
Parker, W. K. 1870 A.
1875 A.
Pavlow, A. 1885 A, p. 113.
Selenka, E. 1869 A.
Shufeldt, R. W. 1885 A.
1886 A.
1886 B.
Steinmann and Döderlein 1890 A, p. 673.
Woodward, A. S. 1886 C, p. 370.
Zittel, K. A. 1890 A, p. 834.

Order CECOMORPHÆ Huxley.

Huxley, T. H. 1867 A.
Cope, E. D. 1889 R, p. 871.
1898 B, p. 95.
Coues, E. 1884 A. p. 732 (Longipennes); p. 787 (Pygopodes).
Gadow, H. 1898 A, p. 31. (Colymbiformes.)
Huxley, T. H. 1872 A, p. 234.
Mivart, St. G. 1892 A, p. 276 (Pygopodiformes); p. 275 (Gaviæ); p. 276 (Tubinares).

Shufeldt, R. W. 1890 B. (Pygopodes.)
1890 C. (Pygopodes.)
1891 C. (Pygopodes.)
1892 C. (Pygopodes.)
Stejneger, L. 1885 A, p. 64.
Zittel, K. A. 1890 A, p. 838 (Podicipitiformes); p. 843 (Tubinares); p. 844 (Charadriiformes, part).

Superfamily *COLYMBOIDEA* Stejneger.

Stejneger, L. 1885 A, p. 66. (Colymboideæ.) | Shufeldt, R. W. 1892 C. (Podicipoideæ.)

COLYMBIDÆ.

Beddard, F. E. 1897 A, p. 468. (Colymbi.)
Coues E, 1886 A. (Colymbus, but not as here recognized.)
1884 A, p. 789 (Colymbidæ); p. 792 (Podicipedidæ).
Edwards, A. M. 1868 A, p. 278. (Colymbides.)
Fürbringer, M. 1888 A, p. 1153. (Colymbidæ, Podicipidæ.)
1890 B. (Colymbidæ, Podicipidæ.)

Fürbringer, M. 1896 A, p. 508. (Podicipidæ.)
Gadow, H. 1891 A.
Magnus, H. 1869 A.
Parker, W. K. 1868 A, p. 148. (Podicipinæ.)
Reichenow, A. 1871 A.
Selenka, E. 1869 A.
Shufeldt, R. W. 1890 A. (Podicipidæ.)
1892 C. (Podicipidæ.)
Stejneger, L. 1885 A, p. 66.

COLYMBUS Linn. Type *C. cristatus* Linn.

Beddard, F. E. 1897 A, p. 468.
Brandt, J. F. 1840 A, p. 161.
Edwards, A. M. 1868 A, p. 278.
Gadow, H. 1891 A, pl. ix, figs. 4–6; pl. xiii, fig. 6. (Podiceps.)
Giebel, C. G. 1866 A, p. 29. (On Podiceps.)
Lydekker, R. 1891 C.
Magnus, H. 1869 A. (On Podiceps.)
Owen, R. 1856 F, p. 212.
Shufeldt, R. W. 1890 A.
1892 C.
1894 B.

Colymbus auritus Linn.

Brandt, J. F. 1840 A, p. 204, pl. xiv. (Podiceps.)
Shufeldt, R. W. 1891 B, p. 818.
1892 D. p. 396.
1897 A, p. 646.
Recent; Northern Hemisphere: Pleistocene; Oregon, Tennessee.

Colymbus holbœlii (Reinh.).

Shufeldt, R. W. 1891 B, p. 818.
1892 D, p. 396.
Recent; North America: Pleistocene (Equus beds); Oregon.

Colymbus nigricollis (Brehm.).

Cope, E. D. 1878 H, p. 389. (Podiceps "near californicus.")
1889 F. (Podiceps californicus).
1889 S, p. 980. (Podiceps californicus.)
Shufeldt, R. W. 1890 A, p. 170, text figs. 1, 2.
1891 B, p. 818. (C. n. californicus.)
1892 D, p. 396. (C. n. californicus.)
Recent; Western North America: Pleistocene (Equus beds); Oregon.

524 FOSSIL VERTEBRATA OF NORTH AMERICA. [BULL. 179.

ÆCHMOPHORUS Coues. Type *Podiceps occidentalis* Lawr.

Coues, E. 1862, Proc. Acad. Nat. Sci. Phila., p. 229.
 1884 A, p. 793.

Æchmophorus occidentalis (Lawr.).

Cope, E. D. 1878 H, p. 389. (Podiceps.)
 1889 F. (Podiceps.)

Cope, E. D. 1889 S, pp. 978, 980. (Podiceps.)
Coues, E. 1884 A, p. 793.
Shufeldt, R. W. 1891 B, p. 818.
 1892 D, p. 396.
Recent; Western North America: Pleistocene;
 Oregon.

PODILYMBUS Lesson. Type *Colymbus podiceps* Linn.

Lesson, 1831, Traité d'Orn., p. 595.
Coues, E. 1884 A, p. 796.
Shufeldt, R. W. 1890, A, p. 173.

Podilymbus podiceps (Linn.).

Cope, E. D. 1878 H, p. 389.
 1889 F

Cope, E. D. 1889 S, p. 980.
Coues, E. 1884 A, p. 797. (P. podicipes.)
Shufeldt, R. W. 1891 B, p. 818.
 1892 D, p. 396.
Recent; North America: Pleistocene; Oregon.

Superfamily ALCOIDEA Stejneger.

Stejneger, L. 1885 A, p. 68. (Alcoideæ.)

ALCIDÆ.

Blyth, E. 1837 A.
Coues, E. 1884 A, p. 797.
Edwards, A. M. 1868 A, p. 278.
Fürbringer, M. 1888 A, p. 1148.
Gadow, H. 1891 A.
Huxley, T. H. 1867 A, p. 458.
Owen, R. 1856 F, p. 212. (Alca.)
Parker, W. K. 1868 A, p. 146. (Alcinæ.)

Reichenow, A. 1871 A.
Selenka, E. 1869 A.
Shufeldt, R. W. 1888 A.
 1889 B.
 1889 C.
 1890 A.
 1891 C.

PLAUTUS Brünnich. Type *Alca impennis* Linn.

Brünnich, 1772, Zool. Fundamenta, p. 78.
 The genus *Plautus* is included by many
 authors in *Alca;* hence references are given to
 the literature of the last-named genus.
Coues, E. 1884 A, p. 819. (Alca.)
Giebel, C. G. 1866 A, p. 29. (Alca.)
Lucas, F. A. 1888 A.
Menzbier, M. 1887 A.
Newton, A. 1875 A, p. 733. (Alca.)
Owen, R. 1856 F, p. 212. (Alca.)
Shufeldt, R. W. 1888 A. (Alca, Plautus.)
 1889 B. (Alca.)
 1889 C. (Alca.)

Plautus impennis (Linn.).

American Naturalist 1869 A. (Alca.)
Blasius, W. 1884 A. (Alca.)
Blyth, E. 1837 A. (Alca.)
Champley, R. 1864 A. (Alca.)
Coues, E. 1869 A. (Alca.)

Coues, E. 1884 A, p. 819, fig. 561. (Alca.)
Lucas, F. A. 1888 A.
 1890 C, p. 493, pls. lxxii, lxxiii. ("Great
 auk.")
Menzbier, M. 1887 A.
Newton, A. 1861 A. (Alca.)
 1870 A. (Alca.)
 1875 A, p. 731. (Alca.)
 1879 A. (Alca.)
Orton, J. 1869 A. (Alca.)
Owen, R. 1864 A. (Alca.)
 1866 C, pls. i–iii. (Alca.)
Preyer, W. 1862 A.
Selenka, E. 1869 A, pl. ix, figs. 1, 2. (Alca.)
Shufeldt, R. W. 1888 A, pl. v, figs. 24, 25.
Steenstrup, J. 1855 A, plate.
Stejneger, L. 1885 A, p. 72, fig. 32.
Wyman, J. 1867 A.
 Recently extinct; once an inhabitant of
 Newfoundland, Iceland, Denmark, etc.

URIA Brisson. Type *Colymbus troile* Linn.

Coues, E. 1884 A, p. 814 (Uria); p. 816 (Lomvia).
Selenka, E. 1869 A, pl. ix, fig. 7; pl. lvii, fig. 10.
Shufeldt, R. W. 1889 B, p. 165, pl. vii; p. 537, text
 figs 1, 2, 4, 5, 6, 8.
 1889 C.

Uria affinis (Marsh).

Marsh, O. C. 1872 J, p. 259. (Catarractes.)
Coues, E. 1872 B, p. 350 (Catarractes.)
 1884 A, p. 825. (Lomvia.)

Marsh, O. C. 1870 A, p. 214. (No name.)
 Pleistocene; New Jersey.

Uria antiqua (Marsh).

Marsh, O. C. 1870 A, p. 213. (Catarractes.)
Cope, E. D. 1869 M (1870), p. 237. (Catarbactes.)
 1875 V, p. 41. (Catarbactes.)
Coues, E. 1872 B, p. 350. (Catarractes.)
 1884 A, p. 825. (Lomvia.)
 Miocene; North Carolina.

Superfamily LAROIDEA Stejneger.

Stejneger, L. 1885 A, p. 74. (Laroideæ.)   ;

LARIDÆ.

Brandt, J. F. 1840 A, p. 165.
Coues, E. 1884 A, p. 733.
Edwards, A. M. 1868 A, p. 302.
Fürbringer, M. 1888 A, p. 1158.
Gadow, H. 1891 A.
    1896 A, p. 35. (Lari.)
Parker, W. K. 1868 A, p. 153. (Larinæ.)

Reichenow, A. 1871 A.
Selenka, E. 1869 A.
Shufeldt, R. W. 1890 A.
    1890 B.
    1891 C.
Stejneger, L. 1885 A, p. 77.
Zittel, K. A, 1890 A, p. 845.

LARUS Linnæus.   Type *Larus canus* Linn.

Coues, E. 1884 A, p. 740.
Edwards, A. M. 1868 A, p. 305.
Huxley, T. H. 1867 A, p. 430, fig. 11.
Owen, R. 1856 F, p. 212.
Selenka, E. 1869 A, pl. xv, fig. 12; pl. xvii, figs. 12, 16.
Shufeldt, R. W. 1888 A, pl. ii, figs. 2-12; pl. iv, fig. 21.
    1889 B, pl. vii, fig. 6; pl. ix, fig. 12; pl. x, fig. 18; pl. xi, fig. 26.
    1890 A, p. 545, figures.
    1890 B, p. 66, with figures.
    1891 C.

Larus argentatus Brünn.
Coues, E. 1884 A, p. 743.
Shufeldt, R. W. 1891 B, p. 818. (L. a. smithsonianus.)
    1892 D, p. 397. (L. a. smithsonianus.)
    Recent; North America: Pleistocene; Oregon.

Larus californicus? Lawr.
Coues, E. 1884 A, p. 745.

Shufeldt, R. W. 1891 B, p. 818.
    1892 D, p. 398.
    Recent; Western North America: Pleistocene; Oregon.

Larus oregonus Shufeldt.
Shufeldt, R. W. 1892 D, p. 398, pl. xv, fig. 3.
    1891 B, p. 818. (No description.)
    Pleistocene; Oregon.

Larus philadelphia (Ord).
Coues, E. 1884 A, p. 751. (Chroicocephalus.)
Shufeldt, R. W. 1891 B, p. 818.
    1892 D, p. 399.
    Recent; North America: Pleistocene; Oregon

Larus robustus Shufeldt.
Shufeldt, R. W. 1891 B, p. 818.
    1892 D, p. 398, pl. xv, figs. 1, 2.
    Pleistocene; Oregon.

XEMA Leach.   Type *Larus sabinii* (Sab.).

Coues, E. 1884 A, p. 753.
Shufeldt, R. W. 1893 A.

Xema sabinii (Sab.).
Coues, E. 1885 A, p. 753.

Shufeldt, R. W. 1891 B, p. 818.
    1892 D, p. 399.
    1893 A.
    Recent; Arctic America and Europe: Pleistocene; Oregon.

HYDROCHELIDON Boie.   Type *Sterna nigra* Linn.

Coues, E. 1884 A, p. 770.

Hydrochelidon nigra (Linn.).
Coues, E. 1884 A, p. 770. (H. lariformis.)

Shufeldt, R. W. 1891 B, p. 818. (H. n. surinamensis.)
    1892 D, p. 399. (H. n. surinamensis.)
    Recent; Old and New Worlds: Pleistocene; Oregon.

STERNA Linn.   Type *Sterna caspia* Pall.

Brandt, J. F. 1840 A, p. 218.
Coues, E. 1884 A, p. 756.
Selenka, E. 1869 A, pl. viii, fig. 13.
Shufeldt, R. W. 1893 A.

Sterna elegans? Gamb.
Coues, E. 1884 A, p. 760.
Shufeldt, R. W. 1891 B, p. 818.
    1892 D, p. 399.

    Recent; New England to California and Brazil: Pleistocene; Oregon.

Sterna forsteri? Nutt.
Coues, E. 1884 A, p. 763.
Shufeldt, R. W. 1891 B, p. 818. (S. fosteri.)
    1892 D, p. 399. (S. fosteri.)
    Recent; North America: Pleistocene; Oregon.

Superfamily PROCELLAROIDEA Stejneger.

Stejneger, L. 1885 A, p. 84.   (Procellaroideæ.)
Beddard, F. E. 1897 A, p. 468.
Coues, E. 1884 A, p. 773.   (Tubinares.)
Fürbringer, M. 1888 A, p. 1162.   (Tubinares.)
Gadow, H. 1898 A, p. 32.   (Procellariæ.)
Mivart, St. G. 1892 A, p. 276.   (Tubinares.)

Nicholson and Lydekker 1889 A, p. 1232.   (Tubinares.)
Pycraft, W. P. 1899 A, p. 381.   (Tubinares.)
Shufeldt, R. W. 1888 E.   (Tubinares.)
Zittel, K. A. 1890 A, p. 843.   (Tubinares.)

PROCELLARIIDÆ.

Coues, E. 1884 A, p. 773.
Fürbringer, M. 1888 A, p. 1162.

Pycraft, W. P. 1899 A, p. 402.
Stejneger, L. 1885 A, p. 88.

PUFFINUS Brisson.   Type *Procellaria puffinus* Brünn.

Coues, E. 1884 A, p. 783.
Pycraft, W. P. 1899 A, p. 382.
Selenka, E. 1869 A, pl. xvii, fig. 15.

Puffinus conradi Marsh.

Marsh, O. C. 1870 A, p. 212.

Cope, E. D. 1869 M (1870), p. 237.
Coues, E. 1872 B, p. 350.
      1884 A, p. 825.
Marsh, O. C. 1870 G.
Miocene; Maryland.

Order GRALLÆ.

Beddard, F. E. 1897 A, p. 469.
Cope, E. D. 1898 B, p. 89.
Coues, E. 1884 A, p. 596 (Limicolæ); p. 665 (Alectorides).
Fürbringer, M. 1888 A, p. 1235 (Fulicariæ); p. 1420 (Limicolæ).
Gadow, H. 1898 A, p. 35.   (Charadriiformes.)
Haeckel, E. 1895 A, p. 414.   (Charadriornithes.)
Huxley, T. H. 1867 A, p. 457.   (Charadriomorphæ, Geranomorphæ.)
      1872 A, p. 234.   (Charadriomorphæ, Geranomorphæ.)

Mivart, St. G. 1892 A, p. 273 (Alectorides); p. 275 (Limicolæ).
Owen, R. 1856 F, p. 208.
Parker, W. K. 1868 A, p. 163.
Selenka, E. 1869 A.   (Limicolæ, etc.)
Shufeldt, R. W. 1883 A.   (> Limicolæ.)
      1884 A.   (> Limicolæ.)
      1894 A.
Stejneger, L. 1885 A, p. 91.

Superfamily SCOLOPACOIDEA Stejneger.

Stejneger, L. 1885 A, p. 94.   (Scolopacoideæ.)
Gadow, H. 1898 A, p. 35.   (Limicolæ.)

Huxley, T. H. 1867 A, p. 457.   (Charadriomorphæ.)

CHARADRIIDÆ.

Coues, E. 1885 A, p. 597.
Edwards, A. M. 1868 A, p. 367.   (Totanides, in part.)
Fürbringer, M. 1888 A.

Selenka, E. 1869 A.
Stejneger, L. 1885 A, p. 98.
Zittel, K. A. 1890 A, p. 844.

CHARADRIUS Linn.   Type *Tringa squatarola* Linn.

Cope, E. D. 1884 O, p. 754.
Coues, E. 1884 A, p. 599.
Edwards, A. M. 1868 A, p. 375.
Huxley, T. H. 1867 A, p. 427, figs. 6–8.
Shufeldt, R. W. 1888 A, pl. ii, fig. 10.
      1890 A, pl. xii, figs. 5, 6.
      1891 C.

Charadrius sheppardianus Cope.

Cope, E. D. 1881 B.
      1884 A, p. 823.
      1884 O, p. 755, pl. lix, fig. 5.
Coues, E. 1884 A, p. 823.
Zittel, K. A. 1890 A, p. 844.
Eocene; Colorado.

SCOLOPACIDÆ.

Coues, E. 1884 A, p. 612 (Phalaropodidæ); p.614 (Scolopacidæ).
Edwards, A. M. 1868 A, p. 367.   (Totanides, part.)
Fürbringer, M. 1888 A.
Gadow, H. 1891 A.
Parker, W. K. 1868 A, p. 155.   (Scolopacinæ.)

Reichenow, A. 1871 A.   (Scolopacidæ, Phalaropidæ.)
Shufeldt, R. W. 1884 A.   (Numenius.)
Stejneger, L. 1885 A, p. 105.
Zittel, K. A. 1890 A, p. 844.

PALÆOTRINGA Marsh.   Type *P. littoralis* Marsh.

Marsh, O. C. 1870 A, p. 208.
    1870 G.
Zittel, K. A. 1890 A, p. 844.

Palæotringa littoralis Marsh.

Marsh, O. C. 1870 A, p. 208.
Cope, E. D. 1869 M (1870), p. 238.
    1875 E, p. 246.
Coues, E. 1872 B, p. 349.
    1884 A, p. 828.   (P. littoralis.)
Marsh, O. C. 1870 G.
    1873 D, p. 229.
    1880 G, p. 199.
    Cretaceous; New Jersey.

Palæotringa vagans Marsh.

Marsh, O. C. 1872 E, p. 365.
Coues, E. 1872 B, p. 349.
    1884 A, p. 828.

Marsh, O. C. 1873 D, p. 229.
    1880 G, p. 200.
    Cretaceous; New Jersey.

Palæotringa vetus Marsh.

Marsh, O. C. 1870 A, p. 209.
Cope, E. D. 1869 M (1870), p. 238.
    1875 E, p. 246.
Coues, E. 1872 B, p. 349.
    1884 A, p. 828.
Harlan, R. 1835 B, p. 280.   ("Scolopax.")
Leidy, J. 1851 H, p. 330.   ("Scolopax.")
    1865 A, p. 2.   ("Scolopax.")
    1866 B, p. 237.   ("Scolopax.")
Marsh, O. C. 1870 G.
    1873 E, p. 229.
    1880 G, p. 200.
Morton, S. G. 1834 A, p. 32.   (No name.)
    Cretaceous; New Jersey.

PHALAROPUS Brisson.   Type *Tringa lobata* Linn.

Coues, E. 1884 A, p. 612 (Steganopus); p. 614 (Lobipes).
Edwards, A. M. 1868 A, p. 375.

Phalaropus lobatus (Linn.)

Coues, E. 1884 A, p. 613.   (Lobipes hyperboreus.)

Shufeldt, R. W. 1891 B, p. 820.
    1892 D, p. 413.
    Recent; Northern Hemisphere: Pleistocene;
    Oregon.

Superfamily *GRUIOIDEA Stejneger.*

Stejneger L. 1885 A, p. 121.   (Gruioideæ.)
Beddard, F. E. 1897 A, p. 469.   (Grues.)
Gadow, H. 1898 A, p. 34.   (Gruiformes.)

Huxley, T. H. 1867 A, p. 457.   (Geranomorphæ.)
Selenka, E. 1869 A.
Shufeldt, R. W. 1894 A.   (Gruioidea.)

GRUIDÆ.

Coues, E. 1884 A, p. 666.
Edwards, A. M. 1871 A, p. 1.   (Gruides.)
Fürbringer, M. 1888 A, p. 1203.
Gadow, H. 1891 A.
Owen, R. 1856 F, p. 213.

Parker, W. K. 1868 A, p. 158.   (Gruinæ.)
Selenka, E. 1869 A.
Shufeldt, R. W. 1894 A.
Stejneger, L. 1885 A, p. 123.
Zittel, K. A. 1890 A, p. 846.

GRUS Pallas.   Type *Ardea grus* Linn.

Coues, E. 1884 A, p. 666.
Edwards, A. M. 1871 A, p. 1.
Huxley, T. H. 1867 A, p. 426, fig. 9.
Owen, R. 1856 F, p. 209.
Parker, W. K. 1868 A, p. 158.
Selenka, E. 1869 A, pl. x, fig. 8, pl. xi, fig. 14.
Shufeldt, R. W. 1894 A.

Grus haydeni Marsh.

Marsh, O. C. 1870 A, p. 214.
Cope, E. D. 1869 M (1870), p. 237.

Coues, E. 1872 B, p. 348.
    1884 A, p. 823.
King, C. 1878 A, p. 430.
Marsh, O. C. 1870 G.
    Pliocene (?); Nebraska.

Grus proavus Marsh.

Marsh, O. C. 1872 J, p. 261.
Coues, E. 1872 B, p. 348.
    1884 A, p. 823.
    Pleistocene; New Jersey.

ALETORNIS Marsh.   Type *A. nobilis* Marsh.

Marsh, O. C. 1872 J, p. 256.
Fürbringer, M. 1888 A.
Zittel, K. A. 1890 A, p. 846.

Aletornis bellus Marsh.

Marsh, O. C. 1872 J, p. 256.
Coues, E. 1872 B, p. 349.
    1884 A, p. 824.

King, C. 1878 A, p. 404.
    Eocene (Bridger); Wyoming.

Aletornis gracilis Marsh.

Marsh O. C. 1872 J, p. 258.
Coues, E. 1872 B. p. 348.
    1884 A, p. 824.
King, C. 1878 A, p. 404.
    Eocene (Bridger); Wyoming.

Aletornis nobilis Marsh.

Marsh, O. C. 1872 J, p. 256.
Coues, E. 1872 B, p. 348.
1884 A, p. 823.
King, C· 1878 A, p. 404.
Eocene (Bridger); Wyoming.

Aletornis pernix Marsh.

Marsh, O. C. 1872 J, p. 256.
Coues, E. 1872 B, p. 348.

Coues, E. 1884 A, p. 824.
King, C. 1878 A, p. 404.
Eocene (Bridger); Wyoming.

Aletornis venustus Marsh.

Marsh, O. C. 1872 J, p. 257.
Coues, E. 1872 B, p. 348.
1884 A, p. 824.
King, C. 1878 A, p. 404.
Eocene (Bridger); Wyoming.

RALLIDÆ.

Coues, E. 1884 A, p. 669.
Edwards, A. M. 1871 A, p. 110.
Fürbringer, M. 1888 A, p. 1235.
Gadow, H. 1891 A.
Parker, W. K. 1868 A, p. 161. (Rallinæ.)

Selenka E. 1869 A.
Shufeldt, R. W. 1888 B. (On Porzana.)
1894 A.
Stejneger, L. 1885 A, p. 127.
Zittel, K. A. 1890 A, p. 846.

CRECCOIDES Shufeldt.   Type *C. osbornii* Shufeldt.

Shufeldt, R. W. 1892 B, p. 253.
1892 D, p. 411, footnote.   (Creccoides.)
1892, in Cope, E. D. 1892 J, p. 125.

Creccoides osbornii Shufeldt.

Shufeldt, R. W. 1892 B, p. 253.

Shufeldt, R. W. 1892, in Cope E. D. 1892 J, p. 125.
Pleistocene; Texas.

TELMATORNIS Marsh.   Type *T. priscus* Marsh.

Marsh, O. C. 1870 A, p. 210.
Fürbringer. M. 1888 A, p. 1235.
Marsh, O. C. 1870 G.
Zittel, K. A. 1890 A, p. 846.

Telmatornis affinis Marsh.

Marsh, O. C. 1870 A, p. 211.
Cope, E. D. 1869 M (1870), p. 238.
1875 E, p. 246.
Coues, E. 1872 B, p. 349.
1884 A, p. 829.
Marsh, O. C. 1870 G.
1873 D, p. 229.

Marsh, O. C. 1880 G, p. 201.
Cretaceous; New Jersey.

Telmatornis priscus Marsh.

Marsh, O. C. 1870 A, p. 210.
Cope, E. D. 1869 M (1870), p. 238.
1875 E, p. 246.
Coues, E. 1872 B, p. 349.
1884 A, p. 829.
Marsh, O. C. 1870 G.
1873 D, p. 229.
1880 G, p. 200.
Cretaceous; New Jersey.

FULICA Linn.   Type *F. atra* Linn.

Coues, E. 1884 A, p. 676.
Fürbringer, M. 1888 A, p. 1235.
Zittel, K. A. 1890 A, p. 846.

Fulica americana Gmel.

Cope, E. D. 1889 F.
1889 S, p. 940.
Coues, E. 1884 A, p. 676.

Shufeldt, R. W. 1891 B, p. 819.
1892 D, p. 411.
Recent; North America: Pleistocene; Oregon.

Fulica minor Shufeldt.

Shufeldt, R. W. 1892 D, p. 412, pl. xvii, fig. 32.
1891 B, p. 820. (No description).
Pleistocene; Oregon.

Order CHENOMORPHÆ Huxley.

Huxley, T. H. 1867 A, p. 460.
Cope, E. D. 1898 B, p. 88.
Coues, E. 1884 A, p. 677. (Lamellirostres.)
Edwards, A. M. 1868 A, I, p. 127. (Lamellirostres.)
Fürbringer, M. 1888 A, p. 1173 (Anseres); p. 1184
(Phœnicopteridæ.)
Gadow, H. 1891 A. (Lamellirostres.)
1898 A, p. 33. (Anseriformes.)

Huxley, T. H. 1872 A, p. 234.   (Chenomorphæ.)
Mivart, St. G. 1892 A, p. 277.   (Lamellirostres.)
Owen, R. 1856 F, p. 211.   (Natatores.)
Selenka, E. 1869 A.   (Lamellirostres.)
Shufeldt, R. W. 1889 A.   (Anseres.)
Stejneger, L. 1885 A, p. 132.
Zittel, K A. 1890 A, p. 836 (Anseriformes): p. 839
(Ciconiiformes).

Superfamily *ANATOIDEA* Stejneger.

Stejneger, L. 1885 A, p. 136. (Anatoideæ.)
Gadow, H. 1898 A, p. 33. (Anseres.)
Giebel, C. G. 1866 A, p. 28. (Anseres.)

Magnus, H. 1869 A. (Natatores.)
Reichenow, A. 1871 A. (Anseres.)
Shufeldt, R. W. 1889 A. (Anseres.)

ANATIDÆ.

Beddard, F. E. 1897 A, p. 466.
Coues, E. 1884 A, p. 679.
Gadow, H. 1891 A.
Owen, R. 1856 F, p. 213.

Parker, W. K. 1868 A, p. 152. (Anatinæ.)
Selenka, E. 1869 A.
Shufeldt, R. W. 1889 A.

LOPHODYTES Reichenbach.   Type *Mergus cucullatus* Linn.

Coues, E. 1884 A, p. 716.

Lophodytes cucullatus (Linn.).

Coues, E. 1884 A, p. 718. (Mergus.)

Shufeldt, R. W. 1891 B, p. 818.
    1892 D, p. 402.
Recent; North America: Pleistocene; Oregon.

ANAS Linn.   Type *A. boschas* Linn.

Coues, E. 1884 A, p. 691.
Gadow, H. 1891 A, pl. iv, fig. 2.
Giebel, C. G. 1866 A, p. 28.
Selenka, E. 1869 A.
Shufeldt, R. W. 1889 A.
Zittel, K. A. 1890 A, p. 637.

Anas americana Gmel.

Coues, E. 1884 A, p. 694. (Mareca.)
Shufeldt, R. W. 1891 B, p. 818.
    1892 D, p. 403.
Recent; North America: Pleistocene; Oregon.

Anas boschas Linn.

Coues, E. 1884 A, p. 691. (A. boscas.)
Selenka, E. 1869 A, pl. ii, fig. 9.
Shufeldt, R. W. 1889 A, p. 283.
    1891 B, p. 818.
    1892 D, p. 403.
Recent; Northern Hemisphere: Pleistocene;
Oregon.

Anas carolinensis Gmel.

Coues, E. 1884 A, p. 696. (Querquedula.)
Shufeldt, R. W. 1891 B, p. 818.
    1892 D, p. 403.
Recent; North America: Pleistocene; Oregon.

Anas cyanoptera ? Vieill.

Coues, E. 1884 A, p. 696. (Querquedula.)
Shufeldt, R. W. 1891 D, p. 818.
    1892 D, p. 404.
Recent; South America, Western North
America: Pleistocene; Oregon.

Anas discors Linn.

Coues, E. 1884 A, p. 696. (Querquedula.)
Shufeldt, R. W. 1889 A.
    1891 B, p. 818.
    1892 D, p. 403.
Recent; North America: Pleistocene; Oregon.

SPATULA Boie.   Type *Anas clypeata* Linn.

Coues, E. 1884 A, p. 696.
Shufeldt, R. W. 1889 A, p. 230.
Zittel, K. A. 1890 A, p. 838.

Spatula clypeata (Linn.).

Coues, E. 1884 A, p. 696.

Selenka, E. 1869 A, pl. x, fig. 5. (Anas.)
Shufeldt, R. W. 1889 A, p. 230, figs. 15, 16, 20, 22–25.
    1891 B, p. 819.
    1892 D, p. 404.
Recent; North America, Europe, Asia: Pleistocene; Oregon.

DAFILA Stephens.

Coues, E. 1884 A, p. 692.

Dafila acuta (Linn.).

Coues, E. 1884 A, p. 692.

Shufeldt, R. W. 1891 B, p. 819.
    1892 D, p. 404.
Recent; Northern Hemisphere: Pleistocene;
Oregon.

AIX Boie.   Type *Anas sponsa* Linn.

Coues, E. 1884 A, p. 697.

Aix sponsa (Linn.).

Coues, E. 1884 A, p. 698.

Shufeldt, R. W. 1891 B, p. 819.
    1892 D, p. 405.
Recent; North America: Pleistocene; Oregon.

ATHYA Boie.   Type *Anas ferina* Linn.

Coues, E. 1884 A, p. 701.   (Fuligula.)

Athya marila.

Coues, E. 1884 A, p. 701.   (Fuligula.)

Shufeldt, R. W. 1891 B, p. 819.
1892 D, p. 406.
Recent; North America: Pleistocene; Oregon.

GLAUCIONETTA Stejneger.   Type *Anas clangula* Linn.

Coues, E. 1884 A, p. 704.   (<Clangula.)
Shufeldt, R. W. 1889 A, p. 235.

Glaucionetta islandica (Gmel.).

Coues, E. 1884 A, p. 704.   (Clangula.)

Shufeldt, R. W. 1889 A, p. 235, figs. 19, 21, 26.
1891 B, p. 819.
1892 D, p. 406.
Recent; North America and Northern Europe:
Pleistocene; Oregon.

CLANGULA Leach.   Type *Anas glacialis* Linn.

Coues, E. 1884 A, p. 704 (Clangula); p. 706 (Harelda).

Clangula hyemalis (Linn.).

Coues, E. 1884 A, p. 706.   (Harelda glacialis.)

Shufeldt, R. W. 1891 B, p. 819.
1892 D, p. 406.
Recent; Northern Hemisphere: Pleistocene;
Oregon.

CHEN Boie.

Coues, E. 1884 A, p. 685.

Chen hyperborea (Pall.).

Coues, E. 1884 A, p. 685.   (C. hyperboreus.)

Shufeldt, R. W. 1891 B, p. 819.
1892 D, p. 409.
Recent; North America: Pleistocene; Oregon.

ANSER Brisson.   Type *Anas anser* Linn.

Coues, E. 1884 A, p. 684.
Owen, R. 1856 F, p. 211.
Selenka, E. 1869 A, pl. xvii, fig. 10.
Shufeldt, R. W. 1889 A.

Anser albifrons (Gmel.).

Cope, E. D. 1878 H, p. 389.
1889 F.
1889 S, p. 980.

Coues, E. 1884 A, p. 684.
Shufeldt, R. W. 1891 B, p. 819.
1892 D p. 406.
Recent, Northern regions: Pleistocene; Oregon.

Anser condoni Shufeldt.

Shufeldt, R. W. 1892 D, p. 406, pl. xvi, figs. 19, 26, 27,
1891 A, p. 819.   (Name only.)
Pleistocene; Oregon.

BRANTA Scopoli.   Type *Anas bernicla* Linn.

Coues, E. 1884 A, p. 686.   (Bernicla.)
Shufeldt, R. W. 1889 A.

Branta canadensis (Linn.).

Cope, E. D. 1878 H, p. 389.   (Anser.)
1889 F.   (Anser.)
1889 S, pp. 978, 980.   (Anser.)
Coues, E. 1884 A, p. 688.   (Bernicla.)
Shufeldt, R. W. 1889 A, p. 248, figs. 27-30.
1891 B, p. 819.
1892 D, p. 406.
Recent; North America: Pleistocene; Oregon.

Branta hypsibata (Cope).

Cope, E. D. 1878 H, p. 387.   (Anser.)
1889 F.

Cope, E. D. 1889 S, p. 980.
Coues, E. 1884 A, p. 824.   (Bernicla hypsibates.)
Shufeldt, R. W. 1891 B, p. 819.
1892 D, p. 407.
Zittel, K. A. 1890 A, p. 838.
Pleistocene; Oregon.

Branta propinqua Shufeldt.

Shufeldt, R. W. 1892 D, p. 407, pl. xv. fig. 17.
Cope, E. D. 1878 H, p. 389.   (Anser, "near nigricans.")
1889 F.   (Anser, "near nigricans.")
1889 S, p. 890.   Anser, "near nigricans.")
Shufeldt, R. W. 1891 B, p. 819.
Pleistocene; Oregon.

OLOR Wagler.   Type *Anas cygnus* Linn.

Coues, E. 1884 A, p. 682.   (Cygnus.)
Selenka, E. 1869 A, pl. x, figs. 2, 3.   (Cygnus.)
Shufeldt, R. W. 1889 A, p. 233.

Olor paloregonus Cope.

Cope, E. D. 1878 H, p. 388.   (Cygnus.)

Cope, E. D. 1889 F.   (Cygnus.)
1889 S, pp. 978, 980.   (Cygnus.)
Coues, E. 1884 A, p. 824.   (Cygnus.)
Shufeldt, R. W. 1891 B, p. 819.
1892 D, p. 409, pl. xvi, figs. 18, 21, 26.
Pleistocene; Oregon.

LAORNIS Marsh.   Type *L. edwardsianus* Marsh.

Marsh, O. C. 1870 A, p. 206.
Fürbringer, M. 1888 A, pp. 1141, 1159, 1174.
Marsh, O. C. 1870 G.
Stejneger, L. 1885 A, p. 30.
Zittel, K. A. 1890 A, p. 837.

Laornis edvardsianus Marsh.

Marsh, O. C. 1870 A, p. 206.
Cope, E. D. 1869 M (1870), p. 237.

Cope, E. D. 1875 E, p. 245.
Coues, E. 1872 B, p. 350.
1884 A, p. 828.
Marsh, O. C. 1870 G.
1873 D, p. 230.
1880 G, p. 199.
Cretaceous; New Jersey.

Superfamily PHŒNICOPTEROIDEA Stejneger.

Stejneger, L. 1885 A, p. 153.   (Phœnicopteroideæ.)
Gadow, H. 1898 A, p. 33.   (Phœnicopteri.)

Huxley, T. H. 1867 A, p. 460.   (Amphimorphæ.)
1872 A, p. 234.   (Amphimorphæ.)

PHŒNICOPTERIDÆ.

Coues, E. 1884 A, p. 678.
Edwards, A. M. 1871 A, p. 53.   (Phœnicopterides.)
Fürbringer, M. 1888 A, p. 1184.
Gadow, H. 1891 A.
Parker, W. K. 1868 A, p. 166.   (Phœnicopterinæ.)

Reichenow, A. 1871 A.
Seebohm, H. 1889 A.   (Phœnicopteri.)
Selenka, E. 1869 A.
Stejneger, L. 1885 A, p. 154.
Zittel, K. A. 1890 A, p. 839.   (Phœnicopteri.)

PHŒNICOPTERUS Linn.   Type *P. ruber* Linn.

Coues, E. 1884 A, p. 678.
Edwards, A. M. 1871 A, p. 53.
Fürbringer, M. 1888 A.
Gadow, H. 1877 A.
1891 A, pl. lvii, fig. 20.
Parker, W. K. 1889 B.
Selenka, E. 1869 A.

Phœnicopterus copei Shufeldt.

Shufeldt, R. W. 1892 D, p. 410, pl. xv, fig. 13; pl. xvii, figs. 28, 29, 38.
1891 B, p. 820.
Pleistocene; Oregon.

Order HERODII.

Beddard, F. E. 1897 A, p. 469.   (Herodiones.)
Cope, E. D. 1898 B, p. 88.
Coues, E. 1884 A, p. 647.   (Herodiones.)
Fürbringer, M. 1888 A, p. 1187.   (Pelargo-Herodii.)
Gadow, H. 1891 A.
Huxley, T. H. 1867 A, p. 461.   (Pelargomorphæ.)
1872 A, p. 234.   (Pelargomorphæ.)

Mivart, St. G. 1892 A, p. 272.   (Herodiones.)
Seebohm, H. 1889 A.   (Herodiones.)
Selenka, E. 1869 A.
Stejneger, L. 1885 A, p. 157.
Zittel, K. A. 1890 A, p. 840.   (Pelargo-Herodii.)

Superfamily ARDEOIDEA Stejneger.

Stejneger, L. 1885 A, p. 162.   (Ardeoideæ.)

Gadow, H. 1898 A, p. 32.   (Ardeæ.)

ARDEIDÆ.

Coues, E. 1884 A, p. 654.
Edwards, A. M. 1871 A, p. 88.   ("Ardéides.")
Fürbringer, M. 1888 A.
Gadow, H. 1891 A.
Giebel, C. G. 1866 A, p. 27.
Huxley, T. H. 1867 A, p. 439.

Magnus, H. 1869 A.
Parker, W. K. 1868 A, p. 163.   (Ardeinæ.)
Reichenow, A. 1871 A.
Seebohm, H. 1889 A.
Selenka, E. 1869 A.
Zittel, K. A. 1890 A, p. 841.

ARDEA Linn.   Type *A. cinerea* Linn.

Coues, E. 1884 A, p. 657.
Edwards, A. M. 1871 A, p. 91.
Gadow, H. 1891 A.
Huxley, T. H. 1867 A, p. 437, fig. 19.
Magnus, H. 1869 A.
Owen, R. 1856 F, p. 209.
Selenka, E. 1869 A, pl. xi, fig. 11; pl. xvii, figs. 8, 9.

Zittel, K. A. 1890 A, p. 841.

Ardea paloccidentalis Shufeldt.

Shufeldt, R. W. 1892 D, p. 411, pl. xvii, fig. 3
1891 B, p. 819.   (No description.)
Pleistocene; Oregon.

Order STEGANOPODES.

Beddard, F. E. 1897 A, p. 468.
Brandt, J. F. 1840 A, p. 91.   (Steganopoda.)
Cope, E. D. 1898 B, p. 88.
Coues, E. 1884 A, p. 718.
Fürbringer, M. 1888 A, p. 1168.
Gadow, H. 1891 A.
    1898 A, p. 32.
Giebel, C. G. 1866 A, p. 29.   (Steganopoda.)
Huxley, T. H. 1867 A, p. 461.   (Dysporomorphæ.)

Mivart, St. G. 1892 A, p. 272.
Pycraft, W. P. 1898 A, p. 82.
Seebohm, H. 1889 A.
Selenka, E. 1869 A.
Shufeldt, R. W. 1888 E.
    1894 C.
    1894 D.
Stejneger, L. 1885 A, p. 179.
Zittel, K. A. 1890 A, p. 841.

Superfamily PELECANOIDEA Stejneger.

Stejneger, L. 1885 A, p. 185.   (Pelecanoideæ.)

Shufeldt, R. W. 1899 C.   (Pelecanoidea.)
    1894 D.   (Pelecanoidea.)

PELECANIDÆ.

Coues, E, 1884 A, p. 721.
Edwards, A. M. 1868 A, i, p. 180.
Fürbringer, A. 1891 A.
Gadow, H. 1891 A.
Huxley, T. H. 1867 A, p. 440.
Mivart, St. G. 1878 A.
Parker, W. K. 1868 A, p. 149.   (Pelecaninæ.)

Reichenow, A. 1871 A.
Selenka, E. 1869 A.
Shufeldt, R. W. 1888 E.
    1894 C.
    1894 D.
Zittel, K. A. 1890 A, p. 841.

PELECANUS Linn.   Type *P. onocrotalus* Linn.

Brandt, J. F. 1840 A, p. 141.
Coues, E. 1884 A, p. 722.
Edwards, A. M. 1868 A, i, p. 183.
Gadow, H. 1891 A, pl. lvii, fig. 17.
Huxley, T. H. 1867 A, p. 438, fig. 20.
Magnus, H. 1869 A.
Mivart, St. G. 1878 A.
Owen, R. 1856 F, p. 212.
Pycraft, W. P. 1898 A, pp. 84, 98.
Seebohm, H. 1889 A.
Selenka, E. 1869 A, pl. viii, fig. 7.

Shufeldt, R. W. 1888 E, p. 310, figs. 40–43.
    1894 B.
    1894 C, p. 337.
    1894 D, p. 161.
Zittel, K. A. 1890 A, p. 842.

Pelecanus erythrorhynchos? Gmel.

Coues, F. 1884 A, p. 722.   (P. trachyrhynchus.)
Shufeldt, R. W. 1891 B, p. 818.
    1892 D, p. 401.
Recent; North America: Pleistocene; Oregon.

CYPHORNIS Cope.   Type *C. magnus* Cope.

Cope, E. D. 1894 G.
    The position of this genus is in doubt

Cyphornis magnus Cope.

Cope, E. D. 1894 G, p. 451, pl. xx, figs. 11–14.
    Eocene?; Vancouvers Island.

SULIDÆ.

Coues, E. 1884 A, p. 720.
Fürbringer, M. 1888 A, p. 1172.   (Sulinæ.)
Gadow, H. 1891 A.
Pycraft, W. P. 1898 A, p. 92.
Reichenow, A. 1871 A.

Seebohm, H. 1889 A.
Selenka, E. 1869 A.
Shufeldt, R. W. 1888 E, p. 287.
Stejneger, L. 1885 A, p. 188.
Zittel, K. A. 1890 A, p. 842.

SULA Brisson.   Type *S. bassana* (Linn.).

Brandt, J. F. 1840 A, p. 139.
Coues, E. 1884 A, p. 720.
Edwards, A. M. 1868 A.
Gadow, H. 1891 A.
Mivart, St. G. 1878 A.
Parker, W. K. 1868 A, p. 150.
Pycraft, W. P. 1898 A, p. 82.
Seebohm, H. 1889 A.
Selenka, E. 1869 A, pl. viii, fig. 8.

Shufeldt, R. W. 1888 E, p. 287, figs. 24–33.
    1894 B.

Sula loxostyla Cope.

Cope, E. D. 1869 M (1870), p. 236, fig. 58.
Coues, E. 1872 B, p. 349.
    1884 A, p. 824.
Miocene; North Carolina.

PHALACROCORACIDÆ.

Coues, E. 1884 A, p. 723.
Fürbringer, M. 1888 A, p. 1172. (Phalacrocoracinæ.)
Gadow, H. 1891 A.
Pycraft, W. P. 1898 A, p. 92. (Phalacrocoracinæ.)
Reichenow, A. 1871 A.

Seebohm, H. 1889 A.
Selenka, E. 1869 A.
Shufeldt, R. W. 1888 E, p. 310.
1894 C, p. 337.
Stejneger, L. 1885 A, p. 190.
Zittel, K. A. 1890 A, p. 842.

PHALACROCORAX Brisson.   Type *Pelecanus carbo* Linn.

Coues, E. 1884 A, p. 726.
Edwards, A. M. 1868 A, p. 186.
Gadow, H. 1891 A.
Huxley, T. H. 1867 A, p. 439, fig. 21.
Jeffries, J. A. 1883 A.
1883 B.
1884 A.
Parker, W. K. 1868 A, p. 149.
Pycraft, W. P. 1898 A, p. 88.
Seebohm, H. 1889 A.
Selenka, E, 1869 A, pl. viii, fig. 5.
Shufeldt, R. W. 1883 B.
1888 E, p. 309, fig. 39.
1894 D, p. 161.
Zittel, K. A. 1890 A, p. 842.

King, C. 1878 A, p. 443. (G. idahoensis.)
Zittel, K. A. 1890 A, p. 842. (P. idahoensis.)
Pleistocene; Idaho.

Phalacrocorax macropus Cope.

Cope, E. D. 1878 H, p. 386. (Graculus.)
1889 F. (Graculus.)
1889 S, p. 980. (Graculus.)
Coues, E. 1884 A, p. 824.
Shufeldt, R. W. 1891 B, p. 818.
1892 D, p. 399, pl. xv, figs. 6–9.
Pleistocene; Oregon.

Phalacrocorax idahensis Marsh.

Marsh, O. C. 1870 A, p. 216. (Graculus.)
Cope, E. D. 1869 M (1870), p. 287. (Graculus.)
Coues, E. 1872 B, p. 349. (Graculus.)
1884 A, p. 824.

Phalacrocorax perspicillatus Pallas.

Coues, E. 1884 A, p. 728.
Lucas, F. A. 1890 B, p. 88, pls. ii-iv.
1896 A.
Stejneger, L. 1885 A, p. 191, fig. 92.
Stejneger and Lucas 1890 A, pls. ii-iv.
Recently extinct; Bering Island.

GRACULAVUS Marsh.   Type *G. velox* Marsh.

Marsh, O. C. 1872 E, p. 363.
Fürbringer M. 1888 A, p. 1168.
Stejneger, L. 1885 A, p. 30.
Zittel, K. A. 1890 A, p. 840.
The systematic position of this genus is doubtful.

Marsh, O. C. 1873 D, p. 229.
1880 G, p. 195.
Cretaceous; New Jersey.

Graculavus velox Marsh.

Marsh, O. C. 1872 E, p. 363.
Coues, E. 1872 B, p. 349.
1884 A, p. 826.
Marsh, O. C. 1873 D, p. 229.
1880 G, p. 194.
Cretaceous; New Jersey.

Graculavus pumilis Marsh.

Marsh, O. C. 1872 E, p. 364.
Coues, E. 1872 B, p. 350.
1884 A, p. 826.

Order GALLINÆ.

Blanchard, É. 1857 A. (Gallidæ.)
Cope, E. D. 1898 B, p. 89.
Coues, E. 1884 A, p. 571.
Edwards, A. M. 1871 A, p. 161. ("Gallinacés.")
Fürbringer, M. 1888 A, p. 1255. (Gallidæ.)
Gadow, H. 1891 A.
1898 A, p. 33. (Galliformes.)
Huxley, T. H. 1867 A, p. 459. (Alectoromorphæ.)
1872 A, p. 234. (Alectoromorphæ.)

Mivart, St. G. 1892 A, p. 274. (Galliformes.)
Owen, R. 1856 F, p. 8.
1866 B. (Rasores.)
Parker, W. K. 1866 B.
1868 A, p. 182.
1870 A.
Selenka, E. 1869 A. (Rasores.)
Shufeldt, R. W. 1888 C. (On Gallus.)
Zittel, K. A. 1890 A, p. 847. (Galliformes.)

Suborder ALECTOROPODES.

Coues, E. 1884 A, p. 573.
Gadow, H. 1898 A, p. 34. (Galli.)

Haeckel, E. 1895 A, p. 415. (Alectornithes.)

TETRAONIDÆ.

Coues, E. 1884 A, p. 576.
Gadow, H. 1891 A.
Huxley, T. H. 1867 A, p. 432, figs. 14. (Tetrao.)

Parker, W. K. 1866 B.
Selenka, E. 1869 A.
Shufeldt, R. W. 1881 B.

TYMPANUCHUS Gloger. Type *Tetrao cupido* Linn.

Coues, E. 1884 A, p. 583. (Cupidonia.)
Shufeldt, R. W. 1881 B, pls. x–xii. (Cupidonia.)

Tympanuchus pallidicinctus Ridgw.

Coues, E. 1884 A, p. 584. (Cupidonia cupido pallidicincta.)

Shufeldt, R. W. 1891 B, p. 820.
1892 D, p. 418.
Recent; Great Plains: Pleistocene; Oregon.

PEDIOCÆTES Baird. Type *Tetrao phasianellus* Linn.

Coues, E. 1884 A, p. 581. (Pediœcetes.)

Pediocætes lucasii Shufeldt.

Shufeldt, R. W. 1892 D, p. 414, pl. xvii, fig. 30.
1891 B, p. 820. (Name; no description.)
Pleistocene; Oregon.

Pediocætes nanus Shufeldt.

Shufeldt, R. W. 1892 D, p. 414, pl. xvii, fig. 37.
1891 B, p. 820. (Name; no description.)
Pleistocene; Oregon.

Pediocætes phasianellus (Linn.).

Coues E. 1884 A, p. 581.
Shufeldt, R. W. 1891 B, p. 820. (P. p. columbianus.)
1892 D, p. 414. (P. p. columbianus.)
Recent; western and northern America:
Pleistocene; Oregon.

PALÆOTETRIX Shufeldt. Type *P. gillii* Shufeldt.

Shufeldt, R. W. 1892 D, p. 415.
1891 B, p. 820.

Palæotetrix gillii Shufeldt.

Shufeldt, R. W. 1892 D, p. 415, pl. xvii, fig. 34.
1891 B, p. 820.
Pleistocene; Oregon.

PHASIANIDÆ.

Coues, E. 1884 A, p. 576. (>Meleagrididæ.)
Fürbringer, M. 1888 A, p. 1255. (Meleagrinæ.)
Gadow, H. 1891 A.
Huxley, T. H. 1867 A, p. 432.

Mivart, St. G. 1892 A, p. 275.
Reichenow, A. 1871 A.
Selenka, E. 1869 A.

MELEAGRIS Linn. Type *M. gallopavo* Linn.

Linnæus, C., 1758, Syst. Nat., ed. x, p. 98.
Cope, E. D. 1869 M (1870), p. 238.
Coues, E. 1884 A, p. 576.
Edwards, A. M. 1871 A, p. 171.

Meleagris antiquus Marsh.

Marsh, O. C. 1871 E, p. 126.
Coues, E. 1872 B, p. 347.
1884 A, p. 823.
King, C. 1878 A, p. 412.
Marsh, O. C. 1897 C, p. 527. .
Shufeldt, R. W. 1897 A, p. 648.
Oligocene (White River); Colorado.

Meleagris celer Marsh.

Marsh, O. C. 1872 J, p. 261.
Coues, E. 1872 B, p. 348.
1884 A, p. 823.
Shufeldt, R. W. 1897 A, p. 649.
Pleistocene; New Jersey.

Meleagris superbus Cope.

Cope, E. D. 1869 M (1870), p. 229 (M. superbus); p.
ii (M. altus).
Amer. Jour. Science 1871 A.
Cope, E. D. 1873 X, p. 298. (M. altus.)
Coues, E. 1872 B, p. 348. (M. altus.)
1884 A, p. 823. (M. altus.)
Marsh, O. C. 1870 H, p. 317. (M. altus; no description.)
1870 I, p. 11. (No description.)
1872 J, p. 260. (M. altus.)
1873 P, p. ix. (M. altus.)
Shufeldt, R. W. 1897 A, p. 648. (M. altus.)
Pleistocene; New Jersey.

Meleagris sp. indet.

Wheatley, C. M. 1871 A.
Meleagris gallopavo has been reported by Dr.
Leidy (1889 H, p. 6) from a cave in Pennsylvania. The great majority of species accompanying it live yet in the vicinity.

GALLINULOIDIDÆ.

Lucas F. A. 1900 B, p. 84.

GALLINULOIDES. Type *G. wyomingensis* Eastman.

Eastman, C. R. 1900 E, p. 54.
Lucas, F. A., 1900 B, p. 79.

Gallinuloides wyomingensis Eastman.

Eastman, C. R. 1900 E, p. 54, pl. iv.
Lucas, F. A., 1900 B, p. 79, pl. i.
    Eocene (Bridger, Green River shales); Wyoming.

Order RAPTORES.

Cope, E. D. 1898 B, p. 88. (Accipitres.)
Coues, E. 1884 A, p. 496.
Edwards, A. M. 1871 A, p. 406.
Fürbringer, M. 1888 A, p. 1294 (Accipitres); p. 1306 (Strigidæ).
Gadow, H. 1891 A.
    1898 A, p, 83. (Falciformes.)
Giebel, C. G. 1866 A, p. 25. (Rapaces.)
Huxley, T. H. 1867 A, p. 462. (Aëtomorphæ.)

Huxley, T. H. 1872 A, p. 234. (Aëtomorphæ.)
Magnus, H. 1869 A.
Mivart, St. G. 1892 A, p. 270.
Newton, A. 1868 A. (Aëtomorpha.)
Owen, R. 1866 F, p. 206.
    1866 B.
Reichenow, A. 1871 A.
Selenka, E. 1869 A.

Suborder CATHARTIDES Coues.

Coues, E. 1884 A, p. 557.
Gadow, H. 1898 A, p. 33. (Cathartes.)

Mivart, St. G. 1892 A, p. 271. (Cathartes.)

CATHARTIDÆ.

Coues, E. 1884 A, p. 557.
Edwards, A. M. 1871 A, p. 406.
Fürbringer, M. 1888 A, p. 1294.
Huxley, T. H. 1867 A, p. 463.

Newton, A. 1868 A.
Reichenow, A. 1871 A.
Shufeldt, R. W. 1883 C.

PALÆOBORUS Coues. Type *Cathartes umbrosus* Cope.

Coues, E. 1884 A, p. 822.
Gadow, H. 1891 A, pl. xii, figs. 7, 8. (Cathartes.)
Huxley, T. H. 1867 A, p. 441, fig. 22. (Cathartes.)

Palæoborus umbrosus (Cope.)

Cope, E. D. 1874 I, p. 151. (Cathartes.)
    1874 O, p. 18. (Cathartes.)

Cope, E. D. 1875 F, p. 73. (Vultur.)
    1875 S. (Vultur.)
    1877 K, p. 287, pls. lxvii, figs. 10–18; pl. lxviii. (Vultur.)
Coues, E. 1884 A, p. 822.
Shufeldt, R. W. 1883 C. (Cathartes.)
    Pliocene; New Mexico.

Suborder ACCIPITRES.

Blanchard, E. 1859 A.
Coues, E. 1884 A, p. 517.
Edwards, A. M. 1871 A, p. 406. (A. diurni.)
Fürbringer, M. 1888 A, p. 1294.
Gadow, H. 1891 A.
    1898 A, p. 33.

Huxley, T. H. 1867 A, p. 462. (Gypaëtidæ.)
Mivart, St. G. 1892 A, p. 270. (Falcones.)
Parker, W. K. 1868 A, p. 166. (Accipitrinæ.)
Selenka, E. 1869 A.
Zittel, K. A. 1890 A, p. 842.

FALCONIDÆ.

Blanchard, E. 1859 A. (Falco.)
Coues, E. 1884 A, p. 519.
Gadow, H. 1891 A.

Parker, W. K. 1873 A, p. 45.
Selenka, E. 1869 A.
Shufeldt, R. W. 1881 A.

AQUILA Brisson.   Type *Falco chrysaëtos* Linn.

Blanchard, É. 1859 A, p. 36.
Coues, E. 1884 A, p. 553.
Nicholson and Lydekker 1889 A, p. 1214, fig. 1106.
Selenka, E. 1869 A, pl. xii, figs. 12–16; pl. xvi, figs. 5,6; pl. xvii, fig. 18.

Aquila danana Marsh.

Marsh, O. C. 1871 E, p. 125.
Coues, E. 1872 B, p. 346.
1884 A, p. 822.

King, C. 1878 A, p. 430.
Pliocene?; Nebraska.

Aquila pliogryps Shufeldt.

Shufeldt, R. W. 1892 D, p. 416, pl. xvii, fig. 33.
1891 B, p. 820.
Pleistocene; Oregon.

Aquila sodalis Shufeldt.

Shufeldt, R. W. 1892 D, p. 417, pl. xv, fig. 5.
1891 B, p. 820.
Pleistocene; Oregon.

Suborder STRIGES.

Coues, E. 1884 A, p. 498.
Edwards, A. M. 1871 A, p. 474.   (Strigides.)
Fürbringer, M. 1888 A, p. 1306.   (Strigidæ.)
Gadow, H. 1898 A, p. 36.

Mivart, St. G. 1892 A, p. 271.
Parker, W. K. 1868 A, p. 166.   (Striginæ.)
Shufeldt, R. W. 1881 D.   (On Speotyto.)

STRIGIDÆ.

Blanchard, É. 1859 A, p. 60.   (Strix.)
Coues, E. 1884 A, p. 502.
Fürbringer, M. 1888 A, p. 1306.
Gadow, H. 1891 A.

Huxley, T. H. 1867 A, p. 462.
Newton, A. 1868 A.
Selenka, E. 1869 A.
Shufeldt, R. W. 1881 D.   (On Speotyto.)

BUBO Duméril.   Type *Strix bubo* Linn.

Blanchard, É. 1859 A, p. 63.
Coues, E. 1884 A, p. 503.
Fürbringer, M. 1888 A, p. 1306.

Bubo leptosteus Marsh.

Marsh, O. C. 1871 E, p. 126.
Coues, E. 1872 B, p. 347.
1884 A, p. 822.
King, C. 1878 A, p. 404.

Leidy, J. 1872 B, p. 365.
Nicholson and Lydekker 1889 A, p. 1241.
Eocene (Bridger); Wyoming.

Bubo virginianus (Gmel.).

Coues, E. 1884 A, p. 503.
Shufeldt, R. W. 1891 B, p. 819.
1892 D, p. 418.
Recent; North America: Pleistocene; Oregon.

Order PICARIÆ.

Blanchard, É. 1859 A, p. 126.
Coues, E. 1884 A, p. 444.
Fürbringer, M. 1888 A.
Gadow, H. 1891 A.
Huxley, T. H. 1867 A, p. 467.   (Celeomorphæ.)
1872 A, p. 234.   (Celeomorphæ.)

Mivart, St. G. 1892 A, p. 266.   (Piciformes.)
Owen, R. 1856 F, p. 207.   (Scansores.)
Parker, W. K. 1868 A, p. 169.
Selenka, E. 1869 A.
Stejneger, L. 1885 A, p. 368.

UINTORNIS Marsh.   Type *U. lucaris* Marsh.

Marsh, O. C. 1872 J, p. 259.
Zittel, K. A. 1890 A, p. 850.
The position of this genus is in doubt.

Uintornis lucaris Marsh.

Marsh, O. C. 1872 J, p. 259.

Allen, J. A. 1878 A, p. 445.
Coues, E. 1872 B, p. 347.
1884 A, p. 822.
King, C. 1878 A, p. 404.
Eocene (Bridger); Wyoming.

Order PASSERES.

Blanchard, É. 1859 A, p. 85.
Cope, E. D. 1898 B, p. 90.
Coues, E. 1884 A, p. 238.
Edwards, A. M. 1871 A, p. 296.

Fürbringer, M. 1888 A, p. 1405.
Gadow, H. 1891 A.
1898 A, p. 37.   (Passeriformes.)
Garrod, A. N. 1876 A.

Giebel, C. G. 1866 A, p. 24.
Haeckel, E. 1895 A, p. 415. (Coracornithes.)
Huxley, T. H. 1867 A, p. 469. (Coracomorphæ.)
Lydekker, R. 1891 A, p. 2.
Mivart, St. G. 1892 A, p. 260.
Owen, R. 1866 F, p. 207. (Insessores.)
Parker, W. K. 1868 A, p. 178. (Passerinæ.)

Parker, W. K. 1875 B.
　1878 B.
Selenka, E. 1869 A.
Shufeldt, R. W. 1881 C. (On Lanius.)
　1881 D. (On Eremophilus.)
Stejneger, L. 1885 A, p. 456.

CORVIDÆ.

Coues, E. 1884 A, p. 414.
Gadow, H. 1891 A.

Parker, W. K. 1872 A.
Shufeldt, R. W. 1888 D.

CORVUS Linn.　Type *C. corax* Linn.

Coues, E. 1884 A, p. 415.
Edwards, A. M. 1871 A, p. 302.
Gadow, H. 1891 A, pl. lvii, figs. 9, 47.
Huxley, T. H. 1867 A, p. 451, fig. 32.
Parker, W. K. 1872 A.
Shufeldt, R. W. 1888 D, pl. xv, fig. 1.

Corvus annectens Shufeldt.
Shufeldt, R. W. 1892 D, p. 419, pl. xv, figs. 14-16.
　1891 B, p. 820. (No description.)
　Pleistocene; Oregon.

ICTERIDÆ.

Coues, E. 1884 A, p. 399.
Shufeldt, R. W. 1888 D.

Stejneger, L. 1885 A, p. 544.

SCOLECOPHAGUS Swainson.　Type *Turdus carolinus* Müll.

Coues, E. 1884 A, p. 411.
Shufeldt, R. W. 1888 D.

Scolecophagus affinis Shufeldt.
Shufeldt, R. W. 1892 D, p. 418, pl. xv, fig. 10.
　1891 B, p. 820. (No description.)
　Pleistocene; Oregon.

FRINGILLIDÆ.

Coues, E. 1884 A, p. 339.

Stejneger, L. 1885 A, p. 545.

PALÆOSPIZA Allen.　Type *P. bella* Allen.

Allen, J. A. 1878 A.
　The relationships of this genus are in doubt.

Palæospiza bella Allen.
Allen, J. A. 1878 A, pl. i, figs. 1, 2.

Cope, E. D. 1881 M.
Coues, E. 1884 A, p. 822.
　Eocene ? (Amyzon shales); Colorado.

CIMOLOPTERYX Marsh.　Type *C. rarus* Marsh.

Marsh, O. C. 1892 G, p. 175.
　1889 D, p. 83. (No description.)
　A genus of unknown position among birds.

Cimolopteryx rarus Marsh.
Marsh, O. C. 1892 G, p. 175, pl. iii, fig. 2.
　1889 D, p. 83. (No description.)

Cimolopteryx retusus Marsh.
Marsh, O. C. 1892 G, p. 175.
　Cretaceous (Laramie); Wyoming.

PALÆONORNIS STRUTHIONOIDES Emmons.

Emmons, E. 1857 A, p. 148, fig. 114.
　Triassic ?; North Carolina.
　A fossil of doubtful relationships.

FOSSIL EGG.

Farrington, O. C. 1891 A, pp. 198-200, pls. xx, xxi.
　Miocene; South Dakota.

ICHNITES.

The imprints of the feet of animals on the rocks have been described under various general names, such as *Ichnites*, *Ornithites*, *Ornichichnites*, *Ornithoidichnites*, *Sauroidichnites*, *Batrachoidichnites*. The first attempt to place under generic names the various forms found in North America was made by President Edward Hitchcock in 1845 (Hitchcock, E. 1845 B). This communication the present writer has made the basis for both the generic and the specific names. Many of the names, both generic and specific, were afterwards changed; but this was done in violation of the rules of nomenclature as now recognized. On account of the impossibility of determining with certainty the systematic positions of the animals which made the tracks described, the "genera" are arranged alphabetically.

Bouvé, T. T. 1859 B.
Cope, E. D. 1867 D.
Deane, J. 1844 A.
    1844 B.
    1844 C.
    1845 A.
    1845 B.
    1845 C.
    1845 D.
Edwards, A. Milne 1869 A.
Field, R. 1860 A.
    1860 B.
    1860 C.
Giebel, C. G. 1847 B, p. 84.
Hitchcock, C. H. 1855 A.
    1869 A.
    1871 A.
    1889 A.
    1889 B.
Hitchcock, E. 1836 A.
    1837 A.
    1837 B.
    1841 A.
    1843 A.
    1844 A.

Hitchcock, E. 1844 B.
    1845 A.
    1845 B.
    1847 A.
    1848 A.
    1858 A.
    1861 A.
    1863 A.
    1865 A.
Lyell, C. 1842 A.
    1843 A.
    1846 B.
Marsh, D. 1848 A.
Mitchell, J. 1895 A.
Mudge, B. F. 1866 A.
    1874 A.
Pictet, F. J. 1853 A, p. 406.
Quenstedt, F. A. 1852 A, p. 81.
Rogers, H. D. 1842 A.
Sickler, F. K. L. 1835 A.
Silliman, B. 1837 A.
Snow, F. H. 1887 A.
Voigt, F. S. 1835 A.
Zittel, K. A. 1890 A, p. 411.

ALLOPUS Marsh.   Type *A. littoralis* Marsh.

Marsh, O. C. 1894 D, p. 83.

Allopus littoralis Marsh.
Marsh, O. C. 1894 D, p. 83.
    1894 I, pl. xi, figs. 4, 4a.

Miller, S. A. 1897 A, p. 793.
Coal-measures; Kansas.

AMBLYPUS E. Hitch.   Type *A. dextratus* E. Hitch.

Hitchcock, E. 1858 A, p. 143.

Amblypus dextratus E. Hitch.
Hitchcock, E. 1858 A, p. 143, pl. xxv, fig. 7; pl. xlviii, fig. 5.

Hitchcock, C. H. 1889 B, p. 119.
Triassic; Massachusetts.

ANCYROPUS E. Hitch.   Type *Sauroidichnites heteroclitus* E. Hitch.

Hitchcock, E. 1845 B, p. 24.
    1848 A, p. 242.
    1858 A, p. 138.

Ancyropus heteroclitus E. Hitch.
Hitchcock E. 1845 B, p. 24 (A. heteroclitus, A. Jacksonianus.)
Hitchcock, C. H. 1889 B, p. 119.

Hitchcock, E. 1841 A, p. 479, pl. xxx, fig. 2 (Sauroidichnites heteroclitus); fig. 3 (Saur. jacksonii).
    1848 A, p. 243, pl. xv, figs. 3–5.
    1858 A. p. 139, pl. xxv, figs. 3, 4; pl. liii, figs. 1, 2.
Triassic; Connecticut.

ANOMŒPUS E. Hitch.   Type *A. scambus* E. Hitch.

Hitchcock, E. 1848 A, p. 220.
Hitchcock, C. H. 1889 A.
    1889 B.
Hitchcock, E. 1858 A, p. 55.
    1861 A, p. 150.
    1863 A, p. 48.
    1865 A, p. 31.
Lydekker, R. 1890 A, p. 221.
Warren, J. C. 1854 A, p. 35.

Anomœpus? culbertsonii (King.)

King, A. F. 1844 A, p. 176, figure. (Ornithichnites.)
    1845 A, fig. 2. (Ornithichnites.)
    1845 C, p. 216, fig. 4. (Ornithichnites.)
Lyell, C. 1846 A, p. 25. (No name.)
    Coal-measures; Pennsylvania.
    Assigned to this genus provisionally.

Anomœpus cuneatus C. H. Hitch.

Hitchcock, C. H. 1889 B, pp. 118, 125.
    Triassic; Massachusetts.

Anomœpus curvatus E. Hitch.

Hitchcock, E. 1863 A, p. 48, fig. 1.
    1863 B, p. 86. (Apatichnus curvatus.)
    1865 A, p. 5, pl. i, fig. 2; pl. xv, fig. 2.
    1889 B, p. 118.
Lydekker, R. 1890 A, p. 221.
    Triassic; Connecticut.

Anomœpus? gallinuloides (King).

King, A. F. 1844 A, p. 176, figure. (Ornithichnites.)
    1845 A, fig. 1. (Ornithoidichnites.)
Lyell, C. 1846 A, p. 25. (No name.)
    Coal-measures; Pennsylvania.
    This track is assigned to *Anomœpus* only provisionally.

Anomœpus gracillimus (E. Hitch).

Hitchcock, E. 1845 B, p. 23. (Eubrontes.)
Deane, J. 1861 A, p. 36, pl. iv, figs. 2, 4. (Grallator.)
Hitchcock, C. H. 1865 A, p. 6. (Anomœpus.)
    1889 B, pp. 118, 123.
Hitchcock, E. 1844 A, p. 306, pl. iii, fig. 4. (Ornithoidichnites.)
    1848 A, p. 175, pl. ii, fig. 3. (Brontozoum.)
    1858 A, p. 78, pl. xiii, fig. 5; pl. xxxii, fig. 2. (Grallator.)
    1865 A, p. 6. (Anomœpus.)

Lydekker, R. 1890 A, p. 221.
Mantell, G. A. 1846 A, p. 38. (Ornithichnites.)
Warren, J. C. 1854 A, p. 27. (Brontozoum.)
    Triassic; Massachusetts, Pennsylvania.

Anomœpus intermedius E. Hitch.

Hitchcock, E. 1865 A, pp. 2, 7, pl. i, fig. 1; pl. xv, fig. 1.
Hitchcock, C. H. 1889 B, p. 118.
Hitchcock, E. 1863 A, pp. 49, 53. (No description.)
    1863 B, pp. 85, 86. (Name only.)
Lydekker, R. 1890 A, p. 221.
    Triassic; Massachusetts, Connecticut.
    This species is based on part of *Brontozoum isodactylum* Edw. Hitchcock, 1858 A, p. 69. See under *Plesiornis minor.*

Anomœpus isodactylus C. H. Hitch.

Hitchcock, C. H. 1889 B, pp. 118, 124.
    Triassic; Massachusetts.

Anomœpus major E. Hitch.

Hitchcock, E. 1858 A, p. 56, pl. viii.
Deane, J. 1861 A, p. 48, pl. xxxi.
Eyermann, J. 1866 A, p. 32.
Hitchcock, C. H. 1871 A, p. xxi, fig. 2.
    1873 A, p. 433, fig. 1.
    1889 B, p. 118.
Woodworth, J. B. 1895 A.
    Triassic; Massachusetts, New Jersey.

Anomœpus minimus E. Hitch.

Hitchcock, E. 1865 A, p. 5, pl. ii, figs. 1, 2.
Hitchcock, C. H. 1889 B, p. 118.
Hitchcock, E. 1863 B, p. 85. (Name only.)
    Triassic; Massachusetts.

Anomœpus scambus E. Hitch.

Hitchcock, E. 1848 A, p. 222, pl. xiii, figs. 1–6.
Dana, J. D. 1896 A, p. 752, fig. 1175.
Deane, J. 1845 C, p. 80, figs. A, B. (No name.)
    1847 A, p. 78, fig. 3. (No name.)
    1861 A, p. 48, pl. xxxii. (A. minor.)
Eyermann, J. 1889 A, p. 32. (A. minor.)
Hitchcock, C. H. 1855 A, p. 394.
    1889 B, p. 118. (A. minor.)
Hitchcock, E. 1858 A, p. 57, pl. ix, figs. 1, 2; pl. xxxiv, fig. 2. (A. minor.)
    Triassic; Massachusetts, New Jersey.

ANTHRACOPUS Leidy.   Type *A. ellangowensis* Leidy.

Leidy, J. 1879 B, p. 164.

Anthracopus ellangowensis Leidy.

Leidy, J. 1879 B, p. 164.

Mason, W. D. H. 1878 A, pp. 717, 725. (No name.)
    Coal-measures; Pennsylvania.

ANTICHEIROPUS E. Hitch.   Type, no species indicated.

Hitchcock, E. 1865 A, p. 10.

Anticheiropus hamatus E. Hitch.

Hitchcock, E. 1865 A, p. 10, pl. ix, figs. 1, 2.
Hitchcock, C. H. 1889 B, p. 118.
    Triassic; Massachusetts.

Anticheiropus pilulatus E. Hitch.

Hitchcock, E. 1865 A, p. 10, pl. ix, fig. 3.
Hitchcock, C. H. 1889 B, p. 118.
    Triassic; Massachusetts.

ANTIPUS E. Hitch.   Type *A. flexiloquus* E. Hitch.

Hitchcock, E. 1858 A, p. 115.

Antipus bifidus E. Hitch.

Hitchcock, E. 1858 A, p. 116, pl. xxxvi, fig. 8; pl. xlviii, fig. 10.
Hitchcock, C. H. 1889 B, p. 119.
Triassic; Massachusetts.

Antipus flexiloquus E. Hitch.

Hitchcock, E. 1858 A, p. 115, pl. xx, fig. 10.
Hitchcock, C. H. 1889 B, p. 119.
Triassic; Massachusetts.

APATICHNUS E. Hitch.   Type *A. circumagens* E. Hitch.

Hitchcock, E. 1858 A, p. 99.
1861 A, p. 150.

Apatichnus bellus E. Hitch.

Hitchcock, E. 1858 A, p.101, pl. xvii, fig. 6; pl. xxxv, fig. 8; pl. xlv, fig. 6.
Dana, J. D. 1896 A, p. 752, fig. 1174.
Hitchcock, C. H. 1889 B, p. 118.
Triassic; Massachusetts.

Apatichnus circumagens E. Hitch.

Hitchcock, E. 1858 A, p. 100, pl. xvii, fig. 5; pl. xxxv, fig. 6.

Deane, J. 1861 A, p. 37, pl. vi, fig. 1; pl. xix; pl. xxxiv, fig. 1, 2.
Hitchcock, C. H. 1889 B, p. 118.
Triassic; Massachusetts.

Apatichnus crassus C. H. Hitch.

Hitchcock, C. H. 1889 B, pp. 118, 122, 124.
Triassic; New Jersey.

Apatichnus holyokensis C. H. Hitch.

Hitchcock, C. H. 1889 B, pp. 118, 123, 124.
Triassic; Massachusetts.

ARACHNICHNUS E. Hitch.   Type *A. dehiscens* E. Hitch.

Hitchcock, E. 1858 A, p. 117.

Arachnichnus dehiscens E. Hitch.

Hitchcock, E. 1858 A, p. 117, pl. xx, figs. 12,13; pl. xxxvii, fig. 2.

Hitchcock, C. H. 1889 B, p. 119.
Triassic; Massachusetts.

ARGOIDES E. Hitch.   No type; *A. isodactylatus* may be taken.

Hitchcock, E. 1845 B, p. 24.
1848 A, p. 184. (Argozoum.)
1858 A, p. 81. (Argozoum.)

Argoides isodactylatus E. Hitch.

Hitchcock, E. 1845 B, p. 24.
Adams, C. B. 1846 A, p. 215. (Argoides minimus?)
Hitchcock, C. H. 1889 B, p. 118. (Argozoum paridigitatum.)
Hitchcock, E. 1836 A, p. 325, fig. 9. (Ornithichnites minimus.)
1837 A, p. 175. (Ornithichnites minimus.)
1841 A, p. 496, pl. xlv, figs. 38, 39. (Ornithoidichnites isodactylus.)
1848 A, p. 187, pl. vi, figs, 4, 5. (Argozoum paridigitatum.)
1858 A, p. 82, pl. xiv, fig. 3; pl. xxxv, fig. 4; pl. xxxix, fig. 1. (Argozoum paridigitatum.)
Warren, J. C. 1854 A, p. 29. (Argozoum paridigitatum.)
Triassic; Massachusetts, Connecticut.

Argoides macrodactylatus (E. Hitch.).

Hitchcock, E. 1845 B, p. 25. (Platypterna.)
Hitchcock, C. H. 1889 B, p. 118. (Argozoum paridigitatum.)
Hitchcock, E. 1841 A, p. 494, pl. xliii, fig. 35. (Ornithoidichnites.)

Hitchcock, E. 1848 A, p. 186, pl. vi, fig. 3. (Argozoum disparidigitatum.)
1858 A, p. 82, pl. xiv, fig. 2. (Argozoum disparidigitatum.)
arren, J. C. 1854 A, p. 29. (Argozoum disparidigitatum.)
Triassic; Massachusetts, Connecticut, New Jersey.

Argoides redfieldianus E. Hitch.

Hitchcock, E. 1845 B, p. 24.
Hitchcock, C. H. 1889 B, p. 118. (Argozoum.)
Hitchcock, E. 1844 A, p. 304, pl. iii, fig. 1. (Ornithoidichnites redfieldii.)
1848 A, p. 185. (Argozoum.)
1858 A, p. 81, pl. xiv, fig. 1. (Argozoum.)
1863 B, p. 85. (Argozoum redfieldianum, doubtful species.)
1865 A, p. 2. (Argozoum.)
Triassic; Massachusetts.

Argoides robustus E. Hitch.

Hitchcock, E. 1845 B, p. 24.
Giebel, C. G. 1847 B, p. 36. (Ornithichnites robustus.)
Hitchcock, E. 1836 A. p. 319, fig. 3. (Ornithichnites ingens minor.)
1837 B, p. 175. (O. robustus.)
Triassic; Massachusetts.

BAROPUS Marsh. Type *B. lentus* Marsh.

Marsh, O. C. 1894 D, p. 83.

Baropus lentus Marsh.

Marsh, O. C. 1894 D, p. 83, pl. ii, fig. 5.
Dana, J. D. 1896 A, p. 684, fig. 1116.

Marsh, O. C. 1894 I, p. 338, pl. xi, fig. 5.
1895 C, p. 492, fig. 11.
Miller, S. A. 1897 A, p. 793.
Coal-measures; Kansas.

BATRACHICHNUS Woodworth. Type *B. plainvillensis* Woodworth.

Woodworth, J. B. 1900 A, p. 458.

Batrachichnus plainvillensis Woodworth.
Woodworth, J. B. 1900 A, p. 453, pl. xl; text-fig. 1.
Carboniferous; Massachusetts.

BATRACHOIDES E. Hitch. Type *Batrachoides nidificans* E. Hitch.

Hitchcock, E. 1858 A, p. 121.
Hitchcock, C. H. 1871 A, p. xxi.
1873 A, p. 436.
Shepard, C. U. 1867 A.
The name Batrachoides is preoccupied by Lacépède for a genus of fishes.

Batrachoides nidificans E. Hitch.
Hitchcock, E. 1858 A, p. 122, pl. xxi, fig. 5; pl. l, figs. 1–4.

Hitchcock, C. H. 1889 B, p. 119.
Shepard, C. U. 1867 A.
Triassic; Massachusetts.
Batrachoides antiquior described in Hitchcock, E. 1858 A, p. 123, pl. xxi, fig. 6; pl. l, fig. 2, was rejected in 1865. (Hitchcock C. H. 1865 A, p. 2.) It is regarded as wholly of inorganic origin.

BATRACHOPUS E. Hitch. Type *B. deweyanus* E. Hitch.

Hitchcock E. 1845 B, p. 25.
Hitchcock, C. H. 1871 A, p. xxi. (Anisichnus, to replace Anisopus preoccupied.)
Hitchcock, E. 1848 A, p. 226. (Anisopus.)
1858 A, pp. 60, 61. (Anisopus.)
Zittel, K. A. 1890, p. 411.

Batrachopus deweyanus E. Hitch.
Hitchcock, E. 1845 B, p. 25.
Dana, J. D. 1896 A, p. 751, fig. 1171. (Anisopus.)
Deane, J. 1845 A, p. 161, figure. (Batrachoidichnites deweyi.)
1861 A, p. 46, pl. xxv, fig. 2; pl. xxvi, fig. 2. (Anisopus.)
Hitchcock, C. H. 1855 A, p. 394. (Anisopus.)
1889 B, p. 119. (Anisichnus.)
Hitchcock, E. 1841 A, p. 489, pl. xxxix, fig. 26; pl. xlviii, fig. 44. (Ornithoidichnites parvulus.)
1843 A, p. 261, pl. xl, fig. 9. (Sauroidichnites deweyi.)
1844 A, p. 305. (Batrachoidichnites parvulus, B. deweyi.)
1848 A, p. 226, pl. xvi, figs. 5, 6. (Anisopus.)

Hitchcock, E. 1858 A, p. 60, pl. ix, fig. 3; pl. xli, fig. 2; pl. xlii, figs. 1, 2; pl. liii, fig. 8; pl. lviii, fig. 11. (Anisopus.)
Triassic; Connecticut.

Batrachopus gracilior (E. Hitch.).
Hitchcock, E. 1863 A, p. 54. (Anisopus.)
Hitchcock, C. H. 1889 B, p. 119. (Anisichnus.)
Hitchcock, E. 1865 A, p. 6, pl. i, fig. 3. (Anisopus.)
Triassic; Massachusetts.

Batrachopus gracilis (E. Hitch.).
Hitchcock, E. 1848 A, p. 228, pl. xvi, fig. 34. (Anisopus.)
Dana, J. D. 1896 A, p. 751, fig. 1172. (Anisopus.)
Deane, J. 1861 A, p. 46, pl. xxvi, fig. 1. (Anisopus.)
Hitchcock, C. H., 1889 B, pp. 119, 122, 123. (Anisichnus.)
Hitchcock, E. 1858 A, p. 61, pl. ix, fig. 4; pl. xxxv, fig. 5; pl. xxxvi, figs. 1–5; pl. lviii, fig. 9. (Anisopus.)
1861 A, p. 155, fig. 4. (Anisopus.)
Lydekker, R. 1890 A, p. 221. (Anisopus.)
Triassic; Massachusetts, New Jersey, Pennsylvania.

CHEIROTHEROIDES E. Hitch. Type *C. pilulatus* E. Hitch.

Hitchcock, E. 1858 A, p. 130.

Cheirotheroides pilulatus E. Hitch.
Hitchcock, E. 1858 A, p. 130, pl. xxiii, figs. 3, 4; pl. xxxvi, fig. 6; pl. liv, fig. 8.

Hitchcock, C. H. 1889 B, p. 119.
Triassic; Massachusetts.

CHELICHNUS Jardine.   Type *Testudo duncani* Owen.

Jardine, W., 1850, Ann. Nat. Hist., vi, p. 205.

Chelichnus wymanianus Lea.

Lea, I. 1856 C, p. 78.

Cope, E. D. 1871 C, p. 242.
Lea, I. 1856 B, p. 123.
Triassic; Pennsylvania.

CHELONOIDES E. Hitch.   Type *C. incedens* E. Hitch.

Hitchcock, E. 1858 A, p. 140.

Chelonoides incedens E. Hitch.

Hitchcock, E. 1858 A, p. 140, pl. xxxi, fig. 3

Hitchcock, C. H. 1889 B, p. 119.
Triassic; Massachusetts.

CHIROTHERIUM Kaup.   Type *C. barthii* Kaup.

Kaup, J. J., 1835, Thierreich, i, p, 246; figure, p. 247.
  The spelling Cheirotherium is usually employed for this genus. The original form is given above.
Cunningham, J. 1839 A, p. 12.
Gaudry, A. 1890 C, p. 171.
Hitchcock, C. H. 1889 A, p, 133.
  1873 A, p. 436.
  1889 A, p. 186.
King, A. T. 1845 A, p. 343. (Thenaropus,[1] type *T. hetcrodactylus.*)
Lortet, L. 1892 A, p. 125.
Lyell, C. 1846 A, p. 25.
Miller, S. A. 1889 A, p. 619.
Neues Jahrbuch Mineral., etc., 1835, pp. 110, 111, 122, 243.
Owen, R. 1860 E, p. 163.

Quenstedt, F. A. 1885 A, p. 121.
Warren, J. C. 1856 A.
Zittel, K. A. 1890 A, p. 4.
  [1] This name had previously been employed by King for tracks of a very different kind.

Chirotherium? heterodactylum (King.)

King, A. T. 1845 A, pp. 343, 348, figs. 7a–7c. (Thenaropus.)
  1845 C, p. 216. (Thenaropus.)
Lea, I. 1853 A, p. 187. (Thenaropus.)
Lyell, C. 1846 A, p. 25. ("Cheirotherium.")
  Coal-measures; Pennsylvania.

Chirotherium? reiteri Moore.

Moore, W. D. 1873 A.
  Coal-measures; Pennsylvania.

COLLETTOSAURUS Cox.   Type *C. indianaensis* Cox.

Cox, E. T. 1874 A, p. 247.
Miller, S. A. 1889 A, p. 620.

Collettosaurus indianaensis Cox.

Cox, E. T. 1874 A, p. 247, plate.
Amer. Jour. Science 1876 A.
  Coal-measures; Indiana.

COMPTICHNUS E. Hitch.   Type *C. obesus* E. Hitch.

Hitchcock, E. 1865 A, p. 9.

Comptichnus obesus E. Hitch.

Hitchcock, E. 1865 A, p. 9, pl. v, fig. 4; pl. xviii, fig. 6.

Hitchcock, C. H. 1889 B, p. 119.
Triassic; Massachusetts.

CORVIPES E. Hitch.   Type *C. lacertoideus* E. Hitch.

Hitchcock, E. 1858 A, p. 98.

Corvipes lacertoideus E. Hitch.

Hitchcock, E. 1858 A, p. 98, pl. xvii. fig. 8; pl. xlvii, fig. 1; pl. xxxv, fig. 7.

Hitchcock, C. H. 1889 B, pp. 118, 122.
Triassic; Massachusetts, Connecticut.

CRUCIPES Butts.   Type *C. parvus* Butts.

Butts, E. 1891 A, p. 19.

Crucipes parvus Butts.

Butts, E. 1891 A, p. 19.
  Coal-measures; Missouri.

CUNICHNOIDES E. Hitch.   Type *C. marsupialoides* E. Hitch.

Hitchcock, E. 1858 A, p. 54.

Cunichnoides marsupialoides E. Hitch.

Hitchcock, E. 1858 A, p. 55, pl. ix, fig. 5; pl. ix, figs. 2, 3, 4.

Hitchcock, E. 1861 A, p. 154.
Triassic; Connecticut.

DROMOPUS Marsh.   Type *D. agilis* Marsh.

Marsh, O. C. 1894 D, p. 82.
Fürbringer, M. 1900 A, pp. 627, 673.

Dromopus agilis Marsh.

Marsh, O. C. 1894 D, p. 182, pl. ii, fig. 3; pl. iii, fig. 3.

Dana, J. D. 1895 A, p. 684, fig. 1115.
Marsh, O. C. 1894 I, p. xi, fig. 3.
    1895 D, p. 492, fig. 10.
Miller, S. A. 1897 A, p. 793.
Coal-measures; Kansas.

EUBRONTES E. Hitch.   No type was mentioned, but *E. giganteus* may be taken.

Hitchcock, E. 1845 B, p. 23.
Hitchcock, C. H. 1878 A, p. 432 (Brontozoum).
    1889 B, p. 118.   (Brontozoum.)
Hitchcock, E. 1847 A, p. 50.   (Brontozoum.)
    1848 A, p. 169.   (Brontozoum.)
    1858 A, p. 63.   (Brontozoum.)
Huxley, T. H. 1876 B, xiii, p. 515.   (Brontozoum.)
Lydekker, R. 1890 A, p. 220.   (Brontozoum.)
Owen, R. 1860 E, p. 290.   (Brontozoum.)
    Brontozoum is a substitute name for the earlier *Eubrontes*.

Eubrontes approximatus (E. Hitch.).

Hitchcock, E. 1865 A, p. 23, pl. x, fig. 2.   (Brontozoum.)
Hitchcock, C. H. in Hitchcock, E. 1865 A, p. 24.   (Brontozoum.)
    1889 B, pp. 118, 121, 122, 123.   (Brontozoum.)
Triassic; Massachusetts, New Jersey.
    This species is based on portions of *Eubrontes giganteus*, but the writer can not determine whether or not any of the figures which have been published to illustrate *E. giganteus* belong here.

Eubrontes dananus E. Hitch.

Hitchcock, E. 1845 B, p. 23.
Cook, G. H. 1885 A, p. 96.   (Brontozoum sillimanium.)
Deane, J., 1861 A, p. 37, pls. vii, ix, x.
Hitchcock, C. H. 1889 B, pp. 118, 122, 123.   (Brontozoum sillimanium.)
Hitchcock, E. 1836 A, p. 318.   (Ornithichnites tuberosus, part.)
    1837 A, p. 175.   (Ornithichnites tuberosus.)
    1841 A, p. 486, in part, pl. xxxvii, fig. 21; pl. xxxviii, fig. 22 (Ornithoidichnites tuberosus);
    p. 488 (Ornithoidichnites cuneatus, in part).
    1843 A, p. 254 (Ornithoidichnites tuberosus);
        p. 256, pl. xi, fig. 2 (O. sillimanii).
    1844 A, p. 300.   (O. tuberosus, part?.)
    1847 A, p. 49.   (Brontozoum sillimanium.)
    1848 A, p. 171, pl. iii, fig. 2.   (B. sillimanium.)
    1858 A, p. 68, pl. xii, fig. 3; pl. xxxiii, figs. 4, 5; pl. xliii, fig. 6.   (B. sillimanium.)
    1865 A, p. 7, pl. v, fig. 1.   (B. sillimanium.)

Lydekker, R., 1890 A, p. 220.   (B. sillimani.)
Redfield, W. C. 1843 A, p. 136.   (Ornithoidichnites tuberosus.)
Warren, J. C. 1854 A, p. 27.   (B. sillimanium.)
    Triassic; Massachusetts, Connecticut, New Jersey, Pennsylvania.

Eubrontes divaricatus (E. Hitch.).

Hitchcock, E. 1865 A, p. 7.   (Brontozoum.)
Cook, G. H. 1885 A, p. 96.   (Brontozoum.)
Hitchcock, C. H. 1889 B, p. 126.   (Brontozoum.)
Hitchcock, E. 1865 A, p. 2.   (B. isodactylum, in part.)
    Triassic; Massachusetts, Connecticut.
    This species is based partly on *Brontozoum isodactylum*, Edw. Hitchcock, 1858 A, p. 69. See under *Plesiornis minor*.

Eubrontes oxpansus E. Hitch.

Hitchcock, E. 1845 B, p. 23.
    1841 A, p. 487, pl. xxxviii, fig. 23; pl. xxxix, fig. 24.   (Ornithoidichnites.)
    1848 A, p. 174, pl. iii, fig. 1.   (Brontozoum.)
Triassic; Massachusetts.
    The materials on which this species was based have evidently been assigned to some other species; which one can not be determined.

Eubrontes exsertus (E. Hitch.).

Hitchcock, E. 1858 A, p. 67, pl. xii, fig. 1; pl. xxxvii-li; pl. lvii, fig. 3.   (Brontozoum.)
Deane, J., 1861 A, p. 39, pl. xiii.   (Brontozoum.)
Hitchcock, C. H. 1889 B.   (Brontozoum.)
    Triassic; Massachusetts.

Eubrontes giganteus E. Hitch.

Hitchcock, E. 1845 B, p. 23.
Cook, G. H. 1885 A, p. 96.   (Brontozoum.)
Dana, J. D. 1895 A, p. 752, fig. 1177.   (Brontozoum.)
Deane, J. 1861 A, p. 42, pl. xv.   (Brontozoum.)
Giebel, C. G. 1847 B, p. 35.   (Ornithichnites.)
Hitchcock, C. H. 1889 B.   (Brontozoum.)
Itchcock, E. 1836 A, p. 317.   (Ornithichnites.)
    1837 A, p. 175.   (Ornithichnites.)
    1841 A, p. 484, pl. xxxvi, fig. 18.   (Ornithoidichnites.)
    1845 A, p. 68.   (Ornithoidichnites.)

Hitchcock, E. 1847 A, p. 57. (Brontozoum.)
1848 A, p. 169, pl. i, fig. 1. (Brontozoum.)
1858 A, p. 64, pl. xxxiii, figs. 1–3; pl. xli,
fig. 1; pl. liii, fig. 7; pl. lvii, fig. 1. (Brontozoum.)
1861 A, p. 149. (Brontozoum.)
1865 A, p. 23, fig. 1. (Brontozoum.)
Lydekker, R. 1890 A, p. 220.
Owen, R. 1860 E, p. 289. (Brontozoum.)
Pictet, F. J. 1853 A, p. 406, pl. xx, fig. 8. (Ornithichnites.)
Quenstedt, F. A. 1852 A, p. 82. (Ornithichnites.)
Warren, J. C. 1854 A, p. 25. (Brontozoum.)
Triassic; Massachusetts, Connecticut, New
Jersey.

Eubrontes minusculus (E. Hitch.)

Hitchcock, E. 1858 A, p. 65, pl. xi, fig. 1; pl. xl, fig.
2; pl. xli, fig. 1; pl. xlii, fig. 3; pl. lvii, fig. 2.
(Brontozoum.)
Cook, G. H. 1885 A, p. 96. (Brontozoum.)
Deane, J. 1861 A, p. 41, pls. xiv, xvi, xvii. (Bronto-
zoum.)
Hitchcock, C. H. 1889 B, pp. 118, 122. (Bronto-
zoum.)
Triassic; Massachusetts, New Jersey.

Eubrontes tuberatus (E. Hitch.)

Hitchcock, E. 1858 A, p. 66, pl. xl, fig. 2; pl. liii,
fig. 7. (Brontozoum.)
Hitchcock, C. H. 1889 B, p. 118. (Brontozoum.)
Triassic; Massachusetts.

Eubrontes tuberosus E. Hitch.

Hitchcock, E. 1845 B, p. 23.
Deane, J. 1861 A, p. 39, pl. xii. (Brontozoum
validum.)
Hitchcock, C. H. 1889 B, p. 118. (B. validum.)
Hitchcock, E. 1836 A, p. 318. (Ornithichnites
tuberosus, part.)
1841 A, p. 486, in part, pl. xxxvii, fig. 20.
(Ornithoidichnites.)
1848 A, p. 172, pl. li, figs. 1, 2. (Brontozoum
loxonyx.)
1858 A, p. 67, pl. xli, fig. 2; pl. xxxviii–li;
pl. lvii, fig. 3. (Brontozoum validum.)
Lydekker, R. 1890 A, p. 220. (B. validum.)
Warren, J. C. 1854 A, p. 27. (B. loxonyx.)
Triassic; Massachusetts.

EUPALAMOPUS nom. nov. Type *Palamopus dananus* E. Hitch. = *P. clarki* E. Hitch.

Hitchcock, E. 1848 A, p. 217. (Palamopus, not of
1845.)
1858 A, p. 127. (Palamopus, not of 1845.)
This generic name is intended as a substitute
for *Palamopus* E. Hitch., 1848. The latter name
had been employed in 1845 to notate a genus
based on *P. anomalus* E. Hitch., later called
Macropterna recta E. Hitch.

Eupalamopus dananus (E. Hitch.).

Hitchcock, E. 1848 A, p. 217, pl. xi, figs. 1, 2. (Pa-
lamopus.)
Hitchcock, C. H. 1889 B, p. 119. (P. clarki.)
Hitchcock, E. 1858 A, p. 127, pl. xxiii, fig. 2; pl.
xliv, fig. 2. (P. clarki.)
Triassic; Massachusetts.

EXOCAMPE E. Hitch. Type *E. arcta* E. Hitch.

Hitchcock, E. 1858 A, p. 142.
Through error this genus is called, on pl.
xxv of Hitchcock's paper of 1858, *Hectocampe.*

Exocampe arcta E. Hitch.

Hitchcock, E. 1858 A, p. 143, pl. xxv, figs. 5, 6, 10;
pl. xlix, fig. 5.
Deane, J. 1861 A, p. 57, pl. xxxix.
Hitchcock, C. H. 1889 B, p. 119.
Triassic; Massachusetts.

Exocampe minima E. Hitch.

Hitchcock, E. 1865 A, p. 11, pl. xviii, fig. 3, and
Appendix.

Hitchcock, C. H. 1889 B, p. 119.
Hitchcock, E. 1863 B, p. 86. (Insufficient descrip-
tion.)
Triassic; Massachusetts.

Exocampe ornata E. Hitch.

Hitchcock, E. 1858 A, p. 143, pl. xxv, fig. 11; pl.
lviii. figs. 1, 6.
Deane, J. 1861 A, p. 57, pl. xxxviii, fig. 1.
Hitchcock, C. H. 1889 B, p. 119.
Triassic; Massachusetts.

FULICOPUS E. Hitch. Type *F. lyellianus* E. Hitch.

Hitchcock, E. 1845 B, p. 23.
1848 A, p. 177. (Æthyopus.)
1858 A, p. 70. (Amblonyx.)
Warren, J. C. 1854 A, p. 28. (Æthyopus.)

Fulicopus giganteus (E. Hitch.).

Hitchcock, E. 1858 A, p. 71, pl. xiii, fig. 1; pl.
xxxviii, figs. 1, 2; pl. lvii, figs. 5. (Amblonyx.)
Hitchcock, C. H. 1889 B, p. 118. (Amblonyx.)
Hitchcock, E. 1865 A, p. 2. (Amblonyx.)
Triassic; Massachusetts.

Fulicopus lyellianus (E. Hitch.).

Hitchcock, E. 1845 B, p. 23.

Hitchcock, C. H. 1889 B, p. 118. (Amblonyx.)
Hitchcock, E. 1843 A, p. 257, pl. xl, fig. 1. (Ornithoidichnites lyellii.)
1848 A, p. 178, pl. iv, fig. 1. (Æthyopus.)
1858 A, p. 71, pl. xiii, fig. 2; pl. xxviii, fig. 2.
(Amblonyx.)
1863 B, p. 85. (A. lyellianus., doubtful
species.)
1865 A, p. 2. (Amblonyx.)
Warren, J. C. 1854 A, p. 28. (Æthyopus.)
Triassic; Massachusetts.

GIGANDIPUS E. Hitch.   Type *G. caudatus* E. Hitch.

Hitchcock, E. 1856 A, p. 97.
Hitchcock, C. H. 1878 A, p. 484, fig. 3. (Gigantitherium.)
Hitchcock, E. 1855 A. p. 416. (Gigadipus; no definition.)
   1858 A, p. 93. (Gigantitherium.)
Warren, J. C. 1856 A.

Gigandipus caudatus E. Hitch.

Hitchcock, E. 1856 A, p. 97, text-figures.
Hitchcock, C. H. 1889 B, p. 118. (Gigantitherium.)

Hitchcock, E. 1855 A, p. 416. (Gigadipus; insufficient description.)
Hitchcock, E. 1858 A, p. 93, pl. xvi, figs. 1, 2; pl. xliv, fig. 4. (Gigantitherium.)
Triassic; Massachusetts.

Gigandipus minor (E. Hitch.).

Hitchcock, E. 1858 A, p. 95, pl. xvii, fig. 1: pl. xli, fig. 2; pl. xlii, fig. 2. (Gigantitherium.)
Hitchcock, C. H. 1889 B, p. 118. (Gigantitherium.)
Triassic; Massachusetts.

GRALLATOR E. Hitch.   No type assigned.

Hitchcock, E. 1858 A, p. 72.
Lydekker R. 1890 A, p. 222.

Grallator cuneatus E. Hitch.

Hitchcock, E. 1858 A, p. 74, pl. xiii, fig. 6: pl. xxxix, figs. 1, 3; pl. xli, figs. 1, 2.
Deane, J. 1861 A, p. 38, pl. viii.
Eyermann, J. 1889 A, p. 32.
Hitchcock, C. H. 1878 A. p. 432.
   1889 B, p. 118, 122.
Hitchcock, E. 1841 A, p. 488, text-fig. 106; pl. xxxix, fig. 25; pl. xlviii, figs. 45, 55. (Ornithoidichnites.)
   1848 A, p. 258. (Ornithoidichnites.)
Lydekker, R. 1890 A, p. 222.
   Triassic; Connecticut, New Jersey.

Grallator cursorius E. Hitch.

Hitchcock, E. 1858 A, p. 72, pl. xiii, fig. 3; pl. lviii, fig. 4.
Hitchcock, C. H., 1889 B, pp. 118, 122.
   Triassic; Massachusetts, New Jersey.

Grallator formosus E. Hitch.

Hitchcock, E. 1858 A, p. 75, text-fig., p. 77.
Cook, G. H. 1885 A, p. 96.
Deane, J. 1861 A, p. 39, pl. xi.
Hitchcock, C. H. 1889 B, pp. 118, 122.
   Triassic; Connecticut, Massachusetts, New Jersey.

Grallator gracilis C. H. Hitch.

Hitchcock, C. H. 1865 A, p. 8, pl. ix, fig. 7.
   1869 A.
   1889 B, p. 118.
   Triassic; Massachusetts, New Jersey.

Grallator parallelus E. Hitch.

Hitchcock, E. 1847 A, p. 50, fig. 2. (Brontozoum)
Cook, G. F. 1885 A, p. 96.
Hitchcock, C. H. 1889 B, pp. 118, 122.
Hitchcock, E. 1836 A, p. 318. (Ornithichnites tuberosus dubius.)
   1837 B, p. 175. (O. parallelus.)
   1848 A, p. 175, pl. iii, figs. 3, 4. (Brontozoum.)
   1863 B, p. 86.
   1865 A, p. 7, pl. iv, fig. 1.
Warren, J. C. 1854 A, p. 27. (Brontozoum.)
   Triassic; Massachusetts, New Jersey.

Grallator tenuis E. Hitch.

Hitchcock, E. 1858 A, p. 73, pl. xiii, fig. 4; pl. liii, fig. 5.
Deane, J. 1861 A, p. 36, pl. iv, fig. 3; pl. xxvii, fig. 2.
Eyermann, J. 1889 A, p. 32.
Hitchcock, C. H, 1889 B, p. 118.
   Triassic; Massachusetts, New Jersey, Connecticut.

HARPEDACTYLUS E. Hitch.   Type *H. tenuissimus* E. Hitch.

Hitchcock, E. 1845 B, p. 24.
   1848 A, p. 206.
   1858 A. p. 112.

Harpedactylus crassus E. Hitch.

Hitchcock, E. 1865 A, p. 12, pl. iii, fig. 1.
Hitchcock, C. H. 1889 B, p. 119.
Hitchcock, E. 1863 B, p. 86. (No description.)
   Triassic; Massachusetts.

Harpedactylus gracilior E. Hitch.

Hitchcock, E. 1865 A, p. 12, pl. iii, fig. 2.
Hitchcock, C. H. 1889 B, p. 119.

Hitchcock, E. 1863 B, p. 86. (No description.)
   Triassic; Massachusetts.

Harpedactylus tenuissimus E. Hitch.

Hitchcock, E. 1845 B, p. 24.
Hitchcock, C. H. 1889 B, p. 119. (H. gracilis.)
Hitchcock, E. 1841 A, p. 482, pl xxxiv, fig. 13. (Sauroidichnites.)
   1848 A, p. 206, pl. xiv, fig. 2. (H. gracilis.)
   1858 A. p. 112, pl. xx, fig. 4; pl. iii, fig. 5. (H. gracilis.)
   Triassic; Massachusetts.

546   FOSSIL VERTEBRATA OF NORTH AMERICA.   [BULL. 179.

HELCURA E. Hitch.   Type *H. littoralis* E. Hitch.

Hitchcock, E. 1848 A, p. 244.
1858 A, p. 141.

Helcura anguinea E. Hitch.

Hitchcock, E. 1856 A, p. 141, pl. xxxvi, fig. 9.
Triassic; Massachusetts.

Helcura littoralis E. Hitch.

Hitchcock, E. 1848 A, p. 244, pl. xv, fig. 1; pl. xxiii,
fig. 3.

Hitchcock, C. H. 1889 B, p. 119.   (H. caudata.),
Hitchcock, E. 1858 A, p. 140, pl. xxxvii, fig. 8; pl.
xl, fig. 1.   (H. caudata.)
Triassic; Massachusetts.

Helcura surgens E. Hitch.

Hitchcock, E. 1858 A, p. 141, pl. xxxvi, fig. 10.
Hitchcock, C. H. 1889 B, p. 119.
Triassic; Massachusetts.

HOPLICHNUS E. Hitch.   Type *H. quadrupedans* E. Hitch.

Hitchcock, E. 1848 A, p. 230.
1858 A, p. 134.
It is doubtful whether or not this genus of
footmarks was produced by a vertebrate animal.

Hoplichnus equus E. Hitch.

Hitchcock, E. 1858 A, p. 134, pl. xxiv, figs. 3–5.
Hitchcock, C. H. 1889 B, p. 119.
Triassic; Massachusetts, Connecticut?

Hoplichnus quadrupedans E. Hitch.

Hitchcock, E. 1848 A, p. 230, pl. xvi, figs. 7, 8; pl.
xxii, fig. 3.
Hitchcock, C. H. 1889, B, p. 119.   (H. poledrus.)
Hitchcock, E. 1858 A, p. 136, pl. xxiv, figs. 6, 7;
pl. xlviii, fig. 9.   (H. poledrus.)
Triassic; Massachusetts.

HYLOPUS Dawson.   Type *H. logani* Dawson.

Dawson, J. W. 1894 A, p. 260.
1895 A, p. 77.

Hylopus caudifer Dawson.

Dawson, J. W. 1895 A, p. 78.
Coal-measures; Nova Scotia.

Hylopus hardingi Dawson.

Dawson, J. W. 1895 A, p. 78.
Subcarboniferous; Nova Scotia.

Hylopus logani Dawson.

Dawson, J. W. 1894 A, p. 260, figure.
1895 A, p. 78.
Subcarboniferous; Nova Scotia.

Hylopus minor Dawson.

Dawson, J. W. 1895 A, p. 78.
Coal-measures; Nova Scotia.

Hylopus? trifidus Dawson.

Dawson, J. W. 1895 A, p. 78.
Coal-measures; Nova Scotia.

Hylopus sp. indet.

Dawson, J. W. 1895 A, p. 78.
Coal-measures; Nova Scotia.

HYPHEPUS E. Hitch.   Type *H. fieldi* E. Hitch.

Hitchcock, E. 1858 A, pp. 97,180.

Hyphepus fieldi E. Hitch.

Hitchcock, E. 1858 A, pp. 97, 189, pl. xvii, fig. 2;
pl. xxxv, fig. 11; pl. xli, fig. 2; pl.xlii, fig. 2.

Hitchcock, C. H. 1889 B, p. 118.
Triassic; Massachusetts.

ISOCAMPE E. Hitch.   Type *I. strata* E. Hitch.

Hitchcock, E. 1858 A, p. 119.

Isocampe strata E. Hitch.

Hitchcock, E. 1858 A, p. 120, pl. xx, fig. 5; pl.
xxxvi, fig. 5.

Hitchcock, C. H. 1889 B, p. 118.
Triassic; Connecticut, Massachusetts.

LANGUNCULAPES E. Hitch.   Type *L. latus* E. Hitch.

Hitchcock, E. 1858 A, p. 132.

Langunculapes latus E. Hitch.

Hitchcock, E. 1858 A, p. 132, pl. xxiv, fig. 1; pl.
xlv, fig. 4.

Hitchcock, C. H. 1889, B, p. 119.
Triassic; Massachusetts.

LEPTONYX E. Hitch.   Type *L. lateralis* E. Hitch.

Hitchcock, E. 1865 A, p. 8.

Leptonyx lateralis E. Hitch.

Hitchcock, E, 1865 A, p. 8, pl. v, fig. 3.

Hitchcock, C. H. 1889 B, p. 118.
Triassic; Massachusetts.

LIMNOPUS Marsh.   Type *L. vagus* Marsh.

Marsh, O. C. 1894 D, p. 82.

Limnopus vagus Marsh.

Marsh, O. C. 1894 D, p. 82, pl. ii, fig. 2; pl. iii, fig. 2.
Dana, J. D. 1896 A, p. 684, fig. 1114.

Marsh, O. C. 1894 I, pl. xi, fig. 2.
1895 D, p. 492, fig. 9.
Miller, S. A. 1897 A, p. 793.
Coal-measures; Kansas.

NANOPUS Marsh.   Type *N. caudatus* Marsh.

Marsh, O. C. 1894 D, p. 82.

Nanopus caudatus Marsh.

Marsh, O. C. 1894 D, p. 82, pl. ii, fig. 1; pl. iii, fig. 1.
Dana, J. D. 1896 A, p. 684, fig. 1113. (Nanopus.)

Marsh, O. C. 1894 I, pl. xi, fig. 1.
1895 D, p. 492, fig. 8.
Miller, S. A. 1897, A, p. 793.
Coal-measures; Kansas.

NOTALACERTA Butts.   Type *N. missouriensis* Butts.

Butts, E. 1891 A, p. 18.

Notalacerta missouriensis Butts.

Butts, E. 1891 A, p. 18, fig. 2.
Coal-measures; Missouri.

NOTAMPHIBIA Butts.   Type *N. magna* Butts.

Butts, E. 1891 B, p. 44.

Notamphibia magna Butts.

Butts, E. 1891 B, p. 44.
Coal-measures; Missouri.

ORTHODACTYLUS E. Hitch.   Type *O. floriferus* E. Hitch.

Hitchcock, E. 1858 A, p. 113.

Orthodactylus floriferus E. Hitch.

Hitchcock, E. 1858 A, p. 114, pl. xx, fig. 7.
Deane, J. 1861 A, p. 44, pl. xxiii.
Hitchcock, C. H. 1889 B, p. 119.
Triassic; Massachusetts.

Orthodactylus introvergens E. Hitch.

Hitchcock, E. 1858 A, p. 114, pl. xx, fig. 8; pl. li, fig. 1.

Hitchcock, C. H. 1889 B, p. 119.
Triassic; Massachusetts.

Orthodactylus linearis E. Hitch.

Hitchcock, E. 1858 A, p. 115, pl. xx, fig. 9; pl. xlviii, fig. 4.
Deane, J. 1861 A, p. 57, pl. xxxviii, fig. 2.
Hitchcock, C. H. 1889 B, p. 119.
Triassic; Massachusetts.

OTOZOUM E. Hitch.   Type *E. moodii* E. Hitch.

Hitchcock, E. 1847 A, p. 54.
Field, R. 1860 B, p. 339.
Hitchcock, C. H. 1878 A, p. 485.
1889 B.
Hitchcock, E. 1848 A, p. 214.
1858 A, p. 123.
Owen, R. 1860 E, p. 165.
Quenstedt, F. A. 1852 A, p. 157.
Warren, J. C. 1854 A, p. 32.

Otozoum caudatum C. H. Hitch.

Hitchcock, C. H. 1871 A, p. xx.
1889 B, pp. 119, 127.
Triassic; Connecticut.

Otozoum moodii E. Hitch.

Hitchcock, E. 1847 A, p. 55, fig. 1.
Dana, J. D. 1896 A, p. 752, fig. 1176.
Deane, J. 1861 A, p. 55, pl. xxxvii.
Hitchcock, C. H. 1871 A, p. xxi.
1889 B, pp. 119, 127.
Hitchcock, E. 1848 A, p. 214, pl. xii, fig. 1.
1858 A, p. 123, pl. xxii; pl. xxiii, fig. 1; pl. xxxiii, figs. 4, 5; pl. xlvi, fig. 5.
1861 A, p. 148, fig. 2.
Triassic; Massachusetts, Connecticut.

Otozoum parvum C. H. Hitch.

Hitchcock, C. H. 1889 B, pp. 119, 122, 127.
Triassic; New Jersey, Pennsylvania.

548   FOSSIL VERTEBRATA OF NORTH AMERICA.   [BULL. 179.

PALÆOSAUROPUS nom. nov.   Type *Sauropus primævus* Lea.

Dawson, J. W. 1868 A, p. 357. (Sauropus.)
  1895 A, p. 77. (Sauropus.)
Lea, I. 1849 A, p. 91. (Sauropus; preoccupied by
  E. Hitchcock, 1845.)
Lesley, J. P., 1890 A, p. 728, figures. (Sauropus.)
Miller, S. A. 1889 A, p. 625. (Sauropus.)
Owen, R. 1860 E, p. 167. (Sauropus.)

Palæosauropus antiquior (Dawson).

Dawson, J. W. 1882 A, p. 652.   Sauropus.)
  1895 A, P. 77. (Sauropus.)
Subcarboniferous; Nova Scotia.

Palæosauropus primævus (Lea).

Lea, I. 1849 A, p. 91, figure. (Sauropus.)
Cope, E. D. 1871 C, p. 241. (Sauropus.)
Dana, J. D. 1896 A, p. 645, fig. 1023. (Sauropus.)
Lea, I. 1850 A. (Sauropus.)
  1852 A, p. 307, pls. xxxi, xxxii. (Sauropus.)
  1855 A. (Sauropus.)
Miller, S. A. 1889 A, p. 625, fig. 1193. (Sauropus.)
Silliman, B. 1852 B. (Sauropus.)

Zittel, K. A. 1890 A, p. 411. (Sauropus.)
  Coal-measures; Pennsylvania.

Palæosauropus sydenensis (Dawson).

Dawson, J. W. 1868 E, p. 431, fig. 1. (Sauropus.)
  1868 A. p. 358, fig. 140. (Sauropus.)
  1876 A, p. 441. (Sauropus.)
  1878 A, p. 358, fig. 140; suppl., p. 63. (Sauropus.)
  1895 A, p. 77. (Sauropus.)
Lesley, J. P. 1890 A, p. 920, figure. (Sauropus.)
  Coal-measures; Cape Breton Island.

Palæosauropus unguifer (Dawson).

Dawson, J. W. 1872 A, p. 251, figure. (Sauropus.)
  1876 A, p. 441. (Sauropus.)
  1878 A, suppl., p. 64. (Sauropus.)
  1882 A, p. 652. (Sauropus.)
  1894 A, p. 267. (Sauropus.)
  1895 A, p. 77. (Sauropus.)
Carboniferous (Millstone Grit); Nova Scotia.

PALAMOPUS E. Hitch.   Type *P. anomalus* E. Hitch.

Hitchcock, E, 1845 B, p. 24.
  1845 B, p. 24. (Sillimanius, in part.)
  1848 A, p. 233. (Macropterna.)
  1858 A, p. 128. (Macropterna.)
Lydekker, R. 1890 A, p. 219. (Macropterna.)
  This is not the *Palamopus* E. Hitch. of 1848 and
subsequently. The latter is here replaced by
Eupalamopus. Since E. Hitchcock's *Macropterna
recta* (=*P. anomalus*) belongs here, it becomes
necessary to refer here also the other species of
Macropterna.

Palamopus anomalus E. Hitch.

Hitchcock, E. 1845 B, p. 24.
Amer. Jour. Science 1847 B, p. 276.
Deane, J. 1847 A, p. 79, fig. 4. (Sauroidichnites
palmatus.)
  1847 B. (Sauroidichnites palmatus.)
Hitchcock, E. 1836 A, p. 324, fig. 15. (Ornithichnites palmatus.)
  1837 B, p. 175. (Sauroidichnites palmatus.)
  1841 A, p. 483, pl. xxxiv, fig. 15. (Sauroidichnites palmatus.)
  1848 A, p. 235, pl. xv, fig. 6. (Macropterna
recta.)
Triassic; Massachusetts.

Palamopus divaricans (E. Hitch.).

Hitchcock, E. 1848 A, p. 237, pl. xv, fig. 7. (Macropterna.)
Dana, J. D. 1896 A, p. 752, fig. 1178. (Macropterna.)
Deane, J. 1861 A, p. 45, pl. xxiv, fig. 1. (Macropterna.)
Hitchcock, C. H. 1889 B, p. 119. (Macropterna.)

Hitchcock. E. 1841 A, p. 483 (part), pl. xxxiv, fig.
16. (Sauroidichnites palmatus.)
  1858 A, p. 129, pl. xxiii, fig. 7. (Macropterna.)
Lydekker, R. 1890 A, p, 219. (Macropterna.)
  Triassic; Massachusetts.

Palamopus gracilipes (E. Hitch.).

Hitchcock, E. 1858 A, p. 129, pl. xxiii, fig. 6; pl.
xxxiv, fig. 1. (Macropterna.)
?Deane, J. 1861 A, p. 47, pl. xxix, fig. 1. (Macropterna.)
Hitchcock, C. H. 1889 B, p. 119. (Macropterna.)
  Triassic; Massachusetts.

Palamopus rogersianus (E. Hitch.).

Hitchcock, E. 1845 B, p. 24. (Sillimanius rogersianus: Argoidbes minimus.)
Deane, J. 1861 A, p. 47, pl. xxviii, fig. 2. (Macropterna vulgaris.)
Hitchcock, C. H. 1889 B, p. 119. (Macropterna
vulgaris.)
Hitchcock, E. 1841 A, p. 497 (in part), pl. xlii, fig.
30. (Ornithoidichnites minimus, not of 1836,
1837.)
  1843 A, p. 256, pl. xl, fig. 7. (Ornithoidichnites rogersi.)
  1848 A, p. 187, pl. vi, fig. 6 (Argozoum
minimum); p. 233, pl. xv. fig. 9 (Macropterna rhynchosauroidea.)
  1858 A, p. 128, pl. xxiii, fig. 5; pl. xxxv, fig.
9; pl. xxxvii, fig. 4; pl. xlviii, fig. 7; pl.
xlix, fig. 6. (M. vulgaris.)
Triassic; Massachusetts, Connecticut.

PLATYPTERNA E. Hitch.   Type, none given; *P. deaniana* may be taken.

Hitchcock, E. 1845 B, p. 25.
Edwards, A. Milne 1868 A, p. 136.
    1869 A, p. 672.
Hitchcock, E. 1845 B, p. 25. (Calopus, type *C. delicatulus.*)
    1848 A, p. 188.
    1858 A, p. 83.
Warren, J. C. 1854 A, p. 29.

Platypterna concamerata (E. Hitch.).

Hitchcock, E. 1848 A, p. 207, pl. xiv, fig. 3. (Harpedactylus.)
Hitchcock, C. H. 1889 B, p. 118. (P. varica.)
Hitchcock, E. 1858 A, p. 85, pl. xiv, fig. 8; pl. xlvii, fig. 4. (P. varica.)
    Triassic; Massachusetts.

Platypterna deaniana E. Hitch.

Hitchcock, E. 1845 B, p. 25.
Hitchcock, C. H. 1889 B, p. 118.
Hitchcock, E. 1841 A, p. 493, pl. xliii, figs. 31, 32. (Ornithoidichnites deanii.)
    1848 A, p. 189, pl. vii, fig. 1.
    1858 A, p. 83, pl. xiv, fig. 4.
    1863 B, p. 85.
    1865 A, p. 2.
    Triassic; Connecticut.

Platypterna delicatula E. Hitch.

Hitchcock, E. 1845 B, p. 25. (Calopus.)
Hitchcock, C. H. 1889 B, p. 118.
Hitchcock, E. 1841 A, p. 497, pl. xlv, fig. 40. (Ornithoidichnites.)

Hitchcock, E. 1848 A, p. 190, pl. vii, fig. 4.
    1858 A, p. 84, pl. xiv, fig. 6; pl. lviii, fig. 8.
    Triassic; Connecticut.

Platypterna digitigrada E. Hitch.

Hitchcock, E. 1858 A, p. 86, pl. xiv, fig. 9; pl. li, fig. 2.
Hitchcock, C. H. 1889 B, p. 118.
    Triassic; Massachusetts.

Platypterna gracillima E. Hitch.

Hitchcock, E. 1858 A, p. 86, pl. xiv, fig. 12.
    1865 A, p. 2.
    Triassic; Massachusetts.
    This species was rejected in the publication last quoted; but no statement is made as to what species the specimens are to be referred.

Platypterna recta E. Hitch.

Hitchcock, E. 1848 A, p. 209, pl. v, fig. 5. (Harpedactylus.)
Hitchcock, C. H. 1889 B, p. 118.
Hitchcock, E. 1858 A, p. 84, pl. xiv, fig. 7; pl. xlvii, fig. 3.
    Triassic; Massachusetts.

Platypterna tenuis E. Hitch.

Hitchcock, E. 1845 B, p. 25.
Hitchcock, C. H. 1889 B, p. 118.
Hitchcock, E. 1841 A, p. 494, pl. xliii, figs. 33, 34. (Ornithoidichnites.)
    1848 A, p. 189, pl. vii, figs. 2, 3.
    1858 A, p. 84, pl. xiv, fig. 5; pl. lviii, fig. 10.
    Triassic; Connecticut.

PLECTROPTERNA E. Hitch.   Type *Plectropus minitans* E. Hitch.

Hitchcock, E. 1858 A, p. 108. (For Plectropus, preoccupied.)
    1845 B, p. 24. (Plectropus.)
    1848 A, p. 198. (Plectropus.)

Plectropterna angusta E. Hitch.

Hitchcock, E. 1858 A, p. 110, pl. xviii, fig. 4: pl. xxxvi, fig. 3.
Hitchcock, C. H. 1889 B, p. 119.
    Triassic; Massachusetts.

Plectropterna elegans E. Hitch.

Hitchcock, C. H. 1889 B, pp. 122, 125.
    Triassic; Connecticut.

Plectropterna gracilis E. Hitch.

Hitchcock, E. 1858 A, p. 109, pl. xviii, fig. 3; pl. xlviii, fig. 2.
Hitchcock, C. H. 1889 B, p. 119.
Hitchcock, E. 1865 A, p. 24.
    Triassic; Massachusetts, Connecticut.

Plectropterna lineans E. Hitch.

Hitchcock, E. 1858 A, p. 110, pl. xviii, fig. 5; pl. xxxv, fig. 10.
Hitchcock, C. H. 1889 B, p. 119.
    Triassic; Connecticut.

Plectropterna minitans E. Hitch.

Hitchcock, E. 1845 B, p. 24. (Plectropus.)
Hitchcock, C. H. 1889 B, p. 119.
Hitchcock, E. 1841 A, p. 481, pl. xxxiii, fig. 11; pl. xlviii, fig. 47 (Sauroidichnites minitans); p. 482, pl. xxxiii, fig. 12; pl. xlviii, fig. 48 (S. longipes).
    1844 A, p. 304. (Sauroidichnites.)
    1845 B, p. 24. (Plectropus longipes.)
    1848 A, p. 198, pl. ix, figs. 2, 3 (Plectropus minitans); p. 199, pl. viii, fig. 4; pl. x, figs. 1-3 (Plectropus longipes).
    1858 A, p. 108, pl. xviii, fig. 2; pl. xix, figs. 10-12.
    Triassic; Connecticut, Massachusetts.

PLESIORNIS E. Hitch.   No type assigned.

Hitchcock, E. 1858 A, p. 102.
1861 A, p. 150.
1865 A, p. 85.

Plesiornis giganteus C. H. Hitch.

Hitchcock, C. H. 1889 B, pp. 118, 122, 126.
Triassic; Connecticut.

Plesiornis minimus (E. Hitch.).

Hitchcock, E. 1845 B, p. 23. (Argoides.)
Hitchcock, C. H. 1889 B, pp. 118, 122. (P. æqualipes.)
Hitchcock, E. 1841 A, p. 497 in part, pl. xlv, fig. 41. (Orinithoidichnites minimus.)
1848 A, p. 187, pl. vi, fig. 6. (Argozoum minimum.)
1858 A, p. 104, text-figure. (P. æqualipes.)
Triassic; Massachusetts, Connecticut.

Argoides minimus of 1845 was based on *Ornithoidichnites minimus* of 1841. This was a composite species, a part of which was assigned to *Palamopus rogersianus* of the present work (*Macropterna vulgaris* of E. Hitchcock.) The remaining part must bear the name given it in 1845.

Plesiornis minor (E. Hitch.).

Hitchcock, E. 1845 B, p. 23. (Fulicopus.)
Cook G. H. 1885 A, p. 96. (Brontozoum isodactylum.)
Deane, J. 1844 A, p. 76. (Ornithichnites fulicoides.)
1861 A, p. 43, pls. xviii, xxxvii. (P. quadripes.)

Eyermann, J. 1889 A, p. 82. (B. isodactylum.)
Field, R. 1860 A. (Ornithichnites fulicoides.)
Hitchcock, C. H. 1889 B, p. 118. (P. quadripes.)
Hitchcock, E. 1843 A, p. 258, pl. xi, fig. 3. (Ornithichnites fulicoides.)
1844 A, p. 297. (Ornithichnites fulicoides.)
1848 A, p. 179, pl. iv, figs. 2, 3. (Æthyopus minor.)
1858 A, p. 69, pl. xii, fig. 3; pl. xl, fig. 1; pl. xlvi, fig. 3; pl. lvii, fig. 4 (Brontozoum isodactylum); 1 p. 102, pl. xvii, fig. 7; pl. xxxv, figs. 1, 2; pl. xlv, fig. 5 (P. quadripes).
Mantell, G. A. 1846 A. (Ornithichnites fulicoides.)
Quenstedt, F. A. 1852 A, p. 82. (Ornithichnites fulicoides.)
Warren, J. C. 1854 A, p. 28. (Æthyopus minor.)
Triassic; Massachusetts.

Either the whole or a part of this species was divided into the two new species *Brontozoum isodactylum* and *Anomœpus intermedius*, which see. The *B. isodactylum* of Cook and Eyermann may not belong here.

Plesiornis mirabilis E. Hitch.

Hitchcock, E. 1865 A, p. 35, pl. xx.
Hitchcock. C. H. 1889 B, p. 118.
Triassic; Massachusetts.

Plesiornis pilulatus E. Hitch.

Hitchcock, E. 1858 A, p. 103, pl. xvii, fig. 8; pl. xxxvi, fig. 4.
Hitchcock, C. H. 1889 B, p. 118.
Triassic; Massachusetts.

POLEMARCHUS E. Hitch.   Type *Sauroidichnites polemarchius* E. Hitch = *P. gigas* E. Hitch.

Hitchcock, E. 1845 B, p. 24.
1848 A, p. 197.
1858 A. p. 107.

Polemarchus gigas E. Hitch.

Hitchcock, E. 1845 B, p. 24.
Hitchcock, C. H. 1889 B, pp. 119, 122.
Hitchcock, E. 1841 A, p. 483, pl. xxxv, fig. 17. (Sauroidichnites polemarchius.)

Hitchcock, E. 1844 A, p. 304. (Sauroidichnites polemarchius.)
1845 A, p. 63. (Sauroidichnites polemarchius.)
1848 A, p. 197, pl. ix, fig. 1.
1858 A, p. 107, pl. xviii, fig. 1; pl. lix, fig. 3.
Lesley, J. P. 1889 A, p. 729.
Triassic; Massachusetts, New Jersey, Pennsylvania.

SAUROPUS E. Hitch.   Type *S. barrattii* E. Hitch.   Not *Sauropus* Lea, 1849.

Hitchcock, E. 1845 B, p. 24.
Hitchcock, C. H. 1871 A, p. xxi. (Chimæerichnus, for Chimæra E. Hitch., preoccupied.)
1886 B, p. 118. (Chimærichnus.)
Hitchcock, E. 1858 A' p. 118. (Chimæra.)

Sauropus barrattii E. Hitch.

Hitchcock, E. 1845 B, p. 24.
Hitchcock, C. H. 1889 B, p. 118. (Chimæerichnus.)
Hitchcock. E. 1841 A, p. 477, pl. xxx, fig. 1. (Sauroidichnites.)

Hitchcock, E. 1848 A, p. 225, pl. xiv, fig. 1. (Anomœpus.)
1858 A. p. 118, pl. xxi, figs. 1–4; pl. xxxvii, fig. 1. (Chimæerichnus.)
Triassic; Connecticut, Massachusetts.

Sauropus ingens (C. H. Hitch.).

Hitchcock, C. H. 1889 B, pp. 118, 122. (Chimæerichnus.)
Triassic; New Jersey.

SELENICHNUS E. Hitch. Type *S. falcatus* E. Hitch.

Hitchcock, E. 1858 A, p. 133.

Selenichnus breviusculus E. Hitch.

Hitchcock, E. 1858 A, p. 134, pl. **xxiii**, fig. 9; pl. lx, fig. 7.
Deane, J. 1861 A, p. 48, pl. **xxx**, fig. 1.
Hitchcock, C. H. 1889 B, p. 119.
Triassic; Massachusetts.

Selenichnus falcatus E. Hitch.

Hitchcock, E. 1858 A, p. 133, pl. **xxiii**, fig. 8; pl. lx, fig. 8.
Deane, J. 1861 A, p. 48, pl. **xxx**, fig. 2.
Hitchcock, C. H. 1889 B, p. 119.
Triassic; Massachusetts.

SHEPARDIA E. Hitch. Type *S. palmipes* E. Hitch.

Hitchcock, E. 1858 A, p. 131.

Shepardia palmipes E. Hitch.

Hitchcock, E. 1858 A, p. 131, pl. **xxiv**, fig. 2.

Hitchcock, C. H. 1889 B, p. 119.
Triassic; Massachusetts.

SILLIMANIUS E. Hitch. Type, none assigned.

Hitchcock, E. 1845 B, p. 24.
1848 A, p. 191. (Ornithopus; no type.)
1858 A, p. 87. (Ornithopus.)
Originally four species, *S. adamsanus*, *S. tetradactylus*, *S. gracilior*, and *S. rogersianus*, were assigned to this genus. The last-named belongs under *Palamopus*. Later (1848) Hitchcock referred the first three species to *Ornithopus*, a new genus. As *Sillimanius* was based on described and figured forms, *Ornithopus* becomes a synonym.

Sillimanius gracilior E. Hitch.

Hitchcock, E. 1845 B, p. 24.
Hitchcock, C. H. 1889 B, p. 118. (Ornithopus.)
Hitchcock, E. 1841 A, p. 498, pl. xlvi, fig. 43 (Ornithoidichnites).
1843 A. (Ornithoidichnites.)
1848 A, p. 193, pl. viii, fig. 2. (Ornithopus.)
1858 A, p. 88, pl. xiv, fig. 11; pl. lviii, fig. 7. (Ornithopus.)
Triassic; Connecticut.

Sillimanius tetradactylus E. Hitch.

Hitchcock, E. 1845 B, p. 24.
Giebel, C. G. 1847 B, p. 36. (Ornithichnites.)
Hitchcock, C. H. 1889 B, p. 118. (Ornithopus gallinaceus.)
Hitchcock, E. 1836 A, p. 323. (Ornithichnites.)
1837 A, p. 175. (Ornithichnites.)
1841 A, p. 497, pl. xlvi, fig. 42. (Ornithoidichnites.)
1848 A, p. 192, pl. viii, fig. 1. (Ornithopus gallinaceus.)
1858 A, p. 87, pl. xiv, fig. 10; pl. lviii, fig. 1. (Ornithopus gallinaceus.)
Pictet, F. J. 1853 A, p. 407. (Ornithichnites.)
Quenstedt, F. A. 1852 A, p. 82. (Ornithichnites.)
Warren, J. C. 1854 A, p. 29. (Ornithopus gallinaceus.)
Triassic; Massachusetts, Connecticut.

STENODACTYLUS E. Hitch. Type *S. curvatus* E. Hitch.

Hitchcock, E. 1858 A, p. 116.

Stenodactylus curvatus E. Hitch.

Hitchcock, E. 1858 A, p. 116, pl. **xx**, fig. 11; pl. **xxiv**, fig. 3.

Hitchcock, C. H. 1889 B, p. 119.
Triassic; Massachusetts.

STEROPOIDES E. Hitch. Type not indicated.

Hitchcock, E. 1845 B, p. 24.
Edwards, A. Milne 1863 A, p. 136. (Tridentipes.)
1869 A, p. 672. (Tridentipes.)
Hitchcock, E. 1848 A, p. 182. (Steropezoum.)
1858 A, p. 88, (Tridentipes, to replace Steropezoum.)
Lydekker, R. 1890 A, p. 222. (Tridentipes.)

Steropoides elegans E. Hitch.

Hitchcock, E. 1845 B, p. 24. (S. elegans, S. elegantior.)
Giebel, C. G. 1847 B, p. 36. (Ornithichnites.)
Hitchcock, C. H. 1889 B, p. 118. (Tridentipes.)

Hitchcock, E. 1836 A, p. 320. (Ornithichnites diversus); p. 321 (O. diversus platydactylus).
1841 A, p. 491, pl. xli, fig. 28, 29; pl. xlviii, figs. 57-62 (Ornithoidichnites elegans); pl. 493, pl. xlii, fig. 30; pl. xlviii, fig. 58 (O. elegantior).
1848 A, p. 183, pl. v, fig. 2 (Steropezoum elegans); p. 184, pl. v, fig. 3 (S. elegantius).
1858 A, p. 90, pl. xv, fig. 2; pl. xiv, fig. 6; pl. lii, figs. 8, 11. (Tridentipes.)
Pictet, F. J. 1853 A, p. 406. (Ornithichnites diversus.)
Triassic; Massachusetts.

Steropoides infelix nom. nov.

Hitchcock, C. H. 1889 B, p. 118. (Tridentipes elegantior.)
Hitchcock, E. 1858 A, p. 90, pl. xv, fig. 3; pl. xlv,
fig. 1. (Tridentipes elegantior, not Steropezoum
elegantius, 1848.)
Triassic; Massachusetts.

Steropoides ingens E. Hitch.

Hitchcock, E. 1845 B, p. 24. (S. ingens, Sillimanius adamsanus.)
Cook, G. H. 1885 A, p. 96. (Tridentipes.)
Giebel, C. G. 1847 B, p. 35 (Ornithichnites ingens);
p. 36 (O. danæ.)
Hitchcock, C. H. 1889 B, pp. 118, 122. (Tridentipes.)
Hitchcock, E. 1836 A, p. 319. (Ornithichnites.)
1841 A, p. 490, pl. xl, fig. 37. (Ornithoidichnites.)
1844 A, p. 306, pl. iii, fig. 5. (O. danæ.)
1848 A, p. 182, pl. v, fig. 1 (Steropezoum
ingens); p. 191, pl. vii, fig. 5 (Ornithopus
adamsanus.)
1858 A, p. 89, pl. xv, fig. 1. (Tridentipes.)
Lesley, J. P. 1890 A, p. 1211, figure. (Tridentipes.)

Pictet, F. J. 1853 A, p. 406. (Ornithichnites.)
Quenstedt, F. A. 1852 A, p. 82. (Ornithichnites.)
Triassic; Massachusetts, New Jersey.

Steropoides loripes (E. Hitch.).

Hitchcock, E. 1848 A, p. 193, pl. vii, fig. 3. (Ornithopus.)
Hitchcock, C. H. 1889 B, p. 118. (Tridentipes insignis.)
Hitchcock, E. 1841 A, p. 495, pl. xliv, figs. 36, 37;
pl. xlviii, fig. 63. (Ornithoidichnites divaricatus.)
1858 A, p. 91, pl. xv, fig. 4; pl. xlv, fig. 3; pl.
xlvii, fig. 2. (Tridentipes insignis.)
Lesley, J. P. 1890 A, p. 1211, figure. (Tridentipes.)
Triassic; Massachusetts.

Steropoides uncus (E. Hitch.).

Hitchcock, E. 1858 A, p. 91, pl. xv, fig. 5; pl. xlvi,
fig. 1. (Tridentipes.)
Hitchcock, C. H. 1889 B, p. 118. (Tridentipes uncus?)
1863 B, p. 85. (T. uncus, doubtful species.)
Hitchcock, E. 1865 A. p. 20. (Tridentipes.)
Triassic; Massachusetts.

TARSODACTYLUS E. Hitch.   Type *T. caudatus* E. Hitch.

Hitchcock, E. 1858 A. p. 98.

Tarsodactylus caudatus E. Hitch.

Hitchcock E. 1858 A, p. 99.
Deane, J. 1861 A, p. 54, pl. xxxv.
Hitchcock, C. H. 1889 B, p. 118.
Triassic; Massachusetts.

Tarsodactylus expansus C H. Hitch.

Hitchcock, C. H. 1866 A, p. 301.
1889 B, p. 118.
Triassic; Massachusetts.

THENAROPUS King.   Type not named; *T. leptodactylus* may be assumed.

King, A. T. 1844 A, p. 177.
Amer. Jour. Science 1845 C. (Spheropezium.)
Hitchcock, E. 1848 A, p. 218.
King, A. T. 1845 A, p. 343. (Spheropezopus.)
1845 C. (Spheropezium.)
Lyell, C. 1846 A, p. 25. (No name.)
Miller, S. A. 1889 A, p. 626. (Sphæropezium,
Thenaropus.)
Zittel, K. A. 1890 B, p. 411.
The tracks assigned to this genus are regarded
by Lyell as being artifacts.

Thenaropus leptodactylus King.

King, A. T. 1844 A, p. 177, fig. 3.
Amer. Jour. Science 1845 C, p. 345. (Spheropezium.)
King, A. T. 1845 A, p. 345, fig. 3. (Spheropezopus.)
1845 C, p. 216, fig. 2. (Spheropezium.)
Coal-measures; Pennsylvania.

Thenaropus ovoidactylus King.

King, A. T. 1844 A, p. 178, fig. 7.

Amer. Jour. Science 1845 C, p. 347. (Spheropezium.)
King, A. T. 1845 A, p. 347, fig. 6. (Spheropezopus.)
Coal-measures; Pennsylvania.

Thenaropus pachydactylus King.

King, A. T. 1844 A, p. 177, fig. 4.
Amer. Jour. Science 1845 C, p. 346. (Spheropezium.)
King, A. T. 1845 A, p. 346, fig. 4. (Spheropezopus.)
1845 C, p. 216, fig. 1. (Spheropezium.)
Coal-measures; Pennsylvania.

Thenaropus sphærodactylus King.

King, A. T. 1844 A, p. 177, figs. 5, 6.
Amer. Jour. Science 1845 C, p. 346. (Spheropezium thærodactylum.)
King, A. T. 1845 A, p. 346, figs. 5a, 5b. (Spheropezopus thærodactylum.)
1845 C, fig. 3. (Spheropezium thærodactylum.)
Coal-measures; Pennsylvania.

THINOPUS Marsh.   Type *T. antiquus* Marsh.

Marsh, O. C. 1896 G. p. 374.

Thinopus antiquus Marsh.

Marsh, O. C. 1896 G, p. 374, figure.
Devonian; Pennsylvania.

TOXICHNUS E. Hitch.   Type *T. inæqualis* E. Hitch.

Hitchcock E. 1865 A, p. 12.

Toxichnus inæqualis E. Hitch.

Hitchcock, E. 1865 A, p. 12, pl. v, fig. 5.

Hitchcock, E. 1889 B, p. 119.
  Triassic; Massachusetts.

TRIÆNOPUS E. Hitch.   Type *T. baileyanus* E. Hitch.

Hitchcock, E. 1845 B, p. 24.
   1848 A, p. 202.
   1858 A, p. 111.
Warren, J. C. 1854 A, p. 30.

Triænopus baileyanus E. Hitch.

Hitchcock, E. 1845 B, p. 24. (T. baileyanus, T. emmonsianus.)
Hitchcock, C. H. 1889 B, pp. 119, 122. (T. leptodactylus.)
Hitchcock, E. 1841 A, p. 479, pl. xxx, figs. 4–6. (Sauroidichnites emmonsii); p. 480, pl. xxxii, figs. 8, 9 (S. baileyi).

Hitchcock, E. 1848 A, p. 203, pl. x, fig. 4 (T. baileyanus); p. 204, pl. x, fig. 5 (T. emmonsianus).
   1858 A, p. 111, pl. xix, figs. 1–9; pl. xx, figs. 1–3; pl. xlv, fig. 8; pl. lii, fig. 1. (T. leptodactylus.)
Warren, J. C. 1854 A, p. 31. (T. baileyanus, T. emmonsianus.)
   Triassic; Connecticut.

TRIHAMUS E. Hitch.   Type *T. elegans* E. Hitch.

Hitchcock, E. 1865 A, p. 9.
  Triassic; Massachusetts.

Trihamus elegans E. Hitch.

Hitchcock, E. 1865 A, p. 9, pl. ii, fig. 3.
Hitchcock, C. H. 1889 B, p. 118.

Hitchcock, E. 1863 B, p. 86. (No description.)
  Triassic; Connecticut.

Trihamus magnus C. H. Hitch.

Hitchcock, C. 1889 B, pp. 118, 126.
  Triassic; Connecticut.

TYPOPUS E. Hitch.   Type *T. abnormis* E. Hitch.

Hitchcock, E. 1845 B, p. 25.
   1848 A, p. 212.
   1856 A, p. 105.

Typopus abnormis E. Hitch.

Hitchcock, E. 1845 B, p. 25.
Hitchcock, C. H. 1889 B, p. 119, 122.
Hitchcock, E. 1844 A, pp. 294, 307, pl. iii, figs. 6–8. (Sauroidichnites.)

Hitchcock, E. 1848 A, p. 212, pl. x, fig. 6.
   1858 A, p. 105, pl. xvii, fig. 9; pl. xlv. fig. 7.
   Triassic; Massachusetts, Connecticut.

Typopus gracilis E. Hitch.

Hitchcock, E. 1858 A, p. 106, pl. xvii, fig. 10.
Hitchcock, C. H. 1889 B, p. 119.
   Triassic; Connecticut.

XIPHOPEZA E. Hitch.   Type *X. triplex* E. Hitch.

Hitchcock, E. 1848 A, p. 239.
   1858 A, p. 113.

Xiphopeza triplex E. Hitch.

Hitchcock, E. 1858 A, p. 113, pl. xx, fig. 6; pl. lii, figs. 3, 4, 6.

Deane, J. 1861 A, p. 45, pl. xxiv, fig. 2.
Hitchcock, C. H. 1889 B, p. 119.
   Triassic; Massachusetts.

Class MAMMALIA Linn.

Linnaeus, C., 1858, Syst. Nat., ed. 10, 1, p. 14.
Albrecht, P. 1886 A.
Allen, H. 1882 A.
Ameghino, F. 1891 D, p. 282.
   1896 A.
   1898 B.
   1899 A, p. 555.
Baird, S. F. 1857 A.
Baur, G. 1884 C.
   1884 D.
   1885 G.
   1885 H.

Baur, G. 1886 A.
   1886 B.
   1889 D.
   1891 C, p. 354.
   1892 D, p. 210.
   1894 A.
Baur and Case 1897 A.
Bemmelen, J. F. 1899 A, p. 162.
Branco, W. 1897 A.
Cope, E. D. 1871 B, p. 196.
   1873 E, p. 645.
   1874 M.

Cope, E. D. 1880 B, p. 58.
1881 J.
1882 A.
1883 M, p. 562.
1884 E.
1884 I, p. 43.
1884 K, p. 328.
1884 L, p. 324.
1884 O, p. 166.
1884 FF.
1884 HH.
1884 II.
1886 A.
1885 G.
1887 S, pp. 241, 317, 341.
1889 C.
1889 R.
1891 N, p. 64.
1893 M.
1898 B, p. 101.
Dawkins, W. B. 1870 A.
1880 A.
Dobson, G. E. 1885 A.
Fleischman, A. 1891 A.
Flower, W. H. 1869 A.
1876 B.
1876 C.
1883 A.
1883 D.
1885 A.
Flower and Lydekker 1891 A.
Frenkel, F. 1873 A.
Fürbringer, M. 1885 B.
Gadow, H. 1889 A, p. 474.
1896 A, p. 42.
1898 A, p. 39.
Gaupp, E. 1894 A.
Gegenbaur, C. 1864 A.
1865 B.
1898 A.
Gill, T. 1871 A.
1872 A.
Goodrich, E. S. 1894 E.
Haacke, W. 1888 A, p. 8.
1893 A, p. 719.
Haeckel, E. 1895 A, p. 419
Hallmann, E. 1837 A.
Hoffmann, C. K. 1879 A, p. 31.
Howes, G. B. 1887 B.
1893 A.
1896 A.
Humphrey, G. M. 1866 A.
Humphreys, J. 1889 A, p. 137.
Huxley, T. H. 1870 A.
1872 A, p. 273.
1876 B, xiv, p. 33.
1880 B.
Kehner, F. A. 1876 A.
Kehrer, G. 1886 A, p. 83.
Kingsley and Ruddick 1900 A, p. 219.
Köstlin, O. 1844 A.
Kollmann, J. 1888 A.
Kükenthal, W. 1890 C.
1892 A.
1892 B.
1896 B.
Landois, — 1881 A, p. 125.

Lataste, F. 1889 A.
Leboucq, H. 1896 A.
Leche, W. 1893 A.
1893 B.
1895 A.
1896 A.
1896 B.
Lemoine 1898 A.
Lydekker, R. 1885 B.
1885 C.
1886 A.
1886 B.
1887 A.
Maggi, L. 1898 B.
1898 C.
1898 D.
Marsh, O. C. 1877 E.
1885 C.
1887 A.
1891 H.
1898 B.
Mehnert, E. 1889 A.
Nicholson and Lydekker 1889 A, p. 1245.
Osborn, H. F. 1887 D.
1888 E.
1888 F.
1892 E.
1892 I.
1893 A.
1893 D.
1893 G.
1897 C.
1896 A.
1898 B.
1898 L.
1898 O, p. 175.
1898 R, p. 309.
1900 J, p. 943.
Owen, R. 1843 A.
1845 B, p. 296.
1851 C.
1858 A.
1868 A.
Parker, W. K. 1868 A, p. 192.
1874 B.
1879 B.
1885 A.
1886 A.
Parsons, F. G. 1899 A.
1900 A.
Pollard, E. C. 1893 A
Reh, L. 1895 A.
Römer, F. 1898 A.
Röse, C. 1892 A.
1892 D.
1892 E.
1893 C.
1895 A.
1896 A.
Roger, O. 1889 A.
1896 A.
Rüttimeyer, L. 1877 A.
1888 A.
1891 A.
1892 A.
Schlosser, M. 1888 A.
1889 A.

Schlosser, M. 1890 A.
    1890 C.
    1890 D.
    1892 A.
    1893 A.
Schmidt, O. 1886 A.
Schwalbe, G. 1894 A.
Scott, W. B. 1891 B, p. 378.
    1891 C.
    1892 A.
    1892 C.
    1899 B, p. 92.
Seeley, H. G. 1888 D, p. 106.
    1888 E.
    1888 H.
    1898 A, p. 357.
Slade, D. D. 1888 A.
Smith, J. 1895 A.
Steinmann and Döderlein 1890 A, p. 678.
Thilenius, G. 1894 A.
Thomas, O. 1888 A.

Thomas, O. 1890 A.
    1892 B.
Thompson, J. A. 1897 A, p. 41.
Tomes, C. S. 1898 A, p. 299.
Tornier, G. 1886 A.
    1888 A.
    1890 A.
Trouessart, E. L. 1889 A.
    1898 A.
    1899 A.
Turner, H. N. 1848 A.
Weber, Max 1887 A.
    1893 A.
    1893 B.
Weyhe, — 1875 A.
Woodward, A. S. 1898 B, p. 245.
Woodward, M. F. 1892 A.
    1893 A.
    1896 A.
Zittel, K. A. 1898 A.
    1893 B.

Subclass PROTOTHERIA Gill.

Gill, T. 1872 B, p. vi (Prototheria); pp. 27, 46
(Ornithodelphia).
Cope, E. D. 1887 Q.
    1889 R, pp. 873, 874.
    1891 N, p. 377.
    1898 B, p. 102.
Flower, W. H. 1883 A.
    1883 D, p. 377.
    1885 A. (Ornithodelphia.)
Flower and Lydekker 1889 A, p. 117.
Gadow, H. 1898 A, p. 39.
Gill, T. 1888 A.

Haeckel, E. 1895 A, p. 468. (Monotrema in part.)
Huxley, T.H. 1872 A, p. 274. (Ornithodelphia.)
    1880 B.
Osborn, H. F. 1893 A, p. 391.
    1899 B, p. 92.
Parker, W. K. 1885 A, p. 25.
Steinmann and Döderlein 1890 A, p. 696.
Tomes, C. S. 1898 A, p. 300.
Wilson and Hill 1897 A, p. 429.
Woodward, A. S. 1898 B, p. 248.
Zittel, K. A. 1898 B, p. 61. (Eplacentalia.)

Order PROTODONTA Osborn.

Osborn, H. F. 1887 E.
Cope, E. D. 1889 R, p. 874.
    1891 N, pp. 64, 66.
    1898 B, p. 102.
Gadow, H. 1898 A, p. 40.

Haeckel, E. 1895 A, p. 470. (Dromatheridæ.)
Hay, O. P. 1899 F, p. 682.
Osborn, H. F. 1888 G, p. 222.
Zittel, K. A. 1898 B, p. 96.

DROMATHERIIDÆ.

Gill, T. 1872 B, p. 27.
Cope, E. D. 1889 R, p. 874.
    1898 B, p. 102.
Marsh, O. C. 1887 A, p. 344.

Nicholson and Lydekker 1889 A, p. 1272.
Osborn, H. F. 1887 F, p. 291.
    1888 G, p. 222.
Steinmann and Döderlein 1890 A, p. 697.

DROMATHERIUM Emmons.   Type *D. sylvestre* Emmons.

Emmons, E. 1857 A, p. 93.
Ameghino, F, 1896 A.
Emmons, E. 1858 A, p. 78. (No description.)
    1860 A, p. 175.
Flower and Lydekker 1891 A, p. 113.
Goodrich, E. S. 1894 A, p. 426.
Leidy, J. 1857 B, p. 150.
    1869 A, p. 410.
    1870 F.
Lydekker, R. 1887 A, p. 269.
Marsh, O. C. 1891 I, p. 237.
Nicholson and Lydekker 1889 A, p. 1272.

Osborn H. F. 1886 B.
    1887 D.
    1887 E.
    1887 F, p. 291.
    1888 B.
    1888 F, p. 1067.
    1888 G, pp. 222, 289.
    1891 C, p. 776.
    1892 I.
    1893 D.
    1895 C. p. 6.
Owen, R. 1860 E, p. 302

Schlosser, M. 1890 C, 242, fig. I, 1.
Scott, W. B. 1892 A, p. 405.
Winge, H. 1893 C, p. 75.
Woodward, A. S. 1898 B, p. 256.
Zittel, K. A. 1890 B, p. 96.
  Authors previously to 1886 included *Microconodon* in this genus.

Dromatherium sylvestre Emmons.

Emmons, E, 1857 A, p. 93, fig. 66.
Cope, E. D. 1875 V, p. 51.
Edwards, A. M. 1895 A. (Identification doubtful.)
Emmons, E. 1860 A, p. 175, fig. 152.
Flower, W. H. 1883 D, p. 375.
Leidy, J. 1859 G.
  1869 A, p. 410.

Leidy, J. 1870 F.
Osborn, H. F. 1886 A.
  1886 B, p. 360.
  1887 E.
  1888 G, pl. ix, fig. 77.
  1891 C, p. 777.
  1895 C, p. 6, pl. AA, fig. 2. ("Dromatherium.")
Trouessart, E. L. 1898 A, p, 1252.
Roger, O. 1896 A, p. 12.
Woodward, A. S. 1890 A, p. 393.
Zittel, K. A. 1893 B, p. 96, fig. 73.
  Triassic; North Carolina, New Jersey?
  Authors writing before 1886 included *Microconodon tenuirostris* Osb. in this species.

MICROCONODON Osborn.   Type *M. tenuirostris* Osborn.

Osborn, H. F. 1886 A.
Ameghino, F. 1896 A.
Cope, E. D. 1888 R.
Goodrich, E. S. 1894 A, p. 429.
Marsh, O. C. 1891 I, p. 612.
Nicholson and Lydekker 1889 A, p. 1273.
Osborn, H. F. 1886 B, p. 362.
  1887 D.
  1887 E.
  1888 F, p. 1073.
  1888 G, p. 223.
  1891 C, p. 777.
  1892 I.
  1893 G, p. 496.
  1895 C, p. 3.
  1898 R, p. 353.

Schlosser, M. 1890 C, p. 242, fig. I, 2.
Winge, H. 1893 C, p. 76.
Woodward, A. S. 1898 B, p. 256.
Zittel, K. A. 1893 B, p. 97.
  Before the year 1886 included in *Dromatherium*.

Microconodon tenuirostris Osborn.

Osborn, H. F. 1886 A.
Cope, E. D. 1888 R, pl. xiv.
Osborn, H. F. 1886 B, p. 362.
  1888 G, p. 223, fig. A, 10, B.
  1898 R, p. 353, fig. 14.
Roger, O. 1896 A, p. 12.
Seeley, H. G. 1895 B, p. 89.
Trouessart, E. L. 1898 A, p. 1252.
  Triassic; North Carolina.

Order ALLOTHERIA Marsh.

Marsh, O. C. 1880 F, p. 239.
  Most of the authors below cited employ Cope's name *Multituberculata* for this group. Citations of such authors are followed by the contraction (Mult.).
Ameghino, F. 1894 A, p. 332. (Mult.)
Cope, E. D. 1884 O, p. 167. (Allotheria.)
  1884 Y, p. 687. (Mult.)
  1888 J. (Mult.)
  1888 K. (Mult.)
  1888 M. (Mult.)
  1888 R. (Mult.)
  1888 CC, p. 302. (Mult.)
  1889 C. (Mult.)
  1889 R, p. 874. (Mult.)
  1891 N, p. 65. (Mult.)
  1898 B, p. 102. (Mult.)
Flower and Lydekker 1891 A, p. 109. (Mult.)
Gadow, H. 1898 A, p. 39.
Goodrich, E. S. 1894 A, p. 429.
Haeckel, E. 1895 A, p. 476.
Hay, O. P. 1899 F, p. 682.
Lydekker, R. 1887 A, p. 195. (Mult.)
Major, Fors. 1893 A, p. 210.
Marsh, O. C. 1887 A, p. 345.
  1889 D, p. 83.

Marsh, O. E. 1891 H.
  1891 I, p. 239.
  1892 B.
Matthew, W. D. 1899 A, p. 28. (Mult.)
Nicholson and Lydekker 1889 A, p. 1266. (Mult.)
Osborn, H. F. 1887 F, pp. 284, 285. (Mult.)
  1888 B, p. 75. (Mult.)
  1888 C, p. 232. (Mult.)
  1888 E, p. 926. (Mult.)
  1888 F, p. 1068. (Mult.)
  1888 G. (Mult.)
  1891 A. (Mult.)
  1891 C, p. 778. (Mult.)
  1891 E, p. 599. (Mult.)
  1893 B, p. 313. (Mult.)
  1893 D. (Mult.)
  1897 C. (Mult.)
  1898 O. (Mult.)
Owen, R. 1884 A. (Mult.)
Roger, O. 1896 A, p. 3. (Allotheria.)
Schlosser, M. 1890 C, p. 276. (Mult.)
Steinmann and Döderlein 1890 A, p. 698. (Mult.)
Trouessart, E. L. 1898 A, p. 1253.
Woodward, A. S. 1898 B, p. 248. (Mult.)
Zittel, K. A. 1893 B, p. 72.

BOLODONTIDÆ.

Osborn, H. F. 1887 F, p. 285.
Cope, E. D. 1887 H, p. 567. (Chirogidæ.)
   1888 J. (Chirogidæ.)
   1888 CC, p. 307. (Chirogidæ.)
   1898 B, p. 102. (Chirogidæ.)
Haeckel, E. 1895 A, p. 470. (Bolodontida.)
Lydekker, R. 1887 A, p. 202.
Marsh, O. C. 1887 A, p. 329. (Bolodon.)
Nicholson and Lydekker 1889 A, p. 1269
Osborn, H. F. 1887 A. (Bolodon.)

Osborn, H. F. 1888 C, p. 233.
   1888 G, pp. 213, 251.
   1891 B, p. 44. (Allodontidæ.)
Owen, R. 1871 A, p. 54. (Bolodon.)
Roger, O. 1896 A, p. 4.
Steinmann and Döderlein 1890 A, p. 699.
Trouessart, E. L. 1898 A, p. 1258.
Woodward, A. S. 1898 B, p. 251. (Bolodon.)
Zittel, K. A. 1893 B, p. 77.

ALLACODON Marsh.   Type *A. lentus* Marsh.

Marsh, O. C. 1889 F, p. 178.
Osborn, H. F. 1891 B.
   1893 B, pp. 314, 315, 317.
Trouessart, E. L. 1896 A, p. 1258.
Zittel, K. A. 1893 B, p. 79.

Allacodon fortis Marsh.

Marsh, O. C. 1892 B, p. 255, pl. vii, fig. 4.
Roger, O. 1896 A, p. 4.
Trouessart, E. L. 1898 A, p. 1258.
Walcott, C. D. 1900 A, p. 23.
Cretaceous (Laramie); Wyoming.

Allacodon lentus Marsh.

Marsh, O. C. 1889 F, p. 178, pl. viii, figs. 17–31.
   1897 C, p. 521, fig. 89.
Osborn, H. F. 1891 A, p. 130, fig. 7.
   1891 B, p. 45.
   1891 E, p. 604, fig. 9.

Roger, O. 1896 A, p. 4.
Trouessart, E. L. 1898 A, p. 1258.
Cretaceous (Laramie); Wyoming.

Allacodon pumilis Marsh.

Marsh, O. C. 1889 F, p. 179.
   1892 B, p. 255, pl. vii, fig. 3.
Osborn, H. F. 1891 A, p. 131.
   1891 E, p. 604.
Roger, O. 1896 A, p. 4.
Trouessart, E. L. 1898 A, p. 1258.
Walcott, C. D. 1900 A, p. 23.
Zittel, K. A. 1893 B, p. 79, fig. 49.
Cretaceous (Laramie); Wyoming.

Allacodon rarus Marsh.

Marsh, O. C. 1892 B, p. 256, pl. vii, fig. 5.
Roger, O. 1896 A, p. 4.
Trouessart, E. L. 1898 A, p. 1258.
Cretaceous (Laramie); Wyoming.

ALLODON Marsh.   Type *A. laticeps* Marsh.

Marsh, O. C. 1881 F, p. 511.
Lydekker, R. 1887 A, p. 203. (=? Bolodon.)
   1893 A. (Syn.? of Bolodon.)
Nicholson and Lydekker 1889 A, p. 1270.
Marsh, O. C. 1887 A, p. 329.
Osborn, H. F. 1887 A.
   1888 C, p. 233.
   1888 G, pp. 218, 251.
Osborn and Earle 1895 A, p. 5.
Schlosser, M. 1890 C, p. 275.
Steinmann and Döderlein 1890 A, p. 699.
Trouessart, E. L. 1898 A, p. 1258.
Winge, H. 1893 C, p. 77. (Allodon, Bolodon.)
Zittel, K. A. 1893 B, pp. 78, 722.

Roger, O. 1896 A, p. 4.
Trouessart, E. L. 1898 A, p. 1258.
Woodward, A. S. 1898 B, p. 250, fig. 147.
Zittel, K. A. 1893 B, p. 78, fig. 48 A, B, C.
Jurassic; Wyoming.

Allodon laticeps Marsh.

Marsh, O. C. 1881 F, p. 511.
   1887 A, p. 331, pl. vii, figs. 1–6.
Osborn, H. F. 1888 G, p. 217, fig. 10.
Roger, O. 1896 A, p. 4.
Trouessart, E. L. 1898 A, p. 1258.
Zittel, K. A. 1893 B, p. 78, fig. 48 D, E, F.
Jurassic; Wyoming.

Allodon fortis Marsh.

Marsh, O. C. 1887 A, p. 331, pl. vii, figs. 7–10.

CHIROX Cope.   Type *C. plicatus* Cope.

Cope, E. D. 1884 K, p. 821.
   1887 H.
   1888 I, p. 10.
   1889 L.
Matthew, W. D. 1897 C, p. 265.
Nicholson and Lydekker 1889 A, p. 1269.
Osborn, H. F. 1887 F, p. 286.
   1888 G p, 219.

Osborn, H. E. 1893 D.
Winge, H. 1893 C, p. 77.
Zittel, K. A. 1893 B, p. 79.

Chirox plicatus Cope.

Cope, E. D. 1884 K, p. 821.
   1887 H.
   1888 I, p. 10, fig. 7.

Cope, E. D. 1888 CC, p. 302.
1889 C, p. 269, fig. 91.
1896 B, p. 104, fig. 42.
Matthew, W. D. 1897 C, p. 263.
1899 A, p. 29.

Osborn, H. F. 1889 G, p. 219, fig. 12.
1891 E, p. 605, fig. 11.
Roger, O. 1896 A, p. 4.
Trouessart, E. L. 1898 A, p. 1256.
Eocene (Torrejon); New Mexico.

PLAGIAULACIDÆ.

Gill, T. 1872 B, p. 27.
Ameghino, F. 1889 A, p. 268. (Plagiaulacoidea, in part; Plagiaulacidæ.)
1891 B, p. 38.
Cope, E. D. 1882 U.
1882 W.
1884 O, p. 167.
1884 Y, p. 687.
1888 J.
1888 CC, p. 307.
1889 R, p. 874.
1896 B, p. 102.
Haeckel, E. 1895 A, p. 470. (Plagiaulacida.)
Lydekker, R, 1887 A, p. 195.
Marsh, O. C. 1879 G, p. 397.
1880 F, p. 239.
1881 F, p. 512.
1887 A, pp. 329, 344.
1889 D, p. 84 (Cimolodontidæ); p. 85 (Dipriodontidæ); p. 86 (Tripriodontidæ).

Marsh, O. C. 1889 F, p. 177. (Cimolomidæ.)
1891 I, p. 239.
Matthew, W. D. 1899 A, p. 28.
Nicholson and Lydekker 1889 A, p. 1257.
Osborn, H. F. 1887 F, p. 285.
1888 G, p. 213.
1891 B. (Plagiaulacidæ, Stereognathidæ.)
1891 C, p. 781.
1892 D.
1893 B.
1893 D.
1893 G, p. 214.
Roger, O. 1896 A, p. 4.
Schlosser, M. 1884 B, p. 640.
1888 A, p. 564.
Steinmann and Döderlein 1890 A, p. 700.
Winge, H. 1893 C, p. 77.
Woodward, A. S. 1898 B, p. 249.
Zittel, K. A. 1893 B, p. 79.

PLAGIAULACINÆ.

Osborn and Earle 1895 A, p. 11.   !

PLAGIAULAX Falconer.   Type *P. becklesii* Falconer.

Falconer, H. 1857 A.
This genus being the type of the family *Plagiaulacidæ*, the following references to its literature are given, although no species are known from our region.
Ameghino, F. 1891 A, p. 40.
Cope, E. D. 1882 U, p. 416.
1882 W, p. 520.
1884 O, p. 168.
1884 Y, p. 691.
Falconer, H. 1862 A.
1868 A, ii, p. 408, pls. xxiii, xxiv.

Lemoine, V. 1882 A.
Lydekker, R. 1887 A, p. 196.
1893 A.
Nicholson and Lydekker 1889 A, p. 1257.
Osborn, H. F. 1888 C.
1888 G, p. 215.
Owen, R. 1871 A, p. 75.
Schlosser, M. 1890 C, p. 277.
Seeley, H. 1888 H, p. 141.
Winge, H. 1893 C, p. 78.
Woodward, A. S. 1898 B, p. 249
Zittel, K. A. 1893 B, p. 80.

CTENACODON Marsh.   Type *C. serratus* Marsh.

Marsh, O. C. 1879 G, p. 396.
Cope, E. D. 1882 U, p. 416.
1882 W, p. 521.
1884 O, p. 168.
1884 Y, p. 691.
1888 J.
Lydekker, R. 1893 A.
Marsh, O. C. 1885 C.
1887 A, p. 332.
Nicholson and Lydekker 1889 A, p. 1257.
Osborn, H. F. 1887 F, p. 285.   .
1888 G, p. 215.
1893 A.
1893 B, p. 314.
Schlosser, M. 1890 C, p. 277.
Trouessart, E. L. 1898 A, p. 1254.
Winge, H. 1893 C, p. 77.

Woodward, A. S. 1898 B, p. 251.
Wortman, J. L. 1885 A, p. 299.
Zittel, K. A. 1893 B, pp. 81, 722.

Ctenacodon nanus Marsh.

Marsh, O. C. 1881 F, p. 512.
Roger, O. 1896 A, p. 4.
Trouessart, E. L. 1898 A, p. 1254.
Jurassic; Wyoming.

Ctenacodon potens Marsh.

Marsh, O. C. 1887 A, p. 333, pl. viii, figs. 2, 3, 7-9.
Osborn, H. F. 1891 E, p. 604, fig. 8a.
Roger, O. 1896 A, p. 4.
Trouessart, E. L. 1898 A, p. 1254.
Woodward, A. S. 1898 B, p. 250, fig. 147.
Jurassic; Wyoming.

Ctenacodon serratus Marsh.

Marsh, O. C. 1879 G, p. 396.
Cope, E. D. 1884 Y, p. 608, fig. 7.
    1888 J, fig. 9*a*.
    1889 C, p. 270, fig. 98*a*.
Marsh, O. C. 1880 F, p. 238, figure.
    1887 A, p. 333, pl. viii, figs. 1, 4–6.

Marsh, O. C. 1897 C, p. 508, fig. 70.
Osborn, H. F. 1888 G, p. 216, fig. 8.
Roger, O. 1896 A, p. 4.
Trouessart, E. L. 1898 A, p. 1254.
Woodward, A. S. 1898 B, p. 250, fig. 147.
Zittel, K. A. 1893 B, p. 80, figs. 53, 54.
    Jurassic; Wyoming.

NEOPLAGIAULAX Lemoine.   Type *N. eocænus* Lemoine.

Lemoine, V. 1883 A.
Cope, E. D. 1887 H, p. 567.
Lemoine, V. 1891 A, p. 289.
Lydekker, R. 1887 A, p. 196.
Marsh, O. C. 1887, A, p. 344.
Matthew, W. D. 1897 C. p. 265.
Nicholson and Lydekker 1889 A, p. 1267.
Osborn, H. F. 1891 C, p. 781.
    1892 I.
Roger, O. 1896 A, p. 6.
Schlosser, M. 1890 C, p. 277.
Winge, H. 1893 C, p. 78.
Zittel, K. A. 1893 B, p. 84.

Neoplagiaulax americanus Cope.

Cope, E. D. 1885 M, p. 493.
    1886 I, p. 451.

Cope, E, D. 1888 CC, p. 302.
Matthew, W. D. 1897 C, p. 263.
    1899 A, p. 28.
Osborn and Earle 1895 A, p. 7.
Roger, O. 1896 A, p. 6.
Trouessart, E. L. 1898 A, p. 1257.
    Eocene (Puerco); New Mexico.

Neoplagiaulax molestus Cope.

Cope, E. D. 1889 I.
    1888 CC, p. 307, pl. v, figs. 10, 11.
Matthew, W. D. 1897 C, p. 263.
    1899 A, p. 28.
Osborn and Earle 1895 A, p. 6. (Chirox.)
Roger, O. 1896 A, p. 6.
Trouessart, E. L. 1898 A, p. 1257.
    Eocene (Torrejon); New Mexico.

PTILODUS Cope.   Type *P. mediævus* Cope.

Cope, E. D. 1881 V.
    1882 U, p. 416.
    1882 W, p. 520.
    1884 O, pp. 168, 172.
    1884 Y, p. 691.
    1888 J, p. 12.
Lydekker, R. 1887 A, p. 196.
Nicholson and Lydekker 1889 A, p. 1267.
Osborn, H. F. 1888 G, p. 217.
    1891 A, p. 134.
    1893 B, p. 314, pl. vii, figs. 1–6, text figs. 1, 2.
    1893 D, p. 30.
Winge, H. 1893 C, p. 78.
Woodward, A. S. 1898 B, p. 252.
Zittel, K. A. 1893 B, p. 83.

Ptilodus mediævus Cope.

Cope, E. D. 1881 V.
    1882 W, p. 522.
    1884 O, p. 173, pl. xxiii *d*, fig. 1.
    1884 Y, p. 696, fig. 8.

Cope, E. D. 1888 CC, p. 302.
    1893 B, p. 104, fig. 41.
Lydekker, R. 1887 A, p. 196, fig. 30.
Matthew, W. D. 1897 C, p. 263.
    1899 A, p. 28.
Osborn and Earle 1895 A, p. 7.
Roger, O, 1896 A, p. 5.
Trouessart, E. L. 1898 A, p. 1256.
    Eocene (Torrejon); New Mexico.

Ptilodus trouessartianus Cope.

Cope, E. D. 1882 AA, p. 686.
    1884 O, p. 737, pl. xxv *f*, fig. 19.
    1885 M, p. 493.
    1888 CC, p. 302.
Matthew, W. D. 1897 C, p. 263.
    1899 A, p. 28.
Osborn, H. F. 1893 B, p. 315, text fig. 1.
Roger O, 1896 A, p. 5.
Trouessart, E. L. 1898 A, p. 1256.
    Eocene (Torrejon); New Mexico.

CIMOLOMYS Marsh.   Type *C. gracilis* Marsh.

Marsh, O. C. 1889 D, P. 84.
Cope, E. D. 1892 U, p. 762. (Cimolodon.)
Marsh, O. C. 1889 D, p. 84 (Cimolodon, type *C. nitidus*); p. 85 (Nanomys, type *N. minutus*).
    1892 B, p. 261. (Nanomyops, to replace Nanomys.)
Nicholson and Lydekker 1889 A, p. 1268.
Osborn, H. F. 1891 A, p. 125.
    1891 B.
    1893 B, p. 314. (Ptilodus.)
Steinmann and Döderlein 1890 A, p. 700.
Trouessart, E. L. 1898 A, p. 1256.
Zittel, K. A. 1893 B, p. 82.

Cimolomys agilis (Marsh).

Marsh, O. C. 1889 B, p. 255, pl. vi, fig. 8. (Cimolodon.)
Roger, O. 1896 A, p. 6.
Trouessart, E. L. 1898 A, p. 1257.
Walcott, C. D. 1900 A, p. 23. (Cimolodon.)
    Cretaceous (Laramie); Wyoming.

Cimolomys bellus Marsh.

Marsh, O. C. 1889 D, p. 84.
    1892 B, p. 262, pl. vii, fig. 2.
Osborn, H. F. 1891 A, p. 125. (Syn. of C. gracilis.)

Osborn, H. F. 1891 B, p. 44.
1891 E, p. 599.
1893 B, p. 316. (Ptilodus.)
Roger, O. 1896 A, p. 6.
Trouessart, E. L. 1898 A, p. 1256. (Syn. of C. gracilis.)
Cretaceous (Laramie); Wyoming.

Cimolomys digona Marsh.

Marsh, O. C. 1889 F, p. 177, pl. vii, figs. 1-4, 13-16.
1897 C, p. 521, fig. 87.
Osborn, H. F. 1891 A, p. 126.
1891 B, p. 44.
1891 E, p. 599.
1893 B, p. 317. (Ptilodus.)
Roger, O. 1896 A, p. 6.
Trouessart, E. L. 1898 A, p. 1256.
Cretaceous (Laramie); Wyoming.

Cimolomys formosus (Marsh).

Marsh, O. C. 1889 F, p. 179, pl. viii, figs. 32-39. (Halodon.)
Osborn, H. F. 1891 A, p. 126, fig. 1,c.
1891 B, p. 44.
1893 B, p. 315. (Ptilodus.)
Roger, O. 1896 A, p. 6.
Trouessart, E. L. 1898 A, p. 1257.
Zittel, K. A. 1893 B, p. 84, fig. 61,C. (Halodon.)
Cretaceous (Laramie); Wyoming.

Cimolomys gracilis Marsh.

Marsh, O. C. 1889 D, p. 84, pl. ii, figs. 1-4.
Osborn, H. F. 1891 A, p. 125, figs. 2,a, 3.
1891 B, p. 44.
1891 E, pp. 599, 601, figs. 3, 4.
1893 B, p. 317. (Ptilodus.)
Roger, O. 1896 A, p. 6.
Trouessart, E. L. 1898 A, p. 1256.
Zittel, K. A. 1893 B, p. 81, fig. 56.
Cretaceous (Laramie); Wyoming.

Cimolomys minutus (Marsh).

Marsh, O. C. 1889 D, p. 85, pl. ii, figs. 9-12. (Nanomys.)
1892 B, p. 261, pl. vi, fig. 2. (Nanomyops.)

Osborn, H. F. 1891 A, p. 126, figs. 2,c, 3.
1891 E, p. 600, figs. 3, 4.
1892 B, p. 44.
1893 B, p. 316.
Roger, O. 1896 A, p. 6.
Trouessart, E. L. 1898 A, p. 1256. (Syn. of C. gracilis.)
Cretaceous (Laramie); Wyoming.

Cimolomys nitidus (Marsh).

Marsh, O. C. 1889 D, p. 84, pl. ii, figs. 5-8. (Cimolodon.) .
Cope, E. D. 1892 U, p. 762. (Cimolodon.)
Marsh, O. C. 1892 B, p. 255, pl. vi, figs. 7, 9. (Cimolodon.)
Osborn, H. F. 1891 A, p. 126, fig. 2,b. (Cimolodon.)
1891 B, p. 44. (Cimolomys.)
1891 E, p. 600, figs. 3, 4.
1893 B, p. 316. (Ptilodus.)
Roger, O. 1896 A, p. 6.
Trouessart, E. L. 1898 A, p. 1257. (Syn. of C. gracilis.)
Zittel, K. A. 1898 B, p. 82, figs. 57, 58.
Cretaceous (Laramie); Wyoming.

Cimolomys parvus (Marsh).

Marsh, O. C. 1892 B, p. 254, pl. vi, fig. 4. (Cimolodon.)
Roger, O. 1896 A, p. 6.
Trouessart, E. L. 1898 A, p. 1257.
Cretaceous (Laramie); Wyoming.

Cimolomys serratus (Marsh).

Marsh, O. C. 1889 D, p. 87, pl. iii, figs. 4-6, 14-17. (Halodon.)
Osborn, H. F. 1891 A, p. 126, figs. 1,b, 5,c. (Halodon.)
1891 B, p. 44.
1891 E, p. 600, fig. 2, 4.
1893 B, pp. 315, 318. (In part to Ptilodus; in part to Meniscoëssus.)
Roger, O. 1896 A, p. 6. (Halodon.)
Trouessart, E. L. 1898 A, p. 1257. (Syn. of C. gracilis.)
Cretaceous (Laramie); Wyoming.

MENISCOËSSUS Cope.   Type *M. conquistus* Cope.

Cope, E. D. 1882 BB.
1884 Y, p. 691.
1888 J.
1889 L.
1892 U, p. 762. (Dipriodon=? Ptilodus.)
Marsh, O. C. 1889 D, p. 85 (Dipriodon, type D. robustus); p. 86 (Selenacodon, type S. fragilis; Tripriodon, type T. cælatus;); p. 87 (Halodon in part, type H. sculptus).
1891 I.
1892 D, p. 82.
1892 G, p. 174.
Nicholson and Lydekker 1889 A, p. 1268.
Osborn, H. F. 1887 F, p. 286.
1888 G, p. 217.
1891 A, p. 128.
1891 B.
1891 C, pp. 779, 782.

Osborn, H. F. 1891 E, pp. 597, 609.
1893 B, pp. 818, 317, 320. (Syn? of Paronychodon.)
1893 D, p. 80.
Schlosser, M. 1890 C, p. 277.
Trouessart, E. L. 1898 A, p. 1254.
Wortmann, J. L. 1883 A, p. 709.
1885 A, p. 296.
Zittel, K. A. 1893 B, p. 83.

Meniscoëssus brevis (Marsh).

Marsh, O. C. 1889 F, p. 177, pl. vii, figs. 9-12. (Selenacodon.)
Osborn, H. F. 1891 A, p. 129, fig. 5,d.
1891 B, p. 45.
1891 E, p. 602, fig. 5.
1893 B, p, 318.
Roger, O. 1896 A, p. 5. (Syn. of M. conquistus.)

Trouessart, E. L. 1896 A, p. 1255. (Syn. of M. conquistus.)
Cretaceous (Laramie); Wyoming.

Meniscoëssus caperatus (Marsh).

Marsh, O. C. 1889 D, p. 86, pl. iii, figs. 18–20. (Tri-
priodon.)
Dana, J. D. 1896 A, p. 853, fig. 1443. (Tripriodon.)
Marsh, O. C. 1892 B, p. 261, pl. v, fig. 2. (Triprio-
don.)
Osborn, H. F. 1891 A, p. 129.
    1891 B, p. 45.
    1891 E, p. 602, fig. 8.
    1896 B, p. 318.
Trouessart, E. L. 1896 A, p. 1255. (Syn. of M. con-
quistus.)
Cretaceous (Laramie); Wyoming.

Meniscoëssus cœlatus (Marsh).

Marsh, O. C. 1889 D, p. 86, pl. ii, figs. 19–21. (Tri-
priodon.)
Dana, J. D. 1896 A, p. 853, fig. 1442. (Tripriodon.)
Major, Fors. 1893 A, p. 211. (Tripriodon.)
Osborn, H. F. 1891 A, p. 129, fig. 4,b.
    1891 B, p. 45.
    1891 E, p. 602, figs. 5, 8.
    1896 B, p. 319.
Roger, O. 1896 A, p. 5. (Syn. of M. conquistus.)
Trouessart, E. L. 1896 A, p. 1255. (Syn. of M. con-
quistus.)
Zittel, K. A. 1893 B, p. 82, fig. 59B. (Tripriodon.)
Cretaceous (Laramie); Wyoming.

Meniscoëssus conquistus Cope.

Cope, E. D. 1882 BB.
    1884 Y, p. 691, fig. 7.
    1888 J.
    1889 C, p. 270, fig. 98 d, f.
    1889 L, p. 490.
Marsh, O. C. 1889 D, p. 82.
Osborn, H. F. 1891 B p. 45.
    1891 E, p. 597 E, fig. 1.
    1891 I.
    1893 B, p. 319.
Osborn and Earle, 1895 A, p. 14.
Roger, O. 1896 A, p. 5.
Trouessart, E. L. 1896 A, p. 1255.
Zittel, K. A. 1893 B, p. 82, fig. 59 A.
Cretaceous (Laramie); Wyoming.

Meniscoëssus fragilis (Marsh).

Marsh, O. C. 1889 D, p. 86, pl. ii, figs. 22–24.
(Selenacodon.)
Major, Fors. 1893 A, p. 211. (Selenacodon.)

Osborn, H. F. 1891 A, p. 128, figs. 4,c, 5,d, 6.
    1891 B, p. 45.
    1891 E, p. 602, figs. 5, 8.
    1893 B, p. 318.
Roger, O. 1896 A, p. 5. (Syn. of M. conquistus.)
Trouessart, E. L. 1896 A, p. 1255. (Syn. of M.
conquistus.)
Cretaceous (Laramie); Wyoming.

Meniscoëssus lunatus (Marsh).

Marsh, O. C. 1889 D, p. 85, pl. ii, figs. 16–18.
(Dipriodon.)
Cope, E. D. 1892 U, p. 762. (?Ptilodus.)
Dana, J. D. 1896 A, p. 858, fig. 1440. (Dipriodon.)
Marsh, O. C. 1892 B, p. 261, pl. v, figs. 6, 7.
(Dipriodon.)
Osborn, H. F. 1891 A, pp. 128, 130, figs. 4,d, 6.
(Dipriodon.)
    1891 E, p. 602, figs. 5, 8.
    1893 B, p. 318.
Roger, O. 1896 A, p. 5. (Syn. of M. conquistus.)
Trouessart, E. L. 1896 A, p. 1255.
Zittel, K. A. 1893 B, p. 83, fig. 60A. (Dipriodon.)
Cretaceous (Laramie); Wyoming.

Meniscoëssus robustus (Marsh).

Marsh, O. C. 1889 D, p. 85, pl. ii, figs. 13–15.
(Dipriodon.)
Major, Fors. 1893 A, p. 211. (Dipriodon.)
Osborn, H. F. 1891 A, pp. 128,130, figs. 4,a. (Di-
priodon.)
    1891 B, p. 45.
    1891 C, p. 782.
    1891 E, p. 601, figs. 5, 7, 8.
    1893 B, p. 318.
Roger, O. 1896 A, p. 5. (Syn. of M. conquistus.)
Trouessart, E. L. 1896 A, p. 1255. (Syn. of M.
conquistus.)
Cretaceous (Laramie); Wyoming.

Meniscoëssus sculptus (Marsh).

Marsh, O. C. 1889 D, p. 87, pl. iii, figs. 7–13.
(Halodon.)
Dana, J. D. 1896 A, p. 858, figs. 1441 a–c. (Halodon.)
Marsh, O. C. 1892 B, p. 261, pl. v, fig. 4. (Halodon.)
Osborn, H. F. 1891 A, p. 126, figs. 1,a, 5,a. (Halo-
don.)
    1891 B, p. 44.
    1891 E, p. 600, fig. 2.
    1893 B, p. 318.
Roger, O. 1896 A, p. 5. (Syn. of M. conquistus.)
Trouessart, E. L. 1896 A, p. 1255. (Syn. of M.
conquistus.)
Cretaceous (Laramie); Wyoming.

ORACODON Marsh. Type *O. anceps* Marsh.

Marsh, O. C. 1889 F, p. 178.
Osborn, H. F. 1891 B.
    1893 B, p. 317. (Syn.? of Meniscoëssus.)
Zittel, K. A. 1893 B, p. 83.

Oracodon anceps Marsh.

Marsh, O. C. 1889 F, p. 178, pl. viii, figs. 13–16.
1892 B, p. 256, pl. vii, figs. 6, 7.

Marsh, O. C. 1897 C, p. 521, fig. 91.
Osborn, H. F. 1891 A, p. 131.
    1891 B, p. 45.
    1891 E, p. 605.
    1893 B, p. 319.
Roger, O. 1896 A, p. 5.
Trouessart, E. L. 1896 A, p. 1255.
Cretaceous (Laramie); Wyoming.

Oracodon conulus Marsh.

Marsh, O. C. 1892 B, p. 256, pl. vii, fig. 8.
Dana, J. D. 1896 A, p. 853, fig. 1439 a, b.
Osborn, H. F. 1893 B, p. 319.
Roger, O. 1896 A, p. 5.

Trouessart, E. L. 1898 A, p. 1256.
Walcott, C. D. 1900 A, p. 29.
Zittel, K. A. 1893 B, p. 84, fig. 63.
Cretaceous (Laramie); Wyoming.

CAMPTOMUS Marsh. Type *C. amplus* Marsh.

Marsh, O. C. 1889 D, p. 87.
Nicholson and Lydekker 1889 A, p. 1268.

Camptomus amplus Marsh.

Marsh, O. C. 1889 D, p. 87, pl. v. figs. 1–4, 18–23.
Osborn, H. F. 1891 A, p. 131.

Osborn, H. F. 1891 B, p. 45.
Osborn, H. F. 1891 E, p. 605.
Trouessart, E. L. 1898 A, p. 1256.
Woodward, A. S. 1898 B, p. 253.
Zittel, K. A. 1893 B, p. 73, fig. 44.
Cretaceous (Laramie); Wyoming.

POLYMASTODONTINÆ.

Osborn and Earle 1896 A, p. 11.
Unless otherwise indicated, the following authorities employ for this group, regarded as of family value, the term *Polymastodontidæ.*
Cope, E. D. 1884 Y, p. 687.
1888 J.
1888 CC, p. 307.
Haeckel, E. 1896 A, p. 470. (Polymastodontida.)
Lydekker, R. 1887 A, p. 199.

Nicholson and Lydekker 1889 A, p. 1269.
Osborn, H. F. 1887 F, p. 286.
1888 C.
1888 G, pp. 213, 221.
Roger, O. 1896 A, p. 6.
Schlosser, M. 1890 C, p. 277.
Trouessart, E. L. 1898 A, p. 1253.
Woodward, A. S. 1898 B, p. 252.
Zittel, K. A. 1893 B, p. 85.

CATOPSALIS Cope. Type *C. foliatus* Cope.

Cope, E. D. 1882 U, p. 416.
1882 W, p. 520.
1884 O, pp. xxxv, 168, 170, 733.

Catopsalis foliatus Cope.

Cope, E. D. 1882 U, p. 416.
1884 O, p. 171, pl. xxiiid, fig. 2.
1884 Y, p. 688, fig. 5. (Polymastodon.)

Cope, E. D. 1888 CC, p. 302. (Polymastodon.)
Matthew, W. D. 1897 C, p. 263.
1899 A, p. 28.
Osborn and Earle 1895 A, p. 11. (Polymastodon.)
Roger, O. 1896 A, p. 7. (Polymastodon.)
Trouessart, E. L. 1898 A, p. 1254. (Polymastodon.)
Eocene (Puerco); New Mexico.

TÆNIOLABIS Cope. Type *T. sulcatus* Cope.

Cope, E. D. 1882 Z, p. 604.
1884 O, p. 193. (Tæniolabis.)
1885 M, p. 493. (Syn? of Polymastodon.)
Zittel, K. A. 1893 B, p. 85. (Syn. of Polymastodon.)
By most authors this genus is included with *Polymastodon*, and to this the student is referred.

Tæniolabis sulcatus Cope.

Cope, E. D. 1882 Z, p. 604.
1884 O, p. 193, pl. xxiii e, fig. 7. (T. scalper.)
1885 M, p. 493. (T. scalper, syn.? of Polymastodon taöensis).

Roger, O. 1896 A, p. 6. (Syn. of Polymastodon taöensis.)
Eocene (Puerco); New Mexico.
After 1884 Prof. Cope regarded this species as based on what was probably an incisor tooth of *Polymastodon taöcnsis* Cope. Should Cope's identification eventually prove correct, the name *Tæniolabis* will supersede *Polymastodon.* The name *scalper* appears to be merely a substitute for the earlier *T. sulcatus.*

POLYMASTODON Cope. Type *P. taöensis* Cope.

Cope, E. D. 1882 AA, p. 684.
1884 O, pp. xxxv, 168, 732.
1884 Y, p. 688.
1888 I, p. 11.
1889 L.
Lydekker, R. 1887 A, p. 200.
Nicholson and Lydekker 1889 A, p. 1269.
Osborn, H. F. 1887 A.
1888 C.
1891 E, p. 610.
1893 B.
1898 E.

Osborn and Earle 1895 A, p. 11.
Steinmann and Döderlein 1890 A, p. 700.
Tomes, C. S. 1898 A, p. 330.
Trouessart, E. L. 1898 A, p. 1253.
Winge, H. 1893 B, p. 78.
Woodward, A. S. 1898 B, p. 252.
Zittel, K. A. 1893 B, p. 85.

Polymastodon attenuatus Cope.

Cope, E. D. 1885 M, p. 494.
1888 CC, p. 302.
Matthew, W. D. 1897 C, p. 263.

Osborn and Earle 1895 A, p. 12, fig. 1A.
Roger, O. 1896 A, p. 7.
Trouessart, E. L. 1898 A, p. 1253.
Woodward, A. S. 1898 B, p. 253, fig. 149.
Eocene (Puerco); New Mexico.

Polymastodon fissidens Cope.

Cope, E. D. 1884 K, p. 322. (Catopsalis.)
1884 Y, p. 688.
1888 CC. p. 302.
Matthew, W. D. 1897 C, p. 263.
Osborn and Earle 1895 A, pp. 12, 14.
Roger, O. 1896 A, p. 7.
Trouessart. E. L. 1898 A, p. 1253.
Eocene (Torrejon); New Mexico.

Polymastodon latimolis Cope.

Cope, E. D. 1885 J, p. 385.
1888 CC, p. 302.
Matthew, W. D. 1899 A, p. 28.
Osborn and Earle 1895 A, pp. 11, 12.
Roger, O. 1896 A, p. 7.
Trouessart, E. L. 1898 A, p. 1253.
Eocene (Puerco); New Mexico.

Polymastodon selenodus Osborn and Earle.

Osborn and Earle 1895 A, pp. 12, 15.

Matthew, W. D. 1897 C, p. 263.
1899 A, p. 28.
Roger, O. 1896 A, p. 7.
Trouessart, E. L. 1898 A, p. 1254.
Eocene (Puerco); New Mexico.

Polymastodon taöensis Cope.

Cope, E. D. 1882 AA, p. 684.
1882 AA, p. 685. (Catopsalis pollux.)
1883 E. (Catopsalis pollux.)
1884 O, p. 732, pl. xxxiii c, fig. 6 (Polymastodon taöensis); p. 734, pl. xxiii c, figs. 1–5 (Catopsalis pollux).
1884 Y, p. 689, figs. 3, 4.
1885 M, p. 493.
1888 I, p. 11, fig. 8.
1888 CC, p. 302.
1889 C, p. 270, fig. 92.
Lydekker, R. 1887 A, p. 200, fig. 32.
Matthew, W. D. 1897 C, p. 263.
1899 A, p. 28.
Osborn and Earle 1895 A, pp. 11, 12, fig. 1, C.
Roger, O. 1896 A, p. 6.
Trouessart, E. L. 1898 A, p. 1253.
Woodward, A. S. 1898 B, p. 253, fig. 149.
Zittel, K. A. 1893 B, p. 86, fig. 66.
Eocene (Puerco); New Mexico.

Subclass EUTHERIA Gill.

Gill, T. 1872 B, p. vi.
Cope, E. D. 1887 Q.
1889 R, p. 874.
1891 N, pp. 65, 66.
1898 B, p. 104.
Flower, W. H. 1883 A.
1883 D, p. 383.
Flower and Lydekker 1891 A, p. 173.
Gill, T. 1888 A.
Haeckel, E. 1895 A, p. 467. (Prodidelphia.)

Osborn, H. F. 1893 A, p. 391.
1894 A. (Cenoplacentalia, Mesoplacentalia.)
1896 L.
1899 B, p. 92.
Osborn and Earle 1895 A, p. 3. (Cenoplacentalia, Mesoplacentalia.)
Schlosser, M. 1898 A. p. 301.
Tomes, C. S. 1898 A, p. 805.
Woodward, A. S. 1898 B, p. 254. (Metatheria.)
Zittel, K. A. 1893 B, p. 117. (>Placentalia.)

Superorder *DIDELPHIA Blainville.*

Cope, E. D. 1889 R, p. 874.
1898 B, p. 104.
Flower W. H. 1883 A. (Metatheria, Didelphia.)
1883 D, p. 378. (Metatheria.)
1885 A.
Flower and Lydekker 1891 A, p. 128. (Metatheria, Didelphia.)
Gadow, H. 1898 A, p. 40. (Metatheria.)
Gill, T. 1872 A, p. 305.
1872 B, pp. 25, 46.

Hay, O. P. 1899 F, p. 682.
Huxley, T. H. 1872 A, p. 276.
1880 B. (Metatheria.)
Nicholson and Lydekker 1889 A, p. 1270.
Osborn, H. F. 1893 A, p. 391. (Metatheria.)
1898 L. (Metatheria.)
Parker, W. K. 1885 A, p. 59. (Metatheria.)
Wilson and Hill 1897 A, p. 429. (Metatheria.)
Woodward, A. S. 1898 B, p. 254. (Metatheria.)

Order MARSUPIALIA.

Baird, S. F. 1857 A, p. 230. (Marsupiata.)
Baur, G. 1885 K, p. 475.
Bensley, B. A. 1900 A, p. 558.
Cope, E. D. 1873 E, p. 644.
1880 B, p. 453.
1884 E.
1884 O, pp. 167, 788.

Cope, E. D. 1884 Y.
1888 D, p. 255.
1889 C.
1889 R, pp. 874, 876 (Pantotheria, in part; Marsupialia.)
1891 N, p. 69 (Marsupialia); p. 72 (Pantotheria).

Flower, W. H. 1869 A, p. 257.
  1876 B, xiii, p. 327.
  1885 A.
Flower and Lydekker 1891 A, p. 128.
Gaudry, A. 1878 B, p. 9.
Gegenbaur, C. 1898 A.
Gill, T. 1872 A, p. 305.
  1872 B, pp. 25, 46.   (Didelphia.)
Gill and Coues 1877 A, p. 1072.
Gruber, W. 1859 A.
Haeckel, E. 1899 A, p. 479.
Huxley, T. H. 1870 A.
  1872 A, p. 273 (Metatheria); p. 276 (Didel-
    phia).
Kollman, J. 1888 A.
Kükenthal,W. 1891 A, p. 368.
  1892 A.
Leche, W. 1893 A, p. 521.
  1893 B.
  1895 A, p. 83.
  1896 A, p. 292.
  1896 B.
Lydekker, R. 1887 A, p. ix (Pantotheria, in part);
  p. 146 (Marsupialia).
  1896 B.
  1900 A, p. 922.
Marsh, O. C. 1877 E.
  1880 F, p. 239.   (Pantotheria, in part.)
  1885 C.   (Pantotheria, in part.)
  1887 A, p. 345.   (Pantotheria, in part; Marsu-
    pialia.)
Nicholson and Lydekker 1889 A, p. 1272.
Osborn, H. F. 1888 G, p. 256.
  1891 A, p. 125.   (Pantotheria, in part.)
  1891 C, p. 780.   (Pantotheria, in part.)
  1891 E, p. 599.   (Pantotheria, in part.)
  1893 A, p. 379.
  1898 D.
  1898 L.
  1900 C, p. 566.
Owen, R. 1838 A, p. 120.
  1839 A.
  1839 D.

Owen, R. 1843 A, p. 72.
  1845 B, p. 273.
  1851 C, p. 733.
  1866 B, p. 328.
  1868 A, p. 285.
  1871 A.
Parker, W. K. 1868 A, p. 195.
Paulli, S. 1899 A, p. 147.
Pollard, E. C. 1893 A.
Reh, L. 1895 A, p. 163.
Roger, O. 1889 A.
Röse, C. 1892 A, p. 638.
  1892 D, p. 508.
  1892 E.
Ryder, J. 1878 A, p. 49.
Schlosser, M. 1886 C.
  1889 A.
  1890 A.
  1890 C.
  1890 D.
  1892 A.
  1898 A.
Schmidt, O. 1886 A, p. 93.
Schwalbe, J. 1894 A.
Slade, D. D. 1888 A, p. 242.
Steinmann and Döderlein 1890 A, p. 700.
Thomas, O. 1888 A.
  1892 B.
Thompson, J. A. 1897 A, p. 43.
Tims, H. W. M. 1896 A.
Tomes, C. S. 1898 A, p. 562.
Tornier, G. 1890 A.
Trouessart, E. L. 1898 A, p. 1155.
Turner, W. 1899 A.
Weyhe, —— 1875 A, p. 107.
Wilson and Hill 1897 A, p. 427.
Winge, H. 1893 C, pp. 61, 86.
Woodward, A. S. 1898 B, p. 255.
Woodward, M. F. 1893 A.
  1896 B.
Wortman, J. L. 1886 A, p. 493.
Zittel, K. A. 1893 B, p. 86 (Marsupialia); pp. 95,
  723, 724 (Pantotheria, in part).

Suborder POLYPROTODONTIA.

Flower and Lydekker 1891 A, p. 133.
Gadow, H. 1898 A, p. 40.
Lydekker, R. 1887 A, p. 254.
Nicholson and Lydekker 1889 A, p. 1272.

Owen, R. 1868 A, p. 293.   ("Polyprotodont.")
  1871 A, p. 88.   ("Polyprotodont section.")
Roger, O. 1896 A, p. 12.
Woodward, A. S. 1898 B, p. 255.

STAGODONTIDÆ.

Marsh, O. C. 1889 F, p. 178.
Cope, E. D. 1892 J, p. 760.   (Thlæodontidæ.)
Marsh, O. C. 1892 B, pp. 256, 257.

Osborn, H. F. 1891 B, p. 45.
  1891 C, p. 780.
  1892 D.

STAGODON Marsh.   Type *S. nitor* Marsh.

Marsh, O. C, 1889 F, p. 178.
Ameghino. F, 1894 A, p. 335.
Cope, E. D. 1892 U, p. 760.
Marsh, O. C. 1891 I, p. 240.
  1892 B, p. 257.
Osborn, H. F. 1891 B, p. 45.
  1891 C, p. 780.
  1893 B, p. 324.

Trouessart, E. L. 1898 A, p. 670.
Zittel, K. A. 1893 B, p. 510.

Stagodon nitor Marsh.

Marsh, O. C. 1889 F, p. 178, pl. vii. figs. 22-25.
  1892 B, p. 262, pl. viii, figs. 1, 2.
Osborn, H. F. 1891 A, p. 131, fig. 8a.
  1891 B, p. 45.

Osborn, H. F. 1891 C, p. 780, fig. 1.
   1891 E, p. 606, fig. 12.
   1896 B, p. 323.
Roger, O. 1896 A, p. 90.
Zittel, K. A. 1898 B, p. 511, fig. 425A.
   Cretaceous (Laramie); Wyoming.

Stagodon tumidus Marsh.

Marsh, O. C. 1889 F, p. 178, pl. vii, figs. 17–21.
Dana, J. D. 1896 A, p. 858, fig. 1488 a, b.
Marsh, O. C. 1892 B, p. 262, pl. viii, fig. 3.
Roger, O. 1896 A, p. 89.
   Cretaceous (Laramie); Wyoming.

Stagodon validus Marsh.

Marsh, O. C. 1892 B, p. 256, pl. viii, figs. 4–7.
Cope, E. D. 1892 U, p. 760.
Dana, J. D. 1896 A, p. 858, fig. 1487 a–c.
Marsh, O. C. 1897 C, p. 521, fig. 92.
Osborn, H. F. 1891 B, p. 45.
Roger, O. 1896 A, p. 90.
Zittel, K. A. 1898 B, p. 511, figs. 425B, 426.
   Cretaceous (Laramie); Wyoming.

THLÆODON Cope.   Type *T. padanacus* Cope.

Cope, E. D. 1892 U, p. 759.
   1892 D.
Osborn, H. F. 1898 A, p. 388.
   1898 B, pp. 324, 328.
   1898 D.
Zittel, K. A. 1898 B, p. 511.

Thlæodon padanicus Cope.

Cope, E. D. 1892 U, p. 759, pl. xxii.
Roger, O. 1896 A, p. 90.
Trouessart, E. L. 1898 A, p. 609.
   Cretaceous (Laramie); Wyoming.

PLATACODON Marsh.   Type *P. nanus* Marsh.

Marsh, O. C. 1889 F, p. 178.
   1892 B, p. 259.

Platacodon nanus Marsh.

Marsh, O. C. 1889 F, p. 178, pl. viii, figs. 4–12.
Hatcher, J. B. 1900 A, p. 719.
Marsh, O. C. 1892 B, p. 262, pl. xi, figs. 6, 7.

Osborn, H. F, 1891 A, p. 132, fig. 8b.
   1891 B, p. 45.
   1891 E, p. 609, fig. 12b.
   1898 B, p. 323.
   Cretaceous (Laramie); Wyoming.
   By Hatcher this genus has been referred to the fishes.

CIMOLESTIDÆ.

Marsh, O. C. 1892 B, p. 257.

Trouessart, E. L. 1898 A, p. 1202. (Epanorthidæ, in part.)

PEDIOMYS Marsh.   Type *P. elegans* Marsh.

Marsh, O. C. 1889 D, p. 89.
Nicholson and Lydekker 1889 A, p. 1277.
Osborn, H. F. 1891 A, p. 131.
   1891 B, p. 45.
   1891 E, p. 609.
   1898 B, p. 324.
Zittel, K. A. 1898 B, p.102.

Pediomys elegans Marsh.

Marsh, O. C. 1889 D, p. 89, pl. iv, figs. 23–25.

Marsh, O. C. 1892 B, p. 262, pl. x.
Osborn, H. F. 1891 A, p. 131.
   1891 B, p. 45.
   1891 E, p. 609.
   1898 B, p. 323.
   Roger, O. 1896 A, p. 15.
Trouessart, E. L. 1898 A, p. 1249.
   Cretaceous (Laramie); Wyoming.

BATODON Marsh.   Type *B. tenuis* Marsh.

Marsh, O. C. 1892 B, p. 256.
Osborn, H. F. 1893 B, p. 324.
Zittel, K. A. 1898 B, p. 106.

Batodon tenuis Marsh.

Marsh, O. C. 1892 B, p.256, pl. x, fig. 6; pl. xi, fig. 2.
Dana, J. D. 1896 A, p. 858, fig. 1486 a–c.

Osborn, H. F. 1898 B. p. 324.
Roger, O. 1896 A, p. 19.
Trouessart, E. L. 1898 A, p. 1204.
Walcott, C. D. 1900 A, p. 23.
Zittel, K. A. 1898 B, p. 106, fig. 92.
   Cretaceous (Laramie); Wyoming.

DIDELPHODON Marsh.   Type *D. vorax* Marsh.

Marsh, O. C. 1889 D, p. 88.
   1889 F, p. 179. (Didelphops, to replace Didelphodon.)
   1889 I, p. 15. (Didelphops.)

Nicholson and Lydekker 1889 A, p. 1277.
Osborn, H. F. 1891 B.
   1891 E, p. 610.
   1898 B, p. 323.

Roger, O. 1896 A, p. 18.   (Didelphops.)
Woodward, M. F. 1896 A, p. 586.
Zittel, K. A. 1893 B, p. 105.

Didelphodon comptus Marsh.

Marsh, O. C. 1889 D, p. 88, pl. iv, figs. 5–7, 20–22.
Dana, J. D. 1896 A, p. 853, fig. 1433 a–e.   (Didelphops.)
Marsh, O. C. 1892 B, p. 262, pl. x, figs. 1, 2.   (Didelphops.)
Osborn, H. F. 1891 A, p. 131.   (Didelphops.)
    1891 B, p. 45.   (Didelphops.)
    1891 E, p. 606.   (Didelphops.)
Roger, O. 1896 A, p. 18.   (Didelphops.)
Cretaceous (Laramie); Wyoming.

Didelphodon ferox Marsh.

Marsh, O. C. 1889 D, p. 88, pl. iv, figs. 4, 26–28.
Dana, J. D. 1896 A, p. 853, fig. 1435 a, b.   (Didelphops.)

Marsh, O. C. 1892 B, p. 262, pl. ix, figs. 7, 8.   (Didelphops.)
Osborn, H. F. 1891 A, p. 131.   (Didelphops.)
    1891 B, p. 45.   (Didelphops.)
    1891 E, p. 606.   (Didelphops.)
Roger, O. 1896 A, p. 18.   (Didelphops.)
Cretaceous (Laramie); Wyoming.

Didelphodon vorax Marsh.

Marsh, O. C. 1889 D, p. 88, pl. iv, figs. 1–3.
Dana, J. D. 1896 A, p. 853, fig. 1434.   (Didelphops.)
Marsh, O. C. 1892 B, p. 262, pl. ix, fig. 1.   (Didelphops.)
Osborn, H. F. 1891 A, p. 131.   (Didelphops.)
    1891 B, p. 45.   (Didelphops.)
    1891 E, p. 606.   (Didelphops.)
    1893 B, p. 323, fig. 42.   (Didelphops.)
Zittel, K. A. 1893 B, p. 105, fig. 87.
Cretaceous (Laramie); Wyoming.

CIMOLESTES Marsh.   Type *C. incisus* Marsh.

Marsh, O. C. 1889 D, p. 89.
    1889 F, p. 179.
Nicholson and Lydekker 1889 A, p. 1277.
Osborn, H. F. 1891 A, p. 131.
    1891 B, p. 45.
    1891 E, p. 610.
    1893 B, p. 325.
Zittel, K. A. 1893 B, p. 105.

Cimolestes curtus Marsh.

Marsh, O. C. 1889 D, p. 89, pl. iv, figs. 8–11.
Osborn, H. F. 1891 A, p. 131.
    1891 B, p. 45.
    1891 E, p. 606.
Roger, O. 1896 A, p. 18.

Trouessart, E. L. 1898 A, p. 1204.
Cretaceous (Laramie); Wyoming.

Cimolestes incisus Marsh.

Marsh, O. C. 1889 D, p. 89, pl. iv, figs. 12–19.
Dana, J. D. 1896 A, p. 853, fig. 1432 a–d.
Marsh, O. C. 1892 B, p. 262, pl. x, fig. 5.
Osborn, H. F. 1891 A, p. 131.
    1891 B, p. 45.
    1891 E, p. 606.
    1893 B, p. 323, fig. 4.
Roger, O. 1896 A, p. 18.
Trouessart, E. L. 1898 A, p. 1204.
Zittel, K. A. 1893 B, p. 106, fig. 89.
Cretaceous (Laramie); Wyoming.

TELACODON Marsh.   Type *T. lævis* Marsh.

Marsh, O. C. 1892 B, p. 258.
Osborn, H. F. 1893, B, p. 324.
Zittel, K. A. 1893 B, p. 106.

Telacodon lævis Marsh.

Marsh, O. C. 1892 B, p. 258, pl. ix, figure; pl. x, figure.
Osborn, H. F. 1893 B, p. 323.
Roger, O. 1896 A, p. 19.
Trouessart, E. L. 1898 A, p. 1204.
Zittel, K. A. 1893 B, p. 106, fig. 90.
Cretaceous (Laramie); Wyoming.

Telacodon præstans Marsh.

Marsh, O. C. 1892 B, p. 258, pl. ix, figs. 2, 3; pl. xi fig. 1.
    1897 C, p. 521, fig. 90.
Roger, O. 1896 A, p. 19.
Trouessart, E. L. 1898 A, p. 1204.
Walcott, C. D. 1900 A, p. 23.
Zittel, K. A. 1893 B, p. 106, fig. 91.
Cretaceous (Laramie); Wyoming.

TRICONODONTIDÆ.

Marsh, O. C. 1887 A, p. 341.
Cope, E. D. 1889 C, p. 226.
Haeckel, E. 1895 A. p. 470.   (Triconodontida.)
Lydekker, R. 1887 A, p. 257.
Marsh, O. C. 1887 A, p. 340.   (Spalacotheriidae, Tinodontidæ.)
Nicholson and Lydekker 1889 A, p. 1277.
Osborn, H. F. 1887 F, p. 287.
    1888 C, p. 235.

Osborn, H. F. 1888 G, pp. 227, 241.
    1893 D.   (Triconodonta.)
    1898 A, p. 147.   (Triconodonta.)
    1898 B, p. 177.   (Triconodonta.)
Roger, O. 1896 A, p. 12.
Steinmann and Döderlein 1890 A, p. 697.
Trouessart, E. L. 1898 A, p. 1249.
Winge, H. 1893 C, p. 75.
Zittel, K. A. 1893 B, p. 97.   (Triconodonta.)

TRICONODONTINÆ nom. nov.

Goodrich, E. S. 1894 A, p. 420.   (Amphilestes.)
Osborn, H. F. 1888 G, p. 228.   (Amphilestinæ.)

Winge, H. 1893 C, p. 76.   (Amphilestidæ.)

TRICONODON Owen. Type *T. mordax* Owen.

Owen, R., 1859, Encyc. Brit., ed. 8., xvii, p. 161.
Cope, E. D. 1888 D, p. 257.
　　1889 C, p. 227.
Leche, W. 1896 B, p. 280.
Lydekker, R. 1887 A, p. 256.
Marsh, O. C. 1879 F, p. 215.
　　1880 A, p. 237.
Nicholson and Lydekker 1889 A, p. 1277..
Osborn, H. F. 1887 F, p. 287.
　　1888 C, p. 235.
　　1888 F, p. 1071.
　　1888 G, p. 228.
　　1893 D, p. 22.
　　1898 R, p. 320.
Owen, R. 1871 A, p. 56.

Schlosser, M. 1890 C, p. 242, fig. I, 5.
Scott, W. B. 1892 A, p. 405.
Steinmann and Döderlein 1890 A, p. 697.
Thomas, O, 1892 B.
Trouessart, E. L. 1898 A, p. 1249.
Winge, H. 1893 C, p. 77.
Zittel, K. A. 1898 B, p. 97.

Triconodon bisulcus Marsh.

Marsh, O. C. 1880 F, p. 237.
　　1887 A, p. 343.
Roger, O. 1896 A, p. 14.
Trouessart, E. L. 1898 A, p. 1250.
　　Jurassic; Wyoming.

PRIACODON Marsh. Type *Tinodon ferox* Marsh.

Marsh, O. C. 1887 A, p. 341.
Nicholson and Lydekker 1889 A, pp. 1275, 1277.
Osborn, H. F. 1887 F, p. 287. (Syn. of Tricono-
　　don.)
　　1888 G, p. 229.
Trouessart, E. L., 1898 A, p. 1250.
Winge, H. 1893 C, p. 76.
Zittel, K. A. 1898 B, p. 98.

Priacodon ferox Marsh.

Marsh, O. C. 1880 F, p. 236. (Tinodon.)

Cope, E. D. 1889 C, p. 227, pl. viii, fig. 4; text fig.
　　51. (Triconodon.)
Marsh, O. C. 1887 A, p. 341, pl. x, fig. 9.
Osborn, H. F. 1888 G, p. 229, fig. 10 a.
Roger, O. 1896 A, p. 13. (Priacodon, Tinodon.)
Trouessart, E. L. 1898 A, p. 1250. (Syn. of Tricon-
　　odon ferox Owen.)
Woodward, A. S. 1898 B, p. 259, fig. 152.
　　Jurassic; Wyoming.

TINODONTINÆ nom. nov.

Goodrich, E. S. 1894 A, p. 417. (Phascolotherium.)
Marsh, O. C. 1879 F, p. 216. (Tinodontidæ.)
　　1887 A, p. 340. (Tinodontidæ.)
Osborn, H. F. 1887 F, p. 288. (Phascolotheriidæ.)
　　1888 G, p. 229. (Phascolotheriinæ.)

Owen, R. 1846 B, p. 61. (On Phascolotherium.)
Trouessart, E. L. 1898 A, p. 1249. (Triconodonti-
　　dæ, in part.)

TINODON Marsh. Type *T. bellus* Marsh.

Marsh, O. C. 1879 F, p. 216.
　　1885 C.
　　1887 A, p. 340.
Osborn, H. F. 1887 F, p. 288.
　　1888 C, p. 235.
　　1888 F, p. 1075.
　　1888 G, p. 229.
Winge, H. 1893 C, p. 76.
Zittel, K. A. 1893 B, p. 98.

Tinodon bellus Marsh.

Marsh, O. C. 1879 F, p. 216, figure.
　　1887 A, p. 340, pl. x, fig. 1.
Osborn, H. F. 1888 G, p. 229, figure.
Roger, O. 1896 A, p. 13.

Trouessart, E. L. 1898 A, p. 1251.
　　Jurassic; Wyoming?

Tinodon lepidus Marsh.

Marsh, O. C. 1879 G, p. 398.
　　1887 A, p. 343.
Roger, O. 1896 A, p. 13.
Trouessart, E. L. 1898 A, p. 1251.
　　Jurassic; Wyoming.

Tinodon robustus Marsh.

Marsh, O. C. 1879 G, p. 397.
　　1887 A, p. 343.
Roger, O. 1896 A, p. 13.
Trouessart, E. L. 1898 A, p. 1251.
　　Jurassic; Wyoming.

SPALACOTHERIINÆ.

Osborn, H. F. 1888 G, p. 229.
Lydekker, R. 1887 A, p. 292. (Spalacotheriidæ.)
Marsh, O. C. 1887 A, p. 340. (Spalacotheriidæ.)

Nicholson and Lydekker 1889 A, p. 1277. (Spala-
　　cotheriidæ.)
Owen, R. 1885 C, p. 426. (On Spalacotherium.)

568   FOSSIL VERTEBRATA OF NORTH AMERICA.   [BULL. 179.

MENACODON Marsh.   Type *M. rarus* Marsh.

Marsh, O. C. 1887 A, p. 340.
Ameghino, F. 1896 A, p. 467.   (Syn. of Tinodon.)
Osborn, H. F. 1888 F, p. 1075.
    1888 G, p. 230.
Winge, H. 1893 C, p. 76.
Zittel, K. A. 1893 B, p. 98.   (Syn. of Tinodon.)

Menacodon rarus Marsh.

Marsh, O. C. 1887 A, p. 340, pl. x, figs. 5, 6.

Cope, E. D. 1889 C, p. 227, fig. 52.
Osborn, H. F. 1888 G, p. 230, fig. 21.
Roger, O. 1896 A, p. 13.
Trouessart, E. L. 1898 A, p. 1251.
Walcott, C. D. 1900 A, p. 23.
Jurassic; Wyoming.

AMPHITHERIIDÆ.

Owen, R. 1846 B, p. 29.
Goodrich, E. S. 1894 A, p. 409.   (Amphitherium.)
Lydekker, R. 1887 A, p. 269.
    1896 B, p. 52.
Marsh, O. C. 1887 A, p. 338.   (Diplocynodontidæ.)
Nicholson and Lydekker 1889 A, 1273.

Osborn, H. F. 1887 F, p. 289.   (Diplocynodontidæ.)
    1888 C, p. 234.
    1888 F, p. 1078.   (Dicrocynodontidæ.)
    1888 G, p. 230.
Owen, R. 1868 A, p. 287.   (Genus Amphitherium.)
Roger, O. 1896 A, p. 14.
Trouessart, E. L. 1898 A, p. 1246.
Zittel, K. A. 1893 B, p. 100.

DICROCYNODON Marsh.   Type *Diplocynodon victor* Marsh.

Marsh, O. C. in Osborn, H. F. 1888 G, p. 268.   (To
replace Diplocynodon, preoccupied.)
    1880 F, p. 235.   (Diplocynodon.)
    1887 A, p. 338.   (Diplocynodon.)
    1889 I, p. 14.
    1897 C, p. 508.   (Diplocynodon.)
Osborn, H. F. 1887 D.   (Diplocynodon.)
    1888 D, p. 17.
    1897 C, p. 1000.
Schlosser, M. 1890 C, p. 242.   (Diacynodon.)
Zittel, K. A. 1893 B, p. 99.

Dicrocynodon victor Marsh.

Marsh, O. C. 1880 F, p. 235, figure.   (Diplocynodon.)
    1887 A, p. 338, pl. x, fig. 3.   (Diplocynodon.)
    1897 C, p. 508, fig. 69.   (Diplocynodon.)
Osborn, H. F. 1888 G, p. 232, fig. 22.   (Diplocyn-
odon.)
Roger, O. 1896 A, p. 14.
Trouessart, E. L. 1898 A, p. 1251.
Woodward, A. S. 1898 B, p. 259, fig. 152
Zittel, K. A. 1893 B, p. 99, fig. 78.
Jurassic; Wyoming.

DOCODON Marsh.   Type *D. striatus* Marsh.

Marsh, O. C. 1881 F, p. 512.
    1887 A, p. 339.
Osborn, H. F. 1888 G, p. 233.
Zittel, K. A. 1893 B, p. 99.

Docodon striatus Marsh.

Marsh, O. C. 1881 F, p. 512.
    1887 A, p. 339, pl. x, fig. 2.

Osborn, H. F. 1888 G, p. 232, fig. 23.
Roger, O. 1896 A, p. 14.
Trouessart, E. L. 1898 A, p. 1252.
Woodward, A. S. 1898 B, p. 259, fig. 152.
Jurassic; Wyoming.

ENNACODON Marsh.   Type *Enneodon crassus* Marsh.

Marsh, O. C. 1899 I, p. 15.   (To replace *Enneodon*
Marsh, preoccupied.)
    1887 A, p. 339.   (Enneodon.)
Nicholson and Lydekker 1889 A, p. 1277.   (En-
neodon.)
Osborn, H. F. 1888 G, p. 232.   (Enneodon.)
Zittel, K. A. 1893 B, p. 99.   (Ennacodon.)

Ennacodon affinis Marsh.

Marsh, O. C. 1887 A, p. 339.   (Enneodon.)
Roger, O. 1896 A, p. 14.
Trouessart, E. L. 1898 A, p. 1252.

Walcott, C. D. 1900 A, p. 23.   (Enneodon.)
Jurassic; Wyoming.

Ennacodon crassus Marsh.

Marsh, O. C. 1887 A, p. 339, pl. x, fig. 4.   (Enneo-
don.)
Osborn, H. F. 1888 G, p. 232, fig. 24.   (Enneodon.)
Roger, O. 1896 A, p. 14.
Trouessart, E. L. 1898 A, p. 1252.
Walcott, C. D. 1900 A, p. 23.   (Enneodon.)
Jurassic; Wyoming.

PAURODONTIDÆ.

Marsh, O. C. 1887 A, p. 341.
Osborn, H. F. 1887 F, p. 289.   (Peralestidæ, Pau-
rodontidæ.)
    1888 G, p. 233.   (Peralestidæ.)

Trouessart, E. L. 1898 A, p. 1246.   (Amphitheriidæ,
in part.)
Woodward, M. F. 1896 A, p. 587.   (On Peralestes.)

PAURODON Marsh.   Type *P. valens* Marsh.

Marsh, O. C. 1887 A, p. 342.
Ameghino, F. 1899 A, p. 571.
Cope, E. D. 1888 D, p. 257.
    1889 C, p. 236.
Osborn, H. F. 1887 F, p. 289.
    1888 G, p. 233.
Winge, H. 1898 C, p. 76.

Paurodon valens Marsh.

Marsh, O. C. 1887 A, p. 342, pl. ix, figs. 7, 8.
Osborn, H. F. 1888 G, p. 233, figs. 25 a, b.
Roger, O. 1896 A, p. 15.
Trouessart, E. L. 1898 A, p. 1247.
Walcott, C. D. 1900 A, p. 23.
Jurassic; Wyoming.

DRYOLESTIDÆ.

Marsh, O. C. 1879 G, p. 397.
Ameghino, F. 1896 A, p. 409.   (Stylacodontidæ.)
Lydekker, R. 1887 A, p. 289.   (Stylodontidæ.)
Marsh, O. C. 1879 E, p. 60.   (Stylodontidæ.)
    1887 A, p. 334.
Nicholson and Lydekker 1889 A, p. 1410.   (Stylodontidæ.)

Osborn, H. F. 1887 F, p. 290.   (Stylodontidæ.)
    1888 A, p. 298.   (Stylacodontidæ.)
    1888 C, p. 232.   (Stylodontidæ.)
    1888 G, p. 236.   (Stylacodontidæ.)
Steinmann and Döderlein 1890 A, p. 698.   (Stylacodontidæ.)
Trouessart, E. L. 1898 A, p. 1246.   (Amphitheriidæ, in part.)

STYLACODON Marsh.   Type *S. gracilis* Marsh.

Marsh, O. C. 1879 E, p. 60.
Ameghino, F. 1896 A, p. 454.
Lydekker, R. 1887 A, p. 289 (Stylacodon); p. 290 (Stylodon).
Marsh, O. C. 1885 C.
    1887 A, p. 335.
Nicholson and Lydekker 1889 A, p. 1276.   (Amblotherium.)
Osborn, H. F. 1887 D.   (Stylodon.)
    1887 F, p. 290.   (Syn.? of Stylodon.)
    1888 A, p. 300.
    1888 G, p. 207 (Stylodon); pp. 236, 243 (Stylacodon).
Owen, R. 1866 D.   (Stylodon, not of Beck, 1837.)
Zittel, K. A. 1898 B, p. 102.   (Syn.? of Dryolestes.)

Stylacodon gracilis Marsh.

Marsh, O. C. 1879 E, p. 60.
    1887 A, pl. ix, fig. 1.
    1897 C, p. 508, fig. 68.
Osborn, H. F. 1888 G, p. 237, fig. 26.
Roger, O. 1896 A, p. 15.
Trouessart, E. L. 1898 A, p. 1248.
Jurassic; Wyoming.

Stylacodon validus Marsh.

Marsh, O. C. 1880 F, p. 236.
Roger, O. 1896, A, p. 15.
Trouessart, E. L. 1898 A, p. 1249.
Jurassic; Wyoming.

DRYOLESTES Marsh.   Type *D. priscus* Marsh.

Marsh, O. C. 1878 D.
    1879 F, p. 215.
    1885 C.
    1887 A, pp. 333, 334.
Nicholson and Lydekker 1889 A, pp. 1276, 1277.
Osborn, H. F. 1888 C, p. 235.
    1888 G, p. 237.   (Syn.? of Phascolestes.)
Schlosser, M. 1890 C, p. 242.
Steinmann and Döderlein 1890 A, p. 698.
Winge, H. 1898 C, p. 76.
Woodward, M. F. 1896 A, p. 567.
Zittel, K. A. 1898 B, p. 102.

Dryolestes arcuatus Marsh.

Marsh, O. C. 1879 G, p. 397.
    1887 A, p. 343.
Roger, O. 1896 A, p. 15.
Trouessart, E. L. 1898 A, p. 1248.
Jurassic; Wyoming.

Dryolestes gracilis Marsh.

Marsh, O. C. 1881 F, p. 513.
    1887 A, p. 343.
    1897 C, p. 526.
Roger, O. 1896 A, p. 15.
Trouessart, E. L. 1898 A, p. 1248.
Jurassic; Wyoming, Colorado.

Dryolestes obtusus Marsh.

Marsh, O. C. 1880 F, p. 237.
    1887 A, p. 343.
Roger, O. 1896 A, p. 15.
Trouessart, E. L. 1898 A, p. 1248.
Jurassic; Wyoming.

Dryolestes priscus Marsh.

Marsh, O. C. 1878 D.
    1879 F, p. 215.
    1887 A, p. 334, pl. ix, fig. 2.
Osborn, H. F. 1888 G, p. 238, fig. 28.
Roger, O. 1896 A, p. 15.
Trouessart, E. L. 1898 A, p. 1248.
Woodward, A. S. 1898 B, p. 259, fig. 152.
Zittel, K. A. 1898 B, p. 102, fig. 81.
Jurassic; Wyoming.

Dryolestes tenax Marsh.

Marsh, O. C. 1889 D, p. 87.
Osborn, H. F. 1891 A, p. 131.
    1891 B, p. 45.
    1891 E, p. 606.
Roger, O. 1896 A, p. 15.
Trouessart, E. L. 1898 A, p. 1248.
Cretaceous (Laramie); Wyoming.

Dryolestes vorax Marsh.

Marsh, O. C. 1879 F, p. 215,
1887 A, p. 334, pl. ix, figs. 3, 4.
Osborn, H. F. 1888 G, p. 237, fig. 27.

Roger, O. 1896 A, p. 15.
Trouessart, E. L. 1898 A, p. 1248.
Woodward, A. S. 1898 B, p. 259, fig. 152.
Jurassic; Wyoming.

ASTHENODON Marsh.   Type *A. segnis* Marsh.

Marsh, O. C. 1887 A, p. 336, pl. ix, figs. 6, 7.
Nicholson and Lydekker 1889 A, p. 1276.
Osborn, H. F. 1888 A, p. 300.
1888 F, p. 1071.
1888 G, p. 238.
Winge, H. 1893 C, p. 76.
Zittel, K. A. 1893 B, p. 102.

Asthenodon segnis Marsh.

Marsh, O. C. 1887 A, p. 336, pl. ix, figs. 6, 7
Osborn, H. F. 1888 G, p. 238, figs. 29, A, B.
Roger, O. 1896 A, p. 15.
Trouessart, E. L. 1898 A, p. 1248.
Woodward, A. S. 1898 B, p. 259, fig. 152.
Jurassic; Wyoming.

LAODON Marsh.   Type *L. venustus* Marsh.

Marsh, O. C. 1887 A, p. 337.
Ameghino, F. 1899 A, p. 571.
Nicholson and Lydekker 1889 A, p. 1276.
Osborn, H. F. 1888 G, p. 238.
Winge, H. 1893 C, p. 76.

Laodon venustus Marsh.

Marsh, O. C. 1887 A, p. 337, pl. ix, fig. 5.

Osborn, H. F. 1888 G, p. 238, fig. 30.
Roger, O. 1896 A, p. 15.
Trouessart, E. L. 1898 A, p. 1248.
Walcott, C. D. 1900 A, p. 23.
Woodward, A. S. 1898 B, p. 259, fig. 152.
Jurassic; Wyoming.

DIDELPHIDÆ.

Ameghino, F. 1889 A, p. 277.
Baird, S. F. 1857 A, p. 231.   (Didelphydæ, Pera-
theridæ.)
Baur, G. 1885 K, p. 475.
Cope, E. D. 1889 R, p. 876.
Flower, W. H. 1883 D, p. 380.
Flower and Lydekker 1891 A, p. 133.
Gegenbaur, C. 1898 A.
Haeckel, E. 1895 A, p. 481.   (Didelphyida.)
Lydekker, R. 1887 A, p. 278.
1896 B, p. 55.
Nicholson and Lydekker 1889 A, p. 1280.

Osborn, H. F. 1893 G, p. 497.
1900 C, p. 566.
Owen, R. 1838 A.
1839 A.
Parker, W. K. 1885 A, p. 59.
Roger, O. 1896 A, p. 18.
Schlosser, M. 1889 A.
Schmidt, O. 1886 A, p. 97.
Tomes, C. S. 1898 A, p. 562.
Winge, H. 1893 C, p. 89.
Wyman, J. 1872 A.
Zittel, K. A. 1893 B, p. 105.

DIDELPHIS[1] Linnæus.   Type *D. marsupialis* Linn.

Linnæus, C., 1758, Syst. Nat., ed. 10, i, p. 54.
Ameghino, F. 1889 A, p. 277.
Aymard, A., 1848, Mem. Soc. Acad. du Puy.
(Peratherium.)
Baird, S. F. 1857 A, p. 232.
Bensley, B. A. 1900 A, p. 559.
Burmeister, H. 1879 B, p. 183.
Cope, E. D. 1873 T, p. 1.   (Herpetotherium, type
H. fugax.)
1873 CC, p. 4.   (Embassis, type *E. alternans,*
Peratherium.)
1874 B, pp. 465, 468 (Embassis, Herpe-
totherium); p. 472 (Didelphys).
1874 P, p. 23.   (Herpetotherium.)
1879 A, p. 45.   (Didelphys.)
1884 O, p. 37 (Didelphys); pp. 260, 268, 788,
789 (Peratherium).
1884 Y, p. 687.   (Peratherium.)
Coues, E. 1872 A.
Emery, C. 1897 C.
Flower, W. H. 1885 A.

Flower and Lydekker 1891 A, p. 134.
Gaudry, A. 1878 B, p. 11.
Kükenthal, W. 1891 B, p. 656, figs. 1–8.
Lydekker, R. 1887 A, p. 278.
Marsh, O. C. 1877 E.
Nicholson and Lydekker 1887 A, p. 1458.
Osborn, H. F. 1892 I.
Owen, R. 1838 A.
1843 A, pp. 57, 72.
1866 A, p. 330, *et seq.*
1868 A, p. 288.
Pictet, F. J. 1853 A, p. 394.
Rösc, C. 1892 E, p. 641.
Schlosser, M. 1887 B, p. 145.   (Peratherium.)
1890 C, p. 244.
1890 D.
Thomas, O. 1892 B.
Trouessart, E. L. 1898 A, p. 1280.
Weil, R. 1899 A, p. 103, pls. ii, iii.
Zittel, K. A. 1893 B, p. 105.

---

[1] This is the original orthography. The form *Didelphys* is usually employed by writers.

Didelphis alternans (Cope).

Cope, E. D. 1873 CC, p. 7. (Embassis.)
1874 B, p. 468. (Embassis.)
1884 O, pp. 794, 796, pl. lxii, figs. 22–24. (Peratherium.)
Matthew, W. D. 1899 A, p. 52.
Roger, O. 1896 A, p. 20.
Trouessart, E. L. 1898 A, p. 1232.
Oligocene (White River); Colorado.

Didelphis comstocki (Cope).

Cope, E. D. 1884 O, p. 269 (Peratherium comstocki); pl. xxv a, fig. 15. P. comstockianum).
Matthew, W. D. 1899 A, p. 31.
Roger, O. 1896 A, p. 19.
Trouessart, E. L. 1898 A, p. 1230.
Eocene (Wasatch); Wyoming.

Didelphis fugax (Cope).

Cope, E. D. 1873 T, p. 1. (Herpetotherium fugax, err. typ.)
1873 CC, pp. 5, 6. (Herpetotherium.)
1874 B, p. 467. (Herpetotherium.)
1874 P, p. 23. (Herpetotherium.)
1884 O, p. 794, pl. lxii, figs. 1–9. (Peratherium.)
1884 Y, p. 687, fig. 1. (Peratherium.)
Matthew, W. D. 1899 A, p. 52.
Roger, O. 1896 A, p. 21.
Trouessart, E. L. 1898 A, p. 1232.
Oligocene (White River); Colorado.

Didelphis huntii (Cope).

Cope, E. D. 1873 CC, p. 5. (Herpetotherium.)
1873 CC, pp. 5, 6 (Herpetotherium stevensoni); p. 8 (Miothen gracile).
1874 B, p. 466 (Herpetotherium huntii); p. 467 (H. stevensoni); p. 470 (Domnina gracilis).
1884 O, pp. 794, 796, pl. lxii, figs. 12–16. (Peratherium.)
1887 Y, p. 687. (Peratherium.)
Matthew, W. D. 1899 A, p. 52.
Roger, O. 1896 A, p. 21.
Trouessart, E. L. 1898 A, p. 1232.
Oligocene (White River); Colorado.

Didelphis marginalis (Cope).

Cope, E. D. 1873 CC, pp. 5, 6. (Herpetotherium.)

Cope, E. D. 1874 B, p. 468. (Embassis.)
1884 O, pp. 794, 796, pl. lxii, figs. 19–21. (Peratherium.)
Matthew, W. D. 1899 A, p. 52.
Roger, O. 1896 A, p. 21.
Trouessart, E. L. 1898 A, p. 1232.
Oligocene (White River); Colorado.

Didelphis pygmæa Scott.

Scott, W. B. 1884 A, p. 442, figure.
Matthew, W. D, 1899 A, p. 52.
Roger, O. 1896 A, p. 21.
Trouessart, E. L. 1898 A, p. 1232.
Oligocene (White River); Colorado.

Didelphis scalaris (Cope).

Cope, E. D. 1873 CC, pp. 5, 7. (Herpetotherium.)
1874 B, p. 467. (Herpetotherium.)
1884 O, pp. 794, 797, pl. lxii, figs. 17, 18. (Peratherium.)
Matthew, W. D. 1899 A, p. 52.
Roger, O. 1896 A, p. 21.
Trouessart, E. L. 1898 A, p. 1232.
Oligocene (White River); Colorado.

Didelphis tricuspis (Cope).

Cope, E. D. 1873 CC, p. 5. (Herpetotherium.)
1874 B, p. 466. (Herpetotherium.)
1884 O, pp. 794, 796, pl. lxii, figs. 10, 11. (Peratherium.)
Roger, O. 1896 A, p. 21.
Trouessart, E. L. 1898 A, p. 1232.
Oligocene; Colorado.

Didelphis virginiana Kerr.

Kerr, Robert, 1792, Animal Kingdom, i, No. 386. (Fide J. A. Allen.)
·Baird, S. F. 1857 A. p. 232.
Coues, E. 1872 A.
Emmons, E. 1858 A, p. 78. ("Insectivorous mammal.")
Leidy, J. 1860 B, p. 116, pl. xxiii, fig. 2.
1869 A, p. 410.
Trouessart, E. L. 1898 A, p. 1233. (Subsp. of D. marsupialis.)
Wyman, J. 1872 A.
Recent; eastern North America: Pleistocene; South Carolina.

Superorder *MONODELPHIA*.

Cope, E. D. 1889 R, p. 874.
1898 B, p. 104.
Flower, W. H. 1883 A.
1883 D, p. 383. (Eutheria, or Monodelphia.)
1885 A, p. 6. (Eutheria, Monodelphia.)
Gadow H. 1898 A, p. 41. (Eutheria.)
Gill, T. 1872 A, p. 295.
1872 B, p. 46.

Haeckel, E. 1895 A, p. 487. (Placentalia.)
Huxley, T. H. 1872 A, p. 281.
1880 B. (Monodelphia, Eutheria.)
Nicholson and Lydekker 1889 A, p. 1289. (Eutheria.)
Osborn, H. F. 1893 A, p. 391. (Eutheria.)
1898 L, p. 456. (Eutheria.)
Woodward, A. S. 1898 B, p. 266. (Eutheria.)

Order BRUTA Linn.

Linnæus, C., 1758, Syst. Nat., ed. 10, i, p. 33.
Unless otherwise indicated, the following authorities employ the name *Edentata* for this order.

Ameghino, F. 1889 A, p. 653.
Baur, G. 1885 K, p. 477.
Blainville, H. M. D. 1839 A. ("Edentés.")
1839 B. ("Edentés.")

Blainville, H. M. D. 1839 C, p. 113.
1864 A, iv, DD. pls. i-vi.
Broom, R. 1897 A.
Burmeister, H. 1866 B, p. 148.
1871 A.
1871 B.
1879 B, p. 275.
Cope, E. D. 1880 B, p. 454.
1884 X.
1887 S, p. 342.
1888 I, p. 4.
1888 K, p. 163.
1889 C.
1889 P.
1889 R, p. 875.
1897 A, p. 323.
1896 B, p. 105.
Cuvier, G, 1834 A, viii, pt. i, pp. 129-270.
Filhol, H. 1894 A.
Flower, W. H. 1869 A, p. 265.
1882 A.
1883 A.
1883 D, p. 383.
1885 A.
Flower and Lydekker 1891 A, p. 176.
Gadow, H. 1898 A, p. 41.
Gaudry, A. 1878 B, p. 192.
Gegenbaur, C. 1898 A.
Gervais, P. 1873 A.
Gill, T. 1872 A, p. 304. (Bruta.)
1872 B, pp. 23, 50. (Bruta.)
Gill and Coues 1877 A, p. 1069.
Gray, J. E. 1869 A, p. 361. (Bruta.)
Haeckel, E. 1895 A, pp. 490, 511.
Hay, O. P. 1899 F, p. 682. (Bruta.)
Huxley, T. H. 1872 A, p. 282. (Edentata, or Bruta.)
Leche, W. 1898 A, p. 527.
1895 A, p. 106.
1897 A.
Leidy, J. 1871 C, p. 365.
Marsh, O. C. 1877 E.
1897 B, p. 144.
Lydekker, R. 1896 B.

Nicholson and Lydekker 1889 A, 1299.
Osborn, H. F. 1888 F, p. 1067.
1898 D.
1893 G, p. 498.
1897 F, p. 611.
Owen, R. 1840 F, p. 85. (Bruta, or Edentata.)
1842 D.
1845 B, p. 317.
1851 C, p. 740.
1880 D.
1866 B, p. 398.
1868 A, p. 272. (Bruta.)
Parker, W. K. 1868 A, p. 199.
1885 A, p. 90.
1896 A.
Parsons, F. G. 1899 A.
1900 A.
Rapp, W. 1852 A.
Reh, L. 1895 A.
Roemer, F. 1892 A.
1898 A.
Röse, C. 1892 D, p. 495.
Roger, O. 1896 A, p. 90.
Schlosser, M. 1890 A.
1890 C.
1890 D.
1892 A.
1898 A.
Schmidt, O. 1886 A, p. 110.
Slade, D. D. 1888 A, p. 242.
1890 A.
Steinmann and Döderlein 1890 A, p. 736.
Thomas, O. 1888 A.
Tomes, C. S. 1898 A, p. 370.
Trouessart, E. L. 1898 A, p. 1094.
Turner, H. N. 1848 A.
Weber, Max 1898 A.
1898 B.
Weyhe, — 1875 A, p. 105.
Woodward, A. S. 1898 B, p. 277.
Wortman, J. L. 1886 A, p. 408. (Bruta.]
1896 B, p. 259.
1897 B.
Zittel, K. A. 1893 B, p. 117.

Suborder TÆNIODONTA Cope.

Cope, E. D. 1876 C.
Ameghino, F. 1897 A, pp. 83, 89.
Cope, E. D. 1877 K, pp. 85, 157.
1882 E, p. 146.
1882 L, p. 72.
1883 I, p. 79.
1884 O, pp. 186, 187, 739.
1887 S, p. 343.
1888 K.
1897 A, p. 322.
1898 B, p. 115.
Marsh, O. C. 1875 C, p. 221. (Tillodontia, in part.)

Marsh, O. C. 1897 B. (>Stylinodontia.)
Osborn, H. F. 1897 F, p. 611. (Ganodonta.)
1898 A, p. 146. (Ganodonta.)
Rütimeyer, L. 1891 A.
Trouessart, E. L. 1898 A, p. 667. (Ganodonta.)
Woodward, A. S. 1898 B, p. 277.
Wortman, J. L. 1885 A, p. 298.
1896 B, p. 259. (Ganodonta.)
1897 A. (Ganodonta.)
1897 B. (Ganodonta.)
Zittel, K. A. 1893 B, p. 508. (Tillodontia, in part.)

CONORYCTIDÆ.

Wortman, J. L. 1896 B, p. 260.
Matthew, W. D. 1897 C, p. 262.

Trouessart, E. L. 1898 A, p. 667.
Wortman, J. L. 1897 B, pp. 95, 109.

CONORYCTES Cope.   Type *C. comma* Cope.

Cope, E. D. 1881 H, p. 486.
    1881 U, p. 829.
    1883 E.
    1883 Q, p. 168.
    1883 EE.
    1884 AA, p. 795.   (Hexodon, type *H. molestus.*)
    1884 O, p. 196.
    1888 CC, p. 316.
Nicholson and Lydekker 1889 A, p. 1453.
Osborn and Earle 1895 A, p. 40.
Pavlow, M. 1887 A.
Schlosser, M. 1885 A, p. 684.
    1886 B, p. 37.
Scott, W. B. 1892 B, p. 323.
    1892 D, p. 80.
Trouessart, E. L. 1898 A, p. 667.
Wortman, J. L. 1896 B, p. 259.

Wortman, J. L. 1897 B, p. 101, 109.
Zittel, K. A. 1893 B, p. 510.

Conoryctes comma Cope.

Cope, E. D. 1881 H, p. 486.
    1881 U, p. 829.
    1884 AA, p. 796, fig. 3.   (Hexodon molestus.)
    1884 O, pp. xxxv, 198, pl. xxiii *e*, figs. 1–5; pl. xxv *c*, figs. 3, 4.
    1888 CC, pp. 308, 317.
Matthew, W. D. 1897 C, p. 264.
    1899 A, p. 29.
Roger, O. 1896 A, p. 89.
Schlosser, M. 1886 B, p. 37.
Trouessart, E. L. 1898 A, p. 667.
Wortman, J. L. 1897 B, p. 101, fig. 36.
    Eocene (Puerco); New Mexico.

ONYCHODECTES Cope.   Type *O. tisonensis* Cope.

Cope, E. D. 1888 CC, p. 317.
Ameghino, F. 1897 A, p. 84.
Matthew, W. D. 1897 C, p. 266.
Osborn and Earle 1895 A, p. 40.
Scott, W. B. 1892 B, p. 323.
    1892 D, p. 80.
Wortman, J. L. 1896 B, p. 258.
    1897 B, pp. 97, 109.
Zittel, K. A, 1893 B, p. 510.

Onychodectes rarus Osb. and Earle.

Osborn and Earle 1895 A, p. 42, fig. 13.
Matthew, W. D. 1897 C, p. 264.
    1899 A, p. 28.

Roger, O. 1896 A, p. 89.
Trouessart, E. L. 1898 A, p. 667.
Wortman, J. L. 1897 B, p. 100.
    Eocene (Puerco); New Mexico.

Onychodectes tisonensis Cope.

Cope, E. D. 1888 CC, p. 318, pl. v, figs. 8, 9.
Matthew, W. D. 1897 C, p. 264.
    1899 A, p 28.
Osborn and Earle 1895 A, p. 40, fig. 12.
Roger, O. 1896 A, p. 89.
Trouessart, E. L. 1898 A, p. 667.
Wortman, J. L. 1897 B, p. 97, figs. 31–34.
    Eocene (Puerco); New Mexico.

STYLINODONTIDÆ.

Marsh, O. C. 1876 C, p. 221.
Ameghino, F. 1897 A, p. 89.
Cope, E. D. 1876 C, p. 39.   (Ectoganidæ, Calamodontidæ.)
    1877 K, p. 157.   (Ectoganidæ.)
    1884 O, p. 188.   (Ectoganidæ, Calamodontidæ.)
    1888 CC, p. 310.   (Hemiganidæ.)
Flower, W. H. 1876 C, p. 124.   (Stylinodontidæ, Ectoganidæ.)

Marsh, O. C. 1876 G, p. 60.
    1877 E.
    1897 B, p. 137.
Matthew, W. D. 1897 C, p. 262.
Nicholson and Lydekker 1889 A, p. 1410.   (Calamodontidæ,)
Roger, O. 1896 A, p. 88.   (Calamodontidæ.)
Wortman, J. L. 1896 B, p. 260.
    1897 B, pp. 95, 109.
Zittel, K. A. 1893 B, p. 506.

WORTMANIA Hay.   Type *Hemiganus otariidens* Cope.

Hay, O. P. 1899 A, p. 593.
Cope, E. D. 1885 L.   (Hemiganus, in part.)
    1888 CC, p. 311.   (Hemiganus, in part.)
Nicholson and Lydekker 1889 A, p. 1453.   (Hemiganus, in part.)
Trouessart, E. L. 1898 A, p. 668.   (Hemiganus.)
Wortman, J. L. 1897 B.   (Hemiganus.)
Zittel, K. A. 1893 B, p. 510.   (Hemiganus, in part.)

Wortmania otariidens (Cope.)

Cope, E. D. 1885 L, p. 493.   (Hemiganus.)

Cope, E. D. 1888 CC, p. 311, pl. iv; pl. v, figs. 1–7.   (Hemiganus.)
Hay, O. P. 1899 A, p. 593.   (Wortmania.)
Matthew, W. D. 1897 C, p. 264.   (Hemiganus.)
    1899 A, p. 28.   (Hemiganus.)
Roger, O. 1896 A, p. 89.   (Hemiganus.)
Trouessart, E. L. 1898 A, p. 668.   (Hemiganus.)
Wortman, J. L. 1897 B, pp. 64, 67, figs. 2, 3.   (Hemiganus.)
Zittel, K. A. 1893 B, p. 510.   (Hemiganus.)
    Eocene (Puerco); New Mexico.

PSITTACOTHERIUM Cope.   Type *P. multifragum* Cope.

Cope, *E. D.* 1882 P, p. 157.
Ameghino, F. 1897 A, p. 89.
Cope, E. D. 1882 E, p. 191.
    1882 CC. (Hemiganus, type *H. vultuosus*.)
    1883 E.
    1883 I, p. 77.
    1884 O, pp. 195, 738 (Psittacotherium); explan. pl. xxiii *c* (Hemiganus).
    1885 L. (Hemiganus.)
    1887 F, p. 469.
    1888 I, p. 4.
    1888 CC, p. 311. (Hemiganus.)
    1889 C.
    1897 A, p. 322.
Nicholson and Lydekker 1889 A, p. 1409 (Psittacotherium); p. 1458 (Hemiganus).
Osborn, H. F. 1897 F, p. 611.
    1898 A, p. 145.
Schlosser, M. 1890 C, p. 252.
    1892 A, p. 225.
Scott, W. B. 1892 B, p. 323. (Hemiganus.)
    1892 D, p. 80. (Hemiganus.)
Steinmann and Döderlein 1890 A, p. 729. (Hemiganus.)
Wortman, J. L. 1896 B. (Psittacotherium, Hemiganus.)
    1897 B, pp. 65, 67 (type of Hemiganus=P. multifragum); p. 109 (Hemiganus, Psittacotherium).
Zittel, K. A. 1898 B, p. 508. (Psittacotherium, Hemiganus.)

Psittacotherium multifragum Cope.

Cope, *E. D.* 1882 P, p. 157.
    1882 E, p. 191 (P. multifragum); p. 192 (P. aspasiæ).
    1882 CC. (Hemiganus vultuosus.)
    1884 O, p. 196, pl. xxiii *c*, figs. 7-12 (Hemiganus vultuosus); pl. xxiv *b*, fig. 2; pl. xxiv *b*, figs. 3, 4 (P. aspasiæ).
    1885 L. (Hemiganus vultuosus.)
    1887 F, p. 469. (P. multifragum, P. megalodus.)
    1888 I, p. 4, fig. 1.
    1888 CC, p. 302 (P. multifragum, P. aspasiæ, P. megalodus); p. 303 (Hemiganus vultuosus).
    1889 C, p. 262.
Hay, O. P. 1899 A, p. 598.
Matthew, W. D. 1897 C, pp. 264, 320.
    1899 A, p. 29.
Osborn and Earle 1895 A, p. 42.
Roger, O. 1896 A, p. 88 (P. multifragum, P. aspasiæ, P. megalodus); p. 89 (Hemiganus vultuosus).
Wortman, J. L. 1896 B, p. 259.
    1897 B, p. 65 *et seq.*, figs. 1, 4, 6-9, 11-13, 15-20; p. 67 (Hemiganus vultuosus, a synonym); p. 71 (P. megalodus and P. aspasiæ, synonyms).
Zittel, K. A. 1898 B, p. 508, fig. 423 (P. multifragum); p. 510 (Hemiganus vultuosus).
Eocene (Puerco); New Mexico.

CALAMODON Cope.   Type *C. simplex* Cope.

Cope, *E, D.* 1874 O, p. 5.
    1875 A.
    1875 C, p. 23.
    1876 C.
    1877 K, p. 162.
    1882 E, p. 146.
    1882 L.
    1884 O, p. 188.
    1888 I, p. 5.
    1889 C, p. 262.
    1894 I, p. 594. (Calamodon, or Conicodon.)
    1897 A, p. 322.
Flower and Lydekker 1891 A, p. 442.
Major, Fors. 1893 A, p. 209.
Marsh, O. C, 1875 A, p. 151.
Nicholson and Lydekker 1889 A, p. 1410.
Osborn, H. F 1897 F.
Rütimeyer, L. 1891 A, p. 126.
Schlosser, M. 1890 C, p. 252.
Steinmann and Döderlein 1890 A, p. 730.
Wortman, J. L. 1897 B, pp. 88, 110.
Zittel, K. A. 1898 B, p. 509.

Calamodon arcamœnus Cope.

Cope, *E. D.* 1874 O, p. 6.

Cope, E. D. 1875 C, p. 24.
    1877 K, p. 163, pl. xli, figs. 13-17; pl. xlii, figs. 1-5; pl. xliv, fig. 1.
Matthew, W. D. 1899 A, p. 32.
Roger, O. 1896 A, p. 89.
Zittel, K. A. 1898 B, p. 509. (Stylinodon.)
Eocene (Wasatch); New Mexico.

Calamodon simplex Cope.

Cope, *E. D.* 1874 O, p. 6.
    1875 C, p. 24.
    1877 K, p. 166, pl. xlii, figs. 6-8; pl. xliii; pl. xliv, figs. 2-5.
    1882 E, p. 147.
    1884 O, p. 189, pl. xxiv *b*, fig. 1;? pl. xxiv *e*, figs. 20-22.
    1888 I, p. 5, fig. 2.
    1889 C, p. 262, fig. 85.
King, C. 1878 A, p. 377.
Matthew, W. D. 1899 A, p. 32.
Osborn and Wortman 1892 A, p. 83.
Roger, O. 1896 A, p. 89.
Wortman, J. L. 1897 B, p. 89, figs. 22-24.
Zittel, K. A. 1898 B, p. 509. (Stylinodon.)
Eocene (Wasatch); Wyoming, New Mexico.

STYLINODON Marsh.   Type *S. mirus* Marsh.

Marsh, *O. C.* 1874 C, p. 532.
Ameghino, F. 1897 A, p. 89.
Cope, E. D. 1875 A.
Flower and Lydekker 1891 A, p. 442.

Marsh, O. C. 1875 A.
    1877 E.
    1897 B.
Nicholson and Lydekker 1889 A, p. 1410.

Osborn, H. F. 1897 F, p. 612.
Wortman, J. L. 1887 F, p. 290.
    1896 B, p. 259.
    1897 A, p. 110.
Zittel, K. A. 1898 B, p. 509.

Stylinodon cylindrifer (Cope).

Cope, E. D. 1881 D, p. 184. (Calamodon.)
    1884 O, p. 192, pl. xxiv a, figs. 15, 16. (Calamodon.)
Matthew, W. D. 1899 A, p. 36.
Roger, O. 1896 A, p. 89. (Calamodon.)
Trouessart, E. L. 1898 A, p. 669.

Wortman, J. L. 1897 B, p. 92, fig. 25.
Zittel K. A. 1898 B, p. 509.
    Eocene (Wind River, Bridger); Wyoming.

Stylinodon mirus Marsh.

Marsh, O. C. 1874 C, p. 582.
King, C. 1878 A, p. 404.
Marsh, O. C. 1897 B, p. 138, figs. 1–7.
Matthew, W. D. 1899 A, p. 43.
Roger, O. 1896 A, p. 89. (Calamodon.)
Trouessart, E. L. 1898 A, p. 669.
Wortman, J. L. 1897 B, p. 93, figs. 26–30.
    Eocene (Bridger); Wyoming.

ECTOGANUS Cope.   Type E. gliriformis Cope.

Cope, E. D. 1874 O, p. 4.
    1875 C, p. 23.
    1876 A, p. 3.
    1877 K, p. 158.
    1897 A, p. 322.
Wortman J. L. 1896 B, p. 259.
    1897 B, p. 88.
Zittel, K. A. 1898 B, p. 510.

Ectoganus gliriformis Cope.

Cope, E. D. 1874 O, p. 4.
    1877 K, p. 160, pl. xli, figs. 1–12.

King, C. 1878 A, p. 377.
Roger, O. 1896 A, p. 89.
    Eocene; New Mexico.

Ectoganus novomehicanus Cope.

Cope, E. D. 1874 O, p. 6. (Calamodon.)
    1875 C, p. 24. (Calamodon.)
    1877 K, p. 159, pl. xl, figs. 34–39.
Matthew, W. D. 1899 A, p. 32. (Calamodon.)
Roger, O. 1896 A, p. 89.
    Eocene (Wasatch); New Mexico.

DRYPTODON Marsh.   Type D. crassus Marsh.

Marsh, O. C. 1876 H, p. 408.
    1877 E.
    1897 B, 188.
Trouessart, E. L. 1898 A, p. 669.
Wortman, J. L. 1897 B, p. 88, footnote 4.
Zittel K. A. 1898 B, p. 510.

Dryptodon crassus Marsh.

Marsh, O. C. 1876 H, p. 408.
King, C. 1878 A, p. 377.
Matthew, W. D. 1899 A, p. 32.
Roger, O. 1896 A, p. 89.
Trouessart, E. L. 1898 A, p. 669.
    Eocene (Wasatch); New Mexico.

Suborder XENARTHRA.

Ameghino, F. 1889 A, p. 661. (Gravigrada.)
Burmeister, H. 1879 B, p. 279 (Phyllophaga); p. 387 (Effodientia).
Cope, E. D. 1889 C. (Xenarthri.)
    1889 P, p. 657.
Flower, W. H. 1883 A. (Pilosa.)
Flower and Lydekker 1891 A, p. 179. (Pilosa.)
Gadow, H. 1898 A, p. 41.
Gill, T. 1872 B, pp. 23, 24. (Vermilingua, Tardigrada, Loricata.)

Haeckel E. 1895 A, p. 490. (Bradytheria.)
Roger, O. 1896 A, p. 91.
Trouessart, E. L. 1898 A, p. 1094.
Wortman, J. L. 1897 B, p. 109.
Zittel, K. A. 1898 B, p. 124.
    Dr. Gill, in Standard Natural History, vol. v, 1884, p. 66, recognized the peculiar character of the vertebræ of this group and employed the adjective term xenarthral; but he did not formally name the group.

MEGATHERIIDÆ.

Ameghino, F. 1889 A, p. 665 (Megatheridæ); p. 690 (Megalonychidæ); p. 740 (Mylodontidæ).
    1894 A, p. 397. (Gravigrada.)
Burmeister, H. 1866 B, p. 149. (<Gravigrada.)
Claypole, E. W. 1891 A.
Cope, E. D. 1889 C.
    1889 P, p. 658.
    1889 R, p. 876.
Flower, W. H. 1882 A.
    1883 D, p. 384.
Flower and Lydekker 1891 A, p. 158.
Gill, T. 1872 B, p. 24.
Lönnberg, E. 1899 A. (Mylodontidæ.)

Lydekker, R. 1887 A, p. 85.
    1894 A, p. 70.
    1896 B, p. 102. (Megalotheriidæ.)
Nicholson and Lydekker 1889 A, p. 1296.
Roger, O. 1896 A, p. 91 (Megatheridæ); p. 93 (Megalonychidæ); p. 97 (Mylodontidæ).
Schlosser, M. 1890 C, p. 251.
Steinmann and Döderlein 1890 A, p. 789.
Trouessart, E. L. 1898 A, p. 1098 (Megalonychidæ); p. 1107 (Megatheriidæ).
Zittel, K. A. 1898 B, p. 130 (Megatheridæ); p. 133 (Megalonychidæ); p. 136 (Mylodontidæ).

MEGATHERIUM Cuvier.   Type *M. americanum* Cuvier.

Cuvier, G., 1796, Mag. Encycl., i.   (*Fide* Agassiz.)
Adam, W. 1834 A.
Agassiz, L. 1842 C.
    1863 A.
    1863 B.
Ameghino, F. 1886 A, p. 178.
    1889 A, p. 666.
Blainville, H. M. D. 1839 A, p. 67.
    1839 C, p. 114.
    1864 A, iv, EE, p. 39, pls. i–iv.
Bru, D. J. B. 1804 A.
Buckland, W. 1837 A, i, p. 112.
Burmeister, H. 1866 B, p. 150.
    1870 A, p. 381.
    1878 A.
    1879 B, p. 285.
    1888 B.
Clift, W. 1835 A.
Cope, E. D. 1889 C, p. 188.
    1889 P, p. 660.
Couper, J. H. 1842 A.
Cuvier, G., 1798, Table Élemént., p. 146.
    1804 B, pp. 376–387, pls. xxiv, xxv.
    1834 A, viii, pt. i, pp. 581–570, pl. ccxvii.
Dawkins, W. B. 1870 A, p. 233.
De Kay, J. E. 1842 C, p. 98.
Desor, E. 1852 A, p. 58.
Flower, W. H. 1882 A.
    1883 D, p. 385.
Flower and Lydekker 1891 A, p. 185.
Gervais, P. 1889 A.
    1873 A, p. 464.
Giebel, C. G. 1847 A, p. 111.
Harlan, R. 1825 A, p. 199.
Holmes, F. S. 1858 B.
Koken, E. 1888 C, p. 5.
Leidy J. 1855 B, pp. 49, 59.
Lund, P. W. 1841 A, p. 73.
    1842 A.
Lydekker R. 1887 A, p. 86.
    1894 A, p. 71.
    1896 B, p. 102. (Megalotherium.)
Nicholson and Lydekker 1889 A, p. 1296.
Owen, R. 1840 B.
    1840 F, pp. 64, 85.
    1842 D.
    1843 B.
    1845 B, p. 333, pls. lxxxiii, lxxxiv.
    1847 C, p. 66.
    1851 C.
    1852 A.
    1855 B.
    1856 B.
    1858 C.
    1860 D.
    1866 B, p. 402.
    1868 A, p. 274.
Pander and d'Alton 1818 A.
Pictet, F. J. 1853 A, p. 264, fig. 216,
Schmidt, O. 1886 A, p. 115.
Steinmann and Döderlein 1890 A, p. 739.
Trouessart, E. L. 1898 A, p. 1108.
Woodward, A. S. 1898 B, p. 279.
Zittel, K. A. 1893 B, p. 131.

Megatherium americanum Cuvier.

Cuvier, G., 1800, in Shaw's "General Zoology," i,
    p. 165.

Ameghino, F. 1889 A, p. 668, pl. xli, fig. 4; pl.
    lxxxi, figs. 1–3.
Blainville, H. M. D. 1864 A.
Burmeister, H. 1866 B, p. 150, pl. v.
    1878 A, pl. xi.
    1879 B, p. 321.
    1889 A, p. 27, pl. xi, figs. 7, 8.
Desmarest, A. G., 1822, Mammalogie, p. 365.   (M.
    cuvieri.)
Flower and Lydekker 1891 A, p. 185, figs. 60–62.
Gervais and Ameghino 1880 A, p. 135.
Giebel, C. G. 1847 A, p. 111. (M. cuvieri.)
Leidy, J. 1877 A, p. 218, pl. xxxiv, figs. 42, 43.
    (With doubt.)
Lydekker, R. 1887 A, p. 86.
    1894 A, p. 73, pls. xlv, xlvi.
Owen, R. 1840 F, p. 100, pls. xxx–xxxii.
    1851 C.
    1855 B, pls. xxii–xxvii.
    1856 B, pls. xxi–xxvi.
Quenstedt, F. A. 1852 A, p. 44, pl. iii, fig. 13.
Reinhardt, J. 1875 A, pl. iv, figs. 6, 7.
Roger, O. 1896 A, p. 91.
Trouessart, E. L. 1898 A, p. 1109.
Zittel, K. A. 1898 B, p. 132, figs. 104, 107, 108.
    This species has been recorded with doubt
    from the phosphate beds of South Carolina. The
    species had previously been known only
    from South America.

Megatherium mirabile Leidy.

Leidy, J. 1852 H, p. 117.
Cooper, W. 1824 A, p. 114, pl. viii. ("Megatheri-
    um.")
    1828 A, p. 267. ("Megatherium.")
Cope, E. D. 1889 P, p. 660.
Couper, J. H. 1842 A, p. 216.
    1846, in Hodson, W. B. 1846 A.
Cuvier, G. 1834 A, viii, p. 338. ("Megatherium.")
De Kay, J. E. 1842 C, p. 98. (M. cuvieri.)
Habersham, J., in Hodson, W. B. 1846 A, p. 25.
    (M. cuvieri.)
Harlan, R. 1825 A, p. 200. (M. cuvieri.)
    1834 B, p. 63. (M. cuvieri.)
    1835 B, p. 269. (M. cuvieri.)
    1842 B, p. 143. ("Megatherium.")
    1843 B, p. 70. (Megatherium.)
Hodson, W. B. 1846 A.
Holmes, F. S. 1849 A, p. 197. ("Megatherium.")
Leidy, J. 1854 A, p. 10.
    1855 B, pp. 49, 59, pl. xv.
    1860 B, p. 111, pl. xx, figs. 8, 8a.
    1869 A, p. 411.
Lydekker, R. 1887 A, p. 90.
Lyell, C. 1843 B, p. 128. ("Megatherium.")
    1844 A, p. 323. ("Megatherium.")
Mitchell, S. L. 1824 A, pl. vi, figs. 12, 13. ("Mega-
    therium.")
Roger, O. 1896 A, p. 92.
Trouessart, E. L. 1898 A, p. 1110.
T. R. D. 1839 A. ("Megatherium.")
Wyman, J. 1856 A.
Zittel, K. A. 1893 B, p. 132.
    Pleistocene; Georgia, South Carolina.

MEGALONYX Jefferson. Type *M. jeffersonii* Desm.

Jefferson, Thos. 1799 A.
Allen, J. A. 1876 A.
Ameghino, F. 1889 A, p. 690.
Amer. Geologist 1891 G, p. 198.
Blainville, H. M. D. 1839 B.
    1839 C, p. 116.
    1864 A, iv, EE, p. 42, pls. iii, iv.
Bronn, H. G. 1838 A, p. 1258.
Burmeister, H. 1866 B, p. 179.
    1879 B, p. 382.
Claypole, E. W. 1891 A.
    1897 A.
Cope, E. D. 1871 I, p. 72.
    1889 C, p. 186.
    1889 P, p. 660, pl. xxxi.
Cuvier, G. 1804 A, pp. 358–376. ("Megalonix.")
    1834 A, viii, pt. i, pp. 304–330.
De Kay, J. E. 1842 C, p. 98.
Gervais, P. 1855 A, p. 44.
    1873 A, p. 466.
Giebel, C. G. 1847 A, p. 111.
Harlan, R. 1825 A, p. 201.
    1835 B, p. 271.
Holmes, F. S. 1858 B.
Leidy, J. 1855 B, pp. 3, 57.
    1857 K.
    1860 A, p. 107.
Lund, P. W. 1841 A, p. 72.
Lydekker, R. 1887 A, p. 111.
    1896 B, p. 105.
Nicholson and Lydekker 1889 A, p. 1299.
Owen, R. 1840 F, p. 64.
    1842 D.
    1843 B.
    1845 B, p. 333.
    1852 A.
    1866 B, p. 411.
Pictet, F. J. 1853 A, p. 258, pl. viii, figs. 1–3.
Reinhardt, J. 1879 A.
Safford, J. M. 1892 A.
    1892 B.
Trouessart, E. L. 1898 A, p. 1106.
Wheatley, C. M. 1871 A.
Wistar, C. 1799 A.
Wortman, J. L. 1897 B.
Wyman, J. 1850 D.
Zittel, K. A. 1893 B, p. 134.

Megalonyx dissimilis Leidy.

Leidy, J. 1855 B, pp. 45, 57, pl. xvi, figs. 8, 15.
Cope, E. D. 1889 P, p. 660.
    1896 F, p. 379. (Syn. of M. jeffersonii.)
Cuvier, G. 1804 A, pl. xxiii, fig. 13. ("Megalonix.")
    1834 A, pl. ccxvi, fig. 13. ("Megalonyx.")
Leidy, J. 1852 H, p. 117. (No description.)
    1854 A, p. 9. (No description.)
    1869 A, p. 412.
Roger, O., 1896 A, p. 96.
Trouessart, E. L. 1898 A, p. 1106.
Wailes, B. L. C. 1854 A, p. 286.
    Pleistocene; Mississippi.
    Prof. Cope (1896 F, p. 379) regards this species, described by Leidy, as synonymous with *M. jeffersonii.*

Megalonyx jeffersonii (Desmarest).

Desmarest, A. G., 1822, Mammalogie, p. 366. (Megatherium.)
Bronn, H. G. 1838 A, p. 1255, pl. xlv, fig. 10.
Burmeister, H. 1866 B, p. 180.
Claypole, E. W. 1891 A. ("Megalonyx"; this species?)
    1897 A. ("Megalonyx"; this species?)
Cooper, W. 1831 A, pp. 171, 206.
    1833 A. ("Megalonyx.")
Cooper, Smith, and De Kay 1831 A, p. 371.
Cope, E. D. 1869 E, p. 172.
    1889 P, p. 660.
    1893 A, p. 50.
    1896 F, p. 379.
    1899 A, p. 210.
Cope and Wortman 1884 A, p. 39, pl. v, figs. 1, 2.
Cuvier, G. 1804 A, pp. 358–376, pl. xxiii, except fig. 13. ("Megalonix.")
    1834 A, viii, pt. i, p. 304; pl. ccxvi, except fig. 13. ("Megalonyx.")
De Kay, J. E. 1842 C, p. 99.
Dickeson, M. W. 1846 A, p. 106.
Harlan, R. 1825 A, p. 201.
    1831 A, p. 269, pls. xii–xiv (M. laqueatus; M. jeffersonii); p. 334 (Aulaxodon s. Pleurodon).
    1831 B, p. 74.
    1834 B, p. 65 (M. jeffersonii); p. 67 (M. laqueatus.)
    1835 A, p. 273. (M. laqueatus.)
    1835 B, p. 271. (M. laqueatus.)
    1835 C, p. 319, pls. xii–xiv; pl. xv, figs. 5–7. (M. laqueatus.)
    1843 A, p. 209. (M. laqueatus, in part.)
Jefferson, Thos. 1799 A. ("Megalonyx.")
Kindle, E. M. 1898 A, p. 485.
Leidy J. 1852 H, p. 117. (M. jeffersonii; M. potens.)
    1854 A, p. 9.
    1854 I, p. 200.
    1855 B, pp. 3, 57, pls. i–xiii.
    1860 A, p. 107, pl. vi, fig. 1.
    1869 A, pp. 411, 412.
    1870 H, p. 13.
    1889 E, p. 27.
Lesley, J. P. 1889 A, p. 383.
Lindahl, J. 1892 A, pl. v, figs. 2, 4, 6.
Lund, 1839 A, p. 578.
Lydekker, R. 1887 A, p. 111.
    1896 B, p. 107.
Lyell, C. 1843 B, p. 128. ("Megalonyx.")
    1844 A, p. 323. ("Megalonyx.")
Mercer, H. C. 1897 A, figs. 1–26.
Nicholson and Lydekker 1889 A, p. 1229.
Orton, E. 1891 A.
Owen, D. D. 1857 A, p. 273.
Owen, R. 1840 F, pp. 64, 99, pl. xvii, fig. 1.
    1842 D.
    1843 B, p. 343.
Quenstedt, F. A. 1852 A, p. 44, pl. i, fig. 1.
    1885 A, p. 61, pl. i, fig. 15.
Rafinesque, C. S., 1832, Atlantic Jour., p. 18. (Aulaxodon speleum.)

Roger, O. 1896 A, p. 96.
Steinmann and Döderlein 1890 A, p. 741.
Troost, G. 1835 A, p. 144.
Trouessart, E. L. 1898 A, p. 1106.
Wailes, B. L. C. 1854 A, p. 286.
Ward, H. A. 1864 A.
Wistar, C. 1799 A, p. 526, pls. i, ii. ("Megalonix.")
Wortman, J. L. 1897 B, p. 78, figs. 5, 10.
Wyman, J. 1850 A, p. 58, figs. 1-8. (M. laqueatus.)
1862 A, p. 422.
Zittel, K. A. 1893 B, p. 134, fig. 110.
Pleistocene; West Virginia, Northern Illinois, Indiana, Ohio, Kentucky, Tennessee, Alabama, Mississippi, Florida.

Megalonyx leidyi Lindahl.

Lindahl, J. 1892 A, pp. 1-10, pls. i-iv; pl. v, figs. 1, 3, 5, 7-9.
Cope, E. D. 1889 P, p. 660, pl. xxxi. ("Megalonyx.")
1896 A, p. 49.
Roger, O. 1896 A, p. 96.
Trouessart, E. L. 1898 A, p. 1106.
Udden, J. A. 1891 A, p. 342.
Williston, S. W. 1897 I, p. 304.
1898 I, p. 98.
Pliocene; Kansas.

Megalonyx leptostomus Cope.

Cope, E. D. 1893 A, p. 49, pl. xiii, figs. 1-3.
Matthew, W. D. 1899 A, p. 75.
Roger, O. 1896 A, p. 96.
Trouessart, E. L. 1898 A, p. 1106.
Pliocene (Blanco); Texas.

Megalonyx loxodon Cope.

Cope, E. D. 1871 I, p. 74, figs. 1, 2.
1889 P, p. 660.
1896 F, p. 379.
1899 A, p. 211.
Lesley, J. P. 1889 A, p. 383, figures.

Roger, O. 1896 A, p. 96.
Trouessart, E. L. 1898 A, p. 1107.
Pleistocene; Pennsylvania.

Megalonyx scalper Cope.

Cope, E. D. 1899 A, p. 218, pl. xviii, figs. 2, 2a.
Pleistocene; Pennsylvania.
According to Cope this may be the same species as Ereptodon priscus.

Megalonyx tortulus Cope.

Cope, E. D. 1871 I, p. 84, fig. 12.
1889 P, p. 660.
1896 F, p. 379.
1899 A, p. 217.
Lesley, J. P. 1889 A, p. 384, figures.
Roger, O. 1896 A, p. 96.
Trouessart, E. L. 1898 A, p. 1107.
Pleistocene; Pennsylvania.

Megalonyx validus Leidy.

Leidy, J. 1868 A, p. 175.
Cope, E. D. 1871 I, p. 80.
Leidy, J. 1869 A, p. 412, pl. xxx, fig. 6.
Roger, O. 1896 A, p. 96.
Trouessart, E. L. 1898 A, p. 1106.
Pleistocene; Texas.

Megalonyx wheatleyi Cope.

Cope, E. D. 1871 I, p. 75, figs. 3-6; fig. 10.
1871 F. (M. dissimilis, not of Leidy.)
1871 I, p. 83, fig. 11 (M. sphenodon); figs. 7-8 (M. dissimilis).
1898 A, p. 50. (M. wheatleyi, M. dissimilis.)
1896 F, p. 378.
1899 A, pp. 210, 211, 212.
Lesley, J. P. 1889 A, p. 384, figures. (M. sphenodon, M. wheatleyi.)
Roger, O. 1896 A, p. 96.
Trouessart, E. L. 1898 A, p. 1107.
Pleistocene; Pennsylvania.

 EREPTODON Leidy.   Type *E. priscus* Leidy.

Leidy, J. 1853 E.
1855 B, pp. 46, 58.
Trouessart, E. L. 1898 A, p. 1106. (Syn.? of Megalonyx.)
Zittel, K. A. 1893 B, p. 134. (Syn.? of Megalonyx.)

Ereptodon priscus Leidy.   •

Leidy, J. 1853 E.

Cope, E. D. 1899 A, p. 219.
Leidy, J. 1855 B, pp. 46, 58, pl. xiv, figs. 9, 10; pl. xvi, fig. 18.
1869 A, p. 412.
Roger, O. 1896 A, p. 96. (Megalonyx.)
Trouessart, E. L. 1898 A, p. 1106. (Megalonyx.)
Wailes, B. L. C. 1854 A, p. 286.
Pleistocene; Mississippi.

MYLODON Owen.   Type *M. harlani* Owen.

Owen, R. 1840 F, p. 63.
Ameghino, F. 1889 A, p. 741.
1898 A, p. 324.
Blainville, H. M. D. 1864 A, iv, EE, p. 39, pls. i-iv.
Bronn, H. G. 1838 A, p. 1256. ("Orycterotherium.")
Burmeister, H. 1865 B.
1866 B, p. 160.
1879 B, p. 345.
Cope, E. D. 1882 FF, p. 923.
1889 C, p. 188.
1889 P, p. 660.

Desor, E. 1852 A, p. 58.
Flower and Lydekker 1891 A, p. 189.
Gervais, P. 1855 A, p. 45.
1873 A, p. 467.
Gervais and Ameghino 1880 A, p. 155.
Giebel, C. G. 1847 A, p. 117.
Harlan, R. 1842 B, p. 141. (Aulaxodon seu Pleurodon, in part.)
1842 C, p. 111. (Orycterotherium, type O. missouriense.)
1843 A.
1843 B. (Orycterotherium.)

Leidy, J. 1858 E. (Eubradys, type *E. antiquus.*)
　　1855 B, pp. 47, 58.
Lönnberg, E. 1899 A.
Lydekker, R. 1887 A, p. 104.
　　1894 A, p. 77.
　　1896 B, p. 105.
Marsh, O. C. 1883 B.
Nicholson and Lydekker 1889 A, p. 1298.
Owen, R. 1840 F, p. 57. (Glossotherium.)
　　1842 D.
　　1843 B.
　　1845 C.
　　1852 A.
　　1855 B.
　　1856 B.
　　1858 C.
　　1860 D.
　　1866 B, p. 400.
Pictet, F. J. 1853 A, p. 269.
Reinhardt, J. 1866 A.
　　1879 A.
Schmidt, O. 1886 A, p. 115.
Steinmann and Döderlein 1890 A, p. 740.
Wortman, J. L. 1897 B.
Zittel, K. A. 1893 B, p. 137.

Mylodon harlani Owen.

Owen, R. 1840 F, pp. 68, 69, pl. xvii, figs. 3, 4.
Carpenter, W. M. 1846 A, p. 249. (Oryctotherium;" identification very doubtful.)
Claypole, E. W. 1891 A.
Cooper, W. 1831 A, p. 172. ("Megalonyx.")
　　1833 A, p. 166. ("Megalonyx.")
Cope, E. D. 1871 I, p. 85.
　　1889 P, p. 666. (M. laqueatus.)
　　1896 H, p. 458, pl. x, figs. 1, 2. (M. harlanii.)
　　1899 A, p. 210. (M.? harlanii.)
Dickeson, M. W. 1846 A. (Megalonyx laqueatus.)
Flower and Lydekker 1891 A, p. 189.
Giebel, C. G. 1847 A, p. 118.
Harlan, R. 1831 B, p. 74, in part; pl. iii, figs. 1–3. (Megalonyx laqueatus.)
　　1835 C, p. 334, pl. xv, figs. 2–4. (Megalonyx laqueatus.)
　　1842 A, p. 392. ("Orycterotherium.")
　　1842 B, p. 141 (Megalonyx laqueatus, Aulaxodon *seu* Pleurodon, in part); p. 142 (Orycterotherium missouriense).
　　1842 C, p. 111. (Orycterotherium missouriense.)
　　1843 A, p. 209 (Megalonyx laqueatus); p. 210 (Orycterotherium missouriense).
　　1843 B. pls. i–iii. (Megalonyx laqueatus, Orycterotherium missouriense.)
Holmes, F. S. 1858 B. ("Megalonyx.")
Leidy, J. 1853 E. (Eubradys antiquus.)
　　1854 A, p. 10.
　　1855 B, pp. 47, 58, pl. xiv, figs. 1–3; pl. xvi, figs. 19, 20.

Leidy, J. 1860 B, p. 111, pl. xx, figs, 7–7b.
　　1869 A, p. 413.
　　1871 C, p. 365.
　　1884 A.
　　1885 B. (M. harlani; ?Orycterotherium missouriense.)
　　1889 F, p. 83, pl. v, figs. 1–4.
Lydekker, R. 1887 A, p. 106.
Lyell, C. 1843 B, p. 128. ('Mylodon.")
　　1844 A, p. 323. (Megalonyx.)
Nicholson and Lydekker 1889 A, p. 1299.
Owen, R. 1842 D.
　　1843 B, p. 341. (M. harlani, Megalonyx laqueatus.)
　　1845 B, p. 335.
　　1845 C.
Perkins, H. C. 1842 A. ("Mylodon": identification doubtful.)
　　1843, in Harlan R. 1843 B, p. 80. (Orycterotherium oregonense.)
　　1844 A, p. 135. (Orycterotherium oregonense.)
Pictet, F. J. 1853 A, p. 270.
Roger, O. 1896 A, p. 98.
Trouessart, E. L. 1898 A, p. 1115.
Wailes, B. L. C. 1854 A, p. 286.
Williston, S. W. 1895 A, p. 175, figs. 1, 2. (Identification doubtful.)
Wilson, T. 1892 A. ("Mylodon.")
Zittel, K. A. 1893 B, p. 138. (M. laqueatus.)
　　In addition to the above a few references, mostly of minor importance, may be found in Leidy, J. 1869 A, pp. 413, 414.
　　Pleistocene; Kentucky, Missouri, Louisiana, Mississippi, South Carolina, Oregon, Kansas?, Pennsylvania?.

Mylodon renidens Cope.

Cope, E. D. 1896 H, p. 460, pl. x, fig. 3; pl. xi, figs. 5, 6.
Roger, O. 1896 A, p. 98.
Trouessart, E. L. 1898 A, p. 1116.
　　Pleistocene; Louisiana.

Mylodon sodalis Cope.

Cope, E. D. 1878 H, p. 385.
　　1889 F.
　　1889 P, p. 660, fig. 1.
　　1889 S, pp. 978, 980.
　　1893 A, p. 87.
Roger, O. 1896 A. p. 98. (Syn. of M. harlani.)
Trouessart, E. L. 1898 A, p. 1116. (Syn. of M. harlani.)
　　Pliocene; Oregon, Texas?.

Mylodon sulcidens Cope.

Cope, E. D. 1896 H, p. 462, pl. x, fig. 4; pl. xi, fig. 7.
Roger, O. 1896 A, p. 98.
Trouessart, E. L. 1898 A, p. 1116.
　　Pleistocene; Louisiana.

MOROTHERIUM Marsh.　　Type *M. gigas* Marsh.

Marsh, O. C. 1874 C, p. 531.
Flower and Lydekker 1891 A, pp. 190, 413. (Syn. of Chalicotherium.)
Marsh, O. C. 1876 G, p. 61.

Marsh, O. C. 1883 B.
　　1892 D, p. 448.
　　1897 B, p. 144.
Zittel, K. A. 1893 B, p. 136.

Morotherium gigas Marsh.

Marsh, O. C. 1874 C, p. 531.
King, C. 1878 A, p. 443.
Marsh, O. C. 1897 B, p. 143, figs. 8, 9.
Roger, O. 1896 A, p. 96.
Trouessart, E. L. 1898 A, p. 1107.
Pliocene; California.

Morotherium leptonyx Marsh.

Marsh, O. C. 1874 C, p. 532.
King, C. 1878 A, p. 443.
Roger, O. 1896 A, p. 96.
Trouessart, E. L. 1898 A, p. 1107.
Pliocene; Idaho.

GLYPTODONTIDÆ.

Ameghino, F. 1883 A. (Glyptodontes.)
    1889 A, p. 758 (Glyptodontia); p. 775 (Glyptodontidæ).
Burmeister, H. 1865 A.
    1865 C.
    1866 B, p. 183. (Biloricata.)
    1871 B.
    1872 A. ("Glyptodonten.")
    1874 A.
    1879 A.
    1879 B, p. 388. (Loricata.)
Cope, E. D. 1889 C, p. 207.
    1889 P, p. 658.
Flower, W. H. 1882 A.
Flower and Lydekker 1891 A, p. 202.

Gadow, H. 1898 A. p. 42.   (Dasypodidæ, part.)
Lydekker, R. 1887 A, p. 114.
    1894 A. ;. 1.
    1896 B. p. 94.
Mercerat, A., 1891, Revista Mus. La Plata, ii, p. 27
    (Hoplophoridæ)
Nicholson and Lydekker 1889 A, p. 1292.
Reh, L. 1895 A, p. 170.
Roger, O. 1896 A, p. 101.
Schlosser, M. 1890 C, p. 251.
Steinmann and Döderlein 1890 A, p. 738.
Trouessart, E. L. 1898 A, p. 1123.
Woodward, A. S. 1898 B, p. 282.
Zittel, K. A. 1893 B, p. 144.

GLYPTODON Owen.   Type *G. clavipes* Owen.

Owen, R., 1838, Parish's Buenos Ayres and La Plata.
    p. 178 B.
Ameghino, F. 1883 A.
    1889 A, p. 775.
    1892 A, p. 22.
Blainville, H. M. D. 1864 A, iv, EE, p. 47, pls. i, ii.
Burmeister, H. 1864 A.
    1865 A.
    1865 C.
    1866 A.
    1866 B, p. 185.
    1871 A.
    1871 B.
    1874 A.
    1879 A.
    1879 B, p. 420.
Cope, E. D. 1889 P, p. 662.
Flower, W. H. 1883 D, p. 387, figs. 39, 40.
Flower and Lydekker 1891 A, p. 202.
Giebel, C. G. 1847 A, p. 109.
Hale, C. S. 1848 A, p. 357.
Holmes, F. S. 1858 B.
Huxley, T. H. 1865 A.
    1872 A, p. 291.
Koken, E. 1888 C.
Lund, P. W. 1839 B, p. 218. (Hoplophorus.)
Lydekker, R. 1887 A, p. 114.
    1894 A, p. 3.
    1896 B, p. 96.
Meyer, H. 1865 A.
Natural Science 1893 A, p. 140, fig. 2.
Nicholson and Lydekker 1889 A, p. 1293.
Owen, R. 1840 B.
    1845 A.
    1847 C, p. 67.
Pictet, F. J. 1853 A, p. 278.
Pouchet, G. 1866 A.
Quenstedt, F. A. 1885 A, p. 63.

Reinhardt, J. 1875 A.
Röse, C. 1892 D. p. 511.
Serres, M. 1865 A.
    1865 B.
Steinmann and Döderlein 1890 A, p. 738.
Trouessart, E. L. 1898 A, p. 1125.
Woodward, A. S. 1898 B, p. 283.
Zittel, K. A. 1893 B, p. 144.

Glyptodon euphractus Lund.

Lund, P. W., 1838, Overs. K. Danske Vid. Selskab.
    Forh. viii, p. iii. (Hoplophorus.)
Ameghino, F. 1883 A, p. 28.
    1889 A, p. 781, pl. liii, fig. 8.
Burmeister, H. 1871 A, pl. viiiA. (Hoplophorus.)
    1879 B, p. 410. (Hoplophorus.)
Gervais and Ameghino 1880 A, p. 202.   (G. euphractus, G. sellowi.)
Leidy, J. 1889 E, p. 25.   (Hoplophorus ornatus?.)
    1892 A, p. 129. (Hoplophorus.)
Lund, P. W. 1839 B, p. 218. (Hoplophorus.)
    1841 A, p. 70, pl. i, fig. 11; pl. xi; pl. xiv; pl. xv, figs. 1-7. (Hoplophorus.)
    1842 A, pl. xxxv, figs. 1-4. (Hoplophorus.)
Lydekker, R. 1887 A, p. 121.
    1894 A, p. 11.
Owen, R. 1868 A, p. 273.
Pouchet, G. 1866 A, p. 337, pls. ix, x. (Hoplophorus.)
Reinhardt, J. 1875 A, p. 165. (Hoplophorus, Schistopleurum.)
Roger, O. 1896 A, p. 101.
Trouessart, E. L. 1898 A, p. 1127.
    Pliocene; South America, Florida.

Glyptodon petaliferus Cope.

Cope, E. D. 1888 N, p. 345.

Cope, E. D. 1889 F.
  1889 P, p. 662.
  1892 AA, p. 180.
Leidy, J. 1889 E, p. 25, pl. iv, fig. 9.
Roger, O. 1896 A, p. 108.
Trouessart, E. L. 1898 A, p. 1127.
Zittel, K. A. 1893 B, p. 147.
  Pliocene: Mexico, Texas, Florida.

Glyptodon reticulatus Owen.

Owen, R., 1845, Cat. Foss. Mam. Aves. Roy. Coll.
  Surg., p. 119.
Ameghino, F. 1883 A, p. 28.   (G. asper.)
  1889 A, p. 784, pl. l, figs. 1-8; pl. li, figs. 1, 2;
    pl. lii, figs. 1-3; pl. liv, figs. 1-11.
Burmeister, H. 1864 A, p. 75.
  1866 B, p. 200, pl. vi; pl. vii, figs. 4-6; pl. viii,
    fig. 6 (G. asper, original description); p.
    205 (G. reticulatus).

Burmeister, H. 1874 A, p. 268, pls. xxiii, xxiv,
  xlxv; pl. xxv, fig. 1; pl. xxviii, fig. 1; pl. xxix,
  figs. 4, 6; pl. xxx, figs. 1, 4; pl. xxxi; pl. xxxii,
  fig. 2; pl. xxxiii, figs. 2, 3, 5, 6; pl. xxxiv, fig.
  2; pl. xxxvii; pl. xl, fig. 1 (G. asper); p. 385
  (G. reticulatus).
  1879 B, p. 422 (G. reticulatus); p. 424 (Schis-
    topleurum asperum).
Gervais and Ameghino 1880 A, p. 196. (G. typus.)
Leidy, J. 1889 E, p. 27, pl. v, figs. 11, 12. (G. asper.)
  1892 A. (G. asper.)
Lydekker, R. 1887 A, p. 117.
  1894 A, p. 5.
Nodot, L. 1855 A. (Schistopleurum.)
Roger, O. 1896 A, p. 102.
Trouessart, E. L. 1898 A, p. 1125.
  Additional synonomy and literature in Ame-
    ghino, F. 1889 A, p. 784.
  Pliocene; South America, Florida.

TOMIOPSIS Cope.   Type *T. ferruminatus* Cope.

Cope, E. D. 1893 E, p. 317.
  The position of this genus is uncertain.

Tomiopsis ferruminatus Cope.
Cope, E. D. 1893 E, p. 317.
  Neocene?; Texas.

DASYPODIDÆ.

Broom, R. 1897 A.
Burmeister, H. 1879 B, p. 426. (Dasypidæ.)
  1883 A. (Genus Eutatus.)
Cope, E. D. 1889 P, p. 658.
Flower, W. H. 1868 A.
  1882 A.
  1883 D, p. 386.
  1885 A.
Flower and Lydekker, 1891 A, p. 194.
Gadow, H. 1898 A, p. 42. (Dasypodidæ, part.)
Giebel, C. G. 1867 A, p. 545. (Dasypus.)
Gill, T. 1872 B, p. 24.
Gray, J. E. 1869 A, p. 377.
Hæckel, E. 1895 A, p. 516. (Dasypodida.)
Huxley, T. H. 1872 A, p. 290.

Lydekker, R. 1887 A, p. 135.
  1894 A, p. 51.
  1896 B, p. 94.
Nicholson and Lydekker 1889 A, p. 1291.
Owen, R. 1882 A. (Dasypodæ.)
  1851 C, p. 743. (Dasypus.)
  1855 B, p. 381.
Roemer, F. 1892 A, p. 525.
  1893 A.
Roger, O. 1896 A, p. 107.
Schlosser, M. 1890 C, p. 251.
Steinmann and Döderlein 1890 A, p. 739.
Weber, M. 1898 A.
  1893 B.
Zittel, R. A. 1893 B, p. 150. (Subord. Dasypoda.)

CHLAMYTHERIUM Lund.   Type *C. humboldtii* Lund.

Lund, P. W., 1838, Overs. K. Danske Vid. Selsk.
  Forh. viii, p. 111.
Ameghino, F. 1883 A. pp. 28, 33.
  1886 A, p. 208.
Flower and Lydekker 1891 A, p. 201.
Gervais and Ameghino 1880 A, p. 211.
Lydekker, R. 1887 A, p. 135.
  1894 A, p. 52.
Lund, P. W. 1839, A, p. 572.
  1839, B, p. 217.
  1841 A, pp. 69, 232.
  1842 A, p. 187, pl. xxxiv; pl. xxxv, fig. 5.
Pictet, F. J. 1853 A, p. 274.
Röse, C. 1892 D, p. 516.
Zittel, K. A. 1893 B, p. 151.
  The name of this genus is usually spelled *Chla-
mydotherium;* but the original form is that here
given. The original description is adequate.

Chlamytherium humboldtii Lund.
Lund, P. W., 1838, Overs. K. Danske Vid. Selsk.
  Forh., viii, p. 111.
Ameghino, F. 1883 A, p. 33.
Cope, E. D. 1889 P, p. 662 ("Glyptodon floridanus[1]
  Leidy.")
  1892 AA, p. 180.
Leidy, J. 1889 B, p. 97. (Glyptodon septentric'
  alis.)
  1889 E, p. 24, pl. iv, figs 3-6.
  1892 A.
Lund, P. W. 1839 B, p. 217.
  1841 A, p. 69, pl. i, figs. 7-10, 12, 13; pl. ii, figs.
    1-3; pl. xii, figs. 1, 6, 7; pl. xiii, figs. 2, 6,
    7-11; pl. xiv, fig. 1.
  1842 A, p. 187, pl. xxxiv; pl. xxxv, fig. 5.
Lydekker, R. 1887 A, p. 136.
  1896 B, p, 96.

[1] The writer has made search in Leidy's papers without finding this name. It occurs in neither
Leidy's nor Cope's lists in Bull. No. 84, U. S. Geol. Survey (Correlation paper). The name is probably due to an error on part of Professor Cope.

Roger, O. 1896 A, p. 108 (Glyptodon septentriona-
lis); p. 108 (C. humboldti).
Trouessart, E. L. 1898 A, p. 1127 (Glyptodon sep-
tentrionalis); p. 1137 (C. humboldti).

Zittel, K. A. 1893 B, p. 147 (Glyptodon floridanus[1]);
p. 152 (C. humboldti).
Pliocene; South America, Florida.

Order SIRENIA.

Allen, J. A. 1881 A.
Baur, G. 1887 A, p. 493.
Brandt, J. F. 1862 A.
  1862 B.
  1868 A.
Cope, E. D. 1880 B, p. 456.
  1887 S, p. 342.
  1889 C, p. 142.
  1889 R, pp. 874, 876.
  1890 F,
  1898 B, p. 109.
Cuvier, G. 1834 A, viii, pt. ii, pp. 1–74, pls. ccxx,
  ccxxi.
Flot, L. 1886 A.
  1886 B.
Flower, W. H. 1876 B, xiii, p. 409.
  1883 A.
  1883 D, p. 389.
  1885 A.
Flower and Lydekker 1891 A, p. 212.
Gadow, H. 1898 A, p. 42.
Gegenbaur, C, 1898 A.
Gervais, P. 1874 B, p. 578.   ("Sirénidea.")
Gill, T. 1872 A, p. 301.
  1872 B, pp. 13, 48, 91.
  1873 A.
Gill and Coues, 1877 A, p. 1041.
Gray, J. E. 1866 A, p. 356.
Hæckel, E. 1896 A, p. 567.
Hay, O. P. 1899 F, p. 682.
Huxley, T. H. 1872 A, p. 330.

Krauss, F. 1858 A.
  1862 A.
  1870 A.
Kükenthal, W. 1890 C.
  1897 A.
Lartet, E. 1866 A.
Lepsius, G. R. 1881 A.   (Sirenia.)
Lydekker, R. 1892 A.
  1896 B,
Marsh, O. C. 1877 E.
Nicholson and Lydekker 1889 A, p. 212.
Owen, R. 1866 B, p. 429.
  1868 A, p. 288.
  1875 A.
  1875 B.
Parker, W. K. 1868 A, p. 218.
Roger, O. 1896 A, p. 247.
Ryder, J. A. 1878 A.
  1885 D.
Schlosser, M. 1889 A.
  1890 D.
Schmidt, O. 1886 A, p. 242.
Seeley, H. G. 1888 N, p. 181.
Slade, D. D. 1888 A, p. 244.
Steinmann and Döderlein 1890 A, p. 708.
Tomes, C. S. 1898 A, p. 385.
Trouessart, E. L. 1898 A, p. 999.
Woodward, H. 1885 D.
Zigno, A. de 1887 A.
Zittel, K. A. 1893 B, pp. 187, 201, 740.

PRORASTOMIDÆ.

Cope, E. D. 1889 R, p. 876.
  1890 F, p. 698.
Flower and Lydekker 1891 A, p. 224.   (Halithe-
riidæ.)
Gadow, H. 1898 A, p. 43.
Haeckel, E. 1895 A, p. 566.   (Halicorida.)
Zittel, K. A. 1893 B, p. 194.
  This family is based on *Prorastomus sirenoides,*

Owen, found in the island of Jamaica. The
following references are given to the literature
of this animal: Owen, R. 1855 C; Owen, R.
1875 B, pls. xxviii, xxix; Leidy, J. 1869 A, p.
414; Trouessart, E. L. 1898 A, p, 999; Thomas
and Lydekker 1897 A, p. 599; Lepsius, G. R.
1881 A; Tomes, C. S. 1898 A, p. 390.

HALITHERIIDÆ.

Gill, T. 1872 B, pp. 13, 91, 92.
Brandt, J. F. 1868 A, p. 344.   (<Halicorida.)
Cope, E. D. 1889 R, p. 876.
  1890 F, p. 698.
  1898 B, p. 110.
Flower and Lydekker 1891 A, p. 222.

Lepsius, G. R. 1881 A.   (Halitherium.)
Lydekker, R. 1887 A, p. 5.
Nicholson and Lydekker 1889 A, p. 1310.
Roger, O. 1896 A, p. 247.
Trouessart, E. L. 1898 A, p. 1004.   (Halicoridæ.)
Zittel, K. A. 1893 B, p. 195.   (<Halicoridæ.)

DIOPLOTHERIUM Cope.   Type *D. manigaultii* Cope.

Cope, E. D. 1883 N.
  1883 X.
  1890 ʋ, p. 698.

Flower, W. H. 1885 A, p. 223.
Flower and Lydekker 1891 A, p. 223.
Nicholson and Lydekker 1889 A, p. 1811.

[1] The writer has made search in Leidy's papers without finding this name. It occurs in neither
Leidy's nor Cope's lists in Bull. No. 84, U. S. Geol. Survey (Correlation paper). The name is probably due to an error on part of Professor Cope.

Zittel, K. A. 1898 B, p. 201.
It is not at all certain that this genus belongs among the Sirenia.

Dioplotherium manigaultii Cope.
Cope, E. D. 1883 N.

Cope, E. D. 1883 X.
    1890 F, p. 696, pl. xxv, figs. 1–5.
Roger, O. 1896 A, p. 249.
Trouessart, E. L. 1898 A, p. 1006.
Tertiary (Phosphate beds); South Carolina.

HEMICAULODON Cope.   Type *H. effodiens* Cope.

Cope, E. D. 1869 E, p. 190.
    1883 N, p. 52.
Zittel, K. A. 1898 B, p. 201.
A genus of uncertain position.

Hemicaulodon effodiens Cope.
Cope, E. D. 1869 E, p. 191, pl. v, fig. 6.

Brandt, J. F. 1873 A, p. 290.
Cope, E. D. 1890 F, p. 699.
Leidy, J. 1869 A, p. 440.
Roger, O. 1896 A, p. 249.
Trouessart, E. L. 1898 A, p. 1006.
Eocene; New Jersey.

DESMOSTYLUS Marsh.   Type *D. hesperus* Marsh.

Marsh, O. C. 1888 B, p. 95.
Flower and Lydekker 1891 A, p. 223.
Nicholson and Lydekker 1889 A, p. 1310.
Zittel, K. A. 1898 B, p. 201.
A genus of doubtful position.

Desmostylus hesperus Marsh.
Marsh, O. C. 1888 B, p. 95, figs. 1–3.
Roger, O. 1896 A, p. 249.
Trouessart, E. L. 1898 A, p. 1006.
Tertiary; California.

TRICHECHIDÆ.

Gill, T. 1872 B, pp. 14, 91.
Blainville, H. M. D. 1864 A, iii, U, pls. i–xi. ("Lamantinos.")
Brandt, J. F. 1868 A, p. 343. (Manatida.)
Cope, E. D. 1890 F, p. 700. (Manatidæ.)
Flower, W. H. 1885 A, p. 215. (Manatidæ.)
Flower and Lydekker 1891 A, p. 215. (Manatidæ.)

Gray, J. E. 1866 A, p. 357. (Manatidæ.)
Haeckel, E. 1895 A, p. 566. (Manatida.)
Nicholson and Lydekker 1889 A, p. 1311. (Manatidæ.)
Roger, O. 1896 A, p. 247. (Manat. idæ.)
Trouessart, E. L. 1898 A, p. 999. (Manatidæ.)
Zittel, K. A. 1898 B, p. 195. (Manatidæ.)

TRICHECHUS Linn.   Type *T. manatus* Linn.

Linnæus, C., 1758, Syst. Nat., ed., 10, i, p. 34.
Unless otherwise indicated, the following writers employ for the genus the name *Manatus.*
Allen, J. H. 1846 A.
Baikie, B. 1857 A.
Blainville, H. M. D. 1864 A, iii, U, plates.
Brandt, J. F 1862 A.
    1862 B.
    1868 A.
Cope, E. D. 1890 F, p. 700.
Cuvier, G. 1834 A, viii, pt. ii, pp. 16–40, pl. coxx, figs. 1, 2. (No generic name.)
Flower, W. H. 1876 B, xiii, p. 410.
    1885 A.
Flower and Lydekker 1891 A, p. 215.
Gill, T. 1873 A.
Gray, J. E. 1857 A.
Harlan, R. 1834 B, p. 73.
    1835 B, p. 278.
    1835 J, p. 385.
Howes and Harrison 1893 A.
Kneeland, S. 1850 A.
Krauss, F. 1858 A.
    1862 A.
Kükenthal, W. 1890 C.
    1891 C.
    1896 A.
    1897 A.

Lepsius, G. R. 1881 A.
Murie, J. J. 1872 A.
Owen, R. 1845 B, p. 371, pl. xcvi.
    1868 A, p. 284.
    1875 A.
    1875 B.
Parker, W. K. 1868 A, p. 218.
Thomas and Lydekker 1897 A, p. 596. (Trichechus.)
Trouessart, E. L. 1898 A, p. 1000.
Ward, H. L. 1887 A.
Woodward, H. 1885 D, p. 424.
Zittel, K. A. 1898 B, p. 195.

Trichechus antiquus (Leidy.)

Leidy, J. 1856 J, p. 165. (Manatus.)
Cope, E. D. 1890 F, p. 700. (M. fossilis.)
Leidy, J. 1860 B, p. 117, pl. xxiv, figs. 5–7. (Manatus.)
    1869 A, p. 414. (Manatus.)
    1877 A, p. 214. (Manatus.)
    1889 E, p. 27. (Manatus.)
    1892 A. (Manatus.)
Roger, O. 1896 A, p. 247. (Manatus.)
Trouessart, E. L. 1898 A, p. 1000. (Manatus.)
Pleistocene; South Carolina: Pliocene (Peace Creek), Florida.
Not improbably the same as *T. manatus.*

Trichechus inornatus (Leidy.)

Leidy, J. 1873 B, p. 336, pl. xxxvii, figs. 16, 17. (Manatus.)
Roger, O. 1896 A, p. 247. (Manatus.)
Trouessart, E. L. 1898 A, p. 1000. (Manatus.)
Tertiary (Phosphate beds); South Carolina.

Trichechus manatus Linn.

Linnæus, C., 1758, Syst. Nat. ed., 10, i, p. 34.
Balkie, B. 1857 A. (M. latirostris.)
Baur, G. 1887 D, p. 840. (Manatus americanus.)
Giebel, C. G. 1847 A, p. 231. (Manatus americanus fossilis.)
Gray, J. E. 1857 A. (M. latirostris, M. americanus.)
Hartlaub, C. 1886 A. (M. latirostris.)
Lepsius, G. R. 1881 A. (Manatus americanus.)
Murie, J. 1872 A. (M. americanus.)
Owen, R. 1866 B, p. 432. (M. americanus.)
Parker, W. K. 1868 A, p. 218, pl. xxix, fig. 21. (M. americanus.)
Trouessart, E. L. 1898 A, p. 1000. (M. latirostris.)
Recent; and probably including many or all

of the undetermined bones of *Manatus* which have been found along the Eastern and Floridan coasts.

Trichechus sp. indet.

The remains referred to in the following citations belong to uncertain and probably indeterminable species. Most of them probably belong to T. manatus.

Allen, J. H. 1846 A, p. 41. (Manatus.)
Cope, E. D. 1867 C, p. 138. (Manatus.)
De Kay, J. C. 1842 C, p. 123. (M. giganteus.)
Gibbes, R. W. 1850 C, p. 67. (Manatus.)
Harlan, R. 1834 B, p. 73. (Manatus.)
  1835 B, p. 278. (Manatus.)
  1835 J, p. 385. (Manatus.)
Leidy, J. 1854 A, p. 10. (Manatus.)
  1869 A, p. 414. (Manatus.)
Pictet, F. J. 1853 A, p. 372. (M. americanus fossilis.)
Smith, J. L. 1844 A. (Manatus.)
Pleistocene; New Jersey, Maryland, Virginia, North Carolina, South Carolina, Florida.

HYDRODAMALIDÆ.

Palmer, T. S. 1895 A, p. 450.
Brandt, J. F. 1849 A, p. 141. (Edentata seu Rhytinea).
  1868 A, p. 100 (Rhytinæ); p. 344 (<Halicorida).
Cope, E. D. 1890 F, p. 701. (Rhytinidæ.)
Flower, W. H. 1885 A, p. 221. (Rhytinidæ.)
Flower and Lydekker 1891 A, p. 221. (Rhytinidæ.)

Gadow, H. 1896 A, p. 45. (Halicoridæ, part.)
Gill, T. 1872 B, p. 14. (Rhytina.)
Haeckel, E. 1895 A, p. 566. (Rhytinida.)
Lepsius, G. R. 1881 A. (Rhytina.)
Lydekker, R. 1887 A, p. 15. (Rhytinidæ.)
Nicholson and Lydekker 1889 A, p. 1311. (Rhytinidæ.)
Zittel, K. A. 1893 B, p. 196. (<Halicoridæ.)

HYDRODAMALIS Retzius.

Retzius, 1894, K. Vet. Acad. Handl. Stockholm, xv, p. 292.
Baer, K. E. 1840 A. (Rytina.)
Brandt, J. F. 1849 A. (Rhytina.)
  1862 A. (Rhytina.)
  1862 B. (Rhytina.)
  1862 C. (Rhytina.)
  1867 D. (Rhytina.)
  1868 A. (Rhytina.)
Cope, E. D. 1890 F, p. 698. (Rhytina.)
Flower and Lydekker 1891 A, p. 221. (Rhytina.)
Gray, J. E. 1866 A, p. 365. (Rytina.)
Lepsius, G. R. 1881 A. (Rhytina.)
Lydekker, R. 1887 A, p. 15. (Rhytina.)
Nordmann, A. 1862 A. (Rhytina.)
Owen, R. 1845 B, p. 316. (Rytina.)
  1875 A. (Rhytina.)
  1875 B. (Rhytina.)
Palmer, T. S. 1895 A.
Weyhe, — 1875 A, p. 100. (Rhytina.)
Woodward, H. 1885 D, p. 415. (Rhytina.)
Zittel, K. A. 1893 B, p. 206. (Rhytina.)

Hydrodamalis gigas (Zimm.).

Zimmermann, 1780, Geogr. Geschichte, ii, p. 426, fig. 5. (Manati). (Fide Trouessart.)
Brandt, J. F. 1849 A. (Rhytina borealis.)
  1866 A, p. 113. (R. borealis.)
  1866 C. (R. borealis.)
  1867 A. (R. borealis.)

Type *H. gigas* (Zimm.).

Brandt, J. F. 1868 A, pls. i-ix. (R. borealis.)
Büchner, E. 1891 A, pl. i. (Rhytina.)
Claudius, M. 1867 A. (R. stelleri.)
Cope, E. D. 1890 F, p. 701. (R. gigas.)
Evermann, B. W. 1893 A. (R. gigas.)
Flower, W. H. 1876 B, xiii, p. 411.
  1885 A, p. 231. (R. stelleri.)
Flower and Lydekker 1891 A, p. 222. (R. stelleri.)
Gray, J. E. 1866 A, p. 365. (R. gigas.)
Hutchinson, H. N. 1893 A, p. 246, pl. xxvi. (R. gigas.)
Lydekker, R. 1887 A, p. 15, fig. 5. (R. gigas.)
Murie, J. 1872 B. (Rhytina.)
Nicholson and Lydekker 1889 A, p. 1311. (R. gigas.)
Nordmann, A. 1862 A. (Rhytina stelleri.)
Palmer, T. S. 1895 A.
Roger, O. 1896 A, p. 249. (Rhytina.)
Stejneger, L. 1883 A, (R. gigas.)
  1884 A. (R. gigas.)
  1886 A, p. 317. ("Sea-cow.")
  1887 A. (R. gigas.)
  1893 A. (R. gigas.)
Steller, G. W. 1751 A. ("Manatus.")
Trouessart, E. L. 1898 A, p. 1008. (Rhytina.)
Woodward, H. 1885 A. (R. stelleri.)
  1885 D, p. 417, fig. 2. (R. gigas.)
Zigno, A de 1887 A. (R. borealis.)
Zittel, K. A. 1893 B, p. 200. (R. stelleri.)
Recently extinct; Commander Islands.

Order CETE Linn.

Linnæus, C., 1758, Syst. Nat., ed. 10, 1, p. 75.
Unless otherwise indicated, the following
authors use the term Cetacea for this order.
Albrecht, P. 1886 A.
Allen, J. A. 1881 A.   (Cete.)
Brandt, J. F. 1873 A.
Cope, E. D. 1867 C, p. 144.
    1880 B, p. 456.
    1887 S, p. 342.
    1889 C, p. 207.
    1890 E.
    1891 N, p. 69.
Duvernoy, G. L. 1851 A.
Flower, W. H. 1864 A.
    1867 A.
    1876 B, xiii, p. 449.
    1883 A.
    1883 D, p. 391.
    1885 A.
Flower and Lydekker 1891 A, p. 225.
Gadow, H. 1898 A. p. 44.
Gegenbaur, C. 1898 A.
Gill, T. 1871 B.
    1872 B, pp. 14, 49, 92.   (Cete.)
    1873 C.
Gill and Coues 1877 A, p. 1043.
Gray, J. E. 1886 A, p. 61.
Gruber, W. 1859 A.
Haeckel, E. 1895 A. p. 562.   (Cetomorpha, part.)
Hay, O. P. 1899 F, p. 682.
Howes, G. B. 1888 A, p. 506.
Huxley, T. H. 1870 A.
    1872 A, p. 333.
Jaekel, O. 1891 D, p. 198.
Kükenthal, W. 1888 A.
    1889 A.
    1890 A.
    1890 B.
    1890 C.
    1891 C.
    1893 B.
    1893 C.
    1895 A.
Leboucq, H. 1887 A.
    1888 A.

Leboucq, H. 1896 A.
Leche, W. 1895 A, p. 119.
Leidy, J. 1871 C, p. 365.
Lydekker, R. 1896 B.
Marsh, O. C. 1877 E.
Murie, J. 1865 A.
Nicholson and Lydekker 1889 A, p. 1300.
Osborn, H. F. 1888 F, p. 1067.
    1893 D.
Owen, R. 1843 A, p. 72.
    1845 B, pp. 311, 345.
    1851 C, p. 739.
    1866 B, p. 415.
    1868 A, p. 276.
Paquier, V, 1894 A, p. 18.
Parker, W. K. 1868 A, p. 217.
Roger, O. 1896 A, p. 75.
Ryder, J. A. 1885 D.
Schlosser, M. 1889 A.
    1890 A.
    1890 C.
Schmidt, O. 1886 A. p. 246.
Slade, D. D. 1888 A, p. 244.
Steinmann and Döderlein 1890 A, p. 705.
Struthers, J. 1871 A.
    1872 A.
    1881 A.
    1888 A.
    1893 A.
    1895 A.
Thomas, O. 1892 B.
Tomes, C. S. 1898 A. p. 375.
Trouessart, E. L. 1898 A. p. 1009.
Van Beneden, P. J. 1868 A.   ("Cétacés.")
    1880 A.
    1882 A.
Van Beneden and Gervais 1880 A.   ("Cétacés.")
Weber, M. 1886 A.
    1887 A.
    1888 A.
Weyhe, — 1875 A, p. 99.
Woodward, A. S. 1898 B, p. 269.
Wortman, J. L. 1886 A, p. 413.
Zittel, K. A. 1893 B, pp. 155, 183, 745.

Suborder ZEUGLODONTES.

Ameghino, F. 1896 A, pp. 393, 409.   (Zeuglodontes.)
Brandt, J. F. 1873 A, p. 291.   (Zeuglodontina.)
Cope, E. D. 1889 R, p. 876.   (Archæoceti.)
    1890 E, p. 601.   (Archæoceti.)
    1898 B, p. 108.   (Archæoceti)
Dames, W. 1894 B.   (Archæoceti.)
Flower, W. H. 1883 A.   (Archæoceti.)
    1883 D, p. 395.   (Archæoceti.)
Flower and Lydekker 1891 A, p. 246.   (Archæoceti.)
Gadow, H. 1898 A, p. 44.   (Archæoceti.)
Gill, T. 1871 A.   (Zeuglodontes.)
    1871 B.   (Zeuglodontia.)
    1872 A, p. 301.   (Zeuglodontia.)
    1872 B, p. v (<Zeuglodontia); p. 14 (<Zeug-
        lodontes).
    1873 C.

Haeckel, E. 1895 A, p. 566.   (Archiceta.)
Huxley, T. H. 1872 A, p. 349.   (<Phocodontia.)
Lydekker, R. 1887 A, p. 49.   (Archæoceti.)
Müller, J. 1847 E, p. 199.   ("Zeuglodonten.")
    1849 A.   ("Zeuglodonten.")
    1851 A.   ("Zeuglodonten.")
Nicholson and Lydekker 1889 A, p. 1304.   (Arch-
        æoceti.)
Pictet, F. J. 1853 A, p. 375.   ("Zeuglodontes.")
Roger, O. 1896 A, p. 75.   (Archæoceti.)
Thompson, D'A. W. 1890 A.
Trouessart, E. L. 1898 A, p. 1009.   (Zeuglodonta.)
Weber, M. 1886 A, p. 235.
Woodward, A. S. 1898 B, p. 270.   (Archæoceti.)
Zittel, K. A. 1893 B, p. 167.

BASILOSAURIDÆ.

Unless otherwise indicated, the following authors use the family name Zeuglodontidæ.

Brandt J. F. 1868 A, p. 351.
    1873 A, pp. 11, 291.
    1874 A, p. 28.
Burmeister, H. 1879 B, p. 580.
Cope, E. D. 1868 I, p. 184. (Basilosauridæ.)
    1890 E, p. 601 (Zeuglodontidæ); p. 614 (Basilosauridæ).
Flower and Lydekker 1891 A, p. 246.
Gill, T. 1871 B, p. 122. (Basilosauridæ.)
    1872 A, p. 301. (Basilosauridæ.) .

Gill, T. 1873 C. (Basilosauridæ.)
Huxley, T. H. 1870 A.
Lydekker, R. 1887 A, p. 49.
    1898 C.
Natural Science 1894 A.
Nicholson and Lydekker 1889 A, pp. 1301, 1304.
Owen, R. 1865 B.
Steinmann and Döderlein 1890 A, p. 706.
Trouessart, E. L. 1898 A, p. 1009.
Van Beneden, P. J. 1865 A, p. 65.
Zittel, K. A. 1898 B, p. 167.

BASILOSAURUS Harlan. Type *Zeuglodon cetoides* Owen.

Harlan, R. 1834 C.
Bartlett, J. 1846 A. (Zeuglodon.)
Blainville, H. M. D. 1864 A, iii, U, p. 134.
Brandt, J. F. 1873 A, p. 295. (Zeuglodon.)
    1874 A, pp. 28, 47. (Zeuglodon.)
Bronn, H. G. 1838 A, p. 1070.
Buckley, S. B. 1843 A. (Zygodon.)
    1846 A. (Basilosaurus.)
Burmeister, H. 1871 C. (Zeuglodon.)
Carus, C. G. 1847 A. (Hydrarchus.)
    1849 A. (Zeuglodon.)
Conrad, T. A. 1840 A.
Cope, E. D. 1867 C, p. 155.
    1889 C, p. 207.
Dames, W. 1883 B. p. 132. (Zeuglodon.)
    1894 A, p. 219. (Zeuglodon.)
    1894 B. (Zeuglodon.)
Dana, J. D. 1875 A. (Zeuglodon.)
    1896 A, p. 931. (Zeuglodon.)
Delfortrie, E. 1874 A. (Zeuglodon.)
Emmons, E. 1845 A.
    1845 B.
    1846 A.
Flower, W. H. 1876 B, xiii, p. 450. (Zeuglodon.)
    1883 A. (Zeuglodon.)
    1883 D, p. 395. (Zeuglodon.)
Flower and Lydekker 1891 A, p. 246. (Zeuglodon.)
Gaudry, A. 1878 B, P. 39.
Geinitz, B. 1847 A.
Gibbes, R. W. 1847 B.
Giebel, C. G. 1847 C, p. 717. (Hydrarchos.)
Hammerschmidt, C. E. 1848 A.
Harlan, R. 1834 B, p. 77.
    1834 C, p. 397.
    1834 E.
    1835 B, p. 282.
    1835 D.
    1835 F.
    1839 A, p. 23.
Huxley, T. H. 1870 A. (Zeuglodon.)
    1870 B, p. xlviii. (Zeuglodon.)
    1872 A, p. 349. (Zeuglodon.)
Jaekel, O. 1891 D, p. 198. (Zeuglodon.)
Koch, A, 1845 A. (Hydrarchos, or Hydrargos.)
    1851 A. (Zeuglodon.)
    1851 B. (Zeuglodon.)
    1857 B. (Zeuglodon.)
Leidy, J. 1869 A, p. 427.

Lister, G. 1846 A. (Zeuglodon.)
Lucas, F. A. 1895 A. (Zeuglodon.)
    1900 C, p. 810.
Lydekker, R. 1887 A, p. 49. (Zeuglodon.)
    1898 C, p. 558. (Zeuglodon.)
Marsh, O. C. 1877 E, p. 372. .
Meyer, H. 1847 A.
Molin, R. 1859 A, p. 125. (Zeuglodon.)
Müller, J. 1847 A. (Basilosaurus.)
    1847 B. (Zeuglodon.)
    1847 C. (Zeuglodon.)
    1847 E. (Zeuglodon.)
    1849 A. (Zeuglodon.)
    1851 A. (Zeuglodon.)
Natural Science 1894 A. (Zeuglodon.)
Nicholson and Lydekker 1889 A, p. 1304.
Osborn, H. F. 1898 D. (Zeuglodon.)
Owen, R. 1839 B, p. 24. (Zeuglodon; to replace Basilosaurus of Harlan.)
    1841 C. (Zeuglodon.)
    1845 B, p. 360. (Zeuglodon.)
    1865 B. (Zeuglodon.)
Probst, J. 1886 A, p. 137. (Zeuglodon.)
Rogers, H. D. 1845 A. (Zeuglodon.)
Schlosser, M. 1890 A, p. 87. (Zeuglodon.)
Thompson, D'A. W. 1890 A.
Trouessart, E. L. 1898 A, p. 1009. (Zeuglodon.)
Van Beneden, P. J. 1861 A. (Zeuglodon.)
    1865 A, p. 65. (Zeuglodon.)
Warren, J. C. 1854 D. (Zeuglodon.)
Weber, M. 1886 A, p. 233. (Zeuglodon.)
Woodward, A. S. 1898 B, p. 270.
Wyman J. 1850 E. (Zeuglodon.)
Zittel, K. A. 1898 B, p. 167.-

Basilosaurus brachyspondylus (Müller).

Müller, J. 1847 B, p. 388. (Zeuglodon.)
Brandt, J. F. 1874 A, p. 47. (Zeuglodon.)
Cope, E. D. 1867 C, p. 155. (Doryodon.)
    1890 E. P. 602. (Zeuglodon.)
Dames, W. 1883 B. p. 132. (Zeuglodon.)
    1894 B. (Zeuglodon.)
Dana, J. D. 1875 A, p. 342. (Zeuglodon.)
Koch, A. 1845 A, p. 676. ("Zygodon," in part.)
    1851 B, pl. vii. (Zeuglodon.)
Eocene; Alabama.

Basilosaurus cetoides (Owen).

Owen, R. 1841 C, p. 69, pls. vii–ix. (Zeuglodon.)
Agassiz L. 1848 A, p. 4. (Zeuglodon.)
Bouvé, T. T. 1859 A, p. 49. ("Zeuglodon.")
Buckley, S. B. 1843 A, p. 409. ("Zygodon.")
    1846 A, p. 125. (Zeuglodon.)
Carus, C. G. 1847 A. (Hydrarchus harlani, in
    part.)
    1849 A, p. 385, pl. xxxix A, fig. 4. (Zeuglo-
    don.)
Cope, E. D. 1867 C, p. 155.
    1868 I, p. 185. (Polyptychodon interruptus
    Emmons a synonym ?.)
    1890 E. p. 602 (Zeuglodon; p. 614 (Basilosaur-
    us).
Dames W. 1883 B. p. 135, (Z. cetoides.)
    1894 B. (Z. macrospondylus.)
Dana, J. D. 1875 A, p. 342. (Z. macrospondylus.)
    1875 B. ("Hydrarchus harlani.")
    1896 A, p. 906, fig. 1532. (Zeuglodon.)
De Kay, J. E. 1842 C, p. 123. (Z. harlani.)
Emmons, E. 1845 A, figures. (Zeuglodon.)
    1845 B. (Zeuglodon.)
    1846 A, pls. i, ii. (Zeuglodon.)
    1851 A. (Zeuglodon).
    1858 B, p. 201, fig. 25. (Zeuglodon.)
Geinitz, H. B. 1847 A. ("Hydrarchus harlani.")
    1847, in Carus, C. G. 1847 A, p. 1.
Gervais, P. 1874 B, p. 58, pl. xix, figs. 6, 7. (Zeug-
    lodon.)
Gibbes, R. W. 1847 B, pl. i, figs. 1–4, 8.
    1850 A.
Giebel, C. G. 1847 D.
Hammerschmidt, C. E. 1848 A. (B. kochii, B.
    harlani.)
Harlan, R. 1834 B, p. 77. ("Basilosaurus.")
    1834 C, p. 397, pl. xx, figs. 1, 2. ("Basilo-
    saurus.")
    1835 D, p. 337. ("Basilosaurus.")
    1835 F, p. 349, pls. xxvi–xxviii. (Basilo-
    saurus.)
    1839 A, p. 23. ("Basilosaurus.")
    1839 B. ("Basilosaurus.")
    1839 C. ("Basilosaurus.")
    1840 A, p. 67. ("Basilosaurus.")

Koch, A. 1845 A. (Zeuglodon macrospondylus,
    Hydrarchus harlani.)
    1845 B, p. 676. ("Zygodon," in part.)
    1851 A. (Zeuglodon macrospondylus.)
    1857 A. ("Zeuglodon.")
    1857 B. (Zeuglodon macrospondylus.)
Leidy, J. 1869 A, p. 427.
    1869 B, p. 95. (Zeuglodon.)
Lucas, F. A. 1895 B, p. 745. (Zeuglodon.)
    1900 C, p. 809. (Basilosaurus.)
    1900 E, p. 327, pls. v–vii. (Basilosaurus.)
Lydekker, R. 1887 A, p. 50. (Zeuglodon.)
Lyell, C. 1846 C, p. 409. ("Zeuglodon.")
Müller, J. 1847 A, p. 421. (Zeuglodon.)
    1847 B. (Zeuglodon macrospondylus.)
    1847 C. (Zeuglodon.)
    1847 D. (Zeuglodon.)
    1847 E, p. 193. (Zeuglodon macrospondy-
    lus.)
    1849 A, plates. (Zeuglodon macrospondy-
    lus.)
    1851 A. (Zeuglodon macrospondylus.)
Owen, R. 1839 B, p. 24. ("Zeuglodon.")
    1845 B, p. 360, pl. xci. (Zeuglodon.)
    1860 E, p. 345, figs. 108, 109. (Zeuglodon.)
Pictet, F. J. 1853 A, p. 378. (Zeuglodon macros-
    pondylus, hydrarchus, trachyspondylus.)
Quenstedt, F. A. 1852 A, p. 78. (Zeuglodon.)
    1885 A, p. 112. (Zeuglodon.)
Roger, O. 1896 A, p. 75. (Zeuglodon.)
Schuchert, C. 1900 A, p. 328.
Trouessart, E. L. 1898 A, p. 1009. (Zeuglodon.)
Tuomey, M. 1847 A, p.283, figures. ("Zeuglodon.")
    1847 B, p. 151. ("Zeuglodon.")
    1847 C. ("Zeuglodon.")
Wailes, B. L. C. 1854 A, p. 277. (Zeuglodon har-
    lani.)
Wyman, J. 1845 A. (Hydrarchos sillimani.)
Zittel, K. A. 1893 B, p. 168, figs. 131, 132. (Zeug-
    lodon.)
    Eocene; Alabama, Mississippi, Louisiana, Ar-
    kansas, Florida, North Carolina.
    For identification of the figures of Müller's
    plates (Müller J. 1849 A.), see Leidy, J. 1869 A, p.
    430. Additional synonymy may also be found
    there.

DORUDON Gibbes.   Type *D. serratus* Gibbes.

Gibbes, R. W. 1845 A.
Agassiz, L. 1848 A.
Cope, E. D. 1867 C, p. 154. (Doryodon.)
    1890 E, p. 602. (Doryodon.)
Gibbes, R. W. 1848 A.
Leidy, J. 1852 C. (Pontogeneus, type *P. priscus.*)
    1869 A, p. 427.
Lucas, F. A. 1900 C, p. 810.
Müller, J. 1849 A, p. 1. (Syn. of Zeuglodon
    1851 A. (Syn. of Zeuglodon.)
Owen, R. 1846 C.
    1846 D.
Zittel, K. A. 1893 B, p. 168.

Dorudon serratus Gibbes.

Gibbes, R. W. 1845 A, p. 254, pl. i.
Agassiz, L. 1848 A, p. 254.

Amer. Jour. Science 1845 A, p. 218. ("Zeuglo-
    don.")
Carus, C. G. 1847 A. (Hydrarchus harlani.)
    1849 A, p. 369, pl. xxxix A, figs. 1–3; pl.
    xxxix B. (Zeuglodon hydrarchos.)
Cope, E. D. 1867 C, p. 155. (Doryodon.)
    1868 I, p. 186. (Pontogeneus priscus; 'Del-
    phinoid.")
    1869 G, p. 6. (Pontogeneus priscus.)
    1890 E, p. 614. (Doryodon serratus.)
Dames, W. 1894 B.
Gibbes, R. W. 1845 A.
    1847 B, p. 5, pl. ii, figs. 1–8; pl. iii, figs. 4–6.
    (Basilosaurus cetoides, in part.)
    1848 A, p. 57.
Koch, A. 1845 B, p. 676. ("Zygodon," in part.)
    1851 B, pl. vii. (Zeuglodon brachyspondy-
    lus.)

Leidy, J. 1852 C, p. 52. (Pontogeneus.)
    1869 A, p. 428, pl. xxix, figs. 2-5.
Lydekker, R. 1887 A, p. 50. (Zeuglodon hydrarchus; syn.? of Z. cétoïdes.)
    1892 B, p. 559. (Z. hydrarchus.)
Müller, J. 1847 C, p. 106. (D. serratus, syn. of Zeuglodon cetoïdes.)
    1847 E,p.193. (Zeuglodon brachyspondylus.)
    1849 A, pl. i, fig. 1 (in part); pls. iii, iv; pl. v, fig. 1; pl. viii, figs. 3, 9, 10; pl. xii, fig. 11?; pl. xiii, figs. 1, 2, 6, 7; pl. xviii; pl. xix,

Müller, J.—Continued.
    figs. 1–5; pls. xx, xxi; pl. xxiii, fig. 4; pl. xxvi; pl. xxvii, figs. 1, 2, 5 (Zeuglodon brachyspondylus).
    1851 A. (Zeuglodon brachyspondylus.)
Reichenbach, H. G. L., in Carus, C. G. 1847 A, p. 13. (Basilosaurus kochii.)
Trouessart, E. L. 1898 A, p. 1010. (Zeuglodon.)
    Eocene; Alabama, South Carolina, Louisiana.
    It is possible that Dorudon serratus and Zeuglodon brachyspondylus are distinct species.

PONTOBASILEUS Leidy.   Type *P. tuberculatus* Leidy:

Leidy, J. 1873 B, p. 337,
Trouessart, E. L. 1898 A, p. 1010.
Zittel, K. A. 1893 B, p. 167. (Syn. of Zeuglodon.)

Pontobasileus tuberculatus Leidy.
    Leidy, J. 1873 B, p. 337, pl. xxxvii, fig. 15.
    Trouessart, E. L. 1898 A, p. 1010.
    Tertiary; Alabama?.

SAUROCETUS Agassiz.   Type *S. gibbesii* Agassiz.

Agassiz, L. 1848 A.
Burmeister, H. 1871 C.
Cope, E. D. 1890 E, p. 615.
Leidy, J. 1869 A, p. 431.
Trouessart, E. L. 1898 A, p. 1010. (Syn. of Zeuglodon.)

Zittel, K. A. 1893 B. p. 171.
    A genus of uncertain position.

Saurocetus gibbesii Agassiz.
    Agassiz, L. 1848 A.
    Eocene; South Carolina.
    For additional references see under the genus.

Suborder ODONTOCETE Flower.

Brandt, J. F. 1873 A, p. 202. (Odontocetoidea.)
Cope, E. D. 1889 R, p. 876. (Odontoceti.)
    1890 E, p. 602. (Odontoceti.)
    1896 B. p. 109. (Odontoceti.)
Dames, W. 1894 B, p. 36. (Odontoceti.)
Flower, W. H. 1864 A, p. 388. (Delphinoidea, or Odontocete.)
    1867 A, p. 111. (Delphinoidea, or Odontocete.)
    1868 A. (Physeter.)
    1871 A. (Odontoceti.)
    1876 B, xiii, p. 450. (Odontoceti.)
    1883 D, p. 396. (Odontoceti.)
    1885 A.
Flower and Lydekker 1891 A, p. 247. (Odontoceti.)
Gadow, H. 1898 A, p. 44. (Odontoceti.)

Gill, T. 1871 B, p. 122. (Denticete.)
    1872 B, pp. 14, 98. (Denticete.)
    1873 C, p. 26. (Denticetes.)
Gray, J. E. 1866 A, pp. 62, 194. (Denticete.)
Haeckel, E. 1895 A, p. 566. (Denticeta, part.)
Howes, G. B. 1888 A, p. 507. (Odontoceti.)
Huxley, T. H. 1872 A, p. 340. (Delphinoidea.)
Lydekker, R. 1887 A, p. 58. (Odontoceti.)
Nicholson and Lydekker 1889 A, p. 1305. (Odontoceti.)
Roger, O. 1896 A, p. 76. (Odontoceti.)
Schlosser M. 1890 C, p. 241.
Van Beneden and Gervais 1880 A, p. 298. ("Cétodontes.")
Woodward, A. S. 1898 B, p. 271.
Zittel, K. A. 1893 B, p. 168. (Odontoceti.)

SQUALODONTIDÆ.

Cope, E. D. 1867 C, p. 144. (Cynorcidæ.)
    1889 R, p. 876.
    1890 E, p. 602.
    1896 A, p. 139.
Dames, W. 1894 B, p. 36. (Squalodontidæ, Mesoceti.)
Flower, W. H. 1883 D, p. 397.
Flower and Lydekker 1891 A, p. 257.

Gill, T. 1872 B, p. 14. (Cynorcidæ.)
Lydekker, R. 1887 A, p. 876.
Nicholson and Lydekker 1889 A, p. 1306.
Owen, R. 1865 B.
Roger, O. 1894 A, p. 76.
Steinmann and Döderlein 1890 A, p. 706.
Van Beneden, P. J. 1865 A. ("Squalodons.")
Zittel, K. A. 1893 B, p. 169.

SQUALODON Grateloup.   Type *Delphinoides grateloupii* Pedroni.

Grateloup, J. P. S. 1840 A, p. 208.
Agassiz, L. 1841 A. (Phocodon, type *P. scillæ.*)
Ameghino, F. 1896 A, p. 510.
Brandt, J. F. 1873 A, pp. 299, 315.
    1874 A, p. 28.

Cope, E. D. 1867 C, p. 150 (Squalodon; Cynorca, type *C. proterva*); p. 151 Colophonodon).
    1868 J, p. 185. (Cynorca.)
Dames, W. 1894 B, p. 19.
Depéret, C. 1897 A, p. 278.

Flower and Lydekker 1891 A, p. 257.
Gervais, P. 1859 A, p. 309.
	1862 A.
	1876 A. (Phocodon.)
Gaudry, A. 1878 B, p. 80.
Gibbes, R. W. 1847 B.
Jourdan, — 1861 A, p. 370 (Rhizoprion, type *R. bariensis*); p. 371 (Squalodon).
Leidy, J. 1853 C, p. 377. (Colophonodon, type *C. holmesii*.)
	1856 K, p. 220. (Macrophoca, type *M. atlantica*.)
	1869 A, p. 416.
Longhi, P. 1899 A, p. 374.
Lydekker, R. 1887 A, p. 75.
Meyer, H. 1840 A.
	1841 A.
	1847 A. (Squalodon, Pachyodon.)
Molin, R. 1859 A, p. 125. (Squalodon, Pachyodon.)
Müller, J. 1847 C. (Squalodon, syn. of Zeuglodon.)
	1849 A, p. 29.
Nicholson and Lydekker 1889 A, p. 1306.
Osborn, H. F. 1898 D, p. 14.
Owen, R. 1845 B, p. 564. (Phocodon.)
	1865 B.
	1866 B, p. 405. (Squalodon, Phocodon.)
Paquier, V. 1894 A.
Pictet, F. J. 1853 A, p. 379.
Probst, J. 1885 A.
	1886 A, p. 105.
Quenstedt, F. A. 1885 A, p. 113.
Roger, O. 1896 A, p. 76.
Steinmann and Döderlein 1890 A, p. 706.
Trouessart, E. L. 1898 A, p. 1011.
Van Beneden, P. J. 1861 A.
	1865 A, p. 64.
Van Beneden and Gervais 1880 A, pp. 426, 519 (Squalodon); p. 451 (Phocodon).
Weber, M. 1886 A, p. 233.
Woodward, A. S. 1898 B, p. 272.
Zittel, K. A. 1893 B, p. 170.

Squalodon atlanticus Leidy.

Leidy, J. 1856 K, p. 220. (Macrophoca.)
Cope, E. D. 1867 B, p. 132.
	1867 C, p. 151. (Colophonodon.)
	1869 N, p. 739.
	1875 H, p. 363.
	1890 E, p. 608.
Leidy, J. 1869 A, p. 416, pl. xxviii, figs. 4–7; pl. xxx, fig. 18.
Lydekker, R. 1887 A, p. 77.
Roger, O. 1896 A, p. 77.

Van Beneden and Gervais 1880 A, p. 440.
	Miocene; New Jersey, Maryland.

Squalodon debilis (Leidy).

Leidy, J. 1856 N, p. 265. (Phoca.)
Allen, J. A. 1880 A, p. 473. (Squalodon?)
Cope, E. D. 1867 C, p, 144.
Leidy, J. 1869 A, p. 415, pl. xxviii, figs. 12, 13. (Phoca.)
Roger, O. 1896 A, p. 74. (Phoca.)
Trouessart, E. L. 1898 A, p. 1014.
Zittel, K. A. 1893 B, p. 684. (Phoca.)
	Eocene; South Carolina.

Squalodon holmesii Leidy.

Leidy, J. 1853 C, p. 377. (Colophonodon.)
Cope, E. D. 1867 C, pp. 144, 153.
	1890 E, p. 608.
Leidy, J. 1869 A, p. 418, pl. xxviii, figs. 15–17; pl. xxix, fig. 9.
Roger, O. 1896 A, p. 77.
Trouessart, E. L. 1898 A, p. 1015.
	Eocene; South Carolina.

Squalodon? modestus (Leidy).

Leidy, J. 1869 A, p. 415, pl. xxviii, fig. 14. (Phoca.)
Allen, J. A. 1880 A, p. 474. ("Squalodont.")
Roger, O. 1896 A, p. 74. (Phoca.)
Zittel, K. A. 1893 B, p. 684. (Phoca.)
	Eocene; South Carolina.

Squalodon pelagius Leidy.

Leidy, J. 1869 A, p. 420, pl. xxix, fig. 1.
Cope, E. D. 1890 E, p. 615.
Roger, O. 1896 A, p. 77.
Trouessart, E. L. 1898 A, p. 1015.
	Eocene; South Carolina.

Squalodon protervus Cope.

Cope, E. D. 1867 C, pp. 144, 152 (Cynorca); p. 151, in part (Squalodon).
	1868 I, p. 185. (Cynorca.)
	1890 E, p. 615.
Leidy, J. 1869 A, pp. 384, 423, pl. xxviii, figs. 18, 19.
Roger, O. 1896 A, p. 77.
Trouessart, E. L. 1898 A, p. 1015.
	Eocene; South Carolina.
	The original description refers almost wholly to a canine tooth of *Dicotyles*.

Squalodon tiedemani Allen.

Allen, J. A. 1887 A, pls. v, vi,
Trouessart, E. L. 1898 A, p. 1015. (S. tiedmanni.)
	Eocene; South Carolina.

## AGOROPHIUS Cope.					Type *Zeuglodon pygmæus* Müller.

Cope, E. D. 1895 A, p. 139.
	1896 J, p. 141.
Trouessart, E. L. 1898 A, p. 1069.

Agorophius pygmæus (Müller).

Müller, J. 1849 A, p. 29, pl. xxxiii, figs. 1, 2. (Zeuglodon.)
Agassiz, L., in Leidy, J. 1869 A, p. 420. (Phocodon holmesii.)
Cope, E. D. 1867 C, p. 155. (Doryodon.)
	1868 I, p. 186. (Zeuglodon.)

Cope, E. D. 1890 E, p. 615. (Squalodon pygæmus errore.)
	1895 A, p. 139.
Dames, W. 1894 B. (Zeuglodon.)
Gervais, P. 1871 A, p. 188. (Squalodon.)
Gibbes, R. W. 1847 B, p. 8. ("Zeuglodon.")
Holmes, F. S. 1849 A, p. 197. ("Zeuglodon," "Basilosaurus.")
Leidy, J. 1854 A, p. 8. (Basilosaurus.)
	1869 A, p. 420, pl. xxix, figs. 7, 8. (Squalodon.)
Lydekker, R. 1892 B, p. 559. (Doryodon.)

590   FOSSIL VERTEBRATA OF NORTH AMERICA.   [BULL. 179.

Müller, J. 1851 A, p. 242.
Roger, O. 1896 A, p. 77.   (Squalodon.)
Trouessart, E. L. 1898 A, p. 1014 (Squalodon); p. 1069 (Agorophius).
Tuomey, M. 1847 A.   ("Zeuglodon.")
 1847 B.   ("Zeuglodon.")
 1847 C.   ("Zeuglodon.")

Tuomey, M. 1848, Rept. Geol. S. C., p. 166.   ("Zeuglodon.")
 1849, Rept. Geol. S. C., p. 69.   ("Zeuglodon.")
Van Beneden and Gervais 1880 A, p. 441.   (Squalodon.)
Eocene; South Carolina.

GRAPHIODON Leidy.   Type *G. vinearius* Leidy.

Leidy, J. 1870 R, p. 122.
 1873 B, p. 337.
Trouessart, E. L. 1898 A, p. 1011.   (Syn. of Squalodon?)

Graphiodon vinearius Leidy.

Leidy, J. 1870 R, p. 122.

Cope, E. D. 1890 E, p. 615.   (Squalodon.)
Leidy, J. 1873 B, p. 337, pl. xxii, fig. 7.
Roger, O. 1896 A, p. 77.   (Squalodon.)
Trouessart, E. L. 1898 A, p. 1015.   (Squalodon?.)
Miocene; Massachusetts.

CETERHINOPS Leidy.   Type *C. longifrons* Leidy.

Leidy, J. 1877 A, p. 280.
A genus of uncertain position.

Ceterhinops longifrons Leidy.

Leidy, J. 1877 A, p. 230, pl. xxxiv, fig. 7.
Tertiary (Phosphate beds); South Carolina.

PLATANISTIDÆ.

Brandt, J. F. 1874 A, p. 17.   (Platanistinæ.)
Cope, E. D. 1869 G, p. 6.
 1890 E, p. 608.
Flower, W. H. 1867 A, p. 113.
 1868 B, p. 349.
Flower and Lydekker 1891 A, p. 257.
Gill, T. 1871 B, p. 123.   (Platanistidæ, Rhabdosteidæ.)
 1872 B, pp. 14, 95 (Platanistidæ); p. 15 (Rhabdosteidæ).

Huxley, T. H. 1872 A, p. 342.
Lydekker, R. 1893 C, p. 564.
Nicholson and Lydekker 1889 A, p. 1306.
Owen, R. 1868 A, p. 282.   (Platanista.)
Paquier, V. 1894 A, p. 11.
Roger, O. 1896 A, p. 77.
Zittel, K. A. 1893 B, p. 171.

PLATANISTINÆ.

Brandt, J. F. 1874 A, p. 17.

Flower, W. H. 1867 A, p. 114.

CHAMPSODELPHIS Gervais.   Type *C. macrogenius* Gervais.

Gervais, P., 1852, Zool. Paléont. Franc., ed. 1, p. 152.
Brandt, J. F. 1873 A, p. 263.
 1874 A, p. 19.
Flower and Lydekker 1891 A, p. 259.
Longhi, P. 1899 A, p. 323.
Lydekker, R. 1887 A, p. 74.
Nicholson and Lydekker 1889 A, p. 1306.
Probst, J. 1886 A, pp. 106, 122.
Van Beneden and Gervais 1880 A, p. 482.

Zittel, K. A. 1893 B, p. 171.

Champsodelphis acutidens (Cope).

Cope, E. D. 1867 C, p. 145.   (Priscodelphinus·)
Brandt, J. F. 1873 A, p. 286.   (Priscodelphinus.)
Leidy, J. 1869 A, p. 433.   (Priscodelphinus.)
Trouessart, E. L. 1898 A, p. 1020.
Van Beneden and Gervais 1880 A, p. 508.
Miocene; Maryland.

LOPHOCETUS Cope.   Type *Delphinus calvertensis* Harlan.

Cope, E. D. 1867 C, p. 146.
 1869 G, p. 6.
 1890 E, p. 606.
Zittel, K. A. 1893 B, p. 173.
A genus of uncertain position.

Lophocetus calvertensis (Harlan).

Harlan, R. 1842 D, p. 195, 3 plates.   (Delphinus.)
Brandt, J. F. 1873 A, p. 288.
Cope, E. D. 1866 H, p. 297.   (Pontoporia.)

Cope, E. D. 1867 C, p. 146.
 1890 E, pp. 606, 615.
Leidy, J. 1869 A, p. 435.
Roger, O. 1896 A, p. 79.
Trouessart, E. L. 1898 A, p. 1022.
Van Beneden and Gervais 1880 A, p. 512.
Zittel, K. A. 1893 B, p. 173.
Miocene; Maryland.

DELPHINODON Leidy.   Type *Squalodon mento* Leidy.

Leidy, J. 1869 A, p. 424.
Cope, E. D. 1890 E, p. 615.
Leidy, J. 1869 A, p. 428.   (Phocageneus, type *P. venustus* Leidy.)
Lydekker, R. 1887 A, p. 75.   (Syn. of Squalodon.)
Trouessart, E. L. 1898 A, p. 1023.
Zittel, K. A. 1893 B, p. 170.   (Syn. of Squalodon.)

Delphinodon mento Cope.

Cope, E. D. 1867 B, p. 132.   (Squalodon.)
   1867 C, pp. 151, 152.   (Squalodon.)
   1890 E, p. 615.
Leidy, J. 1869 A, p. 424, pl. xxx, figs. 7, 8, 9.
Roger, O. 1896 A, p. 79.
Trouessart, E. L. 1898 A, p. 1023.
Wyman, J. 1850 B, p. 230, figs. 5–7.   ("Cetacean.")
   Miocene; Maryland, South Carolina.

Delphinodon venustus (Leidy).

Leidy, J. 1869 A, p. 426, pl. xxix, fig. 10.   (Phocageneus.)
Agassiz, L. in Wyman J. 1850 B, p. 230, fig. 4.   ("Phocodon.")
Cope, E. D. 1867 C, p. 152.   (Squalodon mento, in part.)

Cope, E. D. 1890 E, p. 615.
Roger, O. 1896, A, p. 79.
Trouessart, E. L. 1898 A, p. 1023.
   Miocene; Virginia.

Delphinodon leidyi nom. nov.

Allen, J. A. 1880 A, p. 473.   (D. wymani.)
Cope, E. D. 1867 C, pp. 151, 152.   (Squalodon wymanii, in part.)
   1890 E, p. 615.   (D. wymanii.)
Leidy, J. 1856 N. p. 265.   (Phoca wymani, in part.)
   1869 A, p. 425, pl. xxx, figs. 10–12.   (D. wymanii.)
Roger, O. 1896 A, p. 79.   (D. wymanii.)
Trouessart, E. L. 1898 A, p. 1023.   (D. wymanii.)
Wyman, J. 1850 B, p. 229, in part?.   ("Seal.")
   Miocene; Maryland.

The name *Delphinodon leidyi* is intended to replace *D. wymani* Leidy, which is the same, in part, as *Squalodon wymanii* Cope 1867 C, p. 152, but not same as *S. wymani* Cope 1867 B, p. 132. The type of *D. leidyi* is Leidy, J. 1869 A, pl. xxx, fig. 12.

See references under *Phoca wymani.*

PRISCODELPHINUS Leidy.   Type *P. harlani* Leidy.

Leidy, J. 1851 G, p. 326.
Cope, E. D. 1868 I, pp. 186, 187 (Priscodelphinus); pp. 186, 190 (Tretosphys, type *T. gabbii.*
   1869 G, p. 7.   (Tretosphys.)
   1890 E, pp. 604, 615.
Leidy, J. 1869 A, p. 433 (Priscodelphinus); p. 434 (Tretosphys.)
Trouessart, E. L. 1898 A, p. 1023.
Van Beneden and Gervais 1880 A, pp. 493, 510.
Zittel, K. A. 1893 B, p. 172.

Priscodelphinus gabbii Cope.

Cope, E. D. 1868 I, p. 191.   (Delphinapterus.)
Brandt, J. F. 1873 A, p. 287.   (Tretosphys.)
Cope, E. D. 1869 G, pp. 6, 7, 8.   (Tretosphys.)
   1890 E, p. 615.
Leidy, J. 1869 A, p. 434.   (Tretosphys.)
Roger, O. 1896 A, p. 78.
Trouessart, E. L. 1898 A, p. 1024.
Van Beneden and Gervais 1880 A, p. 512.   (Tretosphys.)
   Miocene; Maryland.

Priscodelphinus grandævus Leidy.

Leidy, J. 1851 G, p. 327.   (Priscodelphinus.)
Brandt, J. F. 1873 A, p. 287.   (Tretosphys.)
Cope, E. D. 1867 C, p. 144.
   1868 I, p. 191.   (Delphinapterus.)
   1869 G, pp. 7, 9.   (Tretosphys.)
   1890 E, p. 604, figs. 2, 3.
Leidy, J. 1858 C, p. 377.
   1854 A, p. 8.
   1869 A, p. 434.   (Tretosphys.)
Roger, O. 1896 A, p. 78.
Trouessart, E. L. 1898 A, p. 1024.
Van Beneden and Gervais 1880 A, p. 511.   (Tretosphys.)

Zittel, K. A. 1893 B, p. 173, fig. 134.
   Miocene; New Jersey.

Priscodelphinus harlani Leidy.

Leidy, J. 1851 G, p. 326.
Brandt, J. F. 1873 A, p. 286.
Cope, E. D. 1867 C, p. 144.
   1868 I, p. 188.
   1869 N, p. 739.
   1875 H, p. 363.
   1890 E, p. 615.   (Syn. of P. grandævus.)
Harlan, R. 1825 B, p. 232.   ("Plesiosaurus.")
   1834 B, p. 77.   ("Plesiosaurus.")
   1835 B, p. 821.   ("Plesiosaurus.")
Leidy, J. 1858 C, p. 377.
   1854 A, p. 8.
   1865 A, p. 1.
   1869 A, p. 433.
Roger, O. 1896 A, p. 78.   (Syn. of P. grandævus.)
Trouessart, E. L. 1898 A, p. 1024.
   Miocene; New Jersey.

Priscodelphinus lacertosus Cope.

Cope, E. D. 1868 I, p. 189 (Tretosphys); p. 190 (Delphinapterus lacertosus, D. hawkinsii.)
   1868 I, p. 190.   (Delphinapterus,Tretosphys.)
   1869 G, p. 7.   (Tretosphys.)
   1869 N, p. 739.   (Delphinapterus lacertosus, D. hawkinsii.)
   1875 H, p. 363.
   1890 E, p. 615.
Leidy, J. 1869 A, p. 434.   (Tretosphys.)
Roger, O. 1896 A, p. 78.
Trouessart, E. L. 1898 A, p. 1024.
Van Beneden and Gervais 1880 A, p. 512.
   Miocene; Maryland.

Priscodelphinus ruschenbergeri Cope.

Cope, E. D. 1868 I, p. 189. (Delphinapterus.)
Brandt, J. F. 1873 A, p. 287. (Tretosphys.)
Cope, E. D. 1869 G, pp. 7, 9. (Tretosphys.)
1890 E, p. 615.
Leidy, J. 1869 A, p. 435. (Tretosphys.)
Roger, O. 1896 A, p. 78.
Trouessart, E. L. 1898 A, p. 1024.
Van Beneden and Gervais 1880 A, p. 512. (Tretosphys.)
Miocene; Maryland.

Priscodelphinus ursæus Cope.

Cope, E. D. 1869 G, pp. 7, 8. (Tretosphys.)
Brandt, J. F. 1873 A, p. 287. (Tretosphys.)
Cope, E. D. 1875 H, p. 363.
1890 E, p. 615.
Leidy, J. 1869 A, p. 435. (Tretosphys.)
Roger, O. 1896 A, p. 78.
Trouessart, E. L. 1898 A, p. 1024.
Van Beneden and Gervais 1880 A, p. 512.
Miocene; New Jersey, Maryland.

· ZARHACHIS Cope.   Type *Z. flagellator* Cope.

Cope, E. D. 1868 I, pp. 186, 189.
1869 G, p. 9.
1890 E, p. 604.
Leidy, J. H. 1869 A, p. 435.
Trouessart, E. L. 1898 A, p. 1024.
Zittel, K. A. 1893 B, p. 172.

Zarhachis flagellator Cope.

Cope, E. D. 1868 I, p. 189, in part.
Brandt, J. F. 1873 A, p. 287.
Cope, E. D. 1869 G, p. 9.
1890 E, p. 615.
Leidy, J. 1869 A, p. 435.
Roger, O. 1896 A, p. 78.
Trouessart, E. L. 1898 A, p. 1024.
Van Beneden and Gervais 1880 A, p. 512
Miocene; Maryland,

Zarhachis tysonii Cope.

Cope, E. D. 1869 G, p. 9.

Brandt, J. F. 1873 A, p. 288.
Cope, E. D. 1868 I, p. 189. (Z. flagellator, in part.)
1890 E, p. 615.
Leidy, J. 1869 A, p. 435.
Roger, O. 1896 A, p. 78.
Trouessart, E. L. 1898 A, p. 1025.
Van Beneden and Gervais 1880 A, p. 512.
Miocene; Maryland.

Zarhachis velox Cope.

Cope, E. D. 1869 G, pp. 9, 10.
Brandt, J. F. 1873 A, p. 288.
Cope, E. D. 1868 I, p. 189. (Z. flagellator, in part.)
1875 H, p. 363.
1890 E, p. 615.
Leidy, J. 1869 A, p. 615.
Roger, O. 1896 A, p. 78.
Trouessart, E. L. 1898 A, p. 1025.
Van Beneden and Gervais 1880 A, p. 512.
Miocene; Maryland.

IXACANTHUS Cope.   Type *I. cœlospondylus* Cope.

Cope, E. D. 1868 G, p. 159.
Brandt, J. F. 1873 A, p. 286.
Cope, E. D. 1868 I, p. 187.
1875 H, p. 364. (Belosphys, type B. spinosus.)
1890 E, pp. 604, 615.
Leidy, J. 1869 A, p. 435.
Trouessart, E. L. 1898 A, p. 1025.
Zittel, K. A. 1898 B, p. 172.

Ixacanthus atropius Cope.

Cope, E. D. 1868 I, p. 188. (Priscodelphinus.)
Brandt, J. F. 1873 A, p. 286. (Priscodelphinus.)
Cope, E. D. 1869 G, p. 6. (Priscodelphinus.)
1875 H, p. 364. (Belosphys.)
1890 E, p. 615.
Leidy, J. 1869 A, p. 433. (Priscodelphinus.)
Roger, O. 1896 A, p. 78.
Trouessart, E. L. 1898 A, p. 1025.
Van Beneden and Gervais 1880 A, p. 511. (Priscodelphinus.)
Miocene; Maryland.

Ixacanthus cœlospondylus Cope.

Cope, E. D. 1868 G, p. 159.
Brandt, J. F. 1873 A, p. 288.
Cope, E. D. 1868 I, p. 187.
1869 N, p. 739.
1875 H, p. 364.
1890 E, p. 604.

Leidy, J. 1869 A, p. 435.
Trouessart, E. L. 1898 A, p. 1025.
Van Beneden and Gervais 1880 A, p. 513.
Miocene; Maryland, New Jersey?.

Ixacanthus conradi (Leidy).

Leidy, J. 1852 B, p. 35. (Delphinus.)
Brandt, J. F. 1873 A, p. 286. (Priscodelphinus.)
Cope, E. D. 1868 I, p. 188. (Priscodelphinus.)
1875 H, p. 364. (Belosphys.)
1890 E, p. 615.
Leidy, J. 1869 A, p. 433. (Priscodelphinus.)
Roger, O. 1896 A, p. 79.
Trouessart, E. L. 1898 A, p. 1025 (I. cœlospondylus); p. 1082 (Delphinus conradi).
Van Beneden and Gervais 1880 A, p. 511. (Priscodelphinus.)
Miocene; Virginia, Maryland.

Ixacanthus spinosus Cope.

Cope, E. D. 1868 I, p. 187. (Priscodelphinus.)
Brandt, J. F. 1873 A, p. 286. (Priscodelphinus.)
Cope, E. D. 1875 H, p. 364. (Belosphys.)
1890 E, p. 603, fig. 1; p. 615.
Leidy, J. 1869 A, p. 433. (Priscodelphinus.)
Roger, O. 1896 A p. 78.
Trouessart, E. L. 1898 A, p. 1025.
Van Beneden and Gervais 1880 A, p. 511. (Priscodelphinus.)
Miocene; Maryland.

Ixacanthus stenus Cope.

Cope, E. D. 1868 I, p. 188.   (Priscodelphinus.)
Brandt, J. F. 1873 A, p. 286.
Cope, E. D. 1875 H, p. 364.   (Belosphys.)
    1890 E, p. 615.

Leidy, J. 1869 A, p. 483.   (Priscodelphinus.)
Roger, O. 1896 A, p. 79.
Trouessart, E. L. 1898 A, p. 1025.
Van Beneden and Gervais 1880 A, p. 511.
    Miocene; Maryland.

AGABELUS Cope.   Type *A. porcatus* Cope.

Cope, E. D. 1875 H, p. 363.
    1890 E, pp. 604, 615.
Flower and Lydekker 1891 A, p. 260.
Trouessart, E. L. 1898 A, p. 1025.
Zittel, K. A. 1893 B, p. 174.

Agabelus porcatus Cope.

Cope, E. D. 1875 H, p. 363.
    1890 E, pp. 604, 615.
Roger, O. 1896 A, p. 79.
Trouessart, E. L. 1898 A, p. 1025.
    Miocene; New Jersey.

RHABDOSTEUS Cope.   Type *R. latiradix* Cope.

Cope, E. D. 1867 B, p. 132.
Brandt, J. F. 1874 A, p. 13.
Cope, E. D. 1⁸⁶⁷ C, p. 145.
    1890 E, p. 604.
Leidy, J. 1869 A, p. 435.
Trouessart, E. L. 1898 A, p. 1025.
Van Beneden and Gervais 1880 A, p. 513.
Zittel, K. A. 1893 B, p. 174.

Rhabdosteus latiradix Cope.

Cope, E. D. 1867 B, p. 132.
Brandt, J. F. 1873 A, p. 288.
Cope, E. D. 1867 C, p. 145.
    1890 E, pp. 606, 607, fig. 4; p. 615.
Leidy, J. 1869 A, p. 435.
Roger, O. 1896 A, p. 79.
Trouessart, E. L. 1898 A, p. 1025.
    Miocene; Maryland.

CETOPHIS Cope.   Type *C. heteroclitus* Cope.

Cope, E. D. 1868 I, p. 184.
    1890 E, p. 604.
Trouessart, E. L. 1898 A, p. 1025.
Zittel, K. A. 1893 B, p. 173.
A genus of doubtful position.

Cetophis heteroclitus Cope.

Cope, E. D. 1868 I, p. 185.
    1890 E, pp. 603, 606.
Leidy, J. 1869 A, p. 431.
Roger, O. 1896 A, p. 79.
Trouessart, E. L. 1898 A, p. 1025.
    Miocene; Maryland,

DELPHINIDÆ.

Albrecht, P. 1886 A, p. 344.
Brandt, J. F. 1873 A, p. 227.
Cope, E. D. 1868 I, p. 186.
    1889 C, p. 207.
    1890 E, p. 609.
Cuvier, G. 1834 A, viii, pp. 75–170, pls. ccxxii, ccxxiv.
Duvernoy, G. L. 1851 A.   ("Dauphins.")
Flower, W. H. 1867 A, p. 113.
    1868 B, p. 349.
    1883 B.
    1883 D, p. 396.
    1885 A, p. 349.
Flower and Lydekker 1891 A, p. 260.

Gill, T. 1871 B, p. 123.
    1872 B, pp. 14, 94.
Gray, J. E. 1866 A, p. 228.
Huxley, T. H. 1872 A, p. 342.
Lydekker, R. 1887 A, p. 78.
Nicholson and Lydekker 1889 A, p. 1307.
Osborn, H. F. 1888 F, p. 1075.
Owen, R. 1866 B, p. 424.
Roger, O. 1896 A, p. 79.
Trouessart, E. L. 1898 A, p. 1026.
True, F. W. 1889 A.
Weyhe, — 1875 A, p. 100.
Zittel, K. A. 1893 B, p. 174.

DELPHININÆ.

Brandt, J. F. 1873 A, p. 243.
Flower, W. H. 1867 A, p. 113.
Gill, T. 1872 B, p. 15.   (Delphininæ, Globiocephalinæ.)

True, F. W. 1889 A, pp. 10, 13, 15.

DELPHINUS Linn.   Type *D. delphis* Linn.

Linnæus, C., 1758, Syst. Nat., ed. 10, i, p. 77.
Brandt, J. F. 1873 A, p. 244.
Cope, E. D. 1868 I, p. 186.

Duvernoy, G. L. 1851 A.   ("Dauphin.")
Flower, W. H. 1883 B, pp. 500, 511.
    1883 D, p. 399.

Flower and Lydekker 1891 A, p. 271.
Gervais, P. 1859 A, p. 301.
Gray, J. E. 1866 A, p. 239.
Nicholson and Lydekker 1889 A, p. 1307.
Pictet, F. J. 1853 A, p. 381.
Trouessart, E. L. 1898 A, p. 1031.
True, F. W. 1889 A, pp. 44, 160.
Zittel, K. A. 1893 B, p. 175.

Delphinus occiduus Leidy.

Leidy, J. 1868 E, p, 197.
Brandt, J. F. 1873 A, p. 286.

Cope, E. D. 1890 E, p. 616.
Leidy, J. 1869 A, p. 431.
    1871 C, p. 365.
Roger, O. 1896 A, p. 79.
Tomes, C. S. 1898 A, p. 376.
Trouessart, E. L. 1898 A, p. 1031.
    Miocene; California.

Delphinus sp. indet.

Wyman, J. 1850 B, p. 231, figs. 5–7.
Leidy, J. 1869 A, p. 432.

DELPHINAPTERINÆ.

Gill, T. 1871 B, p. 124.
    1872 B, p. 15.   .

True, F. W. 1889 A, pp. 10, 11.

DELPHINAPTERUS Lacépède.   Type *D. leucas* Pall.

Lacépède, B. G. E., 1804, Hist. Nat. Cét., p. xli.
Brandt, J. F. 1873 A, p. 234.
Cope, E. D. 1868 I, p. 186.
    1890 E, p. 609.
Dawson, J. W. 1883 A, p. 201.   (Beluga.)
Flower, W. H. 1883 B, p. 505.
    1889 A, p. 146.
Flower and Lydekker 1891 A, p. 262.
Gray, J. E., 1828, Spicelegia Zool., i, p. 2.   (Beluga.)
    1866 A, p. 307.
Lydekker, R. 1887 A, p. 78.
Probst, J. 1886 A, p. 129.   (Beluga.)
True, F. W. 1889 A, p. 146.
Van Beneden and Gervais 1880 A, p. 531.   (Beluga.)
Zittel, K. A. 1893 B, p. 176.   (Beluga.)

Delphinapterus leucas Pall.

Cope, E. D. 1890 E, pl. xxi.
Dawson, J. W. 1883 A, p. 201.   (Beluga catodon.)
    1895 B.   (B. catodon.)
Flower and Lydekker 1891 A, p. 262, fig. 91.
Owen, R. 1868 A, p. 281.
Trouessart, E. L. 1898 A, p. 1050.
True, F. W. 1889 A, p. 187, pl. xlvi, figs. 1, 2.
    Recent; Arctic seas.   Pleistocene; Ontario.

Delphinapterus orcina Cope.

Cope, E. D. 1875 V, p. 49.
    1890 E, p. 610.
Roger, O. 1896 A, p. 80.   (Beluga.)
Trouessart, E. L. 1898 A, p. 1051.
    Miocene; North Carolina.

Delphinapterus vermontana (Thomp.).

Thompson, Z. 1850 A, p. 257, figs. 1–13.   (Delphinus.)
Amer. Naturalist 1871 A.   (Beluga.)
Brandt, J. F. 1873 A, p. 289.   (Beluga.)
Cope, E. D. 1867 C, p. 144.   (Beluga.)
    1890 E, p. 610.
Dawson, J. W. 1878 A, supp., p. 28.   (Beluga.)
    1883 A, p. 201.   (Beluga.)
Gilpin B, 1873 A.   (Beluga vermontana?)
Honeyman, D. 1874 A.   (Beluga.)
    1874 B.   (Beluga.)
Leidy, J. 1869 A, p. 436.   (Beluga.)
Logan, W. E. 1863 A, p. 919.   (Beluga.)
Thompson, Z. 1853 A, p. 15, figs. 1–13.   (Beluga.)
    1859 A, p. 299.   (Beluga.)
Trouessart, E. L. 1898 A, p. 1051.
    Pleistocene; Vermont, Quebec.

MONODON Linn.   Type *M. monoceros* Linn.

Linnæus, C., 1758, Syst. Nat., ed. 10, i, p. 75.
Berthold, A. A. 1850 A.
Brandt, J. F. 1873 A, p. 231.
Clark, J. W. 1871 A.
Flower, W. H. 1883 B.
    1883 D.
Flower and Lydekker 1891 A, p. 260.
Gray, J. E. 1866 A, p. 310.
Jaeger, G. 1851 A.
Leboucq, H. 1887 A, p. 203.
Lydekker, R. 1887 A, p. 78.
Nicholson and Lydekker 1889 A, p. 1307.
Owen, R. 1845 B, p. 347.
    1868 A, p. 279, fig. 220.
Pictet, F. J. 1853 A, p. 384.
Smith, J. 1895 A, p. 331.
Van Beneden and Gervais 1880 A, p. 523.
Zittel, K. A. 1893 B, p. 176.

Monodon monoceros Linn.

Linnæus, C., 1758, Syst. Nat., ed. 10, i, p. 75.

Barton, B. S. 1805 A, p. 98.   ("Monodon.")
Brandt, J. F. 1873 A, p. 232.
Clark, J. W. 1871 A.
Flower, W. H. 1883 D, p. 398, fig. 49.
Flower and Lydekker 1891 A, p. 251, fig. 90.
Gray, J. E. 1866 A, p. 311.
Jaeger, G. 1857 A, p. 571.
Lydekker, R. 1887 A, p. 78.
Nicholson and Lydekker 1889 A, p. 1307.
Owen, R. 1845 B, p. 347, pl. lxxxvii.   ("Monodon.")
Tomes, C. S. 1898 A, p. 381, fig. 174.
Trouessart, E. L. 1898 A, p. 1052.
Turner, H. N. 1872 A.
    1873 A.
    1876 A.
Van Beneden and Gervais 1880 A, p. 523, pl. xliv, figs. 6–9; pl. xlv.
Zittel, K. A. 1893 B, p. 176.
    Recent; Arctic seas.   Pleistocene; Alaska.

PHYSETERIDÆ. ·

Cope, E. D. 1889 C, p. 207.
  1890 E, p. 606.
  1895 A, p. 137. (Choneziphiidæ.)
Duvernoy, G. L. 1851 A. ("Les cachalots.")
Flower, W. H. 1867 A, p. 113.
  1868 B, p. 345.
  1883 D, p. 305.
Flower and Lydekker 1891 A, p. 247.
Gill, T. 1871 B, p. 123.
· 1872 B, pp. 15, 96. (Physeteridæ, Ziphiidæ.)

Gray, J. E. 1866 A, p. 195. (Catodontidæ.)
Huxley, T. H. 1872 A, p. 340.
Lydekker, R. 1887 A, p. 53.
Nicholson and Lydekker 1889 A, p. 1305.
Pouchet and Beauregard 1889 A.
Roger, O. 1896 A, p. 80.
Van Beneden, P. J. 1877 B. (Physeterula.)
Zittel, K. A. 1898 B, p. 176.

PHYSETERINÆ.

Brandt, J. F. 1873 A, p. 205.
Flower, W. H. 1867 A, p. 114.
  1883 D, p. 395.

Gill, T. 1871 B, p. 124.
  1872 B, pp. 15, 96.
Zittel, K. A. 1898 B, p. 177.

PHYSETER Linn.   Type *P. macrocephalus* Linn.

Linnæus, C., 1758, Syst. Nat., ed. 10, i, p. 76.
Brandt, J. F. 1873 A, p. 205.
Cope, E. D. 1890 E, p. 606.
Flower, W. H. 1868 B.
  1885 A.
Flower and Lydekker 1891 A, p. 248.
Gray, J. E. 1866 A, p. 210.
Harlan, R. 1834 B, pp. 74, 75. (Megistosaurus, Nephrosteon, synonyms of Physeter.)
  1835 A, p. 76. (Megistosaurus.)
  1835 B, p. 279. (Megistosaurus, Nephrosteon.)
Huxley, T. H. 1872 A, p. 340.
Leidy, J. 1856 L, p. 255. (Orycterocetus, type *O. quadratidens.*)
  1857 A, p. 271. (Orycterocetus.)
  1869 A, p. 436 (Orycterocetus); p. 444 (Physeter.)
· 1876 A, p. 176. (Dinoziphius.)
Owen, R. 1845 B, p. 353.
Probst, J. 1886 A.
Van Beneden and Gervais 1880 A, p. 303 (Physeter); p. 344 (Dinoziphius.)
Zittel, K. A. 1898 B, p. 177.

Physeter carolinensis (Leidy).

Leidy, J. 1877 A, p. 216, pl. xxxiv, fig. 6. (Dinoziphius.)
Cope, E. D. 1890 E, p. 608. (Hoplocetus; may not refer to P. carolinensis.)

Physeter macrocephalus Linn.

Linnæus, C., 1758, Syst. Nat., ed. 10, i, p. 76.
Flower, W. H. 1868 B, pls. lv–lxi.
Gibbes, R. W. 1850 B. ("Physeter.")

Gray, J. E. 1866 A, p. 212. (Physeter tursio.)
Harlan, R. 1828 A, p. 186.
  1835 A, p. 76.
  1835 B, p. 279. (Megistosaurus and Nephrosteon = Physeter macrocephalus.)
Jaeger, G. 1857 A, p. 573.
Leidy, J. 1869 A, p. 444.
Owen, R. 1845 B, p. 353, pls. lxxxix, lxxxix a.
  1868 A, p. 281.
Rafinesque, C. S., 1833, Atlantic Journal, p. 12. ("Nephrosteon.")
Roger, O. 1896 A, p. 80.
Tomes, C. S. 1898 A, p. 378.
Trouessart, E. L. 1898 A, p. 1056.
  Recent in warm seas: Pleistocene?; Louisiana. "A doubtful fossil."

Physeter vetus (Leidy).

Leidy, J. 1853 C, p. 378. (P. antiquus, preoccupied by Gervais.)
Brandt, J. F. 1873 A, p. 289. (Catodon.)
Cope, E. D. 1869 N, p. 739. (P. antiquus.)
  1875 V, p. 50.
  1890 E, p. 615.
Emmons, E. 1858 B, p. 212, fig. 34. (P. antiquus.)
Gibbes, R. W. 1847 B, p. 11. (P. macrocephalus?.)
  1850 B. (Catodon.)
Leidy, J. 1860 B, p. 117, pl. xxiv, figs. 8, 9. (P. antiquus.)
  1869 A, p. 436. (Catodon.) .
Roger, O. 1896 A, p. 80.
Trouessart, E. L. 1898 A, p. 1065.
  Pleistocene; South Carolina. Miocene; North Carolina, Virginia, New Jersey.

ORYCTEROCETUS Leidy.   Type *O. quadratidens* Leidy.

Leidy, J. 1856 C, p. 378.
Cope, E. D. 1890 E, p. 607. (Syn. of Physeter.)
Leidy, J. 1857 A, p. 271.
  1869 A, p. 436.
Trouessart, E. L. 1898 A, p. 1055. (Syn. of Physeter.)
Zittel, K. A. 1898 B, p. 177. (Syn. of Physeter.)

Orycterocetus cornutidens Leidy.

Leidy, J. 1856 L, p. 255, in part.

Brandt, J. F. 1873 A, p. 289.
Cope, E. D. 1867 C, p. 144. (O. crocodilinus.)
  1875 V, p. 50.
  1890 E, p. 615. (Physeter.)
Emmons, E. 1858 B, p. 211, fig. 33.
Leidy, J. 1857 A, p. 271.
  1869 A, p. 437.
Roger, O. 1896 A, p. 80. (Physeter.)
Trouessart, E. L. 1898 A, p. 1055. (Physeter.)
  Miocene; North Carolina, Maryland.

Orycterocetus quadratidens Leidy.

Leidy, J. 1853 C, p. 378.
Brandt, J. F. 1873 A, p. 289.
Cope, E. D. 1875 V, p. 50.
1890 E, p. 615. (Physeter.)
Emmons, E. 1858 B, p, 210, fig. 32.

Leidy, J. 1856 L, p. 255. (O. cornutidens, in part.)
1857 A, p. 271.
1869 A, p. 436, pl. xxx, figs. 16, 17.
Roger, O. 1896 A, p. 81. (Physeter.)
Trouessart, E. L. 1898 A, p. 1055. (Physeter.)
Miocene; Virginia, North Carolina.

HYPOCETUS Lydekker.   Type *Mesocetus poucheti* Moreno.

Lydekker, R. 1893 D, p. 7 (Hypocetus); p. 8 (Paracetus).
Ameghino, F. 1894 A, p. 437. (Diaphorocetus, to replace Mesocetus Moreno, preoccupied.)
Cope, E. D. 1895 A, p. 135. (Paracetus.)
Lydekker, R. 1896 B, p. 68.
Moreno, F. P. 1892 A, p. 395, pl. x. (Mesocetus, type M. *poucheti;* not Mesocetus of Van Beneden.)

Trouessart, E. L. 1898 A, p. 1052. (Diaphorocetus.)

Hypocetus mediatlanticus (Cope).

Cope, E. D. 1895 A, p. 135. (Paracetus.)
Roger, O. 1896 A, 81. (Paracetus.)
Trouessart, E. L. 1898 A, p. 1053. (Diaphorocetus.)
Miocene; Maryland.

HOPLOCETUS Gervais.   Type *H. crassidens* Gervais.

Gervais, P. 1852, Zool. Paléont. Franc., ed. 1, p. 26.
Cope, E. D. 1890 E, p. 607.
Flower and Lydekker 1891 A, p. 251.
Leidy, J. 1869 A, p. 438.
Lydekker, R. 1887 A, p. 60.
Owen, R. 1846 B, p. 536. (Balænodon, type *B. physaloides;* may not be identical genus.)
Pictet, F. J. 1853 A, p. 388.
Probst, J. 1886 A, p. 106.
Van Beneden and Gervais 1880 A, p. 239.
Zittel, K. A. 1893 B, p. 177.

Hoplocetus obesus Leidy.

Leidy, J. 1868 E, p. 196.
Brandt, J. F. 1873 A, p. 290.
Cope, E. D. 1890 E, p. 608 (H. carolinensis); p. 615 (H. obesus).
Leidy, J. 1869 A, p. 438, pl. xxx, figs. 13–15.
Trouessart, E. L. 1898 A, p. 1054.
Miocene; California.
    Professor Cope's name, *H. carolinensis,* more probably refers to *Physeter carolinensis;* or it may be due to a confusion of the two names.

ONTOCETUS Leidy.   Type *O. emmonsi* Leidy.

Leidy, J. 1859 G, p. 162.
1869 A, p. 440.
A genus of doubtful position.

Ontocetus emmonsi Leidy.

Leidy, J. 1859 G, p. 162.
Brandt, J. F. 1873 A, p. 290.

Cope, E. D. 1867 C, p. 144.
1875 V, p. 51.
Emmons, E. 1860 A, p. 219.
Leidy, J. 1869 A, p. 440.
Trouessart, E. L. 1898 A, p. 1055. (Hoplocetus.)
Miocene; North Carolina.

ZIPHIINÆ.

Brandt, J. F. 1873 A, p. 209.
Flower, W. H. 1867 A, p. 114.
1872 A.
1883 D, p. 396.
Gill, T. 1872 B, p. 15.
Flower and Lydekker 1891 A, p. 251.
Gervais, P. 1850 A. (Ziphioidea.)

Gervais, P. 1859 A, pp. 285, 291. (" Ziphidés.")
Lydekker R. 1887 A, p. 63.
Owen, R. 1889 A. (Ziphius.)
Probst, J. 1886 A, p. 108. (Ziphioidæ.)
Roger, O. 1896 A, p. 81.
Zittel, K. A. 1893 B, p. 178.

CHONEZIPHIUS Duvernoy.   Type *Ziphius planirostris* Cuv.

Duvernoy, G. L. 1851 A, p. 61.
Brandt, J. F. 1873 A, p. 218.
Cope, E. D. 1890 E, pp. 607,608, part.
Flower and Lydekker 1891 A, p. 257.
Gervais, P. 1859 A, p. 288.
Leidy, J. 1877 A, p. 218.
Lydekker, R. 1887 A, p. 64.
Owen, R. 1889 A, p. 35.
Pictet, F. J. 1853 A, p. 385.

Trouessart, E. L. 1898 A, p. 1059, part.
Van Beneden and Gervais 1880 A, p. 413.
Zittel, K. A. 1893 B, p. 178.

Choneziphius cœlops (Leidy).

Leidy, J. 1876 A, p. 81. (Eboroziphius.)
Cope, E. D. 1890 E, p. 616.
Leidy, J. 1877 A, p. 224, pl. xxx, fig. 5; pl. xxxi, fig. 3. (Eboroziphius.)

Roger, O. 1896 A, p. 82.
Trouessart, E. L. 1898 A, p. 1060.
Tertiary (Phosphate beds); South Carolina.

Choneziphius liops Leidy.

Leidy, J. 1876 A, p. 81.
Cope, E. D. 1890 E, p. 616.
Leidy, J. 1877 A, p. 222, pl. xxx, fig. 1; pl. xxxi,
fig. 2.
Roger, O. 1896 A, p. 82.
Trouessart, E. L. 1898 A, p. 1060.
Tertiary (Phosphate beds); South Carolina.

Choneziphius trachops Leidy.

Leidy, J. 1876 A, p. 81.
Cope, E. D. 1890 E, pp. 608,615.
Leidy, J. 1877 A, p. 218, pl. xxx, fig. 2; pl. xxxi,
fig. 1.
Roger, O. 1896 A, p. 82.
Trouessart, E. L. 1898 A, p. 1060.
Tertiary (Phosphate beds); South Carolina.

PROROZIPHIUS Leidy. Type *P. macrops* Leidy.

Leidy, J. 1876 E, p. 87.
1877 A, p. 227.
Trouessart, E. L. 1898 A, p. 1060. (Syn. of Choneziphius.)

Proroziphius chonops Leidy.

Leidy, J. 1876 F, p. 114.
Cope, E. D. 1890 E, p. 616. (Choneziphius.,
Leidy, J. 1877 A, p. 229, pl. xxxii, figs. 3, 4.
Roger, O. 1896 A, p. 82. (Choneziphius.)

Trouessart, E. L. 1898 A, p. 1060. (Choneziphius.)
Tertiary (Phosphate beds); South Carolina.

Proroziphius macrops Leidy.

Leidy, J. 1876 E, p. 87.
Cope, E. D. 1890 E, p. 616,. (Choneziphius.)
Leidy, J. 1877 A, p. 227, pl. xxxii, figs. 1, 2.
Roger, O. 1896 A, p. 82. (Choneziphius.)
Trouessart, E. L. 1898 A, p. 1060. (Choneziphius.)
Tertiary (Phosphate beds); South Carolina.

PELYCORHAMPHUS Cope. Type *P. pertortus* Cope.

Cope, E. D. 1895 A, p. 137.

Pelycorhamphus pertortus Cope.

Cope, E. D. 1895 A, p. 137.

Trouessart, E. L. 1898 A, p. 1057.
Miocene. Locality unknown.

ANOPLONASSA Cope. Type *A. forcipata* Cope.

Cope, E. D. 1869 E, p. 189.
1869 O.
1890 F, p. 700.
1895 A, p. 138.
Trouessart, E. L. 1898 A, pp. 1008, 1057.
Zittel, K. A. 1893 B, p. 179.

Anoplonassa forcipata Cope.

Cope, E. D. 1869 E, p. 189, pl. v, fig. 5.

Brandt, J. F. 1873 A, p. 289.
Cope, E. D. 1869 O.
1890 F, p. 700, fig. 2.
Leidy, J. 1869 A, p. 436.
Roger, O. 1896 A, p. 82.
Trouessart, E. L. 1898 A, p. 103.
Van Beneden and Gervais 1880 A, p. 386, text
figure.
Tertiary (Phosphate beds); South Carolina.

MESOPLODON Gervais. Type *Delphinus sowerbensis.*

Gervais, P. 1850 A, p. 16.
Brandt, J. F. 1873 A, p. 220. (Mesoödon.)
Cope, E. D. 1890 E, p. 607.
Flower, W. H. 1872 A p. 209.
Flower and Lydekker 1891 A, p. 254.
Forbes, H. O. 1893 A.
Gervais, P. 1850 A, p. 16. (Dioplodon.)
1859 A, p. 289 (Dioplodon); p. 290 (Mesoplo-
don.)
Huxley, T. H. 1864 A, p. 388. (Micropteron; Be-
lemnoziphius, type *B. compressus.*)
Leidy, J. 1876 A, p. 81. (Belemnoziphius.)
1877 A, p. 226. (Dioplodon.)
Owen, R. 1889 A, p. 38. (Belemnoziphius.)
Probst, J. 1886 A, p. 113.
Trouessart, E. L. 1898 A, p. 1063.

Van Beneden and Gervais 1880 A, p. 392 (Mesoplodon); pp. 403, 419 (Dioplodon).
Zittel, K. A. 1893 B, p. 179. (Mesoplodon, Dioplodon.)

Mesoplodon prorops (Leidy).

Leidy, J. 1876 A, p 81. (Belemnoziphius.)
Cope, E. D. 1890 E, p. 616.
Leidy, J. 1877 A, p. 226, pl. xxx, figs. 3, 4. (Dio-
plodon.)
Roger, O. 1896 A. p. 82.
Trouessart, E. L. 1898 A, p. 1065.
Van Beneden, P. J. 1880 A. (Ziphius longirostris;
=? M. prorops.)
Tertiary (Phosphate beds); South Carolina.

Suborder MYSTICETE.

Brandt, J. F. 1873 A, p. 17. (Balænoidea.)
Cope, E. D. 1889 R, p. 876. (Mystacoceti.)
   1890 E, p. 610. (Mysticete.)
   1896 B, p. 109. (Mysticete.)
Flower, W. H. 1864 A, p. 388. (Balænoidea, or Mysticete.)
   1867 A, p. 110. (Mystacoceti, or Balænoidea.)
   1876 B, xiii, p. 450. (Mystacoceti.)
   1883 D, p. 394. (Mystacoceti.)
Flower and Lydekker 1891 A, p. 234. (Mystacoceti.)
Gadow, H. 1898 A, p. 44. (Mystacoceti.)
Gill, T. 1871 B, p. 122. (Mysticete.)
   1872 B, pp. 16, 97. (Mysticete.)

Gill, T. 1873 C, p. 25. (Mysticete.)
Gray, J. E. 1866 A, pp. 68, 194. (Mysticete.)
Haeckel, E. 1895 A, p. 566. (Mysticeta.)
Huxley, T. H. 1872 A, p. 336. (Balænoidea.)
Kükenthal, W. 1891 A, p. 365. ("Bartenwale.")
Leboucq, H. 1887 A. ("Mysticétes.")
Murie, J. 1865 A.
Roger, O. 1896 A, p. 83. (Mystacoceti.)
Ryder, J. 1878 A.
Van Beneden, P. J. 1880 B, p. 11. ("Mysticétes.")
Van Beneden and Gervais 1880 A, pp. 29, 239. ("Mysticétes.")
Woodward, A. S. 1898 B, p. 273. (Mystacoceti.)
Zittel, K. A. 1893 B, p. 180. (Mystacoceti.)

BALÆNIDÆ.

Brandt, J. F. 1873 A, p. 18 (Balænidæ); p. 26. (Balænopteridæ). .
Carte and Macalister 1868 A. (>Balænopteridæ.)
Cope, E. D. 1889 R, p. 876.
   1890 E, p. 611.
   1895 A, p. 139.
Cuvier, G. 1834 A, viii, pt. ii, pp. 249–332. ("Baleines.")
Duvernoy, G. L. 1851 A. ("Les Baleines.")
Flower, W. H. 1864 A. (Balænidae, Balænopteridæ.)
   1867 A, p. 115. (>Balænopterinæ.)
   1876 B, xiii, p. 450. (Balænidae, Balænopteridæ.)
   1883 D, p. 394.
Flower and Lydekker 1891 A, p. 234.
Gervais, P. 1871 A.
Gray, J. E. 1866 A, p. 75 (Balænidæ); p. 106 (Balænopteridæ).
Holder, J. B. 1883 A.

Huxley, T. H. 1872 A, p. 339.
Kükenthal, W. 1890 C.
Marsh, O. C. 1877 E.
Murie, J. 1865 A. (Balænopteridæ.)
Nicholson and Lydekker 1889 A, p. 1302.
Owen, R. 1866 B, p. 415.
   1868 A, p. 276.
Roger, O. 1896 A, p. 83 (Bálænidæ); p. 87 (Balænopteridæ).
Struthers, J. 1871 A. (>Balænopteridæ.)
   1872 A. (>Balænopteridæ.)
   1881 A. (>Balænidæ.)
   1893 A. (Balænidae, Balænopteridæ.)
   1895 A. (>Balænidæ.)
Van Beneden P. J. 1872 A. (>Balænidæ.)
   1880 A. (>Balænidæ.)
   1882 A. (Balænidæ, Balænopteridæ.)
Zittel, K. A. 1893 B, p. 181 (Balænidæ); p. 183 (Balænopteridæ).

METOPOCETUS Cope.   Type *M. durinasus* Cope.

Cope, E. D. 1896 J, p. 141.

Trouessart, E. L. 1898 A, p. 1069.
Miocene; Maryland.

Metopocetus durinasus Cope.

Cope, E. D. 1896 J, p. 141, pl. xi, fig. 3.

CEPHALOTROPUS Cope.   Type *C. coronatus* Cope.

Cope, E. D. 1896 J, pp. 141, 143.

Trouessart, E. L. 1898 A, p. 1069.
Miocene; Maryland?.

Cephalotropus coronatus Cope.

Cope, E. D. 1896 J, p. 143, pl. xi, fig. 2.

CETOTHERIUM Brandt.   Type *C. rathkii* Brandt.

Brandt, J. F., 1843, Bull. Acad. Imp. St. Petersb., i, p. 145.
   1844 A.
   1873 A, pp. 61, 142, 143.
Cope, E. D. 1869 G, p. 10. (Eschrichtius.)
   1890 E, p. 612.
   1895 A, p. 140.
   1896 J, p. 141.
Flower and Lydekker 1891 A, p. 245.
Gaudry, A. 1878 B, p. 32. (Plesiocetus.)
Gray, J. E. 1866 A, p. 131, figs. 21, 22. (Eschrichtius.)

Lydekker, R. 1887 A, p. 42.
   1893 D, pt. ii, p. 2.
Pictet, F. J. 1853 A, p. 388.
Van Beneden, P.-J. 1872 A, p. 416.
   1872 B, p. 15 (Plesiocetus); p. 16 (Cetotherium).
   1875 A. (Plesiocetus.)
   1880 B, p. 17. (Plesiocetus.)
Van Beneden and Gervais 1880 A, p. 268. (Cetotherium); p. 274 (Plesiocetus).
Zittel, K. A. 1893 B, p. 181 (Plesiocetus); p. 182 (Cetotherium).

Cetotherium cephalum Cope.

Cope, E. D. 1867 B, p. 131.   (Eschrichtius.)
    1867 C, p. 148.   (Eschrichtius.)
    1869 G, pp. 10, 11.   (Eschrichtius.)
    1890 E, p. 613, pl. xxii, text fig. 7.
    1891 N, fig. 39.
    1896 J, p. 146, pl. xii, fig. 2.
Dana, J. D. 1896 A, p. 912, fig. 1589.
Leidy, J. 1869 A, p. 442.   (Eschrichtius.)
Roger, O. 1896 A, p. 84.
Trouessart, E. L. 1898 A, p. 1071.
    Miocene; Maryland.

Cetotherium davidsonii Cope.

Cope, E. D. 1872 Z, p. 29.   (Eschrichtius.)
Bowers, S. 1889 A.   (Eschrichtius.)
Cope, E. D. 1890 E, p. 616.   (Balænoptera.)
    1896 J, p. 146, pl. xii, fig. 4.
Roger, O. 1896 A, p. 86.   (Balænoptera.)
Trouessart, E. L. 1898 A, p. 1079.   (Referred to
    Balænoptera davidsoni Scammon.)
    Miocene; California.

Cetotherium leptocentrum Cope.

Cope, E. D. 1867 C, p. 147.   (Eschrichtius.)
    1869 G, p. 11.   (Eschrichtius.)
    1890 E, p. 616.
    1896 A, p. 148.   (C. crassangulum.)
    1896 J, p. 145, pl. xii, fig. 1.
Leidy, J. 1869 A, p. 442.
Roger, O. 1896 A, p. 84 (C. leptocentrum); p. 85
    (C. crassangulum.)
Trouessart, E. L. 1898 A, p. 1071.
    Miocene; Virginia.

Cetotherium megalophysum Cope.

Cope, E. D. 1896 A, p. 146.
    1896 J, p. 143, pl. xi, fig. 1.
Roger, O. 1896 A, p. 84.

Trouessart, E. L. 1898 A, p. 1071.
    Miocene; Maryland?.

Cetotherium mysticetoides (Emmons).

Emmons, E. 1858 B, p. 205, fig. 26.   (Balæna.)
Cope, E. D. 1875 V, p. 42.   (Eschrichtius.)
    1890 E, p. 616.
Leidy, J. 1869 A, p. 440.   (Balæna.)
Roger, O. 1896 A, p. 84 (Cetotherium); p. 87 (Ba-
    læna).
Trouessart, E. L. 1898 A, p. 1071 (Cetotherium); p.
    1092 (Balæna).
Van Beneden, P. J. 1882 A, p. 23.
    Miocene; North Carolina.

Cetotherium parvum Trouessart.

Trouessart, E. L. 1898 A, p. 1071.   (To replace C.
    pusillum Cope, not of Nordmann.
Cope, E. D. 1868 G, p. 159.   (Balænoptera pusilla.)
    1868 I, p. 190 (?Delphinapterus tyrannus); p.
    191 (Eschrichtius pusillus.)
    1869 G, p. 7 (?Delphinapterus tyrannus); p.
    11 (Eschrichtius pusillus.)
    1890 E, p. 616.   (C. pusillum.)
    1895 A, p. 145.   (C. pusillum.)
Leidy, J. 1869 A, p. 442.   (Eschrichtius pusillus.)
Roger, O. 1896 A, p. 84.   (C. pusillum.)
    Miocene; Maryland, South Carolina.

Cetotherium polyporum Cope.

Cope, E. D. 1869 I.   (Eschrichtius.)
    1870 F, p. 285.   (Eschrichtius.)
    1875 V, p. 42, pl. v, figs. 3, 3a.   (Eschrich-
    tius.)
    1890 E, p. 616.
    1895 A, p. 149, pl. vi, fig. 7.
Roger, O. 1896 A, p. 84.
Trouessart, E. L. 1898 A, p. 1071.
    Miocene; North Carolina.

SIPHONOCETUS Cope.   Type *Balæna prisca* Leidy.

Cope, E. D. 1895 A, p. 140.
Trouessart, E. L. 1898 A, p. 1074.

Siphonocetus clarkianus Cope.

Cope, E. D. 1895 A, p. 140, pl. vi, fig. 4.
Roger, O. 1896 A, p. 85.
Trouessart, E. L. 1898 A, p. 1074.
    Miocene; Maryland.

Siphonocetus expansus Cope.

Cope, E. D. 1868 I, p. 193.   (Megaptera.)
    1869 G, p. 11.   (Eschrichtius.)
    1890 E, p. 616.   (Cetotherium.)
    1896 A, p. 140, pl. vi, fig. 5.
Leidy, J. 1869 A, p. 442.   (Eschrichtius.)
Roger, O. 1896 A, p. 85.

Trouessart, E. L. 1898 A, p. 1074.
    Miocene; Virginia, Maryland.

Siphonocetus priscus (Leidy).

Leidy, J. 1851 F, p. 308.   (Balæna.)
Cope, E. D. 1867 B, p. 132.   (Balæna.)
    1867 C, p. 144.   (Balænoptera.)
    1868 I, p. 193.   (Balænoptera.)
    1869 G, pp. 10, 11.   (Eschrichtius.)
    1890 E, p. 616.   (Cetotherium.)
    1895 A, p. 140, pl. vi, fig. 3.
Leidy, J. 1869 A, p. 441.   (Eschrichtius.)
Roger, O. 1896 A, p. 85.
Trouessart, E. L. 1898 A, p. 1074.
    Miocene; Virginia.

ULIAS Cope.   Type *U. moratus* Cope.

Cope, E. D. 1895 A, p. 141.
Trouessart, E. L. 1898 A, p. 1074.

Ulias moratus Cope.

Cope, E. D. 1895 A, p. 141, pl. vi, fig. 1.
Roger, O. 1896 A, p. 85.
    Miocene; Maryland?

TRETULIAS Cope.   Type *T. buccatus* Cope.

Cope, *E. D.* 1895 A, p. 143.
Trouessart, E. L. 1898 A, p. 1074.

Tretulias buccatus Cope.

Cope, *E. D.* 1895 A, p. 143, pl. vi, fig. 2.

Roger, O. 1896 A, p. 85.
Trouessart, E. L. 1898 A, p. 1074.
Miocene; Maryland?.

MESOCETUS Van Beneden.   Type *M. longirostris* Van Beneden.

Van Beneden, *P. J.* 1880 B, p. 22.
Cope, *E. D.* 1895 A, p. 153.
Van Beneden, P. J. 1882 B.

Mesocetus siphunculus Cope.

Cope, *E. D.* 1895 A, p. 153.

Cope, *E. D.* 1896 J, p. 146, pl. xii, fig. 6.
Roger, O. 1896 A, p. 85.
Trouessart, E. L. 1898 A, p. 1076.
Miocene; Virginia.

RHEGNOPSIS Cope.   Type *Balæna palæatlanticus* Leidy.

Cope, *E. D.* 1896 J, p. 145.
Leidy, J. 1869 A, p. 440 (Protobalæna, type *B. palæatlanticus;* preocc. by Van Beneden, 1867).
Trouessart, E. L. 1898 A, p. 1077.

Rhegnopsis palæatlanticus (Leidy).

Leidy, *J.* 1851 F, p. 308. (Balæna.)
Cope, E. D. 1867 B, p. 132. (Balæna.)

Cope, E. D. 1867 C, p. 147.   (Balæna.)
    1868 I, p. 198.   (Balænoptera.)
    1890 E, p. 616.   (Balænoptera.)
    1896 J, p. 145, pl. xii, fig. 5.
Leidy, J. 1869 A, p. 440.   (Protobalæna.)
Roger, O. 1896 A, p. 86.   (Balænoptera.)
Trouessart, E. L. 1898 A, p. 1077.
    Miocene; Virginia.

BALÆNOPTERA Lacépède.

Lacépède, *B. G. É.*, 1804, Hist. Nat. d. Cétacés; Tab. ord., p. xxxvi.
Brandt, J. F. 1873 A, p. 27.
Carte and Macalister 1869 A.
Cope, E. D. 1890 E, p. 612.
Flower, W. H. 1864 A, p. 373.
    1885 A.
Flower and Lydekker 1891 A, p. 242.
Gray, J. E. 1866 A, pp. 186, 382.
Lydekker, R. 1887 A, p. 34.
Murie, J. 1865 A.   (Physalus.)
Struthers, J. 1871 A.
Struthers, J. 1872 A.

Struthers, J. 1898 A.
Tomes, C. S. 1898 A, p. 383.
Van Beneden, P. J. 1880 B, p. 14.
    1882 A, p. 61.
Van Beneden and Gervais 1880 A, p. 137.
Zittel, K. A. 1898 B, p. 182.

Balænoptera sursiplana Cope.

Cope, *E. D.* 1895 A, p. 151.
Roger, O. 1896 A, p. 86.
Trouessart, E. L. 1898 A, p. 1077.
Miocene; Maryland.

MEGAPTERA Gray.   Type *M. longimana.*

Gray, *J. E.*, 1846, Zool. "Erebus" and "Terror," p. 16.
Flower, W. H. 1864 A.
    1883 D, p. 394.
Flower and Lydekker 1891 A, P. 241.
Gervais, P. 1871 A.
Gray, J. E. 1866 A, p. 117.
Lydekker, R. 1887 A, p. 31.
Struthers, J. 1885 A.
    1888 A.
Van Beneden, P. J. 1868 A.
    1880 B, P. 13.
    1882 A, p. 26.
Van Beneden and Gervais 1880 A, P. 116.
Zittel, K. A. 1893 B, p. 182.

Megaptera boops (Linn).

Linnæus, *C.*, 1758, Syst. Nat., ed. 10, i, p. 76. (Balæna.)
Dawson, J. W. 1883 A, p. 202.   (M. longimana.)
Flower and Lydekker 1891 A, p. 241, fig. 79.
Gervais, P. 1871 A.
Gray, J. E. 1866 A, p. 119, figs. 14-18.   (M. longimana.)
Struthers, J. 1885 A, p. 124.
    1888 A.
Trouessart, E. L. 1898 A, p. 1085.
Van Beneden and Gervais 1880 A, p. 120, pl. x; pl. xi, figs. 1-8.
    Recent in North Atlantic: Pleistocene; Ontario.

MESOTERAS Cope.   Type *M. kerrianus* Cope.

Cope, *E. D.* 1870 R, p. 128.
    1870 F, p. 286.
    1875 V, p. 44, pl. viii, fig. 1.
    1890 E, p. 613.
Zittel, K. A. 1893 B, p. 183.

Mesoteras kerrianus Cope.

Cope, *E. D.* 1870 R, p. 128.

Cope, E. D. 1870 F, p. 286.
    1875 V, p. 44.
    1890 E, pp. 613, 616.
Roger, O. 1896 A, p. 86.
Trouessart, E. L. 1898 A, p. 1088.
    Miocene; North Carolina.

BALÆNA Linn.   Type *B. mysticetus* Linn.

Linnæus, C., 1758, Syst. Nat., ed. 10, i, p. 75.
Flower, W. H. 1864 A, p. 390.
  1883 D, p. 395.
  1885 A.
Flower and Lydekker 1891 A, p. 286.
Gervais, P. 1871 A.
Gray, J. E. 1866 A, p. 79.
Holder, J. B. 1883 A.
Huxley, T. H. 1872 A, p. 337.
Lydekker R. 1887 A, p. 16.
Nicholson and Lydekker 1889 A, p. 1308.
Owen, R. 1866 B, p. 415.
Parker, W. K. 1868 A, p. 217.
Pictet, F. J. 1853 A, p. 387.
Struthers, J. 1881 A.
  1893 A,
  1895 A.
Roger, O. 1896 A, p. 87.

Trouessart, E. L. 1898 A, p. 1089.
Van Beneden and Gervais 1880 A, p. 32.
Zittel, K. A. 1893 B, p. 183.

Balæna affinis Owen.

Owen, R, 1846 B, p. 530, fig. 221.
Cope, E. D. 1895 A, p. 152.
Emmons, E. 1858 B, p. 206. (Doubtful identification.)
Lydekker, R. 1887 A, p. 17, fig. 7.
Zittel K. A. 1893 B, p. 183.
  Miocene; England, Maryland?, North Carolina.
  For citations to descriptions of uncertain and unidentified remains of Cete see Leidy 1869 A, p. 443.

Order UNGULATA.

Baur, G. 1884 F.
Cope, E. D. 1873 E, p. 561.
  1873 F.
  1875 C, p. 26.
  1881 I.
  1882 Y.
  1883 D.
  1883 FF.
  1884 E.
  1885 O.
  1885 U.
  1887 I.
  1887 N.
  1888 T.
  1889 C.
  1889 R, pp. 875, 876.
Flower, W. H. 1883 D, p. 421.
  1885 A.
Flower and Lydekker 1891 A, p. 273.
Gadow, H. 1898 A, p. 45.
Gegenbaur, C. 1898 A.
Gill, T. 1872 A, p. 298.
  1872 B, pp. 8, 47, 70.
Haeckel, E. 1895 A, p. 524.
Huxley, T. H. 1872 A, p. 292.
Kowalevsky, W. 1873 A, p. 152.
  1874 A.
Leuthardt, F. 1891 A, p. 93.
Lydekker, R. 1885 C.
  1886 A.
  1886 B.
  1896 B.
Marsh, O. C. 1877 E.
  1884 F, p. 173.
Mettam, A. E. 1895 A.
Nicholson and Lydekker 1889 A, p. 1312.

Osborn, H. F. 1890 D.
  1892 H.
  1892 I.
  1893 A, p. 379.
  1893 D.
Owen R. 1845 B, p. 523.
  1851 C, p. 735.
  1868 A, p. 340.
Parker, W. K. 1868 A, p. 219. (Herbivora.)
Paulli, S. 1900 A.
Pavlow, M. 1888 A.
  1890 A.
  1892 A.
Roger, O. 1896 A, p, 142.
Rütimeyer, L. 1888 A.
  1891 A.
Ryder, J. A. 1878 A.
Schlosser, M. 1885 A.
  1886 A.
  1886 B.
  1887 A.
  1890 D.
  1892 A, p. 216.
Schmidt, O. 1886 A, p. 126.
Slade, D. D. 1888 A, p. 244.
  1893 A.
Steinmann and Döderlein 1890 A, p. 746.
Tacker, J. 1892 A.
Tomes, C. S. 1898 A, p. 390.
Trouessart, E. L. 1898 A, p.
Turner, H. N. 1848 A.
Wiedersheim, R. 1881 A, p. 360.
Woodward, A. S. 1898 B, p. 287.
Wortman, J. L. 1886 A, p. 469.
  1893 B, p. 421.
Zittel, K. A. 1893 B, p. 203.

Suborder CONDYLARTHRA Cope.

Cope, E. D. 1881 X, p. 1017.
Ameghino, F. 1891 C, p. 214.
  1896 A, p. 445.
Branco, W. 1897 A, p. 18.

Cope, E. D. 1882 A. (Taxeopoda, in part; Condylarthra.)
  1882 E, p. 178.
  1882 F.

Cope, E. D. 1882 R.
   1882 Y. (Taxeopoda, in part: Condylarthra.)
   1883 D. (Taxeopoda, in part; Condylarthra.)
   1883 E.
   1884 C. (Taxeopoda, in part.)
   1884 O, pp. 378, 382 (Condylarthra); pp. 167, 374, 378 (Taxeopoda, in part).
   1884 AA.
   1885 K, p. 458.
   1885 O.
   1885 U.
   1886 Q, p. 238. (Taxeopoda, in part.)
   1887 I.
   1887 N. (Taxeopoda, in part.)
   1887 8, p. 342. (Taxeopoda, in part; Condylarthra.)
   1888 Q. (Taxeopoda, in part.)
   1888 T.
   1888 CC.
   1889 C. (Taxeopoda, in part; Condylarthra.)
   1891 N, p. 79. (Taxeopoda, in part; Condylarthra.)
   1892 8.
   1894 B, p. 120 (Condylarthra); pp. 106, 120 (Taxeopoda, in part).
   1893 D.
Earle, C. 1893 B.
   1893 D.
Flower and Lydekker 1891 A, p. 438.
Gadow, H. 1898 A, p. 46.
Haeckel, E. 1895 A, pp. 529, 533.
Hay, O. P. 1899 F, p. 682.
Lydekker, R. 1885 E, p. 469. (Taxeopoda, in part.)
   1886 A, p. 172.
Marsh, O. C. 1884 F, p. 171. (Protungulata.)
   1892 D, p. 445. (Mesodactyla.)

Marsh, O. C. 1894 L, p. 26. (Mesodactyla.)
Matthew, W. D. 1897 B, pp. 261, 323 (Protungulata); p. 293 (Condylarthra).
Nicholson and Lydekker 1889 A, p. 1378
Osborn, H. F. 1888 F.
   1890 D, p. 536.
   1892 I.
   1893 A, p. 383.
   1893 D.
   1898 H, p. 175.
Osborn and Earle 1895 A, p. 47.
Parsons, F. G. 1899 A.
   1900 A.
Pavlow, M. 1887 A. (Taxeopoda, in part; Condylarthra.)
   1888 A.
Röse, C. 1896 A, p. 84.
Roger, O. 1896 A, p. 163.
Rütimeyer, L. 1888 A. (Taxeopoda, in part; Condylarthra.)
   1890 A. (Taxeopoda, in part; Condylarthra.)
   1891 A.
Schlosser, M. 1884 B, p. 641.
   1885 A, p. 683.
   1886 B.
   1887 A.
   1888 A, p. 584.
   1889 A.
   1890 C.
   1892 A.
Scott, W. B. 1892 A, p. 427.
Steinmann and Döderlein 1890 A, p. 749.
Tomes, C. S. 1898 A, p. 443.
Woodward, A. S. 1898 B, p. 289.
Wortman, J. L. 1883 A, p. 706. (Taxeopoda, in part.)
   1885 A, p. 297. (Taxeopoda, in part.)
Zittel, K. A. 1893 B, pp. 213, 725.

PHENACODONTIDÆ.

Cope, E. D. 1881 X, p. 1018.
Ameghino, F. 1891 C, p. 215.
Cope, E. D. 1882 E, p. 178.
   1882 F, pp. 96, 96.
   1882 R.
   1883 E.
   1884 O, pp. 384, 386.
   1884 AA, pp. 793, 892.
   1887 N, p. 986.
Matthew, W. D. 1897 C, pp. 294, 299.
   1899 A, pp. 28, 29, 32.
Nicholson and Lydekker 1889 A, p. 218.

Osborn and Earle 1895 A, p. 49.
Pavlow, M. 1887 A.
   1888 A.
Roger, O. 1896 A, p. 165.
Rütimeyer, L. 1888 A, p. 44.
Schlosser, M. 1885 A, p. 683.
   1886 B.
   1888 A, p. 565.
   1890 C.
Steinmann and Döderlein 1890 A, p. 750.
Wortman, J. L. 1882 B, p. 724.
Zittel, K. A. 1893 B, p. 218.

OXYACODON Osb. and Earle.   Type O. apiculatus Osb. and Earle.

Osborn and Earle 1895 A, p. 25.
Matthew, W. D. 1897 C, p. 292.
   1899 A, p. 28.
Trouessart, E. L. 1898 A, p. 72.
   1899 A, p. 1290.
   A genus of uncertain position; placed by Trouessart first in Primates, later in Carnivora; by Matthew in Creodonta incertæ sedis.

Oxyacodon agapetillus (Cope).
Cope, E. D. 1884 K, p. 320. (Anisonchus.)
   1888 CC, p. 305. (Anisonchus.)
Matthew, W. D. 1897 C, pp. 264, 292, fig. 10.

Matthew, W. D. 1899 A, p. 28.
Roger, O. 1896 A, p. 165. (Anisonchus.)
Trouessart, E. L. 1899 A, p. 1290.
Eocene (Puerco); New Mexico.

Oxyacodon apiculatus Osb. and Earle.
Osborn and Earle 1895 A, p. 25, fig. 6.
Matthew, W. D. 1897 C, p. 292.
   1899 A, p. 28.
Roger, O. 1896 A, p. 252.
Trouessart, E. L. 1898 A, p. 72.
   1899 A, p. 1290.
Eocene (Puerco). New Mexico.

PROTOGONODON Scott.   Type *Mioclænus pentacus* Cope.

Scott, W. B. 1892 B, p. 322.
Earle, C. 1893 B.
　　1893 D, p. 51.
Matthew, W. D. 1897 C. pp. 301, 302.
Osborn, H. F. 1894 A, p. 235.
Osborn and Earle 1895 A, pp. 64, 67.
Scott, W. B. 1892 A, p. 427.
Zittel, K. A. 1893 B, p. 586.

Protogonodon pentacus (Cope).

Cope, E. D. 1888 CC, p. 325.   (Mioclænus.)
Earle, C. 1893 B.

Matthew, W. D. 1897 C, pp. 264, 302.
　　1899 A, p. 28.
Osborn and Earle 1895 A, p. 67, figs. 20, 21.
Roger, O. 1896 A, p. 166.
Scott, W. B. 1892 A, p. 427.
　　1892 B, p. 322.
Eocene (Puerco); New Mexico.

Protogonodon stenognathus Matth.

Matthew, W. D. 1897 C, pp. 264, 302.
　　1899 A, p. 28.
Eocene (Puerco); New Mexico.

TETRACLÆNODON Scott.   Type *Mioclænus floverianus* Cope.

Scott, W. B. 1892 B, p. 299.
Cope, E. D. 1881 H, pp. 487, 492.   (Protogonia, type *P. subquadrata*.)
　　1882 E, p. 178.   (Protogonia.)
　　1882 F, p. 96.   (Protogonia.)
　　1884 O, pp. 386, 424.   (Protogonia.)
　　1884 AA, p. 893.   (Protogonia.)
　　1887 I, p. 657.   (Protogonia.)
　　1888 CC, p. 359.   (Protogonia.)
　　1893, in Earle, C. 1893 B, p. 378.   (Euprotogonia, to replace Protogonia preoccupied.)
　　1894 I, p. 594.   (Euprotogonia.)
Earle, C. 1893 D, p. 50.   (Euprotogonia.)
　　1898 A, p. 262.   (Euprotogonia.)
Hay, O. P. 1899 A, p. 593.
Matthew, W. D. 1897 C. pp. 301, 303.   (Euprotogonia.)
　　1899 A, p. 29.
Nicholson and Lydekker 1889 A, p. 1382.   (Protogonia.)
Osborn, H. F. 1894 A, p. 235.   (Euprotogonia.)
　　1898 A, p. 146.   (Euprotogonia.)
　　1898 G, p. 163.   (Euprotogonia.)
Osborn and Earle 1895 A, pp. 64, 66.   (Euprotogonia.)
Pavlow, M. 1887 A.   (Protogonia.)
Rütimeyer, L. 1888 A, p. 47.   (Protogonia.)
Schlosser, M. 1894 A, p. 158.   (Protogonia.)
Scott, W. B. 1892 A, p. 427 (Protogonia); p. 421 (Tetraclænodon).
　　1892 B, p. 299.   (Tetraclænodon, type *Mioclænus floverianus* Cope.)
　　1892 D, p. 78.   (Tetraclænodon.)
Steinmann and Döderlein 1890 A, p. 751.   (Protogonia.)
Trouessart, E. L. 1898 A, p. 724.   (Euprotogonia.)
Zittel, K. A. 1893 B, p. 218 (Protogonia); p. 588 (Tetraclænodon).

Tetraclænodon minor (Matth).

Matthew, W. D. 1897 C, pp. 264, 310.   (Euprotogonia.)
Cope, E. D. 1888 CC, p. 305.   (Protogonia zuniensis, in part.)
Matthew, W. D. 1899 A, p. 29.   (Euprotogonia.)
Eocene (Torrejon); New Mexico.

Tetraclænodon puercensis (Cope).

Cope, E. D. 1881 H, p. 492.   (Phenacodus.)
　　1882 E, p. 180.   (Syn. Phenacodus vortmani.)
　　1882 EE, p. 333.   (Protogonia plicifera.)
　　1883 M, p. 561.   (Phenacodus calceolatus.)

Cope, E. D. 1884 K, p. 309.   (Protogonia plicifera.)
　　1884 O, pp. 433, 488, pl. xxv, figs. 12, 13; pl. lvii *f*, figs. 8, 9 (Phenacodus puercensis); pp. 433, 487, pl. xxiv *g*, fig. 7 (Phenacodus calceolatus); p. 424, pl. xxv *f*, figs. 2, 3 (Protogonia plicifera).
　　1884 AA, p. 893, fig. 14 (Protogonia plicifera); p. 900, fig. 22 (Phenacodus puercensis).
　　1885 J, p. 387.   (Phenacodus.)
　　1887 I, p. 657.   (Phenacodus puercensis.)
　　1888 CC, pp. 305, 359 (Phenacodus puercensis, P. plicifera, P. calceolata); p. 330 (Mioclænus floverianus).
Earle, C. 1893 D, p. 50.   (Euprotogonia.)
Hay, O. P. 1899 A, p. 593.   (Tetraclænodon.)
Matthew, W. D. 1897 C, pp. 264, 304, 320.   (Euprotogonia.)
　　1899 A, p. 29.
Osborn, H. F. 1898 G, p. 163, fig. 4.   (Euprotogonia.)
Osborn and Earle 1895 A, p. 65, fig. 19 (Euprotogonia puercensis); p. 65 (Euprotogonia plicifera).
Pavlow, M. 1887 A, figs. 1, 3.   (Phenacodus.)
　　1888 A, p. 154.   (Phenacodus.)
Roger, O. 1896 A, p. 82 (T. flowerianus); p. 165 (Euprotogonia calceolata, E. puercensis).
Schlosser, M. 1896 B, p. 135, pl. vi, fig. 33 (Phenacodus puercensis); pl. v, fig. 30; pl. vi, fig. 32 (Protogonia plicifera).
Scott, W. B. 1892 B, p. 300.   (Tetraclænodon floverianus.)
Trouessart, E. L. 1898 A, p. 724.   (Euprotogonia.)
Wortman, J. L. 1896 A, p. 106, pl. ii, fig. A, text fig. 16 (Euprotogonia puercensis); pl. ii, fig. F, text fig. 17 (Euprotogonia plicifera).
Zittel, K. A. 1893 B, p. 218, fig. 158.
Eocene (Torrejon); New Mexico.

Tetraclænodon subquadratus Cope.

Cope, E. D. 1881 H, p. 492.   (Protogonia.)
　　1884 O, p. 426, pl. lvii, figs. 11, 12.   (Protogonia.)
　　1884 AA, p. 893.   (Protogonia.)
　　1888 CC, p. 305.   (Syn. of Protogonia puercensis.)
Matthew, W. D. 1897 C, p. 304.   (Syn. of Euprotogonia puercensis.)
　　1899 A, p. 29.   (Syn. of Euprotogonia puercensis.)
Osborn and Earle 1895 A, p. 64.   (Euprotogonia.)
Pavlow, M. 1887 A, fig. 7.   (Protogonia.)
Scott, W. B. 1892 A, p. 427, fig. 7.   (Protogonia.)
Eocene (Torrejon); New Mexico.

PHENACODUS Cope.   Type *P. primævus* Cope.

Cope, E. D. 1873 D, p. 3.
Baur, G. 1890 B.
Branco, W. 1897 A, p. 18.
Cope, E. D. 1875 C, p. 18.
    1877 K, p. 173.
    1881 H, p. 487.
    1881 X.
    1882 E, p. 177.
    1882 F, p. 96.
    1882 Y.
    1883 B, p. 275.
    1883 G.
    1883 BB.
    1884 O, pp. 386, 428.
    1884 AA, p. 893 (Phenacodus); p. 900 (Tris
        pondylus, type P. vortmani).
    1885 K, p. 458.
    1886 L.
    1886 Q.
    1888 T.
    1889 C.
    1889 J, p. 192.
    1892 S, p. 411.
Earle, C. 1893 D.
Flower and Lydekker 1889 A, p. 439.
Leuthardt, F. 1891 A.
Lockwood, S. 1886 A, p. 325.
Lydekker R. 1886 D, p. 326.
    1889 K.
Marsh, O. C. 1892 C, p. 352.   (Syn. of Helohyus.)
Matthew, W. D. 1897 C, pp. 299, 301, 308.
Nicholson and Lydekker 1889 A, p. 1380.
Osborn, H. F. 1890 D, p. 532.
    1898 D.
Owen, R. 1886 B.
Pavlow, M. 1887 A.
    1888 A, p. 135.
    1900 A, p. 46.
Roger, O. 1889 A.
Rütimeyer, L. 1888 A, p. 38.
Schlosser, M. 1885 A, p. 683.
    1886 A, pp. 252, 432.
    1886 B.
    1894 A, p. 158.
Scott, W. B. 1892 A, p. 405.
Steinmann and Döderlein 1890 A, p. 751.
Tomes, C. S. 1898 A, p. 443.
Woodward, A. S. 1898 B, p. 290.
Woodward, H. 1886 B.
Wortman, J. L. 1882 B, p. 725.
    1883 A, p. 708, fig. 125.
    1886 A, p. 472.
Zittel, K. A. 1893 B, p. 219.

Phenacodus brachypternus Cope.

Cope, E. D. 1882 E, pp. 179, 180.
    1884 O, pp. 433, 490, pl. xxv *e*, fig. 14.
    1884 AA, p. 901, fig. 24.
Matthew, W. D. 1899 A, p. 32.
Roger, O. 1896 A, p. 166.
Trouessart, E. L. 1898 A, p. 725.
    Eocene (Wasatch); Wyoming.

Phenacodus hemiconus Cope.

Cope, E. D. 1882 E, p. 179.
    1884 O, pp. 433, 463, pl. xxv *e*, fig. 16.

Matthew, W. D. 1899 A, p. 32.
Roger, O. 1896 A, p. 166.
Trouessart, E. L. 1898 A, p. 725.
    Eocene (Wasatch); Wyoming.

Phenacodus macropternus Cope.

Cope, E. D. 1882 E, pp. 179, 180.
    1884 O, pp. 433, 490, pl. xxv *e*, fig. 15.
Matthew, W. D. 1899 A, p. 32.
Roger, O. 1896 A, p. 166.
Trouessart, E. L. 1898 A, p. 725.
    Eocene (Wasatch); Wyoming.

Phenacodus nuniěnus Cope.

Cope, E. D. 1884 O, pp. 433, 434, pl. lvii *h*, figs. 1, 2.
    1884 AA, p. 898.
Matthew, W. D. 1897 C, p. 398, footnote.
    1899 A, p. 32.
Roger, O. 1896 A, p. 166.
Trouessart, E. L. 1898 A, p. 725.   (P. nunenius.)
    Eocene (Wasatch); Wyoming.

Phenacodus omnivorus Cope.

Cope, E. D. 1874 O, p. 11.
    1875 C, p. 18.
    1877 K, p. 178, pl. xlv, fig. 6.
    1882 E, p. 179.
    1884 O, p. 463.
Matthew, W. D. 1899 A, p. 32.   (Syn. of P.
    primævus.)
    Eocene (Wasatch); New Mexico.

Phenacodus primævus Cope.

Cope, E. D. 1873 D, p. 3.
    1874 B, p. 458.
    1874 O, p. 10.
    1875 C, p. 18.
    1877 K, p. 174, pl. xlv, figs. 1–5.
    1881 D, p. 200.   (P. primævus; P. trilobatus.)
    1882 E, p. 178.
    1883 G, p. 563, pls. i, ii.
    1883 BB, pl. xii.
    1884 O, pp. 433, 435, pls. lvii *b*; lvii *c*; lvii *d*;
        lvii *e*; lvii *h*, figs. 1–16 (P. primævus); p.
        463, pl. lviii, fig. 11 (P. trilobatus).
    1884 AA, p. 893, figs. 15–19, pls. xxviii, xxix.
    1887 I, p. 657.
    1887 N, p. 987, fig. 1.
    1887 S, pp. 270, figs. 47, 48; p. 273, fig. 51; p.
        275, pl. vii, figs. 3, 4.
    1889 U.
King, C. 1878 A, p. 376.
Matthew, W. D. 1899 A, pp. 32, 36.
Osborn, H. F. 1897 B.
    1898 G, pl, xii, text figs. 1–4.
    1898 Q, p. 11.
Osborn and Wortman 1892 A, pp. 83, 86.
Pavlow, M. 1887 A.
Roger, O. 1896 A, p. 166.   (P. primævus, P. trilo-
    batus.)
Schlosser, M. 1886 B, p. 12.   (P. trilobatus.)
Scott, W. B. 1892 A, p. 427, fig. 12.
Trouessart, E. L. 1898 A, p. 725.
Wortman, J. L. 1883 A, p. 708, fig. 127.
Zittel, K. A. 1893 B, p. 219, figs. 151, 153, 159, 160.
    Eocene (Wasatch, Wind River); Wyoming,
    New Mexico.

Phenacodus sulcatus Cope.

Cope, E. D. 1874 O, P. 11.
   1875 C, p. 18.
   1877 K, p. 179, pl. xlv, fig. 7.
   1882 E, p. 179.
Matthew, W. D. 1899 A, p. 32.
Roger, O. 1896 A, p. 166.
   Eocene; New Mexico.

Phenacodus vortmani Cope.

Cope, E. D. 1880 N, p. 747. (Hyracotherium.)
   1881 D, p. 99.
   1882 E, p, 179, 180. (P. vortmani, P. apter-
   nus.)
   1884 O, pp. 433, 464, pls. lvii *g*; lvii *h*, fig. 17;
   lviii, figs. 8–10.

Cope, E. D. 1884 AA, p. 899, text figs. 20, 21, pl. xxx;
   p. 900, footnote. ("Trispondylus.")
   1887 I, p. 657. (P. wortmani.)
   1889 U.
Matthew, W. D. 1899 A, pp. 32, 36.
Nicholson and Lydekker 1889 A, p. 1382.
Osborn and Wortman 1892 A, p. 86.
Pavlow, M. 1887 A.
Roger, O. 1896 A, p. 166. (P. vortmani, P. apter-
   nus.)
Schlosser, M. 1886 B, p. 12.
Scott, W. B. 1892 A, p. 427, fig. 7.
Trouessart, E. L. 1898 A, p. 725. (P. wortmanni.)
Zittel, K. A. 1893 B, p. 219. (P. wortmani.)
   Eocene (Wasatch, Wind River); Wyoming.

ECTOCION Cope.   Type *Oligotomus osbornianus* Cope.

Cope, E. D. 1882 X.
   1884 O, pp. 694, 695.
   1887 N, pp. 997, 1061.
Earle, C. 1892 C, p. 276. (Ectocium.)
Matthew, W. D. 1897 C, p. 301.
Wortman, J. L. 1888 A, p. 710. (Ectocium.)
   1896 A, pp. 81, 83. (Ectocium.)
Zittel, K. A. 1893 B, p. 242. (Syn. of Eohippus.)

Ectocion osbornianum Cope.

Cope, E. D. 1882 E, p. 182. (Oligotomus.)

Cope, E. D. 1882 X.
   1884 O, p. 696, pl. xxv *e*, figs. 9, 10.
   1887 N, p. 1061, fig. 25.
Matthew, W. D. 1899 A, pp. 32, 36. (Phenacodus.)
Roger, O. 1896 A, p. 169. (Eohippus.)
Trouessart, E. L. 1898 A, p. 724.
Zittel, K. A. 1893 B, p. 242, fig. 177. (Eohippus.)
   Eocene (Wasatch, Wind River); Wyoming.

MENISCOTHERIIDÆ.

Cope, E. D. 1882 R, p. 334.
Ameghino, F. 1897 A, p. 84.
Cope, E. D. 1882 F, p. 95.
   1883 E.
   1884 O, p. 384.
   1884 AA, pp. 793, 901.
Matthew, W. D. 1897 C, p. 294.
Nicholson and Lydekker 1889 A, p. 1382.

Osborn, H. F. 1891 D, p. 911.
   1892 C.
   1893 F, p. 127.
Osborn and Earle 1895 A, pp. 47, 49.
Roger, O. 1896 A, p. 167.
Pavlow, M. 1887 A.
Steinmann and Döderlein 1890 A, p. 751.
Zittel, K. A. 1893 B, p. 220.

MENISCOTHERIUM Cope.   Type *M. chamense* Cope.

Cope, E. D. 1874 O, p. 8.
   1877 K, p. 251.
   1881 G, p. 396.
   1882 F, p. 97.
   1882 R, p. 334.
   1884 O, pp. 386, 493.
   1884 AA, p. 901.
   1886 K, p. 21.
   1887 I, p. 657.
   1887 N, p. 1018.
   1892 S, p. 412.
Lydekker, R. 1889 B.
Marsh, O. C. 1892 D, p. 445.
Matthew, W. D. 1897 C, p. 309.
Nicholson and Lydekker 1889 A, pp. 1376, 1382.
Osborn, H. F. 1890 D, p. 582.
   1891 D, p. 911.
   1892 C.
   1892 G, figs. 1–4.
   1893 D, p. 43.
   1893 F, p. 127.
Osborn and Wortman 1893 A, p. 16.
Pavlow, M. 1887 A.

Schlosser, M. 1885 A, p. 683.
   1886 B, p. 21.
Scott, W. B. 1892 A, p. 429.
Wortman, J. L. 1886 A, p. 475.
   1896 B, p. 262.
Zittel, K. A. 1893 B, p. 221.

Meniscotherium chamense Cope.

Cope, E. D. 1874 O, p. 8.
   1877 K, pp. 251, 252, pl. lxvi, fig. 18.
   1884 AA, p. 904.
King, C. 1878 A, p. 376.
Marsh, O. C. 1892 D, p. 445.
Matthew, W. D. 1899 A, p. 32.
Roger, O. 1896 A, p. 167.
Trouessart, E. L. 1898 A, p. 726.
   Eocene (Wasatch); New Mexico.

Meniscotherium tapiacitis Cope.

Cope, E. D. 1882 C, p. 470.
   1884 O, p. 506, pl. xxv *f*, fig. 15.
   1884 AA, p. 905, fig. 26 *b*.
Matthew, W. D. 1899 A, p. 32.

Roger, O. 1896 A, p. 167.
Eocene (Wasatch); New Mexico.

Meniscotherium terrærubræ Cope.

Cope, E. D. 1881 H, p. 493.
1882 C, p. 470.
1884 O. p. 496, pls. xxv *f*, figs. 12–14; pl. xxv *g*.
1884 AA, p. 904, figs. 25–28.

Matthew, W. D. 1899 A, p. 32.
Pavlow, M. 1887 A, figs. 12 *a*, 12 *b*.
Roger, O. 1896 A, p. 167.
Trouessart, E. L. 1898 A, p. 726.
Zittel, K. A. 1893 B, p. 221, fig. 163.
Eocene (Wasatch); New Mexico.

HYRACOPS Marsh.   Type *H. socialis* Marsh.

Marsh, O. C. 1892 D, p. 446.
Cope, E. D. 1892 S, p. 412.   (Syn.? of Meniscotherium.)
Matthew, W. D. 1899 A, p. 32, footnote.
Osborn, H. F. 1892 G, p. 506, figs. 1, 2.   (Syn. of Meniscotherium.)
Zittel, K. A. 1893 B, p. 221.

Hyracops socialis Marsh.

Marsh, O. C. 1892 D, p. 447, figures.

Matthew, W. D. 1899 A, p. 32.
Osborn, H. F. 1892 G, p. 506, figs. 1–4.   (Meniscotherium.)
1893 F, p. 128, figs. 1, 2.   (Meniscotherium.)
Roger, O. 1896 A, p. 167.
Trouessart, E. L. 1898 A, p. 726.
Woodward, A. S. 1898 B, p. 289.
Zittel, K. A. 1893 B, p. 221, fig. 162.
Eocene (Wasatch); New Mexico.

MIOCLÆNIDÆ.

Osborn and Earle 1895 A, pp. 48, 61.
Matthew, W. D. 1897 C, p. 311.

Roger, O, 1896 A, p. 163.
Scott, W. B. 1892 B, p. 322.   (Not named.)

MIOCLÆNUS Cope.   Type *M. turgidus* Cope.

Cope, E. D. 1881 H, pp. 487, 490, 491.
1881 U, p. 830.
1882 E, p. 187.
1882 K, p. 71.
1884 K, p. 312.
1884 O, pp. 259, 324.
1884 S, p. 349.
1888 D.
1888 CC, p. 319.
Earle, C. 1893 B, p. 377.
Major, Fors. 1893 A, p. 199.
Matthew, W. D. 1897 C, p. 312.
Osborn, H. F. 1888 E.
1888 F, p. 1067.
Osborn and Earle 1896 A, p. 48.
Osborn and Wortman 1892 A, p. 118.
Pavlow, M. 1887 A.
Schlosser, M. 1887 B, p. 320.
Scott, W. B. 1892 A, p. 413.
1892 B, p. 321.
1892 D, p. 90.
Steinmann and Döderlein 1890 A, p. 712.
Trouessart, E. L. 1898 A, p. 719.
Zittel, K. A. 1893 B, pp. 565, 588, 589.

Mioclænus acolytus Cope.

Cope, E. D. 1882 C, p. 462.   (Hyopsodus.)
1882 C, p. 468.   (M. minimus.)
1884 K, pp. 309, 312.   (M. minimus.)
1884 O, pp. 235, 238, pl. xxiii *d*, figs. 5, 6 (Hyopsodus); pp. 325, 327, pl. xxv *e*, figs. 22–24 (M. minimus.)
1884 S, p. 349.   (M. minimus.)
1888 CC, p. 335.
Matthew, W. D. 1897 C, p. 317, fig. 18.
1899 A, p. 29.
Roger, O. 1896 A, p. 164.   (M. acolytus, M. minimus.)
Schlosser, M. 1887 B, p. 221.   (M. minimus.)
Scott, W. B. 1892 B, p. 323.

Trouessart, E. L. 1898 A, p. 719.
Eocene (Torrejon); New Mexico.
For Matthew's disposition of *M. minimus* see *Ellipsodon inæquidens.*

Mioclænus lemuroides Matth.

Matthew, W. D. 1892 C, p. 315, figs. 15, 16.
Cope, E. D. 1888 CC, p. 334.   (M. turgidunculus, in part.)
Matthew, W. D. 1899 A, p. 29.
Trouessart, E. L. 1898 A, p. 719.
Eocene (Torrejon); New Mexico.

Mioclænus lydekkerianus Cope.

Cope, E. D. 1888 CC, p. 328.   (Type specimen only, *fide* Matthew.)
Matthew, W. D. 1897 C, pp. 264, 312, 313.
1899 A, p. 29.
Roger, O. 1896 A, p. 166.   (Protogonodon.)
Scott, W. B. 1892 B, p. 322.   (Protogonodon.)
Trouessart, E. L. 1898 A, p. 719.
Eocene (Torrejon); New Mexico.

Mioclænus turgidunculus Cope.

Cope, E. D. 1888 CC, p. 334.   (In part, *fide* Matthew.)
1888 CC, p. 334.
Matthew, W. D. 1897 C, pp. 264, 313.
1899 A, p. 28.
Roger, O. 1896 A, p. 164.   (Protogonodon.)
Scott, W. B. 1892 B, p. 322.
Trouessart, E. L. 1898 A, p. 719.
Eocene (Puerco); New Mexico.

Mioclænus turgidus Cope.

Cope, E. D. 1881 U, p. 830.
1881 H, p. 489.
1884 K, p. 312.
1884 O, p. 325, pl. xxv *e*, figs. 19, 20; pl. lvii *f*, figs. 3, 4.

Cope, *E. D.* 1884 S, p. 349, fig. 15.
　1888 CC, p. 334.　(M. zittelianus.)
Earle, C. 1893 B, p. 379.
Matthew, W. D. 1897 C, pp. 264, 312.
　1899 A, p. 29.
Osborn and Earle 1895 A, p. 50, fig. 17.
Pavlow, M. 1887 A, fig. 8.

Roger, O. 1896 A, p. 163 (M. turgidus); p. 164 (M. zittelianus).
Schlosser, M. 1887 B, p. 221.
Scott, W. B. 1892 B, p. 321.　(M. turgidus, M. zittelianus.)
Trouessart, E. L. 1898 A, p. 719.
　Eocene (Torrejon); New Mexico.

PROTOSELENE Matth. Type *Mioclænus opisthacus* Cope.

Matthew, *W. D.* 1897 C, p. 317.

Protoselene opisthaca (Cope).

Cope, *E. D.* 1882 EE, p. 833.　(Mioclænus.)
　1882 C, p. 467.　(Hemithlæus.)
　1882 EE, p. 853.　(Mioclænus baldwini.)
　1884 K, p. 312.　(Mioclænus opisthacus, M. baldwini.)
　1884 O, p. 407, pl. xxv *f*, figs. 8, 9 (Hemithlæus); pp. 325, 328, pl. xxv *f*, fig. 16 (Mioclænus baldwini).

Cope, E. D. 1888 CC, p. 332.　(Mioclænus.)
Matthew, W. D. 1897 C, pp. 264, 317, figs. 19, 20.
　1899 A, p. 29.
Roger, O. 1896 A, p. 164.　(Mioclænus.)
Schlosser, M, 1887 B, p. 221.　(Mioclænus baldwini.)
Trouessart, E. L. 1898 A, p. 720.
　Eocene (Torrejon); New Mexico.

Suborder PERISSODACTYLA Owen.

Owen, *R.* 1848 A, p. 131.
Branco, W. 1897 A, p. 18.
Burmeister, H. 1867 A, p. 237.　(Imparidigitata.)
Cope, E. D. 1873 E, p. 562.
　1873 F, p. 39.
　1874 M.
　1880 B, p. 457.
　1881 G.
　1881 I, p. 269.
　1881 J.
　1882 A.
　1882 E, p. 177.
　1882 Y.
　1883 D.　(Diplarthra, part.)
　1884 O, pp. 379, 614 (Perissodactyla); pp. 167, 378, 613 (Diplarthra, part).
　1886 C, p. 615.　(Diplarthra, part.)
　1887 N.
　1887 S.
　1888 K, p. 163.　(Diplarthra, part.)
　1889 C, p. 148, *et seq.*
　1889 R, p. 877.
　1891 N, p. 84.　(Diplarthra, part.)
　1892 S.
　1898 B, p. 125.　(Diplarthra, part.)
Cope and Wortman 1884 A, p. 15.
Flower, W. H. 1876 B, xiii, p. 327.
　1883 A.
　1883 D, p. 427.
　1885 A.
Flower and Lydekker 1891 A, p. 368.
Gadow, H. 1898 A, p. 46.
Gegenbaur, C. 1896 A.
Gill, T. 1872 A, p. 299.
　1872 B, pp. 11, 71, 84.
Gruber, W. 1859 A.
Haeckel, E. 1895 A, p. 543.
Hay, O. P. 1899 F, p. 682.
Huxley, T. H. 1870 A, p. 440.
　1872 A, p. 293.
Leidy, J. 1871 C, p. 356.
Leuthardt, F. 1891 A.　(Mesaxonia.)
Lydekker, R. 1885 E, p. 469.　(Diplarthra, part.)
　1886 A, p. 1.

Lydekker, R. 1891 B.
　1896 B.
Marsh, O. C. 1884 F, p. 177.　(Clinodactyla, Mesaxonia.)
　1892 C, p. 349.　(Mesaxonia.)
Nicholson and Lydekker 1889 A, p. 1382.
Osborn, H. F. 1890 D, pp. 505, 541, 559.
　1893 A, p. 379.
　1893 D.
　1898 I, p. 79.
　1900 C, p. 568.
Osborn and Wortman 1892 A.
Owen, R. 1845 B, p. 572.　("Isodactyle Ungulata.")
　1857 A.
　1866 B, p. 444.
　1868 A, p. 352.
Paulli, S. 1900 A, p. 181.
Pavlow, M. 1887 A.
Roger, O, 1896 A, p. 168.
Röse, C. 1896 A.
Rütimeyer, L. 1865 A.　(Imparidigitata.)
　1888 A.　(Diplarthra, part.)
Schlosser, M. 1885 A.
　1886 A, p. 252.
　1886 B.
　1887 A.
　1888 A.
　1889 A.
　1890 C.
　1890 D.
　1892 A.
Schmidt, O. 1886 A, p. 189.
Scott, W. B. 1892 A, p. 430.
Steinmann and Döderlein 1890 A, p. 765.
Tornier, G. 1888 A, p. 299.
Turner, H. N. 1848 A.
Weber, M. 1886 A, p. 226.
Wiedersheim, R. 1881 A, p. 360.　(Perissodactyli.)
Woodward, A. S. 1898 B, p. 319.
Wortman, J. L. 1883 A, p. 706.
　1885 A, p. 297.　(Diplarthra, part.)
　1898 B, p. 208.　(Diplarthra, part.)
Zittel, K. A. 1893 B, pp. 224, 726.

Superfamily *EQUOIDEA*.

Gill, T. 1872 B, p. 87.   (Solidungula.) | Osborn, H. F. 1898 I, p. 79.   (Hippoidea.)

EQUIDÆ.

Ameghino, F. 1889 A, p. 502.
  1891 A.
Boule, M. 1899 A, p. 531.   ("Équidés.")
Branco, W. 1897 A, p. 18.
Cope, E. D. 1874 M.
  1881 G, pp. 378, 380, 400.
  1881 O.   (Equidæ, Anchitheriidæ.)
  1884 O, pp. 615, 715, 716.
  1887 N.
  1889 C.
  1892 S.
  1893 A.
Cope and Wortman 1884 A, p. 26.
Cornevin, C. 1882 A.
Cuvier, G. 1834 A, iii, pp. 193–220, with plates.
Ewart, J. C. 1894 A.
  1894 B.
Farr, M. S. 1896 A.
Flower, W. H. 1876 B, xiii, p. 308.
  1876 C, p. 109.
Flower and Lydekker 1891 A, p. 376.
Gadow, H. 1898 A, p. 47.
George, ——. 1869 A.
Gervais, P. 1874 A.
Gray, J. E. 1869 A, p. 262.
Haeckel, E. 1895 A, p. 547.   (Hippotherida.)
Huxley, T. H. 1870 A, p. 439.
  1870 B, p. xlix.
  1872 A, p. 295.
  1876 B, xiv, p. 34.
  1880 B, p. 649.
Klever, E. 1889 A.
Kowalevsky, W. 1873 B.
Leidy, J. 1869 A, p. 302.   (>Anchitheriidæ.)
  1871 C, p. 359.   (Equidae, Anchitheriidæ.)
Lydekker, R. 1882 A.
  1886 A, p. 49.
  1886 D, p. 326.
  1891 B.
Major, F. 1873 A, p. 109.

Major, F. 1877 A.   ·
  1880 A.
Marsh, O. C. 1877 E.
  1892 C, p. 339.
Matthew, W. D. 1899 A, p. 22, *et seq.*
Nicholson and Lydekker 1889 A, p. 1359.
Osborn, H. F. 1893 D, p. 37.
  1898 I, p. 79.
Osborn and Wortman 1892 A, p. 123.
Owen, R. 1845 B, p. 572.
  1870 B.
  1878 C, p. 218.
Pavlow, M. 1888 A, p. 134.
  1890 A.
Ridewood, W. G. 1895 A.
Roger, O. 1889 A.
  1896 A, p. 168.
Rütimeyer, L. 1862 A.
  1875 A.
  1888 A, p. 16.
Ryder, J. 1877 A.
Schlosser, M. 1886 A.   (Hippidæ.)
  1886 B.
Schmidt, O. 1886 A, p. 201.
Scott, W. B. 1886 B.
  1891 B.
  1895 C, p. 78.
Slade, D. D. 1893 A.
Steinmann and Döderlein 1890 A, p. 778.
Tornier, G. 1888 A, p. 301.
Trouessart, E.-L. 1898 A, p. 783.
Tscherski, J. D. 1892 A, p. 257.
Weyhe, ——. 1875 A, p. 121.
Wilckens, M. 1884 A.
  1884 B.
  1885 A.
  1888 A.
Woodward, A. S. 1898 B, p. 337.
Wortman, J. L. 1882 B.
Zittel, K. A. 1893 B, pp. 231, 259.

HYRACOTHERIINÆ.

Branco, W. 1897 A, p. 18.
Cope, E. D. 1881 G, p. 381.
  1884 O, p. 618.
  1886 K.   (Hyracotheriidæ.)
  1887 N, p. 994.
  1889 C, p. 162 *et seq.*
Haeckel, E. 1895 A, p. 548.   (Hyracotherida.)
Marsh, O. C. 1892 C, p. 352.   (>Orohippidæ.)
Osborn, H. F. 1893 D, p. 37.   (Systemodontinæ, in part.)

Osborn and Wortman 1892 A, p. 98.
Roger, O. 1896 A, p. 168.
Rütimeyer, L. 1888 A, p. 27.
Steinmann and Döderlein 1890 A, p. 769.   (Hyracotheriidæ.)
Wortman, J. L. 1896 A.
Wortman and Earle 1893 A.   (Systemodontinæ, in part.)
Zittel, K. A. 1893 B, p. 239.

Eohippus Marsh.   Type *E. validus* Marsh.

Marsh, O. C. 1876 H, p. 401.
  Since many writers regard Marsh's genus *Eohippus* as a synonym, or at most a subgenus of *Hyracotherium* Owen, and refer to the American species under the latter name, the literature of both genera is included in the following citations. Unless otherwise indicated the authors in this list employ the term *Hyracotherium.*

Ameghino, F, 1896 A, p. 410.
Cope, E. D. 1877 K, p. 258.
  1881 G, p. 381.
  1881 Y.   (Systemodon, type *Hyracotherium tapirinum.*)
  1883 B, p. 275.
  1884 O, pp. 618, 624.   (Hyracotherium, Systemodon.)
  1886 K.   (Hyracotherium, Systemodon.)

Cope, E. D. 1887 I.
  1887 N, p. 994 (Hyracotherium, Systemodon.
  1889 C.
  1892 S, p. 411.
Earle, C. 1893 C p. 393. (Hyracotherium, Systemodon.)
  1896 A.
Flower and Lydekker 1891 A, pp. 373, 374. (Hyracotherium, Systemodon.)
Kowalevsky, W. 1873 A, p. 205.
Landois, — 1881 A, p. 127.
Leuthardt, F. 1891 A. (Hyracotherium.)
Lydekker, R. 1886 A, p. 10.
  1886 D, p. 326.
Marsh, O. C. 1877 E, p. 358.
  1879 D, p. 504.
  1892 C, p. 351. (Eohippus, Hyracotherium.)
  1898 E, p. 568.
Matthew, W. D. 1897 C, p. 309.
  1899 A, p. 33.
Nicholson and Lydekker 1889 A, pp. 1356, 1380. (Hyracotherium, Systemodon.)
Owen, R. 1844 A, p. 226.
  1857 C, p. 65.
Osborn, H. F. 1890 A.
  1890 B, p. 274.
  1890 D, p. 505.
  1893 D.
Osborn and Wortman 1892 A, pp. 83, 98 (Hyracotherium); p. 124 (Systemodon).
Pavlow, M. 1887 A, p. 350.
  1888 A. (Eohippus, Systemodon,)
  1900 A, p. 46.
Rütimeyer, L. 1862 A, p. 577.
  1891 A, p. 45.
Schlosser, M. 1886 A, pp. 252, 432. (Hyracotherium, Systemodon.)
  1886 B. (Hyracotherium, Eohippus, Systemodon.)
  1887 A.
Scott, W. B. 1886 B.
  1892 A, p. 430. (Systemodon.)
Steinmann and Döderlein 1890 A, p. 770.
Tomes, C. S. 1898 A, p. 412. (Hyracotherium.)
Trouessart, E. L. 1898 A, p. 770 (Hyracotherium); p. 771 (Eohippus, subgenus); p. 772 (Pliolophus, subgenus).
Wilckens, M. 1885 B, p. 210.
Woodward, A. S. 1898 B, p. 323.
Wortman, J. L. 1882 B, vi, p. 68.
  1883 A, p. 709.
  1886 A, p. 477. (Hyracotherium, Eohippus.)
  1893 B.
  1896 A. (Hyracotherium, Systemodon.)
Wortman and Earle, 1893 A. (Hyracotherium, Systemodon.)
Zittel, K. A. 1893 B, p. 239 (Hyracotherium); p. 277 (Systemodon).

Eohippus craspedotus (Cope).

Cope, E. D. 1880 N, p. 747. (Hyracotherium.)
  1881 D, p. 199. (Hyracotherium.)
  1882 E, p. 186. (Hyracotherium.)
  1884 O, p. 631, pl. lviii, figs. 1, 2. (Hyracotherium.)
  1887 I, p. 656. (Hyracotherium.)

Matthew, W. D. 1899 A, p. 36. (Hyracotherium.)
Roger, O. 1896 A, p. 169. (Hyracotherium.)
Trouessart, E. L. 1898 A, p. 771. [Hyracotherium (Eohippus).]
Wortman, J. L. 1896 A, p. 97, pl. ii, fig. M, text fig. 7. (Hyracotherium.)
  Eocene (Wind River); Wyoming.

Eohippus cristatus (Wort.).

Wortman, J. L. 1896 A, p. 96, pl. ii, fig. H; text fig. 6. (Hyracotherium.)
Matthew, W. D. 1899 A, p. 33. (Hyracotherium.)
Trouessart, E. L. 1898 A, p. 771. [Hyracotherium (Eohippus).]
  Eocene (Wasatch); Wyoming.

Eohippus index (Cope).

Cope, E. D. 1873 D, p. 4. (Orotherium.)
Allen, H. 1886 B. (Hyracotherium cuspidatum, Pliolophus vintanus.)
Cope, E. D. 1873 U, p. 3. (Orotherium vasacciense.)
  1874 B, p. 459. (Orotherium.)
  1875 C, p. 20 (Orohippus index); p. 18 (Orotherium vintanum); pp. 20, 21 (Orohippus angustidens); p. 22 (O. cuspidatus.)
  1877 K, p. 262 (Orohippus index); p. 255, pl. lxv, figs. 1–12 (Orotherium vintanum); pp. 262, 265, pl. lxvi, figs. 1–6 (Hyracotherium angustidens); pp. 262, 267, pl. lxv, fig. 18 (Hyracotherium cuspidatum); p. 264, pl. lxvi, figs. 7–11 (Hyracotherium vasacciense.)
  1880 N, p. 747. (Hyracotherium angustidens, H. vasacciense.)
  1881 D, p. 198. (Hyracotherium angustidens.)
  1882 E, p. 186 (Hyracotherium angustidens, H. vasacciense, H. venticolum); pp. 179, 181, pl. xxv e, fig. 17, premolars (Phenacodus laticuneus.)
  1884 O, pp. 630, 650 (Hyracotherium index); pp. 630, 648, pl. xlix a, fig. 16 (Hyracotherium angustidens); p. 631 (Hyracotherium cuspidatum); pp. 630, 634 (Hyracotherium vasacciense, in part.)
King, C. 1878 A, p. 377. (Eohippus angustidens, E. cuspidatus.)
Lydekker, R. 1886 A, p. 12, fig. 3. (Hyracotherium angustidens.)
Matthew, W. D. 1899 A, p. 30, footnote; p. 33. (Hyracotherium.)
Pavlow, M. 1888 A, pl. i, fig. 10. (Hyracotherium angustidens.)
Roger, O. 1896 A, p. 169 (Hyracotherium angustidens); p. 171 (Epihippus cuspidatus).
Trouessart, E. L. 1898 A, p. 771 [Hyracotherium (Eohippus)]; p. 776 (Epihippus).
Wortman, J. L. 1896 A, p. 99, pl. ii, fig. L; text figs. 9, 10. (Hyracotherium.)
  Eocene (Wasatch); New Mexico, Wyoming.

Eohippus pernix Marsh.

Marsh, O. C. 1876 H, p. 402.
King, C. 1878 A, p. 377.
Marsh, O. C. 1892 C, p. 349, figs. 14, 15.
Matthew, W. D. 1899 A, p. 33. (Syn. of Hyracotherium index.)

Pavlow, M. 1888 A, p. 145.
Roger, O. 1896 A, p. 169.
Trouessart, E. L. 1898 A, p. 772.   [Hyracotherium (Eohippus).]
Wortman, J. L. 1896 A, p. 82.   (Syn.? of Hyracotherium index.)
Eocene (Wasatch); Wyoming.

Eohippus tapirinus (Cope).

Cope, E. D. 1875 C, p. 20.   (Orohippus.)
    1877 K, p. 263, pl. lxvi, figs. 12–16.   (Hyracotherium.)
    1881 Y.   (Systemodon.)
    1882 E, p. 183.   (Systemodon.)
    1884 O, p. 619, pl. lvi, figs. 1, 2.   (Systemodon.)
    1887 N, p. 999, fig. 8.   (Systemodon.)
Hay, O. P. 1899 A, p. 593.   (Hyracotherium.)
King, C. 1878 A, p. 377.
Matthew, W. D. 1899 A, p. 33.   (Hyracotherium.)
Osborn and Wortman 1892 A, p. 125, fig. 17, only part, if any.   (Systemodon.)
Pavlow, M. 1888 A, p. 174.   (Systemodon.)
Roger, O. 1896 A, p. 183.   (Systemodon.)
Scott, W. B. 1892 A, p. 430.   (Systemodon.)
Trouessart, E. L. 1898 A, p. 771.   [Hyracotherium (Eohippus).]
Woodward, A. S. 1898 B, p. 323, fig. 183.
Wortman, J. L. 1896 A, p. 94, pl. ii, fig. K, text figs. 4, 5.   (Hyracotherium.)

Wortman and Earle 1893 A, p. 170.   (Systemodon.)
Eocene (Wasatch); Wyoming, New Mexico.

Eohippus validus Marsh.

Marsh, O. C. 1876 H, p. 401.
King, C. 1878 A, p. 377.
Matthew, W. D. 1899 A, p. 33.   (Syn. of E. vasacciensis.)
Roger, O. 1896 A, p. 169.
Trouessart, E. L. 1898 A, p. 772.   [Hyracotherium (Eohippus).]
Wortman, J. L. 1896 A, p. 82.   (Syn.? of Hyracotherium vasacciense.)
Eocene (Wasatch); New Mexico.

Eohippus vasacciensis (Cope).

Cope, E. D. 1872 O, p. 474.   (Notharctus.)
    1872 E, p. 471.   (Notharctus vasachiensis.)
    1873 E, pp. 605, 646.   (Orotherium.)
    1873 U, p. 3.   (Orotherium.)
    1875 C, p. 21.   (Orohippus.)
    1884 O, p. 634.   (Hyracotherium.)
King, C. 1878 A, p. 376.   (Orohippus.)
Matthew, W. D. 1899 A, p. 33.   (Hyracotherium.)
Roger, O. 1896 A, p. 169.
Trouessart, E. L. 1898 A, p. 771.   [Hyracotherium(Eohippus).]
Wortman, J. L. 1896 A, p. 98, pl. ii, fig. G; text fig. 8.   (Hyracotherium.)
Eocene (Wasatch); Wyoming.

PLIOLOPHUS Owen.   Type *P. vulpiceps* Owen.

Owen, R. 1857 C, p. 54, pls. ii–iv.
Cope, E. D. 1881 G, p. 381.
    1884 O, pp. 618, 650.
    1887 N, pp. 994, 997.
    1892 S, p. 411.
Earle, C. 1896 A.
Owen, R. 1860 E, p. 326.
Wortman, J. L. 1896 A, pp. 83, 102.
Zittel, K. A. 1893 B, pp. 239, 243.   (Syn. of Hyracotherium.)
    It is doubtful whether the American species belong to this genus or to *Eohippus;* indeed, the genus itself is of doubtful validity.

Pliolophus cristonensis Cope.

Cope, E. D. 1877 K, p. 254, pl lxv, figs. 13, 14.   (Orotherium.)
    1875 C, pp. 20, 21.   (Orohippus major, not of Marsh.)
    1877 K, p. 257, pl. lxv, figs. 15–17.   (Orotherium lœvii.)

Cope, E. D. 1884 O, p. 651.   (P. cristonensis, P. lœvii.)
Matthew, W. D. 1899 A, p. 33.   (Hyracotherium.)
Roger, O. 1896 A, p. 170.   (Pachynolophus.)
Trouessart, E. L. 1898 A, p. 772.   [Hyracotherium (Pliolophus).]
Wortman, J. L. 1896 A, p. 102, pl. ii, figs. i, i', text figs. 11, 12.   [Hyracotherium (Pliolophus).]
Zittel, K. A. 1898 B, p. 243, fig. 180.   (Pachynolophus.)
Eocene (Wasatch); New Mexico, Wyoming.

Pliolophus montanus Wort.

Wortman, J. L. 1896 A, p. 103, pl. ii, fig. J; text fig. 13.   [Hyracotherium (Pliolophus).]
Matthew, W. D. 1899 A, p. 33.
Trouessart, E. L. 1898 A, p. 772.   [Hyracotherium (Pliolophus).]
Eocene (Wasatch); Wyoming.

PROTOROHIPPUS Wortman.   Type *Hyracotherium venticolum* Cope.

Wortman, J. L. 1896 A, pp. 91, 104.
Osborn, H. F. 1897 G. p. 258.

Protorohippus venticolus (Cope).

Cope, E. D. 1881 D, p. 198.   (Hyracotherium.)
    Unless otherwise indicated, the following authors refer this species to *Hyracotherium.*
Cope, E. D. 1884 O, pp. 630, 635, pl. xlix a, figs. 1–15; pls. xlix b, xlix c.

Cope, E. D. 1887 N, p. 995, pl. xxx, text figs. 5–7.
    1887 S, p. 270, pl. vi, figs. 1, 2.
    1888 T.
    1889 C, p. 159.
Matthew, W. D. 1899 A, p. 36.   (Protorohippus.)
Osborn H. F. 1896 C, p. 706, illus. 4.
    1898 Q, pp. 10, 15.
Pavlow, M. 1887 A.
    1888 A, p. 114, pl. i, figs. 1, 9.

Roger, O. 1896 A, p. 168.
Rütimeyer, L. 1888 A, p. 15.
Schlosser, M. 1886 B, pl. v, fig. 32; pl. vi, fig. 34. (Hyracotherium.)
Woodward, A. S. 1898 B, p. 324, fig. 184. (Hyracotherium.)

Wortman, J. L. 1883 A, p. 709, figs. 128–130.
1896 A, p. 105, pl. ii, fig. N; text figs. 14, 15. (Protorohippus.)
Zittel, K. A. 1893 B, p. 240, figs. 174, 175.
Eocene (Wind River); Wyoming, Colorado.

HELOHIPPUS Marsh.	Type *Lophiodon pumilus* Marsh.

Marsh, O. C. 1892 C, p. 353.

Helohippus pumilus Marsh.

Marsh, O. C. 1871 D, p. 38. (Lophiodon.)
Leidy, J. 1872 B, p. 362. (Lophiodon.)

Leidy, J. 1872 C, p. 20. (Lophiodon.)
Marsh, O. C. 1892 C, p. 353.
Matthew, W. D. 1899 A, p. 45.
Roger, O. 1896 A, p. 169.
Eocene (Bridger); Wyoming.

OROHIPPUS Marsh.	Type *O. pumilis* Marsh.

Marsh, O. C. 1872 I, p. 207.
Cope, E. D. 1872 N, p. 466. (Helotherium, type *H. procyoninus.*)
1873 E, p. 606 (Orohippus); p. 607 (Oligotomus).
1873 U, p. 2. (Oligotomus, type *O. cinctus.*)
1874 M. (Syn. of Hipposyus.)
1875 C, pp. 13, 18, 20 (Orotherium); p. 19 (Orohippus).
1877 K, p. 258 (Syn. of Hyracotherium); p. 252 (Orotherium).
1887 N, pp. 997, 1074. (Syn. of Pliolophus.)
1892 S, p. 411.
Dana, J. D. 1896 A, p. 913, fig. 1540.
Farr, M. S. 1896 A, p. 157. (Syn. of Pachynolophus.)
Flower, W. H. 1876 B, xiii, p. 327.
Flower and Lydekker 1891 A, p. 374. (Syn. of Pachynolophus.)
Gaudry, A. 1878 B, p. 142.
Huxley, T. H. 1876 B, xiv, p. 34.
1876 C, p. 294.
Landois, — 1881 A, p. 128.
Leidy, J. 1871 C, p. 358. (Lophiotherium.)
1873 B, p. 69. (Lophiotherium.)
Lydekker, R. 1886 A, p. 10. (Syn. of Hyracotherium.)
Marsh, O. C. 1872 I, p. 217. (Orotherium, type *O. uintanum.*)
1873 H, p. 407.
1874 B, p. 247.
1874 D, p. 66.
1876 G.
1877 E.
1879 D, p. 504.
1892 C, p. 348.
1898 E, p. 568.
Nicholson and Lydekker 1889 A, p. 1356.
Osborn, H. F. 1890 D, p. 506. (Syn. of Pliolophus.)
Osborn, Scott, and Speir 1878 A, p. 17 (Oligotomus); p. 24 (Orohippus).
Schlosser, M. 1886 B, p. 13. (Orotherium.)
Scott, W. B. 1891 B, p. 302.
Tomes, C. S. 1898 A. p. 415.
Weithofer, A. 1888 A, p. 278.
Wortman, J. L. 1883 A, p. 709. (Syn. of Hyracotherium.)
1896 A, pp. 92, 93, 108, fig. 18; pl. ii, fig. E.
Zittel, K. A. 1893 B, p. 242. (Syn. of Pachynolophus.)

Orohippus agilis Marsh.

Marsh, O. C. 1873 H, p. 407.
Cope, E. D. 1875 C, p. 22.
1877 K, p. 257. (Hyracotherium.)
Gaudry, A. 1878 B, p. 143.
King, C. 1878 A, p. 404.
Marsh, O. C. 1874 B, pp. 248, 255, figure.
1892 C, p. 349, figs. 16, 17.
Matthew, W. D. 1899 A, p. 45.
Osborn, H. F. 1890 D, p. 506. (Epihippus.)
Roger, O. 1896 A, p. 171. (Epihippus.)
Trouessart, E. L. 1898 A, p. 777. (Epihippus.)
Eocene (Bridger); Wyoming: Eocene (Uinta); New Mexico.

Orohippus ballardi (Marsh).

Marsh, O. C. 1871 D, p. 39. (Lophiotherium.)
Leidy, J. 1872 B, p. 365. (Lophiotherium.)
Marsh, O. C. 1872 I, p. 217. (Lophiotherium.)
Roger, O. 1896 A, p. 170. (Pachynolophus.)
Eocene (Bridger); Wyoming.

Orohippus cinctus (Cope).

Cope, E. D. 1873 U, p. 2. (Oligotomus.)
1873 E, p. 607. (Oligotomus.)
1884 O, p. 658, pl. xxiv, fig. 26. (Pliolophus.)
Matthew, W. D. 1899 A, p. 45. (Oligotomus.)
Osborn, Scott, and Speir 1878 A, p. 17. (Oligotomus.)
Roger, O. 1896 A, p. 170. (Pachynolophus.)
Trouessart, E. L. 1898 A, p. 774.
Wortman, J. L. 1896 A, p. 108.
Eocene (Bridger); Wyoming.

Orohippus major Marsh.

Marsh, O. C. 1874 B, p. 248.
Matthew, W. D. 1899 A, p. 45.
Osborn, H. F. 1890 D, p. 544, figures. (Pliolophus.)
Osborn, Scott, and Speir 1878 A, p. 24, pl. ix, figs. 1–7. (Identification doubtful.)
Trouessart, E. L. 1898 A, p. 772.
Eocene (Bridger); Wyoming.

Orohippus osbornianus (Cope).

Cope, E. D. 1884 O, pp. 630, 647, pl. xxiv, fig. 23. (Hyracotherium.)
1873 E, p. 607. (Orotherium sylvaticum, not Lophiotherium sylvaticum of Leidy.)

Cope, E. D. 1873 U, p. 2.   (Orotherium sylvaticum.)
    1877 K, p. 262.   (Hyracotherium sylvaticum.)
Matthew, W. D. 1899 A, p. 45.   (Hyracotherium.)
Roger, O. 1896 A, p. 169.   (Hyracotherium.)
Trouessart, E. L. 1898 A, p. 774.
    Eocene (Bridger); Wyoming.

Orohippus procyoninus Cope.

Cope, E. D. 1872 N, p. 466.   (Helotherium.)
    1873 E, p. 606.
    1875 C, pp. 20, 22.
    1877 K, pp. 262, 266.   (Hyracotherium.)
    1884 O, pp. 631, 711, pl. xxiv, fig. 22.   (Lambdotherium.)
    1887 S, p. 45.   (Lambdotherium.)
Marsh, O. C. 1873 K, p. 249.   (Syn.? of O. pumilis.)
Matthew, W. D. 1899 A, p. 45.   (Hyracotherium.)
Roger, O. 1896 A, p. 193.   (Lambdotherium.)
    Eocene (Bridger); Wyoming.

Orohippus pumilis Marsh.

Marsh, O. C. 1872 I, p. 207.
Cope, E. D. 1873 E, p. 606.   (Syn. of O. procyoninus.)
    1875 C, p. 22.   (Syn. of O. procyoninus.)
    1877 K, p. 266.   (Syn.? of Hyracotherium procyoninum.)

Marsh, O. C. 1874 B, p. 249.
Matthew, W. D. 1899 A, p. 45.
Osborn, Scott, and Speir 1878 A, p. 24.
Roger, O. 1896 A, p. 170.   (Pachynolophus.)
Trouessart, E. L. 1898 A, p. 773.
    Eocene (Bridger); Wyoming.

Orohippus sylvaticus (Leidy).

Leidy, J. 1870 T, p. 126.   (Lophiotherium.)
Cope, E. D. 1884 O, pp. 631, 647, 652.   (Pliolophus.)
Leidy, J. 1871 C, p. 358.   (Lophiotherium.)
    1872 B, p. 364.   (Lophiotherium.)
    1873 B, pp. 69, 327, pl. vi, figs. 33–35.   (Lophiotherium.)
Matthew, W. D. 1899 A, p. 45.
Roger, O. 1896 A, p. 170.   (Pachynolophus.)
Trouessart, E. L. 1898 A, p. 773.
    Eocene (Bridger); Wyoming.

Orohippus uintanus (Marsh).

Marsh, O. C. 1872 I, p. 217.   (Orotherium.)
Cope, E. D. 1881 D, p. 199.   (Orotherium.)
    1884 O, p. 651.   (Pliolophus vintanus.)
Matthew, W. D. 1899 A, p. 45.
Roger, O. 1896 A, p. 170.   (Pachynolophus.)
Trouessart, E. L. 1898 A, p. 777, in part.   (Epihippus uintensis.)
    Eocene (Bridger); Wyoming.

EPIHIPPUS Marsh.   Type *E. gracilis* Marsh.

Marsh, O. C., 1878, Pop. Sci. Month., p. 678.
Cope, E. D. 1887 N, pp. 1061, 1063.
    1890 C, p. 470.
    1892 S, p. 411.
Flower and Lydekker 1891 A, p. 374.
Marsh, O. C. 1879 D, p. 504.   (No definition.)
    1892 C, p. 353.
    1898 E, p. 568.
Matthew, W. D. 1899 A, p. 49.
Nicholson and Lydekker 1891 A, p. 1358.
Osborn, H. F. 1890 D, pp. 506, 529.
Schlosser, M. 1886 B, p. 14.
Steinmann and Döderlein 1890 A, p. 780.
Wortman, J. L. 1896 A, p. 92.
Zittel, K. A. 1893 B, p. 244.

Epihippus gracilis Marsh.

Marsh, O. C. 1871 D, p. 38.   (Anchitherium.)
    1878, in King, C. 1878 A, p. 407.
    1874 B, p. 249.   (Orohippus.)

Matthew, W. D. 1899 A, p. 49.
Roger, O. 1896 A, p. 171 (Epihippus); p. 174 (Mesohippus?).
Scott and Osborn 1887 A, p. 259.
Trouessart, E. L. 1898 A, p. 777 (Epihippus); p. 781 (Mesohippus).
    Eocene (Uinta); Utah.

Epihippus uintensis Marsh.

Marsh, O. C. 1875 B, p. 247.   (Orohippus.)
    1878, in King, C. 1878 A, p. 407.
Matthew, W. D. 1899 A, p. 49.
Osborn, H. F. 1890 D, p. 529, pl. xi, figs. 3, 5.
    1895 A, p. 98.
Roger, O. 1896 A, p. 171.
Scott and Osborn 1887 A, p. 259.
Trouessart, E. L. 1898 A, p. 777, in part.
Zittel, K. A. 1893 B, p. 245, fig. 184.
    Eocene (Uinta); Utah.

MIOHIPPUS Marsh.   Type *M. annectens* Marsh.

Marsh, O. C. 1874 B, p. 249.
Ameghino, F. 1896 A, p. 442.   (Mesohippus.)
Cope, E. D. 1874 B, p. 496.   (Anchitherium.)
    1881 G, p. 399.   (Mesohippus.)
    1884 O, p. 714.   (Anchitherium.)
    1887 N, p. 1067.   (Mesohippus.)
    1892 S, p. 412.   (Miohippus.)
    1894 D, p. 791.   (Mesohippus, Miohippus.)
Dana, J. D. 1896 A, p. 913, fig. 1540.   (Miohippus, Mesohippus.)
Farr, M. S. 1896 A, p. 150.   (Mesohippus.)
Huxley, T. H. 1876 C, pp. 292, 294.   (Miohippus, Mesohippus.)

Landois, — 1881 A, p. 128 (Protohippus); p. 29 (Mesohippus).
Lydekker, R. 1886 A, p. 45.   (Mesohippus.)
Marsh, O. C. 1874 D, p. 66.   (Miohippus.)
    1875 B, p. 248.   (Mesohippus, type *Anchitherium bairdi*.)
    1876 G, p. 61.   (Mesohippus.)
    1877 E, p. 359.   (Miohippus, Mesohippus.)
    1879 D, p. 504.   (Mesohippus.)
    1892 C, p. 354.   (Miohippus, Mesohippus.)
    1898 E, p. 568.   (Miohippus, Mesohippus.)
Nicholson and Lydekker 1889 A, p. 1358 (Mesohippus); p. 1359 (Miohippus).

Osborn, H. F. in Scott and Osborn 1890 B, p. 87.
(Mesohippus.)
Osborn and Wortman 1895 A, p. 352. (Mesohippus.)
Pavlow, M. 1888 A. (Mesohippus.)
Schlosser, M 1886 B, p. 14. (Miohippus and Mesohippus, syns. of Anchitherium.)
Scott, W. B. 1886 B. (Miohippus, Mesohippus.)
    1891 B. (Mesohippus.)
    1893 B, p. 660. (Miohippus, Mesohippus.)
    1895 C, p. 79. (Miohippus, Mesohippus.)
Scott and Osborn 1887 B, p. 171. (Mesohippus.)
    1890 B, p. 87. (Mesohippus.)
Steinmann and Döderlein 1890 A, p. 781. (Mesohippus.)
Tomes, C. S. 1898 A. p. 415.
Weithofer, A. 1888 A, p. 278.
Woodward, A. S. 1898 B, p. 328. (Mesohippus.)
Wortman, J. L. 1883 A, p. 710. (Anchitherium.)
    1898 A. (Mesohippus.)
Zittel, K. A. 1893 B, p. 249. (Mesohippus.)

Miohippus agrestis (Leidy).

- Leidy, J. 1873 B, pp. 251, 323, pl. vii, figs. 16, 17. (Anchitherium.)
Cope, E. D. 1874 B, p. 498. (Anchitherium.)
Matthew, W. D. 1899 A, p. 56. (Syn. of Anchippus texanus.)
Roger, O. 1896 A, p. 174. (Mesohippus.)
Trouessart, E. L. 1898 A, p. 781. (Mesohippus.)
    Miocene; Montana.

Miohippus anceps Marsh.

Marsh, O. C. 1874 B, p. 250. (Anchitherium.)
Cope, E. D. 1886 H, p. 368. (Syn.? of Anchitherium equiceps.)
King, C. 1878 A, p. 424.
Marsh, O. C. 1892 C, p. 350, figs. 20, 21.
Matthew, W. D. 1899 A, p. 63. (Mesohippus.)
Roger, O. 1896 A, p. 174.
Scott, W. B. 1893 B, p. 659.
    1895 C, p. 113.
Trouessart, E. L. 1898 A, p. 781.
Zittel, K. A. 1893 B, p. 250, fig. 194. (Anchitherium.)
    Miocene (John Day); Oregon.

Miohippus annectens Marsh.

Marsh, O. C. 1874 B, p. 249.
King, C. 1878 A, p. 424.
Marsh, O. C. 1894 F, p. 91.
Matthew, W. D. 1899 A, pp. 56, 63. (Mesohippus.)
Roger, O. 1896 A, p. 174.
Scott, W. B. 1893 B, p. 659.
    1895 C, p. 80, pl. i, figs. 6–8. (M. annectens.?)
Trouessart, E. L. 1898 A, p. 781.
    Oligocene? (White River); Montana?: Miocene (John Day); Oregon.

Miohippus australis (Leidy).

Leidy, J. 1873 B, pp. 250, 323, pl. xx, fig. 19. (Anchitherium.)
Cope, E. D. 1893 A, p. 23. (Anchitherium.)
Matthew, W. D. 1899 A, p. 69. (Syn. of Anchippus texanus.)
Roger, O. 1896 A, p. 174. (Mesohippus.)
Trouessart, E. L. 1898 A, p. 781. (Mesohippus.)
    Miocene (Loup Fork); Texas.

Miohippus bairdi (Leidy).

Leidy, J. 1850 C, p. 122. (Palæotherium.)
Cope, E. D. 1873 CC, p. 14. (Anchitherium.)
    1874 B, p. 496. (Anchitherium.)
    1879 D, p. 75. (Anchitherium.)
Farr, M. S. 1896 A, pl. xiii, text figs. 1–4. (Mesohippus.)
Hayden, F. V. 1858 B, p. 158. (Anchitherium.)
King, C. 1878 A, p. 412. (Mesohippus.)
Leidy, J. 1852 J, p. 572. (Anchitherium.)
    1852 L, p. 64. (Palæotherium.)
    1854 A, p. 67, pl. x, figs. 14–21; pl. xi. (Anchitherium.)
    1869 A, pp. 333, 402, pl. xx. (Anchitherium.)
    1870 P, p. 112. (Anchitherium.)
    1871 C, p. 362. (Anchitherium.)
    1873 B, pp. 218, 322, pl. vii, fig. 15. (Anchitherium.)
Lydekker, R. 1886 A, p. 49. (Anchitherium.)
Marsh, O. C. 1874 B, p. 250. (Anchitherium.)
    1875 B, p. 248. (Mesohippus.)
Matthew, W. D. 1899 A, p. 55. (Mesohippus.)
Nicholson and Lydekker 1889 A, p. 1359. (Mesohippus.)
Osborn, H. F. 1896 B, p. 174. (Titanotherium.)
Osborn and Wortman 1894 A, p. 213. (Mesohippus.)
    1895 A, p. 353. (Mesohippus.)
Pavlow, M. 1888 A, p. 148, pl. i, figs. 4, 13. (Anchitherium.)
Roger, O. 1896 A, p. 173. (Mesohippus.)
Scott, W. B. 1891 B, pls. xxii, xxiii.
Scott and Osborn 1887 B, p. 171. (Mesohippus.)
    1890 B, p. 87, figs. 10, 11. (Mesohippus.)
Trouessart, E. L. 1898 A, p. 780. (Mesohippus.)
Zittel, K. A. 1893 B, p. 250, fig. 191. (Mesohippus.)
    Oligocene (White River); South Dakota, Nebraska, Colorado, Oregon?.

Miohippus brachylophus (Cope).

Cope, E. D. 1879 D, p. 74. (Anchitherium.)
    1879 B, p. 58. (Anchitherium.)
    1894 D. (Anchitherium or Miohippus.)
Matthew, W. D. '899 A, p. 63. (Mesohippus.)
Roger, O. 1896 A, p. 174.
Trouessart, E. L. 1898 A, p. 781.
    Miocene (John Day); Oregon.

Miohippus celer Marsh.

Marsh, O. C. 1874 B, p. 251. (Anchitherium.)
Cope, E. D. 1874 B, p. 497. (Anchitherium.)
King, C. 1878 A, p. 412. (Mesohippus.)
Marsh, O. C. 1875 B, p. 248. (Mesohippus.)
    1892 C, p. 350, figs. 18, 19. (Mesohippus.)
    1897 C, p. 524, fig. 101.
Matthew, W. D. 1899 A, p. 56. (Mesohippus.)
Osborn and Wortman 1895 A, p. 353. (Mesohippus.)
Roger, O. 1896 A, p. 174.
Trouessart, E. L. 1898 A, p. 780. (Mesohippus.)
Zittel, K. A. 1893 B, p. 250, fig. 192. (Mesohippus.)
    Oligocene (White River); Colorado?.

Miohippus condoni (Leidy).

Leidy, J. 1870 P, p. 112. (Anchitherium.)
Cope, E. D. 1879 D, p. 76. (Anchitherium.)

King, C. 1878 A, p. 424.
Leidy, J. 1871 C, p. 362. (Anchitherium.)
    1871 J. (Anchitherium.)
    1873 B, pp. 218, 323, pl. ii, fig. 5. (Anchitherium.)
Marsh, O. C. 1874 B, p. 249. (Anchitherium.)
Matthew, W. D. 1899 A, p. 63. (Mesohippus.)
Roger, O. 1896 A, p. 174.
Trouessart, E. L. 1898 A, p. 781.
    Miocene (John Day); Oregon.

Miohippus copei (Osb. and Wort.).

Osborn and Wortman 1895 A, pp. 353, 356, fig. 5. (Mesohippus and Anchitherium.)
Farr, M. S. 1896 A, p. 164, fig. 5. (Mesohippus.)
Matthew, W. D. 1899, A, p. 56. (Mesohippus.)
Trouessart, E. L. 1898 A, p. 781. (Mesohippus.)
    Oligocene (White River); South Dakota.

Miohippus cuneatus (Cope).

Cope, E. D. 1873 T, p. 7. (Anchitherium.)
    1873 CC, p. 14. (Anchitherium.)
    1874 B, p. 497. (Anchitherium.)
    1879 D, p. 76. (Anchitherium.)
Matthew, W. D. 1899 A, p. 56. (Syn. of M. bairdi.)
Osborn and Wortman 1895 A, p. 353, footnote. (Mesohippus.)
Roger, O. 1896 A, p. 174. (Mesohippus.)
Scott, W. B. 1895 C, p. 113. (Mesohippus.)
Trouessart, E. L. 1898 A, p. 780. (Mesohippus.)
    Oligocene (White River); Colorado.

Miohippus equiceps (Cope).

Cope, E. D. 1879 D, p. 73. (Anchitherium.)
    1879 B, p. 58. (Anchitherium.)
    1886 H, p. 368. (Anchitherium anceps?)
    1889 C, p. 271, fig. 83. (Anchitherium.)
    1894 D, p. 791. (Anchitherium or Miohippus.)
Matthew, W. D. 1899, A, pp. 56, 63. (Mesohippus.)
Roger, O. 1896 A, p. 174.
Scott, W. B. 1895 C, p. 79.
Trouessart, E. L. 1898 A, p. 781.
    Miocene (John Day); Oregon. Oligocene? (White River?); Montana?:

Miohippus exoletus (Cope).

Cope, E. D. 1874 B, p. 496. (Anchitherium.)
    1879 D, p. 76. (Anchitherium.)
    1894 D, p. 791. (Anchitherium.)
Matthew, W. D. 1899 A, p. 56. (Syn. of M. bairdi.)
Osborn and Wortman 1895 A, p. 353. (Mesohippus.)
Roger, O. 1896 A, p. 174. (Mesohippus.)
Trouessart, E. L. 1898 A, p. 782. (Anchitherium.)
    Oligocene (White River); Colorado.

Miohippus intermedius (Osb. and Wort.).

Osborn and Wortman 1895 A, pp. 353, 354, fig. 4. (Mesohippus.)
Farr, M. S. 1896 A, p. 169, fig. 6. (Mesohippus.)
Matthew, W. D. 1899 A, p. 56. (Mesohippus.)
Trouessart, E. L. 1898 A, p. 781. (Mesohippus.)
    Oligocene (White River); South Dakota. Nebraska.

Miohippus longicristis (Cope).

Cope, E. D. 1879 D, p. 75. (Anchitherium.)
    1879 B, p. 58. (Anchitherium.)
    1886 H, p. 368. (Anchitherium.)
    1894 D, p. 791. (Anchitherium.)
Matthew, W. D. 1899 A, p. 63. (Mesohippus.)
Roger, O. 1896 A, p. 174.
Scott and Osborn 1890 B, p. 88, fig. 12. (Anchitherium.)
Trouessart, E. L. 1898 A, p. 782.
Zittel, K. A. 1893 B, p. 251, fig. 195. (Anchitherium.)
    Miocene (John Day); Oregon.

Miohippus præstans (Cope).

Cope, E. D. 1879 S. (Anchitherium.)
    1887 N, p. 1069, fig. 36. (Anchitherium.)
    1889 B, p. 455. (Anchitherium.)
    1894 D, p. 791. (Anchitherium.)
Farr, M. S. 1896 A, p. 163. (Mesohippus.)
Matthew, W. D. 1899 A, p. 63. (Mesohippus.)
Osborn and Wortman 1895 A, p. 356. (Anchitherium.)
Roger, O. 1896 A, p. 174.
Trouessart, E. L. 1898 A, p. 782. (Anchitherium.)
    Miocene (John Day); Oregon.

Miohippus ultimus (Cope).

Cope, E. D. 1886 B, p. 357. (Anchitherium.)
    1886 H, p. 368. (Anchitherium.)
    1887 N, p. 1068. (Anchitherium.)
Farr, M. S. 1896 A, p. 169. (Anchitherium.)
Matthew, W. D. 1899 A, p. 69. (Mesohippus.)
Roger, O. 1896 A, p. 174.
Trouessart, E. L. 1898 A, p. 782.
    Miocene (Loup Fork); Oregon.

Miohippus westonii (Cope).

Cope, E. D. 1889 I, p. 153. (Anchitherium.)
Ami, H. M. 1891 A. (Anchitherium.)
Cope, E. D, 1891 A, p. 20, pl. xiv, figs. 1, 2. (Anchitherium.)
Matthew, W. D. 1899 A, p. 56. (Mesohippus.)
Roger, O. 1896 A, p. 174. (Mesohippus.)
Trouessart, E. L. 1898 A, p. 781. (Mesohippus.)
    Oligocene (White River); Canada.

ANCHITHERIUM Meyer. · Type *A. aurelianense* (Blainv.).

Meyer, H. 1844 A, p. 298.
Ameghino, F. 1896 A, p. 442.
Blainville, H. M. D. 1864 A, iv, Y, pl. vii. ("Palæotherium hippoides, ou equinum, ou aurelianense.")
Cope, E. D. 1874 B, p. 496.
    1881 G, p. 399.

Cope, E. D. 1884 O, p. 714.
    1887 N, p. 1067.
    1889 C, p. 255 et seq.
    1892 S, p. 412.
    1894 D, p. 791.
Dawkins, W. B. 1870 A.
Depéret, C. 1887 A, p. 213.

Filhol, H. 1891 A, p. 169.
Flower, W. H. 1876 B, xiii, p. 327.
Gaudry, A. 1878 B, p. 124.
Gervais, P. 1859 A, p. 83.
Huxley, T. H. 1870 B, p. xlix.
　　1872 A, p. 306.
　　1876 B, xiv, p. 34.
　　1876 C, pp. 292, 294.
Kowalevsky, W. 1873 B.
Leidy, J. 1854 A, pp. 67, 114.
　　1869 A, p. 108.
　　1871 C, p. 362.
Leuthardt, F. 1891 A.
Lydekker, R. 1886 A, p. 45.
Major, Fors. 1873 A, p. 103.
　　1877 A.
Marsh, O. C. 1874 B, p. 247.
Nicholson and Lydekker 1889 A, p. 1358.
Osborn and Wortman, 1895 A, p. 352.
Pavlow, M. 1888 A.
Pomel, A. 1847 A, pp. 202, 203, 207.
Schlosser, M. 1886 B, p. 14.
Scott, W. B. 1886 B.

Scott, W. B. 1891 B.
　　1891 D, p. 90.
　　1893 B, p. 660.
　　1895 C, p. 94.
Scott and Osborn 1890 B, p. 88.
Steinmann and Döderlein 1890 A, p. 781.
Weitbofer, A. 1888 A, p. 278.
Wilckens, M. 1884 A.
Woodward, A. S. 1898 B, p. 327.
Wortman, J. L. 1882 B, vi, p. 69.
　　1883 A, p. 710.
　　1893 A.
Zittel, K. A. 1893 B, p. 250.

Anchitherium equinum Scott.
Scott, W. B. 1893 B, p. 661.
Cope, E. D. 1894 D, p. 790.
Matthew, W. D. 1899 A, p. 69.
Roger, O. 1896 A, p. 174.
Scott, W. B. 1895 C, p. 94, pl. iii, figs. 23–28; pl. iv, figs. 30, 31.
Trouessart, E. L. 1898 A, p. 782.
　　Miocene (Loup Fork); Montana.

DESMATIPPUS Scott.　　Type *D. crenidens* Scott.

Scott, W. B. 1893 B, p. 661.
Cope, E. D. 1894 D, p. 790. (Syn. of Anchitherium.)
Scott, W. B. 1895 C, p. 84.
Trouessart, E. L. 1898 A, p. 783. (Desmathippus.)

Desmatippus crenidens Scott.
Scott, W. B. 1893 B, p. 661.

Cope, E. D. 1894 D, p. 791. (Anchitherium.)
Matthew, W. D. 1899 A, p. 69. (Anchippus.)
Roger, O. 1896 A, p. 175.
Scott, W. B. 1895 C, p. 84, pl. ii, figs. 9–14.
Trouessart, E. L. 1898 A, p. 783. (Desmathippus.)
　　Miocene (Loup Fork); Montana.

ANCHIPPUS Leidy.　　Type *A. texanus* Leidy.

Leidy, J. 1868 I, p. 232.
Cope, E. D. 1884 O, p. 714.
　　1887 N, p. 1067.
　　1892 W, p. 943.
　　1893 A, p. 22.
Dawkins, W. B. 1870 A.
Leidy, J. 1869 A, p. 312.
　　1871 C, p. 362.
Marsh, O. C. 1874 B, p. 254.
Matthew, W. D. 1899 A, p. 22.
Nicholson and Lydekker 1889 A, p. 1359.

Anchippus brevidens Marsh.
Marsh, O. C. 1874 B, p. 254.
Cope, E. D. 1892 H, p. 326.
　　1892 W, p. 943.

Cope, E. D. 1898 A, p. 23.
King, C. 1878 A, p. 443.
Matthew, W. D. 1899 A, p. 69.
Roger, O. 1896 A, p. 175.
Trouessart, E. L. 1898 A, p. 783.
　　Miocene (Loup Fork); Oregon.

Anchippus texanus Leidy.
Leidy, J. 1868 I, p. 231.
　　1869 A, pp. 312, 403, pl. xxi, fig. 13.
　　1871 C, p. 362.
Marsh, O. C. 1874 B, p. 254.
Matthew, W. D. 1899 A, p. 56.
Roger, O. 1896 A, p. 175.
Trouessart, E. L. 1898 A, p. 783.
　　Miocene (Loup Fork?); Texas, Colorado.

HYPOHIPPUS Leidy.　　Type *H. affinis* Leidy.

Leidy, J. 1858 E, p. 26. (Subgenus.)
Cope, E. D. 1892 H, p. 326.
　　1892 W, p. 943. (Syn. of Protohippus.)
　　1893 A, p. 20. (Syn. of Protohippus.)
Leidy, J. 1869 A, p. 311.
　　1871 C, p. 362.
Marsh, O. C. 1874 B, p. 255.
Nicholson and Lydekker 1889 A, p. 1359.
Schlosser, M. 1886 B, p. 14.
Zittel, K. A. 1893 B, p. 250. (Syn.? of Anchitherium.)

Hypohippus affinis Leidy.
Leidy, J. 1858 E, p. 26. [Anchitherium(Hypohippus).]
Cope, E. D. 1893 A, p. 22.
Hayden, F. V. 1858 B, p. 158.
Leidy, J. 1869 A, pp. 311, 402, pl. xxi, figs. 11, 12.
　　1871 C, p. 362.
Roger, O. 1896 A, p. 174.
Trouessart, E. L. 1898 A, p. 783.
　　Pliocene; South Dakota.

EQUINÆ.

Boule, M. 1899 A, p. 531. ("Équidés.")
Osborn, H. F. 1893 D, p. 137.
Osborn and Wortman 1892 A, p. 93.

Roger, O. 1896 A, p. 175.
Steinmann and Döderlein 1890 A, p. 782.
Zittel, K. A. 1893 B, p. 252.

MERYCHIPPUS Leidy.

Type *M. insignis* Leidy.

Leidy, J. 1856 Q, p. 311.
Unless otherwise indicated, the following authors refer to this genus under the name *Protohippus.*
Ameghino, F. 1891 C, p. 215.
Cope, E. D. 1873 R, p. 420.
1877 K, p. 322.
1881 G, p. 399.
1884 O, p. 714.
1887 N, p. 1068.
1889 C.
1892 H, p. 826. (Parahippus.)
1892 W, p. 943. (Merychippus, Parahippus.)
1893 A, p. 20. (Protohippus; Merychippus a synonym.)
Dana, J. D. 1896 A, p. 913, fig. 1540.
Dawkins, W. B. 1870 A.
Farr, M. S. 1896 A, p. 169.
Flower and Lydekker 1891 A, p. 380.
Gaudry, A. 1877 A. (Merychippus.)
Huxley, T. H. 1876 C, p. 294.
Landois, —— 1881 A, p. 128.
Leidy, J. 1858 E, p. 26 (Protohippus, type *P. perditus)*; p. 27 (Merychippus).
1869 A, p. 275 (Protohippus); p. 292 (Merychippus); p. 313 (Parahippus).
1870 T, p. 127.
1871 C, p. 360 (Protohippus); p. 361 (Merychippus); p. 362 (Parahippus).
Major, Fors. 1873 A, p. 108. (Merychippus.)
Marsh, O. C. 1877 E.
1879 D, p. 504.
1892 C, p. 354.
Matthew, W. D. 1899 A, p. 69.
Nicholson and Lydekker 1889 A, p. 1359 (Parahippus); p. 1360 (Protohippus).
Pavlow, M. 1888 A. (Protohippus, Merychippus.) 1892 A.
Roger, O. 1889 A.
Rütimeyer, L. 1862 A, p. 611, (Merychippus.)
1877 A. (Merychippus.)
Schlosser, M. 1886 B, p. 14 (Merychippus, Parahippus); p. 15 (Protohippus, syn. of Hipparion).
Scott, W. B. 1891 D, p. 90 (Merychippus); p. 93 (Protohippus).
1896 C, p. 79. (Protohippus.)
Steinmann and Döderlein 1890 A, p. 782 (Merychippus); p. 783 (Protohippus).
Weithofer, A. 1888 A, p. 276.
Woodward, A. S. 1896 B, p. 337.
Wortman, J. L. 1882 B, vi, p. 71.
1883 A, p. 711.
Zittel, K. A. 1893 B, p. 252 (Merychippus); p. 255 (Protohippus).

Merychippus avus (Marsh).

Marsh, O. C. 1874 B, p. 253. (Protohippus.)
King, C. 1878 A, p. 443. (Protohippus.)
Marsh, O. C. 1897 C, p. 524, fig. 100. (Protohippus.)

Matthew, W. D. 1899 A, p. 70. (Protohippus.)
Trouessart, E. L. 1898 A, p. 785. (Protohippus.)
Miocene (Loup Fork); Oregon.

Merychippus castilli (Cope).

Cope, E. D. 1885 S, p. 1208, pl. xxxvii, fig. 6. (Protohippus.)
1886 W, p. 150, fig. 2. (Protohippus.)
1893 A, p. 26. (Protohippus.)
Matthew, W. D. 1899 A, p. 70. (Protohippus.)
Roger, O. 1896 A, p. 177. (Protohippus.)
Trouessart, E. L. 1898 A, p. 784. (Protohippus.)
Miocene (Loup Fork); Mexico.

Merychippus cumminsii (Cope).

Cope, E. D. 1893 A, p. 67, pl. xx, fig. 7; pl. xxiii, fig. 1. (Equus.)
1894 A, p. 67. (Equus.)
1899 A, p. 255. (Equus.)
Gidley, J. W. 1901, Bull. Amer. Mus. Nat. Hist., xiv, p. 126. (Protohippus.)
Matthew, W. D. 1899 A, p. 75. (Equus.)
Trouessart, E. L. 1898 A, p. 791. (Equus.)
Pliocene (Blanco); Texas, Oklahoma?.

Merychippus fossulatus (Cope).

Cope, E. D. 1893 A, p. 30, pls. v, vi, vii. (Protohippus.)
Matthew, W. D. 1899 A, p. 70. (Pliohippus.)
Roger, O, 1896 A, p. 177. (Protohippus.)
Trouessart, E. L. 1898 A, p. 784. (Protohippus.)
Miocene (Loup Fork); Texas.

Merychippus insignis Leidy.

Leidy, J. 1856 Q, p. 311,
Cope, E. D. 1874 P, p. 13. (Protohippus.)
1889 B, p. 457. (Protohippus.)
1892 W, p. 944, pl. xxvi, fig. 4 (Protohippus insignis); fig. 5, (P. medius).
1893 A, p. 25. (Syn. in part of Protohippus medius; in part of P. sejunctus.)
Hayden, F. V. 1858 B, p. 158.
King, C. 1878 A, p. 430.
Leidy, J. 1858 E, p. 27.
1869 A, pp. 296, 402, pl. xvii, figs. 3–7.
1871 C, p. 361.
Major, Fors. 1877 A.
Matthew, W. D. 1899 A, p. 69. (Protohippus medius.)
Osborn, H. F. 1898 I, p. 107, fig. 20.
Roger, O. 1896 A, p. 175 (Protohippus insignis); p. 177 (P. medius).
Scott, W. B. 1893 B, p. 659. (Protohippus.)
Trouessart, E. L. 1898 A, p. 784. (Protohippus sejunctus, part; P. medius, part.)
Weithofer, A. 1888 A, p. 278.
Zittel, K. A. 1893 B, p. 252, fig. 196.
Miocene (Loup Fork); South Dakota, Kansas, Oregon, Mexico.

Merychippus labrosus (Cope).

Cope, E. D. 1874 P, p. 13. (Protohippus.)
1874 B, p. 523. (Protohippus.)
1893 A, p. 26. (Protohippus.)
Matthew, W. D. 1899 A, p. 69. (Protohippus labrosus; syn. of P. perditus.)
Roger, O. 1896 A, p. 177. (Protohippus.)
Trouessart, E. L. 1898 A, p. 784. (Protohippus.)
Miocene (Loup Fork); Colorado.

Merychippus lenticularis (Cope).

Cope, E. D. 1893 A, p. 41, pl. xii, figs. 1, 2. (Protohippus.)
Matthew, W. D. 1899 A, p. 75. (Protohippus.)
Roger, O. 1896 A, p. 177. (Protohippus.)
Trouessart, E. L. 1898, p. 784. (Protohippus.)
Pliocene (Palo Duro); Texas.

Merychippus mirabilis Leidy.

Leidy, J. 1858 E, p. 27.
Cope, E. D. 1892 W, p. 943. (Protohippus.)
1898 A, p. 32, pl. viii; pl. ix, fig. 1; pl. x, fig. 47. (Protohippus.)
Hayden, F. V. 1858 F, p. 158.
King, C. 1878 A, p. 430.
Leidy, J. 1869 A, pp. 299, 327, 402, pl. xvii, figs. 8-15; pl. xxvii, fig. 1.
1870 J, p. 67.
1871 C, p. 361.
1873 B, pp. 248, 250, 322, pl. xx, figs. 16, 20. (M. mirabilis?.)
Matthew, W. D. 1899 A, p. 70. (Pliohippus.)
Roger, O. 1896 A, p. 175.
Trouessart, E. L. 1898 A, p. 784. (Protohippus.)
Weithofer, A. 1888 A, p. 278.
Miocene (Loup Fork); Nebraska, Colorado, South Dakota, Texas, Utah?.

Merychippus pachyops (Cope.)

Cope, E. D. 1892 W, p. 944, pl. xxvi, fig. 1. (Protohippus.)
1893 A, p. 26, pl. xi, fig. 1; pl. xii; pl. xvii, figs. 2, 3. (Protohippus.)
Matthew, W. D. 1899 A, p. 70. (Pliohippus.)
Roger, O. 1896 A, p. 177. (Protohippus.)
Trouessart, E. L. 1898 A, p. 783. (Protohippus.)
Miocene (Loup Fork); Texas.

Merychippus parvulus (Marsh.)

Marsh, O. C. 1868 A, p. 374. (Equus.)
Cope, E. D. 1892 H, p. 326. (Protohippus.)
1892 W, p. 943. (Protohippus.)
1898 A, pp. 25, 29, pl. xi figs. 1, 2. (Protohippus parvulus?.)
King, C. 1878 A, p. 430. (Protohippus.)
Leidy, J. 1868 D, p. 195. ("Equus.")
1869 A, p. 400. (Equus.)
1871 C, p. 360. (Equus.)
1873 B, pp. 252, 323, pl. xx, fig. 23. ("Anchitherium?")
Marsh, O. C. 1874 B, p. 251. (Protohippus.)
Matthew, W. D. 1899 A, p. 69. (Syn. of Protohippus placidus.)
Osborn, H. F. 1890 B, p. 89. (Anchitherium?)
Roger, O. 1896 A, p. 174 (Mesohippus?); p. 177 (Protohippus).

Trouessart, E. L. 1898 A, p. 785. (Subsp. of Protohippus placidus.)
Miocene (Loup Fork); Nebraska, Texas?.

Merychippus perditus (Leidy.)

Leidy, J. 1858 E, p. 26. [Equus (Protohippus).]
Burmeister, H. 1875 A, p. 12. (Protohippus.)
Cope, E. D. 1874 B, pp. 524, 528. (Protohippus.)
1874 P, p. 18. (Protohippus.)
1889 B, p. 447, pl. ii, figs. 9, 12. (Protohippus profectus.)
1892 W, p. 943, pl. xxvi, fig. 2. (Protohippus.)
1893 A, pp. 24, 42, pl. x, fig. 2. (Protohippus.)
1894 A, p. 66. (Protohippus.)
Hayden, F. V. 1858 F, p. 158. (Equus perditus, Parahippus cognatus.)
King, C. 1878 A, p. 430. (Protohippus.)
Leidy, J. 1858 E, p. 26. [Anchitherium (Parahippus) cognatus.]
1869 A, pp. 275, 327, 400, pl. xvii, figs. 1, 2; pl. xxvii, fig. 5 (Protohippus perditus); pl. xxi, figs. 7-10 (Parahippus cognatus).
1870 J, p. 67. (Protohippus.)
1870 W, p. 127. (Protohippus.)
1871 C, p. 360 (Protohippus perditus); p. 362 (Parahippus cognatus).
1873 B, pp. 248-250, 322, pl. xx, figs. 16, 20. (Protohippus.)
Matthew, W. D. 1899 A, pp. 69, 75. (Protohippus.)
Roger, O. 1896 A, p. 175 (Parahippus cognatus); p. 177 (Protohippus perditus).
Trouessart, E. L. 1898 A, p. 784. (Protohippus.)
Miocene (Loup Fork); Nebraska, Colorado, Oklahoma, Kansas, Wyoming: Pliocene (Palo Duro); Texas.

Merychippus phlegon Hay.

Hay, O. P. 1899 G, p. 345. (Equus.)
Cope, E. D. 1893 A, pp. 44, 66, 67, pl. xx, fig. 8 (Equus minutus, preoccupied.)
1893 Q, p. 812. (E. minutus.)
1899 A, p. 255. (E. minutus.)
Gidley, J. W. 1901, Bull. Amer. Mus. Nat. Hist., xiv, p. 127, figs. 18, 19. (Protohippus.)
Matthew, W. D. 1899 A, p. 75. (E. minutus.)
Trouessart, E. L. 1898 A, p. 791. (E. minutus.)
Pliocene (Blanco); Texas.

Merychippus placidus (Leidy.)

Leidy, J. 1869 A, pp. 277, 328, 401, pl. xviii, figs. 39-48; pl. xxvii, figs. 6, 7. (Protohippus.)
Cope, E. D. 1873 R, p. 420. (Protohippus.)
1874 B, pp. 524, 528. (Protohippus.)
1874 P, p. 19. (Protohippus.)
1892 H, p. 325. (Protohippus.)
1892 W, p. 942, pl. xxv. (Protohippus.)
1893 A, pp. 26, 34, pl. xi, figs. 3-8. (Protohippus.)
1893 Q, p. 812. (Protohippus.)
Leidy, J. 1870 W, p. 127. (Protohippus.)
1871 C, p. 360. (Protohippus.)
1873 B, pp. 249, 250, 322, pl. xx, figs. 17, 18. (Protohippus.)
Matthew, W. D. 1899 A, p. 69. (Protohippus.)

Roger, O. 1896 A, p. 177. (Protohippus.)
Trouessart, E. L. 1896 A, p. 785. (Protohippus.)
    Miocene (Loup Fork); Nebraska, Colorado,
    South Dakota, Texas,

Merychippus sejunctus (Cope.)

Cope, E. D. 1874 P, p. 15. (Protohippus.)
    1873 R, p. 420. (Protohippus; no descrip-
    tion.)
    1874 B, pp. 523, 524. (Protohippus.)
    1881 I, p. 271, fig. 3. (Protohippus.)
    1884 J, p. 544, fig. 3. (Protohippus.)
    1886 B, p. 359. (Protohippus.)
    1886 H. (Protohippus.)
    1887 N, p. 1071, fig. 39. (Protohippus.)
    1887 S, p. 271, fig. 3. (Protohippus.)
    1889 C, p. 160. (Protohippus.)
    1893 A, pp. 23, 25. (Protohippus.)
Matthew, W. D. 1899 A, p. 69. (Protohippus.)
Roger, O. 1896 A, p. 177. (Protohippus.)
Scott, W. B. 1893 B, p. 659. (Protohippus.)
Trouessart, E. L. 1898 A, p. 784. (Protohippus.)
Woodward, A. S. 1898 B, p. 337, fig. 194. (Proto-
    hippus.)
Wortman, J. L. 1883 A, p. 711, figs. 141, 142. (Pro-
    tohippus.)
Zittel, K. A. 1893 B, p. 256, figs. 200, 201. (Proto-
    hippus.)
    Miocene (Loup Fork); Colorado, Montana,
    Oregon? New Mexico?.

Merychippus supremus (Leidy.)

Leidy, J. 1869 A, pp. 328, 401, pl. xxvii, figs. 3, 4.
    (Protohippus.)
Cope, E. D. 1893 A, p. 25. (Syn. of Protohippus
    mirabilis.)
King, C. 1878 A, p. 430. (Protohippus.)
Leidy, J. 1870 W, p. 127. (Protohippus.)
    1871 C, p. 360. (Protohippus.)
Matthew, W. D. 1899 A, p. 70. (Syn. of Protohip-
    pus mirabilis.)
Roger, O. 1896 A, p. 177. (Protohippus.)
Trouessart, E. L. 1898 A, p. 784. (Syn.? of Proto-
    hippus mirabilis.)
    Miocene (Loup Fork); South Dakota.

Merychippus sp. indet.

Cope, E. D. 1877 K, p. 323, pl. lxxxv, fig. 7. (Pro-
    tohippus sp.)
    Miocene (Loup Fork); New Mexico.

Merychippus sp. indet.

Scott, W. B. 1895 C, p. 98, pl. ii, fig. 17. (Proto-
    hippus sp.)
Trouessart, E. L. 1898 A, p. 784. (Syn.? of Proto-
    hippus mirabilis.)
    Miocene; Montana.

PLIOHIPPUS Marsh. Type *P. pernix* Marsh.

Marsh, O. C. 1874 B, p. 252.
Cope, E. D. 1879 A, p. 47. (Syn. of Hippidium.)
    1880 B, p. 47. (Syn. of Hippidium.)
Dana, J. D. 1896 A, p. 913, fig. 1540.
Flower and Lydekker 1891 A, p. 381.
Landois, ——. 1881 A, p. 128.
Marsh, O. C. 1874 D.
    1877 E, p. 359.
    1879 D, p. 504.
    1892 C, p. 354.
Matthew, W. D. 1899 A, p. 70.
Nicholson and Lydekker 1889 A, p. 1356.
Schlosser, M. 1886 B, p. 15.
Weithofer, A. 1888 A, p. 276.
Zittel, K. A. 1893 B, p. 256.

Pliohippus gracilis Marsh.

Marsh, O. C. 1892 C, p. 347.
    Pliocene; Oregon.

Pliohippus pernix Marsh.

Marsh, O. C. 1874 B, p. 252.
Cope, E. D. 1880 G. (Hippidium.)
King, C. 1878 A, p. 430.

Marsh, O. C. 1897 C, pp. 524, 527, fig. 99.
Matthew, W. D. 1899 A, p. 70.
Roger, O. 1896 A, p. 177.
Trouessart, E. L. 1898 A, p. 785.
    Miocene (Loup Fork); Nebraska, Colorado.

Pliohippus robustus Marsh.

Marsh, O. C. 1874 B, p. 253.
Cope, E. D. 1880 G. (Hippidium.)
King, C. 1878 A, p. 430.
Matthew, W. D. 1899 A, p. 70.
Roger, O. 1896 A, p. 177.
Trouessart, E. L. 1898 A, p. 785.
    Miocene (Loup Fork); Nebraska.

Pliohippus simplicidens (Cope.)

Cope, E. D. 1892 J, p. 124, fig. 1. (Equus.)
    1892 E, p. 252. (Equus.)
    1892 G, p. 228. (Equus.)
    1892 P. (Equus.)
    1893 A, pp. 41, 44, 66, pl. xx, figs. 1–3. (Equus.)
Gidley, J. W., 1901, Bull. Amer. Mus. Nat. Hist.,
    xiv, p. 123.
Trouessart, E. L. 1898 A, p. 785.
    Pliocene (Palo Duro, Blanco); Texas.

HIPPIDION Owen. Type *Equus neogæus* Lund.

Owen, R. 1870 B, p. 572.
    The amended form of this name, *Hippidium*,
    is employed by many authors.
Ameghino, F. 1889 A, p. 513.
    1891 A.
    1891 C, p. 215.
Branco, W. 1897 A, pp. 19, 78.

Burmeister, H. 1875 A.
    1889 A, p. 3.
Cope, E. D. 1879 A, p. 47.
    1880 B, p. 47.
    1881 G, p. 400.
    1884 O, p. 715.
    1887 N, p. 1078.

Cope, E. D. 1893 A, p. 20.
1899 A, p. 256.
Gervais and Ameghino 1880 A, p. 87.
Lydekker, R. 1893 D, pt. iii, p. 74.
Nicholson and Lydekker 1889 A, p. 1361.
Pavlow, M. 1892 A.
Steinmann and Döderlein 1890 A, p. 784.
Wilckens, M. 1884 B, p. 331.
Woodward, A. S. 1898 B, p. 338.
Wortman, J. L. 1883 A, p. 711.
Zittel, K. A. 1893 B, p. 256.

Hippidion interpolatum Cope.

Cope, E. D. 1893 A, p. 42, pl. xii, figs. 3, 4. (Hippidium.)
Matthew, W. D. 1899 A, p. 75. (Pliohippus.)

Trouessart, E. L. 1898 A, p. 786.
Pliocene (Palo Duro); Texas.

Hippidion spectans Cope.

Cope, E. D. 1880 G. (Hippidium.)
1887 N, p. 1072, fig. 41. (Hippidium.)
1892 E, p. 253. (Hippidium.)
1893 A, p. 43, pl. xii, figs. 5, 6. (Hippidium? spectans.)
Matthew, W. D. 1899 A, pp. 70, 75. (Pliohippus.)
Roger, O. 1896 A, p. 178. (Pliohippus.)
Trouessart, E. L. 1892 A.
1898 A, p. 786.
Zittel, K. A. 1893 B, p. 256. (Pliohippus.)
Miocene (Loup Fork); Oregon: Pliocene (Palo Duro); Texas.

HIPPARION Christol.

Christol, J., 1832, Ann. Sci. Indust. Mid. France, i, p. 180.
Christol's name *Hipparion* for this genus antedates Kaup's *Hippotherium*. It is said by Christol (Bull. Soc. Geol. France, ix, 1852, p. 255) to have been defined, but the present writer has not been able to learn whether or not Christol described or designated a type species.
Ameghino, F. 1891 C, p. 215. (Hippotherium.)
Branco, W. 1897 A, pp. 19, 78. (Hipparion.)
Boule, M. 1899 A, p. 541. (Hipparion.)
Bronn, H. G. 1838 A, p. 1192 (Hippotherium); p. 1193 (Hipparion).
Cope, E. D. 1879 D, p. 76. (Stylonus, type S. seversus Cope.)
1881 G, p. 399. (Hippotherium.)
1884 O, p. 714. (Hippotherium.)
1887 N, p. 1068. (Hippotherium.)
1889 B, p. 430. (Hippotherium.)
1889 C, p. 256. (Hippotherium.)
1893 A, pp. 21, 36. (Hippotherium.)
Filhol, H. 1891 A, p. 186, seq. (Hipparion.)
Flower and Lydekker 1891 A, p. 380. (Hipparion.)
Gaudry, A. 1878 B, p. 124. (Hipparion.)
1890 A. (Hipparion.)
Gervais, P. 1859 A, p. 80. (Hipparion.)
Hensel, R. 1860 A. (Hipparion.)
Huxley, T. H. 1870 A, p. 439. (Hipparion.)
1870 B, p. xlix. (Hipparion.)
1872 A, p. 306. (Hipparion.)
Kaup, J. J. 1833 A, p. 327. (Hippotherium, type H. gracilis Kaup.)
1835 A. (Hippotherium.)
Kowalevsky, W. 1873 A, p. 220. (Hipparion.)
Leidy, J. 1854 C, p. 90. (Hippodon, type H. speciosus Leidy.)
1858 E, p. 27.
1869 A, p. 280. (Hipparion.)
1871 C, p. 360. (Hipparion.)
1873 B, p. 247. (Hipparion.)
Leuthardt, F. 1891 A. (Hippotherium.)
Lydekker, R. 1882 A. (Hipparion.)
1886 A, p. 50. (Hipparion.)
1891 A, p. 880. (Hipparion.)
Major, Fors. 1877 A. Hipparion.)
Marsh, O. C. 1874 B, p. 255. (Hipparion.)
1874 D. (Hipparion.)

Nicholson and Lydekker 1889 A, p. 1360. (Hipparion.)
Osborn, H. F. 1893 D, p. 7. (Hipparion.)
Owen, R. 1845 B, p. 573. (Hipparion.)
Pavlow, M. 1888 A. (Hipparion.)
1890 A. (Hipparion.)
1892 A. (Hipparion.)
Pictet. F. J. 1853 A, p. 314. (Hipparion, Hippotherium.)
Quenstedt, F. A. 1850 A. (Hippotherium.)
Rütimeyer, L. 1862 A, p. 646. (Hipparion.)
Schlosser, M. 1886 B, p. 15. (Hipparion.)
Scott, W. B. 1893 B, p. 661. (Hipparion.)
Steinmann and Döderlein 1890 A, p. 782.
Weithofer, A. 1888 A, p. 273. (Hipparion.)
1888 A. (Hipparion.)
Wilckens, M. 1884 A. (Hipparion.)
Woodward, A. S. 1898 B, p. 338. (Hipparion.)
Wortman, J. L. 1882 B, vi, p. 78. (Hippotherium.)
1883 A, p. 711. (Hippotherium.)
1886 A, p. 483. (Hippotherium.)
Zittel, K. A. 1893 B, p. 252. (Hipparion.)

Hipparion affine Leidy.

Leidy, J. 1869 A, pp. 286, 402, pl. xviii, figs. 20-24. (Hipparion.)
Cope, E. D. 1889 B, p. 434, fig. 1. (Hippotherium occidentale, not of Leidy; H. affine.)
1893 A, p. 36. (Hippotherium.)
Leidy, J. 1871 C, p. 361. (Hipparion.)
Lydekker, R. 1882 A, p. 70. (Hippotherium.)
Pavlow, M. 1888 A, pl. i, fig. 26. (Hipparion.)
Roger, O. 1896 A, p. 176. (Hipparion.)
Trouessart, E. L. 1898 A, p. 786. (Hipparion.)
Pliocene (Loup Fork); Nebraska, Texas.

Hipparion calamarium (Cope).

Cope, E. D. 1875 P, p. 259. (Hippotherium.)
1875 F, p. 70. (Hippotherium.)
1877 K, p. 321, pl. lxxv, figs. 1, 2. (Hippotherium.)
1889 B, p. 451, fig 15. (Hippotherium.)
1893 A, p. 22. (Hippotherium.)
Matthew, W. D. 1899 A, p. 70. (Hipparion.)
Roger, O. 1896 A, p. 176. (Hipparion.)
Trouessart, E. L. 1898 A, p. 789. [Hipparion (Stylonus).]
Miocene (Loup Fork); New Mexico, Colorado.

Hipparion eurystylum (Cope).

Cope, E. D. 1893 A, pp. 43, 66, pl. xii, figs 7, 8; pl. xx, fig. 6. (Equus.)

1899 A, p. 255. (Equus.)

Gidley, J. W., 1901, Bull. Amer. Mus. Nat. Hist., xiv, p. 125. (Hipparion.)

Matthew, W. D. 1899 A, p. 75. (Equus.)

Trouessart, E. L. 1898 A, p. 791. (Equus.)

Pliocene (Goodnight?); Texas.

Hipparion gratum Leidy.

Leidy, J. 1869 A, pp. 287, 326, 402, pl. xviii, figs. 25, 30. (Hipparion.)

Cope, E. D. 1889 B, p. 445, figs. 16, 17. (Hippotherium.)

1892 H, p. 325. (Hippotherium.)

1892 W, p. 942. (Hippotherium.)

1893 A, pp. 21, 26. (Syn. of Protohippus placidus.)

Leidy, J. 1871 C, p. 361. (Hipparion.)

1885 A, p. 33, figure. (Hippotherium ingenuum.)

1887 A, p. 304. (Hippotherium.)

1889 B. (Hippotherium ingenuum.)

1889, E, p. 20, pl. iii, figs. 2–4; text figure. (Hippotherium ingenuum.)

1892 A. (Hippotherium ingenuum.)

Leidy and Lucas 1896 A, p. 49, pl. xviii, figs. 15, 16; pl. xix, figs. 9, 10. (Hippotherium.)

Lydekker, R. 1882 A, p. 70. (Hippotherium.)

Matthew, W. D. 1899 A, p. 69. (Syn. of Protohippus placidus.)

Roger, O. 1896 A, p. 176. (Hipparion ingenuum.)

Trouessart, E. L. 1898 A, p. 785. (Syn. of Protohippus placidus.)

Miocene (Loup Fork); Nebraska, Kansas.

Hipparion isonesum (Cope).

Cope, E. D. 1889 B, p. 451, fig. 23. (Hippotherium.)

1886 B, p. 359. (Hippotherium seversum, in part.)

Matthew, W. D. 1899 A, p. 70. (Hipparion.)

Trouessart, E. L. 1898 A, p. 789. [Hipparion (Stylonus).]

Miocene (Ticholeptus); Montana, Oregon, Colorado.

Hipparion minimum (Douglass).

Douglass, E. 1900 A, p. 26.

Miocene; Montana.

Hipparion montezumæ (Leidy).

Leidy, J., 1882, Proc. Acad. Nat. Sci. Phila., p. 291. (Hippotherium.)

Cope, E. D. 1889 B, p. 435. (Hippotherium.)

Leidy, J. 1889 E, p. 21, pl. v, figs. 5–7, 10. (Hippotherium.)

Matthew, W. D. 1899 A, p. 70. (Hipparion.)

Roger, O. 1896 A, p. 176. (Hipparion.)

Trouessart. E. L. 1898 A, p. 787. (Hipparion montezumai.)

Miocene (Loup Fork); Mexico.

Hipparion occidentale Leidy.

Leidy, J. 1856 D, p. 59. (Hipparion.)

Cope, E. D. 1886 B, p. 359. (Hippotherium.)

1886 H. (Hippotherium.)

Cope, E. D. 1892 AA. (Hippotherium.)

1893 A, p. 36. (Hippotherium.)

Hayden, F. V. 1856 B, p. 158.

Leidy, J. 1856 B, p. 422. (Hipparion.)

1856 E, p. 27. (Hipparion.)

1869 A, pp. 281, 326, 401, pl. xviii, figs. 1–5; pl. xxvii, fig. 2.

1871 C, p. 361. (Hipparion.)

Lydekker, R. 1882 A, p. 70. (Hippotherium.)

Matthew, W. D. 1899 A, p. 70. (Hipparion.)

Roger, O. 1896 A, p. 176. (Hipparion.)

Trouessart, E. L. 1898 A, p. 787. (Hipparion occidentale, in part.)

Wortman, J. L. 1883 A, p. 712. (Hippotherium.)

Miocene (Loup Fork); South Dakota, Texas.

Hipparion paniense (Cope).

Cope, E. D. 1874 P, p. 12. (Hippotherium.)

1874 B, p. 522. (Hippotherium.)

1889 B, p. 447, figs. 13, 14. (Hippotherium.)

Lydekker, R. 1882 A, p. 70. (Hippotherium.)

Matthew, W. D. 1899 A, p. 70. (Hipparion.)

Roger, O. 1896 A, p. 176. (Hipparion.)

Trouessart, E. L. 1898 A, p. 788. (Hipparion.)

Miocene (Loup Fork); Colorado.

Hipparion peninsulatum (Cope).

Cope, E. D. 1885 S, pl. xxxvii, fig. 5. (Hippotherium; no description.)

1886 W, p. 150, fig. 1. (Hippotherium.)

1889 B, p. 436, fig. 4. (Hippotherium.)

Matthew, W. D. 1899 A, p. 70. (Hipparion.)

Roger, O. 1896 A, p. 176. (Hipparion.)

Trouessart, E. L. 1898 A, p. 787. (Hipparion.)

Miocene (Loup Fork); Mexico.

Hipparion plicatile (Leidy).

Leidy, J. 1887 A, p. 309, figure. (Hippotherium.)

Cope, E. D. 1889 B, p. 444, fig. 6. (Hippotherium.)

Leidy, J. 1889 B. (Hippotherium.)

1889 E, p. 20. (Hippotherium.)

1892 AA. (Hippotherium.)

Leidy and Lucas 1896 A, p. 50, pl. xviii, figs. 11–14; pl. xix, figs. 1–8. (Hippotherium.)

Roger, O, 1896 A, p. 176. (Hipparion.)

Trouessart, E. L. 1898 A, p. 787. (Hipparion.)

Miocene; Florida.

Hipparion rectidens (Cope).

Cope, E. D. 1886 B, p. 360. (Hippotherium.)

1889 B, pp. 435, 458. (Hippotherium.)

Roger, O. 1896 A, p. 176. (Hipparion.)

Trouessart, E. L. 1898 A, p. 787. (Hipparion.)

Miocene (Loup Fork); Mexico.

Hipparion relictum (Cope).

Cope, E. D. 1889 K, p. 254. (Hippotherium.)

1889 B, p. 449, figs. 19, 20. (Hippotherium.)

Matthew, W. D. 1899 A, p. 70. (Hipparion.)

Roger, O. 1896 A, p. 176. (Hipparion.)

Trouessart, E. L. 1898 A, p. 788. (Hipparion.)

Pliocene (Blanco); Oregon.

Hipparion retrusum (Cope).

Cope, E. D. 1889 B, p. 446, figs. 7, 8. (Hippotherium.)

Trouessart, E. L. 1898 A, p. 788. (Hipparion.)

Miocene (Loup Fork); Kansas.

Hipparion seversum (Cope).

Cope, E. D. 1879 D, p. 76. (Stylonus.)
   1886 H. (Hippotherium.)
   1889 B, p. 457, fig. 24. (Hippotherium.)
Matthew, W. D. 1899 A, p. 71. (Hipparion.)
Trouessart, E. L. 1898 A, p. 779. [Hipparion (Stylonus).]
Wortman, J. L. 1883 A, p. 712. (Hippotherium.)
Miocene? (Ticholeptus); Oregon.

Hipparion sinclairii (Wort.).

Wortman, J. L, 1883 A, p. 712. (Hippotherium.)
Cope, E. D. 1886 B, p. 359. (Hippotherium.)
   1886 H. (Hippotherium.)
   1889 B, p. 434, fig. 2. (Hippotherium.)
Matthew, W. D. 1899 A, p. 70. (Hipparion.)
Roger, O. 1896 A, p. 176. (Hipparion.)
Trouessart, E. L. 1898 A, p. 787. (Hipparion.)
Miocene (Loup Fork); Oregon.

Hipparion speciosum Leidy.

Leidy, J. 1854 C, p. 90. (Hippodon.)
Cope, E. D. 1875 F, p. 71. (Hippotherium.)
   1877 K, p. 322, pl. lxxv, fig. 3. (Hippotherium.)
   1887 N, p. 1070, fig. 38. (Hippotherium.)
   1889 B, p. 436, fig. 5. (Hippotherium.)
Hayden, F. V. 1858 B, p. 158. (Hipparion.)
Leidy, J. 1856 Q, p. 311. [Hipparion (Hippodon).]
   1858 E, p. 27. (Hipparion *seu* Hippotherium.)
   1869 A, pp. 282, 401, pl. xviii, figs. 6–9. (Hipparion.)
   1871 C, p. 361. (Hipparion.)
   1873 B, pp. 247, 248, 322, pl. xx, figs. 14, 15. (Hipparion.)

Lydekker, R. 1882 A, p. 70. (Hippotherium.)
Matthew, W. D. 1899 A, p. 70. (Hipparion.)
Roger, O. 1896 A, p. 176. (Hipparion.)
Trouessart, E. L. 1898 A, p. 787. (Hipparion.)
Wortman, J. L. 1883 A, p. 711, figs. 138, 139. (Hippotherium.)
Zittel, K. A. 1893 B, p. 253, fig. 196. (Hipparion.)
Miocene (Upper); South Dakota, New Mexico. Colorado, Texas, Kansas.

Hipparion sphenodus (Cope).

Cope, E. D. 1889 B, p. 449, figs. 21, 22. (Hippotherium.)
   1874 P, p. 12. (Hippotherium speciosum, not of Leidy.)
   1884 B, p. 522. (Hippotherium speciosum, not of Leidy.)
Matthew, W. D. 1899 A, p. 71. (Hipparion.)
Trouessart, E. L. 1898 A, p. 788. (Hipparion.)
Miocene (Loup Fork); Colorado.

Hipparion venustum Leidy.

Leidy, J. 1860 B, p. 105, pl. xvi, figs. 32, 33. (Hippotherium.)
Cope, E. D. 1889 B, p. 448, fig. 18. (Hippotherium.)
Leidy, J. 1853 E. (Hipparion venustum; no description.)
   1869 A, p. 401. (Hipparion.)
   1871 C, p. 361. (Hipparion.)
   1877 A, p. 212. (Hipparion.)
Lydekker, R. 1882 A, p. 71. (Hippotherium.)
Roger, O. 1896 A, p. 176. (Hipparion.)
Trouessart, E. L. 1898 A, p. 787. (Hipparion.)
Tertiary; South Carolina.

Equus Linn.   Type *E. caballus* Linn.

Linnæus, C., 1758, Syst. Nat., ed. 10, i, p. 73.
Ameghino, F. 1889 A, p. 503.
   1891 A.
Auld, R. C. 1892 A.
Blake, C. C. 1863 A.
Blake, W. P. 1864 A.
Boas, J. E. 1884 A.
Boule, M. 1899 A, p. 531.
Branco, W. 1897 A, pp. 19, 67.
Burmeister, H. 1867 A, p. 238.
   1875 A.
   1889 A, p. 14.
Cope, E. D. 1881 G, p. 400.
   1884 G, p. 9.
   1884 O, p. 715.
   1884 JJ, p. 40.
   1887 N.
   1889 C, p. 257.
   1892 J, p. 125. (Tomolabis, type *E. fraternus* Leidy, part.)
   1899 A, p. 256.
Cuvier, G. 1834 A, iii, p. 193, plates.
Dana, J. D. 1896 A, p. 913, fig. 1540.
Ewart, J. C. 1894 A.
   1894 B.
Flower, W. H. 1881 A.
   1885 A.
Flower and Lydekker 1891 A, p. 381.
Gaudry, A. 1878 B, p. 137.
Gervais, P. 1874 A.

Gray, J. E. 1869 A, p. 263.
Huxley, T. H. 1870 A, p. 439.
   1870 B, p. xlix.
Leidy, J. 1869 A, p. 258.
   1871 C, p. 359.
Leuthardt, F. 1891 A, p. 102.
Lydekker, R. 1882 A.
   1886 A, p. 65.
Marsh, O. C. 1874 B, p. 256.
   1877 E.
   1879 D, p. 499.
Mojsisovics, A. 1889 A.
Nicholson and Lydekker 1889 A, p. 1361.
Owen, R. 1844 A, p. 230.
   1845 B, p. 572.
   1846 B, p. 383.
   1866 B, p. 447, figs. 300, 303–305.
   1868 A, p. 352, figs. 280–285.
Paulli, S. 1900 A, p. 190.
Pictet, F. J. 1853 A, p. 315.
Ridewood, W. G. 1895 A.
Roger, O. 1896 A.
Ryder, J. A. 1877 A.
Rütimeyer, L. 1862 A, p. 671.
Schlosser, M. 1886 A, pp. 253, 432.
Scott, W. B. 1891 B.
Steinmann and Döderlein 1890 A, p. 783.
Struthers, J. 1885 A.
   1893 B.
Trouessart, E. L. 1892 A.

622   FOSSIL VERTEBRATA OF NORTH AMERICA.   [BULL. 179.

Trouessart, E. L. 1898 A, p. 790.
Weithofer, A. 1888 A, p. 273.
Wheatley, C. M. 1871 A.
Wilckens, M. 1884 A.
    1884 B.
    1885 A.
    The North American species of this genus have been greatly confused, and many of the identifications of authors are erroneous. In this work the species are accepted on the authority of Mr. J. W. Gidley, of the American Museum of Natural History, New York, who has made an exhaustive study of them. His paper appeared in the Bulletin of the Museum, vol. xiv, 1901, pp. 92-142. He is not to be held responsible for the citations that are made under the various species.

Equus caballus Linn.

Linnæus, C., 1758, Syst. Nat., ed. 10, i, p. 73.
Agassiz, L. in Holmes, F. S. 1858 A. ("Horse.")
Boule, M. 1899 A, p. 531, figs. 1, 6, 9, 12, 19.
Broadhead, G. C. 1870 A. ("Horse;" species doubtful.)
Buckland, W. 1831 A, p. 595, pl. iii, figs. 14-16. ("Horse;" species doubtful.)
Cope, E. D. 1884 G, pp. 10, 11.
    1887 N, p. 1072, fig. 43.
    1893 A, pp. 66, 82.
Dana, J. D. 1879 A, p. 233.
Emmons, E. 1858 B, p. 196, figs. 18-21. (E. caballus; identification doubtful.)
Gidley, J. W. 1900 A, p. 111, figs. 4, 5, A.
Holmes, F. S. 1858 A. ("Horse.")
Huxley, T. H. 1872 A, p. 295.
Leidy, J. in Holmes, F. S. 1858 A. ("Horse.")
    1860 B, p. 101.
    1869 A, pp. 261, 399. (E. caballus, E. fossilis.)
    1871 C, p. 360. (E. fossilis.)
    1889 D, p. 16.
    1889 E, p. 19.
Leidy, J. in Whitney, J. D. 1879 A, p. 257.
Lydekker, R. 1882 A, p. 71.
    1886 A, pp. 73, 78, 80, 82, 83, 86.
Marsh, O. C. 1879 D.
Owen, R. 1845 B, p. 572, pls. cxxxvi, cxxxvii.
    1868 A, p. 352, figs. 280-285.
    1890 A.
Pavlow, M. 1888 A.
    1890 A.
Richardson, J. 1854 A, p. 17. (E. fossilis.)
Roger, O. 1896 A, p. 180.
Swallow, G. C. 1866 A. ("Horse;" possibly not E. caballus.)
Wilckens, M. 1884 B, p. 337.
    1884 A, p. 16.
Zittel, K. A. 1893 B, pp. 234, 235, figs. 169G, 170F; p. 257, fig. 202.
    Pleistocene: Alaska, California, Kansas, Missouri, North Carolina, South Carolina, Texas.
    Remains supposed to belong to the same species as our domestic horse have been found in various places in North America. In some cases the identifications have been open to question; in other cases the remains may have been derived from the introduced race. The former existence of the species in Alaska and in California appears well established.

Equus complicatus Leidy.

Leidy, J. 1858 C, p. 11.
Blake, C. C. 1863 A, p. 28. (E. nearcticus Blake.)
Carpenter, W. M. 1838 A, p. 202, figs. 1-3. ("Horse.")
Cooper, Wm. 1831 A, p. 207. ("Equus.")
Cooper, Smith and De Kay, 1831 A, p. 371. ("Horse.")
Cope, E. D. 1869 E, p. 176.
    1869 N, p. 741.
    1878 H, p. 389. (E. major.)
    1882 FF, p. 922. (E. major.)
    1884 G, pp. 10, 11. (E. major.)
    1884 JJ, p. 41. (E. major.)
    1889 F. (E. major.)
    1889 S, p. 980. (E. major.)
    1892 E, p. 251. (E. major.)
    1892 J, p. 123. (E. major.)
    1893 A, p. 83. (E. major.)
    1896 H, p. 463, pl. xi, fig. 8; pl. xii. (E. intermedius.)
    1899 A, pp. 255, 259. (E. complicatus, E. intermedius.)
Cragin, F. W. 1896 A, p. 53.
De Kay, J. E. 1842 C, p. 108. E. major; no description, no figures, no types.)
Gibbes, R. W. 1850 B. (E. americanus.)
    1850 C. (E. americanus.)
Gidley, J. W. 1900 A, p. 114. (E. eous.)
Hay, O. P. 1899 A, p. 593. (E. eous, to replace E. intermedius.)
Holmes, F. S. 1850 A. ("Equus.")
Kindle, E. M. 1898 A, p. 485.
Leidy, J. 1847 A, p. 265, pl. ii, figs. 1, 4-6. (E. americanus, preoccupied.)
    1847 C, p. 328. (E. americanus.)
    1851 K, p. 140. (E. americanus.)
    1853 E. (E. americanus.)
    1854 I, p. 200. (E. americanus.)
    1860 B, p. 100, pl. xv, figs. 2-5, 7, 9, 11-15; pl. xvi, figs. 24-26, 30, 31.
    1868 A, p. 175.
    1869 A, pp. 265, 399. (E. major.)
    1871 A, p. 63. (E. major.)
    1871 C, p. 360. (E. major.)
    1871 F, p. 113.
    1873 B, pp. 244, 321, pl. xxxiii, figs. 3-18.
    1877 A, p. 212. (E. major.)
    1884 A. (E. major.)
    1889 F, p. 38, figures. (E. major.)
    1890 A, p. 182. (Hippotherium princeps.)
Leidy and Lucas 1896 A, pp. viii, 49; pl. ii, figs. 12, 13. (E. major.)
Miller, G. S. 1899 B, p. 372. (E. major.)
Roger, O. 1896 A, p. 176 (Hipparion princeps); p. 178 (E. major); p. 179 (E. intermedius).
Skilton, — 1855 A. (E. major.)
Trouessart, E. L. 1898 A, p. 790. (E. major.)
Wailes, B. C. L. 1854 A, p. 286. (E. americanus.)
Wilckens, M. 1884 B, p. 337.
Williston, S. W. 1897 I, p. 303. (E. major.)
    1898 I, p. 92. (E. major.) .
Woodworth, J. B. 1900 B, p. 459. ("Fossil horse.")
    Pleistocene; Mississippi, Missouri, South Carolina, New Jersey, Pennsylvania, New York, Massachusetts, Indiana, Illinois, Kentucky, Texas, Kansas, Louisiana, Oregon, Washington, Nevada.

Equus conversidens Owen.

Owen, R. 1869 A, p. 267.
Cope, E. D. 1884 G, p. 12.
Leidy, J. 1869 A, p. 400.
        1871 C, p. 360.
Owen, R. 1870 B, pl. lxi, figs. 1, 3.
Roger, O. 1896 A, p. 179.
Trouessart, E. L. 1896 A, p. 791.
Wilckens, M. 1884 B, p. 328.
    Pleistocene; Valley of Mexico.
    This species is not known to occur within the
    limits included in this work.

Equus crenidens Cope.

Cope, E. D. 1884 G, pp. 10, 12.
        1885 S, p. 1208.
        1887 N, p. 1072, fig. 42.
        1889 F.
        1899 A, p. 255.
Cope and Wortman 1884 A, p. 40.
Felix and Lenk 1891, p. 131.
Roger, O. 1896 A, p. 178.
Trouessart, E. L. 1896 A, p. 790.
    Pleistocene; Mexico.

Equus excelsus Leidy.

Leidy, J. 1858 E, p. 26.
Amer. Geologist 1892 A.
Cope, E. D. 1884 G, pp. 10, 13.
        1884 JJ, p. 41.
        1885 S.
        1887 N, p. 1076.
        1889 F.
        1889 S, p. 980.
        1891 J.
        1892 C.
        1893 A, p. 80, pl. xx, figs. 4, 5; pl. xxii, figs. 4, 5.
Dana, J. D. 1879 A, p. 233.
Dawkins, W. B. 1870 A.
Felix and Lenk 1891 A, pp. 128, 135, pl. xxx, fig. 7.
Hayden, F. V. 1858 B, p. 158.
Leidy, J. 1869 A, pp. 266, 400, pl. xix, fig. 39; pl.
    xxi, fig. 31.
        1870 J, p. 67.
        1871 C, p. 360.
        1873 B, p. 242.   (Syn. of E. occidentalis.)
Roger, O. 1896 A, p. 178.   (Syn. of E. occidentalis.)
Trouessart, E. L. 1896 A, p. 790.
Whitney, J. D. 1879 A.
Wilckens, M. 1884 B, p. 337.
Williston, S. W. 1897 I, p. 303.
        1896 I, p. 92.
    Pleistocene; South Dakota, Nebraska, Texas,
    Mexico, Oregon?, Kansas.

Equus fraternus Leidy.

Leidy, J. 1860 B, p. 100, pl. xv, figs. 6, 8, 16–18;
    pl. xvi, figs. 23, 28.
Conrad, T. A. 1869 A, p. 359.
        1871 A.
Cooper, Smith, and De Kay 1831 A, p. 370, in part.
    ("Equus.")
Cope, E. D. 1869 N, p. 741.
        1884 JJ, p. 41.
        1885 S, p. 1209.
        1892 J, p. 125.   (Tomolabis.)

Cope, E. D. 1896 H, p. 465.
        1899 A, pp. 255, 257.   (E. fraternus frater-
    nus.)
Harlan, R. 1835 B, p. 267.   (E. caballus.)
Hayes, Seth 1896 A.
Kindle, E. M. 1898 A, p. 485.
Leidy, J. 1847 A, p. 263, in part (E. curvidens);
    pl. ii, figs. 2, 3 (E. americanus).
        1858 C, p. 11.   (Description insufficient.)
        1869 A, pp. 266, 400.
        1871 A, p. 63.
        1871 C, p. 360.
        1877 A, p. 212.
        1889 D, p. 16.
        1889 F, p. 38.
        1892 A.
Lucas, F. A. 1899 A, p. 764.   (E. fraternus?.)
Marsh, O. C. 1874 B, p. 255.
Roger, O. 1896 A, p. 178.
Trouessart, E. L. 1896 A, p. 790.
Wilckens, M. 1884 B, p. 336.
    Pleistocene; South Carolina, North Carolina,
    Mississippi, Georgia, Kentucky, Texas, New
    Jersey, Indiana, Alaska?, Pennsylvania, Louisi-
    ana.

Equus giganteus Gidley.

Gidley, J. W., 1901, Bull. Amer. Mus. Nat. Hist.,
    xiv, p. 137, fig. 27.
Cope, E. D. 1885 S, p. 1208, pl. xxxvii, fig. 4.   (E.
    crenidens.)
    Pleistocene; Texas.

Equus occidentalis Leidy.

Leidy, J. 1865 C, p. 94.
Bowers, S. 1889 A, p. 391.
Cope, E. D. 1878 H, p. 389.
        1884 G, pp. 10, 11.
        1884 JJ, p. 41.
        1885 S.
        1887 N, p. 1076.
        1889 F.
        1889 S, p. 980.
        1892 A, p. 252.
        1893 A, pp. 66, 81.
        1896 H, p. 464.
        1899 A, p. 255.
Dana, J. D. 1879 A, p. 233.
Le Conte, J. 1883 A.   (E. occidentalis, E. pacifi-
    cus.)
Leidy, J. 1869 A, pp. 266, 400.   (E. excelsus, in
    part.)
        1873 B, pp. 239, 322, pl. xxxiii, figs. 1, 2.
Lydekker, R. 1882 A, p. 72.
Roger, O. 1896 A, p. 178.
Trouessart, E. L. 1896 A, p. 790.
Whitney, J. D. 1879 A, p. 257.
Wilckens, M. 1884 B, p. 337.
Williston, S. W. 1897 I, 303.
        1898 I, p. 92.
    Pleistocene; California.

Equus pacificus Leidy.

Leidy, J. 1868 D, p. 32.
Gidley, J. W., 1901, Bull. Amer. Mus. Nat. Hist.
    xiv, p. 116.
Le Conte, J. 1883 A.

Leidy, J. 1869 A, p. 400.
  1871 C, p. 360.
  1871 D, p. 50.
  Pleistocene; Oregon.

Equus pectinatus Cope.

Cope, E. D. 1899 A, pp. 255, 257.   (E. fraternus pectinatus.)
  Pleistocene; Pennsylvania, Illinois?.

Equus scotti Gidley.

Gidley, J. W. 1900 A, p. 111, figs. 1-3, 5A.
  Pleistocene (Equus beds); Texas.

Equus semiplicatus Cope.

Cope, E. D. 1893 A, p. 80, pl. xxiii, figs. 2, 3.
  1889 F, p. 161.   (E. fraternus, not of Leidy.)
  1899 A, p. 255.
Trouessart, E. L. 1898 A, p. 791.
  Pleistocene; Texas.

Equus tau Owen.

Owen, R. 1870 B, p. 565, pl. lxi, fig. 4.
Cope, E. D. 1884 G, p. 12 (E. tau); pp. 10, 15 (E. barcenæi).
  1884 JJ, p. 41.   (E. barcenæi.)
  1885 S.   (E. barcenæi.)
  1889 F.   (E. tau, E. barcenæi.)
  1893 A, p. 79.
  1899 A, p. 255.
Felix and Lenk 1891 A, pp. 127, 136, pl. xxx, figs. 6, 6a (E. barcenai); p. 135 (E. tau).
Leidy, J. 1869 A, p. 401.
Owen, R. 1869 A, p. 268.   (No description.)
Roger, O. 1896 A, p. 178.   (E. barcenæi.)
Trouessart, E. L. 1898 A, p. 790.
Wilckens, M. 1884 B, p. 329.
  Pleistocene; Mexico, Texas.

Superfamily TAPIROIDEA.

Gill, T. 1872 B, p. 12 (Tapiroidea); p. 88 (Lophiodontoidea).

LOPHIODONTIDÆ.

By some authors this group is regarded as a subfamily of the Tapiridæ.
Cope, E. D. 1879 C, pp. 228, 232.
  1879 V, p. 771f.
  1881 G, pp. 378, 379, 381.
  1881 O.
  1884 O, pp. 614, 617.
  1887 N, pp. 991, 993.
  1889 C, p. 158.
Flower, W. H. 1876 B, xiii, p. 327.
  1883 D, p. 428.
Flower and Lydekker 1891 A, p. 373.
Gaudry, A. 1897 A.
Gill, T. 1872 B, p. 12.
Lydekker, R. 1886 A, p. 6.
  1886 D, p. 326.
Nicholson and Lydekker 1889 A, pp. 1354, 1364.

Osborn, H. F. 1893 D, p. 37.
  1898 I, p. 79.
Osborn and Wortman 1892 A, p. 92 (Lophiodontidæ, Lophiodontinæ); p. 127 (Helaletidæ, Helaletinæ).
  1895 A, p. 358.
Rütimeyer, L. 1888 A, p. 27.
  1891 A, p. 21.
Trouessart, E. L. 1898 A, p. 760.
Woodward, A. S. 1898 B, p. 322.
Wortman, J. L. 1882 B, p. 723.
  1883 A, p. 710.
  1886 A, p. 478.
Wortman and Earle 1893 A, p. 173.   (Helaletidæ =Helaletinæ+Colodontinæ.)
Zittel, K. A. 1893 B, p. 274.   (Lophiodontinæ.)

LOPHIODON Cuv.   Type *L. tapiroides* Cuv.

The following references are given to the literature of the genus *Lophiodon*, the type of its family. No American species are here recognized as belonging to this genus, although several have at times been assigned to it by authors.
Cuvier, G., 1822, Oss. Foss., ed. 2, vol. ii, pt. 1, p. 176.
Blainville, H. M. D. 1864 A, iv, Y, p. 80, plates.
Cope, E. D. 1873 E, p. 595.
  1873 H, p. 212.
  1881 G, p. 381.
  1884 O, p. 618.
  1887 N, pp. 994, 998.
Filhol, H. 1885 A.
  1888 A.
Flower and Lydekker 1891 A, p. 373.
Gaudry, A. 1877 A.
  1878 B, p. 63.
Gervais, P. 1859 A, p. 117.

Leidy, J. 1871 C, p. 357.
  1872 B, p. 361.
  1872 C, p. 19.
  1873 B, p. 219.
Lydekker, R. 1886 A, p. 6.
Maack, G. A. 1865 A, p. 6.
Major, Fors. 1878 A, pp. 102, 204.
Nicholson and Lydekker 1889 A p 1354.
Osborn, H. F. 1892 B.
  1898 I, p. 87, fig. 5.
Osborn and Wortman 1892 A, p. 131.
  1895 A, pp. 358, 361.
Owen, R. 1844 A, p. 224.
  1845 B, p. 606.
  1848 A, p. 126.
  1857 C.
  1858 B, p. 78.
  1860 E, p. 323.
Pictet, F. J. 1853 A, p. 804.

Pomel, A. 1847 A, p. 203.
Rütimeyer, L. 1862 A.
 1891 A, p. 20.
Schlosser, M. 1886 B, p. 27.
 1888 A, p. 585.

Steinmann and Döderlein 1890 A, p. 771.
Wilckens, M. 1885 B, p. 209.
Wortman, J. L. 1896 A, p. 85.
Wortman and Earle 1898 A, p. 173.
Zittel, K. A. 1893 B, p. 276.

HEPTODON Cope.   Type *Lophiodon ventorum* Cope.

Cope, E. D. 1882 GG.
Branco, W. 1897 A, p. 20.
Cope, E. D. 1884 O, pp. 618, 658.
 1887 N, p. 994.
Earle, C. 1898 C, p. 398.
Nicholson and Lydekker 1889 A, p. 1356.
Osborn, H. F. 1892 B, p. 763.
Osborn and Wortman 1892 A, pp. 127–182.
 1896 A, p. 358.
Schlosser M. 1886 B, p. 28.
Scott, W. B. 1892 A, p. 431.
Wortman, J. L. 1896 A, p. 85.
Wortman and Earle 1898 A.
Zittel, K. A. 1898 B, p. 275.

Matthew, W. D. 1899 A, p. 33.
Roger, O. 1896 A, p. 181.
Trouessart, E. L. 1898 A, p. 760.
Wortman, J. L. 1896 A, p. 86.
 Eocene (Wasatch); Wyoming.

Heptodon singularis Cope.

Cope, E. D. 1875 C, p. 19.   (Hyrachyus.)
 1877 K, p. 267, pl. lxvi, fig. 17.   (Hyrachyus.)
 1884 O, p. 618, fig. 31.
 1887 N, p. 990, fig. 4.
Matthew, W. D. 1899 A, p. 38.
Roger, O. 1896 A, p. 181 (Heptodon); p. 185 (Hyrachyus).
Schlosser, M. 1886 B, p. 28.   (Pachynolophus.)
Trouessart, E. L. 1898 A, p: 745 (Hyrachyus?); p. 761 (Heptodon?).
Wortman, J. L. 1896 A, p. 86.
 Eocene (Wind River); New Mexico.

Heptodon calciculus Cope.

Cope, E. D. 1880 N, p. 747.   (Lophiodon.)
 1881 D, p. 197.   (Pachynolophus.)
 1882 GG.
 1884 O, p. 656, pl. xxia, fig. 6.
 1887 N, p. 997, fig. 9.
Matthew, W. D. 1899 A, p. 36.
Osborn and Wortman 1892 A, p. 127, fig. 18.
Roger, O. 1896 A, p. 181.
Trouessart, E. L. 1898 A, p. 760.
Wortman, J. L. 1896 A, p. 86.
Wortman and Earle 1898 A, p. 177.
 Eocene (Wind River); Wyoming.

Heptodon ventorum Cope.

Cope, E. D. 1880 N, p. 747.   (Lophiodon.)
 1881 D, p. 197.   (Pachynolophus.)
 1882 E, p. 187.   (Pachynolophus.)
 1882 GG.
 1884 O, p. 654, pl. xxixa, figs. 4, 5.
 1887 N, p. 997, fig. 9.
Matthew, W. D. 1899 A, p. 36.
Roger, O. 1896 A, p. 180.
Trouessart, E. L. 1898 A, p. 760.
Wortman, J. L. 1896 A, p. 86.
Zittel, K. A. 1898 B, p. 275, fig. 217.
 Eocene (Wind River); Wyoming.

Heptodon posticus Cope.

Cope, E. D. 1882 E, p. 187.   (Pachynolophus.)
 1882 GG.
 1884 O, p. 654, pl. lvi, fig. 6.

HELALETES Marsh.   Type *H. boops* Marsh.

Marsh, O. C. 1872 I, p. 218.
Cope, E. D. 1879 A, p. 48.   (Syn. of Tapirulus.)
 1881 G, p. 381.
 1884 O, p. 618 (Helaletes); p. 693 (Desmatotherium).
 1887 N, p. 994 (Helaletes, Dilophodon); p. 1000 (Desmatotherium).
Flower and Lydekker 1891 A, p. 375.
Gaudry, A. 1897 A, p. 321.
Marsh, O. C. 1877 E.
Nicholson and Lydekker 1889 A, p. 1356.
Osborn, H. F. 1890 D, p. 505.
 1892 B, p. 763.
Osborn and Wortman 1892 A, pp. 82, 93, 127, 130.   (Helaletes, Desmatotherium, Dilophodon.)
 1896 A, p. 358, 360.
Schlosser, M. 1886 B, p. 27.
Scott, W. B. 1883 A, p. 46 (Desmatotherium, type *D. guyotii*); p. 51 (Dilophodon, type *D. minusculus*); p. 53 (Helaletes).
 1892 A, p. 431.
Scott and Osborn 1883 A, p. 12.   (Desmatotherium.)
 1887 A, p. 260.   (Desmatotherium.)

Trouessart, E. L. 1898 A, p. 761.   (Helaletes.)
Wortman and Earle 1893 A.
Zittel, K. A, 1898 B, p. 275.

Helaletes boops Marsh.

Marsh, O. C. 1872 I, p. 218.
Cope, E. D. 1873 E, p. 605.   (Hyrachyus.)
 1873 U, p. 6.   (Hyrachyus.)
 1884 O, p. 661.
King, C. 1878 A, p. 404.
Matthew, W. D. 1899 A, p. 45.
Osborn and Wortman 1892 A, p. 131.
 1896 A, p. 360.
Roger, O. 1896 A, p. 181.
Trouessart, E. L. 1898 A, p. 761.
Wortman and Earle 1898 A, p. 177.
 Eocene (Bridger); Wyoming.

Helaletes guyotii (Scott).

Scott, W. B. 1883 A, p. 46, pl. viii.   (Desmatotherium.)
Cope, E. D. 1883 II, p. 970.   (Desmatotherium.)
Matthew, W. D. 1899 A, pp. 45, 49.
Osborn, H. F. 1892 B.

Roger, O. 1896 A, p. 181.
Trouessart, E. L. 1898 A, p. 761.
Eocene (Bridger); Wyoming. (Uinta); Utah.

Helaletes minusculus (Scott).

Scott, W. B. 1883 A, p. 51, pl. viii, fig. 4. (Dilophodon.)
Cope, E. D. 1883 II, p. 970. (Dilophodon.)
Matthew, W. D. 1899 A, p. 45.
Roger, O. 1896 A, p. 181.
Trouessart, E. L. 1898 A, p. 961.
Wortman and Earle 1893 A, p. 180.
Eocene (Bridger); Wyoming.

Helaletes nanus Marsh.

Marsh, O. C. 1871 D, p. 37. (Lophiodon.)

Cope, E. D. 1873 E, p. 605. (Hyrachyus.)
1873 U, p. 6. (Hyrachyus.)
1884 O, p. 661. (Hyrachyus.)
Leidy, J. 1872 B, p. 361 (Hyrachyus); p. 362 (Lophiodon.)
1872 C, p. 20. (Hyrachyus.)
1873 B, pp. 67, 327, pl. ii, fig. 14; pl. vi, fig. 42; pl. xxvi, fig. 11; pl. xxvii, figs. 21, 22. (Hyrachyus.)
Marsh, O. C. 1872 I, p. 218.
Matthew, W. D. 1899 A, p. 45.
Osborn, H. F. 1895 A, p. 505.
Osborn and Wortman 1895 A, p. 360.
Trouessart, E. L. 1898 A, p. 761. (Syn. of H. boops.)
Eocene (Bridger); Wyoming.

COLODON Marsh.   Type *C. luratus* Marsh.

Marsh, O. C. 1890 D, p. 524.
Earle, C. 1893 C, p. 393. (Colodon.)
Gaudry, A. 1897 A. (Colodon.)
Osborn, H. F. 1898 D. (Colodon.)
Osborn and Wortman 1895 A, pp. 358, 362. (Colodon.)
Scott, W. B. 1892 A, p. 430. (Mesotapirus.)
Scott and Orborn 1890 A, p. 524. (Mesotapirus, conjectural.)
Wortman, J. L. 1893 A. (Colodon.)
Wortman and Earle 1893 A, p. 173. (Colodon, Mesotapirus.)
Zittel, K. A. 1893 B, p. 275. (Colodon.)

Colodon dakotensis Osb. and Wort.

Osborn and Wortman 1895 A, p. 362, fig. 7. (Colodon.)
Hatcher, J. B. 1896 B, p. 171, pl. iii, figs. 3, 3 a. (Colodon.)
Matthew, W. D. 1899 A, p. 56. (Colodon.)
Trouessart, E. L. 1898 A, p. 761. (Colodon.)
Oligocene (White River); South Dakota.

Colodon longipes Osb. and Wort.

Osborn and Wortman 1894 A, p. 214. (?Mesohippus.)
Hatcher, J. B. 1896 B, p. 169. (Syn.? of Colodon occidentalis.)
Matthew, W. D. 1899 A, p. 56. (Colodon.)
Osborn and Wortman 1895 A, p. 366. (Colodon ?.)
Roger, O. 1896 A, p. 174. (Mesohippus ?.)
Trouessart, E. L. 1898 A, p. 761. (Colodon.)
Oligocene (White River); South Dakota.

Colodon luxatus Marsh.

Marsh, O. C. 1890 D, p. 524. (Colodon.)
Hatcher, J. B. 1896 B, p. 170. (Syn. of Colodon occidentalis.)

Marsh, O. C. 1893 D, p. 411, pl. x. (Colodon.)
Matthew, W. D. 1899 A, p. 56. (Colodon.)
Roger, O. 1896 A, p. 181. (Syn. of Colodon occidentalis.)
Wortman and Earle 1893 A, p. 174. (Syn. of Colodon occidentalis.)
Oligocene (White River); South Dakota.

Colodon occidentalis (Leidy).

Leidy, J. 1868 I, p. 232. (Lophiodon.)
Hatcher, J. B. 1896 B, p. 170, pl. iii, figs. 2, 6. 7. (Colodon.)
Leidy, J. 1869 A, pp. 239, 391, pl. xxi, figs. 28–30 (Lophiodon.)
1870 P, p. 112. (Lophiodon.)
1871 C, p. 357. (Lophiodon.)
1873 B, pp. 219, 327 (Lophiodon occidentalis), pl. ii, fig. 1 (Lophiodon oregonensis). ·
Matthew, W. D. 1899 A, pp. 56, 63. (Colodon.)
Osborn and Wortman 1895 A, pp. 362, 365. (Colodon.)
Roger, O. 1896 A, p. 181. (C. occidentalis.)
Scott and Osborn, 1890 A, p. 524.
Trouessart, E. L. 1898 A, p. 762.
Wortman and Earle 1893 A, p. 174, figs. 6, 7. (Colodon.)
Oligocene (White River); South Dakota; Miocene (John Day); Oregon.

Colodon procuspidatus Osb. and Wort.

Osborn and Wortman 1895 A, pp. 362, 364, fig. 8.
Hatcher, J. B. 1896 B, p. 170. (Syn.? of C. occidentalis.)
Matthew, W. D. 1899 A, p. 56. (Colodon.)
Trouessart, E. L. 1898 A, p. 170. (Colodon.)
Oligocene (White River); South Dakota.

TAPIRIDÆ.

Blainville, H. M. D. 1864 A, iv, Z, p. 1. pls. i–vi.
Branco, W. 1897 A, p. 19.
Cope, E. D. 1874 M.
1879 C, p. 228.
1881 G, pp. 378, 395.
1881 O.
1884 O, pp. 615, 693.
1887 N.
Cuvier, G. 1834 A, pp. 273–430, 542, with plates.

Earle, C. 1893 C.
1896 B.
Flower, W. H. 1876 B, xiii, p. 328.
Flower and Lydekker 1891 A, p. 370.
Gaudry, A. 1897 A.
Ghigi, A. 1900 A, p. 17.
Gray, J. E. 1867 A, p. 878. (Tapiridæ, Tapirinæ.)
1869 A, p. 252 (Tapiridæ); p. 253 (Tapirinæ).
Hatcher, J. B. 1896 B.

Huxley, T. H. 1872 A, p. 310.
Leidy, J. 1871 C, p. 357.
Lydekker, R. 1886 A, p. 2.
   1896 B.
Nicholson and Lydekker 1889 A, p. 1354.
Osborn, H. F. 1890 D, p. 542.
   1893 D, p. 37. (Tapiridæ, Tapirinæ; Systemodontinæ, in part.)
   1898 l, p. 79.
Osborn and Wortman 1892 A, p. 93. (Systemodontinæ, Tapirinæ.)
Owen. R. 1845 B, p. 604.
Pavlow, M. 1887 A, p. 353.
   1888 A, p. 173.

Roger, O. 1896 A, p. 180.
Rütimeyer, L. 1888 A, p. 14.
Schlosser, M. 1886 A, p. 253.
   1886 B, p. 27.
   1888 A.
Slade, D. D. 1893 A.
Steinmann and Döderlein 1890 A, p. 771.
Trouessart, E. L. 1898 A, p. 760 (Tapiridæ); p. 764 (Tapirinæ).
Woodward, A. S. 1898 B, p. 321.
Wortman and Earle 1893 A. (Tapiridæ, Tapirinæ; Systemodontinæ, in part.)
Zittel, K. A. 1893 B, p. 273 (Tapiridæ); p. 276 (Tapirinæ).

HOMOGALAX Hay.   Type *Systemodon primævus* Wort.

Hay, O. P. 1899 A, p. 593.
Osborn, H. F. 1898 I, p. 87. (Systemodon.)
Trouessart, E. L. 1898 A, p. 765. (Systemodon.)
Wortman, J. L. 1896 A. (Systemodon.)
For the literature of *Systemodon*, which bears more or less on the species here included under *Homogalax*, see under *Eohippus*.

Homogalax primævus (Wort.).

Wortman, J. L. 1896 A, p. 89, fig. 3. (Systemodon.)
Hay, O. P. 1899 A, p. 593.
Matthew, W. D. 1899 A, p. 34. (Systemodon.)
Osborn and Wortman 1892 A, p. 125. (Systemodon tapirinus, in part.)
Trouessart, E. L. 1898 A, p. 765. (Systemodon.)
Eocene (Wasatch); Wyoming.

Homogalax protapirinus (Wort.).

Wortman, J. L. 1896 A, pp. 89, 90. (Systemodon.)

Hay, O. P. 1899 A, p. 593.
Matthew, W. D. 1899 A, p. 34. (Systemodon.)
Trouessart, E. L. 1898 A, p. 765. (Systemodon.)
Eocene (Wasatch); Wyoming.

Homogalax semihians (Cope).

Cope, E. D. 1882 A, p. 184. (Systemodon.)
   1884 O, pp. 619, 622, pl. lvi, figs. 3, 4. (Systemodon.)
Hay, O. P. 1899 A, p. 593.
Matthew, W. D. 1899 A, p. 34. (Systemodon.)
Osborn and Wortman 1892 A, pp. 124, 126. (Systemodon.)
Roger, O. 1896 A, p. 183. (Systemodon.)
Scott, W. B. 1892 A, p. 430. (Systemodon.)
Trouessart, E. L. 1898 A, p. 765. (Systemodon.)
Wortman, J. L. 1896 A, p. 89. (Systemodon.)
Eocene (Wasatch); Wyoming.

ISECTOLOPHUS Scott and Osb.   Type *I. annectens* Scott and Osb.

Scott and Osborn 1887 A, p. 260.
Cope, E. D. 1887 N, p. 994.
   1890 C, p. 470.
Earle, C. 1893 C.
Flower and Lydekker 1891 A, p. 374.
Hatcher, J. B. 1896 B, p. 177.
Nicholson and Lydekker 1889 A, p. 1355.
Osborn, H. F. 1890 D, p. 518.
   1892 B, p. 764.
Osborn, Scott, and Speir 1878 A, p. 54, part. (Helaletes.)
Scott, W. B. 1892 A, p. 430.
Wortman and Earle 1893 A.
Zittel, K. A. 1893 B, p. 277.

Isectolophus annectens Scott and Osb.

Scott and Osborn 1887 A, p. 260.
Hatcher, J. B. 1896 B, p. 177.

Matthew, W. D. 1899 A, p. 49.
Osborn, H. F. 1890 D, p. 520, pl. x
   1892 B, p. 764.
   1895 A, p. 98.
Roger, O. 1896 A, p. 183.
Trouessart, E. L. 1898 A, p. 765.
Wortman and Earle 1893 A, p. 171.
Eocene (Uinta); Wyoming.

Isectolophus latidens Scott and Osb.

Osborn, Scott, and Speir 1878 A, p. 54. (Helaletes.)
Hatcher, J. B. 1896 B, p. 177.
Matthew, W. D. 1899 A, p. 45.
Osborn, H. F. 1890 D, p. 519.
Trouessart, E. L. 1898 A, p. 765.
Wortman and Earle 1893 A, p. 170.
Eocene (Bridger); Wyoming.

PROTAPIRUS Filhol.   Type *P. priscus* Filhol=P. filholi Troues.

Filhol, H. 1877 A, p. 131.
Cope, E. D. 1887 N, pp. 994, 996.
Earle, C. 1893 C.
   1896 B, p. 934.
Filhol, H. 1885 A, p 1.
Gaudry, A. 1897 A, p. 320.
Hatcher, J. B. 1896 B.
Wortman, J. L. 1898 A.

Wortman and Earle 1893 A, p. 161
Zittel, K. A. 1893 B, p. 278.

Protapirus obliquidens Wort. and Earle.

Wortman and Earle 1893 A, p. 162, figs. 1–4.
Earle, C. 1893 C, p. 395.
Hatcher, J. B. 1896 B, p. 162.

Matthew, W. D. 1899 A, p. 56.
Roger, O. 1896 A, p. 183.
Trouessart, E. L. 1898 A, p. 766.
   Oligocene (White River); South Dakota.

Protapirus simplex Wort. and Earle.

Wortman and Earle 1893 A, p. 168, fig. 1, A.
Earle, C. 1893 C, p. 394.
Hatcher, J. B. 1896 B, p. 168, pl. iii, figs. 5, 5a.
Matthew, W. D. 1899 A, p. 56.
Roger, O. 1896 A, p. 183.

Trouessart, E. L. 1898 A, p. 766.
   Oligocene (White River); South Dakota.

Protapirus validus Hatcher.

Hatcher, J. B. 1896 B, p. 162, pl. ii, figs. 1–4; text figs. 1, 2.
Earle, C. 1896 B, p. 934.
Matthew, W. D. 1899 A, p. 56.
Roger, O. 1896 A, p. 183.
Trouessart, E. L. 1898 A, p. 766.
   Oligocene (White River); South Dakota.

        TAPIRAVUS Marsh.   Type *T. validus* Marsh.

Marsh, O. C. 1877 D, p. 252.
Cope, E. D. 1887 N, pp. 996, 1007.
Earle, C. 1893 C, p. 345.
Hatcher, J. B. 1896 B, p. 178.
Marsh, O. C. 1877 E.
Scott, W. B. 1883 A, p. 50.
Zittel, K. A. 1893 B, p. 278.

Tapiravus rarus Marsh.

Marsh, O. C. 1877 D, p. 252.
King, C. 1878 A, p. 430.

Matthew, W. D. 1899 A, p. 71.
Roger, O. 1896 A, p. 183.
Trouessart, E. L. 1898 A, p. 766.
   "Pliocene"; Rocky Mountain region.

Tapiravus validus Marsh.

Marsh, O. C. 1871 F. (Lophiodon.)
Hatcher, J. B. 1896 B, p. 178.
Marsh, O. C. 1877 D, p. 252.
Trouessart, E. L. 1898 A, p. 766.
   Miocene; New Jersey.

        TANYOPS Marsh.   Type *T. undans* Marsh.

Marsh, O. C. 1894 M, p. 348.

Tanyops undans Marsh.

Marsh, O. C. 1894 M, p. 348.
   Miocene; South Dakota.

        TAPIRUS Brisson.   Type *Hippopotamus terrestris* Linn.

Brisson, 1762, Regnum Anim., p. 176. (*Fide* C. H. Merriam.)
Blainville, H. M. D. 1864 A, iv, Z, p. 1, plates.
Branco, W. 1897 A, p. 20.
Cope, E. D. 1873 E, p. 594.
   1873 H, p. 212.
   1881 G, p. 395.
   1887 N, p. 1007.
   1889 C, p. 254.
Cope and Wortman 1884 A, p. 26.
Cuvier, G. 1834 A, iii, pp. 273–430, with plates.
Döderlein, L. 1878 A.
Earle, C. 1893 A.
   1893 C.
Flower and Lydekker 1891 A. p. 370.
Ghigi, A. 1900 A, p. 17.
Gray, J. E. 1867 A, p. 879.
   1869 A, p. 254.
Hatcher, J. B. 1896 A, p. 173.
Huxley, T. H. 1872 A, p. 311.
Kowalevsky, W. 1873 A, pp. 218, 222.
Leuthardt, F. 1891 A, p. 113.
Lydekker, R. 1886 A, p. 3.
Meyer, H. 1867 A.
Nicholson and Lydekker 1889 A, p. 1354.
Owen, R. 1845 B, p. 604.
   1866 B, p. 444, figs. 299, 301.
   1868 A, p. 357.
Paulli, S. 1900 A, p. 182.
Pictet, F. J. 1853 A, p. 301.
Rütimeyer, L. 1862 A.
Scott, W. B. 1892 A, p. 431.
Trouessart, E. L. 1898 A, p. 767.
Wortman and Earle 1893 A.
Zittel, K. A. 1893 B, p. 278.

Tapirus haysii Leidy.

Leidy, J. 1860 B, p. 106, pl. xvii, figs. 4, 7–10.
Cope, E. D. 1869 E, p. 176, pl. iii, fig. 6.
   1869 J.
   1871 I, p. 96.
   1895 F, p. 447.
   1896 G, p. 597.
   1899 A, p. 253.
Cope and Wortman 1884 A, p. 26.
Hays, J. 1852 A. ("Tapir.")
Leidy, J. 1852 G. (No description.)
   1852 I. (No description.)
   1860 B, p. 106, pl. xvii, figs. 4, 7–10.
   1873 B, p. 391.
Meyer, H. 1867 A, p. 169.
Roger, O. 1896 A, p. 184.
Trouessart, E. L. 1898 A, p. 769. [T. (Elasmognathus).]
Wailes, B. C. L. 1854 A, p. 286.
   Pleistocene; South Carolina, Mississippi, Virginia, Pennsylvania, Indiana.

Tapirus terrestris (Linn.).

Linnæus, C., 1758, Syst. Nat., ed. 10, i, p. 74. (Hippopotamus.)
   Unless otherwise indicated, the following authors call the species *T. americanus.*
Blake, W. P. 1868 A. ("Tapir.")
Carpenter, W. M. 1846 A, p. 247, figs. 3, 4. ("Tapir.")
Cope, E. D. 1871 I, p. 96.
Cope and Wortman 1884 A, p. 26. (T. terrestris.)
Earle, C. 1893 A.
Ghigi, A. 1900 A, p. 22.
Gray, J. E. 1867 A, p. 879. (T. terrestris.)

Hatcher, J. B. 1896 B, p. 174.
Holmes, F. S. 1858 B.
   1859 A, p. 183.
Kindle, E. M. 1898 A, p. 485. (T. terrestris?.)
Leidy, J. 1849 A.
   1860 B, p. 106, pl. xvii, figs. 1–3, 6, 11, 12.
   1869 A, p. 391.
   1884 B.
   1889 B.
   1889 E, p. 19.
   1892 A.

Roger, O. 1896 A, p. 185.
Wailes, B. C. L. 1854 A, App., p. 285.
Wheatley, C. M. 1871 A. ("Tapir.")
Whitney, J. D. 1879 A, p. 250.
   Pleistocene; Virginia, Illinois, California, Texas, Louisiana, Kentucky, Mississippi, Indiana, Ohio, South Carolina.
   Prof. E. D. Cope (1899 A, p. 253) refers all specimens of fossil *Tapirus* found in the United States to *T. haysii.*

Superfamily *BRONTOTHERIOIDEA.*

Osborn, H. F. 1898 I, p. 79. (Titanotheroidea.)

BRONTOTHERIIDÆ.

Marsh, O. C. 1873 H, p. 486. (Brontotheridæ.)
Brandt, J. F. 1878 A, pp. 10, 19. (Brontotherinæ.)
Cope, E. D. 1881 G, pp. 378, 379, 397. (Menodontidæ.)
   1884 O, pp. 615, 713. (Menodontidæ.)
   1887 L, p. 926. (Menodontidæ.)
   1887 N. (Menodontidæ.)
   1889 C, p. 158. (Menodontidæ.)
   1898 B, p. 185. (Menodontidæ.)
Earle, C. 1892 C, p. 273. (Titanotheriidæ.)
Flower, W. H. 1876 B, xiii, p. 327. (Titanotheridæ.)
   1876 C, p. 109. (Titanotheriidæ.)
Flower and Lydekker 1891 A, p. 413. (Titanotheriidæ.)
Gervais, P. 1876 B. (Brontotherides.)
Hatcher, J. B. 1893 B, p. 216. (Titanotheriidæ.)
Lydekker, R. 1896 B, pp. 170, 377. (Titanotheriidæ.)
Marsh, O. C. 1874 A, p. 66. (Brontotheridæ.)
   1874 D, p. 81. (Brontotheridæ.)
   1875 B, p. 245. (Brontotheridæ.)
   1876 C. (Brontotheridæ.)
   1876 G, p. 60. (Brontotheridæ.)
   1877 E. (Brontotheridæ.)

Marsh, O. C. 1884 F, p. 190. (Brontotheridæ.)
   1887 B, p. 336. (Brontotheridæ.)
   1889 A, p. 163. (Brontotheridæ.)
   1889 H. (Brontotheridæ.)
   1895 D, p. 498. (Brontotheridæ.)
   1897 D, p. 166. (Brontotheridæ.)
Nature 1874 A.
Nicholson and Lydekker 1889 A, p. 1374. (Titanotheriidæ.)
Osborn, H. F. 1890 D, p. 514. (Titanotheriidæ.)
   1893 D. (Titanotheriidæ.)
   1897 D. (Titanotheriidæ.)
   1898 I, p. 79. (Titanotheriidæ.)
   1900 C, p. 570. ("Titanotheres.")
Osborn and Wortman 1892 A, p. 93. (Titanotheriidæ.)
Roger, O. 1996 A, p. 192. (Titanotheridæ.)
Schlosser, M. 1886 B, p. 31. (Menodontidæ.)
   1888 A, p. 588. ("Brontotherien.")
Scott and Osborn 1887 B, p. 157. (Menodontidæ.)
Steinman and Döderlein 1890 A, p. 777. (Brontotheridæ.)
Woodward, A. S. 1898 B, p, 329. (Titanotheriidæ.)
Zittel, K. A. 1893 B, p. 298. (Titanotheriidæ.)

LAMBDOTHERIINÆ.

Earle, C. 1892 C, p. 274. (Palæosyopinæ.)
Flower and Lydekker 1891 A, p. 413. (Lambdotheriidæ.)
Nicholson and Lydekker 1889 A, p. 1371. (Lambdotheriidæ.)
Osborn, H. F. 1898 D, p. 87.

Osborn and Wortman 1892 A, p. 93.
Roger, O. 1896 A, p. 192.
Steinmann and Döderlein 1890 A, p. 777.
Trouessart, E. L. 1898 A, p. 737.
Zittel, K. A. 1893 B, p. 300.

LAMBDOTHERIUM Cope. Type *L. popoagicum* Cope.

Cope, E. D. 1880 N, p. 746.
   1881 G, p. 396.
   1884 O, pp. 69⁴, 709.
   1887 N.
Earle, C. 1892 C.
Flower and Lydekker 1891 A, p. 413.
Nicholson and Lydekker 1889 A, p. 1372.
Osborn, H. F. 1897 A.
   1897 D.
Schlosser, M. 1886 B, p. 19.
   1887 A, p. 576.
Scott, W. B. 1892 A, p. 432.
Wortman, J. L. 1883 A, p. 710.
   1886 A, p. 480.
Zittel, K. A. 1893 B, p. 300.

Lambdotherium popoagicum Cope.
Cope, E. D. 1880 N, p. 746.
   1881 D, p. 196.
   1884 O, pp. 710, pl. xxixa, fig. 7; pl. lviii, figs. 3–5.
   1887 N, p. 1060, figs. 24, 26.
   1887 S, p. 275, pl. vii, figs. 5, 6.
Matthew, W. D. 1899 A, p. 36.
Osborn, H. F. 1897 A.
   1897 G, p. 256.
Roger, O. 1896 A, p. 192.
Trouessart, E. L. 1898 A, p. 737.
   Eocene (Wind River); Wyoming, Colorado.

LIMNOHYOPS Marsh.   Type *L. laticeps* Marsh.

Marsh, O. C. 1890 D, p. 525.
Cary, A. 1892 A.
Cope, E. D. 1873 E, p. 598. (Limnohyus.)
    1881 G, p. 396. (Limnohyus.)
    1884 O, pp. 697, 705. (Limnohyus.)
    1887 N, p. 1061. (Limnohyus.)
Earle, C. 1891 B, p. 113.
    1892 C, pp. 270, 276, 350.
Leidy, J. 1872 H, p. 241. (Limnohyus, type *L. laticeps*, not Limnohyus Marsh.)
    1873 B, pp. 57, 323. (Limnohyus.)
Marsh, O. C. 1872 G, p. 124. (Palæosyops, type *P. laticeps*, not Palæosyops Leidy, 1870.)
Matthew, W. D. 1899 A, p. 47. (Syn. of Palæosyops.)
Osborn, Scott, and Speir 1878 A, p. 41. (Limnohyus.)
Schlosser, M. 1886 B, p. 18. (Limnohyus.)
Zittel, K. A. 1893 B, p. 303.

Limnohyops fontinalis (Cope).

Cope, E. D. 1873 B, p. 35. (Palæosyops.)
    1873 E, p. 594. (Limnohyus.)
    1884 O, p. 707, pl. xlix, fig. 9; pl. l, fig. 4; pl. lviiia, figs. 4, 5. (Limnohyus.)
Earle, C. 1891 B, p. 115.
    1892 C, p. 368.

Matthew, W. D. 1899 A, p. 47. (Palæosyops.)
Roger, O. 1896 A, p. 193.
Trouessart, E. L. 1898 A, p. 738.
    Eocene (Bridger); Wyoming.

Limnohyops laticeps Marsh.

Marsh, O. C. 1872 G, p. 122. (Palæosyops.)
Cope, E. D. 1884 O, p. 707. (Limnohyus.)
Earle, C. 1891 B, p. 114.
    1892 C, pp. 269, 351, pl. xi, figs. 8, 9; pl. xii, figs. 24-27; pl. xiv, figs. 39-42; p. 330, pl. xii, figs. 14-16; pl. xiv, figs. 43, 44. (Palæosyops minor, not of Marsh.)
Leidy, J. 1873 B, pp. 28-45 (in part); pl. iv, figs. 3-6; pl. v, figs. 10, 11; pl. xxiv, figs. 6, 7 (Palæosyops paludosus); pp. 58, 326, pl. xxiii, fig. 13 (Limnohyus).
Marsh, O. C. 1884 F, fig. 68. (Palæosyops.)
    1890 D, p. 525.
Matthew, W. D. 1897 A, p. 57. (Palæosyops.)
    1899 A, p. 47. (Palæosyops.)
Osborn, Scott, and Speir 1878 A, p. 42. (Limnohyus.)
Roger, O. 1896 A, p. 193.
Trouessart, E. L. 1898 A, p. 738.
    Eocene (Bridger); Wyoming.

PALÆOSYOPS Leidy.   Type *P. paludosus* Leidy.

Leidy, J. 1870 Q, p. 113.
Cope, E. D. 1873 E, p. 591.
    1881 G, p. 396.
    1884 O, pp. 694, 697.
    1887 N, p. 1061.
    1892 R.
Earle, C. 1891 B.
    1892 C.
    1895 A.
Flower, W. H. 1876 B, xiii, p. 327.
Hatcher, J. B. 1895 A, p. 1090. (Manteoceras, type *M. vallidens*=P. manteoceras Osb., not P. vallidens Cope.)
Leidy, J. 1871 C, p. 354.
    1871 G.
    1872 B, p. 358.
    1872 H, p. 241.
    1873 B, p. 27, pl. xxxi, fig. 1.
Marsh, O. C. 1872 G, p. 124. (Limnohyus, type *L. robustus*.)
    1877 E. (Limnohyus.)
Matthew, W. D. 1897 A.
Nicholson and Lydekker 1889 A, p. 1372.
Osborn, H. F. 1883 A, p. 31.
    1890 D, pp. 513, 553.
    1895 A, p. 87. (Manteoceras=Telmatherium.)
    1897 D.
Osborn, Scott, and Speir 1878 A, p. 27.
Osborn and Wortman 1892 A, p. 132.
Schlosser, M. 1886 B, p. 18.
Scott, W. B. 1886 A, p. 307.
    1891 B, p. 371 seq.
    1892 A, p. 432.

Steinmann and Döderlein 1890 A, p. 777.
Woodward, A. S. 1898 B, p. 332.
Zittel, K. A. 1893 B, p. 301.

Palæosyops humilis Leidy.

Leidy, J. 1872 G.
    1872 A.
    1872 H, p. 242. (Limnohyus.)
    1873 B, pp. 58, 326, pl. xxiv, fig. 8.
Matthew, W. D. 1899 A, p. 47.
Roger, O. 1896 A, p. 193.
Trouessart, E. L. 1898 A, p. 739.
    Eocene (Bridger); Wyoming.

Palæosyops junior Leidy.

Leidy, J. 1872 J, p. 277.
    1873 B, pp. 57, 326.
Matthew, W. D. 1899 A, p. 47.
Roger, O. 1896 A, p. 193.
Trouessart, E. L. 1898 A, p. 739.
    Eocene (Bridger); Wyoming.
    Probably belongs to *P. paludosus*.

Palæosyops lævidens Cope.

Cope, E. D. 1873 E, p. 591. (Limnohyus.)
    1884 O, pp. 699, 701, pl. l, figs. 1-3.
Earle, C. 1892 C, p. 317.
Leidy, J. 1872 N, p. 359. (P. paludosus, *fide* Cope.)
Matthew, W. D. 1899 A, p. 47. (Syn. of P. laticeps.)
Roger, O. 1896 A, p. 192.
Trouessart, E. L. 1898 A, p. 738.
    Eocene (Bridger); Wyoming.

Palæosyops longirostris Earle.

Earle, C. 1892 C, p. 338.
Matthew, W. D. 1899 A, p. 47.   (Syn.? of Telmatotherium megarhinum.)
Roger, O. 1896 A, p. 198.
Trouessart, E. L. 1898 A, p. 739.
Eocene (Bridger); Wyoming.

Palæosyops paludosus Leidy.

Leidy, J. 1870 Q, p. 113.
Cope, E. D. 1873 B, p. 35.   (P. lævidens.)
    1873 E, pp. 592, 593.   (P. major, Limnohyus paludosus, L. diaconus.)
    1873 U, p. 4.   (P. diaconus sp. nov.)
    1884 O, p. 105   (P. paludosus); pp. 698, 699, 701, pl. 11. fig. 2; pl. 111, figs. 1, 2 (P. major); p. 706, pl. 11, fig. 3 (L. diaconus).
    1887 N, p. 1082, fig. 27.   (P. major.)
Earle, C. 1891 B, p. 115.   (L. diaconus, invalid.)
    1892 C, p. 279, pl. x, fig. 1; pl. xii, figs. 17–23; pl. xiii; pl. xiv, figs. 45–49, and text figures.
    1895 A, p. 624.
King, C. 1878 A, p. 404.
Leidy, J. 1871 B, p. 373.   (P. major.)
    1871 C, p. 355.
    1871 I.
    1871 K, P. 229.   (P. major; orig. description.)
    1872 B, pp. 359, 365.   (P. major.)
    1872 G, p. 168.

Leidy, J. 1872 H, p. 241.   (P. major.)
    1873 B, pp. 28, 325, pl. iv, figs. 1, 2, 7, 8; pl. v, figs. 4–11; pl. xix, figs. 1–4; pl. xx, figs. 1–7; pl. xxiii, figs. 3–6; pl. xxiv, figs. 6, 7; pl. xxix, fig. 5 (P. paludosus); pp. 45, 326, pl. xx, fig. 8; pl. xxiii, figs. 1, 2, 7–12, 14–16; pl. xxiv, figs. 1–5 (P. major).
    1873 E, p. 592.   (P. major.)
L. P. B. 1885 A, p. 191, fig. 24.   (P. robustus.)
Marsh, O. C. 1872 G, pp. 122, 124 (P. paludosus); p. 124 (P. robustus sp. nov.).
    1885 B, fig. 105.   (P. robustus.)
    1886 F, fig. 69.   (P. robustus.)
Matthew, W. D. 1897 A, p. 57.
    1899 A, p. 46.
Osborn, H. F. 1898 Q, p. 11.
Osborn, Scott, and Speir 1878 A, p. 27, pl. ii, figs. 1–13; pl. iii, fig. 8; pl. v (P. major); p. 34, pl. i; pl. iii, figs. 1–9 (P. paludosus).
Roger, O. 1896 A, p. 192 (P. paludosus); p. 193 (Limnohyops diaconus).
Trouessart, E. L. 1898 A, p. 738.   (P. paludosus, Limnohyops diaconus.)
Zittel, K. A. 1893 B, p. 301, fig. 242 (P. major); p. 302, fig. 244 (P. paludosus).
Eocene (Bridger); Wyoming.

Palæosyops ultimus Matth.

Matthew, W. D. 1897 A, p. 57.   (No description.)
    1899 A, p. 50.   (No description.)
Eocene (Uinta); Utah.

TELMATHERIUM Marsh.   Type *T. validum* Marsh.

Marsh, O. C. 1872 G, p. 123.
Cope, E. D. 1879 W, p. 33.   (Leurocephalus.)
    1881 G, p. 396.   (Leurocephalus.)
    1884 O, p. 694.   (Leurocephalus.)
    1887 N, p. 1061.   (Leurocephalus.)
Earle, C. 1891 B, p. 115.   (Telmatotherium.)
    1892 C, pp. 269, 341.   (Telmatotherium.)
    1895 A, p. 627.   (Telmatotherium.)
Hatcher, J. B. 1895 A.   (Telmatotherium.)
Leidy, J. 1873 B, p. 323.   (Syn. of Palæosyops.)
Marsh, O. C. 1880 H, p. 10.   (Telmatotherium.)
Osborn, H. F. 1895 A, p. 82.   (Telmatotherium.)
    1897 D.   (Telmatotherium.)
Osborn, Scott, and Speir 1878 A, p. 42.   (Leurocephalus, type L. *cultridens*.)
Scott, W. B. 1892 A, p. 432.
Zittel, K. A. 1893 B, p. 303.   (Telmatotherium.)

Telmatherium boreale (Cope).

Cope, E. D. 1880 N, p. 746.   (Palæosyops.)
    1881 D, p. 196 (Palæosyops); p. 196 (Lambdotherium brownianum).
    1884 O, pp. 699, 703, pl. lviii. fig. 3 (Palæosyops); p. 709, pl. lvia, fig. 10 (Lambdotherium brownianum).
Earle, C. 1892 C, pp. 319, 384.   (Palæosyops.)
Matthew, W. D. 1897 A, p. 57.   (Palæosyops.)
    1899 A, p. 36.   (Telmatotherium.)
Osborn, H. F. 1897 D.
Osborn and Wortman 1892 A, p. 132, fig. 19.   (Palæosyops.)

Roger, O. 1896 A, p. 193.   (Palæosyops borealis, Lambdotherium brownianum.
Trouessart, E. L. 1898 A, p. 737 (Lambdotherium brownianum); p. 738 (Palæosyops).
Eocene (Wind River); Wyoming.

Telmatherium cultridens (Scott and Osb.)

Osborn, Scott, and Speir 1878 A, p. 42, pl. iv.   (Leurocephalus.)
Earle, C. 1891 B, p. 116.   (Telmatotherium.)
    1892 C, pp. 269, 343, pl. x, fig. 3; pl. xii, figs. 12, 13.   (Telmatotherium.)
Matthew, W. D. 1897 A, p. 58.   (Telmatotherium.)
    1899 A, p. 47.   (Telmatotherium.)
Osborn, H. F. 1895 A, pp. 83, 95.   (Telmatotherium.)
Osborn and Wortman 1892 A, p. 134.   (Telmatotherium.)
Roger, O. 1896 A, p. 193.
Trouessart, E. L. 1898 A, p. 739.   (Telmatotherium.)
Eocene (Bridger); Wyoming.

Telmatherium diploconum Osborn.

Osborn, H. F. 1895 A, p. 85, fig. 6.   (Telmatotherium.)
Matthew, W. D. 1899 A, p. 50.   (Variety minus.)
Roger, O. 1896 A, p. 193.
Trouessart, E. L. 1898 A, p. 740.   (Telmatotherium.)
Eocene (Uinta); Utah.

Telmatherium hyognathum (Scott and Osb.).

Scott and Osborn 1890 A, p. 513.   (Palæosyops.)
Earle, C. 1891 B, p. 116.   (Palæosyops.)
   1892 C, p. 848, pl. x, figs. 10, 11.   (Telmatotherium.)
Matthew, W. D. 1897 A, p, 57.   (Telmatotherium.)
   1899 A, p. 47 (Syn. of T. validum); p. 50 (Telmatotherium hyognathum).
Osborn, H. F. 1395 A, pp. 83, 87.   (Telmatotherium.)
Roger, O. 1896 A, p. 193.
Trouessart, E. L. 1898 A, p. 740.   (Telmatotherium.)
   Eocene (Bridger); Wyoming: (Uinta); Utah.

Telmatherium megarhinum (Earle).

Earle, C. 1891 A, p. 45.   (Palæosyops.)
   1891 B, p. 117.   (Palæosyops.)
   1892 C, p. 321, pl. x, fig. 2; pl. xi, figs. 4-7 (Palæosyops megarhinum); p. 330, pl. xii, figs. 14-16; pl. xiv, figs. 43, 44 (P. minor, fide Wortman in litt.).
   1895 A, p. 623.   (Palæosyops.)
Leidy, J. 1873 B, pp. 28-45, in part; pl. iv, figs. 3-6; pl. v, figs. 10, 11; pl. xxiv, figs. 6, 7.   (Palæosyops paludosus, fide Earle.)
Matthew, W. D. 1899 A, p. 47.   (Telmatotherium.)
Osborn, H. F. 1895 A, pp. 83, 84.   (Telmatotherium.)
Roger, O. 1896 A, p. 193.

Trouessart, E. L. 1898 A, p. 738, in part (Palæosyops minor); p. 739 (P. megarhinum).
   Eocene (Bridger); Wyoming.

Telmatherium validum Marsh.

Marsh, O. C. 1872 G, p. 123.
Cope, E. D. 1884 O, p. 699, in part; pl. li, fig. 1.   (Palæosyops vallidens.)
Earle, C. 1891 B, p. 116.   (Telmatotherium.)
   1892 C, p. 342.   (Telmatotherium.)
Matthew, W. D. 1899 A, p. 47.   (Telmatotherium.)
Osborn, H. F. 1895 A, pp. 82, 94.   (Telmatotherium.)
Osborn, Scott, and Speir 1878 A, p. 40.   (Palæosyops vallidens.)
Roger, O. 1896 A, p. 193.   (T. validum, T. vallidens.)
Trouessart, E. L. 1898 A, p. 739.   (Telmatotherium.)
   Eocene (Bridger); Wyoming.

Telmatherium vallidens (Cope).

Cope, E. D. 1872 S, p. 487.   (Palæosyops.)
   1873 E, p. 592.   (Palæosyops.)
   1884 O, p. 699, pl. lii, fig. 3; pl. liii, fig. 1 (Palæosyops.)
Earle, C. 1892 C, p. 316.   (Palæosyops.)
Matthew, W. D. 1899 A, p. 47.   (Telmatotherium.)
Trouessart, E. L. 1898 A, p. 739.
   Eocene (Bridger); Wyoming.

DOLICHORHINUS Hatcher.   Type *Telmatotherium cornutum* Osb.

Hatcher, J. B. 1895 A, p. 1090.
Trouessart, E. L. 1898 A, p. 740.

Dolichorhinus cornutus (Osb.).

Osborn, H. F. 1895 A, p. 90, figs. 10, 11.   (Telmatotherium.)

Hatcher, J. B. 1895 A, p. 1090.
Matthew, W. D. 1897 A, p. 58.   (Telmatotherium.)
   1899 A, p. 50.   (Telmatotherium.)
Roger, O. 1896 A, p. 193.   (Telmatotherium.)
Trouessart, E. L. 1898 A, p. 740.
   Eocene (Uinta); Utah.

MANTEOCERAS Hatcher.   Type *Palæosyops manteoceras* Osb.

Hatcher, J. B. 1895 A, p. 1090.

Manteoceras manteoceras (Osb.).

Osborn, H. F. in Matthew, W. D. 1899 A, p. 47.   (Palæosyops.)
Hatcher, J. B. 1895 A, p. 1089, pl. xxxix, fig. 2 (Telmatotherium vallidens, not of Cope); p. 1090 (made type of Manteoceras.)

Matthew, W. D. 1899 A, pp. 47, 50.   (Palæosyops.)
Osborn, H. F. 1895 A, pp. 83, 87, figs. 7-9.   (Telmatotherium vallidens, not of Cope.)
   Eocene (Bridger); Wyoming. (Uinta); Utah.

DIPLACODON Marsh.   Type *D. elatus* Marsh.

Marsh, O. C. 1875 B, p. 246.
Cope, E. D. 1881 G, p. 397.
   1884 O, p. 718.
   1887 N, p. 1063.
   1890 C, p. 470.
Earle, C. 1892 C.
Hatcher, J. B. 1895 A.   (Diplacodon, Protitanotherium.)
Marsh, O. C. 1884 F, p. 172.
   1889 A, p. 165.
Matthew, W. D. 1897 A, p. 58.
Nicholson and Lydekker 1889 A, p. 1374.
Osborn, H. F. 1890 D, pp. 512, 553.
   1897 D.

Schlosser, M. 1886 B, p. 19.
Scott, W. B. 1891 B, p. 371 et seq.
   1892 A, p. 432.
Scott and Osborn 1887 A, p. 263.
Steinmann and Döderlein 1890 A, p. 777.
Zittel, K. A. 1893 B, p. 303.

Diplacodon elatus Marsh.

Marsh, O. C. 1875 B, p. 246.
King, C. 1878 A, p. 407.
Marsh, O. C. 1889 H.
Matthew, W. D. 1899 A, p. 50.
Osborn, H. F. 1890 D, p. 514, pls. viii, ix.
Roger, O. 1896 A, p. 193.

Scott and Osborn 1887 A, p. 262.
Trouessart, E. L. 1898 A, p. 740.
  Eocene (Uinta); Utah.

Diplacodon emarginatus Hatcher.

Hatcher, J. B. 1895 A, p. 1085, pl. xxxviii, figs. 1–4.

Matthew, W. D. 1899 A, p. 50.
Trouessart, E. L. 1898 A, p. 740.
  Eocene (Uinta); Utah.
  Hatcher, as cited, has proposed for this species, under certain contingencies, the new generic name *Protitanotherium*.

DŒODON Cope.   Type *D. shoshonensis* Cope.

Cope, E. D. 1879 D, p. 77.
  1881 G, p. 397.
  1884 O, p. 713.
  1887 N, p. 1063.
Roger, O. 1896 A, p. 194.   (Syn. of Titanotherium.)
Trouessart, E. L. 1898 A, p. 740.   (Syn. of Titanotherium.)
Zittel, K. A. 1893 B, p. 304.   (Daleodon; syn. of Titanotherium.)

Dæodon shoshonensis Cope.

Cope, E. D. 1879 D, p. 77.
  1879 B, p. 58.
Matthew, W. D. 1899 A, p. 63.
Roger, O. 1896 A, p. 194.   (Titanotherium.)
Trouessart, E. L. 1898 A, p. 740.   (Titanotherium.)
  Miocene (John Day); Oregon.

BRONTOTHERIINŒ.[1]

Brandt, J. F. 1878 A, pp. 10, 19.   (Brontotherinæ.)
Earle, C. 1892 C, p. 274.   (Titanotherinæ.)
Osborn, H. F. 1893 D.   (Titanotherinæ.)
  1900 F, p. 94.   ("Titanotheres.")

Osborn and Wortman 1892 A, p. 93.   (Titanotheriinæ.)
Roger, O. 1896 A, p. 194.   (Titanotherinæ.)
Zittel, K. A. 1893 B, p. 304.   (Titanotherinæ.)

MEGACEROPS Leidy.   Type *M. coloradensis* Leidy.

Leidy, J. 1870 D, p. 2.
Cope, E. D. 1873 E, pp. 564, 565.   (Megaceratops.)
  1873 F, p. 66.   (Megaceratops.)
  1873 K, p. 102.   (Megaceratops.)
  1873 CC, p. 14.   (Megaceratops.)
  1887 N, p. 1063.   (Syn. of Menodus.)
  1889 I, p. 153.   (Haplacodon, type *Menodus angustigenis* Cope.)
  1891 A, p. 9.   (Menodus, in part.)
Earle, C. 1892 C, p. 296.   (Haplacodon.)
Leidy, J. 1871 C, p. 352.   (Megaceratops.)
  1873 B, pp. 289, 335.
  1874 B.   (Syn. of Titanotherium.)
Marsh, O. C. 1873 F, p. 296.
  1874 A.
  1875 B, p. 245.
  1876 C, p. 338.
  1887 B, p. 331.   (Allops, type *A. serotinus*.)
  1889 A, p. 165.   (Allops, Megacerops.)
  1890 D, p. 523.   (Diploclonus, type *D. amplus*.)
Nicholson and Lydekker 1889 A, p. 1374.
Osborn, H. F. 1890 D, p. 513, 542.   (Syn. of Titanotherium.)
  1896 B.   (Titanotherium, in part.)
Scott and Osborn 1887 A, pp. 255, 262.   (Menodus, in part.)
  1887 B, p. 157.   (Syn. of Menodus.)
Trouessart, E. L. 1898 A, p. 740.   (Syn. of Titanotherium.)

Megacerops? amplus (Marsh).

Marsh, O. C. 1890 D, p. 523.   (Diploclonus.)

Osborn, H. F. 1896 B, pp. 192, 193.   (Titanotherium.)
Roger, O. 1896 A, p. 195.   (Titanotherium.)
Trouessart, E. L. 1898 A, p. 742.   (Syn. of Titanotherium montanum.)
  Oligocene (White River); South Dakota.

Megacerops angustigenis (Cope).

Cope, E. D. 1886 U, p. 81.   (Menodus.)
Ami, H. M. 1891 A.   (Menodus.)
Cope, E. D. 1887 N, p. 1067.   (Menodus.)
  1889 I, p. 153.   (Type of Haplacodon.)
  1889 O, p. 628.   (Menodus syceras.)
  1891 A, p. 13, pl. v, figs. 1, 2; pl. vi, fig. 2; pl. vii, fig. 1; pl. xi, fig. 1?; pl. xii, fig. 4?; pl. xiii, figs. 1?, 4?, 6?, 8? (Menodus angustigenis); p. 18, pl. vii, fig. 2?; pl. viii, fig. 4? (M. syceras.)
Matthew, W. D. 1899 A, p. 58.   (Titanotherium.)
Osborn, H. F. 1896 B, p. 184 (Titanotherium angustigenis); pp. 178, 193 (T. syceras, a syn.? of T. acer.)
Roger, O. 1896 A, p. 194 (Titanotherium syceras); p. 195 (T. angustigenis).
Trouessart, E. L. 1898 A, p. 741 (Titanotherium angustigenis); p. 743 (T. syceras, a syn. of T. acer.)
  Oligocene (White River); Swift Current Creek, Canada.

Megacerops coloradensis Leidy.

Leidy, J. 1870 D.
Cope, E. D. 1873 E, p. 585.   (Megaceratops coloradoensis.)

[1] In the arrangement of the genera and species of the Brontotheriinæ the writer has had the benefit of the conclusions arrived at by Prof. H. F. Osborn, at the date of going to press, in his studies in the preparation of his forthcoming monograph of the Titanotheres.

Cope, E. D. 1873 F, p. 66. (Megaceratops colorado-ensis.)
1887 N, p. 1063. (Megaceratops colorado-ensis.)
?1891 A, p. 9 (Menodus coloradoensis); p. 10, pl. vi, fig. 1; pl. viii, figs. 1–3; pl. ix, figs. 2, 3; pl. x, fig. 2; pl. xii, fig. 1; pl. xiii, figs. 5, 7 (Menodus? americanus); p. 10, pl. ix, fig. 1; pl. x, fig. 3; pl. xii, figs. 2, 3c; pl. xiii, fig. 3 (Menodus? proutii).
Leidy, J. 1871 C, p. 352.
1873 B, pp. 239, 335, pl. i, figs. 2, 3; pl. ii, fig. 2.
Marsh, O. C. 1873 F, p. 296.
1874 A, p. 81.
1897 C, p. 527.
Matthew, W. D. 1899 A, p. 57. (Titanotherium.)
Osborn, H. F. 1890 D, p. 15, p. 513. (Titanotherium.)
1896 B, p. 171, text figure; p. 175, pl. iii. (Titanotherium.)
Roger, O. 1896 A, p. 194. (Titanotherium.)
Trouessart, E. L. 1898 A, p. 741. (Titanotherium.)
Oligocene (White River); South Dakota.

Megacerops? crassicornis (Marsh).

Marsh, O. C. 1891 E, p. 268. (Allops.)
Oligocene (White River); South Dakota.

Megacerops dispar (Marsh).

Marsh, O. C. 1887 B, p. 328, figs. 7, 8; not figs. 5, 6. (Brontops.)
1891 E, p. 269. (Brontops validus.)
Osborn, H. F. 1896 B, p. 188. (Titanotherium.)
Roger, O. 1896 A, p. 195. (Titanotherium.)
Scott and Osborn 1887 B, p. 158, text figs. 3³, 5³, 6³. (Menodus coloradensis.)
Trouessart, E. L. 1898 A, p. 742. (Titanotherium.)
Zittel K. A. 1893 B, p. 307, figs. 249, 250. (Titanotherium.)
Oligocene (White River); South Dakota.

Megacerops? serotinus (Marsh).

Marsh, O. C. 1887 B, p. 331. (Allops.)
Cope, E. D. 1887 L, p. 926. (Menodus.)
Osborn, H. F. 1896 B, p. 192. (Titanotherium.)
Roger, O. 1896 A, p. 195. (Titanotherium.)
Oligocene (White River); South Dakota.

BRONTOTHERIUM Marsh.   Type *B. gigas* Marsh.

Marsh, O. C. 1873 H, p. 486.
Cope, E. D. 1873 S, p. 2. (Miobasileus, type *M. ophryas*.)
1873 DD. (Miobasileus.)
1874 A, p. 108. (Miobasileus.)
1887 L, p. 926. ("Brontops.")
1887 N, p. 1063. (Syn. of Menodus.)
1891 A, p. 9. (Menodus, in part.)
Flower, W. H. 1876 B, xiii, p. 327.
Flower and Lydekker 1891 A, p. 413. (Syn. of Titanotherium.)
Leidy, J. 1874 B. (Syn. of Titanotherium.)
Marsh, O. C. 1874 A, p. 81.
1874 D, p. 66.
1875 B, p. 245.
1876 C, p. 334. (Brontotherium.)
1887 B, p. 326 (Brontops, type *B. robustus*); p. 330 (Titanops, type *T. curtus*).
1889 A, p. 163 (Brontops); p. 165 (Titanops).
1889 H. (Brontops.)
Nature 1874 A.
Osborn, H. F. 1896 B, p. 175. (Titanotherium, in part.)
Schlosser, M. 1886 A, p. 253.
Scott and Osborn 1887 A, p. 262. (Menodus, in part.)
1887 B, p. 157. (Syn. of Menodus.)
Trouessart, E. L. 1898 A, p. 740. (Syn. of Titanotherium.)
Woodward, A. S. 1898 B, p. 330 (Titanotherium, in part); p. 332 (Brontotherium).

Brontotherium bucco (Cope.)

Cope, E. D. 1873 CC, p. 11. (Symborodon.)
1873 DD. (Symborodon.)
1874 B, p. 485, in part. (Symborodon.)
1874 G, p. 89. (Symborodon.)

Cope, E. D. 1887 N, p. 1063. (Symborodon.)
Roger, O. 1896 A, p. 195. (Titanotherium.)
Trouessart, E. L. 1898 A, p. 742. (Titanotherium.)
Oligocene (White River); Colorado.

Brontotherium curtum (Marsh.)

Marsh, O. C. 1887 B, p. 330, fig. 11. (Titanops.)
Cope, E. D. 1887 L, p. 926. (Syn. of Menodus platyceras.)
1891 D, p. 48. (Menodus peltoceras.)
Marsh, O. C. 1897 C, p. 522, fig. 94. (Titanops.)
Matthew, W. D. 1899 A, p. 58. (Titanotherium curtum, T. peltoceras.)
Osborn, H. F. 1896 B, p. 189, pl. iv. (Titanotherium.)
Roger, O. 1896 A, p. 194 (Titanotherium curtum, a syn. of T. platyceras); p. 195 (T. peltoceras).
Trouessart, E. L. 1898 A, p. 741 (Titanotherium peltoceras); p. 743 (T. curtum, a syn. of T. platyceras).
Oligocene (White River); Colorado, South Dakota.

Brontotherium dolichoceras (Scott and Osb.).

Scott and Osborn 1887 B, p. 160, text figs. 3³, 5³, 6³. (Menodus.)
Cope, E. D. 1887 N, p. 1063. (Menodus.)
Matthew, W. D. 1899 A, p. 58. (Titanotherium.)
Osborn, H. F. 1896 B, p. 185, pl. iv; text fig. p. 169. (Titanotherium.)
Osborn and Wortman 1895 A, p. 349. (Titanotherium.)
Roger, O. 1896 A, p. 194. (Titanotherium.)
Trouessart, E. L. 1898 A, p. 742. (Titanotherium.)
Oligocene (White River); South Dakota.

Brontotherium gigas Marsh.

Marsh, O. C. 1873 H, p. 486.
King, C. 1878 A, p. 412.
Marsh, O. C. 1876 C, pl. xii.
     1887 B, p. 830, fig. 12.   (Titanops elatus.)
     1897 C, p. 527.
Matthew, W. D. 1899 A, p. 58.   (Titanotherium
     elatum.)
Osborn, H. F. 1896 B, p. 175 (Titanotherium gi-
     gas); p. 171, text figure (T. elatum); p. 189, pl.
     iv (T. elatum).
     1896 C, pp. 709, 714, illust. 5.   (Titanothe-
     rium elatum.)
     1896 Q, p. 17.   (Titanotherium elatum.)
Osborn and Wortman 1895 A, p. 349.   (Titanothe-
     rium elatum.
Roger, O. 1896 A, p. 194.   (Titanotherium gigas.)
Trouessart, E. L. 1898 A, p. 742.   (Titanotherium
     gigas; T. elatum, a subsp. of T. montanum.)
     Oligocene (White River); Colorado, South
     Dakota.

Brontotherium hypoceras (Cope).

Cope, E. D. 1874 B, p. 491.   (Symborodon.)
Osborn, H. F. 1896 B, p. 183.   (Titanotherium.)
Trouessart, E. L. 1898 A, p. 742.   (Syn.? of T. in-
     gens.)
     Oligocene (White River); Colorado.

Brontotherium medium (Marsh).

Marsh, O. C. 1891 E, p. 269.   (Titanops.)
     Oligocene (White River); South Dakota.

Brontotherium? ophryas (Cope).

Cope, E. D. 1873 S, p. 3.   (Miobasileus.)
     1873 CC, p. 14.   (Miobasileus.)
     1874 A, p. 108.   (Miobasileus.)
     1874 B, p. 490.   (Symborodon.)
Osborn, H. F. 1896 B, p. 177.   (Titanotherium.)
Trouessart, E. L. 1898 A, p. 743.   (Titanotherium.)
     Oligocene (White River); Colorado.

Brontotherium platyceras (Scott and
Osb.).

Scott and Osborn 1887 B, p. 160, text fig. 4.   (Meno-
     dus.)
Cope, E. D. 1887 L, p. 926.   (Menodus.)
     1887 N, p. 1063.   (Menodus.)
     1891 A, p. 9.   (Menodus.)
Matthew, W. D. 1899 A, p. 58.   (Titanotherium.)
Osborn, H. F. 1896 B, pp. 167, 185, pl. iv, text figs.
     7, 7a, and p. 167.   (Titanotherium.)
Roger, O. 1896 A, p. 194.   (Titanotherium.)
Trouessart, E. L. 1898 A, p. 743.   (Titanotherium.)
Zittel, K. A. 1893 B, p. 309, fig. 251.   (Titanothe-
     rium.)
     Oligocene (White River); South Dakota.

Brontotherium ramosum (Osborn).

Osborn, H. F. 1896 B, pp. 167, 194, pl. iv, figures;
     text fig. 13.   (Titanotherium.)
Matthew, W. D. 1899 A, p. 58.   (Titanotherium.)
Trouessart, E. L. 1898 A, p. 743.   (Titanotherium.)
     Oligocene (White River); South Dakota.

Brontotherium robustum (Marsh).

Marsh, O. C. 1887 B, p. 826.   (Brontops.)
Hatcher, J. B. 1893 B, p. 212.   (Titanotherium.)
Hutchinson, H. N. 1893 A, p. 160, pl. xv.   (Titano-
     therium.)
Marsh, O. C. 1889 A, p. 163, pl. vi.   (Brontops.)
     1889 H, p. 706.   (Brontops.)
Matthew, W. D. 1899 A, p. 58.   (Titanotherium.)
Osborn, H. F. 1896 A, p. 162.   (Titanotherium.)
     1896 B, p. 187, text fig. 8.   (Titanotherium.)
Osborn and Wortman 1895 A, p. 346, pls. viii, ix;
     text figs. 1-3.   (Titanotherium.)
Trouessart, E. L. 1898 A, p. 742.   (Titanotherium.)
Woodward, A. S. 1898 B, p. 331, fig. 190.   (Titano-
     therium.)
Zittel, K. A. 1893 B, p. 306, fig. 247.   (Titanothe-
     rium.)
     Oligocene (White River); Nebraska, Colo-
     rado.

Brontotherium? selwynianum(Cope).

Cope, E. D. 1889 D, p. 628.   (Menodus.)
Ami, H. M. 1891 A.   (Menodus.)
Cope, E. D. 1891 A, p. 17, pl. v, figs. 3–3b.   (Meno-
     dus.)
Matthew, W. D. 1899 A, p. 58.   (Titanotherium.)
Osborn, H. F. 1896 B, p. 193.   (Titanotherium
     selwynianus.)
Roger, O. 1896 A, p. 194.   (Titanotherium.)
Trouessart, E. L. 1898 A, p. 741.   (Titanotherium.)
     Oligocene (White River); Swift Current Creek,
     Canada.

Brontotherium tichoceras (Scott and
Osb.).

Scott and Osborn 1887 A, p. 159, text figs. 3³, 5⁴, 6³.
     (Menodus.)
Cope, E. D. 1887 N, p. 1063.   (Menodus.)
     1891 A, p. 9.   (Menodus tichoceras, M. al-
     tirostris.)
Matthew, W. D. 1899 A, p. 58.   (Titanotherium.)
Osborn, H. F. 1896 B, p. 184, pl. iii, figures; and
     p. 169.   (Titanotherium.)
Roger, O. 1896 A, p. 195.   (Titanotherium.)
Trouessart, E. L. 1898 A, p. 742.   (Titanotherium.)
     Oligocene (White River); South Dakota.

TITANOTHERIUM Leidy.   Type *Palæotherium? proutii* Owen, Norw., Evans.

Leidy, J. 1852 J, p. 552.
Cary, A. 1892 A, p. 309, pl. xviii, fig. 2.
Cope, E. D. 1873 E, pp. 564, 585.
     1874 G, p. 90.
     1874 K, p. 224.
     1881 G, p. 397.   (Menodus.)

Cope, E. D. 1884 O, p. 713.   (Menodus.)
     1887 L, p. 926.   (Menodus.)
     1887 N, p. 1063.   (Menodus.)
     1891 A, p. 8.   (Menodus.)
Flower, W. H. 1876 B, xiii, p. 327.
     1876 C, p. 115.

Flower and Lydekker 1891 A, p. 413. (Titanotherium, in part.)
Gaudry, A. 1878 B, p. 163.
Green, F. V. 1853 A.
Hatcher, J. B. 1893 B.
Hay, O. P. 1899 A, p. 594. (Menodus.)
Leidy, J. 1850 C, p. 122. (Palæotherium.)
1851 A. (Palæotherium.)
1852 A. (Rhinoceros.)
1853 D, p. 392. (Titanotherium; Eotherium, type E. americanum.)
1854 A, pp. 72, 114.
1856 H, p. 92.
1869 A, p. 207.
1871 C, p. 352.
1874 B.
1874 C.
Lydekker, R. 1896 B, pp. 170, 377.
Marsh, O. C. 1873 H, p. 486. (Titanotherium; Menodus preoccupied.)
1874 A.
1875 B, p. 245.
1876 C, p. 388. (Menodus.)
1884 F, p. 225. (Menodus.)
1887 B, p. 328. (Menops, type M. varians.)
1889 A, p. 165. (Menops.)
Nature 1874 A.
Nicholson and Lydekker 1889 A, p. 1374. (Titanotherium, in part.)
Osborn, H. F. 1890 D, pp. 513, 542. (Titanotherium, in part.)
1896 A.
1896 B. (Titanotherium, in part.)
?Pomel, A. 1849, Biblioth. Univ. Genève, x, p. 75. (Menodus.)
Prout, H. A. 1860 A. (Leidyotherium.)
Rütimeyer, L. 1862 A, p. 571.
Schlosser, M. 1886 A, p. 253. (Titanotherium, Menodus.)
1886 B, p. 19.
Scott, W. B. 1891 A, p. 371. (Titanotherium, in part.)
1892 A, p. 432. (Titanotherium, in part.)
Scott and Osborn 1887 A, p. 262. (Menodus.)
1887 B, p. 158. (Menodus.)
Steinmann and Döderlein 1890 A, p. 778. (Menodus.)

Titanotherium heloceras (Cope).

Cope, E. D. 1873 S, p. 4. (Megaceratops heloceras, err. typ.)
1873 CC, p. 14. (Symborodon.)
1873 DD. (Symborodon heloceros.)
1874 A, p. 109. (Symborodon.)
1874 B, pp. 484, 487. (Symborodon.)
1887 N, p. 1063. (Symborodon.)
Hatcher, J. B. 1893 B, p. 219, fig. 3.
Matthew, W. D. 1899 A, p. 58.
Osborn, H. F, 1896 B, p. 179, pl. iii; text fig. p. 171.
Trouessart, E. L. 1898 A, p. 741.
Oligocene (White River); Colorado.

Titanotherium ingens (Marsh).

Marsh, O. C. 1874 A, p. 85, pls. i, iv. (Brontotherium.)

Cope, E. D. 1874 Q. (Syn. of Symborodon trigonoceras.)
1887 N, p. 1063. (Menodus.)
1891 A, p. 91. (Syn. of Menodus americanus.)
Gaudry, A. 1878 B, p. 75, fig. 87. (Brontotherium.)
Hatcher, J. B. 1893 B, p. 219, fig. 2.
King, C. 1878 A, p. 411. (Brontotherium.)
Marsh, O. C. 1876 C, pls. x-xiii, and text figs. 1, 2. (Brontotherium.)
1884 F, figs. 64, 160, 161. (Brontotherium.)
1885 B, figs. 101, 130, 131. (Brontotherium.)
1897 C, p. 522, fig. 93. (Brontotherium.)
Matthew, W. D. 1899 A, p. 58.
Nature 1874 A, figs. 1, 2. (Brontotherium.)
Osborn, H. F. 1896 B, p.182, text fig. p. 171.
Scott and Osborn 1887 B, p. 158. (Syn. of Menodus coloradensis.)
Trouessart, E. L. 1898 A, p. 742.
Zittel, K. A. 1893 B, p. 304, fig. 245.
Oligocene (White River); Colorado, South Dakota.

Titanotherium proutii (Owen, Norw., and Evans).

Owen, Norwood, and Evans 1850 A, p. 66. (Palæotherium?.)
Ami, H. M. 1891 A. (Menodus americanus, M. proutii.)
Cope, E. D. 1874 K.
?1887 N, p. 1063, pl. xxxii. (Menodus giganteus.)
Hayden, F. V. 1858 B, p. 157.
Leidy, J. 1850 C, p. 122. (Palæotherium.)
1851 A. (Palæotherium.)
1852 A, p. 2. (Rhinoceros americanus.)
1852 J, p. 551, pl. ix, figs. 3, 3a?; pl. xii b, figs. 3, 4, 6 (Palæotherium proutii); explan. pl. xii b (P. maximum).
1852 L, p. 64. (Palæotherium.)
1853 D, p. 392. (T. proutii; Eotherium americanum.)
1854 A, pp. 72, 114, pl. xvi; pl. xvii, figs. 1-10 (T. proutii); pp. 78, 114, pl. xvii, figs. 11-13 (Palæotherium giganteum).
1854 E, p. 157.
1856 H, p. 92.
1869 A, pp. 206, 389, pl. xxiv. (T. prouti.)
1871 C, p. 354.
1874 B.
1874 C.
Marsh, O. C. 1870 D.
1873 H, p. 486.
1874 A, p. 81.
1874 H, p. 486. (T. prouti.)
1876 C, p. 138.
Osborn, H. F. 1896 B, p. 158. (T. proutii, T. maximum, T. americanum.)
Osborn and Wortman 1896 A, p. 34, fig. 2.
Pictet, F. J. 1853 A, p. 311. (Palæotherium.)
Prout, H. A. 1846 A, fig. 1. ("Palæotherium.")
1847 A, figs. 1, 2. (Palæotherium?)
Roger, O 1896 A, p. 194. (T. proutii, T. americanum.)
Scott and Osborn 1887 B, p. 158.
Trouessart, E. L. 1898 A, p. 741. (T. proutii, T. americanum.)

Zittel, K. A. 1893 B, p. 306, fig. 246.
Oligocene (White River); South Dakota?.
Reference of specimens to this species by writers since the original description may or may not be correct. More than one species may be included.

Titanotherium trigonoceras (Cope).

Cope, E. D. 1873 CC, p. 13. (Symborodon.)
1874 B, p. 488. (Symborodon.)
1874 G, pp. 89, 90. (Symborodon trigonoceras.)
1874 P, p. 28. (Symborodon.)
1874 Q, p. 2. (Symborodon.)
1887 L, p. 926. (Menops varians=Menodus trigonoceras.)

SYMBORODON Cope.

Cope, E. D. 1873 S, p. 2.
1873 DD.
1874 A, p. 108.
1874 B, p. 480.
1874 G, p. 90.
1874 K, p. 224.
1881 G, p. 397.
1884 O, p. 713.
1887 L, p. 926.
1887 N, p. 1063.
1887 S.
1889 C.
1891 A, p. 9.
Flower, W. H. 1876 B, xiii, p. 327.
Flower and Lydekker 1891 A, p. 413. (Syn. of Titanotherium.)
Leidy, J. 1874 B. (Syn. of Titanotherium.)
Marsh, O. C. 1874 A, p. 81. (Syn. of Brontotherium.)
1875 B, p. 246. (Anisacodon, type A. montanus.)
1876 C, p. 338 (Megacerops, part; Brontotherium, part); p. 339 (Diconodon, to replace Anisacodon Marsh, 1875; preoccupied by Marsh, 1872).
Nicholson and Lydekker 1896 A, p. 1374. (Syn.? of Titanotherium.)
Osborn, H. F. 1890 D, p. 513. (Titanotherium, in part.)
1896 B. (Titanotherium, in part.)
Scott and Osborn 1887 B, p. 157. (Syn.? of Menodus.)
Steinmann and Döderlein 1890 A, p. 778.
Trouessart, E. L. 1898 A, p. 740. (Syn. of Titanotherium.)

Symborodon acer Cope.

Cope, E. D. 1873 S. p. 4. (Megaceratops.)
1873 CC, p. 12 (S. altirostris); p. 13 (S. acer).
1873 DD. (S. acer, S. altirostris.)
1874 A, p. 109.
1874 B, p. 486, pls. v, vi; pl. viii, fig. 1 (S. altirostris); p. 488, pl. vii; pl. viii, fig. 3 (S. acer).
1874 G, p. 89. (S. acer, S. altirostris.)
1887 N, p. 1064, fig. 28; pl. xxxiv, fig. 3 (S. acer); pl. xxxiii, fig. a; pl. xxxiv, fig. 1 (S. altirostris).

Cope, E. D. 1887 N, p. 1065, figs. 29, 30, 31?. (Symborodon.)
1889 C, p. 259, fig. 81. (Symborodon,)
1891 A, p. 9. (Symborodon.)
Marsh, O. C. 1887 B, p. 328, fig. 9; not fig. 10. (Menops varians.)
Matthew, W. D. 1899 A, p. 58.
Osborn, H. F. 1896 B, p. 180, pl. iii, text figs. 501 and 6355 on p. 167; not text figs. 5, 6 (T. trigonoceras); p. 189, pl. iii (T. varians).
Scott and Osborn 1887 B, p. 158. (Syn. of Menodus coloradensis.)
Trouessart, E. L. 1898 A, p. 741. (T. trigonoceras.)
Oligocene (White River); Colorado, South Dakota.

Type S. torvus Cope.

Cope, E. D. 1891 A, p. 9. (S. acer, S. altirostris.)
Hatcher, J. B. 1893 B, p. 219, fig. 1. (Titanotherium.)
Matthew, W. D. 1899 A, p. 58. (Titanotherium.)
Osborn, H. F. 1896 B, pp. 178, 179, pl. iv, and text figs. 3 A, 4. (Titanotherium.)
Roger, O. 1896 A, p. 195. (Titanotherium acer, T. altirostris.)
Trouessart, E. L. 1898 A, p. 743.
Oligocene (White River); Colorado, Nebraska, South Dakota, Canada.

Symborodon montanus (Marsh).

Marsh, O. C. 1875 B, p. 246. (Anisacodon.)
King, C. 1878 A, p. 412. (Diconodon.)
Marsh, O. C. 1876 C, p. 339. (Diconodon.)
Matthew, W. D. 1899 A, p. 58. (Titanotherium.)
Osborn, H. F. 1896 B, p. 183. (Titanotherium.)
Roger, O. 1896 A, p. 195. (Titanotherium.)
Trouessart, E. L. 1898 A, p. 742. (Titanotherium.)
Oligocene (White River); South Dakota.

Symborodon torvus Cope.

Cope, E. D. 1873 S, p. 2. (S. torrus, errore typog.)
1873 CC, p. 14.
1874 A, p. 108.
1874 B, p. 485, pl. ii, fig. 1; pls. iii, iv; pl. viii, fig. 2. (S. bucco.)
1874 G.
1887 N, p. 1063, pl. xxxiii, fig. b; pl. xxxiv, fig. 2. (S. bucco.)
1891 A, pp. 9, 11. (S. bucco, S. proutii.)
Matthew, W. D. 1899 A, p. 58. (Titanotherium.)
Osborn, H. F. 1896 B, p. 176, pl. iii; not text figs. 2, 3 (Titanotherium torvum); pp. 167, figure; p. 179, pl. iii (T. bucco).
1898 Q, p. 9. (Menodus.)
Roger, O. 1896 A, p. 195. (Titanotherium.)
Trouessart, E. L. 1898 A, p. 742. (Syn. of Titanotherium bucco.)
Woodward, A. S. 1898 B, p. 331, fig. 190. (Menodus.)
Oligocene (White River); Colorado, South Dakota, Nebraska.

TELEODUS Marsh.   Type *T. avus* Marsh.

Marsh, O. C. 1890 D, p. 524.
Hatcher, J. B. 1893 B, p. 217.
Osborn, H. F. 1896 B, p. 175 (Syn.? of Diplacodon);
  p. 194 (Syn.? of Titanotherium).
Zittel, K. A, 1893 B, pp. 304, 306.   (Syn. of Titano-
  therium.)

Teleodus avus Marsh.

Marsh, O. C. 1890 D, p. 524.

Matthew, W. D. 1899 A, p. 57.   (Titanotherium.)
Osborn, H. F. 1896 B, p. 194.   (Titanotherium
  avum.)
Roger, O. 1896 A, p. 194.   (Titanotherium.)
Trouessart, E. L. 1898 A, p. 740.   (Diplacodon.)
Oligocene (White River); South Dakota.

Superfamily RHINOCEROTOIDEA.

Gill, T. 1872 B, pp. 12, 85, 87.

Osborn, H. F. 1896 I, pp. 79, 80.

HYRACODONTIDÆ.

Branco, W. 1897 A, p. 20
Cope, E. D. 1879 C, p. 228.
  1879 V, p. 771 a.
  1881 G, pp. 378, 393.
  1881 O.
  1884 O, pp. 615, 691.
  1887 K.
  1887 N.
Osborn, H. F. 1893 D, p. 37.
  1896 I, pp. 79, 80, 94.

Osborn, H. F. 1900 C, p. 570.
Osborn and Wortman 1895 A, p. 367.
Roger, O. 1896 A, p. 185.
Scott and Osborn 1883 A, p. 17.   ("Hyracodon
  series.")
Steinmann and Döderlein 1890 A, p. 772.   (Hyra-
  codontinæ.)
Trouessart, E. L. 1898 A, p. 744.   (Hyracodontinæ.)
Zittel, K. A. 1893 B, p. 284.   (Subfamily.)

HYRACHYINÆ.

Osborn, H. F. 1893 D, p. 40.
  1896 I, p. 85.

Osborn and Wortman 1892 A, p. 93.

HYRACHYUS Leidy.   Type *H. agrarius* Leidy.

Leidy, J. 1871 K, p. 229.
Brandt, J. F. 1878 A, p. 24.
Cope, E. D. 1873 E, p. 594.
  1873 H.
  1881 G, p. 381.
  1883 II, p. 969.
  1884 O, pp. 618, 657.
  1887 N, p. 994.
  1889 C.
Earle, C. 1892 C, p. 329.
Filhol, H. 1885 A.
  1888 A, p. 180.
Flower, W. H. 1876 B, xiii, p. 327.
Flower and Lydekker 1891 A, p. 373.
Gaudry, A. 1877 A.
  1878 B, p. 65.
  1897 A, p. 320.
Leidy, J. 1871 C, p. 357.
  1872 B, p. 360.
  1872 C, p. 19.
  1873 B, pp. 59, 327.
Marsh, O. C. 1875 B, p. 244.
  1877 E.
Nicholson and Lydekker 1889 A, p. 1355.
Osborn, H. F. 1890 D, p. 505.
  1896 I, p. 87, fig. 5.
Osborn, Scott, and Speir 1878 A, p. 49.
Osborn and Wortman 1892 A, p. 129.
Pavlow, M. 1888 A, p. 177.
  1892 A.
Rütimeyer, L. 1891 A, p. 28.

Schlosser, M. 1886 B, p. 23.
Scott, W. B. 1883 A, p. 49.
  1896 B.
Scott and Osborn 1883 A, p. 11.
  1887 A, p. 261.
Steinmann and Döderlein 1890 A, p. 771.
Trouessart, E. L. 1898 A, p. 744.
Wortman and Earle 1893 A.
Zittel, K. A. 1893 B, p. 284.

Hyrachyus affinis (Marsh).

Marsh, O. C. 1871 D, p. 37.   (Lophiodon.)
Leidy, J. 1872 B, p. 362.   Lophiodon.)
  1872 C, p. 20.   (Lophiodon.)
  Eocene (Bridger); Wyoming.
  This species is referred to the genus *Hyra-
  chyus* with doubt.

Hyrachyus agrarius Leidy.

Leidy, J. 1871 K, p. 229.
Cope, E. D. 1873 E, p. 606.
  1873 U, p. 6.   (H. implicatus, specimen 1.)
  1884 O, pp. 660, 675.
  1887 N, p. 997, pl. xxxi, text fig. 10.   (H.
  agrestis.)
  1889 C, p. 257, fig. 80.   (H. agrestis.)
King, C. 1878 A, p. 404.   (H. agrarius, H. bairdianus.)
Leidy, J. 1871 B, p. 373.
  1871 C, p. 357.   (H. agrarius, H. agrestis.)
  1871 K, p. 229.

Leidy, J. 1872 B, p. 361.
1872 C, p. 19.
1873 B, pp. 60, 327, pl. ii, figs. 11, 12; pl. iv,
figs. 9-18; pl. xx, figs. 25, 26.
Marsh, O. C. 1871 D, p. 36. (Lophiodon bairdianus.)
1884 F, p. 64, fig. 71. (H. bairdianus.)
Matthew, W. D. 1899 A, p. 46.
Osborn, H. F. 1890 D, p. 505.
1892 B, p. 764.
1898 I, p. 81, fig. 1; p. 88, fig. 6; p. 89, fig. 8;
p. 128, fig. 33.
1898 Q, p. 9.
Osborn, Scott, and Speir 1878 A, p. 49.
Osborn and Wortman 1895 A, p. 367, figs. 9-11.
Pavlow, M. 1892 A.
Roger, O. 1896 A, p. 185.
Trouessart, E. L. 1898 A, p. 744.
Eocene (Bridger); Wyoming, Utah.

Hyrachyus eximius Leidy.

Leidy, J. 1871 K, p. 229.
Cope, E. D. 1873 E, p. 595.
1873 H, p. 213.
1884 O, pp. 660, 662, pl. xxiiia, fig. 1; pl. liii,
fig. 3; pls. liv, lv, lva; pl. lviiia, figs. 5, 6.
Leidy, J. 1871 B, p. 373.
1872 B, p. 361.
1873 B, pp. 66, 327, pl. iv, figs. 19, 20; pl. xix,
fig. 5; pl. xxvi, figs. 9, 10.
Matthew, W. D. 1899 A, p. 46.
Osborn, H. F. 1892 I, p. 764.
Osborn, Scott, and Speir 1878 A, pp. 49, 52.
Pavlow, M. 1892 A, p. 171.
Roger, O, 1896 A, p. 185.
Trouessart, E. L. 1898 A, p. 744.
Zittel, K. A. 1893 B, p. 284, figs. 224, 225.
Eocene (Bridger); Wyoming.

Hyrachyus imperialis Osb., Scott, and Speir.

Osborn, Scott, and Speir 1878 A, p. 50.
Cope, E. D. 1884 O, p. 660.
Matthew, W. D. 1899 A, p. 46.

Trouessart, E. L. 1898 A, p. 744.
Eocene (Bridger); Wyoming.

Hyrachyus implicatus Cope.

Cope, E. D. 1873 U, p. 5, specimen 2.
1873 E, p. 604.
1884 O, pp. 660, 675, 676, pl. lviii, figs. 6, 7
(H. implicatus); p. 660 (H. intermedius).
Matthew, W. D. 1899 A, p. 46.
Osborn, Scott, and Speir 1878 A, p. 51 (H. impli-
catus); p. 52 (H. intermedius, not of Filhol); p.
52 (H. crassidens).
Roger, O. 1896 A, p. 185.
Trouessart, E. L. 1898 A, p. 744. (H. implicatus,
H. intermedius.)
Eocene (Bridger); Wyoming.

Hyrachyus modestus Leidy.

Leidy, J. 1870 O, p. 109. (Lophiodon.)
Cope, E. D. 1884 O, p. 660.
Leidy, J. 1871 C, p. 357. (Lophiodon.)
1872 B, p. 361.
1872 C, p. 20.
1873 B, pp. 67, 367, pl. ii, figs. 11, 13.
Matthew, W. D. 1899 A, p. 46.
Roger, O. 1896 A, p. 185.
Trouessart, E. L. 1898 A, p. 744.
Eocene (Bridger); Wyoming.

Hyrachyus paradoxus Osb., Scott, and Speir.

Osborn, Scott, and Speir 1878 A, pp. 53, 135.
Matthew, W. D. 1899 A, p. 46.
Eocene (Bridger); Wyoming.

Hyrachyus princeps Marsh.

Marsh, O. C. 1872 G, p. 125.
Cope, E. D. 1873 E, p. 595.
1873 U, p. 6.
1884 O, pp. 660, 661, pl. lii, fig. 4
Matthew, W. D. 1899 A, p. 46.
Roger, O. 1896 A, p. 185.
Trouessart, E. L. 1898 A, p. 744.
Eocene (Bridger); Wyoming.

COLONOCERAS Marsh. Type C. agrestis Marsh.

Marsh, O. C. 1873 H, p. 407.
Brandt, J. F. 1878 A, p. 25.
Cope, E. D. 1879 C, p. 233.
1881 G, p. 381.
1884 O, p. 618.
1887 N, p. 994.
Marsh, O. C. 1877 E.
1884 F, fig. 10.
1897 D, p. 166.
Nicholson and Lydekker 1889 A, p. 1356.
Osborn, H. F. 1890 D, p. 505.
1898 D.

Scott, W. B. 1883 A, p. 53.
Scott and Osborn 1883 A, p. 20.
Zittel, K. A. 1893 B, p. 285.

Colonoceras agrestis Marsh.

Marsh, O. C. 1878 H, p. 407.
1897 D, p. 167, fig. 2.
Matthew, W. D. 1899 A, p. 46.
Roger, O. 1896 A, p. 185.
Trouessart, E. L. 1898 A, p. 745.
Eocene (Bridger); Wyoming.

TRIPLOPODINÆ.

Cope, E. D. 1881 G, pp. 378, 382. (Triplopodidæ.)
1881 O. (Triplopodidæ.)
1884 O, p. 615. (Triplopidæ.)
1887 N, pp. 992, 998. (Triplopodidæ.)

Nicholson and Lydekker 1889 A, p. 1356. (Triplo-
podidæ.)
Osborn, H. F. 1898 I, p. 85.
Osborn and Wortman 1892 A, p. 93.

TRIPLOPUS Cope.   Type *T. cubitalis* Cope.

Cope, E. D. 1880 I.
　　1880 M.
　　1881 G, p. 382.
　　1884 O, p. 678.
　　1890 C, p. 472.
Nicholson and Lydekker 1889 A, pp. 1355, 1356.
　　(Prothyracodon.)
Osborn, H. F. 1890 D, p. 524.
Schlosser, M. 1886 B, p. 23.
Scott, W. B. 1883 A, p. 53.
Scott and Osborn 1883 A, p. 12.
　　1887 A, p. 260.   (Prothyracodon, type *P. in-*
　　termedium=T. obliquidens.)
Zittel, K. A. 1893 B, p. 285.

Triplopus amarorum Cope.

Cope, E. D. 1881 G, p. 389.
1884 O, pp. 660, 687, pl. lv, figs. 6–9; pl. lviii, fig. 2.
Matthew, W. D. 1899 A, p. 46.
Osborn, H. F. 1897 A, p. 57.
Roger, O. 1896 A, p. 185.
Trouessart, E. L. 1898 A, p. 745.
　　Eocene (Bridger); Wyoming.

Triplopus cubitalis Cope.

Cope, E. D. 1880 I.
　　1881 G, p. 383.
　　1884 O, p. 687, pls. lva, figs. 10–12; pl. lvia.
　　1887 N, p. 999, fig. 13.
Matthew, W. D. 1899 A, p. 46.
Osborn, H. F. 1890 D, p. 528.
Roger, O. 1896 A, p. 185.
Scott, W. B. 1896 B, p. 367.
Trouessart, E. L. 1898 A. p. 745.
　　Eocene (Bridger); Wyoming.

Triplopus obliquidens (Scott and Osb.).

Scott and Osborn 1887 A, p. 259.   (Hyrachyus.)
Matthew, W. D. 1899 A, p. 49.
Osborn, H. F. 1890 D, p. 525, pl. xi, figs. 6–10.
Roger, O. 1896 A, p. 185.
Scott and Osborn 1887 A, p. 260.   (Prothyracodon
　　intermedium.)
Trouessart, E. L. 1898 A, p. 745.
　　Eocene (Bridger); Wyoming: (Uinta); Utah.

ANCHISODON Cope.   Type *A. quadriplicatus* Cope.

Cope, E. D. 1879 L, p. 270.
　　1879 C, p. 233.
Scott, W. B. 1883 A, p. 50.

Anchisodon quadriplicatus Cope.

Cope, E. D. 1873 S, p. 1.   (Hyracodon.)
Brandt, J. F. 1878 A, p. 32.   (Aceratherium.)
Cope, E. D. 1873 CC, p. 14.   (Hyracodon.)
　　1874 B, p. 495.   (Aceratherium.)
　　1879 C, p. 234.
　　1879 L.
Matthew, W. D. 1899 A, p. 57.   (Aceratherium.)

Roger, B. 1896 A, p. 185.
Trouessart, E. L. 1898 A, p. 745 (Anchisodon); p.
　　749 (Aceratherium).
　　Oligocene (White River); Colorado.

Anchisodon tubifer Cope.

Cope, E. D. 1879 L, p. 270.
　　1879 C, p. 234.
Matthew, W. D. 1899 A, p. 63.   (Aceratherium.)
Roger, O. 1896 A, p. 185.
Trouessart, E. L. 1898 A, p. 745.
　　Miocene (John Day); Oregon.

HYRACODONTINÆ.

Branco, W, 1897 A, p. 20.   (Hyracodontidæ.)
Osborn, H. F. 1898 I, p. 85.
Osborn and Wortman 1892 A, p. 93.

Steinmann and Döderlein 1890 A, p. 772.
Trouessart, E. L. 1898 A, p. 744.

HYRACODON Leidy.   Type *H. nebraskensis* Leidy.

Leidy, J. 1856 H, p. 91.
Brandt, J. F. 1878 A, p. 22.
Cope, E. D. 1879 C, p. 234.
　　1879 V, pp. 771a, 771d.
　　1884 O, p. 691.
　　1887 L, p. 925.
Dawkins, W. B. 1870 A.
Flower, W. H. 1876 C, p. 107.
Flower and Lydekker 1891 A, p. 412.
Gaudry, A. 1877 A.
Leidy, J. 1869 A, p. 232.
　　1871 C, p. 356.
Lydekker, R. 1886 A, p. 158.
Marsh, O. C. 1875 B, p. 244.
　　1877 E.
Nicholson and Lydekker 1889 A, p. 1356.
Osborn, H. F., in Scott and Osborn 1890 B, p. 89.
　　1890 D, p. 511.
Pavlow, M. 1892 A.

Scott, W. B. 1896 B.
　　1897 A.
Scott and Osborn, 1883 A, pp. 10, 17.
Woodward, A. S. 1898 B, p. 334.
Wortman, J. L. 1893 A.
Zittel, K. A. 1893 B, p. 285.

Hyracodon arcidens Cope.

Cope, E. D. 1873 S, p. 2.
Brandt, J. F. 1878 A, p. 24.
Cope, E. D. 1873 CC, p. 14.
　　1874 B, p. 493.
　　1879 C, p. 234.
Matthew, W. D. 1899 A, p. 57.
Roger, O. 1896 A, p. 186.
Scott and Osborn 1887 B, p. 171.
Trouessart, E. L. 1898 A, p. 746.
　　Oligocene (White River); Colorado.

Hyracodon major Scott and Osb.

Scott and Osborn 1887 B, p. 170.
Matthew, W. D. 1899 A, p. 57.
Roger, O. 1896 A, p. 186.
Trouessart, E. L. 1898 A, p. 746.
Oligocene (White River). Locality unknown.

Hyracodon nebraskensis Lei...

Leidy, J. 1850 C, p. 121. (Rhinoceros.)
Brandt, J. F. 1878 A, p. 28.
Cope, E. D. 1873 CC, p. 14.
    1874 B, p. 493.
    1879 C, p. 234.
    1886 U, p. 83. [Aceratherium (Cænopus) pumilum, in part.]
    1887 N, p. 1008, fig. 17.
    1889 I, p. 154. (Aceratherium pumilum, in part.)
    1891 A, p. 19, pl. iv, fig. 4. (Aceratherium pumilum.)
Hayden, F. V. 1858 B, p. 157. (Rhinoceros.)
King, C. 1878 A, p. 412. (Rhinoceros.)
Kowalevsky, W. 1873 A, p. 138.
Leidy, J. 1851 I, p. 331. (Aceratherium.)
    1852, J, p. 556, pls. xii A, fig. 6; pl. xxii B, fig. 3. (Rhinoceros.)
    1852 L, p. 63. (Rhinoceros.)

Leidy, J. 1854 A, pp. 86, 114, pls. xiv, xv. (Rhinoceros.)
    1854 E, p. 157. (Aceratherium.)
    1856 H, p. 92.
    1865 B, p. 176.
    1869 A, p. 391.
    1871 C, p. 357.
Lydekker, R. 1886 A, p. 159.
Matthew, W. D. 1899 A, p. 57.
Osborn, H. F. 1896 C, pp. 711, 714, illus. 7.
    1898 I, pp. 82, 93, figs. 2, 11.
    1898 Q, p. 18.
Roger, O. 1896 A, p. 187.
Scott, W. B. 1896 B, pls. i-iii.
Scott and Osborn, 1887 B, p. 169.
Trouessart, E. L. 1898 A, p. 746.
Zittel, K. A. 1893 B, p. 286, fig. 226.
Oligocene (White River); South Dakota, Colorado, Nebraska.
The name of this species is commonly spelled *nebrascensis*. The original form is here given.

Hyracodon? planiceps Scott and Osb.

Scott and Osborn 1887 B, p. 170.
Matthew, W. D. 1899 A, p. 57.
Trouessart, E .L. 1898 A, p. 746.
Oligocene (White River); South Dakota?.

AMYNODONTIDÆ.

Scott and Osborn 1883 A, pp. 4, 12.
Branco, W. 1897 A, p. 21. (Amynodontinæ.)
Cope, E. D. 1883 II, p. 969.
Nicholson and Lydekker 1889 A, p. 1365. (Amynodontinæ.)
Osborn, H. F. 1890 D, p. 50. (Amynodontidæ, Amynodontinæ.)
    1895 A, p. 95.

Osborn, H. F. 1898 A, pp. 79, 80.
Osborn and Wortman 1892 A, p. 93. (Amynodontidæ, Amynodontinæ.)
    1894 A, p. 208.
Pavlow, M. 1893 A, p. 41.
Scott and Osborn 1887 B, p. 164.
Zittel, K. A. 1893 B, p. 286.

AMYNODON Marsh.   Type *Diceratherium advenum* Marsh.

Marsh, O. C. 1877 D, p. 251.
Brandt, J. F. 1878 A, pp. 15, 25.
Cope, E. D. 1879 V, p. 771 f.
    1882 II, p. 926. (Orthocynodon.)
    1883 II, p. 969. (Orthocynodon.)
    1887 K.
    1890 C, p. 472.
Flower and Lydekker 1891 A, p. 412.
Marsh, O. C. 1877 E.
    1893 D, p. 409.
Nicholson and Lydekker 1889 A, p. 1365.
Osborn, H. F. 1890 D, pp. 506, 551 (Amynodon); p. 508 (Orthocynodon, as subgenus).
    1892 B, p. 764.
    1893 D.
Pavlow, M. 1892 A, p. 172. (Amynodon, Orthocynodon.)
    1893 A.
Scott and Osborn 1882 A, p. 223 (Amynodon); p. 223 (Orthocynodon. type O. *antiquus*).
    1883 A. (Orthocynodon, Amynodon.)
    1884 A. (Orthocynodon.)
    1887 A, p. 261.
    1887 B, p. 164.
Zittel, K. A, 1893 B, p. 286.

Amynodon advenus Marsh.

Marsh, O. C. 1875 B, p. 244. (Diceratherium.)
Brandt, J. F. 1878 A, p. 32. (Diceratherium.)
King, C. 1878 A, p. 407.
L. P. B. 1885 A, p. 191, fig. 25.
Marsh, O. C. 1877 D, p. 252.
    1884 F, fig. 72.
    1885 B, fig. 106.
    1897 D, p. 167, fig. 3. (Diceratherium.)
Matthew, W. D. 1899 A, p. 50.
Osborn, H. F. 1890 D, p.508. [A. (Orthocynodon).]
Pavlow, M. 1892 A, p. 172. (Orthocynodon.)
Roger, O. 1896 A, p. 186.
Scott, W. B. 1892 A, p. 431.
Scott and Osborn 1882 A, p. 224. (Orthocynodon.)
    1887 A, p. 262.
Trouessart, E. L. 1898 A, p. 746.
Eocene (Uinta); Utah.

Amynodon antiquus (Scott and Osb.).

Scott and Osborn 1883 A, p. 4, pl. i; text fig. 1. (Orthocynodon.)
Matthew, W. D. 1899 A, p. 46.
Osborn, H. F. 1890 D, p. 508. [A. (Orthocynodon).]

Pavlow, M. 1893 A, p. 37.
Roger, O. 1896 A, p. 186.
Trouessart, E. L. 1898 A, p. 746.
Eocene (Bridger): Wyoming.

Amynodon intermedius Osb.

Osborn, H. F. 1890 D, p. 508, pl. x, fig. 10.

Matthew, W. D. 1899 A, p. 50.
Osborn, H. F. 1895 A, p. 95.
Pavlow, M. 1893 A, p. 37, pl. iii.
Roger, O. 1896 A, p. 186.
Trouessart, E. L. 1898 A, p. 746.
Zittel, K. A. 1893 B, p. 287, fig. 227.
Eocene (Uinta): Utah.

METAMYNODON Scott and Osb.   Type *M. planifrons* Scott and Osb.

Scott and Osborn 1887 B, p. 165.
Cope, E. D. 1887 K.
    1887 N.
    1890 C, p. 472.
Flower and Lydekker 1891 A, p. 412.
Nicholson and Lydekker 1889 A, p. 1365.
Osborn, H. F. 1890 D, p. 507.
    1893 D, p. 40.
    1893 E.
Pavlow, M. 1893 A, p. 41.
Wortman, J. L. 1893 A.
Zittel, K. A. 1893 B, p. 287.

Metamynodon planifrons Scott and Osb.

Scott and Osborn 1887 B, p. 165, figs. 7-9.
Cope, E. D. 1887 N, p. 1001, figs. 15, 16.
Matthew, W. D. 1899 A, p. 57.
Osborn, H. F. 1896 C, pp. 705, 710, illus. 6.
    1896 I, p. 83, fig. 3; p. 91, fig. 10; p. 93, fig. 12.
    1898 Q, p. 9.
Osborn and Wortman 1894 A, p. 209.
    1895 A, p. 373, pls. x, xi.
Roger, O. 1896 A, p. 186.
Oligocene (White River): South Dakota.

RHINOCEROTIDÆ.

By some authors the spelling *Rhinoceridæ* is employed.
Blainville, H. M. D. 1864 A, iii, x, pls. i-xiv.
Branco, W. 1897 A, p. 20
Brandt, J. F. 1849 B.
    1878 A.
Cope, E. D. 1879 C.
    1879 V.
    1881 G, pp. 378, 393.
    1881 O.
    1884 O, pp. 615, 691.
    1887 K. (Cænopidæ.)
    1897 N. (Rhinocerotidæ, Cænopidæ.)
Cuvier, G. 1834 A, iii, pp. 1-192, with plates.
Falconer, H. 1868 A, p. 309.
Flower, W. H. 1876 A.
    1876 B, xiii, p. 328.
    1883 D, p. 428.
Flower and Lydekker 1891 A, p. 402.
Gray, J. E. 1867 B.
    1869 A.
Huxley, T. H. 1872 A, p. 307.
Leidy, J. 1869 A, p. 219.
    1871 C, p. 356.
Lydekker, R. 1881 A.
    1886 A, p. 90.

Nicholson and Lydekker 1889 A, p. 1364.
Osborn, H. F., in Scott and Osborn 1890 B, p. 90.
    1893 D.
    1898 I, pp. 79, 80, 94, 95.
    1900 E, p. 885.
    1900 H.
Owen, R. 1846 B, p. 587.
    1866 B, p. 450.
    1868 A, p. 356.
Pavlow, M. 1887 A, p. 358.
    1888 A, p. 173.
    1892 A.
Roger, O. 1889 A.
    1896 A, p. 186.
Schlosser, M. 1886 A, p. 268.
    1886 B, p. 22.
    1888 A.
    1890 C.
Scott, W. B. 1892 A, p. 432.
Scott and Osborn 1884 A.
Slade, D. D. 1893 A.
Steinmann and Döderlein 1890 A. p. 772.
Trouessart, E. L. 1898 A, p. 743.
Weyhe, — 1875 A, p. 121.
Woodward, A. S. 1898 B, p. 332.
Zittel, K. A. 1893 B, p. 281.

DICERATHERIINÆ.

Osborn, H. F. 1898 I, p. 121.
    1900 E, p. 885.
    1900 H, p. 232.

Osborn and Wortman 1892 A, p. 93.
Scott and Osborn 1883 A, p. 20. ("Diceratherium series.")

TRIGONIAS Lucas.   Type *T. osborni* Lucas.

Lucas, F. A. 1900 D, p. 221.
Osborn, H. F. 1900 H, p. 233.

Trigonias osborni Lucas.

Lucas, F. A. 1900 D, p. 221, figs. 1, 2.

Osborn, H. F. 1900 D, p. 767, fig. 1.
Oligocene (White River): South Dakota.

LEPTACERATHERIUM Osb. Type *Aceratherium trigonodum* Osb. and Wort.

Osborn, H. F. 1896 I, p. 182.

Leptaceratherium trigonodum (Osb. and Wort.).

Osborn and Wortman 1894 A, p. 201, pl. i, fig. A; text fig. 1. (Aceratherium.)
Matthew, W. D. 1899 A, p. 57.

Osborn, H. F. 1896 I, pp. 113, 115, 132, pl. xii; pl. xiii, fig. 1; pl. xix, fig. 27; text figs. 25, 26, 35, 44.
Osborn and Wortman 1895 A, p. 373. (Aceratherium.)
Roger, O. 1896 A, p. 188. (Rhinoceros.)
Trouessart, E. L. 1898 A, p. 748. (Aceratherium.)
    Oligocene (White River); South Dakota.

CÆNOPUS Cope. Type *Aceratherium mite* Cope.

Cope, E. D. 1880 M, p. 610.
    1881 G, p. 393.
    1884 O, p. 691.
    1887 K.
Osborn, H. F. 1900 E, p. 233.
Zittel, K. A. 1893 B, pp. 289, 291. (Syn. of Aceratherium.)

Cænopus copei Osb.

Osborn, H. F. 1898 I, pp. 113, 128, 131, 146, pl. xv; text figs. 25, 33, 34, 44. (Aceratherium.)
Matthew, W. D. 1899 A, p. 57. (Aceratherium.)
Osborn, H. F. 1900 H, p. 226.
Osborn and Wortman 1894 A, p. 203, pl. ii, fig. B; text fig. 2, B. (Aceratherium mite, not of Cope.)
    1895 A, p. 371. (No name.)
Trouessart, E. L. 1899 A, p. 1846. (Aceratherium.)
    Oligocene (White River); South Dakota.

Cænopus mitis Cope.

Cope, E. D. 1874 B, p. 493. (Aceratherium.)
Brandt, J. F. 1878 A, p. 30. (Aceratherium.)
Cope, E. D. 1879 C, p. 235. (Aceratherium.)
    1879 P, p. 333. (Aceratherium.)
    1879 V, p. 771c. (Aceratherium.)
    1880 M. (Cænopus.)
    1885 H. (Aceratherium.)
    1886 Q, p. 83. (Aceratherium mite; A. pumilium, in part.)
    1887 K. (Cænopus.)
    1889 I, p. 154. (Aceratherium mite; A. pumilium, in part.)
    1891 A, p. 19, pl. iv, figs. 2, 3 (Cænopus); p. 19, pl. iv, fig. 4. (C. pumilis.)
Lydekker, R. 1881 A, p. 21. (Aceratherium.)
Matthew, W. D. 1899 A, p. 57. (Aceratherium.)
Osborn, H. F. 1898 C. (Aceratherium.)
    1898 I, pp. 89, 136, figs. 8, 36, 37, 39. (Aceratherium.)
Osborn and Wortman 1894 A, p. 203.
    1895 A, p. 371. (Aceratherium.)
Roger, O. 1896 A, p. 187 (Rhinoceros pumilum); p. 188 (R. mite).
Scott, W. B. 1896 B, p. 367. (Aceratherium.)
Trouessart, E. L. 1898 A, p. 748. (Aceratherium mite, A. pumilum.)
Wortman, J. L. 1893 A, p. 104. (Aceratherium.)
    Oligocene (White River); South Dakota, Canada.

Cænopus occidentalis (Leidy).

Leidy, J. 1851 E, p. 276. (Rhinoceros.)
Brandt, J. F. 1878 A, p. 31. (Aceratherium.)
Cope, E. D. 1873 CC, p. 14. (Aceratherium.)
    1874 B, p. 495. (Aceratherium.)

Cope, E. D. 1879 C, p. 235. (Aceratherium.)
    1879 P, p. 333. (Aceratherium.)
    1887 K.
    1887 N, p. 1000, fig. 14.
    1891 A, p. 19. (Aceratherium.)
King, C. 1878 A, p. 412. (Aceratherium.)
Hayden, F. V. 1858 B, p. 157. (Rhinoceros.)
Leidy, J. 1850 B, p. 119. (Rhinoceros; no description.)
    1851 I, p. 331. (Aceratherium.)
    1852 J, p. 552, pl. ix, figs. 1, 2. (Rhinoceros.)
    1852 L, p. 64. (Rhinoceros.)
    1853 D, p. 392. (Rhinoceros.)
    1854 A, pp. 81, 114, pls. xii, xiii. (Rhinoceros.)
    1854 E, p. 157. (Aceratherium.)
    1865 B. (Rhinoceros.)
    1869 A, pp. 220, 390, pl. xxi, fig. 34; pl. xxii; pl. xxiii, figs. 1–3. (Rhinoceros, Aceratherium.)
    1871 C, p. 356. (Rhinoceros.)
Lydekker, R. 1881 A, p. 21. (Aceratherium.)
    1886 A, p. 143. (Rhinoceros.)
Matthew, W. D. 1899 A, p. 57. (Aceratherium.)
Osborn, H. F. 1893 C. (Aceratherium.)
    1898 I, pp. 90, 109, 128, 131, 134, 150, pl. xiii, figs. 5, 6, 7; pl. xvi; pl. xix, fig. 9; text figs. 9, 21, 33, 34, 35. (Aceratherium.)
    1900 H, p. 245. (Cænopus.)
Osborn and Wortman 1894 A, p. 204, pl. ii, fig. C; text fig. 3, A. (Aceratherium.)
    1895 A, p. 373, fig. 12. (Aceratherium.)
Pavlow, M. 1892 A. (Aceratherium.)
Pohlig, H. 1893 A, p. 41, pl. iii. (Aceratherium.)
Roger, O. 1896 A, p. 188. (Rhinoceros.)
Scott and Osborn 1887 B, p. 169. (Aceratherium.)
Trouessart, E. L. 1898 A, p. 747. (Aceratherium.)
    Oligocene (White River); South Dakota, Nebraska, Colorado, Canada.

Cænopus simplicidens Cope.

Cope, E. D. 1891 D, p. 48. (Cænopus.)
Matthew, W. D. 1899 A, p. 57. (Aceratherium.)
Osborn, H. F. 1898 I, p. 145, fig. 43. (Aceratherium.)
Osborn and Wortman 1894 A, 209. (Aceratherium.)
Roger, O. 1896 A, p. 188. (Rhinoceros.)
Trouessart, E. L. 1898 A, p. 748. (Aceratherium.)
    Oligocene (White River); Nebraska.

Cænopus tridactylus Osb.

Osborn, H. F. 1893 C. (Aceratherium.)
Hatcher, J. B. 1894 B. (Diceratherium proavitum.)
    1897 A. (Diceratherium proavitum and Aceratherium tridactylum.)

Matthew, W. D. 1899 A, p. 57. (Aceratherium.)
Osborn, H. F. 1893 E, p. 55. (Aceratherium.)
1898 I, pp. 84, 94, 109, 115, 128, 131, 134, 158,
pl. xliii, fig. 8; pl. xvii; pl. xix, fig. 30; pl.
xx, text figs. 4, 13, 21, 26, 33 34, 35 48, 49
(Aceratherium.)
1898 Q, pp. 13, 15. (Aceratherium.)
1900 H, p. 245. (Cænopus.)
Osborn and Wortman 1894 A, p. 206, pl. ii, fig. E;
pl. iii (Aceratherium tridactylum); p. 208 (Diceratherium proavitum).

Osborn and Wortman 1895 A, p. 373. (Acerathe-
rium.)
Roger, O. 1896 A, p. 189. (Rhinoceros proavitum.)
Scott, W. B. 1896 B, p. 381. (Aceratherium.)
Trouessart, E. L. 1898 A, p. 748 (Aceratherium
tridactytum); p. 752 (Diceratherium proavi-
tum).
Oligocene (White River); South Dakota.
North Dakota.

DICERATHERIUM Marsh.   Type *D. armatum* Marsh.

Marsh, O. C. 1875 B, p. 242.
Branco, W. 1897 A, p. 21.
Brandt, J. F. 1878 A, p. 32.
Cope, E. D. 1879 C, p. 233.
1879 V, pp. 771b, 771h.
1881 G, p. 393.
1884 O, p. 692.
1887 N, p. 1004.
Flower, W. H. 1876 B, xiii, p. 327.
Flower and Lydekker 1891 A, p. 411.
Hatcher, J. B. 1894 B.
Lydekker, R. 1886 A. p. 91. (Syn. of Rhinoceros.)
Marsh, O. C. 1876 G, p. 60.
1877 E.
Matthew, W. D. 1899 A, p. 22.
Nicholson and Lydekker 1889 A, p. 1368.
Osborn, H. F. 1898 I.
1900 H, p. 232.
Schlosser, M. 1886 B, p. 24.
Scott and Osborn 1883 A, p. 21.
Woodward, A. S. 1898 B, p. 335.
Zittel, K. A. 1893 B, p. 292.

Diceratherium annectens Marsh.

Marsh, O. C. 1873 H, p. 409. (Rhinoceros.)
Brandt, J. F. 1878 A, p. 4. (Rhinoceros.)
Marsh, O.C.in King, C. 1878 A, p. 424. (Dicera-
therium.)
Leidy, J. 1873 B, p. 328. (Rhinoceros.)
Matthew, W. D. 1899 A, p. 63. (Aceratherium.)
Trouessart, E. L. 1898 A, p. 748. (Aceratherium.)
Miocene (John Day); Oregon.

Diceratherium armatum Marsh.

Marsh, O. C. 1875 B, p. 242.
Brandt, J. F. 1878 A, p. 32.
King, C. 1878 A, p. 424.
Matthew, W. D. 1899 A, p. 63.
Roger, O. 1896 A, p. 189. (Rhinoceros.)
Trouessart, E. L. 1898 A, p. 752.
Miocene (John Day); Oregon.

Diceratherium hesperium (Leidy).

Leidy, J. 1865 B, p. 177. (Rhinoceros.)
Brandt, J. F. 1878 A, p. 40. (Rhinoceros.)
Cope, E. D. 1879 p. 333. (Syn.? of A. pacificum.)
Dana, J. D. 1879 A, p. 233. (Rhinoceros.)
Leidy, J. 1869 A, pp. 230, 390, pl. xxiii, figs. 11, 12.
(Rhinoceros.)
1870 P, p. 112. (Rhinoceros.)
1871 C, p. 356. (Rhinoceros.)

Leidy, J. 1873 B, pp. 220, 328, pl. ii, figs. 8, 9. (Rhi-
noceros hesperius?.)
Matthew, W. D. 1899 A, p. 63.
Roger, O. 1896 A, p. 189. (Rhinoceros.)
Trouessart, E. L. 1898 A, p. 751. (Aphelops.)
Whitney, J. D. 1879 A, p. 243. (Rhinoceros.)
Miocene; California: (John Day); Oregon?

Diceratherium nanum Marsh.

Marsh, O. C. 1875 B, p. 243.
Brandt, J. F. 1878 A, p. 32.
King, C. 1878 A, p. 424.
Matthew, W. D. 1899 A, p. 63.
Roger, O. 1896 A, p. 189. (Rhinoceros.)
Trouessart, E. L. 1898 A, p. 752.
Miocene (John Day); Oregon.

Diceratherium oregonense (Marsh).

Marsh, O. C. 1873 H, p. 410. (Rhinoceros.)
Brandt, J. F. 1878 A, p. 41. (Rhinoceros.)
Leidy, J. 1873 B, p. 328. (Rhinoceros.)
Matthew, W. D. 1899 A, p. 71. (Aceratherium.)
Roger, O. 1896 A, p. 189. (Rhinoceros.)
Trouessart, E. L. 1898 A, p. 751. (Aphelops.)
Miocene (Loup Fork); Oregon.

Diceratherium pacificum (Leidy).

Leidy, J. 1871 M, p. 248. (Rhinoceros.)
Brandt, J. F. 1878 A, p. 41. (Rhinoceros.)
Cope, E. D. 1879 B, pp. 58, 235.
1879 C, p. 229.
1879 P, p. 333. (Aphelops.)
1879 V, p. 771c. (Diceratherium.)
King, C. 1878 A, p. 424 (Rhinoceros.)
Leidy, J. 1870 P, p. 112. (Rhinoceros occidentalis.)
1873 B, pp. 221, 328, pl. ii, figs. 6, 7; pl. vii.
figs. 24, 25. (Rhinoceros.)
Lydekker, R. 1881 A, p. 21.
Matthew, W. D. 1899 A, p. 63.
Roger, O. 1896 A, p. 188. (Rhinoceros.)
Trouessart, E. L. 1898 A, p. 748.
Miocene (John Day); Oregon.

Diceratherium truquianum (Cope).

Cope, E. D. 1879 P, p. 333. (Aceratherium.)
1879 C, pp. 229, 335. (Aceratherium.)
Lydekker, R. 1881 A, p. 21. (Aceratherium.)
Matthew, W. D. 1899 A, p. 63. (Aceratherium.)
Roger, O, 1896 A, p. 188. (Rhinoceros.)
Trouessart, E. L. 1898 A, p. 748. (A. trucquianum.)
Miocene (John Day); Oregon.

ELASMOTHERIINÆ.

Bonaparte, C. L. 1851, Conspect. Syst. Mastozool., pp. 472–479. (Elasmotheriina.)
Gill, T. 1872 B, pp. 12, 88. (Elasmotheriidæ.)
Osborn and Wortman 1892 A, p. 93. (Aceratheriinæ.)

Osborn and Wortman 1895 A, p. 371. (Aceratheriinæ.)
Osborn, H. F. 1898 I, p. 121. (Aceratheriinæ.)
1900 H, p. 240. (Aceratheriinæ.)
Zittel, K. A. 1893 B, p. 296

ACERATHERIUM Kaup.   Type *Rhinoceros incisivus* Cuv.

Kaup, P. 1832 B, p. 904.
Branco, W. 1897 A, p. 21.
Brandt, J. F. 1878 A, pp. 11, 37.
Bronn, H. G. 1838 A, p. 1213, pl. xlvi, fig. 2.
Cope, E. D. 1874 B, p. 493.
1879 C, pp. 229, 285.
1879 P, p. 333.
1879 V, p. 771a.
1881 G, p. 393.
1883 Y.
1884 O, p. 691. (Aceratherium.)
1887 K.
1887 N, p. 1004.
Duvornoy, G. L. 1855 A, p. 51.
Filhol, H. 1891 A, p. 201.
Flower, W. H. 1876 B, xiii, p. 328.
Flower and Lydekker 1889 A, p. 411. (Syn. of Rhinoceros.)
Gaudry, A. 1878 B, p. 46.
Hatcher, J. B. 1894 C, p. 240. (Aphelops.)
Leidy J. 1851 I, p. 331.
1854 A, p. 81.
Lydekker, R. 1881 A, p. 9.
1884 B, pp. 2, 8.
1886 A, p. 91. (Syn. of Rhinoceros.)
Marsh, O. C. 1877 E.
Mermier, É. 1896 A, p. 225.
Nicholson and Lydekker 1889 A, p. 1367.
Osborn, H. F. 1890 D, pp. 511, 550 (Aceratherium).
p. 542 (Aphelops).
1898 C.
1898 E, p. 52.
1898 I.
1898 P.
1899 A, p. 161, pl. 1.
1900 H, p. 241. (Aceratherium or Aphelops.)

Peters, K. 1870 A.
Pictet, F. J. 1853 A, p. 296.
Pomel, A. 1853 A, p. 76. (Acerotherium.)
Schlosser, M. 1886 A, p. 253.
1886 B, p. 24 (Aphelops); p. 25 (Aceratherium).
Scott, W. B. 1892 A, p. 431.
1895 C, p. 122. (Aphelops.)
1896 B. (Aceratherium, Aphelops.)
Scott and Osborn 1883 A, p. 10 *et seq.*; p. 21. (Aceratherium, Aphelops.)
1887 B, p. 169.
1890 B, p. 89 (Aceratherium); p. 92 (Aphelops).
Steinmann and Döderlein 1890 A, p. 774. (Aceratherium, Peraceras, Aphelops.)
Williston, S. W. 1894 E.
Woodward, A. S. 1898 B, p. 334. (Aceratherium.)
Wortman, J. L. 1893 A.
Zittel, K. A. 1893 B, p. 289.

Aceratherium platycephalum Osb. and Wort.

Osborn and Wortman 1894 A, p. 206, pl. ii, E; text figs. 2, E: 3, D.
Matthew, W. D. 1899 A, p. 57.
Osborn, H. F. 1898 I, pp. 115, 116, 128, 130, 134, pl. xiii, fig. 9: pl. xviii; text figs. 26, 27, 33, 35.
1900 H, pp. 240, 243.
Osborn and Wortman 1895 A, p. 373.
Roger, O. 1896 A, p. 188. (Rhinoceros.)
Trouessart, E. L. 1898 A, p. 748.
Oligocene (White River); South Dakota.

APHELOPS Cope.   Type *A. megalodus* Cope.

Cope, E. D. 1873 C, p. 1.
Branco, W. 1897 A, p. 21. (Aphelops, Peraceras.)
Brandt, J. F, 1878 A, p. 30.
Cope, E. D. 1875 F, p. 71.
1877 K, p. 316.
1879 C, pp. 229, 235.
1879 P, p. 333.
1879 V, p. 771a.
1880 K, p. 540. (Peraceras, type *Aphelops superciliosus.*)
1881 G, p. 393. (Aphelops, Peraceras.)
1884 G, pp. 8, 9.
1884 O, p. 691. (Aphelops, Peraceras.)
1887 N, p. 1004. (Peraceras.)
Flower, W. H. 1876 B, xiii, p. 328.

Aphelops crassus (Leidy).

Leidy, J. 1858 E, p. 28. (Rhinoceros.)
Brandt, J. F. 1878 A, p. 30. (Aceratherium.)

Cope, E. D. 1874 B, p. 521. (Aceratherium.)
1874 P, p. 12.
1877 K, p. 317.
1879 C, p. 237.
1879 V, p. 771j. (Aceratherium.)
Hayden, F. V. 1858 B, p. 157. (Rhinoceros.)
King, C. 1878 A, p. 424. (Diceratherium.)
Leidy, J. 1865 B. (Rhinoceros.)
1869 A, pp. 228, 390, pl. xxiii, figs. 4–9. (Rhinoceros.)
1871 C, p. 356. (Rhinoceros.)
Lydekker, R. 1881 A, p. 20. (Aceratherium.)
Matthew, W. D. 1899 A, p. 71. (Syn.? of Aceratherium megalodum.)
Scott and Osborn 1890 B, p. 92.
Roger, O. 1896 A, p. 188. (Rhinoceros.)
Trouessart, E. L. 1898 A, p. 751.
Miocene (Loup Fork); Nebraska, Kansas, Colorado.

Aphelops jemezanus Cope.

Cope, E. D. 1875 P, p. 260.
1875 F, p. 72.
1877 K, pp. 317, 319, pl. lxxiii, figs. 3, 4; pl. lxxiv, fig. 4.
Lydekker, R. 1881 A, p. 21. (Aceratherium.)
Matthew, W. D. 1899 A, p. 72. (Aceratherium.)
Roger, O. 1896 A, p. 189. (Rhinoceros.)
Trouessart, E. L. 1898 A, p. 751.
Miocene (Loup Fork); New Mexico.

Aphelops malacorhinus Cope.

Cope, E. D. 1878 X, p. 488.
1878 H, p. 383.
1879 C, pp. 229, 237.
1879 V, p. 771f, figs. 7, 8.
1880 K, p. 540. (Peraceras.)
1892 AA. (A. malacorhinus, A. longipes.)
Leidy, J. 1890 A. (Rhinoceros longipes.)
Leidy and Lucas 1896 A, p. 45, pl. x, figs. 11-16; pl. xi, figs. 3, 4, 11; pl. xiii, figs. 6-8; pl. xiv, fig. 2; pl. xvi, figs. 5-10.
Lydekker, R. 1881 A, p. 21. (Aceratherium.)
Matthew, W. D. 1899 A, p. 71. (?Teleoceras.)
Roger, O. 1896 A, p. 189. (Rhinoceros longipes.)
Trouessart, E. L. 1898 A, p. 75.
Miocene (Loup Fork); Kansas, Florida.

Aphelops? matutinus (Marsh).

Marsh, O. C. 1870 F. (Rhinoceros.)
1893 D, p. 411, pl. x, fig. 4. (Rhinoceros.)
Roger, O. 1896 A, p. 188. (Rhinoceros.)
Trouessart, E. L. 1898 A, p. 748. (Aceratherium.)
Miocene; New Jersey.

Aphelops megalodus Cope.

Cope, E. D. 1873 C, p. 1. (Aceratherium; made type of Aphelops.)
Brandt, J. F. 1878 A, p. 30. (Aceratherium.)
Cope, E. D. 1873 CC, p. 14.
1874 B, p. 520. (Aceratherium.)
1874 P, p. 11.
1875 F, p. 71.
1877 K, p. 317.
1878 X.

Cope, E. D. 1879 C, pp. 229, 237.
1879 V, p. 771c, figs. i, 2, 4, 5.
1881 J, p. 270.
1889 C, p. 189 et seq.
Lydekker, R. 1881 A, p. 21. (Aceratherium.)
1886 A, p. 141. (Rhinoceros.)
Matthew, W. D. 1899 A, p. 71. (Aceratherium.)
Osborn, H. F. 1898 C. (Aceratherium.)
1898 C. (Aceratherium.)
1898 I, p. 115, fig. 26. (Aceratherium.)
Pavlow, M. 1892 A, p. 176. (Aceratherium.)
Roger, O. 1896 A, p. 188. (Syn. of Rhinoceros crassus.)
Trouessart, E. L. 1898 A, p. 751.
Zittel, K. A. 1893 B, p. 291, fig. 232. (Rhinoceros.)
Miocene (Loup Fork); Colorado.

Aphelops meridianus (Leidy).

Leidy, J. 1865 B, p. 176. (Rhinoceros.)
Brandt, J. F. 1878 A, p. 40. (Rhinoceros.)
Cope, E. D. 1875 F, p. 71.
1877 K, pp. 316, 317, pl. lxxiii, figs. 1, 2; pl. lxxiv, figs. 1-3.
1879 C, pp. 229, 237.
Leidy, J. 1869 A, pp. 229, 390, pl. xxiii, fig. 10. (Rhinoceros.)
1871 C, p. 356. (Rhinoceros.)
Lydekker, R. 1881 A, p. 21. (Aceratherium.)
Matthew, W. D. 1899 A, p. 71. (Syn.? of A. malacorhinus.)
Roger, O. 1896 A, p. 189. (Rhinoceros.)
Trouessart, E. L. 1898 A, p. 751.
Miocene (Loup Fork); Texas, New Mexico.

Aphelops profectus (Matth.).

Matthew, W. D. 1899 A, p. 71. (Aceratherium.)
Miocene (Loup Fork); Colorado.

Aphelops superciliosus (Cope).

Cope, E. D. 1880 K, p. 540. (Peraceras.)
1887 N, p. 1004, fig. 18. (Peraceras.)
Matthew, W. D. 1899 A, p. 71. (Teleoceras.)
Osborn, H. F. 1898 I, p. 106, fig. 18; p. 109, fig. 21.
Roger, O. 1896 A, p. 189. (Rhinoceros.)
Trouessart, E. L. 1898 A, p. 751.
Miocene (Loup Fork); Nebraska.

TELEOCERATINÆ.

Osborn, H. F. 1900 H, p. 249. (Brachypodinæ.) |

TELEOCERAS Hatcher.   Type *T. major* Hatcher.

Hatcher, J. B. 1894 A.
1894 C, p. 241.
Osborn, H. F. 1898 E, p. 51.
1898 I, p. 98. (Syn. of Aceratherium.)
1900 H.

Teleoceras fossiger (Cope).

Cope, E. D. 1878 H, p. 382. (Aphelops.)
1878 X. (Aphelops.)
1879 C, pp. 229, 237. (Aphelops.)
1879 V, p. 771c, figs. 3, 6. (Aphelops.)
1880 E, p. 141. (Aphelops.)
1883 R. (Aphelops.)
1884 G, p. 8. (Aphelops.)

Cope, E. D. 1884 J, p. 309. (Aphelops.)
1884 Q. (Aphelops.)
1887 N, p. 1006, fig. 22. (Aphelops.)
1892 AA. (Aphelops proterus.)
1893 A, p. 20. (Aphelops.)
Hatcher, J. B. 1894 A. (Teleoceras major.)
1894 C, p. 241, pl. i, fig. 1; pl. ii, figs. 2, 6 (Teleoceras major); pl. ii, figs. 3, 5 (Aphelops fossiger).
Leidy, J. 1885 A, p. 33. (Rhinoceros proterus Leidy.)
1886 A, figs. 1, 2. (?Eusyodon maximus.)
1887 A. (Rhinoceros proterus.)
1890 A. (Rhinoceros proterus.)

Leidy, J. 1892 A. (Rhinoceros proterus.)
Leidy and Lucas 1896 A, p. 41, pls. viii, ix; pl. x,
figs. 1–10; pl. xi, figs. 1, 2, 5–10; pl. xii, figs. 1–14;
pl. xiii, figs. 1–5; pl. xiv, figs. 1–7; pl. xv, figs.
1–8. (Aphelops fossiger, with Rhinoceros proterus Leidy a synonym.)
Lydekker, R. 1881 A, p. 20.
    1884 B, p. 10.
Marsh, O. C. 1887 B, p. 325, figs. 3, 4. (Aceratherium
acutum.)
    1897 C, pp. 525, 527, fig. 102. (Aceratherium
    acutum.)
Matthew, W. D. 1899 A, p. 71.
Osborn, H. F. 1893 C.
    1898 C, fig. 1. (Teleoceras.)

Osborn, H. F. 1898 E, pls. iv, iv A. (Teleoceras.)
    1898 I, pp. 99, 109, 110, figs. 15, 21, 22. (Acer-
    atherium.)
    1898 Q, p. 12.
    1900 H, p. 249.
Pavlow, M. 1892 A, p. 176.
Roger, O. 1896 A, p. 188. (Rhinoceros.)
Scott, W. B. 1893 B, p. 659. (Aphelops.)
Scott and Osborn 1890 B, p. 92, pls. ii, iii, text
    fig. 15. (Aphelops.)
Trouessart, E. L. 1898 A, p. 751. (Aphelops.)
Williston, S. W. 1894 E, pl. viii. (Aphelops.)
Zittel, K. A. 1893 B, p. 292, fig. 233. (Aphelops.)
    Miocene (Loup Fork); Kansas, Nebraska,
Colorado, Florida.

RHINOCEROTINÆ.

Branco, W. 1897 A, p. 21.
Brandt, 1878 A, p. 25.
Osborn, H. F. 1898 I, p. 121. (Ceratorhinæ.)
    1900 H, p. 264. (Ceratorhinæ, Rhinoceroti-
    næ.)
Osborn and Wortman 1892 A, p. 98.
Roger, O. 1896 A, p. 186. (Rhinocerinæ.)
Trouessart, E. L. 1898 A, p. 747.
Zittel, K. A. 1893 B, p. 288. (Rhinocerinæ.)

Although no species found in North America
is here admitted to this subfamily and to the
genus *Rhinoceros*, citations to some of the literature of this subfamily and the genus are given
below, since the genus *Rhinoceros* is the type
of this whole group. For additional bibliography see Osborn 1898 I, pp. 121–125; 1900 H, p.
266.

RHINOCEROS Linn.

Linnæus C., 1758, Syst. Nat., ed. 10, i, p. 56.
Brandt, J. F. 1849 B.
    1878 A, p. 33.
Cope, E. D. 1879 C, p. 229.
    1881 G, p. 393.
    1884 O, p. 692.
    1887 N, p. 1008.
    1889 C, p. 258.
Cuvier, G. 1834 A, iii, pp. 1–185, with plates.
Duvornoy, G. L. 1855 A.
Filhol, H. 1891 A, p. 194.
Flower, W. H. 1876 A.
    1885 A.
Flower and Lydekker 1891 A, p. 402.
Gaudry, A. 1878 B, p. 44.
Gervais, P. 1859 A, p. 87.
Gray, J. E. 1867 B, p. 1008.
    1869 A, p. 300.
Huxley, T. H. 1872 A, p. 307.
Kowalevsky, W. 1873 A, pp. 219, 222.
Leidy, J. 1854 A, pp. 79, 114.
    1871 C, p. 356.
Lydekker, R. 1881 A.
    1886 A, p. 91.
Meyer, H. 1864 A.
Nicholson and Lydekker 1889 A, p. 1368.

Osborn, H. F. 1898 E, p. 52.
    1898 I.
Owen, R. 1844 A, p. 220.
    1845 B, p. 587.
    1846 A, p. 325.
Pavlow, M. 1888 A, p. 173.
    1892 A.
Peters, K. 1870 A.
Roger, O. 1896 A, p. 186.
Trouessart, E. L. 1898 A, p. 752.
Zittel, K. A. 1898 B, p. 293.

Rhinoceroides alleghaniensis Feath-
erst.

Featherstonhaugh, G. W. 1831 A, pl. i.
Blainville, H. M. D. 1864 A, Rhinoceros, p. 172.
    (Rhinoceros.)
Giebel, C. G. 1847 A, p. 185.
Harlan, R. 1834 B, p. 62.
    1835 B, p. 268.
Leidy, J. 1854 A, pp. 10, 79.
    1869 A, p. 444.
    The subject of the above references is now
generally conceded to be merely a worn piece
of sandstone.

Suborder ARTIODACTYLA Owen.

Owen, R. 1848 A, p. 131.
Baur, G. 1884 F.
Branco, W. 1897 A, p. 22.
Burmeister, H. 1867 A, p. 236. (Paridigitata.)
Cope, E. D. 1873 E, p. 562.
    1873 F, p. 39.
    1874 M.
    1880 B, p. 458.

Cope, E. D. 1881 I, p. 269.
    1881 J.
    1882 A.
    1882 Y.
    1883 D. (Diplarthra, part.)
    1884 C.
    1884 H.
    1884 O, pp. 879, 716.

Cope, E. D. 1884 DD.
1885 B.
1886 C, p. 615.
1887 B.
1887 N, p. 985.
1887 S.
1888 K, p. 163.   (Diplarthra, part.)
1888 X.
1889 C, p. 148 *et seq.*
1889 G.
1891 N, p. 84.   (Diplarthra, part.)
1898 B, p. 125.   (Diplarthra, part.)
Cope and Wortman 1884 A, p. 15.
Earle, C. 1893 B.
Flower, W. H. 1873 A.
1874 A.
1876 B, xiii, p. 327.
1883 A.
1883 D, p. 428.
1885 A.
Flower and Lydekker 1891 A, p. 275.
Gadow, H. 1898 A, p. 48.
Gegenbaur, C. 1898 A.
Gill, T. 1872 A, p. 298.
1872 B, pp. 8, 71.
Gill and Coues 1877 A, p. 1020.
Gruber, W. 1859 A.   (Multungulata.)
Haeckel, E. 1895 A, p. 549.
Hay, O. P. 1899 F, p. 682.
Huxley, T. H. 1870 A, p. 440.
1872 A, p. 312.
• Kowalevsky, W. 1874 A.
Leuthardt, F. 1891 A, p. 121.   (Paraxonia.)
Lydekker, R. 1885 C.
1885 E, p. 469.   (Diplarthra, part.)
1896 B.
Marsh, O. C. 1877 E.
1884 F, p. 177.   (Clinodactyla, Paraxonia.)

Nicholson and Lydekker 1889 A, p. 1313.
Osborn, H. F. 1893 A, p. 379 *et seq.*
1893 D.
1900 C, p. 568.
Owen, R. 1845 B, p. 527.   ("Anisodactyle ungulates.")
1857 A.
1866 B, p. 457.
1868 A, p. 343.
Paulli, S. 1900 A, p. 193.
Pavlow, M. 1900 A.
Pomel, A. 1848 B, p. 326.   ("Artiodactyles.")
Röse, C. 1896 A.
Roger, O. 1896 A, p. 198.
Rütimeyer, L. 1865 A.   (Ungulata paridigitata.)
1888 A.   (Diplarthra, part.)
Schmidt, O. 1886 A, p. 137.
Schlosser, M. 1885 A, p. 684.
1886 B.
1888 A.
1889 A.
1890 C.
1890 D.
1892 A.
Scott, W. B. 1892 A, p. 432.
Slade, D. D. 1893 A.
Steinmann and Döderlein 1890 A, p. 784.
Tomes, C. S. 1898 A, p. 392.
Tornier, G. 1888 A.
Turner, H. N. 1848 A.
Weber, M. 1886 A, p. 226.
Weyhe, ——. 1875 A, p. 118.
Wiedersheim, R. 1881 A.
Woodward, A. S. 1898 B, p. 338.
Wortman, J. L. 1883 A, p. 706.
1880 A, p. 297.   (Diplarthra, part.)
1893 B, p. 208.   (Diplarthra, part.)
Zittel, K. A. 1893 B, pp. 315, 727.

Superfamily *PANTOLESTOIDEA* Cope.

Cope, E. D. 1887 B, p. 378. | Cope, E. D. 1898 B, p. 130.   (Trigonolestoidea.)

PANTOLESTIDÆ.

Cope, E. D. 1884 O, p. 719.
1886 C, p. 615.
1887 B, p. 378.
1888 X, p. 1079.
Haeckel, E. 1895 A, p. 554.   (Pantolestida.)

Matthew, W. D. 1899 A, pp. 34, 48.   (Homacodontidæ, part.)
Roger, O. 1896 A, p. 198.
Steinmann and Döderlein 1890 A, p. 780.
Zittel, K. A. 1893 B, p. 324.

PANTOLESTES Cope.   Type *P. longicaudus* Cope.

Cope, E. D. 1872 N, p. 466.
Ameghino, F. 1896 A, p. 444.
Cope, E. D. 1875 C, pp. 13, 15.
1875 L, p. 21.
1877 K, pp. 134, 145.
1882 E, p. 149.
1883 M, p. 547.
1884 H, pp. 21, 25.
1884 O, pp. 215, 717.
1885 B.
1886 C, p. 617.
Marsh, O. C. 1894 L, p. 263.   (Homacodon, in part.)
Matthew, W. D. 1899 A, p. 48.

Osborn, H. F. 1893 D, p. 42.
Pavlow, M. 1900 A, p. 59.
Scott, W. B. 1890 A, p. 486.
1891 A, pp. 9, 45, 66.
1892 A, p. 433.
Schlosser, M. 1886 B, p. 39.
Trouessart, E. L. 1898 A, p. 800.
Wortman, J. L. 1898 A, pp. 101, 135.
Zittel, K. A. 1893 B, p. 324.
In many of the papers here cited this genus includes species which are here assigned to *Trigonolestes*. The systematic position of *Pantolestes longicaudus* is in doubt.

Pantolestes longicaudus Cope.

Cope, E. D. 1872 N, p. 467.
1873 E, p. 549.   (Notharctus.)
1873 U, p. 3.   (Notharctus.)
1882 E, p. 150.
1883 W, p. 191.

Cope, E. D. 1884 O, pp. 717, 725, pl. xxiv, figs. 13–17.
1886 C, p. 617.
Matthew, W. D. 1899 A, p. 48.
Roger, O. 1896 A, p. 198.
Trouessart, E. L. 1898 A, p. 800.
Eocene (Bridger); Wyoming.

TRIGONOLESTES Cope.   Type *Pantolestes brachystomus* Cope.

Cope, E. D. 1894 E, p. 868.
Ameghino, F. 1896 A, p. 444.
Cope, E. D. 1897 B.
Matthew, W. D. 1899 A, p. 34.
Osborn and Earle 1895 A, p. 70.
Scott, W. B. 1899 A, pp. 46, 62.

Trigonolestes brachystomus Cope.

Cope, E. D. 1882 E, p. 187.   (Mioclænus.)
1882 K.   (Mioclænus.)
1883 M, p. 547.   (Pantolestes.)
1884 O, pp. 718, 721, pl. xxiiid, figs. 16–21.
(Pantolestes.)
1886 C, p. 617, fig. 8.   (Pantolestes.)
1888 X, p. 1086, fig. 2.   (Pantolestes.)
1894 E, p. 868.   (Trigonolestes.)
Matthew, W. D. 1899 A, p. 34.
Roger, O. 1896 A, p. 198.
Schlosser, M. 1886 B, p. 135, pl. v, fig. 19; pl. vi,
fig. 29.   (Pantolestes.)
Scott, W. B. 1891 A, p. 46.   (Pantolestes.)
1892 A, pp. 433, 438, fig. 8⁴.   (Pantolestes.)
Trouessart, E. L. 1898 A, p. 800.
Zittel, K A. 1893 B, p. 325, fig. 262.   (Pantolestes.)
Eocene (Wasatch); Wyoming.

Trigonolestes chacensis (Cope).

Cope, E. D. 1875 C, p. 15.   (Pantolestes.)
1877 K, p. 146, pl. xlv, fig. 17.   (Pantolestes.)
1882 E, p. 150.   (Pantolestes.)
1884 O, pp. 717, 719, pl. xxivd, fig. 5.   (Pantolestes.)
Matthew, W. D. 1899 A, p. 34.   (Trigonolestes.)
Roger, O. 1896 A, p. 198.   (Pantolestes.)
Trouessart, E. L. 1898 A, p. 800.   (Pantolestes.)
Eocene (Wasatch); Wyoming.

Trigonolestes etsagicus (Cope).

Cope, E. D. 1882 E, p. 189.   (Mioclænus.)

Cope, E. D. 1883 M, p. 547.   (Pantolestes.)
1884 O, pp. 717, 724, pl. xxve, fig. 21.   (Pantolestes.)
1886 C, p. 618.   (Pantolestes.)
Matthew, W. D. 1899 A, p. 34.   (Trigonolestes.)
Roger, O. 1896 A, p. 198.
Trouessart, E. L. 1898 A, p. 800.   (Pantolestes.)
Wortman, J. L. 1898 A, p. 101, footnote.   (Syn.?
of Eohyus distans.)
Eocene (Wasatch); Wyoming.

Trigonolestes metsiacus (Cope).

Cope, E. D. 1882 E, p. 149.   (Pantolestes.)
1884 O, pp. 717, 719, pl. xxivd, fig. 6.   (Pantolestes.)
Matthew, W. D. 1899 A, p. 34.   (Trigonolestes.)
Roger, O. 1896 A, p. 199.   (Pantolestes.)
Trouessart, E. L. 1898 A, p. 800.   (Pantolestes.)
Eocene (Wasatch); Wyoming.

Trigonolestes nuptus Cope.

Cope, E. D. 1882 E, p. 150.   (Pantolestes.)
1884 O, pp. 718, 720, pl. xxivd, fig. 7.   (Pantolestes.)
Matthew, W. D. 1899 A, p. 32.   (Trigonolestes.)
Roger, O. 1896 A, p. 199.   (Pantolestes.)
Trouessart, E. L. 1898 A, p. 800.   (Pantolestes.)
Eocene (Wasatch); Wyoming.

Trigonolestes secans Cope.

Cope, E. D. 1881 D, p. 187.   (Pantolestes.)
1882 E, p. 150.   (Pantolestes.)
1884 O, pp. 717, 725, pl. xxvd, fig. 6.   (Pantolestes.)
Matthew, W. D. 1899 A, p. 36.   (Trigonolestes?.)
Roger, O. 1896 A, p. 199.   (Pantolestes.)
Trouessart, E. L. 1898 A, p. 800.   (Pantolestes.)
Eocene (Wind River); Wyoming, Colorado.

HOMACODONTIDÆ.

Marsh, O. C. 1894 L, p. 263.
Matthew, W. D. 1899 A, pp. 37, 48, part.

Scott, W. B. 1899 A, p. 84.

HOMACODON Marsh.   Type *H. vagans* Marsh.

Marsh, O. C. 1872 G, p. 126.
Ameghino, F. 1896 A, p. 444.
Cope, E. D. 1873 E, p. 608.
1894 E, p. 868.
Flower and Lydekker 1891 A, p. 294.
Marsh, O. C. 1877 E, p. 364.
1894 L, p. 263.
Pavlow, M, 1900 A, p. 59.
Scott, W. B. 1890 A, pp. 485, 486.

Scott, W. B. 1891 A, pp. 9, 46.
1898 C, p. 81.
1899 A, pp. 46, 65.
Trouessart, E. L. 1898 A, p. 800.   (Syn. of Panto-lestes.)
Woodward, A. S. 1898 B, p. 341.
Wortman, J. L. 1898 A, pp. 94, 98, 101, 102, 135.
Zittel, K. A. 1893 B, p. 361.

Homacodon priscus Marsh.

Marsh, O. C. 1894 L, p. 261, fig. 3.
Roger, O. 1896 A, p. 199.   (Pantolestes.)
Trouessart, E. L. 1898 A, p. 800.   (Pantolestes.)
Eocene; Wyoming, New Mexico.

Homacodon pucillus Marsh.

Marsh, O. C. 1894 L, p. 261, fig. 4.
Roger, O. 1896 A, p. 199.   (Pantolestes.)
Trouessart, E. L. 1898 A, p. 800.   (Pantolestos.)
Eocene; Wyoming, New Mexico.

Homacodon vagans Marsh.

Marsh, O. C. 1872 G, p. 126.
Cope, E. D. 1894 E, p. 868.
King, C. 1878 A, p. 404.
Marsh, O. C. 1894 L, p. 262, figs. 5–8.
Matthew, W. D. 1899 A, p. 48.
Roger, O. 1896 A, p. 199.   (Pantolestes.)
Scott, W. B. 1891 A, p. 486.
Trouessart, E. L. 1898 A, p. 800.   (Pantolestes.)
Woodward, A. S. 1898 A, p. 342, fig. 196.
Eocene (Bridger); Wyoming.

NANOMERYX Marsh.   Type *N. caudatus* Marsh.

Marsh, O. C. 1894 L, p. 263.

Nanomeryx caudatus Marsh.

Marsh, O. C. 1894 L, p. 263, figs. 9, 10.
Cope, E. D. 1894 E, p. 868.

Matthew, W. D. 1899 A, p. 48.
Roger, O. 1896 A, p. 199.
Trouessart, E. L. 1896 A, p. 801.
Eocene (Bridger); Wyoming.

BUNOMERYX Wort.   Type not indicated; *B. elegans*, being figured, may be taken.

Wortman, J. L. 1898 A, p. 97.
Scott, W. B. 1899 A, p. 84.

Bunomeryx elegans Wort.

Wortman, J. L. 1898 A, p. 100, fig. 2.
Matthew, W. D. 1899 A, p. 50.
Trouessart, E. L. 1899 A, p. 1348.
Eocene (Uinta); Utah.

Bunomeryx montanus Wort.

Wortman, J. L. 1898 A, p. 97.
Matthew, W. D. 1899 A, p. 50.
Trouessart, E. L. 1899 A, p. 1348.
Eocene (Uinta); Utah.

HELOHYIDÆ.

Marsh, O. C. 1877 E, p. 364.
Cope, E. D. 1892 S, p. 411.
Marsh, O. C. 1892 C, p. 352.

Matthew, W. D. 1899 A, p. 48.   (Homacodontidæ, in part.)
Scott, W. B. 1890 A, p. 485.

HELOHYUS Marsh.   Type *H. plicodon* Marsh.

Marsh, O. C. 1872 I, p. 207.
Cope, E. D. 1873 D, p. 3.   ("Thinotherium," of Marsh.)
    1892 S, p. 411.
    1894 E, p. 868.   (Syn. of Phenacodus.)
Marsh, O. C. 1872 I, p. 208.   (Thinotherium, type *T. validum;* not Thinotherium Cope, 1870.)
    1877 E.
    1892 C, p. 351.
    1894 L, p. 264.
Schlosser, M. 1886 B, p. 12.   (Syn.? of Phenacodus.)
Scott, W. B. 1889 C, p. 76.
    1890 A, pp. 486, 502.
    1890 B, p. 366.
Wilckens, M. 1885 B, p. 308.
Wortman, J. L. 1898 A, p. 94.
Zittel, K. A. 1893 B, p. 220.   (Thinotherium Marsh.)

Helohyus lentus Marsh.

Marsh, O. C. 1871 D, p. 39.   (Elotherium.)
Leidy, J. 1872 B, p. 365.   (Elotherium.)
    1873 B, p. 124.   (Elotherium.)

Marsh, O. C. 1873, in King, C. 1878 A, p. 404.   (Helohyus.)
    1894 L, p. 265, fig. 16.
Matthew, W. D. 1899 A, p. 48.
Roger, O, 1896 A, p. 167.   (Phenacodus?.)
Trouessart, E. L. 1898 A, p. 726.   (Phenacodus.)
Eocene (Bridger); Wyoming.

Helohyus plicodon Marsh.

Marsh, O. C. 1872 I, p. 207.
    1894 L, p. 264, figs. 11–14, 17.
Matthew, W. D. 1899 A, p. 48.
Roger, O. 1896 A, p. 166.   (Phenacodus?.)
Scott, W. B. 1890 B, pl. xlv, fig. 14.
Eocene (Bridger); Wyoming.

Helohyus validus Marsh.

Marsh, O. C. 1872 I, p. 208.   (Thinotherium.)
    1894 L, p. 265, fig. 15.
Matthew, W. D. 1899 A, p. 48.
Roger, O. 1896 A, p. 166.   (Phenacodus?.)
Eocene (Bridger); Wyoming.

EOHYIDÆ.

Marsh, O. C. 1894 L, p. 260.

Cope, E. D. 1894 E, p. 867.

EOHYUS Marsh.   Type *E. distans* Marsh.

Marsh, O. C. 1894 L, p. 259.
Cope, E. D. 1894 E, p. 868.
Marsh, O. C. 1877 E.   (No description.)
Matthew, W. D. 1899 A, p. 32, footnote.
Wilckens, M. 1885 B, p. 303.   (No description.)

Eohyus distans Marsh.

Marsh, O. C. 1894 L, p. 259, fig. 1.
Matthew, W. D. 1899 A, p. 32.
Roger, O. 1896 A, p. 199.

Trouessart, E. L. 1898 A, p. 801.
Wortman, J. L. 1898 A, p. 101, footnote.
   Eocene (Wasatch); New Mexico.

Eohyus robustus Marsh.

Marsh, O. C. 1894 L, p. 260.
Matthew, W. D. 1899 A, p. 32.
Roger, O. 1896 A, p. 199.
Trouessart, E. L. 1898 A, p. 801.
   Eocene (Wasatch); New Mexico.

Superfamily ANTHRACOTHERIOIDEA *Gill.*

Gill, T. 1872 B, pp. 11, 83.
Cope, E. D. 1898 B, p. 130.

Owen, R. 1848 A, p. 116.   (Anthracotheria.)

ANTHRACOTHERIIDÆ.

Blainville, H. M. D. 1864 A, iv, Y, p. 121, with plates.
Branco, W. 1897 A, p. 23.
Cope, E. D. 1884 H, p. 26.   (Hyopotamidæ.)
   1885 B, p. 489.   (Hyopotamidæ.)
   1888 X, pp. 1080, 1085.
Flower and Lydekker 1891 A, p. 292.
Gill, T. 1872 B, pp. 11, 83.
Haeckel, E. 1895 A, p. 554.   (Anthracotherida.)
Kowalevsky, W. 1873 A.   (Anthracotheriidæ, Hyopotamidæ.)
   1874 A.   (Hyopotamidæ.)
Leidy, J. 1869 A, p. 202.
   1871 C, p. 355.
Lydekker, R. 1883 A, p. 147.
   1885 C, p. 215.
Marsh, O. C. 1894 F, p. 92.
   1894 K, p. 178.   (Ancodontidæ.)

Meyer, H. 1856 B, pp. 61–66.
Nicholson and Lydekker 1889 A, p. 1324.
Osborn and Wortman 1894 A, p. 219.
Pavlow, M. 1900 A, p. 18.
Pomel, A. 1848 B, p. 321.   ("Les Anthracotheriums.")
Roger, O. 1896 A, p. 199.
Rütimeyer, L. 1857 A.   ("Anthracotherien.")
Schlosser, M. 1886 A, p. 255.
   1886 B, p. 79.
Scott, W. B. 1890 B, p. 392.   (Hyopotamidæ.)
Steinmann and Döderlein 1890 A, p. 792.
Teller, F. 1884 A.
Trouessart, E. L. 1898 A, p. 801.
Woodward, H. 1898 B, p. 348.
Wortman, J. L. 1886 A, p. 485.
Zittel, K. A. 1893 B, p. 325.

ANTHRACOTHERIINÆ.

Roger, O. 1896 A, p. 199.
Trouessart, E. L. 1898 A, p. 801.

Zittel, K. A. 1893 B, p. 327.

ANTHRACOTHERIUM Cuv.   Type *A. magnum* Cuv.

Cuvier, G. 1822, Oss. Foss., iii, p. 396.
Bayle, E. 1855 A.
Blainville, H. M. D. 1864 A, iv, Y, p. 121, pls. i–iii.
Bronn, H. G. 1838 A, p. 1225; pl. xlvi, fig. 4.
Cuvier, G. 1834 A, p. 464, pl. 161.
Filhol, H. 1891 B, p. 89.
Flores, E. 1897 A, p. 92.
Gaudry, A. 1878 B, pp. 42, 94.
Gervais, P. 1859 A, p. 189.
Hoernes, R. 1875 A.
   1876 A.
   1877 A.
Kowalevsky, W. 1873 A.
Lydekker, R. 1883 A, p. 148.
   1885 C, p. 235.
Marsh, O. C. 1894 C, p. 409.   (Heptacodon, type *H. curtus*.)
   1894 F, p. 92.   (Heptacodon.)
   1894 K, p. 178.   (Heptacodon.)

Meyer, H. 1834 A, p. 59.
   1854 A.
Nicholson and Lydekker 1889 A, p. 1324.
Osborn and Wortman 1894 A, p. 222.
Owen, R. 1845 B, p. 567.
   1848 A, p. 109.
Pavlow, M. 1900 A, p. 21.
Pictet, F. J. 1853 A, p. 332.
Pomel, A. 1847 A, pp. 208, 207.
   1847 C, p. 381.
   1848 B, p. 324.
   1853 A, p. 89.
Renevier, E. 1879 A.
Rütimeyer, L. 1857 A.
   1857 B.
   1857 C.
Teller, F. 1884 A.
Woodward, A. S. 1898 B, p. 348.
Zittel, K. A. 1893 B, p. 327.

Anthracotherium curtum (Marsh.)

Marsh, O. C. 1894 C, p. 409.   (Heptacodon.)
1894 K, p. 176, fig. 1.   (Heptacodon.)
Matthew, W. D. 1899 A, p. 58.
Osborn and Wortman 1894 A, p. 221, figs. 6, 7 (A. curtum?); p. 223 (A. occidentale).
Roger, O. 1896 A, p. 200.   (A. curtum, A. occidentale.)
Trouessart, E. L. 1898 A, p. 803.   (A. curtum, A. occidentale.)
Oligocene (White River); South Dakota.

Anthracotherium gibbiceps (Marsh).

Marsh, O. C. 1894 K, p. 175, fig. 2.   (Heptacodon.)
Matthew, W. D. 1899 A, p. 59.
Roger, O. 1896 A, p. 201.
Trouessart, E. L. 1898 A, p. 803.
Oligocene (White River); South Dakota?.

Anthracotherium karense Osb. and Wort.

Osborn and Wortman 1894 A, p. 222, fig. 8.
Matthew, W. D. 1899 A, p. 59.
Roger, O. 1896 A, p. 200.
Trouessart, E. L. 1898 A, p. 803.
Oligocene (White River); South Dakota.

OCTACODON Marsh.   Type *O. valens* Marsh.

Marsh, O. C. 1894 F, p. 92.
1894 K, p. 178.
Trouessart, E. L. 1898 A, P. 801.   (Syn. of Anthracotherium.)

Octacodon valens Marsh.

Marsh, O. C. 1894 F, p. 92, fig. 1.
1894 K. p. 178, fig. 6.

Matthew, W. D. 1899 A, p. 59.   (Syn. of Anthracotherium karense.)
Roger, O. 1896 A, p. 201.   (Anthracotherium.)
Trouessart, E. L. 1898 A, p. 103.   (Anthracotherium.)
Oligocene (White River); South Dakota.

ELOMERYX Marsh.   Type *Heptacodon armatus* Marsh.

Marsh, O. C. 1894 K, p. 176.

Elomeryx armatus Marsh.

Marsh O. C. 1894 F, p. 93, fig. 2.   (Heptacodon.)
1894 K. p. 176, figs. 3, 4.
Matthew, W. D. 1899, A, p. 59.   (Syn. of Anthracotherium karense.)
Roger, O. 1896 A, p. 201.
Trouessart, E. L. 1898 A, p. 804.
Oligocene (White River); South Dakota.

Elomeryx mitis Marsh.

Marsh, O. C. 1894 K, p. 177, fig. 5.
Matthew, W. D. 1899 A, p. 59.   (Anthracotherium.)
Roger, O. 1896 A, p. 201.
Trouessart, E. L. 1898 A, p. 804.
Oligocene (White River); South Dakota?.

ANCODON Pomel.   Type *A. velaunum* Pomel.

Pomel A., 1847, Arch. Sci. Phys. et Nat., Genève, v, p. 207.
Unless otherwise indicated, the following authors use the name *Hyopotamus* for this genus. *Ancodon* antedates it. *Ancodon* was afterwards emended into *Ancodus*.
Aymard, A., 1848, Ann. Soc. Agri. et Sci., du Puy, xii.   (Bothriodon, type *B. platyrhinchus*.)
Cope, E. D. 1887 A, p. 383.
1888 X, p. 1086.
1889 C, p. 163.
Filhol, H. 1882 A, p. 85.   (Ancodus.)
Gaudry, A. 1878 B, p. 94.
Gervais, P. 1859 A, p. 191.
Kowalevsky, W. 1874 A.
Leidy, J. 1869 A, p. 202.
1871 C, p. 355.
Lydekker, R. 1883 A, p. 154.
1885 A, p. 218.
Major, Fors. 1873 A, p. 101.
Marsh, O. C. 1894 K, p. 178.   (Ancodus.)
Nicholson and Lydekker 1889 A, p. 1325.
Owen, R. 1848 A, p. 104.   (Hyopotamus, type *H. bovinus*.)
Pavlow, M. 1900 A, p. 19.

Pomel, A. 1848 B, p. 322.   (Ancodus.)
1853 A, p. 91.   (Ancodus.)
Rütimeyer, L. 1891 A, p. 56.
Schlosser, M. 1886 B, p. 79.
Scott, W. B. 1890 A, p. 488,
1894 C.
1894 D.   (Ancodus.)
1894 E.   (Ancodus.)
Steinmann and Döderlein 1890 A, p. 794.
Trouessart, E. L. 1898 A, p. 804.
Woodward, A. S. 1898 B, p. 349.
Wortman, J. L. 1898 A.
Zittel, K. A. 1893 B, p. 329.   (Ancodus.)

Ancodon americanus (Leidy).

Leidy, J. 1856 D, p. 59.   (Hyopotamus.)
Hayden, F. V. 1858 B, p. 157.   [Chæropotamus (Hyopotamus).]
King, C. 1878 A, p. 411.   (Hyopotamus.)
Leidy, J. 1856 B, p. 422.   (Hyopotamus.)
1857 C, p. 89.   [Chæropotamus (Hyopotamus).]
1869 A, pp. 202, 389, pl. xxi, figs. 1–6.   (Hyopotamus.)
1871 C, p. 355.   (Hyopotamus.)

Marsh, O. C. 1890 D, p. 524. (Hyopotamus deflectus.)
　　1894 K, p. 178, fig. 7. (Ancodus deflectus.)
Matthew, W. D. 1899 A, p. 59. (Hyopotamus americanus, H. deflectus.)
Osborn and Wortman 1894 A, p. 219, fig. 6, A. (Hyopotamus.)
Roger, O. 1896 A, p. 202. (Hyopotamus americanus, H. deflectus.)
Scott, W. B. 1894 E, p. 461, pl. xxxiii, figs. 2, 3 (Ancodus americanus); p. 461 (A. deflectus).
Trouessart, E. L. 1898 A, p. 806. (Hyopotamus americanus, H. deflectus.)
　　Oligocene (White River); South Dakota.
An examination of the type of Professor Marsh's *Hyopotamus deflectus* shows that the character on which the species was based holds good on only one side.

Ancodon brachyrhynchus (Osb. and Wort.).

Osborn and Wortman 1894 A, p. 220, fig. 6 B. (Hyopotamus.)
Matthew, W. D. 1899 A, p. 59. (Hyopotamus.)
Roger, O. 1896 A, p. 202. (Hyopotamus.)
Scott, W. B. 1894 D, p. 492. (Ancodus.)
　　1894 E, pl. xxiii, fig. 1; pl. xxiv, figs. 5–8. (Ancodus.)
Trouessart, E. L. 1898 A, p. 806. (Hyopotamus.)
Woodward, A. S. 1898 B, p. 350, fig. 200. (Hyopotamus.)
　　Oligocene (White River); South Dakota.

Ancodon rostratus Scott.

Scott, W. B. 1894 E, appendix, p. 536. (Ancodus.)
　　Oligocene (White River); South Dakota.

Superfamily SUOIDEA.

Cope, E. D. 1885 B, p. 482. (Bunodonta.)
　　1887 B, p. 378. (Bunodonta.)
　　1898 B, p. 130.
Flower, W. H. 1873 A. (Suina.)
　　1874 A. (Suina.)
　　1876 B, xiii, p. 328. (Suina.)
　　1883 D, p. 430. (Bunodonta.)
　　1885 A. (Suina.)
Flower and Lydekker 1891 A, p. 278. (Suina.)

Gray, J. E. 1868 A, p. 22. (Suina.)
　　1869 A, p. 238. (Suina.)
Kowalevsky, W. 1873 A, p. 152. (Bunodonta, in part.)
Lydekker, R. 1883 A. (Suina, Bunodontia, in part.)
　　1884 C. (Suina, Bunodontia, in part.)
Marsh, O. C. 1877 E. (Bunodonta.)
Tomes, C. S. 1898 A, p. 392. (Suina.)

SUIDÆ.

Baur, G. 1884 F, p. 596.
Blainville, H. M. D. 1864 A, iv, AA, p. 105.
Branco, W. 1897 A, p. 23.
Cope, E. D. 1887 B, pp. 378, 384.
　　1888 X, pp. 1080, 1090.
Cope and Wortman 1884 A, p. 16.
Coues, E. 1878 A.
Cuvier, G. 1834 A, p. 221 et seq., with plates.
Flower, W. H. 1873 A.
Flower and Lydekker 1891 A, p. 281.
Garrod, A. H. 1877 A.
Gray, J. E. 1868 A.
　　1869 A, p. 327.
Haeckel, E. 1895 A, p. 555. (Suida.)
Huxley, T. H. 1870 A, p. 440.
　　1872 A, p. 813.
Leidy, J. 1871 C, p. 352.
Lydekker, R. 1884 C, p. 49.
　　1885 C, p. 250.
Nicholson and Lydekker 1889 A, p. 1318.
Osborn, H. F. 1883 A, p. 34.

Owen, R. 1845 B, p. 543.
Parker, W. K. 1874 B.
Roger, O. 1896 A, p. 202.
Rütimeyer, L. 1857 B. ("Schweine.")
　　1862 B. ("Schweine.")
Schlosser, M. 1886 A, p. 255.
　　1886 B, p. 82.
　　1890 C, p. 273.
Schmidt, O. 1886 A, p. 137.
Slade, D. D. 1893 A.
Smith, J. 1895 A, p. 340.
Stehlin, H. G. 1899 A.
Steinmann and Döderlein 1890 A, p. 794.
Taeker, J. 1892 A.
Trouessart, E. L. 1898 A, p. 808.
Turner, H. N. 1848 A.
Weyhe, —— 1875 A, p. 120.
Wilckens, M. 1885 B.
Woodward, A. S. 1898 B, p. 341.
Zittel, K. A. 1893 B, p. 331.

ACHÆNODONTINÆ.

Cope, E. D. 1887 B, p. 378. (Elotheriidæ.)
　　1888 X, p. 1089. (Elotheriinæ.)
Marsh, O. C. 1893 D, p. 410. (Elotheriidæ.)
　　1894 B, p. 408. (Elotheriidæ.)
　　1894 H. (Elotheriidæ.)
Osborn, H. F. 1893 D, p 41. (Elotheriidæ.)
Roger, O. 1896 A, p. 203.

Stehlin, H. G. 1899 A, p. 123. ("Elotherien.")
Trouessart, E. L. 1898 A, p. 808. (Achænodontinæ.)
Zittel, K. A. 1893 B, p. 334. (Achænodontinæ.)
　　Possibly the subfamily name *Elotheriinæ* has priority.

ACHÆNODON Cope.   Type *A. insolens* Cope.

Cope, E, D. 1873 D, p. 2.
Ameghino, F. 1896 A, p. 144.
Cope, E. D. 1874 B, p. 457.
  1883 I, p. 77.
  1883 II, p. 969.
  1884 O, pp. 259, 342, 738.
  1884 Z, p. 718.
  1885 K, p. 469.
  1887 S. p. 261.
Lydekker, R. 1885 E, p. 470.
Nicholson and Lydekker 1889 A, p. 1323.
Osborn, H. F. 1883 A, p. 26.
Schlosser, M. 1885 A, p. 684.
  1886 B, p. 39. *
  1887 B, p. 220.
Zittel, K. A. 1893 B, p. 234.

Achænodon insolens Cope.

Cope, E. D. 1873 D, p. 2.
  1874 B, pp. 457, 463.

Cope, E. D. 1874 M, p. 78, fig. 5.
  1882 JJ, p. 584.
  1884 O, p. 343, pls. lvii, lviia.
  1885 K, p. 470, fig. 17.
Matthew, W. D. 1899 A, pp. 48, 50.
Osborn; H. F. 1883 A, p. 24.
  1895 A, p. 105.
Roger, O. 1896 A, p. 208.
Trouessart, E. L. 1898 A, p. 808.
  Eocene (Bridger); Wyoming: (Uinta); Utah.

Achænodon robustus Osb.

Osborn, H. F. 1883 A, p. 24, pl. vi.
Cope, E. D. 1885 K, p. 471, fig. 18.
Matthew, W. D. 1899 A, p. 48.
Osborn, H. F. 1895 A, p. 103.
Roger, O. 1896 A, p. 208.
Trouessart, E. L. 1898 A, p. 808.
Zittel, K. A. 1893 B, p. 335, fig. 270.
  Eocene (Bridger); Wyoming.

PARAHYUS Marsh.   Type *P. vagus* Marsh.

Marsh, O. C. 1876 H, p. 402.
Cope, E. D. 1894 E, p. 868.
Marsh, O. C. 1877 E.
Nicholson and Lydekker 1889 A, p. 1323.
Osborn, H. F. 1883 A, pp. 23, 24:
  1893 D, p. 41.
Trouessart, E. L. 1898 A, p. 808. (Syn. of Achænodon.)
Wilckens, M. 1885 B, p. 305.
Zittel, K. A. 1893 B, p. 334. (Syn.? of Achænodon.)

Parahyus aberrans Marsh.

Marsh, O. C. 1894 L, p. 261, fig. 2.
Matthew, W. D. 1899 A, p. 34.
Roger, O. 1896 A, p. 208. (Achænodon.)

Trouessart, E. L. 1898 A, p. 808. (Achænodon.)
  Eocene (Bridger); Wyoming.

Parahyus vagus Marsh.

Marsh, O. C. 1876 H, p. 402.
King, C. 1878 A, p. 377. (P. vagans.)
Matthew, W. D. 1899 A, p. 34.
Osborn, H. F. 1883 A, p. 24 [Achænodon (Parahyus).]
Roger, O. 1896 A, p. 208. (Achænodon.)
Scott, W. B. 1898 A, p. 323. (Achænodon.)
Trouessart, E. L. 1898 A, p. 808. (Achænodon.)
Wilckens, M. 1885 B, p. 305.
  Eocene (Bridger); Wyoming.

PROTELOTHERIUM Osb.   Type *Elotherium uintense* Osb.

Osborn, H. F. 1895 A, p. 105.

Protelotherium uintense Osb.

Osborn, H. F. 1895 A, p. 102, figs. 16, 17 (Elotherium); p. 105 (Protelotherium).

Matthew, W. D. 1899 A. p. 50.
Roger, O. 1896 A, p. 208. (Elotherium.)
Scott, W. B. 1898 A, p. 323. (Achænodon.)
Trouessart, E. L. 1898 A, p. 808.
  Eocene (Uinta); Utah.

ELOTHERIUM Pomel.   Type *E. magnum* Pomel.

Pomel, A. 1847 B, p. 307.
Aymard, A., 1848, Sci. et Bell. Lett. du Puy, xii, p. 240. (Entelodon.)
Cope, E. D. 1874 B, p. 504. (Pelonax, type *P. ramosus.*)
  1879 B, p. 59. (Boöchœrus, type *B. humerosus.*)
  1884 H, p. 22.
  1885 B, p. 483.
  1888 X, p. 1088.
  1889 C. (Elotherium, Boöchœrus.)
Filhol, H. 1882 A, p. 190.
Flower, W. H. 1873 A.
Flower and Lydekker 1891 A, p. 292.
Gaudry, A, 1877 A. (Entelodon.)
  1878 B, p. 92. (Entelodon.)

Gervais, P. 1859 A, p. 194. (Entelodon.)
Greene, F. V. 1858 A. (Archæotherium.)
Grinnell and Dana 1876 A.
Hoernes, R. 1892 A. (Entelodon.)
Kowalevsky, W. 1873 A, p. 258. (Entelodon.)
  1876 A, p. 415. (Entelodon.)
Leidy, J. 1850 A, p. 90. (Archæotherium, type *A mortoni* Leidy.)
  1851 J. (Arctodon.)
  1852 J, p. 558. (Archæotherium.)
  1852 L, p. 64. (Archæotherium.)
  1854 A, pp. 57, 114. (Archæotherium.)
  1869 A, pp. 174, 388.
  1871 C, p. 358.
Lydekker, R. 1885 C, p. 249.
  1896 B, p. 161.

Marsh, O. C. 1877 E.
1894 B, pl. ix.
1894 H.
Nicholson and Lydekker 1889 A, p. 1323.
Osborn, H. F. 1883 A, p. 23. (Entelodon.)
Pictet, F. J. 1853 A, p. 828.
Pomel, A., 1847, Bull. Soc. Géol. France, iv, p. 1083.
1848 B, p. 325.
1853 A, p. 88.
Schlosser, M. 1886 B, p. 80. (Entelodon.)
Scott, W. B. 1884 B.
1896 A.
1898 A.
Stehlin, H. G. 1899 A, pp. 121, 122.
Steinmann and Döderlein 1890 A, p. 793. (Entelodon.)
Trouessart, E. L. 1898 A, p. 808.
Wilckens, M. 1885 B, pp. 210, 216, 217.
Woodward, A. S. 1898 B, p. 343.
Zittel, K. A. 1893 B, p. 335 (Elotherium); p. 337 (Boöchœrus).

Elotherium clavum Marsh.

Marsh, O. C. 1893 D, p. 409, pl. ix, fig. 1.
1897 C, p. 523, fig. 97. (Entelodon.)
Matthew, W. D. 1899 A, p. 59.
Roger, O. 1896 A, p. 204.
Trouessart, E. L. 1898 A, p. 810.
Oligocene (White River); South Dakota?, Colorado.

Elotherium coarctatum Cope.

Cope, E. D. 1889 O, p. 629.
1886 U, p. 84. (E. mortoni.)
1891 A, p. 20, pl. xiv, fig. 3. (E. arctatum.)
Matthew, W. D. 1899 A, p. 59. (E. arctatum.)
Roger, O. 1896 A, p. 204. (E. arctatum.)
Oligocene (White River); Northwest Territory, Canada.

Elotherium crassum Marsh.

Marsh, O. C. 1873 H, p. 487,
Cope, E. D. 1874 B, p. 504. (Pelonax?.)
King, C. 1878 A, p. 411.
L. P. B. 1885 A, p. 193, fig. 27.
Lydekker, R. 1896 B, p. 162, fig. 35. ·
Marsh, O. C. 1884 F, fig. 75.
1885 B, fig. 106.
1893 D, p. 408, pl. viii.
1894 B, pl. ix.
1894 H, pl. x.
1897 C, pl. xxx. (Entelodon.)
Matthew, W. D. 1899 A, p. 59.
Roger, O. 1896 A, p. 204.
Stehlin, H. G. 1899 A, p. 123.
Trouessart, E. L. 1898 A, p. 809.
Woodward, A. S. 1898 B, p. 344, fig. 197.
Zittel, K. A. 1893 B, p. 337, fig. 373.
Oligocene (White River) ; South Dakota, Colorado?.

Elotherium humerosum (Cope).

Cope, E. D. 1879 B, p. 59. (Boöchœrus.)
1879 H. p. 131. (Boöchœrus.)
Matthew, W. D. 1899 A, p. 64. (Boöchœrus.)
Roger, O. 1896 A, p. 204. (Boöchœrus.)
Trouessart, E. L. 1898 A, p. 810. (Boöchœrus.)

Wortman, J. L. 1896 A, p. 120, footnote.
Miocene (John Day); Oregon.
Probably identical with E. imperator.

Elotherium imperator Leidy.

Leidy, J. 1873 B, pp. 217, 320 (E. imperator); pl. ii, figs. 3, 4 (E. superbum); pl. vii, fig. 2 (E. imperator).
Cope, E. D. 1874 B, p. 505. (E. superbum.)
1879 B, p. 58.
1888 X, p. 1090.
Leidy, J. 1870 P, p. 112. (E. superbum.)
Matthew, W. D. 1899 A, p. 63.
Scott, W. B. 1898 A.
Trouessart, E. L. 1898 A, p. 809. (Syn. of E. superbum.) ~
Miocene (John Day); Oregon.

Elotherium ingens Leidy.

Leidy, J. 1856 J, p. 164. (Entelodon.)
1857 D, p. 175. (Elotherium.)
1858 B, p. 157. (Entelodon.)
1869 A, pp. 192, 388, pl. xxvii, figs. 8-11.
1870 P, p. 112.
1871 C, p. 353.
1873 B, p. 320.
Matthew, W. D. 1899 A, p. 59.
Scott, W. B. 1896 A.
1898 A, pl. xvii; pl. xviii, figs. 3-12.
Trouessart, E. L. 1898 A, p. 809. (Syn. E. robustum.)
Oligocene (White River); South Dakota.

Elotherium mortoni Leidy.

Leidy, J. 1850 A, p. 90. (Archæotherium.)
Ami, H. M. 1891 A.
Cope, E. D. 1885 H. (Entelodon.)
1886 U, p. 84.
1888 X, p. 1089.
1889 I, p. 155.
Dawkins, W. B. 1870 A. (Archæotherium.)
Greene, F. V. 1853 A, p. 66. ("Archæotherium.")
Hayden, F. V. 1858 B, p. 157.
Hoernes, R. 1892 A. (Entelodon.)
King, C. 1878 A, p. 411.
Leidy, J. 1851 J, p. 278. ("Arctodon.")
1852 J, p. 558, pl. x, figs. 1-3; pl. xi, fig. 1 (Archæotherium mortoni); p. 572 (A. robustum).
1852 L, p. 64 (Archæotherium mortoni); p. 65 (Arctodon vetustum).
1853 D, p. 392. (Entelodon.)
1854 A, pp. 57, 114, pls. viii, ix; pl. x, figs. 1-7 (Archæotherium mortoni); pp. 66,114, pl. x, figs. 8-13 (A. robustum).
1854 E, p. 157. (Entelodon.)
1857 D, p. 175.
1869 A, pp. 175, 388, pl. xvi.
1871 C, p. 353.
1873 B, pp. 125, 320, pl. vii, figs. 28, 29.
Matthew, W. D. 1899 A, p. 59.
Roger, O. 1896 A, p. 204. (E. mortoni, E. robustum.)
Matthew, W. B., 1898 A, pl. xviii, figs. 1, 2.
Scott and Osborn 1887 B, p. 156. (Entelodon.)
Trouessart, E. L. 1898 A, p. 809.
Wortman, J. L. 1893 A, p. 97.

Zittel, K. A. 1893 B, p. 336, fig. 271.
Oligocene (White River); South Dakota, Nebraska, Colorado.

Elotherium ramosum Cope.

Cope, E. D. 1874 P, p. 27.
1874 B, pp. 463, 504.   (Pelonax.)
1883 F, p. 217.
1888 X, p. 1089.
Matthew, W. D. 1899 A, p. 59.
Osborn, H. F. 1896 C, pp. 713, 715, illus. 9.
1898 Q, p. 16.
Roger, O. 1896 A, p. 204.
Trouessart, E. L. 1898 A, p. 809.
Oligocene (White River); Colorado, North Dakota.

Elotherium superbum Leidy.

Leidy, J. 1868 B.
Cope, E. D. 1874 B, p. 505.
Dana, J. D. 1879 A, p. 283.
King, C. 1878 A, p. 411.
Leidy, J. 1869 A, p. 388.
1871 C, p. 353.
Roger, O. 1896 A, p. 204.
Trouessart, E. L. 1898 A, p. 809.
Whitney, J. D. 1879 A, p. 244.
Miocene; California.

AMMODON Marsh.   Type *A. leidyanum* Marsh.

Marsh, O. C. 1893 D, p. 409.
Trouessart, E. L. 1898 A, p. 808.   (Syn. of Elotherium.)

Ammodon bathrodon Marsh.

Marsh, O. C. 1874 C, p. 534.   (Elotherium.)
King, C. 1878 A, p. 411.   (Elotherium.)
Marsh, O. C. 1893 D, p. 410, pl. ix, fig. 4.
Matthew, W. D. 1899 A, p. 59.   (Elotherium.)
Roger, O. 1896 A, p. 204.   (Elotherium.)
Trouessart, E. L. 1898 A, p. 808.   (Elotherium.)
Oligocene (White River); South Dakota?.

Ammodon leidyanus Marsh.

Marsh, O. C. 1893 D, p. 409, pl. ix.
Cope, E. D. 1869 N, p. 740.   (Elotherium.)
Leidy, J. 1869 A, p. 388.   (Elotherium.)

Marsh, O. C. 1871 F, p. 10.   (Elotherium; no description.)
1874 C, p. 534.   (Elotherium; no description.)
Roger, O. 1896 A, p. 204.   (Elotherium.)
Oligocene (White River); Colorado?.
Trouessart, E. L. 1898 A, p. 808.   (Elotherium.)
Miocene; New Jersey.
Not any of the literature of this species previous to 1893 contains an adequate description.

Ammodon potens Marsh.

Marsh, O. C. 1893 D, p. 410.
1897 C, p. 527.
Matthew, W. D. 1899 A, p. 59.   (Elotherium.)
Roger, O. 1896 A, p. 204.   (Elotherium.)
Trouessart, E. L. 1898 A, p. 808.   (Elotherium.)

HYOTHERIINÆ.

Cope, E. D. 1894 E, p. 869.   (Leptochœridæ.)
Marsh, O. C. 1894 L, p. 273.   (Leptochœridæ.)
Roger, O. 1896 A, p. 204.

Trouessart, E. L. 1898 A, p. 810.
Zittel, K. A. 1893 B, p. 337.

HYOTHERIUM Meyer.   Type *H. sömeringii* Meyer.

Meyer, H. 1834 A, p. 43.
Bronn, H. G. 1838 A, p. 1222, pl. xlvi, fig. 7.
Cope and Wortman, 1884 A, p. 17.
Filhol, A. 1881 A.
1891 A, p. 207.
Gaudry, A. 1878 B, p. 70.
Gervais, P. 1859 A, p. 181.
Lydekker, R. 1884 C, p. 91.
1885 C, p. 253.
Meyer, H. 1841 A.
Peters, K. 1869 B.
Pictet, F. J. 1853 A, p. 333.
Stehlin, H. G. 1899 A, pp. 44, 185, 235, 313.
Wilckens, M. 1885 B, p. 337.
Woodward, A. S. 1898 B, p. 343.
Zittel, K. A. 1893 B, p. 341.

Hyotherium americanum Scott and Osb.

Scott and Osborn 1887 B, p. 155.
Cope, E. D. 1888 E, p. 66.
Matthew, W. D. 1899 A, p. 60.   (Bothrolabis.)
Roger, O. 1896 A, p. 207.   (Bothrolabis.)
Stehlin, H. G. 1899 A, p. 113.
Trouessart, E. L. 1898 A, p. 815.   (Bothrolabis.)
Oligocene (White River); South Dakota?

Hyotherium platyops Cope.

Cope, E. D. 1881 C, p. 174.
Lydekker, R. 1884 C, p. 93.
Matthew, W. D. 1899 A, p. 60.   [Bothrolabis
(Palæochœrus).]
Roger, O. 1896 A, p. 207.   (Bothrolabis.)
Trouessart, E. L. 1898 A, p. 815.   (Bothrolabis.)
Oligocene (White River); South Dakota.

PERCHŒRUS Leidy.   Type *Palæochœrus probus* Leidy.

Leidy, J. 1869 A, p. 194.
Cope, E D. 1879 A, p. 44. (Thinohyus.)
    1879 E, p. 374. (Thinohyus.)
    1884 O, p. 37. (Thinohyus.)
    1887 B, p. 384. (Thinohyus.)
    1888 E, p. 62. (Thinohyus.)
    1894 E, p. 869. (Thinohyus.)
Cope and Wortman 1884 A, p. 17. (Thinohyus.)
Leidy, J. 1871 C, p. 353.
Marsh, O. C. 1875 B, p. 248. (Thinohyus type *T. lentus* Marsh.)
    1877 E. (Perchœrus, Thinohyus.)
    1894 L, p. 271.
Nicholson and Lydekker 1889 A, p. 1322. (Perchœrus, Thinohyus.)
Schlosser, M. 1886 B, p. 88.
Scott, W. B. 1892 A, p. 438. (Perchœrus, Thinohyus.)
Stehlin, H. G. 1889 A, p. 112. (Perchœrus); p. 118 (Tinohyus).
Wilckens, M. 1885 B, p. 304.
Zittel, K. A. 1898 B, p. 342. (Perchœrus, Thinohyus.)

Perchœrus antiquus Marsh.

Marsh, O. C. 1893 D, p. 411. pl. x, fig. 1. [Perchœrus (Dicotyles).]
    1870 I, p. 11. (Dicotyles antiquus; no description.)
    1894 L, p. 271, fig. 28. (Thinohyus.)
Matthew, W. D. 1899 A, p. 60. (Perchœrus.)
Roger, O. 1896 A, p 207. (Thinohyus.)
Stehlin, H. G. 1899 A, p. 114. (Tinohyus.)
Trouessart, E. L. 1898 A, p. 815 (Perchœrus); p. 816 (Thinohyus.)
    Miocene; New Jersey.

Perchœrus lentus (Marsh).

Marsh, O. C. 1875 B, p. 248 (Thinohyus.)
Cope, E. D. 1879 E p. 375 (Thinohyus.)
King, C. 1871 A, p. 424. (Thinohyus.)
Marsh, O. C. 1894 L, p. 271, fig. 26. (Thinohyus.)
Matthew, W. D. 1899 A p. 64. (Thinohyus.)
Roger, O. 1896 A, p. 207. (Thinohyus.)
Scott, W. B. 1892 A, p. 438, fig 8³. (Thinohyus.)

Stehlin, H. G. 1899 A, p. 114 (Tinohyus); p. 193. (Thinohyus.)
Trouessart, E. L. 1898 A, p. 815. (Thinohyus.)
    Miocene (John Day); Oregon.

Perchœrus nanus (Marsh).

Marsh, O. C. 1894 L, p. 271, fig. 28. (Thinohyus.)
Matthew, W. D. 1899 A, p. 60. (Thinohyus.)
Roger, O. 1896 A, p. 207. (Thinohyus.)
Stehlin, H. G. 1899 A, p. 113 (Tinohyus); pp. 193, 273.
Trouessart, E. L. 1898 A, p. 816. (Thinohyus.)
    Oligocene (White River); South Dakota.

Perchœrus probus Leidy.

Leidy, J. 1856 J, p. 165. (Palæochœrus.)
Hayden, F. V. 1858 B, p. 157. (Palæochœrus.)
King, C. 1878 A, p. 411.
Leidy, J. 1869 A. pp. 194, 389, pl. xxi, figs. 20–27.
    1871 C, p. 353.
Matthew, W. D. 1899 A, p. 60.
Roger, O. 1896 A, p. 207.
Scott, W. B. 1892 A, p. 438, fig. 8³.
Stehlin, H. G. 1899 A, pp. 112, 193.
Wilckens, M. 1885 B, p. 305.
    Oligocene (White River); South Dakota.

Perchœrus robustus (Marsh).

Marsh, O. C. 1894 F, p. 94. (Thinohyus.)
Matthew, W. D. 1899 A, p. 60. (Thinohyus.)
Stehlin, H. G. 1899 A, p. 114 (Tinohyus); p. 193
Trouessart, E. L. 1898 A, p. 816. (Thinohyus.)
    Oligocene (White River); South Dakota.

Perchœrus socialis (Marsh).

Marsh, O. C. 1875 B, p. 249. (Thinohyus.)
Cope, E. D. 1879 B, p. 58. (Palæochœrus.)
King, C. 1878 A, p. 424. (Thinohyus.)
Marsh, O. C. 1894 L, p. 271, fig. 25. (Thinohyus.)
Matthew, W. D. 1899 A, p. 64. (Thinohyus.)
Stehlin, H. G. 1899 A, p. 113. (Tinohyus.)
Trouessart, E. L. 1898 A, p. 816. (Thinohyus.)
    Miocene (John Day); Oregon.

LEPTOCHŒRUS Leidy.   Type *L. spectabilis* Leidy.

Leidy, J. 1856 I, p. 88.
Cope, E. ³). 1⁸⁄₉⁴ E, p. 869.
Leidy, J. 1869 A, p. 197.
    1871 C, p. 354.
Nicholson and Lydekker 1889 A, p. 1323.
Schlosser, M. 1886 B, p. 86.
Scott, W. B. 1891 A, p. 46.
Zittel, K. A. 1893 B, p. 341.

Leptochœrus gracilis Marsh.

Marsh, O. C. 1894 L, p. 272, figs. 29, 30.
Matthew, W. D. 1899 A, p. 60.
Roger, O 1896 A, p. 206.
    Oligocene (White River); South Dakota.

Leptochœrus spectabilis Leidy.

Leidy, J. 1856 I, p. 88.
Hayden, F. V. 1858 B, p. 157.
King, C. 1878 A, p. 411.
Leidy, J. 1869 A, pp. 197, 389, pl. xxi, figs. 14–19.
    1871 C, p. 354.
Matthew, W. D. 1899 A. p. 59.
Roger, O 1896 A, p. 206.
Trouessart, E. L. 1898 A, p. 814.
Wilckens, M. 1885 B, p. 306.
    Oligocene (White River); South Dakota.

NANOHYUS Leidy.   Type *N. porcinus* Leidy.

Leidy, J. 1869 B, p. 65.
1869 A, p. 200.
1871 C, p. 354.
Scott, W. B. 1892 A, p. 441.
Stehlin, H. G. 1899 A, p. 221.

Nanohyus porcinus Leidy.

Leidy, J. 1869 B, p. 65.
1869 A, pp. 200, 389, pl. xxix, figs. 11, 12.

Leidy, J. 1871 C, p. 354.
Matthew, W. D. 1899 A, p. 60.
Roger, O. 1896 A, p. 207.
Stehlin, H. G. 1899 A, p. 113.
Trouessart. E. L. 1898 A, p. 815.
Wilckens, M. 1885 B, p. 306.
    Oligocene (White River); South Dakota?.

CHÆNOHYUS Cope.   Type *C. decedens* Cope.

Cope, E. D. 1879 E, p. 373.
1887 B, p. 384.
1888 E, p. 63.
1888 X, p. 1088.
Cope and Wortman 1884 A, p. 17.
Flower and Lydekker 1895 A, p. 291.
Nicholson and Lydekker 1889 A, p. 1321.
Stehlin, H. G. 1899 A, pp. 113, 193.
Zittel, K. A. 1893 B, p. 342.

Chænohyus decedens Cope.

Cope. E. D. 1879 E, p. 373.
1888 E, p. 63.
Matthew, W. D. 1899 A, p. 64.
Roger, O. 1896 A, p. 207.
Stehlin, H. G. 1899 A, pp. 114, 193.
Trouessart, E. L. 1898 A, p. 815.
    Miocene (John Day); Oregon.

BOTHROLABIS Cope.   Type *B. rostratus* Cope.

Cope, E. D. 1888 E, pp. 63, 66.
1888 X, p. 1088.
Flower and Lydekker 1891 A, p. 291.
Stehlin, H. G. 1899 A, pp. 113, 193, 272, 320.
Steinmann and Döderlein 1890 A, p. 796.
Zittel, K. A. 1893 B, p. 342.

Bothrolabis pristinus (Leidy).

Leidy, J. 1873 B, pp. 216, 319, pl. vii, figs. 13, 14.
    (Dicotyles.)
Cope, E. D. 1879 B, p. 58.   (Palæochœrus.)
1888 E, pp. 66, 70.
Cope and Wortman, 1884 A, pp. 18, 19.   (Dicotyles.)
Leidy, J. 1870 P, p. 112.   ("Peccary.")
Matthew, W. D. 1899 A, p. 64.
Reinhardt, J. 1880 A, p. 30.   (Dicotyles.)
Roger, O. 1896 A, p. 207.
Stehlin, H. G. 1899 A, pp. 114, 193, 273, 320.
Wilckens, M. 1885 B, p. 306.   (Dycotyles.)
    Miocene (John Day); Oregon.

Bothrolabis rostratus Cope.

Cope. E. D. 1888 E, pp. 66, 77.
Matthew, W. D. 1899 A, p. 64.

Roger, O. 1896 A, p. 207.
Stehlin, H. G. 1899 A, pp. 114, 193, 320.
Trouessart, E. L. 1898 A, p. 815.
    Miocene (John Day); Oregon.

Bothrolabis subsequans Cope.

Cope, E. D. 1879 E, p. 374.   (Palæochœrus.)
1888 E, pp. 66, 67.
Lydekker, R. 1884 C, p. 93.   (Hyotherium)
Matthew, W. D. 1899 A, p. 64.
Roger, O. 1896 A, p. 207.
Stehlin, H. G. 1899 A, pp. 114, 193, 273, 320.
Trouessart, E. L. 1898 A, p. 815.
    Miocene (John Day); Oregon.

Bothrolabis trichænus Cope.

Cope, E. D. 1879 E, p. 373.   (Thinohyus.)
1879 B, p. 58.   (Palæochœrus condoni.)
1888 E, pp. 66, 74.
Matthew, W. D. 1899 A, p. 64.
Roger, O. 1896 A, p. 207.
Stehlin, H. G. 1899 A, pp. 114, 193, 273, 320.
Trouessart, E. L. 1898 A, p. 815.
    Miocene (John Day); Oregon.

TAYASSUINÆ.

The name hitherto employed for this subfamily has been *Dicotylinæ*.

Branco. W. 1897 A, p. 23.
Cope, E. D. 1888 E.
1888 X, p 1088.
1889 G, p. 134.
Cope and Wortman 1884 A, pp. 16, 17.
Flower and Lydekker 1891 A, p. 289.   (Dicotylidæ.)

Gill, T. 1872 B, pp. 11, 83.   (Dicotylidæ.)
Gray, J. E. 1868 A, p. 43.   (Dicotylidæ.)
1869 A, p. 350.   (Dicotylidæ.)
Leidy, J. 1853 G.
Roger, O. 1896 A, p. 207.
Trouessart. E. L. 1898 A, p. 816.
Zittel, K. A. 1893 B, p. 342.

TAYASSU Fischer.   Type *Dicotyles torquatus* Cuv.

Fischer, G. 1814. Zoognosia, iii, pp. 284–289.
Unless otherwise indicated, the following
authors use for this genus the name *Dicotyles*.
Ameghino, F. 1889 A, p. 573.
Baird, S. F. 1859 A, p. 627.
Burmeister, H. 1867 A, p. 236.   ("Dicotyle Cuv.")
Cope, E. D. 1875 L, p. 21.
    1887 B, p. 384.
    1888 E, p. 62.
    1888 X, p. 1088.
    1889 G, p. 134.   (Dicotyles; Mylohyus, type
        D. nasutus.)
    1899 A. p. 259.   (Mylohyus.)
Cope and Wortman 1884 A, p. 17.
Cuvier, G., 1817, Règne Anim, i, p. 237.
Flower, W. H. 1873 A.
Flower and Lydekker 1891 A, p. 289.
Gervais and Ameghino 1880 A, p. 111.   ("Dico-
    tyle.")
Gray, J. E. 1868 A, p. 45.
    1869 A, p. 351.
Holmes, F. S. 1858 B.
Klippart, J. H. 1875 A.
Leidy, J. 1853 G.
    1857 J.
    1871 C, p. 352.
Lydekker, R. 1885 C, p. 250.
Nicholson and Lydekker 1889 A, p. 1322.
Owen, R. 1845 B, p. 559.
    1848 A, p. 123.
    1866 B, p. 458.
Palmer, T. S. 1899 A, p. 494.   (Tayassu.)
Paulli, S. 1900 A, p. 198.
Reinhardt, J. 1880 A.
Rütimeyer, L. 1862 A, pp. 578, 584.
Schlosser, M. 1885 A, p. 685.
    1886 A, p. 255.
    1886 B, p. 89.
Scott, W. B. 1890 B.
    1891 A, p. 48, *et seq.*
    1892 A, p. 438.
Scott and Osborn 1890 B, p. 76.
Stehlin, H. G. 1899 A, pp. 111, 186, 208, 270, 318, 332.
Steinmann and Döderlein 1890 A, p. 796.
Trouessart, E. L. 1898 A, p. 817.
Turner, H. N. 1848 A, p. 70.
Wilckens, M. 1885 B, pp. 217, 307.
Woodward, A. S. 1898 B, p. 346.
Zittel, K. A. 1893 B, p. 342.

Tayassu hesperius (Marsh.)

Marsh, O. C. 1871 D, p. 42.   (Dicotyles.)
Cope, E. D. 1879 E, p. 373.   (Chœnohyus.)
Cope and Wortman 1884 A, p. 18.   (Dicotyles)
King, C. 1878 A, p. 443.   (Dicotyles.)
Stehlin, H. G. 1899 A. pp. 115, 194.   (Dicotyles.)
Trouessart, E. L. 1898 A, p. 815.   (Bothrolabis
    subæquans.)
Wilckens, M. 1885 B, p. 306.   (Dicotyles.)
    Pliocene; Oregon.

Tayassu nasutus (Leidy.)

Leidy, J. 1868 I, p. 230.   (Dicotyles.)
Conrad, T. A. 1869 A, p. 363.   (Dicotyles.)

Cope, E. D. 1869 E, p. 176.   (Dicotyles.)
    1869 N, p. 741.   (Dicotyles.)
    1888 X, p. 1089.   (Dicotyles.)
    1889 G. p. 134.   (Dicotyles, Mylohyus.)
    1899 A. p. 263.   (Mylohyus.)
Cope and Wortman 1884 A, p. 18.   (Dicotyles.)
Kindle, E. M. 1898 A, p. 485.   (Dicotyles.)
Leidy, J. 1860 C.   ("Peccary.")
    1869 A, pp. 385, 445, pl. xxviii, figs. 1, 2.
        (Dicotyles.)
    1871 C, p. 352.   (Dicotyles.)
Reinhardt, J. 1880 A, p. 29.   (Dicotyles.)
Roger, O. 1896 A, p. 208.   (Dicotyles.)
Stehlin, H. G. 1899 A, pp. 115, 190, 273, 319.   (Dico-
    tyles.)
Trouessart, E. L. 1898 A, p. 817.   (Dicotyles.)
    Pleistocene; Indiana, New Jersey.

Tayassu pennsylvanicus (Leidy.)

Leidy, J. 1889 H, p. 8, pl. ii, figs. 3–6.   (Dicotyles.)
Cope, E. D. 1899 A, pp. 259, 262.   (Mylohyus.)
Leidy, J. 1880 A, p. 347.   (Dicotyles nasutus.)
Mercer, H. C. 1894 A, p. 98.   (Dicotyles.)
    Pleistocene; Pennsylvania.

Tayassu serus (Cope.)

Cope, E. D. 1878 C, p. 224.   (Dicotyles.)
    1889 G. p. 134.   (Dicotyles.)
Cope and Wortman 1884 A, p. 19.   (Dicotyles.)
Matthew, W. D. 1899 A, p. 72.   (Dicotyles.)
Reinhardt, J. 1880 A, p. 28.   (Dicotyles.)
Roger, O. 1896 A, p. 208.   (Dicotyles.)
Stehlin, H. G. 1899 A, pp. 115, 189, 319.   (Dico-
    tyles.)
Trouessart, E. L. 1898 A, p. 817.   (Dicotyles.)
    Miocene (Loup Fork); Kansas.

Tayassu tajacu (Linn.).

Linnæus, C., 1758, Syst. Nat., ed. 10, i, p. 50.   (Sus.)
Ameghino, F. 1889 A, p. 574.   (Dicotyles.)
Baird, S. F. 1857 A, p. 627, pl. lxxxvii.   (Dico-
    tyles.)
Blainville, H. M. D. 1864 A, iv, AA, p. 138, pls. ii,
    iii, v–viii.   (Sus torquatus.)
Cope, E. D. 1867 C, p. 151 (Cynorca proterva, in
    part); p. 155 (Dicotyles torquatus.)
    1868 I, p. 185.   ("Dicotyles.")
    1889 G. p. 134.   (Dicotyles tajassus.)
Cope and Wortman 1884 A, pp. 18, 19.   (Dicotyles
    torquatus.)
Flower and Lydekker 1891 A, p. 290, fig. 109.
    (Dicotyles.)
Kindle, E. M. 1898 A, p. 485.   (Dicotyles torqua-
    tus.)
Le Conte, J. L. 1848 A, p. 104.   (Dicotyles torqua-
    tus.)
Leidy, J. 1860 B, p. 108, pl. xvii, figs. 13, 14.   (Dico-
    tyles fossilis.)
    1869 A, p. 384.   (Dicotyles lenis.)
    1871 C, p. 352.   (Dicotyles lenis.)
Lydekker, R. 1885 C, p. 252.   (Dicotyles.)

Marsh, O. C. 1885 B, p. 293, fig. 112. (Dicotyles torquatus.)
Reinhardt, J. 1880 A. (Dicotyles torquatus, D. lenis.)
Roger, O. 1896 A, p. 208. (D. torquatus.)
Scott, W. B 1890 B, p. 385. (Dicotyles torquatus.)
Stehlin, H. G. 1899 A, pp. 111, 187, 208, 318, 334. (Dicotyles torquatus.)
Trouessart, E. L. 1898 A, p. 817. (Dicotyles.)
Wyman, J. 1862 A, p. 422. (Dicotyles.)
Zittel, K. A. 1893 B, p. 342, fig. 280. (D. torquatus.)
Living: South America, Central America, North America to Arkansas: Pleistocene; South Carolina, Maryland, Virginia, Wisconsin, Illinois, Iowa, Kentucky.

Tayassu tetragonus (Cope).

Cope, E. D. 1899 A, pp. 259, 260, pl. xxi, figs. 3–3b. (Mylohyus.)
Pleistocene; Pennsylvania.

Tayassu sp. indet.

Leidy, J. 1869 A, p. 387. p.. xxviii, figs.
tyles.)
Pliocene?; Niobrara River.

Tayassu sp. indet.

Scott, W. B. in Scott and Osborn 1890 B, p. 76, fig. 6. (Dicotyles.)
Stehlin, H. G. 1899 A, pp. 20, 115, 195. (Dicotyles.)
Williston, S. W. 1894 C, p. 24. (Platygonus.)
Miocene (Loup Fork); Nebraska?

PLATYGONUS Le Conte.   Type *P. compressus* Le Conte.

Le Conte, J. L. 1848 A, p. 103. (Platigonus.[1])
Cope, E. D. 1887 B, p. 384.
    1889 G, p. 134.
Le Conte, J. L. 1848 A, p. 104 (Hyops, type H. depressifrons); p. 105 (Protochœrus, type P. prismaticus=Platygonus compressus).
    1848 B, p. 257.
Leidy, J. 1853 G. (Platygonus; Euchœrus, type E. macrops=P. compressus.)
    1869 A, p. 383.
    1871 C, p. 353.
    1889 G, p. 41.
Marsh, O. C. 1877 E.
Nicholson and Lydekker 1889 A, p. 1322.
Pictet, F. J. 1853 A, p. 303 (Platygonus); p. 326 (Hyops).
Reinhardt, J. 1880 A, p. 29.
Schlosser, M. 1886 B, p. 89.
Stehlin, H. G. 1899 A, pp. 116, 190, 210, 319.
Steinmann and Döderlein 1890 A, p. 796.
Wilckens, M. 1885 B, p. 270 (Hyops); p. 304 (Platygonus).
Williston, S. W. 1894 C.
Zittel, K. A. 1893 B, p. 342.

Platygonus bicalcaratus Cope.

Cope, E. D. 1893 A, p. 68, fig. 5.
Matthew, W. D. 1899 A, p. 75.
Stehlin, H. G. 1899 A, p. 117.
Trouessart, E. L. 1898 A, p. 816.
Williston, S. W 1894 C, p. 24.
Pliocene (Blanco); Texas.

Platygonus compressus Le Conte.

Le Conte, J. L. 1848 A, p. 103, figs. 1, 2.
Cope, E. D. 1869 E, p. 176. (Dicotyles.)
    1884 G, p. 15.
    1888 X, p. 1092, fig. 6.
    1889 F.
Cope and Wortman, 1884 A, p. 20.
Harlan, R. 1825 A, p. 222. ("Extinct peccari.")
Kindle, E. M. 1898 A, p. 485.
Klippart, J. H. 1875 A. (Dicotyles.)
L. P. B. 1885 A, p. 193, fig. 28.
Le Conte, J. L. 1848 A, p. 104 (Hyops depressifrons); p. 105 (Protochœrus prismaticus).
    1848 B, pls 1–iv.

Le Conte, J. L. 1852 A, p. 3 (Dicotyles depressifrons); p. 5 (Protochœrus prismaticus).
    1852 B. (Dicotyles costatus.)
    1852 D. (Hyops depressifrons, P. compressus.)
Leidy, J. 1853 G, p. 323, pl. xxxvii, figs. 12–16; pl. xxviii, figs. 2, 3 (P. compressus); p. 342, pl. xxxviii, fig. 1 (Dicotyles depressifrons); p. 342, pl. xxxvii, fig. 18 (Protochœrus prismaticus); p. 342 (Dicotyles costatus); p. 340, pl. xxxvi, figs. 1, 2; pl. xxxvii, figs. 5–8 (Euchœrus macrops).
    1856 R. (Dicotyles.)
    1857 J, pl. vi, figs. 2–7. (Dicotyles.)
    1862 A. (Dicotyles.)
    1868 I, p. 231. (Dicotyles.)
    1869 A, pp. 200, 383.
    1870 H, p. 13.
    1871 C, p. 353.
    1889 G, p. 41, pl. viii, fig. 1.
    1889 H, p. 12.
Marsh, O. C. 1884 F, fig. 761
    1885 B, fig. 109.
Miller, G. S. 1899 B, p. 372.
Newberry, J. S. 1874 F, p. 77. (Dicotyles.)
    1875 A, p. vi. (Dicotyles.)
Pictet, F. J. 1853 A, p. 303 (P. compressus); p. 335 (Protochœrus prismaticus).
Reinhardt, J. 1880 A, p. 29. (Dicotyles.)
Roger, O. 1896 A, p. 208.
Stehlin, H. G. 1899 A, pp. 117, 190, 273.
Trouessart, E. L. 1898 A, p. 817.
Williston, S. W. 1894 C, p. 24.
    1897 I, p. 303.
    1898 I, p. 92.
Wyman, J. 1862 A, p. 422. ("Dicotyles.")
Pleistocene; Illinois, Indiana, Kentucky, Ohio, New York, Missouri, Iowa, Kansas.

Platygonus condoni Marsh.

Marsh, O. C. 1871 D, p. 41.
Cope, E. D. 1886 B, p. 359.
    1886 H.
Cope and Wortman 1884 A, pp. 18, 19. (Dicotyles.)
King, C. 1878 A, p. 443.
Reinhardt, J. 1880 A, p. 30.
Roger, O. 1896 A, p. 208.
Stehlin, H. G. 1899 A, p. 118.

[1] This, the original spelling of this generic name, was probably due to a printer's error.

Trouessart, E. L. 1898 A, p. 816.
Wilckens, M. 1885 B, p. 306.
Williston, S. W. 1894 C, p. 24.
Pliocene; Oregon.

Platygonus leptorhinus Willist.

Williston, S. W. 1894 F.
Stehlin, H. G. 1899 A, pp. 117, 191, 273, 333.
Trouessart, E. L, 1898 A, p. 816.
Williston, S. W. 1894 C, p. 27, pls. vii, viii; text figs. 1-6.
Pliocene; Kansas.

Platygonus rex Marsh.

Marsh, O. C. 1894 L, p. 273, figs. 31, 32.
Roger, O. 1896 A, p. 208.
Stehlin, H. G. 1899 A, pp. 21, 115. (Dicotyles.)
Trouessart, E. L. 1898 A, p. 816.
Pliocene; Oregon.

Platygonus striatus Marsh.

Marsh, O. C. 1871 D, p. 41.
King, C. 1878 A, p. 430.
Matthew, W. D. 1899 A, p. 72.
Roger, O. 1896 A, p. 208.
Stehlin, H. G. 1899 A, p. 118.
Trouessart, E. L. 1898 A, p. 816.
Williston, S. W. 1894 C, p. 24.
Miocene (Loup Fork); Nebraska.

Platygonus vetus Leidy.

Leidy, J. 1882 A, p. 301.
Cope, E. D. 1888 X, p. 1089.
Cope and Wortman 1884 A.
Dugès, A. 1891 A, p. 16, pls. i, ii. (P. alemanii.)
Leidy, J. 1889 G, p. 49.
1889 H, p. 12, pl. ii, figs. 1, 2.
Lesley, J. P. 1889 A, p. 690, figure.
Mercer. H. C. 1894 A, p. 98. (Dycotyles pennsylvanicus.)
Roger, O. 1896 A, p. 208.
Stehlin, H. G. 1899 A, p. 117.
Trouessart, E. L. 1898 A, p. 816.
Williston, S. W. 1894 C.
Pleistocene; Pennsylvania.

Platygonus ziegleri Marsh.

Marsh, O. C. 1871 D, p. 40.
Cope and Wortman 1884 A, p. 21.
Leidy, J. 1872 B. p. 365.
Matthew, W. D. 1899 A, p. 48.
Roger, O. 1896 A, p. 207.
Stehlin, H. G. 1899 A, pp. 20, 118, 191.
Trouessart, E. L. 1898 A, p. 816.
Williston, S. W. 1894 C, p. 24.
Eocene (Bridger); Wyoming.
The generic position of this species is doubtful.

THINOTHERIUM Cope. Type *T. annulatum* Cope. Not *Thinotherium* Marsh, 1872.

Cope, E. D. 1870 F, p. 292.
A genus of which little is known. Regarded at the time of description as belonging to the Hippopotamidæ.

Thinotherium annulatum Cope.

Cope, E. D. 1870 F, p. 292.
Roger, O. 1896 A, p. 207.
Miocene?; Virginia.

SUINÆ.

Roger, O. 1896 A, p. 209.

Zittel, K. A. 1893 B, p. 343.

SUS Linn. Type *S. scrofa* Linn.

Linnæus, C., 1758. Syst. Nat., ed. 10, i, p. 49.
Blainville, H. M. D. 1864 A, iv, AA, pls. i-ix.
Flower and Lydekker 1891 A, p. 281.
Hensel, R. 1875 A.
Huxley, T. H. 1872 A, p. 313.
Owen, R. 1845 B, p. 544.
1866 B, p. 458.
1868 A, p. 344, fig. 273.
Parker, W. K. 1874 B, pls. xxviii-xxxvii.
Paulli, S. 1900 A, p. 193.
Pomel, A. 1848 A.
Roger, O. 1896 A, p. 209.
Rütimeyer, L. 1862 B.
1867, Verb. naturf. Gesellsch. Basel, iv, pp. 139-176.
Stehlin, H. G. 1899 A.
Trouessart, E. L. 1898 A, p. 820.
Woodward, A. S. 1898 B, 345.
Zittel, K. A. 1893 B, p. 344.

Sus scrofa Linn.

Linnæus, C., 1758. Syst. Nat., ed. 10, i, p. 49.
Agassiz, L. in Holmes F. S. 1858 A ("Hog.")
Blainville H. M. D. 1864 A, iv, AA, p. 113, pls. ii, iii, vi-viii.

Cope, E. D. 1870 E, p. 284. (Sus vagrans.)
1873 E (S. vagrans.)
1873 F, p. 291. (Sus vagrans.)
1873 L, p. 207. (S. scrofa.)
Emmons, E. 1858 B. p. 198, fig. 22.
Holmes, F. S. 1850 B. p. 203. ("Hog.")
1858 A ("Hog.")
Leidy, J. 1860 B, p. 109, pl. xix, figs. 4, 5.
1869 A, p. 443.
1873 D.
Owen, R. 1866 B, pp. 458, 467, figs. 313-315.
Paulli, S. 1900 A, p. 193, figs. 7, 8.
Rafinesque, C. S., 1831, Enum. some remark, nat. objects. (Aper pecari.)
Roger, O. 1896 A, p. 209.
Rütimeyer, L. 1862 B, p. 26.
Stehlin, H. G. 1899 A.
Trouessart, E. L. 1898 A, p. 821.
This species has been reported from various parts of the United States, North Carolina, South Carolina, New Jersey, Nebraska, and from deposits ranging from the Miocene to the present; but it is most probable that all the remains have been derived from the introduced race.

Superfamily *CAMELOIDEA* Gill.

Gill, T. 1872 B, p. 76.
Following the suggestions of Dr. W. B. Scott, this group is made to include the Agriochœridæ and their allies. Unless otherwise indicated, the following authorities use for this group, in the restricted sense, the name Tylopoda.
Burmeister, H. 1867 A, p. 233.
Cope, E. D. 1898 B, p. 132.
Flower, W. H. 1874 A.
    1885 A.
Flower and Lydekker 1891 A. p. 295.

Rütimeyer, L. 1867 B.   (Camelina.)
Schlosser, M. 1886 A, p. 255.
Scott, W. B. 1890 B, p. 387.
    1891 A.
    1891 C, p. 63.
    1892 C, p. 15.
    1898 C, p. 75.
    1899 A, p. 111.
Slade, D. D. 1893 A.
Woodward, A. S. 1898 B. p. 358.
Wortman, J. L. 1898 A, p. 94.

AGRIOCHŒRIDÆ.

Unless otherwise indicated, the following authors use the family name Oreodontidæ.
Ameghino, F. 1889 A, p. 576.
Bettany, G. T. 1873 A.
    1876 A.
Cope, E. D. 1874 B, p. 463.
    1884 H, p. 26.
    1884 M.
    1884 T.
    1884 DD.
    1885 B, p. 487.
    1887 B, pp. 379, 386.
    1887 N.
    1888 X. pp. 1080, 1092.
    1889 C, p. 171, *et seq.*
Flower and Lydekker 1891 A, p. 293.   (Cotylopidæ)
Gill, T. 1872 B, pp. 10, 81.
Leidy, J. 1869 A, pp. 71, 379 (Oreodontidæ); p. 131 (Agriochœridæ).
    1871 C, p. 345 (Oreodontidæ); p. 348 (Agriochœridæ).
    1873 B, p. 318.

Major, Fors. 1873 A, p. 104.
Marsh, O. C. 1877 E.
Nicholson and Lydekker 1889 A, p. 1326.   (Cotylopidæ.)
Osborn and Wortman 1894 A, p. 215.
Roger, O. 1896 A, p. 212.
Schlosser, M. 1886 A, p. 255.
    1886 B, p. 46.
    1888 A.
    1890 C.
Scott, W. B. 1885 B.
    1889 C.
    1890 A, p. 490.
    1890 B.
    1891 A, p. 71.
    1892 A. p. 433.
    1895 C, p. 164.
    1898 C, p. 74.
    1899 A, p. 85.
Steinmann and Döderlein 1890 A, p. 799.
Trouessart, E. L. 1898 A, p. 835.
Woodward, A. S. 1898 B, p. 354.
Zittel, K. A. 1893 B, pp. 348, 356.

AGRIOCHŒRINÆ.

Cope, E. D. 1884 M, p. 571.
Flower and Lydekker 1891 A, p. 293.
Leidy, J. 1869 A, p. 131.   (Agriochœridæ.)
    1871 C, p. 348.   (Agriochœridæ.)
Lydekker. R. 1896 B, p. 375.   (Agriochœridæ.)
Major, Fors. 1873 A, p. 104.   (Agriochœridæ.)
Osborn, H. F. 1893 F, p. 132.   (Artionychia.)
    1893 J, p. 611.   (Artionychia.)

Osborn and Wortman 1893 A, p. 4.   (Artionychia.)
Roger, O. 1896 A, p. 212.
Scott, W. B. 1889 C, p. 75.
    1890 B, pp. 320, 358.
    1892 A, p. 434.
    1899 A, p. 101.   (Agriochœridæ.)
Trouessart, E. L. 1898 A, p. 833.
Zittel, K. A. 1893 B, p. 351.

PROTAGRIOCHŒRUS Scott.   Type *P. annectens* Scott.

Scott, W. B. 1899 A, p. 100.

Protagriochœrus annectens Scott.

Scott, W. B. 1899 A, p. 100, pl. iv, figs. 26–28.
    Eocene (Uinta): Utah.

AGRIOCHŒRUS Leidy.   Type *A. antiquus* Leidy.

Leidy, J. 1850 C, p. 121.
Cope, E. D. 1879 E, p. 375.   (Coloreodon, type *C. ferox.*)
    1879 K, p. 197.   (Merycopater, type *Hyopotamus guyotianus.*)
    1884 M, pp. 504, 559 (Agriochœrus); pp. 504, 570 (Coloreodon).
    1887 B. p. 398 (Coloreodon); p. 398 (Agriochœrus).

Cope, E. D. 1888 X, p. 1094.
    1889 C, p. 48.
    1889 G, p. 119.
    1894 E, p. 869.   (Coloreodon.)
Leidy, J. 1852 L, p. 64.
    1854 A, pp. 24, 113.
    1869 A, p. 132.
    1871 C, p. 349.
    1873 B, p. 216.

Marsh, O. C. 1854 L, p. 270. (Agriomeryx, type
  A. migrans.)
Nicholson and Lydekker 1889 A, p. 1327. (Agri-
  ochœrus, Coloreodon.)
Osborn, H. F. 1893 D, p. 44. (Agriochœrus, Ar-
  tionyx.)
    1893 E. (Artionyx.)
    1893 F. (Artionyx.)
    1893 J, p. 611. (Artionyx.)
    1893 K. (Artionyx.)
    1898 A, p. 146.
Osborn and Wortman 1893 A, p. 5. (Artionyx,
  type *A. gaudryi.*)
Schlosser, M. 1885 A, p. 685.
    1886 B, p. 47.
Scott, W. B. 1889 C, p. 75.
    1890 A.
    1890 B, p. 358 (Agriochœrus): p. 361 (Colo-
      reodon).
    1891 A, p. 48.
    1891 D, p. 90.
    1892 A, p. 434.
    1894 B.
    1894 D.
    1894 E.
    1899 A, p. 88, 108.
Steinmann and Döderlein 1890 A, p. 799.
Trouessart, E. L. 1898 A, p. 834.
Woodward, A. S. 1898 B, p. 356.
Wortman, J. L. 1893 A. (Artionyx.)
    1895 A, p. 145.
Zittel, K. A. 1893 B, p. 352. (Agriochœrus, Colo-
  reodon.)

Agriochœrus antiquus Leidy.

Leidy, J. 1850 C, p. 122.
Cope, E. D. 1879 E, p. 375. (Merycopater.)
    1884 M, p. 560.
Hayden, F. V. 1858 B, p. 157.
King, C. 1878 A, p. 411.
Leidy, J. 1852 J, p. 571.
    1852 L, p. 65.
    1853 D, p. 392.
    1854 A, p. 24, pl. l, figs. 5–10.
    1854 E, p. 157.
    1869 A, pp. 132, 381, pl. xiii, fig. 4.
    1870 P, p. 112.
    1871 C, p. 349.
    1873 B, pp. 216, 319.
Matthew, W. D. 1899 A, p. 60.
Roger, O. 1896 A, p. 212.
Trouessart, E. L. 1895 A, p. 833.
Wortman, J. L. 1895 A, p. 145, pl. l; text figs. 8–15,
    17 (A. latifrons, not of Leidy); p. 177 (A. an-
    tiquus).
    Oligocene (White River); South Dakota.

Agriochœrus auritus Leidy.

Leidy, J. 1852 J, p. 563, pl. xv, figs. 1, 2. (Eucro-
  taphus.)
Cope, E. D. 1884 M, p. 560. (Eucrotaphus.)
Leidy, J. 1854 A, pp. 56, 113, pl. vii, figs. 1–3.
    (Eucrotaphus.)
    1869 A, p. 381. (Syn?. of A. major.)
Trouessart, E. L. 1898 A, p. 834. (Syn?. of A. ma-
  jor.)
    Oligocene (White River); South Dakota.
    If this species and *A. major* are identical, *A.*
    auritus will stand as the correct name.

Agriochœrus ferox (Cope.)

Cope, E. D. 1879 E, p. 375. (Coloreodon.)
    1884 M, pp. 505, 570, figure. (Coloreodon.)
    1884 T, p. 282. (Coloreodon.)
Matthew, W. D. 1899 A, p. 64. (Agriochœrus.)
Roger, O. 1896 A, p. 213. (Coloreodon.)
Scott, W. B. 1890 B, pl. xiv, fig. 13; pl. xv, fig. 22.
    (Coloreodon.)
Trouessart, E. L. 1898 A, p. 834. [Agriochœrus
    (Coloreodon).]
Wortman, J. L. 1895 A, p. 178. (Agriochœrus.)
    Miocene (John Day); Oregon.

Agriochœrus gaudryi (Osb. and Wort.).

Osborn and Wortman 1893 A, p. 5, figs. 1, 3, 4, 5 B.
    (Artionyx.)
Ameghino, F. 1894 A, p. 319. (Artionyx.)
Osborn, H. F. 1893 F, p. 131, fig. 4. (Artionyx.)
Roger, O. 1896 A, p. 212.
Scott, W. B. 1894 R.
Trouessart, E. L. 1898 A, p. 834. [Agriochœrus
    (Coloreodon).]
Wortman, J. L. 1895 A, p. 170, figs. 1, 19, 20 (Ag-
    riochœrus major); p. 178 (Agriochœrus gaud-
    ryi?).
    Oligocene (White River); South Dakota.

Agriochœrus guyotianus Cope.

Cope, E. D. 1879 D, p. 77. (Hyopotamus.)
    1879 B, p. 58. (Merycopater.)
    1879 E, p. 375. (Merycopater guiotianus.)
    1879 K, p. 197. (Merycopater.)
    1884 M, pp. 560, 563.
    1884 T, p. 281.
    1889 G, pl. iii.
Lydekker, R. 1883 A, p. 157. (Merycopater.)
Roger, O. 1896 A, p. 212.
Scott, W. B. 1890 B, pl. xiv, figs. 11, 12.
Trouessart, E. L. 1898 A, p. 834.
Wortman, J. L. 1895 A, figs. 2–7, 16–18.
    Miocene (John Day); Oregon.

Agriochœrus latifrons Leidy.

Leidy, J. 1869 A, pp. 135, 381, pl. xiii, figs. 1–3.
Cope, E. D. 1884 M, p. 560.
    1884 T, p. 281.
Leidy, J. 1867 D, p. 32.
    1871 C, p. 349.
    1873 B, pp. 216, 319.
Lydekker, R. 1896 B, p. 374, fig. 78.
Marsh, O. C. 1875 B, p. 250.
Matthew, W. D. 1899 A, p. 60.
Roger, O. 1896 A, p. 212.
Scott, W. B. 1890 B, text figs. vii, x³.
    1891 A, p. 49.
Trouessart, E. L. 1898 A, p. 834.
Woodward, A. S. 1898 B, p. 357, fig. 204.
Wortman, J. L. 1895 A, p. 177 (in part, not figs. 8–
    15, 17).
Zittel, K. A. 1893 B, p. 352, fig. 287.
    Oligocene (White River); South Dakota.

Agriochœrus macrocephalus (Cope.)

Cope, E. D. 1879 E, p. 376. (Coloreodon.)
    1884 M, p. 570. (Coloreodon.)
    1884 T, p. 281. (Coloreodon.)
Matthew, W. D. 1899 A, p. 64. (Coloreodon.)
Roger, O. 1896 A, p. 213. (Coloreodon.)

664   FOSSIL VERTEBRATA OF NORTH AMERICA.   [BULL. 179

Trouessart, E. L. 1898 A, p. 835. [Agriochœrus (Coloreodon).]
Wortman, J. L. 1895 A, p. 178.
Miocene (John Day); Oregon.

Agriochœrus major Leidy.

Leidy, J. 1856 J, p. 164.
Cope, E. D. 1884 M. p. 560.
  1884 T, p. 281.
Hayden, F.V. 1858 B, p. 157.
Leidy, J. 1869 A, pp. 134, 381.
  1871 C, p. 349.
Matthew, W. D. 1899 A, p. 60.
Roger, O. 1896 A, p. 212.
Trouessart, E. L. 1898 A, p. 834.
Wortman, J. L. 1895 A, p. 178.
Oligocene (White River); South Dakota.
It is not unlikely that this species is identical with *A. auritus* Leidy. If this view is accepted, the name *auritus* will supersede *major*.

Agriochœrus migrans (Marsh).

Marsh, O. C. 1894 L, p. 270, fig. 24. (Agriomeryx.)

Matthew, W. D. 1899 A, p. 60. (Agriochœrus.)
Trouessart, E. L. 1898, A, p. 834. [Agriochœrus (Coloreodon).]
Oligocene (White River); South Dakota.

Agriochœrus ryderanus Cope.

Cope, E. D. 1881 C, p. 173. (Coloreodon.)
  1884 M, pp. 560, 566.
  1884 T, p. 281.
Roger, O. 1896 A, p. 212.
Trouessart, E. L. 1898 A, p. 834.
Wortman, J. L. 1895 A, p. 178.
Miocene (John Day); Oregon.

Agriochœrus trifrons Cope.

Cope, E. D. 1884 M, pp. 560, 561.
  1884 T, p. 281.
Matthew, W. D. 1899 A, p. 64.
Roger, O. 1896 A, p. 212.
Trouessart, E. L. 1898 A, p. 834.
Wortman, J. L. 1895 A, p. 177.
Miocene (John Day); Oregon.

PROTOREODONTINÆ.

Scott, W. B. 1890 A, p. 503.
Flower and Lydekker 1891 A, p. 293.
Marsh, O. C. 1894 L, p. 267. (Eomericidæ.)
Roger, O. 1896 A, p. 212.

Scott, W. B. 1890 B, pp. 320, 361.
  1899 A, p. 85. (Oreodontidae. in part.)
Zittel, K. A. 1893 B, p. 351.

PROTOREODON Scott and Osb.   Type *P. parvus* Scott and Osb.

Scott and Osborn 1887 A, p. 257.
Berg. C., 1899, Commun. Mus. Nac. Buenos Aires, i, p. 79. (Chorotherium, to replace Agriotherium, preoccupied.)
Cope, E. D. 1888 X, p. 1082.
  1890 C, p. 471.
Flower and Lydekker 1891 A, p. 293.
Marsh, O. C. 1894 L, p. 268. (Syn. of Eomeryx.)
Nicholson and Lydekker 1889 A, p. 1328.
Pavlow, M. 1900 A, pp. 34, 44.
Scott, W. B. 1889 C, p. 75.
  1890 A, p. 487, pl. vii, figs. 1–8.
  1890 B, p. 361, pl. xiii, figs. 1–3; pl. xvi, figs. 23–26.
  1892 A, p. 433.
  1894 E.
  1898 C. p. 79. (Agriotherium, type *A. paradoxicum*.)
  1899 A, p. 85.
Wortman, J. L. 1898 A, p. 96.
Zittel, K. A. 1893 B, p. 351.

Matthew, W. D. 1899 A, p. 50. (Agriotherium.)
Scott, W. B. 1899 A, pp. 86, 95, pl. iv, figs. 24, 25. (Protoreodon.)
Eocene (Uinta); Utah.

Protoreodon parvus Scott and Osb.

Scott and Osborn 1887 A, p. 257, figure.
Cope, E. D. 1890 C, p. 470.
Marsh, O. C. 1894 L, p. 268. (Syn. of Eomeryx pumilis.)
Matthew, W. D. 1899 A, p. 50. (Eomeryx.)
Pavlow, M. 1900 A, p. 34.
Roger, O. 1896 A, p. 212.
Scott, W. B. 1890 A, p. 487, pl. vii, figs. 1–8.
  1890 B, pl. xiii, figs. 1–3; pl. xvi, figs. 23–26.
  1899 A, pp. 89, 95, pl. iii, figs. 19–21.
Trouessart, E. L. 1898 A, p. 833.
Wortman, J. L. 1898 A, p. 96.
Zittel, K. A. 1893 B, p. 351, fig. 286.
Eocene (Uinta); Utah.

Protoreodon minor Scott.

Scott, W. B. 1899 A, p. 95, pl. iii, fig. 23.
Eocene (Uinta); Utah.

Protoreodon paradoxicus Scott.

Scott, W. B. 1898 C, p. 79. (Agriotherium.)

Protoreodon sp. indet.

Scott, W. B. 1899 A, p. 95, pl. iii, fig. 22.
Eocene (Uinta); Utah.

EOMERYX Marsh.   Type *E. pumilis* Marsh.

Marsh, O. C. 1894 L, p. 266.
Cope, E. D. 1894 F, p. 869. (Syn. of Protoreodon.)
Marsh, O. C. 1877 E. (Inadequate description; no species.)

Pavlow, M. 1900 A, p. 34.
Scott, W. B. 1890 A, p. 488.
  1899 A, pp. 85, 86. (Syn.? of Protoreodon.)
Scott and Osborn 1887 A, p. 259.

Wortman, J. L. 1898 A, p. 96.
Zittel, K. A. 1893 B, p. 351.   (Syn.? of Protoreodon.)

Eomeryx pumilis Marsh.

Marsh, O. C. 1875 B, p. 250.   (Agriochœrus.,
Cope, E. D. 1884 M, p. 560.   (Agriochœrus.)
King, C. 1878 A, p. 407.   (Agriochœrus.)
Marsh, O. C. 1894 L, p. 266, figs. 18–22.

Matthew, W. D. 1899 A, p. 50.
Pavlow, M. 1900 A, p. 34.
Scott, W. B. 1890 A, p. 487.   (Agriochœrus.)
    1899 A, pp. 86, 96.   (Protoreodon.)
Scott and Osborn 1887 A, p. 258.   (Protoreodon?.)
Trouessart, E. L. 1898 A, p. 833.   (Syn. of Proto-
reodon parvus.)
Wortman, J. L. 1898 A, p. 96.
Eocene (Uinta); Utah.

HYOMERYX Marsh.   Type *H. breviceps* Marsh.

Marsh, O. C. 1894 L, p. 145.
Scott, W. B. 1898 C, p. 81.
    1899 A, pp. 96, 99.
Wortman, J. L. 1895 A, p. 97.

Hyomeryx breviceps Marsh.

Marsh, O. C. 1894 L, p. 268, fig. 19.

Matthew, W. D. 1899 A, p. 50.
Roger, O. 1896 A, p. 212.
Trouessart, E. L. 1898 A, p. 833.
Wortman, J. L. 1895 A, p. 145.
Eocene (Uinta); Utah.

MERYCOIDODONTINÆ.

Branco, W. 1897 A, p. 24.   (Oreodontinæ.)
Cope, E. D. 1884 M, p. 571.   (Oreodontinæ.)
Flower and Lydekker 1891 A, p. 298.   (Cotylopi-
næ.)
Gill, T. 1872 B, p. 81.   (Oreodontinæ.)

Lydekker, R. 1896 B, p. 373.   (Cotylopidæ.)
Schlosser, M. 1890 D, p. 722.   ("Oreodonten.")
Scott, W. B. 1889 C. p. 75.   (Oreodontinæ.)
    1892 A, p. 434.   (Oreodontinæ.)
Zittel, K. A. 1893 B, p. 352.   (Oreodontidæ.)

MERYCOIDODON Leidy.   Type *M. culbertsonii* Leidy.

Leidy, J. 1848 B, p. 47.
    The generic name *Merycoidodon* clearly has
priority over both *Oreodon* and *Cotylops*, not-
withstanding the fact that it has been discarded
by Leidy himself and succeeding writers.   Un-
less otherwise indicated, the following authors
employ the name *Oreodon*.
Bettany, G. T. 1873 A.
    1876 A.
Bruce, A. T. 1883 A, p. 36, figs. 1–4.   (Oreodon.)
Cope, E. D. 1884 H, p. 22.
    1884 M, p. 505.
    1885 B, p. 483.
    1887 B, p. 386.
    1887 S, p. 85.
    1888 X, p. 1098.
Dawkins, W. B. 1870 A.
Flower, W. H. 1873 A.
Flower and Lydekker 1891 A, p. 298.   (Cotylops.)
Gaudry, A. 1878 B, p. 82.
Grinnell and Dana 1876 A.
Hay, O. P. 1899 A, p. 594.   (Merycoidodon.)
Leidy, J. 1851 D, p. 238 (Oreodon); p. 239 (Coty-
lops, type *O. speciosa*).
    1852 J, pp. 540, 548.
    1852 L, p. 64.   (Oreodon, Merycoidodon.)
    1854 A, pp. 29, 47, 113.
    1869 A, pp. 72, 379.
    1870 P, p. 111.
    1871 C, p. 34ᵇ.
    1873 B, p. 201.
Marsh, O. C. 1873 H, p. 409.
    1875 B, p. 250.
    1877 E.
    1884 F, p. 172.
Nicholson and Lydekker 1889 A, p. 1326.   (Coty-
lops.)

Schlosser, M. 1886 B, p. 47.
Scott, W. B. 1884 C.
    1885 B.
    1889 C, p. 75.
    1890 A.
    1890 B, p. 320.
    1891 A, p. 53.
    1894 D.
    1894 H.
    1895 C, p. 125, *et seq.*
    1899 A, p. 87.
Steinmann and Döderlein 1890 A, p. 799.
Trouessart, E. L. 1898 A, p. 835.
Woodward, A. S. 1898 B, p. 354.
Wortman, J. L. 1895 A, p. 173.
Zittel, K. A. 1893 B, p. 352.

Merycoidodon affinis (Leidy).

Leidy, J. 1869 A, pp. 105, 380, pl. ix, fig. 3.   (Oreo-
don.)
Cope, E. D. 1884 M, p. 512.   (Oreodon.)
    1884 T, p. 281.   (Oreodon.)
Leidy, J. 1869 A, p. 96, No. 10.   (Oreodon.)
    1871 C, p. 346.   (Oreodon.)
    1873 B, p. 212.   (Oreodon.)
Roger, O. 1896 A, p. 213.   (Oreodon.)
Trouessart, E. L. 1898 A, p. 835.   (Oreodon.)
Oligocene (White River); South Dakota.

Merycoidodon bullatus (Leidy).

Leidy, J. 1869 A, pp. 106, 380.   (Oreodon.)
Cope, E. D. 1884 M, p. 517.   (Syn. of Eucrotaphus
jacksoni.)
King, C. 1878 A, p. 411.   (Eporeodon.)
Leidy, J. 1871 C, p. 346.   (Oreodon.)
    1873 B, p. 212 (Oreodon bullatus); p. 318
    (Oreodon culbertsonii).

Marsh, O. C. 1875 B, p. 250.   (Eporeodon.)
Matthew, W. D. 1899 A, p. 60.   (Oreodon.)
Osborn and Wortman 1894 A, p. 218, fig. 5c.   (Oreodon.)
Roger, O. 1896 A, p. 213.   (O. jacksoni.)
Scott, W. B. 1890 B, pl. xiii, fig. 6.   (Oreodon.)
Trouessart, E. L. 1898 A, p. 836.   (Eucrotaphus.)
   Oligocene (White River); South Dakota.

Merycoidodon culbertsonii Leidy.

Leidy, J. 1848 B, p. 47, plate.
   Unless otherwise indicated, the following authors refer to this species under the generic name Oreodon.
Bettany, G. T. 1873 A.
Cope, E. D. 1884 M, pp. 511, 512, 513.   (O. culbertsonii and O. culbertsonii periculorum.)
   1888 X, p. 1094.
Farr, M. S. 1896 A, p. 157.
Gaudry, A. 1878 B, p. 81, fig. 90.
Hayden, F. V. 1858 B, p. 157.
King, C. 1878 A, p. 411.
Leidy, J. 1851 D, p. 239 (Cotylops speciosa); p. 238 (O. priscum).
   1851 E, p. 276.   (Oreodon robustum.)
   1852 J, p. 548, pl. x, figs. 4–6; pl. xiii, figs. 3, 4.
   1852 L, p. 64.   (Oreodon priscum, M. culbertsonii.)
   1853 D, p. 392.
   1854 A, pp. 45, 113, pls. ii, iii; pl. iv, figs. 1–5; pl. v, figs. 1, 2; pl. vi, figs. 8–11.
   1854 E.
   1869 A, pp. 86, 379, pl. vi, fig. 1; pl. vii, fig. 2; pl. ix, fig. 1.
   1870 J, p. 67.
   1870 P, p. 112.
   1871 C, p. 345.
   1873 B, pp. 211, 318.
Lydekker, R. 1885 C, p. 207.
Marsh, O. C. 1870 D.
Matthew, W. D. 1899 A, p. 60.
Osborn and Wortman 1894 A, p. 215, fig. 5 A.
Roger, O. 1896 A, p. 213.
Scott, W. B. 1890 B, pl. xii; pl. xiii, fig. 4; pl. xvi, figs. 27–29; text figs. ii, iii, iv, x.
Stewart, A. 1897 A, pl. i.

Trouessart, E. L. 1898 A, p. 835.
Zittel, K. A. 1893 B, p. 353, fig. 288.
   Oligocene (White River); South Dakota.

Merycoidodon gracilis Leidy.

Leidy, J. 1851 D, p. 239.   (Oreodon.)
Cope, E. D. 1874 B, p. 498.   (Oreodon.)
   1884 M, pp. 511, 512.   (Oreodon, with varieties gracilis and coloradoensis.)
   1884 T, p. 281.   (Oreodon.)
Hayden, F. V. 1858 B, p. 157.   (Oreodon.)
King, C. 1878 A, p. 411.   (Oreodon.)
Leidy, J. 1852 J, p. 550, pl. xi, figs. 2, 3; pl. xiii. figs. 5, 6.   (Oreodon.)
   1852 L, p. 64.   (Oreodon.)
   1853 D, p. 392.   (Oreodon.)
   1854 A, p. 53, pl. v, figs. 3, 4; pl. vi, figs. i-7.   (Oreodon, Merycoidodon.)
   1854 E, p. 157.   (Oreodon.)
   1869 A, pp. 94, 379, pl. vi, figs. 2, 3.   (Oreodon.)
   1871 C, p. 346.   (Oreodon.)
   1873 B, p. 211.   (Oreodon.)
Osborn and Wortman 1894 A, p. 216, fig. 5, B.   (Oreodon.)
Roger, O. 1896 A, p. 213.   (Oreodon.)
Scott, W. B. 1890 B, pl. xiii, fig. 5; text figs. i. ix, x, 5.   (Oreodon.)
Scott and Osborn 1887 B, p. 154.   (Oreodon.)
Trouessart, E. L. 1898 A, p. 835.   (Oreodon.)
Zittel, K. A. 1893 B, p. 353.   (Oreodon.)
   Oligocene (White River); South Dakota, Nebraska, Colorado.

Merycoidodon hybridus (Leidy).

Leidy, J. 1869 A, pp. 105, 380, pl. ix, fig. 4.   (Oreodon.)
   1871 C, p. 346.   (Oreodon.)
   1873 B, p. 212.   (Oreodon.)
Roger, O. 1896 A, p. 213.
Trouessart, E. L. 1898 A, p. 835.
   Oligocene (White River); South Dakota.

Merycoidodon minor Cope.

Cope, E. D. 1888 X, p. 1094.   (Name only.)

EUCROTAPHUS Leidy.   Type E. jacksoni Leidy.

Leidy, J. 1850 A, p. 90.
Cope, E. D. 1884 H, p. 23.
   1884 M, pp. 504, 513.
   1884 T.
   1885 B, p. 484.
   1887 B, p. 386.
   1888 X, p. 1093.
Leidy, J. 1851 D, p. 239.
   1852 J, p. 562.
   1852 L, p. 64.
   1854 A, pp. 56, 113.
   1854 C, p. 157.
   1873 B, p. 212.
Nicholson and Lydekker 1889 A, p. 1327.
Scott, W. B. 1890 B, p. 320.   (Syn. of Oreodon.)

Wortman, J. L. 1893 A.
Zittel, K. A. 1893 B, pp. 352, 855.

Eucrotaphus jacksoni Leidy.

Leidy, J. 1850 A, p. 90.
Cope, E. D. 1884 M, pp. 514, 517 (E. jacksoni. p. 519 (subsp. leptacanthus).
Leidy, J. 1852 L, p. 65.
   1854 A, pp. 56, 113, pl. vii, figs. 4–6.
   1869 A, pp. 134, 381.
Lydekker, R. 1877 A, p. 334.   (Eporeodon.)
Roger, O. 1896 A, p. 213.   (Oreodon.)
Scott, W. B. 1890 B, p. 372, fig. x, 2.   (Oreodon.)
Trouessart, E. L. 1898 A, p. 836.
   Oligocene (White River); South Dakota.

EPOREODON Marsh. Type *Oreodon occidentalis* Marsh.

Marsh, O. C. 1875 B, p. 249.
Lydekker, R. 1885 C, p. 208.
Marsh, O. C. 1877 E, p. 365.
Nicholson and Lydekker 1889 A, p. 1327.
Osborn and Wortman 1894 A, p. 218.
Schlosser, M, 1886 B, p. 47.
Scott, W. B. 1890 B, p. 339.
　　1895 C, p. 125.
Trouessart, E. L. 1898 A, p. 836. (Syn. of Eucrotaphus.)
Zittel, K. A. 1893 B, p. 355.

Eporeodon longifrons (Cope).

Cope, E. D. 1884 M, pp. 519, 520. (Eucrotaphus major longifrons.)
Matthew, W. D. 1899 A, p. 64. (Eporeodon major, var. longifrons.)
　Miocene (John Day); Oregon.

Eporeodon? major (Leidy).

Leidy, J. 1854 A, p. 55, pl. iv, fig. 6. (Oreodon, Merycoidodon.)
Cope, E. D. 1884 M, pp. 519, 520. (Eucrotaphus.)
　　1884 T, p. 281. (Eucrotaphus.)
Hayden, F. V. 1858 B, p. 157. (Oreodon.)
King, C. 1878 A, p. 411.
Leidy, J. 1858 D. (Oreodon, no description.)
　　1854 E. (Syn. of Oreodon culbertsoni.)
　　1856 J, p. 164. (Oreodon.)
　　1869 A, pp. 99, 380, pl. vii, fig. 1; pl. viii. (Oreodon.)
　　1871 C, p. 346. (Oreodon.)
Marsh, O. C. 1875 B, p. 250.
　　1897 C, p. 524, fig. 98.
Matthew, W. D. 1899 A. pp. 60, 64.
Osborn and Wortman 1894 A, p. 218, fig. 5, D.
Roger, O. 1896 A, p. 213.
Scott, W. B. 1890 B, p. 339.
Scott and Osborn 1887 B, p. 155. (Eucrotaphus.)
Trouessart, E. L. 1898 A, p. 836. (Eucrotaphus.)
　Oligocene (White River); South Dakota, North Dakota, Colorado, Nebraska.

Eporeodon occidentalis Marsh.

Marsh, O. C. 1873 H, p. 409. (Oreodon.)
Cope, E. D. 1879 B, p. 59. (Eucrotaphus.)
　　1884 M, p. 517. (Syn. of Eucrotaphus jacksoni.)
King, C. 1878 A, p. 424.
Leidy, J. 1873 B, p. 318 (Syn.? of Oreodon bullatus); ?pl. vii, fig. 12 (Oreodon culbertsoni).
Marsh, O. C. 1875 B, p. 250.
　　1884 F, figs. 73, 162, 163.
Matthew, W. D. 1899 A, p. 64.
Trouessart, E. L. 1898 A, p. 836. (Syn. of Eucrotaphus jacksoni.)
　Miocene (John Day); Oregon.

Eporeodon pacificus (Cope).

Cope, E. D. 1884 M, p. 519. (Eucrotaphus jacksoni pacificus.)
　　1884 N, p. 23. (Eucrotaphus.)
　　1885 B, p. 484. (Eucrotaphus.)
　　1888 X, p. 1094. (Eucrotaphus.)
Matthew, W. D. 1899 A, p. 64. (E. occidentalis var. pacificus.)
Trouessart, E. L. 1898 A, p. 836. (Eucrotaphus.)
　Miocene (John Day); Oregon.

Eporeodon socialis Marsh.

Marsh, O. C. 1885 B, figs. 128, 129.
Matthew, W. D. 1899 A, p. 64.
Roger, O. 1896 A, p. 213.
　Miocene (John Day); Oregon.

Eporeodon trigonocephalus (Cope).

Cope, E. D. 1884 M, p. 511. (Eucrotaphus.)
　　1884 T, p. 281. (Eucrotaphus.)
Matthew, W. D. 1899 A, p. 64.
Roger, O. 1896 A, p. 213.
Trouessart, E. L. 1898 A, p. 836. (Eucrotaphus.)
　Miocene (John Day); Oregon.

MESOREODON Scott. Type *M. chelonyx* Scott.

Scott, W. B. 1893 B, p. 661.
　　1895 C, pp. 125, 164.

Mesoreodon chelonyx Scott.

Scott, W. B. 1893 B, p. 661.
Matthew, W. D. 1899 A, p. 60.
Roger, O. 1896 A, p. 214.
Scott, W. B. 1895 C, p. 125, pl. iii, fig. 29; pl. iv, figs. 32–34; pl. v, figs. 35–42; pl. vi, figs. 46, 47.
Trouessart, E. L. 1898 A, p. 836.
　Miocene (Deep River); Montana.

Mesoreodon intermedius Scott.

Scott, W. B. 1893 B, p. 661.
Matthew, W. D. 1899 A, p, 60.
Roger, O. 1896 A, p. 214.
Scott, W. B. 1895 C, p. 145, pl. v, figs. 43, 44.
Trouessart, E. L. 1898 A, p. 836.
　Miocene (Deep River); Monta:

MERYCOCHŒRUS Leidy. Type *M. proprius* Leidy.

Leidy, J. 1858 F, p. 24.
Bettany, G. T. 1876 A.
Cope, E. D. 1884 H, p. 23.
　　1884 M, pp. 504, 520.
　　1884 T.

Cope, E. D. 1885 B, p. 484.
　　1888 X, p. 1093.
　　1889 C.
　　1889 G, p. 113.
Leidy, J. 1869 A, p. 110.

Leidy, J. 1870 P, p. 111.
    1871 C, p. 347.
    1873 B, pp. 199, 202.
• Marsh, O. C. 1877 E.
Nicholson and Lydekker 1889 A, p. 1327.
Osborn, H. F. 1901 A, p. 46.
Scott, W. B. 1889 C, p. 75.
    1890 B, p. 340.
    1891 D, p. 90.
    1892 A, p. 435.
    1894 E.
    1895 C, p. 165.
Scott and Osborn 1890 B, p. 73.
Steinmann and Döderlein 1890 A, p. 800.
Trouessart, E. L. 1898 A, p. 837.
Zittel, K. A. 1893 B, p. 355.

Merycochœrus chelydra Cope.

Cope, E. D. 1884 M. pp. 521, 523.
    1884 T, p. 281.
Matthew, W. D. 1899 A, p. 64.
Roger, O. 1896 A, p. 214.
Trouessart, E. L. 1898 A, p. 838.
Miocene (John Day); Oregon.

Merycochœrus cœnopus Scott.

Scott, W. B. 1890 B, p. 346, pl. xvi, figs. 33, 34.
Matthew, W. D. 1899 A, p. 72. (M. cenopus.)
Roger, O. 1896 A, p. 214.
Scott and Osborn 1890 B, p. 73. (M. cenopus.)
Trouessart, E. L. 1898 A, p. 838.
Miocene (Loup Fork); Nebraska.

Merycochœrus laticeps Douglass.

Douglass, E. 1900 B, p. 428 figs. 1–3.
Miocene (Loup Fork); Montana.

Merycochœrus leidyi Bettany.

Bettany, G. T. 1876 A, P. 270, pl. xviii.
Cope, E. D. 1879 B, p. 59.
    1884 M, pp. 521, 523.
Matthew, W. D. 1899 A, p. 64.
Roger, O. 1896 A, p. 214.
Trouessart, E. L. 1898 A, p. 838.
Miocene (John Day); Oregon.

Merycochœrus macrostegus Cope.

Cope, E. D. 1884 M, pp. 521, 526.
    1884 T, p. 281.
Matthew, W. D. 1899 A, p. 64.
Roger, O. 1896 A, p 214.
Scott, W. B. 1890 B, pl. xiv, figs. 8, 9.
Trouessart, E. L. 1898 A, p. 838.
Miocene (John Day); Oregon.

Merycochœrus montanus Cope.

Cope, E. D. 1884 M, pp. 521, 531.
    1884 H, p. 23.
    1884 T, p. 282.
    1885 B, p. 484.
    1886 B, p. 359.
    1886 H.
    1889 G, p. 113, fig. 9.
    1895 C, p. 151.

Matthew, W. D. 1899 A, p. 72.
Roger, O. 1896 A, p. 214.
Scott, W. B 1890 B, p. 342, text figs. v, v¹.
    1893 B, p. 659.
Trouessart, E. L. 1898 A, p. 838.
Miocene (Loup Fork); Montana.

Merycochœrus obliquidens Cope.

Cope, E. D. 1886 H, p. 368.
    1886 B, p. 359.
Matthew, W. D. 1899 A, p. 72.
Roger, O. 1896 A, p. 214.
Trouessart, E. L. 1898 A, p. 838.
Miocene (Loup Fork); Oregon.

Merycochœrus proprius Leidy.

Leidy, J. 1858 E, p. 21.
Cope, E. D. 1884 M, pp. 521, 535.
    1884 T, p. 282.
Hayden, F. V. 1858 B, p. 157.
King, C. 1878 A, p. 411.
Leidy, J. 1869 A, pp. 110, 380, pl. x.
    1870 O, p. 109.
    1871 C, p. 347.
    1873 B, p. 201.
Matthew, W. D. 1899 A, p. 72.
Trouessart, E. L. 1898 A, p. 838.
Miocene (Loup Fork); Wyoming.

Merycochœrus rusticus Leidy.

Leidy, J. 1870 O p. 109.
Cope, E. D, 1884 M, pp. 521, 535.
    1884 T, p. 282.
Leidy, J. 1871 C, p. 347.
    1873 B, pp. 199, 319, pl. iii, figs. 1–3; pl. vii,
    figs. 1–5; pl. xx, figs. 9–11.
Matthew, W· D. 1899 A, p. 72.
Roger, O. 1896 A, p. 214.
Scott, W. B. 1890 B, pp. 341, 348. (Merychyus.)
Trouessart, E. L. 1898 A, p. 838.
Miocene (Loup Fork); Wyoming.

Merycochœrus superbus (Leidy).

Leidy, J. 1870 P, p. 111. (Oreodon.)
Bettany, G. T. 1876 A, p. 269, pl. xvii. (M. temporalis.)
Cope, E. D. 1879 B, p. 59. (Eucrotaphus.)
    1884 M, pp. 521, 522.
    1884 T, p. 281.
    1888 X, p. 1094, pl. xxvi.
King, C. 1878 A, p. 424. (Eporeodon.)
Leidy, J. 1871 C, p. 346. (Oreodon.)
    1873 B, pp. 211, 319, pl. i, fig. 1; pl. ii, fig. 16;
    pl. vii, figs. 7–11. (Oreodon.)
Marsh, O. C 1875 B, p. 250. (Eporeodon.)
Matthew, W. D. 1899 A, p. 64.
Roger, O. 1896 A, p. 214.
Scott, W. B. 1890 B, p. 339, pl. xiv, fig. 10.
Trouessart, E. L. 1898 A, p. 838.
Miocene (John Day); Oregon.

Merycochœrus sp. indet.

Leidy, J. 1873 B, p. 208, pl. xx, figs. 9–11.

MERYCHYUS Leidy.   Type *M. elegans* Leidy.

Leidy, J. 1858 E, p. 25.
Bettany, G. T. 1876 A.
Cope, E. D. 1878 H, p. 380. (Ticholeptus, type *T. zygomaticus.*)
     1878 P. (Ticholeptus.)
     1884 M, pp. 504, 535.
     1884 T.
     1885 B, p. 484.
     1887 B, p. 386.
     1887 S, p. 85.
     1888 X, p. 1093.
Dawkins, W. B. 1870 A, p. 232.
Hayden, F. V. 1858 B, p. 157.
Leidy, J. 1869 A, p. 115.
     1871 C, p. 347.
     1873 B, pp. 202, 213.
Marsh, O. C. 1877 E.
Nicholson and Lydekker 1889 A, p. 1327.
Scott, W. B. 1889 C, p. 75.
     1890 B, p. 347, pl. xiv, fig. 36.
     1891 A, p. 53.
     1891 D, p. 90.
     1892 A, p. 435.
     1895 C, p. 165.
Steinmann and Döderlein 1890 A, p. 800.
Trouessart, E. L. 1898 A, p. 837.
Zittel, K. A. 1893 B, p. 355.

Merychyus arenarum Cope.

Cope, E. D. 1884 M, pp. 536, 537, 540. (With subsp. leptorhynchus.)
     1884 H, p. 23.
     1884 T, p. 282.
     1885 B, p. 484.
     1888 X, p. 1080, pl. xxvi.
Matthew, W. D. 1899 A, p. 72.
Roger, O. 1896 A, p. 214.
Scott, W. B. 1890 B, pl. xiii, fig. 7; pl. xvi, figs. 30-32.
Trouessart, E. L. 1898 A, p. 837.
     Miocene (Loup Fork): Wyoming.

Merychyus elegans Leidy.

Leidy, J. 1858 E, p. 24.
Bettany, G. T. 1876 A.
Cope, E. D. 1874 B, p. 529.
     1874 P, p. 19.
     1884 M, pp. 536, 545.
     1884 T, p. 282.
King, C. 1878 A, p. 430.
Leidy, J. 1869 A, pp 118, 380, pl. xi, figs. 1-11.
     1871 C, p. 348.
     1873 B, p. 201.
Matthew, W. D. 1899 A, p. 72.
Roger, O. 1896 A, p. 214.
Scott and Osborn 1890 B, p. 72.
Trouessart, E. L. 1898 A, p. 837.
     Miocene (Loup Fork). Nebraska.

Merychyus major Leidy.

Leidy, J. 1858 E, p. 26.
Cope, E. D. 1874 B, p. 529.
     1874 P, p. 19.

Cope, E. D. 1884 M, pp. 536, 545.
     1884 T, p. 282.
Hayden, F. V. 1858 B, p. 157.
Leidy, J. 1869 A, pp. 121, 380, pl. xi, figs. 15, 16.
     1870 O, p. 109.
     1871 C, p. 348.
     1873 B, p. 201.
Matthew, W. D. 1899 A, p. 72.
Roger, O. 1896 A, p. 214.
Trouessart, E. L. 1898 A, p. 837.
Zittel, K. A. 1893 B, p. 355.
     Miocene (Loup Fork): Nebraska, Colorado.

Merychyus medius Leidy.

Leidy, J. 1858 E, p. 26.
Cope, E. D. 1875 F, p. 70.
     1877 K, p. 324.
     1884 M, pp. 536, 545.
     1884 T, p. 282.
Dawkins, W. B. 1870 A, p. 232.
Hayden, F. V. 1858 B, p. 157.
Leidy, J. 1869 A, pp. 119, 121, 380, pl. xi, figs. 12-14.
     1870 O, p. 109.
     1871 C, p. 348.
     1873 B, p. 201.
Matthew, W. D. 1899 A, p. 72.
Roger, O. 1896 A, p. 214.
Trouessart, E. L. 1898 A, p. 837.
     Miocene (Loup Fork): Nebraska, New Mexico, Colorado.

Merychyus pariogonus Cope.

Cope, E. D. 1884 M, pp. 536, 542.
     1884 T, p. 282.
     1886 B, p. 359.
     1886 H.
Matthew, W. D. 1899 A, p. 72.
Roger, O. 1896 A, p. 214.
Scott, W. B. 1893 B, p. 659.
     1895 C, p. 148.
Trouessart, E. L. 1898 A, p. 837.
     Miocene (Loup Fork); Wyoming, Montana.

Merychyus zygomaticus Cope.

Cope, E. D. 1878 P, p. 129. (Ticholeptus.)
     1878 H, p. 380. (Ticholeptus.)
     1884 M, pp. 536, 545.
     1884 T, p. 282.
     1886 B, p. 359.
     1886 H.
Matthew, W. D. 1899 A, p. 72.
Roger, O. 1896 A, p. 214.
Scott, W. B. 1893 B, p. 659.
     1895 C, p. 146, pl. v, fig. 45.
Trouessart, E. L. 1898 A, p. 837.
     Miocene (Loup Fork): Montana.
     Dr. W. D. Matthew (1899 A, p. 72) has attributed to Cope a species, *Merychyus euryops*, but search has failed to discover the name in any of Cope's papers. It is probably a manuscript name only.

LEPTAUCHENIA Leidy.   Type *L. decora* Leidy.

Leidy, J. 1856 I, p. 88.
Cope, E. D. 1884 M, pp. 504, 546.
    1887 B, p. 386.
    1888 X, p. 1093.
Leidy, J. 1858 E, p. 25.
    1869 A, p. 122.
    1871 C, p. 348.
Matthew, W. D. 1899 A, p. 24.
Nicholson and Lydekker 1889 A, p. 1327.
Schlosser, M. 1886 B, p. 49.
Scott, W. B. 1890 B, p. 353.
    1892 A, p. 435.
    1896 C, p. 166.
    1899 A, p. 96.
Trouessart, E. L. 1898 A, p. 838.
Zittel, K. A. 1893 B, p. 355.

Leptauchenia decora Leidy.

Leidy, J. 1856 I, p. 88.
Cope, E. D. 1884 M, p. 546.
Hayden, F. V. 1858 B, p. 157.
Leidy, J. 1869 A, pp. 127, 381, pl. xii, figs. 21, 22.
    1871 C, p. 348.
Matthew, W. D. 1899 A, p. 61.
Roger, O. 1896 A, p. 215.
Trouessart, E. L. 1898 A, p. 838.
    Oligocene (White River); South Dakota, Colorado.

Leptauchenia major Leidy.

Leidy, J. 1856 J, p. 163.
Cope, E. D. 1884 M, p. 546.
Hayden, F. V. 1858 B, p. 157.
King, C. 1878 A, p. 411.
Leidy, J. 1869 A, pp. 124, 380, pl. xii, figs. 1–5.
    1871 C, p. 348.
Matthew, W. D. 1899 A, p. 61.
Roger, O. 1896 A, p. 215.
Scott, W. B. 1890 B, pl. xv, figs. 15, 16.
Trouessart, E. L. 1898 A, p. 838.
    Oligocene (White River); South Dakota.

Leptauchenia nitida Leidy.

Leidy, J. 1869 A, pp. 129, 381, pl. xii, figs. 21, 22.
Cope, E. D. 1884 M, p. 546.
    1884 T, p. 282.
Leidy, J. 1871 C, p. 348.
Matthew, W. D. 1899 A, p. 61.
Roger, O. 1896 A, p. 215.
Scott, W. B. 1890 B, pl. xvi, fig. 35.
Trouessart, E. L. 1898 A, p. 838.
    Oligocene (White River); South Dakota.

CYCLOPIDIUS Cope.   Type *C. simus* Cope.

Cope, E. D. 1878 C, p. 221.
    1878 C, p. 219 (Pithecistes, type *P. brevifacies*); p. 220 (Brachymeryx, type *B. feliceps*=Cyclopidius simus, juv.).
    1878 L. (Cyclopidius, Brachymeryx, Pithecistes.)
    1884 M. pp. 504, 546 (Cyclopidius); pp. 504, 557 (Pithecistes).
    1884 T.
    1886 B, p. 359.
    1887 B, p. 386. (Cyclopidius, Pithecistes.)
    1888 X, p. 1093. (Cyclopidius, Pithecistes.)
Matthew, W. D. 1899 A, pp. 24, 73.
Nicholson and Lydekker 1889 A, p. 1327. (Cyclopidius, Pithecistes.)
Scott, W. B. 1890 B, p. 356 (Cyclopidius); p. 357 (Pithecistes).
    1891 B, p. 370. (Pithecistes.)
    1891 D, p. 90. (Cyclopidius, Pithecistes.)
    1892 A, p. 435. (Cyclopidius, Pithecistes.)
    1895 C, p. 162 (Cyclopidius); p. 164 (Pithecistes).
Steinmann and Döderlein 1890 A, p. 800.
Zittel, K. A. 1893 B, p. 356. (Cyclopidius, Pithecistes.)

Cyclopidius brevifacies (Cope).

Cope, E. D. 1878 C, p. 219. (Pithecistes.)
    1884 M, p. 558. (Pithecistes.)
    1884 T, p. 212. (Pithecistes.)
    1886 B, p. 359. (Pithecistes.)
    1886 H. (Pithecistes.)
Matthew, W. D. 1899 A, p. 73. (Cyclopidius.)
Roger, O. 1896 A, p. 215. (Pithecistes.)

Scott, W. B. 1890 B, pl. xv, figs. 20, 21. (Pithecistes.)
    1893 B, p. 659. (Pithecistes.)
Trouessart, E. L. 1898 A, p. 839. (Pithecistes.)
    Miocene (Loup Fork); Montana.

Cyclopidius decedens (Cope).

Cope, E. D. 1884 M, p. 558. (Pithecistes.)
    1884 T, p. 282. (Pithecistes.)
    1886 B, p. 359. (Pithecistes.)
    1886 H. (Pithecistes.)
Matthew, W. D. 1899 A, p. 73. (Cyclopidius.)
Roger, O. 1896 A, p. 215. (Pithecistes.)
Scott, W. B. 1893 B, p. 659. (Pithecistes.)
Trouessart, E. L. 1898 A, p. 839. (Pithecistes.)
    Miocene (Loup Fork); Montana.

Cyclopidius emydinus Cope.

Cope, E. D. 1884 M, p. 553.
    1884 T, p. 282.
    1886 B, p. 359.
    1886 H.
    1888 X, pl. xxviii.
Roger, O. 1896 A, p. 215.
Scott, W. B. 1890 B, pl. xv, figs. 17–19
    1893 B, p. 659.
Trouessart, E. L. 1898 A, p. 839.
Zittel, K. A. 1893 B, p. 356, fig. 291.
    Miocene (Deep River); Montana.

Cyclopidius heterodon Cope.

Cope, E. D. 1878 C, p. 222. (Cyclopidius.)
    1884 M, p. 559. (Pithecistes.)
    1884 T, p. 282. (Pithecistes.)

Cope, E. D. 1886 B, p. 359. (Pithecistes.)
1886 H. (Pithecistes.)
Matthew, W. D. 1899 A, p. 73. (Cyclopidius.)
Roger, O. 1896 A, p. 215. (Pithecistes.)
Scott, W. B. 1893 B, p. 659. (Pithecistes.)
Trouessart, E. L. 1898 A, p. 839. (Pithecistes.)
Miocene (Loup Fork); Montana.

Cyclopidius incisivus Scott.

Scott, W. B. 1893 B, p. 659.
Matthew, W. D. 1899 A, p. 73.
Roger, O. 1896 A, p. 215.
Scott, W. B. 1895 C, p. 163.

Trouessart, E. L. 1898 A, p. 839.
Miocene (Loup Fork); Montana.

Cyclopidius simus Cope.

Cope, E. D. 1878 C, p. 221.
1878 C, p. 220. (Brachymeryx feliceps.)
1884 M, p. 547.
1884 T, p. 282.
1886 H.
Roger, O. 1896 A, p. 215.
Scott, W. B. 1893 B, p. 659.
Trouessart, E. L. 1898 A, p. 839.
Miocene (Deep River); Montana.

HADROHYUS Leidy. Type *H. supremus* Leidy.

Leidy, J. 1871 M, p. 248.
1873 B, p. 222.
A genus of uncertain position; founded on unsatisfactory material.

Hadrohyus supremus Leidy.

Leidy, J. 1871 M, p. 248.
1873 B, pp. 222, 331, pl. xvii, fig. 26.
Miocene; Oregon.

LEPTOMERYCINÆ.

Cope, E. D. 1879 B, p. 66. (Hypertragulidæ, in part.)
1889 G, p. 121. (Hypisodontidæ.)
Roger, O. 1896 A, p. 226.

Scott, W. B. 1899 A, p 15. (Leptomerycidæ.)
Trouessart, E. L. 1898 A, p. 863.
Zittel, K. A. 1893 B, p. 389.

LEPTOMERYX Leidy. Type *L. evansi* Leidy.

Leidy, J. 1853 D, p. 394.
Cope, E. D. 1873 T, p. 8. (Trimerodus, type *T. cedrensis*.)
1874 E, p. 110.
1874 P, p. 26.
1879 A, p. 44.
1884 H, p. 23.
1884 DD.
1885 B, p. 484.
1887 B, p. 389.
1889 G, p. 121.
Farr, M. S. 1896 A, p. 163.
Flower and Lydekker 1891 A, p. 307.
Leidy, J. 1869 A, p. 165.
1871 C, p. 351.
Lydekker, R. 1887 A, p. 332.
Nicholson and Lydekker 1889 A, p. 1333.
Osborn and Wortman 1892 B.
Rütimeyer, L. 1883 A, p. 98.
Scott, W. B. 1891 A, p. 3.
1891 B, p. 342.
1891 C, p. 63.
1892 A, p. 436.
1895 A.
1898 C, p. 77.
1899 A, pp. 15–111.
Steinmann and Döderlein 1890 A, p. 802.
Wortman, J. L. 1893 A.
Zittel, K. A. 1893 B, p. 389.

Leptomeryx esulcatus Cope.

Cope, E. D. 1889 I, p. 154.
1891 A, p. 22, pl. xiv, fig. 5.
Matthew, W. D. 1899 A, p. 61.
Roger, O. 1896 A, p. 226.
Trouessart, E. L. 1898 A, p. 863.
Oligocene (White River); Canada.

Leptomeryx evansi Leidy.

Leidy, J. 1853 D, p. 394.
Cope, E. D. 1873 T, p. 8. (Trimerodus cedrensis.)
1873 CC, p. 14. (Trimerodus cedrensis.)
1874 B, pp. 464, 503.
1874 P, p. 27.
1879 B, p. 66.
1879 D, p. 63.
1884 H, p. 23.
1885 B, p. 484.
1889 G, p. 122.
Hayden, F. V. 1858 B, p. 157.
King, C. 1878 A, p. 411.
Leidy, J. 1854 E, p. 157.
1857 D, p. 176. (Dorcatherium.)
1869 A, pp. 165, 383, pl. xiv, figs 1–8.
1870 P, p. 112.
1871 C, p. 351.
1873 B, pp. 216, 317. (Except John Day specimens.)
Lydekker, R. 1887 A, p. 333.
Major, Fors. 1873 A, p. 105.
Matthew, W. D. 1899 A, p. 61.
Roger, O. 1896 A, p. 226.
Scott, W. B. 1891 B, p. 343, figs. D–I.
1891 A, p. 15, pl. i, figs. 1, 2.
Scott and Osborn 1887 B, p. 156.
Trouessart, E. L. 1898 A, p. 863.
Oligocene (White River): South Dakota, Colorado, Nebraska.

Leptomeryx mammifer Cope.

Cope, E. D. 1886 U, p. 84.
Ami, H. M. 1891 A.
Cope, E. D. 1885 H, p. 163. (No description.)
1889 I, p. 154.

Cope, E. D. 1891 A, p. 22, pl. xiv, figs. 6, 7.
Matthew, W. D. 1899 A, p. 61.
Roger, O. 1896 A, P. 226.
Trouessart, E. L. 1898 A, p. 863.
Oligocene (White River); Canada, Colorado.

Leptomeryx semicinctus Cope.

Cope, E. D. 1889 I, p. 154.
1891 A, p. 23, pl. xiv, fig. 8.
Matthew, W. D. 1899 A, p. 61.
Roger, O. 1896 A, p. 226.
Trouessart, E. L. 1896 A, p. 863.
Oligocene (White River); Canada.

HYPISODUS Cope.　　Type *H. ringens* Cope.

Cope, E. D. 1873 CC, p. 5.
1873 R, p. 419.
1874 B, p. 501.
1874 E.
1884 H, p. 25.
1884 DD, p. 1036.
1885 B, p. 486.
1887 B, p. 389.
1889 G, p. 121.
Scott, W. B. 1891 B, p. 351.
1899 A, pp. 19, 112.
Zittel, K. A. 1893 B, p. 389.

Hypisodus minimus Cope.

Cope, E. D. 1873 T, p. 8. (Leptauchenia.)

Cope, E. D. 1873 R, p. 419.
1873 CC, p. 7 (H. ringens); p. 14 (Leptauchenia minima.)
1874 B, p. 501. (H. minimus, with H. cingens as synonym.)
1874 P, p. 26.
1884 H, p. 25.
1885 B, p. 486.
1889 G, p. 122.
Matthew, W. D. 1899 A, p. 61.
Roger, O. 1896 A, p. 226.
Scott and Osborn 1887 B, p. 157.
Trouessart, E. L. 1898 A, p. 863. (With H. cingens as synonym.)
Oligocene (White River); Colorado.

OROMERYX Marsh.　　Type *O. plicatus* Marsh.

Marsh, O. C. 1894 L, p. 269.
Cope, E. D. 1894 E, p. 869.
Marsh, O. C. 1877 E, p. 364. (Insufficient definition; no species.)
Scott, W. B. 1898 C, pp. 77, 78.
1899 A, pp. 83, 114.
Scott and Osborn 1887 A, p. 259. (Name only.)
Zittel, K. A. 1893 B, p. 361.

Oromeryx plicatus Marsh.

Marsh, O. C. 1894 L, p. 269, fig. 23.
Matthew, W. D. 1899 A, p. 50.
Roger, O. 1896 A, p. 215.
Trouessart, E. L. 1898 A, p. 839.
Eocene (Uinta); Utah.

CAMELOMERYX Scott.　　Type *C. longiceps* Scott.

Scott, W. B. 1898 C, p. 77.
1899 A, pp. 67, 82, 114.

Camelomeryx longiceps Scott.

Scott, W. B. 1898 C, p. 78.

Matthew, W. D. 1899 A, p. 50.
Scott, W. B. 1899 A, p. 67, pl. iii, figs. 15-18.
Wortman, J. L. 1898 A, p. 97. (Bunomeryx montanus, in part.)
Eocene (Uinta); Wyoming.

LEPTOREODON Wort.　　Type *L. marshi* Wort.

Wortman, J. L. 1898 A, p. 95.
Scott, W. B. 1898 C, p. 75. (Merycodesmus, type M. gracilis Scott.)
1899 A, pp. 51, 64, 113.

Leptoreodon gracilis Scott.

Scott, W. B. 1898 C. p. 76. (Merycodesmus.)
Matthew, W. D. 1899 A, p. 50. (Syn. of L. marshi.)
Scott, W. B. 1899 A, pp. 52, 67, pl. ii, figs. 10-14.
Eocene (Uinta); Utah.

Leptoreodon marshi Wort.

Wortman, J. L. 1898 A, p. 95, fig. 1.
Matthew, W. D. 1899 A, p. 50.
Scott, W. B. 1899 A, pp. 52, 67.
Trouessart, E. L. 1899 A, p. 1348.
Eocene (Uinta); Utah.

STIBARUS Cope.　　Type *S. obtusilobus* Cope.

Cope, E. D. 1873 T, p. 3.
1874 B, p. 503.
1886 C, pp. 614, 619.
1893 G, p. 148.
Zittel, K. A. 1893 B, p. 361.
The systematic position of this genus is doubtful.

Stibarus obtusilobus Cope.

Cope, E. D. 1873 T, p. 3.

Cope, E. D. 1873 CC, p. 9.
1874 B, p. 503.
1886 C. p. 618. ("Stibarus.")
Matthew, W. D. 1899 A, p. 61.
Roger, O. 1896 A, p. 215.
Trouessart, E. L. 1898 A, p. 840.
Oligocene (White River); Colorado.

PROTOCERATINÆ.

Zittel, K. A. 1893 B, p. 405.
Lydekker, R. 1896 B, p. 376. (Protoceratidæ.)
Marsh, O. C. 1891 A, p. 82. (Protoceratidæ.)
1897 D, p, 165. (Protoceratidæ.)
Osborn, H. F. 1892 B, p. 353. (Protoceratidæ.)

Osborn, H. F. 1893 I, p. 322. (Protoceratidæ.)
Roger, O. 1896 A, p. 237.
Scott, W. B. 1899 A, p. 19.
Woodward, A. S. 1898 B, p. 361. (Protoceratidæ.)

PROTOCERAS Marsh. Type *P. celer* Marsh.

Marsh, O. C. 1891 A, p. 81.
Cope, E. D. 1893 G, p. 147.
Lydekker, R. 1896 B, p. 376.
Marsh, O. C. 1897 D, p. 165.
Osborn, H. F. 1893 D, p. 42.
1893 E.
1893 I, p. 321.
Osborn and Wortman 1892 B, p. 351.
Scott, W. B. 1895 A.
1896 C. p. 77.
1899 A, pp. 19-112.
Woodward, A. S. 1898 B, p. 362.
Wortman, J. L. 1893 A.
1896 A, p. 99.
Zittel, K. A. 1893 B, p. 405.

Protoceras celer Marsh.

Marsh, O. C. 1891 A, p. 81.
Cope, E. D. 1893 G, p. 147, pls. i, ii.
Dean, B. 1893 A.
Marsh, O. C. 1893 D,.p 407. pl. vii.
1897 D, p. 169, pls. ii-v; pl. vi, fig. 1; pl. vii,
figs. 3, 4; text figs. 4, 6, 7.
Matthew, W. D. 1899 A, p. 61.

Osborn, H. F. 1893 I, p. 321, figs. 1-3.
1893 L, p. 128.
1896 C, pp. 712, 715, illus. 8.
1898 Q. p. 16.
Osborn and Wortman 1892 B, pp. 353-371, figs. 1-6.
Roger, O. 1896 A, p. 236.
Scott, W. B. 1895 A, pls. xx-xxii.
Trouessart, E. L. 1898 A, p. 864.
Woodward, A. S. 1898 B, p. 362, fig. 206.
Zittel, K. A. 1893 B, p. 405, fig. 388.
Oligocene (White River); South Dakota.

Protoceras comptus Marsh.

Marsh, O. C. 1894 F, p. 93.
1897 D, p. 172, pl. vi, fig. 2.
Matthew, W. D. 1899 A, p. 61.
Roger, O. 1896 A, p. 237.
Trouessart, E. L. 1898 A, p. 864.
Oligocene (White River); South Dakota.

Protoceras nasutus Marsh.

Marsh, O. C. 1897 D, p. 168; fig. 5.
Matthew, W. D. 1899 A, p. 61.
Oligocene (White River); South Dakota.

CALOPS Marsh. Type *C. cristatus* Marsh.

Marsh, O. C. 1894 F, p. 94.
1897 D, p. 174.

Calops consors Marsh.

Marsh, O. C. 1897 D, p. 175, pl. vi, figs. 1, 2.
Matthew, W. D. 1899 A, p. 61.
Oligocene (White River); South Dakota.

Calops cristatus Marsh.

Marsh, O. C. 1894 F, p. 94.
1894 L, p. 273.
Matthew, W. D. 1899 A, p. 61.
Roger, O. 1896 A, p. 237.
Trouessart, E. L. 1898 A, p. 864.
Oligocene (White River); South Dakota.

CAMELIDÆ.

Ameghino, F. 1889 A, p. 579.
Baur, G. 1885 I, p. 196.
Branco, W. 1897 A, p. 24.
Cope, E. D. 1874 B, p. 464.
1874 M.
1875 Q.
1876 G, p. 146.
1877 K, p 342.
1881 J, p. 646.
1884 G, p. 16.
1884 H, p. 26.
1886 C.
1887 B, pp. 379, 391.
1887 S, p. 222.
1888 X, p. 1080.
1889 C,
1889 G, p. 119.
Flower, W. H. 1876 B, xiii, p. 328.
Flower and Lydekker 1891 A, p. 295.
Gill T. 1872 B, p. 71.
Huxley, T. H. 1872 A, p. 328.
Leidy, J. 1871 C, p. 349.

Lydekker, R. 1885 C, p. 139.
1896 B, p. 163.
Marsh, O. C. 1877 E.
Matthew, W. D. 1899 A, p. 22, *et seq.*
Nicholson and Lydekker 1889 A, p. 1334.
Osborn, H. F. 1900 C, p. 568.
Owen, R. 1845 B, p. 528.
1868 A, p. 349.
1870 C.
Roger, O. 1896 A, p. 215.
Rütimeyer, L. 1877 A.
1878 A, p. 19.
1883 A, p. 9. (Camelina.)
Schmidt, O. 1886 A, p. 154.
Scott, W. B. 1891 A.
1899 A, pp. 22, 116.
Steinmann and Döderlein 1890 A, p. 800.
Tornier, G. 1888 A.
Wilckens, M. 1885 C.
Wortman, J. L. 1898 A.
Zilliken, J. E. 1879 A.
Zittel, K. A. 1893 B, p. 357.

Bull. 179——43

LEPTOTRAGULINÆ.

Cope, E. D. 1879 B, p. 66. (Hypertragulidæ, in part.)
Roger, O. 1896 A, p. 215.

Trouessart, E. L. 1898 A, p. 839.
Zittel, K. A. 1893 B, p. 361.

LEPTOTRAGULUS Scott and Osb.

Scott and Osborn 1887 A, p. 258.
Cope, E. D. 1889 G, p. 119.
    1890 C, p. 471.
    1894 E, p. 869.
Flower and Lydekker 1891 A, p. 304.
Nicholson and Lydekker 1889 A, p. 1334.
Osborn, H. F. 1893 D, p. 42.
Scott, W. B. 1890 A, p. 479.
    1891 A, pp. 10, 47.
    1892 A, p. 435.
    1898 C, p. 75.
    1899 A, pp. 47, 113.
Steinmann and Döderlein 1890 A, p. 800.
Trouessart, E. L. 1898 A, p. 840.

Type L. proavus Scott and Osb.

Wortman, J. L. 1898 A, p. 103. (Parameryx, in part.)
Zittel, K. A. 1893 B, p. 361.

Leptotragulus proavus Scott and Osb.

Scott and Osborn 1887 A, p. 258.
Matthew, W. D. 1899 A, p. 50.
Roger, O. 1896 A, p. 215.
Scott and Osborn 1890 A, p. 479, pl. vii, figs. 9–16.
Trouessart, E. L. 1898 A, p. 840.
Wortman, J. L. 1898 A, p. 103. (Parameryx.)
Zittel, K. A. 1893 B, p. 361.
    Eocene (Uinta); Wyoming.

PARAMERYX Marsh.

Marsh, O. C. 1894 L, p. 269.
Cope, E. D. 1894 E, p. 869. (Syn. of Leptotragulus.)
Marsh, O. C. 1877 E, p. 364. (Insufficient definition; no species.)
Scott, W. B. 1890 A, p. 488.
    1898 C, pp. 74, 77.
    1899 A, p. 47. (Syn.? of Leptotragulus.)
Scott and Osborn 1887 A, p. 259.
Wilckens, M. 1885 C, p. 418.
Wortmann, J. L. 1898 A, p. 103.

Type P. lævis Marsh.

Cope, E. D. 1894 E, p. 869.
Matthew, W. D. 1899 A, p. 50.
Roger, O. 1896 A, p. 215. (Leptotragulus.)
Trouessart, E. L. 1898 A, p. 840. (Leptotragulus.)
    Eocene (Uinta); Wyoming.

Parameryx sulcatus Marsh.

Marsh, O. C. 1894 L, p. 269.
Matthew, W. D. 1899 A, p. 50.
Roger, O. 1896 A, p. 215. (Leptotragulus.) ·
Trouessart, E. L. 1898 A, p. 840. (Leptotragulus.)
    Eocene (Uinta); Wyoming.

Parameryx lævis Marsh.

Marsh, O. C. 1894 L, p. 269, figs. 20, 21.

ITHYGRAMMODON Osb., Scott, and Speir.

Osborn, Scott, and Speir 1878 A, p. 56.
Cope, E. D. 1886 C, p. 613.
    1887 B, p. 391.
Scott, W. B. 1891 A, p. 9.
Zittel, K. A. 1893 B, p. 361.

Type I. cameloides Osb., Scott, and Speir.

Cope, E. D. 1886 C, p. 613.
    1887 B, p. 391.
Matthew, W. D. 1899 A, p. 47.
Roger, O. 1896 A, p. 215.
Trouessart. E. L. 1898 A, p. 840.
    Eocene (Bridger); Wyoming.

Ithygrammodon cameloides Osb., Scott, and Speir.

Osborn, Scott, and Speir 1878 A, p. 57, pl. x, figs. 1–4.

HYPERTRAGULUS Cope.

Cope, E. D. 1873 R, p. 419.
    1874 B, p. 502.
    1874 E.
    1874 P, p. 26.
    1884 H, p. 24.
    1885 B, p. 485.
    1887 B, p. 389.
    1889 G, p. 120.
Flower and Lydekker 1891 A, p. 307.
Nicholson and Lydekker 1889 A, p. 1334.
Rütimeyer, L. 1883 A, p. 99.
Schlosser, M. 1886 B, p. 75.
Scott, W. B. 1891 D, p. 90.
    1895 A.
    1899 A, pp. 17–112.
Zittel, K. A. 1893 B, p. 389.

Type H. calcaratus Cope.

Hypertragulus calcaratus Cope.

Cope, E. D. 1873 T, p. 7. (Leptauchenia.)
    1873 R, p. 419. (H. calcaratus. H. tricostatus.)
    1873 CC, p. 14. (Leptauchenia.)
    1874 B, p. 502 (H. calcaratus); p. 503 (H. tricostatus.)
    1874 P, p. 26 (H. calcaratus); p. 27 (H. tricostatus.)
    1879 B, p. 66.
    1879 D, p. 63. (In part.)
    1884 H, p. 24. (In part.)
    1885 B, p. 485. (In part.)
Matthew, W. D. 1899 A, p. 61. (H. calcaratus, H. tricostatus.)

HAY.] CATALOGUE. 675

Roger, O. 1896 A, p. 226. (H. calc., H. tricostatus.)
Scott, W. B. ?1893 B, p. 659.
  ?1895 C, p. 167.
  1899 A, p. 17. (In part.)
Trouessart, E. L. 1898 A, p. 863. (H. calc., H. tricostatus.)
  Oligocene (White River); Colorado, South Dakota, Montana, Canada: (Deep River beds?) Montana?

Hypertragulus hesperius sp. nov.

Cope, E. D. 1879 B, p. 66. (H. calcaratus and Leptomeryx evansi, John Day specimens.)
  1879 D, p. 63. (H. calcaratus, in part.)
  1884 H, p. 24. (H. calcaratus, in part.)
  1885 B, p. 485. (H. calcaratus, in part.)
  1889 G, p. 122, pl. vi. (H. calcaratus.)

Leidy, J. 1873 B, pp. 216, 317. (Leptomeryx evansi, John Day specimens.)
Scott, W. B. ?1893 B, p. 659. (H. calcaratus.)
  ?1895 C, p. 167. (H. calcaratus.)
  1899 A, p. 17, in part; pl. i, figs. 3, 4. (H. calcaratus.)
  Miocene (John Day); Oregon: (Deep River?); Montana?

Hypertragulus transversus Cope.

Cope, E. D. 1889 I, p. 154.
  1891 A, p. 42, pl. xiv, fig. 4.
Matthew, W. D. 1899 A, p. 61.
Roger, O. 1896 A, p. 226.
Trouessart, E. L. 1898 A, p. 864.
  Oligocene (White River); Canada.

POËBROTHERIINÆ.

Unless otherwise indicated, the following writers regard this group as a family, the Poebrotheriidæ.
Cope, E. D. 1874 P, p. 26.
  1874 B, p. 464.
  1884 M, p. 508.
  1886 C, p. 613.
  1887 B, pp. 379, 390.

Cope, E. D. 1888 X, p. 1080.
  1889 G, p. 119.
Lydekker, R. 1885 C, p. 149.
Nicholson and Lydekker 1889 A, p. 1334.
Roger, O. 1896 A, p. 216. (Poëbrotherinæ.)
Trouessart, E. L. 1898 A, p. 840. (Poëbrotherinæ.)
Zittel, K. A. 1893 B, p. 361. (Poëbrotherinæ.)

PROTYLOPUS Wort.   Type *P. petersoni* Wort.

Wortman, J. L. 1898 A, pp. 104, 136.
Scott, W. B. 1899 A, pp. 22–113.

Protylopus petersoni Wort.

Wortman, J. L. 1898 A, p. 104, figs. 3–6.

Matthew, W. D. 1899 A, p. 50.
Scott, W. B. 1899 A, p. 22, pl. ii, figs. 5–9.
Trouessart, E. L. 1899 A, p. 1349.
  Eocene (Uinta); Utah.

POËBROTHERIUM Leidy.   Type *P. wilsoni* Leidy.

Leidy, J. 1847 B, p. 322.
Baur, G. 1885 H, p. 196.
Branco, W. 1897 A, p. 24.
Bruce, A. T. 1883 A, p. 39.
Cope, E. D. 1874 B, p. 498.
  1874 E.
  1874 P, p. 26.
  1875 Q, p. 262.
  1876 G, p. 146.
  1884 H, pp. 25, 26.
  1885 B, pp. 482, 486.
  1886 C, p. 618.
  1887 B, p. 390.
Flower and Lydekker 1891 A, p. 304.
Leidy, J. 1848 A.
  1852 J, p. 571. (Pœbrotherium.)
  1852 L, p. 64. (Pœluotherium, misprint.)
  1854 A, p. 113.
  1869 A, p. 141.
  1871 C, p. 350.
Lydekker, R. 1885 C, p. 149.
Marsh, O. C. 1877 E.
Nicholson and Lydekker 1889 A, p. 1334.
Schlosser, M. 1886 B, p. 48.
Scott, W. B. 1890 A, p. 483.

Scott, W. B. 1891 A.
  1891 D, p. 90.
  1892 A, p. 435.
  1895 C, p. 179.
  1898 C, p. 74. (Parameryx.)
  1899 A, p. 23 *et seq.*
Steinmann and Doderlein 1898 A, p. 800.
Trouessart, E. L. 1898 A, p. 840.
Wilckens, M. 1885 C, p. 419.
Woodward, A. S. 1898 B, p. 358.
Wortman, J. L. 1893 A.
  1898 A, pp. 110, 136.
Zittel, K. A. 1893 B, p. 362.

Poëbrotherium eximium sp. nov.

Wortman, J. L. 1898 A, p. 111, fig. 7. (P. wilsoni, not of Leidy.)
  Oligocene (White River); South Dakota.

Poëbrotherium labiatum[1] Cope.

Cope, E. D. 1881 I, p. 271, fig. 4.
  1881 J, p. 541, fig. 4.
  1886 C, p. 618, fig. 7.
  1887 S, p. 271, fig. 4.
Matthew, W. D. 1899 A, p. 60.

[1] Trouessart refers to Bull. U. S. Geog. and Geol. Survey, vol. I, p. 26, as the place of original description of this species; but it is not mentioned there. Cope's reference to ' Hayden iv, pl. cxv," is to an unpublished work.

Roger, O. 1896 A, p. 216.
Scott, W. B. 1891 A, pl. i, fig. 7; pl. ii, figs. 11-39;
pl. iii, figs. 40-42, 48-51.
Trouessart, E. L. 1898 A, p. 840.
Wortman, J. L. 1898 A, p. 113, pl. xi, figs. C, D.
Zittel, K. A. 1893 B, pp. 359, 362, fig. 292.
   Oligocene (White River); Colorado, South
   Dakota.

Poëbrotherium wilsoni Leidy.

Leidy, J. 1847 B, p. 322.
Cope, E. D. 1874 B, p. 500.   *
   1874 P, p. 24.
   1886 C, p. 618, figs. 3, 9.
   1889 G, 114, fig. 12. (P. vilsoni.)
Hayden, F. V. 1858 B, p. 157.
King, C. 1878 A, p. 411.
Leidy, J. 1848 A.
   1852 J, p. 571. (Pœbrotherium.)

Leidy, J. 1852 L, p. 64. (Pœluotherium, misprint.)
   1854 A, p. 19. pl. i, figs. 1-4.
   1854 E, p. 157.
   1869 A, pp. 141, 381, pl. xiii, figs. 5-7. (Pœbrotherium.)
   1871 C, p. 350.
Matthew, W. D. 1899 A, p. 61.
Roger, O. 1896 A, p. 216.
Scott, W. B. 1890 A, p. 483.
   1891 A, pl. i, figs. 1-3, 8, 9; pl. ii, fig. 1: pl.
   iii, figs. 44-47.
Trouessart, E. L. 1898 A, p. 840.
Wilckens, M. 1885 C, p. 419.
Wortman, J. L. 1898 A, p. 111, figs. 8-10; pl. xi,
   fig. B.
Zittel, K. A. 1893 B, p. 362, fig. 294.
   Oligocene (White River); South Dakota, Nebraska.

PROTOMERYX Leidy.   Type *P. halli* Leidy.

Leidy, J. 1856 J, p. 163.
Cope, E. D. 1886 C, p. 618 (Gomphotherium, type
G. sternbergii; preoccupied by Gomphotherium
Burmeister).
Leidy, J. 1869 A, p. 160.
   1871 C, p. 350.
Nicholson and Lydekker 1889 A, p. 1334. (Gom-
photherium.)
Scott, W. B. 1891 A. (Protomeryx, Gomphothe-
rium.)
   1899 A, pp. 17, 46, 113. (Gomphotherium.)
Wilckens, M. 1885 C, p. 421.
Wortman, J. L. 1898 A, pp. 114, 137. (Gomphothe-
rium.)
Zittel, K. A. 1893 B, p. 362. (Gomphotherium.)
   In case future discoveries shall prove that
Cope's *Gomphotherium sternbergii* is generically
different from Leidy's *Protomeryx halli* a new
generic name must replace *Gomphotherium*.

Protomeryx cameloides (Wort.)

Wortman, J. L. 1898 A, p. 115, pl. xi, fig. F; text
figs. 15-19. (Gomphotherium.)
Douglass, E. 1900 A, p. 14. (Gomphotherium.)
Matthew, W. D. 1899 A, p. 64.
Trouessart, E. L. 1899 A, p. 1349. (Gomphothe-
rium.)
   Miocene (John Day); Oregon.

Protomeryx halli Leidy.

Leidy, J. 1856 J, p. 163.

Cope, E. D. 1874 P, p. 23. (Poëbrotherium.)
Hayden, F. V. 1858 B, p. 157.
King, C. 1878 A, p. 411.
Leidy, J. 1869 A, pp. 160, 382, pl. xv, figs. 8, 9.
   1871 C, p. 351.
Matthew, W. D. 1899 A, p. 61.
Roger, O. 1896 A, p. 216.
Trouessart, E. L. 1898 A, p. 841.
   Oligocene (White River); South Dakota, Colo-
   rado.

Protomeryx serus (Douglass).

Douglass, E. 1900, p. 12, pl. i, fig. 1. (Gomphotherium.)
   Miocene: Montana.

Protomeryx sternbergii (Cope.)

Cope, E. D. 1879 B, p. 59. (Poëbrotherium.)
   1884 H, p. 25. (Poebrotherium.)
   1885 B, p. 486. (Poëbrotherium.)
   1886 C, p. 618, fig 10. (Gomphotherium.)
Matthew, W. D. 1899 A p. 64. (Protomeryx.)
Roger, O. 1896 A, p. 216. (Gomphotherium.)
Trouessart, E. L. 1898 A, p. 840. (Gomphotherium.)
Wortman, J. L. 1898 A, p. 115, pl. xi, fig. E, text
figs. 11-14.
Zittel, K A. 1893 B, p. 362, fig. 295. (Gomphotherium.)
   Miocene (John Day); Oregon.

MIOLABINÆ nom. nov.

Cope, E. D. 1884 G, p. 16. (Protolabidæ.)
   1886 C, p. 613. (Protolabidæ.)
   1887 B, pp. 379, 391. (Protolabidæ.)
   1888 X, p. 1080. (Protolabidæ.)

Roger, O. 1896 A, p. 217. (Protolabinæ.)
Trouessart, E. L. 1898 A, p. 841. (Protolabinæ)
Zittel, K. A. 1893 B, p. 363. (Protolabinæ.)

MIOLABIS Hay.   Type *Protolabis transmontanus* Cope.

Hay, O. P. in Matthew, W. D. 1899 A, pp. 24, 74.
   1899 A, p. 593.
Matthew, W. D. 1899 A, pp. 24, 74.
Wortman, J. L. 1898 A, p. 120. (Protolabis.)

Miolabis transmontanus (Cope).

Cope, E. D. 1879 B, p. 67. (Protolabis)
   1879 N, p 131. (Protolabis.)
   1886 B, p. 359. (Protolabis.)

Cope, E. D. 1886 C, p. 620, fig. 11.   (Protolabis.)
1886 H.   (Protolabis.)
1887 S. p. 223, fig. 13.   (Protolabis.)
Hay, O. P. 1899 A. p. 593.
Matthew, W. D. 1899 A, p. 74.

Roger, O. 1896 A, p. 216.   (Protolabis.)
Trouessart, E. L. 1898 A, p. 841.   (Protolabis.)
Wortman, J. L. 1898 A, p. 122.   (Protolabis.)
Zittel, K. A. 1893 B, p. 363, fig. 296.   (Protolabis.)
Miocene (Loup Fork); Oregon.

PROCAMELUS Leidy.   Type *P. occidentalis* Leidy.

Leidy, J. 1858 E, p 23.
Baur, G. 1885 I.
Branco, W. 1897 A, p. 24.
Cope, E. D. 1873 R, p. 420.
1874 P, p. 20.
1875 F, p. 75.
1875 Q, p. 262.
1876 G, p. 144.   (Procamelus; Protolabis, type *P. heterodontus*.)
1877 K, p. 325.   (Procamelus, Protolabis.)
1884 G, p. 16.
1886 C, p. 612 (Procamelus); p. 613 (Protolabis).
1887 B, p. 391 (Protolabis); p. 393 (Procamelus).
1889 C.
1893 A, p. 36.
Gaudry, A. 1878 B, p. 123.
Leidy, J. 1859 A, p. 147 (Procamelus); p. 158 (Homocamelus, type *H. caninus*).
1871 C, p. 350.   (Homocamelus.)
1873 A, p. 312.   (Protocamelus.)
1873 B, p. 258 (Procamelus, or Protocamelus).
1873 C, p. 15.   (Protocamelus.)
Marsh, O. C. 1877 E.
Nicholson and Lydekker 1889 A, p. 1334.   (Procamelus, Protolabis.)
Osborn, Scott, and Speir 1878 A, pl. x, fig. 6.   (Protolabis.)
Schlosser, M. 1886 B, p. 48 (Procamelus, Protolabis); p. 49 (Homocamelus).
Scott, W. B. 1891 A, p. 49 (Procamelus, Protolabis); p. 50 (Homocamelus).
1891 D, p. 90.   (Procamelus, Protolabis.)
1892 A, p. 486.
1895 C, p. 179.   (Protolabis.)
Steinmann and Doderlein 1890 A, p. 800 (Protolabis); p. 801 (Procamelus).
Wilckens M. 1885 C, p. 421 (Procamelus); p. 424 (Homocamelus).
Woodward, A. S. 1898 B, p. 359.
Wortman, J. L. 1898 A, pp. 120, 122, 139.
Zittel, K. A. 1893 B, p. 363.   (Protolabis, Procamelus.)

Procamelus altus Marsh.

Marsh, O. C. 1894 L, p. 274, figs. 33, 34.
Cope, E. D. 1894 E, p. 869.   ("Procamelus.")
Matthew, W. D. 1899 A, p. 73.   (Syn. of P. robustus.)
Roger, O. 1896 A, p. 216.
Trouessart, E. L. 1898 A, p. 841.
Miocene; Oregon, Colorado.

Procamelus angustidens Cope.

Cope, E. D. 1874 P, p. 20.
1874 B, p. 529.
1877 K, p. 327.

Cope, E. D. 1893 A, p. 37.
Matthew, W. D. 1899 A, p. 74.
Roger, O. 1896 A, p. 216.
Scott, W. B. 1891 A, p. 57, pl. i, fig. 6.
Trouessart, E. L. 1898 A, p. 842.
Wortman, J. L. 1898 A, p. 124.   (Syn. of P. occidentalis.)
Miocene (Loup Fork); Colorado, Kansas.

Procamelus fissidens Cope.

Cope, E. D. 1876 G, p. 145.
1877 K, p. 327.
1893 A, p. 37.
Matthew, W. D. 1899 A, p. 74.   (Miolabis?.)
Roger, O. 1896 A, p. 216.
Trouessart, E. L. 1898 A, p. 841.
Wilckens, M. 1885 C, p. 423.
Miocene (Loup Fork); Colorado.

Procamelus gracilis Leidy.

Leidy, J. 1858 F, p. 89.
Cope, E. D. 1875 F, p. 70.
1877 K, p. 328 (P. gracilis); p. 329, pl. lliii, fig. 2; pl. lxxvi; pl. lxxvii, figs. 1-3; pl. lxxviii, figs. 1-9; pl. lxxix (P. occidentalis, *fide* Cope 1893 A, p. 37.)
1893 A, p. 37.
Hayden, F. V. 1858 B, p. 157.
Leidy, J. 1869 A, pp. 155, 382, pl. xiv, fig. 15.
1871 C, p. 350.
1873 A, p. 312.   (Protocamelus.)
1873 C, p. 15.   (Protocamelus.)
Leidy and Lucas 1891 A, p. 54.
Matthew, W. D. 1899 A, p. 73.
Trouessart, E. L. 1898 A, p. 841.
Wilckens, M. 1885 C. p. 422.
Wortman, J. L. 1898 A, p. 124, fig. 20.
Miocene (Loup Fork); Nebraska, New Mexico, Texas.

Procamelus heterodontus Cope.

Cope; E. D. 1873 R, p. 420.
1871 B, p. 530.
1874 P, p. 20.
1876 G, p. 145.   (Protolabis.)
1877 K, p. 325.
1879 B, p. 68.   (Protolabis.)
1886 C, p. 620, fig. 11.
Hay, O. P. 1899 A, p. 593.
Matthew, W. D. 1899 A, p. 74.   (Protolabis.)
Roger, O. 1896 A, p. 216.
Trouessart, E. L. 1898 A, p. 841.   (Protolabis.)
Wilckens, M. 1885 C, p. 422.
Wortman, J. L. 1898 A, pp. 120, 123.   (Syn. of P. robustus.)
Miocene (Loup. Fork); Colorado.

Procamelus lacustris Douglass.

Douglass, E. 1900 A, p. 18, pl. i, fig. 2.
Miocene; Montana.

Procamelus leptognathus Cope.

Cope, E. D. 1893 A, p. 37.
Matthew, W. D. 1899 A, p. 73.
Roger, O. 1896 A, p. 216.
Trouessart, E. L. 1898 A, p. 842.
Miocene (Loup Fork); Texas.

Procamelus madisonius Douglass.   [1]

Douglass, E. 1900 A, p. 15, pl. ii.
Miocene; Montana.

Procamelus major (Leidy).

Leidy, J. 1886 B, p. 11.   (Auchenia.)
Cope, E. D. 1892 AA.   (Pliauchenia.)
Leidy, J. 1887 A.   (Auchenia.)
1889 A.   (Auchenia.)
1892 A.   (Auchenia.)
Leidy and Lucas 1896 A, pp. viii, 53; pl. xv, fig. 9; pl. xvii, figs. 1-16; pl. xviii, figs. 1-4.
Roger, O. 1896 A, p. 219.   (Auchenia.)
Trouessart, E. L. 1898 A, p. 845.   (Lama.)
Pliocene; Florida.

Procamelus minimus (Leidy).

Leidy, J. 1886 B, p. 11.   (Auchenia.)
Cope, E. D. 1892 AA.   (Auchenia.)
Leidy, J. 1889 D, p. 17, pl. iii, fig. 5.   (Auchenia.)
1892 A.   (Auchenia.)
Leidy and Lucas 1896 A, pp. viii, 53, pl. xvi, figs. 1-4.
Roger, O. 1896 A, p. 218.   (Auchenia.)
Trouessart, E. L. 1898 A, p. 845.   (Lama.)
Pliocene; Florida.

Procamelus minor (Leidy).

Leidy, J. 1886 B, p. 12.   (Auchenia.)
Cope, E. D. 1892 AA.   (Pliauchenia media.)
Leidy, J. 1889 A.   (Auchenia.)
1892 A.   (Auchenia.)
Leidy and Lucas 1896 A, pp. viii, 53, pl. xv, fig. 10; pl. xvii, figs. 17, 18; pl. xviii, figs. 5, 6.   (P. medius, error for P. minor.)
Roger, O. 1896 A, p. 218.   (Auchenia.)
Trouessart, E. L. 1898 A, p. 846.   (Lama.)
Pliocene; Florida.

Procamelus montanus (Douglass).

Douglass, E. 1900 A, p. 13, pl. iii.   (Protolabis.)
Miocene; Montana.

Procamelus occidentalis Leidy.

Leidy, J. 1858 E, p. 23.
Baur, G. 1885 I.
Cope, E. D. 1874 B, p. 531.
1874 P, p. 22.
1875 F, p. 70, pl. ii.
1876 G, pp. 144, 146.
1877 K, p. 329,[1] pl. liii, fig. 2; pl. lxxvi; pl. lxxv'i, figs. 1-3; pl. lxxviii, figs. 1-9; pl. lxxix.
1877 M.

Cope, E. D. 1881 J, p. 547, fig. 9.
1886 C, p. 621, figs. 2, 12, 14.
1887 S, p. 223, fig. 14.
1889 G, p. 114, fig. 10.
1893 A, p. 37.
Hayden, F. V. 1858 B, p. 157.
Leidy, J. 1869 A, pp. 151, 382, pl. ix, fig. 5; pl. xv, figs. 5-7 (P. occidentalis); pp. 158, 382, pl. xiv, figs. 16, 17 (Homocamelus caninus).
1871 C, p. 350.   (P. occidentalis, Homocamelus caninus.)
1873 A, p. 312.   (Protocamelus.)
1873 B, pp. 258, 317, pl. xx.
1873 C, p. 15.   (Protocamelus.)
Matthew, W. D. 1899 A, p. 73.
Osborn, Scott, and Speir 1878 A, pl. x, fig. 5.
Owen, R. 1870 C, p. 74.
Roger, O. 1896 A, p. 216 (P. occidentalis); p. 217 (Homocamelus caninus).
Scott, W. B. 1891 A, pp. 51, 57, pl. i, figs. 4, 5; pl. iii, fig. 52.
Trouessart, E. L. 1898 A, p. 841 (P. occidentalis). p. 842 (Homocamelus caninus).
Wilckens, M. 1885 C, p. 422 (P. occidentalis); p. 424 (Homocamelus caninus).
Woodward, A. S. 1898 B, p. 359, fig. 205.
Wortman, J. L. 1898 A, pp. 116, 123, 124, pl. xi, fig. G.
Zittel, K. A. 1893 B, p. 363, fig. 279.
Miocene (Loup Fork); South Dakota, Nebraska, New Mexico, Texas.

Procamelus robustus Leidy.

Leidy, J. 1858 F, p. 89.
Cope, E. D. 1881 C, p. 175.   (Protolabis prehensilis.)
1893 A, p. 37.
Hay, O. P. 1899 A, p. 593.
Hayden, F. V. 1858 B, p. 157.
King, C. 1878 A, p. 430.
Leidy, J. 1869 A, pp. 148, 381, pl. xv, figs. 1-4.
1871 C, p. 350.
1873 B, pp. 258, 317.
Matthew, W. D. 1899 A, p. 73.
Roger, O. 1896 A, p. 216.   (Protolabis robustus, P. prehensilis.)
Trouessart, E. L. 1898 A, p. 841.
Wilckens, M. 1885 C, p. 423.
Wortman, J. L. 1898 A, pp. 120, 123.
Miocene (Loup Fork); Nebraska, Kansas, Texas, New Mexico.

Procamelus virginiensis Leidy.

Leidy, J. 1873 C, p. 15.   (Protocamelus.)
1873 A, p. 311.   (Protocamelus.)
1873 B, pp. 259, 317, pl. xxvii, figs. 26-29.
Roger, O. 1896 A, p. 216.
Trouessart, E. L. 1898 A, p. 842.
Wilckens, M. 1885 C, p. 422.
Miocene; Virginia.

[1] This description and the accompanying figures are referred later by Professor Cope to *P. gracilis*. Dr. Wortman (1898 A, p. 124) appears to regard the New Mexican specimens as *P. occidentalis*.

CAMELOPS Leidy.   Type *C. kansanus* Leidy.

Leidy, J. 1854 F, p. 172.
Cope, E. D. 1884 G, p. 16. (Holomeniscus, type *Auchenia hesterna*.)
    1886 C, p. 621. (Holomeniscus.)
    1887 B, p. 392. (Holomeniscus.)
Cragin, F. W. 1892 A, p. 257. (Holomeniscus.)
Leidy, J. 1858 E. p. 24. (Megalomeryx, type *M. niobrarensis*.)
    1871 C, p. 349 (Camelops); p. 351 (Megalomeryx).
Scott, W. B. 1891 A, p. 52. (Holomeniscus.)
    1892 A, p. 436. (Holomeniscus.)
Steinmann and Döderlein 1890 A, p. 801. (Holomeniscus.)
Whitney, J. D. 1879 A, p. 250. (Megalomeryx.)
Wilckens, M. 1885 C, p. 425. (Megalomeryx.)
Wortman, J. L. 1898 A, p. 128.
Zittel, K. A. 1893 B, p. 364 (Camelops); p. 365 (Holomeniscus).

Camelops californicus (Leidy).

Leidy, J. 1870 T, p. 126. (Auchenia.)
Bowers, S. 1889 A, p. 391. (Holomeniscus.)
Cope, E. D. 1884 G, p. 18. (Auchenia.)
Cragin, F. W. 1892 A, p. 257. (Auchenia.)
Leidy, J. 1871 C, p. 349. (Auchenia.)
    1873 B, p. 255. (Auchenia.)
Roger, O. 1896 A, p. 217. (Holomeniscus.)
Trouessart, E. L. 1898 A, p. 843. (Subsp. of Holomeniscus hesternus.)
Whitney, J. D. 1879 A, p. 248. (Auchenia.)
Wilckens, M. 1885 C, p. 431. (Auchenia.)
Zittel, K. A. 1893 B, p. 365. (Holomeniscus.)
Pleistocene; California.

Camelops hesternus (Leidy).

Leidy, J. 1873 F. (Auchenia.)
Cope, E. D. 1884 G, p. 17. (Holomeniscus.)
    1889 F. (Holomeniscus.)
    1889 S, pp. 978, 980. (Holomeniscus.)
    1892 E, p. 251. (Holomeniscus.)
Cragin, F. W. 1892 A, p. 257. (Auchenia.)
Dana, J. D. 1879 A, p. 233. (Auchenia.)
Leidy, J. 1873 B, pp. 255, 317, pl. xxxvii, figs. 1–3. (Auchenia.)
Roger, O. 1896 A, p. 217. (Holomeniscus.)
Trouessart, E. L. 1898 A, p. 843. (Holomeniscus.)
Whitney, J. D. 1879 A, p. 250. (Auchenia.)
Wilckens, M. 1885 C, p. 431. (Auchenia.)
Wortman, J. L. 1898 A, p. 130. (Syn. of Camelops kansanus.)
Yates, L. G. 1874 B, p. 18. (Auchenia.)
Pleistocene: California, Oregon, Mexico.
From this species the Texan and New Mexican specimens are excluded. Not improbably the species is identical with *C. kansanus*.

Camelops kansanus Leidy.

Leidy, J. 1854 F, p. 172.
Cope, E. D. 1893 A, p. 72 (Auchenia huerfanensis); p. 84, pl. xxiii, fig. 4 (Holomeniscus sulcatus).
Cragin, F. W. 1892 A, p. 258. (Auchenia huerfanensis.)
    1896 A, p. 53. (A. huerfanensis.)
Hayden, F. V. 1858 B, p. 157. (Megalomeryx niobrarensis.)
King, C. 1878 A, p. 430. (Megalomeryx niobrarensis.)
Leidy, J. 1856 T, p. 166, pl. xvii, figs. 8–10.
    1858 E, p. 24. (Megalomeryx niobrarensis.)
    1869 A, p. 382 (C. kansanus); pp. 161, 382, pl. xiv, figs. 12–14 (Megalomeryx niobrarensis.)
    1871 C, p. 349 (C. kansanus); p. 351 (Megalomeryx niobrarensis.)
    1873 B, p. 260, pl. xxvii, figs. 21, 25 (Megalomeryx niobrarensis?); p. 317 (Procamelus? niobrarensis).
Owen, R. 1870 C, p. 75.
Roger, O. 1896 A, p. 216 (Procamelus niobrarensis); p. 217 (Camelops kansanus, Holomeniscus sulcatus); p. 218 (Auchenia huerfanensis).
Trouessart, E. L. 1898 A, p. 842 (C. kansanus); p. 845 (Lama huerfanensis).
Wilckens, M. 1885 C, p. 425.
Williston, S. W. 1897 I, p. 303 (C. kansanus); p. 304 (Auchenia huerfanensis).
    1898 I, p. 92. (Auchenia kansanus, A. huerfanensis.)
Wortman, J. L. 1898 A, p. 130.
Pleistocene; Kansas.

Camelops macrocephalus (Cope).

Cope, E. D. 1893 A, p. 85, pl. xxiii, fig. 5. (Holomeniscus.)
Roger, O. 1896 A, p. 217. (Holomeniscus.)
Trouessart, E. L. 1898 A, p. 843. (Holomeniscus.)
Pliocene (Equus beds); Texas.

Camelops vitakerianus (Cope).

Cope, E. D. 1878 H, p. 380. (Auchenia.)
    1884 G, p. 17. (Holomeniscus.)
    1889 F. (Holomeniscus.)
    1889 S, pp. 978, 980. (Holomeniscus.)
    1893 A, p. 86. (Holomeniscus.)
Roger, O. 1896 A, p. 217. (Holomeniscus.)
Trouessart, E. L. 1898 A, p. 843. (Holomeniscus.)
Wortman, J. L. 1898 A, p. 132.
Pliocene; Oregon.

PLIAUCHENIA Cope.   Type *P. humphreysiana* Cope.

Cope, E. D. 1875 P, p. 258.
    1875 F, pp. 69, 75.
    1875 Q, p. 262.
    1876 G.
    1877 K, p. 340.

Cope, E. D. 1886 C, p. 612.
    1887 B, p. 392.
    * 1893 A, p. 70.
Flower and Lydekker 1891 A, p. 304.
Nicholson and Lydekker 1889 A, p. 1331.

Schlosser, M. 1886 B, p. 49.
Scott, W. B. 1891 A, p. 51.
Wilckens, M. 1885 C, p. 428.
Wortman, J. L. 1898 A, p. 126.
Zittel, K. A. 1893 B, p. 364.

Pliauchenia humphreysiana Cope.

Cope, E. D. 1875 P, p. 258.
1875 F, p. 69.
1877 K, p. 344, pl. lxxvii, fig. 4.
1886 C, p. 621.
1892 AA.
Matthew, W. D. 1899 A, p. 74.
Roger, O. 1896 A, p. 217.
Trouessart, E. L. 1898 A, p. 842.　(P. humphresi-
ana.)
Wilckens, M. 1885 C, p. 430.
Wortman, J. L. 1898 A, p. 127.　(P. humphresi-
ana.)
Zittel, K. A. 1893 B, p. 364.　(P. humphriesiana.)
Miocene (Loup Fork); New Mexico, Kansas.

Pliauchenia minima Wort.

Wortman, J. L. 1898 A, p. 127.
Matthew, W. D. 1899 A, p. 74.
Miocene (Loup Fork); Kansas.

Pliauchenia spatula Cope.

Cope, E. D. 1893 A, p. 70. pl. xxi, figs. 1, 2.
Matthew, W. D. 1899 A, p. 75.
Roger, O. 1896 A, p. 217.
Trouessart, E. L. 1898 A, p. 842.
Wortman, J. L. 1898 A, p. 127.
Pliocene (Blanco); Texas.

Pliauchenia vulcanorum Cope.

Cope, E. D. 1875 P, p. 259.
1875 F, p. 70.
1877 K, p. 345, pl. lxxvii, fig. 5.
1892 AA.
Matthew, W. D. 1899 A, p. 74.
Roger, O. 1896 A, p. 217.
Trouessart, E. L. 1898 A, p. 842.
Wilckens, M. 1885 C, p. 430.
Wortman, J. L. 1898 A, p. 127.
Miocene (Loup Fork); New Mexico.
As Dr. Wortman has pointed out, the generic
position of this species is doubtful.

CAMELINÆ.

Cope, E. D. 1887 B, pp. 379, 394.　(Eschatiidæ.)
1888 X, p. 1080.　(Eschatiidæ.)
Roger, O. 1896 A, p. 217.

Trouessart, E. L. 1898 A, p. 843.
Zittel, K. A. 1893 B, p. 364.

ESCHATIUS Cope.　Type *E. conidens* Cope.

Cope, E. D. 1884 G, pp. 16, 18.
1886 C, p. 622.
1891 L, p. 1118.
Nicholson and Lydekker 1889 A, p. 1335.
Scott, W. B. 1891 A, p. 52.
1892 A, p. 436.
Steinmann and Döderlein 1890 A, p. 801.
Zittel, K. A. 1893 B, p. 365.

Eschatius conidens Cope.

Cope, E. D. 1884 G, p. 19.
1889 F.
1889 S, pp. 978, 980.

Roger, O. 1896 A, p. 217.
Trouessart, E. L. 1898 A, p. 843.
Wortman, J. L. 1898 A, p. 134.
Pliocene; Oregon, Mexico.

Eschatius longirostris Cope.

Cope, E. D. 1884 G, p. 20.
1889 F.
1889 S, pp. 978, 980.
Roger, O. 1896 A, p. 217.
Trouessart, E. L. 1898 A, p. 843.
Wortman, J. L. 1898 A. p. 135.
Pliocene; Oregon.

CAMELUS Linn.　Type *C. dromedarius* Linn.

Linnæus, C., 1758, Syst. Nat., ed. 10, 1, p. 65.
Baur, G. 1885 l, p. 196.
Blainville, H., M. D. 1864 A, iv, CC, p. 61, pls. i-v.
Cope, E. D. 1886 C. p. 612.
Flower and Lydekker 1891 A, p. 296.
Lydekker, R. 1885 C, p. 139.
Nicholson and Lydekker 1889 A, p., 1335.
Owen, R. 1866 B, p. 470, figs. 307, 318.
1868 A, p. 349, fig. 278.
Paulli, S. 1900 A, p. 203.

Roger, O. 1896 A, p. 219.
Scott, W. B. 1891 A.
Trouessart, E. L. 1898 A, p. 846.
Woodward, A. S. 1898 B, p. 360.
Zittel, K. A. 1893 B, p. 764.

Camelus americanus Wort.

Wortman, J. L. 1898 A, p. 133, fig. 7.
Trouessart, E. L. 1899 A, p. 1349.
Pleistocene; Nebraska.

TELEOPTERNUS Cope.　Type *T. orientalis* Cope.

Cope, E. D. 1899 A, p. 263.
Professor Cope expressed himself as being un-
able to determine whether this genus was a
camelid or a cervid.

Teleopternus orientalis Cope.

Cope, E. D. 1899 A, p. 264, pl. xxi, figs. 4, 4a.
Pleistocene; Pennsylvania.

Superfamily *TRAGULOIDEA.*

Citations to the literature of this superfamily and of the family Tragulidæ are here presented because by some authors the Leptomerycinæ, here placed in the Cameloidea, have been assigned to the Tragulidæ and may yet be retained there. Unless otherwise indicated, the following authors use the name Tragulina in referring to this group.
Cope, E. D. 1885 B.
Flower, W. H. 1874 A.

Flower, W. H. 1885 A.
Flower and Lydekker 1891 A, p. 305.
Gill, T. 1872 B, pp. 9, 73, 80.   (Traguloidea.)
Osborn and Wortman 1892 B, p. 370.
Rütimeyer, L. 1867 B.
　1877 A, p. 21.
　1878 A, p. 21.
　1883 A, p. 15.
　1884 A, p. 403.
Scott, W. B. 1892 A, p. 436.

TRAGULIDÆ.

Boas, J. E. V. 1890 A.
Branco W. 1897 A, p. 25.
Cope, E. D. 1884 H, p. 27.
　1887 B, pp. 379, 388.
　1888 X, p. 1081.
　1889 G, p. 120.
Edwards, A. M. 1864 A.
Flower, W. H. 1873 A.
Flower and Lydekker 1889 A, p. 305.
Gill, T. 1872 B, pp. 9, 73, 80.
Gray, J. E. 1869 C.
Huxley, T. H. 1870 A, p. 439.
　1872 A, p. 326.

Kowalevsky, W. 1873 A, p. 186.
Lydekker, R. 1885 C, p. 150.
　1896 B, p. 164.
Nicholson and Lydekker 1889 A, p. 1333.
Osborn and Wortman 1892 B, p. 370.
Roger, O. 1896 A, p. 224.
Schlosser, M. 1886 A, p. 254.
　1886 B, p. 72.
Scott, W. B. 1890 B, p. 388.
Steinmann and Döderlein 1890 A, p. 802.
Trouessart, E. L. 1898 A, p. 857.
Woodward, A. S. 1898 B, p. 360.
Zittel, K. A. 1893 B, p. 381.

Superfamily *BOÖIDEA.*

Boas, J. E. V. 1890 A.   (Ruminantia.)
Brandt, E. 1888 A.   (Ruminantia.)
Cope, E. D. 1884 H, p. 27.   (Selenodonta.)
　1884 DD.   (Selenodonta.)
　1885 B, p. 483.   (Selenodonta.)
　1887 B, p. 304 (Pecora); p. 378 (Selenodonta).
　1898 B, p. 130.   (Boöidea.)
Cuvier, G. 1834 A, vi.   (Ruminantia.)
Flower, W. H. 1873 A.   (Ruminantia.)
　1874 A.   (Pecora.)
　1883 D, p. 430.   (Pecora.)
　1885 A.   (Pecora.)
Flower and Lydekker 1891 A, p. 307.   (Pecora.)
Gaudry, A. 1878 B, p. 77.   ("Les Ruminants.")
Gill, T. 1872 B, pp. 9, 76.
Grüber, W. 1859 A.   (Ruminantia.)
Huxley, T. H. 1872 A, p. 322.   (Ruminantia.)
Kowalevsky, W. 1873 A, p. 171.   (Selenodonta.)
　1874 A.   (Ruminantia.)
Lydekker, R. 1883 A.   (Ruminantia.)

Marsh, O. C. 1877 E.   (Selenodonta.)
Mayo, F. 1888 A.   (Ruminantia.)
Mettam, A. E. 1895 A.   (Ruminantia.)
Osborn and Wortman 1892 B, p. 370.   (Pecora.)
Owen, R. 1845 B, p. 527.   (Ruminantia.)
　1857 A.   (Ruminantia.)
Parker, W. K. 1885 A, p. 201.
Roger, O. 1889 A.   (Ruminantia.)
Rütimeyer, L. 1862 A, p. 603.   (Ruminantia.)
　1865 A.
　1867 B.
　1868 A.
　1884 A, p. 425.   (Selenodonta.)
Schlosser, M. 1889 A.
Schmidt, O. 1886 A, p. 150 (Ruminantia); p. 173 (Cavicornia).
Scott, W. B. 1899 A, p. 118.   (Pecora.)
Steinmann and Döderlein 1890 A, p. 797.
Turner, H. N. 1848 A.
Zittel, K. A. 1893 B, p. 413.   (Cavicornia.)

CERVIDÆ.

Ameghino, F. 1889 A, p. 599.
Baird, S. F. 1857 A, p. 629.
Baur, G. 1884 F. p. 599.
Branco, W. 1897 A, p. 25.   (Cervicornia.)
Brandt, E. 1888 A.
Brooke, V. 1878 A, p. 883.
Cope, E. D. 1887 B, pp. 379, 394, 398.
　1888 X, p. 1081.
　1889 G, p. 124.
Cope and Wortman 1884 A, p. 22.
Coues, E. 1878 B.
Cuvier, G. 1834 A, vi, p. 43, plates.
Flower, W. H. 1873 A.

Flower, W. H. 1874 A.
　1883 D, p. 431.
　1885 A.
Flower and Lydekker 1891 A, p. 31.
Gill, T. 1872 B, pp. 9, 73.
Leidy, J. 1871 C, p. 351.
Leuthardt, F. 1891 A, p. 126.
Lydekker, R. 1885 C, p. 73.
Lydekker, R. 1896 B, p. 164.
　1898 A.
Marsh, O. C. 1877 E.
Nicholson and Lydekker 1889 A, p. 1336.
Nitsche, H. 1899 A.

Owen, R. 1866 B.
Pohlig, H. 1892 A.
Roger, O. 1887 A.
Rütimeyer, L. 1867 B.   (Cervicornia.)
1877 A.
1878 A, p. 25.
1881 A.
1882 A.

Rütimeyer, L. 1883 A, p. 20.
Schmidt, O. 1886 A, p. 158.
Steinmann and Döderlein 1890 A, p. 804.
Trouessart, E. L. 1898 A, p. 864.
Weyhe, ——. 1875 A, p. 118.
Woodward, A. S. 1898 B, p. 363.
Zittel, K. A. 1893 B, p. 390.   (Cervicornia.)

CERVULINÆ.

Brooke, V. 1874 A.
Cope, E. D. 1887 B, pp. 394, 395.   (Cosorycinæ,
Cosorycidæ.)
Flower, W. H. 1874 A.
Roger, O. 1896 A, p. 227.
Rütimeyer, L. 1881 A, p. 23.

Rütimeyer, L. 1882 A, p. 16.   (Cervulina.)
1884 A, p. 445.
Steinmann and Döderlein 1890 A, p. 805.
Trouessart, E. L. 1898 A, p. 865.
Zittel, K. A. 1893 B. 396.

PALÆOMERYX Meyer.   Types P. bojani and P. kaupii.

Meyer, H. 1834 A, p. 92.
Blainville, H. M. D. 1864 A, iv, BB, p. 135.
Depéret, C. 1887 A, p. 210.
Douglass, E. 1900 A. p. 20.
Flower and Lydekker, 1891 A, p. 330.
Hofmann, A., 1888, Jahrb. k. k. geol. Reichsanst.,
xxxviii, p. 551.
Lydekker, R., 1883, Mem. Geol. Surv. India, ii, pp.
173, 174.   (Propalæomeryx.)
Nicholson and Lydekker 1889 A, p. 1338.
Pictet, F. J. 1853 A. p. 350.
Rütimeyer, L. 1883 A. p. 79.

Schlosser, M. 1886 C, p. 294.
Steinmann and Döderlein 1890 A, p. 805.
Trouessart, E. L. 1898 A, p. 867.
Zittel, K. A. 1893 B, p. 397.

Palæomeryx americanus Douglass.
Douglass, E. 1900 A, p. 20, pl. iv, figs. 2, 3.
Miocene (Loup Fork); Montana.

Palæomeryx madisonius Douglass.
Douglass, E. 1900 A, p. 23.
Miocene (Loup Fork); Montana.

BLASTOMERYX Cope.   Type Merycodus gemmifer Cope.

Cope, E. D. 1877 K, p. 350.
1875 F, p. 68.   (Dicrocerus, not of Lartet.)
1875 O, p. 257.   (Dicrocerus.)
1878 C, p. 222.
1884 H, p. 27.
1885 B, p. 488.
1887 B, p. 396.
1889 C, p. 206.
1889 G, p. 125.
Nicholson and Lydekker 1889 A, p. 1342.
Scott, W. B. 1891 D, p. 90.
1895 C, p. 167.
Scott and Osborn 1890 B, p. 76, figs. 7–9.
Zittel, K. A. 1893 B, p. 399.

Blastomeryx antilopinus Scott.
Scott, W. B. 1893 B, p. 662.
Matthew, W. D. 1899 A, p. 74.
Roger, O. 1896 A, p. 230.
Scott, W. B. 1895 C, p. 168, pl. vi, figs. 48–51.
Trouessart, E. L. 1898 A, p. 870.
Miocene (Loup Fork); Montana.

Blastomeryx borealis Cope.
Cope, E. D. 1878 C, p. 222.
1878 H, p. 382.
1878 L, p. 58.
1884 DD, p. 1035.
1886 B, p. 359.

Cope, E. D. 1886 H.
1889 G, p. 128, figs. 16, 19.
Matthew, W. D. 1899 A, p. 74.
Roger, O. 1896 A, p. 230.
Scott, W. B. 1891 D, p. 91.
1893 B, p. 664.
1895 C, p. 168.
Trouessart, E. L. 1898 A, p. 870.
Miocene (Loup Fork); Montana, Oregon.

Blastomeryx gemmifer Cope.
Cope, E. D. 1874 B, p. 531.   (Merycodus.)
1874 P, p. 22.   (Merycodus.)
1875 F, p. 68.   (Dicrocerus.)
1875 O, p. 257.   (Dicrocerus.)
1877 K, pp. 350, 360, pl. lxxxii, fig. 13.
[Dicrocerus (Blastomeryx).]
1884 DD, p. 1035.   (Cosoryx.)
1885 B, p. 488.
1893 A, p. 39.
Matthew, W. D. 1899 A, p. 74.
Roger, O. 1896 A, p. 230.
Scott and Osborn 1890 B, p. 76.
Trouessart, E. L. 1898 A, p. 870.
Wilckens, M. 1885 C, p. 425.   (Cosoryx.)
Zittel, K. A. 1893 B, p. 399, fig. 330.
Miocene (Loup Fork); Colorado, Texas, New
Mexico, Nebraska.

MERYCODUS Leidy.   Type *M. necatus* Leidy.

Leidy, J. 1854 C, p. 90.
Cope, E. D. 1875 F, p. 69.   (Dicrocerus, not of
   Lartet.)
   1875 O. (Dicrocerus.)
   1877 K, p. 346. (Dicrocerus.)
   1879 A, p. 46. (Procervulus, Merycodus, and
      Cosoryx, "nomina nuda.")
   1880 B, p. 46. (Syn. of Procervulus Gaudry.)
   1884 H, p. 27. (Cosoryx.)
   1884 Z, p. 718. (Cosoryx.)
   1884 DD. (Cosoryx.)
   1885 B, p. 488. (Cosoryx.)
   1887 B, p. 396. (Cosoryx.)
   1889 C. (Cosoryx.)
   1889 G, p. 125. (Cosoryx.)
Hay, O. P. 1899 A, p. 594.
Hensel, R. 1859 A. (Dicrocerus Lartet, not of
   Cope.)
Leidy, J. 1858 E, p. 23. (Merycodus.)
   1869 A, pp. 162, 382 (Merycodus); pp. 173, 383
      (Cosoryx, type *C. furcatus*.
   1871 C, p. 350 (Merycodus); p. 351 (Cosoryx).
   1873 B, p. 318. (Merycodus.)
Marsh, O. C. 1877 E. (Cosoryx.)
Nicholson and Lydekker 1889 A, p. 1345. (Coso-
   ryx.)
Rütimeyer, L. 1878 A, p. 90. (Cosoryx.)
Trouessart, E. L. 1898 A, p. 870. (Cosoryx.)
Wilckens, M. 1885 C, p. 425.
Zittel, K. A. 1893 B, p. 399. (Cosoryx.)

Merycodus agilis (Douglass).

Douglass, E. 1900 A, p. 23, pl. iv, fig. 1. (Cosoryx.)
   Miocene; Montana.

Merycodus furcatus (Leidy).

Leidy, J. 1869 A, pp. 173, 383, pl. xxviii, fig. 8.
   (Cosoryx.)
Cope, E. D. 1874 I, p. 148. (Cosoryx.)
   1875 O, p. 257. (Dicrocerus.)
   1877 K, p. 350, pls. lxxx, lxxxi, fig. 1; pl.
      lxxxii, fig. 1. (Dicrocerus.)
   1880 E, p. 141. (Procervulus.)
   1881 J, p. 547, fig. 10. (Cosoryx.)
   1887 S, p. 271, fig. 5.
   1889 C. (Cosoryx.)
   1889 G, p. 114, fig. 11. (Cosoryx.)
Dawkins, W. B. 1870 A, p. 232. (Cosoryx.)
King, C. 1878 A, p. 430. (Cosoryx.)
Leidy, J. 1871 C, p. 352. (Cosoryx.)
Matthew, W. D. 1899 A, p. 74. (Cosoryx.)
Roger, O. 1896 A, p. 230. (Cosoryx.)
Scott and Osborn 1890 B, p. 82, pl. i. (Cosoryx.)
Trouessart, E. L. 1898 A, p. 870. (Cosoryx.)
Zittel, K. A. 1893 B, p. 399, fig. 331. (Cosoryx.)
   Miocene (Loup Fork); Nebraska, New Mexico.

Merycodus necatus Leidy.

Leidy, J. 1854 C, p. 90. (Merycodus.)
Cope, E. D. 1875 F, p. 69. (Dicrocerus.)
   1875 O, p. 257. (Dicrocerus.)
   1877 K, pp. 350, 353, pl. lxxxi, figs. 2–6; pl.
      lxxxii, figs. 2, 3. (Dicrocerus.)
   1880 E, p. 141. (Procervulus.)
   1889 G, p. 128, fig. 18. (Cosoryx.)

Hayden, F. V. 1858 B, p. 157. (M. necatus, Cer-
   vus warreni.)
King, C. 1878 A, p. 430. (Cervus warreni, M. ne-
   catus.)
Leidy, J. 1854 E, p. 157.
   1858 E, p. 23. (M. necatus, Cervus warreni.)
   1869 A, pp. 162, 382, pl. xiv, figs. 9, 10 (M.
      necatus); pp. 172, 379, pl. xxvii, fig. 12
      (Cervus warreni).
   1871 C, p. 350 (M. necatus); p. 351 (Cervus
      warreni).
   1873 B, p. 318.
Matthew, W. D. 1899 A, p. 74. (Cosoryx.)
Roger, O. 1896 A, p. 230.
Trouessart, E. L. 1898 A, p. 870.
Wilckens, M. 1885 C, p. 425.
   Pliocene; South Dakota, Wyoming Nebraska,
      New Mexico.

Merycodus ramosus (Cope).

Cope, E. D. 1874 I, p. 148, in part. (Cosoryx.)
   1874 O, p. 16. (Cosoryx.)
   1875 F, p. 69. (Dicrocerus.)
   1875 O, p. 257. (Dicrocerus.)
   1877 K, pp. 350, 356. (Syn., in part, of Dicro-
      cerus furcatus.)
   1880 E, p.141. (Procervulus.)
   1889 C, p. 205. Cosoryx.)
   1889 G, p. 128, fig. 18. (Cosoryx.)
Matthew, W. D. 1899 A, p. 74. (Syn. of Cosoryx
   furcatus.)
Trouessart, E. L. 1898 A, p. 870. (Syn. of Cosoryx
   furcatus.)
   Pliocene; New Mexico.

Merycodus tehuanus (Cope).

Cope, E. D. 1877 K, pp. 350, 359, pl. lxxxii, figs.
   10–?12. (Dicrocerus.)
Matthew, W. D. 1899 A, p. 74. (Cosoryx.)
Roger, O. 1896 A, p. 230. (Cosoryx.)
Trouessart, E. L. 1898 A, p. 870. (Syn. of Cosoryx
   trilateralis.)
   Pliocene; New Mexico.
   A doubtful species.

Merycodus teres (Cope).

Cope, E. D. 1874 I, p. 148. (Cosoryx.)
   1874 O, p. 16. (Cosoryx.)
   1875 F, p. 69. (Dicrocerus.)
   1875 O, p. 257. (Dicrocerus.)
   1877 K, pp. 347, 356, pl. lxxxi, fig. 7; pl.
      lxxxii, fig. 6. (Dicrocerus.)
Matthew, W. D. 1899 A, p. 74. (Cosoryx.)
Roger, O. 1896 A, p. 230. (Cosoryx.)
Scott and Osborn 1890 B, p. 83. (Cosoryx.)
Trouessart, E. L. 1898 A, p. 870. (Cosoryx.)
   Miocene (Loup Fork); New Mexico.

Merycodus trilateralis (Cope).

Cope, E. D. 1877 K, pp. 357, pl. lxxxi, fig. 8; pl.
   lxxxii, figs. 7–9. (Dicrocerus.)
Matthew, W. D. 1899 A, p. 74. (Cosoryx.)
Roger, O. 1896 A, p. 230. (Cosoryx.)
Scott and Osborn 1890 B, p. 83. (Cosoryx.)
Trouessart, E. L. 1898 A, p. 870. (Cosoryx.)
   Miocene (Loup Fork); New Mexico.

CERVINÆ.

Baird, S. F. 1857 A, p. 630.
Flower and Lydekker 1891 A, p. 316.
Gill, T. 1872 B, p. 9.
Lydekker, R. 1898 A, p. 32.
Nicholson and Lydekker 1889 A, p. 1342. ·
Roger, O. 1887 A. (Cervina.)

Roger, O. 1896 A, p. 230.
Rütimeyer, L. 1882 A, p. 22. (Cervina.)
    1883 A, p. 20. (Cervina.)
    1884 A, p. 408. (Cervina.)
Steinmann and Döderlein 1890 A, p. 806.
Zittel, K. A. 1893 B, p. 400.

DAMA Zimm.   Type *D. virginiana* Zimm.

Zimmermann, E., 1777, Zool. Geog., p. 532.
Baird, S. F. 1857 A, p. 637. (Cervus, in part.)
Cope, E. D. 1887 B, p. 398. (Cariacus.)
    1889 G, p. 132. (Cariacus.)
Flower and Lydekker 1891 A, p. 329. (Cariacus.)
Gray, J. E. 1850. (Cariacus.)
Hensel, R. 1859 A, p. 257. (Cariacus.)
Lydekker, R. 1885 C, p. 82. (Cervus.)
    1898 A, p. 243. (Mazama.)
Nicholson and Lydekker, 1889 A, p. 1342. (Caria-
    cus.)
Rafinesque, C. 1832-33, Atlantic Jour., p. 109, fig-
    ure. (Odocoileus.)
Roger, O. 1887 A. (Cariacus.)
    1896 A, p. 230. (Cervus.)
Rütimeyer, L. 1881 A, p. 47. (Cariacus.)
    1882 A, p. 26. (Cariacus.)
    1884 A, p. 410. (Cariacus.)
Zittel, K. A. 1893 B, p. 401. (Cariacus.)

Dama dolichopsis (Cope).

Cope, E. D. 1878 GG, p. 189. (Cariacus.)
    1878 H, p. 379. (Cariacus.)
Cope and Wortman 1884 A, p. 22, pl. ii. (Caria-
    cus.)
Kindle, E. M. 1898 A, p. 485. (Cariacus.)
Pleistocene; Indiana.

Dama ensifer (Cope).

Cope, E. D. 1889 F, p. 163. (Cariacus.)
Roger, O. 1896 A, p. 232. (Cervus.)
Trouessart, E. L. 1898 A, p. (Cariacus.)
Pleistocene (Equus beds); Oregon.

Dama lævicornis (Cope).

Cope, E. D. 1896 F, p. 393. (Cariacus.)
    1899 A, p. 265, pl. xxi, fig. 5. (Cariacus.)
Trouessart, E. L. 1898 A, p. 892. (Cariacus.)
Pleistocene; Pennsylvania.

Dama virginiana Zimm.

Zimmermann, E., 1777, Zool. Geog., p. 532.
Allen, J. A. 1876 A, p. 48. (Cervus.)
Baird, S. F. 1857 A, p. 643, text fig. 13. (Cervus.)
Cooper, W. 1831 A, p. 207. (Cervus.)
Cope, E. D. 1869 A, p. 176. (Cariacus.)
    1869 N, p. 742. (Cariacus.)
    1871 K, p. 176. (Cariacus.)
    1899 A, p. 266. (Cariacus.)
Cuvier, G. 1834 A, vi, p. 63, pl. 166. (Cervus.)
Emmons, E. 1858 B, p. 200. (Cervus.)
Erxleben, C. P., 1777, Syst. Reg. Anim., Mamm.,
    p. 312. (Cervus dama americanus; americanu
    not nomenclatorial.)
Flower and Lydekker, 1891 A, p. 326, fig. 132
    (Cariacus.)
Hensel, R. 1859 A, p. 257, pl. xi, fig. 6. (Cariacus.
    Cervus.)
Kindle, E. M, 1898 A, p. 485. (Cariacus.)
Leidy, J. 1854 I, p. 200. (Cervus.)
    1860 B, p. 109, pl. xx, figs. 1–4. (Cervus.)
    1869 A, p. 376. (Cervus virg.)
    1889 B, p. 96. ("Deer.")
    1889 E, p. 22. (Cervus.)
    1889 H, p. 6. (Cervus.)
    1892 A. (Cervus.)
Lydekker, R. 1898 A, p. 249, pl. xx; text figs. 68, 69
Rafinesque, C. 1832–33, Atlantic Jour. p. 109, fig-
    ure. (Odocoileus speleus.)
Roger, O. 1896 A, p. 232. (Cervus.)
Trouessart, E. L. 1898 A, p. 892. (Cariacus.)
Walles, B. L. C. 1854 A, p. 286. (Cervus.)
Winchell, A. 1864 A, p. 223. ("Deer.")
Wyman, J. 1862 A, p. 421. (Cervus.)
    Pleistocene and Recent. Fossil from South
    Carolina, New Jersey, Indiana, Kentucky, Vir-
    ginia, Pennsylvania, Illinois, Mississippi.

Dama whitneyi (Allen).

Allen, J. A. 1876 A, p. 49. (Cervus.)
    Pleistocene; lead region of Illinois, Iowa,
    and Wisconsin.

CERVUS Linn.   Type *C. elaphus* Linn.

Linnæus, C., 1758, Syst. Nat., ed. 10, i, p. 66.
    By many of the following authors the generic
    name *Cervus* is employed in a more comprehen-
    sive sense than is here given it.
Baird, S. F. 1857 A, p. 637, in part.
Brooke, V. 1878 A.
Cope, E. D. 1887 B, p. 398.
Cuvier, G. 1834 A, vi, p. 43, plates.
Flower and Lydekker 1891 A, p. 319.
Leidy, J. 1871 C, p. 351.

Lydekker, R. 1885 C, p. 78.
    1898 A.
Nicholson and Lydekker 1889 A, p. 1338.
Nitsche, H. 1891 A.
Owen, R. 1866 B, p. 477.
Paulli, S. 1900 A, p. 210.
Pictet, F. J. 1853 A, p. 358.
Rütimeyer, L. 1884 A, p. 408.
Trouessart, E. L. 1898 A, p. 880.
Zittel, K. A. 1893 B, p. 401. (Elaphus.)

Cervus canadensis Erxl.

Erxleben, C. P., 1777, Syst. Reg. Anim., Mamm., p. 305. (Cervus elaphus canadensis.)
Allen, J. A. 1876 A, p. 48.
Baird, S. F. 1857 A, p. 638, text figs. 9–11.
Cooper, W. 1831 A, p. 207.
Cope, E. D. 1869 E, p. 178.
   1869 N, p. 742.
Cuvier, G. 1834 A, vi, p. 49, pl. 164, figs. 13–22.
De Kay, J. E. 1842 C, p. 120, pl. xxix, fig. 2. (Elaphus americanus.)
Flower and Lydekker 1891 A, p. 322, fig. 129.
Hall, J. 1846 B, p. 391. (Elaphus.)
   1887 A, p. 39. (Elaphus.)
Leidy, J. 1869 A, p. 377.
   1889 H, p. 6.
Logan, W. E. 1863 A, p. 914.
Lydekker, R. 1898 A, p. 96, figs. 24, 25.

Putnam, F. W., 1900, 33d Report Peabody Mus. Amer. Archæl, Ethnol., Harvard Univ., p. 275. ("Elk.")
Roger, O. 1896 A, p. 235. (C. americanus fossilis.)
Scott, W. B. 1885 C, text fig. 1.
Trouessart, E. L. 1898 A, p. 883.
Tscherski, J. D. 1892 A, p. 222. (Cervus canadensis maral; in Russia.)
Winchell, A, 1864 A, p. 224. ("Elk.")
Zittel, K. A. 1893 B, p. 402.
   Pleistocene and Recent. Fossil in New York, New Jersey, Kentucky, North Carolina, Florida, Ontario; also fossil in Europe and Siberia.

Cervus fortis Cope.

Cope, E. D. 1878 C, p. 223.
Roger, O. 1896 A, p. 235.
Trouessart, E. L. 1898 A, p. 882.
   Pliocene (Loup Fork); Oregon.

CERVALCES Scott.   Type *Cervus americanus* Harlan.

Scott, W. B. 1885 A, p. 420.
Cope, E. D. 1887 B, p. 398.
   1889 G, p. 132.
Flower and Lydekker 1891 A, p. 327.
Lydekker, R. 1898 A, pp. 51, 60. (Alces.)
Nicholson and Lydekker 1889 A, p. 1341.
Scott, W. B. 1885 D, p. 4.
Zittel, K. A. 1893 B, p. 404.

Cervalces americanus (Harlan)

Harlan, R., 1825 A, p. 245. (Cervus.)
Cooper, W. 1831 A, p. 174. ("Cervus.")
Cooper, Smith, and De Kay 1831 A, p. 371. (Cervus alces?.)
Cope, E. D. 1889 G, p. 134.
De Kay, J. E. 1842 C, p. 120, in part. (Elaphus.)
Flower and Lydekker, 1891 A, p. 327.
Godman, J. D. 1828 A, ii, pl. opp. p. 197; iii, p. 242. ("Fossil elk.")

Harlan, R. 1834 B, p. 70. (Cervus.)
   1835 B, p. 275. (Cervus.)
Leidy, J. 1869 A, p. 379. (Cervus.)
   1884 B. (Cervus.)
Lydekker, R. 1898 A, p. 60, fig. 14. (Alces scotti.)
Osborn, H. F. 1896 N, p. 845. (Cervalces.)
   1898 Q, p. 7, plate.
Roger, O. 1896 A, p. 236. (Cervus.)
Scott, W. B. 1885 A, p. 420, with 2 figures.
   1885 C, p. 174, pl. ii, text figs. 2, 5, 7.
   1885 D, p. 4, plate.
Trouessart, E. L. 1898 A, p. 888 (Cervus canadensis, in part); p. 886 (Cervalces americanus.)
Williston, S. W. 1897 I, p. 303. (Cervalces? americanus?.)
Wistar, C. 1818 A, p. 375, pl. x, figs. 4, 5. ("Cervus.")
   Pleistocene; Kentucky, New Jersey.

ALCES Jardine.   Type *Cervus alces* Linn.

Baird, S. F. 1857 A, p. 631. (Alce.)
Cope, E. D. 1887 B, p. 378.
Flower and Lydekker 1891 A, p. 326.
Lydekker, R. 1898 A, p. 49.
Roger, O. 1887 A.
Rütimeyer, L. 1881 A, p. 54.
   1882 A, p. 28.
   1884 A, p. 414.
Scott, W. B. 1885 C.
Zittel, K. A. 1893 B, p. 404.

Alces americanus Jardine.

Jardine, W., 1837, Nat. Library, iii, p. 125.
Baird, S. F. 1857 A, p. 631, text figs. 1, 2. (Alce.)
Cooper, W. 1831 A, p. 207. (Cervus alces.)
Flower and Lydekker 1891 A, p. 326, fig. 133. (Alces machlis, in part.)
Hensel, R. 1859 A, pl. xi, fig. 2. (C. alces of Europe.)
Leidy, J. 1869 A, p. 377. (Cervus alces.)
Lydekker, R. 1885 C, p. 78. (A. machlis, in part.)
   1898 A, p. 52, pl. ii, text fig. 11. (A. machlis.)

Meyer, H. 1832 A, pl. xxxvii. (Cervus alces fossilis, not A. americanus.)
Nitsche, H. 1891 A. (C. alces, of Europe.)
Richardson, J. 1854 A, p. 20, pls. xx, xxi; pl. xxii, fig. 1; pl. xxiv (Cervus alces); p. 102 (Alces muswa).
Scott, W. B. 1885 C, p. 181; figs. 3, 4. (A. machlis.)
Trouessart, E. L. 1898 A, p. 887.
Williston, S. W. 1898 I, p. 92. (A. americanus?.)
Zittel, K. A. 1893 B, p. 404.
   Recent in northern portions of North America: Pleistocene; Kentucky, Kansas?, Alaska.

Alces brevitrabalis Cope.

Cope, E. D. 1889 F, p. 162, February.
Trouessart, E. L. 1898 A, p. 887.
   Pleistocene; Washington.

Alces semipalmatus Cope.

Cope, E. D. 1889 F, p. 162, February.
Roger, O. 1896 A, p. 237. (Cervus.)
Trouessart, E. L. 1898 A, p. 887.
   Pleistocene; Washington.

RANGIFER H. Smith.   Type *Cervus tarandus* Linn.

Smith. H., 1827, Griffith's Anim. Kingd., v, p. 304.
Baird, S. F. 1857 A, p. 633.
Brooke, V. 1878 A, p. 927.
Cope, E. D. 1887 B, p. 398.
Flower and Lydekker 1891 A, p. 324.
Lydekker, R. 1885 C, p. 79.
　1898 A, p. 33.
Paulli, S. 1900 A, p. 216.
Roger, O. 1887 A.
Rütimeyer, L. 1881 A, p. 51.
　1882 A, p. 27.
　1884 A, p. 413.
Zittel, K. A. 1893 B, p. 404.

Rangifer caribou (Gmelin).

Baird, S. F. 1857 A, p. 633, text figs. 3–6.
Cooper, W. 1831 A, p. 207.   (Cervus tarandus.)
Cope, E. D. 1869 N, p. 740.   (R. grœnlandicus.)
Dana, J. D. 1875 C, p. 354.   (R. tarandus.)
Fisher, G. J. 1859 A, p. 194.   ("Reindeer.")
Leidy, J. 1859 H.   ("Reindeer.")
　1869 A, p. 377, pl. xxviii, fig. 9.   (Cervus tarandus.)
　1879 A.   (R. caribou, Cervus muscatinensis.)
　1880 A.
　1889 H, p. 5, pl. ii, fig. 22.
Lesley, J. P. 1889 A, p. 855, figure.   ("Reindeer.")
Lydekker, R. 1898 A, p. 42.   (R. tarandus caribou.)

Mercer, H. C. 1894 A, p. 98.
Shaler, N. S. 1871 B, p. 167.   (Tarandus rangifer.)
Trouessart, E. L. 1898 A, p. 888.
　Recent; Newfoundland to Alaska; not on Arctic coasts.   Pleistocene: New England, New York, New Jersey, Pennsylvania, Kentucky, Iowa, Alaska.

Rangifer tarandus (Linn.).

Linnæus, C., 1758, Syst. Nat., 10, i, p. 67.   (Cervus.)
Brandt, J. F. 1867 B.
Buckland, W. 1831 A, pp. 595, 597, 605, pl. iii, figs. 11–13.   ("Reindeer.")
Hensel, R. 1859 A, pl. xi fig. 1.   (Cervus.)
Lydekker, R. 1885 C, p. 79, fig. 7.
　1898 A, p. 37.
Richardson, J. 1854 A, pp. 20, 115, pl. xxii, fig. 2; pl. xxiii.
Struckmann, C. 1880 A.   (Cervus.)
Trouessart, E. L. 1898 A, p. 887.
Tscherski, J. D. 1892 A, p. 199.
Zittel, K. A. 1893 B, p. 404.
　Recent in northern Europe and northern America.   Pleistocene in Alaska.   Has been reported from New England, New York, New Jersey, and Kentucky, but the remains probably belong to R. caribou.

ANTILOCAPRIDÆ.

Flower and Lydekker 1891 A, p. 333.
Gadow, H. 1898 A, p. 49.   (Cervidæ.)
Murie, J. 1870 A.

Nicholson and Lydekker 1889 A, p. 1345.
Trouessart, E. L. 1898 A, p. 903.

ANTILOCAPRA Ord.   Type *A. americana* Ord.

Baird, S. F. 1857 A, p. 665.
Flower and Lydekker 1891 A, p. 333.
Nitsche, H. 1899 A, p. 186.
Rütimeyer, L. 18.   (Dicranocerus.)
Zittel, K. A. 1893 B, p. 417.

Antilocapra americana Ord.
Allen, Harrison, 1889 A.

Allen, J. A. 1876 A, p. 48.
Baird, S. F. 1857 A, p. 666, pl. xxv.
Murie, J. 1870 A.
Richardson, J. 1854 A, p. 131, pls. xvi–xix.
　Recent; Mexico and California to British America.   Pleistocene in lead region of Illinois, Iowa, and Wisconsin.

BOVIDÆ.

Baird, S. F. 1857 A, p. 664.   (Cavicornia.)
Baur, G. 1890 B.
Branco, W. 1897 A, p. 27.
Brandt, F. 1888 A.   (Cavicornia.)
Brooke, V. 1878 A, p. 884.
Cope, E. D. 1887 B, pp. 379, 394.
　1888 X, p. 1081.
　1889 C, p. 162.
Cope and Wortman 1884 A, pp. 22, 24.
Cuvier, G. 1834 A, vi, p. 217.
Flower W. H. 1874 A.
　1883 D, p. 432.
Flower and Lydekker 1891 A, p. 334.
Gadow, H. 1898 A, p. 49.
Gill, T. 1872 B, pp. 8, 76.
Lydekker, R. 1885 C, p. 1.
　1896 B.
　1898 B.
Nicholson and Lydekker 1889 A, p. 1345.

Nitsche, H. 1899 A, p. 186.
Owen, R. 1866 B.
　1868 A, p. 348.
Richardson, J. 1854 A, p. 28.
Rütimeyer, L. 1862 B.
　1865 A.
　1867 B.   (Cavicornia.)
　1877 A.
　1878 A.
　1883 A, p. 10.   (Cavicornia.)
Sanson, A. 1878 A.
Steinmann and Döderlein 1890 A, p. 808.
Tomes, C. S. 1898 A, p. 407.
Trouessart, E. L. 1898 A, p. 904.
Turner, H. N. 1850 A.
Weyhe, — 1875 A, p. 119.
Wilckens, M. 1885 E.
Zittel, K. A. 1893 B, p. 425.   (Cavicornia.)

ANTILOPINÆ.

Baird, S. F. 1857 A, p. 664.
Flower and Lydekker 1891 A, p. 349. ("Rupica-
prine section.")
Nicholson and Lydekker 1889 A, p. 1346. ("An-
telopes.")

Rütimeyer, L. 1867 B. (Antilopina.)
1877 A, p. 17. (Antilopina.)
1878 A, p. 78. ("Antilopen.")
Zittel, K. A. 1893 B, p. 415.

NEMORHŒDUS H. Smith.

Smith, H., 1827, Griffith's Anim. Kingd., iv, p. 277.
Flower and Lydekker 1891 A, p. 350.
Rütimeyer, L. 1877 A, p. 43. (Nemorhedus.)
1878 A, p. 90. (Kemas.)
Trouessart, E. L. 1898 A, p. 964.

Nemorhœdus palmeri Cragin.

Cragin, F. W. 1899 A, p. 610, pl. lvii.
1900 A, p. 1.
Pleistocene; Colorado.

CAPRINÆ.

Baird, S. F. 1857 A, p. 664. (Ovinæ.)
Flower and Lydekker 1891 A, p. 351. ("Caprine
section.").
Gill, T. 1872 B, pp. 9, 77. (Caprinæ, Ovinae.)

Owen, R. 1868 A, p. 348. (Ovidæ.)
Stehlin, H. G. 1893 A.
Trouessart, E. L. 1898 A, p. 968.
Zittel, K. A. 1893 B, p. 422. (Ovinæ.)

OVIS Linn. Type *O. aries* Linn.

Linnæus, C., 1758, Syst. Nat., ed, 10, i, p. 70.
Flower, W. H. 1885 A.
Flower and Lydekker 1891 A, p. 354.
Owen, R. 1866 B, p. 474, figs. 323, 324.
1868 A, p. 349, fig. 279.
Stehlin, H. G. 1893 A, p. 71.
Trouessart, E. L. 1898 A, p. 975.
Zittel, K. A. 1893 B, p. 424.

Ovis mamillaris Foster.

Foster, J. W. 1837 A, p. 82, figs. 19 a, b.
De Kay, J. E. 1842 C, p. 112.
Leidy, J. 1869 A, p. 375.
Roger, O. 1896 A, p. 244.
Trouessart, E. L. 1898 A, p. 978.
Pleistocene?; Ohio.
A doubtful fossil.

OVIBOS Blainv. Type *O. moschatus* (Zimm.).

Blainville, H. M. D., 1816, Bull. Soc. Philom., p. 76.
Agassiz, L. 1853 A.
Baird, S. F. 1857 A, p. 680.
Dawkins, W. B. 1872 A.
Flower and Lydekker 1891 A, p. 357.
Leidy, J. 1852 F, p. 71. (Boötherium.)
1853 A, p. 12. (Boötherium.)
1870 M, p. 97. (Boötherium.)
Lönnberg, E. 1900 A, p. 686.
Marsh, O. C. 1877 E.
Nicholson and Lydekker 1889 A, p. 1350.
Owen, R. 1856 C, p. 124.
Richardson, J. 1854 A, p. 120. (Bubalus.)
Rütimeyer, L. 1868 A.
1898 B, p. 139.
Zittel, K. A. 1893 B, p. 424.

Ovibos appalachicolus Rhoads.

Rhoads, S. N. 1895 A, p. 248. (Bison.)
Lucas, F. A. 1899 A, p. 756.
Rhoads, S. N. 1897 A, p. 492.
Pleistocene; Pennsylvania.

Ovibos bombifrons (Harlan).

Harlan, R. 1825 A, p. 271. (Bos.)
Cooper, W. 1831 A, pp. 173, 206. (Bos.)
Cooper, Smith, and De Kay 1831 A, p. 371. (Bos.)
Cope and Wortman 1884 A, p. 24.
Dawkins, W. B. 1870 A, p. 233. (Ovibos.)
1872 A, p. 29. (Boötherium.)

De Kay, J. E. 1828 A, p. 286. (Bos.)
1842 C, p. 110. (Bos.)
Harlan, R. 1834 B, p. 71. (Bos.)
1835 B, p. 275. (Bos.)
Leidy, J. 1852 F, p. 71. (Boötherium.)
1853 A, p. 17, pl. iv, fig. 2; pl. v, figs. 1, 2.
(Boötherium.)
1854 G, p. 210. (Boötherium.)
1869 A, p. 374.
Lydekker, R. 1885 C, p. 39.
Meyer, H. 1835 A, p. 143. (Bos.)
Pictet, F. J. 1853 A, p. 366.
Roger, O. 1896 A, p. 244. (O. priscus.)
Rütimeyer, L. 1865 A, p. 328. (O. priscus.)
1868 A, pp. 9, 17. (O. priscus.)
Trouessart, E. L. 1898 A, p. 984, in part only.
Wistar, C. 1818 A, p. 379, pl. xi, figs. 10, 11. (No
systematic name.)
Pleistocene; Kentucky.

Ovibos cavifrons Leidy.

Leidy, J. 1852 F, p. 71. (Boötherium or Ovibos.)
Agassiz, L, 1851 B, p. 179. (Bos bombifrons.)
Cooper, W. 1831 A, pp. 173, 206. (Bos pallasii.)
Cope and Wortman 1884 A, p. 24.
Dawkins, W. B. 1870 A, p. 233.
1872 A, p. 29. (Boötherium.)
De Kay, J. E. 1828 A, p. 291, pl. vi. (Bos pallasii.)
1842 C, p. 110. (Bos pallasii.)
Giebel, C. G. 1847 A, p. 154. (B. pallasii.)

Godman, J. D. 1828 A, p. 244. (" De Kay's fossil ox.")
Harlan, R. 1834 B, p. 72. (Bos pallasii.)
1835 B, p. 276. (B. pallasii.)
Leidy, J. 1853 A, p. 12, pl. iii, figs. 1, 2; pl. iv, fig. 1. (Boötherium.)
1854 G, pp. 209, 210. (Boötherium.)
1869 A, p. 374.
1870 L.
1870 M, p. 97. (Boötherium.)
Lydekker, R. 1885 C, p. 40. (O. bombifrons, in part.)
1898 B, p. 148. (O. bombifrons, in part.)
McGee, W J 1887 A, p. 217.
Meyer, H. 1835 A, p. 155. (B. pallasii.)
Pictet, F. J. 1853 A, p. 366. (Bos pallasii.)
Richardson, J. 1854 A, pp. 25, 120, pl. xi, figs. 2-4. (Ovibos maximus.)
Rütimeyer, L. 1865 A, p. 327. (O. priscus.)
St. John, S. 1851 A, p. 235. (Bos bombifrons.)
Trouessart, E. L. 1898 A, p. 986. (O. bombifrons, in part.)
Wailes, B. L. C. 1854 A, p. 286. (Boötherium.)
Zittel, K. A. 1893 B, p. 425. (O. priscus.)
Pleistocene; Indian Territory, Missouri, Kentucky, Ohio, Iowa, Alaska.

Ovibos moschatus (Zimm.).

Zimmermann, E., 1780, Geog.Gesch., ii, p. 86. (Bos.)
Baird, S. F. 1857 A, p. 681.
Buckland, Wm. 1831 A, pp. 597, 605. (" Musk-ox.")
Cuvier, G. 1834 A, vi, p. 311, pl. clxxi, figs. 15-17. (Bos.)
Dall, W. H. 1869 A. ("Musk-ox.")
Dawkins, W. B. 1872 A.
1883 A.
Gratacap, L. P. 1896 A, p. 994.
Leidy, J. 1854 G, p. 210.
1869 A, p. 373.
Lönnberg, E. 1900 A, p. 686, figures.
Lydekker, R. 1888 C, p. 38.
1898 B, p. 142.
Owen, R. 1856 C, p. 124, figs. 3-6. (Bubalus.)
Richardson, J. 1854 A, pp. 22, 66, 119, pls. ii-v; pl. xi, figs. 1-5; pl xiv, fig. 5.
Roger, O. 1896 A, p. 244.
Rütimeyer, L. 1868 A, p. 6, figs. 1, 2.
Trouessart, E. L. 1898 A, p. 984.
Tscherski, J. D. 1892 A, p. 153, pl. iv, figs. 3-6.
Woodward, H. 1894 A, l.
Recent; northern North America, Greenland. Pleistocene; Alaska, Greenland, northern Europe, and Asia.

BOVINÆ.

Flower and Lydekker 1891 A, p. 360. ("Bovine section.")
Gill, T. 1872 B, pp. 8, 77.
Roger, O. 1896 A, p. 244.
Stehlin, H. G. 1893 A.
Trouessart, E. L. 1898 A, p. 985.
Zittel, K. A. 1893 B, p. 425.

Bison II. Smith.   Type *Bos bison* Linn.

Smith, H., 1827, in Griffith's Anim. Kingd., v, p. 373.
Allen, J. A. 1876 B.
Baird, S. F. 1857 A, p. 682. (Bos.)
Brandt, F. 1867 C, p. 150.
Cope, E. D. 1889 G, p. 131.
Desor, E. 1852 A, p. 58. (Harlanus.)
Flower and Lydekker 1891 A, p. 360. (Bos.)
Leidy, J. 1871 C, p. 344.
Lucas, F. A. 1897 A.
1899 A.
Lydekker, R. 1898 B, p. 5 (Bos.); p. 50 (Bison, as subgenus).
Marsh, O. C. 1877 E.
Owen, R. 1846 A. p. 94. (Harlanus, type H. americanus=B. latifrons.)
1847 D, p. 18. (Harlanus.)
1856 C, p. 129.
Pictet, J. F. 1853 A, p. 303. (Harlanus.)
Rütimeyer, L. 1865 A, p. 335.
Trouessart, E. L. 1898 A, p. 994.
Tscherski, J. D. 1892 A, p. 75.
Zittel, K. A. 1893 B, p. 430.

Bison alleni Marsh.

Marsh, O. C. 1877 D, p. 252.
Blake, W. P. 1898 A, pp. 71, 72.
Cope, E. D. 1877 Y, p. 629.
1894 A, p. 68. (Bos crampianus; no description.)

Cope, E. D. 1894 H, p. 456, pl. xxii, figs. 1-4. (Bos crampianus.)
King, C. 1878 A, p. 430.
Lucas, F. A. 1897 A,
1899 A, pp. 756, 765, pls. lxxvii-lxxx; text fig. 2; text fig. 1[8].
Matthew, W. D. 1899 A, p. 74.
Rhoads, S. N. 1897 A, p. 488, pl. xii, fig. 5. (Bos crampanius.)
Roger, O. 1896 A, p. 246.
Trouessart, E. L. 1898 A, p. 995.
Williston, S. W. 1897 I, p. 302. (B. crampanius, B. alleni.)
1898 I, p. 92. (B. crampianus, B. alleni.)
Miocene (Loup Fork; Kansas, Oklahoma, Idaho.

Bison antiquus Leidy.

Leidy, J. 1852 H, p. 117.
Allen, J. A. 1876 B, p. 21,? pl. iii, figs. 1, 3, 6, 8, 10.
Cope, E. D. 1869 E, p. 176. (Bos antiquus.?)
1894 H, p. 457. (Syn. of Bos americanus.)
Leidy, J. 1853 A, p. 11, pl. ii, fig. 1.
1854 G, p. 210.
1867 A, p. 85. (B. latifrons?.)
1869 A, p. 371. (B. latifrons, in part.)
1870 H, p. 13. (Bos.)
1873 B, pp. 253, 318, in part; pl. xxviii, figs. 4, 5, 6,? 7.? (B. latifrons.)

Lucas, F. A. 1899 A, p. 759, pl. lxvii-lxx.
  1899 B, p. 17.
Lydekker, R. 1896 B, p. 61. (Syn. of Bos priscus.)
Middleton and Moore 1900 A, p. 178, plate.
Rhoads, S. N. 1897 A, p. 484 (B. antiquus); p. 501,
  pl. xii, fig. 2 (B. californicus).
Richardson, J. 1854 A, p. 139.
Rütimeyer, L. 1865 A, p. 336.
  1868 A, p. 60. (B. priscus.)
Zittel, K. A. 1893 B, p. 430.
  Pleistocene; Kentucky, Virginia?, Illinois?,
  California.

Bison bison (Linn.).

Linnæus, C., 1758, Syst. Nat., ed. 10, i, p. 72. (Bos.)
  Unless otherwise indicated, the following au-
  thors refer to this species under the name *Bison
  americanus.*
Allen, J. A. 1876 B, pl. xi, fig. 2; pl. iii, figs. 2, 4, 7,
  9, 11; pl. v, figs. 1-8; pl. vi, figs. 1-8; pl. vii, figs.
  1-6, 9, 10; pl. viii, figs. 1-11; pl. ix, figs. 1-8; pl.
  x, figs. 1-12; pl. xi, figs. 1-10; pl. xii, figs. 1-6.
Baird, S. F. 1857 A, p. 682. (Bos.)
Cooper, W. 1831 A, p. 43. (Bos americanus.)
Cooper, Smith, and De Kay 1831 A, p. 371. (Bos
  americanus.)
Cope and Wortman 1884 A, p. 24.
De Kay, J. E. 1842 C, p. 110.
Knight, ,— 1885 A, p. 166. (" Buffalo.")
Leidy, J. 1853 A, p. 5.
  1854 G, p. 210.
  1869 A, p. 371.
  1870 K, p. 69. (Bos americanus.)
  1880 A, p. 347.
  1889 E, p. 22.
  1889 H, p. 5.
  1892 A.
Lydekker, R. 1898 B, p. 79. (Bos.)
Mercer, H. C. 1894 A, p. 98.
Packard, A. S. 1867 A, pp. 243, 245, 246, 257, 261.
Rhoads, S. N. 1897 A, p. 484, pl. xii, fig. 1.
Richardson, J. 1854 A, p. 139, pl. vi, figs. 3, 4; pl.
  vii, figs. 4, 5; pl. viii, figs. 9-12.
Rütimeyer, L. 1865 A, p. 335.
  1867 B, p. 99.
  1868 A, p. 58.
Shaler N. S. 1876 A, p. 234.
Trouessart, E. L. 1898 A, p. 995.
Underwood, L. 1890 A, p. 968.
Williston, S. W. 1897 I, p. 302.
  1898 I, p. 91.
Wyman, J. 1862 A, p. 421. (" Bos.")
  Within recent times occupying a large por-
  tion of N. America. Pleistocene; Illinois, Iowa,
  Wisconsin, Kansas, Colorado, Pennsylvania,
  New York, South Carolina, Kentucky.

Bison crassicornis Richardson.

Richardson, J. 1854 A, pp. 40, 139, pl. ix; pl. xi, fig.
  6; pl. xii, figs. 1-4; pl. xiii, figs. 1, 2; pl. xv, figs.
  1-4.
Allen, J. A. 1876 B, pp. 14, 15, 21, ?pl. viii, fig. 12.
  (Syn. of B. antiquus.)
Buckland, W. 1831 A, p. 595, pl. iii, figs. 1-7.
  (Bos urus.)
Cope, E. D. 1894 H, p. 457. (Bos.)

Leidy, J. 1854 G, p. 210. (Bison priscus, B. crassicornis.)
  1869 A, p. 371 (Bison priscus); p. 372 (Syn.
  of Bison latifrons).
Lucas, F. A. 1897 A.
  1899 A, p. 760, pls. lxxiii-lxxvi; ?pl. lxxiii;
  text figs. 1, 2.
  1899 B, p. 18.
Lydekker, R. 1885 C, pp. 25, 26. (Bos bonasus, var.
  priscus.)
  1896 B, p. 61. (Syn. of Bos priscus.)
Rhoads, S. N. 1897 A, p. 500, pl. xii, figs. 3, 6. (B.
  alaskensis.)
Richardson, J. 1854 A, pp. 38, 139, pl. vi, figs. 5, 6;
  pl. vii, fig. 1; pl. x, figs. 1-6. (B. priscus?.)
Woodward, H. 1894 A. (B. priscus.)
  Pleistocene; Alaska.

Bison ferox Marsh.

Marsh, O. C. 1877 D, p. 252.
Blake, W. P. 1898 A, pp. 71, 72.
Cope, E. D. 1877 Y, p. 629.
King, C. 1878 A, p. 430.
Lucas, F. A. 1899 A, p. 766, pl. lxxxi.
Matthew, W. D. 1899 A, p. 74.
Rhoads, S. N. 1897 A, pp. 489, 500.
Roger, O. 1896 A, p. 246.
Trouessart, E. L. 1898 A, p. 995.
  Miocene (Loup Fork); Nebraska.

Bison latifrons (Harlan.)

Harlan, R. 1825 A, p. 273. (Bos.)
Allen, J. A. 1876 B, p. 7, pl. i; pl. ii, figs. 1-4; pl.
  vii, fig. 11.
Blainville, H. M. D. 1864 A, iv AA, p. 187. (Sus
  americanus.)
Blake, W. P. 1898 A, pp. 66, 71 (Bos arizonica);
  p. 69 (Bison latifrons).
    1898 B, p. 247. (B. latifrons, B. arizonica.)
Bojanus, L. H. 1827 A, p. 427. (Syn. of Urus pris-
  cus.)
Brandt, F. 1867 C.
Broadhead G. C. 1870 B. (Bos.)
Carpenter, W. 1846 A, p. 246, figs. 1, 2. (" Ex-
  tinct ox.")
Collett, J. in Cope and Wortman 1884 A, p. 24,
  footnote.
Cooper, W. 1831 A, p. 174. (Bos.)
Cope, E. D. 1884 G, p. 21. (Bos.)
  1889 F.
  1869 G, p. 131.
  1894 H, p. 457. (Bos.)
Cope and Wortman 1884 A, p. 24.
Couper, J. H. 1842 A, p. 216. (" Bos," Sus ameri-
  canus.)
Cuvier, G. 1806 B, p. 382, pl. xxxiv, fig. 2.
  (" Aurochs.")
  1834 A, vi, p. 287, pl. clxxiii, fig. 2.
  (" Aurochs.")
De Kay, J. E. 1828 A, p. 286. (Bos.)
  1842 C, p. 110. (Bos.)
Faujas, St. F. 1802 A, p. 190, pl. xliii. (" Bœuf
  fossil.")
Flower and Lydekker 1891 A, p. 364. (Bos.)
Giebel, C. G. 1847 A, p. 153 (Bos priscus, in part);
  p. 174 (Sus americanus).

Godman, J. D. 1828 A, iii, p. 243, pls. opp. pp. 242, 245. ("Latifrons.")
Harlan, R. 1834 B, p. 71. (Bos.)
    1835 B, p. 276. (Bos.)
    1842 B, p. 143, pl. iii, fig. 1. (?Bus americana.)
Leidy, J. 1852 H, p. 117.
    1853 A, p. 8, pls. i, ii, figs. 2–7.
    1854 C, p. 89.
    1854 G, p. 210.
    1860 B, p. 109, pl. xvii, figs. 15, 16. (Identification doubtful.)
    1867 A, p. 85.
    1869 A, p. 371.
    1871 C, p. 345, in part.
    1889 C, p. 12.
Lucas, F. A. 1899 A, p. 767, pls. lxxxii, lxxxiii.
Lydekker, R. 1885 C, p. 27.
    1898 B, p. 92. (Bos.)
Meyer, H. 1835 A, p. 141. (Syn. of Bos priscus.)
Owen, R. 1846 A, p. 94. (Harlanus americanus.)
    1847 D, p. 20, pl. vi. (H. americanus.)
Peale, R. 1803 A, p. 325, pl. vi. ("Great Indian buffalo.")
Pictet, F. J. 1853 A, p. 366. (Bos.)
Rafinesque, C. S. 1831, Enum. remark, nat. objects (Taurus latifrons, T. gigas); 1832–33, Atlantic Jour., p. 28 (Taurus latifrons).
Rhoads, S. N. 1897 A, p. 486, pl. xii, fig. 4.
Roger, O. 1896 A, p. 208 (Harlanus americanus.) p. 246 (Bison latifrons.)
Rütimeyer, L. 1865, p. 336.
    1868 A, p. 60. (B. priscus.)
Sanson, A. 1878 A, p. 757. (Bos.)

Shaler, N. S. 1876 A, p. 235. (Bos.)
Smith, H. P. 1887 A, p. 19, pl. i.
Stehlin, H. G. 1899 A, pp. 21, 118. (Platygonus ziegleri.)
Trouessart, E. L. 1898 A, p. 817 (Harlanus americanus); p. 995 (Bison latifrons).
Wailes, B. L. C. 1854 A, p. 286.
Whitney, J. D. 1879 A, p. 247.
Wilckens, M. 1885 B, p. 270. (Bus americana.)
Zittel, K. A. 1893 B, p. 430.
    Pleistocene; Kentucky, South Carolina?, Mississippi, Ohio, Georgia?, Texas, Florida, Arizona, Mexico.

Bison occidentalis Lucas.

Lucas, F. A. 1898 D, p. 678.
Allen, J. A. 1876 B, p. 34, pl. iv. (B. antiquus.)
?Dall, W. H. 1869 A, p. 136. ("Buffalo.")
?Hay, R. 1885 A, p. 98. ("Bison.")
?Leidy, J. 1870 K, p. 69. (Bos americanus.)
Lucas, F. A. 1899 A, p. 758, pls. lxv, lxvi.
    1899 B, p. 17, pls. viii, ix.
Stewart, A. 1897 B, p. 127, pl. xvii; text fig. 1. (B. antiquus.)
Williston, S. W. 1897 I, p. 302. (B. antiquus.)
    1898 I, p. 91. (B antiquus.)
    Pleistocene; Alaska, Kansas.
    Bos bovis.—This species has been reported from Ashley River, South Carolina, by F. S. Holmes (1850 B, p. 203). Leidy (1869 A, p. 443), characterizes it as a doubtful fossil. See also Holmes, F. S. 1858 A, in which paper are found extracts from letters written by Louis Agassiz and Dr. Leidy.

Order ANCYLOPODA Cope.

Cope, E. D. 1889 I, p. 158.
Ameghino F. 1894 A, p. 311.
Cope, E. D. 1889 C.
    1891 N, p. 76.
    1898 B, p. 120. (Chalicotheria.)
Gadow, H. 1898 A, p. 47.
Hay, O. P. 1899 F, p. 682.
Marsh, O. C. 1892 D, p. 448. (Chalicotheria.)
    1897 B, p. 144. (Chalicotheria.)
Osborn, H. F. 1898 D, p. 48.

Osborn, H. F. 1898 F.
    1898 J. (Perissonychia.)
    1898 K.
    1898 A, p. 146.
Osborn and Wortman 1893 A, p. 4. (Ord. Ancylopoda; subord. Perissonychia.)
Roger, O. 1896 A, p. 196.
Trouessart, E. L. 1898 A, p. 693.
Woodward, A. S. 1898 B, p. 307.

Superfamily CHALICOTHERIOIDEA.

Gill, T. 1872 B, p. 71.
Osborn, H. F. 1898 I, p. 79.

Scott and Osborn 1890 B, p. 99.

CHALICOTHERIIDÆ.

Gill, T. 1872 B. pp. 8, 72.
Ameghino, F. 1891 D, p. 284.
    1894 A, p. 316.
Cope, E D. 1879 C, p. 228.
    1881 G, pp. 378, 379, 395.
    1881 O.
    1884 O, pp. 615, 694.
    1887 N.
    1891 L, p. 1117.
Earle, C. 1892 C, p. 276.
Flower and Lydekker 1891 A, p. 413.
Haeckel, E. 1895 A, p. 546. (Chalicotherida.)
Lydekker, R. 1886 A, p. 161.
Marsh, O. C. 1877 D, p. 249. (Moropodidæ.)

Marsh, O. C. 1877 E. (Moropodidæ.)
    1897 B, p. 145. (Moropodidæ.)
Nicholson and Lydekker 1889 A, p. 1372.
Osborn, H. F. 1891 D.
    1892 C.
    1898 D, p. 48.
Osborn and Wortman 1893 A, p. 13.
Roger, O. 1896 A, p. 196.
Schlosser, M. 1886 A, p. 258.
    1886 B, p. 18.
Steinmann and Döderlien 1890 A, p. 776.
Wortman, J. L. 1882 B, p. 723.
    1883 A, p. 709.
Zittel, K. A. 1898 B, p. 308.

CHALICOTHERIUM Kaup.   Type *C. goldfussi* Kaup.

Kaup, J. J. 1833, Oss. Foss. Darmstadt, pt. ii, p. 4.
Ami, H. M. 1891 A.
Bronn, H. G. 1838 A, p. 1201, pl. xlvi, fig. 2.
Cope, E. D. 1884 O, p. 695.
    1887 N, p. 1061.
    1889 P, p. 658.
    1891 A, p. 7.
Depéret, C. 1892 A, p. 61.
Earle, C. 1892 C, p. 276.
Filhol, H. 1891 A, p. 294.
Flower and Lydekker 1891 A, p. 413.
Gaudry, A. 1875 B.
    1877 A.
Kowalevsky, W. 1873 A, p. 175.
Lydekker, R. 1886 A, p. 162.
Marsh, O. C. 1874 A, p. 82.
    1877 E.
    1884 F, p. 172.
    1897 B, p. 143.
Nicholson and Lydekker 1889 A, p. 1373.
Osborn, H. F. 1888 D.
    1891 D, p. 911.
    1892 C.
    1892 G.

Osborn, H. F. 1893 D, pp. 6, 43.
    1893 F, fig. 1.
    1893 K.
Osborn and Wortman 1893 A.
Pictet, F. J. 1853 A, p. 337.
Schlosser, M. 1883 A.
    1886 B, p. 19.
    1893 B, p. 659.
Scott, W. B. 1891 A, p. 3.
    1891 D, p. 90.
Scott and Osborn 1890 B, p. 99.
Steinmann and Döderlein 1890 A, p. 778.
Trouessart, E. L. 1898 A, p. 697.
Wortman, J. L. 1883 A, p. 710.
Zittel, K. A. 1893 B, p. 312.

Chalicotherium bilobatum Cope.

Cope, E. D. 1889 I, p. 151.
    1891 A, p. 8, pl. iv, fig. 1.
Matthew, W. D. 1899 A, p. 58.
Osborn and Wortman 1893 A, p. 2.
Roger, O. 1896 A, p. 197.
Trouessart, E. L. 1898 A, p. 698.
    Oligocene (White River); Canada.

MOROPUS Marsh.   Type *M. distans* Marsh.

Marsh, O. C. 1877 D, p. 249.
Flower and Lydekker 1891 A, pp. 190, 413.
Marsh, O. C. 1877 E, p. 251.
    1892 D, p. 448.
    1897 B, p. 144.
Matthew, W. D. 1899 A, p. 72.
Nicholson and Lydekker 1889 A, p. 1373.
Osborn, H. F. 1891 D, p. 911.
    1893 D, p. 43.
    1893 F, p. 122.
Osborn and Wortman 1893 A, p. 2.   (Syn. of Chalicotherium.)
Scott, W. B. 1893 B, p. 1.
Wortman, J. L. 1897 B, p. 60.
Zittel, K. A. 1893 B, p. 314.

Moropus distans Marsh.

Marsh, O. C. 1877 D, p. 249.
King, C. 1878 A, p. 424.
Matthew, W. D. 1899 A, p. 68.
Osborn, H. F. 1893 F, p. 122.
Roger, O. 1896 A, p. 197.
Trouessart, E. L. 1898 A, p. 698.
    Miocene (John Day); Oregon.

Moropus elatus Marsh.

Marsh, O. C. 1877 D, p. 250.
King, C. 1878 A, p. 430.
Matthew, W. D. 1899 A, p. 72.
Osborn, H. F. 1893 F, p. 122.
Roger, O. 1896 A, p. 197.
Scott and Osborn 1890 B, p. 99, fig. 18.   (Chalicotherium.)
Trouessart, E. L. 1898 A, p. 698.
Zittel, K. A. 1893 B, p. 315, fig. 255.
    Miocene (Loup Fork); Nebraska.

Moropus senex Marsh.

Marsh, O. C. 1877 D, p. 250.
King, C. 1878 A, p. 424.
Matthew, W. D. 1899 A, p. 68.
Osborn, H. F. 1893 F, p. 122.
Roger, O. 1896 A, p. 197.
Trouessart, E. L. 1898 A, p. 698.
    Miocene (John Day); Oregon.

SPHENOCŒLUS Osb.   Type *S. uintensis* Osb.

Osborn, H. F. 1895 A, p. 98.
A genus of doubtful position.

Sphenocœlus uintensis Osb.

Osborn, H. F. 1895 A, p. 98. figs. 12–15.

Matthew, W. D. 1899 A, p. 50.
Roger, O. 1896 A, p. 197.
    Eocene (Uinta); Utah.

Order AMBLYPODA Cope.

Cope, E. D. 1875 C, p. 28.
1875 M.
1877 K, p. 179.
1877 L, p. 619. (Protencephala, in part.)
1877 S. (Protoencephala, in part; Amblypoda.)
1881 J.
1882 A (Amblypoda); p. 447 (Hyodonta.)
1882 E, pp. 165, 177.
1882 Y.
1883 D.
1884 E.
1884 O, pp. 167, 378, 607.
1884 EE.
1885 F.
1885 O.
1886 C, p. 616.
1886 K, p. 721,
1887 N. (Amblypoda, Hyodonta.)
1887 S, p. 459.
1888 F.
1889 C.
1889 R, p. 877.
1891 N, p. 84.
1898 B, p. 124.
Cope and Wortman 1884 A, p. 12.
Earle, C. 1892 B.
Flower, W. H. 1876 B, xiii, p. 388.
1883 D, p. 426.
Flower and Lydekker 1891 A, p. 437.
Gadow, H. 1898 A, p. 45.
Haeckel, E. 1895 A, p. 541.
Hay, O. P. 1899 F, p. 682.
Lydekker, R. 1885 E, p. 469.

Lydekker, R. 1886 A, p. 175.
1896 B, p. 173.
Marsh, O. C. 1884 F, pp. 173, 193. (Amblydactyla.)
1892 C, p. 323. (Amblydactyla.)
1893 C, p. 324. (Amblydactyla.)
Nicholson and Lydekker 1889 A, p. 1383.
Osborn, H. F. 1888 F.
1890 D, p. 538.
1893 A, pp. 383, 459.
1893 D.
1898 A, p. 146.
1898 H.
1900 C, p. 567.
Roger, O. 1896 A, p. 155.
Rütimeyer, L. 1890 A, p. 16.
Science 1885 A, pp. 489, 490. (Amblypoda vs.
Amblydactyla.)
Schlosser, M. 1887 A.
1888 A, p. 586.
1889 A.
1890 C.
1890 D.
Scott, W. B. 1886 A, p. 303.
1892 A, p. 439.
Steinmann and Döderlein 1890 A, p. 762.
Tomes, C. S. 1898 A, p. 440.
Trouessart, E. L. 1898 A, p. 712.
Weithofer, K. A. 1888 B.
Woodward, A. S. 1898 B, p. 292.
Wortman, J. L. 1882 B, p. 721.
1885 A, p. 297.
Zittel, K. A. 1893 B, pp. 207, 432, 725.

[Genera of uncertain position, possibly ancestral to the *Amblypoda*.]

SYNCONODON Osb.   Type *S. sexcuspis* Osb.

Osborn, H. F. 1898 H, p. 171.

Synconodon sexcuspis Osb.

Osborn, H. F. 1898 H, pp. 171, 182, fig. 1 C.

Osborn, H. F. 1898, p. 320, pl. viii, fig. C. (No
name.)
Trouessart, E. L. 1899 A, p. 1346.
Cretaceous (Laramie) Wyoming.

ECTOCONODON Osb.   Type *E. petersoni* Osb.

Osborn, H. F. 1898 H, p. 171.

Ectoconodon petersoni Osb.

Osborn, H. F. 1898 H, p. 171, 182, fig. 1 F.

Osborn, H. F. 1898, p. 320, pl. viii, fig. F. (No
name.)
Trouessart, E. L. 1899 A, p. 1346.
Cretaceous (Laramie) Wyoming.

PROTOLAMBDA Osb.   Type *P. hatcheri* Osb.

Osborn, H. F. 1898 H, p. 172.

Protolambda hatcheri Osb.

Osborn, H. F. 1898 H, pp. 172, 182, fig. 1 A.

Osborn, H. F. 1898, p. 320, pl. viii, fig. A. (No
name.)
Trouessart, E. L. 1899 A, p. 1346.
Cretaceous (Laramie) Wyoming.

Suborder TALIGRADA Cope.

Cope, E. D. 1883 Z.
1884 O, p. 600.
1884 EE, p, 1111.
1889 R, p. 877.
1898 B, p. 124.

Flower and Lydekker 1891 A, p. 436.
Nicholson and Lydekker 1889 A, p. 1385.
Osborn, H. F. 1898 H, pp. 177, 180, 181.
Osborn and Earle 1895 A, p. 43.
Wortman, J. L. 1885 A, p. 298.

PERIPTYCHIDÆ.

Cope. E. D. 1882 D.
    1882 A, pp. 446, 447.
    1883 O.
    1884 O, pp. 384, 385.
    1884 AA, pp. 793, 794.
    1887 N, p. 986.
    1888 K, p. 162.
    1889 C, p. 177.
    1897 B.
Earle, C. 1893 D, p. 51.
Haeckel, E. 1895 A, p. 584. (Periptychida.)
Lydekker, R. 1886 A, p. 172.
Matthew, W. D. 1897 C, pp. 262, 294, 295.
    1899 A, p. 29.

Nicholson and Lydekker 1889 A, pp. 1323, 1379.
Osborn, H. F. 1898 A, p. 146.
    1898 H, p. 181.
Osborn and Earle 1895 A, pp. 47, 52, 61.
Pavlow, M. 1887 A, p. 354.
Roger, O. 1896 A, p. 164.
Schlosser, M. 1886 B.
    1888 A, p. 585.
    1890 C.
Scott, W. B. 1892 A, p. 428.
Steinmann and Döderlein 1890 A, p. 750.
Zittel, K. A. 1893 B, p. 216.

PERIPTYCHINÆ.

Osborn and Earle 1895 A, p. 52.
Matthew, W. D. 1897 C, pp. 295, 296.

Osborn, H. F. 1898 H, p. 182.
Trouessart, E. L. 1898 A, p. 722.

PERIPTYCHUS Cope.   Type *P. carinidens* Cope.

Cope, E. D. 1881 H, p. 484.
    1881 H, p. 487. (Catathlæus, type *C. rhab-*
     dodon.)
    1881 N.
    1881 U. (Catathlæus.)
    1882 A, p. 447.
    1882 E, p. 158.
    1882 F, p. 96. (Periptychus, Catathlæus.)
    1882 DD.
    1883 G.
    1883 O.
    1883 AA.
    1884 E.
    1884 O, pp. 382, 387.
    1884 AA, pp. 794, 796.
    1884 EE, p. 1120.
    1887 I, p. 658.
    1894 E, p. 868.
Earle, C. 1893 D.
Flower and Lydekker 1891 A, p. 439.
Lydekker, R. 1886 A, p. 172.
Matthew, W. D. 1899 A, p. 28.
Nicholson and Lydekker 1889 A, p. 1379.
Osborn, H. F. 1888 H.
    1890 D, p. 532.
    1893 B, p. 326.
    1898 H, pp. 181, 186.
Osborn and Earle 1895 A, p. 53.
Pavlow, M. 1887 A, p. 354.
Schlosser, M. 1886 B.
Scott, W. B. 1892 A, p. 428.
Steinmann and Döderlein 1890 A, p. 750.
Wortman, J. L. 1886 A, p. 470.
Zittel, K. A. 1893 B, p. 217.

Periptychus carinidens Cope.

Cope, E. D. 1881 N, p. 337.
    1881 H, p. 484.
    1883 M, p. 561.
    1884 O, pp. 403, pl. xxiiid, figs. 14, 15; pl.
     xxiv, fig. 5; pl. xxva, fig. 16.
    1884 AA, p. 801.

Cope, E. D. 1888 CC, p. 305 (P. carinidens); pp.
    305, 354 (P. brabensis).
Matthew, W. D. 1897 C, pp. 264, 297.
    1899 A, p. 29.
Osborn, H. F. 1898 H, p. 182.
Osborn and Earle 1895 A, p. 55. (P. brabensis.)
Roger, O. 1896 A, p. 165.
Trouessart, E. L. 1898 A, p. 722.
    Eocene (Torrejon); New Mexico.

Periptychus coarctatus Cope.

Cope, E. D. 1883 Q, p. 168.
    1884 K, p. 309.
    1884 O, explan., pl. xxixd, figs. 7, 8.
    1884 AA, p. 802, fig. 10.
    1885 J, p. 387.
    1888 CC, pp. 305, 354. (P. coarctatus, C.
     brabensis.)
Matthew, W. D. 1897 C, p. 296.
    1899 A, p. 28.
Osborn, H. F. 1898 H, p. 182.
Osborn and Earle 1895 A, p. 54.
Roger, O. 1896 A, p. 165. (P. coarctatus, P.
    brabensis.)
    Eocene (Puerco); New Mexico.

Periptychus rhabdodon Cope.

Cope, E. D. 1881 U, p. 829. (Catathlæus.)
    1881 H, p. 487. (Catathlæus.)
    1882 C, p. 465.
    1883 G, p. 564, pl. li, figs. 2, 3.
    1884 K, p. 309.
    1884 O, p. 391, pl. xxiif; pl. xxiig, figs.
     1–11; pl. lviif, figs. 1, 2.
    1884 AA, p. 794, figs. 2, 6–9.
    1887 I, p. 658.
    1887 S, p. 268, fig. 44.
    1888 CC, p. 305.
Lydekker, R. 1886 A, p. 174, figs. 24, 25.
Matthew, W. D. 1897 C, pp. 264, 296, 320.
    1899 A, p. 29.
Osborn, H. F. 1890 D, p. 533, diagrams 6, 7.

Osborn, H. F. 1898 H, p. 182.
Osborn and Earle 1895 A, p. 53.
Pavlow, M. 1887 A, p. 359, figs. 2, 5a.
Roger. O. 1896 A, p. 165.

Schlosser, M. 1886 B, p. 35.
Trouessart, E. L. 1898 A, p. 722.
Zittel, K. A. 1893 B, p. 218, figs. 152, 153, 156, 157.
Eocene (Torrejon); New Mexico.

ECTOCONUS Cope. Type *Periptychus ditrigonus* Cope.

Cope, E. D. 1884 AA, p. 796.
1884 E, p. 313.
1884 O, p. xxxv.
1888 CC, p. 355.
Osborn, H. F. 1898 B, p. 325.
1898 H, p. 173.
Osborn and Earle 1895 A, p. 56.
Schlosser, M. 1886 B, p. 37.
Zittel, K. A. 1893 B, p. 218.

Ectoconus ditrigonus Cope.

Cope, E. D. 1882 C, p. 465. (Periptychus.) ·
1883 Q, p. 168. (Conoryctes.)
1883 EE, p. 968. (Conoryctes.)

Cope, E. D. 1884 A, p. 796. (Ectoconus.)
1884 O, p. xxxv (Ectoconus); p. 404, pl.
xxxg, fig. 12 (Periptychus): explan. pl.
xxixd, figs. 2-6 (Conoryctes).
1884 AA, p. 797, figs. 4, 5.
1888 CC, p. 356.
Matthew, W. D. 1897 C, pp. 264, 320
1899 A, p. 28.
Osborn, H. F. 1898 H, p. 182.
Osborn and Earle 1895 A, p. 56.
Roger, O. 1896 A, p. 165.
Schlosser, M. 1886 B, p. 37, pl. vi. figs. 30, 31.
Trouessart, E. L. 1898 A, p. 723.
Eocene (Puerco); New Mexico.

ANISONCHINÆ.

Osborn and Earle 1895 A, p. 52.
Matthew, W. D. 1897 C, pp. 295, 297.

Osborn, H. F. 1898 H, p. 182.
Trouessart, E. L. 1898 A, p. 720.

HAPLOCONUS Cope. Type *H. lineatus* Cope.

Cope, E. D. 1882 V, p. 417.
1882 C, p. 466.
1882 F, p. 96.
1884 O, pp. 386, 415.
1884 AA, p. 795.
1887 I, p. 657.
Lydekker, R. 1896 A, p. 175.
Matthew, W. D. 1897 C, p. 298.
Osborn and Earle 1895 A, p. 58.
Pavlow, M. 1887 A.
Schlosser, M. 1886 B, p. 37.
Steinmann and Döderlein 1890 A, p. 750.
Zittel, K. A. 1893 B, p. 216.

Haploconus angustus Cope.

Cope, E. D. 1881 U, p. 831. (Miochænus.)
1881 H, pp. 490, 491. (Miochænus.)
1882 V, p. 418.
1884 O, p. 416, pl. lvi f, fig. 6.
1887 I.
1888 CC, p. 304.
Matthew, W. D. 1897 C, p. 298. (Syn. of H. lineatus.)
Roger, O. 1896 A, p. 164 (H. angustus): p. 250 (Tricentes angustus).
Trouessart, E. L. 1898 A, p. 720. (Syn. of H. lineatus.)
Eocene (Torrejon); New Mexico.

Haploconus corniculatus Cope.

Cope, E. D. 1888 CC, p. 349.

Matthew, W. D. 1897 C, p. 298.
1899 A, p. 29.
Roger, O. 1896 A, p. 165.
Trouessart, E. L. 1898 A, p. 720.
Eocene (Torrejon); New Mexico.

Haploconus lineatus Cope.

Cope, E. D. 1882 V, p. 417.
1882 C, p. 466. (H. xiphodon.)
1884 K, p. 309.
1884 O, pp. 416, 417, pl. xxve, figs. 1-4 (H.
lineatus); pp. 416, 420, pl. xxve, figs. 5, 6
(H. xiphodon).
1884 AA, p. 804, fig. 13a, 13b (H. lineatus)
fig. 13c (H. xiphodon).
1887 I, p. 658.
1888 CC, p. 304. (H. lineatus, H. xiphodon.)
Lydekker, R. 1886 A, p. 175.
Matthew, W. D. 1897 C, pp. 264, 298.
1899 A, p. 29.
Osborn, H. F. 1893 B, p. 325.
Osborn and Earle 1895 A, p. 59.
Roger, O. 1896 A, p. 165. (H. lineatus, H. xiphodon.)
Schlosser, M. 1886 B, p. 37, pl. v, fig. 27.
Scott, W. B. 1892 A, p. 429. (H. xiphodon.)
Zittel, K. A. 1893 B, p. 216, fig. 154.
Eocene (Torrejon); New Mexico.
It is not improbable that this species will
prove to be identical with *H. angustus.*

ANISONCHUS Cope. Type *A. sectorius* Cope.

Cope, E. D. 1881 H, pp. 487, 488.
1882 E, p. 178.
1882 F, p. 96.
1883 AA, p. 408.
1884 E, p. 314.

Cope, E. D. 1884 O, pp. 386, 408.
1884 AA, pp. 795, 808.
1887 I, p. 657.
Matthew, W. D. 1897 C, p. 298.
Nicholson and Lydekker 1889 A, p. 1380.

Osborn and Earle 1895 A, p. 60.
Pavlow, M. 1887 A.
Schlosser, M. 1886 B, p. 37.
Scott, W. B. 1892 A, p. 428.
Trouessart, E. L. 1898 A, p. 721.
Zittel, K. A. 1893 B, p. 216.

Anisonchus gillianus Cope.

Cope, E. D. 1882 C, p. 467.　(Haploconus.)
　　1882 AA, p. 686.　(Haploconus.)
　　1884 O, pp. 409, 411, pl. xxve, fig. 7 (A. gillianus, A. apiculatus); pl. xv/, figs.10, 11 (H. gillianus).
　　1884 AA, p. 803.
　　1888 CC, pp. 305, 352.　(Hemithlæus apiculatus, A. gillianus.)
Matthew, W. D. 1897 C, p. 299.
　　1899 A, p. 28.
Roger, O. 1896 A, p. 165.　(A gillianus, Hemithlæus apiculatus.)
Scott, W. B. 1892 A, p. 428.
Trouessart, E. L. 1898 A, p. 721.
Eocene (Puerco); New Mexico.

Anisonchus sectorius Cope.

Cope, E. D. 1881 H, p. 488.
　　1881 H, p. 490.　(Mioclænus mandibularis.)
　　1881 U, p. 831.　(Mioclænus sectorius, M. mandibularis.)
　　1882 C, p. 467.
　　1884 K, p. 309.
　　1884 O, pp. 325, 339, pl. lvii/, fig. 7 (Mioclænus mandibularis); pp. 409, 413, pl. xxvc, figs. 5, 6, 8 (Anisonchus sectorius).
　　1884 AA, p. 803, fig. 12.
　　1884 CC, p. 305, 351.　(A. sectorius, A. mandibularis.)
Matthew, W. D. 1897 C, pp. 298, 299.
　　1899 A, p. 29.
Osborn and Earle 1895 A, p. 61.　(A. sectorius, A. mandibularis.)
Pavlow, M. 1887 A, fig. 9.
Roger, O. 1896 A, p. 164.　(A. sectorius, A. mandibularis.)
Schlosser, M. 1886 B, p. 37, pl. vi, fig. 26.
　　1887 B, p. 221.　(Mioclænus mandibularis.)
Scott, W. B. 1892 A, p. 428.
Trouessart, E. L. 1898 A, p. 721.
Eocene (Torrejon); New Mexico.

ZETODON Cope.　Type *Z. gracilis* Cope.

Cope, E. D. 1883 Q, p. 169.
　　1883 EE.
　　1884 O, explan. pl. xxixd, fig. 9.
　　1884 AA, pp. 795, 804.
Matthew, W. D. 1897 C, p. 299.
Osborn, H. F. 1898 B, p. 325.
Pavlow, M. 1887 A, p. 302.
Schlosser, M. 1886 B, p. 37.
Zittel, K. A. 1893 B, p. 218.
　　According to Matthew this genus is probably founded on crushed jaws of small specimens of *Anisonchus.*

Zetodon gracilis Cope.

Cope, E. D. 1883 Q, p. 169.
　　1883 EE.
　　1884 O, explan. pl. xxixd, fig. 9.
　　1884 AA, pp. 802, 805, fig. 11 d, d'.
　　1888 CC, p. 305.
Matthew, W. D. 1897 C, p. 299.
　　1899 A, p. 28.
Roger, O. 1896 A, p. 165.
Eocene (Puerco); New Mexico.

HEMITHLÆUS Cope.　Type *H. kowalevskianus* Cope.

Cope, E. D. 1882 DD, p. 832.
　　1884 O, pp. 386, 405.
　　1884 AA, pp. 795, 803.
　　1884 K, p. 309.
　　1887 I, p. 657.
　　1888 CC, p. 352.
Nicholson and Lydekker 1889 A, p. 1380.
Osborn and Earle 1895 A, p. 60.
Pavlow, M. 1887 A.
Schlosser, M. 1886 B, p. 37.
Scott, W. B. 1892 A, p. 428.
Zittel, K. A. 1893 B, p. 217.

Hemithlæus kowalevskianus Cope.

Cope, E. D. 1882 DD, p. 832.
　　1882 DD, p. 832.　(Anisonchus coniferus, in part.)

Cope, E. D. 1884 O, p. 405, pl. xxv, figs. 6, 7 (H. kowalevskianus); p. 409, in part,? pl. xxivg, fig. 6a-c (Anisonchus coniferus).
　　1884 AA, p. 802, fig. 11c.
　　1888 CC, pp. 305, 353, fig. 11,
Matthew, W. D. 1897 C, pp. 264, 297, fig. 11.
　　1899 A, p. 28.
Osborn and Earle 1895 A, p. 60, fig. 18.
Roger, O. 1896 A, p. 164.　(H. kowalevskianus, Anisonchus coniferus.)
Scott, W. B. 1892 A, p. 428.　(Anisonchus coniferus.)
Trouessart, E. L. 1898 A, p. 721.　(H. kowalevskianus.)
Zittel, K. A. 1898 A, p. 721, fig. 155.　(H. kowalewskyanus.)
Eocene (Puerco); New Mexico.

CONACODON Matth.　Type *Haploconus entoconus* Cope.

Matthew, W. D. 1897 C, p. 298.
Trouessart, E. L. 1898 A, p. 721.

Conacodon cophater (Cope).

Cope, E. D. 1884 K, p. 321.　(Anisonchus.)
　　1888 CC, pp. 304, 349.　(Haploconus.)

Matthew, W. D. 1897 C, pp. 264, 298.
　　1899 A, p. 28.
Osborn, H. F. 1898 H, p. 182.
Roger, O. 1896 A, p. 164.　(Haploconus.)
Trouessart, E. L. 1898 A, p. 721.
Eocene (Puerco); New Mexico.

Conacodon entoconus (Cope.)

Cope, E. D. 1882 AA, p. 686. (Haploconus.)
 1882 DD, p. 832. (Anisonchus coniferus, in part.)
 1884 O, pp. 416, 421, pl. xxxvf, figs. 4, 5 (Haploconus); p. 409, pl. xxivg, fig. 6 (Anisonchus coniferus).
 1884 AA, p. 804 (Haploconus); p. 808, fig. 12c (Anisonchus coniferus).
 1888 CC, p. 304 (Haploconus); p. 305 (Anisonchus coniferus).

Matthew, W. D. 1897 C, pp. 264, 298.
 1899 A, p. 28.
Osborn, H. F. 1898 H, p. 182.
Pavlow, M. 1887 A, fig. 10. (Haploconus.)
Roger, O. 1896 A, p. 165.
Schlosser, M. 1886 B, p. 37, pl. vi, fig. 28. (Haploconus.)
Scott, W. B. 1892 A, p. 428. (Haploconus.)
 Eocene (Puerco); New Mexico.

PANTOLAMBDIDÆ.

Cope, E. D. 1883 Z.
 1884 O, p. 601.
 1888 K, p. 162.
 1897 B, p. 336.
Haeckel, E. 1895 A, pp. 530, 542. (Pantolambdina.)
Matthew, W. D. 1897 C, p. 294.

Nicholson and Lydekker 1889 A, p. 1385.
Osborn, H. F. 1898 H, p. 182.
Roger, O. 1896 A, p. 155.
Steinmann and Döderlein 1890 A, p. 762.
Zittel, K. A. 1893 B, p. 434.

PANTOLAMBDA Cope.   Type *P. bathmodon* Cope.

Cope, E. D. 1882 V, p. 418.
Ameghino, F. 1896 A, p. 445.
Cope, E. D. 1882 F, p. 96.
 1883 Z.
 1884 O, p. 601.
 1884 EE, pp. 1111, 1120.
 1885 A, p. 478.
 1888 CC, p. 360.
 1889 C.
Earle, C. 1892 A, pp. 150, 158.
 1892 B.
Matthew, W. D. 1897 C, pp. 308, 319.
Nicholson and Lydekker 1889 A, p. 1385.
Osborn, H. F. 1888 H.
 1893 E.
 1897 C, p. 1013, fig. 14.
 1898 A, p. 146.
 1898 F, p. 91.
 1898 H, pp. 173, 183, 184.
 1898 M, pp. 585, 587.
Osborn and Earle 1895 A, p. 43.
Osborn and Wortman 1892 A, p. 122.
Schlosser, M. 1888 A, p. 585.
Scott, W. B. 1892 A, p. 439.
Zittel, K. A. 1893 B, p. 435.

Pantolambda bathmodon Cope.

Cope, E. D. 1882 V, p. 418.
 1883 M, p. 557.

Cope, E. D. 1883 Z, p. 406.
 1884 O, p. 603, pl. xxixb; pl. xxixc, figs. 3–7.
 1884 EE, p. 1111, figs. 3–5.
 1885 F, p. 55.
 1887 S, p. 275, pl. vii, fig. 2.
 1888 F, p. 82, fig. 1.
 1888 CC, pp. 305, 360.
 1889 C, p. 237.
Matthew, W. D. 1897 C, p. 320.
 1899 A, p. 29.
Osborn, H. F. 1898 H, pp. 181, 182, 184, figs. 8–13.
Osborn and Earle 1895 A, p. 43, figs. 14–16.
Roger, O. 1896 A, p. 155.
Trouessart, E. L. 1898 A, p. 712.
 Eocene (Torrejon); New Mexico.

Pantolambda cavirictus Cope.

Cope, E. D. 1888 Q, p. 159.
 1883 EE.
 1884 O, explan. pl. xxixd, fig. 1.
 1884 EE, p. 1113, fig. 6.
 1888 F, p. 83, fig. 2.
 1888 CC, pp. 305, 360.
Matthew, W. D. 1897 C, p. 320.
 1899 A, p. 29.
Osborn, H. F. 1898 H, pp. 172, 179, 183, figs. 2, 7, 13.
Roger, O. 1896 A, p. 155.
Trouessart, E. L. 1898 A, p. 712.
 Eocene (Torrejon); New Mexico.

Suborder PANTODONTA Cope.

Cope, E. D. 1873 F, pp. 40, 67.
 1875 C, p. 28.
 1875 M.
 1877 K, p. 182.
 1877 S.
 1882 A.
 1882 E, p. 165.
 1883 M, p. 558.
 1883 Z.
 1884 O, pp. 379, 510, 514, 601.
 1884 EE.

Cope, E. D. 1885 O.
 1898 B, p. 124.
Lydekker, R. 1886 A, p. 176. (Coryphodontia.)
Marsh, O. C. 1884 F, p. 193. (Coryphodontia.)
 1893 C, p. 323. (Coryphodontia.)
Nicholson and Lydekker 1889 A, p. 1385. (Coryphodontia.)
Osborn, H. F. 1893 A, p. 459.
 1898 H, pp. 177, 190, 188.
Science 1885 A, p. 490.
Scott, W. B. 1886 A, p. 303.

CORYPHODONTIDÆ.

Marsh, O. C. 1876 D, p. 428.
Cope, E. D. 1872 EE, p. 279. (Bathmodontidæ.)
    1873 E, pp. 563, 585. (Bathmodontidæ.)
    1873 K. (Bathmodontidæ.)
    1882 E, p. 165.
    1882 M.
    1884 E.
    1884 L, p. 324.
    1884 O, p. 515.
    1884 EE, p. 1192.
    1888 F, p. 80.
    1898 H.
Earle, C. 1892 B.
Flower and Lydekker 1889 A, p. 438.
Haeckel, E. 1895 A, p. 542. (Coryphodontida.)

Lydekker, R. 1886 A, p. 176.
Marsh, O. C. 1877 A.
    1877 E.
    1893 C, p. 323.
    1893 H.
Matthew, W. D. 1899 A, p. 33.
Nicholson and Lydekker 1889 A, p. 1386.
Osborn, H. F. 1888 H.
    1896 H, p. 182.
Roger, O. 1896 A, p. 155.
Rütimeyer, L. 1888 A, p. 13.
Schlosser, M. 1888 A, p. 586.
Steinmann and Döderlein 1890 A, p. 763.
Trouessart, E. L. 1898 A, p. 712.
Zittel, K. A. 1893 B, p. 435.

CORYPHODON Owen.   Type *C. eocænus* Owen.

Owen, R. 1845 B, p. 607 pl. cxxxv.
Ameghino, F. 1896 A, p. 445.
Amer. Jour. Science 1873 A.
Cope, E. D. 1872 A. (Bathmodon.)
    1872 G, p. 350 (Bathmodon): p. 353 (Loxo-
        lophodon).
    1872 J, p. 417 (Bathmodon, type *B. radians*
        Cope); p. 420 (Loxolophodon, type *Bath-
        modon semicinctus*).
    1872 S. (Loxolophodon, in part.)
    1872 T, p. 542. (Metalophodon, type *M.
        armatus* Cope.)
    1872 AA. (Bathmodon.)
    1872 LL. (Metalophodon.)
    1873 A. (Bathmodon.)
    1873 D, p. 3. (Bathmodon, Metalophodon.)
    1873 E, p. 586 (Bathmodon); pp. 586, 589
        (Metalophodon.)
    1873 F, p. 67 (Bathmodon): pp. 67, 71 (Meta-
        lophodon.)
    1873 K. (Bathmodon, Metalophodon.)
    1874 M. (Bathmodon.)
    1875 C, pp. 24, 29. (Bathmodon.)
    1875 M. (Bathmodon.)
    1876 D, p. 64. (Bathmodon.)
    1877 K, p. 187.
    1877 L.
    1877 S.
    1878 T.
    1882 E, p. 145 (Coryphodon); pp. 165, 175
        (Metalophodon); pp. 165, 167 (Ectacodon,
        type *E. cinctus*); pp. 165, 167 (Manteodon,
        type *M. subquadratus*).
    1882 H, p. 294. (Coryphodon, Bathmodon.)
    1882 M. (Coryphodon, Metalophodon, Man-
        teodon.)
    1883 U.
    1883 Z. (Bathmodon.)
    1884 O, pp. 513, 517, 521 (Coryphodon); pp.
        517, 544 (Bathmodon): pp. 517, 554 (Meta-
        lophodon); p. 517 (Manteodon); pp. 517,
        519 (Ectacodon).
    1884 EE, pp. 1117, 1194 (Coryphodon); p.
        1194 (Ectacodon, Manteodon.)
    1885 A, p. 478. (Bathmodon.)
    1887 S. (Bathmodon, Metalophodon.)
    1888 F, p. 85. (Coryphodon, Bathmodon.)

Cope, E. D. 1889 C.
    1893 H.
Earle, C. 1892 A, pp. 150, 155 (Metalophodon =
    Coryphodon); p. 165 (Manteodon): p. 166 (Ec-
    tacodon).
    1892 B. (Coryphodon, Manteodon.)
Flower and Lydekker 1891 A, p. 437.
Gervais, P. 1859 A, p. 127.
Hébert, Ed. 1856 A.
Huxley, T. H. 1870 A.
L. P. B. 1885 A, p. 190.
Lydekker, R. 1886 A, p. 177.
Maack, G. A. 1865 A, p. 7.
Marsh, O. C. 1876 D.
    1877 A.
    1877 H.
    1878 H.
    1884 F, p. 225. (Loxolophodon = Corypho-
        don.)
    1893 C.
    1893 H, p. 321.
Nicholson and Lydeckker 1889 A, p. 1387. (Co-
    ryphodon, Ectacodon, Manteodon.)
Osborn, H. F. 1890 D, p. 538.
    1897 C, p. 1014.
    1898 F.
    1898 H, pp. 173, 191.
    1898 M, p. 585.
Osborn and Wortman 1892 A, pp. 118, 120, 121,
    figures.
Owen, R. 1846 B, p. 299.
    1848 A, p. 126.
    1858 B, vol. i, p. 78.
    1860 E, p. 322.
    1878 D, p. 216, pl. xi.
    1886 A.
Pictet, F. J. 1853 A, p. 304.
Rütimeyer, L. 1862 A.
Scott, W. B. 1892 A, p. 439.
Steinmann and Döderlein 1890 A, p. 763.
Trouessart, E. L. 1898 A, p. 712.
Vasseur, G. 1875 A.
Wortman, J. L. 1893 B, p. 426.
Zittel, K. A. 1893 B, p. 438. (Coryphodon, Ectaco-
    don, Manteodon.)

Coryphodon anax Cope.

Cope, E. D. 1882 E, p. 168.
1882 H, p. 294. (Bathmodon pachypus.)
1883 U, p. 68. (Bathmodon pachypus.)
1884 O, pp. 523, 537, pl. xlivb; pl. xlive; pl. xlivf (C. anax); p. 549, pl. xlivd; pl. xlive, figs. 7–13; pl. xlivg (Bathmodon pachypus).
1884 EE, p. 1116, figs. 9, 23 (C. anax); p. 1194, figs. 14, 15 (Bathmodon pachypus).
1893 H, p. 251.
Earle, C. 1892 A, pp. 151, 152, 155, 156, 164, fig. 2, H.
Matthew, W. D. 1899 A, p. 33. (Syn. of C. lobatus.)
Osborn, H. F. 1898 H, pp. 191, 197, 198, 204, in part. (Syn. of C. lobatus.)
Osborn and Wortman 1892 A, pp. 83, 119, fig. 15.
Roger, O. 1896 A, p. 155.
Trouessart, E. L. 1898 A, p. 714.
Woodward, A. S. 1898 B, p. 293.
Zittel, K. A. 1893 B, p. 438, fig. 359 (Coryphodon); p. 438 (Ectacodon).
Eocene (Wasatch); Wyoming.
See remarks under C. lobatus.

Coryphodon armatus (Cope).

Cope, E. D. 1872 T, p. 543. (Metalophodon.)
1872 LL. (Metalophodon.)
1873 E, p. 589. (Metalophodon.)
1873 F, p. 71. (Metalophodon.)
1873 K. (Metalophodon.)
1877 K, p. 250. (Metalophodon.)
1884 O, p. 555, pl. xlix, figs. 1–7. (Metalophodon.)
1884 EE, p. 1202. (Metalophodon.)
1893 H, p. 251. (Metalophodon.)
Earle, C. 1892 A, p. 151 (Metalophodon); p. 159 ("Metalophodon").
Matthew, W. D. 1899 A, p. 33.
Osborn, H. F. 1898 H, pp. 191, 194, 197, 208, figs. 15, 16, 17, 25.
Roger, O. 1896 A, p. 156.
Trouessart, E. L. 1898 A, p. 713.
Eocene (Wasatch); Wyoming.

Coryphodon cinctus (Cope).

Cope, E. D. 1882 E, p. 167. (Ectacodon.)
1882 M. (Ectacodon.)
1884 O, p. 520, pl. xliva, fig. 6. (Ectacodon.)
1884 EE, p. 1197, fig. 16. (Estacodon.)
1888 F, p. 84, fig. 3. (Ectacodon.)
1889 C, p. 238, figs. 63, 64. (Ectacodon.)
Earle, C. 1892 A, p. 166, fig. 2, G. (Ectacodon.)
Matthew, W. D. 1899 A, p. 33.
Osborn, H. F. 1898 H, pp. 191, 197, 198, 202, figs. 15, 16.
Roger, O. 1896 A, p. 156. (Ectacodon.)
Trouessart, E. L. 1898 A, p. 714. (Ectacodon.)
Zittel, K. A. 1893 B, p. 438, fig. 360.
Eocene (Wasatch); Wyoming.

Coryphodon curvicristis Cope.

Cope, E. D. 1882 E, pp. 170, 172.
1884 O, pp. 524, 533, pl. xlivc.
1884 EE, pp. 1116, 1201, figs. 9, 22.
Earle, C. 1892 A, pp. 151, 156, 163.
Matthew, W. D. 1899 A, p. 33.
Osborn, H. F. 1898 H, pp. 191, 197, 199, 207.

Roger, O. 1896 A, p. 155.
Trouessart, E. L. 1898 A, p. 713.
Eocene (Wasatch); Wyoming.

Coryphodon cuspidatus Cope.

Cope, E. D. 1875 C, pp. 29, 30. (Bathmodon.)
1877 K, p. 206, pl. xlvi, fig. 1.
1881 D, p. 194.
1882 E, p. 170.
1884 O, pp. 523, 525.
1884 EE, pp. 1116, 1202, fig. 9.
1893 H, p. 251.
Earle, C. 1892 A, pp. 155, 156, 161 (C. cuspidatus). p. 162, "No. 276" (C. obliquus).
Matthew, W. D. 1899 A, p. 33. (Syn. of C. elephantopus.)
Osborn, H. F. 1898 H, pp. 191, 197, 199, 205.
Roger, O. 1896 A, p. 155.
Trouessart, E. L. 1898 A, p. 713.
Eocene (Wasatch); New Mexico, Wyoming

Coryphodon elephantopus Cope.

Cope, E. D. 1874 O, p. 10. (Bathmodon.)
1875 C, p. 29. (Bathmodon.)
1875 F, p. 95, pls. v, vi. (Bathmodon.)
1877 K, pp. 206, 217, pl. l, figs. 5, 6; pls. li, lii, liii; pl. liv, fig. 1.
1882 E, pp. 170, 171.
1884 O. pp. 523, 531, pl. xxixc.
1884 EE, p. 1193, figs. 12, 13, 20, 21.
1887 S, p. 269, figs. 45, 46.
1889 C, p. 159, figure.
1891 L, p. 1118.
1893 H, p. 251.
Earle, C. 1892 A, pp. 151, 155, 160, fig. 2, D. 1892 B.
King, C. 1878 A, p. 377.
Lydekker, R. 1886 A, p. 177, fig. 27.
Matthew, W. D. 1899 A, p. 33.
Osborn, H. F. 1898 H, pp. 191, 196, 197, 199, figs. 19, 20.
Osborn and Wortman 1892 A, pp. 83, 119.
Roger, O. 1896 A, p. 155.
Scott, W. B. 1886 A, p. 304.
Trouessart, E. L. 1898 A, p. 713.
Zittel, K. A. 1893 B, p. 436, fig. 357.
Eocene (Wasatch); New Mexico, Wyoming.

Coryphodon hamatus Marsh.

Marsh, O. C. 1876 D, p. 426, figs. 1, 2.
Cope, E. D. 1891 L.
Earle, C. 1892 A, pp. 151, 155, 156, 162.
King, C. 1878 A, p. 377.
Lydekker, R. 1886 A, p. 176.
Marsh, O. C. 1877 A, p. 81, pl. iv.
1885 B, figs. 92, 93, 103, 132, 133.
1893 C, p. 325, pl. vi.
1893 H, pl. xviii, text fig. 1.
Matthew, W. D. 1899 A, p. 33.
Osborn, H. F. 1898 H, pp. 191, 197, 199, 214, in part.
Roger, O. 1896 A, p. 155.
Trouessart, E. L. 1898 A, p. 713.
Woodward, A. S. 1898 B, p. 293, figs. 166–168.
Zittel, K. A. 1893 B, p. 437, figs. 356, 358.
Eocene (Wasatch); Wyoming.

Coryphodon latidens Cope.

Cope, E. D. 1875 C, p. 29.   (Bathmodon.)
   1877 K, pp. 206, 214.
   1882 E, p. 170.
   1884 O, p. 523.
   1884 EE, p. 1200, fig. 22.
   1887 Q.
   1888 F, p. 86, fig. 4.
   1889 C, p. 241, fig. 25.
   1893 H, p. 251.
Earle, C. 1892 A, pp. 151, 160.   (Syn.? of C. elephantopus.)
King, C. 1878 A, p. 377,
Matthew, W. D. 1899 A, p. 33.
Osborn, H. F. 1898 H, pp. 191, 197, 199, 206.
Roger, O. 1896 A, p. 155.
Trouessart, E. L. 1896 A, p. 713.   (Syn.? of C. elephantopus.)
   Eocene (Wasatch); New Mexico, Wyoming.

Coryphodon latipes Cope.

Cope, E. D. 1873 F, pp. 68, 70.   (Bathmodon.)
   1873 E, pp. 587, 588.   (Bathmodon.)
   1882 E, p. 170.
   1884 O, pp. 523, 526, pls. xxixa, figs. 4, 5; pl. xlviii, figs. 7–14.
   1893 H, p. 251.
Earle, C. 1892 A, pp. 150, 151.
Matthew, W. D. 1899 A, p. 33.
Osborn, H. F. 1898 H, pp. 191, 197, 199.
Roger, O. 1896 A, p. 155.
Trouessart, E. L. 1896 A, p. 713.
   Eocene; New Mexico.

Coryphodon lobatus Cope.

Cope, E. D. 1877 K, pp. 205, 209, pl. xlvi, figs. 2–10.
   1882 E, p. 170.
   1884 O, p. 523.
Earle, C. 1892 A, pp. 151, 155, 164.   (Syn. of C. anax.)
Matthew, W. D. 1899 A, p. 33.
Osborn, H. F. 1898 H, pp. 191, 197, 198; p. 204 in part.
Trouessart, E. L. 1896 A, p. 714.   (Syn. of C. anax.)
   Eocene; New Mexico, Wyoming.
   Earle (1892 A, p. 164) makes this species a synonym of *C. anax.* Should it result that the specimens described as *C. anax, C. pachypus,* and *C. lobatus* belong to one species, this must bear the name *C. lobatus,* this name having priority of publication. This course has been followed by Osborn 1898 H, p. 204.

Coryphodon marginatus Cope.

Cope, E. D. 1882 E, pp. 170, 174.
   1884 O, pp. 524, 535, pl. xlive, fig. 5.
Earle, C. 1892 A, pp. 151, 163.   (Syn.? of C. curvicristis.)
Matthew, W. D. 1899 A, p. 33.
Osborn, H. F. 1898 H, pp. 191, 197, 198, 210.
Roger, O. 1896 A, p. 155.
Trouessart, E. L. 1896 A, p. 714.
   Eocene (Wasatch); Wyoming.

Coryphodon molestus Cope.

Cope, E. D. 1874 D, p. 9.   (Bathmodon.)

Cope, E. D. 1874 D, p. 8 (Bathmodon simus); p. 9 (B. lomas).
   1875 C, p. 29.   (Bathmodon molestus, simus, lomas.)
   1875 F, p. 95.   (Bathmodon molestus, lomas, simus.)
   1877 K, pp. 206, 229, pl. lvi; pl. lvii, figs. 1, 2   (C. molestus); pp. 206, 225, pl. lv   (C. simus).
   1882 E, pp. 170, 171.   (C. molestus, C. simus.)
   1884 O, p. 524   (C. molestus); pp. 524, 532   (C. simus.)
   1884 EE, p. 1197.   (C. molestus.)
   1893 H, p. 251.   (C. molestus, C. simus.)
Earle, C. 1892 A, pp. 151, 155, 160.   (Syn. of C. elephantopus.)
Matthew, W. D. 1899 A, p. 33.   (Syn. of C. latidens.)
Osborn, H. F. 1898 H, pp. 191, 197, 199, 208, fig. 25.   (Syn. of C. armatus.)
Trouessart, E. L. 1896 A, p. 714.   (Syn. of C. elephantopus.)
   Eocene (Wasatch); New Mexico.

Coryphodon obliquus Cope.

Cope, E. D. 1877 K, pp. 205, 207, pl. xlvii, figs. 1–7.
   1882 E, p. 170.
   1884 O, p. 523.
   1893 H, p. 251.
Earle, C. 1892 A, pp. 151, 155, 156, 162.   (Except specimen No. 276.)
Matthew, W. D. 1899 A, p. 33.   (Syn. of C. elephantopus.)
Osborn, H. F. 1898 H, pp. 191, 196, 197, 199.   (Syn. of C. elephantopus.)
Osborn and Wortman 1892 A, pp. 83, 119.
Roger, O. 1896 A, p. 155.
Trouessart, E. L. 1896 A, p. 713.
   Eocene (Wasatch); New Mexico.

Coryphodon radians Cope.

Cope, E. D. 1872 J, p. 418.   (Bathmodon.)
   1872 G, p. 351 (Bathmodon radians); p. 352 (B. semicinctus).
   1872 AA.   (Bathmodon.)
   1873 E, p. 587.   (Bathmodon.)
   1873 F, p. 68.   (Bathmodon.)
   1875 C, p. 29.   (Bathmodon.)
   1877 K, pp. 206, 211.   (Coryphodon.)
   1881 D, p. 194.   (Coryphodon.)
   1884 O, p. 544, pls. xlv, xlvi; pl. xlvii, figs. 1–6; pl. xlviii, figs. 1–6.   (Bathmodon.)
   1888 F, p. 85.   (Bathmodon.)
   1889 C, p. 239, fig. 64.
Earle, C. 1892 A, pp. 151, 155, 159, fig. 2, B, C.
   1892 B.
King, C. 1878 A, p. 377.
Marsh, O. C. 1876 D, p. 425.
Matthew, W. D. 1899 A, p. 33.
Osborn, H. F. 1898 F, pl. x.
   1898 H, pp. 191, 197, 196, 213.
   1898 M, p. 585, fig. 1.
   1898 Q, p. 12.
Osborn and Wortman 1892 A, pp. 83, 119.
Roger, O. 1896 A, p. 155.
Trouessart, E. L. 1896 A, p. 712.
   Eocene (Wasatch); Wyoming, New Mexico.

Coryphodon repandus Cope.

Cope, E. D. 1882 C, pp. 170, 171.
 1884 O, pp. 524, 532, pl. xlive, figs. 1-4.
Earle, C. 1892 A, pp. 151, 155, 156.   (Syn. of C. radians.)
Matthew, W. D. 1899 A, p. 33.   (Syn. of C. testis.)
Osborn, H. F. 1898 H, pp. 191, 197, 198, 201.
Trouessart, E. L. 1898 A, p. 712.   (Syn. of C. radians.)
 Eocene (Wasatch); Wyoming.

Coryphodon semicinctus Cope.

Cope, E. D. 1872 J, p. 420.   (Bathmodon, Loxolophodon.)
 1872 G, p. 352.   (Bathmodon.)
 1872 S, p. 488.   (Loxolophodon.)
 1873 E, pp. 567, 587, 588.   (Bathmodon.)
 1873 F, pp. 68, 70.   (Bathmodon.)
 1884 O, p. 544.   (Syn.? of C. radians.)
Earle, C. 1892 A, p. 151.   (Bathmodon.)
Marsh, O. C. 1872, 8.   (Loxolophodon.)
 1873 B, p. 117.   (Loxolophodon.)
 1873 K, p. 147.   (Loxolophodon.)
 1876 D, p. 425.
Osborn, H. F. 1898 H, p. 191.
Trouessart, E. L. 1898 A, p. 712.   (Syn.? of C. radians.)
 Eocene (Wasatch); Wyoming.

Coryphodon singularis Osb.

Osborn, H. F. 1898 H, pp. 191, 197, 199, 214, fig. 215.
Matthew, W. D. 1899 A, p. 36.
Trouessart, E. L. 1899 A, p. 1346.
 Eocene (Wind River); Wyoming.

Coryphodon subquadratus (Cope).

Cope, E. D. 1882 E, p. 166.   (Manteodon.)
 1882 M, p. 73.   (Manteodon.)
 1884 O, p. 518, pl. xliva, figs. 1-5.   (Manteodon.)
 1884 EE, p. 1117, fig. 9/.   (Manteodon quadratus.)
Earle, C. 1892 A, pp. 151, 165, fig. 2F.   (Manteodon.)
Matthew, W. D. 1899 A, p. 33.   (Syn. of C. hamatus.)

Osborn, H. F. 1898 H, pp. 191, 197, 214.   (Syn.? of C. hamatus.)
Roger, O. 1896 A, p. 156.
Trouessart, E. L. 1898 A, p. 714.   (Manteodon.)
 Eocene (Wasatch); Wyoming.

Coryphodon testis (Cope).

Cope, E. D. 1882 E, p. 175.   (Metalophodon.)
 1882 M.   (Metalophodon.)
 1884 O, p. 557, pl. xliva, fig. 13.   (Metalophodon.)
 1884 EE, p. 1197, fig. 16.   (Metalophodon.)
 1888 E, p. 84, fig. 3.   (Metalophodon.)
 1889 C, p. 238, fig. 63.   (Metalophodon.)
Earle, C. 1892 A, pp. 151, 155, 156, 158, fig. 2, E. 1892 B.
Matthew, W. D. 1899 A, p. 33.
Osborn, H. F. 1898 H, p. 174, 175, 189, 190, 197, 202, figs. 4, 6, 14, 15, 16, 17A, 18C, 18a, 21-23, 29.
 1900 F, p. 93, fig. 7.
Roger, O. 1896 A, p. 155.
Trouessart, E. L. 1898 A, p. 713.
 Eocene (Wasatch); Wyoming.

Coryphodon ventanus Osb.

Osborn, H. F. 1898 H, pp. 191, 197, 210, figs. 16, 17, 26D.
Matthew, W. D. 1899 A, p. 36.
Trouessart, E. L. 1899 A, p. 1346.
 Eocene (Wind River); Wyoming, Colorado.

Coryphodon wortmani Osb.

Osborn, H. F. 1898 H, pp. 191, 197, 199, 212, figs. 18B, 27.
Matthew, W. D. 1899 A, p. 36.
Trouessart, E. L. 1899 A, p. 1346.
 Eocene (Wind River); Wyoming.

The following citations are given to descriptions of specifically undetermined remains of *Coryphodon.*
Cope, E. D. 1877 K, p. 237, pl. lvii, figs. 8-7 pl. lviii; p. 241, pl. lviii, fig. 9; pls. lix, 1 , p. 247, pl. lxi; p. 248, pl. lxii; pl. lxiii, figs. 1-4.
Osborn, H. F. 1897 G, p. 256.

BATHYOPSIDÆ.

Osborn, H. F. 1898 H, p. 182.

BATHYOPSIS Cope.   Type *B. fissidens* Cope.

Cope, E. D, 1881 D, p. 194.
 1881 CC, p. 75.
 1884 O, pp. 561, 596.
 1885 F, pp. 44, 58.
 1885 N.
Nicholson and Lydekker 1889 A, p. 1388.
Osborn, H. F. 1881 A, p. 17.
 1898 H, pp. 173, 215, 217, fig. 3.
Scott, W. B. 1886 A, p. 304.
Trouessart, E. L. 1898 A, p. 717.   (Syn. of Uintatherium.)
Zittel, K. A. 1893 B, p. 447.

Bathyopsis fissidens Cope.

Cope, E. D. 1881 D, p. 182.

Cope, E. D. 1881 CC. p. 75.
 1882 E, p. 176.
 1884 O, p. 597, pl. xxixa, figs. 1-3; pl. lviiia, fig. 1.
 1884 EE, pp. 1111, 1115, fig. 7.
 1885 F, p. 53, fig. 35.
Marsh, O. C. 1884 F, p. 220, figs. 196, 197.   (Uintatherium.)
Matthew, W. D. 1899 A, p. 36.
Roger, O. 1896 A, p. 158.
Trouessart, E. L. 1898 A, p. 718.   (Uintatherium.)
 Eocene (Wind River), Wyoming.

Suborder DINOCEREA Marsh.

Marsh, O. C. 1872 S.
    Unless otherwise indicated, the following authorities refer to this suborder under the amended form of the name *Dinocerata.*
Baur, G. 1890 C.
Brandt, J. F. 1878 A, p. 16.
Cope, E. D. 1873 E, p. 564.
    1873 F, pp. 40, 42.
    1872 J, p. 11.
    1873 X.
    1875 C, p. 28.
    1875 M.
    1877 K, p. 182.
    1882 A, p. 445.
    1884 E.
    1884 O, pp. 510, 512, 559, 587.
    1884 EE, p. 1111.
    1885 F, p. 40.
    1885 N.
    1885 O.
    1886 F.
    1888 T.
    1898 B, p. 125.
Filhol, H. 1884 A.
Flower, W. H. 1876 B, xiii, p. 308.
Gadow, H. 1898 A, p. 45. (Amblypoda, part.)
Garrod, A. H. 1873 A.
    1873 B.
Gaudry, A. 1885 A.
Haeckel E. 1895 A, p. 542.
Hay, O. P. 1899 F, p. 682.
Hill, F. C. 1881 A.

L. P. B. 1885 A.
Lydekker, R. 1886 A, p. 179.
Marsh, O. C. 1873 B.
    1873 F.
    1873 G.
    1873 I.
    1873 J.
    1873 K.
    1873 L.
    1873 M.
    1873 P.
    1874 A, p. 84.
    1874 D.
    1876 A.
    1876 G, p. 60.
    1884 F.
    1885 B.
Nicholson and Lydekker 1889 A, p. 1387.
Osborn, H, F. 1893 D.
    1898 H, pp. 177, 180.
    1900 F, p. 89.
Osborn, Scott, and Speir 1878 A, p. 62.
Owen, R. 1876 A, p. 401.
Rütimeyer, L. 1888 A, p. 13.
Schlosser, M. 1888 A, p. 587.
    1890 C, p. 250.
Science 1885 A, p. 488.
Scott, W. B. 1892 A, p. 439.
Steinmann and Döderlein 1890 A, p. 764.
Tomes, C. S. 1898 A, p. 440.
Zittel, K. A. 1893 B, p. 439. (Amblypoda, part.)

TINOCERIDÆ.

Marsh, O. C. 1872 M, p. 323.
Cope, E. D. 1873 E, pp. 563, 564. (Eobasilidæ.)
    1873 F, p. 41. (Eobasilidæ.)
    1875 K. (Eobasilidæ.)
    1873 X, p. 292. (Eobasilidæ.)
    1888 F, pp. 80, 86. (Uintatheriidæ.)
    1889 R, p. 877. (Uintatheriidæ.)
Flower, W. H. 1876 B, xiii, p. 387. (Uintatheriidæ.)
    1876 C, p, 114. (Uintatheriidæ.)
Haeckel E. 1895 A, p. 580. (Dinoceratida.)

! Lydekker, R. 1886 A. p. 179. (Uintatheriidæ.)
    1896 B. (Uintatheriidæ.)
Marsh, O. C. 1873 F, p. 295. (Tinoceratidæ.)
    1884 F, p. 191. (Tinoceratidæ.)
Nicholson and Lydekker 1889 A, p. 1387. (Uintatheriidæ.)
Osborn, H. F. 1898 H, p. 182. (Uintatheriidæ.)
Roger, O. 1896 A, p. 156. (Dinoceratidæ.)
Trouessart, E. L. 1898 A, p. 715. (Uintatheriidæ.)
Zittel, K. A. 1893 B, p. 439. (Dinoceratidæ.)

UINTATHERIUM Leidy.   Type *U. robustum* Leidy.

Leidy, J. 1872 G, p. 169.
Cope, E. D. 1873 E, pp. 564, 580.
    1873 F, p. 60.
    1873 K.
    1874 M.
    1877 K, p. 186.
    1882 H, p. 294.
    1883 U.
    1884 O, pp. 561, 587.
    1884 EE, p. 1117.
    1885 F, pp. 44, 53.
    1885 N, p. 594 (Uintatherium; Ditetrodon, type *Uintatherium segne.*)
    1888 F, p. 88, figure.
    1889 C.
Flower, W. H. 1876 B, xiii, p. 387.
    1876 C, p. 118.
Flower and Lydekker 1891 A, p. 436.

Leidy, J. 1872 A, p. 240. (Uintatherium, Uintamastix.)
    1872 G, p. 169. (Uintamastix, type *U. atrox.*)
    1872 H, p. 241.
    1873 B, pp. 93, 331.
L. P. B. 1885 A.
Marsh, O. C. 1873 F.
    1874 D.
    1881 G.
    1884 F, p. 191.
Nicholson and Lydekker 1889 A, p. 1388.
Osborn, H. F. 1881 A.
    1890 D, p. 538.
    1895 A, p. 82.
    1897 G, pp. 252, 258.
    1898 F, p. 91.
    1898 H, p. 174.
    1898 M, p. 587.

Osborn, Scott, and Speir 1878 A, p. 62.
Science 1885 A, p. 490.
Scott, W. B. 1891 B, pp. 375, 385.
Trouessart, E. L. 1898 A, p. 717.
Woodward, A. S. 1898 B, p. 299.
Zittel, K. A. 1898 B, p. 446.

Uintatherium alticeps Scott.

Scott, W. B. 1886 A, p. 307, fig. 4.
Matthew, W. D. 1899 A, p. 44.
Roger, O. 1896 A, p. 157.   (Tinoceras.)
Eocene (Bridger); Wyoming.

Uintatherium latifrons Marsh.

Marsh, O. C. 1884 F, p. 220, figs. 11, 125–128.
    1885 B, pp. 249, 262, fig. 48.
Matthew, W. D. 1899 A, p. 44.
Roger, O. 1896 A, p. 157.
Trouessart, E. L. 1898 A, p. 718.
Eocene (Bridger); Wyoming.

Uintatherium leidianum Osb., Scott, and Speir.

Osborn, Scott, and Speir 1878 A, p. 68, pls. vi–viii.
Cope, E. D. 1884 O, pp. 593, 599, fig. 26.
    1884 EE, p. 1117, fig. 10.
    1885 F, p. 53, fig. 27.
Hill, F. C. 1881 A, p. 524.
Marsh, O. C. 1884 F, p. 221, fig. 198.
Matthew, W. D. 1899 A, p. 44.
Osborn, H. F. 1881 A, pp. 18, 19, 23, pl. ii.
Roger, O. 1896 A, p. 158.
Trouessart, E. L. 1898 A, p. 718.
Eocene (Bridger); Wyoming.

Uintatherium princeps Osb., Scott, and Speir.

Osborn, Scott, and Speir 1878 A, p. 81, pl. vii, fig. 2.
Cope, E. D. 1884 O, p. 593.
Marsh, O. C. 1884 F, p. 223.
Matthew, W. D. 1899 A, p. 44.
Eocene (Bridger); Wyoming.

Uintatherium robustum Leidy.

Leidy, J. 1872 G, p. 169.
Cope, E. D. 1873 E, pp. 561, 563.
    1873 F, p. 64.
    1873 K, p. 102.
    1873 X, p. 291.
    1882 H, p. 295.
    1884 O, p 589, pl. xxxvi, figs. 1, 2.
    1885 F, p. 53.
King, C. 1878 A, p. 403.
Leidy, J. 1872 A, p. 240.   (U. robustum, Uinta-
    mastix atrox.)
    1872 G, p. 169.   (Uintamastix atrox, orig-
        inal description.)
    1872 H, p. 241.   (U. robustum, Uintamastix
        atrox.)
    1873 B, pp. 96, 333, pl. xxv, figs. 6–12; pl.
        xxvi, figs. 1–3; pl. xxvii, figs. 30–34.
Marsh, O. C. 1872 S.
    1873 B, p. 117.
    1873 K, p. 147.
    1873 L, p. 218.
    1884 F, p. 219, figs. 61, 62, 195.
    1885 B, fig. 98, 99.
Matthew, W. D. 1899 A, p. 43.
Osborn, Scott, and Speir 1878 A, p. 82.
Roger, O. 1896 A, p. 157.
Trouessart, E. L. 1898 A, p. 717.
Eocene (Bridger); Wyoming.

Uintatherium segne Marsh.

Marsh, O. C. 1884 F, p. 222, figs. 4, 42, 101, 102, 199,
    200.
Cope, E. D. 1885 N, p. 594.   (Ditetrodon.)
L. P. B. 1885 A, p. 186, fig. 16.
Marsh, O. C. 1885 B, figs. 78, 79.
Matthew, W. D. 1899 A, p. 44.
Roger, O. 1896 A, p. 158.
Trouessart, E. L. 1898 A, p. 718.
Eocene (Bridger); Wyoming.

Uintatherium sp. indet. Cope.

Cope, E. D. 1884 O, p. 593, pls. xxxiv, xxxv.
Eocene (Bridger); Wyoming.

DINOCERAS Marsh.   Type *D. mirabilis* Marsh.

Marsh, O. C. 1872 R, p. 343.
Barbour, E. H. 1890 B, p. 395.
Cope, E. D. 1873 E, p. 580.   (Syn. of Uintatherium.)
    1877 S.   (Syn. of Uintatherium.)
    1884 O, p. 587.   (Syn.? of Uintatherium.)
    1885 F, pp. 44, 53.   (Octotomus, type *D. lati-
        ceps*.)
    1885 O.
Flower, W. H. 1876 B, xiii, p. 387.
Flower and Lydekker 1891 A, p. 437.   (Syn. of
    Uintatherium.)
Garrod, A. H. 1873 A, p. 74.
Gaudry, A. 1878 B, p. 74.
    1889 A.
L. P. B. 1885 A.
Leidy, J. 1878 B, pp. 96, 332.   (Syn. of Uintathe-
    rium.)
Lydekker, R. 1886 A, p. 182.
Marsh, O. C. 1873 F.
    1873 G.

Marsh, O. C. 1874 D.
    1876 A, p. 162.
    1877 H.
    1881 G.
    1884 F, p. 200.   (Paroceras, type *D. laticeps*.)
Nicholson and Lydekker 1889 A, p. 1388.
Osborn, H. F. 1898 M, p. 587.
Osborn, Scott, and Speir 1878 A, p. 62.
Owen, R. 1876 A, p. 401.
Trouessart, E. L. 1898 A, p. 715.
Wiedersheim, R. 1881 A, p. 368.
Woodward, A. S. 1898 B, p. 296.
Zittel, K. A. 1898 B, p. 446.

Dinoceras agreste Marsh.

Marsh, O. C. 1884 F, pp. 19, 197, fig. 15.
Matthew, W. D. 1899 A, p. 43.
Roger, O. 1896 A, p. 156.
Trouessart, E. L. 1898 A, p. 715.
Eocene (Bridger); Wyoming.

Dinoceras cuneum Marsh.

Marsh, O. C. 1884 F, pp. 77, 197, figs. 93, 94, 170, 171.
  1885 B, figs. 118, 119.
Matthew, W. D. 1899 A, p. 43.
Roger, O. 1896 A, p. 156.
Trouessart, E. L. 1898 A, p. 715.
  Eocene (Bridger); Wyoming.

Dinoceras distans Marsh.

Marsh, O. C. 1884 F, pp. 13, 15, 16, 29, 199, figs. 4, 8, 10, 31.
L. P. B. 1885 A, p. 181, fig. 8.
Marsh, O. C. 1885 B, figs. 41, 45, 47, 66.
Matthew, W. D. 1899 A, p. 43.
Roger, O. 1896 A, p. 156.
  Eocene (Bridger); Wyoming.

Dinoceras laticeps Marsh.

Marsh, O. C. 1873 I, p. 301.
Cope, E. D. 1885 F, p. 53, fig. 34.   (Octotomus.)
King, C. 1878 A, p. 408.
L. P. B. 1885 A, figs. 18, 23.
Marsh, O. C. 1876 A, p. 168, pl. v.
  1884 F, p. 200, text figs. 14, 22, 27, 33, 47, 50, 57, 58, 112, 136, 138-140; pls. x-xiv, xliii.
  [Dinoceras (Paroceras).]
  1885 B, figs. 51, 57, 62, 68, 71, 72, 84, 85, 87, 94, 95.
Matthew, W. D. 1899 A, p. 43.
Osborn and Speir 1879 A, p. 304.   (Uintatherium.)
Roger, O. 1896 A, p. 156.
Trouessart, E. L. 1898 A, p. 715.
  Eocene (Wyoming); Wyoming.

Dinoceras lucare Marsh.

Marsh, O. C. 1873 H, p. 408.
King, C. 1878 A, p. 408.
L. P. B. 1885 A, p. 176.
Leidy, J. 1873 B, p. 334.
Marsh, O. C. 1884 F, p. 200, pl. ix; text figs. 46, 108-110, 172, 173.
  1885 B, figs. 83, 124, 125.
Matthew, W. D. 1899 A, p. 43.
Roger, O. 1896 A, p. 156.
Scott, W. B. 1886 A, p. 307.
Trouessart, E. L. 1898 A, p. 715.
  Eocene (Wyoming); Wyoming.

Dinoceras mirabile Marsh.

Marsh, O. C. 1872 R.
Cope, E. D. 1873 D, p. 1.   (Uintatherium.)
  1873 E, pp. 581, 584.   (Uintatherium.)
  1873 F, p. 65.   (Uintatherium.)
  1873 X, p. 292.   (Uintatherium.)
  1874 B, p. 456.   (Uintatherium.)
  1884 O, p. 589.   (Uintatherium.)
  1885 F, p. 53, figs. 25, 26, 33.   (Uintatherium.)
  1889 C, p. 158, figure.   (Uintatherium.)
Emerton, J. H. 1887 A.
King, C. 1878 A, p. 403.
L. P. B. 1885 A, with figures.
Leidy, J. 1872 H, p. 241.   (Syn.? of Uintatherium robustum.)
  1873 B, pp. 97, 108, 333.   (Syn.? of U. robustum.)
Lydekker, R. 1886 A, p. 182.
Marsh, O. C. 1872 S.
  1873 B, p. 118, pls. i, ii.
  1873 K, pls. i, ii.
  1873 L, p. 218.
  1876 A, p. 165, pls. ii-iv.
  1881 G, pl. ii.
  1884 F, pls. i-vii, xx-xliii, xliv-lv; text figs. 3, 7, 13, 26, 34, 35, 39, 40, 43-45, 63, 89, 90, 91, 92, 95, 96, 99, 100, 106-109, 111, 113, 114, 118-123, 129-132, 143-146, 152, 153.
  1885 B, figs. 38-40, 44, 50, 69, 70, 76, 80-82, 100, 116, 117, 120-123, 134-136.
Matthew, W. D. 1899 A, p. 43.   (Uintatherium.)
Nature 1873 A.
  1873 B.
Osborn, H. F. 1881 A, p. 24.   (Uintatherium.)
  1900 F, p. 93.   (Uintatherium.)
Roger, O. 1896 A, p. 156.
Trouessart, E. L. 1898 A, p. 715.
Woodward, A. S. 1898 B, p. 296, fig. 169.
Zittel, K. A. 1893 B, p. 440, figs. 361, 362, 364-367.
  Eocene (Bridger); Wyoming.

Dinoceras reflexum Marsh.

Marsh, O. C. 1884 F, p. 201, fig. 174.
Matthew, W. D. 1899 A, p. 43.
Roger, O. 1896 A, p. 156.
Trouessart, E. L. 1898 A, p. 715.
  Eocene (Bridger); Wyoming.

TINOCERAS Marsh.   Type *Titanotherium? anceps* Marsh.

Marsh, O. C. 1872 L, p. 322.   (Issued in advance, Aug. 19.)
Barbour, E. H. 1890 B, p. 395.   .
Cope, E. D. 1872 S. p. 487.   (Loxolophodon [1], in part.)
  1872 T, p. 542.   (Loxolophodon.)
  1872 W.   (Lefalophodon, error for Loxolophodon; issued Aug. 19.)
  1872 II.   (Loxolophodon.)
  1873 E, pp. 564, 565.   (Loxolophodon.)
  1873 F, pp. 40-43.   (Loxolophodon.)
  1873 K.   (Loxolophodon.)
  1873 X, p. 293.   (Loxolophodon.)

Cope, E. D. 1877 K, p. 148.   (Loxolophodon.)
  1879 Q.   (Loxolophodon.)
  1884 O, pp. 561, 569.   (Loxolophodon.)
  1885 F, p. 44.   (Loxolophodon.)
  1885 N.   (Loxolophodon; Tetheopsis, type *Tinoceras stenops.*)
  1886 F.   (Tetheopsis.)
Filhol, H. 1884 A, pl. ix.   (Loxolophodon.)
Flower and Lydekker 1891 A, p. 437.   (Syn. of Uintatherium.)
Hill, F. C. 1881 A.   (Loxolophodon.)
L. P. B. 1885 A.
Leidy, J. 1873 B, p. 332.   (Syn. of Uintatherium.)

---

[1] The generic title *Loxolophodon* was first employed by Professor Cope for *Bathmodon semicinctus*, now regarded as a species of *Coryphodon*. Later Professor Cope used the name for the genus here called *Tinoceras*. In this sense the type was *L. cornutus* (Cope, 1884 O, p. 572).

Lydekker, R. 1886 A, p. 180.
   1893 E, p. 142.   (Tinoceras, "miscalled.")
Marsh, O. C. 1872 G, p. 504.
   1873 F, p. 296.
   1873 G.
   1873 J.
   1873 L, p. 216.
   1873 M.
   1873 P, p. ii.   (" Lefalophodon.")
   1874 D.
   1877 A, p. 81.
   1881 G.
   1884 F, p. 191.
   1885 B.
Nicholson and Lydekker 1889 A, p. 1388.
Osborn, H. F. 1881 A, pls. iii, iv.   (Loxolophodon.)
Osborn, Scott, and Speir 1878 A, p. 62.
Osborn and Speir 1879 A, p. 304, pl. ii.   (Loxolophodon.)
Science 1885 A, p. 490.
Steinmann and Döderlein 1890 A, p. 765.
Trouessart, E. L. 1898 A, p. 716.   (Loxolophodon.)
Woodward, A. S. 1898 B, p. 299.
Zittel, K. A. 1893 B, p. 446.

Tinoceras affine Marsh.

Marsh, O. C. 1884 F, p. 204, figs, 16, 178, 179.
Matthew, W. D. 1899 A, p. 43.
Roger, O. 1896 A, p. 157.
Trouessart, E. L. 1898 A, p. 716.   (Loxolophodon.)
Eocene (Bridger); Wyoming.

Tinoceras anceps Marsh.

Marsh, O. C. 1871 D, p. 35.   (Titanotherium?.)
Cope, E. D. 1873 F, p. 41.   (Titanotherium.)
   1873 J.   (Titanotherium.)
   1884 O, p. 572.
King, C. 1878 A, p. 403.
Leidy, J. 1872 B, p. 365.   (Titanotherium.)
   1873 B, p. 334.   (Uintatherium.)
Marsh, O. C. 1872 G, p. 123 (Mastodon anceps),
   p. 504 (Tinoceras).
   1872 L.
   1873 B, p. 117.
   1873 H p. 408.
   1873 K, p. 147.
   1873 L, p. 218.
   1873 N, p. 51.
   1884 F, figs. 97, 98, 175-177.
Matthew, W. D. 1899 A, p. 43.
Roger, O. 1896 A, p. 157.
Trouessart, E. L. 1898 A, p. 716.   (Loxolophodon.)
   Eocene (Bridger); Wyoming.

Tinoceras annectens Marsh.

Marsh, O. C. 1884 F, p. 205, figs. 6, 21, 36, 87.
L. P. B. 1885 A, fig. 4.
Marsh, O. C. 1885 B, figs. 43, 56, 73, 74.
Matthew, W. D. 1899 A, p. 43.
Roger, O. 1896 A. p. 157.
Trouessart, E. L. 1898 A, p. 716.   (Loxolophodon.)
   Eocene (Bridger); Wyoming.

Tinoceras cornutum (Cope).

Cope, E. D. 1872 W.   (Lefalophodon dicornutus,
   error for Loxolophodon cornutus.)
Barbour, E. H. 1890 B, p. 395, fig. 8.

Cope, E. D. 1872 S, p. 488.   (Loxolophodon.)
   1872 II.   (Loxolophodon.)
   1872 KK.   (Eobasileus.)
   1873 E, pp. 568, 575, pls. i-iv.   (Loxolophodon.)
   1873 F, pp. 45, 54.   (Loxolophodon.)
   1873 W, p. 151.   (Eobasileus.)
   1873 X, p. 298, pls. i, ii.   (Loxolophodon.)
   1873 Y, p. 49.   (Eobasileus.)
   1873 BB.   (Loxolophodon.)
   1884 O, pp. 573, 574, pls. xxxvii-xliii.   (Loxolophodon.)
   1885 F, p. 45, pl. i, text figs. 24, 29, 31, 32.   (Loxolophodon.)
   1885 O.   (Loxolophodon.)
   1887 S, p. 277, pl. viii.   (Uintatherium.)
Leidy, J. 1873 B, p. 333.   (Loxolophodon.)
Lydekker, R. 1886 A, p. 180.
Marsh, O. C. 1873 F, p. 294 (T. grande); p. 296 (T. cornutum).
   1873 G.
   1873 J.   (T. grande.)
   1873 L, p. 216 (T. cornutum); p. 217 (T. grande).
   1873 M, p. 306.   (T. cornutum, T. grande.)
   1873 P, p. ii.
   1884 F, p. 206, figs. 180, 181.
Matthew, W. D. 1899 A, p. 44.   (Uintatherium.)
Nature 1873 C.   (Loxolophodon.)
Osborn, H. F. 1896 C, p. 706, illus. 2.   (Uintatherium.)
   1898 Q, p. 17.   (Uintatherium.)
   1900 F, p. 90, fig. 2.   (Uintatherium.)
Osborn and Speir 1879 A, p. 304, pl. i.   (Loxolophodon.)
Roger, O. 1896 A, p. 157.
Trouessart, E. L. 1898 A, p. 716.   (Loxolophodon.)
Zittel, K. A. 1893 B, p. 447, fig. 368.
   Eocene (Bridger); Wyoming.

Tinoceras crassifrons Marsh.

Marsh, O. C. 1884 F, p. 208, figs. 29, 182.
L. P. B. 1885 A, fig. 12.
Marsh, O. C. 1885 B, fig. 25.
Matthew, W. D. 1899 A, p. 43.
Roger, O. 1896 A, p. 157.
Trouessart, E. L. 1898 A, p. 716.   (Loxolophodon.)
   Eocene (Bridger); Wyoming.

Tinoceras galeatum Cope.

Cope, E. D. 1873 D, p. 1.   (Eobasileus.)
   1874 B, p. 456, pl. i.   (Eobasileus, Loxolophodon.)
   1874 T.   (Eobasileus.)
   1884 O, pp. 573, 585, pls. xliii, xliv.   (Loxolophodon.)
   1885 F, p. 46.
Marsh, O. C. 1884, F, p. 209, figs. 183, 184.
Matthew, W. D. 1899 A, p. 44.   (Uintatherium.)
Osborn, H. F. 1881 A, pp. 21, 22.   (Loxolophodon.)
Roger, O. 1896 A, p. 157.
Trouessart, E. L. 1898 A, p. 716.   (Loxolophodon.)
   Eocene (Bridger); Wyoming.

Tinoceras grande Marsh.

Marsh, O. C. 1872 M, p. 323.
Cope, E. D. 1873 E, p. 575.   (Loxolophodon cornutus.)
   1873 F, pp. 54, 61.

L. P. B. 1885 A, fis. 31, 32.
Marsh, O. C. 1873 F, p. 294.
   1873 G.
   1873 L, pp. 217, 218.
   1884 F, p. 210, figs. 20, 49, 87, 88, 185.
   1885 B, figs. 55, 86, 113-115.
Matthew, W. D. 1899 A, p. 44.
Roger, O. 1896 A, p. 157.
Trouessart, E. L. 1898 A, p. 716.   (Loxolophodon.)
   Eocene (Bridger); Wyoming.

Tinoceras hians Marsh.

Marsh, O. C. 1884 F, p. 210, figs. 32, 186.
   1885 B, fig. 67.
Matthew, W. D. 1899 A, p. 44.   (Uintatherium.)
Roger, O. 1896 A, p. 157.
Trouessart, E. L. 1898 A, p. 717.   (Loxolophodon.)
   Eocene (Bridger); Wyoming.

Tinoceras ingens Marsh.

Marsh, O. C. 1884 F, p. 211, figs. 9, 17, 23, 25, 28, 51, 115-117, 124, 141, 142.
Barbour, E. H. 1890 B, p. 394, fig. 7.
Hutchison, H. N. 1893 A, p. 150, pl. xiv.
L. P. B. 1885 A, figs. 6. 10, 22.
Lydekker, R. 1886 A, p. 181, fig. 30.
Marsh, O. C. 1885 B, figs. 46, 52, 58, 63, 88, 96, 97, 126, 127, 137.
Matthew, W. D. 1899 A, p. 44.   (Uintatherium.)
Osborn, H. F. 1900 F, p. 94.   (Uintatherium.)
Roger, O. 1896 A, p. 157.
Trouessart, E. L. 1898 A, p. 717.   (Loxolophodon.)
   Eocene (Bridger); Wyoming.

Tinoceras jugum Marsh.

Marsh, O. C. 1884 F, p. 212, fig. 187.
Matthew, W. D. 1899 A, p. 44.   (Uintatherium.)
Roger, O. 1896 A, p. 157.
Trouessart, E. L. 1898 A, p. 717.   (Loxolophodon.)
   Eocene (Bridger); Wyoming.

Tinoceras lacustre Marsh.

Marsh, O. C. 1872 R, p. 344.   (Dinoceras.)
Cope, E. D. 1873 E, pp. 581, 584.   (Uintatherium.)
   1873 F, pp. 61, 66.   (Uintatherium.)
   1884 O, pp. 589, 591, pl. xxxvi, figs. 3-8.   (Uintatherium.)
   1886 F.   (Uintatherium.)
King, C. 1878 A, p. 403.   (Dinoceras.)
Marsh, O. C. 1873 L, p. 218.   (Dinoceras.)
   1884 F, p. 212, fig 188.
Matthew, W. D. 1899 A, p. 44.   (Uintatherium.)
Roger, O. 1896 A, p. 157.
Trouessart, E. L. 1898 A, p. 717.   (Loxolophodon.)
   Eocene (Bridger); Wyoming.

Tinoceras latum Marsh.

Marsh, O. C. 1884 F, p. 213, figs. 189, 190.
Matthew, W. D. 1899 A, p. 44.   (Uintatherium.)
Roger, O. 1896 A, p. 157.
Trouessart, E. L. 1898 A, p. 717.   (Loxolophodon.)
   Eocene (Bridger); Wyoming.

Tinoceras longiceps Marsh.

Marsh, O. C. 1884 F, p. 214, figs. 38, 48, 191, 192.
   1885 B, pp. 275, 279, figs. 75, 85.
Matthew, W. D. 1899 A, p. 44.   (Uintatherium.)
Roger, O. 1896 A, p. 157.
Trouessart, E. L. 1898 A, p. 717.   (Loxolophodon.)
   Eocene (Bridger); Wyoming.

Tinoceras pugnax Marsh.

Marsh, O. C. 1884 F, p. 215, figs. 5, 18, 19, 24, 25, 29, 67.
L. P. B. 1885 A, fig. 9.
Marsh, O. C. 1885 B, figs. 42, 53, 54, 59, 64, 89, 104.
Matthew, W. D. 1899 A, p. 44.   (Uintatherium.)
Roger, O. 1896 A, p. 157.
Trouessart, E. L. 1898 A, p. 717.   (Loxolophodon.)
   Eocene (Bridger); Wyoming.

Tinoceras speirianum Osb.

Osborn, H. F. 1881 A, p. 20, pl. i.   (Loxolophodon.)
Cope, E. D. 1884 O, pp. 573, 599, fig. 28.   (Loxolophodon.)
   1885 F, p. 46, fig. 28.
Marsh, O. C. 1884 F, p. 216, fig. 193.
Matthew, W. D. 1899 A, p. 44.   (Uintatherium.)
Roger, O. 1896 A, p. 157.
Trouessart, E. L. 1898 A, p. 717.   (Loxolophodon.)
   Eocene (Bridger); Wyoming.

Tinoceras stenops Marsh.

Marsh, O. C. 1884 F, p. 217, figs. 53, 54, 194.
Cope, E. D. 1885 N, p. 594.   (Tetheopsis.)
   1886 F, p. 155.
L. P. B. 1885 A, figs. 20, 21.
Lydekker, R. 1886 A, p. 179, fig. 28.
Marsh, O. C. 1885 B, figs. 90, 91.
Matthew, W. D. 1899 A, p. 44.   (Uintatherium.)
Roger, O. 1896 A, p. 157.
Trouessart, E. L. 1898 A, p. 717.   (Loxolophodon.)
   Eocene (Bridger); Wyoming.

Tinoceras vagans Marsh.

Marsh, O. C. 1884 F, p. 218, fig. 12.
   1885 B, fig. 49.
Matthew, W. D. 1899 A, p. 44.   (Uintatherium.)
Roger, O. 1896 A, p. 157.
Trouessart, E. L. 1898 A, p. 717.   (Loxolophodon.)
   Eocene (Bridger); Wyoming.

ELACHOCERAS Scott.   Type E. parvum Scott.

Scott, W. B. 1886 A, p. 304.
Nicholson and Lydekker 1889 A, pp. 1384, 1388.
Zittel, K. A. 1893 B, p. 447.

Matthew, W. D. 1899 A, p. 44.   (Uintatherium.)
Roger, O. 1896 A, p. 156.
Trouessart, E. L. 1898 A, p. 715.
   Eocene (Bridger); Wyoming.

Elachoceras parvum Scott.

Scott, W. B. 1886 A, p. 305, figs. 2, 3.

· EOBASILEUS Cope.   Type "*E. cornutus*" = *E. pressicornis*.

Cope, E. D. 1872 R, p. 485.   (Aug. 20, 1872 as Palæont. Bull. No. 6.)
    1872 S, p. 488.   (Loxolophodon, in part.)
    1872 W.   (Lefalophodon, in part.)
    1872 II.   (Loxolophodon, in part.)
    1873 E, pp. 564, 567, 575, 645.
    1873 F, p. 54.
    1873 J.
    1873 K.
    1873 W.
    1873 Z.
    1884 O, p. 561.
    1885 F, pp. 44, 45.
Leidy, J. 1873 B, p. 332.   (Syn. of Uintatherium.)
Marsh, O. C. 1873 B, pp. 118, 122.
    1873 F.
    1873 K, p. 151.
Nicholson and Lydekker 1889 A, p. 1388.
Science 1885 A, p. 490.
Trouessart, E. L. 1898 A, p. 718.
Zittel, K. A. 1893 B, pp. 446, 447.
    A genus of doubtful validity; probably equivalent to *Tinoceras*.

Eobasileus furcatus Cope.

Cope, E. D. 1872 W.   (Lefalophodon bifurcatus, error for Loxolophodon furcatus.)
    1872 S, p. 488.   (Loxolophodon.)
    1872 II.   (Loxolophodon.)
    1872 KK, p. 774.
    1873 E, p. 580.
    1873 F, p. 59.
    1873 W, p. 159.   (Uintatherium.)

Cope, E. D. 1873 X, p. 291.
    1884 O, p. 565 (Eobasileus); pls. xxxii xxxiii   (Uintatherium).
    1885 F, p. 45.
Marsh, O. C. 1873 B, p. 118.
    1873 F, p. 295.
    1873 K, p. 151.
    1873 L, p. 217.
    1884 F, p. 223.   (Tinoceras.)
Matthew, W. D. 1899 A, p. 44.   (Uintatherium.)
Trouessart, E. L. 1898 A, p. 718.
    Eocene (Bridger); Wyoming.

Eobasileus pressicornis Cope.

Cope, E. D. 1872 W.   (Lefalophodon excressicornis, error for Loxolophodon pressicornis.)
    1872 R, p. 485.   (E. cornutus, not Loxolophodon cornutus of Cope, E. D. 1872 W.)
    1872 S, p. 488.   (Loxolophodon.)
    1872 II, p. 580.   (Loxolophodon.)
    1872 KK, p. 774.
    1873 E, p. 575.
    1873 F, p. 54.
    1873 W, p. 159.   (Uintatherium.)
    1884 O, p. 562, pl. xxx, figs. 1–5; pl. xxxi.
Marsh, O. C. 1873 F, p. 295.
    1873 L, p. 217.
    1884 F, p. 223.   (Tinoceras.)
Matthew, W. D. 1899 A, p. 44.   (Uintatherium)
Roger, O, 1896 A, p. 158.
Trouessart, E. L. 1898 A, p. 718.
    Eocene (Bridger); Wyoming

Order PROBOSCIDEA.

Baur, G. 1889 K.
Cope, E. D. 1873 E, pp. 561–568.
    1873 F, p. 38.
    73 J, p. 11.
    .874 M.
    1880 B, p. 457.
    1882 A, p. 444.
    1882 Y.
    1883 D.
    1884 O, pp. 167, 378, 379.
    1887 S, p. 342.
    1888 D, p. 257.
    1889 C, 142 *et seq.*
    1889 J.
    1889 R, pp. 875, 877.
Cope and Wortman 1884 A, p. 13.
Flower, W. H. 1876 B, xiii, p. 350.
    1883 D, p. 423.
    1885 A.
Flower and Lydekker 1891 A, p. 418.
Gadow, H. 1898 A, p. 45.
Gaudry, A. 1878 B, p. 169.
Gill, T. 1872 A, p. 301.
    1872 B, pp. 13, 48, 89.
Gill and Coues 1877 A, p. 1087.
Grant, E. 1842 A, p. 769.
Gray, J. E. 1869 A, p. 358.
Haeckel, E. 1895 A, p. 533.
Hay, O. P. 1899 F, p. 682.
L dekker, R. 1884 B, p. 16.
    1886 B.

Lydekker, R. 1896 B.
Marsh, O. C. 1877 E.
    1884 F, p. 172 *et seq.*
Nicholson and Lydekker 1889 A, p. 1390.
Osborn, H. F. 1890 D, p. 539.
    1898 A.
    1898 D.
    1898 A, p. 146.
    1900 C, p. 567.
    1900 F, p. 89.
Owen, R. 1845 B, p. 613 ("Proboscidians".)
    1848 A, p. 132.
    1866 B, p. 437.
    1868 A, p. 359.
Paulli, S. 1900 B, p. 235.
Roger, O. 1896 A, p. 158.
Ryder, J. A. 1878 A.
Schlosser, M. 1890 A.
    1890 C.
    1890 A.
Scott, W. B. 1892 A, p. 439.
Slade, D. D. 1898 A.
Steinmann and Döderlein 1890 A, p. 754
Thomas, O. 1890 A.
Tomes, C. S. 1898 A, p. 424.
Weithofer, K. A. 1888 B.
    1890 A.
    1893 A.
Woodward, A. S. 1898 B, p. 299.
Wortman, J. L. 1885 A, p. 297.
    1886 A, p. 488.
Zittel, K. A. 1893 B, pp. 206, 447, 743.

ELEPHANTIDÆ.

Adams, A. L. 1877 B.
Blainville, H. M. D. 1864 A, iii, 8, pls. i-xvii.
Brandt, J. F. 1869, p. 28.
Cope, E. D. 1873 E, p. 563.
　1875 C, p. 25.
　1889 J, p. 182.
　1893 A.
Cuvier, G. 1799 A. ("Elephans.")
　1806 A. ("Les elephans.")
　1834 A, i, p. 463; ii, p. 1–873, pls. 1–29.
Edwards, A. M. 1865 A.
Falconer, H. 1868.A.
Flower, W. H. 1876 B, xiii, p. 351.
Flower and Lydekker 1891 A, p. 423.
Gadow, H. 1898 A, p. 46.
Gill and Coues 1877 A, p. 1087.
Gray, J. E. 1869 A, p. 358.
Lortet and Chantre 1879 A.

Lydekker, R. 1880 A, p. 197.
　1884 B, p. 17.
　1886 B, p. 13.
Naumann, E. 1879 A.
Nicholson and Lydekker 1889 A, p. 1394.
Roger, O. 1889 A.
　1896 A, p. 156.
Scott, W. B. 1891 C, p. 66.
Sirodot, S. 1876 A.
　1876 B.
Steinmann and Döderlein 1890 A, p. 757.
Thomas, O. 1890 A.
Trouessart, E. L. 1898 A, p. 700.
Weithofer, K. A. 1890 A.
　1893 A.
Weyhe, —— 1875 A, p. 121.
Zittel, K. A. 1893 B, p. 458.

MAMMUT Blumenbach.　Type *M. ohioticum* Blumenbach=*M. americanus* Kerr.

Blumenbach, J. F., 1799, Naturgeschichte, ed. vi, p. 698.

Unless otherwise indicated, the following writers refer to this genus under the name *Mastodon.*

Ameghino, F. 1889 A, p. 633.
Blumenbach, J. F. 1803, Naturgeschichte, ed. vii, p. 723. (Mammut.)
Brandt, J. F. 1869 A, p. 31.
Bronn, H. G. 1838 A, p. 1232 (Tetracaulodon); p. 1233 (Mastodon).
Burmeister, H. 1867 A, p. 287.
Cope, E. D. 1877 C, p. 584. (Cænobasileus, type *C. tremontigerus.*)
　1884 G, p. 2. (Mastodon, Dibelodon, Tetrabelodon.)
　1889 J, pp. 193, 204. (Dibelodon, Tetrabelodon.)
Cope and Wortman 1884 A, p. 31.
Cuvier, G. 1806 A. ("Mastodonte.")
　1806 B, pp. 270–312.
　1817, Règne Anim., p. 233.
　1834 A, ii, pp. 247–273.
Falconer, H. 1857 B (Mastodon); p. 316 (Trilophodon, Tetralophodon.)
　1865 A.
Falconer and Cautley 1846 A.
Flower, W. H. 1876 B, xiii, p. 351.
　1883 D, p. 425.
Flower and Lydekker 1891 A, p. 431.
Gaudry, A. 1878 B, p. 170.
　1891 B.
Gervais, P. 1855 A.
Godman, J. D. 1825 A.
　1830 A. (Tetracaulodon.)
Grant, E. 1842 A, p. 771. (Mastodon, Tetracaulodon.)
Harlan, R. 1825 A, p. 209.
　1834 B, p. 51. (Tetracaulodon.)
　1835 B, p. 257. (Tetracaulodon.)
　1842 B, p. 142 (Tetracaulodon); p. 143 (Hippopotamus).
　1843 A, p. 210. (Tetracaulodon.)
Hays, I, 1834 A.
　1842 A. (Tetracaulodon.)

Hays, I. 1842 B, p. 275. (Missourium, Tetracaulodon, Mastodon.)
　1843 A.
　1844 A. (Tetracaulodon.)
Horner, W. E. 1841 A.
　1841 B. (Mastodon, Tetracaulodon.)
　1841 C. (Tetracaulodon.)
Kaup, J. J. 1832 A. (Tetracaulodon.)
　1843 A. (Tetracaulodon.)
Koch, A. C. 1839 B. (Missourian.)
　1842 A, p. 714. (Tetracaulodon.)
　1843 A. (Leviathan.)
　1845 A. (Mastodon.)
Leidy, J. 1869 A, pp. 240, 392.
　1871 C, p. 358.
　1873 G, p. 415.
Lortet and Chantre 1879 A.
Lydekker, R. 1880 A, p. 202. (Mastodon, Trilophodon.)
　1886 B, p. 14.
Marsh, O. C. 1877 E.
Meyer, H. 1867 B.
Nasmyth, A. 1842 A, p. 775. (Mastodon, Missourium, Tetracaulodon.)
Naumann, E. 1879 A.
Nicholson and Lydekker 1889 A, p. 1395.
Owen, R. 1842 E, p. 689.
　1844 A, p. 219.
　1845 B, p. 613.
Pavlow, M. 1894 A.
Pictet, F. J. 1853 A, p. 286.
Plummer, J. T. 1843 A.
Pohlig, H. 1892 B, p. 313.
Quenstedt, F. A. 1852 A, p. 53.
Richardson, J. 1854 B.
Scott, W. B. 1887 A.
　1891 D, p. 90.
　1892 A, p. 439.
Steinmann and Döderlein 1890 A, p. 758.
Thomas, O. 1890 A.
Trouessart, E. L. 1898 A, p. 700.
Warren, J. C. 1846 A.
　1852 C.
Woodward, A. S. 1898 B, p. 303.
Woodward, H. 1891 A.
Zittel, K. A. 1893 B, p. 459.

Mammut americanum (Kerr).

Kerr, R. 1792, Anim. Kingdom, p. 116. (Elephas americanus.)
    Unless otherwise indicated, the following writers call this species *Mastodon americanus.*
Agassiz, L. 1862 B. (M. giganteus.)
Allen, H. 1886 B.
Amer. Geologist 1891 C, p. 335. ("Mastodon.")
Amer. Jour. Science 1828 A, p. 187. ("Mastodon.")
    1835 A. ("Mastodon.")
    1846 A, p. 263. (Mastodon.)
    1846 B, p. 131, fig. 1. (M. giganteus.)
    1852 A. (M. giganteus.)
    1853 A. (M. giganteus.)
    1875 A, p. 222, (M. ohioticus.)
    1882 A, p. 294. ("Mastodon.")
Ami, H. M. 1898 A, p, 80.
Andrews, E. 1875 A, p. 32. ("Mastodon.")
Annan, R. 1793 A. (No name.)
Anonymous 1834 A. ("Mastodon.')
    1837 A. ("Mastodon.")
Atwater. C. 1820 A, p. 245, pl. ii, figs. A, B. ("Mastodon.")
Barton, B. S. 1805 A, ("Mammoth.")
Bell, R. 1898 A. (M. giganteus.)
Bensted, W. H. 1861 A. ("Missourium.")
Blainville, H. M. D. 1864 A, iii, S, 232, pls. xvi, xvii (M. ohioticus); iv, Z, p. 44 (Tapirus mastodontoideus).
Blake, W. P. 1855 A. (M. giganteus.)
    1868 A, p. 381. ("Mastodon.")
Blatchley, W. S. 1898 A, p. 90.
?Brevoort, J. C. 1859 A. ("Mastodon.")
Britton, N. L. 1885 A. ("Mastodon.")
Bronn, H. G. 1838 A, p. 1235, pl. xliv, fig. 6. (M. giganteus.)
Camper, P. 1788 A, pls. viii, ix. ("Mamonteum," not as generic name.)
Carpenter, W. M. 1838 A, p. 202. ("Mastodon.")
    1839 A. ("Mastodon.")
    1846 A, pp. 244, 249 ("Mastodon"); p 250 ("Dinotherium").
Chaloner, A. D. 1843 A. ("Mastodon.")
Chapman, E. J. 1858 A. (M. ohioticus?.)
Cheney, T. A. 1872 A. (M. giganteus.)
Christie, W. J. 1856 A. ("Mastodon.")
Claypole, E. W. 1895 B. ("Mastodon.")
Collett, J. 1882 A. (M. giganteus.)
Cooper, W. 1831 A, p. 160. (M. maximus.)
    1843 A, p. 33. (M. giganteum.)
Cooper, Smith, and De Kay 1831 A, p. 370. ("Mastodon.")
Cope, E. D. 1868 L, p. 251. (Tetracaulodon ohioticus, Trilophodon ohioticus.)
    1869 N, p. 740. (Trilophodon ohioticus.)
    1871 F. (Trilophodon ohioticus.)
    1871 I, p. 95.
    1874 J, p. 221. (M. ohioticus.)
    1875 F, p. 66. (M. ohioticus.)
    1875 V, p. 51. (Trilophodon ohioticus)
    ?1877 C, p. 584. (Cænobasileus tremontigerus.)
    1882 O. (M. giganteus.)
    1884 G, pp. 5, 8.
    1884 V, p. 525. (M. ohioticus.)
    1885 S.
    1887 S, p. 56. (Trilophodon ohioticus.)

Cope, E. D. 1889 F, p. 164. (Mastodon.)
    1889 J, p. 206, text fig. 3, pl. xv, fig. 2.
    1899 A, p. 252. (Mastodon.)
Cope and Wortman 1884 A, p. 33, pl. iii, figs. 1, 2, pl. vi, fig. 1.
Couper, J. H. 1842 A, p. 216. ("Hippopotamus," "Mastodon.")
    1846 in Hodson, W. B. 1846 A, p. 31. (M. giganteus.)
Croom, H. B. 1835 A, p. 170. (M. giganteus.)
Cuvier, G. 1799 A, p. 19. (Animal de l'Ohio.)
    1806 A, p. 55. ("Mastodonte.")
    1806 B, pp. 270, 312, pls. xlix-lvi. ("Mastodonte.")
    1806 C, pls. lxvi-lxix. ("Mastodonte.")
    1834 A, ii, pp. 247-273, x, p. 477 (M. maximus); p. 478, pls. xix-xxv (M. americanus).
Daily State Journal 1870 A. ("Mastodon.")
Dana, J. D. 1875 A. (M. giganteus.)
    1879 A, p. 233.
Dawkins, W. B. 1870 A, p. 232. ("Mastodon.")
Dawson, J. W. 1868 A, figs. 23, 24. ("Mastodon.";
De Kay, J. E. 1824 A. (M. giganteus.)
    1842 C, p. 102. (M. giganteus.)
Depéret, C. 1887 A, p. 175. (M. ohioticus.)
Desor, E. 1849 A, p. 207. ("Mastodon.")
    1852 A, p 58. ("Mastodon.")
Dickeson, M. W. 1846 A, p. 106. (M. giganteus.)
Dometier, W. 1803 A. ("Mammoth.")
Falconer. H. 1857 B. [M. (Trilophodon) ohioticus.]
    1868 A, i, p. 49, pl. vii, fig. 2. (M. ohioticus)
Falconer and Cautley 1846 A, p. 16, pl. iii, fig. 9: pl. xl, fig. 16; pl. xliv, fig. 4. (M. ohioticus.)
Fischer, — 1808, Programme d'Invitation, p. 19. (Harpagmotherium canadense.)
Foster, J. W. 1838 A, p. 79. (M. maximus.)
    1839 A, fig. 1. (M. giganteum.)
    1849 A. (M. giganteum.)
    1857 A. (M. giganteum.)
Gaudry, A. 1891 B, p. 4, pl. ii, fig. 7.
Gazley, S. 1830 A. ("Mastodon.")
Giebel, C. G. 1847 A, p. 197 (Tapirus mastodontoides); p. 202 (M. giganteum).
Gilbert, G. K. 1871 A, p. 220. ("Mastodon.")
Goddard, — 1841 A. ("Mastodon.")
Godman, J. D. 1825 A, pl. ii. (M. giganteus.)
    1828 A, p. 205, pls. opp. pp. 205, 225, 281, 233, 236. (M. giganteum.)
    1830 A, pls. xvii, xviii. (Tetracaulodon mastodontoideum.)
Gray, A. 1846 A. ("Mastodon.")
Hall, J. 1843 A, p. 363, plate, fig. No. 74; text fig. 173. (M. maximus.)
    1871 A, pls. iii-vii. (M giganteus.)
Hallowell. E. 1847 A. ("Mastodon.")
Harlan, R 1825 A, p 211 (M. giganteus), p. 224 (Tapirus mastodontoides, M. tapiroides).
    1834 B p. 47 (M. giganteum, M. maximus) pp. 57, 59 (Tapirus mastodontoides, M. tapiroides).
    1835 B, p 254 (M. giganteum), p. 262 (M. tapiroides) p 265 (Tapirus mastodontoides).
    1842 E. (M. giganteus.)
    1843 A, p. 210. (" Mastodon ")
Hartt, C F 1871 A. (' Mastodon)
Hayden, F. V. 1866 A. ("Mastodon.")

Hayes, S. 1895 A. ("Mastodon.")
Hays, I. 1834 A (M. giganteum); p. 334, pl. xxiv
(M. cuvieri): p. 334, pl. xxv (M. jeffersonii): p.
334, pl. xxviii (M. collinsii): p. 334, pl. xxix
(Tetracaulodon godmani).
   1841 A, pp. 102, 106. ("Tetracaulodon," M.
   giganteum.)
   1842 A, p. 183. (Tetracaulodon, M. giganteum.)
   1842 B, p. 265. (M. giganteum, Tetracaulodon godmanii, T. tapiroides, T. kochii.)
   1843 A, p. 46 ("Tetracaulodon"); p. 48 (M.
   giganteum.)
   1846 A, p. 269. ("Tetracaulodon.")
Hazeltine, J. 1835 A. ("Mastodon.")
Hicks, L. E. 1873 A. p. 79. ("Mastodon.")
Hitchcock, E. 1872 A. ("Mastodon.")
   1885 A. ("Mastodon.")
Hodson, W. B. 1846 A, p. 12. (M. giganteus.)
Holmes, F. S. 1849 A, p. 197. ("Dinotherium.")
   1850 B, p. 203. ("Dinotherium," "Mastodon.")
   1858 B, p. 442. ("Mastodon.")
Holmes, N. 1857 A. (M. giganteus.)
Horner, W. E. 1840 A, p. 281 (M. giganteum); p.
282 (Missourium kochii).
   1841 A. ("Mastodon.")
   1841 B, p. 308. ("Tetracaulodon.")
   1842 A, p. 53. ("Mastodon.")
Horner and Hays 1842 A, pls. i–iv. ("Masto-
don.")
Hoy, P. R. 1871 A, p. 147. (M. giganteus.)
Hunt. J. G. 1874 A. ("Mastodon.")
Hunter, W. 1769 A, pl. iv, figs. 1, 3, 5. ("Pseud-
elephant," "animal incognitum."
Hutchinson, H. N. 1893 A, p. 217, pl. xxi.
Jackson, J. B. S. 1845 A. (M. giganteus.)
Kaup, J. J. 1843 A, p. 172 (M. cuvieri, M. jeffersoni, M. giganteus): p. 173 (Tetracaulodon
bucklandi, T. godmani): p. 174 (M. collinsii).
Kindle, E. M. 1898 A, p. 485.
Klippart, J. H. 1875 B. ("Mastodon.")
Koch, A. C. 1839 A. ("Mammoth.")
   1839 B. ("Mastodon," "Missourian.")
   1840 A. (Missourium kochii,)
   1841, Description of the Missourian. (Leviathan missourii, Tetracaulodon tapy-
   roides, T. osagei, T. kochii.)
   1842 A, p. 715. (Tetracaulodon godmani,
   T. kochi, T. tapiroides.)
   1843 A. (Leviathan missouriensis, Missourium theristocaulodon.)
   1845 A. (Mastodon rugatum, Tetracaulodon godmani, T. haysii, T. kochii, Missourium theristocaulodon.)
   1851 A. (M. giganteum.)
   1857 A. (M. giganteum.)
   1858 A. (M. giganteum.)
Lapham, I. A. 1855 A. (M. giganteus.)
Lathrop, S. P. 1851 A. ("Mastodon.")
Leidy, J. 1849 A, p. 182.
   1858 D. ("Mastodon.")
   1860 B, p. 108, pl. xix, figs. 1–3. (M. ohioticus.)
   1868 A, p. 175.
   1869 A, pp. 240, 392.
   1870 M. p. 96.
   1871 C, p. 358.

Leidy, J. 1871 D.
   1871 F, p. 113.
   1873 B, pp. 237, 255, 330, pl. xxii, figs. 5, 6;
   pl. xxviii, fig. 9.
   1877 A, p. 213.
   1884 A.
   1884 B.
   1889 B, p. 9.
Leidy and Lucas 1896 A, p. 16.
Lesley, J. P. 1889 A, p. 380.
Lockwood, S. 1883 A.
Lortet and Chantre 1879 A, p. 291.
Lydekker, R. 1886 B, p. 15, fig. 3.
Lyell, C. 1843 B. (M. giganteum.)
McCallie, S. W. 1892 A. ("Mastodon.")
Marsh, O. C. 1867 A. (M. ohioticus.)
   1875 F. ("Mastodon.")
   1877 E. ("Mastodon.")
   1884 F, fig. 74.
   1885 B, fig. 107.
   1892 I, p. 350, pl. viii.
   1893 E, pl. viii.
   1897 C, pl. xxxi.
Maxwell, J. B. 1845 A. ("Mastodon.")
Miller, G. S. 1899 B, p. 373.
Mitchell, S. L. 1818 A, pl. vi, figs. 1–4: pl. viii,
figs. 1–3. ("Mastodon.")
Moore, J. 1897 A, p. 277, plate.
Nasmyth, A. 1842 A, p. 776 (M. giganteum); p.
777 (Tetracaulodon godmani, T. kochii, T.
tapiroides).
Newberry, J. S. 1870 A, p. 77. (M. giganteum.)
Nicholson and Lydekker 1889 A, p. 1397.
Osborn, H. F. 1898 Q, p. 19.
   1899 G, p. 539. (Mastodon.)
Owen, R. 1842 E, p. 689. (M. giganteum.)
   1845 B, p. 616, pl. cxliv, figs. 1–11, 13, 14. (M
   giganteus.)
   1846 A, p. 94. (M. giganteus.)
   1846 B, pp. 273–290, fig. 102. (M. giganteus.)
   1847 D. (M. giganteus.)
   1860 E, p. 353. (M. ohioticus.)
   1866 B, p. 441, fig. 297. (M. giganteus.)
Packard, A. S. 1868 A, p. 33. (M. giganteus.)
Panton, J. H. 1891 A. (M. giganteus.)
   1892 A. (M. giganteus.)
Pavlow M. 1894 A, pl. i; pl. ii, fig. 2. (M. ohioticus.)
Peale, R. 1802 A. ("Mammoth.")
   1802 B, pl. v, fig. 1. ("Mammoth.")
   1803 C. ("Mammoth.")
Pictet, F. J. 1853 A, p. 302. (Tapirus mastodontoides.)
Prime, A. J. 1845 A, pl. iv. ("Mastodon.")
Putnam, F. W. 1884 A, p. 112. ("Mastodon.")
   1885 A. ("Mastodon.")
Quart. Jour. Geol. Soc., 1845, i, p. 566. (M. gigan-
teus.)
Quenstedt, F. A. 1852 A, p. 53, pl. iv, figs. 7, 8
(M. giganteum.)
   1885 A, p. 73, pl. v, figs. 7, 8. (M. giganteum.)
Rice, F. P. 1895 A.
Richardson, J. 1854 A, p. 101 (Elephas rupertianus); p. 141 (M. rupertianus).
   1851 B. (M. giganteus.)
   1855 A, p. 131. (M. giganteus.)
Roger, O. 1896 A, p. 160.

Safely, R. 1866 A, p. 426, ("Mastodon.")
Savage, J. 1878 A, p. 1011 (Mastodon ohioticus.)
Shaler, N. S. 1871 A, p. 162. (M. ohioticus.)
Shurtleff, N. B. 1846 A. (M. giganteus.)
Silliman, B., sr. 1835 A, p. 165. ("Mastodon.")
Silliman, B., jr. 1868 A. ("Mastodon:" may be some other species.)
Smith, J. A. 1846 A, p. 19. ("Mastodon.")
Steinmann and Döderlein 1890 A, p. 759.
Stewart, T. P. 1828 A.
Todd, A. 1876 A. (M. augustidens.)
Troost, G. 1835 A, pp. 139, 236. ("Mastodon.")
Trouessart, E. L. 1898 A, p. 704.
Turner, G. 1799 A. ("Mastodon", "Incognitum.")
T. R. J. 1889 A. ("Mastodon.")
Van Rensselaer, J. 1826 A. (M. giganteum.)
  1827 A. ("Mastodon.")
  1828 A. (M. giganteum.)
Wailes, B. L. C. 1854 A, pp. 284, 286. (M. giganteus.)
Warren, J. C. 1846 A. (M. giganteus.)
  1852 B. (M. giganteus.)
  1852 C, pls. i-xxvi+i. (M. giganteus.)
  1853 A, figs. 1-3. (M. giganteus.)
  1855 A. (M. giganteus.)
  1855 C. (M. giganteus.)
  1856 A. (M. giganteus.)
Wheatley, C. M. 1871 A, p. 236.
Wheeler, W. 1878 A, p. 11 ("Mastodon.")
Whipple, S. H. 1844 A. ("Mastodon.")
Whitfield, R. P. 1891 A. ("Mastodon.")
Whitney, J. D. 1879 A, p. 252.
Whittlesey, C. 1848 A, p. 215. ("Mastodon.")
Wilder, B. G. 1871 A, p. 58. ("Mastodon.")
Williston, S. W. 1897 I. p. 301.
  1898 I, p. 91.
Winchell, A. 1864 A. ("Mastodon.")
? Winslow, C. 1857 A. (M. giganteus.)
  1868 A, p. 407. ("Mastodon.")
Wistar, C. 1818 A. ("Mastodon.")
Wizlizenius, A. 1858 A. (M. giganteus.)
Woolworth, S. 1847 A, p. 31, fig. 2. ("Mastodon.")
Worthen, A. H. 1871 A. ("Mastodon.")
Wyman, J. 1850 A, p. 57. (M. giganteus.)
  1853 B. ("Mastodon.")
  1862 A, p. 422. ("Mastodon.")
Yates, L. G. 1874 B.
Zittel, K. A. 1893 B, p. 464.
  Pleistocene: Virginia, Ohio, Michigan, New Jersey, Connecticut, Massachusetts, Maryland, South Carolina, Texas, Kansas, Colorado, California, Tennessee, Illinois, Indiana, Pennsylvania, New York, Mississippi, Manitoba, Cape Breton Island, Ontario, etc.

Mammut brevidens (Cope).

Cope, E. D. 1889 J, pp. 199, 200, 201, fig. 5. (Tetrabelodon.)
  1884 V, p. 525. (Mastodon proavus.)
  1886 B, p. 359. (Mastodon proavus.)
  1886 H. (Mastodon proavus.)
Matthew, W. D. 1899 A, p. 68. (Tetrabelodon.)
Roger, O. 1896 A, p. 160. (Mastodon.)
Scott, W. B. 1893 B, p. 659. (Mastodon proavus.)
  1895 C, p. 179. (Mastodon proavus.)
Trouessart, E. L. 1898 A, p. 701. (Mastodon.)
Zittel, K. A. 1893 B, p. 465. [Mastodon (Trilophodon).]
  Miocene (Loup Fork); Montana.

Mammut campestre (Cope).

Cope, E. D. 1878 C, p. 225. (Tetralophodon.)
  1878 N, p. 128. (Mastodon.)
  1884 G, p. 5. (Mastodon.)
  1884 V, pp. 524, 526. (Mastodon.)
  1889 J, pp. 195, 204, pls. ix, x. (Tetrabelodon.)
  1893 A, p. 61. (Tetrabelodon.)
Lydekker, R. 1886 B, p. xi. (Mastodon.)
Matthew, W. D. 1899 A, p. 68. (Tetrabelodon.)
Roger, O. 1896 A, p. 160. (Mastodon.)
Trouessart, E. L. 1898 A, p. 702. (Mastodon.)
Zittel, K. A. 1893 B, p. 465. [Mastodon (Tetralophodon).]
  Miocene (Loup Fork); Kansas, Nebraska.

Mammut chapmani (Cope).

Cope, E. D. 1874 J, p. 222. (Mastodon.)
  1875 F, p. 73. (Mastodon.)
  1877 K, pp. 314, 315. (Mastodon.)
Hays, I. 1884 A, explan. pl. xxii, figs. 3, 4. (No specific name.)
Leidy, J. 1869 A, pp. 248, 396. (Mastodon.)
Trouessart, E. L. 1898 A, p. 701. (Syn. of Mastodon obscurus.)
  The origin of the specimen on which this species rests is unknown. It is not certainly known to be American.

Mammut euhypodon (Cope).

Cope, E. D. 1884 V, pp. 524, 525. (Mastodon.)
  1884 G, p. 5. (Mastodon.)
  1889 J, pp. 195, 202, pl. xiii, text fig. 7. (Tetrabelodon.)
Lydekker, R. 1886 B, p. xi. (Mastodon.)
Matthew, W. D. 1899 A, p. 68. (Tetrabelodon.)
Roger, O. 1896 A, p. 160. (Mastodon.)
Trouessart, E. L. 1898 A, p. 702. (Mastodon euhyphodon.)
Wagner, G. 1899 A, p. 103. (Syn.? of M. shepardi.)
Zittel, K. A. 1893 B, p. 465. (Mastodon euhyphodon.)
  Miocene (Loup Fork); Kansas.

Mammut floridanum (Leidy).

Leidy, J. 1886 B, p. 11. (Mastodon.)
Cope, E. D. 1889 J, p. 205. (Tetrabelodon floridanus; syn.? of T. serridens.)
  1892 AA. (Syn.? of Mastodon serridens.)
Leidy, J. 1886 C. (Mastodon.)
  1887 A. (Mastodon.)
  1890 C. (Mastodon.)
  1892 A. (Mastodon.)
Leidy and Lucas 1892 A, pp. viii, 15, pls. i-vii (Mastodon.)
Trouessart, E. L. 1898 A, p. 702. (Mastodon.)
Wagner, G. 1899 A, p. 103. (Syn.? of Tetracaulodon shepardii.)
Zittel, K. A. 1893 B, p. 465. (Mastodon.)
  Pliocene; Florida.

Mammut humboldii (Cuvier).

Cuvier G. 1825, Oss. Foss., ed. 2, vol. v, pt. ii, p. 527. (Mastodon.)
  The specific name is usually spelled humboldtii. The original form is here adopted.
Ameghino, F. 1889 A, p. 645. (Mastodon.)
Blainville, H. M. D. 1864 A (1845), iii, pp. 249-285, 302. (Mastodon.)

Burmeister, H. 1867 A, p. 288, pl. xiv. (Masto-
don.)
    1889 A, p. 33. (Mastodon.)
Cope, E. D. 1884 G, pp. 5, 7. (Dibelodon.)
    1884 V, p. 525. (Mastodon.)
    1889 J, p. 195. (Dibelodon.)
    1892 BB, p. 1059. (Dibelodon.)
    1893 A, p. 60, pl. xvi, figs. 2–4; pl. xvii, fig. 4.
    (Dibelodon.)
Cuvier, G. 1806 C, p. 413. ("Mastodonte hum-
boldien.")
    1812, Oss. Foss., ed. 1, vol. ii, art. xi, p. 13.
    ("Mastodonte humboldien."
Gervais, P. 1855 A, p. 18, pl. v, figs. 9, 10. (M.
    humboldii.)
Gervais and Ameghino 1880 A, p. 109. (Masto-
don.)
Lydekker R. 1886 B, p. 41, fig. 10. (Mastodon.)
Matthew, W. D. 1899 A, p. 75. (Dibelodon.)
Meyer, H. 1867 B, p. 64, pl. vi. (Mastodon.)
Roger, O. 1896 A, p. 161.
Trouessart, E. L. 1898 A, p. 704. (Mastodon.)
Wyman, J. 1855 C, p. 279, pl. xiii. (Mastodon.)
Zittel, K. A. 1893 B, p. 464. (Mastodon.)
    Pliocene (Blanco); South America, Texas.

Mammut mirificum (Leidy).

Leidy, J. 1858 B. (Mastodon.)
    The following authors refer to this species
    under the name *Mastodon mirificus.*
Cope, E. D. 1874 J, p. 222.
    1877 K, p. 313.
    1878 C, pp. 226, 227.
    1884 V, p. 526.
    1889 J, pp. 196, 206.
    1892 G, p. 228.
    1893 A, pp. 61, 62.
Hayden, F. V. 1858 B, p. 157.
King, C. 1878 A, p. 412.
Leidy, J. 1858 E, p. 28. [M. (Tetralophodon).]
    1869 A, pp. 249, 396, pl. xxv, figs. 1, 2.
    1870 J, p. 67.
    1871 A, p. 64.
    1871 C, p. 358.
    1873 B, pp. 237, 330.
Lydekker, R. 1886 B, p. xi.
Matthew, W. D. 1899 A, p. 69. (Dibelodon.)
Roger, O. 1896 A, p. 160.
Trouessart, E. L. 1898 A, p. 705.
Zittel, K. A. 1893 B, p. 465.
    Miocene (Loup Fork); Nebraska.

Mammut obscurum (Leidy).

Leidy, J. 1869 A, p. 396, pl. xxvii, figs. 13, 15?, 16?.
    Unless otherwise indicated, the following
    writers call this species *Mastodon obscurus.*
Cope, E. D. 1872 DD.
    1874 J, p. 221.
    1875 V, p. 51.
    1877 K, p. 315.
    1880 S.
    1884 V, p. 526.
    1889 J, pp. 196, 200, 208. (Tetrabelodon.)
?Croom, H. B. 1835 A, p. 170. (M. angustidens.)
Dana, J. D. 1879 A, p. 233.
Emmons, E. 1858 B, p. 198, fig. 23. (M. giganteus.)
Gibbes, R. W. 1850 E. (M. angustidens.)
Harlan, R. 1842 B, p. 143. (M. longirostris.)

Leidy, J. 1858 D, p. 12. ("Mastodon.")
    1870 M, p. 99.
    1871 A, p. 64.
    1871 C, p. 358.
    1873 B, pp. 231, 330 (in part), pl. xxii,
    figs. 1–4.
Lydekker, R. 1884 B, p. 28.
    1886 B, pp. x, 30.
Lyell, C. 1843 B, p. 128. (M. longirostris.)
    1845 A, p. 427. (M. longirostris.)
Marsh, O. C. 1897 C, p. 527.
Roger, O. 1896 A, p. 160.
Trouessart, E. L. 1898 A, p. 701.
Warren, J. C. 1898 A, p. (M. angustidens.)
    1852 A. (M. angustidens.)
    1852 C, p. 78, pl. xxvi. (M. angustidens.)
Whitfield, R. P. 1888 A, p. 252.
Whitney, J. D. 1879 A, p. 252.
Zittel, K. A. 1893 B, p. 465.
    Founded on a cast of a tooth said to have come
    from the Miocene of Maryland; also reported
    from North Carolina, South Carolina, Colorado
    (Pliohippus beds).

Mammut præcursor (Cope).

Cope, E. D. 1893 A, p. 64, pls. xviii, xix. (Dibelo-
don.)
    1892 BB, p. 1059. (Mastodon præcursor; no
    description.)
    1893 C, p. 203. (Mastodon.)
Matthew, W. D. 1899 A, pp. 69, 75. (Dibelodon.)
Roger, O. 1896 A, p. 160.
Trouessart, E. L. 1898 A, p. 702. (Mastodon.)
    Miocene (Loup Fork); Texas. Pliocene
    (Blanco); Texas.

Mammut proavus (Cope).

Cope, E. D. 1873 CC, p. 10. (Mastodon.)
    Unless otherwise indicated, the following
    writers call this species *Mastodon proavus.*
Cope, E. D. 1874 B, p. 531.
    1874 J, p. 222.
    1874 P, p. 22.
    1884 G, p. 5. (M. angustidens.)
    1884 V, p. 525. (M. angustidens var.)
    1889 J, pp. 195, 202, pl. xl; text fig. 6. (Te-
    trabelodon angustidens proavus.)
    1892 E, p. 253. ("M. angustidens type."
Lydekker, R. 1884 B, p. 17.
    1886 B, p. x.
Matthew, W. D. 1899 A, p. 66. (Tetrabelodon
    angustidens.)
Roger, O. 1896 A, p. 160.
Scott, W. B. 1893 B, p. 659.
Trouessart, E. L. 1898 A, p. 702.
Zittel, K. A. 1893 B, p. 465.
    Miocene (Loup Fork); Kansas, Nebraska,
    Colorado, New Mexico.

Mammut productum Cope.

Cope, E. D. 1874 J, p. 221. (Mastodon.)
    Unless otherwise indicated, the following
    authors call this species *Mastodon productus.*
Amer. Jour. Science 1875 A, p. 222.
Cope, E. D. 1875 F, p. 72.
    1877 K, pp. 24, 306, pls. lxx, lxxii.
    1884 G, p. 5.
    1884 V, p. 524.

Cope, E. D. 1889 J, pp. 195, 204.   (Tetrabelodon.)
  1893 A, p. 58.
  1893 C, p. 203.
Leidy, J. 1873-B, pp. 235, 330 (in part), pl. xxii,
  figs. 1-4.   (M. obscurus.)
Lydekker, R. 1886 B, p. x.
Matthew, W. D. 1899 A, p. 68.   (Tetrabelodon.)
Roger, O. 1896 A, p. 160.
Trouessart, E. L. 1898 A, p. 702.
Zittel, K. A. 1893 B, p. 465.
  Miocene (Loup Fork); New Mexico.

Mammut rugosidens Leidy.

Leidy, J. 1890 B.   (Mastodon.)
Roger, O. 1896 A, p. 161.   (Mastodon.)
Trouessart, E. L. 1898 A, p. 702.   (Mastodon.)
Zittel, K. A. 1893 B, p. 465.   (Mastodon.)
  Upper Pliocene; South Carolina.

Mammut serridens (Cope).

Cope, E. D. 1884 V, pp. 524, 525, pl. iii, figs. 2, 3.
  (Mastodon.)
  1884 G, pp. 5, 7.   (Mastodon.)
  1885 S.   (Mastodon.)
  1889 F, p. 164.   (Mastodon.)
  1889 J, pp. 200, 205, fig. 8.   (Tetrabelodon.)
  1892 AA.   (Mastodon.)
  1893 A, p. 18, pl. iii, figs. 2, 3.   (Tetrabelodon
  serridens cimarronis.)
Leidy and Lucas 1896 A, p. 18.   (Mastodon.)
Lydekker, R. 1886 B, p. x.   (Mastodon.)
Matthew, W. D. 1899 A, p. 68.   (Tetrabelodon.)
Roger, O. 1896 A, p. 160.   (Mastodon.)
Trouessart, E. L. 1898 A, p. 702.   (Mastodon.)
Zittel, K. A. 1893 B, p. 465.   (Mastodon.)
  Miocene (Loup Fork); Texas.

Mammut shepardi (Leidy).

Leidy, J. 1870 M, p. 98.   (Mastodon.)
Bowers, S. 1889 A.   (Mastodon.)
Cope, E. D. 1874 J, p. 222.   (Mastodon.)
  1884 G, p. 5.   (Dibelodon.)
  1884 V. pp. 524, 525.   (Mastodon.)
  1889 F.   (Dibelodon.)

Cope, E. D. 1889 J, p. 204.   (Dibelodon shepardi,
  D. tropicus.)
  1892 E, p. 258.   (Dibelodon.)
  1892 G, p. 228.   (Mastodon.)
  1893 A, p. 57, pl. xv.   (Tetrabelodon.)
  1893 C.   (Tetrabelodon.)
  1893 P, p. 473, pl. xii.   (Tetrabelodon.)
  1893 Q. p. 812.   (Tetrabelodon.)
Felix and Lenk 1891 A, p. 132, pl. xxx, fig. 1.
  (Mastodon.)
Leidy, J. 1871 A, p. 64.   (Mastodon.)
  1871 C, p. 358.   (Mastodon.)
  1871 J.   (Mastodon.)
  1872 K, p. 142.   (M. obscurus.)
  1873 B, pp. 231, 330 (in part), pl. xxi, figs.
  1-4.   (M. obscurus.)
Matthew, W. D. 1899 A, p. 75.   (Tetrabelodon.)
Roger, O. 1896 A, p. 161.   (Mastodon.)
Trouessart, E. L. 1898 A, p. 702.   (Mastodon.)
Wagner, G. 1899 A, p. 99, pls. xxiv, xxv.   (Te-
  tracaulodon.)
Zittel, K. A. 1898 B, p. 465.   (Mastodon.)
  Pliocene (Blanco); California, Texas, Kansas,
  Mexico.

Mammut tropicum (Cope).

Cope, E. D. 1884 G, p. 7.   (Dibelodon.)
  1889 J. pp. 195, 198.   (Dibelodon.)
  1892 G, p. 227.   (Mastodon successor.)
  1892 J, p. 123.   (Mastodon sp.)
  1892 BB, p. 1059.   (Dibelodon.)
  1893 A, pp. 58, 59, 62, pl. xvi, fig. 1; pl. xvii.
  figs. 1-3.   (Dibelodon.)
  1893 C, p. 203.   (Dibelodon.)
Felix and Lenk 1891 A, p. 133.   (Mastodon.)
Lydekker, R. 1886 B, p. xi.   (Mastodon.)
Matthew, W. D. 1899 A, p. 75.   (Dibelodon.)
Meyer, H. 1867 B, p. 64, pl. vi.   (Mastodon humboldtii.)
Roger, O. 1896 A, p. 161.   (Mastodon successor.)
Trouessart, E. L. 1898 A, p. 703.   (Mastodon.)
Zittel, K. A. 1893 B, p. 46.   (Syn. of Mastodon hum-
  boldtii.)
  Pliocene (Blanco); Texas, Mexico, South
  America.

ELEPHAS Linn.   Type *E. maximus* Linn.

Linnæus, C. 1758 Syst. Nat., ed. 10, i, p. 33.
Adams, A. L. 1874 A.
  1877 B.
  1879 A.
  1881 A.
Blainville, H. M. D. 1867 A, iii, 8, plates.
Brandt, J. F. 1869 A, p. 84.
Cope, E. D. 1889 J, p. 194.
  1893 M, p. 1015.
Cuvier, G. 1806 A.
  1834 A, i, p. 455, ii, pp. 1-373.
Falconer, H. 1857 B.   (Elephas, with subgenera
  Stegodon, Loxodon, and Euelephas.)
  1865 A.
  1868 A.
Falconer and Cautley, 1846 A.
Flower, W. H. 1876 B, xiii, p. 351.
  1883 D, p. 425.
  1885 A.
Flower and Lydekker, 1891 A, p. 424.

Gaudry, A. 1878 B, p. 175.
Grant, E. 1842 A, p. 771.
Gray, J. E. 1869 A, p. 359.
Leuthardt, F. 1891 A.
Lydekker, R. 1880 A, pp. 197, 256.
  1886 B, p. 78.
Marsh, O. C. 1877 E.
Möbius, K. 1892 A.
Naumann, E. 1879 A.
Nicholson and Lydekker, 1889 A, p. 1401.
Owen, R. 1844 A, p. 208.
  1845 B, p. 625.
  1868 A, p 360, figs. 289-242.
Paulli, S. 1900 A, p. 236.
Pictet, F. J 1853 A, p. 280.
Pohlig. H 1889 A.
  1892 B.
Quenstedt, F. A. 1885 A, p. 67.
Röse, C. 1893 A.
Rütimeyer, L. 1888 A, p. 11.

Scott, W. B. 1887 A.
Steinmann and Döderlein 1890 A, p. 760.
Thomas, O. 1890 A.
Turner, H. N. 1848 A, p. 71.
Weber, M. 1898 A, p. 185.
Welthofer, K. A. 1890 A.
　　1893 A.
Woodward, A. S. 1898 B, p. 305.
Zittel, K. A. 1893 B, p. 467.

Elephas columbi Falconer.

Falconer, H. 1857 B, p. 319.　[E. (Euelephas).]
Bell, R. 1898 A, p. 371, fig. 1.
Blake, C. C. 1862 A, pl. iv.　(E. texianus.) ·
　　1862 B.　(E. texianus.)
Briggs and Foster in Billings, E. 1863 A, figs. 1–5.
　(E. jacksoni.)
Cope, E. D. 1874 J, p. 221.　(E. primigenius co-
　lumbi.)
　　1889 J, p. 209, pl. xiv.　(E. primigenius co-
　lumbi.)
　　1894 A, p. 68.　(E. primigenius var.)
Cragin, F. W. 1896 A, p. 53.　(E. imperator?.)
Dall, W. H. 1891 A.
Dawkins, W. B. 1870 A, p. 232.
Donald, J. F. 1871 A.　(E. primigenius var. jack-
　soni.)
Edwards, A. M. 1865 A.
Falconer, H. 1868 A, ii, p. 212, pl. x.
Felix and Lenk 1891 A, p. 131.
Gervais and Ameghino 1880 A, p. 213.
Hayden, F. V. 1858 B, p. 157.
?Kindle, E. M. 1898 A, p. 485.　(E. americanus.)
Leidy, J. 1858 B, p. 10.　(E. imperator.)
　　1858 E, p. 29.　[E. (Euelephas) imperator.]
　　1869 A, p. 397.　(E. americanus in part.)
　　1870 M, p. 97.
　　1871 C, p. 359.　(E. americanus, part.)
　　1873 B, pp. 238, 329.　(E. americanus, part.)
　　1873 G.
　　1875 A, p. 121.　(E. americanus, part.)
　　1877 A, p. 213.　(E. americanus.)
　　1889 D, p. 17, pl. iii, figs. 8, 9.　(E. america-
　nus, or E. columbi.)
　　1892 A, p. 129.
Logan, W. E. 1863 A, p. 914, figs. 495–498.　(Euele-
　phas jacksoni.)
Lucas, F. A. 1900 A, p. 349.
Lydekker, R. 1886 B, pp. xi, 171.
Mather, W. W. 1838 A, p. 362.　(E. jacksoni.)
Miller, G. S. 1899 B, p. 373.
Nicholson and Lydekker 1889 A, p. 1406,
Owen, R. 1859, Rep. Brit. Assoc. Adv. Sci. for 1858,
　p. lxxxvi.　(E. texianus, without description.)
Pohlig, H. 1887 A, p.1 17.　(E. trogontheri?, E. col-
　umbi.)
　　1889 A, p. 247.　(E. columbi, E. imperator.)
　　1892 B, p. 327.
Trouessart, E. L. 1898 A, p. 711.
Zittel, K. A. 1893 B, p. 473.
　　Pleistocene; Georgia, South Carolina, Ala-
　bama, Texas, Colorado, Oregon, New Mexico,
　California, New York?, Florida, Mexico, North-
　west Territory, Hudson Bay.

Elephas primigenius Blumenbach.

Blumenbach, J. F., 1803, Handb. Naturg., 1st
　French ed., vol. ii, p. 407.　(*Fide* Lydekker.)

Adams, A. L. 1879 A.
　　1881 A.
Agassiz, L. 1850 A, p. 100.　("Elephant.")
Amer. Geologist 1891 F.
Amer. Jour. Science 1853 B, p. 146.
Anonymous 1834 A.　("Elephant.")
· 　　1837 A.　("Elephant.")
Baer, K. E. 1866 A.
　　1867 A.
Barbour, E. H. 1890 A
Barton, B. S. 1805 A, p. 98.　("Elephant.")
Bell, R. 1898 A.
Beyer, S. W. 1899 A, p. 211.
Blainville, H. M. D. 1864 A, iii 8, pls. iii, v, vi,
　viii, x.
Blake, W. P. 1855 A.
　　1864 A.　("Elephas.")
　　1867 B.
　　1900 A, p. 257.
Blatchley, W. S. 1898 A, p. 89.
Brandt, J. F. 1848 A.
　　1866 A.
　　1866 B.
　　1867 A.
Buckland, W. 1831 A, p. 594, pls. i, ii.
Carpenter, W. M. 1846 A, p. 249.　("Elephant ")
Chaloner, A. D. 1843 A, p. 321.　("Elephant.")
Charlton, O. C. 1890 A.
Cooper, W. 1831 A, p. 168.
　　1843, Proc. Geol. Soc. Lond., p. 33.　("Mam-
　moth.")
Cooper, Smith, and De Kay 1831 A, p. 371.　("Ele-
　phant.")
Cope, E. D. 1884 G, p. 8.
　　1888 B, p. 293.
　　1889 F.
　　1889 J, p. 207, pl. xvi, fig. 20.
　　1889 S.
　　1893 A, p. 87.
　　1894 A, p. 68.
　　1894 H, p. 453.
Cope and Wortman 1884 A, p. 32, pl. vi, figs. 2–5.
Cottle, T. 1852 A.
　　1853 A.
Couper, J. H. 1842 A, p. 217.　("Mammoth.")
　　1846 in Hodson, W. B. 1846 A, p. 45.
Cuvier, G. 1799 A, p. 15, pls. v, vi.　("Mammoth.")
　　1806 A, pp. 45, 149, 264.
　　1834 A, ii, pp. 1–246, with plates.
C. D. 1838 A.
Dall, W. H. 1869 A.
　　1873 A.
Davison, C. 1894 A.
Dawkins, W. D. 1870 A, p. 233.
Dawson, G. M. 1894 A.
De Kay, J. G. 1842 C, p. 100 (E. primigenius); p.
　101, pl. xxxii (E. americanus).
Diffenberger, F. R. 1873 A.　(E. americanus.)
Dupont, E. 1867 A.
Duralde, M. 1804 A.
Edwards, A. M. 1865 A.
Edwards, T. 1793 A.　(No name.)
Emmons, E. 1858 B, p. 200, fig. 24.
Falconer, H. 1857 B.　[E. (Euelephas) primige-
　nius.]
　　1865 A.
Falconer and Cautley 1846 A, pl. i, fig. 1; pl. xiii.
　A, figs. 1–3; pl. xiii B, figs. 1–3.

Felix and Lenk 1891 A, pp. 126, 131.
Flower, W. H. 1876 B, xiii, 351.
  1883 C.
Flower and Lydekker 1891 A, p. 428, figs. 184, 185.
Foster, J. W. 1838 A, p. 79. (E. primogenius.)
  1889 A, p. 190, fig. 2. ("Elephant.")
  1857 A.
  1872 A, p. 143. (E. mississippiensis.)
Gaudry, A. 1872 A.
  1873 A.
Gazley, S. 1830 A. ("Elephant.")
Geinitz, H. B. 1885 A, p. 66, pl. iii.
Gibbes, R. W. 1850 D.
Giebel, C. G. 1847 A, p. 208.
Godman, J. D. 1828 A, ii, pp. 255-265.
Harlan, R. 1823 A. ("Elephant.")
  1825 A, p. 207.
  1828 A, p. 189.
  1834 B, p. 57.
  1835 B, p. 263.
  1835 G, p. 359. ("Elephant.")
  1842 E.
Haymond, R. 1844 A, p. 294. ("Megatherium.")
Hays, I. 1844 A, p. 43. ("Elephant.")
Higley, W. K. 1866 A, plate.
Holmes, F. S. 1850 B, p. 203. ("Elephant.")
Horner, W. E. 1840 A, p. 299. ("Elephant."
Howorth, H. W. 1870 A.
  1881 A.
  1881 B.
  1892 A.
  1893 A.
  1894 A.
Hutchinson, H. N. 1893 A, p. 193, pl. xx.
Kindle, E. M. 1898 A, p. 485.
Lambe, L. M. 1898 A, p. 136.
Landois —, 1871 A, p. 47, pl. ix.
Leidy, J. 1854 A, p. 9. (E. americanus.)
  1860 B, p. 108, pl. xviii. (E. americanus.)
  1869 A, pp. 251, 397. (E. primigenius, E. americanus.)
  1870 K, p. 69. (E. americanus.)
  1871 C, p. 359 (E. americanus.)
  1875 A. (E. americanus, part.)
  1877 A, p. 212. (E. americanus.)
  1889 D, p. 17, pl. iii, figs. 6-9.
  1889 E, p. 22.
Lucas, F. A. 1899 A, p. 764.
  1900 A, p. 349, figures.
Lydekker, R. 1886 B, p. 175, figs. 30-32.
Lyell, C. 1843 C, p. 33. ("Elephant.")
Madison, ——. 1806 A, p. 486.
Marsh, O. C. 1877 E, p. 367. (E. americanus.)
Meyer, H. 1852 A, pl. xiv, figs. 1-4.
Mitchell, S. L. 1818 A, pl. vi, figs. 2, 3, 5, 6. ("Elephant.")
Möbius, K. 1892 A.
Nature 1871 A.
Naumann, E. 1879 A, pls. i-vii.
Newberry, J. G. 1870 A, p. 77. (E. primogenius.)

Newberry, J. G. 1871 A, p. 241. (E. americanus.)
Nicholson and Lydekker 1889 A, pp. 1404, 1406.
Owen, R. 1844 A, p. 210.
  1845 B, p. 628, pl. cxlviii, figs. 6-8.
  1846 B, p. 217, figs. 85, 86, 90-95.
  1847 A.
Packard, A. S. 1868 A, p. 33. (E. americanus.)
  1886 A.
Panton, J. H. 1891 A.
  1892 A.
Parker, H. W. 1884 A.
Pictet, F. J. 1853 A, p. 283.
Pohlig, H. 1887 A, p. 117.
  1887 B, p. 254.
  1889 A, p. 248. (E. primigenius, E. americæ.)
  1892 B, p. 327. (E. primigenius, E. americæ.)
Quenstedt, F. A. 1885 A, p. 68, pl. iv, figs. 11-19.
Richardson, J. 1854 A, p. 11 (Mastodon giganteus);
  p. 141 (E. primigenius).
  1855 A, p. 132.
Rogers, H. D. 1854 A.
Schmidt, F. 1872 A.
Shaler, N. S. 1871 A, p. 148.
  1871 B, p. 167.
Steinmann and Döderlein 1890 A, p. 761.
Steitz, A. 1870 A.
Stirrup, M. 1893 A.
  1894 A.
Thompson, Z. 1850 A, p. 256. ("Elephant.")
  1853 A, p. 14.
Toll, E. 1895 A. ("Mammoth.")
Troost, G. 1835 A, p. 143. (Elaphus primigenius)
Trouessart, E. L. 1898 A, p. 711.
Turner, G. 1799 A. ("Mammoth.")
Van Rensselaer, J. 1828 A, ("Elephant.")
Walles, B. L. C. 1854 A, p. 286.
Warren, J. C. 1852 C. p. 142.
Welthofer, C. A. 1893 A.
Whitney, J. D. 1879 A, p. 254.
Williston, S. W. 1898 I, p. 91.
Winchell, A. 1863 A.
Winslow, C. F. 1857 A.
  1868 A.
Wistar, C. 1818 A. ("Elephant.")
Woodhull, A. A. 1872 A.
Woodward, H. 1869 A.
Woolworth, S. 1847 A, p. 31, fig. 1. (E. americanus.)
Worthen, A. H. 1871 A.
Wylie, T. A. 1859 A.
Wyman, J. 1853 B.
  1857 A. ("Elephant.")
Yates, L. G. 1874 B, p. 18.
Zittel, K. A. 1893 B, p. 469, figs. 387-390.
  Pleistocene; New York, Vermont, New Jersey, Pennsylvania, Alaska, British Columbia, Ontario, Alberta, Mexico, Arizona, Montana, Colorado, Michigan, Kentucky, Ohio, Missouri, Illinois, Iowa, Indiana, Kansas, Georgia, North Carolina, South Carolina, Texas, Oklahoma, California, Oregon.

Order TILLODONTIA Marsh.

Marsh, *O. C.* 1875 C, p. 221.
Cope, E. D. 1875 C, p. 23.
    1876 A, pp. 3, 4. (Tillodonta.)
    1877 K, pp. 85, 370. (Tillodonta.)
    1882 E, p. 146.
    1882 L.
    1883 I, p. 79.
    1884 O, pp. 186, 194, 739. (Tillodonta.)
    1887 S, p. 343.
    1889 C, p. 144.
    1891 L, p. 1117.
    1891 N, p. 73.
    1893 M, p. 1015.
    1896 B, p. 115.
Flower, W. H. 1876 B, xiii, p. 514.
Flower and Lydekker 1891 A, p. 441.
Gadow, H. 1898 A, p. 42.
Haeckel, E. 1895 A, pp. 501, 502.
Hay, O. P. 1899 F, p. 682.
Lydekker, R. 1887 A, p. 1.

Lydecker, R. 1896 B, p. 378.
Marsh, O. C. 1876 B, p. 249.
    1876 G, p. 60.
    1877 E.
    1892 D, p. 443.
    1895 D, p. 498.
Nicholson and Lydekker 1889 A, p. 1408. (Tillodontia, Anchippodontia.)
Osborn, H. F. 1888 F.
    1893 D.
    1898 A, p. 146.
Osborn and Earle 1895 A, p. 40.
Roger, O. 1896 A, p. 88.
Schlosser, M. 1890 C, p. 250.
Steinmann and Döderlein 1890 A, p. 728.
Tomes, C. S. 1898 A, p. 446.
Woodward, A. S. 1898 B, p. 374.
Wortman, J. L. 1885 A, p. 298.
    1897 B, p. 61.
Z'ttel, K. A. 1893 B, pp. 508, 725.

ESTHONYCHIDÆ.

Cope, *E. D.* 1889 R, p. 876.
Haeckel, E. 1895 A, p. 502. (Esthonychida.)
Matthew, W. D. 1899 A, p. 42.
Nicholson and Lydekker 1889 A, p. 1409. (Platychœropidæ.)

Roger, O. 1896 A, p. 88.
Steinmann and Döderlein 1890 A, p. 729.
Trouessart, E. L. 1898 A, p. 665. (Esthonycidæ.)
Zittel, K. A. 1893 B, p. 6.

ESTHONYX Cope.   Type *E. bisulcatus* Cope.

Cope, *E. D.* 1874 O, p. 6.
    1876 C.
    1877 K, p. 158.
    1882 E, p. 146.
    1882 L.
    1883 I, p. 77.
    1884 O, pp. 194, 196, 202, 738.
    1884 S, pp. 351, 479.
    1885 T.
    1885 X, p. 1208.
    1888 I, p. 4.
    1888 K, p. 163.
    1889 C, p. 221.
Lydekker, R. 1885 D. (Syn. of Platychœrops.)
    1887 A, p. 3.
Nicholson and Lydekker 1889 A, p. 1409.
Schlosser, M. 1890 C, p. 252.
Scott, W. B. 1892 D, p. 80.
Tomes, C. S, 1898 A, p. 446.
Woodward, A. S. 1898 B, p. 375.
Wortman, J. L. 1885 A, p. 299.
    1886 A, p. 425.
    1897 B, pp. 62, 63.
Zittel, K. A. 1893 B, p. 506.

Esthonyx acer Cope.

Cope, *E. D.* 1874 O, p. 7.
    1875 C, p. 24. (Syn. of E. bisulcatus.)
    1877 K, p. 154. (Syn. of E. bisulcatus.)
    1882 E, p. 148.
    1884 O, p. 204.
    1884 S, p. 480.

Matthew, W. D. 1899 A, p. 31.
Roger, O. 1896 A, p. 88.
Trouessart, E. L. 1898 A, p. 665.
    Eocene (Wasatch); New Mexico.

Esthonyx acutidens Cope.

Cope, *E. D.* 1881 D, p. 185.
    1882 E, p. 148.
    1884 O, pp. 204, 210, pl. xxiva, figs. 17-21.
Matthew, W. D. 1899 A, p. 36.
Roger, O. 1896 A, p. 88.
Trouessart, E. L. 1898 A, p. 665.
Wortman, J. L. 1897 B, p. 62.
    Eocene (Wind River); Wyoming.

Esthonyx bisulcatus Cope.

Cope, *E. D.* 1874 O, p. 6.
    1875 C, p. 24.
    1877 K, p. 154, pl. xl, figs. 27-33.
    1884 O, p. 204.
King, C. 1878 A, p. 377.
Matthew, W. D. 1899 A, p. 31.
Roger, O. 1896 A, p. 88.
Trouessart, E. L. 1898 A, p. 665.
    Eocene (Wasatch); New Mexico.

Esthonyx burmeisterii Cope.

Cope, *E. D.* 1874 O, p. 7.
    1875 C, p. 24.
    1877 K, p. 156, pl. xl, fig. 26.
    1882 E, p. 147.
    1884 O, p. 204, pl. xxivc, figs. 1-10.

Cope, E. D. 1884 S, p. 479, fig. 23.
Lydekker, R. 1887 A, p. 3.
Matthew, W. D. 1899 A, p. 31.
Nicholson and Lydekker 1889 A, p. 1409, fig. 1290.
Roger, O. 1896 A, p. 88.
Trouessart, E. L. 1898 A, p. 665.
Woodward, A. S. 1898 B, p. 375, fig. 212.
Wortman, J. L. 1886 A, p. 425, fig. 207.
Zittel, K. A. 1893 B, p. 506, fig. 419.
Eocene (Wasatch); New Mexico, Wyoming.

Esthonyx spatularius Cope.
Cope, E. D. 1880 O, p. 908.
1881 D, p. 187.
1882 E, p. 148.
1884 O, pp. 204, 211, pl. xxiva, figs. 22-25.
1884 S, p. 479.
Matthew, W. D. 1899 A, pp. 31, 36.
Roger, O. 1896 A, p. 88.
Trouessart, E. L. 1898 A, p. 665.
Eocene (Wasatch, Wind River); Wyoming.

ANCHIPPODONTIDÆ.

Gill, T. 1872 B, p. 87.
Flower, W. H. 1876 C, p. 22. (Tillotheriidæ.)
Gadow, H. 1898 A, p. 42. (Tillotheriidæ.)
Haeckel, E. 1895 A, p. 502. (Tillotherida.)
Lydekker, R. 1887 A, p. 1. (Anchippodontidæ.)
1896 B, p. 878. (Anchippodontidæ.)
Marsh, O. C. 1875 C, p. 221. (Tillotheriidæ, Anchippodontidæ.)

Marsh, O. C. 1876 G, p. 60. (Tillotheriidæ.)
1877 E. (Tillotheriidæ.)
Nicholson and Lydekker 1889 A, p. 1408. (Anchippodontidæ.)
Roger, O. 1896 A, p. 88. (Tillotheriidæ.)
Steinmann and Döderlein 1890 A, p. 730. (Tillotheriidæ.)
Zittel, K. A. 1893 B, p. 506. (Tillotheriidæ.)

TILLOTHERIUM Marsh.   Type T. hyracoides Marsh.

Marsh, O. C. 1873 H, p. 485.
Cope, E. D. 1882 E, p. 146.
1892 L.
1884 O, p. 195.
1888 I.
1888 L.
Flower, W. H. 1876 B, xiii, p. 514.
1876 C, p. 123.
Flower and Lydekker 1891 A, p. 441.
Lydekker, R. 1887 A, p. 1.
Marsh, O. C. 1875 B, p. 241.
1875 C.
1876 B, pls. viii, ix.
1876 G, p. 60.
1877 E.
1895 D, p. 498.
Nicholson and Lydekker 1889 A, p. 1408.
Osborn, H. F. 1897 G, p. 252.
Schlosser, M. 1890 C, p. 275.
1892 A, p. 226.
Steinmann and Döderlein 1890 A, p. 730.
Tomes, C. S. 1898 A, p. 416.
Woodward, A. S. 1898 B, p. 376.
Wortman, J. L. 1886 A, p. 433.
1897 B, pp. 61, 63.
Zittel, K. A. 1893 B, p. 507.

Lydekker, R. 1887 A, p. 2, fig. 2.
Marsh, O. C. 1876 B, p. 250, pl. viii; pl. ix, figs. 1-3; text fig. 1.
1895 D, p. 498, fig. 48.
Matthew, W. D. 1899 A, p. 42.
Nicholson and Lydekker 1889 A, p. 1408, fig. 1289.
Osborn, H. F. 1897 G, p. 257.
Roger, O. 1896 A, p. 88.
Trouessart, E. L. 1898 A, p. 666.
Woodward, A. S. 1898 B, p. 376, fig. 213.
Wortman, J. L. 1897 B, p. 63.
Zittel, K. A. 1893 B, p. 507, fig. 420.
Eocene (Bridger); Wyoming, Colorado.

Tillotherium hyracoides Marsh.
Marsh, O. C. 1873 H, p. 485.
King, C. 1878 A, p. 404.
Matthew, W. D. 1899 A, p. 42.
Roger, O. 1896 A, p. 88.
Trouessart, E. L. 1898 A, p. 666.
Eocene (Bridger); Wyoming.

Tillotherium fodiens Marsh.
Marsh, O. C. 1875 B, p. 241.
Flower, W. H. 1876 C, p. 123, fig. 3.
Flower and Lydekker 1891 A, p. 441, fig. 193.
King, C. 1878 A, p. 404.

Tillotherium latidens Marsh.
Marsh, O. C. 1874 C, p. 533.
1876 B, explan. pl. ix, fig. 4.
Matthew, W. D. 1899 A, p. 42.
Roger, O. 1896 A, p. 88.
Trouessart, E. L. 1898 A, p. 666.
Wortman, J. L. 1897 B, p. 63.
Zittel, K. A. 1893 B, p. 507, fig. 421.
Eocene (Bridger); Wyoming.

ANCHIPPODUS Leidy.   Type A. riparius Leidy.

Leidy, J. 1868 I, p. 232
Cope, E. D. 1873 E, p. 605.
1873 G, p. 208.
1875 J, p. 3.
1876 A, p. 3.
1876 C.
1884 O, p. 195.
1888 L.
Flower, W. H. 1876 C, p. 122.

Flower and Lydekker 1891 A, p. 441.
Leidy, J. 1871 F, p. 114. (Trogosus, type T. castoridens.)
1872 B, p. 359. (Trogosus.)
1873 B, p. 328.
Lydekker, R. 1887 A, p. 1.
Marsh, O. C. 1874 B, p 255.
1875 B, p. 241.
1877 E.

Nicholson and Lydekker 1889 A, pp. 1408, 1410.
Wortman, J. L. 1897 B, p. 61.
Zittel, K. A. 1893 B, p. 506.

Anchippodus minor (Marsh).

Marsh, O. C. 1871 D, p. 36. (Palæosyops.)
Cope, E. D. 1873 E, p. 605.
Flower, W. H. 1876 B, xiii, p. 252.
King, C. 1878 A, p. 404.
Leidy, J. 1871 F, p. 114. (Trogosus castoridens.)
  1871 I. (Palæosyops minor.)
  1872 B, p. 360. (Trogosus castoridens.)
  1872 D, p. 37. (A. riparius.)
  1873 B, p. 71, pl. v, figs. 1–3. (Trogosus castoridens); p. 328 (A. riparius).
Lydekker, R. 1887 A, p. 2. (Syn. of A. riparius.)
Marsh, O. C. 1873 H, p. 485. (Anchippodus.)
  1876 B, p. 252, pl. ix, fig. 5. (Anchippodus.)
Matthew, W. D. 1899 A, p. 42.
Roger, O. 1896 A, p. 88 (Syn. of A. riparius); p. 192 (Palæosyops).
Trouessart, E. L. 1898 A, p. 666 (A. riparius, in part); p. 738 (Palæosyops minor, in part).
Wortman, J. L. 1897 B, p. 63. (Tillotherium.)
Zittel, K. A. 1893 B, p. 508, fig. 422.

Eocene (Bridger); Wyoming.

Anchippodus riparius Leidy.

Leidy, J. 1868 I, p. 232.
Conrad, T. A. 1869 A, p. 363.
Cope, E. D. 1869 N, p. 740.
Flower, W. H. 1876 B, xiii, p. 514.
  1876 C, p. 120.
Leidy, J. 1869 A, p. 408, pl. xxx, figs. 45, 46.
  1872 D, p. 37.
  1873 B, pp. 72, 328, in part.
Lydekker, R. 1887 B, p. 2, in part.
Matthew, W. D. 1899 A, p. 42.
Roger, O. 1896 A, p. 88.
Trouessart, E. L. 1898 A, p. 666.
Zittel, K. A. 1893 B, p. 508.

Eocene; New Jersey.

Anchippodus vetulus Leidy.

Leidy, J. 1871 K, p. 229. (Trogosus.)
  1872 B, p. 360. (Trogosus.)
  1873 B, p. 75 (Trogosus): p. 329, pl. vi, fig. 43. (Anchippodus).
Matthew, W. D. 1899 A, p. 42.
Roger, O. 1896 A, p. 88.
Trouessart, E. L. 1898 A, p. 666.

Eocene (Bridger); Wyoming.

Order GLIRES Linn.

Linnæus, C., 1758, Syst. Nat., ed. 10, i, p. 56. (Exclusive of Rhinoceros.)
  Unless otherwise indicated, the following authors denominate this group Rodentia.
Adloff, P. 1898 A. ("Rodentien.")
Allen, J. A. 1877 A.
Alston, E. R. 1876 A. (Glires.)
Baird, S. F. 1857 A, p. 235.
Brandt, J. F. 1855 A.
Cope, E. D. 1874 B, p. 474.
  1880 B. 455.
  1881 E.
  1881 J, p. 545.
  1883 S.
  1884 O, pp. 166, 812.
  1887 S, p. 342.
  1888 D, p. 257.
  1888 I.
  1889 C, p. 143.
  1889 R, pp. 875, 876. (Glires.)
  1893 M, p. 1015.
  1896 B, p. 111. (Glires.)
Cope and Wortman 1884 A, p. 36.
Coues, E. 1877 B.
Cuvier, G. 1884 A, viii, pt. i, pp. 1–128, with plates.
Fleischmann, A. 1888 A.
Flower, W. H. 1883 A.
  1883 D, p. 415.
  1885 A.
Flower and Lydekker 1891 A, p. 443.
Gadow, H. 1898 A, p. 42.
Gill, T. 1872 A, p. 303. (Glires.)
  1872 B, p. 20. (Glires.)
Gill and Coues 1877 A, p. 1061.
Gruber, W. 1859 A. (Glires.)
Haeckel, E. 1895 A, p. 506.
Hay, O. P. 1899 F, p. 682. (Glires.)
Huxley, T. H. 1872 A, p. 369.
Leidy, J. 1871 C, p. 363.
Lydekker, R. 1896 B.

Major, Fors. 1893 A, p. 202.
Marsh, O. C. 1877 E.
Nicholson and Lydekker 1889 A, p. 1411.
Osborn, H. F. 1900 C, p. 567.
Owen, R. 1845 B, p. 398.
  1851 C.
  1866 B, p. 364.
  1868 A, p. 294.
Palmer, T. S. 1897 A, p. 103.
Parker, W. K. 1868 A, p. 207.
  1885 A, p. 192.
Parsons, F. G. 1899 A.
  1900 A.
Paulli, S. 1900 B, p. 539.
Roger, O. 1896 A, p. 110.
Schlosser, M. 1884 A.
  1884 B.
  1888 A.
  1889 A.
  1890 A.
  1890 C.
  1890 D.
Scott, W. B. 1892 A, p. 426.
Slade, D. D. 1888 A, p. 243.
  1890 A. ,
  1892 B.
Smith J. 1895 A, p. 341.
Steinmann and Döderlein 1890 A, p. 731.
Thomas, O. 1896 A, p. 1012.
Tomes, C. S. 1898 A, p. 448.
Tomes, J. 1350 A, p. 529.
Trouessart, E. L. 1898 A, p. 388.
Tullberg, T. 1899 A.
Turner, H. N. 1848 A.
Weyhe, — 1875 A, p. 111.
Winge, H. 1888 A, p. 103.
Woodward, A. S. 1898 B, p. 373.
Woodward, M. F. 1894 A.
Wortman, J. L. 1886 A, p. 466.
Zittel, K. A. 1893 B, pp. 512, 553, 727.

Suborder SIMPLICIDENTATA.

Alston, E. R. 1876 A.
Flower and Lydekker 1891 A, p. 448.
Nicholson and Lydekker 1889 A, p. 1413.
Palmer, T. S. 1897 A.

Schlosser, M. 1884 A, p. 123. (Miodonta.)
Thomas, O. 1896 A, p. 1015.
Tullberg, T. 1899 A, pp. 57, 341.
Woodward, A. S. 1898 B, p. 377.

Superfamily SCIUROMORPHA.

Alston, E. R. 1876 A.
Brandt, J. F. 1855 A.
Cope, E. D. 1881 E, p. 361.
　1883 S, p. 43.
　1884 O, p. 812.
Flower, W. H. 1883 D, p. 417.
Flower and Lydekker 1891 A, p. 448.
Gadow, H. 1898 A, p. 42.
Haeckel, E. 1895 A, p. 509.
Lydekker, R. 1896 B.
Nicholson and Lydekker 1889 A, p. 1419.

Palmer, T. S. 1897 A.
Schlosser, M. 1884 A, p. 98.
　1890 C, p. 250.
Slade, D. D. 1888 A, p. 243.
　1892 B.
Roger, O. 1896 A, p. 115.
Steinmann and Döderlein 1890 A, p. 732.
Thomas, O. 1896 A, p. 1015.
Tullberg, T. 1899 A, pp. 283, 461. (Sciuromorphi.)
Zittel, K. A. 1893 B, p. 521 (Protrogomorpha); p. 527 (Sciuromorpha).

SCIURIDÆ.

Allen, J. A. 1877 A, xi, p. 637.
Alston, E. R. 1876 A.
Baird, S. F. 1857 A, p. 240.
Brandt, J. F. 1855 A.
Cope, E. D. 1888 I, p. 9.
　1889 C, p. 267.
　1898 M, p. 1015.
Flower and Lydekker 1891 A, p. 450.
Leidy, J. 1871 C, p. 363.
Major, Fors. 1873 A.
　1898 A, p. 186.
Marsh, O. C. 1877 D, p. 253. (Allomyidæ.)

Nicholson and Lydekker 1889 A, p. 1420.
Roger, O. 1896 A, p. 115.
Schlosser, M. 1890 C, p. 250.
Steinmann and Döderlein 1890 A, p. 733.
Thomas, O. 1896 A, p. 1015.
Tomes, J. 1850 A, p. 536.
Trouessart, E. L. 1898 A, p. 395.
Tullberg, T. 1899 A, pp. 290, 468.
Winge, H. 1888 A, p. 185. (Including Sciurini, Castorini.)
Zittel, K. A. 1893 B, p. 528.

PTEROMYINÆ.

Major, Fors. 1898 A, p. 187.
Marsh, O. C. 1877 D, p. 253. (Allomyidæ.)

Trouessart, E. L. 1898 A, p. 395.
Winge, H. 1888 A, p. 115. (Allomyini.)

ALLOMYS Marsh.　Type A. nitens Marsh.

Marsh, O. C. 1877 D, p. 253.
Cope, E. D. 1879 D, p. 67. (Meniscomys, type M. hippodus.)
　1881 E, p. 365. (Meniscomys.)
　1883 S, p. 51. (Meniscomys.)
　1884 O, p. 826. (Meniscomys.)
　1888 B, p. 291 (Meniscomys=Sciurodon, fide Schlosser.)
Hay, O. P. 1899 O, p. 253.
Major, Fors. 1893 A, p. 192.
Marsh, O. C. 1877 E.
Nicholson and Lydekker 1889 A, p. 1421. (Meniscomys.)
Riggs, E. S. 1899 A, p. 183. (Protogaulus, type Meniscomys hippodus.)
Schlosser, M. 1884 A, p. 136.
Winge, H. 1888 A, pp. 114, 164.
Zittel, K. A. 1893 B, p. 529.

Allomys cavatus (Cope).

Cope, E. D. 1881 E, p. 366. (Meniscomys.)
　1881 Q, p. 586. (Meniscomys.)
　1885 S, p. 52, fig. 7. (Meniscomys.)

Cope, E. D. 1884 O, pp. 827, 830, pl. lxiii, figs. 12–15 (Meniscomys.)
Major, Fors. 1898 A, p. 192.
Matthew, W. D. 1899 A, p. 62.
Roger, O. 1896 A, p. 116.
Trouessart, E. L. 1898 A, p. 396.
Miocene (John Day); Oregon.

Allomys hippodus (Cope).

Cope, E. D. 1879 D, p. 67. (Meniscomys.)
　1880 C, p. 55. (Meniscomys.)
　1881 E, p. 365. (Meniscomys.)
　1881 Q, p. 586. (Meniscomys.)
　1883 S, p. 52, fig. 6. (Meniscomys.)
　1884 O, pp. 827, 828, pl. lxii, figs. 7–10 (Meniscomys.)
Hay, O. P. 1899 C, p. 253.
Matthew, W. D. 1899 A, p. 62.
Riggs, E. S. 1899 A, p. 183, text figure, p. 185. (Protogaulus.)
Roger, O. 1896 A, p. 116.
Trouessart, E. L. 1898 A, p. 396.
Miocene (John Day); Oregon.

Allomys liolophus (Cope).

Cope, E. D. 1881 E, p. 366. (Meniscomys.)
　　　1881 Q, p. 586. (Meniscomys.)
　　　1883 S, p. 52, fig. 53. (Meniscomys.)
　　　1884 O, pp. 827, 829, pl. lxiii, fig. 11. (Meniscomys.)
Matthew, W. D. 1899 A, p. 62.
Roger, O. 1896 A, p. 116.
Trouessart, E. L. 1898 A, p. 396.
　　Miocene (John Day); Oregon.

Allomys multiplicatus (Cope).

Cope, E. D. 1879 B, p. 55. (Meniscomys.)
　　　1879 D, p. 68. (Meniscomys.)
　　　1883 S, p. 52, fig. 8. (Meniscomys.)
　　　1884 O, pp. 827, 832. (Syn. of Meniscomys nitens.)
Matthew, W. D. 1899 A, p. 62.

Roger, O. 1896 A, p. 116.
Trouessart, E. L. 1898 A, p. 396.
　　Miocene (John Day); Oregon.

Allomys nitens Marsh.

Marsh, O. C. 1877 D, p. 253.
Cope, E. D. 1881 E, p. 366. (Meniscomys.)
　　　1881 Q, p. 586. (Meniscomys.)
　　　1883 S, p. 52, fig. 8. (Meniscomys.)
　　　1884 O, pp. 827, 832, pl. lxiii, figs. 16, 17. (Meniscomys.)
King, C. 1878 A, p. 424.
Major, Fors. 1893 A, p. 193.
Matthew, W. D. 1899 A, p. 62.
Roger, O. 1896 A, p. 116.
Trouessart, E. L. 1898 A, p. 396.
Zittel, K. A. 1893 B, p. 529, fig. 440.
　　Miocene (John Day); Oregon.

SCIURIN.E.

Baird, S. F. 1857 A, p. 240.
Major, Fors. 1893 A.
Thomas, O. 1896 A, p. 1015.

Trouessart, E. L. 1898 A, p. 403.
Winge, H. 1888 A, pp. 109, 190. (<Sciurini.)

SCIURUS [1] Linn. Type *S. vulgaris* Linn.

Linnæus, C., 1758, Syst. Nat., ed. 10, I, p. 63.
Allen, J. A. 1877 A, xi, p. 666.
Baird, S. F. 1857 A, p. 243.
Cope, E. D. 1881 E, p. 363.
　　　1883 S, p. 49.
　　　1884 O, p. 816.
Flower and Lydekker 1891 A, p. 450.
Nicholson and Lydekker 1889 A, p. 1421.
Owen, R. 1868 A, p. 299.
Parker, W. K. 1868 A, p. 207, pl. xxv, figs. 12-15.
Tomes, J. 1850 A, p. 535.
Trouessart, E. L. 1898 A, p. 410.
Zittel, K. A. 1893 B, p. 329.

Sciurus ballovianus Cope.

Cope, E. D. 1881 C, p. 177.
　　　1881 E, p. 363.
　　　1881 Q.
　　　1883 S, p. 50, fig. 4, a–d.
　　　1884 O, p. 818, pl. lxiii, figs. 5, 6.
Matthew, W. D. 1899 A, p. 62.
Roger, O. 1896 A, p. 117.
Trouessart, E. L. 1898 A, p. 425.
　　Miocene (John Day); Oregon.

Sciurus calycinus Cope.

Cope, E. D. 1871 I, p. 86.
Allen, J. A. 1877 A, xi, p. 931.
Cope, E. D. 1883 S, p. 373.
　　　1899 A, p. 199.
Roger, O. 1896 A, p. 117.
Trouessart, E. L. 1898 A, p. 425.
Wheatly, C. M. 1871 B.
　　Pleistocene; Pennsylvania.

Sciurus panolius Cope.

Cope, E. D. 1869 E, p. 174, pl. iii, figs. 5, 5a.
Allen, J. A. 1877 A, xi, p. 932.
Cope, E. D. 1883 S, p. 373.
Leidy, J. 1869 A, p. 404.
Roger, O. 1896 A, p. 117.
Trouessart, E. L. 1898 A, p. 404.
　　Pleistocene; Virginia.

Sciurus relictus Cope.

Type, E. D. 1878 CC, p. 3. (Paramys.)
Allen, J. A. 1877 A, xi, p. 932.
Cope, E. D. 1869 J, p. 3. (No description.)
　　　1874 B, p. 475.
　　　1881 E, p. 363.
　　　1881 Q.
　　　1883 S, p. 50, fig. 4, e, f.
　　　1884 O, p. 817, pl. lxv, fig. 35.
Matthew, W. D. 1899 A, p. 52.
Roger, O. 1896 A, p. 117.
Trouessart, E. L. 1898 A, p. 425.
　　Oligocene (White River); Colorado.

Sciurus vortmani Cope.

Cope, E. D. 1879 E, p. 370.
　　　1884 E, p. 363.
　　　1881 Q.
　　　1883 S, p. 50, fig. 4, g, h.
　　　1884 O, p. 816, pl. lxiii, fig. 4.
Matthew, W. D. 1899 A, pp. 52, 62.
Trouessart, E. L. 1898 A, p. 155.
　　Miocene (John Day); Oregon, Colorado?.

[1] Besides the species below recorded, Dr. Leidy has reported (1889 H, p. 5) *S. carolinensis* from a cave in Pennsylvania.

TAMIAS Illiger.   Type *Sciurus striatus* Linn.

Illiger, 1811, Prod. Syst. Mamm., p. 83.
Allen, J. A. 1877 A, xi, p. 779.
Baird, S. F. 1857 A, p. 291.
Flower and Lydekker 1891 A, p. 452.
Nicholson and Lydekker 1889 A, p. 1421.
Tomes, J. 1850 A, p. 538.
Trouessart, E. L. 1898 A, p. 429.
Winge, H. 1888 A, p. 195.
Zittel, K. A. 1893 B, p. 530.

Tamias lævidens Cope.

Cope, E. D. 1869 E, p. 174, pl. iii, fig. 4.
Allen, J. A. 1877 A, xi, p. 932.
Cope, E. D. 1883 S, p. 373.
Leidy, J. 1869 A, p. 404.
Roger, O. 1896 A, p. 117.
Trouessart, E. L. 1898 A, p. 433.
Pleistocene; Virginia.
     T. striatus has been reported by Dr. Leidy
(1889 H, p. 5.) from a cave in Pennsylvania.

ADJIDAUMO Hay.   Type *Gymnoptychus minutus* Cope.

Hay, O. P. 1899 C, p. 253.
Cope, E. D. 1874 B, p. 476. (Gymnoptychus; not
 of 1873, the type not the same.)
    1881 E, p. 364. (Gymnoptychus.)
    1883 S, p. 50. (Gymnoptychus.)
    1884 O, p. 819. (Gymnoptychus.)
    1888 B, p. 291. (Gymnoptychus.)
Nicholson and Lydekker 1889 A, p. 1421. (Gym-
 noptychus.)
Scott, W. B. 1895 E, p. 286. (Gymnoptychus.)
Tullberg, T. 1899 A, p. 480. (Gymnoptychus.)
Zittel, K. A. 1893 B, p. 532. (Gymnoptychus.)

Adjidaumo minutus (Cope.)

Cope, E. D. 1873 T, p. 6. (Gymnoptychus.)
Allen, J. A. 1877 B, p. 945. (Gymnoptychus
 minutus, G. nasutus.)
Cope, E. D. 1873 T, p. 6. (G. nasutus.)
    1873 CC, p. 3. (G. nasutus, G. minutus.)
    1874 B, p. 476. (G. nasutus, a syn. of G. tri-
 lophus.)
    1881 E, p. 364. (Gymnoptychus.)
    1881 Q. (Gymnoptychus.)

Cope, E. D. 1883 S, p. 51, fig. 5. (Gymnoptychus.)
    1884 O, p. 822, pl. lxv, figs. 19-30. (Gym-
 noptychus.)
Hay, O. P. 1899 C, p. 253.
Matthew, W. D. 1899 A, p. 52. (Gymnoptychus.)
Roger, O. 1896 A, p. 119. (Gymnoptychus.)
Trouessart, E. L. 1898 A, p. 578. (Gymnoptychus.)
Oligocene (White River); Colorado.

Adjidaumo trilophus (Cope).

Cope, E. D. 1873 T, p. 6. (Gymnoptychus.)
Allen, J. A. 1877 B, p. 945. (Gymnoptychus.)
Cope, E. D. 1873 CC, p. 3. (Gymnoptychus.)
    1874 B, p. 476. (Gymnoptychus.)
    1881 E, p. 364. (Gymnoptychus.)
    1881 Q. (Gymnoptychus.)
    1883 S, p. 51, fig. 5. (Gymnoptychus.)
    1884 O, p. 826, pl. lxv, figs. 31-34. (Gymnop-
 tychus.)
Hay, O. P. 1899 C, p. 253.
Matthew, W. D. 1899 A, p. 52. (Gymnoptychus.)
Roger, O. 1896 A, p. 120. (Gymnoptychus.)
Oligocene (White River); Colorado.

ARCTOMYS Schreber.   Type not designated.

Schreber, J. C., 1780, Säugeth., iv, p. 721.
Allen, J. A. 1877 A, xi, p. 909.
Alston, E. R. 1876 A.
Baird, S. F. 1857 A, p. 338.
Blainville, H. M. D. 1864 A, iv, EE, p. 11, plate.
Cope, E. D. 1869 E, p. 172. (Stereodectes, type
 S. tortus.)
    1869 J, p. 3. (Stereodectes.)
Flower and Lydekker 1891 A, p. 454.
Giebel, C. G. 1859 A.
Nicholson and Lydekker 1889 A, p. 1420.
Owen, R. 1866 B, p. 382.
Palmer, T. S. 1897 A.
Parker, W. K. 1868 A, p. 207, pl. xxiv, figs. 14-16.
Tomes, J. 1850 A, p. 538.
Zittel, K. A. 1893 B, p. 529.

Arctomys monax (Linn).

Linnæus, C., 1758, Syst. Nat., ed. 10, i, p. 60. (Mus.)
Allen, J. A. 1877 A, xi, p. 911.
Baird, S. F. 1857 A, p. 359, pl. xlix, fig. 1.
Cope, E. D. 1869 E, p. 173.
    1883 S, p. 373.
Flower and Lydekker 1891 A, p. 450, fig. 198.
Giebel, C. G. 1859 A.

Leidy, J. 1862 A, p. 423.
    1869 A, p. 404.
    1889 H, p. 5.
Trouessart, E. L. 1898 A, p. 445.
Zittel, K. A. 1893 B, p. 529, fig. 439.
Recent; United States to Hudson Bay: Pleis-
 tocene; Illinois, Virginia.

Arctomys tortus (Cope).

Cope, E. D. 1869 E, p. 172, pl. iii, figs. 3, 3a. (Stereo-
dectes.)
    1869 J. (Stereodectes.)
Leidy, J. 1869 A, pp. 404, 446.
Trouessart, E. L. 1898 A, p. 445. (A. monax.)
Pleistocene; Virginia.

Arctomys vetus Marsh.

Marsh, O. C. 1871 E, p. 121.
Allen, J. A. 1877 A, xi, p. 933.
King, C. 1878 A, p. 430.
Matthew, W. D. 1899 A, p. 66.
Roger, O. 1896 A, p. 116.
Trouessart, E. L. 1898 A, p. 445.
Miocene (Loup Fork); Nebraska.

MYLAGAULIDÆ.

Cope, E. D. 1881 Q, p. 586.
    1881 E, p. 373.

Matthew, W. D. 1899 A, p. 66.
Riggs, E. S. 1899 A, p. 181.

MYLAGAULUS Cope.   Type *M. sesquipedalis* Cope.

Cope, E. D. 1878 H, p. 384.
    1881 E, p. 373.
    1883 S, p. 56.
Nicholson and Lydekker 1889 A, p. 1415.
Riggs, E. S. 1899 A, p. 181.
Zittel, K. A. 1893 B, p. 581.

Mylagaulus monodon Cope.

Cope, E. D. 1881 E, p. 374.
    1881 Q. p. 586.
    1883 S, p. 56, fig. 12.
Matthew, W. D. 1899 A, p. 66.
Riggs, E. S. 1899 A, p. 184.

Roger, O. 1896 A, p. 119.
Trouessart, E. L. 1898 A, p. 448.
Miocene (Loup Fork); Nebraska, Colorado.

Mylagaulus sesquipedalis Cope.

Cope, E. D. 1878 H, p. 384.
    1881 E, p. 375.
    1881 Q. p. 586.
    1883 S, p. 56, fig. 12.
Matthew, W. D. 1899 A, p. 66.
Roger, O. 1896 A, p. 119.
Trouessart, E. L. 1898 A, p. 448.
Miocene (Loup Fork); Nebraska.

MESOGAULUS Riggs.   Type *M. ballensis* Riggs.

Riggs, E. S. 1899 A, p. 181.

Mesogaulus ballensis Riggs.

Riggs, E. S. 1899 A, p. 181, text figures.
Miocene (Deep River); Montana.

CASTORIDÆ.

Allen, J. A. 1877 A, p. 431,
Alston, E. R. 1876 A, p. 78.
Baird, S. F. 1857 A, p. 350. (Castorinæ.)
Brandt, J. F. 1855 A.
Flower and Lydekker 1891 A, p. 457.
Leidy, J. 1871 C, p. 363.
Merriam, J. C. 1896 A, p. 364.
Nicholson and Lydekker 1889 A, p. 1419.

Roger, O. 1896 A, p. 118.
Steinmann and Döderlein 1890 A, p. 733.
Thomas, O. 1896 A, p. 1015.
Trouessart; E. L. 1898 A, p. 447.
Tullberg, T. 1899 A, pp. 306, 472.
Winge, H. 1888 A, p. 137. (Castorini.)
Zittel, K. A. 1893 B, p. 528.

CASTOR Linn.   Type *C. fiber* Linn.

Linnæus, C., 1858, Syst. Nat., ed. 10, i, p. 58.
Allen, J. A. 1877 A, p. 482.
Baird, S. F. 1857 A, p. 355.
Blainville, H. M. D. 1864 A, iv, EE, p. 13, pls. i, ii.
Brandt, J. F. 1855 A.
Cope, E. D. 1881 E, p. 368.
    1884 O, p. 838.
    1888 J, p. 10.
Flower, W. H. 1885 A.
Flower and Lydekker 1891 A, p. 457.
Nicholson and Lydekker 1889 A, p. 1419.
Owen, R. 1866 B, pp. 364, 374, figs. 230, 240.
Pictet, F. J. 1853 A, p. 250.
Tomes, J. 1850 A, p. 739.
Trouessart, E. L. 1898 A, p. 447.
Wyman, J. 1850 D.
Zittel, K. A. 1893 B, p. 531.

Castor canadensis Kuhl.

Kuhl, H., 1820, Beiträge zur Zool., p. 64.
Allen, J. A. 1877 A, p. 433. (C. fiber.)
Baird, S. F. 1857 A, p. 355, pl. xlviii, fig. 1.

Brandt, J. F. 1855 A, p. 64, pls. i-iii. (C. americanus.)
Cope, E. D. 1869 E, p. 173. (C. fiber.)
    1878 H, p. 389. (C. fiber.)
    1881 E, p. 370. (C. fiber.)
    1884 O, p. 839. (C. fiber.)
    1889 F, p. 160. (C. fiber.)
    1889 B, p. 980. (C. fiber.)
    1899 A, p. 200. (C. fiber.)
Leidy, J. 1860 B, p. 111, pl. xxi, figs. 1, 2.
    1869 A, p. 405.
    1889 H, p. 5, pl. i, fig. 8. (C. fiber.)
Logan, W. E. 1863 A, p. 914.
Lydekker, R. 1885 B, p. 263.
Roger, O. 1896 A, p. 119.
Trouessart, E. L. 1898 A, p. 447.
Wyman, J. 1850 A, p. 61, fig. 4. (C. fiber americanus.)
    Recent; North America. Pleistocene; New York, Pennsylvania, Virginia, Tennessee, New Jersey, South Carolina, Oregon, Ontario, Canada.

EUCASTOR Leidy.   Type *E. tortus* Leidy.

Leidy, J. 1858 E, p. 23. (Subgenus of Castor.)
Allen, J. A. 1877 A, p. 449.
Cope, E. D. 1881 E, p. 369.
    1883 S, p. 55.
    1884 O, p. 839.
Flower and Lydekker 1891 A, p. 458.
Nicholson and Lydekker 1889 A, p. 1419.
Zittel, K. A. 1893 B, p. 581.

Eucastor tortus Leidy.

Leidy, J. 1858 E, p. 23. [Castor (Eucastor).]
Allen, J. A. 1877 A, p. 449.

Cope, E. D. 1881 E, p. 369.
    1881 Q, p. 586.
    1883 S, p. 55.
Hayden, F. V. 1858 B, p. 158.
Leidy, J. 1869 A, pp. 341, 405, pl. xxvi, figs. 21, 22.
    (Castor.)
    1871 C, p. 363. (Castor.)
Matthew, W. D. 1899 A, p. 66.
Merriam, J. C. 1896 A.
Roger, O. 1896 A, p. 119.
Trouessart, E. L. 1898 A, p. 448.
    Miocene (Loup Fork); Nebraska.

STENEOFIBER E. Geoffroy.   Type not indicated.

Geoffroy, E., 1833, Revue Encyclop.
Cope, E. D. 1877 K, p. 297.
    1881 E, p. 369.
    1884 O, p. 838. (Syn. of Castor.)
Flower and Lydekker 1891 A, p. 458. (Chalicomys, Palæocastor.)
Leidy, J. 1869 A, p. 306. (Palæocastor, type *Steneofiber nebrascensis.*)
    1871 C, p. 363. (Palæocastor.)
Nicholson and Lydekker 1889 A, p. 1420. (Chalicomys, Palæocastor.)
Pictet, F. J. 1853 A, p. 252.
Pomel, A. 1847 C, p. 381.
    1853 A, p. 20.
Schlosser, M. 1884 A, p. 21.
Scott, W. B. 1895 C, p. 76.
Tullberg, T. 1899 A, p. 474.
Zittel, K. A. 1893 B, p. 530.

Steneofiber gradatus Cope.

Cope, E. D. 1879 D, p. 63.
    1879 B, p. 55.
    1881 E, p. 370. (Castor.)
    1881 Q. p. 586. (Castor.)
    1883 S, p. 55. (Castor.)
    1884 O, pp. 839, 844, pl. lxiii, fig. 22. (Castor.)
Matthew, W. D. 1899 A, p. 62.
Roger, O. 1896 A, p. 118.
Trouessart, E. L. 1898 A, p. 449.
    Miocene (John Day); Oregon.

Steneofiber montanus Scott.

Scott, W. B. 1893 B, p. 680.
Matthew, W. D. 1899 A, p. 52.
Roger, O. 1896 A, p. 119.
Scott, W. B. 1895 C, p. 76.
Trouessart, E. L. 1898 A, p. 450.
    Oligocene (White River); Montana.

Steneofiber nebrascensis Leidy.

Leidy, J. 1856 I, p. 89.
Allen, J. A. 1877 A, p. 451.

Cope, E. D. 1877 K, pp. 297, 300.
    1881 E, p. 370. (Castor.)
    1881 Q, p. 586. (Castor.)
    1883 S, p. 55. (Castor.)
    1884 O, p. 839. (Castor.)
Hayden, F. V. 1858 B, p. 158.
King, C. 1878 A, p. 412. (Palæocastor.)
Leidy, J. 1857 D, p. 176. (Chalicomys.)
    1869 A, pp. 338, 406, pl. xxvi, figs. 7–11. (Palæocastor.)
    1871 C, p. 363. (Palæocastor.)
Matthew, W. D. 1899 A, p. 52.
Roger, O. 1896 A, p. 118.
Trouessart, E. L. 1898 A, p. 449.
    Oligocene (White River); South Dakota, Nebraska, North Dakota.

Steneofiber pansus Cope.

Cope, E. D. 1874 J, p. 222.
Allen, J. A. 1877 A, p. 453.
Cope, E. D. 1875 F, p. 73.
    1877 K, p. 297, pl. lxix, figs. 4–14.
    1881 E, p. 370.
    1881 K. (Castor.)
    1883 S, p. 55, fig. 11. (Castor.)
    1884 O, pp. 839, 840. (Castor.)
Matthew, W. D. 1899 A, p. 66.
Roger, O. 1896 A, p. 118.
Trouessart, E. L. 1898 A, p. 449.
    Miocene (Loup Fork); New Mexico, Nebraska.

Steneofiber peninsulatus (Cope).

Cope, E. D. 1881 E, p. 370. (Castor.)
    1879 B, p. 55. (Steneofiber nebrascensis?.)
    1881 Q, p. 586. (Castor.)
    1883 S, p. 55, figs. 9, 10. (Castor.)
    1884 O, p. 840, pl. lxiii, figs. 18–21. (Castor.)
Matthew, W. D. 1899 A, p. 62.
Roger, O. 1896 A, p. 118.
Trouessart, E. L. 1898 A, p. 449.
    Miocene (John Day); Oregon.

SIGMOGOMPHIUS Merriam.   Type *S. lecontei* Merriam.

Merriam, J. C. 1896 A.

Sigmogomphius lecontei Merriam.

Merriam, J. C. 1896 A, figs. 1, 2.

Matthew, W. D. 1899 A, p. 66.
Trouessart, E. L. 1898 A, p. 450.
    Pliocene; California.

ISCHYROMYIDÆ.

Alston, E. R. 1876 A, p. 78.
Allen, J. A. 1877 B, p. 944.
Cope, E. D. 1881 E, p. 368.
    1884 O, pp. 37, 834 (Protomyidæ); p. 835
    (Ischyromyidæ).
Nicholson and Lydekker 1889 A, p. 1420.

Roger, O. 1896 A, p. 110.
Trouessart, E. L. 1898 A, p. 451.
Tullberg, T 1899 A, p. 463.
Winge, H. 1888 A, p. 114.
Zittel, K. A. 1893 B, p. 521.

PARAMYS Leidy. Based on three species, *P. delicatus, delicatior, delicatissimus.*

Leidy, J. 871 L, p. 231.
    As the species here included in *Paramys* have,
    by Cope and others, been referred to the genus
    Plesiarctomys, citations are given to the litera-
    ture of the latter genus also.
Allen, J. A. 1877 A, xi, p. 933 (Paramys); p. 935
    (Sciuravus).
    1877 B, p. 946. (Pseudotomus.)
Alston, E. R. 1876 A, p. 78. (Pseudotomus.)
Cope, E. D. 1872 N, p. 467. (Pseudotomus, type
    P. hians.)
    1873 E, p. 610. (Pseudotomus.)
    1877 K, p. 170. (Plesiarctomys.)
    1883 S, p. 45. (Plesiarctomys.)
    1884 O, p. 175. (Plesiarctomys.)
Flower and Lydekker 1891 A, p. 457. (Plesiarc-
    tomys.)
Leidy, J. 1871 L, p. 230 (Sciuravus); p. 231 (Para
    mys).
    1872 B, p. 357.
    1873 B, pp. 109, 335 (Paramys); p. 113 (Sci-
    uravus).
Major, Fors. 1893 A, p. 202. (Plesiarctomys.)
Marsh, O. C. 1871 E, p. 122. (Sciuravus, type *S.
    nitidus.*)
    1873 G, p. 246. (Sciuravus.)
Matthew, W. D. 1897 C, p. 266. (Plesiarctomys.)
Nicholson and Lydekker 1889 A, p. 1420. (Plesi-
    arctomys, Paramys.)
Osborn, H. F. 1895 A, p. 81.
Osborn, Scott, and Speir 1878 A, p. 83.
Scott, W. B. 1890 A, p. 474. (Plesiarctomys.)
    1892 A, p. 426.
Trouessart, E. L. 1898 A, p. 451.
Winge, H. 1888 A, pp. 114, 164.
Zittel, K. A. 1893 A, p. 521 (Paramys); p. 528
    (Plesiarctomys.)

Paramys buccatus (Cope).

Cope, E. D. 1877 K, p. 171, pl. lxiv, fig. 8. (Plesi-
    arctomys.)
    1881 D, p. 184. (Plesiarctomys.)
    1882 E, p. 146. (Plesiarctomys.)
    1883 S, p. 47. (Plesiarctomys.)
    1884 O, p. 179, pl. xliva, fig. 14. (Plesiarc-
    tomys.)
Matthew, W. D. 1899 A, p. 30, 39.
Roger, O. 1896 A, p. 110.
Trouessart, E. L. 1898 A, p. 451.
    Eocene (Wasatch); New Mexico, Wyoming:
    (Bridger); Wyoming.

Paramys delicatior Leidy.

Leidy, J. 1871 L, p. 231.
Allen, J. A. 1877 A, xi, p. 934.

Cope, E. D. 1873 E, p. 610.
    1873 U, p. 4.
    1877 K, p. 172, pl. xliv, figs. 10, 11. (Plesi-
    arctomys.)
    1881 D, p. 184. (Plesiarctomys.)
    1882 E, p. 146. (Plesiarctomys.)
    1883 S, p. 47. (Plesiarctomys.)
    1884 O, p. 182, pl. xxiva, figs. 11-13. (Plesi-
    arctomys.)
Leidy, J. 1872 B, p. 357.
    1873 B, pp. 110, 335, pl. vi, figs. 26, 27; pl.
    xxvii, figs. 16-18.
Matthew, W. D. 1899 A, pp. 30, 35, 38.
Osborn, H. F. 1897 G, p. 256. (Plesiarctomys deli-
    catior?.)
Roger, O. 1896 A, p. 110.
Trouessart, E. L. 1898 A, p. 451.
    Eocene (Wind River, Bridger); Wyoming
    (Wasatch); New Mexico, Wyoming.

Paramys delicatissimus Leidy.

Leidy, J. 1871 L, p. 231.
Allen, J. A. 1877 A, xi, p. 935.
Cope, E. D. 1873 E, p. 610.
    1873 U, p. 4.
    1877 K, p. 172, pl. xxiv, figs. 9, 12. (Plesiarc-
    tomys.)
    1881 D, p. 184. (Plesiarctomys.)
    1883 S, p. 47, figs. 1, 2. (Plesiarctomys.)
    1884 O, p. 179, pl. xxiva, figs. 1-10. (Plesiarc-
    tomys.)
Leidy, J. 1872 B, p. 357.
    1873 B, pp. 111. 335, pl. vi, figs. 28, 29.
Matthew, W. D. 1899 A, pp. 30, 34, 39.
Roger, O. 1896 A, p. 110.
Trouessart, E. L. 1898 A, p. 451.
Zittel, K. A. 1893 B, p. 522, fig. 430.
    Eocene (Bridger); Wyoming: (Wasatch?);
    New Mexico, Wyoming,

Parymys delicatus Leidy.

Leidy, J. 1871 L, p. 231.
Allen, J. A. 1877 A, xi, p. 934.
Cope, E. D. 1873 E, p. 610.
    1873 U, p. 4.
King, C. 1878 A, p. 404.
Leidy, J. 1872 B, p. 357.
    1873 B, pp. 110, 335, pl. vi, figs. 23-25.
Matthew, W. D. 1899 A, p. 38.
Roger, O. 1896 A, p. 110.
Trouessart, E. L. 1898 A, p. 451.
    Eocene (Bridger); Wyoming.

Paramys hians (Cope.)

Cope, E. D. 1872 N, p. 467. (Pseudotomus.)
Allen, J. A. 1877 B, p. 946. Pseudotomus.)

Cope, E. D. 1873 E, p. 611. (Pseudotomus.)
  1883 S, p. 47. (Plesiarctomys.)
  1884 O, p. 183, pl. xxiv, figs. 3–5. (Plesiarctomys.
  tomys.
Matthew, W. D. 1899 A, p. 39.
Roger, O. 1896 A, p. 110.
Trouessart, E. L. 1898 A, p. 451.
  Eocene (Bridger); Wyoming.

Paramys leptodus Cope.

Cope, E. D. 1873 U, p. 8.
Allen, J. A. 1877 A, xi, p. 934.
Cope, E. D. 1873 E, p. 609.
  1883 S, p. 47. (Plesiarctomys.)
  1884 O, p. 183, pl. xxiv, fig. 1. (Plesiarc-
  tomys.)
Matthew, W. D. 1899 A, p. 39.
Roger, O. 1896 A, p. 111.
Trouessart, E. L. 1898 A, p. 451.
  Eocene (Bridger); Wyoming.

Paramys nitidus (Marsh.)

Marsh, O. C. 1871 E, p. 122. (Sciuravus.)
Allen, J. A. 1877 A, xi, p. 935.
King, C. 1878 A, p. 404.
Leidy, J. 1872 B, p. 358. (Sciuravus.)
  1873 B, p. 113. (Sciuravus.)
Matthew, W. D. 1899 A, p. 39.
Roger, O. 1896 A, p. 111.
Trouessart, E. L. 1898 A, p. 451.
  Eocene (Bridger); Wyoming.

Paramys parvidens (Marsh).

Marsh, O. C. 1872 I, p. 220. (Sciuravus.)
Allen, J. A. 1877 A, xi, p. 936. (Sciuravus.)
Matthew, W. D. 1899 A, p. 39.
Roger, O. 1896 A, p. 111.
Trouessart, E. L. 1898 A, p. 451.
  Eocene (Bridger); Wyoming.

Paramys robustus Marsh.

Marsh, O. C. 1872 I, p. 218.
Allen, J. A. 1877 A, xi, p. 934.
Matthew, W. D. 1899 A, p. 39.
Osborn, H. F. 1895 A, p. 81.

Roger, O. 1896 A, p. 111.
Trouessart, E. L. 1898 A, p. 451.
  Eocene (Bridger); Wyoming.

Paramys sciuroides (Scott and Osb.).

Scott and Osborn 1887 A, p. 256. (Plesiarctomys.)
Major, Fors. 1893 A, p. 202. (Plesiarctomys.)
Matthew, W. D. 1899 A, p. 49.
Scott and Osborn 1890 A, p. 475, pl. xi, figs. 1, 2;
  text fig. 1. (Plesiarctomys.)
Roger, O. 1896 A, p. 111.
Trouessart, E. L. 1898 A, p. 451.
Zittel, K. A. 1893 B, p. 522, fig. 429.
  Eocene (Uinta); Utah.

Paramys superbus Osb., Scott, and Speir.

Osborn, Scott, and Speir 1878 A, p. 84.
  Eocene (Bridger); Wyoming.

Paramys uintensis Osborn.

Osborn, H. F. 1895 A, p. 81.
Matthew, W. D. 1899 A, p. 49.
Roger, O. 1896 A, p. 111.
Trouessart, E. L. 1898 A, p. 451.
  Eocene (Uinta); Utah.

Paramys undans (Marsh).

Marsh, O. C. 1871 E, p. 122. (Sciuravus.)
Allen, J. A. 1877 A, xi, p. 965. (Sciuravus.)
Cope, E. D. 1873 E, p. 610.
  1873 U, p. 4.
  1883 S, p. 47. (Plesiarctomys.)
Leidy, J. 1872 B, p. 358. (Sciuravus.)
  1873 B, p. 113. (Sciuravus.)
Matthew, W. D. 1899 A, p. 39.
Roger, O. 1896 A, p. 111.
Trouessart, E. L. 1898 A, p. 451.
  Eocene (Bridger); Wyoming.

Paramys sp. indet.

Leidy, J. 1873 B, pp. 113, 335, pl. vi, fig. 30.
  (Sciuravus.)
Allen, J. A. 1877 A, xi, p. 936.
  Eocene (Bridger); Wyoming.

TILLOMYS Marsh.   Type *T. senex* Marsh.

Marsh, O. C. 1872 I, p. 219.
Allen, J. A. 1877 A, xi, p. 938.
Nicholson and Lydekker 1889 A, p. 1420.
  (=? Ischyromys.)
Schlosser, M. 1884 A, p. 130. (=? Ischyromys.)
Zittel, K. A. 1893 B, p. 522.

Tillomys parvus Marsh.

Marsh, O. C. 1872 I, p. 219.
Allen, J. A. 1877 A, xi, p. 939.
Matthew, W. D. 1899 A, p. 39.
Roger, O. 1896 A, p. 111.

Trouessart, E. L. 1898 A, p. 451.
  Eocene (Bridger); Wyoming

Tillomys senex Marsh.

Marsh, O. C. 1872 I, p. 219.
Allen, J. A. 1877 A, xi, p. 938.
King, C. 1878 A, p. 404.
Matthew, W. D. 1899 A, p. 39.
Roger, O. 1896 A, p. 111.
Trouessart, E. L. 1898 A, p. 451.
  Eocene (Bridger); Wyoming.

TAXYMYS Marsh.   Type *T. lucaris* Marsh.

Marsh, O. C. 1872 I, p. 219.
Allen, J. A. 1877 A, xi, p. 938.
Marsh, O. C. 1880 H, p. 10. (Tachymys.)
Zittel, K. A. 1893 B, p. 522. (Toxymys.)

Taxymys lucaris Marsh.

Marsh, O. C. 1872 I, p. 219.

Allen, J. A. 1877 A, xi, p. 938.
King, C. 1878 A, p. 404.
Matthew, W. D. 1899 A, p. 39.
Roger, O. 1896 A, p. 111.
Trouessart, E. L. 1898 A, p. 452.
  Eocene (Bridger); Wyoming.

COLONYMYS Marsh.   Type *C. celer* Marsh.

Marsh, O. C. 1872 I, p. 220.
Allen, J. A. 1877 A, xi, p. 938.
Marsh, O. C. 1878 G, p. 246.
Matthew, W. D. 1899 A, p. 39.   (Colonomys.)
Nicholson and Lydekker 1889 A, p. 1419.
Zittel, K. A. 1898 B, p. 522.

Colonymys celer Marsh.

Marsh, O. C. 1872 I, p. 220.
Allen, J. A. 1877 A, xi, p. 938.
Matthew, W. D. 1899 A, p. 39.
Roger, O. 1896 A, p. 111.
Trouessart, E. L. 1898 A, p. 452.
Eocene (Green River); Wyoming.

APATEMYS Marsh.   Type *A. bellus* Marsh.

Marsh, O. C. 1872 I, p. 221.
1877 E.
A genus of uncertain position.

Apatemys bellulus Marsh.

Marsh, O. C. 1872 I, p. 221.
Matthew, W. D. 1899 A, p. 39.
Roger, O. 1896 A, p. 89.
Trouessart, E. L. 1898 A, p. 669.
Eocene (Bridger); Wyoming.

Apatemys bellus Marsh.

Marsh, O. C. 1872 I, p. 221.
King, C. 1878 A, p. 404.
Matthew, W. D. 1899 A, p. 39.
Roger, O. 1896 A, p. 89.
Trouessart, E. L. 1898 A, p. 669.
Eocene (Bridger); Wyoming.

ISCHYROMYS Leidy.   Type *I. typus* Leidy.

Leidy, J. 1856 I, p. 89.
Allen, J. A. 1877 B, p. 944 (Ischyromys); p. 945 (Gymnoptychus).
Cope, E. D. 1873 S, p. 1.   (Colotaxis, type *C. cristatus*=I. typus.)
    1873 T, p. 5.   (Gymnoptychus, type *G. chrysodon*=I. typus.)
    1874 B, p. 477.
    1881 E, p. 366.
    1883 S, p. 47.
    1884 O, p. 883.
    1888 B, p. 291.   (=Sciuromys, *fide* Schlosser.)
Hay, O. P. 1899 C, p. 253.
Leidy, J. 1871 C, p. 563.
Nicholson and Lydekker 1889 A, p. 1420.
Schlosser, M. 1884 A, p. 130.
Trouessart, E. L. 1898 A, p. 452.
Tullberg, T. 1899 A, p. 464.
Winge, H. 1888 A, pp. 115, 164.
Zittel, K. A. 1898 B, p. 522.

Ischyromys typus Leidy.

Leidy, J. 1856 I, p. 89.
Allen, J. A. 1877 A, xi, p. 931.   (Gymnoptychus chrysodon.)

Allen, J. A. 1877 B, p. 944 (I. typus); p. 945 (Gymnoptychus crysodon); p. 948 (Colotaxis cristatus).
Cope, E. D. 1873 S, p. 1.   (Colotaxis cristatus.)
    1873 T, p. 5.   (Gymnoptychus chrysodon.)
    1873 CC, p. 3.   (Colotaxis cristatus.)
    1874 B, p. 477.
    1881 Q.
    1883 S, p. 47, fig. 3.
    1884 O, p. 885, pl. lxvii, figs. 1–12.
    1888 I, p. 8, fig. 5.
    1889 C, p. 267, fig. 88.
Hay, O. P. 1899 C, p. 253.
Hayden, F. V. 1856 B, p. 158.
King, C. 1878 A, p. 412.
Leidy, J. 1869 A, pp. 335, 405, pl. xxvi, figs. 1–6.
    1871 C, p. 563.
Matthew, W. D. 1899 A, p. 52.
Roger, O. 1896 A, p. 111.
Schlosser, M. 1884 A, p. 136.
Trouessart, E. L. 1898 A, p. 452.
Tullberg, T. 1899 A, p. 463.
Zittel, K. A. 1898 B, p. 522, fig. 431.
Oligocene (White River); South Dakota, Nebraska, Colorado.

MYSOPS Leidy.   Type *M. minimus* Leidy.

Leidy, J. 1871 L, p. 232.
Allen, J. A. 1877 A, xi, p. 936.
Cope, E. D. 1883 S, p. 47.   (Syllophodus; no type indicated.)
    1884 O, p. 846.   (Syllophodus.)
Flower and Lydekker 1891 A, p. 484.
Leidy, J. 1873 B, p. 111.
Nicholson and Lydekker 1889 A, p. 1416 (Syllophodus); p. 1420 (Mysops).
Zittel, K. A. 1898 B, p. 523.

Mysops fraternus Leidy.

Leidy, J. 1873 B, pp. 112, 336, pl. xxvii, figs. 14, 15.
Allen, J. A. 1877 A, xi, p. 937.
Cope, E. D. 1883 S, p. 47.   (Syllophodus.)
Matthew, W. D. 1899 A, p. 39.

Roger, O. 1896 A, p. 111.
Trouessart, E. L. 1898 A, p. 452.
Eocene (Bridger); Wyoming.

Mysops minimus Leidy.

Leidy, J. 1871 L, p. 232.
Allen, J. A. 1877 A, p. 937.   (M. minutus.)
Cope, E. D. 1883 S, p. 47.   (Syllophodus.)
King, C. 1878 A, p. 404.
Leidy, J. 1872 B, p. 357.
    1873 B, pp. 111, 336, pl. vi, figs. 31, 32.
Matthew, W. D. 1899 A, p. 39.
Roger, O. 1896 A, p. 111.
Trouessart, E. L. 1898 A, p. 452.
Eocene (Bridger); Wyoming.

Superfamily MYOMORPHA Brandt.

Brandt, J. F. 1855 A.
Alston, E. R. 1876 A.
Cope, E. D. 1881 E, p. 361.
    1883 S, p. 43.
    1884 O, p. 812.
    1889 P, p. 659.
Flower and Lydekker 1891 A, p. 459.
Gadow, H. 1898 A, p. 43.
Haeckel, E. 1896 A, p. 510.
Lydekker, R. 1896 B.

Nicholson and Lydekker 1889 A, p. 1416.
Roger, O. 1896 A, p. 120.
Schlosser, M. 1884 A, p. 98.
Slade, D. D. 1888 A.
    1892 B.
Thomas, O. 1896 A, p. 1016.
Trouessart, E. L. 1898 A, p. 453.
Tullberg, T. 1899 A, pp. 151, 386. (Myomorphi.)
Zittel, K. A. 1893 B, p. 533.

MURIDÆ.

Adloff, P. 1896 A. ("Muriden.")
Alston, E. R. 1876 A, p. 80.
Baird, S. F. 1857 A, p. 427.
Burmeister, H. 1879 B, p. 197. (Murini.)
Coues, E. 1877 B, i, p. 1.
Flower, W. H. 1883 D, p. 418.
Flower and Lydekker 1891 A, p. 461.
Nicholson and Lydekker 1889 A, p. 1416.

Parker, W. K. 1868 A, p. 207.
Thomas, O. 1896 A, p. 1017.
Tomes, J. 1850 A, p. 542.
Trouessart, E. L. 1898 A, p. 458.
Tullberg, T. 1899 A, pp. 242, 432 (Hesperomyidæ;
    pp. 251, 445 (Muridæ).
Winge, H. 1888 A, p. 123.
Zittel, K. A. 1893 B, p. 537.

CRICETINÆ.

Baird, S. F. 1857 A, p. 445. (Sigmodontes.)
Coues, E. 1877 B, i, p. 7. (Sigmodontes, in part.)
Flower and Lydekker 1891 A, p. 463.
Merriam, C. H. 1894 A, p. 226.
Nehring, A. 1898 A. (Genus Cricetus.)
Palmer, T. S. 1897 A.

Roger, O. 1896 A, p. 120. (Cricetidæ.)
Thomas, O. 1888 B, p. 132. (Cricetl.)
    1896 A, p. 1019. (Sigmodontinæ.)
Trouessart, E. L. 1898 A, pp. 505, 511.
Tullberg, T. 1899 A, pp. 219, 430.
Zittel, K. A. 1893 B, p. 534. (Cricetidæ.)

Eumys Leidy.   Type E. elegans Leidy.

Leidy, J. 1856 I, p. 90.
Cope, E. D. 1874 B, p. 474.
    1877 K, p. 300.
    1881 E, p. 376.
    1883 S, p. 165.
    1884 O, p. 848.
Leidy, J. 1871 C, p. 364.
Lydekker, R. 1887 A, p. 322.
Nicholson and Lydekker 1889 A, p. 1418.
Zittel, K. A. 1893 B, p. 535.

Eumys elegans Leidy.

Leidy, J. 1856 I, p. 90.
Allen, J. A. 1877 B, p. 946.
Cope, E. D. 1874 B, p. 474.

Cope, E. D. 1881 E, p. 376.
    1881 Q.
    1883 S, p. 165, figs. 13, 14.
    1884 O, p. 849, pl. lxv, figs 1-14.
Hayden, F. V. 1858 B, p. 158.
King, C. 1878 A, p. 412.
Leidy, J. 1869 A, pp. 342, 407, pl. xxvi, figs. 12, 13.
    1871 C, p. 364
Lydekker, R. 1887 A, p. 322.
Matthew, W. D. 1899 A, p. 58.
Roger, O. 1896 A, p. 121.
Trouessart, E. L. 1898 A, p. 511.
Zittel, K. A. 1893 B. p. 535, fig. 448.
    Oligocene (White River); South Dakota,
    Colorado, Nebraska.

Peromyscus Gloger.   Type P. arboreus Gloger.

Unless otherwise indicated the following authorities call this genus Hesperomys.
Alston, E. R. 1876 A, p. 84.
Baird, S. F. 1857 A, p. 453.
Burmeister, H. 1879 B, p. 205.
Cope, E. D. 1881 E, p. 376.
    1884 O, p. 852.
Coues, E. 1877 B, i, p 43 (Hesperomys); p. 45 (Subg. Vesperimus).
Flower and Lydekker 1891 A, p. 463. (Cricetus.)
Lydekker, R. 1885 B, p 229.
Nicholson and Lydekker 1889 A, p. 1418.
Parker, W. K. 1868 A, p. 207, pl. xxv, figs. 6, 7. (Cricetus.)

Thomas, O. 1888 B, p. 133. (Cricetus.)
    1896 A, p. 1019.
Zittel, K. A. 1893 B, p. 585.

Peromyscus leucopus (Raf.).

Baird, S. F. 1857 A, p. 459. (Hesperomys.)
Cope, E. D. 1869 E, p. 173. (Hesperomys.)
    1871 I, p. 87. (Hesperomys.)
    1881 E, p. 377. (Hesperomys.)
    1899 A, p. 201. (Hesperomys leucopus.?)
Coues, E. 1877 B, i, p. 50, pl. iii, figs. 18-21. [Hesperomys (Vesperimus).]
Leidy, J. 1889 H, p. 6. (Hesperomys.)
Roger, O. 1896 A, p 122. (Vesperomys.)

Trouessart, E. L. 1896 A, p. 512.
Wheatley, C. M. 1871 B. (Hesperomys.)
Zittel, K. A. 1893 B, p. 535. (Vesperomys leucopus fossilis.)
　Recent; eastern United States from Carolinas to New York. Pleistocene?; caves of Virginia and Pennsylvania, with doubts as to identifications.

Peromyscus loxodon (Cope).

Cope, E. D. 1874 I, p. 148. (Hesperomys.)
Allen, J. A. 1877 B, p. 946. (Eumys.)
Cope, E. D. 1874 O, p. 17. (Hesperomys.)
　1875 F, p. 73. (Eumys.)
　1877 K, p. 300, pl. lxix, fig. 15. (Eumys.)
　1881 E, p. 376. (Hesperomys.)
　1881 Q. (Hesperomys.)

Cope, E. D. 1883 S, p. 165. (Hesperomys.)
Matthew, W. D. 1899 A, p. 66.
Roger, O. 1896 A, p. 122. (Hesperomys.)
Trouessart, E. L. 1898 A, p. 512.
　Miocene (Loup Fork); New Mexico.

Peromyscus nematodon (Cope).

Cope, E, D. 1879 E, p. 370. (Hesperomys.)
　1881 C, p. 176. (Eumys.)
　1881 Q. (Hesperomys.)
　1883 S, p. 165, fig. 15. (Hesperomys.)
　1884 O, p. 852, pl. lxvi, fig. 33. (Hesperomys)
Matthew, W. D. 1899 A, p. 62. (Hesperomys.)
Roger, O. 1896 A, p. 122. (Hesperomys.)
Trouessart, E. L. 1898 A, p. 512.
　Miocene (John Day); Oregon.

PACICULUS Cope.　　Type *P. insolitus* Cope.

Cope, E. D. 1879 E, p. 370.
　1881 E, p. 377.
　1883 S, p. 166.
　1884 O, p. 853.
Nicholson and Lydekker 1889 A, p. 540.
Scott, W. B. 1895 D.
　1896 E, p. 286.
Zittel, K. A. 1893 B, p. 536.

Paciculus insolitus Cope.

Cope, E. D. 1879 E, p. 371.
　1881 E, p. 377.
　1881 Q.
　1883 S, p. 166, fig. 15.
　1884 O, p. 854, pl. lxvi, figs. 31, 32.

Matthew, W. D. 1899 A, p. 62.
Roger, O. 1896 A, p. 124.
Trouessart, E. L. 1898 A, p. 540.
　Miocene (John Day); Oregon.

Paciculus lockingtonianus Cope.

Cope, E. D. 1881 C, p. 176. (Eumys.)
　1881 E, p. 377.
　1881 Q.
　1883 S, p. 166.
　1884 O, p. 854, pl. lxiv, fig. 10.
Matthew, W. D. 1899 A, p. 62.
Roger, O. 1896 A, p. 124.
Trouessart, E. L. 1898 A, p. 540.
　Miocene (John Day); Oregon.

NEOTOMINÆ.

Merriam, C. H. 1894 A, p. 228.
Baird, S. F. 1857 A, p. 445. (Sigmodontes, in part.)
Coues, E. 1877 B, i, p. 7. (Sigmodontes, in part.)

Thomas, O. 1896 A, p. 1020.
Trouessart, E. L. 1898 A, p. 540.

NEOTOMA Say and Ord.　　Type *N. floridana* Say and Ord.

Say and Ord, 1825, Jour. Acad. Nat. Sci., Phila., iv, pp. 345, 346.
Alston, E. R. 1876 A, p. 85.
Baird, S. F. 1857 A, p. 486.
Coues, E. 1877 B, i, p. 7.
Flower and Lydekker 1891 A, p. 464.
Merriam, C. H. 1894 A, p. 239.
Nicholson and Lydekker 1889 A, p. 1418.
Trouessart, E. L. 1898 A, p. 541.
Zittel, K. A. 1893 B, p. 536.

Neotoma floridana Say and Ord.

Say and Ord, 1818, Bull. Soc. Philom., Paris, p. 181.
Baird, S. F. 1857 A, p. 487, pl. liii, fig. 2.
Coues, E. 1877 B, i, p. 14, pl. i, figs. 1–4.
Leidy, J. 1869 A, p. 407.
　1880 A, p. 347.
　1889 H, p. 6.
Merriam, C. H. 1894 A, p. 244.
Rhoads, S. N. 1894 A, p. 346.
Roger, O. 1896 A, p. 124.

Say and Ord, 1825, Jour. Acad. Nat. Sci., Phila., iv, p. 346, pls. xxi, xxii.
Trouessart, E. L. 1898 A, p. 541.
　Recent; southeastern United States. Pleistocene?; Pennsylvania?

Neotoma magister Baird.

Baird, S. F. 1857 A, p. 498, pl. liii, fig. 4; pl. liv, fig. 978.
Cope, E. D. 1869 E, p. 173. (N. magister, N. floridana?)
　1883 S, p. 373.
Coues, E. 1877 B, i, p. 29, pl. ii, fig. 13.
Leidy, J. 1869 A, p. 407.
　1880 A, p. 347.
Mercer, H. C. 1897 A.
Merriam, C. H. 1894 A, p. 244.
Rhoads, S. N. 1894 A, p. 213.
Roger, O. 1896 A, p. 124.
Trouessart, E. L. 1898 A, p. 542.
　Pleistocene; Pennsylvania, Virginia, Kentucky, Missouri.

MICROTINÆ.

Alston, E. R. 1876 A, p. 85. (Arvicolinæ.)
Baird, S. F. 1857 A, p. 506. (Arvicolinæ.)
Blackmore, H. P. 1874 A. (Arvicolidæ.)
Brandt, J. F. 1885 A. (Arvicolini.)
Coues, E. 1877 B, i, p. 131. (Arvicolinæ.)
Flower, W. H. 1863 D, p. 419. (Arvicolinæ.)
Flower and Lydekker 1891 A, p. 465.
Lataste, F., 1883, Le Nat., ii, pp. 328, 332, 343, 347. (Arvicolinæ.)
     1884, Ann. Mus. Civic. Genova, xx, p. 255. (Arvicolinæ.)

Merriam, C. H. 1894 A, p. 227. (Arvicolinæ.)
Miller, G. S. 1896 A, p. 8.
Nehring, A. 1875 A.
Roger, O. 1896 A, p. 124. (Arvicolidæ.)
Thomas, O. 1896 A, p. 1020.
Trouessart, E. L. 1898 A, p. 545.
Tullberg, T. 1899 A, pp. 227, 430. (Arvicolidæ.)
Zittel, K, A. 1893 B, p. 536. (Arvicolidæ.)

ANAPTOGONIA Cope.   Type *A. hiatidens* Cope.

Cope, E. D. 1871 I, p. 91. [Arvicola (Anaptogonia).]
     1883 S, p. 373. [Arvicola (Anaptogonia).]
     1896 F, p. 379.
     1899 A, p. 201.
Miller, G. S. 1896 A, p. 74. [Microtus (Anaptogonia).]
Trouessart, E. L. 1898 A, p. 551. (Anaptogenia.)

Anaptogonia cloacina Cope.

Cope, E. D. 1896 A, p. 380.
Trouessart, E. L. 1898 A, p. 551.
Pleistocene; Pennsylvania.

Anaptogonia hiatidens Cope.

Cope, E. D. 1871 I, p. 91, fig. 18. [Arvicola (Anaptogonia).]

Allen, J. A. 1877 B, p. 947. [Arvicola (Anaptogonia).]
Cope, E. D. 1883 S, p. 373, fig. 24, no. 18. [Arvicola (Anaptogonia):]
     1896 F, p. 379.
     1899 A, p. 201 (A. hiatidens); p. 206 (Schistodelta sulcata[1]).
Miller, G. S. 1896 A, p. 74. [Microtus (Anaptogonia).]
Roger, O. 1896 A, p. 125. (Arvicola.)
Trouessart, E. L. 1898 A, p. 551.
Wheatley, C. M. 1871 B.
     Pleistocene; Pennsylvania.

SYCIUM Cope.   Type *S. cloacinum* Cope.

Cope, E. D, 1899 A, p. 203.

Sycium cloacinum Cope.

Cope, E. D. 1899 A, p. 203, text figure, p. 204.
     Pleistocene; Pennsylvania.

MICROTUS Schrank.   Type *M. arvalis* (Pall.).

Schrank, F. 1798, Fauna Boica, i, p. 72.
     Unless otherwise indicated, the following authors refer to this genus under the name *Arvicola* Lacépède, 1801
Alston, E. R. 1876 A, p. 85.
Baird, S. F. 1857 A, p. 509.
Blackmore, H. P. 1874 A.
Cope, E. D. 1871 I, p. 87. (Isodelta, subgenus, type *Arvicola speothen*.)
     1896 F, p. 379. (Microtus.)
     1899 A, pp. 201, 204. (Microtus.)
Coues, E. 1877 B, i, p. 149.
Flower and Lydekker 1891 A, p. 466.
Lydekker, R. 1885 B, p. 230.
Miller, G. S. 1896 A, pp. 15, 44 (Microtus); p. 58 (subgenus Pitymys).
Nehring, A. 1875 A.
Owen, R. 1868 A, p. 298.
Parker, W. K. 1868 A, p. 208, pl. xxiv, figs. 9-13.
Scott, W. B. 1892 A, p. 407.
Thomas, O. 1896 A, p. 1021.
Tomes, J. 1850 A, p. 550.
Trouessart, E. L. 1898 A, p. 552. (Microtus.)
Zittel, K. A. 1893 B, p. 536.

Microtus didelta Cope.

Cope, E. D. 1871 I, p. 89, fig. 15 [Arvicola (Pitymys)]: p. 90 [A. (Pitymys) sigmodus].
Allen, J. A. 1877 B, p. 947. [Arvicola (Pitymys) didelta, sigmoda.]
Cope, E. D. 1883 S, p. 373, fig. 24, no. 15. (Arvicola.)
     1899 A, pp. 205, 207.
Roger, O. 1896 A, p. 125. (Arvicola.)
Trouessart, E. L. 1898 A, p. 556 [Microtus (Pitymys)]: p. 562 (M. sigmodus).
Wheatley. C. M. 1871 B. (A. sigmodus.)
     Pleistocene; Pennsylvania.

Microtus diluvianus Cope.

Cope, E. D, 1896 F, p. 381.
     1899 A, p. 205, text figure.
Trouessart, E. L. 1898 A, p. 561.
     Pleistocene; Pennsylvania.

Microtus involutus Cope.

Cope, E. D. 1871 I, p. 89, fig. 16. [Arvicola (Pitymys).]

[1] This is regarded by Cope as possibly different from *A. hiatidens.*

Allen, J. A. 1877 B, p. 947. (Arvicola.)
Cope, E. D. 1883 S, p. 374, fig. 25.
　　　1899 A, p. 205.
Roger, O. 1896 A, p. 125. (Arvicola.)
Trouessart, E. L. 1898 A, p. 562.
Wheatley, C. M. 1871 B.
　　　Pleistocene; Pennsylvania.

Microtus pennsylvanicus (Ord).

Baird, S. F. 1857 A, p. 523. (Arvicola.)
Cope, E. D. 1871 I, p. 87. (Arvicola riparia.)
Coues, E. 1877 B, i, p. 156, pl. iv, figs. 42–49. (Arvicola riparius.)
Leidy, J. 1862 A, p. 423. ("Arvicola.")
　　　1869 A, p. 406. (Arvicola riparia?.)
　　　1889 H, p. 6. (Arvicola riparius.)
Miller, G. S. 1894 A, p. 190, pl. iii, fig. 2; pl. iv, fig. 1. (Arvicola riparius.)
　　　1896 A, p. 63, fig. 33, c. [Microtus (Microtus).]
Trouessart, E. L. 1898 A, p. 562. [Microtus (Microtus).]
　　　Recent; eastern United States. Pleistocene; Illinois, Virginia.

Microtus pinetorum (Le Conte).

Le Conte, J. L., 1829, Ann. Lyc. Nat. Hist. N. Y., iii, p. 132, pl. ii. (Psammomys.)
Baird, S. F. 1857 A, p. 544, pl. liv, No. 1719. [Arvicola (Pitymys).]
Cope, E. D. 1871 I, p. 87. [Arvicola (Pitymys).]
　　　1899 A, p. 205.

Coues, E. 1877 B, i, p. 219, pl. v, figs. 70–73. [Arvicola (Pitymys).]
Miller, G. S. 1896 A, p. 59, pl. i, fig. 2; text fig. 31, a. [Microtus (Pitymys).]
Trouessart, E. L. 1898 A, p. 555. [Microtus (Pitymys).]
Wheatley, C. M. 1871 B.
　　　Recent; United States. Pleistocene?; Pennsylvania. Not given in Cope 1899 A, as found in Port Kennedy cave.

Microtus speothen Cope.

Cope, E. D. 1871 I, p. 87, fig. 13. (Arvicola.)
Allen, J. A. 1877 B, p. 947. [Arvicola (Isodelta) speothen, A. (Pitymys) tetradelta.]
Cope, E. D. 1871 I, p. 88, fig. 14. (Arvicola tetradelta.)
　　　1883 S, p. 373, fig. 24, no. 13 [Arvicola (Isodelta)]; p. 373, fig. 24, no. 14 (A. tetradelta).
　　　1896 F, p. 383.
　　　1899 A, pp. 205, 206.
Miller, G. S. 1896 A, p. 75. [Microtus (Isodelta).]
Roger, O. 1896 A, p. 125. (Arvicola.)
Trouessart, E. L. 1898 A, p. 561.
Wheatley, C. M. 1871 B. (Arvicola.)
　　　Pleistocene; Pennsylvania.

Microtus sp. indet.

Cope, E. D. 1889 F, p. 161. (Arvicola.)
　　　Pleistocene (Equus beds); Oregon.

FIBER Cuv.　Type *Castor zibethicus* Linn.

Cuvier, G., 1800, Leçons d Anat. Comp., i, tabl. i.
Baird, S. F. 1857 A, p. 560.
Brandt, J. F. 1855 A, p. 171. (Ondatra.)
Cope, E. D. 1889 C, p. 249.
Coues, E. 1877 B, i, p. 251.
Flower and Lydekker 1891 A, p. 470.
Miller, G. S. 1896 A, pp. 14, 71.
Nicholson and Lydekker 1889 A, p. 1417.
Tomes, J. 1850 A, p. 550.
Tullberg, T. 1899 A, p. 235.
Zittel, K. A. 1893 B, p. 587.

Coues, E. 1877 B, i, p. 254.
Flower and Lydekker 1891 A, p. 471, fig. 209.
Holmes, F. S. 1850 B, p. 201. ("Muskrat.")
Leidy, J. 1860 B, p. 113, pl. xxii, figs. 1–4.
　　　1869 A, p. 406.
Miller, G. S. 1896 A, p. 72, pl. ii, fig. 13, text figs. 37–39.
Owen, R. 1866 B, p. 375, fig. 241.
Roger, O. 1896 A, p. 126.
Trouessart, E. L. 1898 A, p. 566.
Tullberg, T. 1899 A, p. 235.
Wheatley, C. M. 1871 A. (Fiber sp. indet.)
　　　Recent; North America north of Mexico. Pleistocene; South Carolina, New Jersey, Pennsylvania?.

Fiber zibethicus (Linn.).

Linnæus, C., 1766, Syst. Nat. ed. xii, p. 79. (Castor.)
Baird, S. F. 1857 A, p. 561, pl. liv.
Brandt, J. F. 1855 A, p. 171, pl. iii, figs. 9–16. (Ondatra.)

HELISCOMYS Cope.　Type *H. vetus* Cope.

Cope, E. D. 1873 CC, p. 3.
Allen, J. A. 1877 A, xi, p. 936.
Cope, E. D. 1874 B, p. 475.
　　　1881 E, p. 375.
　　　1883 S, p. 57.
　　　1884 O, p. 845.
Nicholson and Lydekker 1889 A, p. 1419.
Zittel, K. A. 1893 B, p. 582.
　　　A genus of uncertain position.

Heliscomys vetus Cope.

Cope, E. D. 1873 CC, p. 3.

Allen, J. A. 1877 A, xi, p. 936.
Cope, E. D. 1874 B, p. 475.
　　　1881 E, p. 375.
　　　1881 Q.
　　　1883 S, p. 56, fig. 13, a–d.
　　　1884 O, p. 847, pl. lxv, figs. 14–18.
Matthew, W. D. 1899 A, p. 52.
Roger, O. 1896 A, p. 120.
Trouessart, E. L. 1896 A, p. 578.
　　　Oligocene (White River); Colorado.

GEOMYIDÆ.

Alston, E. R. 1876 A, p. 87.
Baird, S. F. 1857 A. p. 364.   (Saccomyidæ.)
Brandt, J. F. 1855 A, p. 188.   (Sciurospalacoides.)
Cope, E. D. 1875 B.   (Saccomyidæ.)
Coues, E. 1875 B, p. 275.
    1875 C, p. 130.
    1877 B, viii, p. 488 (Saccomyidæ); p. 492
    (Geomyidæ); x, p. 607 (Geomyidæ).
Flower, W. H. 1883 D, p. 419.
Flower and Lydekker 1891 A, p. 478.

Merriam, C. H. 1895 A.
Nicholson and Lydekker 1889 A, p. 1416.
Roger, O. 1896 A, p. 119.
Scott, W. B. 1895 D, p. 283.
Thomas, O. 1896 A, p. 1022.
Trouessart, E. L. 1898 A, p. 571.
Tullberg, T. 1899 A, pp. 312, 475 (Geomyidæ); p.
    465 (Protoptychidæ).
Winge, H. 1888 A, p. 138.   (Saccomyidæ.)
Zittel, K. A. 1893 B, p. 582.

PROTOPTYCHUS Scott.   Type *P. hatcheri* Scott.

Scott, W. B. 1895 E, p. 269.
    1896 D, p. 923.
Tullberg, T. 1899 A, p. 465.

Protoptychus hatcheri Scott.
Scott, W. B. 1895 E, p. 269, figs. 1–6.

Matthew, W. D. 1899 A, p. 49.
Roger, O. 1896 A, p. 119.
Scott, W. B. 1896 D, p. 923.
Trouessart, E. L. 1898 A, p. 577.
Tullberg, T. 1899 A, p. 464.
    Eocene (Uinta); Utah.

GEOMYS Raf.   Type *Mus tuza* Ord.

Rafinesque, C. S., 1817, Amer. Month. Mag., ii, p. 45.
Alston, E. R. 1876 A, p. 87.
Baird, S. F. 1857 A, p. 366.
Coues, E. 1875 C, p. 130.
    1877 B, x, p. 611.
Flower and Lydekker 1891 A, p. 478.
Merriam, C. H. 1895 A, p. 109.
Nicholson and Lydekker 1889 A, p. 1416.
Tomes, J. 1850 A, p. 549.
Trouessart, E. L. 1898 A, p. 571.

Geomys bisulcatus Marsh.
Marsh, O. C. 1871 E, p. 121.
Allen, J. A. 1877 B, p. 947.
King, C. 1878 A, p. 430.
Matthew, W. D. 1899 A, p. 66.
Roger, O. 1896 A, p. 120.
Trouessart, E. L. 1898 A, 571.
    Miocene (Loup Fork); Nebraska.

Geomys bursarius (Shaw).
Shaw, G., 1800, Trans. Linn. Soc. v, p. 227, pl. viii.
    (Mus.)
Baird, S. F. 1857 A, p. 372, pl. l, fig. 2.
Cope, E. D. 1869 E, p. 173.
Coues, E. 1875 C, p. 131.
    1877 B, x, p. 613.
Leidy, J. 1862 A.   (Pseudostoma.)
    1867 B.
    1869 A, p. 406.
Merriam, C. H. 1895 A, p. 120, pl. i; pl. ix, figs. 8, 9:
    pl. x, fig. 6; pl. xiii, fig. 11; pl. xiv, fig. 2; pl. xv,
    fig. 11; pl. xvii, fig. 3; pl. xviii, fig. 1; pl. xix,
    fig. 3; text figs. 1, 8, 11, 55; with synonomy.
Trouessart, E. L. 1898 A, p. 571.
Williston, S. W. 1897 I, p. 304.
    Recent; Mississippi Valley. Pleistocene; Illi-
    nois, Nebraska.

THOMOMYS Wied.   Type *T. rufescens* Wied.

Wied., 1839, N. Act. Acad. Leop. Carol., xix, p. 377.
Allen, J. A. 1893, A, pp. 58, 64.
Alston, E. R. 1876 A, p. 87.
Baird, S. F. 1857 A, p. 388.
Coues, E. 1875 C, p. 134.
    1877 B, x, p. 621.
Flower and Lydekker 1891 A, p. 478.
Merriam, C. H. 1898 A, p. 198,
Nicholson and Lydekker 1889 A, p. 1416.
Trouessart, E. L. 1898 A, p. 575.

Thomomys bulbivorus (Richardson).
Richardson, J., 1829, Fauna Bor.-Amer., i, p. 206,
    pl. 18B.
Allen, J. A. 1893 A, p. 56, pl. i, fig. 14.
Baird, S. F. 1857 A, p. 389, pl. lii, fig. 1.
Brandt, J. F. 1855 A, p. 187.   ((Tomomys.)
Cope, E. D. 1883 S, p. 374.
    1889 S, p. 980.

Coues, E, 1875 C, p. 136.   (T. talpoides bulbivorus.)
    1877 B, x, p. 626.   (T. talpoides bulbivorus.)
Merriam, C. H. 1895 A, p. 198, figs 68–71.
Trouessart, E. L. 1898 A, p. 575.
Tullberg, T. 1899 A, p. 328.
    Recent; northwestern United States. Pleisto-
    cene; Oregon.

Thomomys clusius Coues.
Coues, E. 1875 C, p. 138.
Allen, J. A. 1893 A, pp. 61, 62, pl. i, fig. 2.
Cope, E. D. 1878 H, p. 389.   ("Thomomys near
    clusius.")
    1889 F.
    1889 S, p. 980.
Coues, E. 1877 B, x, p. 629, pl. vii.
Trouessart, E. L. 1898 A p. 576.
    Recent; Wyoming, Idaho.   Pleistocene;
    (Equus beds); Oregon.

Thomomys talpoides Richardson.

Richardson, J., 1828, Zool. Jour., iii, p. 518.
Allen, J. A. 1898 A, p. 66.
Cope, E. D. 1878 H, p. 389.
    1889 F.
Coues, E. 1875 C, pp. 134, 135.

Coues, E. 1877 B, x, p. 628.
Roger, O. 1896 A, p. 120.
Trouessart, E. L. 1898 A, p. 576.
    Recent; Canada to Upper Missouri River,
    Pleistocene; Oregon.

PLEUROLICUS Cope.   Type *P. sulcifrons* Cope.

Cope, E. D. 1879 D, P. 66.
    1879 B, p. 55.
    1881 E, p. 380.
    1883 S, p. 166.
    1884 O, p. 866.
Flower and Lydekker 1891 A, p. 479.
Nicholson and Lydekker 1889 A, p. 1416,
Scott, W. B. 1895 D.
    1895 E, pp. 270, 286,
Zittel, K. A. 1898 B, p. 533.

Pleurolicus diplophysus Cope.

Cope, E. D. 1881 E, p. 381.
    1881 Q.
    1883 S, p. 167, fig. 16, c, d.
    1884 O, pp. 867, 869, pl. lxiv, fig. 9.
Matthew, W. D. 1899 A, p. 62.
Roger, O. 1896 A, p. 120.
Scott, W. B. 1895 E, p. 284, figs. 5, 6.
Trouessart, E. L. 1898 A, p. 578.
    Miocene (John Day); Oregon.

Pleurolicus leptophrys Cope.

Cope, E. D. 1881 E, p. 381.
    1881 Q.
    1883 S, p. 167, fig. 16, a, b.
    1884 O, p. 868, pl. lxiv, figs. 7, 8.
Matthew, W. D. 1899 A, p. 62.
Roger, O. 1896 A, p. 120.
Trouessart, E. L. 1898 A, p. 578.
    Miocene (John Day); Oregon.

Pleurolicus sulcifrons Cope.

Cope, E. D. 1879 D, p. 66.
    1879 B, p. 55.
    1881 E, p. 381.
    1881 Q.
    1883 S, p. 167,
    1884 O, p. 867, pl. lxiv, fig. 6.
Matthew, W. D. 1899 A, p. 62.
Roger, O. 1896 A, p. 120.
Trouessart, E. L. 1898 A, p. 578.
    Miocene (John Day); Oregon.

ENTOPTYCHUS Cope.   Type *E. cavifrons* Cope.

Cope, E. D. 1879 D, p. 64.
    1881 E, p. 378.
    1883 S, p. 168.
    1884 O, p. 855.
Flower and Lydekker 1891 A, p. 479.
Nicholson and Lydekker 1889 A, p. 1416.
Scott, W. B. 1895 D.
    1895 E, p. 286.
Zittel, K. A. 1898 B, p. 533.

Entoptychus cavifrons Cope.

Cope, E. D. 1879 D, p. 64.
    1879 B, p. 55.
    1881 E, p. 380.
    1881 Q.
    1883 S, p. 169.
    1884 O, pp. 856, 862, pl. lxiv, fig. 4.
Matthew, W. D. 1899 A, p. 62.
Roger, O. 1896 A, p. 120.
Trouessart, E. L. 1896 A, p. 578.
    Miocene (John Day); Oregon.

Entoptychus crassiramis Cope.

Cope, E. D. 1879 D, p. 65.
    1879 B, p. 55.
    1881 E, p. 380.
    1881 Q.
    1883 S, p. 168, fig. 17.
    1884 O, pp. 858, 864, pl. lxiv, fig. 5.
Matthew, W. D. 1899 A, p 62.
Roger, O. 1896 A, p. 120.
Trouessart, E L 1898 A, p. 579.
    Miocene (John Day), Oregon.

Entoptychus lambdoideus Cope.

Cope, E. D. 1881 E, p. 380.
    1881 Q.
    1883 S, p. 169.
    1884 O, pp. 858, 860, pl. lxiv, fig. 2.
Matthew, W. D. 1899 A, p. 62.
Roger, O. 1896 A, p. 120.
Trouessart, E. L. 1898 A, p. 578.
    Miocene (John Day); Oregon.

Entoptychus minor Cope.

Cope, E. D. 1881 E, p. 379.
    1881 Q.
    1883 S, p. 170.
    1884 O, pp 858, 861, pl. lxiv, fig. 3.
Matthew, W. D. 1899 A, p. 62.
Roger, O. 1896 A, p. 120.
Trouessart, E. L. 1898 A, p. 578.
    Miocene (John Day), Oregon.

Entoptychus planifrons Cope.

Cope, E D. 1879 D, p. 65.
    1879 B, p. 55.
    1881 E, p 379.
    1881 Q.
    1883 S, p. 167, fig. 18.
Matthew, W. D 1899 A, p 62.
Roger, O. 1896 A p. 120.
Trouessart E. L 1898 A p 578.
Zittel K A 1898 B p. 533 fig. 445.
    Miocene (John Day). Oregon

DIPODIDÆ.

Allen, J. A. 1900 A, p. 199.
Baird, S. F. 1857 A, p. 428.   (Dipodinæ.)
Flower, W. H. 1883 D, p. 419.
Flower and Lydekker 1891 A, p. 479.
Nicholson and Lydekker 1889 A, p. 1416.

Thomas, O. 1896 A, p. 1023.
Tullberg, T. 1899 A, pp. 182, 408.
Winge, H. 1888 A, p. 118.
Zittel. K. A. 1898 B, p. 526.

ZAPODINÆ.

Coues, E., 1875, Bull. U. S. Geol. Surv., ser. 2, no. 5, p. 253.   (Zapodidæ.)
Alston, E. R. 1876 A, p. 89.   (Jaculinæ.)
Coues, E. 1877 B, vii, p. 454.   (Zapodidæ.)

Flower and Lydekker 1891 A, p. 480.
Thomas, O. 1896 A, p. 1023.
Trouessart, E. L. 1898 A, p. 590.

ZAPUS Coues.   Type *Dipus hudsonius* Zimm.

Coues, E., 1875, Bull. U. S. Geol. Surv. Terr., ser. 2, no. 5, p. 253.
Allen, J. A. 1900 A, p. 200.
Baird, S. F. 1857 A, p. 429.   (Jaculus.)
Brandt, J. F. 1855 A.   (Jaculus.)
Coues, E. 1876 A.
    1877 B, vii, p. 465.
Flower and Lydekker 1891 A, p. 480.
Nicholson and Lydekker 1889 A, p. 1416.
Palmer, T. S. 1897 A.
Preble, E. A. 1899 A, p. 13.
Thomas, O. 1896 A, p. 1023, footnote.

Baird, S. F. 1857 A, p. 430.   (Jaculus.)
Cope, E. D. 1871 I, p. 86.   (Jaculus.)
    1883 S, p. 373.   (Meriones.)
    1899 A, p. 200.
Coues, E. 1876 A.
    1877 B, vii, p. 465.
Flower and Lydekker 1891 A, p. 480.   (Z. hudsonianus.)
Nicholson and Lydekker 1889 A, p. 1416.
Preble, E. A. 1899 A, p. 15, pl. i, figs. 3, 3a.
Trouessart, E. L. 1898 A, p. 590.
Tullberg, T. 1899 A, p. 185.
Wheatley, C. M. 1871 B.   (Jaculus.)
Zittel, K. A. 1898 B, p. 539.

Zapus hudsonius (Zimm.)

Zimmermann, E., 1780, Geog. Gesch., ii, p. 358.   (Dipus.)
Allen, J. A. 1900 A, p. 201.

Recent; Alaska to Hudson Bay, south to New Jersey, west to Iowa and Missouri. Pleistocene; Pennsylvania.

Superfamily HYSTRICOMORPHA.

Alston, E. R. 1876 A, p. 91.
Brandt, J. F. 1855 A.
Cope E. D. 1881 E, p. 361.
    1883 S, p. 43.
    1884 O, p. 812.
Flower and Lydekker 1891 A, p. 480.
Gadow, H. 1898 A, p. 48.
Haeckel, E. 1895 A, p. 510.
Lydekker, R. 1896 B.

Nicholson and Lydekker 1889 A, p. 1413.
Schlosser, M. 1884 A, p. 98.
    1890 C, p. 250.
Slade, D. D. 1888 A, p. 243.
    1892 B.
Steinmann and Döderlein 1890 A, p. 733.
Thomas, O. 1896 A, p. 1024.
Tullberg, T. 1899 A, pp. 82, 363.
Zittel, K. A. 1898 B, p. 539.

HYSTRICIDÆ.

Allen, J. A. 1877 A, iii, p. 385.
Alston, E. R. 1876 A, p. 93.
Baird, S. F. 1857 A, p. 565.
Cope, E. D. 1875 C, p. 26.
Flower, W. H. 1883 D, p. 420.
Flower and Lydekker 1891 A, p. 484.
Leidy, J. 1871 C, p 364.
Nicholson and Lydekker 1889 A, p. 1415.

Roger, O. 1896 A, p. 127.
Trouessart, E. L. 1898 A, p. 614.
Tullberg, T. 1899 A, pp 83, 366.
Turner, H. N. 1848 A.
Winge, H. 1888 A, p. 126.
Zittel, K. A. 1898 B, p. 539.
    No species of this family, as its limits are here recognized, are found in North America.

ERETHIZONTIDÆ.

Thomas, O. 1896 A, p 1025.
Allen, J. A 1877 A, iii, p 385.   (Hystricidæ, in part.)
Baird, S F. 1857 A. p 565   (Hystricidæ, in part.)

Flower W H 1883 D, p. 420.   (Hystricidæ, in part.)
Trouessart, E L 1898 A, p 619   (Cœndidæ.)
Tullberg, T. 1899 A, pp. 108, 368.

ERETHIZONTINÆ.

Thomas, O 1896 A, p. 1025.

ERETHIZON F. Cuv.   Type *Hystrix dorsata* Linn.

Cuvier, F., 1822, Mém. de Muséum, Paris, ix, p. 426.
Allen, J. A. 1877 A, iii, p. 386.
Alston, E. R. 1876 A, p. 94. (Erythizon.)
Flower, W. H. 1883 D, p. 420.
Flower and Lydekker 1891 A, p. 484.
Leidy, J. 1858 E, p. 22. (Hystricops, subgenus.)
Nicholson and Lydekker 1889 A, p. 1415.
Winge, H. 1888 A, p. 129.
Zittel, K. A. 1893 B, p. 540.

Erethizon cloacinus Cope.

Cope, E. D. 1871 I, p. 93, fig. 19.
Allen, J. A. 1877 A, iii, p. 398.
Cope, E. D. 1883 S, p. 379.
    1899 A, p. 199.
Roger, O. 1896 A, p. 127.
Trouessart, E. L. 1898 A, p. 620.
Wheatley, C. M. 1871 B.
    Pleistocene; Pennsylvania.

Erethizon dorsatus (Linn.).

Linnæus, C., 1758, Syst. Nat., ed. 10, i, p. 57. (Hystrix.)

Allen, J. A. 1877 A, iii, p. 568.
Cope, E. D. 1899 A, p. 198.   (E. dorsatum?)
Leidy, J. 1889 H, p. 5.
Trouessart, E. L. 1898 A, p. 620.
Tullberg, T. 1899 A, p. 109.
    Recent; northeastern U. S.; Canada, west to
    Saskatchewan. Pleistocene; Pennsylvania.

Erethizon venustus (Leidy).

Leidy, J. 1858 E, p. 22. [Hystrix (Hystricops).]
Allen, J. A. 1877 A, iii, p. 397. (Hystrix? venustus.)
Cope, E. D. 1881 Q. (Hystrix.)
Hayden, F. V. 1858 B, p. 407. (Hystrix.)
King, C. 1878 A, p. 430. (Hystrix.)
Leidy, J. 1869 A, pp. 343, 407, pl. xxvi, figs. 23, 24.
    (Hystrix.)
    1871 C, p. 364. (Hystrix.)
Roger, O. 1896 A, p. 127.
Trouessart E. L. 1898 A, p. 620.
Zittel, K. A. 1893 B, p. 540.
    Pleistocene; Nebraska.

CASTOROIDIDÆ.

Allen, J. A. 1887 A, v, p. 419.
Flower and Lydekker 1891 A, p. 488.
Moore, J. 1891 A.
Nicholson and Lydekker 1889 A, p. 1415.

Roger, O. 1896 A, p. 136.
Trouessart, E. L. 1898 A, p. 630.
Zittel, K. A. 1893 B, p. 547.

CASTOROIDES Foster.   Type *C. ohioensis* Foster.

Foster, J. W. 1838 A, p. 81,
Allen, J. A. 1877 A, v, p. 419.
Alston, E. R. 1876 A, p. 78.
Baird, S. F. 1857 A, p. 363.
Cope, E. D. 1883 S, p. 370.
Flower and Lydekker 1891 A, p. 488.
Moore, J. 1893 A, p. 67.
Nicholson and Lydekker 1889 A, p. 1415.
Pictet, F. J. 1853 A, p. 252.
Scott, W. B. 1892 A, p. 410.
Winge, H. 1888 A, p. 139.
Wyman, J. 1846 B, p. 391.
    1850 A, p. 56.
Zittel, K. A. 1893 B, p. 547.

Castoroides ohioensis Foster.

Foster, J. W. 1838 A, p. 81, figs. A–D.
Agassiz, L. 1851 B.
Allen, J. A. 1877 A, v, p. 423.
Alston, E. R. 1876 A, p. 79.
Baird, S. F. 1857 A, p. 363.
Cope, E. D. 1883 S, p. 370, figs. 22, 23.
    1888 I, p. 6, figs. 3, 4.
    1889 C, p. 264.
Cope and Wortman 1884 A, p. 37, pl. iv, figs. 1–3.
De Kay, J. E. 1842 C, p. 75, pl. xi, fig. 3 [Castor
    (Trogotherium) ohionis].
Foster, J. W. 1837 A, p. 80, figs. 15–18. (No generic
    or specific name.)
Hall, J. 1846 A.
    1846 B.

Kindle, E. M. 1896 A, p. 485.
Langdon, F. W. 1883 A.
Le Conte, J. L. 1852 C.
Leidy, J. 1860 B, p. 114, pl. xxii, figs. 5–8.
    1867 C, p. 97.
    1869 A, p. 405.
    1880 A, p. 347.
    1889 H, p. 14, pl. ii, figs. 7–20.
Mercer, H. C. 1894 A, p. 98.
Miller, G. S. 1899 B, p. 375.
Moore, J. 1890 A.
    1890 B.
    1891 A.
    1893 A.
    1900 A, p. 171, pls. i, ii.
Newberry, J. S. 1870 A, p. 83.
    1874 A, p. 92.
    1875 A, p. vi.
Pictet, F. J. 1853 A, p. 253.
Pomel, A. 1848 D, p. 165.
Roger, O. 1896 A, p. 136.
Trouessart, E. L. 1898 A, p. 630.
Whittlesey, C. 1848 A, p. 215.
Wyman, J. 1846 A.
    1846 B, pls. 37–39.
    1850 A, pp. 62–64, fig. 5.
    1850 D, p. 280.
Tullberg, T. 1899 A, p. 474.
    Pleistocene; Ohio, Indiana, Illino's, Tennes-
    see, New York, South Carolina, Pennsylvania,
    Texas, Michigan.

DASYPROCTIDÆ.

Adloff, P. 1898 A, p. 388. (Genus Dasyprocta.)
Alston, E. R. 1876 A, p. 95.
Flower and Lydekker 1889 A, p. 488.
Nicholson and Lydekker 1889 A, p. 1414.
Parker, W. K. 1868 A, p. 207, pl. xxiv, fig. 8.
(Genus Dasyprocta.)

Roger, O. 1896 A, p. 128.
Trouessart, E. L. 1898 A, p. 633.
Tullberg, T. 1899 A, pp. 90, 370. (Caviidæ, in
part.)
Winge, H. 1888 A, p. 135. (Dasyproctini.)
Zittel, K. A, 1893 B, p. 540.

AGOUTI Lacépède.   Type *Mus paca* Linn.

Lacépède, B. G., 1799, Tab. ord., gen. mamm., p. 9.
Cuvier, F.,1807, Ann. Mus., Paris, x, p. 208. (Cœlogenys.)
Flower and Lydekker 1891 A, p. 489. (Cœlogenys.)
Harlan, R. 1825 A, p. 126. (Osteopera, type *O. platycephala*=C. paca.)
Palmer, T. S. 1897 A.
Thomas, O. 1896 A, p. 1026. (Cœlogenys.)
Trouessart, E. L. 1898 A, p. 635. (Cœlogenys.)
Tullberg, T. 1899 A, p. 91. (Cœlogenys.)
Agouti paca (Linn.).
Linnæus, C., 1766, Syst. Nat., ed. 12, p. 81. (Mus.)

Harlan, R. 1825 A, p. 126. (Osteopera platycephala.)
Leidy, J. 1869 A, p. 444. (Cœlogenys.)
Palmer, T. S. 1897 A.
Pictet, F. J. 1853 A, p. 254. (Cœlogenys.)
Roger, O. 1896 A, p. 129. (Cœlogenys.)
Trouessart, E. L. 1898 A, p. 635. (Cœlogenys.)
Tullberg, T. 1899 A, p. 91.

Recent; Mexico to Paraguay. Reported by
Harlan as a fossil found in vicinity of Philadelphia; but doubtless transported thither.

CAVIIDÆ.

Adloff, P. 1898 A, p. 387. (Genus Cavia.)
Alezais, H. 1898 A. (Genus Cavia.)
1899 A. (Genus Cavia.)
Flower and Lydekker 1891 A, p. 489.

Thomas, O. 1896 A, p. 1026.
Trouessart, E. L. 1898 A, p. 636.
Tullberg, T. 1899 A, pp. 90, 370. (In part).
Zittel, K. A. 1893 B, p. 544.

HYDROCHŒRUS Brisson.   Type *Sus hydrochæris* Linn.

Brisson, A. D., 1762, Reg. Anim., pp. 12, 80.
Burmeister, H. 1879 B, p. 263.
Flower and Lydekker 1891 A, p. 490.
Leidy, J. 1853 E, p. 241. (Oromys, type *O. æsopi;* no description.)
Nicholson and Lydekker 1889 A, p. 1414.
Owen, R. 1868 A, p. 296, fig. 235.
Schlosser, M. 1890 C, p. 251.
Thomas, O. 1896 A, p. 1026.
Tomes, J. 1850 A, p. 557.
Tullberg, T. 1899 A, p. 105.
Zittel, K. A. 1893 B, p. 546.

Hydrochœrus æsopi Leidy.
Leidy, J. 1856 J, p. 165.
Allen, J. A. 1877 B, p. 949.
Cope, E. D. 1883 S, p. 379, fig. 30.
1888 I, p. 9, fig. 6.
Leidy, J. 1853 E, p. 241. (Oromys æsopi; no description.)

Leidy, J. 1860 B, p. 112, pl. xxi, figs. 3-6.
1869 A, p. 407.
Roger, O. 1896 A, p. 186.
Trouessart, E. L. 1898 A, p. 643.
Pleistocene; South Carolina.

Hydrochœrus robustus Leidy.
Leidy, J. 1888 B, p. 276, fig. 1.
1890 B, p. 184. (H.? magnus Gerv. and
Amegh.)
Roger, O. 1896 A, p. 186.
Pleistocene, Nicaragua, Florida?.
There is doubt about the identification of the
Florida specimen.
Dr. Jeffries Wyman has reported the finding
of a skull of the capybara, *H. hydrochæris*, on
the southwestern frontier of the United States.
Proc. Bost. Soc. Nat. Hist., vii, 1860, p. 350.

Suborder DUPLICIDENTATA.

Alston, E. R. 1876 A.
Flower and Lydekker 1891 A, p. 491.
Nicholson and Lydekker 1889 A, p. 1412.
Palmer, T. S. 1897 A.

Schlosser, M. 1884 A, p. 123. (Pliodonta.)
Thomas, O. 1896 A, p. 1026.
Tullberg, T. 1899 A, pp. 42, 338.
Woodward, A. S. 1898 B, p. 377.

Superfamily *LAGOMORPHA.*

Brandt, J. F. 1855 A.
Cope, E. D. 1881 E, p. 361.
    1883 S, p. 43.
    1884 O, p. 812.
Gadow, H. 1898 A. p. 42.
Lydekker, R. 1896 B.
Major, Fors. 1893 A, p. 208.
    1899 A, p. 433.

Roger, O. 1896 A, p. 140.
Schlosser, M. 1884 A, p. 98.
Slade, D. D. 1892 B.
Steinmann and Döderlein 1890 A, p. 785.
Thomas, O. 1896 A, p. 1026.
Zittel, K. A. 1893 B, p. 549.

OCHOTONIDÆ.

Trouessart, E. L. 1898 A, p. 644.
Allen, J. A. 1877 A, iv, p. 405. (Lagomyidæ.)
Flower, W. H. 1883 D, p. 421. (Lagomyidæ.)
Flower and Lydekker 1891 A, p. 491. (Lagomyidæ.)

Major, Fors. 1899 A, p. 463. (Lagomyidæ.)
Schlosser, M. 1884 A, p. 9. (Lagomyidæ.)
Tullberg, T. 1899 A, p. 53. (Lagomyidæ.)
Zittel, K. A. 1893 B, p. 551. (Lagomyidæ.)

OCHOTONA Link.   Included *O. pusillus, alpinus,* and *ochotona.*

Link H. F., 1795, Zool. Beytr., i, pt. ii, p. 74.
Allen, J. A. 1877 A, p. 407. (Lagomys.)
Cope, E. D. 1871 I, p. 93. (Praotherium.)
Cuvier, G., 1798, Tabl. Elém., p. 132. (Lagomys.)
Flower and Lydekker 1891 A, p. 491. (Lagomys.)
Major, Fors. 1899 A, p. 433. (Lagomys.)
Meyer, H. 1845 A. (Lagomys.)
Thomas, O. 1896 A, p. 1026.
Trouessart, E. L. 1898 A, p. 645 (Ochotona); p. 649 (Praotherium).
Winge, H. 1888 A, p. 112. (Lagomys.)

Zittel, K. A. 1893 B, p. 553. (Lagomys.)

Ochotona palatinus (Cope).

Cope, E. D. 1871 I, pp. 94, 102, fig. 20. (Praotherium.)
    1883 S, p. 379. (Lagomys.)
    1899 A, p. 209. (Lagomys.)
Roger, O. 1896 A, p. 140. (Praotherium.)
Trouessart, E. L. 1898 A, p. 649. (Praotherium.)
    Pleistocene; Pennsylvania.

LEPORIDÆ.

Allen, J. A. 1877 A, ii, p. 265.
Alston, E. R. 1876 A, p. 97.
Baird, S. F. 1857 A, p. 572.
Brandt, J. F. 1855 A.
Cope, E. D. 1881 J, p. 545.
Flower and Lydekker 1891 A, p. 492.
Giebel, C. G. 1880 A.
Hemstedt, R. 1870 A.
Hilgendorf, F. 1866 A, p. 673.
Howes, G. B. 1887 B.
Leidy, J. 1871 C, p. 363.
Major, Fors. 1899 A, p. 465.

Marsh, O. C. 1877 E.
Nicholson and Lydekker 1889 A, p. 1412.
Parker, W. K. 1868 A, p. 207.
Roger, O. 1896 A, p. 140.
Steinmann and Döderlein 1890 A, p. 785.
Thomas, O. 1896 A, p. 1027.
Tomes, J. 1850 A, p. 560.
Trouessart, E. L. 1898 A, p. 648.
Tullberg, T. 1899 A, p. 50.
Winge, H. 1888 A, p. 110.
Woodward, M. F. 1894 A.
Zittel, K. A. 1893 B, p. 550.

PALÆOLAGUS Leidy.   Type *P. haydeni* Leidy.

Leidy, J. 1856 I, p. 89.
Cope, E. D. 1873 T, p. 4. (Tricium, type *T. avunculus* = P. haydeni.)
    1874 B, p. 477.
    1881 E, p. 381.
    1883 S, p. 170.
    1884 O, p. 870.
Leidy, J. 1869 A, p. 331.
    1871 C, p. 363.
Lydekker, R. 1887 A, p. 325.
Major, Fors. 1899 A, pp. 443, 470.
Nicholson and Lydekker 1889 A, p. 1412.
Winge, H. 1888 A, p. 113.
Zittel, K. A. 1893 B, p. 550.

Palæolagus haydeni Leidy.

Leidy, J. 1856 I, p. 89.

Cope, E. D. 1873 S, p. 1. (P. agapetillus.)
    1873 T, p. 4 (Tricium avunculus); p. 5 (T. agapetillum).
    1873 CC, p. 4. (Tricium agapetillum, T. annæ, T. avunculus.)
    1874 B, p. 478. (P. haydeni, P. agapetillus.)
    1879 B, p. 55.
    1879 D, p. 63.
    1881 E, p. 384. (P. haydeni, P. agapetillus.)
    1881 Q. (P. haydeni, P. agapetillus.)
    1883 S, p. 173, fig. 20.
    1884 O, pp. 874, 875, pl. lxvi, figs. 1–27.
Hayden, F. V. 1858 B, p. 158.
King, C. 1878 A, p. 412.
Leidy, J. 1869 A, pp. 331, 40⁴, pl. xxvi, figs. 14–20.
    1871 C, p. 363.
Lydekker, R. 1887 A, p. 326.

Major, Fors. 1899 A, p. 443, pl. xxxvi, fig. 36.
Matthew, W. D. 1899 A, 53.
Roger, O. 1896 A, p. 140.
Trouessart, E. L. 1898 A, p. 648.
Zittel, K. A. 1893 B, p. 550, fig. 464.
Oligocene (White River); South Dakota, Nebraska, Colorado, Oregon?.

Palæolagus intermedius Matth.

Matthew, W. D. 1899 A, p. 53.
Oligocene (White River); Colorado.

Palæolagus leporinus (Cope).

Cope, E. D. 1873 T, p. 5. (Tricium.)
1873 CC, p. 4. (Tricium.)
Oligocene (White River); Colorado.
This is probably a synonym of *P. haydeni*. It was not recognized by Professor Cope after 1873; but I do not find that he referred it to any of his other species.

Palæolagus triplex Cope.

Cope, E. D. 1873 T, p. 4.
1873 CC, p. 4.
1874 B, p. 479.
1881 E, p. 384.
1883 S, p. 173.
1884 O, pp. 874, 881, pl. lxvi, fig. 28.

Major, Fors. 1899 A, p. 443.
Matthew, W. D. 1899 A, p. 53.
Roger, O. 1896 A, p. 140.
Trouessart, E. L. 1898 A, p. 648.
Oligocene (White River); Colorado.

Palæolagus turgidus Cope.

Cope, E. D. 1873 T, p. 4. (P. turgidus, Tricium paniense.)
1873 CC, p. 4. (P. turgidus, Tricium paniense.)
1874 B, p. 479. (P. turgidus, Tricium paniense.)
1881 E, p. 384.
1881 Q.
1883 S, p. 173.
1884 O, pp. 874, 882, pl. lxvi, fig. 28; pl. lxvii, figs. 13-27.
1885 H.
1886 U, p. 80.
1889 I.
1891 A, p. 5, pl. xlv, fig. 9.
Lydekker, R. 1887 A, p. 326.
Matthew, W. D. 1899 A, p. 52.
Roger, O, 1896 A, p. 140.
Trouessart, E. L. 1898 A, p. 848.
Oligocene (White River); Colorado, western Canada.

PANOLAX Cope.   Type *P. sanctæfidei* Cope.

Cope, E. D. 1874 I, p. 151.
1874 O, p. 17.
1877 K, p. 296.
1881 E, p. 381.
1883 S, p. 173.
Nicholson and Lydekker 1889 A, p. 1412.
Zittel, K. A. 1893 B, p. 550.

Panolax sanctæfidei Cope.

Cope, E. D. 1874 I, p. 151.

Cope, E. D. 1874 O, p. 17.
1875 F, p. 73.
1877 K, p. 296, pl. lxix, figs. 16-22. (P. sanctæfidæl.)
1881 Q. p. 586.
1883 S, p. 173.
Matthew, W. D. 1899 A, p. 67.
Roger, O. 1896 A, p. 140.
Trouessart, E. L. 1898 A, p. 648.
Miocene (Loup Fork); New Mexico.

LEPUS Linn.   Type *L. timidus* Linn.

Linnæus, C., 1758, Syst. Nat., ed. 10, i, p. 57.
Adloff, P. 1898 A, p. 392.
Allen, J. A. 1877 A, ii, p. 282.
Baird, S. F. 1857 A, p. 573.
Cope, E. D. 1881 E, p. 384.
1883 S, p. 173.
1884 O, p. 885.
1889 C, p. 269.
Flower, W. H. 1885 A.
Flower and Lydekker 1891 A, p. 492.
Giebel, C. G. 1880 A.
Hilgendorf, F. 1866 A, p. 673.
Howes, G. B. 1896 A.
Huxley, T. H. 1872 A, p. 371.
Major, Fors. 1899 A, pp. 433, 463.
Owen, R. 1866 B, pp. 364, 367, figs. 229, 233.
Parker, W. K. 1868 A, p. 207, pl. xxv, figs. 1-5.
Tomes, J. 1850 A, p. 560.
Trouessart, E. L. 1898 A, p. 649.
Zittel, K. A. 1893 B, p. 550.

Lepus ennisianus Cope.

Cope, E. D. 1881 E, p. 385.
1881 Q.
1883 S, p. 173, fig. 21.

Cope, E. D. 1884 O, pp. 881, 886, pl. lxiv, fig. 11; pl. xlvi, fig. 29.
Matthew, W. D. 1899 A, p. 62.
Roger, O. 1896 A, p. 140.
Trouessart, E. L. 1898 A, p. 649.
Miocene (John Day); Oregon.

Lepus sylvaticus Bachm.

Bachman, J., 1837, Jour. Acad. Nat. Sci., Phila., vii, p. 403.
Allen, J. A. 1877 A, ii, p. 327.
Baird, S. F. 1857 A, p. 597, pl. lviii, fig. 1.
Cope, E. D. 1869 E, p. 175.
1871 I, p. 93.
1899 A, p. 209.
Holmes, F. S. 1858 B.
Leidy, J. 1860 B, p. 113, pl. xxii, fig. 1.
1862 A.
1869 A, p. 403.
1889 H, p. 6.
Trouessart, E. L. 1898 A, p. 658.
Wheatley, C. M. 1871 B.
Recent; eastern United States. Pleistocene; Illinois, South Carolina, Pennsylvania, Virginia.

ORDER INSECTIVORA.

Allen, H. 1880 A.
Baur, G. 1885 K, p. 459.
Blainville, H. M. D. 1864 A, i, H, pls. i-xi.
. Cope, E. D. 1874 B, pp. 464, 472.
　　1876 A.
　　1877 K, pp. 85, 158.
　　1880 B, p. 455.
　　1883 I, p. 78.
　　1884 O, pp. 185, 197, 739.
　　1885 K.
　　1887 S, p. 348.
　　1888 D.
　　1891 N, p. 72.
　　1898 B, p. 115.
Dobson, G. E. 1883 A.
Flower, W. H. 1876 B, xiii, p. 514.
　　1883 A.
　　1883 D, p. 400.
　　1885 A.
Flower and Lydekker 1891 A, p. 610.
Gadow H. 1898 A, p. 51.
Gill, T. 1872 A, p. 303.
　　1872 B, pp. 18, 49.
　　1875 A.
　　1883 B.
Gill and Coues 1877 A, p. 1058.
Haeckel, E. 1895 A, pp. 578, 580.
Hay, O. P. 1899 F, p. 682.
Huxley, T. H. 1870 A.
　　1872 A, p. 375.
Leche, W. 1880 A.
　　1893 A.
　　1896 A, p. 286.
　　1897 A.
　　1897 B.
Leidy, J. 1871 C, p. 364.
Lydekker, R. 1885 B, p. 14.
　　1896 B.
Marsh, O. C. 1877 E.
Meyer, O. 1885 A, p. 229.

Mivart, St. G. 1867 B.
　　1867 C.
　　1868 A.
　　1871 A.
Nicholson and Lydekker 1889 A, p. 1454.
Osborn, H. F. 1888 G, p. 260.
　　1893 D.
　　1898 A, p. 146.
　　1899 B, p. 94.
　　1900 C, p. 566.
Owen, R. 1843 A, p. 57.
　　1845 B, p. 412.
　　1866 B, p. 385.
　　1868 A, p. 301.
Parker, W. K. 1868 A, p. 210.
　　1885 A, p. 125.
　　1886 A.
Parsons, F. G. 1899 A.
　　1900 A.
Paulli, S. 1900 B, p. 484.
Pomel, A. 1848 C, p. 246.
Ryder, J. A. 1878 A, p. 49.
Schlosser, M. 1886 C.
　　1887 B, p. 80.
　　1888 A.
　　1890 A.
　　1890 C.
　　1890 D.
Scott, W. B. 1892 A, p. 422.
Slade, D. D. 1888 A, p. 242.
　　1892 A.
Steinmann and Döderlein 1890 A, p. 709.
Tomes, C. S. 1898 A, p. 481.
Tornier, G. 1890 A.
Trouessart, E. L. 1898 A, p. 166.
Weyhe, — 1875 A, p. 104.
Woodward, M. F. 1896 A.
Wortman, J. L. 1886 A, p. 417.
Zittel, K. A. 1898 B, pp. 557, 571, 727.

TALPIDÆ.

Baird, S. F. 1857 A, p. 57.
Bate, C. S. 1867 A, p. 377.　(Genus Talpa.)
Blainville, H. M. D. 1864 A, i, H, pls. i, v, vii, viii,
　　ix.　(Genus Talpa.)
Cope, E. D. 1883 I, p. 80.
Flower, W. H. 1883 D, p. 403.
　　1885 A.
Flower and Lydekker 1891 A, p. 628.
Dobson, G. E. 1883 A, p. 126.
Gill, T. 1872 B, p. 18.
　　. 1875 A.
Kober, J. 1882 A.
　　1884 A.
Leche, W. 1895 A, p. 50.
Mivart, St. G. 1867 B.

Mivart, St. G. 1867 C, i, p. 286; ii, p. 150.
　　1871 A, p. 75.
Moseley and Lankester 1868 A.
Nicholson and Lydekker 1889 A, p. 1458.
Owen, R. 1868 A, p. 303.
Parker, W. K. 1868 A, p. 210, pl. xxvii, figs. 1-17.
　　1885 B, p. 378.
Pomel, A. 1848 C, p. 246.　(Talpina.)
Roger, O. 1896 A, p. 23.
Schlosser, M. 1887 B, p. 125.
Tomes, C. S. 1898 A, p. 491.　(Genus Talpa.)
Trouessart, E. L. 1898 A, p. 208.
True, F. W. 1896 A, p. 18.
Woodward, M. F. 1896 A, p. 575.
Zittel, K. A. 1898 B, p. 563.

Bull. 179——47

SCALOPS [1] Cuvier.   Type *Sorex aquaticus* Linn.

Baird, S. F. 1857 A, p. 58.
Dobson, G. E. 1883 A, p. 159.
Flower and Lydekker 1891 A, p. 630.
Leche, W. 1895 A, p. 50.
Mivart, St. G. 1867 B, p. 281.
    1867 C, i, p. 311; ii, p. 151.
    1871 A, p. 77.

Owen, R. 1868 A, p. 308.
True, F. W. 1896 A, pp. 7, 19.
Zittel, K. A. 1893 B, p. 565.

Scalops sp. indet.

Cope, E. D. 1871 I, p. 95.
    Pleistocene; Pennsylvania.

TALPAVUS Marsh.   Type *T. nitidus* Marsh.

Marsh, O. C. 1872 G, p. 128.
    1877 E.
Nicholson and Lydekker 1889 A, p. 1458.
Schlosser, M. 1887 B, p. 143.
Zittel, K. A. 1893 B, p. 564.

King, C. 1878 A, p. 403.
Matthew, W. D. 1899 A, p. 42.
Roger, O. 1896 A, p. 23.
Trouessart, E. L. 1898 A, p. 208.
    Eocene (Bridger); Wyoming.

Talpavus nitidus Marsh.

Marsh, O. C. 1872 G, p. 128.

SORICIDÆ.

Baird, S. F. 1857 A, p. 7.
Blainville, H. M. D. 1864 A, i, H, pls. ii, v, vii,
    viii, ix, x. (Genus Sorex.)
Brandt, E. 1868 A, p. 76, pls i-vi. (Genus Sorex.)
    1870 A, p. 1. (Genus Sorex.)
Flower, W. H. 1883 D, p. 403.
Flower and Lydekker 1891 A, p. 621.
Gervais, P. 1859 A, p. 54.
Giebel, C. G. 1852 A, p. 222. ("Spitzmäuse.")
Gill, T. 1872 B, p. 18.
    1875 A.
Gill and Coues 1877 A, p. 1058.
Leche, W. 1895 A, p. 47.

Merriam, C. H. 1895 C.
Mivart, St. G. 1867 C, p. 289. (Genus Sorex.)
Nicholson and Lydekker 1889 A, p. 1457.
Owen, R. 1868 A, p. 305.
Parker, W. K. 1868 A, p. 210, pl. xxviii, fig. 1-7.
    1885 B, p. 378.
Peters, W. 1852 A, p. 220. ("Spitzmäuse.")
Pomel, A. 1848 C, p. 248. ("Soriciens.")
Roger, O. 1896 A, p. 25.
Schlosser, M. 1887 B, p. 120.
Wortman, J. L. 1886 A, p. 426.
Zittel, K. A. 1893 B, p. 567.

PROTOSOREX Scott.   Type *P. crassus* Scott.

Scott, W. B. 1894 A, p. 446.

Protosorex crassus Scott.

Scott, W. B. 1894 A, p. 446.

Matthew, W. D. 1899 A, p. 55.
Roger, O. 1896 A, p. 25.
Trouessart, E. L. 1898 A, p. 186.
    Oligocene (White River); South Dakota.

BLARINA Gray.   Type *Sorex talpoides* Gapper = *S. brevicaudus* Say.

Gray, J. E., 1837, Proc. Zool. Soc. Lond., p. 124.
Baird, S. F. 1857 A, p. 36.
Merriam, C. H. 1895 B, pp. 5, 9.
    For synonymy of this genus see Merriam as
quoted.

Blarina simplicidens Cope.

Cope, E. D. 1899 A, p. 219, text figure.
    Pleistocene; Pennsylvania.

LEPTICTIDÆ.

Cope, E. D. 1882 E, pp. 156, 157.
    1884 O, p. 259.
    1884 S, p. 347.
    1886 L, p. 966.
Gill, T. 1872 B, p. 19.
Matthew, W. D. 1899 A, p. 55.
Nicholson and Lydekker 1889 A, p. 1459.

Roger, O. 1896 A, p. 22. (Ictopsidæ.)
Schlosser, M. 1886 C.
    1887 B, p. 140. (Ictopsidæ.)
Steinmann and Döderlein 1890 A, p. 710. (Ictopsidæ.)
Trouessart, E. L. 1898 A, p. 216. (Ictopsidæ.)
Zittel, K. A. 1893 B, p. 561. (Ictopsidæ.)

[1] *Scalops aquaticus* has been reported by Dr. Leidy (1889 H, p. 5) from a cave in Pennsylvania in
which a very large proportion of the animals found belong to species yet living in the vicinity.

ICTOPS Leidy.   Type *I. dakotensis* Leidy.

Leidy, J, 1868 J, p. 316.
Cope, E. D. 1881 D, p. 192.
  1882 E, p. 157.
  1884 O, pp. 259, 265, 801.
  1884 S, pp. 352, 478.
Leidy, J. 1871 C, p. 364.
Schlosser, M. 1887 B, p. 141.
Scott, W. B. 1892 A, p. 422.
Trouessart, E. L. 1898 A, p. 216.
Zittel, K. A. 1893 B, p. 216.

Ictops bullatus Matth.

Matthew, W. D. 1899 A, p. 55.
  Oligocene (White River); South Dakota.

Ictops dakotensis Leidy.

Leidy, J. 1868 J, p. 316.
Cope, E. D. 1884 O, pp. 266, 800.

King, C. 1878 A, p. 412.
Leidy, J. 1869 A, pp. 351, 408, pl. xxvi.
  1871 C, p. 364.
Matthew, W. D 1899 A, p. 55.
Roger, O. 1896 A, p. 22.
Scott, W. B. 1892 A, p. 423.
Trouessart, E. L. 1898 A, p. 216.
  Oligocene (White River); South Dakota.

Ictops didelphoides Cope.

Cope, E. D. 1881 D, p. 192.
  1884 O, p. 268, pl. xxva, fig. 9.
Matthew, W. D. 1899 A, p. 35.   (Palæictops?.)
Roger, O. 1896 A, p. 22.
Schlosser, M. 1887 B, p. 142.
Trouessart, E. L. 1898 A, p. 216.
  Eocene (Wind River); Wyoming.

PALÆICTOPS Matth.   Type *Ictops bicuspis* Cope.

Matthew, W. D. 1899 A, p. 31.

Palæictops bicuspis (Cope).

Cope, E. D. 1880 N, p. 746.   (Stypolophus.)
  1881 D, p. 192.   (Ictops.)
  1882 E, p. 160.   (Ictops.)
  1884 O, p. 266, pl. lviii *f*, figs. 2, 3.   (Ictops.)

Cope, E. D. 1884 S, p. 478, fig. 21.   (Ictops.)
Matthew, W. D. 1899 A, p. 31.
Roger, O. 1896 A, p. 22.   (Ictops.)
Trouessart, E. L. 1898 A, p. 216.   (Ictops.)
Zittel, K. A. 1893 B, p. 561, fig. 470.   (Ictops.)
  Eocene (Wasatch); Wyoming.

LEPTICTIS Leidy.   Type *L. haydeni* Leidy.

Leidy, J. 1868 J, p. 315.
Bruce, A. T. 1883 A, P. 43, fig. 9.
Cope, E. D. 1874 B, p. 472.
  1882 E, p. 157.
  1884 O, pp. 260, 801.
  1884 S, p. 352.
  1886 L, p. 966.
Leidy, J. 1871 C, P. 364.
Schlosser, M. 1887 B, p. 141.
Scott, W. B. 1892 A, p. 422.
Wortman, J. L. 1893 A.
Zittel, K. A. 1893 B, p. 562.

Leptictis haydeni Leidy.

Leidy, J. 1868 J, p. 315.
Cope, E. D. 1884 O, p. 800.
  1884 S, p.478, fig. 22.
King, C. 1878 A, p. 412.
Leidy, J. 1869 A, pp. 345, 408, pl. xxvi, figs. 25–28.
  1871 C, p. 364.
Matthew, W. D. 1899 A, p. 55.
Roger, O. 1896 A, p. 22.
Schlosser, M. 1887 B, p. 141.
Zittel, K. A. 1893 B, p. 562, fig. 471.
  Oligocene (White River); South Dakota.

MESODECTES Cope.   Type *M. caniculus* Cope.

Cope, E. D. 1875 C, p. 30.
  1873 T, p. 3.   (Isacis, preoccupied.)
  1873 CC, P. 5.   (Isacis.)
  1874 B, pp. 465, 470.   (Isacis.)
  1882 E, p. 157.
  1884 O, pp. 259, 788, 801.
  1884 S, p. 352.
Matthew, W. D. 1899 A, p. 55.
Scott, W. B. 1892 A, p. 423.
Zittel, K. A. 1893 B, p. 562.

Mesodectes caniculus Cope.

Cope, E. D. 1873 T, p. 3.   (Isacis.)
  1873 CC, p. 8.   (Isacis.)
  1874 B, p. 473.   (Isacis.)
  1874 P, p. 23.   (Isacis.)
  1884 O, p. 805, pl. lxii, figs. 33–50.
  1884 S, p. 478.
Matthew, W. D. 1899 A, p. 55.
Roger, O. 1896 A, p. 22.
Trouessart, E. L. 1898 A, p. 216.
  Oligocene (White River); Colorado.

GEOLABIS Cope.   Type *G. rhynchæus* Cope.

Cope, E. D. 1884 O, p. 807.
Schlosser, M. 1887 B, p. 142.
Zittel, K. A. 1893 B, p. 562.

Geolabis rhynchæus Cope.

Cope, E. D. 1884 O, p. 808, pl. lxii, figs. 30–32.

Cope, E. D. 1874 B, p. 479, fig. 22.   (Domnina sp.)
Matthew, W. D. 1899 A, p. 55.
Roger, O. 1896 A, p. 22.
Trouessart, E. L. 1898 A, p. 217.
  Oligocene (White River); Colorado.

DIACODON Cope.   Type *D. alticuspis* Cope.

Cope, E. D. 1875 C, p. 11.
    1876 A, p. 3.
    1877 K, p. 133.
    1882 E, p. 157.
    1884 O, pp. xxxv. 260, 713.
    1884 S, p. 349.
Schlosser, M. 1887 B, p. 142.   (Syn. of Centetodon
    Marsh.)
Zittel, K. A. 1893 B, p. 562.

Diacodon alticuspis Cope.

Cope, E. D. 1875 C, p. 12.
    1877 K, p. 132, pl. xlv, fig. 19.
    1884 S, p. 350, fig. 18a.

Matthew, W. D. 1899 A, p. 31.
Roger, O. 1896 A, p. 22.
Trouessart, E. L. 1898 A, p. 217.
    Eocene (Wasatch); New Mexico.

Diacodon celatus Cope.

Cope, E. D. 1875 C, p. 12.,
    1877 K, p. 133, pl. xlv, fig. 20.
    1884 S, p. 350, fig. 18b.
Matthew, W. D. 1899 A, p. 31.
Roger, O. 1896 A, p. 22.
Trouessart, E. L. 1893 B, p. 217.
    Eocene (Wasatch); New Mexico.

CENTETODON Marsh.   Type *C. pulcher* Marsh.

Marsh, O. C. 1872 I, p. 209.
Schlosser, M. 1887 B, p. 144.
Zittel, K. A. 1893 B, p. 562.

Centetodon altidens Marsh.

Marsh, O. C. 1872 I, p. 214.
Matthew, W. D. 1899 A, p. 41.
Roger, O. 1896 A, p. 23.
Trouessart, E. L. 1898 A, p. 217.
    Eocene (Bridger); Wyoming.

Centetodon pulcher Marsh.

Marsh, O. C. 1872 I, p. 209.
King, C. 1878 A, p. 408.
Matthew, W. D. 1899 A, p. 41.
Roger, O. 1896 A, p. 22.
Trouessart, E. L. 1898 A, p. 217.
    Eocene (Bridger); Wyoming.

PASSALACODON Marsh.   Type *P. litoralis* Marsh.

Marsh, O. C. 1872 I, p. 208.
Schlosser, M. 1887 B, p. 143.
Zittel, K. A. 1887 B, p. 562.

Passalacodon litoralis Marsh.

Marsh, O. C. 1872 I, p. 208.

King, C. 1878 A, p. 408.
Matthew, W. D. 1899 A, p. 42.
Roger, O. 1896 A, p. 23.
Trouessart, E. L. 1893 B, p. 217.
    Eocene (Bridger); Wyoming.

ANISACODON Marsh.   Type *A. elegans* Marsh.

Marsh, O. C. 1872 I, p. 209.
Schlosser, M. 1887 B, p. 143.
Zittel, K. A. 1893 B, pp. 304, 308, 562.

Anisacodon elegans Marsh.

Marsh, O. C. 1872 I, p. 209.

Matthew, W. D. 1899 A, p. 42.
Roger, O. 1896 A, p. 23.
Trouessart, E. L. 1898 A, p. 217.
    Eocene (Bridger); Wyoming.

CENTRACODON Marsh.   Type *C. delicatus* Marsh.

Marsh, O. C. 1872 I, p. 215.
Schlosser, M. 1887 B, p. 143.
    A genus of uncertain position.

Centracodon delicatus Marsh.

Marsh, O. C. 1872 I, p. 215.

Matthew, W. D. 1899 A, p. 42.
Roger, O. 1896 A, p. 23.
Trouessart, E. L. 1898 A, p. 218.
    Eocene (Bridger); Wyoming.

ENTOMODON Marsh.   Type *E. comptus* Marsh.

Marsh, O. C. 1872 I, p. 214.
Schlosser, M. 1887 B, p. 143.
Zittel, K. A. 1893 B, p. 562.

Entomodon comptus Marsh.

Marsh, O. C. 1872 I, p. 214.

Matthew, W. D. 1899 A, p. 42.
Roger, O. 1896 A, p. 23.
Trouessart, E. L. 1898 A, p. 218.
    Eocene (Bridger); Wyoming.

ENTOMACODON.   Type *E. minutus* Marsh.

Marsh, O. C. 1872 I, p. 214.
  1877 E.
Schlosser, M. 1887 B, p. 143.
Zittel, K. A. 1893 B, p. 562.

Entomacodon angustidens Marsh.

Marsh, O. C. 1872 I, p. 214.
King, C. 1878 A, p. 408.
Matthew, W. D. 1899 A, p. 42.
Roger, O, 1896 A, p. 23.

Trouessart, E. L. 1898 A, p. 218.
  Eocene (Bridger); Wyoming.

Entomacodon minutus Marsh.

Marsh, O. C. 1872 I, p. 222.
Matthew, W. D. 1899 A, p. 42.
Roger, O. 1896 A, p. 23.
Schlosser, M. 1887 B, p. 143.
Trouessart, E. L. 1898 A, p. 218.
  Eocene (Bridger); Wyoming.

EURYACODON Marsh.   Type *E. lepidus* Marsh.

Marsh, O. C. 1872 I, p. 223.
Schlosser, M. 1887 B, p. 218.

Euryacodon lepidus Marsh.

Marsh, O. C. 1872 I, p. 223.

Matthew, W. D. 1899 A, p. 42.
Roger, O. 1896 A, p. 23.
Trouessart, E. L. 1898 A, p. 218.
  Eocene (Bridger); Wyoming.

ANOMODON Le Conte.   Type *A. snyderi* Le Conte.

Le Conte, J. L. 1848 A, p. 106.
  The position of this genus is doubtful. Roger
  and Trouessart (*loc. cit. infra*) assigned it to
  the Erinaceidæ.

Anomodon snyderi Le Conte.

Le Conte, J. L. 1848 A, p. 106, fig. 3.
Cope, E. D. 1869 E, p. 175.

Leidy, J. 1856 T, p. 171, pl. xvii, figs. 25, 26.
  1862 A.
  1869 A, p. 408.
  1870 H, p. 13.
Pictet, F. J. 1853 A, p. 179.
Roger, O. 1896 A, p. 27.
Trouessart, E. L. 1898 A, p. 178.
  Pleistocene; northern Illinois.

DOMNINA Cope.   Type *D. gradata* Cope.

Cope, E. D. 1873 T, p. 1.
  1873 CC, p. 4 (Domnina); p. 5 (Miothen,
  type *M. crassigenis*).
  1874 B, pp. 465, 469, 470.
  1884 O, pp. 788, 810.
Matthew, W. D. 1899 A, p. 55.
Nicholson and Lydekker 1889 A, p. 1463.
Schlosser, M. 1887 D, p. 78.
Zittel, K. A. 1893 B, p. 562 (Syn.? of Geolabis);
  p. 578.
  The relationships of this genus are doubtful.

Domnina crassigenis Cope.

Cope, E. D. 1873 CC, p. 8. (Miothen.)
  1874 B, p. 470.

Cope, E. D. 1884 O, p. 811, pl. lxii, figs. 24–29.
Matthew, W. D. 1899 A, p. 55.
Roger, O. 1896 A, p. 30.
Trouessart, E. L. 1898 A, p. 106.
  Oligocene (White River); Colorado.

Domnina gradata Cope.

Cope, E. D. 1873 T, p. 1.
  1873 CC, p. 8.
  1874 B, p. 469.
  1884 O, p. 810, pl. lxii, figs. 25, 26.
Matthew, W. D. 1899 A, p. 55.
Roger, O. 1896 A, p. 30.
Trouessart, E. L. 1898 A, p. 106.
  Oligocene (White River); Colorado.

Order CHIROPTERA.

Allen, H. 1880 A.
  1882 A.
  1886 A.
  1898 A.
Blainville, H. M. D. 1864 A, i, G, pls. i–xv.
Cope, E. D. 1873 E, p. 644.
  1880 B, p. 455.
  1884 O, p. 167.
  1888 D, p. 255.
  1889 C, p. 143.
  1889 R, pp. 875, 876.
Dobson, G. E. 1878 A.
Flower, W. H. 1876 B, xiii, p. 514.
  1883 A.
  1883 D, p. 405.
  1885 A.

Flower and Lydekker 1891 A, p. 641.
Gadow, H. 1898 A, p. 52.
Gill, T. 1872 A, p. 302.
  1872 B, pp. 16, 49.
Gill and Coues 1877 A, p. 1056.
Gruber, W. 1859 A.
Haeckel, E. 1895 A, pp. 593, 595.
Hay, O. P. 1899 F, p. 682.
Huxley, T. H. 1870 A, p. 440.
  1872 A, p. 385.
Leboucq, H. 1896 A.
Leche, W. 1877 A.
  1895 A, p. 74.
Lydekker, R. 1896 B.
Marsh, O. C. 1877 E.
Nicholson and Lydekker 1889 A, p. 1459.

Osborn, H. F. 1893 A, p. 387.
1893 D, p. 12.
Owen, R. 1843 A, p. 56.
1845 B, p. 424.
Parker, W. K. 1868 A, p. 214.
1885 A, p. 188.
1886 A.
Roger, O. 1896 A, p. 28.
Ryder, J. A. 1878 A, p. 48.
Schlosser, M. 1887 B, p. 55.

Schlosser, M. 1889 A.
1890 A.
1890 C, p. 269.
Slade, D. D. 1888 A, p. 242.
Tomes, C. S. 1898 A, p. 496.
Trouessart, E. L. 1898 A, p. 77.
Weyhe, —1875 A, p. 102.
Winge, H. 1893 A.
1893 B.
Zittel, K. A. 1893 B, pp. 572, 729.

VESPERTILIONIDÆ.

Allen, H. 1893 A, p. 53.
Blainville, H. M. D. 1864 A, i, G, p. 71.
Flower, W. H. 1883 D, p. 410.
Flower and Lydekker 1891 A, p. 660.
Miller, G. S. 1897 A.
Nicholson and Lydekker 1889 A, p. 1461.

Roger, O. 1896 A, p. 28.
Trouessart, E. L. 1898 A, p. 103.
Winge, H, 1893 A, p. 34.
1893 B, p. 82.
Zittel, K. A. 1893 B, p. 577.

PIPISTRELLUS Kaup.   Type *Vespertilio pipistrellus* Schreber.

Allen, H. 1893 A, p. 121. (Vesperugo.)
Cope, E. D. 1884 O, p. 373. (Vesperugo.)
Dobson, G. E. 1878 A, p. 183. (Vesperugo, part.)
Flower and Lydekker 1891 A, p. 661. (Vesperugo, part.)
Miller, G. S. 1897 A, p. 87. (Pipistrellus.)
Nicholson and Lydekker 1889 A, p. 1461. (Vesperugo.)
Zittel, K. A. 1893 B, p. 578. (Vesperugo.)

Pipistrellus anemophilus (Cope).

Cope, E. D. 1880 N, p. 745. (Vesperugo.
1881 D, p. 184. (Vesperugo.)
1884 O, p. 374. (Vesperugo.)
Matthew, W. D. 1899 A, p. 35. (Vesperugo.)
Roger, O. 1896 A, p. 29. (Vesperugo.)
Trouessart, E. L. 1898 A, p. 115. (Vesperugo.)
Eocene (Wind River); Wyoming.

VESPERTILIO Linn.   Type *V. murinus* Linn.

Linnæus, C., 1858, Syst. Nat., ed. 10, i, p. 31.
Allen, H. 1893 A, p. 111. (Adelonycteris.)
Dobson, G. E. 1878 A, p. 183. (Vesperugo, part.)
Flower and Lydekker 1891 A, p. 661. (Vesperugo, in part.)
Miller, G. S. 1897 A, p. 96.

Dobson, G. E. 1878 A, p. 193. [Vesperugo (Vesperus.)]
Leidy, J. 1889 H, p. 5[1].
Mercer, C. H. 1897 A. (Adelonycteris.)
Miller, G. S. 1897 A, p. 96, fig. 24, b.
Trouessart, E. L. 1898 A, p. 106. [Vesperugo (Vesperus.)]
Recent; Georgia to California and northward.
Pleistocene?; cave in Tennessee.

Vespertilio fuscus Beauvois.

Beauvois, 1796, Cat. Peale's Mus., p. 14.
Allen, H. 1893 A, p. 112. (Adelonycteris.)

NYCTITHERIUM Marsh.   Type *N. velox* Marsh.

Marsh, O. C. 1872 G, p. 127.
1877 E.
Miller, G. S. 1897 A, p. 16.
Nicholson and Lydekker 1889 A, p. 1461. (Syn. of Vesperugo.
Zittel, K. A. 1893 B, p. 578.

Trouessart, E. L. 1898 A, p. 106.
Eocene (Bridger); Wyoming.

Nyctitherium velox Marsh.

Marsh, O. C. 1872 G, p. 127.
King, C. 1878 A, p. 403.
Matthew, W. D. 1899 A, p. 42.
Miller, G. S. 1897 A, p. 37.
Roger, O. 1896 A, p. 30.
Trouessart, E. L. 1898 A, p. 106.
Eocene (Bridger); Wyoming.

Nyctitherium priscum Marsh.

Marsh, O. C. 1872 G, p. 128.
King, C. 1878 A, p. 403.
Matthew, W. D. 1899 A, p. 42.
Miller, G. S. 1897 A, p. 33.

NYCTILESTES Marsh.   Type *N. serotinus* Marsh.

Marsh, O. C. 1872 I, p. 215.
1877 E.
Miller, G. S. 1897 A, p. 16.
Nicholson and Lydekker 1889 A, p. 1461.
Zittel, K. A. 1893 B, p. 578.

King, C. 1878 A, p. 403.
Matthew, W. D. 1899 A, p. 42.
Miller, G. S. 1897 A, p. 35.
Roger, O. 1896 A, p. 30.
Trouessart, E. L. 1898 A, p. 106.
Eocene (Bridger); Wyoming.

Nyctilestes serotinus Marsh.

Marsh, O. C. 1872 I, p. 215.

[1] *V. subulatus* has been reported by Dr. Leidy (loc. cit.) from the same cave.

Order FERÆ Linnæus.

Linnæus, C., 1758, Syst. Nat., ed. 10, i, p. 37.
Unless otherwise indicated, the following
authors apply the name *Carnivora* to this order.
Albrecht, P. 1879 A.
Allen, J. A. 1880 A.
Ameghino, F. 1889 A, p. 295.
Baird, S. F. 1857 A, p. 78.
Baur, G. 1890 B.
Blainville, H. M. D. 1864 A, ii, with 117 plates.
Cope, E. D. 1873 E, p. 646.
    1875 C, p. 25.
    1876 A.
    1879 G.
    1880 B, p. 455.
    1881 J, p. 545.
    1884 E.
    1884 O, pp. 263, 888.
    1887 S, pp. 264, 343.
    1888 D.
    1888 I, p. 6.
    1889 C.
    1889 R, p. 875.
Cuvier, G. 1834 A, v, p. 486; vii.
Filhol, H. 1883 A.
Flower, W. H. 1869 A, p. 269.
    1869 B, p. 4.
    1876 B, xiii, p. 487.
    1888 D, p. 432.
    1885 A.
Flower and Lydekker 1891 A, p. 496.
Gadow, H. 1898 A, p. 50.
Gill, T. 1872 A, p. 297. (Feræ.)
    1872 B, pp. 47, 56. (Feræ.)
Gaudry, A. 1878 B, p. 209.
Gray, J. E. 1869 A, p. 2.
Gruber, W. 1859 A.
Haeckel, E. 1895 S, pp. 578,584.
Hay, O. P. 1899 F, p. 682.
Humphreys, J. 1849 A, p. 142.
Huxley, T. H. 1872 A, p. 350.

Kittl, E. 1887 A.
Kollmann, J. 1888 A.
Lydekker, R. 1885 B, p. 20.
    1896 B.
Marsh, O. C. 1877 E.
Mivart, St. G. 1881 A, p. 473.
Osborn, H. F. 1900 C, p. 568.
    1892 I.
Owen, R. 1845 B, p. 473.
    1851 C, p. 729.
    1866 B, p. 487.   *
    1868 A, p. 327.
Parker, W. K. 1868 A, p. 215.
    1885 A, p. 198.
Parsons, F. G. 1899 A.
    1900 A.
Paulli, S. 1900 B, p. 489.
Scheidt, P. 1894 A.
Schlosser, M. 1886 C.
    1888 B.
    1889 A.
    1890 A.
    1890 C.
    1890 D.
    1892 A, p. 217.
Scott, W. B. 1888 A.
    1892 A, p. 424.
Slade, D. D. 1888 A, p. 243.
Steinmann and Döderlein 1890 A, p. 716.
Tims, H. W. M. 1896 B, p. 461.
Tomes, C. S., 1898 A, p. 459.
Turner, H. N. 1848 A.
Waterhouse, G. R. 1839 A.
Weber, M. 1886 A, p. 227.
Weyhe, — 1875 A, p. 116.
Winge, H. 1895 B. (Carnivora vera.)
Woodward, A. S. 1896 H, p. 380.
Wortman, J. L. 1886 A, p. 448.
Wortman and Matthew 1899 A, p. 109.
Zittel, K. A. 1893 B, pp. 578, 606, 678, 731.

Suborder CREODONTA Cope.

Cope, E. D. 1875 A, p. 3.
Ameghino, F. 1889 A, pp. 285, 298.
    1891 C, p. 214.
Cope, E. D. 1875 C, p. 23. (Bunotheriidæ, in part.)
    1876 E. (Bunotheria, in part.)
    1877 K, p. 72 (Bunotheria, in part); pp. 85, 87 (Creodonta).
    1877 S. (Bunotheria, in part.)
    1880 C.
    1881 S.
    1882 E, p. 156.
    1883 I (Bunotheria, in part); p. 78 (Creodonta).
    1884 E.
    1884 L.
    1884 O, pp. 185, 251, 738 (Creodonta); pp. 167, 185, 737, 800 (Bunotheria, in part).
    1884 S.
    1885 U.
    1886 L.
    1887 M.
    1887 S, p. 343.

Cope, E. D. 1888 D.
    1888 K.
    1888 CC, p. 308.
    1889 C. (Bunotheria, in part; Creodonta.)
    1889 R. (Bunotheria, in part; Creodonta.)
    1891 N, p. 72. (Bunotheria, in part; Creodonta.)
    1898 B, p. 114 (Creodonta); pp. 105, 114 (Bunotheria).
Earle, C. 1898 A, p. 261.
Flower and Lydekker 1891 A, p. 606.
Gadow, H, 1898 A, p. 50.
Hay, O. P. 1899 F, p. 682.
Ihering, H.1891 A, p. 209.
Lemoine, V, 1880 A, p. 586.
Lydekker, R. 1885 E, p. 471.
    1887 A, p. viii. (Bunotheria, in part.)
    1900 A, p. 924.
Matthew, W. D. 1899 A, p. 28.
Nicholson and Lydekker 1889 A, p. 1449.
Osborn, H. F. 1888 F.
    1892 I.

Osborn, H. F. 1898 D.
    1894 A, p. 236.   (Bunotheria, in part.)
    1898 A, p. 146.
    1898 H, p. 174.
    1899 B, p. 98.
    1900 C, p. 567.
Osborn and Wortman 1892 A, p. 108.
Roger, O. 1896 A, p. 30.
Rütimeyer, L. 1888 A.
    1891 A, p. 98.
Schlosser, M. 1886 C.
    1887 A.
    1887 B, p. 162.
    1888 A, p. 584.
    1889 A.
    1890 A.
    1890 C.
    1890 D.

Schlosser, M. 1892 A.
Scott, W. B. 1888 B.
    1890 A.
    1891 B, p. 366.
    1892 A, p. 418.
    1892 B.
    1892 D.
    1895 F, p. 719.
Slade, D. D. 1890 A.
Steinmann and Döderlein 1890 A, p. 711.
Trouessart, E. L. 1898 A, p. 219.
Winge, H. 1895 B, p. 46.   (Carnivora primitiva.)
Woodward, A. S. 1898 B, p. 380.
Wortman, J. L. 1885 A, p. 296.   (Bunotheria, in
    part; Creodonta.)
Wortman and Matthew 1899 A, p. 109.
Zittel, K. A. 1893 B, pp. 579, 605, 725.

OXYCLÆNIDÆ.

Scott, W. B. 1892 B, p. 294.
Earle, C. 1898 A, p. 261.   (Chriacidæ.)
Matthew, W. D. 1897 C, p. 265 (Chriacidæ); p. 267
    (Oxyclænidæ.)
    1899 A, pp. 28, 29.
Osborn and Earle 1895 A, p. 20.

Roger, O. 1896 A, p. 32.
Scott, W. B. 1892 D, p. 77.
Trouessart, E. L. 1898 A, p. 219.   (Includes Chria-
    cidæ.)
Zittel, K. A. 1893 B, p. 584.

OXYCLÆNUS Cope.   Type *Mioclænus cuspidatus* Cope.

Cope, E. D. 1884 K, p. 312.
    1888 CC, p. 320, footnote.
Matthew, W. D. 1897 C, pp. 268, 276.
Scott, W. B. 1892 A, p. 419.
    1892 B, p. 296.
    1892 D, p. 77.
Zittel, K. A. 1893 B, p. 584.

Oxyclænus cuspidatus Cope.

Cope, E. D. 1884 K, p. 312.   (Mioclænus, Oxyclæ-
    nus.)
    1888 CC, p. 321.   (Mioclænus.)
Matthew, W. D. 1897 C, p. 276.
    1899 A, p. 28.
Osborn and Earle 1895 A, p. 23.   (Protochriacus
    simplex?.)

Roger, O. 1896 A, p. 32.
Scott, W. B. 1892 B, p. 296.
Trouessart, E. L. 1898 A, p. 219.
    Eocene (Puerco); New Mexico.

Oxyclænus simplex Cope.

Cope, E. D. 1884 K, pp. 314, 315.   (Chriacus.)
    1888 CC, p. 336.   (Chriacus.)
Matthew, W. D. 1897 C, p. 277, fig. 8.
    1899 A, p. 28.
Roger, O. 1896 A, p. 249.   (Protochriacus.)
Scott, W. B. 1892 B, p. 296.   (Protochriacus.)
    Eocene (Puerco); New Mexico.

CHRIACUS Cope.   Type *Pelycodus pelvidens* Cope.

Cope, E. D. 1883 I, p. 80.
    1884 K, p. 312.
    1884 O, p. 740.
    1884 S, P. 352.
    1888 CC, p. 336.
Earle, C. 1898 A, p. 261.
Matthew, W. D. 1897 C, p. 272.
Osborn and Earle 1895 A, p. 21.
Schlosser, M. 1887 B, p. 49.
Scott, W. B. 1892 A, p. 421, fig. 4.
    1892 B, p. 295 (Chriacus); p. 296 (Epichria-
    cus, type C. *schlosserianus*).
    1892 D, p. 77.
Steinmann and Döderlein 1890 A, p. 712.
Zittel, K. A. 1893 B, p. 585.   (Chriacus, Epichria-
    cus.)

Chriacus angulatus Cope.

Cope, E. D. 1875 C, p. 14.   (Pelycodus.)
    1877 K, p. 144, pl. xxxix, fig. 15.   (Tomi-
    therium.)
    1882 E, p. 151.   (Pelycodus.)
    1883 I, p. 80.   (Chriacus.)
    1884 K, p. 314.
    1884 O, pp. 225, 231, pl. xxivd, fig. 4 (Pelyco-
    dus); p. 740 (Chriacus).
    1884 S, p. 358, fig. 14.
Matthew, W. D. 1899 A, p. 30.   (Pelycodus.)
Roger, O. 1896 A, p. 250.
Scott, W. B. 1892 B, p. 296.
Trouessart, E. L. 1898 A, p. 220.
    Eocene (Wasatch); New Mexico, Wyoming.

Chriacus baldwini Cope.

Cope, E. D. 1882 C, p. 463.   (Deltatherium.)
1884 K, p. 314, in part.   (C. pelvidens.)
1884 O, p. 282, pl. xxiiid, fig. 12 (Deltatherium);   p. 225, in part, pl. xxivd, fig. 3
(Pelycodus pelvidens).
1888 CC, pp. 304, 340, in part (C. baldwini);
p. 304, in part (C. pelvidens).
Matthew, W. D. 1897 C, pp. 208, 274, figs. 5, 6.
1899 A, p. 29.
Osborn and Earle 1895 A, p. 21.
Roger, O. 1896 A, p. 250.
Scott, W. B. 1892 B, p. 296.
Trouessart, E. L. 1898 A, p. 220.
Eocene (Torrejon); New Mexico.

Chriacus pelvidens Cope.

Cope, E. D. 1881 Z, p. 1019.   (Lipodectes.)
1882 C, p. 462.   (Pelycodus.)
1882 E, p. 151.   (Pelycodus.)
1883 I, p. 80.
1884 K, p. 313, part.
1884 O, p. 225, in part (Pelycodus); p. 740
(Chriacus).
1888 CC, pp. 304, 336, in part.
Matthew, W. D. 1897 C, pp. 263, 273, fig. 4.
1899 A, p. 29.
Roger, O. 1896 A, p. 250.
Trouessart, E. L. 1898 A, p. 220.
Eocene (Torrejon); New Mexico.

Chriacus schlosserianus Cope.

Cope, E. D. 1888 CC, p. 338.
Matthew, W. D. 1897 C, pp. 263, 268, 275.
1899 A, p. 29.
Roger, O. 1896 A, p. 249.   (Epichriacus.)
Scott, W. B. 1892 A, p. 421.   (Epichriacus.)
1892 B, p. 296.   (Epichriacus.)
1892 D, p. 77.   (Epichriacus.)
Trouessart, E. L. 1898 A, p. 220.   (Epichriacus.)
Eocene (Torrejon); New Mexico.

Chriacus stenops Cope.

Cope, E. D. 1888 CC, p. 341.
Matthew, W. D., 1899 A, p. 29.   (Syn. of C. pelvidens.)
Roger, O. 1896 A, p. 250.
Scott, W. B. 1892 A, p. 421.
1892 B, p. 296.
Trouessart, E. L. 1898 A, p. 220.
Eocene (Torrejon); New Mexico.

Chriacus truncatus Cope.

Cope, E. D. 1884 K, p. 313.
1888 CC, pp. 304, 336, fig. 8.
Matthew, W. D. 1897 C, pp. 263, 275, fig. 7.
1899 A, p. 29.
Roger, O. 1896 A, p. 250.
Trouessart, E. L. 1898 A, p. 220.
Eocene (Torrejon); New Mexico.

PENTACODON Scott.   Type *Chriacus inversus* Cope.

Scott, W. B. 1892 B, p. 296.
1892 D, p. 77.
Zittel, K. A. 1893 B, p. 585.

Pentacodon inversus (Cope).

Cope, E. D. 1888 CC, p. 342.   (Chriacus.)

Roger, O. 1896 A, p. 250.
Scott, W. B. 1892 B, p. 297.
Trouessart, E. L. 1898 A, p. 220.
Eocene (Puerco); New Mexico.

LOXOLOPHUS Cope.   Type *L. adapinus* Cope=*Chriacus hyattianus* Cope.

Cope, E. D. 1885 J, p. 386.
Hay, O. P. 1899 A, p. 593.
Matthew, W. D. 1897 C, p. 268.   (Protochriacus.)
Osborn and Earle 1895 A, p. 22.
Scott, W. B. 1892 B, p. 296 (Protochriacus, type
P. priscus); p. 297 (Loxolophus).
1892 D, p. 77.   (Protochriacus, Loxolophus.)
Zittel, K. A. 1893 B, p. 585.   (Protochriacus, Loxolophus.)

Loxolophus attenuatus (Osb. and Earle).

Osborn and Earle 1895 A, p. 22.   (Protochriacus.)
Matthew, W. D. 1897 C, pp. 263, 269.   (Protochriacus.)
1899 A, p. 28.
Roger, O. 1896 A, p. 249.   (Protochriacus.)
Trouessart, E. L. 1898 A, p. 220.   (Protochriacus.)
Eocene (Puerco); New Mexico.

Loxolophus hyattianus (Cope).

Cope, E. D. 1885 J, p. 385 (Chriacus hyattianus);
p. 386 (L. adapinus).

Cope, E. D. 1888 CC, pp. 304, 336.   (Chriacus.)
Matthew, W. D. 1897 C, pp. 263, 268, 269, fig. 2.
(Protochriacus.)
1899 A, p. 28.   (Protochriacus hyattianus,
Loxolophus adapinus.)
Roger, O. 1896 A, p. 249 (Protochriacus); p. 250
(Loxolophus).
Scott, W. B. 1892 B, p. 297.
Trouessart, E. L. 1898 A, p. 220.
Eocene (Puerco); New Mexico.

Loxolophus priscus (Cope).

Cope, E. D. 1888 CC, p. 337, text fig. 6.   (Chriacus.)
Matthew, W. D. 1897 C, pp. 263, 269.   (Protochriacus.)
1899 A, p. 28.
Osborn and Earle 1895 A, p. 22.   (Protochriacus.)
Roger, O. 1896 A, p. 249.   (Protochriacus.)
Scott, W. B. 1892 B, p. 296.   (Protochriacus.)
Trouessart, E. L. 1898 A, p. 220.   (Protochriacus.)
Zittel, K. A. 1893 B, p. 585, fig. 489.   (Protochriacus.)
Eocene (Puerco); New Mexico.

TRICENTES Cope. Type *T. crassicollidens* Cope.

Cope, E. D. 1884 K, p. 315.
  1884 O, p. xxxv.
  1884 R, p. 60.
  1884 S, p. 353.
  1885 K, p. 464.
Matthew, W. D. 1897 C, p. 270.
Osborn and Earle 1895 A, p. 23.
Scott, W. B. 1892 B, p. 297.
  1892 D, p. 77.
Zittel, K. A. 1893 B, p. 565.

Tricentes crassicollidens Cope.

Cope, E. D. 1884 K, p. 315.
  1884 R, p. 61.
  1888 CC, p. 304.
Matthew, W. D. 1897 C, pp. 263, 272.
  1899 A, p. 29.
Roger, O. 1896 A, p. 250.
Trouessart, E. L. 1898 A, p. 221.
  Eocene (Torrejon); New Mexico.

Tricentes subtrigonus Cope.

Cope, E. D. 1881 H, p. 491 (Mioclænus subtrigonus); p. 492 (Phenacodus zuniensis).
  1883 M, p. 555. (Mioclænus subtrigonus, M. bucculentus.)

Cope, E. D. 1884 K, pp. 312, 315, 316.
  1884 O, pp. 325, 328, pl. xxiv/, fig. 4; pl. lvii/, fig. u (M. subtrigonus); pp. 325, 341, pl. xxivg, fig. 2 (M. bucculentus): pp. 433, 491, pl. lvii/, fig. 10 (Phenacodus zuniensis).
  1884 R, p. 61. (Tricentes subtrigonus, T. bucculentus.)
  1884 S, p. 349, fig. 17. (M. subtrigonus.)
  1888 CC, pp. 303, 321 (M. subtrigonus); pp. 303, 336, fig. 7 (T. bucculentus); pp. 304, 340 (Chriacus baldwini, in part); pp. 305, 359 (Protogonia zuniensis, in part).
Matthew, W. D. 1897 C, pp. 263, 270, fig. 3; p. 310.
  1899 A, p. 29.
Osborn and Earle 1895 A, p. 24. (Tricentes bucculentus.)
Roger, O. 1896 A, p. 165 (Euprotogonia zuniensis); p. 250 (Tricentes bucculentus, T. subtrigonus).
Schlosser, M. 1887 B, p. 221. (Mioclænus subtrigonus, M. bucculentus.)
Scott, W. B. 1892 B, pp. 297, 299.
Trouessart, E. L. 1898 A, p. 221. (T. bucculentus, T. subtrigonus.)
  Eocene (Torrejon); New Mexico.
The above synonomy is based on Matthew. W. D., 1897 C.

ELLIPSODON Scott. Type *Tricentes inæquidens* (Cope.)

Scott, W. B. 1892 B, p. 298.
  1892 D, p. 77.

Ellipsodon inæquidens (Cope).

Cope, E. D. 1884 K, p. 317. (Tricentes.)
  1888 CC, pp. 304, 336, fig. 8 (Tricentes); p. 335 (Mioclænus minimus).

Matthew, W. D. 1897 C, pp. 263, 315, fig. 17. (Mioclænus.)
Roger, O. 1896 A, p. 250.
Scott, W. B. 1892 B, p. 298.
Trouessart, E. L. 1898 A, p. 221 (Ellipsodon); P. 719 (Mioclænus).
  Eocene (Torrejon); New Mexico.

ARCTOCYONIDÆ.

Cope, E. D. 1877 K, p. 90.
  1880 C, p. 78.
  1882 E, pp. 156, 157.
  1883 I, p. 78.
  1884 O, p. 259.
  1885 K, p. 469.
Flower and Lydekker 1891 A, p. 609.
Gill, T. 1872 B, pp. 59, 68.
Matthew, W. D. 1897 C, p. 288.

Osborn and Wortman 1892 A, p. 115.
Roger, O. 1896 A, p. 32.
Schlosser, M. 1887 B, p. 219.
Scott, W. B. 1892 A, p. 420.
  1892 B, pp. 295, 298.
  1892 D, pp. 77, 78.
Trouessart, E. L. 1898 A, p. 222.
Winge, H. 1895 B, pp. 46, 51.
Zittel, K. A. 1893 B, p. 586.

ARCTOCYON Blainv. Type *A. primævus* Blainv.

Blainville, H. M. D. 1864 A (1841), ii, L, p. 73.
  No American species belonging to *Arctocyon* are known. The following citations are, however, given to the literature of the genus, since it is the type of the family.
Cope, E. D. 1880 C, p. 79.
  1882 E, p. 157.
  1884 O, p. 259.
Gaudry, A, 1878 B, p. 22.
Gervais, P. 1859 A, p. 220.

Lemoine, V. 1878 A.
  1882 B.
  1891 A, p. 272.
Osborn, H. F. 1888 A, p. 32.
  1890 C, p. 59, fig. 4.
Schlosser, M. 1887 B, p. 221.
Trouessart, E. L. 1898 A, p. 222.
Winge, H. 1895 B, pp. 46, 91.
Woodward, A. S. 1898 B, p. 881.
Zittel, K. A. 1893 B, p. 586.

CLÆNODON Scott.   Type *Mioclænus ferox* Cope.

Scott, W. B. 1892 B, p. 298.
Matthew, W. D. 1897 B, p. 232.
 1897 C, p. 288.
 1898 A, p. 880.
Osborn and Earle 1895 A, p. 26.
Scott, W. B. 1892 A, p. 420.
 1892 D, p. 78.
Winge, H. 1895 B, pp. 46, 100, 101.
Zittel, K. A. 1893 B, p. 587.

Clænodon corrugatus (Cope).

Cope, E. D. 1883 M, p. 557. (Mioclænus.)
 1883 W, p. 191. (M. meniscus, no description.)
 1884 K, p. 312 (Mioclænus); p. 313 (Oxyclænus).
 1884 O, pp. 325, 341, pl. xxiv*f*, fig. 5. (Mioclænus corrugatus); pl. xxiv*g*, fig. 8 (M. ferox).
 1884 S, p. 349. (Mioclænus.)
 1887 S, p. 275, pl. vii, fig. 1. (Mioclænus.)
 1888 CC, p. 332. (Mioclænus.)
 1889 C, p. 349. (Mioclænus).
Matthew, W. D. 1897 C, pp. 264, 290, 320.
 1899 A, p. 29.
Roger, O. 1896 A, p. 32.
Schlosser, M. 1887 B, p. 221.
Scott, W. B. 1892 B, p. 299.
Trouessart, E. L. 1898 A, p. 223.
Zittel, K. A. 1893 B, p. 587, fig. 493.
 Eocene (Torrejon); New Mexico.

Clænodon ferox (Cope).

Cope, E. D. 1883 M, p. 547. (Mioclænus.)
 1884 K, p. 312 (Mioclænus): p. 313 (Oxyclænus.)
 1884 O, pp. 325, 328, pl. xxiv*f*, figs. 6, 7, 9–15 (not fig. 8, *fide* Cope). (Mioclænus.)
 1884 S, p. 350. (Mioclænus.)
 1888 CC, pp. 303, 321. (Mioclænus.)
Matthew, W. D. 1897 C, pp. 264, 289, 320.
 1899 A, p. 29.
Osborn and Earle 1895 A, p. 27.
Roger, O. 1896 A, p. 32.
Schlosser, M. 1887 B, p. 221.
Trouessart, E. L. 1898 A, p. 222.
 Eocene (Torrejon); New Mexico.

Clænodon protogonioides (Cope).

Cope, E. D. 1882 EE, p. 833. (Mioclænus.)
 1884 K, p. 312. (Mioclænus.)
 1884 O, pp. 325, 340, pl. xxv, fig. 17. (Mioclænus.)
 1888 CC, p. 329. (Mioclænus.)
Matthew, W. D. 1897 C, pp. 264, 291.
 1899 A, p. 29.
Roger, O. 1896 A, p. 32.
Schlosser, M. 1887 B, p. 221. (Mioclænus.)
Scott, W. B. 1892 A, p. 420, fig. 4.
 1892 B, p. 299.
Trouessart, E. L. 1898 A, p. 223.
 Eocene (Puerco and Torrejon); New Mexico.

ANACODON Cope.   Type *A. ursidens* Cope.

Cope, E. D. 1882 E, pp. 178, 181.
 1882 F, p. 96.
 1884 O, pp. 386, 427.
Osborn, H. F. 1898 D.
Osborn and Earle 1895 A, p. 27.
Osborn and Wortman 1892 A, p. 115.
Pavlow, M. 1887 A.
Scott, W. B. 1892 B, p. 300.
 1892 D, p. 78.
Wortman, J. L. 1882 B, p. 775.
Zittel, K. A. 1898 B, p. 588.

Anacodon ursidens Cope.

Cope, E. D. 1882 E, p. 181.
 1884 O, p. 427, pl. xxv*r*, fig. 11.
 1884 AA, p. 900, fig. 23.
Matthew, W. D. 1899 A, p. 31.
Osborn and Wortman 1892 A, p. 115, fig. 131.
Roger, O. 1896 A, p. 33.
Trouessart, E. L. 1898 A, p. 223.
 Eocene (Wasatch); Wyoming.

TRIISODONTIDÆ.

Scott, W. B. 1892 B, pp. 294, 300.
Matthew, W. D. 1897 C, p. 277.
 1899 A, pp. 28, 29.
Roger, O. 1896 A, p. 33.
Scott, W. B. 1892 A, p. 421.

Scott, W. B. 1892 D, pp. 77, 78.
Trouessart, E. L. 1898 A, p. 223.
Wortman, J. L. 1899 A, p. 146.   (Subfamily of Mesonychidæ.)
Zittel, K. A. 1898 B, p. 588.

TRIISODON Cope.   Type *T. quivirensis* Cope.

Cope, E. D. 1881 S, p. 667.
 1881 H, p. 484.
 1882 E, p. 157.
 1882 H, p. 299.
 1884 K, p. 311.
 1884 O, pp. 260, 270.
 1884 S, p. 349.
 1887 Q.

Matthew, W. D. 1897 C, p. 278.
Scott, W. B. 1892 A, p. 420.
 1892 B, p. 300.
 1892 D, p. 78.
Schlosser, M. 1887 B, p. 209.
Winge, H. 1895 B, pp. 46, 51.
Wortman, J. L. 1899 A, p. 147.
Zittel, K. A. 1898 B, p. 588.

Triisodon gaudrianus (Cope).

Cope, E. D. 1888 CC, pp. 303, 326.   (Mioclænus.)
Matthew, W. D. 1897 C, p. 280.
    1899 A, p. 28.
Roger, O. 1896 A, p. 33.   (T. gaudryanus.)
Scott, W. B. 1892 B, p. 302.   (Goniacodon.)
Trouessart, E. L. 1898 A, p. 224.   (Goniacodon.)
Eocene (Puerco); New Mexico.

Triissodon heilprinianus Cope.

Cope, E. D. 1882 E, p. 198.
    1882 C, p. 468.   (Conoryctes crassicuspis.)
    1884 K, p. 310 (T. rusticus); p. 311 (T. heil-
        prinianus).
    1884 O, p. 201, pl. xxiiid, fig. 6 (Conoryctes
        crassicuspis); pp. 271, 273, pl. xxviiia, fig.
        2 (T. heilprinianus.)
    1885 J, p. 386.   (Sarcothraustes coryphæus.)
    1888 CC, pp. 303, 321 [Mioclænus (Goniaco-
        don) heilprinianus]; pp. 303, 321 (Mioclæ-
        nus bathygnathus, M. rusticus); pp. 303,
        323 (Mioclænus crassicuspis); pp. 304, 343
        (T. biculminatus); p. 323, figs. 1-4 (Sarco-
        thraustes coryphæus).
Matthew, W. D. 1897 C, pp. 263, 279, 320.
    1899 A, p. 28.
Osborn and Earle 1895 A, p. 28, fig. 7 (T. bicul-
    minatus); p. 29 (Sarcothraustes coryphæus, S.
    crassicuspis).

Roger, O. 1896 A, p. 33 (T. heilprinranus, T. bicul-
    minatus, T. rusticus, Sarcothraustus coryphæus,
    S. bathygnathus); p. 34 (S. crassicuspis).
Schlosser, M. 1886 B, p. 37.   (Zetodon crassicus-
    pis.)
Scott, W. B. 1892 B, p. 300 (T. heilprinianus); p. 302
    (Goniacodon rusticus); p. 303 (Sarcothraustes
    crassicuspis, S. bathygnathus).
    1897 A, p. 3.   (Miacis bathygnathus.)
Trouessart, E. L. 1898 A, p. 224 (T. heilprinianus);
    p. 223 (T. biculminatus); p. 224 (Sarcothraustes
    coryphæus); p. 225 (S. bathygnathus; S. crassi-
    cuspis).
Eocene (Puerco); New Mexico.
    The above synonomy is based on Matthew,
    W. D., 1897 C.

Triisodon quivirensis Cope.

Cope, E. D. 1881 S. p. 667.
    1881 G, p. 485.
    1884 K, p. 311.
    1884 O, pp. 271, 272, pl. xxvc, fig. 2.
    1884 S, pp. 257, 350, fig. 1.
    1888 CC, p 304.
Matthew, W. D. 1897 C, p. 279.
    1899 A, p. 28.
Roger, O. 1896 A, p. 33.
Scott, W. B. 1892 B, p. 301.
Trouessart, E. L. 1898 A, p. 223.
Wortman, J. L. 1886 A, p. 423 fig. 205.
Eocene (Puerco); New Mexico.

GONIACODON Cope.   Type G. levisanus Cope.

Cope, E. D. 1888 CC, pp. 320, 321.   (Subgenus.)
Matthew, W. D. 1897 C, pp. 278, 281.
    1899 A, p. 29.
Scott, W. B. 1892 B, p. 301.
    1892 D, p. 78.
Zittel, K. A. 1893 B, p. 589.

Cope, E. D. 1884 O, pp. 271, 273, pl. xxivf, fig. 3.
    (Triisodon.)
    1888 CC, pp. 303, 321.   [Mioclænus (Goniaco-
        don).]
Matthew, W. D. 1897 C, pp. 263, 282.
Roger, O. 1896 A, p. 33.
Scott, W. B. 1892 B, p. 302.
Trouessart, E. L. 1898 A, p. 224.
Eocene (Torrejon); New Mexico.

Goniacodon levisanus (Cope).

Cope, E. D. 1883 M, p. 546.   (Triisodon.)
    1884 K, p. 311.   (Triisodon.)

MICROCLÆNODON Scott.   Type Triisodon assurgens Cope.

Scott, W. B. 1892 B, p. 302.
    1892 D, p. 78.
Zittel, K. A. 1893 B, p. 589.

Cope, E. D. 1888 CC. pp. 304, 321.   [Mioclænus
    (Goniacodon).]
Roger, O. 1896 A, p. 33.
Scott, W. B. 1892 B, p. 302.
Trouessart, E. L. 1898 A, p. 224.
Eocene (Puerco); New Mexico.

Microclænodon assurgens (Cope).

Cope, E. D. 1884 K, p. 311.   (Triisodon.)
    1884 S, p. 350.   (Diacodon.)

SARCOTHRAUSTES Cope.   Type S. antiquus Cope.

Cope, E. D. 1882 E, pp. 156, 193.
    1884 O, pp. 260, 346.
    1884 S, p. 267.
    1888 D, p. 255.
Matthew, W. D. 1897 C, p. 282.
Osborn and Earle 1895 A, p. 28.
Schlosser, M. 1887 B, p. 207.
Scott, W. B. 1892 A, p. 421.
    1892 B, p. 302.
    1892 D, p. 78.

Winge, H. 1895 B, pp. 46, 51.
Zittel, K. A. 1893 B, p. 589.

Sarcothraustes antiquus Cope.

Cope, E. D. 1882 E, p. 193.
    1882 H, p. 207.   (Triisodon conidens.)
    1884 K, p. 311.   (Triisodon conidens.)
    1884 O, pp. 271, 274, pl. xxiiid, figs. 9, 10
        (Triisodon conidens); p. 347, pl. xxivd,
        figs. 19-22 (S. antiquus).

Cope, E. D. 1884 S, p. 267 (S. antiquus); p. 350
(Diacodon conidens).
　　1888 CC, pp. 303, 320. (Mioclænus antiquus,
　　　M. conidens.)
Matthew, W. D. 1897 C, pp. 263, 283.
　　1899 A, p. 29.

Osborn and Earle 1895 A, p. 29.
Roger, O. 1896 A, p. 33.
Scott, W. B. 1893 B, p. 303.
Trouessart, E. L. 1898 A, p. 224.
Wortman, J. L. 1899 A, p. 146.
　　Eocene (Torrejon); New Mexico.

MESONYCHIDÆ.

Cope, E. D. 1876 A, p. 3.
　　1880 C, pp. 78, 79.
　　1882 E, pp. 157, 158.
　　1884 O, pp. 259, 260, 355.
　　1884 S, pp. 260, 261.
　　1888 CC, p. 310.
Matthew, W. D. 1897 C, p. 284.
Nicholson and Lydekker 1889 A, p. 1458.
Osborn, H. F. 1898 D.
Osborn and Wortman 1892 A, p. 112.
Roger, O. 1896 A, p. 34.

Schlosser, M. 1886 C.
Scott, W. B. 1888 B, p. 164.
　　1890 A, p. 472.
　　1892 A, p. 419.
　　1892 B, pp. 294, 303.
　　1892 D, pp. 77, 78.
Steinmann and Döderlein 1890 A, p. 712.
Winge, H. 1895 A, pp. 46, 49, 50. (Mesonychini.)
Wortman, J. L. 1899 A. p. 146. (With subfamilies
　　Mesonychinæ and Triïssodontinæ.)
Zittel, K. A. 1893 B, p. 589.

DISSACUS Cope. Type *D. navajovius* Cope.

Cope, E. D. 1881 Z, p. 1019.
　　1882 E, p. 158.
　　1884 K, p. 324.
　　1884 O, pp. 260, 344, 741.
　　1884 S, p. 267.
　　1888 CC, p. 343.
Earle, C. 1898 A, p. 261.
Lemoine, V. 1891 A, p. 271.
Matthew, W. D. 1897 C, pp. 284, 308.
Nicholson and Lydekker 1889 A. p. 1458.
Osborn, H. F. 1898 B, p. 325.
Osborn and Earle 1895 A, p. 30.
Osborn and Wortman 1892 A, p. 112.
Schlosser, M. 1887 B, p. 209.
Scott, W. B. 1892 A, pp. 418, 420.
　　1892 B, p. 303.
　　1892 D, p. 78.
Winge, H. 1895 B, pp. 46, 49, 91, 100, 101.
Wortman, J. L. 1899 A, p. 147.
Zittel, K. A. 1893 B, p. 589.

Dissacus leptognathus Osb. and Wort.

Osborn and Wortman 1892 A, p. 112 (D. leptog-
nathus); fig. 10 [? Dissacus (Pachyæna) leptog-
nathus].
Matthew, W. D. 1899 A, p. 31.
Roger, O. 1896 A, p. 34.
Scott, W. B. 1892 B, p. 304.
Trouessart, E. L. 1898 B, p. 225.
　　Eocene (Wasatch); Wyoming.

Dissacus navajovius Cope.

Cope, E. D. 1881 H, p. 484. (Mioclænus.)
　　1881 Z, p. 1019.
　　1882 E, p. 165.
　　1882 EE, p. 334. (D. carnifex.)
　　1884 O, p. 345. pl. xxvc, fig. 1 (D. navajovius);
　　　pp. 345, 741, pl. xxivg, figs. 3, 4 (D. carni-
　　　fex).
　　1884 S, p. 267, fig. 11.
　　1888 CC, pp. 304, 344 (D. navajovius); p. 304
　　　(D. carnifex).
Matthew, W. D. 1897 C, pp. 263, 284, 320.
　　1899 A, pp. 29, 31, footnote.
Osborn, H. F. 1898 H, p. 174, fig. 5. (D. carnifex.)
Roger, O. 1896 A, p. 34. (D. navajovius, D. carnifex.)
Trouessart, E. L. 1898 A, p. 225. (D. carnifex, D.
navajovius.)
Wortman, J. L. 1886 A, p. 420, fig. 202.
　　Eocene (Torrejon); New Mexico.

Dissacus saurognathus Wort.

Wortman, J. L. in Matthew, W. D. 1897 C, pp. 263,
285, 320, fig. 9.
Matthew, W. D. 1899 A, p. 29.
Osborn and Earle 1895 A, p. 30, figs. 8, 9. (D. car-
nifex, not of Cope.)
　　Eocene (Torrejon); New Mexico.

PACHYÆNA Cope. Type *P. ossifraga* Cope.

Cope, E. D. 1874 O, p. 13.
　　1877 K, p. 94.
　　1881 Z, p. 1018.
　　1882 E, p. 165.
Earle, C. 1898 A, p. 261.
Matthew, W. D. 1897 C, p. 284.
　　1899 A, p. 31, footnote.
Osborn and Earle 1895 A, p. 38.
Scott, W. B. 1888 B, p. 165.
　　1892 A, p. 420.
　　1892 B, p. 304.

Scott, W. B. 1892 D, p. 79.
Winge, H. 1895 B, pp. 46, 49, 91, 100, 101.
Zittel, K. A. 1898 B, p. 590.

Pachyæna gigantea Osb. and Wort.

Osborn and Wortman 1892 A, p. 113, figs. 11, 12.
Matthew, W. D. 1899 A, p. 31.
Roger, O. 1896 A, p. 34.
Trouessart, E. L. 1898 A, p. 225.
Zittel, K. A. 1893 B, p. 590, fig. 495.
　　Eocene (Wasatch); Wyoming.

Pachyæna intermedia Wort.

Wortman, J. L. 1899 A, p. 147.
Matthew, W. D. 1899 A, p. 31. (No description.)
Eocene (Wasatch); Wyoming.

Pachyæna ossifraga Cope.

Cope, E. D. 1874 O, p. 13.
 1875 C, p. 11.
 1875 F, p. 94.
 1877 K, p. 94, pl. xxxix, fig. 10.
 1881 Z, p. 1018. (Mesonyx.)
 1882 E, p. 165. (Mesonyx.)
 1882 S, p. 334.

Cope, E. D. 1884 O, p. 362, pl. xxive, fig. 14–19; pl.
 xxviiia, fig. 1; pl. xxviiib; pl. xxviiic; pl.
 xxviiid. (Mesonyx.)
 1884 S, p. 264, figs. 2, 6, 8, 10. (Mesonyx.)
King, C. 1878 A, p. 376.
Matthew, W. D. 1899 A, p. 31.
Osborn and Wortman 1892 A, p. 112.
Roger, O. 1896 A, p. 34.
Schlosser, M. 1887 B, p. 206.
Steinmann and Döderlein 1890 A, p. 714.
Trouessart, E. L. 1898 A, p. 225.
Wortman, J. L. 1886 A, p. 418, figs. 200,201. (Mesonyx.)
 1899 A, p. 147.
Eocene (Wasatch); New Mexico, Wyoming.

MESONYX Cope. Type *M. obtusidens* Cope.

Cope, E. D. 1872 M, p. 460.
Allen, H. 1886 B.
Cope, E. D. 1872 R, p. 483. (Synoplotherium, type
 S. lanius.)
 1872 HH.
 1873 E, p. 550 (Mesonyx); p. 554 (Synoplotherium).
 1873 G, p. 198 (Mesonyx); p. 203 (Synoplotherium).
 1875 C, p. 5 (Mesonyx); p. 9 (Synoplotherium).
 1875 L, p. 21.
 1876 A, p. 1 (Synoplotherium); p. 3 (Mesonyx).
 1880 C, p. 79.
 1881 Z.
 1882 E, p. 158.
 1882 S, p. 334.
 1884 O, pp. 260, 348.
 1884 S, p. 263.
 1887 S, pp. 251, 278.
Earle, C. 1898 A, p. 261.
Flower, W. H. 1876 B, xiii, p. 514.
Flower and Lydekker 1891 A, p. 609.
Major, Fors. 1893 A, p. 200.
Marsh, O. C. 1876 H, p. 403. (Dromocyon, type D.
 vorax.)
 1877 E. (Dromocyon, Mesonyx.)
Matthew, W. D. 1897 C, p. 284.
Nicholson and Lydekker 1889 A, p. 1458.
Osborn, H. F. 1893 D, p. 6.
Osborn and Earle 1895 A, p. 37.
Schlosser, M. 1887 B, p. 206 (Synoplotherium); p.
 207 (Mesonyx).
 1890 B, p. 64.
Scott, W. B. 1888 B, p. 164 (Mesonyx); p. 165 (Syno-
 plotherium).
 1890 A, p. 472.
 1892 A, p. 420.
 1892 B, p. 306.
 1892 D, p 79.
Steinmann and Döderlein 1890 A, p. 713.
Wilckens, M. 1885 D, p. 523. (Dromocyon.)
Winge, H. 1896 B, pp. 46, 49, 91.
Woodward, A. S. 1898 B, p. 382.
Wortman, J. L. 1886 A, p. 417.
Zittel, K. A. 1893 B, p. 590.

Mesonyx? dakotensis Scott.

Scott, W. B. 1892 B, p. 306.
Trouessart, E. L. 1898 A, p. 226.
Oligocene (White River); South Dakota.

Mesonyx lanius Cope.

Cope, E. D. 1872 R, p. 483. (Synoplotherium.)
 1873 E, p. 557, pls. v, vi. (Synoplotherium.)
 1873 G, p. 207. (Synoplotherium.)
 1877 K, p. 94. (Synoplotherium.)
 1884 O, pp. 264, 348, 355, pl. xxvii, figs. 25–28;
 pl. xxviii; pl. xxix, figs. 1–6; pl. xixa,
 fig. 1.
 1884 S, p. 265, fig. 7.
Matthew, W. D. 1899 A, p. 41.
Osborn and Wortman 1892 A, p. 113.
Roger, O. 1896 A, p. 34.
Trouessart, E. L. 1898 A, p. 226.
Eocene (Bridger); Wyoming.

Mesonyx obtusidens Cope.

Cope, E. D. 1872 M, p. 460.
 1873 E, p. 552.
 1873 G, p. 201.
 1884 O, pp. 348, 355, pl. xxvi, figs. 3–12; pl.
 xxvii, figs. 1–24.
 1884 S, p. 264.
 1887 M.
Matthew, W. D. 1899 A, pp. 41, 49.
Osborn, H. F. 1896 C, p. 107, illus. 3.
 1898 Q, p. 18.
Roger, O. 1896 A, p. 34.
Scott, W. B. 1888 B, p. 155, pl. v; pl. vi, fig. 1;
 pl. vii, figs. 1–3.
 1892 B, p. 307.
Trouessart, E. L. 1898 A, p. 226.
Eocene (Bridger); Wyoming: (Uinta); Utah.

Mesonyx uintensis Scott.

Scott, W. B. 1888 B (1887), p. 168. (M.? uintensis.)
Cope, E. D. 1887 M, p. 927. (M. uintaensis.)
Matthew, W. D. 1899 A, p. 49.
Osborn, H. F. 1895 A, p. 79, fig. 4.
Roger, O. 1896 A, p. 34.
Scott, W. B. 1890 A, p. 471, pl. x, fig. 9.
Scott and Osborn 1887 A, p. 255. (Name only.)

Trouessart, E. L. 1898 A, p. 226.
Eocene (Uinta): Utah.

Mesonyx vorax (Marsh).

Marsh, O. C. 1876 H, p. 408. (Dromocyon.)
King, C. 1878 A, p. 408.
Matthew, W. D. 1899 A, p. 41.

Schlosser, M. 1890 B, p. 63. (Dromocyon.)
Scott, W. B. 1892 B, p. 306. (Syn. of Mesonyx obtusidens.)
Trouessart, E. L. 1898 A, p. 228. (Syn. of M. obtusidens.)
Eocene (Bridger); Wyoming.

PROVIVERRIDÆ.

Schlosser, M. 1886 C, p. 298.
Flower and Lydekker 1891 A, p. 608.
Lydekker, R. 1887 A, p. 307.
Matthew, W. D. 1899 A, pp. 31, 35, 40.
Nicholson and Lydekker 1889 A, p. 1452.
Roger, O. 1896 A, p. 34.
Scott, W. B. 1892 B, pp. 294, 307.

Scott, W. B. 1892 D, pp. 77, 79.
1894 F, p. 501.
Steinmann and Döderlein 1890 A, p. 712.
Trouessart, E. L. 1898 A, p. 226.
Winge, H. 1895 B, pp. 46, 48. (Proviverrini.)
Zittel, K. A. 1893 B, p. 592.

DELTATHERIUM Cope.   Type *D. fundaminis* Cope.

Cope, E. D. 1881 N, p. 337.
1881 H, p. 486.
1881 Z, p. 1019. (Lipodectes, type *L. penetrans* Cope.)
1882 E, p. 157. (Deltatherium, Lipodectes.)
1884 O, pp. 260, 277.
1884 S, p. 352.
1888 D, p. 255.
Lydekker, R. 1887 A, p. 309.
Nicholson and Lydekker 1889 A, p. 1453.
Osborn and Earle 1895 A, p. 39.
Schlosser, M. 1887 A, p. 219.
1888 B, pp. 16, 23.
Scott, W. B. 1892 A, p. 418.
· 1892 B, p. 308.
1892 D, p. 79.
Trouessart, E. L. 1898 A, p. 227.
Winge, H. 1895 B, pp. 46, 49, 91, 101.
Zittel, K. A. 1893 B, p. 592.

Deltatherium fundaminis Cope.

Cope, E. D. 1881 N, p. 337.
1881 H, p. 486.
1881 Z, p. 1019. (Lipodectes penetrans.)
1882 C, p. 463.

Cope, E. D. 1882 X, p. 522.
1884 O, p. 278, pl. xxiiie, figs. 8–11; pl. xxva, fig. 10; pl. xxvd, fig. 8.
1884 S, p. 352, fig. 20.
1888 D.
1888 CC, p. 304.
1889 C, p. 229, fig. 54.
Lydekker, R. 1887 A, p. 309.
Matthew, W. D. 1899 A, p. 29.
Osborn and Earle 1895 A, p. 40, figs. 10, 11.
Roger, O. 1896 A, p. 35.
Trouessart, E. L. 1898 A, p. 227.
Zittel, K. A. 1893 A, p. 598, fig. 497.
Eocene (Puerco); New Mexico.

Deltatherium? interruptum Cope.

Cope, E. D. 1882 C, p. 463.
1884 O, p. 282, pl. xxiiid, fig. 13.
1888 CC, p. 303, 328. (Mioclænus.)
Matthew, W. D. 1899 A, p. 28. (Mioclænus.)
Roger, O. 1896 A, p. 164.
Scott, W. B. 1892 D, p. 323. (Genus not determined.)
Trouessart, E. L. 1898 A, p. 720.
Eocene (Puerco); New Mexico.

STYPOLOPHUS Cope.   Type *S. pungens* Cope.

Cope, E. D. 1872 N, p. 466.
1872 E, p. 469. (Triacodon, not of Marsh.)
1873 E, p. 559 (Stypolophus); pp. 560, 611 (Triacodon.)
1874 O, p. 13. (Prototomus, type *P. viverrinus.*)
1875 C, pp. 5, 9. (Prototomus.)
1875 L, p. 21. (Stypolophus.)
1876 A. (Stypolophus.)
1877 K, pp. 87, 109, 112. (Stypolophus.)
1879 A, p. 43. (Stypolophus.)
1880 C, p. 79. (Stypolophus.)
1882 E, p. 158 (Stypolophus); p. 164 (Prototomus.)
1884 E. (Stypolophus.)
1884 O, pp. 35, 260, 285 (Stypolophus); p. 35 (Prototomus.)

Cope, E. D. 1884 S, p. 350. (Stypolophus.)
1887 S, p. 251. (Stypolophus.)
1888 D. (Stypolophus.)
1889 C, p. 232. (Stypolophus.)
Leidy, J. 1871 F, p. 115. (Sinopa, type *S. rapax;* description insufficient.)
1872 B, p. 365.[1] (Sinopa.)
1873 B, p. 116. (Sinopa.)
Marsh, O. C. 1872 G, p. 126. (Limnocyon, type *L. verus.*)
1877 E.
Matthew, W. D. 1899 A, p. 31. (Sinopa.)
Nicholson and Lydekker, 1889 A, p. 1437. (Stypolophus=Proviverra; Limnocyon.)
Osborn, Scott, and Speir 1878 A, p. 18. (Sinopa.)
Schlosser, M. 1887 B, p. 218. (Stypolophus.)
1888 B, p. 16. (Stypolophus.)

[1] In case this paper was issued before Cope 1872 N ("Aug. 3, 1872") *Sinopa* will replace *Stypolophus.*

Schlosser, M. 1890 C, p. 244.   (Stypolophus.)
Scott, W. B. 1892 A, p. 419.   (Sinopa.)
    1892 B, p. 306.   (Sinopa.)
    1892 D, p. 79.   (Sinopa.)
Trouessart, E. L. 1898 A, p. 227.
Wilckens, M. 1885 D, p. 524.   (Sinopa.)
Winge, H. 1895 B, pp. 46, 47, 52, 91, 100, 101.
Woodward, A. S. 1898 B, p. 383.
Zittel, K. A. 1893 B, pp. 592.   (Sinopa.)

Stypolophus aculeatus Cope.

Cope, E. D. 1872 M, p. 460.   (Triacodon.)
    1872 E, p. 469.   (Triacodon.)
    1873 E, pp. 560, 611.   (Triacodon.)
    1873 U, p. 3.   (Triacodon.)
    1877 K, p. 112.
    1882 E, p. 161.
    1884 O, pp. 290, 299, pl. xxiv, figs. 6, 7; p. xxvii, figs. 1, 2.
Matthew, W. D. 1899 A, p. 31.   (Sinopa.)
Roger, O. 1896 A, p. 35.
Scott, W. B. 1892 B, p. 310.   (Sinopa.)
Trouessart, E. L. 1898 A, p. 227.   (Sinopa.)
    Eocene (Wasatch); New Mexico.

Stypolophus agilis (Marsh).

Marsh, O. C. 1872 I, p. 204.   (Limnocyon.)
Matthew, W. D. 1899 A, p. 40.   (Sinopa.)
Roger, O. 1896 A, p. 35 (S. agilis); p. 41 (Miacis? agilis).
Schlosser, M, 1890 B, p. 63.   (Limnocyon.)
Scott, W. B. 1892 B, p. 310.   (Sinopa.)
Trouessart, E. L. 1898 A, p. 227 (Sinopa); p. 236 (Miacis?).
    Eocene (Bridger); Wyoming.

Stypolophus brevicalcaratus Cope.

Cope, E. D. 1872 E, p. 469.   (S. brevicolcarabus, error.)
    1873 E, p. 560.
    1884 O, pp. 290, 291, pl. xxiv, fig. 9.
Matthew, W. D. 1899 A, p. 40.   (Syn. of Sinopa agilis.)
Roger, O, 1896 A, p. 35.
Scott, W. B. 1892 B, p. 310.   (Sinopa.)
rouessart, E. L. 1898 A, p. 227.   (Sinopa.)
    Eocene (Bridger), Wyoming.

Stypolophus hians Cope.

Cope, E. D. 1877 K, pp. 72, 118, pl. xxxviii, figs. 12-20.
    1876 A, p. 1.   (No description.)
    1884 O, p. 290.   (Prototomus.)
Matthew, W. D. 1899 A, p. 31.   (Sinopa.)
Roger, O. 1896 A, p. 36.
Scott, W. B. 1892 B, p. 310.   (Sinopa.)
Trouessart, E. L. 1898 A, p. 228.   (Sinopa.)
    Eocene (Wasatch); New Mexico, Wyoming.

Stypolophus insectivorus Cope.

Cope, E D. 1872 E, p. 469.
    1873 E, p. 559.
    1884 O, p, 290, pl. xxiv, figs. 10, 11.
Matthew, W. D. 1899 A, p. 40.
Roger, O. 1896 A, p 35
Scott, W. B. 1892 B, p. 310   (Sinopa.)
Trouessart, E. L. 1898 A, p. 227   (Sinopa.)
    Eocene (Bridger), Wyoming.

Stypolophus multicuspis Cope.

Cope, E. D. 1875 C, p. 10.   (Prototomus.)
    1877 K, pp. 112, 116, pl. xxxix, figs. 12-14.
    1884 O, p. 290.
Matthew, W. D. 1899 A, p. 31.   (Sinopa.)
Roger, O. 1896 A, p. 36.
Scott, W. B. 1892 B, p. 310.   (Sinopa.)
Trouessart, E. L. 1898 A, p, 227.   (Sinopa.)
    Eocene (Wasatch); New Mexico.

Stypolophus pungens Cope.

Cope, E. D. 1872 N, p. 466.
    1873 E, p. 559.
    1884 O, pp. 290, 291, pl. xxiv, fig. 8.
Matthew, W. D. 1899 A, p. 40.   (Sinopa.)
Roger, O. 1896 A, p. 35.
Scott, W. B. 1892 B, p. 310.   (Sinopa.)
Trouessart, E. L. 1898 A, p. 227.   (Sinopa.)
    Eocene (Bridger); Wyoming.

Stypolophus rapax (Leidy).

Leidy, J. 1872 B, p. 356.   (Sinopa.)
King, C. 1878 A, p. 403.   (Sinopa.)
Leidy, J. 1871 F, p. 115.   (Sinopa; insufficient description.)
    1873 B, pp. 116, 118, 316, pl. vi, fig. 44.   (Sinopa.)
Matthew, W. D. 1899 A, p. 40.   (Sinopa.)
Roger, O. 1896 A, p. 35.
Schlosser, M. 1887 B, p. 35.   (Sinopa.)
Scott, W. B. 1892 B, p. 311.   (Sinopa.)
Trouessart, E. L. 1898 A, p. 227.   (Sinopa.)
    Eocene (Bridger); Wyoming.

Stypolophus secundarius Cope.

Cope, E. D. 1875 C, p. 9.   (Prototomus.)
    1877 K, p. 115.
    1884 O, p. 290.
Matthew, W. D. 1899 A, p. 31.   (Sinopa.)
Roger, O. 1896 A, p. 35.
Scott, W. B. 1892 B, p. 310.   (Sinopa.)
Trouessart, E. L. 1898 A, p. 227.   (Sinopa.)
    Eocene (Wasatch); New Mexico.

Stypolophus strenuus Cope.

Cope, E. D. 1875 C, p. 10.   (Prototomus.)
    1877 K, p. 117, pl. xxxix, fig. 11.
    1880 N, p. 746.
    1882 E, p. 161.   (Syn. of S. whitese.)
    1884 O, p. 290.
Matthew, W. D. 1899 A, p. 31.   (Sinopa.)
Roger, O. 1896 A, p. 36.
Scott, W. B. 1892 B, p. 310.   (Sinopa.)
Trouessart, E. L. 1898 A, p. 228.   (Sinopa.)
    Eocene (Wasatch), New Mexico, Wyoming.

Stypolophus verus (Marsh).

Marsh, O. C. 1872 G, p. 126.   (Limnocyon.)
Matthew, W. D. 1899 A, p. 40.   (Sinopa.)
Roger, O. 1896 A, p. 35 (S. verus); p. 4i (Miacis vetus).
Schlosser, M. 1890 B, pp. 62, 64.   (Limnocyon vetus.)
Scott, W. B. 1892 B, p. 310.   (Sinopa.)
Trouessart, E. L. 1898 A, p. 227 (Sinopa); p. 236 (Miacis? vetus).
    Eocene (Bridger); Wyoming.

Stypolophus viverrinus Cope.

Cope, E. D. 1874 O, p. 13.   (Prototomus.)
    1875 C, p. 9.   (Prototomus.)
    1877 K, p. 112, pl. xxxviii, figs. 1–11.
    1884 O, p. 290.   (Prototomus.)
Matthew, W. D. 1899 A, pp. 31, 35.   (Sinopa.)
Roger, O. 1896 A, p. 35.
Scott, W. B. 1892 B, p. 310.   (Sinopa.)
Trouessart, E. L. 1897 A, p. 227.   (Sinopa.)
    Eocene (Wasatch, Wind River); New Mexico,
    Wyoming.

Stypolophus whitese Cope.

Cope, E. D. 1882 E, p. 161.

Cope, E. D. 1881 D, p. 192.   (S. strenuus, not of
    Cope, 1875.)
    1884 O, pp. 290, 292, pl. xxvb, figs. 8–14; pl.
    xxvd, figs. 1, 2.
    1884 S, p. 347, fig. 13.
    1889 C, p. 232, fig. 56.
Lydekker, R. 1887 A, p. 306, fig. 46.   (Proviverra.)
Matthew, W. D. 1899 A, pp. 31, 35.   (Sinopa.)
Roger, O. 1896 A, p. 36.
Scott, W. B. 1892 B, p. 310.   (Sinopa.)
Trouessart, E. L. 1897 A, p. 228.   (Sinopa.)
Woodward, A. S. 1898 B, p. 383, fig. 214.
Wortman, J. L. 1886 A, p. 422, fig. 204.
    Eocene (Wasatch and Wind River); Wyo-
    ming.

TRIACODON Marsh.   Type *T. fallax* Marsh.

Marsh, O. C. 1871 E, p. 123.   (Not *Triacodon* of
    Cope.)
Schlosser, M. 1887 B, p. 144.
Zittel, K. A. 1893 B, p. 603.
    The systematic position of this genus is un-
    certain.

Triacodon fallax Marsh.

Marsh, O. C. 1871 E, p. 123.
Cope, E. D. 1873 E, p. 612.
Leidy, J. 1872 B, p. 357.
    1873 B, p. 123.
Matthew, W. D. 1899 A, p. 41.
Schlosser, M. 1887 B, p. 144.
    Eocene (Bridger); Wyoming.

Triacodon grandis Marsh.

Marsh, O. C. 1872 I, p. 222.
Cope, E. D. 1873 U, p. 3.
Matthew, W. D. 1899 A, p. 41.
Schlosser, M. 1887 B, p. 144.
    Eocene (Bridger); Wyoming.

Triacodon nanus Marsh.

Marsh, O. C. 1872 I, p. 223.
Matthew, W. D. 1899 A, p. 41.
Schlosser, M. 1887 B, p. 144.
    Eocene (Bridger); Wyoming.

PROVIVERRA Rütimeyer.   Type *P. typica* Rütimeyer.

Rütimeyer, L., 1862, Schweiz. Gesell. Naturwiss.,
    p. 80.
Cope, E. D. 1880 C, p. 79.
    1882 E, p. 158.
    1884 O, p. 260.
Flower and Lydekker 1891 A, p. 608.
Gaudry, A, 1878 B, p. 20.
Lydekker, R. 1887 A, p. 308.
Nicholson and Lydekker 1889 A, p. 1452.
Schlosser, M. 1887 B, p. 213.
Scott, W. B. 1892 A, p. 419.
    1892 B, p. 311.

Winge, H. 1895 B, p. 91.
Zittel, K. A. 1893 B, p. 593.

Proviverra americana Scott.

Scott, W. B. 1892 B, p. 311.
Matthew, W. D. 1899 A, p. 40.
Osborn, Scott, and Speir 1878 A, p. 18.   (Sinopa
    rapax, not of Leidy.)
Roger, O. 1896 A, p. 36.
Trouessart, E. L. 1897 A, p. 228.
    Eocene (Bridger); Wyoming.

DIDELPHODUS Cope.   Type *Deltatherium absarokæ* Cope.

Cope, E. D. 1882 X, p. 522.
    1884 O, pp. 260, 283.
    1884 S, p. 351.
    1888 D, p. 255.
Nicholson and Lydekker 1889 A, p. 1458.
Schlosser, M. 1887 B, p. 214.
    1888 B, pp. 16, 23.
    1890 C, p. 244.
Scott, W. B. 1892 B, p. 311.
    1892 D, p. 79.
Winge, H. 1895 B, pp. 46, 49, 91, 101.
Zittel, K. A. 1893 B, p. 595.

Didelphodus absarokæ Cope.

Cope, E. D. 1881 S, p. 669.   (Deltatherium.)
    1882 E, p. 161.   (Deltatherium.)
    1884 X, p. 522.
    1884 O, p. 284, pl. xxive, fig. 13.
    1884 S, p. 351, fig. 19.
Matthew, W. D. 1899 A, p. 31.
Roger, O. 1896 A, P. 36.
Scott, W. B. 1892 B, p. 311.
Trouessart, E. L. 1897 A, p. 229.
    Eocene (Wasatch); Wyoming.

PALÆOSINOPA Matth.   Type *P. veterrima* Matth.

Matthew, W. D. 1899 A, p. 31.   (Palæosinopa
"Wortman"; no description.)
For description of this genus and its type species
see Bull. Amer. Mus. Nat. Hist., xiv, 1901, p. 22.

Palæosinopa veterrima Matth.

Matthew, W. D. 1899 A, p. 31.   (P. veterrima
"Wortman": no description.)
Eocene (Wasatch); Wyoming.

VIVERRAVIDÆ.

Wortman and Matthew 1899 A, p. 136.

VIVERRAVUS Marsh.   Type *V. gracilis* Marsh.

Marsh, O. C. 1872 G, p. 127.
Cope, E. D. 1873 E, p. 560.
   1875 C, pp. 5, 11.   (Didymictis, type *D. pro-
    tenus.*)
   1876 A.   (Didymictis.)
   1877 K, pp. 87, 123.   (Didymictis.)
   1880 C, p. 79.   (Didymictis.)
   1882 C, p. 470.   (Didymictis.)
   1882 E, pp. 157, 158.   (Didymictis.)
   1884 O, pp. 260, 304.   (Didymictis.)
   1884 S, p. 484.   (Didymictis.)
   1887 M, p. 92⁷.   (Didymictis.)
   1888 K, p. 162.   (Didymictis.)
Matthew, W. D. 1897 C, p. 286.
   1899 A, pp. 30, 35.
Nicholson and Lydekker 1889 A, p. 1437.   (Didymictis, Viverravus.)
Osborn, H. F. 1888 F, p. 1076.   (Didymictis.)
Osborn and Wortman 1892 A, p. 111.   (Didymictis.)
Schlosser, M. 1890 B, pp. 1, 61.   (Didymictis.)
   1899 A, p. 86.   (Didymictis.)
Scott, W. B. 1890 A, p. 473.   (Didymictis.)
   1890 C, p. 265.   (Didymictis.)
   1892 A, p. 419.   (Didymictis.)
   1892 B, p. 318 (Didymictis); p. 321 (Viverra-
    vus).
   1892 D, p. 80.   (Didymictis.)
Steinmann and Döderlein 1890 A, p. 716.   (Didy-
   mictis.)
Trouessart, E. L. 1897 A, p. 235.
Winge, H. 1895 B, pp. 46, 52, 53, 91.
Wortman, J. L. 1886 A, p. 430.
Zittel, K. A. 1893 B, p. 602 (Didymictis); p. 603
   (Viverravus).

Viverravus altidens (Cope).

Cope, E. D. 1880 N, p. 746.   (Didymictis.)
   1881 D, p. 190.   (Didymictis.)
   1882 E, p. 159.   (Didymictis.)
   1884 O, pp. 305, 307, pl. xxva, figs. 13, 14.
    (Didymictis.)
Earle C. 1898 A, p. 261.   (Didymictis.)
Matthew, W. D. 1897 C, p. 287.   (Didymictis.)
   1899 A, p. 35.   (Viverravus.)
Osborn, H. F. 1897 G, p. 255.   (Didymictis.)
Roger, O. 1896 A, p. 40   (Didymictis.)
Scott, W. B. 1888 B, p. 169.   (Didymictis.)
   1892 B, p. 320.   (Didymictis.)
Trouessart, E. L. 1897 A, p. 235.   (Didymictis.)
   Eocene (Wasatch, Wind River, Bridger); Wy-
   oming.

Viverravus curtidens (Cope).

Cope, E. D. 1882 E, p. 160.   (Didymictis.)
   1884 O, pp. 306, 313, pl. xxivd, fig. 10.   (Didy-
    mictis.)
   1884 S, p. 484.   (Didymictis curtus, error? for
    D. curtidens.)
Matthew, W. D. 1897 C, p. 287.   (Didymictis.)
   1899 A, p. 30.   (Viverravus.)
Osborn and Wortman 1892 A, pp. 83, 111.   (Didy-
   mictis.)
Roger, O. 1896 A, p. 40.   (Didymictis.)
Scott, W. B. 1892 B, P. 320.   (Didymictis.)
Trouessart, E. L. 1897 A, p. 235.   (Didymictis.)
   Eocene (Wasatch); Wyoming.

Viverravus gracilis Marsh.

Marsh, O. C. 1872 G, p. 127.
Cope, E. D. 1881 D, p. 191.   (Didymictis dawkinsianus.)
   1882 E, pp. 158, 159.   (Didymictis dawkins-
    ianus.)
   1884 O, pp. 306, 310, pl. xxva, fig 11.   (Didy-
    mictis dawkinsianus.)
   1884 S, p. 485, fig. 30.   (Didymictis dawkins-
    ianus.)
Matthew, W. D. 1897 C, pp. 264, 286.   (Didymictis
   dawkinsianus.)
   1899 A, pp. 35, 40.
Osborn, H. F. 1897 G, p. 255.   (Didymictis dawkinsianus.)
Osborn and Wortman 1892 A, pp. 83, 111.   (Didymictis dawkinsianus.)
Roger, O. 1896 A, p. 40 (Didymictis dawkinsianus);
   p. 41 (V. gracilis).
Schlosser, M. 1890 B, p. 63.
Scott, W. B. 1892 B, p. 320 (Didymictis dawkinsianus); p. 321 (V. gracilis).
Trouessart, E. L. 1897 A, p. 235 (Didymictis daw-
   kinsianus); p. 236 (V. gracilis).
Wortman and Matthew 1899 A, p. 136.
Zittel, K. A. 1893 B, p. 603, fig. 506 B.   (Didymictis
   dawkinsianus.)
   Eocene (Wasatch, Wind River, Bridger), Wy-
   oming, Colorado.

Viverravus haydenianus (Cope).

Cope, E. D. 1882 C, p. 464.   (Didymictis.)
   1884 K, 309.   (Didymictis primus.)
   1884 O, pp. 305, 306, pl. xxiiic, figs. 12, 13.
    (Didymictis)
   1884 S, p. 484.   (Didymictus.)

Cope, E. D. 1888 CC, p. 304.   (D. haydenianus, D.
    primus.)
Matthew, W. D. 1897 C, pp. 286, 287.   (Didymictis.)
    1899 A, p. 29.   (Viverravus.)
Roger, O. 1896 A, p. 40.   (Didymictus haydeni-
    anus, D. primus.)
Scott, W. B. 1892 B, p. 320.   (Didymictus.)
Trouessart, E. L. 1897 A, p. 234 (D. haydenianus);
    p. 235 (D. primus).
Zittel, K. A. 1893 B, p. 608, fig. 506 A.   (Didymic-
    tis.)
    Eocene (Torrejon); New Mexico.

Viverravus leptomylus (Cope).

Cope, E. D. 1880 O, p. 908.   (Didymictis.)
    1881 D, p. 191.   (Didymictis.)
    1882 E, p. 159.   (Didymictus.)
    1884 O, pp. 305, 309, pl. xxvα, fig. 12.   (Didy-
    mictis.)
Matthew, W. D. 1897 C, p. 287.   (Didymictis.)
    1899 A, p. 30, footnote; p. 35.   (Viverravus.)
Osborn and Wortman 1892 A, pp. 83, 111.   (Didy-
    mictis.)
Roger, O. 1896 A, p. 40.   (Didymictis.)
Scott, W. B. 1892 B, 320.   (Didymictis.)
Trouessart, E. L. 1897 A, p. 235.   (Didymictis.)
    Eocene (Wind River); Wyoming.

Viverravus massetericus (Cope).

Cope, E. D. 1882 E, p. 159.   (Didymictis.)
    1884 O, pp. 306, 312, pl. xxive, fig. 11.   (Didy-
    mictis.)
    1884 S, p. 484, fig. 29.   (Didymictis.)
Matthew, W. D. 1897 C, p. 287.   (Didymictis.)

Matthew, W. D. 1899 A, p. 30.   (Viverravus.)
Roger, O. 1896 A, p. 40.
Scott, W. B. 1892 B, p. 320.   (Didymictis.)
Trouessart, E. L. 1897 A, p. 235.   (Didymictis.)
    Eocene (Wasatch); Wyoming.

Viverravus nitidus Marsh.

Marsh, O. C. 1872 I, p. 205.
Matthew, W. D. 1899 A, p. 40.
Roger, O. 1896 A, p. 41.
Schlosser, M. 1887 B, p. 143.   (Entomodon?.)
    1890 B, p. 63.
Scott, W. B. 1892 B, p. 321.
Trouessart, E. L. 1897 A, p. 236.
    Eocene (Bridger); Wyoming.

Viverravus protenus (Cope).

Cope, E. D. 1874 O, p. 15.   (Limnocyon.)
    1875 C, p. 11.   (Didymictis.)
    1875 L, p. 21.   (Didymictis.)
    1877 K, p. 123, pl. xxxix, figs. 1–9.   (Didy-
    mictis.)
    1882 E, p. 159.   (Didymictis.)
    1884 O, pp. 305, 311, pl. xxvd, figs. 4, 5.
    (Didymictis.)
    1887 S, p. 245.   (Didymictis.)
Matthew, W. D. 1897 C, p. 286.   (Didymictis.)
    1899 A, pp. 30, 35.   (Viverravus.)
Osborn, H. F. 1897 G, p. 255.   (Didymictis.)
Roger, O. 1896 A, p 40.   (Didymictis.)
Scott, W. B. 1892 B, p. 320.   (Didymictis.)
Trouessart, E. L. 1897 A, p. 235.   (Didymictis.)
    Eocene (Wasatch); New Mexico, Wyoming,
    Colorado.

TELMATOCYON Marsh.   Type *Limnocyon riparius* Marsh.

Marsh, O. C. 1899 B, p. 397.

Telmatocyon riparius Marsh.

Marsh, O. C. 1872 I, p. 203.   (Limnocyon.)
    1899 B, p. 397.
King, C. 1878 A, p. 403.   (Limnocyon.)

Matthew, W. D. 1899 A, pp. 35, 40.   (Viverravus.)
Roger, O. 1896 A, p. 41.   (Viverravus.)
Schlosser, M. 1890 B, pp. 63, 64.   (Limnocyon.)
Scott, W. B. 1891 B, pp. 310, 321.   (Viverravus.)
Trouessart, E. L. 1897 A, p. 236.   (Viverravus.)
    Eocene (Wind River, Bridger); Wyoming.

AMBLOCTONIDÆ.

Cope, E. D. 1877 K, p. 89.
    1880 C, pp. 78, 79.   (Amblyctonidæ.)
    1882 E, pp. 157, 158.   (Amblyctonidæ.)
    1884 O, pp. 259, 260, 739.   (Amblyctonidæ.)
Haeckel, E. 1895 A, pp. 578, 583.   (Palæonictida.)
Osborn, H. F. 1893 D, p. 83.   (Palæonictidæ.)
Osborn and Wortman 1892 A, p. 103.   (Palæonic-
    tidæ.)

Roger, O. 1896 A, p. 37.   (Palæonictidæ.)
Schlosser, M 1886 C.
Scott, W. B. 1892 B, pp 294 311.   (Palæonictidæ.)
    1892 D, pp. 77, 79.   (Palæonictidæ.)
Trouessart, E. L. 1897 A, p. 229   (Palæonictidæ.)
Winge, H 1895 B, pp. 47, 52, 91   (Palæonictidæ.)
Wortman, J. L. 1899 A, p. 140   (Palæonictidæ.)
Zittel, K. A. 1893 B, p 595.   (Palæonictidæ.)

PALÆONICTIS Blainv.   Type *P. gigantea* Blainv.

Blainville, H. M. D. 1864 A (1841), ii, N, p. 79
Cope, E. D. 1876 A, p. 3.
    1877 K, p. 88.   (Palæonyctis.)
    1884 O, p. 260.   (Palæonyctis.)
Gervais, P. 1859 A, p. 225.
Gaudry, A. 1878 B, p. 18.
Osborn, H. F. 1892 A.
    1893 E.

Osborn, H. F. 1893 H.
    1893 I, p. 321.
Osborn and Wortman 1892 A, pp. 81, 96, 100, 104.
Pictet, F. J. 1858 A, p. 212.
Schlosser, M. 1887 B, p. 211.
Scott, W. B. 1892 A, p. 418.
    1892 B, p. 311.
    1892 D, p. 79.

Winge, H. 1895 B, pp. 47, 52, 91, 93, 100, 101.
Wortman, J. L. 1894 A, p. 156.
Zittel, K. A. 1893 B, p. 596.

Palæonictis occidentalis Osb. and Wort.

Osborn and Wortman 1892 A, p. 104, pl. iv.

Matthew, W. D. 1899 A, p. 31.
Osborn, H. F. 1893 H, p. 435, fig. 1.
Roger, O. 1896 A, p. 37.
Trouessart, E. L. 1897 A, p. 230.
Zittel, K. A. 1893 B, p. 595, fig. 499.
Eocene (Wasatch); Wyoming.

AMBLOCTONUS Cope.   Type *A. sinosus* Cope.

Cope, E. D. 1875 C, pp. 5, 7.
1875 A.
1877 K, pp. 87, 90.
1880 C, p. 79. (Amblyctonus.)
1882 E, p. 158. (Amblyctonus.)
1884 O, p. 260. (Amblyctonus.)
1887 S, p. 251. (Amblyctonus.)
1889 C, p. 190. (Amblyctonus.)
Flower and Lydekker 1891 A, p. 609. (Amblyctonus.)
Nicholson and Lydekker 1889 A, p. 1453. (Amblyctonus.)
Osborn, H. F. 1893 H, p. 486.
Osborn and Wortman 1892 A, pp. 81, 99, 104.
Schlosser, M. 1887 B, p. 206.

Scott, W. B. 1892 B, p. 312.
1892 D, p. 80.
Zittel, K. A. 1893 B, p. 596. (Amblyctonus.)

Ambloctonus sinosus Cope.

Cope, E. D. 1875 C, p. 8.
1877 K, p. 91, pl. xxxiii.
1881 J, p. 546, fig. 5. (Amblyctonus.)
1884 O, p. 361. (Amblyctonus.)
1884 S, p. 265, fig. 4. (Amblyctonus.)
Matthew, W. D. 1899 A, p. 31. (Amblyctonus.)
Osborn and Wortman 1892 A, p. 106, fig. 8.
Roger, O. 1896 A, p. 37.
Trouessart, E. L. 1897 A, p. 230.
Eocene (Wasatch); New Mexico.

HYÆNODONTIDÆ.

Cope, E. D. 1875 L, p. 22.
1877 K, p. 89. (Oxyænidæ.)
1880 C, pp. 78, 79. (Oxyænidæ.)
1882 E, pp. 156, 158. (Oxyænidæ.)
1884 O, pp. 259, 260. (Oxyænidæ.)
1884 S, pp. 261, 480. (Oxyænidæ.)
1886 L, p. 965.
Flower and Lydekker 1891 A, p. 608.
Gill, T. 1872 B, pp. 59, 68.
Haeckel, E. 1895 A, pp. 578, 583. (Hyænodontida.)
Leidy, J. 1871 C, p. 342.
Lydekker, R. 1884 A, p. 348.
1885 B, p. 20.
1885 E.
Nicholson and Lydekker 1889 A, p. 1450.
Osborn, H. F. 1893 D.
1900 I, p. 276. (Oxyænidæ.)

Osborn and Wortman 1892 A, p. 104.
Roger, O. 1896 A, p. 37.
Schlosser, M. 1886 C. (Oxyænidæ, Hyænodontidæ.)
Scott, W. B. 1888 B, p. 184.
1892 A, p. 419.
1892 B, pp. 294, 318.
1892 D, pp. 77, 79.
1894 F.
1895 F, p. 720.
Steinmann and Döderlein 1890 A, p. 714.
Trouessart, E. L. 1897 A, p. 230.
Winge, H. 1895 B, pp. 46, 48, 50, 91. (Hyænodontidæ.)
Wortman, J. L. 1899 A, p. 139. (Oxyænidæ, Hyænodontidæ.)
Zittel, K. A. 1893 B, p. 596.

OXYÆNINÆ.

Osborn, H. F. 1900 I, p. 276. (Oxyænidæ.)
Schlosser, M. 1886 C. (Oxyænidæ.)

Wortman, J. L. 1899 A, p. 139. (Oxyænidæ.)

OXYÆNA Cope.   Type *O. lupina* Cope.

Cope, E. D. 1874 O, p. 11.
1875 C, p. 9,
1875 L.
1876 A.
1877 K, pp. 87, 95.
1880 C, p. 79.
1882 E, p. 158.
1882 S, p. 480.
1884 E.
1884 O, pp. 260, 313.
1887 S, p, 251.
Flower and Lydekker 1891 A, p. 608.
Lydekker, R. 1885 B, p. 35.
Matthew, W. D. 1896 A, p. 880.

Nicholson and Lydekker 1889 A, p. 1451. (Oxyæna.)
Osborn, H. F. 1900 I, p. 269.
Osborn and Wortman 1892 A, p. 108.
Schlosser, M. 1887 B, p. 204.
1890 B, pp. 36, 71.
Scott, W. B. 1888 B, p. 184.
1892 A, p. 418.
1892 B, p. 314.
1892 D, p. 79.
Winge, H. 1895 B, pp. 47, 53, 91, 93, 101.
Wortman, J. L. 1894 A, p. 152.
Zittel, K. A. 1893 B, p. 597.

Oxysena forcipata Cope.

Cope, E. D. 1874 O, p. 12.
    1875 C, p. 9.
    1876 A.
    1877 K, pp. 72, 105, pls. xxxv, figs. 7–12; pl. xxxvi; pl. xxxvii, figs. 1–5.
    1882 E, p. 164.
    1884 O, p. 318, pl. xxivb, figs. 11–15; pl. xxivc, figs. 1–18.
    1884 S, pp. 260, 482, figs. 3, 25.
King, C. 1878 A, p. 376.
Matthew, W. D. 1899 A, p. 31.
Osborn, H. F. 1897 G, p. 255.
Osborn and Wortman 1892 A, p. 109.
Roger, O. 1896 A, p. 37.
Trouessart, E. L. 1897 A, p. 230.
    Eocene (Wasatch); New Mexico, Wyoming.
    From a remark by Dr. Wortman (1899 A, p. 140) we may infer that a part, or most probably all, of the remains described by Cope in 1884 O, p. 318, as *O. forcipata* belong to *O. lupina.*

Oxysena huerfanensis Osb.

Osborn, H. F. 1897 G, p. 255.
Matthew, W. D. 1899 A, p. 35.
Trouessart, E. L. 1899 A, p. 1290.
    Eocene (Wind River); Colorado.

Oxysena lupina Cope.

Cope, E. D. 1874 O, p. 11.
    1875 C.
    1877 K, p. 101, pl. xxxiv, figs. 14–17; pl. xxxv, figs. 1–4.
King, C. 1878 A, p. 376.
Matthew, W. D. 1899 A, p. 31.
Osborn, H. F. 1900 I, p, 276.
Osborn and Wortman 1892 A, p. 108, fig. 9.
Roger, O. 1896 A, p. 37.
Trouessart, E. L. 1897 A, p. 597, fig. 601.
Wortman, J. L. 1899 A, p. 140, pl. vii; text figs. 1, 2.
    Eocene (Wasatch); New Mexico, Wyoming.

Oxysena morsitans Cope.

Cope, E. D. 1874 O, p. 12.
    1875 C, p. 9.
    1877 K, p. 98, pl. xxxiv, figs. 1–13.
    1881 J, p. 546, fig. 6.
    1884 S, p. 263, fig. 5.
    1889 C, p. 190.
Matthew, W. D. 1899 A, p. 31.
Roger, O. 1896 A, p. 37.
Trouessart, E. L. 1897 A, p. 230.
    Eocene (Wasatch); New Mexico.

PATRIOFELIS Leidy.   Type *P. ulta* Leidy.

Leidy, J. 1870 G, p. 11,
Cope, E. D. 1880 C, p. 79.
    1880 N, p. 745. (Protopsalis, type *P. tigrinus* Cope.)
    1881 D, p. 193. (Protopsalis.)
    1882 E, p. 158. (Patriofelis, Protopsalis.)
    1884 O, p. 260 (Patriofelis); pp. 260, 321 (Protopsalis).
    1884 S, p. 480 (Protopsalis); p. 483 (Patriofelis).
Leidy, J. 1871 C, p. 344.
    1873 B, p. 114.
Marsh, O. C. 1872 I, p. 202. (Limnofelis, type *L. ferox.*)
    1872 O, p. 406. (Oreocyon, type *O. latidens.*)
    1877 E. (Orocyon, Limnofelis.)
Nicholson and Lydekker 1889 A, p. 1452. (Protopsalis.)
Osborn, H. F. 1893 H, p. 436.
    1896 C, pp. 705, 710, illus. 1.
    1898 Q, p. 15.
    1900 I, p. 269.
Schlosser, M. 1887 B, p. 210. (Patriofelis, Protopsalis.)
    1890 B, p. 36. (Protopsalis.)
Scott, W. B. 1892 B, p. 313 (Patriofelis); p. 316 (Protopsalis).
    1892 D, p. 79 (Protopsalis); p. 80 (Patriofelis.)
    1894 F, p. 583.
Winge, H. 1895 B, p. 97.
Woodward, A. S. 1898 B, p. 384.
Wortman, J. L. 1894 A.
Zittel, K. A. 1893 B, p. 596 (Patriofelis); p. 597 (Protopsalis).

Patriofelis ferox (Marsh).

Marsh, O. C. 1872 I, p. 202. (Limnofelis.)

King, C. 1878 A, p. 403.
Lydekker, R. 1896 B, p. 372, fig. 77.
Matthew, W. D. 1899 A, p. 41.
Osborn, H. F. 1898 Q, p. 10.
    1900 I, p. 276, figs. 2, 4–8.
Roger, O. 1896 A, p. 37.
Schlosser, M. 1890 B, p. 63. (Limnofelis.)
Scott, W. B. 1892 B, p. 316. (Syn.? of Protopsalis tigrinus.)
Trouessart, E. L. 1897 A, p. 230.
Woodward, A. S. 1898 B, p. 384, fig. 215.
Wortman, J. L. 1894 A, pl. i, text figs. 1–5.
    Eocene (Bridger); Wyoming.

Patriofelis latidens (Marsh).

Marsh, O. C. 1872 I, p. 203. (Limnofelis.)
    1872 O, p. 406. (Oreocyon.)
Matthew, W. D. 1899 A, p. 41.
Schlosser, M. 1890 B, p. 63. (Limnofelis.)
Scott, W. B. 1892 B, p. 313. (Syn. of P. ulta.)
Trouessart, E. L. 1897 A, p. 230. (Syn. of P. ulta.)
Wortman, J. L. 1894 A, p. 129.
    Eocene (Bridger); Wyoming.
    A doubtful species.

Patriofelis tigrina (Cope).

Cope, E. D. 1880 N, p. 745. (Protopsalis.)
    1881 D, p. 193. (Protopsalis.)
    1884 O, pp. 260, 322, pl. xxvb, figs. 1–7. (Protopsalis.)
    1884 S, p. 483, fig. 27. (Protopsalis.)
Matthew, W. D. 1899 A, p. 35.
Osborn, H. F. 1900 I, p. 277, fig. 7 B. (Protopsalis.)
Roger, O. 1896 A, p. 38. (Protopsalis.)
Schlosser, M. 1887 B, p. 210. (Protopsalis.)
Scott, W. B. 1888 B, p. 174. (Protopsalis.)
    1892 B, 316. (Protopsalis.)
Trouessart, E. L. 1897 A, p. 231. (Protopsalis.)

Wortman, J. L. 1894 A, p. 180.
  Eocene (Wind River); Wyoming.

Patriofelis ulta Leidy.

Leidy, J. 1870 G, p. 11.
  1871 C, p. 344.
  1872 B, p. 355.
  1873 B, pp. 114, 316, pl. ii, fig. 10.
Matthew, W. D. 1899 A, p. 41.

Osborn, H. F. 1897 G, P. 25.
  1900 I, P. 279.
Osborn and Wortman 1892 A, p. 97.
Roger, O. 1896 A, p. 37.
Scott, W. B. 1892 B, p. 313.
Trouessart, E. L. 1897 A, p. 230.
Wortman, J. L. 1894 A, pp. 129, 130. 164.
  Eocene (Bridger); Wyoming.

HYÆNODONTINÆ.

Schlosser, M. 1886 C.   (Hyænodontidæ.)
Winge, H. 1895 B, p. 46.   (Hyænodontini.)

Wortman, J. L. 1899 A, p. 139.   (Hyænodontidæ.)

HYÆNODON Laizer and Parieu.   Type *H. leptorhynchus* Laizer and Parieu.

Laizer and Parieu 1839 A.
Blainville, H. M. D. 1864 A, u, L, p. 65, pl. xii, figs.
  1-5.   (Taxotherium.)
Cope, E. D. 1875 C, p. 5.
  1883 I, p. 79.
  1884 O, p. 739.
  1884 S, p, 344.
  1886 L, p. 966.
  1888 D.
Farr, M. S. 1896 A, p. 163.
Filhol, H, 1876 A. p. 169.
Flower, W. H. 1876 B, xiii, p. 514.
Flower and Lydekker 1891 A, p. 608.
Gaudry, A. 1878 B, p. 13.
Gervais, P. 1859 A, p. 232.
Giebel, C. G. 1847 A, p. 43.
Leidy, J. 1869 A, p. 38.
  1871 C, p. 342.
Lydekker, R. 1884 A, p. 348.
  1885 B, p. 21.
  1900 A, p. 927.
Marsh, O. C. 1877 E.
Nicholson and Lydekker 1889 A, p. 1450.
Osborn and Wortman 1894 A, p. 224.
Owen, R. 1852 B.
  1868 A, p. 340.
Pictet, F. J. 1853 A, p. 196.
Pomel, A. 1853 A, p. 115.
Schlosser, M. 1887 B, p. 173.
Scott, W. B. 1888 B, pp. 164, 175.
  1892 A, p. 419.
  1892 B, p. 317.
  1892 D, p. 79.
  1894 F.
  1895 B.
Trouessart, E. L. 1897 A, p. 232.
Vasseur, G. 1874 A.
Wilckens, M. 1885 D, p. 490.
Winge, H. 1895 B, pp. 46, 50, 91, 100, 101.
Woodward, A. S. 1898 B, p. 385.
Wortman, J. L. 1894 A, p. 155.
Zittel, K. A. 1893 B, p. 599.

Hyænodon crucians Leidy.

Leidy, J. 1853 D, p. 393.
Cope, E. D. 1874 B, p. 505.
Hayden, F. V. 1858 B, p. 158.
King, C. 1878 A, p. 411.

Leidy, J. 1869 A, pp. 48, 369, pl, ii.
  1871 C, p. 342.
Matthew, W. D. 1899 A, p. 53.
Osborn and Wortman 1894 A, p. 224.
·Roger, O. 1896 A, p. 40.
Scott, W. B. 1888 B, p. 175.
  1892 B, p. 318.
  1894 F.
Trouessart, E. L. 1897 A, p. 234.
  Oligocene (White River); South Dakota.

Hyænodon cruentus Leidy.

Leidy, J. 1853 D, p. 393.
Hayden, F. V. 1858 B, p. 158.
King, C. 1878 A, p. 411.
Leidy, J. 1869 A, pp. 47, 369, pl. v, figs. 10, 11.
  1871 C, p. 342.
Matthew, W. D. 1899 A, p. 53.
Osborn and Wortman 1894 A, p. 224.
Roger, O. 1896 A, p. 40.
Scott, W. B. 1888 B, p. 175, pl. vi, fig. 2.
  1894 F, figs. 1-4, 6, 7, 10.
  1895 B, pl. xii A.
Trouessart, E. L. 1897 A, p. 234.
Woodward, A. S. 1898 B, p. 387, fig. 217.
  Oligocene (White River); South Dakota, Colo-
  rado.

Hyænodon horridus Leidy.

Leidy, J. 1853 D, p. 392.
Cope, E. D. 1874 B, p. 505.
  1884 S, p. 346, fig. 12.
Flower, W. H. 1876 C, p. 120.
Hayden, F. V. 1858 B, p. 158.
King, C. 1878 A, p. 411.
Leidy, J. 1869 A, pp. 39, 369, pl. iii.
  1871 C, p. 342.
Matthew, W. D. 1899 A, p. 53.
Osborn and Wortman 1894 A, p. 224.
Roger, O. 1896 A, p. 40.
Scott, W. B. 1888 B, p. 175, pl. vii, figs. 4-6.
  1894 F, fig. 5.
  1895 B, p. 442.
Scott and Osborn 1887 B, p. 151.
Trouessart, E. L. 1897 A, p. 234.
Wortman, J. L. 1886 A, p. 421, fig. 203.
Zittel, K. A. 1893 B, p. 601, fig. 505.
  Oligocene (White River); South Dakota.

Hyænodon leptocephalus Scott.

Scott, W. B. in Scott and Osborn 1887 B, p. 152, text fig. 1.
Matthew, W. D. 1899 A, p. 58.
Osborn and Wortman 1894 A, p. 224.
Roger, O. 1896 A, p. 40.
Scott, W. B. 1888 B, p. 175.
1894 F.
Trouessart, E. L. 1897 A, p. 234.
Oligocene (White River); South Dakota.

Hyænodon mustelinus Scott.

Scott, W. B. 1894 F, p. 499.
Oligocene (White River); South Dakota.

Hyænodon paucidens Osb. and Wort.

Osborn and Wortman 1894 A, p. 223.
Matthew, W. D. 1899 A, p. 53.
Roger, O. 1896 A, p. 40.
Scott, W. B. 1894 F.
Trouessart, E. L. 1897 A, p. 234.
Oligocene (White River); South Dakota.

HEMIPSALODON Cope. Type *H. grandis* Cope.

Cope, E. D. 1885 H, p. 163.
1891 A, p. 6.
Nicholson and Lydekker 1889 A, p. 1452.
Scott, W. B. 1892 B, p. 316.
1892 D, p. 79.
1894 F, p. 585.
Zittel, K. A. 1893 B, p. 598.

Hemipsalodon grandis Cope.

Cope, E. D. 1885 H, p. 163.

Ami, H. M. 1891 A.
Cope, E. D. 1886 U, p. 80.
1889 I, p. 151.
1891 A, p. 6, pls. ii, iii.
Matthew, W. D. 1899 A, p. 53.
Roger, O. 1896 A, p. 38.
Scott, W. B. 1892 B, p. 317.
Trouessart, E. L. 1897 A, p. 231.
Oligocene (White River); Northwest Territory, Canada.

OXYÆNODON Matth. Type *O. dysodus* Matth.

Matthew, W. D. 1899 A, p. 49. (Reference to figured specimen.)
Osborn, H. F. 1900 I, p. 278.
Wortman, J. L. 1899 A, p. 145.

Oxyænodon dysodus Matth.

Matthew, W. D. 1899 A, p. 49. (Reference to figured specimen.)
Osborn, H. F. 1896 A, p. 78, fig. 3. ("? Hyænodon.")
Eocene (Uinta); Utah.

Oxyænodon dysclerus nom. nov.

Wortman, J. L. 1899 A, p. 145, fig. 3. (O. dysodus, not of Matthew.)
Eocene (Uinta); Utah.
It is by no means certain that this species belongs to the genus *Oxyænodon*, as represented by the type jaw.

UINTACYONIDÆ.

The following authors call this family *Miacidæ*.
Cope, E. D. 1880 C, pp. 78, 79.
1882 E, pp. 156, 157.
1884 O, pp. 259, 260.
1884 S, p. 483.
1886 L, p. 966.
1888 K, p. 163.
1888 EE, p. 163.
Matthew, W. D. 1897 C, p. 286.
1899 A, p. 49. (Canidæ, in part.)
Nicholson and Lydekker 1889 A, p. 1437.

Roger, O. 1896 A, p. 40.
Schlosser, M. 1886 C.
Scott, W. B. 1888 A.
1888 B, p. 173.
1892 A, p. 419.
1892 B, pp. 294, 318.
1892 D, pp. 77, 80.
1895 F, p. 720.
Steinmann and Döderlein 1890 A, p. 716.
Trouessart, E. L. 1897 A, p. 234.
Zittel, K. A. 1893 B, p. 602.

UINTACYON Leidy. Type *U. edax* Leidy.

Leidy, J. 1872 J, p. 277.
Ameghino, F. 1896 A, p. 454. (Miacis, in part.)
Cope, E. D. 1880 C, p. 79.
1882 E, p. 157.
1884 O, pp. 260, 301. (Miacis, in part.)
1884 S, p. 483. (Miacis, in part.)
Leidy, J. 1873 B, p. 118. (Uintacyon.)
Matthew, W. D. 1897 C, p. 286. (Miacis, in part.)
Nicholson and Lydekker 1889 A, p. 1437. (Miacis, in part.)

Osborn, H. F. 1892 I. (Miacis in part.)
Osborn and Wortman 1892 A, pp. 95, 110. (Miacis, in part.)
Schlosser, M. 1888 B, p. 58 (Miacis); p. 61 (Uintacyon).
1890 B, p. 60. (Miacis, in part.)
Scott, W. B. 1888 B, p. 173. (Miacis, in part.)
1890 A, p. 473. (Miacis, in part.)
1892 B, p. 320. (Miacis, in part.)

Scott, W. B. 1892 D, p. 80. (Miacis, in part.)
1898 B, p. 363. (Miacis, in part.)
Trouessart, E. L. 1897 A, p. 235. (Miacis, in part.)
Winge, H. 1896 B, pp. 46, 52, 91, 92. (Miacis, in part.)
Wortman and Matthew 1899 A, p. 110.
Zittel, K. A. 1898 B, p. 602. (Miacis, in part.)

Uintacyon bathygnathus (Scott).

Scott, W. B. 1888 B, p. 172. (Miacis.)
Matthew, W. D. 1899 A, p. 40.
Roger, O. 1896 A, p. 41. (Miacis.)
Schlosser, M. 1887 B, p. 59. (Miacis.)
Scott, W. B. 1889 B, p. 220. (Miacis.)
1892 B, p. 320. (Miacis.)
Trouessart, E. L. 1897 A, p. 235. (Miacis.)
Eocene (Bridger); Wyoming.

Uintacyon brevirostris (Cope.)

Cope, E. D. 1881 D, p. 190. (Miacis.)
1882 E, p. 158. (Miacis.)
1884 O, pp. 302, 303. (Miacis.)
1884 S, p. 484, fig. 28. (Miacis.)
Matthew, W. D. 1899 A, pp. 30, 35.
Osborn and Wortman 1892 A, pp. 83, 111. (Miacis.)
Roger, O. 1896 A, p. 41. (Miacis.)
Scott, W. B. 1892 B, p. 320. (Miacis.)
Trouessart, E. L. 1897 A, p. 236. (Miacis.)
Wortman and Matthew 1899 A, p. 112.
Eocene (Wind River, Wasatch); Wyoming.

Uintacyon canavus (Cope).

Cope, E. D. 1881 D, p. 189. (Miacis.)
1882 E, p. 158. (Miacis.)
1884 O, p. 302. (Miacis.)
Matthew, W. D. 1899 A, pp. 30, 35. (Uintacyon.)
Osborn and Wortman 1892 A, p. 110. (Miacis.)
Roger, O. 1896 A, p. 41. (Miacis.)
Scott, W. B. 1892 B, p. 320. (Miacis.)
Trouessart, E. L. 1897 A, p. 236. (Miacis.)
Wortman and Matthew 1899 A, p. 112.
Eocene (Wasatch and Wind River); Wyoming.

PRODAPHÆNUS Matth.

Matthew, W. D. 1899 A, p. 49. (April 8.)
Wortman and Matthew 1899 A, p. 114. (June 21.)

Prodaphænus scotti Wort. and Matth.

Wortman and Matthew 1899 A, p. 115.
Matthew, W. D. 1899 A, p. 49. (Name only.)
Eocene (Uinta); Wyoming.

PROCYNODICTIS Wort. and Matth.

Wortman and Matthew 1899 A, p. 121.
Put in this family provisionally.

Uintacyon edax Leidy.

Leidy, J. 1872 J, p. 277.
Cope, E. D. 1884 O, p. 303. (Miacis.)
1887 Q. (Miacis.)
King, C. 1878 A, p. 403. (Uintacyon.)
Leidy, J. 1873 B, pp. 118, 316, pl. xxvii, figs. 6–10. (Uintacyon.)
Matthew, W. D. 1899 A, p. 40. (Uintacyon.)
Osborn and Wortman 1892 A, pp. 83, 110. (Miacis.)
Roger, O. 1896 A, p. 40. (Miacis.)
Schlosser, M. 1888 B, p. 61.
Scott, W. B. 1892 B, p. 320. (Miacis.)
Trouessart, E. L. 1897 A. p. 235. (Miacis.)
Wortman and Matthew 1899 A, p. 114.
Eocene (Bridger); Wyoming.

Uintacyon promicrodon Wort. and Matth.

Wortman and Matthew 1899 A, p. 111.
Eocene (Wasatch); Wyoming.

Uintacyon pugnax Wort. and Matth.

Wortman and Matthew 1899 A, p. 114.
Eocene; Wyoming.

Uintacyon vorax Leidy.

Leidy, J. 1872 J, p. 277.
Cope, E. D. 1884 O, p. 302. (Miacis.)
Leidy, J. 1873 B, pp. 120, 316, pl. xxvii, figs. 11–13. (Uintacyon.)
Matthew, W. D. 1899 A, pp. 35, 40. (Uintacyon.)
Roger, O. 1896 A, p. 40. (Miacis.)
Schlosser, M. 1888 B, p. 61.
Trouessart, E. L. 1897 A, p. 235. (Miacis.)
Wortman and Matthew 1899 A, p. 113.
Eocene (Wind River, Bridger); Wyoming.

Uintacyon vulpinus Scott and Osb.

Scott and Osborn 1887 A, p. 255. (Amphicyon.)
Matthew, W. D. 1899 A, p. 49. (Miacis.)
Roger, O. 1896 A, p. 41. (Miacis.)
Scott, W. B. 1890 A, p. 474. (Miacis.)
1892 B, p. 321. (Miacis.)
Trouessart, E. L. 1897 A, p. 236. (Miacis.)
Eocene (Uinta); Utah.

Type P. uintensis (Osb.).

Prodaphænus uintensis (Osb.).

Osborn, H. F. 1895 A, p. 77, fig. 2. (Miacis.)
Matthew, W. D. 1899 A, p. 49.
Roger, O. 1896 A, p. 41. (Miacis.)
Scott, W. B. 1898 B, p. 363, fig. 1?. (Miacis.)
Trouessart, E. L. 1897 A, p. 236. (Miacis.)
Eocene (Uinta); Utah.

Type P. vulpiceps Wort. and Matth.

Procynodictis vulpiceps Wort. and Matth.

Wortman and Matthew 1899 A, p. 121, figs. 7, 8.
Matthew, W. D. 1899 A, p. 49. (No description.)
Eocene (Uinta); Utah.

THINOCYON Marsh.   Type *T. velox* Marsh.

Marsh, O. C. 1872 I, p. 204.
Schlosser, M. 1890 B, p. 64.
Scott, W. B. 1892 B, p. 321.
Zittel, K. A. 1893 B, p. 602.   (Syn.? of Miacis.)

Thinocyon velox Marsh.

Marsh, O. C. 1872 I, p. 204.

Matthew, W. D. 1899 A, p. 40.
Roger, O. 1896 A, p. 41.
Schlosser, M. 1890 B, p. 63.
Scott, W. B. 1892 B, p. 321.
Trouessart, E. L. 1897 A, p. 237.
    Eocene (Bridger); Wyoming.

VULPAVUS Marsh.   Type *V. palustris* Marsh.

Marsh, O. C. 1871 E, p. 124.
Cope, E. D. 1872 E, p. 470. (Miacis, type *M. parvivorus.*)
    1878 E, p. 561. (Miacis, Viverravus.)
    1880 C, p. 79. (Miacis.)
    1882 E, p. 157. (Miacis.)
    1884 O, pp. 260, 301. (Miacis, in part.)
Matthew, W. D. 1899 A, p. 39.
Nicholson and Lydekker 1889 A, p. 1437. (Miacis, in part.)
Osborn and Wortman 1892 A, pp. 95, 110. (Miacis, in part.)
Schlosser, M. 1899 B, p. 60 (Miacis); p. 64. (Vulpavus.)
Scott, W. B. 1888 B, p. 173. (Miacis, in part.)
    1890 A, p. 473. (Miacis, in part.)
    1892 B, p. 320. (Miacis, in part.)
    1892 D, p. 80. (Miacis in part.)
    1898 B, p. 363. (Miacis, in part.)
Trouessart, E. L. 1897 A, p. 236. (Syn.? of Miacis.)
Winge, H. 1895 B, pp. 46, 52, 91, 92. (Miacis, in part.)
Wortman and Matthew 1899 A, p. 110.
Zittel, K. A. 1893 B, p. 602.   (=? Miacis.)

Vulpavus palustris Marsh.

Marsh, O. C. 1871 E, p. 124.
King, C. 1878 A, p. 403.
Leidy, J. 1872 B, p. 356.
    1873 B, p. 118.
Matthew, W. D. 1899 A, p. 39.
Roger, O. 1896 A, p. 41. (Miacis?.)
Schlosser, M. 1890 B, p. 62.
Trouessart, E. L. 1897 A, p. 236. (Miacis?.)
Wortman and Matthew 1899 A, p. 118, figs. 4–6.
    Eocene (Bridger); Wyoming.

Vulpavus parvivorus (Cope).

Cope, E. D. 1872 E, p. 470. (Miacis.)
    1873 E, p. 560. (Viverravus.)
    1873 U, p. 3. (Viverravus.)
    1884 O, pp. 302, 304, pl. xxiv, fig. 12. (Miacis.)
Matthew, W. D. 1899 A, p. 39. (Syn. of Vulpavus palustris.)
Roger, O. 1896 A, p. 40. (Miacis.)
Scott, W. D. 1892 B, p. 320. (Miacis.)
Trouessart, E. L. 1897 A, p. 235. (Miacis.)
Wortman and Matthew 1899 A, pp. 119, 120.
    Eocene (Bridger); Wyoming.

ZIPHACODON Marsh.   Type *Z. rugatus* Marsh.

Marsh, O. C. 1872 I, p. 216.
    1877 E. (Name only.)
Matthew, W. D. 1899 A, p. 40.
Schlosser, M. 1890 B, p. 64.
Zittel, K. A. 1893 B, p. 603.

Ziphacodon rugatus Marsh.

Marsh, O. C. 1872 I, p. 216.
Roger, O. 1896 A, p. 42.
Schlosser, M. 1890 B, p. 64.
Trouessart, E. L. 1897 A, p. 237.
    Eocene (Bridger); Wyoming.

HARPALODON Marsh.   Type *H. sylvestris* Marsh.

Marsh, O. C. 1872 I, p. 216.
Schlosser, M. 1890 B, p. 64.
Zittel, K. A. 1893 A, p. 603.

Harpalodon sylvestris Marsh.

Marsh, O. C. 1872 I, p. 216,
Matthew, W. D. 1899 A, p. 40.
Roger, O. 1896 A, p. 42.
Schlosser, M. 1890 B, p. 64.
Trouessart, E. L. 1897 A, p. 231.
    Eocene (Bridger); Wyoming.

Harpalodon vulpinus Marsh.

Marsh, O. C. 1872 I, p. 217.
Matthew, W. D. 1899 A, p. 40.
Roger, O. 1896 A, p. 42.
Schlosser, M. 1890 B, p. 64.
Trouessart, E. L. 1897 A, p. 237.
    Eocene (Bridger); Wyoming

CARCINODON Scott.   Type *Mioclænus filholianus* Cope.

Scott, *W. B.* 1892 B, p. 323.
Matthew, W. D. 1897 C, p. 293.
  1899 A, p. 28.
Zittel, K. A. 1893 B, p. 586.
  A genus of uncertain position placed by Matthew in Creodonta, "incertæ sedis."

Carcinodon filholianus (Cope).
Cope, *E. D.* 1888 CC, pp. 303, 329.   (Mioclænus.)
Matthew, W. D. 1897 C, pp. 264, 293.
  1899 A, p. 28.
Roger, O. 1896 A, p. 41.
Scott, W. B. 1892 B, p. 323.
Trouessart, E. L. 1897 A, p. 237.
Eocene (Puerco); New Mexico.

PARADOXODON Scott.   Type *Chriacus rütimeyeranus* Cope.

Scott, *W. B.* 1892 B, p. 322.
  1892 D, p. 80.
Zittel, K. A. 1893 B, p. 586.
  A genus of uncertain position.

Parodoxodon rütimeyeranus (Cope).
Cope, *E. D.* 1888 CC, pp. 304, 323.   (Chriacus.)

Matthew, W. D. 1899 A, p. 28.   (Mioclænus.)
Roger, O. 1896 A, p. 41.
Scott, W. B. 1892 B, p. 323.
Trouessart, E. L. 1897 A, p. 237.
Eocene (Puerco); New Mexico.

Suborder FISSIPEDIA.

Cope, E. D. 1882 D, pp. 471, 473.
  1888 W, p. 1019.   (Carnivora vera.)
  1889 R, p. 876.
  1898 B, p. 118.
Flower, W. H. 1876 B, xiii, p. 487.   (Carnivora vera.)
  1883 D, p. 433.
Gill, T. 1872 A, p. 297.
  1872 B, p. 56.

Gill and Coues 1877 A, p. 1005.   [Feræ (Fissipedia).]
Huxley, T. H. 1872 A, p. 351.
Lydekker, R. 1896 B.
Nicholson and Lydekker 1889 A, p. 1424.   (Carnivora vera.)
Weber, M. 1886 A, p. 227.
Woodward, A. S. 1898 B, p. 389.   (Carnivora vera.)
Zittel, K. A. 1893 B, pp. 606, 678.

Superfamily *ARCTOIDEA*.

Flower, W. H. 1869 B, p. 15.
Cope, E. D. 1882 D, p. 473.   (Hypomycteri, in part.)
  1884 O, pp. 890, 892.   (Hypomycteri, in part.)
Flower, W. H. 1883 D, p. 439.
Flower and Lydekker 1891 A, p. 556.

Huxley, T. H. 1872 A, p. 858.
Mivart, St. G. 1881 A, p. 474.
  1885 B, p. 340.
Schlosser, M. 1888 B, p. 6.
Tims, H. W. M. 1896 B, p. 462.
Tomes, C. S. 1898 A, p. 472.

URSIDÆ.

Albrecht, P. 1879 A.
Baird, S. F. 1857 A, p. 206.
Blainville, H. M. D. 1864 A, ii, K, pls. i–xviii.
Cope, E. D. 1882 D, p. 473.
  1884 O, p. 892.
Cuvier, G. 1834 A, ii, p. 171.
Delbos, J. 1858 A.
  1860 A.
  1860 B.
Flower, W. H. 1869 A, p. 277.
  1876 B, xiii, p. 488.
  1883 D, p. 441.
  1885 A.
Flower and Lydekker 1891 A, p. 557.
Gadow, H. 1898 A, p. 50.
Gill, T. 1872 B, pp. 58, 66.
Gray, J. E. 1869 A, p. 215.
Haeckel, E. 1895 A, p. 585.   (Ursida.)
Lydekker, R. 1884 A, p. 202.
  1885 B p 106, in part.
Mivart, St. G. 1885 B p 394
Nicholson and Lydekker 1889 A, p. 1430.

Owen, R. 1843 A, p. 62.
Owen, R. 1845 B, p. 501.
  1866 B, p. 490.
  1868 A, p. 335.
Pictet, F. J. 1853 A, p. 188.
Roger, O. 1896 A. p. 52.
Schlosser, M. 1888 A.
  1888 B, pp. 3, 14, 15.
  1899 B, p. 95.
Scott, W. B. 1888 A.
  1892 A.
Steinmann and Döderlein 1890 A, p. 721.
Tornier, G. 1888 A.
  1890 A.
Trouessart, E. L. 1897 A, p. 238.
Turner, H. N. 1848 A.
Waterhouse, G. R. 1839 A.
Weyhe. — 1876 A, p. 117.
Winge, H. 1895 B, pp. 47, 63, 91.   (Ursidæ, Ursini)
Woodward, A S 1898 B, p. 394.
Zittel K. A 1893 B, p. 639.

URSUS Linn.   Type *U. arctos* Linn.

Linnæus, C., 1758, Syst. Nat., ed. 10, i, p. 47.
Allen, H. 1888 A.
Baird, S. F. 1857 A, p. 216.
Blainville, H. M. D. 1864 A, ii, K, pls. i-xviii.
Burmeister, H. 1879 B, p. 169.
Delbos, J. 1858 A.
   1860 A.
   1860 B.
Flower, W. H. 1869 B. p. 6.
   1883 D, p. 441.
   1885 A.
Flower and Lydekker 1891 A, p. 557.
Gray, J. E. 1869 A, p. 218.
Lydekker, R. 1884 A, p. 206.
   1885 B, p. 159.
Merriam, C. H. 1896 A.
Middendorff, A. T. 1850 A.
Mivart, St. G. 1885 B, p. 389.
Nicholson and Lydekker 1889 A, p. 1431.
Owen, R. 1843 A, p. 62.
   1846 B, p. 77.
   1866 B, p. 490.
   1868 A, p. 335.
Pictet, F. J. 1853 A, p. 185.
Schlosser, M. 1888 B, p. 88.
   1899 B, p. 95.
   1900 A, p. 261.
Scott, W. B. 1892 A, p. 425.
Steinmann and Döderlein 1890 A, p. 722
Trouessart, E. L. 1897 A, p. 238.
Turner, H. N. 1848 A, p. 75.
Zittel, K. A. 1893 B, p. 641.

Ursus americanus Pallas.

Allen, J. A. 1876 C, p. 333.
Baird, S. F. 1857 A, p. 225, pl. xliii, figs. 10, 13.

Cope, E. D. 1869 E, p. 176.
   1896 F, p. 447.
Harlan, R. 1835 C, p. 329.
Leidy, J. 1853 B, p. 303.   ("Ursus.")
   1856 T, p. 169.
   1859 F, p. 111.
   1869 A, p. 369.
Merriam, C. H. 1896 A.
Miller, G. S. 1899 A, p. 54.   (U. americanus fossilis.)
Pictet, F. J. 1853 A, p. 189.
Trouessart, E. L. 1897 A, p. 243.
Wailes, B. L. C. 1854 A, p 286.   (U. americanus fossilis.)
   Recent; North America. Pleistocene; Kentucky, Mississippi, Pennsylvania.

Ursus amplidens Leidy.

Leidy, J. 1853 B, p. 303.
Cope, E. D. 1869 E, p. 176.
Forshey, C. G. 1846 A, p. 163.   ("Polar bear;" species problematical.)
Leidy, J. 1856 T, p. 168, pl. xvii, figs. 13-16.
   1869 A, p. 370.
Merriam, C. H. 1896 A.
Miller, G. S. 1899 A, p. 54.
Roger, O. 1896 A, p. 55.   (Arctotherium.)
Trouessart, E. L. 1897 A, p. 243.
Wailes, B. L. C. 1854 A, p. 286.
Wortman, J. L. 1882 A.   (U. amplidens=U. ferox.)
Zittel, K. A. 1893 B, p. 643.
   Pleistocene; Mississippi, Virginia.

Ursus procerus Miller.

Miller, G. S. 1899 A, p. 55.
   Pleistocene; Ohio.

ARCTODUS Leidy.   Type *A. pristinus* Leidy.

Leidy, J. 1854 C, p. 90.
Ameghino, F. 1889 A, p. 315.   (Arctotherium.)
Bravard, 1860, Cat. Foss. Amer. Merid., p. 8.   (Arctotherium.)
Cope, E. D. 1871 I, p. 96.   (Ursus.)
   1899 A, p. 221.   (Tremarctos.)
Flower and Lydekker 1891 A, p. 561.   (Arctotherium.)
Gervais, P., 1855, Hist. Nat. Mamm., ii, p. 20.   (Tremarctos, type *T. ornatus.*)
Gervais and Ameghino 1880 A, p. 23.   (Arctotherium.)
Lydekker, R. 1884 A, p. 236.   (Arctotherium.)
   1885 B, p. 157.   (Arctotherium.)
Nicholson and Lydekker 1889 A, p. 1429 (Arctodus); 1432 (Arctotherium.)
Steinmann and Döderlein 1890 A, p. 722.   (Arctotherium.)
Zittel, K. A. 1893 B, p. 641.   (Arctotherium.)

Arctodus haplodon (Cope).

Cope, E. D. 1896 F, p. 383.   [Ursus (Tremarctus).]
   1871 F, p. 15.   (Ursus pristinus, not of Leidy.)
   1871 I, p. 96.   (Ursus pristinus.)
   1871 P, p. 58.   (Ursus pristinus.)
   1896 F, p. 447.   (Arctotherium pristinum.)

Cope, E. D. 1899 A, p. 220, pl. xix, figs. 2, 2a.   (Ursus.)
Miller, G. S. 1899 A, p. 54.   (Ursus.)
Trouessart, E. L. 1897 A, p. 245.   (Tremarctos.)
Wheatley, C. M. 1871 A.   (Ursus pristinus.)
   Pleistocene (Port Kennedy Cave); Pennsylvania.

Arctodus pristinus Leidy.

Leidy, J. 1854 C, p. 90.   (Arctodus.)
Cope, E. D. 1896 F, p. 383.   (Ursus.)
   1899 A, p. 221.   (Ursus.)
Leidy, J. 1860 B, p. 115, pl. xxiii, figs. 3, 4.   (Arctodus.)
   1869 A, p. 370.   (Arctodus.)
Miller, G. S. 1899 A, p. 54.   (Arctodus.)
Roger, O. 1896 A, p. 55.   (Arctodus.)
Trouessart, E. L. 1897 A, p. 246.   (Tremarctos.)
   Pleistocene; South Carolina.

Arctodus simus (Cope).

Cope, E. D. 1879 U, p. 791.   (Arctotherium.)
   1891 K, p. 997, pl. xxi.   (Arctotherium.)
   1899 A, p. 221.   (Ursus.)
Miller, G. S. 1899 A, p. 54.   (Arctotherium.)
Roger, O. 1896 A, p. 54.   (Arctotherium.)
Trouessart, E. L. 1897 A, p. 245.   (Tremarctos.)
   Pleistocene; California.

PROCYONIDÆ.

Albrecht, P. 1879 A.
Allen, H. 1880 A.
Cope, E. D. 1880 D, p. 883.
    1881 C, p. 165.
    1882 D, p. 473.
    1884 E.
    1884 O, p. 892.
Flower, W. H. 1869 B, p. 5.
    1883 D, p. 440.
Flower and Lydekker 1891 A, p. 562.
Gill, T. 1872 B, pp. 58, 67.
Gray, J. E. 1869 A, p. 242.
Haeckel, E. 1895 A, p. 585.   (Procyonida.)

Huxley, T. H. 1880 A.
Mivart, St. G. 1885 B, p. 398.
Nicholson and Lydekker 1869 A, 1429.
Roger, O. 1896 A, p. 55.
Schlosser, M. 1888 B, p. 3 (Subursi); p. 12 (Procyonidæ).
    1890 B, p. 23.   (Subursi.)
    1899 A, p. 94.   ("Subursen.")
Scott, W. B. 1892 A, p. 426.
Trouessart, E. L. 1897 A, p. 248.
Winge, H. 1895 B, pp. 47, 63, 91.
Wortman and Matthew 1899 A, p. 129.
Zittel, K. A. 1893 B, p. 644.

CERCOLEPTINÆ.

Blainville, H. M. D. 1864 A, ii, L, p. 20, pls. v, vii–x.
Cope, E. D. 1882 D, p. 473.   (Cercoleptidæ.)
    1884 O, p. 892.   (Cercoleptidæ.)

Gill, T. 1872 B, pp. 58, 67 (Cercoleptidæ); pp. 59, 67 (Bassarididæ).
Trouessart, E. L. 1897 A, p. 245.

LEPTARCTUS Leidy.   Type *L. primus* Leidy.

Leidy, J. 1856 Q, p. 311.
    1871 C, p. 344.
Lydekker, R. 1884 A, p. 204.
Nicholson and Lydekker 1889 A, p. 1429.
Schlosser, M. 1899 A, p. 84.
Wortman, J. L. 1894 B, p. 229.
Zittel, K. A. 1893 B, p. 645.

King, C. 1878 A, p. 430.
Leidy, J. 1869 A, pp. 70, 370, pl. i, figs. 15, 16.
    1871 C, p. 344.
Matthew, W. D. 1894 A, p. 67.
Roger, O. 1896 A, p. 55.
Schlosser, M. 1899 A, p. 84.
Trouessart, E. L. 1897 A, p. 248.
Wortman, J. L. 1894 B, p. 229.
Zittel, K. A. 1893 B, p. 645.
    Miocene (Loup Fork); South Dakota.

Leptarctus primus Leidy.

Leidy, J. 1856 Q, p. 311.
Hayden, F. V. 1858 B, p. 158.

BASSARISCUS Coues.   Type *Bassaris astuta* Licht.

Coues, E., 1887, Science (2), ix, p. 516.
Allen, J. A. 1879 A, p. 331.   (Bassaris.)
Baird, S. F. 1857 A, p. 147.   (Bassaris.)
Flower, W. H. 1869 B, p. 31.   (Bassaris.)
Mivart, St. G. 1885 B, p. 361.   Bassaris.)
Schlosser, M. 1890 B, p. 4.   (Bassaris.)
    1890 C, p. 267.   (Bassaris.)
    1899 A, p. 86.   (Bassaris.)
Winge, H. 1895 B, pp. 47, 64, 102, 103.   (Bassaris.)
Wortman and Matthew 1899 A, p, 134.

Bassariscus astutus (Licht.).

Allen, J. A. 1879 A, p. 325.   (Bassaris.)
Baird, S. F. 1857 A, p. 147, pl. lxxiv, fig. 2.   (Bassaris.)
Cope, E. D. 1895 F, p. 447.
Flower, W. H. 1869 B, p. 10, fig. 3a.   (Bassaris.)
Trouessart, E. L. 1897 A, p. 249.
    Recent; Mexico to Oregon. Pleistocene (Port Kennedy); Pennsylvania, *fide* Cope.

PROCYONINÆ.

PROCYON Storr.   Type *Ursus lotor* Linn.

Storr, G. C. 1780, Prod. Meth. Mamm., p. 35.
Baird, S. F. 1857 A, p. 207.
Blainville, H. M. D. 1864 A, ii, L, p. 13, pls. iii, vi.
Flower, W. H. 1869 B, p. 9.
    1883 D, p. 440.
Flower and Lydekker 1891 A, p. 564.
Giebel, C. G. 1857 A, p. 349.
Gray, J. E. 1869 A, p. 242.
Huxley, T. H. 1880 A.
Mivart, St. G. 1885 B, p. 346.
Nicholson and Lydekker 1889 A, p. 1429.
Owen, R. 1868 A, p. 334.
Schlosser, M. 1890 B, p. 23.
Scott, W. B. 1892 A, p. 426.

Winge, H. 1895 B, pp. 47, 65, 91.
Wortman and Matthew 1899 A, p. 133.
Zittel, K. A. 1893 B, p. 644.

Procyon lotor (Linn.).

Linnæus, C., 1758, Syst. Nat., ed. 10, i, p. 48. (Ursus.)
Allen, J. A. 1876 C, p. 333.
Baird, S. F. 1857 A, p. 209.
Cope, E. D. 1869 E, p. 176.
Flower, W. H. 1869 B, p. 10, fig. 3.
Giebel, C. G. 1857 A, p. 349.
Leidy, J. 1860 B, p. 115, pl. xxiii, fig. 1.
    1869 A, p. 370.
    1889 H, p. 5.

Trouessart, E. L. 1897 A, p. 251.
　　Recent; Central and North America. Pleis-
　　tocene; South Carolina, Virginia, Pennsylva-
　　nia, New Jersey.

Procyon priscus Le Conte.

Le Conte, J. L. 1848 A, p. 106.
Cope, E. D. 1869 E, p. 176.

Leidy, J. 1856 T, p. 169, pl. xvii, figs. 17-24.
　　1862 A.
　　1869 A, p. 370.
　　1870 H, p. 13.
Pictet, F. J. 1853 A, p. 191.
Roger, O. 1896 A, p. 55.
Trouessart, E. L. 1897 A, p. 251.
　　Pleistocene; Galena, Illinois.

PHLAOCYON Matth.　Type *P. leucosteus* Matth.

Matthew, W. D. 1899 A, p. 54.
Wortman and Matthew 1899 A, p. 129.

Phlaocyon leucosteus Matth.

Matthew, W. D. 1899 A, p. 54.
Wortman and Matthew 1899 A, pp. 129, 131, pl. vi.
　　Oligocene (White River); Colorado.

MYXOPHAGUS Cope.　Type *M. spelæus* Cope.

Cope, E. D. 1869 E, p. 176.

Myxophagus spelæus Cope.

Cope, E. D. 1869 E, p. 176, pl. iii, figs. 2, 2a.
　　1869 J, p. 3. (No description.)

Leidy, J. 1869 A, p. 445.
Roger, O. 1896 A, p. 55.
Trouessart, E. L. 1897 A, p. 252.
　　Pleistocene; Virginia.

MUSTELIDÆ.

Albrecht, P. 1879 A.
Baird, S. F. 1857 A, p. 148.
Blainville, H. M. D. 1864 A, ii M, pls. i-xiv.
Cope, E. D. 1875 L, p. 21.
　　1882 D, p. 473.
　　1884 O, p. 892.
Coues, E. 1877 A, pp. 1, i.
Flower, W. H. 1869 B.
　　1883 D, p. 439.
Flower and Lydekker 1891 A, p. 567.
Gill, T. 1872 B, pp. 58, 64.
Gray, J. E. 1865 A, p. 100.
　　1869 A, p. 79.
Haeckel, E. 1895 A, p. 587. (Mustelida.)
Humphreys, J. 1889 A, p. 158.
Lydekker, R. 1884 A, p. 178.
Mayer, P. 1886 A, p. 276.

Mivart, St. G. 1885 B, p. 398.
Nicholson and Lydekker 1889 A, p. 1426.
Owen, R. 1845 B, p. 494.
　　1868 A, p. 333.
Roger, O. 1896 A, p. 56.
Schlosser, M. 1888 B, pp. 3, 14, 107.
　　1890 C, p. 265.
Scott, W. B. 1892 A, p. 426.
Steinmann and Döderlein 1890 A, p. 728.
Tomes, C. S, 1898 A, p. 472.
Trouessart, E. L. 1897 A, p. 252.
Turner, H. N. 1848 A.
Waterhouse, G. R. 1839 A.
Winge, H. 1895 B, pp. 47, 65, 69, 91.
Woodward, A. S. 1898 B, p. 396.
Zittel, K. A. 1893 B, pp. 645, 654.

MELINÆ.

Blainville, H. M. D, 1864 A, ii, L, p. 2, pls. ii, vi,
　　xii. (Genus Meles.)
Baird, S. F. 1857 A, p. 190.
Coues, E. 1877 A, pp. 5, 9, 10. (Melinæ, Mephit-
　　inæ.)
Flower, W. H. 1883 D, p. 439.
Flower and Lydekker 1891 A, p. 572.
Gill, T. 1872 B, p. 64 (Melinæ); p. 65 (Mephitinæ).

Mivart, St. G. 1885 B, p. 394.
Moseley and Lankester 1868 A. (Genus Meles.)
Owen, R. 1868 A, p. 333. (Melidæ.)
Roger, O. 1896 A, p. 60.
Trouessart, E. L. 1897 A, p. 252.
Winge, H. 1895 B, p. 47. (Melini.)
Zittel, K. A. 1893 B, p. 650.

TAXIDEA Storr.　Type *Ursus taxus* Schreber.

Storr, G. C. 1780, Prod. Meth. Mamm., p. 34.
Baird, S. F. 1857 A, p. 201.
Flower, W. H. 1869 B, p. 11.
　　1883 D, p. 440.
Flower and Lydekker 1891 A, p. 576.
Gray, J. E. 1865 A, p. 140.
　　1869 A, p. 129.
Mivart, St. G. 1885 B, p. 367.
Owen, R. 1868 A, p. 333.
Waterhouse, G. R., 1838, Trans. Zool. Soc. Lond., ii,
　　pp. 343.
Winge, H. 1895 B, p. 68.

Taxidea taxus (Schreber).

Baird, S. F. 1857 A, p. 202. (T. americana.)

Cope, E. D. 1899 A, p. 239. (T. americana.)
Gray, J. E. 1865 A, p. 141. (T. americana.)
Mivart, St. G. 1885 B, p. 367. (T. americana.)
　　Recent. Ohio to Oklahoma, north to Hudson
　　Bay. Pleistocene; Pennsylvania.

Taxidea sulcata Cope.

Cope, E. D. 1878 C, p. 227.
　　1889 F, p. 162. (Syn. of T. americana.)
Roger, O. 1896 A, p. 61.
Trouessart, E. L. 1897 A, p. 252.
　　Pleistocene (Equus beds); State of Washing-
　　ton.

MEPHITIS Cuvier. Type "*Les Moufettes*."

Cuvier, G 1800, Leçons d'Anat. Comp., i, tabl. i.
Ameghino, F. 1889 F, p. 822.
Baird, S. F. 1857 A, p. 191.
Blainville, H. M. D. 1864 A, ii, M, p. 19, pls. i, lx–xiv.
Burmeister, H. 1879 B, p. 162.
Cope, E. D. 1899 A, p. 232.
Coues, E. 1875 A.
Filhol, H. 1891 A, p. 109.
Flower, W. H. 1869 B, p. 11.
  1883 D, p. 439.
Flower and Lydekker 1891 A, p. 572.
Gray, J. E. 1865 A, p. 147.
  1869 A, p. 136.
Mivart, St. G. 1885 B, p. 370.
Schlosser, M. 1888 B, pp. 113, 114, 140.
Trouessart, E. L. 1897 A, p. 259.
Winge, H. 1895 B, pp. 68, 79.
Zittel, K. A. 1893 B, p. 652.

Mephitis fossidens Cope.

Cope, E. D. 1896 F, p. 386.
  1899 A, pp. 231, 232, 233, pl. xviii, fig. 7.
Trouessart, E. L. 1897 A, p. 259.
Pleistocene; Pennsylvania.

Mephitis frontata Coues.

Coues, E. 1875 A, p. 7, fig. 1.
Allen, J. A. 1876 C, p. 333. (Syn. of M. mephitica.)
Coues, E. 1877 A, p. 193.

Trouessart, E. L. 1897 A, p. 260. (M. mephitica frontata.)
Pleistocene; Pennsylvania.

Mephitis leptops Cope.

Cope, E. D. 1899 A. pp. 232, 235, pl. xviii, figs. 9, 9a.
Pleistocene; Pennsylvania.

Mephitis mephitica (Shaw).

Allen, J. A. 1876 C, p. 333.
Baird, S. F. 1857 A, p. 195, pl. lx, fig. 1.
Cope, E. D. 1895 F, p. 447.
  1899 A, p. 232.
Coues, E. 1875 A, p. 8.
  1877 A, p. 195. (Synonymy and literature.)
Leidy, J. 1889 H, p. 5.
Trouessart, E. L. 1897 A, p. 259.
  Recent; Louisiana northward to Nova Scotia and Saskatchewan. Pleistocene; Pennsylvania, *fide* Allen.

Mephitis obtusata Cope.

Cope, E. D. 1899 A, p. 236.
Pleistocene; Pennsylvania.

Mephitis orthrostica Cope.

Cope, E. D. 1896 F, p. 389.
  1899 A, p. 234, pl. xviii, figs. 8, 8a.
Trouessart, E. L. 1897 A, p. 259.
Pleistocene; Pennsylvania.

SPILOGALE Gray. Type *Mephitis interrupta* Raf.

Gray, J. E. 1865 A, p. 150.
Coues, E. 1877 A, p. 192. (Subgenus.)
Flower and Lydekker 1891 A, p. 574.
Merriam, C. H., 1890, N. A. Fauna, no. 4, p. 7.
Trouessart, E. L. 1897 A, p. 262.

Spilogale perdicida Cope.

Cope, E. D. 1869 E, p. 177, pl. iii, fig. 1. (Galera.)
  1869 J, p. 3. (Hemiacis.)

Cope, E. D. 1896 F, p. 386. (Galera.)
Coues, E. 1875 A, p. 8. (Mephitis.)
  1877 A, p. 18. (Spilogale putorius?.)
Leidy, J. 1869 A, p. 445. (Galera.)
Roger, O. 1896 A, p. 60. (Galictis.)
Trouessart, E. L. 1897 A, p. 262. (Syn. of S. putorius.)
Pleistocene; Virginia.

OSMOTHERIUM Cope. Type *O. spelæum* Cope.

Cope, E. D. 1896 F, p. 385.
  1899 A, p. 230.

Osmotherium spelæum Cope.

Cope, E. D. 1896 F, p. 385.

Cope, E. D. 1899 A, p. 231, pl. xviii, fig. 6.
Trouessart, E. L. 1897 A, p. 260.
Pleistocene; Pennsylvania.

PELYCICTIS Cope. Type *P. lobulatus* Cope.

Cope, E. D. 1896 F, p. 390.
  1899 A, p. 237.

Pelycictis lobulatus Cope.

Cope, E. D. 1896 F, p. 390.

Cope, E. D. 1899 A, p. 237, pl. xviii, fig. 10; text figure.
Trouessart, E. L. 1897 A, p. 260.
Pleistocene; Pennsylvania.

MUSTELINÆ.

Baird, S. F. 1857 A, p. 148. (Martinæ.)
Flower, W. H. 1883 D, p. 440.
Flower and Lydekker 1891 A, p. 579.
Gill, T. 1872 B, p. 64.
Lydekker, R. 1884 A, p. 178.
Mivart, St. G. 1885 B, p. 394.

Roger, O. 1896 A, p. 56.
Trouessart, E. L. 1897 A, p. 263.
Turner, H. N. 1848 A.
Winge, H. 1895 B, pp. 47, 66, 69.
Zittel, K. A. 1893 B, p. 646.

BUNÆLURUS Cope.   Type *B. lagophagus* Cope.

Cope, E. D. 1873 CC. p. 8.
    1874 B, p. 507.
    1884 O, p. 946.
Trouessart, E. L. 1897 A, p. 264 (Bunælurus); p. 268
    (Syn. of Palæogale).

Bunælurus lagophagus Cope.

Cope, E. D. 1873 CC, p. 8.
    1873 CC, p. 10. (Canis osorum.)
    1874 B, p. 507 (Canis osorum); p. 508 (B.
    lagophagus).

Cope, E. D. 1884 O, p. 946, pl. lxviia, figs. 13, 14 (Bun-
    ælurus; Plesiogale, on plate); p. 947, pl.
    lxviia, fig. 12 (B. osorum).
Matthew, W. D. 1899 A, p. 54.
Roger, O. 1896 A, p. 44 (Daphænos osorum); p. 59
    (Bunælurus lagophagus).
Scott, W. B. 1893 A.
Trouessart, E. L. 1897 A, p. 264 (B. lagophagus, B.
    osorum); p. 296 (Daphænos osorum).
Zittel, K. A. 1893 B, p. 649.
    Oligocene (White River); Colorado.

GALICTIS Bell.   Type *Viverra vittata* (Schreber).

Bell, T., 1826, Zool. Journ., ii, p. 551.
Burmeister, H. 1879 B, p. 156.
Cope, E. D. 1867 C, p. 156. (Galera.)
Coues, E. 1877 A, p. 17.
Flower, W. H. 1869 B, p. 12. (Galera.)
Flower and Lydekker 1891 A, p. 579.
Gray, J. E. 1865 A, p. 121. (Galera.)
Mivart, St. G. 1885 B, p. 376.
Nehring, A. 1886 A, p. 177.
Nicholson and Lydekker 1889 A, p. 1428.
Schlosser, M. 1888 B, p. 114.
Zittel, K. A. 1893 B, p. 650.

Galictis macrodon Cope.

Cope, E. D. 1867 C, p. 155. (Galera.)
Cope and Wortman 1884 A, p. 5.
Coues, E. 1877 A, p. 17. (Galera.)
Leidy, J. 1869 A, p. 369, pl. xxx, figs. 1–3. (Galera.)
Roger, O. 1896 A, p. 60.
Trouessart, E. L. 1897 A, p. 264.
Wortman, J. L. 1883 B, p. 1001. (Putorius.)
    Pleistocene; Maryland.

STENOGALE Schlosser.   Type, none designated.

Schlosser, M. 1888 B, pp. 110, 115, 151.
Cope, E. D. 1890 I, p. 950.
Zittel, K. A. 1893 B, p. 647.

Stenogale robusta Cope.

Cope, E. D. 1890 I, p. 950.
Matthew, W. D. 1899 A, p. 68.
Trouessart, E. L. 1897 A, p. 267.
    Miocene (Loup Fork); Nebraska.

PARICTIS Scott.   Type *P. primævus* Scott.

Scott, W. B. 1893 A, p. 658. (Parictis, error.)

Parictis primævus Scott.

Scott, W. B. 1893 A, p. 658. (Parietis princeous,
    error.)

Matthew, W. D. 1899 A, p. 63.
Roger, O. 1896 A, p. 57.
Trouessart, E. L. 1897 A, p. 267.
    Miocene (John Day); Oregon.

MUSTELA Linn.   Type *M. martes* Linn.

Linnæus, C., 1758, Syst. Nat. ed. 10, i, p. 45.
Baird, S. F. 1857 A, p. 149.
Blainville, H. M. D. 1864 A, ii, M, plates.
Coues, E. 1877 A, p. 59.
Filhol, H. 1891 A, p. 94.
Flower, W. H. 1869 B, p. 13.
Flower and Lydekker 1891 A, p. 94.
Gray, J. E. 1865 A, p. 110.
Lydekker, R. 1885 B, p. 176.
Mivart, St. G. 1885 B, p. 378.
Nicholson and Lydekker 1889 A, p. 1428.
Owen, R. 1868 A, p. 333.
Zittel, K. A. 1893 B, p. 648.

Mustela americana Turton.

Allen, J. A. 1876 C, p. 333.
Baird, S. F. 1857 A, p. 152, pl. xxxvii, fig. 1.
Coues, E. 1877 A, p. 81, pl. v. (Synonymy and lit-
    erature.)
Gray, J. E. 1865 A, p. 106. (Martes.)
Trouessart, E. L. 1897 A, p. 272.
    Recent; North America. Pleistocene or Re-
    cent, in caves; Pennsylvania, *fide* Allen.

Mustela diluviana Cope.

Cope, E. D. 1899 A, p. 229, pl. xviii, figs. 5, 5a.
    Pleistocene; Pennsylvania.

Mustela parviloba (Cope).

Cope, E. D. 1873 C, p. 1. [Ælurodon mustelinus;
    not Mustela (Plesiogale) mustelina of Pomel.]
    1874 B, p. 520. (Martes mustelinus.)
    1874 P, p. 11. (Martes mustelinus.)
Coues, E. 1877 A, p. 16. (Mustela mustelina.)
Matthew, W. D. 1899 A, p. 68.
Roger, O. 1896 A, p. 59. (Mustela mustelina.)
Scott and Osborn 1890 B, p. 71. ("M. parviloba
    Cope.")
Trouessart, E. L. 1897 A, p. 272. (M. parviloba.)
    Miocene (Loup Fork); Colorado, Nebraska.

Mustela pennantii Erxleben.

Erxleben, C. P., 1777, Syst. Regn. Anim., p. 470.
Allen, J. A. 1876 C, p. 333.
Baird, S. F. 1857 A, p. 149, pl. xxxvi, fig. 1.
Coues, E. 1877 A, p. 62, pl. ii. (Synonymy and
    literature.)
Gray, J. E. 1865 A, p. 107. (Martes.)
Trouessart, E. L. 1897 A, p. 272.

PUTORIUS[1] Cuv.   Type *Mustela putorius* Linn.

Baird, S. F. 1867 A, p. 159.
Coues, E. 1877 A, p. 96.
Flower, W. H. 1883 D, p. 440.
Flower and Lydekker 1891 A, p. 585.
Gervais, P. 1859 A, p. 251.
Gray, J. E. 1865 A, p. 109.
   1869 A, p. 87.
Mivart, St. G. 1885 B, p. 379.
Owen, R. 1843 A, p. 70.
Schlosser, M. 1888 B, p. 111.
Scott, W. B. 1888 A.
Trouessart, E. L. 1897 A, p. 278.
Zittel, K. A. 1898 B, p. 649.

Putorius nambianus Cope.

Cope, E. D. 1874 I, p. 147. (Martes.)

Cope, E. D. 1874 O, p. 16.   (Martes.)
   1875 F, p. 68.   (Mustela.)
   1877 K, p. 305, pl. lxix, fig. 3.
Coues, E. 1877 A, p. 16.   (Mustela.)
Matthew, W. D. 1899 A, p. 68.
Roger, O. 1896 A, p. 59.   (Mustela).
Trouessart, E. L. 1897 A, p. 275.
   Miocene (Loup Fork); New Mexico.

Putorius vison (Schreber).

Allen, J. A. 1876 C, p. 333.
Baird, S. F. 1857 A, p. 177, pl. xxxvii, figs. 2, 3.
Coues E. 1877 A, p. 160. (Synonymy and literature.)
Trouessart, E. L. 1897 A, p. 274.
   Recent; North America.   In caves of Pennsylvania, *fide* Allen.

GULO Storr.   Type *Ursus luscus* Linn.

Baird, S. F. 1857 A, p. 180.
Blainville, H. M. D. 1864 A, ii M, p. 21, pls. iii, vii, ix-xiv.
Flower, W. H. 1869 B.
Flower and Lydekker 1891 A, p. 591.
Gervais, P. 1859 A, p. 247.
Gray, J. E. 1865 A, p. 120.
Mivart, St. G. 1885 B, p. 381.
Zittel, K. A. 1898 B, p. 649.

Gulo luscus (Linn).

Linnæus, C., 1758, Syst. Nat., ed. 10, 1, p. 47 (Ursus.)
Baird, S. F. 1857 A, p. 181.
Cope, E. D. 1899 A, p. 229.
Flower and Lydekker 1891 A, p. 591.
   Recent; northern parts of North America. Pleistocene; Pennsylvania.

LUTRINÆ.

Albrecht, P. 1879 A. (Lutridæ.)
Baird, S. F. 1857 A, p. 183.
Coues, E. 1877 A, pp. 5, 298.
Flower, W. H. 1883 D, p. 439.
Flower and Lydekker 1891 A, p. 567.
Gill, T. 1872 B, pp. 65, 66.
Gray, J. E. 1865 A, p. 123. (Lutrina.)

Lydekker, R. 1884 A, p. 187.
Mivart, St. G. 1885 B, p. 394.
Nicholson and Lydekker 1889 A, p. 1426.
Roger, O. 1896 A, p. 62.
Trouessart, E. L. 1897 A, p. 281.
Winge, H. 1895 B, pp. 47, 70. (Lutrini.)
Zittel, K. A. 1898 B, p. 652.

POMATOTHERIUM E. Geoffroy.   Type *P. valetoni* E. Geoffroy.

Geoffroy, E., 1833, Rev. Encyc., lix, p. 80.
Cope, E. D. 1890 I, p. 951. (Brachypsalis, type *B. pachycephalus*.)
   1896 F, p. 385.
   1899 A, p. 230.
Lydekker, R. 1885 B, p. 190. (Syn. of Lutra.)
Meyer, H. 1847, p. 182. (Stephanodon.)
Nicholson and Lydekker 1889 A, p. 1427.
Pomel, A. 1847 C, p. 380. (Lutrictis, type *Lutra valetoni*.)
   1853 A, p. 46. (Lutrictis.)
Schlosser, M. 1888 B, p. 112.

Zittel, K. A. 1898 B, p. 652. (Pomatotherium, Brachypsalis.)

Pomatotherium pachycephalum Cope.

Cope, E. D. 1890 I, p. 951. (Brachypsalis.)
   1896 F, p. 385. (Pomatotherium?.)
Matthew, W. D. 1899 A, p. 68. (Brachypsalis pachygnathus.)
Roger, O. 1896 A, p. 62. (Brachypsalis.)
Trouessart, E. L. 1897 A, p. 281.
   Miocene (Loup Fork); Nebraska.

LUTRA Brisson.   Type *Mustela lutra* Linn.

Brisson, A. D., 1762, Regn. Anim., p. 201.
Baird, S. F. 1857 A, p. 184.
Barnston, G. 1863 A.
Blainville, H. M. D. 1864 A, ii, M, p. 26, pls. v, viii, ix-xiv.
Burmeister, H. 1879 B, p. 165.
Coues, E. 1877 A, p. 294.

Flower, W. H. 1869 B.
   1883 D, p. 439.
Flower and Lydekker 1891 A, p. 467.
Gray, J. E. 1865 A, p. 125.
   1869 A, p. 108.
Lydekker, R. 1884 A, p. 187.
   1885 B, p. 190.

[1] *Putorius erminius* has been reported by Leidy (1889 H, p. 5) from a cave in Pennsylvania, in which by far the greater number of animals belong to species now living in the vicinity.

Martin, W. 1836 A
Mivart, St. G. 1885 B, p. 383.
Nicholson and Lydekker 1889 A, p. 1427.
Owen, R. 1846 B, p. 119.
　　1866 B, p. 491.
Schlosser, M. 1888 B, p. 112.
Winge, H. 1896 B, pp. 69, 91.
Zittel, K. A. 1898 B, p. 658.

Lutra canadensis Schreber.

Allen, J. A. 1876 C, p. 333.
Baird, S. F. 1857 A, p. 184, pl. xxxviii.
Coues, E. 1877 A, p. 295, pl. xvii. (Synonymy and literature.)
Trouessart, E. L. 1897 A, p. 285.
　Recent; North America. In caves of Pennsylvania, *fide* Allen.

Lutra piscinaria Leidy.

Leidy, J. 1873 B, p. 316, pl. xxxi, fig. 4; p. 230.
Cope, E. D. 1878 H, p. 389. (Lutra, "near piscinaria.")
　1889 F.
　1889 S, p. 980.
Coues, E. 1877 A, p. 823.
Roger, O. 1896 A, p. 62.
Trouessart, E. L. 1897 A, p. 285.
　Pliocene; Idaho.

Lutra rhoadsii Cope.

Cope, E. D. 1896 F, p, 391.
　1899 A, p. 238, pl. xviii, fig. 11.
Trouessart, E. L. 1897 A, p. 285.
　Pleistocene; Pennsylvania.

LUTRICTIS Cope.　Type *L. lycopotamicus* Cope.

Cope, E. D. 1879 B, p. 66.

Lutrictis lycopotamicus Cope.

Cope, E. D. 1879 B, p. 66.
Matthew, W. D. 1899 A, p. 68.

Roger, O. 1896 A, p. 63. (Lutra.)
Trouessart, E. L. 1897 A, p. 285. (Lutra.)
Zittel, K. 1893 B, p. 654. (Lutra.)
　Miocene (Loup Fork); Oregon.

MEGENCEPHALON Osb., Scott, and Speir.　Type *M. primævus* Osb., Scott, and Speir.

Osborn, Scott, and Speir 1878 A, p. 20.
Cope, E. D. 1879 W, p. 38.
Scott, W. B. 1888 B.

Megencephalon primævus Osb., Scott, and Speir.

Osborn, Scott, and Speir 1878 A, p. 20, pl. ix, fig. 8?.

Bruce, A. T. 1883 A, p. 39, pl. vii, fig. 6.
Cope, E. D. 1883 II, p. 970.
Roger, O. 1896 A, p. 63.
Trouessart, E. L. 1897 A, p. 287. (Megencephalum.)
　Eocene; Wyoming,

Superfamily CYNOIDEA Flower.

Flower, W. H. 1869 B, p. 24.
Cope, E. D. 1882 D, p. 473. (Hypomycteri, in part.)
　1884 O, pp. 890, 892. (Hypomycteri, in part).
Flower, W. H. 1883 D, p. 437.
Flower and Lydekker 1891 A, p. 544.
Gill, T. 1872 B, p. 63.

Huxley, T. H. 1872 A, p. 358.
Schlosser, M. 1888 B, p. 6.
Tims, H. W. M. 1896 B, p. 462.
Tomes, C. S. 1898 A, p. 464.
Zittel, K. A. 1893 B, p. 618.

CANIDÆ.

Baird, S. F. 1857 A, p. 102.
Blainville, H. M. D. 1864 A, ii, P, pls. i–xiv.
Cope, E. D. 1875 L, p. 21.
　1879 F.
　1881 C.
　1881 F.
　1882 D, p. 478.
　1883 T.
　1884 O, pp. 268, 892.
　1888 D.
Cuvier, G. 1884 A, vii, p. 464.
Filhol, H. 1883 A.
Flower, W. H. 1869 B, p. 24.
　1871 B.
　1883 D, p. 437.
Flower and Lydekker 1891 A, p. 544.
Gadow, H. 1896 A, p. 50.
Gill, T. 1872 B, pp. 57, 63.

Gray, J. E. 1868 B.
　1869 A, p. 178.
Haeckel, E. 1895 A, p. 585. (Canida.)
Humphreys, J. 1889 A, p. 155.
Huxley, T. H. 1872 A, p. 358.
　1880 A.
Lydekker, R. 1884 A, p. 240.
　1885 B, p. 106.
Mivart, St. G. 1881 A, pp. 479, 491.
Nicholson and Lydekker 1889 A, p. 1435.
Noack, T. 1894 A.
Owen, R. 1845 B, p. 475.
　1866 B.
　1868 A, p. 330.
Parsons, F. G, 1899 A.
　1900 A.
Roger, O. 1896 A, p. 42.
Schlosser, M. 1888 B, pp. 3, 14, 18,

Schmidt, O. 1886 A, p. 259.
Scott, W. B, 1890 C.
    1891 A, p. 6.
    1891 B, p. 373.
    1891 C, p. 66.
    1897 A.
    1898 B.
Steinmann and Döderlein 1890 A, p. 719.
Tims, H. W. M. 1896 A.
    1896 B, p. 445.

Tornier, G. 1888 A.
Trouessart, E. L. 1897 A, p. 288.
Turner, H. N. 1848 A.
Waterhouse, G. R. 1839 A.
Weyhe, — 1875 A, p. 116.
Wilckens, M. 1885 D.
Winge, H. 1895 B, pp. 47, 63. (Canini.)
Woodward, A. S. 1898 B, p. 389.
Wortman and Matthew 1899 A, p. 129.
Zittel, K. A. 1898 B, p. 619.

SIMOCYONINÆ.

Gill, T. 1872 B, pp. 59, 67. (Simocyonidæ.)
Roger, O. 1896 A, p. 50.

Trouessart, E. L. 1897 A, p. 291.

ENHYDROCYON Cope.   Type *E. stenocephalus* Cope.

Cope, E. D. 1879 B, p. 56.
    1879 F, pp. 179, 185.
    1879 H, p. 181.
    1881 C, p. 178.
    1883 T, p. 245.
    1884 O, pp. 898, 935.
Flower and Lydekker 1891 A, p. 562.
Nicholson and Lydekker 1889 A, p. 1433.
Schlosser, M. 1888 B, p. 61.
Scott, W. B. 1898 B, p. 414.
Wortman and Matthew 1899 A, p. 130.

Zittel, K. A. 1898 B, p. 633.

Enhydrocyon stenocephalus Cope.

Cope, E. D. 1879 B, p. 56.
    1883 T, p. 245, fig. 12.
    1884 O, p. 935, pl. lxix, figs. 3-5.
Matthew, W. D. 1899 A, p. 63.
Roger, O. 1896 A, p. 51.
Trouessart, E. L. 1897 A, p. 291.
Miocene (John Day); Oregon.

HYÆNOCYON Cope.   Type *H. basilatus* Cope.

Cope, E. D. 1879 E, p. 372.
    1881 C, pp. 179, 181.
    1881 P.
    1883 T, p. 245.
    1884 O, pp. 893, 942.
Flower and Lydekker 1891 A, p. 562.
Nicholson and Lydekker 1889 A, p. 1433.
Schlosser, M. 1888 B, p. 61.
    1890 B, p. 26.
Wortman and Matthew 1899 A, p. 130.
Zittel, K. A. 1893 B, p. 633.

Hyænocyon basilatus Cope.

Cope, E. D. 1879 B, p. 57, part. (Enhydrocyon.)
    1879 E, p. 372, in part.
    1881 C, p. 181, in part.
    1883 T, p. 246, fig. 8c.
    1884 O, p. 942, pl. lxxv, fig. 3.

Matthew, W. D. 1899 A, p. 63.
Roger, O. 1896 A, p. 50.
Schlosser, M. 1890 B, p. 25.
Trouessart, E. L. 1897 A, p. 292.
Miocene (John Day); Oregon.

Hyænocyon sectorius Cope.

Cope, E. D. 1883 T, p. 246, fig. 13d.
    1879 B, p. 57, in part. (Enhydrocyon basil-
    atus.)
    1879 E, p. 372, in part. (E. basilatus.)
    1881 C, p. 181, in part. (E. basilatus.)
    1884 O, p. 943, pl. lxx, fig. 1.
Matthew, W. D. 1899 A, p. 63.
Roger, O. 1896 A, p. 51.
Schlosser, M. 1890 B, p. 25.
Trouessart, E. L. 1897 A, p. 292.
Miocene (John Day); Oregon.

OLIGOBUNIS Cope.   Type *Ictocyon crassivultus* Cope.

Cope, E. D. 1881 P, p. 497.
    1883 T, p. 246.
    1884 O, pp. 893, 939.
Nicholson and Lydekker 1889 A, p. 1433.
Schlosser, M. 1888 B, pp. 18, 60, 105.
Scott, W. B. 1898 B, p, 401.
Wortman and Matthew 1899 A, p. 131.
Zittel, K. A. 1893 B, p. 633.

Oligobunis crassivultus Cope.

Cope, E. D. 1879 F, p. 190. (Icticyon.)

Cope, E. D. 1881 C, p. 181. (Icticyon.)
    1881 P, p. 497.
    1883 T, p. 246, fig. 14.
    1884 O, p. 940, pl. lxix, figs. 1, 2, text fig. 34.
Matthew, W. D. 1899 A, p. 63.
Roger, O. 1896 A, p. 51.
Schlosser, M. 1888 B, p. 105.
Trouessart, E. L. 1897 A, p. 292.
Zittel, K. A. 1898 B, p. 633, fig. 530.
Miocene (John Day); Oregon.

CANINÆ.

Cope, E. D. 1879 F.
Gill, T. 1872 B, p. 63.
Lydekker, R. 1884 A, p. 240.
Roger, O. 1896 A, p. 42.

Scott, W. B. 1890 C.
Trouessart, E. L. 1897 A, p. 293.
Zittel, K. A. 1893 B, p. 622.

CYNODICTIS Bravard and Pomel.	Type *C. parisiensis* Pomel.

Bravard and Pomel, 1850, Notice Oss. foss. Débruge, p. 5.
Filhol, H. 1876 A, p. 66.
1883 A.
Flower and Lydekker 1891 A, p. 555.
Gervais, P. 1859 A, p. 216.
Huxley, T. H. 1880 A.
Lydekker, R. 1884 A, p. 243.
1885 B, p. 107.
Nicholson and Lydekker 1889 A, p. 1437.
Pomel, A. 1853 A, p. 66.
Roger, O. 1896 A, p. 42.
Schlosser, M. 1888 B, pp. 16, 23, 40.
1890 B, p. 45.
Scott, W. B. 1888 A.
1889 A, p. 21.
1889 B, p. 211.
1890 C, p. 38.	(Hesperocyon.)
1896 C, p. 72.
1898 B, p. 364.
1899 B, p. 137.
Scott and Osborn 1887 B, p. 152.
Trouessart, E. L. 1897 A, p. 293.
Wilckens, M. 1885 D, p. 497.
Wortman and Matthew 1899 A, pp. 118, 130.
Woodward, A. S. 1898 B, p. 390.
Zittel, K. A. 1893 B, p. 622.

Cynodictis gregarius (Cope).

Cope, E. D. 1873 T, p. 3.	(Canis.)
Bruce, A. T. 1883 A, p. 41, fig. 7.	(Galecynus.)
Cope, E. D. 1873 CC, p. 9.	(Canis.)
1874 B, p. 506.	(Canis.)
1879 B, p. 58.	(Canis.)
1879 D, p. 63.	(Canis.)
1881 C, p. 181.	(Galecynus.)
1883 F, p. 217.	(Galecynus.)
1883 T, p. 241.	(Galecynus.)
1884 O, pp. 915, 916, pl. lxviia, figs. 7–11; pl. lxviii, figs. 5–8.	(Galecynus.)
Matthew, W. D. 1899 A, p. 54.
Roger, O. 1896 A, p. 45.	(Galecynus.)

Scott, W. B. 1898 B, p. 367, pl. xix, figs. 11–13; pl. xx, figs. 23–24.
Trouessart, E. L. 1897 A, p. 298.	(Galecynus.)
Wortman and Matthew 1899 A, pp. 123, 130.
Oligocene (White River); Colorado, Nebraska, South Dakota: (John Day); Oregon, *fide* Cope.

Cynodictis hylactor Hay.

Hay, O. P. 1899 C, p. 254.
Cope, E. D. 1884 O, p. 916.	(Syn. of Galecynus gregarius.)
Hayden, F. V. 1858 B, p. 158.	(Amphicyon gracilis.)
Leidy, J. 1856 A, p. 90.	(Amphicyon gracilis, not of Pomel.)
1869 A, pp. 36, 369, pl. i, fig. 7; pl. v, figs. 6–9.	(Amphicyon gracilis.)
1871 C, p. 342.	(Amphicyon gracilis.)
Matthew, W. D. 1899 A, p. 54.	(Syn. of C. lippincottianus.)
Roger, O. 1896 A, p. 44.	(Daphænos gracilis.)
Schlosser, M. 1888 B, p. 85.	(Amphicyon gracilis.)
Scott and Osborn 1887 B, p. 152.	(C. gracilis.)
Trouessart, E. L. 1897 A, p. 296.	(Daphænos gracilis.)
Oligocene (White River); South Dakota.

Cynodictis lippincottianus (Cope).

Cope, E. D. 1873 CC, p. 9.	(Canis.)
1874 B, p. 506.	(Canis.)
1879 B, p. 58.	(Canis.)
1879 D, p. 63.	(Canis.)
1884 O, p. 919, pl. lxvii, figs. 5, 6.	(Galecynus.)
Matthew, W. D. 1899 A, p. 54.
Roger, O. 1896 A, p. 45.	(Galecynus.)
Scott, W. B. 1898 B, p. 400.
Wortman and Matthew 1899 A, pp. 123, 128, 130.
Oligocene (White River); Colorado, Oregon.

Cynodictis temnodon Wort. and Matth.

Wortman and Matthew 1899 A, p. 130.
Oligocene (White River); South Dakota.

NOTHOCYON[1] Matth.

Matthew, W. D. 1899 A, p. 62.	(April 8.)
Wortman and Matthew 1899 A, pp. 124, 130, pl. vi. (June 21.)

Nothocyon geismarianus (Cope).

Cope, E. D. 1879 D, p. 71.	(Canis.)
1879 B, p. 58.	(Canis.)
1881 C, p. 181.	(Galecynus.)
1883 T, p. 240, figs. 5, 6.	(Galecynus.)
1884 O, pp. 915, 920, pl. lxx, figs. 2, 3; pl. lxxa.	(Galecynus.)
1889 C, p. 233, fig. 59.	(Cynodictis.)
Matthew, W. D. 1899 A, p. 62.
Roger, O. 1896 A, p. 45.	(Galecynus.)
Scott, W. B. 1898 B, p. 365.	(Canis.)
Trouessart, E. L. 1897 A, p. 299.	(Galecynus.)

Woodward, A. S. 1898 B, p. 391.
Wortman and Matthew 1899 A, pp. 127, 130.
Zittel, K. A. 1893 B, p. 626, fig. 524.	(Galecynus.)
Miocene (John Day); Oregon.

Nothocyon latidens (Cope).

Cope, E. D. 1881 C, p. 181.	(Galecynus.)
1883 T, p. 241.	(Galecynus.)
1884 O, pp. 915, 930, pl. lxx, figs. 4, 5.	(Galecynus.)
Matthew, W. D. 1899 A, p. 62.
Roger, O. 1896 A, p. 45.	(Galecynus.)
Scott, W. B. 1898 B, p. 365.	(Canis.)
Trouessart, E. L. 1897 A, p. 299.	(Galecynus.)
Wortman and Matthew 1899 A, pp. 127, 130.
Miocene (John Day); Oregon.

[1] Originally the three species here included were assigned to this genus by Matthew. No type was designated, but *Canis geismarianus* may be taken. Later, Wortman and Matthew include in the genus also *Canis parvidens* Mivart.

Nothocyon lemur (Cope).

Cope, E. D. 1879 E, p. 371.   (Canis.)
   1881 C, p. 181.   (Galecynus.)
   1883 T, p. 241, fig. 7.   (Galecynus.)
   1884 O, pp. 915, 931, pl. lxx, figs. 6–8.   (Galecynus.)

Matthew, W. D. 1899 A, p. 62.
Roger, O. 1896 A, p. 45.   (Galecynus.)
Scott, W. B. 1898 B, p. 365.   (Canis.)
Trouessart, E. L. 1897 A, p. 299.   (Galecynus.)
Wortman and Matthew 1899 A, pp. 127, 130.
   Miocene (John Day); Oregon.

DAPHŒNUS Leidy.   Type *D. vetus* Leidy.

Leidy, J. 1853 D, p. 393.
   This word is usually spelled *Daphænus*. The original form is given.
Cope, E. D. 1884 O, p. 894.   (Syn. of Amphicyon.)
Eyerman, J. 1896 A, p. 279.
Flower and Lydekker 1891 A, p. 556.   (Daphænus.)
Nicholson and Lydekker 1889 A, p. 1435.
Roger, O. 1896 A, p. 44.   (Daphænos.)
Scott, W. B. 1889 B, p. 211.
   1890 C, p. 37.
   1891 A, p. 6.
   1897 A, p. 1.
   1898 B, p. 326.
Trouessart, E. L. 1897 A, p. 296.   (Daphænos.)
Winge, H. 1895 A, pp. 46, 52, 91, 92.
Wortman and Matthew 1899 A, pp. 115, 118, 129.   (Daphænus.)
Zittel, K. A. 1893 B, p. 625.   (Daphænos.)

Daphœnus angustidens (Marsh).

Marsh, O. C. 1871 E, p. 124.   (Amphicyon.)
King, C. 1878 A, p. 411.   (Amphicyon.)
Lydekker, R. 1884 A, p. 247.   (Amphicyon.)
Matthew, W. D. 1899 A, p. 54.   (Syn. of Cynodictis lippincottianus.)
Roger, O. 1896 A, p. 44.
Schlosser, M, 1890 B, pp. 62, 64.   (Amphicyon.)
Trouessart, E. L. 1897 A, p. 296.   (Daphænos.)
   Oligocene (White River); Nebraska.

Daphœnus dodgei Scott.

Scott, W. B. 1898 B, p. 362, pl. xix, figs. 6, 7.   (Daphænus.)
Wortman and Matthew 1899 A, p. 129.   (Daphænus.)
   Oligocene (White River); Nebraska.

Daphœnus felinus Scott.

Scott, W. B. 1898 B, p. 361, pl. xx, figs. 15–17.   (Daphænus.)
Wortman and Matthew 1899 A, p. 129.   (Daphænus.)
   Oligocene (White River); Nebraska.

Daphœnus hartshornianus (Cope).

Cope, E. D. 1873 CC, p. 9.   (Canis.)

Cope, E. D. 1874 B, p. 505.   (Canis.)
   1881 C, p. 178.   (Canis.)
   1883 T, p. 237, figs. 2a, 3.   (Amphicyon.)
   1884 O, p. 896, 906.   (Amphicyon.)
Leidy, J. 1869 A, p. 32, in part; pl. i, figs. 3, 4, 6.   (Amphicyon vetus=A. hartshornianus, *fide* Cope.)
Lydekker, R. 1884 A, p. 247.   (Amphicyon.)
Matthew, W. D. 1899 A, pp. 52, 62.   (Daphænus.)
Roger, O. 1896 A, p. 44.
Schlosser, M. 1888 B, p. 85.   (Amphicyon,)
Scott, W. B. 1890 C, p. 37.   (Daphænus.)
   1898 B, p. 361, pl. xix, figs. 1–5; pl. xx, figs. 19–21a; p. 363, text fig. A, 2.   (Daphænus.)
Trouessart, E. L. 1897 A, p. 296.   (Daphænos.)
Wortman and Matthew 1899 A, p. 129.   (Daphænus.)
   Oligocene (White River); Colorado, South Dakota.

Daphœnus robustus Scott.

Scott, W. B. 1897 A, p. 2.   (Daphænus.)
   Oligocene (White River); Colorado.

Daphœnus vetus Leidy.

Leidy, J. 1853 D, p. 393.
Cope, E. D. 1881 C, p. 178.   (Amphicyon.)
   1883 T, p. 237.   (Amphicyon.)
   1884 O, p. 894.   (Amphicyon.)
Hayden, F. V. 1858 B, p. 158.   (Amphicyon.)
King, C. 1878 A, p. 411.   (Amphicyon.)
Leidy, J. 1854 E, p. 157.   (Amphicyon.)
   1869 A, pp. 32, 369, pl. i, figs. 1–6.   (Amphicyon; see D. hartshornianus.)
   1871 C, p. 342.   (Amphicyon.)
Lydekker, R. 1884 A, p. 248.   (Amphicyon.)
Matthew, W. D. 1899 A, p. 53.   (Daphænus.)
Roger, O. 1896 A. p. 44.   (Daphænos.)
Schlosser, M. 1888 B, p. 85.   (Amphicyon.)
Scott, W. B. 1890 C, p. 37.   (Daphænus.)
   1897 A, p. 1.   (Daphænus.)
   1898 B, pl. xix, figs. 8–10.   (Daphænus.)
Trouessart, E. L. 1897 A, p. 296.   (Daphænos.)
Wortman and Matthew, 1899 A, p. 129.
   Oligocene (White River); South Dakota, Colorado.

PARADAPHÆNUS Matth.   Type *Canis cuspigerus* Cope.

Matthew, W. D. 1899 A, p. 62.   (April 8.)
Wortman and Matthew 1899 A, p. 129.   (June 21.)

Paradaphænus cuspigerus (Cope).

Cope, E. D. 1879 D, p. 70.   (Canis.)
   1879 B, p. 58.   (Canis.)
   1879 E, p. 372.   (Amphicyon entoptychi.)
   1881 C, p. 178.   (Amphicyon.)
   1883 T, p. 237, fig. 1.   (Amphicyon.)
   1884 O, p. 898, pl. lxviii, figs. 1–4.   (Amphicyon.)
Lydekker, R. 1884 A, p. 247.   (Amphicyon cuspigerus, A. entoptychi.)

Matthew, W. D. 1899 A, p. 62.
Roger, O. 1896 A, p. 44.   (Daphænos.)
Scott, W. B. 1890 C, p. 37.   (Daphænus.)
Trouessart, E. L. 1897 A, p. 296.   (Daphænos.)
Wortman and Matthew 1899 A, p. 129.
   Miocene (John Day); Oregon.

Paradaphænus transversus Wort. and Matth.

Wortman and Matthew 1899 A, p. 129.
   Miocene (John Day); Oregon.

TEMNOCYON Cope.   Type *T. altigenis* Cope.

Cope, E. D. 1879 D, p. 68.
    1879 F, pp. 178, 180.
    1881 C, pp. 178, 179.
    1883 T, p. 237.
    1884 O, pp. 898, 902.
Eyerman, J, 1896 A, p. 267.
Flower and Lydekker 1891 A, p. 555.
Nicholson and Lydekker 1889 A, p. 1436.
Schlosser, M. 1888 B, pp. 18, 23, 24, 56.
Scott, W. B. 1890 C, p. 38.
    1897 A, p. 3.
    1898 B.
Wortman and Matthew 1899 A, pp. 115, 130.
Zittel, K. A. 1893 B, p. 626.

Temnocyon altigenis Cope.

Cope, E. D. 1879 D, p. 68.
    1879 B, p. 58.
    1881 C, p. 179.
    1883 T, p. 238, figs. 2, 3.
    1884 O, p. 903, pl. lxviii, fig. 9; pl. lxx, fig. 11.
Matthew, W. D. 1899 A, p. 62.·
Roger, O. 1896 A, p. 44.

Trouessart, E. L. 1897 A, p. 296.
Wortman and Matthew 1899 A, p. 130.
Zittel, K. A. 1893 B, p. 625, fig. 528, C, D.
    Miocene (John Day); Oregon.

Temnocyon ferox Eyerman.

Eyerman, J. 1894 A.
    1896 A.
Matthew, W. D. 1899 A, p. 62.
Trouessart, E. L. 1897 A, p. 297.
Wortman and Matthew 1899 A, p. 117, fig. 3, p. 130.
    Miocene (John Day); Oregon.

Temnocyon wallovianus Cope.

Cope, E. D. 1881 C, p. 179.
    1883 T, p. 239.
    1884 O, pp. 903, 905, pl. lxx, fig. 10.
Matthew, W. D. 1899 A, p. 62.
Roger, O. 1896 A, p. 44.
Trouessart, E. L. 1897 A, p. 296.
Wortman and Matthew 1899 A, p. 130.
Zittel, K. A. 1893 B, p. 625, fig. 528 B.
    Miocene (John Day); Oregon.

MESOCYON Scott.   Type *Temnocyon coryphæus* Cope.

Scott, W. B. 1890 C, p. 38.
Eyerman, J. 1894 A. (Hypotemnodon, type *Temnocyon coryphæus*.)
    1896 A, p. 284. (Hypotemnodon.)
Hay, O. P. 1899 C, p. 254.
Scott, W. B. 1897 A, p. 3. (Hypotemnodon.)
    1898 B. (Hypotemnodon.)
Matthew, W. D. 1899 A. p. 63. (Hypotemnodon.)
Trouessart, E. L. 1897 A, p. 297. (Hypotemnodon.)
Wortman and Matthew 1899 A, pp. 118, 130. (Hypotemnodon.)

Mesocyon coryphæus (Cope).

Cope, E. D. 1879 F, p. 180. (Temnocyon.)
    1879 B, p. 58. (Canis hartshornianus.)
    1881 C, p. 179. (Temnocyon.)
    1883 T, p. 239, fig. 4. (Temnocyon.)
    1884 O, pp. 903, 905, pl. lxxi; pl. lxxia, figs. 1-7; pl. lxxiia, figs. 4-7. (Temnocyon.)
Eyerman, J. 1894 A. (Hypotemnodon.)

Eyerman, J. 1896 A, p. 284. (Hypotemnodon.)
Hay, O. P. 1899 C, p. 253.
Matthew, W. D. 1899 A, p. 63. (Hypotemnodon.)
Scott, W. B. 1890 C, p. 39. (Mesocyon.)
Trouessart, E. L. 1897 A, p. 297. (Hypotemnodon.)
Wortman and Matthew 1899 A, p. 130. (Hypotemnodon.)
Zittel, K. A. 1893 B, p. 625, fig. 528 A.
    Miocene (John Day) Oregon.

Mesocyon josephi (Cope).

Cope, E. D. 1881 C, p. 179. (Temnocyon.)
    1883 T, p. 239. (Temnocyon.)
    1884 O, pp. 903, 912, pl. lxx, fig. 9. (Temnocyon.)
Matthew, W. D. 1899 A, p. 63.
Roger, O. 1896 A, p. 44.
Trouessart, E. L. 1897 A, p. 297. (Temnocyon.)
    Miocene (John Day); Oregon.

CYNODESMUS Scott.   Type *C. thooides* Scott.

Scott, W. B. 1893 B, p. 660.
    1895 C, p. 63.
    1897 A, p. 3.
Wortman and Matthew 1899 A, p. 118.

Cynodesmus thooides Scott.

Scott, W. B. 1893 B, p. 660.

Matthew, W. D. 1899 A, p. 54.
Roger, O. 1896 A, p. 44.
Scott, W. B. 1895 C, p. 63, pl. i, figs. 1–5.
Trouessart, E. L. 1897 A, p. 297.
    Oligocene (White River); Montana?.

ÆLURODON Leidy.   Type *A. ferox* Leidy=*A. sævus* Leidy.

Leidy, J. 1858 E, p. 22.
Cope, E. D. 1875 N, p. 256.
    1881 F, p. 387.
    1883 T, p. 243.
    1884 O, p. 893.

Flower and Lydekker 1891 A, p. 562.
Leidy, J. 1858 E, p. 21. (Subgenus Epicyon, type *Canis haydeni.*)
    1869 A, p. 68.
    1871 C, p. 344.

Nicholson and Lydekker 1889 A, p. 1433.
Schlosser, M. 1888 B, pp. 24, 61.
    1890 B, p. 25. (Prohyæna, type *Canis wheelerianus* Cope.)
Scott, W. B. 1888 A.
    1891 D, p. 90.
    1892 A, p. 425.
    . 1898 B, p. 402.
Scott and Osborn 1890 B, p. 66.
Trouessart, E. L. 1897 A, p. 297.
Zittel, K. A. 1893 B, p. 626.

Ælurodon compressus Cope.

Cope, E. D. 1890 J, p. 1067.
Matthew, W. D. 1899 A, p. 67.
Roger, O. 1896 A, p. 45.
Scott and Osborn 1890 B, p. 67, fig. 2 A. (*Æ. hyænoides*, not of Cope.)
Trouessart, E. L. 1897 A, p. 297.
    Miocene (Loup Fork); Nebraska.

Ælurodon haydeni (Leidy).

Leidy, J. 1858 E, p. 21. [Canis (Epicyon).]
Cope, E. D. 1874 P, p. 11. (Canis.)
    1875 N, p. 256. (Amphicyon.)
    1883 T, p. 242. (Canis.)
Hayden, F. V. 1858 B, p. 158. [Canis (Epicyon).]
Leidy, J. 1869 A, pp. 30, 368, pl. i, fig. 10. [Canis (Epicyon).]
    1871 C, P. 341. (Canis.)
Matthew, W. D. 1899 A, p. 67.
Roger, O. 1896 A, p. 45.
Scott and Osborn 1890 B, p. 66, fig. 2 B.
Trouessart, E. L. 1897 A, p. 297.
    Miocene (Loup Fork); Nebraska.

Ælurodon hyænoides Cope.

Cope, E. D. 1881 F, p. 388.
    1881 BB, p. 1028.
    1883 T, p. 244, fig. 11.
    1884 O, p. 945, fig. 35c.
Matthew, W. D. 1899 A, p. 67.
Roger, O. 1896 A, p. 45.
Schlosser, M. 1890 B, p, 25.
Trouessart, E. L. 1897 A, p. 298.
    Miocene (Loup Fork); Kansas or Nebraska.

Ælurodon meandrinus Hatcher.

Hatcher, J. B. 1894 C, p. 239.
Matthew W. D. 1899 A, p. 67.
Roger, O. 1896 A, p. 45.
Trouessart, E. L. 1897 A, p. 298.
    Miocene (Loup Fork); Nebraska.

Ælurodon sævus (Leidy).

Leidy, J. 1858 E, p. 21. (Canis.)
Cope, E. D. 1874 B, p. 519. (Canis.)
    1874 P, p. 11. (Canis.)
    1875 F, p. 68. (Canis.)

CANIMARTES Cope. Type *C. cumminsii* Cope.

Cope E. D. 1892 Y, p. 1029.
    1892 I, p. 327.
    1893 A, p. 51.

Canimartes cumminsii Cope.

Cope, E. D. 1892 Y, p. 1029.

Cope, E. D. 1877 K, p. 302. (Canis lupus.)
    1881 F, p. 387.
    1883 T, p. 243, fig. 9.
    1884 O, p. 945, text fig. 36.
    1888 W, p. 1020.
    1889 C, p. 234, fig, 60.
    1890 J, p. 1067, pl. xxxii.
    1898 B, p. 100, fig. 37.
Hayden, F. V. 1858 B, p. 158. (Ælurodon ferox. Canis sævus.)
King, C. 1878 A, p. 430. (Canis.)
Leidy, J. 1858 E, p. 21 (Canis sævus): p. 22 (Ælurodon ferox).
    1869 A, pp. 28, 368, pl. i, fig. 9 (Canis sævus); pp. 68, 367, pl. i, figs. 13, 14 (Ælurodon ferox).
    1870 V, p. 65. (Ælurodon ferox?.)
    1871 C, p. 341 (Canis sævus); p. 344 (Ælurodon ferox).
Matthew, W. D. 1899 A, p. 67.
Roger, O. 1896 A, p. 45.
Schlosser, M. 1888 B, pp. 24, 61. (Canis.)
Scott and Osborn 1890 B, p. 66, figs. 1, 2.
Trouessart, E. L. 1897 A, p. 297.
    Miocene(Loup Fork); Nebraska, New Mexico.

Ælurodon taxoides Hatcher.

Hatcher, J. B. 1894 C, p. 236, pl. i, fig. 2.
Matthew, W. D. 1899 A, p. 67.
Roger, O. 1896 A, p. 45.
Trouessart, E. L. 1897 A, p. 298.
    Miocene (Loup Fork); Nebraska.

Ælurodon ursinus (Cope).

Cope, E. D. 1875 N, p. 256. (Canis.)
    1875 F, p. 68. (Canis.)
    1877 K, p. 304, pl. lxix, fig. 1. (Canis.)
    1879 A, p. 46. (Amphicyon?.)
    1881 C, p. 178. (Canis.)
    1883 T, p. 242. (Canis.)
    1884 O, p. 39. (Amphicyon.)
Matthew, W. D. 1899 A, p. 67.
Roger, O. 1896 A, p. 45.
Scott and Osborn 1890 B, p. 68.
Trouessart, E. L. 1897 A, p 298.
    Miocene(Loup Fork); New Mexico, Nebraska.

Ælurodon wheelerianus Cope.

Cope, E. D. 1877 K, p. 302, pl. lxix, fig. 2. (Canis.)
    1881 F, p. 388.
    1883 T, p. 244, fig. 11.
Matthew, W. D. 1899 A, p. 67.
Roger, O. 1896 A, p. 45.
Schlosser, M. 1890 B, p. 25. (Prohyæna.)
Scott and Osborn 1890 B, p. 67.
Trouessart, E. L. 1897 A, p. 297.
Zittel, K. A. 1893 B, p. 626.
    Miocene(Loup Fork); New Mexico, Nebraska.

Cope, E. D. 1892 I, p. 327.
    1893 A, p. 52, pl. xiv, figs. 12–14.
Matthew, W. D. 1899 A, p. 75.
Roger, O. 1896 A, p. 59.
Trouessart, E. L. 1897 A, p. 298.
    Pliocene (Blanco); Texas.

TOMARCTUS Cope. Type *T. brevirostris* Cope.

Cope, E. D. 1873 T, p. 2.
    1874 B, p. 519.
    1879 F, pp. 179, 185.
    1883 T, p. 243.
Nicholson and Lydekker 1889 A, p. 1433.

Tomarctus brevirostris Cope.
Cope, E. D. 1873 T, p. 2.

Cope, E. D. 1873 CC, p. 9.
    1874 B, p. 520.
    1874 P, p. 11.
    1883 T, p. 243.
Matthew, W. D. 1899 A, p. 68.
Roger O. 1896 A, p. 55.
    Miocene (Loup Fork); Colorado.

CANIS Linn. Type *C. familiaris* Linn.

Linnæus, C., 1758, Syst. Nat., ed. 10, i, p. 38.
Baird, S. F. 1857 A, p. 104.
Blainville, H. M. D. 1864 A, ii, P, plates.
Cope, E. D. 1877 K, p. 300.
    1879 F, pp. 178, 184.
    1881 C, p. 178.
    1884 O, p. 893.
Cuvier, G. 1834 A, vii, p. 465.
Filhol, H. 1883 A.
Flower, W. H. 1869 B, p. 24.
    1871 B.
    1876 B, xiii, p. 487.
    1883 D, p. 438.
    1885 A.
Flower and Lydekker 1891 A, p. 546.
Gray, J. E. 1868 B, p. 508.
    1869 A, p. 193.
Huxley, T. H. 1872 A, p. 353.
    1880 A.
Leche, W. 1895 A, p. 60.
Lydekker, R. 1884 A, p. 252.
    1885 B, p. 123.
Maggi, L. 1896 D.
    1896 E, figs. 1-20.
Nicholson and Lydekker 1889 A, p. 1435.
Owen, R. 1868 A, p. 331, fig. 262.
Paulli, S. 1900 B, p. 489.
Pictet, F. J. 1853 A, p. 202.
Schlosser, M. 1888 B, p. 20.
Shufeldt, R. W. 1900 A, p. 395. (Vulpes.)
Steinmann, and Döderlein 1890 A, p. 720.
Trouessart, E. L. 1897 A, p. 299.
Wilckens, M. 1885 A, p. 518.
Winge, H. 1896 B, pp. 60, 91, 102, 103.
Wortman, J. L. 1886 A, p. 397.
Zittel, K. A. 1893 B, p. 627.

Canis? anceps Scott.
Scott, W. B. 1893 B, p. 660.
Matthew, W. D. 1899 A, p. 67.
Roger, O. 1896 A, p. 46.
Scott, W. B. 1895 C, p. 75.
Trouessart, E. L. 1897 A, p. 302.
    Miocene (Loup Fork); Montana.

Canis brachypus Cope.
Cope, E. D. 1881 F, p. 389.
    1883 T, p. 242, fig. 8.
    1888 W.
Matthew, W. D. 1899 A, p. 67.
Roger, O. 1896 A, p. 48.
    Miocene (Loup Fork); Wyoming.

Canis cinereoargentatus Schreber.
Allen, J. A. 1876 C, p. 333. (Vulpes virginianus.)

Baird, S. F. 1857 A, p. 139, pl. xxxv, fig. 1.
Cope, E. D. 1895 F, p. 447. (Vulpes.)
Leidy, A. p. 368. (C. virginianus.)
    1889 H, p. 5. (C. virginianus.)
Redfield, W. C. 1850 A, p. 255. ("Vulpes"; may be not this species.)
Trouessart, E. L. 1897 A, p. 313. (Urocyon.)
    Pleistocene; lead region of Illinois, New York?, Pennsylvania.

Canis dirus Allen.
Allen, J. A. 1876 A, p. 48.
Cope, E. D. 1877 K, p. 301.
Roger, O. 1896 A, p. 48.
Trouessart, E. L. 1897 A, p. 302.
    Pleistocene; lead region of Upper Mississippi River.

Canis indianensis Leidy.
Leidy, J. 1869 A, p. 368.
Allen, J. A. 1876 A, p. 48 (C. indianensis, C. primævus); p. 49 (C. mississipiensis).
Cope, E. D. 1894 A, p. 453, pl. xxi, figs. 14-16.
    1899 A, p. 227.
Cope and Wortman 1884 A, p. 9. (C. lupus.)
Leidy, J. 1854 I, p. 200. (C. primævus; name preoccupied.)
    1856 T, p. 167, pl. xvii, figs. 11, 12. (C. primævus.)
    1873 B, pp. 230, 315, pl. xxxi, fig. 2.
    1873 F, p. 260.
Roger, O. 1896 A, p. 48. (C. indianensis, C. mississippiensis.)
Trouessart, E. L. 1897 A, p. 302. (C. indianensis, C. mississippiensis.)
Whitney, J. D. 1879 A, p. 246.
Yates, L. G. 1874 B, p. 18.
    Pleistocene; Indiana, lead region of Upper Mississippi River; California (Leidy).

Canis latidentatus (Cope).
Cope, E. D. 1899 A, p. 228, pl. xviii, figs. 4, 4a. (Vulpes.)

Canis latrans Say.
Say, T., 1823, Long's Exped. R. Mts., i, p. 168.
Baird, S. F. 1857 A, p. 113, pl. lxxvi.
Cope, E. D. 1878 H, p. 389.
    1883 T, p. 242.
    1889 F, p. 161.
    1889 S, p. 980.
Cope and Wortman 1884 A, p. 7.
Dana, J. D. 1879 A, p. 233.
Haeckel, E. 1895 A, p. 586.
Leidy, J. 1869 A, p. 369.

Trouessart, E. L. 1897 A, p. 304.
Whitney, J. D. 1879 A, p. 246.
Wyman, J. 1862 A, p. 422.
    Recent; northern part of Mississippi Valley to
    Alberta. Pleistocene; Illinois, Indiana, Oregon.

Canis? marshii Hay.

Hay, O. P. 1899 C, p. 253.
Leidy, J. 1872 B, p. 356. (C. montanus.)
Lydekker, R. 1884 A, p. 258. (C. montanus.)
Marsh, O. C. 1871 E, p. 123. (C. montanus, not of
    Pearson.)
Matthew, W. D. 1899 A, pp. 40, 67. (C. montanus.)
Roger, O. 1896 A, p. 46. (C. montanus.)
Schlosser, M. 1890 B, pp. 62, 64. (C. montanus.)
Trouessart, E. L. 1897 A, p. 307. (C. montanus?.)
    Eocene (Bridger); Wyoming.
    The generic position of this species is uncer-
tain, but the specific title *montanus* is not
available.

Canis occidentalis Richardson.

Richardson, J., 1829, Fauna Bor.-Amer., p. 60.
Baird, S. F. 1857 A, p. 104, pl. xxxi.
Cope, E. D. 1883 T, p. 242. (C. lupus.)
Leidy, J. 1869 A, p. 368.
    1889 H, p. 5. (C. lupus.)
Lydekker, R. 1884 A, p. 264. (C. lupus?.)
Trouessart, E. L. 1897 A, p. 302.
Williston, S. W. 1898 I, p. 92. (C. lupus.)
Wyman, J. 1862 A, p. 422.
    Recent; North America. Pleistocene; lead
region of Upper Mississippi River.

Canis priscolatrans Cope.

Cope, E. D. 1899 A, p. 227 pl. xviii, figs. 3–8g.
    Pleistocene; Pennsylvania.

Canis temerarius Leidy.

Leidy, J. 1858 E, p. 21.
Cope, E. D. 1883 T, p. 242.
    1884 O, p. 915.
Hayden, F. V. 1858 B, p. 158.
King, C. 1878 A, p. 430.
Leidy, J. 1869 A, pp. 29, 368, pl i, fig. 12.
    1871 C, p. 341.
Matthew, W. D. 1899 A, p. 67.
Roger, O. 1896 A, p. 48.
Trouessart, E. L. 1897 A, p. 302.
    Miocene (Loup Fork); Nebraska, Colorado.

Canis vafer Leidy.

Leidy, J. 1858 E, p. 21.
Cope, E. D. 1875 F, p. 68.
    1877 K, pp. 301, 302.
    1884 O, p. 915.
Hayden, F. V. 1858 B, p. 158.
Leidy, J. 1869 A, pp. 29, 368, pl. i, fig. 11.
    1871 C, p. 341.
    1873 B, p. 315.
Matthew, W. D. 1899 A, p. 67.
Roger, O. 1896 A, p. 48.
Schlosser, M. 1888 B, p. 28.
Scott and Osborn 1890 B, p. 69, fig. 3. (C. vafer?.)
Trouessart, E. L. 1897 A, p. 310. (Vulpes.)
    Miocene (Loup Fork); Nebraska.

PACHYCYON Allen. Type *P. robustus* Allen.

Allen. J. A. 1885 A, p. 1.

Pachycyon robustus Allen.

Allen, J. A. 1885 A, p. 1, pls. i–iii.

Cope, E. D. 1886 G.
Trouessart, E. L. 1897 A, p. 302. (Canis.)
    Pleistocene; Virginia.

HYÆNIDÆ.

Blainville, H. M. D. 1864 A, ii, Q, pls. i–viii.
Cope, E. D. 1882 D, p. 474.
    1884 O, p. 892.
    1888 D, p. 256.
Flower and Lydekker 1891 A, p. 540.
Gill, T. 1872 B, p. 57.
Gray, J. E. 1869 A, p. 211.
Haeckel, E. 1895 A, p. 578. (Hyænida.)
Lydekker, R. 1884 A, p. 274.
    1885 B, p. 68.
Mivart, St. G. 1881 A.
    1882 A.
Nicholson and Lydekker 1889 A, p. 1440.

Owen, R. 1845 B, p. 482.
Roger, O. 1896 A, p. 65.
Schlosser, M. 1888 B, pp. 4, 14.
    1890 B, p. 24.
    1890 C, p. 265.
Tomes, C. S. 1898 A, p. 467.
Trouessart, E. L. 1897 A, p. 317.
Turner, H. N. 1848 A.
Waterhouse, G. R. 1839 A.
Winge, H. 1895 B, pp. 47, 59.
Woodward, A. S. 1898 B, p. 397.
Zittel, K. A. 1893 B, p. 660.

BOROPHAGUS Cope. Type *B. diversidens* Cope.

Cope, E. D. 1892 Y, p. 1028.
    1892 I, p. 326.
    1898 A, p. 52.

Borophagus diversidens Cope.

Cope, E. D. 1892 Y, p. 1028.

Cope, E. D. 1892 I, p. 326.
    1893 A, p. 54, pl. xiii, fig. 4.
Matthew, W. D. 1899 A, p. 75.
Roger, O. 1896 A, p. 66. [Hyaena (Borophagus).]
Trouessart, E. L. 1897 A, p. 320.
    Pliocene (Blanco); Texas.

Superfamily *ÆLUROIDEA* Flower.

Flower, W. H. 1869 B, p. 22.
Cope, E. D. 1882 D, p. 473.   (Epimycteri.)
   1884 O, pp. 890, 892.   (Epimycteri.)
Flower, W. H. 1883 D, p. 434.
Flower and Lydekker 1891 A, p. 501.
Huxley, T. H. 1872 A, p. 358.

Mivart, St. G. 1881 A, p. 474.
   1882 A.
Schlosser, M. 1888 B, p. 6.
   1890 C, p. 265.
Tims, H. W. M. 1896 B, p. 462.

FELIDÆ.

Adams, G. I. 1896 A.
Baird, S. F. 1857 A, p. 80.
Blainville, H. M. D. 1864 A, ii, O, pls. i–xix.
Cope, E. D. 1879 F.
   1880 D, pp. 834, 851.
   1881 C.
   1882 D, p. 474.
   1884 O, pp. 263, 892.
   1888 D.
Cuvier, G. 1884 A, vii, p. 361.
Flower, W. H. 1876 B, xiii, p. 488.
   1883 D, p. 434.
Flower and Lydekker 1891 A, p. 502.
Gill, T. 1872 B, p. 57.
Gray, J. E. 1869 A, p. 5.
Haeckel, E. 1895 A, p. 587.   (Felida.)
Lydekker, R. 1884 A, p. 313.
Mivart, St. G. 1881 A, p. 439.
   1882 A.
Nicholson and Lydekker 1889 A, p. 1443.

Osborn and Wortman 1892 A, p. 94.
Owen, R. 1845 B, p. 486.
Parsons, F. G. 1899 A.
   1900 A.
Roger, O. 1889 A.
   1896 A, p. 66.
Scheidt, P. 1894 A.
Schlosser, M. 1888 B, pp. 4, 14.
   1890 B, p. 34.
Scott, W. B. 1889 B, p. 235.
Slade, D. D. 1890 A.
Steinmann and Döderlein 1890 A, p. 725.
Tomes, C. S. 1898 A, p. 469.
Turner, H. N. 1848 A.
Waterhouse, G. R. 1839 A.
Weyhe, —, 1875 A, p. 116.
Williston, S. W. 1898 K.
Winge, H. 1895 B, pp. 47, 58.
Woodward, A. S. 1898 B, p. 398.
Zittel, K. A. 1893 B, pp. 663, 677.

MACHAIRODONTINÆ.

Adams, G. I. 1896 A, p. 443.   (Machærodontinæ.)
Cope, E. D. 1880 D, p. 835.   (Nimravidæ.)
   1881 C.   (Nimravidæ.)
   1882 D, p. 474.   (Nimravidæ.)
   1884 O, pp. 892, 947.   (Nimravidæ.)
Gill, T. 1872 B, pp. 59, 60.   (Machairodontinæ.)
Lydekker, R. 1885 E, p. 469.   (Nimravidæ.)
Mivart, St. G. 1881 A, p. 439.   (Nimravidæ.)
Nicholson and Lydekker 1889 A, p. 1443.   (Nimravidæ.)
Osborn and Wortman 1892 A.   (Nimravidæ.)
Roger, O. 1896 A, p. 67.   (Machairodinæ.)
Schlosser, M. 1890 B, p. 38.   (Nimravidæ.)

Scott, W. B. 1888 A.   (Nimravidæ.)
   1889 B, p. 238.   (Nimravidæ.)
   1892 A, p. 425.   (Nimravidæ.)
   1898 B, p. 404.
Scott and Osborn 1887 B, p. 153.   (Nimravidæ.)
Steinmann and Döderlein 1890 A, p. 726.   (Nimravinæ.)
Trouessart, E. L. 1897 A, p. 343.   (Machærodinæ.)
Winge, H. 1895 B, pp. 47, 55, 56.
Woodward, A. S. 1898 B, p. 399.   (Nimravidæ, or Machærodontidæ.)
Zittel, K. A. 1893 B, p. 667.

MACHAIRODUS Kaup.   Type *M. neogæus* Lund.

Kaup, J., 1833 Oss. foss. Mus. Darmstadt.
   The name of this genus is commonly spelled *Machærodus*.
Adams, G. I. 1896 A, p. 432.
   1897 A, p. 147.
Brown, H. G. 1838 A, p. 1277, pl. xlv, fig. 4.
Burmeister, H. 1866 B, p. 123.
   1879 B, p. 106.
Cope, E. D. 1879 F, p. 171.
Dawkins and Sanford 1872 A, p. 184.
Filhol H. 1891 A, p. 47.
Gervais, P. 1859 A, p. 230.
Hartmann, R. 1893 A.
Kittl, E. 1887 A, p. 321.
Leidy, J. 1854 A, pp. 96, 115.

Lydekker, R. 1884 A, p. 332.
   1885 B, p. 41.
March, O. C. 1877 E.
Nicholson and Lydekker 1889 A, p. 1449.
Owen, R. 1845 B, p. 490.
   1846 B, p. 174.
   1847 C, p. 67.
   1868 A, p. 339.
Pictet, F. J. 1853 A, p. 230.
Quenstedt, F. A. 1885 A, p. 42.
Steinmann and Döderlein 1890 A, p. 727.
Trouessart, E. L. 1897 A, p. 347.
Winge, H. 1895 B, pp. 47, 56, 91.
Zittel, K. A. 1893 B, p. 673.

Machairodus gracilis Cope.

Cope, E. D. 1880 D, p. 857.   (Smilodon.)
  1895 F, p. 448.   (Smilodon.)
  1899 A, p. 240, pl. xx, fig. 1.
Cope and Wortman 1884 A, p. 5.   (Smilodon.)
Cragin, F. W. 1892 B, p. 17.   (Machærodus.)
Lydekker, R. 1884 A, p. 333.   (Machærodus.)
Roger, O. 1896 A, p. 70.   (Machairodus.)
Trouessart, E. L. 1897 A, p. 849.
  Pleistocene; Pennsylvania.

Machairodus mercerii Cope.

Cope, E. D. 1895 F, p. 448.   (Uncia.)
  1895 F, p. 392.   (Uncia mercerii, in part.)
  1899 A, p. 240 (Machærodus); p. 245, pl. xx, figs. 2-2c (Smilodon).
Roger, O. 1896 A, p. 72.   (Felis.)
Trouessart, E. L. 1897 A, p. 351.   (Felis.)
  Pleistocene; Pennsylvania.

ÆLUROTHERIUM Adams.   Type *Patriofelis leidyana* Osb. and Wort.

Adams, G. I. 1896 A, p. 442.

Ælurotherium leidyanum (Osb. and Wort.).

Osborn and Wortman 1892 A, p. 98, fig. 5.   (Patriofelis.)
Adams, G. I. 1896 A, p. 442.

Matthew, W. D. 1899 A, p. 41.
Roger, O. 1896 A, p. 37.   (Patriofelis.)
Scott, W. B. 1892 B, p. 313.   (Genus doubtful.)
Trouessart, E. L. 1897 A, p. 231 (Patriofelis); p. 344 (Ælurotherium).
Wortman, J. L. 1894 A, p. 164.   (Not Patriofelis.)
  Eocene (Bridger); Wyoming.

EUSMILUS Gerv.   Type *E. bidentatus* (*Filhol*).

Gervais, 1875, Jour. Zool., iv, p. 419.
Cope, E. D. 1884 O, p. 948.
Flower, W. H. 1883 D, p. 435.
Flower and Lydekker 1891 A, p. 524.
Hatcher, J. B. 1895 A.
Lydekker, R. 1884 A, p. 382.
Mivart, St. G. 1881 A, p. 487.
Schlosser, M. 1890 B, p. 48.
Trouessart, E. L. 1897 A, p. 846.

Zittel, K. A. 1893 B, p. 672.

Eusmilus dakotensis Hatcher.

Hatcher, J. B. 1895 B.
Adams, G. I. 1896 B, p. 48, pl. ii, fig. 3.
Matthew, W. D. 1899 A, p. 55.
Trouessart, E. L. 1897 A, p. 346.
  Oligocene (White River); South Dakota.

HOPLOPHONEUS Cope.   Type *Machærodus oreodontis* Cope.

Cope, E. D. 1874 B, p. 509.
Adams, G. I. 1896 A.
  1896 B.
Cope, E. D. 1879 F, pp. 170, 171.
  1880 D, pp. 836, 849.
  1881 C, p. 167.
  1884 O, pp. 948, 992.
Flower and Lydekker 1891 A, p. 524.
Leidy, J. 1869 A, p. 53.   (Drepanodon.)
  1871 C, p. 343.   (Drepanodon.)
Mivart, St. G. 1881 A, pp. 433, 439.
Nicholson and Lydekker 1889 A, p. 1446.
Osborn and Wortman 1894 A, p. 227.
Schlosser, M. 1890 B, p. 48.
Scott, W. B. 1889 B.
  1898 B, p. 411.
Scott and Osborn 1887 B, p. 153, text fig. 2.
Steinmann and Döderlein 1890 A, p. 727.
Williston, S. W. 1895 A, p. 170.   (Dinotomius, type *D. atrox* Willist.
Winge, H. 1895 B, pp. 47, 55, 91.
Wortman, J. L. 1893 A.
Zittel, K. A. 1893 B, p. 672.

Hoplophoneus catocopis (Cope).

Cope, E. D. 1887 P, p. 1019.   (Machærodus.)
Adams, G. I. 1896 A, p. 433.
Cope, E. D. 1892 AA.   (Machærodus.)
Matthew, W. D. 1899 A, p. 68.   (Machærodus.)
Roger, O. 1896 A, p. 70.   (Machairodus.)
Trouessart, E. L. 1897 A, p. 346.
  Miocene (Loup Fork); Kansas.

Hoplophoneus cerebralis Cope.

Cope, E. D. 1880 F, p. 143.   (Machærodus.)
Adams, G. I. 1896 A, p. 429, pl. xi, fig. 1.
  1896 B, p. 50, pl. i, fig. 1.
Cope, E. D. 1880 D, p. 850.
  1884 O, pp. 976, 997, pl. lxxva, figs. 3-5.
Matthew, W. D. 1899 A, p. 63.
Roger, O. 1896 A, p. 68.
Trouessart, E. L. 1897 A, p. 345.
  Miocene (John Day); Oregon.

Hoplophoneus crassidens (Cragin).

Cragin, F. W. 1892 B, p. 17.   (Machærodus.)
Adams, G. I. 1896 A, p. 433.
Matthew, W. D. 1899 A, p. 55.
Trouessart, E. L. 1897 A, p. 346.
Williston, S. W. 1895 A, p. 174, pl. xix.   (Machærodus.)
  Miocene? (Loup Fork?); Kansas.

Hoplophoneus insolens Adams.

Adams, G. I. 1896 A, p. 429, pl. xi, fig. 5.
  1896 B, p. 48, pl. i, fig. 5.
Matthew, W. D. 1899 A, p. 55.
Osborn and Wortman 1894 A, p. 227.   (H. occidentalis, not of Leidy.)
Trouessart, E. L. 1897 A, p. 846.
  Oligocene (White River); South Dakota.

Hoplophoneus occidentalis (Leidy).

Leidy, J. 1869 A, pp. 63, 367, pl. v, fig. 5.   (Drepanodon.)

Adams, G. I. 1896 A, p. 428, pl. xi, fig. 6.
   1896 B, p. 47, pl. i, fig. 6; pl. ii, figs. 1, 2.
Cope, E. D. 1884 O, p. 998.
Leidy, J. 1866 A, p. 345. (Drepanodon occiden-
   talis; no description.)
   1871 C, p. 343. (Drepanodon.)
Matthew, W. D. 1899 A, p. 54.
Riggs, E. S. 1896 A, pl. 1.
Roger, O. 1896 A, p. 68.
Scott and Osborn 1887 B, p. 154.
Trouessart, E. L. 1897 A, p. 346.
Williston, S. W. 1895 A, p. 170, pl. xviii. (Dino-
   tomius atrox.)
   1898 K, p. 349, figure.
   Oligocene (White River); South Dakota.

Hoplophoneus oreodontis Cope.

Cope, E. D. 1873 CC, p. 9. (Machærodus.)
Adams, G. I, 1896 A, p. 429, pl. xi, fig. 2.
   1896 B, p. 50, pl. i, fig. 2.
Bruce, A. T. 1883 A, p. 42, fig. 8.
Cope, E. D. 1874 B, p. 509.
   1874 P, p. 23.
   1880 D, p. 850.
   1884 O, pp. 976, 993, pl. lxviia, fig. 17; pl.
   lxxva, figs. 1, 2.
Matthew, W. D. 1899 A, p. 55.
Mivart, St. G. 1881 A, p. 434, fig. 185.
Roger, O. 1896 A, p. 68.
Trouessart, E. L. 1897 A, p. 345.
   Oligocene (White River); Colorado.

Hoplophoneus primævus (Leidy).

Leidy, J. 1851 H, p. 329. (Machairodus.)
Adams, G. I. 1896 A, p. 419, pl. x; pl. xi, fig. 3.
   1896 B, p, 49, pl. i, fig. 4.
Cope, E, D. 1880 D, p. 850.
   1884 O, p. 993.
Hayden, F. V. 1858 B, p. 158.
King, C. 1878 A, p. 411. (Drepanodon.)

Leidy, J. 1852 J, p. 564, pl. xii A, fig. 5. (Machairodus.)
   1858 D, p. 392. (Machairodus.)
   1854 A, pp. 94, 115, pl. xviii. (Machairodus.)
   1854 E, p. 156. (Machairodus.)
   1857 D, p. 176. (Drepanodon.)
   1869 A, pp. 54, 367, pl. i, fig. 8. (Drepanodon.)
   1871 C, p. 343. (Drepanodon.)
Matthew, W. D. 1899 A, p. 75.
Osborn, H. F. 1898 Q, p. 11.
Roger, O. 1896 A, p. 68.
Scott and Osborn, 1887 B, p. 153, pl. i.
Trouessart, E. L. 1897 A, p. 345.
   Oligocene (White River); South Dakota.

Hoplophoneus robustus Adams.

Adams, G. I. 1896 B, p. 49, pl. i, fig. 4.
   1896 A, p. 428, pl. xi, fig. 4.
Leidy, J. 1869 A, p. 54 (in part); pl. iv, fig. 1.
Matthew, W. D. 1899 A, p. 55.
Osborn and Wortman 1894 A, p. 227. (H. pri-
   mævus, not of Leidy.)
Trouessart, E. L. 1897 A, p. 346.
   Oligocene (White River); South Dakota, Ne-
   braska.

Hoplophoneus strigidens Cope.

Cope, E. D. 1879 D, p. 71. (Machærodus.)
Adams, G. I. 1896 A, p. 429.
Cope, E. D. 1879 B, p. 56. (Machærodus.)
   1879 X, p. 798a. (Machærodus.)
   1880 D, p. 351.
   1884 O, p. 1001, pl. lxxva, fig. 6.
Matthew, W. D. 1899 A, p. 63.
Roger, O. 1896 A, p. 68.
Trouessart, E. L. 1897 A, p. 346.
   Miocene (John Day); Oregon.
   The generic position of this species is doubt-
   ful.

DEINICTIS Leidy.   Type *D. felina* Leidy.

Leidy, J. 1856 H, p. 91.
   Most authors have adopted the spelling *Dinictis.* The original name is *Deinictis.*
Adams, G. I. 1896 A.
Allen, H. 1888 A.
Cope, E. D. 1873 T, p. 2. (Daptophilus, type *D.
   squalidens* Cope.)
   1874 B, p. 508. (Daptophilus.)
   1879 F, pp. 169, 170.
   1880 D, pp. 836, 845 (Dinictis); pp. 836, 847
   (Pogonodon).
   1880 F, p. 143 (Dinictis; Pogonodon, type *P.
   platycopis).*
   1881 C, p. 167. (Dinictis, Pogonodon.)
   1884 O, pp. 948, 973 (Dinictis); pp. 948, 981
   (Pogonodon).
Flower and Lydekker 1891 A, p. 523 (Dinictis);
   p. 524 (Pogonodon).
Gaudry, A. 1878 B, p. 220. (Dinictis.)
Leidy, J. 1854 D, p. 127. (Dinictis; no description.)
   1869 A, p. 64. (Dinictis.)

Leidy, J. 1871 C, p. 343. (Dinictis.)
Mivart, St. G. 1881 A, pp. 435, 439 (Dinictis); pp.
   437, 439 (Pogonodon).
Nicholson and Lydekker 1889 A, p. 1444 (Dinic-
   tis); p. 1446 (Pogonodon).
Schlosser, M. 1890 B, p. 42.
Scott, W. B. 1892 A, p. 425.
Steinmann and Döderlein, 1890 A, p. 726 (Dinic-
   tis); p. 727 (Pogonodon).
Trouessart, E. L. 1897 A, p. 344.
Williston, S. W. 1895 A, p. 173, figure.
Winge, H. 1895 B, pp. 47, 55, 91.
Zittel, K. A. 1893 B, p. 669 (Dinictis); p. 671 (Po-
   gonodon).

Deinictis bombifrons Adams.

Adams, G. I. 1895 A, p. 577, pl. xxvi. (Dinictis.)
Matthew, W. D. 1899 A, p. 54. (Dinictis.)
Roger, O. 1896 A, p. 68. (Dinictis.)
Trouessart, E. L. 1897 A, p. 345. (Dinictis.)
   Oligocene (White River); South Dakota?.

Deinictis brachyops (Cope).

Cope, E. D. 1879 D, p. 72, in part.   (Machærodus.)
Adams, G. I. 1896 A, p. 431, pl. xii, fig. 5.   (Dinictis.)
Cope, E. D. 1879 B, p. 56.   (Hoplophoneus.)
    1879 K, p. 197.   (Hoplophoneus.)
    1880 D, pp. 843, 849, fig. 11.   (Pogonodon.)
    1884 O, pp. 972, 967, pl. lxxiv, figs. 3–10.
        (Pogonodon.)
Matthew, W. D. 1899 A, p. 63.   (Pogonodon.)
Roger, O. 1896 A, p. 68.   (Pogonodon.)
Trouessart, E. L. 1897 A, p. 345.   (Dinictis.)
    Miocene (John Day); Oregon.

Deinictis cyclops Cope.

Cope, E. D. 1879 F, p. 176.   (Dinictis.)
Adams, G. I. 1896 A, p. 430, pl. xii, fig. 2.   (Dinictis.)
Cope, E. D. 1880 D, p. 846, fig. 8.   (Dinictis.)
    1884 O, p. 974, pl. lxxv, fig. 1.   (Dinictis.)
Matthew, W. D. 1899 A, p. 63.
Mivart, St. G. 1881 A, p. 435.   (Dinictis.)
Roger, O. 1896 A, p. 67.
Trouessart, E. L. 1897 A, p. 344.   (Dinictis.)
Zittel, K. A. 1893 B, p. 670, fig. 560.   (Dinictis.)
    Oligocene (John Day); Oregon.

Deinictis felina Leidy.

Leidy, J. 1856 H, p. 91.
Adams, G. I. 1896 A, p. 430, pl. xii, fig. 3.   (Dinictis.)
Cope, E. D. 1879 F, p. 176.   (Dinictis.)
    1880 D, p. 846.   (Dinictis.)
    1884 O, pp. 974, 978.   (Dinictis.)
Gaudry, A. 1878 B, p. 220, fig. 292.
Hayden, F. V. 1858 B, p. 158.
King, C. 1878 A, p. 411.   (Dinictis.)
Leidy, J. 1854 D, p. 127.   (Dinictis; no description.)
    1869 A, pp. 64, 368, pl. v, figs. 1–4.   (Dinictis.)
    1871 C, p. 343.   (Dinictis.)
Matthew, W. D. 1899 A, p. 54.   (Dinictis.)
Roger, O. 1896 A, p. 67.   (Dinictis.)
Schlosser, M. 1890 B, p. 41.
Scott, W. B. 1889 A.   (Dinictis.)
Scott and Osborn 1887 B, p. 153.   (Dinictis.)
Trouessart, E. L. 1897 A, p. 344.   (Dinictis.)
Zittel, K. A. 1893 B, p. 669, fig. 559.   (Dinictis.)
    Oligocene (White River); South Dakota,
    Colorado.

Deinictis fortis Adams.

Adams, G. I. 1896 A, p. 574, text fig. 1.   (Dinictis.)
    1896 A, pp. 430, 436, pl. xii, fig. 4.   (Dinictis.)
Matthew, W. D. 1899 A, p. 54.   (Dinictis.)
Roger, O. 1896 A, p. 68.   (Dinictis.)
Trouessart, E. L. 1897 A, p. 345.   (Dinictis.)
    Oligocene (White River); Colorado.

Deinictis major Lucas.

Lucas, F. A. 1898 A, p. 399.   (Dinictis.)
    Oligocene (White River); Nebraska.

Deinictis paucidens Riggs.

Riggs, E. S. 1896 B, p. 237.   (Dinictis.)
Adams, G. I. 1896 A, p. 436, footnote.   (Syn. of D. fortis.)
Matthew, W. D. 1899 A, p. 54.   (Dinictis.)
    Oligocene (White River).

Deinictis platycopis (Cope).

Cope, E. D. 1879 X, p. 798b.   (Hoplophoneus.)
Adams, G. I. 1896 A, p. 431, pl. xii, fig. 6.   (Dinictis.)
Cope, E. D. 1879 E, p. 373.   (Hoplophoneus.)
    1880 D, p. 847, fig. 9.   (Pogonodon.)
    1880 F, p. 143.   (Pogonodon.)
    1884 O, p. 982, text fig. 38, pl. lxxiva.   (Pogonodon.)
Matthew, W. D. 1899 A, p. 63.   (Pogonodon.)
Mivart, St. G. 1881 A, p. 189.   (Pogonodon.)
Roger, O. 1896 A, p. 68.   (Pogonodon.)
Trouessart, E. L. 1897 A, p. 345.   (Dinictis.)
Zittel, K. A. 1893 B, p. 671, fig. 562.   (Pogonodon platycopsis.)
    Miocene (John Day); Oregon.

Deinictis squalidens Cope.

Cope, E. D. 1873 T, p. 2.   (Daptophilus.)
Adams, G. I. 1896 A, p. 430, pl. xii, fig. 1.   (Dinictis.)
Cope, E. D. 1873 CC, p. 9.   (Daptophilus.)
    1874 B, p. 508.   (Daptophilus.)
    1879 F, p. 176.   (Dinictis.)
    1880 D, p. 847,   (Dinictis.)
    1884 O, pp. 974, 979, pl. lxviia, figs. 15, 16.   (Dinictis.)
Matthew, W. D. 1899 A, p. 54.   (Dinictis.)
Roger, O. 1896 A, p. 68.   (Dinictis.)
Trouessart, E. L. 1897 A, p. 344.   (Dinictis.)
    Oligocene (White River); Colorado.

SMILODON Lund.   Type *S. populator* Lund.

Lund, P. W. 1842 A, p. 198.
Adams, G. I. 1896 A, p. 433.
    1897 A, p. 147.
Ameghino, F. 1889 A, p. 333.
Blainville, H. M. D. 1864 A, iv, EE, p. 7, pl. xx.
Cope, E. D. 1880 D, p. 854.
    1881 C, p. 166.
    1896 A, p. 240.
Filhol, H. 1888 B, p. 129.   (On Machairodus bidentatus.)
Gervais P. 1878 A.
Gervais and Ameghino 1880 A, p. 11.
Leidy, J. 1868 A, p. 175.   (Trucifelis, as subgenus. with *Felis fatalis* as type.)
    1869 A, p. 366.   (Trucifelis.)

Lydekker, R. 1885 B, p. 41.   (Syn. of Machærodus.)
Nicholson and Lydekker 1889 A, p. 1449.
Owen, R. 1847 C, p. 67.
Zittel, K. A. 1893 B, p. 675.   (Syn. of Machairodus.)

Smilodon fatalis (Leidy).

Leidy, J. 1868 A, p. 175.   [Felis (Trucifelis).]
Adams, G. I. 1896 A, p. 433.
Cope, E. D. 1880 D, p. 857.
    1899 A, p. 240.
Cope and Wortman 1884 A, p. 5.
Leidy, J. 1869 A, p. 366, pl. xxvii, figs. 10, 11.   (Trucifelis.)
    1873 R, p. 260.   (Trucifelis.)
Lydekker, R. 1884 A, p. 333.   (Machærodus.)

Roger, O. 1896 A, p. 70. (Machairodus.)
Trouessart, E. L. 1897 A, p. 349.
　　Pleistocene; Texas.

Smilodon floridanus (Leidy).

Leidy, J. 1889 A, p. 29. (Drepanodon or Machairodus.)
Adams, G. I. 1896 A, p. 433.

Cope, E. D. 1892 AA. (Machærodus.)
Leidy, J. 1889 D, p. 14, pl. iii, fig. 1. (Machairodus.)
　　1892 A. (Machairodus.)
Roger, O. 1896 A, p. 70. (Machairodus.)
Trouessart, E. L. 1897 A, p. 349.
Zittel, K. A. 1893 B, p. 675. (Machairodus.)
　　Pleistocene; Florida.

DINOBASTIS Cope.　Type *D. serus* Cope.

Cope, E. D. 1893 K, p. 896.
Adams, G. I. 1896 A, pp. 433, 435.
　　1897 A, p. 145. (Syn. of Smilodon.)
Cope, E. D. 1894 H, p. 454.

Dinobastis serus Cope.

Cope, E. D, 1893 K, p. 896.

Cope, E. D. 1894 A, p. 68.
　　1894 H, p. 454, pl. xxi, figs. 1–13.
Roger, O. 1896 A, p. 70.
Trouessart, E. L. 1897 A, p. 349.
　　Pleistocene; Kansas, Oklahoma.

ARCHÆLURUS Cope.　Type *A. debilis* Cope.

Cope, E. D. 1879 X, p. 798a.
Adams, G. I. 1896 A, p. 438.
Allen, H. 1888 A.
Cope, E. D. 1879 E, p. 372.
　　1880 D, p. 836, 841.
　　1881 C, p. 167.
　　1884 O, pp. 948, 952.
Flower and Lydekker 1891 A, p. 524.
Mivart, St. G. 1881 A, pp. 436, 439.
Nicholson and Lydekker 1889 A, p. 1446.
Schlosser, M. 1890 B, p. 40.
Winge, H. 1895 B, pp. 47, 55, 91.
Zittel, K. A. 1893 B, p. 672.

Archælurus debilis Cope.

Cope, E. D. 1879 X, p. 798a.

Cope, E. D. 1879 D, p. 72. (Part of Machærodus brachyops, "female skull.")
　　1879 E, p. 372.
　　1880 D, p. 842, figs. 3, 4, 7.
　　1881 J, p. 546, fig. 7.
　　1884 O, p. 953, pl. lxxia, figs. 8–16.
　　1889 C, p. 190.
Matthew, W. D. 1899 A, p. 63.
Mivart, St. G. 1881 A, p. 436, fig. 188.
Roger, O. 1896 A, p. 68.
Trouessart, E. L. 1897 A, p. 343. (Exclusive of Pseudælurus intrepidus.)
　　Miocene (John Day); Oregon.

NIMRAVUS Cope.　Type *N. gomphodus* Cope.

Cope, E. D. 1879 F, pp. 169, 170.
Adams, G. I. 1896 A, p. 438.
　　1897 A, p. 146.
Cope, 1880 D, pp. 836, 843.
　　1881 C, p. 167.
　　1884 O, pp. 948, 963.
Flower and Lydekker 1891 A, p. 524.
Lydekker, R. 1887 A, p. 313.
Mivart, St. G. 1881 A, pp. 435, 439.
Nicholson and Lydekker 1889 A, p. 1446.
Winge, H. 1895 B, pp. 47, 55, 91.
Zittel, K. A. 1893 B, p. 671.

Nimravus confertus Cope.

Cope, E. D. 1880 D, p. 844, fig. 10.
Adams, G. I. 1897 A, p. 149, fig. 5.
Cope, E. D. 1881 C, p. 172.
　　1884 O, p. 972, pl. lxxia, fig. 17.
Matthew, W. D. 1899 A, p. 63.

Roger, O. 1896 A, p. 68.
Trouessart, E. L. 1897 A, p. 344.
　　Miocene (John Day); Oregon.

Nimravus gomphodus Cope.

Cope, E. D. 1880 D, p. 844, figs. 6, 7.
Adams, G. I. 1897 A, p. 149, fig. 4.
Cope, E. D. 1879 F, p. 170. (N. brachyops; not Deinictis brachyops.)
　　1881 C, p. 171.
　　1881 J, p. 546, fig. 8.
　　1884 O, p. 964, pl. lxxiia, figs. 1–3; pl. lxxiii.
Matthew, W. D. 1899 A, p. 63.
Mivart, St. G. 1881 A, p. 434, fig. 186. (N. brachyops.)
Roger, O. 1896 A, p. 68.
Trouessart, E. L. 1897 A, p. 344.
Zittel, K. A. 1893 B, p. 673, fig. 563.
　　Miocene (John Day); Oregon.

FELINÆ.

Adams, G. I. 1896 A, p. 444.
Gill, T. 1872 B, p. 59.
Roger, O, 1896 A, p. 66 (Prœlurinæ.); p. 67 (Machairodinæ, in part.)

Trouessart, E. L. 1897 A, p. 350.
Zittel, K. A. 1893 B, p. 676.

PSEUDÆLURUS Gervais.   Type *Felis quadridentata* Blainv.

Gervais, P., 1848–1852, Zool. et Paléont. franç., 1st
  edition, p. 127.
Adams, G. I. 1897 A, p. 145.
Cope, E. D. 1879 F, p. 170.
  1881 C, p. 167.
  1884 O, p. 948.
Depéret, C. 1892 A, p. 20.
Filhol, H. 1872 A, p. 3.
  1876 A, p. 158.
  1891 A, p. 73.
Gervais, P. 1859 A, p. 232.
Leidy, J. 1869 A, p. 52.
  1871 C, p. 343.
Lydekker, R. 1884 A, p. 314.
  1885 B, p. 64.
Mivart, St. G. 1881 A, p. 434.
Schlosser, M. 1890 B, p. 39.
Trouessart, E. L. 1897 A, p. 350.

Winge, H. 1895 B, pp. 47, 54, 78, 91.
Zittel, K. A. 1893 B, p. 666.

Pseudælurus intrepidus Leidy.

Leidy, J. 1858 E, p. 22.   [Felis (Pseudælurus).]
Cope, E. D. 1879 X, p. 798a.
Hayden, F. V. 1858 B, p. 158.   [Felis (Pseudæ-
  lurus).]
King, C. 1878 A, p. 411.   (Drepanodon.)
Leidy, J. 1869 A, pp. 52, 367, pl. i, fig. 8.
  1871 C, p. 343.
Matthew, W. D. 1899 A, p. 68.
Roger, O. 1896 A, p. 67.
Schlosser, M. 1890 B, p. 40.
Scott and Osborn 1890 B, p. 71.
Trouessart, E. L. 1897 A, p. 343.   (Syn. of Archæ-
  lurus debilis.)
  Miocene (Loup Fork); Nebraska, Colorado.

FELIS [1] Linn.   Type *F. catus* Linn.

Linnæus, C., 1758, Syst. Nat., ed. 10, i, p. 41.
Baird, S. F. 1857 A, p. 81.
Blainville, H. M. D. 1864 A, ii, O, p. 1, pl. xix.
Cope, E. D. 1879 F, pp. 170, 171.
  1899 A, p. 247.   (Felis, Uncia.)
Flower, W. H. 1869 B, p. 15.
  1883 D, p. 434.
  1885 A.
Flower and Lydekker 1891 A, p. 502.
Leche, W. 1895 A, p. 56.
Mivart, St. G. 1881 A.
  1882 A.
Nicholson and Lydekker 1889 A, p. 1447.
Owen, R. 1866 B
  1868 A.
Pictet, F. J. 1853 A, p. 225.
Trouessart, E. L. 1897 A, p. 351.
Williston, S. W. 1898 K.
Winge, H. 1895 B, pp. 54, 91, 100, 101.
Zittel, K. A. 1893 B, p. 676.

Felis atrox Leidy.

Leidy, J. 1853 F, p. 34.
Cope, E. D. 1880 D, p. 858.   (Uncia.)
  1899 A, p. 249.   (Uncia.)
Cope and Wortman 1884 A, p. 5.
Leidy, J. 1860 B, p. 103.
  1869 A, p. 365.
Roger, O. 1896 A, p. 72.
Trouessart, E. L. 1897 A, p. 351.
Wailes, B. L. C. 1854 A, p. 286.

Felis augustus Leidy.

Leidy, J. 1872 E, p. 39.
Cope, E. D. 1880 D, p. 858.   (Uncia.)
Leidy, J. 1873 B, pp. 227, 315, pl. vii, figs. 18, 19;
  pl. xx, fig. 24.
Matthew, W. D. 1899 A, p. 68.   (Felis?.)
Roger, O. 1896 A, p. 73.   (F. angusta.)
Trouessart, E. L. 1897 A, p. 351.
Zittel, K. A. 1893 B, p. 677.   (F. angustus.)
  Miocene (Loup Fork); Nebraska.

Felis calcaratus Cope.

Cope, E. D. 1899 A, p. 250, pl. xxi, figs, 2, 2a.
  1895 F, p. 448.   (F. rufus, not of authors.)
  Pleistocene; Pennsylvania.

Felis concolor Linn.

Linnæus, C., 1771, Mantissa, p. 522.
Baird, S. F. 1857 A, p. 83, pl. lxxi.
Giebel, C. G. 1872 A, p. 431.
Leidy, J. 1888 A, p. 9.
Trouessart, E. L. 1897 A, p. 352.
True, F. W. 1891 A, p. 591.
  Recent; Mexico to Canada. Pleistocene; Il-
  linois.

Felis eyra Desm.

Baird, S. F. 1857 A, p. 88, pl. lxii, fig. 1.
Cope, E. D. 1895 F, p. 449.
  1899 A, p. 250.
Giebel, C. G. 1872 A, p. 431.
  Recent; Guiana to Mexico. Pleistocene;
  Pennsylvania.

Felis hillianus Cope.

Cope, E. D. 1892 Y, p. 1029.
  1892 I, p. 327.
  1893 A, p. 55, pl. xiv, figs. 1–11.
Matthew, W. D. 1899 A, p. 75.
Roger, O. 1896 A, p. 72.
Trouessart, E. L. 1897 A, p. 351.
  Pliocene (Blanco); Texas.

Felis imperialis Leidy.

Leidy, J. 1873 F, p. 259.
Adams, G. I. 1896 A, p. 434.
Leidy, J. 1873 B, pp. 228, 315, pl. xxxi, fig. 3.
Roger, O. 1896 A, p. 73.
Trouessart, E. L. 1897 A, p. 351.
Whitney, J. D. 1879 A, p. 246.
Yates, L. G. 1874 B, p. 18.
  Pleistocene; California.

[1] Besides the species below recorded, Dr. Leidy (1889 H, p. 5) has reported *F. canadensis* from a cave
in Pennsylvania.

Felis inexpectatus (Cope).

Cope, E. D. 1895 F, p. 449. (Crocuta.)
1896 F, p. 392. (Uncia mercerii, in part.)
1899 A, p. 247, pl. xxi, figs. 1–1f. (Uncia.)
Lydekker, R., 1896, Zool. Record for 1895, Mamm.,
. p. 28. (Nimravus?.)
Roger, O. 1896 A, p. 66 (Hyaena crocuta inex-
pectata).
Trouessart, E. L. 1897 A, p. 351. (Syn. of Felis
mercerii.)
Pleistocene; Pennsylvania.

Felis? maximus Scott and Osb.

Scott and Osborn 1890 B, p. 70, figs. 4, 5.

Adams, G. I. 1896 A, p. 434.   (Genus uncertain.)
Cragin, F. W. 1892 B, p. 17.
Matthew, W. D. 1899 A, p. 68.   (Felis?.)
Roger, O. 1896 A, p. 72.
Trouessart, E. L. 1897 A, p. 351.
Williston, S. W. 1898 K, p. 349.
Miocene (Loup Fork); Kansas.

Felis ruffus Güldenst.

Allen, J, A. 1876 C, p. 333.   (Lynx rufus.)
Baird, S. F. 1857 A, p. 90.   (Lynx rufus.)
Trouessart, E. L. 1897 A, p. 368.   (F. rufa.)
Recent; United States to Mexico. Pleisto-
cene?; caves of Pennsylvania, fide Allen.

Suborder PINNIPEDIA.

Allen, J. A. 1880 A.
Cope, E. D. 1888 CC, p. 310.
1889 R, p. 876.
1896 B, pp. 118, 120.
Flower, W. H. 1883 D, p. 442.
1885 A.
Flower and Lydekker 1891 A, p. 592.
Gadow, H. 1898 A, p. 50.
Gill, T. 1866 A.
1872 A.
1872 B, pp. 56, 68.
Gray, J. E. 1866 A.
Gruber, W. 1859 A.
Haeckel, E. 1895 A, p. 588.
Huxley, T. H. 1872 A, p. 859.
Kükenthal, W. 1890 C.
Leboucq, H. 1888 A, p. 859.
Lydekker, R. 1896 B.
Mivart, St. G. 1885 A.

Murie, J. 1874 A.
Nicholson and Lydekker 1889 A, p. 1422.
Osborn, H. F. 1893 D.
Paulli, S. 1900 B, p. 505.
Reh, L. 1894 A.
Roger, O. 1896 A, p. 73.
Schlosser, M. 1889 A.
1890 C, p. 266.
1890 D.
Smith, J. 1895 A, p. 342.
Steinmann and Döderlein 1890 A, p. 728.
Thompson, D'A. W. 1890 A.
Tomes, C. S. 1898 A, p. 475.
Tornier, G. 1890 A, p. 449.
Trouessart, E. L. 1897 A, p. 369.
Van Beneden, P. J. 1877 A.
Weber, M. 1886 A, p. 227.
Wortman, J. L. 1894 A, pp. 157, 162.
Zittel, K. A. 1893 B, p. 680.

OTARIIDÆ.

Brookes, 1828, Cat. Anat. Zool. Mus., p. 36. (Ota-
riadæ, fide Allen.)
Allen, J. A. 1880 A, p. 187.
Flower and Lydekker 1898 A, p. 593.
Gill, T. 1866 A, p. 7.

Gill, T. 1872 B, pp. 68, 69.
Trouessart, E. L. 1897 A, p. 369.
Winge, H. 1895 B, pp. 47, 70.
Zittel, K. A. 1893 B, p. 683.   (Otaridæ.)

EUMETOPIAS Gill.   Type *Arctocephalus monteriensis* Gray = *Otaria stelleri* Lesson.

Gill, T. 1866 A, pp. 7, 11.
Allen, J. A. 1880 A, p. 231.
Flower and Lydekker 1891 A, p. 593. (Otaria, in
part.)
Trouessart, E. L. 1897 A, p. 370.

Eumetopias stelleri (Lesson).

Allen, J. A. 1880 A, p. 232, fig. 37. (Synonymy and
literature.)

Bowers, S. 1889 A.
Gray, J. E. 1866 A, p. 60. (Otaria.)
1872 A, figs. 4, 5.
Trouessart, E. L. 1897 A, p. 370.
Recent; Pacific coast from California to
Japan. Pleistocene; California, fide Bowers.

ODOBENIDÆ.

This name is usually spelled *Odobænidæ*.
Unless otherwise indicated, the following au-
thors use the family name *Trichechidæ*.
Allen, J. A, 1880 A, pp. 3, 5.
Flower, W. H. 1883 D, p. 443.
Flower and Lydekker 1891 A, p. 596.
Gadow, H. 1898 A, p. 51.

Gill, T. 1866 A, pp. 7, 11. (Rosmaridæ.)
1872 B, pp. 27, 69, 70. (Rosmaridæ.)
Gray, J. E., 1821, Lond. Med. Repos., p. 303.
1825, Ann. Philos., p. 340.
1866 A, p. 33. (Trichechina.)
Haeckel, E, 1895 A, p. 590. (Trichechida.)
Huxley, T. H. 1872 A, p. 360.

Mivart, St. G. 1885 A. (Trichechina.)
Nicholson and Lydekker 1889 A, pp. 1422, 1424.
Reh, L. 1894 A, p. 12.
Roger, O. 1896 A, p. 75.
Turner, H. N. 1848 A, pp. 85, 88. (Trichecina, subfamily.)

Van Beneden, P. J. 1877 A, p. 39.
Winge, H. 1896 B, pp. 47, 73. (Trichechini.)
Wortman, J. L. 1886 A, p. 451.
Zittel, K. A. 1893 B, p. 685.

ODOBENUS Brisson.   Type *Phoca rosmarus* Linn.

Brisson, A. D. 1762, Regnum Anim., ed. ii, pp. 12, 30–31.
Unless otherwise indicated, the following authors refer to this genus under the name *Trichechus*.
Allen, J. A. 1880 A, p. 14. (Odobænus.)
Blainville, H. M. D. 1864 A, ii, J. (Trichechus.)
Boyd, C. H. 1881 A.
Flower and Lydekker 1891 A, p. 597.
Gill, T. 1866 A, p. 13. (Rosmarus.)
Gratiolet, P. 1858 A. (Odobenotherium, type *O. lartetianus.*)
Gray, J. E. 1866 A, p. 35.
Kükenthal, W. 1894 A, p. 77.
Mivart, St. G. 1885 A, p. 493.
Murie, J. 1871 A.
Nicholson and Lydekker, 1889 A, p. 1424.
Owen, R. 1845 B, p. 510.
Palmer, T. S. 1899 A, p. 494.
Pictet, F. J. 1853 A, p. 233.
Rhoads, S. N. 1898 A, p. 200. (Rosmarus.)
Slade, D. D. 1888 A, p. 243.
Smith, J. 1895 A, p. 339, fig. 7.
Stannius, H. 1842 B.
Tomes, C. S. 1898 A, p. 478.
Trouessart, E. L. 1897 A, p. 375.
Turner, H. N. 1848 A.
Van Beneden, P. J. 1877 A, p. 39.
Wiegmann, F. A. 1838 A, p. 113. (Odobenus.)
Zittel, K. A. 1893 B, p. 685.

Odobenus rosmarus (Linn.).

Linnæus, C., 1758, Syst. Nat., ed. 10, i p. 38. (Phoca.)
Unless otherwise indicated, the following authors refer this species to *Trichechus.*
Allen, J. A. 1880 A, p. 23, figs. 15, 16, 18, 20, 23, 25, 26, 28, 30, 32, 34. (Synonymy and literature.)
Blainville, H. M. D. 1864 A, ii, J, pls. i, iv, vii, ix. (Trichechus.)
Gervais, P. 1859 A, p. 275. (Trichechus.)
Giebel, C. G. 1847 A, p. 222.
Goethe, J. W. 1831 A, p. 8, pl. iv. ("Walross.")
Gray, J. E. 1866 A, p. 37, fig. 12.
Gratiolet, P., 1858 A, pl. v. (Odobenotherium lartetianus.)
Jaeger, G. 1844 A. (Wallross.)
Kersten, H. 1821 A.
Kükenthal, W. 1894 A, p. 77.
Malmgren, A. J. 1864 A, p. 505, pl. vii. (Odobænus.)
1865 A. (Odobænus.)

Murie, J. 1871 A.
Owen, R. 1853 A.
1866 B, p. 498, fig. 339.
1868 A, p. 338.
Peters, W. 1864 A, plate. (Odobænus rosmarus.)
Reh, L. 1894 A, p. 12.
Rhoads, S. N. 1898 A, p. 200. (Rosmarus.)
Roger, O. 1896 A, p. 75.
Stannius, H. 1842 B.
Trouessart, E. L. 1897 A, p. 375.
Van Beneden, P. J. 1877 A, p. 39, pl. viii, figs. 7–9.
Wiegmann, F. A. 1838 A, p. 113. (Odobenus.)
Recent; shores of Northern Atlantic and Arctic oceans from Labrador to Siberia. Pleistocene; Labrador.

Odobenus virginianus (De Kay).

De Kay, J. E. 1842 C, p. 56, pl. xix, fig. 1. (Trichechus.)
Agassiz, L. 1851 A, p. 252. ("Walrus.")
Allen, J. A. 1880 A, p. 57. (Odobænus.)
Amer. Naturalist 1871 B, p. 316.
Barton, B. S. 1805 A, p. 96. (Trichechus rosmarus.)
Cope, E. D. 1869 N, p. 740. (Trichechus rosmarus.)
1874 T. ("Walrus.")
1878 BB. (Trichechus rosmarus.)
Harlan, R. 1834 B, p. 72. (Trichechus rosmarus.)
1835 B, p. 277. (Trichechus rosmarus.)
Leidy, J. 1857 G, p. 83, pl. iv; pl. v, fig. 1. (Trichechus.)
1869 A, p. 416. (Trichechus rosmarus.)
1876 A, p. 80. ("Walrus.")
1877 A, p. 214, pl. xxx, fig. 6. (Rosmarus obesus.)
Lyell, C. 1843 C, p. 32. (Trichechus rosmarus.)
Mitchell, Smith, and Cooper 1828 A. ("Walrus.")
Newberry, J. S. 1870 C, p. 75. (Trichechus rosmarus.)
1874 D, p. 71. (Trichechus rosmarus.)
Packard, A. S. 1867 A, pp. 243, 246. (Rosmarus obesus.)
Rhoads, S. N. 1898 A, p. 196.
Pleistocene; Maine, Nova Scotia, Massachusetts, New Jersey, Virginia, South Carolina.
Some of the remains described by the above authors may belong to *O. rosmarus;* indeed, it is not certain that any are distinct.

PHOCIDÆ.

Allen, J. A. 1880 A, p. 412.
Blainville, H. M. D. 1864 A, ii, I, p. 5; J, p. 1, pls. i–ix.
Cuvier, G. 1834 A, viii, pl. i, pp. 381–458, with plates.

Flower, W. H. 1883 D, p. 443.
Flower and Lydekker 1891 A, p. 600.
Gill, T. 1866 A, p. 5.
1872 B, pp. 68, 69.

Gray, J. E. 1866 A, p. 1.
Huxley, T. H. 1872 A, p. 360.
Kükenthal, W. 1894 A, p. 97.
Leche, W. 1893 B.
Mivart, St. G. 1885 A.
Nicholson and Lydekker 1889 A, p. 1422.
Owen, R. 1845 B, p. 505.
    1868 A, p. 336.

Reh, L. 1894 A, p. 35.
Roger, O. 1896 A, p. 73.
Scott, W. B. 1892 A, p. 424.
Turner, H. N. 1848 A.
Van Beneden, P. J. 1877 A.
Winge, H. 1895 B, pp. 47, 72, 74, 76, 91.
Zittel, K. A. 1893 B, p. 683.

CYSTOPHORA Nilsson.   Type *C. borealis* Nilsson=*Phoca cristata* Erxl.

Nilsson, S., 1820, Skand. Fauna, i, p. 382.
Allen, J. A. 1880 A, p. 723.
Flower and Lydekker 1891 A, p. 605.
Owen, R. 1868 A, p. 337.

Cystophora cristata (Erxl.).

Erxleben, C. P., 1777, Syst. Regn. Anim., p. 590.
    (Phoca.)

Allen, J. A. 1880 A, pp. 470, 724, figs. 52-56. (Synonymy and literature.)
Leidy, J. 1869 A, p. 416.
Lyell, C. 1843 C, p. 32. (C. proboscidea.)
Owen, R. 1866 B, p. 496.
Trouessart, E. L. 1897 A, p. 378.
    Recent; along shores of Northern Atlantic; said by Lyell to have been found in Tertiary strata of Martha's Vineyard.

LOBODON Gray.   Type *Phoca carcinophaga* Homb., Jacq.

Gray, J. E., 1844, Zool. Erebus and Terror, i, p. 5.
Flower and Lydekker 1891 A, p. 605.
Gray, J. E. 1866 A, p. 8, fig. 2.
Trouessart, E. L. 1897 A, p. 381. (Subgenus of Ogmorhinus.)

Lobodon vetus Leidy.

Leidy, J. 1853 C, p. 377, figure. (Stenorhynchus.)

Allen, J. A. 1880 A, p. 475.
Cope, E. D. 1869 N, p. 740.
Leidy, J. 1865 A, p. 2. (Stenorhynchus.)
    1869 A, p. 415.
Trouessart, E. L. 1897 A, p. 381. [Ogmorhinus (Lobodon).]
    Miocene?; New Jersey.

PHOCA Linn.   Type *P. vitulina* Linn.

Linnæus, C., 1758, Syst. Nat., ed. 10. i, p. 37.
Allen, J. A. 1880 A, p. 462.
Flower, W. H. 1883 D. p. 443.
    1885 A.
Flower and Lydekker 1891 A, p. 601.
Gray, J. E. 1866 A, p. 31.
Huxley, T. H. 1872 A, p. 361.
Mivart, St. G. 1885 A, p. 486.
Owen, R. 1866 B, p. 494.
Pictet, F. J. 1853 A, p. 232.
Toula, F. 1897 A.
Van Beneden, P. J. 1877 A.
Winge, H. 1895 B, pp. 47, 91.
Zittel, K. A. 1893 B, p. 684.

Phoca grœnlandica.

Allen, J. A. 1880 A, pp. 470, 630. (Synonymy and literature.)
Dawson, J. W. 1877 B.

Leidy, J. 1856 G, p. 90, pl. iii.
Leidy, J. 1869 A, p. 415.
Logan, W. E. 1863 A, p. 920.
Trouessart, E. L. 1897 A, p. 387. [Phoca (Pagophilus).]
    Recent in Northern Atlantic and in Arctic Ocean around the poles. Pleistocene; Canada, Maine.

Phoca wymani Leidy.

Leidy, J. 1854 A, p. 8.
Allen, J. A. 1880 A, p. 471.
Cope, E. D. 1867 B, p. 132, in part.
    1867 C, pp. 151, 152, in part.
Leidy, J. 1869 A, p. 415, in part.
Roger, O. 1896 A, p. 74.
Wyman, J. 1850 B, p. 229, figs. 1-3 (?).
    Miocene; Virginia.
    See citations under *Delphinodon leidyi*.

Order PRIMATES.

Blainville, H. M. D. 1864 A, i, A, B, F, pls. i-xi.
Cope, E. D. 1873 E, pp. 546, 644.
    1880 B, p. 456. (Quadrumana.)
Flower, W. H. 1882 B.
    1883 A.
    1883 D, p. 444.
    1885 A.
Flower and Lydekker 1891 A, p. 680.
Fry, E. 1846 A.

Gaudry, A. 1878 B, p. 223. ("Les quadrumanes.")
Gill, T. 1872 A, p. 296.
    1872 B, pp. 1, 47, 50.
Gill and Coues 1877 A.
Hay, O. P. 1899 F, p. 682.
Hubrecht, A. A. W. 1896 A.
Huxley, T. H. 1872 A, p. 388.
Lydekker, R. 1885 B.
    1896 B.

Marsh, O. C. 1872 N. (Quadrumana.)
1872 S. (Quadrumana.)
1877 E.
Mivart, St. G. 1864 A.
1865 A.
1867 A.
1868 B.
1873 A.
1875 A.
Nicholson and Lydekker 1889 A, p. 1468.
Osborn, H. F. 1892 I.
1893 G, p. 498.
1898 A, p. 146.
Osborn and Earle 1895 A, p. 15.
Owen, R. 1835 A.
1843 A, p. 55. (Quadrumana.)
1845 B, p. 433. (Quadrumana.)
1859 C.
1866 B, p. 511. (Quadrumana.)

Owen, R. 1868 A, p. 313. (Quadrumana.
Parsons, F. G. 1899 A.
1900 A.
Paulli, S. 1900 B, p. 542.
Roger, O. 1896 A, p. 249.
Schlosser, M. 1887 B, p. 7. (Quadrumana.)
1889 A.
1890 D.
Scott, W. B. 1892 A, p. 417.
Slade, D. D. 1891 A.
1891 B.
Steinmann and Döderlein 1890 A, p. 741.
Topinard, P. 1892 A.
Tornier, G. 1888 A.
Trouessart, E. L. 1897 A, p. 1.
Weyhe, ——. 1875 A, p. 114.
Winge, H. 1895 A.
Zittel, K. A. 1898 B, pp. 685, 747.

Suborder PROSIMIÆ.

Branco, W. 1897 A, p. 30. ("Halbaffen.")
Cope, E. D. 1876 E, p. 86. (>Mesodonta.)
1877 K, pp. 85, 134. (>Mesodonta.)
1880 B, p. 456.
1882 A, p. 446. (Mesodonta.)
1882 E, p. 151.
1883 I, pp. 77, 79. (Prosimiæ, Mesodonta.)
1884 O, pp. 239, 739 (Prosimiæ; pp. 185, 211, 738 (Mesodonta).
1884 R, p. 60. (Lemuroidea.)
1885 K. (Lemuroidea.)
1888 EE, p. 164. (>Mesodonta.)
Flower, W. H. 1876 B, xiv, p. 11. (Lemuroidea.)
1882 B. (Lemuroidea.)
Flower and Lydekker 1891 A, p. 682. (Lemuroidea.)
Gill, T. 1872 A, p. 296.
Hubrecht, A. A. W. 1896 A, p. 150. (Lemures.)
Leche, W. 1897 C.

Lemoine, V. 1880 A, p. 587.
Lydekker, R. 1896 B. (Lemuroidea.)
Mivart, St. G. 1868 B.
Nicholson and Lydekker 1889 A, p. 1465. (Lemuroidea.)
Osborn, H. F. 1888 F. p. 1068. (Lemuroidea.)
1890 C, p. 55. (Mesodonta.)
1900 C, p. 567. (Mesodonta.)
Osborn and Wortman 1892 A, p. 102. (Lemuroidea, Mesodonta.)
Roger, O. 1896 A, p. 249.
Rütimeyer, L. 1888 A, p. 29. (>Mesodonta.)
Schlosser, M. 1887 B, p. 19. (>Pseudolemuridæ.)
Schmidt, O. 1886 A. p. 294.
Steinmann and Döderlein 1890 A, p. 742.
Trouessart, E. L. 1897 A.
Winge, H. 1895 A, p. 12. (Lemuroidei.)
Zittel, K. A. 1893 B, pp. 686, 711, 725.

MIXODECTIDÆ.

Cope, E. D. 1883 I, p. 80. (Anaptomorphidæ, Myxodectidæ.)
* 1884 K, pp. 318, 319. (Anaptomorphidæ, Myxodectidæ.)
1884 O, p. 240. (Anaptomorphidæ, Myxodectidæ.)
1884 R, p. 60. (Anaptomorphidæ, Myxodectidæ.)
1885 K, pp. 459, 464. (Anaptomorphidæ, Myxodectidæ.)
Haeckel, E. 1895 A, pp. 600, 604. (Necrolemures, in part.)

Matthew, W. D. 1899 A, p. 29.
Osborn and Earle 1895 A, p. 16.
Osborn and Wortman 1892 A, pp. 101, 102.
Roger, O. 1896 A, p. 253. (Anaptomorphidæ.)
Trouessart, E. L. 1897 A, p. 66. (Tarsiidæ, in part.)
Winge, H. 1896 A, p. 12. (Tarsiidæ, in part; Tarsiinæ.)
Zittel, K. A. 1893 B, p. 696.

MIXODECTES Cope.   Type M. pungens Cope.

Cope, E. D. 1883 M, p. 558.
Branco, W. 1897 A, p. 31.
Cope, E. D. 1884 K, p. 319.
1884 O, pp. xxxv, p. 240.
1885 K, p. 465.
Matthew, W. D. 1897 B, pp. 231, 265.
1897 C, p. 265.
1898 B, p. 369.

Nicholson and Lydekker 1889 A, p. 1467.
Osborn, H. F. 1898 A, p. 145.
Schlosser, M. 1887 B, pp. 40, 49.
Zittel, K. A. 1893 B, p. 697.

Mixodectes crassiusculus Cope.

Cope, E. D. 1883 M, p. 560.
1884 O, p. 242, pl. xxiv/, fig. 2.

Cope, E. D. 1888 CC, p. 304.
Matthew, W. D. 1899 A, p. 29.
Roger, O. 1896 A, p. 254.
Trouessart, E. L. 1897 A, p. 67.
Eocene (Torrejon); New Mexico.

Mixodectes pungens Cope.

Cope, E. D. 1883 M, p. 559.

Cope, E. D. 1884 O, p. 241, pl. xxiv/, fig. 1.
1885 K, p. 465, fig. 9.
1888 CC, p. 304.
Matthew, W. D. 1897 C, p. 266, fig. 1.
1899 A, p. 29.
Roger, O. 1896 A, p. 254.
Trouessart, E. L. 1897 A, p. 67.
Eocene (Torrejon); New Mexico.

ANAPTOMORPHUS Cope.   Type *A. æmulus* Cope.

Cope, E. D. 1872 U, p. 554.
Branco, W. 1897 A, p. 31.
Cope, E. D. 1873 E, p. 549.
1877 K, p. 135.
1882 F, p. 152.
1882 N, p. 73.
1884 K, p. 319.
1884 O, pp. xxxv, 214, 240, 245.
1884 R, p. 60.
1885 K, p. 465.
1888 G, p. 8.
Filhol, H. 1874 A, p. 1.
1883 B, p. 3 (Anaptomorphus); p. 15 (Washakius).
1885 A.
Flower, W. H. 1882 B, p. 445.
Flower and Lydekker 1891 A, p. 697.
Hubrecht, A. A. W. 1896 A, p. 162.
Leidy, J. 1873 B, p. 123. (Washakius, type *W. insignis.*)
Nicholson and Lydekker 1889 A, p. 1467.
Osborn, H. F. 1888 F, p. 1077.
1892 I.
1895 C, p. 7, pl. AA, figs. 7, 9.
Osborn and Wortman 1892 A, p. 102.
Schlosser, M. 1887 B, p. 34 (Washakius=?Opisthotomus); pp. 39, 48 (Anaptomorphus).
Science 1885 A, p. 489.
Scott, W. B. 1892 A, p. 418.
Steinmann and Döderlein 1890 A, p. 744.
Trouessart, E. L. 1897 A, p. 67.
Winge, H. 1895 A, p. 15.
Zittel, K. A. 1893 B, p. 697.

Anaptomorphus æmulus Cope.

Cope, E. D. 1872 U, p. 554.

Cope, E. D. 1873 E, p. 549.
1884 O, p. 248, pl. xxv, fig. 10.
1885 K, p. 465, fig. 11.
Marsh, O. C. 1875 B, p. 239. (Syn. of Antiacodon nanus.)
Matthew, W. D. 1899 A, p. 38.
Nicholson and Lydekker 1889 A, p. 1468, fig. 1351.
Osborn and Wortman 1892 A, p. 103. (Antiacodon.)
Roger, O. 1896 A, p. 253. (A. aemulus, with "Washakius fallax" as a synonym.)
Trouessart, E. L. 1897 A, p. 67.
Eocene (Bridger); Wyoming.

Anaptomorphus homunculus Cope.

Cope, E. D. 1882 E, p. 152.
1882 N, p. 73.
1884 K, p. 466, fig. 12.
1884 O, p. 249, pl. xxive, fig. 1.
Matthew, W. D. 1899 A, p. 30. (Omomys.)
Nicholson and Lydekker 1889 A, p. 1467, fig. 1350.
Osborn, H. F. 1895 C, p. 11, fig D.
Osborn and Wortman 1892 A, p. 102, fig. 6.
Roger, O. 1896 A, p. 254.
Trouessart, E. L. 1897 A, p. 67.
Zittel, K. A. 1893 B, p. 697, fig. 576.
Eocene (Wasatch); Wyoming.

Anaptomorphus insignis (Leidy).

Leidy, J. 1873 B, pp. 123, 336, pl. xxvii, figs. 3, 4. (Washakius.)
Matthew, W. D. 1899 A, p. 38. (Syn. of A. æmulus.)
Trouessart, E. L. 1897 A, p. 67. (Anaptomorphus.)
Eocene (Bridger); Wyoming.

ANTIACODON Marsh.   Type *A. venustus* Marsh.

Marsh, O. C. 1872 I, p. 210.
Cope, E. D. 1873 C, pp. 13, 17.
1875 I, p. 21.
Filhol, H. 1883 B, p. 17.
Marsh, O. C. 1875 B, p. 239.
1877 E.
Matthew, W. D. 1899 A, p. 38. (=?Omomys.)
Schlosser, M. 1887 B, p. 38.
Zittel, K. A. 1893 B, p. 697. (Syn. of Anaptomorphus.)

Antiacodon nanus Marsh.

Marsh, O. C. 1872 I, p. 213. (Hemiacodon.)
1875 B, p. 239. (Antiacodon Anaptomorphus æmulus, a synonym.)

Matthew, W. D. 1899 A, p. 38. (Hemiacodon? nanus.)
Trouessart, E. L. 1897 A, p. 74. (Hemiacodon.)
Eocene (Bridger); Wyoming.
Marsh's *Antiacodon nanus* seems to be his earlier *Hemiacodon nanus*.

Antiacodon venustus Marsh.

Marsh, O. C. 1872 I, p. 210.
Cope, E. D. 1873 E, p. 607. (Syn. of A. pygmæus.)
1873 U, p. 1. (Syn. of A. pygmæus.)
Matthew, W. D. 1899 A, p. 38.
Roger, O. 1896 A, p. 253. (Sarcolemur.)
Trouessart, E. L. 1897 A, p. 67 (Syn. of A. æmulus); p. 73 (Sarcolemur).
Eocene (Bridger); Wyoming.

OMOMYS Leidy.   Type *O. carteri* Leidy.

Leidy, J. 1869 B, p. 63.
Cope, E. D. 1875 C, p. 13.
    1877 K, p. 134.
    1884 O, pp. 215, 240.
Leidy, J. 1869 A, p. 408.
    1871 C, p. 364.
Nicholson and Lydekker 1889 A, p. 1467.
Osborn and Wortman 1892 A, p. 102.
Schlosser, M. 1887 B, p. 35.
Zittel, K. A. 1893 B, p. 697.

Omomys carteri Leidy.

Leidy, J. 1869 B, p. 63.
    1869 A, p. 408, pl. xxix, figs. 13, 14.
    1871 C, p. 365.
    1872 B, p. 356.
    1873 B, pp. 120, 336.
Matthew, W. D. 1899 A, p. 38.
Roger, O. 1896 A, p. 254.
Trouessart, E. L. 1897 A, p. 67.
    Eocene (Bridger); Wyoming.

CYNODONTOMYS Cope.   Type *C. latidens* Cope.

Cope, E. D. 1882 E, p. 151.
    1883 M, p. 559.
    1884 K, p. 319.
    1884 O, pp. xxxv, pp. 239, 240, 243.
    1885 K, p. 465.
Nicholson and Lydekker 1889 A, p. 1467.
Schlosser, M. 1887 B, pp. 40, 48.
Zittel, K. A. 1893 B, p. 698.

Cynodontomys latidens Cope.

Cope, E. D. 1882 E, p. 151.
    1884 O, p. 244, pl. xxive, fig. 2.
    1885 K, p. 465, fig. 10.
Matthew, W. D. 1899 A, p. 30.
Roger, O. 1896 A, p. 254.
Trouessart, E. L. 1897 A, p. 68.
    Eocene (Wasatch); Wyoming.

PROSINOPA Trouessart.   Type *Sinopa erimia* Leidy.

Trouessart, E. L. 1897 A, p. 68.
Scott, W. B. 1892 B, p. 310. (Sinopa.)

Prosinopa eximia (Leidy).

Leidy, J. 1873 B, pp. 118, 316, pl. xvi, fig. 4b. (Sinopa.)

Matthew, W. D. 1899 A, p. 38.
Roger, O. 1896 A, pp. 35, 254. (Sinopa?.)
Scott, W. B. 1892 B, p. 310. (Sinopa?.)
Trouessart, E. L. 1897 A, p. 68.
    Eocene (Bridger); Wyoming.

LIMNOTHERIIDÆ.

Marsh, O. C. 1875 B, p. 239.
Bronn, H. G. 1838 A, p. 1224, pl. xlvi, fig. 6. (Adapis.)
Cope, E. D. 1884 K, p. 318.
    1884 O, pp. 213, 240.
    1885 K, p. 459.
    1888 EE, p. 163.
Filhol, H. 1883 B, p. 19. (Adapis.)
Flower, W. H. 1882 B, p. 444. (Adapis.)
Gaudry, A. 1878 B, p. 226. (Adapis.)
Haeckel, E. 1895 A, pp. 600, 604. (Necrolemures, in part.)
Leche, W. 1897 C, p. 147. (Adapis.)

Marsh, O. C. 1875 B, p. 240. (Lemuravidæ.)
Marsh, O. C. 1876 G, p. 60. (Limnotheriidæ, Lemuravidæ.)
    1877 E.
    1895 D, p. 498. (Limnotheriidæ, Lemuravidæ.)
Nicholson and Lydekker 1889 A, p. 1466. Adapidæ, Hyopsodontidæ.)
Schlosser, M. 1887 B, p. 21 (Adapidæ); pp. 21, 54 (Hyopsodidæ.)
Steinmann and Döderlein 1890 A, p. 743. (Adapidæ, Hyopsodidæ.)
Zittel, K. A. 1893 B, p. 689. (Pachylemuridæ.)

TOMITHERIUM Cope.   Type *T. rostratum* Cope.

Cope, E. D. 1872 E, p. 470.
    1873 E, pp. 546, 645.
    1876 E.
    1877 K, p. 134.
    1884 O, pp. 216, 218.
    1885 K, pp. 460, 461.
Flower and Lydekker 1891 A, p. 698.
Lockwood, S. 1886 A, p. 324.
Matthew, W. D. 1899 A, p. 37. (Syn.? of Notharctus.)
Nicholson and Lydekker 1889 A, p. 1466.
Osborn, Scott, and Speir 1878 A, p. 18.
Schlosser, M. 1887 B, p. 22.
Zittel, K. A. 1893 B, p. 698.

Tomitherium affine (Marsh).

Marsh, O. C. 1872 I, p. 207. (Limnotherium.)
Cope, E. D. 1884 O, p. 223. (Limnotherium.)
Marsh, O. C. 1875 B, p. 239. (Limnotherum; T. rostratum, a syn.)
Matthew, W. D. 1899 A, p. 37. (Notharctus.)
Trouessart, E. L. 1897 A, p. 70. (Syn. of T. rostratum.)
    Eocene (Bridger); Wyoming.
    Possibly identical with *T. rostratum* Cope.

Tomitherium rostratum Cope.

Cope, E. D. 1872 E, p. 470.

Cope, E. D. 1873 E, p. 548.
   1873 U, p. 3.
   1876 E.
   1884 O, p. 221, pl. xxv, figs. 1–9.
   1885 K, p. 464, figs. 5–7.
Marsh, O. C. 1875 B, p. 239. (Syn. of Limnotherium affine.)

Matthew, W. D. 1899 A, p. 37. (Syn. of Notharctus tenebrosus.)
Osborn, Scott, and Speir 1878 A, p. 13.
Roger, O. 1896 A, p. 251.
Trouessart, E. L. 1897 A, p. 70.
   Eocene (Bridger); Wyoming.

LIMNOTHERIUM Marsh. Type *L. tyrannus* Marsh.

Marsh, O. C. 1871 D, p. 43.
Cope, E. D. 1884 O, p. 220.
   1887 S, p. 253.
Marsh, O. C. 1872 N.
   1875 B, p. 239.
   1877 E.
   1895 D, p. 498.
Matthew, W. D. 1899 A, p. 37. (Syn.? of Notharctus.)
Nicholson and Lydekker 1889 A, p. 1372.
Schlosser, M. 1887 B, p. 36. (Syn. of Tomitherium.)
Trouessart, E. L. 1897 A, p. 70. (Syn. of Tomitherium.)

Zittel, K. A. 1893 B, p. 693. (Syn. of Tomitherium.)

Limnotherium tyrannus Marsh.

Marsh, O. C. 1871 D, p. 43.
Cope, E. D. 1872 E, p. 471. (Notharctus.)
King, C. 1878 A, p. 403.
Leidy, J. 1872 B, p. 364.
Matthew, W. D. 1899 A, p. 37. (Notharctus.)
Roger, O. 1896 A, p. 251. (Tomitherium.)
Trouessart, E. L. 1897 A, p. 70. (Tomitherium.)
   Eocene (Bridger); Wyoming.

NOTHARCTUS Leidy. Type *N. tenebrosus* Leidy.

Leidy, J. 1870 Q, p. 114.
Cope, E. D. 1872 E, p. 471.
   1873 E, p. 548.
   1877 K, p. 135.
   1879 A, p. 48. (Syn. of Adapis.)
   1884 O, p. 34.
   1885 K, p. 460,
Filhol, H. 1883 B, p. 19.
Flower and Lydekker 1891 A, p. 698.
Leidy, J. 1871 C, p. 344.
   1873 B, p. 86.
Marsh, O. C. 1875 B, p. 239.
   1894 L, p. 263.
Matthew W. D. 1899 A, p. 37.
Nicholson and Lydekker 1889 A, p. 1466.

Schlosser, M. 1887 B, p. 22.
Zittel, K. A. 1893 B, p. 694.

Notharctus tenebrosus Leidy

Leidy, J. 1870 Q, p. 114.
Cope, E. D. 1872 E, p. 471.
   1885 K, p. 461, fig. 4.
Leidy, J. 1871 C, p. 344.
   1872 B, p. 364.
   1873 B, pp. 86, 329, pl. vi, figs. 36, 37.
Matthew, W. D. 1899 A, p. 37.
Roger, O. 1896 A, p. 252.
Trouessart, E. L. 1897 A, p. 70.
   Eocene (Bridger); Wyoming.

LEMURAVUS Marsh. Type *L. distans* Marsh.

Marsh, O. C. 1875 B, p. 239.
Cope, E. D. 1877 K, p. 148.
Marsh, O. C. 1876 G, p. 60.
   1877 E.
   1895 D, p. 498.
Nicholson and Lydekker 1889 A, p. 1468.
Zittel, K. A. 1893 B, p. 691.

Lemuravus distans Marsh.

Marsh, O. C. 1875 B, p. 238.
King, C. 1878 A, p. 403.
Matthew, W. D. 1899 A, p. 37. (Syn. of Hyopsodus paulus.)
Roger, O. 1896 A, p. 250.
Trouessart, E. L. 1897 A, p. 70.
   Eocene (Bridger); Wyoming.

PELYCODUS Cope. Type *Prototomus jarrovii* Cope.

Cope, E. D. 1875 C, p. 13.
   1875 L, p. 21.
   1882 E, p. 151.
   1884 O, pp. 216, 224.
   1885 K, pp. 459, 467.
Flower and Lydekker 1891 A, p. 699.
Nicholson and Lydekker 1889 A, p. 1465.
Osborn, H. F. 1888 A, p. 1076.
Rütimeyer, L. 1888 A, p. 34.
   1891 A, p. 115.

Schlosser, M. 1887 B, p. 21.
   1894 A, p. 159.
Steinmann and Doderlein 1890 A, p. 743.
Zittel, K. A. 1893 B, p. 691.

Pelycodus frugivorus Cope.

Cope, E. D. 1875 C, p. 14.
   1877 K, p. 144, pl. xxxix, fig. 16. (Tomitherium.)
   1881 D, p. 187. (P. nunienum.)

Cope, E. D. 1882 E, p. 151.
    1884 O, pp. 182, 225, 230, pl. xxva, figs. 4, 5.
Earle, C. 1898 A, p. 262.
Matthew, W. D. 1899 A, pp. 30, 34.
Roger, O. 1896 A, p. 250. (P. frugivorus, P. nunienum.)
Rütimeyer, L. 1888 A, p. 35·
Trouessart, E. L. 1897 A, p. 71.
Zittel, K. A. 1893 B, p. 692, fig. 570.
    Eocene (Wasatch, Wind River); New Mexico, Wyoming.

Pelycodus jarrovii Cope.

Cope, E. D. 1874 O, p. 14. (Prototomus.)
    1875 C, p. 14.
    1876 E, p. 88. (Tomitherium.)
    1877 K, pp. 137, 141, pl. xxxix, figs. 17, 18; pl. xl, figs. 1–15. (Tomitherium.)
    1881 D, p. 187.
    1882 E, p. 151.
    1884 O, pp. 225, 228.

Matthew, W. D. 1899 A, p. 30.
Roger, O. 1896 A, p. 250.
Trouessart, E. L. 1897 A, p. 70.
    Eocene (Wasatch); New Mexico, Wyoming.

Pelycodus tutus Cope.

Cope, E. D. 1877 K, p. 141, pl. xxxix, fig. 19; pl. xl, figs. 16–25. (Tomitherium.)
    1876 E, p. 88. (No description.)
    1881 D, p. 187.
    1882 E, p. 151.
    1884 O, pp. 224, 225, 228, pl. xxva, figs. 1–3.
    1885 K, p. 468, figs. 13, 16.
Matthew, W. D. 1899 A, pp. 30, 34.
Roger, O. 1896 A, p. 250.
Rütimeyer, L. 1888 A, p. 35.
Trouessart, E. L. 1897 A, p. 71.
Zittel, K. A. 1893 B, p. 692, fig. 569.
    Eocene (Wasatch, Wind River); New Mexico, Wyoming, Colorado.

*Hyopsodus Leidy.   Type *H. paulus* Leidy.

Leidy, J. 1870 O, p. 110.
Cope, E. D. 1875 C, pp. 13, 18.
    1875 L, p. 21.
    1877 K, pp. 135, 150.
    1882 GG, p. 1029. (Diacodexis, type D. laticuneus.)
    1884 O, pp. 216, 234 (Hyopsodus); pp. 386, 492 (Diacodexis).
    1885 K, p. 460.
    1885 X, p. 1208.
    1887 S, p. 257.
Earle, C. 1898 A, p. 262.
Flower and Lydekker 1891 A, p. 697.
Gaudry, A. 1877 A.
Leidy, J. 1871 C, p. 354.
    1872 B, p. 362.
    1872 C, p. 20.
    1873 B, p. 75.
Lydekker, R. 1885 A.
Marsh, O. C. 1871 D, p. 42. (Hypsodus.)
    1875 B, p. 239.
    1877 E.
Nicholson and Lydekker 1889 A, p. 1465.
Osborn, H. F. 1888 E.
Osborn, Scott, and Speir 1878 A, p. 15.
Pavlow, M. 1887 A, p. 369. (Diacodexis.)
    1888 A, p. 135.
Rütimeyer, L. 1888 A, p. 34.
    1891 A, p. 118.
Schlosser, M. 1887 B, p. 21.
    1894 A, p. 159.
Scott, W. B. 1892 A, p. 418.
Steinmann and Doderlein 1890 A, p. 743.
Zittel, K. A. 1893 B, p. 220 (Diacodexis); p. 693 (Hyopsodus).

Hyopsodus gracilis Marsh.

Marsh, O. C. 1871 D, p. 42. (Hypsodus.)
Cope, E. D. 1872 E, p. 471. (Notharctus.)
King, C. 1878 A, p. 407.
Leidy, J. 1872 B, p. 368.
    1872 C, p. 20.
Matthew, W. D. 1899 A, pp. 37, 49.
Roger, O. 1896 A, p. 251.
Scott, W. B. 1890 A, p. 471.

Scott and Osborn 1887 A, p. 255.
Trouessart, E. L. 1897 A, p. 71.
    Eocene (Bridger, Uinta); Wyoming.

Hyopsodus laticuneus (Cope).

Cope, E. D. 1882 E, pp. 179, 181. (Phenacodus laticuneus, in part.)
    1882 GG, p. 1029. (Diacodexis.)
    1884 O, p. 492 pl. xxve, figs. 17, molars; p. 18 (Diacodexis).
    1884 AA, p. 901, figs. 24b, c. (Diacodexis.)
Matthew, W. D. 1899 A, p. 30.
Pavlow, M. 1887 A, p. 369. (Diacodexis.)
Roger, O. 1896 A, p. 167. (Diacodexis?.)
Trouessart, E. L. 1897 A, p. 726.
    Eocene (Wasatch); Wyoming.
    Dr. Matthew (op. cit.) states that the molars, upper and lower, of Cope's D. laticuneus belong to Hyopsodus, the premolars to Hyracotherium index.

Hyopsodus lemoinianus Cope.

Cope, E. D. 1882 E, p. 148,
    1884 O, pp. 235, 236, pl. xxiv, figs. 8, 9.
Matthew, W. D. 1899 A, pp. 30, 34.
Roger, O. 1896 A, p. 251.
Trouessart, E. L. 1897 A, p. 71.
    Eocene (Wasatch, Wind River); Wyoming.

Hyopsodus minusculus Leidy.

Leidy, J. 1873 B, pp. 81, 320, pl. xxvii, fig. 5.
Cope, E. D. 1875 C, p. 18. (Syn. of H. vicarius.)
    1880 O, p. 908. (Syn. ? of H. vicarius.)
    1884 O, p. 237. (Syn. of H. vicarius.)
King, C. 1878 A, p. 403,
Matthew, W. D. 1899 A, p. 37.
Roger, O. 1896 A, p. 251.
Trouessart, E. L. 1897 A, p. 71.
    Eocene (Bridger); Wyoming.
    May be identical with H. vicarius.

Hyopsodus miticulus Cope.

Cope, E. D. 1874 O, p. 8. (Esthonyx.)
    1875 C, p. 18.
    1877 K, p. 150, pl. xlv, figs. 10–12.

Cope, E. D. 1882 E, p. 149.
  1884 O, p. 235.
Matthew, W. D. 1899 A, pp. 30, 34.
Roger, O. 1896 A, p. 251.
Trouessart, E. L. 1897 A, p. 71.
  Eocene (Wasatch, Wind River); New Mexico.

Hyopsodus paulus Leidy.

Leidy, J. 1870 O, p. 110.
Cope, E. D. 1873 E, p. 609.
  1873 U, p. 18.
  1875 C, p. 18.
  1881 D, p. 186.
  1882 E, p. 148.
  1884 O, pp. 235, 237.
  1885 K, p, 460, fig. 2.
  1887 I, p. 657.
King, C. 1878 A, p. 403.
Leidy, J. 1871 C, p. 354.
  1872 B, p. 363.
  1873 B, pp. 75, 320, pl. vi, figs. 1-9, 18-20.
Marsh, D. C. 1871 D, p. 42. (Hypsodus.)
Matthew, W. D. 1899 A, pp. 30, 34, 37.
Osborn, Scott, and Speir 1878 A, p. 15.
Pavlow, M. 1887 A, p. 371.
Roger, O. 1896 A, p. 251.
Trouessart, E. L. 1897 A, p. 71.
  Eocene (Wasatch, Wind River, Bridger); Wyoming.

MICROSYOPS Leidy.   Type *M. gracilis* Leidy.

Leidy, J. 1872 C, p. 20.
Cope, E. D. 1875 C, p. 13.
  1877 K, p. 134.
  1884 K, p. 319.
  1884 O, pp. xxxv, 215, 216, 240.
  1885 K, p. 465.
Flower and Lydekker 1891 A, p. 698.
Gaudry, A. 1877 A.
Leidy, J. 1873 B, p. 82.
Marsh, O. C. 1875 B, p. 239.
Matthew, W. D. 1897 C, p. 265.
Nicholson and Lydekker 1889 A, p. 1466
Schlosser, M. 1887 A, pp. 40, 49.
Zittel, K. A. 1893 B, p. 693.

Microsyops elegans (Marsh).

Marsh, O. C. 1871 D, p. 43. (Limnotherium.)
Cope, E. D. 1884 O, p. 217.
  1887 S, p, 257, fig. 30.
King, C. 1878 A, p. 403.
Leidy, J. 1872 B, p. 364. (Limnotherium.)
  1873 B, pp. 83, 320.
Matthew, W. D. 1899 A, p. 37.
Roger, O. 1896 A, p. 251 (syn. of M. gracilis); p. 252 (Notharctus).
Trouessart, E. L. 1897 A, p. 72.
  Eocene (Bridger); Wyoming.

Microsyops gracilis Leidy.

Leidy, J. 1872 C, p. 20.
Cope, E. D. 1884 O, p. 216.
  1885 K, p. 465, fig. 2.

Hyopsodus powellianus Cope.

Cope, E. D. 1884 O, p. 235, pl. xxiiid, figs. 3, 4.
  1882 E, pp. 179, 180. (Phenacodus zuniensis.)
Matthew, W. D. 1897 C, p. 270.
  1899 A, p. 30.
Osborn, H. F. 1897 G, p. 256.
Roger, O. 1896 A, p. 251.
Trouessart, E. L. 1897 A, p. 71.
  Eocene (Wasatch); Wyoming.

Hyopsodus vicarius Cope.

Cope, E. D. 1873 U, p. 1. (Microsyops.)
  1873 E, p. 609. (Microsyops.)
  1875 C, p. 18.
  1880 O, p. 908.
  1881 D, p. 186.
  1882 E, p. 149.
  1884 O, pp. 235, 237, pl. xxiv, figs. 20, 21; pl. xxva, fig. 7.
  1885 K, p. 460, fig. 2.
Lydekker, R. 1885 E, p, 470.
Matthew, W. D. 1899 A, pp. 30, 34, 37.
Nicholson and Lydekker 1889 A, p. 1465, fig. 1348
Roger, O. 1896 A, p. 251.
Schlosser, M. 1887 B, p. 21.
Trouessart, E. L. 1897 A, p. 71.
  Eocene (Wasatch, Wind River, Bridger); Wyoming.

Leidy, J. 1873 B, p. 83, pl. vi, figs. 14-17.
Marsh, O. C. 1881 D, p. 188.
Matthew, W. D. 1899 A, pp. 30, 34, 37.
Roger, O. 1896 A, p. 251.
  Eocene (Wasatch, Wind River, Bridger); Wyoming.

Microsyops scottianus Cope.

Cope, E. D. 1881 D, p. 188.
  1884 O, p. 217, pl. xxiva, fig. 2.
Matthew, W. D. 1899 A, p. 34.
Roger, O. 1896 A, p. 251.
Trouessart, E. L. 1897 A, p. 72.
  Eocene (Wind River); Wyoming.

Microsyops speirianus Cope.

Cope, E. D. 1880 O, p. 908.
  1881 D, p. 188.
  1882 C, p. 463. (M. spierianus.)
  1884 O, p. 216, pl. xxva, fig, 8. (M. spierianus.)
Matthew, W. D. 1899 A, p. 30.
Roger, O. 1896 A, p. 251.
Trouessart, E. L. 1897 A, p. 72.
  Eocene (Wind River); Wyoming.

Microsyops uintensis Osborn.

Osborn, H. F. 1895 A, p. 77, fig. 1.
Matthew, W. D. 1899 A, p. 49.
Roger, O. 1896 A, p. 251.
Trouessart, E. L. 1897 A, p. 72.
  Eocene (Uinta); Wyoming.

MICROSUS Leidy.   Type *M. cuspidatus* Leidy.

Leidy, J. 1870 Q, P. 11.
    1871 C, p. 354.
    1872 B, p. 363.
    1873 B, p. 81.
Trouessart, E. L. 1897 A, p. 72.   (Syn. of Microsyops.)
Zittel, K. A. 1893 B, pp. 688, 693.

Microsus cuspidatus Leidy.

Leidy, J. 1870 Q, p. 113.
    1871 C, p. 354.
    1872 B, p. 363.
    1873 B, pp. 81, 321, pl. vi, figs. 10, 11.
Matthew, W. D. 1899 A, p. 37.
Trouessart, E. L. 1897 A, p. 72.   (Syn. of Microsyops gracilis.)
    Eocene (Bridger); Wyoming.

MENOTHERIUM Cope.   Type *M. lemurinum* Cope.

Cope, E. D. 1873 R, p. 419.
    1874 B, p. 510.
    1875 N, p. 256.
    1884 O, pp. 788, 808.
Marsh, O. C. 1875 B, p. 240.   (Laopithecus, type *L. robustus*.)
    1877 E.
Matthew, W. D. 1899 A, p. 59.   (Syn. of Leptochœrus.)
Scott, W. B. 1890 A, p. 471.   (Menotherium, Laopithecus.)
Zitttel, K. A. 1893 B, p. 696.

Menotherium lemurinum Cope.

Cope, E. D. 1873 R, p. 419.
    1874 B, p. 510.
    1874 P, p. 22.

Cope, E. D. 1884 O, p. 808, pl. lxvi, figs. 34-36.
Matthew, W. D. 1899 A, p. 60.   (Leptochœrus.)
Trouessart, E. L. 1897 A, p. 72.
    Oligocene (White River); Colorado.

Menotherium robustum (Marsh).

Marsh, O. C. 1875 B, p. 240.   (Laopithecus.)
King, C. 1878 A, p. 411.   (Laopithecus.)
Marsh, O. C. 1893 D, p. 412, pl. x, fig. 5.   (Laopithecus.)
Matthew, W. D. 1889 A, p. 59.   (Syn. of Leptochœrus spectabilis.)
Roger, O. 1896 A, p. 252.   (Laopithecus.)
Trouessart, E. L. 1897 A, p. 72.   (Syn. of M. lemurinum.)
    Oligocene (White River); 30 miles south of Black Hills, South Dakota.

INDRODON Cope.   Type *I. malaris* Cope.

Cope, E. D. 1884 K, pp. 318, 319.
    1884 O, p. xxxv.
    1884 R, p. 60.
    1885 K, p. 465.
Flower and Lydekker 1891 A, p. 699.
Matthew, W. D. 1897 C, p. 265.
Osborn and Earle 1895 A, p. 16.
Zittel, K. A. 1893 B, p. 696.

Indrodon malaris Cope.

Cope, E. D. 1884 K, p. 318.

Cope, E. D. 1884 R, p. 61.
    1885 K, p. 465.
    1888 CC, p. 304.
Matthew, W. D. 1897 C, p. 320.
    1899 A, p. 29.
Osborn and Earle 1896 A, p. 16, figs. 3-5.
Roger, O. 1896 A, p. 252.
Trouessart, E. L. 1897 A, p. 72.
    Eocene (Torrejon); New Mexico.

OPISTHOTOMUS Cope.   Type *O. astutus* Cope.

Cope, E. D. 1875 C, pp. 13, 15.
    1877 K, pp. 135, 151.
    1884 O, p. 216.
    1885 K, p. 460.
Osborn, Scott, and Speir 1878 A, p. 17.
Schlosser, M. 1887 B, p. 34.   (=?Washakius.)
Zittel, K. A. 1893 B, p. 696.

Opisthotomus astutus Cope.

Cope, E. D. 1875 C, p. 16.
    1877 K, p. 152, pl. xlv, fig. 9.

Osborn, Scott, and Speir 1878 A, p. 17.
Roger, O. 1896 A, p. 252.
Trouessart, E. L. 1897 A, p. 73.
    Eocene; New Mexico, Wyoming.

Opisthotomus flagrans Cope.

Cope, E. D. 1875 C, p. 16.
    1877 K, p. 152, pl. xlv, fig. 8.
Roger, O. 1896 A, p. 252.
Trouessart, E. L. 1897 A, p. 73.
    Eocene; New Mexico.

APHELISCUS Cope.   Type *A. insidiosus* Cope.

Cope, E. D. 1875 C, pp. 13, 16.
    1877 K, pp. 135, 137.
    1884 O, p. 216.
    1885 K, p. 460.
Zittel, K. A. 1893 B, p. 696.

Apheliscus insidiosus Cope.

Cope, E. D. 1874 O, p. 14.   (Prototomus.)

Cope, E. D. 1875 C, p. 17.
    1877 K, p. 147.
    1885 K, p, 460.
Roger, O. 1896 A, p 253.
Trouessart, E. L. 1897 A, p. 73.
    Eocene; New Mexico.

SARCOLEMUR Cope.   Type *Antiacodon furcatus* Cope.

Cope, E. D. 1875 N, p. 256.
  1877 K, pp. 134, 147.
  1884 O, pp. 216, 233.
  1885 K, p. 460.
Trouessart, E. L. 1898 A, p. 73.
Zittel, K. A. 1893 B, p. 73.

Sarcolemur crassus Cope.

Cope, E. D. 1875 C, p. 17. (Antiacodon.)
  1877 K, p. 149, pl. xlv, fig. 16.
Roger, O. 1896 A, p. 253.
Trouessart, E. L. 1897 A, p. 73.
  Eocene (Wasatch); New Mexico.

Sarcolemur mentalis Cope.

Cope, E. D. 1875 C, p. 17. (Antiacodon.)
  1875 N, p. 256.
  1877 K, p. 149, pl. xlv, fig. 15.

Roger, O. 1896 A, p. 253.
Trouessart, E. L. 1897 A, p. 73.
  Eocene (Wasatch); New Mexico.

Sarcolemur pygmæus Cope.

Cope, E. D, 1872 M, p. 2 of Palæont. Bull. (Lophiotherium.)
  1872 M (1873), p. 461. (Hyopsodus.)
  1873 E, p. 607 (Antiacodon pygmæus); p. 608
   (A. furcatus).
  1873 U, p. 1. (Antiacodon pygmæus, Antiacodon furcatus.)
  1875 N, p. 256. (S. furcatus, S. pygmæus.)
  1877 K, p. 148. (S. pygmæus, S. furcatus.)
  1884 O, p. 233, pl. xxiv, figs. 18, 19.
Matthew, W. D. 1899 A, p. 38.
Roger, O. 1896 A, p. 253.
Trouessart, E. L. 1897 A, p. 73.
  Eocene (Bridger); Wyoming.

HIPPOSYUS Leidy.   Type *H. formosus* Leidy.

Leidy, J. 1872 D, p. 37.
Cope, E. D. 1874 M, p. 82.
  1887 S, p. 258.
Leidy, J. 1873 B, p. 90.
Marsh, O. C. 1875 B, p. 239.
  1877 E.
Schlosser, M. 1887 B, p. 36.
Zittel, K. A. 1893 B, p. 696.

Hipposyus formosus Leidy.

Leidy, J. 1872 D, p. 37.
  1873 B, pp. 90, 321, pl. vi, fig. 41; pl. xxvii,
  figs. 1, 2.

Matthew, W. D. 1899 A, p. 37.
Roger, O. 1896 A, p. 253.
Trouessart, E. L. 1897 A, p. 73.
  Eocene (Bridger); Wyoming.

Hipposyus robustior Leidy.

Leidy, J. 1872 D, p. 364. (Notharctus.)
Cope, E. D. 1872 E, p. 471. (Notharctus.)
Leidy, J. 1873 B, pp. 93, 321, pl. vi, fig. 40.
Matthew, W. D. 1899 A, p. 37.
Roger, O. 1896 A, p. 253.
Trouessart, E. L. 1897 A, p. 74.
  Eocene (Bridger); Wyoming.

BATHRODON Marsh.   Type *B. typus* Marsh.

Marsh, O. C. 1872 I, p. 211.
Cope, E. D. 1882 E, p. 165.
Marsh, O. C. 1875 B, p. 238.
  1877 A, p. 81.
  1877 E.
Schlosser, M. 1887 B, p. 37.
Zittel, K. A. 1893 B, p. 696.

Bathrodon annectens Marsh.

Marsh, O. C. 1872 I, p. 211.
Matthew, W. D. 1899 A, p. 37.

Roger, O. 1896 A, p. 253.
Trouessart, E. L. 1897 A, p. 74.
  Eocene (Bridger); Wyoming.

Bathrodon typus Marsh.

Marsh, O. C. 1872 I, p. 211.
Matthew, W. D. 1899 A, p. 37.
Roger, O. 1896 A, p. 253.
Trouessart, E. L. 1897 A, p. 74.
  Eocene (Bridger); Wyoming.

TELMALESTES Marsh.   Type *T. crassus* Marsh.

Marsh, O. C. 1872 I, p. 206.
Cope, E. D. 1877 K, p. 148. (Telmatolestes.)
Marsh, O. C. 1872 N, p. 405. (Telmatolestes.)
  1875 B, p. 239. (Telmatolestes.)
  1877 E. (Telmatolestes.)
Nicholson and Lydekker 1889 A, p. 1466. (Tel-
  matolestes.)
Schlosser, M. 1887 B, p. 37. (Telmatolestes.)
Trouessart, E. L. 1897 A, p. 74. (Telmatolestes.)

Zittel, K. A. 1893 B, p. 696. (Telmatolestes.)

Telmalestes crassus Marsh.

Marsh, O. C. 1872 I, p. 206.
Matthew, W. D. 1899 A, p. 37.
Roger, O. 1896 A, p. 253. (Telmatolestes.)
Schlosser, M. 1887 B, p. 37.
Trouessart, E. L. 1897 A, p. 74.
  Eocene (Bridger); Wyoming.

HEMIACODON Marsh.   Type *H. gracilis* Marsh.

Marsh, O. C. 1872 I, p. 212.
   1877 E.
Schlosser, M. 1887 B, p. 37.
Zittel, K. A. 1893 B, p. 688.

Hemiacodon gracilis Marsh.

Marsh, O. C. 1872 I, p. 212.
Roger, O. 1896 A, p. 253.
Trouessart, E. L. 1897 A. p. 74.
   Eocene (Bridger); Wyoming.

Hemiacodon pucillus Marsh.

Marsh, O. C. 1872 I, p. 213.
Matthew, W. D. 1899 A, p. 38.
Roger, O. 1896 A, p. 253.   (H. pusillus.)
Trouessart, E. L. 1897 A, p. 74.   (H. pusillus.)
   Eocene (Bridger); Wyoming.

THINOLESTES Marsh.   Type *T. anceps* Marsh.

Marsh, O. C. 1872 I, p. 205.
   1872 N.
   1875 B, p. 239.
   1877 E.
Nicholson and Lydekker 1889 A, p. 1466
Schlosser, M. 1887 B, p. 37.
Zittel, K. A. 1893 B, p. 696.

Thinolestes anceps Marsh.

Marsh, O. C. 1872 I, p. 205.
Matthew, W. D. 1899 A, p. 37.
Roger, O. 1896 A, p. 253.
Trouessart, E. L. 1897 A, p. 74.
   Eocene (Bridger); Wyoming.

STENACODON Marsh.   Type *S. rarus* Marsh.

Marsh, O. C. 1872 I, p. 210.
Schlosser, M. 1887 B, p. 37.
Zittel, K. A. 1893 B, p. 696.

Stenacodon rarus Marsh.

Marsh, O. C. 1872 I, p. 210.

Marsh, O. C. 1894 L, p. 265.
Matthew, W. D. 1899 A, p. 48.
Roger, O. 1896 A, p. 253.
Trouessart, E. L. 1897 A, p. 74.
   Eocene (Bridger); Wyoming.

MESACODON Marsh.   Type *M. speciosus* Marsh.

Marsh, O. C. 1872 I, p. 212.
   1875 B, p. 239.
   1877 E.
Schlosser, M. 1887 B, p. 37.
   1890 B, p. 64.
Zittel, K. A. 1893 B, p. 696.

Mesacodon speciosus Marsh.

Marsh, O. C. 1872 I, p. 212.
Matthew, W. D. 1899 A, P. 38.
Roger, O. 1896 A, p. 253.
Trouessart, E. L. 1897 A, p. 74.
   Eocene (Bridger); Wyoming.

PALÆACODON Leidy.   Type *P. verus* Leidy.

Leidy, J. 1872 C, p. 21.
   1873 B, p. 122,
Marsh, O. C. 1872 I, p. 224.
   1875 B, p. 239.
   1877 E.
Schlosser, M. 1887 B, pp. 35, 144.
Zittel, K. A. 1893 B, p. 688

Palæacodon vagus Marsh.

Marsh, O. C. 1872 I, p. 224.
King, C. 1878 A, p. 403.
Matthew, W. D. 1899 A, p. 38.
Roger, O. 1896 A, p. 253.

Trouessart, E. L. 1897 A, p. 74.
   Eocene (Bridger): Wyoming.

Palæacodon verus Leidy.

Leidy, J. 1872 C, p. 21.
King, C. 1878 A, p. 403.
Leidy, J. 1872 B, p. 356.
   1873 B, pp. 122, 336, pl. vi, fig. 46.
Marsh, O. C. 1872 I, p. 224.
Matthew, W. D. 1899 A, p. 38.
Roger, O. 1896 A, p. 253.
Trouessart, E. L. 1897 A, p. 74.
   Eocene (Bridger); Wyoming

ADDENDA ET CORRIGENDA.

Page 131, second column, line 15. For "Lazier," read "Laizer."

Page 230. Under Trouessart, E. L., 1897 A, instead of "pp. 665–1264," read "pp. 1–664." Wherever after page 716 of the present Bibliography and Catalogue Trouessart, E. L., 1898 A, is quoted, "1897" must be substituted for "1898."

Page 274. For "*EUSELACHII*," read "*ARISTOSELACHII*." *Euselachii* has recently been employed by Parker and Haswell in a different sense.

Page 304. Under *Otodus divaricatus* insert "Leidy, J." before "1873 B," etc.

Page 361. Above ' Osteolepidæ" insert the following: **Spermatodus pustulosus** Cope. *Cope, E. D* 1894 F, p. 438, text fig. 4, Permian, Texas. Both genus and species are new.

Page 387. Under *Empo nepaholica*, after "Stewart, A. 1900 A. p. 332," insert "pl lix, figs. 1–9; pl. lxi figs. 2–5."

Page 393. Under the genus *Anogmius* insert "Cope, E. D., 1877 G, p. 584."

Page 399. Under *Syllæmus latifrons*, after "1877 X. p. 27," insert "pl. xxiii, fig. 1."

Page 442. Under *Lytoloma platyops*, after "Cope, E. D., 1869 M, p. 149," insert "pl. vi."

Page 548. Under *Palæosauropus primævus*, instead of "Silliman, B," read "Amer. Jour. Science."

Page 620. *Hipparion minimum* was orginally described as *Anchitherium minimum*, and ought to have · been arranged under *Anchitherium*.

Page 659. As type of *Tayassu* substitute " *T. pecari* Fischer."

795

INDEX.